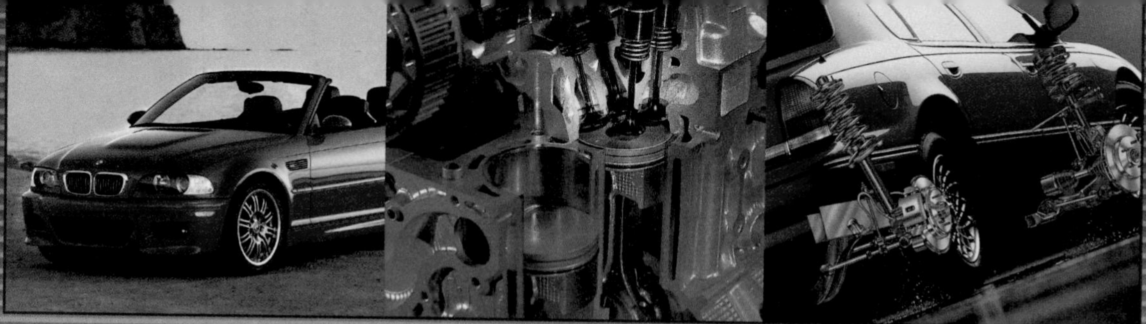

AUTOMOTIVE TECHNOLOGY

A SYSTEMS APPROACH

4TH EDITION

Jack Erjavec

DELMAR
CENGAGE Learning™

Australia • Brazil • Japan • Korea • Mexico • Singapore • Spain • United Kingdom • United States

DELMAR
CENGAGE Learning™

Automotive Technology
Fourth Edition
Jack Erjavec

Vice President, Technology and Trades SBU: **Alar Elken**

Editorial Director: **Sandy Clark**

Senior Acquisitions Editor: **Dave Boelio**

Developmental Editor: **Christopher Shortt**

Marketing Director: **Dave Garza**

Channel Manager: **Bill Lawrensen**

Marketing Coordinator: **Mark Pierro**

Production Director: **Mary Ellen Black**

Production Editor: **Barbara L. Diaz**

Art/Design Specialist: **Rachel Baker**

Technology Project Manager: **Kevin Smith**

Technology Project Specialist: **Linda Verde**

Editorial Assistant: **Kevin Rivenburg**

Library of Congress Control Number:

ISBN-13: 978-1-4018-4831-6

ISBN-10: 1-4018-4831-1

Delmar Cengage Learning
5 Maxwell Drive
Clifton Park, NY 12065-2919
USA

Cengage Learning products are represented in Canada by Nelson Education, Ltd.

For your lifelong learning solutions, visit **delmar.cengage.com**

Visit our corporate website at **www.cengage.com**

Notice to the Reader

Publisher does not warrant or guarantee any of the products described herein or perform any independent analysis in connection with any of the product information contained herein. Publisher does not assume, and expressly disclaims, any obligation to obtain and include information other than that provided to it by the manufacturer. The reader is expressly warned to consider and adopt all safety precautions that might be indicated by the activities described herein and to avoid all potential hazards. By following the instructions contained herein, the reader willingly assumes all risks in connection with such instructions. The publisher makes no representations or warranties of any kind, including but not limited to, the warranties of fitness for particular purpose or merchantability, nor are any such representations implied with respect to the material set forth herein, and the publisher takes no responsibility with respect to such material. The publisher shall not be liable for any special, consequential, or exemplary damages resulting, in whole or part, from the readers' use of, or reliance upon, this material.

Printed in the United States of America
6 7 8 9 X 11 10 09 08

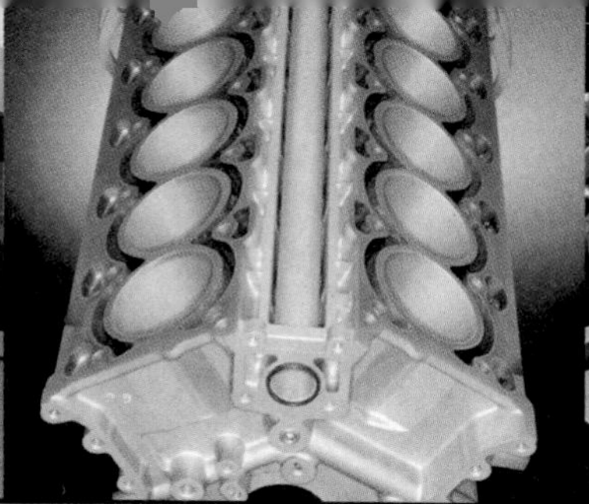

CONTENTS

SECTION 1 AUTOMOTIVE TECHNOLOGY

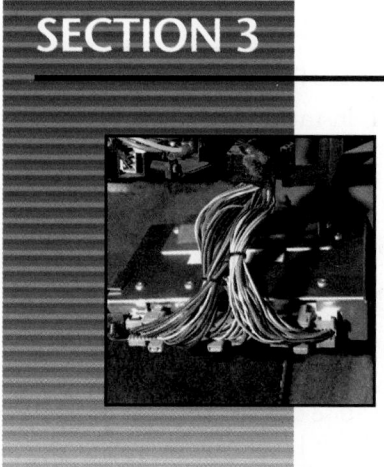

SECTION 4

ENGINE PERFORMANCE

SECTION 5

MANUAL TRANSMISSIONS AND TRANSAXLES

SECTION 6 AUTOMATIC TRANSMISSIONS AND TRANSAXLES

SECTION 7

SUSPENSION AND STEERING SYSTEMS

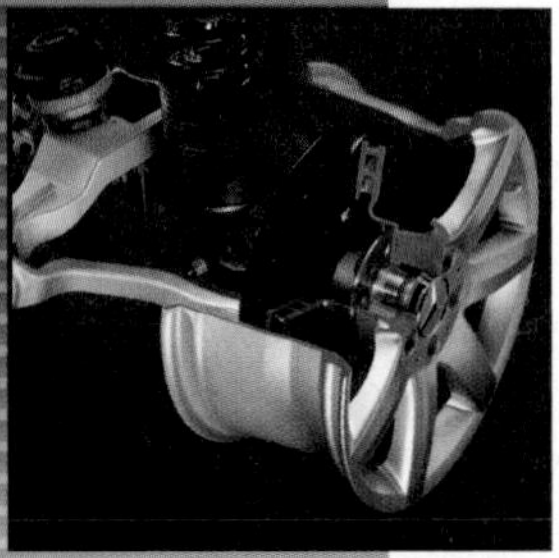

SECTION 8

BRAKES

PHOTO SEQUENCES

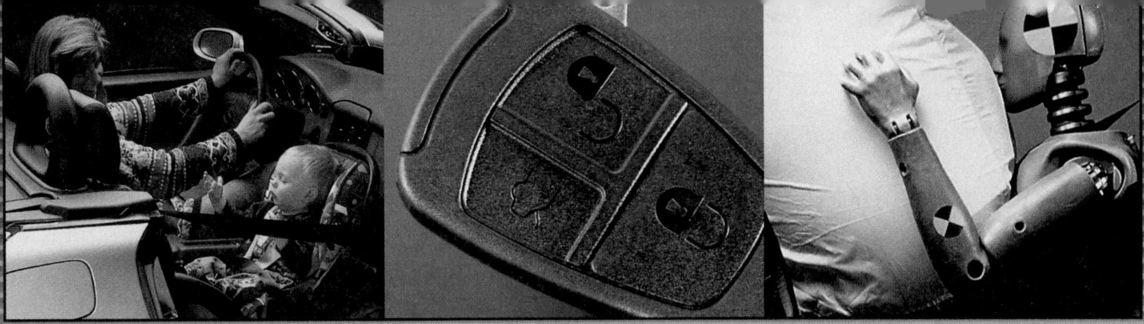

PREFACE

About the Book

All of the changes to the various systems of an automobile and the integration of those systems has made becoming a successful technician more challenging than ever before. This book, *Automotive Technology: A System's Approach,* was designed and written to prepare students for those challenges. The emphasis of this book is on those things students need to know about the vehicles of yesterday, today, and tomorrow.

This does not mean the pages are filled with fact after fact. Rather, each topic is explained in a logical way, slowly but surely. After many years of teaching, I have a good sense of how students read and study technical material. I also know what things draw their interest into a topic, and keep it there. This knowledge has been incorporated in the writing and the features of this book.

This new edition of *Automotive Technology: A System's Approach* represents the many changes that have taken place in the automotive industry over the past few years. With each new edition, a new challenge (for me) presents itself. What should I include and what should I delete? I hope that I made the right choices. Of course if I did, I give much of the credit to the feedback I have received from users of the third edition and those individuals who reviewed this new edition while it was in the making. They all did a fantastic job and showed they are truly dedicated to automotive education.

New to This Edition

This new edition is not the previous edition with a new cover and some new chapters. Although information from the third edition was retained, each chapter has been updated in response to the changing industry. In addition, some new chapters should be helpful for students and their instructors.

The chapters that were dedicated to carburetors are gone. Carburetors, themselves, have been gone for many years. They have also been totally eliminated from standardized testing, such as ASE, for quite a few years. It seems contradictory to include carburetors in any book that claims to cover current technology.

The first section of chapters, which gives an overview of the automotive industry, careers, working as a technician, tools, diagnostic equipment, and basic automotive systems, has a totally new look. Three chapters are new and one other has been totally revamped. Chapter 1 explores the career opportunities in the automotive industry. This discussion has been expanded to include more about an alternative career, a parts counterperson. A new chapter, Chapter 2, discusses workplace skills. This chapter goes through the process of getting a job and keeping it. It also covers some of the duties common to all automotive technicians.

Chapter 6, another new chapter, covers the special and diagnostic tools required for working in each of the eight primary certification areas. The tools discussed include all of the required tools for each area as defined by NATEF.

Also, another new chapter, Chapter 7 presents the science and math principles that are the basis for the operating principles of an automobile. Too often we as instructors, assume our students know these basics. I have included this chapter to serve as a reference for those students who want to be good technicians and to do that, need a better understanding of why things happen the way they do. Appendix C is a chart that leads the student to the section of Chapter 7 that pertains to the information covered in other chapters of the book.

The remainder of Section One has been updated with more coverage of current trends, safety issues (including blood-borne pathogens), shop equipment, and preventive maintenance services.

Section 2, which includes the chapters about engines, has been changed to include more coverage on the latest engine designs and technologies. There is more coverage on theory, diagnosis, and service to alloy engines and overhead camshaft engines. There are discussions on the latest trends, including variable valve timing and lift and variable compression ratios.

It seems everyone appreciated the coverage of basic electricity and electronics in the previous edition. As a result, nothing was deleted from those chapters. A new chapter covering troubleshooting and repairing electrical and electronic systems has been added. Coverage of all the major electrical systems has been increased to include new technologies, including 42-volt systems, new exterior lighting systems, new restraint systems, adaptive systems (such as cruise control), and many new accessories. The rest of the Electricity section has been brought up-to-date with additional coverage on body computers and the use of lab scopes and graphing meters.

The entire Engine Performance section has been updated. New to this edition is a thorough discussion of fuels and alternative energy sources. In this new chapter, Chapter 26, the benefits and disadvantages of the various alternative fuels are discussed, as are hybrid vehicles. Added emphasis on diagnostics was the main goal of the revision of the remainder of this section.

Sections 5 and 6 cover transmissions and drive lines. All of the chapters in these sections have been updated to include more coverage on electronic controls. There is also more coverage on six-speed transmissions, automatic manual transmissions, new differential designs, and electronic automatic transmissions and transaxles.

The Suspension and Steering Systems section has increased coverage on electronic controls and systems, including the new magneto-rheological shock absorbers and four-wheel steering systems. The chapter about wheel alignment has been updated to include the latest techniques for performing a four-wheel alignment.

The Brakes section has also been updated to reflect current technology, which includes the latest antilock brake, stability control, and traction control systems.

Heating and air-conditioning systems are discussed in Section 9. The safety equipment and accessories that were part of this section in the previous edition have been moved to the electricity chapter, which is more appropriate for the contents. The content in Chapters 50 and 51 was totally revised for the third edition and received only minor changes in this edition.

Organization and Goals of This Edition

This edition is still a comprehensive guide to the service and repair of our contemporary automobiles. It is divided into nine sections that relate to the specific automotive systems. Within each section, the chapters describe the various subsystems and individual components. Diagnostic and service procedures unique to different automobile manufacturers are also included in these chapters. Because many automotive systems are integrated, the chapters spend time explaining these important relationships.

Effective diagnostic skills begin with learning to isolate the problem. By identifying the system that contains the problem, the exact cause is easier to pinpoint. Learning to think logically about troubleshooting problems is crucial to mastering this essential skill. Therefore, logical troubleshooting techniques are discussed throughout this text. Each chapter describes ways to isolate the problem system and then individual components of that system.

This *systems approach* gives the student important preparation for the ASE certification exams. These exams are categorized by the automobile's major systems. The book's sections are outlined to match the ASE test specifications and competency task lists. The review questions at the end of every chapter give students practice answering ASE-style questions.

More importantly, a systems approach allows students to have a better understanding of the total vehicle. With this understanding, they have a good chance for a successful career as an automotive technician. That is the single most important goal of this text.

Acknowledgments

I would like to acknowledge and thank the following dedicated and knowledgeable educators for their comments, criticisms, and suggestions during the review process for this new edition:

Anthony Allegro
Lincoln Technical Institute
Union, NJ

Jim Clarke
Lincoln Technical Institute
Melrose Park, IL

Ken Egan
Louisiana Technical College
Winnfield, LA

Tom Fearfield
Lincoln Technical Institute
Union, NJ

John Goldbeck
Porter and Chester
Stratford, CT

Troy Mennis
Lincoln Technical Institute
Grand Prairie, TX

Clifford Murray
Holland College
Charlottetown, Prince Edward Island

Daniel Ortolivo
Sequoia Institute
Fremont, CA

Al Playter
Centennial College
Scarborough, Ontario

Roger Trenkle
Lincoln Technical Institute
Union, NJ

Manuscript Review

Jim Baum
Sequoia Institute
Fremont, CA

Douglas Borders
Sequoia Institute
Fremont, CA

Joe Casolary
Sequoia Institute
Fremont, CA

David DeMasi
Sequoia Institute
Fremont, CA

Rick Errington
Sequoia Institute
Fremont, CA

Jeff Fowler
Sequoia Institute
Fremont, CA

Victor Gilbert
Sequoia Institute
Fremont, CA

Andrew J. Knight
Sequoia Institute
Fremont, CA

Don Moore
Sequoia Institute
Fremont, CA

James Raybourn
Sequoia Institute
Fremont, CA

We would like to acknowledge the following educators for their insightful suggestions and comments on ways to improve the text prior to the development of the fourth edition.

Stephen Belitsos
Vermont Technical College
Randolph Center, VT

C. Neel Flannagan
Aiken Technical College
Aiken, SC

Ralph Greer
John M Patteron State Technical College
Montgomery, AL

Ed Hester
Cedar Valley College
Lancaster, TX

Carl Hinkley
Central Maine Technical College
Auburn, ME

Dave Hostert
Morton College
Cicero, IL

Robert Huetl
Ivy Tech State College
Southbend, IN

Ken Pickerill
Ivy Tech State College
Sellersburg, IN

Ray Taylor
Flint River Technical College
Auburn, ME

John Thorp
Illinois Central College
East Peoria, IL

Rex Weber
Northwest Iowa Community College
Sheldon, IA

Mark Wharton
Pennco Tech
Bristol, PA

We would like to thank those individuals who reviewed our supplement package:

Steve Bilger
Valdosta Technical College
Valdosta, GA

Drew Goddard
Hennepin Technical College
Hopkins, MN

Damian Hall
Nova Scota Community College
Halifax, MN

Randy Olsen
Lake Regional State College
Devils Lake, ND

Dan Rowland
Victor Valley College
Victorville, CA

Contributing Companies

We would like to thank the following companies who provided technical information and art for this edition and the third edition:

Accu Industries
Actron Manufacturing Company
Adaptive Activity Network
AlliedSignal Automotive Aftermarket
American Honda Motor Co., Inc.
American Isuzu Motors, Inc.
Atlas Engineering and Manufacturing, Inc.
Audi of America, Inc.
Autolite
Automatic Transmission Parts, Incorporated
Automatic Transmission Rebuilders Association (ATRA)
Automotive Lift Institute
Battery Council International
Bear Automotive Service Equipment Company
Bendix Brakes by AlliedSignal
Binks Manufacturing Company
BMW of North America, Incorporated
Brodhead Garrett
C&M Cleaning Systems
Central Tools, Inc.
Champion Spark Plug Company
Clevite Engine Parts
Cooper Automotive
CRC Industries, Inc.
DaimlerChrysler Corp.
Dalloz Safety
Dana Corporation
Delco-Remy Co.
Delphi Corporation
Denso Sales California, Inc.
Detroit Gasket
DuPont Company
Echlin Manufacturing Company
Edge Diagnostic Systems
EIS—Standard Motor Products
Emhart Fastening Teknologies
Environmental Systems Products, Inc.
Everco Industries, Inc.
Exide Batteries
Federal-Mogul Corporation
Fel-Pro, Inc.
Fluke Corporation
Ford Motor Company
FRAM
Fred V. Fowler Co., Inc.
Frontline Equipment Company
Gates Rubber Company
GFI Control Systems, Inc.
Goodson Shop Supplies
Goodyear Tire Company
Gutman Advertising Agency
Haldex Traction Systems
Hall-Toledo, Inc.
Hennessy Industries, inc.
Hickok Inc.
Hunter Engineering Company
Hyundai Motor America
Infiniti Motor Company
ITW DeVilbiss
Jaspar Engine and Transmission Exchange, Inc.
John Bean Company
Johnson Matthey, Catalytic Systems Division
Kansas Instruments
K-Line Industries
Lexus of America Co.
Lincoln Automotive
Lisle Corp.
Loctite Corp.
Lubriplate Division Fiske Brothers Refining Co.
Luk Automotive Systems
Mac Tools
Mazda Motor of America, Inc.
Melling Engine Parts
Mercedes-Benz of N.A., Inc.
Midtronics Inc.
Mitchell Repair Inc.
Mitsubishi Motor Sales of America, Inc.
Moog Automotive, Inc.
National Institute for Automotive Service Excellence (ASE)

Neway Manufacturing, Inc.
Nissan North America, Inc.
Old World Industries
OTC Tool and Equipment, Division of SPX Corporation
Perfect Circle/Dana Corporation
Permatex, Inc.
Pollution Control Products Co.
Porsche Cars, North America Inc.
Prestone Products
Progressive Diagnostics—WaveFile AutoPro
Pullman/Holt Corporation
Quaker State Corporation
Robert Bosch Corp.
Robinair, SPX Corporation
RTI Technologies, Inc.
Saab Cars USA
SAE International
Securall Safety Storage Equipment
Service Solutions, SPX Corporation
Siemans VDO Automotive
Sioux Tools, Inc.
Snap-on Tools Company
Society of Automotive Engineers, Inc.
SPX Corporation, Aftermarket Tool and Equipment Group
SPX Genisys Tools
SPX Service Solutions
SPX-OTC/Automotive Electronic Diagnostic Tools
Stanley Tools, a Product Group of The Stanley Works, New Britain, CT
Stant Manufacturing Inc.
Storm Vulcan Mattoni
Subaru of America, Inc.
Sun Electric Corporation
Sunnen Products Company
Super Ford Magazine
Superlift Suspension Systems
Texaco's Magazine, LUBRICATION
The L.S. Starrett Co.
Tracer Products
Transtar Industries Inc.
TRW, Incorporated
U.S. Environmental Protection Agency
Vette Brakes and Products, Incorporated
Visteon Corporation
Volkswagen of America, Inc.
Volvo Car Corporation
Wagner Brake Products
Zexel Torsen Inc.

Portions of materials contained herein have been reprinted with permission of General Motors Corporation.

ABOUT THE AUTHOR

Jack Erjavec has become a fixture in the automotive textbook publishing world. Jack has many years of experience as a technician, educator, author, and editor. Jack has authored or co-authored more than thirty automotive textbooks and training manuals. He has also written technical articles for trade magazines. Jack has a Master of Arts degree in Vocational and Technical Education from Ohio State University. He spent twenty years at Columbus State Community College as an instructor and administrator. Jack has also been a long-time affiliate of the North American Council of Automotive Teachers, including serving on the board of directors and as executive vice-president. He is very active in the industry and has served as a consultant for several automotive manufacturers. Jack is also associated with ATMC, SAE, ASA, ATRA, AERA, and other automotive professional associations.

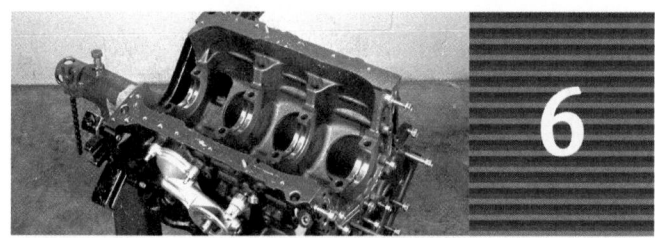

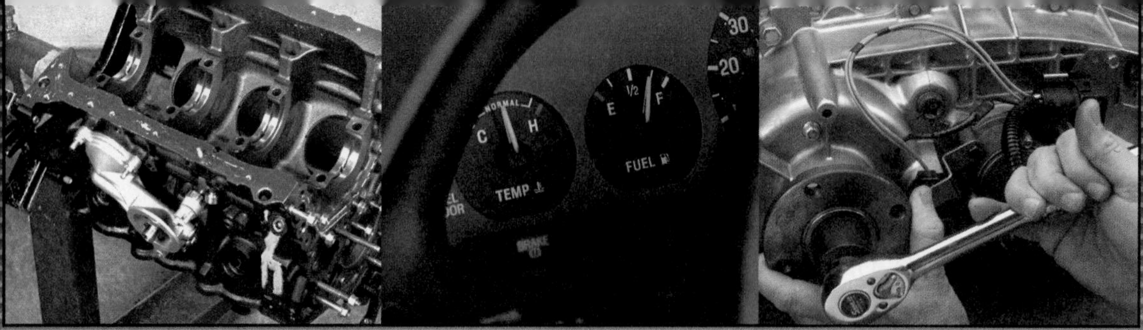

PHOTO SEQUENCES

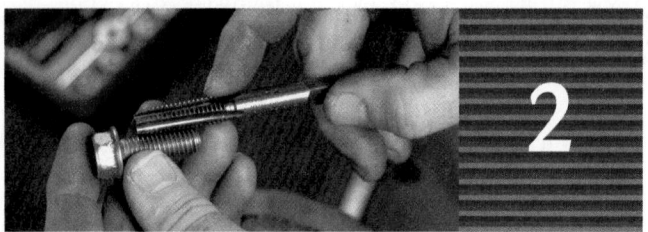

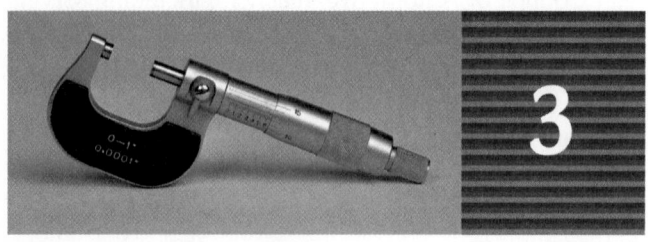

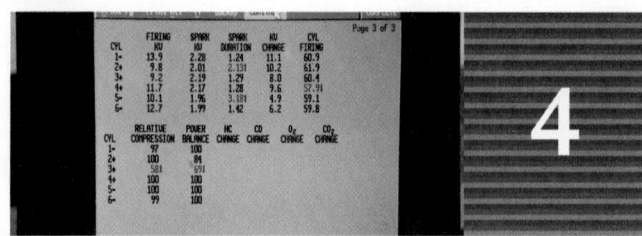

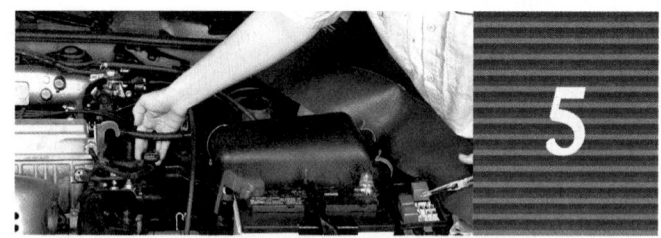

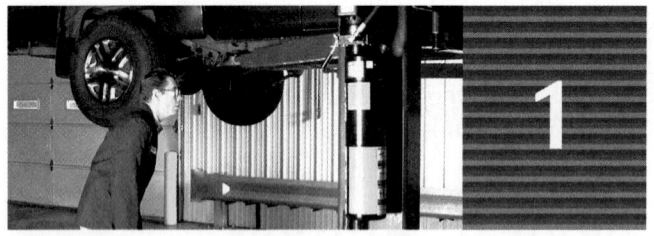

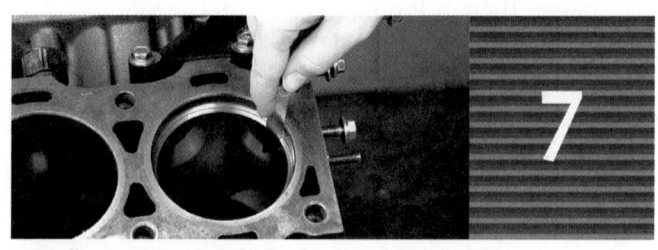

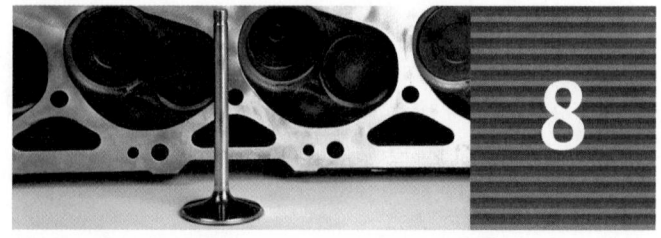

Replacing a Timing Belt on an OHC Engine ■ *348*

Performing a Voltage Drop Test ■ *410*

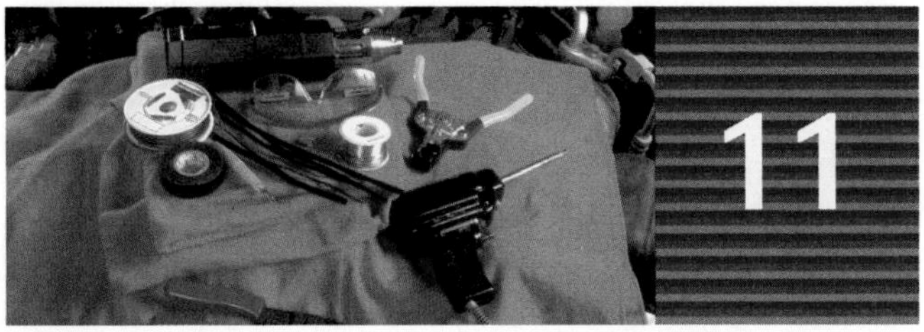

Soldering Two Copper Wires Together ■ *430*

Typical Procedure for Cleaning a Battery Case, Tray, and Cables ■ *448*

Voltage Drop Testing of a Starter Circuit ■ *472*

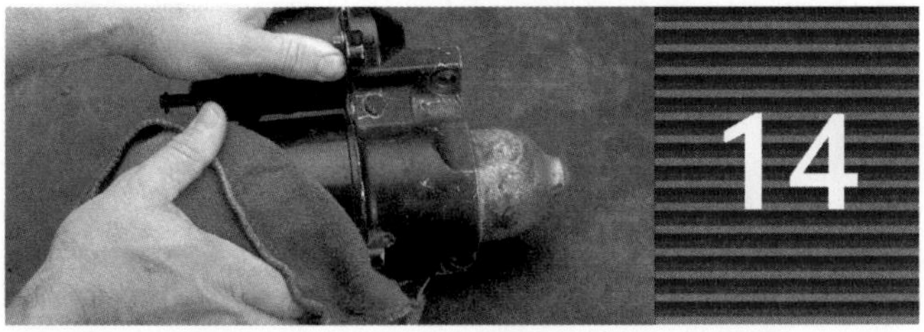

Typical Procedure for Disassembling a Delco-Remy Starter ■ *478*

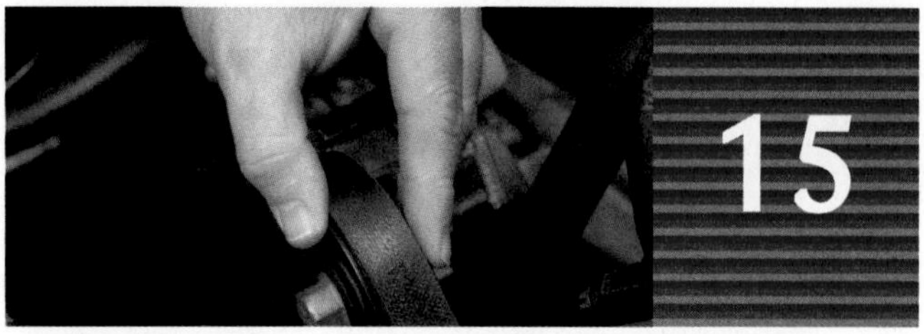

Typical Procedure for Inspecting, Removing, Replacing, and Adjusting a Drive Belt ■ *496*

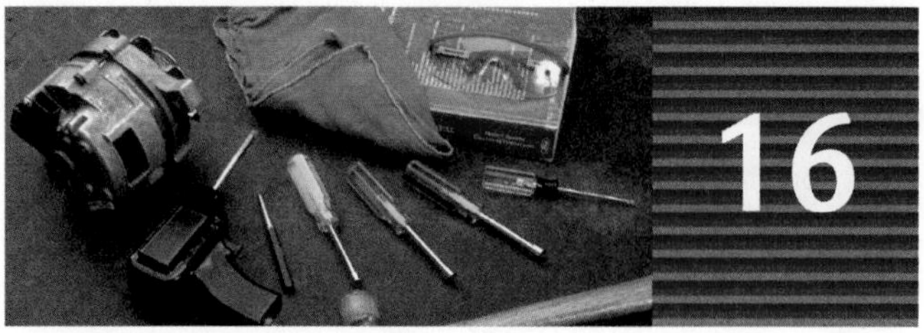

Typical Procedure for Disassembling a Ford IAR AC Generator ■ *500*

Removing a Multifunction Switch ■ *522*

Bench Testing a Fuel Gauge Sending Unit ■ *538*

Typical Procedure for Replacing a Servo Unit ■ *557*

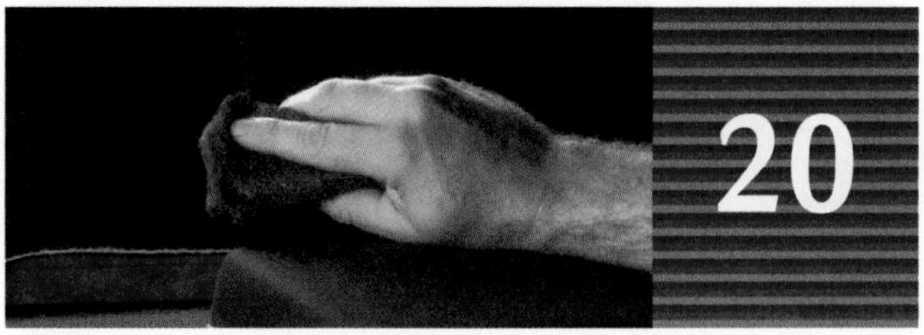

Typical Procedure for Grid Wire Repair ■ *568*

21

22

23

24

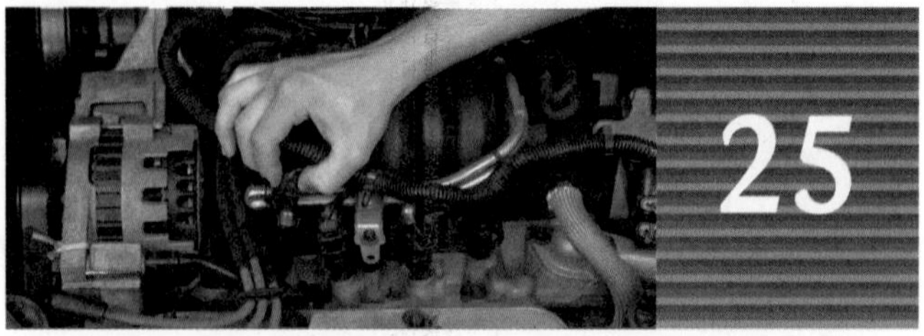
25

26

27

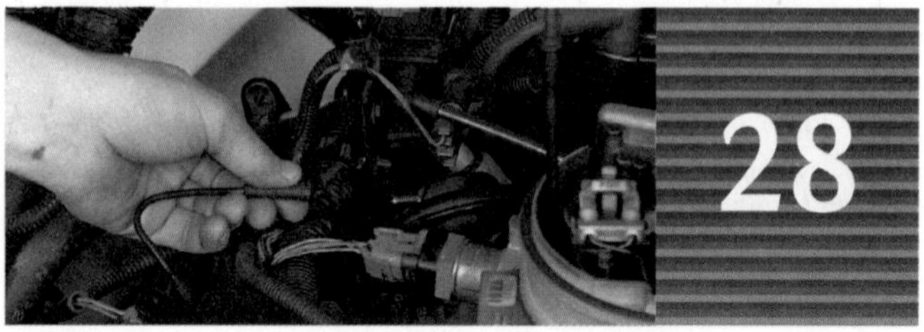
28

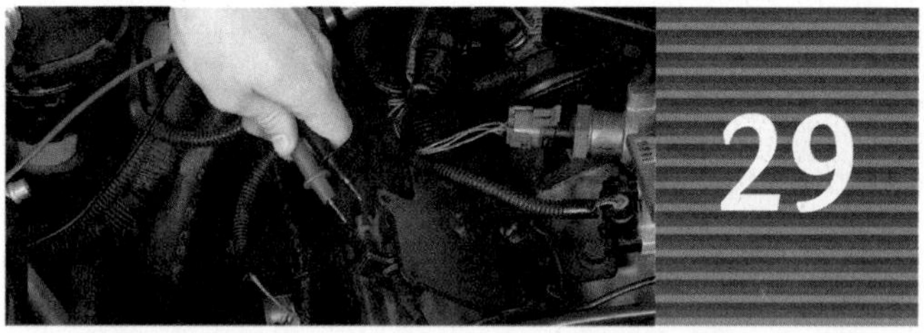
29

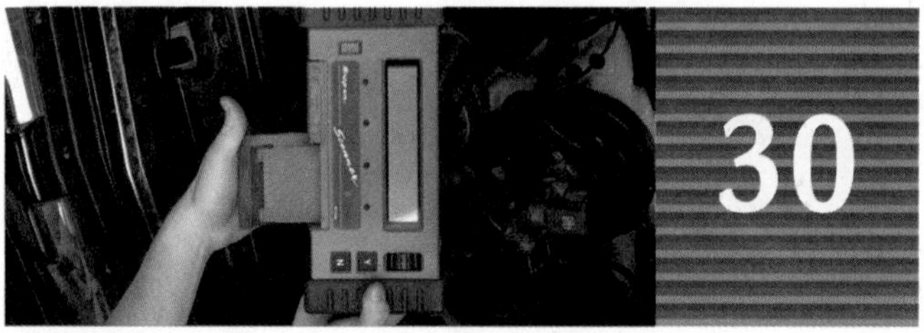
30

31

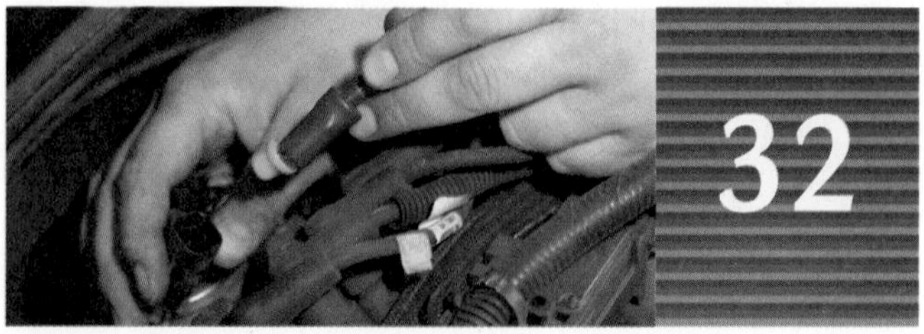
32

Testing a Ford MAP Sensor ■ 891

Installing and Aligning a Clutch Disc ■ 915

Disassembling a Typical Transaxle ■ 950

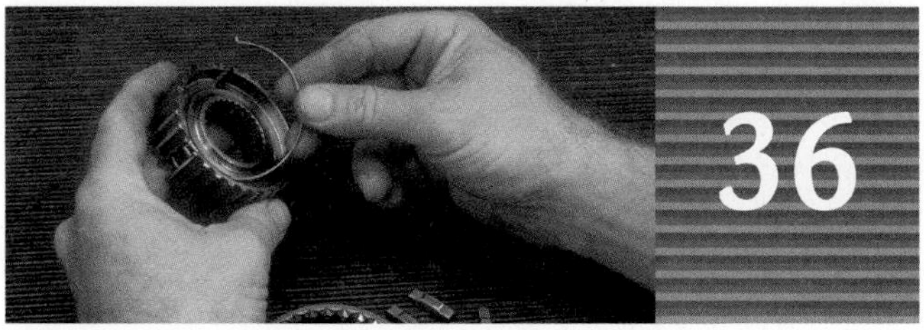

Reassembling a Typical Transaxle ■ 956

Removing and Replacing a CV Joint Boot ■ 968

Disassembling a Single Universal Joint ■ 976

Reassembling a Single Universal Joint ■ 977

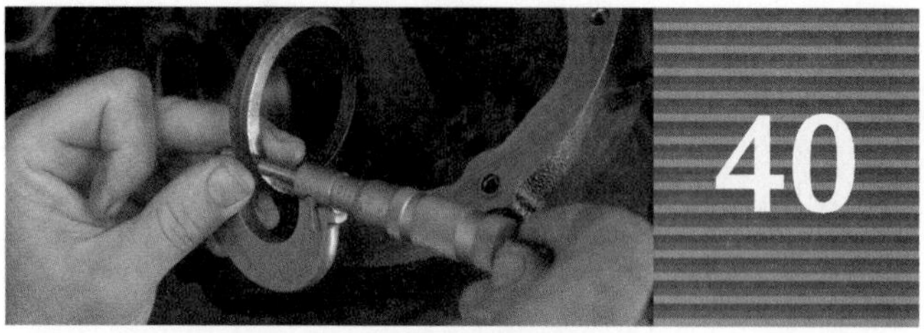

Measuring and Adjusting Backlash and Side-Bearing Preload on a Final Drive Assembly with a Shim Pack ■ 989

Measuring and Adjusting Backlash and Side-Bearing Preload on a Final Drive Assembly with Adjusting Nuts ■ 990

Typical Procedure for Overhauling a 4T60E Transaxle ■ 1066

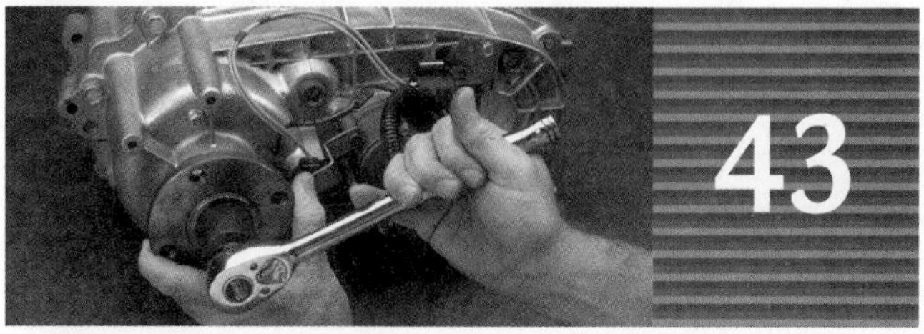

Typical Procedure for Disassembling a Warner 13-56 Transfer Case ■ 1084

Typical Procedure for Reassembling a Warner 13-56 Transfer Case ■ 1086

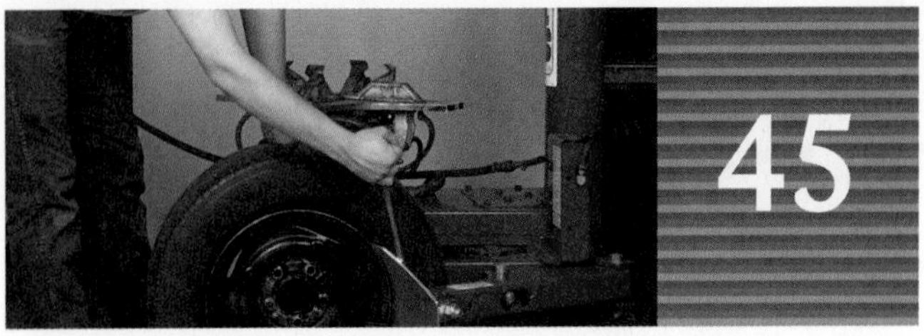
45

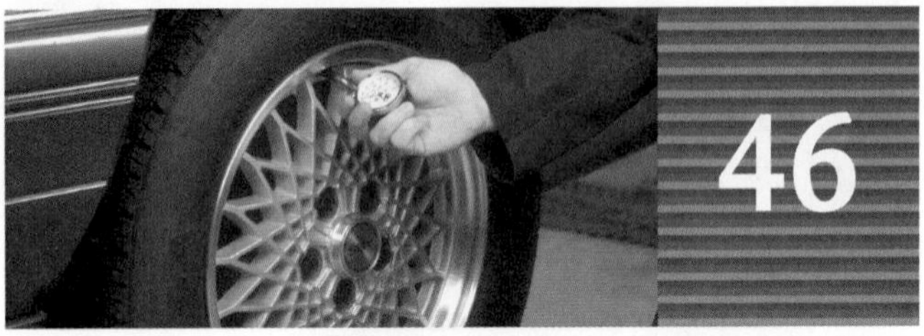
46

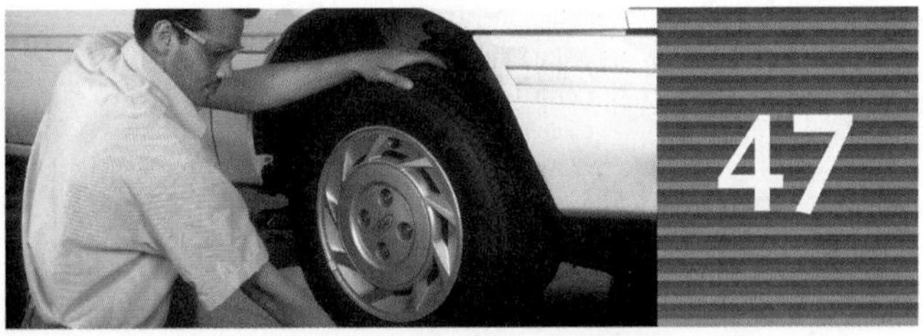
47

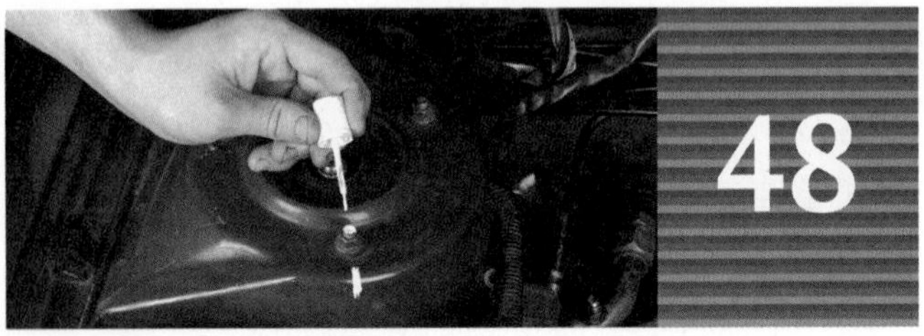
48

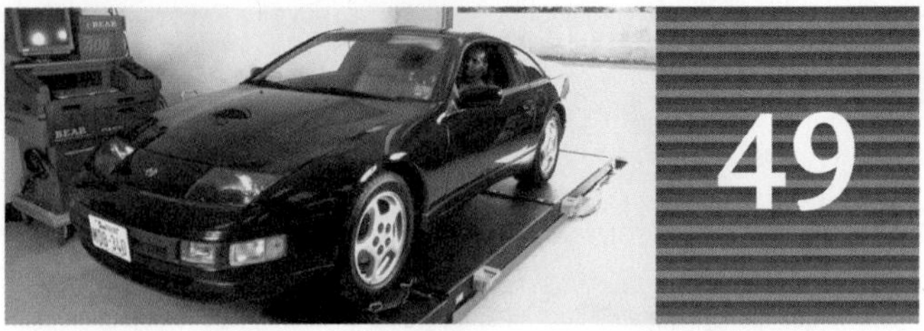
49

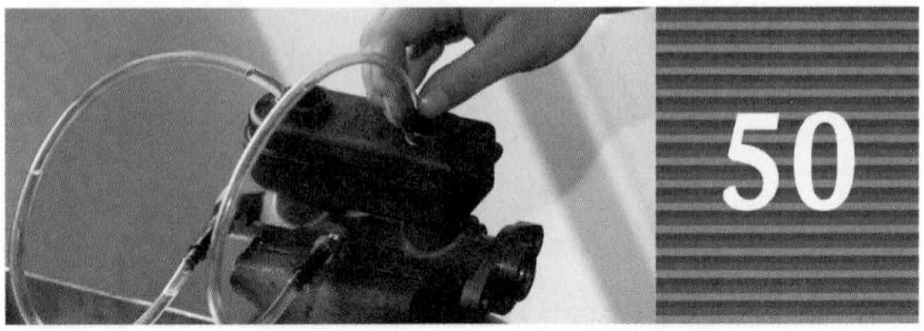
50

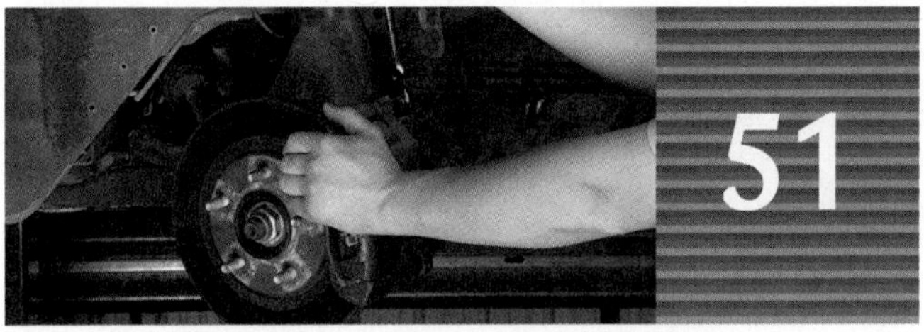
51

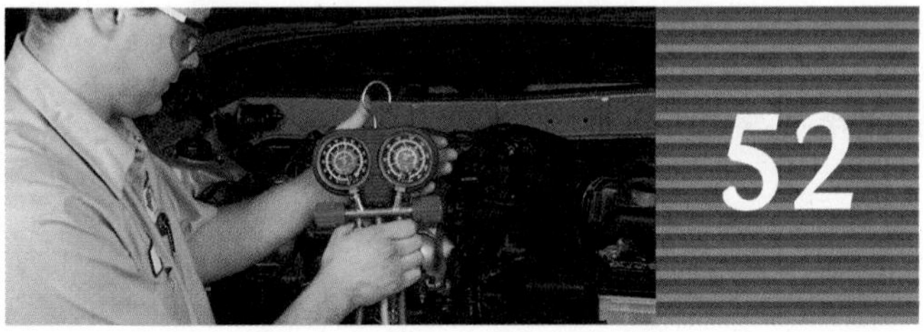
52

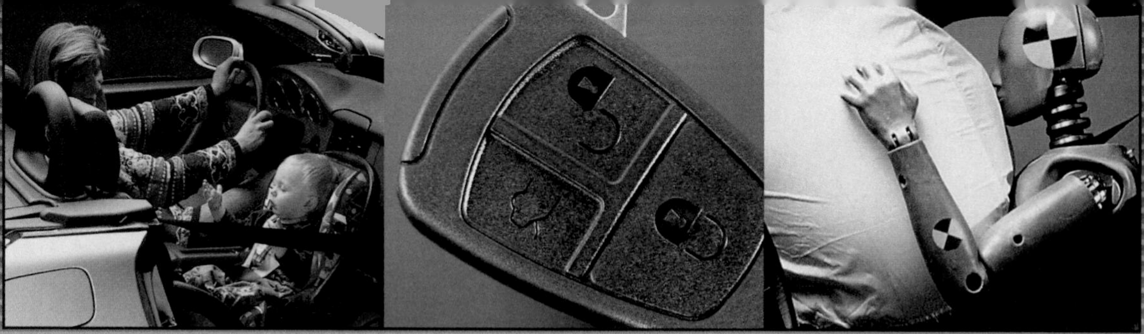

FEATURES OF THE TEXT

Learning how to maintain and repair today's automobiles can be a daunting endeavor. To guide the readers through this complex material, we have built in a series of features that will ease the teaching and learning processes.

OBJECTIVES

Each chapter begins with the purpose of the chapter, stated in a list of objectives. Both cognitive and performance objectives are included in the lists. The objectives state the expected outcome that will result from completing a thorough study of the contents in the chapters.

45

WHEEL ALIGNMENT

OBJECTIVES

■ Explain the benefits of accurate wheel alignment. ■ Explain the importance of correct wheel alignment angles. ■ Describe the different functions of camber and caster with regard to the vehicle's suspension. ■ Identify the purposes of steering axis inclination. ■ Explain why toe is the most critical tire wear factor of all the alignment angles. ■ Identify the purposes of turning radius or toe-out. ■ Explain the condition known as tracking. ■ Perform a prealignment inspection. ■ Describe the various types of equipment that can be used to align the wheels of a vehicle. ■ Describe how alignment angles can be changed on a vehicle. ■ Understand the importance of rear-wheel alignment. ■ Know the difference between two-wheel and four-wheel alignment procedures.

Swift and sure steering responses are needed at today's driving speeds. To accomplish this, the wheels must be in alignment. Wheel alignment allows the wheels to roll without scuffing, dragging, or slipping on different types of road conditions. Proper alignment of both the front and the rear wheels ensures greater safety in driving, easier steering, longer tire life, reduction in fuel consumption, and less strain on the parts that make up the steering and suspension systems of the vehicle (Figure 45–1).

Figure 45–1 Technicians who specialize in wheel alignment are always in demand. Courtesy of Hunter Engineering Company

There are a multitude of angles and specifications that the automotive manufacturers must consider when designing a car. The multiple functions of the suspension system complicate things a great deal for design engineers. They must take into account more than basic geometry. Durability, maintenance, tire wear, available space, and production cost are all critical elements. Most elements contain a degree of compromise in order to satisfy the minimum requirements of each.

Most technicians do not need to be concerned with such requirements. All they need to do is restore the vehicle to the condition the design engineer specified. To do this, however, the technician must be totally familiar with the purpose of basic alignment angles.

The alignment angles are designed in the vehicle to properly locate the vehicle's weight on moving parts and to facilitate steering. If these angles are not correct, the vehicle is misaligned. The effects of misalignment are given in Table 45–1. It is important to remember that alignment angles, when specified in the text, are those specific angles that should exist when the system is being measured under a given set of conditions. During regular performance, these angles change as the traveling surface and vehicle driving forces change.

An important thing to remember when doing wheel alignment and diagnosis is that roads are not designed to be flat. They are designed to allow rain to flow off the surface rather than to accumulate. To accomplish this drainage, the road's surface is paved at a slight angle.

1177

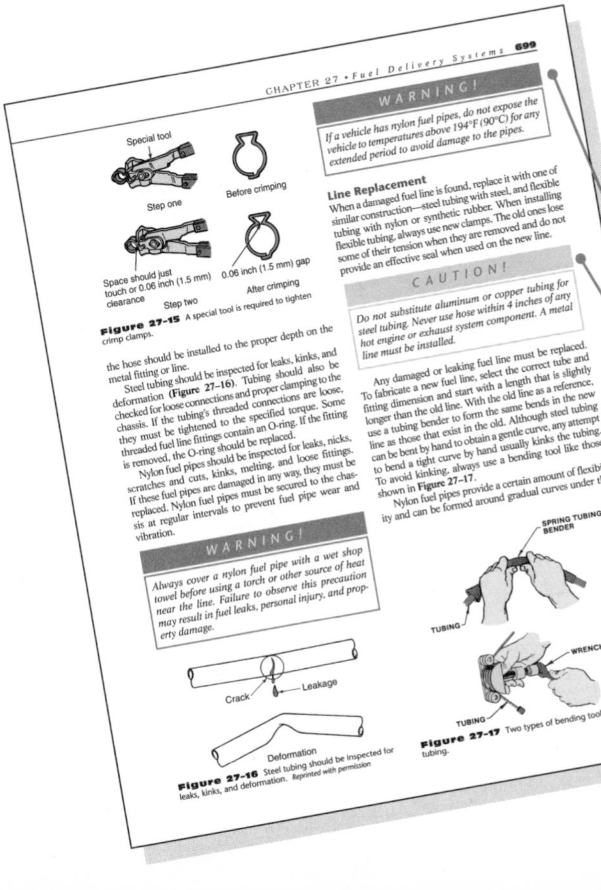

CHAPTER 27 • Fuel Delivery Systems 699

Special tool

Before crimping

Step one

Space should just touch or 0.06 inch (1.5 mm) clearance

0.06 inch (1.5 mm) gap

After crimping

Step two

Figure 27–15 A special tool is required to tighten crimp clamps.

the hose should be installed to the proper depth on the metal fitting or line.

Steel tubing should be inspected for leaks, kinks, and deformation (Figure 27–16). Tubing should also be checked for loose connections and proper clamping to the chassis. If the tubing's threaded connections are loose, they must be tightened to the specified torque. Some threaded fuel line fittings contain an O-ring. If the fitting is removed, the O-ring should be replaced.

Nylon fuel pipes should be inspected for leaks, nicks, scratches and cuts, kinks, melting, and loose fittings. If these fuel pipes are damaged in any way, they must be replaced. Nylon fuel pipes must be secured to the chassis at regular intervals to prevent fuel pipe wear and vibration.

WARNING!

Always cover a nylon fuel pipe with a wet shop towel before using a torch or other source of heat near the line. Failure to observe this precaution may result in fuel leaks, personal injury, and property damage.

Crack

Leakage

Deformation

Figure 27–16 Steel tubing should be inspected for leaks, kinks, and deformation. Reprinted with permission

WARNING!

If a vehicle has nylon fuel pipes, do not expose the vehicle to temperatures above 194°F (90°C) for any extended period to avoid damage to the pipes.

Line Replacement

When a damaged fuel line is found, replace it with one of similar construction—steel tubing with steel, and flexible tubing with nylon or synthetic rubber. When installing flexible tubing, always use new clamps. The old ones lose some of their tension when they are removed and do not provide an effective seal when used on the new line.

CAUTION!

Do not substitute aluminum or copper tubing for steel tubing. Never use hose within 4 inches of any hot engine or exhaust system component. A metal line must be installed.

Any damaged or leaking fuel line must be replaced. To fabricate a new fuel line, select the correct tube and fitting dimension and start with a length that is slightly longer than the old line. With the old line as a reference, use a tubing bender to form the same bends in the new line as those that exist in the old. Although steel tubing can be bent by hand to obtain a gentle curve, any attempt to bend a tight curve by hand usually kinks the tubing. To avoid kinking, always use a bending tool like those shown in Figure 27–17.

Nylon fuel pipes provide a certain amount of flexibility and can be formed around gradual curves under the

SPRING TUBING BENDER

TUBING

WRENCH

TUBING

Figure 27–17 Two types of bending tools for steel tubing.

CAUTIONS AND WARNINGS

Instructors often tell us that shop safety is their most important concern. Cautions and warnings appear frequently in every chapter to alert students to important safety concerns.

SHOP TALK

These features are sprinkled throughout each chapter to give practical, commonsense advice on service and maintenance procedures.

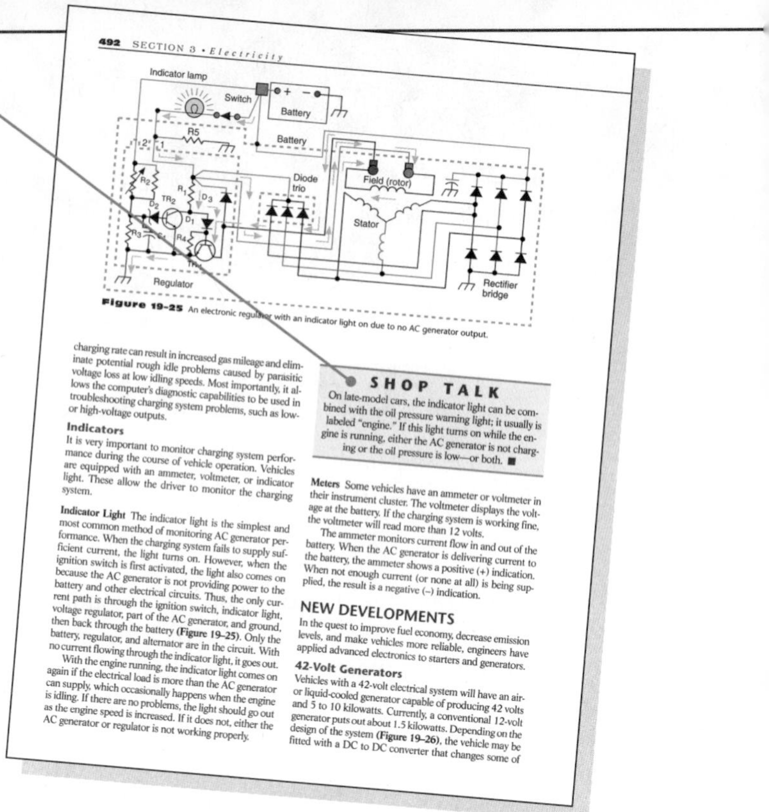

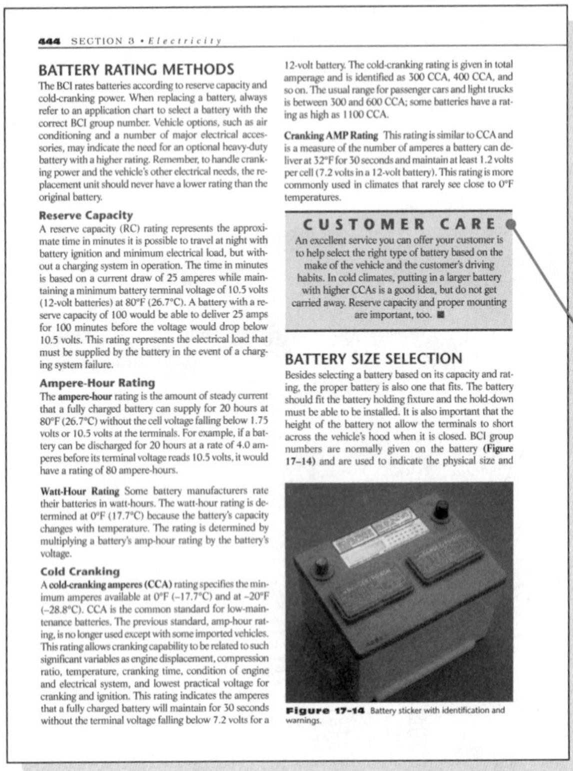

CUSTOMER CARE

Creating a professional image is an important part of shaping a successful career in automotive technology. The customer care tips were written to encourage professional integrity. They give advice on educating customers and keeping them satisfied.

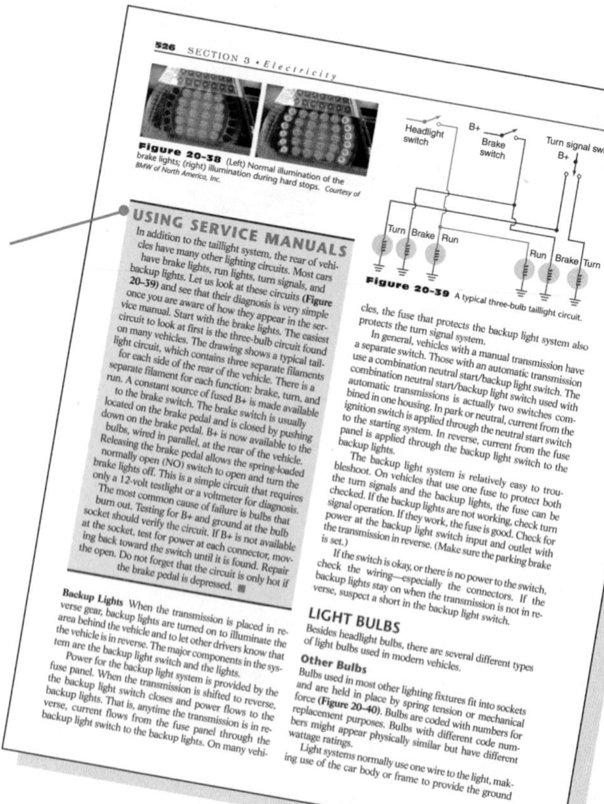

USING SERVICE MANUALS

Learning to use service manuals can be a confusing and time-consuming task. Therefore, we have included this feature to familiarize the reader with the information available in this important resource. Tips and instructions on how to locate the information are found throughout the text. Students will practice recording and analyzing service manual data in the *Tech Manual* that accompanies the textbook.

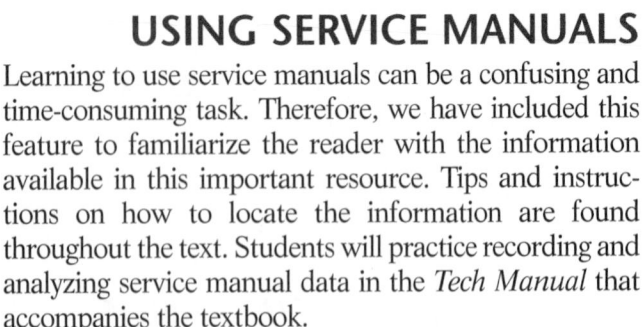

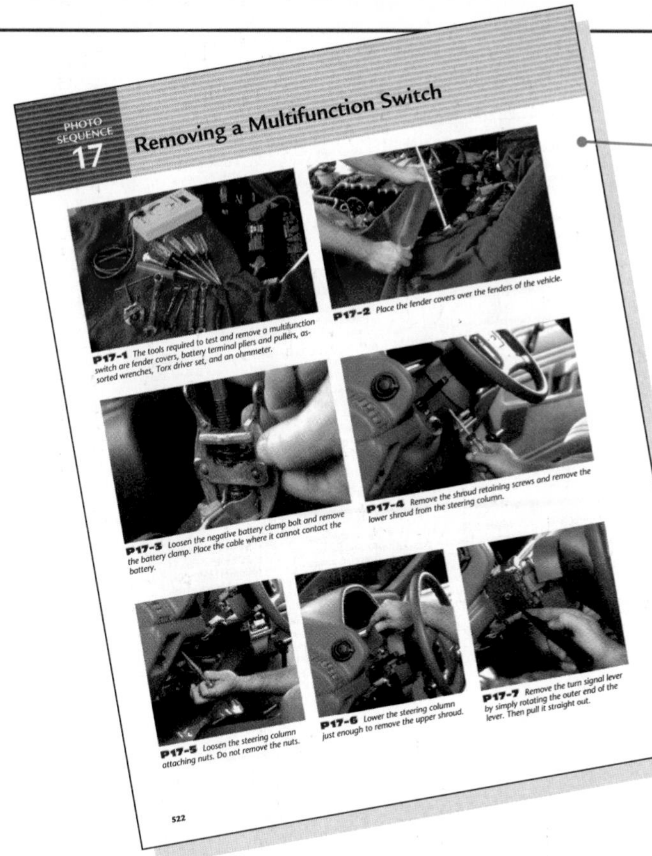

PHOTO SEQUENCES

Step-by-step photo sequences illustrate practical shop techniques. The photo sequence focus on techniques that are common, need-to-know service and maintenance procedures. These photo sequences give students a clean, detailed image of what to look for when they perform these procedures. This was a popular feature of the previous editions, so we now have a total of 52.

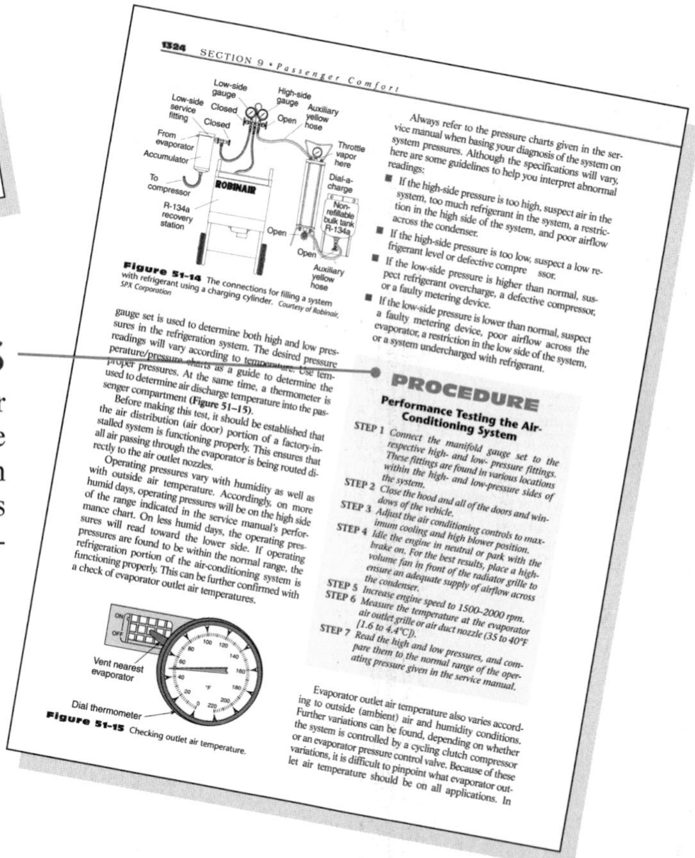

PROCEDURES

This feature gives detailed, step-by-step instructions for important service and maintenance procedures. These hands-on procedures appear frequently and are given in great detail because they help to develop good shop skills and help to meet competencies required for ASE certification.

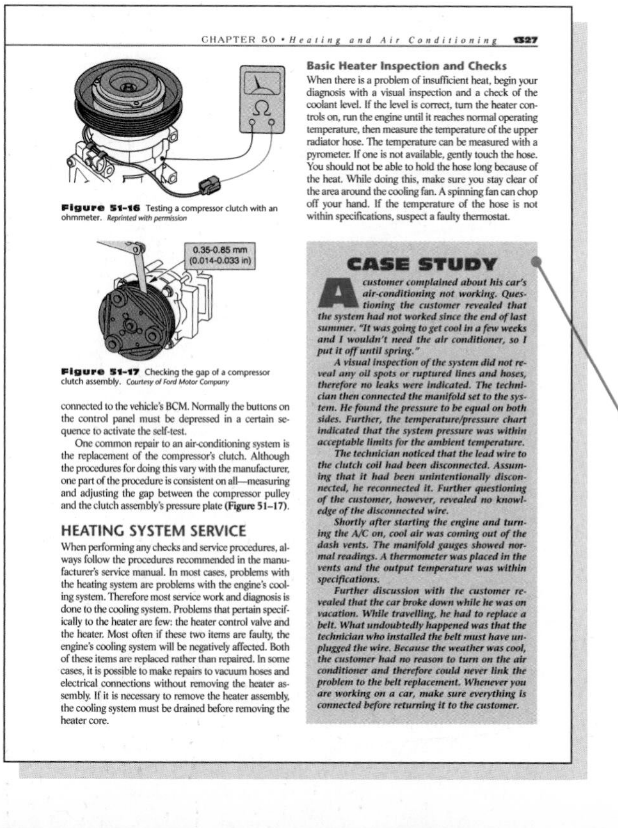

CASE STUDIES

Case Studies highlight our emphasis on logical troubleshooting. A service problem is outlined at the end of the chapters, and then a technician's solution is described. This gives the student a practical example of logical troubleshooting. The *Tech Manual* provides many open-ended case studies for more practice.

KEY TERMS

Each chapter ends with a list of the terms that were introduced in the chapter. These terms are highlighted in the text when they are first used, and many are defined in the glossary.

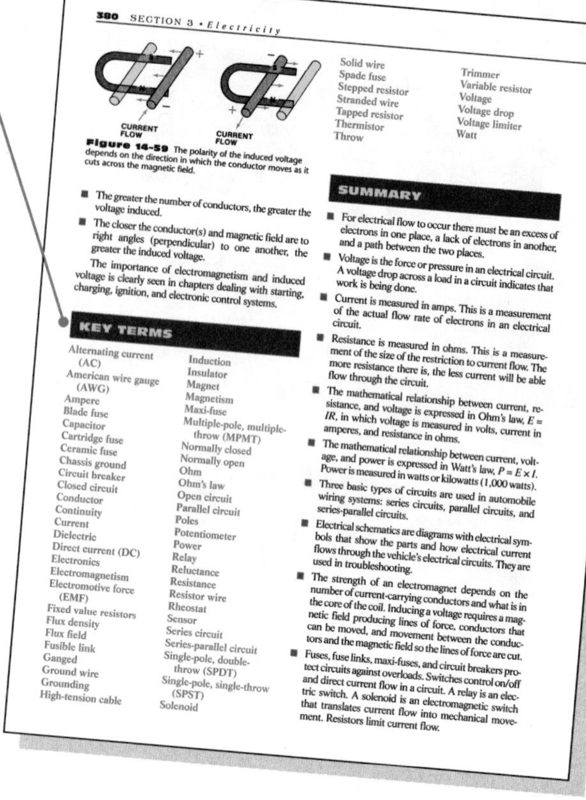

CURRENT FLOW CURRENT FLOW

Figure 14-59 The polarity of the induced voltage depends on the direction in which the conductor moves as it cuts across the magnetic field.

- The greater the number of conductors, the greater the voltage induced.
- The closer the conductor(s) and magnetic field are to right angles (perpendicular) to one another, the greater the induced voltage.

The importance of electromagnetism and induced voltage is clearly seen in chapters dealing with starting, charging, ignition, and electronic control systems.

KEY TERMS

Alternating current (AC), American wire gauge (AWG), Ampere, Blade fuse, Capacitor, Cartridge fuse, Ceramic fuse, Chassis ground, Circuit breaker, Closed circuit, Conductor, Continuity, Current, Dielectric, Direct current (DC), Electronics, Electromagnetism, Electromotive force (EMF), Fixed value resistors, Flux density, Flux field, Fusible link, Ganged, Ground wire, Grounding, High-tension cable, Induction, Insulator, Magnet, Magnetism, Maxi-fuse, Multiple-pole, multiple-throw (MPMT), Normally closed, Normally open, Ohm, Ohm's law, Open circuit, Parallel circuit, Poles, Potentiometer, Power, Relay, Reluctance, Resistance, Resistor wire, Rheostat, Sensor, Series circuit, Series-parallel circuit, Single-pole, double-throw (SPDT), Single-pole, single-throw (SPST), Solenoid, Solid wire, Spade fuse, Stepped resistor, Stranded wire, Tapped resistor, Thermistor, Throw, Trimmer, Variable resistor, Voltage, Voltage drop, Voltage limiter, Watt

SUMMARY

- For electrical flow to occur there must be an excess of electrons in one place, a lack of electrons in another, and a path between the two places.
- Voltage is the force or pressure in an electrical circuit. A voltage drop across a load in a circuit indicates that work is being done.
- Current is measured in amps. This is a measurement of the actual flow rate of electrons in an electrical circuit.
- Resistance is measured in ohms. This is a measurement of the size of the restriction to current flow. The more resistance there is, the less current will be able flow through the circuit.
- The mathematical relationship between current, resistance, and voltage is expressed in Ohm's law, $E = IR$, in which voltage is measured in volts, current in amperes, and resistance in ohms.
- The mathematical relationship between current, voltage, and power is expressed in Watt's law, $P = E \times I$. Power is measured in watts or kilowatts (1,000 watts).
- Three basic types of circuits are used in automobile wiring systems: series circuits, parallel circuits, and series-parallel circuits.
- Electrical schematics are diagrams with electrical symbols that show the parts and how electrical current flows through the vehicle's electrical circuits. They are used in troubleshooting.
- The strength of an electromagnet depends on the number of current-carrying conductors and what is in the core of the coil. Inducing a voltage requires a magnetic field producing lines of force, conductors that can be moved, and movement between the conductors and the magnetic field so the lines of force are cut.
- Fuses, fuse links, maxi-fuses, and circuit breakers protect circuits against overloads. Switches control on/off and direct current flow in a circuit. A relay is an electric switch. A solenoid is an electromagnetic switch that translates current flow into mechanical movement. Resistors limit current flow.

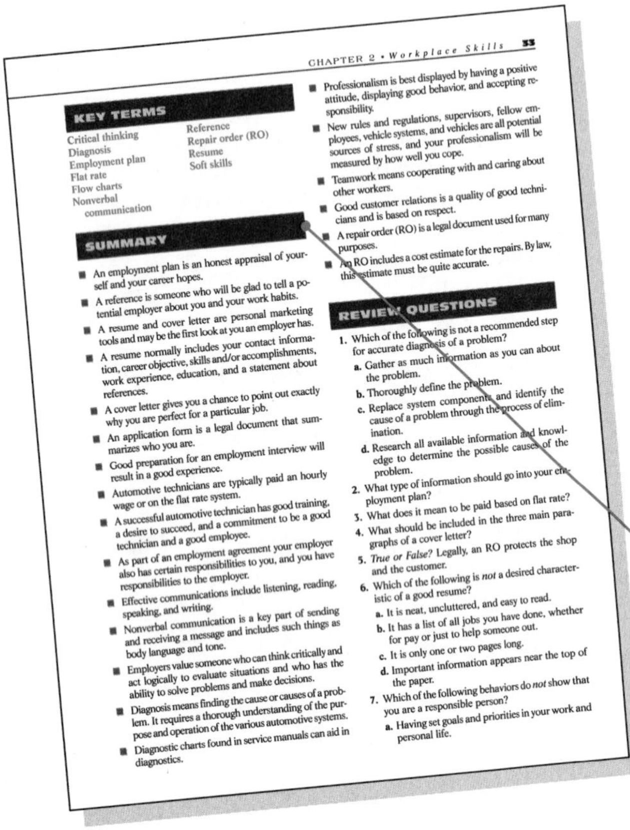

KEY TERMS

Critical thinking, Diagnosis, Employment plan, Flat rate, Flow charts, Nonverbal communication, Reference, Repair order (RO), Resume, Soft skills

SUMMARY

- An employment plan is an honest appraisal of yourself and your career hopes.
- A reference is someone who will be glad to tell a potential employer about you and your work habits.
- A resume and cover letter are personal marketing tools and may be the first look at you an employer has.
- A resume normally includes your contact information, career objective, skills and/or accomplishments, work experience, education, and a statement about references.
- A cover letter gives you a chance to point out exactly why you are perfect for a particular job.
- An application form is a legal document that summarizes who you are.
- Good preparation for an employment interview will result in a good experience.
- Automotive technicians are typically paid an hourly wage or on the flat rate system.
- A successful automotive technician has good training, a desire to succeed, and a commitment to be a good technician and a good employee.
- As part of an employment agreement your employer also has certain responsibilities to you, and you have responsibilities to the employer.
- Effective communications include listening, reading, speaking, and writing.
- Nonverbal communication is a key part of sending and receiving a message and includes such things as body language and tone.
- Employers value someone who can think critically and act logically to evaluate situations and who has the ability to solve problems and make decisions.
- Diagnosis means finding the cause or causes of a problem. It requires a thorough understanding of the purpose and operation of the various automotive systems.
- Diagnostic charts found in service manuals can aid in diagnostics.
- Professionalism is best displayed by having a positive attitude, displaying good behavior, and accepting responsibility.
- New rules and regulations, supervisors, fellow employees, vehicle systems, and vehicles are all potential sources of stress, and your professionalism will be measured by how well you cope.
- Teamwork means cooperating with and caring about other workers.
- Good customer relations is a quality of good technicians and is based on respect.
- A repair order (RO) is a legal document used for many purposes.
- An RO includes a cost estimate for the repairs. By law, this estimate must be quite accurate.

REVIEW QUESTIONS

1. Which of the following is not a recommended step for accurate diagnosis of a problem?
   a. Gather as much information as you can about the problem.
   b. Thoroughly define the problem.
   c. Replace system components and identify the cause of a problem through the process of elimination.
   d. Research all available information and knowledge to determine the possible causes of the problem.
2. What type of information should go into your employment plan?
3. What does it mean to be paid based on flat rate?
4. What should be included in the three main paragraphs of a cover letter?
5. *True or False?* Legally, an RO protects the shop and the customer.
6. Which of the following is *not* a desired characteristic of a good resume?
   a. It is neat, uncluttered, and easy to read.
   b. It has a list of all jobs you have done, whether for pay or just to help someone out.
   c. It is only one or two pages long.
   d. Important information appears near the top of the paper.
7. Which of the following behaviors do *not* show that you are a responsible person?
   a. Having set goals and priorities in your work and personal life.

SUMMARY

Highlights and key bits of information from the chapter are listed at the end of each chapter. This listing is designed to serve as a refresher for the reader.

REVIEW QUESTIONS

A combination of short-answer essay, fill-in-the-blank, multiple-choice, and ASE-style questions make up the end-of-chapter review questions. Different question types are used to challenge the reader's understanding of the chapter's contents. The chapter objectives are used as the basis for the review questions.

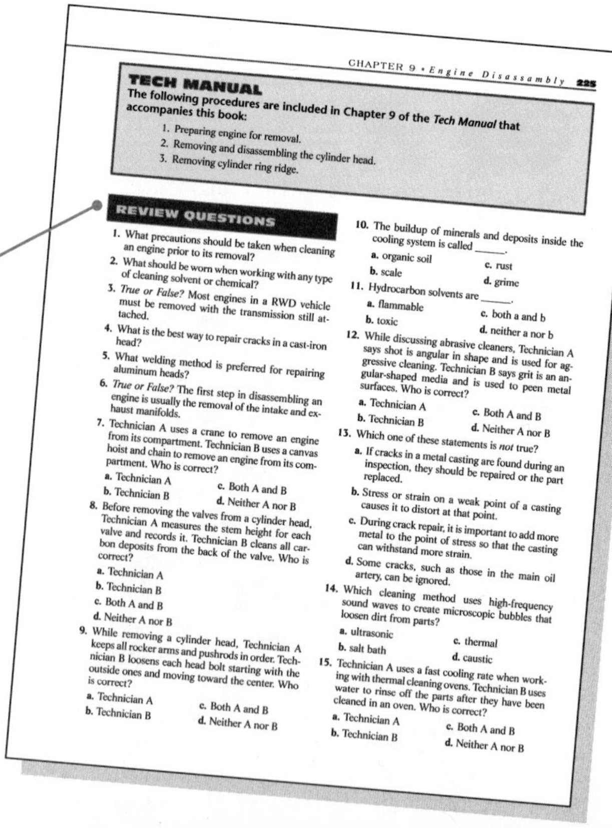

TECH MANUAL

The following procedures are included in Chapter 9 of the *Tech Manual* that accompanies this book:

1. Preparing engine for removal.
2. Removing and disassembling the cylinder head.
3. Removing cylinder ring ridge.

REVIEW QUESTIONS

1. What precautions should be taken when cleaning an engine prior to its removal?
2. What should be worn when working with any type of cleaning solvent or chemical?
3. *True or False?* Most engines in a RWD vehicle must be removed with the transmission still attached.
4. What is the best way to repair cracks in a cast-iron head?
5. What welding method is preferred for repairing aluminum heads?
6. *True or False?* The first step in disassembling an engine is usually the removal of the intake and exhaust manifolds.
7. Technician A uses a crane to remove an engine from its compartment. Technician B uses a canvas hoist and chain to remove an engine from its compartment. Who is correct?
   a. Technician A c. Both A and B
   b. Technician B d. Neither A nor B
8. Before removing the valves from a cylinder head, Technician A measures the stem height for each valve and records it. Technician B cleans all carbon deposits from the back of the valve. Who is correct?
   a. Technician A
   b. Technician B
   c. Both A and B
   d. Neither A nor B
9. While removing a cylinder head, Technician A keeps all rocker arms and pushrods in order. Technician B loosens each head bolt starting with the outside ones and moving toward the center. Who is correct?
   a. Technician A c. Both A and B
   b. Technician B d. Neither A nor B
10. The buildup of minerals and deposits inside the cooling system is called _____.
    a. organic soil c. rust
    b. scale d. grime
11. Hydrocarbon solvents are _____.
    a. flammable c. both a and b
    b. toxic d. neither a nor b
12. While discussing abrasive cleaners, Technician A says shot is angular in shape and is used for aggressive cleaning. Technician B says grit is an angular-shaped media and is used to peen metal surfaces. Who is correct?
    a. Technician A c. Both A and B
    b. Technician B d. Neither A nor B
13. Which one of these statements is *not* true?
    a. If cracks in a metal casting are found during an inspection, they should be repaired or the part replaced.
    b. Stress or strain on a weak point of a casting causes it to distort at that point.
    c. During crack repair, it is important to add more metal to the point of stress so that the casting can withstand more strain.
    d. Some cracks, such as those in the main oil artery, can be ignored.
14. Which cleaning method uses high-frequency sound waves to create microscopic bubbles that loosen dirt from parts?
    a. ultrasonic c. thermal
    b. salt bath d. caustic
15. Technician A uses a fast cooling rate when working with thermal cleaning ovens. Technician B uses water to rinse off the parts after they have been cleaned in an oven. Who is correct?
    a. Technician A c. Both A and B
    b. Technician B d. Neither A nor B

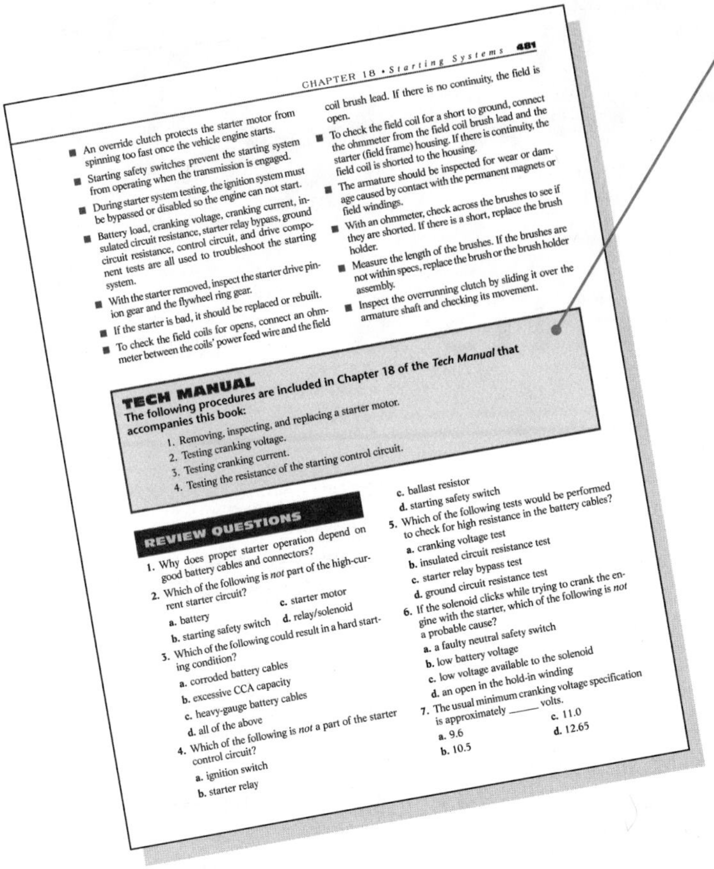
CHAPTER 18 • Starting Systems **481**

- An override clutch protects the starter motor from spinning too fast once the vehicle engine starts.
- Starting safety switches prevent the starting system from operating when the transmission is engaged.
- During starter system testing, the ignition system must be bypassed or disabled so the engine can not start.
- Battery load, cranking voltage, cranking current, insulated circuit resistance, starter relay bypass, ground circuit resistance, control circuit, and drive component tests are all used to troubleshoot the starting system.
- With the starter removed, inspect the starter drive pinion gear and the flywheel ring gear.
- If the starter is bad, it should be replaced or rebuilt.
- To check the field coils for opens, connect an ohmmeter between the coils' power feed wire and the field coil brush lead. If there is no continuity, the field is open.
- To check the field coil for a short to ground, connect the ohmmeter from the field coil brush lead and the starter (field frame) housing. If there is continuity, the field coil is shorted to the housing.
- The armature should be inspected for wear or damage caused by contact with the permanent magnets or field windings.
- With an ohmmeter, check across the brushes to see if they are shorted. If there is a short, replace the brush holder.
- Measure the length of the brushes. If the brushes are not within specs, replace the brush or the brush holder assembly.
- Inspect the overrunning clutch by sliding it over the armature shaft and checking its movement.

TECH MANUAL
The following procedures are included in Chapter 18 of the *Tech Manual* that accompanies this book:
1. Removing, inspecting, and replacing a starter motor.
2. Testing cranking voltage.
3. Testing cranking current.
4. Testing the resistance of the starting control circuit.

REVIEW QUESTIONS
1. Why does proper starter operation depend on good battery cables and connectors?
2. Which of the following is *not* part of the high-current starter circuit?
   a. battery
   b. starting safety switch
   c. starter motor
   d. relay/solenoid
3. Which of the following could result in a hard starting condition?
   a. corroded battery cables
   b. excessive CCA capacity
   c. heavy-gauge battery cables
   d. all of the above
4. Which of the following is *not* a part of the starter control circuit?
   a. ignition switch
   b. starter relay
   c. ballast resistor
   d. starting safety switch
5. Which of the following tests would be performed to check for high resistance in the battery cables?
   a. cranking voltage test
   b. insulated circuit resistance test
   c. starter relay bypass test
   d. ground circuit resistance test
6. If the solenoid clicks while trying to crank the engine with the starter, which of the following is *not* a probable cause?
   a. a faulty neutral safety switch
   b. low battery voltage
   c. low voltage available to the solenoid
   d. an open in the hold-in winding
7. The usual minimum cranking voltage specification is approximately ____ volts.
   a. 9.6
   b. 10.5
   c. 11.0
   d. 12.65

TECH MANUAL REFERENCES

Each service-related chapter ends with a reference to the *Tech Manual* that accompanies this text. Each of these references lists the procedures that are detailed in the *Tech Manual*.

Size

Automotive spark plugs are available in either 14-mm or 18-mm diameters. The 14-mm variety can have either a flat seat that requires a gasket or a tapered seat that does not. The latter is the most commonly used. All 18-mm plugs feature tapered seats that match similar seats in the cylinder head and need no gasket. All spark plugs have a hex-shaped shell that accommodates a socket wrench for installation and removal. The 14-mm, tapered seat plugs have shells with a ⅝-inch (47.7-mm) hex; 14-mm gasketed and 18-mm tapered seat plugs have shells with a 13⁄16-inch (20.67-mm) hex.

METRIC EQUIVALENTS

Throughout the text, all measurements are given in UCS and metric increments.

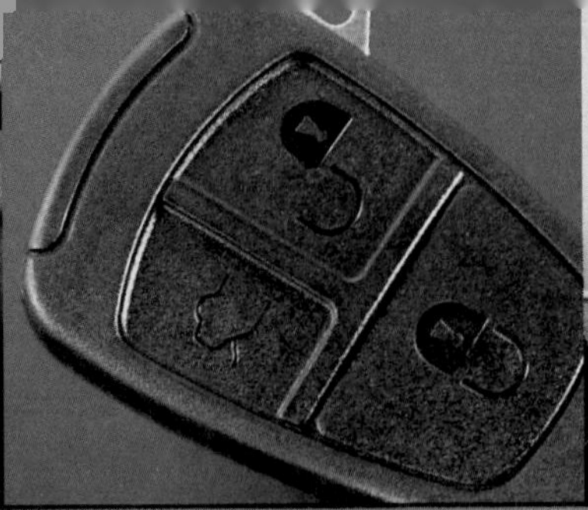
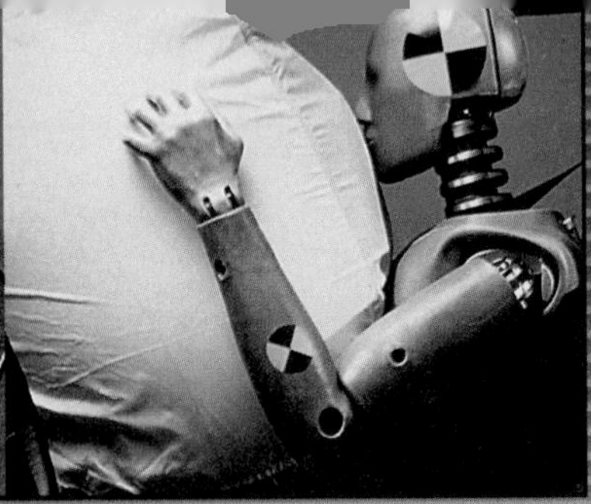

SUPPLEMENTS

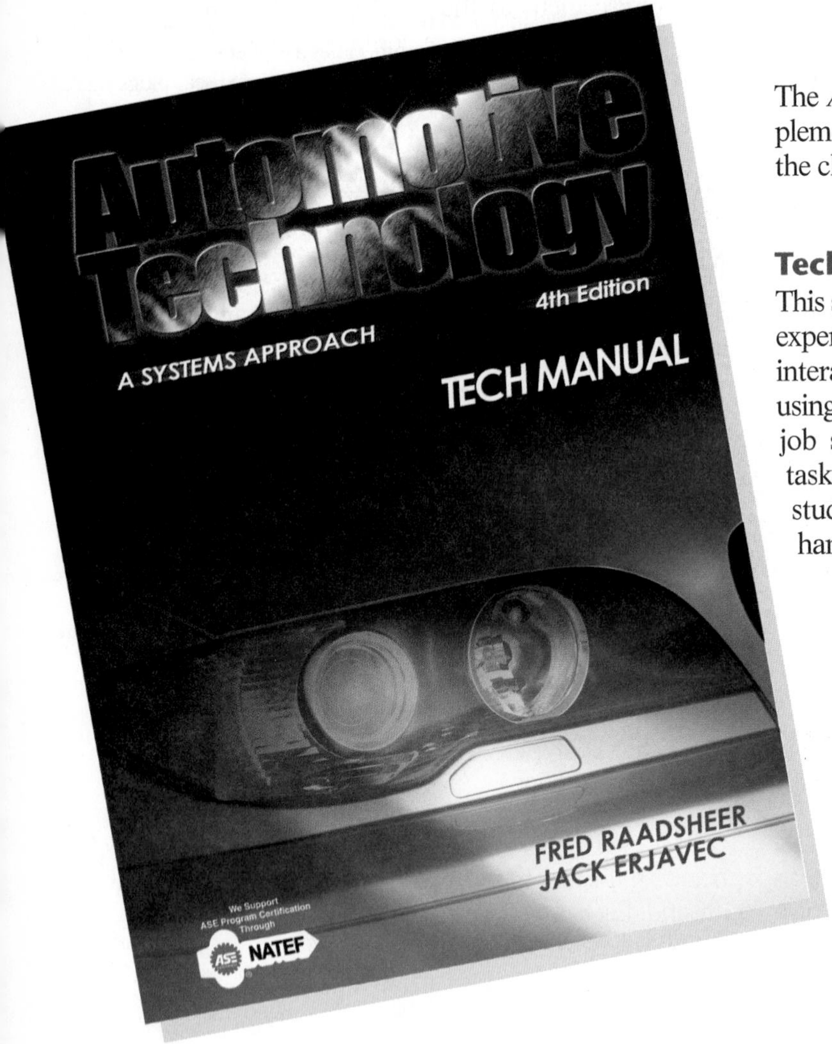

The *Automotive Technology* package offers a full complement of supplements to keep instructors up to date in the classroom.

Tech Manual

This student *Tech Manual* gives hands-on, practical shop experience. It contains hundreds of shop activities and interactive job sheets, with practice in troubleshooting, using diagnostic charts, and using service manuals. Many job sheets are directly correlated to specific NATEF tasks. Service manual report sheets, open-ended case studies, review questions, and ASE prep tests reinforce hands-on learning.

Instructor's Manual

This comprehensive guide provides lecture outlines with teaching hints, answers to review questions from the textbook and answers to *Tech Manual* questions, as well as guidelines for using the *Tech Manual*. A correlation chart to the ASE and NATEF task lists provides references to topic coverage in both the text and *Tech Manual*.

Instructor's e.resource

The new instructor's e.resource on CD-ROM for the fourth edition of *Automotive Technology* is a true asset to all instructors. It has been created as an improvement and replacement to the third edition's instructor's resource kit. The e.resource contains: **PowerPoint (PPT) Slides** for each chapter, providing lectures for all book topics—more than a *thousand PPTs* in total. Selected text images included within the PPTs reinforce concepts. The **Computerized Test Bank** offers an average of 40 test questions per chapter. Question formats include fill in the blank, multiple choice, true/false, and ASE style. An **Image Library,** new to this edition, contains nearly 2400 searchable full color images for classroom display. The entire **Instructor's Manual** plus **NATEF and ASE Correlation Charts** are available as PDFs on the e.resource. Also included is a **Create-Your-Own-Jobsheet Template.** The **Challenging Concepts** from the Student CD are provided for use in the classroom. Questions, video clips, and online resources support each Challenging Concept.

Student CD-ROM

A Student CD-ROM is generally a CD with test questions that provides little more than the same material that they work with in the classroom every day. With the new Erjavec Student CD-ROM at the back of this book, students have a tool that is useful to them because it gives them what they need—instructional reinforcement of automotive concepts and principles that they often struggle with.

The challenging concepts included in the Student CD-ROM were taken from a large cross-section of the automotive market. Automotive instructors from various areas of the country provided a list of concepts that have challenged their students for years. Delmar Learning took those concepts and created an interactive CD-ROM tool that offers students several different ways of grasping concepts that they typically have trouble understanding the first time around. Each concept includes an introductory narrated explanation of the key points, a video clip that clearly illustrates the procedure associated with the concept, a set of theory questions that help students know if they have mastered the basic principles behind the concept, and a set of practical application questions to help students determine their ability to apply the theory they learn. In addition, there are interactive games applicable to each automotive area as well as online support that includes numerous links to websites containing pertinent automotive information.

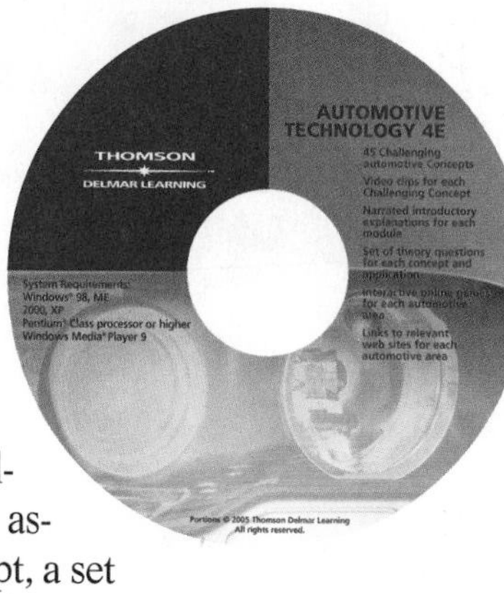

Delmar's ASE Test Preparation Series

This package of ASE test preparation booklets covers A1–A8, L1, P2, X1, and C1. These books are intended for any automotive technician who is preparing to take an ASE examination. Each book combines refresher materials with an abundance of sample test questions, as well as a wealth of information regarding test-taking strategies and the ASE exam style. This series includes the history of the ASE, basic statistics, how the tests are determined, who is expected to pass or fail, and scoring of the tests.

SECTION 1

Automotive Technology

Servicing today's vehicles is much different than it was just a few years ago. Demands on the automotive industry to build more reliable, cleaner, safer, and more fuel-efficient vehicles has greatly changed the way products are built, operated, and serviced.

Today's automotive technicians must keep up with these changes. Automobile mechanics who have not kept up with these changes have left or been forced out of the industry. Many wanting to enter the service industry find it difficult to gain the required knowledge without education, and they last only a short while in the industry. These factors have resulted in a shortage of qualified technicians. This shortage comes at a time when more cars are on the road and, therefore, more cars need to be serviced. As a result, there are more excellent career opportunities for skilled and certified service technicians than ever before. There are also many related careers available for those who have the required knowledge and skills.

The material in Section 1 explores these career opportunities. It also presents what it takes to be a professional technician and a quick look at the major systems of an automobile. Individual chapters present how to get started in an automotive career, some of the important safety practices, common hand and special tools, shop and diagnostic equipment, and the math and science skills needed to be a successful automotive technician.

WE ENCOURAGE PROFESSIONALISM

ASE

THROUGH TECHNICIAN CERTIFICATION

1

CAREERS IN THE AUTOMOTIVE INDUSTRY

OBJECTIVES

■ Describe the reasons why today's automotive industry is considered a global industry. ■ Explain how computer technology has changed the way vehicles are built and serviced. ■ Explain why the need for qualified automotive technicians is increasing. ■ Describe the major types of businesses that employ automotive technicians. ■ List some of the many job opportunities available to people with a background in automotive technology. ■ Describe the different ways a student can gain work experience while attending classes. ■ Describe the requirements for ASE certification as an automotive technician and as a master auto technician.

SERVICING TODAY'S VEHICLES

When the first automobile rolled down a street over one hundred years ago, life changed. Only the elite owned one of these early horseless carriages, which were a sign of wealth and status. Today, an automobile is a necessity. Most Americans would have a difficult time surviving without an automobile. We need our cars and we need the automotive industry. Each year millions of new cars and light trucks are produced and sold in North America **(Figure 1–1)**. The automotive industry's part in the total economy of the United States is second only to the food industry. Manufacturing, selling, and servicing these vehicles is an incredibly large, diverse, and expanding industry.

Thirty years ago, America's "big three" automakers—General Motors Corporation, Ford Motor Company, and Chrysler Corporation—dominated the auto industry. That is no longer true. The industry is now a global industry **(Table 1–1)**. Automakers from Japan, Korea, Germany, Sweden, and other European and Asian countries compete with U.S. companies for domestic and foreign sales; in fact the two best-selling cars in the United States are from Japanese car companies.

Several foreign manufacturers, such as Honda, Toyota, and BMW, operate assembly plants in the United States and Canada. Chrysler Corporation has merged with Mercedes-Benz to form a new company—DaimlerChrysler. No longer is Chrysler only a domestic automobile company; this merger has made it a global company. Many smaller auto manufacturers have been bought by larger companies to form larger global automobile companies. Most often the ownership of a car company is not readily identifiable by the brand name, an example of which is Ford Motor Company; Ford brands include Ford, Mercury, Lincoln, Jaguar, Volvo, Mazda, Aston Martin, and Rover. Many more mergers and acquisitions in the future will continue to create global automobile manufacturers. A number of vehicles are built jointly by the United States and foreign manufacturers. These vehicles are built in North America to be sold here or exported to other countries. Some of these joint ventures manufacture automobiles overseas and import the vehicles into North America.

This cooperation between manufacturers and the public acceptance of imported vehicles has resulted in an extremely wide selection of vehicles from which cus-

Figure 1–1 Ford's F-150 pickup has been the best-selling vehicle in America for many years.

TABLE 1–1 WORLDWIDE UNIT SALES OF PASSENGER CARS, LIGHT-, MEDIUM-, AND HEAVY-DUTY TRUCKS

Manufacturer	Country of Origin	Approx. Units Sold Annually	Notes
General Motors Corp.	U.S.	8.5 million	Includes Holden, Hummer, Vauxhall, Opel, and Saab
Ford Motor Co.	U.S.	7.0 million	Includes Land Rover, Volvo Car Corp., Jaguar, and Aston Martin
Toyota Motor Corp.	Japan	6.1 million	Includes Lexus, Daihatsu, and Hino
Volkswagen AG	Germany	4.9 million	Includes Audi, Bentley, Bugatti, Lamborghini, Rolls-Royce, Skoda, and Seat
DaimlerChrysler AG	Germany	4.5 million	Includes Chrysler Group, Freightliner, Mercedes-Benz, Setra, Smart, Sterling, Thomas Bull Buses, and Western Star
PSA Peugeot Citroen	France	3.2 million	
Hyundai Group	Korea	2.9 million	Includes Kia
Honda Motor Co.	Japan	2.8 million	Includes Acura
Nissan Motor Co.	Japan	2.7 million	
Renault SA	France	2.4 million	Includes Dacia and Renault-Samsung Motors
Fiat Auto S.p.A	Italy	2.1 million	Includes Ferrari, Alfa Romeo, Iveco, Lancia, and Maserati
Mitsubishi Motors Corp.	Japan	1.8 million	
Suzuki Motor Co.	Japan	1.7 million	
BMW Group	Germany	1.0 million	Includes Mini
Mazda Motor Co.	Japan	965 thousand	Owned by Ford Motor Co.
AutoVaz	Russia	640 thousand	
Fuji Heavy Industries Ltd.	Japan	550 thousand	Includes Subaru
Isuzu Motors Ltd.	Japan	456 thousand	
First Auto Works	China	356 thousand	
Daewoo Motor Co.	Korea	317 thousand	
Proton	Malaysia	215 thousand	
TELCO	India	205 thousand	
AutoGaz	Russia	194 thousand	
Ssangyong Motor Co.	Korea	161 thousand	
Volvo Truck Co	Sweden	157 thousand	
Paccar	U.S.	92 thousand	Includes DAF, Kenworth, Leyland, Peterbilt, and Foden
Navistar International	U.S.	88 thousand	
Mahindra & Mahindra	India	70 thousand	
Porsche AG	Germany	55 thousand	
MAN Nutzfahrzeuge Group	Germany	54 thousand	
Scania	Sweden	44 thousand	
Ashok Leland	India	33 thousand	
Nissan Diesel Motor Co.	Japan	29 thousand	
Hindustan Motors	India	23 thousand	
Eicher Motors	India	12 thousand	
Bajaj Tempo	India	9 thousand	

tomers may choose. This variety has also created new challenges for automotive technicians based on one simple fact: Along with the different models come different systems.

The Importance of Automotive Technicians

The automobile started out as a simple mechanical beast. It moved people and things with little regard to the environment, safety, and comfort. Through the years these concerns have been the impetus for design changes. A technical area that has affected vehicle design the most is the same one that has greatly influenced the rest of our lives—electronics. Today's automobiles are sophisticated, electronically controlled machines. To provide comfort and safety while still being friendly to the environment, these new machines use the latest developments of many different technologies—mechanical and chemical engineering, hydraulics, refrigeration, pneumatics, physics, and, of course, electronics.

Because electronics play an important part in the operation of all automotive systems, an understanding of electronics is a must for all automotive technicians. The needed level of understanding is not that of an engineer; instead, technicians need a practical understanding of electronics. In addition to mastering the mechanical skills needed to remove, repair, and replace faulty or damaged components, today's technician also must be able to diagnose and service electronic systems.

Computers and electronic devices are used to control the engine and its support systems. Because of these controls, today's automobiles use less fuel, perform better, and run cleaner than those in the past **(Figure 1–2)**.

Electronic controls also are used to activate shifting in transmissions, eliminate brake lockup in antilock braking systems (ABS), improve handling by controlling steering and suspension systems, and provide passenger

Figure 1-2 Today's cars use less fuel, perform better, and run clearner than they did a few years ago. *Used with permission from Nissan North America, Inc.*

protection and comfort. The amount of electronics used on cars and trucks is increasing with each model year. Consider these facts:

- About 80 % of all functions on vehicles is controlled by electronics.
- Electronically controlled antilock braking systems were once optional on some vehicles but are now standard equipment on many vehicles.
- Some vehicles use twelve or more different computers to control their different systems. Separate computers are used to control the engine, the transmission/transaxle, security systems, instrumentation, climate control, the suspension system, steering system, and antilock brakes.
- Vehicle diagnostic systems anticipate breakdowns, contact emergency road service, and guide technicians through the repair process.
- Brake lights vary in size and brightness according to the pressure put on the brake pedal.
- Headlights have moveable reflectors that allow the lights to follow the curves in the road.
- Some vehicles are now available with an infrared system that provides improved vision at night and in bad weather.
- In the near future, some model vehicles will have multiple video cameras to view the area all around the vehicle, eliminating blind spots.
- Complex electronic circuitry is and will be used to control engine systems so that the engine's exhaust contains very low amounts or zero pollutants.
- Intelligent cruise control devices combine speed control with braking. The vehicle's brakes will be applied automatically to maintain safe distances between moving vehicles. The distance will be monitored by radar.
- Global navigation and satellite tracking systems allow the driver to avoid adverse traffic and road conditions by giving detailed travel routing.

There are many reasons for the increasing incorporation of electronics into automobiles. Electronics are based on electricity and electricity moves at the speed of light. This means systems can be monitored and their mode of operation changed very quickly. Because they have no moving parts, electronic components do not wear. This means they last a long time and do not require periodic adjustments. Electronic components are also very light. Reducing vehicle weight means improved performance and fuel mileage.

Right now, only the creativity of the designers limits the future use of electronics in the automotive industry. That creativity will shape the vehicles of the future. However, other factors will also influence the shape of the future and the use of electronics. One of the main factors has been and will be legislation. Throughout recent history, car manufacturers have responded to laws designed to make automobiles safer and run cleaner.As the manufacturers respond to legislation, new systems and components are introduced. With these come new learning requirements for technicians. Anyone desiring to be a good technician must update his or her skills to keep up with the technology.

Many states have passed laws that require vehicle owners to have their cars' exhaust tested on an annual basis. Most states require that vehicles pass an I/M-240 or similar test. An I/M-240 test focuses on the inspection and maintenance of emission controls. That is why it is called an **Inspection/Maintenance (I/M)** test. This test, and the laws that tell the owners their vehicles must pass the test, affects the work of a technician. The cause of test failures must be found and corrected.

The Need for Quality Service

Vehicles will continue to become more complex; therefore, the need for good technicians will continue to grow. Currently there is a great shortage of qualified automotive technicians. This means there are, and will be, excellent career opportunities for good technicians. Good technicians are able to diagnose problems in both the simple and the complex systems of today's automobiles. Of course, after the cause of a problem has been identified, the system must be properly serviced or repaired **(Figure 1–3)**.

With the increase in the price of new vehicles came increased public demand for very reliable vehicles. The public also demands that when things do go wrong, they should be corrected the first time they take the vehicle back to the dealership—they expect the problem to be "fixed right the first time." This feeling also carries through to older vehicles, those out of warranty and serviced by someone other than the dealership. Paying for repairs and parts that do not fix the problem is not something consumers want, nor should we expect them to. It is also not something that helps the reputation of technicians or the manufacturer of the vehicle.

The primary reason some technicians are unable to fix a particular problem is simply that they cannot find the cause of the problem. Today's vehicles are complex, which means that a great amount of knowledge and understanding is required to diagnose them. Today's technicians must have good **diagnostic skills**. Individuals who can identify and solve problems the first time the vehicle is brought into the shop are wanted by the industry. For them, there are many excellent opportunities.

The high cost of electronic components and many mechanical parts has made the hit-or-miss method of repair too expensive. Too often, mechanics who do not understand how to properly troubleshoot an electronic system automatically replace its most expensive component—the computer, which often results in a very expensive wrong guess. Computers are very reliable. Normally, the cause of a problem in a computer system is the failure of an inexpensive switch or sensor, a poor electrical connection, or a bad mechanical part within the system.

The Need for Ongoing Service

The use of electronic controls has not eliminated the need for routine service and scheduled maintenance **(Figure 1–4)**. In fact, it has made it more important than ever.

Figure 1-3 Good technicians are able to follow a specific manufacturer's diagnostic charts and interpret the results of diagnostic tests.

Figure 1-4 Regular preventive maintenance is important for keeping electronic control systems operating correctly. A common part of PM is changing the engine's oil and filter.

Although the computer systems can make adjustments to cover up some problems, a computer cannot replace parts that wear. A computer cannot tighten loose belts, change weak or dirty coolant, or change dirty engine oil. Simple problems such as these can set off a chain of unwanted events in an engine control system. Electronic controls are designed to help a well-maintained vehicle operate efficiently. They are not designed to repair systems.

The computer, through its control devices, may attempt to compensate for a problem by making adjustments to the engine's systems. As a result, the engine will run reasonably well, but its overall performance and efficiency will be lowered.

Various maintenance procedures usually are performed according to a schedule recommended by the vehicle's manufacturer. These maintenance procedures are referred to as **preventive maintenance (PM)** because they are designed to prevent problems. Scheduled preventive maintenance normally includes oil and filter changes, coolant and lubrication services, replacement of belts and hoses, and replacement of spark plugs, filters, and worn electrical parts **(Figure 1–5)**.

If the vehicle's owner fails to follow the recommended maintenance schedule, the vehicle's warranty might not cover problems that result. For example, if the engine fails during the period of time covered by the warranty, the warranty may not cover the engine if the owner does not have proof that the engine's oil was changed according to the recommended schedule.

Warranties A **warranty** is an agreement by the auto manufacturer to have its authorized dealers repair, replace, or adjust certain parts if they become defective. This agreement normally lasts until the vehicle has been driven a certain number of miles, typically 36,000 (58,000 km), and/or until the vehicle has been owned for a certain length of time, typically three years. In order for the warranty to cover the cost of the repair, the problem must occur during the time or miles covered by the warranty. There are basically two types of warranties: those offered by the manufacturer and those ordered by federal and state laws.

The details of most manufacturer warranties normally vary, depending on the manufacturer, vehicle model, and year. Most manufacturers provide several levels of warranty coverage. There is often a basic warranty that covers the complete vehicle for the first year or first 12,000 miles (19,200 km), whichever comes first. Additional warranties may cover the engine, transmission, drive axles, powertrain, battery, safety restraint systems, the body, or other parts of the vehicle. These warranties extend the warranty time for these items. Sometimes on these warranties, the owner must pay a certain amount of money, called the **deductible**. The manufacturer pays for all repair costs over the deductible amount. Battery warranties are often prorated, which means that the amount of the repair bill covered by the warranty decreases over time. Some of these warranties are held by a third party, such as the manufacturer of tires. Although the manufacturer sold the vehicle with the tires already installed, the warranty of the tires is the responsibility of the tire manufacturer.

The two government-mandated warranties are the Federal Emissions Defect Warranty and the Federal Emissions Performance Warranty. The Federal Emissions Defect Warranty ensures that the vehicle meets all required emissions regulations and that the vehicle's emission control system works as designed and will continue to do for two years or 24,000 miles. Typically covered by this warranty are the following systems:

- Air induction
- Fuel metering
- Ignition
- Exhaust
- Positive crankcase ventilation
- Fuel evaporative control
- Emission control system sensors

The warranty does not cover malfunctions caused by accidents, floods, misuse, modifications, poor maintenance, or the use of leaded fuels. The Federal Emissions Performance Warranty covers catalytic converter and engine control module for a period of 8 years or 80,000 miles. If the owner properly maintains the vehicle and it fails an emissions test approved by the Environmental Protection Agency (EPA), the manufacturer's dealer will repair those emission-related parts covered by the warranty, free of charge. Some states, such as California, require the manufacturers to offer additional or extended warranties.

All warranty information can be found in the vehicle's owner's manual. Whenever there are questions about the warranties, carefully read that section in the owner's manual. If you are working on a vehicle and know the part or system is covered under a warranty, make sure to tell the customer before proceeding with your work. Doing this will save the customer money and you will earn his or her trust.

Increased Vehicle Age Like the price of everything else, the price of a new car has risen sharply. To purchase a new car, many people have taken out loans and have contracted to make car payments for up to 7 years. These long-term loans are the only way many can afford a new vehicle.

Another way people are able to afford a new car is through leasing. When you lease a vehicle, you never really own it. Normally you use the vehicle for 2 to 5 years,

USING THE MAINTENANCE SCHEDULE

GM wants to help keep these vehicles in good working condition. Because of all the different ways people use their GM vehicles, maintenance needs vary. More frequent checks and replacement than you will find in the schedules in this section may be needed. Read this section, and keep in mind the customer's driving habits.

The proper fluids and lubricants to use are listed in Part D. Use the proper fluids and lubricants whenever servicing these vehicles.

The schedules are for vehicles that:

- carry passengers and cargo within recommended limits. Refer to "Vehicle Certification Label" in this section.
- are driven on reasonable road surfaces within legal driving limits.
- are driven off-road in the recommended manner. Refer to the Owner's Manual.
- use the recommended unleaded fuel.

SELECTING THE RIGHT SCHEDULE

Schedule I Definition

Follow Maintenance Schedule I if any one of these are true:

- Most trips are less than 5 to 10 miles (8 to 16 km). This is particularly important when outside temperatures are below freezing.
- Most trips include extensive idling (such as frequent driving in stop-and-go traffic).
- The vehicle is operated in dusty areas or off-road frequently.
- Trailer towing or using a carrier on top of the vehicle frequently.

Schedule I should also be followed if the vehicle is used for delivery service, police, taxi, or other commercial applications.

Schedule II Definition

Follow Schedule II ONLY if none of the conditions from Schedule I are true.

Schedule I Intervals

Every 3,000 Miles (4,800 km) or 3 Months
Engine Oil and Filter Change
Chassis Lubrication
Drive Axle Service

At 6,000 Miles (9,600 km)
Tire Rotation

Every 15,000 Miles (24,000 km)
Air Filter Inspection,
if driving in dusty conditions
Front Wheel Bearing Repack (two-wheel drive only) (or at each brake relining, whichever occurs first).

Every 30,000 Miles (48,000 km)
Air Filter Replacement
Fuel Filter Replacement

Every 50,000 Miles (80,000 km)
Automatic Transmission Service (severe conditions only).

Every 60,000 Miles (96,000 km)
Engine Accessory Drive Belt Inspection
Engine Timing Check
Fuel Tank, Cap, and Lines Inspection

Every 100,000 Miles (160,000 km)
Spark Plug Replacement
Spark Plug Wire Inspection
Positive Crankcase Ventilation (PCV) Valve Inspection

Every 150,000 Miles (240,000 km)
Cooling System Service (or every 60 months, whichever occurs first)

Schedule II Intervals

Every 7,500 Miles (12,000 km)
Engine Oil and Filter Change (or every 12 months, whichever occurs first)
Chassis Lubrication (or every 12 months, whichever occurs first)
Drive Axle Service
Tire Rotation

Every 30,000 Miles (48,000 km)
Fuel Filter Replacement
Air Filter Replacement
Front Wheel Bearing Repack (two-wheel drive only) (or at each brake relining, whichever occurs first)

Every 50,000 Miles (80,000 km)
Automatic Transmission Service (severe conditions only)

Every 60,000 Miles (96,000 km)
Engine Accessory Drive Belt Inspection
Fuel Tank, Cap, and Lines Inspection
Engine Timing Check

Every 100,000 Miles (160,000 km)
Spark Plug Wire Inspection
Spark Plug Replacement
Positive Crankcase Ventilation (PCV) Valve Inspection

Every 150,000 Miles (240,000 km)
Cooling System Service (or every 60 months, whichever occurs first)

Figure 1-5 A typical preventive maintenance schedule.

then give it back to the dealership. During the time you use it, you have a payment and are responsible for the maintenance of the vehicle. At the end of the lease, the car can be bought for its residual value. The **residual value** is simply the vehicle's projected worth at the end of the lease. Leasing is attractive to some because the monthly payments are based on the selling price minus the residual value of the vehicle.

The average age of on-the-road automobiles is 7 years. There are no signs that the trend of keeping cars longer will stop. Older vehicles provide most of the major repair and overhaul work performed in dealerships and independent garages. However, this does not mean that these cars are not complex. Most of the cars on the road have electronic controls, and servicing them requires training and specialized equipment.

Career Opportunities

Automotive service technicians can enjoy careers in many different types of automotive businesses. Because of the skills required to be a qualified technician, there are also career opportunities for those who do not want to repair automobiles the rest of their lives. There are also many opportunities for good technicians who want to change careers. The knowledge required to be a good service technician can open many doors of opportunity.

Dealerships New car dealerships **(Figure 1–6)** serve as the link between the vehicle manufacturer and the customer. They are privately owned businesses. Most dealerships are franchised operations, which means the owners have signed a contract with particular auto manufacturers and have agreed to sell and service their vehicles.

The manufacturer usually sets the sales and service policies of the dealership. Most warranty repair work is done at the dealership. The manufacturer then pays the dealership for making the repair. The manufacturer also provides the service department at the dealership with the training, special tools, equipment, and information needed to repair its vehicles. The manufacturers also help the dealerships get service business. Often, their commercials stress the importance of using their replacement parts and promote their technicians as the most qualified to work on their products.

Working for a new car dealership can have many advantages. Technical support, equipment, and the opportunity for ongoing training are usually excellent. At a dealership, you have a chance to become very skillful in working on the vehicles you service. However, working on one or two types of vehicles does not appeal to everyone. Some technicians want diversity.

Independent Service Shops **Independent shops (Figure 1–7)** may service all types of vehicles or may specialize in particular types of cars and trucks or specific systems of a car. Independent shops outnumber dealerships by six to one. As the name states, an independent service shop is not associated with any particular automobile manufacturer. Many independent shops are started by technicians eager to be their own boss and run their own business.

Figure 1–6 Dealerships sell and service vehicles made by specific auto manufacturers.

Figure 1–7 Full-service gasoline stations are not as common as they used to be, but they are a good example of an independent service shop.

An independent shop may range in size from a two-bay garage with two to four technicians, to a multiple-bay service center with twenty to thirty technicians. A **bay** is simply a work area for a complete vehicle. The amount of equipment in an independent shop varies; however, most are well equipped to do the work they do best. Working in an independent shop may help you develop into a well-rounded technician.

Specialty shops specialize in areas such as engine rebuilding, transmission/transaxle overhauling, and air conditioning, brake, exhaust, cooling, emissions, and electrical work. A popular type of specialty shop is the "quick lube" shop, which takes care of the preventive maintenance of vehicles. It hires lubrication specialists who change fluids, belts, and hoses, in addition to checking certain safety items on the vehicle.

The number of specialty shops that service and repair only one or two systems of the automobile has steadily increased over the past 10 to 20 years. Technicians employed by these shops have the opportunity to become very skillful in one particular area of service.

Franchise Repair Shop A great number of jobs are available at service shops that are run by large companies, such as Firestone, Goodyear, Midas, and Procare. These shops do not normally service and repair all of the systems of the automobile. However, their customers do come in with a variety of service needs. Technicians employed by these shops have the opportunity to become very proficient in many areas of service and repair.

Some independent shops may look like they are part of a franchise but are actually independent. Good exam-

Figure 1-8 NAPA service centers are good examples of independent repair shops that have affiliated with a large business. In these arrangements, the shops are still run independently.

ples of this type of shop are the NAPA service centers **(Figure 1–8)**. These centers are not controlled by NAPA, nor are they franchises of NAPA. They are called NAPA service centers because the facility has met NAPA's standards of quality and the owner has agreed to use NAPA as his primary source of parts and equipment.

Store-Associated Shops Other major employers of auto technicians are the service departments of department stores. Many large stores that sell automotive parts often offer certain types of automotive services, such as brake, exhaust system, and wheel and tire work.

Fleet Service and Maintenance Any company that relies on several vehicles to do its business faces an ongoing vehicle service and preventive maintenance problem. Small fleets often send their vehicles to an independent shop for maintenance and repair. Large fleets, however, usually have their own preventive maintenance and repair facilities and technicians.

Utility companies (such as electric, telephone, or cable TV), car rental companies, overnight delivery services, and taxicab companies are good examples of businesses that usually have their own service departments. These companies normally purchase their vehicles from one manufacturer. Technicians who work on these fleets have the same opportunities and benefits as technicians in a dealership. In fact, the technicians of some large fleets are authorized to do warranty work for the manufacturer. Many good career opportunities are available in this segment of the auto service industry.

JOB CLASSIFICATIONS

The automotive industry offers numerous types of employment for people with a good understanding of automotive systems.

Service Technician

A **service technician** assesses vehicle problems, performs all necessary diagnostic tests, and competently repairs or replaces faulty components. The skills to do this job are based on a sound understanding of auto technology, on-the-job experience, and continuous training in new technology as it is introduced by auto manufacturers.

Individuals skilled in automotive service are called technicians, not mechanics. There is a good reason for this. *Mechanic* stresses the ability to repair and service mechanical systems. While this skill is still very much needed, it is only part of the technician's overall job. Today's vehicles require mechanical knowledge plus an understanding of other technologies, such as electronics, hydraulics, and pneumatics.

A technician may work on all systems of the car or may become specialized. Specialty technicians concentrate on servicing one system of the automobile, such as electrical, brakes **(Figure 1–9)**, or transmission. These specialties require advanced and continuous training in that particular field.

Shop Foreman

The **shop foreman** is the one who helps technicians with more difficult tasks and serves as the quality control expert. In some shops, this is the role of the **lead tech**. For the most part, both jobs are the same. Some shops have technician teams. On these teams, there are several technicians, each with a different level of expertise. The lead tech is sort of the shop foreman of the team. Lead techs and shop foremen have a good deal of experience and excellent diagnostic skills.

Figure 1-9 Specialty technicians work on only one vehicle system, such as brakes.

Service Advisor

The person who greets customers at a service center is the **service advisor**, sometimes called a service writer or consultant. Service advisors need to have an understanding of all major systems of an automobile and be able to identify all major components and their locations. They also must be able to describe the function of each of those components and be able to identify related components. A good understanding of the recommended service and maintenance intervals and procedures is also required. With this knowledge they are able to explain the importance and complexity of each service and are able to recommend other services.

A thorough understanding of warranty policies and procedures is also a must. Service advisors must be able to explain and verify the applicability of warranties, service contracts, service bulletins, and campaign/recalls procedures.

Service advisors also serve as the liaison between the customer and the technician in most dealerships. They have responsibility for explaining the customer's concerns and/or requests to the technician plus keeping track of the progress made by the technician so the customer can be informed. This monitoring is also important because it impacts the completion of service on the vehicles of other customers.

Often automotive technicians or students of automotive service programs realize a need to change career choices but desire to stay in the service industry. Becoming a service writer, advisor, or consultant is a good alternative. This job is good for those who have the technical knowledge but lack the desire or physical abilities to physically work on automobiles.

Many of the requirements for being a successful technician apply to being a successful service consultant. However, being a service consultant requires greater skill levels in customer relations, internal communication and relations, and sales. Service consultants must communicate well with customers, over the telephone or in person, in order to satisfy their needs or concerns. Most often this satisfaction involves the completion of a repair order, which contains customer information, instructions to the technicians, and a cost estimate.

Accurate estimates are not only highly appreciated by the customer, but they are also required by law in most states. Writing an accurate estimate requires a solid understanding of the automobile, good communications with the customers and technicians, and good reading and math skills.

Most shops use computers to generate the repair orders and estimates and to schedule the shop's workload. Therefore, having solid computer skills is an asset for service advisors.

Service Manager

The service manager is responsible for the operation of the entire service department at a large dealership or independent shop. Normally, customer concerns and complaints are handled by the service manager. Therefore, a good service manager has good people skills in addition to organizational skills and a solid automotive background.

In a dealership, the service manager makes sure the manufacturers' policies on warranties, service procedures, and customer relations are carried out. The service manager also arranges for technician training and keeps all other shop personnel informed and working together.

Parts Counterperson

A parts counterperson **(Figure 1–10)** can have several different duties and is commonly called a parts person or specialist. Parts specialilsts are found in nearly all automotive dealerships and auto parts retail and wholesale stores. They sell auto parts directly to customers and issue materials and supplies to auto repair specialists working in automotive services facilities and body shops. A parts counterperson must be friendly, professional, and efficient when working with all customers, both on the phone and in person.

Depending on the parts store or department, duties may also include delivery of parts, purchasing a variety of automotive parts, maintaining inventory levels, and issuing parts to customers and technicians. Responsibilities include preparing purchase orders, scheduling deliveries, assisting in the receipt and storage of parts and supplies, and maintaining contact with vendors. An understanding of automotive terminology and systems is a must for good parts counterpersons.

This career is an excellent alternative for those who know about cars but would rather not work on them. Much of the knowledge required to be a technician is also required for a parts person. However, a parts specialist

Figure 1–10 A parts counterperson has an important role in the operation of a store or dealership.

requires a different set of skills. Most automotive parts specialists acquire the sales and customer service skills needed to be successful primarily thorugh on-the-job experience and training. They may also gain the necessary technical knowledge on the job or through educational programs and/or experience. To better understand the world of the parts industry refer to **Figure 1–11**, which defines the common terms used by parts personnel.

ACCOUNTS RECEIVABLE Money due from a customer.

ALPHANUMERIC A numbering system commonly used in parts catalogs and price listings. This system uses a combination of letters and numbers. They are placed in order starting from the left digit and working across to the right.

BACK ORDER Parts ordered from a supplier that have not been shipped to the store or shop because supplier has none in its inventory.

BILL OF LADING A shipping document acknowledging receipt of goods and stating terms of delivery.

CATALOGING The process of looking up the needed parts in a parts catalog.

CORE CHARGE A charge that is added when a customer buys a remanufactured part. Core charges are refunded to the customer when he or she returns a rebuildable part.

CORRECTION BULLETIN A bulletin that corrects catalog errors due to printing errors or inaccurately assigned part numbers.

CUSTOMER RELATIONS A description of how a salesperson interacts with the customer.

DEALERS The jobber's wholesale customers, such as service stations, garages, and vehicle dealers, who install parts in their customers' vehicles.

DISCOUNT The amount of savings being offered to a customer, normally expressed as a percentage.

DISTRIBUTOR A large-volume parts-stocking business that sells to wholesalers.

FREIGHT CHARGE A charge added to special order parts to cover their transportation to the store.

GROSS PROFIT The selling price of a part minus its cost.

HIGH-VOLUME Describes a popular item, which is sold in large numbers.

INDIVIDUALLY PRICED The condition of having each part of a display priced for customer convenience.

INVENTORY The parts a store or shop has in its possession for resale.

INVENTORY CONTROL A method of determining amounts of merchandise to order based on supplies on hand and past sales of the item.

INVOICE The record of a sale to a customer.

JOBBER The owner or operator of an auto parts store usually wholesaling products to volume purchasers such as dealers, fleet owners, and businesses. They also may sell retail to do-it-yourselfers.

LIST PRICE The suggested selling price for an item.

MARGIN Same as gross profit.

MARKUP The amount a business charges for a part above the actual cost of the part.

NET PRICE A business's profit after deducting the cost of all of its merchandise and all expenses involved in operating the business.

NO-RETURN POLICY A store policy that certain parts cannot be returned after purchase. It is common to have a no-return policy on electrical and electronic parts.

ON HAND The quantity of an item that the store or shop has in its possession.

PERPETUAL INVENTORY A method of keeping a continuous record of stock on hand through sales receipts and/or invoices.

PHYSICAL INVENTORY The process whereby each part is manually counted and the number on hand is written on a form or entered into a computer.

PROFIT The amount received for goods or services above the shop's or store's cost for the part or service.

PURCHASE ORDER A form giving someone the authority to purchase goods or services for a company.

REMANUFACTURED PART A part that has been reconditioned to its original specifications and standards.

RESTOCKING FEE The fee charged by a store or supplier for having to handle a returned part.

RETAIL Selling merchandise to walk-in trade (do-it-yourselfers).

RETURN POLICY A policy regarding the return of unwanted and unneeded parts. Return policies may include restocking fees or prohibit the return of certain parts.

SELLING PRICE The price at which a part is sold. This price will vary according to the type of customer (retail or wholesale) that is purchasing the part.

SPECIAL ORDER An order placed whenever a customer purchases an item not normally kept in stock.

STOCK ORDER A process by which the store orders more stock from its suppliers in order to maintain its inventory.

STOCK ROTATION Selling the older stock on hand before selling the newer stock.

SUPERSESSION BULLETIN A bulletin sent by the parts supplier that lists part numbers that now replace (supersede) previous part numbers.

TURNOVER The number of times each year that a business buys, sells, and replaces a part.

VENDOR The supplier.

WARRANTY RETURN A defective part returned to the supplier due to failure during its warranty period.

WAREHOUSE DISTRIBUTOR The jobber's supplier who is the link between the manufacturer and the jobber.

WHOLESALE The business's price to large-volume customers.

Figure 1-11 Some of the common terms used by parts personnel.

Parts Manager

The parts manager is in charge of ordering all replacement parts for the repairs the shop performs. The ordering and timely delivery of parts is extremely important for the smooth operation of the shop. Delays in obtaining parts or omitting a small but crucial part from the initial parts order can cause frustrating holdups for both the service technicians and customers.

Most dealerships and large independent shops keep an inventory of commonly used parts, such as filters, belts, hoses, and gaskets. The parts manager is responsible for maintaining this inventory.

An understanding of automotive systems and their parts, thoroughness, attention to detail, and the ability to work with people face-to-face and over the phone are essential for a parts manager.

RELATED CAREER OPPORTUNITIES

In addition to careers in automotive service, there are many other job opportunities directly related to the automotive industry.

Parts Distribution

The **aftermarket** refers to the network of businesses **(Figure 1–12)** that supplies replacement parts to independent service shops, car and truck dealerships, fleet operations, and the general public.

Vehicle manufacturers and independent parts manufacturers sell and supply parts to approximately a thousand warehouse distributors throughout the United States. These **warehouse distributors (WDs)** carry substantial inventories of many part lines.

Warehouse distributors serve as large distribution centers. WDs sell and supply parts to parts wholesalers, commonly known as jobbers.

Jobbers sell parts and supplies to shops and do-it-yourselfers. Jobbers often have a delivery service that gets the desired parts to a shop shortly after they ordered them. Some parts stores focus on individual or walk-in customers. These businesses offer the do-it-yourselfers repair advice, and some even offer testing of old components. Selling good parts at a reasonable price and offering extra services to their customers are the characteristics of successful parts stores. Many jobbers operate machine shops that offer another source of employment for skilled technicians. Jobbers or parts stores can be independently owned and operated. They can also be part of a larger national chain **(Figure 1–13)**. Auto manufacturers have also set up their own parts distribution systems to their dealerships and authorized service outlets. Parts manufactured by the original vehicle manufacturer are called **original equipment manufacturer (OEM)** parts.

Opportunities for employment exist at all levels in the parts distribution network, from warehouse distributors to the counterpeople at local jobber outlets.

Marketing and Sales

Companies that manufacture equipment and parts for the service industry are constantly searching for knowledgeable people to represent and sell their products. For example, a sales representative working for an aftermarket parts manufacturer should have a good knowledge of the company's products. The sales representative also works with WDs, jobbers, and service shops to make sure the parts are being sold and installed correctly. They also help coordinate training and supply information so that everyone using their products is properly trained and informed.

Other Opportunities

Other career possibilities for those trained in automotive service include automobile and truck recyclers, insurance

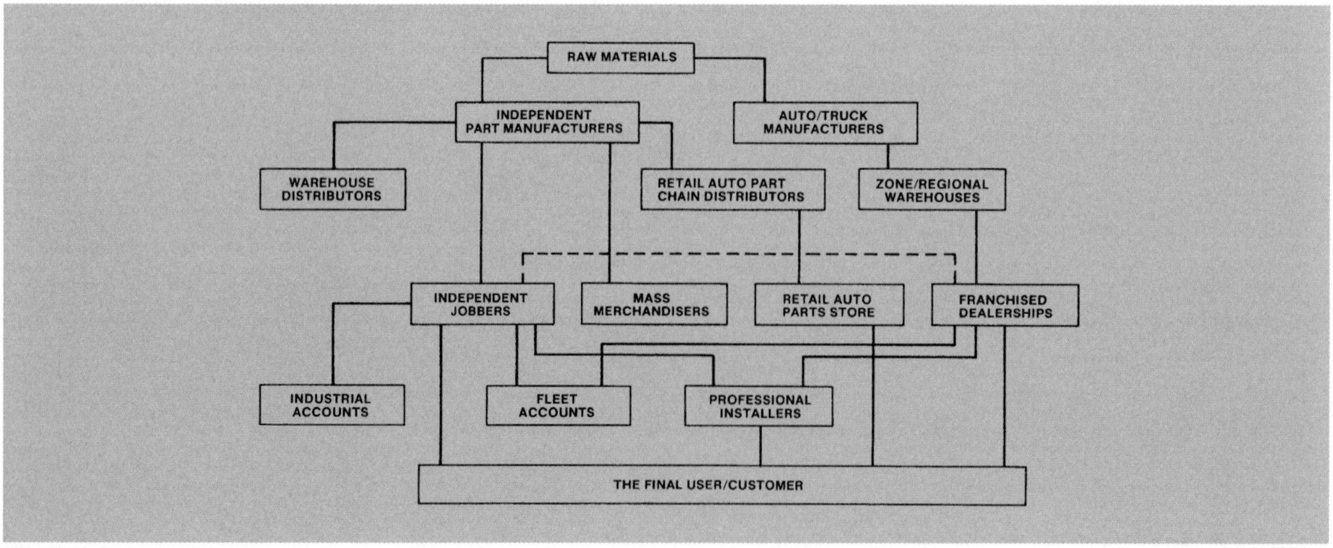

Figure 1–12 The auto parts supply network.

Figure 1-13 Many parts stores are part of a national corporation with stores located across the country.

company claims adjusters, auto body shop technicians, and trainers for the various manufacturers or instructors for an automotive program. The latter two careers require solid experience and a thorough understanding of the automobile. It is not easy being an instructor or trainer; however, passing on knowledge can be very rewarding. Undoubtedly, there is no other career that can have as much impact on the automotive service industry as that of a trainer or instructor.

TRAINING FOR A CAREER IN AUTOMOTIVE SERVICE

Those interested in a career in auto service can receive training in formal school settings—secondary, post-secondary, and vocational schools, and technical or community colleges, both private and public.

Student Work Experience

There are many ways to gain work experience while you are a student. You may already be involved in one of the following; if not, consider becoming involved in one of these possibilities.

Job Shadowing Program In this program you follow an experienced technician or service writer. The primary program objective is to expose you to the "real world," to see what it takes to be a successful technician or service writer. By job shadowing, you will also become familiar with the total operation of a service department.

Mentoring Program This program has the lowest participation rate of all these programs but can be one of the most valuable. In a mentoring program, you have someone who is successful to use as an expert. Your mentor has agreed to stay in contact with you, to answer questions, and to encourage you. When you have a good mentor, you have someone who may be able to explain things a little differently than the way things were explained in class. A mentor may also be able to give real life examples of why some of the things you need to learn are important.

Cooperative Education This type of program is typically 2 years in length. One year is spent at school and the other in a dealership or service facility. This does not mean 1 solid year is spent at school; rather you spend 8 to 12 weeks at school, and then work for 8 to 12 weeks. The switching back and forth continues for 2 years. Not only do you earn an hourly wage while you are working, you also earn credit toward your degree or diploma. While at work, you get a chance to practice and perfect what you learned in school. Your experiences at work are carefully coordinated with your experiences at school; therefore, it is called a cooperative program—industry cooperates with education. Examples of this type of program are the Chrysler CAPS, Ford ASSET, GM ASEP, and Toyota T-Ten (in Canada these are called T-TEP) programs.

Apprenticeship Program Similar to a cooperative education program, an apprenticeship program combines work experiences with education. The primary difference between the two programs is that in an apprenticeship program students attend classes in the evening after completing a day's work. During this rigorous training program, you receive a decent hourly wage and plenty of good experience. You start the program as a helper to an experienced technician and can begin to do more on your own as you progress through the program. Most apprenticeship programs take 2 years to complete. Automobile manufacturers and dealers often sponser these programs.

Part-Time Employment The success of this experience depends on you and your drive to learn. Working part-time will bring you good experience, some income, and a good start in getting a great full-time position after you have completed school. The best way to approach this is to find a position and service facility that will allow you to grow. You need to start at a right level and be able to take on more difficult tasks when you are ready. The most difficult challenge when working part-time is to keep up with your education while you are working. Many times work may get in the way, but if you truly want to learn, you will find a way to fit your educational needs around your work schedule.

Postgraduate Education A few manufacturer programs are designed for graduates of postsecondary schools. These programs train individuals to work on particular vehicles. For example, BMW's Service Technician Education Program (STEP) is a scholarship program for the top graduates of automotive postsecondary schools. Students in the program apply what they learned in their 2-year program and learn to diagnose and service BMW products. BMW says this program is the most respected and intense training program of its kind in the world. For more information go to http://www.bmwstep.com.

Canada's Automotive Apprenticeship Program

Canada has an extremely well-defined apprenticeship program for nearly fifty different trades, one of which is automotive service. The programs are based on the tasks performed by journeymen in the trade. The apprenticeship programs are 2 to 5 years in length with 85% of the time working in the occupation and the rest of the time spent at a school. Since most of Canada and other countries require that automotive technicians be certified or licensed in the area in which they are working, participating in an apprenticeship program is one way individuals can get a start in an automotive-related career. This certification is unique to Canada and does not include ASE certification.

Canada's Red Seal Program **(Figure 1–14)** was designed to make it easier for skilled workers in many different trades to move across the country and obtain jobs. Through the program, apprentices who have completed their training and become certified journeypersons are able to obtain a Red Seal endorsement on the Certificate of Qualification and Apprenticeship by successfully completing an Interprovincial Standards Examination. The red seal allows qualified automotive technicians to work in the field in any province or territory in Canada without needing to take further examinations. The red seal is a mark of high achievement and, in most cases, is not a requirement for employment.

The Need for Continuous Learning

Training in automotive technology and service does not end with graduation. Nor does the *need to read* end. A professional technician constantly learns and keeps up to date. In order to maintain your image as a professional and to keep your knowledge and skills up to date, you need to do what you can to learn new things. You need to commit yourself to lifelong learning. There are many ways you can keep up with the changing times and technology. Short courses on specific systems or changes are often available from manufacturers, training groups, associations, and schools. It is wise to participate in these classes as soon as you can. If you wait too long, other changes will have occurred and you may have a difficult time catching up.

Besides attending classes, you can also be updated by reading automotive magazines or the latest editions of automotive textbooks. As soon as you realize what subjects you need refreshing or updating on, respond as soon as you can. A good technician takes advantage of every opportunity to learn.

STANDARD
CANADA
We support the red seal program of Canada

Figure 1–14 Canada's Red Seal is a mark of high achievement.

ASE CERTIFICATION

The National Institute for **Automotive Service Excellence (ASE)** has established a voluntary certification program for automotive, heavy-duty truck, auto body repair, and engine machine shop technicians. In addition to these programs, ASE also offers individual testing in the areas of automotive and heavy-duty truck parts, service consultant, alternate fuels, advanced engine performance, and a variety of other areas. This certification system combines voluntary testing with on-the-job experience to confirm that technicians have the skills needed to work on today's more complex vehicles. ASE recognizes two distinct levels of service capability—the automotive technician and the master automotive technician. The master automotive technician is certified by ASE in all major automotive systems. The automotive technician may have certification in only several areas.

To become ASE certified, a technician must pass one or more tests that stress system diagnosis and repair procedures. The eight basic certification areas in automotive repair follow:

1. Engine repair
2. Automatic transmission/transaxle
3. Manual transmissions and drive axles
4. Suspension and steering
5. Brakes
6. Electrical systems
7. Heating and air conditioning
8. Engine performance (driveability)

After passing at least one exam and providing proof of 2 years of hands-on work experience, the technician becomes ASE-certified. Retesting is necessary every 5 years to remain certified. A technician who passes one examination receives an automotive technician shoulder patch. The master automotive technician patch is

Figure 1-15 ASE certification shoulder patches worn by (*left*) automotive technicians and (*right*) master automotive technicians.

awarded to technicians who pass all eight of the basic automotive certification exams **(Figure 1–15)**.

ASE also offers advanced-level certification in some areas. The most commonly sought advanced certification for automobile technicians is the L1 or Advanced Engine Performance. Individuals seeking this certification must be certified in Electricity and Engine Performance before taking this exam.

ASE also offers specialist certifications. For example, to become a certified Undercar Specialist, you must have certification in Suspension and Steering, Brake, and Exhaust Systems (a speciality test). Certification is also available for Parts Counterperson and Service Consultants.

As mentioned, ASE certification requires that you have 2 years of full-time, hands-on working experience as an automotive technician. You may receive credit toward this 2-year experience requirement by completing formal training in one or a combination of the following:

- High school training
- Post–high school training
- Short courses
- Apprenticeship programs

Each certification test consists of forty to eighty multiple-choice questions. The questions are written by a panel of technical service experts, including domestic and import vehicle manufacturers, repair and test equipment and parts manufacturers, working automotive technicians, and automotive instructors. All questions are pretested and quality-checked on a national sample of technicians before they are included in the actual test. Many test questions force the student to choose between two distinct repair or diagnostic methods.

For further information on the ASE certification program, go to their Web site at http://www.ase.com.

KEY TERMS

Aftermarket
Automotive Service Excellence (ASE)
Bay
Deductible
Diagnostic skills
Independent shops
Inspection/Maintenance (I/M)
Jobbers
Lead tech
Original Equipment Manufacturer (OEM)
Preventive maintenance (PM)
Residual value
Service advisor
Service technician
Shop foreman
Warehouse distributors (WDs)
Warranty

SUMMARY

- The modern auto industry is a global industry involving vehicle and parts manufacturers from many countries.
- Electronic computer controls are found on many auto systems, such as engines, ignition systems, transmissions, steering systems, and suspensions. The use of electronics in automobiles is increasing rapidly.
- The increasing complexity of vehicles, the increasing age of vehicles on the road, and the need to comply with federal laws concerning emission control and mileage are three reasons the need for quality service technicians is increasing.
- Preventive maintenance is extremely important in keeping today's vehicles in good working order.
- New car dealerships, independent service shops, specialty service shops, fleet operators, and many other businesses are in great need of qualified service technicians.
- A solid background in auto technology may be the basis for many other types of careers within the industry. Some examples are parts management, collision damage appraisal, sales, and marketing positions.
- Training in auto technology is available from many types of secondary, vocational, and technical schools. Auto manufacturers also have cooperative programs with schools to ensure that graduates understand modern systems and the equipment to service them.
- The National Institute for Automotive Service Excellence (ASE) actively promotes professionalism within the industry. Its voluntary certification program for automotive technicians and master auto technicians helps guarantee a high level of quality service.
- The ASE certification process involves both written tests and credit for on-the-job experience. Testing is available in many areas of auto technology.

REVIEW QUESTIONS

1. Give a brief explanation of why electronics are so widely used on today's vehicles.
2. Explain the basic requirements for becoming a successful automotive technician.
3. List at least five different types of businesses that hire service technicians. Describe the types of work these businesses handle and the advantages and disadvantages of working for them.
4. Name four ways that you can gain work experience while you are a student.
5. *True or False?* An apprentice has a prescribed set of tasks he or she should complete during on-the-job experiences.
6. Repair work performed on vehicles still under the manufacturer's warranty is usually performed by ______.
   a. independent service shops
   b. dealerships
   c. specialty shops
   d. either a or b
7. Which of the following businesses perform work on only one or two automotive systems?
   a. dealerships
   b. independent service shops
   c. specialty shops
   d. fleet service departments
8. Normally whose job is it to prepare a repair cost estimate for a customer?
   a. service manager
   b. parts manager
   c. master automotive technician
   d. service advisor
9. Technician A says that after an individual passes a particular ASE certification exam, he or she is certified in that test area. Technician B says that all of the questions on an ASE exam force the test taker to choose between two distinct repair methods. Who is correct?
   a. Technician A
   b. Technician B
   c. Both A and B
   d. Neither A nor B
10. To be successful, today's automotive technician must have:
    a. an understanding of electronics.
    b. the ability to repair and service mechanical systems.
    c. the dedication to always be learning something new.
    d. all of the above.
11. A technician must have a minimum of ______ year(s) of hands-on work experience to get ASE certification.
    a. 1
    b. 2
    c. 3
    d. 4
12. A technician who passes all eight basic ASE automotive certification tests is certified as a(n) ______.
    a. automotive technician
    b. master automotive technician
    c. service manager
    d. parts manager
13. Technician A says battery warranties are often prorated. Technician B says prorated warranties have a deductible. Who is correct?
    a. Technician A
    b. Technician B
    c. Both A and B
    d. Neither A nor B
14. Wholesale auto parts stores that sell aftermarket parts and supplies to service shops and the general public are called ______.
    a. warehouse distributors
    b. mass merchandisers
    c. jobbers
    d. free-lancers
15. Ongoing technical training and support is available from ______.
    a. aftermarket parts manufacturers
    b. auto manufacturers
    c. jobbers
    d. all of the above

2

WORKPLACE SKILLS

OBJECTIVES

■ Develop a personal employment plan. ■ Seek and apply for employment. ■ Prepare a resume and cover letter. ■ Prepare for an employment interview. ■ Accept employment. ■ Understand how automotive technicians are compensated. ■ Understand the proper relationship between an employer and an employee. ■ Explain the key elements of on-the-job communications. ■ Be able to use critical thinking and problem-solving skills. ■ Explain how you should look and act to be regarded as a professional. ■ Explain how fellow workers and customers should be treated. ■ Describe the information that should be included on a repair order. ■ Explain how repair costs can be estimated.

This chapter gives an overview of what you should do to get a job and how to keep it. The basis for this discussion is respect; respect for yourself, your employer, fellow employees, your customers, and everyone else. Also included in this discussion are the key personal characteristics required of all seeking to be successful automotive technicians and employees.

SEEKING AND APPLYING FOR EMPLOYMENT

Becoming employed, especially in the field in which you want a career, involves many steps. As with many things in life, you must be adequately prepared before taking the next step toward employment. This discussion suggests ways you can prepare and what to expect while taking these steps.

Employment Plan

An **employment plan** is nothing more than an honest appraisal of yourself and your career hopes. It includes your specific job goals, and when you expect to reach them, and how your attitudes, interests, aptitudes, and skills match the requirements of your target job. An employment plan should contain your short-term goals (4 to 6 months) and long-term goals, as well as a prioritized list of potential employers or types of employers. You may need to present your employment plan to someone while you are seeking employment, so make sure it is complete. Even if no one else will see it, you should be thorough because it will help you make good career choices and will serve as the basis for your personal marketing tools.

Think about the type of job you want and do some research to determine the requirements for that job. Evaluate yourself against those requirements and determine if you are ready for the job. Also consider the conditions in which you would work with that type of job. Ask yourself if you are willing and able to be a productive worker in those conditions. If not, find a job that is similar to your desire and pursue that type of job.

To begin the self-appraisal part of your employment plan, think about what your interests and skills are. Ask yourself:

- Why am I looking for a job?
- What specifically do I hope to gain by having a job?
- What do I like to do?
- What am I good at?
- Which of my skills would I like to use in my job?

By honestly answering these questions, you should be able to identify the jobs that will help you meet your goals. If you are just seeking a job to pay bills or buy a car and have no intention of turning this job into a career, be honest with yourself and your potential employer. If you are hoping to begin a successful career, realize you will probably start at the bottom of the ladder to success. You must also realize that how quickly you climb the ladder is your responsibility. An employer's responsibility is merely to give you a fair chance to climb it.

Honestly evaluate yourself, and your life, to determine what skills you have. Even if you have never had a job, you still have skills and talents that can be offered to an employer. Think about your life and make a list of all of the things you have learned from your school, friends, family, and through television, volunteering, books, hobbies, and so on. You may be surprised by the number of skills you really have. Identify these skills as being either technical or personal skills.

Technical skills include things you can do well and enjoy, such as:

- Use a computer
- Work with tools, machines, or equipment
- Play video games
- Do math problems
- Maintain or fix things
- Figure out how things work
- Make things with your hands
- Work with ideas and information
- Solve puzzles or problems
- Study or read
- Do experiments or research a topic
- Express yourself through writing

Personal skills are also called **soft skills** and are things that are part of your personality. These are things you are good at or enjoy doing, such as:

- Working with people
- Caring for or help people
- Working as a member of a team and independently
- Leading or supervising others
- Following orders or instructions
- Persuading people
- Negotiating with others

By identifying these skills you will have created your personal skills inventory. From the inventory you match your skills and personal characteristics to the needs and desires of potential employers. The inventory will also come in handy when marketing yourself for a job, such as when preparing your resume and cover letter and during an interview.

References

A **reference** is someone who will be glad to tell a potential employer about you and your work habits. A reference can be anyone who knows you other than a family member or close friend. Employers contact references to verify or complete their picture of who you are. Make a list of three to five people, with their contact information, that you can use as references. If you do not present references to a potential employer, he or she may assume that you cannot find anyone who has anything nice to say about you. You probably will not be considered for the job.

Choose your references wisely. Teachers (past and present), coaches, and administrators of your school are good examples of who you can ask to be a reference. People you have worked for or have helped are also good references. Try also to get someone whose opinion is respected, such as a priest, minister, or elder in your church or someone you know well that is in a high position. Always talk to your references first, if possible, and get permission to give their names and telephone numbers to an employer. If they do not seem comfortable with giving you a reference, take the hint and move on to someone else. If someone is willing to provide you with a written reference, make several copies of the recommendation so you can attach them to your resume and/or the applications you fill out.

Application forms often have a section for personal references. Make sure you have your list with you when applying for a job. Your list of references does not need to be included in your resume; merely state that references are available on request. Make sure you give copies of your resume to your references.

Identifying Job Possibilities

One of the things you identified in your employment plan was your preferred place to work. This may have been a specific business but probably was a type of business, such as a new car dealership or independent shop. Now your task is to identify the companies in that type of business that are looking for someone. To do this, you can look through the help-wanted section in the newspaper **(Figure 2–1)**. You can also check your school's job post-

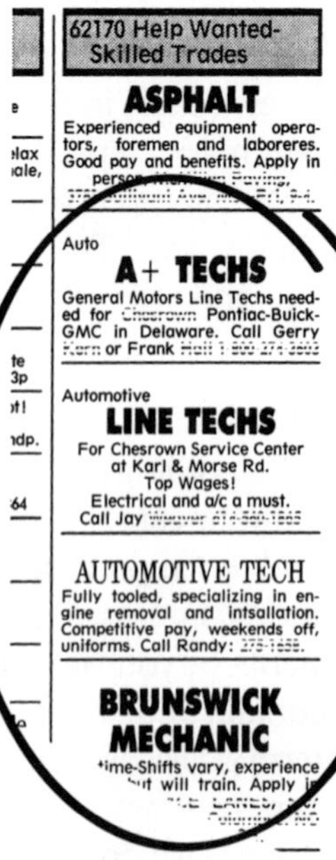

Figure 2-1 Check the help-wanted ads in your local newspaper for businesses that are looking for technicians.

ing board or ask people you know that already work in the business. If there is nothing available in the business you prefer, look for openings in the type of business that was second on your priority list.

Carefully look at the description of the job. Make sure you meet the qualifications for the job before you apply. For example, if you have a drug problem and the ad states that all applicants will be drug tested, you should not bother applying and should concentrate on breaking the habit. Even if the ad says nothing about testing for drug use, you should know that there is no place for drugs at work and continued drug use will only jeopardize your career.

Preparing Your Resume

Your **resume** and cover letter are your own personal marketing tools and may be the first look at you an employer has. Although not all employers require a resume, you should prepare one for those that do. Preparing a resume also forces you to look at your qualifications for a job. That alone justifies having a resume.

A resume must be neatly typewritten. If you do not have access to a computer or a typewriter, your local library probably has them available for public use.

Keep in mind that although you may spend hours writing and refining your resume, an employer may only take a minute or two from his or her busy schedule to look it over. With this in mind, put together a resume that tells the employer who you are in such a way that he or she wants to interview you.

A resume normally includes your contact information, career objective, skills and/or accomplishments, work experience, education, and a statement about references. There are different formats you can follow when designing your resume. If you have limited work experience, make sure the resume emphasizes your skills and accomplishments rather than work history. Even if you have no work experience, you can sell yourself by highlighting some of the skills and attributes you identified in your employment plan.

When listing or mentioning your attributes and skills, express them in a way that shows how they relate to the job you are seeking. For instance, if you practice every day at your favorite sport so you can make the team, you may want to describe yourself as being persistent, determined, motivated, and goal-oriented. Another example is if you have ever pulled an all-nighter to get an assignment done on time, it can mean that you work well under pressure and always get the job done. Another example would be if you keep your promises and do what you said you would do, you may want to describe yourself as reliable, a person who takes commitment seriously.

Identifying your skills may be a difficult task, so have your family and/or friends help you. Keep in mind that you have qualities and skills that employers want. You need to recognize them, put them in a resume, and tell them to your potential employer. Do not put the responsibility of figuring out who you are on the employers—tell them.

Figure 2–2 is an example of a basic resume for an individual seeking an entry-level position as a technician. Here are some guidelines to follow when you are designing your resume:

- Make sure it is neat, uncluttered, and easy to read
- Use quality white paper.
- Keep it short—a maximum of one or two pages.
- Use dynamic words to describe your skills and experience, such as accomplished, achieved, communicated, completed, created, delivered, designed, developed, directed, established, founded, instructed, managed, operated, organized, participated, prepared, produced, provided, repaired, and supervised.
- Chose your words carefully; remember that the resume is a look at you.
- Make sure all information is accurate.
- Make sure the information you think is the most important stands out and is positioned near the top of the page.
- Design your resume with a clean letter type (font) and wide margins (1-½ inches on both sides is good) so that it is easy on the eyes.
- Only list the "odd" jobs you had if they are related to the job you are applying for.
- Do not repeat information.
- Proofread the entire resume to catch spelling and grammatical errors. If you find them, fix them and print a new, clean copy.
- Do not make hand-written corrections or use correction fluid to cover mistakes.
- Make sure your resume is not dirty and wrinkled when you deliver it.

Preparing Your Cover Letter

A cover letter **(Figure 2–3)** should be presented with every resume you mail, e-mail, fax, or personally deliver. A cover letter gives you a chance to point out exactly why you are perfect for the job. Make sure you address the letter to a person, not just a title. If you do not know the person's name, call the company and ask for the correct spelling of the person's name and his or her title.

Make sure the words you choose for the letter are upbeat. Use a natural writing style, keeping it professional but friendly. Try hard not to start every sentence with "I"; make some "you" statements. Make sure you check the

Jack Erjavec
1234 My Street
Somewhere, OZ 99902
123-456-7890

Performance-oriented student, with an excellent reputation as a responsible and hard-working achiever, seeking a position as an entry-level automotive technician in a new car dealership.

Skills and Attributes

- People oriented
- Motivated
- Committed
- Strong communication and teamwork skills
- Honest
- Reliable
- Organized
- Methodical
- Creative problem-solver
- Good hand skills

Work Experience

2002–2004 Somewhere Soccer Association (Assistant coach)

- Instructed and supervised junior team
- Performed administrative tasks as the Coach required

2000–2002 Carried out various odd jobs within the community

- Washing and waxing cars, picking up children from school, raking leaves, cutting grass

Education

Somewhere Senior High School, graduated in 2003
Somewhere Community College, currently enrolled in the Automotive Technology Program

Extracurricular Activities

1999–2003 Active member of the video game club
1999–2003 Member of the varsity soccer team

Hobbies and Activities

Reading auto-related magazines, going to races, doing puzzles, working on cars with family and friends.

References

Available upon request.

Figure 2-2 A sample of a resume for someone who has little work experience.

letter for spelling and grammatical errors, and make sure that it is on quality paper and is neat and clean.

Typically a cover letter has three paragraphs, each with their purpose. In the first one, you tell the employer you are interested in working for that company, the position you are interested in, and why. This paragraph also includes how you found out about the open position, which could be a reference to a help-wanted ad, a job posting at school, and/or a referral by someone who works for the company. Make sure this paragraph shows that you know about the company and what the job involves.

In the second paragraph, sell yourself by addressing one or two of your qualifications for the job and describing them in more detail than you did in your resume. Make sure you expand on the material in your resume rather than simply repeat it. Point out any special train-

Jack Erjavec
1234 My Street
Somewhere, OZ 99902

March 24, 2004

Mrs. Need Someone
Service Manager, Exciting New Cars
56789 Big Dealer Avenue
Somewhere, OZ 99907

Re: application for an entry-level automotive technician position

Dear Mrs. Someone:

Your ad in the March 14 edition of the *Dogpatch* for an automotive technician greatly interested me, as this position is very much in line with my immediate career objective—a career position as an automotive technician in a new car dealership. Because of the people and cars featured at your dealership, I know working there would be exciting.

I have tinkered with cars for most of my life and am currently enrolled in the Automotive Technology program at Somewhere Community College. I chose this program because you are on the advisory council and I knew it must be a good program. I have good hand skills and work hard to be successful. My being on the varsity soccer team for four years should attest to that. I also enjoy working with people and have developed excellent communication skills. The position you have open is a perfect fit for me. A resume detailing my skills and work experience is attached for your review.

I would appreciate an opportunity to meet with you to further discuss my qualifications. In the meantime, many thanks for your consideration, and I look forward to hearing from you soon. I can be reached by phone at 123-456-7890, most weekdays after 2:30 PM. If I am unable to answer the phone when you call, please leave a message and I will return your call as soon as I can. Thanks again.

Sincerely,

Jack Erjavec

encl.

Figure 2-3 An example of a cover letter that can be sent with the resume in Figure 2–2.

ing or experience you have that directly relates to the job. When doing this, give a summary without listing the places and dates. This information is listed in your resume so simply refer to the resume for details. This summary is another opportunity for you to let the employers know you understand what they do and what the job involves.

The third paragraph is typically the end or closing. Make sure you thank the employer for taking the time to review your resume and ask him or her to contact you to make an appointment for an interview. Make sure you give a phone number where you can be reached. If you have particular times when it is best to contact you, put those times in this paragraph. Finally, make sure you sign the letter before sending it.

Make sure you have a clear and understandable message on your telephone's answering machine, just in case you miss a potential employer's call. Also have an organized work area around the phone so you accurately schedule any interview appointments.

You should not send out the same cover letter to all potential employers. Adjust the letter to match the company and position you are applying for. Yes, this means a little more work, but it will be worth it. For example, if you state in one letter that you have always been a "Chevy nut" that may help at a Chevrolet dealership but will not at a Toyota or Ford dealership.

Contacting Potential Employers

Unless the help-wanted ad or job posting tells you otherwise, it is best to drop your resume and cover letter off in person (preferably to the person who does the hiring). When you are doing this, make sure you tell the employer who you are and the job you want. Make sure you are prepared for what happens next. You may be given an interview right then. You may be asked to fill out an application. If so, fill it out.

Before you leave, thank the employer and ask if you can call back in a few days if you do not hear from him. If you do not hear back within a week, call to make sure the employer received your resume, reminding him of who you are and what job you applied for. If he tells you that the job is filled or that no jobs are available, politely thank him for considering you and tell him you will stay in touch in case there is a future job opening.

Applications

An application form is a legal document that summarizes who you are. It is also another marketing tool for you. Filling out the application is the first task the employer has asked you to do, so do it thoroughly and carefully. Make sure you are prepared to fill out an application before you go. Take your own pen and a paperclip so that you can attach your resume to the application. Make sure you have your reference list with you. When filling out the application, neatly print your answers.

Read over the entire application before filling it out. Make sure you follow the directions carefully. Too often applicants try to rush through the application and make mistakes or provide the wrong information. Also by reading through the application before you fill in the blanks, you have a better chance of filling it out neatly. A messy application or one with crossed out or poorly erased information tells employers you may not care about the quality of your work.

By following the directions and providing the employer with the information asked for, you are demonstrating that you have the ability to read, understand, and follow written instructions, rules, and procedures. When answering the questions in the application, be honest.

Make sure you completely fill out the application. Doing this shows the employer that you can complete a task. Answer every question. Write N/A (nonapplicable) if a question does not apply to you. If lines are left blank, the employer may think you do not pay attention to the details of a job or are a bit lazy. When you have completed the application, sign it and attach your cover letter and resume to it.

The Interview

Typically if employers are interested in you, you will be contacted to come in for an interview. This is a good sign. If they were not impressed with what they know of you so far, they probably would not ask for an interview. Knowing this should give you some confidence as you prepare for the interview.

Although an interview does not last very long, it is a time when you can either get the job or lose it. Whether or not you realize it, you have a variety of qualities you can sell.

Get ready for the interview by taking some time to learn as much as you can about the company. Think of some of the reasons the company should hire you. When doing this, think of how both of you would benefit. Think of questions you might ask the interviewer to show you are interested in the job and the business. Then make a list of questions that you think the employer might ask. Think about how you should answer each of them and practice the answers with your family and friends. Some of the more common interview questions include:

- What can you tell me about yourself?
- Why are you interested in the job?
- What are your strengths and weaknesses?
- If we offer you a job, what can you offer us?
- Do you have any questions about the job?

Make sure you are on time (early is good) for the interview. If you are not exactly sure how to get to the business or what types of problems you may face getting there (such as traffic jams or construction), make a trip there at the time of the interview, but one or two days before. If you must be late, or if you cannot make it to the interview, call the employer as soon as possible and explain why. Ask if you can arrange for a new interview time.

Determine the days and hours you can work and when you can start to work before you go to the interview. Make sure you take your Social Security card (or SIN card), extra copies of your resume, a list of your references and their contact information, as well as copies of any letters of recommendation you have. Do not be surprised if your interviewer takes notes during the interview. You should also take paper and a pen so that you can take notes as well.

Here are some things you should do to have a successful interview:

- Show up looking neat and professional. Wear something more formal than what you would wear on the job.
- Try to relax before the interview.
- When you are greeted by the interviewer, introduce yourself and be ready to shake hands. Do it firmly!
- Listen closely to the interviewer and look at the interviewer while he or she talks.
- Answer all questions carefully and honestly. If you do not have an immediate answer, think about it before you open your mouth. If you do not understand the question, restate the question in the way you understand it. The interviewer will then know what question you are answering.
- Never answer questions with a simple yes or no. Answer all questions with examples or explanations that show your qualities or skills.
- Market yourself but do not lie about or exaggerate your abilities.
- Show your desire and enthusiasm for the job, but try to be yourself—not too shy, not too aggressive.
- Never say anything negative about other people or past employers.
- Do not be overly familiar with the interviewer and do not use slang during the interview, even if the interviewer does.
- Restate your interest in the job and summarize your good points at the end of the interview.
- Ask the interviewer if you can call back in a few days.

After the Interview

When you have left the business after the interview, go to a quiet place and reflect on what just took place. Think about what you did well and what you could have done better. Write these down so you can refer to them when you are preparing for your next interview.

Within three days after the interview, write a letter to the interviewer thanking him or her for their time. Make sure you remind them of your interest and qualifications. Take advantage of this additional chance to market yourself but do not be overly aggressive when doing this.

Remember, finding a job takes time and seldom do you land a job on your first attempt. If you do not get a job offer as a result of a first interview, do not give up. Do your best not to feel depressed or dejected. Simply realize that, although you are qualified, someone with more experience was chosen. Send a thank you letter anyway; this may prompt the interviewer to think of you the next time a similar job becomes available.

Review your cover letter, resume, and interview experience. Identify anything that can improve your marketing tools. Do not feel shy about asking the employer who did not hire you what you could have done better. Discuss your job hunt with your family and friends who will provide support and encouragement. Keep in touch with people you know who are working and who may have job leads. Explore other options. Do not rule out volunteering or job shadowing as a means of connecting with the workplace.

If you do get a job offer, do not be afraid to discuss the terms and conditions before accepting. Find out, or confirm, things such as what you will be doing, the hours you will be working, how you will be paid, and what to do when you report to work the first day. If you have any concerns, do not hesitate to share them with someone whose opinion you respect before committing yourself to the job. Do not commit to the job and then change your mind a few days later. If you have any doubts about the job, think seriously about it before you accept or decline.

ACCEPTING EMPLOYMENT

When you accept the job, you are entering into an agreement with the employer. That agreement needs to be honored. Make sure you are ready to start to work. You need to have transportation to and from work and the required tools and clothes for the job.

You will also need a Social Security (or, in Canada, a social insurance) number. If you do not already have one, you need one quickly. In the United States, you can apply for a social security number if you are a legal citizen or if you have a nonimmigrant visa status and have permission to work in the United States or are required to do so by federal, state, or local regulations.

To apply for a social security number, you must appear in person at a Social Security office to complete the application form. You must take your birth certificate or valid passport with the necessary cards and authorizations to be employed. Once the forms are completed and submitted, it may take more than two weeks for you to receive a card with your number on it.

Typically before you begin to work, or at least before you get paid, you will fill out state and federal income tax forms. These forms give the company authorization to deduct income taxes from your wages. When you are an employee for a company, the company must deduct those taxes. One form you will fill out is the employees withholding allowance certificate form, called the W-4. This form tells the employer how much, according to a scale, should be deducted from your pay for taxes. Basically the form asks how many exemptions you would like to claim. What you should claim depends on many things, and it is best that you seek advice from someone before you fill this in. In fact, do this well before you arrive to fill out the form.

Compensation

Part of your agreement with an employer is compensation. Basically the employer agrees to pay you in exchange for your work. Along with the pay, the employer may offer benefits it pays for or provides and some that you pay part of. Make sure you understand the benefits and seek help in choosing which you should participate in.

Keep in mind that when you accept employment you accept the terms of compensation offered to you. Do not show up on the first day of work demanding more. After you have worked, progressed on the job, and made the company money, you can ask for more.

Automotive technicians are typically paid according to their abilities. Most often, new or apprentice technicians are paid by the hour. While being paid they are learning the trade and the business. Time is usually spent working with a master technician or doing low-skilled jobs. As an apprentice learns more, he or she can earn more and take on more complex jobs. Once technicians have demonstrated a satisfactory level of skills, they may go on flat rate.

Flat rate is a pay system in which a technician is paid for the amount of work he or she does. Each job has a flat rate time. Pay is based on that time, regardless of how long it took to complete the job. To explain how this system works, suppose a technician is paid $15.00 per hour flat rate. If a job has a flat rate time of 3 hours, the technician will be paid $45.00 for the job, regardless of how long it took to complete it. Experienced technicians beat the flat rate time, nearly all of the time. Their weekly pay is based on the time turned in, not on the time spent. If the technician turns in 60 hours of work in a 40-hour workweek, he or she actually earned $22.50 each hour worked. However, if he or she turned in only 30 hours in the 40-hour week, the hourly pay is $11.25.

The flat rate system favors good technicians that work in a shop that has a large volume of work. Although this pay plan offers excellent wages, it is not a recommended pay plan for new and inexperienced technicians. The use of flat rate times allows for more accurate repair estimates to the customers. It also rewards productive technicians. Many good shops do not pay on the flat rate system; rather, they pay a good hourly rate to their productive technicians. Some even have bonus plans that allow technicians to make more when they are highly productive.

WORKING AS AN AUTOMOTIVE TECHNICIAN

Landing a job is only the beginning of your career. Once you have the job, you will have to keep it. Your performance during the first few weeks will determine how long you will stay employed and how soon you will get a raise or a promotion. During the first weeks on the job and the rest of time you have the job, make sure you arrive on time to work. If you are going to be late or absent, call the employer as soon as you can. Once you are at work, be cheerful and cooperative with those around you, but do not spend a lot of time talking when you should be working.

Find out what is expected of you, such as how many hours you are expected to work and when breaks are allowed. Show that you are willing to learn and to help out in emergencies. Make sure you ask about anything you are not sure of, but try to think things out for yourself whenever you can.

A successful automotive technician has good training, a desire to succeed, and a commitment to be a good technician and employee. A good employee works well with others and strives to make the business successful. The required training is not just in the automotive field. Because good technicians spend a great deal of time working with service manuals, good reading skills are a must. Technicians must also be able to accurately describe what is wrong to customers and the service advisor. Often these descriptions are done in writing; therefore, a technician also needs to be able to write well.

Technicians also should have a basic knowledge of computers and basic keyboarding skills. Computers not only control the major systems of today's vehicles, they are also used for diagnostics, tracking customers, and recordkeeping, and as sources for information. If you have little or no experience with computers, take a computer course and spend time with a computer. If you do not have access to one, go to your local library.

Employer-Employee Relationships

Being a good employee requires more than job skills. When you become an employee, you sell your time, skills, and efforts. In return, your employer pays you for these resources.

As part of the employment agreement, your employer also has certain responsibilities:

- *Instruction and supervision.* You should be told what is expected of you. A supervisor should observe your work and tell you if it is satisfactory and offer ways to improve your performance.
- *Clean, safe place to work.* An employer should provide a clean and safe work area and a place for personal cleanup.
- *Wages.* You should know how much you are to be paid before accepting a job. You should understand what your pay will be based on. Will you be paid by the hour, by the amount of work completed, or by a combination of these two? Your employer should pay you on designated paydays.
- *Fringe benefits.* When you are hired, you should be told what benefits, in addition to wages, you can ex-

pect. Fringe benefits usually include paid vacations and employer contributions to health insurance and retirement plans.

- *Opportunity and fair treatment*. Opportunity means you are given a chance to succeed and possibly advance within the company. Fair treatment means all employees are treated equally, without prejudice or favoritism.

On the other side of this business transaction, employees have responsibilities to their employers. Your obligations as an employee include the following:

- *Regular attendance*. A good employee is reliable. Businesses cannot operate successfully unless their workers are on the job. One of the first things a potential employer will ask an instructor is about the student's attendance.
- *Following directions*. As an employee, you are part of a team. Doing things your way may not serve the best interests of the company.
- *Responsibility*. Be willing to answer for your behavior and work habits.
- *Productivity*. Remember that you are paid for your time as well as your skills and effort. You have a duty to be as effective as possible when you are at work.
- *Loyalty*. Loyalty is expected by any employee. Being loyal means that you act in the best interests of your employer, both on and off the job.

COMMUNICATIONS

Employers value employees that can communicate. Effective communications include listening, reading, speaking, and writing. Communication is a two-way process. It involves sending a message and receiving a response and possibly clarifying that the response has been received and understood. Because communication is based on a message, the role of listening and reading is receiving that message.

You should carefully follow all oral and written directions that pertain to your job. If you do not fully understand them, ask for clarification. You also need to be a good listener. Like other things in life, messages can appear to be good, bad, or have little worth to you. Regardless of how you rate the message, you should show respect to the person giving the message. Look at the person while they speak and listen to their message before you respond. Try to fully comprehend the message by asking questions about it and gathering as many details as possible. Try to put yourself in the other person's shoes and listen without bias.

Obviously, when you read something you are receiving a message without the advantage of seeing the message sender. Therefore you must take what you read at face value. This is important because being able to read and understand the information and specifications given in service information is a must for automotive technicians **(Figure 2–4)**.

The purpose of speaking and writing is to send a message. Do your best to think through the words you use to convey the message. Pay attention to how the intended receiver of the information is listening and adjust your words and mannerisms accordingly. This consideration is also important when you write out your message. Think about who the message is going to and adjust your words to match the abilities and attitudes of the reader. Also, keep in mind that more than one person may read it, so think of others' needs as well.

Obviously, working in an automotive facility requires speaking to your supervisors, fellow employees, and customers. Always keep in mind that communication is a two-way street; do not try to totally control the conversation, and give listeners a chance to speak.

GENERAL SPECIFICATIONS	
DISPLACEMENT	1.9L
NUMBER OF CYLINDERS	1–4
BORE AND STROKE	
1 9L	82 × 88 (3.23 × 3.46)
FIRING ORDER	1-3-4-2
OIL PRESSURE (HOT 2000 RPM)	240–450 kPa (35–65 psi)
DRIVE BELT TENSION	178–311 (40–70 Lb-Ft)
CYLINDER HEAD AND VALVE TRAIN 1 2	
COMBUSTION CHAMBER VOLUME (oc)	EFI-HO 55 ± 1.6
VALUE GUIDE BORE DIAMETER	EFI 39.9 ± 0.8
Intake	13.481-13 519 mm (0.531-0.5324 in.)
Exhaust	13.481-13.519 mm (0.531-0.532 in.)
VALVE GUIDE I.D.	
Intake and Exhaust	8.063–8.094 mm (.3174–.3187 in.)
Width — Intake & Exhaust	1.75–2.32 mm (0.069–0.091 in.)
Angle	45°
Runout (T.I.R.)	0.076 mm (0.003 in.) MAX.
Bore Diameter (Insert Counterbore Diameter)	
Intake	(EFI-HO) 43.763 mm (1.723 in.) MIN.
	43.788 mm (1.724 in.) MAX.
	(EFI) 39.940 mm (1.572 in.) MIN.
	39.965 mm (1.573 in.) MAX.
Exhaust	(EFI-HO) 38.263 mm (1.506 in.) MIN.
	38.288 mm (1.507 in.) MAX.
	(EFI) 34.940 mm (1.375 in.) MIN.
	39.965 mm (1.573 in) MAX.
GASKETS SURFACE FLATNESS	0.04 mm (0.0016 in.)/26 mm (1 in.)
	0.08 mm (0.003 in.)/156 mm (6 in.)
	0.15 mm (0.006 in.) Total
HEAD FACE SURFACE FINISH	0.7/2.5 0.8 (28/100 .030)
VALVE STEM TO GUIDE CLEARANCE	
Intake	0.020–0.069 mm (0.0008–0.0027 in.)
Exhaust	0.046–0.095 mm (0.0018–0.0037 in.)
VALVE HEAD DIAMETER	
Intake	42.1–41.9 mm (1.66–1.65 in.)
Exhaust	37.1–36.9 mm (1.50–1.42 in.)
VALVE FACE RUNOUT	
LIMIT	Intake & Exhaust 05 mm (0.002 in.)
VALVE FACE ANGLE	45.6°
VALVE STEM DIAMETER (Std.)	
Intake	8.043–8.025 mm (0.3167–0.3159 in.)

Figure 2-4 Being able to read and understand the information and specifications given in service information is a must for automotive technicians. *Courtesy of Ford Motor Company*

Proper telephone etiquette is also important. Most businesses will tell you how to answer the phone, typically involving the name of the company followed by your name. Make sure you listen carefully to the person calling. When you are the one making the call, make sure you introduce yourself and state the overall purpose of the phone call. Again the key to proper phone etiquette is respect.

You will also be required to write things, such as warranty reports and work orders. You may also need to either speak with or write to customers, parts suppliers, and supervisors to clarify an issue. Take your time and write clear, concise, complete, and grammatically correct sentences and paragraphs. Doing this will not only help get your message across but will also make you a more prized employee.

Nonverbal Communication

In any communication some of the true meaning is lost in the simple transmission of a message from a sender to a receiver. In many cases, the heard message is often far different than the one intended, because the words spoken are not always understood or are interpreted wrongly because of personal feelings. Therefore it is important to realize that a major part of communication is nonverbal. **Nonverbal communication** is a key part of sending and receiving a message. Pay attention to your nonverbal communication as well as that of others.

Nonverbal communication includes such things as body language and tone. Body language includes facial expression, eye movement, posture, and gestures. All of us read people's faces for ways to interpret what they say or feel, such as looking for a nod of a head. We also look at posture to provide insights about how the other person feels about the message. Posture can indicate self-confidence, aggressiveness, fear, guilt, or anxiety. Similarly, we look at gestures such as how they place their hands or give a handshake.

Posture and other aspects of body language have been identified as important keys to communication. Many scholars have studied and classified them and defined what they indicate. Some divide postures into two basic groups: open/closed and forward/back.

Open/closed is the most obvious. People with their arms folded, legs crossed, and bodies turned away are signaling that they are rejecting or are closed to messages. People fully facing you with open hands and both feet planted on the ground are saying they are open to and accepting the message.

Forward/back indicates whether people are actively or passively reacting to the message. When they are leaning forward and pointing toward you, they are actively accepting or rejecting the message. When they are leaning back, looking at the ceiling, doodling on a pad, or cleaning their glasses, they are either passively absorbing or ignoring the message.

You can alter the meaning of words significantly by changing the tone of your voice. Think of how many ways you can say "no"; you could express mild doubt, terror, amazement, anger, and other emotions.

SOLVING PROBLEMS AND CRITICAL THINKING

Employers value someone who can think critically and act logically to evaluate situations. They also value employees with the ability to solve problems and make decisions. **Critical thinking** is the art of being able to judge or evaluate something without bias or prejudice. When diagnosing an automotive problem, critical thinkers are able to locate the cause of the problem because they respond to what is known, not what is supposed.

Good critical thinkers begin their process of problem solving by careful observation of what is or what is not happening. Based on these observations, they declare something as a fact. For example, it is a fact that the right headlamp does not light, and it is a fact that the left headlamp does light. Based on these facts, a critical thinker is quite sure that the source of the problem is related to the right headlamp and not the left one. Therefore the focus of any testing of the system is centered on the right headlamp. A critical thinker then studies the circuit and determines the test points. Prior to conducting any test, he or she knows what to test for and what the possible results are. Further, he or she knows what those results would indicate.

Critical thinkers solve problems in an orderly way and do not depend on chance or guesswork. They come to conclusions based on a sound reason. They also understand that if a specific problem exists only during certain conditions, there are a limited number of solutions. They further understand the relationship between how often the problem occurs and the probability of accurately predicting the problem. Also, they understand that one problem may cause other problems and know how to identify the connection between the problems.

Solving problems is something we do every day of our lives. Often the problems are trivial, such as deciding what to watch on television. Other times they are critical and demand much thought. At these times thinking critically will really pay off. It is impossible to guarantee that critical thinking will lead to the correct decision; however, it will lead to good decisions and solutions.

Diagnosis

The word ***diagnosis*** is commonly used to define a primary duty of an automotive technician. Diagnosis is not guessing, and it is more than following a series of interrelated steps in order to find the solution to a specific problem. Diagnosis is a way of looking at systems that are not functioning properly and finding out why. Through an understanding of the purpose and operation of the system, you can accurately diagnose problems.

In service manuals there are diagnostic aids given for many different systems. These are either symptom based or flow charts. **Flow charts** or decision trees **(Figure 2–5)** guide you through a step-by-step process. As you answer the questions given at each step, you are told what your next step should be. Symptom-based diagnostic charts **(Figure 2–6)** focus on a solid definition of the problem and offer a list of possible causes of the problem. Sometimes the diagnostic aids are a combination of the two, a flow chart based on clearly defined symptoms.

When these diagnostic aids are not available or prove to be ineffective, most good technicians conduct a good visual inspection and then take a logical approach to solv-

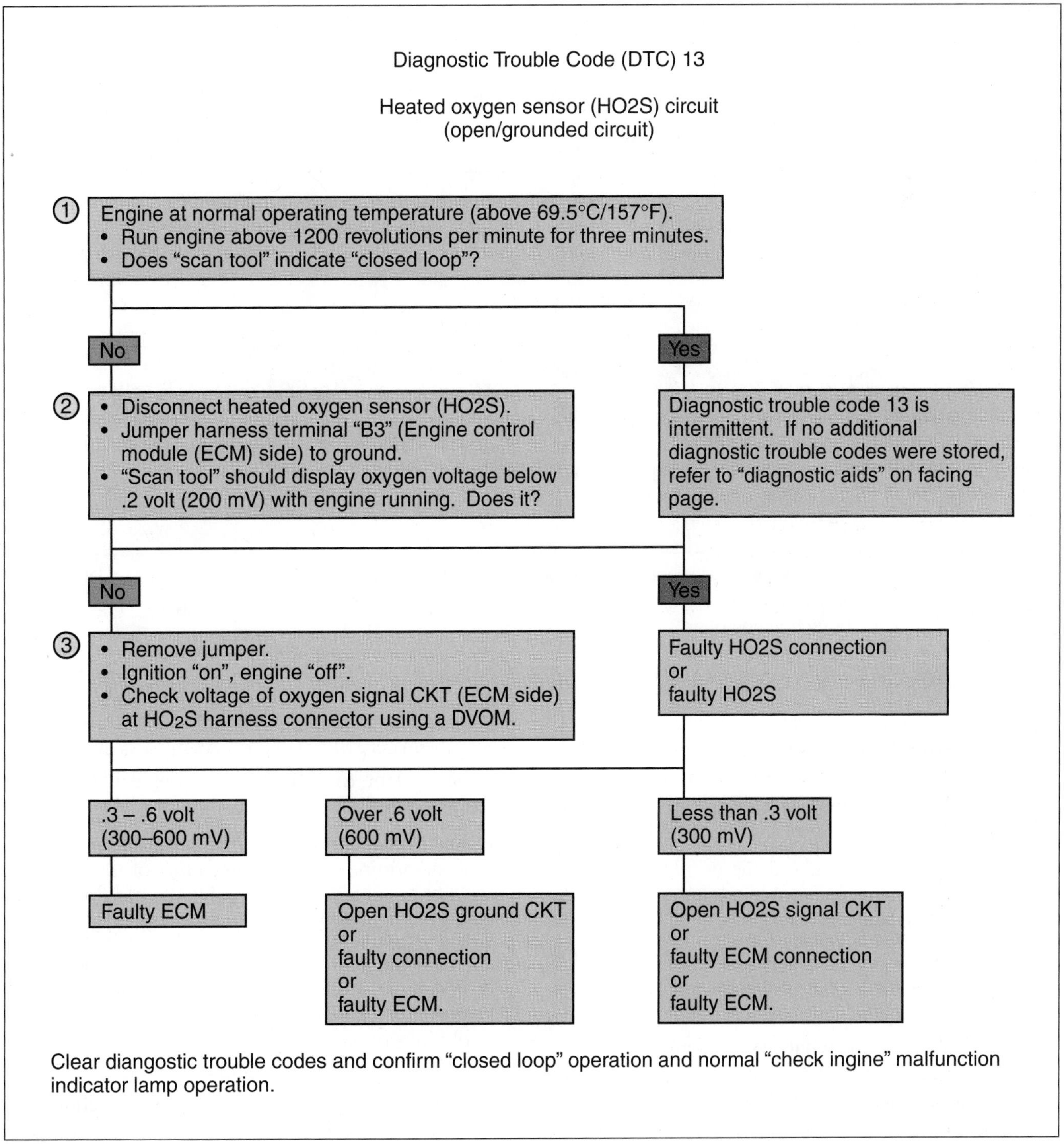
Diagnostic Trouble Code (DTC) 13
Heated oxygen sensor (HO2S) circuit (open/grounded circuit)
① Engine at normal operating temperature (above 69.5°C/157°F).
• Run engine above 1200 revolutions per minute for three minutes.
• Does "scan tool" indicate "closed loop"?
No
Yes
② • Disconnect heated oxygen sensor (HO2S).
• Jumper harness terminal "B3" (Engine control module (ECM) side) to ground.
• "Scan tool" should display oxygen voltage below .2 volt (200 mV) with engine running. Does it?
Diagnostic trouble code 13 is intermittent. If no additional diagnostic trouble codes were stored, refer to "diagnostic aids" on facing page.
No
Yes
③ • Remove jumper.
• Ignition "on", engine "off".
• Check voltage of oxygen signal CKT (ECM side) at HO2S harness connector using a DVOM.
Faulty HO2S connection or faulty HO2S
.3 – .6 volt (300–600 mV)
Over .6 volt (600 mV)
Less than .3 volt (300 mV)
Faulty ECM
Open HO2S ground CKT or faulty connection or faulty ECM.
Open HO2S signal CKT or faulty ECM connection or faulty ECM.
Clear diangostic trouble codes and confirm "closed loop" operation and normal "check ingine" malfunction indicator lamp operation.

Figure 2-5 A typical decision tree for diagnostics.

DTC B3138 DOOR LOCK CIRCUIT (HIGH)

STEP	ACTION	YES	NO
1	Was BCM diagnostic check performed?	Go to step 2	See BCM diagnostics
2	* Check for current DTCs with scan tool. * Disconnect power to LH door lock switch. * Does scan tool display B3138 as a current code?	Go to step 3	Go to step 6
3	* Check for current DTCs with scan tool. * Disconnect power to RH door lock switch. * Does scan tool display B3138 as a current code?	Go to step 4	Go to step 7
4	* Disconnect the brown BCM (C1) connector. * Backprobe connectors with a digital multimeter. * Measure voltage between A4 (LT BLU) and ground. * Does multimeter show battery voltage?	Go to step 5	Go to step 8
5	Locate and repair short to battery voltage in CKT 195 (LT BLU) between the BCM and the LOCK relay, or the left or right front door switches.	Go to step 9	—
6	Replace the LH power door lock switch. Is the repair complete?	Go to step 9	—
7	Replace the RH power door lock switch. Is the repair complete?	Go to step 9	—
8	* Replace the BCM. * Program the BCM with proper calibrations. * Perform the learn procedure. Is the repair complete?	Go to step 9	—
9	* Reconnect all disconnected components. * Clear the DTCs. Is the action complete?	System OK	—

Figure 2-6 A symptom-based diagnostic chart.

ing the problem. This approach relies on critical thinking skills, as well as system knowledge. Logical diagnosis follows these steps:

1. Gather information about the problem. Find out when and where the problem happens and what exactly happens.
2. Verify that the problem exists. Take the vehicle for a road test and try to duplicate the problem, if possible.
3. Thoroughly define what the problem is and when it occurs. Pay strict attention to the conditions present when the problem happens. Also pay attention to the entire vehicle; another problem may be evident to you that was not evident to the customer.
4. Research all available information and knowledge to determine the possible causes of the problem. Try to match the exact problem with a symptoms chart or think about what is happening and match a system or some components to the problem.
5. Isolate the problem by testing. Narrow down the probable causes of the problem by checking the obvious or easy-to-check items.
6. Continue testing to pinpoint the cause of the problem. Once you know where the problem should be, test until you find it!

7. Locate and repair the problem, then verify the repair. Never assume that your work solved the original problem. Make sure the problem is history before returning it to the customer.

PROFESSIONALISM

The key to effective communications was respect. Communication was defined as a two-way process. So is respect. You should respect others and others should respect you. Respect cannot be commanded, it must be earned. As a technician, you can earn the respect of others in many ways. All of these are the result of the amount of professionalism you display. Professionalism carries through all aspects of life and is best shown by having a positive attitude, displaying good behavior, and accepting responsibility.

A successful automotive technician is a highly skilled and knowledgeable individual. A professional automotive technician has the same skills and knowledge but also dresses and acts appropriately. A professional demonstrates positive behaviors, attitudes, and responsibility.

Demonstrations of positive behaviors and attitudes include evidence of:

- Self-esteem, pride, and confidence
- Honesty, integrity, and personal ethics
- A positive attitude toward learning, growth, and personal health
- Initiative, energy, and persistence to get the job done
- Respect for others
- A display of initiative and assertiveness

You show your employer and the rest of the world that you are responsible by displaying:

- The ability to set goals and priorities in work and personal life
- The ability to plan and manage time, money, and other resources to achieve goals
- The willingness to follow rules, regulations, and policies
- The willingness to fulfill the responsibilities of your job
- Assuming responsibility and accountability for your decisions and actions
- The ability to apply ethical reasoning

Coping with Change

Demonstrations of your professionalism are also evident by how you react to change. Unfortunately, work environments never stay the same. New rules and regulations, supervisors, fellow employees, vehicle systems, and vehicles are all potential sources of stress. Rather than focusing on the negatives of the changes, you should identify the positives. Doing this will help you minimize stress. If you feel stress, do what you can to relieve it. Activities such as walking, running, or playing sports help reduce stress. Sharing the stress with others also helps as long as you show respect to the person you want to listen. When you are stressed, whether it is caused by something at work or in your personal life, it is difficult to be a productive worker. Therefore do your best to put things in perspective and do some critical thinking to identify what you can do to change the situation that is causing the stress.

When the source of stress is related to your job, you may be in a situation that will not change or one that you are no longer willing to cope with. In these cases, it may be wise to find employment elsewhere. This stress can be the result of change or can be caused by your realization that you do not like what you are doing or are not capable of doing what is required of you.

If you decide that leaving your job is the best solution to the problem, do it professionally. Do not simply stop showing up for work or walk up to the employer and say "I quit!" The best way to quit a job is to write a letter of resignation and present it personally to the employer. The letter should state your reason for leaving the company. When doing this, be careful not to attack the business, the employer, or fellow workers. You can simply say you are looking at other opportunities or have found another job. Bad-mouthing the business is a sure way of losing a good work reference, one that you may need to land your next job. The letter should also include the last day you intend to work. Your last day should be approximately two weeks after you notify the employer. At the end of your letter of resignation, thank the employer for the opportunity to work for him or her and for the experiences they provided for your personal growth.

INTERPERSONAL RELATIONSHIPS

As an employee, you have certain responsibilities toward your fellow workers. You will be a member of a team. Teamwork means cooperating with and caring about other workers. All members of the team (the business that employs you) should understand and contribute to the goals of the business. Keep in mind that if the business does not make money, you may not have a job in the future. Your responsibility is more than simply doing your job. You should also:

- Suggest improvements that may make the business more successful
- Display a positive attitude

- Work with team members to achieve common goals
- Respect the thoughts and opinions of your fellow workers and your employer
- Exercise give-and-take for the benefit of the business
- Value individual diversity
- Respond to praise or criticism in a professional way
- Provide constructive praise or criticism
- Channel and control emotional reactions to situations
- Resolve conflicts in a professional way
- Identify and react to any intimidation or harassment

Customer Relations

Good customer relations begin at the technician level. Learn to listen and communicate clearly. Be polite and organized, particularly when dealing with customers on the telephone. Always be as honest as you possibly can.

Look like and present yourself as a professional, which is what automotive technicians are. Professionals are proud of what they do and they show it. Always dress and act appropriately and watch your language, even when you think no one is near.

Respect the vehicles on which you work. They are important to the lives of your customers. Always return the vehicle to the owner in a clean, undamaged condition. Remember, a car is the second largest expense a customer has. Treat it that way. It does not matter if you like the car. It belongs to the customer; treat it respectfully.

Explain the repair process to the customer in understandable terms. Whenever you are explaining something to a customer, make sure you do this in a simple way without making the customer feel stupid. Always show customers respect and be courteous to them. Not only is this the right thing to do but it also leads to loyal customers. Make repair estimates as precise as possible. No one likes surprises, particularly when substantial amounts of money are involved.

To help develop your customer relations skills, special customer care tips appear throughout this text. They contain sound advice you can share with customers on personal vehicle care. They also give you advice on how to conduct business in a courteous and professional manner.

REPAIR ORDERS

A **repair order (RO)** is written for every vehicle brought into the shop for service. Repair orders may also be called service or work orders and contain information about the customer, the vehicle, the customer's concern or request, and an estimate of the cost of the services and when the services should be completed **(Figure 2–7)**. Repair orders are legal documents that are used for many other purposes, such as payroll and general record keeping **(Figure 2–8)**. Legally, an RO protects the shop and the customer. An RO is signed by the customer, who in doing so authorizes the service and accepts the terms noted on the RO. The customer, however, is protected against being charged more than the estimate given on the RO, unless he or she later authorizes a higher amount. Some states allow shops to be within 10% of the estimate, while others hold the shop to the amount that was estimated.

Today, most service facilities have shop management software in their computers. The information for the completion of an RO is input on the computer's keyboard. The software package also helps in the estimation of repair costs. The software also takes information from the RO and saves it in various files, each defined by its purpose.

Guidelines for Estimating Repair Costs

For legal reasons and to establish good customer relations, projected repair costs must be calculated with as much accuracy as possible. To do this, do the following.

1. Make sure you have the correct information about the vehicle.
2. Always use the correct labor and parts guide or database for that specific vehicle.
3. Locate the exact service for that specific vehicle in the guide or database.
4. Using the guidelines provided in the guide or database, choose the proper time allocation listed for the service.
5. Multiply the allocated time by the shop's hourly flat rate.
6. Using the information given in the guide or database, identify the parts that will be replaced for that service.
7. Locate the cost of the parts in the guide or database or in the catalogs used by the shop.
8. Repeat the process for all other services required or requested by the customer.
9. Multiply the time allocations by the shop's hourly flat rate.
10. Add all of the labor costs together; this sum is the labor estimate for those services.
11. Add the cost of all the parts together; this sum is the estimate for the parts required for the services.
12. Add the total labor and parts costs together. If the shop charges a standard fee for shop supplies, add it to the labor and parts total. This sum is the cost estimate to present to the customer.

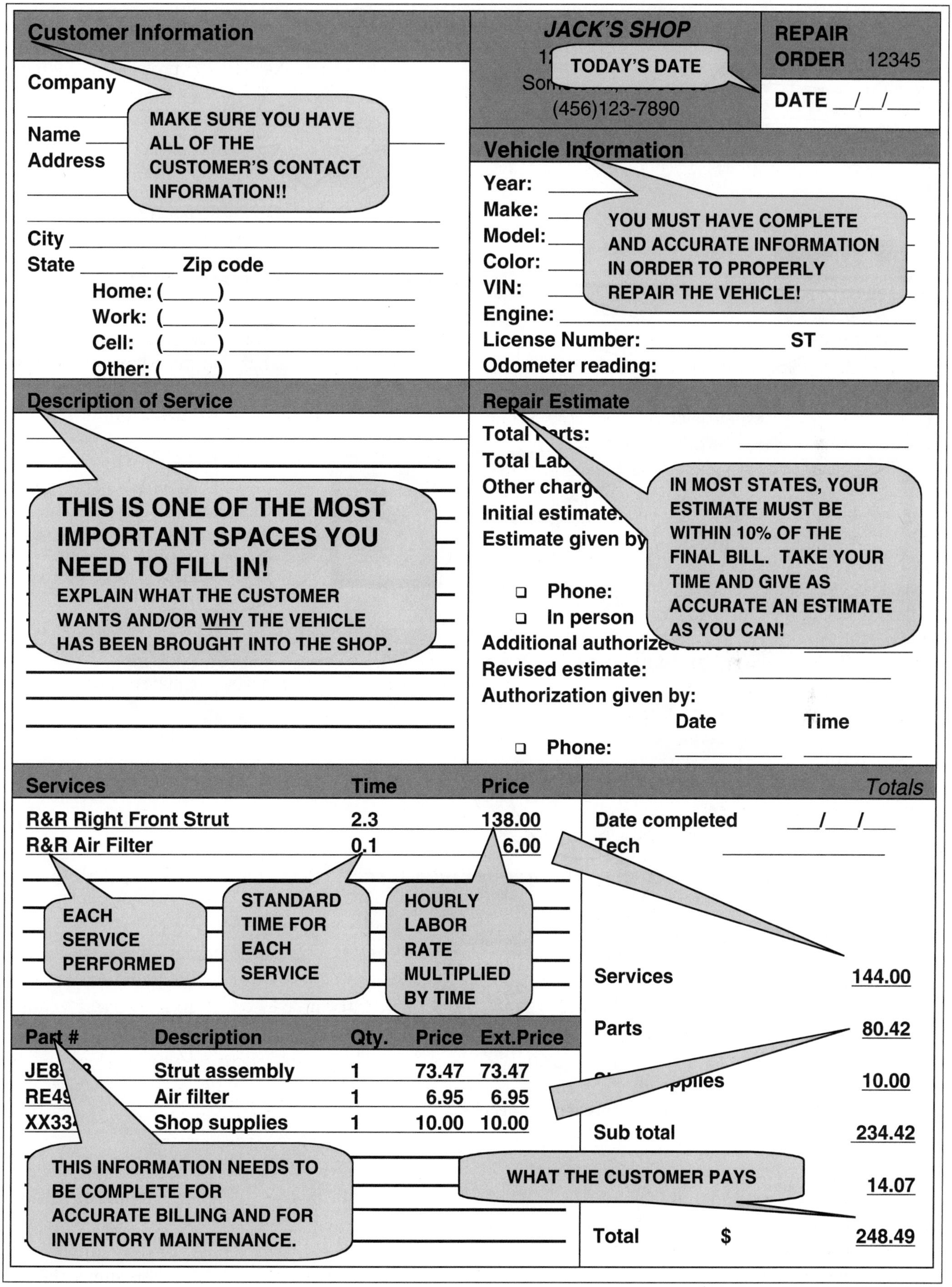
Customer Information
Company
Name
Address
City
State Zip code
Home: ( )
Work: ( )
Cell: ( )
Other: ( )
MAKE SURE YOU HAVE ALL OF THE CUSTOMER'S CONTACT INFORMATION!!
JACK'S SHOP
(456)123-7890
TODAY'S DATE
REPAIR ORDER 12345
DATE __/__/__
Vehicle Information
Year:
Make:
Model:
Color:
VIN:
Engine:
License Number: ST
Odometer reading:
YOU MUST HAVE COMPLETE AND ACCURATE INFORMATION IN ORDER TO PROPERLY REPAIR THE VEHICLE!
Description of Service
THIS IS ONE OF THE MOST IMPORTANT SPACES YOU NEED TO FILL IN!
EXPLAIN WHAT THE CUSTOMER WANTS AND/OR WHY THE VEHICLE HAS BEEN BROUGHT INTO THE SHOP.
Repair Estimate
Initial estimate:
Phone:
In person
Additional authorized
Revised estimate:
Authorization given by:
Date Time
Phone:
IN MOST STATES, YOUR ESTIMATE MUST BE WITHIN 10% OF THE FINAL BILL. TAKE YOUR TIME AND GIVE AS ACCURATE AN ESTIMATE AS YOU CAN!
Services Time Price
R&R Right Front Strut 2.3 138.00
R&R Air Filter 0.1 6.00
EACH SERVICE PERFORMED
STANDARD TIME FOR EACH SERVICE
HOURLY LABOR RATE MULTIPLIED BY TIME
Totals
Date completed __/__/__
Tech
Services 144.00
Parts 80.42
10.00
Sub total 234.42
14.07
Total $ 248.49
Part # Description Qty. Price Ext.Price
Strut assembly 1 73.47 73.47
Air filter 1 6.95 6.95
Shop supplies 1 10.00 10.00
THIS INFORMATION NEEDS TO BE COMPLETE FOR ACCURATE BILLING AND FOR INVENTORY MAINTENANCE.
WHAT THE CUSTOMER PAYS

Figure 2-7 A completed repair order.

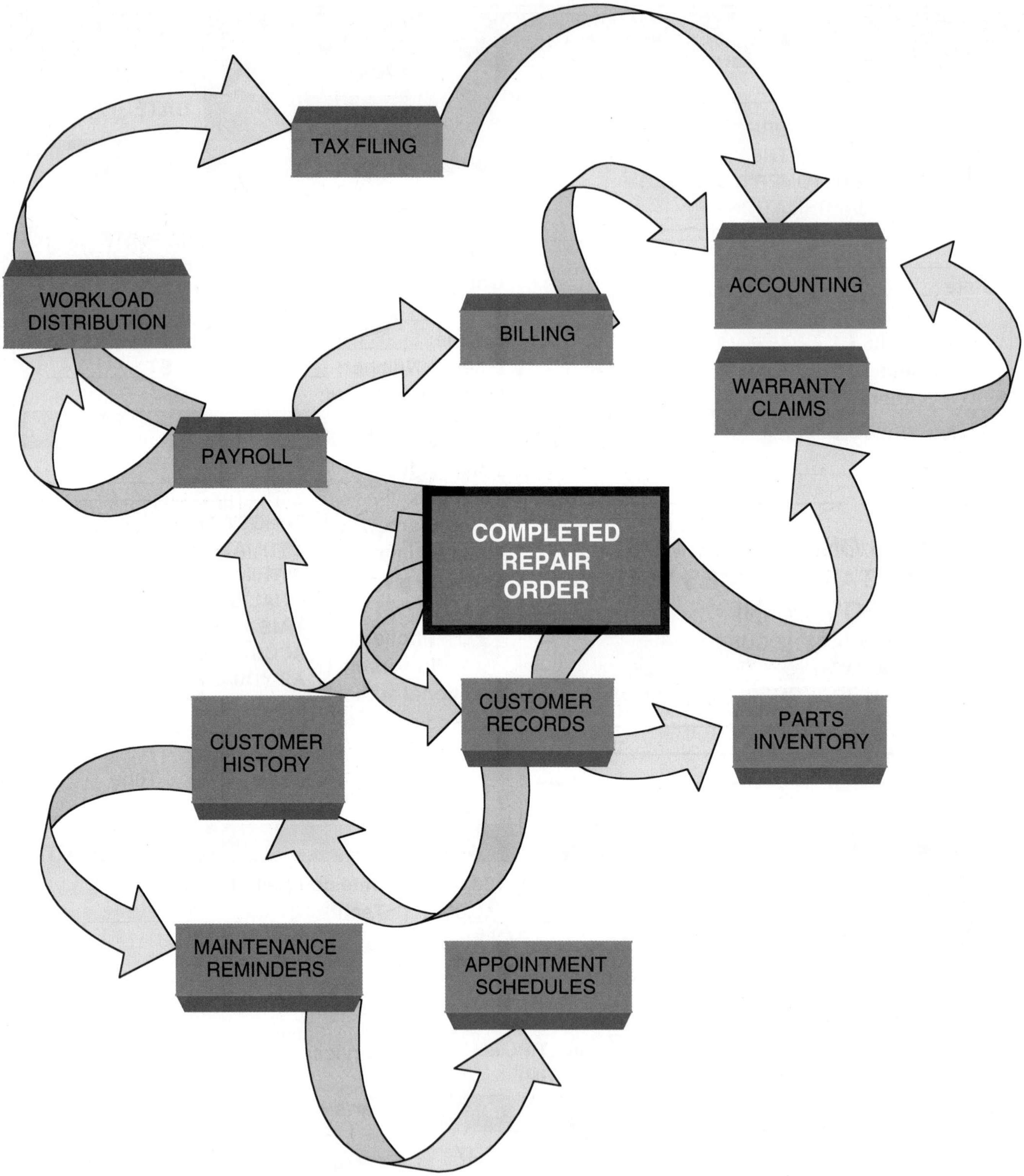

Figure 2-8 Where the information from a repair order goes.

KEY TERMS

Critical thinking
Diagnosis
Employment plan
Flat rate
Flow charts
Nonverbal communication
Reference
Repair order (RO)
Resume
Soft skills

SUMMARY

- An employment plan is an honest appraisal of yourself and your career hopes.
- A reference is someone who will be glad to tell a potential employer about you and your work habits.
- A resume and cover letter are personal marketing tools and may be the first look at you an employer has.
- A resume normally includes your contact information, career objective, skills and/or accomplishments, work experience, education, and a statement about references.
- A cover letter gives you a chance to point out exactly why you are perfect for a particular job.
- An application form is a legal document that summarizes who you are.
- Good preparation for an employment interview will result in a good experience.
- Automotive technicians are typically paid an hourly wage or on the flat rate system.
- A successful automotive technician has good training, a desire to succeed, and a commitment to be a good technician and a good employee.
- As part of an employment agreement your employer also has certain responsibilities to you, and you have responsibilities to the employer.
- Effective communications include listening, reading, speaking, and writing.
- Nonverbal communication is a key part of sending and receiving a message and includes such things as body language and tone.
- Employers value someone who can think critically and act logically to evaluate situations and who has the ability to solve problems and make decisions.
- Diagnosis means finding the cause or causes of a problem. It requires a thorough understanding of the purpose and operation of the various automotive systems.
- Diagnostic charts found in service manuals can aid in diagnostics.
- Professionalism is best displayed by having a positive attitude, displaying good behavior, and accepting responsibility.
- New rules and regulations, supervisors, fellow employees, vehicle systems, and vehicles are all potential sources of stress, and your professionalism will be measured by how well you cope.
- Teamwork means cooperating with and caring about other workers.
- Good customer relations is a quality of good technicians and is based on respect.
- A repair order (RO) is a legal document used for many purposes.
- An RO includes a cost estimate for the repairs. By law, this estimate must be quite accurate.

REVIEW QUESTIONS

1. Which of the following is not a recommended step for accurate diagnosis of a problem?
   a. Gather as much information as you can about the problem.
   b. Thoroughly define the problem.
   c. Replace system components and identify the cause of a problem through the process of elimination.
   d. Research all available information and knowledge to determine the possible causes of the problem.
2. What type of information should go into your employment plan?
3. What does it mean to be paid based on flat rate?
4. What should be included in the three main paragraphs of a cover letter?
5. *True or False?* Legally, an RO protects the shop and the customer.
6. Which of the following is *not* a desired characteristic of a good resume?
   a. It is neat, uncluttered, and easy to read.
   b. It has a list of all jobs you have done, whether for pay or just to help someone out.
   c. It is only one or two pages long.
   d. Important information appears near the top of the paper.
7. Which of the following behaviors do *not* show that you are a responsible person?
   a. Having set goals and priorities in your work and personal life.

b. Showing a willingness to follow rules, regulations, and policies.

c. Showing a willingness to share the consequences of your mistakes with others.

d. Using ethical reasoning when making decisions.

8. *True or False?* If you decide that leaving your job is the best way to relieve stress at work, you should stop showing up for work and send your employer a letter stating why you left the company.

9. *True or False?* When you are filling out an application, do not attempt to answer the questions that do not pertain to you and your situation.

10. Which of the following is not the right thing to do when you are being interviewed for a job?

    a. Show up looking neat and in the clothing you would wear on the job.

    b. Never hesitate with an answer to a question, because hesitation may indicate you are a shy person.

    c. To avoid saying too much or offending the interviewer, answer as many questions as you can with a simple yes or no.

    d. Listen closely to the interviewer and look at the interviewer while he or she talks.

11. Which of the following is *not* a characteristic of a good employee?

    a. Reliable c. Overly sociable

    b. Responsible d. Loyal

12. Applicant A goes to a quiet place immediately after an interview and reflects on what just took place. Applicant B sends a letter of thanks to the interviewer if he has not heard back from the employer within two weeks. Who is doing the right thing?

    a. A only c. Both A and B

    b. B only d. Neither A nor B

13. Technician A always looks at people while they are speaking and listens to their message before responding. Technician B always looks at customers when they are speaking to show she is interested in what they are saying. Who is doing the right thing?

    a. A only c. Both A and B

    b. B only d. Neither A nor B

14. Technician A always speaks to customers with his arms folded across his chest because he does not know what else to do with them. Technician B always tries to fully comprehend the message by asking questions about it and gathering as many details as possible. Who is doing the right thing?

    a. A only c. Both A and B

    b. B only d. Neither A nor B

15. Which of the following would *not* be considered a soft skill?

    a. Enjoying solving puzzles or problems

    b. Caring for or helping people

    c. The ability to work independently

    d. Taking care to follow orders or instructions

3

WORKING SAFELY IN THE SHOP

OBJECTIVES

■ **Understand the importance of safety and accident prevention in an automotive shop.** ■ **Explain the basic principles of personal safety, including protective eye wear, clothing, gloves, shoes, and hearing protection.** ■ **Explain the procedures and precautions for safely using tools and equipment.** ■ **Explain the precautions that need to be followed to safely raise a vehicle on a lift.** ■ **Explain what should be done to maintain a safe working area in a shop, including running the engines of vehicles in the shop and venting the exhaust gases.** ■ **Describe the purpose of the laws concerning hazardous wastes and materials, including the right-to-know laws.** ■ **Describe your rights, as an employee and/or student, to have a safe place to work.**

Working on automobiles can be dangerous. It can also be fun and very rewarding. To keep the fun and rewards rolling in, you need to try to prevent accidents by working safely. In an automotive repair shop, there is great potential for serious accidents, simply because of the nature of the business and the equipment used. When there is carelessness, the automotive repair industry can be one of the most dangerous occupations.

Unless you want to get hurt or want your fellow students or employees to get hurt, you should strive to work safely. Shop accidents can cause serious injury, temporary or permanent disability, and death. Think about these facts:

- Vehicles, equipment, and many parts are very heavy; their weight can cause severe injuries.
- Many parts of a car become very hot and can cause severe burns.
- High fluid pressures can build up inside the cooling system, fuel system, or battery; these can spray dangerous fluids on you and especially into your eyes.
- Batteries contain highly corrosive and potentially explosive acids; these can cause bad skin burns or blindness.
- Fuels and commonly used cleaning solvents are flammable.
- Exhaust fumes are poisonous and can be deadly.
- During some repairs, technicians can be exposed to harmful dust particles and vapors that can cause chronic or terminal diseases.

All of these can be enough to scare you away from working on cars. However, the chances of your being injured while working on a car are close to nil if you learn to work safely and use common sense. Shop safety is the responsibility of everyone in the shop—you, your fellow students or employees, and your employer or instructor. Everyone must work together to protect the health and welfare of all who work in the shop.

This chapter contains many safety guidelines concerning personal, work area, tool and equipment, and hazardous material safety. In addition to this chapter, special warnings have been used throughout this book to alert you to situations in which carelessness could result in personal injury. Finally, when working on cars, always follow safety guidelines given in service manuals and other technical literature. They are there for your protection.

PERSONAL SAFETY

Personal safety simply involves those precautions you take to protect yourself from injury. This includes wearing protective gear, dressing for safety, working professionally, and correctly handling tools and equipment.

Personal Safety Precautions

As you work on cars, there are many parts and fluids that you do not want to keep. Some of the common wastes, as well as how to dispose of them, are given here.

Eye Protection Your eyes can become infected or permanently damaged by many things in a shop.

Some procedures, such as grinding, result in tiny particles of metal and dust that are thrown off at very high speeds. These metal and dirt particles can easily get into your eyes, causing scratches or cuts on your eyeball. Pressurized gases and liquids escaping a ruptured hose or hose fitting can spray a great distance. If these chemicals get into your eyes, they can cause blindness. Dirt and sharp bits of corroded metal can easily fall down into your eyes while you are working under a vehicle.

Eye protection should be worn whenever you are exposed to these risks. To be safe, you should wear **safety glasses** whenever you are working in the shop. There are many types of eye protection available **(Figure 3–1)**. To provide adequate eye protection, safety glasses have lenses made of safety glass. They also offer some sort of side protection. Regular prescription glasses do not offer sufficient protection and therefore should not be worn as a substitute for safety glasses. When prescription glasses are worn in a shop, they should be fitted with side shields.

Wearing safety glasses at all times is a good habit to get into. To help develop this habit, wear safety glasses that fit well and feel comfortable.

Some procedures may require that you wear other eye protection in addition to safety glasses. For example, when you are working around air conditioning systems, you should wear splash goggles and, when cleaning parts with a pressurized spray, you should wear a face shield. The face shield not only gives added protection to your eyes but it also protects the rest of your face.

If chemicals such as battery acid, fuel, or solvents get into your eyes, flush them continuously with clean water. Have someone call a doctor and get medical help immediately.

Many shops have eye wash stations or safety showers **(Figure 3–2)** that should be used whenever you or someone else has been sprayed or splashed with a chemical.

Clothing Your clothing should be well fitted and comfortable but made with strong material. Loose, baggy clothing can easily get caught in moving parts and machinery. Neckties should not be worn. Some technicians prefer to wear coveralls or shop coats to protect their personal clothing. Your work clothing should offer you some protection but should not restrict your movement.

Hair and Jewelry Long hair and loose, hanging jewelry can create the same type of hazard as loose-fitting clothing. They can get caught in moving engine parts and machinery. If you have long hair, tie it back or tuck it under a cap.

Never wear rings, watches, bracelets, or neck chains. These can easily get caught in moving parts and cause serious injury.

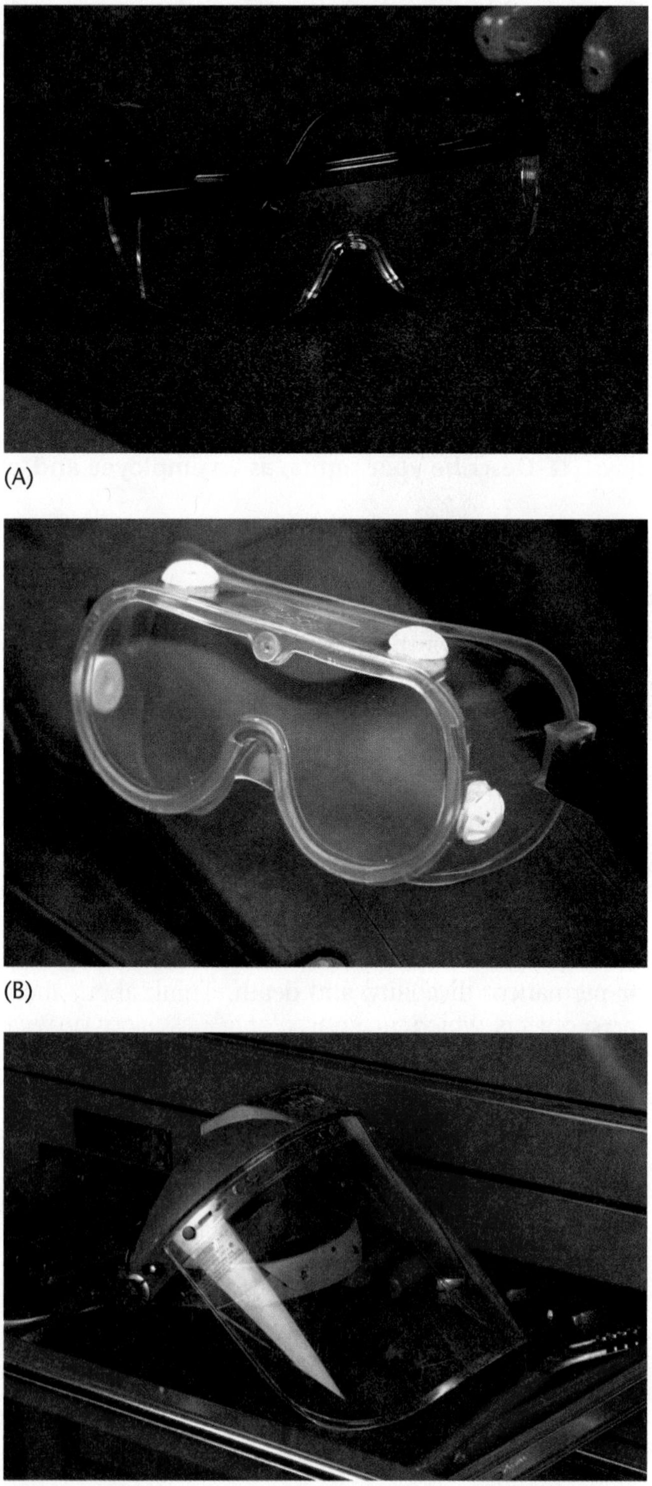

(A)

(B)

(C)

Figure 3–1 (A) Safety glasses; (B) safety (splash) goggles; and (C) a face shield. *Courtesy of Goodson Shop Supplies*

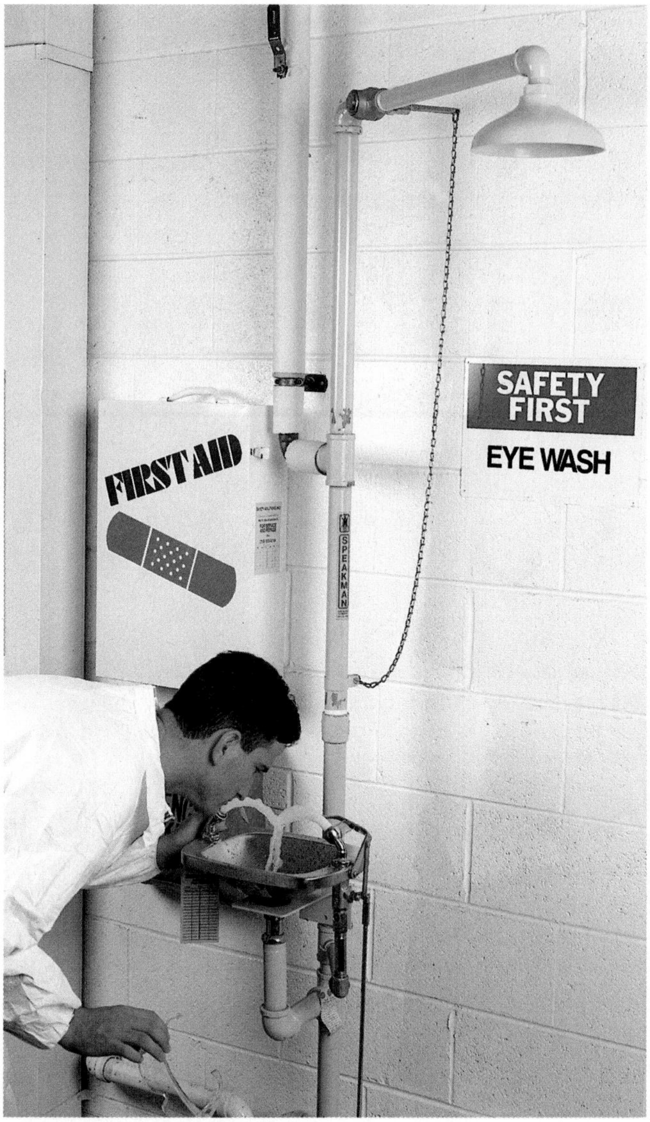

Figure 3-2 A combination eye wash and safety shower. *Courtesy of DuPont Company*

Shoes Automotive work involves the handling of many heavy objects, which can be accidentally dropped on your feet or toes. Always wear shoes that are made of leather or a similar material or boots with nonslip soles. Steel-tipped safety shoes can give added protection to your feet. Jogging or basketball shoes, street shoes, and sandals are inappropriate in the shop.

Gloves Good hand protection is often overlooked. Gloves are worn to protect your hands from disease, injury, and to keep them clean. Many different types of gloves are worn by technicians. A well-fitted pair of heavy work gloves should be worn during operations such as grinding and welding or when handling high-temperature components. Polyurethane or vinyl gloves should be worn when handling strong and dangerous caustic chemicals. These chemicals can easily burn your skin. Latex surgical-type **(Figure 3–3)** and nitrile gloves are worn as protection against disease to keep grease from building up under and around your fingernails. Latex gloves are more comfortable to wear but weaken when they are exposed to gas, oil, and solvents. Nitrile gloves are not as comfortable as latex gloves but they are not affected by gas, oil, and solvents. Your choice of hand protection should be based on what you are doing.

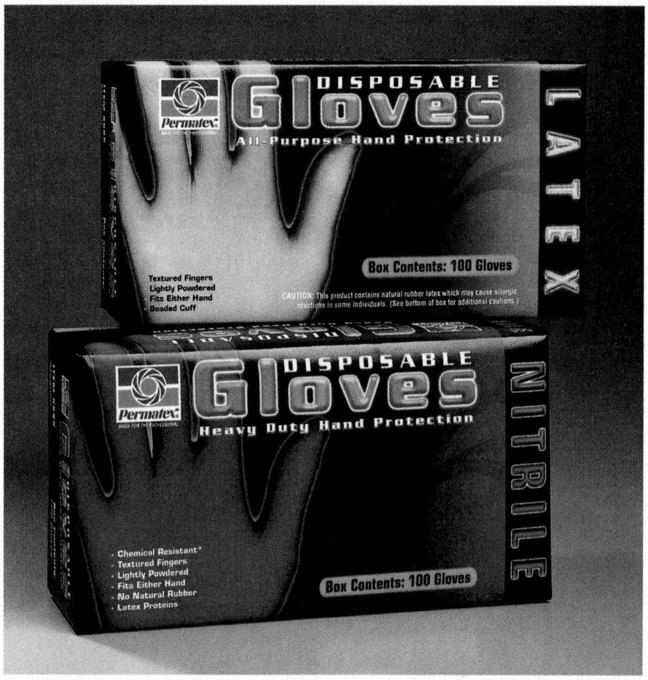

Figure 3-3 Disposable latex gloves. *Courtesy of Permatex, Inc.*

Disease Prevention When you are ill with something that may be contagious, see a doctor and do not go to work or school until the doctor says there is little chance of someone else contracting the illness from you. Doing this will protect others and if others do this, you will be protected.

You should also be concerned with and protect yourself and others from bloodborne pathogens. **Bloodborne pathogens** are pathogenic microorganisms that are present in human blood and can cause disease in humans. These pathogens include, but are not limited to, hepatitis B virus (HBV) and human immunodeficiency virus (HIV). For everyone's protection, any injury that causes bleeding should be dealt with as a threat to others. You should avoid contact with the blood of another. If you need to administer some form of first aid, make sure you wear hand protection before you do so. You should also wear gloves and other protection when handling the item that caused the cut. This item should be sterilized immediately. Most importantly, like all injuries, report the accident to your instructor or supervisor.

Ear Protection Exposure to very loud noise levels for extended periods of time can lead to hearing loss. Air wrenches, engines running under a load, and vehicles running in enclosed areas can all generate annoying and harmful levels of noise. Simple earplugs or earphone-type protectors **(Figure 3–4)** should be worn in environments that are constantly noisy.

Respiratory Protection It is not uncommon for a technician to work with chemicals that have toxic fumes. **Respirators** or respiratory masks **(Figure 3–5)** should be worn whenever you will be exposed to toxic fumes or excessive amounts of dust. Cleaning parts with solvents and painting are the most common activities during which respiratory masks should be worn.

High-efficiency respiratory masks should also be worn when handling parts that have asbestos dust on them or when handling hazardous materials. The proper handling of these materials is covered in great detail later in this chapter.

(A)

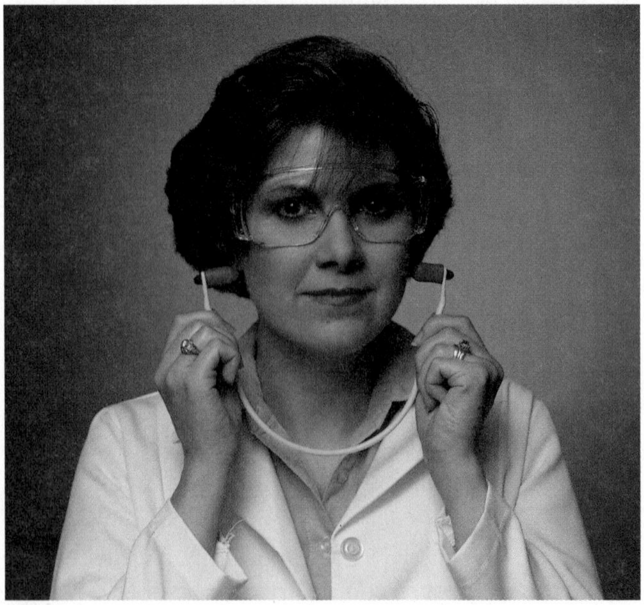

(B)

Figure 3–4 Typical (A) earmuffs and (B) earplugs. *Courtesy of Wilson Safety Company*

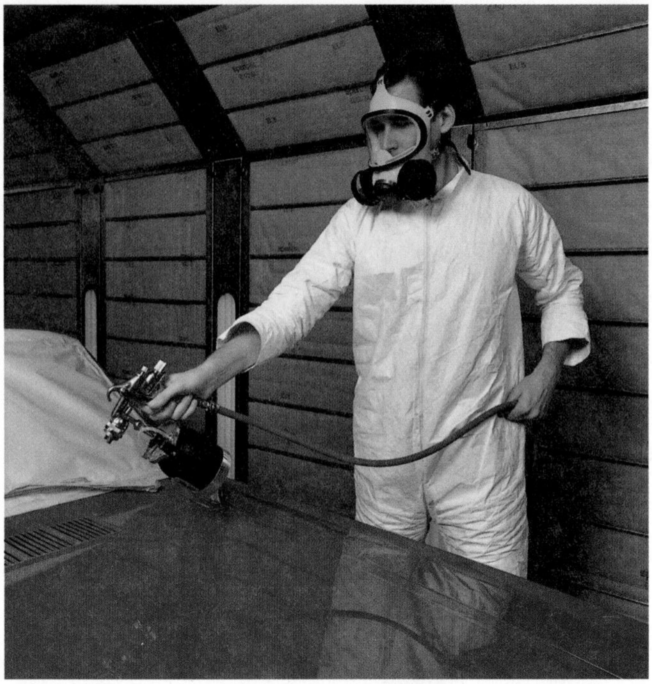

Figure 3–5 A respirator should be worn whenever you will be exposed to toxic fumes or excessive dust. *Photo courtesy of Binks Manufacturing Company*

Lifting and Carrying Knowing the proper way to lift heavy objects is important. You should also use back-protection devices when you are lifting a heavy object. Always lift and work within your ability and ask others to help when you are not sure whether you can handle the size or weight of an object. Even small, compact parts can be surprisingly heavy or unbalanced. Think about how you are going to lift something before beginning. When lifting any object, follow these steps:

1. Place your feet close to the object. Position your feet so that you will be able to maintain a good balance.
2. Keep your back and elbows as straight as possible. Bend your knees until your hands reach the best place to get a strong grip on the object **(Figure 3–6)**.
3. If the part is in a cardboard box, make sure the box is in good condition. Old, damp, or poorly sealed boxes will tear and the part will fall out.
4. Firmly grasp the object or container. Never try to change your grip as you move the load.
5. Keep the object close to your body, and lift it up by straightening your legs. Use your leg muscles, not your back muscles.
6. If you must change your direction of travel, never twist your body. Turn your whole body, including your feet.
7. When placing the object on a shelf or counter, do not bend forward. Place the edge of the load on

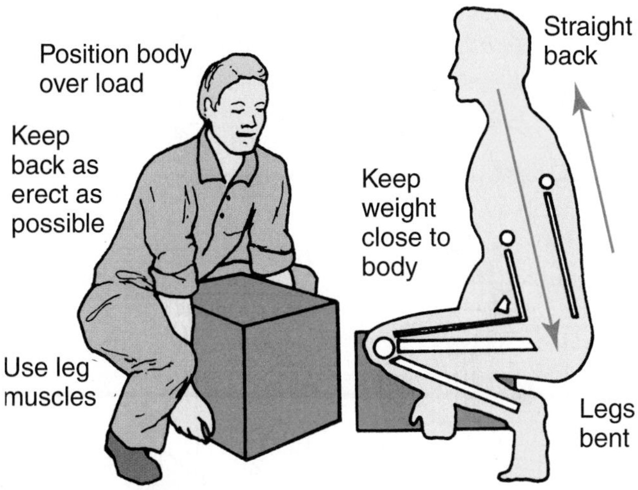

Figure 3-6 Use your leg muscles—*never* your back—to lift heavy objects.

the shelf and slide it forward. Be careful not to pinch your fingers.

8. When setting down a load, bend your knees and keep your back straight. Never bend forward. This strains the back muscles.
9. When lowering something heavy to the floor, set the object on blocks of wood to protect your fingers.

Professional Behavior

Accidents can be prevented simply by the way you act. The following are some guidelines to follow while working in a shop. This list does not include everything you should or should not do; it merely gives you some things to think about:

- Never smoke while working on a vehicle or while working with any machine in the shop.
- To prevent serious burns, keep your skin away from hot metal parts such as the radiator, exhaust manifold, tailpipe, catalytic converter, and muffler.
- Always disconnect electric engine cooling fans when working around the radiator. Many of these will turn on without warning and can easily chop off a finger or hand. Make sure you reconnect the fan after you have completed your repairs.
- When working with a hydraulic press, make sure the pressure is applied in a safe manner. It is generally wise to stand to the side when operating the press.
- Properly store all parts and tools by putting them away in a place where people will not trip over them. This practice not only cuts down on injuries, but also reduces time wasted looking for a misplaced part or tool.

TOOL AND EQUIPMENT SAFETY

An automotive technician must adhere to the following shop safety guidelines when working with tools and equipment.

Hand Tool Safety

Careless use of simple hand tools such as wrenches, screwdrivers, and hammers causes many shop accidents that could be prevented.

Keep all hand tools grease-free and in good condition. Tools that slip can cause cuts and bruises. If a tool slips and falls into a moving part, it can fly out and cause serious injury.

Use the proper tool for the job. Make sure the tool is of professional quality. Using poorly made tools or the wrong tools can damage parts or the tool itself, or could cause injury. Never use broken or damaged tools.

Be careful when using sharp or pointed tools. Do not put sharp tools or other sharp objects into your pockets. They can stab or cut your skin, ruin automotive upholstery, or scratch a painted surface. If a tool is supposed to be sharp, make sure it is sharp. Dull tools can be more dangerous than sharp tools.

Power Tool Safety

Power tools are operated by an outside source of power, such as electricity, compressed air, or hydraulic pressure. Safety around power tools is very important. Serious injury can result from carelessness. Always wear safety glasses when using power tools.

If the tool is electrically powered, make sure it is properly grounded. Check the wiring for cracks in the insulation, as well as for bare wires, before using it. Also, when using electrical power tools, never stand on a wet or damp floor. Disconnect the power source before doing any work on the machine or tool. Before plugging in any electric tool, make sure its switch is in the off position. When you have finished using the tool, turn it off and unplug it. Never leave a running power tool unattended.

When using power equipment on a small part, never hold the part in your hand. Always mount the part in a bench vise or use vise grip pliers. Never try to use a machine or tool beyond its stated capacity or for operations requiring more than the rated power of the tool.

When working with larger power tools, such as a bench or floor grinding wheel, check the machine and the grinding wheels for signs of damage before using them. If the wheels are damaged, they should be replaced and not used. Check the speed rating of the wheel and make sure it matches the speed of the machine. Never spin a grinding wheel at a speed higher than it is rated for. Be sure to place all safety guards in position **(Figure 3–7)**. A safety guard is a protective cover that is placed over a moving part. Although the safety guards are designed to

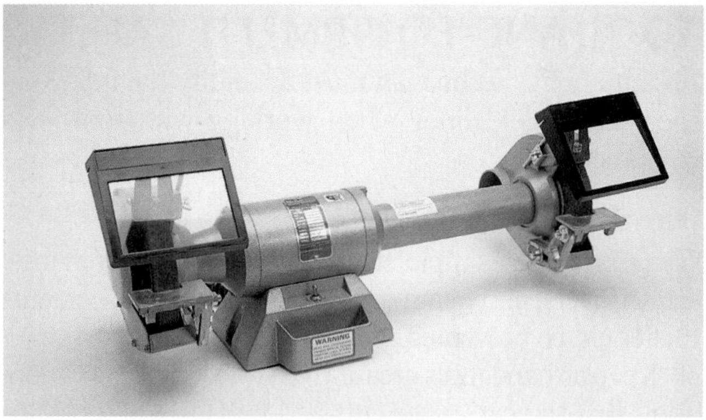

Figure 3-7 A bench grinder with its safety shields in place. *Courtesy of Snap-on Tools Company*

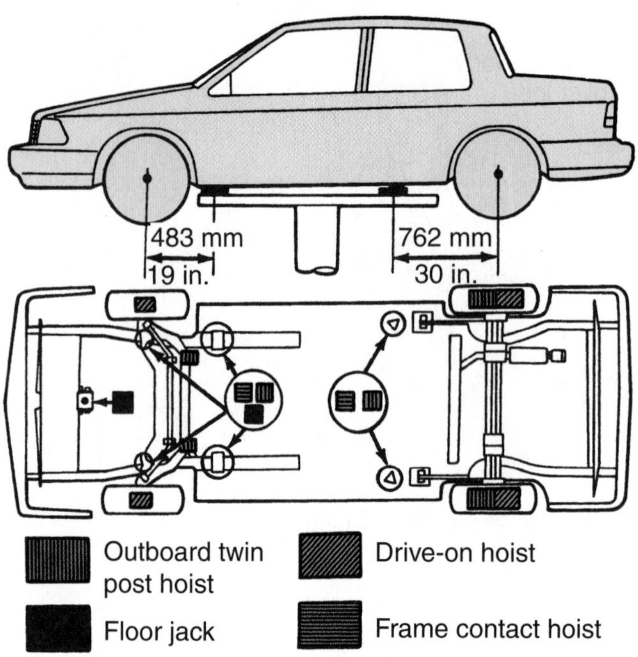

Figure 3-8 Hoisting and lifting points for a typical unibody car. *Courtesy of DaimlerChrysler Corporation*

prevent injury, you should still wear safety glasses and/or a face shield while using the machine. Make sure there are no people or parts around the machine before starting it. Keep your hands and clothing away from the moving parts. Maintain a balanced stance while using the machine.

Compressed Air Equipment Safety

Tools that use compressed air are called **pneumatic tools**. Compressed air is used to inflate tires, apply paint, and drive tools. Compressed air can be dangerous when it is not used properly.

When using compressed air, safety glasses and/or a face shield should be worn. Particles of dirt and pieces of metal, blown by the high-pressure air, can penetrate your skin or get into your eyes.

Before using a compressed air tool, check all hose connections. Always hold an air nozzle or air control device securely when starting or shutting off the compressed air. A loose nozzle can whip suddenly and cause serious injury. Never point an air nozzle at anyone. Never use compressed air to blow dirt from your clothes or hair. Never use compressed air to clean the floor or workbench.

Never spin bearings with compressed air. If the bearing is damaged, one of the steel balls or rollers might fly out and cause serious injury.

Finally, pneumatic tools must always be operated at the pressure recommended by the manufacturer.

Lift Safety

Always be careful when raising a vehicle on a lift or a hoist. Adapters and hoist plates must be positioned correctly on twin post- and rail-type lifts to prevent damage to the underbody of the vehicle. There are specific lift points. These points allow the weight of the vehicle to be evenly supported by the adapters or hoist plates. The correct lift points can be found in the vehicle's service manual. **Figure 3–8** shows typical locations for unibody and frame cars. These diagrams are for illustration only. Always follow the manufacturer's instructions. Before operating any lift or hoist, carefully read the operating manual and follow the operating instructions.

Once you feel the lift supports are properly positioned under the vehicle, raise the lift until the supports contact the vehicle. Then, check the supports to make sure they are in full contact with the vehicle. Shake the vehicle to make sure it is securely balanced on the lift, then raise the lift to the desired working height.

WARNING!

Before working under a car, make sure the lift's locking device is engaged.

The Automotive Lift Institute (ALI) is an association concerned with the design, construction, installation, operation, maintenance, and repair of automotive lifts. Their primary concern is safety. Every lift approved by ALI has the label shown in **Figure 3–9**. It is a good idea to read through the safety tips included on that label before using a lift.

Jack and Jack Stand Safety

A vehicle can be raised off the ground by a hydraulic jack **(Figure 3–10)**. A handle on the jack is moved up and down to raise part of a vehicle and a valve is turned to release the hydraulic pressure in the jack to lower the part. At the end of the jack is a lifting pad. The pad must be positioned under an area of the vehicle's frame or at one of the manufacturer's recommended lift points. Never

AUTOMOTIVE LIFT

SAFETY TIPS

Post these safety tips where they will be a constant reminder to your lift operator. For information specific to the lift, always refer to the lift manufacturer's manual.

1. Inspect your lift daily. Never operate if it malfunctions or if it has broken or damaged parts. Repairs should be made with original equipment parts.
2. Operating controls are designed to close when released. Do not block open or override them.
3. Never overload your lift. Manufacturer's rated capacity is shown on nameplate affixed to the lift.
4. Positioning of vehicle and operation of the lift should be done only by trained and authorized personnel.
5. Never raise vehicle with anyone inside it. Customers or by-standers should not be in the lift area during operation.
6. Always keep lift area free of obstructions, grease, oil, trash, and other debris.
7. Before driving vehicle over lift, position arms and supports to provide unobstructed clearance. Do not hit or run over lift arms, adapters, or axle supports. This could damage lift or vehicle.
8. Load vehicle on lift carefully. Position lift supports to contact at the vehicle manufacturer's recommended lifting points. Raise lift until supports contact vehicle. Check supports for secure contact with vehicle. Raise lift to desired working height. CAUTION: If you are working under vehicle, lift should be raised high enough for locking device to be engaged.
9. Note that with some vehicles, the removal (or installation) of components may cause a critical shift in the center of gravity and result in raised vehicle instability. Refer to the vehicle manufacturer's service manual for recommended procedures when vehicle components are removed.
10. Before lowering lift, be sure tool trays, stands, etc. are removed from under vehicle. Release locking devices before attempting to lower lift.
11. Before removing vehicle from lift area, position lift arms and supports to provide an unobstructed exit (See Item #7).

These "Safety Tips," along with "Lifting it Right," a general lift safety manual, are presented as an industry service by the Automotive Lift Institute. For more information on this material, write to: ALI, P.O. Box 1519, New York, NY 10101.

Look For This Label on all Automotive Service Lifts.

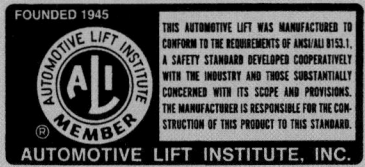

FOUNDED 1945

AUTOMOTIVE LIFT INSTITUTE ALI MEMBER ®

THIS AUTOMOTIVE LIFT WAS MANUFACTURED TO CONFORM TO THE REQUIREMENTS OF ANSI/ALI B153.1, A SAFETY STANDARD DEVELOPED COOPERATIVELY WITH THE INDUSTRY AND THOSE SUBSTANTIALLY CONCERNED WITH ITS SCOPE AND PROVISIONS. THE MANUFACTURER IS RESPONSIBLE FOR THE CONSTRUCTION OF THIS PRODUCT TO THIS STANDARD.

AUTOMOTIVE LIFT INSTITUTE, INC.

Figure 3-9 Automotive lift safety tips. *Courtesy of Automotive Lift Institute*

place the pad under the floorpan or under steering and suspension components, because they can easily be damaged by the weight of the vehicle. Always position the jack so that the wheels of the vehicle can roll as the vehicle is being raised.

WARNING!

Never use a lift or jack to move something heavier than it is designed for. Always check the rating before using a lift or jack. If a jack is rated for 2 tons, do not attempt to use it for a job that requires a 5-ton jack. It is dangerous for you and the vehicle.

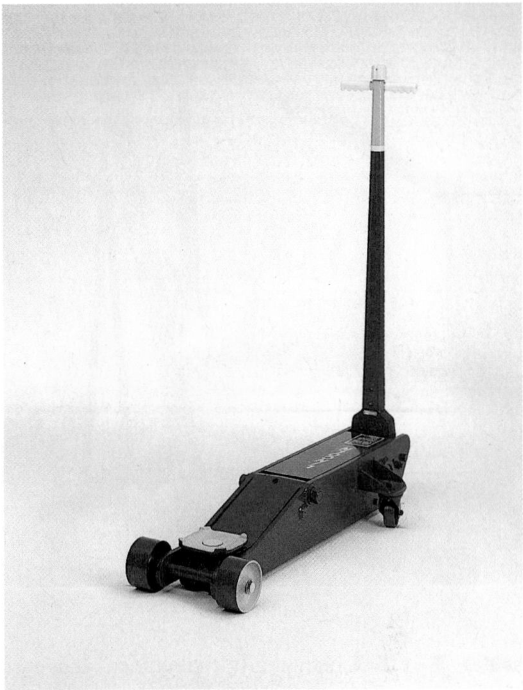

Figure 3-10 Typical hydraulic jack. *Courtesy of Lincoln Automotive Company*

Safety stands, also called **jack stands (Figure 3–11)**, are supports of various heights that sit on the floor. They are placed under a sturdy chassis member, such as the frame or axle housing, to support the vehicle. Once the safety stands are in position, the hydraulic pressure in the jack should be slowly released until the weight of the vehicle is on the stands. Like jacks, jack stands also have a capacity rating. Always use a jack stand of the correct rating.

Never move under a vehicle when it is supported by only a hydraulic jack. Rest the vehicle on the safety stands before moving under the vehicle.

The jack should be removed after the jack stands are set in place. This eliminates a hazard, such as a jack

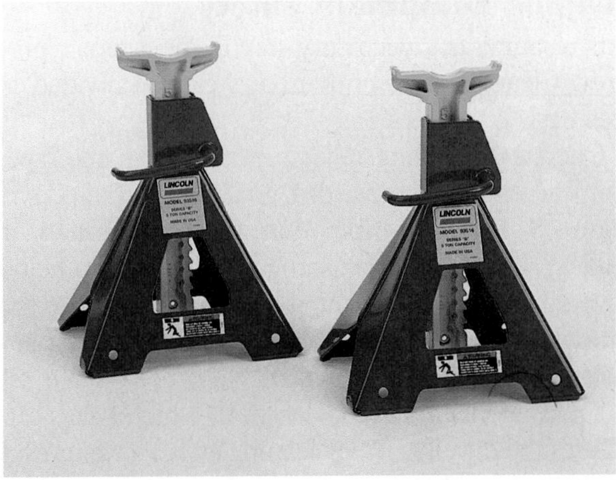

Figure 3-11 Jack stands should be used to support the vehicle after it has been raised by a jack. *Courtesy of Lincoln Automotive Company*

Figure 3-12 A heavy-duty chain hoist. *Courtesy of Lincoln Automotive Company*

handle sticking out into a walkway. A jack handle that is bumped or kicked can cause a tripping accident or cause the vehicle to fall.

Chain Hoist and Crane Safety

Heavy parts of the automobile, such as engines, are removed by using chain hoists **(Figure 3–12)** or cranes. Another term for a chain hoist is chain fall. Cranes often are called cherry pickers.

To prevent serious injury, chain hoists and cranes must be properly attached to the parts being lifted. Always use bolts with enough strength to support the object being lifted. After you have attached the lifting chain or cable to the part that is being removed, have your instructor check it. Place the chain hoist or crane directly over the assembly. Then, attach the chain or cable to the hoist.

Cleaning Equipment Safety

Parts cleaning is a necessary step in most repair procedures. Cleaning automotive parts can be divided into three basic categories.

Chemical cleaning relies primarily on some type of chemical action to remove dirt, grease, scale, paint, or rust **(Figure 3–13)**. A combination of heat, agitation, mechanical scrubbing, or washing may be used to help remove dirt. Chemical cleaning equipment includes small parts washers, hot/cold tanks, pressure washers, spray washers, and salt baths.

Thermal cleaning relies on heat, which bakes off or oxidizes the dirt. Thermal cleaning leaves an ash residue on the surface that must be removed by an additional cleaning process, such as airless shot blasting or spray washing.

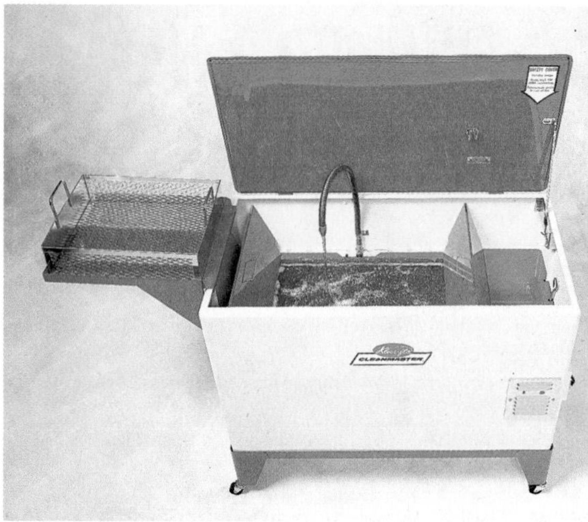

Figure 3-13 An automotive parts washer. *Courtesy of Broadhead-Garrett*

Abrasive cleaning relies on physical abrasion to clean the surface. This includes everything from a wire brush to glass bead blasting, airless steel shot blasting, abrasive tumbling, and vibratory cleaning. Chemical in-tank solution sonic cleaning might also be included here because it relies on the scrubbing action of ultrasonic sound waves to loosen surface contaminants.

Vehicle Operation

When a customer brings a vehicle in for service, certain driving rules should be followed to ensure your safety and the safety of those working around you. For example, before moving a car into the shop, buckle your safety belt. Make sure no one is near, the way is clear, and there are no tools or parts under the car before you start the engine.

Check the brakes before putting the vehicle in gear. Then, drive slowly and carefully in and around the shop.

When road testing the car, obey all traffic laws. Drive only as far as is necessary to check the automobile and verify the customer's complaint. Never make excessively quick starts, turn corners too quickly, or drive faster than conditions allow.

If the engine must be kept running while you are working on the car, block the wheels to prevent the vehicle from moving. Place the transmission in park for automatic transmissions or in neutral for manual transmissions. Set the parking (emergency) brake. Never stand directly in front of or behind a running vehicle.

When parking a vehicle in the shop, always roll the windows down. This allows for access if the doors accidentally lock.

Run the engine only in a well-ventilated area to avoid the danger of poisonous **carbon monoxide (CO)** in the engine exhaust. CO is an odorless but deadly gas. Most shops have an exhaust ventilation system **(Figure 3–14)**; always use it. Connect the hose from the vehicle's tailpipe

Figure 3-14 When running an engine in a shop, make sure the vehicle's exhaust is connected to the shop's exhaust ventilation system.

Figure 3-15 Flammable liquids should be stored in safety-approved containers. *Courtesy of Gutman Advertising Agency*

to the intake for the vent system. Make sure the vent system is turned on before running the engine. If the work area does not have an exhaust venting system, use a hose to direct the exhaust out of the building.

WORK AREA SAFETY

Your work area should be kept clean and safe. The floor and bench tops should be kept clean, dry, and orderly. Any oil, coolant, or grease on the floor can make it slippery. Slips can result in serious injuries. To clean up oil, use a commercial oil absorbent. Keep all water off the floor. Water causes smooth floors to become slippery, and it readily conducts electricity. Aisles and walkways should be kept clean and wide enough for you to easily move through them. Make sure the work areas around machines are large enough for you to safely operate the machines.

Make sure all drain covers are snugly in place. Open drains or covers that are not flush to the floor can cause toe, ankle, and leg injuries.

Keep an up-to-date list of emergency telephone numbers clearly posted next to the telephone. The numbers of a doctor, hospital, and fire and police departments should be included. Also, the work area should have a first-aid kit for treating minor injuries and eye-flushing kits readily available. You should know where these items are kept.

Gasoline is a highly **flammable** volatile liquid. Something that is flammable catches fire and burns easily. A **volatile liquid** is one that vaporizes very quickly. Flammable volatile liquids are potential firebombs. Always keep gasoline or diesel fuel in an approved safety can **(Figure 3–15)**, and never use gasoline to clean your hands or tools.

Handle all solvents (or any liquids) with care to avoid spillage. Keep all solvent containers closed, except when pouring. Proper ventilation is very important in areas where volatile solvents and chemicals are used. Solvents and other combustible materials must be stored in approved and designated storage cabinets or rooms **(Figure 3–16)**. Storage rooms should have adequate ventilation.

Be extra careful when transferring flammable materials from bulk storage. Static electricity can build up to the point where it creates a spark that could cause an

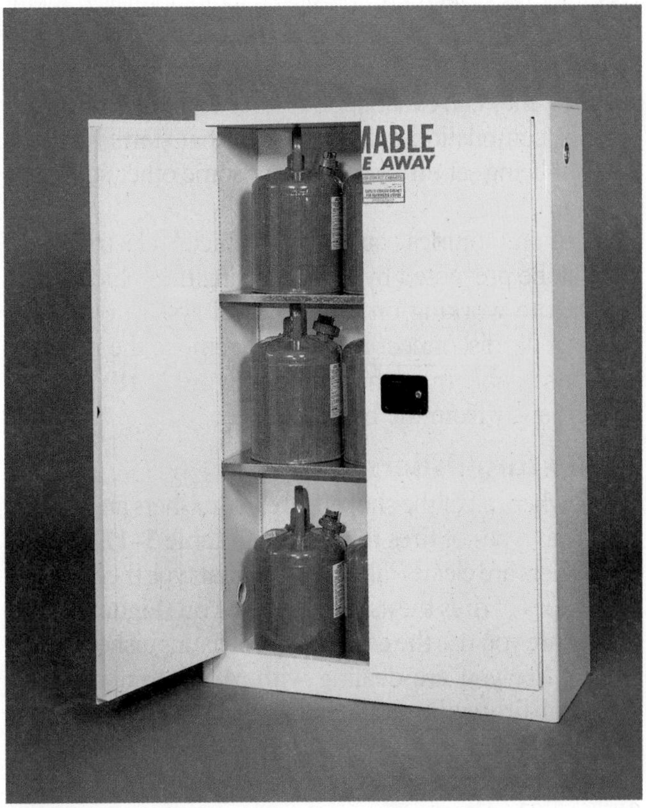

Figure 3-16 Store combustible materials in approved safety cabinets. *Courtesy of Securall Safety Storage Equipment*

Figure 3–17 Dirty rags and towels should be kept in an approved container. *Courtesy of DuPont Company*

explosion. Discard or clean all empty solvent containers. Solvent fumes in the bottom of these containers are very flammable. Never light matches or smoke near flammable solvents and chemicals, including battery acids.

Oily rags should also be stored in an approved metal container **(Figure 3–17)**. When these oily, greasy, or paint-soaked rags are left lying about or are not stored properly, they can cause spontaneous combustion. Spontaneous combustion results in a fire that starts by itself, without being set off by a match or some other source of ignition.

Fires and injuries caused by a vehicle's electrical system can be prevented by disconnecting the vehicle's battery before working on the electrical system or before welding. To disconnect the battery, remove the negative or ground cable from the battery **(Figure 3–18)** and position it away from the battery.

Fire Extinguishers

Know where all of the shop's fire extinguishers are located and what types of fires they put out **(Table 3–1)**. Fire extinguishers are clearly labeled as to what type they are and what types of fires they should be used on **(Figure 3–19)**. Make sure you use the correct type of extinguisher for the type of fire you are dealing with. A multipurpose dry chemical fire extinguisher will put out ordinary combustibles, flammable liquids, and electrical fires. Never put water on a gasoline fire. The water will just spread the fire. The proper fire extinguisher will smother the flames. Remember, during a fire, never open doors or

Figure 3–18 Before doing any electrical work or working around the battery, disconnect the negative lead of the battery.

windows unless it is absolutely necessary; the extra draft will only make the fire worse. Make sure the fire department is contacted before or during your attempt to extinguish a fire.

To extinguish a fire, stand 6 to 10 feet (2 to 3 meters) from the fire. Hold the extinguisher firmly in an upright position. Aim the nozzle at the base and use a side-to-side motion, sweeping the entire width of the fire. Stay low to avoid inhaling the smoke. If it gets too hot or too smoky, get out. Remember, never go back into a burning build-

Figure 3–19 Know the location and types of fire extinguishers available in the shop. *Courtesy of DuPont Company*

TABLE 3–1 GUIDE TO EXTINGUISHER SELECTION

	Class of Fire	Typical Fuel Involved	Type of Extinguisher
Class A Fires (green)	**For Ordinary Combustibles** Put out a class A fire by lowering its temperature or by coating the burning combustibles.	Wood Paper Cloth Rubber Plastics Rubbish Upholstery	Water*[1] Foam* Multipurpose dry chemical[4]
Class B Fires (red)	**For Flammable Liquids** Put out a class B fire by smothering it. Use an extinguisher that gives a blanketing, flame-interrupting effect; cover whole flaming liquid surface.	Gasoline Oil Grease Paint Lighter fluid	Foam* Carbon dioxide[5] Halogenated agent[6] Standard dry chemical[2] Purple K dry chemical[3] Multipurpose dry chemical[4]
Class C Fires (blue)	**For Electrical Equipment** Put out a class C fire by shutting off power as quickly as possible and by always using a nonconducting extinguishing agent to prevent electric shock.	Motors Appliances Wiring Fuse boxes Switchboards	Carbon dioxide[5] Halogenated agent[6] Standard dry chemical[2] Purple K dry chemical[3] Multipurpose dry chemical[4]
Class D Fires (yellow)	**For Combustible Metals** Put out a class D fire of metal chips, turnings, or shavings by smothering or coating with a specially designed extinguishing agent.	Aluminum Magnesium Potassium Sodium Titanium Zirconium	Dry powder extinguishers and agents only

**Cartridge-operated water, foam, and soda-acid types of extinguishers are no longer manufactured. These extinguishers should be removed from service when they become due for their next hydrostatic pressure test.*

Notes:

(1) Freezes in low temperatures unless treated with antifreeze solution, usually weighs more than 20 pounds (9 kg), and is heavier than any other extinguisher mentioned.

(2) Also called ordinary or regular dry chemical (sodium bicarbonate)

(3) Has the greatest initial fire-stopping power of the extinguishers mentioned for class B fires. Be sure to clean residue immediately after using the extinguisher so sprayed surfaces will not be damaged (potassium bicarbonate)

(4) The only extinguishers that fight A, B, and C classes of fires. However, they should not be used on fires in liquefied fat or oil of appreciable depth. Be sure to clean residue immediately after using the extinguisher so sprayed surfaces will not be damaged (ammonium phosphates)

(5) Use with caution in unventilated, confined spaces.

(6) May cause injury to the operator if the extinguishing agent (a gas) or the gases produced when the agent is applied to a fire is inhaled.

ing for anything. To help remember how to use an extinguisher, remember the word "PASS."

Pull the pin from the handle of the extinguisher.

Aim the extinguisher's nozzle at the base of the fire.

Squeeze the handle.

Sweep the entire width of the fire with the contents of the extinguisher.

MANUFACTURER'S WARNINGS AND GOVERNMENT REGULATIONS

A typical shop contains many potential health hazards for those working in it. These hazards can cause injury, sickness, health impairments, discomfort, and even death. Hazards can be classified as chemical, wastes, physical, and ergonomic:

- **Chemical hazards** are caused by high concentrations of vapors, gases, or solids in the form of dust.
- **Hazardous wastes** are those substances that are the result of a service.
- **Physical hazards** include excessive noise, vibration, pressures, and temperatures.
- **Ergonomic hazards** are conditions that impede normal and/or proper body position and motion.

Many government agencies are charged with ensuring safe work environments for all workers. Federal agencies include the **Occupational Safety and Health Administration (OSHA)**, Mine Safety and Health Administration (MSHA), and National Institute for Occupational Safety and Health (NIOSH). These agencies, as well as state and local governments, have instituted regulations that must be understood and followed. Everyone in a shop has the responsibility for adhering to these regulations.

OSHA

In 1970, OSHA was formed by the federal government to "assure safe and healthful working conditions for working men and women; by authorizing enforcement of the standards developed under the Act; by assisting and encouraging the States in their efforts to assure safe and healthful working conditions by providing research, information, education, and training in the field of occupational safety and health."

Safety standards have been established that are consistent across the country. It is the employers' responsibility to provide a place of employment that is free from all recognized hazards and that will be inspected by government agents knowledgeable in the law as it applies to working conditions. All safety and health issues of the automotive industry are controlled by OSHA.

In the United States, OSHA regulates the use of many potentially hazardous materials. The Environmental Protection Agency (EPA) regulates their disposal. As a result, the regulations from these two authorities are best treated as adjoining laws that deal with hazardous materials and waste. OSHA and the EPA have other strict rules and regulations that help to promote safety in the auto shop. These are described throughout this text whenever they are applicable. Maintaining a vehicle involves handling and managing a wide variety of materials and wastes. Some of these wastes can be toxic to fish, wildlife, and humans when improperly managed. No matter the amount of waste produced, it is to the shop's legal and financial advantage to manage the wastes properly and, even more importantly, to prevent the pollution of our natural resources.

RIGHT-TO-KNOW LAW

An important part of a safe work environment is the employees' knowledge of potential hazards. Every employee in the shop is protected by **right-to-know laws**. These laws were put into effect when OSHA's Hazard Communication Standard was published in 1983. This document was originally intended for chemical companies and manufacturers that require employees to handle potentially hazardous materials in the workshop. Since then, most states have enacted their own right-to-know laws. The federal courts have decided that these regulations should apply to all companies, including auto repair shops.

The general intent of right-to-know laws is for employers to provide their employees with a safe working place when hazardous materials are involved. Specifically, there are three areas of employer responsibility.

Primarily, all employees must be trained about their rights under the legislation, the nature of the hazardous chemicals in their workplace, and the contents of the labels on the chemicals. All of the information about each chemical must be posted on **material safety data sheets (MSDS)** and must be accessible. The manufacturer of the chemical must give these sheets to its customers, if they are requested to do so **(Figure 3–20)**. They detail the chemical composition and precautionary information for all products that can present a health or safety hazard. The Canadian equivalents to MSDS are called **workplace hazardous materials information systems (WHMIS)**.

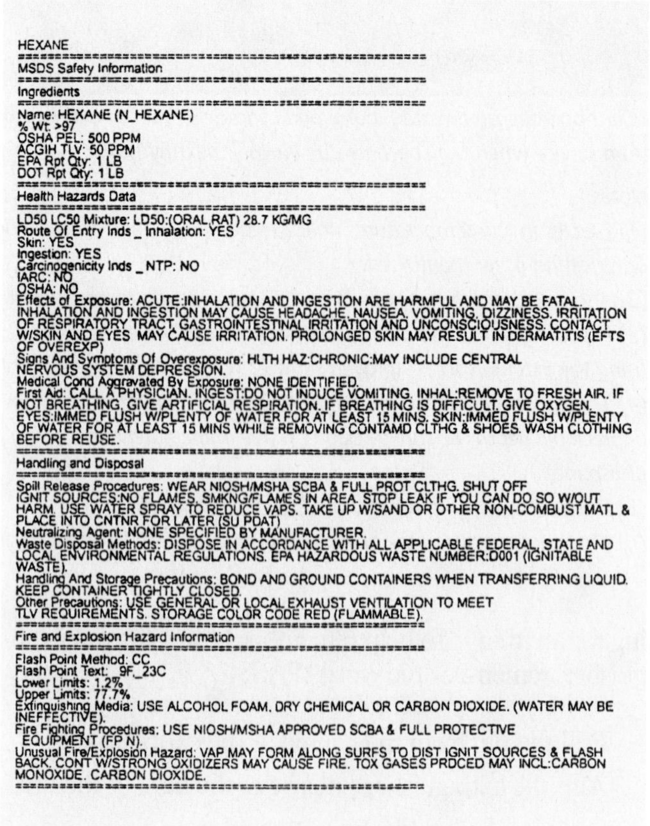

HEXANE

MSDS Safety Information

Ingredients

Name: HEXANE (N_HEXANE)
% Wt: >97
OSHA PEL: 500 PPM
ACGIH TLV: 50 PPM
EPA Rpt Qty: 1 LB
DOT Rpt Qty: 1 LB

Health Hazards Data

LD50 LC50 Mixture: LD50:(ORAL,RAT) 28.7 KG/MG
Route Of Entry Inds _ Inhalation: YES
Skin: YES
Ingestion: YES
Carcinogenicity Inds _ NTP: NO
IARC: NO
OSHA: NO
Effects of Exposure: ACUTE:INHALATION AND INGESTION ARE HARMFUL AND MAY BE FATAL. INHALATION AND INGESTION MAY CAUSE HEADACHE, NAUSEA, VOMITING, DIZZINESS, IRRITATION OF RESPIRATORY TRACT, GASTROINTESTINAL IRRITATION AND UNCONSCIOUSNESS. CONTACT W/SKIN AND EYES MAY CAUSE IRRITATION. PROLONGED SKIN MAY RESULT IN DERMATITIS (EFTS OF OVEREXP)
Signs And Symptoms Of Overexposure: HLTH HAZ:CHRONIC:MAY INCLUDE CENTRAL NERVOUS SYSTEM DEPRESSION.
Medical Cond Aggravated By Exposure: NONE IDENTIFIED.
First Aid: CALL A PHYSICIAN. INGEST:DO NOT INDUCE VOMITING. INHAL:REMOVE TO FRESH AIR. IF NOT BREATHING, GIVE ARTIFICIAL RESPIRATION. IF BREATHING IS DIFFICULT, GIVE OXYGEN. EYES:IMMED FLUSH W/PLENTY OF WATER FOR AT LEAST 15 MINS. SKIN:IMMED FLUSH W/PLENTY OF WATER FOR AT LEAST 15 MINS WHILE REMOVING CONTAMD CLTHG & SHOES. WASH CLOTHING BEFORE REUSE.

Handling and Disposal

Spill Release Procedures: WEAR NIOSH/MSHA SCBA & FULL PROT CLTHG. SHUT OFF IGNIT SOURCES:NO FLAMES, SMKNG/FLAMES IN AREA. STOP LEAK IF YOU CAN DO SO W/OUT HARM. USE WATER SPRAY TO REDUCE VAPS. TAKE UP W/SAND OR OTHER NON-COMBUST MATL & PLACE INTO CNTNR FOR LATER (SU PDAT)
Neutralizing Agent: NONE SPECIFIED BY MANUFACTURER.
Waste Disposal Methods: DISPOSE IN ACCORDANCE WITH ALL APPLICABLE FEDERAL, STATE AND LOCAL ENVIRONMENTAL REGULATIONS. EPA HAZARDOUS WASTE NUMBER:D001 (IGNITABLE WASTE).
Handling And Storage Precautions: BOND AND GROUND CONTAINERS WHEN TRANSFERRING LIQUID. KEEP CONTAINER TIGHTLY CLOSED.
Other Precautions: USE GENERAL OR LOCAL EXHAUST VENTILATION TO MEET TLV REQUIREMENTS. STORAGE COLOR CODE RED (FLAMMABLE).

Fire and Explosion Hazard Information

Flash Point Method: CC
Flash Point Text: _9F,_23C
Lower Limits: 1.2%
Upper Limits: 77.7%
Extinguishing Media: USE ALCOHOL FOAM, DRY CHEMICAL OR CARBON DIOXIDE. (WATER MAY BE INEFFECTIVE).
Fire Fighting Procedures: USE NIOSH/MSHA APPROVED SCBA & FULL PROTECTIVE EQUIPMENT (FP N).
Unusual Fire/Explosion Hazard: VAP MAY FORM ALONG SURFS TO DIST IGNIT SOURCES & FLASH BACK. CONT W/STRONG OXIDIZERS MAY CAUSE FIRE. TOX GASES PRDCED MAY INCL:CARBON MONOXIDE, CARBON DIOXIDE.

Figure 3-20 Material safety data sheets are an important part of employee training and should be readily accessible.

An MSDS must include the following information about the product:

- The trade and chemical name of the product
- The manufacturer of the product
- All of the ingredients of the product
- Health hazards such as headaches, skin rashes, nausea, and dizziness
- The product's physical description; this information may include the product's color, odor, permissible exposure limit (PEL), threshold limit value (TLV), specific gravity, boiling point, freezing point, evaporation data, and volatility rating
- The product's explosion and fire data, such as flash point
- The reactivity and stability data
- The product's weight compared to air
- Protection data, including first aid and proper handling

Employees must become familiar with the general uses, protective equipment, accident or spill procedures, and any other information regarding the safe handling of the hazardous material. This training must be given to employees annually and provided to new employees as part of their job orientation.

Furthermore, all hazardous material must be properly labeled, indicating what health, fire, or reactivity hazard it poses and what protective equipment is necessary when handling each chemical. The manufacturer of the hazardous waste materials must provide all warnings and precautionary information, which must be read and understood by the user before application. Attention to all label precautions is essential for the proper use of the chemical and for prevention of hazardous conditions. A list of all hazardous materials used in the shop must be posted for the employees to see.

Finally, shops must maintain documentation on the hazardous chemicals in the workplace, proof of training programs, records of accidents or spill incidents, satisfaction of employee requests for specific chemical information via the MSDS, and a general right-to-know compliance procedure manual utilized within the shop.

HAZARDOUS MATERIALS

WARNING!

When handling any hazardous waste material, be sure to wear the proper safety equipment covered under the right-to-know law. Follow all required procedures correctly. This includes the use of approved respirator equipment.

As mentioned before, some of the materials used in auto repair shops can be dangerous. The solvents and other chemical products used in an auto shop carry warnings and caution information that must be read and understood by all who use them.

Many repair and service procedures generate hazardous wastes. Dirty solvents and cleaners are good examples of hazardous wastes. Something is classified as a hazardous waste by the Environmental Protection Agency if it is on the EPA list of known harmful materials. A complete EPA list of hazardous wastes can be found in the *Code of Federal Regulations*. It should be noted that no material is considered a hazardous waste until the shop is finished using it and is ready to dispose of it.

Regulations on hazardous waste handling and generation have led to the development of equipment now commonly found in shops. Examples of these are thermal cleaning units, close-loop steam cleaners, waste oil furnaces, oil filter crushers, refrigerant recycling machines, engine coolant recycling machines, and highly absorbent cloths.

WARNING!

The shop is ultimately responsible for the safe disposal of hazardous wastes, even after the waste leaves the shop. Only licensed waste removal companies should be used to dispose of the waste. Make sure you know what the company is planning to do with the waste. Make sure you have a written contract stating what is supposed to happen to the waste. Leave nothing to chance. In the event of an emergency hazardous waste spill, contact the National Response Center (1-800-424-8802) immediately. Failure to do so can result in a $10,000 fine, a year in jail, or both.

Many shops use full-service haulers to remove hazardous waste from the property. Besides hauling the hazardous waste away, the hauler also takes care of all the paperwork, deals with the various government agencies, and advises the shop on how to recover the disposal costs.

Guidelines for Handling Shop Wastes

To protect yourself, you should consider the following.

Oil Recycle oil. Set up equipment, such as a drip table or screen table with a used-oil collection bucket, to collect oil that drips off parts. Place drip pans underneath vehicles that are leaking fluids onto the storage area. Do not mix other wastes with used oil, except as allowed by your recycler. Used oil generated by a shop (and/or oil received from household do-it-yourself generators) may be burned on site in a commercial space heater. Also, used oil may be burned for energy recovery. Contact state and local authorities to determine requirements and to obtain necessary permits.

Oil Filters Drain for at least 24 hours, crush **(Figure 3–21)** and recycle used oil filters.

Batteries Recycle batteries by sending them to a reclaimer or back to the distributor. Keeping shipping receipts can demonstrate that you have recycled. Store batteries in a watertight, acid-resistant container. Inspect batteries for cracks and leaks when they come in. Treat a dropped battery as if it were cracked. Acid residue is hazardous because it is corrosive and may contain lead and other toxins. Neutralize spilled acid by covering it with baking soda or lime, and dispose of all hazardous material.

Metal Residue from Machining Collect metal filings when machining metal parts. Keep separate and recycle if possible. Prevent metal filings from falling into a storm sewer drain.

Refrigerants Recover and/or recycle refrigerants during the servicing and disposal of motor vehicle air conditioners and refrigeration equipment. It is not allowable to knowingly vent refrigerants into the atmosphere. Recovery and/or recycling during servicing must be performed by an EPA-certified technician using certified equipment and following specified procedures.

Figure 3-21 A hydraulic single oil filter crusher. *Courtesy of SPX/OTC Service Solutions*

Solvents Replace hazardous chemicals with less toxic alternatives that have equal performance. For example, substitute water-based cleaning solvents for petroleum-based solvent degreasers. To reduce the amount of solvent used when cleaning parts, use a two-stage process: dirty solvent followed by fresh solvent. Hire a hazardous waste management service to clean and recycle solvents. (Some spent solvents must be disposed of as hazardous waste, unless recycled properly.) Store solvents in closed containers to prevent evaporation. Evaporation of solvents contributes to ozone depletion and smog formation. In addition, the residue from evaporation must be treated as a hazardous waste. Properly label spent solvents and store on drip pans or in diked areas and only with compatible materials.

Containers Cap, label, cover, and properly store aboveground outdoor liquid containers and small tanks within a diked area and on a paved impermeable surface to prevent spills from running into surface or ground water.

Other Solids Store materials such as scrap metal, old machine parts, and worn tires under a roof or tarpaulin to protect them from the elements and to prevent potential contaminated runoff. Consider recycling tires by retreading them.

Liquid Recycling Collect and recycle coolants from radiators. Store transmission fluids, brake fluids, and solvents containing chlorinated hydrocarbons separately, and recycle or dispose of them properly.

Shop Towels/Rags Keep waste towels in a closed container marked "contaminated shop towels only." To reduce costs and liabilities associated with disposal of used towels, which can be classified as hazardous wastes, investigate using a laundry service that is able to treat the wastewater generated from cleaning the towels.

Hiring a Hauler Hire a reputable, financially stable, and state-approved hauler, who will dispose of your shop wastes legally. If hazardous waste is dumped illegally, your shop may be held responsible.

Waste Storage Always keep hazardous waste separate, properly labeled and sealed in the recommended containers. The storage area should be covered and may need to be fenced and locked if vandalism could be a problem. Select a licensed hazardous waste hauler after seeking recommendations and reviewing the firm's permits and authorizations.

Asbestos

Asbestos has been identified as a health hazard. **Asbestos** is a term used to describe a number of naturally occurring fibrous materials. It has been identified as a carcinogen and has been shown to cause a number of diseases that result in cancer. Asbestos-caused cancer, or mesothelioma, is a form of lung cancer. When breathed in, the asbestos fibers cause scarring of the lungs and/or cause damage to the lung's air passages. The injuries and scars in the lung become an effective holding place for the asbestos. Obviously, you want to avoid breathing in asbestos dust and fibers. When working with asbestos materials, such as brake pads, clutch discs, and some engine gaskets, there are certain guidelines you should follow. All asbestos waste must be disposed of in accordance with OSHA and EPA asbestos regulations.

Generally, the EPA does not regulate the removal of asbestos brakes unless debonding or grinding of asbestos brake pads constitutes more than 50% of the shop's work. At such facilities, the asbestos materials are regulated as a hazardous waste and handled accordingly; they are stored in an enclosed container and sent to a hazardous waste hauler. However, even when asbestos wastes are not regulated as hazardous wastes, the EPA recommends that shops capture asbestos from brake shoes in a separate container. Use a low-pressure/wet-cleaning method, an OSHA-preferred method of compliance. *Never* blow brake dust and never use an air hose for cleaning.

One asbestos cleaning method is the use of a **high-efficiency particulate-arresting (HEPA)** vacuum cleaner. This vacuum cleaner captures the asbestos in a special filter. When used on a brake system, the vacuum cleaner completely houses the brake assembly. Built-in gloves allow a technician to clean the assembly with compressed air through a window and without direct contact with the assembly. The dust is drawn in by the vacuum cleaner.

WARNING!

Make sure the enclosure of the vacuum cleaner fits tightly around the brake or clutch assembly before using compressed air to clean it.

Once the filter in the HEPA vacuum cleaner is full, it must be wetted with a mist of water before it is removed from the cleaner. The filters must be placed in an impermeable container, labeled, and disposed of in the manner prescribed by local or state laws.

Another approved asbestos cleaning method is to use water mixed with an organic solvent or wetting agent. It is important that the wetting agent be allowed to flow through the brake drum or around the brake disc before the brakes are disassembled for further cleaning. Position a catch basin under the brake assembly to capture any contaminated liquid. As the assembly is being disassembled, the parts should be misted with the wetting agent.

To minimize the risks of working around asbestos, follow these simple personal hygiene guidelines:

- Do not smoke while or after working with the materials.
- Thoroughly wash yourself before eating.
- Shower after work.
- Change into work clothes when you arrive at work, and change out of your work clothes after work. Do not take work clothing home.

For more information on work environment safety, contact the United States EPA Office of Compliance at *http://es.inel.gov* or the Coordinating Committee for Automotive Repair (CCAR)–Greenlink at *http://www.ccar-greenlink.org* or 888-476-5465.

KEY TERMS

Abrasive cleaning
Asbestos
Bloodborne pathogens
Carbon monoxide (CO)
Chemical cleaning
Chemical hazards
Ergonomic hazards
Flammable
Hazardous wastes
High efficiency particulate arresting (HEPA)
Jack stands
Material safety data sheets (MSDS)
Occupational Safety and Health Administration (OSHA)
Physical hazards
Pneumatic tools
Power tools
Respirators
Right-to-know laws
Safety glasses
Safety stands
Thermal cleaning
Volatile liquid
Workplace hazardous materials information systems (WHMIS)

SUMMARY

- Dressing safely for work is very important. Wear snug-fitting clothing, eye and ear protection, protective gloves, steel-toed shoes, and caps to cover long hair.
- When choosing eye protection, make sure it has safety glass and offers side protection.
- A respirator should be worn whenever you are working around toxic fumes or excessive dust.
- When shop noise exceeds safe levels, protect your ears by wearing earplugs or earmuffs.
- Safety while using any tool is essential, and even more so when using power tools. Before plugging in a power tool, make sure the power switch is off. Disconnect the power before servicing the tool.

- Always observe all relevant safety rules when operating a vehicle lift or hoist. Jacks, jack stands, chain hoists, and cranes can also cause injury if not operated safely.
- Use care whenever it is necessary to move a vehicle in the shop. Carelessness and playing around can lead to a damaged vehicle and serious injury.
- Carbon monoxide (CO) gas is a poisonous gas present in engine exhaust fumes. Exhaust must be properly vented from the shop using tailpipe hoses or other reliable methods.
- Adequate ventilation is also necessary when working with any volatile solvent or material.
- Gasoline and diesel fuel are highly flammable and should be kept in approved safety cans.
- Never light matches near any combustible materials.
- It is important to know when to use each of the various types of fire extinguishers. When fighting a fire, aim the nozzle at the base and use a side-to-side sweeping motion.
- Right-to-know laws came into effect in 1983 and are designed to protect employees who must handle hazardous materials and wastes on the job.
- Material safety data sheets (MSDS) contain important chemical information and must be furnished to all employees annually. New employees should be given the sheets as part of their job orientation.
- All hazardous and asbestos waste should be disposed of according to OSHA and EPA regulations.

REVIEW QUESTIONS

1. What is the correct way to dispose of used oil filters?
2. Where in the shop should a list of emergency telephone numbers be posted?
3. When should eye protection be worn?
4. How should a class B fire be extinguished?
5. Where can complete EPA lists of hazardous wastes be found?
6. Which of the following statements about safety glasses is true?
   - **a.** They should offer side protection.
   - **b.** The lenses should be made of a shatterproof material.
   - **c.** Some service operations require that additional eye protection be worn with safety glasses.
   - **d.** All of the above statements are true.
7. Gasoline is ______.
   - **a.** highly volatile
   - **b.** highly flammable
   - **c.** dangerous, especially in vapor form
   - **d.** all of the above
8. Technician A says it is recommended that you wear shoes with nonslip soles in the shop. Technician B says steel-toed shoes offer the best foot protection. Who is correct?
   - **a.** A only
   - **b.** B only
   - **c.** Both A and B
   - **d.** Neither A nor B
9. Technician A says used engine coolant should be collected and recycled. Technician B says all oil-based waste materials can be collected in the same container if an approved waste disposal company is hired to rid the shop of the oil. Who is correct?
   - **a.** Technician A
   - **b.** Technician B
   - **c.** Both A and B
   - **d.** Neither A nor B
10. Which method for cleaning parts may leave a residue that must be removed by further cleaning?
    - **a.** chemical
    - **b.** abrasive
    - **c.** thermal
    - **d.** all of the above
11. Federal right-to-know laws concern ______.
    - **a.** auto emission standards
    - **b.** hazards associated with chemicals used in the workplace
    - **c.** employee benefits
    - **d.** hiring practices
12. Which of the following is/are important when working in an automotive shop?
    - **a.** Using the proper tool for the job.
    - **b.** Avoiding loose-fitting clothes.
    - **c.** Wearing steel-toed shoes.
    - **d.** All of the above.
13. When a material reacts violently with water or other materials, it is said to have high ______.
    - **a.** corrosivity
    - **b.** volatility
    - **c.** ignitability
    - **d.** reactivity
14. Which of the following is *not* recommended for use when trying to extinguish flammable liquid fires?
    - **a.** foam
    - **b.** carbon dioxide
    - **c.** water
    - **d.** dry chemical
15. Technician A says some machines can be routinely used beyond their stated capacity. Technician B

says a power tool can be left running unattended if the technician puts up a power-on sign. Who is correct?

a. Technician A **c.** Both A and B
b. Technician B **d.** Neither A nor B

16. What is the correct procedure for using a fire extinguisher to put out a fire?

17. Technician A ties his long hair behind his head while working in the shop. Technician B covers her long hair with a brimless cap. Who is correct?

a. Technician A **c.** Both A and B
b. Technician B **d.** Neither A nor B

18. Technician A uses compressed air to blow dirt from his clothes and hair. Technician B uses compressed air to clean off the top of a work bench. Who is correct?

a. Technician A **c.** Both A and B
b. Technician B **d.** Neither A nor B

19. Heavy protective gloves should be worn when ______.

a. welding
b. grinding metal
c. working with caustic cleaning solutions
d. all of the above

20. Proper disposal of oil filters includes ______.

a. recycling used filters
b. draining them for at least 24 hours
c. crushing them
d. all of the above

4

AUTOMOTIVE SYSTEMS AND PREVENTIVE MAINTENANCE

OBJECTIVES

■ Explain the major events that have influenced the development of the automobile during the last 35 years. ■ Explain the difference between unitized and body-over-frame vehicles. ■ Describe the manufacturing process used in a modern automated automobile assembly plant. ■ List the basic systems that make up an automobile and name their major components and functions. ■ Explain the importance of preventive maintenance, and list at least six examples of typical preventive maintenance.

HISTORICAL BACKGROUND

The automobile has changed quite a bit since the first horseless carriage went down an American street. In 1896, both Henry Ford and Ransom Eli Olds test drove their first gasoline-powered vehicles. This same year is credited as the beginning of the automotive industry, not because of what Ford or Olds did, but because the Duryea Brothers by 1896 had made thirteen cars in their factory, which was the first to make cars for customers.

In the beginning, the automobile looked like the horse-drawn carriage it was designed to replace **(Figure 4–1)**. In fact, for many years most cars looked like carriages. In 1919, 90% of the cars had carriagelike open bodies. Although body styles changed, cars continued to have carriagelike features. It was not until 1939 that running boards began disappearing.

These early cars had rear-mounted engines and very tall tires. They were designed to move people down dirt roads. The automobile changed to meet new conditions: Roads were improved and became paved, more people owned cars, manufacturers tried to sell more cars, concern for safety and the environment grew, and new technology was developed. Because of all of these changes, the automobile became more practical, more affordable, safer, more comfortable, more dependable, and faster. Although many improvements have been made to the original design, the basic structure of the automobile has changed very little.

Nearly all of today's cars still use gasoline engines to drive two or more wheels. A steering system is used to control the direction of the car. A brake system is used to slow down and stop the car. A suspension system is used to absorb road shocks and help the driver maintain control on bumpy roads. The parts of these major systems are mounted on steel frames and the frame is covered with body panels. These panels give the car its shape and protect those inside from the weather and dirt. The body panels also offer some protection for the passengers if the automobile is in an accident.

Although the basics of an automobile have changed little in the past 100 years, the parts and the control systems have changed greatly. The entire automobile is technologically light-years ahead of Ford's and Olds's early models **(Figure 4–2)**. The use of new technology has changed the slow, unreliable, user-hostile vehicles of the early 1900s into vehicles that travel at very high speeds,

Figure 4–1 The 1886 Benz Patent Motor Wagen, one of the first automobiles made. *Courtesy of Mercedes-Benz of N.A., Inc.*

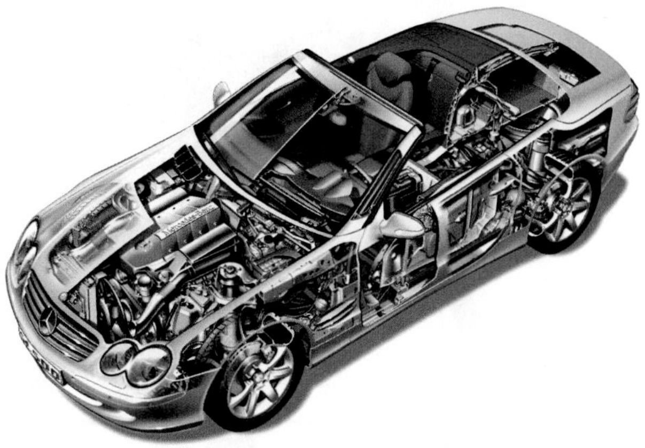

Figure 4-2 A cutaway of a late-model car showing some of the technology of today's cars. *Courtesy of DaimlerChrysler Corporation*

operate trouble-free for thousands of miles, and provide comforts that even the rich had not dreamed of in 1896.

The most dramatic changes have occurred during the past 40 years. In 1965, legislation was passed limiting the levels of exhaust emissions. Although there was little immediate effect on the industry because of this legislation, automobile manufacturers needed to focus on the future. They needed to build cleaner-burning engines. In the following years, stricter emissions laws were passed and manufacturers were required to develop systems to control emissions.

World events in the 1970s continued to shape the development of today's automobile. An oil embargo by Arab nations in 1973 caused the price of gasoline to quickly increase to four times its normal price. This event caused most Americans to realize that the supply of gasoline and other nonrenewable resources was limited. Buyers wanted cars that were not only kind to the environment but that also used a smaller amount of fuel.

In 1975, Congress passed the **Corporate Average Fuel Economy (CAFE)** standards, which require auto makers not only to manufacture clean-burning engines but also to equip their vehicles with engines that burn gasoline efficiently. Under the CAFE standards, different models from each manufacturer are tested for the number of miles they can be driven on a gallon of gas. The fuel efficiencies of these vehicles are averaged together to arrive at a corporate average. The CAFE standards have increased many times since they were first established. A manufacturer that does not meet CAFE standards for a given model year faces heavy fines.

MODERN POWER PLANTS

In trying to produce more efficient vehicles, American manufacturers put four-cylinder and other small engines into their vehicles, instead of large eight-cylinder engines. Some basic engine systems like carburetors and ignition breaker points were replaced by electronic fuel injection and electronic ignition systems.

By the mid-1980s, the American automobile had gained a measure of self-control over emissions and fuel efficiency through the use of computers and other electronics. Fuel and air were carefully monitored and consumed in proportions that maximized the performance of the smaller engines while minimizing the production of harmful pollutants.

After a prolonged period of economic growth in the 1980s, the demand for good performance was once again a shaping force in automotive design. Electronic sensors are now used to monitor engine functions. Computerized engine control systems control air and fuel delivery, ignition timing, emission systems operation, and a host of other related operations. The result is a clean-burning, fuel-efficient, and powerful engine **(Figure 4–3)**.

DESIGN EVOLUTION

Not too long ago, nearly every car and truck built in America was built with body-over-frame construction, rear-wheel drive, and symmetrical designs. Today, most cars do not have a separate frame; instead, the frame and body are built as a single unit, called a unibody. In 1977, when most cars were built on a frame, the average weight of a car was 4,500 pounds (16,800 kg). Today, because of unibody construction and changes in materials, the average weight is 3,000 pounds (1,120 kg). Most trucks are still built on a frame.

Another major influence on design was the switch from rear-wheel drive to front-wheel drive. Making this

Figure 4-3 A late-model lightweight V-12 engine. *Courtesy of DaimlerChrysler Corporation*

switch accomplished many things. The most notable benefits of front-wheel drive are improved traction for the drive wheels, increased interior space, shorter hood lines, and a very compact driveline. Because of the weight and loads pickup trucks are designed to move, they remain rear-wheel drive.

Perhaps the most obvious design change through the years has been in body styles. Body styles have changed in response to the other design considerations and to trends of the day. For example, in the 1950s America had a strange preoccupation with the unknown, outer space; this led to cars that had rocketlike fins. Since then fins have disappeared and body styles have become more rounded to reduce air drag.

Unitized Construction

A **unibody** has no separate frame **(Figure 4–4)**. It is a stressed hull structure in which each of the body parts supplies structural support and strength to the entire vehicle.

The major advantage of unibody vehicles is that they tend to be more tightly constructed because the major parts are all welded together. This design characteristic helps protect the occupants during a collision. However, it causes damage patterns that differ from those of body-over-frame vehicles. Rather than localized damage, the stiffer sections used in unibody design tend to transmit and distribute impact energy throughout more of the vehicle.

Nearly all unibodies are constructed from steel. A few cars, such as the Audi A8, use aluminum instead. An aluminum car body and frame can weigh up to 40% less than an identical body made of steel.

Body-Over-Frame Construction

In body-over-frame construction, the frame is the vehicle's foundation. The body and all major parts of the vehicle are attached to the frame. The frame must also be strong enough to keep the other parts in alignment should a collision occur.

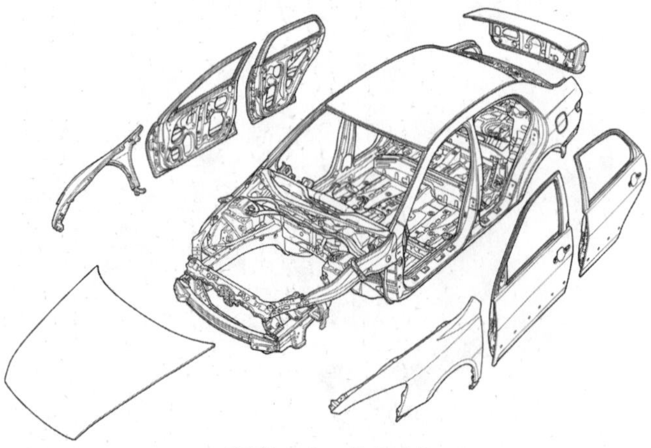

Figure 4–4 The structure of a unibody car. *Courtesy of American Honda Motor Co., Inc.*

Figure 4–5 A typical hydroformed frame for a pickup truck. *Courtesy of Dana Corporation*

The frame is an independent, separate component that is not welded to any of the major units of the body. The body is generally bolted to the frame **(Figure 4–5)**. Large, specially designed rubber mounts are placed between the frame and body structure to reduce noise and vibration from entering the passenger compartment. Quite often, two layers of rubber are used in the mounting pads to provide a smoother ride. Body-over-frame designs are still used on many of today's pickup trucks, full-size vans, and a few full-size passenger cars.

The frame rails are made of stamped steel and are welded together. Some frames are made by a **hydroforming** process, which uses high-pressure water to shape the steel into the desired shape.

BODY SHAPES

Various methods of classifying vehicles exist. Vehicles may be classified by engine type, body/frame construction, fuel consumption structure, type of drive, or the classifications most common to consumers, which are body shape, seat arrangement, and number of doors. Seven basic body shapes are used today:

1. **Sedan.** A vehicle with front and back seats that accommodates four to six persons is classified as either a two- or four-door sedan **(Figure 4–6)**. Often, a two-door sedan is called a coupe **(Figure 4–7)**. If the vehicle's B pillars do not extend up through the side windows, the car is called a hardtop. It can also be classified as either a two- or four-door hardtop.

Figure 4-6 This Toyota Camry is an example of a typical late-model sedan. *Reprinted with permission.*

Figure 4-9 Two-passenger convertibles are called sports cars. Vipers are among the best-known sports cars. *Courtesy of DaimlerChrysler Corporation*

2. **Convertible.** After an absence from the domestic market for several years, many manufacturers have offered convertible cars since 1985. Convertibles have vinyl roofs that can be raised or lowered **(Figure 4–8)**. Like a hardtop, the B pillar stops at the belt line of the car. It can be available in two- and four-door models. Some convertibles have both front and rear seats. Those without rear seats are commonly referred to as sports cars **(Figure 4–9)**.
3. **Liftback** or **hatchback.** The distinguishing feature of this vehicle is its rear luggage compartment, which is an extension of the passenger compartment. Access to the luggage compartment is gained through an upward opening hatch-type door **(Figure 4–10)**. A car of this design can be a three- or five-door model. The third or fifth door is the rear hatch.
4. **Station wagon.** A station wagon **(Figure 4–11)** is characterized by its roof, which extends straight

Figure 4-7 This Acura CL Type-S concept car is a modified coupe. *Courtesy of American Honda Motor Co., Inc.*

Figure 4-10 The distinguishing feature of a hatchback is its rear luggage compartment, which is an extension of the passenger compartment. *Courtesy of Saab Cars Inc.*

Figure 4-8 A BMW M3 convertible. *Courtesy of BMW of North America, Incorporated*

Figure 4-11 A late-model Subaru Legacy station wagon. *Courtesy of Subaru of America, Inc.*

back, allowing a spacious interior luggage compartment in the rear. The rear door, which can be opened in various ways depending on the model, provides access to the luggage compartment. Station wagons come in two- and four-door models and have space for up to nine passengers.

5. **Pickups.** Pickup truck body designs have an open cargo area behind the driver's compartment. There are many varieties available today; there are compact, medium-sized **(Figure 4–12)**, full-sized, and heavy-duty pickups. They can also be had in two-, three-, or four-door models. Some have extended cab areas with seats in back of the front seat. They are available in two-wheel drive, four-wheel drive (4×4), or all-wheel drive **(Figure 4–13)**.
6. **Vans.** The van body design **(Figure 4–14)** has a tall roof and a totally enclosed large cargo or passenger area. Vans can seat from two to twelve passengers, depending on size and design. Basically there are two sizes of vans: mini- and full-size.

Figure 4–12 This Nissan Frontier is an example of a medium sized pickup truck with four doors. *Courtesy of Nissan Motor Company*

Figure 4–13 A full-size Dodge Ram pickup with four-wheel drive. *Courtesy of DaimlerChrysler Corporation.*

Figure 4–14 This Chrysler Voyager is an example of a late-model minivan. Full-size vans are also available. *Courtesy of DaimlerChrysler Corporation*

7. **Sport utility vehicles (SUVs).** This classification of vehicles covers a range of body designs. SUVs are best described as multipurpose vehicles and depending on their size and design can carry a wide range of passengers. A good majority of SUVs have four-wheel drive, although some do not. The classification of SUV implies that the vehicles are designed to do well off-the-road. This is not always the case. Buyers may choose SUVs for status, size, utility, and/or off-road play. Most small SUVs are based on an automobile platform and take on many different looks and features **(Figure 4–15)**. Mid-size SUVs are larger and typically offer more features and comfort **(Figure 4–16)**. There are many large SUVs available **(Figure 4–17)**. These vehicles can seat up to nine adults and tow up to six tons.

Crossover vehicles are a new trend in automotive offerings. These vehicles are a mixture of a station wagon

Figure 4–15 This Honda Element is considered a small SUV and has many unique features including anytime four-wheel drive. *Courtesy of American Honda Motor Co., Inc.*

Figure 4–16 This Lexus GX470 is a midsize SUV with a V-8 engine and many luxuries. *Courtesy of Lexus of America Co.*

Figure 4–17 A late-model Nissan Armada is a good example of a large SUV. *Used with permission from Nissan North America, Inc.*

and an SUV. They have SUV features but are not quite the size of an SUV.

TECHNOLOGICAL ADVANCES

Perhaps the thing that has brought about the greatest change in the automotive industry is the computer. Not only are engine support systems controlled by computers, nearly every other major system on a car has some sort of electronic control. Initially, electronic controls were added to help maintain low emissions levels from vehicles. As these controls became more sophisticated, they improved engine performance. Electronic controls have done so much for engine efficiency that some of the early emission control systems have been eliminated.

The use of electronics has and will continue to change the automobile, as will advances made in other technologies. New composite materials are being used for engine parts. Soon most of the engine may be made with plastic-based or ceramic materials. Steel body panels are being replaced with aluminum and plastic parts that are bonded to a frame with special adhesives. The remaining steel body parts are thin, high-strength steel. Steel, aluminum, and composites are being used to reduce vehicle weight. Reduced weight results in better performance and fuel economy, especially when the weight of moving engine parts is reduced. A loss of weight here reduces the frictional drag of the engine, resulting in great increases in power output.

Vehicles powered by electric motors have also been introduced by the manufacturers. Some use a battery as the power source for the motor; others use a small automotive-type engine to charge a battery and to act as a supplemental power source to the electric motor. There are even vehicles that use hydrogen to feed a fuel cell that is capable of generating electricity for the electric motor.

The way automobiles of today and tomorrow look and run is being shaped by the constant influx of new technology. Automotive technicians must stay abreast of these changes and be able to diagnose and service the new systems.

THE BASIC ENGINE

The engine provides the power to drive the wheels of the vehicle. All automobile engines, both gasoline and diesel, are classified as internal combustion engines because the combustion or burning that creates energy takes place inside the engine. **Combustion** is the burning of an air/fuel mixture. As a result of combustion, large amounts of pressure are generated in the engine. This pressure or energy is used to power the car. The engine must be built strong enough to hold the pressure and temperatures formed by combustion.

The following sections cover the basic parts and the major systems of a gasoline engine.

Cylinder Block

The biggest part of the engine is the **cylinder block**, which is also called an **engine block (Figure 4–18)**. The cylinder block is a large casting of metal (cast iron or aluminum) that is drilled with holes to allow for the passage of lubricants and coolant through the block and to provide spaces for movement of mechanical parts. The block contains the cylinders, which are round passageways fitted with pistons. The block houses or holds the major mechanical parts of the engine.

Cylinder Head

The **cylinder head** fits on top of the cylinder block to close off and seal the top of the cylinders. The **combustion chamber** is an area into which the air/fuel mixture is compressed and burned. The cylinder head contains all or most of the combustion chamber. The cylinder head also contains **ports**, which are passageways through which the air/fuel mixture enters and burned gases exit the cylinder. A cylinder head can be made of cast iron or aluminum.

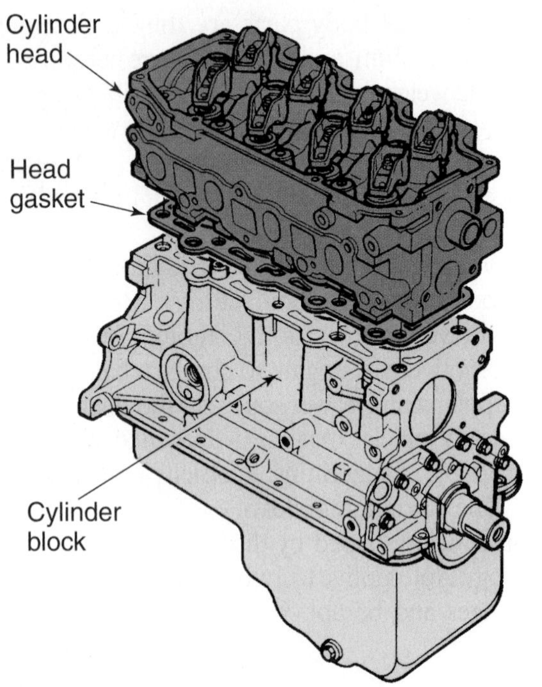

Figure 4–18 The two major units of an engine, the cylinder block and the cylinder head, are sealed together with a gasket and are bolted together. *Courtesy of Federal-Mogul Corporation*

Piston

The burning of air and fuel takes place between the cylinder head and the top of the piston. The **piston** is a can-shaped part closely fitted inside the cylinder **(Figure 4–19)**. In a four-stroke cycle engine, the piston moves through four different movements or strokes to complete one cycle. These four are the intake, compression, power, and exhaust strokes. On the intake stroke, the piston moves downward, and a charge of air/fuel mixture is introduced into the cylinder. As the piston travels upward, the air/fuel mixture is compressed in preparation for burning. Just before the piston reaches the top of the cylinder, ignition occurs and combustion starts. The pressure of expanding gases forces the piston downward on its power stroke. When it reciprocates, or moves upward again, the piston is on the exhaust stroke. During the exhaust stroke, the piston pushes the burned gases out of the cylinder.

Connecting Rods and Crankshaft

The reciprocating motion of the pistons must be converted to rotary motion before it can drive the wheels of a vehicle. This conversion is achieved by linking the piston to a **crankshaft** with a **connecting rod**. As the piston is pushed down on the power stroke, the connecting rod pushes on the crankshaft causing it to rotate. The end of the crankshaft is connected to the transmission to continue the power flow through the drivetrain and to the wheels.

Valve Train

A **valve train** is a series of parts used to open and close the intake and exhaust ports. A valve is a movable part that opens and closes a passageway. A camshaft controls the movement of the valves **(Figure 4–20)**, causing them to open and close at the proper time. Springs are used to close the valves.

Manifolds

A **manifold** is a metal ductwork assembly used to direct the flow of gases to or from the combustion chambers. Two separate manifolds are attached to the cylinder head **(Figure 4–21)**. The **intake manifold** delivers a mixture of air and fuel to the intake ports. The **exhaust manifold** mounts over the exhaust ports and carries exhaust gases away from the cylinders.

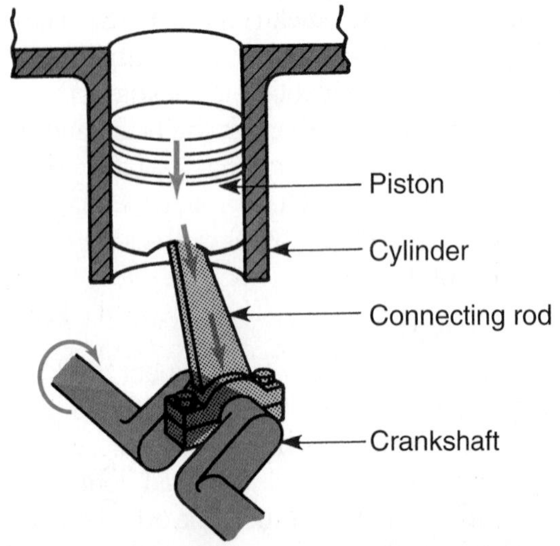

Figure 4–19 The engine's pistons fit tightly in the cylinders and are connected to the engine's crankshaft with connecting rods.

Figure 4–20 The valve train for one cylinder. Notice that this engine uses two intake and two exhaust valves in each cylinder. *Reprinted with permission*

Figure 4-21 The blue manifold is the intake manifold and the red manifold is for the exhaust. *Courtesy of Ford Motor Company*

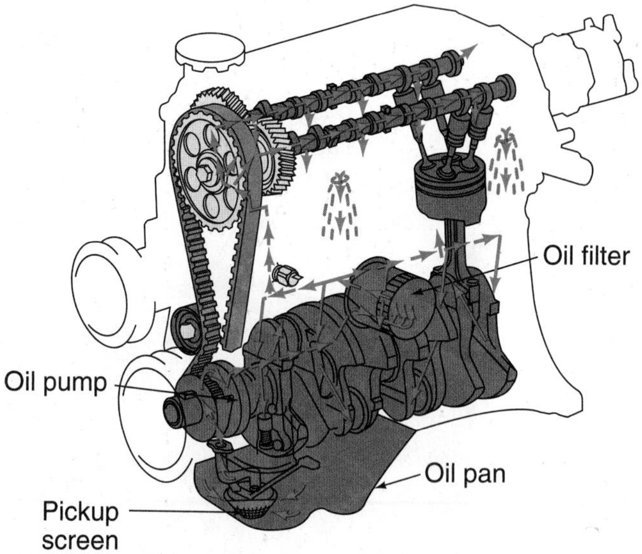

Figure 4-22 Oil flow in a typical engine's lubrication system.

ENGINE SYSTEMS

The following sections present a brief explanation of the systems that help an engine run and keep running.

Lubrication System

The moving parts of an engine need constant lubrication. Lubrication limits the amount of wear and reduces the amount of friction in the engine. **Friction**, which occurs when two objects rub against each other, generates heat.

Motor or engine oil is the fluid used to lubricate the engine. Several quarts of oil are stored in an **oil pan** bolted to the bottom of the engine block. The oil pan is also called the crankcase or **oil sump**. When the engine is running, an oil pump draws oil from the pan and forces it through oil galleries, which are small passageways that direct the oil to the moving parts of the engine.

Oil from the pan passes through an oil filter before moving through the engine **(Figure 4–22)**. The filter removes dirt and metal particles from the oil. Premature wear and damage to parts can result from dirt in the oil. Regular replacement of the oil filter and oil is an important step in a preventive maintenance program.

Cooling System

The burning of the air/fuel mixture in the combustion chambers of the engine produces large amounts of heat. This heat must not be allowed to build up and must be reduced or it can easily damage and warp the metal parts of an engine. To prevent this, engines have a cooling system **(Figure 4–23)**.

The most common way to cool an engine is to circulate a liquid coolant through passages in the engine block and cylinder head. An engine can also be cooled by passing air over and around the engine. Few air-cooled engines are used in automobiles today because it is very difficult to maintain a constant temperature at the cylinders. If the engine is kept at a constant temperature, it will run more efficiently. A liquid cooling system also has a supply of hot coolant available to operate a heater for the passenger compartment. The cooling system is designed to cool the engine, not the passengers inside the car. Cooling the passengers is the responsibility of the air-conditioning system.

A typical cooling system relies on a **water pump** that circulates the coolant through the system. The pump is typically driven by the engine. The coolant, a mixture of water and antifreeze, is pushed through passages called **water jackets** in the cylinder block and head to remove heat from the area around the cylinders' combustion chambers. The heat picked up by the coolant is sent to the **radiator** that transfers the coolant's heat to the outside air as the coolant flows through its tubes. To help remove the heat from the coolant, a cooling fan is used to pull cool outside air in through the fins of the radiator.

To raise the boiling point of the coolant, the cooling system is pressurized. To maintain this pressure, a radiator or **pressure cap** is fitted to the radiator. A **thermostat** is used to block off circulation in the system until a preset temperature is reached. This allows the engine to warm up faster. The thermostat also keeps the engine temperature at a predetermined level. Since parts of the cooling system are located in various spots under the vehicle's hood, hoses are used to connect these parts and keep the system sealed.

Fuel and Air System

The fuel and air system is designed to supply the correct amount of fuel mixed with the correct amount of air to

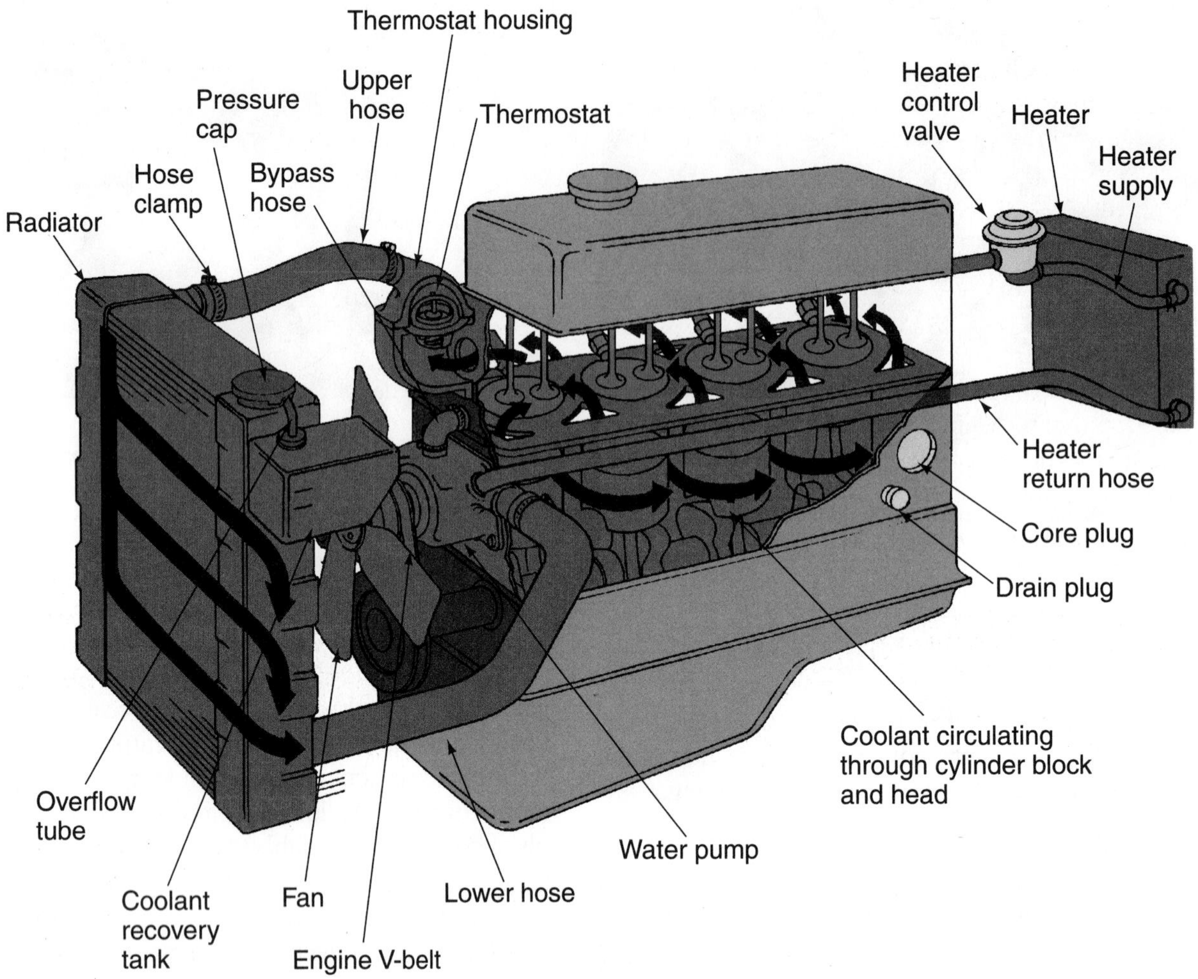

Figure 4-23 A typical cooling system. *Courtesy of Gates Rubber Co.*

the cylinders of the engine. This system also

- stores the fuel for later use.
- collects and cleans the outside air.
- delivers fuel to a device that controls the amount of fuel going to the engine.
- breaks down the fuel into very fine droplets (atomizes it) and mixes the fuel with air to form a vapor.
- changes the fuel and air ratios to meet the needs of the engine during different operating conditions.

The fuel system is made up of several different parts. A fuel tank stores the liquid gasoline. Fuel lines carry the liquid from the tank to the other parts of the system. A pump moves the gasoline from the tank through the lines. A filter removes dirt or other particles from the fuel. A fuel pressure regulator keeps the pressure below a specified level. An air filter cleans the outside air before it is delivered to the cylinders. Fuel injectors or a carburetor mix the liquid gasoline with air for delivery to the cylinders. An intake manifold directs the air/fuel mixture to each of the cylinders **(Figure 4–24)**.

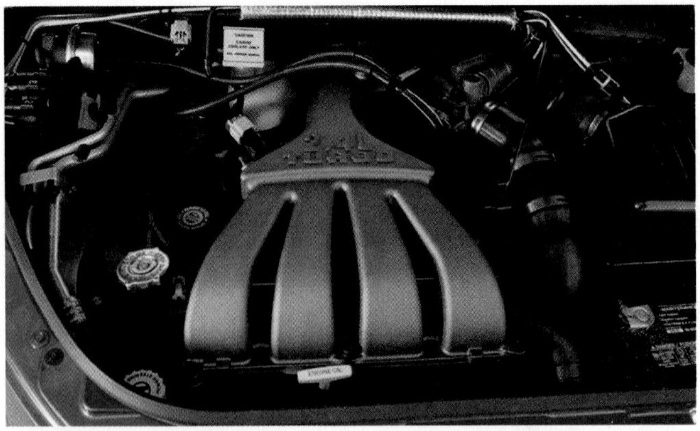

Figure 4-24 The intake manifold for this four cylinder is quite dominant when looking under the hood on this vehicle. *Courtesy of DaimlerChrysler Corporation*

Emission Control System

In the past, one of the chief contributors to air pollution was the automobile. For some time now, engines have been engineered to emit very low amounts of certain pollutants. The pollutants that have been drastically reduced are **hydrocarbons (HC)**, carbon monoxide (CO), and **ox-**

ides of nitrogen (NO_x). The Environmental Protection Agency establishes emissions standards that limit the amount of these pollutants a vehicle can emit.

To meet these standards, many changes have been made to the engine itself. Moreover, there have been systems developed and added to the engines to reduce the pollutants they emit. A list of the most common pollution-control devices follows:

- ***Positive crankcase ventilation*** **(PCV)** system. This system reduces HC emissions by drawing fuel and oil vapors from the crankcase and sending them into the intake manifold, where they are delivered to and burned in the cylinders. This system prevents the pressurized vapors from escaping the engine and entering the atmosphere.
- *Evaporative emission control system.* This system reduces HC emissions by drawing fuel vapors from the fuel system and releasing them into the intake air to be burned. This system stops the vapors from leaking into the atmosphere.
- ***Exhaust gas recirculation*** **(EGR)** *system.* This system introduces exhaust gases into the intake air to reduce the temperatures reached during combustion. This reduces the chances of NO_x forming during combustion.
- *Catalytic converter.* Located in the exhaust system, the catalytic converter allows for the burning or converting of HC, CO, and NO_x into harmless substances, such as water.
- *Air injection system.* This system reduces HC emissions by introducing fresh air into the exhaust stream to cause minor combustion of the HC in the engine's exhaust.

Exhaust System

During the exhaust stroke, the engine's pistons move up and push the burned air/fuel mixture, or exhaust, out of the combustion chamber and into the exhaust manifold. From the manifold, the gases travel through the other parts of the exhaust system until they are expelled into the atmosphere **(Figure 4–25)**. The exhaust system is designed

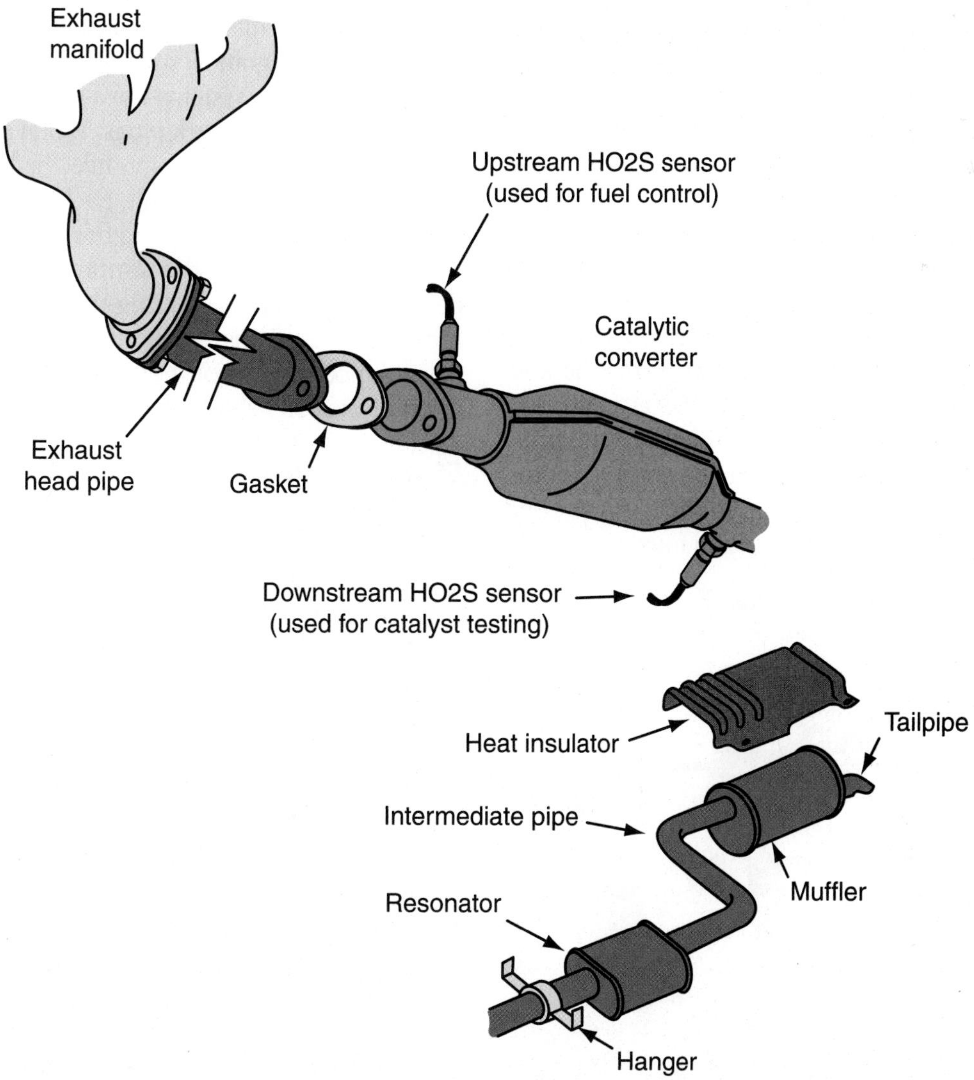

Figure 4-25 A typical exhaust system on a late-model car.

to direct toxic exhaust fumes away from the passenger compartment, to quiet the sound of the exhaust pulses, and to burn or catalyze pollutants in the exhaust. A typical exhaust system contains the following components:

- Exhaust manifold and gasket
- Exhaust pipe, seal, and connector pipe
- Intermediate pipes
- Catalytic converter(s)
- Muffler
- Resonator
- Tailpipe
- Heat shields
- Clamps, gaskets, and hangers

ELECTRICAL AND ELECTRONIC SYSTEMS

Automobiles have many circuits that carry electrical current from the battery to individual components. The total electrical system includes such major subsystems as the ignition system, starting system, charging system, and the lighting and other electrical systems.

Ignition System

After the air/fuel mixture has been delivered to the cylinder and compressed by the piston, it must be ignited. A gasoline engine uses an electrical spark to ignite the mixture. Generating this spark is the role of the ignition system.

The **ignition coil** generates the electricity that creates this spark **(Figure 4–26)**. The coil transforms the low voltage of the battery into a burst of 30,000 to 100,000 volts. This burst is what ignites the mixture. The mixture must be ignited at the proper time in order for complete combustion to occur. Although the exact proper time varies with engine design, ignition must occur at a point before the piston has completed its compression stroke.

Figure 4-26 An ignition module and coil assembly for four cylinders.

On most engines, the motion of the piston and the rotation of the crankshaft are monitored by a **crankshaft position sensor**. The sensor electronically tracks the position of the crankshaft and relays that information to an ignition control module. Based on input from the crankshaft position sensor, and, in some systems, the electronic engine control computer, the ignition control module then turns the battery current to the coil on and off at precisely the right time so that the voltage surge arrives at the cylinder at the right time.

The voltage surge from the coil must be distributed to the correct cylinder, since only one cylinder is fired at a time. In earlier systems, this was the job of the **distributor**. A distributor is driven by a gear on the camshaft at one-half of the crankshaft speed. It transfers the high-voltage surges from the coil to the spark plug wires in the correct firing order. The spark plug wires then deliver the high voltage to the spark plugs, which are screwed into the cylinder head. The voltage jumps across a space between two electrodes on the end of each **spark plug** and causes a spark. This spark ignites the air/fuel mixture.

Today, most ignition systems do not have a distributor. Instead, these systems have several coils—typically one for each pair of spark plugs. When a coil is activated by the electronic control module, high voltage is sent through the spark plug circuit. Each spark plug circuit includes two spark plugs, which fire at the same time. One spark plug fires during the compression stroke of a cylinder and the other fires during the exhaust stroke of another cylinder and is wasted. In this way, the electronic control module controls both the timing and the distribution of the coil's spark-producing voltage.

Starting and Charging Systems

The starting system is responsible for getting the engine started **(Figure 4–27)**. When the ignition key is turned to the start position, a small amount of current flows from the battery to a **solenoid** or relay. This activates the solenoid or relay and closes another electrical circuit that allows full battery voltage to reach the starter motor. The starter motor then rotates the flywheel mounted on the rear of the crankshaft. As the crankshaft turns, the pistons move through their strokes. At the correct time for each cylinder, the ignition system provides the spark to ignite the air/fuel mixture. If good combustion takes place, the engine will now rotate on its own without the need of the starter motor. The ignition key is now allowed to return to the on position. From this point on, the engine will continue to run until the ignition key is turned off.

The electrical power for the engine and the rest of the car comes from the car's battery. The battery is especially

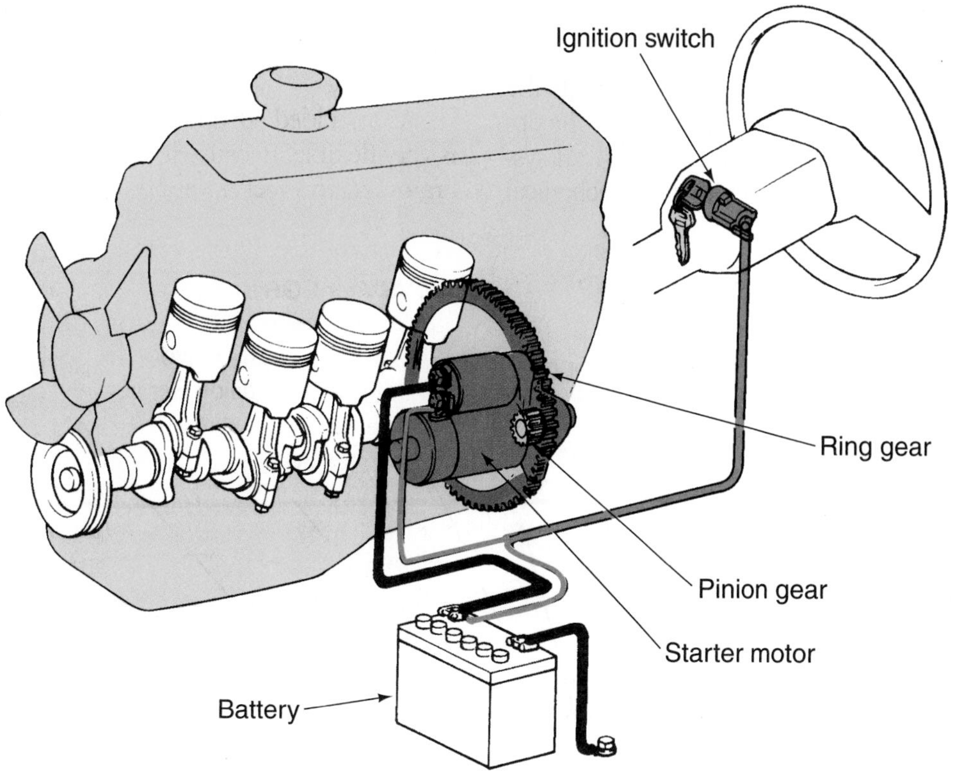

Figure 4-27 A typical starting system.

important for the operation of the starting system. While the starter is rotating the crankshaft, it uses a lot of electricity. This tends to lower the amount of power in the battery. Therefore, a system is needed to recharge the battery so that engine starts can be made in the future.

The charging system is designed to recharge and maintain the battery's state of charge. It also provides electrical power for the ignition system, air conditioner, heater, lights, radio, and all electrical accessories when the engine is running.

The charging system includes an AC generator (alternator), voltage regulator, indicator light, and the necessary wiring **(Figure 4–28)**. Rotated by the engine's crankshaft through a drive belt, the **AC generator** converts mechanical energy into electrical energy. When the output or electrical current from the charging system flows back to the battery, the battery is being charged. When the current flows out of the battery, the battery is said to be discharging.

Electronic Engine Controls

Nearly all vehicles on the road have an electronic engine control system. This is a system comprised of many electronic and electromechanical parts. The system is designed to continuously monitor the operation of the engine and to make adjustments that will cause the engine to run more efficiently. Electronic engine control systems have dramatically improved fuel mileage, engine performance, and driveability and have greatly reduced exhaust emissions.

Electronic control systems have three main types of components: input sensors, a computer, and output

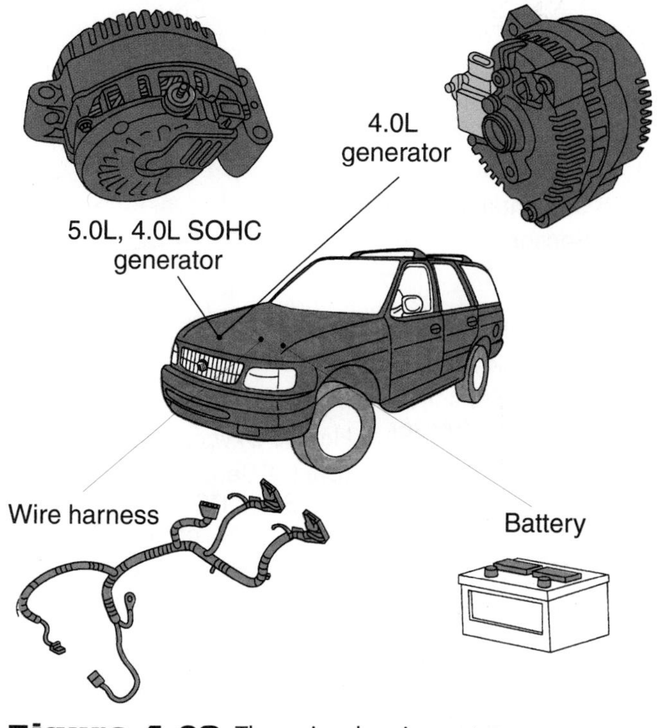

Figure 4-28 The major charging system components for a late-model vehicle. *Courtesy of Ford Motor Company*

devices **(Figure 4–29)**. The computer analyzes data from the input sensors. Then, based on the inputs and the instructions held in its memory, the computer directs the output devices to make the necessary changes in the operation of some engine systems. Electronic control systems have fewer moving parts than old-style mechanical and vacuum controls. Therefore, the engine and other support systems can maintain their calibration almost indefinitely.

As an added advantage, an electronic control system is very flexible. Because it uses computers, it can be programmed to meet a variety of different vehicle engine

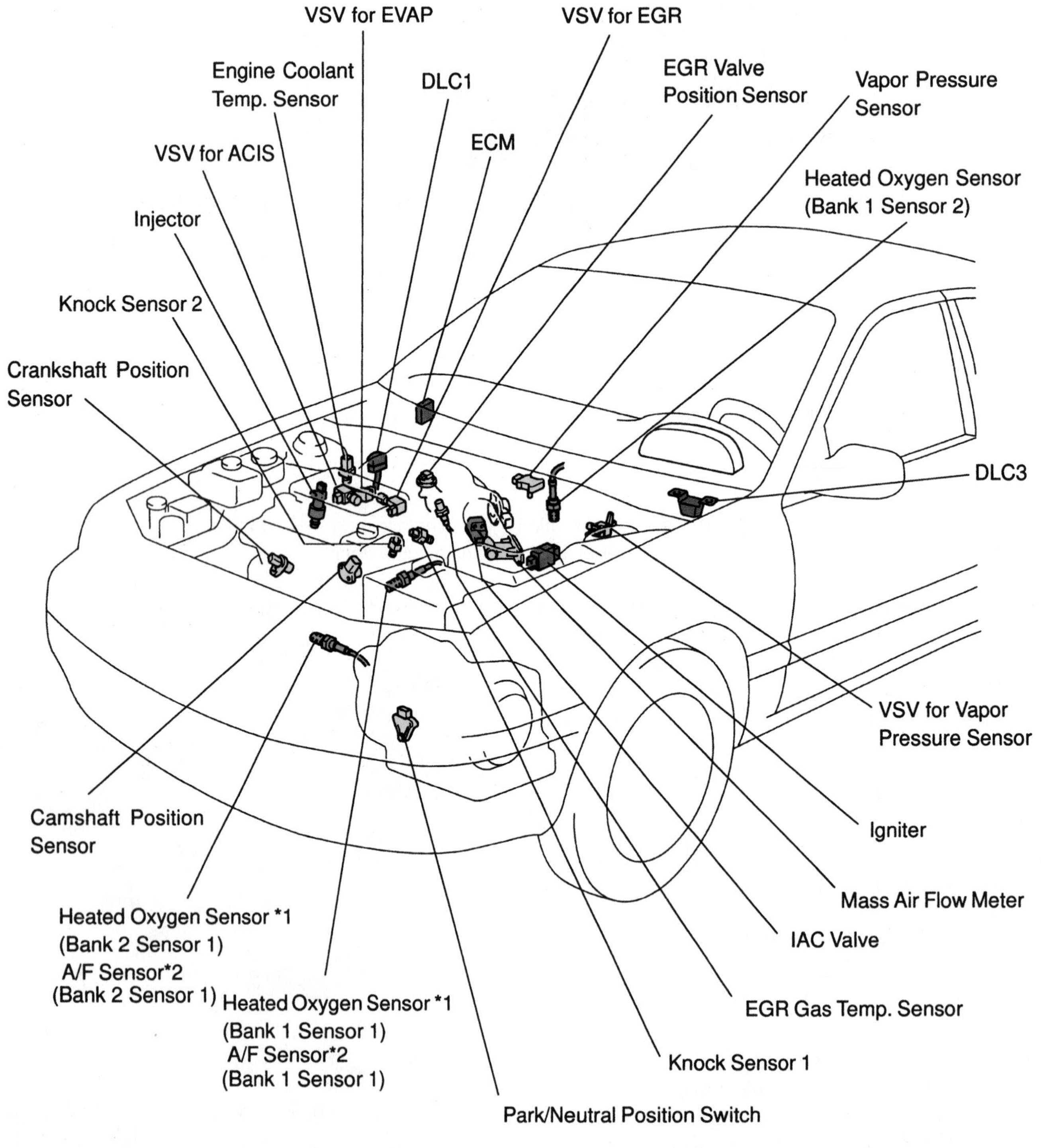

*1 : Except California Specification vehicles
*2 : Only for California Specification vehicles

Figure 4-29 Late-model electronic engine control systems are made up of many different sensors and actuators and a central computer or control module.

combinations or calibrations. Critical quantities that determine an engine's performance can be changed easily by changing data that is stored in the computer's memory.

On-Board Diagnostics

Today's engine control systems are **on-board diagnostic (OBD II)** second generation systems. These systems were developed to ensure proper emission control system operation for the vehicle's lifetime by monitoring emission-related components and systems for deterioration and malfunction. This monitoring also includes a check of the tank ventilation system for vapor leaks. The OBD system consists of the engine and transmission control modules, their sensors and actuators, along with the diagnostic software.

The computer can detect system problems even before the driver notices a driveability problem because many problems that affect emissions can be electrical or even chemical in nature.

When the OBD system determines that a problem exists, a corresponding diagnostic trouble code is stored in the computer's memory. The computer also illuminates a yellow dashboard light indicating "check engine" or MIL (malfunction indicator lamp) or displays an engine symbol. This light informs the driver of the need for service, not of the need to stop the vehicle.

A blinking or flashing dashboard lamp indicates a rather severe level of engine misfire. When this occurs, the driver should reduce engine speed and load and have the vehicle serviced as soon as possible. After the problem has been fixed, the dashboard lamp will turn off automatically or must be reset.

DRIVETRAIN

The **drivetrain** is made up of all components that transfer power from the engine to the driving wheels of the vehicle. The exact components used in a vehicle's drivetrain depend on whether the vehicle is equipped with rear-wheel drive, front-wheel drive, or four-wheel drive.

Today, most cars are front-wheel drive (FWD). Some larger luxury and performance cars are rear-wheel drive (RWD). Most pickup trucks, minivans, and SUVs are also RWD vehicles. Power flow in a RWD vehicle passes through the **clutch** or **torque converter**, manual or automatic transmission, and the driveline (drive shaft assembly). Then it goes through the rear differential, the rear-driving axles, and onto the rear wheels.

Power flow through the drivetrain of FWD vehicles passes through the clutch or torque converter, then moves through the transaxle, the driving axles, and onto the front wheels.

Four-wheel drive (4WD) or all-wheel drive (AWD) vehicles combine features of both rear- and front-wheel drive systems so that power can be delivered to all wheels either on a permanent or an on-demand basis. Typically, if a truck, pickup, or SUV has 4WD, the system is based on a RWD and a front drive axle is added. When a car has AWD or 4WD, the drivetrain is a modified FWD system. Modifications include a rear drive axle and an assembly that transfers some of the power to the rear axle.

Clutch

A clutch is used with manual transmissions/transaxles. It mechanically connects the engine's flywheel to the transmission/transaxle input shaft **(Figure 4–30)**. This is accomplished by a special friction plate that is splined to the input shaft of the transmission. When the clutch is engaged, the friction plate contacts the flywheel and transfers power to the input shaft.

When stopping, starting, and shifting from one gear to the next, the clutch is disengaged by pushing down on the clutch pedal. This moves the clutch plate away from the flywheel, stopping the power flow to the transmission. The driver can then shift gears without damaging the transmission or transaxle. Releasing the clutch pedal re-engages the clutch and allows power to flow from the engine to the transmission.

Manual Transmission

A manual or standard transmission is one in which the driver manually selects the gear of choice. Proper gear selection allows for good driveability and requires some driver education.

Whenever two or three gears have their teeth meshed together, a gearset is formed. The movement of one gear in the set will cause the others to move. If any of the gears in the set are a different size than the others, the gears will move at different speeds. The size ratio of a gearset is called the **gear ratio** of that gearset.

A manual transmission houses a number of individual gearsets, which produce different gear ratios **(Figure 4–31)**. The driver selects the desired operating gear or gear ratio. A typical manual transmission has four or five forward gear ratios, neutral, and reverse.

Automatic Transmission

An automatic transmission does not need a clutch pedal and shifts through the forward gears without the control of the driver. Instead of a clutch, it uses a torque converter to transfer power from the engine's flywheel to the transmission input shaft. The torque converter allows for smooth transfer of power at all engine speeds **(Figure 4–32)**.

Shifting in an automatic transmission is controlled by a hydraulic and/or electronic control system. In a hydraulic system, an intricate network of valves and other components uses hydraulic pressure to control the operation of planetary gearsets. These gearsets provide the three or four forward speeds, neutral, park, and reverse

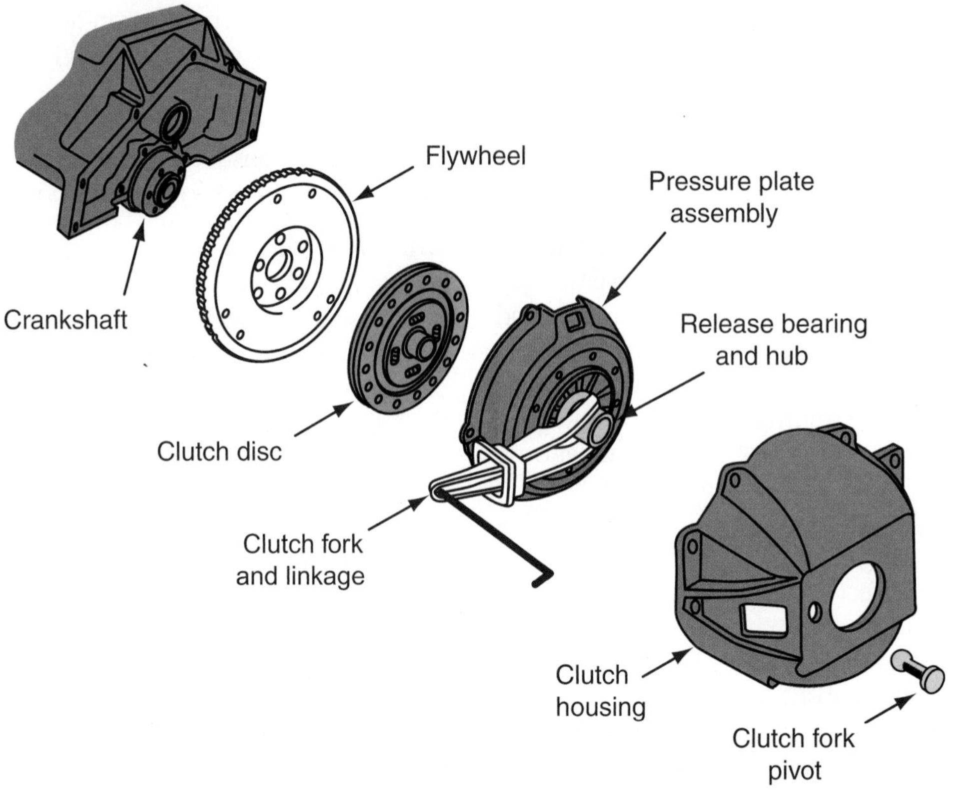

Figure 4-30 The major components of a clutch assembly for a manual transmission.

gears normally found in automatic transmissions. Newer electronic shifting systems use electric solenoids to control shifting mechanisms. Electronic shifting is precise and can be varied to suit certain operating conditions.

Driveline

Drivelines are used on RWD vehicles and 4WD vehicles. They connect the output shaft of the transmission to the gearing in the rear axle housing. They are also used to connect the output shaft to the front and rear drive axles on a 4WD vehicle.

A driveline consists of a hollow drive or propeller shaft that is connected to the transmission and drive axle differential by universal joints (U-joints). These U-joints allow the drive shaft to move with the movement of the rear suspension, preventing damage to the shaft.

Differential

On RWD vehicles, the drive shaft turns perpendicular to the forward motion of the vehicle. The differential gear-

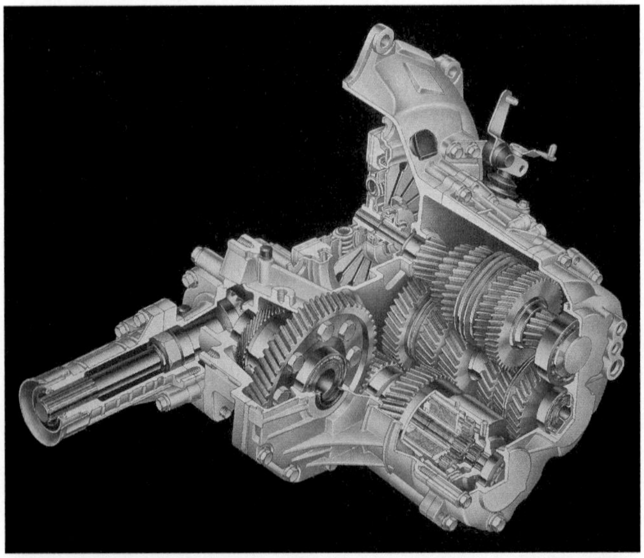

Figure 4-31 A typical manual transaxle. *Courtesy of DaimlerChrysler Corporation*

Figure 4-32 Cutaway of a six-speed automatic transmission shown with the torque converter in the housing. *Courtesy of BMW of North America, Incorporated*

Figure 4-33 The driveline connects the output from the transmission to the differential unit and drive axles. *Courtesy of BMW of North America, Incorporated*

ing in the rear axle housing is designed to turn the direction of the power so it can be used to drive the wheels of the vehicle. The power flows into the **differential**, where it changes direction, then flows to the rear axles and wheels **(Figure 4–33)**.

The gearing in the differential also multiplies the torque it receives from the drive shaft by providing a final gear reduction. Also, it divides the torque between the left and right driving axles and wheels so that a differential wheel speed is possible. This means one wheel can turn faster than the other when going around turns.

Driving Axles

Driving axles are solid steel shafts that transfer the torque from the differential to the driving wheels. A separate axle shaft is used for each driving wheel. In a RWD vehicle, the driving axles and part of the differential are enclosed in an axle housing that protects and supports these parts. Some rear drive axle units are mounted to an independent suspension and the drive axle assembly is similar to that of a FWD vehicle.

Each drive axle is connected to the side gears in the differential. The inner ends of the axles are splined to fit into the side gears. As the side gears are turned, the axles to which they are splined turn at the same speed.

The drive wheels are attached to the outer ends of the axles. The outer end of each axle has a flange mounted to it. A **flange** is a rim for attaching one part to another part. The flange, fitted with studs, at the end of an axle holds the wheel in place. **Studs** are threaded shafts, resembling bolts without heads. One end of the stud is screwed or pressed into the flange. The wheel fits over the studs and a nut, called the **lug nut**, is tightened over the open end of the stud. This holds the wheel in place.

The differential carrier supports the inner end of each axle. A bearing inside the axle housing supports the outer end of the axle shaft. This bearing, called the axle bearing, allows the axle to rotate smoothly inside the axle housing.

Transaxle

A **transaxle** is used on FWD vehicles. It is made up of a transmission and differential housed in a single unit **(Figure 4–34)**. The gearsets in the transaxle provide the required gear ratios and direct the power flow into the differential. The differential gearing provides the final gear reduction and splits the power flow between the left and right drive axles.

The drive axles extend from the sides of the transaxle. The outer ends of the axles are fitted to the hubs of the drive wheels. **Constant velocity (CV) joints** mounted on each end of the drive axles allow for changes in length and angle without affecting the power flow to the wheels.

Four-Wheel-Drive System

4WD or AWD vehicles combine the features of rear-wheel-drive transmissions and front-wheel-drive transaxles. Additional **transfer case** gearing splits the power flow between a differential driving the front wheels and a rear differential that drives the rear wheels. This transfer case can be a housing bolted directly to the transmission/transaxle, or it can be a separate housing mounted somewhere in the driveline. Most RWD-based four-wheel-drive vehicles have a drive shaft connecting the output of the transmission to the rear axle and another connecting the output of the transfer case to the front drive axle. Typically, AWD cars have a center differential that splits the torque between the front and rear drive axles.

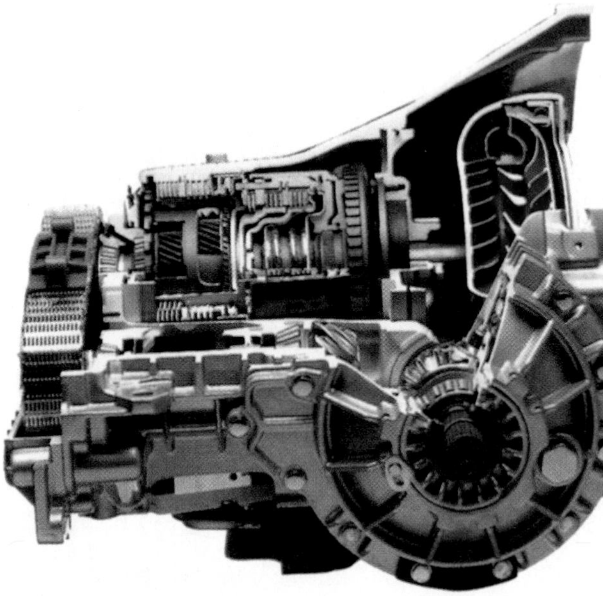

Figure 4-34 A cutaway of an automatic transaxle. *Courtesy of DaimlerChrysler Corporation*

RUNNING GEAR

The **running gear (Figure 4–35)** of a vehicle includes those parts that are used to control the vehicle, which includes the wheels and tires and the suspension, steering, and brake systems.

Suspension System

The suspension system **(Figure 4–36)** includes such components as the springs, shock absorbers, MacPherson struts, torsion bars, axles, and connecting linkages. These components are designed to support the body and frame, the engine, and the drivelines. Without these systems, the comfort and ease of driving the vehicle would be reduced.

Springs or **torsion bars** are used to support the axles of the vehicle. The two types of springs commonly used are the coil spring and the leaf spring. Torsion bars, which are long spring steel rods, are also used. One end of the rod is connected to the frame, while the other end is connected to the movable parts of the axles. As the axles move up and down, the rod twists and acts as a spring.

Shock absorbers dampen the upward and downward movement of the springs. This is necessary to limit the vehicle's reaction to a bump in the road.

Figure 4-35 The running gear in a typical late-model FWD car. *Courtesy of DaimlerChrysler Corporation*

Steering System

The steering system allows the driver to control the direction of the vehicle. It includes the steering wheel, steering gear, steering shaft, and steering linkage.

Two basic types of steering systems are used today: the **rack-and-pinion** and **recirculating ball** systems **(Figure 4–37)**. The rack-and-pinion system is commonly used in passenger cars. The recirculating ball system is normally used only on pickup trucks, SUVs, and full-size luxury cars.

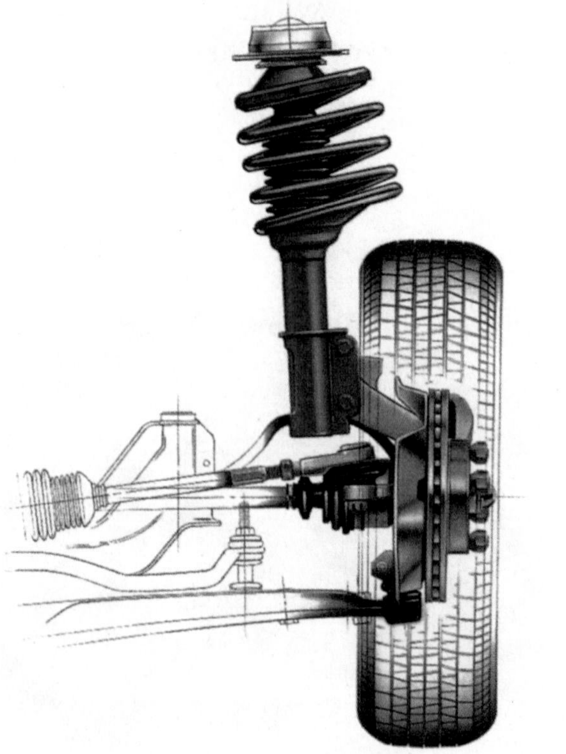

Figure 4-36 A strut assembly of a typical suspension system. *Courtesy of Ford Motor Company*

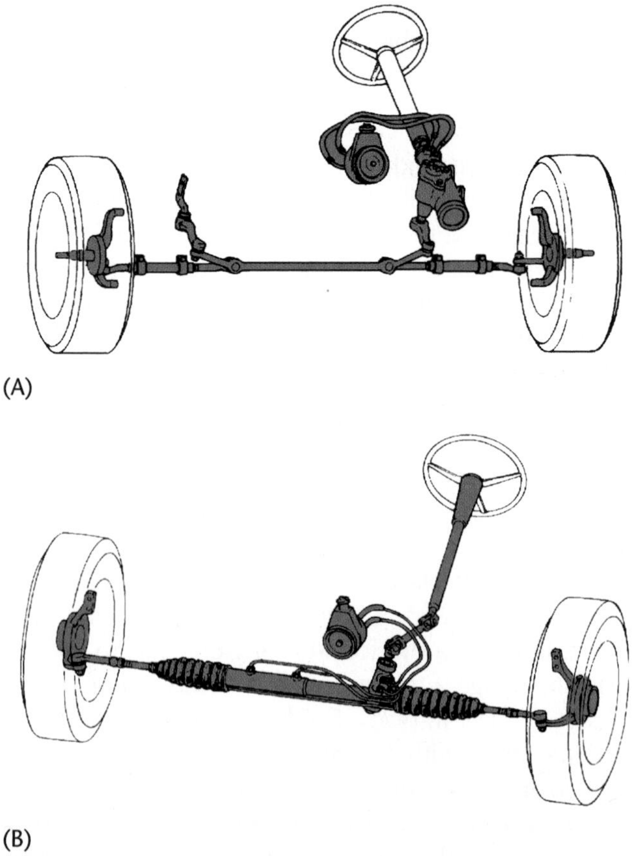

Figure 4-37 (A) A parallelogram-type steering system. (B) A rack and pinion steering system. *Courtesy of Moog Automotive, Inc.*

Steering gears provide a gear reduction to make changing the direction of the wheels easier. On most vehicles, the steering gear is also power assisted to ease the effort of turning the wheels. In a power-assisted system, a pump provides hydraulic fluid under pressure to the steering gear. Pressurized fluid is directed to one side or the other of the steering gear to make it easier to turn the wheels.

Some vehicles are equipped with speed-sensitive power-steering systems. These systems change the amount of power assist according to vehicle speed. The greatest amount of power assist occurs when the vehicle is moving slowly and it decreases as speed increases.

Brakes

Obviously, the brake system is used to slow down and stop a vehicle **(Figure 4–38)**. Brakes, located at each wheel, use friction to slow and stop a vehicle.

The brakes are activated when the driver presses down on the brake pedal. The brake pedal is connected to a plunger in a **master cylinder**, which is filled with hydraulic fluid. As pressure is put on the brake pedal, a force is applied to the hydraulic fluid in the master cylinder. This force is increased by the master cylinder and transferred through brake hoses and lines to the four brake assemblies.

Two types of brakes are used—**disc brakes** and **drum brakes**. Many vehicles use a combination of the two types: disc brakes at the front wheels **(Figure 4–39)** and drum brakes at the rear wheels; others have disc brakes at all wheels.

Most vehicles have power-assisted brakes. Many vehicles use a vacuum **brake booster** to increase the pressure applied to the plunger in the master cylinder. Others use hydraulic pressure from the power steering pump to increase the pressure on the brake fluid. Both of these systems lessen the amount of pressure that must be applied to the brake pedal and increase the responsiveness of the brake system.

Figure 4-38 A typical brake system with antilock disc brakes at the front and rear wheels. *Reprinted with permission*

Figure 4-39 A disc brake unit with a wheel speed sensor for ABS. *Courtesy of DaimlerChrysler Corporation*

Nearly all late-model vehicles have an **antilock brake system (ABS)**. The purpose of ABS is to prevent skidding during hard braking, which gives the driver more control of the vehicle during hard stops.

Wheels and Tires

The only contact a vehicle has with the road is through its tires and wheels. Tires are made of forms of rubber and other materials to give them strength, and are filled with air. Wheels are made of metal and are bolted to the axles or spindles **(Figure 4–40)**. Wheels hold the tires in

Figure 4-40 An alloy wheel with high-performance tires. *Courtesy of Mercedes-Benz of N.A., Inc.*

place. Wheels and tires come in many different sizes. Their sizes must be matched to one another and to the vehicle.

VEHICLE IDENTIFICATION

Before any service is done to a vehicle, it is important for you to know exactly what type of vehicle you are working on. The best way to do this is to refer to the **vehicle identification number (VIN)**. The VIN is given on a plate behind the lower corner of the driver's side of the windshield, as well as other locations on the vehicle. The VIN is made up of seventeen characters and contains all pertinent information about the vehicle. The use of the seventeen number and letter code became mandatory beginning with 1981 vehicles and is used by all manufacturers of both domestic and foreign vehicles.

Each character of a VIN has a particular purpose. The first character identifies the country where the vehicle was manufactured; for example:

- 1 or 4—United States
- 2—Canada
- 3—Mexico
- J—Japan
- K—Korea
- S—England
- W—Germany

The second character identifies the manufacturer; for example:

- A—Audi
- B—BMW
- C—Chrysler
- D—Mercedes Benz
- F—Ford
- G—General Motors
- H—Honda
- N—Nissan
- T—Toyota

The third character identifies the vehicle type or manufacturing division (passenger car, truck, bus, and so on). The fourth through eighth characters identify the features of the vehicle, such as the body style, vehicle model, and engine type.

The ninth character is used to identify the accuracy of the VIN and is a check digit. The tenth character identifies the model year; for example:

- S—1995
- T—1996
- V—1997
- W—1998
- X—1999
- Y—2000
- 1—2001
- 2—2002
- 3—2003
- 4—2004
- 5—2005

The eleventh character identifies the plant where the vehicle was assembled, and the twelfth to seventeenth characters identify the production sequence of the vehicle as it rolled off the manufacturer's assembly line.

The specifics needed for decoding the characters of the VIN can be found in a service manual for the vehicle.

PREVENTIVE MAINTENANCE

Preventive maintenance (PM) involves performing certain services to a vehicle on a regularly scheduled basis, before there is any sign of trouble. Regular inspection and routine maintenance can prevent major breakdowns and expensive repairs. They also keep cars and trucks running efficiently and safely.

A recent survey of 2,375 vehicles conducted during National Car Care Month found that more than 90% of the cars looked at needed some form of service. The cars were inspected for exhaust emissions, fluid levels, tire pressure, and other safety features. The results indicated that 34% of the cars had restricted air filters, 27% had worn belts, 25% had clogged PCV filters, 14% had worn hoses, and 20% had defective batteries, battery cables, or terminals.

During the fluid and cooling system inspection, 39% failed due to bad or contaminated transmission or power steering fluid, 36% had worn out or dirty engine oil, 28% had inadequate cooling system protection, and 8% had a faulty radiator cap.

In the safety category, 50% failed due to worn or improperly inflated tires, 32% had inoperative headlights or brake lights, and 14% had worn wipers.

Professional technicians constantly remind their customers about the need for preventive maintenance. The schedule for a particular vehicle's PM program is outlined in the owner's manual, which lists the recommended service intervals and the specifications for the various fluids to be used.

A typical preventive maintenance schedule recommends particular service at mileage or time intervals. Driving habits and conditions should also be used to determine the frequency of PM service intervals. For example, vehicles that frequently are driven for short distances in city traffic may require more frequent oil changes due to the more rapid accumulation of condensation and unburned fuel in the oil. Most manufacturers also specify more frequent service intervals for vehicles that are used to tow a trailer or for those that operate in extremely dusty or unusual conditions.

Typical PM Services

Preventive maintenance involves many different service operations. Often they are no more than making a visual inspection and checking fluids. The results of these checks tell you what to do next. Many of these basic checks and maintenance services are covered in this section. Additional PM services are covered throughout this book.

Engine Oil Perhaps the PM service that is best known by the public is changing the engine's oil and filter. Since oil is the lifeblood of an engine, it is critical that the oil be changed on a regular basis. Photo Sequence 1 shows the steps involved in changing the engine oil and the oil filter. Whenever this procedure is done, make sure the new engine oil has the correct rating for the vehicle.

Cooling System Whenever you change an engine's oil, you should also do a visual inspection of many of the parts and systems under the hood of the vehicle, including the cooling system. Check the level of the coolant in the coolant recovery tank **(Figure 4–41)**. If the level is too low, more coolant should be added through the mouth of the tank, not through the radiator. Visually inspect all of the hoses in the cooling system for signs of leakage and/or damage. If any of the hoses are swollen or cracked or show signs of leakage, they should be replaced.

CAUTION!

Never remove the radiator cap when the coolant is hot. Because the system is pressurized, the coolant can be hotter than boiling water and will cause severe burns. Wait until the top radiator hose is not too hot to touch. Then press down on the cap and slowly turn it until it hits the first stop. Now slowly let go of the cap. If there is any built-up pressure in the system, it will be released when the cap is loosened. After all the pressure has been exhausted, turn the radiator cap to remove it.

Drive Belts Check the condition of all of the drive belts on the engine. Carefully look to see if they are cracked or glazed **(Figure 4–42)**. A belt that is damaged in any way should be replaced.

Battery Visually inspect the battery. Check for signs of corrosion on the battery, the battery cables, battery tray, and the battery hold-down fixtures **(Figure 4–43)**. Normally this corrosion can be cleaned off with a mixture of baking soda and water. However, if there is heavy corrosion, the battery should be removed and the battery cleaned.

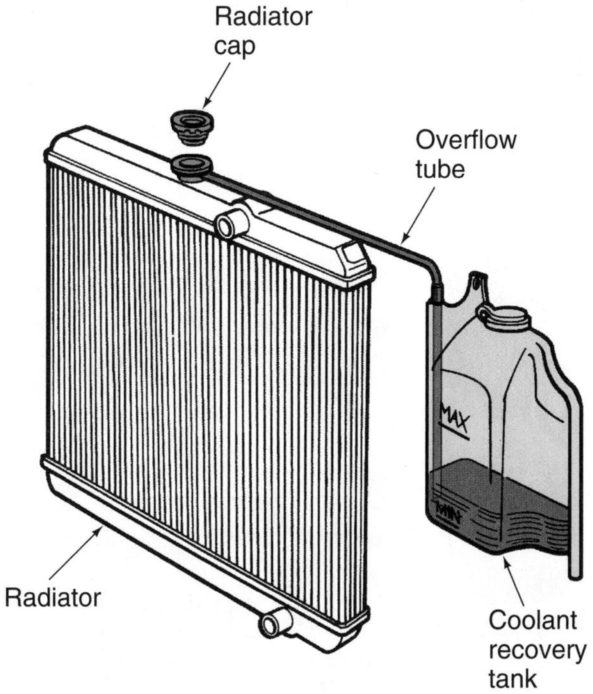

Figure 4–41 The level of coolant in the cooling system should be checked at the coolant recovery tank. *Courtesy of DaimlerChrysler Corporation*

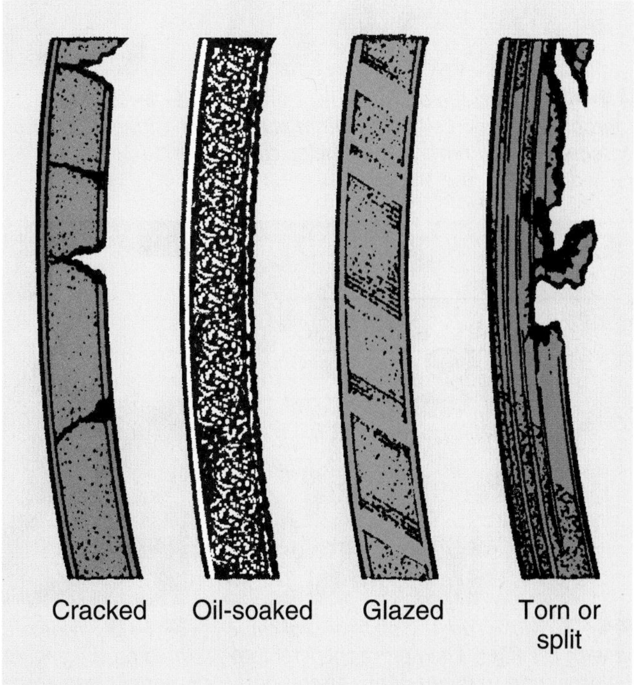

Figure 4–42 Drive belts should be inspected. *Courtesy of DaimlerChrysler Corporation*

PHOTO SEQUENCE 1

Changing the Oil and Oil Filter

P1-1 *Always make sure the vehicle is positioned safely on a lift or supported by jack stands before working under it. Before raising the vehicle, allow the engine to run awhile. After it is warm, turn off the engine.*

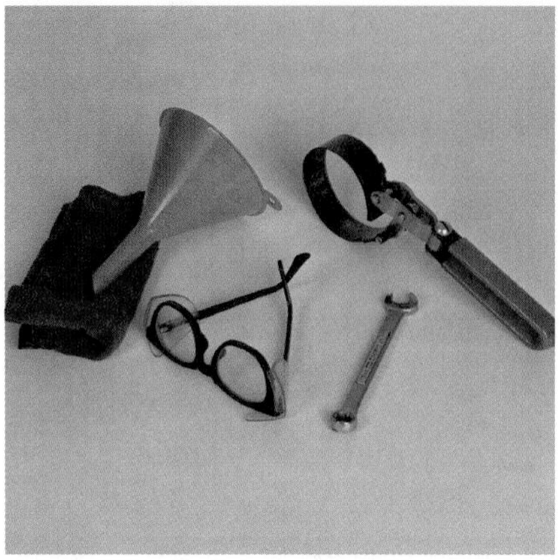

P1-2 *The tools and other items needed to change the engine's oil and oil filter are rags, a funnel, an oil filter wrench, safety glasses, and a wrench for the drain plug.*

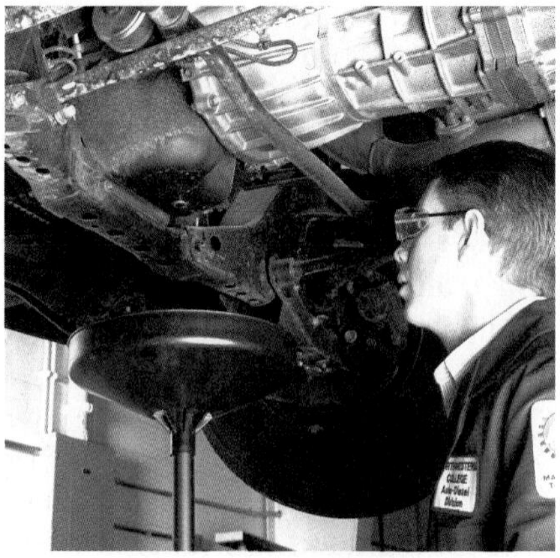

P1-3 *Place the oil drain pan under the drain plug before beginning to drain the oil.*

P1-4 *Loosen the drain plug with the appropriate wrench. After the drain plug is loosened, quickly remove it so the oil can freely drain from the oil pan.*

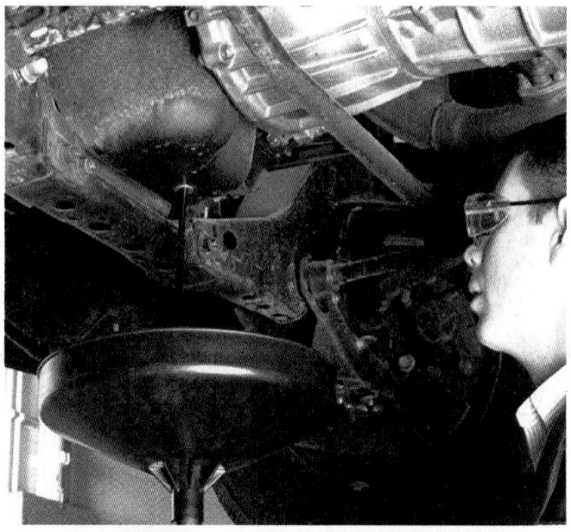

P1-5 *Make sure the drain pan is positioned so it can catch all of the oil.*

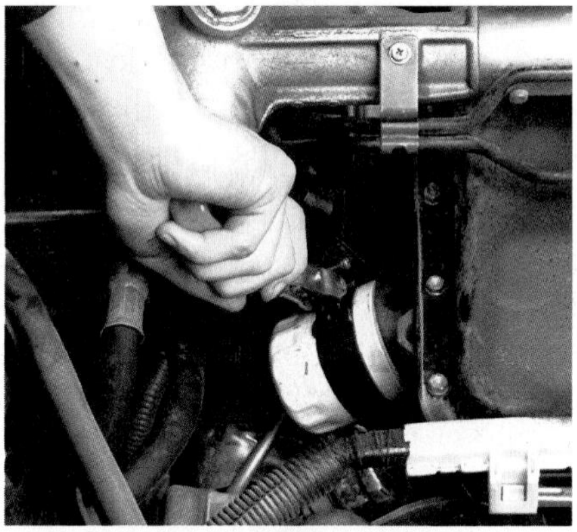

P1-6 *While the oil is draining, use an oil filter wrench to loosen and remove the oil filter.*

P1-7 *Make sure the oil filter seal came off with the filter. Then place the filter into the drain pan so it can drain. After it has completely drained, discard the filter according to local regulations.*

P1-8 *Wipe off the oil filter sealing area on the engine block. Then apply a coat of clean engine oil onto the new filter's seal.*

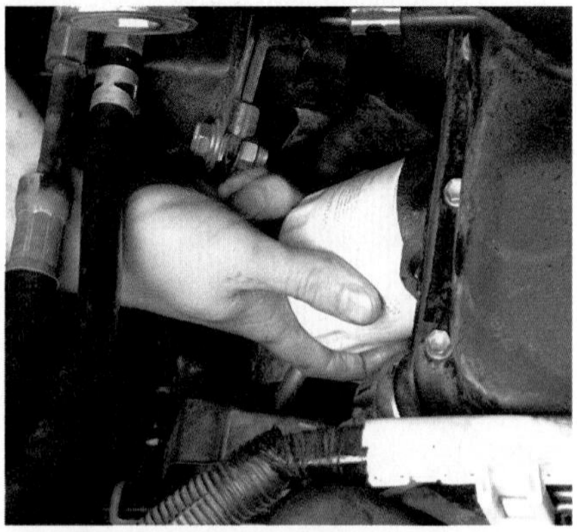

P1-9 *Install the new filter and hand-tighten it. Oil filters should be tightened according to the directions given on the filter.*

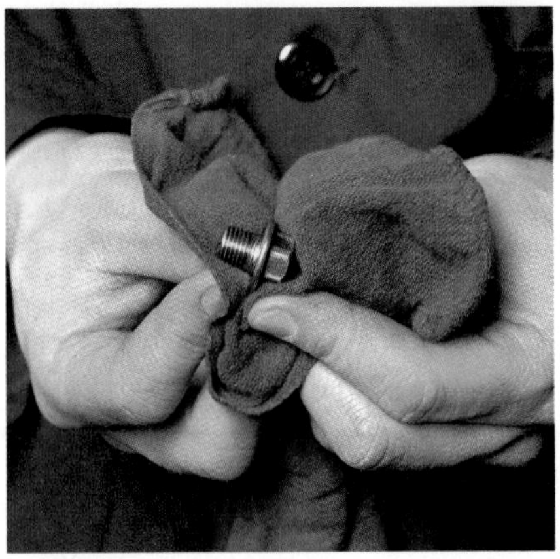

P1-10 *Prior to installing the drain plug, wipe off its threads and sealing surface with a clean rag.*

P1-11 *The drain plug should be tightened according to the manufacturer's recommendations. Overtightening can cause thread damage, while undertightening can cause an oil leak.*

P1-12 *With the oil filter and drain plug installed, lower the vehicle and remove the oil filler cap.*

P1-13 *Carefully pour the oil into the engine. The use of a funnel usually keeps oil from spilling on the engine.*

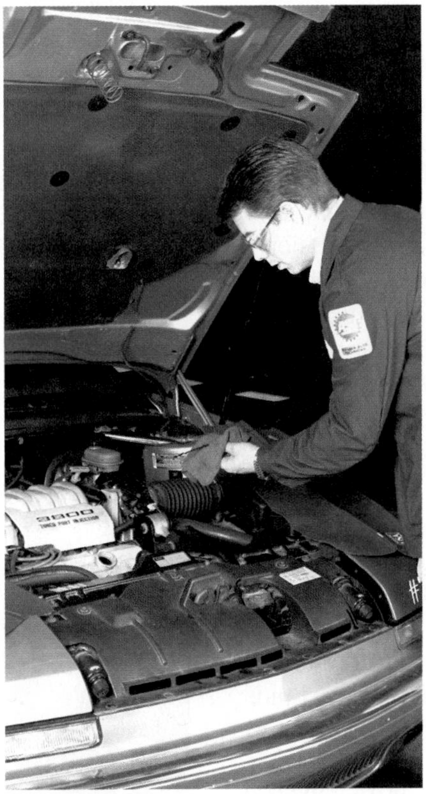

P1-14 *After the recommended amount of oil has been put in the engine, check the oil level.*

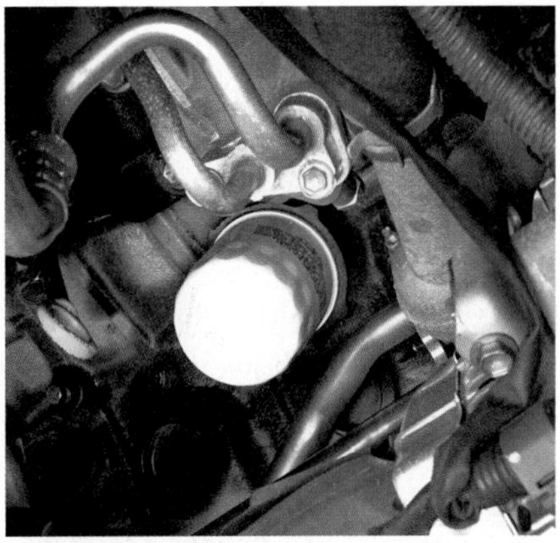

P1-15 *Start the engine and allow it to reach normal operating temperature. While the engine is running, check the engine for oil leaks, especially around the oil filter and drain plug. If there is a leak, shut down the engine and correct the problem.*

P1-16 *After the engine has been turned off, recheck the oil level and correct it as necessary.*

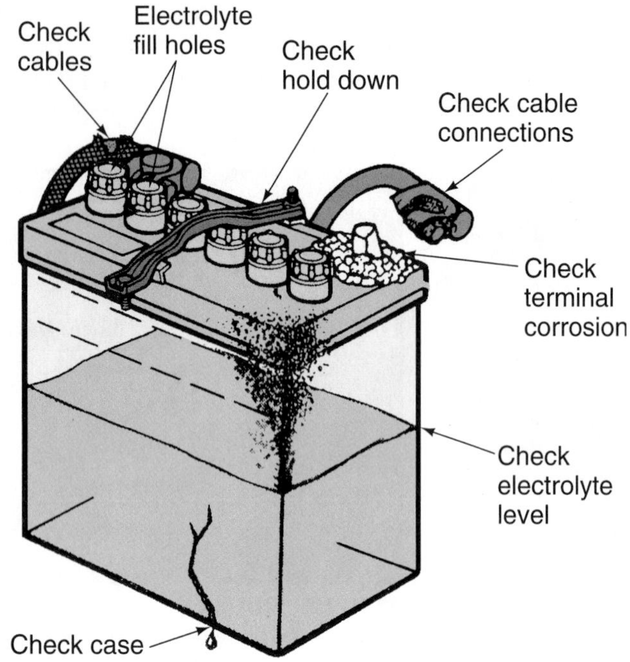

Figure 4-43 Batteries should be carefully checked for damage, dirt, and corrosion.

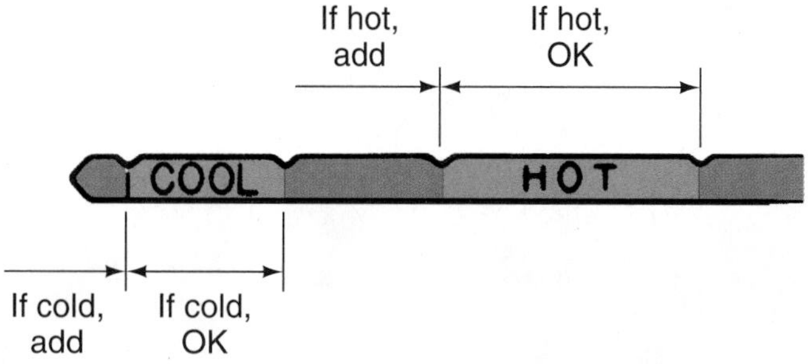

Figure 4-44 Automatic transmission fluid should be checked regularly. Normally the level is checked when the engine is warm. The normal cold level is well below the normal hot level.

Transmission Fluid The oil used in automatic transmissions is called automatic transmission fluid (ATF). This special fluid is dyed red so that it is not easily confused with engine oil. Make sure the engine is warm and the vehicle is level. Then set the parking brake and allow the engine to idle. Sometimes the manufacturer recommends that the ATF level be checked when the transmission is placed into Park; however, some may require some other gear. Make sure you follow those requirements. Locate the fluid dipstick (normally located to the rear of the engine) and pull it out of its tube. Check the level of the fluid on the dipstick **(Figure 4–44)**. If the level is low, add only enough to bring the level to full. Make sure you only use the fluid recommended by the manufacturer.

Manual transmissions, transaxles, and drive axle units require the use of specific lubricants or oils and the levels need to be checked according to the manufacturer's recommended service intervals. Some manufacturers recommend that the fluids be changed periodically. Most repair shops have an air-operated dispenser for these fluids; others rely on a hand-operated oil pump **(Figure 4–45)**.

Power-Steering Fluid Now locate the power-steering pump. The level of power-steering fluid is checked with the engine off. The filler cap on the power-steering pump normally has a dipstick. Unscrew the cap and check the level. The level of the fluid is normally checked when the engine is warm. If the fluid is cold, it will read lower than normal. Add fluid as necessary.

Brake Fluid The next fluid level to check is the brake fluid. Brake fluid levels are checked at the master cylinder. There are basically two designs of master cylinders and therefore two basic ways to check brake fluid. It is important to clean the area around the master cylinder caps before removing them to prevent dirt from falling into the reservoir.

One master cylinder design is found mostly on older vehicles. These are made of cast iron or aluminum and have a metal bail that snaps over the master cylinder cover to hold it in place. Normally the bail can be moved in only one direction. Once moved out of the way, the master cylinder cap can be removed.

The other master cylinder design has a plastic reservoir mounted to the top of it. This reservoir holds the brake fluid. Most often the caps on this type of master cylinder are screwed on **(Figure 4–46)**. To remove them, simply unscrew them. The caps of some plastic reservoirs have snaps to hold them. Unsnap the cap to check the fluid.

A rubber diaphragm attached to the inside of the master cylinder cap is designed to stop dirt, moisture, and air from entering the reservoir. Make sure the diaphragm is not damaged. Check the fluid level and add as necessary. Make sure the fluid is of the correct type and is fresh and clean. Brake fluid tends to absorb moisture, so old

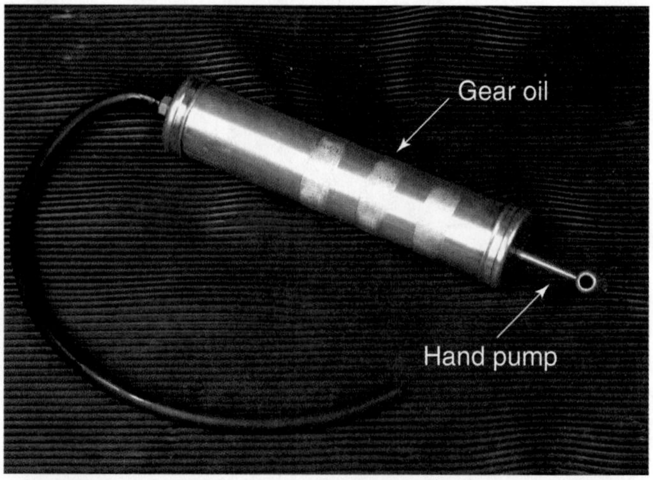

Figure 4-45 A hand-operated pump used to fill transmissions and drive axles with lubricant.

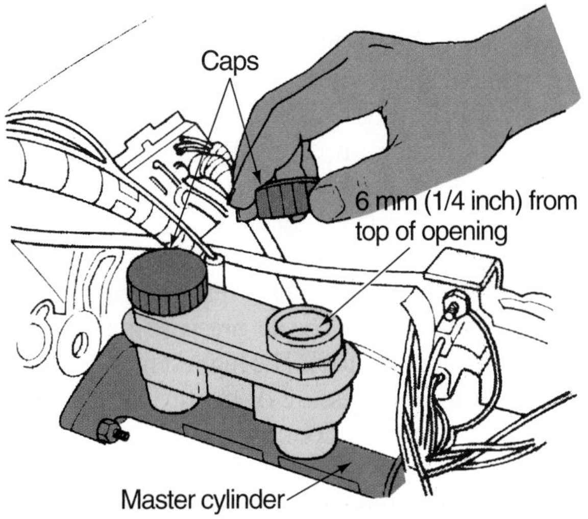

Figure 4–46 Some master cylinders have twist-off caps.

brake fluid is most likely to be contaminated. When reinstalling the cap, make sure you do not damage the diaphragm.

On some vehicles with manual transmissions, there is another but smaller master cylinder close to the brake master cylinder. This is the clutch master cylinder. Its fluid level needs to be checked, which is done in the same way as for brake fluid. In most cases, the clutch master cylinder uses the same type of fluid as the brake master cylinder. However, check this out before adding any fluid.

Windshield Washer Fluid The last fluid level to check is that of the windshield washer **(Figure 4–47)**. Visually check the level and add fluid as necessary. Do not add straight water to the washer tank, especially in cold weather. The water can freeze and crack the tank or clog the washer hoses and nozzles. Always use windshield washer fluid.

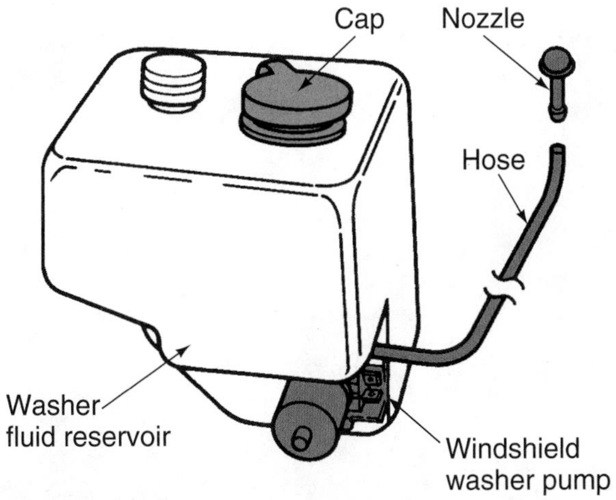

Figure 4–47 Check the level of the windshield washer fluid at the fluid reservoir. *Courtesy of Volkswagen of America, Inc.*

Windshield Wipers Several things should be checked after the hood of the vehicle is closed. One of them is the condition of the windshield wipers. Wiper blades can become dull, torn, or brittle. If any of these conditions are evident, the wiper blades should be replaced. Before replacing just the blades, check the condition of the wiper arms. Look for signs of distortion or damage. Also check the condition of the spring on the arm. This spring is designed to keep the wiper blade fairly tight against the windshield. If the spring is weak or damaged, the blade will not do an acceptable job of cleaning the glass.

Most wiper blade assemblies have replaceable blades or inserts **(Figure 4–48)**. Sometimes, when blades are not available or when the arm is damaged, it is necessary to replace the complete assembly. To replace just the wiper blades, grab hold of the assembly and pivot it away from the windshield. Once the arm is moved to its maximum position, it should stay there until it is pivoted back to the windshield. Doing this will allow you to easily replace the wiper blades without damaging the vehicle's paint or glass.

There are three basic designs of wiper blade inserts **(Figure 4–49)**. To determine which one will work on your vehicle, carefully look at your wiper assembly. Remove the old insert and install the new one. After installation, pull on the insert to make sure it is secured properly. If the insert comes loose while the wipers are moving across the windshield, the wiper arm could scratch the glass.

Most often wiper blades are replaced as an assembly. There are several different methods used to secure the blades to the wiper arm **(Figure 4–50)**. Most replacement blades come with the necessary adapters to secure the blade to the arm.

When it is necessary to replace the wiper arm or the entire assembly, the assembly must be removed from the vehicle. The wiper arms are either mounted onto a threaded shaft and held in place by a nut, or they are pressed over a splined shaft. Most often this type mount is reinforced by a clip that must be released before the

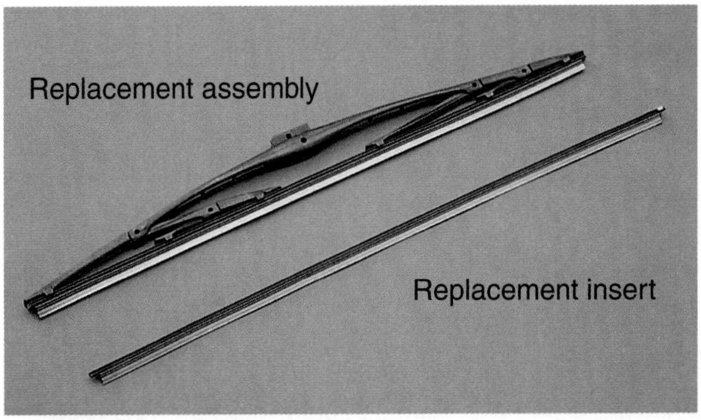

Figure 4–48 Windshield wiper blades are replaced as a complete assembly, or blade inserts are fitted into the blade.

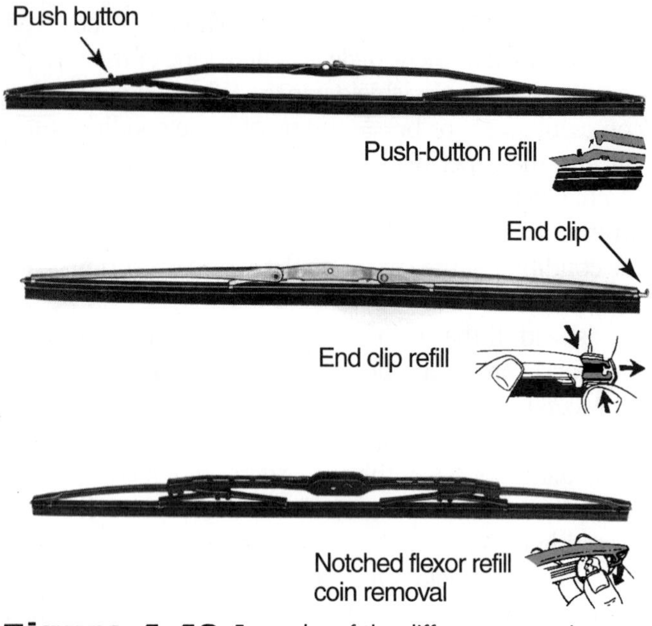

Figure 4-49 Examples of the different ways wiper blade inserts are secured to the blade assembly. *Courtesy of Cooper Automotive, Inc.*

arm can be pulled from the shaft. When wiper arms are reinstalled on their mounts, make sure they are positioned so the blades cannot hit the frame of the windshield while the wipers are operating. When checking the placement and operation of the wipers, wet the windshield before turning on the wipers. The water will serve as a lubricant for the wipers.

Tires The last simple check is that of the tires. They should be checked for damage and wear. Tires should have at least 1/16″ (1.59 mm) of tread remaining. Any less and the tire should be replaced. Tires have "tread wear indicators" molded into them. When the wear bar shows across the width of the tread, the tire is worn beyond its limits. Also check the tires for bulges, nails, tears, and other damage. All of these indicate that the tire should be replaced.

Check the inflation of the tires. To do this, use a tire-pressure gauge **(Figure 4–51)**. Press the gauge firmly onto the tire's valve stem. The air pressure in the tire will push the scale out of the tool. The highest number shown on the scale is the air pressure of the tire. Compare this reading with the specifications for the tire.

The correct tire pressure is listed in the vehicle's owner's manual or on a decal stuck on the driver's side doorjamb. The air pressure rating on the tire is not the amount of pressure the tire should have; it only indicates the maximum pressure the tire should ever have when it is cold.

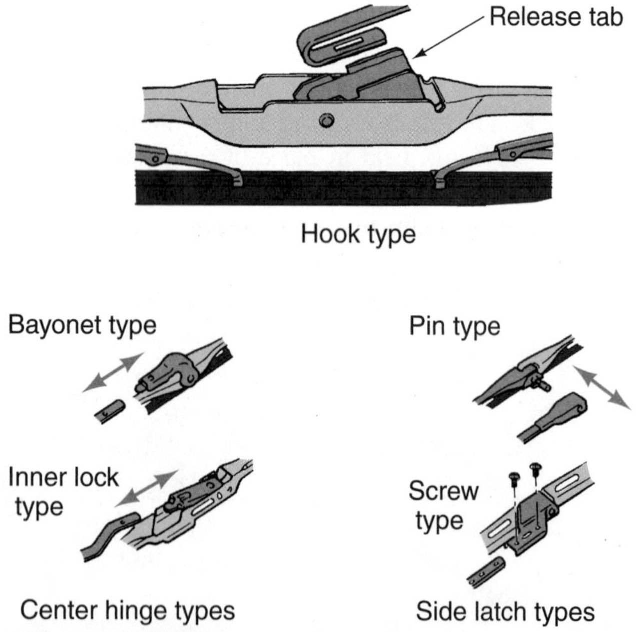

Figure 4-50 Examples of the different ways windshield wiper blades are secured to the wiper arm. *Courtesy of DaimlerChrysler Corporation*

Additional PM Checks

The following PM checks are in addition to those items specified by the manufacturer. These should be performed at the suggested time intervals to help ensure safe and dependable vehicle operation:

Time: Whenever refueling

- Check the engine oil level.
- Check the windshield washer fluid level.
- Look for low or underinflated tires.

Time: At least monthly

- Check the tire pressure. (Check tires when cold, not after a long drive.)
- Check the coolant in the coolant recovery reservoir.
- Check the operation of all exterior lights, including the brake lights, turn signals, and hazard warning flashers.

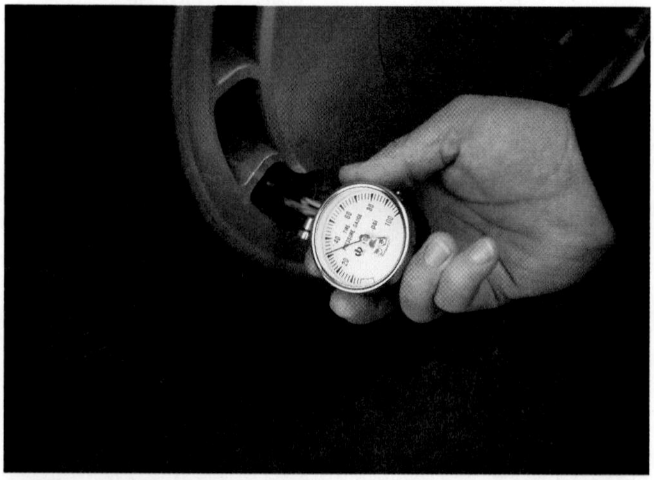

Figure 4-51 Check the tires and wheels for damage and proper inflation.

Time: At least twice a year

- Check windshield washer spray and wiper operation.
- Check and replace worn wiper blades.
- Check for worn tires and loose wheel lug nuts.
- Check pressure in spare tire.
- Check headlight alignment.
- Check muffler, exhaust pipes, and clamps.
- Check the brake fluid level.
- Inspect the lap/shoulder belts for wear.
- Check radiator, heater, and air-conditioning hoses for leaks or damage.

Time: At least once a year

- Lubricate all hinges and all outside key locks.
- Lubricate the rubber weatherstrips for the doors.
- Clean the body's water drain holes.
- Check the power-steering fluid level.
- Check the connections at the battery.
- Lubricate the transmission controls and linkage.

Time: While operating the vehicle

- Pay attention to and note any changes in the sound of the exhaust or any smell of exhaust fumes in the vehicle.
- Check for vibrations in the steering wheel. Notice any increased steering effort or looseness in the steering wheel.
- Notice if the vehicle constantly turns slightly or pulls to one side of the road.
- When stopping, listen and check for strange sounds, pulling to one side, increased brake pedal travel, or hard-to-push brake pedal.
- If any slipping or changes in the operation of the transmission occur, check the transmission fluid level.
- Check for fluid leaks under the vehicle. (Water dripping from the air-conditioning system after use is normal.)
- Check the automatic transmission's park function.
- Check the parking brake.

KEY TERMS

AC generator
Antilock brake system (ABS)
Brake booster
Clutch
Combustion
Combustion chamber
Connecting rod
Constant velocity (CV) joint
Convertible
Corporate Average Fuel Economy (CAFE)
Crankshaft
Crankshaft position sensor
Cylinder block
Cylinder head
Differential
Disc brakes
Distributor
Drivetrain
Drum brakes
Engine block
Exhaust gas recirculation (EGR)
Exhaust manifold
Flange
Friction
Gear ratio
Hatchback
Hydrocarbons (HC)
Hydroforming
Ignition coil
Intake manifold
Liftback
Lug nut
Manifold
Master cylinder
Oil pan
Oil sump
On-board diagnostic (OBD II)
Oxides of nitrogen (NO_x)
Pickup
Piston
Port
Positive crankcase ventilation (PCV)
Pressure cap
Rack and pinion
Radiator
Recirculating ball
Running gear
Sedan
Shock absorber
Solenoid
Spark plug
Sport utility vehicle (SUV)
Spring
Station wagon
Steering gear
Stud
Thermostat
Torque converter
Torsion bar
Transaxle
Transfer case
Unibody
Valve train
Van
Vehicle identification number (VIN)
Water jacket
Water pump

SUMMARY

- Dramatic changes to the automobile have occurred over the last 40 years, including the addition of emission control systems, more fuel-efficient and cleaner-burning engines, and lighter body weight.
- In addition to being lighter than body-over-frame vehicles, unibodies offer better occupant protection by distributing impact forces throughout the vehicle.
- Today's computerized engine control systems regulate such things as air and fuel delivery, ignition timing, and emissions. The result is an increase in overall efficiency.
- All automotive engines are classified as internal combustion, because the burning of the fuel and air occurs inside the engine. Diesel engines share the same

major parts as gasoline engines, but they do not use a spark to ignite the air/fuel mixture.

- The cooling system maintains proper engine temperatures. Liquid cooling is more efficient than air cooling and more commonly used.
- The lubrication system distributes motor oil throughout the engine. This system also contains the oil filter necessary to remove dirt and other foreign matter from the oil.
- The fuel system is responsible not only for fuel storage and delivery, but also for atomizing and mixing it with the air in the correct proportion.
- The exhaust system has three primary purposes: to channel toxic exhaust away from the passenger compartment, to quiet the exhaust pulses, and to burn the emissions in the exhaust.
- The electrical system of an automobile includes the ignition, starting, charging, and lighting systems. Electronic engine controls regulate these systems very accurately through the use of computers.
- Modern automatic transmissions use a computer to match the demand for acceleration with engine speed, wheel speed, and load conditions. It then chooses the proper gear ratio and, if necessary, initiates a gear change.
- The running gear is critical to controlling the vehicle. It consists of the suspension system, braking system, steering system, and wheels and tires.
- Preventive maintenance involves regularly scheduled service on a vehicle to keep it operating efficiently and safely. Professional technicians should stress the importance of PM to their customers.

TECH MANUAL

The following procedure is included in Chapter 4 of the *Tech Manual* that accompanies this book:

1. Basic preventive maintenance inspection

REVIEW QUESTIONS

1. Under the CAFE standards, for what are vehicles tested?
2. Describe the information found in a VIN.
3. Define internal combustion.
4. In addition to the battery, what does the charging system include?
5. What preventive maintenance checks should be made whenever the vehicle is refueled?
6. Which of the following is *not* a typical emission control system?
   a. EGR **c.** EPA
   b. PCV **d.** EFE
7. Automatic transmissions use a ______ instead of a clutch to transfer power from the flywheel to the transmission's input shaft.
   a. differential
   b. U-joint
   c. torque converter
   d. constant velocity joint
8. Technician A stresses the need to follow the manufacturer's recommendations for preventive maintenance when servicing his customer's car. Technician B says the proper PM service intervals depend on the customer's driving habits and typical driving conditions. Who is correct?
   a. Technician A **c.** Both A and B
   b. Technician B **d.** Neither A nor B
9. Which of the following is *not* one of the strokes of a four-cycle engine?
   a. compression **c.** intake
   b. exhaust **d.** combustion
10. Technician A says the PCV system is designed to limit CO emissions. Technician B says catalytic converters reduce HC emissions at the tailpipe. Who is correct?
    a. Technician A **c.** Both A and B
    b. Technician B **d.** Neither A nor B
11. Which type of engine is classified as internal combustion?
    a. gasoline **c.** both a and b
    b. diesel **d.** neither a nor b

12. What does the valve train do?

a. It delivers fuel to a device that controls the amount of fuel going to the engine.

b. It houses the major parts of the engine.

c. It converts a reciprocating motion to rotary motion.

d. It opens and closes the intake and exhaust ports of each cylinder.

13. Technician A says that a liquid-cooled engine maintains a constant operating temperature. Technician B says that oil is circulated through the cooling system to remove heat from the engine's parts. Who is correct?

a. Technician A **c.** Both A and B

b. Technician B **d.** Neither A nor B

14. An engine will not start and no spark is found at the spark plugs when the engine is turned over by the starter. Technician A says the problem is probably the battery. Technician B says the ignition system is most likely at fault. Who is correct?

a. Technician A **c.** Both A and B

b. Technician B **d.** Neither A nor B

15. Which emission control system introduces exhaust gases into the intake air to reduce the formation of NO_x in the combustion chamber?

a. evaporative emission controls

b. exhaust gas recirculation

c. air injection

d. early fuel evaporation

16. Technician A says many vehicles use an AC generator as the charging unit. Technician B says many vehicles use an alternator as the charging unit. Who is correct?

a. Technician A **c.** Both A and B

b. Technician B **d.** Neither A nor B

17. Technician A says a transaxle delivers torque to the front and the rear drive axles. Technician B says a transaxle is most commonly found in 4WD pickups and SUVs. Who is correct?

a. Technician A **c.** Both A and B

b. Technician B **d.** Neither A nor B

18. Which of the following is *not* part of the running gear?

a. differential **c.** suspension

b. steering **d.** brakes

19. Which of the following PM checks should be made most often?

a. coolant in the coolant recovery reservoir

b. operation of all exterior lights

c. engine oil level

d. spare tire pressure

20. Technician A says tires should have a tread depth of at least 1⁄16 inch. Technician B says tires have a tread wear indicator built into them. Who is correct?

a. Technician A **c.** Both A and B

b. Technician B **d.** Neither A nor B

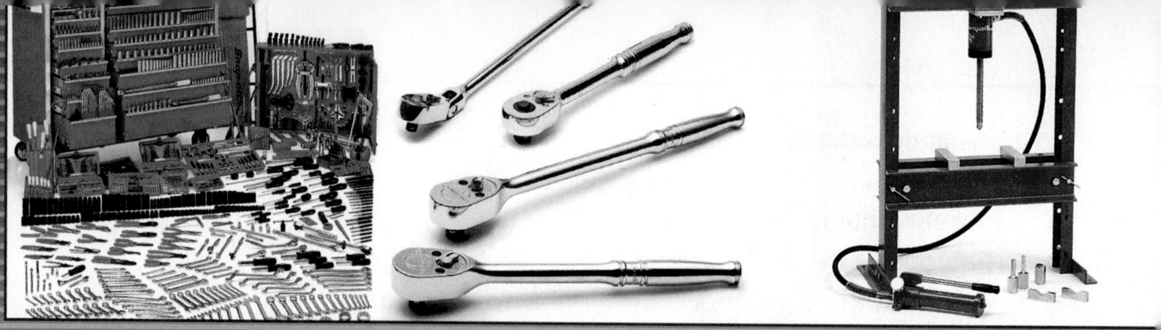

5

HAND TOOLS AND SHOP EQUIPMENT

OBJECTIVES

■ List the basic units of measure for length, volume, and mass in the two measuring systems. ■ Describe the different types of fasteners used in the automotive industry. ■ List the various mechanical measuring tools used in the automotive shop. ■ Describe the proper procedure for measuring with a micrometer. ■ List some of the hand tools used in auto repair. ■ List the common types of shop equipment and state their purpose. ■ Describe the use of common pneumatic, electrical, and hydraulic power tools found in an automotive service department. ■ Describe the different sources for service information that are available to technicians.

Repairing the modern automobile requires the use of various tools. Many of these tools are common hand and power tools used every day by a technician. Other tools are very specialized and are only for specific repairs on specific systems and/or vehicles. This chapter presents some of the more commonly used hand and power tools with which every technician must be familiar. Because units of measurement play such an important part in tool selection and in diagnosing automotive problems, this chapter begins with a presentation of measuring systems. Prior to the discussion on tools, there is a discussion on another topic that relates very much to measuring systems—fasteners.

MEASURING SYSTEMS

Two systems of weights and measures now exist side by side in the United States—the British Imperial (U.S.) system and the international or metric system.

The basic unit of linear measurement in the Imperial system is the inch. The basic unit of linear measurement in the metric system is the meter. The meter is easily broken down into smaller units, such as the centimeter (1/100 meter) and millimeter (1/1,000 meter).

All units of measurement in the metric system are related to each other by a factor of 10. Every metric unit can be multiplied or divided by the factor of 10 to get larger units (multiples) or smaller units (submultiples). This makes the metric system much easier to use, with less chance of math errors than when using the Imperial system **(Figure 5–1)**.

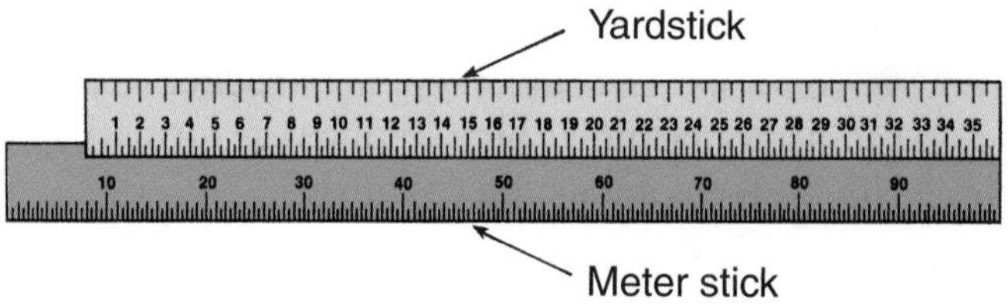

Figure 5–1 A meter stick is made of 1,000 increments known as millimeters and is slightly longer than a yardstick.

The United States passed the Metric Conversion Act in 1975 in an attempt to get American industry and the general public to use the metric system, as the rest of the world does. While the general public has been slow to drop the customary measuring system of inches, gallons, and pounds, many industries, led by the automotive industry, have now adopted the metric system for the most part.

Nearly all vehicles are now built to metric standards. Technicians must be able to measure and work with both systems of measurement. The following are some common equivalents in the two systems:

Linear Measurements

1 meter (m) = 39.37 inches (in.)

1 centimeter (cm) = 0.3937 inch

1 millimeter (mm) = 0.03937 inch

1 inch = 2.54 centimeters

1 inch = 25.4 millimeters

1 mile = 1.6093 kilometers

Square Measurements
1 square inch = 6.452 square centimeters
1 square centimeter = 0.155 square inches

Volume Measurements
1 cubic inch = 16.387 cubic centimeters
1,000 cubic centimeters = 1 liter (l)
1 liter (l) = 61.02 cubic inches
1 gallon = 3.7854 liters

Weight Measurements
1 ounce = 28.3495 grams
1 pound = 453.59 grams
1,000 grams = 1 kilogram
1 kilogram = 2.2046 pounds

Temperature Measurements
1°Fahrenheit (F) = $\frac{9}{5}$C + 32°
1°Celsius (C) = $\frac{5}{9}$(F – 32°)

Pressure Measurements
1 pound per square inch (psi) = 0.07031 kilograms (kg) per square centimeter
1 kilogram per square centimeter = 14.22334 pounds per square inch
1 bar = 14.504 pounds per square inch
1 pound per square inch = 0.06895 bars
1 atmosphere = 14.7 pounds per square inch

Torque Measurements
10 foot pounds (lb) = 13.558 Newton (N) meters
1 Nm = 0.7375 ft. lb
1 ft. lb. = 0.138 m kg
1 cm kg = 7.233 ft. lb
10 cm kg = 0.98 N-m

FASTENERS

Fasteners are those things used to secure or hold parts of something together. Many types and sizes of fasteners are used by the automotive industry. Each fastener is designed for a specific purpose and condition. One type of fastener most commonly used is the threaded fastener. Threaded fasteners include bolts, nuts, screws, and similar items that allow a technician to install or remove parts easily **(Figure 5–2)**.

Threaded fasteners are available in many sizes, designs, and threads. The threads can be either cut or rolled into the fastener. Rolled threads are 30% stronger than cut threads. They also offer better fatigue resistance because there are no sharp notches to create stress points.

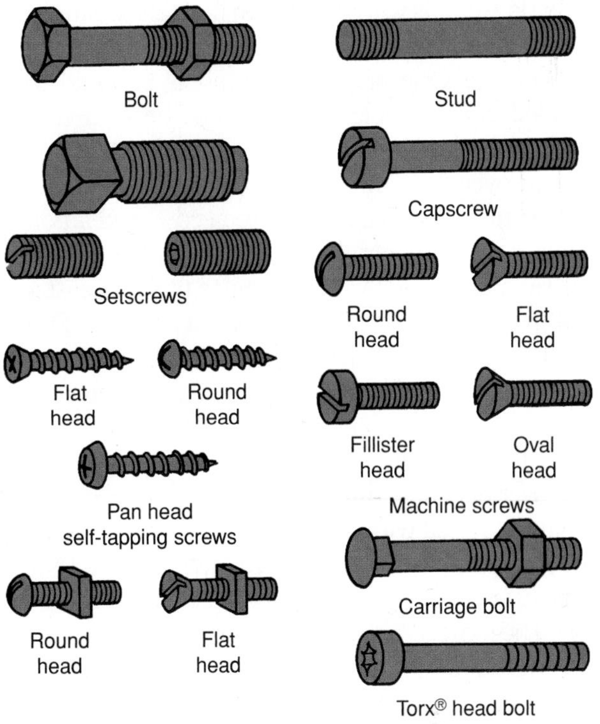

Figure 5–2 Common automotive threaded fasteners.

Fasteners are made to Imperial or metric measurements. There are four classifications for the threads of Imperial fasteners: Unified National Coarse (UNC), Unified National Fine (UNF), Unified National Extrafine (UNEF), and Unified National Pipe Thread (UNPT or NPT). Metric fasteners are also available in fine and coarse threads.

Coarse threads are used for general-purpose work, especially where rapid assembly and disassembly is required. Fine-threaded fasteners are used where greater holding force is necessary. They are also used where greater resistance to vibration is desired.

Bolts have a head on one end and threads on the other. Bolts are identified by defining the head size, shank diameter, thread pitch, length, **(Figure 5–3)** and its grade. Bolts have a shoulder below the head and the threads do not travel all the way from the head to the end of the bolt.

Cap screws are similar to bolts; however, cap screws have no shoulder. The threads travel from the head to the end of the bolt. It is important that you never use a cap screw in place of a bolt.

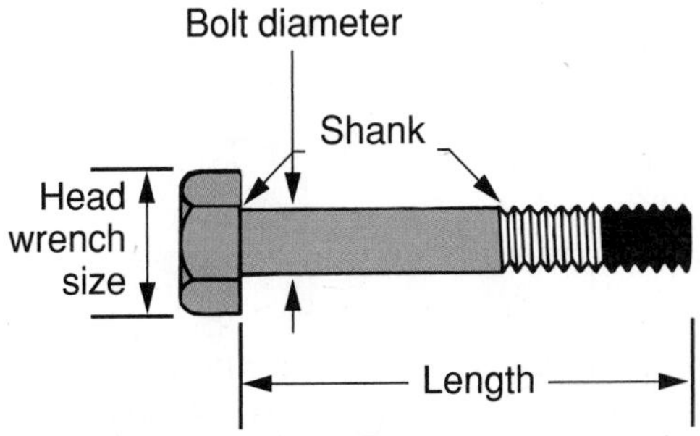

Figure 5–3 Basic terminology for bolt identification.

Studs are rods with threads on both ends. Most often, the threads on one end are coarse while the other end has fine threads. One end of the stud is screwed into a threaded bore. A hole in the part to be secured is fitted over the stud and held in place with a nut that is screwed over the stud. Studs are used when the clamping pressures of a fine thread are needed and a bolt will not work. If the material the stud is being screwed into is soft (such as aluminum) or granular (such as cast iron), fine threads will not withstand a great amount of pulling force on the stud. Therefore, a coarse thread is used to secure the stud in the work piece and a fine-threaded nut is used to secure the other part to it. Doing this results in having the clamping force of fine threads and the holding power of coarse threads.

Nuts are used with other threaded fasteners when the fastener is not threaded into a piece of work. Nuts of many different designs are found on today's cars **(Figure 5–4)**. The most common one is the hex nut, which is used with studs and bolts and is tightened with a wrench.

Setscrews are used to prevent rotary motion between two parts, such as a pulley and shaft. Setscrews are either headless or have a square head. Headless setscrews require an Allen wrench or screwdriver to loosen and tighten them.

Machine screws are similar to cap screws but have a flat point. Machine screws can have a round, flat, Torx®, oval, or fillister head.

Self-tapping screws are used to fasten sheet-metal parts or to join light metal, wood, or plastic parts together. These screws form their own threads in the material they are screwed into.

Bolt Identification

The **bolt head** is used to loosen and tighten the bolt. A socket or wrench fits over the head and is used to screw the bolt in or out. The size of the bolt head varies with the diameter of the bolt and is available in Imperial and metric wrench sizes. Many confuse the size of the head with the size of the bolt. The size of a bolt is determined by the diameter of its shank. The size of the bolt head determines what size wrench is required to screw it. **Table 5–1** lists the most common bolt head sizes. Notice that the sizes are listed as fractions of an inch or as millimeters.

Bolt diameter is the measurement across the major diameter of the threaded area or across the **bolt shank**. The length of a bolt is measured from the bottom surface of the head to the end of the threads.

The **thread pitch** of a bolt in the Imperial system is determined by the number of threads that are in one inch of the threaded bolt length and is expressed in number of threads per inch. A UNF bolt with a 3/8-inch (9.54 mm) diameter would be a 3/8 × 24 bolt. It would have 24 threads per inch. Likewise a 3/8-inch (9.54 mm) UNC bolt would be called a 3/8 × 16.

The distance, in millimeters, between two adjacent threads determines the thread pitch in the metric system. This distance will vary between 1.0 and 2.0, and depends on the diameter of the bolt. The lower the number, the closer the threads are placed and the finer the threads are.

The bolt's tensile strength, or grade, is the amount of stress or stretch it is able to withstand before it breaks. The type of material the bolt is made of and the diameter of the bolt determines its grade. In the Imperial system, the tensile strength of a bolt is identified by the number of radial lines (**grade marks**) on the bolt's head. More lines mean higher tensile strength **(Figure 5–5)**. Count the number of lines and add two to determine the grade of a bolt.

A property class number on the bolt head identifies the grade of metric bolts. This numerical identification is comprised of two numbers. The first number represents the tensile strength of the bolt. The higher the number, the greater the tensile strength. The second number

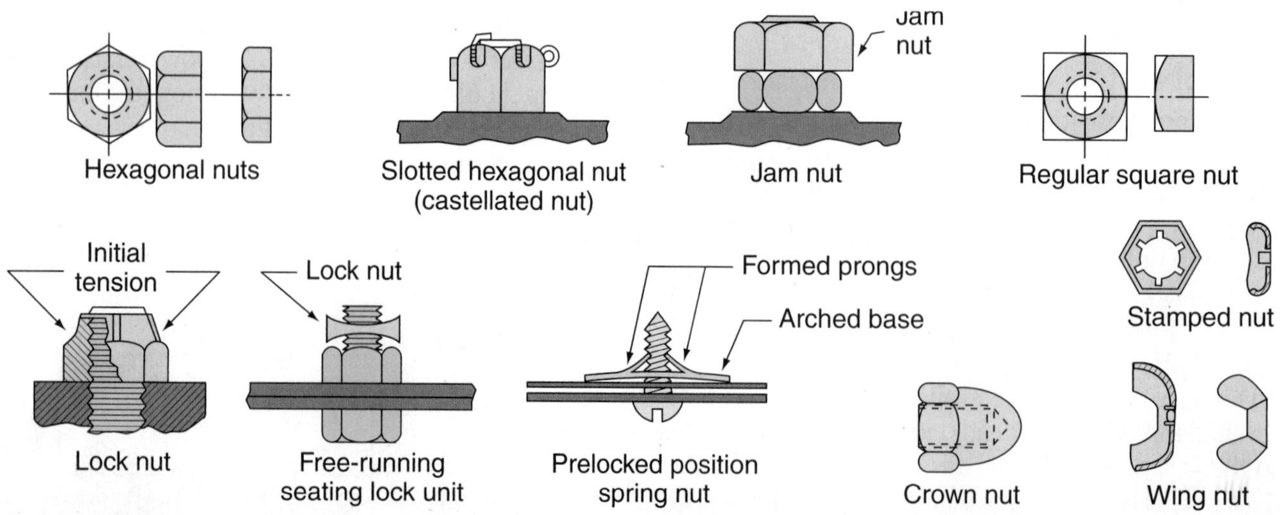

Figure 5-4 Many different types of nuts are used on automobiles. Each type has a specific purpose.

TABLE 5–1 STANDARD BOLT HEAD SIZES

Common English (U.S. Customary) Head Sizes	Common Metric Head Sizes
Wrench Size (inches)*	**Wrench Size (millimeters)***
3/8	9
7/16	10
1/2	11
9/16	12
5/8	13
11/16	14
3/4	15
13/16	16
7/8	17
15/16	18
1	19
1 1/16	20
1 1/8	21
1 3/16	22
1 1/4	23
1 5/16	24
1 3/8	26
1 7/16	27
1 1/2	29
	30
	32

**This does not suggest equivalency.*

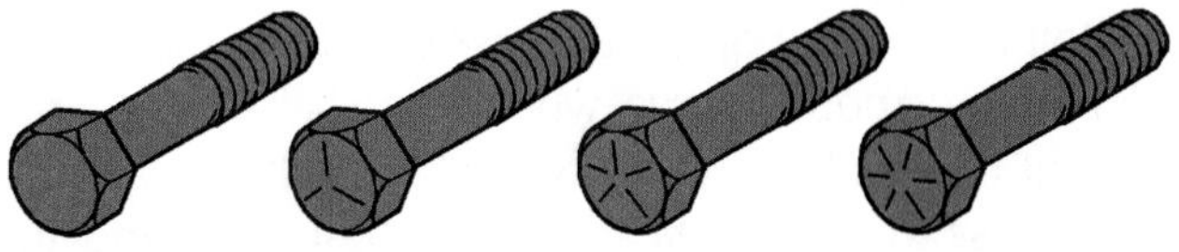

Customary (inch) bolts—identification marks correspond to bolt strength—increasing numbers represent increasing strength.

Metric bolts—identification class numbers correspond to bolt strength—increasing numbers represent increasing strength.

Figure 5-5 Bolt grade markings. *Courtesy of General Motors Corporation*

represents the yield strength of the bolt. This number represents how much stress the bolt can take before it is unable to return to its original shape without damage. The second number represents a percentage rating. For example, a 10.9 bolt has a tensile strength of 1,000 MPa (145,000 psi) and a yield strength of 900 MPa (90% of 1,000). A 10.9 metric bolt is similar in strength to an SAE grade 8 bolt.

Nuts are graded to match their respective bolts **(Table 5–2)**. For example, a grade 8 nut must be used with a grade 8 bolt. If a grade 5 nut were used, a grade 5 connection would result. Grade 8 and critical applications require the use of fully hardened flat washers. These will not dish out when torqued, as soft washers will.

TABLE 5–2 STANDARD NUT STRENGTH MARKINGS

Inch System		Metric System	
Grade	**Identification**	**Class**	**Identification**
Hex Nut Grade 5	 3 Dots	Hex Nut Property Class 9	 Arabic 9
Hex Nut Grade 8	 6 Dots	Hex Nut Property Class 10	 Arabic 10
Increasing dots represent increasing strength.		Can also have blue finish or paint dab on hex flat. Increasing numbers represent increasing strength.	

Bolt heads can pop off because of **fillet** damage. The fillet is the smooth curve where the shank flows into the bolt head **(Figure 5–6)**. Scratches in this area introduce stress to the bolt head, causing failure. Removing any burrs around the edges of holes can protect the bolt head. It is also a good practice to place flat washers with their rounded, punched side against the bolt head and their sharp side to the work surface.

Fatigue breaks are the most common type of bolt failure. A bolt becomes fatigued from working back and forth when it is too loose. Undertightening the bolt causes this problem. Bolts can also be broken or damaged by overtightening, being forced into a nonmatching thread, or bottoming out, which happens when the bolt is too long.

Tightening Bolts

Any fastener is near worthless if it is not as tight as it should be. When a bolt is properly tightened, it will be "spring loaded" against the part it is holding. This spring effect is caused by the stretch of the bolt when it is tightened. Normally a properly tightened bolt is stretched to 70% of its elastic limit. The elastic limit of a bolt is that point of stretch from which the bolt will not return to its original shape when it is loosened. Not only will an overtightened or stretched bolt not have sufficient clamping force, it will also have distorted threads. The stretched threads will make it more difficult to screw and unscrew the bolt or a nut on the bolt.

Washers

Many different types of washers are used with fasteners. The type of washer it is defines the purpose of the washer. Flat washers are used to spread out the load of tightening a nut or bolt. This stops the bolt head or nut from digging into the surface as it is tightened. Soft, flat washers, sometimes called compression washers, are also used to spread the load of tightening and help seal one component to another. Copper washers are often used with oil pan bolts to help seal the pan to the engine block.

Lock washers are used to lock the head of a bolt or nut to the work piece to keep it from coming loose and to prevent damage to softer metal parts.

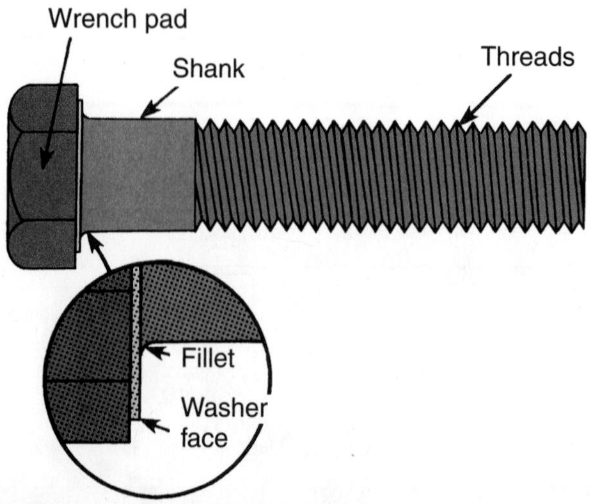

Figure 5-6 Bolt fillet detail.

Thread Lubricants and Sealants

Often manufacturers recommend that the threads of a bolt or stud be coated with a sealant or lubricant. The most commonly used lubricant is antiseize compound. Antiseize compound is used where a bolt might become difficult to remove after a period of time, for example in an aluminum engine block. Thread lubricants introduce the possibility of a hydrostatic lock, where oil is trapped in a blind hole. When the bolt contacts the oil, it cannot compress it; therefore the bolt cannot be properly tightened and a cracked part may result.

Thread sealants are used on bolts that are tightened into an oil cavity or coolant passage. The sealant prevents the liquid from seeping past the threads. Another commonly used thread chemical, called threadlocker **(Figure 5–7)**, prevents a bolt from working loose as the engine or another part vibrates.

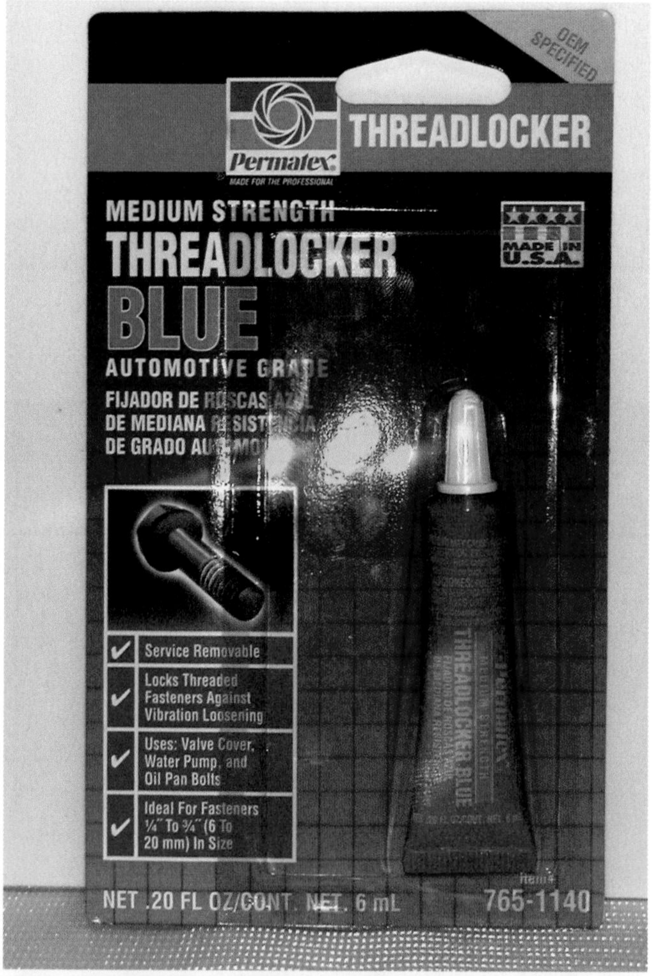

Figure 5-7 A container of thread locker. *Courtesy of Permatex, Inc.*

Screw Pitch Gauge

The use of a **screw pitch gauge (Figure 5–8)** provides a quick and accurate method of checking the thread pitch of a fastener. The leaves of this measuring tool are marked with the various pitches. To check the pitch of threads, simply match the teeth of the gauge with the threads of the fastener. Then, read the pitch from the leaf.

Screw pitch gauges are available for the various types of fastener threads used by the automotive industry: Unified National Coarse and Fine threads, metric threads, International Standard threads, and Whitworth threads.

Taps and Dies

The hand **tap** is a small tool used for hand cutting internal threads **(Figure 5–9)**. An internal thread is cut on the inside of a part, such as a thread on the inside of a nut. Taps are also available that only clean and restore threads that were previously cut. Taps are selected by size and thread pitch. Photo Sequence 2 goes through the correct procedure for repairing damaged threads with a tap.

When tapping a bore, rotate the tap in a clockwise direction. Then, turn the tap counterclockwise about a quarter turn to break off any metal chips that may have accumulated in the threads. These small metal pieces can damage the threads as you continue to tap. These metal chips are gathered in the tap's flutes, which are recessed areas between the cutting teeth of the tap **(Figure 5–10)**. After backing off the tap, continue rotating the tap clockwise. Remember to back off the tap periodically and make sure all of the existing threads in the bore have been recut by the tap.

Hand-threading **dies** are the opposite of taps because they cut external (outside) threads on bolts, rods, and pipes rather than internal threads. Dies are made in various sizes and shapes, depending on the particular work for which they are intended. Dies may be solid (fixed size), split on one side to permit adjustment, or have two halves held together in a collet that provides for individual adjustments. Dies fit into holders called die stocks.

Threaded Inserts

When the threads in a bore are excessively damaged, it is better to replace them than try to tap them. A thread insert can be used to restore the original threads. Inserts require drilling the bore to a larger diameter and tapping that bore to allow the insert to be screwed into it. The inner threaded diameter of the insert will provide fresh threads for the bolt **(Figure 5–11)**.

Spark Plug Thread Repair Sometimes when spark plugs are removed from a cylinder head, the threads have traces

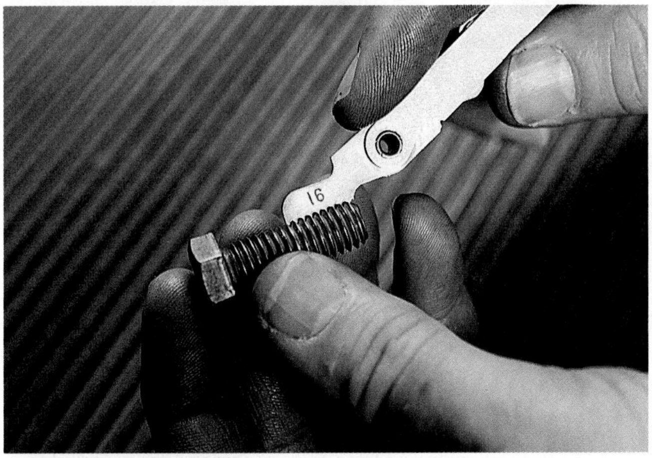

Figure 5–8 Using a screw pitch gauge to check the thread pitch on a bolt.

Figure 5–9 A tap and die set. *Courtesy of Snap-on Tools Company*

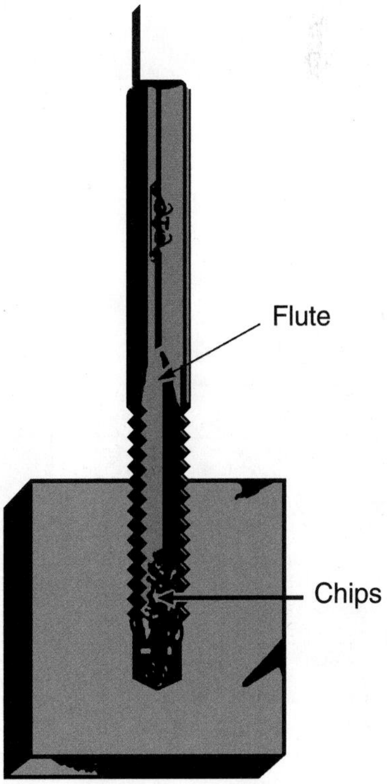

Figure 5–10 Metal chips are gathered into the flutes of a tap.

PHOTO SEQUENCE 2

Repairing Damaged Threads with a Tap

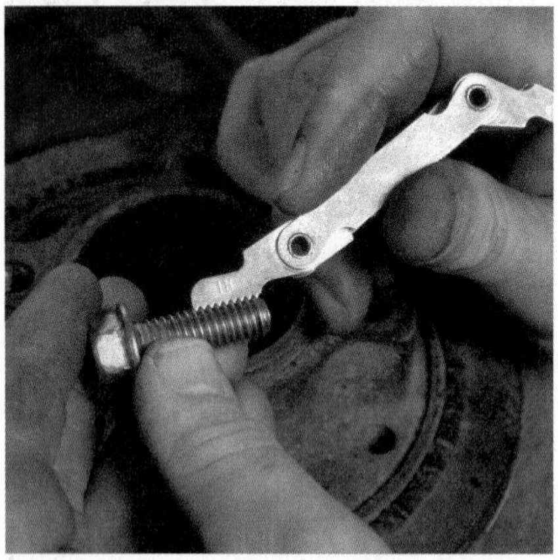

P2-1 *Using a thread pitch gauge, determine the thread size of the fastener that should fit into the damaged internal threads.*

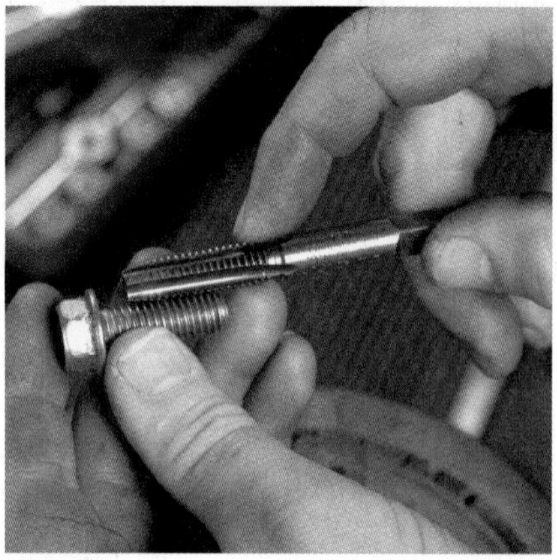

P2-2 *Select the correct size and type of tap for the threads and bore to be repaired.*

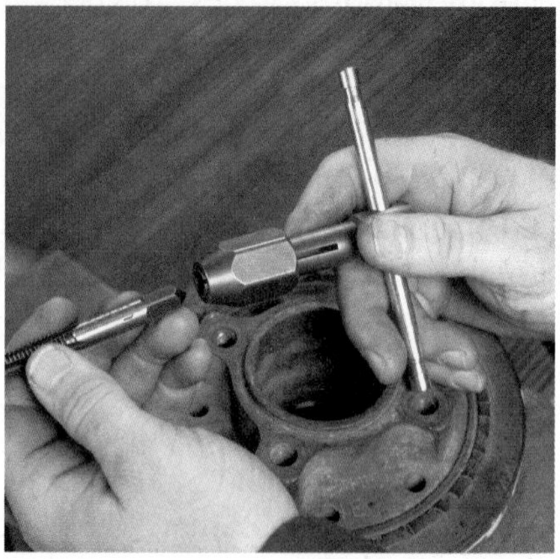

P2-3 *Install the tap into a tap wrench.*

P2-4 *Start the tap squarely in the threaded hole using a machinist square as a guide.*

P2-5 *Rotate the tap clockwise into the bore until the tap has run through the entire length of the threads.*

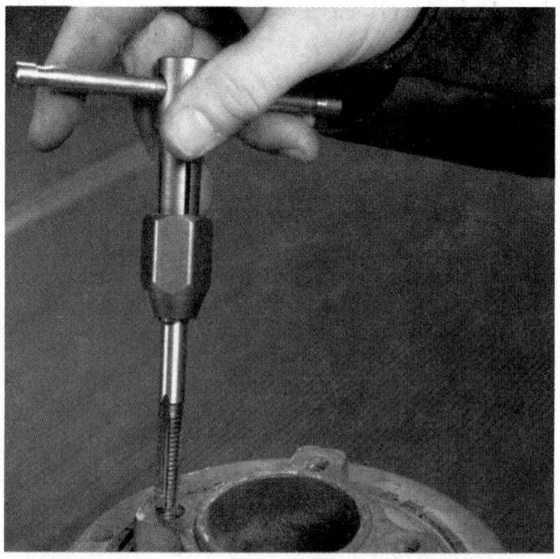

P2-6 *Drive the tap back out of the hole by turning it counterclockwise.*

P2-7 *Clean the metal chips left by the tap out of the hole.*

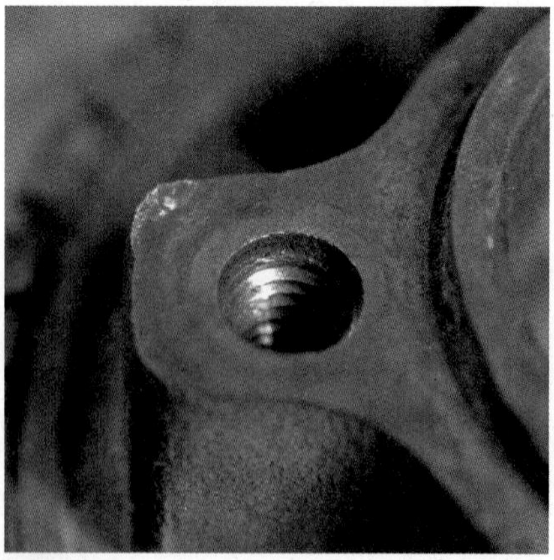

P2-8 *Inspect the threads left by the tap to be sure they are acceptable.*

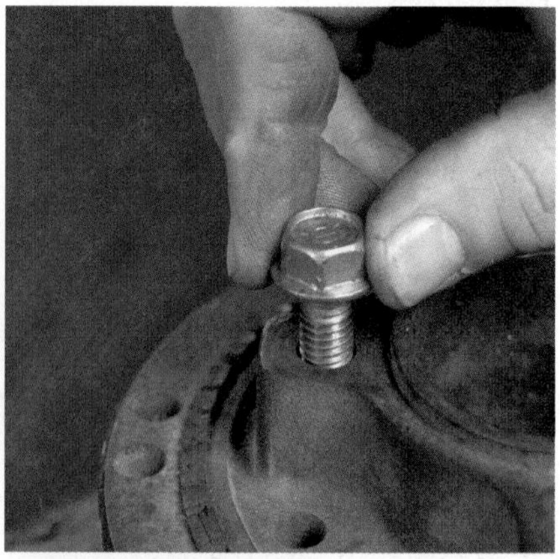

P2-9 *Test the threads by threading the correct fastener into the threaded hole.*

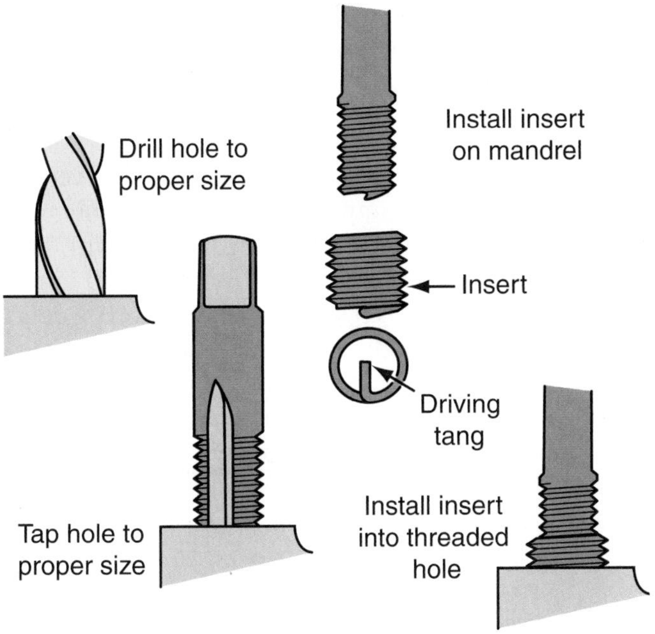

Figure 5-11 Using a threaded insert (heli-coil ®) to repair damaged threads. *Courtesy of Emhart Fastening Teknologies*

of metal on them. This happens more often with aluminum heads. When this occurs, the spark plug bore must be corrected by installing thread inserts.

> **SHOP TALK**
>
> Never change spark plugs when the cylinder head is hot. The bores for the plugs can take on an oval shape as the cylinder head cools without spark plugs in the bores. ■

When installing spark plugs, if the plugs cannot be installed easily by hand, the threads in the cylinder head may need to be cleaned with a thread-chasing tap. There are special taps for spark plug bores, simply called **spark plug thread taps**. Be especially careful not to cross-thread the plugs when working with aluminum heads. Always tighten the plugs with a torque wrench and the correct spark plug socket, following the vehicle manufacturer's specifications. Also, when changing spark plugs in aluminum heads, the temperature of the heads should be ambient temperature before attempting to remove the plugs.

MEASURING TOOLS

Some service work, such as engine repair, requires very exact measurements, often in ten-thousandths (0.0001) of an inch or thousandths (0.001) of a millimeter. Accurate measurements with this kind of precision can only be made by using precise measuring devices.

Measuring tools are precise and delicate instruments. In fact, the more precise they are, the more delicate they are. They should be handled with great care. Never pry, strike, drop, or force these instruments. They may be permanently damaged.

Precision measuring instruments, especially micrometers, are extremely sensitive to rough handling. Clean them before and after every use. All measuring should be performed on parts that are at room temperature to eliminate the chance of measuring something that has contracted because it was cold or has expanded because it was hot.

> **SHOP TALK**
>
> Check measuring instruments regularly against known good equipment to ensure that they are operating properly and are capable of accurate measurement. Always refer to the appropriate material for the correct specifications before performing any service or diagnostic procedures. The close tolerances required for the proper operation of some automotive parts make using the correct specifications and taking accurate measurements very important. Even the slightest error in measurement can be critical to the durability and operation of an engine and other systems. ■

Machinist's Rule

The **machinist's rule** looks very much like an ordinary ruler. Each edge of this basic measuring tool is divided into increments based on a different scale. As shown in **Figure 5–12**, a typical machinist's rule based on the

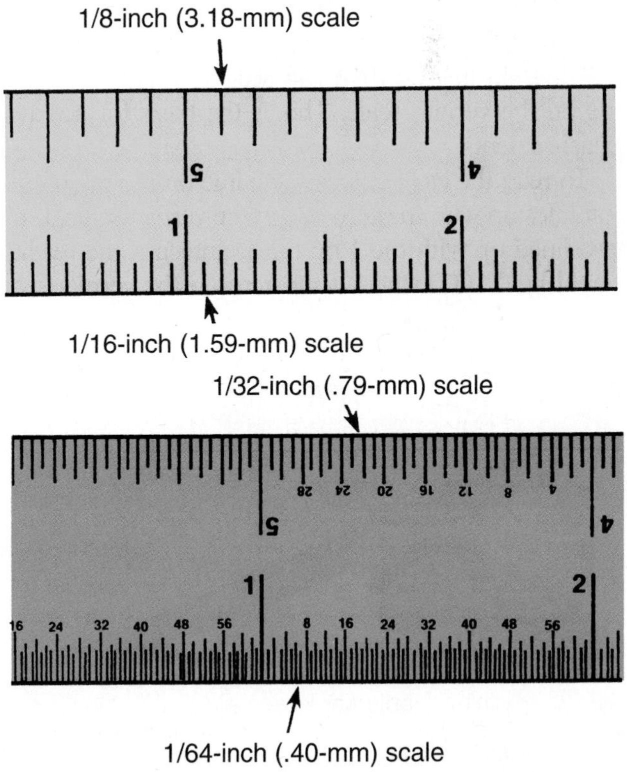

Figure 5-12 Graduations on a typical machinist's rule.

Imperial system of measurement may have scales based on ⅛-, 1/16-, 1/32-, and 1/64-inch (3.18-, 1.59-, .79-, and .40-mm) intervals. Of course, metric machinist rules are also available. Metric rules are usually divided into 0.5-mm and 1-mm increments.

Some machinist's rules may be based on decimal intervals. These are typically divided into 1/10-, 1/50-, and 1/1,000-inch (0.1-, 0.03-, and 0.01-mm) increments. Decimal machinist's rules are very helpful when measuring dimensions specified in decimals; they make such measurements much easier.

Vernier Caliper

A **vernier caliper** is a measuring tool that can make inside, outside, or depth measurements. It is marked in both British Imperial and metric divisions called a vernier scale. A vernier scale consists of a stationary scale and a movable scale, in this case the vernier bar to the vernier plate. The length is read from the vernier scale.

A vernier caliper has a movable scale that is parallel to a fixed scale **(Figure 5–13)**. These precision measuring instruments are capable of measuring outside and inside diameters and most will even measure depth. Vernier calipers are available in both Imperial and metric scales. The main scale of the caliper is divided into inches; most measure up to 6 inches. Each inch is divided into 10 parts, each equal to 0.100 inch. The area between the 0.100 marks is divided into four. Each of these divisions is equal to 0.025 inches.

The vernier scale has 25 divisions, each one representing 0.001 inch. Measurement readings are taken by combining the main and vernier scales. At all times, only one division line on the main scale will line up with a line on the vernier scale. This is the basis for accurate measurements.

To read the caliper, locate the line on the main scale that lines up with the zero (0) on the vernier scale. If the zero lined up with the 1 on the main scale, the reading would be 0.100 inches. If the zero on the vernier scale does not line up exactly with a line on the main scale, then look for a line on the vernier scale that does line up with a line on the main scale.

Dial Caliper

The **dial caliper (Figure 5–14)** is an easier-to-use version of the vernier caliper. Imperial calipers commonly measure dimensions from 0 to 6 inches (0 to 150 mm). Metric dial calipers typically measure from 0 to 150 mm in increments of 0.02 mm. The dial caliper features a depth scale, bar scale, dial indicator, inside measurement jaws, and outside measurement jaws.

The main scale of an British Imperial dial caliper is divided into one-tenth (0.1) inch graduations. The dial indicator is divided into one-thousandth (0.001) inch graduations. Therefore, one revolution of the dial indicator needle equals one-tenth inch on the bar scale.

A metric dial caliper is similar in appearance; however, the bar scale is divided into 2-mm increments. Additionally, on a metric dial caliper, one revolution of the dial indicator needle equals 2 mm.

Both English and metric dial calipers use a thumb-operated roll knob for fine adjustment. When you use a dial caliper, always move the measuring jaws backward and forward to center the jaws on the object being measured. Make sure the caliper jaws lay flat on or around the object. If the jaws are tilted in any way, you will not obtain an accurate measurement.

Although dial calipers are precision measuring instruments, they are only accurate to plus or minus two-thousandths (±0.002) of an inch. Micrometers are preferred when extremely precise measurements are desired.

Micrometers

The **micrometer** is used to measure linear outside and inside dimensions. Both outside and inside micrometers are

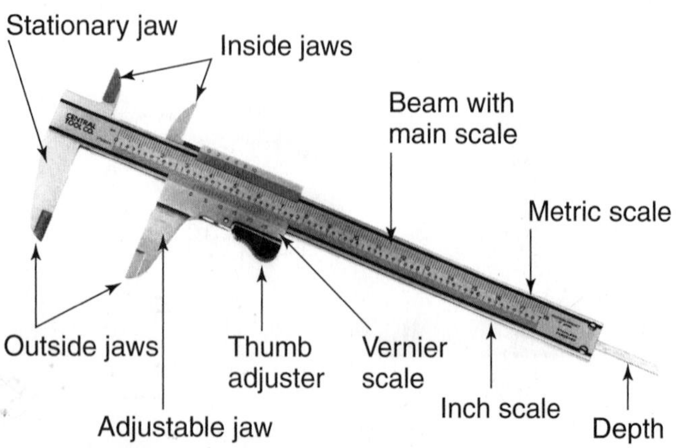

Figure 5–13 A vernier caliper. *Courtesy of Central Tools, Inc.*

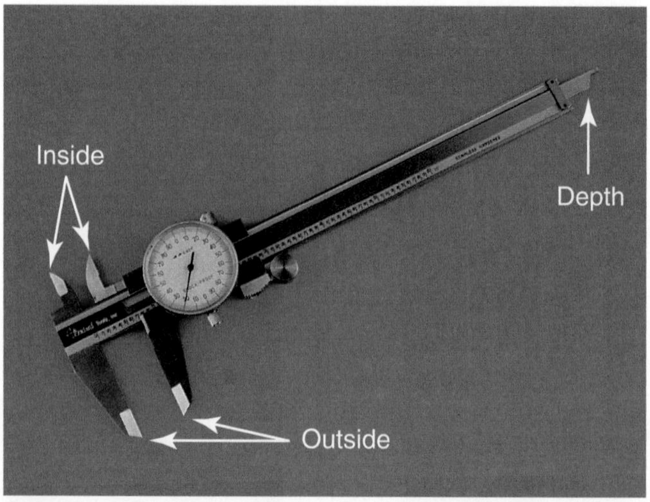

Figure 5–14 A dial vernier caliper. *Courtesy of Central Tools, Inc.*

calibrated and read in the same manner. Measurements on both are taken with the measuring points in contact with the surfaces being measured.

The major components and markings of a micrometer include the frame, anvil, spindle, locknut, sleeve, sleeve numbers, sleeve long line, thimble marks, thimble, and ratchet **(Figure 5–15)**. Micrometers are calibrated in either inch or metric graduations and are available in a range of sizes. The proper procedure for measuring with an inch-graduated outside micrometer is outlined in Photo Sequence 3.

Most micrometers are designed to measure objects with accuracy to 0.001 (one-thousandth) inch. Micrometers are also available to measure in 0.0001 (ten-thousandths) of an inch. This type of micrometer should be used when the specifications call for this much accuracy.

Reading a Metric Outside Micrometer The metric micrometer is read in the same manner as the inch-graduated micrometer, except the graduations are expressed in the metric system of measurement. Readings are obtained as follows:

- Each number on the sleeve of the micrometer represents 5 millimeters (mm) or 0.005 meter (m) **(Figure 5–16A)**.
- Each of the 10 equal spaces between each number, with index lines alternating above and below the horizontal line, represents 0.5 mm or five-tenths of a mm. One revolution of the thimble changes the reading one space on the sleeve scale or 0.5 mm **(Figure 5–16B)**.
- The beveled edge of the thimble is divided into 50 equal divisions with every fifth line numbered: 0, 5, 10, . . . 45. Since one complete revolution of the thimble advances the spindle 0.5 mm, each graduation on the thimble is equal to one hundredth of a millimeter **(Figure 5–16C)**.
- As with the inch-graduated micrometer, the three separate readings are added together to obtain the total reading **(Figure 5–17)**.

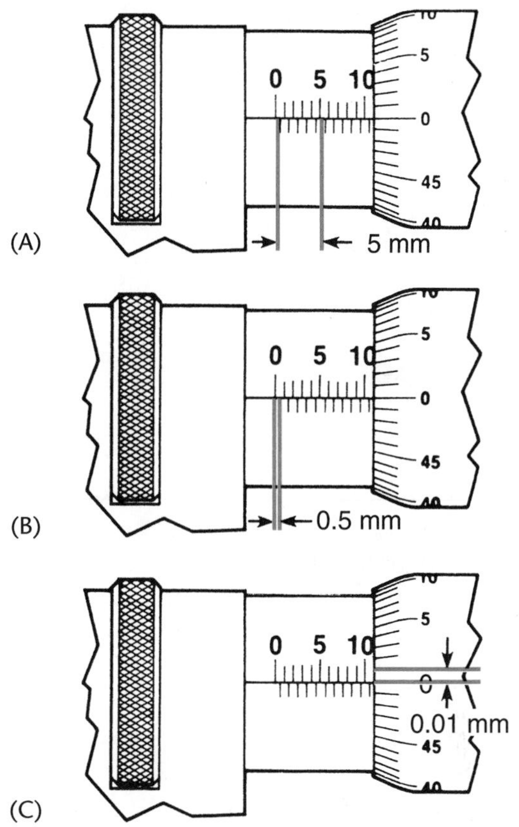

Figure 5-16 Reading a metric micrometer: (A) 5 mm plus (B) 0.5 mm plus (C) 0.01 mm equals 5.51 mm.

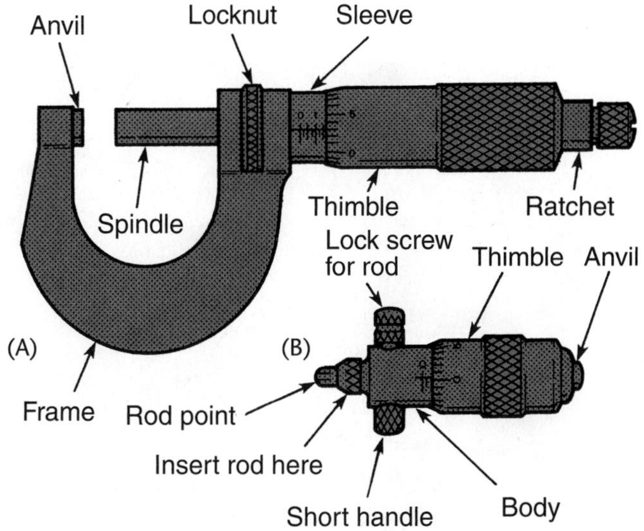

Figure 5-15 Major components of (A) an outside and (B) an inside micrometer.

Using an Outside Micrometer To measure small objects using an outside micrometer, open the jaws of the tool and slip the object between the spindle and anvil. While holding the object against the anvil, turn the thimble using

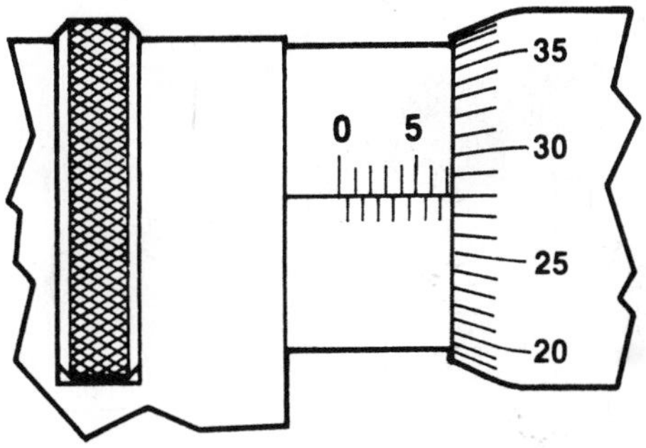

Figure 5-17 The total reading on this micrometer is 7.28 mm.

PHOTO SEQUENCE

3 Using a Micrometer

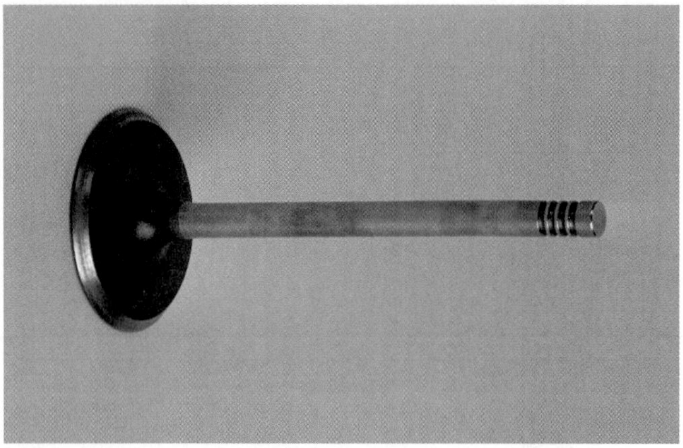

P3-1 *Micrometers can be used to measure the diameter of many different objects. By measuring the diameter of a valve stem in two places, the wear of the stem can be determined.*

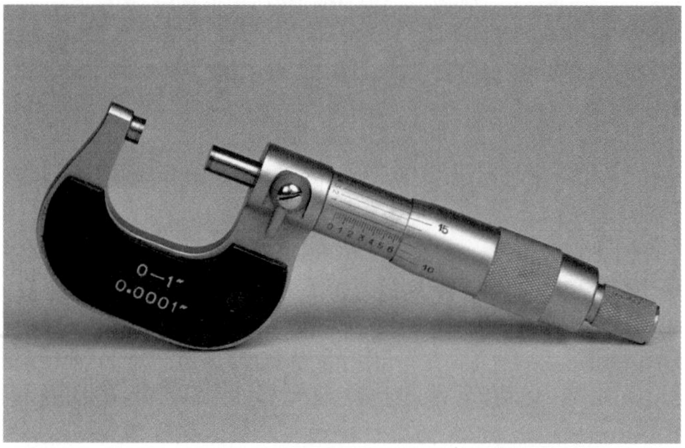

P3-2 *Because the diameter of a valve stem is less than one inch, a 0-to-1-inch (0-to-25-mm) outside micrometer is used.*

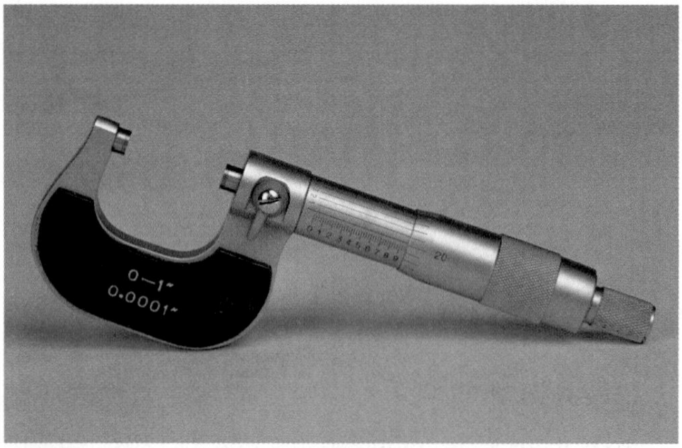

P3-3 *The graduations on the sleeve each represent 0.025 inch. To read a measurement on a micrometer, begin by counting the visible lines on the sleeve and multiplying them by 0.025.*

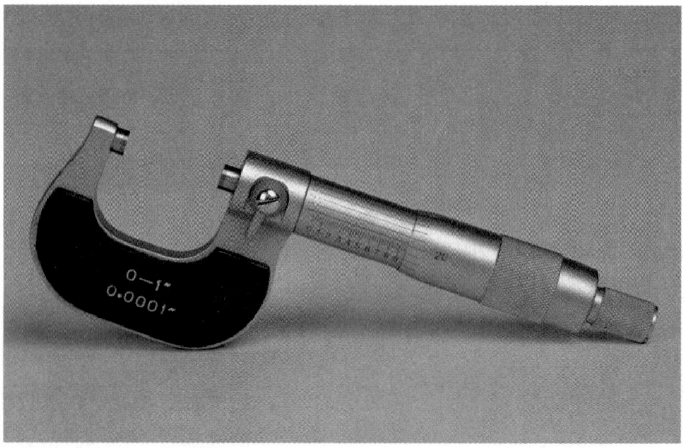

P3-4 *The graduations on the thimble assembly define the area between the lines on the sleeve. The number indicated on the thimble is added to the measurement shown on the sleeve.*

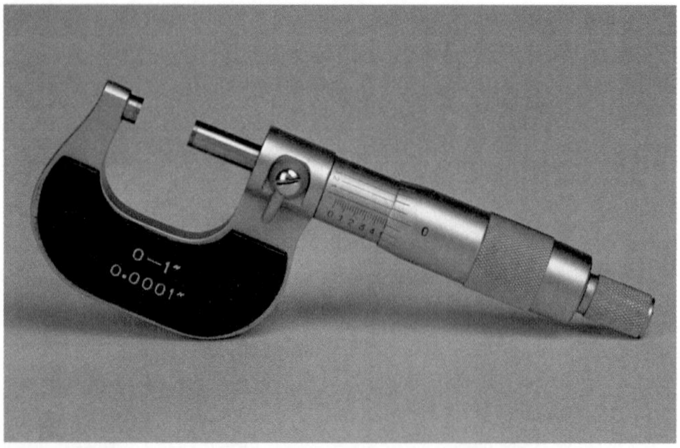

P3-5 *A micrometer reading of 0.500 inch (12.70 mm).*

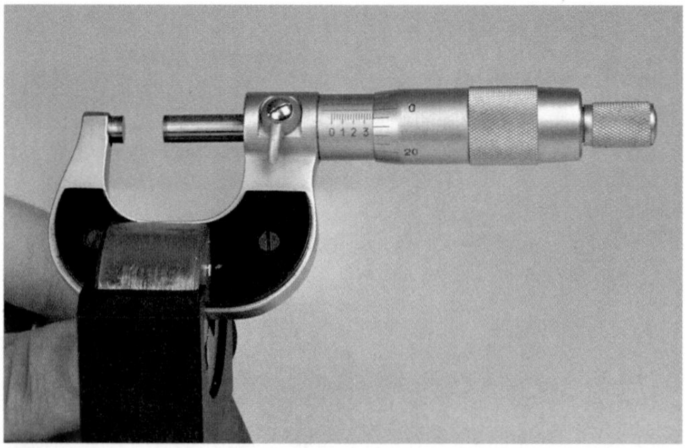

P3-6 *A micrometer reading of 0.375 inch (9.53 mm).*

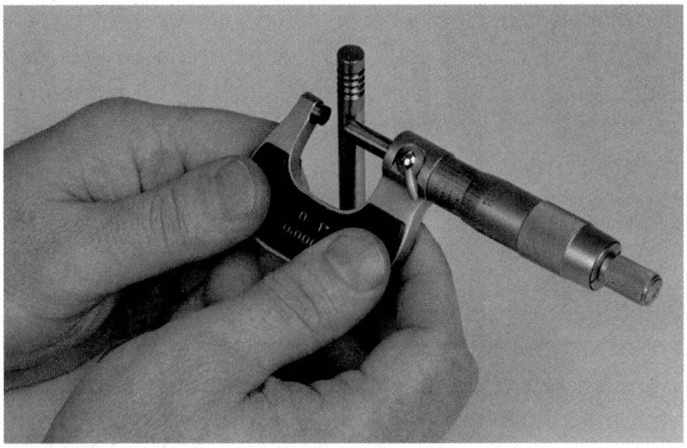

P3-7 *Normally, little stem wear is evident directly below the keeper grooves. To measure the diameter of the stem at that point, close the micrometer around the stem.*

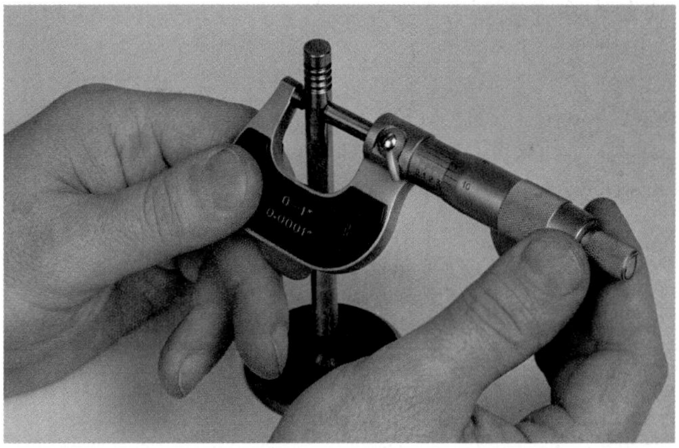

P3-8 *To get an accurate reading, slowly close the micrometer until a slight drag is felt while passing the valve in and out of the micrometer.*

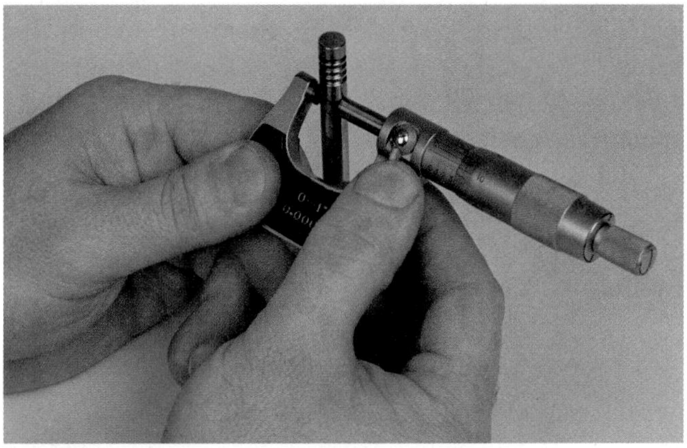

P3-9 *To prevent the reading from changing while you move the micrometer away from the stem, use your thumb to activate the lock lever.*

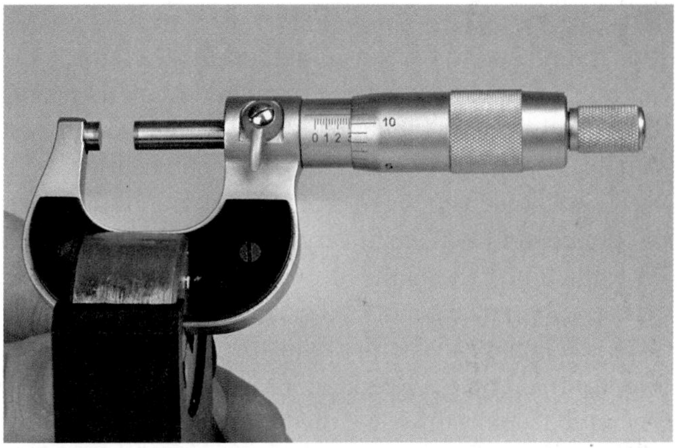

P3-10 *This reading (0.311 inch [7.89 mm]) represents the diameter of the valve stem at the top of the wear area.*

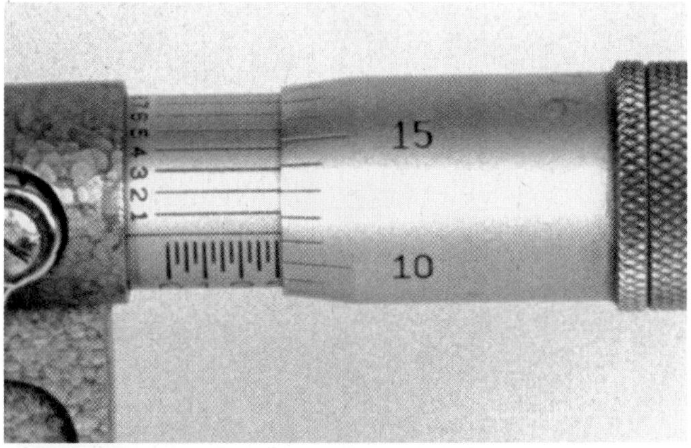

P3-11 *Some micrometers are able to measure in 0.0001 (ten-thousandths) of an inch. Use this type of micrometer if the specifications call for this much accuracy. Note that the exact diameter of the valve stem is 0.3112 inch (7.90 mm).*

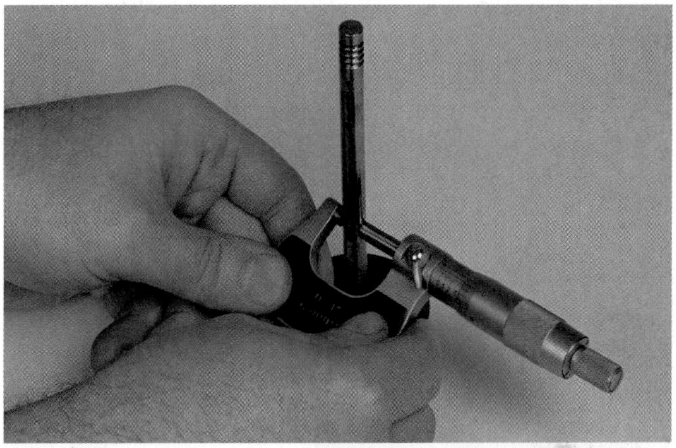

P3-12 *Most valve stem wear occurs above the valve head. The diameter here should also be measured. The difference between the diameter of the valve stem just below the keepers and just above the valve head represents the amount of valve stem wear.*

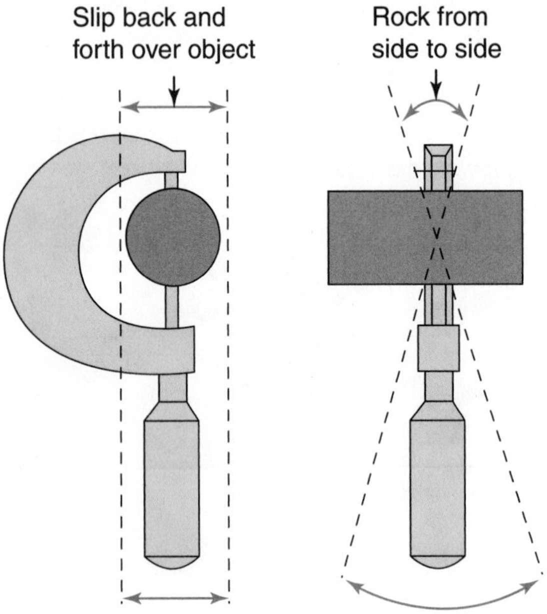

Figure 5-18 Slip the micrometer over the object and rock it from side to side.

your thumb and forefinger until the spindle contacts the object. Never clamp the micrometer tightly. Use only enough pressure on the thimble to allow the work to just fit between the anvil and spindle. To get accurate readings, you should slip the micrometer back and forth over the object until you feel a very light resistance, while at the same time rocking the tool from side to side to make certain the spindle cannot be closed any further **(Figure 5–18)**. When a satisfactory adjustment has been made, lock the micrometer. Read the measurement scale.

Some technicians use a digital micrometer, which is easier to read. Since these tools do not have the various scales, the measurement is displayed and read directly off the micrometer. **Figure 5–19** shows a digital outside micrometer.

To measure a larger object such as a piston, select a micrometer of the proper size. Micrometers are available in a number of different sizes. The size is dictated by the smallest measurement it can make to the largest. Examples of these sizes are the 0-to-1-inch (0-to-25-mm), 1-to-2-inch (25-to-50-mm), 2-to-3-inch (50-to-75-mm), and 3-to-4-inch (75-to-100-mm) micrometers.

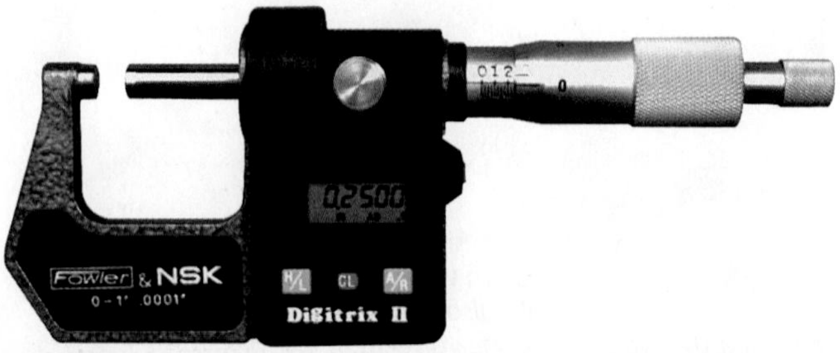

Figure 5-19 A digital micrometer. *Courtesy of Fred V. Fowler Co., Inc.*

Reading an Inside Micrometer Inside micrometers are used to measure the inside diameter of a bore or hole. To do this, place the tool inside the bore and extend the measuring surfaces until each end touches the bore's surface. If the bore is large, it might be necessary to use an extension rod to increase the micrometer's range. These extension rods come in various lengths. The inside micrometer is read in the same manner as an outside micrometer.

To obtain a precise measurement, keep the anvil firmly against one side of the bore and rock the micrometer back and forth and side to side to ensure that the micrometer is in the center of the bore. As with the outside micrometer, this procedure will require a little practice until you get the feel of the correct resistance on both ends of the tool.

Reading a Depth Micrometer A depth micrometer **(Figure 5–20)** is used to measure the distance between two parallel surfaces. The sleeves, thimbles, and ratchet screws operate in the same way as other micrometers. Likewise, depth micrometers are read in the same way as other micrometers.

If a depth micrometer is used with a gauge bar, it is important to keep both the bar and the micrometer from rocking. Any movement of either part will result in an inaccurate measurement.

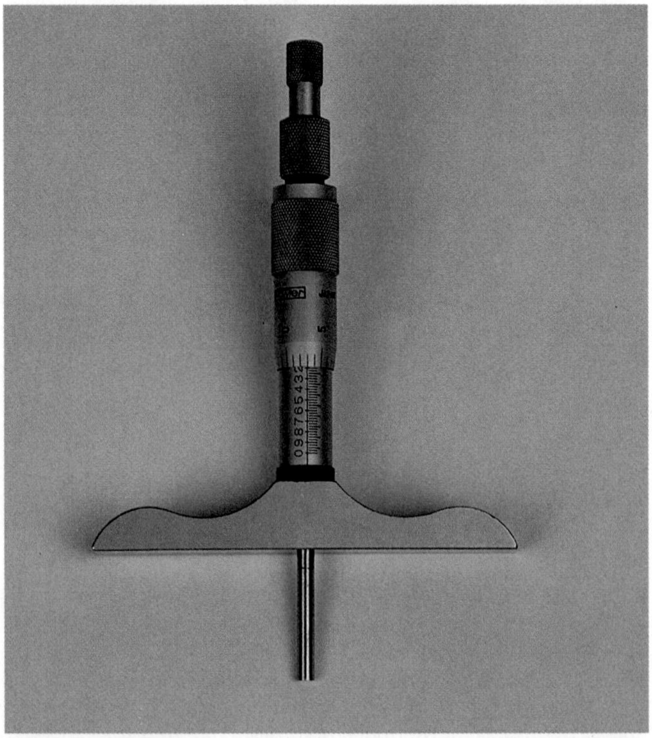

Figure 5-20 A depth micrometer. *Courtesy of Central Tools, Inc.*

SHOP TALK

Measurements with any micrometer will be reliable only if the micrometer is calibrated correctly. To calibrate a micrometer, close the micrometer over a micrometer standard. If the reading differs from that of the standard, the micrometer should be adjusted according to the instructions provided by the tool manufacturer. Proper care of a micrometer is also important to ensure accurate measurements. This care includes:

- Always clean the micrometer before using it.
- Do not touch the measuring surfaces.
- Store the tool properly. The spindle face should not touch the anvil face; a change in temperature might spring the micrometer.
- Clean the micrometer after use. Wipe it clean of any oil, dirt, or dust using a lint-free cloth.
- Never use the tool as a clamp or tighten the jaws too tightly around an object.
- Do not drop the tool.
- Check the calibration weekly. ■

Telescoping Gauge

Telescoping gauges (Figure 5–21) are used for measuring bore diameters and other clearances. They may also be called **snap gauges**. They are available in sizes ranging from fractions of an inch through 6 inches (150 mm). Each gauge consists of two telescoping plungers, a handle, and a lock screw. Snap gauges are normally used with an outside micrometer.

To use the telescoping gauge, insert it into the bore and loosen the lock screw. This will allow the plungers to snap against the bore. Once the plungers have expanded, tighten the lock screw. Then, remove the gauge and measure the expanse with a micrometer.

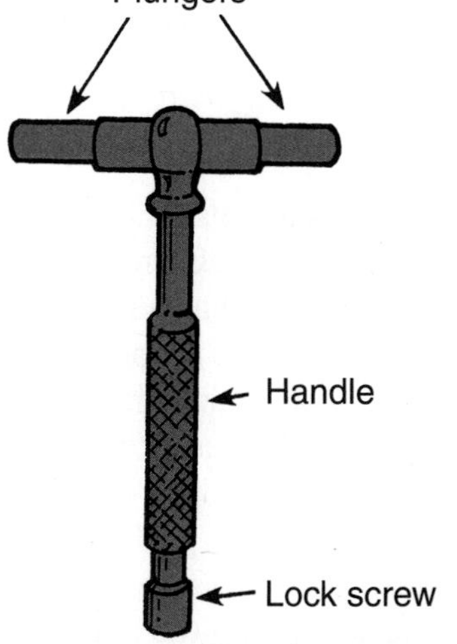

Figure 5-21 Parts of a telescoping gauge.

Small Hole Gauge

A small hole or **ball gauge** works just like a telescoping gauge. However, it is designed to be used on small bores. After it is placed into the bore and expanded, it is removed and measured with a micrometer **(Figure 5–22)**.

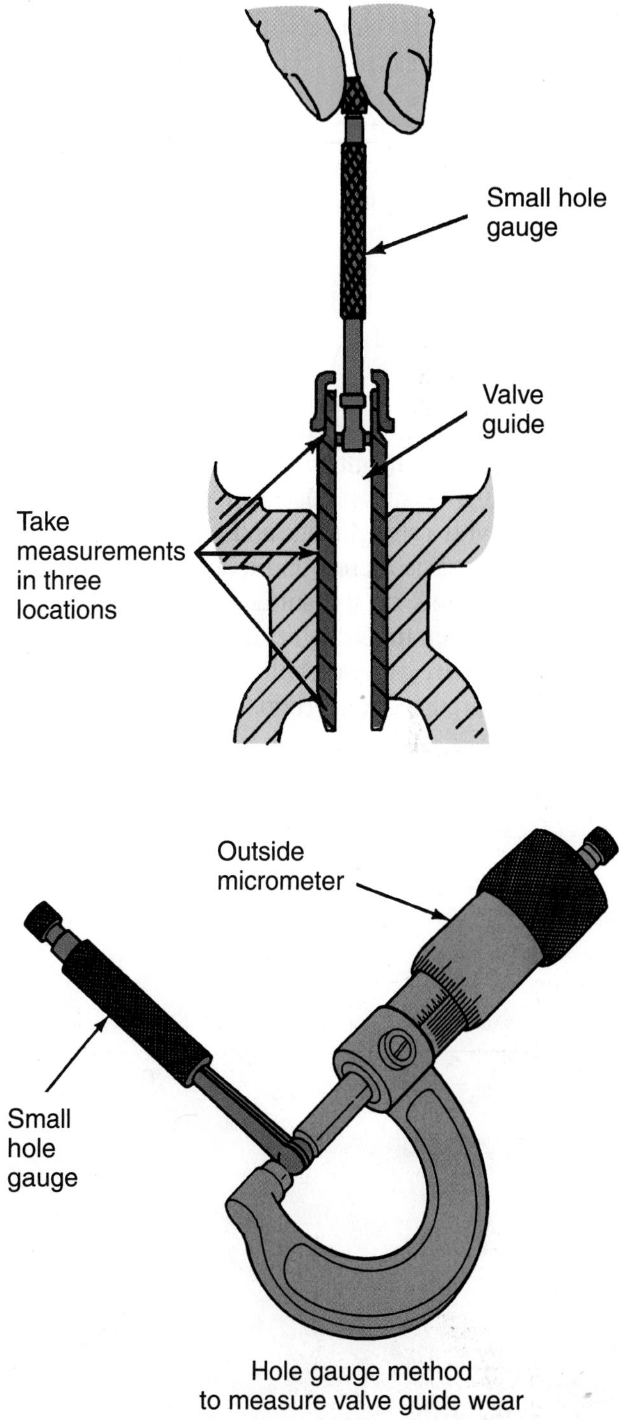

Figure 5-22 Insert the ball gauge into the bore to be measured. Then expand it, lock it, and remove it. Now measure it with an outside micrometer. *Courtesy of Ford Motor Co.*

Like the telescoping gauge, the small hole gauge consists of a lock, handle, and an expanding end. The end expands or retracts by turning the gauge handle.

Feeler Gauge

A **feeler gauge** is a thin strip of metal or plastic of known and closely controlled thickness. Several of these metal strips are often assembled together as a feeler gauge set that looks like a pocket knife **(Figure 5–23)**. The desired thickness gauge can be pivoted away from others for convenient use. A steel feeler gauge pack usually contains strips or leaves of 0.002- to 0.010-inch (0.508- to .2540-mm) thickness (in steps of 0.001 inch [.0254 mm]) and leaves of 0.012- to 0.024-inch (.3048- to .6096-mm) thickness (in steps of 0.002 inch [.0508 mm]).

A feeler gauge can be used by itself to measure piston ring side clearance, piston ring end gap, connecting rod side clearance, crankshaft endplay, and other distances.

Round wire feeler gauges are often used to measure spark plug gap. The round gauges are designed to give a better feel for the fit of the gauge in the gap.

Straightedge

A **straightedge** is no more than a flat bar machined to be totally flat and straight, and to be effective it must be flat and straight. Any surface that should be flat can be checked with a straightedge and feeler gauge set. The straightedge is placed across and at angles on the surface. At any low points on the surface, a feeler gauge can be placed between the straightedge and the surface **(Figure 5-24)**. The size gauge that fills in the gap indicates the amount of warpage or distortion.

Dial Indicator

The **dial indicator (Figure 5–25)** is calibrated in 0.001-inch (.0254-mm) (one-thousandth inch) increments.

Figure 5-23 Typical feeler gauge set.

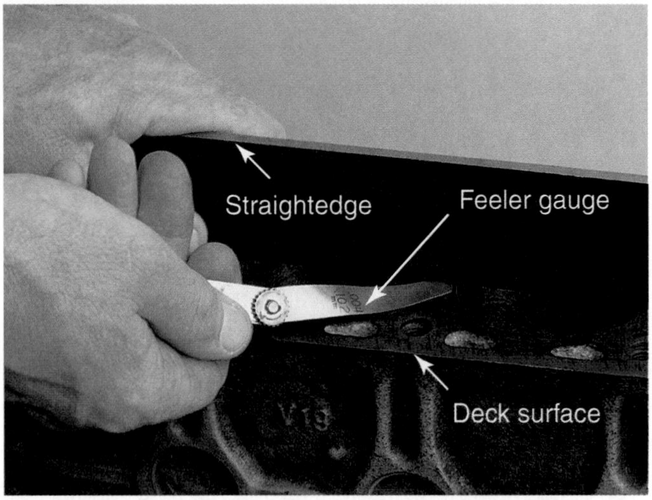

Figure 5-24 Using a feeler gauge and precision straightedge to check for warpage.

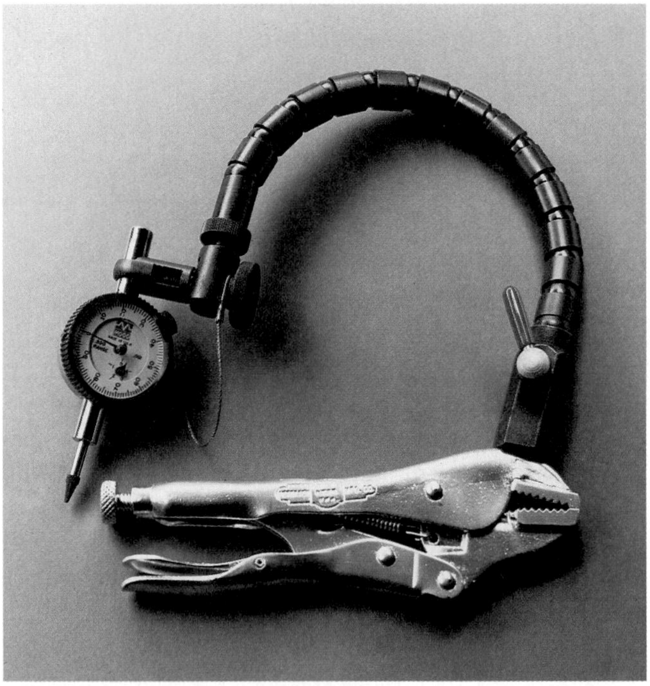

Figure 5-25 A dial indicator with a highly adaptive holding fixture. *Courtesy of Moog Automotive, Inc.*

Metric dial indicators are also available. Both types are used to measure movement. Common uses of the dial indicator include measuring valve lift, journal concentricity, flywheel or brake rotor runout, gear backlash, and crankshaft end play. Dial indicators are available with various face markings and measurement ranges to accommodate many measuring tasks.

To use a dial indicator, position the indicator rod against the object to be measured. Then, push the indicator toward the work until the indicator needle travels far enough around the gauge face to permit movement to

Figure 5-26 This dial indicator setup will measure the amount this axle can move in and out.

be read in either direction **(Figure 5–26)**. Zero the indicator needle on the gauge. Always be sure the range of the dial indicator is sufficient to allow the amount of movement required by the measuring procedure. For example, never use a 1-inch (25-mm) indicator on a component that will move 2 inches (50.8 mm).

HAND TOOLS

Most service procedures require the use of hand tools. Therefore, technicians need a wide assortment of these tools. Each has a specific job and should be used in a specific way. Most service departments and garages require their technicians to buy their own hand tools. A set of master technician's hand tools and a tool chest are shown in **Figure 5–27**.

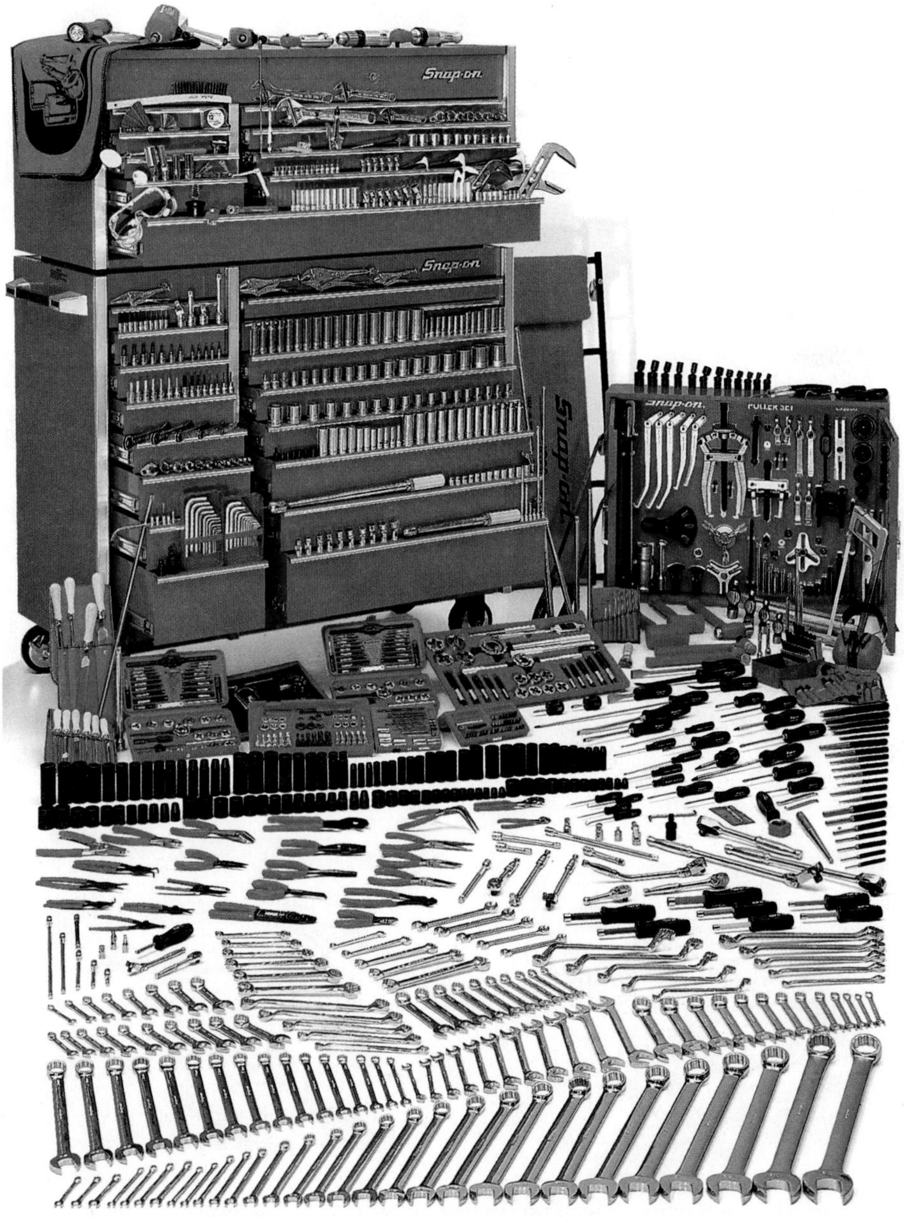

Figure 5-27 A typical set of automotive hand tools. *Courtesy of Snap-on Tools Company*

Wrenches

The word *wrench* means twist. A wrench is a tool for twisting and/or holding bolt heads or nuts. Nearly all bolt heads and nuts have six sides; the jaw of a wrench fits around these sides to turn the bolt or nut. All technicians should have a complete collection of **wrenches**. This includes both metric and SAE wrenches in a variety of sizes and styles **(Figure 5–28)**. The width of the jaw opening determines its size. For example, a ½-inch (12.70-mm) wrench has a jaw opening (from face to face) of ½ inch (12.70 mm). The size is actually slightly larger than its nominal size so the wrench fits around a nut or bolt head of equal size.

Figure 5-28 A technician needs many different sets of wrenches. *Courtesy of Snap-on Tools Company*

SHOP TALK

Metric and SAE wrenches are not interchangeable. For example, a 9⁄16-inch wrench is 0.02 inch larger than a 14-millimeter nut. If the 9⁄16-inch wrench is used to turn or hold a 14-millimeter nut, the wrench will probably slip. This may cause rounding of the points of the nut and possibly skinned knuckles as well. ■

The following is a brief discussion of the types of wrenches used by automotive technicians.

Open-End Wrench The jaws of the open-end wrench **(Figure 5–29)** allow the wrench to slide around two sides of a bolt or nut head where there might be insufficient clearance above or on one side of the nut to accept a box wrench.

Box-End Wrench The end of the box-end wrench is boxed or closed rather than open. The jaws of the wrench fit completely around a bolt or nut, gripping each point on the fastener. The box-end wrench is not likely to slip off a nut or bolt. It is safer than an open-end wrench. Box-end wrenches are available as 6 point and 12 point **(Figure 5–30)**. The 6-point box end grips the screw more securely than a 12-point box-end wrench can and avoids damage to the bolt head.

Combination Wrench The combination wrench has an open-end jaw on one end and a box-end on the other. Both ends are the same size. Every auto technician should have two sets of wrenches: one for holding and one for turning. The combination wrench is probably the best choice for the second set. It can be used with either open-end or box-end wrench sets and can be used as an open-end or box-end wrench.

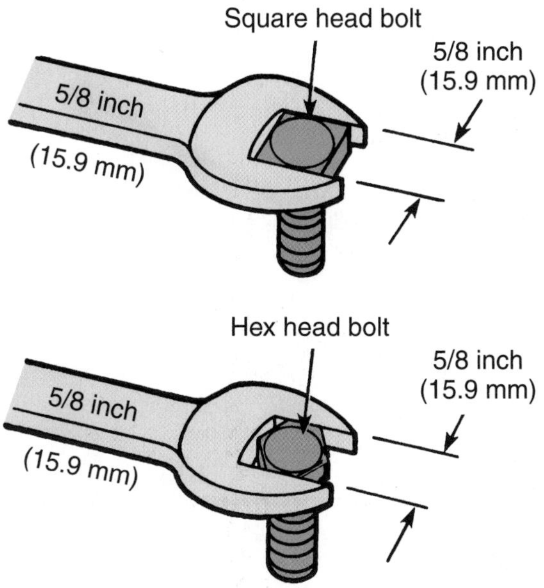

Figure 5-29 An open-end wrench grips only two sides of a fastener.

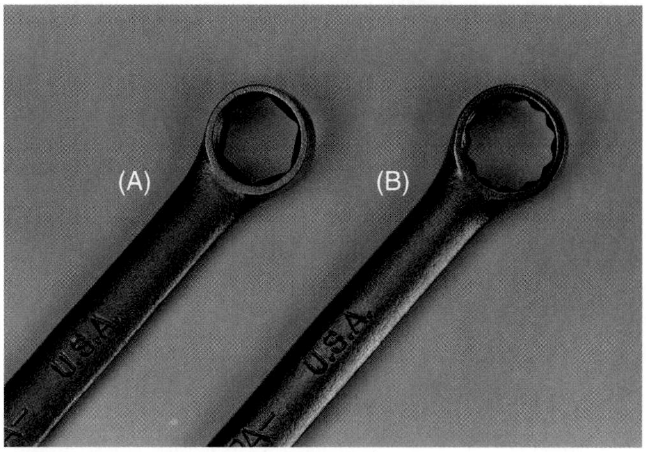

Figure 5-30 (A) Six-point and (B) twelve-point box-end wrenches are available.

Flare Nut (Line) Wrenches Flare nut or line wrenches should be used to loosen or tighten brake line or tubing fittings. Using open-ended wrenches on these fittings tends to round the corners of the nut, which are typically made of soft metal and can distort easily. Flare nut wrenches surround the nut and provide a better grip on the fitting. They have a section cut out so that the wrench can be slipped around the brake or fuel line and dropped over the flare nut.

Allen Wrench Setscrews are used to fasten door handles, instrument panel knobs, engine parts, and even brake calipers. A set of fractional and metric hex head wrenches, or Allen wrenches **(Figure 5–31)**, should be in every technician's toolbox. An Allen wrench can be L-shaped or can be mounted in a socket driver and used with a ratchet.

Adjustable-End Wrench An adjustable-end wrench (commonly called a crescent wrench) has one fixed jaw and one movable jaw. The wrench opening can be adjusted by rotating a helical adjusting screw that is mated to teeth in the lower jaw. Because this type of wrench does not firmly grip a bolt's head, it is likely to slip. Adjustable wrenches should be used carefully and *only* when it is absolutely necessary. Be sure to put all of the turning pressure on the fixed jaw.

Sockets and Ratchets

A set of Imperial and metric sockets combined with a ratchet handle and a few extensions should be included in your tool set. The ratchet allows you to turn the socket in one direction with force and in the other direction without force, which allows you to tighten or loosen a bolt

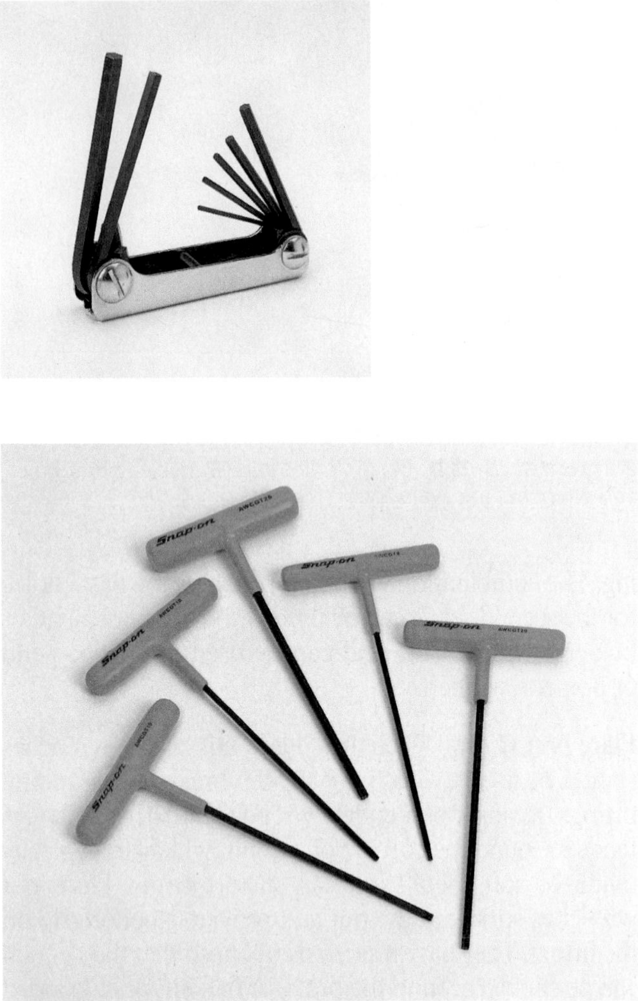

Figure 5-31 Top: A handy tool containing many different Allen wrenches. Bottom: Tee-handle Allen wrenches designed for better gripping and easier torque application. *Courtesy of Snap-on Tools Company*

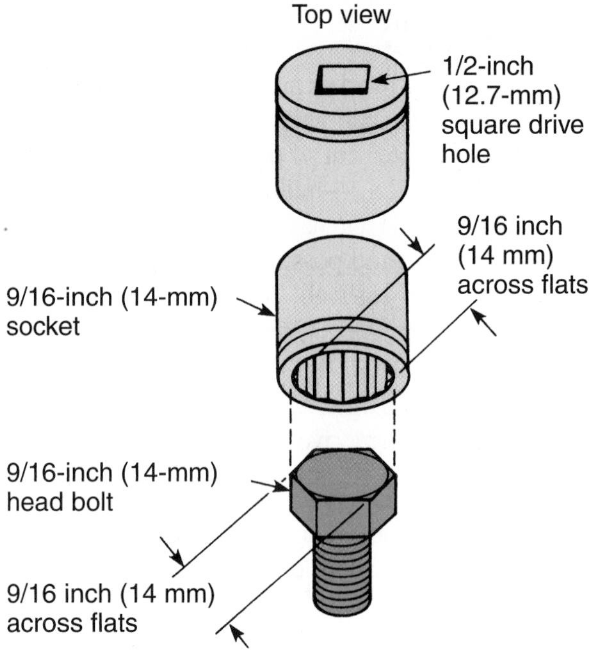

Figure 5-32 The size of the correct socket is the same size as the size of the bolt head or nut.

without removing and resetting the wrench after you have turned it. In many situations, a socket wrench is much safer, faster, and easier to use than any other wrench. In fact, sometimes it is the only wrench that will work.

The basic socket wrench set consists of a ratchet handle and several barrel-shaped sockets. The socket fits over and around a bolt or nut **(Figure 5–32)**. Inside, it is shaped like a box-end wrench. Sockets are available in 6, 8, or 12 points. A 6-point socket has stronger walls and improved grip on a bolt compared to a normal 12-point socket. However, 6-point sockets have half the positions of a 12-point socket. Six-point sockets are mostly used on fasteners that are rusted or rounded. Eight-point sockets are available to use on square nuts or square-headed bolts. Some axle and transmission assemblies use square-headed plugs in the fluid reservoir.

The top side of a socket has a square hole that accepts a square lug on the socket handle. This square hole is the drive hole. The size of the hole and handle lug (1/4 inch, 3/8 inch, 1/2 inch, [6.35 mm, 9.54 mm, 12.70 mm] and so on) indicates the drive size of the socket wrench. One handle fits all the sockets in a set. On better-quality handles, a spring-loaded ball in the square drive lug fits into a depression in the socket. This ball holds the socket to the handle. An assortment of socket (ratchet) handles is shown in **Figure 5–33**.

Not all socket handles are ratcheting. Some, called breaker bars, are simply long arms with a swivel drive used to provide extra torque onto a bolt to help loosen it.

Figure 5-33 An assortment of ratchets. *Courtesy of Snap-on Tools Company*

These are available in a variety of lengths and drive sizes. Sometimes nut drivers are used. These handles look like screwdrivers and have a drive shaft on the end of the shaft. Sockets and/or various attachments are inserted on the drive lug. These drivers are only used when bolt tightness is low.

Sockets are available in various sizes, lengths, and bore depths. Both standard SAE and metric socket wrench sets are necessary for automotive service. Normally, the larger the socket size, the longer the socket or the deeper the well. Deep-well sockets **(Figure 5–34)** are made extra long to fit over bolt ends or studs. A spark plug socket is an example of a special purpose deep-well socket. Deep-well sockets are also good for reaching nuts or bolts in limited-access areas. Deep-well sockets should not be used when a regular-size socket will do the job. The longer socket develops more twist torque and tends to slip off the fastener.

Heavier-walled sockets are designed for use with an impact wrench and are called impact sockets. Most sockets are chrome-plated, except for impact sockets, which are not.

WARNING!

Never use a nonimpact socket with an impact wrench.

Special Sockets Screwdriver (including Torx® driver) and Allen wrench attachments are also available for use with a socket wrench. **Figure 5–35** shows a typical set of screwdriver attachments and three specialty sockets. These socket wrench attachments are very handy when

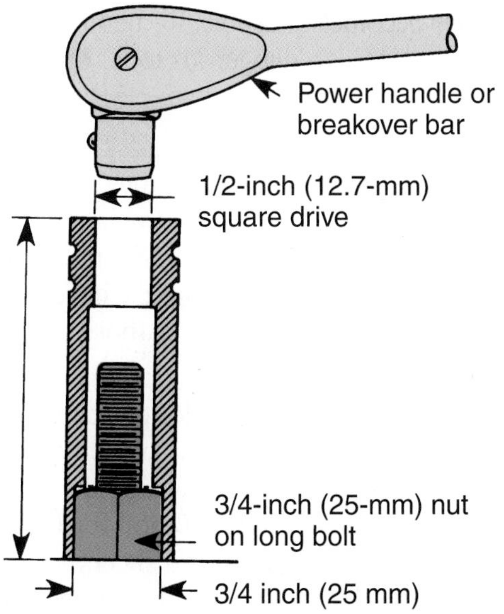

Figure 5-34 Deep-well sockets fit over the ends of bolts and studs.

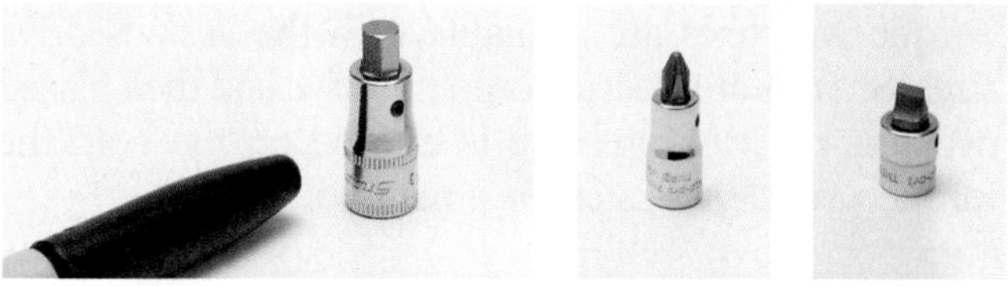

(A) (B) (C)

Figure 5-35 Typical screwdriver attachment set including (A) a hex driver, (B) a Phillips driver, and (C) slot tip driver. *Courtesy of Snap-on Tools Company*

a fastener cannot be loosened with a regular screwdriver. The leverage given by the ratchet handle is often just what it takes to break a stubborn screw loose.

Swivel sockets are also available. These sockets are fitted with a flexible joint that accommodates odd angles between the socket and the ratchet handle. These sockets are often used to work bolts that are difficult to reach.

Although crowfoot sockets are not really sockets, they are used with a ratchet or breaker bar. These sockets are actually the end of an open-end or line wrench made with a drive bore, which allows a ratchet to move the socket.

Extensions An extension is commonly used to separate the socket from the ratchet or handle. The extension moves the handle away from the bolt and makes the use of a ratchet more feasible. Extensions are available in all common drive sizes and in a variety of lengths. The most common lengths are 1 inch, 3 inches, 6 inches, and 10 inches; however 2- and 3-foot extensions are also quite common.

Wobble extensions allow a socket to pivot slightly at the drive connection. This type of extension provides for a more positive connection to the socket than swivel joints but only allows approximately 16 degrees of flexibility.

Socket Adapters When sockets of a different drive size must be used with a particular ratchet or handle, an adapter can be inserted between the socket and the drive on the handle. An example of a common adapter is one that allows for the use of a ¼-inch drive socket on a ⅜-inch drive ratchet.

Torque Wrenches

Torque wrenches measure how tight a nut or bolt is. Many of car's nuts and bolts should be tightened to a certain amount and have a torque specification that is expressed in foot-pounds (USCS) or Newton-meters (metric). A foot-pound is the work or pressure accomplished by a force of 1 pound through a distance of 1 foot. A Newton-meter is the work or pressure accomplished by a force of 1 kilogram through a distance of 1 meter.

A torque wrench is basically a ratchet or breaker bar with some means of displaying the amount of torque exerted on a bolt when pressure is applied to the handle.

Torque wrenches are available with the various drive sizes. Sockets are inserted onto the drive and then placed over the bolt. As pressure is exerted on the bolt, the torque wrench indicates the amount of torque.

SHOP TALK

Following torque specifications is critical. However, there is a possibility that the torque spec is wrong as printed. (In other words, someone made a mistake.) If the torque spec seems way too tight or loose for the size of bolt, find the torque spec in a different source. If the two specs are the same, use it. If they are different, use the one that seems right. ■

There are four basic types of torque wrenches **(Figure 5–36)** available. These types are available with inch-pound and foot-pound increments.

- A beam torque wrench. On this type, the beam points to the torque reading. This torque wrench is not highly accurate.
- A "click"-type torque wrench. On this type of wrench, the desired torque reading is set on the handle. When the torque reaches that level, the wrench clicks.
- A dial torque wrench. This type of torque wrench has a dial that indicates the torque exerted on the wrench. Some designs of this type have a light or buzzer that turns on when the desired torque is reached.
- A digital readout type. This style of torque wrench displays the torque digitally and is commonly used to measure turning effort as well as to tighten bolts. Some designs of this type have a light or buzzer that turns on when the desired torque is reached. Any good torque wrench allows a technician to tighten a bolt or nut to the correct torque.

The correct torque provides the tightness and stress that the manufacturer has found to be the most desirable and reliable. For example, engine-bearing caps that are too tight distort the bearings, causing excessive wear and incorrect oil clearance. This often results in rapid wear of other engine parts due to decreased oil flow. Insufficient torque can result in out-of-round bores and subsequent failure of the parts.

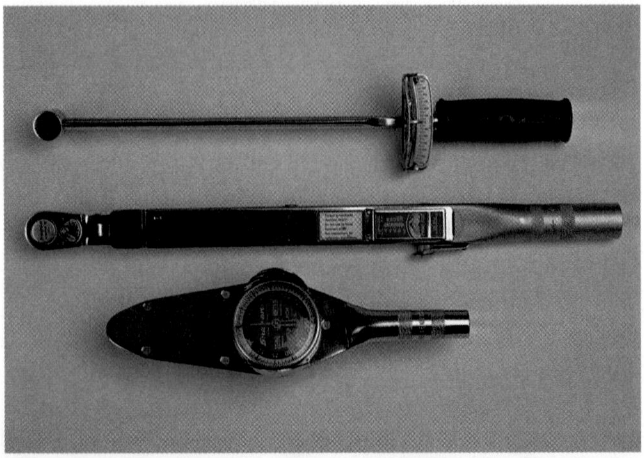

Figure 5-36 The basic types of torque wrenches.

When using a torque wrench, follow these steps to get an accurate reading:

1. Locate the torque specs and procedures in a service manual
2. Mentally divide the torque specification by three.
3. Hold the wrench so that it is at a 90-degree angle from the fastener being tightened.
4. Tighten the bolt or nut to one-third of the specification.
5. Then, tighten the bolt to two-thirds of the spec.
6. Now tighten the bolt to within 10 foot-pounds of the spec.
7. Tighten the bolt to the specified torque.
8. Recheck the torque.

Screwdrivers

A screwdriver drives a variety of threaded fasteners used in the automotive industry. Each fastener requires a specific kind of screwdriver, and a well-equipped technician has several sizes of each.

SHOP TALK

A screwdriver should not be used as a chisel, punch, or pry bar. Screwdrivers were not made to withstand blows or bending pressures. When misused in such a fashion, the tips will wear, become rounded, and tend to slip out of the fastener. At that point, a screwdriver becomes unusable. Remember a defective tool is a dangerous tool. ■

Screwdrivers are defined by their sizes, their tips **(Figure 5–37)**, and the types of fasteners they should be used with. Your tool set should include both blade and Phillips drivers in a variety of lengths from 2-inch "stubbies" to 12-inch screwdrivers. You also should have an assortment of special screwdrivers, such as those with a Torx® head design.

- *Standard Tip Screwdriver:* A slotted screw accepts a screwdriver with a standard or blade-type tip. The standard tip screwdriver is probably the most common type **(Figure 5–38)**. It is useful for turning carriage bolts, machine screws, and sheet metal screws. The width and thickness of the blade determine the size of a standard screwdriver. Always use a blade that fills the slot in the fastener.

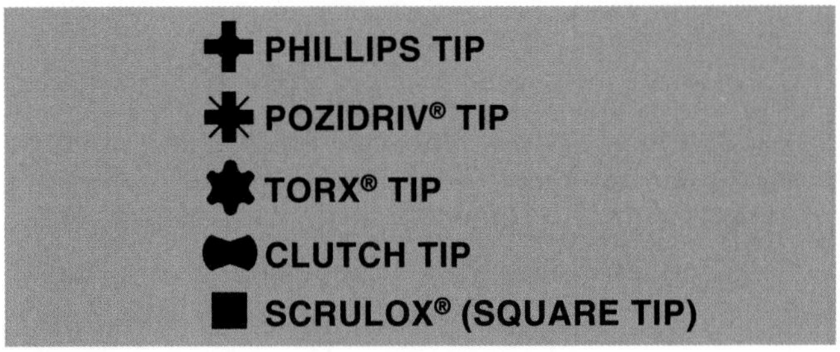

Figure 5-37 The various screwdriver tips that are available. *Courtesy of Snap-on Tools Company*

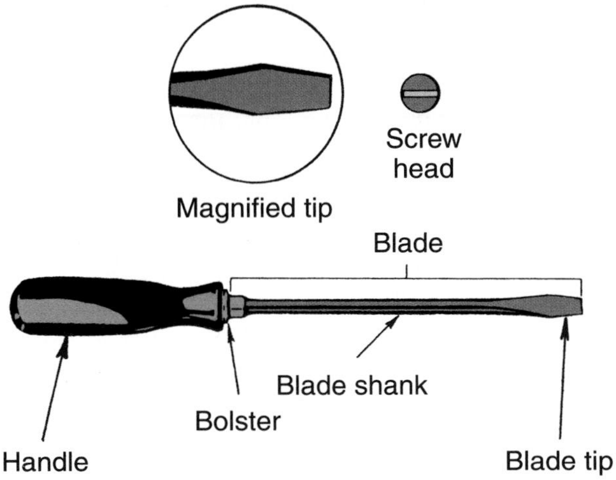

Figure 5-38 The standard tip screwdriver is used with slotted head fasteners.

- *Phillips Screwdriver:* The tip of a **Phillips screwdriver** has four prongs that fit the four slots in a Phillips head screw **(Figure 5–39)**. The four surfaces enclose the screwdriver tip so it is less likely that the screwdriver will slip out of the fastener. Phillips screwdrivers come in sizes #0 (the smallest), #1, #2, #3, and #4 (the largest).
- *Reed and Prince Screwdriver:* The tip of a Reed and Prince screwdriver is like a Phillips except that the prongs come to a point rather than to a blunt end.

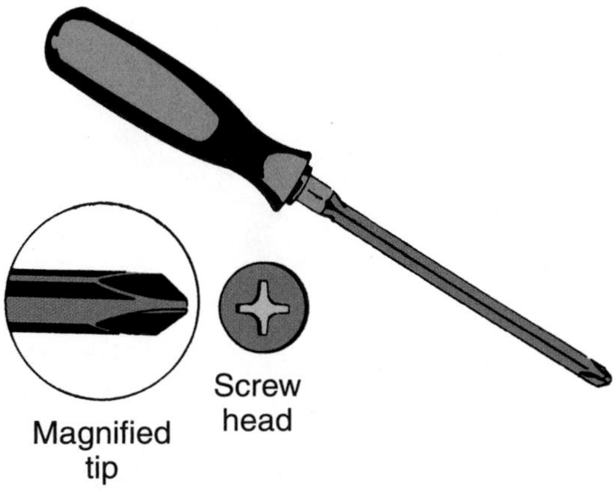

Figure 5-39 The tip of a Phillips screwdriver has four prongs that provide a good grip on the fastener.

- *Pozidriv® Screwdriver:* The **Pozidriv screwdriver** is like a Phillips but its tip is flatter and blunter. The squared tip grips the screw's head and slips less than a Phillips screwdriver.
- *Torx® Screwdriver:* The **Torx** screwdriver is used to secure headlight assemblies, mirrors, and luggage racks. Not only does the six-prong tip provide greater turning power and less slippage, but the Torx fastener also provides a measure of tamper resistance. Torx drivers come in sizes T15 (the smallest), T20, T25, and T27 (the largest).
- *Clutch Driver:* Fasteners that require a clutch driver are normally used in non-load-bearing places. Clutch head fasteners offer a degree of tamper resistance and offer less slippage than a standard slot screw. The clutch head design has been called a butterfly or figure-eight. Automotive technicians do not often use these drivers.
- *Scrulox® Screwdriver:* The Scrulox screwdriver has a square tip. The tip fits into a square recess in the top of a fastener. This type of fastener is commonly used on truck bodies, campers, and boats.

Impact Screwdriver

An impact screwdriver is used to loosen stubborn screws. Impact screwdrivers have interchangeable heads and bits that allow the handles of the tools to be used with various screw head designs.

To use an impact screwdriver **(Figure 5–40)**, select the correct bit and insert it into the driver's head. Then hold the bit against the screw slot while firmly twisting the handle in the desired direction. Strike the handle with a hammer. The force of the hammer will exert a downward force on the screw and, at the same time, exert a twisting force on the screw.

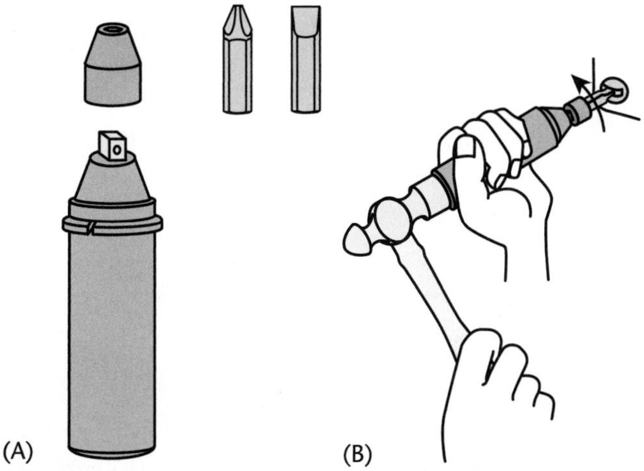

Figure 5-40 (A) An impact screwdriver set. (B) An impact screwdriver automatically tries to rotate the screw when it is struck with a hammer.

Pliers

Pliers are gripping tools used for working with wires, clips, and pins. At a minimum, an auto technician should own several types: standard pliers for common parts and wires, needle nose for small parts, and large, adjustable pliers for large items and heavy-duty work. A brief discussion on the different types of pliers follows:

- *Combination pliers* **(Figure 5–41)** are the most common type of pliers and are frequently used in many kinds of automotive repair. The jaws have both flat and curved surfaces for holding flat or round objects. Also called slip-joint pliers, the combination pliers have many jaw-opening sizes. One jaw can be moved up or down on a pin attached to the other jaw to change the size of the opening.
- *Adjustable pliers,* commonly called *channel locks* **(Figure 5–42)**, have a multiposition slip joint that allows for many jaw-opening sizes.
- *Needle nose pliers* have long, tapered jaws **(Figure 5–43)**. They are great for holding small parts or for reaching into tight spots. Many needle nose pliers also have wire-cutting edges and a wire stripper. Curved needle nose pliers allow you to work on a small object around a corner.
- *Locking pliers*, or *vise grips*, are similar to the standard pliers, except they can be tightly locked around an object **(Figure 5–44).** They are extremely useful for

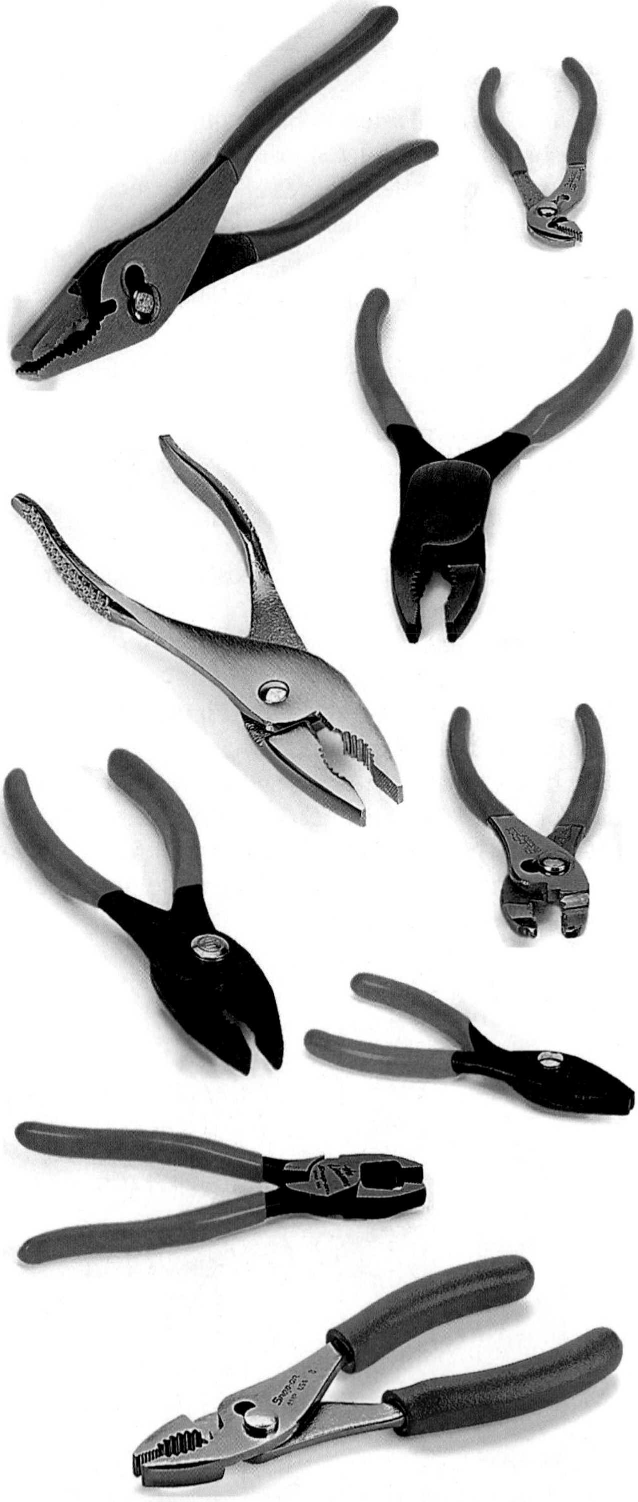

Figure 5-41 Various combination pliers. *Courtesy of Snap-on Tools Company*

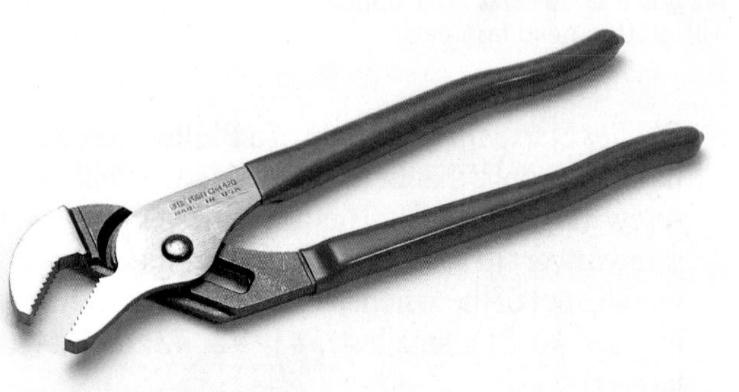

Figure 5-42 A typical adjustable pliers. *Courtesy of Snap-on Tools Company*

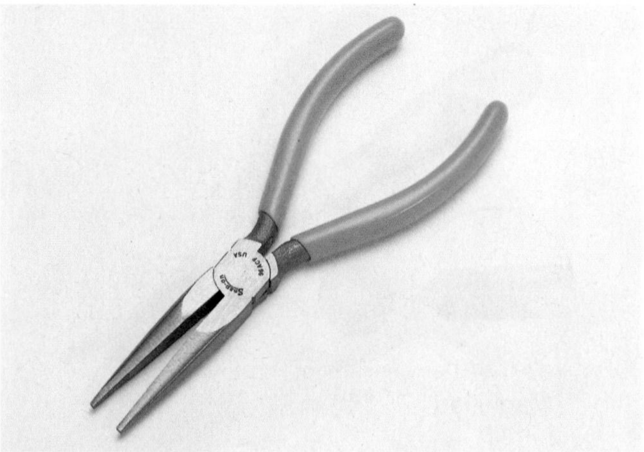

Figure 5-43 A typical needle nose pliers. *Courtesy of Snap-on Tools Company*

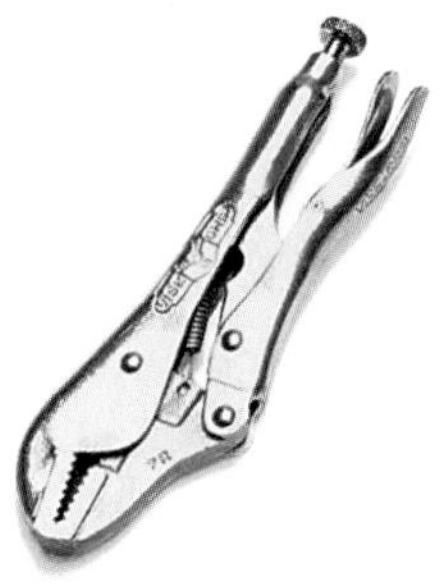

Figure 5-44 Locking pliers, commonly called vise grips. *Courtesy of Snap-on Tools Company*

Figure 5-45 Various diagonal-cutting pliers. *Courtesy of Snap-on Tools Company*

holding parts together. They are also useful for getting a firm grip on a badly rounded fastener that is impossible to turn with a wrench or socket. Locking pliers come in several sizes and jaw configurations for use in many auto repair jobs.

- *Diagonal-cutting pliers*, or cutters, are used to cut electrical connections, cotter pins, and wires on a vehicle. Jaws on these pliers have extra-hard cutting edges that are squeezed around the item to be cut **(Figure 5–45)**.
- *Snap or lock ring pliers* are made with a linkage that allows the movable jaw to stay parallel throughout the range of opening **(Figure 5–46)**. The jaw surface is usually notched or toothed to prevent slipping.
- *Retaining ring pliers* are identified by their pointed tips that fit into holes in retaining rings. Retaining ring pliers come in fixed sizes but are also available in sets with interchangeable jaws.

Hammers

Hammers are identified by the material and weight of the head. There are two groups of hammer heads: steel and soft faced **(Figures 5–47** and **5–48)**. Your tool set should include at least three hammers: two ball-peen hammers, one 8-ounce and one 12- to 16-ounce hammer, and a small sledgehammer. You should also have a plastic and lead or brass-faced mallet. The heads of steel-faced hammers are made from high-grade alloy steel. The steel is deep forged and heat treated to a suitable degree of hardness. Soft-faced hammers have a surface that yields when it strikes an object. Soft-faced hammers should be used on machined surfaces and when marring a finish is undesirable. For example, a brass hammer should be used to strike gears or shafts because it will not damage them.

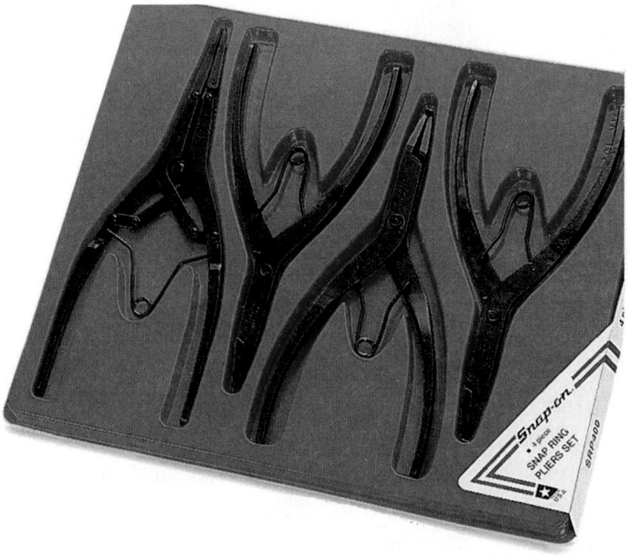

Figure 5-46 Snap ring pliers. *Courtesy of Snap-on Tools Company*

Figure 5-47 Various steel-faced hammers. *Courtesy of Snap-on Tools Company*

Figure 5-48 Soft-faced hammers. *Courtesy of Snap-on Tools Company*

Chisels and Punches

Chisels are used to cut metal by driving them with a hammer. Automotive technicians use a variety of chisels for cutting sheet metal, shearing off rivet and bolt heads, splitting rusted nuts, and chipping metal. A variety of chisels are available, each with a specific purpose, including flat, cape, round-nose cape, and diamond point chisels.

Punches are used for driving out pins, rivets, or shafts; aligning holes in parts during assembly; and marking the starting point for drilling a hole. Punches are designated by their point diameter and punch shape. Drift punches are used to remove drift and roll pins. Some drifts are made of brass; these should be used whenever you are concerned about possible damage to the pin or to the surface surrounding the pin. Tapered punches are used to line up bolt holes. Starter or center punches are used to make an indent before drilling to prevent the drill bit from wandering.

Removers

Rust, corrosion, and prolonged heat can cause automotive fasteners, such as cap screws and studs, to become

Figure 5-49 Stud installation/removal tool.

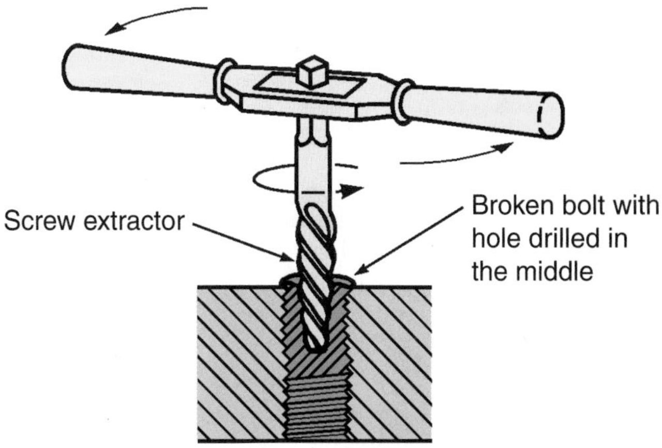

Figure 5-51 Using a screw extractor to remove a broken bolt.

stuck. A box wrench or socket is used to loosen cap screws. A special gripping tool is designed to remove studs. However, if the fastener breaks off, special extracting tools and procedures must be employed.

One type of stud remover is shown in **Figure 5–49**. These tools are also used to install studs. Stud removers have hardened, knurled, or grooved eccentric rollers or jaws that grip the stud tightly when operated. Stud removers/installers are turned by a socket wrench drive handle, a socket, or wrench.

Extractors are used on screws and bolts that are broken off below the surface. Twist drills, fluted extractors, and hex nuts are included in a screw extractor set **(Figure 5–50)**. This type of extractor lessens the tendency to expand the screw or stud that has been drilled out by providing gripping power along the full length of the stud.

Screw extractors are often called easy outs. To use an extractor, the bolt must be drilled and the extractor forced into that bore. The teeth of the extractor grip the inside of the drilled bore and allow the bolt to be turned out **(Figure 5–51)**. Easy outs typically have the size of the required drill bit stamped on one side.

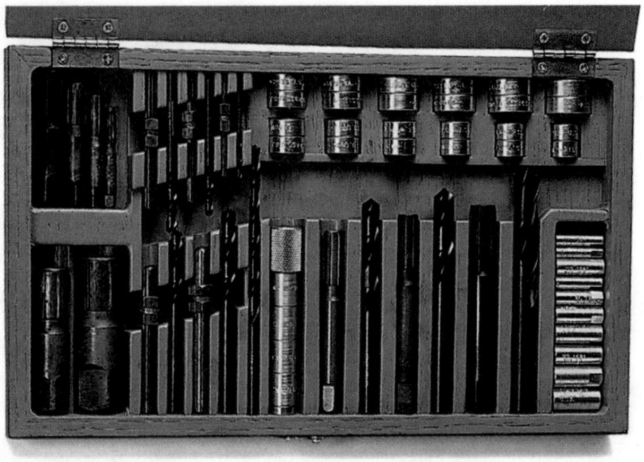

Figure 5-50 Screw extractor set. *Courtesy of Snap-on Tools Company*

At times a broken bolt can be loosened and removed from its bore by driving it in a counterclockwise direction with a chisel and hammer. A bolt broken off above the surface may be able to be removed with locking pliers.

Hacksaws

A hacksaw is used to cut metal **(Figure 5–52)**. The blade only cuts on the forward stroke. The teeth of the blade should always face away from the saw's handle. The number of teeth on the blade determines the type of metal the saw can be used on. A fine-toothed blade is best for thin sheet metal, whereas a coarse blade is used on thicker metals.

When using a hacksaw, never bear down on the blade while pulling it toward you; this will dull the blade. Use the entire blade while cutting.

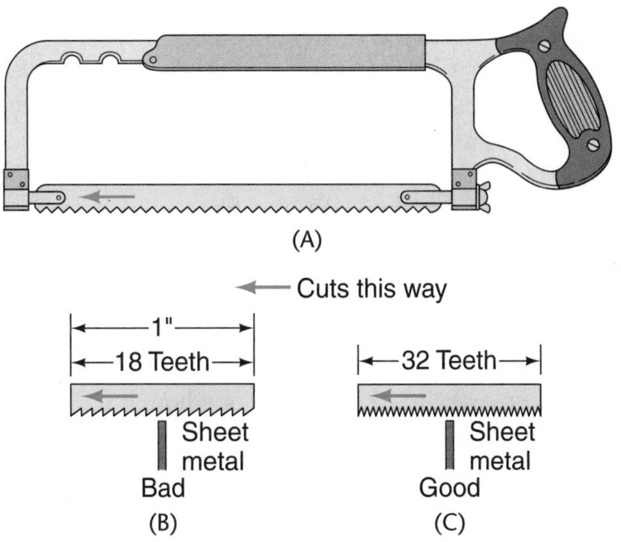

Figure 5-52 (A) The teeth on the blade in a hacksaw should face forward. (B) A coarse blade should not be used with sheet metal. (C) A fine blade will work well with sheet metal.

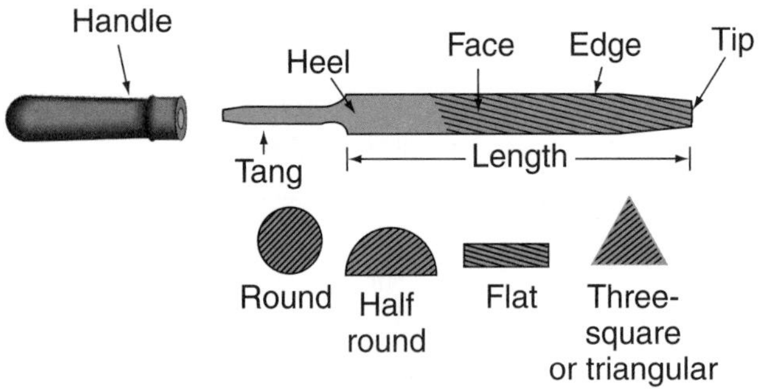

(A)

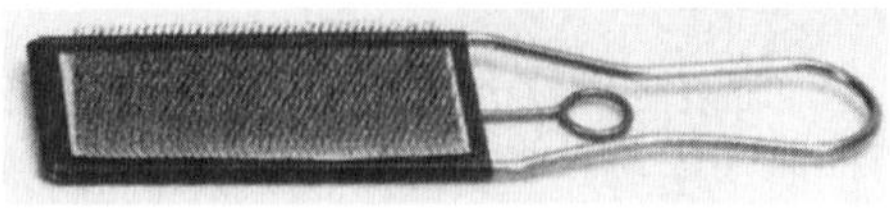

(B)

Figure 5-53 (A) Files come in a variety of shapes. *Courtesy of Ford Motor Company* (B) A file card. *Courtesy of Snap-on Tools Company*

Files

Files are commonly used to shape or smooth metal edges. Files typically have square, triangular, rectangular (flat), round, or half-round shapes **(Figure 5–53)**. They also vary in size and coarseness. The most commonly used files are the half-round and flat with either single-cut or double-cut designs. A single-cut file has its cutting grooves lined up diagonally across the face of the file. The cutting grooves of a double-cut file run diagonally in both directions across the face. Double-cut files are considered first cut or roughening files because they can remove large amounts of metal. Single-cut files are considered finishing files because they remove small amounts of metal.

To avoid personal injury, files should always be used with a plastic or wooden handle. Like hacksaws, files only cut on the forward stroke. Coarse files are used for soft metals, and smoother, or finer, files are used to work steel and other hard metals.

Keep files clean, dry, and free of oil and grease. To clean filings from the teeth of a file, use a special tool called a file card.

Gear and Bearing Pullers

Many precision gears and bearings have a slight interference fit (**press fit**) when installed on a shaft or housing. For example, the inside diameter of a bore is 0.001 inch smaller than the outside diameter of a shaft. When the shaft is fitted into the bore it must be pressed in to overcome the 0.001-inch interference fit. This press fit prevents the parts from moving on each other. The removal of gears and bearings must be done carefully. Prying or hammering can break or bind the parts. A puller with the proper jaws and adapters should be used when applying force to remove gears and bearings. Using proper tools, the force can be applied with a slight and steady motion.

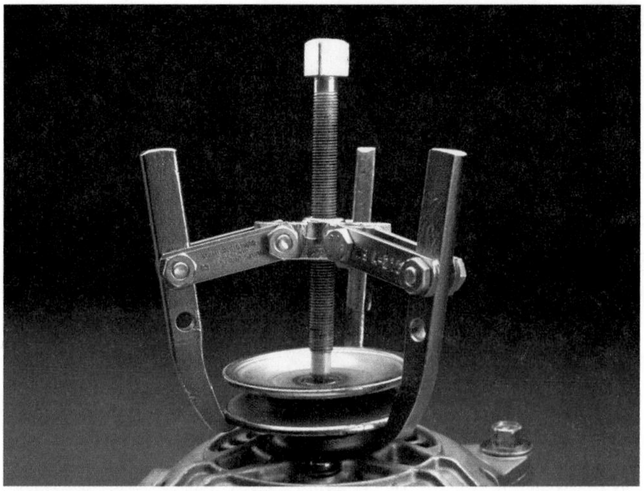

Figure 5-54 The jaws on this puller are reversible to allow for inside and outside pulls.

Many pullers are designed to accommodate inside and outside pulls **(Figure 5–54)**. These pullers may come with various jaw lengths and shapes to allow them to work in a number of different situations.

Some pullers are fitted to the end of a slide hammer **(Figure 5–55)** to remove slightly press fit items. After the jaws of the puller are secure in or around the object to be removed, the weight on the tool's hammer is slid back with force against the handle of the tool, generating a pulling force and jerking the object out of its bore.

Bushing and Seal Pullers and Drivers

Another commonly used group of special tools includes the various designs of bushing and seal drivers and pullers. Pullers are either threaded or slide hammer-type tools. Always make sure you use the correct tool for the job; bushings and seals are easily damaged if the wrong tool or procedure is used. Car manufacturers and specialty tool companies work closely together to design and manufacture special tools required to repair cars. Most

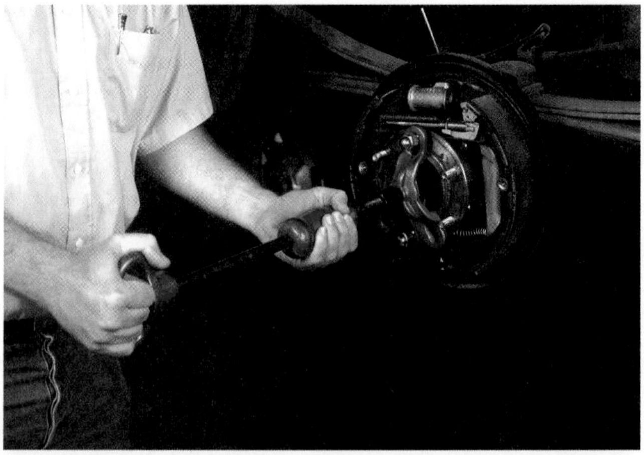

Figure 5-55 Using a slide hammer-type puller to remove a drive axle.

of these special tools are listed in the appropriate service manuals.

Seal drivers are designed to fit squarely against the seal case and inside the seal lip. A soft hammer is used to tap the seal driver and drive the seal straight into the housing. Some tool manufacturers market a seal driver kit with drivers to fit many common seals.

Trouble Light

Adequate light is necessary when working under and around automobiles. A **trouble light** may be battery powered (like a flashlight) or may need to be plugged into a wall socket. Some shops have trouble lights that pull down from a reel suspended from the ceiling. Trouble lights use either an incandescent bulb or a fluorescent tube. Because incandescent bulbs can pop and burn, it is highly recommended that you use only fluorescent tubes. Take extra care when using a trouble light. Make sure the cord does not get caught in a rotating object. The bulb or tube is surrounded by a cage or enclosed in clear plastic to prevent accidental breaking and burning.

SHOP EQUIPMENT

Some tools and equipment are supplied by the service facility and few technicians have these as part of their tool assortment. These tools are commonly used but there is no need for each technician to own them. Many shops have one or two of each.

Bench Vises

Often repair work is completed with a part or assembly removed from the vehicle. The repairs are typically safely and quickly made by securing the assembly. Small parts are usually secured with a bench vise. The vise is bolted to a workbench to give it security. The object to be held is placed into the tool's jaws and the jaws are tightened around the object. If the object could be damaged or marred by the jaws, brass jaw caps are installed over the jaws before the object is placed between them.

Bench Grinder

This electric power tool is generally bolted to a workbench. The grinder should have safety shields and guards. Always wear face protection when using a grinder. A bench grinder is classified by wheel size. Six- to ten-inch wheels are the most common in auto repair shops. Three types of wheels are available with this bench tool.

1. Grinding wheel, for a wide variety of grinding jobs from sharpening cutting tools to deburring
2. Wire wheel brush, for general cleaning and buffing, removing rust, scale, and paint, deburring, and so forth
3. Buffing wheel, for general purpose buffing, polishing, and light cutting

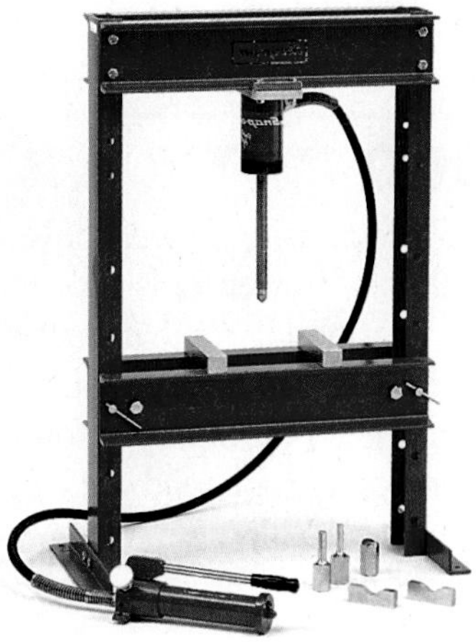

Figure 5-56 A floor-mounted hydraulic press. *Courtesy of Snap-on Tools Company*

Presses

Many automotive jobs require the use of powerful force to assemble or disassemble parts that are press-fit together. Removing and installing piston pins, servicing rear axle bearings, pressing brake drum and rotor studs, and performing transmission assembly work are just a few examples. Presses can be hydraulic, electric, air, or hand driven. Capacities range up to 150 tons of pressing force, depending on the size and design of the press. Smaller arbor and C-frame presses can be bench or pedestal mounted, while high-capacity units are freestanding or floor mounted **(Figure 5–56)**.

WARNING!

Always wear safety glasses when using a press.

Grease Guns

Some shops are equipped with air-powered grease guns, while in others, technicians use a manually operated grease gun. Both types can force grease into a grease fitting. Hand-operated grease guns are often preferred because the pressure of the grease can be controlled by the technician. However, many shops use low air pressure to activate a pneumatic grease gun. The suspension and steering system may have several grease or zerk fittings.

POWER TOOLS

Power tools make a technician's job easier. They operate faster and with more torque than hand tools. However, power tools require greater safety measures. Power tools do not stop unless they are turned off. Power is furnished by air (pneumatic), electricity, or hydraulic fluid. *Power tools should only be used for loosening nuts and/or bolts.*

SHOP TALK

Safety is critical when using power tools. Carelessness or mishandling of power tools can cause serious injury. Do not use a power tool without obtaining permission from your instructor. Be sure you know how to operate the tool properly before using it. Prior to using a power tool, read the instructions carefully. ■

Impact Wrench

An **impact wrench (Figure 5–57)** is a portable hand-held reversible wrench. A heavy-duty model can deliver up to 450 foot-pounds (607.5 Nm) of torque. When triggered, the output shaft, onto which the impact socket is fastened, spins freely at 2,000 to 14,000 rpm, depending on the wrench's make and model. When the impact wrench meets resistance, a small spring-loaded hammer situated near the end of the tool strikes an anvil attached to the drive shaft onto which the socket is mounted. Each impact moves the socket around a little until torque equilibrium is reached, the fastener breaks, or the trigger is released. Torque equilibrium occurs when the torque of the bolt equals the output torque of the wrench. Impact wrenches can be powered either by air or by electricity.

SHOP TALK

When using an air impact wrench, it is important that only impact sockets and adapters be used. Other types of sockets and adapters, if used, might shatter and fly off, endangering the safety of the operator and others in the immediate area. ■

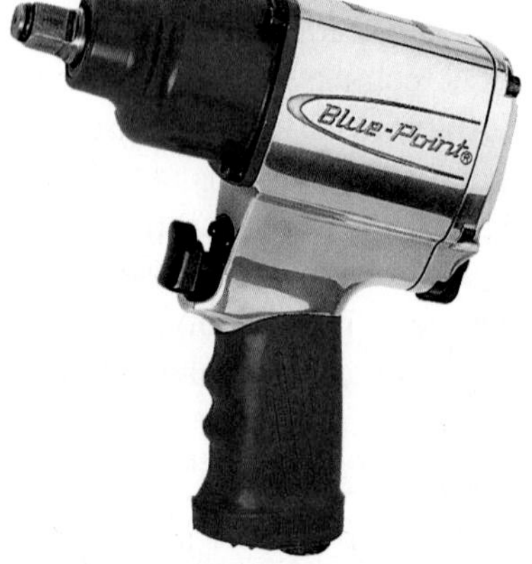

Figure 5-57 A typical air impact wrench. *Courtesy of Snap-on Tools Company*

An impact wrench uses compressed air or electricity to hammer or impact a nut or bolt loose or tight. Light-duty impact wrenches are available in three drive sizes—¼, ⅜, and ½ inch—and two heavy-duty sizes—¾ and 1 inch.

WARNING!

Impact wrenches should not be used to tighten critical parts or parts that may be damaged by the hammering force of the wrench.

Air Ratchet

An air ratchet, like the hand ratchet, has a special ability to work in hard-to-reach places. Its angle drive reaches in and loosens or tightens where other hand or power wrenches just cannot work **(Figure 5–58)**. The air ratchet looks like an ordinary ratchet but has a fat handgrip that contains the air vane motor and drive mechanism. Air ratchets usually have a ⅜-inch drive. Air ratchets are not torque sensitive; therefore, a torque wrench should be used on all fasteners after snugging them up with an air ratchet.

Air Drill

Air drills are usually available in ¼-, ⅜-, and ½-inch (6.35-, 9.54-, and 9.54-mm) sizes. They operate in much the same manner as an electric drill, but are smaller and lighter. This compactness makes them a great deal easier to use for drilling operations in auto work.

Blowgun

Blowguns are used for blowing off parts during cleaning. Never point a blowgun at yourself or anyone else. A blowgun **(Figure 5–59)** snaps into one end of an air hose and directs airflow when a button is pressed. Always use an OSHA-approved air blowgun. Before using a blowgun, be sure it has not been modified to eliminate air-bleed holes on the side.

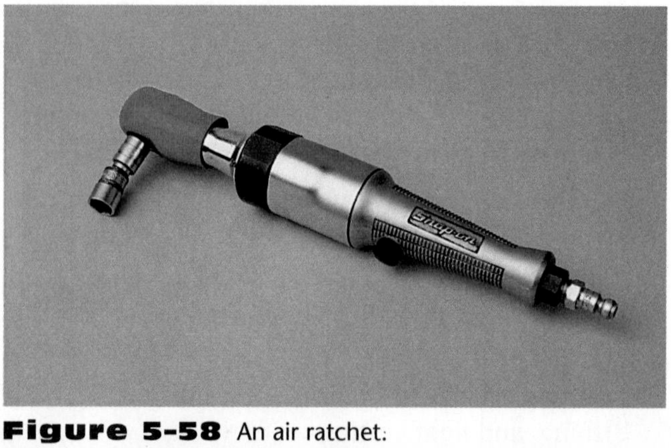

Figure 5-58 An air ratchet.

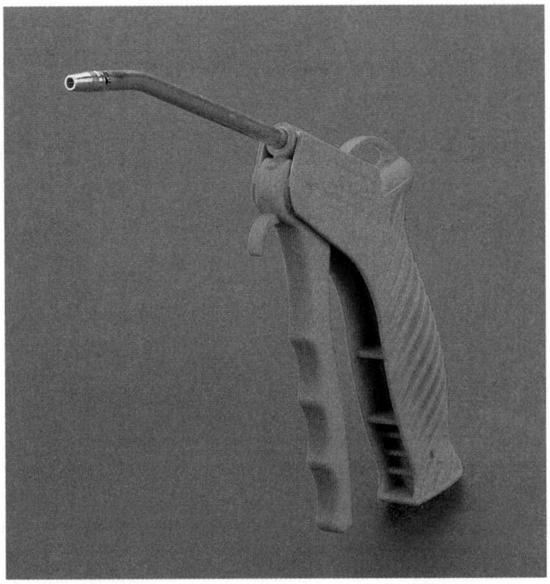

Figure 5-59 An OSHA-approved air blowgun. *Courtesy of ITW DeVilbiss Co.*

JACKS AND LIFTS

Jacks are used to raise a vehicle off the ground and are available in two basic designs and in a variety of sizes. The most common jack is a hydraulic floor jack, which is classified by the weights it can lift: 1 ½, 2, and 2 ½ tons, and so on. These jacks are controlled by moving the handle up and down. The other design of portable floor jack uses compressed air. Pneumatic jacks are operated by controlling air pressure at the jack.

CAUTION!

Before lifting a vehicle with air suspension, turn off the system. The switch is usually in the trunk.

When a vehicle is raised by a jack, it should be supported by safety stands **(Figure 5–60)**. Never work under a car with only a jack supporting it; always use safety stands. Hydraulic seals in the jack can let go and allow the vehicle to drop.

Figure 5-60 Whenever you have raised a vehicle with a floor jack, the vehicle should be supported with jack stands.

The hydraulic floor lift is the safest lifting tool and is able to raise the vehicle high enough to allow you to walk and work under it. Various safety features prevent a hydraulic lift from dropping if a seal does leak or if air pressure is lost. Before lifting a vehicle, make sure the lift is correctly positioned.

Floor Jack

A floor jack is a portable unit mounted on wheels. The lifting pad on the jack is placed under the chassis of the vehicle, and the jack handle is operated with a pumping action. This forces fluid into a hydraulic cylinder in the jack, and the cylinder extends to force the jack lift pad upward and to lift the vehicle. Always be sure that the lift pad is positioned securely under one of the car manufacturer's recommended lifting points. To release the hydraulic pressure and lower the vehicle, the handle or release lever must be turned slowly.

The maximum lifting capacity of the floor jack is usually written on the jack decal. Never lift a vehicle that exceeds the jack lifting capacity. This action may cause the jack to break or collapse, resulting in vehicle damage or personal injury.

Lift

A lift is used to raise a vehicle so the technician can work under the vehicle. The lift arms must be placed under the car manufacturer's recommended lifting points prior to raising a vehicle. There are three basic types of lifts: frame contact **(Figure 5–61)**, wheel contact, and axle engaging. These categories define where the frame contact points align with the vehicle.

Twin posts are used on some lifts **(Figure 5–62)**, whereas other lifts have a single post **(Figure 5–63)**. Some lifts have an electric motor, which drives a hydraulic pump to create fluid pressure and force the lift upward. Other lifts use air pressure from the shop air supply to force the lift upward. If shop air pressure is used for this purpose, the air pressure is applied to fluid in the lift cylinder. A control lever or switch is placed near the lift. The control lever supplies shop air pressure to the lift cylinder, and the switch turns on the lift pump motor. Always be sure that the safety lock is engaged after the lift is raised **(Figure 5–64)**. When the safety lock is released, a release lever is operated slowly to lower the vehicle.

The arms of a lift are fitted with **foot pads (Figure 5–65)** or adapters that can be lifted up to contact the vehicle's lift points to add clearance between the arms and the vehicle. This clearance allows for secure lifting without damaging any part of the body or underbody of the vehicle.

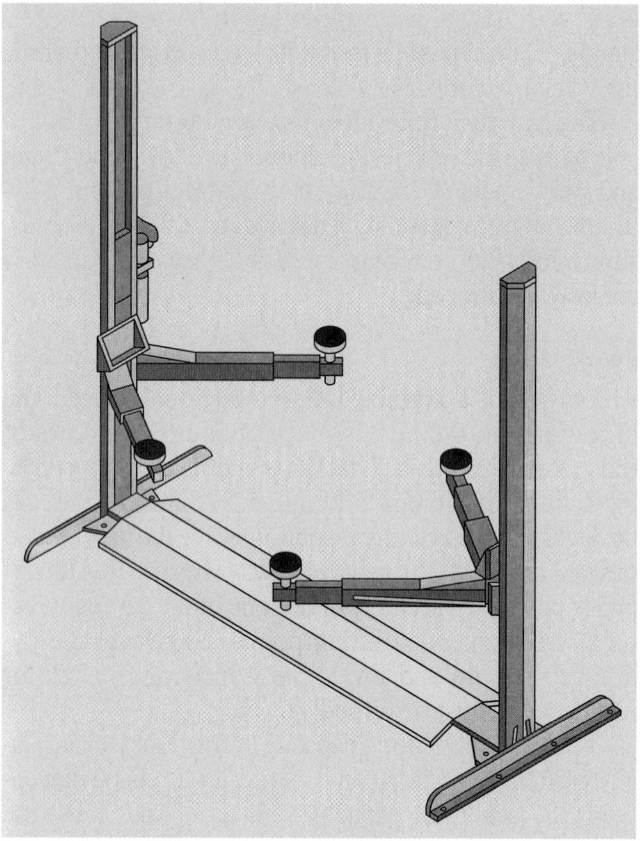

Figure 5-61 An aboveground or surface mount frame-contact lift. *Courtesy of Automotive Lift Institute*

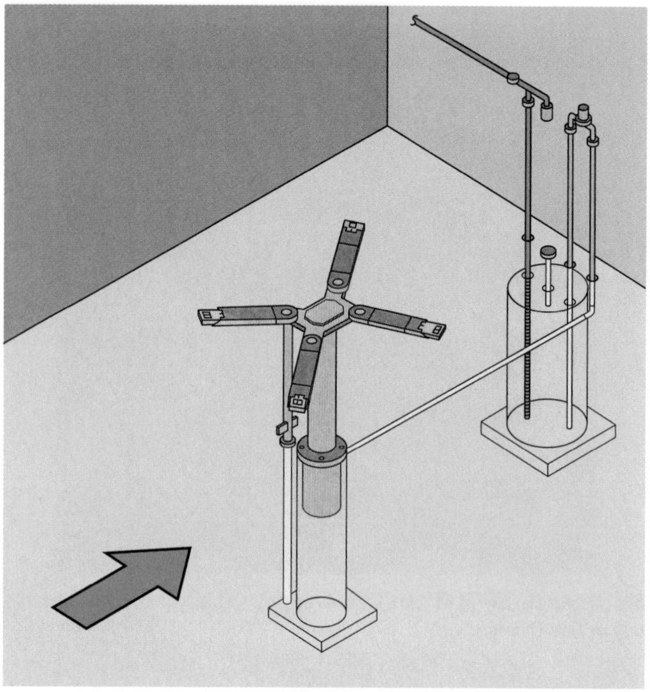

Figure 5-63 The typical setup for a single post lift. *Courtesy of Automotive Lift Institute*

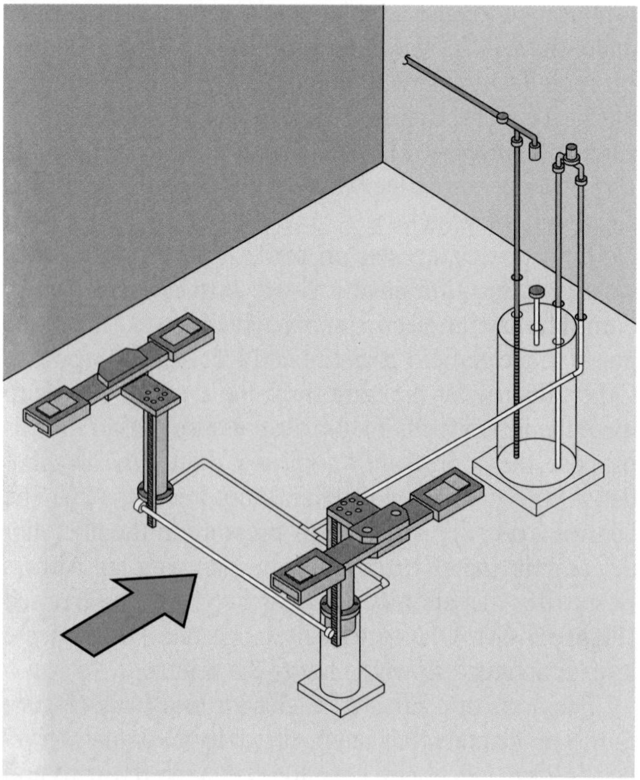

Figure 5-62 The typical setup for a twin post lift. *Courtesy of Automotive Lift Institute*

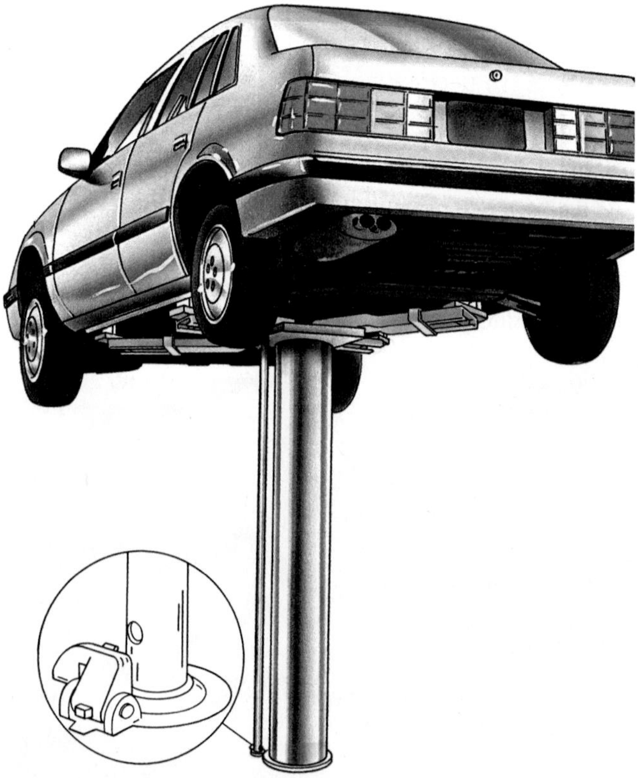

Figure 5-64 Make sure the locking device or safety is fully engaged after the vehicle has been raised to the desired height. *Courtesy of Automotive Lift Institute*

Figure 5-65 Foot pads on the arms of a lift. *Courtesy of Automotive Lift Institute*

Portable Crane

To remove and install an engine, a portable crane, frequently called a cherry picker, is used. To lift an engine, attach a pulling sling or chain to the engine. Some engines have eye plates for use in lifting. If they are not available, the sling must be bolted to the engine. The sling attaching bolts must be large enough to support the engine and must thread into the block a minimum of 1½ times the bolt diameter. Connect the crane to the chain. Raise the engine slightly and make sure the sling attachments are secure. Carefully lift the engine out of its compartment.

Lower the engine close to the floor so the transmission and torque converter or clutch can be removed from the engine, if necessary.

Engine Stands/Benches

After the engine has been removed, use the crane to raise the engine. Position the engine next to an engine stand. Most stands use a plate with several holes or adjustable arms. The engine must be supported by at least four bolts that fit solidly into the engine. The engine should be positioned so that its center is in the middle of the engine's stand adapter plate. The adapter plate can swivel in the stand. By centering the engine, the engine can be easily turned to the desired working positions.

Some shops have engine mounts bolted to the top of workbenches. The engine is suspended off the side of the workbench. These have the advantage of a good working space next to the engine, but they are not mobile and all engine work must be done at that location.

After the engine is secured to its mount, the crane and lifting chains can be removed and disassembly of the engine can begin.

SERVICE INFORMATION

Perhaps the most important tools you will use are service manuals **(Figure 5–66)**. There is no way a technician can remember all of the procedures and specifications needed to repair an automobile correctly. Thus, a good technician relies on service manuals and other sources for this information. Good information plus knowledge allows a technician to fix a problem with the least amount of frustration and at the lowest cost to the customer.

Figure 5-66 One of your most important tools is a service manual.

Auto Manufacturers' Service Manuals

The primary source of repair and specification information for any car, van, or truck is the manufacturer. The manufacturer publishes service manuals each year, for every vehicle built. These manuals are written for professional technicians.

Because of the enormous amount of information, some manufacturers publish more than one manual per year per car model. They may be separated into sections such as chassis, suspension, steering, emission control, fuel systems, brakes, basic maintenance, engine, transmission, body, and so on **(Figure 5–67)**.

When complete information with step-by-step testing, repair, and assembly procedures is desired, nothing can match auto manufacturers' repair manuals. They cover all repairs, adjustments, specifications, detailed diagnostic procedures, and special tools required. They can be purchased directly from the automobile manufacturer.

To help you learn how to effectively use service manuals, you will find service manual references and tips throughout this text.

Since many technical changes occur on specific vehicles each year, manufacturers' service manuals need to be constantly updated. Updates are published as service bulletins (often referred to as Technical Service Bulletins or TSBs) that show the changes in specifications and repair procedures during the model year. These changes do not appear in the service manual until the next year. The car

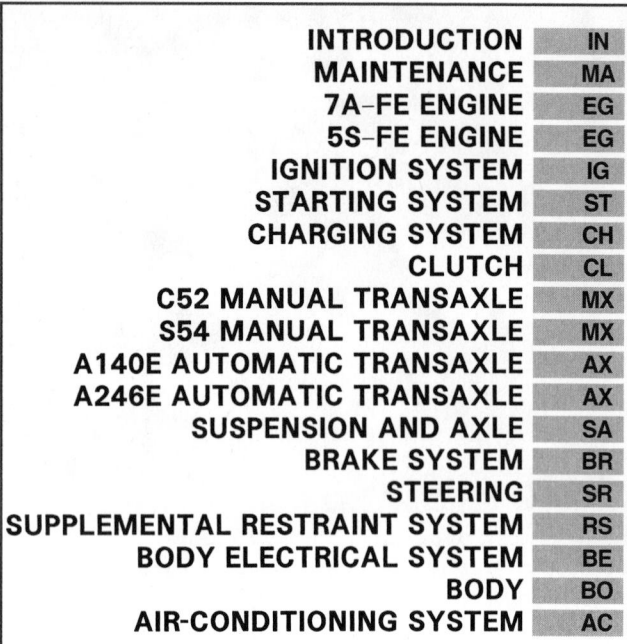

INTRODUCTION	IN
MAINTENANCE	MA
7A-FE ENGINE	EG
5S-FE ENGINE	EG
IGNITION SYSTEM	IG
STARTING SYSTEM	ST
CHARGING SYSTEM	CH
CLUTCH	CL
C52 MANUAL TRANSAXLE	MX
S54 MANUAL TRANSAXLE	MX
A140E AUTOMATIC TRANSAXLE	AX
A246E AUTOMATIC TRANSAXLE	AX
SUSPENSION AND AXLE	SA
BRAKE SYSTEM	BR
STEERING	SR
SUPPLEMENTAL RESTRAINT SYSTEM	RS
BODY ELECTRICAL SYSTEM	BE
BODY	BO
AIR-CONDITIONING SYSTEM	AC

Figure 5-67 The main index of a factory service manual showing that the manual is divided by major vehicle systems. *Reprinted with permission*

manufacturer provides these bulletins to dealers and repair facilities on a regular basis.

Automotive manufacturers also publish a series of technician reference books. The publications provide general instructions about the service and repair of the manufacturers' vehicles and also indicate their recommended techniques.

General and Specialty Repair Manuals

Service manuals are also published by independent companies rather than the manufacturers. However, they pay for and get most of their information from the car makers. The manuals contain component information, diagnostic steps, repair procedures, and specifications for several makes of automobiles in one book. Information is usually condensed and is more general than the manufacturers' manuals. The condensed format allows for more coverage in less space and, therefore, is not always specific. They may also contain several years of models as well as several makes in one book.

Finding Information in Service Manuals

Although the manuals from different publishers vary in presentation and arrangement of topics, all service manuals are easy to use after you become familiar with their organization. Most shop manuals are divided into a number of sections, each covering different aspects of the vehicle. The beginning sections commonly provide vehicle identification and basic maintenance information. The remaining sections deal with each different vehicle system in detail, including diagnostic, service, and overhaul procedures. Each section has an index indicating more specific areas of information.

To obtain the correct system specifications and other information, you must first identify the exact system you are working on. The best source for vehicle identification is the VIN. The code can be interpreted through information given in the service manual. The manual may also help you identify the system through identification of key components or other identification numbers and/or markings.

To use a service manual:

1. Select the appropriate manual for the vehicle being serviced.
2. Use the table of contents to locate the section that applies to the work being done.
3. Use the index at the front of that section to locate the required information.
4. Carefully read the information and study the applicable illustrations and diagrams.
5. Follow all of the required steps and procedures given for that service operation.
6. Adhere to all of the given specifications and perform all measurement and adjustment procedures with accuracy and precision.

Aftermarket Suppliers' Guides and Catalogs

Many of the larger parts manufacturers have excellent guides on the various parts they manufacture or supply. They also provide updated service bulletins on their products. Other sources for up-to-date technical information are trade magazines and trade associations.

Lubrication Guides

These specially designed service manuals contain information on lubrication, maintenance, capacities, and underhood service. The lubrication guide includes lube and maintenance instructions, lubrication diagrams and specifications, vehicle lift points, and preventive maintenance mileage/time intervals. The capacities listed include cooling system, air conditioning, cooling system air bleed locations, wheel and tire specifications, and wheel lug torque specifications. The underhood information includes specifications for tune-up; mechanical, electrical, and fuel systems; diagrams; and belt tension.

Owner's Manuals

An owner's manual comes with the vehicle when it is new. It contains operating instructions for the vehicle and its accessories. It also contains valuable information about checking and adding fluids, safety precautions, a complete list of capacities, and the specifications for the various fluids and lubricants for the vehicle.

Flat-Rate Manuals

Flat-rate manuals contain standards for the length of time a specific repair is supposed to require. Normally, they also contain a parts list with approximate or exact prices of parts. They are excellent for making cost estimates and are published by the manufacturers and independents.

Computer-Based Information

The same information that is available in service manuals and bulletins is also available electronically on **compact disks (CD-ROMs) (Figure 5–68)**, digital video disks (DVDs), and the Internet. A single compact disk can hold a quarter million pages of text, eliminating the need for a huge library to contain all of the printed manuals. Using electronics to find information is also easier and quicker. The disks are normally updated monthly and not only contain the most recent service bulletins but also engineering and field service fixes. DVDs can hold more information than CDs; therefore, fewer disks are needed with systems that use DVDs. The CDs and DVDs are inserted into a computer. All a technician needs to do is enter vehicle information and then move to the appropriate part or system. The appropriate information will then appear on the computer's screen **(Figure 5–69)**. Online data can be updated instantly and requires no space for physical storage. These systems are easy to use and the information is quickly accessed and displayed. The computer's keyboard, mouse, and/or light pen are used to make selections from the screen's menu. Once the information is retrieved, a tech can read it off the screen or print it out and take it to the service bay.

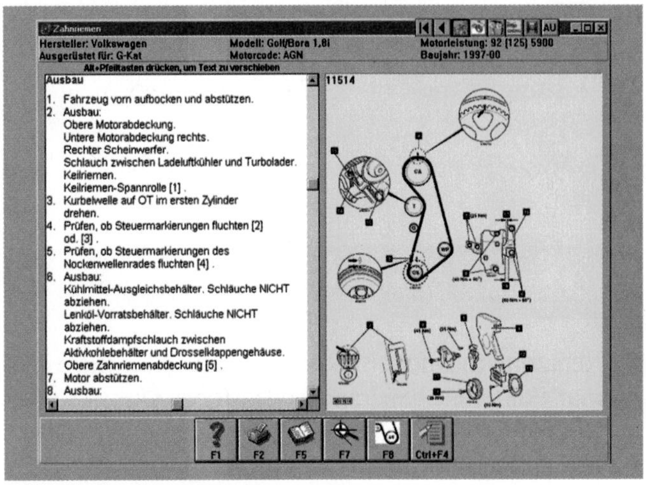

Figure 5–69 All of the relevant information to perform a service is displayed on the computer screen. *Courtesy of Robert Bosch*

Hotline Services

Hotline services provide answers to service concerns by telephone. Manufacturers provide help by telephone for technicians in their dealerships. There are subscription services for independents to be able to get repair information by phone. Some manufacturers also have a phone modem system that can transmit computer information from the car to another location. The vehicle's diagnostic link is connected to the modem. The technician in the service bay runs a test sequence on the vehicle. The system downloads the latest updated repair information on that particular model of car. If that does not repair the problem, a technical specialist at the manufacturer's location will review the data and propose a repair.

iATN

The International Automotive Technician's Network (iATN) is comprised of a group of thousands of professional automotive technicians from around the world. The technicians in this group exchange technical knowledge and information with other members. The web address for this group is http://www.iatn.net.

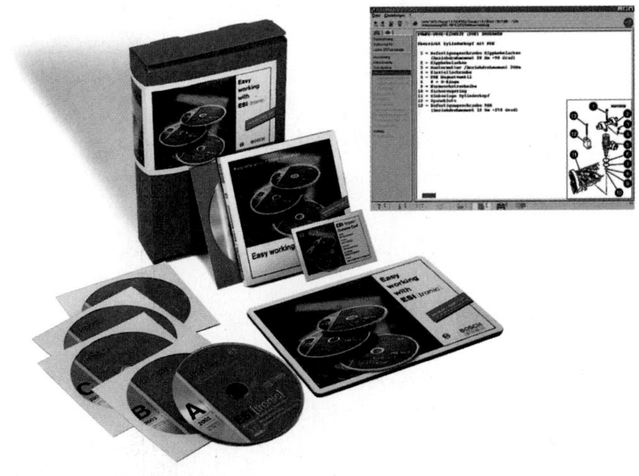

Figure 5–68 The use of CD-ROMs and a computer makes accessing information quick and easy. *Courtesy of Robert Bosch*

KEY TERMS

Ball gauge
Blowgun
Bolt diameter
Bolt head
Bolt shank
Compact disk-read only memory (CD-ROM)
Chisel
Dial caliper
Dial indicator
Die
Extractor
Feeler gauge
Fillet
Flat-rate
Foot pads
Grade marks
Impact wrench
Machinist's rule
Micrometer
Phillips screwdriver
Pliers
Pozidriv screwdriver
Press fit
Punch
Screw pitch gauge
Snap gauge
Spark plug thread tap

Straightedge
Tap
Telescoping gauge
Thread pitch
Torque wrench
Torx screwdriver
Trouble light
Vernier caliper
Wrench

SUMMARY

- Repairing the modern automobile requires the use of many different hand and power tools. Units of measurement play a major role in tool selection. Therefore, it is important to be knowledgeable about the Imperial and the metric systems of measurement.
- Measuring tools must be able to measure objects to a high degree of precision. They should be handled with care at all times and cleaned before and after every use.
- A micrometer can be used to measure the outside diameter of shafts and the inside diameter of holes. It is calibrated in either inch or metric graduations.
- Telescoping gauges are designed to measure bore diameters and other clearances. They usually are used with an outside micrometer. Small hole gauges are used in the same manner as the telescoping gauge, usually to determine valve guide diameter.
- The screw pitch gauge provides a fast and accurate method of measuring the threads per inch (pitch) of fasteners. This is done by matching the teeth of the gauge with the fastener threads and reading the pitch directly from the leaf of the gauge.
- It is crucial to use the proper amount of torque when tightening nuts or cap screws on any part of a vehicle, particularly the engine. A torque-indicating wrench makes it possible to duplicate the conditions of tightness and stress recommended by the manufacturer.
- Metric and SAE size wrenches are not interchangeable. An auto technician should have a variety of both types.
- A screwdriver, no matter what type, should never be used as a chisel, punch, or pry bar.
- The hand tap is used for hand cutting internal threads and for cleaning and restoring previously cut threads. Hand-threading dies cut external threads and fit into holders called die stocks.
- Carelessness or mishandling of power tools can cause serious injury. Safety measures are needed when working with such tools as impact and air ratchet wrenches, blowguns, bench grinders, lifts, hoists, and hydraulic presses.
- The primary source of repair and specification information for any vehicle is the manufacturer's service manual. Updates are published as service bulletins and include changes made during the model year, which will not appear in the manual until the following year.
- Flat-rate manuals are ideal for making cost estimates. Published by manufacturers and independent companies, they contain figures showing how long specific repairs should take to complete, as well as a list of the necessary parts and their prices.

TECH MANUAL

The following procedures are included in Chapter 5 of the *Tech Manual* that accompanies this book:

1. Repairing and/or replacing damaged threads.
2. Cutting with a chisel.
3. Proper use of a vehicle hoist.
4. Proper use of a hydraulic floor jack.
5. Proper use of a stationary grinder.
6. Proper use of a service manual.
7. Proper use of electronic media.

REVIEW QUESTIONS

1. How often should the calibration of a micrometer be checked?
2. List some common uses of the dial indicator.
3. What determines the size of a wrench?
4. *True or False?* The same information available in service manuals and bulletins is also available electronically: on compact disks (CD-ROMs), digital video disks (DVDs), and the Internet.
5. How do manufacturers inform technicians about changes in vehicle specifications and repair procedures during the model year?
6. Which of the following wrenches is the best choice for turning a bolt?

   a. open-end **c.** combination
   b. box-end **d.** none of the above

7. Technician A says a vernier caliper can be used to measure the outside diameter of something. Technician B says a vernier caliper can be used to measure the inside diameter of a bore. Who is correct?

   a. Technician A **c.** Both A and B
   b. Technician B **d.** Neither A nor B

8. Technician A says a tap cuts external threads. Technician B says a die cuts internal threads. Who is correct?

   a. Technician A **c.** Both A and B
   b. Technician B **d.** Neither A nor B

9. Which of the following screwdrivers is like a Phillips but has a flatter and blunter tip?

   a. standard
   b. Torx®
   c. Pozidriv®
   d. clutch head

10. Which of the following types of pliers is best for grasping small parts?

    a. adjustable **c.** retaining ring
    b. needle nose **d.** snap ring

11. Technician A uses a punch to align holes in parts during assembly. Technician B uses a punch to drive out rivets. Who is correct?

    a. Technician A **c.** Both A and B
    b. Technician B **d.** Neither A nor B

12. An extractor is used for removing broken ______.

    a. seals **c.** pistons
    b. bushings **d.** bolts

13. Which of the following statements about items that are press fit is *not* true?

    a. Many precision gears and bearings have a slight interference fit when installed on a shaft or housing.
    b. The press fit allows slight motion between the parts and therefore prevents wear.
    c. The removal of gears and bearings that are press fit must be done carefully to avoid breaking or binding the parts.
    d. A puller with the proper jaws and adapters should be used when applying force to remove press-fit gears and bearings.

14. Technician A uses a blowgun to blow off parts during cleaning. Technician B uses a blowgun to clean off his uniform after working. Who is correct?

    a. Technician A **c.** Both A and B
    b. Technician B **d.** Neither A nor B

15. Technician A says flare nut or line wrenches should be used to loosen or tighten brake line or tubing fittings. Technician B says open-end wrenches will surround the fitting's nut and provide a positive grip on the fitting. Who is correct?

    a. Technician A **c.** Both A and B
    b. Technician B **d.** Neither A nor B

16. Technician A uses a dial caliper to take inside and outside measurements. Technician B uses a dial caliper to take depth measurements. Who is correct?

    a. Technician A **c.** Both A and B
    b. Technician B **d.** Neither A nor B

17. For a measurement that must be made within one ten-thousandth of an inch, Technician A uses a machinist's rule. For the same accuracy, Technician B uses a standard micrometer. Who is correct?

    a. Technician A **c.** Both A and B
    b. Technician B **d.** Neither A nor B

18. Technician A says portable floor jacks are operated by hydraulics. Technician B says portable floor jacks may be operated by compressed air. Who is correct?

    a. Technician A **c.** Both A and B
    b. Technician B **d.** Neither A nor B

19. Technician A uses a 9⁄16-inch wrench to turn a 14-millimeter nut. Technician B uses a 9⁄16-inch wrench to hold a 14-millimeter nut. Who is correct?

a. Technician A
b. Technician B
c. Both A and B
d. Neither A nor B

20. When using an air impact wrench, Technician A uses impact sockets and adapters. Technician B uses chrome-plated sockets. Who is correct?

a. Technician A
b. Technician B
c. Both A and B
d. Neither A nor B

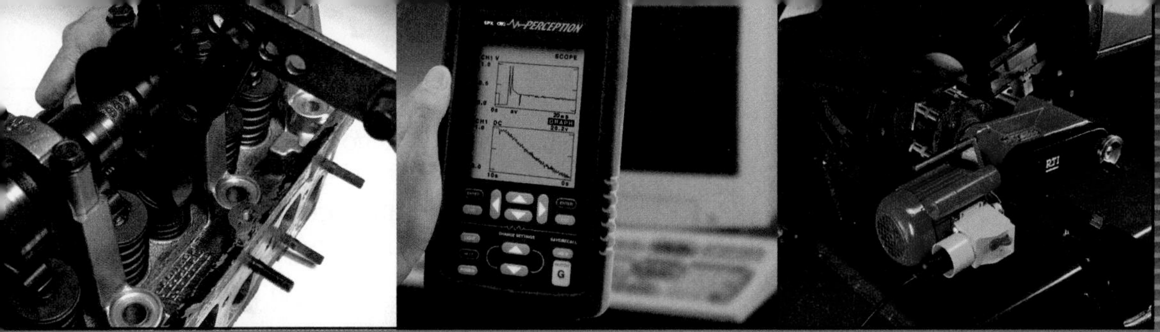

6

DIAGNOSTIC EQUIPMENT AND SPECIAL TOOLS

OBJECTIVES

■ Describe the various diagnostic tools used to check an engine and its related systems. ■ Describe the common tools used to service an engine and its related systems. ■ Describe the various diagnostic tools used to check electrical and electronic systems. ■ Describe the common tools used to service electrical and electronic systems. ■ Describe the various diagnostic tools used to check a vehicle's drivetrain. ■ Describe the common tools used to service a vehicle's drivetrain. ■ Describe the various diagnostic tools used to check a vehicle's running gear for wear and damage. ■ Describe the common tools used to service a vehicle's running gear. ■ Describe the various diagnostic tools used to check a vehicle's heating and air-conditioning system. ■ Describe the common tools used to service a vehicle's heating and air-conditioning system.

Diagnosing and servicing the various systems of an automobile require many different tools. Tools that are used to check the performance of a system or component are commonly referred to as diagnostic tools. Tools designed for a particular purpose or system are referred to as special tools. This chapter looks at the common diagnostic and special tools required to service the different systems of a vehicle.

ENGINE REPAIR TOOLS

Engine repair and diagnostic tools are discussed in the following paragraphs. This discussion does not cover all of the tools you may need: only the most commonly used are discussed. Details of when and how to use these tools are presented in Section 2 of this book.

Compression Testers

Engines depend on the compression of the air/fuel mixture for power output. The compression stroke of the piston compresses the air/fuel mixture within the combustion chamber. If the combustion chamber leaks, some of the air/fuel mixture will escape while it is being compressed, resulting in a loss of power and a waste of fuel. The leaks can be caused by burned valves, a blown head gasket, worn rings, a slipped timing belt or chain, worn valve seats, a cracked head, and more.

If a symptom suggests that the cause of a problem may be poor compression, a compression test is performed.

A **compression gauge** is used to check cylinder compression. The dial face on the typical compression gauge indicates pressure in both pounds per square inch **(psi)** and metric kilopascals **(kPa)**. The range is usually 0 to 300 psi and 0 to 2,100 kPa. There are two basic types of compression gauges: the push-in gauge **(Figure 6–1)** and a screw-in gauge.

The push-in type has a short stem that is either straight or bent at a 45-degree angle. The stem ends in a tapered rubber tip that fits any size spark plug hole. After the spark plugs have been removed, the rubber tip is placed in the spark plug hole and held there while the engine is cranked through several compression cycles. Although simple to use, the push-in gauge may give inaccurate readings if it is not held tightly in the hole.

The screw-in gauge has a long, flexible hose that ends in a threaded adapter. This type of compression tester is often used because its flexible hose can reach into areas that are difficult to reach with a push-in-type tester. The threaded adapters are changeable and come in several thread sizes to fit 10-mm, 12-mm, 14-mm, and 18-mm diameter holes. The adapters screw into the spark plug holes in place of the spark plugs.

Most compression gauges have a vent valve that holds the highest pressure reading on its meter. Opening the valve releases the pressure when the test is complete.

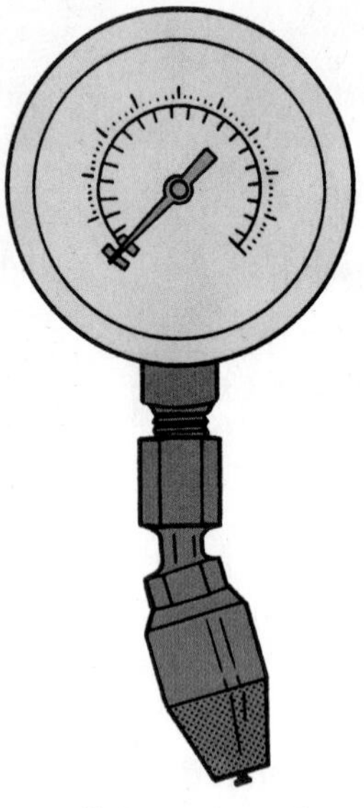

Figure 6-1 A push-in compression gauge.

Cylinder Leakage Tester

If a compression test shows that any of the cylinders are leaking, a cylinder leakage test can be performed to measure the percentage of compression lost and help locate the source of leakage.

A **cylinder leakage tester (Figure 6–2)** applies compressed air to a cylinder, with the piston at the top of its bore, through the spark plug hole. A threaded adapter on the end of the air pressure hose screws into the spark plug hole. A pressure regulator in the tester controls the pressure applied to the cylinder. A gauge registers the percentage of air pressure lost from the cylinder when the compressed air is applied. The scale on the dial face reads 0 to 100%.

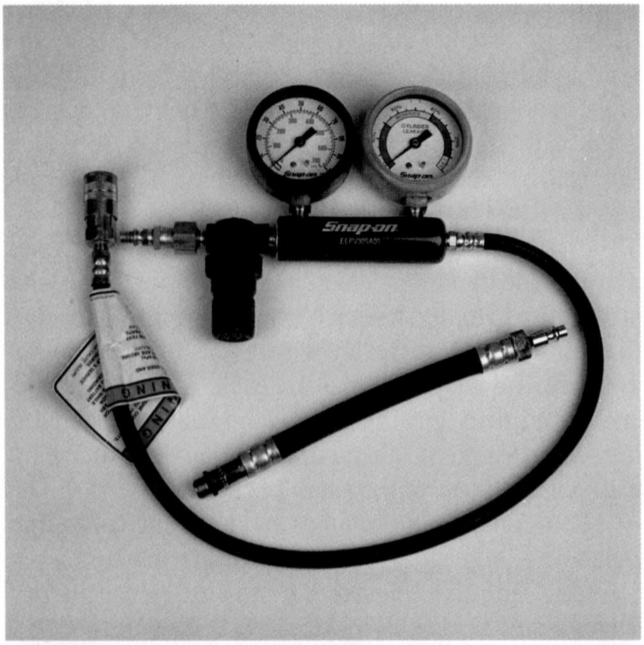

Figure 6-2 A cylinder leakage tester.

A zero reading means there is no leakage in the cylinder. A reading of 100% indicates that the cylinder will not hold any pressure. The location of the compression leak can be found by listening and feeling around various parts of the engine.

Oil Pressure Gauge

Checking the engine's oil pressure gives information about the condition of the oil pump, the pressure regulator, and the entire lubrication system. Lower-than-normal oil pressures can be caused by excessive engine bearing clearances. Oil pressure is checked at the sending unit passage with an externally mounted mechanical oil pressure gauge. Various fittings are usually supplied with the oil pressure gauge to fit different openings in the lubrication system.

Stethoscope

A **stethoscope** is used to locate the source of engine and other noises. The stethoscope pickup is placed on the suspected component, and the stethoscope receptacles are placed in the technician's ears **(Figure 6–3)**. Some sounds can be heard easily without using a listening device, but others are impossible to hear unless amplified, which is what a stethoscope does. It can also help you distinguish between normal and abnormal noise. The best results, however, are obtained with an electronic listening device. With this tool you can tune into the noise, which allows you to eliminate all other noises that might distract or mislead you.

Transaxle Removal and Installation Equipment

The engines of some FWD vehicles are removed by lifting them from the top. Others must be removed from the bottom and the procedure requires different equipment. Make sure you follow the instructions given by the manufacturer and use the appropriate tools and equipment. The required equipment varies with manufacturer and vehicle model; however, most accomplishes the same thing.

To remove the engine from under the vehicle, the vehicle must be raised. A crane and/or support fixture is

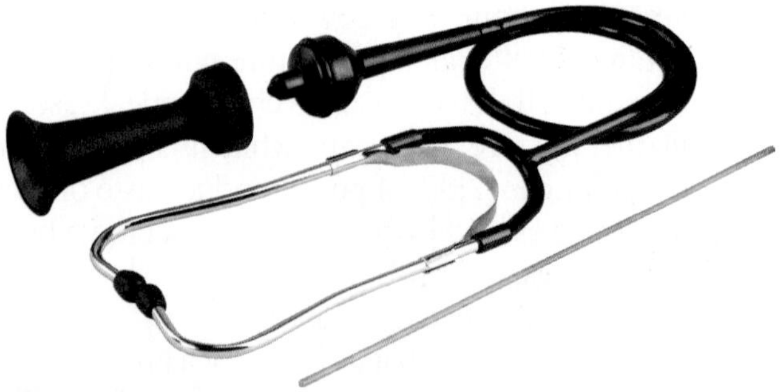

Figure 6-3 A stethoscope. *Courtesy of SPX-OTC Automotive Electronic Diagnostic Tools*

used to hold the engine and transaxle assembly in place while the engine is being readied for removal. Once the engine is ready, the crane is used to lower the engine onto an engine cradle. The cradle is similar to a hydraulic floor jack and is used to lower the engine further so it can be rolled out from under the vehicle.

Often a transverse-mounted engine is removed with the transaxle as a unit. The transaxle can be separated from the engine once it has been lifted out of the vehicle. In this case, the drive axles must be disconnected from the transaxle before removing the unit.

Ridge Reamer

After many miles of use, a ridge is formed at the top of the engine's cylinders. Because the top piston ring stops traveling before it reaches the top of the cylinder, a ridge of unworn metal is left. This ridge must be removed to push the pistons out of the block without damaging them. This ridge is removed with a **ridge reamer**. The tool is adjusted for the bore and then inserted into it. Rotate the tool clockwise with a wrench to remove the ridge. Remove just enough metal to allow the piston assembly to slip out of the bore without causing damage to the surface of the bore or to the piston. If the ridge is too large, the top rings will hit it and possibly break the ring lands.

After the ridge-removing operation, wipe all the metal cuttings out of the cylinder. Use an oily rag to wipe the cylinder. The cuttings will stick to it.

Ring Compressor

A **ring compressor (Figure 6–4)** is used to install a piston with piston rings into a cylinder bore. The compressor wraps around the rings to make their outside diameter smaller than the inside diameter of the bore. With the compressor tool adjusted properly, the piston assembly can be pushed easily into the bore without damaging the bore or piston.

Ring Expander

To prevent damage to the piston rings during removal and installation, a **ring expander** should be used. To install a piston ring, the ring must be made large enough to fit over the piston. The piston rings fit into the jaws of the expander and the handle of the tool is squeezed to expand the ring. Expand the rings only to the point at which they can fit over the piston.

Figure 6–4 A piston ring compressor. *Courtesy of Snap-on Tools Company*

Ring Groove Cleaner

Before installing piston rings onto a piston, the ring grooves should be cleaned. The carbon and other debris that may be present in the back of the groove will not allow the rings to compress evenly and completely into the grooves. Piston ring grooves are best cleaned with a **ring groove cleaner**. This tool is adjustable to fit the width and depth of the groove. Make sure it is properly adjusted before using it and make sure you do not damage the piston while cleaning it.

Dial Bore Indicator

Cylinder bore taper and out-of-roundness can be measured with a micrometer and a telescoping gauge. However, most shops use a **dial bore gauge**. This gauge typically consists of a handle, guide blocks, a lock, an indicator contact, and an indicator. They also come with extensions that make them adaptable to various size bores. As the dial bore gauge is moved inside the bore, the indicator will show any change in the bore's diameter.

Cylinder Deglazer

The proper surface finish on a cylinder wall acts as a reservoir for oil to lubricate the piston rings and prevent piston and ring scuffing. If the inspection and measurements of the cylinder wall show that surface conditions, taper, and out-of-roundness are within acceptable limits, the cylinder walls only need to be deglazed. Combustion heat, engine oil, and piston movement combine to form a thin residue on the cylinder walls that is commonly called glaze.

The common types of cylinder deglazers or **glaze breakers** use an abrasive with about 220 or 280 grit. The glaze breaker is installed in a slow-moving electric drill or in a honing machine. Many deglazers use round stones that extend on coiled wire from the center shaft. This type of deglazer may also be used to lightly hone the bore. Various sizes of resilient-based hone-type brushes are available for honing and deglazing.

Cylinder Hone

You should hone a cylinder whenever there are minor problems with the bore. Honing sands the walls to remove imperfections. A **cylinder hone** usually consists of two or three stones. The hone rotates at a selected speed and is moved up and down the cylinder's bore. The stones have outward pressure on them and remove some metal from the bore as they rotate within it. Honing oil flows over the stones and onto the cylinder wall to control the temperature and flush out any metallic and abrasive residue. The correct stones should be used to ensure that the finished walls have the correct surface finish. Honing stones are classified by grit size; usually the lower the grit number, the coarser the stone.

Cylinder honing machines are available in manual and automatic models. The major advantage of the automatic type is that it allows the technician to dial in the exact crosshatch angle needed.

When cylinder surfaces are badly worn or excessively scored or tapered, a **boring bar** is used to cut the cylinders for oversize pistons or sleeves. A boring bar leaves a pattern on the cylinder wall similar to uneven screw threads; therefore, you should hone the bore to the correct finish after it has been bored.

Cam Bearing Driver Set

The camshaft is supported by several friction-type bearings, or bushings. They are designed as one piece and are typically pressed into the camshaft bore in the cylinder head or block; however, some overhead camshaft (OHC) engines use split bearings to support the camshaft. Camshaft bearings are normally replaced during engine rebuilding. Cam bearings are normally press fit into the block or head using a bushing driver and hammer.

V-Blocks

The various shafts in an engine must be straight and not distorted. Visually it is impossible to see any distortions unless the shaft is severely damaged. Warped or distorted shafts will cause many problems, including premature wear of the bearings they ride on. The best way to check a shaft is to place the ends of the shaft onto V-blocks. These blocks will support the shaft and allow you to rotate the shaft. Place the plunger of a dial indicator on the journals of the shaft and rotate the shaft. Any movement of the indicator's needle suggests a problem.

Valve and Valve Seat Resurfacing Equipment

Whenever the valves have been removed from the cylinder head, the valve heads and valve seats should be resurfaced. The most critical sealing surface in the valve train is between the face of the valve and its seat in the cylinder head. Leakage between these surfaces reduces the engine's compression and power and can lead to valve burning. To ensure proper seating of the valve, the seat area on the valve face and seat must be the correct width, at the correct location, and concentric with the guide. These conditions are accomplished by renewing the surface of the valve face and seat.

Valve and valve seat grinding or refacing is done by machining with a grinding stone or metal cutters to achieve a fresh, smooth surface on the valve faces and stem tips. Valve faces suffer from burning, pitting, and wear caused by opening and closing millions of times during the life of an engine. Valve stem tips wear because of friction from the rocker arms or actuators.

Valve Guide Repair Tools

The amount of valve guide wear can be measured with a ball gauge and micrometer. If wear or taper is excessive, the guide must be machined or replaced. If the original guide can be removed and a new one inserted, press out the old valve guide by placing the properly sized driver into the guide. Then press out the guide. To install a new guide, use a press and the same driver that was used to remove the old guide. Align the new guide and press straight down, not at an angle.

Knurling is one of the fastest ways to restore the inside diameter (ID) dimensions of a worn valve guide. The process raises the surface of the guide ID by plowing tiny furrows through the surface of the metal. As the knurling tool cuts into the guide, metal is raised or pushed up on either side of the indentation, effectively decreasing the ID of the guide hole. A burnisher is used to make the ridges flat and produce the proper-sized hole to restore the correct guide-to-stem clearance.

Reaming is used to repair worn guides by increasing the ID of a guide to take an oversize valve stem or by restoring the guide to its original diameter after installing inserts. When reaming a guide, limit the amount of metal removed and always reface the valve seat after the valve guide has been reamed. Some valve guide liners or inserts are not precut to length and the excess must be milled off before finishing.

Valve Spring Compressor

To remove the valves from a cylinder head, first the valve spring assemblies must be removed. To do this, the valve spring must be compressed enough to remove the valve keepers, then the retainer. Many types of **valve spring compressors** are available. Some designs allow valve spring removal while the cylinder head is still on the engine block; other designs are only used when the cylinder head is removed.

The pry bar-type compressor is used when installing valve oil seals when the cylinder head is still mounted to the block. With the cylinder's piston at top dead center (TDC), shop air is fed into the cylinder to hold the valve up and prevent it from falling into the cylinder. The pry bar is then used to compress the valve spring so the valve keepers can be removed.

Some OHC engines require the use of a special spring compressor **(Figure 6–5)**. Often these special tools can be used when the cylinder head is attached to the block and when it is on a bench. These compressors bolt to the cylinder head and have a threaded plunger that fits onto the retainer. As the plunger is tightened down on the retainer, the spring compresses.

C-clamp-type valve compressors can only be used on cylinder heads after they have been removed. This type of compressor usually is a universal tool with interchangeable jaws. The spring is compressed either pneumatically or manually after the compressor is in place. One end of the clamp is positioned on the valve head and

Figure 6-5 A typical spring compressor for OHC valves.

the other on the valve's retainer. After the compressor is adjusted, the compressor is activated to squeeze down on the spring. Once the spring is compressed, the valve keepers can be removed. Then the tension of the compressor is slowly released and the valve retainer and spring can be removed.

Valve Spring Tester

Before valve springs are reused, they should be checked to make sure they are within specifications. This checking should include their freestanding height and squareness. If those two dimensions are good, the spring should be checked with a **valve spring tester**. A valve spring tester checks each valve's open and close pressure. Correct close pressure guarantees a tight seal. The open pressure overcomes valve train inertia and closes the valve when it should close. The tester's gauge reflects the pressure of the spring when it is compressed to the installed or valve-closed height. Read the pressure on the tester and compare this reading to specifications. Any pressure outside the pressure range given in the specifications indicates the spring should be replaced.

Torque Angle Gauge

Most manufacturers recommend the torque-angle method for tightening cylinder head bolts, which requires the use of a **torque angle gauge**. Typically two steps are involved: tighten the bolt to the specified torque, and then tighten the bolt an additional amount. The latter is expressed in degrees. To accurately measure the number of degrees added to the bolt, a torque angle gauge **(Figure 6–6)** is attached to the wrench. The additional tightening will stretch the bolt and produce a very reliable clamp load that is much higher than can be achieved just by torquing.

Oil Priming Tool

Prior to starting a freshly rebuilt engine, the oil pump must be primed. There are several ways to prelubricate, or prime, an engine. One method is to drive the oil pump with an electric drill. With some engines, it is possible to make a drive that can be chucked in an electric drill motor to engage the drive on the oil pump. Insert the fabricated oil pump drive extension into the oil pump through the distributor drive hole. To control oil splash, loosely set the valve cover(s) on the engine. After running the oil pump for several minutes, remove the valve cover and see whether there is any oil flow to the rocker arms. If oil reached the cylinder head, the engine's lubrication system is full of oil and is operating properly. If no oil reached the cylinder head, there is a problem either with the pump, with the alignment of an oil hole in a bearing, or perhaps with a plugged gallery.

Figure 6-6 A torque angle gauge is attached to the drive lug of a torque wrench. *Courtesy of SPX-OTC Automotive Electronic Diagnostic Tools*

Using a prelubricator **(Figure 6–7)**, which consists of an oil reservoir attached to a continuous air supply, is the

Figure 6-7 An engine preluber kit. *Courtesy of SPX-OTC Automotive Electronic Diagnostic Tools*

best method of prelubricating an engine without running it. When the reservoir is attached to the engine and the air pressure is turned on, the prelubricator will supply the engine's lubrication system with oil under pressure.

Cooling System Pressure Tester

A cooling system pressure tester **(Figure 6–8)** contains a hand pump and a pressure gauge. A hose is connected from the hand pump to a special adapter that fits on the radiator filler neck. This tester is used to pressurize the cooling system and check for coolant leaks. Additional adapters are available to connect the tester to the radiator cap. With the tester connected to the radiator cap, the pressure relief action of the cap may be checked.

Coolant Hydrometer

A coolant **hydrometer** is used to check the amount of antifreeze in the coolant. This tester contains a pickup hose, a coolant reservoir, and a squeeze bulb. The pickup hose is placed in the radiator coolant. When the squeeze bulb is squeezed and released, coolant is drawn into the reservoir. As coolant enters the reservoir, a float moves upward with the coolant level. A pointer on the float indicates the freezing point of the coolant on a scale located on the reservoir housing.

Coolant Recovery and Recycle System

A coolant recovery and recycle machine **(Figure 6–9)** typically can drain, recycle, fill, flush, and pressure test a cooling system. Usually additives are mixed into the used coolant during recycling. These additives either bind to contaminants in the coolant so they can be easily removed, or they restore some of the chemical properties in the coolant.

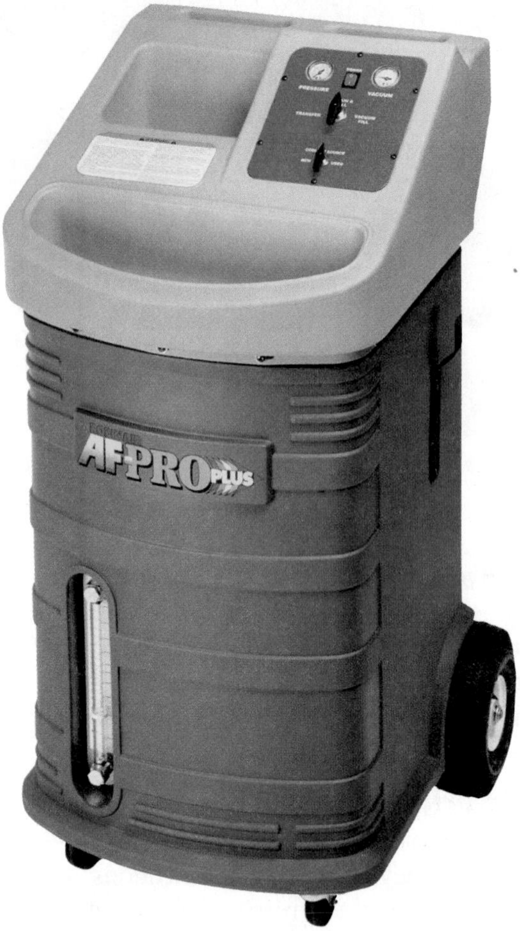

Figure 6-9 A coolant recycling machine that drains, back flushes, and fills the cooling system. *Courtesy of Robinair, SPX Corporation*

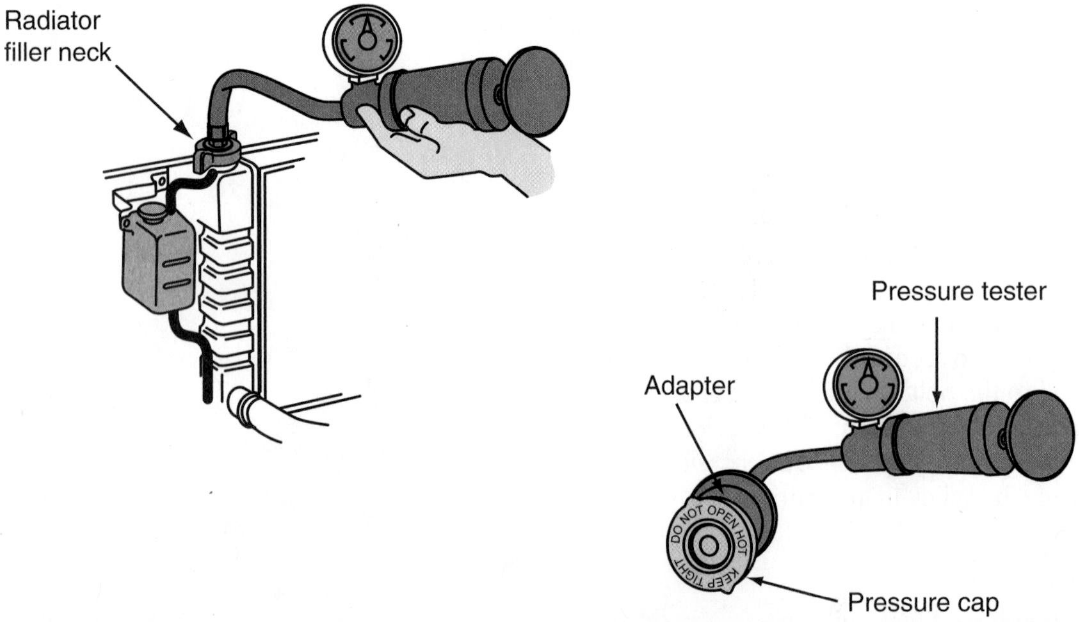

Figure 6-8 A cooling system pressure tester.

ELECTRICAL/ELECTRONIC SYSTEM TOOLS

Electrical system service and diagnostic tools are discussed in the following paragraphs. This discussion does not cover all of the tools you may need; rather, these tools are the most commonly used by the service industry. Many automotive systems are electrically controlled and operated; therefore, these tools are also used in those systems. Details of when and how to use these tools are presented in Section 3 of this book, as well as in the sections that discuss the various other automotive systems.

Computer Memory Saver

Memory savers are nothing more than power sources to maintain the memory circuits in electronic accessories and the engine, transmission, and body computers.

Two types of memory savers are available or can be made for satisfactory use on late-model vehicles. For a power source, you can use a 12-volt automotive battery **(Figure 6–10)** or a 12-volt dry-cell lantern battery. A cigarette lighter adapter plug with wire leads and large alligator clips is attached to the auxiliary battery.

Circuit Tester

Circuit testers **(Figure 6–11)** are used to check for voltage in an electrical circuit. Low-voltage testers are used to troubleshoot 6- to 12-volt circuits. High-voltage circuit testers diagnose primary and secondary ignition circuits.

A circuit tester, commonly called a **test light**, looks like a stubby ice pick. Its handle is transparent and contains a light bulb. A probe extends from one end of the handle and a ground clip and wire from the other end. When the ground clip is attached to a good ground and the probe touched to a live connector, the bulb in the handle will light up. If the bulb does not light, voltage is not available at the connector.

Figure 6–10 A 12-volt memory saver. *Courtesy of Snap-on Tools Company*

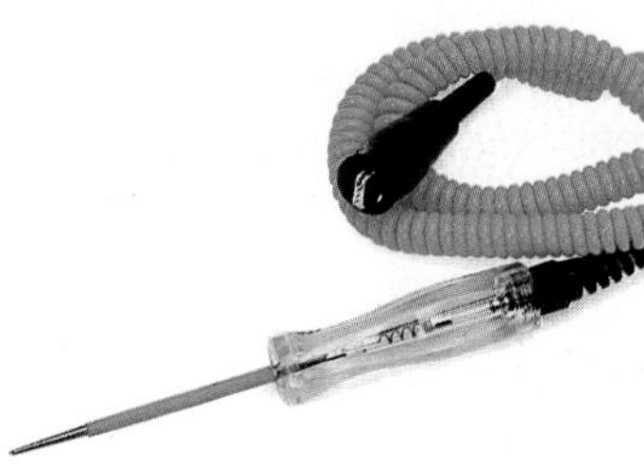

Figure 6–11 A typical circuit tester, commonly called a test light. *Courtesy of Snap-on Tools Company*

WARNING!

Do not use a conventional 12-V test light to diagnose components and wires in computer systems. The current draw of these test lights may damage computers and computer system components. High-impedance test lights are available for diagnosing computer systems.

A self-powered test light is called a **continuity tester**. Used on open circuits, it looks like a regular test light, except that it has a small internal battery. When the ground clip is attached to one end of the wire or circuit and the probe is touched to the other end, the lamp will light if there is continuity in the circuit. If an open circuit exists, the light will not be illuminated.

WARNING!

Do not use any type of test light or circuit tester to diagnose automotive air bag systems. Use only the vehicle manufacturer's recommended equipment on these systems.

Voltmeter

A **voltmeter** has two leads: a red positive lead and a black negative lead. The red lead should be connected to the positive side of the circuit or component. The black should be connected to ground or to the negative side of the component. Voltmeters should be connected across the circuit being tested.

A voltmeter measures the voltage available at any point in an electrical system. A voltmeter can also be used to test voltage drop across an electrical circuit, component, switch, or connector. A voltmeter can also be used to check for proper circuit grounding.

Ohmmeter

An **ohmmeter** measures resistance to current flow in a circuit. In contrast to the voltmeter, which operates by the voltage available in the circuit, an ohmmeter is battery powered. The circuit being tested must have no power applied. If the power is on in the circuit, the ohmmeter will be damaged.

The two leads of the ohmmeter are placed across or in parallel with the circuit or component being tested.

The red lead is placed on the positive side of the circuit and the black lead is placed on the negative side of the circuit. The meter sends current through the component and determines the amount of resistance based on the voltage dropped across the load. The scale of an ohmmeter reads from 0 to infinity (∞). A 0 reading means there is no resistance in the circuit and may indicate a short in a component that should show a specific resistance. An infinity reading indicates a number higher than the meter can measure, which usually indicates an open circuit.

Ammeter

An **ammeter** measures current flow in a circuit. Current is measured in amperes. Unlike the voltmeter and ohmmeter, the ammeter must be placed into the circuit or in series with the circuit being tested. Normally, this requires disconnecting a wire or connector from a component and connecting the ammeter between the wire or connector and the component. The red lead of the ammeter should always be connected to the side of the connector closest to the positive side of the battery and the black lead should be connected to the other side.

It is much easier to test current using an ammeter with an inductive pickup **(Figure 6–12)**. The pickup clamps around the wire or cable being tested. The ammeter determines amperage based on the magnetic field created by the current flowing through the wire. This type of pickup eliminates the need to separate the circuit to insert the meter.

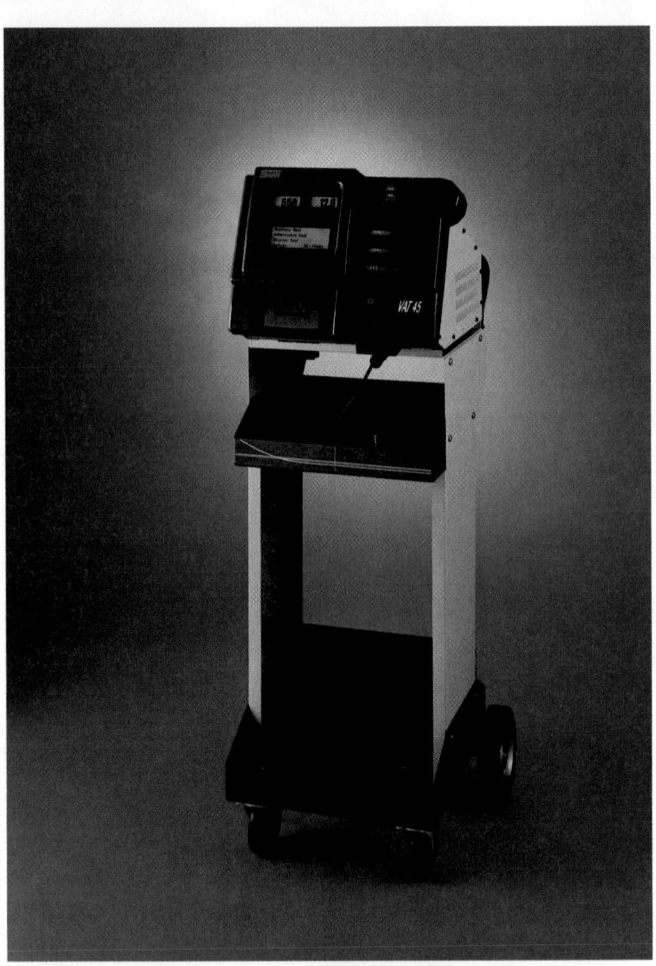

Figure 6–13 The volt/ampere tester is used for testing batteries and the starting and charging systems. *Courtesy of Snap-on Tools Company*

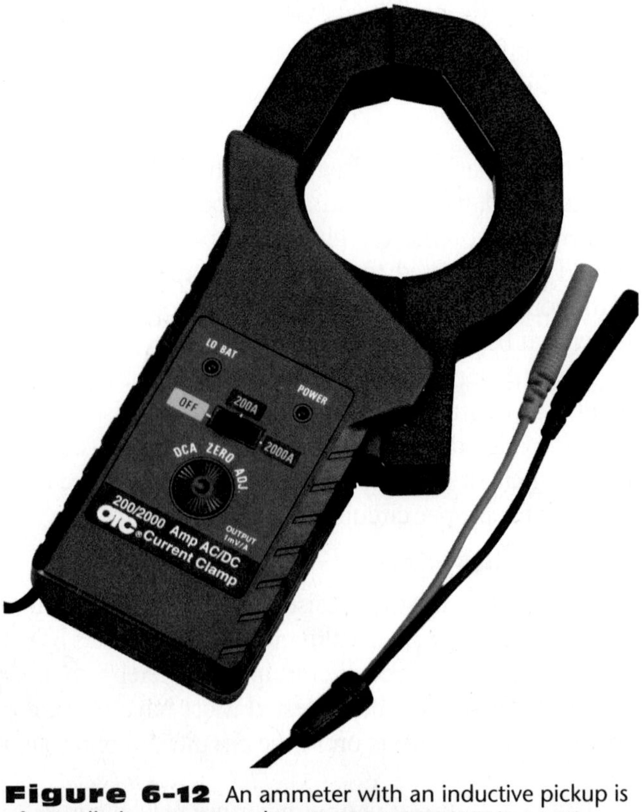

Figure 6–12 An ammeter with an inductive pickup is often called a current probe. *Courtesy of SPX/OTC/Automotive Electronic Diagnostic Tools*

Volt/Ampere Tester

A volt/ampere tester **(VAT)**, shown in **Figure 6–13**, is used to test batteries, starting systems, and charging systems. The tester contains a voltmeter, ammeter, and carbon pile. The carbon pile is a variable resistor. When the tester is attached to the battery and turned on, the carbon pile draws current out of the battery. The ammeter will read the amount of current draw. The maximum current draw from the battery, with acceptable voltage, is compared to the rating of the battery to see if the battery is okay. A VAT also measures the current draw of the starter and current output from the charging system.

Multimeters

A **multimeter** is a must for diagnosing the individual components of an electrical system. Multimeters have different names, depending on what they measure and how they function. A volt-ohm-milliamp meter is referred to as a VOM or DVOM, if it is digital. A **digital multimeter**

Figure 6–14 A typical multifunctional, low-impedance multimeter. *Courtesy of OTC Tool and Equipment, Division of SPX Corporation*

(DMM) can measure many more things than volts, ohms, and low current.

Most multimeters **(Figure 6–14)** measure direct current (dc) and alternating current (ac) amperes, volts, and ohms. More advanced multimeters may also measure diode continuity, frequency, temperature, engine speed, and dwell, and/or duty cycle.

Multimeters are available with either digital or analog displays. DMMs provide great accuracy by measuring volts, ohms, or amperes in tenths, hundredths, or thousandths of a unit. Several test ranges are usually provided for each of these functions. Some meters have multiple test ranges that must be manually selected; others are autoranging.

Analog meters use a sweeping needle against a scale to display readings and are not as precise as digital meters. Analog meters have low input impedance and should not be used on sensitive electronic circuits or components. Digital meters have high impedance and can be used on electronic circuits as well as electrical circuits.

Lab Scopes

An oscilloscope or **lab scope** is a visual voltmeter **(Figure 6–15)**. A lab scope converts electrical signals to a visual image representing voltage changes over a specific period of time. This information is displayed in the form of a continuous voltage line called a **waveform** or **trace**. With a

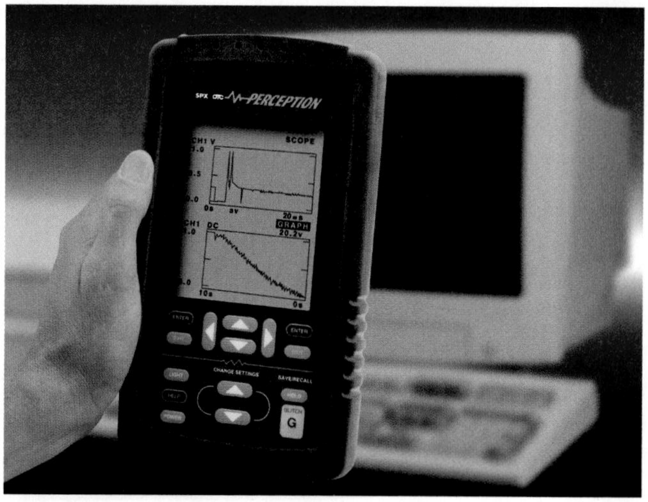

Figure 6–15 A hand-held dual-trace lab scope. *Courtesy of OTC Tool and Equipment, Division of SPX Corporation*

scope, precise measurement is possible. A scope displays any change in voltage as it occurs.

An upward movement of the voltage trace on an oscilloscope indicates an increase in voltage, and a downward movement of this trace represents a decrease in voltage. As the voltage trace moves across an oscilloscope screen, it represents a specific length of time.

The size and clarity of the displayed waveform is dependent on the voltage scale and the time reference selected. Most scopes are equipped with controls that allow voltage and time interval selection. It is important when choosing the scales to remember that a scope displays voltage over time.

Dual-trace scopes can display two different waveform patterns at the same time. This type scope is especially important for diagnosing intermittent problems.

Battery Hydrometer

On unsealed batteries, the specific gravity of the electrolyte can be measured to give a fairly good indication of the battery's state of charge. A hydrometer **(Figure 6–16)** is used to perform this test. A battery hydrometer consists of a glass tube or barrel, rubber bulb, rubber tube, and a glass float or hydrometer with a scale built into its upper stem. The glass tube encases the float and forms a reservoir for the test electrolyte. Squeezing the bulb pulls electrolyte into the reservoir.

When filled with test electrolyte, the sealed hydrometer float bobs in the electrolyte. The depth to which the glass float sinks in the test electrolyte indicates its relative weight compared to water. The reading is taken off the scale by sighting along the level of the electrolyte.

Wire and Terminal Repair Tools

Many automotive electrical problems can be traced to faulty wiring. Loose or corroded terminals; frayed,

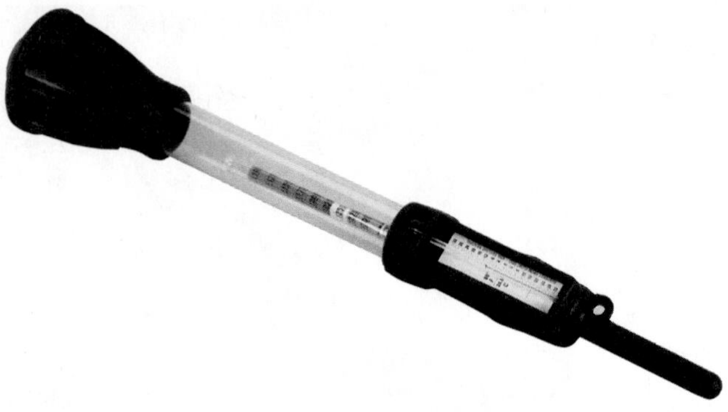

Figure 6–16 A battery hydrometer is used to measure the specific gravity of a battery's electrolyte. *Courtesy of SPX-OTC Automotive Electronic Diagnostic Tools*

broken, or oil-soaked wires; and faulty insulation are the most common causes.

Wires and connectors are often repaired or replaced. Sometimes an entire length of wire is replaced; other times only a section is. In either case, the wire must have the correct terminal or connector to work properly in the circuit. Wire cutters, stripping tools, terminal crimpers, and **connector picks** are the most commonly used tools for wire repair. Also, soldering equipment is used to provide the best electrical connection for a wire to another wire and for a wire to a connector.

Headlight Aimers

Headlights must be kept in adjustment to obtain maximum illumination. Sealed beams that are properly adjusted cover the correct range and afford the driver the proper nighttime view. Headlights can be adjusted using headlamp-adjusting tools or by shining the lights on a chart. Headlight-aiming tools give the best results with the least amount of work. Many late-model vehicles have levels built into the headlamp assemblies that are used to correctly adjust the headlights.

Most headlight aimers use mirrors with split images, like split-image finders on some cameras, and spirit levels to determine exact adjustment. When using any headlight-aiming equipment, follow the instructions provided by the equipment manufacturer.

ENGINE PERFORMANCE TOOLS

Diagnostic and special tools for the air, fuel, ignition, emission, and engine-control systems are discussed in the following paragraphs. This discussion does not cover all of the tools you may need; rather, these tools are the most commonly used by the service industry. Some are also used when diagnosing or servicing the controls of other automotive systems. Details of when and how to use these tools are covered in Section 4 of this book, as well as in other sections where necessary.

Scan Tools

The introduction of computer-controlled ignition and fuel systems brought with it the need for tools capable of troubleshooting electronic engine control systems. A variety of computer scan tools are available today that do just that. A **scan tool (Figure 6–17)** is a microprocessor designed to communicate with the vehicle's computer. Connected to the computer through diagnostic connectors, a scan tool can access trouble codes, run tests to check system operations, and monitor the activity of the system. Trouble codes and test results are displayed on an LCD (liquid crystal display) screen or printed out on the scanner printer.

The scan tool is connected to specific diagnostic connectors on the vehicle. Some manufacturers have one diagnostic connector that connects the data wire from each computer to a specific terminal in this connector. Other vehicle manufacturers have several different diagnostic connectors on each vehicle, and each of these connectors may be connected to one or more computers. The scan tool must be programmed for the model year, make of vehicle, and type of engine.

With OBD-II, the diagnostic connectors are located in the same place on all vehicles. Also, any scan tool designed for OBD-II will work on all OBD-II systems; therefore, the need to have designated scan tools or cartridges is eliminated. The OBD-II scan tool has the ability to run diagnostic tests on all systems and has freeze frame capabilities.

As automotive computer systems become more complex, the diagnostic capabilities of scan testers continue to expand. Many scan testers now have the capability to store, or "freeze," data into the tester during a road test **(Figure 6–18)**, and then play back this data when the vehicle is returned to the shop.

Figure 6–17 A typical scan tool shown with various test leads, adapters, and memory card. *Courtesy of SPX Genisys Tools*

Figure 6-18 Using a scan tool during a road test.
Courtesy of OTC Tool and Equipment, Division of SPX Corporation

Engine Analyzers

When performing a complete engine performance analysis, an engine analyzer is used. An engine analyzer houses all of the necessary test equipment. Although the term engine analyzer is often loosely applied to any multipurpose test meter, a complete engine analyzer incorporates most, if not all, of the test instruments needed to diagnose driveability problems. Photo Sequence 4 shows some of the test modes of a typical engine analyzer. Most engine analyzers have the following diagnostic tools: a compression gauge, an exhaust analyzer, pressure and vacuum gauges, a voltmeter, an ohmmeter, a vacuum pump, an ammeter, a tachometer, an oscilloscope, a timing light and probe, and, of course, a scan tool

With an engine analyzer, you can perform tests on the battery, starting system, charging system, primary and secondary ignition circuits, electronic control systems, fuel system, emissions system, and engine assembly. The analyzer is connected to these systems by a variety of leads, inductive clamps, probes, and connectors. The data received from these connections is processed by several computers within the analyzer.

Most engine analyzers have both manual and automatic test modes. In the manual modes, any single test, such as cylinder compression or generator output, can be performed. The manual test mode is useful when there is little need to do a complete test. The automatic test mode is useful when looking at general performance. When the automatic test mode is selected, specific tests are automatically performed in a specific sequence.

The analyzer may compare all the test results to the vehicle manufacturer's specifications. When the test series is completed, the analyzer prints a report indicating those readings that were not within specifications. Many analyzers also provide diagnostic assistance for the problems indicated by the readings that were not within specifications. Vehicle specifications may be updated simply by obtaining a new disk from the equipment manufacturer.

A phone modem is contained in some engine analyzers to provide networking capabilities. This phone modem allows the technician to unload all technical reports and pattern reports of a specific problem vehicle to off-location technical support teams.

Fuel Pressure Gauge

A fuel **pressure gauge** is essential for diagnosing fuel injection systems. These systems rely on very high fuel pressures, from 35 to 70 psi. A drop in fuel pressure reduces the amount of fuel delivered to the injectors and results in a lean air/fuel mixture.

A fuel pressure gauge is used to check the discharge pressure of fuel pumps, the regulated pressure of fuel injection systems, and injector pressure drop. This test can identify faulty pumps, regulators, or injectors and can identify restrictions present in the fuel delivery system. Restrictions are typically caused by a dirty fuel filter, collapsed hoses, or damaged fuel lines.

Some fuel pressure gauges also have a valve and outlet hose for testing fuel pump discharge volume. The manufacturer's specification for discharge volume is given as a number of pints or liters of fuel that should be delivered in a certain number of seconds.

CAUTION!

While testing fuel pressure, be careful not to spill gasoline. Gasoline spills may cause explosions and fires, resulting in serious personal injury and property damage.

Injector Balance Tester

The injector balance tester **(Figure 6–19)** is used to test the injectors in a port fuel injected engine for proper

Figure 6-19 A fuel injection balance tester.

PHOTO SEQUENCE 4

Diagnosing Engine, Ignition, Electrical, and Fuel Systems with an Engine Analyzer

P4-1 *Connect the analyzer leads and hoses to the engine, according to the directions given by the tester's manufacturer.*

P4-2 *With the engine at normal operating temperature, enter the necessary information regarding the vehicle being tested in the analyzer.*

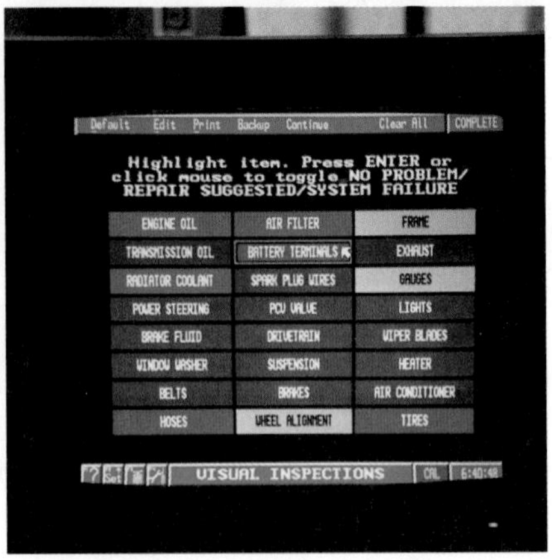

P4-3 *Perform a visual inspection of the vehicle according to the menu on the tester. Then enter the results into the analyzer.*

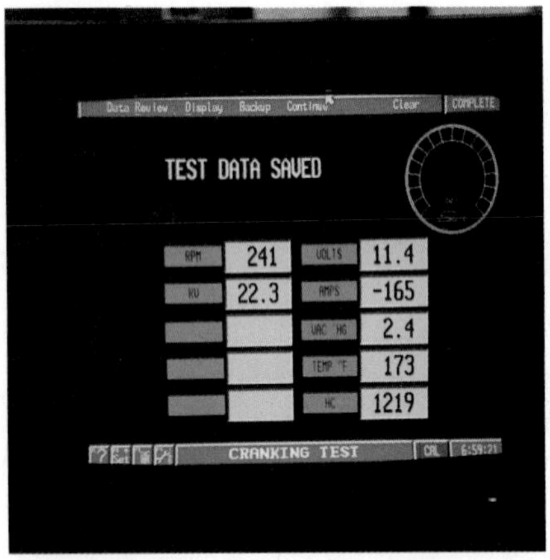

P4-4 *Perform battery and cranking tests, and observe the results on the screen.*

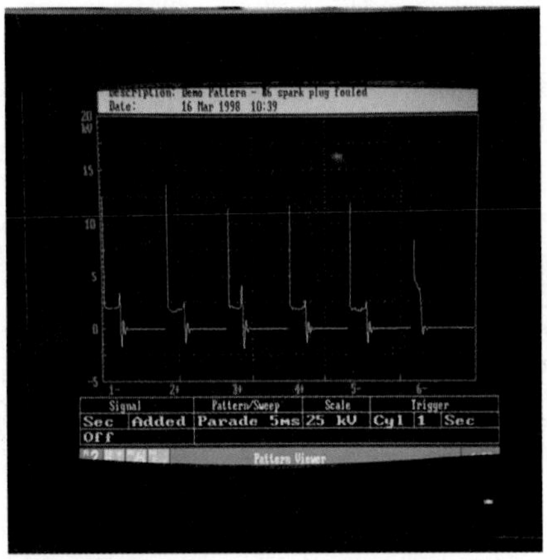

P4-5 *Perform primary ignition circuit tests and check secondary kV, and observe the results on the screen.*

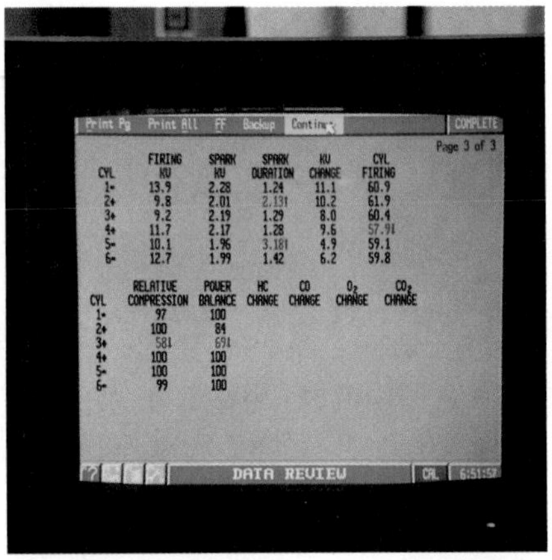

P4-6 *Perform a cylinder performance test and observe the results on the analyzer screen. Look for imbalance problems.*

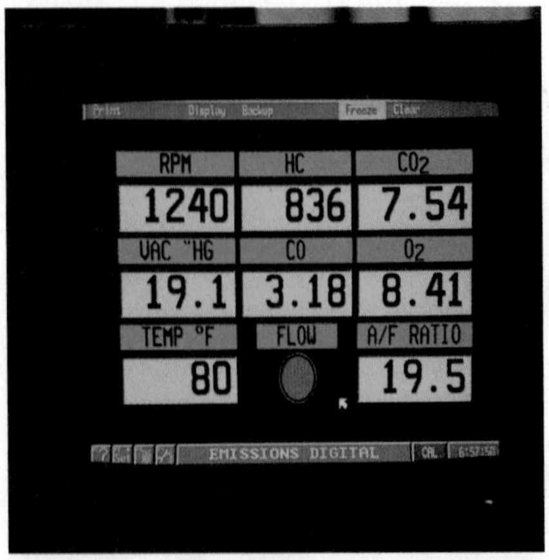

P4-7 *Perform a complete exhaust gas analysis and observe the results. Determine if previously noted faults would cause the readings.*

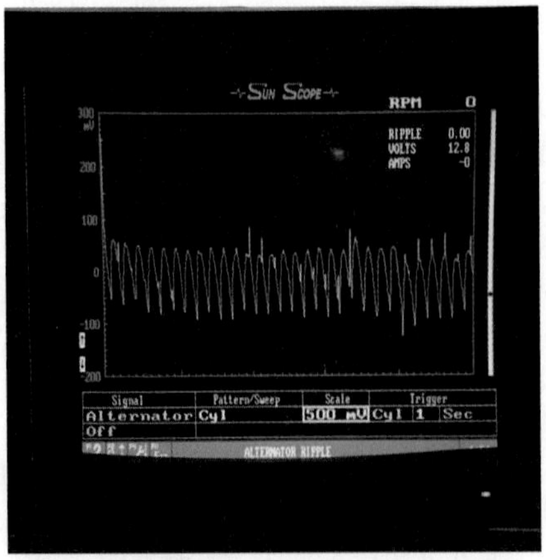

P4-8 *Display the charging system's waveform and check the generator's condition.*

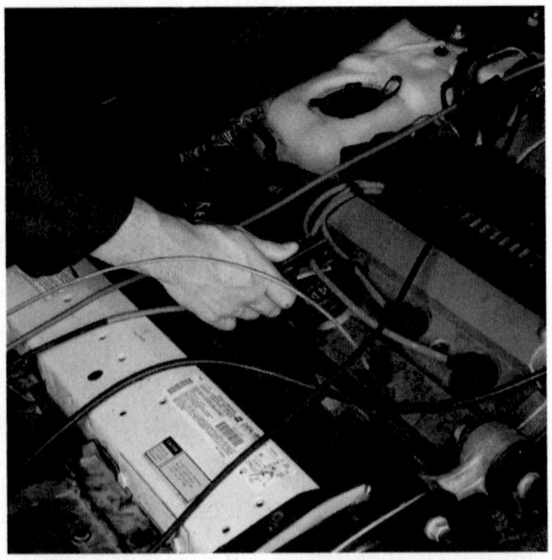

P4-9 *Turn off the engine and disconnect the analyzer leads and hoses. Summarize the results, including all that is good and all that needs further testing.*

operation. A fuel pressure gauge is also used during the injector balance test. The injector balance tester contains a timing circuit, and some injector balance testers have an off-on switch. A pair of leads on the tester must be connected to the battery with the correct polarity. The injector terminals are disconnected, and a second double lead on the tester is attached to the injector terminals.

The fuel pressure gauge is connected to the Schrader valve on the fuel rail, and the ignition switch should be cycled two or three times until the specified fuel pressure is indicated on the pressure gauge. When the tester push button is depressed, the tester energizes the injector winding for a specific length of time, and the technician records the pressure decrease on the fuel pressure gauge. This procedure is repeated on each injector.

If the pressure drops very little, or if there is no pressure drop, the injector's orifice is restricted or the injector is faulty. If there is an excessive amount of pressure drop, the injector plunger is sticking open. Sticking injector plungers may result in a rich air/fuel mixture.

WARNING!

Electronic fuel injection systems are pressurized, and these systems require depressurizing prior to fuel pressure testing and other service procedures.

Injector Circuit Test Light

A special test light called a **noid light** can be used to determine if a fuel injector is receiving its proper voltage pulse from the computer. The wiring harness connector is disconnected from the injector and the noid light is plugged into the connector. After disabling the ignition to prevent starting, the engine is turned over by the starter motor. The noid light will flash rapidly if the voltage signal is present. No flash usually indicates an open in the power feed or ground circuit to the injector.

Fuel Injector Cleaners

Fuel injectors spray a certain amount of fuel into the intake system. If the fuel pressure is low, not enough fuel will be sprayed. Low pressure may also occur if the fuel injector is dirty. Normally, clogged injectors are the result of inconsistencies in gasoline detergent levels and the high sulfur content of gasoline. When these sensitive fuel injectors become partially clogged, fuel flow is restricted. Spray patterns are altered, causing poor performance and reduced fuel economy.

The solution to a sulfated and/or plugged fuel injector is to clean it, not replace it. There are two kinds of fuel injector cleaners. One is a pressure tank. A mixture of solvent and unleaded gasoline is placed in the tank, following the manufacturer's instructions for mixing, quantity, and safe handling. The vehicle's fuel pump is disabled and, on some vehicles, the fuel line must be blocked between the pressure regulator and the return line. Then, the hose on the pressure tank is connected to the service port in the fuel system. The in-line valve is then partially opened and the engine is started. It should run at approximately 2,000 rpm for about 10 minutes to clean the injectors thoroughly.

An alternative to the pressure tank is a pressurized canister **(Figure 6–20)** in which the solvent solution is premixed. Use of the canister-type cleaner is similar to this procedure but does not require mixing or pumping. The canister is connected to the injection system's servicing fitting, and the valve on the canister is opened. The engine is started and allowed to run until it dies. Then, the canister is discarded.

Fuel Line Tools

Many vehicles are equipped with quick-connect line couplers. These work well to seal the connection but are nearly impossible to disconnect if the correct tools are not used. There is a variety of quick-connect fittings and tools.

Pinch-Off Pliers

The need to pinch off a rubber hose is common during diagnostics and service. Special pliers are designed to do this without damaging the hose **(Figure 6–21)**. These pliers are much like vise-grip pliers in that they hold their position until they are released. The jaws of the pliers are

Figure 6–20 A fuel injector cleaner using the shop's compressed air. *Courtesy of OTC Tool and Equipment, Division of SPX Corporation*

Figure 6-21 Pinch-off pliers closing off a vacuum hose.

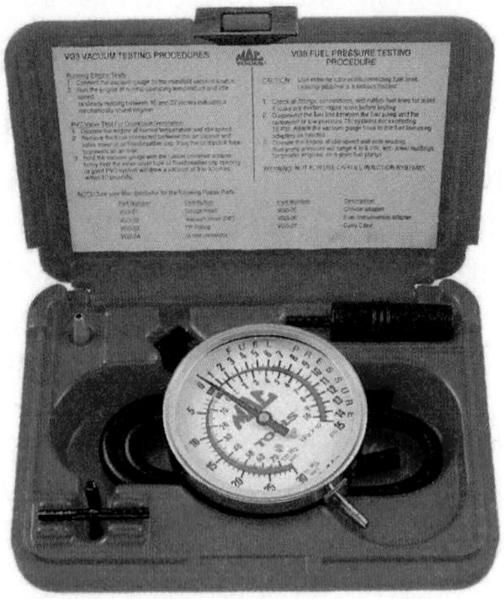

Figure 6-22 A vacuum/pressure tester used to measure engine vacuum. *Courtesy of Mac Tools*

flat and close in a parallel motion. Both of these features prevent damage to the hose.

Vacuum Gauge

Measuring intake manifold vacuum is another way to diagnose the condition of an engine. Manifold vacuum is tested with a vacuum gauge. **Vacuum** is formed on a piston's intake stroke. As the piston moves down, it lowers the pressure of the air in the cylinder—if the cylinder is sealed. This lower cylinder pressure is called engine vacuum. If there is a leak, atmospheric pressure will force air into the cylinder and the resultant pressure will not be as low. The reason atmospheric pressure enters is simply that whenever there is a low and high pressure, the high pressure always moves toward the low pressure. Vacuum is measured in inches of mercury (in. Hg) and in kilopascals (kPa).

To measure vacuum, a flexible hose on the vacuum gauge **(Figure 6–22)** is connected to a source of manifold vacuum, either on the manifold or at a point below the throttle plates. The test is made with the engine cranking or running. A good vacuum reading is typically at least 16 in. Hg. However, a reading of 15 to 20 in. Hg (50 to 65 kPa) is normally acceptable. Since the intake stroke of each cylinder occurs at a different time, the production of vacuum occurs in pulses. If the amount of vacuum produced by each cylinder is the same, the vacuum gauge will show a steady reading. If one or more cylinders are producing different amounts of vacuum, the gauge will show a fluctuating reading.

Vacuum Pump

There are many vacuum operated devices and vacuum switches on cars. These devices use engine vacuum to cause a mechanical action or to switch something on or off. The tool used to test vacuum-actuated components is the vacuum pump **(Figure 6–23)**. There are two types of vacuum pumps: an electrical-operated pump and a hand-held pump. The hand-held pump is most often used for diagnostics. A hand-held vacuum pump consists of a hand pump, a vacuum gauge, and a length of rubber hose used to attach the pump to the component being tested. Tests with the vacuum pump can usually be performed without removing the component from the vehicle.

When the handles of the pump are squeezed together, a piston inside the pump body draws air out of the component being tested. The partial vacuum created by the pump is registered on the pump's vacuum gauge. While forming a vacuum in a component, watch the action of

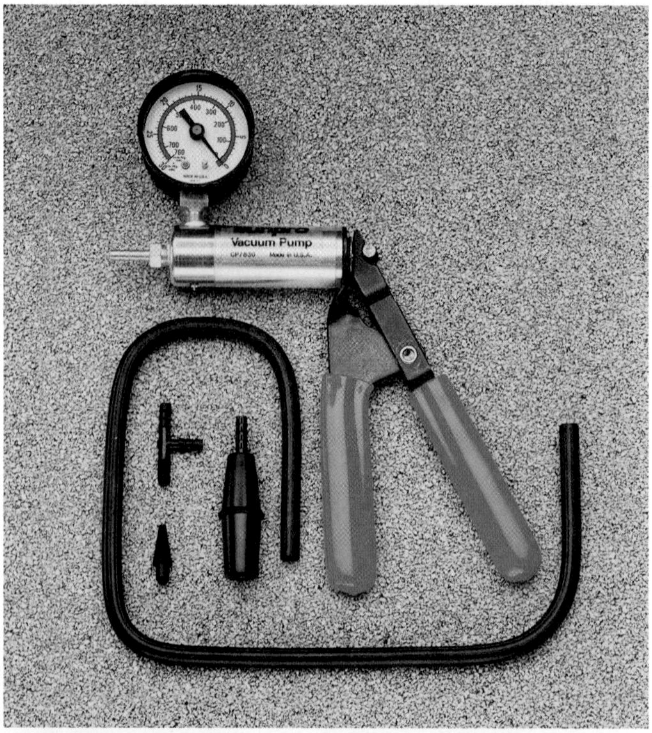

Figure 6-23 A typical vacuum pump with accessories. *Courtesy of Actron Manufacturing Company*

the component. The vacuum level needed to actuate a given component should be compared to the specifications given in the factory service manual.

The vacuum pump is also used to locate vacuum leaks by connecting the vacuum pump to a suspect vacuum hose or component and applying vacuum. If the needle on the vacuum gauge begins to drop after the vacuum is applied, a leak exists somewhere in the system.

Vacuum Leak Detector

A vacuum or compression leak might be revealed by a compression check, a cylinder leakage test, or a manifold vacuum test. However, finding the location of the leak can often be very difficult.

A simple, but time-consuming way to find leaks in a vacuum system is to check each component and vacuum hose with a vacuum pump. Simply apply vacuum to the suspected area and watch the gauge for any loss of vacuum. A good vacuum component holds the vacuum applied to it.

Another method of leak detection is done by using an ultrasonic leak detector **(Figure 6–24)**. Air rushing through a vacuum leak creates a high-frequency sound higher than the range of human hearing. An ultrasonic leak detector is designed to hear the frequencies of the leak. When the tool is passed over a leak, the detector responds to the high-frequency sound by emitting a warning beep. Some detectors also have a series of light emitting diodes (LEDs) that light up as the frequencies are received. The closer the detector is moved to the leak, the more LEDs light up or the faster the beeping occurs, allowing the technician to zero in on the leak. An ultrasonic leak detector can sense leaks as small as 1/500 inch and accurately locate the leak to within 1/16 inch.

An ultrasonic leak detector can also be used to detect the source of compression leaks, bearing wear, and electrical arcing. It can also be used to diagnose fuel injector operation.

Tachometer

A **tachometer** is used to measure engine speed. Like other meters, tachometers are available in analog and digital types. Digital meters are the most common. Tachometers are connected to the ignition system to monitor ignition pulses, which are then converted to engine speed by the meter.

Several types of inductive pickup tachometers that simplify rpm testing are available. An inductive tachometer simply clamps over the number 1 spark plug wire. The digital display gives the engine rpm, based on the magnetic pulses created by the secondary voltage in the wire. This type of tachometer is suitable for distributorless ignition systems.

Timing Light

A **timing light (Figure 6–25)** is essential for checking the ignition timing in relation to crankshaft position. Two

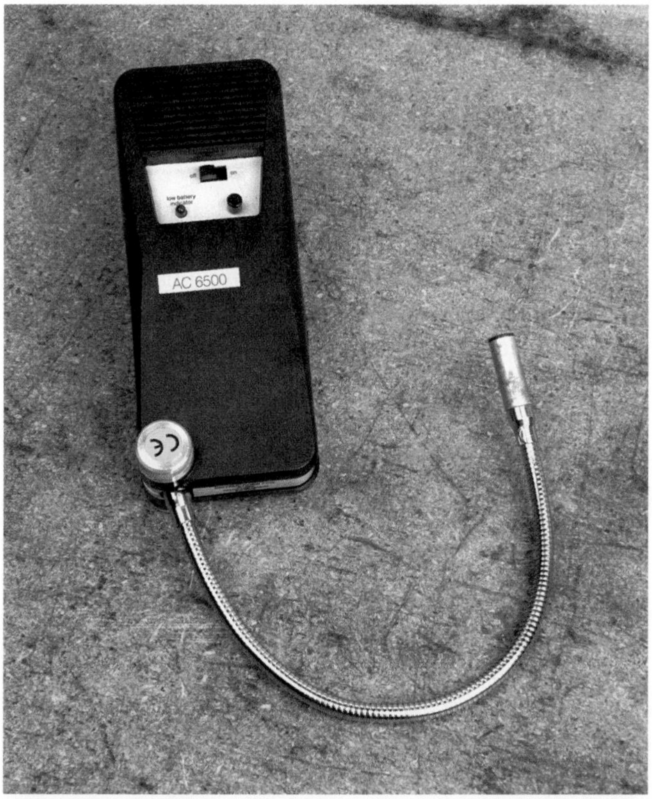

Figure 6-24 An ultrasonic vacuum leak detector.

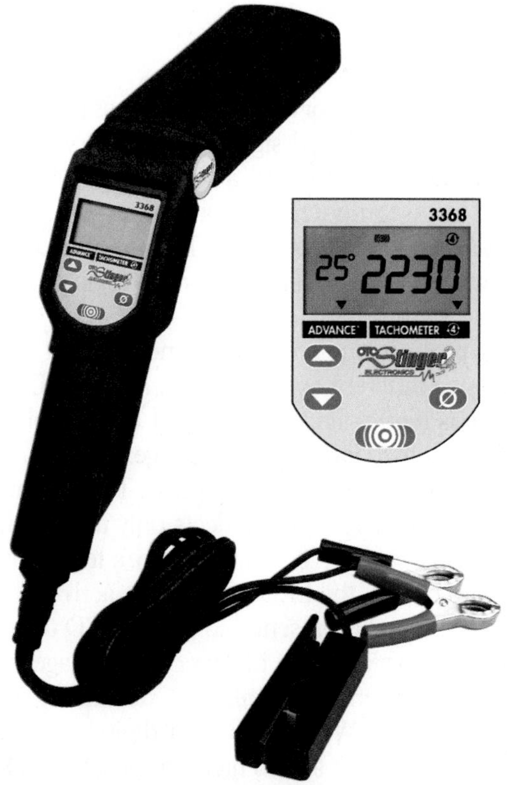

Figure 6-25 A digital timing light. *Courtesy of SPX-OTC Automotive Electronic Diagnostic Tools*

leads on the timing light must be connected to the battery terminals with the correct polarity. Most timing lights have an inductive clamp that fits over the number 1 spark plug wire. Older timing lights have a lead that goes in series between the number 1 spark plug wire and the spark plug. A trigger on the timing light acts as an off-on switch. When the trigger is pulled with the engine running, the timing light emits a beam of light each time the spark plug fires.

Many timing lights have a timing advance knob that can be used to check spark advance. A more versatile advanced timing light is the digital timing light. This type of timing light electronically measures timing advance as the engine rpm is increased and displays timing advance on an LED display. This light flashes only when a trigger is squeezed. When the trigger is released, the LED displays engine rpm. This combined feature eliminates the need for a separate tachometer when setting timing.

Spark Tester

A **spark tester** is a fake spark plug. The tester is constructed like a spark plug but does not have a ground electrode. In place of the electrode there is a grounding clamp. Using test spark plugs is an easy way to determine if the ignition problem is caused by something in the primary or secondary circuit.

The spark tester is inserted in the spark plug end of an ignition cable. When the engine is cranked, a spark should be seen from the tester to a ground. Experience with these testers will also help you determine the intensity of the spark.

Logic Probes

In some circuits, pulsed or digital signals pass through the wires. These on-off digital signals either carry information or provide power to drive a component. Many sensors, used in a computer-control circuit, send digital information back to the computer. To check the continuity of the wires that carry digital signals, a logic probe can be used.

A **logic probe** is similar in appearance to a test light. It contains three different-colored LEDs. A red LED lights when there is high voltage at the point being probed. A green LED lights to indicate low voltage. A yellow LED indicates the presence of a voltage pulse. The logic probe is powered by the circuit and reflects only the activity at the point being probed. When the probe's test leads are attached to a circuit, the LEDs display the activity.

If a digital signal is present, the yellow LED turns on. When there is no signal, the LED is off. If voltage is present, the red or green LEDs will light, depending on the amount of voltage. When there is a digital signal and the voltage cycles from low to high, the yellow LED will be lit and the red and green LEDs will cycle, indicating a change in the voltage.

Figure 6–26 A heated oxygen sensor socket. *Courtesy of SPX-OTC Automotive Electronic Diagnostic Tools*

Sensor Tools

Oxygen sensors are replaced as part of the preventive maintenance program and when they are faulty. Because they are shaped much like a spark plug with wires or a connector coming out of the top, ordinary sockets do not fit well. For this reason, tool manufacturers provide special sockets for these sensors **(Figure 6–26)**.

Special sockets are also available for other sending units and sensors.

Static Strap

Because electronic components are sensitive to voltage, static electricity can destroy them. Static straps are available for technicians to wear while working on or around electronic components. These straps typically are worn around a wrist and connected to a known good ground on the vehicle. The straps send all static electricity to the ground of the vehicle, thereby eliminating the chance of this electricity going to the electronic components.

Pyrometers

The converter should be checked for its ability to convert CO and HC into CO_2 and water by doing a delta temperature test. To conduct this test, use a hand-held digital **pyrometer (Figure 6–27)**. By touching the pyrometer probe or placing it near to the exhaust pipe just ahead of and just behind the converter, there should be an increase of at least 100°F or 8% above the inlet temperature reading as the exhaust gases pass through the converter. If the outlet temperature is the same or lower, nothing is happening inside the converter. A pyrometer can also be used to measure the temperature of the coolant at various stages of its travel.

Spark Plug Sockets

Special sockets are available for the installation and removal of spark plugs **(Figure 6–28)**. These sockets are deep sockets with a hex nut drive at the end to allow a technician to turn them with a ratchet or an open-end wrench. The sockets are available in the common sizes of

Figure 6-27 A hand-held digital infrared temperature sensor. *Courtesy of SPX-OTC Automotive Electronic Diagnostic Tools*

spark plugs (5/8-inch and 13/16-inch) and have a 3/8-inch drive. The socket is built with a rubber sleeve that surrounds the insulator part of the spark plug to prevent cracking or other damage to the plug while it is being removed or installed.

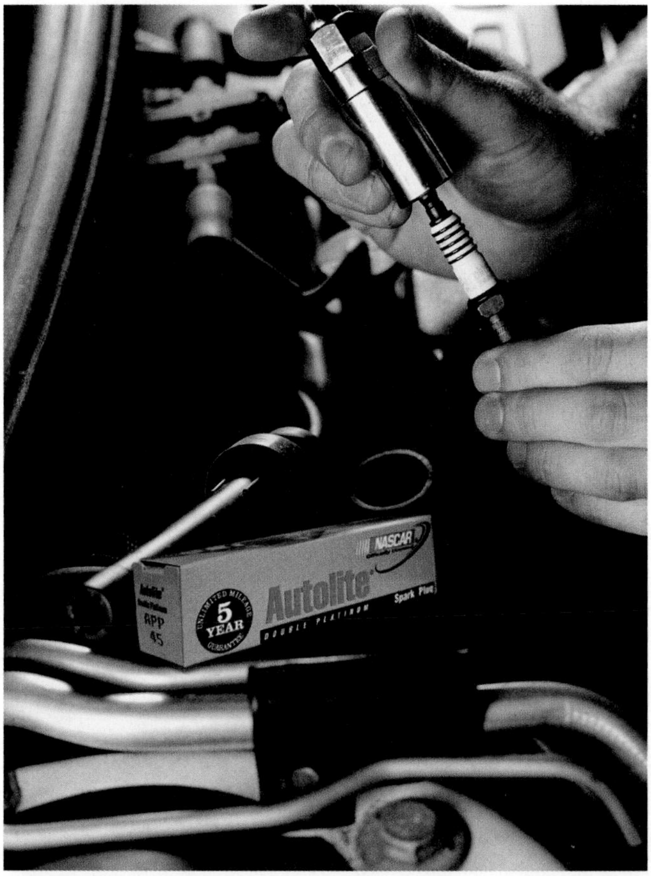

Figure 6-28 A spark plug socket. *Courtesy of Autolite*

Exhaust Analyzers

Federal laws require that new cars and light trucks meet specific emissions levels. State governments have also passed laws requiring that owners maintain their vehicles so that the emissions remain below an acceptable level. Most states require an annual emissions inspection to meet that goal. Many shops have an exhaust analyzer for inspection purposes.

Exhaust analyzers **(Figure 6–29)** are also very valuable diagnostic tools. By looking at the quality of an engine's exhaust, a technician is able to look at the effects of the combustion process. Any defect can cause a change in exhaust quality. The amount and type of change serves as the basis of diagnostic work.

Early emission analyzers measured the amount of hydrocarbons (HC) and carbon monoxides (CO). Exhaust analyzers normally measure HC in parts per million (ppm) or grams per mile (g/mi). CO is measured as a percent of the total exhaust.

Many of the emission control devices that have been added to vehicles over the past 30 years—especially catalytic converters—have decreased the amount of HC and CO in the exhaust. These devices alter the contents of the exhaust. Therefore, checking the HC and CO contents in the exhaust may not be a true indication of the operation of an engine.

The manufacturers of exhaust analyzers have altered their machines so that they can look at the efficiency of an engine in spite of the effectiveness of the emission controls. These machines are four-gas exhaust analyzers. In addition to measuring HC and CO levels, a four-gas exhaust analyzer also monitors carbon dioxide (CO_2) and oxygen (O_2) levels in the exhaust. The latter two gases are changed only slightly by the emission controls and

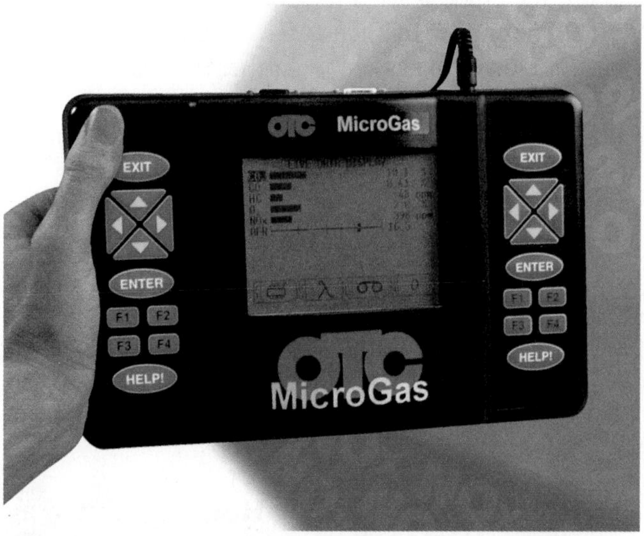

Figure 6-29 A MicroGas five-gas exhaust analyzer. *Courtesy of OTC Tool and Equipment, Division of SPX Corporation*

therefore can be used to check engine efficiency. Many exhaust analyzers are also available that measure a fifth gas, oxides of nitrogen (NO_x).

By measuring NO_x, CO_2, and O_2, in addition to HC and CO, a technician gets a better look at the efficiency of the engine. Keep in mind an exhaust analyzer is an excellent diagnostic tool and is not just for comparing emissions levels against standards. There is a desired relationship among the five gases. Any deviation from this relationship can be used to diagnose a driveability problem.

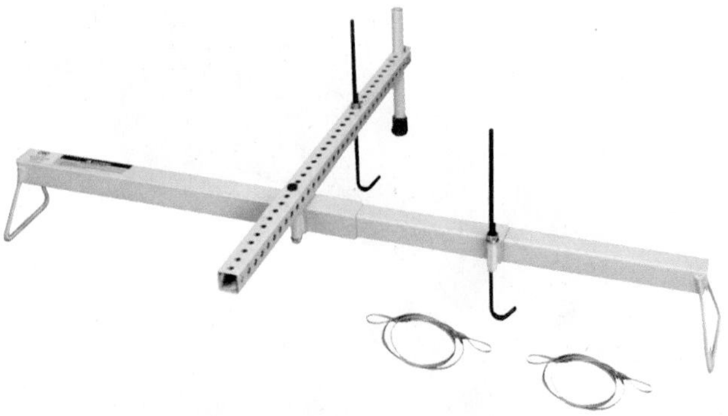

Figure 6-30 An engine support is used on many FWD vehicles to hold the engine in place while the transaxle is removed. *Courtesy of SPX-OTC Automotive Electronic Diagnostic Tools*

TRANSMISSION AND DRIVELINE TOOLS

The repair and diagnostic tools for manual and automatic transmissions, as well as those required for driveline service, are discussed in the following paragraphs. This discussion does not cover all of the tools you may need; rather, these tools are the most commonly used by the service industry. Details of when and how to use these tools are covered in Sections 5 and 6 of this book.

Transaxle Removal and Installation Equipment

The removal and replacement (R&R) of transaxles mounted to transversely mounted engines may require different tools than those needed to remove a transmission from a RWD vehicle. Make sure you follow the manufacturer's instructions before attempting to remove the engine and/or transaxle from a FWD vehicle.

To remove the engine and transmission from under the vehicle, the vehicle must be raised. A crane and/or support fixture is used to hold the engine and transaxle assembly in place while the assembly is being readied for removal. When everything is set for removal of the assembly, the crane is used to lower the assembly onto a cradle. The cradle is similar to a hydraulic floor jack and is used to lower the assembly further so it can be rolled out from under the vehicle. The transaxle can be separated from the engine once it has been removed from the vehicle.

When the transaxle is removed as a single unit, the engine must be supported while it is in the vehicle before, during, and after transaxle removal. Special fixtures **(Figure 6–30)** mount to the vehicle's upper frame or suspension parts. These supports have a bracket that is attached to the engine. With the bracket in place, the engine's weight is on the support fixture and the transmission can be removed.

Transmission/Transaxle Holding Fixtures

Special holding fixtures should be used to support the transmission or transaxle after it has been removed from the vehicle. These holding fixtures may be stand-alone units or may be bench mounted. They allow the transmission to be easily repositioned during repair work.

Transmission Jack

A transmission jack **(Figure 6–31)** is designed to help you while removing a transmission from under the vehicle. The weight of the transmission makes it difficult and un-

Figure 6-31 A typical transmission jack.

safe to remove it without much assistance and/or a transmission jack. These jacks fit under the transmission and are usually equipped with hold-down chains, which are used to secure the transmission to the jack. The transmission's weight rests on the jack's saddle.

Transmission jacks are available in two basic styles. One is used when the vehicle is raised by a hydraulic jack and sitting on jack stands. The other style is used when the vehicle is raised on a lift.

Axle Pullers

Axle pullers are used to pull rear axles in RWD vehicles. Most rear axle pullers are slide hammer-type.

Special Tool Sets

Vehicle manufacturers and specialty tool companies work closely together to design and manufacture special tools required to repair transmissions. Most of these special tools are listed in the appropriate service manuals and are part of each manufacturer's "essential tool kit."

Clutch Alignment Tool

To keep the clutch disc centered on the flywheel while assembling the clutch, a clutch alignment tool is used. The tool is inserted through the input shaft opening of the pressure plate and is passed through the clutch disc. The tool then is inserted into the pilot bushing or bearing. The outside diameter (OD) of the alignment tool that goes into the pilot must be only slightly smaller than the ID of the pilot bushing. The OD of the tool that holds the disc in place must likewise be only slightly smaller than the ID of the disc's splined bore. The effectiveness of this tool depends on its diameter, so it is best to have various sizes of clutch alignment tools **(Figure 6–32)**.

Clutch Pilot Bearing/Bushing Puller/Installer

To remove and install a clutch pilot bearing or bushing, special tools are needed. These tools not only make the job easier but also prevent damage to the bore in the flywheel.

Figure 6-32 A clutch alignment tool set with various sizes of pilots, adapters, and alignment cones. *Courtesy of SPX-OTC Automotive Electronic Diagnostic Tools*

Figure 6-33 A universal joint bearing press with adapters. *Courtesy of Snap-on Tools Company*

Figure 6-34 An angle gauge used to check the angle of the drive shaft. *Courtesy of Snap-on Tools Company*

Universal Joint Tools

Although servicing universal joints can be done with hand tools and a vise, many technicians prefer the use of specifically designed tools. One such tool is a C-clamp modified to include a bore that allows the joint's caps to slide in while tightening the clamp over an assembled joint to remove it **(Figure 6–33)**. Other tools are the various drivers used with a press to press the joint in and out of its yoke.

Drive Shaft Angle Gauge

Critical to the durability of universal joints and vibration-free vehicle operation is the angle of the drive shaft. The angle of the drive shaft at the transmission should equal its angle at the drive axle. There are many ways to measure the angle; one way involves the use of an **inclinometer** or drive shaft angle gauge **(Figure 6–34)**.

Hydraulic Pressure Gauge Set

A common diagnostic tool for automatic transmissions is a hydraulic pressure gauge. A pressure gauge measures pressure in pounds per square inch (psi) and/or kilopascals (kPa). The gauge is normally part of a kit that contains various fittings and adapters.

SUSPENSION AND STEERING TOOLS

Suspension and steering repair and diagnostic tools as well as wheel alignment tools and equipment are discussed in the following paragraphs. This discussion does not cover all of the tools you may need; rather, these tools are the most commonly used by the service industry. Details of when and how to use these tools are covered in Section 7 of this book.

Tire Tread Depth Gauge

A tire tread depth gauge measures tire tread depth. This measurement should be taken at three or four locations around the tire's circumference to obtain an average tread depth. This gauge is used to determine the remaining life of a tire as well as for comparing wear of one tire to the other tires. It is also used when making tire warranty adjustments.

Power Steering Pressure Gauge

A power steering pressure gauge is used to test the power steering pump pressure. This test is also important when checking hydraulic boost brake systems. Because the power steering pump delivers extremely high pressure during this test, the recommended procedure in the vehicle manufacturer's service manual must be followed.

A pressure gauge with a shutoff valve is installed between the pump and the steering gear. Adapters are used to make good connections with the vehicle's power steering system.

Control Arm Bushing Tools

A variety of control arm bushing tools are available to remove and replace control arm bushings. Old bushings are pressed out of the control arm. A C-clamp tool can be used to remove the bushing. The C-clamp is installed over the bushing. An adapter is selected to fit on the bushing and push the bushing through the control arm. Turning the handle on the C-clamp pushes the bushing out of the control arm.

New bushings can be installed by driving or pressing them in place. Adapters are available for the C-clamp tool to install the new bushings. After the correct adapters are selected, position the bushing and tool on the control arm. Turning the C-clamp handle pushes the bushing into the control arm.

Tie-Rod End and Ball Joint Puller

Some car manufacturers recommend a tie-rod end and ball joint puller to remove tie-rod ends and pull ball joint studs from the steering knuckle. A tie-rod end remover is a safer and easier way of separating ball joints than a pickle fork **(Figure 6–35)**.

Ball joint removal and pressing tools are designed to remove and replace pressed-in ball joints on front suspension systems. Often these tools are used in conjunction with a hydraulic press. The size of the removal and pressing tool must match the size of the ball joint.

Some ball joints are riveted to the control arm and the rivets are drilled out for removal.

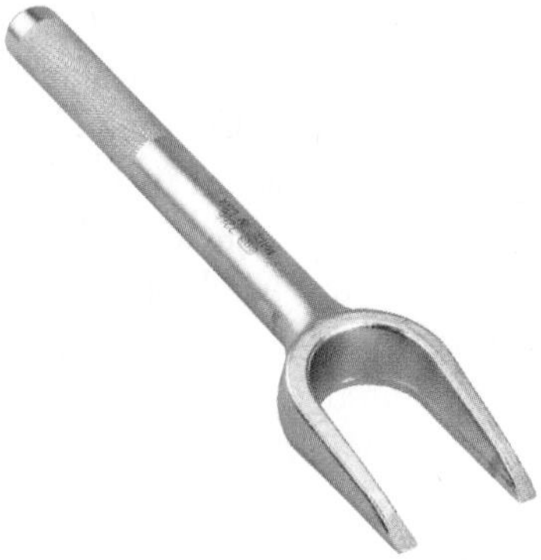

Figure 6-35 A separator tool pickle fork. *Courtesy of SPX-OTC Automotive Electronic Diagnostic Tools*

Front Bearing Hub Tool

Front bearing hub tools are designed to remove and install front wheel bearings on FWD cars. These bearing hub tools are usually designed for a specific make of vehicle and the correct tools must be used for each application. Failure to do so may result in damage to the steering knuckle or hub. Also, the use of the wrong tool will waste quite a bit of your time.

Pitman Arm Puller

A pitman arm puller is a heavy-duty puller designed to remove the pitman arm from the pitman shaft **(Figure 6–36)**. These pullers can also be used to separate tie-rod ends and ball joints.

Tie-Rod Sleeve-Adjusting Tool

A tie-rod sleeve-adjusting tool **(Figure 6–37)** is required to rotate the tie-rod sleeves and perform some front wheel adjustments. Never use anything except a tie-rod adjusting tool to adjust the tie-rod sleeves. Tools such as pipe wrenches will damage the sleeves.

Steering Column Special Tool Set

A wheel puller is used to remove the steering wheel from its shaft. Mount the puller over the wheel's hub after the horn button and air bag have been removed. Make sure

Figure 6-36 A pitman arm puller is designed to remove the pitman arm from the pitman shaft. *Courtesy of SPX-OTC Automotive Electronic Diagnostic Tools*

Figure 6–37 A tie-rod sleeve-adjusting tool. *Courtesy of SPX-OTC Automotive Electronic Diagnostic Tools*

Figure 6–38 A cradle-type coil spring compressor. *Courtesy of RTI Technologies, Inc.*

you exactly follow the recommendations for air bag module removal. Screw the bolts into the threaded bores in the steering wheel. Then tighten the puller's center bolt against the steering wheel shaft until the steering wheel is free.

Special tools are also required to service the lock mechanism and ignition switch.

Shock Absorber Tools

Often shock absorbers can be removed with regular hand tools, but there are times when special tools may be necessary. The shocks are under the vehicle and are subject to dirt and moisture, which may make it difficult to loosen the mounting nut from the stud of the shock. Wrenches are available to hold the stud while attempting to loosen the nut. There are also tools for pneumatic chisels that help to work off the nut.

Spring/Strut Compressor Tool

Many types of coil spring compressor tools are available to the automotive service industry. These tools are designed to compress the coil spring and hold it in the compressed position while removing the strut from the coil spring **(Figure 6–38)**, removing the spring from a short-long arm (SLA) suspension, or performing other suspension work. Various types of spring compressor tools are required on different types of front suspension systems.

One type of spring compressor uses a threaded compression rod that fits through two plates, an upper and lower ball nut, a thrust washer, and a forcing nut. The two plates are positioned at either end of the spring. The compression rod fits through the plates with a ball nut at either end. The upper ball nut is pinned to the rod. The thrust washer and forcing nut are threaded onto the end of the rod. Turning the forcing nut draws the two plates together and compresses the spring.

There is a tremendous amount of energy in a compressed coil spring. Never disconnect any suspension component that will suddenly release this tension. This action could result in serious personal injury and vehicle or property damage.

Power Steering Pump Pulley Special Tool Set

When a power steering pump pulley must be replaced, it should never be hammered off or on. Doing so will cause internal damage to the pump. Normally the pulley can be removed with a gear puller, although special pullers are available **(Figure 6–39)**. To install a pulley, a special tool is used to press the pulley on without a press or the need to drive the pulley in place.

Figure 6–39 A power steering pump pulley service kit. *Courtesy of SPX-OTC Automotive Electronic Diagnostic Tools*

Brake Pedal Depressor

A brake pedal depressor must be installed between the front seat and the brake pedal to apply the brakes while checking some front wheel alignment angles to prevent the vehicle from moving.

Wheel Alignment Equipment—Four Wheel

Many automotive shops are equipped with a computerized four-wheel alignment machine **(Figure 6–40)** that can check all front- and rear-wheel alignment angles quickly and accurately.

After vehicle information is keyed into the machine and the wheel units are installed, the machine must be compensated for wheel runout. When compensation is complete, alignment measurements are instantly displayed. Also displayed are the specifications for that vehicle. In addition to the normal alignment specifications, the screen may display asymmetric tolerances, different left- and right-side specifications, and cross specifications. (A difference is allowed between left and right sides.) Graphics and text on the screen show the technician where and how to make adjustments. As the adjustments are made on the vehicle, the technician can observe the center block slide toward the target. When the block aligns with the target, the adjustment is within half the specified tolerance.

Tire Changer

Tire changers are used to demount and mount tires. A wide variety of tire changers are available, and each one has somewhat different operating procedures. Always follow the procedure in the equipment operator's manual and the directions provided by your instructor.

Figure 6–40 A computerized four-wheel alignment setup. *Courtesy of RTI Technologies, Inc.*

Figure 6–41 A wheel balancer. *Courtesy of Snap-on Tools Company*

Wheel Balancer—Electronic Type

The most commonly used wheel balancer requires that the tire/wheel assembly be taken off and mounted on the balancer's spindle **(Figure 6–41)**. Weights are added to balance the tire/wheel assembly. The wheel assembly is rotated at high speed and the machine indicates the amount of weight to be added and the location where the weights should be placed.

Several electronic dynamic/static balancer units are available that permit balancing while the wheel and tire are on the vehicle. Often a strobe light flashes at the heavy point of the tire and wheel assembly.

Wheel Weight Pliers

Wheel weight pliers are actually combination tools designed to install and remove clip-on lead wheel weights **(Figure 6–42)**. The jaws of the pliers are designed to hook into a hole in the weight's bracket. The pliers are then moved toward the outside of the wheel and the weight is pried off. On one side of the pliers is a plastic hammer head used to tap the weights onto the rim.

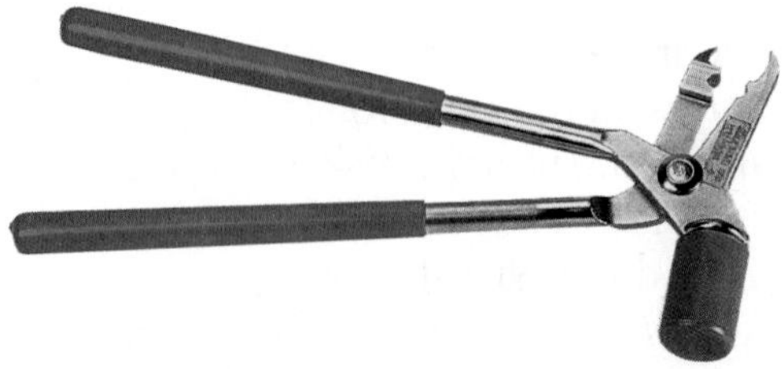

Figure 6–42 Wheel weight pliers. *Courtesy of Snap-on Tools Company*

BRAKE SYSTEM TOOLS

The repair and diagnostic tools for brake service are discussed in the following paragraphs. This discussion does not cover all of the tools you may need; rather, these tools are the most commonly used by the service industry. Details of when and how to use these tools are presented in Section 8 of this book.

Cleaning Equipment and Containment Systems

The following systems and methods are used to safely contain asbestos in the workplace and one or more of these should be used whenever you are doing brake work.

With negative-pressure enclosure and HEPA vacuum systems, cleaning and inspecting brake assemblies is performed inside a tightly sealed protective enclosure that covers and contains the entire brake assembly. The enclosure prevents the release of asbestos fibers into the air and is designed so that you can clearly see the work in progress. A vacuum pump and an HEPA filter keep the enclosure under negative pressure as work is done.

Low-pressure wet cleaning systems wash dirt from the brake assembly and catch the contaminated cleaning agent in a basin. The reservoir contains water with an organic solvent or wetting agent. To prevent any asbestos-containing brake dust from becoming airborne, the flow of liquid should be controlled so that the brake assembly is gently flooded.

Hold-down Spring and Return Spring Tools

Brake shoe return springs used on drum brakes are very strong and require special tools for removal and installation. Most return spring tools have special sockets and hooks to release and install the spring ends. Some are built like pliers **(Figure 6–43)**.

Hold-down springs for brake shoes are much lighter than return springs, and many such springs can be released and installed by hand. A hold-down spring tool **(Figure 6–44)** looks like a cross between a screwdriver and a nut driver. A specially shaped end grips and rotates the spring retaining washer.

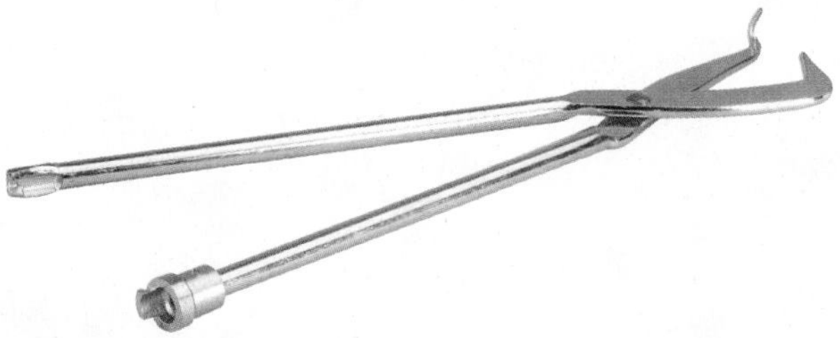

Figure 6–43 Brake spring pliers. *Courtesy of SPX-OTC Automotive Electronic Diagnostic Tools*

Figure 6–44 A hold-down spring compressor tool. *Courtesy of SPX-OTC Automotive Electronic Diagnostic Tools*

Boot Drivers, Rings, and Pliers

Dust boots attach between the caliper bodies and pistons of disc brakes to keep dirt and moisture out of the caliper bores. A special driver is used to install a dust boot with a metal ring that fits tightly on the caliper body. The circular driver is centered on the boot placed against the caliper and then hit with a hammer to drive the boot into place. Other kinds of dust boots fit into a groove in the caliper bore before the piston is installed. Special rings or pliers are then needed to expand the opening in the dust boot and let the piston slide through it for installation.

Caliper Piston Removal Tools

A caliper piston can usually be slid or twisted out of its bore by hand. Rust and corrosion (especially where road salt is used in the winter) can make piston removal difficult. One simple tool that will help with the job is a set of special pliers that grips the inside of the piston and lets you move it by hand with more force. These pliers work well on pistons that are only mildly stuck.

For a severely stuck caliper piston, a hydraulic piston remover can be used. This tool requires that the caliper be removed from the car and installed in a holding fixture. A hydraulic line is connected to the caliper inlet and a hand-operated pump is used to apply up to 1,000 psi of pressure to loosen the piston. Because of the danger of spraying brake fluid, always wear eye protection when using this equipment.

Drum Brake Adjusting Tools

Although almost all drum brakes built during the past 30 years have some kind of self-adjuster, the brake shoes still require an initial adjustment after they are installed. The star wheel adjusters of many drum brakes can be adjusted with a flat-blade screwdriver. Brake adjusting spoons **(Figure 6–45)** and wire hooks designed for this specific purpose can make the job faster and easier, however.

Figure 6–45 A drum brake adjustment tool. *Courtesy of Snap-on Tools Company*

Brake Cylinder Hones

Cylinder hones are used to clean light rust, corrosion, pits, and built-up residue from the bores of master cylinders, wheel cylinders, and calipers. A hone can be a very useful—sometimes necessary—tool when you have to overhaul a cylinder. A hone will not, however, save a cylinder with severe rust or corrosion.

The most common cylinder hones have two or three replaceable abrasive stones at the ends of spring-loaded arms. Spring tension usually is adjustable to maintain proper stone pressure against the cylinder walls. The other end of the hone is mounted in a drill motor for use, and the hone's flexible shaft lets the motor turn the hone properly without being precisely aligned with the cylinder bore.

Another kind of hone is the **brush** or **ball hone**. It has abrasive balls attached to flexible metal brushes that are, in turn, mounted on the hone's flexible shaft. In use, centrifugal force moves the abrasive balls outward against the cylinder walls; tension adjustment is not required. A brush hone provides a superior surface finish and is less likely to remove too much metal than a stone hone.

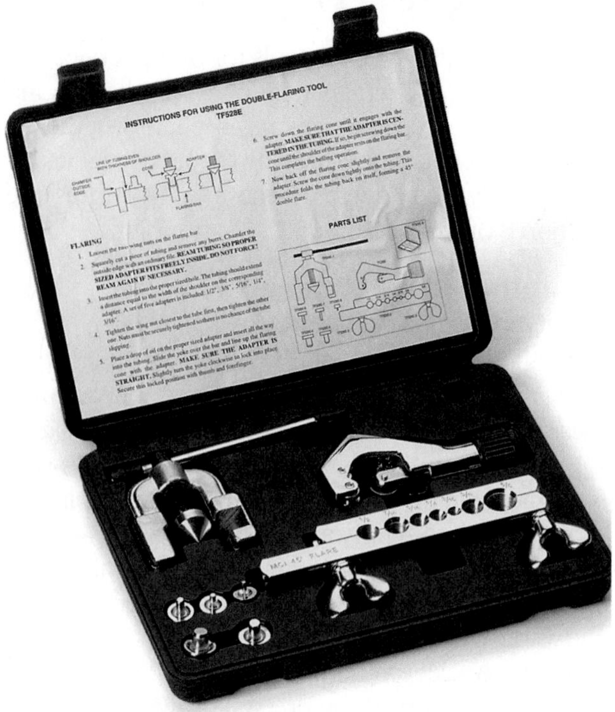

Figure 6–46 A typical tubing tool set. *Courtesy of Snap-on Tools Company*

Tubing Tools

The rigid brake lines, or pipes, of the hydraulic system are made of steel tubing to withstand high pressure and to resist damage from vibration, corrosion, and work hardening. Brake lines often can be purchased in preformed lengths to fit specific locations on specific vehicles. Straight brake lines can also be purchased in many lengths and several diameters and bent to fit specific vehicle locations. Even with prefabricated lines available, you probably will have many occasions to cut and bend steel lines and form flared ends for installation. The common tools **(Figure 6–46)** you should have are:

- A tubing cutter and reamer
- Tube benders
- A double flaring tool for SAE flares
- An International Standards Organization (ISO) flaring tool for European-style ISO flares

Figure 6–47 A brake disc micrometer. *Courtesy of Snap-on Tools Company*

Brake Disc Micrometer

A special micrometer should be used to check the thickness of a rotor accurately. A brake disc micrometer **(Figure 6–47)** has pointed anvils that allow the tip to fit into grooves worn on the rotor. This type of micrometer is read in the same way as other micrometers but is made with a range from 0.300 to 1.300 inches. Digital calipers are also used to measure disc brake thickness **(Figure 6–48)**.

Figure 6–48 A digital caliper for measuring brake disc thickness. *Courtesy of Bendix Brakes by Allied Signal*

Drum Micrometer

A drum micrometer is a single-purpose instrument used to measure the inside diameter of a brake drum. A drum micrometer has two movable arms on a shaft. One arm has a precision dial indicator; the other arm has an outside anvil that fits against the inside of the drum. In use, the arms are secured on the shaft by lock screws that fit into grooves every 1/8 inch (0.125) on the shaft. The dial indicator is graduated in 0.005-inch increments.

Metric drum micrometers work the same way except that the shaft is graduated in 1-cm major increments and the lock screws fit in notches every 2 mm.

Brake Shoe Adjusting Gauge (Calipers)

A brake shoe adjusting gauge is an inside-outside measuring device **(Figure 6–49)**. This gauge is often called a brake shoe caliper. During drum brake service, the inside part of the gauge is placed inside a newly surfaced drum and expanded to fit the drum diameter. The lock screw is then tightened and the gauge moved to the brake shoes installed on the backing plate. The brake shoes are then adjusted until the outside part of the gauge just slips over them. This action provides a rough adjustment of the brake shoes. Final adjustment must still be done after the drum is installed, but the brake shoe gauge makes the job faster.

Brake Lathes

Brake lathes are special power tools used only for brake service. They are used to turn and resurface brake rotors and drums. Turning involves cutting away very small amounts of metal to restore the surface of the rotor or drum. The traditional brake lathe is an assembly mounted on a stand or workbench. The bench lathe requires that the drum or rotor be removed from the vehicle and mounted on the lathe for service.

As the drum or rotor is turned on the lathe spindle, a carbide steel cutting bit is passed over the drum or rotor **friction** surface to remove a small amount of metal. The cutting bit is mounted rigidly on a lathe fixture for precise control as it passes across the friction surface.

An on-car lathe **(Figure 6–50)** is bolted to the vehicle suspension or mounted on a rigid stand to provide a stable mounting point for the cutting tool. The rotor may be turned by either the vehicle's engine or drive train (for a FWD vehicle) or by an electric motor and drive attachment on the lathe. As the rotor is turned, the lathe cutting tool is moved across both surfaces of the rotor to refinish it. An on-car lathe not only has the obvious advantage of speed, it also rotates the rotor on the vehicle wheel bearings and hub so that these sources of runout, or wobble, are compensated for during the refinishing operation.

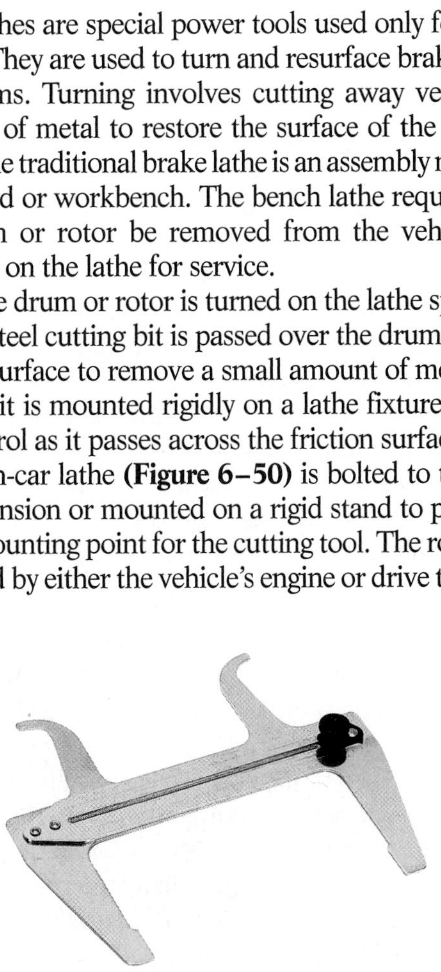

Figure 6-49 A drum brake shoe adjusting gauge. *Courtesy of Snap-on Tools Company*

Figure 6-50 An on-vehicle disc brake lathe. *Courtesy of RTI Technologies, Inc.*

Bleeder Screw Wrenches

Special bleeder screw wrenches often are used to open bleeder screws. Bleeder screw wrenches are small, 6-point box wrenches with strangely offset handles for access to bleeder screws in awkward locations. The 6-point box end grips the screw more securely than a 12-point box wrench can and avoids damage to the screw.

Pressure Bleeders

Removing the air from the closed hydraulic brake system is very important. This removal is done by bleeding the system. Bleeding can be done manually, with a vacuum pump, or with a pressure bleeder **(Figure 6–51)**. The latter is preferred because it is quick and very efficient.

Pressure bleeding is a fast and efficient way to bleed a brake system for two reasons. First, the master cylinder does not have to be refilled several times, and, second, the job can be done by one person. Pressurized brake fluid flows into the master cylinder and out through the brake lines, quickly forcing air out of the lines.

Other tools used in brake bleeding operations include a large rubber syringe, used to remove fluid from the master cylinder on some systems; master cylinder bleeder tubes, used to return fluid to the master cylinder reservoir from the outlet ports during bench bleeding; and assorted line and port plugs, used to close lines and valves temporarily during service and keep out dirt and moisture.

Figure 6–51 A diaphragm-type pressure bleeder. *Courtesy of Snap-on Tools Company*

Figure 6–52 An R-134a manifold gauge set. *Courtesy of Robinair, SPX Corporation*

HEATING AND AIR-CONDITIONING TOOLS

The repair and diagnostic tools for the heating, ventilation, and air-conditioning systems are discussed in the following paragraphs. This discussion does not cover all of the tools you may need; rather, these tools are the most commonly used by the service industry. Details of when and how to use these tools are covered in Section 9 of this book.

Manifold Gauge Set

A **manifold gauge set (Figure 6–52)** is used when discharging, charging, evacuating, and for diagnosing trouble in an A/C system. With the new legislation on handling refrigerants, all gauge sets are required to have a valve device to close off the end of the hose so that the fitting not in use is automatically shut.

The low-pressure gauge is graduated into pounds of pressure from 1 to 120 (with cushion to 250) in 1-pound graduations, and, in the opposite direction, in inches of vacuum from 0 to 30. This is the gauge that should always be used in checking pressure on the low-pressure side of the system. The high-pressure gauge is graduated from 0 to 500 pounds pressure in 10-pound graduations. This gauge is used for checking pressure on the high-pressure side of the system.

The gauge manifold is designed to control refrigerant flow. When the manifold test set is connected into the system, pressure is registered on both gauges at all times.

Because R-134a is not interchangeable with R-12, separate sets of hoses, gauges, and other equipment are required to service vehicles. All equipment used to service R-134a and R-12 systems must meet SAE standard J1991. The service hoses on the manifold gauge set must have manual or automatic backflow valves at the service port connector ends to prevent the refrigerant from being released into the atmosphere during connection and disconnection. Manifold gauge sets for R-134a can be identified by labels on the gauges and/or have a light blue color on the face of the gauges.

For identification purposes, R-134a service hoses must have a black stripe along their length and be clearly labeled. The low-pressure hose is blue with a black stripe. The high-pressure hose is red with black stripe and the center service hose is yellow with a black stripe. Service hoses for one type of refrigerant will not easily connect into the wrong system, as the fittings for an R-134a system are different than those used in an R-12 system.

Service Port Adapter Set

To connect a manifold gauge set to an A/C system, adapters are sometimes needed **(Figure 6–53)**. The high-side fitting on many vehicles with an R-12 system may require the use of a special adapter to connect the manifold gauge set to the service port. The service hoses of some manifold gauge sets are not equipped with a Schrader valve-depressing pin. Therefore, when connecting this type hose to a Schrader valve, an adapter must be used. The manifold and gauge sets for R-12 and R-134a are not interchangeable; therefore, there are no suitable or allowable adapters for using R-12 gauges on an R-1134a system or vice versa.

Figure 6–53 An A/C port adapter set. *Courtesy of SPX-OTC Automotive Electronic Diagnostic Tools*

Electronic Leak Detector

An electronic leak detector **(Figure 6–54)** is safe and effective and can be used with all types of refrigerants. A hand-held battery-operated electronic leak detector contains a test probe that is moved about 1 inch per second in areas of suspected leaks. Since refrigerant is heavier than air, the probe should be positioned below the test point. An alarm or a buzzer on the detector indicates the presence of a leak. On some models, a light flashes when refrigerant is detected.

Fluorescent Leak Tracer

To find a refrigerant leak using the fluorescent tracer system, first introduce a fluorescent dye into the air-conditioning system with a special infuser included with the detector equipment. Run the air conditioner for a few minutes, giving the tracer dye fluid time to circulate and penetrate. Wear the tracer protective goggles and scan the system with a black-light glow gun. Leaks in the system will shine under the black light as a luminous yellow-green **(Figure 6–55)**.

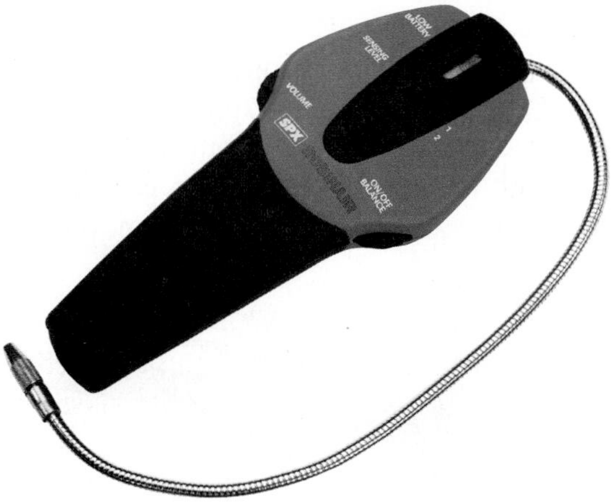

Figure 6–54 An electronic leak detector. *Courtesy of Robinair, SPX Corporation*

Figure 6–55 With a fluorescent tracer system, refrigerant leaks shine a luminous yellow-green under a black light. *Courtesy of Tracer Products*

Refrigerant Identifier

A refrigerant identifier **(Figure 6–56)** is used to identify the type of refrigerant present in a system. This test should be done before any service work. The tester is used to identify the purity and quality of the refrigerant sample taken from the system.

Refrigerant Charging Station

A charging station **(Figure 6–57)** removes, recycles, evacuates, and recharges an air-conditioning system. The proper amount of refrigerant is adjusted through controls on the station.

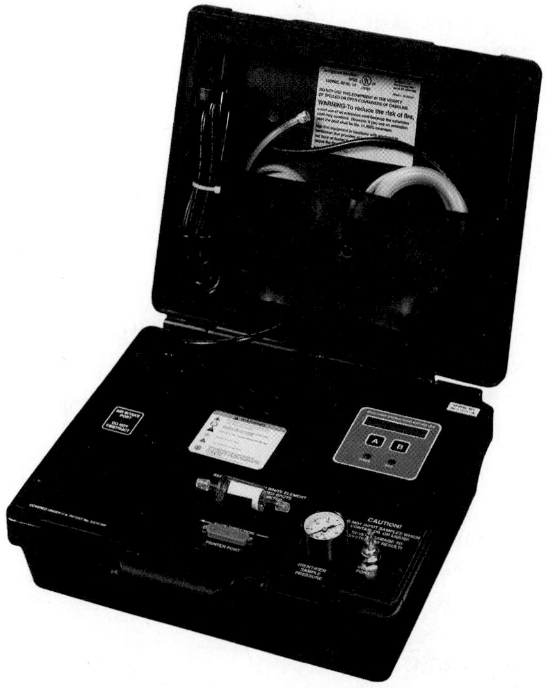

Figure 6–56 A refrigerant identifier. *Courtesy of RTI Technologies, Inc.*

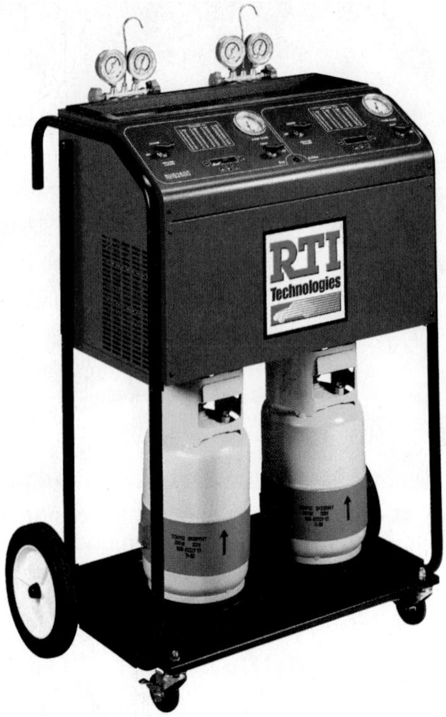

Figure 6-57 A dual (R-12 and R-143a) charging station. *Courtesy of RTI Technologies, Inc.*

A **charging cylinder** is designed to meter out a desired amount of refrigerant by weight. Compensation for temperature variations is accomplished by reading the pressure on the gauge of the cylinder and dialing the plastic shroud. The calibrated chart on the shroud contains corresponding pressure readings for the refrigerant being used.

Refrigerant Recovery and Recycling System

There are currently two types of refrigerant recycling machines, the single-pass and the multipass. Both have the ability to draw the refrigerant from the vehicle, filter and separate the oil from it, remove moisture and air from it, and store the refrigerant until it is reused.

In a single-pass system **(Figure 6–58)**, the refrigerant goes through each stage before being stored. In multipass systems, the refrigerant may go through all stages or some of the stages before being stored. Either system is acceptable if it has the underwriters' laboratory (UL)-approved label.

All recycled refrigerant must be safely stored in DOT CFR Title 49 or UL-approved containers. Containers specifically made for R-134a should be so marked. Before any container of recycled refrigerant can be used, it must be checked for noncondensable gases.

Thermometer

A digital readout or dial-type **thermometer (Figure 6–59)** is often used to measure the air temperature at the vent

Figure 6-58 A single-pass refrigerant recovery and recycling machine. *Courtesy of RTI Technologies, Inc.*

outlets, which indicates the overall performance of the system. The thermometer can also be used to check the temperature of refrigerant lines, hoses, and components while diagnosing a system. While doing the latter, an electronic pyrometer works best and is often used.

Compressor Tools

Although compressors are usually replaced when they are faulty, certain service procedures for them are standard practice. Most of these procedures focus on compressor clutch and shaft seal service and they require special tools. Clutch plate tools are required to gain access to the shaft seal. They are also needed to reinstall the clutch plate after service.

Many different tools are required to perform these services to a compressor. Typically to replace a shaft seal, you will need an adjustable or fixed spanner wrench **(Figure 6–60)**, clutch plate installer/remover, ceramic seal installer/remover, seal assembly installer/remover, seal seat

Figure 6-59 A dial-type thermometer used to check A/C operation. *Courtesy of Robinair, SPX Corporation*

Figure 6-60 An A/C clutch spanner wrench. *Courtesy of SPX-OTC Automotive Electronic Diagnostic Tools*

installer/remover, shaft seal protector, snap ring pliers, O-ring remover, and O-ring installer. Some of these tools are for a specific model compressor; others are universal fit or have interchangeable parts to allow them to work on a variety of compressors.

Hose and Fitting Tools

An A/C system is a closed system, meaning outside air should never enter the system and the refrigerant in the system should never exit to the outside. To maintain this closed system, special fittings and hoses are used. Often, special tools, such as the spring-lock coupling tool set, are required when servicing the system's fittings and hoses. Without this tool, it is impossible to separate the connector and not damage it.

KEY TERMS

Ammeter
Ball hone
Boring bar
Brush hone
Charging cylinder
Compression gauge
Connector pick
Continuity tester
Cylinder hone
Cylinder leakage tester
Dial bore gauge
Digital multimeter (DMM)
Dual-trace
Friction
Glaze breaker
Hydrometer
Inclinometer
kPa
Lab scope
Logic probe
Manifold gauge set
Multimeter
Noid light
Ohmmeter
Pressure gauge
psi
Pyrometer
Ridge reamer
Ring compressor
Ring expander
Ring groove cleaner
Scan tool
Spark tester
Stethoscope
Tachometer
Test light
Thermometer
Timing light
Torque angle gauge
Trace
Vacuum
Valve spring compressor
Valve spring tester
VAT
Voltmeter
Waveform

SUMMARY

- Common diagnostic tools used to check an engine and its related systems include a compression gauge, cylinder leakage tester, oil pressure gauge, stethoscope, dial bore indicator, valve spring tester, cooling system pressure tester, coolant hydrometer, engine analyzers, fuel pressure gauge, injector balance tester, injector circuit test light, vacuum gauge, vacuum pump, vacuum leak detector, spark tester, logic probes, pyrometers, and exhaust analyzers.
- Common tools used to service an engine and its related systems include transaxle removal and installation equipment, ridge reamer, ring compressor, ring expander, ring groove cleaner, cylinder deglazer, cylinder hone, boring bar, cam bearing driver set, V-blocks, valve and valve seat resurfacing equipment, valve guide repair tools, valve spring compressor, torque angle gauge, oil priming tool, a coolant recovery and recycle system, fuel injector cleaners, fuel line tools, pinch-off pliers, timing light, and spark plug sockets.
- Some of the common diagnostic tools for electronic and electrical systems include a test light, continuity tester, voltmeter, ohmmeter, ammeter, volt/ampere tester, DMM, lab scope, scan tools, and battery hydrometer.
- Common electrical and electronic system service tools include a computer memory saver, wire and terminal repair tools, headlight aimers, static straps, and sensor tools.
- Diagnostic tools for a vehicle's drive train include a drive shaft angle gauge and hydraulic pressure gauge set.
- Tools required to service the drive train include transaxle removal and installation equipment, transmission/transaxle holding fixtures, transmission jack, axle pullers, special tool sets, clutch alignment tool, clutch pilot bearing/bushing puller/installer, and universal joint tools.
- The various diagnostic tools used on a vehicle's running gear include a tire tread depth gauge, power steering pressure gauge, wheel alignment equipment, brake disc micrometer, and drum micrometer.
- Some of the common tools used to service a vehicle's running gear include control arm bushing tools, tie-rod end and ball joint pullers, front bearing hub tool, pitman arm puller, tie-rod sleeve adjusting tool, steering column special tool set, shock absorber tools, spring/strut compressor tool, power steering pump pulley special tool set, brake pedal depressor, tire changer, wheel balancer, wheel weight pliers, brake cleaning equipment and containment systems, hold-

down spring and return spring tools, boot drivers and pliers, caliper piston removal tools, drum brake adjusting tools, brake cylinder hones, tubing tools, brake shoe adjusting gauge, brake lathes, bleeder screw wrenches, and pressure bleeders.

- Common tools used to check a vehicle's heating and air-conditioning system include manifold gauge set, service port adapter set, electronic leak detector, fluorescent leak tracer, and thermometer.
- Tools used to service air-conditioning systems include a refrigerant identifier, refrigerant charging station, refrigerant recovery and recycling system, compressor tools, and hose and fitting tools.

TECH MANUAL

The following procedures are included in Chapter 6 of the Tech Manual that accompanies this book:

1. Checking the pressure in a cooling system.
2. Understanding DMM controls.
3. Identifying special drivetrain tools.
4. Dismounting and mounting a tire from a wheel.
5. Measuring brake rotor thickness.
6. Checking the clearance (air gap) of an air-conditioning compressor's clutch.

REVIEW QUESTIONS

1. What are the two types of test lights and how do they differ?

2. *True or False?* Knurling is used to repair worn valve guides by increasing the inside diameter of the guide.

3. Name the two basic types of compression gauges.

4. What tool is used to test engine manifold vacuum?

5. Which of the following statements is *not* true?

a. Exhaust analyzers allow a technician to look at the effects of the combustion process in an engine.

b. Most exhaust analyzers measure HC in parts per million or grams per mile.

c. CO is measured as a percent of the total exhaust.

d. Emission controls greatly alter HC and CO_2 emissions.

6. Technician A says a brake disc micrometer has pointed anvils. Technician B says a brake drum micrometer is a large inside micrometer that is read like any other micrometer. Who is correct?

a. Technician A **c.** Both A and B

b. Technician B **d.** Neither A nor B

7. Technician A says a pyrometer measures temperature. Technician B says a thermometer measures temperature. Who is correct?

a. Technician A **c.** Both A and B

b. Technician B **d.** Neither A nor B

8. *True or False?* A lab scope is a visual voltmeter that shows voltage over a period of time.

9. When using a voltmeter, Technician A connects it across the circuit being tested. Technician B connects the red lead of the voltmeter to the more positive side of the circuit. Who is correct?

a. Technician A **c.** Both A and B

b. Technician B **d.** Neither A nor B

10. Technician A uses a digital volt/ohmmeter to test voltage. Technician B uses the same tool to test resistance. Who is correct?

a. Technician A **c.** Both A and B

b. Technician B **d.** Neither A nor B

11. Technician A says a charging station removes, recycles, and recharges an A/C system. Technician B says a charging cylinder meters out the desired amount of refrigerant by weight. Who is correct?

a. Technician A **c.** Both A and B

b. Technician B **d.** Neither A nor B

12. Which of the following statements about manifold gauge sets is *not* true?

a. An adapter is required for using R-12 gauges on an R-1134a system.

b. A manifold gauge set is used when discharging, charging, and evacuating and for diagnosing trouble in an A/C system.

c. The gauge manifold is designed to control refrigerant flow. When the manifold test set is connected into the system, pressure is registered on both gauges at all times.

d. R-134a service hoses have a black stripe along their length, the low-pressure hose is blue, and the high-pressure hose is red.

13. *True or False?* A brake shoe adjusting gauge is an inside-outside measuring device used to initially adjust the expanse of brake shoes before the brake drum is installed.

14. When conducting an oil pressure test: Technician A says lower than normal pressure can be caused by a burned intake valve. Technician B says lower than normal oil pressure can be caused by excessive engine bearing clearances. Who is correct?

a. Technician A
b. Technician B
c. Both A and B
d. Neither A nor B

15. Which of the following conditions can be revealed by fuel pressure readings?

a. faulty fuel pump

b. restricted fuel injectors

c. restricted fuel delivery system

d. all of the above

16. Technician A uses an ultrasonic leak detector to check bearing wear. Technician B uses it to detect electrical arcing. Who is correct?

a. Technician A
b. Technician B
c. Both A and B
d. Neither A nor B

17. Technician A says a sulfated and plugged fuel injector is caused by electrical problems. Technician B says a sulfated and plugged fuel injector can be cleaned. Who is correct?

a. Technician A
b. Technician B
c. Both A and B
d. Neither A nor B

18. The tests conducted by a computer scan tool can be also done by some

a. fuel injector pulse testers

b. exhaust analyzers

c. engine analyzers

d. digital volt/ohmmeters

19. When using a fuel injector pulse tester, Technician A says that little or no pressure drop indicates a plugged or defective injector. Technician B says that no pressure drop indicates an overly rich condition. Who is correct?

a. Technician A
b. Technician B
c. Both A and B
d. Neither A nor B

20. It is much easier to test current using an ammeter equipped with a(n) __________.

a. continuity tester
b. carbon pile
c. inductive pickup
d. tachometer

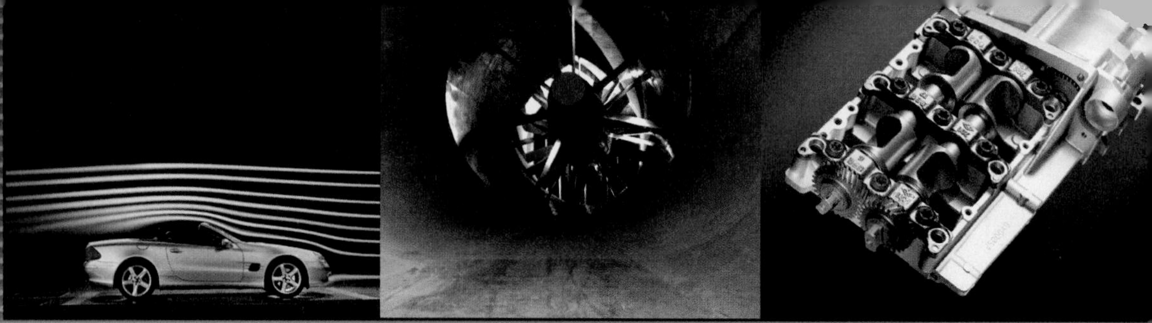

7 BASIC THEORIES AND MATH

OBJECTIVES

■ Describe how all matter exists. ■ Explain what energy is and how energy is converted. ■ Calculate the volume of a cylinder. ■ Explain the forces that influence the design and operation of an automobile. ■ Describe and apply Newton's laws of motion to an automobile. ■ Define friction and describe how it can be minimized. ■ Describe the various types of simple machines. ■ Explain the difference between torque and horsepower. ■ Differentiate between a vibration and a sound. ■ Explain Pascal's law and give examples of how it applies to an automobile. ■ Explain the behavior of gases. ■ Explain how heat affects matter. ■ Describe what is meant by the chemical properties of a substance. ■ Explain the difference between oxidation and reduction. ■ Describe the origin and practical applications of electromagnetism.

This chapter contains many of the subjects you have learned or will learn in other courses. This material is not intended to take the place of those other courses but, rather, to emphasize the knowledge you will need to gain employment and be successful in an automotive career. Many of the facts presented in this chapter are addressed later in greater detail according to the topic. Make sure you understand the contents of this chapter. If you have difficulty answering the chapter review questions, study the appropriate content in the chapter until you clearly understand and are able to answer the questions correctly.

MATTER

Matter is anything that occupies space. All matter exists as a gas, liquid, or solid. Gases and liquids are considered fluids because they move or flow easily and easily respond to pressure. A gas has neither shape nor volume of its own and tends to expand without limits. A liquid takes a shape and has volume. A solid is matter that does not flow.

Atoms and Molecules

All matter consists of countless tiny particles called **atoms** and **molecules**. An atom may be defined as the smallest particle of an element in which all the chemical characteristics of the element are present. Atoms are so small they cannot be seen with an electron microscope, which magnifies millions of times. A substance with only one type of atom is referred to as an **element**. More than 100 elements are known to exist at present and of the known elements, 92 occur naturally. The remaining elements have been manufactured in laboratories.

Small, positively charged particles called protons are located in the center, or nucleus, of each atom. In most atoms, neutrons are also located in the nucleus. Neutrons have no electrical charge, but they add weight to the atom. The positively charged protons tend to repel each other, and this repelling force could destroy the nucleus. The presence of the neutrons with the protons in the nucleus cancels the repelling action of the protons and keeps the nucleus together. Electrons are small, very light particles with a negative electrical charge. Electrons move in orbits around the nucleus of an atom.

A proton is approximately 1,840 times heavier than an electron, and this makes the electron much easier to move than a proton. Because the electrons are orbiting around the nucleus, centrifugal force tends to move the electrons away from the nucleus. However, the attraction between the positively charged protons and the negatively charged electrons holds the electrons in their orbits. The atoms of the different elements have different numbers of protons, electrons, and neutrons. Some of the lighter elements have the same number of protons and neutrons in the nucleus, but many of the heavier elements have more neutrons than protons.

The simplest atom is the hydrogen (H) atom, which has one proton in the nucleus and one electron orbiting

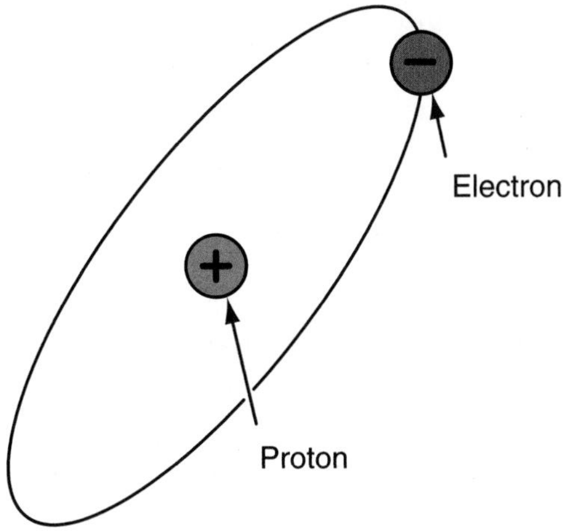

Figure 7–1 A hydrogen atom.

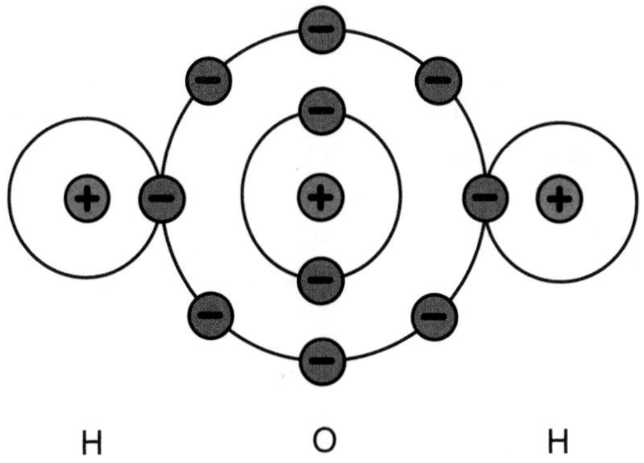

Figure 7–3 A molecule of water.

around the nucleus **(Figure 7–1)**. The nucleus of a copper (CU) atom contains 29 protons and 34 neutrons, while 29 electrons orbit in 4 different rings around the nucleus. Since 2, 8, and 18 electrons are the maximum number of electrons on the first 3 electron rings next to the nucleus, the fourth ring must have 1 electron **(Figure 7–2)**. The outer ring of an atom is called the valence ring, and the number of electrons on this ring determines the electrical characteristics of the element. Elements are listed on the atomic scale, or periodic table, according to their number of protons and electrons. For example, hydrogen is number 1 on this scale and copper is number 29.

For some elements, a single atom does not exist. An example of this is oxygen, which has a chemical symbol of O. Pure oxygen exists as pairs of oxygen atoms and has a symbol of O_2, which is a molecule of oxygen. A molecule is the smallest particle of an element or compound that can exist and still retain the characteristics of the element or compound. Some materials contain only one type of atom, whereas a compound may be described as a liquid, solid, or gas that contains two or more types of atoms. An oxygen atom readily combines with another oxygen atom or atoms of many other elements to form a compound. Many other atoms also have this characteristic.

Water is a compound that contains oxygen and hydrogen atoms. The chemical symbol for water is H_2O.

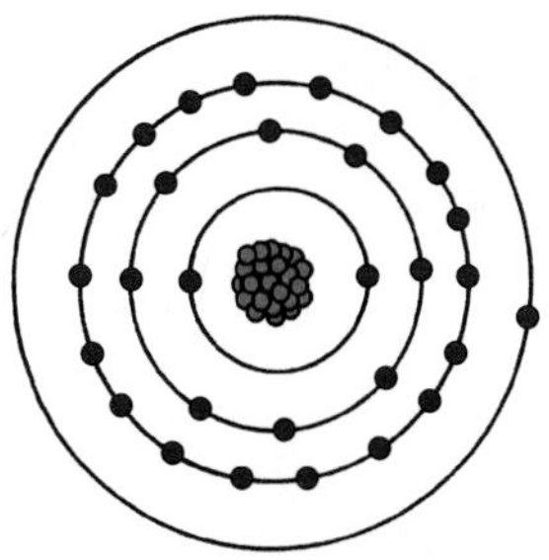
Figure 7–2 A copper atom.

This chemical symbol indicates that each molecule of water contains two atoms of hydrogen and one oxygen atom **(Figure 7–3)**.

States of Matter

The particles of a solid are held together in a rigid structure. When a solid dissolves into a liquid, its particles break away from this structure and mix evenly in the liquid, forming a **solution**. When heated, most liquids **evaporate**, which means that the atoms or molecules of which they are made break free from the body of the liquid to become gas particles. If all of the liquid in a solution has evaporated, the solid is left behind. The particles of the solid normally arrange in a structure called a crystal.

Absorption and Adsorption Not all solids dissolve in a liquid; rather, the liquid will be either absorbed or adsorbed. The action of a sponge serves as the best example of absorption. When a dry sponge is put into water, the water is absorbed by the sponge. The sponge does not dissolve; the water merely penetrates into the sponge and the sponge becomes filled with water. There is no change to the atomic structure of the sponge, nor does the structure of the water change. If we put a glass into water, the glass does not absorb the water. However, the glass still gets wet, as a thin layer of water adheres to the glass. This is adsorption. Materials that *absorb* fluids are **permeable** substances. **Impermeable** substances *adsorb* fluids. Some materials are impermeable to most fluids while others are impermeable to just a few.

ENERGY

Energy may be defined as the ability to do work. Because all matter consists of atoms and molecules that are in constant motion, all matter has energy. Energy is not matter, but it affects the behavior of matter. Everything that happens requires energy, and energy comes in many forms.

Each form of energy can change into other forms. However, the total amount of energy never changes; it can only be transferred from one form to another, not created or destroyed. This is known as the principle of the conservation of energy.

Engine efficiency is a measure of the relationship between the amount of energy put into the engine and the amount of available energy from the engine. Engine efficiency is expressed in a percentage. The formula for determining efficiency is: (output energy ÷ input energy) × 100.

Other aspects of the engine are expressed in efficiencies, including mechanical efficiency, volumetric efficiency, and thermal efficiency. They are expressed as a ratio of input (actual) to output (maximum or theoretical). Efficiencies are always less than 100%. The difference between the efficiency and 100% is the percentage lost during the process. For example, if there were 100 units of energy put into the engine and 28 units were used to power the vehicle, the efficiency would be equal to 28%. This would mean that 72% of the energy received was wasted or lost.

Kinetic and Potential Energy

When energy is released to do work, it is called **kinetic energy**. This type of energy may also be referred to as energy in motion. Stored energy may be called **potential energy**.

Many components and systems have potential energy and, at times, kinetic energy. The ignition system is a source for high electrical energy. The heart of the ignition system is the ignition coil, which has much potential energy. When it is time to fire a spark plug, that energy is released and becomes kinetic energy as it creates a spark across the gap of a spark plug.

Energy Conversion

Energy conversion occurs when one form of energy is changed to another form. Since energy is not always in the desired form, it must be converted to a form we can use. Some of the most common automotive energy conversions are listed here.

- *Chemical to Thermal Energy.* Chemical energy in gasoline or diesel fuel is converted to thermal energy when the fuel burns in the engine cylinders.
- *Chemical to Electrical Energy.* The chemical energy in a battery **(Figure 7–4)** is converted to electrical energy to power many of the accessories on an automobile.
- *Electrical to Mechanical Energy.* In the automobile, the battery supplies electrical energy to the starting motor, and this motor converts the electrical energy to mechanical energy to crank the engine.
- *Thermal to Mechanical Energy.* The thermal energy that results from the burning of the fuel in the engine is converted to mechanical energy, which is used to move the vehicle.
- *Mechanical to Electrical Energy.* The generator is driven by mechanical energy from the engine. The generator converts this energy to electrical energy, which powers the electrical accessories on the vehicle and recharges the battery.
- *Electrical to Radiant Energy.* Radiant energy is light energy. In the automobile, electrical energy is converted to thermal energy, which heats up the inside of light bulbs so they illuminate and release radiant energy.

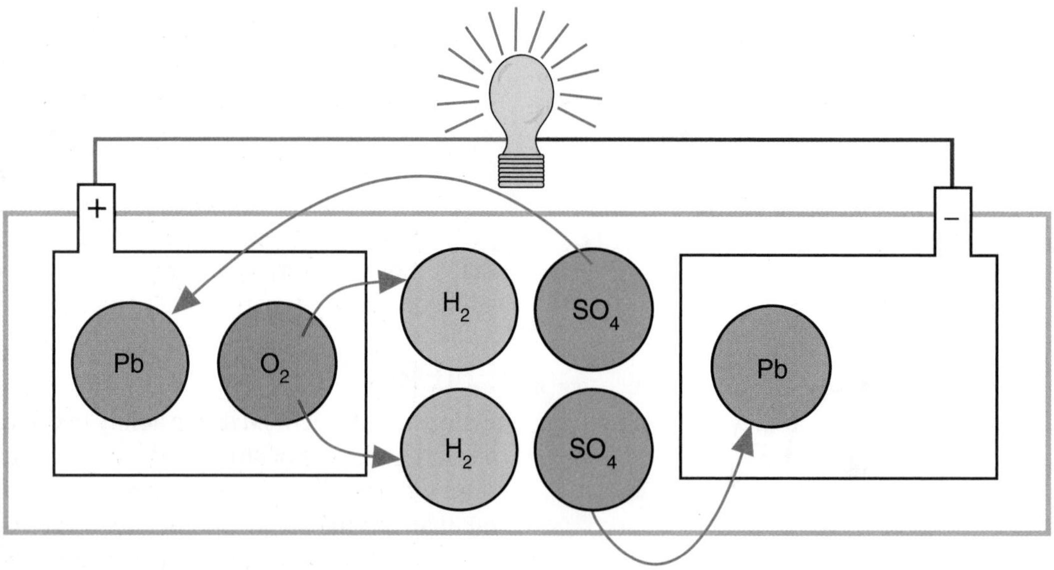

Figure 7-4 Chemical energy is converted to electrical energy in a battery.

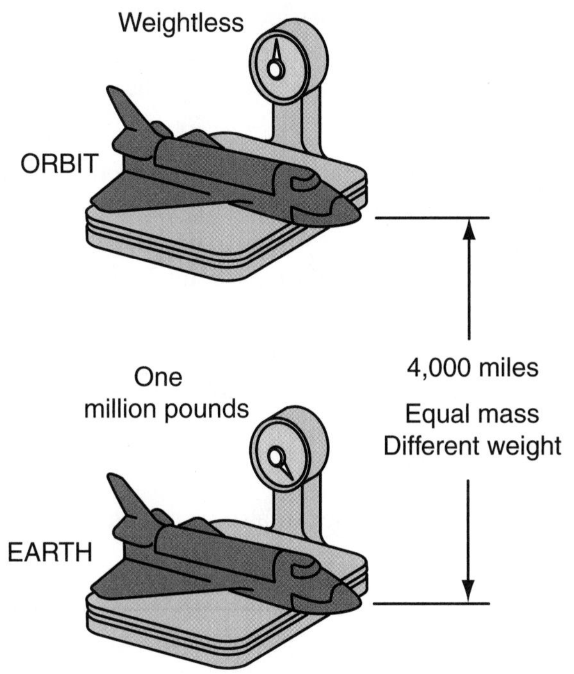

Figure 7-5 The difference in weight of a space shuttle on Earth and in space.

Mass and Weight

Mass is the amount of matter in an object. **Weight** is a force and is measured in pounds or kilograms. Gravitational force gives the mass its weight. As an example, a spacecraft can weigh 500 tons (one million pounds) here on earth where it is affected by the earth's gravitational pull. In outer space, beyond the earth's gravity and atmosphere, the spacecraft is nearly weightless **(Figure 7–5)**.

To convert kilograms into pounds, simply multiply the weight in kilograms by 2.2046. For example, if something weighs 5 kilograms it weighs 11.023 pounds (5 × 2.2046). To express the answer in pounds and ounces, convert the .023 pounds into ounces. Since there are 16 ounces in a pound, multiply 16 by 0.023 (16 × 0.023 = 0.368 ounces). Therefore 5 kilograms is equal to 11 pounds 0.368 ounces.

Size

The size of something is related to its mass. The size of an object defines how much space it occupies. Size dimensions are typically stated in terms of length, width, and height. Length is a measurement of how long something is from one end to another. Width is a measurement of how wide something is from one side to another. Obviously height is a measurement of the distance from something's bottom to its top. All three of these dimensions are measured in inches, feet, yards, and miles in the English system and meters in the metric system.

To convert a meter into feet, multiply the number of meters by 3.281. To convert the feet into inches, simply multiply the answer in feet by 12. To convert 0.01 millimeter into inches, begin by converting 0.01 millimeter into meters. Because 1 millimeter (mm) is equal to 0.001 meters, you need to multiply 0.01 by 0.001 (0.001 × 0.01 = 0.00001). Then multiply 0.00001 meters by 3.281 (0.00001 × 3.281 = 0.00003281 feet). Now convert feet into inches by multiplying by 12 (0.00003281 × 12 = 0.00039372 inches).

An easier way to do this conversion would be to use the conversion factor that states 1 millimeter is equal to 0.03937 inches. To use this conversion factor, multiply 0.01 millimeter by 0.03937 (0.01 × 0.03937 = 0.0003937 inches).

Sometimes distance measurements are made with a rule that has fractional rather than decimal increments. Most automotive specifications are given decimally; therefore, fractions need to be converted into decimals. It is also easier to add and subtract dimensions if they are expressed in decimal form rather than in fractions. For example, suppose you want to find the rolling circumference of a tire and you know the diameter of the tire to be 20⅜ inches. The distance around the tire is the circumference and it is equal to the diameter multiplied by a constant called pi (π). Pi is equal to approximately 3.14; therefore, the circumference of the tire is equal to 20⅜ inches multiplied by 3.14. This calculation is much easier if you convert the 20⅜ inches into a whole number and a decimal. To convert the ⅜ to a decimal, divide the 3 by 8 (3 ÷ 8 = 0.375). Therefore, the diameter of the tire is 20.375 inches. Now multiply the diameter by π (20.375 × 3.14 = 63.98). The circumference of the tire is nearly 64 inches.

VOLUME

Volume is also a measurement of size and is related to mass and weight. Volume is the amount of space occupied by an object in three dimensions: length, width, and height. For example, a pound of gold and a pound of feathers both have the same weight, but the pound of feathers occupies a much larger volume. In the English system, volume is measured in cubic inches, cubic feet, cubic yards, or gallons. The measurement for volume in the metric system is cubic centimeters or liters **(Figure 7–6)**.

The volume of a container is basically calculated by multiplying the measured length, width, and height of an object. For example, if a box has a length of 2 inches, a width of 3 inches, and a height of 4 inches, it has a volume of 24 cubic inches (2 × 3 × 4 = 24). Different shapes have different formulas for calculating volume but all consider the three basic dimensions of the object.

The volume of an engine's cylinders determines its size, expressed as displacement. This size does *not* reflect

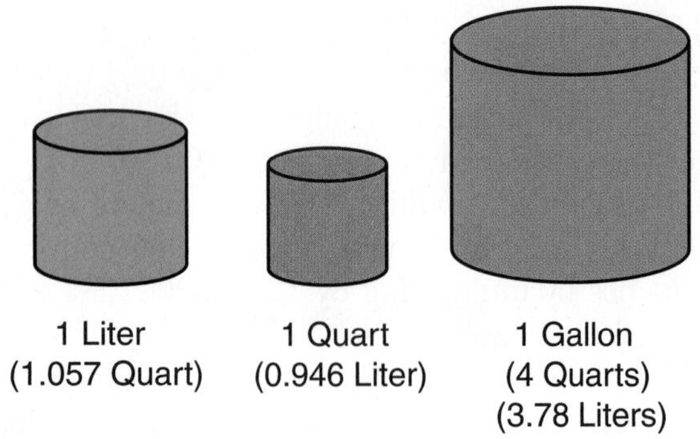

Figure 7-6 A comparison of metric and English units of volume.

the external (length, width, and height) of the engine. Cylinder **displacement** is the volume of a cylinder between when the cylinder's piston is at its lowest point of travel, or bottom dead center (BDC), and its highest point of travel (TDC). This is called the stroke of the piston **(Figure 7–7).** Displacement is usually measured in cubic inches, cubic centimeters, or liters. The total displacement of an engine (including all cylinders) is a rough indicator of its power output. Total displacement is the sum of displacements for all cylinders in an engine. Engine cubic inch displacement (CID) may be calculated as follows:

$$CID = \pi \times R^2 \times L \times N$$

where $\pi = 3.1416$
R = radius of the cylinder opening or the diameter (bore) ÷ 2
L = length of stroke
N = number of cylinders in the engine

Example: Calculate the CID of a six-cylinder engine with a 3.7-in. bore and 3.4-in. stroke.

$CID = 3.1416 \times 1.85^2 \times 3.4 \times 6$

$CID = 219.66$

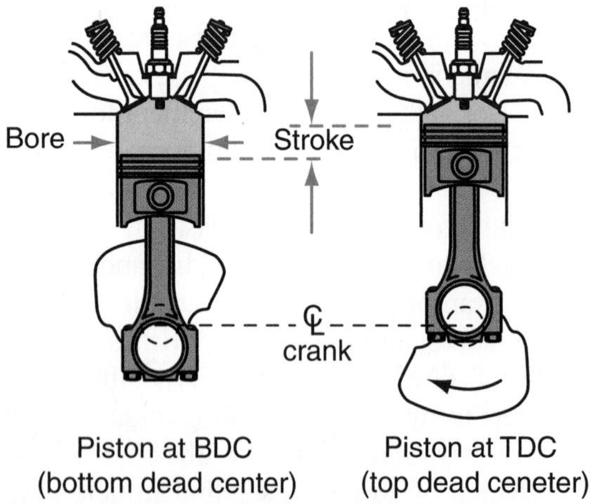

Figure 7-7 The bore and stroke of an engine.

Most of today's engines are described by their metric displacement. Cubic centimeters and liters are determined by using metric measurements in the displacement formula.

Example: Calculate the metric displacement of a four-cylinder engine with a 78.9-mm stroke and a 100-mm bore. Before you use the formula to find the displacement in cubic centimeters, you must convert the millimeter measurements to centimeters. 78.9 mm = 7.89 cm and 100 mm = 10 cm.

Displacement = $3.1416 \times 5^2 \times 7.89 \times 4$

Displacement = 2479 cubic centimeters (cc) or approximately 2.5 liters (L)

Engine displacement can also be calculated by using this formula:

$$0.7854 \times \text{Bore} \times \text{Bore} \times \text{Stroke} \times \text{Number of cylinders} = \text{Displacement}$$

Ratios

Often automotive features are expressed as ratios. A ratio expresses the relationship between two things. If something is twice as large as some other thing, there is a ratio of 2:1. Sometimes ratios are used to compare the movement of an object. For example, if a gear with a 2-inch diameter drives a gear with a 4-inch diameter, the ratio of the gears is 1:2.

The **compression ratio** of an engine expresses how much the air/fuel mixture will be compressed as the piston in a cylinder moves from BDC to TDC of the cylinder. The compression ratio is defined as the ratio of the volume in the cylinder above the piston when the piston is at the bottom of its travel to the volume in the cylinder above the piston when the piston is at its uppermost position **(Figure 7–8).** The formula for calculating the compression ratio is as follows:

volume above the piston at BDC ÷
volume above the piston at TDC

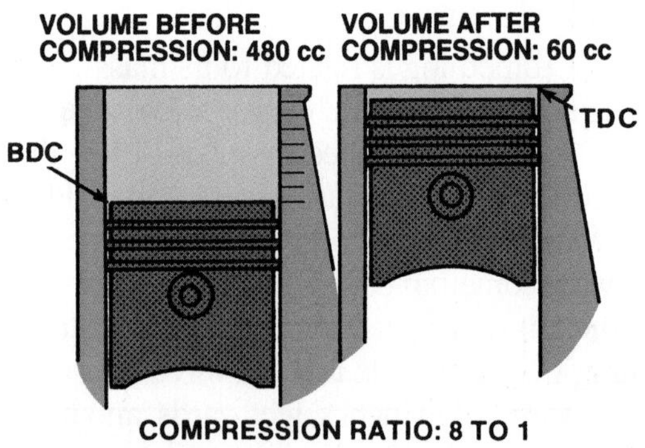

Figure 7-8 An engine's compression ratio indicates the amount the air/fuel mixture is compressed during the compression stroke.

or

total cylinder volume ÷
total combustion chamber volume

In many engines, the top of the piston is even or level with the top of the cylinder block at TDC. The combustion chamber is in the cavity in the cylinder head above the piston. This is modified slightly by the shape of the top of the piston. The volume of the combustion chamber must be added to each volume in the formula to get an accurate calculation of compression ratio.

Example: Calculate the compression ratio if the total piston displacement is 45 cubic inches and the combustion chamber volume is 5.5 cubic inches.

$45 + 5.5 \div 5.5 = 9.1$

Therefore, the compression ratio is 9.1 to 1 or 9.1:1

Proportions

Ratios can also be used to express the correct mixture for something. An example of this would be the amount of engine coolant that should be mixed with water before the engine's cooling system is refilled **(Figure 7–9)**. Typically specifications call for 50% coolant and 50% water, or a ratio of 1:1. This mixture allows for maximum hot and cold protection. To apply this ratio, suppose a cooling system has a capacity of 9.5 liters. Because most engine coolant is sold in gallon containers, to determine the amount of coolant that should be put in the system we must first convert the liter capacity to gallon capacity. One gallon equals 3.7854 liters, so we need to divide 9.5 liters by 3.7854 ($9.5 \div 3.7854 = 2.5097$). Now we know the total capacity of the cooling system is a little more than 2.5 gallons. To determine how much coolant or antifreeze to put in the system, we divide the total capacity by 2, which gives the quantity equal to 50% of the capacity ($2.5 \div 2 = 1.25$). Therefore, to obtain the correct mixture, $1\frac{1}{4}$ gallons of coolant should be mixed with $1\frac{1}{4}$ gallons of water.

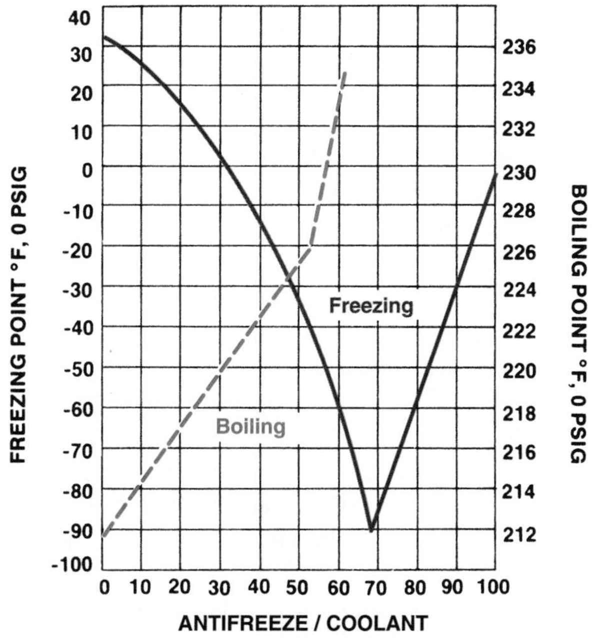

Figure 7-9 The relationship of the percentage of antifreeze to the freezing and boiling points of the engine's coolant.

FORCE

A **force** is a push or pull and can be large or small. Force can be applied to objects by direct contact or from a distance. Gravity and electromagnetism are examples of forces that are applied from a distance. Forces can be applied from any direction and with any intensity. For example, if a pulling force on an object is twice that of the pushing force, the object will be pulled at one-half the pulling force. When two or more forces are applied to an object, the combined force is called the resultant. The resultant is the sum of the size and direction of the forces. For example, when a mass is suspended by two lengths of wire, each wire should carry half the weight of the mass. If we move the attachment of the wires so they are at an angle to the mass, the wires now carry more force. The wires carry the force of the mass plus a force that pulls against the other wire.

Automotive Forces

When a vehicle is at rest, gravity exerts a downward force on the vehicle. The ground exerts an equal and opposite upward force and supports the vehicle. When the engine is running and its power output is transferred to the vehicle's drive wheels, the wheels exert a force against the ground in a horizontal direction. This force causes the vehicle to move but is opposed by the mass of the vehicle **(Figure 7–10)**. To move the vehicle faster, the force supplied by the wheels must increase beyond the opposing forces. As the vehicle moves faster, it pushes against the air as it travels. This push becomes a growing opposing force, and the force at the drive wheels must overcome

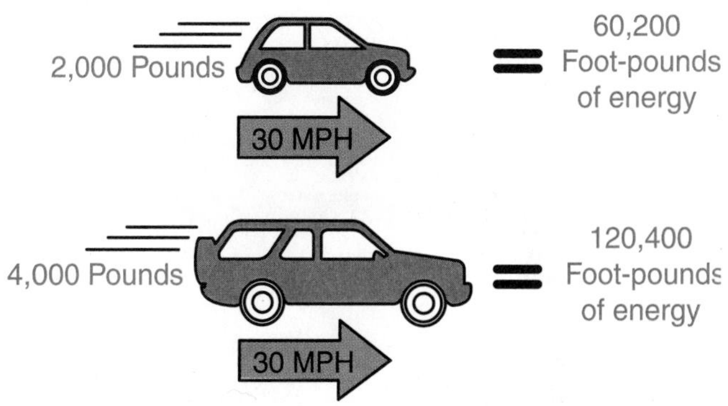

Figure 7-10 The amount of energy required to move a vehicle depends on its mass.

the force in order for the vehicle to increase speed. After the vehicle has achieved the desired speed, no additional force is required at the drive wheels.

Balanced and Unbalanced Force When the applied forces are balanced and there is no overall resultant force, the object is said to be in **equilibrium**. An object sitting on a solid flat surface is in equilibrium, because its weight is supported by the surface and there is no resultant force. If the surface is put on an angle, the object will tend to slide down the surface. If the surface is at a slight angle, the force will cause the object to slowly slide down the surface. If the surface is at a severe angle, the downward force will cause the object to quickly slide down the slope. In both cases, the surface is still supplying the force needed to support the object but the pull of gravity is greater and the resultant force causes the object to slide down the slope.

Turning Forces Forces can cause rotation as well as straight line motion. A force acting on an object that is free to rotate has a turning effect, or turning force. This force is equal to the size of the force multiplied by the distance of the force from the turning point around which it acts.

Forces on Tires and Wheels

If you roll a cone-shaped piece of metal on a smooth surface, the cone does not roll in a straight line. Rather, it moves toward the direction of the tilt on the cone. The weight of the cone is applied to the surface, but part of the weight at the large end of the cone is applied at an angle to the small end of the cone **(Figure 7–11)**. Riding a bicycle is another example. When you want to make a left turn, it is easier if you tilt the bicycle to the left. The reason for this action is that a tilted, rolling wheel tends to move in the direction of the tilt. Similarly, if a tire and wheel on an automobile are tilted, the tire and wheel will tend to move in the direction of the tilt. This principle is used in front wheel alignment.

Figure 7-11 A tire at an angle will roll in the same way as a cone would.

While riding a bicycle, the force applied to the bicycle is projected through the bicycle's front fork to the road surface by your weight. The centerline of the front fork is tilted rearward in relation to the vertical centerline of the wheel. When the handle bars are turned, the tire pivots on the vertical centerline of the wheel. Since the tire's pivot point is behind the front fork centerline where your weight is projected against the surface of the road, the front wheel tends to return to the straight-ahead position after a turn. The wheel also tends to remain in the straight-ahead position as the bicycle is driven. This principle of resultant forces is also the basis for the theories applied during front wheel alignment.

Centrifugal/Centripetal Forces

When an object moves in a circle, its direction is continuously changing, and all changes in direction require a force. The forces required to maintain circular motion are called **centripetal** and **centrifugal force**. The size of these forces depends on the size of the circle and the mass and speed of the object.

Centripetal force tends to pull the object toward the center of the circle, whereas centrifugal force tends to push the object away from the center. The centripetal force that keeps an object whirling around on the end of a string is caused by **tension** in the string. If the string breaks, there is no longer string tension and centripetal force and the object will fly off in a straight line because of the centrifugal force on it. Gravity is the centripetal force that keeps the planets orbiting around the sun. Without this centripetal force, the Earth would move in a straight line through space.

Wheel and Tire Balance

When the weight of a wheel and tire assembly is distributed equally around the center of wheel rotation, the wheel and tire have proper static balance. Being statically balanced, the wheel and tire assembly will not tend to rotate by itself, regardless of the wheel position. If the weight is not distributed equally, the wheel and tire assembly is statically unbalanced. As the wheel and tire rotate, centrifugal force acts on this static unbalance and causes the wheel to "tramp" or "hop" **(Figure 7–12)**.

Dynamic balance exists when the weight thrown to the sides of the tire and wheel assembly is equal when the assembly is rotating **(Figure 7–13)**. To illustrate this balance, assume we have a bar with a ball attached by string to both ends of the bar. If we cause the bar to rotate, the balls will turn with the bar and centripetal and centrifugal force will keep the balls in an orbit around the rotat-

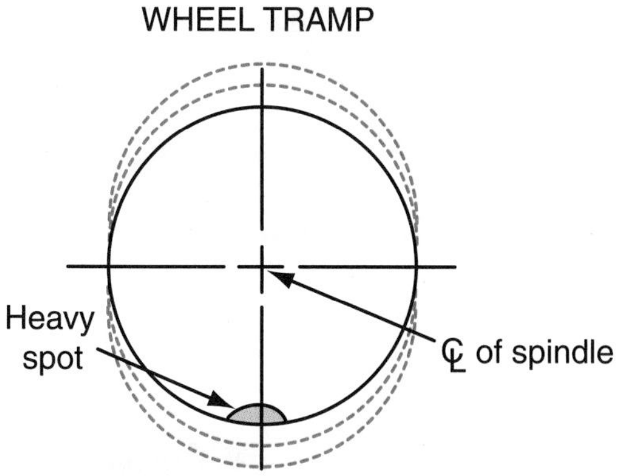

Figure 7-12 Wheel tramp is the result of a tire and wheel assembly being statically unbalanced.

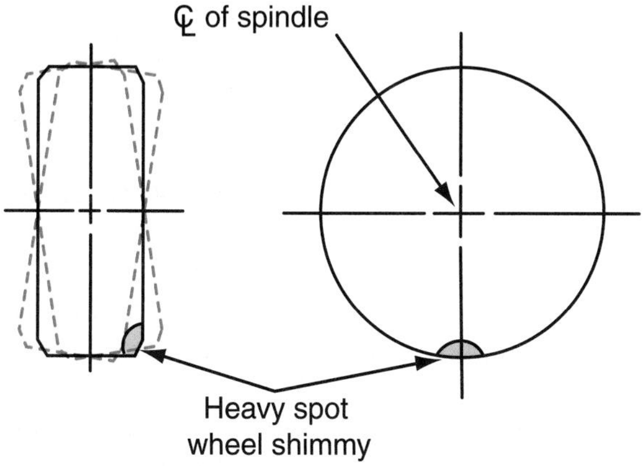

Figure 7-13 Dynamic imbalance causes wheel shimmy.

ing bar. If the two balls weigh the same and are at an equal distance from the bar, the bar will rotate smoothly. However, if one of the balls is heavier than the other the bar will wobble as it rotates. The greater the difference in the weight of the balls, the greater the wobble. The wobble can eventually destroy the mechanism used to rotate the bar.

Now, if we add some weight to the end of the bar that has the lighter ball, the weights and forces can be equalized and the wobble removed. This principle illustrates how we dynamically balance a wheel and tire assembly **(Figure 7–14)**.

When we think of all the parts of an automobile that rotate, it is easy to see why proper balance is important. Improper balance can cause premature wear or destruction of parts.

Pressure

Pressure is a force applied against an object and is measured in units of force per unit of surface area (pounds per square inch or kilograms per square centimeter). Mathematically, pressure is equal to the applied force divided by the area over which the force acts. Consider two

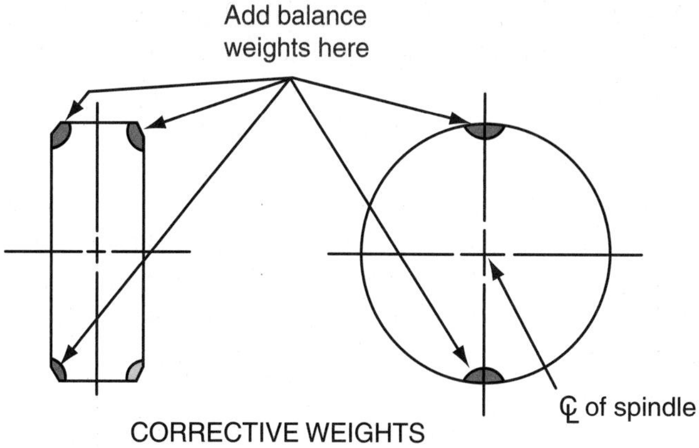

Figure 7-14 Adding a weight to counteract with the heavy spot of a tire and wheel assembly.

10 pound weights sitting on a table; one occupies an area of 1 square inch and the other an area of 4 square inches. The pressure exerted by the first weight would be 10 pounds per 1 square inch or 10 psi. The other weight, although it weighs the same, will exert only 2.5 psi (10 pounds per 4 square inches = 10 ÷ 4 = 2.5). This illustrates an important concept: A force acting over a large area exerts less **pressure** than the same force acting over a small area.

Because pressure is a force, all principles of force apply to pressure. If more than one pressure is applied to an object, the object will respond to the resultant force. Also, all matter (liquids, gases, and solids) tends to move from an area of high pressure to a low-pressure area.

TIME

The word **time** is used to mean many things. For our look into science, time will be defined as a measurement of the duration of an event that has happened, is happening, or will happen. Time is measured by the increments of a clock: seconds, minutes, and hours. Often an automotive technician is concerned with how long something occurs, such as the length of time a spark plug fires to cause combustion in an engine's cylinder. This time, called spark duration, is typically about 3 milliseconds (0.003 seconds) and is measured with a lab scope because it would be very difficult to measure that short a time period with a clock.

Technicians also monitor how many times a cycle is repeated within a period of time, such as a minute. A tachometer, which measures engine revolutions per minute, is an often-used diagnostic tool.

MOTION

When the forces on an object do not cancel each other out, they will change the motion of the object. The object's speed, direction of motion, or both will change. The greater the mass of an object, the greater the force needed to change its motion. This resistance to change in motion is called **inertia**. Inertia is the tendency of an object at rest to remain at rest, or the tendency of an object in motion to stay in motion. The inertia of an object at rest is called static inertia, whereas dynamic inertia refers to the inertia of an object in motion. Inertia exists in liquids, solids, and gases. When you push and move a parked vehicle, you overcome the static inertia of the vehicle. If you catch a ball in motion, you overcome the dynamic inertia of the ball.

When a force overcomes static inertia and moves an object, the object gains momentum. **Momentum** is the product of an object's weight times its speed. Momentum is a type of mechanical energy. An object loses momentum if another force overcomes the dynamic inertia of the moving object.

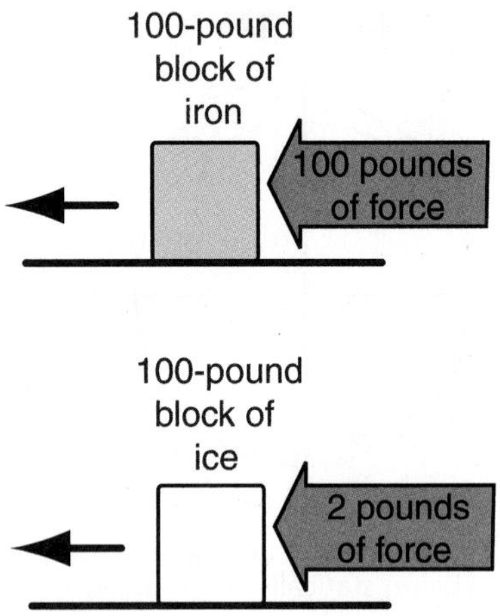

Figure 7–15 Sliding ice across a surface produces less friction than sliding a rougher material, such as iron, across a surface.

Rates

Speed is the distance an object travels in a set amount of time. It is calculated by dividing distance covered by time taken. We refer to the speed of a vehicle in terms of miles per hour (mph) or kilometers per hour (km/h). **Velocity** is the speed of an object in a particular direction. **Acceleration**, which only occurs when a force is applied, is the rate of increase in speed. Acceleration is calculated by dividing the change in speed by the time it took for that change. **Deceleration** is the reverse of acceleration, as it is the rate of a decrease in speed.

Newton's Laws of Motion

How forces change the motion of objects was first explained three centuries ago by Sir Isaac Newton. These explanations are known as Newton's laws. Newton's first law of motion is called the law of inertia. It states that an object at rest tends to remain at rest and an object in motion tends to remain in motion, unless some force acts on it. When a car is parked on a level street, it remains stationary unless it is driven or pushed.

Newton's second law states that when a force acts on an object the motion of the object will change. This change in motion is equal to the size of the force divided by the mass of the object on which it acts. Trucks have a greater mass than cars. Since a large mass requires a larger force to produce a given acceleration, a truck needs a larger engine than a car.

Newton's third law says that for every action there is an equal and opposite reaction. A practical application of this law occurs when the wheel on a vehicle strikes a bump in the road surface. This action drives the wheel and suspension upward with a certain force, and a specific amount of energy is stored in the spring. After this action occurs, the spring forces the wheel and suspension downward with a force equal to the initial upward force caused by the bump.

Friction

Friction is a force that slows or prevents motion in two moving objects or surfaces that touch. Friction may occur in solids, liquids, and gases. The joining or bonding of the atoms at each of the surfaces causes the friction. When you attempt to pull an object across a surface, the object will not move until these bonds have been overcome. Smooth surfaces produce little friction; therefore, only a small amount of force is needed to break the bonds between the atoms. Rougher surfaces produce a larger friction force because stronger bonds are made between the two surfaces **(Figure 7–15)**. To move an object over a rough surface, such as sandpaper, a great amount of force is required.

Friction is put to good use in disc brakes **(Figure 7–16)**. The friction force between the disc and brake pad slows down the rotation of the wheel, reducing the vehicle's speed. In doing so, it converts the kinetic energy of the vehicle into heat.

Lubrication Friction can be reduced in two main ways: by lubrication or by the use of rollers. The presence of oil or another fluid between two surfaces keeps the surfaces apart. Because fluids (liquids and gases) flow, they allow movement between surfaces. The fluid keeps the surfaces apart, allowing them to move smoothly past one another **(Figure 7–17)**.

Rollers Rollers placed between two surfaces keep the surfaces apart. An object placed on rollers will move smoothly if pushed or pulled. Rollers actually use friction to grip the surfaces and produce rotation. Instead of sliding against one another, the surfaces produce turning forces, which cause each roller to spin, leaving very little friction to oppose motion. Bearings are a type of roller used to reduce the friction between moving parts such as

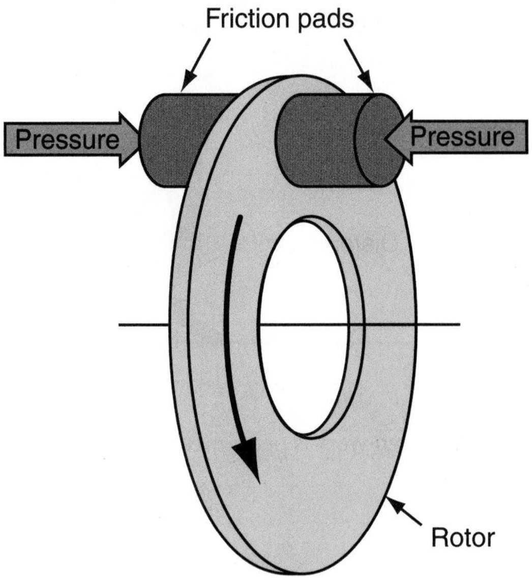

Figure 7-16 As pressure is applied to the friction pads, the pads attempt to stop the rotor to which the tire and wheel are attached.

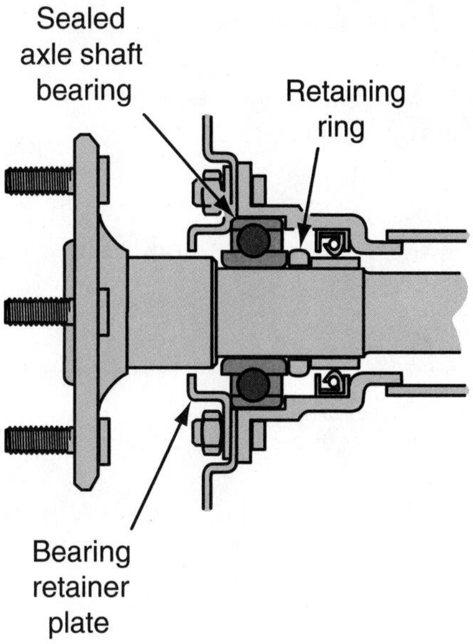

Figure 7-18 A typical ball bearing assembly for an axle shaft.

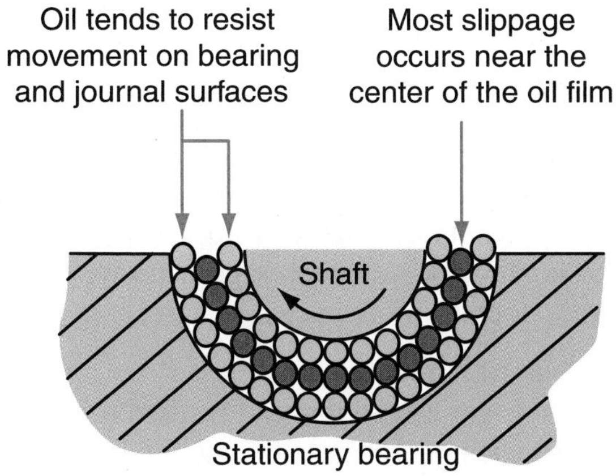

Figure 7-17 Oil separates the rotating shaft from the stationary bearing.

a wheel and its axle **(Figure 7–18)**. As the wheel turns on the axle, the balls in the bearing roll around inside the bearing, drastically reducing the friction between the wheel and axle.

Air Resistance

When a car is driven down the road, resistance occurs between the air and the car's surface. This resistance or friction opposes the momentum, or mechanical energy, of the moving vehicle. The mechanical energy from the engine must overcome the vehicle's inertia and the friction of the air striking the vehicle. The faster an object moves, the greater the air resistance.

Body design, obviously, affects the amount of friction developed by the air striking the vehicle. The total resistance to motion caused by friction between a moving vehicle and the air is referred to as coefficient of drag (Cd). At 45 miles per hour (72 kilometers per hour), half of the engine's mechanical energy can be used to overcome air resistance. Therefore, reducing a vehicle's Cd can be a very effective method of improving fuel economy. Coefficient of drag is also called aerodynamic drag.

Aerodynamics is the study of the effects of air on a moving object **(Figure 7–19)**. The basics of this science are fairly easy to understand. The larger the air area facing the moving air, the more force will be put on that area by the air. The air tends to hold back or resist the forward motion of the object moving against it.

The less air a vehicle pushes out of its way as it moves, the less power it will need to move at a given speed. If engineers want to make a vehicle that uses less fuel and emits fewer pollutants, they simply do whatever it takes to make the engine work less hard. Aerodynamics is one of the things used to accomplish these goals.

Figure 7-19 The movement of air as it goes over a car. *Courtesy of DaimlerChrysler Corporation*

Figure 7–20 A wind tunnel can generate winds as high as 150 miles per hour. *Courtesy of DaimlerChrysler Corporation*

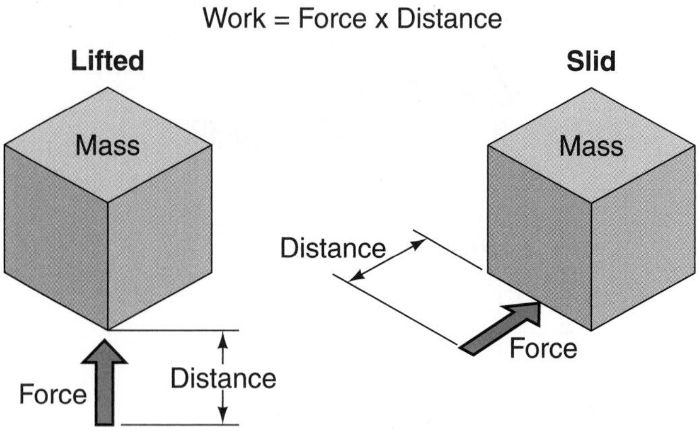

Figure 7–21 When work is performed, a mass is moved a certain distance.

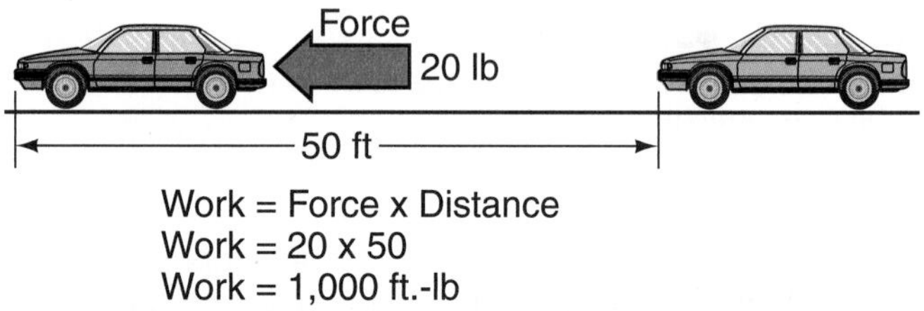

Figure 7–22 One thousand foot-pounds of work.

Most aerodynamic design work is done initially on a computer, then the design is checked and modified by placing a vehicle with that design in a wind tunnel **(Figure 7–20)**. A wind tunnel is a carefully constructed facility with a large fan at one end. Inside the tunnel, the movement of air over, under, and around the vehicle is studied.

Ideally, the air that is moved by the vehicle will follow the contours of the vehicle, which prevents the air from doing funny things as it is pushed away. If the air that moves under the vehicle has a place to push up, the vehicle will tend to lift. This situation creates poor handling, which can be very unsafe. Air can also be trapped under the vehicle, which increases the air drag of the vehicle. If the air moving over the top pushes against the vehicle, there is an increase in air drag. To help direct the air and make some of the air useable, air dams and spoilers or wings are used.

WORK

When a force moves a certain mass a specific distance, **work** is done. When work is accomplished, the mass may be lifted or slid on a surface against a resistance or opposing force **(Figure 7–21)**. Work is equal to the applied force multiplied by the distance the object moved (force × distance = work) and is measured in foot-pounds **(Figure 7–22)**, watts, or Newton-meters. For example, if a force moves a 3,000 pound car 50 feet, 150,000 foot-pounds of work was done.

During work, a force acts on an object to start, stop, or change the direction of the object. It is possible to apply a force to an object and not move the object. For example, you may push with all your strength on a car stuck in a ditch and not move the car. Under this condition, no work is done. Work is only accomplished when an object is started, stopped, or redirected by a force.

Simple Machines

A machine is any device that can be used to transmit a force and, in doing so, change the amount of force and/or its direction. A common example of a simple machine that does both is a valve rocker arm. One end of a rocker arm is pushed up by the action of the engine's camshaft. When this happens, the other end of the rocker arm pushes down on a valve to open it. A rocker arm also is designed to change the size of the force applied to it. Rocker arms provide more movement on the valve side or output than on the input side, a condition referred to as the ratio of the rocker arm. If a rocker arm has a ratio of 1.5:1, one end will move 1.5 times more than the other **(Figure 7–23)**. For example, if the camshaft causes one end of the rocker arm to move ½ inch, the other end will move ¾ of an inch.

The force applied to a machine is called the effort, while the force it overcomes is called the **load**. The effort is often smaller than the load, because a small effort can overcome a heavy load if the effort is moved a larger distance. The machine is then said to give a mechanical advantage. Although the effort will be smaller when using a machine, the amount of work done, or energy used, will be equal to or greater than that without the machine.

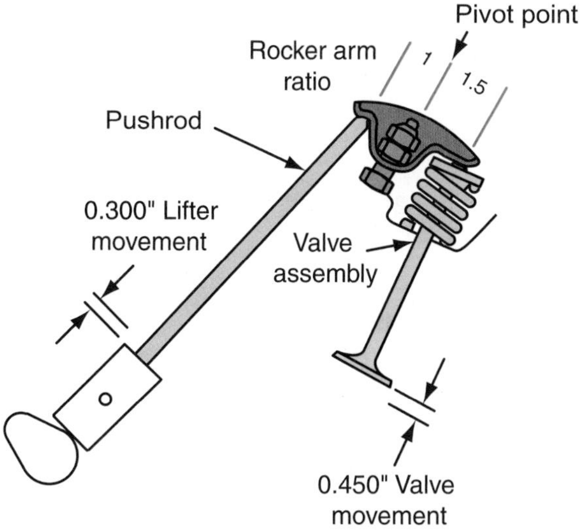

Figure 7-23 A rocker arm with a ratio of 1:1.5.

Figure 7-24 It takes less energy to pull a mass up an inclined plane than to lift the mass vertically.

Inclined Plane The force required to drag an object up a slope **(Figure 7–24)** is less than that required to lift it vertically. However, the overall distance moved by the object is greater when pulled up the slope than if it were lifted vertically. A screw is like an inclined plane wrapped around a shaft. The force that turns the screw is converted to a larger one, which moves a shorter distance and drives the screw in.

Pulleys A **pulley** is a wheel with a grooved rim in which a rope, belt, or chain runs to raise something by pulling on the other end of the rope, belt, or chain. A simple pulley changes the direction of a force but not its size. Also, the distance the force moves does not change. By using several pulleys connected together as a block and tackle, the size of the force can be changed too, so that a heavy load can be lifted using a small force. With a double pulley, the applied force required to move an object can be reduced by one-half but the distance the force must be moved is doubled. A quadruple pulley can reduce the force by four times but the distance will be increased by four times. Pulleys of different sizes can change the amount of required applied force as well as the speed or distance the pulley needs to travel to accomplish work **(Figure 7–25)**.

Levers A **lever** is a device made up of a bar turning about a fixed pivot point called the fulcrum that uses a force applied at one point to move a mass on the other end of the bar. Types of levers are divided into classes. In a class one lever, the fulcrum is between the effort and the load **(Figure 7–26)**. The load is larger than the effort, but it moves

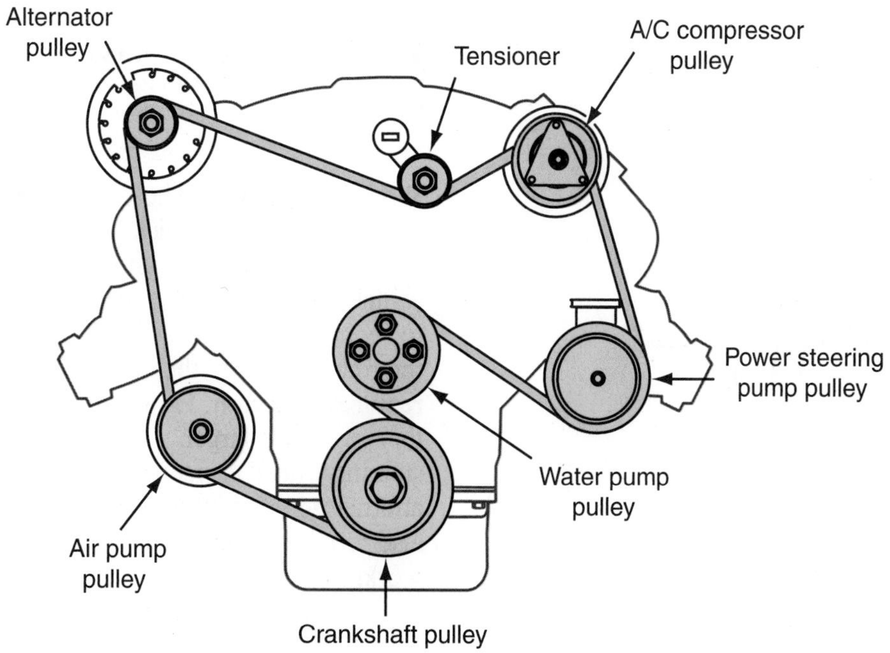

Figure 7-25 Accessories are driven by a common drive belt, but they rotate at different speeds because of the differences in pulley size.

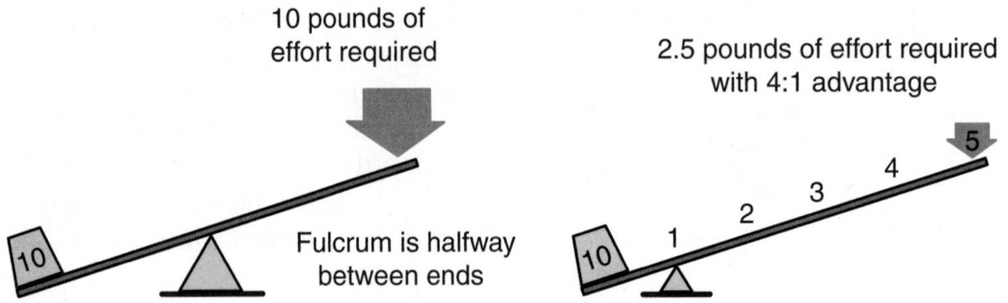

Figure 7-26 A mechanical advantage can be gained with a class one lever.

through a smaller distance. A pair of pliers is an example of a class one lever. In a class two lever, the load is between the fulcrum and the effort. Here again, the load is greater than the effort and moves through a smaller distance **(Figure 7–27)**. In a class three lever, the effort is between the fulcrum and the load. In this case, the load is less than the effort but it moves through a greater distance.

Gears A **gear** is a toothed wheel that becomes a machine when it is meshed with another gear. The action of one gear is that of a rotating lever and moves the other gear meshed with it. Based on the size of the gears in mesh, the amount of force applied from one gear to the other can be changed. Keep in mind this change in gear size does not change the amount of work performed by the gears, although as the force changes so does the distance of travel **(Figure 7–28)**. The relationship of force and distance is inverse. Gear ratios express the mathematical relationship (diameter and number of teeth) of one gear to another.

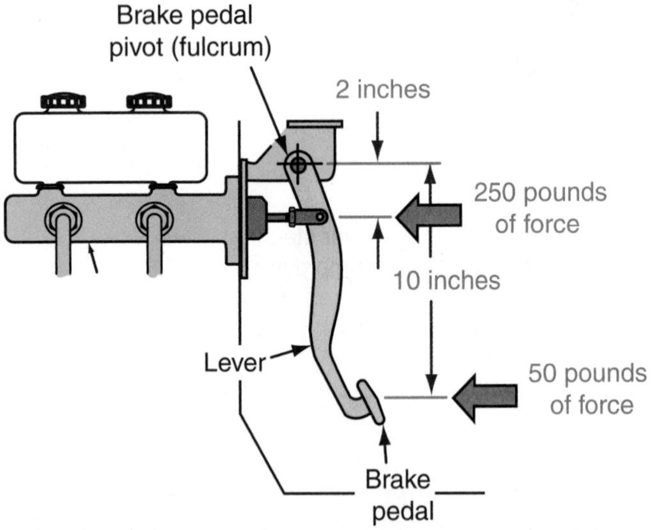

Figure 7-27 A brake pedal assembly is an example of a class two lever.

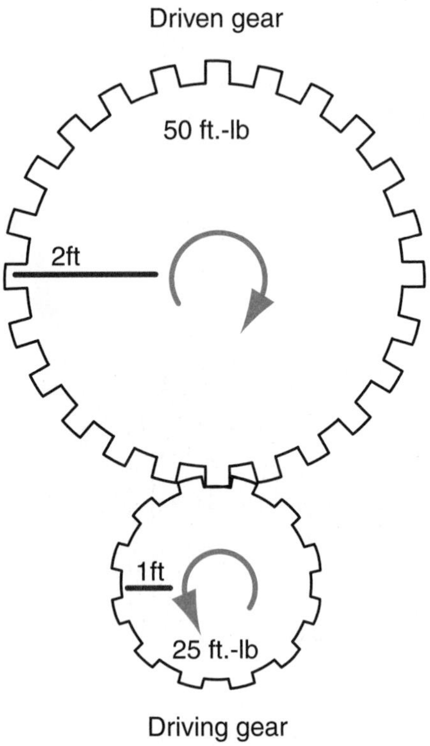

Figure 7-28 When a small gear drives a larger gear, the larger gear turns with more force but travels less; therefore, the amount of work stays the same.

Wheels and Axles The most obvious application of a wheel and axle is a vehicle's tires and wheels. These units revolve around an axle and limit the amount of area of a vehicle that contacts the road. Wheels function as rollers to reduce the amount of friction between a vehicle and the road. Basically, the larger the wheel, the less force is required to turn it. However, the wheel must move farther as it gets larger. An example of this principle is a steering wheel. A steering wheel that is twice the size of another will require one-half the force to turn it but will also require twice the distance to accomplish the same work.

Torque

Torque is a force that tends to rotate or turn things and is measured by the force applied and the distance traveled. The technically correct unit of measurement for torque is pounds per foot (lb-ft.). However, it is common to see torque stated in terms of foot-pounds (ft.-lb). In the metric or SI system, torque is stated in Newton-meters (Nm) or kilogram-meters (kg-m).

An engine creates torque and uses it to rotate the crankshaft. The combustion of gasoline and air in the cylinder creates pressure against the top of a piston. That pressure creates a force on the piston and pushes it down. The force is transmitted from the piston to the connecting rod and from the connecting rod to the crankshaft. The engine's crankshaft rotates with a torque that is transmitted through the drivetrain to turn the drive wheels of the vehicle.

Torque is force times leverage, the distance from a pivot point to an applied force. Torque is generated any time a wrench is turned with force. If the wrench is a foot long and you put 20 pounds of force on it, 20 pounds per foot are being generated. To generate the same amount of torque while exerting only 10 pounds of force, the wrench needs to be 2 feet long **(Figure 7–29)**. To have torque it is not necessary to have movement. When you pull a wrench to tighten a bolt, you supply torque to the bolt. If you pull on a wrench to check the torque on a bolt and the bolt torque is sufficient, torque is applied to the bolt but no movement occurs. If the bolt turns during torque application, work is done. When a bolt does not rotate during torque application, no work is accomplished.

Torque Multiplication When gears with different numbers of teeth mesh, each rotates at a different speed and force. Torque is calculated by multiplying the force by the distance from the center of the shaft to the point where the force is exerted.

The distance from the center of a circle to its outside edge is called the radius. On a gear, the radius is the distance from the center of the gear to the point on its teeth where force is applied.

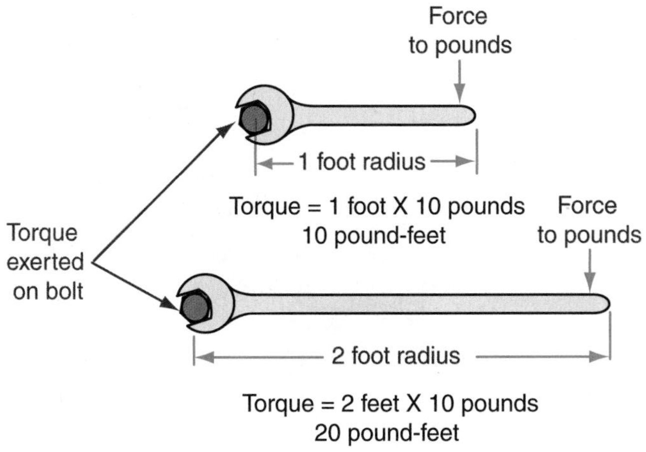

Figure 7-29 The amount of torque applied to a wrench is changed by the length of the wrench.

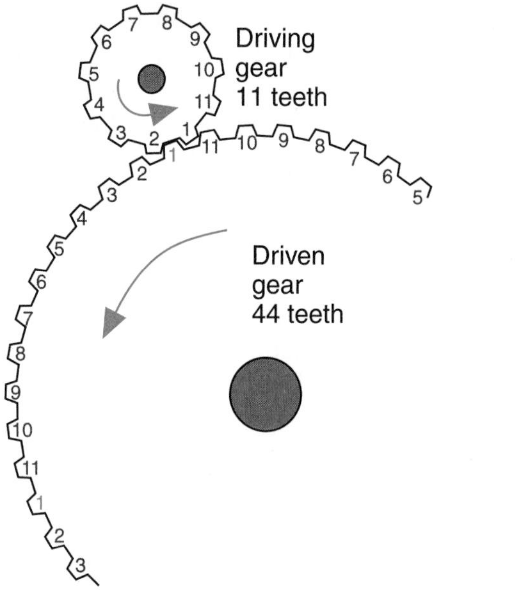

Figure 7-30 The driving gear must rotate four times to rotate the driven gear once.

If a tooth on the driving gear is pushing against a tooth on the driven gear with a force of 25 pounds and the force is applied at a distance of 1 foot (the radius of the driving gear), a torque of 25 ft.-lb is applied to the driven gear. The 25 pounds of force from the teeth of the smaller (driving) gear is applied to the teeth of the larger (driven) gear. If that same force were applied at a distance of 2 feet from the center, the torque on the shaft at the center of the driven gear would be 50 ft.-lb **(see Figure 7–30)**. The same force is acting at twice the distance from the shaft center.

The amount of torque that can be applied from a power source is proportional to the distance from the center at which it is applied. If a fulcrum or pivot point is placed closer to object being moved, more torque is available to move the object, but the lever must move farther than if the fulcrum were farther away from the object. The same principle is used for gears in mesh: A small gear will drive a large gear more slowly but with greater torque.

A drivetrain consisting of a driving gear with 11 teeth and a radius of 1 inch and a driven gear with 44 teeth and a radius of 4 inches will have a torque multiplication factor of 4 and a speed reduction of one-fourth. Thus, the larger gear will turn with four times the torque but one-fourth the speed **(Figure 7–30)**. The radii between the teeth of a gear act as levers; therefore, a gear that is twice the size of another has twice the lever arm length of the other.

Gear ratios express the mathematical relationship of one gear to another. Gear ratios can be varied by changing the diameter and number of teeth of the gears in mesh. A gear ratio also expresses the amount of torque multiplication between two gears. The ratio is obtained by dividing the diameter or number of teeth of the driven gear by the diameter or teeth of the drive gear. If the smaller driving gear has 11 teeth and the larger gear has 44 teeth, the ratio is 4:1.

Power

Power is a measurement for the rate, or speed, at which work is done. The metric unit for power is the watt. A watt is equal to one Newton-meter per second. You can multiply the amount of torque in Newton-meters by the rotational speed to determine the power in watts. Power is a unit of speed combined with a unit of force. For example if you were pushing something with a force of 1 N and it moved at a speed of 1 meter per second, the power output would be 1 watt.

In electrical terms, one watt is equal to the amount of electrical power produced by a current of one ampere across a potential difference of one volt. Mathematically this is expressed as Power (P) = Voltage (E) × Current (I) or $P = E \times I$.

Horsepower

Horsepower is the rate at which torque is produced. James Watt is credited with being the first person to calculate horsepower and power. He measured the amount of work that a horse could do in a specific time. He found that a horse could move 330 pounds 100 feet in one minute **(Figure 7–31)**. Therefore, he determined that one horse could do 33,000 foot-pounds of work in one minute. Thus, one horsepower is equal to 33,000 foot-pounds per minute, or 550 foot-pounds per second. Two horsepower could do this same amount of work in one-half minute. If you push a 3,000-pound (1,360-kilogram) car for 11 feet (3.3 meters) in one-quarter minute, you produce four horsepower.

An engine that produces 300 pound-feet of torque at 4,000 rpm produces 228 horsepower at 4,000 rpm. This is based on the formula that horsepower is equal to torque multiplied by engine speed and that sum divided by 5,252 ([torque × engine speed] ÷ 5252 = horsepower). The constant, 5,252, is used to convert the rpm for torque and horsepower into revolutions per second.

WAVES AND OSCILLATIONS

An **oscillation** is any single swing of an object back and forth between the extremes of its travel. When that motion travels through matter or space, it becomes a **wave**. A mass suspended by a spring, for example, is acted upon by two forces: gravity and the tension in the spring. At the point of equilibrium, the resultant of these forces is zero; they cancel each other out. When the mass is given an initial downward push, the tension of the spring exceeds the weight of the mass. The resultant upward force accelerates the mass back up toward its original position, by which time it has momentum, carrying it farther upward. When the weight exceeds the tension in the spring, the mass is pulled down again and the oscillation repeats itself until the mass is at equilibrium. As the mass oscillates toward the equilibrium position, the size of the oscillation decreases, unless it is again pushed downward. The air around the mass is upset as the mass oscillates and becomes an air wave.

Vibrations

When an object oscillates, it vibrates **(Figure 7–32)**. To prevent the vibration of one mass from causing a vibration in other masses, the oscillating mass must be isolated from other objects. This task is often difficult. For example, consider the engine in an automobile. Even the best-running engine vibrates as it runs. To reduce the transfer of engine vibrations to the vehicle, the engine is held in place by special mounts. The materials used in the mounts must keep the engine in its location but also be elastic

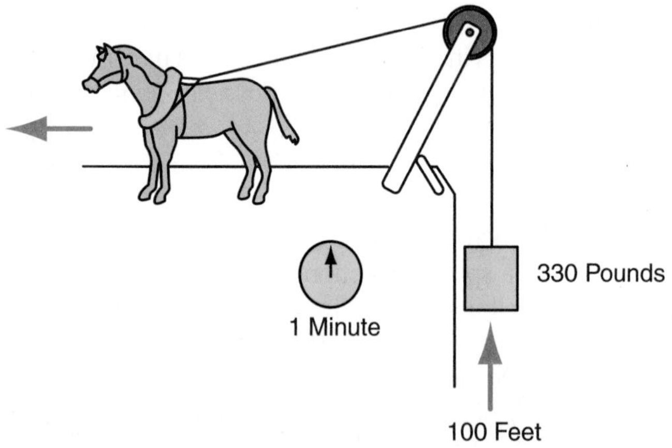

Figure 7–31 James Watt defined one horsepower as the power a horse exerts pulling 330 pounds 100 feet in 1 minute.

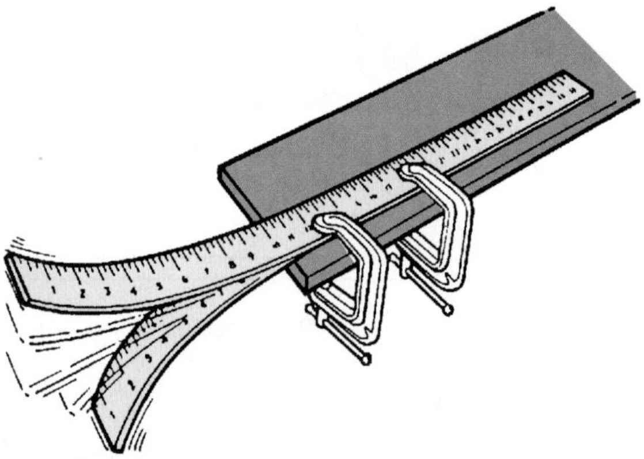

Figure 7–32 Vibrations happen in cycles.

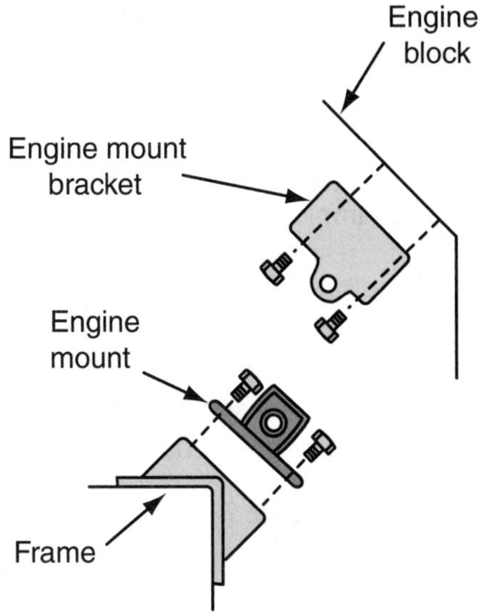

Figure 7-33 An engine mount holds the engine in place and isolates engine vibrations from the rest of the vehicle.

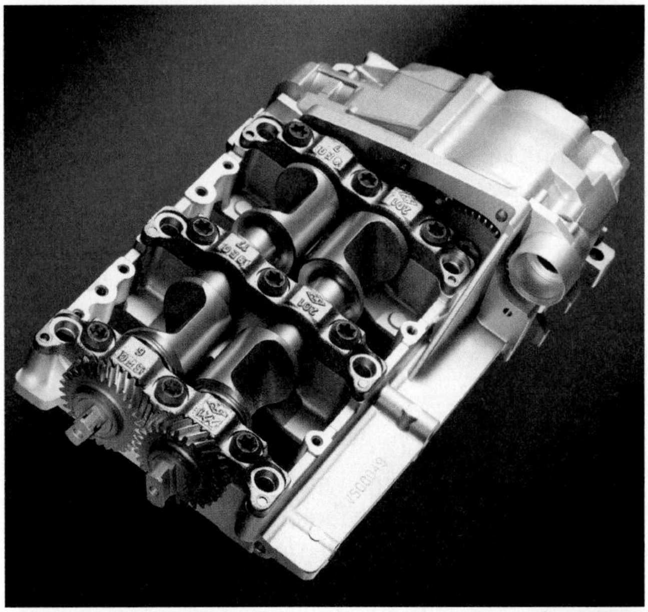

Figure 7-34 Balance shafts are driven by the crankshaft and work to counter crankshaft pulses and vibrations by acting with an equal force but in the opposite direction. *Courtesy of BMW of North America, Incorporated*

enough to absorb the engine's vibrations **(Figure 7–33)**. If the engine were mounted solidly to the vehicle, the vibrations would be felt throughout the vehicle.

In an automobile, many parts vibrate as they operate. Through careful engineering, these vibrations can be isolated or insulated, thereby reducing the chances of the vibrations moving through the vehicle. Vibration control is also important for the reliability of components. If the vibrations are not controlled, the object could shake itself to destruction. Vibration control is the best justification for always mounting components in the way they were designed to be mounted.

Unwanted and uncontrolled vibrations are typically the result of one component vibrating at a different frequency than another part. When two waves or vibrations meet, they add up or interfere. This is called the principle of superposition and is common to all waves. Making unwanted vibrations tolerable may involve canceling out each with an equal and opposite vibration. This approach to vibration reduction is best illustrated by the use of balance shafts in an engine. These shafts are designed to counter the vibrations that result from the rotation of the engine's crankshaft and pistons **(Figure 7–34)**. The balance shaft smoothes the engine by operating at an equal but opposite vibration to the crankshaft.

To cancel a vibration, the vibration must be defined. How many times the vibration occurs in one second is called its **frequency**. Frequency **(Figure 7–35)** is most often expressed in **hertz (Hz)**. One hertz is equal to one cycle per second. The hertz is named in honor of Heinrich Hertz, an early German investigator of radio wave transmission. The **amplitude** and velocity must also be defined. The amplitude of a vibration is its intensity or strength **(Figure 7–36)**. The velocity of a vibration is the result of its amplitude and its frequency. Since every material has a unique resonant or natural vibration frequency, everything on an automobile can vibrate at its own frequency.

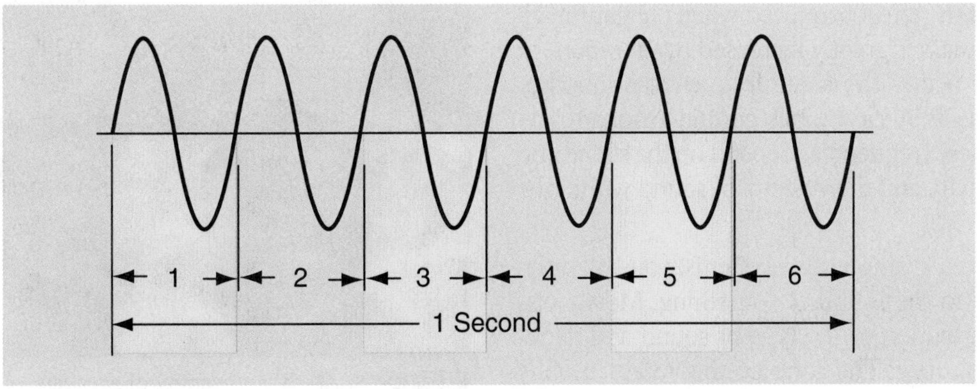

Figure 7-35 Frequency is a statement of how many cycles occur in a second.

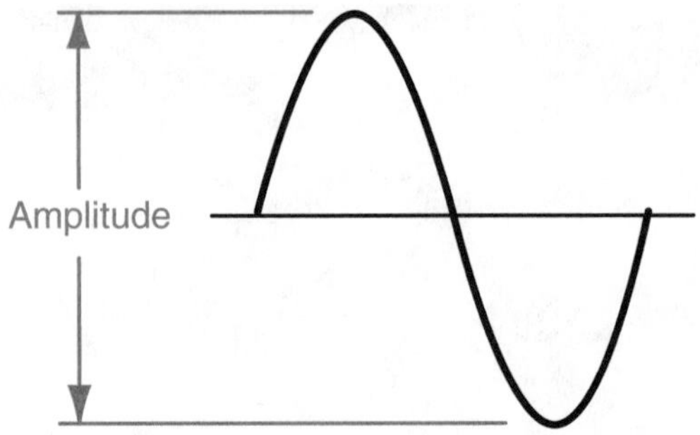

Figure 7-36 Amplitude is a measurement of a vibration's intensity. *Courtesy of General Motors Corporation*

Sound

Vibration is very common and results in the phenomenon of sound. In air, the vibrations that cause sound are transmitted as a wave between air molecules; many other substances transmit sound in a similar way. A vibrating object causes variations in pressure in the surrounding air. Areas of high and low pressure, known as compressions and rarefactions, move through the air as sound waves. Compression makes the sound waves denser, whereas rarefaction makes them less dense. The distance between each compression of a sound wave is called its **wavelength**. Sound waves with a short wavelength have a high frequency and a high pitched sound.

When the rapid variations in pressure occur between about 20 Hz and 20 kHz, sound is potentially audible. Audible sound is the sensation (as detected by the ear) of very small rapid changes in the air pressure above and below atmospheric pressure.

Certain terms are used to describe sound:

- The pitch of a sound is based on its frequency. The greater the frequency, the higher the pitch.
- A decibel is a numerical expression of the relative loudness of a sound.
- Intensity is the amount of energy in a sound wave.
- An overtone is a higher tone that is heard with a tone because of the air waves of the original tone.
- Harmonics are the result of the presence of two or more tones at the same time.
- Resonance is the effect produced when the natural vibration of a mass is greatly increased by vibrations at the same or nearly the same frequency of another source or mass. A cavity has certain resonant frequencies. These frequencies depend on the shape and size of the cavity and the velocity of sound within the cavity.

When diagnosing automotive systems, you will often be told to listen to the sound of something. Mostly you will be paying attention to the type of sound and its intensity and frequency. The tone of the sound usually indicates the type of material causing the noise. If there is high pitch, you know that the source of the sound is something that can vibrate quickly. This means the source is less rigid than something that vibrates at a low pitch. Although pitch depends on the frequency of a sound, the frequency itself can identify the possible sources of the sound. For example, if a sound from an engine increases with an increase in engine speed, you know the source of the sound must be something that is moving faster as a result of the increase in engine speed. If the frequency of the sound appears to be at one-half the speed of the engine, you know the source of the sound is something that is rotating at that speed, such as the camshaft.

Speakers A speaker **(Figure 7–37)** converts electrical energy into sound energy or waves. A constantly changing electrical signal is fed to the coil of a speaker, which lies within the magnetic field of a permanent magnet. The signal in the coil causes it to behave like an electromagnet, making it push against the field of the permanent magnet. The speaker cone is then pushed in and out by the coil in time with the electrical signal. As the cone moves forward, the air immediately in front of it is compressed, causing a slight increase in air pressure. It then moves back past its rest position and causes a reduction in the air pressure (rarefaction). This process continues so that a wave of alternating high and low pressure is radiated away from the speaker cone at the speed of sound. These changes in air pressure are actually sound. The sound from a speaker may be amplified by the space or cavity that surrounds the speaker cone. The room or area that the speaker sits in also works to amplify the sound.

Noise

Noise is any unwanted signal or sound. It can be random or periodic. To identify the source of a noise, it is important to remember that sound or noise is a vibration and the vibration may be traveling through other components. Therefore, the source of the noise is not always where it

Figure 7-37 A variety of speakers. *Courtesy of Robert Bosch*

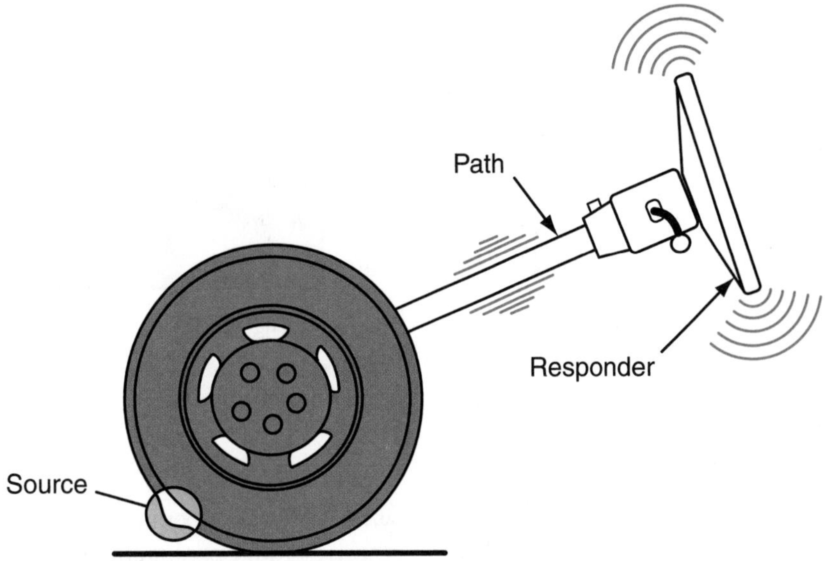

Figure 7–38 A vibration and/or noise will easily move through components so that it appears that the responder (in this case, the steering wheel) is the cause of the noise or vibration.

may appear **(Figure 7–38)**. There are many potential sources of noise in an automobile and manufacturers work hard to prevent noise.

Noise can be prevented or reduced by three different approaches. The most effective way is to intervene at the design stage to redesign a noisy component so that it produces less noise. A relatively new technique for noise reduction is antinoise or active noise control, which involves producing a sound that is similar to, but out of phase with, the noise. This sound effectively cancels the original noise. More obvious methods of noise reduction, or passive noise control, involve the use of filters, insulation, and noise barriers.

A filter is an electrical circuit that allows signals in certain frequency ranges to pass through and that blocks all other frequencies. Sound insulation prevents sound from traveling from one place to another. Heavy materials like concrete are the most effective materials for sound insulation. Sound insulation or deadening materials are placed strategically throughout a modern vehicle. Some sound-deadening materials actually absorb sounds. These materials must be able to vibrate without creating sound. Sound insulators rarely absorb sound.

LIGHT

Light is a form of electromagnetic radiation. In free space, it travels in a straight line at 300 million meters per second. When a beam of light meets an object, a proportion of the rays may be reflected. Some light may be absorbed and some transmitted. Without reflection, we would only be able to see objects that give out their own light. Light always reflects from a surface at the same angle at which it strikes. Therefore, parallel rays of light reflecting off a very flat surface remain parallel. A beam of light reflecting from an irregular surface scatters in all directions. Light that passes through an object is bent or refracted. The angle of refraction depends on the angle at which the light meets the object and the material it passes through. Lenses and mirrors can cause light rays to diverge or converge. When light rays converge, they can reach a point of focus.

These principles are the basis for fiber-optic lighting. With fiber optics, the light from a single lamp moves through one or more fiber cables to illuminate a point remote from the source lamp. The fiber cables are designed to allow the light to travel without losing intensity because of reflection. The light beam is bent through refraction as it travels through the cable and can be delivered to many locations at the same time.

Photo Cells

Radiation is produced in the sun's core during its nuclear reactions and is the source of most of the Earth's energy. A transfer of energy from electromagnetic radiation to electrical energy takes place in a photovoltaic (photo) cell, or solar cell. When no sunlight falls on it, it can supply no electricity.

LIQUIDS

A fluid is something that does not have a definite shape; therefore, liquids and gases are fluids. A characteristic of all fluids is that they will conform to the shape of their container. A major difference between a gas and a liquid is that a gas always fills a sealed container, whereas a liquid may not. A gas also readily expands or compresses according to the pressure exerted on it. Liquids are basically incompressible, which gives them the ability to transmit force **(Figure 7–39)**. The pressure applied to a liquid in a sealed container is transmitted equally in all directions and to all areas of the system and acts with equal force on all areas. As a result, liquids can provide great increases in the force available to do work. They also always seek a common level. A liquid under pressure may also change from a liquid to a gas in response to temperature changes.

Liquids exert pressure on immersed objects, resulting in an upward force called upthrust. The upthrust is equal to the weight of the liquid displaced by the immersed object. If the upthrust on an object is greater than the weight of the object, then the object will float. Large ships float because they displace huge amounts of water, producing a large upthrust.

Laws of Hydraulics

Hydraulics is the study of liquids in motion. Liquids predictably respond to pressures exerted on them. Their reaction to pressure is the basis of all hydraulic applications. This fact allows hydraulics to do work. A simple hydraulic system has liquid, a pump, lines to carry the liquid, control valves, and an output device. The liquid must be available from a continuous source, such as an oil pan or sump. An oil pump is used to move the liquid through the system. The lines that carry the liquid may be pipes, hoses, or a network of internal bores or passages in a single housing. Control valves are used to regulate hydraulic pressure and direct the flow of the liquid. The output device is the unit that uses the pressurized liquid to do work.

More than 300 years ago a French scientist, Blaise Pascal, determined that if you had a liquid-filled container with only one opening and applied force to the liquid through that opening, the force would be evenly distributed throughout the liquid. This explains how pressurized liquid is used to operate and control systems, such as the brake system and automatic transmissions.

Pascal constructed the first known hydraulic device, which consisted of two sealed containers connected by a tube. The pistons inside the cylinders seal against the walls of each cylinder, preventing the liquid from leaking out of the cylinder and air from entering into the cylinder. When the piston in the first cylinder has a force applied to it, the pressure moves everywhere within the system. The force is transmitted through the connecting tube to the second cylinder. The pressurized fluid in the second cylinder exerts force on the bottom of the second piston, moving it upward and lifting the load on the top of it **(Figure 7–40)**. By using this device, Pascal found he could increase the force available to do work **(Figure 7–41)**, just as could be done with levers or gears.

Pascal determined that force applied to liquid creates pressure or the transmission of the force through the liquid. These experiments revealed two important aspects of a liquid when it is confined and put under pressure.

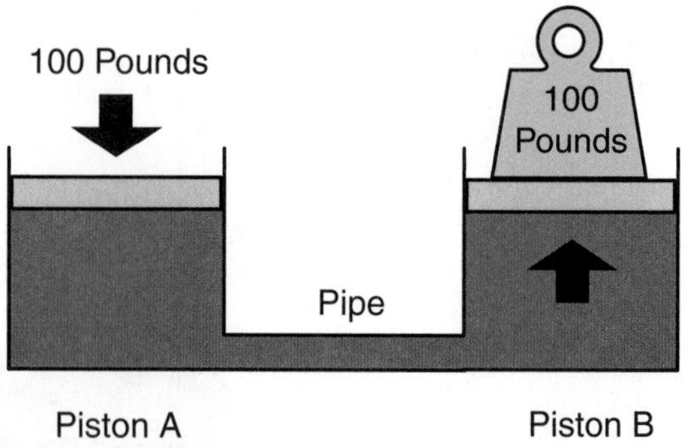

Figure 7–40 In a hydraulic circuit, pressure is transferred equally throughout the system.

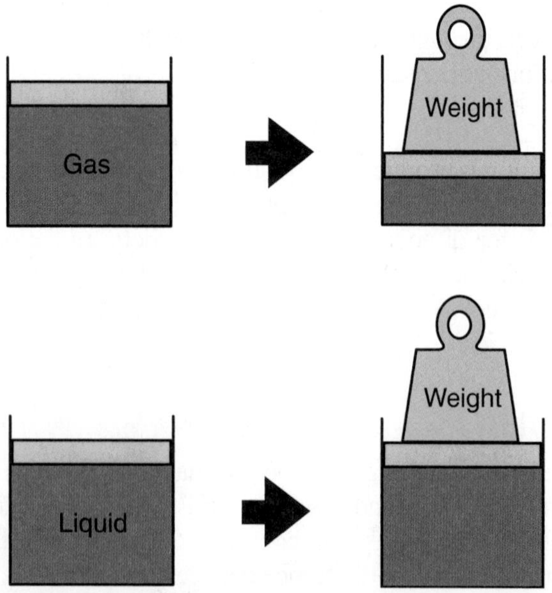

Figure 7–39 Gases compress; liquids (fluids) do not.

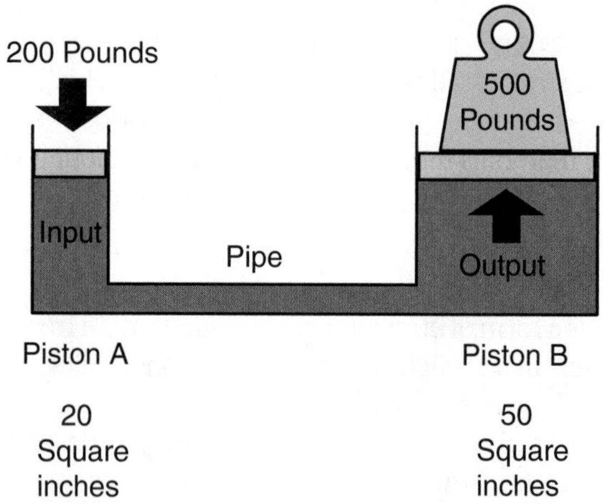

Figure 7–41 The force available to do work can be increased by increasing the size of the piston doing the work.

The pressure applied to it is transmitted equally in all directions, and this pressure acts with equal force at every point in the container. If a liquid is confined and a force is applied, pressure is produced. In order to pressurize a liquid, the liquid must be in a sealed container. Any leak in the container will decrease the pressure.

Mechanical Advantage with Hydraulics

Hydraulics are used to do work in the same way a lever or gear does. All of these systems transmit energy. Because energy cannot be created or destroyed, these systems only redirect energy to perform work and do not create more energy. If a hydraulic pump provides 100 psi, there will be 100 pounds of pressure on every square inch of the system **(Figure 7–42)**. If the system includes a piston with an area of 50 square inches, each square inch receives 100 pounds of pressure. This means there will be 5,000 pounds of force applied to that piston **(Figure 7–43)** but the output's travel will decrease proportionally. The use of the larger piston gives the system a **mechanical advantage**, or increase in the force available to do work. The multiplication of force through a hydraulic system is directly proportional to the difference in the piston sizes throughout the system.

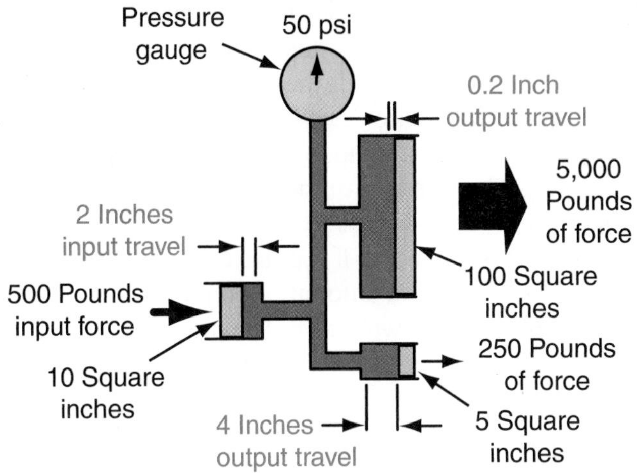

Figure 7–43 Hydraulic systems can provide an increase in force (mechanical advantage), but the output's travel will decrease proportionally.

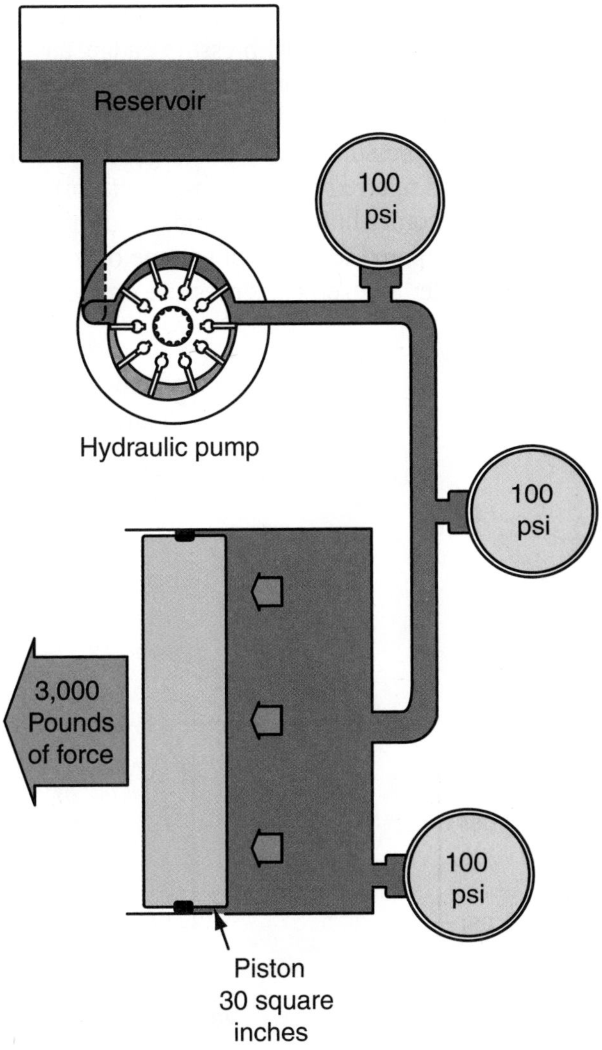

Figure 7–42 A pressure applied to a liquid is transmitted equally and acts with equal force at every point within the hydraulic circuit.

By changing the size of the pistons in a hydraulic system force is multiplied, and as a result low amounts of force can be used to move heavy objects. The mechanical advantage of a hydraulic system can be further increased by the use of levers to increase the force applied to a piston.

Although the force available to do work is increased by using a larger piston in one cylinder, the total movement of the larger piston is less than that of the smaller one. A hydraulic system with two cylinders, one with a 1-inch piston and the other with a 2-inch piston, will double the force at the second piston; however, the total movement of the larger piston will be half the distance of the smaller one.

The use of hydraulics to gain a mechanical advantage is similar to the use of levers or gears. Hydraulics is preferred when the size and shape of the system is of concern. In hydraulics, the force applied to one piston transmits through the fluid and the opposite piston has the same force on it. The distance between the two pistons in a hydraulic system does not affect the force in a static system. Therefore, the force applied to one piston can be transmitted without change to another piston located somewhere else.

A hydraulic system responds to the pressure or force applied to it. The mere presence of different-sized pistons does not always result in fluid power. Either the pressure applied to the pistons must be different or the size of the pistons must be different in order to cause fluid power. If an equal amount of pressure is exerted onto both pistons in a system and both pistons are the same size, neither piston will move and the system is balanced or is at equilibrium. The pressure inside the hydraulic system is called **static pressure** because there is no fluid motion.

When an unequal amount of pressure is exerted on the pistons, the piston with the least amount of pressure on it will move in response to the difference between the two pressures. Likewise, if the size of the two pistons is different and an equal amount of pressure is exerted on the pistons, the fluid will move. The pressure of the fluid while it is in motion is called **dynamic pressure**.

GASES

A gas comprises independent particles—atoms or molecules—in random motion. This means that a gas will fill any container into which it is placed. The random movement of gas particles also ensures that any two gases sharing the same container will totally mix. This phenomenon is **diffusion**.

The kinetic energy in atoms and molecules increases as the temperature increases. A decrease in temperature reduces the kinetic energy. Molecules in solids move slowly compared to those in liquids or gases. Gas molecules move quickly compared to liquid molecules. Because gas molecules are in constant motion, they spread out to fill all the space available. At higher temperatures gas molecules spread out more, whereas at lower temperatures gas molecules move closer together. The bombardment of particles against the sides of the container produces pressure.

Behavior of Gases

Three simple laws describe the predictable behavior of gases: Boyle's law, the pressure law, and Charles's law. Each of these laws describes a relationship among the pressure, volume, and temperature of a gas.

Boyle's law states that the volume and pressure of a mass of gas at a fixed temperature is inversely proportional. If the pressure on a gas increases, its volume will decrease; likewise, if the volume is increased the pressure will decrease.

The pressure law states that the volume of a mass of gas depends on its temperature. The higher the temperature, the greater the volume. If the volume cannot change, the pressure of the gas will. Therefore the pressure and temperature of a gas are also directly related. If you increase one, you also increase the other. This explains why cold air is denser than warm air.

Air Pressure

Because air is gaseous matter with mass and weight, it exerts pressure on the Earth's surface. A one-square-inch column of air extending from the Earth's surface to the outer edge of the atmosphere weighs 14.7 psi at sea level.

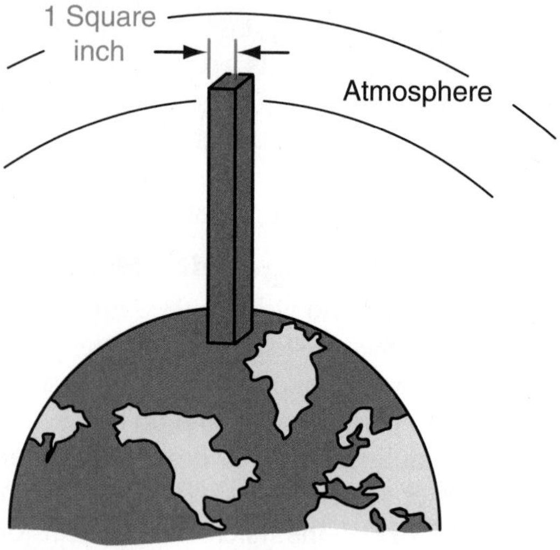

Figure 7–44 One square inch of air equals 14.7 pounds per square inch at sea level.

Therefore, atmospheric pressure is 14.7 psi at sea level **(Figure 7–44)**. **Atmospheric pressure** may be defined as the total weight of the Earth's atmosphere. Pressure greater than atmospheric pressure may be measured in psi gauge (psig). Using a standard pressure gauge, air pressure is compared to that of normal atmospheric pressure. When the actual pressure is 19.7 psi, the gauge will read 5 psi showing the pressure differential **(Figure 7–45)**. The actual pressure is referred to as psi absolute (psia).

When air becomes hotter, it expands, and this hotter air is lighter compared to an equal volume of cooler air. This hotter, lighter air exerts less pressure on the Earth's

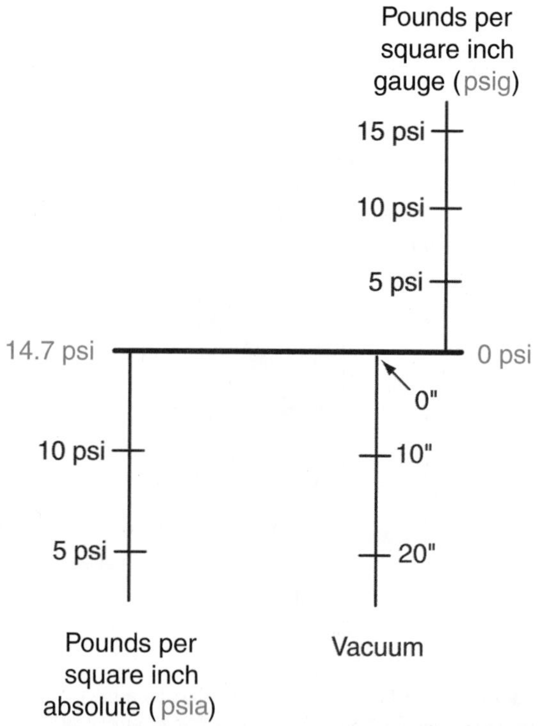

Figure 7–45 The relationship between psia and psig.

surface compared to cooler air, which means the weight of the atmosphere changes with weather. This change is rather slight. As the weight changes, so does the atmospheric pressure. The change in atmospheric pressure is measured with a barometer and is called **barometric pressure**. Barometric pressure at normal atmospheric pressure is 29.92 inches of mercury. The increments for measuring barometric pressure are based on the increments of a barometer. A barometer is a J-shaped tube with mercury in it. One end of the tube is exposed to normal atmospheric pressure and the other end to current atmospheric pressure. When the current atmospheric pressure equals normal atmospheric pressure, the level of the mercury will rise 29.92 inches up the tall part of the J. When the current atmospheric pressure is lower than normal, the normal atmospheric pressure pushes the mercury down. Likewise, when current atmospheric pressure is higher than normal it pushes the mercury up the tube. The amount of mercury movement reflects the difference in the two pressures. This corresponds with a universal law that states that a high pressure always moves toward a lower pressure.

Although the pressure of the atmosphere only changes slightly, the impact of these changes can be critical to the overall operation of an engine. The combustion process depends on having the correct amount of air enter into the cylinders. If the calibrations for the air and the accompanying amount of fuel did not consider the changes in atmospheric pressure, the engine would most often not receive the correct mixture of air and fuel. Today's engines are equipped with a sensor to monitor barometric pressure.

To further consider the law that states that a high pressure always moves to a lower pressure, consider what happens when a nail punctures an automotive tire. The high-pressure air in the tire leaks out until the pressure inside the tire is equal to atmospheric pressure outside the tire. When the tire is repaired and inflated, air with a pressure higher than atmospheric is forced into the tire.

When you climb above sea level, atmospheric pressure decreases. The weight of a column of air is less at an elevation of 5,000 feet (1,524 meters) than it is at sea level. As altitude continues to increase, atmospheric pressure and weight continue to decrease. At an altitude of several hundred miles above sea level the Earth's atmosphere ends, and there is no pressure beyond that point.

Vacuum Scientifically, **vacuum** is defined as the absence of atmospheric pressure. However, it is commonly used to refer to any pressure less than atmospheric pressure. Vacuum may also be referred to as low pressure simply because it is a pressure less than atmospheric pressure.

Vacuum could be measured in psig or psia, but inches of mercury (in. Hg) are most commonly used for this measurement **(Figure 7–46)**. Let us assume that a plastic U tube is partially filled with mercury and atmospheric pressure is allowed to enter one end of the tube. If vacuum is supplied to the other end of the U tube, the mercury is forced downward by the atmospheric pressure. When this movement occurs, the mercury also moves upward on the side where the vacuum is supplied. If the mercury moves downward 10 inches, or 25.4 centimeters (cm), where the atmospheric pressure is supplied and upward 10 inches (25.4 cm) where the vacuum is supplied, 20 in. Hg is supplied to the U tube. The highest possible, or perfect, vacuum is approximately 29.9 in. Hg.

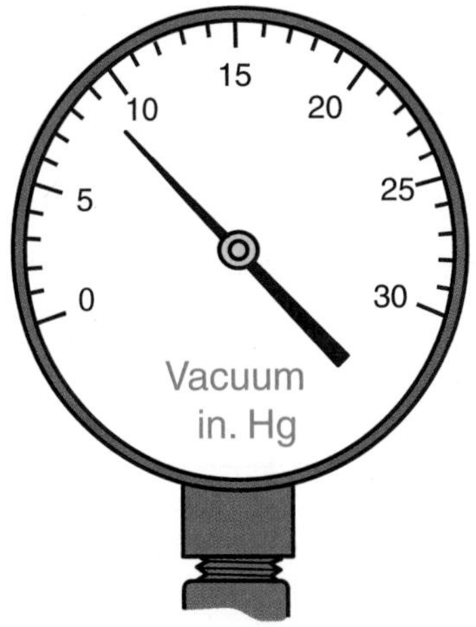

Figure 7-46 A vacuum gauge measures pressures below atmospheric pressure in units of inches of mercury.

HEAT

Heat is a form of energy and is used in many ways. The main sources of heat are the Sun, the Earth, chemical reactions, electricity, friction, and nuclear energy. Heat is the result of the kinetic energy present in all matter; therefore, everything has heat. Cold objects have low kinetic energy because their atoms and molecules are moving very slowly, whereas hot objects have more kinetic energy because their atoms and molecules are moving fast.

Temperature is an indication of an object's kinetic energy. Temperature is measured with a thermometer, which has either a Fahrenheit (F) or Celsius (C) (Centigrade) scale. At absolute zero (–273°C, also referred to as 0 Kelvin) particles of matter do not vibrate, but at all other temperatures particles have motion. The temperature of an object is a statement of how well the object will transfer heat or kinetic energy to or from another object. Heat and temperature are not the same thing; heat is the movement of kinetic energy from one object to another. Temperature is an indication of the amount of kinetic energy something has. Energy from something hot always moves to an object that is colder, until both are at the same temperature. The greater the difference in temperature between the two objects, the faster the heat will flow from one to the other.

Heat is measured in British thermal units (BTUs) and calories. One BTU is the amount of heat required to heat 1 pound of water by 1°F. One calorie is equal to the amount of heat needed to raise the temperature of 1 gram of water 1°C.

Heat Transfer

Heat transfers between two substances at different temperatures through convection, conduction, or radiation. **Convection** is the transfer of heat by the movement of a heated object. Convection can be easily seen by watching a pot of water on a hot stove. The water on the bottom of the pot is the first to be heated by the stove. As the water at the bottom becomes hotter, it expands and becomes lighter than the water at the top of the pan, causing the heavier water to sink toward the bottom and push the warmer water up. This action continues until all of the water in the pan is the same temperature.

Conduction is the movement of heat through a material. Much of the heat generated by a running engine is not used to drive the vehicle. Rather, it is "wasted" as it heats the parts of the engine. The immense heat that results from combustion is absorbed by the engine and is used to push the pistons down. The engine's cooling system uses conduction to move the heat off the parts to help cool the engine. Because the energy from something hot moves toward something colder, the heat from the engine moves to the engine's coolant circulating throughout the system. The object that receives conducted heat can be liquid, gas, or solid.

Radiation does not rely on another material to transfer heat. The moving atoms and molecules within an object create waves of radiant energy called infrared rays. Hot objects give off more infrared rays than colder objects; therefore, a hot object gives off infrared rays to anything around it that is colder. No movement is necessary to transfer this heat. You can feel radiation in action by simply putting your hand near something hot. This radiation can heat up other objects. The hot object cools as it radiates its heat energy. In an engine's cooling system, the radiator uses radiation to transfer heat from the coolant into the surrounding air.

The Effects of Temperature Change

Any time the temperature of an object has changed, a transfer of heat has occurred. A transfer of heat may also cause the object to change size or its state of matter. The amount of heat required to raise the temperature of 1 gram of mass 1°C is called the specific heat capacity. Every substance has its own specific heat capacity and this factor is assigned to material based on its difference from water, which has a specific heat capacity of 1. For example, the temperature of 1 gram of water will increase by 10°C if 10 calories of heat are transferred to it. But if 10 calories of heat were added to 1 gram of copper, the temperature would increase by 111°C because copper has a specific heat capacity of only 0.09 as compared to the 1.0 specific heat capacity of water.

As heat moves in and out of a mass, the movement of atoms and molecules in that mass increases or slows down. With an increase in motion, the size of the mass tends to get bigger or expand through a process called **thermal expansion**. **Thermal contraction** takes place when a mass has heat removed from it and the atoms and molecules slow down. All gases and most liquids and solids expand when heated, with gases expanding the most. Solids, because they are not fluid, expand and contract at a much lower rate. It is important to realize that all materials do not expand and contract at the same rate. For example, an aluminum component will expand at a faster rate than a similar iron component, which explains why aluminum cylinder heads have unique service requirements and procedures when compared to iron cylinder heads.

Thermal expansion takes place every time fuel and air are burned in an engine's cylinders. The sudden temperature increase inside the cylinder causes a rapid expansion of the gases, which pushes the piston downward and causes engine rotation.

Typically, when heat is added to a mass the temperature of the mass increases. This does not always happen, however. In some cases, the additional heat causes no increase in temperature but causes the mass to change its state (solid to liquid or liquid to gas). For example, if we heat an ice cube to 32°F (0°C), it will begin to melt **(Figure 7–47)**. As heat is added to the ice cube, the temperature of the ice cube will not increase until it becomes a liquid. The heat added to the ice cube that did not raise its temperature but caused it to melt is called **latent heat** or the heat of fusion. Each gram of ice at 0°C requires 80 calories of heat to melt it to water at 0°C. As more heat is added to the 0°C water, the water's temperature will once again increase. This process continues until the temperature of the water reaches 212°F (100°C), the boiling temperature of water. At this point, any additional heat applied to the water is latent heat causing the water to change its state to that of a gas. This added heat is called the heat of evaporation.

To change the water gas back to liquid water, the same amount of heat required to change the liquid to a gas must be removed from the gas. At that point the gas condenses to a liquid. As additional heat is removed, the temperature will drop until enough heat is removed to bring its temperature back down to freezing (melting in reverse) point. At that time latent heat must be removed from the liquid before the water turns to ice again.

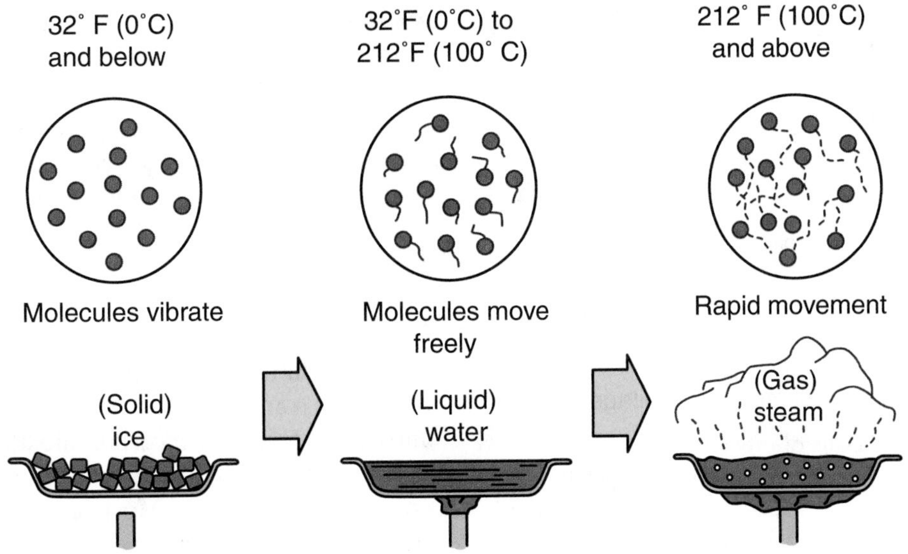

Figure 7-47 Water can exist in three different states of matter.

Controlling Heat

There is a particular temperature range in which all parts of an automobile will operate best. Engineers strive to control those temperatures to ensure reliable and efficient operation. This is a major task as parts that do not perform well when hot are often mounted close to something that is very hot. High heat could transfer to the heat-sensitive parts if insulation or passing outside air were not present. Although some parts tolerate extreme temperatures, they must still be protected from overheating. The combustion inside a cylinder can generate temperatures as great as 2,500°F. If that heat transferred uncontrollably to the metal parts of an engine, those parts would expand to the point where they could no longer move or could melt. This is why the engine's cooling system is so important. The cooling system is a controlled heat transfer system designed to protect the engine and allow it to run more efficiently.

CHEMICAL PROPERTIES

The properties of something are those characteristics used to identify or describe it. Physical properties are readily observable characteristics, such as color, size, luster, and smell. Physical changes are those changes that do not result in the production of a new substance, such as melting an ice cube.

Chemical properties are only observable during a chemical reaction and they describe how one type of matter reacts with another type of matter to form a new and different substance. Chemical properties are quite different from physical properties. An example of a chemical property is a substance's ability to burn. A chemical property of some metals is the ability to combine with oxygen to form rust (iron and oxygen) or tarnish (silver and sulfur). Another example is hydrogen's ability to combine with oxygen to form water.

A solution is a mixture of two or more substances in varying amounts. Most solutions are liquids, but solutions of gases and solids are possible. An example of a gas solution is the air we breathe; it is composed of mostly oxygen and nitrogen. Brass is a good example of a solid solution, as it is composed of copper and zinc. The liquid in a solution is called the **solvent** and the substance added is the solute. If both are liquids, the one present in the smaller amount is usually considered the solute. Solutions can vary widely in terms of how much of the dissolved substance is actually present. A heavily diluted (much water) acid solution has very little acid and may not be harmful or even noticeably acidic.

Specific Gravity

Specific gravity is the heaviness or relative density of a substance compared to that of water. If something is 3.5 times as heavy as an equal volume of water, its specific gravity is 3.5. Its density is 3.5 grams per cubic centimeter, or 3.5 kilograms per liter. Because it is a ratio of two quantities that have the same dimensions (mass per unit volume), specific gravity has no dimension.

Specific gravity checks of a battery's electrolyte are an indication of the battery's state of charge **(Figure 7–48)**.

To calculate the heaviness or **density** of a material, divide the mass of the material in grams by its volume in cubic centimeters.

Chemical Reactions

Chemical changes, or chemical reactions, are changes that result in the production of another substance, such

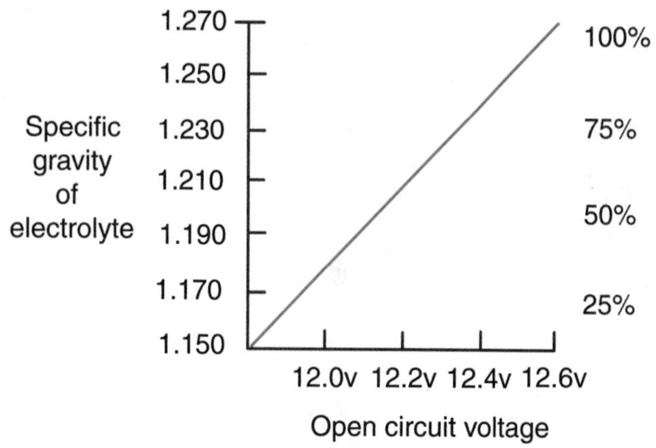

Figure 7-48 Specific gravity checks of a battery's electrolyte are an indication of the battery's state of charge.

as wood turning to carbon after it has burned completely. A chemical reaction is always accompanied by a change in energy. This means energy is given off or taken in during the reaction. Some reactions that release energy need some energy to start the reaction. A reaction takes place when two or more molecules interact and one of the following happens:

- A chemical change occurs.
- Single reactions occur as part of a large series of reactions.
- Ions, molecules, or pure atoms are formed.

Catalysts and Inhibitors

Reactions need a certain amount of energy to happen. If they do not have it, the reaction probably cannot happen. A **catalyst** lowers the amount of energy needed to make a reaction happen. A catalyst is any substance that affects the speed of a chemical reaction without itself being consumed or changed. Catalysts tend to be highly specific, reacting with one substance or a small set of substances. In a car's catalytic converter, the platinum catalyst converts unburned hydrocarbons and nitrogen compounds to products that are harmless to the environment **(Figure 7–49)**. Water, especially salt water, catalyzes oxidation and corrosion. An inhibitor is the opposite of a catalyst and stops or slows the rate of reaction.

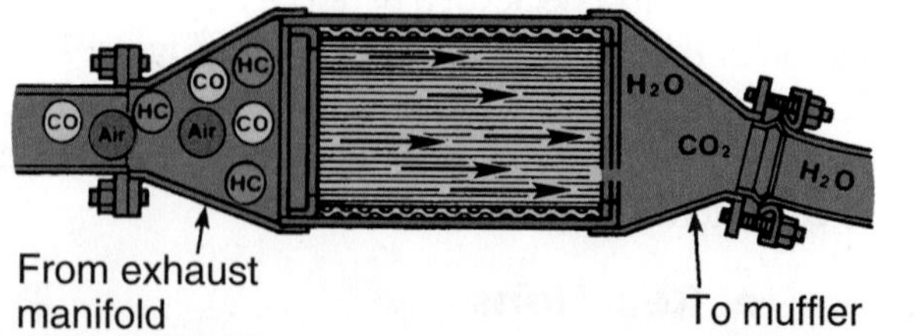

Figure 7-49 A basic catalytic converter that changes pollutants into chemicals that are good for the environment.

Acids/Bases

An ion is an atom or group of ions with one or more positive or negative electric charges. Ions are formed when electrons are added to or removed from neutral molecules or other ions. Many crystalline substances are composed of ions held in regular geometric patterns by the attraction of oppositely charged particles for each other. Ions are what make something an **acid** or a **base**.

Acids are compounds that break into hydrogen (H^+) ions and another compound when placed in an aqueous (water) solution. They have a sour taste, are corrosive, react with some metals to produce hydrogen, react with carbonates to produce carbon dioxide, change the color of litmus from blue to red, and become less acidic when combined with alkalis. Most acids are slow reacting, especially if they are weak acids. Acids also react with bases to form salts.

Alkalis (bases) are compounds that release hydroxide ions (OH^-) and react with hydrogen ions to produce water, thus neutralizing each other. Most substances are neutral (not an acid or a base). Alkalis feel slippery, change the color of litmus from red to blue, and become less alkaline when they are combined with acids.

A hydroxide is any compound made up of one atom each of hydrogen and oxygen, bonded together and acting as the hydroxyl group or hydroxide anion (OH^-). An oxide is any chemical compound in which oxygen is combined with another element. Metal oxides typically react with water to form bases or with acids to form salts. Oxides of nonmetallic elements react with water to form acids or with bases to form salts.

A salt is a chemical compound formed when the hydrogen of an acid is replaced by a metal. Typically, an acid and a base react to form a salt and water.

pH The **pH scale** is used to measure how acidic or basic a solution is. Its name comes from the fact that pH is the absolute value of the power of the hydrogen ion concentration. The scale goes from 0 to 14. Distilled (pure) water is 7. Acids are from 0 and 7 and bases are from 7 to 14. When the pH of a substance is low, the substance has many H^+ ions. When the pH is high, the substance has many OH^- ions. The pH value helps inform scientists and technicians of the nature, composition, or extent of reactivity of substances.

The pH of something is typically checked with litmus paper. Litmus is a mixture of colored organic compounds obtained from several species of lichen. Lichen is a type of plant that is actually a combination of a fungus and an alga.

Reduction and Oxidation

Oxidation is a chemical reaction in which a substance combines with oxygen. Rapid oxidation produces heat fast enough to cause a flame. When fuel burns, substances in the fuel combine with oxygen to form other compounds. This chemical reaction is combustion, which produces heat and fire. The rusting of iron is also an example of oxidation. Unlike fire, rusting occurs so slowly that little heat and no flames are produced.

The addition of hydrogen atoms or electrons is **reduction**. Oxidation and reduction always occur simultaneously: One substance is oxidized by the other, which it reduces. During oxidation, a molecule provides electrons. During reduction, a molecule accepts electrons. Oxidation and reduction reactions are usually called redox reactions. Redox is any chemical reaction in which electrons are transferred. Batteries, also known as voltaic cells, produce an electrical current at a constant voltage through redox reactions.

An oxidizing agent is a substance that accepts electrons and oxidizes something else while being reduced in the process. A reducing agent is a substance that provides electrons and reduces something else being oxidized in the process.

Every atom or ion has a prescribed oxidation number. This value compares the number of protons in an atom and the number of electrons assigned to that atom. In many cases, the oxidation number reflects the actual charge on the atom, but there are many cases in which it does not. The oxidation number is reduced in reduction by adding electrons. The oxidation number of an atom is increased during oxidation by removing electrons. All free, uncombined elements have an oxidation number of zero. Hydrogen, in all its compounds except hydrides, has an oxidation number of +1. Oxygen, in all its compounds except peroxides, has an oxidation number of –2.

Iron combines with oxygen to form rust. The rate of this reaction depends on several factors: temperature, surface area (more iron exposed for oxygen to get at), and catalysts (speed up a reaction but do not react and change themselves). Rusting is just like the burning of other substances.

Corrosion is the wearing away of metal due to chemical reactions, mainly oxidation. It occurs whenever a gas or liquid chemically attacks an exposed surface. It is accelerated by warm temperatures and by acids and salts. Some materials resist corrosion naturally; others can be treated to protect them, such as through painting, coatings, galvanizing, or anodizing.

Galvanizing involves the coating of zinc onto iron or steel to protect them against exposure to the atmosphere and rusting. If galvanizing is properly applied, it can protect the metals for 15 to 30 years or more. If the coating is damaged the iron or steel continues to be protected by sacrificial corrosion, a phenomenon in which atmospheric oxidation spares the iron and affects the zinc.

Anodizing is a method of coating metal for corrosion resistance, electrical insulation, thermal control, abrasion resistance, sealing, improved paint adhesion, and decorative finishing. Anodizing consists of electrically depositing an oxide film from an aqueous solution onto the surface of a metal, often aluminum. During the most common anodizing process, dyes can be added to the oxidation process to give the material a colored surface.

Metallurgy

Metallurgy is the art and science of extracting metals from their ores and modifying the metals for a particular use. It also concerns the chemical, physical, and atomic properties and structures of metals and the principles by which metals are combined to form alloys. An **alloy** is a mixture of two or more metals. Steel, for example, is an alloy of iron plus carbon and other elements.

A metal can be defined as any substance that has one or more of the following properties:

- Good heat and electric conduction
- Malleability—it can be hammered, pounded, or pressed into a shape without breaking
- Ductility—it can be stretched, drawn, or hammered without breaking
- High light reflectivity—it can make light bounce off its surface.
- The capacity to form positive ions in a solution and hydroxides rather than acids when its oxides meet water

About three-quarters of the elements are metals. The most abundant metals are aluminum, iron, calcium, sodium, potassium, and magnesium.

The hardness of something is a statement of its resistance to scratching. **Hardening** is a process that increases the hardness of a metal, deliberately or accidentally, by hammering, rolling, carburizing, heat treating, tempering, or other physical processes. All of these methods deform the metal, but also compact the atoms or molecules to make the material denser. The first few deformations imposed by such treatment weaken the metal, but because of the crystalline structure of metal its strength increases with continued deformations.

Carburizing is a method used to surface-harden steel by heat or mechanical means to increase the hardness of the outer surface while leaving the core relatively soft. This combination of a hard surface and soft interior withstands very high stress and fatigue and also offers low cost and superior flexibility in manufacturing. To carburize, the steel parts are placed in a carbonaceous environment (with charcoal, coke, and carbonates, or carbon dioxide, carbon monoxide, methane, or propane) at a high temperature for several hours. The carbon diffuses into the surface of the steel, altering the crystal structure of the metal. Gears, ball and roller bearings, and piston pins are often carburized.

Heat treating is the changing of the properties of a metal (including iron, steel, aluminum, copper, and titanium) by using heat. It is used to harden metals that have different crystal structures at low and high temperatures. The metal is heated and then quenched (cooled rapidly) to retain the high-temperature constituent. Mid-heating (tempering) may then be used to attain the desired hardness. Heating followed by slow cooling (annealing) is used to soften metals.

Tempering is the heat treating of metal alloys, particularly steel to result in specific properties. For example, raising the temperature of hardened steel from 752°F (4,000°C) and holding it for a time before quenching it in oil decreases its hardness and brittleness and produces strong steel. Quench and temper heat treating is applied at many different cooling rates, holding times, and temperatures and is a very important means of controlling the properties of steel.

Solids under Tension

The atoms of a solid are closely packed, giving it a greater density than most liquids and gases. A solid's rigidity is the result of the strong attraction between its atoms. A force pulling on a solid moves these atoms farther apart, creating an opposing force called tension. If a force pushes on a solid, the atoms move closer together, creating compression. These principles explain how springs function. Springs are used in many automotive systems, most obviously in suspension systems **(Figure 7–50)**.

An elastic substance is a solid that gets larger under tension, gets smaller under compression, and returns to its original size when no force is acting on it. Most solids show some elastic behavior, but there is usually a limit to the force from which recovery is possible. Stresses beyond its elastic limit cause a material to yield, or flow, and the result is permanent distortion or breakage. This limit depends on the material's internal structure; for example, steel, although strong, has a low elastic limit and can be extended only about 1% of its length, whereas rubber can be elastically extended to about 1,000%. Another factor involved in elasticity is the cross-sectional area of the material involved.

Tensile strength is the ratio of the maximum load a material can support without breaking when being stretched and is dependent on the cross-sectional area of the material. When stresses less than the tensile strength are removed, the material completely or partially returns to its original size and shape. As the stress approaches that of the tensile strength, the material forms a narrow, constricted region that is easily broken. Tensile strengths are measured in units of force per unit area.

Electrochemistry

Electrochemistry is a branch of chemistry concerned with the relationship between electricity and chemical change. Many spontaneous chemical reactions release electrical energy, and some of these reactions are used in batteries and fuel cells to produce electric power. The basis for electricity is the movement of electrons from one atom to another.

Electrolysis is an electrochemical process in which electric current is passed through a substance, causing a chemical change, usually the gaining or losing of electrons. It is carried out in an electrolytic cell consisting of separated positive and negative electrodes immersed in an electrolyte solution containing ions.

An **electrolyte** is a substance or compound that conducts electric current as a result of dissociation of its molecules into positive and negative ions. Dissociation is the breaking up of a chemical compound into simpler parts as a result of added energy. The most familiar electrolytes are acids, bases, and salts, which ionize when dissolved in solvents such as water and alcohol. Ions drift to the electrode of opposite charge in an electric field and are the conductors of current in electrolytic cells.

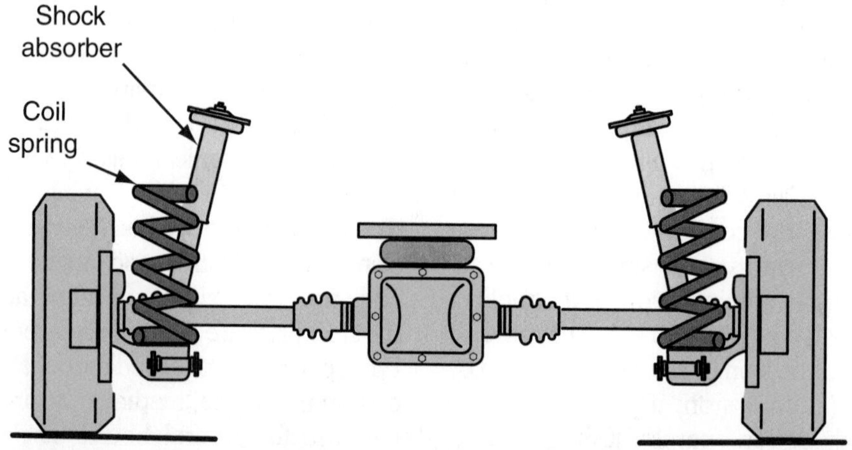

Figure 7-50 The placement of springs in an independent rear suspension.

ELECTRICITY AND ELECTROMAGNETISM

All electrical effects are caused by electric charges. There are two types of electric charge: positive and negative. These charges exert electrostatic forces on each other due to the strong attraction of electrons to protons. An electric field is the area on which these forces have an effect. In atoms protons carry positive charge, while electrons carry negative charge. Atoms are normally neutral, having an equal number of protons and electrons. However, an atom can gain or lose electrons, for example, by being rubbed. It then becomes a charged atom, or ion. Electricity has many similarities with magnetism. For example, the lines of the electric fields between charges take the same form as the lines of magnetic force, so magnetic fields can be said to be equivalent to electric fields. Charges of the same type repel, while charges of a different type attract **(Figure 7–51)**.

Electricity

An electric circuit is simply the path along which an electric current flows. Electrons carry negative charge and can be moved around a circuit by electrostatic forces. A circuit usually consists of a conductive material, such as a metal, in which the electrons are held very loosely to their atoms, thus making movement possible. The strength of the electrostatic force is the voltage. The resulting movement of the electric charge is called an electric current. The higher the voltage the greater the current will be, but the current also depends on the thickness, length, temperature, and nature of the materials that conduct it. The resistance of a material is the extent to which it opposes the flow of electric current. Good conductors have low resistance, which means that a small amount of voltage will produce a large current. In batteries, the dissolving of a metal electrode causes the freeing of electrons, resulting in their movement to another electrode and the formation of a current.

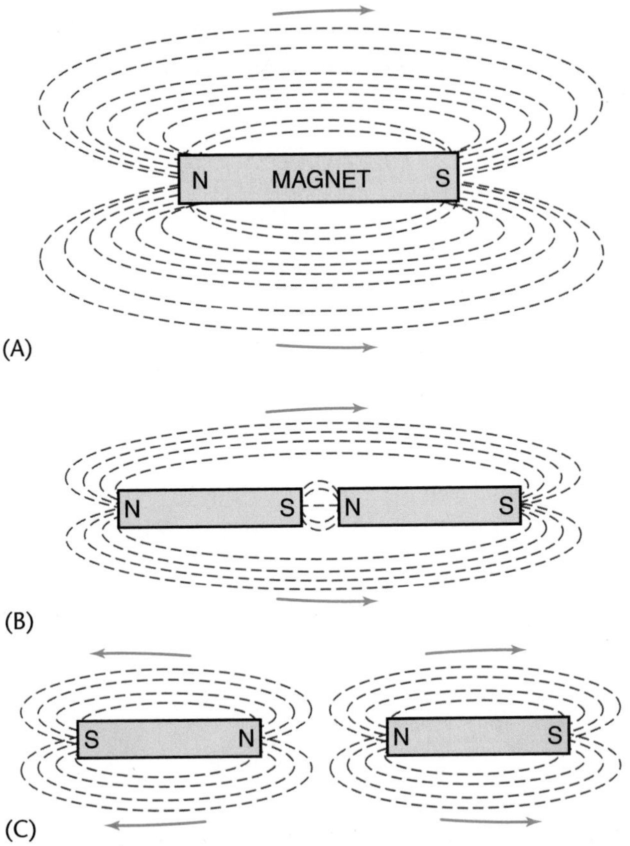

Figure 7-51 (A) In a magnet, lines of force emerge from the north pole and travel to the south pole before passing through the magnet back to the north pole. (B) Unlike poles attract, while (C) similar poles repel each other.

Magnets

Some materials are natural magnets; however, most magnets are produced. The materials typically used to make a permanent magnet are called ferromagnetic materials. These materials consist of mostly iron compounds that are heated. The heat causes the atoms to shift direction, and once they all point in the same direction the metal becomes a magnet. Two distinct poles, called the north and south poles, are at the ends of the magnet, and there is an attraction between the north pole and the south pole. This attraction, or force, set up by a magnet can be observed, but the type of force is not known.

The lines of a magnetic field form closed lines of force from the north to the south. If another iron or steel object enters into the magnetic field, it is pulled into the magnet. If another magnet is introduced into the magnetic field, it will either move into the field or push away from it as a result of the natural attraction of a magnet from north to south. If the north pole of one magnet is introduced to the north pole of another, the two poles will oppose each other and will push away. If the south pole of a magnet is introduced to the north pole of another, the two magnets will join together because the opposite poles are attracted to each other.

The strength of the magnetic force is uniform all around the outside of the magnet. The force is strongest at the surface of the magnet and weakens with distance. If you double the distance of an object from a magnet, the force is reduced by ⅛.

The strength of a magnetic field is typically measured with devices known as magnetometers and in units of Gauss (G).

Electromagnetism

Any electrical current produces magnetism that affects other objects in the same way as permanent magnets. The arrangement of force lines around a current-carrying conductor, its magnetic field, is circular. The magnetic effect of electrical current is increased by making the current-carrying wire into a coil **(Figure 7–52)**.

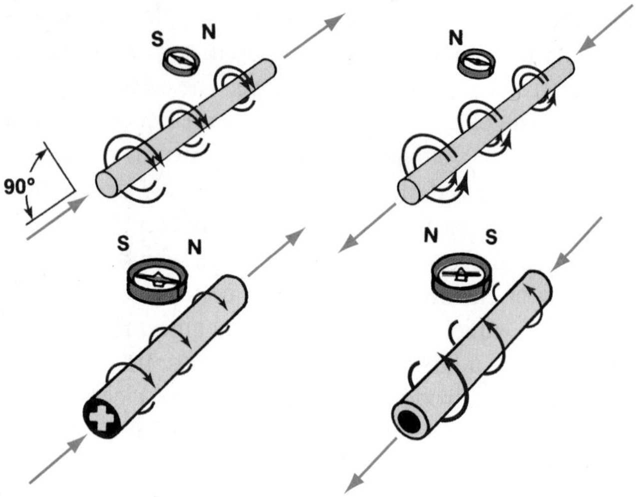

Figure 7-52 When current is passed through a conductor such as a wire, magnetic lines of force are generated around the wire at right angles to the direction of the current flow.

When a coil of wire is wrapped around an iron bar, it is called an **electromagnet**. The magnetic field produced by the coil magnetizes the iron bar, strengthening the overall effect. A field like that of a bar magnet is formed by the magnetic fields of the wires in the coil. The strength of the magnetism produced depends on the number of coils and the size of the current flowing in the wires. Electromagnetic coils and permanent magnets are arranged inside an electric motor so that the forces of electromagnetism create rotation of the armature.

Producing Electrical Energy

There are many ways to generate electricity. The most common way is to use coils of wire and magnets in a generator. Whenever a wire and magnet are moved relative to each other, a voltage is produced **(Figure 7–53)**. In a generator, the wire is wound into a coil. The more turns in the coil and the faster the coil moves, the greater the voltage. The coils or magnets spin around at high speed, typically turned by steam pressure. The steam is usually generated by burning coal or oil, a process that creates pollution. Renewable sources of electricity, such as hydroelectric power, wind power, solar energy, and geothermal power, produce only heat as a pollutant. In automobiles, the generator is spun by a belt driven by the engine's crankshaft. In a generator, the kinetic energy of a spinning object is converted into electrical energy.

A solar cell converts the energy of sunlight directly into electrical energy using layers of semiconductors. Electricity is produced by causing electrons to leave the atoms in the semiconductor material. Each electron leaves behind a hole or gap. Other electrons move into the hole, leaving holes in their atoms. This process continues all the way around a circuit. The moving chain of electrons is an electrical current.

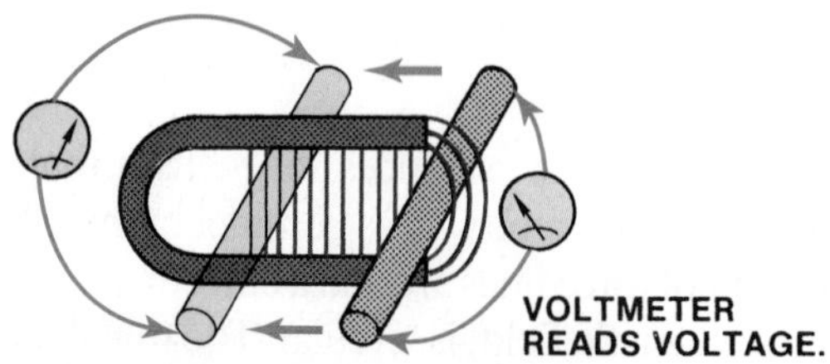

Figure 7-53 Moving a conductor across magnetic lines of force induces a voltage in the conductor.

Radio Waves

Electricity and magnetism are directly related. A changing electric field produces a changing magnetic field, and vice versa. Whenever an electric charge, such as that carried by an electron, accelerates, it gives out energy in the form of electromagnetic radiation. For example, electrons moving up and down a radio antenna produce a type of radiation known as radio waves. Electromagnetic radiation consists of oscillating electric and magnetic fields. There is a wide range of different types of electromagnetic radiation, called the electromagnetic spectrum, extending from low-energy radio waves to high-energy, short wavelength gamma rays, including visible light and X-rays.

KEY TERMS

Acceleration
Acid
Aerodynamics
Alloy
Amplitude
Atmospheric pressure
Atoms
Barometric pressure
Base
Carburizing
Catalyst
Centrifugal force
Centripetal force
Compression ratio
Conduction
Convection
Deceleration
Density
Diffusion
Displacement
Dynamic pressure
Electrolysis
Electrolyte
Electromagnet
Element
Engine efficiency
Equilibrium
Evaporate
Force
Frequency
Friction
Gear
Hardening
Heat
Heat treating
Hertz (Hz)
Horsepower
Impermeable
Inertia
Kinetic energy
Latent heat
Lever
Load
Mass
Matter
Mechanical advantage
Molecules
Momentum
Oscillation
Oxidation

Permeable
pH scale
Potential energy
Power
Pressure
Pulley
Radiation
Reduction
Solution
Solvent
Specific gravity
Speed
Static pressure
Temperature
Tempering
Tension
Tensile strength
Thermal contraction
Thermal expansion
Time
Torque
Vacuum
Velocity
Volume
Wave
Wavelength
Weight
Work

SUMMARY

- Matter is anything that occupies space and it exists as a gas, liquid, or solid.
- All matter consists of countless tiny particles called atoms and molecules.
- When a solid dissolves into a liquid, a solution is formed. Not all solids dissolve in a liquid; rather, the liquid is either absorbed or adsorbed.
- Materials that *absorb* fluids are permeable substances. Impermeable substances *adsorb* fluids.
- Energy is the ability to do work, and all matter has energy.
- The total amount of energy never changes; it can only be transferred from one form to another, not created or destroyed.
- When energy is released to do work, it is called kinetic energy. Stored energy may be called potential energy.
- Energy conversion occurs when one form of energy is changed to another form.
- Mass is the amount of matter in an object. Weight is a force and is measured in pounds or kilograms. Gravitational force gives the mass its weight.
- Volume is the amount of space occupied by an object.
- The volume of an engine's cylinders determines it size, expressed as displacement.
- The compression ratio of an engine is defined as the ratio of the volume in the cylinder above the piston when the piston is at the bottom of its travel to the volume in the cylinder above the piston when the piston is at its uppermost position.
- A force is a push or pull, can be large or small, and can be applied to objects by direct contact or from a distance.
- When an object moves in a circle its direction is continuously changing, and all changes in direction require a force. The forces required to maintain circular motion are called centripetal and centrifugal force.
- Pressure is a force applied against an opposing object and is measured in units of force per unit of surface area (pounds per square inch or kilograms per square centimeter).
- The greater the mass of an object, the greater the force needed to change its motion. Inertia is the tendency of an object at rest to remain at rest, or the tendency of an object in motion to stay in motion.
- When a force overcomes static inertia and moves an object, the object gains momentum. Momentum is the product of an object's weight times its speed.
- Speed is the distance an object travels in a set amount of time. Velocity is the speed of an object in a particular direction. Acceleration, which only occurs when a force is applied, is the rate of increase in speed. Deceleration, the rate of decrease in speed, is the reverse of acceleration.
- Newton's laws of motion state that an object at rest tends to remain at rest and an object in motion tends to remain in motion, unless some force acts on it; when a force acts on an object, the motion of the object will change; and for every action there is an equal and opposite reaction.
- Friction is a force that slows or prevents motion of two moving objects that touch.
- Friction can be reduced in two main ways: by lubrication or by the use of rollers.
- Aerodynamics is the study of the effects of air on a moving object.
- When a force moves a certain mass a specific distance, work is done.
- A machine is any device that can be used to transmit a force and, in doing so, change the amount of force and/or its direction. Examples of simple machines are inclined planes, pulleys, levers, gears, and wheels and axles.
- Torque is a force that tends to rotate or turn things and is measured by the force applied and the distance traveled.
- Gear ratios express the mathematical relationship of one gear to another.
- Power is a measurement of the rate at which work is done and is measured in watts.
- Horsepower is the rate at which torque is produced.
- An oscillation is any single swing of an object back and forth between the extremes of its travel. When

that motion travels through matter or space, it becomes a wave.

- How many times a vibration occurs in one second is called frequency. Frequency is most often expressed in hertz (Hz), which is equal to one cycle per second. The amplitude of a vibration is its intensity or strength.
- Noise is any unwanted signal or sound and can be random or periodic.
- Light is a form of electromagnetic radiation, which travels in a straight line at 300 million meters per second.
- A gas always fills a sealed container, whereas a liquid may not. A gas also readily expands or compresses according to the pressure exerted on it. Liquids are basically incompressible, which gives them the ability to transmit force.
- Hydraulics is the study of liquids in motion.
- Pascal constructed the first known hydraulic device and established what is known as Pascal's law of hydraulics.
- The pressure inside the hydraulic system is called static pressure because there is no fluid motion. The pressure of the fluid while it is in motion is called dynamic pressure.
- Boyle's law states that the volume and pressure of a mass of gas at a fixed temperature are inversely proportional.
- The pressure law states that the pressure exerted by a gas at constant volume increases as the temperature of the gas is increased.
- Charles's law states that the volume of a mass of gas depends on its temperature.
- Atmospheric pressure is the total weight of the Earth's atmosphere. Pressure greater than atmospheric pressure may be measured in psi gauge (psig); actual pressure is measured in psi absolute (psia).
- Scientifically, vacuum is defined as the absence of atmospheric pressure; however, it is commonly used to refer to any pressure less than atmospheric pressure.
- Heat is a form of energy caused by the movement of atoms and molecules and is measured in British thermal units (BTUs) and calories.
- Temperature is an indication of an object's kinetic energy and is measured with a thermometer that has either a Fahrenheit (F) or Celsius (C) (Centigrade) scale.
- Convection is the transfer of heat by the movement of a heated object.
- Conduction is the movement of heat through a material.
- Through radiation, heat is transferred by radiant energy.
- As heat moves in and out of a mass, the size of the mass tends to change.
- Sometimes, additional heat causes no increase in temperature but causes the mass to change its state; this heat is called latent heat.
- The liquid in a solution is called the solvent, and the substance added is the solute.
- Specific gravity is the heaviness or density of a substance compared to that of water.
- A catalyst is any substance that affects a chemical reaction without itself being consumed or changed.
- An ion is an atom or group of ions with one or more positive or negative electric charges. Ions are formed when electrons are added to or removed from neutral molecules or other ions. Ions are what make something an acid or a base.
- The pH scale is used to measure how acidic or basic a solution is.
- Oxidation is a chemical reaction in which a substance combines with oxygen. The addition of hydrogen atoms or electrons is reduction.
- Hardening is a process that increases the hardness of a metal, deliberately or accidentally, by hammering, rolling, carburizing, heat treating, tempering, or other physical processes.
- An elastic substance is a solid that gets larger under tension, gets smaller under compression, and returns to its original size when no force is acting on it.
- Tensile strength is the ratio of the maximum load a material can support without breaking when being stretched and is dependent on the cross-sectional area of the material.
- Electrolysis is an electrochemical process in which electric current is passed through a substance, causing a chemical change, usually the gaining or losing of electrons.
- An electrolyte is a substance or compound that conducts electric current as a result of dissociation of its molecules into positive and negative ions.
- Any electrical current produces magnetism. When a coil of wire is wrapped around an iron bar, it is called an electromagnet.
- The most common way to produce electricity is to use coils of wire and magnets in a generator.

TECH MANUAL

The following procedures are included in Chapter 7 of the *Tech Manual* that accompanies this book:

1. Measuring engine bore and stroke and calculating displacement.
2. Identifying the proper lubricants.
3. Calculating overall gear ratios.
4. Applying Pascal's Law.

REVIEW QUESTIONS

1. Describe Newton's first law of motion and give an application of this law in automotive theory.

2. Explain Newton's second law of motion and give an example of how this law is used in automotive theory.

3. Describe six different forms of energy.

4. Describe four different types of energy conversion.

5. Explain why a rotating tilted wheel moves in the direction of the tilt.

6. Describe the effects of static and dynamic balance.

7. Describe the effect of temperature on the volume of a gas.

8. The nucleus of an atom contains ____________ and ____________ .

9. Work is calculated by multiplying ____________ × ____________ .

10. Energy may be defined as the ability to do ____________ .

11. When one object is moved across the surface of another object, the resistance to motion is called ____________ .

12. Weight is the measurement of the Earth's ____________ ____________ on an object.

13. Torque is a force that does work with a ____________ action.

14. How are engines mounted in a vehicle and why?

15. Vacuum is defined as the absence of ____________ ____________ .

16. While discussing different types of energy, Technician A says when energy is released to do work it is called potential energy. Technician B says stored energy is referred to as kinetic energy. Who is correct?

a. A only
b. B only
c. Both A and B
d. Neither A nor B

17. While discussing friction in matter, Technician A says that friction creates heat. Technician B says that friction occurs in liquids, solids, and gases. Who is correct?

a. A only
b. B only
c. Both A and B
d. Neither A nor B

18. While discussing mass and weight, Technician A says that mass is the measurement of an object's inertia. Technician B says that mass and weight may be measured in cubic inches. Who is correct?

a. A only
b. B only
c. Both A and B
d. Neither A nor B

19. When applying the principles of work and force,

a. work is accomplished when force is applied to an object that does not move.

b. in the metric system the measurement for work is cubic centimeters.

c. no work is accomplished when an object is stopped by mechanical force.

d. if a 50-pound object is moved 10 feet, 500 ft.-lb of work are produced.

20. All these statements about energy and energy conversion are true, *except*:

a. thermal energy may be defined as light energy.

b. chemical to thermal energy conversion occurs when gasoline burns.

c. mechanical energy is defined as the ability to do work.

d. mechanical to electrical energy conversion occurs when the engine drives the generator.

SECTION 2

Engines

The engine of a car or truck is a complex piece of precision-built machinery. You should have a clear understanding of how the internal combustion process of an engine generates usable power before you attempt any engine work.

Engine repair and rebuilding procedures are among the most demanding jobs performed by technicians. Components must be handled with great care and assembled to exacting tolerances. Absolute cleanliness must be maintained. Bolt torque and sealing requirements must be met.

The information found in Section 2 of this textbook explains all of the engine's systems and components, including disassembly, inspection, repair, and reassembly procedures. It corresponds to the materials covered on the ASE certification test for engine repair.

Delmar's Engine Diagnosis and Repair Video Series and CD-ROM Courseware

WE ENCOURAGE PROFESSIONALISM

ASE

THROUGH TECHNICIAN CERTIFICATION

8

AUTOMOTIVE ENGINE DESIGNS AND DIAGNOSIS

OBJECTIVES

■ Describe the various ways in which engines can be classified. ■ Explain what takes place during each stroke of the four-stroke cycle. ■ Outline the advantages and disadvantages of the in-line and V-type engine designs. ■ Define important engine measurements and performance characteristics, including bore and stroke, displacement, compression ratio, engine efficiency, torque, and horsepower. ■ Explain how to evaluate the condition of an engine. ■ List and describe nine abnormal engine noises. ■ Outline the basics of diesel, stratified, and Miller-cycle engine operation.

INTRODUCTION TO ENGINES

The engine provides the power to drive the vehicle's wheels. All automobile engines, both gasoline and diesel, are classified as internal-combustion engines because the combustion or burning that creates energy takes place inside the engine.

The biggest part of the engine is the cylinder block **(Figure 8–1)**. The cylinder block is a large casting of metal that is drilled with holes to allow for the passage of lubricants and coolant through the block and provide spaces for movement of mechanical parts. The block contains the cylinders, which are round passageways fitted with pistons. The block houses or holds the major mechanical parts of the engine.

The cylinder head fits on top of the cylinder block to close off and seal the top of the cylinder **(Figure 8–2)**. The combustion chamber is an area into which the air/fuel mixture is compressed and burned. The cylinder head contains all or most of the combustion chamber. The cylinder head also contains ports through which the air/fuel mixture enters and burned gases exit the cylinder and the bore for the spark plug.

The valve train is a series of parts used to open and close the intake and exhaust ports. A valve is a movable part that opens and closes the ports. A camshaft controls the movement of the valves. Springs are used to help close the valves.

The up-and-down motion of the pistons must be converted to rotary motion before it can drive the wheels of a vehicle. This conversion is achieved by linking the piston to a crankshaft with a connecting rod. The upper end of the connecting rod moves with the piston. The lower end of the connecting rod is attached to the crankshaft and moves in a circle. The end of the crankshaft is connected to the flywheel.

Engine Construction

Modern engines are highly engineered power plants. These engines are designed to meet the performance and fuel efficiency demands of the public. Modern engines are made of lightweight engine castings and stampings; non-iron materials (for example, aluminum, magnesium, fiber-reinforced plastics); and fewer and smaller fasteners to hold things together. These fasteners are made

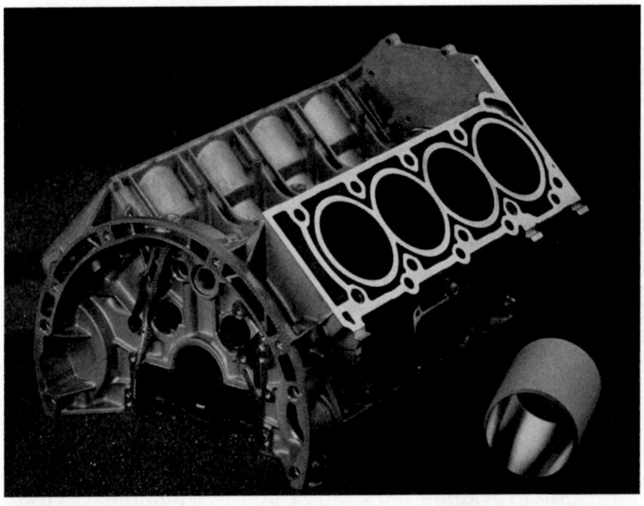

Figure 8–1 A cylinder block and cylinder liner for a late-model aluminum V-8 engine. *Courtesy of DaimlerChrysler Corporation*

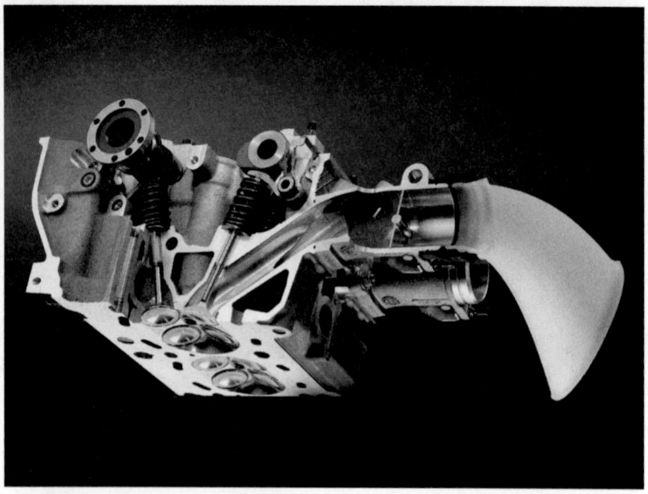

Figure 8-2 A cylinder head assembly for a late-model in-line six-cylinder engine. *Courtesy of BMW of North America, Incorporated*

Figure 8-3 A typical late-model engine. *Courtesy of American Honda Motor Co., Inc.*

possible through computerized joint designs that optimize loading patterns. Each of these newer engine designs has its own distinct personality, based on construction materials, casting configurations, and design **(Figure 8–3)**.

These modern engine-building techniques have changed how engine repair technicians make a living. Before these changes can be explained, it is important to explain the "basics" of engine design and operation.

ENGINE CLASSIFICATIONS

Today's automotive engines can be classified in several ways depending on the following design features:

- Operational cycles. Most technicians will generally come in contact with only four-stroke engines. However, a few older cars have used and some cars in the future will use a two-stroke engine.
- Number of cylinders. Current engine designs include 4-, 5-, 6-, 8-, 10-, and 12-cylinder engines.
- Cylinder arrangement. An engine can be flat (opposed), in-line, or V-type. Other more complicated designs have also been used.
- Valve train type. Engine valve trains can be either the **overhead camshaft (OHC)** type or the camshaft in-block **overhead valve (OHV)** type. Some engines separate camshafts for the intake and exhaust valves. These are based on the OHC design and are called **dual overhead camshaft (DOHC)** engines. V-type DOHC engines have four camshafts—two on each side.
- Ignition type. There are two types of ignition systems: spark and compression. Gasoline engines use a spark ignition system. In a spark ignition system, the air/fuel mixture is ignited by an electrical spark. Diesel engines, or compression ignition engines, have no spark plugs. An automotive diesel engine relies on the heat generated as air is compressed to ignite the air/fuel mixture for the power stroke.
- Cooling systems. There are both air-cooled and liquid-cooled engines in use. Nearly all of today's engines have liquid-cooling systems.
- Fuel type. Several types of fuel currently used in automobile engines include gasoline, natural gas, methanol diesel, and propane. The most commonly used is gasoline although new fuels are being tested.

Four-Stroke Gasoline Engine

In a passenger car or truck, the engine provides the rotating power to drive the wheels through the transmission and driving axles. All vehicle engines, both gasoline and diesel, are classified as internal combustion because the combustion or burning takes place inside the engine. These systems require an air/fuel mixture that arrives in the combustion chamber at the correct time and an engine constructed to withstand the temperatures and pressures created by the burning of thousands of fuel droplets.

The **combustion chamber** is the space between the top of the piston and the cylinder head. It is an enclosed area in which the fuel and air mixture is burned. The piston fits into a hollow metal tube, called a cylinder. The piston moves up and down in the cylinder.

This reciprocating motion must be converted to a rotary motion before it can drive the wheels of a vehicle. This change of motion is accomplished by connecting the piston to a crankshaft with a connecting rod **(Figure 8–4)**. The upper end of the connecting rod moves with the piston as it moves up and down in the cylinder. The lower end of the connecting rod is attached to the crankshaft and moves in a circle. The end of the crankshaft is con-

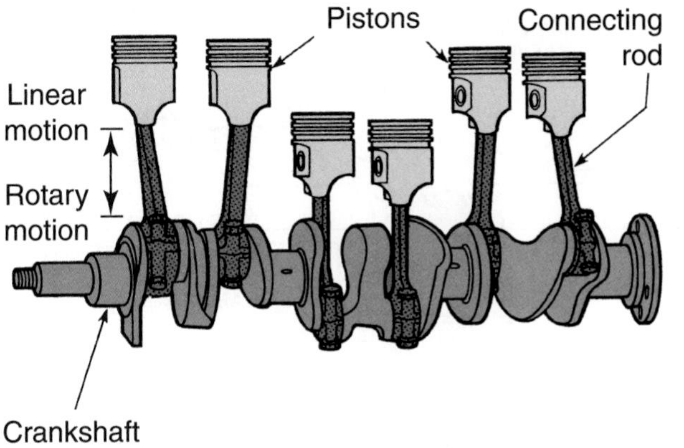

Figure 8–4 The reciprocating motion of the pistons is converted to rotary motion by the crankshaft.

nected to the flywheel, which transfers the engine's power through the drivetrain to the wheels.

In order to have complete combustion in an engine, the right amount of fuel must be mixed with the right amount of air. This mixture must be compressed in a sealed container, then shocked by the right amount of heat at the right time. When these conditions exist, all the fuel that enters a cylinder is burned and converted to power, which is used to move the vehicle. Automotive engines have more than one cylinder. Each cylinder should receive the same amount of air, fuel, and heat, if the engine is to run efficiently.

Although the combustion must occur in a sealed cylinder, the cylinder must also have some means of allowing heat, fuel, and air into it. There must also be a means to allow the burnt air/fuel mixture out so a fresh mixture can enter and the engine can continue to run. To accommodate these requirements, engines are fitted with valves.

There are at least two valves at the top of each cylinder. The air/fuel mixture enters the combustion chamber through an intake valve and leaves (after having been burned) through an exhaust valve. The valves are accurately machined plugs that fit into machined openings. A valve is said to be seated or closed when it rests in its opening. When the valve is pushed off its seat, it opens.

A rotating camshaft, driven by the crankshaft, opens and closes the intake and exhaust valves. **Cams** are raised sections of a shaft that have high spots called **lobes**. As the camshaft rotates, the lobes rotate and lift the valve open by pushing it away from its seat. Once the cam lobe rotates out of the way, the valve, forced by a spring, closes. The camshaft can be located either in the cylinder block or in the cylinder head.

When the action of the valves and the spark plug is properly timed to the movement of the piston, the combustion cycle takes place in four strokes of the piston: the intake stroke, the compression stroke, the power stroke, and the exhaust stroke. A stroke is the full travel of the piston either up or down in the cylinder bore.

The up-and-down movement of the piston on all four strokes is converted to a rotary motion by the crankshaft. It takes two full revolutions of the crankshaft to complete the four-stroke cycle.

The piston moves by the pressure produced during combustion only about half a stroke or one-quarter of crankshaft revolution. This explains why a flywheel is needed. The flywheel stores some of the power produced by the engine to keep the pistons in motion during the rest of the four-stroke cycle. This power is also used to compress the air/fuel mixture just before combustion.

Intake Stroke The first stroke of the cycle is the intake stroke. As the piston moves away from top dead center **(TDC)**, the intake valve opens **(Figure 8–5A)**. The downward movement of the piston increases the volume of the cylinder above it, reducing the pressure in the cylinder. This reduced pressure, commonly referred to as engine vacuum, causes the atmospheric pressure to push a mixture of air and fuel through the open intake valve. (Some engines are equipped with a super- or turbo-charger that pushes more air past the valve.) As the piston reaches the bottom of its stroke, the reduction in pressure stops, causing the intake of air/fuel mixture to slow down. It does not stop because of the weight and movement of the air/fuel mixture. It continues to enter the cylinder until the intake valve closes. The intake valve closes after the piston has reached bottom dead center **(BDC)**. This delayed closing of the valve increases the volumetric efficiency of the cylinder by packing as much air and fuel into it as possible.

Compression Stroke The compression stroke begins as the piston starts to move from BDC. The intake valve closes, trapping the air/fuel mixture in the cylinder **(Figure 8–5B)**. The upward movement of the piston compresses the air/fuel mixture, thus heating it up. At TDC, the piston and cylinder walls form a combustion chamber in which the fuel will be burned. The volume of the cylinder with the piston at BDC compared to the volume of the cylinder with the piston at TDC determines the compression ratio of the engine.

Power Stroke The power stroke begins as the compressed fuel mixture is ignited **(Figure 8–5C)**. With the valves still closed, an electrical spark across the electrodes of a spark plug ignites the air/fuel mixture. The burning fuel rapidly expands, creating a very high pressure against the top of the piston. This drives the piston down toward BDC. The downward movement of the piston is transmitted through the connecting rod to the crankshaft.

Exhaust Stroke The exhaust valve opens just before the piston reaches BDC on the power stroke **(Figure 8–5D)**. Pressure within the cylinder causes the exhaust gas to

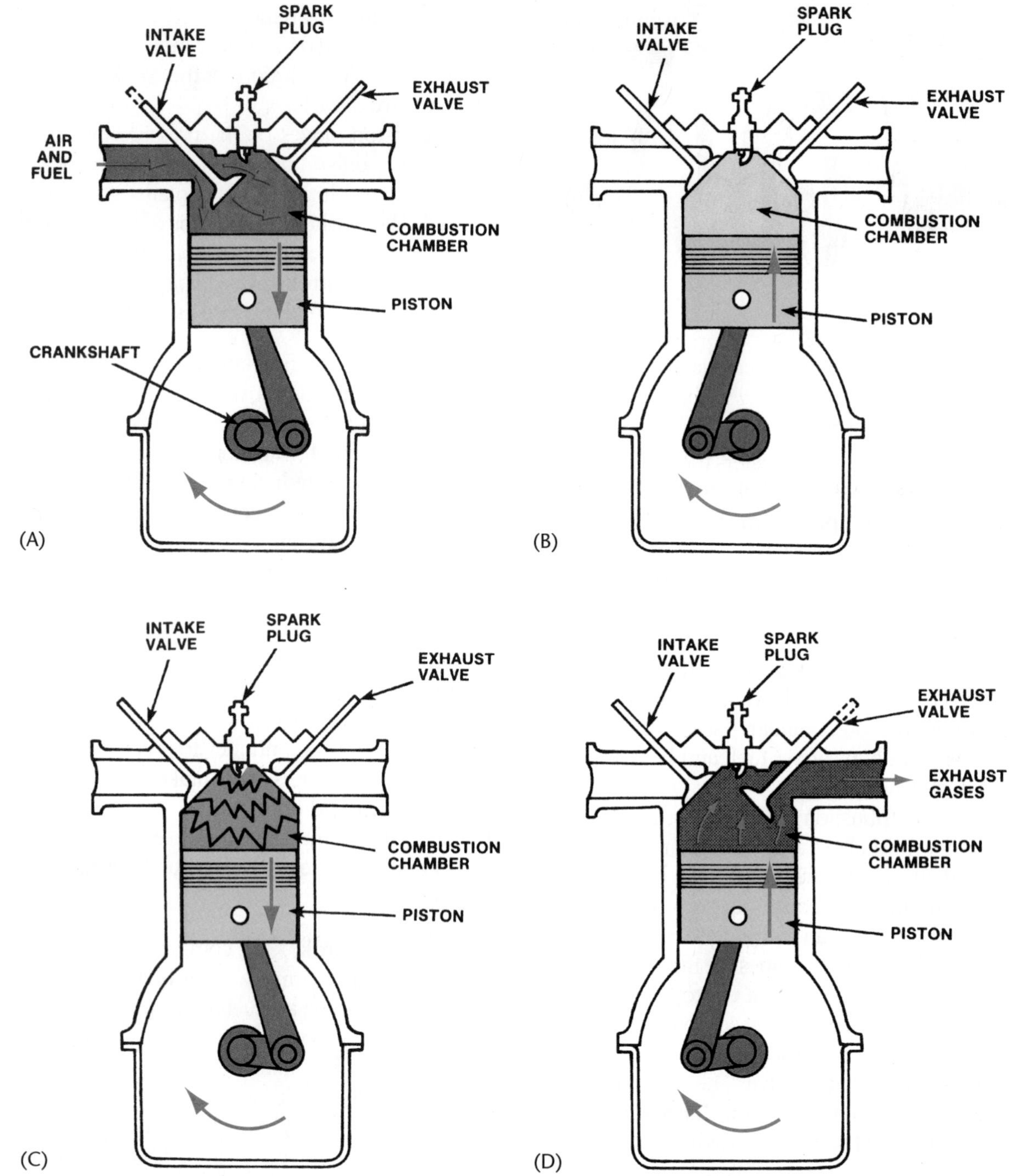

Figure 8-5 (A) Intake stroke, (B) compression stroke, (C) power stroke, and (D) exhaust stroke.

rush past the open valve and into the exhaust system. Movement of the piston from BDC pushes most of the remaining exhaust gas from the cylinder. As the piston nears TDC, the exhaust valve begins to close as the intake valve starts to open. The exhaust stroke completes the four-stroke cycle. The opening of the intake valve begins the cycle again. This cycle occurs in each cylinder and is repeated over and over, as long as the engine is running.

Two-Stroke Gasoline Engine

In the past, several imported vehicles have used two-stroke engines. As the name implies, this engine requires only two strokes of the piston to complete all four operations: intake, compression, power, and exhaust. This is accomplished as follows:

1. Movement of the piston from BDC to TDC completes both intake and compression.

2. When the piston nears TDC, the compressed air/fuel mixture is ignited, causing an expansion of the gases. During this time, the intake and exhaust ports are closed.
3. Expanding gases in the cylinder force the piston down, rotating the crankshaft.
4. With the piston at BDC, the intake and exhaust ports are both open, allowing exhaust gases to leave the cylinder and air/fuel mixture to enter.

Although the two-stroke-cycle engine is simple in design and lightweight because it lacks a valve train, it has not been widely used in automobiles. It tends to be less fuel efficient and releases more pollutants in the exhaust than four-stroke engines. Oil is often in the exhaust stream because these engines require constant oil delivery to the cylinders to keep the piston lubricated. Some of these engines require a certain amount of oil to be mixed with the fuel.

In recent years, however, thanks to a revolutionary pneumatic fuel injection system, there has been some interest in the two-stroke engine. The injection system uses compressed air to flow highly atomized fuel directly into the top of the combustion chamber. The system may be the long-sought-after answer to the fuel economy and emissions problems of the conventional two-stroke engine. This fuel injection system is the basis for the orbital two-stroke direct-injection piston engine, which may be used in vehicles in the future.

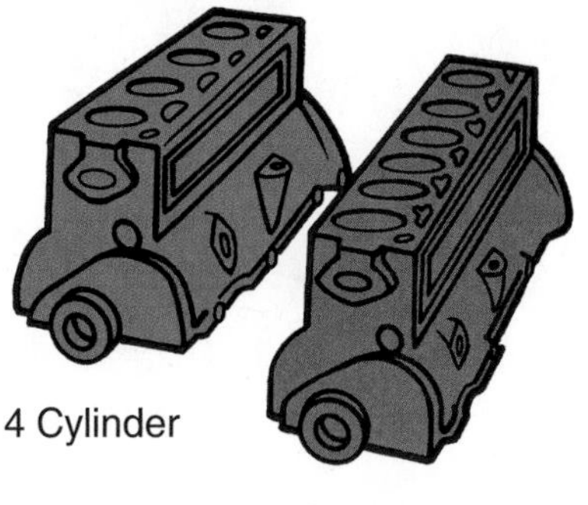

Figure 8-6 In-line engine designs.

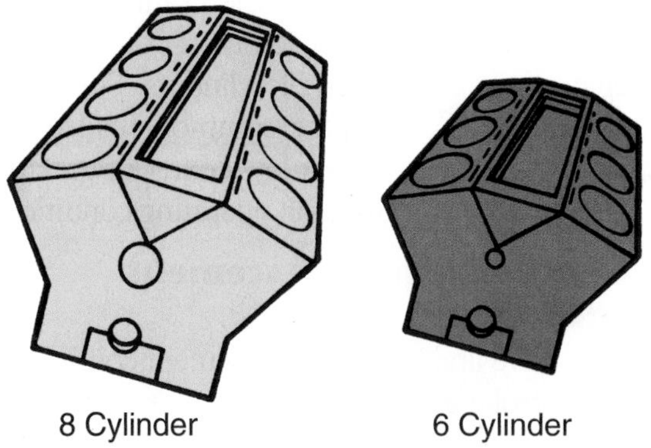

Figure 8-7 V-type engine designs.

Characteristics of Four-Stroke Engine Design

Depending on the vehicle, either an in-line, V-type, slant, or opposed cylinder design can be used. The most popular designs are in-line and V-type engines.

In-Line Engine In the in-line engine design **(Figure 8–6)**, the cylinders are all placed in a single row. There is one crankshaft and one cylinder head for all of the cylinders. The block is cast so that all cylinders are located in an upright position.

In-line engine designs have certain advantages and disadvantages. They are easy to manufacture and service. However, because the cylinders are positioned vertically, the front of the vehicle must be higher. This affects the aerodynamic design of the car. Aerodynamic design refers to the ease with which the car can move through the air. When equipped with an in-line engine, the front of a vehicle cannot be made as low as it can with other engine designs.

V-Type Engine The V-type engine design has two rows of cylinders **(Figure 8–7)** located 60 to 90 degrees away from each other. A V-type engine uses one crankshaft, which is connected to the pistons on both sides of the V. This type of engine has two cylinder heads, one over each row of cylinders.

One advantage of using a V-configuration is that the engine is not as high or long as one with an in-line configuration. The front of a vehicle can now be made lower. This design improves the outside aerodynamics of the vehicle. If eight cylinders are needed for power, a V-configuration makes the engine much shorter, lighter, and more compact. Many years ago, some vehicles had an in-line eight-cylinder engine. The engine was very long and its long crankshaft also caused increased torsional vibrations in the engine.

A variation of the V-type engine is the W-type engine. These engines are basically two V-type engines joined together at the crankshaft. This design makes the engine more compact. They are commonly found in late-model Volkswagens.

Slant Cylinder Engine Another way of arranging the cylinders is in a slant configuration **(Figure 8–8A)**. This arrangement is much like an in-line engine, except the entire block has been placed at a slant. The slant engine was designed to reduce the distance from the top to the bottom of the engine. Vehicles using the slant engine can be designed more aerodynamically.

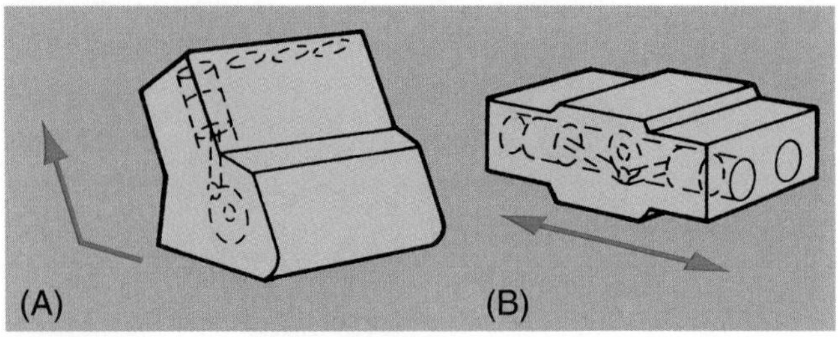

Figure 8-8 (A) Slant cylinder and (B) opposed cylinder engine designs.

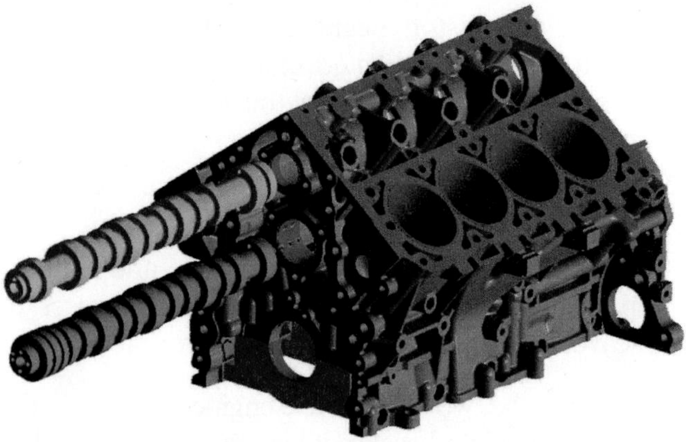

Figure 8-10 An engine block with two camshafts. *Courtesy of General Motors Corporation*

Opposed Cylinder Engine In this design, two rows of cylinders are located opposite the crankshaft **(Figure 8–8B)**. These engines have a common crankshaft and a cylinder head on each bank of cylinders. Porsches and Subarus use this style of engine, commonly called a boxer engine. Boxer engines have a low center of gravity and tend to run smoothly during all operating conditions.

Valve and Camshaft Placement Configurations

Two basic valve and camshaft placement configurations of the four-stroke gasoline engines are used in vehicles.

Overhead Valve (OHV) As the name implies, the intake and exhaust valves on an overhead valve engine are mounted in the cylinder head and are operated by a camshaft located in the cylinder block. This arrangement requires the use of valve lifters, pushrods, and rocker arms to transfer camshaft rotation to valve movement **(Figure 8–9)**. The intake and exhaust manifolds are attached to the cylinder head.

A recent development by GM incorporates two camshafts in the engine block **(Figure 8–10)**. There is a separate camshaft for the intake valves and another for the exhaust valves. This setup allows for variable intake valve operation while keeping the exhaust valves dictated by their camshaft. Having the camshafts placed in the engine block eliminates the need for long valve timing belts and/or chains.

Overhead Cam (OHC) An overhead cam engine also has the intake and exhaust valves located in the cylinder head. But as the name implies, the cam is located in the cylinder head. In an overhead cam engine, the valves are operated directly by the camshaft or through cam followers or tappets **(Figure 8–11)**. Some engines have separate camshafts for the intake and the exhaust valves. These are called dual overhead camshaft (DOHC) engines.

Valve and Camshaft Operation

In OHV engines with the camshaft in the block **(Figure 8–12)**, the valves are operated by valve lifters and

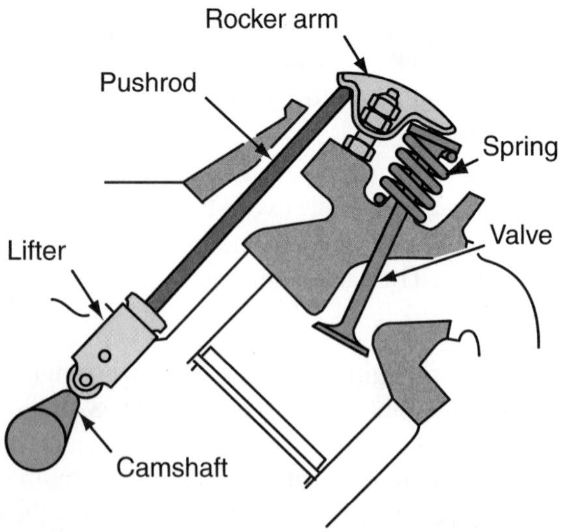

Figure 8-9 The basic valve train for an overhead valve engine.

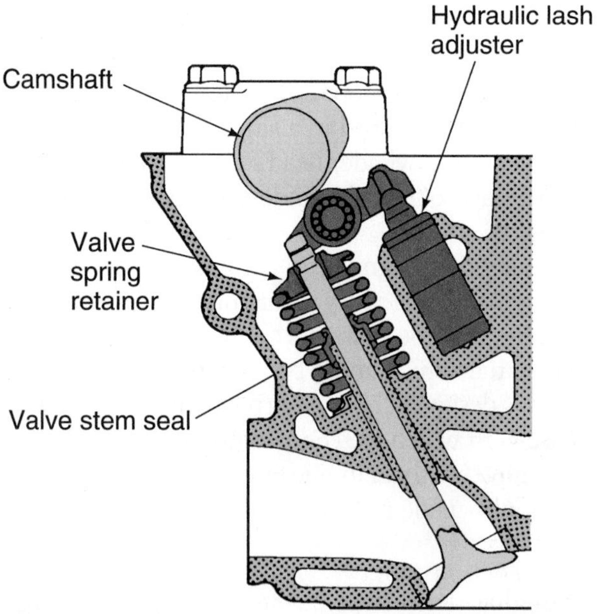

Figure 8-11 Basic valve and camshaft placement in an overhead camshaft engine. *Courtesy of Hyundai Motor America*

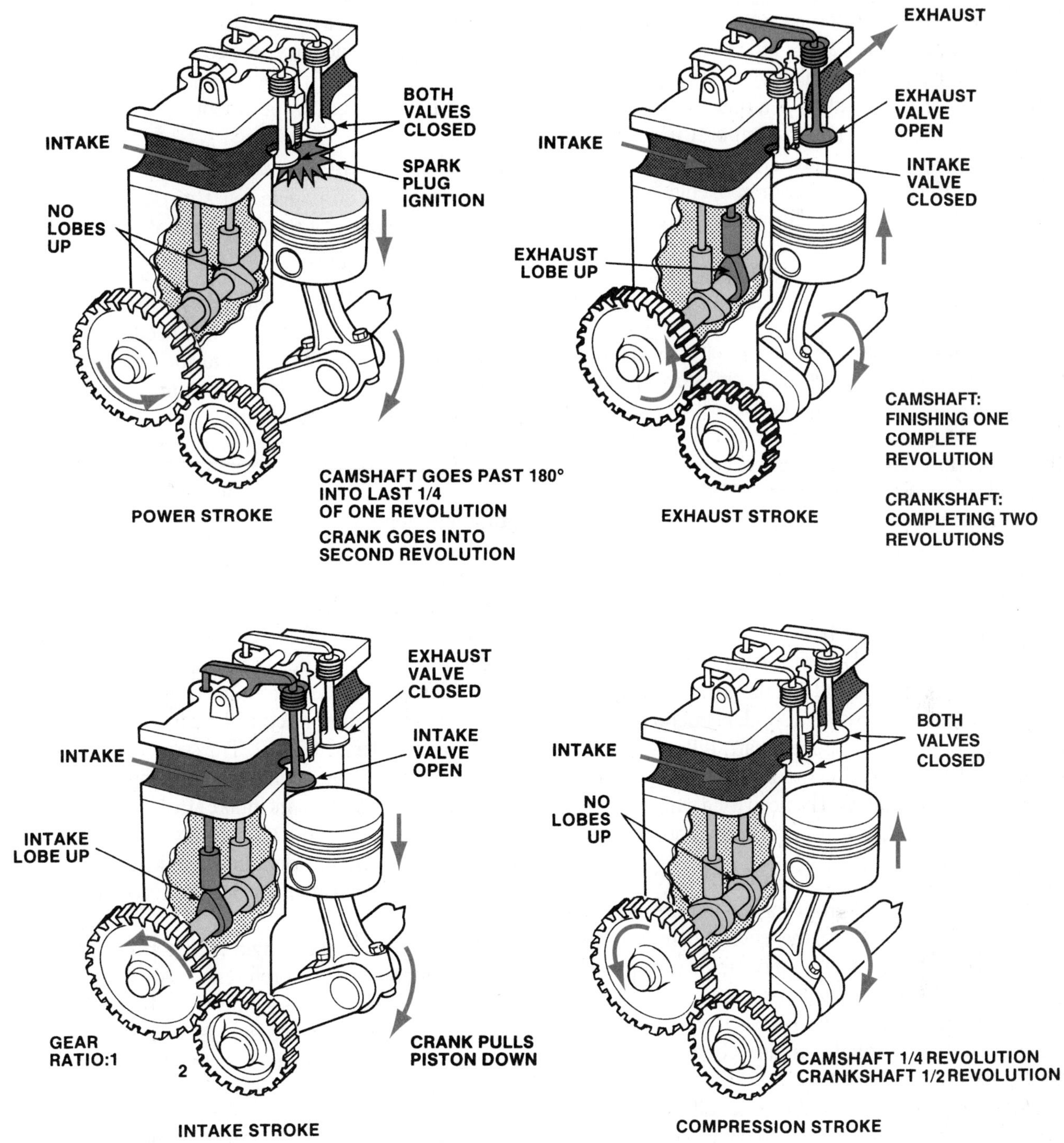

Figure 8-12 Valve operation in an overhead valve engine.

pushrods that are actuated by the camshaft. On overhead cam engines, the cam lobes operate the valves directly and there is no need for pushrods. (Lifters may be used between the camshaft and valves.)

Cam lobes are oval shaped. The placement of the lobe on the shaft determines when the valve will open. Design of the lobe determines how far the valve will open and how long it will remain open in relation to piston movement.

The camshaft is driven by the crankshaft through gears, or sprockets, and a cogged belt, or timing chain. The camshaft turns at half the crankshaft speed and rotates one complete turn during each complete four-stroke cycle.

Engine Location

The engine is usually placed in one of three locations. In most vehicles, it is located at the front of the vehicle, in front of the passenger compartment. Front-mounted engines can be positioned either longitudinally or transversely with respect to the vehicle.

The second engine location is a mid-mount position between the passenger compartment and rear suspension. Mid-mount engines are normally transversely mounted. The third, and least common, engine location is the rear of the vehicle. The engines are typically opposed-type engines.

Each of these engine locations offers advantages and disadvantages.

Front Engine Longitudinal In this type of vehicle, the engine, transmission, front suspension, and steering equipment are installed in the front of the body, and the differential and rear suspension are installed in the rear of the body. Most front engine longitudinal vehicles are rear-wheel drive. Some front-wheel-drive cars with a transaxle have this configuration, and most four-wheel-drive vehicles are equipped with a transfer case and have the engine mounted longitudinally in the front of the vehicle.

Total vehicle weight can be evenly distributed between the front and rear wheels with this configuration. This lightens the steering force and equalizes the braking load. With this design, it is possible to independently remove and install the engine, propeller shaft, differential, and suspension. Longitudinally mounted engines require large engine compartments. The need for a rear-drive propeller shaft and differential also cuts down on passenger compartment space.

Front Engine Transverse Front engines that are mounted transversely sit sideways in the engine compartment. They are used with transaxles that combine transmission and differential gearing into a single compact housing, fastened directly to the engine. Transversely mounted engines reduce the size of the engine compartment and overall vehicle weight.

Transversely mounted front engines allow for downsized, lighter vehicles with increased interior space. However, most of the vehicle weight is toward the front of the vehicle. This provides for increased traction by the drive wheels. The weight also places a greater load on the front suspension and brakes.

Mid-Engine Transverse In this design, the engine and drivetrain are positioned between the passenger compartment and rear axle. Mid-engine location is used in smaller, rear-wheel-drive, high-performance sports cars for several reasons. The central location of heavy components results in a center of gravity very near the center of the vehicle, which vastly improves steering and handling. Since the engine is not under the hood, the hood can be sloped downward, improving aerodynamics and increasing the driver's field of vision. However, engine access and cooling efficiency are reduced. A barrier is also needed to reduce the transfer of noise, heat, and vibration to the passenger compartment.

Gasoline Engine Systems

The operation of an engine relies on several other systems. The efficiency of these systems affects the overall operation of the engine.

Air/Fuel System This system makes sure the engine gets the right amount of both air and fuel needed for efficient operation. For many years air and fuel were mixed in a carburetor, which supplied the resulting mixture to the cylinder. Today, most late-model automobiles have a fuel injection system, which replaces the carburetor but performs the same basic function.

Ignition System This system delivers a spark to ignite the compressed air/fuel mixture in the cylinder at the end of the compression stroke. An engine's **firing order** indicates the order at which an engine's pistons are on their compression stroke and therefore the order in which the cylinders' spark plugs fire. The firing order also indicates the position of all of the pistons in an engine when a cylinder is firing. For example, consider a four-cylinder engine with a firing order of 1-3-4-2. The sequence begins with piston 1 on the compression stroke. During that time, piston 3 is moving down on its intake stroke, 4 is moving up on its exhaust stroke, and 2 is moving down on its power stroke. These events are identified by thinking about what needs to happen in order for 3 to be ready to fire next, and so on.

The firing order of an engine is determined by its design and manufacturer preference. An engine's firing order can be found on the engine or on the engine's emissions label and in service manuals. **Figure 8–13** shows some of the common cylinder arrangements and their associated firing orders.

Lubrication System This system supplies oil to the various moving parts in the engine. The oil lubricates all parts that slide in or on other parts, such as the piston, bearings, crankshaft, and valve stems. The oil reduces friction and enables the parts to move easily so little power is lost and wear is kept to a minimum. The lubrication system also helps transfer heat from one part to another for cooling.

Cooling System This system is also extremely important. Coolant circulates in jackets around the cylinder and in the cylinder head and intake manifold. The cooling system removes part of the heat produced by combustion and prevents the engine from being damaged by overheating.

①②③④⑤⑥ FIRING ORDER 1-5-3-6-2-4 6 CYLINDER	①②③④ RIGHT BANK ⑤⑥⑦⑧ LEFT BANK FIRING ORDER 1-5-4-8-6-3-7-2 V-8
②④⑥⑧ RIGHT BANK ①③⑤⑦ LEFT BANK FIRING ORDER 1-8-4-3-6-5-7-2 V-8	①②③④ FIRING ORDER 1-3-4-2 1-2-4-3 4 CYLINDER
⑤③① RIGHT BANK ⑥④② LEFT BANK FIRING ORDER 1-4-5-2-3-6 V-6	②④⑥ RIGHT BANK ①③⑤ LEFT BANK FIRING ORDER 1-6-5-4-3-2 V-6
②④⑥⑧ RIGHT BANK ①③⑤⑦ LEFT BANK FIRING ORDER 1-8-7-2-6-5-4-3 V-8	①②③④ RIGHT BANK ⑤⑥⑦⑧ LEFT BANK FIRING ORDER 1-5-4-2-6-3-7-8 V-8

Figure 8-13 Common cylinder firing orders.

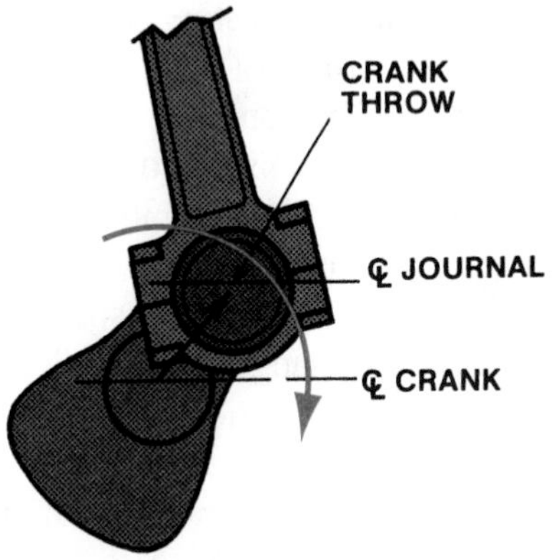

Figure 8-14 The stroke of an engine is equal to twice the crank throw.

Exhaust System This system removes the burned gases from the combustion chamber and limits the noise produced by the engine. It also carries deadly CO away from the passenger compartment to the rear of the vehicle.

Emission Control System Several control devices designed to reduce the amount of pollutants released by the engine have been added to the engine. Engine design changes, such as reshaped combustion chambers and altered valve timing, have also been part of the manufacturers' attempt to reduce emission levels.

ENGINE MEASUREMENT AND PERFORMANCE

Many of the engine measurements and performance characteristics a technician should be familiar with were discussed in Chapter 7. What follows are some of the important facts of each.

Bore and Stroke

The bore of a cylinder is simply its diameter measured in inches (in.) or millimeters (mm). The stroke is the length of the piston travel between TDC and BDC. Between them, bore and stroke determine the displacement of the cylinders. When the bore and stroke are of equal size, the engine is called a *square engine*. Engines that have a larger bore than stroke are called oversquare and engines with a larger stroke than bore are referred to as being undersquare. **Oversquare** engines offer the opportunity to fit larger valves in the combustion chamber and use longer connecting rods, which means oversquare engines are capable of running at higher engine speeds. But because of the size of the bore, the engines tend to be physically larger than undersquare engines. **Undersquare** engines have short connecting rods that aid in the production of more power at lower engine speeds. A square engine is a compromise between the two designs.

The **crank throw** is the distance from the crankshaft's main bearing centerline to the crankshaft throw centerline. The stroke of any engine is twice the crank throw **(Figure 8–14)**.

Displacement

A cylinder's displacement is the volume of the cylinder when the piston is at BDC. An engine's displacement is the sum of the displacements of each of the engine's cylinders **(Figure 8–15)**. Typically, an engine with a larger displacement produces more torque than a smaller displacement engine; however, many other factors influence an engine's power output. Engine displacement can be changed by changing the size of the bore and stroke of an engine.

Compression Ratio

An engine's stated compression ratio is a comparison of a cylinder's volume when the piston is at BDC to the

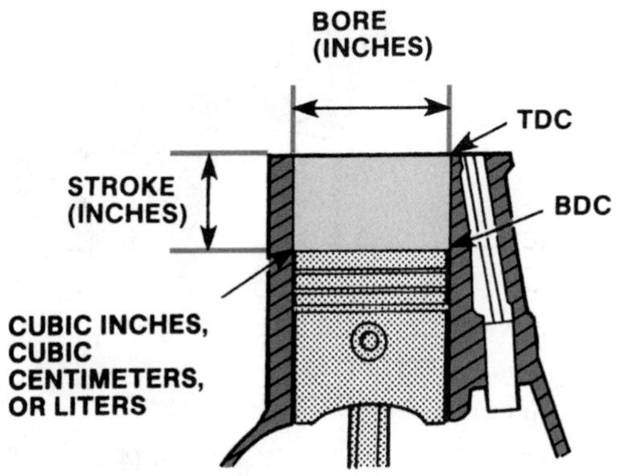

Figure 8-15 Displacement is the volume the cylinder holds between TDC and BDC.

cylinder's volume when the piston is at TDC. The compression ratio is a statement of how the air/fuel mixture is compressed during the compression stroke. It is important to keep in mind that this ratio can change through wear and carbon and dirt buildup in the cylinders. For example, if a great amount of carbon collects on the top of the piston and around the combustion chamber, the volume of the cylinder changes. This buildup of carbon will cause the compression ratio to increase because the volume at TDC will be smaller.

The higher the compression ratio, the more power an engine theoretically can produce. Also, as the compression ratio increases, the heat produced by the compression stroke also increases. Gasoline with a low-octane rating burns fast and may explode rather than burn when introduced to a high-compression ratio, which can cause preignition. The higher a gasoline's octane rating, the less likely it is to explode.

As the compression ratio increases, the octane rating of the gasoline should also be increased to prevent abnormal combustion.

Variable Compression Ratio Engines A system that varies the compression ratio of an engine according to operating conditions was developed by Saab. The engine in this system, called Saab Variable Compression (SVC), has a cylinder head with integrated cylinders. The compression ratio is altered by changing the slope of the engine in relation to the engine block. This action changes the volume of the combustion chamber **(Figure 8–16)**. The cylinder head is pivoted at the crankshaft by a hydraulic actuator and can be moved up to 4 degrees. The engine management system adjusts the angle (and therefore the compression ratio) in response to engine speed, engine load, and fuel quality. The cylinder head is sealed to the engine block by a rubber bellows **(Figure 8–17)**.

Figure 8–17 The hydraulic actuator and rubber bellows of an SVC engine. *Courtesy of Saab Cars USA*

Engine Efficiency

One of the dominating trends in automotive design is increasing an engine's efficiency. Efficiency is simply a measure of the relationship between the amount of energy put into an engine and the amount of energy available from the engine. Other factors, or efficiencies, affect the overall efficiency of an engine.

Volumetric Efficiency Volumetric efficiency describes the engine's ability to have its cylinders filled with air/fuel mixture. If the engine's cylinders are able to be filled with

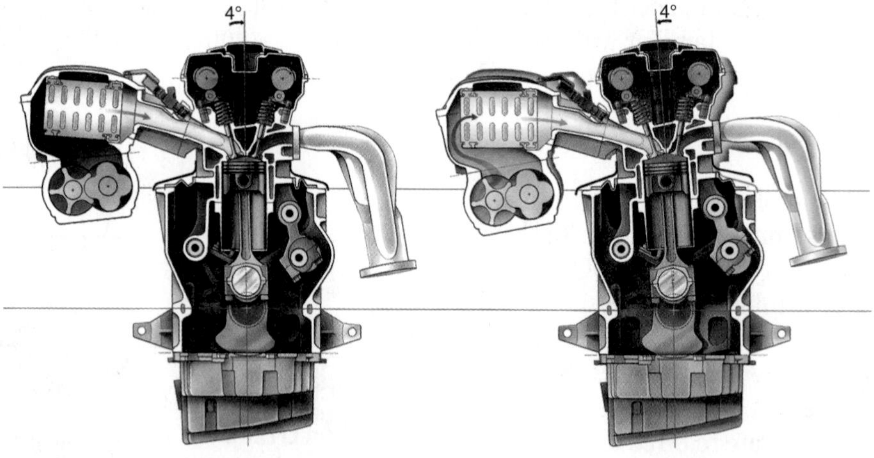

Figure 8–16 The SVC can vary the engine's compression ratio from 8:1 to 14:1. *Courtesy of Saab Cars USA*

air/fuel mixture during its intake stroke, the engine has a volumetric efficiency of 100%. Typically, engines have a volumetric efficiency of 80 to 100% if they are not equipped with a turbo- or supercharger. Basically, an engine becomes more efficient as its volumetric efficiency is increased.

Thermal Efficiency Thermal efficiency is a measure of how much of the heat formed during the combustion process is available as power from the engine. Typically only one-third of the heat is used to power the vehicle. The rest is lost to the surrounding air and engine parts and to the engine's coolant. Obviously, when less combustion heat is lost, the engine is more efficient.

Mechanical Efficiency Mechanical efficiency is a measure of how much power is available once it leaves the engine compared to the amount of power that was exerted on the pistons during the power stroke. Power losses occur because of the friction generated by the moving parts. Minimizing friction increases mechanical efficiency.

Torque versus Horsepower

Torque is a twisting or turning force. Horsepower is the rate at which torque is produced. An engine produces different amounts of torque based on the rotational speed of the crankshaft and other factors. A mathematical representation, or graph, of the relationship between the horsepower and torque of an engine is shown in **Figure 8–18**.

This graph shows that torque begins to decrease when the engine's speed reaches about 1,700 rpm. Brake horsepower increases steadily until about 3,500 rpm. Then it drops. The third line on the graph indicates the horsepower needed to overcome the friction or resistance created by the internal parts of the engine rubbing against each other.

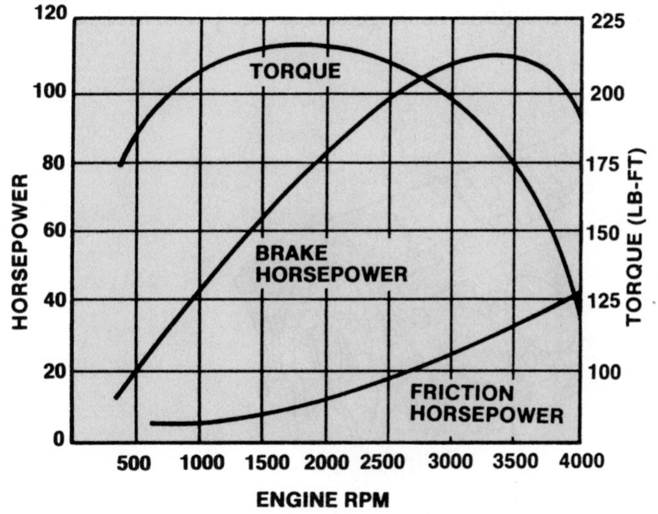

Figure 8-18 The relationship between horsepower and torque.

ENGINE IDENTIFICATION

To find the correct specifications for an engine, a technician must know how to use the vehicle identification number (VIN). The VIN is a code of seventeen numbers and letters stamped on a metal tab that is riveted to the instrument panel close to the windshield **(Figure 8–19)**. From this number much information about the vehicle can be found.

> **USING SERVICE MANUALS**
>
> Normally, information used to identify the size of an engine is given in service manuals at the beginning of the section covering that particular manufacturer. ■

The adoption of the seventeen-number-and-letter code became mandatory beginning with 1981 vehicles. The standard VIN of the United States National Highway Transportation and Safety Administration Department of Transportation is being used by all manufacturers of vehicles, both domestic and foreign.

By referring to the VIN, much information about the vehicle can be determined **(Figure 8–20)**. Identification numbers are also found on the engine. Some manufacturers use tags or stickers attached at various places, such as the valve cover or oil pan. Blocks often have a serial number stamped into them **(Figure 8–21)**. Service manuals typically give the location of the code for a particular engine. The engine code is generally found beside the serial number. A typical engine code might be DZ or MO. These letters indicate the horsepower rating of the engine, whether it was built for an automatic or manual transmission, and other important details. The engine code will help you determine the correct specifications for that particular engine.

Figure 8-19 The VIN is visible through the driver's side of the windshield.

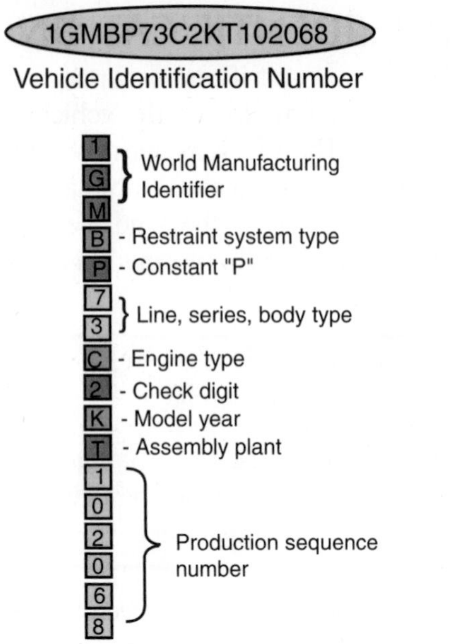

Figure 8-20 The VIN provides a great deal of information.

Casting numbers are often mistaken for serial numbers and engine codes. Manufacturers use a casting number to identify major engine parts on the assembly line. They seldom can be used to identify the type of engine. Normally casting numbers are raised from the metal while ID numbers are usually stamped.

Underhood Label Vehicles produced since 1972 have an underhood emission control label that contains such useful information as ignition timing specifications, emission control devices, engine size, vacuum hose routing, and valve adjustment specifications.

ENGINE DIAGNOSTICS

As the trend toward the integration of ignition, fuel, and emission systems progresses, diagnostic test equipment must also keep up with these changes. New tools and techniques are constantly being developed to diagnose electronic engine control systems. However, not all engine performance problems are related to electronic control systems; therefore, technicians still need to understand basic engine tests. These tests are an important part of modern engine diagnosis.

Compression Test

Internal combustion engines depend on compression of the air/fuel mixture to maximize the power produced by the engine. The upward movement of the piston on the compression stroke compresses the air/fuel mixture within the combustion chamber. The air/fuel mixture gets hotter as it is compressed. The hot mixture is easier to ignite, and when ignited it generates much more power than the same mixture at a lower temperature.

If the combustion chamber leaks, some of the air/fuel mixture will escape when it is compressed, resulting in a loss of power and a waste of fuel. The leaks can be caused by burned valves, a blown head gasket, worn rings, slipped timing belt or chain, worn valve seats, a cracked head, and more.

An engine with poor compression (lower compression pressure due to leaks in the cylinder) will not run correctly and cannot be tuned to factory specifications. If a symptom suggests that the cause of a problem may be poor compression, a compression test is performed.

A compression gauge is used to check cylinder compression. The dial face on the typical compression gauge indicates pressure in both pounds per square inch (psi) and metric kilopascals (kPa). Most compression gauges have a vent valve that holds the highest pressure reading on its meter. Opening the valve releases the pressure when the test is complete. The steps for conducting a cylinder compression test are shown in Photo Sequence 5.

Cylinder Leakage Test

If a compression test shows that any of the cylinders are leaking, a cylinder leakage test can be performed to mea-

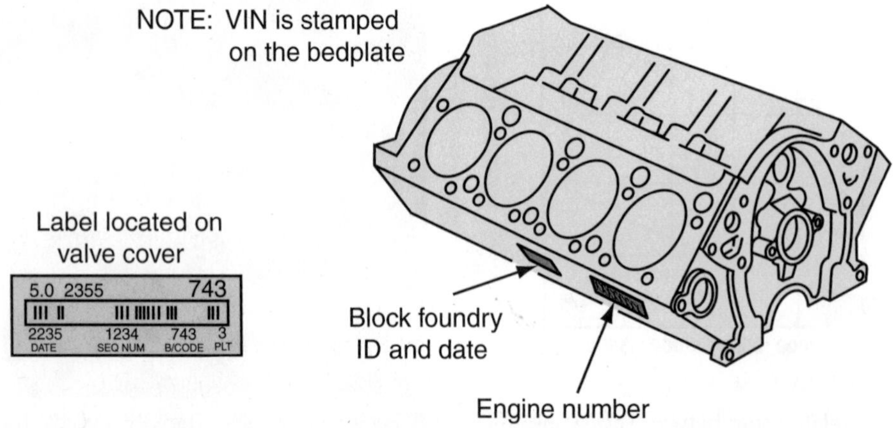

Figure 8-21 Examples of the various identification numbers found on an engine.

sure the percentage of compression lost and help locate the source of leakage. A cylinder leakage tester applies compressed air to a cylinder through the spark plug hole. Before the air is applied to the cylinder, the piston of that cylinder must be at TDC on its compression stroke to ensure that the valves of that cylinder are closed.

A threaded adapter on the end of the air pressure hose screws into the spark plug hole. The source of the compressed air is normally the shop's compressed air system. A pressure regulator in the tester controls the pressure applied to the cylinder. A gauge registers the percentage of air pressure lost from the cylinder when the compressed air is applied. The scale on the gauge typically reads 0 to 100%.

A zero reading means there is no leakage in the cylinder. Readings of 100% indicate that the cylinder will not hold any pressure. Any reading that is more than 0% indicates some leakage. The location of the compression leak can be found by listening and feeling around various parts of the engine. If air is felt or heard leaving the throttle plate assembly, a leaking intake valve is indicated. If a bad exhaust valve is responsible for the leakage, air can be felt leaving the exhaust system during the test. If air is felt or heard coming from the spark plug hole of the cylinder next to the one being tested, a bad head gasket or cylinder head problem is indicated. Air leaving the radiator indicates a faulty head gasket or a cracked block or head. If the piston rings are bad, air will be heard leaving the valve cover's breather cap or the oil dipstick tube.

Most engines, even new ones, experience some leakage around the rings. Up to 20% is considered acceptable during the leakage test. When the engine is actually running, the rings will seal much better and the actual percent of leakage will be lower. However, there should be no leakage around the valves or the head gasket.

Cylinder Power Balance Test

The cylinder power balance test is used to see if all of the engine's cylinders are producing the same amount of power. Ideally, all cylinders will produce the same amount. To check an engine's power balance, the spark plugs for individual cylinders are shorted out one at a time and the change in engine speed is recorded. If all of the cylinders are producing the same amount of power, engine speed will drop the same amount as each plug is shorted. Unequal cylinder power balance can mean a problem in the cylinders themselves, as well as in the rings, valves, intake manifold, head gasket, fuel system, or ignition system.

A power balance test is performed quickly and easily using an engine analyzer, because the firing of the spark plugs can be automatically or manually controlled by pushing a button. Some vehicles have a power balance test built into the engine control computer. This test is either part of a routine self-diagnostic operating mode or must be activated by the technician.

WARNING!

On some computer-controlled engines, certain components must be disconnected before attempting the power balance test. Always check the service manual for appropriate procedures. Be careful not to run the engine with a shorted cylinder for more than 15 seconds. The unburned fuel in the exhaust can build up in the catalytic converter and create an unsafe situation. Also run the engine for at least 10 seconds between cylinder shortings.

Override the controls of the electric cooling fan by using jumper wires to make the fan run constantly. If the fan control cannot be bypassed, disconnect the fan.

Connect the engine analyzer's leads, then turn the engine on and allow it to reach normal operating temperature. Set the engine speed at 1,000 rpm and connect a vacuum gauge to the intake manifold. As each cylinder is shorted, note and record the rpm drop and the change in vacuum.

As each cylinder is shorted, a noticeable drop in engine speed should be noted. Little or no decrease in speed indicates a weak cylinder. If all of the readings are fairly close to each other, the engine is in good condition. If the readings from one or more cylinders differ from the rest, there is a problem. Further testing may be required to identify the exact cause of the problem.

Vacuum Tests

Measuring intake manifold vacuum is another way to diagnose the condition of an engine. Manifold vacuum is tested with a vacuum gauge. Vacuum is formed by the downward movement of the pistons during their intake stroke. If the cylinder is sealed, a maximum amount will be formed. High vacuum improves volumetric efficiency.

Vacuum gauge readings **(Figure 8–22)** can be interpreted to identify many engine conditions, including the ability of the cylinder to seal, the timing of the opening and closing of the engine's valves, and ignition timing.

Ideally each cylinder of an engine will produce the same amount of vacuum; therefore, the vacuum gauge reading should be steady and give a reading of at least 17 inches of mercury (in. Hg). If one or more cylinders produce more or less vacuum than the others, the needle of the gauge will fluctuate. The intensity of the fluctuation indicates the severity of the problem. For example, if the reading on the vacuum gauge fluctuates between 10 and 17 inches of mercury we should look at the rhythm of the needle. If the needle seems to stay at 17 most of the time but drops to 10 and quickly rises, we know the reading

PHOTO SEQUENCE 5

Conducting a Cylinder Compression Test

P5-1 *Before conducting a compression test, disable the ignition and the fuel injection system (if the engine is so equipped).*

P5-2 *Prop the throttle plate into a wide-open position to allow an unrestricted amount of air to enter the cylinders during the test.*

P5-3 *Remove all of the engine's spark plugs.*

P5-4 *Connect a remote starter button to the starter system.*

P5-5 *Many types of compression gauges are available. The screw-in type tends to be the most accurate and easiest to use.*

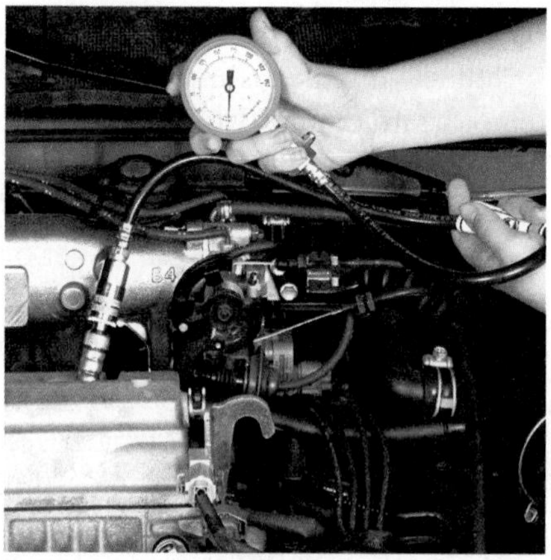

P5-6 *Carefully install the gauge into the spark plug hole of the first cylinder.*

P5-7 *Connect a battery charger to the car to allow the engine to crank at consistent and normal speeds needed for accurate test results.*

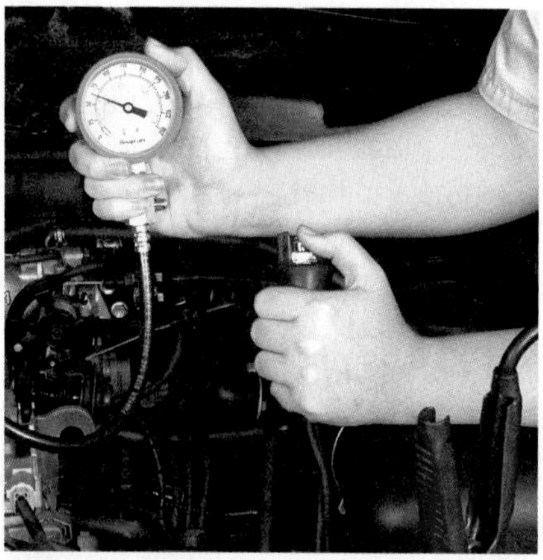

P5-8 *Depress the remote starter button and observe the gauge's reading after the first engine revolution.*

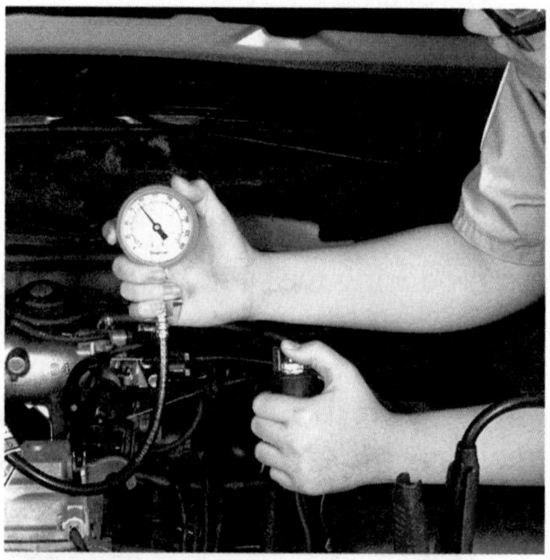

P5-9 *Allow the engine to turn through four revolutions, and observe the reading after the fourth. The reading should increase with each revolution.*

P5-10 *Readings observed should be recorded. After all cylinders have been tested, a comparison of cylinders can be made.*

P5-11 *Before removing the gauge from the cylinder, release the pressure from it using the release valve on the gauge.*

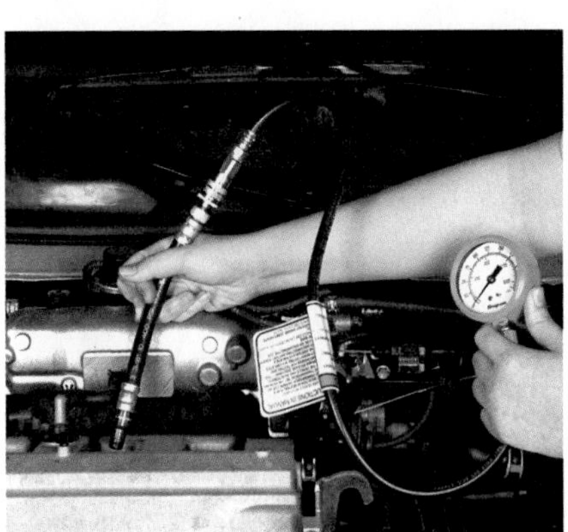

P5-12 *Each cylinder should be tested in the same way.*

P5-13 *After completing the test on all cylinders, compare them. If one or more cylinders is much lower than the others, continue testing those cylinders with the wet test.*

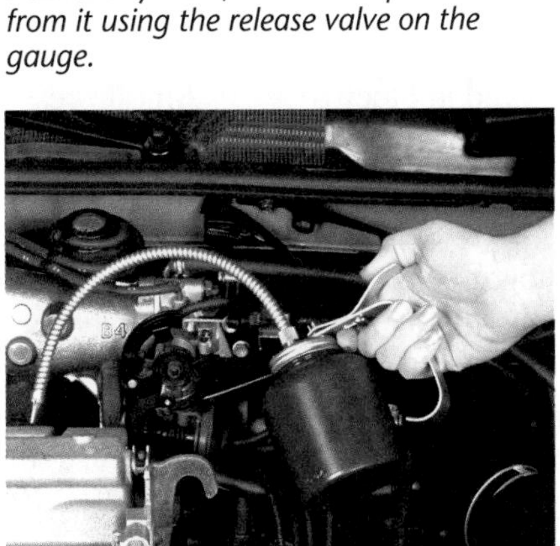

P5-14 *Squirt a small amount of oil into the weak cylinder(s).*

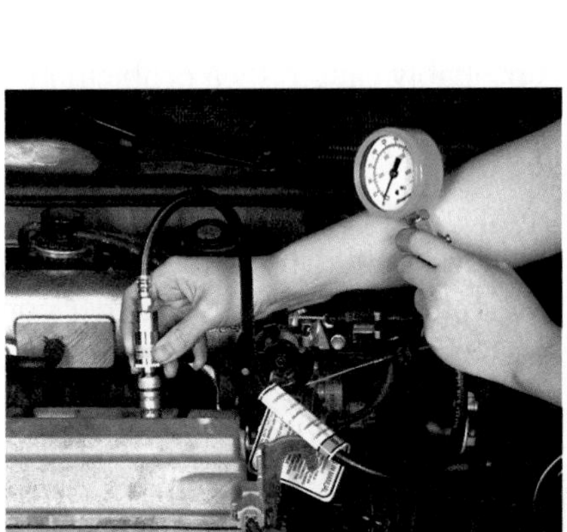

P5-15 *Reinstall the compression gauge into that cylinder and conduct the test.*

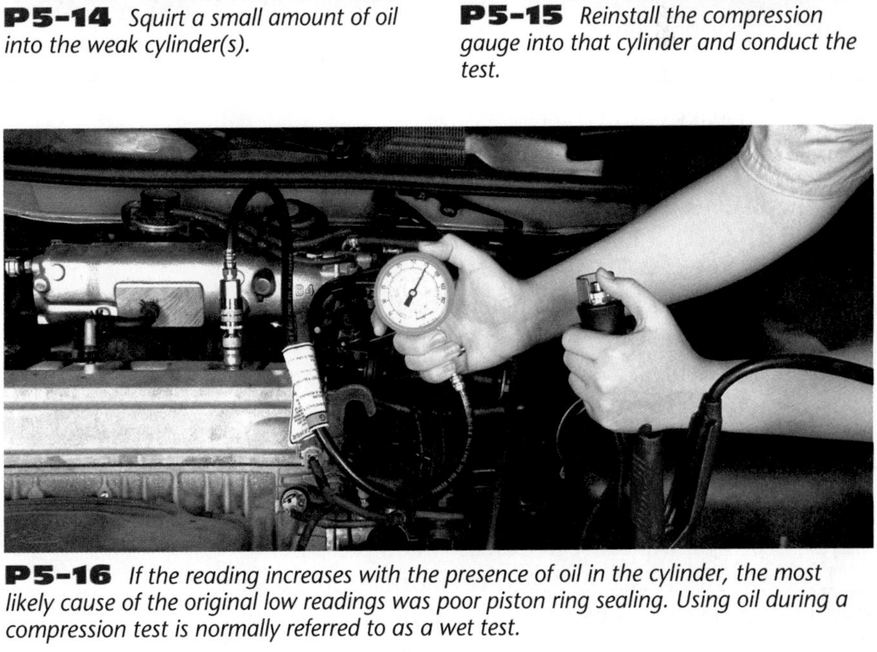

P5-16 *If the reading increases with the presence of oil in the cylinder, the most likely cause of the original low readings was poor piston ring sealing. Using oil during a compression test is normally referred to as a wet test.*

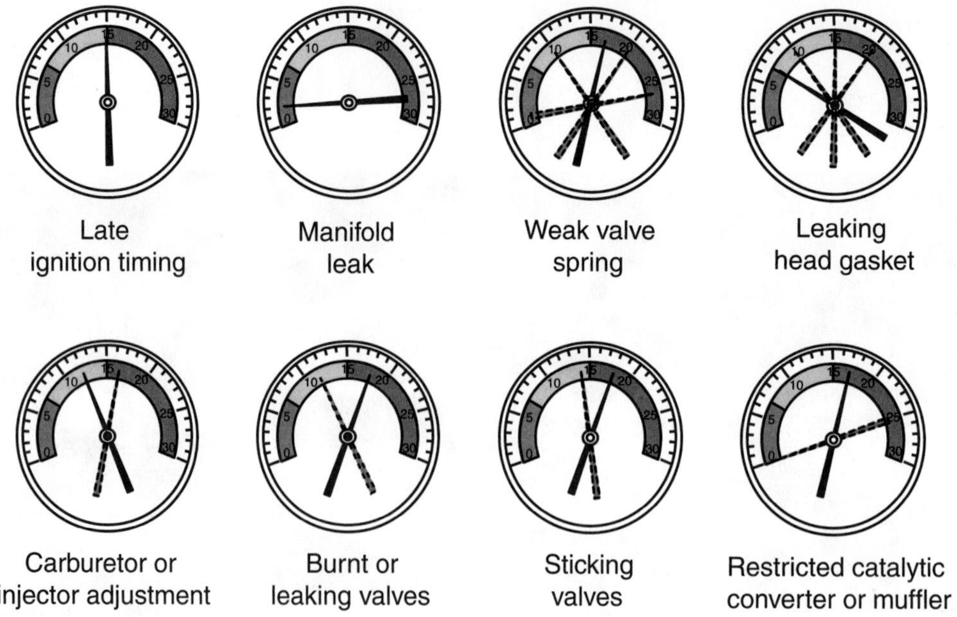

Figure 8–22 Vacuum gauge readings and the engine condition indicated by each.

is probably caused by a problem in one cylinder. Fluctuating or low readings can indicate many different problems. For example, a low, steady reading might be caused by retarded ignition timing or incorrect valve timing. A sharp vacuum drop at regular intervals might be caused by a burned intake valve. Other conditions that can be revealed by vacuum readings follow:

- Stuck or burned valves
- Improper valve or ignition timing
- Weak valve springs
- Faulty PCV, EGR, or other emission-related system
- Uneven compression
- Worn rings or cylinder walls
- Leaking head gaskets
- Vacuum leaks
- Restricted exhaust system
- Ignition defects

Oil Pressure Testing

An oil pressure test is used to determine the wear of an engine's parts. The oil pressure test is performed with an oil pressure gauge, which measures the pressure of the oil as it circulates through the engine. Basically, the pressure of the oil depends on the efficiency of the oil pump and the clearances through which the oil flows. Excessive clearances, most often caused by wear between a shaft and its bearings, will cause a decrease in oil pressure.

Loss of performance, excessive engine noise, and poor starting can be caused by abnormal oil pressure. When the engine's oil pressure is too low, premature wear of its parts will result.

An oil pressure tester is a gauge with a high-pressure hose attached to it. The scale of the gauge typically reads from 0 to 100 psi (0 to 690 kPa). Using the correct fittings and adapters, the hose is connected to an oil passage in the engine block. Normally, the engine's oil pressure sensor is removed and the hose is connected to that port **(Figure 8–23)**.

To conduct the test, simply follow the guidelines given in the service manual and observe the gauge. The pressure is read when the engine is at normal operating temperatures and at a fast idle speed. To get accurate results from the test, make sure you follow the manufacturer's recommendations and compare your findings to specifications. Excessive bearing clearances are not the only possible causes for low oil pressure readings; others are pump-related problems, a plugged oil pickup screen, a

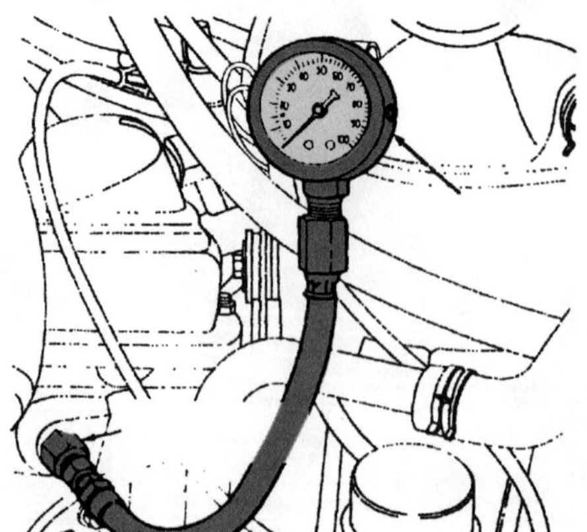

Figure 8–23 The oil pressure gauge is installed into the oil pressure sending unit's bore in the engine block. *Courtesy of DaimlerChrysler Corporation*

weak or broken oil pressure relief valve, low oil level, contaminated oil, or low oil viscosity. Too much oil, cold oil, high oil viscosity, restricted oil passages, and a faulty pressure regulator can cause high oil pressure readings.

EVALUATING THE ENGINE'S CONDITION

Once the compression, cylinder leakage, vacuum, and power balance tests are performed, a technician is ready to evaluate the engine's condition. For example, an engine with good relative compression but high cylinder leakage past the rings is typical of a high-mileage worn engine. This engine would have these symptoms: excessive blowby, lack of power, poor performance, and reduced fuel economy.

If these same compression and leakage test results are found on an engine with comparatively low mileage, the problem is probably stuck piston rings that are not expanding properly. If this is the case, try treating the engine with a combustion chamber cleaner, oil treatment, or engine flush. If this fails to correct the problem, an engine overhaul is required.

A cylinder that has poor compression but minimal leakage indicates a valve train problem. Under these circumstances, a valve might not be opening at the right time, might not be opening enough, or might not be opening at all. This condition can be confirmed on engines with a pushrod-type valve train by pulling the rocker covers and watching the valves operate while the engine is cycled. If one or more valves fail to move, either the lifters are collapsed or the cam lobes are worn. If all of the cylinders have low compression with minimal leakage, the most likely cause is incorrect valve timing.

If compression and leakage are both good, but the power balance test reveals weak cylinders, the cause of the problem is outside the combustion chamber. Assuming there are no ignition or fuel problems, check for broken, bent, or worn valve train components, collapsed lifters, leaking intake manifold, or excessively leaking valve guides. If the latter is suspected, squirt some oil on the guides. If they are leaking, blue smoke will be seen in the exhaust.

NOISE DIAGNOSIS

More often than not, malfunction in the engine will reveal itself first as an unusual noise. This can happen before the problem affects the driveability of the vehicle. Problems such as loose pistons, badly worn rings or ring lands, loose piston pins, worn main bearings and connecting rod bearings, loose vibration damper or flywheel, and worn or loose valve train components all produce telltale sounds. Unless the technician has experience in listening to and interpreting engine noises, it can be very hard to distinguish one from the other.

CUSTOMER CARE

When attempting to diagnose the cause of abnormal engine noise, it may be necessary to temper the enthusiasm of a customer who thinks they have pinpointed the exact cause of the noise using nothing more than their own two ears. While the owner's description may be helpful (and should always be asked for), it must be stressed that one person's "rattle" can be another person's "thump." You are the professional. The final diagnosis is up to you. If customers have been proved correct in their diagnosis, make it a point to tell them so. Everyone feels better about dealing with an automotive technician who listens to them. ■

When correctly interpreted, engine noise can be a very valuable diagnostic aid. For one thing, a costly and time-consuming engine teardown might be avoided. Always make a noise analysis before doing any repair work. This way, there is a much greater likelihood that only the necessary repair procedures will be done. Careful noise diagnosis also reduces the chances of ruining the engine by continuing to use the vehicle despite the problem.

Using a Stethoscope

Some engine sounds can be easily heard without using a listening device, but others are impossible to hear unless amplified. A stethoscope or rubber hose (as mentioned earlier) is very helpful in locating engine noise by amplifying the sound waves. It can also distinguish between normal and abnormal noise. The procedure for using a stethoscope is simple. Use the metal prod to trace the sound until it reaches its maximum intensity. Once the precise location has been discovered, the sound can be better evaluated. A sounding stick, which is nothing more than a long, hollow tube, works on the same principle, though a stethoscope gives much clearer results.

The best results, however, are obtained with an electronic listening device. With this tool you can tune into the noise. Doing this allows you to eliminate all other noises that might distract or mislead you.

WARNING!

Be very careful when listening for noises around moving belts and pulleys at the front of the engine. Keep the end of the hose or stethoscope probe away from the moving parts. Physical injury can result if the hose or stethoscope is pulled inward or flung outward by moving parts.

Common Noises

Following are examples of abnormal engine noises, including a description of the sound, its likely cause, and ways of eliminating it. An important point to keep in mind is that insufficient lubrication is the most common cause of engine noise. For this reason, always check the oil level first before moving on to other areas of the vehicle. Some noises are more pronounced on a cold engine because clearances are greater when parts are not expanded by heat. Remember that aluminum and iron expand at different rates as temperatures rise. For example, a knock that disappears as the engine warms up is probably piston slap or knock. An aluminum piston expands more than the iron block, allowing the piston to fit more closely as engine temperature rises.

Ring Noise This sound can be heard during acceleration as a high-pitched rattling or clicking in the upper part of the cylinder. It can be caused by worn rings or cylinders, broken piston ring lands, or insufficient ring tension against the cylinder walls. Ring noise is corrected by replacing the rings, pistons, or sleeves or reboring the cylinders.

Piston Slap This sound is commonly heard when the engine is cold and often gets louder when the vehicle accelerates. When a piston slaps against the cylinder wall, the result is a hollow, bell-like sound. Piston slap is caused by worn pistons or cylinders, collapsed piston skirts, misaligned connecting rods, excessive piston-to-cylinder wall clearance, or lack of lubrication, resulting in worn bearings. Correction requires either replacing the pistons, reboring the cylinder, replacing or realigning the rods, or replacing the bearings. Shorting out the spark plug of the affected cylinder might quiet the noise.

Piston Pin Knock Piston pin knock is a sharp, metallic rap that can sound more like a rattle if all the pins are loose. It originates in the upper portion of the engine and is most noticeable when the engine is idling and the engine is hot. Piston pin knock sounds like a double knock at idle speeds. It is caused by a worn piston pin, piston pin boss, piston pin bushing, or lack of lubrication, resulting in worn bearings. To correct it, either install oversized pins, replace the boss or bushings, or replace the piston.

Ridge Noise This noise is less common but very distinct. As a piston ring strikes the ridge at the top of the cylinder, the result is a high-pitched rapping or clicking noise that becomes louder during deceleration **(Figure 8–24)**.

There can be more than one reason for the ridge interfering with the ring's travel. For one thing, if new rings are installed without removing the old ridge, the new rings will contact the ridge and make a noise. Also, if the piston pin is very loose or the connecting rod has a loose or burned-out bearing, the piston will go high enough in the cylinder for the top ring to contact the ridge. Thus, in order to eliminate ridge noise, remove the old ring ridge and replace the piston pin or piston.

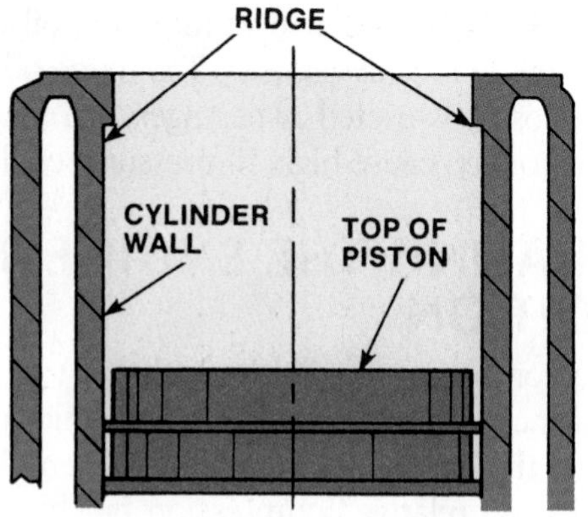

Figure 8-24 When the piston strikes the ridge at the top of the cylinder, a high-pitched rapping or clicking sound is heard.

Rod-Bearing Noise The result of worn or loose connecting rod bearings, this noise is heard at idle as well as at speeds over 35 mph. Depending on how badly the bearings are worn, the noise can range from a light tap to a heavy knock or pound. Shorting out the spark plug of the affected cylinder can lessen the noise, unless the bearing is extremely worn. In this case, shorting out the plug will have no effect. Rod-bearing noise is caused by a worn bearing or crankpin, a misaligned rod, or lack of lubrication, resulting in worn bearings. To correct it, service or replace the crankshaft, realign or replace the connecting rods, and replace the bearings.

Main or Thrust Bearing Noise A loose crankshaft main bearing produces a dull, steady knock, while a loose crankshaft thrust bearing produces a heavy thump at irregular intervals. The thrust bearing noise might only be audible on very hard acceleration. Both of these bearing noises are usually caused by worn bearings or crankshaft journals. To correct the problem, replace the bearings or crankshaft.

Tappet Noise Tappet noise is characterized by a light, regular clicking sound that is more noticeable when the engine is idling. It is the result of excessive clearance in the valve train. The clearance problem area is located by inserting a feeler gauge between each lifter and valve, or between each rocker arm and valve tip, until the noise subsides. Tappet noise can be caused by improper valve adjustment, worn or damaged parts, dirty hydraulic lifters, or lack of lubrication. To correct the noise, adjust the valves, replace any worn or damaged parts, or clean or replace the lifters.

Abnormal Combustion Noises Preignition and detonation noises are caused by abnormal engine combustion.

For instance, **detonation** knock or **ping** is a noise most noticeable during acceleration with the engine under load and running at normal temperature. Excessive detonation can be very harmful to the engine. It is often caused by advanced ignition timing or substantial carbon buildup in the combustion chambers that increases combustion pressure. Carbon deposits that get so hot they glow will also preignite the air/fuel mixture, causing detonation. Another possible cause is fuel with octane that is too low. Detonation knock can usually be cured by removing carbon deposits from the combustion chambers with a rotary wire brush as well as recommending the use of a higher octane gasoline. A malfunctioning EGR valve can also cause detonation.

Sometimes abnormal combustion combines with other engine parts to cause noise. For example, rumble is a term used to describe the knock or noise resulting from another form of abnormal ignition. Rumble is a vibration of the crankshaft and connecting rods that is caused by multisurface ignition. Rumble is a form of preignition in which several flame fronts occur simultaneously from overheated deposit particles. Multisurface ignition causes a tremendous sudden pressure rise near TDC. It has been reported that the rate of pressure rise during rumble is five times the rate of normal combustion.

A loose vibration damper causes a heavy rumble or thump in the front of the engine that is more apparent when the vehicle is accelerating from idle under load or is idling unevenly. A loose flywheel causes a heavy thump or light knock at the back of the engine, depending on the amount of play and the type of engine. Both of these problems are corrected either by tightening or replacing the damper or flywheel.

OTHER ENGINE DESIGNS

The gasoline-powered, internal-combustion piston engine has been the primary automotive power plant for many years and probably will remain so for years to come. Present-day social requirements and new technological developments, however, have necessitated searches for ways to modify or replace this time-proven workhorse. This portion of the chapter takes a brief look at the most likely contenders, and how they work.

Diesel Engine

Diesel engines represent tested, proven technology with a long history of success. Invented by Dr. Rudolph Diesel, a German engineer, and first marketed in 1897, the diesel engine is now the dominant power plant in heavy-duty trucks, construction equipment, farm equipment, buses, and marine applications.

During the late 1970s and early 1980s, many predicted small diesel engines would replace gasoline engines in passenger vehicles. However, stabilized gas prices and other factors dampened the enthusiasm for diesels in these markets.

Diesel engines **(Figure 8–25)** and gasoline-powered engines share several similarities. They have a number of components in common, such as the crankshaft, pistons, valves, camshaft, and water and oil pumps. Both of them are available in four-stroke combustion cycle models. However, the diesel engine and four-stroke compression-ignition engine are easily recognized by the absence of an ignition system. Instead of relying on a spark for ignition, a diesel engine uses the heat produced by compressing air in the combustion chamber to ignite the fuel. The systems used in diesel-powered vehicles are essentially the same as those used in gasoline vehicles.

Figure 8–26 shows the four strokes of a diesel engine. Fuel injection is used on all diesel engines. Injectors spray pressurized fuel into the cylinders just as the piston is completing its compression stroke. The heat of the compressed air ignites the fuel and begins the power stroke.

Glow plugs are used only to warm the combustion chamber when the engine is cold. Cold starting is impossible without these plugs because even the high-compression ratios cannot heat cold air enough to cause combustion.

Diesel combustion chambers are different from gasoline combustion chambers because diesel fuel burns differently. Three types of combustion chambers are used in diesel engines: open combustion chamber, precombustion chamber, and turbulence combustion chamber. The

Figure 8-25 The high output Cummins turbo diesel I-6 engine used in Dodge Ram heavy-duty trucks. *Courtesy of DaimlerChrysler Corporation*

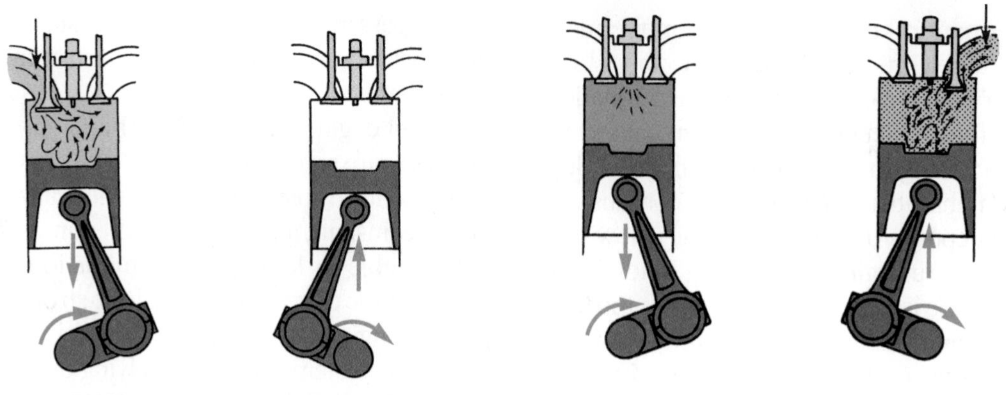

Figure 8-26 A four-stroke diesel engine cycle.

open combustion chamber has the combustion chamber located directly inside the piston. Diesel fuel is injected directly into the center of the chamber. The shape of the chamber and the quench area produces turbulence. The precombustion chamber is a smaller, second chamber connected to the main combustion chamber. On the power stroke, fuel is injected into the small chamber. Combustion is started there and then spreads to the main chamber. This design allows lower fuel injection pressures and simpler injection systems on diesel engines.

Table 8–1 compares the gasoline and diesel four-stroke-cycle engines. Diesel engines are also available in two-stroke-cycle models **(Figure 8–27)**.

TABLE 8–1 COMPARISON BETWEEN GASOLINE AND DIESEL ENGINES

	Gasoline	Diesel
Intake	Air/fuel	Air
Compression	8–10 to 1 130 psi (896 kPa) 545°F	13–25 to 1 400–600 psi (2758–4137 kPa) 1,000°F
Air/fuel mixing point	In intake manifold near the intake valve.	In cylinder near TDC by injection
Combustion	Spark ignition	Compression ignition
Power	464 psi (3199 kPa)	1,200 psi (13,790 kPa)
Exhaust	1,300°–1,800°F (704–982°C) CO = 1%	700°–900°F (371–482°C) CO = 0.5%
Efficiency	22–28%	32–38%

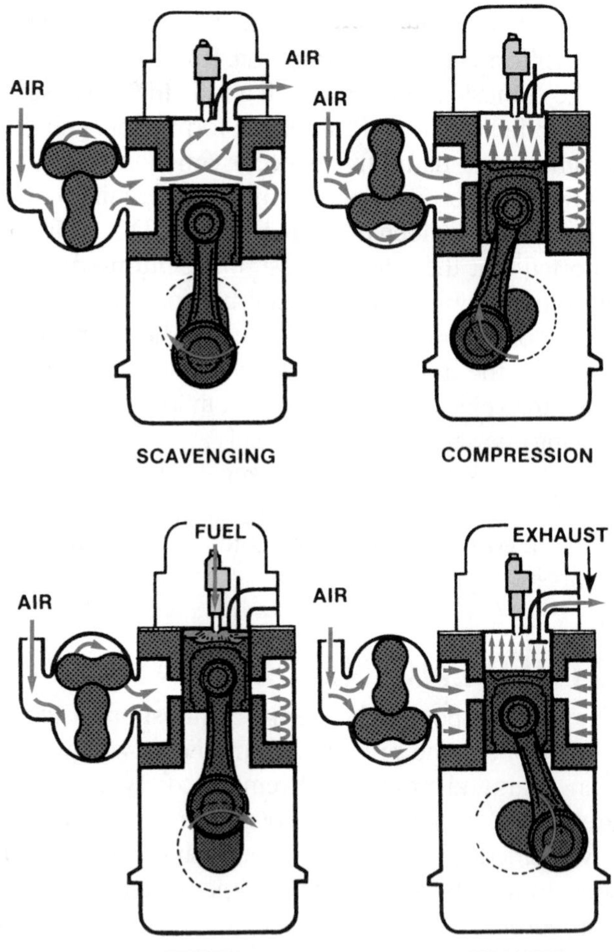

Figure 8-27 A two-stroke diesel engine cycle.

Rotary Engine

The **rotary** engine, or **Wankel** engine, is somewhat similar to the standard piston engine in that it is a spark ignition, internal-combustion engine. Its mechanical design, however, is quite different. For one thing, the rotary engine uses a rotating motion rather than a reciprocating motion. In addition, it uses ports rather than valves for controlling the intake of the air/fuel mixture and the exhaust of the combusted charge.

The heart of a rotary engine is a roughly triangular rotor that "walks" around a smaller, rigidly mounted gear. The rotor is connected to the crankshaft through additional gears in such a manner that, for every rotation of the rotor, the crankshaft revolves three times. The tips of the triangular rotor move within the housing and are in constant contact with the housing walls. As the rotor moves, the volume between each side of the rotor and the housing walls continually changes.

Referring to **Figure 8–28**, when the side of the rotor is in (A), the intake port is uncovered and the air/fuel mixture is entering the upper chamber. As the rotor moves to (B), the intake port closes and the upper chamber reaches its maximum volume. When full compression has reached (C), the 2 spark plugs fire, one after the other, to start the power stroke. At (D), rotor side A uncovers the exhaust port and exhaust begins. This cycle continues until (A) is reached where the chamber volume is at minimum and the intake cycle starts once again.

The fact that the rotating combustion chamber engine is small and light for the amount of power it produces makes it attractive for use in automobiles. Using this small, lightweight engine can provide the same performance as a larger engine. However, the rotary engine, at present, cannot compete with the piston gasoline engine in terms of durability, exhaust emission control, and economy.

After a few years of not offering a rotary engine vehicle, Mazda has released a new version of the engine, called the Renesis. This engine is similar to those used in the past but now produces fewer emissions. The reduction in emissions is the result of many minor changes, most notable of which is the elimination of the peripheral exhaust ports into which the apex seals used to sweep unburned gasoline from the intake port. The new engine has its exhaust ports located very low on the side of each rotor chamber, thereby eliminating the overlap with the intake ports.

Stratified Charge Engine

The stratified charge engine **(Figure 8–29)** combines the features of both the gasoline and diesel engines. It differs from the conventional gasoline engine in that the air/fuel mixture is deliberately stratified to produce a small, rich mixture at the spark plug while providing a leaner, more efficient and cleaner-burning main mixture. In addition, the air/fuel mixture is swirled for more complete combustion.

Referring to (A) in **Figure 8–30**, a large amount of very lean mixture is drawn through the main intake valve on the intake stroke to the main combustion chamber. At the same time, a small amount of rich mixture is drawn through the auxiliary intake valve into the precombustion chamber. At the end of the compression stroke in (B), the spark plug fires the rich mixture in the precombustion chamber. As the rich mixture ignites, it in turn ignites the lean mixture in the main chamber. The lean mixture minimizes the formation of carbon monoxide during the power stroke (C). In addition, the peak temperature stays low enough to minimize the formation of oxides of nitrogen, and the mean temperature is maintained high enough and long enough to reduce hydrocarbon emissions. During the exhaust stroke (D) the hot gases exit through the exhaust valve.

A great deal of automobile engineering research, especially by Japanese and European manufacturers, is being done on these engines. In fact, the Honda CVCC

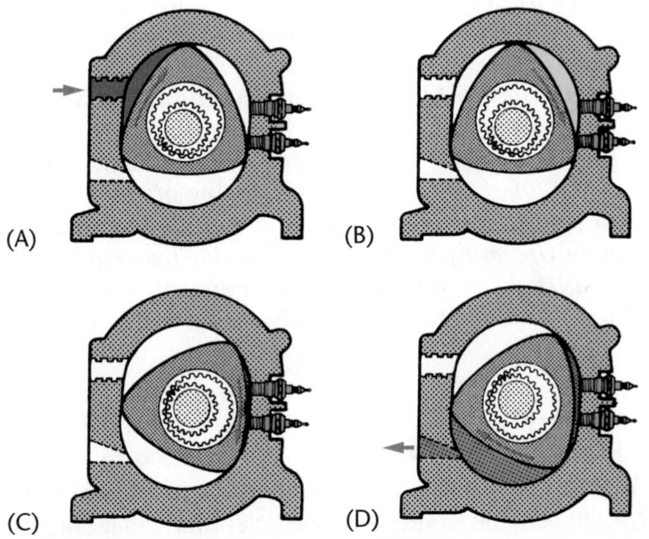

Figure 8-28 A rotary engine cycle.

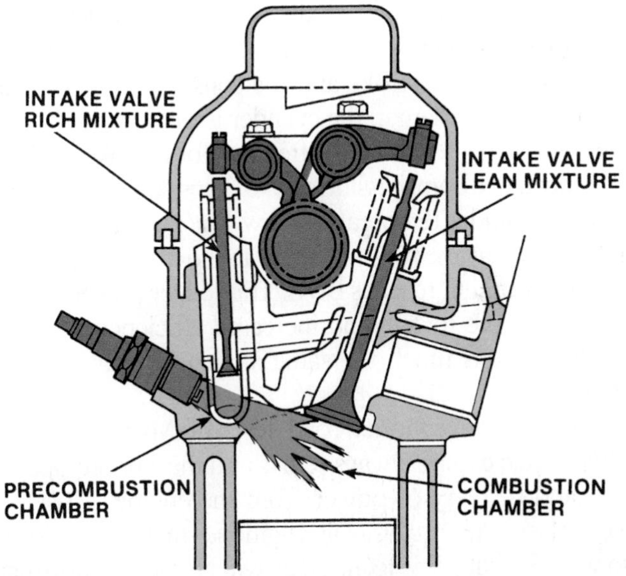

Figure 8-29 A typical stratified charge engine.

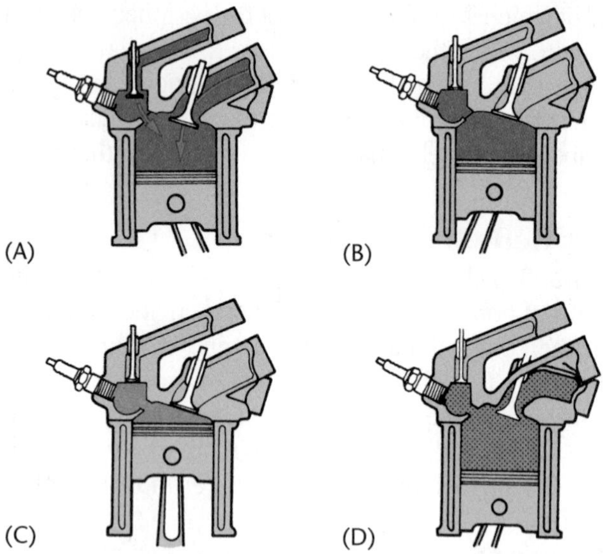

Figure 8-30 The four-stroke cycle of a stratified charge engine.

engine uses a stratified charge design. This engine uses a third valve to release the initial charge. The stratified charge combustion chamber has three important advantages. It produces good part-load fuel economy, can run efficiently on low-octane fuel, and has low exhaust emissions.

Miller-Cycle Engines

A version of the Miller-cycle engine is the base powerplant in late-model Mazda Millennias. This engine design is a modification of the four-stroke engine. During the intake stroke, a supercharger feeds highly compressed air to an intercooler. This cooled, but compressed, air is fed directly into the cylinders. During the intake stroke, the intake valve remains open for a longer-than-normal time, which prevents compression from occurring until the piston has moved one-fifth of the way toward TDC. Then the valve closes and compression occurs. Because of the lengthened intake stroke and the constant supply of air from the supercharger, the cylinder is filled with air and fuel. This does not happen often in a four-stroke engine. The shortened compression stroke keeps compression ratios and cylinder temperatures low. The power stroke begins as soon as the piston is ready to move in its bore and continues until it reaches BDC. This longer power stroke provides more torque and increased efficiency. The exhaust stroke is much the same as that in a four-stroke engine.

Those who conduct road tests of Mazdas with a Miller-cycle engine always remark on the smooth operation and delivery of power from the engine. Most reviewers are surprised that an engine so small can perform so well. Perhaps this technology will be used in more car models in the future.

CASE STUDY

A customer complains that the engine has a rough idle and does not have as much power as it used to. The customer also says that the engine is difficult to start when it is cold and runs better while cruising at highway speeds. On examining the car, it is found that it has a four-cylinder engine with nearly 65,000 miles on it. Driving the car verifies the customer's complaint.

Diagnosis begins with a visual inspection that reveals only that the car has been well maintained. Because there is an endless list of possible causes for these problems, the first test is for engine vacuum. With the vacuum gauge connected to the manifold and the engine at idle speed, the gauge's needle constantly drops from a normal reading to 10 in. Hg. The rhythm of the drop matches the rhythm of the idle. The behavior of the gauge indicates a probable problem in one cylinder.

To verify this diagnosis, a cylinder power balance test is conducted. The results show an engine speed drop of 100 to 125 rpm when cylinders 1, 2, and 4 are shorted, and a drop of only 10 rpm when cylinder 3 is shorted. Based on this test, cylinder 3 is identified as having the problem.

To identify the exact fault, further testing is required. The spark plugs are removed and inspected. All look normal, including the one from cylinder 3. Next, a compression test is taken. All cylinders have normal readings. A cylinder leakage test is then conducted and it too shows normal conditions. The results of the power balance, compression, and cylinder leakage tests lead to the conclusion that the cause has to be in the valve train. Something is preventing a valve from opening.

Removing the cam cover for a visual inspection leads to the discovery of the fault: The intake lobe for cylinder 3 on the camshaft is severely worn. A replacement of the camshaft and matching lifter will correct the problem. The worn lobe only affects the opening of the valve and does not prevent it from sealing, which is why the compression and cylinder leakage test results were normal. Cylinder power and vacuum are affected by the valve not opening fully.

Hybrid Vehicles

A **hybrid vehicle** is one that uses an electric motor and a gasoline engine to move the vehicle. Many manufacturers are building hybrid vehicles. The first hybrids for sale

Figure 8-31 The Toyota Prius is a hybrid vehicle.
Reprinted with permission

in North America were the Honda Insight and the Toyota Prius **(Figure 8–31)**. These cars use an electric motor that also operates as a generator and a small gasoline engine. The motor is placed between the engine and the transaxle and both can be used to power the car. The Prius engine is an Atkinson-cycle four cylinder. The Atkinson-cycle engine is similar to a Miller-cycle engine. The engine gets the car going. While the engine is running, the motor turns and works like a generator to charge the batteries. Once the **nickel metal hydride (NiMH)** batteries are charged, the engine shuts down and the electric motor continues to move the car. The car is designed to have the engine kick on and off to meet the demands of the batteries. Whenever the batteries are fully charged, there is no need to have the engine running.

Combining electric with gasoline power provides low emissions, long driving ranges, and smooth and quiet operation. It also allows the Prius to get 80 to 106 miles (128 to 170 km) per gallon of fuel. Other manufacturers are also working on hybrid vehicles. They seem to be an immediate answer to problems related to electric vehicles.

KEY TERMS

BDC
Cam
Combustion chamber
Crank throw
Detonation
Diesel
Dual overhead camshaft (DOHC)
Firing order
Glow plug
Hybrid vehicle
Lobe
Nickel metal hydride (NiMH)
Overhead camshaft (OHC)
Overhead valve (OHV)
Oversquare
Ping
Rotary
TDC
Torque
Undersquare
Wankel

SUMMARY

- Automotive engines are classified by several different design features such as operational cycles, number of cylinders, cylinder arrangement, valve train type, valve arrangement, ignition type, cooling system, and fuel system.
- The basis of automotive gasoline engine operation is the four-stroke cycle. This includes the intake stroke, compression stroke, power stroke, and exhaust stroke. The four strokes require two full crankshaft revolutions.
- The most popular engine designs are the in-line (in which all the cylinders are placed in a single row) and V-type (which features two rows of cylinders). The slant design is much like the in-line, but the entire block is placed at a slant. Opposed cylinder engines use two rows of cylinders located opposite the crankshaft.
- The two basic valve and camshaft placement configurations currently in use on four-stroke engines are the overhead valve and overhead cam.
- Bore is the diameter of a cylinder, and stroke is the length of piston travel between top dead center (TDC) and bottom dead center (BDC). Together these two measurements determine the displacement of the cylinder.
- Compression ratio is a measure of how much the air and fuel are compressed during the compression stroke. The higher the compression ratio is, the more power an engine can produce.
- Horsepower is the rate at which torque is produced by an engine. The torque is then transmitted through the drivetrain to turn the driving wheels of the vehicle.
- The vehicle identification number, or VIN, is used to identify correct engine specifications. It is stamped on a metal tab, which is riveted to the top of the instrument panel.
- A compression test is conducted to check a cylinder's ability to seal and therefore its ability to compress the air/fuel mixture inside the cylinder.
- A cylinder leakage test is performed to measure the percentage of compression lost and to help locate the source of leakage.
- A cylinder power balance test reveals whether all of an engine's cylinders are producing the same amount of power.
- Vacuum gauge readings can be interpreted to identify many engine conditions, including the ability of the engine's cylinders to seal, the timing of the opening and closing of the engine's valves, and ignition timing.
- An oil pressure test measures the pressure of the engine's oil as it circulates throughout the engine. This

test is very important because abnormal oil pressures can cause a host of problems, including poor performance and premature wear.

- An engine malfunction often reveals itself as an unusual noise. When correctly interpreted, engine noise can be a very helpful diagnostic aid.
- Instead of relying on a spark for ignition, diesel engines use the heat produced by compressing air in the combustion chamber to ignite the fuel.
- Features of both the gasoline and the diesel engine are found in the stratified charge engine. Its major advantages are good part-load fuel economy, low exhaust emissions, and an ability to operate on low-octane fuel.
- In addition to the diesel and the stratified charge, other automotive engines include the rotary or Wankel, the Miller-cycle, and the Atkinson-cycle. The future may bring many electric-powered and hybrid vehicles.

TECH MANUAL

The following procedures are included in Chapter 8 of the *Tech Manual* that accompanies this book:

1. Performing a cylinder compression test.
2. Performing cylinder leakage test.
3. Measuring engine oil pressure.

REVIEW QUESTIONS

1. What occurs in the combustion chamber of a four-stroke engine?
2. Name the four strokes of a four-stroke-cycle engine.
3. As an engine's compression ratio increases, what should happen to the octane rating of the gasoline?
4. What test can be performed to check the efficiency of individual cylinders?
5. Describe tappet noise.
6. Which of the following statements about engines is *not* true?
   a. The engine provides the rotating power to drive the wheels through the transmission and driving axle.
   b. Only gasoline engines are classified as internal combustion.
   c. The combustion chamber is the space between the top of the piston and the cylinder head.
   d. For the combustion in the cylinder to take place completely and efficiently, air and fuel must be combined in the right proportions.
7. Which stroke in the four-stroke cycle begins as the compressed fuel mixture is ignited in the combustion chamber?
   a. power stroke
   b. exhaust stroke
   c. intake stroke
   d. compression stroke
8. What is compression ratio?
   a. diameter of the cylinder
   b. cylinder arrangement
   c. the amount the air/fuel mixture is compressed
   d. none of the above
9. When a customer refers to the engine component that opens and closes the intake and exhaust valves, Technician A believes the customer is referring to the camshaft. Technician B thinks the component in question is the intake manifold. Who is correct?
   a. Technician A
   b. Technician B
   c. Both A and B
   d. Neither A nor B
10. Which engine design would have the highest compression ratio?
    a. diesel engine
    b. Miller-cycle engine
    c. two-stroke-cycle engine
    d. stratified charge engine
11. Technician A says if an engine had good results from a compression test, it will have good results from a cylinder leakage test. Technician B says if an engine had good results from a cylinder leakage test, it will have good results from a compression test. Who is correct?

a. Technician A
b. Technician B
c. Both A and B
d. Neither A nor B

12. What is piston slap?

a. grooves on the side of the piston
b. force applied to the piston
c. noise made by the piston when it contacts the cylinder wall
d. a high-pitched clicking that becomes louder during deceleration

13. Which engine system removes burned gases and limits noise produced by the engine?

a. exhaust system
b. emission control system
c. ignition system
d. air/fuel system

14. In a four cylinder engine with a firing order of 1-3-4-2, in what stroke is the piston for cylinder #3 while the piston for cylinder #1 is traveling upward on its compression stroke?

15. The stroke of an engine is ______ the crank throw.

a. half
b. twice
c. four times
d. equal to

16. *True or False?* During a cylinder leakage test, if air is felt leaving the oil dipstick tube, worn piston rings are indicated.

17. While looking at the results of an oil pressure test, Technician A says higher than normal readings can be caused by a defective pressure regulator. Technician B says higher pressure readings can be expected on a cold engine. Who is correct?

a. Technician A
b. Technician B
c. Both A and B
d. Neither A nor B

18. A vehicle is producing a sharp, metallic, rapping sound originating in the upper portion of the engine. It is most noticeable during idle. Technician A diagnoses the problem as piston pin knock, while Technician B says the problem is most likely a loose crankshaft thrust bearing. Who is correct?

a. Technician A
b. Technician B
c. Both A and B
d. Neither A nor B

19. Which of the following engine problems may be indicated by good results from a compression test and a cylinder leakage test but poor results from a cylinder power balance test?

a. incorrect valve timing
b. collapsed lifter
c. burnt intake valve
d. worn piston rings

20. Which of the following is an expression of how much of the heat formed during the combustion process is available as power from the engine?

a. mechanical efficiency
b. engine efficiency
c. volumetric efficiency
d. thermal efficiency

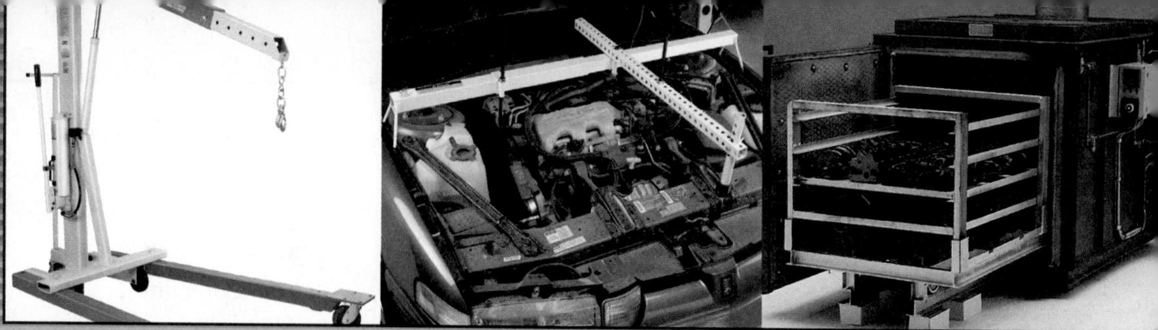

9

ENGINE DISASSEMBLY

OBJECTIVES

■ Prepare an engine for removal. ■ Remove an engine from a FWD and RWD vehicle. ■ Describe how to disassemble and inspect an engine. ■ Name the three basic cleaning processes. ■ Identify the types of cleaning equipment. ■ Describe the common ways to repair cylinder head cracks.

Careful diagnostics will determine if a starting or running problem is caused by the engine itself or by one of its systems, such as the ignition or air/fuel system. When the engine is the source of the problem, or its parts are broken or excessively worn, it normally needs to be rebuilt or overhauled. While some engine repairs can be made with the engine still in the vehicle, most require its removal.

Before removing the engine, clean it and the area around it. Protect the fuel injection and ignition systems while cleaning. Cover them with plastic bags and try not to spray directly at the systems. Also, check the service manual, as procedures and precautions vary from model to model.

REMOVING AN ENGINE

Make sure you have the tools and equipment required for the job before you begin. In addition to hand tools and some special tools, you will need a crane **(Figure 9–1)** and a jack.

The basic procedures for engine removal vary depending on whether the engine is removed from the bottom of the vehicle or through the hood opening. Many FWD vehicles require removal of the engine from the bottom, while most RWD vehicles require the engine to come out from the hood opening. The engine exit point is something to keep in mind while you are disconnecting and removing items in preparation for engine removal.

If you are working on a RWD vehicle, a drive-on type hoist will work best. For FWD vehicles, a frame contact lift is recommended. Make sure to block the wheels so the vehicle does not move while you are working. Open the hood and put fender covers on both front fenders.

> **CUSTOMER CARE**
>
> Make sure your hands, shoes, and clothing are clean before getting into a customer's car. Disposable seat and floor coverings should be used to help protect the interior. ■

Once the vehicle is in position, relieve the pressure in the fuel system using the procedures outlined by the manufacturer. Then disconnect the negative battery cable and place it away from the battery. Then remove the positive cable and the battery. Remember to install a memory

Figure 9–1 To pull an engine out of a vehicle, the chain on the lifting crane is attached to another chain secured to the engine. *Courtesy of OTC Tool and Equipment, Division of SPX Corporation*

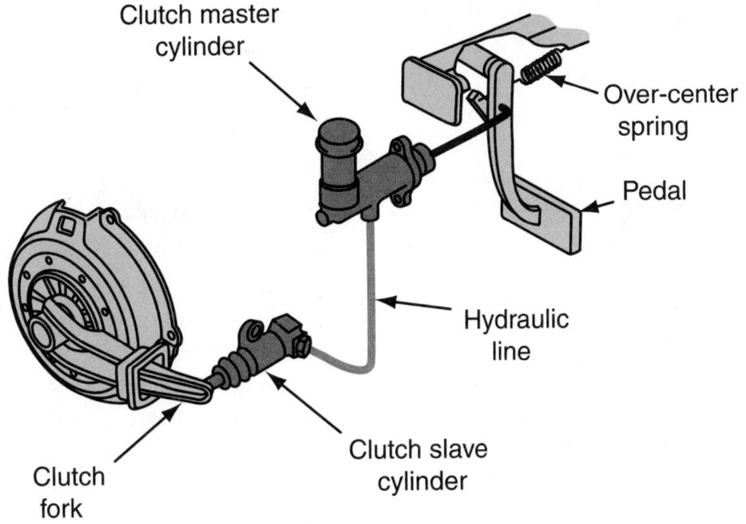

Figure 9-2 If the transmission will be removed with the engine, disconnect or remove the clutch slave cylinder.

saver before you disconnect the battery to prevent the vehicle's computers and other devices from losing what they have stored in their memory.

If the hood must be removed, mark the location of the hinges on the hood. Then unbolt and remove the hood with the help of someone else and place it in a safe place.

Drain the engine's oil and its coolant from the radiator and engine block, if possible. To increase the flow of the coolant out of the system, remove the radiator cap. Make sure the engine is cool before opening the coolant drain and before removing the radiator cap.

If the transmission will be removed with the engine, drain its fluid. While you are under the vehicle, disconnect the shift linkage, transmission cooling lines, all electrical connections, vacuum hoses, and clutch linkages from the transmission. If the clutch is hydraulically operated, unbolt the slave cylinder and set it aside if possible. If this is not possible, disconnect and plug the line to the cylinder **(Figure 9–2)**.

Remove the air intake ducts and air cleaner assembly. Disconnect and plug the fuel line at the fuel rail. If the engine is equipped with a return fuel line from the fuel pressure regulator, disconnect that as well **(Figure 9–3)**. Make sure all fuel lines are closed off with pinch pliers or the appropriate plug or cap. Most late-model fuel lines have quick-connect fittings that are separated by squeezing the

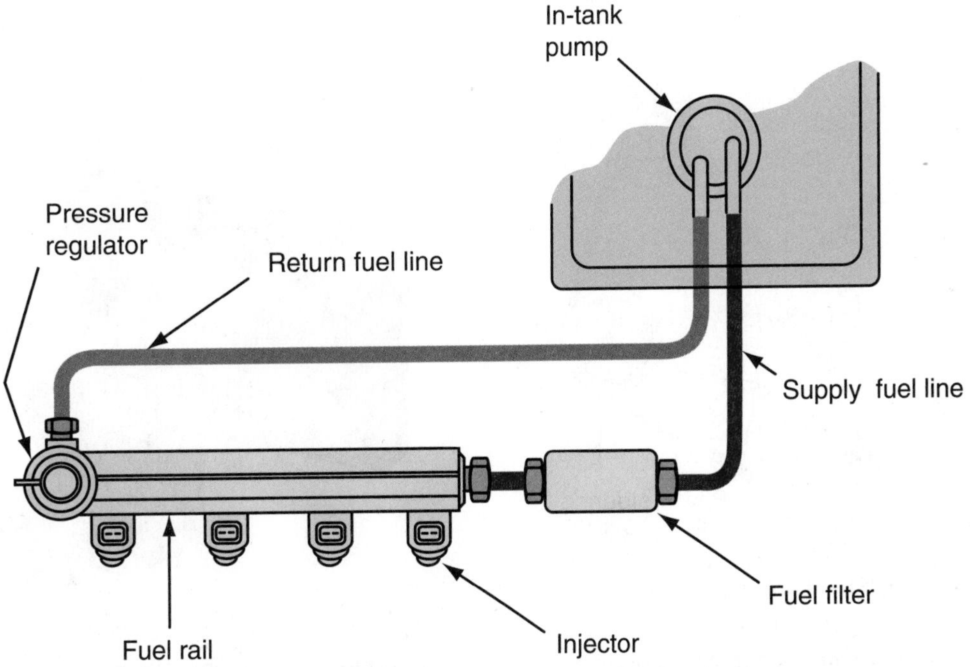

Figure 9-3 Disconnect and plug the fuel lines at the fuel rail and the pressure regulator.

retainer tabs together and pulling the fitting off the fuel line nipple. Now disconnect the throttle linkage at the throttle body and the electrical connector to the throttle position (TP) sensor.

SHOP TALK

Some technicians use instant cameras or video recording cameras to help recall the locations of underhood items by taking pictures before work is started. This technique can be quite valuable considering how complex the underhood systems of current cars have become. ■

Remove all drive belts **(Figure 9–4)**. Unbolt and move the power steering pump and air-conditioning compressor out of the way; do not disconnect lines unless it is necessary. If the A/C compressor needs to be disconnected from its pressure hoses, first the system must be evacuated and the refrigerant captured with a refrigerant reclaimer/recycling machine. Make sure to plug the lines and the connections at the compressor to prevent dirt and moisture from entering.

Remove or move the A/C compressor bracket, power steering pump, air pump, and any other components attached to the engine. Disconnect and plug all transmission and oil cooler lines. Unplug all electrical wires between the engine and the vehicle. Use masking tape as a label to identify all wires and hoses that are disconnected and cap all hoses.

SHOP TALK

When removing the fasteners, pay close attention to their size and type. Many brackets used to secure the engine or accessories use several different size fasteners. Mark and organize the fasteners so their proper location can be easily determined during reassembly. It is a good idea to store the fasteners in several different containers, one for each system or section of the engine. ■

Some engines have a crankshaft position sensor attached above the flywheel or flex plate. This sensor must be removed before separating the engine from the bell housing.

Disconnect the heater inlet and outlet hoses. Then disconnect the upper and lower radiator hoses and any vacuum hose that may get in the way during engine removal. Disconnect the exhaust system, attempt to do this at the exhaust manifold. Now carefully check under the hood to find and remove anything that may interfere with engine removal.

If the radiator is fitted with a fan shroud **(Figure 9–5)**, carefully remove it along with the cooling fan. If the vehicle is equipped with an electric cooling fan, disconnect the wiring to the cooling fan. Unbolt and remove the radiator mounting brackets and remove the radiator. Normally the electric cooling fan assembly and radiator can be removed as a unit **(Figure 9–6)**.

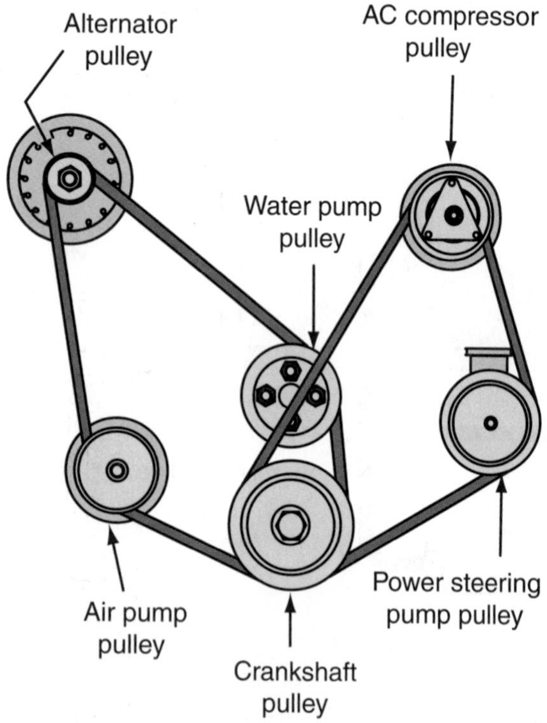

Figure 9-4 Before removing the drive belts, pay attention to what is driven by each belt.

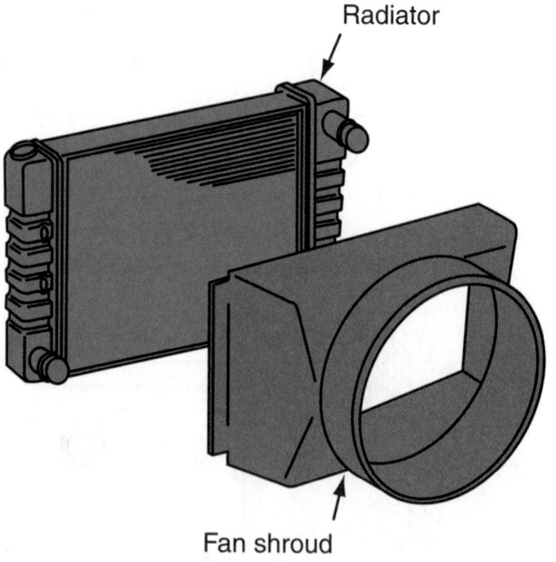

Figure 9-5 If the radiator is fitted with a fan shroud, remove it before attempting to remove the radiator.

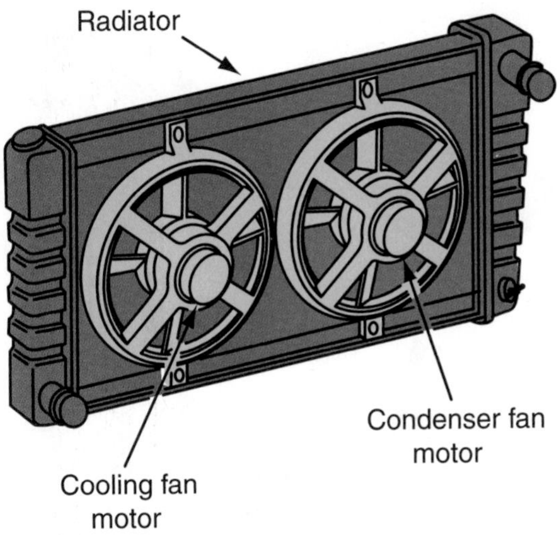

Figure 9–6 The electric cooling fans usually can be removed as a unit with the radiator.

Removing the engine from a RWD vehicle is generally more straightforward than removing one from a FWD model, as there is usually easy access to the cables, wiring, and bell housing bolts. Engines in FWD cars, because of their limited space, can be more difficult to remove as you may need to disassemble or remove large assemblies such as engine cradles, suspension components, brake components, splash shields, or other pieces that would not usually affect RWD engine removal.

FWD Vehicles

Before removing the engine, identify any special tool needs and precautions recommended by the manufacturer. Most often the engine in a FWD vehicle is removed through the bottom of the vehicle. Special tools may be required to hold the transaxle and/or engine in place as it is being disconnected from the vehicle **(Figure 9–7)**. Always refer to the service manual before proceeding to remove the transaxle. You will waste much time and energy if you do not check the manual first.

When the engine is removed through the bottom of the vehicle, use an engine cradle and dolly to support the engine. If the manufacturer recommends engine removal through the hood opening, use an engine hoist. Regardless of the method of removal, the engine and transaxle are usually removed as a unit. The transaxle can be separated from the engine once it has been lifted out of the vehicle.

Make sure the engine ground strap is disconnected, preferably at the engine. Using a large breaker bar, loosen and remove the axle shaft hub nuts. It is recommended that these nuts be loosened with the vehicle on the floor and the brakes applied. Doing so makes the job easier and reduces the chance of damaging the constant velocity (CV) joints and wheel bearings.

Raise the vehicle so you can comfortably work under it. Then remove the wheel and tire assemblies for the

Figure 9–7 A transverse engine support bar provides the necessary support when removing an engine from a FWD vehicle. *Courtesy of OTC Tool and Equipment, Division of SPX Corporation*

front wheels. Tap the splined CV joint shaft with a soft-faced hammer to see if it is loose. Most will come loose with a few taps. Many Ford FWD cars use an interference fit spline at the hub and you will need a special puller for this type of CV joint **(Figure 9–8)**. The tool pushes the shaft out, and on installation pulls the shaft back into the hub.

Disconnect all suspension and steering parts that need to be removed according to the service manual. Index the parts so wheel alignment will be close after reassembly. Normally the lower ball joint must be separated from the steering knuckle. The ball joint will either be bolted to the lower control arm or the ball joint will be held into the knuckle with a pinch bolt **(Figure 9–9)**. Once the ball joint is loose, the control arm can be pulled down and the knuckle can be pushed outward to allow the CV joint shaft to slide out of the hub.

Figure 9–8 Sometimes a puller is required to separate the axle shaft from the hub.

Figure 9-9 Before the axle shaft can be removed, the lower ball joint must be separated from the steering knuckle.

The inboard joint can then be either pried out or will slide out. Some transaxles have retaining clips that must be removed before the inner joint can be removed. Others have a flange-type mounting. These must be unbolted for removal of the shafts. In some cases, the flange-mounted drive shafts may be left attached to the wheel and hub assembly and only unbolted at the transmission flange. The free end of the shafts should be supported and placed out of the way.

Pull the drive axles out of the transaxle. While removing the axles, make sure the brake lines and hoses are not stressed. Suspend them with wire to relieve the weight on the hoses and to keep them out of the way.

Disconnect all electrical connectors and the speedometer cable at the transaxle. Working under the hood, disconnect the shift linkage or cables and the clutch cable. Now, the shift linkages, electrical connections, and speedometer cables should be disconnected.

The exhaust system may also need to be lowered or partially removed. Remove any heat shields that may be in the way.

Now remove the starter. The starter wiring may be left connected or you can completely remove the starter from the vehicle to get it totally out of the way. The starter should never be left to hang by the wires attached to it. The weight of the starter can damage the wires or, worse, break the wires and allow the motor to fall, possibly on you or someone else. Always securely support the starter and position it out of the way after you have unbolted it from the engine.

Removing the Engine Through the Hood Opening Connect the engine sling or lifting chains to the engine. Use the lifting hooks on the engine **(Figure 9–10)** or fasten the sling to the points given in the service manual. Connect the sling to the crane and raise the crane just enough to support the engine.

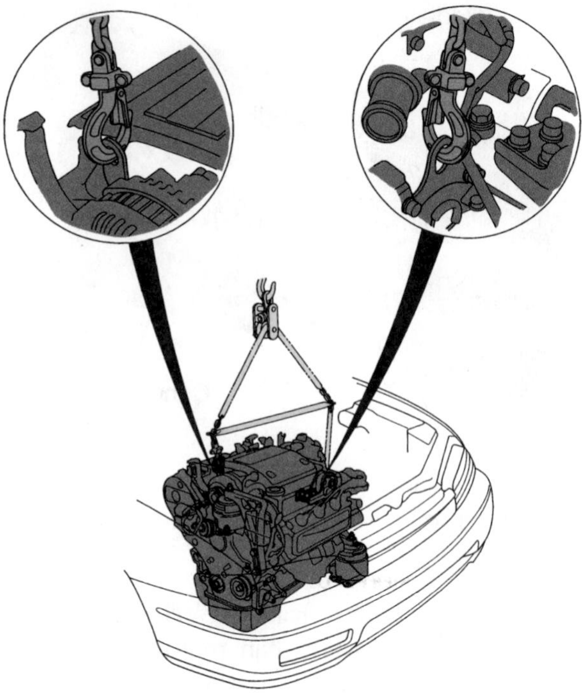

Figure 9-10 Some engines are equipped with eye plates to which the hoist can be safely attached. *Courtesy of American Honda Motor Co., Inc.*

From under the vehicle, remove the cross member. Then remove the mounting bolts for the engine at the engine and transmission mounts. With the transmission jack supporting the transmission, remove the transaxle mounts.

From under the hood, remove all remaining mounts. Raise the engine slightly to free the engine from the mounts. Then slowly raise the engine from the engine compartment. Guide the engine around all wires and hoses to make sure nothing gets damaged. Once the engine is cleared from the vehicle, prepare to separate it from the transaxle.

Removing the Engine from Under the Vehicle Position the engine cradle and dolly under the engine. Adjust the pegs of the cradle so they fit into the recesses on the bottom of the engine and secure the engine.

Remove all engine and transmission mount bolts. If required, remove the frame member from the vehicle. It may also be necessary to disconnect the steering gear from the frame. Double-check to ensure all wires and hoses are disconnected from the engine. With the transmission jack supporting the transmission, remove the transaxle mounts.

Slowly raise the vehicle, lifting it slightly away from the engine. As the vehicle is lifted, the engine remains on the cradle. During this process, continually check for interference with the engine and the body of the vehicle. Also watch for any wires and hoses that may still be attached to the engine. Once the vehicle is clear of the engine, prepare to separate the engine from the transaxle.

RWD Vehicles

On most RWD vehicles, the engine is removed through the hood opening, requiring the use of an engine hoist. Refer to the service manual to determine the proper engine lift points. Never lift the engine by intake manifold bolts. If the transmission is being removed with the engine, it may be easier if you locate the hook of the engine hoist to the chain in such a manner that the engine tips a little toward the transmission. Remember to lift the engine enough to clear the vehicle; it may be necessary to adjust the length of the hoist boom and legs.

Before beginning the final steps for removing the engine, check for anything behind and under the engine that should be disconnected. Loosen and remove all clutch (bell) housing bolts. If the vehicle has an automatic transmission, remove the torque converter mounting bolts.

If the transmission is being removed with the engine, place a drain pan under the transmission and drain the fluid from the transmission. Once the fluid is out, move the drain pan under the rear of the transmission. Use chalk to index the alignment of the rear U-joint and the pinion flange. Then remove the drive shaft. Disconnect the parts of the exhaust system that may get in the way. Disconnect all electrical connections **(Figure 9–11)** and the speedometer cable at the transmission. Make sure you place these away from the transmission so they are not damaged during transmission removal or installation.

Disconnect and remove the transmission and clutch linkage. It is best to do this by disconnecting as little as possible.

Use a floor jack to support the transmission and unbolt the motor mounts. If the engine is removed with its transmission, the front of the engine must come straight up as the transmission moves away from the bottom of the vehicle. Remove the transmission mount and cross member.

To lift the engine out of the vehicle, attach a pulling sling or chain (see Figure 9–1) to the engine. Some engines have eye plates for use in lifting. If they are not available, the sling must be bolted to the engine at the recommended lifting points. The sling attaching bolts must be large enough to support the engine and must thread into the block a minimum of 1½ times the bolt diameter.

Center the boom of the crane directly over the engine and raise the engine slightly. Make sure the engine is securely fastened to the chain and that nothing else is still attached to the engine. Continue raising the engine while pulling it forward. Make sure that the engine does not bind or damage any compartment component during this procedure. When the engine is high enough to clear the

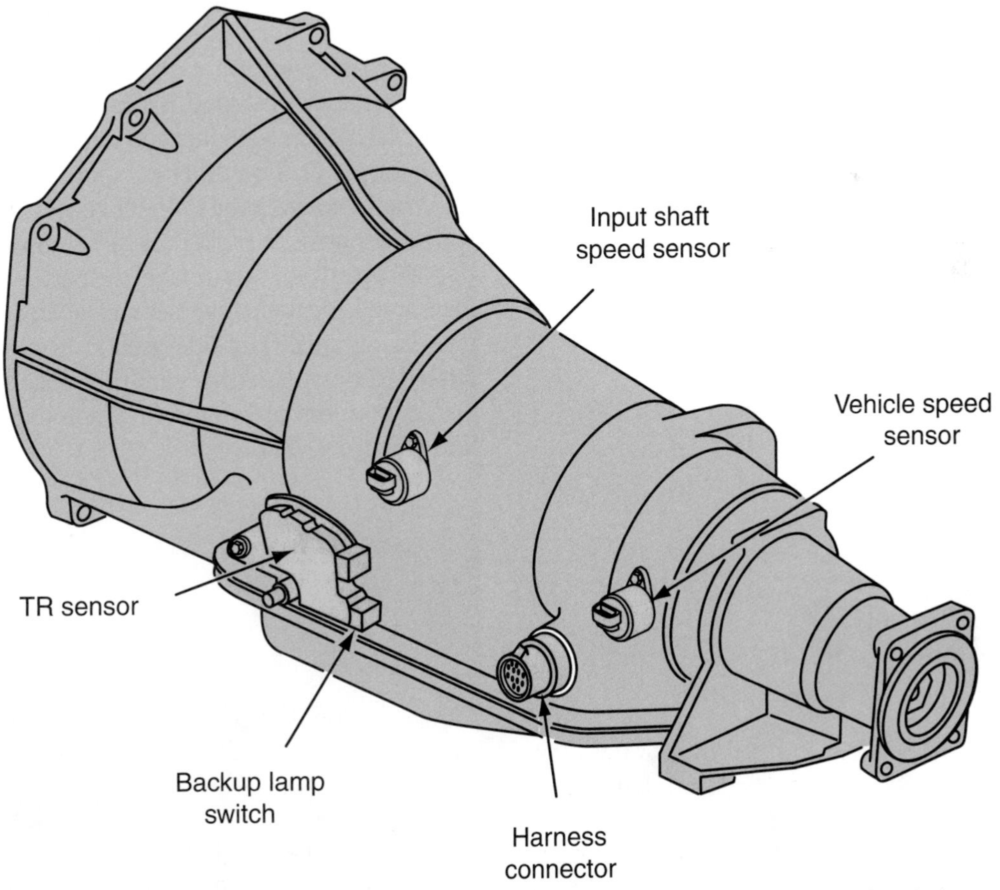

Figure 9-11 If the transmission will be removed with the engine, make sure all electrical connections are disconnected. Move the wiring harnesses to the side so they are not damaged during removal.

radiator, roll the crane and engine straight out and away from the vehicle.

Lower the engine close to the floor so it can be transported to the desired location. If the transmission was removed with the engine, remove the bell housing bolts and inspection plate bolts. On vehicles with an automatic transmission, remove the torque converter-to-flex plate bolts. Use a C-clamp or other brace to prevent the torque converter from falling. Also mark the location of the torque converter in relation to the flex plate for reference during installation.

ENGINE DISASSEMBLY AND INSPECTION

Raise the engine and position it next to an engine stand **(Figure 9–12)**. Mount the engine to the engine stand with bolts. Most stands use a plate with several holes or adjustable arms. The engine must be supported by at least four bolts that fit solidly into the engine. The engine should be positioned so that its center is in the middle of the engine stand's adapter plate to ensure that the engine is not too heavy when rotated on the engine stand.

Once the engine is securely mounted to the engine stand, remove the sling or lifting chain. The engine can now be disassembled and cleaned. Always refer to the service manual before you start to disassemble an engine.

Slowly disassemble the engine and visually inspect each part for any signs of damage. Look for excessive wear on the moving parts. Check all parts for signs of overheating, unusual wear, and chips. Look for signs of gasket and seal leakage.

Figure 9-12 A typical engine stand. *Courtesy of SPX Corporation, Aftermarket Tool and Equipment Group*

The following engine teardown of both cylinder head and block can be considered typical. Exact details will vary slightly depending on the style and type of engine. For instance, in some engines, the overhead camshaft is mounted directly in the cylinder head. In other engines, it is located in a separate housing that is mounted on the cylinder head.

USING SERVICE MANUALS

Look up the specific model car and engine prior to disassembling the engine. ■

Cylinder Head Removal

The first step in disassembly of an engine is usually the removal of the intake and exhaust manifolds. On some in-line engines, the intake and exhaust manifolds are often removed as an assembly.

SHOP TALK

It is important to let an aluminum cylinder head cool completely before removing it. ■

To start cylinder head removal, remove the valve cover or covers and disassemble the rocker arm components. When removing the rocker assembly, check the manufacturer's manual for specific instructions.

After removing the rocker arm and pushrods, check the rocker area for sludge. Excessive buildup can indicate a poor oil change schedule and is a signal to look for similar wear patterns on other components.

When removing the cylinder head, keep the pushrods and rocker arms or rocker arm assemblies in exact order. Use an organizing tray or label the parts with a felt-tipped marker to keep them together and labeled accurately. This type of organization aids greatly in diagnosing valve-related problems. Also, carefully check the lifters for dished bottoms or scratches, which indicate poor rotation **(Figure 9–13)**.

Figure 9-13 Two examples of wear that results when a lifter fails to rotate on the camshaft lobe. *Courtesy of TRW, Incorporated*

The cylinder head bolts are loosened one or two turns each, working from the outside of the cylinder head inward **(Figure 9–14)**. This procedure prevents the distortion that can occur if bolts are all loosened at once. The bolts are then removed, again following the outward to center sequence. With the bolts removed, the cylinder head can be lifted off **(Figure 9–15)**. The cylinder head gasket should be saved to compare with the new head gasket during reassembly.

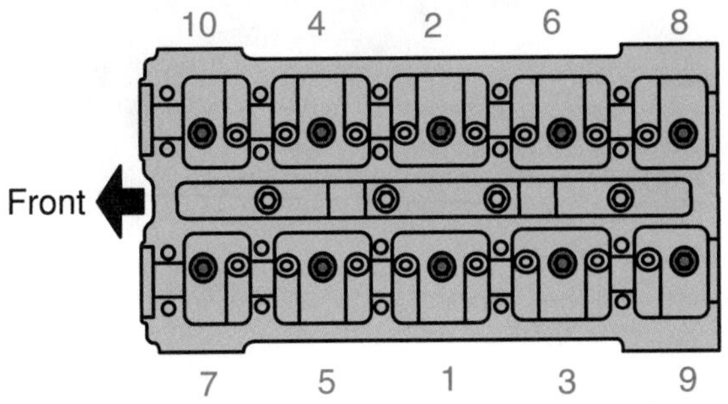

Figure 9-14 When loosening cylinder head bolts, follow the reverse of the specified tightening sequence.

On overhead valve engines, remove the timing cover. On some engines, this may require removing the oil pan. The harmonic balancer or vibration damper usually must also be removed. This normally requires a special puller.

The camshaft can now be carefully removed. Support the camshaft during removal to avoid dragging lobes over bearing surfaces, which would damage bearings and lobes. Do not bump cam lobe edges; this can cause chip-

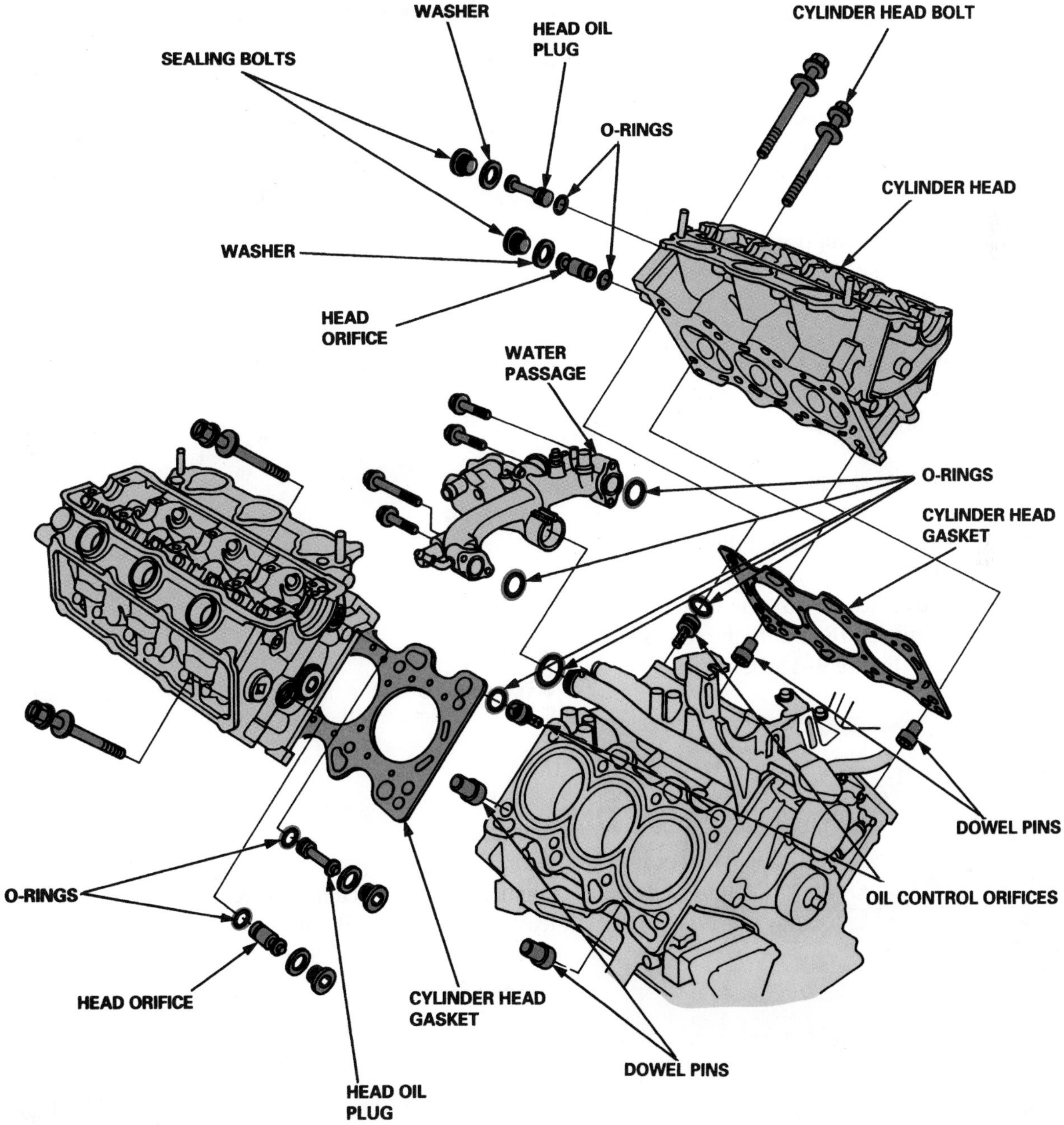

Figure 9-15 The cylinder heads of a typical late-model V6 engine. *Courtesy of American Honda Motor Co., Inc.*

ping. Some engines might require the removal of the thrust plate before taking out the camshaft.

After the camshaft has been removed, visually examine it for any obvious defects—rounded lobes, edge wear, galling, and the like. If either the camshaft or lifters are worn, both should be replaced. Do not install old valve lifters with a new camshaft. Do not install new valve lifters on a used camshaft.

Cylinder Head Disassembly

On overhead cam engines, the camshaft must be removed before the cylinder head can be disassembled. Follow the specified order for loosening the camshaft bearing caps. Also, draw a diagram of the arrangement of the cam follower assemblies and mark each part. This will help ensure that each part is returned to the same position.

If the cam has springs beneath it, there is 60 to 80 psi (414 to 551 kPa) pressure against any spot with the cam lobe against the follower. Random removal of bearing caps can cause the cam to break or spring up, possibly springing out and hitting you. Be sure to follow the service manual for the correct procedure because it is different for the bucket-type lifter design and lash adjuster design.

Next, use a valve spring compressor to begin disassembling the valve train. This compressor allows the technician to compress the valve springs and remove the keepers **(Figure 9–16)**. With the valves still in the cylinder head, measure the stem height for each valve and record it **(Figure 9–17)**. This measurement will be needed during reassembly to set the installed valve stem height.

Next, remove the valve oil seals and the valves. If a valve seems to be stuck in the guide, the tip might be peened over and have a ridge around it. If this is the case, do not drive the valve through the guide. It could score or crack the valve guide or head. Raise the stem and file off the excess metal **(Figure 9–18A)** until the stem slides through the guide easily **(Figure 9–18B)**.

Figure 9-16 Using a valve spring compressor. *Courtesy of Fel-Pro, Inc.*

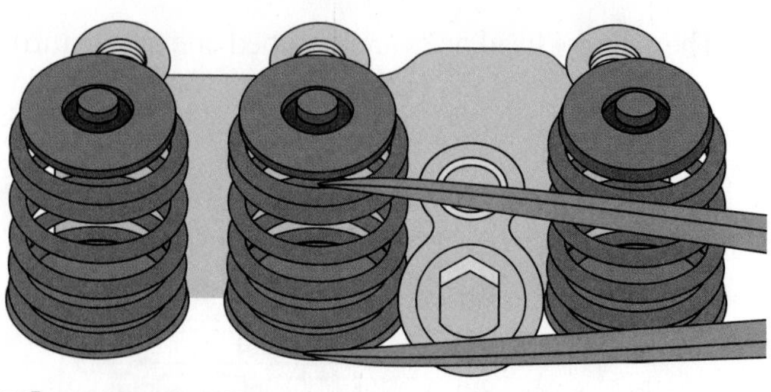

Figure 9-17 Measure valve spring height before disassembling the cylinder head. *Courtesy of Ford Motor Company*

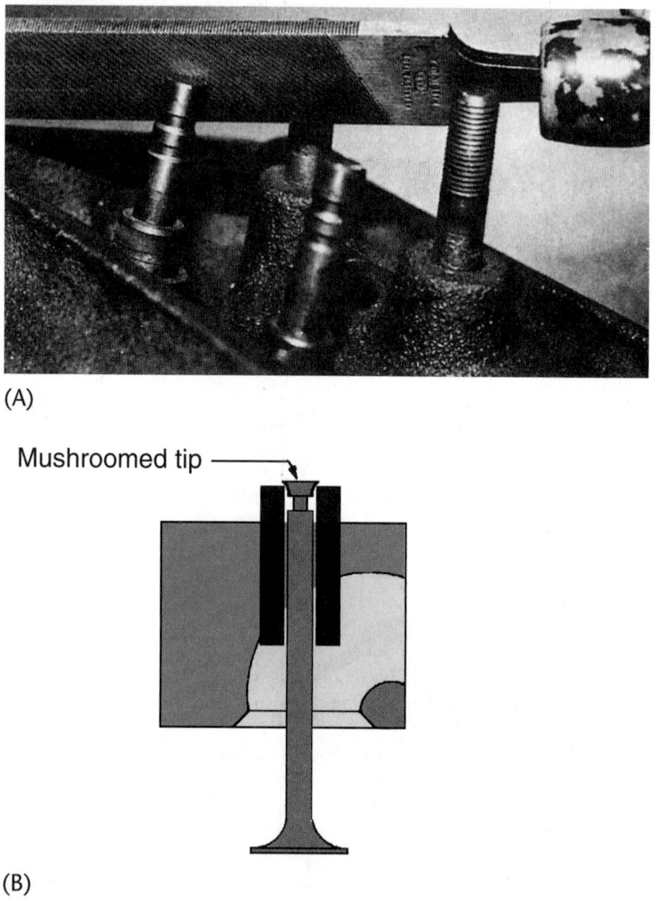

Figure 9-18 A mushroomed valve tip should be (A) filed before removing the valve from (B) its guide.

While removing valves, look for signs of burning, pitting, cracks, grooves, scores, necking, or other signs of wear. These wear patterns signal other problems in the engine. Valves that cannot be refaced without leaving at least a $\frac{1}{32}$-inch (.79-mm) valve margin must be discarded. Also, discard any valve that is badly burned, cracked, pitted, or shows signs of valve stem wear, bent valve stems, or damaged keeper grooves. Examine the backside of the intake valves. A black, oily, carbon buildup on the neck and stem area indicates oil is entering the cylinder through the intake valve guide **(Figure 9–19)**.

Figure 9–19 An oily soot or carbon buildup on the back of the valve indicates bad valve seals.

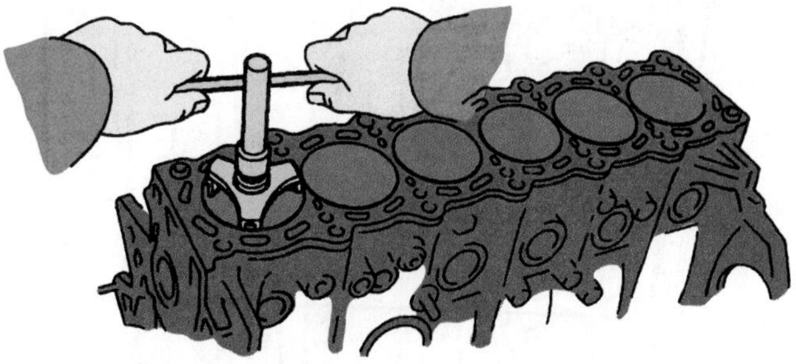

Figure 9–20 Use a ridge reamer to remove the ridge and/or all carbon from the top of the cylinder. *Reprinted with permission*

CUSTOMER CARE

When working on an engine, it is good practice to keep the customer informed of any unexpected problems. ■

Cylinder Block Disassembly

After the cylinder head has been removed, the cylinder block can be torn down. First, remove the oil pan if it was not removed previously. Then, remove the oil pump as directed in the service manual.

Continue the disassembly by removing the timing components. There are three different types of timing component assemblies: the chain and sprocket, the gear, and the timing belt and sprocket.

Often the chain and sprocket assembly and the timing belt and sprocket assembly both have tensioners and guides. All three types of timing mechanisms should be replaced as complete assemblies during an engine overhaul. The tensioners and guides wear and should be replaced as well.

Some OHC engines have complicated drive systems; so if you are unsure of the removal process, consult the proper engine repair manual. Inspect the sprockets for wear, cracks, and broken teeth. Inspect the timing gears for excessive backlash. Check the chain for slackness and wear. Timing belts should always be replaced during an engine overhaul.

Carefully remove the cylinder ridge with a ridge reamer tool. Rotate the tool clockwise with a wrench to remove the ridge **(Figure 9–20)**. Do not cut too deeply; an indentation may be left in the bore. Remove just enough metal to allow the piston assembly to slip out of the bore without causing damage to the bore. If the ridge is too large, the top rings will hit it and possibly break the ring lands.

The ridge is formed at the top of the cylinder **(Figure 9–21)**. Because the top ring stops traveling before it reaches the top of the cylinder, a ridge of unworn metal is left. Carbon also builds up above this ridge, adding to the problem. If the ridge is not removed, the piston's ring lands may be damaged as the piston is driven out of its bore.

After the ridge-removing operation, wipe all the metal cuttings out of the cylinder. Use an oily rag to wipe the cylinder. The cuttings will stick to it.

Prior to removal of the piston and rod assembly, check all connecting rods and main bearing caps for correct position and numbering. If the numbers are not visible, use a center punch or number stamp to number them **(Figure 9–22)**. Caps and rods should be stamped on the external flat surface. Remember, the caps and rods must remain as a set.

To remove the piston and rod assemblies, position the crankshaft throw at the bottom of its stroke. Remove the connecting rod nuts and cap. Tap the cap lightly with a soft hammer or wood block to aid in cap removal. Cover

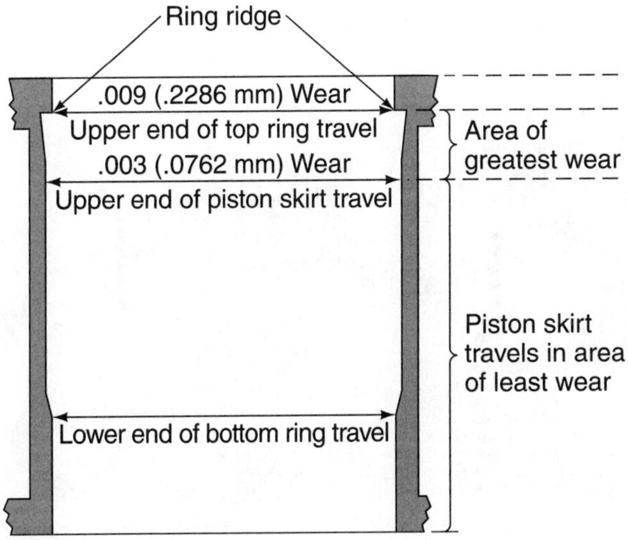

Figure 9–21 Normal cylinder wear. *Courtesy of Dana Corporation*

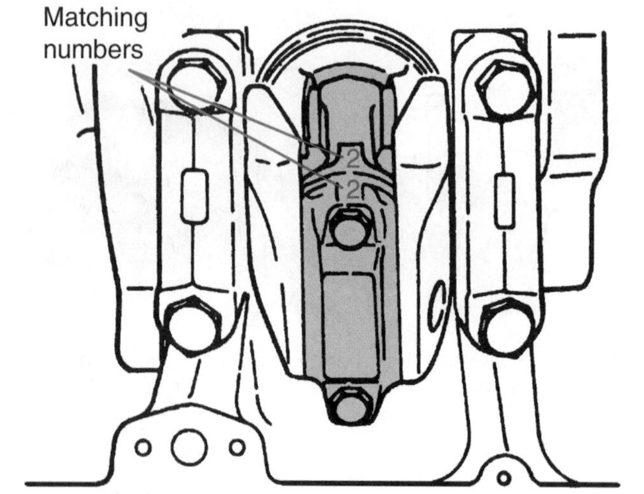

Figure 9–22 Mark the connecting rod caps prior to disassembling the engine. *Courtesy of General Motors Corporation*

Figure 9–24 Inspect each crankshaft journal for damage and wear. *Courtesy of Dana Corporation*

the rod bolts with protectors to avoid damage to the crankshaft journals **(Figure 9–23)**. Carefully push the piston and rod assembly out with the wooden hammer handle or wooden drift and support the piston by hand as it comes out of the cylinder. Be sure the connecting rod does not damage the cylinder during removal. With the bearing inserts in the rod and cap, replace the cap (numbers on same side) and install the nuts. Repeat the procedure for all other piston and rod assemblies.

Remove the flywheel or flex plate. Mark the position of the flywheel on the crankshaft. This aids the reassembly of the engine. In the specified order, loosen and remove the main bearing cap bolts and main bearing cap. Keeping the main bearing caps in order is very important. The location and position of each main bearing cap should be marked.

After removing the main bearing caps, carefully take out the crankshaft by lifting both ends equally to avoid bending and damage. Store the crankshaft in a vertical position to avoid damage.

Remove the main bearings and rear main oil seal from the block and the main bearing caps. Examine the bearing inserts for signs of abnormal engine conditions such as embedded metal particles, lack of lubrication, antifreeze contamination, oil dilution, uneven wear, and wrong or undersized bearings. Inspect them for any unusual wear signs, and inspect the main bearing inserts for indications that they are oversized or undersized. Carefully inspect the main journals on the crankshaft for damage **(Figure 9–24)**.

The block cannot be thoroughly cleaned unless all core plugs **(Figure 9–25)** and oil gallery plugs are removed. It is imperative that all plugs be removed to allow for a thorough cleaning. To remove cup-type "freeze"/core plugs, drive them in and then use a pair of channel lock pliers to pull them out. Flat-type plugs can be removed by drilling a hole near the center and then inserting a slide hammer to pull out the plug. In some engines, the cup-type plug can be removed easily by using a slide hammer or by driving the plug out from the backside with a long rod.

Sometimes removing threaded front and rear oil gallery plugs can be difficult. Using a drill and screw extractor can help.

Figure 9–23 Install rod bolt protectors before removing the rod and piston assembly from the engine block. *Courtesy of Dana Corporation*

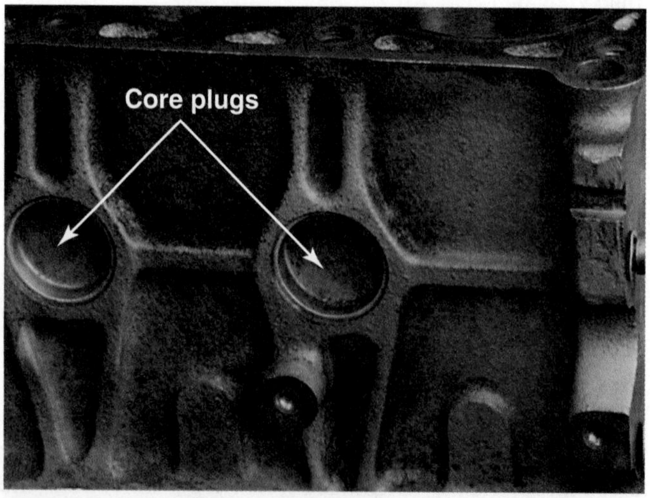

Figure 9–25 Core plugs.

After disassembly, the cylinder head and block and their parts must be visually checked for cracks or other damage before they are cleaned as described in the next section. While inspecting the parts, check the bearings for undersizes and to see if the bearings are the original ones. If the engine has been previously torn down, use caution because a mistake could have been made during the last rebuild. Check for correct sizing before ordering any parts.

CLEANING ENGINE PARTS

When the block or cylinder head parts have been removed, they must be thoroughly cleaned **(Figure 9–26)**. The cleaning method depends on the component to be cleaned and the type of equipment available. An incorrect cleaning method or agent can often be more harmful than no cleaning at all. For example, using caustic soda to clean an aluminum part will dissolve the part.

Only after all components have been thoroughly and properly cleaned can an effective inspection be made or proper machining be done.

Types of Soil Contaminants

Being able to recognize the type of dirt you are to clean will save you time and effort. Basically, there are four types of dirt.

Water-Soluble Soils The easiest dirt to clean is **water-soluble soil**, which includes dirt, dust, and mud.

Organic Soils **Organic soils** contain carbon and cannot be effectively removed with plain water. There are three distinct groupings of organic soils:

- Petroleum by-products derived from crude oil, including tar, road oil, engine oil, gasoline, diesel fuel, grease, and engine oil additives.
- By-products of combustion, including carbon, varnish, gum, and sludge **(Figure 9–27)**.

(A)

(B)

Figure 9-26 From (A) grime to (B) shine.

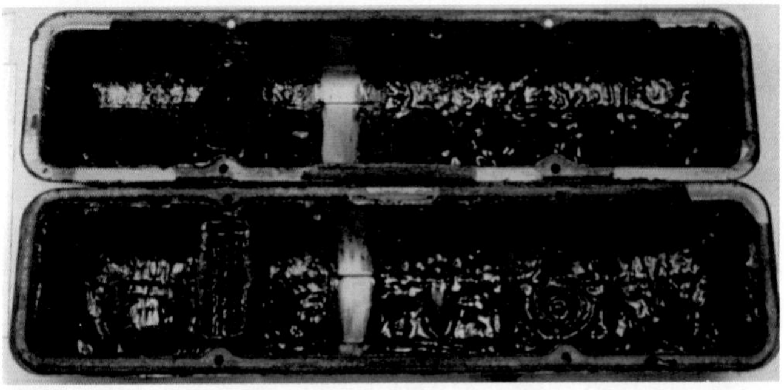

Figure 9-27 Water and by-products of combustion combine with engine oil to form sludge. *Courtesy of Texaco's magazine, LUBRICATION*

- Coatings, including such items as rust-proofing materials, gasket sealers and cements, paints, waxes, and sound-deadener coatings.

Rust **Rust** is the result of a chemical reaction that takes place when iron and steel are exposed to oxygen and moisture. Corrosion, like rust, results from a similar chemical reaction between oxygen and metal containing aluminum. If left unchecked, both rust and corrosion can physically destroy metal parts quite rapidly. In addition to metal destruction, rust also acts to insulate and prevent proper heat transfer inside the cooling system.

Scale When water containing minerals and deposits is heated, suspended minerals and impurities tend to dissolve, settle out, and attach to the surrounding hot metal surfaces. This buildup of minerals and deposits inside the cooling system is known as **scale**. Over a period of time, scale can accumulate to the extent that passages become blocked, cooling efficiency is compromised, and metal parts start to deteriorate.

Cleaning with Chemicals

There are three basic processes for cleaning automotive engine parts. The first process discussed is chemical cleaning.

This method of cleaning uses chemical action to remove dirt, grease, scale, paint, and/or rust.

CAUTION!

When working with any type of cleaning solvent or chemical, wear protective gloves and goggles and work in a well-ventilated area. Prolonged immersion of the hands in a solvent can cause a burning sensation. In some cases a skin rash might develop. There is one caution to mention about all manufactured cleaning materials that cannot be overemphasized: Read the labels carefully before mixing or using. Before using, learn how to operate the equipment.

Unfortunately, the most traditional line of defense against soils involves the use of cleaning chemicals. Chlorinated hydrocarbons and mineral spirits may have some health risks associated with their use through skin exposure and inhalation of vapors. Hydrocarbon cleaning solvents are also flammable. The use of water-based nontoxic chemicals can eliminate such risks.

WARNING!

Prior to using any chemical, read through all the information given on the material safety data sheet (MSDS) or the Canadian workplace hazardous materials information systems sheets (WHMIS) for that chemical. Become aware of the health hazards presented by the various chemicals.

Hydrocarbon solvents are labeled hazardous or toxic and require special handling and disposal procedures. The makers of many water-based cleaning solutions claim their products are biodegradable. Once the cleaning solution has become contaminated with grease and grime, it too becomes a hazardous or toxic waste that can be subject to the same disposal rules as a hydrocarbon solvent.

Some manufacturers offer waste-handling and solvent recycling services. The old solvent is recycled by a distillation process to separate the sludge and contaminants. The solvent is then returned to service and the contaminants disposed of. Independent services for maintaining hot tanks and spray washers are also available.

Chemical Cleaning Machines

Parts Washers. Parts washers (often called **solvent tanks**) are one of the most widely used and inexpensive methods of removing grease, oil, and dirt from the metal surfaces of a seemingly infinite variety of automotive components and engine parts. A typical washer setup **(Figure 9–28)** might consist of a tank to hold a given volume of cleaning solution and some method of applying the solution. These methods include soaking, soaking and agitation, solvent streams, and spray gun applicators.

Soak Tanks. There are two types of soak tanks: cold and hot. Cold soak tanks are commonly used to clean carburetors, throttle bodies, and aluminum parts. A typical cold soak unit consists of a tank to hold the cleaner and a basket to hold the parts to be cleaned. After soaking with or without gentle agitation is complete, the parts are removed, flushed with water, and blown dry with compressed air.

Cleaning time is short, about 20 to 30 minutes, when the chemical cleaner is new. The time becomes progressively longer as the chemical ages. Agitation by raising and lowering the basket (usually done mechanically) will reduce the soak period to about 10 minutes. Some more elaborate tanks are agitated automatically.

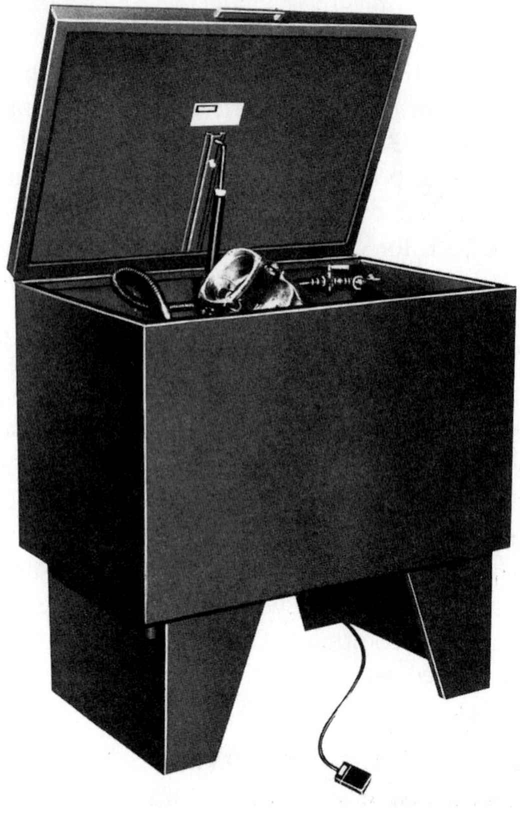

Figure 9-28 A typical parts washer. *Courtesy of Kansas Instruments*

Hot soak tanks are actually heated cold tanks. The source of heat is either electricity, natural gas, or propane. The solution inside the hot tanks usually ranges from 160 to 200°F (71 to 93°C). Most tanks are generally large enough to hold an entire engine block and its related parts.

Hot tanks use a simple immersion process that relies on a heated chemical to lift the grease and grime off the surface. Liquid or parts agitation may also be used to speed up the job. Agitation helps shake the grime loose and also helps the liquid penetrate blind passageways and crevices in the part. Generally speaking, it takes one to several hours to soak most parts clean.

Hot Spray Tanks. Because of the Environmental Protection Agency (EPA) regulations against open steam cleaning, the spray tank has become more popular. The hot spray tank reassembles a large automatic dishwasher and is designed to remove organic and rust soils from a variety of automotive parts **(Figure 9–29)**. In addition to parts being bathed and soaked as in the hot soak method, spray washers add the benefit of moderate pressure cleaning.

Using a hot jet spray washer can cut cleaning time to less than 10 minutes. Normally, a strong soap solution is used as the cleaning agent. The speed of this system, along

Figure 9-29 A compact spray cleaning machine. *Courtesy of Sunnen Products Company*

with lower operating costs, makes it popular with many machine shop owners.

SHOP TALK

Caustic soda, also known as sodium hydroxide, has been used as a cleaner and can be a very dangerous irritant to the eyes, skin, and mucous membranes. These chemicals should be used and handled with care. Because of the accumulation of heavy metals, it is considered a hazardous waste material and must be disposed of in accordance with EPA guidelines. ■

Spray washers are often used to preclean engine parts prior to disassembly. A pass-through spray washer is fully automatic once the parts have been loaded, and the cabinet prevents the runoff from going down the drain or onto the ground (which is not permitted in many areas because of local waste disposal regulations). Spray washers are also useful for postmachining cleanup to remove machine oils and metal chips.

Thermal Cleaning

The second basic process for cleaning engine parts is thermal cleaning. This process relies on heat to bake off or oxidize dirt and other contaminants.

Thermal cleaning ovens **(Figure 9–30)**, especially the pyrolytic type, have become increasingly popular. The main advantage of thermal cleaning is a total reduction of all oils and grease on and in blocks, heads, and other

Figure 9-30 A cleaning furnace. *Courtesy of Pollution Control Products Co.*

parts. The high temperature inside the oven (generally 650 to 800°F [343 to 426°C]) oxidizes all the grease and oil, leaving behind a dry, powdery ash on the parts. The ash must then be removed by shot blasting or washing. The parts come out dry, which makes subsequent cleanup with shot blast or glass beads easier because the shot will not stick.

SHOP TALK

A slow cooling rate is recommended to prevent distortion that could be caused by unequal cooling rates within complex castings. ■

One of the major attractions of cleaning ovens is that they offer a more environmentally acceptable process than chemical cleaning. But, although there is no solvent or sludge to worry about with an oven, the ash residue that comes off the cleaned parts must still be handled according to local disposal regulations.

The maintenance procedure given in the owner's manual must be followed if the ovens are to operate properly.

Abrasive Cleaners

The third process used to clean engine parts involves the use of abrasives. Most abrasive cleaning machines are used in conjunction with other cleaning processes rather than as a primary cleaning process itself. Parts must be dry and grease-free when they go into an abrasive blast machine. Otherwise, the shot or beads will stick.

Abrasive Cleaning Methods

Abrasive Blaster. Airless shot and grit blasters are best used on parts that will be machined after they have been cleaned **(Figure 9–31)**. Two basic types of media are available: shot and grit. Shot is round; grit is angular in shape. Steel shot and glass beads are used primarily for cleaning operations where etching or material removal is not desired. Steel shot and glass beads are also used for peening the surfaces of certain parts. Peening is a process of hammering on the surface of a part. This packs the molecules tightly, thereby increasing the part's resistance to fatigue and stress.

Grit is used primarily for aggressive cleaning jobs or where the surface of the material needs to be etched to improve paint adhesion. Because grit cuts into the surface as it cleans, it removes dirt and scale faster than shot blasting or glass beading. But it also removes metal, leading to some change in tolerances. The beneficial effect of grit blasting is that it roughens the surface, leaving a matte finish to which paint or other surface treatments will stick better than a peened or polished surface. However, grit blasting is an abrasive process that chews out pits in the surface into which pollutants and blast residue can settle. This leads to stress corrosion unless the surface is painted or treated. The tiny crevices also focus surface stresses in the metal, which can lead to cracking in highly loaded parts. Because of that, grit should never be used for peening.

The type of media used for a given job depends on the job itself and the type of equipment. Steel shot is normally used with airless wheel blast equipment, which hurls the shot at the part with the centrifugal force of the spinning wheel. Used with air blast equipment, glass beads are blown through a nozzle by compressed air.

Parts Tumbler. A cleaning alternative that can save considerable labor when cleaning small parts such as engine valves is a tumbler. Various cleaning media can be used in a tumbler to scrub the parts clean. This saves considerable hand labor and eliminates dust. In some tumblers, all parts are rotated and tilted at the same time.

Figure 9–32 A vibratory parts cleaner. *Courtesy of C&M Cleaning Systems*

Vibratory Cleaning. Shakers, as they are frequently called, use a vibrating tub filled with ceramic steel, porcelain, or aluminum abrasive to scrub parts clean **(Figure 9–32)**. Most shakers flush the tub with solvent to help loosen and flush away the dirt and grime. The solvent drains out the bottom and is filtered to remove the sludge.

Cleaning by Hand. Some hand cleaning is inevitable. Regardless of the cleaning process used, it is usually necessary to remove gallery plugs and hand clean the oil galleries **(Figure 9–33)**. Another often neglected area is between the heat shield and the bottom of the intake manifold, where carbon and oil deposits collect. This shield should be removed before cleaning the manifold. Any residual dirt left in the engine after cleaning can lead to failure. Therefore, proper cleaning of all engine components is vital in the rebuilding process. Remove any sur-

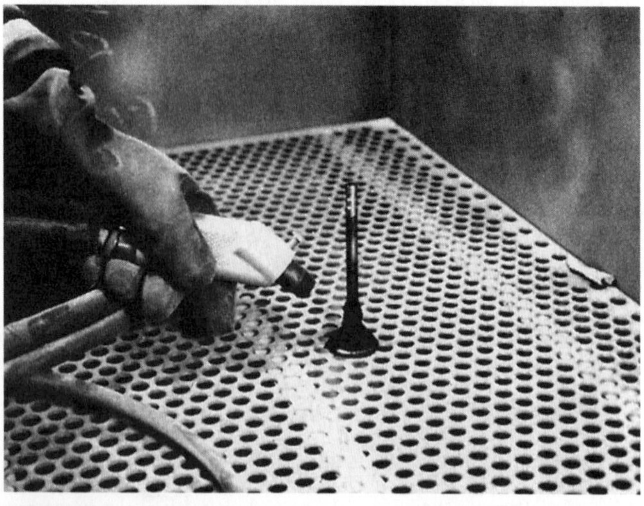

Figure 9–31 Using a blast nozzle to clean the back side of a valve.

Figure 9–33 It is often necessary to remove the gallery plugs and hand clean the oil galleries.

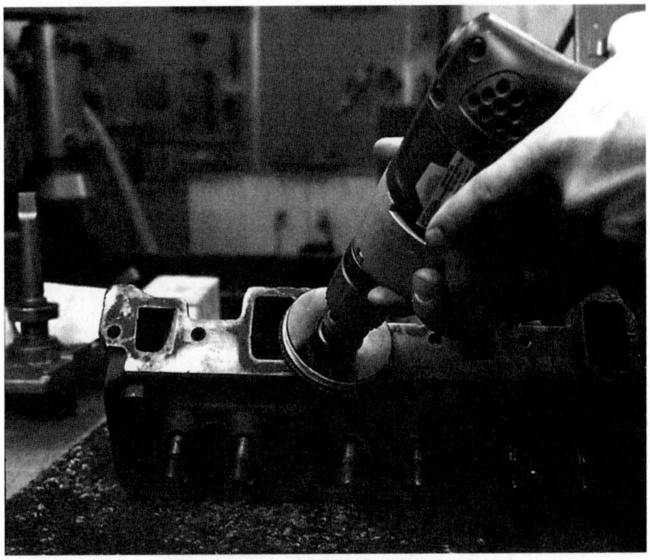

Figure 9–34 Using a power scraper pad will prevent any metal from being removed. *Courtesy of Goodson Shop Supplies*

face irregularities with very fine, emory cloth. Make sure to keep any dirt out of the cylinder bores. The special power scraper pad shown in action in **Figure 9–34** is guaranteed not to remove any metal.

Carbon can be removed from parts using a twist-type wire brush driven by an electric or air drill motor. Using brushes can often be a time-consuming job. Some shops use a wire brush in addition to another cleaning method. Moving the drill motor in a light circular motion against the carbon helps to crack and dislodge the carbon for easier wire brush cleaning.

Alternative Cleaning Methods Three of the most popular alternatives to traditional chemical cleaning systems are ultrasonic cleaning, citrus chemicals, and salt baths.

Ultrasonic Cleaning. This cleaning process has been used for a number of years to clean small parts like jewelry, dentures, and medical instruments. Recently, however, the use of larger ultrasonic units has expanded into small engine parts cleaning. **Ultrasonic cleaning** utilizes high-frequency sound waves to create microscopic bubbles that burst into energy to loosen soil from parts. Because the tiny bubbles do all the work, the chemical content of the cleaning solution is minimized, making waste disposal less of a problem.

Citrus Chemicals. Some chemical producers have developed citrus-based cleaning chemicals as a replacement for the hazardous solvent and alkaline-based chemicals. Because of their citrus origin, these chemicals are safer to handle, easier to dispose of, and even smell good.

Salt Bath. The **salt bath** is a unique process that uses high-temperature molten salt to dissolve organic materials, including carbon, grease, oil, dirt, paint, and some gaskets. For cast iron and steel, the salt bath operates at about 700 to 850°F (371 to 454°C). For aluminum or combinations of aluminum and iron, a different salt solution is used at a lower temperature (about 600°F [315°C]). The contaminants precipitate out of the solution and sink to the bottom of the tank, from which they must be removed periodically. The salt bath itself lasts indefinitely as long as the salt is maintained properly. Like a hot tank, the temperature of the salt bath is maintained continuously.

CRACK DETECTION AND REPAIR

Once the engine parts have been cleaned and given a thorough visual inspection, actual repair work begins. If cracks in the metal casting were discovered during the inspection, they should be repaired or the part replaced.

Cracks in metal castings are the result of stress or strain in a section of the casting. This stress or strain finds a weak point in that section of the casting and causes it to distort or separate at that point **(Figure 9–35)**. Such stresses or strains in castings can develop from the following:

- Pressure or temperature changes during the casting procedure may cause internal material structure defects, inclusion, or voids.
- Fatigue may result from fluctuating or repeated stress cycles. It might begin as a small crack and progress to a larger one under the action of the stress.
- Flexing of the metal may result due to its lack of rigidity.
- Impact damage may occur by a solid, hard object hitting a component.
- Constant impacting of a valve against a hardened seat may produce vibrations that could possibly lead to fracturing a thin-walled casting.

Figure 9–35 Examples of stress cracks.

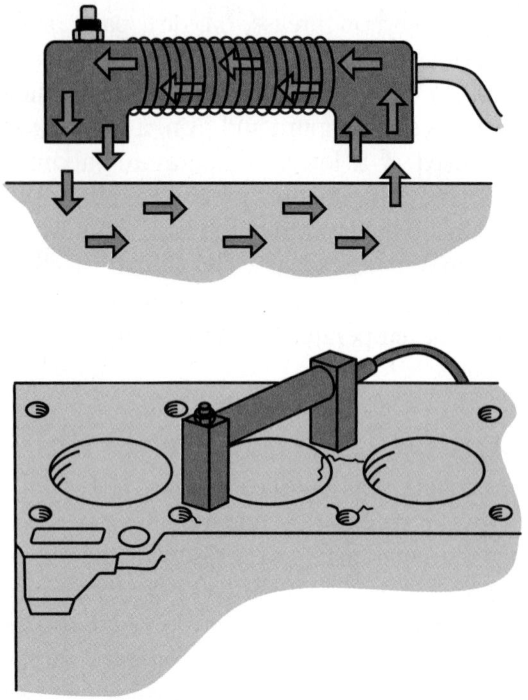

Figure 9–36 MPI testing passes a magnetic field though the iron item being checked.

- Chilling of a hot engine by a sudden rush of cold water or air over the surface may happen.
- Excessive overheating is possible due to improper operation of an engine system.

Cracks can be found by visual inspection; however, many are not easily seen. Therefore, engine rebuilders use special equipment to detect cracks, especially if there is reason to suspect a crack.

Magnetic particle inspection (MPI) uses a permanent magnet or an electromagnet to create a magnetic field in a cast-iron unit **(Figure 9–36)**. When the legs of the detector tool are placed on the metal, the magnetic field travels through the metal. Iron filings are sprinkled on the surface to detect a secondary magnetic field resulting from a crack **(Figure 9–37)**. Because the secondary magnetic field will not form if the crack is in the same direction as the magnet, the magnet must be rotated and the metal checked in both directions.

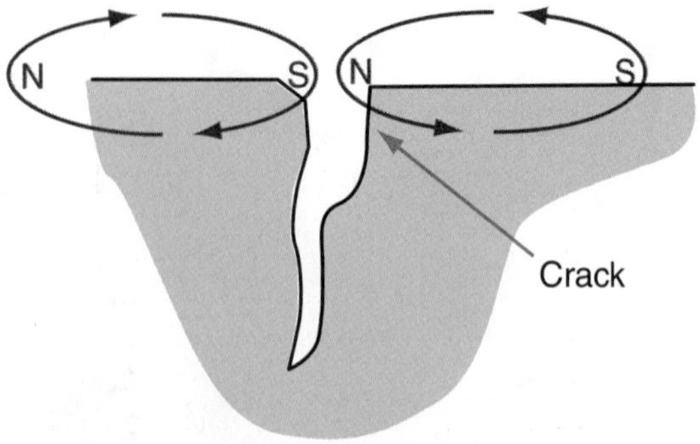

Figure 9–37 A crack causes two opposing magnetic poles to form on each side. The iron filings used with the magnet will show these fields.

Another common way to detect cracks is called magnetic fluorescent crack detection. This method does not rely on a magnet; rather, it uses a magnetic, fluorescent paste. The paste is spread over an area. The magnetic paste will flow over the metal. A black light is used to look at the paste. Wherever the paste has created a line, it has filled in a crack. This method is great for finding small cracks.

No matter what caused the crack, it is important to relieve the stress at the point of distortion or cracking and then add more metal and close the crack. This can be accomplished by the cold process of pinning or the hot process of welding.

Furnace Welding Crack Repairs

Furnace welding is considered by many people to be the best way to repair cracks in a cast-iron head. By preheating the entire casting, the problem of stress cracks forming during the cooling-off period is eliminated. Heat welding, however, requires a good heat source and proficient welding skills. As a rule, this repair is conducted only by a specialist.

Repairing Aluminum Heads

Aluminum heads have become popular in recent years, primarily because of the weight savings they offer. However, there are many problems associated with an aluminum cylinder head. The typical shop is most likely to encounter the following problems:

- Cracks in the aluminum between the valve seat rings **(Figure 9–38)**. These cracks, usually quite small, require close inspection to find. They very seldom leak and can be closed by a light peening. Some shops make no repairs to them.
- Bottomside cracks coming from the coolant passages. These cracks can be repaired by veeing out the damaged area and welding with an aluminum filler rod.

Figure 9–38 Cracks between the center two valve seats are common with aluminum heads.

Figure 9-39 The topside oil artery crack appeared when an oxyacetylene flame was passed over the casting. Carbon in the flame was trapped in the crack, highlighting it.

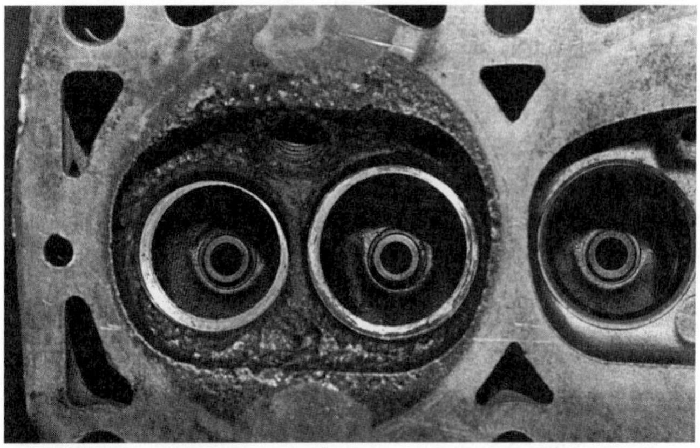

Figure 9-41 Severe coolant related damage can sometimes be repaired, but normally a damaged cylinder head is replaced.

- Topside cracks across the main oil artery. These cracks, although not too common, are usually very visible **(Figure 9–39)**. Most authorities recommend replacing the head completely if such a crack is found. The length of time required to make the repair is not reasonable. The labor cost is not worth the risk of possible failure and is higher than the cost of purchasing an uncracked core.
- Detonation damage can occur on any cylinder **(Figure 9–40)**. Repairs can be made by welding and freehand machining with a rotary burr in a die grinder.
- Meltdown damage is a somewhat common occurrence on the high-swirl combustion chamber heads. Again, repairs can be made by welding and freehand machining.
- Coolant-related metal erosion. If damage around coolant passages is excessive, if the side of any valve seat has been exposed, or if the combustion chamber shows erosion, the head must be repaired or replaced **(Figure 9–41)**. Coolant erosion can be easily fixed by welding and resurfacing.

Figure 9-40 The effect of detonation on a combustion chamber.

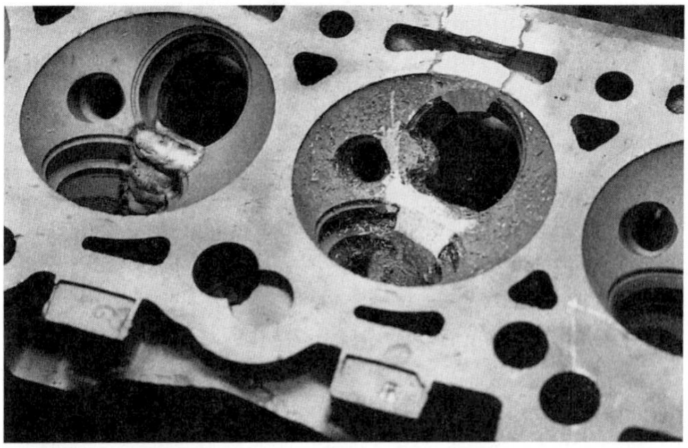

Figure 9-42 TIG-welded aluminum head repair.

Tungsten inert gas (TIG) welding is the preferred repair technique for aluminum heads **(Figure 9–42)**. Welding aluminum is often considered difficult because it welds differently than iron or steel. When exposed to air, aluminum forms an oxide coating on the surface that helps protect the metal against further corrosion. The oxide layer makes welding difficult because it interferes with fusing and weakens the weld. Cleaning the surface can remove the oxide. However, as soon as the metal is heated, oxide reforms (unless the weld is bathed in a constant supply of inert gas).

CASE STUDY

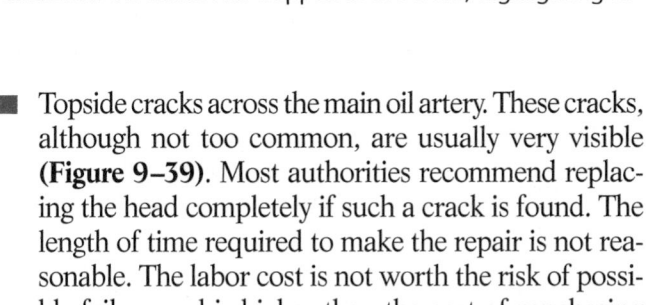

A four-cylinder engine is brought into the shop. The customer complains of excessive oil consumption and oil leaks. Compression and cylinder leakage tests indicate that the cylinders are sealing well and a power balance test indicates all of the cylinders are producing about the same amount of

power. Based on these results, the technician assumes that the problem is leaking valve seals. The initial plan is to replace the seals and regasket the engine.

It is odd that the engine has both of these problems. It has less than 50,000 miles (80,000 km) on it. Not really sure if the problems are related, the technician proceeds to disassemble the engine. Upon removing the valve cover, large amounts of sludge are evident throughout the valve train, normally a sign that the engine has been neglected. However, a review of the files indicates that the oil has recently been changed. In fact, the car has been well-maintained. Is the sludging related to the oil consumption and leaks?

The oil pan is removed and additional sludge is found. The cylinder head is then removed from the block. The piston tops and the combustion chamber are covered with a thick, black carbon coating. Is this buildup related to other problems?

The cylinder head is disassembled, and each of the valve seals is found to be deteriorated. What could cause the deterioration of rubber parts, leaking gaskets, sludging, and carbon buildup in the cylinders? After careful thought, the technician pays attention to the parts taken off the engine during initial disassembly. A thorough inspection is made of the PCV system, and it is discovered that the hose connecting the valve to the manifold is plugged solid. The valve is also found to be plugged.

The PCV system is designed to remove crankcase fumes and pressure from the crankcase. These fumes can cause rapid sludging of the oil and deterioration of rubber parts. Excessive crankcase pressure can cause leaks, as the pressure seeks to relieve itself. A faulty PCV valve can cause all of the problems exhibited by this engine. In fact, it is the cause of the problems.

The engine is resealed and new valve stem seals are installed. The engine is then installed with a new PCV valve and hose. Not only is the customer's complaint taken care of, but so is the cause of the problem.

KEY TERMS

Organic soil
Rust
Salt bath
Scale
Solvent tank
Tungsten inert gas (TGI)
Ultrasonic cleaning
Water-soluble soil

SUMMARY

- When preparing an engine for removal and disassembly, always follow the specific service manual procedures for the particular vehicle being worked on.
- A hoist and chain or crane are needed to lift an engine out of its compartment. Mount the engine to an engine stand with a minimum of four bolts, or set it securely on blocks.
- While an engine teardown of both the cylinder head and block is a relatively standard procedure, exact details vary among engine types and styles. The vehicle's service manual should be considered to be the final word.
- An understanding of specific soil types can save time and effort during the engine cleaning process. The main categories of contaminants include water-soluble and organic soils, rust, and scale.
- Protective gloves and goggles should be worn when working with any type of cleaning solvent or chemical. Read the label carefully before using, as well as all of the information provided on material safety data sheets.
- Parts washers, or solvent tanks, are a popular and inexpensive means of cleaning the metal surfaces of many automotive components and engine parts. Regardless of the type of solvent used, it usually requires some brushing, scraping, or agitation to increase the cleaning effectiveness.
- Alternatives to chemical cleaning have emerged in recent years, including thermal cleaning, ultrasonic cleaning, salt baths, and citrus chemical cleaning.
- Steel shot and glass beads are used for cleaning operations in which etching or material removal is not desired. Grit, the other type of abrasive blaster, is used for more aggressive cleaning jobs.
- Some degree of manual cleaning is necessary in any engine rebuilding job. Very fine, abrasive paper should be used to remove surface irregularities. A twist-type wire brush driven by an electric or air drill motor is also helpful, though it can be time-consuming to work with.
- TIG welding is the preferred repair technique for aluminum cylinder heads. Because it reacts differently to heat than iron or steel, aluminum is considered a challenge to repair by welding.

TECH MANUAL

The following procedures are included in Chapter 9 of the *Tech Manual* that accompanies this book:

1. Preparing engine for removal.
2. Removing and disassembling the cylinder head.
3. Removing cylinder ring ridge.

REVIEW QUESTIONS

1. What precautions should be taken when cleaning an engine prior to its removal?
2. What should be worn when working with any type of cleaning solvent or chemical?
3. *True or False?* Most engines in a RWD vehicle must be removed with the transmission still attached.
4. What is the best way to repair cracks in a cast-iron head?
5. What welding method is preferred for repairing aluminum heads?
6. *True or False?* The first step in disassembling an engine is usually the removal of the intake and exhaust manifolds.
7. Technician A uses a crane to remove an engine from its compartment. Technician B uses a canvas hoist and chain to remove an engine from its compartment. Who is correct?
   - **a.** Technician A
   - **b.** Technician B
   - **c.** Both A and B
   - **d.** Neither A nor B
8. Before removing the valves from a cylinder head, Technician A measures the stem height for each valve and records it. Technician B cleans all carbon deposits from the back of the valve. Who is correct?
   - **a.** Technician A
   - **b.** Technician B
   - **c.** Both A and B
   - **d.** Neither A nor B
9. While removing a cylinder head, Technician A keeps all rocker arms and pushrods in order. Technician B loosens each head bolt starting with the outside ones and moving toward the center. Who is correct?
   - **a.** Technician A
   - **b.** Technician B
   - **c.** Both A and B
   - **d.** Neither A nor B
10. The buildup of minerals and deposits inside the cooling system is called ______.
    - **a.** organic soil
    - **b.** scale
    - **c.** rust
    - **d.** grime
11. Hydrocarbon solvents are ______.
    - **a.** flammable
    - **b.** toxic
    - **c.** both a and b
    - **d.** neither a nor b
12. While discussing abrasive cleaners, Technician A says shot is angular in shape and is used for aggressive cleaning. Technician B says grit is an angular-shaped media and is used to peen metal surfaces. Who is correct?
    - **a.** Technician A
    - **b.** Technician B
    - **c.** Both A and B
    - **d.** Neither A nor B
13. Which one of these statements is *not* true?
    - **a.** If cracks in a metal casting are found during an inspection, they should be repaired or the part replaced.
    - **b.** Stress or strain on a weak point of a casting causes it to distort at that point.
    - **c.** During crack repair, it is important to add more metal to the point of stress so that the casting can withstand more strain.
    - **d.** Some cracks, such as those in the main oil artery, can be ignored.
14. Which cleaning method uses high-frequency sound waves to create microscopic bubbles that loosen dirt from parts?
    - **a.** ultrasonic
    - **b.** salt bath
    - **c.** thermal
    - **d.** caustic
15. Technician A uses a fast cooling rate when working with thermal cleaning ovens. Technician B uses water to rinse off the parts after they have been cleaned in an oven. Who is correct?
    - **a.** Technician A
    - **b.** Technician B
    - **c.** Both A and B
    - **d.** Neither A nor B

16. Parts must be ______ when they go into an abrasive blast machine.

a. wet
b. dry
c. grease-free
d. both b and c

17. While discussing cleaning engine parts, Technician A says the cleaning method used depends on the component to be cleaned and the type of cleaning equipment available. Technician B says sometimes it is best to clean parts by hand with soap and warm water. Who is correct?

a. Technician A
b. Technician B
c. Both A and B
d. Neither A nor B

18. All of the following bolts need to be loosened in a particular order to prevent distortion of the parts they secure, *except* ______.

a. cylinder head bolts
b. main bearing cap bolts
c. camshaft bearing cap bolts
d. connecting rod cap bolts

19. An engine block should be mounted to an engine stand using a minimum of ______ bolts.

a. four
b. six
c. three
d. five

20. Technician A uses light peening to repair small cracks between aluminum head valve seat rings. Technician B repairs bottomside coolant passage cracks by welding. Who is correct?

a. Technician A
b. Technician B
c. Both A and B
d. Neither A nor B

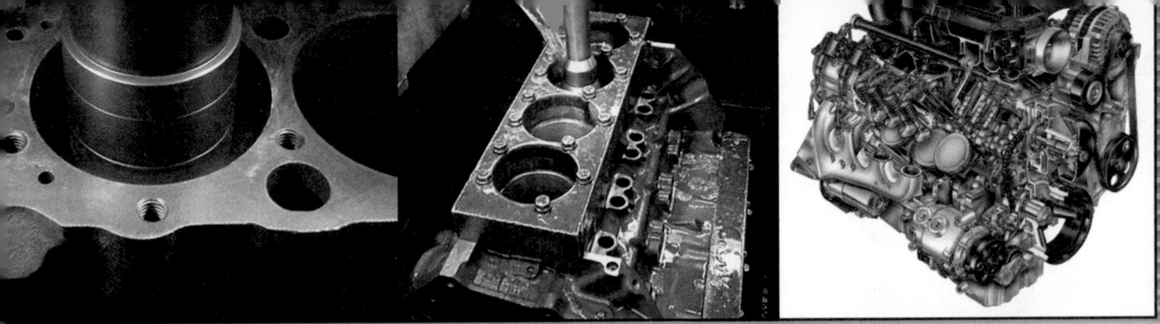

10

SHORT BLOCKS

OBJECTIVES

■ List the parts that make up a short block and briefly describe their operation. ■ Describe the major service and rebuilding procedures performed on cylinder blocks. ■ Describe the purpose, operation, and location of the camshaft. ■ Describe the four types of camshaft drives. ■ Inspect the camshaft and timing components. ■ Describe how to install a camshaft and its bearings. ■ Explain crankshaft construction, inspection, and rebuilding procedures. ■ Explain the function of engine bearings, flywheels, and harmonic balancers. ■ Explain the common service and assembly techniques used in connecting rod and piston servicing. ■ Explain the purpose and design of the different types of piston rings. ■ Describe the procedure for installing pistons in their cylinder bores.

An engine is made up of many parts, each with its own purpose. When there is a major engine failure, shops either rebuild or replace the engine. Most often the **short block** is repaired or replaced as an assembly. A basic short block assembly consists of a cylinder block, crankshaft, crankshaft bearings, connecting rods, pistons and rings, and oil gallery and core plugs. Parts related to the short block but not necessarily included with it are the flywheel and **harmonic balancer**. A short block may also include the engine's camshaft and timing gear. A **long block** is basically a short block with cylinder heads.

CYLINDER BLOCK

The cylinder block makes up the lower section of the engine. It houses the areas where combustion of the air/fuel mixture takes place **(Figure 10–1)**. The upper section of the engine, known as the cylinder head, bolts to the top of the cylinder block. The head is also part of the combustion chamber and contains the valve train components.

The cylinder block **(Figure 10–2)** is normally a one-piece casting, machined so that all the parts contained in it fit properly. Blocks may be cast from several different materials: iron, aluminum, or possibly, in the future, plastic. Some late-model engines are made of two pieces, an upper piece which contains the cylinders and a lower unit that surrounds the crankshaft **(Figure 10–3)**.

The word **cast** refers to how the block is made. To cast is to form molten metal into a particular shape by pouring or pressing it into a mold. This molded piece must then undergo a number of machining operations to make sure all the working surfaces are smooth and true. The top of the block must be perfectly smooth so that the cylinder head can seal it. The base or bottom of the block is also machined to allow for proper sealing of the oil pan.

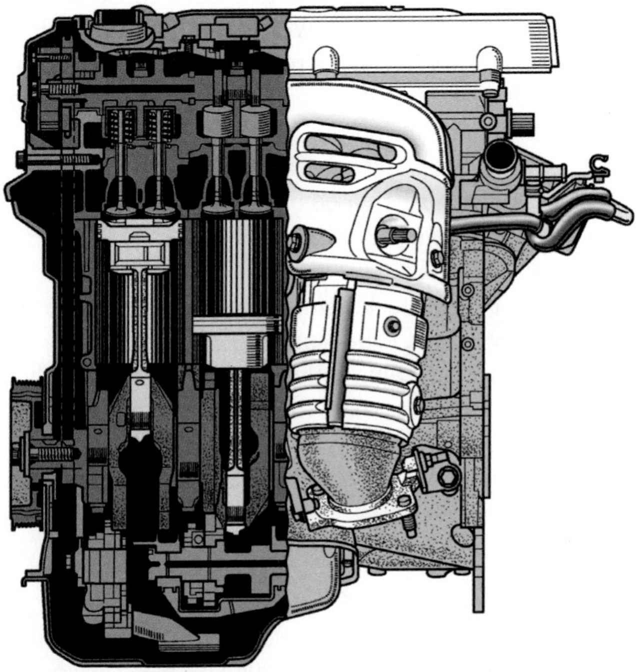

Figure 10-1 A cutaway showing the fit of the piston assemblies and crankshaft in an engine block. *Reprinted with permission*

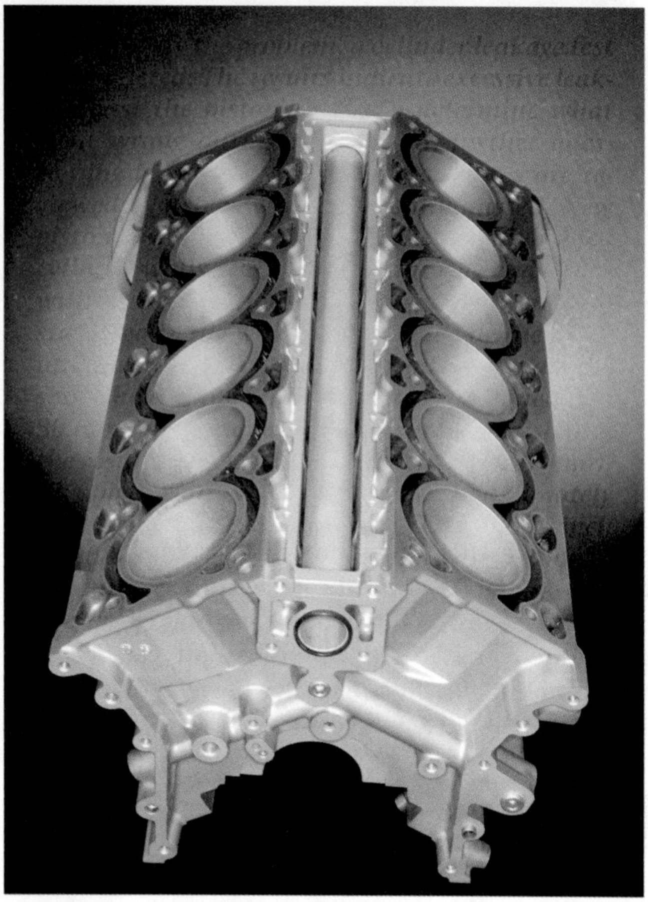

Figure 10-2 An engine block for a 12-cylinder engine. *Courtesy of BMW of North America, Incorporated*

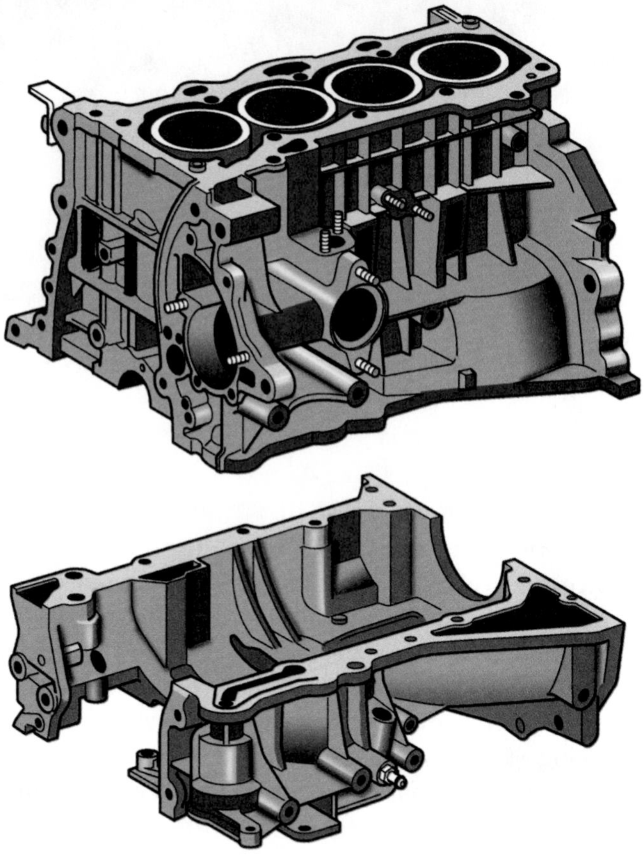

Figure 10-3 Aluminum engine blocks are often two-piece units. *Reprinted with permission*

The cylinder bores must be smooth and have the correct diameter to accept the pistons.

The main bearing area of the block must be align bored to a diameter that will accept the crankshaft. Camshaft bearing surfaces must also be aligned. The word *bore* means to drill or machine a hole. Align boring cuts a series of holes in a straight line.

Cast-iron blocks offer great strength and controlled warpage. With the increased concern for improved gasoline mileage, however, car manufacturers are trying to make the vehicle lighter. One way to do this is to reduce the weight of the block. Aluminum is often used to reduce this weight. Certain materials are added to aluminum to make the aluminum stronger and less likely to warp from the heat of combustion. Aluminum blocks normally have a sleeve or steel liner placed in them to serve as cylinder walls. Steel liners are placed in the mold before the metal is poured. After the metal is poured, the steel liner cannot be removed.

Lubrication and Cooling

A cylinder block contains a series of oil passages that allows engine oil to be pumped through the block and crankshaft and on to the cylinder head. The oil lubricates, cools, seals, and cleans engine components **(Figure 10–4)**.

Some of the heat generated by an engine is absorbed by the block and cylinder heads. This absorbed heat is wasted power and must be removed before it damages the engine. This is the job of the cooling system. Water jackets are also cast in the block around the cylinder

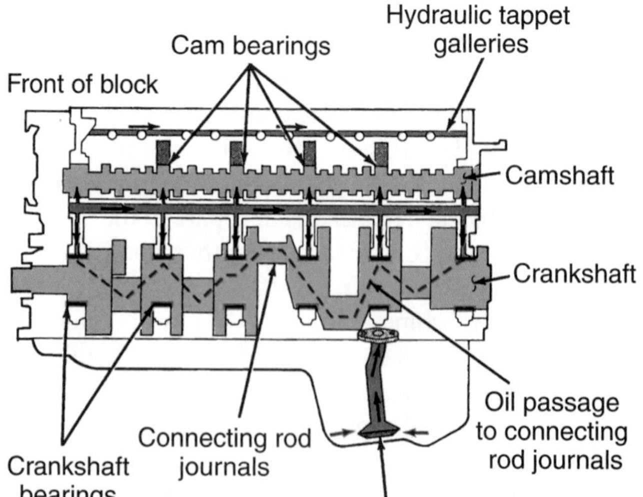

Figure 10-4 Direction of oil flow through this late-model V-10 engine. *Courtesy of DaimlerChrysler Corporation*

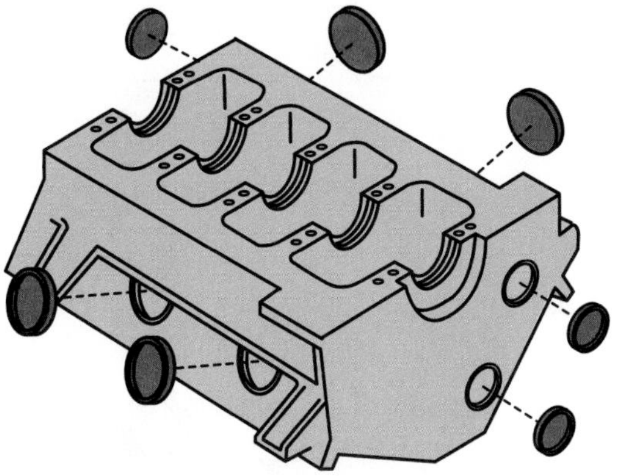

Figure 10-5 Typical core plug locations.

bores. Coolant circulates through these jackets to transfer heat away.

Core Plugs

All cast cylinder blocks use **core plugs**. These are also called expansion plugs. During the manufacturing process, sand cores are used. These cores are partly broken and dissolved when the hot metal is poured into the mold. However, holes must be placed in the block to get the sand out after the block is cast. These core holes are machined and core plugs are placed into them to seal them **(Figure 10–5)**.

Cylinder Sleeves

Some engines, such as those made with aluminum, have **cylinder sleeves (Figure 10–6)**. Some engines have sleeves that can be replaced if the cylinder walls are damaged. Most blocks must be bored out and larger pistons or standard-sized sleeves must be installed.

There are two types of sleeves: wet and dry. Both types are pressed into the block. The dry sleeve is supported from top to bottom by the block. The wet sleeve is supported only at the top and bottom. Coolant touches the center part of a wet sleeve.

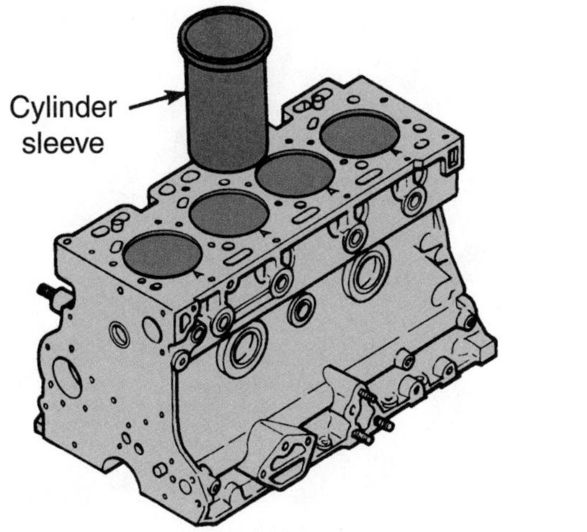

Figure 10-6 A cylinder sleeve is fitted into the bore to serve as a wear surface for the piston rings.

CYLINDER BLOCK RECONDITIONING

Before any reconditioning or rebuilding work is started, threaded holes should be cleaned with the correct-size tap to remove any and all burrs or dirt to allow for proper bolt torquing. Use a bottoming tap in any blind holes. **Chamfering** or counterboring will eliminate thread pulls and jagged edges. If there is damage to the threads, they should be repaired. To restore damaged threads in an aluminum part, a threaded insert should be installed in the bore.

Deck Flatness

The top of the engine block where the cylinder head mounts, is called the **deck**. To check deck warpage, use a precision straightedge and feeler gauge. With the straightedge positioned diagonally across the deck, the amount of warpage is determined by the size of feeler gauge that fits into the gap between the deck and the straightedge **(Figure 10–7)**.

Some engines have special deck flatness requirements. Always refer to the manufacturer's specifications. If specifications are not available, use 0.003 inch (.0762 mm) per 6 inches (152.4 mm), and no more than 0.006-inch (.1524-mm) maximum on any length. If the block has more than one deck surface (such as a V-type engine), each deck should be machined to the same height. This allows for uniform compression and manifold alignment. If the deck is warped and not corrected, valve seat distortion will occur when the head is tightened to the block. Coolant and combustion leakage can also occur.

Cylinder Walls

Inspect the cylinder walls for scoring, roughness, or other signs of wear. Ring and cylinder wall wear can be accel-

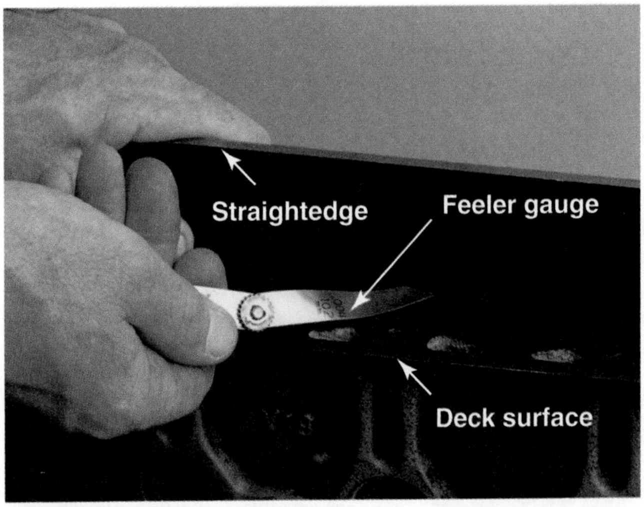

Figure 10-7 Checking for deck warpage with a straightedge and feeler gauge.

erated by dirt. Dirt can get caught in the piston rings and can grind away at the metal surfaces.

Scuffed or scored pistons, rings, and cylinder walls can act as passages for oil to bypass the rings and enter the combustion chamber. Scuffing and scoring occur when the oil film on the cylinder wall is ruptured, allowing metal-to-metal contact of the piston rings on the cylinder wall. Cooling system hot spots, oil contamination, and fuel wash are typical causes of this problem.

Cylinder Bore Inspection

Normally the most cylinder wear occurs at the top of the ring travel area. Pressure on the top ring is at a peak and lubrication at a minimum when the piston is at the top of its stroke. A ridge of unworn material will remain above the upper limit of ring travel. Below the ring travel area, wear is negligible because only the piston skirt contacts the cylinder wall.

A properly reconditioned cylinder must have the correct diameter, have no taper or out-of-roundness, and the surface finish must be such that the piston rings will seat to form a seal that will control oil and minimize blowby.

Taper is the difference in diameter between the bottom of the cylinder bore and the top of the bore just below the ridge **(Figure 10–8)**. Subtracting the smaller diameter from the larger one gives the cylinder taper. Some taper is permissible, but normally not more than 0.006 inch (.1524 mm). If the taper is less than that, reboring the cylinder is not necessary.

Cylinder out-of-roundness is the difference of the cylinder's diameter when measured parallel with the crank and then perpendicular to the crank **(Figure 10–9)**. Out-of-roundness is measured at the top of the cylinder just below the ridge. Typically, the maximum allowable out-

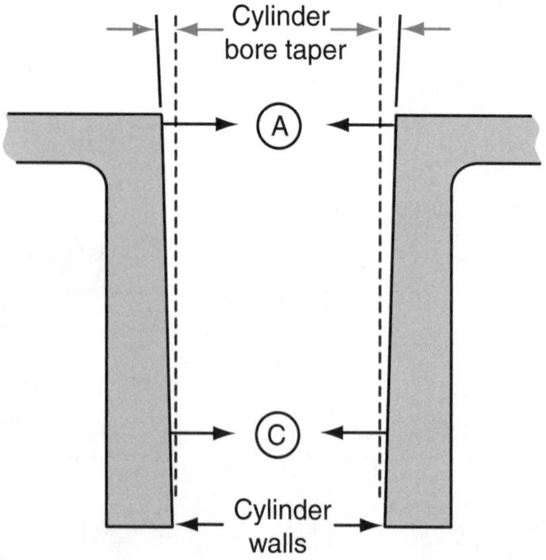

Figure 10–8 To check for taper, measure the diameter of the cylinder at A and C. The difference between the two readings is the amount of taper.

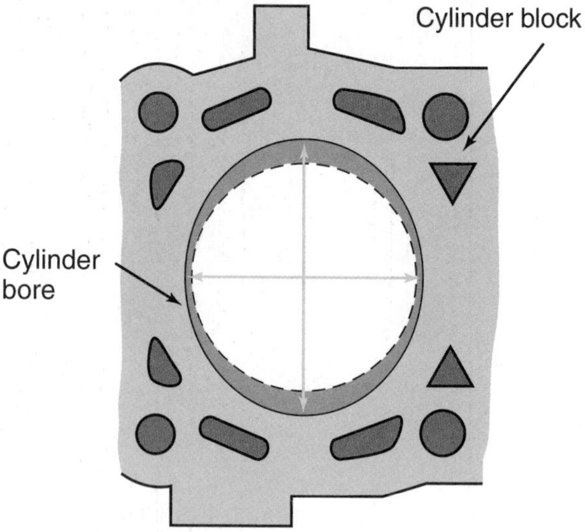

Figure 10–9 To check cylinder out-of-roundness, measure the bore in different locations. *Courtesy of Ford Motor Company*

of-roundness is 0.0015 inch (.0381 mm). Normally a cylinder bore is checked for out-of-roundness with a dial bore gauge **(Figure 10–10)**. However, a telescoping gauge can also be used.

When using a dial bore gauge or a telescoping gauge to check a cylinder's bore, make sure the measuring arms are parallel to the plane of the crankshaft. The best way

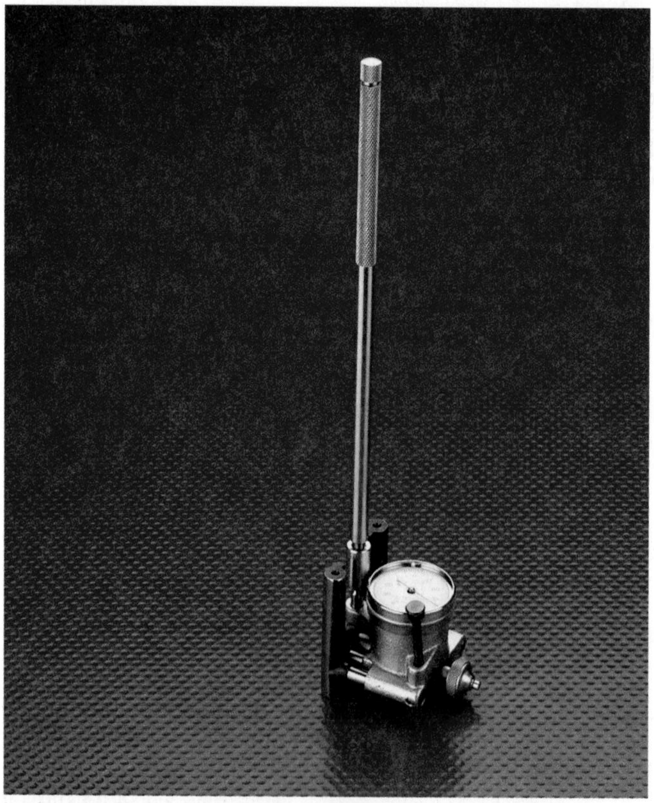

Figure 10–10 A dial bore gauge can be used to check a cylinder's diameter and to check for taper and out-of-roundness. *Courtesy of L.S. Starrett Company*

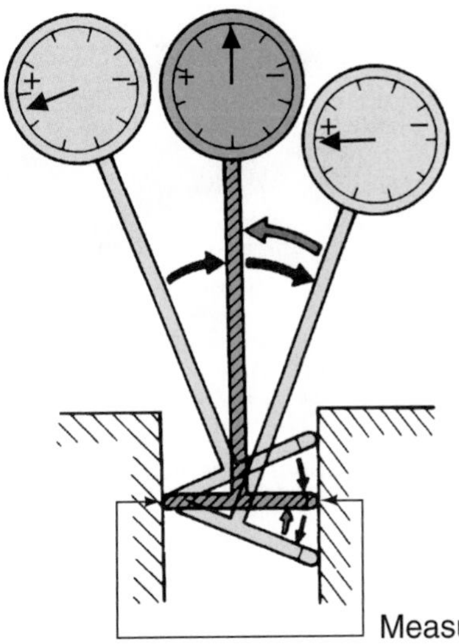

Figure 10-11 To get an accurate reading, rock the gauge until the smallest reading is obtained.

Figure 10-13 The desired cylinder wall finish for most types of piston rings.

to do this is to rock the gauge until the smallest reading is obtained **(Figure 10–11)**.

Cylinder Bore Surface Finish

The proper surface finish on a cylinder wall acts as a reservoir for oil to lubricate the piston rings and prevent piston and ring scuffing. Piston ring faces can be damaged and experience premature wear if the cylinder wall is too rough. A surface that is too smooth will not hold enough oil and will not allow the rings to seat properly. Obtaining the correct cylinder wall finish is important.

The desired cylinder wall finish is comprised of many small crisscross grooves **(Figure 10–12)**. Ideally, these grooves cross at 50- to 60-degree angles, although anything in the 20- to 60-degree range is acceptable. This finish leaves millions of tiny diamond-shaped areas, which serve as lubricant reservoirs **(Figure 10–13)**. This finish also leaves small flat areas or plateaus on the surface. A film of oil adheres to these areas to act as a bearing surface for the piston rings.

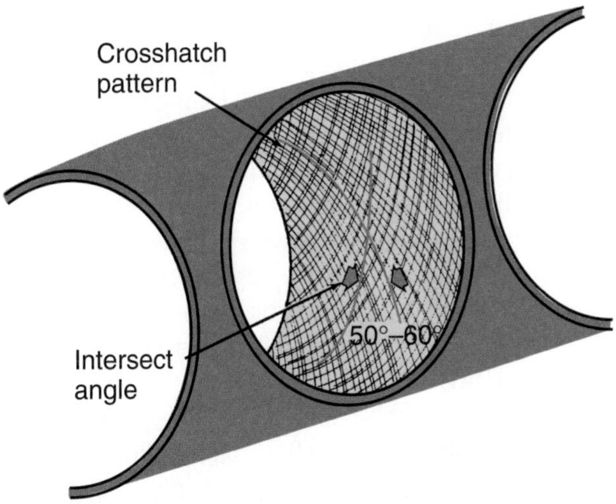

Figure 10-12 Ideal crosshatch pattern for cylinder walls. *Courtesy of DaimlerChrysler Corporation*

If the angle of the crosshatch is too steep, the oil film will be too thin, resulting in ring and cylinder scuffing. If the angle is too flat, the pistons may hydroplane and excessive oil consumption will result.

Cylinder Deglazing If the inspection and measurements of the cylinder wall show that surface conditions, taper, and out-of-roundness are within acceptable limits, the cylinder walls may only need to be deglazed. Combustion heat, engine oil, and piston movement combine to form a thin residue on the cylinder walls commonly called **glaze**.

A glazed cylinder wall allows the piston rings to slide over the walls of the cylinder preventing a positive seal between the two.

It is easy to confuse glaze with the polished surface that appears on the walls of the cylinder after the engine has some miles on it. Often true glazing can be removed by wiping down the cylinders with denatured alcohol or lacquer thinner. Fine honing stones, often referred to as deglazing and honing stones, also remove the glaze and leave the walls with the desired finish. Most often, technicians use a ball hone to deglaze **(Figure 10–14)** and create the desired pattern on the walls of the cylinder.

Cylinder Boring When cylinder surfaces are badly worn or excessively scored or tapered, a **boring bar** or boring machine **(Figure 10–15)** is used to cut the cylinders for oversize pistons or sleeves. A boring bar leaves a pattern on the cylinder wall similar to uneven screw threads. These marks need to be removed because they can cause poor oil control and excessive blowby. Therefore you should hone the bore after it has been bored.

Figure 10–14 Using a resilient-based, hone-type brush, commonly called a ball hone.

Figure 10–15 A cylinder being bored. *Courtesy of Jaspar Engine and Transmission Exchange, Inc.*

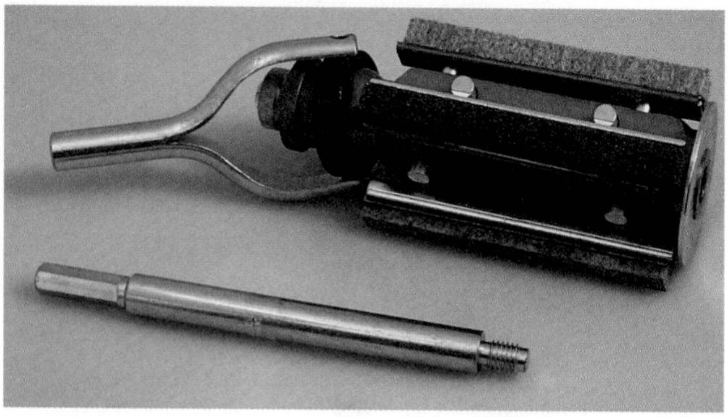

Figure 10–16 A cylinder hone. *Courtesy of Lisle Corp.*

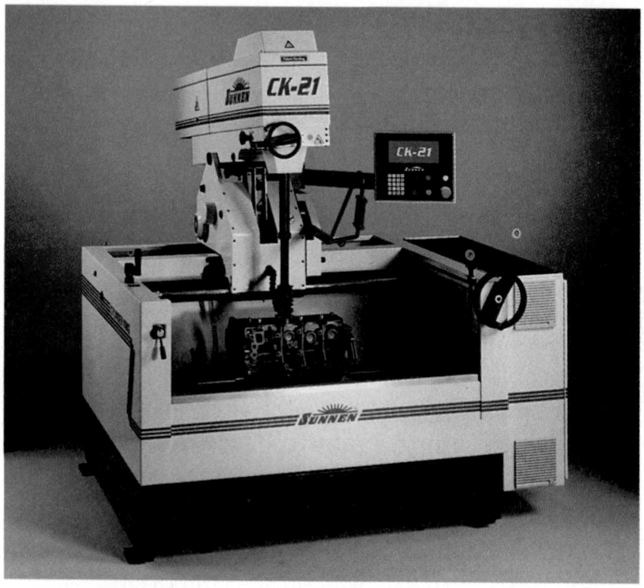

Figure 10–17 An automatic cylinder hone machine. *Courtesy of Sunnen Products Company*

Cylinder Honing A cylinder **hone** usually consists of three or four stones **(Figure 10–16)**. The hone is spun and moved up and down the cylinder's bore. The stones have outward pressure on them and remove some metal from the bore as they rotate within it. The grade of stones is typically specified by the manufacturer of the piston rings. Honing stones are classified by grit size; usually the lower the grit number, the coarser the stone.

Cylinder honing machines are available in manual and automatic models **(Figure 10–17)**. The major advantage of the automatic type is that it allows the technician to dial in the exact crosshatch angle needed. When honing a cylinder by hand, make sure you use a slow-speed (200–450-rpm) electric drill. Mount the honing tool into the drill and insert it into the bore. Adjust the stones so they fit snugly to the narrowest section of the cylinder. Move the drill and the hone up and down in the bore with short strokes. Never remain in one spot too long. Squirt some honing oil on the walls and occasionally stop honing and clean the stones before restarting. Honing oil is used to control the temperature of the cylinder walls and stones and to flush out any metallic and abrasive residue. Continue until the desired results are achieved.

Torque plates simulate the weight and structure of a cylinder head. They are used by engine rebuilding shops and are fastened to the cylinder block to equalize or prevent twist and distortion when honing or boring a cylinder **(Figure 10–18)**.

After resurfacing the cylinder walls, use plenty of hot, soapy water, a stiff bristle brush, and a soft, lint-free cloth to clean the residue **(Figure 10–19)**. Then rinse the block with water and dry it thoroughly. Lightly coat the bore with clean, light engine oil to prevent rust.

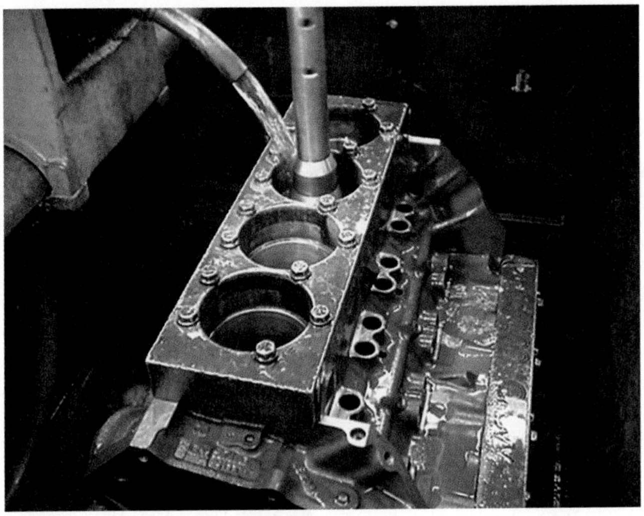

Figure 10-18 Torque plates are fastened to the block during cylinder boring and honing to prevent block distortion during the machining process. *Courtesy of Jaspar Engine and Transmission Exchange, Inc.*

CAUTION!

Always wear eye protection when operating deglazing, honing, or boring equipment.

Lifter Bores

Carefully examine each of the bores for the valve lifters. Look for cracks and evidence of excessive wear. Oblong or egg-shaped bores indicate wear. Typically if these bores exceed allowable wear limits or are damaged, the engine block is replaced. If the bores are rusted, glazed, or have burrs and high spots, they can be honed with a brake wheel cylinder hone. Be careful not to remove more than 0.0005 inch of metal while honing.

Checking Crankshaft Saddle Alignment

Figure 10–20 is an exaggerated illustration of crankcase housing bores that are out of alignment. If the block is warped and its main bearing bores are out of alignment, the crankshaft will inflict heavy loads on one side of the main bearings. Engine blocks that are not severely warped can be repaired by an operation called **line boring**, a machining operation in which the main bearing housing bores are rebored to standard size and alignment **(Figure 10–21)**. Badly warped blocks must be replaced.

The alignment of the crankshaft saddle bore can be checked with a precisely ground arbor placed into the bearing bores. The arbor is rotated in the bores. The effort required to rotate it determines the alignment of the bores.

If a proper arbor is not available, saddle alignment can be checked with a straightedge **(Figure 10–22)**. Place the straightedge in the saddles as shown. Using a feeler gauge that is half the maximum specified oil clearance, try to slide the feeler under the straightedge. If this can be done at any saddle, the saddles are out of alignment and the block must be line-bored. Repeat this procedure at two other parallel positions in the saddles.

Out-of-roundness of the saddles can also be checked by bolting on the main bearing caps and checking each

Figure 10-19 After honing, clean the cylinder with soapy water. *Courtesy of Federal-Mogul Corporation*

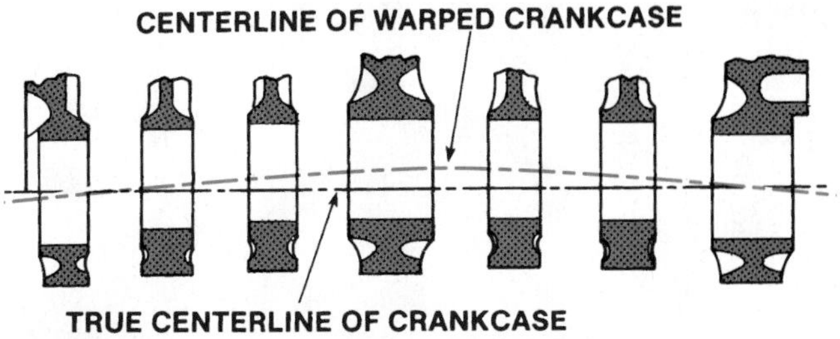

Figure 10-20 An exaggerated view of crankcase housing misalignment.

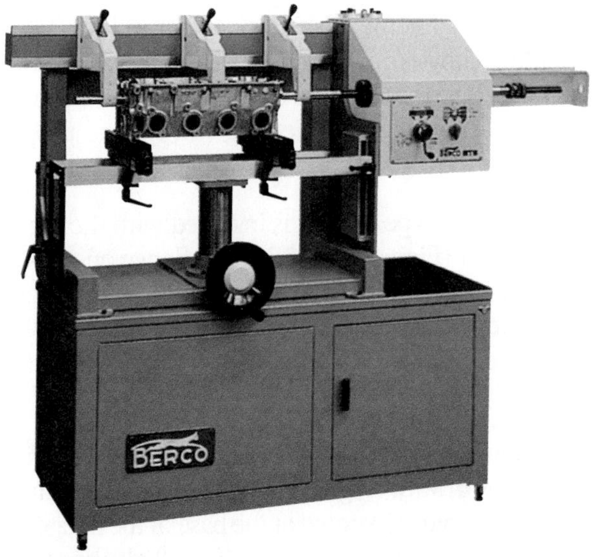

Figure 10-21 A line boring machine for correct crankshaft saddle alignment. *Courtesy of Frontline Equipment Company*

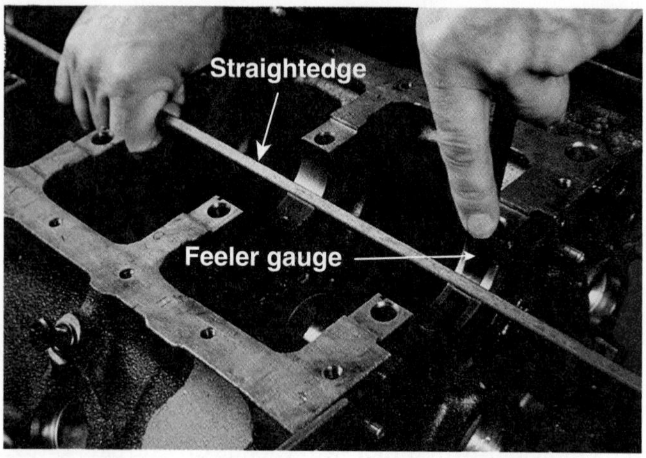

Figure 10–22 Checking bore alignment with a straightedge and feeler gauge. *Courtesy of Federal-Mogul Corporation*

bore with a dial bore gauge or an out-of-roundness indicator.

Installing Core Plugs

Old core and oil gallery plugs are normally removed and replaced during cylinder block reconditioning. When installing new core plugs, make sure they are the correct size and type.

The plugs' bore should be inspected for any damage that would interfere with the proper sealing. If the bore is damaged, it should be bored out for an oversized plug. Oversize (OS) plugs are identified by the OS stamped on the plug.

Coat the plug or bore lightly with a nonhardening oil-resistant (oil gallery) or water-resistant (cooling jacket) sealer. The three basic core plugs are installed as follows.

Disc- or Dished-Type This type fits in a recess in the engine casting with the dished side facing out **(Figure 10–23A).** With a hammer, hit the disc in the center of the crown and drive the plug into the bore until just the crown becomes flat. In this way the plug will expand properly and give a good tight fit.

Cup-Type This type of plug is installed with the flanged edge outward **(Figure 10–23B)**. The flange on cup-type plugs flares outward with the largest diameter at the outer (sealing) edge. The flanged (trailing) edge must be below the chamfered edge of the bore to effectively seal the plugged bore.

Expansion-Type This type of plug is installed with the flanged edge inward **(Figure 10–23C).** The maximum diameter of this plug is located at the base of the flange with the flange flaring inward. When installed, the trailing (maximum) diameter must be below the chamfered edge of the bore to effectively seal.

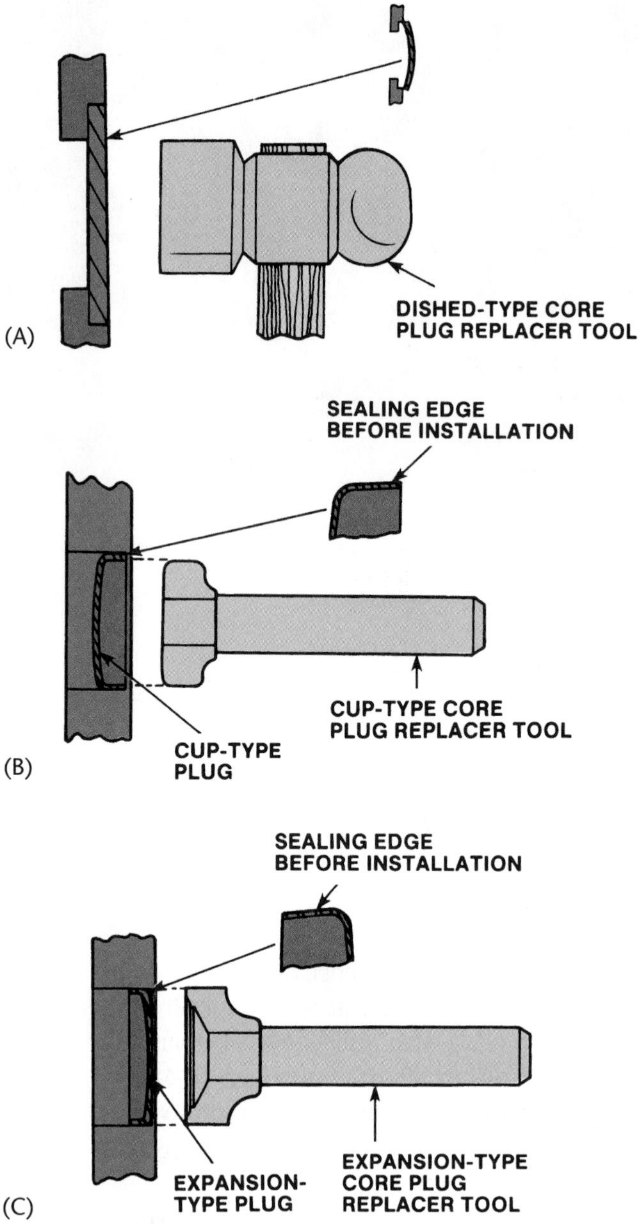

Figure 10–23 Core plug installation methods: (A) dished, (B) cup, and (C) expansion.

CAMSHAFT

A camshaft is a shaft **(Figure 10–24)** with a cam for each exhaust and intake valve, each one placed to allow for the proper timing of each valve. A cam is a device that changes rotary motion into reciprocating motion. Each cam has a high spot or lobe that controls the opening of the valves. The height of the lobe is proportional to the amount the valve will open. Camshafts in older engines had a lobe to operate the fuel pump and a gear to drive the distributor and oil pump. Some diesel engines have cam lobes for fuel injectors, fuel injection pumps, and/or air starting valves.

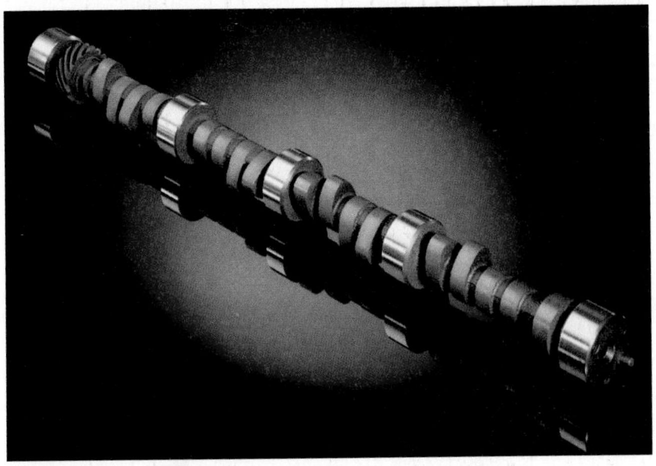

Figure 10-24 A camshaft for a V8 engine. *Courtesy of Melling Engine Parts*

The camshaft can be located in either the cylinder block or cylinder head(s). The camshaft fits into a bore next to the crankshaft on most in-line engines, unless the engine has overhead camshafts. On V-type engines, the camshaft lies in bore above the crankshaft at the center of the block. When the camshaft is in the block **(Figure 10–25)**, the valves are opened through lifters, pushrods, and rocker arms. As the cam lobe rotates, it pushes up on the lifter, which lifts up the pushrod, moving one end of the rocker arm up while the other end pushes the valve down to open it. As the cam rotates, the valve spring closes the valve and maintains the contact between the valve and the rocker arm, thereby keeping the pushrod and the lifter in contact with the rotating cam.

Overhead camshaft engines have the camshaft mounted above the cylinders, in or on the cylinder head **(Figure 10–26)**. OHC engines have no need for pushrods. As the camshaft rotates, the cams ride directly above the valves. The lobes open the valves by either directly depressing the valve or by depressing the valve through the use of a cam follower, rocker arm, or bucket-type tappet. The closing of the valves is still the responsibility of the valve springs.

Service to the camshaft(s) in an OHC engine is usually done when reconditioning the engine's cylinder head. In OHV engines, the camshaft and related parts are inspected and serviced while the short block is being reconditioned.

Timing Mechanisms

A camshaft is driven by the crankshaft at half its speed. This is accomplished through the use of a camshaft drive gear or drive sprocket that is twice as large as the crankshaft sprocket. For every two complete turns of the crankshaft, the camshaft turns once. During that full rotation, the intake and exhaust valves open and close once.

To synchronize the opening and closing of the valves with the position and movement of the pistons, the

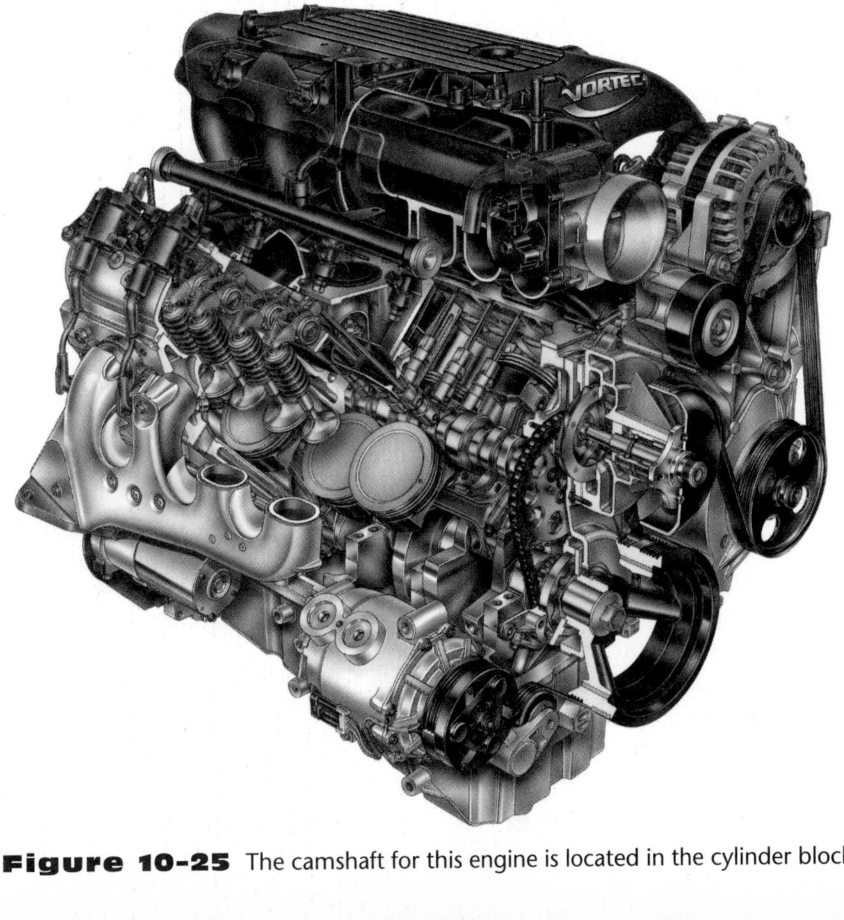

Figure 10-25 The camshaft for this engine is located in the cylinder block.

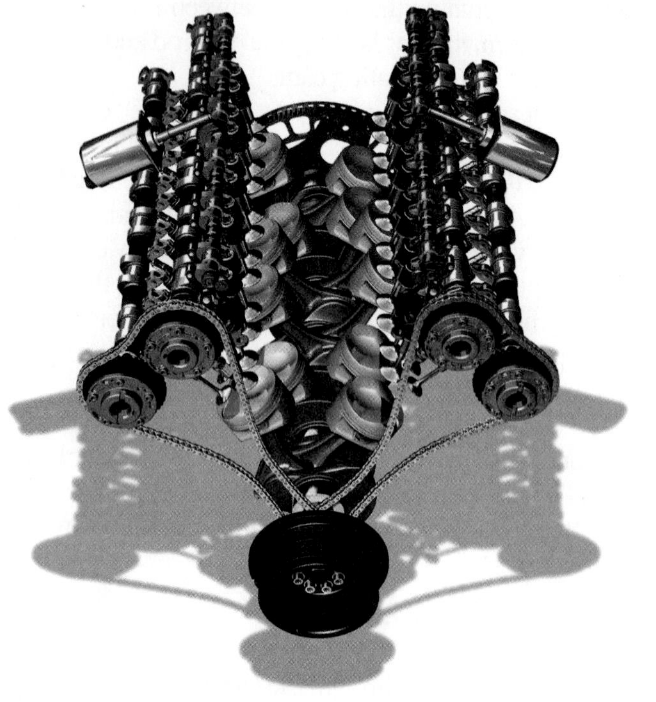

Figure 10-26 The camshaft(s) are located above the cylinders in OHC engines. *Courtesy of BMW of North America, Incorporated*

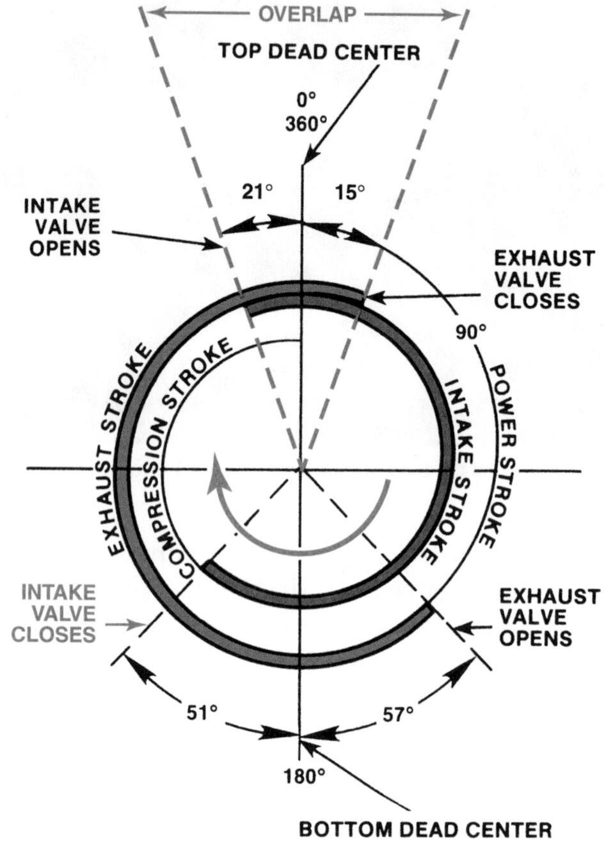

Figure 10-27 Typical valve timing diagram.

camshaft is timed to the crankshaft. In the typical valve timing diagram, **Figure 10–27**, valve action is shown in relation to crankshaft rotation. The intake valve starts to open at 21 degrees before the piston has reached TDC and remains open until it has traveled 51 degrees past BDC. The number of degrees between the valve's opening and closing is called intake valve duration time.

The exhaust stroke begins at 53 degrees before BDC and continues until 15 degrees after TDC, or a total exhaust valve duration time of 200 degrees of crankshaft rotation. Every engine design has its own valve timing requirements; therefore, have camshafts machined with different valve open and close specs. The period of time when both the exhaust and intake valves are open is known as valve **overlap**.

Overlap is critical to exhaust gas scavenging. A camshaft with a long overlap helps empty the cylinders at high engine speeds for improved efficiency. However, because both valves are open for a longer period of time, low rpm cylinder pressure tends to drop. The amount of overlap, because it has an effect on cylinder pressure, affects overall engine efficiency and exhaust emissions. Valve overlap also helps get the intake mixture moving into the cylinder. As the exhaust gases move out of the cylinder, a low pressure is present in the cylinder that causes atmospheric pressure to push the intake charge into the cylinder.

The following are the basic configurations for driving the camshaft.

Gear Drive A gear on the crankshaft meshes directly with another gear on the camshaft **(Figure 10–28A)**. The gear on the crankshaft is usually made of steel. The gear on the camshaft may be steel for heavy-duty applications, or it may be aluminum or pressed fiber when quiet operation is a major consideration. The gears are helical in design because helical gears are stronger and also tend to push the camshaft backward during operation to help prevent the camshaft from walking out of the block.

Chain Drive Sprockets on the camshaft and the crankshaft are linked by a continuous chain **(Figure 10–28B)**. The sprocket on the crankshaft is usually made of steel. The sprocket on the camshaft may be steel for heavy-duty applications. When quiet operation is a major consideration, an aluminum sprocket with nylon covering on the teeth is used. Nearly all OHV engines use a chain drive system. Chain drives are also used on many OHC engines, especially DOHCs. Often multiple chains are used and arranged in an elaborate fashion. These chain arrangements use a chain tensioner to maintain proper tightness and different silencing pads to reduce the noise of the chain.

Belt Drive Sprockets on the crankshaft and the camshaft are linked by a continuous neoprene belt **(Figure 10–28C)**. The belt has square-shaped internal teeth that

(A) OHC engine with belt driven camshaft

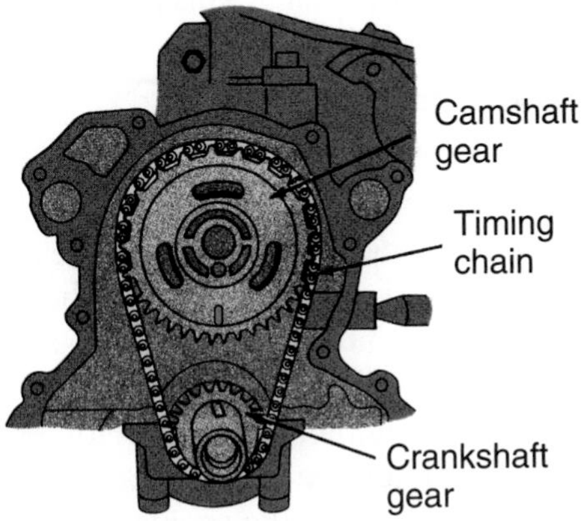

(B) OHV engine with timing chain and gears

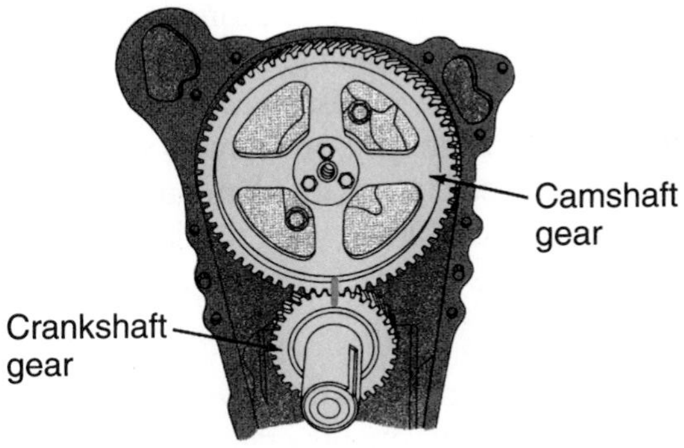

(C) OHV engine with gear driven camshaft

Figure 10-28 The different timing drive mechanisms used today. *Courtesy of Ford Motor Company*

mesh with teeth on the sprockets. The timing belt is reinforced with nylon or fiberglass to give it strength and prevent stretching. This drive configuration is limited to overhead camshaft engines.

The camshaft and crankshaft must always remain in the same relative position to each other. To obtain the correct initial relationship of the components, timing marks on the crankshaft and camshaft are aligned during assembly **(Figure 10–29)**.

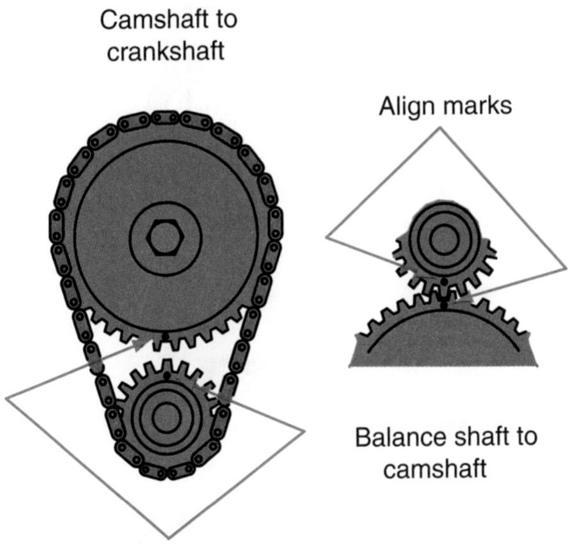

Figure 10-29 Camshaft-to-crankshaft and camshaft-to-balance shaft timing marks.

Valve Lifters

Valve lifters, sometimes called **cam followers** or tappets, follow the contour or shape of the cam lobe. Lifters are either mechanical (solid) or hydraulic. Solid valve lifters provide for a rigid connection between the camshaft and the valves. Hydraulic valve lifters provide the same connection but use oil to absorb the shock that results from the movement of the valve train.

Hydraulic lifters **(Figure 10–30)** are designed to automatically compensate for the effects of engine temperature. Changes in temperature cause valve train components to expand and contract. Hydraulic lifters are designed to automatically maintain a direct connection between valve train parts.

Solid lifters **(Figure 10–31)** do not have this built-in feature and require a clearance between the parts of the valve train. This clearance allows for expansion of the components as the engine gets hot. Periodic adjustment of this clearance must be made. Excessive clearance

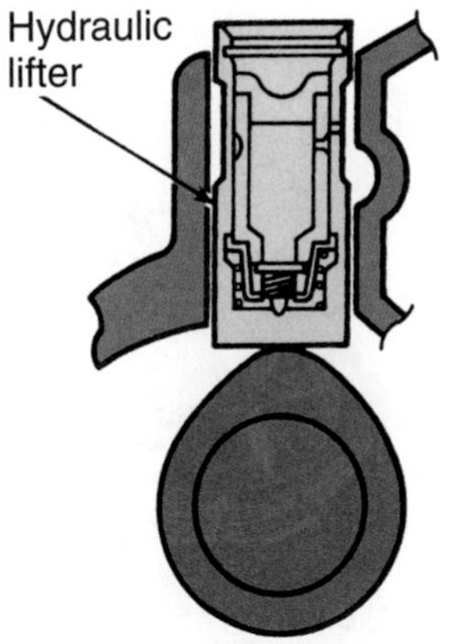

Figure 10-30 A hydraulic valve lifter.

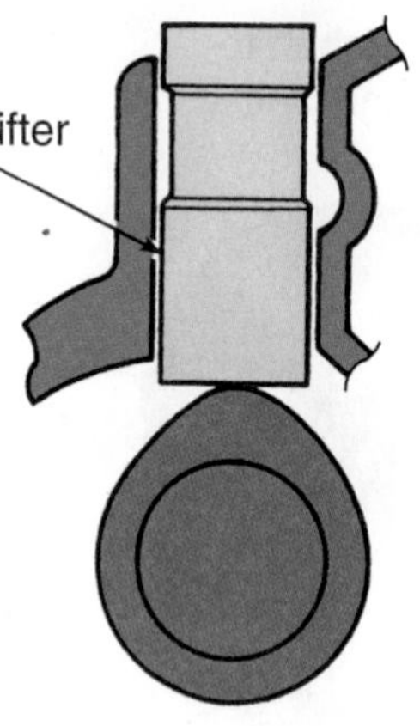

Figure 10-31 A solid valve lifter.

might cause a clicking noise. This clicking noise is also an indication of the hammering of valve train parts against one another, which will result in reduced camshaft and lifter life.

In an effort to reduce the friction—and the resulting power loss—from the lifter rubbing against the cam lobes, engine manufacturers often use roller-type hydraulic lifters. Roller lifters **(Figure 10–32)** are manufactured with a large roller on the camshaft end of the lifter. The roller acts like a wheel and allows the lifter to follow the cam lobe contour better than a flat-type lifter with reduced friction between the two contacting surfaces. Friction is reduced because the lifter rolls along the surface of the cam lobe as opposed to rubbing against it.

Operation of Hydraulic Valve Lifters A typical hydraulic lifter contains a plunger, an oil-metering valve, a pushrod seat, a check valve spring, and a plunger return spring housed in a hardened iron body.

When the lifter is resting on the basic circle of the cam, the valve is closed and the lifter maintains a zero clearance in the valve train. Oil is fed to the lifter through feed holes in the lifter bore. The pressure of the oil seals the oil in the lifter by forcing down the check valve inside the lifter. The oil between the plunger and the check valve forms a rigid connection between the lifter and the pushrod. Whenever there is some clearance in the valve train, a spring between the plunger and the lifter body pushes the plunger up to eliminate the clearance. As the cam lobe turns and opens a valve, the lifter's oil feed hole moves away from the oil feed in the lifter bore. Then no new oil can enter the lifter and the pressure on the plunger pushes it down in the lifter, which allows a small amount of oil to leak out. This leaking out of oil is called **leak-down**. Once the cam rotates and the lifter returns to the base of the cam, oil can again fill the lifter.

If a hydraulic lifter is not able to leak down or does not fill with oil, a noise will be heard from the engine. Non-roller-type lifters must also be able to rotate in their bore when the engine is running. This prevents wear on the bottom of the lifter.

Camshaft Bearings

The camshaft and balance shafts are supported by several friction-type bearings, or bushings. They are designed as one piece and are typically pressed into the camshaft bore in the engine block **(Figure 10–33)**. The bearings are made of either aluminum or of steel with a lining of babbitt. **Babbitt** is a soft slippery material made of mostly lead and tin. Alloys of aluminum are often found in late-model engines. Aluminum bearings have a longer service life in newer engines because the engines run at higher temperatures. Aluminum bearings are harder than babbitt and therefore are more susceptible to damage from dirt and poor lubrication.

Balance shaft and camshaft bearings are normally replaced during engine rebuilding. The old bearings should be inspected for signs of unusual wear that may indicate an oiling or bore alignment problem. Some OHC camshafts use machined bores in the aluminum cylinder head as a bearing surface. Often when this bore is damaged, the cylinder head is replaced.

Balance Shafts

Many late-model engines are fitted with one or more balance (silence) shafts to smooth engine operation. An engine's crankshaft is one of the main sources of engine

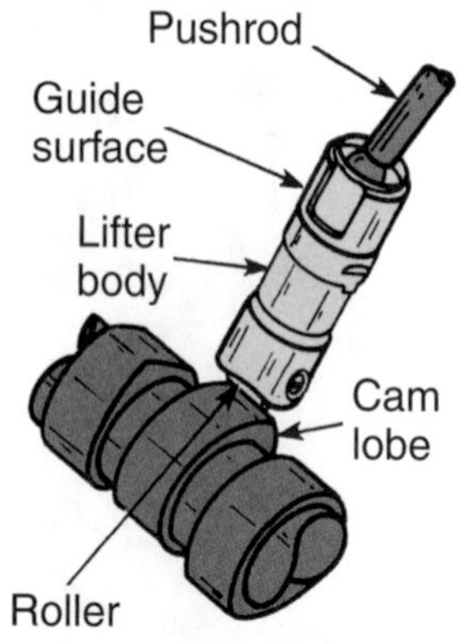

Figure 10-32 A typical roller lifter.

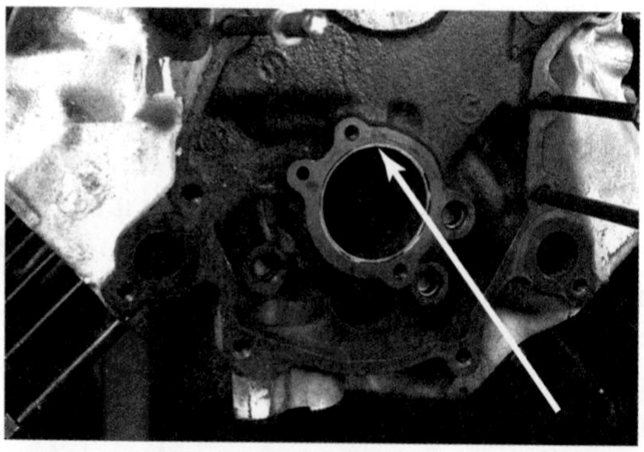

Figure 10-33 The typical camshaft bearing is a full round design.

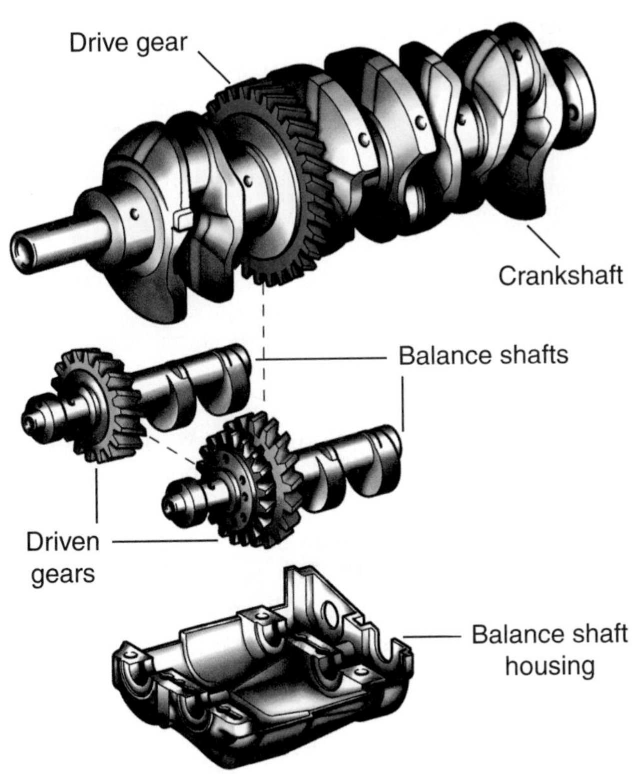

Figure 10-34 Balance shaft assemblies for a four-cylinder engine. *Reprinted with permission*

vibration because its shape makes it inherently out of balance as it spins. Balance shafts are designed to cancel out these vibrations.

In its basic form, a balance shaft is fitted with counterweights designed to mirror the throws of the crankshaft. These weights are rotated in the opposite direction as the crankshaft. As the engine turns, the opposing weights mutually cancel out any vibrations. To do this, the balance shafts rotate at twice the speed of the crankshaft and are synchronized or timed to the rotation of the crankshaft. If the balance shaft(s) are not timed to the crankshaft, the engine may vibrate more than it would without the balance shaft assembly.

Balance shafts are located in the engine block to the right and left side of the crankshaft **(Figure 10–34)** or in the camshaft bore of OHC engines. The shafts are supported by full-round bearings pressed into the block.

Balance shafts are typically inspected and serviced as part of reconditioning or building a short block. The service procedures for balance shafts are the same as those for servicing camshafts. The shaft journals and bearings need to be checked for wear, damage, and proper oil clearances. The runout of the shafts and drive gears should also be checked.

INSPECTION OF CAMSHAFT AND RELATED PARTS

As the engine is being disassembled, all parts should be carefully inspected, including the camshaft and timing gears. The bearings for the camshaft should also be carefully inspected.

Timing Components

The timing belt or chain and crankshaft/camshaft gears (sprockets) should be inspected or replaced if damaged or worn. This inspection should include a timing chain deflection check. To conduct this test, simply depress the chain at its midway point between the gears and measure the amount that the chain can be deflected. If the deflection measurement exceeds specifications, the timing chain and gears should be replaced.

A gear with cracks, spalling, or excessive wear on the tooth surface is an indication of improper **backlash** (either insufficient or excessive). With excessive backlash, operation will be noisy because the teeth will make violent impact contact. When coupled with the normal valve train loads this overloading causes accelerated tooth wear and often breakage. Insufficient backlash places a bind on the gears. Also, it generates high contact forces that can rupture the lubrication film between the teeth, causing spalling and wear.

To measure gear backlash, install a dial indicator and bracketry on the cylinder block **(Figure 10–35)**. Check the backlash between the camshaft gear and the crankshaft gear with a dial indicator at six equally spaced teeth. Hold the gear firmly against the block while making the check. Refer to specifications in the vehicle's service manual for the backlash limits.

Lifters

When inspecting mechanical lifters, carefully check their bottoms and pushrod sockets. Wear, scoring, or pitting

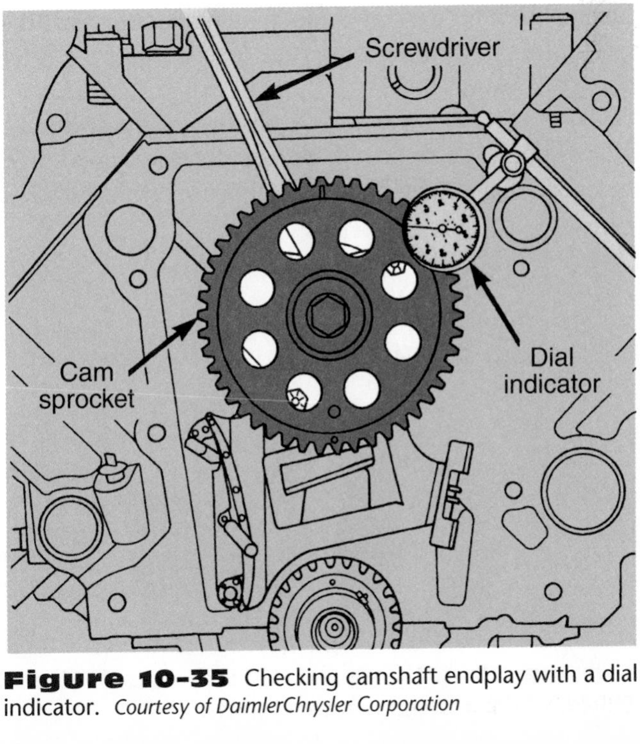

Figure 10-35 Checking camshaft endplay with a dial indicator. *Courtesy of DaimlerChrysler Corporation*

Figure 10-36 A worn lifter and camshaft. *Courtesy of TRW, Incorporated*

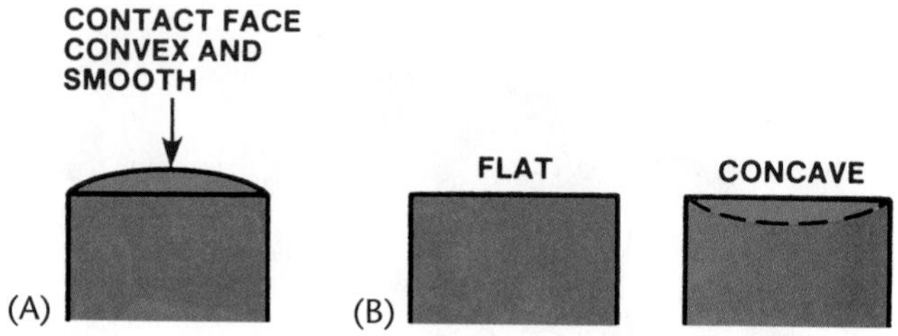

Figure 10-38 (A) Acceptable and (B) unacceptable valve lifter bottoms.

makes their replacement necessary. Any lifter showing pitting or having its contact face worn flat or concave must also be replaced.

Technically, the normal wear of the valve lifters is referred to as adhesive or galling wear. This wear is a result of two solid surfaces (camshaft lobe and lifter face) rubbing against each other, considered normal wear between the cam lobe and lifter. Fortunately, proper lubrication retards this process. However, excessive loading negates the beneficial effects of the lubricant and accelerates the wear process **(Figure 10–36)**. Examples of excessive loading would be incorrectly matched valve springs (too much spring pressure), old lifters on a new camshaft, or new lifters on an old camshaft.

If a camshaft and lifters are going to be reused, the lifters must remain with their respective lobes. Worn valve lifters and improper camshaft installation are common causes of camshaft/lifter failure.

The normal wear path is off center with no edge contact between the lifter and the lobe. The taper on the cam lobe and the spherical radius of the lifter bottom are specifically designed to result in an offset contact pattern, causing the lifter to rotate. The spinning lifter reduces the sliding friction and equalizes the load around the lifter bottom **(Figure 10–37)**.

Whenever the valve train is disturbed, the hydraulic lifters should be removed, disassembled, cleaned, and checked. They should be kept in sequence during removal so that they can be put back in the same place. Lifters should be replaced if the bottoms are worn or pitted or if a new camshaft is installed. The bottoms of new lifters are generally spherical **(Figure 10–38)**.

Any time hydraulic lifters are removed, the varnish and deposits should be carefully removed from the lifter bores in the engine block, and the galleries should be flushed with pressurized oil to clear any dirt from the holes that feed the lifters.

After cleaning, check the lifter's leakdown with a leakdown tester. Lifter (tappet) leakdown rate is important. If the tappets leak down too quickly, noisy operation will result. When diagnosis indicates no cause for noisy tappet operation, the condition can sometimes be remedied by checking the lifter leakdown rate and replacing all lifters that are outside specifications.

Camshaft

After the camshaft has been cleaned, check each lobe **(Figure 10–39)** for scoring, scuffing, fractured surface, pitting, and signs of abnormal wear. Also check for plugged oil passages.

Premature lobe and lifter wear is generally caused by metal-to-metal contact between the cam lobe and lifter bottom due to inadequate lubrication. The nose will be worn from the cam lobes, and the lifter bottoms will be worn to a concave shape or may be worn completely

Figure 10-37 The possible wear patterns of a lifter that does not spin in its bore. *Courtesy of TRW, Incorporated*

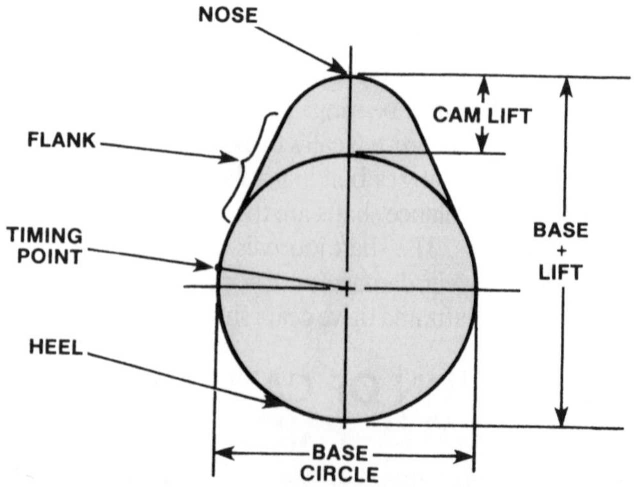

Figure 10-39 Cam lobe nomenclature.

away. This type of failure usually begins within the first few minutes of operation. It is the result of insufficient lubrication.

There are several ways to check cam lobes for wear, but the two most popular are the dial indicator and outside micrometer.

With the camshaft in the engine, use a dial indicator to check the lift of each cam lobe **(Figure 10–40)**. Make sure the pushrod is in the valve lifter socket. Install the dial indicator so that the cup-shaped adapter fits into the end of the pushrod and is in the same plane as the pushrod movement. Connect a remote starter switch into the starting circuit. With the ignition switch off, bump the crankshaft over until the lifter is on the base circle of the camshaft lobe. At this point, the pushrod will be in its lowest position. Set the dial indicator at zero. Continue to rotate the crankshaft slowly until the pushrod is in its fully raised position (highest indicator reading). Compare the total lift recorded on the indicator with specifications. If the lift on the lobe is below the specified service limits, the camshaft and lifters must be replaced.

With the camshaft removed from the engine, cam lobe height can be measured with an outside micrometer. Place the micrometer so it can measure from the heel to the nose of the lobe. Record the measurement for each intake and exhaust lobe. Any variation in height indicates wear. Also check the measurements taken against the manufacturer's cam lobe heights.

Measure each camshaft journal in several places with a micrometer to determine if it is worn excessively. If any journal is 0.001 inch (0.0254 mm) or more below the manufacturer's specifications, it should be replaced.

The camshaft should also be checked for straightness with a dial indicator. Place the camshaft on V-blocks. Position the dial indicator on the center bearing journal and slowly rotate the camshaft. If the dial indicator shows runout (a 0.002-inch [0.0508-mm] deviation), the camshaft is not straight and must be replaced.

If the engine has a worn or damaged camshaft, identify and fix the cause of the damage before installing a new camshaft, lifters and/or followers.

INSTALLING THE CAMSHAFT AND RELATED PARTS

Before installing the camshaft and balance shafts with their bearings, make sure the engine is thoroughly cleaned. Hot water and detergent are best for cleaning blocks, crankshafts, and camshafts to remove grit from honing, grinding, and polishing. Once the parts have been cleaned, blow them dry and immediately coat them with oil to prevent rusting. Also make sure all oil passages are free of dirt and foreign particles.

Coat all parts with a quality assembly lubricant. A good lube is one that has an extreme pressure (EP) lubricant rating and excellent adhesion quality to help prevent scuffing and galling during initial startup. The adhesion quality also prevents the lubricant from draining off the components during engine reassembly.

Camshaft Bearings

Although the installation of camshaft bearings can be done after the rest of the short block is assembled, it may be easier to align the oil holes of the bearings when the crankshaft is not yet installed. Keep in mind that any engine block that needed to have its main bearing bore alignment corrected due to distortion is likely to have camshaft bearing bore misalignment problems.

Cam bearings are normally press-fit into the block or head using a bushing driver and hammer **(Figure 10–41)**. The camshaft journals may have different diameters, with the smallest being on the rear of the block and each journal being progressively larger. Therefore, the bearing at the rear of the block should be installed first.

The new bearing is fit over the expanding mandrel of the tool and the length of the tool is set into the block. A

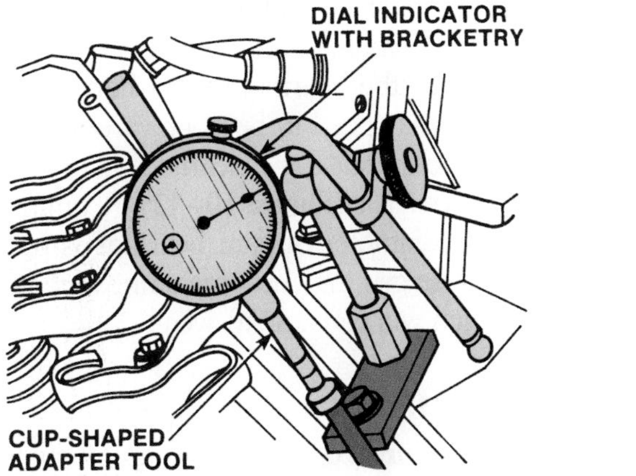

Figure 10–40 Checking a camshaft lobe using a dial indicator.

Figure 10–41 Cam bearings are normally press fit into the block or head using a bushing driver and hammer.
Courtesy of Lisle Corp.

guide cone on the tool is used to keep the tool centered in the bore. Once the bearing is at the outside of its bore, rotate the bearing to align the oil hole in the bearing with the oil hole in the block.

On blocks with grooves behind the bearings, the bearing should be installed with the oil hole at the 2 o'clock position as viewed from the front for normal clockwise camshaft rotation. This position introduces oil into the clearance space outside the loaded area and allows shaft rotation to build an oil film ahead of the load.

While holding the centering cone against the outside bore, drive the bearing into its bore. If the cone and tool are allowed to move while inserting the bearings, the bearing can be damaged. While driving the bearings into their bore, be careful not to shave metal off the backs of the bearings. This may cause a buildup of metal between the outside of the bearing and the housing bore, which will result in a reduction of clearance. To prevent galling, check the housing bores for proper lead-in chamfer before installing the bearings.

After the bearing is fully seated in its bore, double-check the alignment of the bearing's oil hole with the oil hole in the block by inserting a wire through the holes or by squirting oil into the holes. If the oil does not run out, the holes are misaligned. This procedure should be repeated with each bearing.

WARNING!

The use of a standard camshaft bearing driver and hammer is not recommended for aluminum heads because the aluminum bearing supports are very easily damaged or broken, which can result in expensive head replacement.

Camshaft

To install the camshaft, wipe off each cam bearing with a lint-free cloth, then thoroughly coat the camshaft lobes, bearing journals, and distributor drive gear (if there is one) with assembly lube. Also lubricate the lifters. Most premature cam wear develops within the first few minutes of operation. Prelubrication helps to prevent this when the engine is started the first time. Special prelubricants can be used only if specifically recommended by the manufacturer.

The camshaft should be carefully installed to avoid damaging the bearings with the edge of a cam lobe or journal. Be especially careful to keep it straight to prevent it from cutting or grooving the bearings. A threaded bolt in the front of the camshaft can be helpful in guiding the cam in place. Some technicians temporarily install the timing gear onto the camshaft to aid in installing the camshaft into the engine block. An alternative is to install

Figure 10-42 A camshaft being ground. *Courtesy of Jaspar Engine and Transmission Exchange, Inc.*

the camshaft while the block rests on its end. When the camshaft is in place, install the thrust plate and the timing gear.

A camshaft timing gear may need to be pressed off and a replacement pressed on the camshaft prior to installing the camshaft into the block. Be sure to align the thrust plate with the woodruff key during removal to prevent damage to the thrust plate. Both the thrust plate and timing gear must then be aligned with the woodruff key for assembly. Never hammer a gear or sprocket into the shaft. Heat all metal and aluminum gears on a hot plate heated to between 200° and 300°F. To ensure ease of installation, install the gear while it is still hot. This step does not apply to fiber gears. These gears should be carefully installed. Press the camshaft into the gear and be sure to keep the gear square and aligned with the keyway at all times.

Once the shaft is completely in the block, the shaft should be able to be turned by hand. Binding can be caused by a damaged bearing, a nick on the cam's journal, or slight misalignment of the block journals. The cause of the problem should be identified. If the bearing clearance is too small, some technicians ream away a slight amount of the bearing; others hone the bearing. Both of these tasks need to be done carefully. The most practical way to increase the clearance is to grind down the camshaft journals **(Figure 10–42)**. Reaming or honing the inside diameter of cam bearings is not recommended because grit may become embedded in bearing surfaces, which will cause shaft wear.

CRANKSHAFT

Crankshafts **(Figure 10–43)** are generally made of cast iron, forged cast steel, or nodular iron, and then machined. At the centerline of the crankshaft are the main

Figure 10–43 A crankshaft.

bearing journals **(Figure 10–44)**. These journals must be machined to a very close tolerance because the weight and movement of the crankshaft is supported at these points. The number of main bearings is determined by the design of the engine. V-block engines generally have fewer main bearings than an in-line engine with the same number of cylinders because the V-block engine uses a shorter crankshaft.

Offset from the crankshaft centerline are the connecting rod bearing journals. The degree of offset and the number of journals are determined by engine design. An in-line 6-cylinder engine has six connecting rod journals. A V8 engine has only four. Each journal has two connecting rods attached to it, one from each side of the V. The connecting rod journal is also called the crank pin.

Since the connecting rod journals are offset from the centerline of the crankshaft, the weight and pressure from the pistons are also offset from the center of the crankshaft. This could create an imbalanced condition. However, to allow for smooth engine operation, counterweights are added to the crankshaft. These weights are part of the crankshaft and are positioned opposite the connecting rod journals **Figure 10–45)**.

The machining of the main and rod bearing journals must have a very smooth surface at the bearing area. The bearings must fit tightly enough to eliminate noise but must also have enough clearance between them and their

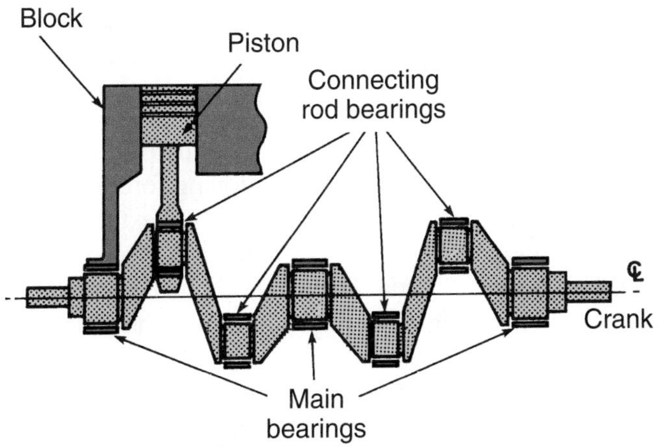

Figure 10–44 Crankshaft bearing and journal locations.

Figure 10–45 A crankshaft with the pistons and connecting rods attached. Notice the counterweights below the connecting rod journals. *Courtesy of BMW of North America, Incorporated*

bearings to allow an oil film of 0.0015 inch (.0381 mm) and 0.002 inch (.0508 mm) to form.

The crankshaft rotates on this film of oil. The oil is supplied by the engine's oil pump. If the crankshaft journals become out of round, tapered, or scored, the oil film will not form properly and the journal will contact the bearing surface, which causes early bearing or crankshaft failure. The main and rod bearings are generally made of lead-coated copper or tin and aluminum. Both of these are softer materials than that used to make the crankshaft. By using the soft material, any wear will appear first on the bearings. Early diagnosis of bearing failure most often will spare the crankshaft and only the bearings will need to be replaced.

The bearings are fed oil under pressure. In order for the oil to reach these bearings, oil passages must be drilled into the crankshaft. Each main bearing journal has a hole drilled into it with a connecting hole or holes leading to one or more rod bearing journals. In this way, all bearing journals receive oil to protect both the bearing and the journal.

The crankshaft configuration determines the engine block design, or the positioning of the connecting rod journals around the centerline of the crankshaft **(Figure 10–46)**.

A crankshaft has two distinct ends. One is called the flywheel end and, as its name implies, this is where the flywheel is attached. The front end or belt drive end of the crankshaft contains a threaded snout or is drilled and tapped. This is for attaching a vibration damper.

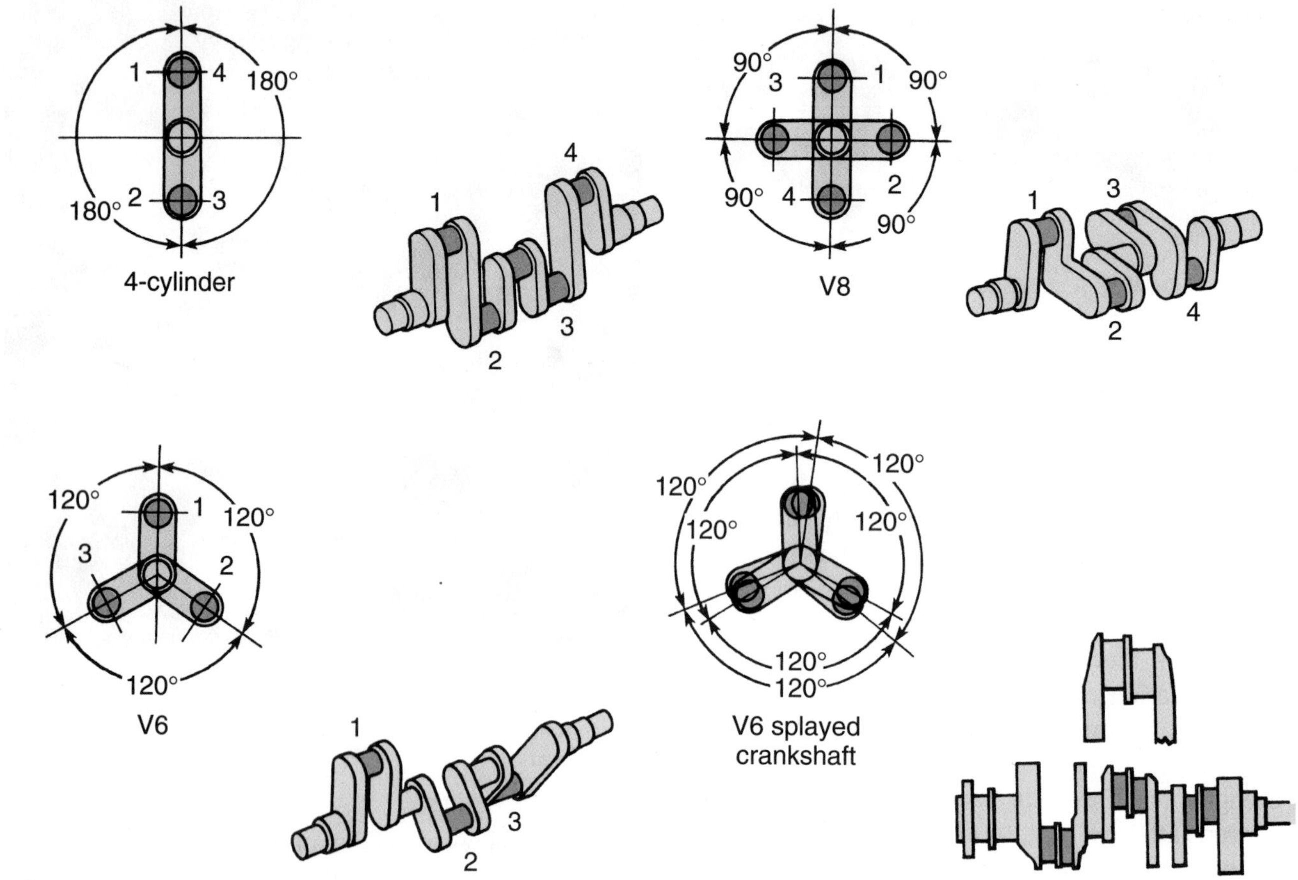

Figure 10-46 Various crankshaft configurations.

Vibration Damper

The purpose of the vibration damper is to dampen crankshaft vibration. The crankshaft can experience a force of more than two tons each time a cylinder fires, causing the crank to momentarily twist and snap back. The center section of the vibration damper **(Figure 10–47)** is attached to the crankshaft. Surrounding the center section is a strip of rubber-like material. Attached to the material is a grooved counterweight. As the crankshaft twists, the center section applies a force to the material. The material must then apply this force to the counterweight. The weight is snapped in the direction of the crankshaft rotation to counterbalance the crankshaft connecting rod journal snapping back against the force due to combustion. The back-and-forth movement of the crankshaft is counterbalanced by the back-and-forth movement of the vibration damper. The vibration damper is sometimes called a harmonic balancer.

Figure 10-47 A vibration damper harmonic balancer.

Flywheel

The **flywheel** also helps to make the engine run more smoothly by applying a constant moving force to carry the crankshaft from one firing stroke to the next. Once the flywheel starts to rotate, its weight tends to keep it rotating. This is called inertia. The flywheel's inertia keeps the crankshaft rotating smoothly in spite of the pulses of power from the pistons.

Because of its large diameter, the flywheel also makes a convenient point for the starter to connect to the engine. The large diameter supplies good gear reduction for the starter, making it easy for the starter to turn the engine against its compression. The surface of a flywheel may be used as part of the clutch. On an engine that drives an automatic transmission, a **flex plate** is used. The automatic transmission torque converter provides the weight required to attain flywheel functions.

CRANKSHAFT INSPECTION AND REBUILDING

Examine the crankshaft carefully. Check for the following:

- Are the vibration damper and flywheel mounting surfaces eroded or fretted?
- Are there indications of damage from previous engine failures?
- Do any of the journal diameters show signs of heat checking or discoloration from high operating temperatures?
- Are any of the sealing surfaces deeply worn, sharply ridged, or scored?
- Are there any signs of surface cracks or hardness distress?

If any or all of these conditions are present, the parts need to be repaired or replaced.

To measure the diameter of a rod journal, use an outside micrometer **(Figure 10–48)**. Measure the journals for size, out-of-roundness, and taper **(Figure 10–49)**. Taper is measured from one side of the journals to the other. The maximum taper is 0.001 inch (.0254 mm).

Compare these measurements to specifications to determine if the crankshaft needs to be reground or replaced. If the journals are within specifications, the journal area needs only to be cleaned up.

Figure 10–48 Measure the diameters of crank journals with an outside micrometer. *Courtesy of Jaspar Engine and Transmission Exchange, Inc.*

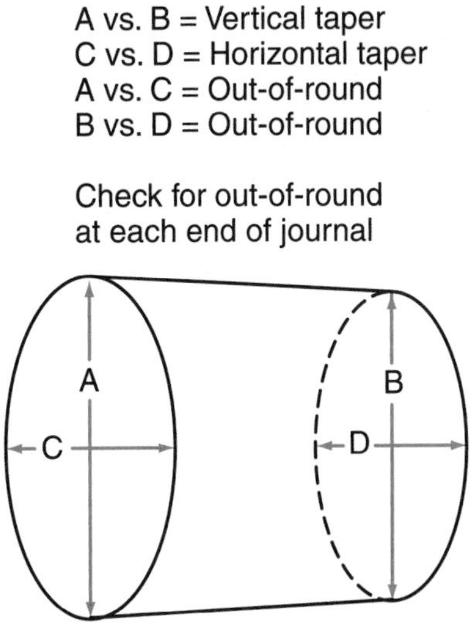

Figure 10–49 Checking crankshaft journals for out-of-roundness and taper.

> **USING SERVICE MANUALS**
>
> Crankshaft specifications can be found in the engine specification section of a service manual. ■

Crankshaft Reconditioning

If the crankshaft is severely damaged, it should be replaced. A crankshaft that has journal taper, grooves the journal surfaces, burnt marks, or small nicks in the journal surface may be reusable after the journals are refinished. This process grinds away some of the metal on the journals to provide an even and mar-free surface. When a crankshaft has been ground, oversize bearings are fitted to the crankshaft to provide for the proper oil clearances.

At times, minor damage to the journals can be corrected by polishing the journals with a very fine crocus cloth. A polishing tool rotates a long loop of emory cloth against the journals as the crankshaft is rotated by a stand. The constant movement of the crocus cloth and the rotation of the crankshaft prevent the creation of any flat spots on the surface. You should also polish the journals after they have been ground.

Checking Crankshaft Straightness

To evaluate the straightness of the crankshaft, the shaft should be supported by V-blocks positioned on the end main bearing journals. Position a dial indicator at the three o'clock position on the center main bearing journal.

Set the indicator at 0 (zero) and turn the crankshaft through one complete rotation. The total deflection of the indicator, the amount greater than zero plus the amount less than zero, is the **total indicator reading (TIR)**. Bow is 50% of the TIR **(Figure 10–50)**. Compare

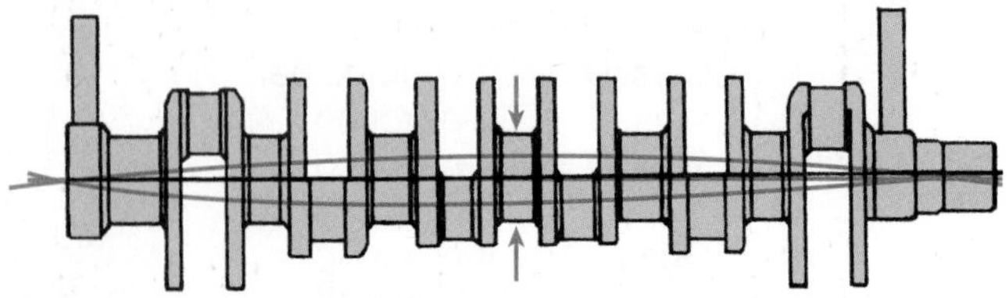

Figure 10-50 Evaluating alignment bow.

the bow of the crankshaft to the acceptable alignment/bow specifications.

A special machine is designed to straighten crankshafts but is only found in serious engine rebuilding shops. In most cases, if the crankshaft is warped, it is replaced.

Flywheel Inspection

Check the runout of the flywheel and carefully inspect its surface. Replacement or resurfacing may be required. When checking the flywheel in a vehicle with an automatic transmission, make sure the torque converter is bolted to the flex plate. Excessive flywheel runout can cause vibrations, poor clutch action, and clutch slippage. With both manual shift and automatic transmissions, inspect the flywheel for a damaged or worn ring gear.

Crankshaft Bearings

Bearings are used to carry the critical loads created by crankshaft movement. They are a major wear item in the engine and require close inspection. Main bearings support the crankshaft journals. Connecting rod bearings are installed between the crankshaft and connecting rods.

Modern crankshaft bearings are known as insert bearings. There are two basic designs of insert bearings **(Figure 10–51)**. A **full round** (one-piece) **bearing** is used in bores that allow the shaft's journals to be inserted into the bearing, such as a camshaft. A **split** (two halves) **bearing** is used where the bearing must be assembled around the journal with the bearing housing being of two parts also, including a cap that holds the assembly together. Crankshaft bearings are typically the split type.

Many crankshafts are fitted with a main bearing that has flanged sides. This type bearing is typically called a thrust bearing and is used to control any horizontal movement or endplay of the shaft. The flange bearing is used in the thrust position of the block. Most thrust main bearings are doubled flanged **(Figure 10–52)**.

Some late-model engines do not use separate main bearing caps; instead they are fitted with a lower engine block assembly. This assembly works like a bridge and contains the lower half of the bore for the main bearings. The assembly is torqued to the engine block and holds the crankshaft in place.

The main bearing caps and lower block assemblies on some engines are given additional strength through the use of additional bolts. Sometimes each main cap is held in place by four bolts, two on each side of the bearing. Other designs may use side bolts that fasten the side of the bearing cap to the engine block. Regardless of the number and position of the bolts, proper tightening sequences **(Figure 10–53)** must be followed.

Bearing Materials

To ensure crankshaft and bearing durability, care must be taken to choose the proper bearing material. Most bearing manufacturers normally meet OEM specifications but also offer alternatives to address unique problems or special needs. Bearings can be made of aluminum, aluminum alloys, copper and lead alloys, and steel backings coated with babbitt. Each alloy has advantages in terms of resistance to corrosion, rate of wear, and fatigue strength. Aluminum alloy bearings are the most commonly used design. These bimetal aluminum bearings contain silicon, which helps to reduce bearing wear. Some bearings use

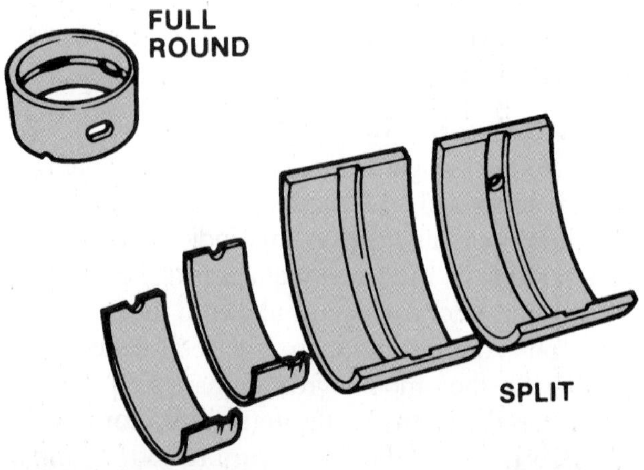

Figure 10-51 Full-round and split insert bearings.

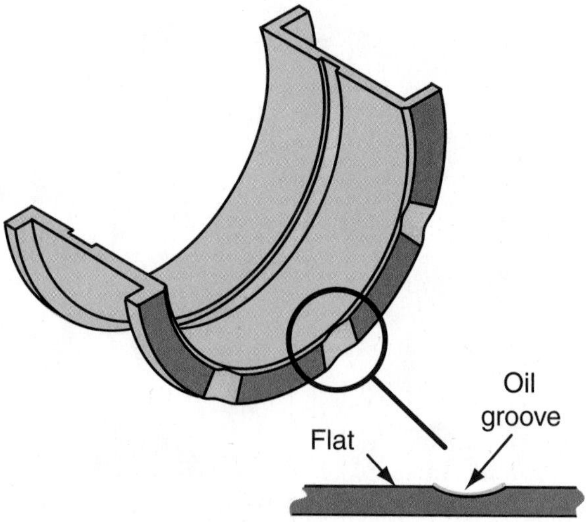

Figure 10-52 A thrust bearing with grooves cut into its flange to provide for better lubrication. *Courtesy of Federal-Mogul Corporation*

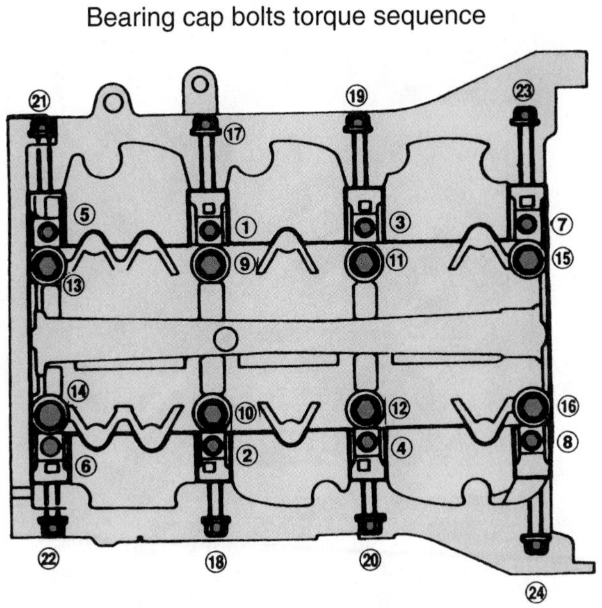

Figure 10–53 Six bolts secure each main cap in this engine. Each bolt must be tightened in correct sequence and to the correct torque. *Courtesy of American Honda Motor Co., Inc.*

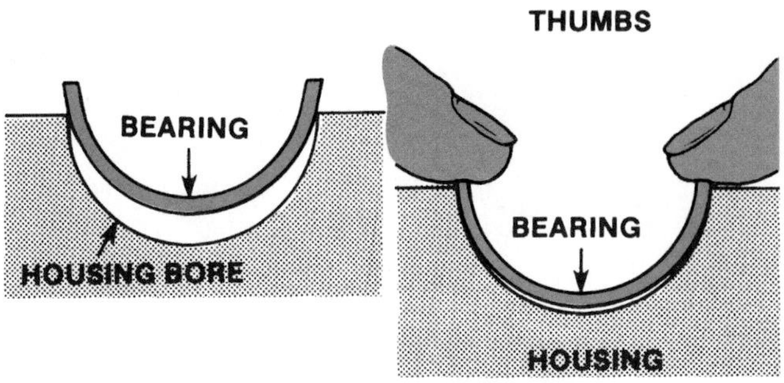

Figure 10–55 Spread requires a bearing to be lightly snapped into place.

a combination of metals to form the bearing, such as a layer of copper-lead alloy on a steel backing followed by a thin coating of babbitt **(Figure 10–54)**. This design takes advantage of the excellent properties of each metal.

Bearing Spread

Most main and connecting rod bearings are manufactured with spread. Bearing spread means that the distance across the outside parting edges of the bearing insert is slightly greater than the diameter of the housing bore. To position a bearing half that has spread, it must be snapped into place by a light forcing action **(Figure 10–55)**. This ensures positive positioning against the inside of the bore and helps to keep the bearings in place during assembly.

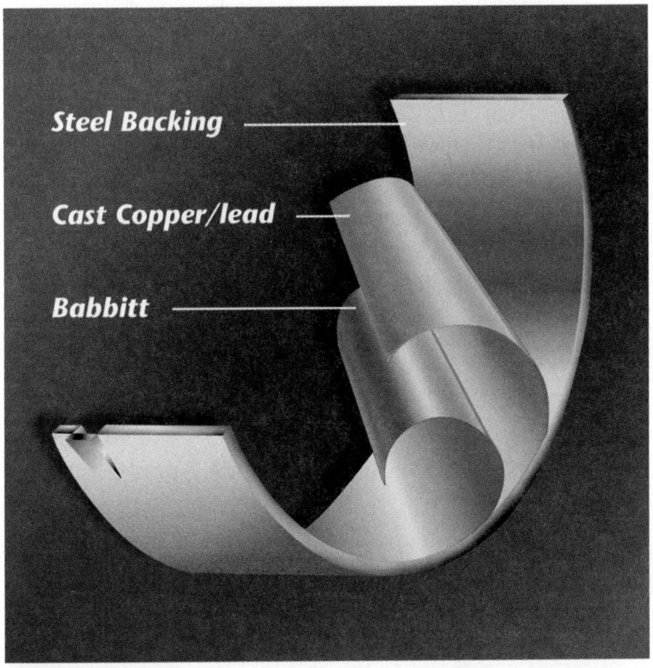

Figure 10–54 The basic construction of a bearing composed of three metals. *Courtesy of Dana Corporation*

Bearing Crush

Each half of a split bearing is made so that it is slightly greater than an exact half. This can be seen quite easily when a half is snapped into place in its housing. The parting faces extend a little beyond the seat **(Figure 10–56)**. This extension is called **crush**.

When the two bearing halves are assembled and the housing cap tightened, the crush sets up a radial pressure on the bearing halves so they are forced tightly into the housing bore.

Bearing crush increases the surface contact between the bearing and its bore, allowing for better heat transfer and compensation for slight bore distortion.

Bearing Locating Devices

Engine bearings must be provided with some means to keep them from rotating or shifting sideways in their housings. Many different methods have been used by manufacturers to keep the bearings in place. The most common way is the use of a locating lug. As shown in **Figure 10–57**, this consists of a protrusion at the parting face of the bearing. The lug fits into a slot in the bearing's bore.

Oil Grooves

Providing an adequate oil supply to all parts of the bearing surface, particularly in the load area, is an absolute

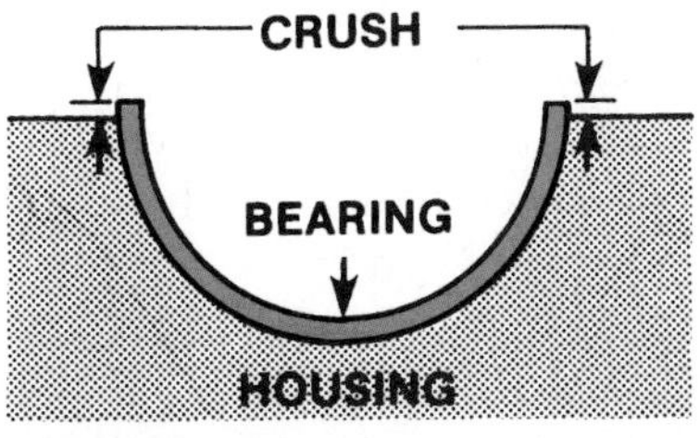

Figure 10–56 Crush ensures good contact between the bearing and the housing.

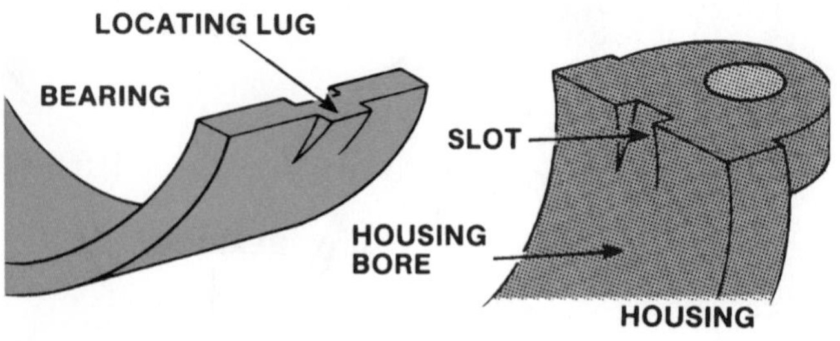

Figure 10-57 The locating lug fits into the slot in the housing.

necessity. In many cases, this is accomplished by the oil flow through the bearing oil clearance. In other cases, however, engine operating conditions are such that this oil distribution method is inadequate. When this occurs, some type of oil groove must be added to the bearing. Some oil grooves are used to ensure an adequate supply of oil to adjacent engine parts by means of oil throw-off.

Most OEM bearings have a full groove around the entire circumference of the bearing and others have a half groove in which only the upper bearing half is grooved.

Oil Holes

Oil holes allow for oil flow through the engine block galleries and into the bearing oil clearance space. Connecting rod bearings receive oil from the main bearings by means of oilways in the crankshaft. Oil holes are also used to meter the amount of oil supplied to other parts of the engine. For example, oil squirt holes in connecting rods are often used to spray oil onto the cylinder walls. When the bearing has an oil groove, the oil hole normally is in line with the groove.

The size and location of oil holes is critical. Therefore, when installing bearings, you must make sure the oil holes in the block line up with holes in the bearings.

Oil Clearance

There must be a gap or clearance between the outside diameter of the crankshaft journals and the inside diameter of its bearings. This clearance allows for the building and maintenance of the oil film. During an engine rebuild, if there is little or no wear on the journals, the proper oil clearance can be restored with the installation of standard size replacement bearings. However, if the crankshaft is worn to the point where the installation of standard-size bearings will result in excessive oil clearance space, a bearing with a thicker wall must be used. Although these bearings are thicker, they are known as undersize because the journals and crank pins of the crankshaft are smaller in diameter. In other words, they are under the standard size.

Undersize bearings are available in 0.001-inch (.0254 mm) or 0.002-inch (.0508 mm) sizes for shafts that are uniformly worn by that amount. Undersize bearings are also available in thicker sizes, such as 0.010 inch

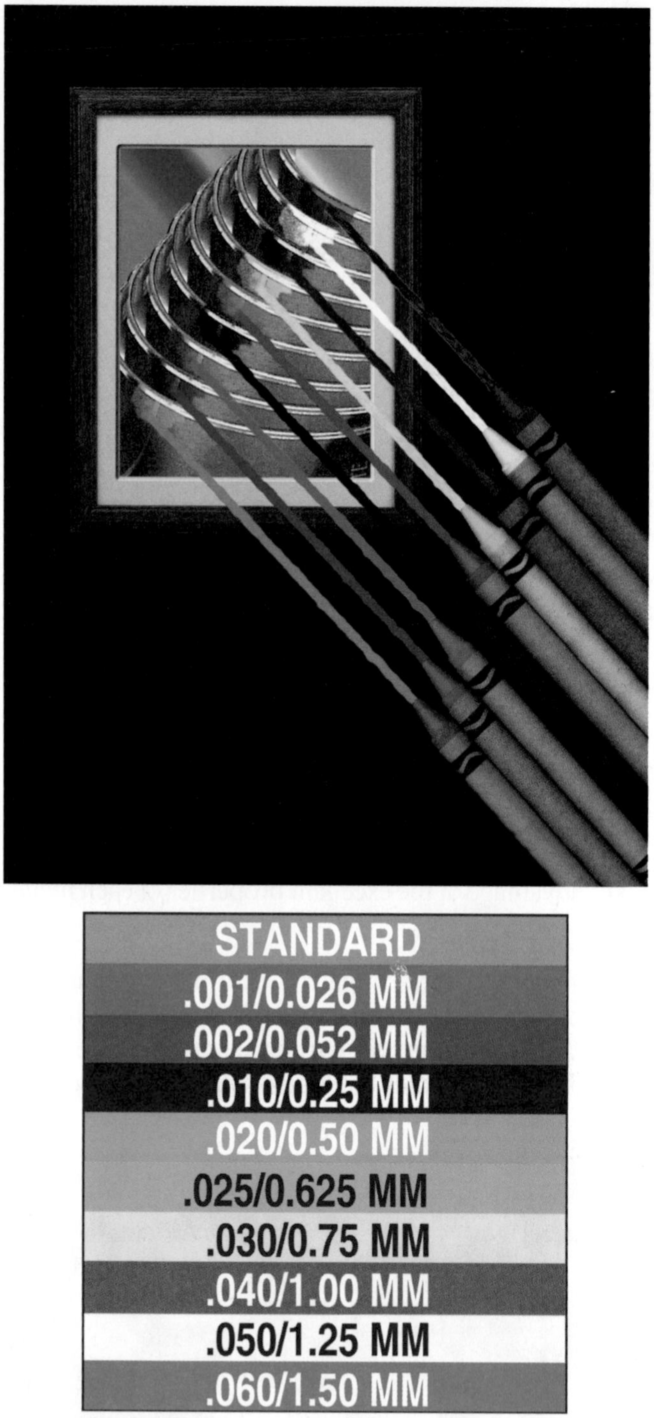

Figure 10-58 Some bearings are color coded to indicate their size. *Courtesy of Dana Corporation*

(.2540 mm), 0.020 inch (.5080 mm), and 0.030 inch (.7620 mm), for use with crankshafts that have been refinished (or reground) to one of these standard undersizes. The difference in thickness of the bearing is normally stamped onto the backside of the bearing. Bearings may also be color-coded to indicate their size **(Figure 10–58).**

Often engines are manufactured with other than standard journal sizes. The manufacturer uses color codes or stamped numbers to indicate which bearing size to use **(Figure 10–59).**

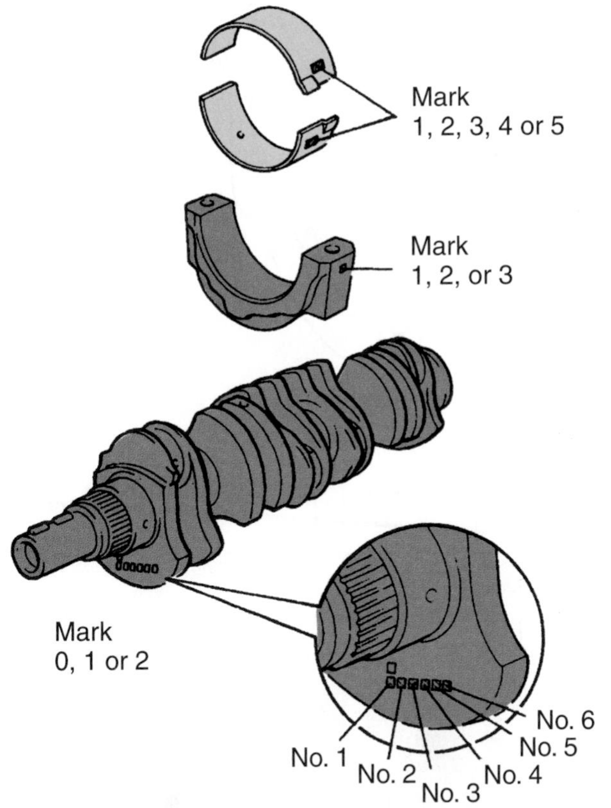

Figure 10–59 Size marking on a crankshaft, connecting rod, and rod bearing. *Reprinted with permission*

SHOP TALK

If the journals measure within specifications but pitting and gouges exist, polish the worst journal to determine whether grinding is necessary. If polishing the journal achieves smoothness, then grinding is probably not necessary. If the crankshaft does not have to be reground, check it for straightness. ■

Bearing Failure and Inspection

As shown in **Figure 10–60**, bearings can fail for a variety of reasons. Oil starvation and dirt are the major reasons for bearing failure. Problems in other engine components, such as bent or twisted crankshafts or connecting rods, or out-of-shape journals, can also cause bearings to wear irregularly.

INSTALLING MAIN BEARINGS AND CRANKSHAFT

When selecting new main bearings, make sure they match the crankshaft journal diameters and main bearing bores. If the crankshaft has been ground undersize, the main bearings must also be undersize. Similarly, if the housing

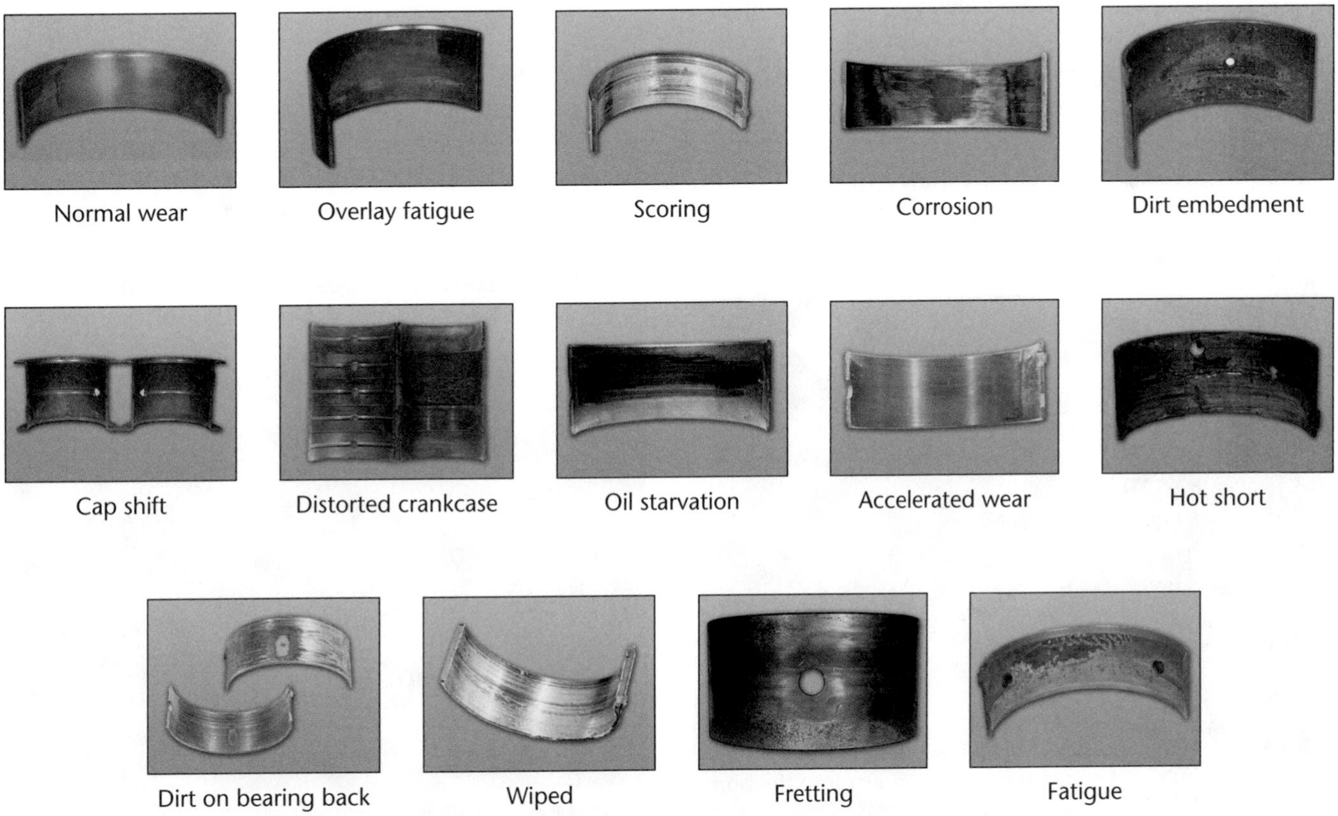

Figure 10–60 Common forms of bearing distress. *Courtesy of Dana Corporation*

bores have been machined oversize by align boring or align honing, the bearings must take up this space. Bearing size is usually marked on the bearing box and on the back of the bearing.

When the bearings are ready to be installed in the main bearing bores, make sure the bore is clean and dry before installing the bearing halves into place. Use a clean, lint-free cloth to wipe the bearing back and bore surface.

Put the new main bearing inserts into each of the main bearing caps **(Figure 10–61)** and into the bearing bores in the cylinder block housings. Make sure all holes align. The backs of the main bearing inserts should never be oiled or greased. Place the crankshaft in the block on the main bearing inserts and arrange the main bearing caps in the correct order and direction over the crankshaft. Follow the factory markings or use those made during disassembly.

The next step is to measure the oil clearance between the crankshaft and the main bearing. Proper lubrication and cooling of the bearing depend on correct crankshaft oil clearances. Scored bearings, worn crankshaft, excessive cylinder wear, stuck piston rings, and worn pistons can result from too small an oil clearance. If the oil clearance is too great, the crankshaft might pound up and down, overheat, and weld itself to the insert bearings.

Plastigage is fine, plastic string used to measure the oil clearance between the bearing and the crankshaft. The procedure for checking bearing clearance with plastigage is outlined in Photo Sequence 6 included in this chapter.

One side of the plastigage's package has stripes for inch measurements, the other side has stripes for metric measurements **(Figure 10–62)**. The string can be purchased to measure different clearance ranges. Usually, only the smallest clearance range is necessary for reassembly work.

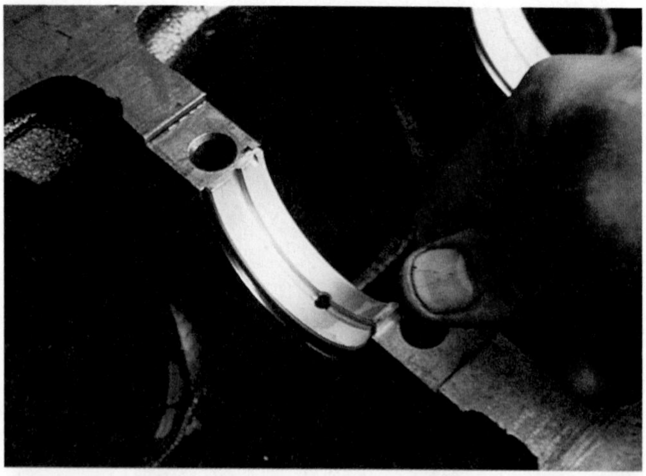

Figure 10-61 Place the bearing inserts into the bore, make sure the locating lugs fit into their recess. *Courtesy of Federal-Mogul Corporation*

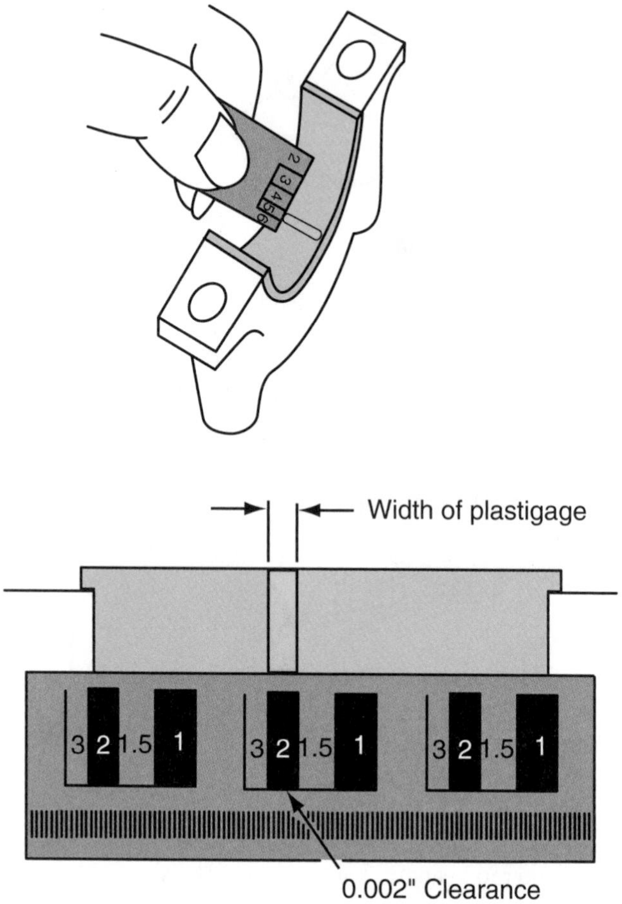

Figure 10-62 Plastigage is available in a variety of ranges and the packing is color coded according to the range. Green has a range of 0.001 to 0.003 inch (0.025 to 0.075 mm).

Crankshaft Endplay

Crankshaft endplay can be measured with a feeler gauge by prying the crankshaft rearward and measuring the clearance between the thrust bearing flange and a machined surface on the crankshaft. Insert the feeler gauge at several locations around the rear thrust bearing face **(Figure 10–63)**. Or position a dial indicator so that

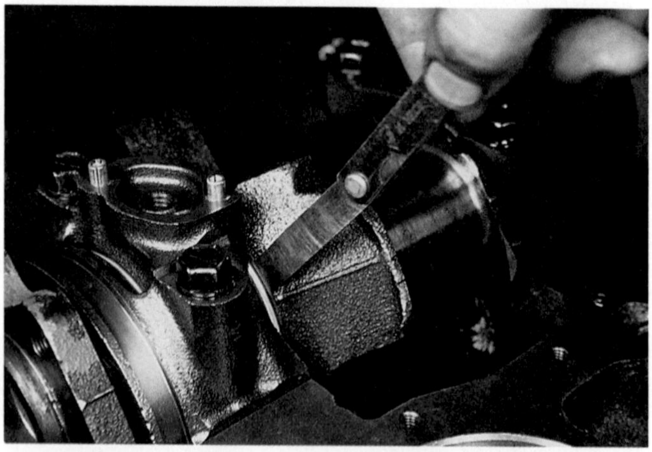

Figure 10-63 Crankshaft endplay can be checked with a feeler gauge. *Courtesy of Federal-Mogul Corporation*

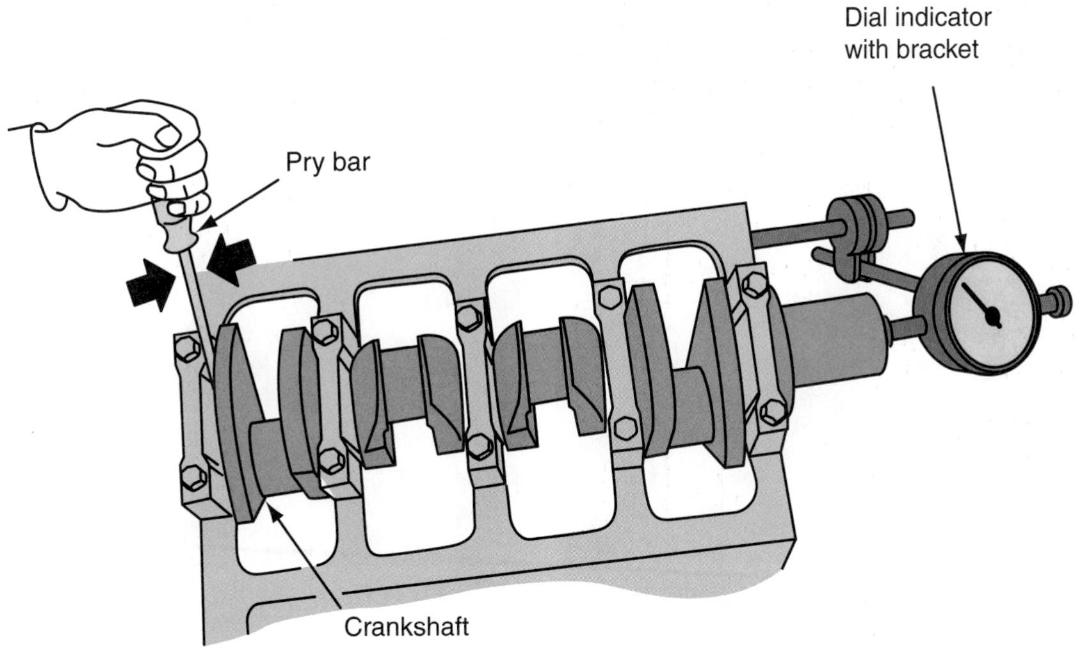

Figure 10-64 Crankshaft endplay can be checked with a pry bar and dial indicator.

the fore and aft movement of the crankshaft can be measured **(Figure 10–64)**.

If the endplay is less than or greater than the specified limits, the main bearing with the thrust surface must be exchanged for one with a thicker or thinner thrust surface. If the engine has thrust washers or shims, thicker or thinner washers or shims must be used.

Most engines require the installation of main bearing seals **(Figure 10–65)** during the final installation of the crankshaft.

Connecting Rod

The connecting rod is used to transmit the pressure applied on the piston to the crankshaft **(Figure 10–66)**. The rod must be very strong and at the same time be kept as light as possible. Connecting rods are generally forged from high-strength steel or are made of nodular steel or cast iron. The center section is made in the form of an "I" for maximum strength with minimum weight.

Figure 10-65 A typical crankshaft seal.

The small end or piston pin end is made to accept the piston pin, which connects the piston to the connecting rod. The piston pin can be pressed-fit in the piston and free fit in the rod. When this is the case, the small end of the rod will be fitted with a bushing. The pin can also be

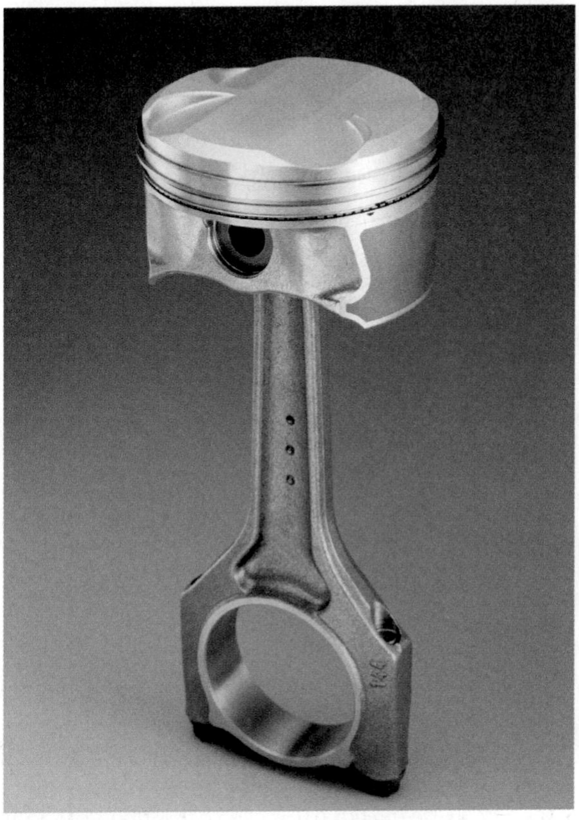

Figure 10-66 A piston and connecting rod assembly.

Checking Main Bearing Clearance with Plastigage

P6-1 *Checking main bearing clearance begins with mounting the engine block upside down on an engine stand.*

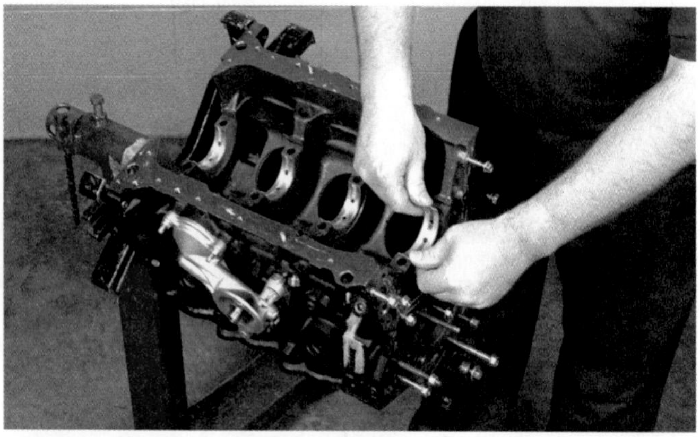

P6-2 *Install main bearings into bores, being careful to properly seat them. Wipe the bearings with a clean lint-free rag.*

P6-3 *Carefully install the crankshaft into the bearings. Try to keep the crankshaft from moving on the bearing surfaces.*

P6-4 *Wipe the crankshaft journals with a clean rag.*

P6-5 *Place a piece of plastigage on the journal. The piece should fit between the radii of the journal.*

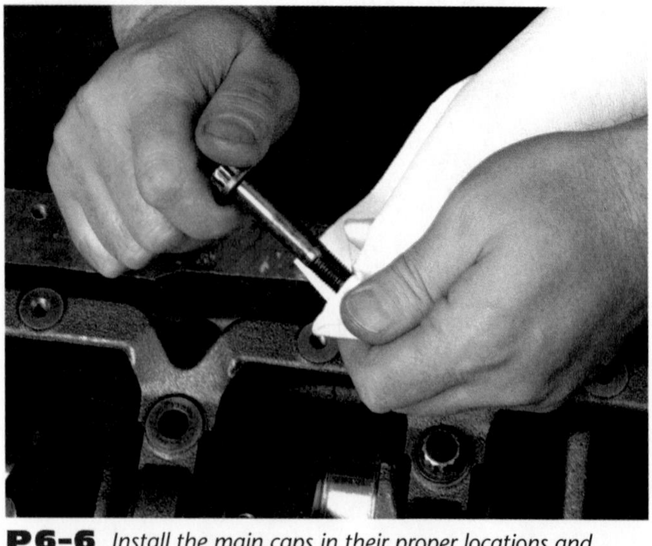

P6-6 *Install the main caps in their proper locations and directions. Wipe the threads of the cap bolts with a clean rag.*

P6-7 *Install the cap bolts and tighten them according to the manufacturer's recommendations.*

P6-8 *Remove the main caps and observe the spread of the plastigage. If the gage did not spread, try again with a larger gage.*

P6-9 *Compare the spread of the gage with the scale given on the plastigage container. Compare the clearance with the specifications.*

P6-10 *Carefully scrape the plastigage off the journal surface.*

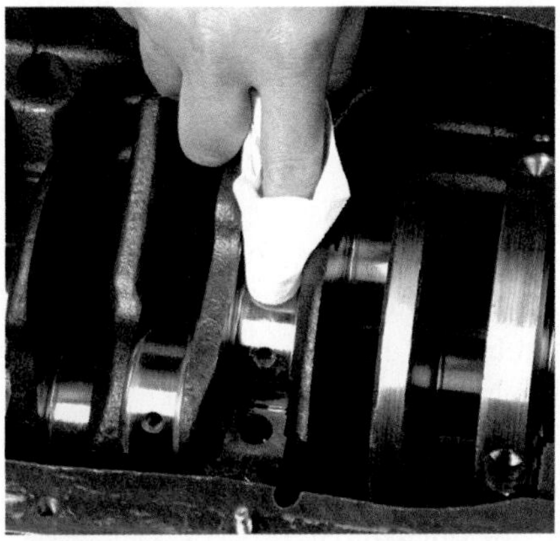

P6-11 *Wipe the journal clean with a rag.*

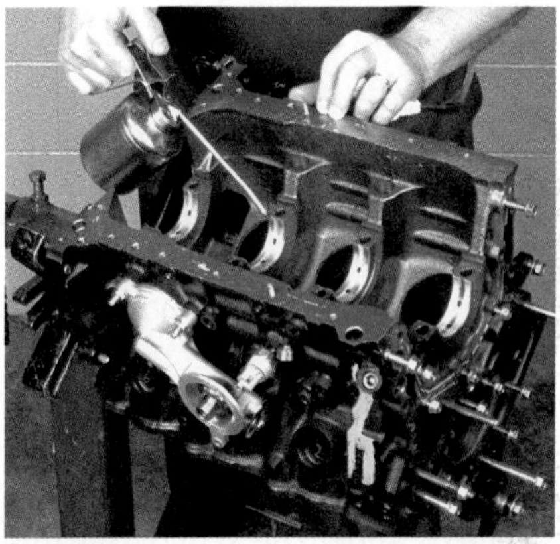

P6-12 *If the clearance was within the specifications, remove the crankshaft and apply a good coat of fresh engine oil to the bearings.*

P6-13 *Reinstall the crankshaft and apply a coat of oil to the journal surfaces.*

P6-14 *Reinstall the main caps and tighten according to specifications.*

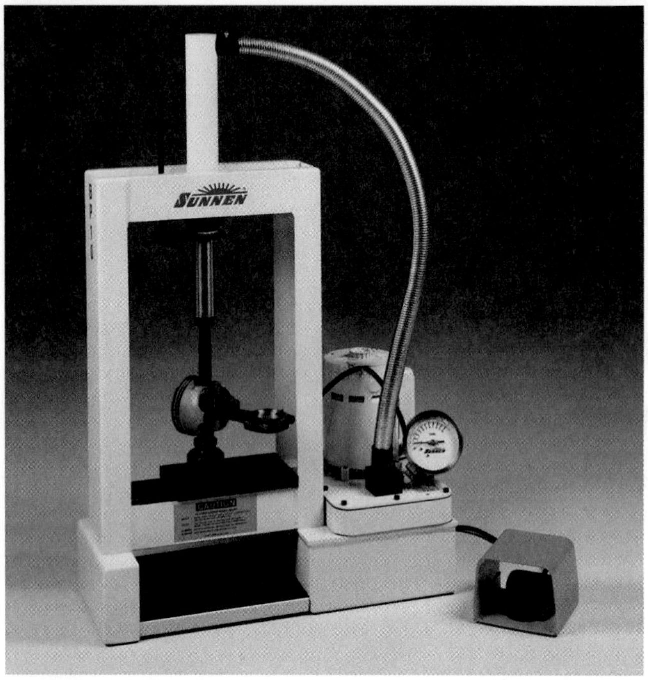

Figure 10-67 A pin press. *Courtesy of Sunnen Products Company*

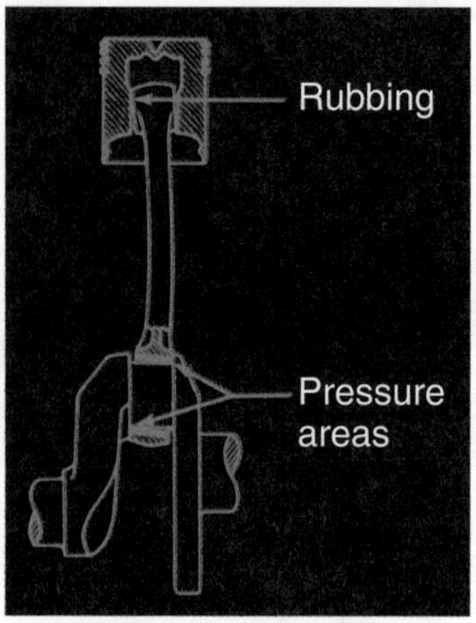

Major misalignment

Bent and twisted connecting rod

Figure 10-68 The effects of rod misalignment on its bearings. *Photos courtesy of Clevite Engine Parts*

a free fit in the piston and pressed-fit in the rod. In this case no bushings are used. The pin simply moves in the piston using the piston hole as a bearing surface. A pin press **(Figure 10–67)** is used to separate and attach pressed-fit piston pins and pistons from their connecting rods. A third mounting allows the pin to move freely in both the piston and the rod. This design requires the use of clips or caps to prevent the pin from moving out against the cylinder walls.

The larger or "big" end of the rod is used to attach the connecting rod to the crankshaft. This end is made in two pieces. The upper half is part of the rod. The lower half is called the rod cap and is bolted to the rod. The connecting rod and its cap are manufactured as a unit and must always be kept together. During production, the rod caps are either machined off the rod or are scribed and broken off. The big end of the rod is fitted with bearing inserts made of the same material as the main bearings. Some connecting rods have a hole drilled through the big end to the bearing area. The bearing insert might have a hole, which will align with this drilling. This hole is used to supply oil for lubricating and cooling the piston skirt. When the rod is properly installed, the oil hole should be pointing to the major thrust area of the cylinder wall.

Closely examine all piston skirts and bearings for unusual wear patterns that may indicate a twisted rod **(Figure 10–68)**. Rods suspected of being bent or distorted can be checked with a rod alignment checker. Normally damaged rods are replaced, although equipment is available to straighten them and to rebore the small and big ends.

PISTON AND PISTON RINGS

The piston forms the lower portion of the combustion chamber. The pressures from combustion are exerted against the top of the piston. This force pushes the piston down in the cylinder. Pistons must be strong enough to face this pressure; however, they should also be as light as possible. This is why most modern pistons are made of aluminum or aluminum alloys.

The top of the piston is called the **head** or **dome**. Just below the dome on the side of the piston is a series of grooves. The grooves are used to contain the piston rings. The high parts between the grooves are called **ring lands**. Below the grooves, as shown in **Figure 10–69**, there is a bore, which is used for the **piston pin**, sometimes called the **wrist pin**. This hole is not always centered in the piston. It can be offset toward the major thrust side of the piston, the side that will contact the cylinder wall during the power stroke. By offsetting the pin, piston slap is eliminated.

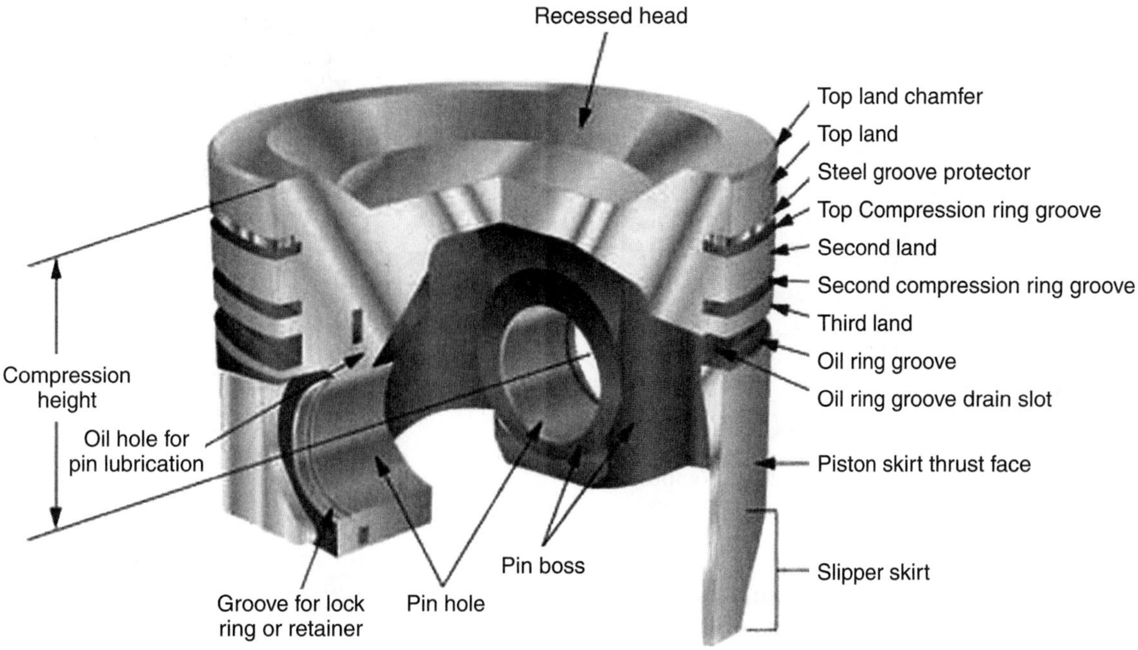

Figure 10-69 The features and terminology used to describe a piston. *Courtesy of Federal-Mogul Corporation*

SHOP TALK

The term *piston slap* is used to describe the noise made by the piston when it contacts the cylinder wall. This noise is usually heard only in older, high-mileage engines that have worn pistons or cylinder walls. The noise is most noticeable when the engine is cold or under a load. ■

To ensure that the piston is installed correctly and has the correct offset, the top of the piston has a mark. The most common mark is a notch, machined into the top edge of the piston. Always refer to the appropriate service manual to determine the correct direction and position of the mark. It is important that the front of the piston match the front of the connecting rod **(Figure 10–70)**.

The base of the piston, the area below the piston pin, is called the piston skirt. The area from just below the bottom ring groove to the tip of the skirt is the piston thrust surface. There are two basic types of piston skirts: the slipper type and the full skirt. The full skirt is used primarily in truck and commercial engines. The slipper type is used for automobile engines and allows the piston enough thrust surface for normal operation. The slipper skirt design also allows the piston to be lighter and reduces piston expansion because there is less material to hold heat.

When an engine is designed, piston expansion determines how much piston clearance will be needed in the cylinder bore. Too little clearance will cause the piston to bind at operating temperatures. Too much will cause piston slap. The normal piston clearance for an engine is about 0.001 to 0.002 inch (.0254 mm to .0508 mm). This clearance is measured between the piston skirt and the cylinder wall.

Piston Rings

Piston rings are used to fill the gap between the piston and cylinder wall.

Piston rings must serve three functions. They seal the combustion chamber at the piston. They remove oil from the cylinder walls to prevent oil from entering into the combustion chamber. They also carry heat from the piston to the cylinder walls to help cool the piston.

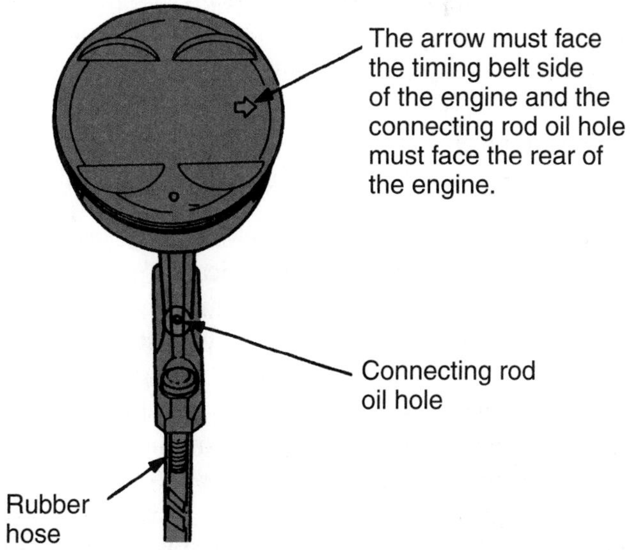

Figure 10-70 Always make sure the markings on the piston and connecting rod are in the correct relationship to each other and that they face the correct direction. *Courtesy of American Honda Motor Co., Inc.*

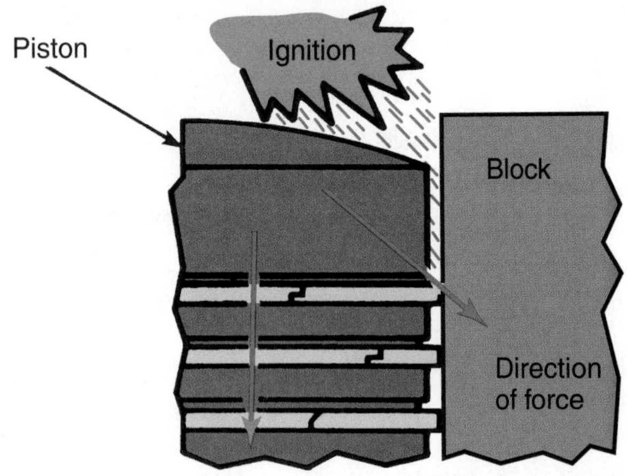

Figure 10-71 The compression rings use combustion pressure to push themselves against the cylinder walls and form a better seal.

There are two basic ring families: compression rings and oil control rings. In most engines, pistons are fitted with two compression rings and one oil control ring. The compression rings are found in the two upper grooves closest to the piston head. The oil ring is fitted to the groove just above the wrist pin.

Compression Rings The compression rings form the seal between the piston and cylinder walls **(Figure 10–71)**. They are designed to use combustion pressure to force them against the cylinder wall. During the power stroke, the pressure caused by the expanding air/fuel mixture is applied between the inside of the ring and the piston groove. This forces the ring into full contact with the cylinder walls. The same force is applied to the top of the ring, forcing it against the bottom of the ring groove. These two actions help to form a tight ring seal.

Compression rings are generally made of cast iron. Most compression rings have a coating on their faces, which aids the wear-in process. Wear-in is the time needed for the rings to conform to the shape and surface of the cylinder wall. Typical soft coatings are graphite, phosphate, iron oxide, and molybdenum. Some compression rings have a hard coating, such as chromium.

Oil Control Rings Oil is constantly being applied to the cylinder walls. The oil is used to lubricate, clean the cylinder wall of carbon and dirt particles, and help cool the piston. Controlling this oil is the primary function of the oil ring. The two most common types of oil rings are the segmented oil ring and the cast-iron oil ring. Both types of rings are slotted so that excess oil from the cylinder wall can pass through the ring. The oil ring groove of the piston is also slotted. After the oil passes through the ring, it can then pass through the slots in the piston and return to the oil sump through the open section of the piston.

Segmented oil rings are made of three pieces: upper and lower scraper rails and an expander. The end gaps of the three separate pieces must be staggered to prevent oil from escaping into the cylinder.

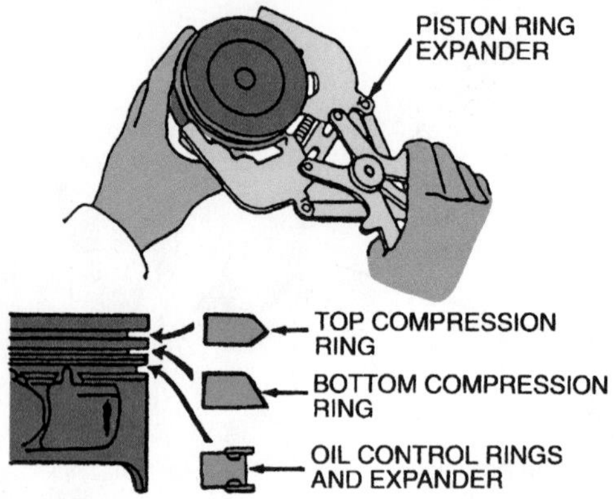

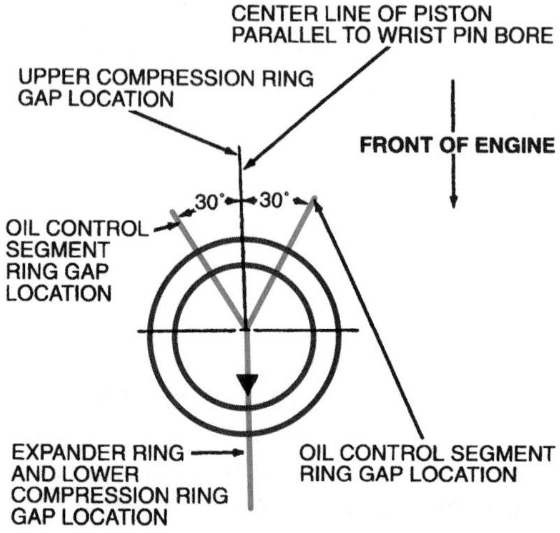

Figure 10-72 Install the piston rings onto the piston with a ring expander. Also make sure the ring end gaps are arranged according to specifications. *Courtesy of Ford Motor Company*

INSTALLING PISTONS AND CONNECTING RODS

Once the crankshaft is in place, the piston and connecting rod assemblies are installed next. Check the marks on the connecting rod caps and the connecting rods to make sure they are a match.

The insert bearings for the connecting rods must be the correct size. If the crankshaft has been machined undersize, matching rod bearing inserts must be installed. The size of the bearing inserts is printed on the box they come in and is stamped on the backs of the bearings or they are color coded.

Snap the new connecting rod bearing inserts into the connecting rods and rod caps. Make sure the tang on the bearing fits snugly into the matching notch.

The piston and rod can be assembled in the block according to the procedure shown in Photo Sequence 7. Re-

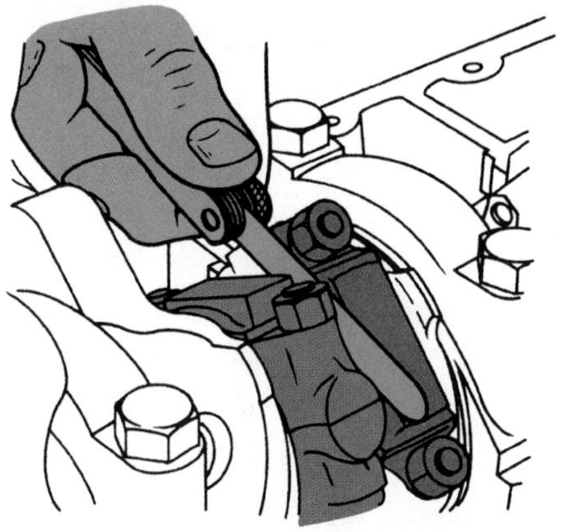

Figure 10–73 Measuring connecting rod side clearance.

member that connecting rods are numbered for easy identification and proper assembly. Also make sure the end gaps of the piston rings are positioned according to the manufacturer's recommendations **(Figure 10–72)** prior to installing the piston assembly in the cylinder bore.

When all the pistons and rods have been installed, connecting rod side clearance can be measured **(Figure 10–73)**. Side clearance is the amount of clearance between the crankshaft and the side of the connecting rod. Side clearance is measured with a feeler gauge. If the clearance is not correct, the rods may need to be machined or replaced.

Be sure to coat the crankshaft assembly with clean lubricant or engine oil. After each piston assembly is installed in the block, rotate the crankshaft and check its freedom of movement. If the crankshaft is hard to rotate after a piston has been installed, remove it and look for signs of binding.

CRANKSHAFT AND CAMSHAFT TIMING

USING SERVICE MANUALS

Normally, camshaft timing marks are shown in the engine section of a service manual, under the heading of Timing Belt or Chain R&R. ■

During most engine rebuilds, a new timing assembly is installed. If wear exists on any component, replacement of the entire assembly is necessary. Wear in the chain, gears, or sprockets results in poor engine performance.

The camshaft drive must be installed so that the camshaft and crankshaft are in time with each other. Both sprockets are held in position by a key or possibly a pin. There are factory timing marks on the crankshaft gear or sprocket and on the camshaft gear or sprocket. The timing marks on all the gears must be positioned according to the manufacturer's instructions.

The chain is installed on the crankshaft gear first, then around the camshaft sprocket. Never wind a chain onto the gears or use a screwdriver, pry bar, or hammer to force a chain into position. Prying or pounding on the chain damages the links, causing the chain to stretch and fail. Carefully place the entire assembly as a unit onto the shafts by pressing both gears evenly, keeping the keyways aligned.

Camshaft End Play

Before proceeding to the next step in the reassembly process, check to be sure that the clearance between the camshaft boss (or gear) and the backing plate is within manufacturer's specifications. Make this check with a feeler gauge. Install shims behind the thrust plate or reposition the camshaft gear and retest the end play. In some cases, adjustment is made by replacing the thrust plate. Some engines limit the endplay of the camshaft through the use of a cam button. A nylon button or Torrington bearing sets on a spring and is installed between the camshaft timing gear and the timing chain cover. To check end play, use a dial indicator setup. Be sure the camshaft end play is not more than recommended in the service manual.

Lifters

It is important to remember that because of the relationship between the camshaft lobe and the lifter, it is extremely important that new lifters be used with a new camshaft. Before installing the lifters, prefill hydraulic lifters with oil. To do this, place them in a container of clean oil and let them sit there overnight to allow the air trapped inside to seep out. Then pump the plunger in the lifter a few times to bleed the remaining air out. Before setting the lifters into their bore in the block, coat the bores with assembly lube. After they have been installed, rotate the camshaft to check for binding or misalignment.

CASE STUDY

***A** car is brought into the shop just 2,000 miles (3,200 km) after the owner rebuilt the engine. The shop is asked to diagnose the cause of excessive blue exhaust smoke, oil consumption, and fouled spark plugs. Responding to questions, the customer indicates that she has rebuilt many engines before and never had this problem. She also has receipts for the parts she installed and the machining she had done to the engine. All of the parts were high quality, and the machining was done by a reputable machine shop.*

PHOTO SEQUENCE 7

Installing a Piston and Rod Assembly

P7-1 *Insert a new piston ring into the cylinder. Use the head of the piston to position the ring so that it is square with the cylinder wall. Use a feeler gauge to check the end gap. Compare end-gap specifications with the measured gap. Correct as needed. Normally, the gaps of the piston rings are staggered to prevent them from being in line with each other. Piston rings are installed easily with a ring expander.*

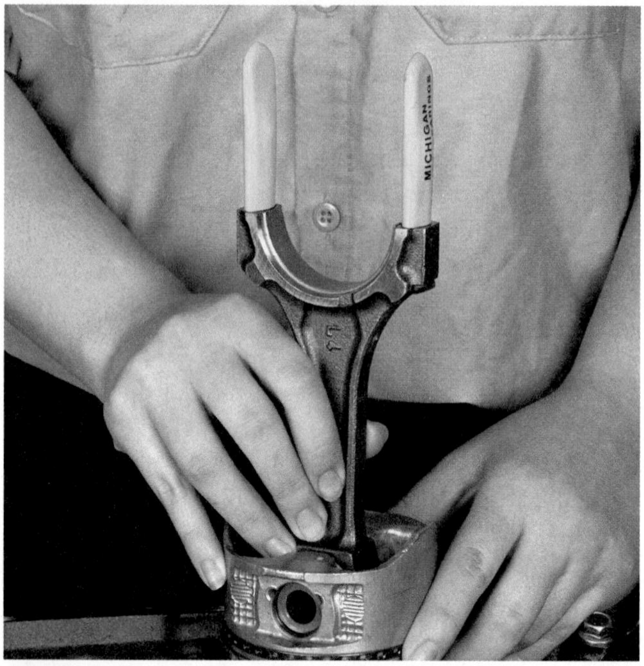

P7-2 *Before attempting to install the piston and rod assembly into the cylinder bore, place rubber or aluminum protectors or boots over the threaded section of the rod bolts. This will help to prevent bore and crankpin damage.*

P7-3 *Lightly coat the piston, rings, rod bearings, cylinder wall, and crankpin with an approved assembly lubricant or a light engine oil. Some technicians submerge the piston in a large can of clean engine oil before it is installed.*

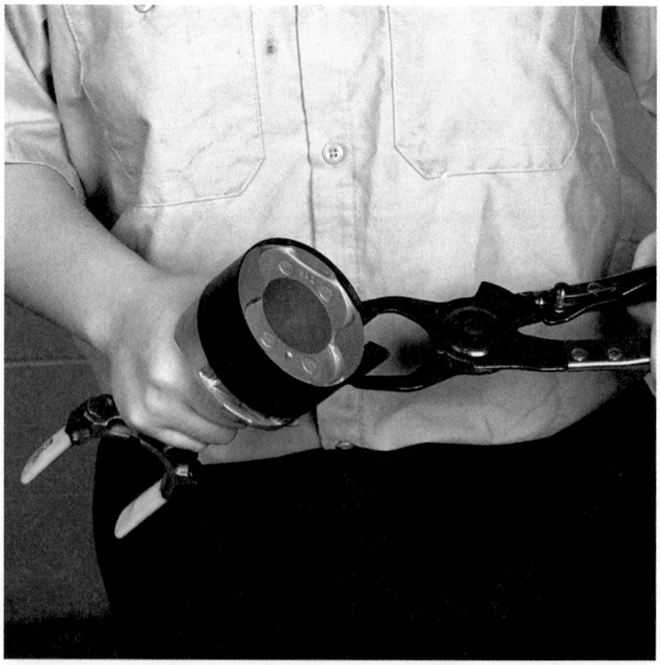

P7-4 *Stagger the ring end gaps and compress the rings with the ring compressor. This tool is expanded to fit around the piston rings. It is tightened to compress the piston rings. When the rings are fully compressed, the tool will not compress any further. The piston will fit snugly, but not tightly.*

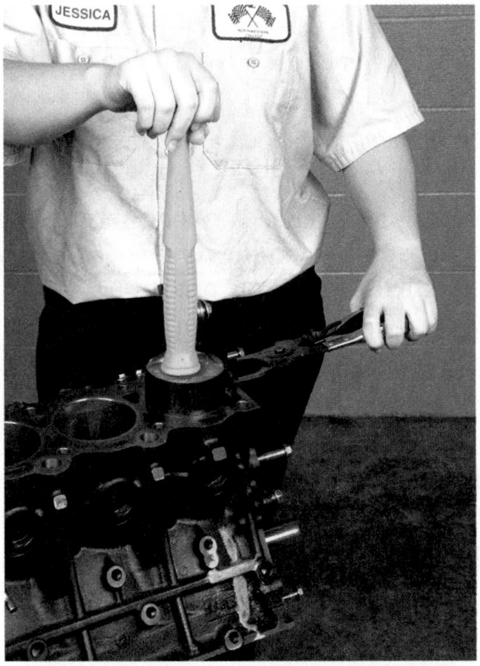

P7-5 *Rotate the crankshaft until the crankpin is at its lowest level (BDC). Then place the piston/rod assembly into the cylinder bore until the ring compressor contacts the cylinder block deck. Make sure that the piston reference mark is in correct relation to the front of the engine. Also, when installing the assembly, make certain that the rod threads do not touch or damage the crankpin.*

P7-6 *Lightly tap on the head of the piston with a mallet handle or block of wood until the piston enters the cylinder bore. Push the piston down the bore while making sure the connecting rod fits into place on the crankpin. Remove the protective covering from the rod bolts.*

P7-7 *Position the matching connecting rod cap and finger tighten the rod nuts. Make sure the connecting rod blade and cap markings are on the same side. Gently tap each cap with a plastic mallet as it is being installed to properly position and seat it. Torque the rod nuts to the specifications given in the service manual. Repeat the piston/rod assembly procedure for each assembly.*

To verify the problem, a cylinder leakage test is conducted. The results indicate excessive leakage past the piston rings. To determine what went wrong during the rebuild, further questioning is needed. Assembly procedures are reviewed. No steps were apparently missed. New piston rings were sized to the freshly-honed cylinder walls and the resultant ring end gap is within specifications. The ring gaps were also staggered in the way recommended by the manufacturer. The work order from the machine shop showed that the cylinders were honed to a 30 room mean square (RMS) surface finish and the resultant size was 3.751 inches (95.27 mm). The pistons are measured to be approximately 3.749 inches (95.22 mm). The 0.002-inch (.0508-mm) clearance is within specifications.

Further examination of the receipts reveals the problem. The customer had originally intended to install cast-iron piston rings. But while the block was in the machine shop, she bought a set of molybdenum rings. Although these rings have advantages over cast-iron rings, they are not best for this block. Molybdenum rings require a smooth cylinder wall finish; typically, a finish of 10 to 15 RMS is recommended. The rougher cylinder wall finish intended for cast-iron rings wore off the molybdenum coating, allowing oil to leak past the rings. This is the cause of the smoke, oil consumption, and spark plug fouling.

KEY TERMS

Babbit
Back lash
Boring bar
Cam follower
Cast
Chamfering
Core plug
Crush
Cylinder sleeve
Deck
Dome
Flex plate
Flywheel
Full round bearing
Glaze
Harmonic balancer
Head
Hone
Leakdown
Line boring
Long block
Overlap
Piston pin
Piston rings
Plastigage
Ring lands
Short block
Split bearing
Taper
Total indicator reading (TIR)
Valve lifters
Wrist pin

SUMMARY

- The basic short block assembly consists of the cylinder block, crankshaft, crankshaft bearings, connecting rods, pistons and rings, oil gallery, and core plugs. On OHV engines the camshaft and its bearings are also included.
- The cylinder block houses the areas in which combustion occurs.
- A properly reconditioned cylinder must be of the correct diameter, have no taper or runout, and have a surface finish that allows the piston rings to seal.
- Glaze is the thin residue that forms on cylinder walls due to a combination of heat, engine oil, and piston movement.
- Core plugs and oil gallery plugs are normally removed and replaced as part of cylinder block reconditioning. The three basic core plugs are the disc- or dished-type, cup-type, and expansion-type.
- A cam changes rotary motion into reciprocating motion. The part of the cam that controls the opening of the valves is the cam lobe. The closing of the valves is the responsibility of the valve springs.
- The camshaft is supported in the cylinder block by friction-type bearings, or bushings, which are typically pressed into the camshaft bore in the block or head. Camshaft bearings are normally replaced during engine rebuilding.
- Solid valve lifters provide for a rigid connection between the camshaft and the valves. Hydraulic valve lifters do the same but use oil to absorb the shock resulting from movement of the valve train. Roller lifters are used to reduce friction and power loss.
- The camshaft is driven by the crankshaft at half its speed.
- The most common ways to measure cam lobe wear are with a dial indicator or an outside micrometer. The dial indicator test should be conducted with the camshaft in the engine. When using an outside micrometer, the camshaft must be out of the engine.
- Most premature cam wear develops within the first few minutes of operation.
- The crankshaft turns on a film of oil trapped between the bearing surface and the journal surface. The journals must be smooth and highly polished. The flywheel adds to an engine's smooth running by applying a constant moving force to carry the crankshaft from one firing stroke to the next. The flywheel surface may be used as part of the clutch.

- Important crankshaft checks include saddle alignment, straightness, clearance, and endplay.
- Bearings carry the critical loads created by crankshaft movement. Most bearings used today are insert bearings.
- Maintaining a specific oil clearance is critical to proper bearing operation. Bearings are available in a variety of undersizes.
- Today's aluminum pistons are lightweight, yet strong enough to withstand combustion pressure.
- Piston rings are used to fill the gap between the piston and cylinder wall. Most of today's vehicle engines are fitted with two compression rings and one oil control ring.
- When installing a piston and connecting rod assembly, various markings can be used to make sure the installation is correct. Always check the service manual for exact locations.
- Connecting rod side clearance determines the amount of oil throw-off from the bearings and is measured with a feeler gauge.
- During most engine rebuilds, a new timing assembly is installed. When installing the timing gears, make sure they are aligned to specifications.

TECH MANUAL

The following procedures are included in Chapter 10 of the *Tech Manual* that accompanies this book:

1. Measuring cylinder bore.
2. Measuring crankshaft journals.
3. Measuring crankshaft endplay.
4. Installing pistons and connecting rods.
5. Measuring camshaft lobes/cam and bearing journals.
6. Installing camshaft bearings.
7. Inspecting and replacing camshaft drives.
8. Testing for worn cam lobes.

REVIEW QUESTIONS

1. Name the two most common ways to measure cam lobe wear.

2. What is the deck?

3. Where does maximum cylinder bore wear occur?

4. What type of valve lifter automatically compensates for the effects of engine temperature?

a. hydraulic
b. solid
c. roller
d. all of the above

5. What is the function of compression rings?

6. Most pistons used today are made from ______.

a. cast iron
b. aluminum
c. ceramic
d. none of the above

7. Core plugs ______.

a. are also called expansion plugs
b. are used in all cast-iron cylinder blocks
c. are a possible source for coolant leaks
d. all of the above

8. Technician A says the maximum amount of cylinder out-of-roundness allowed in most cases is 0.0015 inch (.0381 mm). Technician B says that a cylinder can have a taper of up to 0.010 inch (.2540 mm) and be acceptable. Who is correct?

a. Technician A
b. Technician B
c. Both A and B
d. Neither A nor B

9. After installing cam bearings, Technician A checks that the oil holes in the bearings are properly aligned with those in the block by squirting oil into the holes. Technician B checks for proper alignment by inserting a wire through the holes. Who is correct?

a. Technician A
b. Technician B
c. Both A and B
d. Neither A nor B

10. Technician A says that a cylinder wall with too smooth a surface will prevent the piston rings from seating properly. Technician B says a cylinder wall should have a crosshatch honing pattern. Who is correct?
    - **a.** Technician A
    - **b.** Technician B
    - **c.** Both A and B
    - **d.** Neither A nor B
11. Technician A uses a pry bar to stretch the timing chain onto its sprockets. Technician B cools the timing sprockets so that the timing chain will slip over the teeth. Who is correct?
    - **a.** Technician A
    - **b.** Technician B
    - **c.** Both A and B
    - **d.** Neither A nor B
12. Technician A installs a cup-type core plug with its flanged edge outward. Technician B installs a dish-type core plug with the dished side facing inward. Who is correct?
    - **a.** Technician A
    - **b.** Technician B
    - **c.** Both A and B
    - **d.** Neither A nor B
13. Each half of a split bearing is made so that it is slightly greater than an exact half. What is this extension called?
    - **a.** spread
    - **b.** crush
    - **c.** both a and b
    - **d.** neither a nor b
14. The connecting rod journal is also called the _____.
    - **a.** balancer shaft
    - **b.** vibration damper
    - **c.** plastigage
    - **d.** crank pin
15. Technician A uses a micrometer to measure the connecting rod journal for taper. Technician B uses a micrometer to measure the connecting rod journal for out-of-roundness. Who is correct?
    - **a.** Technician A
    - **b.** Technician B
    - **c.** Both A and B
    - **d.** Neither A nor B
16. Technician A says piston ring end gaps should be the same for each ring on a piston. Technician B says piston ring end gaps should be staggered before installing the piston into its bore. Who is correct?
    - **a.** Technician A
    - **b.** Technician B
    - **c.** Both A and B
    - **d.** Neither A nor B
17. Technician A checks crankshaft endplay or end clearance with a feeler gauge, while Technician B uses a dial indicator. Who is correct?
    - **a.** Technician A
    - **b.** Technician B
    - **c.** Both A and B
    - **d.** Neither A nor B
18. What device in the valve train changes rotary motion into reciprocating motion?
    - **a.** eccentric
    - **b.** cam
    - **c.** bushing
    - **d.** mandrel
19. Technician A says counterweights are added to crankshafts to offset the weight of the connecting rods and pistons. Technician B says the connecting rod bearings are fed a fresh supply of oil through holes drilled in the crankshaft. Who is correct?
    - **a.** Technician A
    - **b.** Technician B
    - **c.** Both A and B
    - **d.** Neither A nor B
20. Which type of oil ring is slotted so that excess oil can pass through it?
    - **a.** cast-iron
    - **b.** segmented
    - **c.** both a and b
    - **d.** neither a nor b

11

CYLINDER HEADS, CAMSHAFTS, AND VALVE TRAINS

OBJECTIVES

■ Describe the purpose of an engine's cylinder head, valves, and related valve parts. ■ Describe the types of combustion chamber shapes found on modern engines. ■ Explain the procedures involved in reconditioning cylinder heads, valve guides, valve seats, and valve faces. ■ Explain the steps in cylinder head and valve reassembly.

CYLINDER HEAD

The cylinder head **(Figure 11–1)** is made of cast iron or aluminum. On overhead valve engines, the cylinder head contains the valves, valve seats, valve guides, valve springs, rocker arm supports, and a recessed area that makes up the top portion of the combustion chamber. On overhead cam engines, the cylinder head contains these items, plus the supports for the camshaft and camshaft bearings **(Figure 11–2)**.

Both overhead valve and overhead cam cylinder heads contain passages that match passages in the cylinder block. These passages allow coolant to circulate in the head and allow oil to drain back into the oil pan. Pressurized oil also moves through some of the passages to lube the camshaft and valve train. The cylinder head also contains tapped holes in the combustion chamber to accept the spark plugs.

The sealing surface of the head must be flat and smooth. However, if the surface is too smooth, the gasket may be able to move. The ideal surface for a cylinder head has a slight texture to it. This finish will grip the gasket and stop it from shifting around. This area must form a tight seal and contain the pressures formed during combustion. To aid in the sealing, a gasket is placed between the head and block. This gasket, called the head gasket, is made of special material that can withstand high temperatures, high pressures, and the expansion of the metals around it. The head also serves as the mounting point for the intake and exhaust manifolds and contains the intake and exhaust ports.

Cylinder head design is one of the most influential factors that affects the overall performance of an engine. The size and shape of the intake and exhaust ports affect

Figure 11–1 A typical late-model cylinder head.

Figure 11–2 An OHC cylinder head. *Courtesy of DaimlerChrysler Corporation*

the velocity and volume of the mixture entering and leaving the cylinders. Most aspects of cylinder head design are carefully tested and calibrated by the manufacturers to ensure optimal performance and fuel economy for the intended application.

Large openings in the cylinder head allow coolant to pass through the head. Coolant must circulate throughout the cylinder head to remove excess heat. The coolant flows from passages in the cylinder block through the head gasket and into the cylinder head. The coolant then passes back to other parts of the cooling system.

COMBUSTION CHAMBER

The performance of an engine, its fuel efficiency, and the level of pollutants in the exhaust all depend to a large extent on the shape of the combustion chamber. An efficient combustion chamber must be compact to minimize the surface area of the walls through which heat is lost to the engine's cooling system. The point of ignition (the nose of the spark plug) should be at the center of the combustion chamber to minimize the flame path, or the distance from the spark to the furthermost point in the chamber. The shorter the flame path, the more evenly the air/fuel mixture will burn.

Manufacturers have designed several different shapes of combustion chambers. Before looking at the popular combustion chamber designs, two terms should be defined.

1. **Turbulence** is a very rapid movement of gases. Turbulence causes better combustion because the air and fuel are mixed better.
2. **Quenching** is the mixing of gases by pressing them into a thin area. The area in which gases are thinned is called the quench area.

Wedge Chamber

In the **wedge-type combustion chamber**, the spark plug is located at the wide part of the wedge **(Figure 11–3)**. As the piston comes up on the compression stroke, the air/fuel mixture is squashed in the quench area. The **quench area** causes the air and fuel to be mixed thoroughly before combustion, which helps to improve the combustion efficiency of the engine. Spark plugs are positioned to allow for rapid and even combustion. When the spark occurs, a flame front moves from the spark plug outward. The wedge-shaped combustion chamber is also called a turbulence- or swirl-type combustion chamber. On newer model cars, the quench area has been reduced, which helps reduce exhaust emissions.

Hemispherical Chamber

The **hemispherical combustion chamber** gets its name from its basic shape. Hemi is defined as half, and spherical means circle. The combustion chamber is shaped like a half circle. This type of cylinder head is also called the **hemi-head**. The piston top forms the base of the hemisphere, and the valves are inclined at an angle of 60 to 90 degrees to each other, with the spark plug positioned between them **(Figure 11–4)**.

This design has several advantages. The flame path from the spark plug to the piston head is short, which gives efficient burning. The cross-flow arrangement of the inlet and exhaust valves allows for a relatively free flow of gases in and out of the chamber. The result is that the engine can breathe deeply, meaning that it can draw in a large volume of mixture for the space available and give a high power output.

The hemispherical combustion chamber is considered a nonturbulence-type combustion chamber. Little turbulence is produced in this type chamber. The air/fuel mixture is compressed evenly on the compression stroke. The spark plug is located directly between the valves. Combustion radiates evenly from the spark plug, completely burning the air/fuel mixture.

One of the more important advantages of the hemispherical combustion chamber is that air and fuel can enter the chamber very easily. The wedge combustion chamber restricts the flow of air and fuel to a certain ex-

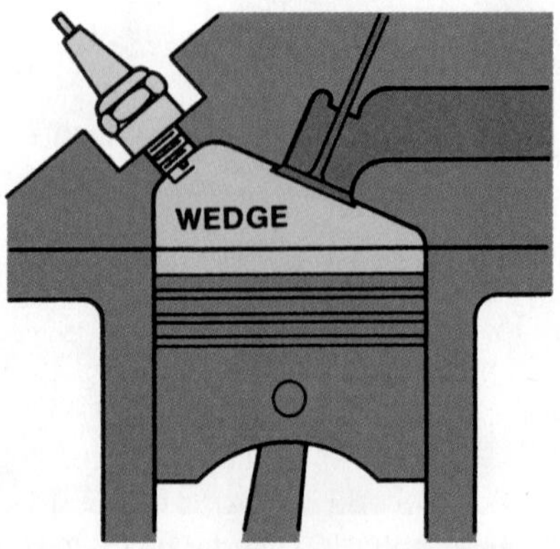

Figure 11-3 A typical wedge combustion chamber.

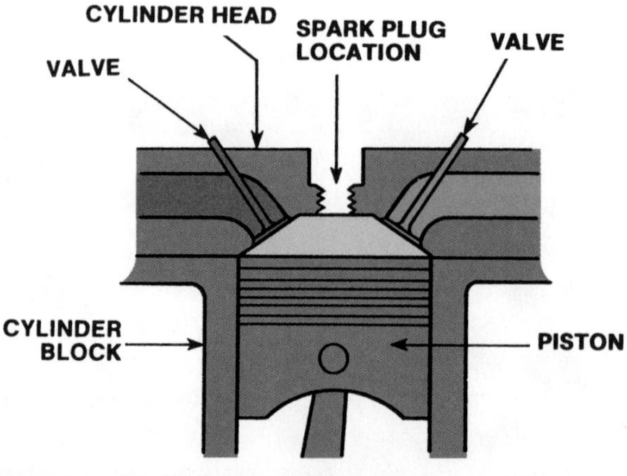

Figure 11-4 A typical hemispherical combustion chamber.

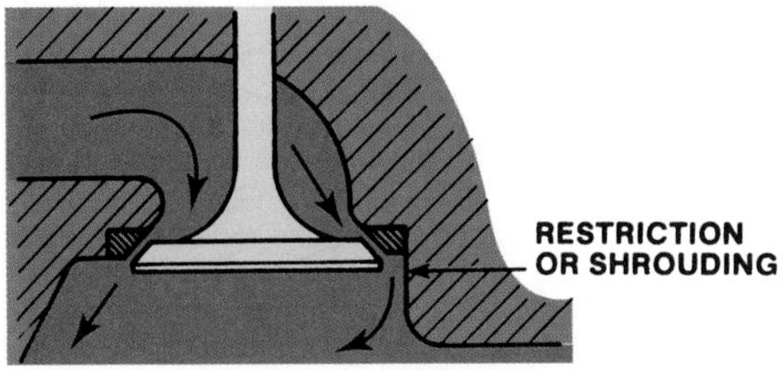

Figure 11-5 Shrouding is a restriction in the flow of intake gases caused by the shape of the combustion chamber.

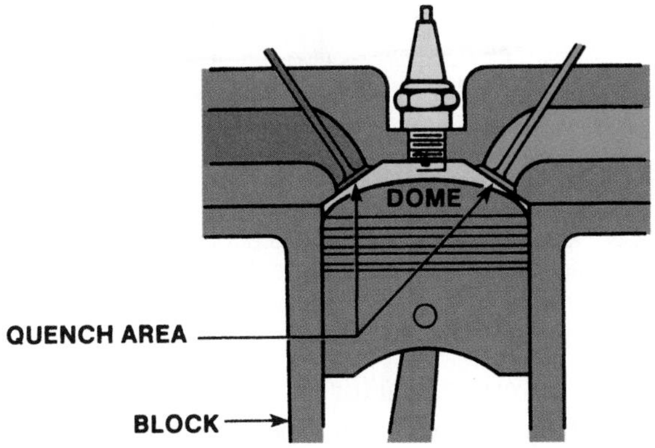

Figure 11-6 Domed pistons improve turbulence by producing a quench area.

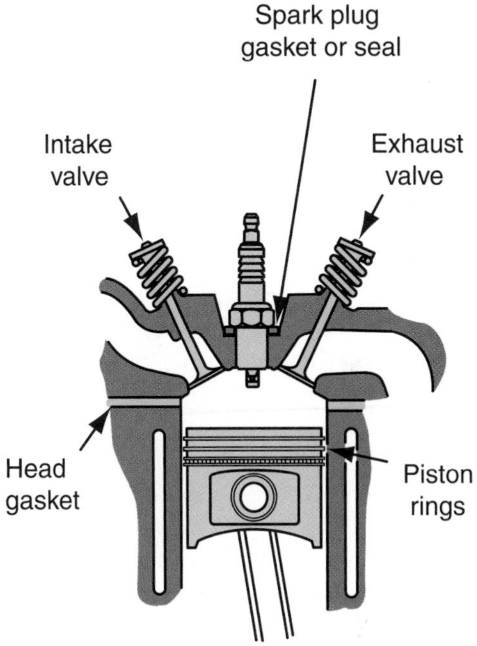

Figure 11-7 The sealing points of a typical engine.

tent. This is called **shrouding**. **Figure 11–5** shows the valve very close to the side of the combustion chamber, which causes the flow of air and fuel to be restricted. Volumetric efficiency is reduced. Hemispherical combustion chambers do not have this restriction. Hemispherical combustion chambers are used on many high-performance applications, especially when large quantities of air and fuel are needed in the cylinder.

Some engines use a dome piston. This type of piston has a quench area to improve turbulence **(Figure 11–6)**. Several variations of this design are used today.

INTAKE AND EXHAUST VALVES

Every cylinder of a four-stroke cycle engine contains at least one intake valve to permit the air/fuel mixture to enter the cylinder and one exhaust valve to allow the burned exhaust gases to escape. The intake and exhaust valves, along with the spark plug gasket and the cylinder head gasket, must also seal the combustion chamber **(Figure 11–7)**.

The type of valve used in automotive engines is called a **poppet**. This is derived from the popping action of the valve as it opens and closes. A poppet valve has a round head with a tapered face, a stem that is used to guide the valve, and a slot that is machined at the top of the stem for the valve spring retainers.

Most valves are made from a special hardened steel or stainless steel. The metals must be able to withstand high temperatures. Other metals are often used in the construction of high-performance valves. Titanium valves, for example, are less affected by combustion temperatures and weigh considerably less than steel valves. Currently ceramic valves are being tested for future use. Ceramic materials weigh less than half of what a comparable size steel valve weighs **(Figure 11–8)** and can withstand extreme temperatures without weakening or becoming deformed.

The head of the valve is the large diameter end and is used to seal the intake or exhaust port. This seal is made

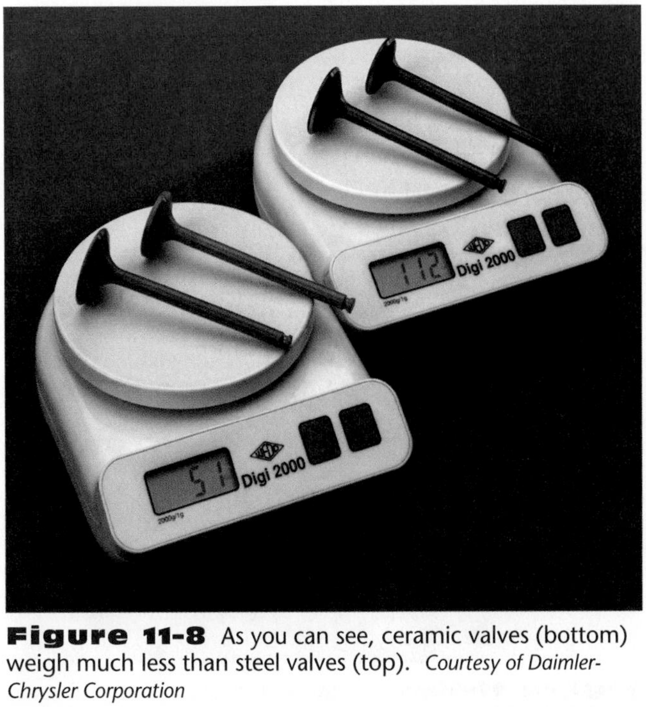

Figure 11-8 As you can see, ceramic valves (bottom) weigh much less than steel valves (top). *Courtesy of DaimlerChrysler Corporation*

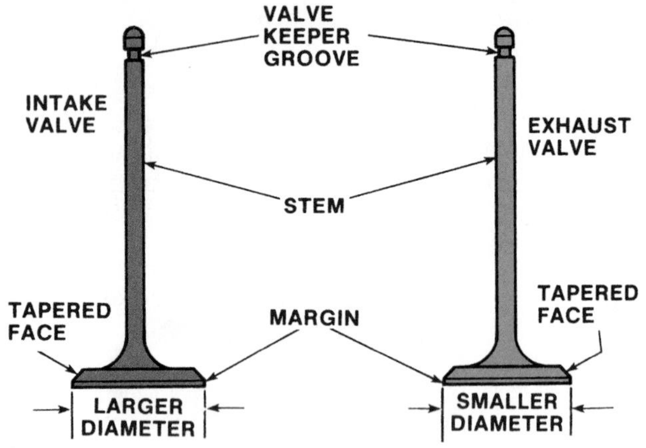

Figure 11-9 Valve nomenclature.

by the **valve face** contacting the valve seat. The valve face is the tapered area machined on the head of the valve. The angle of this taper is determined by the design and manufacturer of the engine. The taper will vary from one engine to another and may vary between intake and exhaust valves in the same engine. The area between the valve face and the head of the valve is called the **margin**.

The intake and exhaust valve heads are different diameters. The intake valve is the larger of the two **(Figure 11–9)**. The size or diameter of the valves is determined by the engine design. The exhaust valve does not need to be as large as the intake because the exhaust gases are pressurized and the intake charge is not, therefore the exhaust gases can exit easier.

As mentioned, the stem guides the valve during its up-and-down movement and connects the valve to its spring through its valve spring retainers and keepers. The stem rides in a guide that is either machined into the head (integral type) or pressed into the head (insert type) as a separate replaceable part.

The valve seat **(Figure 11–10)** is the area of the cylinder head contacted by the face of the valve. The seat may be machined in the head (integral type) or it may be pressed in (insert type) like the valve guide. The valves found in today's engines are highly heat resistant. Heat resistance is very critical for exhaust valves because they must withstand working temperatures between 1,500 and 4,000°F (815 and 2,204°C).

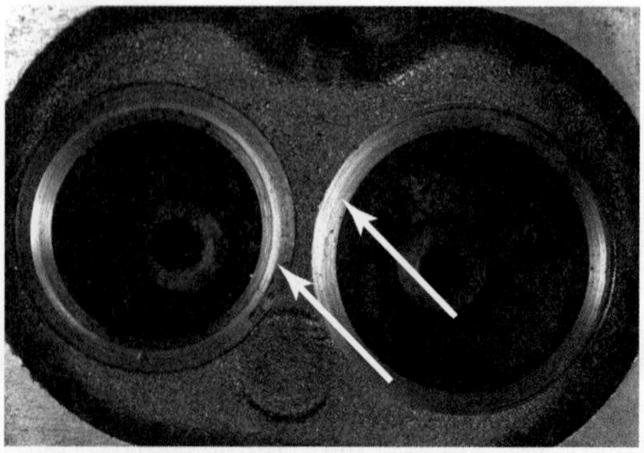

Figure 11-10 Valve seats.

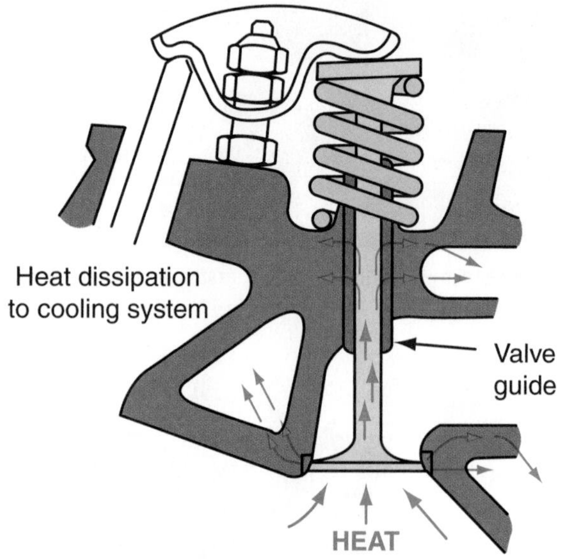

Figure 11-11 Exhaust valves cool by transferring heat to the liquid passages in the cylinder head.

There are two ways for the exhaust valve to cool. First, when the valve face is in contact with its seat, the heat from the valve will be transferred to the cylinder head, which is liquid cooled. The second is through the valve stem to the valve guide and again to the cylinder head **(Figure 11–11)**. To aid in this second method of heat transfer, some exhaust valve stems are hollow. This hollow section is filled with sodium. Sodium is a silver-white alkaline metallic chemical element that transfers heat much better than steel.

WARNING!

Never cut open any sodium-filled valves. Sodium will burn violently when it contacts water.

The valve seat area must be hard enough to withstand the constant closing of the valve and supply good heat transfer. Due to corrosive products found in the exhaust gas, the seats must be highly resistant to corrosion. When the cylinder head material meets these requirements, the seats are machined directly into it. When it does not, the seats are then made of material that will meet the requirements and the seats are pressed into the head.

Important Valve Components of Four-Stroke Engines

Valve Guides Valve guides are the parts that support the valves in the head. They are machined to a fit of a few thousandths of an inch clearance with the valve stem.

This close clearance is important for the following reasons:

- It keeps the engine's oil from being drawn into the combustion chamber past the intake valve stem during the intake stroke and oil leaking out to the exhaust port during times when the pressure in the exhaust port is lower than the pressure in the crankcase.
- It keeps exhaust gases from leaking into the crankcase area past the exhaust valve stems during the exhaust stroke.
- It keeps the valve face in perfect alignment with the valve seat.

Valve guides can be cast integrally with the head **(Figure 11–12A)**, or they can be removable **(Figure 11–12B)**. Removable valve guides usually are press-fit into the head.

Valve Springs, Retainers, and Seals The valve assembly is completed by the spring, retainer, and seal **(Figure 11–13)**. Before the spring and the retainer fit into place, a seal is placed over the valve stem. The seal acts like an umbrella to keep oil from running down the valve stem and into the combustion chamber. The spring, which keeps the valve in a normally closed position, is held in place by the retainer. The retainer locks onto the valve stem with two wedge-shaped parts that are called **valve keepers**. Some engines utilize a single valve spring per valve. Others use two or three springs. Often the second or third spring is a flat spring called a **damper spring**, which is designed to control vibrations.

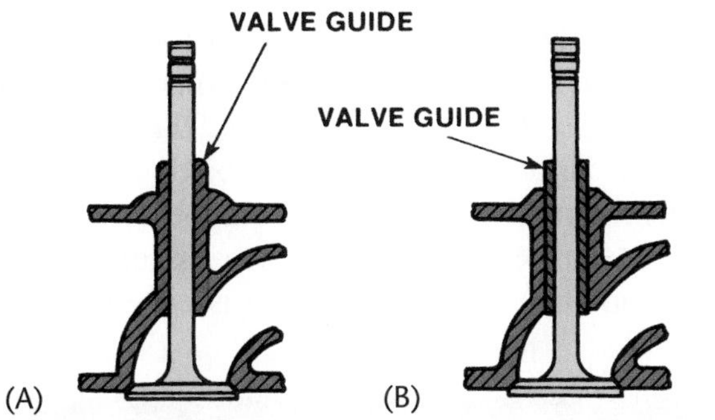

Figure 11-12 (A) Integral and (B) removable valve guides.

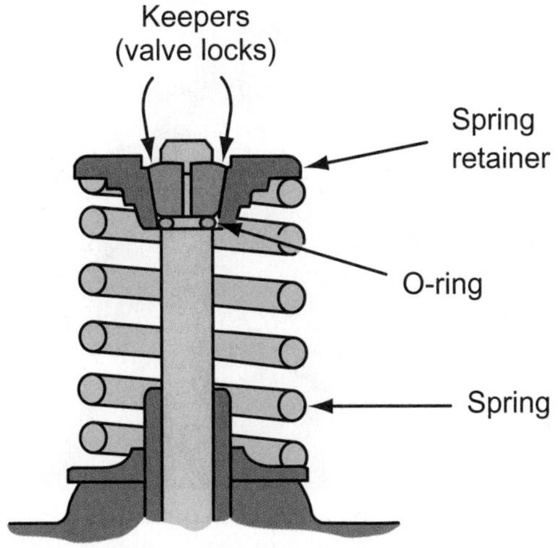

Figure 11-13 A valve assembly with spring, retainer, seals, and keepers.

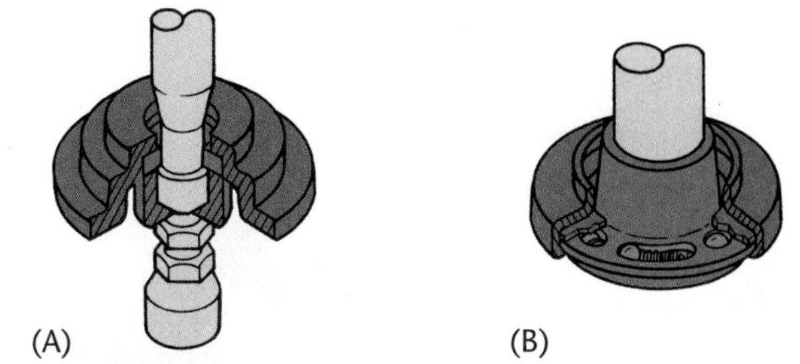

Figure 11-14 (A) Release and (B) positive valve rotators.

Valve Rotators Many engines are equipped with mechanisms that cause the exhaust valves to rotate **(Figure 11–14)**. Their purpose is to keep carbon from building up between the valve face and seat. Carbon buildup can hold the valve partially open, causing it to burn.

Camshaft Bearings The camshaft is part of the cylinder head assembly in all OHC-type engines. The unit that holds the camshaft(s) may be a separate unit bolted to the cylinder head, or the bore for the camshaft is machined into the upper part of the head. The most common design is similar to that of the crankshaft and main bearings. These cylinder heads are machined to accept one or two camshafts above the valves and have caps that secure the camshaft **(Figure 11–15)**. The camshafts are supported by split bearings. When the camshaft is held in a single structure, the camshaft is supported by one-piece insert bearings pressed into the camshaft bore.

Pushrods Pushrods are designed to be the connecting link between the rocker arm and the valve lifter; they are either solid or hollow. The pushrod fits between the valve lifter and the rocker arm to transmit cam action to the valves. Hollow pushrods allow oil to pass from the hydraulic lifter to the rocker arm assembly **(Figure 11–16)**.

Pushrod Guide Plates On some engines, pushrod **guide plates** are used to limit the side movement of the pushrods. The plates hold the pushrods in alignment with the rocker arms. When the pushrods pass through holes in the cylinder head or intake manifold, guide plates are not needed. This is true only when the holes for the pushrods are small enough to limit the sideways movement of the pushrod.

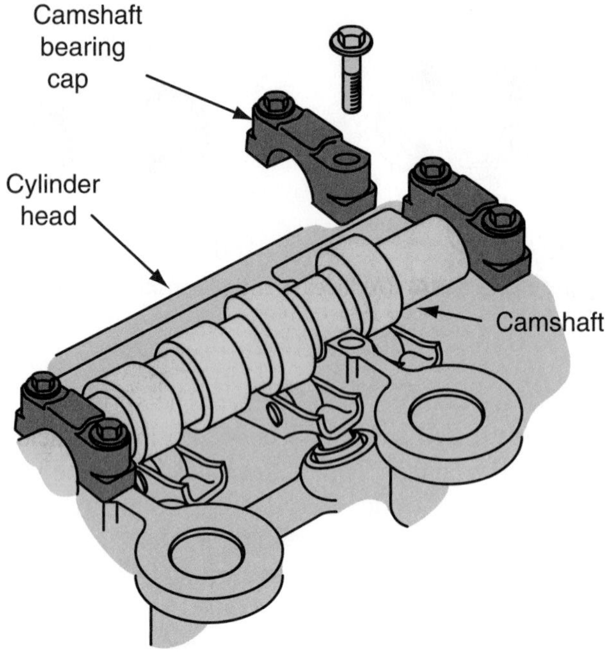

Figure 11-15 Many OHC cylinder heads are machined to accept one or two camshafts above the valves and have bearing caps that secure the camshaft.

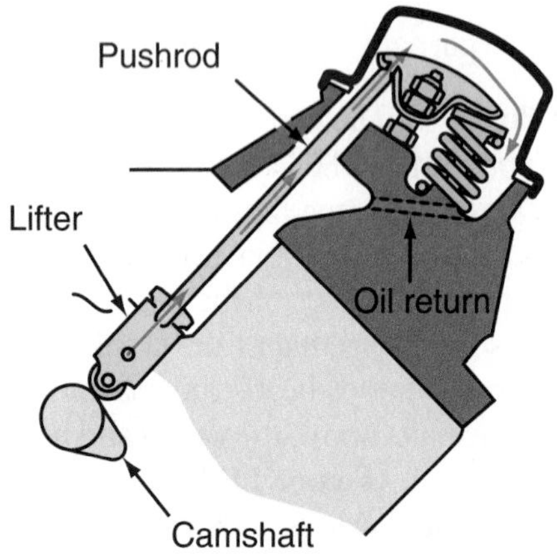

Figure 11-16 Most pushrods have a hole in the center to allow oil to pass from the hydraulic lifter to the rocker arm assembly.

Rocker Arms Rocker arms are designed to do two things: They change the direction of the cam's lifting force and they provide a certain mechanical advantage during valve lifting. As the lifter and pushrod move upward, the rocker arm pivots at the center point, causing a change in direction on the valve side. This change in direction causes the valve to move downward and open. Rocker arms also permit valves to be angled.

On some engines, the valve will open more than the actual lift of the cam lobe. This is done by changing the distance from the pivot point to the ends of the rocker arm. The difference in length from the valve end of the rocker arm and the center of the pivot point (shaft or stud) compared to the pushrod or cam end of the rocker arm and the pivot point (shaft or stud) is expressed as a ratio. Usually, rocker arm ratios range from 1:1 to 1:1.75. A ratio larger than 1:1 results in the valve opening farther than the actual lift of the cam lobe.

Rocker arms are designed and mounted in several ways. Springs, washers, individual rocker arms, and bolts are used in this type of assembly. Other rocker arms are placed on studs that are mounted directly in the cylinder head.

Some overhead camshaft engines use rocker arms in such a way that the camshaft rides directly on top of the rocker arm. One end of the rocker arm fits over a cam follower or lifter and the other end is directly over the valve stem **(Figure 11–17)**. Often OHC cylinder heads have a complex arrangement of rocker arms and pushrods **(Figure 11–18)**. Other overhead camshaft engines do not use rocker arms and the camshaft rides directly on top of the valves.

Rocker arms are made of stamped steel, cast aluminum, or cast iron. Cast adjustable rocker arms are attached to a rocker arm shaft that is mounted on the head by rocker arm brackets. Although a cast rocker arm can be resurfaced, a worn, stamped nonadjustable rocker arm must be replaced.

Cast-iron rockers are used in large, low-speed engines. They normally pivot on a common shaft. Aluminum rockers are generally used on high-performance applications and often pivot on needle bearings. The use of needle bearings reduces the friction at the pivot.

Most domestic engines are equipped with an independent stamped steel rocker arm assembly for each valve, mounted to a stud, which is either pressed or

Figure 11-17 The rocker arm moves with the lobes of the camshaft, and the rocker arm's movement is dampened by a spring assembly mounted next to the rocker arm. *Courtesy of DaimlerChrysler Corporation*

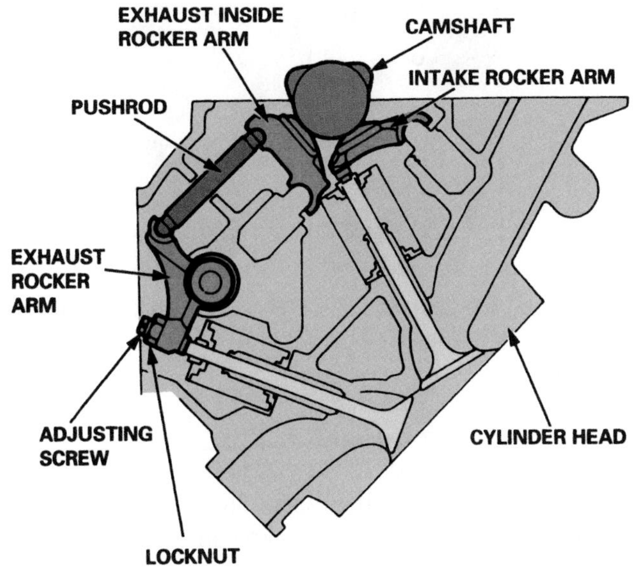

Figure 11–18 A camshaft, rocker arm, and pushrod assembly for an OHC engine. *Courtesy of American Honda Motor Co., Inc.*

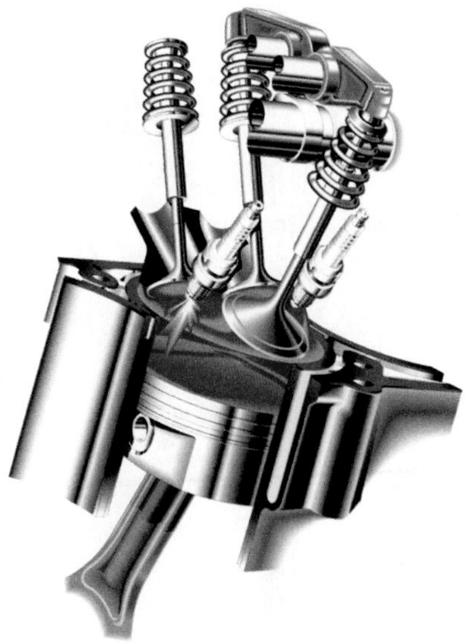

Figure 11–19 An engine with two spark plugs, two intake valves, and one exhaust valve for each cylinder. *Courtesy of Mercedes-Benz of N. A., Inc.*

threaded into the cylinder head. On some engines, the studs are drilled to serve as an oil passage to the rocker arms. Make sure oil can pass through before installing the cylinder head on the block. Replacement press-in studs are available in standard sizes to replace damaged or worn studs and oversized to replace loose studs.

Multivalve Engines

Many newer engines use multivalve arrangements. Automotive engineers have long been obsessed with the idea of additional valves in the cylinder head. It all started in 1918 with the dual-valve Pierce Arrow, which was one of the first cars to use four valves per cylinder as a way to enhance gas flow and increase horsepower.

The basic idea behind using more than one intake and/or exhaust valve is simple—better efficiency. To improve efficiency, engineers need to improve the flow into the combustion chamber and the flow of the gases out of the chamber, which can be done in a number of ways. In the past, the common way was to make the valves larger and to change valve timing. Today, many engines use multiple valves and variable cam timing to improve the efficiency. Larger valves allowed more air in and more exhaust out, but the bigger valves weighed more than smaller ones and therefore required stronger springs to close them. The stronger springs held the valves tighter when they were closed but required more engine power to open them. This fact somewhat diminished the gains of using a larger valve. Also when the engine runs at low speeds, the air moving past a large valve has a lower velocity than it would have if it flowed past a small valve, which reduces engine torque at low engine speeds.

Although two small valves weigh as much or more than one valve, each valve weighs less and therefore the spring tension on each is less, which means the power required to open the valves is also less. Therefore, the gain in efficiency is not offset by the power required to open the valves and the net gain from the engine is realized. Also the velocity of the air in and out at low engine speeds is quicker than it would be with large valves.

Today, multivalve engines can have three valves per cylinder **(Figure 11–19)**, four valves per cylinder **(Figure 11–20)**, or five valves per cylinder **(Figure 11–21)**. The most common arrangement is four valves per cylinder with two intake and two exhaust valves. The different arrangements result from different manufacturer priorities and other features of the engine. In all multivalve engines, the heads are of the cross-flow design.

Using two intake and one or two exhaust valves increases the volume of the intake and exhaust ports. More mixture can enter the cylinders. Thus, multivalve engines have a more complete combustion, which reduces the chances of misfire and detonation. This results in

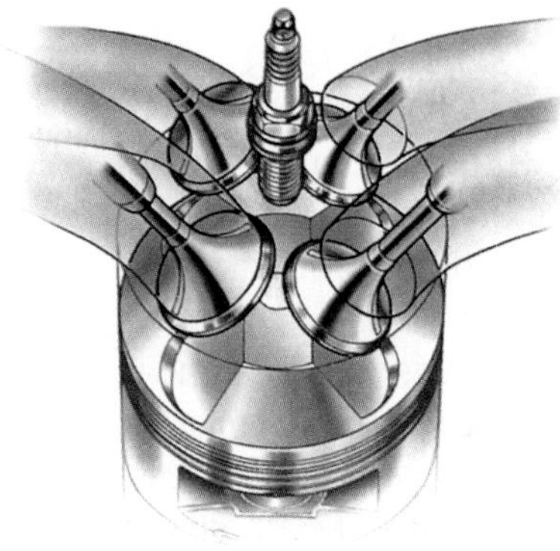

Figure 11–20 A typical layout for a cylinder with four valves. *Courtesy of American Honda Motor Co., Inc.*

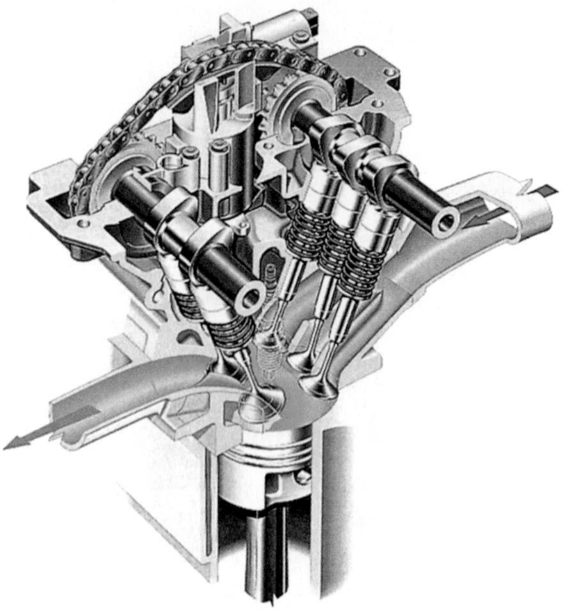

Figure 11-21 This five valve per cylinder arrangement has three intake valves and two exhaust valves. *Courtesy of Audi of America, Inc.*

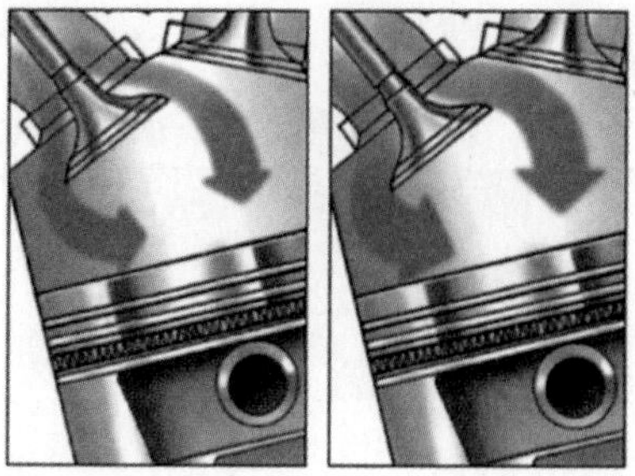

Figure 11-22 Honda's VTEC system changes valve timing and valve lift according to need. *Courtesy of American Honda Motor Co., Inc.*

enhanced fuel efficiency, cleaner exhaust, and increased power output. The velocity of the intake air is higher with small multiple ports than a large single passage. The smaller valves naturally have less mass than big ones, so mechanical inertia is reduced, making a higher engine speed possible before valve float occurs. The more times the cylinder can be filled and evacuated per second, the more horsepower can be obtained.

Because increased air velocity is the main benefit of a multivalve cylinder head, the technology works best at high engine speeds.

The benefits of multivalves are offset to some extent, however, by a more complicated camshaft arrangement. The easiest way to actuate four valves per cylinder is with dual overhead camshafts. These are sometimes difficult to lubricate. The cam drive is even more complicated with V-power plants. Many use a single overhead cam per cylinder bank, with some kind of lever arm actuating the opposite bank of valves.

Variable Valve Timing

How much and when it is best for the valves to open and close varies with engine rpm and the load on the engine. Normally the timing and lift of the valves are controlled by fixed lobes on the camshaft. Engineers design these lobes to meet the engine's needs at the anticipated normal operating speeds and loads and to achieve maximum fuel economy and minimum emission levels during those operating conditions. Basically, power output is usually sacrificed for the sake of economy and emissions levels.

Manufacturers seeking improved driveability and lower fuel consumption and emission levels have designed several methods to vary valve timing and lift in response to the needs of the engine. Many of the systems only vary the timing of the intake camshaft; some vary the timing of the intake and exhaust valves, and a few vary the timing and lift of all the engine's valves **(Figure 11–22)**. Obviously, total control of the valves produces the best results, but these systems are expensive to build and are quite complex.

When cam timing is altered, it is adjusted in its relationship to the crankshaft. At low engine speeds, the intake camshaft is retarded, causing the valves to open later. At moderate speeds, the intake camshaft is advanced and the valves open early. The camshaft is typically retarded again during high-speed operations.

One design uses a hydraulic plunger controlled by the engine's computer to move against the guides for the intake camshaft belt or chain drive. By changing the path of the drive, the intake cam can be advanced for high-speed operation and retarded to normal timing when the engine is at low speeds.

Another design uses pressurized oil to change valve opening, allowing the engine to operate as a three-valve engine at low speeds. At high speeds, the engine operates on four valves, which allows for quicker delivery of, and larger amounts of, intake charge. Below 2,500 rpm, each intake valve follows a separate camshaft lobe. The primary valve opens and closes normally, while the secondary intake opens just enough to keep the engine running smoothly. As the engine reaches 2,500 rpm, pressurized engine oil moves small pins to lock each pair of rocker arms together, causing both intake valves to follow the normal cam lobe. When the engine slows down, the pressurized oil is bled off and the pin releases, separating the two rocker arms.

Another design shifts the phase of the intake cam lobes independently of the exhaust cam lobes and changes the timing of the intake. The exhaust lobes are machined on the outside of the camshaft, while the intake lobes are mated to an internally splined shaft by toothed insets. At the end of the camshaft, near the drive pulley, a phasing mechanism takes signals from the PCM and shifts the

Figure 11-23 The controller/actuator assembly on the intake camshaft of the Lexus VVT-I engine. *Courtesy of Lexus of America Co.*

Figure 11-24 A cutaway of an engine showing cam phasers on both the intake and exhaust camshafts. *Courtesy of BMW of North America, Incorporated*

lobes in accordance with rpm. The phaser can vary the lobes continuously through all driving conditions, boosting fuel economy and driveability.

Phasers can be electronically or hydraulically controlled. In a hydraulically controlled system, the camshafts are controlled through the electronics of the PCM. The system uses a controller/actuator installed on the camshaft **(Figure 11–23)**. The controller contains a vane assembly that is moved by hydraulic pressure. A solenoid controls the hydraulic pressure in the assembly. The solenoid is pulsed by the PCM. The movement of the vane allows the camshaft to rotate on its drive gear, advancing or retarding the valve timing instantaneously. Changes are made continuously as the engine runs.

Some systems adjust a phaser with a solenoid, which controls the flow of pressurized engine oil into a chamber in a two-piece hub in the camshaft sprocket. When the computer closes the valve, oil pressure builds and pushes a geared piston that moves the camshaft to advance valve timing. Although valve lift and overlap remain the same, more power is produced at high engine speeds. When the hub is in the normal position, the engine has a smooth idle and low emissions. On a few engines, phasers are connected to both the intake and exhaust camshafts **(Figure 11–24),** therefore altering the timing of both. By altering the timing of both the intake and exhaust valves, valve overlap is also changed.

A unique setup for controlling intake and exhaust valve timing and lift relies on a conventional camshaft ground for high performance with ultra-high-speed valves to bleed off the fluid in the hydraulic lifters. Based on inputs to the PCM, the fluid pressure in the lifter is changed to delay valve openings, change the amount of time the valve is opened, or even prevent valves from opening at all, which makes it possible to shut down cylinders until more power is needed. The pressure in the lifters is varied within milliseconds to control valve lift and timing during a single rotation of the camshaft. Solenoids are used to control the flow of oil into a piston in each lifter; this effectively determines the tappet's height. At low engine speeds, it can bleed off pressure inside the tappet to prevent all of the camshaft lobe's action from being transferred to the valve. When more power is called for, full oil pressure is applied to the tappet, all of the motion of the lobe is passed along to the valve, and the resultant high-lift, long-duration valve action boosts output.

This basic system can have a different twist. To control the action of the lifters, two spiral-shaped cams are fitted to act on a sliding prism on top of the lifters. By changing the phase between the two cams, the amount of time and the distance that a valve is opened can be instantly and smoothly changed.

Another system operates on an entirely different principle. PCM-controlled solenoids are used to bring rocker arms into contact with a second set of high-performance camshaft lobes. At low speed, the valves are opened and closed by the standard lobes on the camshaft. At high speeds, the solenoids are activated, raising the rocker arms slightly and bringing them into contact with a second set of lobes on the camshaft. These lobes are machined to increase valve lift, duration, and overlap. The result is more power.

One multivalve engine design has groups of three lobes for each pair of valves. The center lobe of the three is shaped for more valve lift and different open and close times. All three lobes actuate rocker arms, but at lower speeds only the outer lobes' rocker arms operate the valves. At high speed, a solenoid valve opens to allow pressurized engine oil to flow through the rocker shaft to a piston in the outer rocker, pushing it partly into the center rocker arm. This forces the center rocker arm's piston partly into the other outer rocker arm, locking the three rockers together. The valves now open according to the shape of the center lobe. When the solenoid valve closes, a spring pushes back the pistons in the rockers, and the engine runs with normal valve timing. This system allows the engine to have better low-end torque and lower emissions, plus better breathing at high speeds.

Soon valves will be controlled by magnetic forces instead of camshaft lobes and springs. A computer will control and pulse electric solenoids, attached to each valve, in response to all of the inputs it receives. With this system, each valve can be individually controlled to ensure the best valve timing and lift for the conditions that exist in a particular cylinder. The result of such a system would be drastically improved power outputs and increased engine speeds, in addition to lower fuel consumption and emission levels.

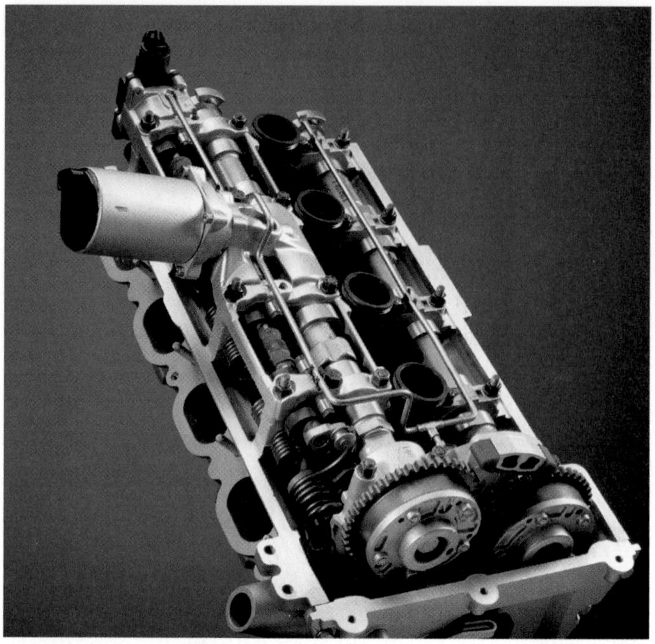

Figure 11-25 The secondary eccentric shaft and stepper motor assembly for the Valvetronic system. *Courtesy of BMW of North America, Incorporated*

Valvetronic System Valvetronic is a system used by BMW, along with variable intake and exhaust valve timing, to regulate the air flow into the cylinders by controlling valve lift. By doing this, the engine has no need for a throttle plate. In fact, this feature is one of the biggest advantages of the system. A throttle plate has a tendency to rob the engine of power, especially at low engine speeds.

The amount of air required to support combustion in a cylinder varies according to the load on the engine at that particular moment. Obviously, when the driver pushes on the throttle pedal, more power is requested. To provide more power, the throttle allows more air to enter the cylinders.

In conventional engines, the throttle plate regulates the flow of incoming air, while the lift and open duration of the intake valve remains constant. During low speeds, the throttle plate is nearly closed and blocks most of the air available for the cylinders. Although this is how the engine remains at a low speed, it also results in a pumping loss. Pumping loss is a term used to describe the difficulty a piston has in moving air into the cylinder as it moves down on its intake stroke. Pumping losses are a major reason why conventional engines consume more fuel in city driving.

Valvetronic solves this problem by controlling the flow of incoming air directly, at the intake valves. This system uses a conventional camshaft with a secondary eccentric shaft and a series of levers and roller followers that are activated by a stepper motor **(Figure 11–25)**. A computer changes the phase of the eccentric cam to change the action of the valves.

At high engine speeds, the system dials in maximum lift, opening the ports for maximum flow to guarantee rapid filling of the combustion chamber **(Figure 11–26)**. At low engine speeds, the system reverts to minimal valve

Figure 11-26 The position of the eccentric shaft to provide maximum lift in a Valvetronic system. *Courtesy of BMW of North America, Incorporated*

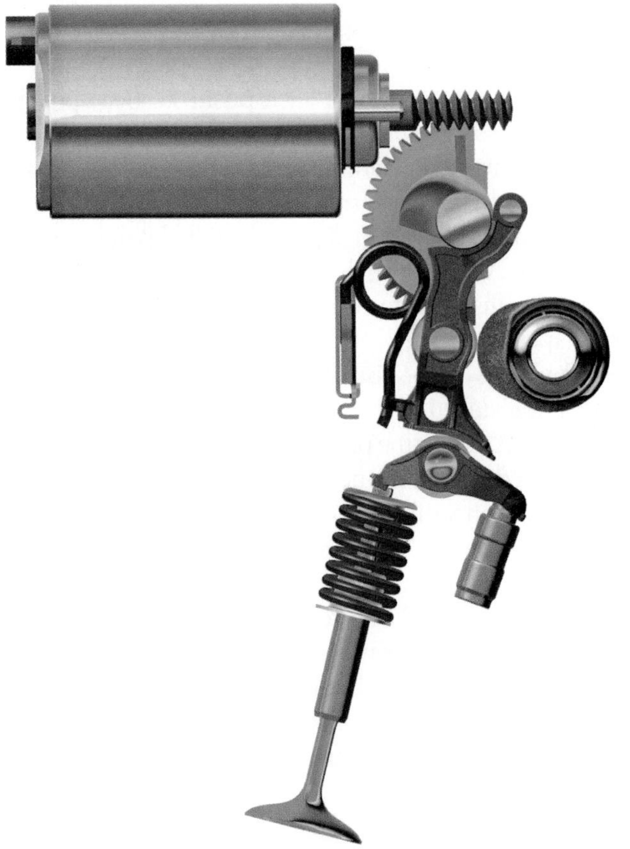

Figure 11-27 The position of the eccentric shaft to provide minimum lift in a Valvetronic system. *Courtesy of BMW of North America, Incorporated*

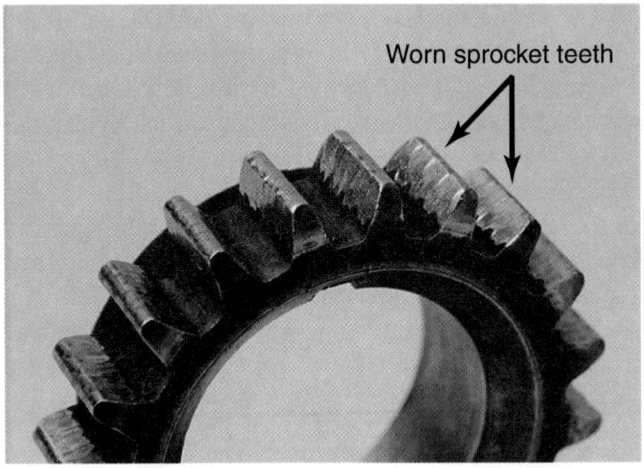

Figure 11-28 Worn teeth on a timing sprocket.

lift **(Figure 11–27)**. This reduces the amount of air/fuel mixture entering the cylinder.

INSPECTION OF CYLINDER HEAD AND VALVE TRAIN

A valve will operate only as well as its actuating parts allow. When inspecting the valve train, each part should be carefully checked. Use the following guidelines when inspecting the components.

Timing Components

The timing belt (or chain) and crankshaft/camshaft sprockets should be inspected or replaced if damaged or worn **(Figure 11–28)**. Cracks, spalling, or excessive wear on the tooth surface of a gear is an indication of improper backlash. Stripped/broken rubber belt failure is often due to insufficient tensioning, extended service life, abusive operation, or worn tensioners. Most manufacturers recommend that a timing belt be replaced every 60,000 miles. Replacing them at this interval can prevent belt breakage. Loose timing belts will jump across the teeth of the timing sprockets, causing shearing of the belt teeth. Localized tensile overloads from overrevving an engine can lead to belt breakage. Also, belts on those engines equipped with adjustable tensioners should be checked for wear whenever a belt is retensioned. Also check the condition and operation of the tensioner. In addition, check for cord separation and cracks on all surfaces **(Figure 11–29)**. If the belts are damaged, they should be replaced. It is wise to replace the belt any time the engine is overhauled.

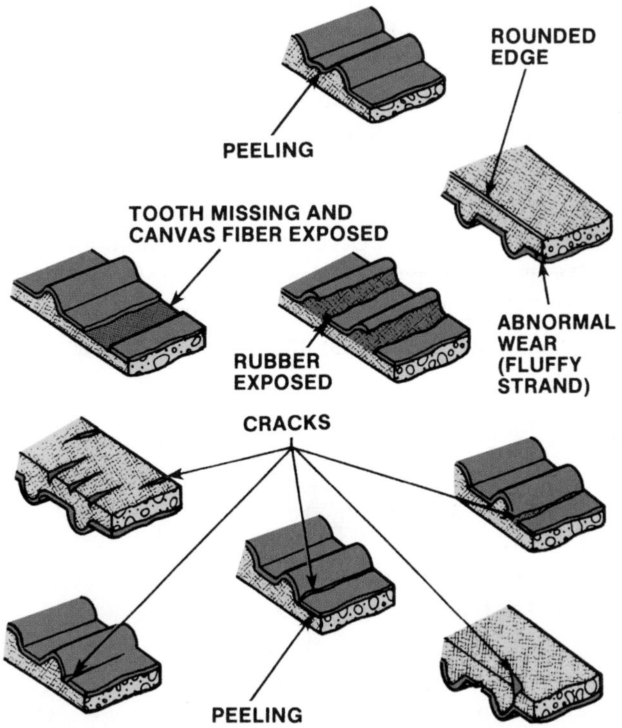

Figure 11-29 Various forms of timing belt wear.

On many engines, severe engine damage will result from a broken timing belt. When a timing belt breaks, the camshaft no longer turns with the crankshaft. As the camshaft slows, the valves may open while the piston is moving to TDC. Because the intake valves are larger than the exhaust valves, they are more likely to hit the top of the pistons, causing damaged pistons and/or bent or broken valves. Therefore, it is absolutely necessary that the belt be replaced according to the manufacturers' recommended mileage intervals and whenever it is removed from the engine.

SHOP TALK

The condition of a timing belt can be checked by pressing your fingernail into the hardened backside of the belt. If no impression is left, the belt is too hard and should be replaced. ■

Rocker Arms

Inspect the rocker shaft assembly for wear, especially at points that contact the valve stem and pushrod. The fit between a cast rocker arm and a rocker shaft is checked by measuring the outside diameter of the shaft and comparing it to the inside diameter of the rocker arm. Excessive clearance requires replacement of the rocker arm or the rocker shaft, or both. Another wear point that should be checked is the pivot area of the rocker arm. Also check for loose mounting studs and nuts or bolts. Other rocker arm wear points are shown in **Figure 11–30**.

Excessive wear on the valve pad occurs when the rocker arm repeatedly strikes the valve tip in a hammer-like fashion. It strikes the valve tip in this way when valve train clearance, or lash, is excessive. Excessive valve lash can occur in several ways. For example, it occurs when mechanical lifters are not adjusted properly or when hydraulic lifters are not working properly. In addition, worn rocker arm valve pads can result from insufficient lubrication. Proper lubrication transfers heat away from the valve pad and reduces the metal-to-metal contact. Also make sure the oil feed in each rocker arm is clear and not plugged with dirt.

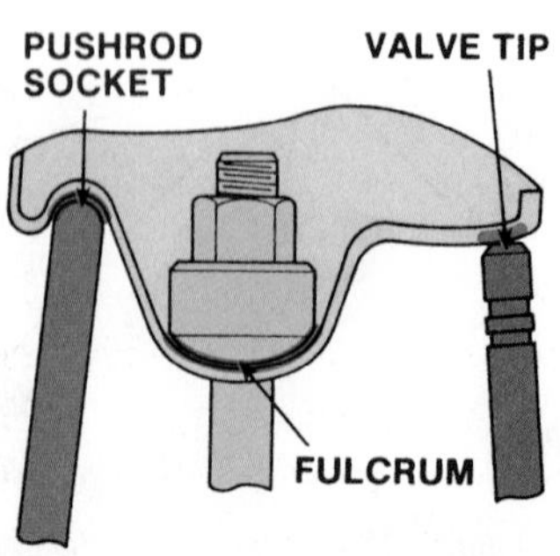

Figure 11-30 Rocker arm wear spots.

Pushrods

Check the straightness of each pushrod. Bent pushrods can be caused by sticking valves and improper valve timing or adjustment. Bent or broken pushrods indicate interference in the valve train. Common causes are the use of incorrect valve springs or an installed height less than specified. Also, insufficient valve-to-piston clearance can cause a collision between the valve and piston at high engine speeds.

Pushrods can be visually checked for straightness while they are installed in the engine by rotating them with the valve closed. With the pushrods out of the engine, they can be checked for straightness by rolling them over a flat surface such as a surface plate. If a pushrod is not straight, it will appear to hop as it is rolled. However, the most accurate way to check for straightness is by using a dial indicator. If more than 0.003 total indicated reading (TIR) is found, the pushrod is not straight and should be replaced.

Hollow pushrods should be looked through to make sure there are no blockages in the rod. No oil will get to the rocker arms if these pushrods are plugged.

The ends of the pushrods should be checked for nicks, grooves, roughness, or signs of excessive wear. Replace any damaged or worn pushrod.

When replacing pushrods, make sure they are the correct length for the application. Also, keep in mind that pushrods that are held in alignment by slots in the cylinder head or guide plates must be hardened.

Cam Followers and Lash Adjusters

Overhead cam follower arm and lash adjuster assemblies should be carefully checked for broken or severely damaged parts. If pads are used to adjust the valve lash, the cups and the shim pads must be carefully checked. A soft shim will not hold the valve at its correct lash, and therefore, each shim should be checked to make sure it is hard. To do this, place the shim on the base circle of the camshaft and press down on the shim with your hand. You should feel no give.

Retainers and Keepers

Valve spring retainers and valve keepers hold the valve spring and valve in place. A worn retainer will allow the spring to move away from the centerline of the valve. This will affect valve operation because spring tension on the valve will not be evenly distributed. Each retainer should be carefully inspected for cracks, because a cracked retainer may result in serious damage to the engine. The inside shape of most retainers is a cone that matches the outside shape of the keepers. This must be a good fit in order for the keepers to stay in their grooves on the valve stem.

The valve stem grooves should match the inside shape of the keepers. Some valves have multiple keeper grooves. Others have only one. All of the valve stem grooves should be inspected for damage and fit by inserting a keeper in them. Both the retainers and keepers should be carefully inspected for damage. If a defect is found, they should be replaced.

Valve Rotators

Most rotators impart positive rotation to the valve during each valve cycle and improve valve life two to five times, in some cases even more. In normal operation, rotators will continue to function for more than 100,000 miles and require no maintenance. However, when valves are refaced or replaced at high mileage, the rotators should be replaced because they cannot be visually inspected accurately. Whether or not they rotate when held in the hand is no indication of their function in the engine. While rotation can only be checked in a running engine, uneven wear patterns develop at the valve stem tip if the rotators are not functioning properly.

Valve Springs

Valve spring assemblies, including the damper springs, should be checked for signs of cracks, breaks, and damage.

Cylinder Heads

Cylinder heads should be carefully inspected after they are cleaned. Any severe damage to the sealing areas indicates the head should be replaced. Also use the appropriate method for detecting cracks. Make sure the cracks are properly repaired or replace the head. Also check the heads for dents, scratches, and corrosion around water passages, especially on aluminum heads.

Camshaft and Bearings

The old camshaft bearings should be inspected for signs of unusual wear that may indicate an oiling or bore alignment problem. On some engines, the bores in the aluminum casting serve as a bearing surface; therefore, the bore should be carefully examined.

Each lobe of the camshaft should be checked for wear, scoring, scuffing, fractured surface, pitting, and signs of abnormal wear. Also check for plugged oil passages.

The camshaft should also be checked for straightness. If the engine has a worn or damaged camshaft, identify the cause of the damage and fix the problem before installing a new camshaft or followers.

Valves

Each valve face should be carefully checked for evidence of burning **(Figure 11–31)** and each stem should be checked for wear **(Figure 11–32)**. Also check the stem for signs of distortion and excessive wear **(Figure 11–33)**. Replace any valves that are badly burned, worn, or bent.

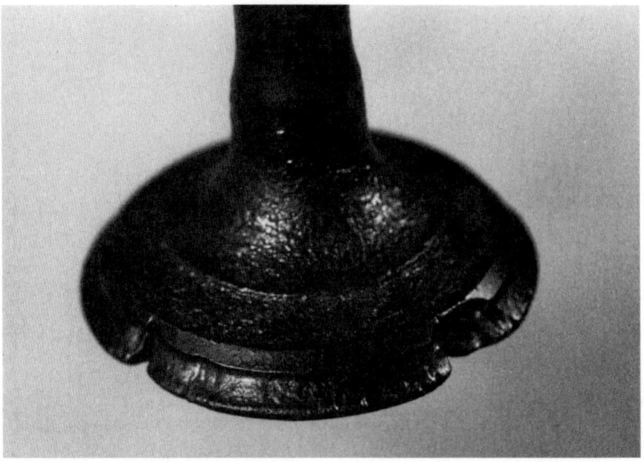

Figure 11-31 A severely burnt valve.

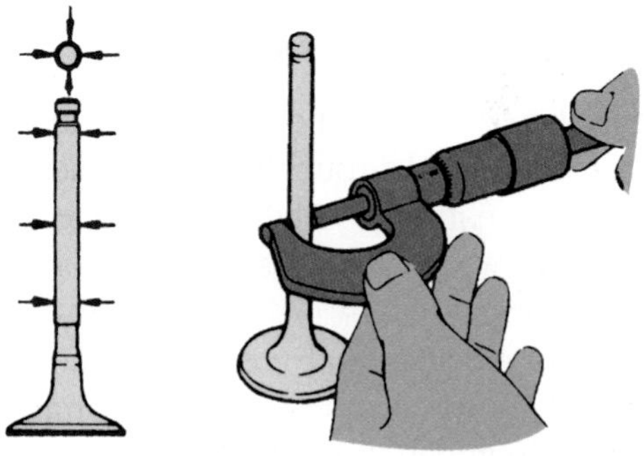

Figure 11-32 Checking a valve stem for wear.

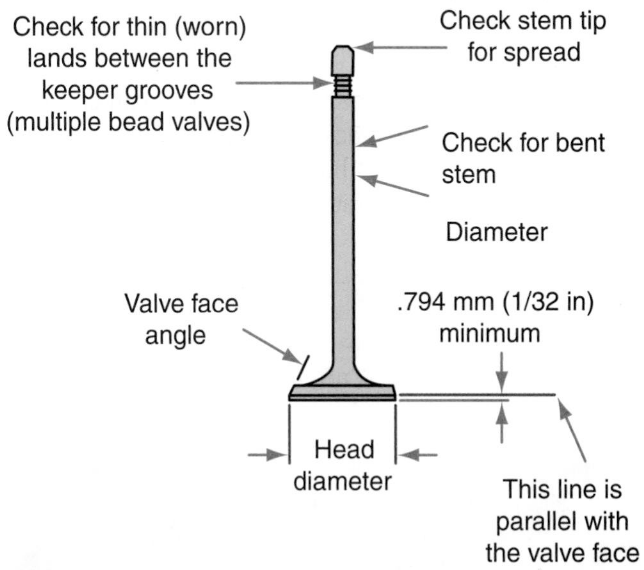

Figure 11-33 Parts of a valve that should be checked during your inspection.

Reusable valves are cleaned by soaking them in solvent, which will soften the carbon deposits **(Figure 11–34)**. The deposits are then removed with a wire buffing wheel. Once the deposits are removed, the valve can be resurfaced.

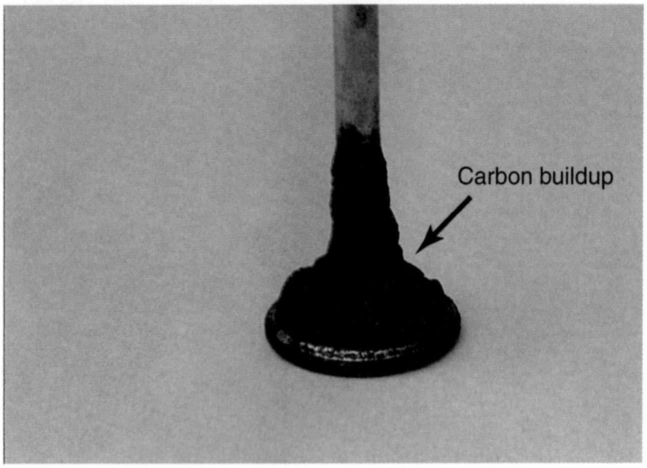

Figure 11-34 Carbon buildup on the back of a valve.

ALUMINUM CYLINDER HEADS

Aluminum heads are commonly used because a typical aluminum head weighs roughly half as much as a cast-iron head. Eliminating anywhere from 20 to 40 pounds (9 to 18 kg) of weight is a plus for fuel economy, but it has its drawbacks.

Aluminum expands and contracts almost twice as much as cast iron in response to temperature changes. This creates a number of problems. When an aluminum head is mated to an iron block, the difference in thermal expansion between head and block creates a great deal of scrubbing stress on the head gasket. Unless the gasket is engineered to take such punishment, leakage and premature gasket failure can result.

Increased thermal expansion and stress can also lead to cracking. The most crack-prone areas in the head are usually the areas around the valve seats **(Figure 11–35)**. High combustion temperatures and the constant pounding of the valve against its seat often cause cracking between the intake and exhaust seats or just under the exhaust seat.

The differing rates of thermal expansion between aluminum and iron creates a lot of stress throughout the head. The head wants to expand in all directions at once as it heats up, but the head bolts keep it from going sideways or lengthwise. The only place left to go is up, so the head tends to bow up in the middle.

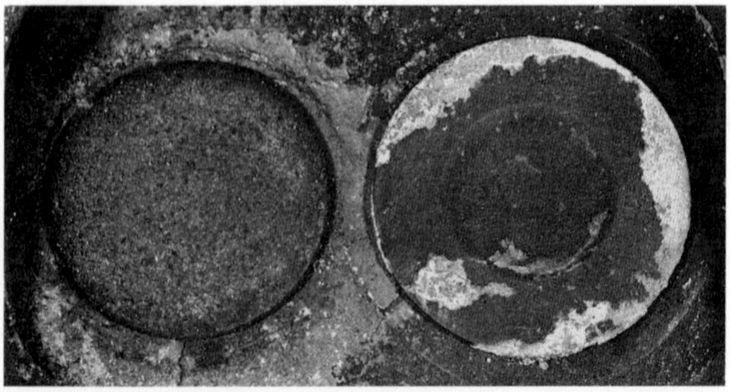

Figure 11-35 Lightweight aluminum heads are prone to cracking, which can lead to a recessed exhaust valve seat.

Aluminum is not as strong as cast iron. Consequently, the head provides less top end support for the block. This can allow more distortion in the upper cylinder bore area, affecting combustion sealing, blowby, and ring life. Using deck plates when boring the block can help minimize some of the distortion that will occur after the head is torqued down. Aluminum cylinder head bolts should never be loosened or tightened when the metal is hot. Doing so may cause the cylinder head to warp due to the torque changes.

Aluminum makes a fairly good bearing material. It is soft and provides good imbedability to foreign particles. But it lacks the support and rigidity of a conventional steel-backed bearing in an iron saddle. If the head overheats and warps, alignment through the cam bores is destroyed.

Aluminum has another drawback—**porosity**. The casting process sometimes leaves microscopic pores in the metal, which can weep oil or coolant. In most instances, the problem can be repaired.

Reconditioning Aluminum Cylinder Heads

Warpage in an aluminum cylinder head is usually the result of overheating (low coolant, uneven coolant circulation within the head, a too-lean fuel mixture, and incorrect ignition timing).

Alignment of the cam bores in an overhead cam head must be checked with either a straightedge and feeler gauge or with a dial indicator. If off by more than 0.002 to 0.003 inch (0.508 to .0762 mm), corrective action is required.

SHOP TALK

Although specifications for the maximum amount of cylinder head surface warpage vary, traditionally, the maximum acceptable limit for cast-iron heads is 0.005 inch (.1270 mm). Aluminum is not as forgiving, so 0.002 inch to 0.003 inch (.0508 mm to .0762 mm) is a more realistic upper limit. ■

Removing metal from the face of any OHV head also alters valve train geometry, which limits the amount of metal that can be removed.

Aluminum cylinder heads can be straightened through the use of heat and special clamping fixtures. Some manufacturers recommend that warped heads not be straightened but instead be replaced. Always follow the recommendations of the manufacturer.

RESURFACING CYLINDER HEADS

There are three reasons for resurfacing the deck surface of a cylinder head:

1. To make the surface flat so the gasket seals properly
2. To raise the compression ratio
3. To square the deck to the main bores

As engines undergo heating and cooling cycles over their life span, certain components tend to warp, especially cylinder heads. By using a precision straightedge or flatness bar and feeler gauge, the amount of warpage can be easily measured. Check the manufacturer's recommendations for the maximum allowable warpage for the particular engine you are working on.

The deck surface should be checked both across the head as well as lengthwise. Be sure to also check flatness of the intake and exhaust manifold mounting surface on the head **(Figure 11–36)**. In general, maximum deformation allowed here is 0.004 inch (.1016 mm).

Heads that are deformed beyond specifications must be surfaced. The finished surface, however, should not be too smooth. It must be rough enough to provide "bite," but not enough to cause a poor seal and leakage.

Surface Finish

No cylinder head surface, no matter how it appears, is perfectly smooth. When viewed in cross section, the surface consists of a series of peaks and valleys. A special instrument, called a profilometer is used to check surface roughness and to measure the distance between the peaks and valleys.

Resurfacing Machines For proper head gasket seating, the finish should consist of shallow scratches and small projections that allow for gasket support and sealing of voids.

Four different types of machines—belt surfacer, milling, broaching, and grinding—can be used in resurfacing operations.

WARNING!

Before attempting to operate any surfacing machine, be sure to become familiar with and follow all the cautions and warnings given in the machine's operation manual. Also, when operating these machines, you must wear safety glasses, goggles, or a face shield.

Belt Surfacers. Belt surfacers resemble belt sanders. These machines are easy to set up and operate. An operator merely places the part to be surfaced on the belt. A restraint rail helps keep the part positioned **(Figure 11–37)**. Some machines have air-operated holddown fixtures.

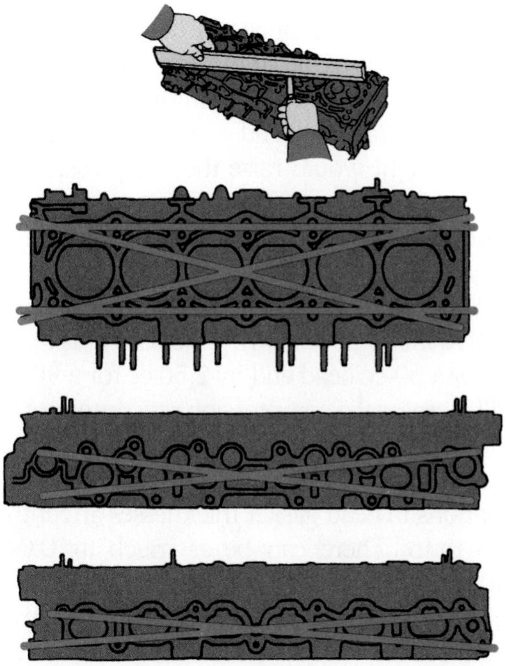

Figure 11–36 Checking a cylinder head for warpage.

(A)

(B)

Figure 11–37 (A) A typical belt surfacer and (B) how it is used.

Figure 11–38 Milling an aluminum cylinder head.

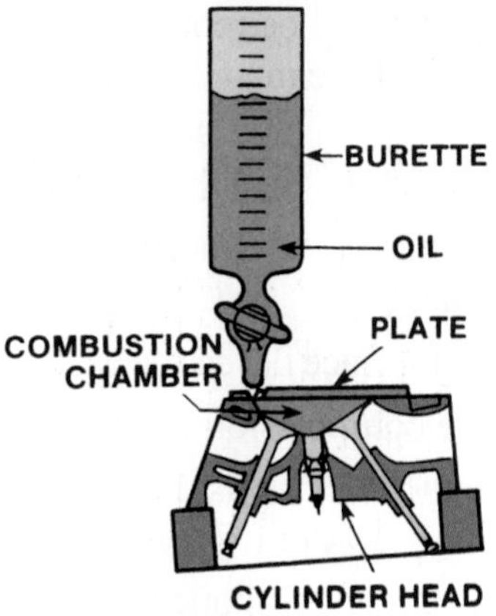

Figure 11–39 CC-ing a cylinder head to find the combustion volume.

Resurfacing quality depends on operator skill and factors such as belt condition, machine horsepower, and the holddown pressure applied.

Milling Machines. **Milling** machines cut away thin layers of metal to create a level, properly finished surface **(Figure 11–38)**. Cutters remove up to 0.050 inch (1.27 mm) per pass. Both rough and finish cuts are usually made to create the desired finish.

Broaching Machines. **Broaching** machines use an underside rotary cutter or broach. A block, cylinder head, or intake manifold is held in an inverted position as the broach passes underneath.

Surface Grinders. Surface Grinders. Surface grinders use a grinding wheel to remove metal stock. They set up and operate similarly to milling machines.

Stock Removal Guidelines The amount of stock removed from the head gasket surface must be limited. Excessive surfacing can lead to problems in the following areas.

Combustion Chamber. It might be necessary to measure and adjust the volume of an engine's combustion chambers. The combustion chamber is equal to the volume of the combustion chamber in the head plus the volume of the cylinder when the piston is at TDC.

Measuring this volume is called **cc-ing** the cylinder head. This is done with the valves and spark plugs installed. The cylinder head is mounted upside down, and a glass or plastic plate is installed over the combustion chamber. A graduated container called a burette is used to fill the combustion chamber with thin oil. The oil is poured through a hole in the plate, as shown in **Figure 11–39**. The amount of oil that enters the combustion chamber is equal to the volume (in cubic centimeters) for this cylinder.

The volume of a combustion chamber can be adjusted in several ways. It may be reduced by surfacing the cylinder head. This, of course, reduces the volumes for all of the chambers in that cylinder head. Individual volumes can be increased by grinding the valve seats to sink the valves and by grinding and polishing metal from the combustion chamber surface. Either method can be used to equalize all the chambers and adjust them to the manufacturer's specifications.

Compression Ratio. Combustion chamber volume directly affects an engine's compression ratio. Boring the block oversize changes the swept volume and the compression ratio. Generally speaking, the compression ratio increases at the same percentage the displacement increased. Boring an engine 0.060 inch (1.52 mm) oversize will increase the displacement 9 cubic inches (147.48 cc) or slightly more than 3%. Assuming the replacement piston has the same compression height as the original, the initial 9.0:1 compression ratio will be increased by 3.0% to 9.27:1. As a rule of thumb, there is a 2% increase at 0.030 inch (.7620 mm) oversize, a 4% increase at 0.060 inch (1.524 mm), and a 6% increase at 0.125 inch (3.175 mm).

Decking the block changes the compression ratio. Removing 0.010 inch (.2540 mm) from the deck surface of an engine with a 4-inch (101.6-mm) bore, 76-cc head, 0.060-inch (1.524 mm) head gasket, 0.080-inch (2.032-mm) deck height would raise the compression ratio by 0.14:1.

Resurfacing the head also increases the compression ratio. Though the effect varies for every type and size of chamber, a good rule of thumb is that when a head is surfaced 0.010 inch (.2540 mm), the chamber is reduced by 1.50 cc for a 60-cc head and by 2.50 cc for a 90-cc head. This will increase the compression ratio by about 0.141:1 to 0.20:1, depending on the specific head configuration and actual swept volume.

Variations in head gasket thicknesses affect the compression ratio. There can be as much as 0.040-inch (1.0160-mm) difference between various types and brands of gaskets. For instance, changing from a soft-faced to a steel or copper shim gasket can increase the compression ratio by as much as 0.50:1.

Fortunately, most aftermarket suppliers either deck or destroke their oversize pistons to enable the technician to reduce or maintain the compression ratio instead of increasing it. These pistons should be used whenever available.

On many OHC engines, it is necessary to restore the distance between the camshaft gear and the crankshaft gear after material has been removed from the head. Special shims are used to raise the camshaft. If 0.030 inch (.7620 mm) was removed from the head surface, the camshaft must be moved up 0.030 inch (.7620 mm). If this distance is not properly restored, valve timing will be altered.

Piston/Valve Interference and Misalignment. When the block or head is surfaced, the piston-to-valve clearance during the valve overlap period becomes less. To prevent the valves from making contact with the piston, a minimum of 0.070 inch (1.778 mm) piston-to-valve clearance is recommended.

Surfacing can also cause valve tips, rocker arms, and pushrods to be dimensioned closer to the camshaft. This causes a change in rocker arm geometry and can also cause hydraulic lifters to bottom out. To correct this problem, pushrods of a different length may need to be installed.

When metal is removed from the block or heads on a V-type OHV engine, the heads will be positioned closer to the crankshaft. This downward movement causes the intake manifold to fit differently between the heads. As a result, ports might be mismatched and manifold bolts might not line up. In order to return the intake manifold to its original alignment, corrective machining on the manifold is required.

GRINDING VALVES

Whenever the valves have been removed from the cylinder head, the valve heads and valve seats should be resurfaced. The most critical sealing surface in the valve train is between the face of the valve and its seat in the cylinder head when the valve is closed. Leakage between these surfaces reduces the engine's compression and power and can lead to valve burning. To ensure proper seating of the valve, the seat area on the valve face and seat must be the correct width, at the correct location, and concentric with the guide. These are accomplished by renewing the surface of the valve face and seat.

Valve grinding or refacing is done by machining a fresh, smooth surface on the valve faces and stem tips. Valve faces suffer from burning, pitting, and wear caused by opening and closing millions of times during the life of an engine. Valve stem tips wear because of friction from the rocker arms or actuators. Valve tips are machined after the valve face is refinished.

Figure 11-40 Grinding a valve face. *Courtesy of Sioux Tools, Inc.*

Figure 11-41 A cutter-type valve resurfacer. *Courtesy of Neway Manufacturing, Inc.*

Valves can be refaced on either grinding **(Figure 11–40)** or cutting **(Figure 11–41)** machines.

USING SERVICE MANUALS

Specifications for valve angles are normally listed in the engine specifications section of a service manual. ■

To start the valve grinding operation, chuck the valve as close as possible to the valve head to eliminate stem flexing from wheel pressure. Set the grinding angle according to the desired angle. Take light cuts using the full grinding wheel width. Make sure coolant is striking the contact point between the valve face and grinding wheel. Remove only enough metal to clean up the valve face. A knifelike edge will heat up and burn easily or might cause preignition **(Figure 11–42)**. The width of the edge of a valve head between the top of the valve and the edge of the face is the valve margin. As a general rule, it is not advisable to grind a valve face to a point where the margin

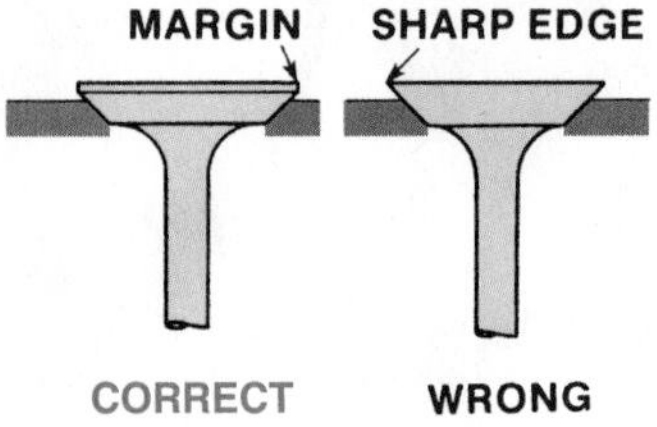

Figure 11-42 A sharp edge on the valve face is not recommended.

is reduced by more than 25% or to where it is less than 0.045 inch (1.143 mm) on the exhaust and 0.030 inch (.7620 mm) on the intake valves.

After grinding, check valve head runout. Use a dial indicator on the valve margin and rotate the valve while it is still in the chuck. Valve runout should not exceed 0.002 inch (.0508 mm) TIR. The face should not show any chatter marks or unground areas. After grinding, examine the valve face for cracks. Sometimes fine cracks are visible only after grinding. Sometimes they occur during grinding due to inadequate coolant flow or excessive wheel pressure.

CAUTION!

Always wear eye protection when operating any type of grinding equipment.

If an interference angle is to be used, the grinder is set ½ to 1 degree less than the standard 30- or 45-degree face angle. Always consult the manufacturer's specifications to determine whether an interference angle is to be used **(Figure 11–43)**. Grinding an interference angle produces a face angle close to 29½ or 44½ degrees. For the valve to seat properly, the face angle cannot be less than 29 or 44 degrees.

Removing metal from the valve face and/or seat will set the valve deeper into the port. As a result, more of the

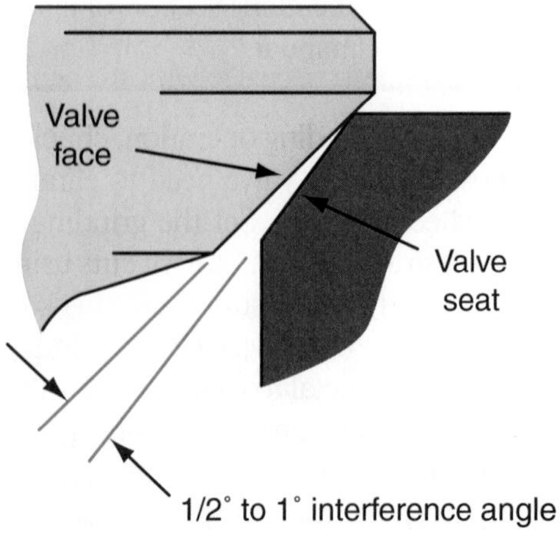

Figure 11-43 The difference between the seat angle and the valve face angle is known as the interference angle.

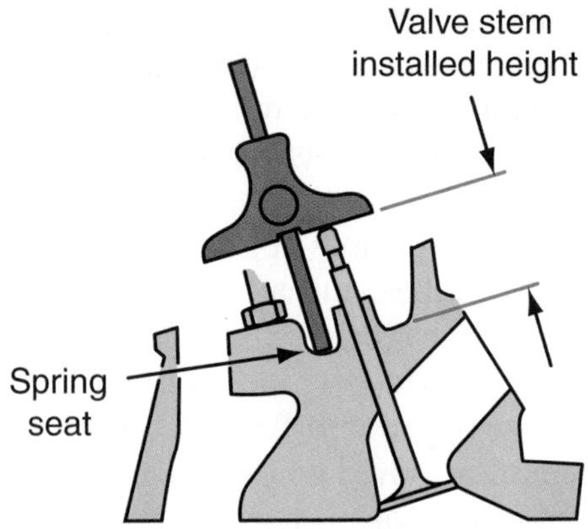

Figure 11-44 Measuring valve stem installed height.

valve stem extends from the other side of the head. If the stem height is greater than that specified by the manufacturer, the valve stem tip must be ground down to bring the overall height of the stem back into specs. In some cases, if the stem height is excessive, the valve and/or valve seat must be replaced. Valve stem installed height is measured from the valve spring pad on the cylinder head to the top of the valve **(Figure 11–44)**. Never remove more material from the tip of the stem than the amount allowed by the manufacturer. When the installed height is excessive, valve train geometry can be thrown off or there can be valve lash problems.

The valve tip is ground so that it is exactly square with the stem. Because valve tips have hardened surfaces up to 0.030 inch (.7620 mm) in depth, only 0.010 inch (.2540 mm) can be removed during grinding. If more than 0.010 inch (.2540 mm) is removed from the tip, the valve must be replaced. Follow the manufacturer's specifications for the allowable limits.

VALVE GUIDE RECONDITIONING

Valve guide problems can be lumped into one of three basic categories: inadequate lubrication, valve geometry problems, and wrong valve stem-to-guide clearance.

Inadequate lubrication can be caused by oil starvation in the upper valve train due to low oil pressure, obstructed oil passages, improper operation of pushrods, and using the wrong type of valve seal. Insufficient lubrication results in stem scuffing, rapid stem and guide wear, possible valve sticking, and ultimately valve failure due to poor seating and overheating.

Geometry problems include an incorrectly installed valve height, off-square springs, and rocker arm tappet screws or rocker arms that push the valve sideways every time it opens. This causes uneven guide wear, leaving an egg-shaped hole. The wear leads to increased stem-to-guide clearance, poor valve seating, and premature valve failure.

As for valve stem-to-guide clearance, a certain minimum amount is needed for lubrication and thermal expansion of the valve stem. Exhaust valves require more clearance than intakes because they run hotter. Clearance should also be close enough to prevent a buildup of varnish and carbon deposits on the stems, which could cause sticking. Insufficient clearance, however, can lead to rapid stem and guide wear, scuffing, and sticking, which prevents the valve from seating fully. This, in turn, causes the valve to run hot and burn.

The amount of valve guide wear can be measured with a ball (small-bore) gauge and micrometer. Insert and expand the ball gauge at the top of the guide. Lock it to that diameter, remove it from the guide and measure the ball gauge with an outside micrometer. Repeat this process with the ball gauge in the middle and the bottom of the guide **(Figure 11–45)**. Compare your measurements to the specifications for valve guide inside diameter. Compare these to specifications. Then compare your measurements against each other. Any difference in reading shows a taper or wear inside the guide.

Another way to check for excessive guide wear is with a dial indicator. The accuracy of this check is directly dependent on the amount the valve is open during the check. Some manufacturers specify this amount or provide special spacers that are installed over the valve stem to ensure the proper height. Attach the dial indicator to the cylinder head and position it so the plunger is at a right angle to the valve stem being measured **(Figure 11–46)**. With the plunger in contact with the valve head, move the valve toward the indicator and set the dial indicator to zero. Now move the valve from the indicator. Observe the reading on the dial while doing this. The reading on the indicator is the total movement of the valve and is indicative of the guide's wear. Compare the reading to specifications.

If the clearance is too great, oil can be drawn past both the intake and exhaust guides. Though oil consumption is more of a problem with sloppy or worn intake guides because the guides are constantly exposed to vacuum, oil can also be pulled down the exhaust guides by suction created in the exhaust port. The outflow of hot exhaust creates a venturi effect as it exits the exhaust port, creating enough vacuum to draw oil down a worn guide **(Figure 11–47)**.

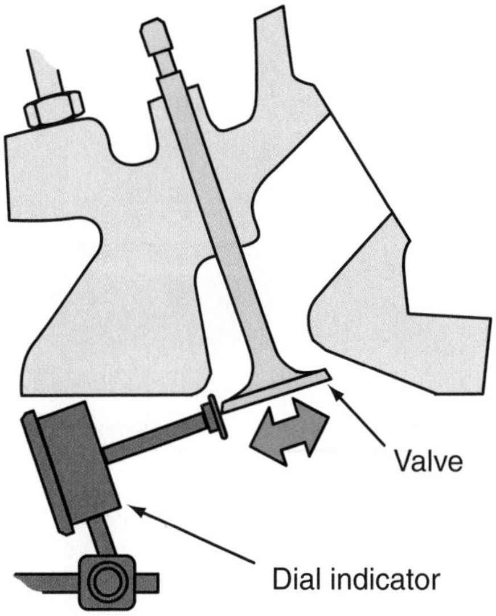

Figure 11-46 Checking for valve guide wear with a dial indicator.

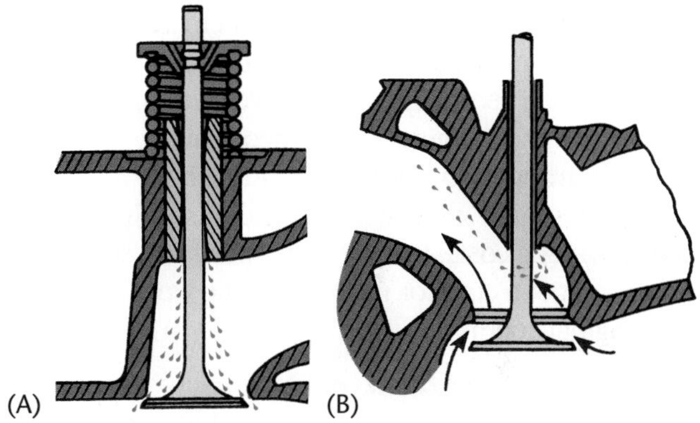

Figure 11-47 (A) Worn intake guides allow the intake vacuum to suck oil down the guide, and (B) worn exhaust guides can do the same.

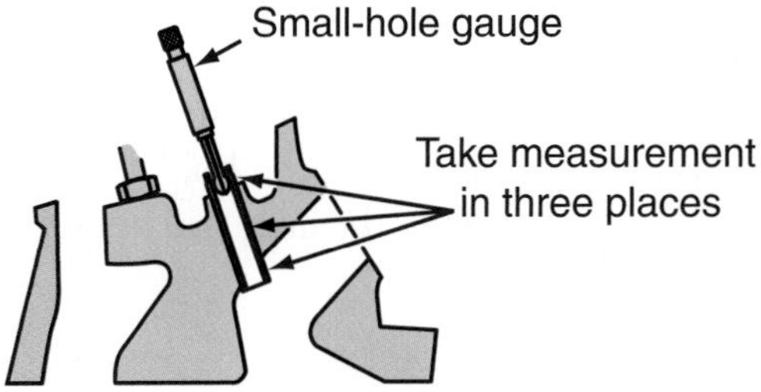

Figure 11-45 Valve guide wear can be measured with a small-hole gauge and a micrometer.

Because it retains oil well, the antiseize and antiwear characteristics of bronze allow a bronze guide to last two to five times longer than a cast-iron guide. However, bronze guides are expensive, so their use in original equipment applications is limited. Bronze is also more difficult to machine. Because of these drawbacks, bronze guides are commonly used only as aftermarket replacements.

Knurling

Knurling is one of the fastest techniques for restoring the inside diameter (ID) dimensions of a worn valve guide. The process raises the surface of the guide ID by plowing tiny furrows through the surface of the metal **(Figure 11–48)**. As the knurling tool cuts into the guide, metal is raised or pushed up on either side of the indentation. This effectively decreases the ID of the guide hole. A burnisher is used to press the ridges flat and is then used to shave

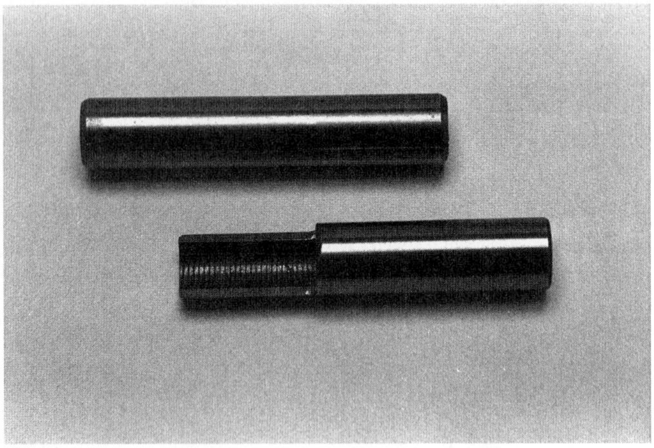

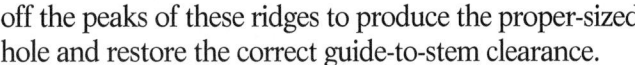

Figure 11-48 Knurling restores the ID dimensions of a worn valve guide by raising the inside surface of the guide ID by plowing tiny furrows through the surface of the metal.

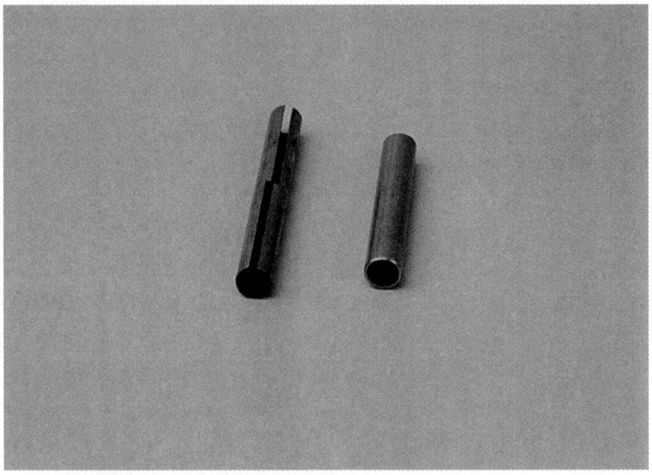

Figure 11-49 A thin wall valve guide liner.

off the peaks of these ridges to produce the proper-sized hole and restore the correct guide-to-stem clearance.

One of the main advantages of knurling is that it does not change the centerline of the valve stem appreciably, so it reduces the amount of work necessary to reseat the valve. Knurling also allows a rebuilder to reuse the old valve if wear is within acceptable limits, helping to reduce rebuilding costs. In spite of its speed and simplicity, knurling is not a cure for restoring badly worn guides to their original condition.

Reaming and Oversized Valves

Reaming is used to repair worn guides by increasing the guide hole size to take an oversize valve stem or by restoring the guide to its original diameter after installing inserts or knurling.

When reaming, limit the amount of metal removed per pass. Always reface the valve seat after the valve guide has been reamed and use a suitable scraper to break the sharp corner (ID) at the top and bottom of the valve guide.

The advantage of reaming for an oversized valve is that the finished product is totally new. The guide is straight, the valve is new, and the clearance is accurate. The use of oversized valve stems is generally considered to be superior to knurling. Yet, like knurling, it is relatively quick and easy. The only tool required is a reamer. The valve centerline is maintained so the work required to finish the seat is reduced. However, since reaming requires the use of new valves, it can be more expensive on an engine with many worn guides. Its use is also limited to heads in which the guides are not worn beyond the limits of the maximum oversize valve that is available. Because of these limitations, many technicians prefer more cost-effective alternatives such as guide liners, inserts, or replacement guides.

Thin-Wall Guide Liners

The thin-wall guide liners **(Figure 11–49)** repair technique offers a number of important advantages and is also popular with many production engine rebuilders, as well as smaller shops. It provides the benefits of a bronze guide surface. It can be used with either integral or replaceable guides. It is faster, easier, and cheaper than installing new guides in heads with replaceable or integral guides, and it maintains guide centering with respect to the seats.

Thin-wall guide liners are manufactured from a phosphor-bronze or silicon-aluminum-bronze material. These liners can be cut to almost any length. They are designed for a 0.002- to 0.0025-inch (.0508- to .0635-mm) press fit. A tight fit is essential for proper heat transfer to the head and to prevent the liner from working loose.

The liners are installed by first boring out the original guides to 0.030 inch (.7620 mm) oversize with a special piloted boring tool pressed into the guide using a driver and air hammer. On guides not precut to length, the excess must be milled off before finishing. The liner is then wedged in place and sized in a single operation by passing a ball broach down through it. This eliminates the need to ream it to size and ensures a tight fit between the liner and guide. If a ream finish is desired, it can be obtained by lubricating the reamer with a bronze-lube and then running it through the guide. For closer than normal stem-to-guide clearance, spiraling or knurling is suggested for added lubrication.

The only trick to using liners is to make sure the hole is round and the correct size. If the hole is distorted excessively or if it is too large, the liner will not fit properly and will cause problems.

Valve Guide Replacement

Replacing the entire valve guide is another repair option possible on cylinder heads with replaceable guides.

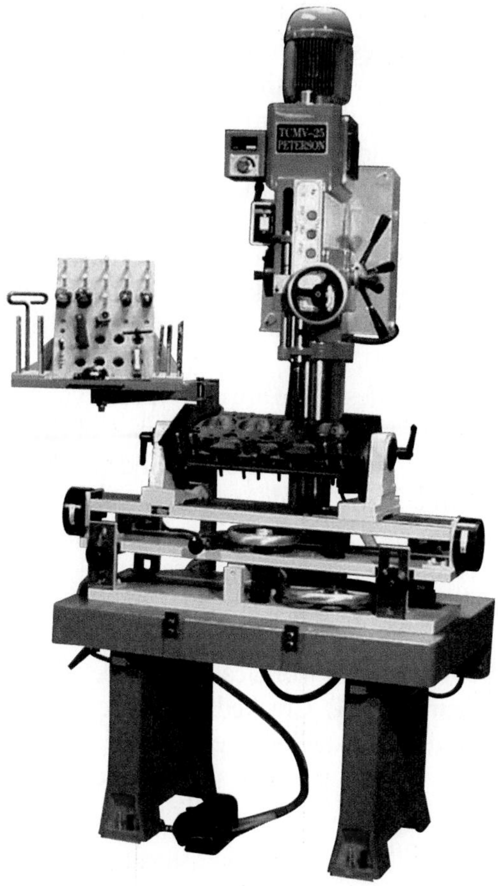

Figure 11–50 A valve seat and guide machine. *Courtesy of Frontline Equipment Company*

However, pressing out the old guides and installing new ones can be difficult with some aluminum heads. Cracking the head or galling the guide hole is always a risk.

Integral Guides To replace integral guides, bore the old guide out and drive a thin-wall replacement guide into the hole. Many shops use a seat and guide machine for this process **(Figure 11–50)**, although it can be done with portable equipment. Drive the replacement guides in cold with approximately 0.002-inch (.0508-mm) press fit. Use an assembly lube to prevent galling. It is necessary to keep the centerline of the guide concentric with the valve seat so that the rocker arm-to-valve stem contact area is not disturbed **(Figure 11–51)**.

Occasionally, a new guide will not be concentric with the valve seat. Install a new seat to correct the problem and check the concentricity of the valve seat with a concentricity gauge.

Insert Guides To remove an old valve guide, place a proper-sized driver so that its end fits snugly into the guide. The shoulder on the driver must also be slightly smaller than the OD of the guide, so it will go through the cylinder head. Use a heavy ball peen hammer or air hammer to drive the pressed guide out of the cylinder head.

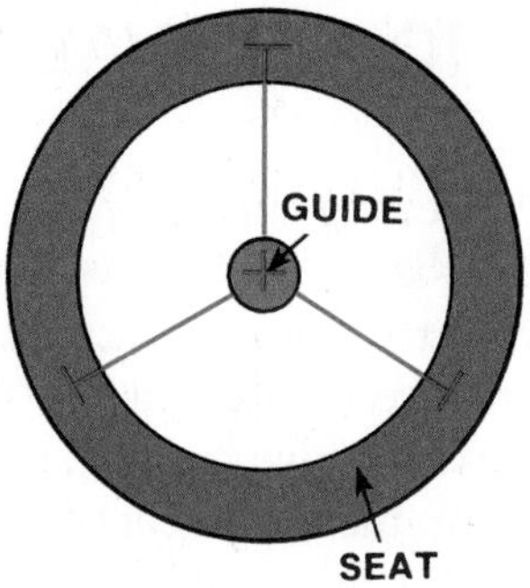

Figure 11–51 The centerline of the guide should be concentric with the seat.

Pressing or driving out and installing new guides is not difficult, but there is always the danger of breaking the guide or tearing up the guide hole in the head. Cast-iron guides in particular have a tendency to gall aluminum heads. Once the hole is damaged, it must be bored out and an oversized guide installed—assuming one is available to fit the application. New guides should be chilled prior to installation because of the needed interference fit between the guide and head. Chilling them in a freezer or with dry ice works well. Lubricant also helps to prevent galling.

When installing new guides, be careful not to damage them. Use a press and the same driver that was used to remove the old guide. Align the new guide and press straight down, not at an angle. An air hammer and special driver can also be used to install new guides.

If the guides are cut off at an angle on the combustion chamber end or cut at an angle at one end or the other, do not press or drive against the angled end.

Find the correct amount of guide protrusion **(Figure 11–52)**. Guide height is important to avoid interference with the valve spring retainer. The guide must also fit the hole tightly or it can work loose. The manufacturer's specifications give the correct valve guide installed height, but it is good practice to measure how far the old guides stick out of the cylinder head and to use this measurement as a reference. As each guide is installed, insert a valve. Check for any stem interference.

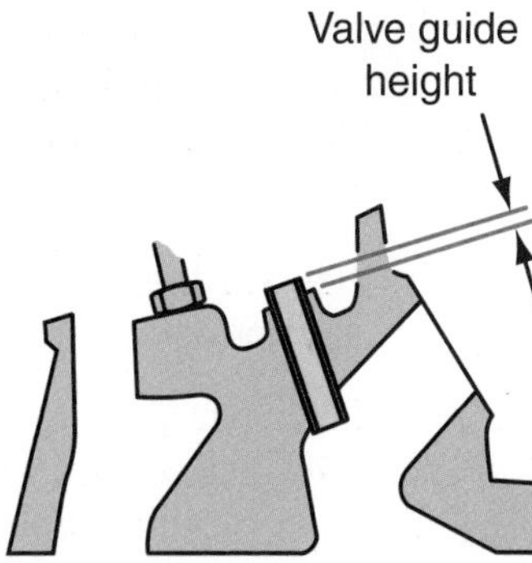

Figure 11–52 Make sure to check the installed height (protrusion) of the new valve guide and correct it if necessary.

RECONDITIONING VALVE SEATS

The most critical sealing surface in the valve train assembly is between the face of the valve and its seat in the cylinder head when the valve is closed. Leakage between these surfaces reduces the engine's compression and power and can lead to valve burning. To ensure proper seating of the valve, the valve seat must be the correct width **(Figure 11–53)**, in the correct location on the valve face, and concentric with the guide (less than 0.002-inch [.0508-mm] runout).

The ideal seat width for automotive engines is 1⁄16 inch for intake valves and 3⁄32 inch (2.38 mm) for exhaust valves. Maintaining this width is important to ensure proper sealing and heat transfer. However, when an existing seat is refinished to make it smooth and concentric, it also becomes wider.

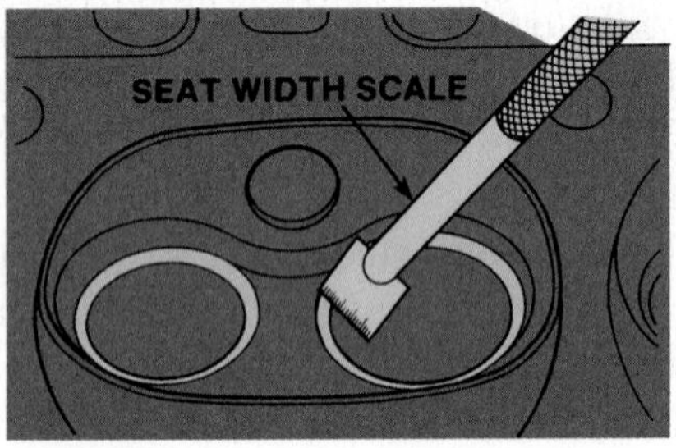
SEAT WIDTH SCALE

Figure 11-53 Checking valve seat width.

Wide seats cause problems. Seating pressure drops as seat width increases. Less force is available to crush carbon particles that stick to the seats, and seats run cooler, allowing deposits to build up on them.

The seat should contact the valve face 1⁄32 inch (.79 mm) from the margin of the valve. When the engine reaches operating temperature, the valve expands slightly more than the seat. This moves the contact area down the valve face. Seats that make too low a contact with the valve face might lose partial contact at normal operating temperatures.

Like valve guides, there are two types of valve seats—integral and insert. Integral seats are part of the casting. Insert seats are pressed into the head and are always used in aluminum cylinder heads.

Valve seats can be reconditioned or repaired by one of two methods, depending on the seat type—machining a counterbore to install an insert seat, or grinding, cutting, or machining an integral seat.

Before starting seat work, carefully check the seats for cracks **(Figure 11–54)**. Cracked integral seats sometimes can be repaired by installing inserts, if the crack is not too deep. Cracked insert seats must be replaced. Check insert seats for looseness with a small pry bar. Replace them if any movement is noted.

Installing Valve Seat Inserts

The following steps outline a typical procedure for valve seat insert removal and replacement.

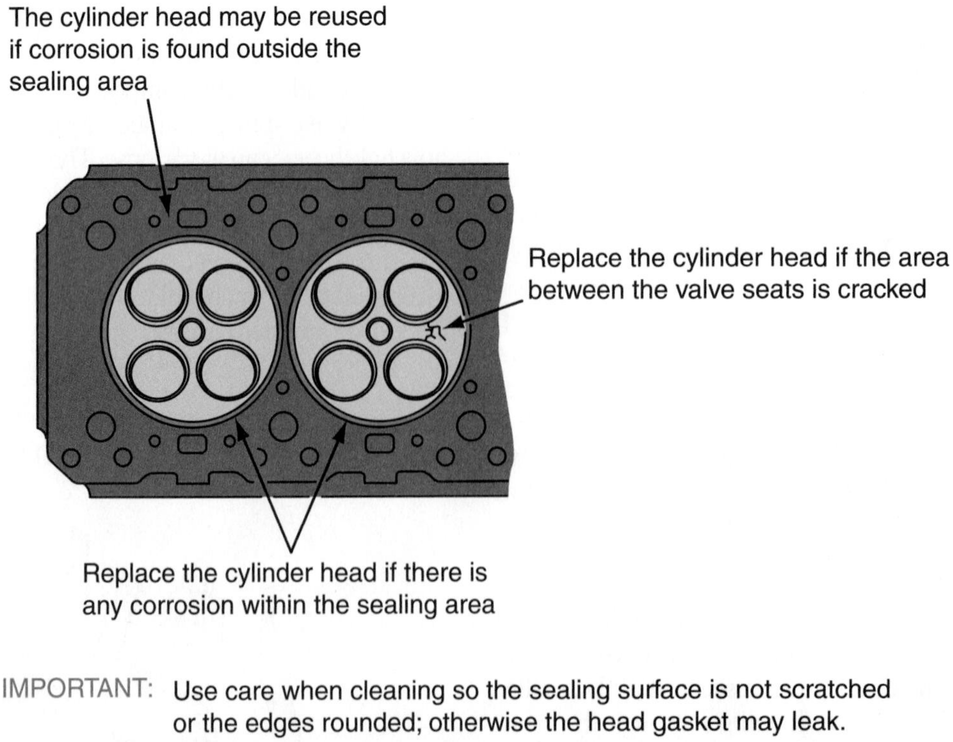
The cylinder head may be reused if corrosion is found outside the sealing area
Replace the cylinder head if the area between the valve seats is cracked
Replace the cylinder head if there is any corrosion within the sealing area
IMPORTANT: Use care when cleaning so the sealing surface is not scratched or the edges rounded; otherwise the head gasket may leak.

Figure 11-54 A careful inspection of a cylinder head includes checking for cracks in the areas between the valve seats. *Courtesy of General Motors Corporation*

PROCEDURE

Insert Valve Seat Removal and Replacement

STEP 1 *To remove the damaged insert, use a puller or a pry bar* ***(Figure 11–55)****.*

STEP 2 *After removal, clean up the counterbore or recut it to accommodate oversized inserts.*

STEP 3 *Insert the counterboring pilot into the valve guide. Then mount the base and ball shaft assembly to the gasket face angle of the cylinder head.*

STEP 4 *Use an outside micrometer to accurately expand the cutterhead to a predetermined size of the counterbore* ***(Figure 11–56)****. Remember that the counterbore should have a slightly smaller ID than the OD of the insert to provide for an interface fit.*

STEP 5 *Place the valve insert counterboring tool over the pilot and ball-shaft assembly. Preset the depth of the valve seat insert at the feed screw.*

STEP 6 *Cut the insert by turning the stop-collar until it reaches the present depth. Use a lubricant on the cutters for a smoother finish.*

STEP 7 *To install the insert, heat it in a parts cleaning oven to approximately 350 to 400°F (176 to 204°C). Chill the insert in a freezer or with dry ice before installation.*

STEP 8 *Press the seat with the proper interference fit using a driver.*

STEP 9 *When the installation is complete, the edge around the outside of the insert is staked as shown in* ***Figure 11–57****. By doing so, the insert will be secured more effectively in the counterbore.*

CAUTION!

Wear the proper gloves when handling dry ice.

Reconditioning Integral Seats

The average valve seat width is 0.060 inch (1.524 mm) and the average seat begins 0.030 inch (.7620 mm) from the valve margin. A properly reconditioned seat has three angles: top, 30 or 15 degrees; seat, 45 or 30 degrees; and throat, 60 degrees. Typically, the 45-degree angle wedges tighter than the 30-degree seat, so it is used more often. Using three angles maintains the correct seat width and sealing position on the valve face **(Figure 11–58)**. Correct sealing pressure and heat transfer from the valve through the seat are also affected. Always check service manual specifications for valve seat angles and valve face-to-seat contact amounts.

Integral valve seats can be reconditioned by grinding, cutting, or machining.

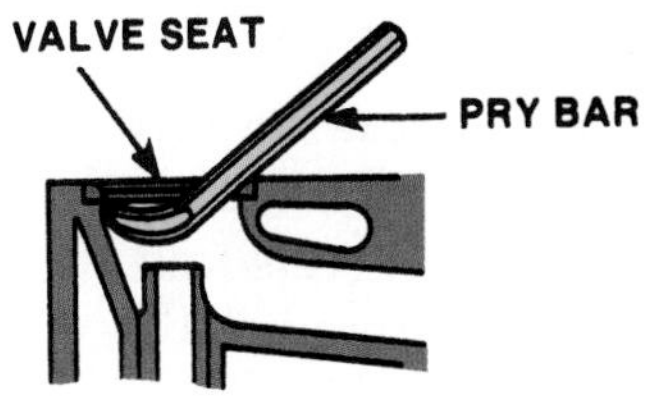

Figure 11–55 Using a pry bar to remove a damaged insert seal.

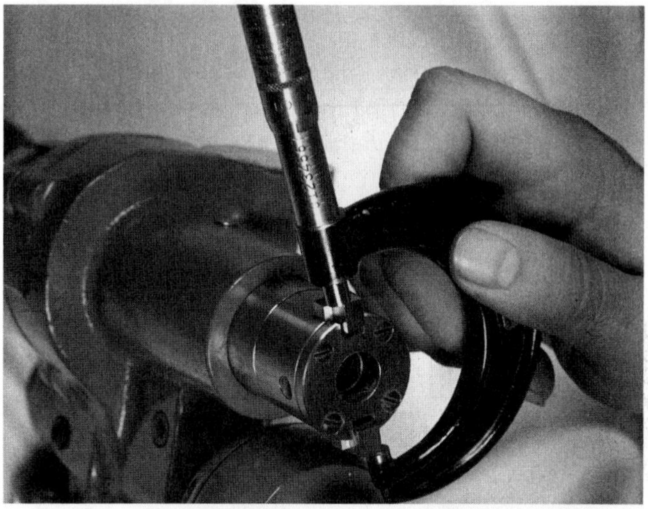

Figure 11–56 Using an outside micrometer to expand the cutter head to allow for an interference fit between the bore and the new valve seat. *Courtesy of Hall-Toledo, Inc.*

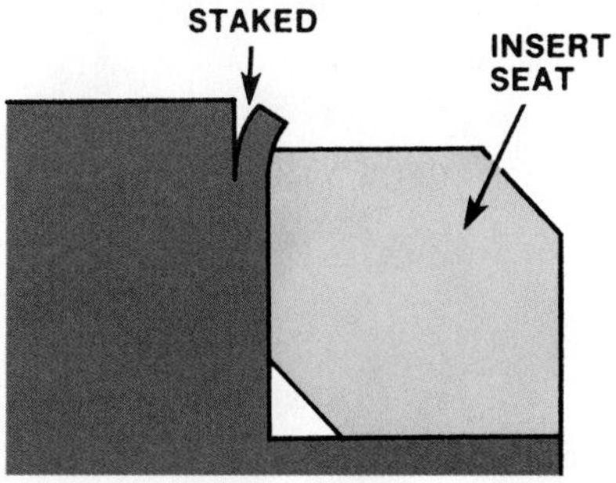

Figure 11–57 Staking the valve seat to the head can be done with a sharp chisel.

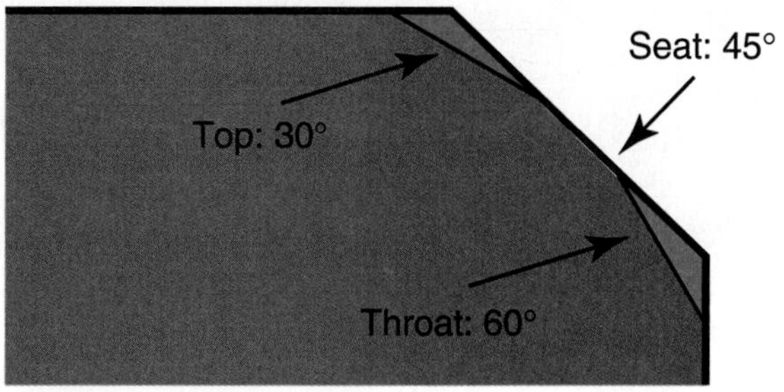

Figure 11–58 The three angles of a properly finished seat.

Grinding Valve Seats

When grinding a valve seat, it is very important to select and use the correct size pilot and grind stone. Hard seats use a soft stone and soft seats (cast iron) use a harder stone. The stone must be properly dressed and cutting oil used to aid in grinding.

SHOP TALK

Before grinding, many technicians clean the seats by placing a piece of fine emery cloth between the stone and the seat and giving the surface a hard rub. This will help prevent contamination of the seat grinding stone with any oil or carbon residue that might be present on the valve seat. Such contamination could cause glazing. ■

The grinding wheel is positioned and centered by inserting a properly sized pilot shaft into the valve guide **(Figure 11–59)**. All valve guide service must be completed before installing the pilot.

Figure 11-59 Grinding the seat.

The seat is ground by continually and quickly raising and lowering the grinder unit on and off the seat. Grinding should only continue until the seat is clean and free of defects.

After the seat is ground, valve fit is checked using machinist dye. The valve face is coated with dye, installed in its seat, and slightly rotated. The valve is then removed and the dye pattern on the valve face and valve seat inspected.

If the valve face and seat are not contacting each other evenly, or if the contact line is too high, the seat must be reground with the same stone used initially to correct the condition. If the line is too low or the width is not correct, the seat must be reground with stones of different angles **(Figure 11–60)**.

Cutting Valve Seats

Cutting valve seats differs from grinding only in the equipment used. Hardened valve seat cutters replace grinding wheels for seat finishing. The basic seat cutting procedures are the same as those for grinding.

Machining Valve Seats

As stated earlier in this chapter, a valve system rebuilding machine can be used to install valve guides and to machine valve seats. Some have optional seat cutters that make three-angle cuts **(Figure 11–61)**. The cutters are set to the proper diameter. Once set, they machine the seat as well as the top and throat angles. Two primary advantages of these cutters over other methods are high speed and precision.

Figure 11-61 A valve seat being cut at three angles at the same time. *Courtesy of Jaspar Engine and Transmission Exchange, Inc.*

Figure 11-60 The fit of the valve face in the seat should be checked carefully.

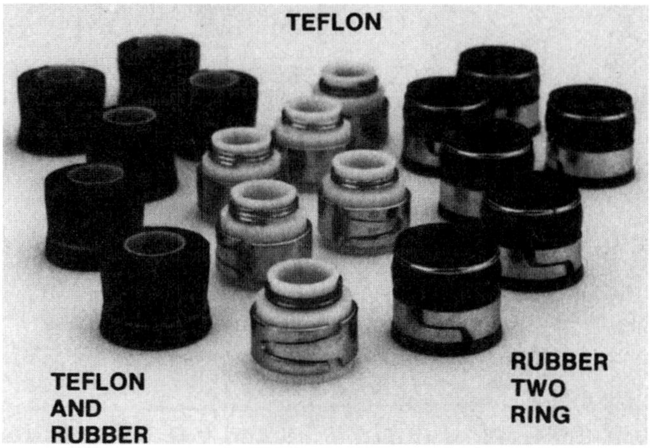

Figure 11-62 Various designs of positive valve seals. *Courtesy of Goodson Shop Supplies*

VALVE STEM SEALS

Valve stem seals are used on many engines to control the amount of oil allowed between the valve stem and guide. The stems and guides will scuff and wear excessively if they do not have enough lubrication. Too much oil produces heavy deposits that build up on the intake valve and hard deposits at the head end of the exhaust valve stem. Worn valve stem seals can increase the oil consumption by as much as 70%.

There are basically three types of seals. **Positive seals** fit tightly around the top of the stem and scrape oil off the valve as it moves up and down **(Figure 11–62)**. Deflector, splash, or **umbrella-type seals** ride up and down on the valve stems to deflect oil away from the guides. O-ring seals installed over the valve stem are also used to prevent oil from moving into the guide when the valve is open.

The ultimate effectiveness of the valve stem seal depends entirely on the way it is secured to the guide. Many guides require machining to accept the stem seals. This must be done using the proper tools. A special valve guide machining tool is available for valve seal cutting. Such a tool is made up of a cutter and pilot, with sizes that vary according to the valve stem diameter and desired guide OD. The pilot is inserted into the guide, and the cutting tool machines the top of the guide.

Installing Positive Valve Seals

To install a positive valve seal **(Figure 11–63)**, place the plastic sleeve in the kit over the end of the valve stem to

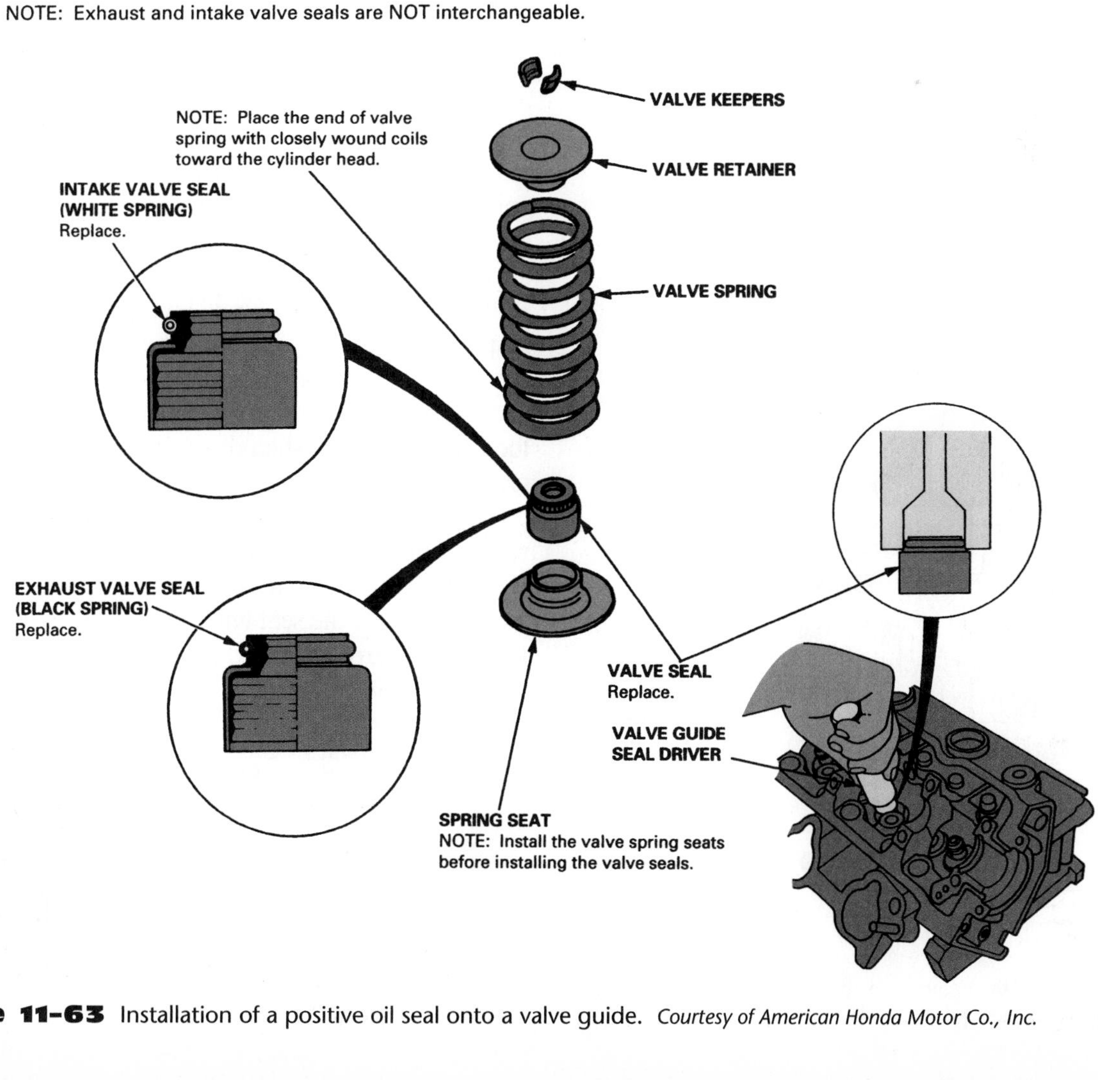

Figure 11-63 Installation of a positive oil seal onto a valve guide. *Courtesy of American Honda Motor Co., Inc.*

protect the seal as it slides over the keeper grooves. Lightly lubricate the sleeve. If it extends more than 1⁄16 inch (1.59 mm) below the lower keeper groove, you might want to remove the sleeve and cut off the excess length for easier removal. Carefully place the seal on the cap over the valve stem and push the seal down until it touches the top of the valve guide. At this point, the installation cap can be removed and placed on the next valve. A special installation tool can be used to finish pushing the seal over the guide until the seal is flush with the top of the guide.

Installing Umbrella-Type Valve Seals

An umbrella-type seal is installed on the valve stem before the spring is installed. It is pushed down on the valve stem until it touches the valve guide boss **(Figure 11–64)**. It will be positioned correctly when the valve first opens.

Installing O-Rings

When installing O-rings, use engine oil to lightly lubricate the O-ring. Then install it in the lower groove of the lock section of the valve stem **(Figure 11–65)**. Make sure the O-ring is not twisted.

Valve Springs

The valve spring performs two functions: It closes the valve and it maintains valve train contact during the opening and closing of the valve. Insufficient spring pressure can lead to valve bounce and breakage. Too much pressure will cause premature camshaft lobe wear and can also lead to valve breakage.

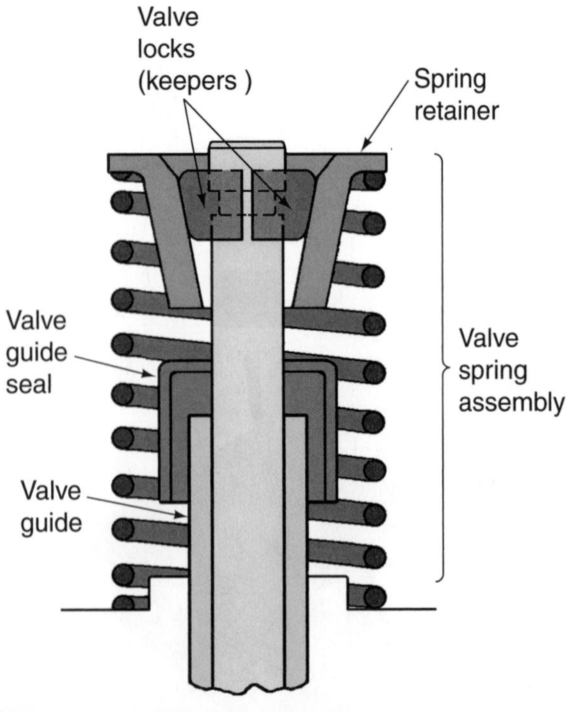

Figure 11-64 Valve assembly with an umbrella-type oil seal. *Courtesy of Federal-Mogul Corporation*

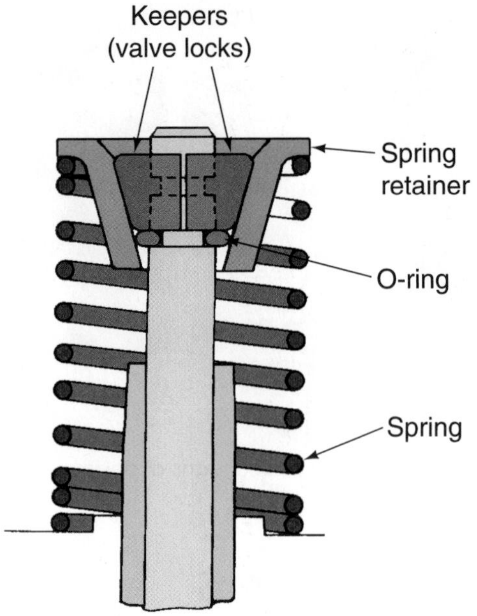

Figure 11-65 Valve assembly with an O-ring valve seal. *Courtesy of Federal-Mogul Corporation*

The common designs of valve springs are illustrated in **Figure 11–66**. A problem that valve springs might have is **spring surge**. As the name implies, spring surge is the violent extending motion of the coils resulting in abnormal oscillation. Always install the closely wound coils of a basket coil-type spring toward the head end of the valve. Mechanical surge and vibration dampers should also be installed toward the head end of the valve. To dampen spring vibrations and increase total spring pressure, some engine manufacturers use a reverse wound secondary spring inside the main spring.

Spring surge can occur when the springs are weak, the installed spring height is improper, or engine speeds are excessive. Whatever the cause, the occurrence of spring surge is visually apparent. The ends of the springs will look smooth or polished due to their rotation during operation. If left alone and not corrected, spring surge can cause damage to the valve train. For example, the self-rotation of the valve springs causes a grinding action between the valve face and seat. As a result, the face will wear down and the seat will recess. Continued operation with spring surge can also cause the springs to break.

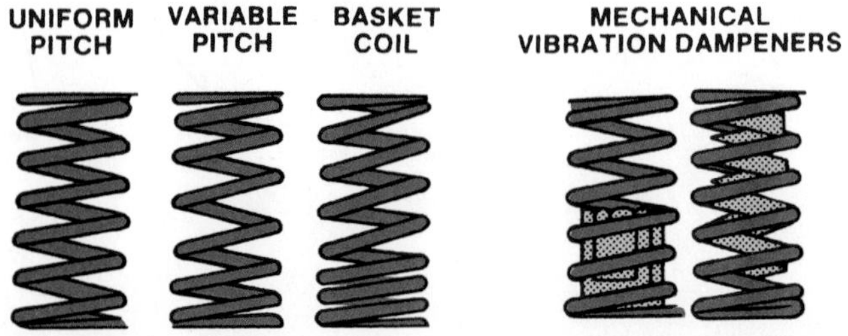

Figure 11-66 Common valve spring designs.

The high stresses and temperatures imposed on valve springs during operation cause them to weaken and sometimes break. Rust pits also cause valve spring breakage. To determine if the spring can be reused, the following tests should be performed.

Freestanding Height Test Line up all the springs on a flat surface and place a straightedge across the tops. All springs should be the same height and free length should be within 1⁄16 inch (1.59 mm) of OE specifications. Throw away any spring that is not within specifications.

Spring Squareness Test A spring that is not square will cause side pressure on the valve stem and abnormal wear. To check squareness, set a spring upright against a square **(Figure 11–67)**. Turn the spring until a gap appears between the spring and the square. Measure the gap with a feeler gauge. If the gap is more than 0.060 inch (1.524 mm), the spring should be replaced.

Open/Close Spring Pressure Test Use a spring tester to check for open and close spring pressure. Close pressure guarantees a tight seal. The open pressure overcomes valve train inertia and closes the valve when it should close. Service manuals list spring pressure specifications according to spring height **(Figure 11–68)**. The proper procedure for testing valve spring tension is given in Photo Sequence 8. Low spring pressure may allow the valve to float during high-speed operation. Excessive spring tension may cause premature valve train wear. Any spring that does not meet specifications should be replaced.

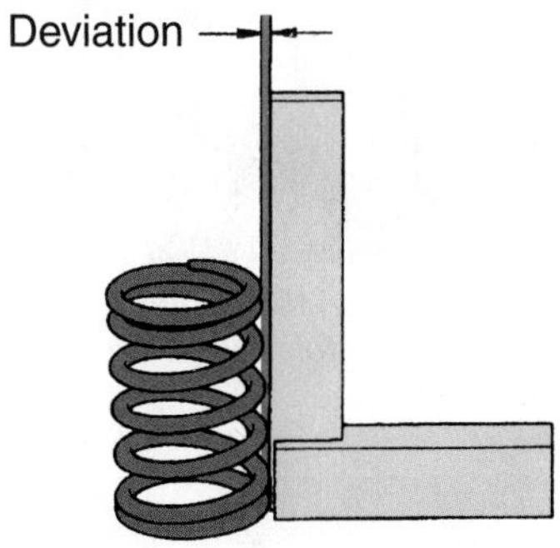

Figure 11-67 A spring squareness test.

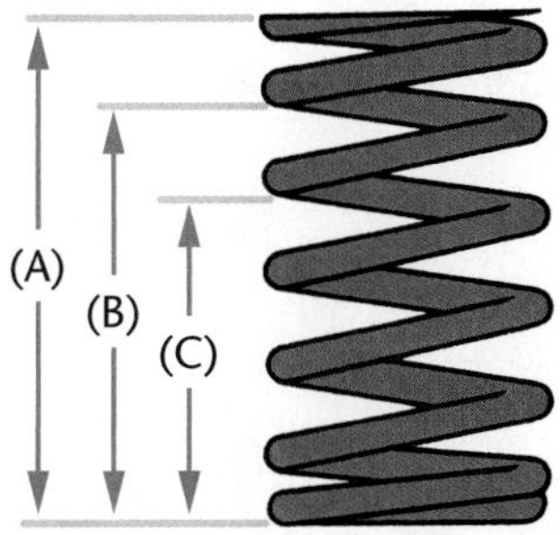

Figure 11-68 Valve spring height terminology: (A) free height, (B) valve closed spring height, and (C) valve open spring height.

ASSEMBLING THE CYLINDER HEAD

Before a cleaned or reconditioned cylinder head is reassembled and installed, two critical measurements must be carefully checked: the installed stem height and the installed spring height.

Installed stem height is determined by measuring the distance between the spring seat and stem tip. Since this measurement directly influences rocker arm geometry and installed spring height, accuracy and precision are important. This is especially true when the valve or valve seat has been ground. A number of tools can be used to obtain an accurate stem height reading including a depth micrometer, vernier caliper, and telescoping gauge.

USING SERVICE MANUALS

Stem height specifications are often unavailable in service manuals. As a guide for assembly, record the stem heights for all valves during disassembly. ■

Another specification can be used that corresponds directly to installed stem height and that is installed spring height. Installed spring height is measured from the spring seat to the underside of the retainer when it is assembled with keepers and held in place. This measurement, which can be made by using a set of dividers or scales, telescoping gauge, or spring height gauge, should be taken only after valve and seat work is completed, valves are installed in their guides, and retainers and keepers are assembled.

If the spec for installed spring height for an exhaust valve is 1.600 inches (40.64 mm), and the measurement is 1.677 inches (42.54 mm), the increase in height is 0.077 inch (1.95 mm). This means the installed stem height has also been increased by 0.077 inch (1.95 mm).

Adjustments to valve spring height can be made with valve spring inserts, otherwise known as spring shims. Even though valve shims come in only three standard thicknesses—0.060, 0.030, and 0.015 inch (1.52, .7620, and .3810 mm)—using combinations of different shims gives the correct amount of compensation (within 0.005 or 0.010 inch [.1270 or .2540 mm]).

SHOP TALK

Valve keepers should be replaced in pairs. If a new keeper is mated with a used one, the spring retainer may cock and break off the valve tip or allow the assembly to come apart. ■

PHOTO SEQUENCE 8

Measuring and Fitting Valve Springs

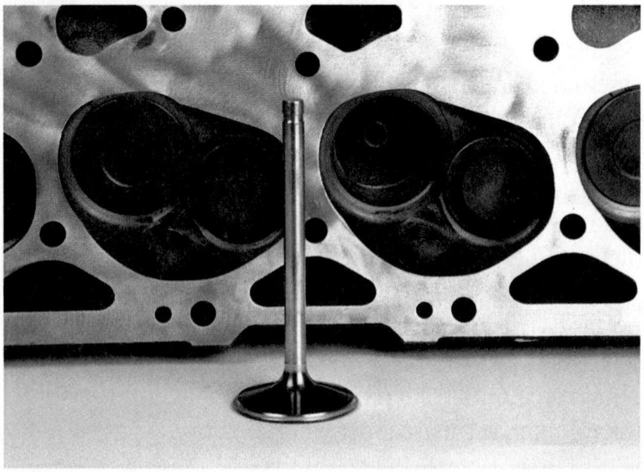

P8-1 *Prior to installing the valve and fitting valve springs, all other head work should be completed.*

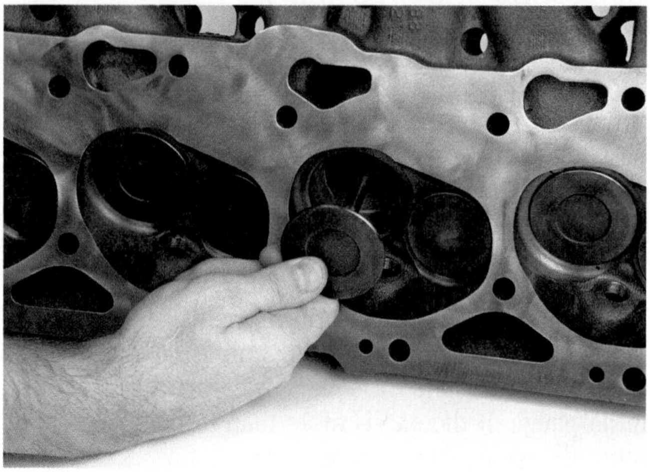

P8-2 *Install the valve into its proper valve guide.*

P8-3 *Install the valve retainer and keepers. Without the spring, these must be held in place.*

P8-4 *While pulling up on the retainer, measure the distance between the bottom of the retainer and the spring pad on the cylinder head with a divider.*

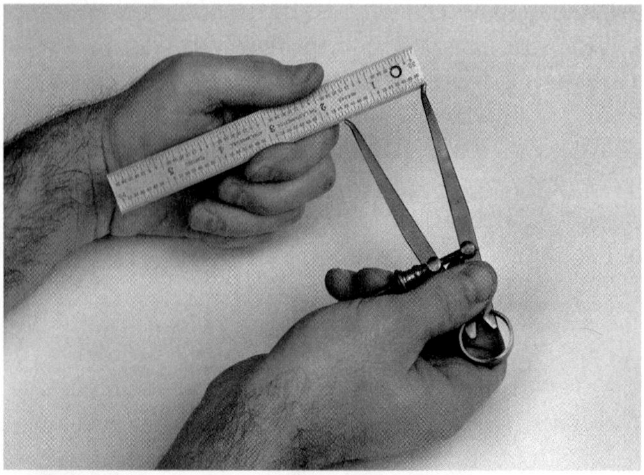

P8-5 *Use a scale to determine the measurement expressed by the divider.*

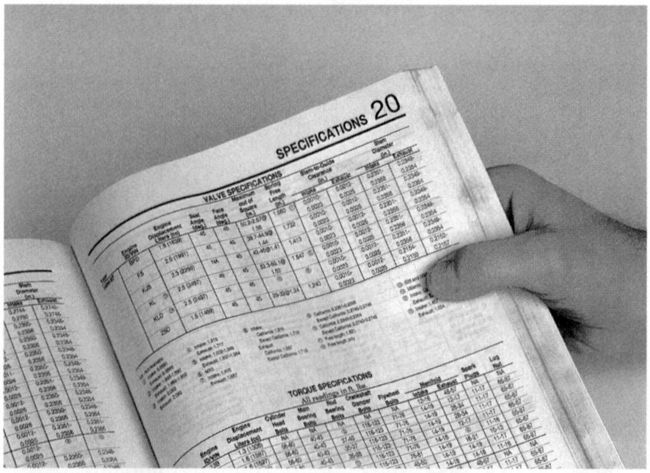

P8-6 *Compare this measurement with the specifications given in the service manual for installed spring height.*

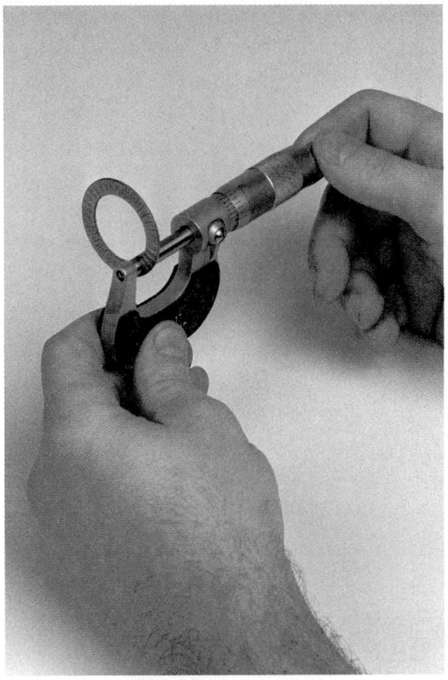

P8-7 *If measured installed height is greater than the specifications, a valve shim must be placed under the spring to correct the difference.*

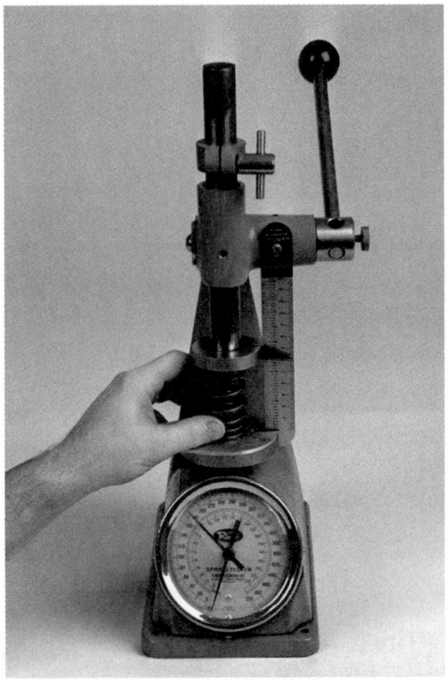

P8-8 *Spring tension must be checked at the installed spring height; therefore, if a shim is to be used, insert it under the spring on the valve spring tension gauge.*

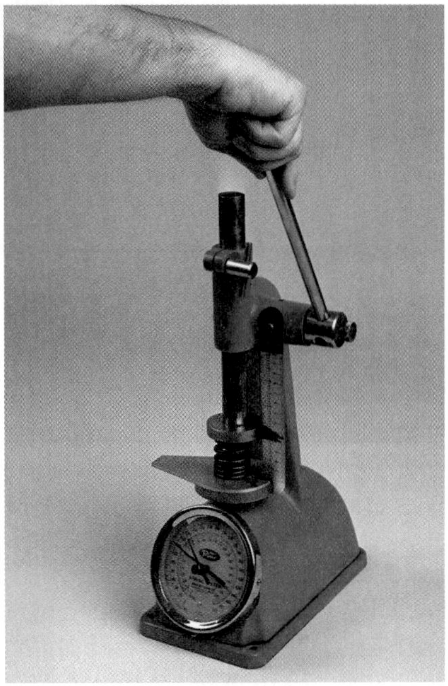

P8-9 *Compress the spring into the installed height by pressing down on the tester's lever.*

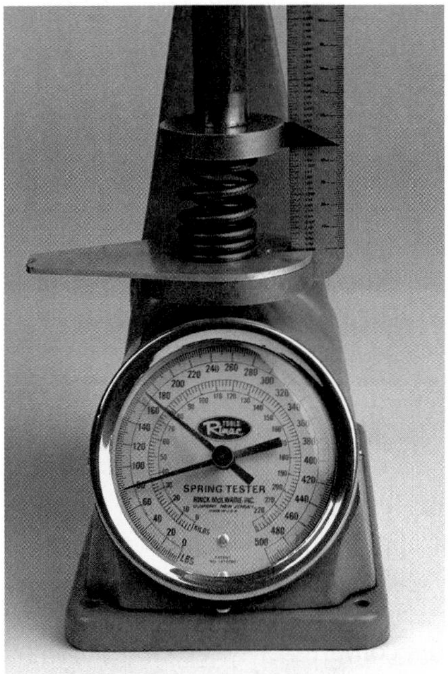

P8-10 *The tension gauge will reflect the pressure of the spring when compressed to the installed or valve closed height. Compare this reading to the specifications.*

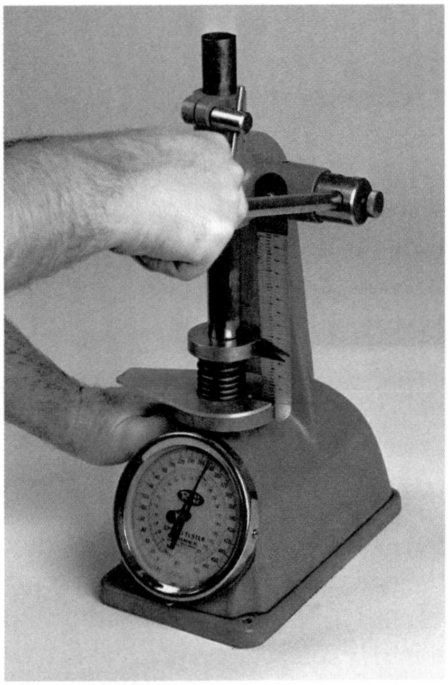

P8-11 *Now compress the spring to the open height specification. Use the rule on the gauge or a scale to measure the compressed height.*

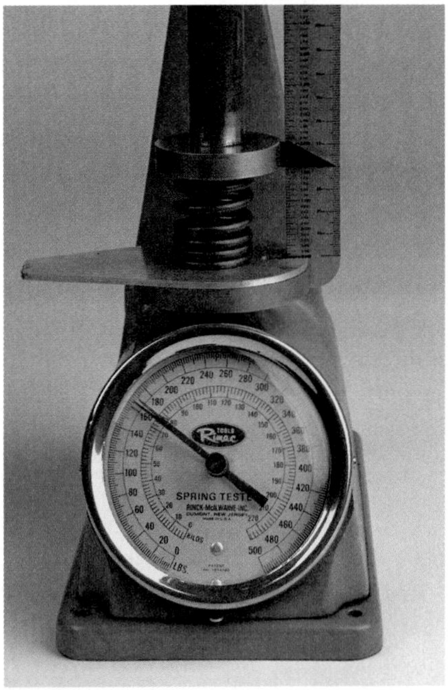

P8-12 *Compare this reading to specifications. Any pressure outside the pressure range given in the specifications indicates the spring should be replaced. After the tension and height have been checked, the spring can be installed on the valve stem.*

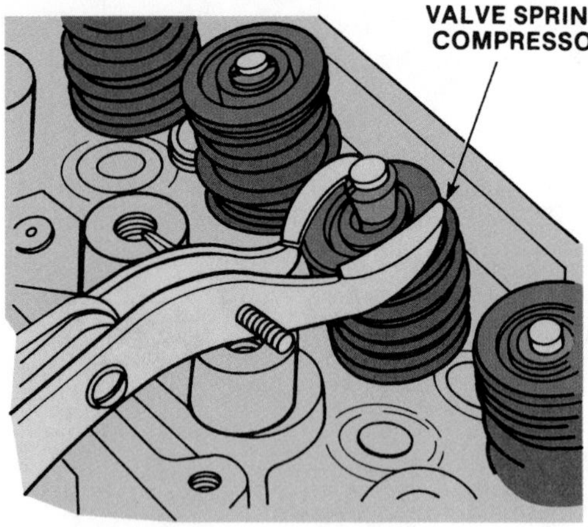

Figure 11-69 Compress the spring just enough to install the keepers.

By comparing spring height to specifications, the desired amount of spring tension correction can be easily determined. For example, if spring height is 0.180 inch (4.59 mm) and the specifications call for 0.149 inch (3.78 mm), a 0.030 inch shim (0.149 inch + 0.030 inch = 0.179 inch [3.78 mm + .7620 mm = 4.54 mm]) would be needed. If more than one shim is required, place the thickest one next to the spring, not on the head. If one side of the shim is serrated or dimpled, place that side over the valve stem and onto the spring seat.

With the valve inserted into its guide, position the valve spring inserts, valve spring, and retainer over the valve stem. Using a valve spring compressor, compress the spring just enough to install the valve keepers into their grooves **(Figure 11–69)**. Excess pressure may cause the retainer to damage the oil seal. Release the spring compressor and tap the valve stem with a rubber mallet to seat the keepers. When doing this, the valve will open slightly. To prevent damage to the valves, never tap the stems with the cylinder head lying flat on the bench. Turn the head on its side or raise it off the bench.

CAUTION!

If the keepers are not fully seated, the spring assembly could fly apart and cause personal injury (or serious damage to the engine if it occurs while the engine is running). For these reasons, it is good practice to assemble the valves with the retainers facing a wall and to wear eye protection.

CASE STUDY

While traveling, the engine in a family's late-model car overheats. Water is added to the radiator. Then the car is taken to a service station to be diagnosed and repaired. Simple tests reveal a heater hose has split and is leaking coolant. The hose is replaced and new coolant added to the system. With the car repaired, the family continues on their trip.

After driving only a short time, smoke is noticed coming out of the tailpipe. The engine shakes violently whenever it is placed in a load condition. The driver turns the car around and heads back to the shop that fixed the leak. Upon arrival, the driver says the technician must have knocked something loose or broken something while replacing the hose. The technician doubts this, but agrees to take another look.

A cylinder power balance test would verify this, but it cannot be conducted because of the erratic idle. Instead, the spark plugs are removed and inspected. All look normal except for plug #2, which looks newer than the rest. A comparison test is taken and cylinder #2 has very low dry and wet readings. Then a cylinder leakage test is conducted. Cylinder #2 has excessive leakage and air can be felt leaving the exhaust pipe.

These test results lead to the conclusion that the engine is running on one less cylinder because an exhaust valve is not sealing. The customer still feels the technician did something wrong but agrees to allow the technician to remove the cylinder head for further diagnosis. As soon as the head is off the block, the problem is identified. The exhaust valve seat for cylinder #2 came loose from the head and is preventing the valve from closing. The technician then explains that this undoubtedly happened when the engine overheated, due to the expansion of the metal. The customer is satisfied with the explanation and allows the technician to correct the problem. The family is then able to continue their trip.

Valve spring compressors are available in different designs for different applications. The most commonly used compressor is hand operated. There are two different hand-operated designs: a universal tool with interchangeable jaws and a compressor specifically designed for OHC engines **(Figure 11–70)**. This compressor allows for positive contact on the retainer without removing the camshaft. Another type of spring compressor is air operated. Air-operated spring compressors are typically found in high-volume engine rebuilding shops.

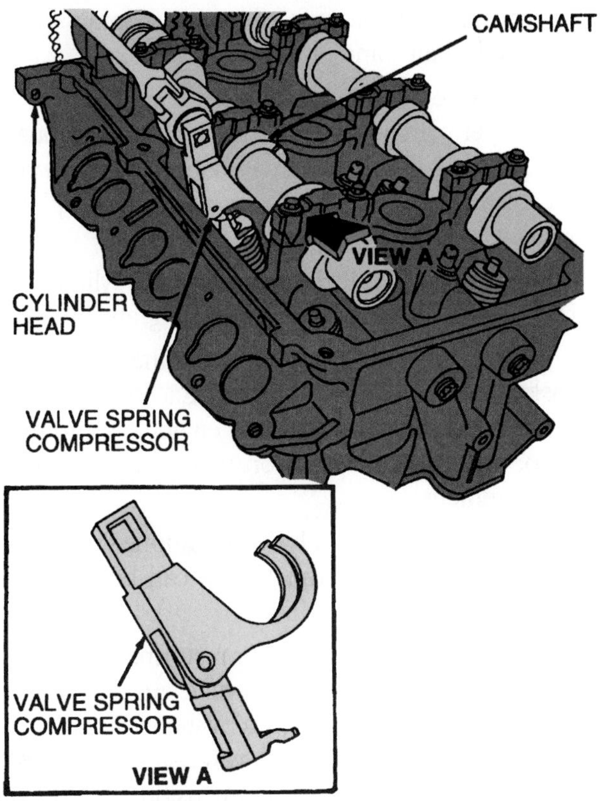

Figure 11–70 A special valve spring compressor for OHC engines. *Courtesy of Ford Motor Company*

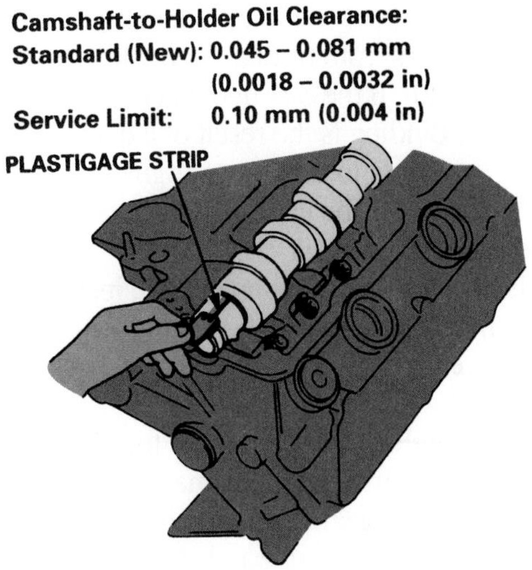

Figure 11–71 OHC camshaft bearing clearances can be checked with plastigage. *Courtesy of American Honda Motor Co., Inc.*

OHC Engines

After the valves are installed and the cylinder head is assembled, the camshaft can be installed in the cylinder head. Some engines have a separate camshaft housing that bolts to the cylinder head. This should be installed with the proper seals and gaskets. Make sure the seals are properly seated in their grooves before tightening the housing to the head.

Service to the camshaft bearings can now be done. Full-round insert bearings are pressed into the bores in the cylinder head. Special tools are designed to make this job easier. Never use a standard camshaft bearing driver and hammer to install these bearings. The hammering can easily break or damage the bearing supports. After each bearing is fully seated in its bore, double-check the alignment of the bearing's oil hole with the oil hole in the head.

Some overhead camshafts do not have bearing inserts. The bores in the aluminum casting serve as a bearing surface. These surfaces can be cleaned up and/or align bored if needed.

Most late-model OHC engines use split bearings and bearing caps or have a separate housing for the camshaft. Working with split camshaft bearings is like working with crankshaft main bearings. This includes checking bearing clearances with plastigage **(Figure 11–71)**.

Now gather the rocker arms, lash adjusters, pushrods, lifters, and other parts that transfer the motion of the camshaft to the valve stem. Coat all of these parts with clean engine oil. The order in which these parts should be installed is given in the service manual. Some parts are installed before the camshaft is installed; others are installed after. Install the required components into their bores in the cylinder head. Make sure they fit securely and do not bind in their bores.

Before installing the camshaft, wipe off each cam bearing with a lint-free cloth, and then thoroughly coat the camshaft lobes and bearing journals with assembly lube **(Figure 11–72)**. Special prelubricants can be used only if specifically recommended by the manufacturer.

Install the camshaft. If the assembly has bearing caps, place them in their correct position and tighten them according to specifications. Once the shaft is in place in the cylinder head, the shaft should be easily turned. On some engines, turning the camshaft causes the valves to open.

Figure 11–72 Coating the camshaft with lubricant. *Courtesy of Dana Corporation*

If this is the case, stand the cylinder head on its end to prevent the valves from hitting the workbench while you turn the camshaft. If the cam does not turn, binding might be the cause. Binding is the result of a damaged bearing, a nick on the cam's journal, or a slight misalignment of the block journals. The cause of the problem should be identified and corrected.

Rocker arm assemblies can now be installed by turning the cam until the cam lobe for the valve faces away from a valve stem. Typically the rocker arms can be slipped into position by depressing the valve spring slightly. Follow the same procedure for all of the valves. Install the camshaft sprockets, make sure the keyways are aligned, then torque the retaining bolt to specifications. Now adjust the valve lash (clearance) before installing the head on the engine.

KEY TERMS

Broaching
CC-ing
Damper spring
Guide plate
Hemi-head
Hemispherical combustion chamber
Knurling
Margin
Milling
Poppet
Porosity
Positive seal
Push rod
Quench area
Quenching
Reaming
Shrouding
Spring surge
Turbulence
Umbrella-type seal
Valve face
Valve keeper
Wedge-type combustion chamber

SUMMARY

- Pushrods are the connecting link between the rocker arm and the valve lifter.
- The rocker arm converts the upward movement of the valve lifter into a downward motion to open the valve. It also permits the valves to be angled.
- Aluminum cylinder heads are used on late-model engines because of their light weight. The thermal expansion characteristics of aluminum can lead to problems such as leaking and cracking.
- An efficient combustion chamber must be compact in order to minimize heat loss. Popular combustion chamber designs include the wedge, hemispherical, swirl, and fast-burn varieties.
- Every cylinder of a four-stroke engine contains at least one intake valve and one exhaust valve.
- Multivalve engines feature either three, four, or five valves per cylinder, which means better combustion and reduced misfire and detonation. These benefits are offset to some extent by a more complicated camshaft arrangement.
- The means of resurfacing the deck of a cylinder head include grinding, milling, belt surfacing, and broaching.
- The amount of stock removed from the cylinder head gasket surface must be limited. Excessive surfacing can create problems with the engine's compression ratio, not to mention piston/valve interference and misalignment.
- The two surfaces of a valve reconditioned by grinding are the face and the tip. Valves can be refaced on grinding or cutting machines.
- One of the fastest methods for restoring the inside diameter dimensions of a worn valve guide is knurling. Reaming repairs worn guides by increasing the guide hole size to take an oversize valve stem or by restoring the guide to its original dimension after knurling or installing inserts.
- Pressing out an old valve guide to install a new one can be difficult on some aluminum heads where the interference fit is considerable.
- To ensure proper seating of a valve, the seat must be the correct width, in the correct location on the valve face, and concentric with the guide.
- When grinding a valve seat, choosing the correct size pilot and stone is important. For soft seats such as cast iron, use a hard stone. For hard seats, a soft stone is needed.
- Valve stem seals are used to control the amount of oil between the valve stem and guide. Too much oil produces deposits, while insufficient lubrication leads to excessive wear.
- The valve spring closes the valve and also maintains the valve train contact during the opening and closing of the valve. To determine if a spring needs to be replaced, three tests are valuable: free-standing height, spring squareness, and open/close spring pressure.
- Two critical measurements that must be made before a cylinder head is reassembled and installed are installed stem height and installed spring height. The first of these is determined by measuring the distance between the spring seat and stem tip. The latter is measured by the spring seat to the underside of the retainer when it is assembled with keepers and held in place.

TECH MANUAL

The following procedures are included in Chapter 11 of the *Tech Manual* that accompanies this book:

1. Inspecting cylinder head for wear.
2. Inspecting and testing valve springs for squareness, pressure, and free height comparison.
3. Inspecting valve spring retainers, locks, and valve lock grooves.
4. Replacing valve stem seals in vehicle.
5. Reconditioning valve faces.
6. Inspecting valve lifters, pushrods, and rocker arms.
7. Reconditioning valve seats.

REVIEW QUESTIONS

1. What happens when valve spring tension is too low?

2. Define valve margin.

3. What usually causes warpage in an aluminum cylinder head?

4. What are the two ways pushrods can be checked for straightness?

5. Why do some technicians not consider knurling a long-term repair?

6. Which of the following can be reconditioned by grinding?

a. valve face
b. valve tip
c. both a and b
d. neither a nor b

7. Which of the following is *not* true of knurling?

a. It is one of the fastest techniques for restoring the ID dimensions of a worn valve guide.
b. It reduces the amount of work necessary to re-seat the valve.
c. It is useful for restoring badly worn guides to their original condition.
d. None of the above.

8. What needs to be done to ensure proper valve timing after many OHC cylinder heads have been resurfaced?

9. To ensure proper seating of the valve, the valve seat must be ______.

a. the correct width
b. in the correct location on the valve face
c. concentric with the guide
d. all of the above

10. When grinding valve seats, ______.

a. a pilot shaft is inserted into the valve guide
b. a hard stone should be used on a hard seat
c. a soft stone should be used on a soft seat
d. all of the above

11. If the valve face and valve seat do not contact each other evenly after grindings, ______.

a. regrind with the same stone
b. regrind with stones of different angles
c. discard cylinder head
d. none of the above

12. Technician A says bronze valve guides retain oil better than cast-iron ones. Technician B says cast-iron valve guides are easier to machine than bronze guides. Who is correct?

a. Technician A
b. Technician B
c. Both A and B
d. Neither A nor B

13. Technician A says positive valve stem seals fit tightly around the top of the stem. Technician B says positive stem seals scrape oil off the valve as it moves up and down. Who is correct?

a. Technician A
b. Technician B
c. Both A and B
d. Neither A nor B

14. When fitting a freshly ground valve into a freshly ground seat, Technician A says the seat should be ground more with the same stone if the margin is too small. Technician B says the seat should be ground more with the same stone if the valve seat is too high. Who is correct?

a. Technician A
b. Technician B
c. Both A and B
d. Neither A nor B

15. Multiple valve engines tend to be more efficient than two valve per cylinder engines because they ______.

a. allow for increased port areas

b. have smaller valves

c. provide less restriction to the airflow

d. all of the above

16. Technician A says low spring pressure may allow a valve to float. Technician B says excessive spring tension may cause premature valve train wear. Who is correct?

a. Technician A **c.** Both A and B

b. Technician B **d.** Neither A nor B

17. Which type of surfacing machine uses underside rotary cutters?

a. milling **c.** belt

b. broaching **d.** grinding

18. Which of the following statements is *not* true?

a. Normally, the desired valve face-to-seat contact area for intake valves is 1⁄16 inch.

b. Normally, the desired valve face-to-seat contact area for exhaust valves is 1⁄16 inch.

c. The average valve seat width is 0.060 inch and the average seat begins 0.030 inch from the valve margin.

d. A properly reconditioned seat has the correct seat width and sealing position on the valve face.

19. ______ is the cooling of gases by pressing them into a thin area.

a. Turbulence

b. Shrouding

c. Reaming

d. Quenching

20. While discussing the Valvetronic system, Technician A says the system has no need for a throttle plate. Technician B says the system alters the duration and the lift of the intake valves. Who is correct?

a. Technician A **c.** Both A and B

b. Technician B **d.** Neither A nor B

12

LUBRICATING AND COOLING SYSTEMS

OBJECTIVES

■ Name and describe the components of a typical lubricating system. ■ Inspect, service, and install an oil pump. ■ Describe the purpose of a crankcase ventilation system. ■ Explain oil service and viscosity ratings. ■ List and describe the major components of the cooling system. ■ Describe the operation of the cooling system. ■ Describe the function of the water pump, radiator, radiator cap, and thermostat in the cooling system. ■ Test and service the cooling system.

The life of an engine depends largely on its lubricating and cooling systems. If an engine does not have a supply of oil or does not cool itself, the engine will quickly be destroyed.

LUBRICATION

An engine's lubricating system does several important things. It holds an adequate supply of oil to cool, clean, lubricate, and seal the engine. It also removes contaminants from the oil and delivers oil to all necessary areas of the engine **(Figure 12–1)**.

Oil Types

Engine oil is a clean or refined form of **crude oil**. Crude oil, when taken out of the ground, is dirty and does not work well as a lubricant for engines. Crude oil must be refined to meet industry standards. Engine oil (often called motor oil) is just one of the many products that comes from crude oil. Engine oil is specially formulated so that it has the following properties:

- Prompt circulation through the engine's lubrication system
- The ability to lubricate without foaming
- The ability to reduce friction and wear
- The ability to prevent the formation of rust and corrosion
- The ability to cool the engine parts it flows on
- The ability to keep internal engine parts clean

To provide these properties, engine oil contains many additives. Because of these additives, choosing the correct oil for each engine application can be a difficult task. However, the **American Petroleum Institute (API)** has developed service ratings for motor oil that greatly simplifies oil selection.

The API classifies engine oil as standard or S-class for passenger cars and light trucks and as commercial or C class for heavy-duty commercial applications. Additionally, various grades of oil within each class are further classified alphabetically according to their ability to meet the engine manufacturers' warranty specifications.

Currently these ratings progress from SA to SL for gasoline engines. SA oil is pure mineral oil with no additives and is designed for very light-duty applications. SB to SL oils have been modified to meet the requirements of current engines. SL oils are recommended for today's

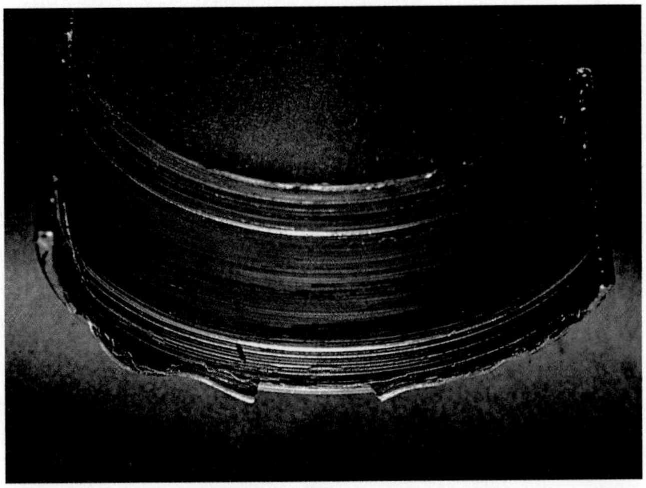

Figure 12-1 Damage to this main bearing was caused by a lack of oil. *Photo courtesy of Clevite Engine Parts*

TABLE 12–1 ENGINE OIL SERVICE RATINGS

Rating	Comments
SA	Straight mineral oil (no additives); not suitable for use in any engine
SB	Nondetergent oil with additives to control wear and oil oxidation
SC	Obsolete since 1964
SD	Obsolete since 1968
SE	Obsolete since 1972
SF	Obsolete since 1980
SG	Obsolete since 1988
SH	Obsolete since 1993
SJ	Obsolete since 1997
SL	Started in 2001

engines. The progression from SB to SL shows how oil engineers have worked with automobile manufacturers to provide efficient and durable engines **(Table 12–1)**.

In addition to oil additives, oil **viscosity** is equally important in selecting an engine oil. The ability of an oil to flow is its viscosity. Viscosity is affected by temperature. For example, hot oil flows faster than cold oil. The rate of oil flow is important to the life of an engine. Because an engine operates under a wide range of temperatures, selecting the correct viscosity becomes even more important.

To standardize oil viscosity ratings, the **Society of Automotive Engineers (SAE)** has established an oil viscosity classification system that is accepted throughout the industry. This system is a numeric rating in which the higher viscosity, or heavier-weight oils, receive the higher numbers. For example, an oil classified as an SAE 50 weight oil is heavier and flows slower than SAE 10 weight oil. Heavyweight oils are best suited for use in high-temperature regions. Low-weight oils work best in low-temperature operations.

Although single-viscosity oils are available, most engine oils are **multiviscosity oils**. These oils carry a combined classification such as 10W-30. Basically this rating says the oil has the viscosity of both a 10- and a 30-weight oil. The "W" after the 10 notes that the oil's viscosity was tested at 0°F (–18°C). This is commonly referred to as the "winter-grade." Therefore, the 10W means the oil has a viscosity of 10 when cold. The 30 rating is the hot rating. This rating was the result of testing the oil's viscosity at 212°F (100°C). To formulate multiviscosity oils, polymers are blended into the oil. Polymers expand when heated. With the polymers, the oil maintains its viscosity to the point where it is equal to a 30-weight oil.

The SAE classification **(Figure 12–2)** and the API rating are usually indicated on oil containers **(Figure 12–3)**. Selecting oils that specifically meet or exceed the manufacturer's recommendations and changing the oil on a regular basis will allow the owner to get the maximum service life from an engine.

ISLAC Oil Ratings The International Lubrication Standardization and Approval Committee (ISLAC) has developed an oil rating that combines SAE viscosity ratings and the API service rating. If an engine oil meets the standards, a sunburst symbol is displayed on the container.

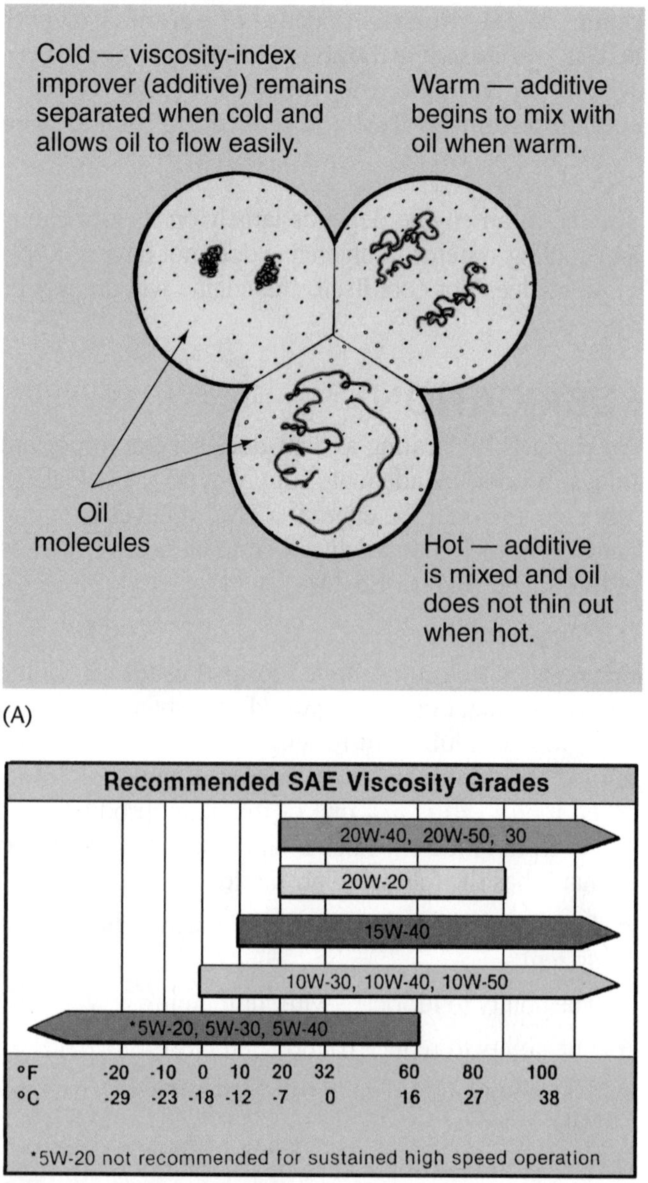

(B)

Figure 12–2 (A) How the polymers affect viscosity. (B) Recommended oil grades according to climate. *Courtesy of Allied Signal Automotive Aftermarket*

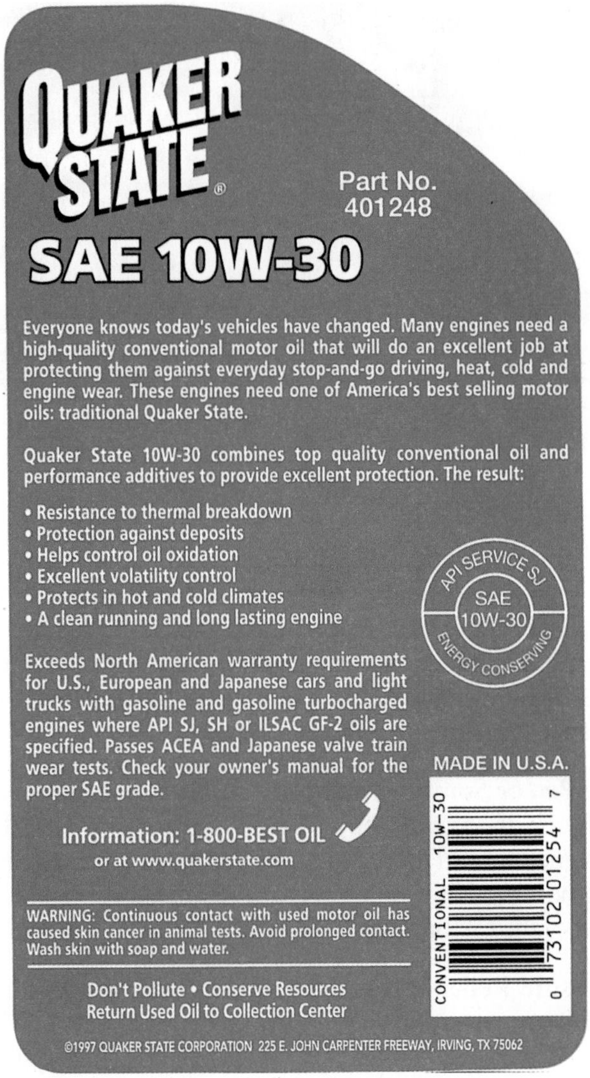

Figure 12-3 API designation and SAE rating. *Courtesy of Quaker State Corporation*

This means the oil is suitable for use in nearly any gasoline engine.

Engine oils can be classified as **energy-conserving** (fuel-saving) **oils**. These are designed to reduce friction, which in turn reduces fuel consumption. Friction modifiers and other additives are used to achieve this.

Synthetic Oils Synthetic oils are considered synthetic because the finished product does not occur naturally and it was made through a chemical, not a natural, process. The introduction of synthetic oils dates back to World War II. Synthetic oils have many advantages over mineral oils, including better fuel economy and engine efficiency by reducing friction. They have low viscosity in low temperatures and a higher viscosity in warm temperatures and tend to have a longer useful life. Synthetic oils cost much more than mineral oils, which is the biggest drawback for using them. Engine oils are available that are blends of mineral oils and synthetics to keep the cost down, but these offer many of the advantages of synthetic oil. Never mix synthetic oils with petroleum-based oils. Also, it is best not to switch from petroleum oils to synthetic oils on an engine that has many miles on it.

CUSTOMER CARE

When changing oil or doing any work on an automobile, use fender covers and do not leave fingerprints on the exterior of the car. If oil or grease gets on the car, clean it off. ■

Oil Consumption

Excessive oil consumption can result from external and internal leaks, faulty accessories, piston rings, and valve guides. Internal leaks **(Figure 12–4)**, which usually result in oil burning, are more difficult to diagnose.

To diagnose an engine that seems to use too much oil, begin by checking the engine for external leaks. These leaks can occur at the valve cover gasket, camshaft expansion plug, oil filter, front and rear oil seals, oil pan gasket, fuel pump gasket, and timing gear cover.

Even the smallest oil leak can cause excessive oil consumption. Losing three drops of oil every 100 feet (30.48 M) equals 3 quarts (2.8 liters) of oil lost every thousand miles. External leaks can occur under normal and abnormal crankcase pressures.

Normal crankcase pressure will cause oil leaks at gaskets or metal-to-metal joints that are in direct contact with oil. Worn seals, faulty gaskets, and loose cover or housing bolts could also cause the problem. Fresh oil on the clutch housing, oil pan, fuel pump, edges of valve covers, external oil lines, distributor shaft housing, base or crankcase filler tube, or at the bottom of the timing gear or chain cover usually indicates that the leak is close to or above that point.

When crankcase pressure is abnormally high, oil is forced out through joints that normally would not leak. Pressure develops when blowby becomes excessive or

Figure 12-4 An indication of an internal oil leak. *Courtesy of Dana Corporation*

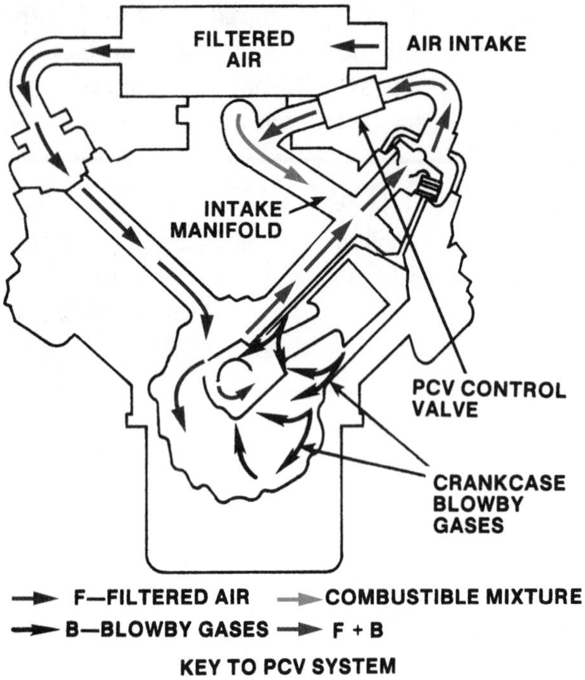

Figure 12-5 The operation of a PCV system.

Figure 12-6 An oil pump with its drive shaft. *Courtesy of Melling Engine Parts*

when a positive crankcase ventilation (PCV) system is not working properly. Blowby is a term used to describe the gases that escape the combustion chamber and enter the crankcase. These gases leak between the piston rings and the cylinder walls when a total seal is not there. Blowby gases are normally pressurized air/fuel mixture and/or pressurized exhaust gases. The PCV system provides a continuous flow of fresh air through the crankcase to relieve the pressure and prevent the formation of corrosive contaminants **(Figure 12–5)**. If the PCV valve or connecting hoses become clogged, excessive pressure will develop in the crankcase. This might force oil into the air cleaner or cause it to be drawn into the intake manifold.

SHOP TALK

Internal leaks are frequently the result of aluminum intake manifolds on V6 and V8 engines because aluminum may warp due to heat. If an unacceptable amount of warpage is found, removing and milling the manifold is necessary to get it to seal. ■

LUBRICATING SYSTEMS

The main components of a typical lubricating system are described here.

Oil Pump The oil pump **(Figure 12–6)** is the heart of the lubricating system. Just as the heart in a human body circulates blood through veins, an engine's oil pump circulates oil through passages in the engine.

Oil Pump Pickup The **oil pump pickup** is a line from the oil pump to the oil stored in the oil pan **(Figure 12–7)**. It usually contains a filter screen, which is submerged in the oil at all times. The screen serves to keep large particles from reaching the oil pump. This screen should be cleaned any time the oil pan is removed.

Oil Pan or Sump The oil pan attaches to the crankcase or block. It serves as the reservoir for the engine's oil. It is designed to hold the amount of oil needed to lubricate the engine when it is running, plus a reserve. The oil pan helps to cool the oil through its contact with the outside air.

Pressure Relief Valve Since the oil pump is a **positive displacement pump**, an oil **pressure relief valve** is included in the system to prevent excessively high system pressures from occurring as engine speed is increased. Once oil pressure exceeds a preset limit, the spring-loaded pressure relief valve opens and allows the excess oil to bypass the rest of the system and return directly to the sump.

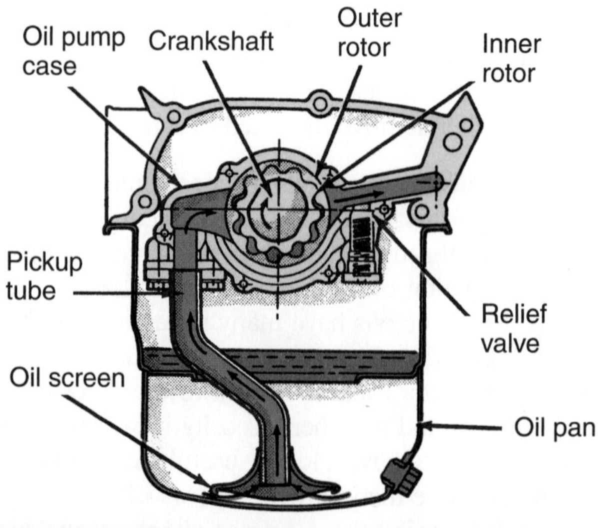

Figure 12-7 Oil pump pickup. *Courtesy of DaimlerChrysler Corporation*

FRAM
TOUGH GUARD

Figure 12-8 An oil filter being installed on an engine block. *Courtesy of FRAM*

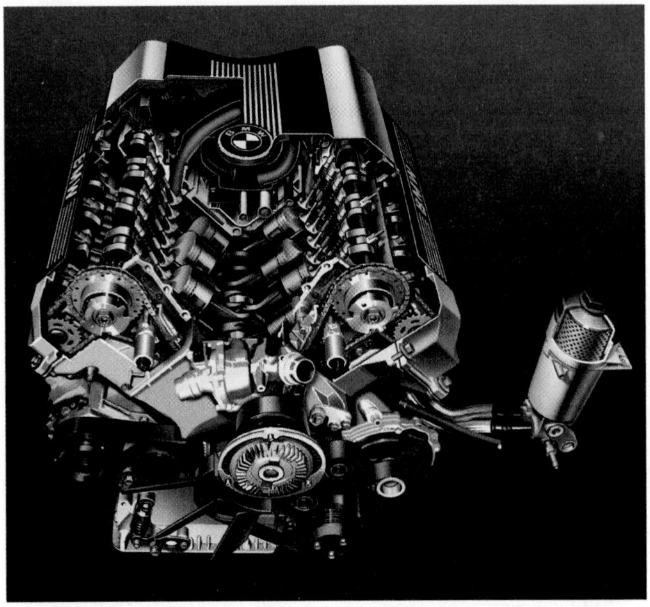

Figure 12-9 Oil lines carry the engine oil in and out of this remote filter before it moves through the engine to lubricate parts. *Courtesy of BMW of North America, Incorporated*

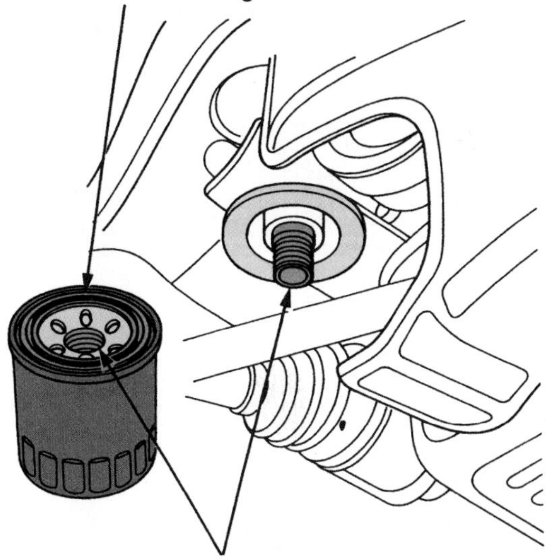
Apply oil to rubber seal before installing.
Inspect threads and rubber seal surface.

Figure 12-10 An oil filter and its seat on the engine. *Courtesy of American Honda Motor Co., Inc.*

Oil Filter All of the oil that leaves the oil pump is directed to the oil filter **(Figure 12–8).** This ensures that very small particles of dirt and metal suspended in the oil will not reach the close-fitting engine parts, causing premature wear. An oil filter is a disposable metal container filled with a special type of treated paper or other filter substance (cotton, felt, or the like) that catches and removes impurities from the oil.

Oil from the engine's oil pump enters the filter and passes through the element of the filter. From the element, the oil flows back into the engine's main oil gallery **(Figure 12–9).**

The oil filter is usually mounted on and sealed to an adapter that bolts to the engine block **(Figure 12–10).** However, it may be attached to the timing cover or remotely mounted with oil lines connecting the filter mount to the oil galleries in the engine block.

Some oil filters have an anti-drainback valve that prevents oil drainage from the filter when the engine is not running. This allows for a supply of filtered oil to the engine as soon as the engine is started and has oil pressure.

Today's engines have a full-flow oil filtration system. All of the oil going to the engine's bearings goes through the filter first. However, should the filter become plugged, a relief valve in the filter opens and allows oil to bypass and go directly to the bearings **(Figure 12–11).** This provides the bearings and the rest of the engine with necessary, though unfiltered, lubrication.

WARNING!

Used oil and oil filters must be disposed of properly and in accordance to local, state, and federal laws.

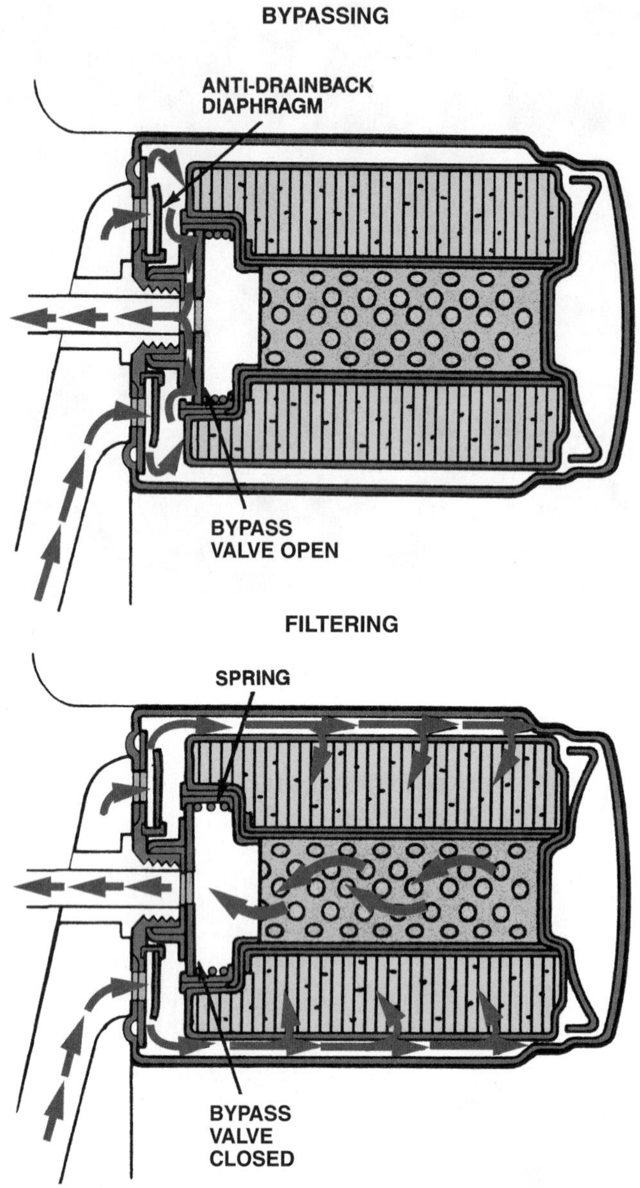

Figure 12-11 Oil flow through the filter. *Courtesy of Ford Motor Company*

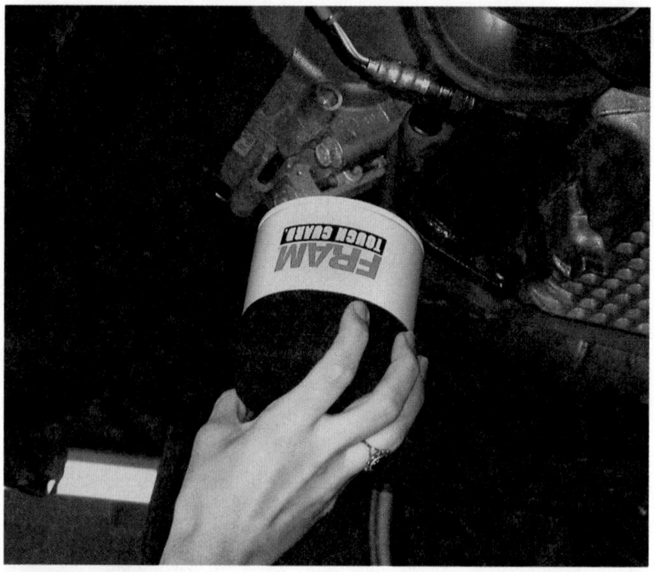

Figure 12-12 Oil filters should be changed when the engine oil is changed. *Courtesy of FRAM*

There are several designs of oil filters. Always use the type of filter recommended for the vehicle you are working on **(Figure 12–12)**.

Engine Oil Passages or Galleries From the filter, the oil flows into the engine oil galleries. These galleries consist of interconnecting passages that have been drilled completely through the engine block during manufacturing. The outside ends of the passages are blocked off so the oil can be routed through these galleries to various parts of the engine. The crankshaft also contains oil passages (oilways) to route the oil from the main bearings to the connecting rod bearing surfaces.

Engine Bearings Since oil is delivered to the engine bearings by an oil gallery, an oil hole is machined in the bearing for alignment with the oil gallery in the engine block. Oil also flows to the connecting rod bearings. Once the oil has been used by the bearing, it flows out of the oil clearance space and is replenished with a fresh supply of oil under pressure from the oil pump. This oil is then thrown off the bearing surface by the spinning motion of the crankshaft. The splashed oil then lubricates other parts of the engine, such as the cylinder walls and pistons.

Oil Pressure Indicator The driver can monitor oil pressure by looking at a gauge, which indicates the engine oil pressure at all times, or it can be a warning light that will come on whenever the engine is running with insufficient oil pressure **(Figure 12–13)**. The warning light is the most common oil pressure indicator.

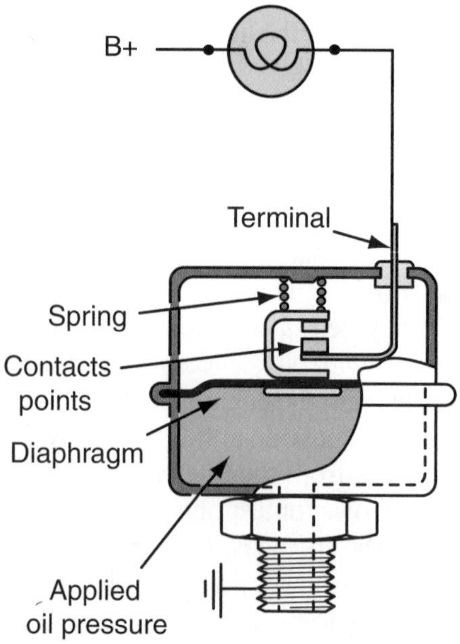

Figure 12-13 An oil pressure sensor for a warning gauge.

Oil Seals and Gaskets These are used throughout the engine to prevent both external and internal oil leaks. The most common materials used for sealing are synthetic rubber, soft plastics, fiber, and cork. In critical areas, these materials might be bonded to a metal.

Dipstick The dipstick is used to measure the level of oil in the oil pan. The end of the stick is marked to indicate when the engine oil level is correct. It also has a mark to indicate the need to add oil to the system.

OIL PUMP

The oil pump is usually located in the oil pan **(Figure 12–14)**. Its purpose is to supply oil to the various moving parts in the engine. To make sure the parts are lubricated, an adequate amount of oil must be delivered to the parts. The amount of oil flow through the engine depends on the volume of oil available, the pressure of the oil, and the clearance or space through which the oil must flow. Engine bearing clearances have a great effect on oil flow.

Increased clearances reduce the resistance to oil flow and, consequently, increase the volume of oil circulating through the engine. This decreased resistance and increased volume lowers the pressure of the oil. The ability of an oil pump to deliver more than the required volume of oil is a safety measure to ensure lubrication of vital parts as the engine wears. Too much oil pressure is seldom a problem. Too little oil pressure can cause poor oil circulation. Poor circulation can result in poor lubrication of some vital parts.

An engine's oil pressure is also determined by the viscosity of the oil and the temperature of the oil. A high-viscosity oil has more flow resistance than a low-viscosity oil. As already mentioned, viscosity decreases as the temperature increases. For this reason, oil pressure is higher in a cold engine than it is when the engine reaches its normal operating temperature.

Figure 12–14 An oil pump mounted to the engine block.

Types of Oil Pumps

The most commonly used oil pumps are the rotor and gear types. Both are positive displacement pumps; that is, a fixed volume of oil passes through the pump with each revolution of its drive shaft. This is because the gears or rotors form a near perfect mechanical seal as they mesh. They trap fixed volumes of oil inside the pump and push it out. Output volume is proportional to pump speed. As engine rpm increases, pump output also increases. Oil pumps are driven, directly or indirectly, by the camshaft or by a gear at the front of the crankshaft.

A typical **rotor-type oil pump (Figure 12–15)** has an inner rotor and an outer rotor, which is driven by the inner rotor. The number of lobes on the rotors varies with the manufacturer of the oil pump. However, the outer rotor always has one more lobe than the inner. When the rotors turn and the rotors' lobes unmesh, oil is drawn into the space. As the rotors continue to turn, the oil becomes trapped between the lobes, cover plate, and top of the pump cavity. Then it is forced out of the pump body by the meshing of the lobes. The meshing of the lobes squeezes the oil out and directs it through the engine. The rate and amount of oil forced out of the pump depends on the diameter and thickness of the pump's rotors.

Gear-type pumps (Figure 12–16) use a drive gear connected to the input shaft and a driven gear. The drive gear turns the driven gear. Both gears trap oil between their teeth and the pump cavity wall. As the gears rotate, the oil is forced out as the gear teeth unmesh. The output volume per revolution depends on the length and depth of the gear teeth. Another style of gear-type oil pump uses an idler gear with internal teeth that spins around the drive gear. In this style of pump, often called a crescent or trochoidal type, the gears are eccentric; that is, as the larger gear turns, it walks around the smaller, moving the oil in the space between.

The rotor type moves a greater volume of oil than a gear type because the space in the open lobe of the outer rotor is greater than the space between the teeth of the gears of a gear-type pump.

Some oil pumps have an intermediate or drive shaft that is driven by a gear on the camshaft. Other oil pumps are driven indirectly by the crankshaft through gears or use an auxiliary shaft meshed with the camshaft. Many oil pumps are an integral part of the timing cover and are driven directly by the crankshaft.

Pressure Regulation

The faster an oil pump turns, the greater its output pressure becomes. Therefore, a pressure-regulating valve is needed to control the maximum oil pressure from the pump. Excessive oil pressure can lead to poor lubrication due to the oil blowing past parts rather than flowing over them. A pressure regulator valve is loaded with a closely

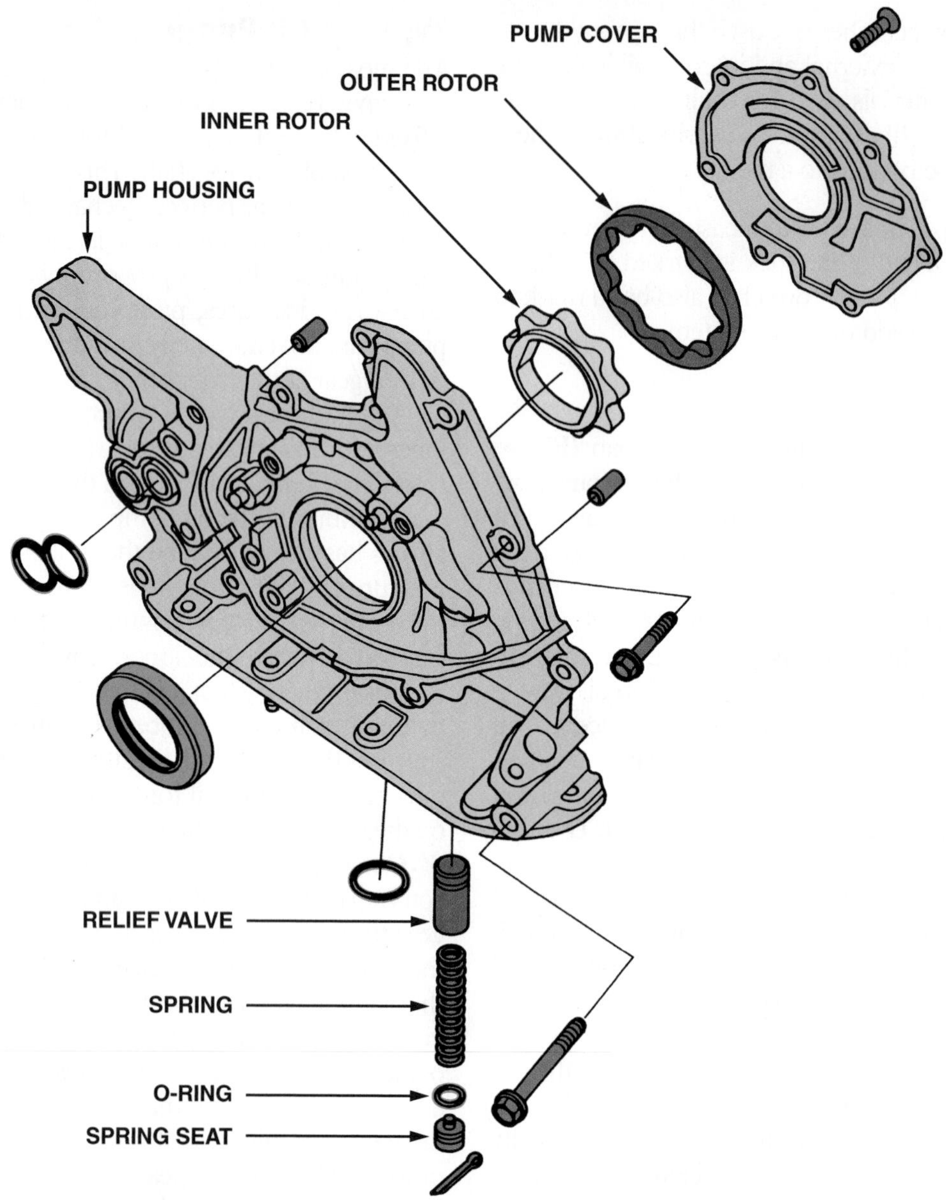

Figure 12-15 A rotor-type oil pump. *Courtesy of American Honda Motor Co., Inc.*

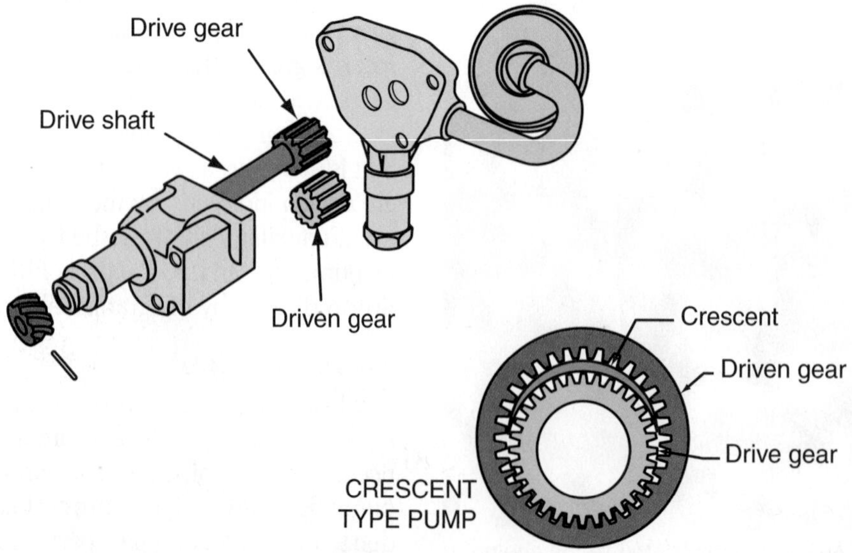

Figure 12-16 Two popular types of gear-driven oil pumps.

Figure 12–17 An oil pressure regulator valve.

calibrated spring that allows oil to bleed off at a given pressure. If the engine's manufacturer decides that 50 psi (344.75 kPa) of oil pressure is desirable in the engine, the pressure-regulating valve **(Figure 12–17)** will not allow the pressure to go beyond 50 psi (344.75 kPa). When the pressure on the output side of the pump reaches this point, it presses against either a check valve, a ball, or a plunger, forcing it to unseat and allow oil to bypass and return to the inlet side of the pump.

OIL PUMP INSPECTION AND SERVICE

Many technicians install a new or rebuilt oil pump on each engine they rebuild. Whenever an engine is being overhauled, the oil pump should be carefully inspected and thoroughly cleaned.

Although the oil pump is probably the best-lubricated part of the engine, it is lubricated before the oil passes through the filter. Therefore, it can experience premature failure because of dirt or other materials entering the pump. Foreign particles can cause three kinds of trouble in a pump:

1. Fine abrasive particles gradually wear the surfaces, causing a reduction in efficiency.
2. Hard particles larger than the clearances can cause scoring and raising of metal as they pass through, finally resulting in seizure.
3. Large particles that cannot pass through will physically lock up the pump.

Of course, when the pump seizes or locks up, the intermediate or drive shaft will also be twisted or sheared off.

During normal operation the pickup bypass valve seats on the supporting cross strap **(Figure 12–18)**. When there is a demand for a large amount of oil and the oil is cold and thick, the valve will unseat and allow the oil to bypass the screen and go directly into the pump. If the screen becomes plugged, the bypass valve will be open during most of the time the engine is running. As the oil

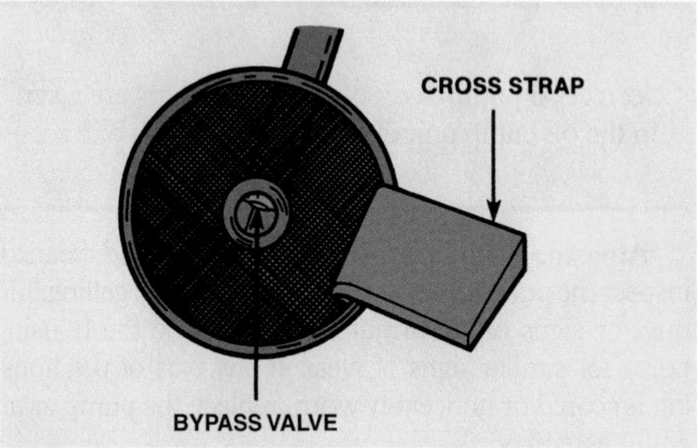

Figure 12–18 A pickup screen with the cross strap bent out of position to show the bypass valve.

rushes through the bypass valve, it can cause a swirling motion or vortex in the oil pan. This could draw up dirt that is either floating in the oil or lying in the bottom of the pan.

To thoroughly inspect the oil pump, it must be disassembled. Carefully remove the pressure relief valve and note the direction in which it is pointing so it can be reinstalled in its proper position. If the relief valve is installed backwards, the pump will not be able to build up pressure.

Before disassembling the pump, mark the gear teeth so they can be reassembled with the same tooth indexing **(Figure 12–19)**. Some pumps have the gears or rotors marked when they are manufactured. Once all the serviceable parts have been removed, clean them and dry them off with compressed air.

SHOP TALK

Use a paint stick or other nonviolent means to mark the gears, if possible. If a center punch is used, make sure you file down the raised material around the indent before reassembling the pump. The raised material that results from making the indent may cause interference and wear if not removed. ■

Figure 12–19 Mark the gear teeth so they can be reassembled with the same indexing.

USING SERVICE MANUALS

Correct oil pump disassembly instructions are given in the oil pump unit of the engines section of a service manual. ■

After the pump has been disassembled and cleaned, inspect the pump gears or rotors for chipping, galling, pitting, or signs of abnormal wear. Examine the housing bores for similar signs of wear. If any part of the housing is scored or noticeably worn, replace the pump as an entire assembly.

Check the mating surface of the pump cover for wear. If it is worn, scored, or grooved, replace the pump. Use a feeler gauge and straightedge to check the flatness of the cover. The service manual gives the maximum and minimum acceptable feeler gauge thicknesses for the cover. If the cover is excessively worn, grooved, or scratched, it should be replaced.

Use an outside micrometer to measure the diameter and thickness of the outer rotor **(Figure 12–20)**. The inner rotor's thickness should also be checked. If these dimensions are less than the specified amount, the rotors must be replaced.

With rotor pumps, assemble the rotors back into the pump body. Use a feeler gauge to check the clearance between the outer rotor and pump body **(Figure 12–21)**. If the manufacturer's specifications are not available, replace the pump or rotors if the measured clearance is greater than 0.012 inch (.3048 mm).

After checking the outer rotor-to-pump housing clearance, position the inner and outer rotor lobes so they face each other. Measure the clearance between them with a feeler gauge **(Figure 12–22)**. A clearance of more than 0.010 inch (.2540 mm) is unacceptable.

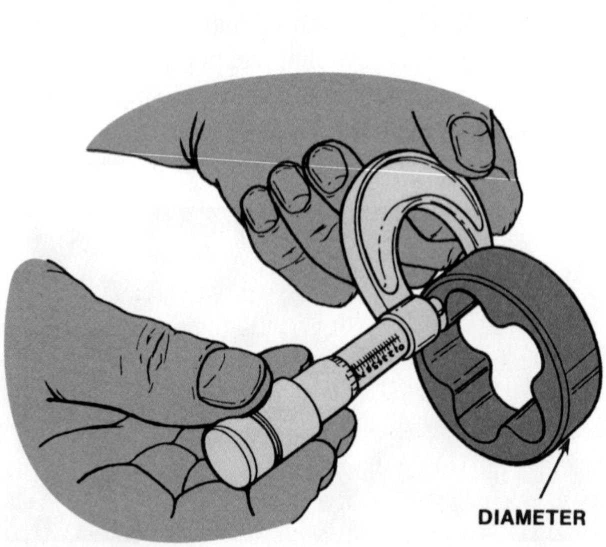

Figure 12-20 Measuring the outer rotor with an outside micrometer.

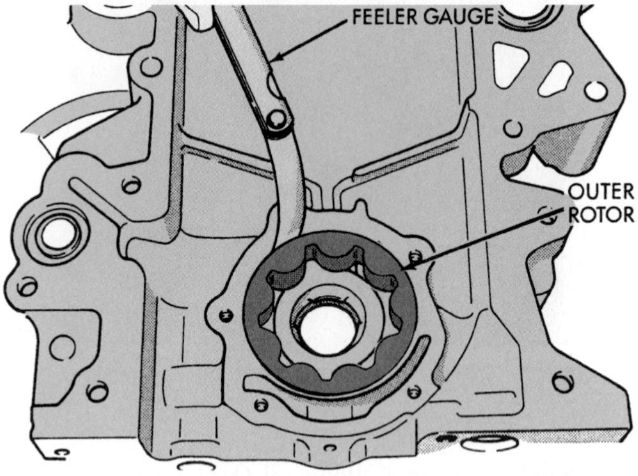

Figure 12-21 Checking clearance between the outer rotor and the pump body. *Courtesy of DaimlerChrysler Corporation*

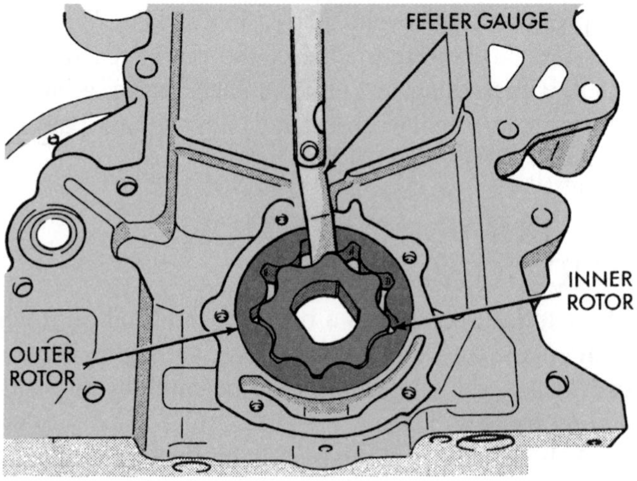

Figure 12-22 Measuring clearance between the inner and outer rotor lobes. *Courtesy of DaimlerChrysler Corporation*

On a gear-type pump, it is important to measure the clearance between the gear teeth and pump housing. Take several measurements at various locations around the housing **(Figure 12–23)** and compare the readings. If the clearance at any point exceeds 0.005 inch (.0762 mm), replace the pump as an assembly.

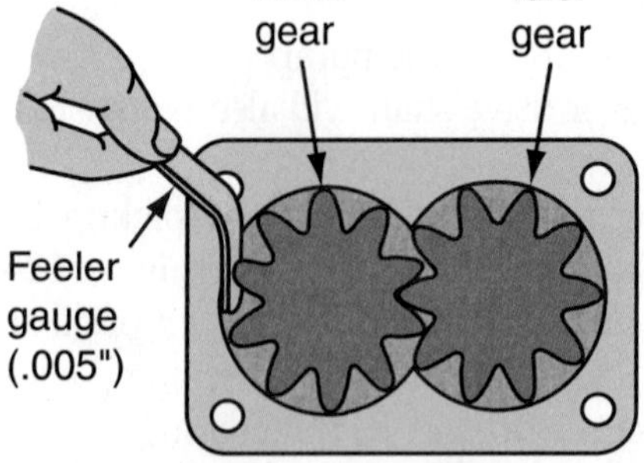

Figure 12-23 Taking several measurements around the housing.

Figure 12-24 Measuring clearance between a straightedge and gears. *Courtesy of General Motors Corporation*

On both gear or rotor oil pumps, place a straightedge across the pump housing and measure the clearance between the straightedge and gears **(Figure 12–24)**. To ensure an accurate reading, make sure the housing surface is clean and free of residual gasket material and that the gears are bottomed in the bore. The desired end play clearance should not exceed 0.003 inch (.1270 mm).

If the pump uses a hexagonal drive shaft, inspect the pump drive and shaft to make sure the corners are not rounded. Check the drive shaft-to-housing bearing clearance by measuring the OD of the shaft and the ID of the housing bearing.

The gasket used to seal the end housing is also designed to provide the proper clearance between the gears and end plate. Consequently, do not substitute another gasket or make a gasket to replace the original one. If a precut gasket was not originally used, seal the end housing with a thin bead of anaerobic sealing material.

Inspect the relief valve spring for signs of collapsing or wear. Check the relief valve spring tension according to specifications. Also, check the relief valve piston for scores and free operation in its bore.

The pickup screen and pump drive **(Figure 12–25)** should be replaced when an engine is rebuilt. The screen and drive must be properly positioned. This is important to avoid oil pan interference and to ensure that the pickup is always submerged in oil. To make the oil pump pickup tube installation easier, several types of drivers are available that are suitable for use with air-powered equipment or with a light mallet.

Make sure the oil pump pickup tube is properly installed and staked, if required, and be sure to use new gaskets and seals **(Figure 12–26)**. Air leaks on the suction side of the oil pump can cause the pressure relief valve to hammer back and forth. Over a period of time, this will cause the valve to fail. Air leaks can also cause oil aeration, foaming, marginal lubrication, and premature engine wear. Care should be taken to make sure all parts on the suction side of the pump fit tightly and there is no place for air leakage. Air leakage often comes from cracked seams in the pickup tube.

On integral pumps, the timing case and gear thrust plate might be worn also. Wear here will limit pump efficiency due to excess clearance. Replace them as necessary.

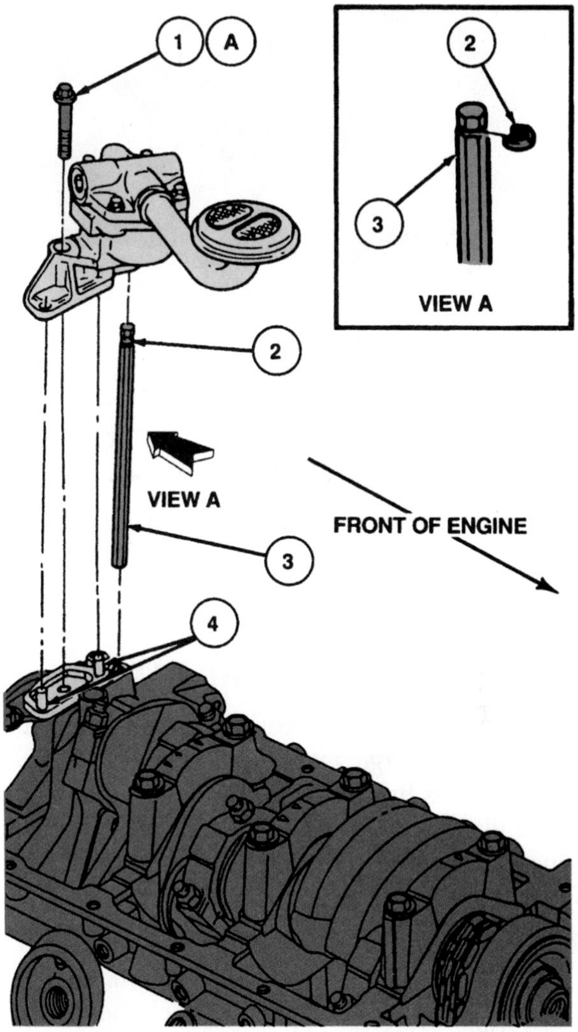

Item	Description
1	Bolt
2	Oil Pump Intermediate Shaft Retaining Ring
3	Oil Pump Intermediate Shaft
4	Dowel
A	Tighten to 40-55 N·m (30-40 Lb-Ft)

Figure 12-25 Installation of an oil pump and its intermediate shaft. *Courtesy of Ford Motor Company*

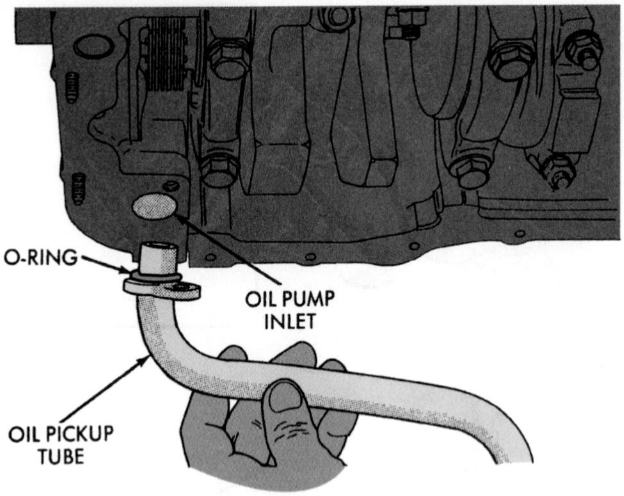

Figure 12-26 Use new gaskets and seals when installing the oil pump pickup tube. *Courtesy of DaimlerChrysler Corporation*

Figure 12-27 The oil pump for this engine is part of the front cover assembly. *Courtesy of General Motors Corporation*

INSTALLING THE OIL PUMP

The pump should be primed before assembly. This can be done by submerging it in clean engine oil. Make sure the inlet port is fully in the oil. Then turn the pump by hand until you see oil flow from the outlet of the pump. To install an older style, distributor-driven oil pump, do so in the following manner:

1. Position the intermediate drive shaft into the distributor socket. With the shaft firmly seated, the stop on the shaft should touch the roof of the crankcase. Remove the shaft and position the stop as necessary.
2. With the stop properly positioned, insert the intermediate drive shaft into the oil pump. Install the pump and shaft as an assembly. Do not attempt to force the pump into position if it will not seat readily. The drive shaft hex might be misaligned with the distributor shaft. To align, rotate the intermediate drive shaft into a new position. Tighten the oil pump attaching screws to torque specifications.
3. Clean and install the oil pump inlet tube and screen assembly.

The installation of a typical camshaft-driven oil pump is done as follows **(Figure 12–27)**:

1. Apply a suitable sealant to the pump and block.
2. Install the pump to its full depth and rotate it back and forth slightly to ensure proper positioning and alignment through the full surface of the pump and the block machined interface surfaces.
3. Once installed, tighten the bolts or screws. The pump must be held in a fully seated position while installing bolts or screws.

SHOP TALK

The instructions here for the installation of either type of pump are general. Specific installation directions, as well as oil priming instructions (if necessary), can be found in the service manual. ■

There are some components that can be considered part of the lubrication system that help to increase engine performance. In newer engines, the baffle assembly **(Figure 12–28)** is one of these components. It is used to

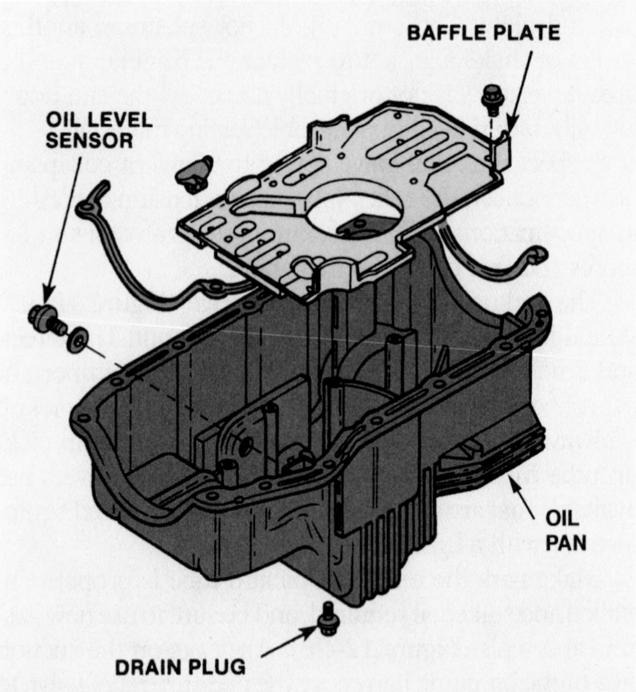

Figure 12-28 An oil pan assembly with baffle plate.

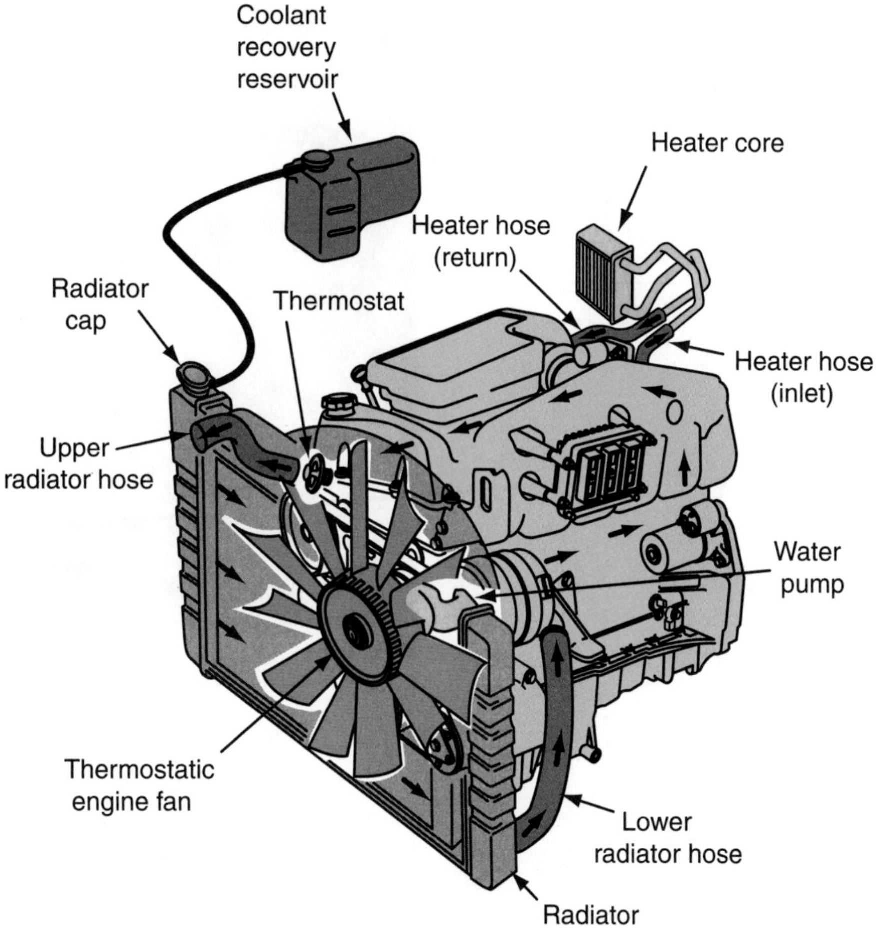

Figure 12–29 The major components of a liquid-cooled system. Arrows indicate the coolant flow.

restrict oil movement in the oil sump. Uncontrolled movement of the oil can allow the oil to mix with air.

COOLING SYSTEMS

Today's internal combustion engines generate a tremendous amount of heat. This heat is created when the air/fuel mixture is ignited and expands inside the engine combustion chamber. Metal temperatures around the combustion chamber can run as high as 1,000°F (537.7°C). To prevent the overheating of cylinder walls, pistons, valves, and other engine parts, it is necessary to dispose of this heat.

Two basic types of cooling systems have been used by automotive manufacturers: liquid-cooled and air-cooled systems. Air-cooled engines have not been used for quite a few years because they had a difficult time maintaining desired engine temperatures and provided poor passenger heating. In a liquid-cooled system **(Figure 12–29)**, heat is removed from around the combustion chambers by a heat-absorbing liquid (coolant) circulating inside the engine. This liquid is pumped through the engine and, after absorbing the heat of combustion, flows into the radiator where the heat is transferred to the atmosphere. The cooled liquid is then returned to the engine to repeat the cycle. These systems are designed to keep engine temperatures within a range where they provide peak performance.

Coolant

Engine **coolant** is actually a mixture of water and antifreeze/coolant. Water alone has a boiling point of 212°F (100°C) and a freezing point of 32°F (0°C) at sea level. A mixture of 67% antifreeze and 33% water will raise the boiling point of the mixture to 235°F (113°C) and lower the freezing point to –92°F (–69°C). As can be seen in **Figure 12–30**, antifreeze in excess of 67% will actually raise the freezing point of the mixture. This chart also indicates why antifreeze that is stored outside might be very hard to pour in temperatures below 0°F. The typical recommended mixture is a 50/50 solution of water and antifreeze/coolant. Some coolant suppliers offer a mixture of water and antifreeze that can be used to top off a cooling system when the level is low **(Figure 12–31)**.

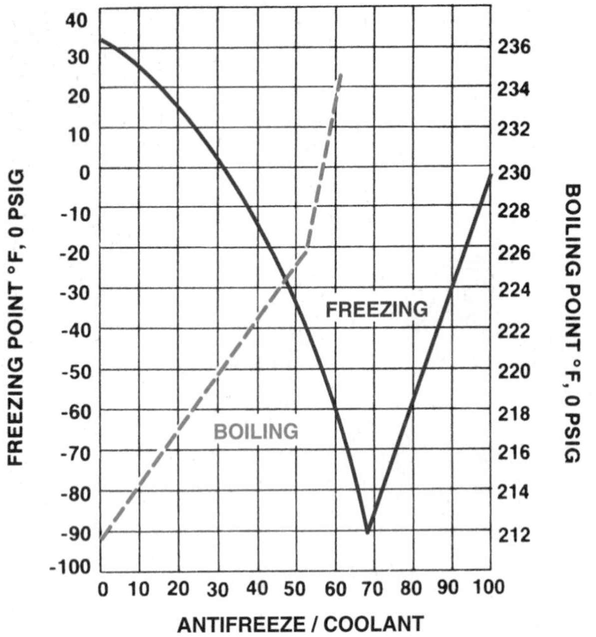

Figure 12-30 The relationship of the percentage of antifreeze in the coolant to the coolant's freezing and boiling points.

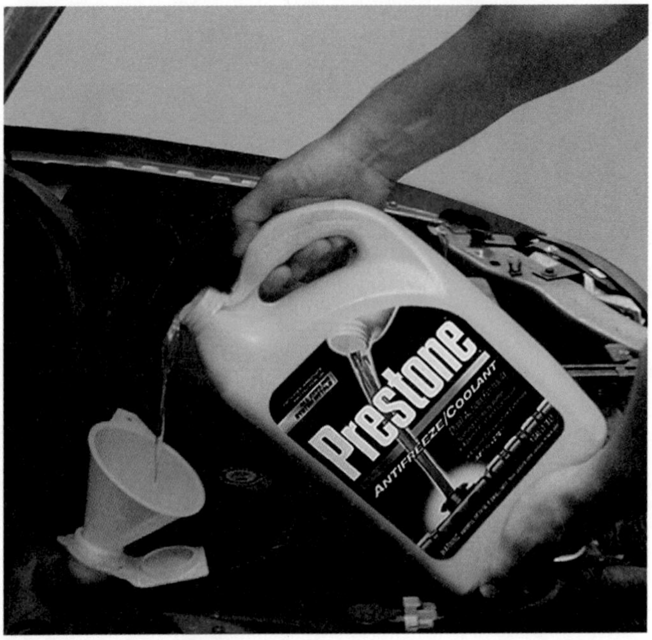

Figure 12-32 The most commonly used antifreeze/coolant is ethylene glycol-based. *Courtesy of Prestone Products*

SHOP TALK

Heat is removed from the engine by the antifreeze/coolant, but is released by the water in the mixture. This is why a 50/50 mixture is typically recommended. ■

The most commonly used antifreeze/coolant is ethylene glycol-based. This type of coolant is green in color and provides good protection regardless of climate **(Figure 12–32)**, but it is poisonous. The coolant has a sweet taste that attracts animals and children and can kill them if it is ingested. Propylene glycol-based coolant has the same basic characteristics as ethylene glycol-based coolant but is not sweet tasting and is less harmful to animals and children. Propylene glycol-based coolants should not be mixed with ethylene glycol.

Other safe coolants **(Figure 12–33)** are also available such as phosphate-free ethylene glycol-based, organic acid technology (OAT) that contains zero phosphates or

Figure 12-31 Topping off a cooling system with a mixture of water and coolant. *Courtesy of Prestone Products*

Figure 12-33 An environmentally safe antifreeze. *Courtesy of Old World Industries*

silicones and is orange in color, and hybrid organic acid technology (HOAT) that is similar to OAT but has additives that are not abrasive to water pumps. Antifreeze formulas have also been developed to prevent or reduce corrosion of aluminum engine parts. Always refer to the service manual when selecting coolant for a particular engine.

The proper mixture of water and antifreeze reduces the amount of rust and lime deposits in the system. These deposits tend to insulate the walls of the water jackets from the coolant. As a result, the coolant is less able to absorb the engine's heat at the points where there is scale. This causes engine hot spots that result in increased component wear and make overheating more likely.

Regardless of the mixture of the coolant or the type of antifreeze used, some lime, rust, and scale will always build in a cooling system. Any deposit on the walls of the water jackets will affect engine cooling. Changes in engine temperature cause the engine parts to expand and contract. Some of these deposits then break off and become suspended in the coolant. The coolant then becomes contaminated and the deposits may collect at a narrow passage, making the passage narrower. This restriction would further lessen the effectiveness of the cooling system. For these reasons and others, the engine's coolant should be replaced and the cooling system flushed every one or two years.

CAUTION!

Never leave coolant out and lying around. Both children and animals will drink it because of its sweet taste. The coolant is poisonous and can cause death.

Chemical Treatments Chemicals are available to clean rust, scale, and other deposits from a cooling system. These chemical cleaners are commonly called radiator flushes. These chemicals are not a substitute for flushing; rather, they are used prior to flushing the system. After the cleaner is added to the cooling system, the engine should be run at the speed and for the amount of time specified on the cleaner's container. Then the system should be totally drained, flushed, and refilled with fresh antifreeze and distilled water. Because distilled water contains fewer minerals than water from a tap, it contributes less to the buildup of scale.

Chemicals are also available to plug small leaks in the cooling system **(Figure 12–34)**. These chemicals work to seal coolant leaks in the radiator and engine (metal components). They do not seal leaking hoses and hose connections. These products are commonly called stop leaks or sealers and are added to the coolant.

Figure 12-34 Stop leak being added to the cooling system. *Courtesy of Prestone Products*

Thermostat

The **thermostat** controls the minimum operating temperature of the engine. The maximum operating temperature is controlled by the amount of heat being produced by the engine at the time and the cooling system's ability to dissipate the heat.

The technical definition of a thermostat is a temperature-responsive control valve. The thermostat controls the temperature and amount of coolant entering the radiator. While the engine is cold, the thermostat remains closed, allowing coolant to circulate only inside the engine. This allows the engine to warm up uniformly and eliminates hot spots. When the coolant reaches the opening temperature of the thermostat, the thermostat begins to allow some flow of coolant to the radiator. The hotter the coolant gets, the more the thermostat opens, allowing more coolant to flow through the radiator. Once the coolant has passed through the radiator and has given up its heat, it re-enters the water pump. Here it is again pushed through the passages surrounding the combustion chambers to pick up heat and start the cycle once again.

Today's thermostat is composed of a specially formulated wax and powdered metal pellet, which is tightly contained in a heat-conducting copper cup equipped with a piston inside a rubber boot. Heat causes the wax pellet to expand, forcing the piston outward, which opens the valve of the thermostat. Today's thermostats are also designed to slow down coolant flow when they are open. This helps to prevent overheating that can result from the coolant moving too quickly through the engine, reducing its effectiveness in absorbing heat.

The most common location of the thermostat is at the front top of the engine block **(Figure 12–35)**. The heat element fits into a recess in the block where it is exposed to hot coolant. The top of the thermostat is then covered

Figure 12–35 A typical thermostat located in the water outlet.

by the water outlet housing, which holds it in place and provides a connection to the upper radiator hose.

The thermostat permits fast warm-up of the engine **(Figure 12–36A)**. Slow warm-up causes moisture condensation in the combustion chambers, which finds its way into the crankcase and causes sludge formation. Most engines are equipped with a coolant bypass, either outside the engine block or built into the casting. Some thermostats are equipped with a bypass valve that shuts off the engine bypass after warm-up, forcing all coolant to flow to the radiator.

Thermostats must start to open at a specified temperature **(Figure 12–36B)**—normally 3°F (1.6°C) above or below its temperature rating. It must be fully opened at about 20°F (–6.6°C) above the start-to-open temperature. They must also permit the passage of a specified amount of coolant when fully open and leak no more than a specified amount when fully closed.

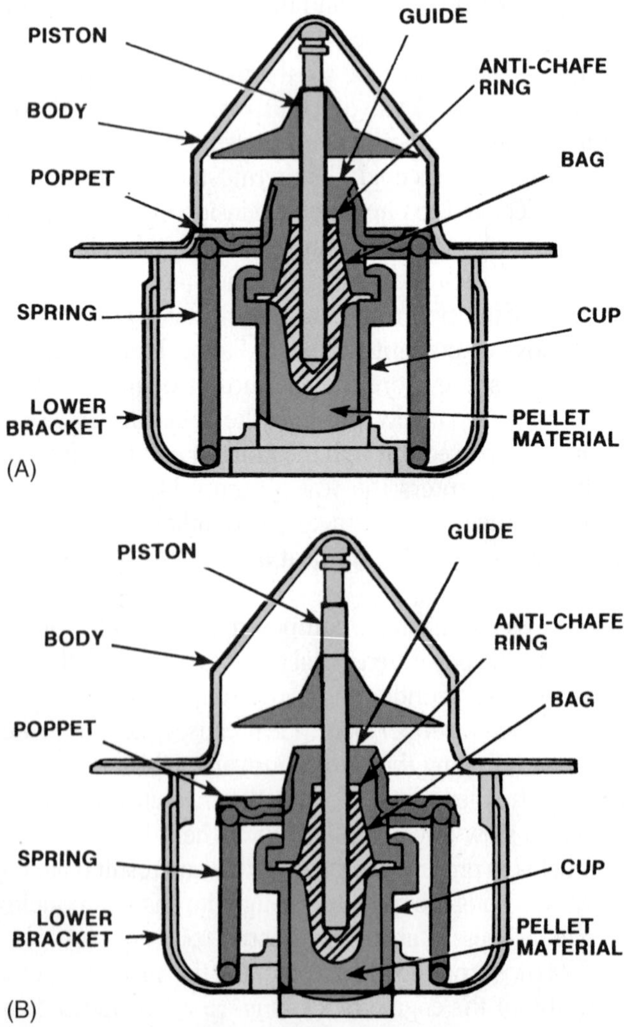

Figure 12–36 (A) Thermostat closed; (B) thermostat open.

Effects on Driveability Many vehicles fail the emissions test because they are operating at too low a temperature. Also, in order to reduce HC exhaust emissions, engine temperatures are kept high. Coolant temperature is raised by using a high-temperature thermostat, a high-pressure radiator cap, and a smaller radiator.

When replacing the thermostat, be sure to use the same temperature thermostat as was used on the original equipment. Using a thermostat with a different opening temperature will cause many driveability problems. The PCM makes adjustments based on operating temperature. If the engine runs cooler than normal, the engine will set fuel injection and ignition timing for a cold engine, which will result in poor gas mileage and excessive emissions.

Water Pump

The heart of the cooling system is the water pump. Its job is to move the coolant through the cooling system. Typically the water pump is driven by the crankshaft through pulleys and a drive V-belt **(Figure 12–37)**. Some pumps may be driven off the camshaft. No matter how they are driven, they all basically work the same way. The pumps are centrifugal-type pumps **(Figure 12–38)** with a rotating paddle-wheel-type impeller to move the coolant. The shaft is mounted in the water pump housing and rotates on bearings. At the drive end, the exposed end, a pulley is mounted to accept the belt. The pulley is driven by the crankshaft. The pump housing usually includes the mounting point for the lower radiator hose.

When the engine is started, the impeller pushes the water from its pumping cavity into the engine block. When the engine is cold, the thermostat is closed. This stops the coolant from reaching the top of the radiator. In order for the water pump to circulate the coolant through the engine during warm-up, a bypass passage is added below the thermostat. This passage must be kept free to eliminate hot spots in the engine. It also allows hot coolant to pass through the valve, which opens the thermostat when it reaches the proper temperature.

Figure 12-37 The front cover of an engine with the water pump bolted to it. *Courtesy of General Motors Corporation*

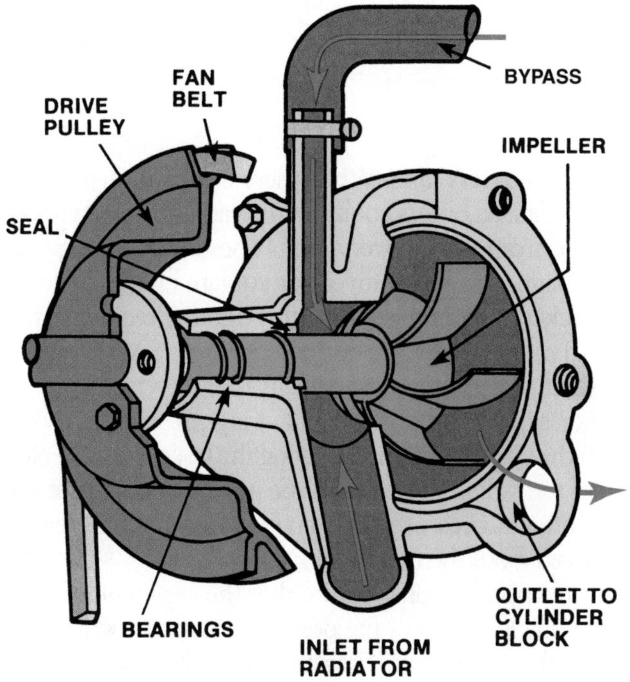

Figure 12-38 An impeller-type water pump.

Radiator

The **radiator** is basically a heat exchanger, transferring heat from the engine to the air passing through it. The radiator itself is a series of tubes and fins that expose the heat from the coolant to as much surface area as possible, thus maximizing the potential of heat being transferred to the passing air.

Factors influencing the efficiency of the radiator are the basic design of the radiator, the area and thickness of the radiator core, the amount of coolant going through the radiator, and the temperature of the cooling air. It is not desirable to have an overly efficient radiator with today's engines. Overefficiency would keep the engine at a low operating temperature, which would increase emission levels.

The radiator is usually based on one of these two designs: cross flow or down flow. In a cross-flow radiator **(Figure 12–39)**, coolant enters on one side, travels through tubes, and collects on the opposite side. In a down-flow radiator, coolant enters the top of the radiator and is drawn downward by gravity. Cross-flow radiators are seen most often on late-model cars because all the coolant flows through the fan airstream, and the design allows for lower hood profiles on body designs.

Most radiators feature petcocks or plugs that allow a technician to drain coolant from the system. Coolant is added to the system at the radiator cap or the recovery tank.

Oil Cooler Radiators used in vehicles with automatic transmissions have a sealed heat exchanger, or form of radiator, located in the coolant outlet tank of the regular radiator. Metal or rubber hoses carry hot automatic transmission fluid to the heat exchanger. The coolant passing over the sealed heat exchanger cools the fluid, which is then returned to the transmission. Cooling the transmission fluid is essential to the efficiency and durability of an automatic transmission.

Radiator Pressure Cap

At one time, the radiator cap was simply designed to keep the coolant (water and alcohol combination in the early days of motoring) from splashing out of the radiator. Today, it still serves that purpose, but it also does much more. Today, radiator caps are equipped with pressure springs and vents. The cap allows for an increase in pressure in the radiator, which raises the boiling point of the coolant. For every pound of pressure put on the coolant, the boiling point is raised about 3¼°F (1.8°C). Today's caps normally are designed to hold between 14 and 17 psi (96.53 and 117.21 kPa). When pressures exceed this level, the seal between the cap and the radiator filler neck opens and allows the excessive pressure to vent into a coolant recovery tank.

There are three basic types of radiator pressure caps: constant pressure, pressure vent, and closed system. The closed system type radiator cap is found on today's cars; the others are found on older model vehicles. The constant pressure type has a lower seal or pressure valve that is held closed until the coolant gets hot enough to build enough pressure to open the valve within the preset

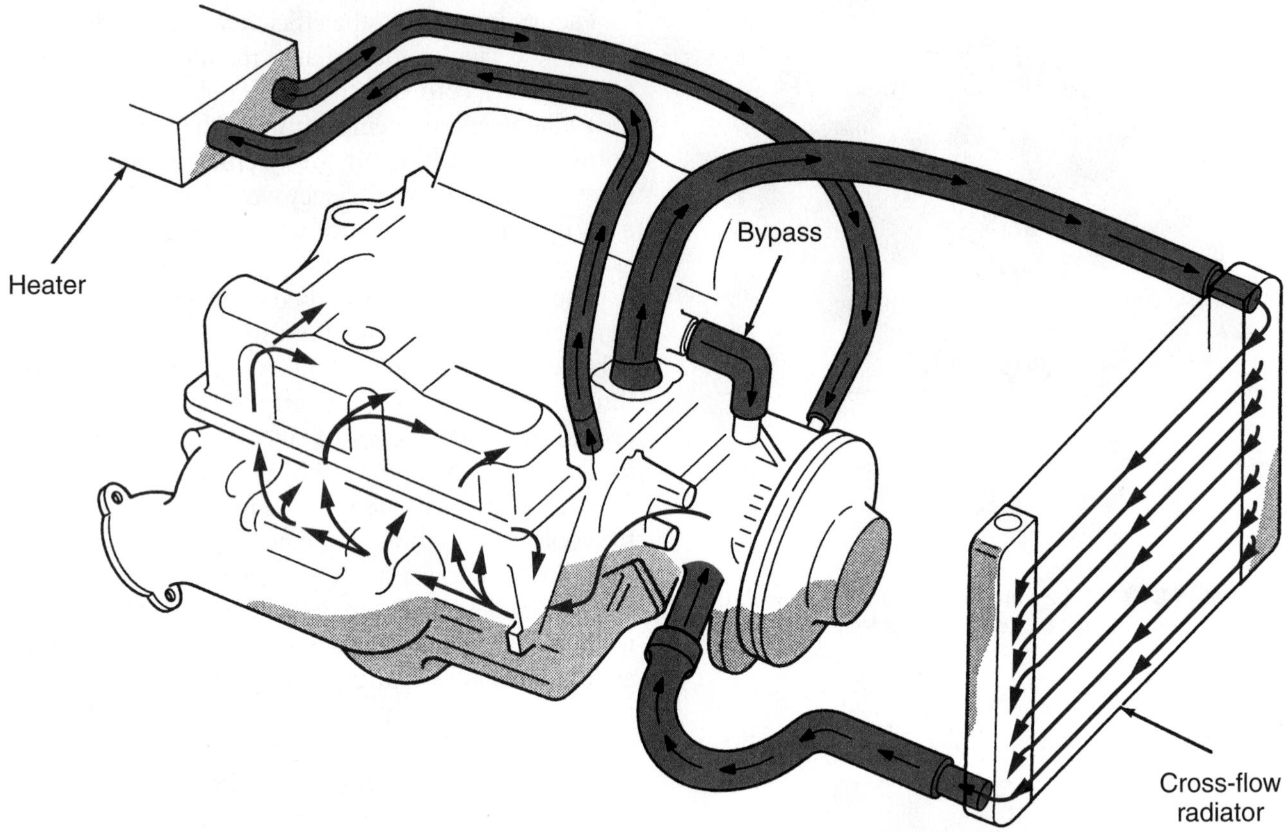

Figure 12-39 The routing of coolant in a cross-flow radiator. *Courtesy of DaimlerChrysler Corporation*

pressure range. The pressure vent-type cap is similar to the constant pressure type. However, it has a vacuum-release valve that is opened by a weight and is kept open to the atmosphere until the pressure is great enough to move the weight and close the valve to prevent atmospheric pressure from entering into the radiator. Like the constant pressure cap, this cap opens to release pressure when it builds to the specified amount.

The closed system type **(Figure 12–40)** works in the same way as the constant pressure cap, except it is designed to keep the radiator full at all times. When the specified pressure is reached, some coolant is released

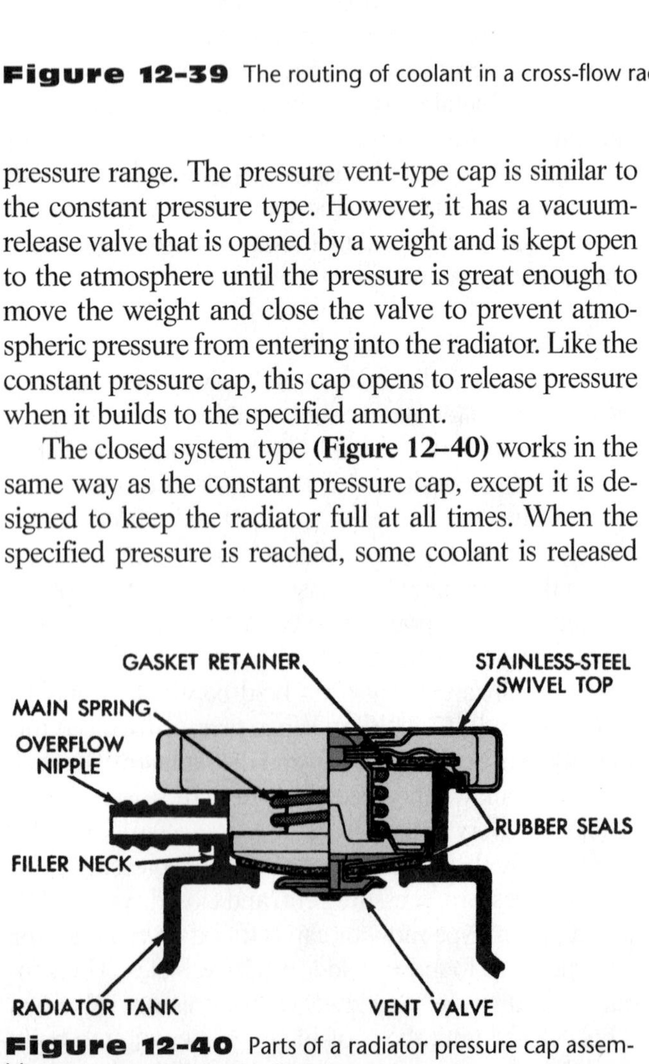

Figure 12-40 Parts of a radiator pressure cap assembly. *Courtesy of DaimlerChrysler Corporation*

into the recovery tank. When there is a vacuum in the radiator (caused by less coolant), the vacuum is used to pull coolant from the recovery tank. These radiator caps are not designed to be removed for coolant checks. Coolant is checked and fluid is added through the recovery tank.

All radiator caps are designed to meet SAE standards for safety. These standards specify that there shall be a detent or safety stop position, allowing pressure to escape from the system without allowing the hot coolant to blow out of the radiator's neck into the person opening the cap. Only after all pressure has been relieved should the cap be removed from the filler neck.

Cap specifications require that the cap must not leak below the low limit of the pressure range and must open above the high limit. Pressure caps should always be tested for the proper pressure release level and checked for gasket cracking, brittleness, or deterioration each time the antifreeze is changed or when any routine cooling system maintenance is performed **(Figure 12–41)**.

Radiator pressure caps are marked indicating the amount of pressure held in the cooling system by the pressure valve's spring. For domestic vehicles, the pressure is stated in pounds per square inch (psi).

Radiator caps for older imported vehicles may be marked "0.9", which indicates that the pressure rating of the cap is 0.9 times normal atmospheric pressure. Since atmospheric pressure is 14.7 psi, a 0.9 cap has a pressure

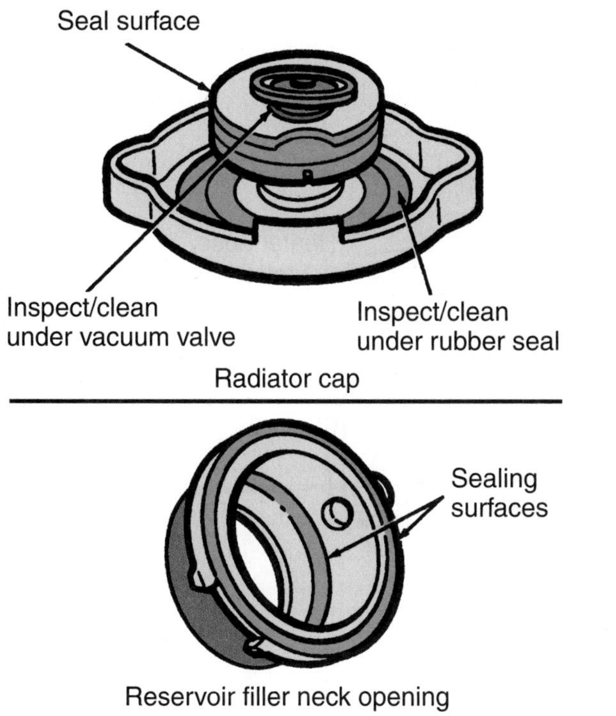

Figure 12–41 Radiator cap inspection. *Courtesy of Ford Motor Company*

rating of about 13.2 psi (14.7 × 0.9). Another common rating is 100. The "100" indicates that the pressure rating of the cap is 100% of atmospheric pressure, or 14.7 psi. Therefore a 15 psi cap would be a good substitute for a 100 cap.

SHOP TALK

Always refer to application charts or a service manual when replacing a pressure cap to make sure the new cap has the same pressure range as the original cap. ■

Expansion Tank Most cooling systems have an **expansion** or **recovery tank**. Expansion tanks are designed to catch and hold any coolant that passes through the pressure cap when the engine is hot. As the engine warms up, the coolant expands. This eventually causes the pressure cap to release. The coolant passes to an expansion tank. When the engine is shut down, the coolant begins to shrink. Eventually, the vacuum spring inside the pressure cap opens and the coolant in the expansion tank is drawn back into the cooling system.

There are marks on most recovery tanks that show where coolant levels should be when the car is running and when it is not **(Figure 12–42)**. To check coolant levels on a car without a recovery tank, remove the radiator cap (when the engine is cold) and see if the coolant is up to where it should be. If there are no markings, make sure

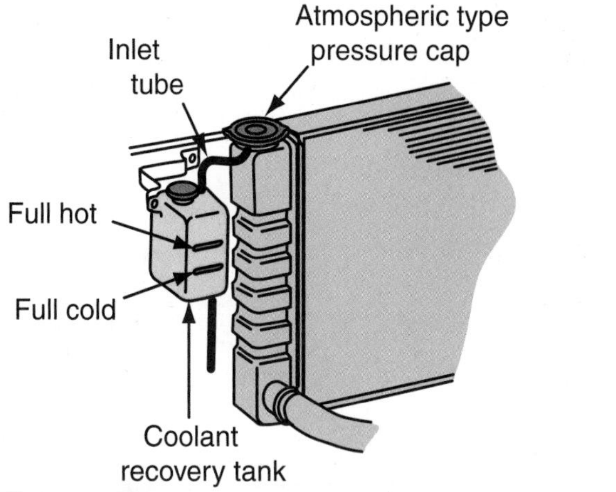

Figure 12–42 A coolant recovery tank with coolant levels marked on the tank.

the coolant is covering the radiator core. If the coolant level is low after repeated filling, there is probably a leak in the cooling system.

WARNING!

When working on the cooling system, remember that at operating temperature the coolant is extremely hot. Touching the coolant or spilling the coolant can cause serious body burns. Never remove the radiator cap when the engine is hot.

Hoses

Coolant flows from the engine to the radiator and from the radiator to the engine through radiator hoses. The radiator is solidly mounted to the vehicle and the engine sits on rubber mounts, which means the engine can move independently of the chassis and the radiator cannot. If the engine were connected solidly to the radiator, the radiator would soon break because of the vibrations and stress. The use of butyl or neoprene rubber hoses cushions the radiator from these vibrations and prevents radiator damage.

A hose is typically made up of three parts: an inner rubber tube, some reinforcement material, and an outer rubber cover. Different covers and reinforcements are used depending on the application of the hose. Hose construction differs based on where it is located and what amounts of temperature and pressure it will face. Cooling system hoses must be able to endure heavy vibrations and be resistant to oil, heat, abrasion, weathering, and pressure.

Most vehicles have at least four hoses in the cooling system; some have five or more **(Figure 12–43)**. Two small diameter hoses send hot coolant from the water pump to the heater core and back. Two larger diameter hoses move the coolant from the water pump to the

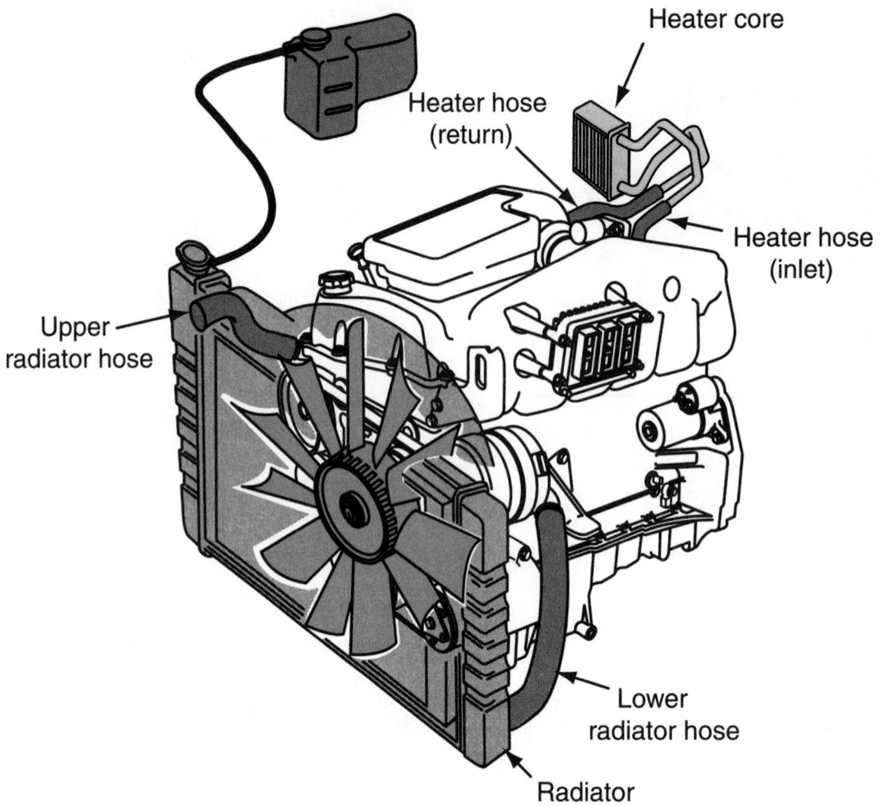

Figure 12-43 A typical cooling system hose arrangement. *Courtesy of American Honda Motor Co., Inc.*

radiator and back into the engine block. The fifth hose, a small diameter bypass hose, allows coolant to circulate within the engine when the thermostat is closed. This hose is not required on all engines because the bypass feature is built into the engine block or cylinder head.

Hoses are sized according to their inside diameter. For example, common heater hoses are $\frac{5}{8}$ or $\frac{3}{4}$ inch. Radiator hoses are larger and have reinforcements that allow them to withstand about six times the normal operating pressure of the cooling system. Lower radiator hoses are normally reinforced with wire to prevent them from collapsing due to the suction of the water pump.

Radiator hoses are seldom straight tubes. They typically must bend or curve around parts to make a good connection without kinking. Straight hoses are not used because bending causes them to collapse at the bend, causing a restriction. Most original equipment radiator hoses are molded to a specific shape to fit specific applications. Often molded hoses are available in a variety of lengths. The hose is then cut to fit a particular application. Some have cutoff marks printed on them to show where they should be cut to fit different applications. Others should be compared to the old hose for a cut reference.

Nearly all original equipment radiator hose is of the molded, curved design. Aftermarket products may be this type or a wire-inserted flex type. The flex-type hose allows greater vehicle coverage per part number but may not be designed for some cars that require radical bends and shapes. Flexible radiator hoses are available in different lengths and diameters. This design can flex or bend into most required shapes without causing a restriction.

Heater hoses are made with reinforcements to help keep their shape. Some applications require a molded shape due to complex routing or curves. Rather than replacing heater hoses with specific molded hoses, formable heater hoses are available. This hose design has a wire spine that allows the hose to bend into a curve without collapsing at the bend. Once the desired shape is obtained, the hose is cut to length and then installed.

Water Outlet The water outlet is the connection between the engine and the upper radiator hose. The water outlet has been called a gooseneck, elbow, inlet, outlet, or thermostat housing. Generally, it covers and seals the thermostat and, in some cases, includes the thermostat bypass.

Most water outlets are made of cast iron, cast aluminum, or stamped steel. Internal corrosion contributes to the failure of water outlets. Cast-iron water outlets are more resistant to this type of corrosion than stamped steel or cast aluminum outlets. A more common cause of failure for a water outlet is the uneven torquing down of the water outlet mounting bolts, which can cause a mounting ear to break off. When this happens, the outlet will not seal and must be replaced.

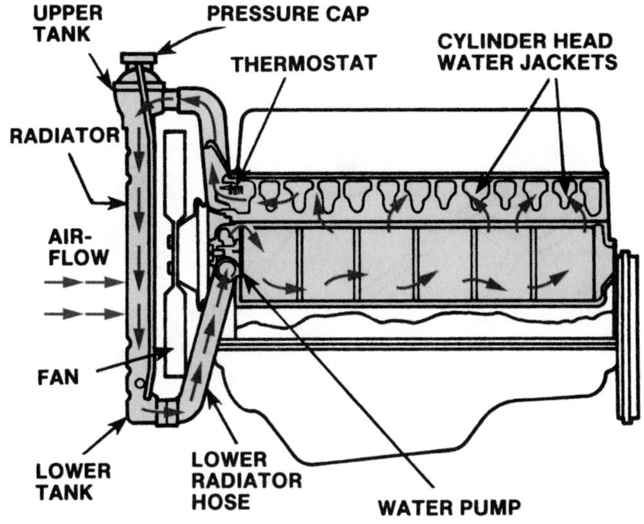

Figure 12-44 The cooling system circulates coolant through the engine's water jackets.

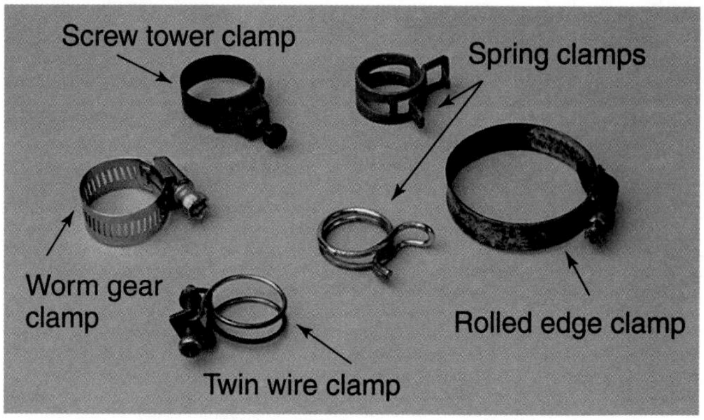

Figure 12-45 Common types of hose clamps.

Water Jackets Hollow passages in the block and cylinder heads surround the areas closest to the cylinders and combustion chambers **(Figure 12–44)**.

Included in the water jackets are soft (core) plugs and a block drain plug. The soft plugs and drain are usually removed during engine teardown. New ones are installed during reassembly. Core plugs are prone to rust and corrosion, and, therefore, will weep coolant or rust through completely. When this happens, the core plugs should be replaced.

Hose Clamps

Hoses are attached to the engine and radiator with clamps. Hose clamps are designed to apply clamping pressure around the outside of the hose at the point where it connects to the inlet and outlet connections at the radiator, engine block, water pump, or heater core. The pressure exerted on this connection is important in making and maintaining a seal at that point.

Original equipment clamps are usually spring steel wires that must be removed and replaced with special pliers. Replacement clamps may use a twin wire with a screw, worm drive screw, or screw tower **(Figure 12–45)**. The worm drive hose clamp is most often used as a replacement clamp for many reasons. This type of clamp provides even pressure around the outside diameter of the hose. It is easy to install and requires no special tools.

Rather than using steel clamps on the hoses, some technicians prefer to use thermoplastic clamps **(Figure 12–46)**. These heat-sensitive clamps are installed on the hose ends and a heat gun is used to shrink the clamp. The shrinking of the clamp tightens the connection. As the engine runs, the heat of the coolant further tightens the connection.

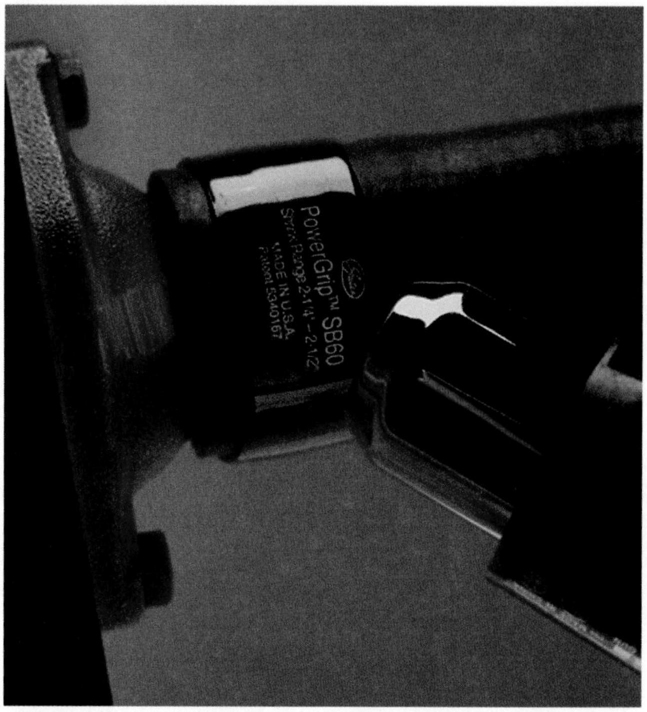

Figure 12-46 Thermoplastic clamps are tightened with a heat gun. *Courtesy of Gates Rubber Company*

Belt Drives

Belt drives have been used for many years. V-belts and V-ribbed (serpentine) belts are used to drive water pumps, power steering pumps, air-conditioning compressors, generators, and emission control pumps **(Figure 12–47)**. Because the belts are flexible, they absorb some shock loads and cushion shaft bearings from excessive loads.

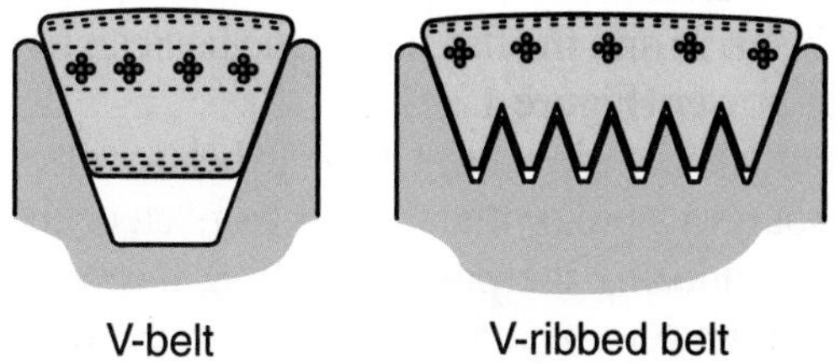

Figure 12-47 A V-belt rides in a single groove whereas a V-ribbed belt rides in several grooves.

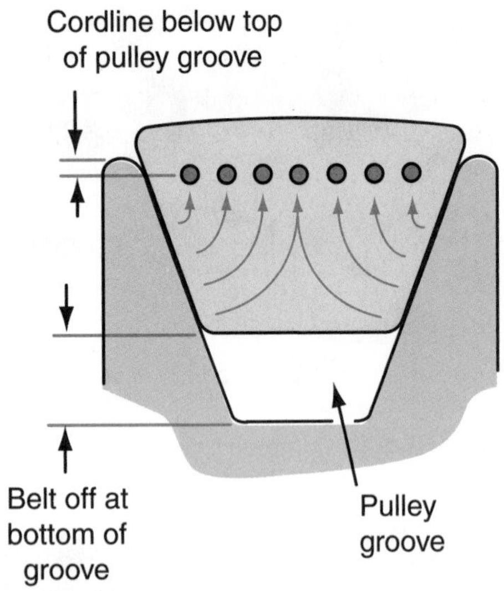

Figure 12-48 The sides of a V-belt contact the grooves of the drive pulley.

The biggest problem with belt drives is that they are neglected. Typically, no one makes an effort to check these vital links until they begin to make noise or break.

V-belts are designed to ride in a matching groove in the pulleys of the system. The angled sides of the belt contact the inside of the pulley grooves **(Figure 12–48)**. This point of contact is where motion is transferred.

Drive belts can be used to drive a single part or a combination of parts. V-belts are typically used to drive a single component. An engine can have three or more V-belts **(Figure 12–49)**. Each of the belts can be a different size and replacement belts must be an exact replacement in length, width, and material. In some cases, two matched belts are used on the same pulley set. This increases the strength of the belt and pulley connection and provides redundancy in case a belt breaks. Matched belts should always be replaced in pairs so they wear together, thus maintaining the same length to prevent slippage and problems.

Most late-model vehicles use a **serpentine belt** to drive all or most accessories. Serpentine belts are long and follow a complex path that weaves around the various pulleys. Though tension is important for all drive belts, it is critical on a serpentine belt because of this complex routing. Serpentine belts are flat on the outside and have a series of continuous ribs on the inside. These ribs are designed to fit into matching grooves in the pulleys. Both the ribbed side and the flat side of the belt can be used to transfer power **(Figure 12–50)**.

Insufficient tension may allow the belt to roll off a pulley or not turn a pulley due to poor contact, or the pulley may slip, reducing the power reaching the component. Excessive tension may put unwanted forces on the pulleys and the shafts they are attached to, leading to belt breakage, glazing, and damage to the driven components.

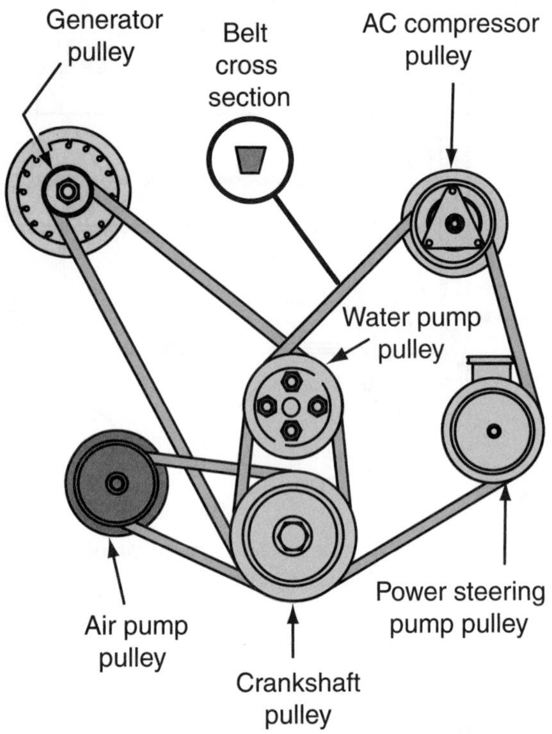

Figure 12-49 An engine can have three or more V-belts to drive different accessories.

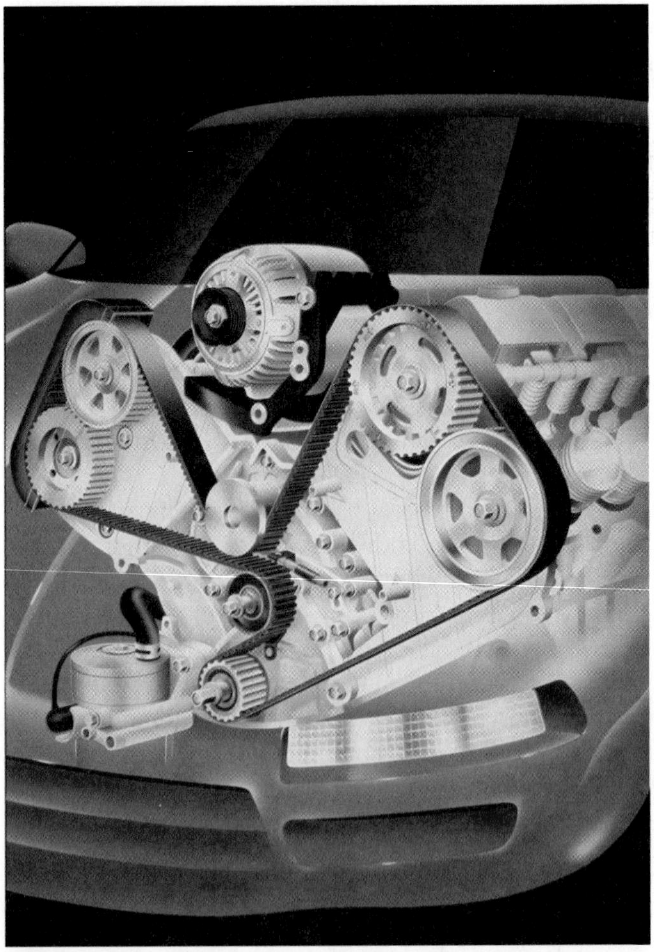

Figure 12-50 A serpentine drive belt. *Courtesy of Gates Rubber Co.*

Over time, serpentine belts will stretch and lose their tension. To compensate for belt stretch and to keep a proper amount of tension on the belt, most serpentine belt systems have a belt tensioner pulley.

A tensioner pulley is typically a spring-loaded pulley **(Figure 12–51)** or wheel that exerts a predetermined amount of the pressure on the belt. This pressure keeps the belt at the desired tension, providing the tensioner pulley was adjusted properly when the belt was installed.

USING SERVICE MANUALS

Proper belt tightening procedures and specifications are given in the specification section of most service manuals. ■

Heat has an adverse effect on drive belts. Belts tend to harden and crack because of excessive heat. Excessive heat usually comes from slippage, which can be attributed to the lack of proper belt tension or oily conditions. When slippage occurs, heat can travel through the drive pulley and down the shaft to the support bearing. These bearings can become damaged. As a V-belt wears, it begins to ride deeper in the pulley groove, reducing its tension and promoting slippage. As this is a normal occurrence, periodic adjustment of belt tensions should be expected.

Cooling Fans

As mentioned earlier, the efficiency of the cooling system is based on the amount of heat that can be removed from the system and transferred to the air. The system needs air. At highway speeds, the ram air through the radiator should be sufficient to maintain proper cooling. At low speeds and idle, the system needs additional air. This air is delivered by a fan. The fan may be driven by the engine, via a belt, or driven by an electric motor.

Figure 12–51 A belt tensioner for a serpentine belt. *Courtesy of Gates Rubber Co.*

The design of the fan found on a vehicle depends on the air requirements of the engine's cooling system. Diameter, pitch, and the number of blades can be varied to attain the needed flow. A fan placed more than 3 inches from the radiator becomes ineffective. It merely recirculates the hot air around the fan blades. For this reason, some radiators are equipped with shrouds. A shroud is a large, circular piece of plastic, metal, or cardboardlike material that extends outward from the radiator to enclose the fan and increase its effectiveness. These shrouds should always be kept intact and should not be modified.

A belt-driven fan is bolted to a pulley on the water pump and turns constantly with the engine. Thus, belt-driven fans always draw air through the radiator from the rear. The power pulley on the crankshaft drives the belt. The fan has several blades made of steel, nylon, or fiberglass attached to a metal hub. Any damage or distortion to the fan will cause it to be out-of-balance, and the fan should be replaced. An out-of-balance fan can cause major problems, including rapid and excessive water pump bearing and seal wear or damage to the radiator if the fan blades hit the radiator.

Since fan air is usually only necessary at idle and low-speed operation, various design concepts are used to limit the fan's operation at higher speeds. Horsepower is required to turn the fan. Therefore, the operation of a cooling fan reduces the available horsepower to the drive wheels, as well as the fuel economy of the vehicle. Fans are also very noisy at high speeds, adding to driver fatigue and total vehicle noise.

To eliminate this power drain during times when fan operation is not needed, many of today's belt-driven fans operate only when the engine and radiator heat up. This is accomplished by a **fan clutch (Figure 12–52)**. When the engine and fan clutch are cold, the fan moves independently from the fan clutch and moves little air. When the engine warms up, the fan clutch engages, locking the fan in and moving a large amount of air. The clutch unit is located between the water pump pulley and the fan. The clutch assemblies rely on a thermostatic spring or silicone fluid. In both cases, the clutch locks the fan to the fan hub when the temperature of the air around the fan reaches a particular point. In most cases, the clutch slips at high speeds; therefore, it is not turning at full engine speed.

Some vehicle manufacturers use flexible blades or **flex blades** that bend or change pitch based on engine speed. That is, at slower speeds, the blade pitch is at the maximum. As engine speed increases the blade pitch decreases, as do the horsepower losses and noise levels.

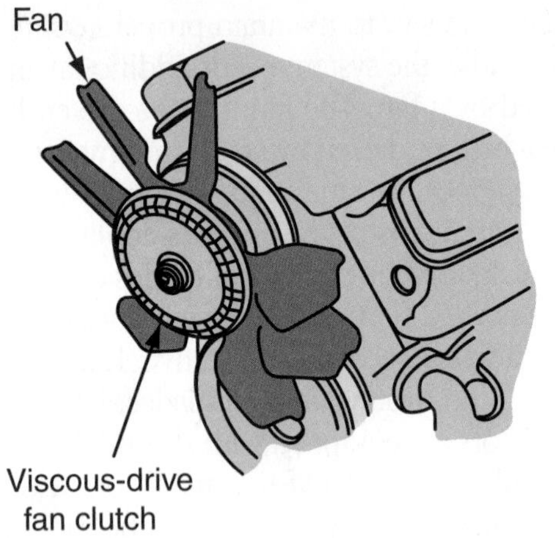

Figure 12-52 A viscous-type fan clutch.

Should the engine coolant temperature exceed approximately 230°F (110°C), the engine coolant temperature switch closes. This energizes the fan relay coil, which in turn closes the relay contacts. The contacts provide 12 volts to the fan motor if the ignition switch is in the on position. The 12-volt supply for the relay coil circuit is independent of the 12-volt supply for the fan motor circuit. The coil circuit extends from the on terminal of the ignition switch, through a fuse in the fuse panel, and to ground through the relay coil and temperature sensor.

Should the air conditioner select switch be turned to any cool position, regardless of engine temperature, a circuit is completed through the relay coil to ground through

Electric Cooling Fans In most late-model applications, to save power and reduce the noise level, the conventional belt-driven, water-pump-mounted engine cooling fan has been replaced with an electrically driven fan **(Figure 12–53)**. This fan and motor are mounted to the radiator shroud and are not connected mechanically or physically to the engine. The 12-volt, motor-driven fan is electrically controlled by either, or both, of two methods: an engine coolant temperature switch or sensor and the air-conditioner switch.

As the schematic in **Figure 12–54** shows, the cooling fan motor is connected to the 12-volt battery supply through a normally open (NO) set of contacts in the cooling fan relay. During normal operation, with the air conditioner off and the engine coolant below a predetermined temperature of approximately 215°F (101.6°C), the relay contacts are open and the fan motor does not operate.

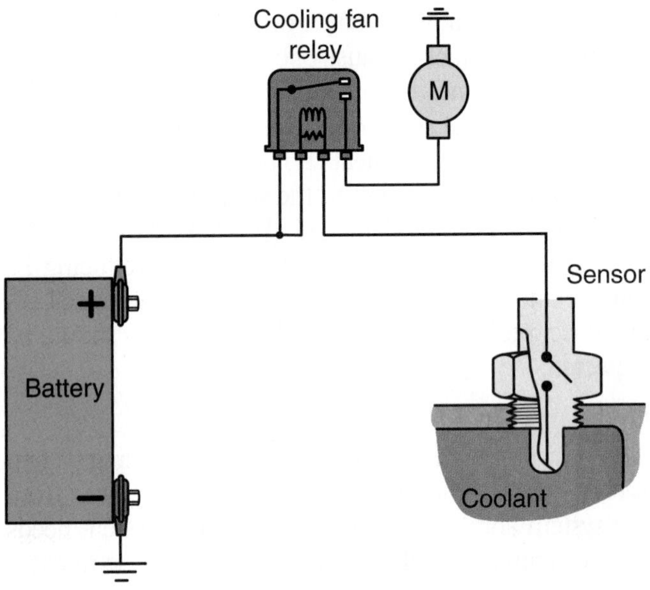

Figure 12-54 A simple schematic for an electric cooling fan.

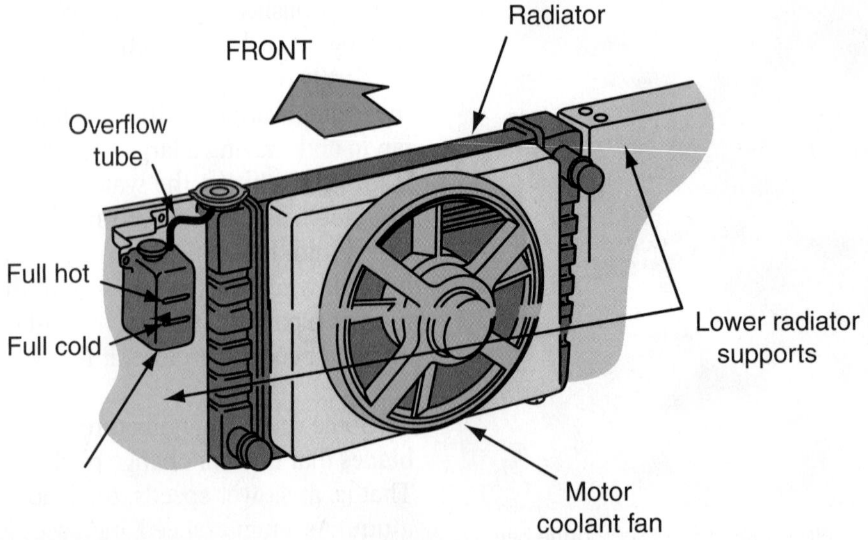

Figure 12-53 An electric cooling fan assembly.

the select switch. This action closes the relay contacts to provide 12 volts to the fan motor. The fan then operates as long as the air conditioner and ignition switches are on.

There are many variations of electric cooling fan operation. Some provide a cool-down period whereby the fan continues to operate after the engine has been stopped and the ignition switch is turned off. These systems may have a second temperature sensor that controls the fan when the engine is off. The fan stops only when the engine coolant falls to a predetermined safe temperature, usually about 210°F (98.8°C). In some systems, the fan does not start when the air conditioner select switch is turned on unless the high side of the air-conditioning (A/C) system is above a predetermined safe temperature.

Some late-model cars control the cooling fan by completing the ground through the engine control computer. Check the service manual to see how an electric cooling fan is controlled before working with it.

CAUTION!

The engine electric cooling fan can come on at any time without warning even if the engine is not running. For this reason, it is always wise to remove the negative terminal from the battery or the electric cooling fan connector while working around an electric fan. Make sure you reconnect the connector before giving the car back to the customer.

Temperature Indicators

Coolant temperature indicators are mounted in the dashboard to alert the driver of an overheating condition. It consists of a temperature gauge and/or a light. A temperature sensor is screwed into a threaded hole in the water jacket **(Figure 12–55)**. Besides indicating coolant temperatures to the driver, temperature sensors supply some important information to today's computer-controlled engine control systems.

Figure 12-55 A coolant temperature sender or sensor.

Temperature Sensors Proper electric cooling fan operation depends on the operation of a temperature sensor. A temperature sensor responds to changes in temperature. Some vehicles use more than one sensor to control the fans and to send engine temperature readings to the Powertrain Control Module (PCM). Based on this information, the PCM will adjust the fuel injection and ignition systems to provide efficient engine operation.

Heater System

A hot liquid passenger compartment heater is part of the engine's cooling system. Heated coolant flows from the engine through heater hoses and a heater control valve to a smaller heater core, or radiator, located in a hollow container on either side of the fire wall. Air is directed or blown over the hot heater core, and the heated air flows into the passenger compartment. Movable doors can be controlled to blend cool air with heated air for more or less heat.

COOLING SYSTEM SERVICING

The cooling system must operate, be inspected, and be serviced as a system. Replacing one damaged part while leaving others dirty or clogged will not increase system efficiency. Service the entire system to ensure good results.

Service involves both a visual inspection of the parts and connections and pressure testing. Pressure testing is used to detect internal or external leaks.

The following is a list of places to check for a suspected leak in the cooling system. Although this list is not complete, it does present the most common areas where a leak can occur.

External Leakage

- Radiator
- Loose hose clamp
- Hose
- Faulty radiator cap
- Dented radiator hose connector for the inlet or outlet hose
- Heater connection
- Water pump, through weep hole
- Cracked or porous water pump housing
- Heater core
- Loose core hole plug in cylinder block
- Cracked thermostat housing
- Water temperature sending unit
- Cylinder head bolts loose or tightened unevenly
- Warped or cracked cylinder head
- Heater control valve

- Cracked cylinder block
- Damaged gasket or dry gasket if engine has been stored
- Coolant reservoir or hose

Internal Leakage

- Faulty head gasket
- Cracked head
- Cracked block
- Transmission fluid cooler

Coolant Condition

A coolant hydrometer is used to check the amount of antifreeze in the coolant. This tester contains a pickup hose, coolant reservoir, and squeeze bulb. The pickup hose is placed in the radiator coolant. When the squeeze bulb is squeezed and released, coolant is drawn into the reservoir. As coolant enters the reservoir, a pivoted float moves upward with the coolant level. A pointer on the float indicates the freezing point of the coolant on a scale located on the reservoir housing.

Testing the Cooling System for Leaks

The most common tool used to test a cooling system is the radiator pressure tester. A radiator pressure tester is really no more than a hand pump with a pressure gauge. The pressure tester is extremely handy for identifying the location of any leak within the cooling system. To use the tester, connect it to the radiator filler neck **(Figure 12–56)**. Run the engine until it is warm, then pump the handle of the tester until its gauge reads the same pressure noted on the radiator cap. Watch the gauge. If the pressure drops, carefully check the hoses, radiator, heat core, and water pump for leaks. Often the leak will initially be obvious because coolant will spray out of the leak. If the pressure drops but there are no external leaks, suspect an internal leak.

The source of an external coolant leak can also be found by a thorough visual inspection or the use of a dye penetrant. Visually, the point of the leak may be wet or have a light gray color, the result of the coolant evaporating at that point. Another common way to identify the source of a leak is to use a dye penetrant and a black light. The dye is poured into the cooling system and the engine is run until it reaches operating temperature. With the engine turned off, the engine and cooling system are inspected with the black light. Where the dyed coolant leaks, a bright or fluorescent green color will be seen.

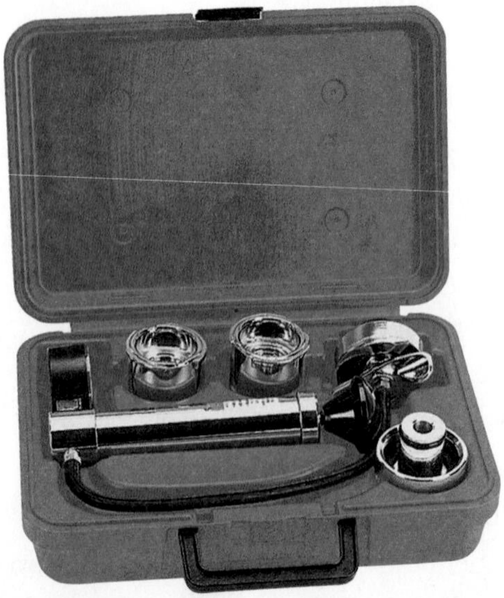

Figure 12-56 Testing the cooling system for leaks with a pressure tester.

Repairing Radiators

Most radiator leak repairs require the removal of the radiator from the vehicle. The coolant must be drained and all hoses and oil cooler lines disconnected. Bolts holding the radiator are then loosened and removed.

The actual radiator repair procedures depend on the material of which it is made and the type of damage. Most radiator repairs are made by radiator specialty shops that employ technicians with knowledge of such work. If the radiator is badly damaged, it should be replaced and a new one should be installed as directed by the manufacturer.

Many of today's radiators have plastic tanks, which are not repaired. If these tanks leak, they are replaced.

Testing the Radiator Pressure Cap

Apply the proper cap testing adapter and radiator pressure cap to the tester head. Pump the tester **(Figure 12–57)** until the pressure valve of the cap releases pressure. The cap should hold pressure in its range as indicated on the tester gauge dial for one minute. If it does not, replace it. Remove the cap from the tester and visually inspect the condition of the cap's pressure valve and upper and lower sealing gaskets. If the gaskets are hard, brittle, or deteriorated, the cap may leak when exposed to hot, pressurized coolant. It should be replaced with a new cap in the same pressure range.

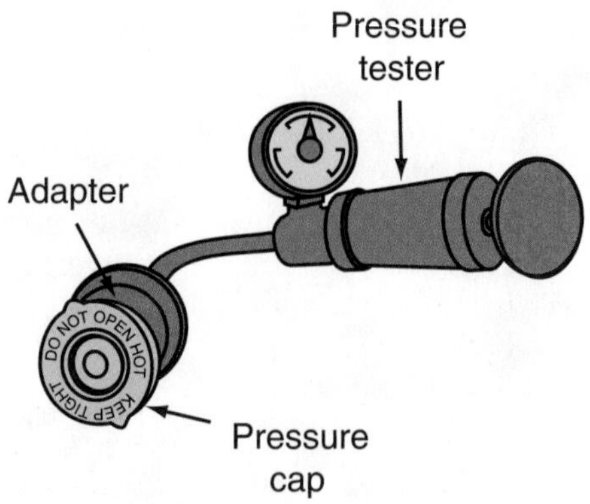

Figure 12-57 Testing a radiator cap with a pressure tester.

CAUTION!

The radiator cap should never be removed when the cap or radiator is hot to the touch. When the pressure in the radiator is suddenly released, the coolant's boiling temperature is reduced, causing the coolant to immediately boil. Because coolant is a thick liquid, it will stick to your skin and cause severe burns if it is hot. The radiator should first be allowed to cool, or force cool it by lightly spraying water on the radiator core. When the cap is cool to the touch and the engine is shut off, use a cloth over the cap and turn it counterclockwise one-quarter turn to the filler neck safety stop. Carefully watch for any liquid or steam loss around the ring of the cap and from the radiator overflow tube. Let the cap remain in this position until all pressure subsides. When evidence of discharge is no longer seen, use a cloth to cover the cap, press it down to pass the cap's ears over the safety stops, and continue to turn counterclockwise to remove the cap.

Testing the Thermostat

Thermostats are often the cause of overheating problems. If a thermostat is stuck closed, no coolant will flow through the radiator. Thermostats are also the cause of poor heater and engine performance. If the thermostat is stuck open, the coolant may not reach the desired temperature because the coolant is cooled before it gets hot.

A properly working thermostat will be closed when the engine is cold and the coolant has a temperature lower than the thermostat's rating. Once the coolant reaches the opening temperature of the thermostat, a waxlike pellet expands inside the thermostat to open it and allow the coolant to flow to the radiator. The hotter the coolant gets, the more the thermostat opens, allowing more coolant to flow through the radiator.

Electronic engine control systems are very dependent on the control and regulation engine coolant temperatures. A faulty thermostat will not only cause overheating or poor heater performance but may cause an increase in fuel consumption and poor engine performance. The PCM's engine management system is programmed to deliver the ideal air/fuel mixture and ignition timing according to the engine's operating conditions. One of the conditions monitored by the PCM is engine temperature.

There are several ways to test the opening temperature of a thermostat. One method does not require that the thermostat be removed from the engine. Remove the radiator pressure cap from a cool radiator and insert a thermometer into the coolant. Start the engine and let it warm up. Watch the thermometer and the surface of the coolant. When the coolant begins to flow, it indicates the thermostat has started to open. The reading on the thermometer indicates the opening temperature of the thermostat. If the engine is cold and coolant circulates, it indicates that the thermostat is stuck open and must be replaced.

The other way to test a thermostat is to remove it. Suspend the thermostat completely submerged in a small container of water so it does not touch the bottom. Place a thermometer in the water so it does not touch the container and only measures water temperature. Heat the water. When the thermostat valve barely begins to open, read the thermometer. This is the opening temperature of this particular thermostat. If the valve stays open after the thermostat is removed from the water, the thermostat is defective and must be replaced.

Several types of commercial testers are available. When using such a tester, be sure to follow the manufacturer's instructions.

Markings on the thermostat normally indicate which end should face toward the radiator. Regardless of the markings, the sensored end must always be installed toward the engine.

When replacing the thermostat, also replace the gasket that seals the thermostat in place and is positioned between the water outlet casting and the engine block **(Figure 12–58)**. Generally, these gaskets are made of a composition fiber material and are die-cut to match the thermostat opening and mounting bolt configuration of the water outlet. Thermostat gaskets generally come with or without an adhesive backing. The adhesive backing of gaskets holds the thermostat securely centered in the mounting flange, leaving both hands of the technician free to align and bolt the thermostat securely in place.

Figure 12-58 Positioning a thermostat into an engine block.

Checking and Replacing Hoses

Carefully check all cooling hoses for leakage, swelling, and chafing. Also change any hose that feels mushy or extremely brittle when squeezed firmly **(Figure 12–59)**. When a hose becomes soft, it is deteriorating and should be replaced before more serious problems result. When a hose is hard, it is brittle and should be replaced. Hard hoses resist flexing and may crack rather than bend. The result is a leak.

Normally, hoses begin to deteriorate from the inside. Pieces of deteriorated hose will circulate through the system until they find a place to rest. This place is usually the radiator core, causing clogging. Deterioration can also cause leaks. Any external bulging or cracking of hoses is a definite sign of failure. When one hose fails, all of the others should be carefully inspected.

The upper radiator hose is subjected to the roughest service life of any hose in the cooling system. It must absorb more engine motion than any of the other hoses. It is exposed to the coolant at its hottest stage, and it is insulated by the hood during hot soak periods. These conditions make the upper hose the most probable to fail.

Be especially watchful for signs of splits when hoses are squeezed. These splits have a habit of bursting wide open under pressure. Also look for rust stains around the clamps. Rust stains indicate that the hose is leaking, possibly because the clamp has eaten into the hose. Loosen the clamp, slide it back, and check for cuts.

The primary cause of coolant hose failure has been identified as an electrochemical attack on the rubber compound in the hose. This is known as **electrochemical degradation (ECD)**. It occurs because the hose, engine coolant, and the engine/radiator fittings form a galvanic (battery) cell. This chemical reaction causes very small cracks in the hose, allowing the coolant to attack and weaken the reinforcement in the hose. ECD can cause pinhole leaks or hose rupture under normal operating pressures. The effects of ECD are accelerated by high temperatures and vibrations.

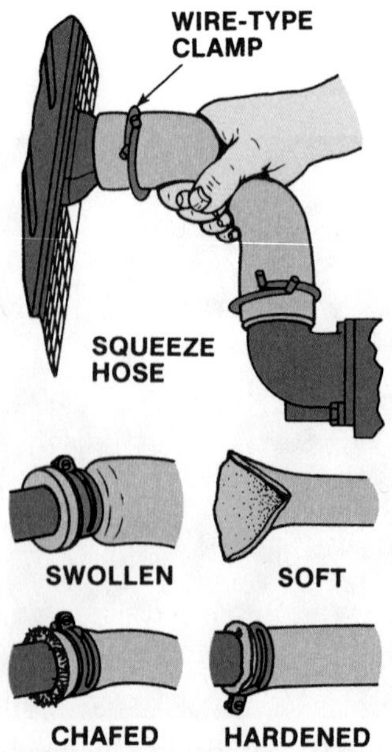

Figure 12-59 Defects in cooling hoses.

The best way to check hoses for the effects of ECD is to squeeze the hose near the clamps or connectors. ECD occurs within two inches of the ends of the hose—not in the middle. Compare the feel of the hose between the middle and the ends. Gaps can be felt along the length of the hose where it has been weakened by ECD. If the ends are soft and feel mushy, chances are the hose is under attack by ECD and the hose should be replaced.

ECD can occur in any cooling system hose and will cause the most damage where the temperature is hottest and air is present with the coolant, which is why upper radiator hoses tend to fail first.

Oil is another enemy to rubber hoses. A hose damaged by oil is swollen, soft, and sticky. If the oil leak is external, eliminate the oil leak or try to reroute the hose away from the oil leak to prevent oil damage to a new hose. At times, the oil damage occurs inside the hose. This damage can be caused by transmission fluid leaking into the coolant or by an internal engine oil leak.

Do not overlook the small bypass hose on some models. It is located between the water pump and engine block. Also, check the lower radiator hose very carefully. This hose contains a coiled wire lining to keep it from collapsing during operation. If the wire loses tension, the hose can partially collapse at high speed and restrict coolant flow, which results in a very elusive overheating problem.

CUSTOMER CARE

Technicians should do their customers a favor and remind them that all cooling hoses should be replaced every two to four years to prevent breakdowns. ■

All cooling system hoses are basically installed the same way. The hose is clamped onto inlet-outlet nipples on the radiator, water pump, and heater.

Replacement radiator hoses must be the correct diameter, length, and shape. Each has a part number, which is often printed on the old hose and on the hose package.

When replacing a hose, drain the coolant system below the level being worked on. Loosen or carefully cut the old clamp. Then, using a knife, carefully cut the end of the old hose **(Figure 12–60)** so it can slide off its fitting. If the hose is stuck, do not pry it off. You could possibly damage the inlet/outlet nipple or the attachment

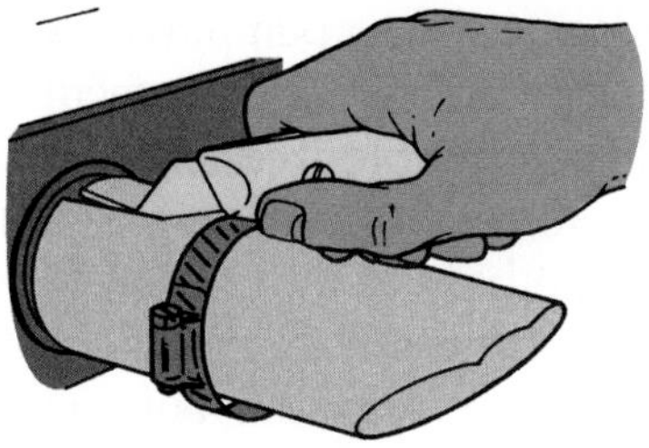

Figure 12-60 Cutting off an old hose.

between the end of the hose and the bead. Simply cut it more so it can come off.

Always clean the neck of the hose fitting or nipple with a wire brush or emery cloth after the old hose has been removed. Burrs or sharp edges could cut into the hose tube and lead to premature failure, and dirt will prevent a good seal.

Dip the ends of the hose in coolant to lubricate it and slip the clamps over each end. Do not reuse old spring-type clamps, even if they look good. Slip the hose over its fittings, engine end first. In cold weather, the hose may be stiff; it can be soaked in warm water to make it more flexible. If the hose does not fit properly, remove it and reverse the ends. Then, slide the clamps to about ¼ inch from the end of the hose after it is properly positioned on the fitting **(Figure 12–61)**. Tighten the clamp securely but do not overtighten.

It is a good idea to readjust the clamp of a newly installed coolant hose after a brief run-in period. The hose end does not contract and expand at the same rate as the metal of the inlet/outlet nipple it is attached to. Rubber coolant hose, warmed by the hot coolant and hot engine, will expand. The clamp compresses the rubber around the hose end and sets it. When the engine cools, the fitting contracts more than the rubber, and the hose will not be as secure, which can result in cold leaks of coolant at the inlet/outlet nipple when the engine is cool. Retightening the clamp eliminates the problem.

Water Pump Service

The majority of water pump failures are attributed to leaks of some sort. When the pump seat fails, coolant will begin to seep out of the weep hole in the casting **(Figure 12–62)**. This is an early indicator of trouble. The seals may simply wear out due to abrasives in the cooling system, or some types of seals crack due to thermal shock such as adding cold water to an overheated engine. This shock could also cause other internal parts to fail.

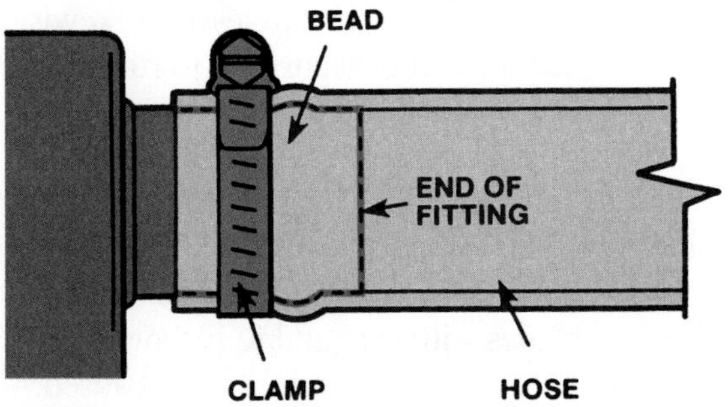

Figure 12-61 New clamps should be placed immediately after the bead of the fitting.

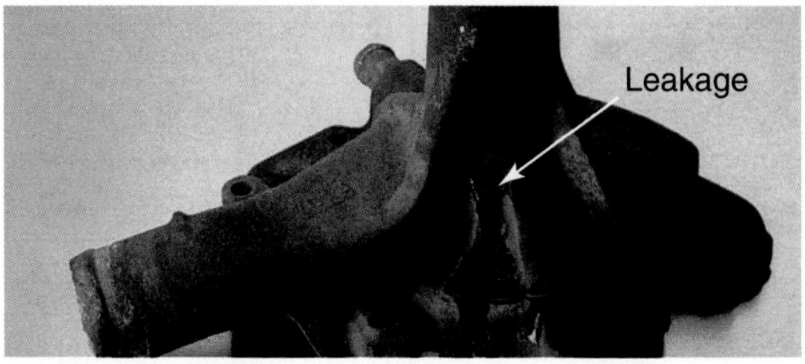

Figure 12-62 Signs of leakage from the water pump vent means the pump's seal is bad. *Courtesy of Federal-Mogul Corporation*

Other failures can be attributed to bearing and shaft problems and an occasional cracked casting. Water pump bearing or seal failure can be caused by surprisingly small out-of-balance conditions that are difficult to spot. Look for the following:

- A bent fan. A single bent blade will cause problems.
- A piece of fan missing.
- A cracked fan blade. Even a small crack will prevent proper flexing.
- Fan mounting surfaces that are not clean or flush.
- A worn fan clutch.

To check a water pump, start the engine and listen for a bad bearing, using a mechanic's stethoscope or rubber tubing. Place the stethoscope or hose on the bearing or pump shaft. If a louder than normal noise is heard, the bearing is defective.

CAUTION!

Whenever working near a running engine, keep your hands and clothing away from the moving fan, pulleys, and belts. Do not allow the stethoscope or rubber tubing to be caught by the moving parts.

There is another test that can be performed on vehicles with an engine-driven fan. With the engine off and the fan belt and shroud removed, grasp the fan and attempt to move it in and out and up and down. More than $\frac{1}{16}$ inch (1.58 mm) of movement indicates worn bearings that require water pump replacement.

To determine whether the water pump is allowing for good circulation, warm up the engine and run it at idle speed. Squeeze the upper hose connection with one hand

and accelerate the engine with the other hand. If a surge on the hose is felt, the pump is working.

Any air being sucked into the cooling system is certain to have a detrimental effect. It cuts down pumping efficiency and causes both rusting and wear at a rate approximately three times above normal. To test for aeration, have the engine fully warmed up, all hose connections tight, and the coolant level up to normal. Attach one end of a small hose to the radiator overflow pipe and put the other end into a jar of water. Run the engine at a fast idle. If a steady stream of bubbles appears in the jar of water, air is getting into the cooling system.

Check first for a cylinder gasket leak by running a compression test. If two adjacent cylinders test low, the gasket is bad. Otherwise, there is an air leak somewhere else in the cooling system.

Replacing the Water Pump When replacing a water pump, it is necessary to drain the cooling system. Any components—belts, fan, fan shroud, shaft spacers, or viscous drive clutch—should be removed to make the pump accessible. Some pumps are attached to the cylinder block as shown in **Figure 12–63**. Loosen and remove the bolts in a crisscross pattern from the center outward. Insert a rag into the block opening and scrape off any remains of the old gasket.

CAUTION!

When working on the coolant system (for example, replacing the water pump or thermostat), a certain amount of coolant will spill on the floor. The antifreeze in the coolant causes it to be very slippery. Always immediately wipe up any coolant that spills to reduce or eliminate the chance of injury.

Figure 12–63 Installing a water pump. *Courtesy of Dana Corporation*

When replacing a water pump, always follow the procedures recommended by the manufacturer. Most often a coating of good waterproof sealer should be applied to a new gasket before it is placed into position on the water pump. Coat the other side of the gasket with sealer, and position the pump against the engine block until it is properly seated. Install the mounting bolts and tighten them evenly in a staggered sequence to the torque specifications with a torque wrench. Careless tightening could cause the pump housing to crack. Check the pump to make sure it rotates freely.

The water pumps on many late-model OHC engines are driven by the engine's timing belt. When replacing the water pump on these engines, always replace the timing belt. Make sure all pulleys and gears are aligned according to specifications when installing the belt.

Checking Fans and Fan Clutches

Fan operation can be checked by spinning the fan by hand. A noticeable wobble or any blade that is not in the same plane as the rest indicates that replacement is in order. The fan can also be checked by removing it and laying it on a flat surface. If it is straight, all the blades should touch the surface. Never attempt to straighten a damaged cooling fan. Bending it back into shape might seem easier (and cheaper) than replacing it, but doing so is risky. Whenever metal is bent, it is weakened.

Fan clutches use a fluid-filled chamber (usually silicone) to turn the fan. Obviously, loss of the drive fluid will render the fan useless.

One of the simplest checks is to visually inspect the fan clutch for signs of fluid loss. Oily streaks radiating outward from the hub shaft mean fluid has leaked out past the bearing seal.

Most fan clutches offer a slight amount of resistance if turned by hand when the engine is cold. They offer drag when the engine is hot. If the fan freewheels easily hot or cold, replace the clutch.

Another check that should be made is to push the tip of a fan blade in and out. Any visible looseness in the shaft bearing means the fan clutch should be replaced.

Fan blades are balanced at the time of manufacture but can be bent if handled carelessly. Likewise, fan clutches are machined very accurately to run true. However, rough handling at the time of pump replacement causes nicks and dents on the mounting faces. This can cause the fan blade to be installed crookedly. Serious trouble might follow soon after the car is back in service. Therefore, technicians should be cautioned to handle fan clutches and blades with care and to file away any nicks, burrs, or dents someone else might have caused. Obviously, if any fins have been broken off the fan clutch, it must be replaced.

> **SHOP TALK**
>
> Damaged fan shrouds can usually be repaired if the damage is not severe. Polyethylene and polypropylene plastics can be hot-air or airless welded. Fiberglass can be glued with epoxy. The important rule here is never leave a fan shroud off. The shroud is needed to help the fan work at peak efficiency and leaving it off will affect the fan's efficiency. ■

Checking Belts

If a belt breaks, at best the fan stops spinning and the coolant does not cool down efficiently. At worst, the water pump stops, the coolant does not circulate, and, eventually, the engine overheats.

Belts, like hoses, are made of elastic rubber compounds. Although they are extremely sturdy, they are primarily designed for transmitting power. Even the best belts last only an average of four years. Advise the customer to replace all belts every four years, regardless of how they look.

Fortunately, belt problems are easily discovered either by visual inspection for cracks, splits, glazing, or oil soakage, or by the screech of slippage A glazed, cracked, or damaged belt should always be replaced. In servicing a multibelt setup, it is very important to replace all of the belts when one belt is bad.

Also inspect the grooves of the drive pulleys for rust, oil, wear, and other damage. If a pulley is damaged, it should be replaced. In many cases, rust, dirt, and oil can be cleaned off the pulley and should be removed before installing a new belt. Also check the alignment of the pulleys.

Misalignment of the V-pulleys reduces the belt's service life and brings about rapid V-pulley wear, which causes thrown belts and screech. Undesirable side or end thrust loads can also be imposed on pulley or pump shaft bearings. Check alignment with a straightedge. Pulleys should be in alignment within 1/16 inch (1.59 mm) per foot of the distance across the face of the pulleys.

Belt Replacement V-belt replacement is a straightforward procedure but care must be taken to install the new belt correctly and under the correct amount of tension. Before removing the old drive belt, disconnect the electric cooling fan at the radiator, if the vehicle has one. Remove the old drive belt by loosening the components that have adjusting slots for belt tension. Then slip the old belt off.

Always use the exact size of replacement belt. The size of a new belt is typically given, along with the part number, on the belt container **(Figure 12–64)**. You can verify that the new belt is a replacement for the old by physically comparing the two. This comparison, however, does not allow for any belt stretch that may have occurred. Therefore, use this comparison only as verification. The best way to select the correct replacement belt is through the catalog and/or by matching the numbers on the old belt to the numbers on the new belt.

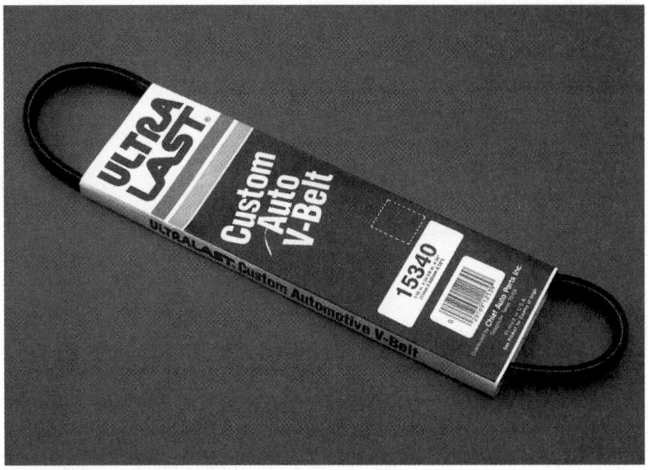

Figure 12-64 A new drive belt.

Place the new belt around the pulleys. Once in place, tighten the component mounting bolts that were loosened during belt removal. Then adjust the tension of the belt and retighten all mounting hardware.

When installing a serpentine belt, make sure it is fed in and around the accessories properly. Service manuals show the proper belt routing. Also make sure the belt tensioner or idler pulley is working properly. This pulley may be a spring-loaded tensioner **(Figure 12–65)** or an adjustable pulley. If the adjustable tensioner pulley is not spring loaded, adjustments are made with a jackscrew or with an off-center pulley bolt. To loosen the tension of the belt, turn the bolt in a counterclockwise direction.

Before removing a serpentine belt, locate a belt-routing diagram in a service manual or on an underhood decal. Compare the diagram with the routing of the old belt. If the actual routing is different from the diagram, draw the existing routing on a piece of paper. To install the new belt, wrap it according to instructions. Make sure the ribs of the belt are seated in the matched grooves on the pulleys. Once the belt is fully routed, put tension on the belt and adjust it to specifications.

Belt Tension Correct belt tension is essential for long belt life and quiet operation. Loose belts may slip on the pulleys and not drive the components at their proper speed. This situation can cause numerous problems, such as low generator output and noise. Excessive tension can also result in noise as the driven components rotate under stress. This stress can lead to premature bearing and bushing failure in water pumps, generators, and power steering pumps.

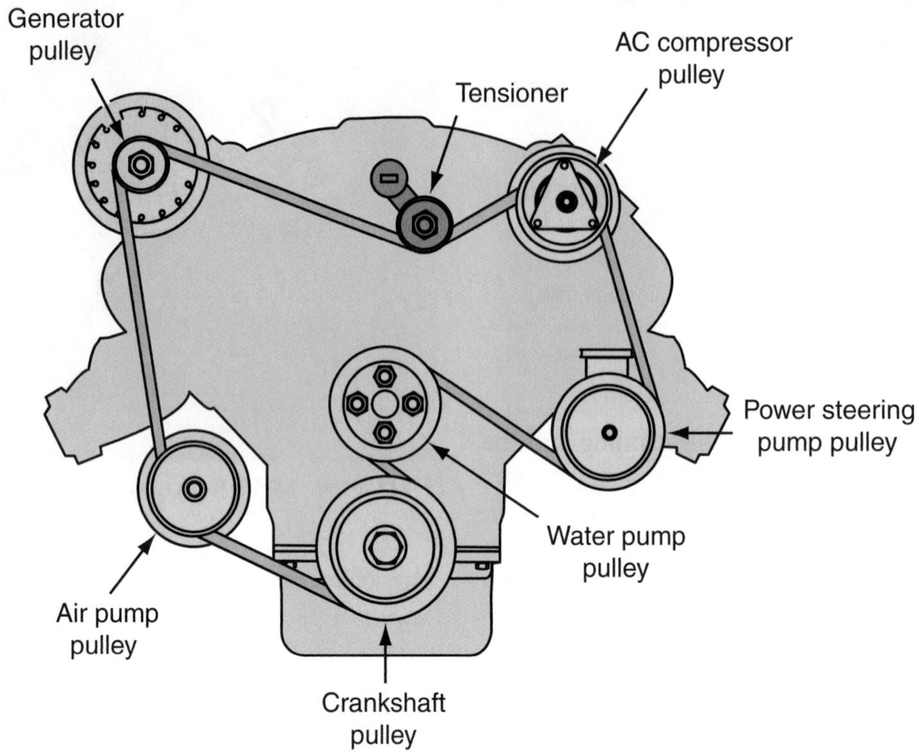

Figure 12-65 The tension of a serpentine belt is typically adjusted with a belt tensioner.

The mounting brackets on generators, power-steering pumps, and air compressors are designed to be adjustable so that proper tension can be maintained on these belts. Some of these brackets have a hole or slot to allow the use of a pry bar or wrench when adjusting. Some automobiles require the fan, fan pulley, and other accessory drive belts to be removed to gain access to belts needing replacement.

After replacing a belt, make sure it is adjusted to specifications. While adjusting the tension, be careful not to damage the part you are prying against while tightening the belt. Tighten all bolts and nuts to keep the tension. Then check the belt's tension with a belt tension gauge **(Figure 12–66)**.

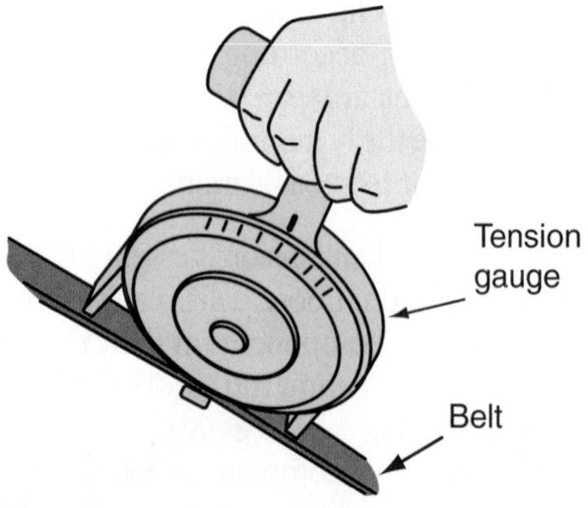

Figure 12-66 The tension of a belt should be checked with a belt tension gauge.

SHOP TALK

It is never advisable to pry a belt onto a pulley. Obtain enough slack so the belt can be slipped on without damaging either the V-belts or a pulley. Some power-steering pumps have a ½-inch drive socket to aid in adjusting belts to the proper tension without prying against any accessory. ■

The water pump on many late-model OHC engines is driven by the engine's timing belt. When replacing the water pump on these engines, always replace the timing belt. Make sure all pulleys and gears are aligned according to specifications when installing the belt.

After installation of the new belts, the engine should be run for 10 to 15 minutes to allow belts to seat and reach their initial stretch condition. Modern steel-strengthened V-belts do not stretch much after the initial run-in, so the retensioning should be done very carefully with an accurate gauge. Recheck the tension again after 5,000 miles.

Coolant Recovery and Recycle System

Whenever the coolant must be drained to service the cooling system, the used coolant should be drained and recycled by a coolant recovery and recycle machine. Typically additives are mixed into the used coolant during recycling. These additives either bind to contaminants in the coolant so they can be easily removed, or they restore some of the chemical properties in the coolant.

Flushing Cooling Systems

Rust and scale will inevitably form in any cooling system. When they do, there are a few vulnerable places they can attack. One happens to be the main seal in the water pump, which keeps coolant away from the bearing and its lubricant. If grit is allowed to erode the seal, the bearing will be the next item to go.

Whenever coolant is changed, and especially before a water pump is replaced, a thorough reverse flushing or back flushing should be performed. Before this flushing is done, chemical cleaners can be added to the cooling system to help dissolve rust and scale deposits.

Reverse flushing is the procedure of forcing clean liquid backwards through the cooling system **(Figure 12–67)**. This carries away rust, scale, corrosion, and other contaminants. A flushing gun that operates on compressed air is used to force clean water and air through the system.

This method of flushing the system is not recommended on systems that use plastic and aluminum radiators. Check the service manual for the proper way of cleaning out the cooling system on vehicles with those kinds of radiators.

Refilling and Bleeding

After the cooling system has been cleaned, refill the system with new coolant mixed to the recommended

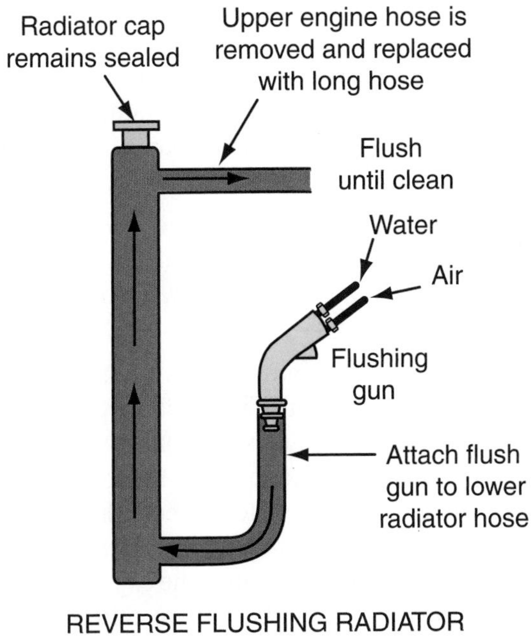

Figure 12-67 The typical setup for backflushing a cooling system.

CASE STUDY

***A** late-model minivan's engine overheats while going up a mountain road. It is a very hot day so water is added to the radiator and the trip continues. After driving 30 miles (48 km), the engine overheats again. Again, water is added. The engine is allowed to cool for an hour while the owner eats. After about 30 miles (48 km) of driving, the engine overheats again. This time the van is taken to a service station to be diagnosed and repaired. It is late in the day and the owner agrees to stay overnight at a local motel. A fresh supply of water is added to the radiator and the owner drives about a mile to the motel.*

In the morning the owner adds water to the system. The van is brought back to the service station for diagnosis. It is obvious that the engine is overheating and there is a loss of fluid. There is no need to verify the customer's complaint. A visual inspection shows little water in the radiator, but no signs of external leakage are found. The oil level is found to be a little high and the oil appears to be thin.

A radiator tester is attached to the system and there is no sign of pressure or coolant leakage. The operation of the water pump is to be checked next. However, this test cannot be done because the engine will not start. The starter cannot turn the engine over. A quick check of the battery reveals it is fully charged. Careful thought suggests the problem is hydrostatic lock-up. This is a condition in which a fluid fills the cylinder. Since a liquid cannot be compressed, the liquid prevents the piston from moving up.

To verify this, the spark plugs are removed. All plugs looked normal except the one for cylinder #5. It is wet with water. With the plugs out, the engine is cranked by the starter motor and water comes gushing out of the cylinder. The exact cause of the internal leak will be best identified by tearing down the engine. During disassembly, a crack in the engine block is found. This crack allowed the coolant to pass from a coolant passage in the block into the cylinder.

This leak is not detected by the radiator pressure test because when the engine is cold, little, if any, coolant leaks through the crack. The crack grows larger as the engine becomes hotter, allowing the cylinder to rapidly fill when it is hot. A new short block is put in the van and the problem is corrected.

strength. Be sure all the flushing water is out of the radiator and evacuated from the engine block before refilling.

The cooling system must be bled to ensure there are no air pockets in the coolant. Air in the system can reduce cooling ability and lead to water pump and other component damage.

Each vehicle has its own specific bleeding procedure. In many cases, the system is filled and the radiator cap is left off. The engine is then run until the thermostat opens and the coolant circulates. Trapped air bubbles escape through the opened radiator.

Other bleeding procedures are more involved. They may require the connection of special bleeder hoses to air bleed valves located on the radiator or other components. Air is bled through the valves as the system is filled. Always follow the bleeding steps outlined in the vehicle's service manual.

CUSTOMER CARE

Many additives, inhibitors, and quick-fix remedies are available for use in the automotive cooling system. These include, but are not limited to, stop leak, water pump lubricant, engine flush, and acid neutralizers. Explain to your customers that extreme caution should be exercised when using any additive in the cooling system. Tell them to read the label directions and precautions in order to know in advance the end results of any additive used. For example, caustic solutions must never be used in aluminum radiators. Alcohol-based remedies should never be used in any cooling system. ■

KEY TERMS

API (American Petroleum Institute)
Coolant
Crude oil
Electrochemical degradation (ECD)
Energy-conserving oil
Expansion tank
Fan clutch
Flex blades
Gear-type pump
Multiviscosity oil
Oil pump pickup
Positive displacement pump
Pressure relief valve
Radiator
Recovery tank
Rotor-type oil pump
Serpentine belt
Society of Automotive Engineers (SAE)
Thermostat
Viscosity

SUMMARY

- An engine's lubrication system has several important purposes: to hold an adequate supply of oil to cool, clean, lubricate, and seal the engine; to remove contaminants from the oil; and to deliver oil to all necessary areas of the engine. The API has developed two basic classifications for engine oils: S class for passenger cars and light trucks and C class for heavy-duty commercial applications. The SAE has standardized oil viscosity ratings. API and SAE classifications are given on the oil container.
- Excessive oil consumption can be a result of external and internal leaks, faulty accessories, piston rings, and valve guides. Internal leaks, which usually result in oil burning, are usually more difficult to diagnose.
- The main components of a typical lubrication system are: an oil pump, oil pump pickup, oil pan, pressure relief valve, oil filter, engine oil passages, engine bearings, crankcase ventilation, oil pressure indicator, oil seals and gaskets, dipstick, and oil coolers.
- The purpose of the oil pump is to supply oil to the various moving parts in the engine. The most commonly used stock pumps are the rotor and gear type. Both are positive displacement pumps. Because the faster the pump turns the greater the pressure becomes, a pressure-regulating valve is installed to control the maximum oil pressure.
- All automotive vehicles are equipped with either an oil pressure gauge or a low-pressure indicator light. The gauges are either mechanically or electrically operated.
- All oil leaving the oil pump is directed to the oil filter. The filter is a disposable metal container filled with a special type of treated paper or other filter substance that catches and holds the oil's impurities.
- After an engine is rebuilt, a new or rebuilt oil pump is often installed. If the old pump is to be reused, it should be carefully inspected for wear and thoroughly cleaned.
- The fluid used as coolant today is a mixture of water and antifreeze/coolant. Closed-cooling systems are cooling systems with an expansion or recovery tank. The function of the water pump is to move the coolant efficiently through the system. The radiator transfers heat from the engine to the air passing through it. The thermostat attempts to control the engine's operating temperature by routing the coolant either to the radiator or through the bypass, or sometimes a combination of both.

- V-belts and V-ribbed belts (called serpentine belts) are used to drive water pumps and to power steering pumps, air-conditioning compressors, generators, and emission control pumps. The fan delivers additional air to the radiator to maintain proper cooling at low speeds and idle. Since fan air is usually only necessary at idle and low-speed operation, various design concepts are used to limit the fan's operation at higher speeds.
- Hollow passages in the block and cylinder heads allow coolant to flow through them. Included in the water jackets are soft plugs and a block drain plug. A temperature indicator is mounted in the dashboard to alert the driver to an overheating condition. A hot-liquid passenger heater is part of the engine cooling system. Radiators for vehicles with automatic transmissions have a sealed heat exchanger located in the coolant outlet tank.
- The basic procedure for testing a vehicle's cooling system includes inspecting the radiator filler neck, inspecting the overflow tube for dents and other obstructions, testing for external leaks, and testing for internal leaks. Most radiator leak repairs require removing the radiator from the vehicle.
- The pressure cap should hold pressure in its range, as indicated on the tester gauge dial, for one minute. A thermostat can be tested in the engine or after it is removed.
- Hoses should be checked for leakage, swelling, and chafing. Any hose that feels mushy or extremely brittle or shows signs of splitting when it is squeezed should be replaced. The majority of water pump failures are attributed to leaks of some sort. Other failures can be attributed to bearing and shaft problems and an occasional cracked casting.
- Fan operation can be checked by spinning the fan by hand. A noticeable wobble or any blade that is not in the same plane as the rest indicates replacement is in order. The fan can also be checked by removing it and laying it on a flat surface. If it is straight, all the blades will touch the surface. Never attempt to repair a damaged fan; replace it. One of the simplest ways to check a fan clutch is to visually inspect it for signs of fluid loss.
- Belt problems are easily discovered by visual inspection or by the screech of slippage. When a new belt is installed, it should be properly adjusted.
- Whenever coolant is changed, a thorough flushing should be performed. The old coolant should be captured and recycled.

TECH MANUAL

The following procedures are included in Chapter 12 of the *Tech Manual* that accompanies this book:

1. Servicing and installing oil pump and oil pan.
2. Inspecting, replacing, and adjusting drive belts and pulleys.
3. Cleaning, inspecting, testing, and replacing electric cooling fans and cooling system-related temperature sensors.

REVIEW QUESTIONS

1. List some of the things that influence a radiator's efficiency.
2. What type of radiator pressure cap is found on today's vehicles?
3. Describe the simple test used to determine whether the water pump is causing good circulation.
4. What typically drives an oil pump?
5. What is the name of the component in the lubrication system that prevents excessively high system pressures from occurring as engine speed increases?
6. Which of the following is a function of the engine's lubrication system?
   - **a.** to hold an adequate supply of oil
   - **b.** to remove contaminants from the oil
   - **c.** to deliver oil to all necessary areas of the engine
   - **d.** all of the above
7. Technician A says the gear pump pumps a greater volume of oil than a rotor pump. Technician B says

engine oil pressure is dependent upon oil viscosity. Who is correct?

a. Technician A
b. Technician B
c. Both A and B
d. Neither A nor B

8. Technician A says many oil filters have a check valve that holds oil in the filter when the engine is not running. Technician B says most engines made in recent years have an oil filtering system somewhere on the input side of the pump. Who is correct?

a. Technician A
b. Technician B
c. Both A and B
d. Neither A nor B

9. What is ECD and what can result from it?

10. *True or False?* To eliminate the drain of engine power during times when cooling fan operation is not needed, many of today's belt-driven fans have a fan clutch that prevents the operation of the fan when the engine and radiator are heated up.

11. When servicing an oil pump, Technician A uses a feeler gauge and straightedge to determine the pump cover flatness. Technician B uses an outside micrometer to measure the diameter and thickness of the outer rotor. Who is correct?

a. Technician A
b. Technician B
c. Both A and B
d. Neither A nor B

12. Technician A says the API classification S stands for standard passenger cars. Technician B says that the API classification C stands for vehicles with compression ignition. Who is correct?

a. Technician A
b. Technician B
c. Both A and B
d. Neither A nor B

13. Technician A says the American Petroleum Institute has established an oil viscosity classification system. Technician B says higher viscosity oils receive the higher rating numbers. Who is correct?

a. Technician A
b. Technician B
c. Both A and B
d. Neither A nor B

14. For every pound of pressure put on engine coolant, the boiling point of the coolant is raised about ______.

a. 2°F (1.1°C)
b. 3°F (1.6°C)
c. 4°F (2.2°C)
d. 5°F (2.7°C)

15. In most automotive applications, the water pump is driven by the ______.

a. gear train
b. camshaft
c. crankshaft
d. impeller

16. Which of the following is not of concern when checking a rotor-type oil pump?

a. cover flatness
b. rotor thickness
c. inner rotor to outer rotor clearance
d. inner rotor to pump housing clearance

17. Which of the following will not result from insufficient drive belt tension?

a. belt slippage on pulleys
b. components will be driven at reduced speeds
c. belt noise
d. premature bearing and bushing failure in water pumps, generators, and power steering pumps

18. When must a thermostat be fully opened?

a. 3°F (1.6°C) above its temperature rating
b. 3°F (1.6°C) above or below its temperature rating
c. 20°F (7°C) above the start-to-open temperature
d. 20°F (7°C) below the start-to-open temperature

19. While discussing the consequences of an air leak on the suction side of an oil pump, Technician A says the engine's oil pressure may be abnormally high. Technician B says the oil may become aerated. Who is correct?

a. Technician A
b. Technician B
c. Both A and B
d. Neither A nor B

20. While installing an oil pump in an engine, Technician A packs the pump with petroleum jelly. Technician B submerges the pump in clean oil and rotates the pump shaft to fill the pump body. Who is correct?

a. Technician A
b. Technician B
c. Both A and B
d. Neither A nor B

13

ENGINE SEALING AND REASSEMBLY

OBJECTIVES

■ Explain the purpose of the various gaskets used to seal an engine. ■ Identify the major gasket types and their uses. ■ Explain general gasket installation procedures. ■ Describe the methods used to seal the timing cover and rear main bearing. ■ Reassemble an engine including core plugs, bearings, crankshaft, camshaft, pistons, connecting rods, timing components, cylinder head, valvetrain components, oil pump, oil pan, and timing covers. ■ Explain the ways to prelubricate a rebuilt engine. ■ Reinstall an engine and observe the correct starting and break-in procedures.

Proper sealing of an engine keeps the low-pressure liquids in the cooling system away from the cylinders and lubricating oil. It also keeps the high pressure of combustion in the cylinders. It prevents both internal and external oil leaks and suppresses and muffles noise.

TORQUE PRINCIPLES

All metals are elastic. **Elasticity** means a bolt can be stretched and compressed up to a certain point. This elastic, springlike property is what provides the clamping force when a bolt is threaded into a tapped hole or when a nut is tightened. As the bolt is stretched a few thousandths of an inch, clamping force or holding power is created due to the bolt's natural tendency to return to its original length **(Figure 13–1)**.

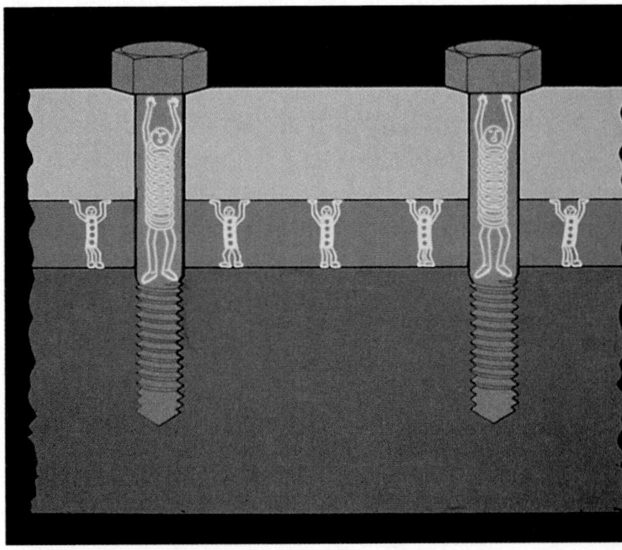

Figure 13-1 Clamping force results from bolt stretch. *Courtesy of Fel-Pro, Incorporated*

Like a spring, the more a bolt is stretched, the tighter it becomes. However, a bolt can be stretched too far. This is obvious when the grip on the wrench feels "mushy." At this point, the bolt can no longer safely carry the load it was designed to support, called the modulus of elasticity.

If a bolt is stretched into **yield**, it takes a permanent set and never returns to normal **(Figure 13–2)**. The bolt will continue to stretch more each time it is used, just like a piece of taffy that is stretched until it breaks.

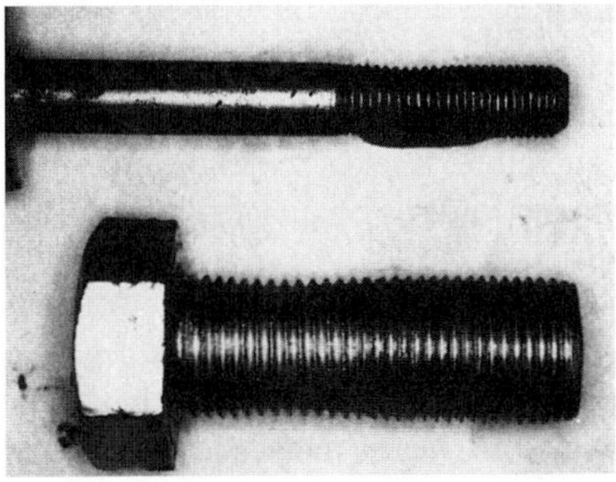

Figure 13-2 These bolts have been torqued past their yield points. Note the soda bottle effect.

Proper use of torque will avoid this yield condition. Torque values are calculated with a 25% safety factor below the yield point. However, some fasteners are intentionally torqued just barely into a yield condition, although not far enough to distort the bolt. This type of fastener, known as a **torque-to-yield (T-T-Y)** bolt, will provide 100% of its intended strength, compared to 75%

TABLE 13–1 STANDARD BOLT AND NUT TORQUE SPECIFICATIONS

Size Nut or Bolt	Torque (foot-pounds) (Newton-meters)	Size Nut or Bolt	Torque (foot-pounds) (Newton-meters)	Size Nut or Bolt	Torque (foot-pounds) (Newton-meters)
1/4–20	7–9 (9.45–12.15 Nm)	7/16–20	57–61 (77–82 Nm)	3/4–10	240–250 (324–338 Nm)
1/4–28	8–10 (10.8–13.5 Nm)	1/2–13	71–75 (96–101 Nm)	3/4–16	290–300 (392–405 Nm)
5/16–18	13–17 (17.55–22.95 Nm)	1/2–20	83–93 (112–126 Nm)	7/8–9	410–420 (554–567 Nm)
5/16–24	15–19 (20.25–25.65 Nm)	9/16–12	90–100 (122–135 Nm)	7/8–14	475–485 (644–655 Nm)
3/8–16	30–35 (40.5–47.25 Nm)	9/16–18	107–117 (144–158 Nm)	1–8	580–590 (783–797 Nm)
3/8–24	35–39 (47.25–52.65 Nm)	5/8–11	137–147 (185–198 Nm)	1–14	685–695 (925–938 Nm)
7/16–14	46–50 (62–67.5 Nm)	5/8–18	168–178 (227–240 Nm)		

when torqued to normal values. These fasteners, however, should not be reused. Some aftermarket gasket manufacturers include new T-T-Y bolts in their gasket sets.

Table 13–1 gives standard bolt and nut torque specifications. If the manufacturer's torque specifications are available, follow them precisely.

The grade and surface condition, whether plated or nonplated, dry or lubricated, oil or antiseize, cut threads or rolled, straight shank or reduced, affects the torque/tension relationship and causes the performance of the connection to change. Because torque is actually a combination of both tension and torsion, it is also a function of friction. The bolt head or nut, whichever is being rotated, produces friction as it is turned, as do the threads when they gall together under the pressure of being in tension. Tests have proven that 90% of work energy is consumed by friction. Friction must first be overcome before any true work is done. To compensate for surface variations, the following formula may be used to approximate the required torque.

$$T = FDC \div 12$$

in which T = Torque in pound-feet
F = Friction factor (torque coefficient)
D = Bolt diameter in inches
C = Bolt tension required in pounds

Nonplated bolts have a rougher surface than plated finishes. Therefore, it takes more torque to produce the same clamping force as on a plated bolt, even with one-third less friction. Add a lubricant and the torque might be as much as two-thirds lower.

Most printed torque values are for dry, plated bolts. Lubricants are beneficial when working with engines. They provide smoother surfaces and more consistent and evenly loaded connections. They also help reduce thread galling.

Reusing a dry nut will produce a connection with decreasing clamp force each time it is used. Nut threads are designed to collapse slightly to carry the bolt load. If dry nuts are reused, increased thread galling will result each time the nuts are reused at the same torque.

Lubrication of fasteners is recommended for consistency **(Table 13–2)**. However, be sure to lubricate all the bolts with the same lubricant. Some lubricants are more slippery than others, which affects torque values. Lubricate the bolt, never the hole. Otherwise, the bolt may merely be tightening against the oil in the hole.

If a bolt with a reduced shank diameter (for example, a connecting rod bolt) is specified by the OEM, never replace it with a standard, straight shank bolt. A reduced shank diameter bolt looks "dog-boned." Its function is to

TABLE 13–2 FRICTION FACTORS (F) AND TORQUE REDUCTIONS FOR LUBRICATED SURFACES ON ALLOY STEEL BOLTS

Lubricant	Friction Factor	Percentage of Torque Reduction Required
Colloidal copper	0.11	Reduce standard torque 45%.
Never-seize	0.11	Reduce standard torque 45%.
Grease	0.12	Reduce standard torque 40%.
Moly-cote (molybdenum disulphite)	0.12	Reduce standard torque 40%.
Heavy oils	0.12	Reduce standard torque 40%.
Graphite	0.14	Reduce standard torque 30%.
White lead	0.15	Reduce standard torque 25%.

reduce the stress on the threads by transferring it to the shank. A standard bolt under similar conditions would break very quickly at the threads.

Keep the following points in mind:

1. Visually inspect the bolts.
   - Threads must be clean and undamaged. Discard all bolts that are not acceptable.
   - Use liquid sealant or engine oil on the threads and seating face of the cylinder head bolts to prevent seizure from rust and corrosion. This is particularly important for an aluminum block or head. Use sealant on those bolts that hold in coolant or oil.
   - Install bolts in their proper holes.
   - Run a nut over the bolt's threads by hand. Discard it if any binding occurs.
   - Clean bolt and cylinder block threads with a thread chaser or tap **(Figure 13–3)**.
2. Apply a light coat of 10W engine oil to threads and bottom face of bolt head. A sealer is required for a bolt that enters a water jacket. This will stop coolant seepage around the bolt threads. Seeping coolant could get in the oil or cause corrosion that might damage parts, resulting in engine failure.
3. Tighten bolts in the recommended sequence. This is important to prevent warpage of the cylinder head or other parts.
   - Use an accurate torque wrench.
   - Tighten bolts to the recommended torque in steps and proper sequence.
4. If bolt heads are not tight against the surface, the bolts should be removed and washers installed.
5. Make sure the bolt is the proper length (not too long).

Figure 13–3 Cleaning bolt holes with a tap. *Used with Permission of Detroit Gasket—Copyright © 1998*

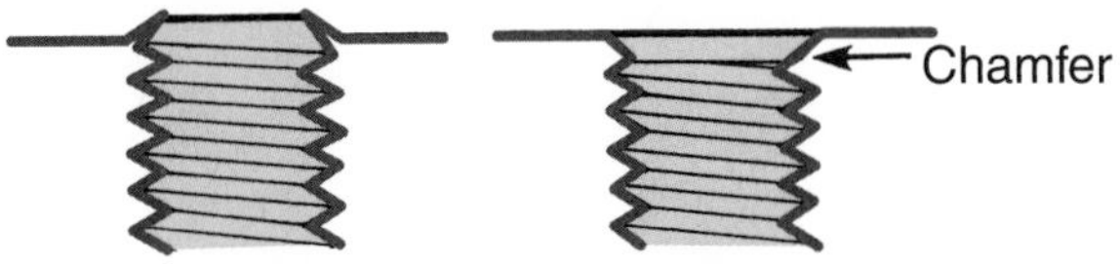

Figure 13–4 Bolt hole threads can pull up, leaving a raised edge. Also, if the block has been resurfaced, the threads may run up to the surface. In either case, the hole must be chamfered.

Bolt hole threads in the engine block often pull up, leaving a raised edge around the hole **(Figure 13–4)**. If the block has been resurfaced, the threads might run up to the surface. In either case, the bolt holes should be tapered at the surface by chamfering and the threads cleaned with an appropriate size bottoming tap. Always repair damaged threads to ensure proper bolt performance.

Many different nuts are used by the automotive industry. Beware of hexagon (hex) nut rotation when using power wrenches. It is deceptively easy to place a nut into a yield condition within seconds. Impact wrenches are the worst offenders. Friction is needed to prevent the nut from spinning. If the nut is lubricated, there is no friction left to stop the impact wrench from hammering the nut past the bolt's yield point or stripping the threads.

SHOP TALK

Impact wrenches should only be used to loosen nuts and bolts. Use other power or hand tools to tighten them. Final tightening should always be done with a torque wrench. ■

Smaller air-powered ratchets do not produce the severe force of impact wrenches and are much safer to use.

Thread Repair

A common fastening problem is threads stripping inside an engine block, cylinder head, or other structure. This problem is usually caused by overtorquing or by incorrectly threading the bolt into the hole. Rather than replacing the block or cylinder head, the threads can be replaced by the use of threaded inserts. The helically coiled insert is the most commonly used.

GASKETS

Gaskets are used predominantly to prevent gas or liquid leakage between two parts that are bolted together **(Figure 13–5)**. Gaskets also serve as spacers, wear insulators, and vibration dampers.

As a spacer, the gasket serves as a shim between two joined engine components, such as the fuel pump lever and cam. It also helps compensate for manufacturing or

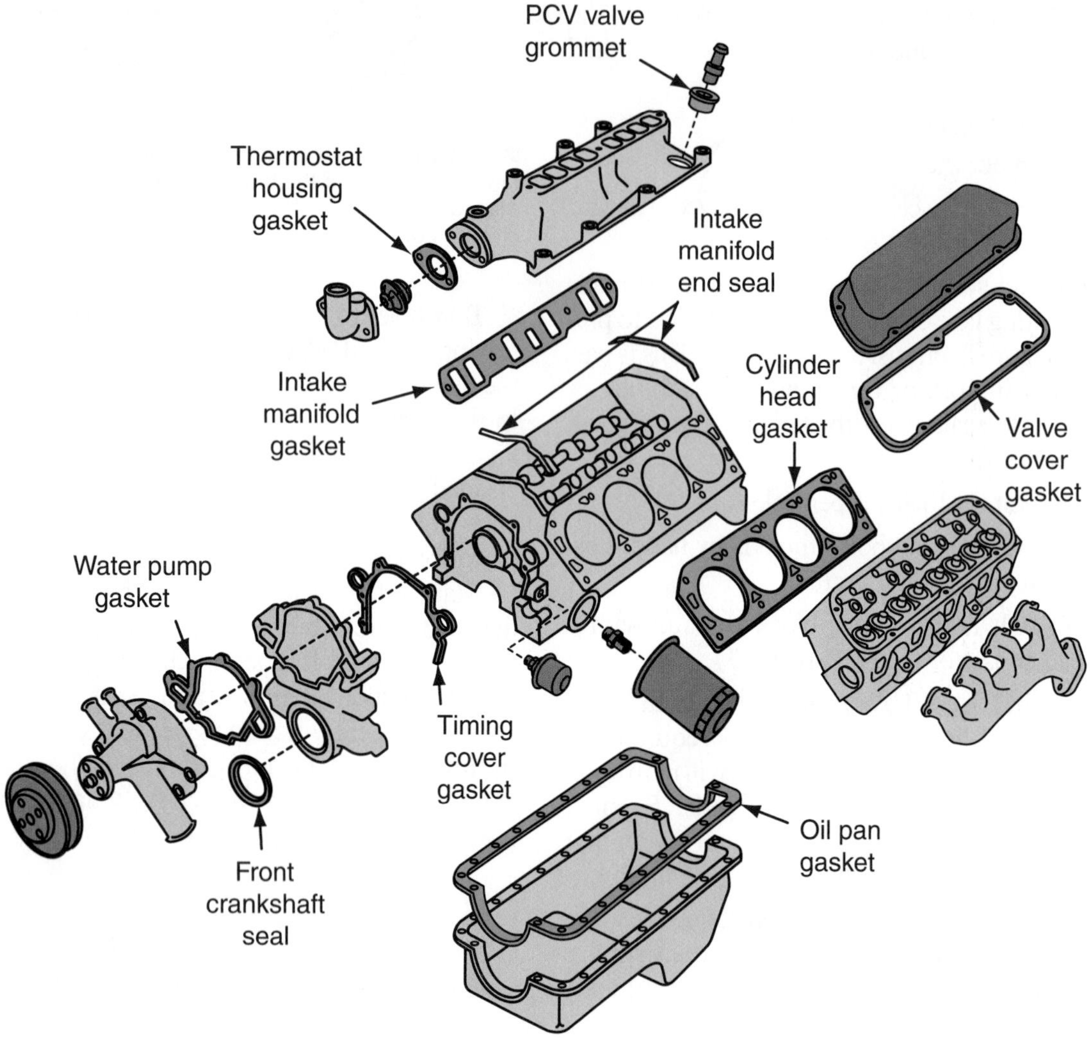

Figure 13-5 Typical engine gasket and seal locations.

rebuilding tolerances in the cylinder head. In the wear insulator function, the gasket material is softer than the separated components. This allows for alternating expansion and contraction with less friction, thus preventing fretting or brinelling of machined surfaces. Keep in mind that the shock absorberlike function does not apply to liquid sealants that are used without a gasket and allow a metal-to-metal seal.

There are four basic classifications of engine gaskets.

- Hard gaskets are steel, stainless steel, copper, or a combination metal enclosure with a compressible and heat-resistant clay/fiber compound sandwiched inside **(Figure 13–6)**. Gaskets in this class are cylinder head gaskets, exhaust manifold gaskets, and some intake manifold gaskets.

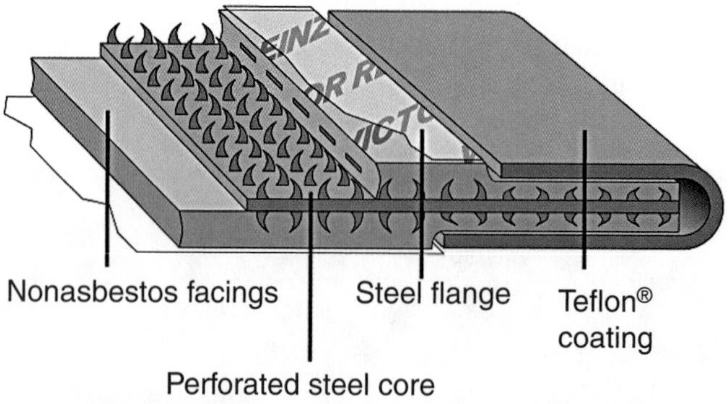

Figure 13-6 Composition of a Teflon-coated perforated steel core gasket used mainly as head and intake manifold gaskets. *Courtesy of Dana Corporation*

- Soft gaskets consist of soft, flexible materials such as cork, rubber, a combination of cork/rubber **(Figure 13–7)**, rubber-coated steel **(Figure 13–8)**, paper, and other compressed fiber materials. Newer materials, sometimes called gaskets-in-a-tube, include silicone gaskets and anaerobics. Soft gaskets are often used on valve covers, oil pans, rocker covers, pushrod covers, water pumps, thermostat housing, timing covers, and inspection plates.

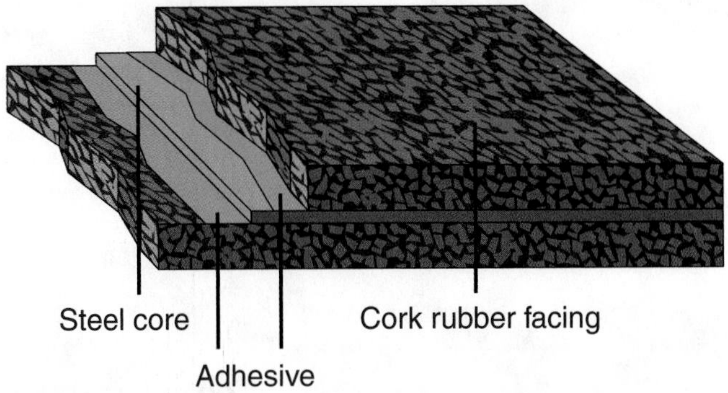

Figure 13-7 The composition of a cork/rubber gasket used mainly as valve cover, oil pan, and timing cover gaskets. *Courtesy of Dana Corporation*

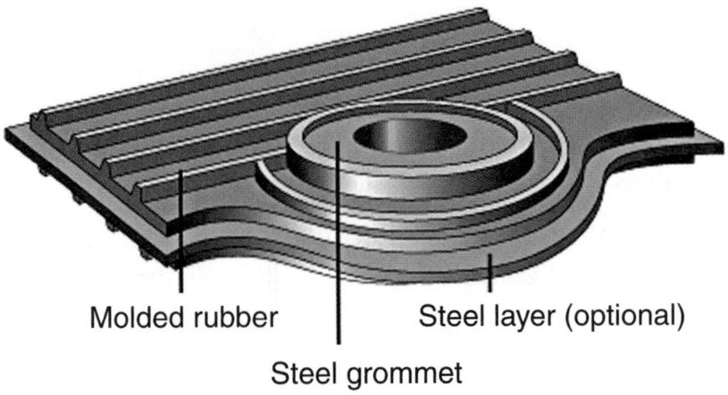

Figure 13-8 The composition of a molded rubber gasket with steel grommets used mainly as valve cover and oil pan gaskets. *Courtesy of Dana Corporation*

- **Silicone gasketing** is available in three grades. The black is for general purpose; blue is for special applications; and red is for high-temperature requirements. Most mechanics use silicone gasketing to aid sealing in corners, notches, or dovetails.
- **Sealants** and **adhesives** generally come in a liquid form to be applied to all metal gaskets such as beaded steel-type cylinder head gaskets and intake manifold gaskets. They dry to flexible, nonhardening sealants to help prevent oil, water, gasoline, air pressure, or vacuum leaks. They are also useful as an **antiseize** for threads on bolts, studs, and fittings.

Gasket Materials

Historically, with heavy-wall, all-cast-iron engines, most gasket materials and constructions were similar, if not identical. One key to the auto industry's success in achieving late-model engine performance goals has been the development of gaskets specifically designed to seal the new engines. These new sealing products have been highly engineered for virtually all critical areas of the different engine designs. Inappropriate substitution of non-OE-equivalent products while servicing an engine may lead to premature failure.

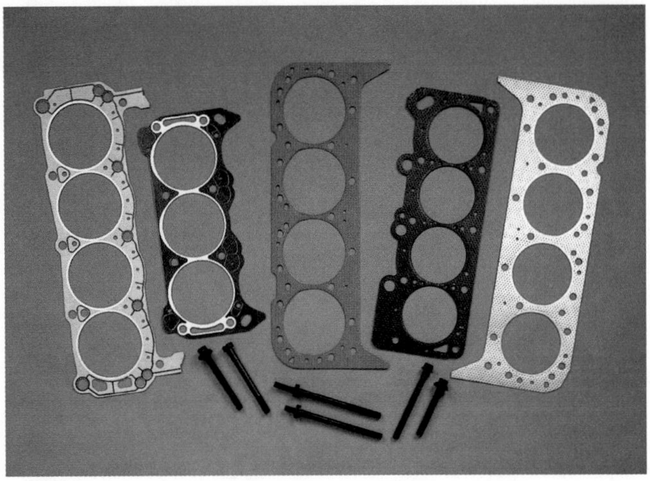
Figure 13-9 Assorted cylinder head gaskets (utilizing various construction and coatings) and cylinder head bolts. *Used with Permission of Detroit Gasket—Copyright © 1998*

Most late-model engines are equipped with gaskets made of molded silicone rubber and cut plastic gaskets with silicone sealing beads. Many gasket design innovations are used **(Figure 13–9)** by different OE suppliers to reach the same goal. That goal is long-term, leak-free joints. The new gaskets have entirely different properties from traditional gasketing materials (asbestos, rubber, cork, and the like) that influence many critical design features, including gasket thickness, compressibility, no-retorque characteristics, near-zero creep, greatly improved corrosion resistance, and superior thermal properties in sealing at temperatures exceeding 1,000°F (538°C).

General Gasket Installation Procedures

The following general instructions will serve as a helpful guide for installing gaskets. Because there are many different gasket materials and designs, it is impossible to list in this chapter directions for every type of installation. Always follow any special directions provided on the instruction sheets packed with most gasket sets. These will give you additional tips on replacing the gaskets of an engine.

1. Never reuse old gaskets. Even if the old gasket appears to be in good condition, it will never seal as well as a new one.
2. Handle new gaskets carefully. Be careful not to damage the new gaskets before placing them on the engine.

USING SERVICE MANUALS

Always refer to the specific engine and engine part section of the service manual for the recommended procedures for using sealants. ■

3. Use gasket sealants only when they are absolutely necessary.

SHOP TALK

Some technicians tend to use too much sealant on gaskets. Do not make this mistake. Because sealants have less strength than gasket materials, they create weaker joints. They can also prevent some gasket material from doing what it is supposed to do, which is to soak up oil and swell to make a tight seal. ■

4. Cleanliness is essential. New gaskets seal best when used on clean surfaces. Thoroughly clean all mating surfaces of dirt, oil deposits, rust, old sealer, and gasket material **(Figure 13–10)**.
5. Use the right gasket in the right position. Always compare the new gasket to the component mating surfaces to make sure it is the right gasket. Check that all bolt holes, dowel holes, pushrod openings, coolant, and lubrication passages line up perfectly with the gasket. Some gaskets will have directions such as "top," "front," or "this side up" stamped on one surface **(Figure 13–11)**. An upside-down or reversed gasket can easily cause loss of oil pressure, overheating, and engine failure.

Cylinder Head Gaskets

Cylinder head gaskets are the most sophisticated type of gasket. They also have the most demanding job. When first starting an engine in cold weather, parts near the combustion chamber are very cold. Then, after only a few minutes of engine operation, these same parts might reach 400°F (204°C). The inner edges of the cylinder head gasket are exposed to combustion flame temperatures from 2,000 to 3,000°F (1,093 to 1,648°C)

Pressures inside the combustion chamber also vary tremendously. On the intake stroke, a vacuum or low pressure exists in the cylinder. Then, after combustion, pressure peaks of approximately 1,000 psi (6,895 kPa) occur. This extreme change from suction to high pressure happens in a fraction of a second.

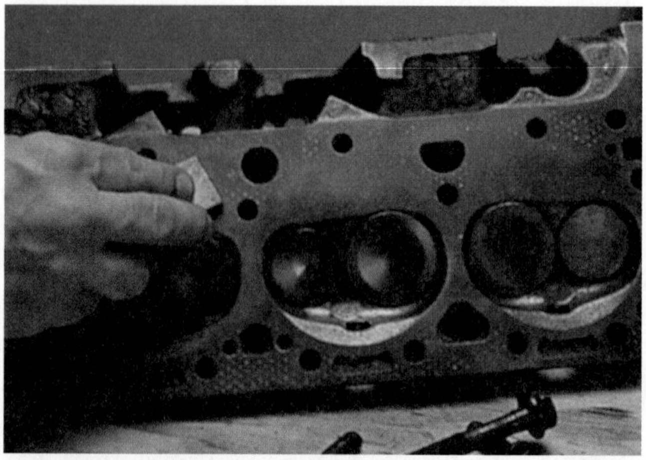

Figure 13–10 Thoroughly clean the sealing surface. *Used with Permission of Detroit Gasket—Copyright © 1998*

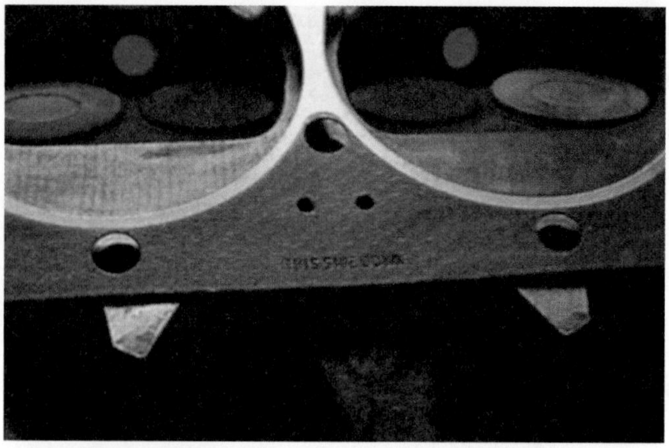

Figure 13–11 Some gaskets have installation directions stamped on them. *Used with Permission of Detroit Gasket—Copyright © 1998*

Cylinder head gaskets, under these conditions, must simultaneously do the following:

- Seal intake stroke vacuum, combustion pressure, and the heat of combustion.
- Prevent coolant leakage, resist rust and corrosion, and, in many cases, meter coolant flow.
- Seal oil passages through the block and head while resisting chemical action.
- Allow for lateral and vertical head movement as the engine heats and cools.
- Be flexible enough to seal minor surface warpage while being stiff enough to maintain adequate gasket compression.
- Fill small machining marks that could lead to serious gasket leakage and failure.
- Withstand forces produced by engine vibration.

Bimetal Engine Requirements Another problem that head gaskets must face is the differing expansion rates of aluminum/cast-iron combination engines **(Figure 13–12)**. Aluminum has a coefficient of thermal expansion two to three times greater than that of steel, depending on the alloy. This creates a scrubbing action on a head gasket as the engine goes from cold to hot and back again. If the surface of the aluminum head has been roughly machined, it can grab and tear the head gasket. To prevent this, slippery nonstick surface coatings and other materials are applied by the manufacturer to the gaskets. This coating provides the needed lubrication between the head and gasket. **Graphite** is a natural lubricant, so graphite-faced gaskets are not coated.

The latest generation of aftermarket gaskets includes a bead sealant that increases clamping pressure around troublesome areas. Another common and desirable

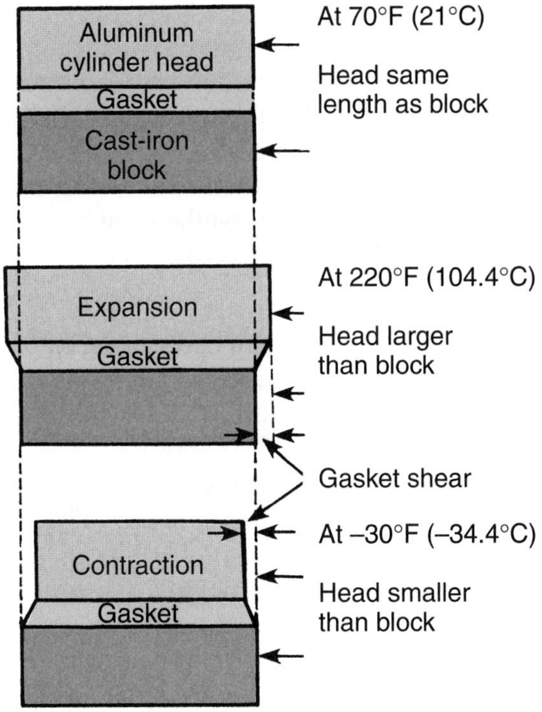

Figure 13–12 Thermal growth characteristics of bimetal engines.

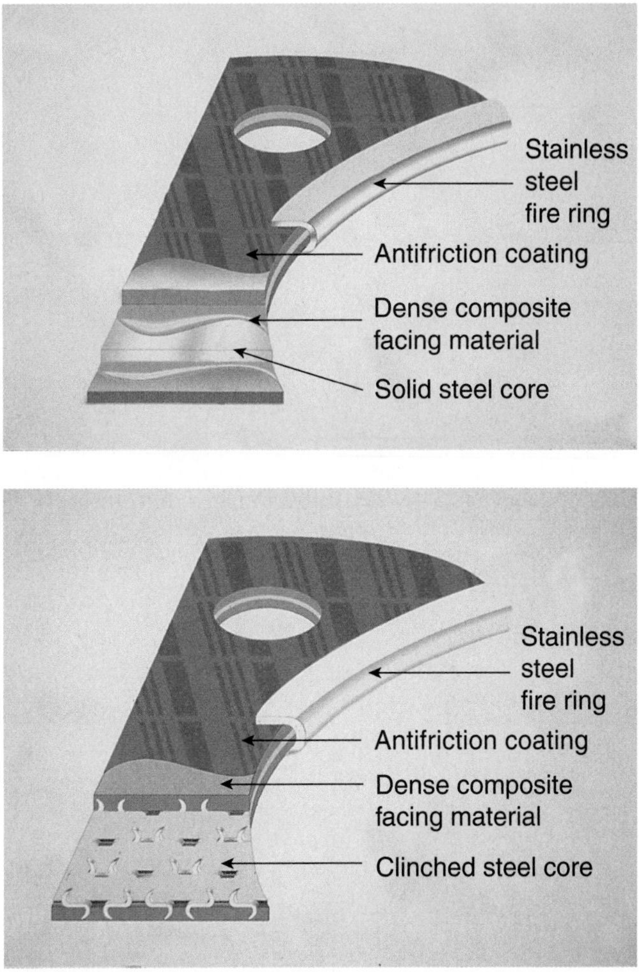

Figure 13–13 Construction of no-retorque head gaskets. *Courtesy of Fel-Pro, Inc.*

gasket design today is the no-retorque type **(Figure 13–13)**. Older gasket designs, such as the steel-faced sandwich and perforated core, require the cylinder head of an engine to be retorqued 300 to 500 miles (480 km to 800 km) after it has been rebuilt. These gaskets take a set after initial engine operation and relax to the point that retorquing is needed to restore the clamping force. The extra labor and expense of retorquing a cylinder head gasket makes the repair more expensive than no-retorque designs in the long run. The no-retorque design retains a higher level of clamping force. Therefore, no retorquing is needed.

Manifold Gaskets

There are three types of manifold gaskets—the intake manifold, exhaust manifold, and the combination intake and exhaust. Each type of manifold gasket has its own sealing characteristics and problems **(Figure 13–14)**. Therefore, be sure to follow the manufacturer's instructions when installing them.

> **USING SERVICE MANUALS**
>
> Refer to the manifold unit in the engine section of the service manual for specific instructions on the installation of manifold gaskets. ■

Valve Cover Gaskets

Valve cover gaskets have an increasingly difficult job in today's high-temperature engines. They must seal between a steel or aluminum (or molded plastic) stamping and a cylinder head surface that might not be machined.

To provide an effective seal, a valve cover gasket must conform to flange distortion. At the same time, it must provide good crush and split resistance to limit torque loss and provide chemical and heat resistance for

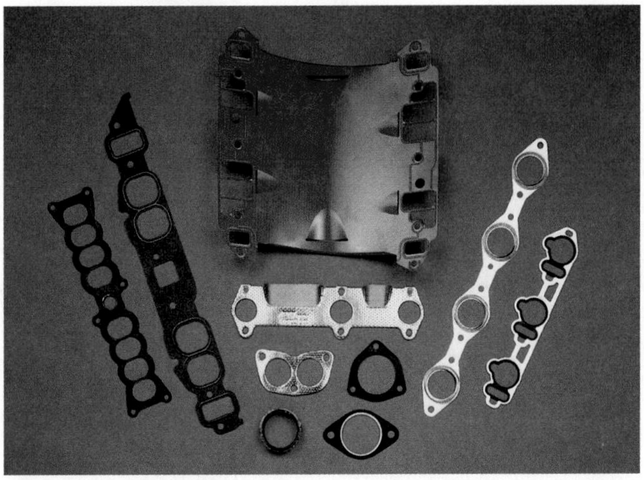

Figure 13–14 Assorted intake and exhaust manifold gaskets for diverse applications. *Used with Permission of Detroit Gasket—Copyright © 1998*

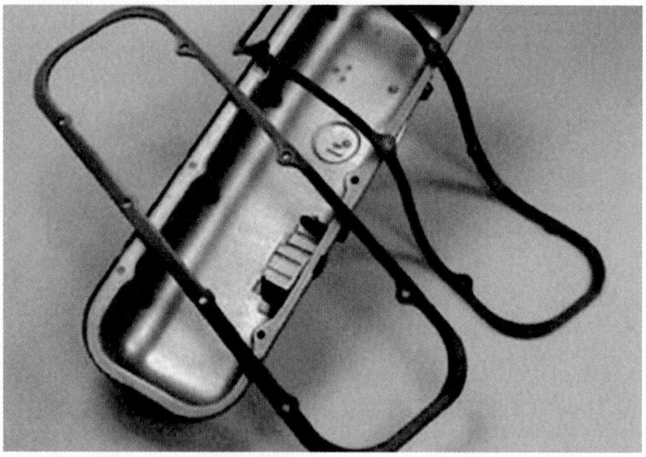

Figure 13-15 (left) A valve cover gasket with a rigid core. (right) A valve cover gasket without reinforcement. *Courtesy of Federal-Mogul Corporation*

long-term sealing. The most commonly used material is a blend of cork and rubber particles. The quality of the materials is crucial when using cork/rubber gaskets, because a low-quality cork/rubber mixture can develop leakage after a fairly short period of operation. Another popular valve cover material is made of synthetic rubber. Whether it is a cork/rubber or synthetic rubber gasket, it must be a perfect fit **(Figure 13–15)**.

Oil Pan Gaskets

An oil pan gasket seals the joint between the oil pan and the bottom of the block **(Figure 13–16)**. The oil pan gasket might also seal the bottom of the timing cover and the lower section of the rear main bearing cap.

Like valve cover gaskets, the oil pan gasket must resist hot, thin engine oil. Oil pans are usually made of

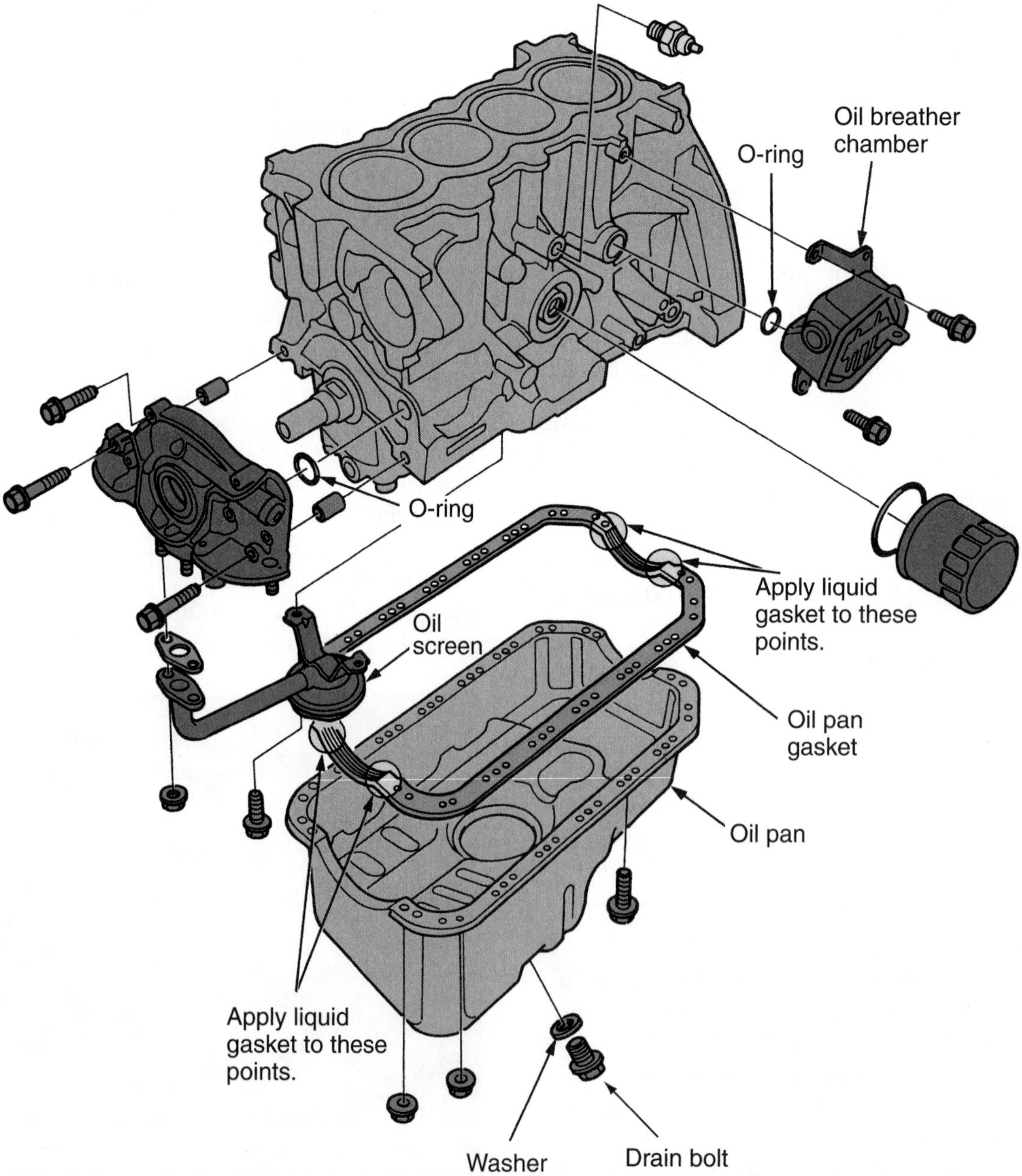

Figure 13-16 Oil pan gasket installation. *Courtesy of American Honda Motor Co., Inc.*

stamped steel, cast iron, or cast aluminum. Because of the added weight and splash of crankcase oil, the pan has many assembly bolts closely spaced. As a result, the clamping force on the oil pan gasket is great. The gasket is thinner and must resist crushing.

Oil pan gaskets are made of several types of material. A commonly used material is synthetic rubber, known for its long-term sealing ability. It is tough, durable, and resists hot engine oil. Synthetic rubber gaskets are easy to remove, so the sealing surfaces need less cleanup.

Carefully follow the recommendations from the gasket manufacturer. Take note of any of the original equipment manufacturer's recommendations that could affect engine sealing. Before installing the oil pan, make sure its flange is flat. The gasket should be mounted with a few dabs of quick-drying contact adhesive. Carefully align the gasket before the adhesive dries. Wait until the adhesive is dry before installing the pan. Tighten the oil pan bolts to the recommended torque specification and sequence given in the service manual.

ADHESIVES, SEALANTS, AND OTHER CHEMICAL SEALING MATERIALS

A number of chemicals can be used to reduce labor and ensure a good seal. Many gasket sets include a label with the proper chemical recommendation for use with that gasket set. Some even include sealers in the sets when the original equipment manufacturer used a sealer to replace a gasket and a gasket cannot be manufactured for that application. They also include sealers in some sets when gasket unions need a sealant to ensure a good seal.

SHOP TALK

Chemical adhesives and sealants give added holding power and sealing ability where two parts are joined. Sealants usually are added to threads where fluid contact is frequent. Chemical thread retainers are either **aerobic** (cures in the presence of air) or **anaerobic** (cures in the absence of air). These chemical products are used in place of lock washers. ■

Adhesives

Gasket adhesives form a tough bond when used on clean, dry surfaces. Adhesives do not aid the sealing ability of the gasket. They are meant only to hold gaskets in place during component assembly. Use small dabs; they will dry quicker for fast installation. Do not assemble components until the adhesive is completely dry. Most adhesives are ideal for use on gasket applications such as valve covers, pushrod covers, manifold and manifold end seals, and oil pan and oil pan end seals **(Figure 13–17)**.

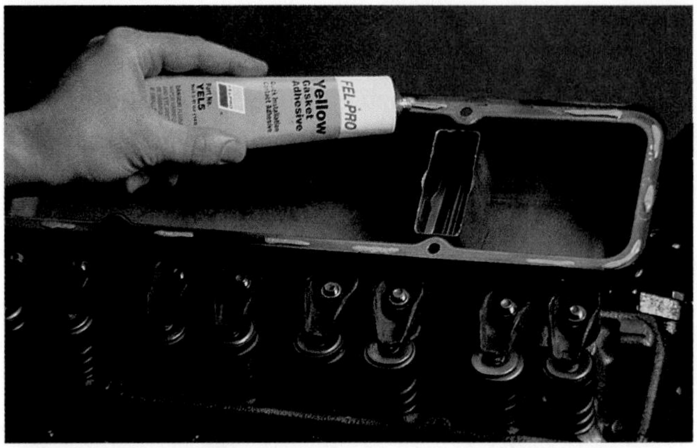

Figure 13–17 An adhesive is often used to hold a gasket in place during assembly. *Courtesy of Fel-Pro, Inc.*

Sealants

General-Purpose Sealants General-purpose sealants come in liquid form and are available in a brush type (known as brush tack) and an aerosol type (known as spray tack). General purpose sealers **(Figure 13–18)** form a tacky, flexible seal when applied in a thin, even coat that aids in gasket sealing by helping to position the gasket during assembly.

WARNING!

Make sure every sealant you use on today's engines is oxygen-sensor safe.

WARNING!

Never use a hard-drying sealant (such as shellac) on gaskets. It will make future disassembly extremely difficult and might damage the gasket.

Flexible Sealants Flexible sealants are most often used on threads of bolts that go into fluid passages. They are

Figure 13–18 Applying gasket sealer with a brush. *Courtesy of Fel-Pro, Inc.*

nonhardening sealers that fill voids, preventing the fluid from running up the threads. They resist the chemical attack of lubricants, synthetic oils, detergents, antifreeze, gasoline, and diesel fuel.

Silicone Formed-in-Place Sealants Silicone gasketing can be used to replace conventional paper, cork, and cork/rubber gaskets. It is generally for use on oil pans, valve covers, thermostat housing, timing covers, water pumps, and other such installations. **Room temperature vulcanizing (RTV)** silicone sealing products are the best known of the formed-in-place (FIP) gasket products.

WARNING!

Be careful not to use excessive amounts of RTV. If too much is applied, it can loosen up and get into the oil system, where it can clog an oil or coolant passage and cause severe engine damage and/or engine overheating.

Today's RTV aerobic silicone formulations are impervious to most automotive fluids, extremely resistant to oil, oxygen-sensor safe, exhibit outstanding flexibility (a necessary feature on modern bimetal engines), and adhere well to a broad range of materials that include plastics, metal, and glass.

To use RTV silicone, make sure the mating surfaces are free from dirt, grease, and oil. Apply a continuous $\frac{1}{8}$-inch (3.18-mm) bead on one surface only (preferably the cover side). Make sure to circle all bolt holes **(Figure 13–19)**. Adjust the shape before a skin forms (in about 1 minute) as shown in **Figure 13–20**. Remove excess RTV silicone with a dry towel or paper towel. Press the parts together. Do not slide the parts together; this will disturb the bead. Tighten all retaining bolts to the manufacturer's specified torque. Cure time is approximately 1 hour for metal-to-metal joints and can take up to 24 hours for $\frac{1}{8}$-inch (3.18-mm) gaps.

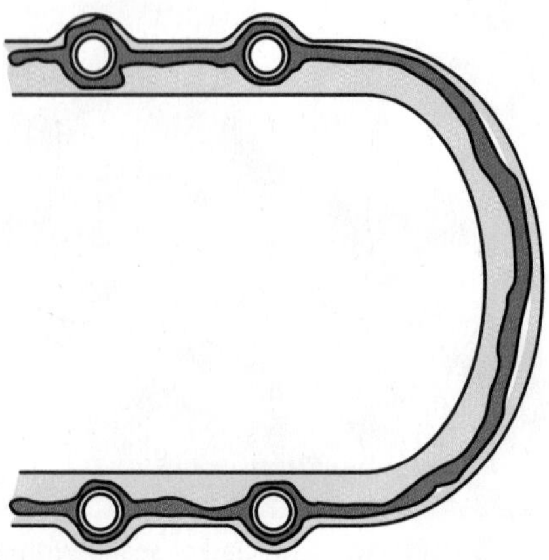

Figure 13-19 When applying RTV, make sure the bolt holes are encircled.

Figure 13-20 Applying a bead of RTV gasket material on a water pump. *Courtesy of Loctite*

CAUTION!

The uncured rubber contained in RTV silicone gasketing irritates the eyes. If any gets in your eyes, immediately flush with clean water or eyewash. If the irritation continues, see a doctor.

Anaerobic Formed-in-Place Sealants These materials are used for thread locking as well as gasketing. As a retaining compound, they are mostly used to hold sleeves, bearings, and locking screw nuts in place where there is a high exposure to vibration.

WARNING!

Never use a sealant or formed-in-place gasket material on exhaust manifolds.

The major difference between aerobic and anaerobic sealants, other than their method of curing, is their gap-filling ability. Typically, 0.050 inch (1.27 mm) ($\frac{3}{64}$ inch) is the absolute limit of any anaerobic's gap-filling ability. Some are only designed to seal 0.005- to 0.010-inch (.1270-mm to .2540-mm) gaps. Anaerobic sealers are intended to be used between the machined surfaces of rigid castings, not on flexible stampings.

SHOP TALK

Once hardened, a good anaerobic bond is unbelievably tenacious and can withstand high temperatures. Therefore, care must be taken in selection. They tend to be highly specialized and not readily interchangeable. For example, there are various levels of thread-locking products that range from medium-strength antivibration agents to high-strength, weldlike retaining compounds. The inadvertent use of the wrong product could make future disassembly an impossibility. Check the label to be certain that anaerobic material will suit the purpose of the application. ■

Antiseize Compounds

Antiseize compounds prevent dissimilar metals from reacting with one another and seizing. This chemical-type material is used on many fasteners, especially those used with aluminum parts. Always follow the manufacturer's recommendations when using this compound.

OIL SEALS

The job of seals is to keep oil and other vital fluids from escaping around a rotating shaft. There are three basic oil seal designs: the fiber packing, the two-piece split lip design, and the one-piece radial design **(Figure 13–21)**.

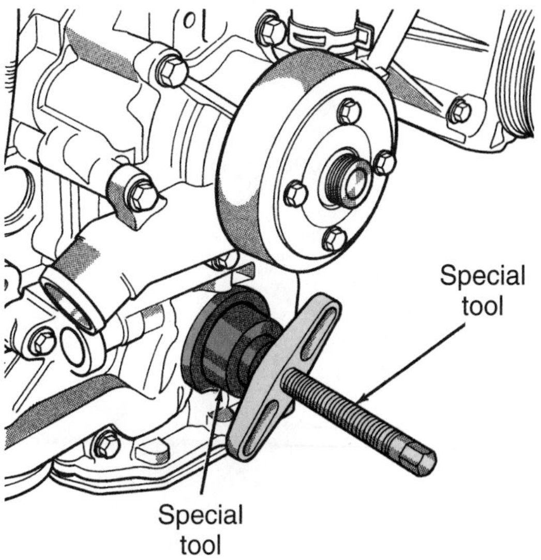

Figure 13-22 The installation of the timing cover oil seal requires the use of a special tool or driver. *Courtesy of DaimlerChrysler Corporation*

SHOP TALK

Whenever installing an oil seal, make sure its lip seal is lubricated with a light coating of grease. Also, make sure the lip portion of the seal is facing the direction that oil is coming against. ■

Timing Cover Oil Seals

An oil seal in the timing cover prevents oil from leaking around the crankshaft. The installation of this seal often requires the use of a special tool **(Figure 13–22)** or a driver. It is important that the seal be positioned squarely in the bore of the timing cover and the crankshaft be positioned in the center of the seal.

Rear Main Bearing Seals

Rear main bearing seals keep oil from leaking at the crankshaft around the rear main bearing. There are two basic types of constructions: wick- or rope-type packing and molded synthetic rubber.

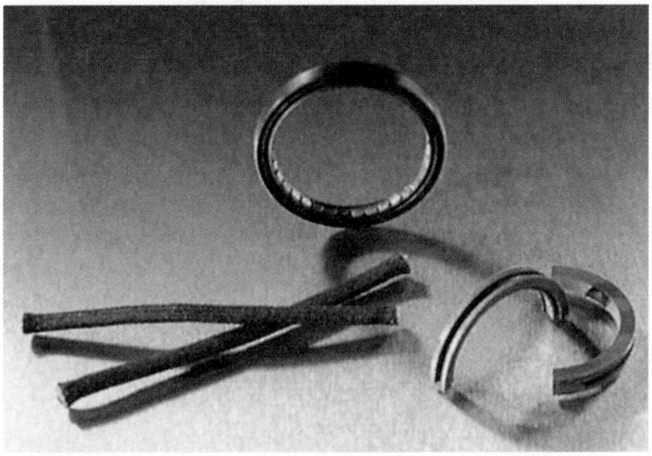

Figure 13-21 The three basic oil seal designs. *Used with permission of Detroit Gasket—Copyright © 1998*

Wick- or rope-type packings are common on many older engines. Molded synthetic rubber lip-type seals are used on many newer engines. They do a good job of sealing even when there is some eccentricity in the shaft, as long as the surface of the shaft is very smooth. Synthetic rubber seals may sometimes be retrofitted to some older engines that have wick seals, but only if the seals are offered as an option by the sealing manufacturer.

Three types of synthetic rubber are used for rear main bearing seals. **Polyacrylate** is commonly used because it is tough and abrasion resistant, with moderate temperature resistance to 350°F (177°C). Silicone synthetic rubber has a greater temperature range, but it has less resistance to abrasion and is more fragile than polyacrylate. Silicone seals must be handled carefully during installation to avoid damage. **Viton** has the abrasion resistance of polyacrylate and the temperature range of silicone, but it is the most expensive of the synthetic types. The synthetic rubber seals may be one piece **(Figure 13–23)** or two pieces **(Figure 13–24)**.

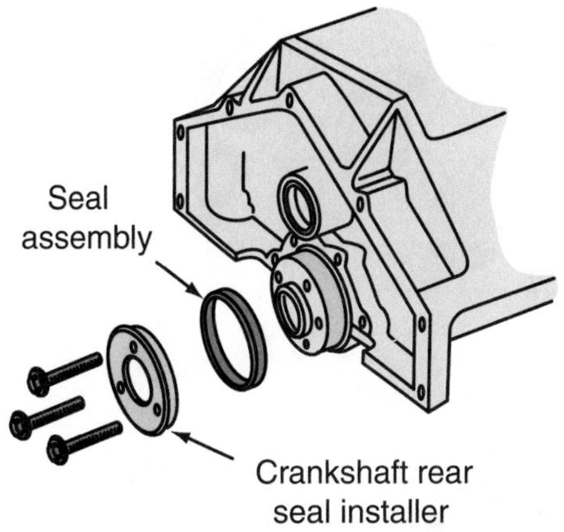

Figure 13-23 Installing a one-piece rubber rear crankshaft seal.

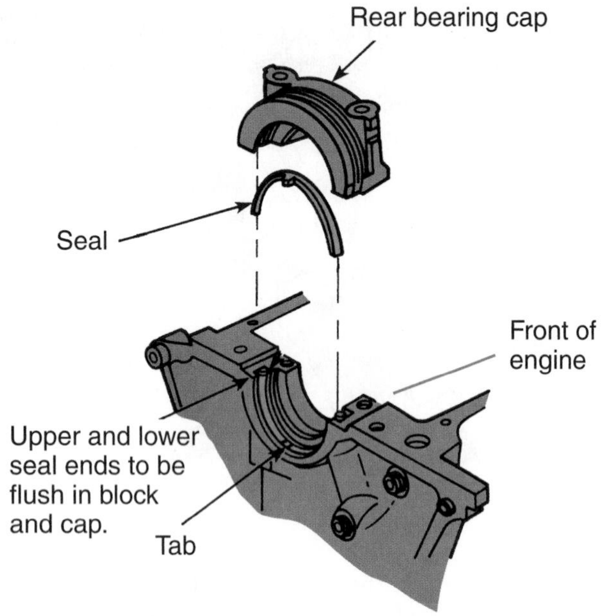

Figure 13-24 A typical two-piece rubber rear crankshaft seal.

No matter what the construction of the seal is, always check the shaft for smoothness. Shafts should be free of nicks and burrs to ensure long oil seal life. Carefully remove any roughness with a very fine emery cloth and then clean the shaft thoroughly. The shaft should have a highly polished appearance and a smooth feel. Also, be sure to check and clean the oil slinger and oil return channel in the bearing cap.

Other Seals

Different parts of the engine have different sealing requirements. Some gaskets must seal pressure; others must seal vacuum. Some seal hot oil **(Figure 13–25)** and others hot antifreeze. Some gaskets seal joints under flexible flanges; others seal joints between fairly rigid flanges held together with large mounting bolts. The construction of each gasket matches its sealing requirement. Choose the correct gasket or sealing material to ensure that proper sealing is achieved.

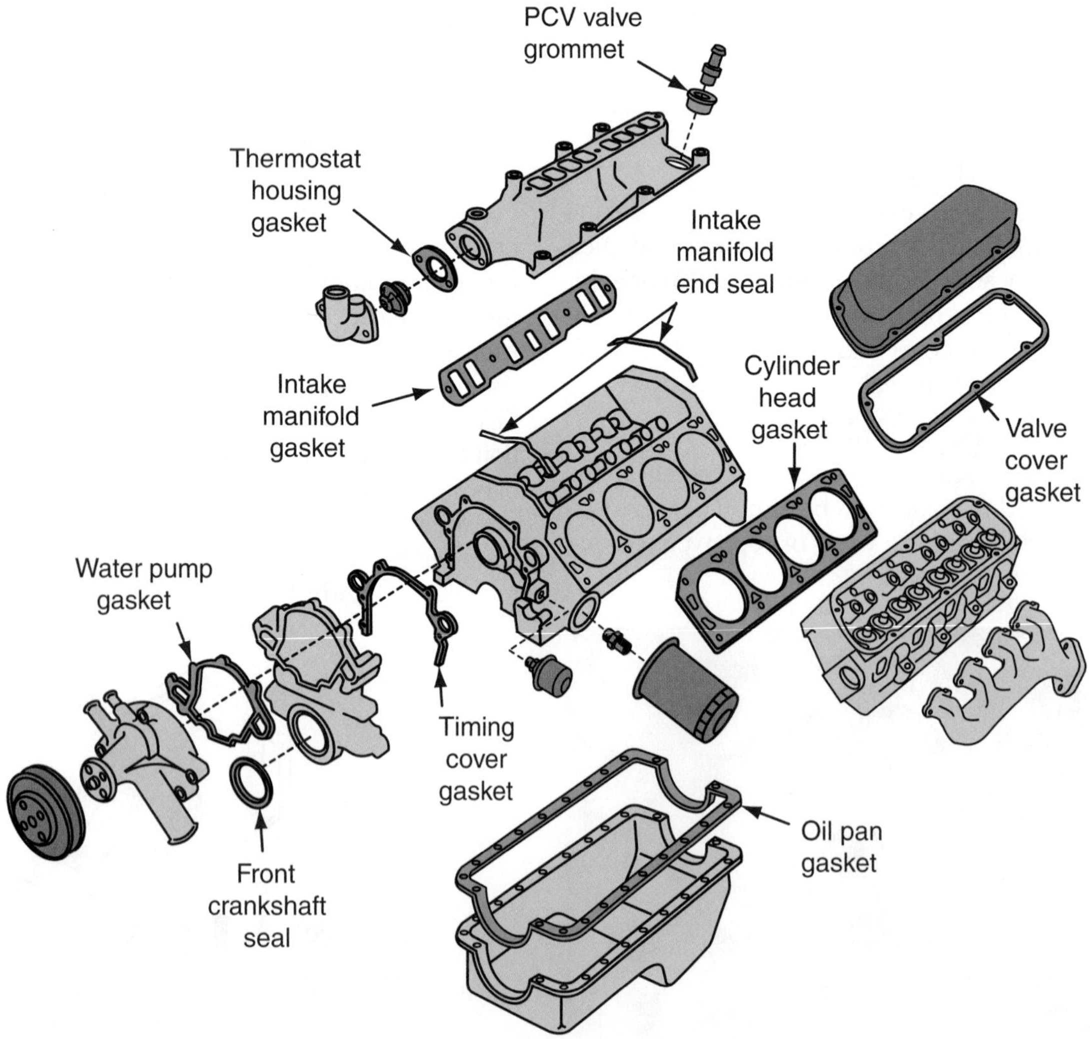

Figure 13-25 The seals and gaskets used to prevent oil leaks in a typical camshaft housing assembly. *Courtesy of General Motors Corporation*

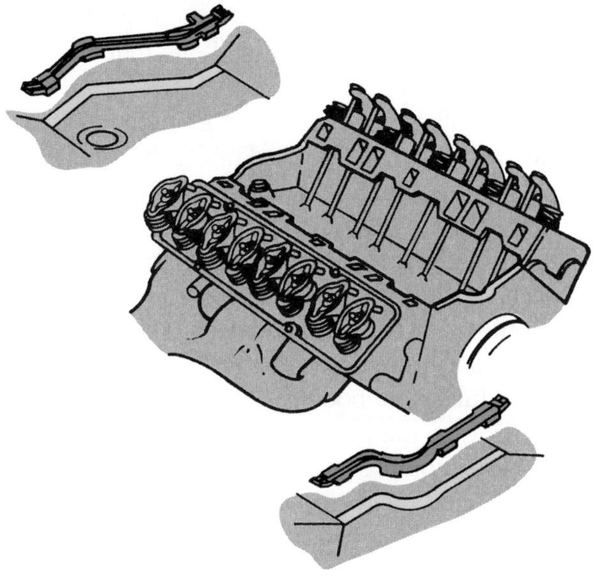

Figure 13-26 Intake manifold end strip seals.

Most gasket sets for V-type engines include manifold end strip seals. They seal the joint between the manifold ends and the block **(Figure 13–26)**. Molded rubber and cork/rubber are materials often used for end strip seals.

The exhaust gas recirculation (EGR) valve takes a sample of the exhaust gases and introduces it back into the intake manifold. This reduces combustion temperatures and reduces NO_x. The EGR valve should be carefully inspected before it is installed on the engine. Check the sealing surface and use a file to remove any minor imperfections that may prevent the valve from sealing properly **(Figure 13–27)**. Also make sure the new gasket is the correct size; some applications have specifically sized holes in the gasket that are used to regulate exhaust flow **(Figure 13–28)**. The use of the wrong gasket could change how the engine performs

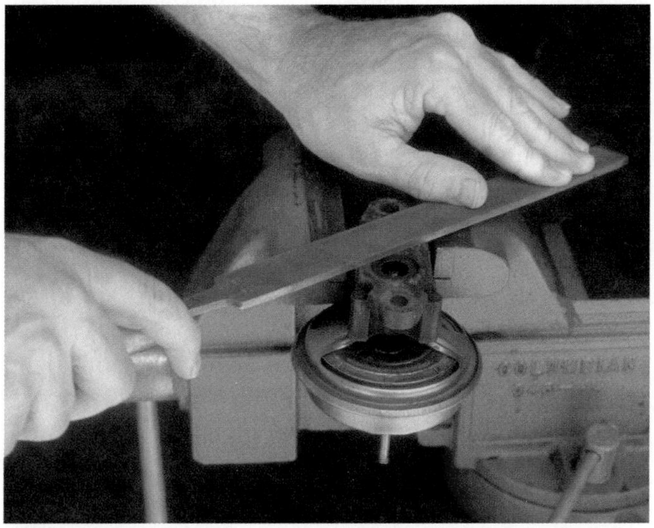

Figure 13-27 Smoothing the sealing surface of an EGR valve.

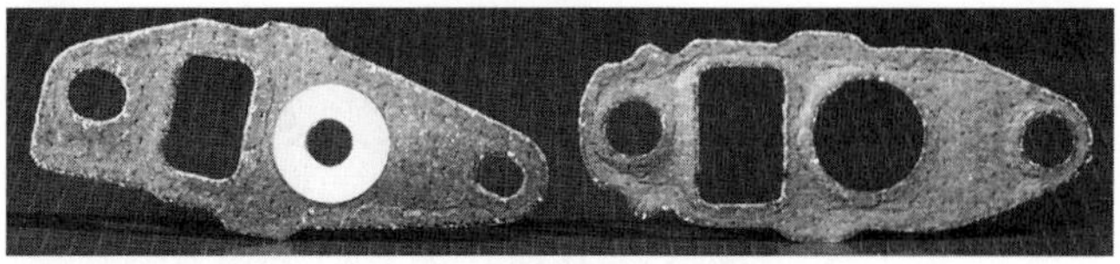

Figure 13-28 Typical EGR valve gaskets. *Courtesy of Fel-Pro, Inc.*

ENGINE REASSEMBLY

When reassembling an engine, the assembly sequence is essentially the reverse of the tear-down sequence outlined in a previous chapter. Details for assembling the major engine parts can be found in the chapters covering that part.

Installing the Cylinder Head and Valve Train

Before installing a cylinder, sort out the cylinder head bolts. Many engines use head bolts of different lengths **(Figure 13–29)**. The service manual usually identifies where the long and short bolts go. The threads of the bolts should be thoroughly cleaned and then lightly lubricated with oil or antiseize compound. Make sure you locate and adhere to the wet torque specs when tightening a bolt that has been lubricated with oil or any other liquid.

The head gasket should be positioned on the block and checked for fit. The cylinder head is then placed in position on the block; the bolts are inserted into the bolt

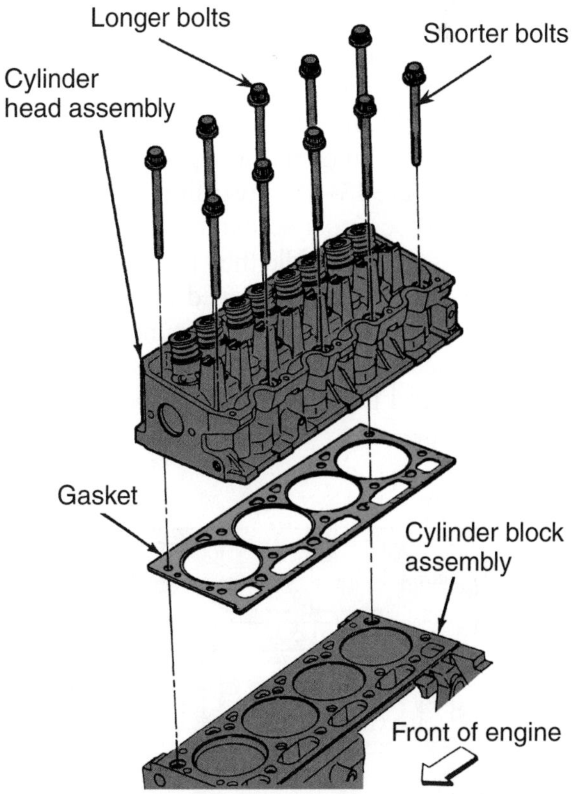

Figure 13-29 Head bolts are different lengths in some engines. *Courtesy of Ford Motor Company*

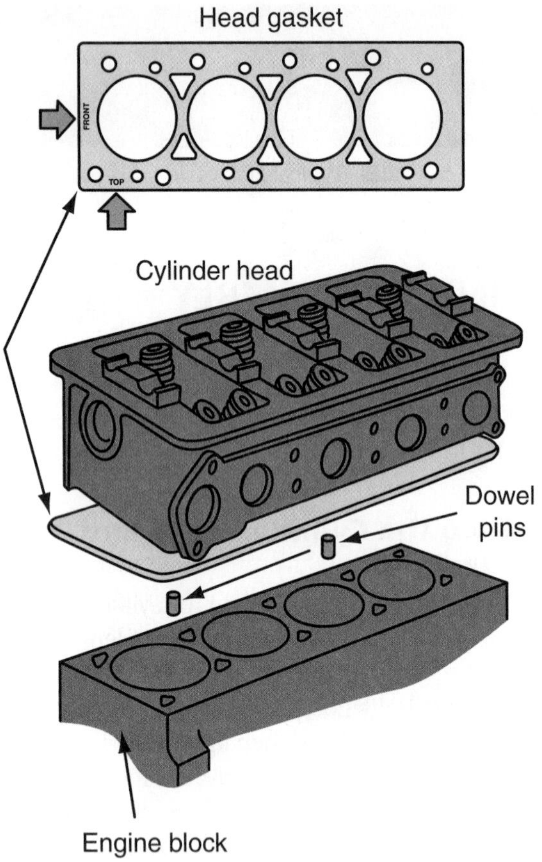

Figure 13-30 When installing a cylinder head onto an engine block, make sure the dowel pins are securely installed in the block and that the head gasket is in the correct position.

holes. Prior to doing this, make sure the dowel pins **(Figure 13–30)** are in place.

Cylinder head bolts must be tightened in the correct order and to the proper amount of torque. A typical tightening sequence **(Figure 13–31)** is usually provided in the service manual. Most cylinder heads are tightened in a sequence that starts in the middle then moves out to the ends. The bolts are generally tightened in two or three stages. If the final torque is 100 foot-pounds, the bolts may first be tightened at 35 foot-pounds. Some manufacturers recommend that the head bolts be retorqued after the engine has been run and warmed up. *Note:* Some manufacturers use torque-to-yield bolts that can not be reused and must be replaced.

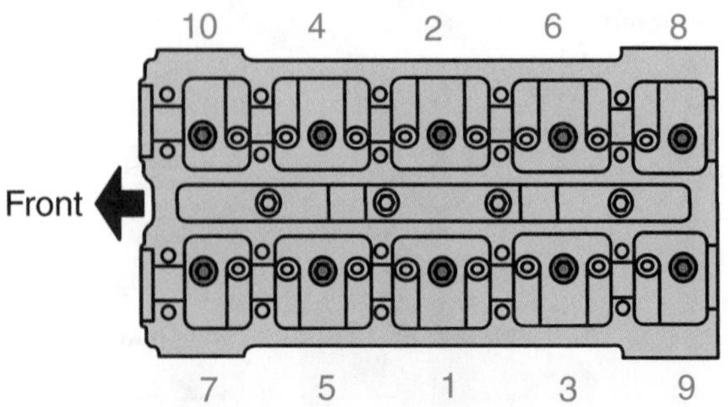

Figure 13-31 Always follow the specified tightening sequence when torquing cylinder head bolts.

When installing head bolts in aluminum heads, hardened steel washers must be used under the bolt heads to prevent galling of the aluminum and to help distribute the load. Make sure the washers are positioned with their rounded or chamfered side up and that there is no debris or burrs under the washers.

The only time sealer should be applied to a head gasket is when one or both sides are totally metal. When the gasket is a full metal body gasket, apply sealer to both sides (except for multilayered steel designs). One-sided metal gaskets require sealer only on the metal side.

Liberally coat the rocker arms with assembly lube or clean engine oil. Then install the pushrods and the rocker arms. Many engines use positive stop rocker arm adjustments. This means that when torquing the rocker arms to spec, the plunger of the hydraulic lifter is properly positioned, giving the correct lifter adjustment.

If the metal was removed from the sealing surface of either the block or head, the geometry of the valve train will be affected. On engines with nonadjustable rocker arms, removal of more than 0.010 inch of material must be compensated for by shimming up the rocker arm assembly supports or by using shorter pushrods.

Torque Angle Gauge Some manufacturers recommend the torque angle method for tightening cylinder head bolts, which requires the use of a torque angle gauge. Torque-to-yield bolts must be tightened according to manufacturer's recommendations. Typically this involves two steps: tighten the bolt to the specified torque, and then tighten the bolt an additional amount. The latter is expressed in degrees. To accurately measure the number of degrees added to the bolt, a torque angle gauge is attached to the wrench. The additional tightening will stretch the bolt, producing a very reliable clamp load that is much higher than can be achieved just by torquing.

Timing Belts

The camshaft and valve train work together to open and close the intake and exhaust valve of the engine. To obtain the correct initial relationship of the components, the corresponding marks are aligned at the time of assembly **(Figure 13–32)**.

Many engines with chain-driven camshafts and all engines with belt-driven camshafts use a tensioner. The tensioner pushes against the belt or chain to keep it tight, which keeps it from slipping on the sprockets, provides more precise valve timing, and compensates for component stretch and wear. The tensioner may have provisions for adjustment or may be self-adjusting. Check the service manual to identify the adjustment procedures **(Figure 13–33)**.

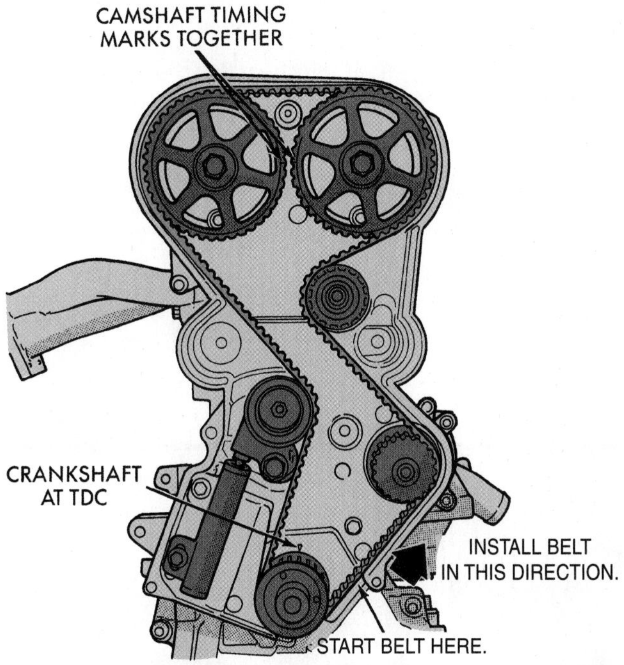

Figure 13-32 Crankshaft and camshaft timing belt installation for a dual overhead cam engine. *Courtesy of DaimlerChrysler Corporation*

USING SERVICE MANUALS

Normally, camshaft timing marks are shown in the engine section of a service manual, under the heading of Timing Belt or Chain R&R. ■

When installing a timing belt, be sure to align the timing marks properly and adjust belt tension to specifications. For more information, see Photo Sequence 9. Timing belts can be replaced with the engine in place in the vehicle and when it is on an engine stand.

Adjusting Valves

All engines that have mechanical lifters use some method of adjustment to bring valve lash (clearance) back into specification. **Valve lash** is checked by inserting a feeler gauge between the valve tip and the rocker arm or cam lobe. Installed valve stem height is critical to avoid incorrect valve train geometry and to correctly set valve lash.

Static valve adjustment is required after an engine has been rebuilt to ensure proper engine starting and prevent

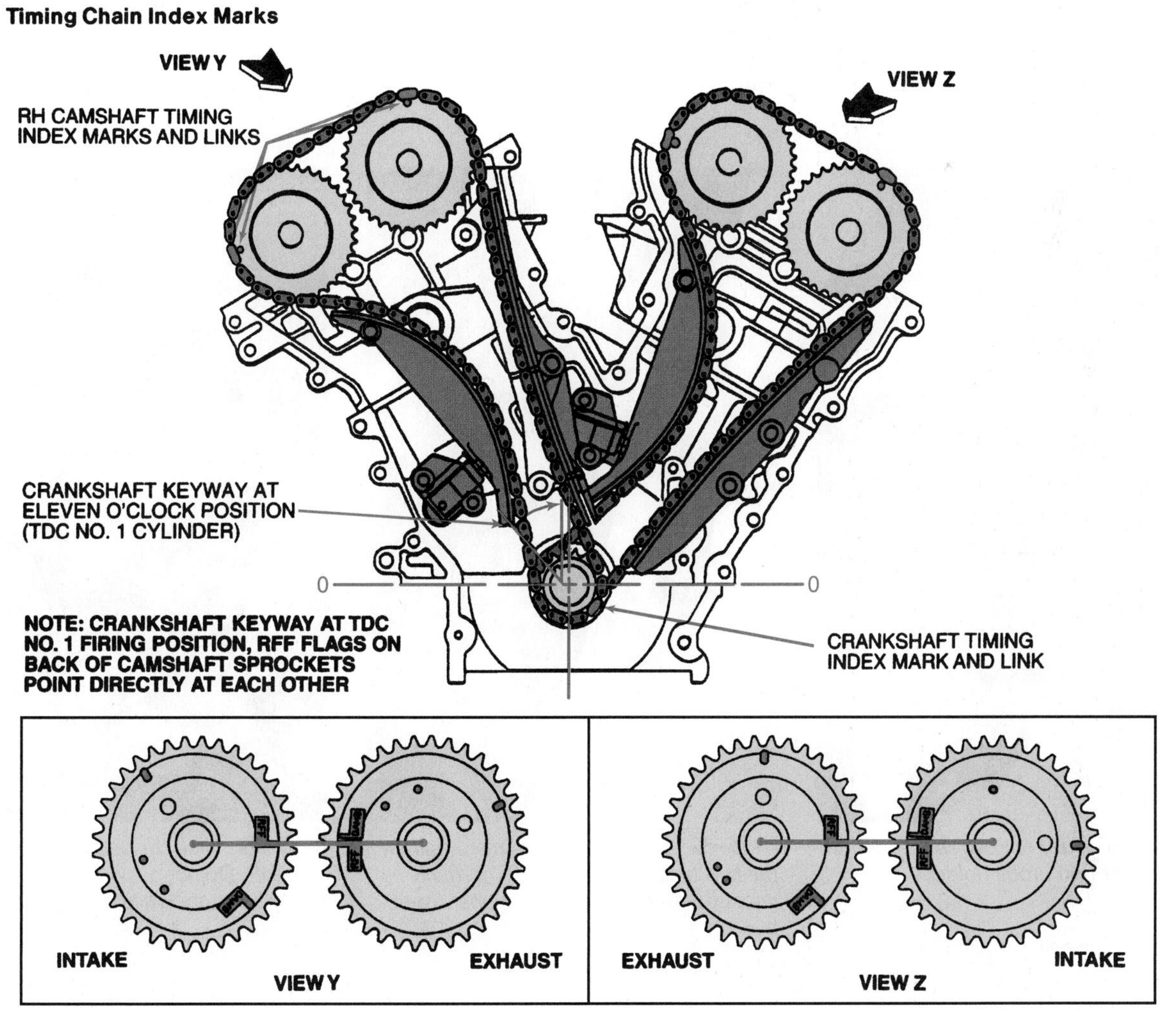

Figure 13-33 Note the timing marks and position of the chain tensioners on this DOHC engine. *Courtesy of Ford Motor Company*

Replacing a Timing Belt on an OHC Engine

P9-1 *Disconnect the negative cable from the battery prior to removing and replacing the timing belt.*

P9-2 *Carefully remove the timing cover. Be careful not to distort or damage it while pulling it up. With the cover removed, check the immediate area around the belt for wires and other obstacles. If some are found, move them out of the way.*

P9-3 *Align the timing marks on the camshaft's sprocket with the mark on the cylinder head. If the marks are not obvious, use a paint stick or chalk to clearly mark them.*

P9-4 *Carefully remove the crankshaft timing sensor and probe holder.*

P9-5 *Loosen the adjustment bolt on the belt tensioner pulley. It is normally not necessary to remove the tensioner assembly.*

P9-6 *Slide the belt off the crankshaft sprocket. Do not allow the crankshaft pulley to rotate while doing this.*

P9-7 *To remove the belt from the engine, the crankshaft pulley must be removed. Then the belt can be slipped off the crankshaft sprocket.*

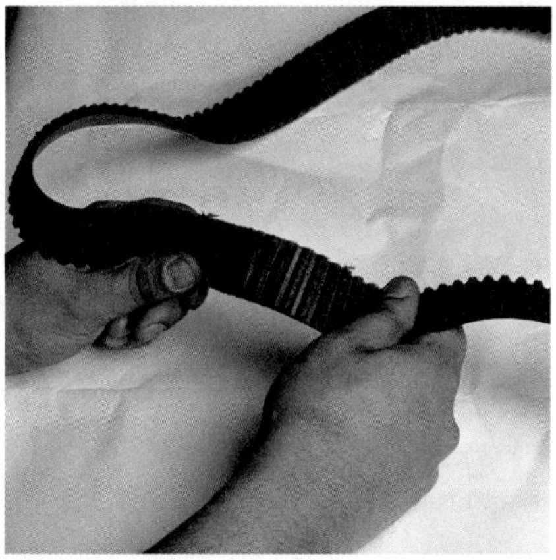

P9-8 *After the belt has been removed, inspect it for cracks and other damage. Cracks will become more obvious if the belt is twisted slightly. Any defects in the belt indicate it must be replaced.*

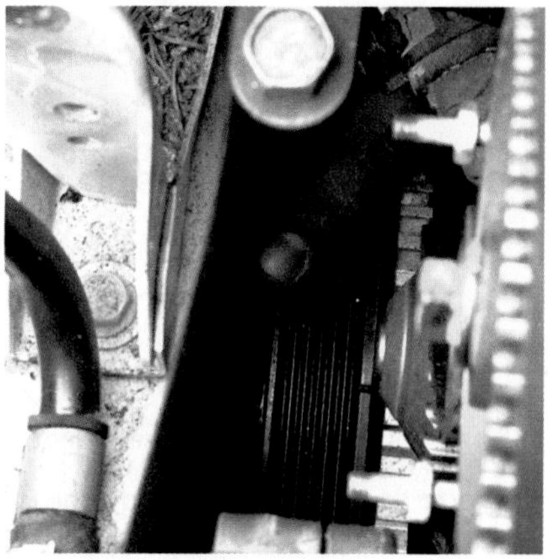

P9-9 *To begin reassembly, place the belt around the crankshaft sprocket. Then reinstall the crankshaft pulley.*

P9-10 *Make sure the timing marks on the crankshaft pulley are lined up with the marks on the engine block. If they are not, carefully rock the crankshaft until the marks are lined up.*

P9-11 *With the timing belt fitted onto the crankshaft sprocket and the crankshaft pulley tightened in place, the crankshaft timing sensor and probe can be reinstalled.*

P9-12 *Align the camshaft sprocket with the timing marks on the cylinder head. Then wrap the timing belt around the camshaft sprocket and allow the belt tensioner to put a slight amount of pressure on the belt.*

P9-13 *Adjust the tension as described in the service manual. Then rotate the engine two complete turns. Recheck the tension.*

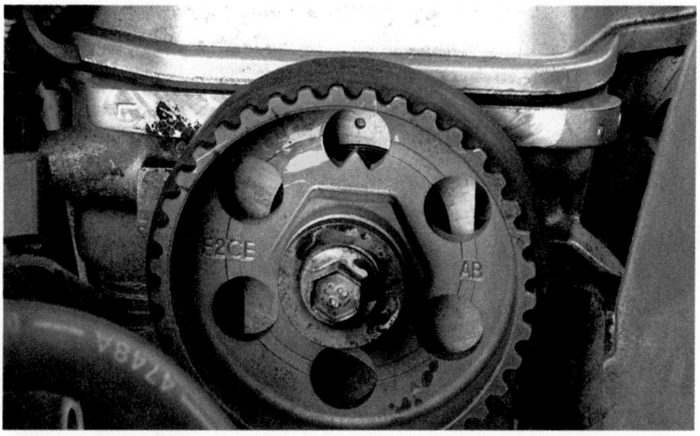

P9-14 *Rotate the engine through two complete turns again, then check the alignment marks on the camshaft and the crankshaft. Any deviation needs to be corrected before the timing cover is reinstalled.*

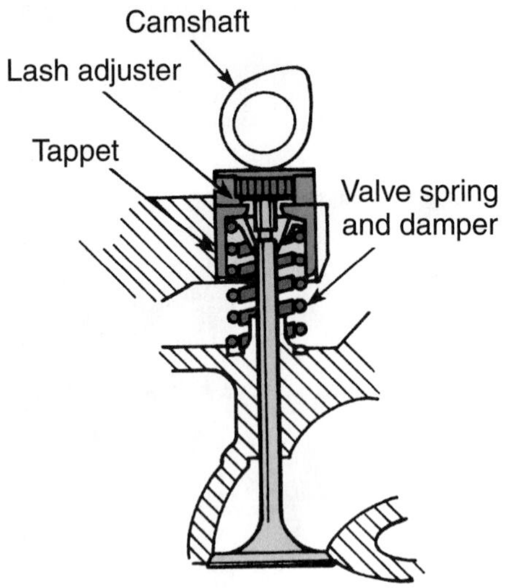

Figure 13-34 Some overhead cam engines feature a cam follower with an adjustment screw.

damage to the valves caused by the pistons hitting the valves. Normally, if the valves are adjusted accurately, no further valve adjustment is required unless the cylinder heads are retightened. All valve adjustments should be made only within the limits prescribed in the manufacturer's vehicle service manual.

There are four basic methods for lash adjustment: rocker arm with adjustable pivots, adjustable pushrods, rocker arms with adjustable screws, and adjustable cam follower (using some type of adjustable screw or replaceable shim). Of these four adjustment types, the first two methods are typically associated with OHV engines. The other two adjustment procedures—rocker arms with adjustment screws and adjustable cam followers **(Figure 13-34)—are commonly found on OHC designs.**

PROCEDURE

Adjusting the valves on an engine that uses shims for valve clearance.

STEP 1 *With the timing belt cover and camshaft covers removed, make sure piston 1 is at TDC* ***(Figure 13–35A)****.*

STEP 2 *Check the camshaft alignment marks. If they are not aligned, rotate the crankshaft one full turn* ***(Figure 13–35B)*** *until they are aligned.*

STEP 3 *Using a feeler gauge, measure and record the valve lash of the valves that are totally closed. Refer to the service manual to identify the closed valves* ***(Figure 13–35C)****.*

STEP 4 *Rotate the crankshaft one full turn with piston 1 again at TDC.*

STEP 5 *Measure and record the valve lash on the valves not measured in the previous step. Compare all measured clearances to specifications.*

STEP 6 *For any valves that do not have the proper lash, follow the rest of this procedure.*

STEP 7 *Rotate the camshaft so that the cam lobe on the valve that needs adjustment is facing up.*

STEP 8 *Using a screwdriver, rotate the notch of the valve lifter and shim assembly so that it is to the side of the camshaft.*

STEP 9 *While holding the camshaft in place, depress the valve lifter assembly.*

STEP 10 *Using a small screwdriver and a magnetic finger, remove the adjusting shim* ***(Figure 13–35D)****.*

STEP 11 *Measure the thickness of the shim with a micrometer* ***(Figure 13–35E)****.*

STEP 12 *Calculate the size of the desired shim by adding the measured clearance to the size of the old shim. Then subtract the desired clearance from that total. To correct excessive clearance, a thicker disc or shim is added. If reduced clearance is needed, a thinner disc or shim must be installed.*

STEP 13 *Install the new shim and recheck the valve lash. Then move to the next valve that needs adjustment and repeat the process.*

STEP 14 *When all valves have been adjusted, reinstall the camshaft covers, timing belt cover, and anything else that has been removed.*

To adjust the valves with adjustable screws, make sure the valve is fully closed and loosen the adjuster locknut on the valve to be adjusted. Then turn the screw to achieve the proper clearance. While holding the screw and preventing it from turning, tighten the locknut. Follow the same procedure for each valve that needs to be adjusted.

Final Reassembly Steps

The steps in final engine assembly involve installing various engine covers and installing manifolds and related items that mount directly to the engine assembly.

Install the Timing Cover When replacing the timing cover, remove the old gaskets and seals from the timing cover and engine block.

Install the new seal using a press, seal driver, or hammer and a clean block of wood. When installing the seal, be sure to support the cover underneath to prevent damage. Apply a light coating of adhesive on the timing cover and position the gasket on the cover. Some manufactur-

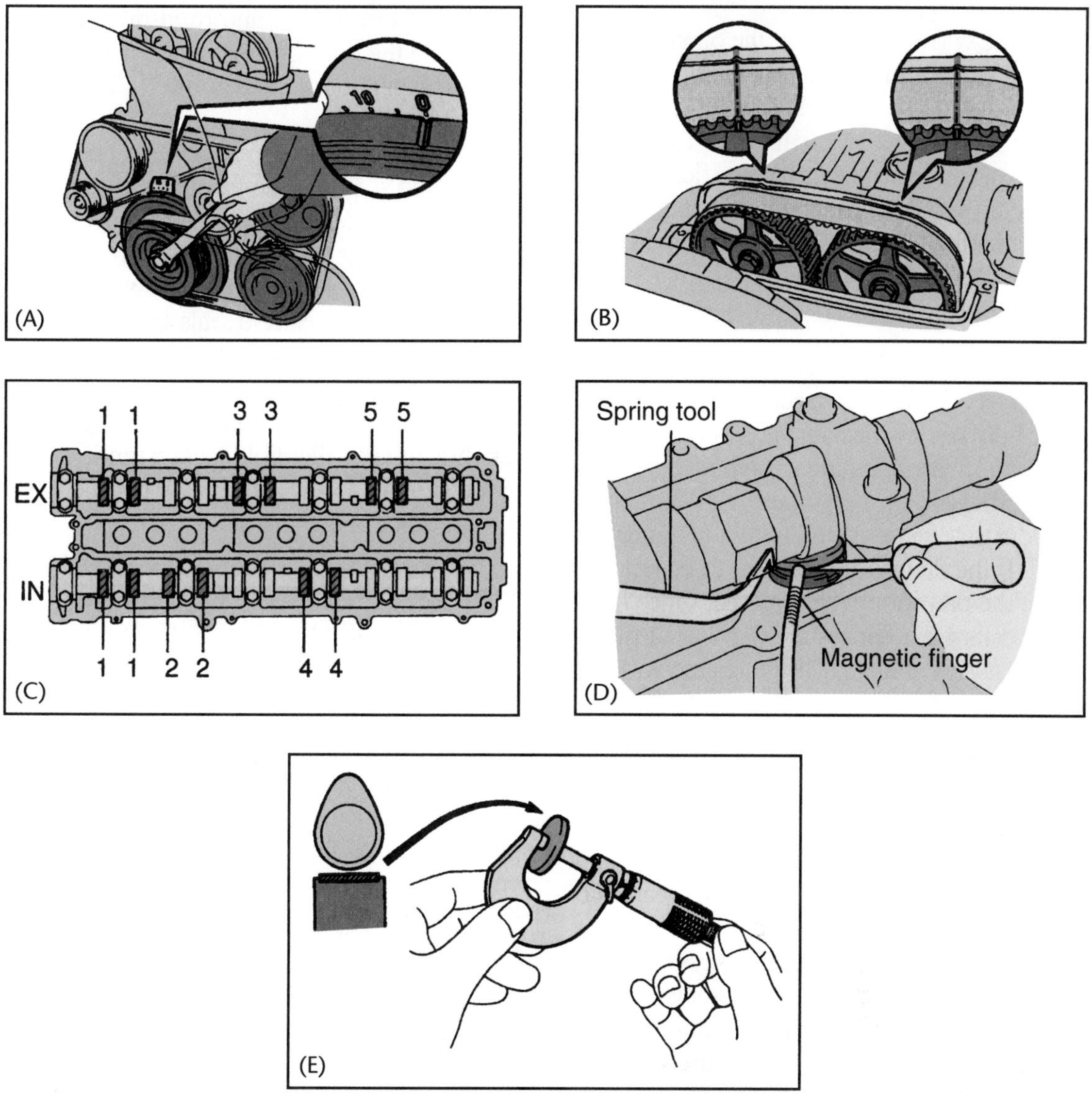

Figure 13-35 To adjust valve lash: (A) Rotate the crankshaft so that it is at TDC, (B) check the camshaft alignment marks, (C) measure and record the valve lash of the valves that are totally closed, (D) use a small screwdriver and a magnetic finger to remove the adjusting shim, and (E) measure the thickness of the shim with a micrometer to determine the correct shim to install. *Reprinted with permission*

ers recommend a sealant be used on both surfaces. Check the service manual and gasket instructions. Finally, mount the timing cover and torque the bolts to specifications.

Install the Vibration Damper Install the vibration damper (harmonic balancer) by using a special installation tool. In most cases, the damper is installed until it bottoms out against the oil slinger and the timing sprocket. Make sure the woodruff key is in place.

Some vibration dampers are not pressed-fit on the crankshaft. Be sure to install the large washer behind the damper-retaining bolt on these engines. Otherwise, the damper might fly off, causing damage and a safety hazard.

Install the Valve Cover To install the valve cover, first make sure the cover's sealing flange is flat, then apply contact adhesive to the valve cover's sealing surfaces in small dabs. Mount the valve cover gasket on the valve cover and align it in position. If the gasket has mounting tabs, use them in tandem with the contact adhesive. Allow the adhesive to dry completely before mounting the valve cover on the cylinder head. Torque the mounting bolts to specifications.

SHOP TALK

Many technicians install the water pump at this point or wait until the engine is back in the vehicle. When installing a water pump, apply a coating of good waterproof sealer to a new gasket and place it in position on the water pump. Coat the other side of the gasket with sealer and position the pump against the engine block until it is properly seated. Install the mounting bolts and tighten them with a torque wrench in a staggered sequence to specifications. Careless tightening could cause the pump housing to crack. After tightening, check the pump to make sure it rotates freely. ■

Install Oil Pan Before installing the oil pan gasket, check the flanges of a steel oil pan for warpage. Use a straightedge **(Figure 13–36)** or lay the pan, flange side down, on a flat surface with a flashlight underneath it to spot uneven edges. Carefully check the flange around bolt holes. Minor distortions can be corrected with a hammer and block of wood. If the flanges are too bent to be repaired in this manner, the pan should be replaced. Once it has been determined that the flanges are flat, install the oil pan with a new gasket. Finally, fill the engine with the proper grade oil.

Install Intake Manifold The intake manifold gasket seals the joint between the intake manifold and the cylinder head. As with cylinder head gaskets, it is important to properly prepare the sealing surfaces of an intake manifold before installing the gaskets. Thoroughly clean all of the sealing surfaces, bolt holes, and bolts. Inspect the surfaces for damage and repair or replace as necessary. Check the gaskets for any markings or installation instructions that may be stamped on them. Check the manufacturer's instructions for recommendations on the use of a supplementary sealant. Most intake manifold gaskets should be coated with a nonhardening sealer.

When installing an intake manifold, it is wise to use guide bolts. These guides make sure the gaskets and the manifold are perfectly aligned before tightening them in place. They also aid in setting the sealer by preventing the manifold from shifting and rupturing the sealant. When tightening the bolts or nuts, make sure you tighten them to the proper torque and in the order specified by the manufacturer.

On a V6 or V8 engine, there may be rubber or cork-rubber end seals for the front and the rear of the manifold. Before installing these seals, thoroughly clean all oil from the mating surfaces. Apply adhesive to the surface to hold the seals in place during installation. Once the intake manifold gaskets and seals are in place, apply a small bead (approximately ⅛-inch [3.18-mm]) of silicone RTV to the point where the seals meet the gasket.

Some V6 and V8 engines have a large one-piece combination intake manifold gasket and manifold splash guard **(Figure 13–37)**. These are installed with the same care as other intake manifold gaskets.

Some intake gaskets purposely block off coolant passages to enable the engine coolant to flow in a predetermined path through the engine. If these ports are not blocked off, the coolant will not flow properly and the engine can overheat.

Prelubrication There are several ways to prelubricate, or prime, an engine. One method is to drive the oil pump with an electric drill. This method is often used on engines that drive the oil pump via the distributor. On most engines, it is possible to make a drive that can be chucked in an electric drill motor to engage the drive on the oil pump. Insert the fabricated oil pump drive extension into the oil pump through the distributor drive hole. Drive the oil pump with the electric drill. To control oil splash, loosely set the valve cover(s) on the engine. After running the oil pump for several minutes, remove the valve cover and see whether there is any oil flow to the rocker arms. If oil reached the cylinder head, the engine's lubrication system is full of oil and is operating properly. If no oil reached the cylinder head, there is a problem either with

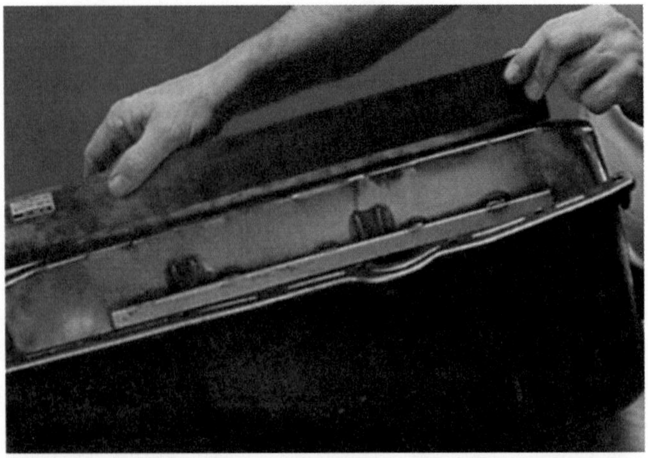

Figure 13-36 Checking the flatness of an oil pan flange. *Used with permission of Detroit Gasket—Copyright © 1998*

Figure 13-37 A one-piece combination intake manifold gasket and manifold splash guard. *Used with permission of Detroit Gasket—Copyright © 1998*

the pump, with the alignment of an oil hole in a bearing, or perhaps with a plugged gallery.

Using a **prelubricator**, which consists of an oil reservoir attached to a continuous air supply, is the best method of prelubricating most late-model engines. When the reservoir is attached to the engine and the air pressure is turned on, the prelubricator will supply the engine's lubrication system with oil under pressure.

Install the Thermostat and Water Outlet Housing Install the thermostat with the temperature sensor facing into the block. If the thermostat is installed upside down, the engine will overheat.

Install Exhaust Manifold When installing the exhaust manifold(s), tighten the bolts in the center of the manifold first to prevent cracking it. If there are dowel holes in the exhaust manifold that align with dowels in the cylinder head, make sure these holes are larger than the dowels. If the dowels do not have enough clearance because of the buildup of foreign material, the manifold will not be able to expand properly, and may crack.

Install Flywheel or Flex Plate Reinstall the engine sling. Raise the engine into the air on a suitable lift, and remove the engine stand mounting head. Set the assembled engine on the floor and support it with blocks of wood while attaching the flywheel or flex plate. Be sure to use the right flywheel bolts and lock washers, and make certain the flywheel is in the correct position. These bolts have very thin heads and the lock washers are thin. If normal bolts or washers are used, they may cause interference with the clutch disc or the torque converter. Make sure the bolts are properly torqued.

Install Clutch Parts If the vehicle has a manual transmission, install the clutch **(Figure 13–38)**. Make sure the transmission's pilot bushing or bearing is in place in the rear of the crankshaft and that it is in good condition.

Using a clutch-aligning tool or an old transmission input shaft, align the clutch disc. Then, tighten the disc and clutch pressure plate to the flywheel. Make sure the disc is installed in the right direction. There should be a marking on it that says "flywheel side." Then install the bell housing if it was removed from the transmission.

Install Torque Converter On cars equipped with automatic transmissions, it is a good practice to replace the transmission's front pump seal. If the transmission was removed from the car with the engine, reinstall it on the engine now.

Install the torque converter, making sure it is correctly engaged with the transmission's front pump. The drive lugs on the converter should be felt engaging the transmission front pump gear. Failure to correctly install the converter can result in damage to the transmission's front pump.

Install Motor Mounts The motor mount bolts may now be installed loosely on the block. The bolts are left loose during engine installation so the mounts can be easily aligned with the front mount brackets. Make sure the mounts are in good condition.

INSTALLING THE ENGINE

Installing a computer-controlled engine can be a complex task requiring special procedures. Referring to the vehicle's service manual is absolutely essential for this

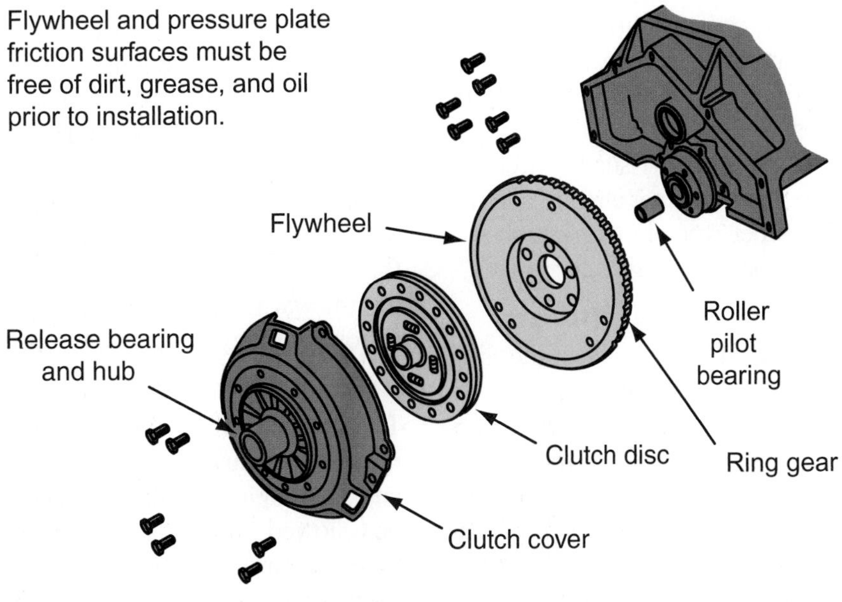

Figure 13–38 A typical clutch assembly

procedure. Typically the procedure is the reverse of the removal procedure.

Installing an Engine into a FWD Vehicle

If the engine will be installed through the top, connect the engine to a sling and then connect the sling to the crane. Slowly lower the engine into the engine compartment. Guide the engine around all wires and hoses to make sure nothing gets damaged. As the engine approaches its position in the engine compartment, align the engine and transmission mounts. Then, lower the engine so you can install the bolts into the mounts. Now raise the vehicle to a good working height.

If the engine will be installed from under the car, install the engine onto the engine cradle and dolly. Lift the vehicle on a hoist or lift and position the engine under the vehicle. Slowly lower the vehicle over the engine while guiding all wires and hoses out of the way. As the vehicle gets close to the engine, align the engine and transmission mounts. Once the mounts are aligned, lower the vehicle so you can install the bolts into the mounts. Now raise the vehicle to a good working height.

While working under the vehicle, install the axle shafts. Then install the remaining engine and transaxle mounts and braces. Connect the exhaust manifold to the exhaust system. Install any heat shields that were removed when the engine was removed. Connect all linkages, lines, hoses, and electrical wiring to the transmission. Now reconnect all suspension and steering parts that were disconnected or removed and install the wheels and tires.

Lower the vehicle to the ground and tighten the axle hub nuts. Connect any disconnected fuel lines and heating system hoses. Now connect the engine ground strap and all electrical connectors and wires. Connect all vacuum hoses. Then connect the throttle linkage and adjust it if necessary.

Install the radiator and the cooling fan(s). Connect the rest of the hoses for the cooling system. Now install the air induction system and connect any remaining items, including the battery cables.

Fill the radiator with coolant and visually check for leaks. If the engine does not have oil in it already, add the specified amount of the proper type of oil. Prime the oil pump of the engine and prepare the engine for startup.

Installing an Engine in a RWD Vehicle

Connect the engine to a sling and then connect the sling to the crane. Place a transmission jack under the transmission to hold it in position. Now slowly lower the engine into the engine compartment. Guide the engine around all wires and hoses to make sure nothing gets damaged. As the engine approaches its position in the engine compartment, align the engine to the input shaft of the transmission or the torque converter hub into the front pump. Carefully wiggle the engine until the input shaft slides through the clutch disc splines or the torque converter seats fully into the transmission. Install and tighten the transmission to engine bolts. Start the engine mount bolts into their bores; you may need to wiggle the engine some to do this. Once the mount bolts are in place, tighten them and remove the transmission jack and engine sling.

Raise the vehicle to a good working height and install all remaining engine and transmission mounts. Connect the exhaust manifold to the exhaust system. Install any heat shields that were removed when the engine was removed. Reconnect the fuel line from the fuel tank to the engine. Lower the vehicle so you can work under the hood. Connect any remaining disconnected fuel lines and heating system hoses. Connect the engine ground strap. Connect all electrical connectors and wires. Connect all vacuum and other hoses. Now connect the throttle linkage and adjust it if necessary.

Install the radiator and the cooling fan(s) and connect the rest of the hoses for the cooling system. Now install the air induction system and connect any remaining items, including the battery cables.

Fill the radiator with coolant and visually check for leaks. If the engine does not have oil in it already, add the specified amount of the proper type of oil. Prime the oil pump of the engine and prepare the engine for startup. Install and align the hood.

Starting Procedure

Set the ignition timing as accurately as possible before starting the engine. The engine can be fine-tuned by using an engine analyzer or diagnostic tester after it has been started and goes through the **break-in** test. However, it is wise to time the ignition distributor before attempting to start the engine even though it is not the final adjustment.

Fill the gasoline tank with several gallons of fresh gasoline. Crank the engine. It will take some time for fuel to be pumped from the tank to the engine if the car has a mechanical fuel pump. When the engine gets fuel, it will try to start. Once it does start, set the throttle to an engine speed of approximately 1,500 rpm. When the engine coolant reaches normal operating temperature, turn off the engine. Look for signs of coolant or oil leaks. After these checks, run the engine at 1,200–1,500 rpm during the warm-up period to ensure adequate initial lubrication for the piston rings, pistons, and camshaft.

Break-In Procedure

To prevent engine damage after it has been rebuilt or completely overhauled and to ensure good initial oil control and long engine life, the proper break-in procedure must be followed. Make a test run at 30 mph (48 km/h) and accelerate at full throttle to 50 mph (80 km/h). Repeat the acceleration cycle from 30 to 50 mph (48 to 80 km/h) at least ten times. No further break-in is necessary. If traffic conditions will not permit this procedure, accelerate the engine rapidly several times through the intermediate gears during the road test. The objective is to apply a load to the engine for short periods of time and in rapid succession soon after the engine warms up. This action thrusts the piston rings against the cylinder wall with increased pressure and results in accelerated ring seating.

Relearn Procedures

The computer in most late-model vehicles must undergo a relearn procedure after the engine has been rebuilt. This procedure allows the computer to learn the condition of the engine and make adjustments according to the engine's restored condition. The last time the engine was run, the computer made adjustments based on the engine faults present. The relearn procedure teaches the computer that those faults were corrected. Always follow the manufacturer's relearn procedures as outlined in the service manual.

CUSTOMER CARE

After the engine has been totally checked over, return it to the owner with the following instructions:

1. Drive the vehicle normally but avoid sustained high speed during the first 500 miles (break-in period).
2. Avoid periods of extended engine idling.
3. Check the oil level frequently during the break-in period. It is not unusual to use 1 or 2 quarts of oil during this time.
4. The oil and oil filter will need to be changed at the end of the break-in time.
5. The cylinder head and intake manifold bolts may need to be retorqued.
6. Some adjustments, such as valve adjustments and ignition timing, will also need to be checked. ■

CASE STUDY

A four-cylinder DOHC engine has just been rebuilt. During the initial running of the engine, it runs quite rough. This is normal until everything gets seated. However, the condition becomes worse the more the engine runs. Slight adjustments to the ignition timing do not help the condition. A visual inspection reveals nothing loose or disconnected. Each hose and wire is traced to make sure it is connected to the proper fitting and terminal.

Because it seems that the engine is running on only three cylinders, a cylinder leakage test is performed. The results indicate excessive leakage past the #4 intake valve. In an attempt to visually locate the problem, the intake cam cover is removed. A look at the camshaft reveals that the #4 intake lobe is worn. This was a new camshaft. Was it defective? Was it not hardened properly? Did something else cause this?

Further visual inspection reveals that the shim used to set valve lash was not fully seated in its cup. The edge of the shim was working like a knife, cutting off the metal from the lobe. The initial rough running of the engine was caused by the shim preventing the valve from fully closing. The condition worsened as the lobe was cut away.

The oil was changed and a new camshaft and shim were installed to correct the problem.

KEY TERMS

Adhesives
Aerobic
Anaerobic
Antiseize
Break-in
Elasticity
Gasket
Graphite
Polyacrylate
Prelubricator
Room temperature vulcanizing (RTV)
Sealants
Silicone gasketing
Torque-to-yield (T-T-Y)
Valve lash
Viton
Yield

SUMMARY

- Elasticity means a bolt can be stretched a certain amount and, when the stretching load is reduced, return to its original size. Yield means a stretched bolt takes a permanent set and never returns to normal. Proper use of torque will prevent a yield condition.
- Gaskets serve as sealers, spacers, wear insulators, and vibration dampeners. Engine gaskets are generally classified as either hard or soft.
- General recommendations for installing gaskets include the following: never reuse old ones; handle new ones carefully, especially the composition-type; use sealants properly; thoroughly clean all mating surfaces; and use the right gasket in the right position.
- Cylinder head gaskets on today's bimetal engines have a demanding job. The no-retorque head gasket retains a high level of clamping force.

- Contact adhesive bonds cork, rubber, fiber, and metal gaskets in place. It does not aid in sealing. Its only purpose is to hold the gasket securely during component assembly.
- General-purpose sealers aid in gasket sealing without upsetting the designed performance of most mechanical gaskets. Hard-drying sealants, such as shellac, should never be used on gaskets because they will make future disassembly very difficult.
- Flexible sealant is often used on bolt threads that go into fluid passages. Silicone gasketing, of which RTV is the best known, is used on oil pans, valve covers, thermostat housing, timing covers, and water pumps. Anaerobic formed-in-place sealants are used for both thread locking and gasketing.
- Oil seals keep oil and other vital fluids from escaping around a rotating shaft. Oil seals should always be replaced during engine rebuilding to ensure against costly do-overs.
- All engines with mechanical lifters have some method of adjustment to bring valve lash (clearance) back into specification. Rocker arms with adjustable pivots and adjustable pushrods are found on OHV engines. Rocker arms with adjustable screws and adjustable cam followers are part of OHC engines.
- The steps in final engine assembly involve installing various engine covers, prelubing the engine, and installing manifolds and related items that mount directly to the engine assembly. The best method of prelubricating an engine under pressure without running it is to use a prelubricator, which consists of an oil reservoir attached to a continuous air supply.
- A proper break-in procedure is necessary to ensure good initial oil contact and long engine life.

TECH MANUAL

The following procedures are included in Chapter 13 of the *Tech Manual* that accompanies this book:

1. Applying RTV silicone sealant.
2. Adjusting valves on an OHC engine.

REVIEW QUESTIONS

1. *True or False?* Make sure you locate and adhere to the "wet" torque specs when tightening a bolt that has been lubricated with oil or any other liquid.
2. What does it mean when a bolt is stretched into yield?
3. Name some applications of hard gaskets.
4. Where are flexible sealers most often used?
5. What are the major differences between aerobic and anaerobic sealants?
6. Which of the following statements is incorrect?
   a. Cylinder head bolts must be tightened to the proper amount of torque.
   b. Most cylinder head bolts are tightened in a sequence that starts on one end.
   c. On some engines, the head bolts are retorqued after the engine has been run and is hot.
   d. Many engines use head bolts of different lengths.
7. *True or False?* There are four basic methods for lash adjustment: rocker arms with adjustable pivots, adjustable pushrods, rocker arms with adjustable screws, and adjustable cam follower.
8. Technician A says T-T-Y bolts can be used after they were removed if there are no signs of distortion or stretching. Technician B says head bolts that pass through a coolant passage should be coated with a nonhardening sealer prior to installing them. Who is correct?
   a. Technician A **c.** Both A and B
   b. Technician B **d.** Neither A nor B
9. Technician A uses adhesives to hold gaskets in place while installing them. Technician B uses RTV sealant on all surfaces that are prone to leaks. Who is correct?
   a. Technician A **c.** Both A and B
   b. Technician B **d.** Neither A nor B
10. Technician A uses soft gaskets on valve covers. Technician B uses soft gaskets on water pumps. Who is correct?
   a. Technician A **c.** Both A and B
   b. Technician B **d.** Neither A nor B

11. Technician A says silicone gasketing can be used to replace conventional paper, cork, and cork/rubber gaskets. Technician B always uses new gaskets and never reuses old ones. Who is correct?

a. Technician A **c.** Both A and B

b. Technician B **d.** Neither A nor B

12. Which of the following statements about the purpose of cylinder head gaskets is *not* true?

a. Seal intake stroke vacuum, seal combustion pressure, and seal the heat of combustion.

b. Prevent coolant leakage, resist rust and corrosion, and, in many cases, meter coolant flow.

c. Allow for lateral and vertical head movement as the engine heats and cools.

d. Meter lubricating oil onto the engine's cylinder walls.

13. *True or False?* On engines with positive stop rocker arm adjustments, the hydraulic lifters must be hand primed before installation to prevent damage to valve train components.

14. Graphite is ______.

a. an anaerobic **c.** both a and b

b. an RTV **d.** neither a nor b

15. Technician A says damaged bolt threads can be repaired with a tap and die set. Technician B says if the bolt heads are not tight against the surface after they have been properly torqued, a washer should be installed under the bolt head. Who is correct?

a. Technician A **c.** Both A and B

b. Technician B **d.** Neither A nor B

16. Silicone RTV gasketing can be used to replace conventional ______.

a. paper gaskets **c.** cork/rubber gaskets

b. cork gaskets **d.** all of the above

17. Which of the following materials is not typically used to form a rear main bearing oil seal?

a. Polyacrylate

b. RTV

c. Silicone synthetic rubber

d. Viton

18. Technician A says aluminum expands at a different rate than cast iron. Technician B says cast iron has a coefficient of thermal expansion two or three times greater than aluminum. Who is correct?

a. Technician A **c.** Both A and B

b. Technician B **d.** Neither A nor B

19. While discussing proper engine break-in, Technician A says the engine should run at idle speed for at least 2 hours. Technician B says the engine should be accelerated at full throttle to 50 mph (80 km/h). This cycle should be repeated at least 10 times. Who is correct?

a. Technician A **c.** Both A and B

b. Technician B **d.** Neither A nor B

20. Which of the following statements about inspecting bolts before assembling an engine is *not* true?

a. Discard all bolts that are not acceptable.

b. Identify what bolts go into the specific bores.

c. Run a nut over the bolt's threads by hand. If any binding occurs, tap the nut.

d. Clean bolt and cylinder block threads with a thread chaser or tap.

SECTION 3

Electricity

The electrical system is perhaps the most important support system of a car or truck. Without an electrical system, gasoline engines would not run. Electricity provides the needed spark for combustion, plus the power needed for starting, lighting and signaling, instrumentation, safety devices, and many other accessories. Advances made in the field of electronics and brought into the automotive industry have allowed our vehicles to be more efficient. Technicians need to be able to diagnose and service both electrical and electronic systems. Although test equipment can simplify working with these systems, understanding electrical principles, component operation, circuit design, and testing procedures is absolutely necessary to be successful as an automotive technician.

Delmar's Automotive Electricity and Electronics Video Series and CD-ROM Courseware. Delmar's Automotive Batteries, Starting and Charging Systems Video Series and CD-ROM Courseware

WE ENCOURAGE PROFESSIONALISM

ASE

THROUGH TECHNICIAN CERTIFICATION

The material covered in Section 3 matches the content areas of the ASE certification test on Electrical Systems.

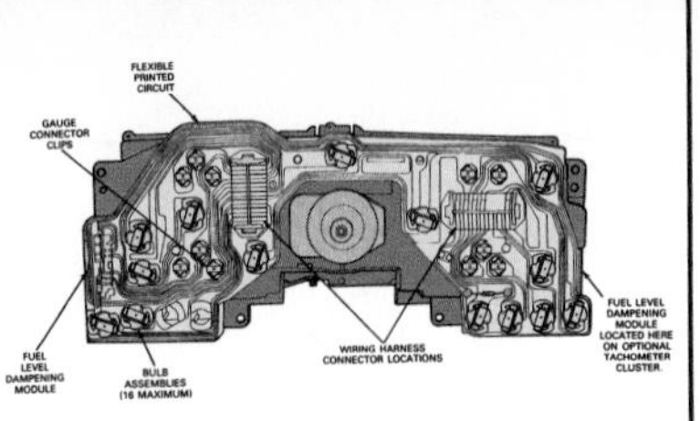

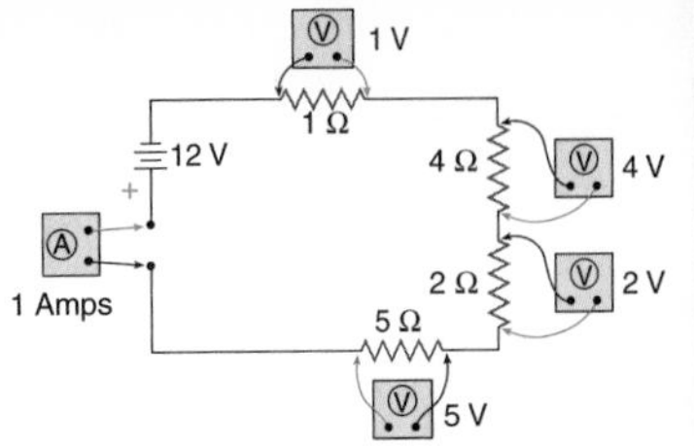

14

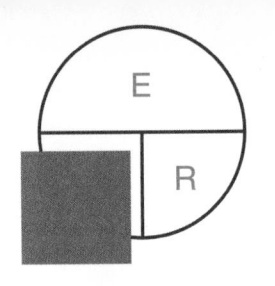

BASICS OF ELECTRICAL SYSTEMS

OBJECTIVES

■ Explain the basic principles of electricity. ■ Define the terms normally used to describe electricity. ■ Use Ohm's law to determine voltage, current, and resistance. ■ List the basic types of electrical circuits. ■ Describe the differences between a series and a parallel circuit. ■ Name the various electrical components and their uses in electrical circuits. ■ Describe the different kinds of automotive wiring. ■ Explain the principles of magnetism and electromagnetism.

There is often confusion concerning the terms electrical and electronic. In this book, electrical and electrical systems refer to wiring and electrical parts, such as generators, lights, and voltage regulators. **Electronics** means computers and other black box-type items used to control engine and vehicle systems.

A good understanding of electrical principles is important to proper diagnosis of any system that is monitored, controlled, or operated by electricity **(Figure 14–1)**.

BASICS OF ELECTRICITY

Perhaps the one reason why some people find it difficult to understand electricity is that they cannot see it. By actually knowing what it is and what it is *not*, you can easily understand it. Electricity is not magic! It is something that takes place or can take place in everything you know. It not only provides power for lights, TVs, stereos, and refrigerators, it is also the basis for the communications between our brain and the rest of our bodies. Although

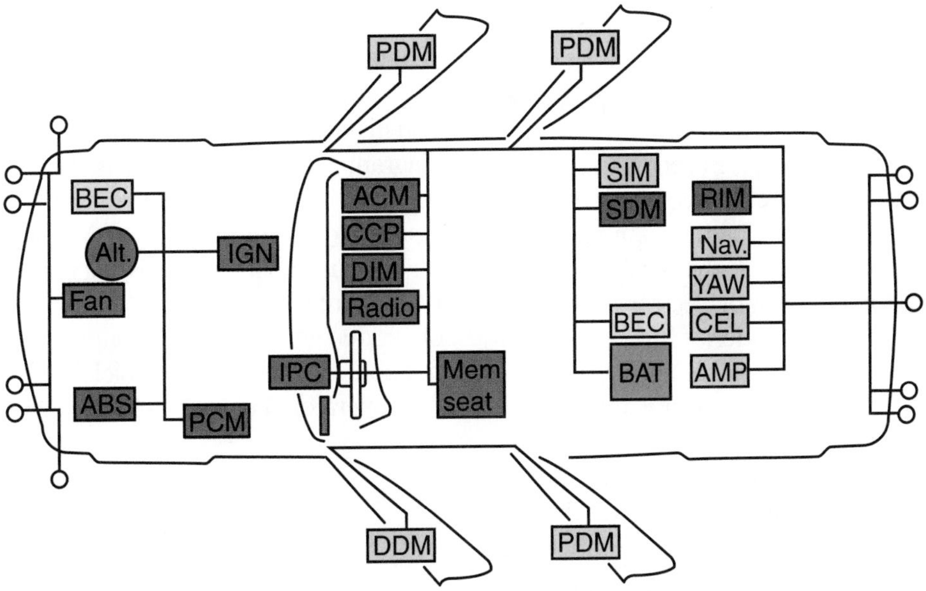

Figure 14–1 Basic overview of the electrical system on a late-model car.

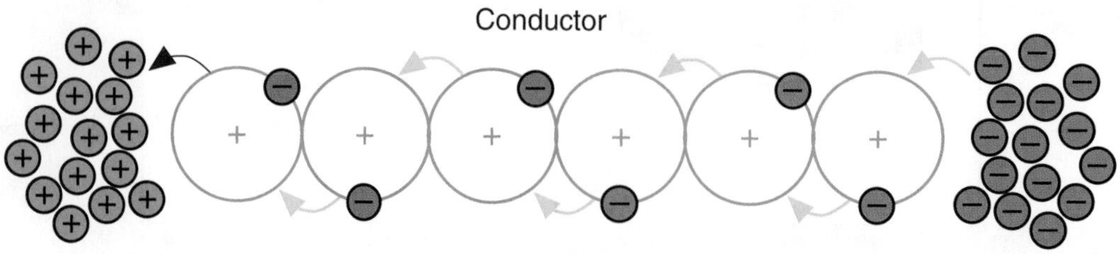

Figure 14-2 Electricity is the flow of electrons from one atom to another.

electricity cannot be seen, the effects of it can be seen, felt, heard, and smelled. One of the most common displays of electricity is a lightning bolt. Lightning is electricity—a large amount of electricity. The power of lightning is incredible. Using the power from much smaller amounts of electricity to perform some work is the basis for an automobile's electrical system. Electricity cannot be seen because it results from the movement of extremely small objects that move at close to the speed of light (186,000 miles [299,000 kilometers] per second).

Flow of Electricity

Electricity is the flow of electrons from one atom to another **(Figure 14–2)**. The release of energy as one electron leaves the orbit of one atom and jumps into the orbit of another is electrical energy. The key to creating electricity is to provide a reason for the electrons to move to another atom.

There is a natural attraction of electrons to protons. Electrons have a negative charge and are attracted to something with a positive charge. When an electron leaves the orbit of an atom, the atom then has a positive charge. An electron moves from one atom to another because the atom next to it appears to be more positive than the one it is orbiting around.

An electrical power source provides for a more positive charge and, to allow for a continuous flow of electricity, it supplies free electrons. To have a continuous flow of electricity, three things must be present: an excess of electrons in one place, a lack of electrons in another place, and a path between the two places.

Two power or energy sources are used in an automobile's electrical system. These are based on a chemical reaction and **magnetism**. A car's battery is a source of chemical energy **(Figure 14–3)**. A chemical reaction in the battery provides for an excess of electrons and a lack of electrons in another place. Batteries have two terminals, a positive and a negative. Basically, the negative terminal is the outlet for the electrons and the positive terminal is the inlet for the electrons to get to the protons. The chemical reaction in a battery causes a lack of electrons at the positive (+) terminal and an excess at the negative (–) terminal. This creates an electrical imbalance, causing the electrons to flow through the path provided by a wire.

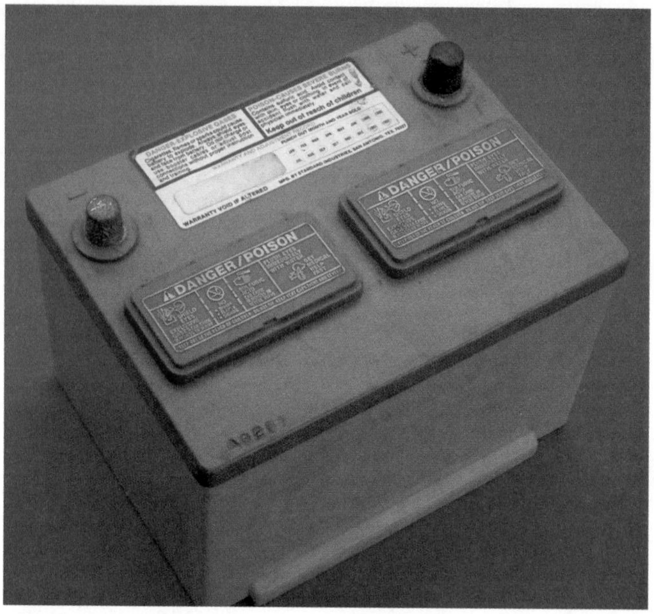

Figure 14-3 An automotive battery.

The chemical process in the battery continues to provide electrons until the chemicals become weak. At that time, either the battery has run out of electrons or all the protons are matched with an electron. When this happens, there is no longer a reason for the electrons to want to move to the positive side of the battery. To the electrons, it no longer looks more positive. Fortunately, the vehicle's charging system continuously restores the battery's supply of electrons. This allows the chemical reaction in the battery to continue indefinitely. In an electrical diagram, a battery is drawn as shown in **Figure 14–4**.

Electricity and magnetism are interrelated. One can be used to produce the other. Moving a wire (a conductor) through an already existing magnetic field (such as a permanent magnet) can produce electricity. This process of producing electricity through magnetism is called **induction**. The heart of a vehicle's charging system is the AC generator **(Figure 14–5)**. A magnetic field, driven by

Figure 14-4 The symbol for a battery.

Figure 14-5 A late-model generator. *Courtesy of Robert Bosch*

the crankshaft via a drive belt, rotates through a coil of wire, producing electricity. The amount of electricity produced depends on a number of things: the strength of the magnetic field, the number of wires that are passed by the magnetic field, and the speed at which the magnetic field moves past the wires.

Electricity is also produced by chemical, photoelectrical, thermoelectrical, and piezoelectrical reactions. These sources of electricity are used throughout a modern automobile. Most are the basis of operation for electronic sensors and are discussed as those sensors are introduced in this book. The two most common ways of producing electricity are through electromagnetic induction and chemical reaction. These are main topics of the chapters in this section of the book.

ELECTRICAL TERMS

Electrical **current** describes the movement or flow of electricity. The greater the number of electrons flowing past a given point in a given amount of time, the more current the circuit has. The unit for measuring electrical current is the **ampere**, usually called an amp. The term ampere was assigned to units of current in honor of Andre Ampere, who studied the relationship between electricity and magnetism. The instrument used to measure electrical current flow in a circuit is called an ammeter.

In the flow of electricity, millions of electrons are moving past any given point at the speed of light. The electrical charge of any one electron is extremely small. It takes millions of electrons to make a charge that can be measured. For these reasons, 1 ampere of current means that 6.28 billion billion electrons are flowing past a given point in 1 second.

There are two types of current: **direct current (DC)** and **alternating current (AC)**. In direct current, the electrons flow in one direction only. In alternating current, the electrons change direction at a fixed rate. Typically, an automobile uses DC, while the current in homes and buildings is AC. Some components of the automobile generate or use AC. These are discussed in later chapters.

There are many theories about the direction of current flow. The conventional theory states that current flows from positive to negative. The electron theory says current moves from negative to positive. And the hole-flow theory basically says something is moving in both directions. Remember that a *theory* is *not* a *fact*. It is a concept that is yet to be proved wrong or right. Therefore, only one theory about current flow is correct and the rest are wrong. For the purposes of this book and for your own understanding of electricity, current flow **(Figure 14–6)** is described as moving from a point of higher potential (voltage) to a point of lower potential (voltage). This statement may not be absolutely correct, but it is sound and is based on what can be observed.

Voltage is electrical pressure. It is the force developed by the attraction of the electrons to protons. The more positive one side of the circuit is, the more voltage is present in the circuit. Voltage does not flow; it is the pressure that causes current flow. To have current flow, some force is needed to move the electrons between atoms. This force is the pressure that exists between a positive and negative point within an electrical circuit. This force, also called **electromotive force (EMF)**, is measured in units called volts. One volt is the amount of pressure required to move 1 ampere of current through a resistance of 1 ohm. Voltage is measured by an instrument called a voltmeter. The unit of measurement for electrical pressure was so named to honor Alessandro Volta who, in 1800, made the first electrical battery.

When any substance flows, it meets **resistance**. The resistance to electrical flow can be measured. The resistance to current flow produces heat. This heat can be measured to determine the amount of resistance. A unit of measured resistance is called an **ohm**. The common symbol for an ohm is shown in **Figure 14–7**. Resistance can be measured by an instrument called an ohmmeter.

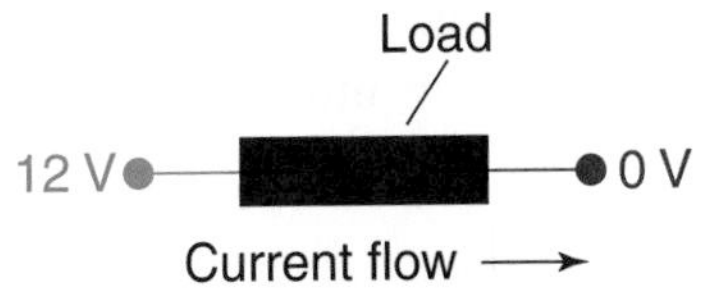

Figure 14-6 Current moves from a point of higher potential to a point of lower potential.

Ω

Figure 14-7 The common symbol for an ohm.

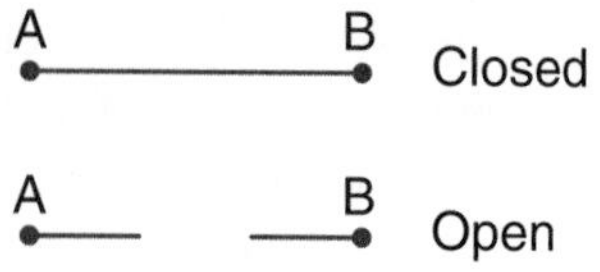

Figure 14-8 Conductors are drawn as lines from one point to another.

Circuit Terminology

An electrical circuit is considered complete when there is a path that connects the positive and negative terminals of the electrical power source. A completed circuit is called a **closed circuit**, whereas an incomplete circuit is called an **open circuit**. When a circuit is complete, there is **continuity**. Conductors are drawn in electrical diagrams as a line connecting two points, as shown in **Figure 14–8**.

Most automotive electrical circuits use the chassis as the path from the negative side of the battery. Electrical components have a lead that connects them to the chassis called the **chassis ground** connections. These connections can be made through a wire or through the mounting of the component. Chassis ground connections are drawn to show which type of connection is normal for that part **(Figure 14–9)**. When the ground is made through the mounting of the component, the connection is represented with the drawing A. When the ground is made by a wire that connects to the chassis, the connection is shown as B.

In a complete circuit, the flow of electricity can be controlled and applied to do useful work, such as light a headlamp or turn over a starter motor. Components that use electrical power put a load on the circuit and consume electrical energy. These components are often referred to as electrical loads. Loads are drawn in electrical diagrams as a symbol representing the part or as a resistor. The typical drawing of a resistor is shown in **Figure 14–10**.

The amount of current that flows in a circuit is determined by the resistance in that circuit. As resistance goes up, the current goes down. The total resistance in a circuit determines how much current will flow through the circuit. The energy used by a load is measured in volts. Amperage stays constant in a circuit but the voltage drops as it powers a load. Measuring voltage drop tells how much energy is being consumed by the load.

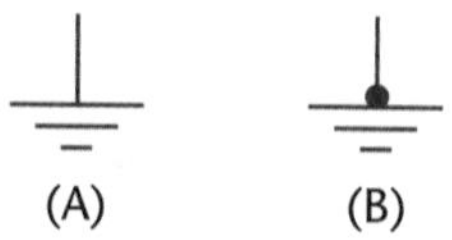

Figure 14-9 Symbols for grounds: (A) made through the component's mounting, and (B) made by a remote wire.

Figure 14-10 The symbol for a resistor.

The amount of electricity consumed by a load is normally called electrical power usage or watts. One watt is equal to one volt multiplied by one ampere. The formula for determining the amount of power consumed by a load is the amount of current through the load multiplied by the voltage drop across the load.

Power Sources

Today's vehicles operate on 12-volt electrical systems. However, the battery stores about 14 volts, although it is rated at 12 volts, and the charging system puts out 14 to 15 volts while the engine is running. The primary source of electrical power when the engine is running is the charging system, so it is fair to say that an automobile's electrical system is a 14-volt system.

In 1954, Cadillacs were built with a 12-volt system. Prior to 1954, vehicles had 6-volt systems. The electrical demands of accessories, such as power windows and seats, put a severe strain on the 6-volt battery. With a 12-volt system, the charging system had to work less hard and there was plenty of electrical power for the electrical accessories.

The increase in voltage also allowed wire sizes to decrease because the amperage required to power things was reduced. To explain this, consider an accessory that required 20 amps to operate in a 6-volt system (120 watts). When the voltage was increased to 12 volts, the system only drew 10 amperes.

Today we are faced with the same situation. The use of computers and the need to keep their memories fresh has put a drain on the battery even when the engine is not running. Plus, the number of electrical accessories has and will continue to grow.

Today's vehicles are very sensitive to voltage change. In fact, their overall efficiency depends on a constant voltage. The demands of new technology make it difficult to maintain a constant voltage and engineers have determined system voltage must be increased to meet those demands. As vehicles evolve, emissions, fuel economy, comfort, convenience, and safety features will put more of a drain on the electrical system. This increased demand is the result of converting purely mechanical systems into electromechanical systems, such as steering, suspension, and braking systems. It has been estimated that in a few years the continuous electrical power demand will be 3,000 to 7,000W. Current 14-volt systems are rated at 800 to 1,500W.

To meet these demands there are two possible solutions: increase the amperage capacity of the battery and charging system or increase system voltage.

Larger capacity batteries and generators are only a band aid solution. Because the generator is driven by engine power, more power from the engine will be required to keep the higher capacity battery charged. Therefore, overall efficiency will decrease.

By moving to a higher system voltage, the battery will need to be larger and heavier. However, because system amperage will be lower, wire size will be smaller and perhaps the weight gain at the battery will be offset by the decreased weight of the wiring.

All of the advantages of moving from 6- to 12-volt systems apply to the move from 12 volts to 42 volts. But why 42 volts? Forty-two volts represent three 12-volt batteries. Engineers decided to take advantage of the fact that a 12-volt battery actually holds a 14-volt charge (3 times 14 volts equals 42 volts). Forty-two-volt systems are also desirable for safety reasons. Sixty volts can stop a person's heart from beating; therefore, 42-volt systems allow for a margin of safety.

During the transition from 14-volt to 42-volt systems, vehicles will be fitted with dual voltage batteries that provide 12 and 36 volts **(Figure 14–11)**. This split voltage system will provide 42 volts to high-power applications such as power steering, traction control, brake, and engine cooling systems. The 14-volt system will power low-load systems, such as lights, air conditioning, power door locks, radios, and navigation systems.

Ohm's Law

In 1827, a German mathematics professor, Georg Ohm, published a book that included his explanation of the behavior of electricity. His thoughts have become the basis for a true understanding of electricity. He found it takes 1 volt of electrical pressure to push 1 ampere of electrical current through 1 ohm of resistance. This statement is the basic law of electricity. It is known as **Ohm's law**.

A simple electrical circuit is a load connected to a voltage source by conductors. The resistor could be a fog light, the voltage source could be a battery, and the conductor could be a copper wire **(Figure 14–12)**.

In any electrical circuit, current (I), resistance (R), and voltage (E) are mathematically related. This relationship is expressed in a mathematical statement of Ohm's law. Ohm's law can be applied to the entire circuit or to any part of a circuit. When any two factors are known, the third factor can be found by using Ohm's law. Using the circle shown in **Figure 14–13**, you can easily find the formula for calculating the unknown element. By covering the element you need to find, the necessary formula is shown in the circle.

To find voltage, cover the E **(Figure 14–14)**. The voltage (E) in a circuit is equal to the current (I) in amperes multiplied by the resistance (R) in ohms.

To find current, cover the I **(Figure 14–15)**. The current (amperage) in a circuit equals the voltage divided by the resistance (in ohms).

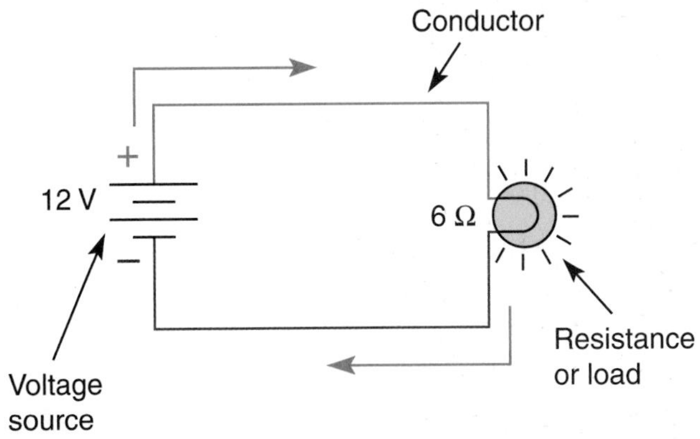

Figure 14-12 A simple circuit consists of a voltage source, conductors, and a resistance or load.

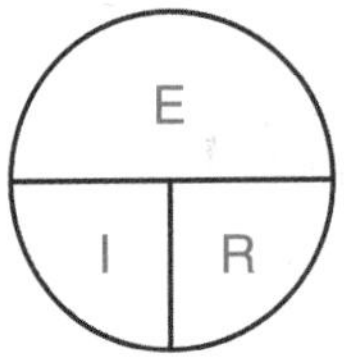

Figure 14-13 Ohm's law.

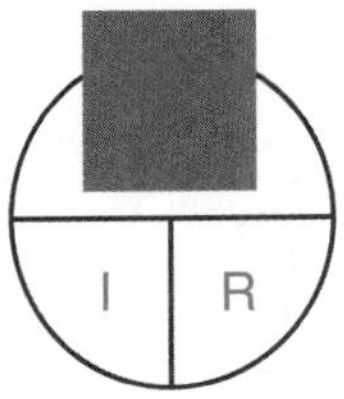

Figure 14-14 To find voltage, cover the *E* and use the exposed formula.

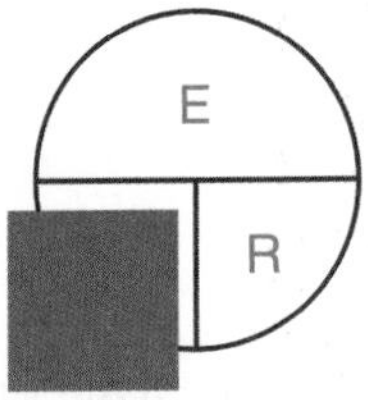

Figure 14-15 To find current, cover the *I* and use the exposed formula.

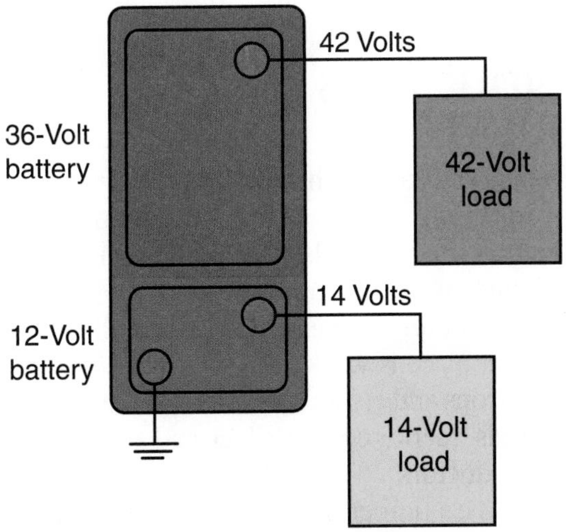

Figure 14-11 A split voltage arrangement providing 14 and 42 volts from a 36-volt battery.

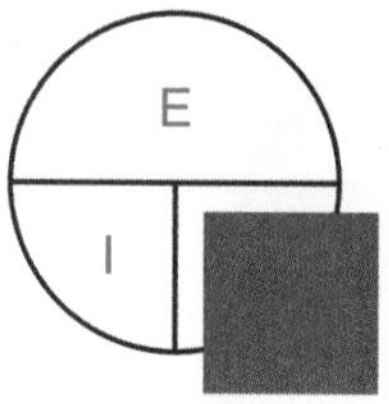

Figure 14-16 To find resistance, cover the *R* and use the exposed formula.

To find resistance, cover the *R* **(Figure 14–16)**. The resistance of a circuit (in ohms) equals the voltage divided by the current (in amperes).

It is very important for technicians to understand Ohm's law. It explains how an increase or decrease in voltage, resistance, or current affects a circuit.

For example, if the fog light in Figure 14–12 has a 6-ohm resistance, how many amperes does it use to operate? Since cars and light trucks have a 12-volt battery and you know two of the factors in the fog light circuit, it is simple to solve for the third.

$$I\text{ (unknown)} = \frac{E\text{ (12 volts)}}{R\text{ (6 ohms)}}$$

$$I = \frac{12}{6}$$

$$I = 2\text{ amperes}$$

In a clean, well-wired circuit, the fog lights will draw 2 amperes of current. What would happen if resistance in the circuit increases due to corroded or damaged wire or connections? If the corroded connections add 2 ohms of resistance to the fog light circuit, the total resistance is 8 ohms **(Figure 14–17)**. The amount of current flowing through the circuit for the lights decreases.

$$I = \frac{12}{6+2} = \frac{12}{8}$$

$$I = 1.5\text{ amperes}$$

If the lights are designed to operate at 2 amperes, this decrease to 1.5 amperes causes them to burn dimly. Cleaning the corrosion away or installing new wires and connectors eliminates the unwanted resistance; the correct amount of current will flow through the circuit, allowing the lamp to burn as brightly as it should.

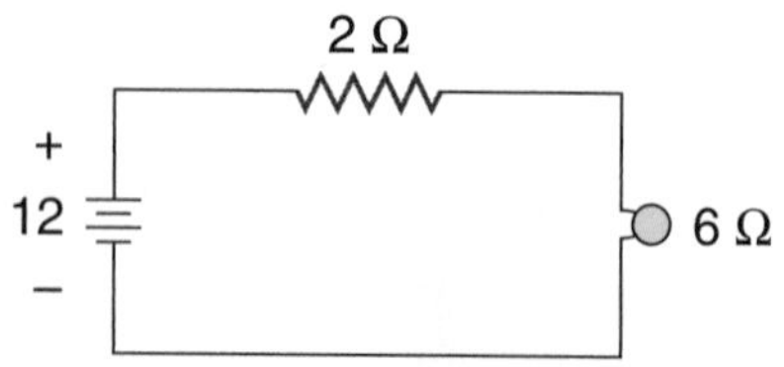

Figure 14-17 This one is the same circuit as **Figure 14–12** but it has a corroded wire, represented by the additional resistor.

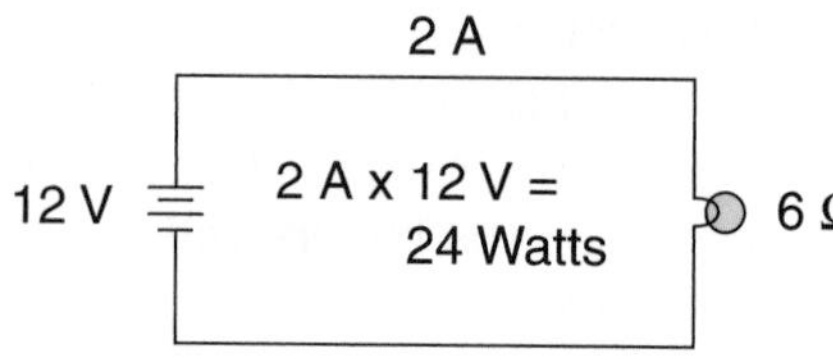

Figure 14-18 Electrical power is calculated by multiplying voltage by current.

Power

Electrical **power**, or the rate of performing work, is found by multiplying the amount of electrical pressure by the amount of current flow (power = voltage × amperage). Power is measured in **watts**. Although power measurements are rarely, if ever, needed in automotive service, knowing the power requirements of light bulbs, electric motors, and other components is sometimes useful when troubleshooting electrical systems.

Another useful formula is one used to find the power of an electrical circuit expressed in watts.

$$P = E \times I$$

That is, power (P) in watts equals the voltage (E) multiplied by the current (I) in amperes **(Figure 14–18)**. This relationship is known as Watt's law. Looking back at the example of the fog light circuit, we can calculate the amount of power used or heat generated by the fog light.

$$P = 12\text{ volts} \times 2\text{ amperes}$$
$$\text{P} = 24\text{ watts}$$

The normal fog light generates 24 watts of power, while the corroded fog light circuit produces the following.

$$P = 12\text{ volts} \times 1.5\text{ amperes}$$
$$P = 18\text{ watts}$$

This reduction in power or heat explains the decrease in bulb brightness.

CONDUCTORS AND INSULATORS

Controlling and routing the flow of electricity requires the use of materials known as conductors and insulators. **Conductors** are materials with a low resistance to the flow of current. If the number of electrons in the outer shell or ring of an atom is less than four, the force holding them in place is weak. The voltage needed to move these electrons and create current flow is relatively small. Most metals, such as copper, silver, and aluminum are excellent conductors.

When the number of electrons in the outer ring is greater than four, the force holding them in orbit is very strong and very high voltages are needed to move them.

These materials are known as **insulators**. They resist the flow of current. Thermal plastics are the most common electrical insulators used today. They can resist heat, moisture, and corrosion without breaking down.

CAUTION!

Your body is a good conductor of electricity. Remember this when working on a vehicle's electrical system. Always observe all electrical safety rules.

Copper wire is by far the most popular conductor used in automotive electrical systems. Wire wound inside electrical units, such as ignition coils and generators, usually has a very thin baked-on insulating coating. External wiring is normally covered with a plastic-type insulating material that is highly resistant to environmental factors like heat, vibration, and moisture. Where flexibility is required, the copper wire is made of a large number of very small strands of wire woven together.

The resistance of a uniform, circular cross-section copper wire depends on the length of the wire, the diameter of the wire, and the temperature of the wire. If the length is doubled, the resistance between the wire ends is doubled. The longer the wire, the greater the resistance. If the diameter of a wire is doubled, the resistance for any given length is cut in half. The larger the wire's diameter, the lower the resistance.

In any circuit, the smallest wire that will not cause excessive voltage drop is used to minimize cost.

The other important factor affecting the resistance of a copper wire is temperature. As the temperature increases, the resistance increases. The effects of temperature are very important in the design of electrical equipment. Excessive resistance caused by normal temperature increases can hurt the performance of the equipment.

Heat is developed in any wire carrying current because of the resistance in the wire. If the heat becomes excessive, the insulation will be damaged. Resistance occurs when electrons collide as current flows through the conductor. These collisions cause friction, which in turn generates heat.

Circuits

A complete electrical circuit exists when electrons flow along a path between two points. In a complete circuit, resistance must be low enough to allow the available voltage to push electrons between the two points. Most automotive circuits contain four basic parts.

1. Power sources, such as a battery or alternator, that provide the energy needed to cause electron flow.
2. Conductors, such as copper wires, that provide a path for current flow.
3. Loads, which are devices that use electricity to perform work, such as light bulbs, electric motors, or resistors.
4. Controllers, such as switches or relays, that control or direct the flow of electrons.

There are also three basic types of circuits used in automotive electrical systems: series circuits, parallel circuits, and series-parallel circuits. Each circuit type has its own characteristics regarding amperage, voltage, and resistance.

Series Circuits A **series circuit** consists of one or more resistors connected to a voltage source with only one path for electron flow. For example, a simple series circuit consists of a resistor (2 ohms in this example) connected to a 12-volt battery **(Figure 14–19A)**. The current can be determined by applying Ohm's law.

$$I = \frac{E}{R} = \frac{12}{2} = 6 \text{ amperes}$$

Another series circuit may contain a 2-ohm resistor and a 4-ohm resistor connected to a 12-volt battery **(Figure 14–19B)**. The word *series* is given to a circuit in which the same amount of current is present throughout the circuit. The current that flows through one resistor also flows through other resistors in the circuit. As that amount of current leaves the battery, it flows through the conductor to the first resistor. At the resistor, some electrical energy or voltage is consumed as the current flows through it. The decreased amount of voltage is then applied to the next resistor as current flows to it. By the time the current is flowing in the conductor leading back to the battery, all voltage has been dropped. All of the source voltage available to the circuit is dropped by the resistors in the circuit.

In a series circuit, the total amount of resistance in the circuit is equal to the sum of all the individual resistors. In the circuit in **Figure 14–19B**, the total circuit resistance

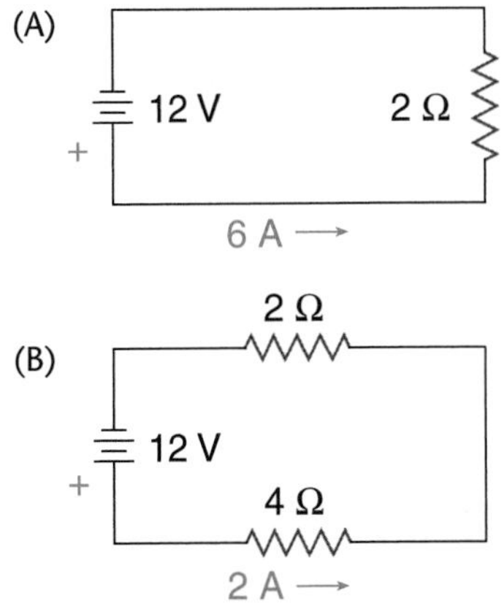

Figure 14–19 In a series circuit, the same amount of current flows through the entire circuit.

is 4 + 2 = 6 ohms. Based on Ohm's law, current is $I = E/R = \frac{12}{6} = 2$ amperes. In a series circuit, current is constant throughout the circuit. Therefore, 2 amps of current flows through the conductors and both resistors.

Ohm's law can be used to determine the voltage drop across parts of the circuit. For the 2-ohm resistor, $E = IR = 2 \times 2 = 4$ volts. For the 4-ohm resistor, $E = 2 \times 4 = 8$ volts. These values are called **voltage drops**. The sum of all voltage drops in a series circuit must equal the source voltage, or 4 + 8 = 12 volts.

An ammeter connected anywhere in this circuit will read 2 amperes, and a voltmeter connected across each of the resistors will read 4 volts and 8 volts, as shown in **Figure 14–20**.

All calculations for a series circuit work in the same way no matter how many resistors there are in series. Consider the circuit in **Figure 14–21**. This circuit has four resistors in series with each other. The total resistance is 12 ohms (5 ohms + 2 ohms + 4 ohms + 1 ohm). Using Ohm's law, we can see that the circuit current is 1 amp ($I = E/R = \frac{12}{12} = 1$ amp). We can also use Ohm's law to determine the voltage drop across each resistor in the circuit. For example, since the circuit current is 1 amp, 4 volts are dropped by the 4-ohm resistor ($E = I \times R = 1$ amp $\times$ 4 ohms = 4 volts).

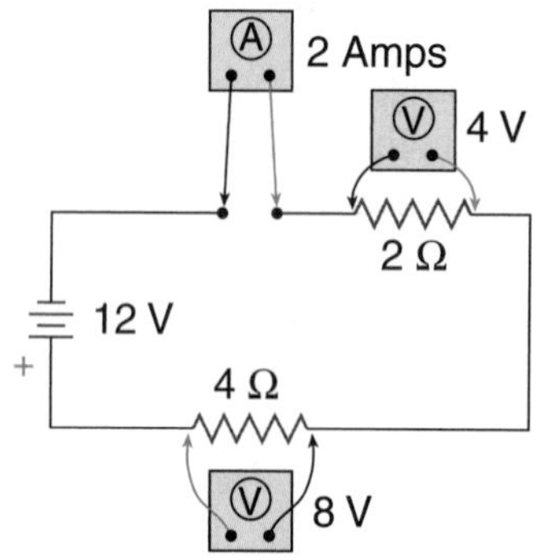

Figure 14-20 Measuring the current and voltage drops in the circuit.

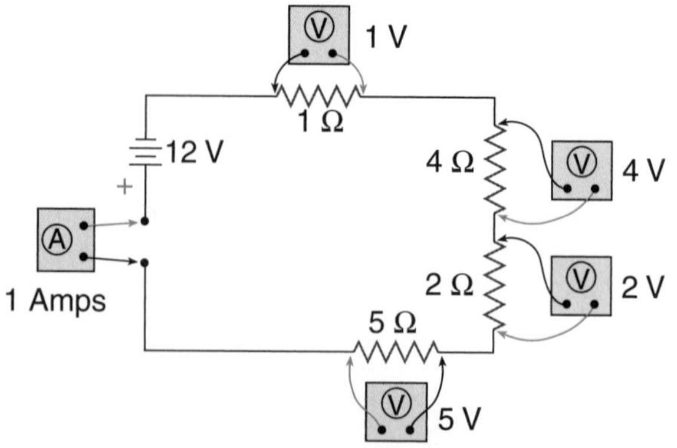

Figure 14-21 Values in the series circuit.

A series circuit is characterized by the following four facts:

1. The circuit's current is determined by the total amount of resistance in the circuit; it is constant throughout the circuit.
2. The voltage drops across each resistor are different if the resistance values are different.
3. The sum of the voltage drops equals the source voltage.
4. The total resistance is equal to the sum of all resistances in the circuit.

Parallel Circuits A **parallel circuit** provides two or more different paths for the current to flow through. Each path has separate resistors (loads) and can operate independently of the other paths. The different paths for current flow are commonly called the legs of a parallel circuit.

A parallel circuit is characterized by the following facts:

1. Total circuit resistance is always lower than the resistance of the leg with the lowest total resistance.
2. The current through each leg will be different if the resistance values are different.
3. The sum of the current on each leg equals the total circuit current.
4. The voltage applied to each leg of the circuit will be dropped across the legs if there are no loads in series with the parallel circuit.

Consider the circuit shown in **Figure 14–22**. Two 3-ohm resistors are connected to a 12-volt battery. The resistors are in parallel with each other, since the battery voltage (12 volts) is applied to each resistor and they have a common negative lead. The current through each resistor or leg can be determined by applying Ohm's law. For the 3-ohm resistors, $I = E/R = \frac{12}{3} = 4$ amperes. Therefore, the total circuit current supplied by the battery is 4 + 4 = 8 amperes. Using Ohm's law, we find that 12 volts are dropped by both resistors **(Figure 14–23)**.

Resistances are not added up to calculate the total resistance in a parallel circuit. Rather, they can be determined by dividing the product of their ohm values by the

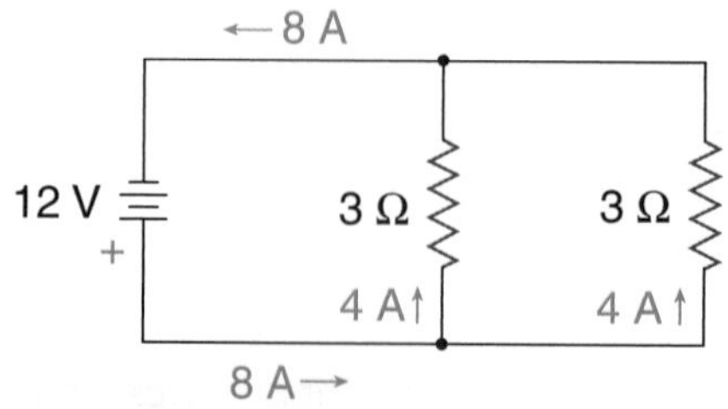

Figure 14-22 A simple parallel circuit.

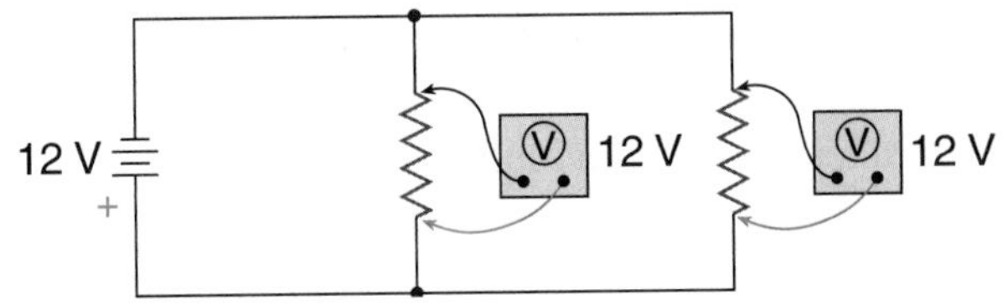

Figure 14–23 A parallel circuit with voltage drops shown.

sum of their ohm values. This formula works when the circuit has two parallel legs.

$$\frac{3 \text{ ohms} \times 3 \text{ ohms}}{3 \text{ ohms} + 3 \text{ ohms}} = \frac{9}{6} = 1.5 \text{ ohms}$$

Total resistance can also be calculated by using Ohm's law if you know the total circuit current and the voltage ($R = E/I = 12/8 = 1.5$ ohms).

Consider another parallel circuit, **Figure 14–24**. In this circuit there are two legs and four resistors. Each leg has two resistors in series. One leg has a 4-ohm and a 2-ohm resistor. The total resistance on that leg is 6 ohms. The other leg has a 1-ohm and a 2-ohm resistor. The total resistance of that leg is 3 ohms. Therefore, we have 6 ohms in parallel with 3 ohms.

Current flow through the circuit can be calculated by different methods. Using Ohm's law, we know that $I = E/R$. If we take the total resistance of each leg and divide it into the voltage, we then know the current through that leg. Since total circuit current is equal to the sum of the current flows through each leg, we simply add the current across each leg together. This will give us total circuit current.

Leg 1: $I = E/R = 12/6 = 2$ amps

Leg 2: $I = E/R = 12/3 = 4$ amps

2 amps + 4 amps = 6 amps = total circuit current

Circuit current can also be determined by finding the total resistance of the circuit. To do this, the product-over-sum formula is used. By dividing this total into the voltage, total circuit current is known.

$$\frac{\text{Leg 1} \times \text{Leg 2}}{\text{Leg 1} + \text{Leg 2}} = \frac{6 \times 3}{6 + 3} = \frac{18}{9} = 2 \text{ ohms}$$

since $I = E/R$, $I = 12/2$, $I = 6$ amps (total circuit current)

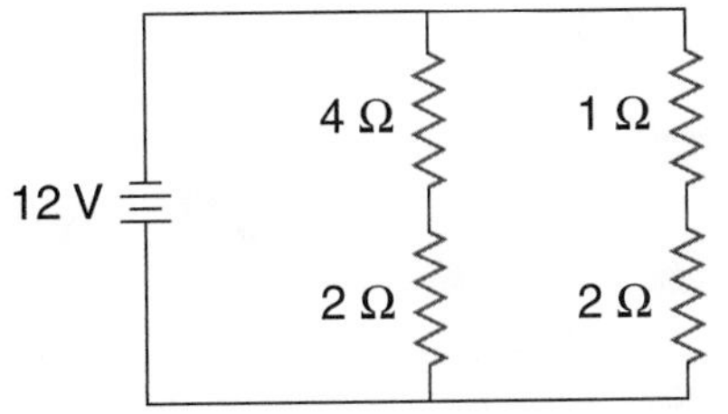

Figure 14–24 Series circuits within a parallel circuit.

When a circuit has more than two legs, the reciprocal formula should be used to determine total circuit resistance. The formula follows:

$$\frac{1}{\frac{1}{R_1} + \frac{1}{R_2} + \frac{1}{R_3} + \ldots \frac{1}{R_n}}$$

To demonstrate how to use this formula, consider the circuit in **Figure 14–25**. Here is a parallel circuit with four legs. The resistances across each leg are 4 ohms, 3 ohms, 6 ohms, and 4 ohms. Using the reciprocal formula, we will find that the total resistance of the circuit is 1 ohm. (Note that the total resistance is lower than the leg with the lowest resistance.)

$$\frac{1}{\frac{1}{4} + \frac{1}{3} + \frac{1}{6} + \frac{1}{4}} =$$

$$\frac{1}{\frac{3}{12} + \frac{4}{12} + \frac{2}{12} + \frac{3}{12}} = \frac{1}{\frac{12}{12}} = \frac{1}{1} = 1$$

The total of this circuit could also have been found by calculating the current across each leg then adding them together to get the total circuit current. Using Ohm's law, if you divide the voltage by the total circuit current, you will get total resistance.

Leg 1: $I = E/R = 12/4 = 3$ amps
Leg 2: $I = E/R = 12/3 = 4$ amps
Leg 3: $I = E/R = 12/6 = 2$ amps
Leg 4: $I = E/R = 12/4 = 3$ amps

Total circuit current = 3 + 4 + 2 + 3 = 12 amps

then,

$$R = E/I = 12/12 = 1 \text{ ohm}$$

Series-Parallel Circuits In a **series-parallel circuit**, both series and parallel combinations exist in the same circuit. If you are faced with the task of calculating the values in a series-parallel circuit, determine all values of the parallel circuit(s) first. By looking carefully at a series-parallel circuit, you will find that it is nothing more than one or more parallel circuits in series with each other or in series with some other resistance.

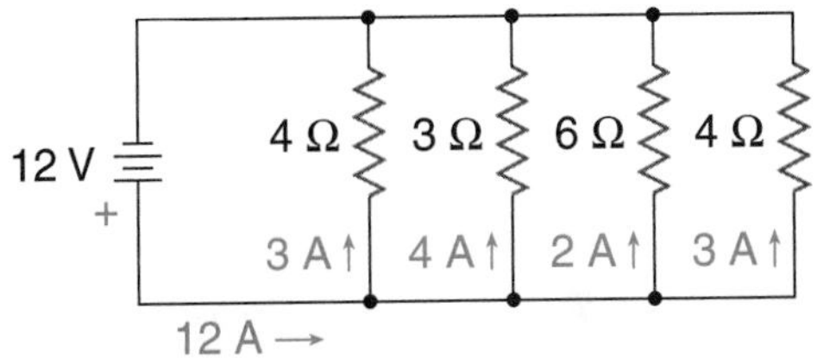

Figure 14–25 A parallel circuit with four legs.

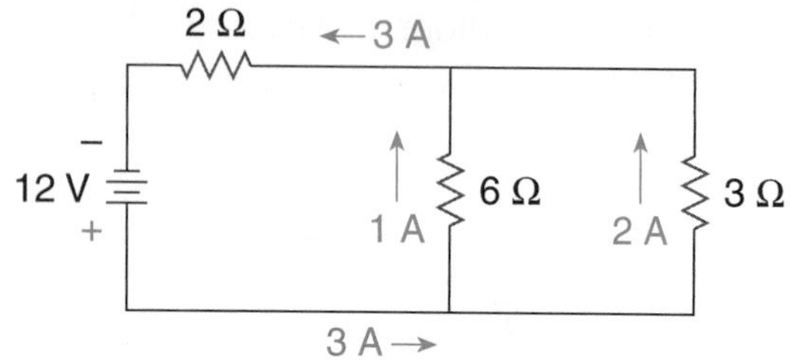

Figure 14-26 In a series-parallel circuit, the sum of the currents through the legs must equal the current through the series part of the circuit.

A series-parallel circuit is illustrated in **Figure 14–26**. The 6- and 3-ohm resistors are in parallel with each other and together are in series with the 2-ohm resistor.

The total current in this circuit is equal to the voltage divided by the total resistance. The total resistance can be determined as follows. The 6- and 3-ohm parallel resistors in **Figure 14–27** are equivalent to a 2-ohm resistor, since (6 × 3)/(6 + 3) = 2. This equivalent 2-ohm resistor is in series with the other 2-ohm resistor. To find the total resistance, add the two resistance values together. This gives a total circuit resistance of 4 ohms (2 + 2 = 4 ohms). The total current, therefore, is $I = {}^{12}\!/\!_4 = 3$ amperes. This means that 3 amps of current is flowing through the 2-ohm resistor in series and 3 amps are divided between the resistors in parallel. In series-parallel circuits, the sum of the currents, flowing in the parallel legs, must equal that of the series resistors' current.

To find the current through each of the resistors in parallel, find the voltage drop across those resistors first. With 3 amperes flowing through the 2-ohm resistor, the voltage drop across this resistor is $E = IR = 3 \times 2 = 6$ volts, leaving 6 volts across the 6- and 3-ohm resistors. The current through the 6-ohm resistor is $I = E/R = {}^{6}\!/\!_6 = 1$ ampere, and through the 3-ohm resistor is $I = {}^{6}\!/\!_3 = 2$ amperes. The sum of these two current values must equal the total circuit current and it does, 1 + 2 = 3 amperes **(Figure 14–27)**. The sum of the voltage drops across the parallel part of the circuit and the series part must also equal source voltage.

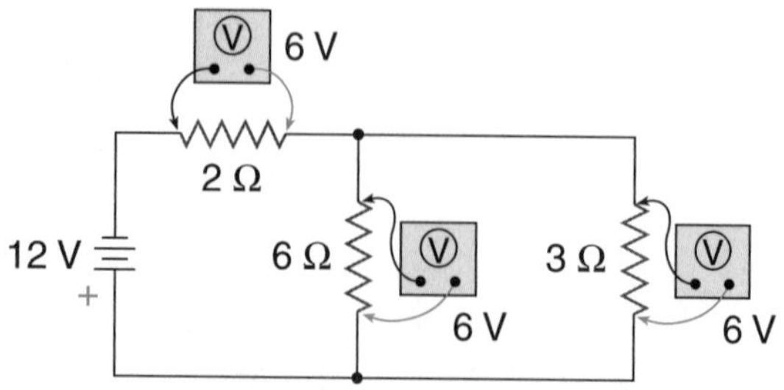

Figure 14-27 The circuit in **Figure 14–26** with voltage drops shown.

Grounding the Load

In the illustrations used to explain series, parallel, and series-parallel circuits, the return wire from the load or resistor connects directly to the negative terminal of the battery **(Figure 14–28)**. If this were the case in an actual vehicle, there would be literally hundreds of wires connected to the negative battery terminal.

To avoid this, auto manufacturers use a wiring style that involves using the vehicle's metal frame components as part of the return circuit. Using the chassis as the negative wire is often referred to as **grounding**. The wire or metal mounting that serves as the contact to the chassis is commonly called the **ground wire** or lead. As shown in **Figure 14–29**, the load is grounded directly to the metal frame. The metal frame then acts as the return wire

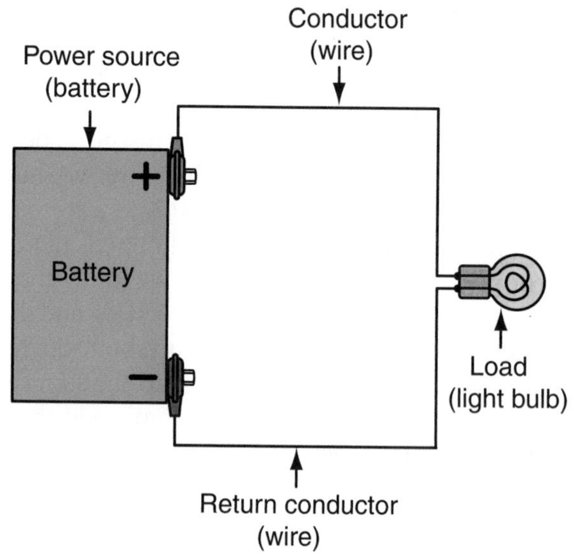

Figure 14-28 Electricity results from the flow of electrons from the negative side of the battery to the positive side.

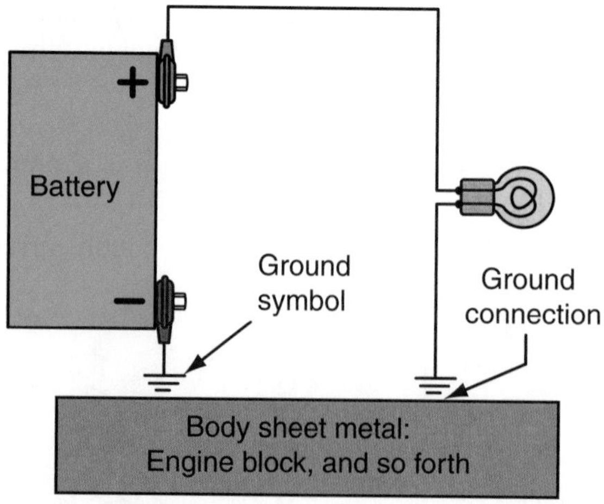

Figure 14-29 The same circuit as in **Figure 14–28** but with a chassis ground.

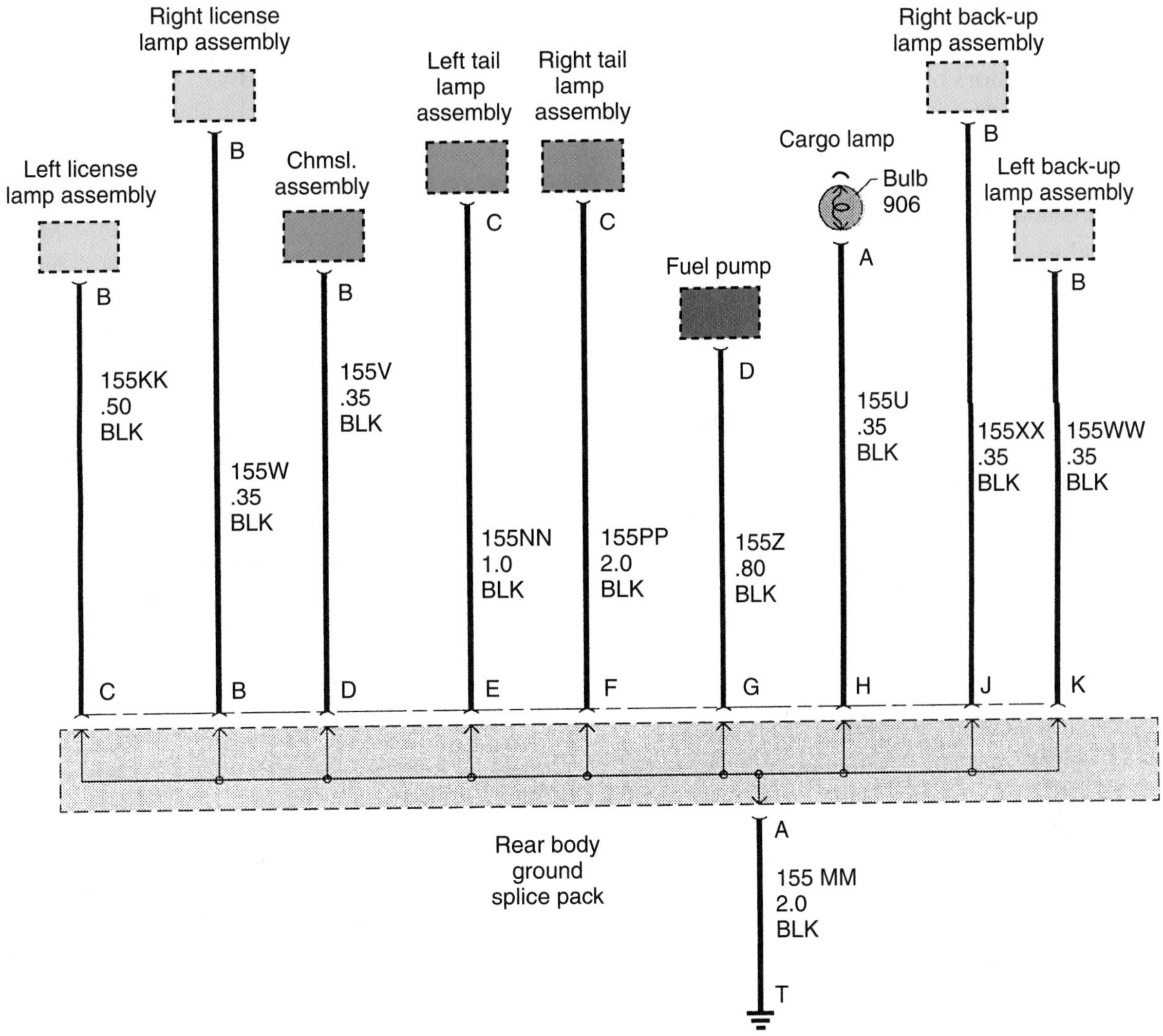

Figure 14–30 Some vehicles use a network of grounding wires and terminals to ensure a good ground in all electrical circuits.

in the circuit. Current passes from the battery, through the load and into the frame. The frame is connected to the negative terminal of the battery through the battery's ground wire.

An electrical component, such as an alternator, may be mounted directly to the engine block, transmission case, or frame. This direct mounting effectively grounds the component without the use of a separate ground wire. In other cases, however, a separate ground wire must be run from the component to the frame or another metal part to ensure a good connection for the return path. The increased use of plastics and other nonmetallic materials in body panels and engine parts has made electrical grounding more difficult. To ensure good grounding back to the battery, some manufacturers now use a network of common grounding terminals and wires **(Figure 14–30)**.

CIRCUIT COMPONENTS

Automotive electrical circuits contain a number of different types of electrical devices. The more common components are outlined in the following sections.

Resistors

As shown in the explanation of simple circuit design, resistors are used to limit current flow (and thereby voltage) in circuits where full current flow and voltage are not needed or desired. Resistors are devices specially constructed to put a specific amount of resistance into a circuit. In addition, some other components use resistance to produce heat and even light. An electric window defroster is a specialized type of resistor that produces heat. Electric lights are resistors that get so hot they produce light.

Automotive circuits typically contain these types of resistors: fixed value, stepped or tapped, and variable.

Fixed value resistors are designed to have only one rating, which should not change. These resistors are used to decrease the amount of voltage applied to a component, such as an ignition coil. Often manufacturers use a special wire, called **resistor wire**, to limit current flow and voltage in a circuit. This wire looks much like normal wire but is not a good conductor and is marked as a resistor.

Tapped or **stepped resistors** are designed to have two or more fixed values, available by connecting wires to the several taps of the resistor. Heater motor resistor packs, which provide for different fan speeds, are an example of this type of resistor **(Figure 14–31)**.

Variable resistors are designed to have a range of resistances available through two or more taps and a control. Two examples of this type of resistor are **rheostats** and **potentiometers**. Rheostats **(Figure 14–32)** have two connections, one to the fixed end of a resistor and one to a sliding contact with the resistor. Moving the control moves the sliding contact away from or toward the fixed end tap, increasing or decreasing the resistance. Potentiometers **(Figure 14–33)** have three connections, one at each end of the resistance and one connected to a sliding contact with the resistor. Moving the control moves the sliding contact away from one end of the resistance but toward the other end. These are called potentiometers because different amounts of potential or voltage can be sent to another circuit. As the sliding contact moves, it

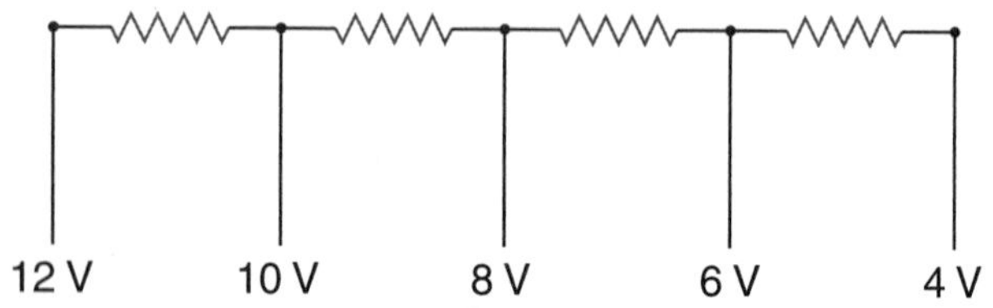

Figure 14-31 A stepped resistor.

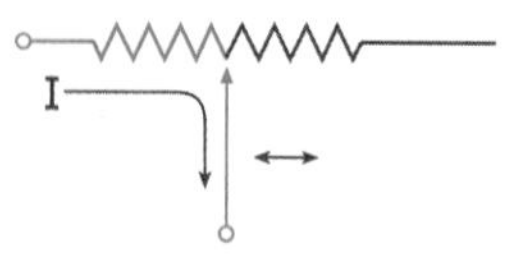

Figure 14-32 A rheostat.

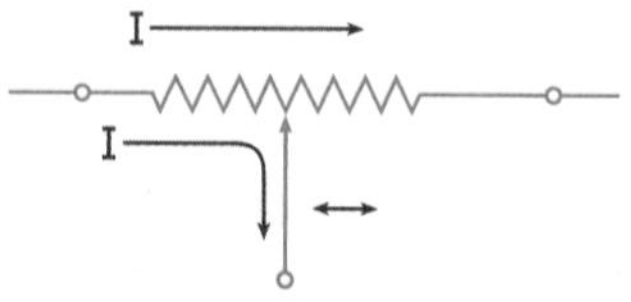

Figure 14-33 A potentiometer.

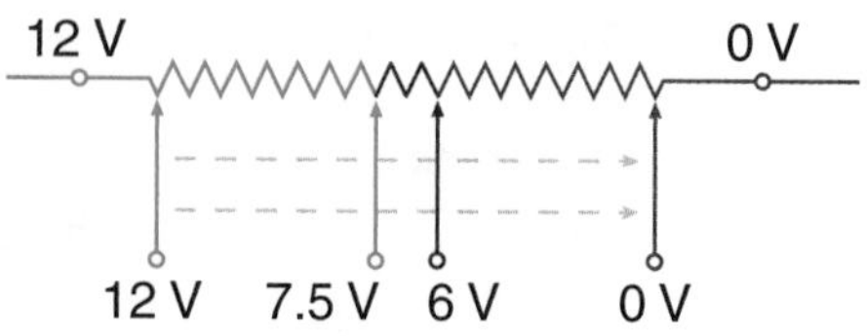

Figure 14-34 Voltage across a potentiometer.

picks up a voltage equal to the source voltage minus the amount dropped by the resistor, so far **(Figure 14–34)**.

Another type of variable resistor is the **thermistor**. This type of resistor is designed to change its resistance value as its temperature changes. Although most resistors are carefully constructed to maintain their rating within a few ohms through a range of temperatures, the thermistor is designed to change its rating. Thermistors are used to provide compensating voltage in components or to determine temperature. As a temperature sensor, the thermistor is connected to a voltmeter calibrated in degrees. As the temperature rises or falls, the resistance also changes, and so does the voltage in the circuit. These changes are read on the temperature gauge. Thermistors are also commonly used to sense temperature and send a signal back to a control unit. The control unit interprets the signal as a temperature value **(Figure 14–35)**.

Circuit Protective Devices

When overloads or shorts in a circuit cause too much current to flow, the wiring in the circuit heats up, the insulation melts, and a fire can result, unless the circuit has some kind of protective device. Fuses, fuse links, maxi-fuses, and circuit breakers are designed to provide protection from high current. They may be used singly or in combination. Typical symbols for protection devices are shown in **Figure 14–36**.

WARNING!

Fuses and other protection devices normally do not wear out. They go bad because something went wrong. Never replace a fuse or fusible link, or reset a circuit breaker, without finding out why it went bad.

Fuses There are three basic types of fuses in automotive use: cartridge, blade, and ceramic **(Figure 14–37)**. The **cartridge fuse** is found on most older domestic cars and a few imports. It is composed of a strip of low temperature melting metal enclosed in a transparent glass or plastic tube. Late-model domestic vehicles and many imports use **blade** or **spade fuses**. The **ceramic fuse** is used on many European imports. The core is a ceramic insulator with a conductive metal strip along one side.

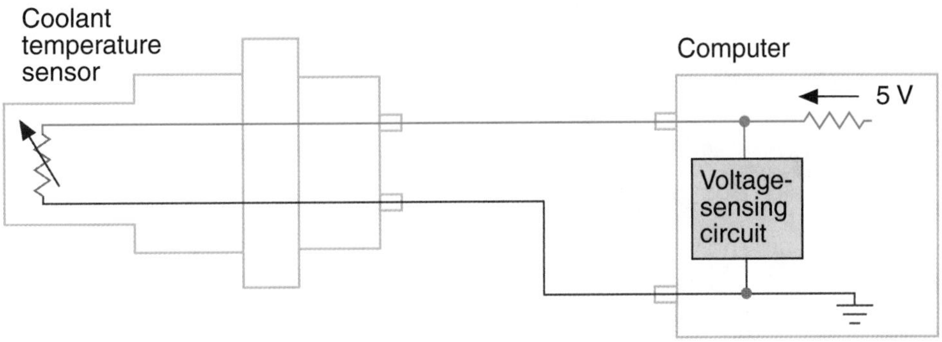

Figure 14-35 A thermistor is used to measure temperature. The sensing unit measures the change in resistance and translates it into a temperature value.

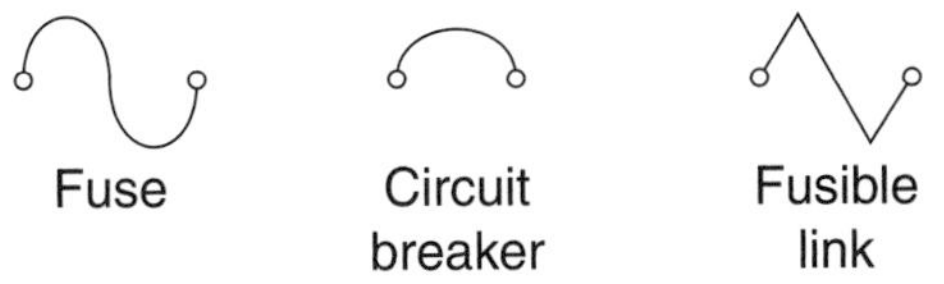

Figure 14-36 Electrical symbols for common circuit protection devices.

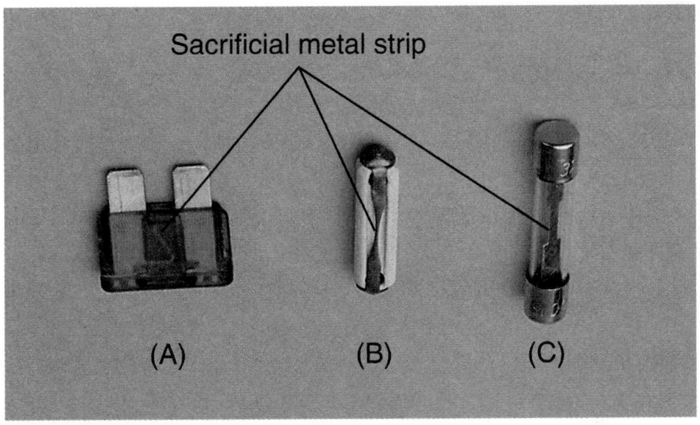

Figure 14-37 (A) A blade fuse, (B) a ceramic fuse, and (C) a cartridge fuse.

Fuses are rated by the current at which they are designed to blow. A three-letter code is used to indicate the types and sizes of fuses. Blade fuses have codes ATC or ATO. All glass SFE fuses have the same diameter, but the length varies with the current rating. Ceramic fuses are available in two sizes, code GBF (small) and the more common code GBC (large). The amperage rating is also embossed on the insulator. Codes, such as AGA, AGW, and AGC, indicate the length and diameter of the fuse. Fuse lengths in each of these series are the same, but the current rating can vary. The code and the current rating are usually stamped on the end cap.

The current rating for blade fuses is indicated by the color of the plastic case **(Table 14–1)**. In addition, it is usually marked on the top. The insulator on ceramic fuses is color coded to indicate different current ratings.

Three basic types of blade fuses are found on today's vehicles: the standard blade, the minifuse, and the maxi-

TABLE 14–1 TYPICAL COLOR CODING OF PROTECTIVE DEVICES

Blade Fuse Color Coding

Ampere Rating	Housing Color
4	pink
5	tan
10	red
15	light blue
20	yellow
25	natural
30	light green

Fuse Link Color Coding

Wire Link Size	Insulation Color
20 GA	blue
18 GA	brown or red
16 GA	black or orange
14 GA	green
12 GA	gray

Maxi-Fuse Color Coding

Ampere Rating	Housing Color
20	yellow
30	light green
40	amber
50	red
60	blue

Figure 14-38 A typical fuse box or panel.

fuse. The minifuse is the commonly used circuit protection device. Minifuses are available in ratings from 5 to 30 amps. The maxi-fuse is a serviceable replacement for a fusible link cable. It is used in circuits that have high operating current. Maxi-fuses are available in 2–100 amp ratings; the most common is the 30-amp.

Fuses are located in a box or panel **(Figure 14–38)**, usually under the dashboard, behind a panel in the foot well, or in the engine compartment. Fuses are generally numbered, and the main components abbreviated. On late-model cars, there may be icons or symbols indicating which circuits they serve. This identification system is covered in more detail in the owner's and service manuals.

Fuse Links Fuse or **fusible links** are used in circuits when limiting the maximum current is not extremely critical. They are often installed in the positive battery lead to the ignition switch and other circuits that have power with the key off. Fusible links are normally found in the engine compartment near the battery. Fusible links are also used when it would be awkward to run wiring from the battery to the fuse panel and back to the load.

A fuse link **(Figure 14–39)** is a short length of small-gauge wire installed in a conductor. Because the fuse link is a lighter gauge of wire than the main conductor, it melts and opens the circuit before damage can occur in the rest of the circuit. Fuse link wire is covered with a special insulation that bubbles when it overheats, indicating that the link has melted.

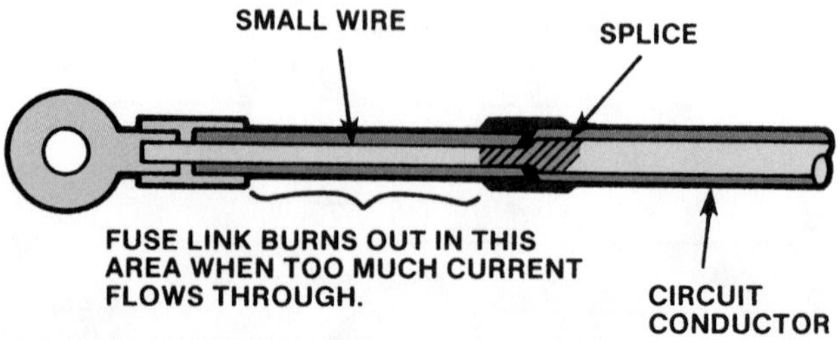

Figure 14-39 A typical fuse link.

CAUTION!

Always disconnect the battery ground cable prior to servicing any fuse link.

Maxi-Fuses Many late-model vehicles use **maxi-fuses** instead of fusible links. Maxi-fuses look and operate like two-prong, blade, or spade fuses, except they are much larger and can handle more current. (Typically, a maxi-fuse is four to five times larger.) Maxi-fuses are usually located in their own underhood fuse block.

Maxi-fuses allow the vehicle's electrical system to be broken down into smaller circuits that are easy to diagnose and repair. For example, in some vehicles a single fusible link controls one-half or more of all circuitry. If it burns out, many electrical systems are lost. By replacing this single fusible link with several maxi-fuses, the number of systems lost due to a problem in one circuit is drastically reduced. This makes it easy to pinpoint the source of trouble.

Circuit Breakers Some circuits are protected by **circuit breakers** (abbreviated c.b. in the fuse chart of a service manual). They can be fuse panel mounted or in-line. Like fuses, they are rated in amperes.

Each circuit breaker conducts current through an arm made of two types of metal bonded together (bimetal arm). If the arm starts to carry too much current, it heats up. As one metal expands faster than the other, the arm bends, opening the contacts. Current flow is broken. A circuit breaker can be cycling **(Figure 14–40)**, or must be manually reset.

In the cycling type, the bimetal arm begins to cool once the current to it is stopped. Once it returns to its original shape, the contacts are closed and power is restored. If the current is still too high, the cycle of breaking the circuit is repeated.

Two types of noncycling or resettable circuit breakers are used. One is reset by removing the power from the circuit. There is a coil wrapped around a bimetal arm **(Figure 14–41A)**. When there is excessive current and the contacts open, a small current passes through the coil.

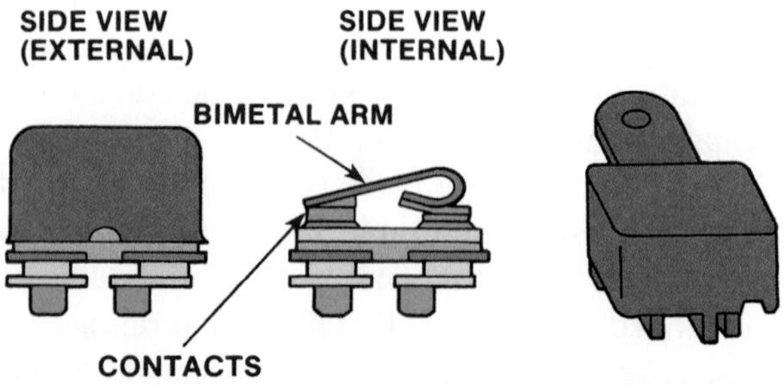

Figure 14-40 A cycling circuit breaker.

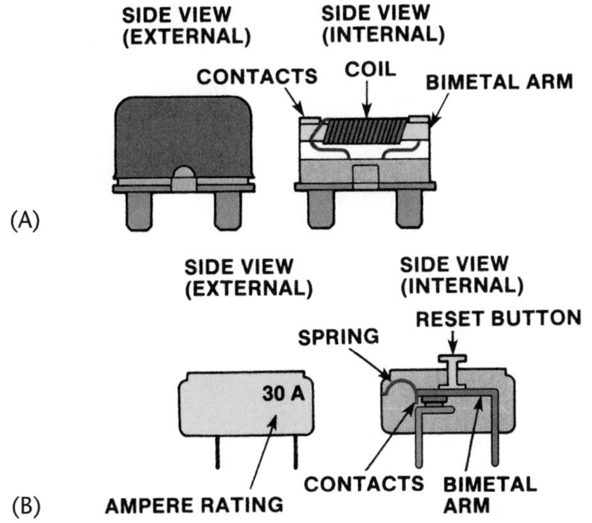

Figure 14–41 Resetting noncycling circuit breakers by (A) removing power from the circuit and (B) depressing a reset button.

This current through the coil is not enough to operate a load, but it does heat up both the coil and the bimetal arm. This keeps the arm in the open position until power is removed. The other type is reset by depressing a reset button. A spring pushes the bimetal arm down and holds the contacts together **(Figure 14–41B)**. When an overcurrent condition exists and the bimetal arm heats up, the bimetal arm bends enough to overcome the spring and the contacts snap open. The contacts stay open until the reset button is pushed, which snaps the contacts together again.

42-Volt Systems New 42-volt systems present a unique problem to circuit protection. Most circuit protection devices used in 12-volt systems are actually rated at 32 volts. If these protection devices were used in a 42-volt system, problems such as severe damage to the vehicle's wiring and electrical components could result. The burning of the components and wiring could also cause a fire.

Protection devices for the 42-volt systems are rated at 55 volts. This rating allows protection during times of high voltage spikes. These protection devices and their receptacles have a unique design to prevent the installation of the wrong type of fuse.

Voltage Limiter Some instrument panel gauges are protected against voltage fluctuations that could damage the gauges or give erroneous readings. A **voltage limiter** restricts voltage to the gauges to a particular amount. The limiter contains a heating coil, a bimetal arm, and a set of contacts. When the ignition is in the on or accessory position, the heating coil heats the bimetal arm, causing it to bend and open the contacts. When the arm cools down to the point that the contacts close, the cycle is repeated. The rapid opening and closing of the contacts produces a pulsating voltage at the output terminal averaging about 5 volts. A voltage limiter is also called an instrument voltage regulator (IVR).

Switches

Electrical circuits are usually controlled by some type of switch. Switches do two things. They turn the circuit on or off, or they direct the flow of current in a circuit. Switches can be under the control of the driver or can be self-operating through a condition of the circuit, the vehicle, or the environment.

Contacts in a switch can be of several types, each named for the job they do or the sequence in which they work. A hinged-pawl switch **(Figure 14–42)** is the simplest type of switch. It either makes (allows for) or breaks (opens) current flow in a single conductor or circuit. This type of switch is a **single-pole**, **single-throw (SPST)** switch. The **throw** refers to the number of output circuits, and the **pole** refers to the number of input circuits made by the switch.

Another type of SPST switch is a momentary contact switch **(Figure 14–43)**. The spring-loaded contact on this switch keeps it from closing the circuit except when pressure is applied to the button. A horn switch is this type of switch. Because the spring holds the contacts open, the switch has a further designation: **normally open**. In the case where the contacts are held closed except when the button is pressed, the switch is designated **normally closed**. A normally closed momentary contact

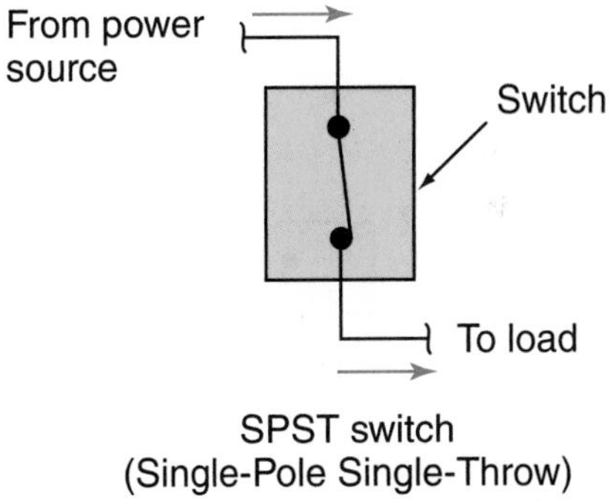

Figure 14–42 An SPST hinged-pawl switch diagram.

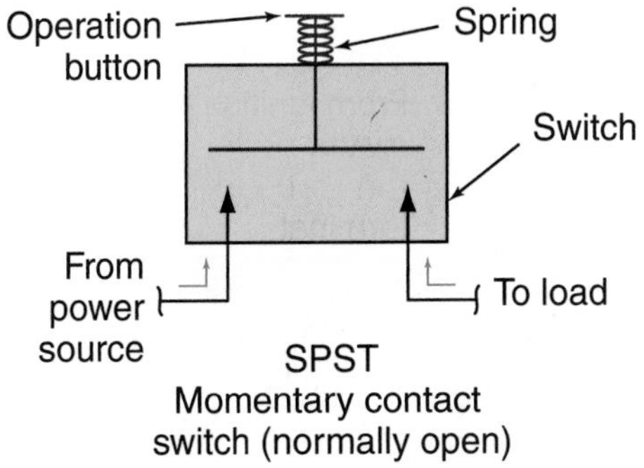

Figure 14–43 An SPST momentary contact switch diagram.

switch is the type of switch used to turn on the courtesy lights when one of the vehicle's doors is opened.

Single-pole, double-throw (SPDT) switches have one wire in and two wires out. **Figure 14–44** shows an SPDT hinged-pawl switch that feeds either the high-beam or low-beam headlight circuit. The dotted lines in the symbol show movement of the switch pawl from one contact to the other.

Switches can be designed with a great number of poles and throws. The transmission neutral start switch shown in **Figure 14–45**, for instance, has two poles and six throws and is referred to as a **multiple-pole, multiple-throw (MPMT)** switch. It contains two movable wipers that move in unison across two sets of terminals. The dotted line shows that the wipers are mechanically linked, or **ganged**. The switch closes a circuit to the starter in either P (park) or N (neutral) and to the back-up lights in R (reverse).

Most switches are combinations of hinged-pawl and push-pull switches, with different numbers of poles and throws. Some special switches are required, however, to satisfy the circuits of modern automobiles. A mercury switch is sometimes used to detect motion in a component, such as the one used in the engine compartment to turn on the compartment light.

Mercury is a very good conductor of electricity. In the mercury switch, a capsule is partially filled with mercury **(Figure 14–46)**. In one end of the capsule are two electrical contacts. The switch is attached to the hood or luggage compartment lid. Normally, the mercury is in the end opposite to the contacts. When the lid is opened, the mercury flows to the contact end and provides a circuit between the electrical contacts.

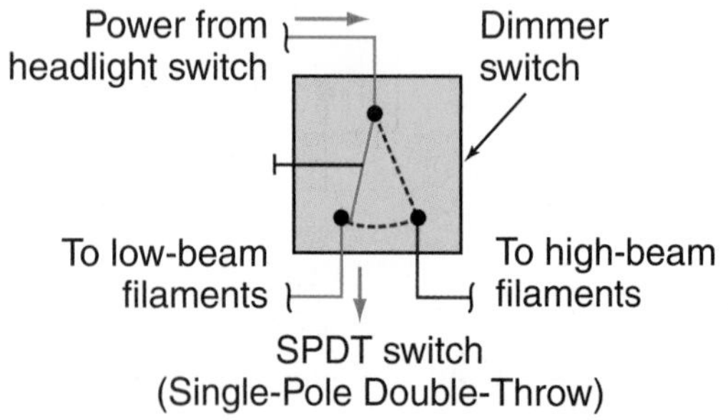

Figure 14–44 An SPDT headlight dimmer switch.

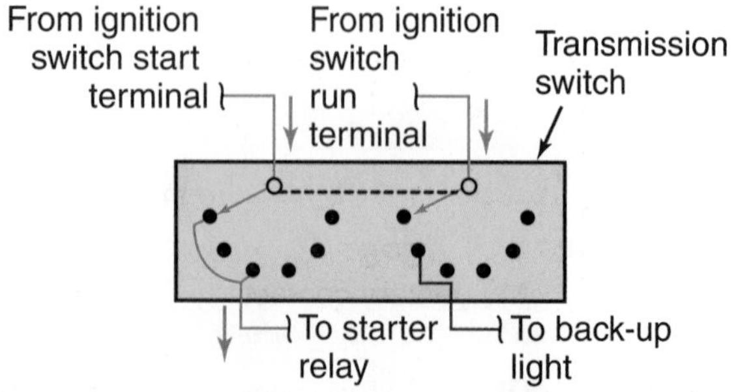

Figure 14–45 An MPMT neutral start safety switch.

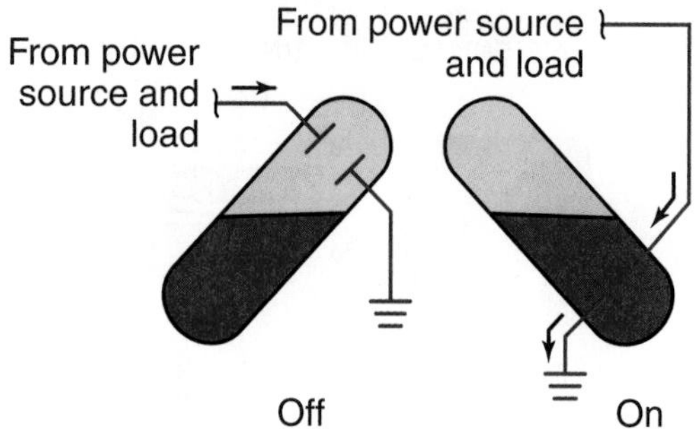

Figure 14–46 A typical mercury switch.

A temperature-sensitive switch usually contains a bimetallic element heated either electrically or by some component in which the switch is used as a **sensor**. When engine coolant is below or at normal operating temperature, the engine coolant temperature sensor is in its normally open condition. If the coolant exceeds the temperature limit, the bimetallic element bends the two contacts together and the switch is closed to the indicator or the instrument panel. Other applications for heat-sensitive switches are time-delay switches and flashers.

Relays

A **relay** is an electric switch that allows a small amount of current to control a high-current circuit **(Figure 14–47)**. When the control circuit switch is open, no current flows to the coil of the relay, so the windings are de-energized. When the switch is closed, the coil is energized, turning the soft iron core into an electromagnet and drawing the armature down. This closes the power circuit contacts, connecting power to the load circuit **(Figure 14–48)**. When the control switch is opened, current stops flowing in the coil and the electromagnet disappears. This releases the armature, which breaks the power circuit contacts.

Solenoids

Solenoids are also electromagnets with movable cores used to change electrical current flow into mechanical movement. They can also close contacts, acting as relays at the same time.

Capacitors

Capacitors (condensers) are constructed from two or more sheets of electrically conducting material with a nonconducting or **dielectric** (antielectric) material placed between them. Conductors are connected to the two sheets. Capacitors are devices that oppose a change of voltage.

If a battery is connected to a capacitor, as shown in **Figure 14–49**, the capacitor will be charged when current flows from the battery to the plates. This current

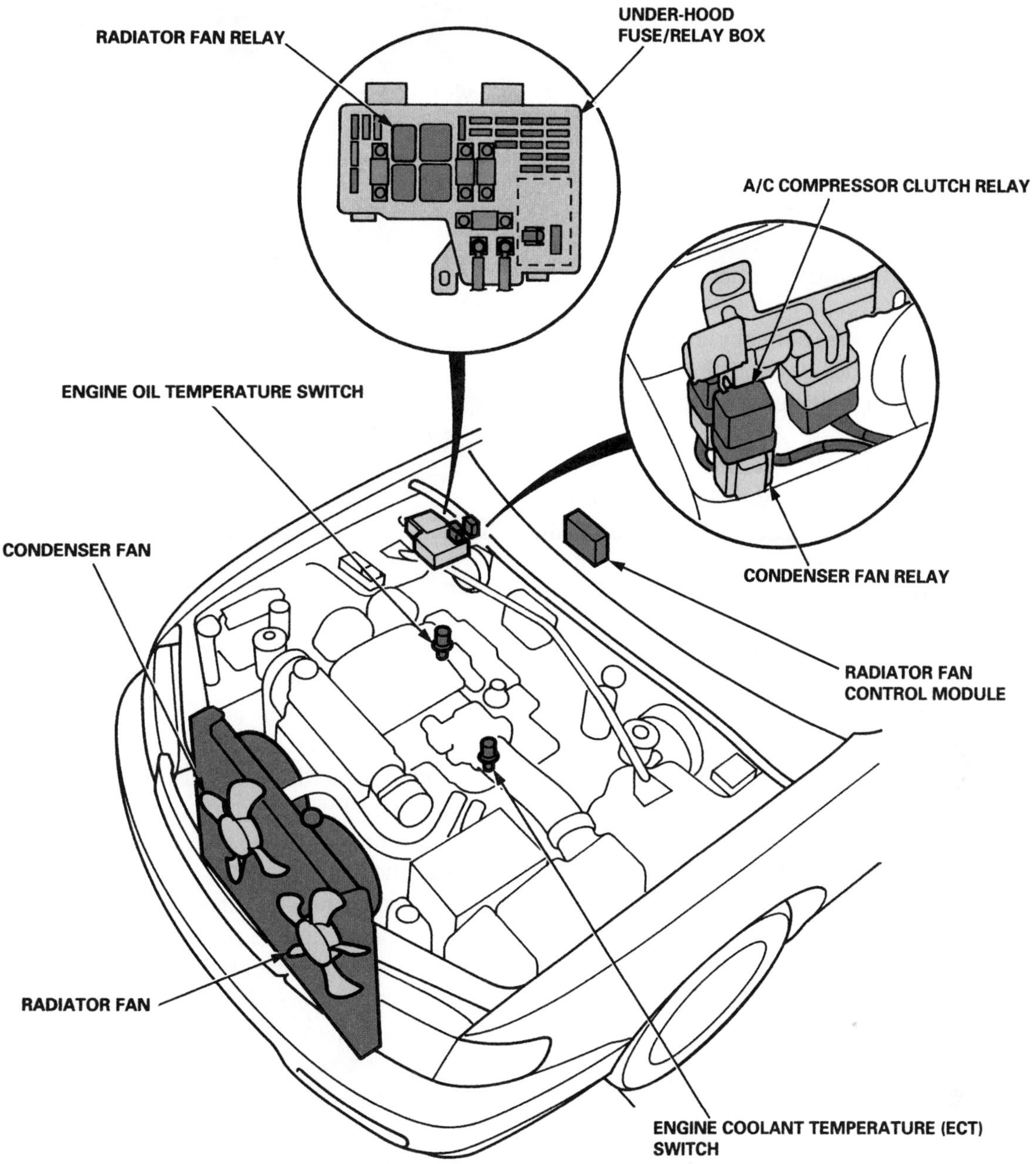

Figure 14-47 An electric cooling fan circuit with control modules and relays. *Courtesy of American Honda Motor Co., Inc.*

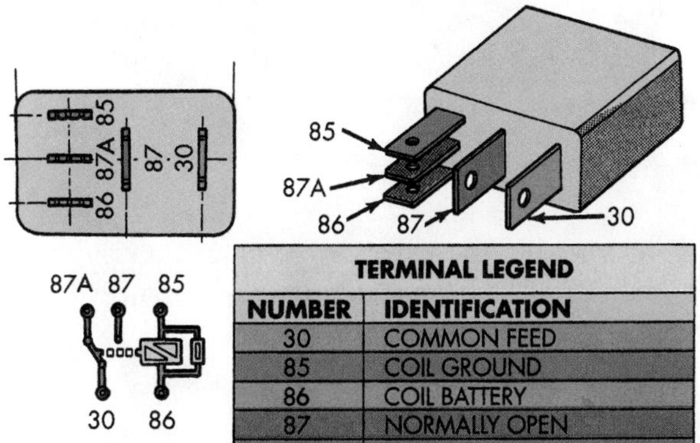

TERMINAL LEGEND	
NUMBER	IDENTIFICATION
30	COMMON FEED
85	COIL GROUND
86	COIL BATTERY
87	NORMALLY OPEN
87A	NORMALLY CLOSED

Figure 14-48 Typical electrical relay inner workings and connections. *Courtesy of DaimlerChrysler Corporation*

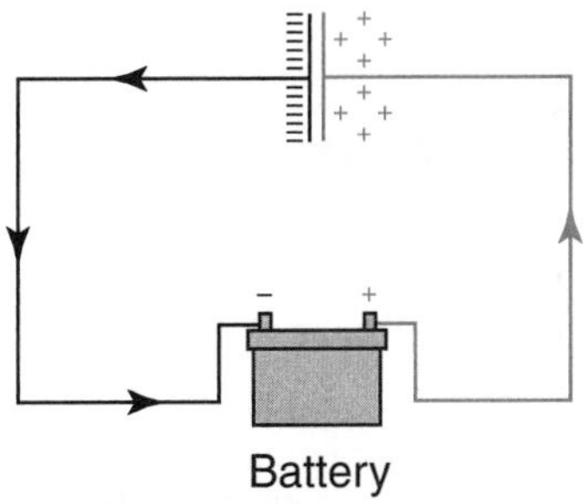

Figure 14-49 Charging a capacitor.

flow will continue until the plates have the same voltage as the battery. At this time, the capacitor is charged.

The capacitor remains charged until a circuit is completed between the two plates. If the charge is routed through a voltmeter, the capacitor will discharge with the same voltage as the battery that charged it. Capacitors are commonly used to filter or clean up voltage signals, such as sound from a stereo. Current can only flow during the period of time that a capacitor is either charging or discharging.

Automotive capacitors are normally encased in metal. The grounded case provides a connection to one set of conductor plates and an insulated lead connects to the other set.

Variable capacitors are called **trimmers** or tuners and are rated very low in capacity because of the reduced size of their conducting plates. For this reason, they are only used in very sensitive circuits, such as radios and other electronic applications.

Wiring

Electrical wires are used to conduct electricity to operate the electrical and electronic devices in a vehicle. There are two basic types of wires used: solid and stranded. **Solid wires** are single-strand conductors. **Stranded wires** are made up of a number of small solid wires twisted together to form a single conductor. Stranded wires are the most commonly used type of wire in an automobile. Electronic units, such as computers, use specially shielded, twisted cable for protection from unwanted induced voltages that can interfere with computer functions **(Figure 14–50)**. In addition, some solid state components use printed circuits.

The current-carrying capacity and the amount of voltage drop in an electrical wire are determined by its length and gauge (size). The wire sizes are established by the Society of Automotive Engineers (SAE), which is the **American wire gauge (AWG)** system. Sizes are identified by a numbering system ranging from number 0 to 20, with number 0 being the largest and number 20 the smallest in a cross-sectional area **(Table 14–2)**. Most automotive wiring ranges from number 10 to 18 and the battery cables are normally at least number 4 gauge. Battery cables are large-gauge wires capable of carrying the high currents for the starter motor.

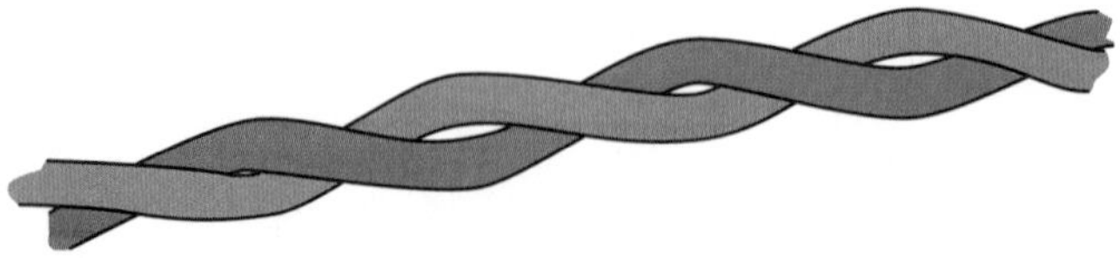

Figure 14-50 Electronic units use specially shielded, twisted cable for protection from unwanted induced voltages that can interfere with computer functions

TABLE 14–2 WIRE GAUGE SIZES

Metric Size (mm^2)	Wire Size	Ampere Capacity
0.5	20	4
0.8	18	6
1.0	16	8
2.0	14	15
3.0	12	20
5.0	10	30
8.0	8	40
13.0	6	50
19.0	4	60

Automotive wiring can also be classified as primary or secondary. Primary wiring carries low voltage to all the electrical systems of the vehicle except to secondary circuits of the ignition system. Secondary wire, also called **high-tension cable**, has extra thick insulation to carry high voltage from the ignition coil to the spark plugs. The conductor itself is designed for low currents.

Wires are commonly grouped together in harnesses. A single-plug harness connector may form the connections for four, six, or more circuits. Harnesses and harness connectors help organize the vehicle's electrical system and provide a convenient starting point for tracking and testing many circuits. Most major wiring harness connectors are located in a vehicle's dash or fire wall area **(Figure 14–51)**.

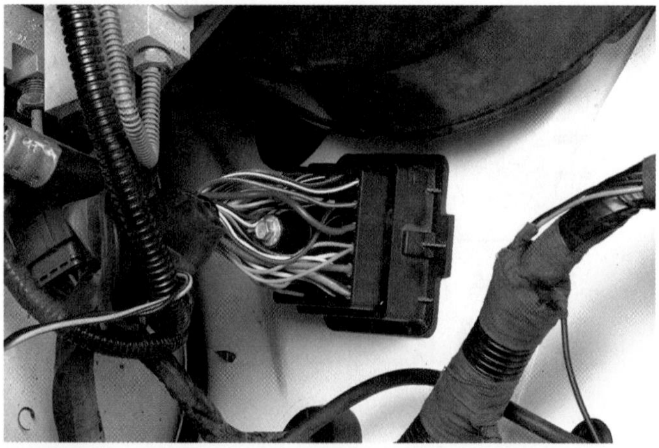

Figure 14-51 A typical front wiring harness bulkhead connector.

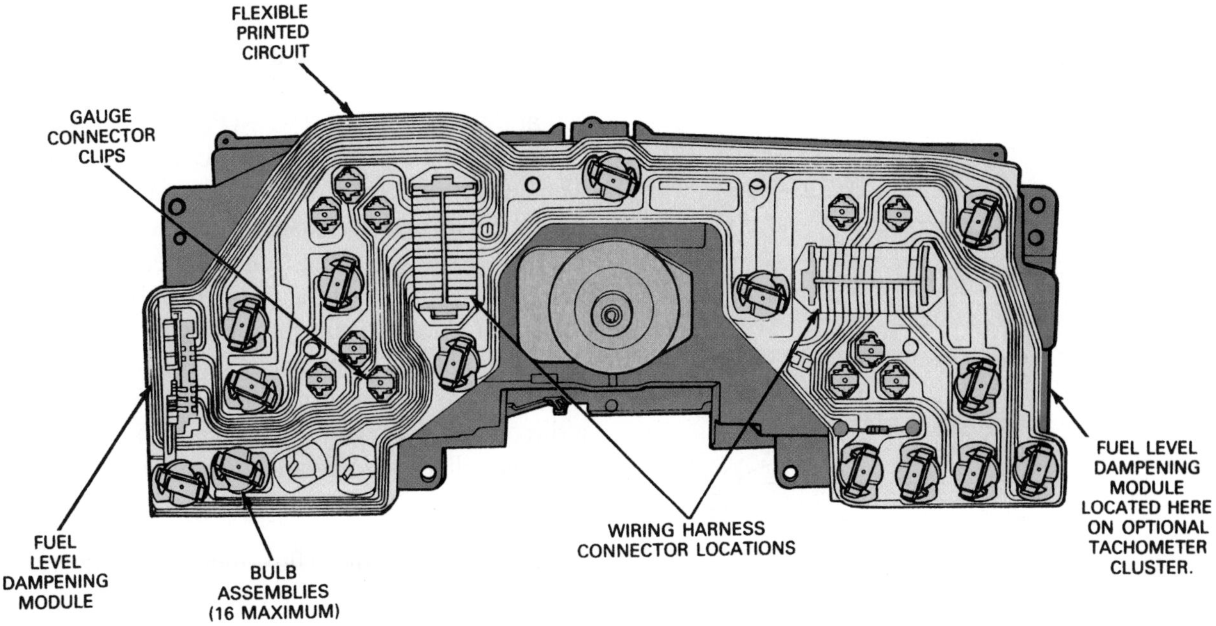

Figure 14–52 A typical printed circuit board.

Flat Wiring As the number of electrical and electronic devices installed in vehicles increases, so does the need for more wiring. More wiring leads to more weight and more potential problems. Also, the size and number of wiring harnesses also increases and spots to carefully tuck them away are limited. For example, a large wiring harness has a difficult time fitting between the roof and the head liner of the vehicle. To run wiring from the front of the vehicle to the rear via the roof, the harnesses are made of small groups of wires, which means there are more harnesses traveling along the roof. Another solution is to make the wiring harnesses flat.

Flat wiring reduces the bulge or thickness of a harness. The copper conductors inside these wires are flattened and no longer round in appearance. In a wiring harness, several flat wires are laid out next to each other and are covered with a plastic insulating material. The plastic offers protection and isolation to the conductors and keeps the harness flat and flexible. In the future, flat wiring may also have electronic components embedded in it. With this design, the wiring harness is not only easier to hide in body panels but also serves as a flexible printed circuit able to be located nearly anywhere in the vehicle.

Printed Circuits Many late-model vehicles use flexible printed circuits **(Figure 14–52)** and printed circuit boards. Both types of printed circuits allow for complete circuits to many components without having to run dozens of wires. Printed circuit boards are typically contained in a housing, such as the engine control module. These boards are not serviceable and in some cases not visible. When these boards fail, the entire unit is replaced.

A flexible printed circuit saves weight and space. It is made of thin sheets of nonconductive plastic onto which conductive metal, such as copper, has been deposited. Parts of the metal are then etched or eaten away by acid. The remaining metal lines form the conductors for the various circuits on the board. The printed circuit is normally connected to the power supply or ground wiring through the use of plug-in connectors mounted on the circuit sheet.

ELECTROMAGNETISM BASICS

Electricity and magnetism are related. One can be used to create the other. Current flowing through a wire creates a magnetic field around the wire. Moving a wire through a magnetic field creates current flow in the wire.

Many automotive components, such as alternators, ignition coils, starter solenoids, and magnetic pulse generators operate using these principles of **electromagnetism**.

Fundamentals of Magnetism

A substance is said to be a **magnet** if it has the property of magnetism—the ability to attract such substances as iron, steel, nickel, or cobalt. These are called magnetic materials.

A magnet has two points of maximum attraction, one at each end of the magnet. These points are called **poles**, with one being designated the north pole and the other the south pole **(Figure 14–53A)**. When two magnets are brought together, opposite poles attract **(Figure 14–53B)**, while similar poles repel each other **(Figure 14–53C)**.

A magnetic field, called a **flux field**, exists around every magnet. The field consists of imaginary lines along

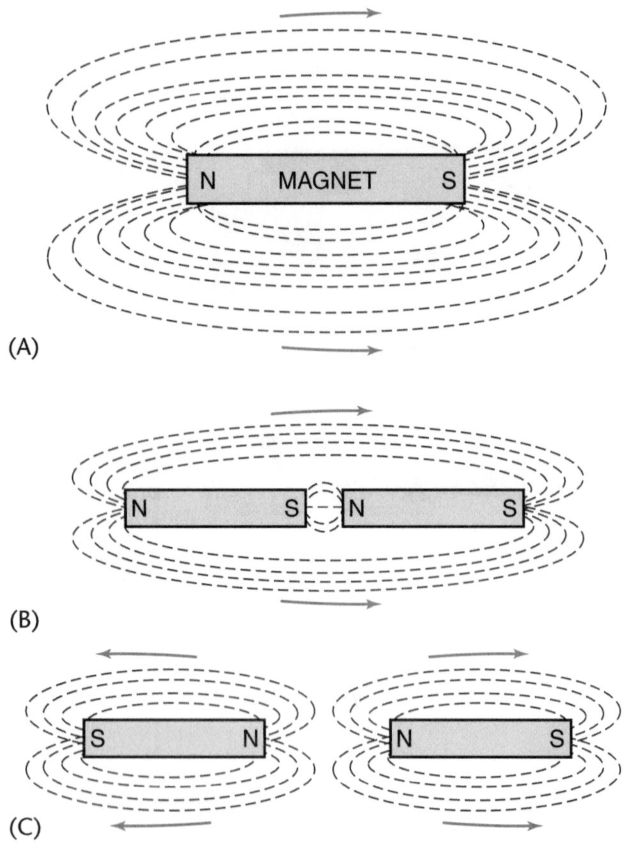

Figure 14-53 (A) In a magnet, lines of force emerge from the north pole and travel to the south pole before passing through the magnet back to the north pole. (B) Unlike poles attract, while (C) similar poles repel each other.

which the magnetic force acts. These lines emerge from the north pole and enter the south pole, returning to the north pole through the magnet itself. All lines of force leave the magnet at right angles to the magnet. None of the lines cross each other. All lines are complete.

Magnets can occur naturally in the form of a mineral called magnetite. Artificial magnets can also be made by inserting a bar of magnetic material inside a coil of insulated wire and passing direct current through the coil. This principle is very important in understanding certain automotive electrical components. Another way of creating a magnet is by stroking the magnetic material with a bar magnet. Both methods force the randomly arranged molecules of the magnetic material to align themselves along north and south poles.

Artificial magnets can be either temporary or permanent. Temporary magnets are usually made of soft iron. They are easy to magnetize but quickly lose their magnetism when the magnetizing force is removed. Permanent magnets are difficult to magnetize. However, once magnetized they retain this property for very long periods.

The earth is a very large magnet, having a north and south pole, with lines of magnetic force running between them. This is why a compass always aligns itself to straight north and south.

In 1820, a simple experiment revealed the existence of a magnetic field around a current-carrying wire. When a compass was held over the wire, its needle aligned itself at right angles to the wire **(Figure 14–54)**. The lines of magnetic force are concentric circles around the wire. The density of these circular lines of force is very heavy near the wire and decreases farther away from the wire. As is also shown in the same figure, the polarity of a current-carrying wire's magnetic field changes depending on the direction the current is flowing through the wire.

Remember, these magnetic lines of force or flux lines do not move or flow around the wire. They simply have a direction, as shown by their effect on a compass needle. These lines of force are always at right angles to the conducting wire.

Flux Density The more flux lines, the stronger the magnetic field at that point. Increasing current increases **flux density**. Also, two conducting wires lying side by side carrying equal currents in the same direction create a magnetic field equal in strength to one conductor carrying twice the current. Adding more wires also increases the magnetic field **(Figure 14–55)**.

Coils Looping a wire into a coil concentrates the lines of force inside the coil. The resulting magnetic field is the sum of all the single-loop magnetic fields added together **(Figure 14–56)**. The overall effect is the same as placing many wires side by side, each carrying current in the same direction.

Magnetic Circuits and Reluctance

Just as current can only flow through a complete circuit, the lines of flux created by a magnet can only occupy a closed magnetic circuit. The resistance that a magnetic

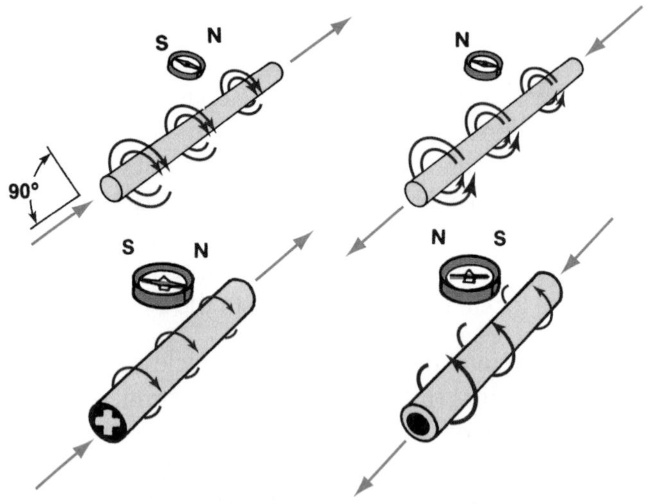

Figure 14-54 When current is passed through a conductor such as wire, magnetic lines of force are generated around the wire at right angles to the direction of the current flow.

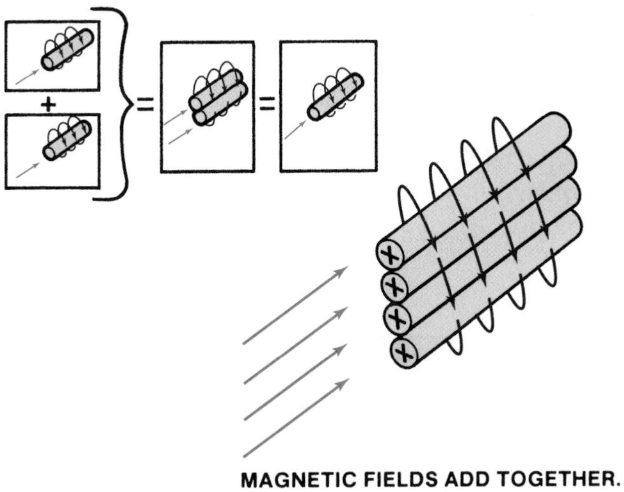

Figure 14-55 Increasing the number of conductors carrying current in the same direction increases the strength of the magnetic field around them.

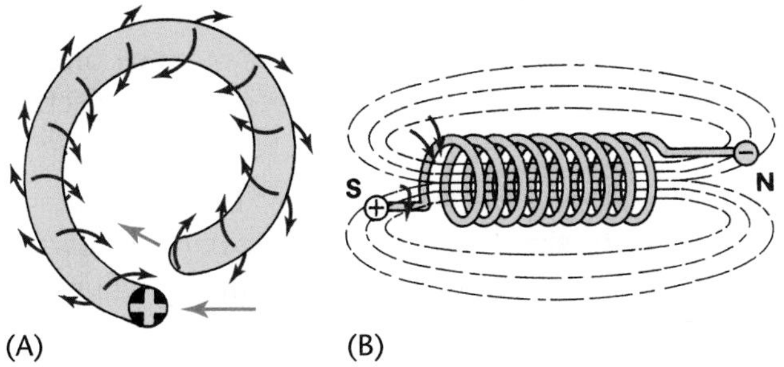

Figure 14-56 (A) Forming a wire loop concentrates the lines of force inside the loop. (B) The magnetic field of a wire coil is the sum of all the single-loop magnetic fields.

circuit offers to a line of flux is called **reluctance**. Magnetic reluctance can be compared to electrical resistance.

Reconsider the coil of wire shown in **Figure 14–56**. The air inside the coil has very high reluctance and limits the strength of the magnetic field that can be produced. However, if an iron core is placed inside the coil, the strength of the magnetic field increases tremendously because the iron core has a very low reluctance **(Figure 14–57)**.

When a coil is wound around an iron core in this manner, it becomes a usable electromagnet. The strength of the magnetic poles in an electromagnet is directly proportional to the number of turns of wire and the current flowing through them.

The equation for an electromagnetic circuit is similar to Ohm's law for electrical circuits. It states that the number of magnetic lines of flux produced is proportional to the ampere-turns divided by the reluctance. To summarize:

- Field strength increases if current through the coil increases.
- Field strength increases if the number of coil turns increases.
- If reluctance increases, field strength decreases.

Induced Voltage

Now that we have explained how current can be used to generate a magnetic field, it is time to examine the opposite effect of how magnetic fields can produce electricity.

Figure 14–58 shows a straight piece of wire with the terminals of a voltmeter attached to both ends. If the wire is moved across a magnetic field, the voltmeter registers a small voltage reading. A voltage has been induced in the wire.

It is important to realize that the wire must cut across the flux lines to induce a voltage. Moving the wire parallel to the lines of flux does not induce voltage.

Voltage can also be induced by holding the wire still and moving the magnetic field at right angles to the wire. This is the exact setup used in a vehicle's alternator. A magnetic field is made to cut across stationary conductors to produce voltage.

The wire becomes a source of electricity and has a polarity or distinct positive and negative end. However, this polarity can be switched depending on the relative direction of movement between the wire and magnetic field **(Figure 14–59)**. This is why charging devices produce alternating current.

The amount of voltage induced depends on four factors:

1. The stronger the magnetic field, the stronger the induced voltage.
2. The faster the field is being cut, the more lines of flux are cut and the stronger the voltage induced.

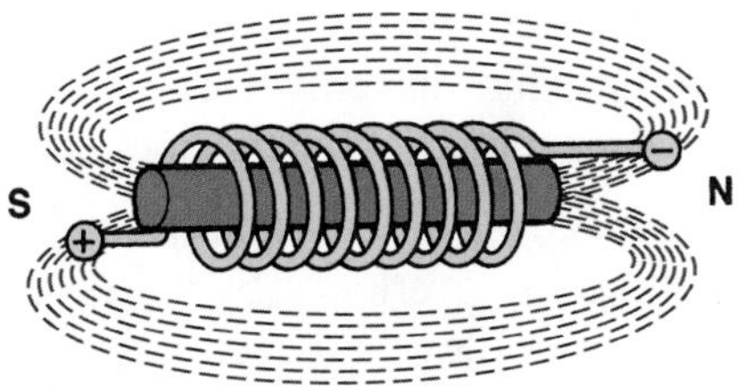

Figure 14-57 Placing a soft iron core inside a coil greatly reduces the reluctance of the coil and creates a usable electromagnet.

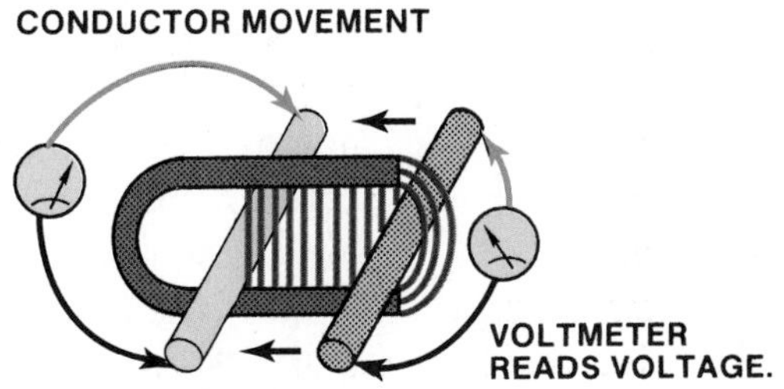

Figure 14-58 Moving a conductor so it cuts across the magnetic lines of force induces a voltage in the conductor.

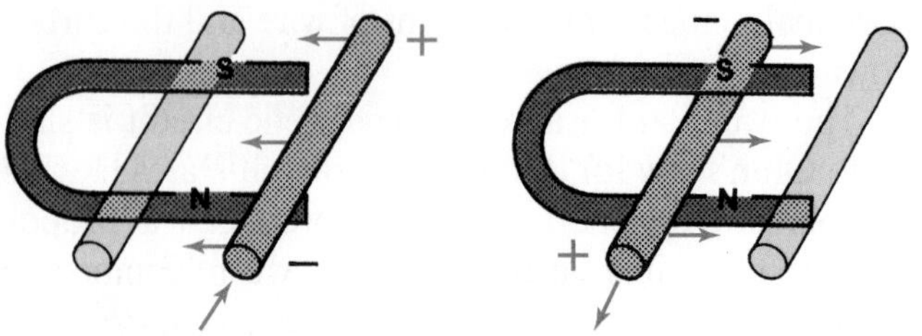

Figure 14-59 The polarity of the induced voltage depends on the direction in which the conductor moves as it cuts across the magnetic field.

3. The greater the number of conductors, the greater the voltage induced.
4. The closer the conductor(s) and magnetic field are to right angles (perpendicular) to one another, the greater the induced voltage.

The importance of electromagnetism and induced voltage is clearly seen in chapters dealing with starting, charging, ignition, and electronic control systems.

KEY TERMS

Alternating current (AC)
American wire gauge (AWG)
Ampere
Blade fuse
Capacitor
Cartridge fuse
Ceramic fuse
Chassis ground
Circuit breaker
Closed circuit
Conductor
Continuity
Current
Dielectric
Direct current (DC)
Electronics
Electromagnetism
Electromotive force (EMF)
Fixed value resistors
Flux density
Flux field
Fusible link
Ganged
Ground wire
Grounding
High-tension cable
Induction
Insulator
Magnet
Magnetism
Maxi-fuse
Multiple-pole, multiple-throw (MPMT)
Normally closed
Normally open
Ohm
Ohm's law
Open circuit
Parallel circuit
Poles
Potentiometer
Power
Relay
Reluctance
Resistance
Resistor wire
Rheostat
Sensor
Series circuit
Series-parallel circuit
Single-pole, double-throw (SPDT)
Single-pole, single-throw (SPST)
Solenoid
Solid wire
Spade fuse
Stepped resistor
Stranded wire
Tapped resistor
Thermistor
Throw
Trimmer
Variable resistor
Voltage
Voltage drop
Voltage limiter
Watt

SUMMARY

- For electrical flow to occur there must be an excess of electrons in one place, a lack of electrons in another, and a path between the two places.
- Voltage is the force or pressure in an electrical circuit. A voltage drop across a load in a circuit indicates that work is being done.
- Current is measured in amps. This is a measurement of the actual flow rate of electrons in an electrical circuit.
- Resistance is measured in ohms. This is a measurement of the size of the restriction to current flow. The more resistance there is, the less current will be able flow through the circuit.
- The mathematical relationship between current, resistance, and voltage is expressed in Ohm's law, $E = IR$, in which voltage is measured in volts, current in amperes, and resistance in ohms.
- The mathematical relationship between current, voltage, and power is expressed in Watt's law, $P = E \times I$. Power is measured in watts or kilowatts (1,000 watts).
- Three basic types of circuits are used in automobile wiring systems: series circuits, parallel circuits, and series-parallel circuits.
- Electrical schematics are diagrams with electrical symbols that show the parts and how electrical current flows through the vehicle's electrical circuits. They are used in troubleshooting.
- The strength of an electromagnet depends on the number of current-carrying conductors and what is in the core of the coil. Inducing a voltage requires a magnetic field producing lines of force, conductors that can be moved, and movement between the conductors and the magnetic field so the lines of force are cut.
- Fuses, fuse links, maxi-fuses, and circuit breakers protect circuits against overloads. Switches control on/off and direct current flow in a circuit. A relay is an electric switch. A solenoid is an electromagnetic switch that translates current flow into mechanical movement. Resistors limit current flow.

REVIEW QUESTIONS

1. What is the name for the formula $E = I \times R$?
2. *True or False?* In a series circuit, circuit current is the same throughout the circuit.
3. What are the two types of wires and which is the most commonly used in automobiles?
4. What four factors determine how much voltage is induced by a magnet?
5. Variable capacitors are called ______.
6. What is the process called in which a conductor cuts across a magnetic field and produces a voltage?
7. What happens in an electrical circuit when the resistance increases?
8. What is the difference between voltage and current?
9. *True or False?* The strength of the magnetic poles in an electromagnet decreases with an increase in the number of turns of wire and the current flowing through them.
10. What is an SPST switch?
11. What is a normally closed switch?
12. Technician A says magnetism is a source of electrical energy in an automobile. Technician B says chemical reaction is a source of electrical energy in an automobile. Who is correct?

    a. Technician A   **c.** Both A and B
    b. Technician B   **d.** Neither A nor B

13. While discussing resistance, Technician A says current increases with a decrease in resistance. Technician B says current decreases with an increase in resistance. Who is correct?

    a. Technician A   **c.** Both A and B
    b. Technician B   **d.** Neither A nor B

14. What is a thermistor?
15. Which type of resistor is commonly used in automotive circuits?

    a. fixed value   **c.** variable
    b. stepped   **d.** all of the above

16. Which of the following is a characteristic of all parallel circuits?

    a. Total circuit resistance is always higher than the resistance of the leg with the lowest total resistance.
    b. The current through each leg is different if the resistance values are different.
    c. The sum of the resistance on each leg equals the total circuit resistance.
    d. The voltage applied to each leg of the circuit will be dropped across the legs if there are loads in series with the parallel circuit.

17. What is the current in a 12-volt circuit with two 6-ohm resistors connected in parallel?

    a. 2 amps   **c.** 6 amps
    b. 4 amps   **d.** 12 amps

18. While discussing a 12-volt circuit with three resistors (a 3 ohm, a 6 ohm, and a 2 ohm) in parallel, Technician A says the total resistance of the circuit is 1 ohm. Technician B says the current flow through the 6-ohm resistor is 12 amps. Who is correct?

    a. Technician A   **c.** Both A and B
    b. Technician B   **d.** Neither A nor B

19. Which of the following does *not* affect the resistance of a uniform, circular copper wire?

    a. The length of the wire
    b. The diameter of the wire
    c. The location of the wire
    d. The temperature of the wire

20. Technician A says a 12-volt light bulb that draws 12 amps has a power output of 1 watt. Technician B says a motor that has 1 ohm of resistance has a power rating of 144 watts if it is connected to a 12-volt battery. Who is correct?

    a. Technician A   **c.** Both A and B
    b. Technician B   **d.** Neither A nor B

15

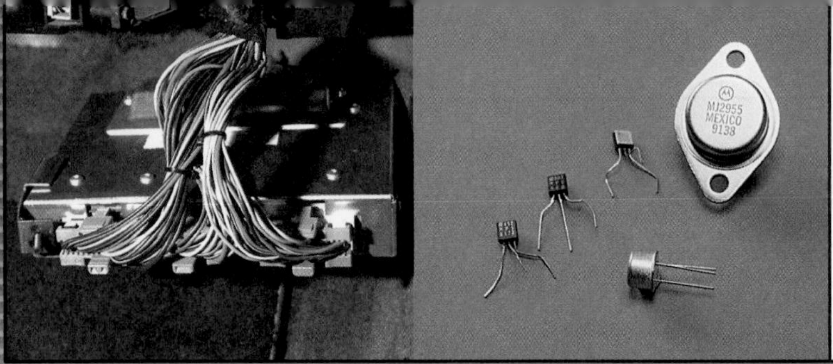

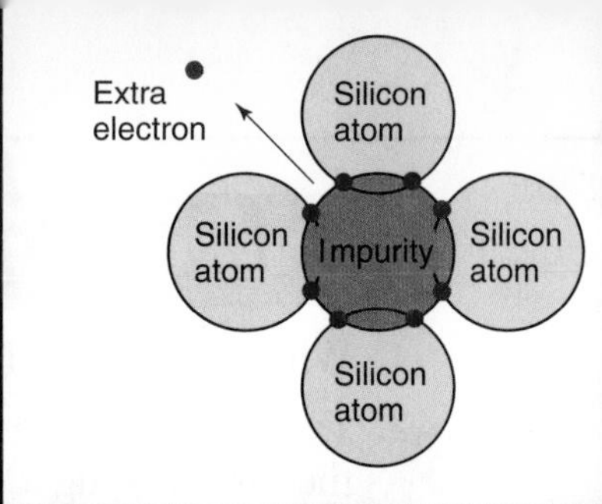

BASICS OF ELECTRONICS

OBJECTIVES

■ Describe how semiconductors, diodes, and transistors work. ■ Explain the principles of operation for common electronic circuits. ■ Explain the principle of multiplexing. ■ Describe the basic function of the central processing unit (CPU). ■ List and describe the functions of the various sensors used by computers. ■ Describe the principle of analog and digital signals. ■ Explain the principle of computer communications. ■ Summarize the function of a binary code. ■ Name the various memory systems used in automotive microprocessors. ■ List and describe the operation of output actuators. ■ Describe the basic electronic logic circuits.

Computerized engine controls and other features of today's cars would not be possible if it were not for electronics. For purposes of clarity, let us define electronics as the technology of controlling electricity. Electronics has become a special technology beyond electricity. Transistors, diodes, semiconductors, integrated circuits, and solid-state devices are all considered to be part of electronics rather than just electrical devices. Keep in mind that all the basic laws of electricity apply to all electronic controls.

Although it is not necessary to understand all of the concepts of computer operation to service the systems they control, a good knowledge will allow you to be more productive.

COMPUTERS

A computer is an electronic device that stores and processes data. It is also capable of operating other computers. The operation of a computer is divided into four basic functions:

1. **Input:** A signal sent from an input device. The device can be a sensor or a button activated by the driver, technician, or mechanical part.
2. **Processing:** The computer uses the input information and compares it to programmed instructions. This information is processed by logic circuits in the computer.
3. **Storage:** The program instructions are stored in the computer's memory. Some of the input signals are also stored in memory for processing later.
4. **Output:** After the computer has processed the sensor input and checked its programmed instructions, it will issue commands to various output devices. These output devices may be instrument panel displays or output actuators. The output of one computer may also be an input to another computer.

Semiconductors

A semiconductor is a material or device that can function as either a conductor or an insulator, depending on how its structure is arranged. Semiconductor materials have less resistance than an insulator but more resistance than a conductor. Some common semiconductor materials include silicon (Si) and germanium (Ge).

In semiconductor applications, materials have a crystal structure. This means that their atoms do not lose and gain electrons as the atoms in conductors do. Instead, the atoms in these semiconductor materials share outer electrons with each other. In this type of atomic structure, the electrons are tightly held and the element is stable.

Because the electrons are not free, crystals cannot conduct current. These materials are called **electrically inert** materials. To function as semiconductors, a small amount of a trace element must be added. The addition of these elements, called **impurities**, allows the material to function as a semiconductor. The type of impurity added determines the type of semiconductor produced.

N-Type Semiconductors N-type semiconductors have loose, or excess, electrons. They have a negative charge.

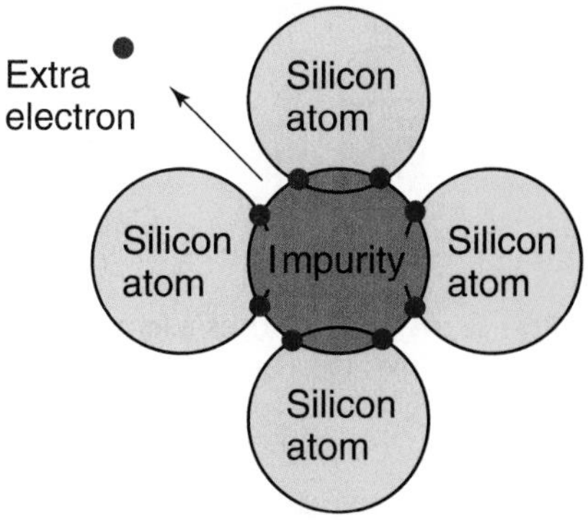

Figure 15-1 Atomic structure of N-type silicon semiconductor.

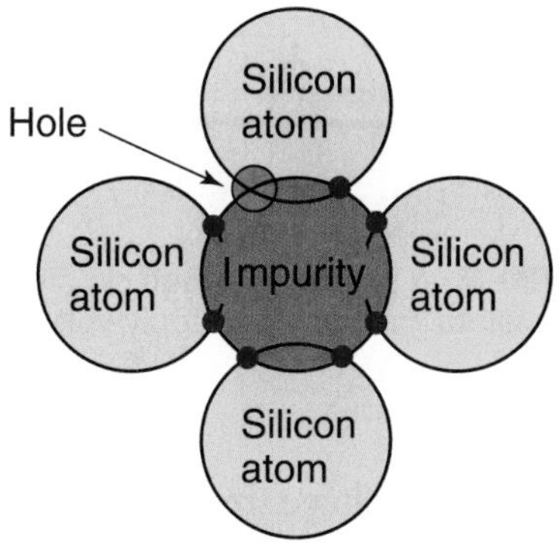

Figure 15-2 Atomic structure of P-type silicon semiconductor.

This enables them to carry current. N-type semiconductors are produced by adding an impurity with five electrons in the outer ring (called pentavalent atoms). Four of these electrons fit into the crystal structure, but the fifth is free. This surplus of electrons produces the negative charge. **Figure 15–1** shows an example.

P-Type Semiconductors P-type semiconductors are positively charged materials, which enables them to carry current. P-type semiconductors are produced by adding an impurity with three electrons in the outer ring (trivalent atoms). When this element is added to silicon or germanium, the three outer electrons fit into the pattern of the crystal, leaving a hole where a fourth electron would fit. This hole is actually a positively charged empty space that carries the current in the P-type semiconductor. **Figure 15–2** shows an example of a P-type semiconductor.

Hole Flow. Understanding how semiconductors carry current without losing electrons requires understanding the concept of hole flow. The holes in a P-type semiconductor, being positively charged, attract electrons. Although the electrons cannot be freed from their atom, they can rearrange their pattern and fill a hole in a nearby atom. Whenever this happens, the moving electron creates another hole. This hole in turn is filled by another electron, and the process continues. The electrons move toward the positive side of the structure, and the holes move to the negative side. This is the principle by which semiconductors carry current.

Semiconductor Uses. Because semiconductors have no moving parts, they seldom wear out or need adjustment. Semiconductors, or solid-state devices, are also small, require little power to operate, are reliable, and generate very little heat. For all these reasons, semiconductors are being used in many applications. It is important to note that all semiconductors must be placed in a circuit with a current-limiting resistance.

Diodes and Transistors Because a semiconductor can function as both a conductor and an insulator, it is very useful as a switching device. How a semiconductor functions depends on the way current flows (or tries to flow) through it. Two common semiconductor devices are diodes and transistors.

Diodes are used for isolation of components or circuits, clamping, or rectification of AC to DC. Transistors are used for amplification or switching.

Diodes. The diode is the simplest semiconductor device. The most commonly used diodes are regular diodes, LEDs, zener diodes, clamping diodes, and photo diodes.

A **diode** allows current to flow in one direction **(Figure 15–3)** but not in the opposite direction. Therefore, it can function as a switch, acting as either conductor or insulator, depending on the direction of current flow. Diodes turn on when the polarity of the current is correct and turn off when the flow has the wrong polarity.

One application of diodes is in the alternator, where they function as one-way valves for current flow. All charging systems, whether AC generators (alternators) or DC generators, produce alternating current. In a DC generator, current is rectified (changed from AC to DC) by a rotating commutator and a set of brushes. In an AC generator, current is rectified by the use of diodes. The diodes

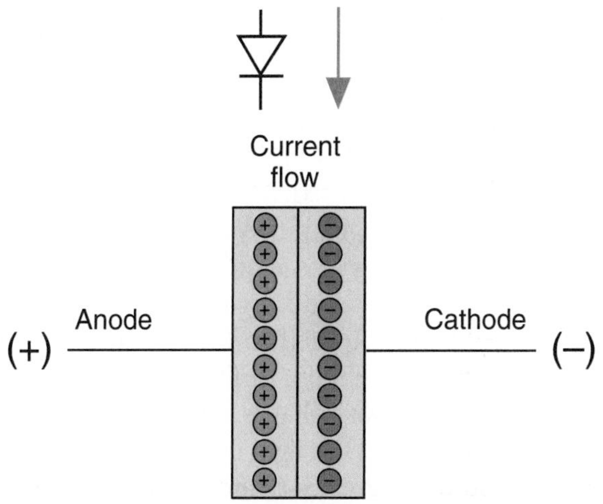

Figure 15-3 A diode and its schematical symbol.

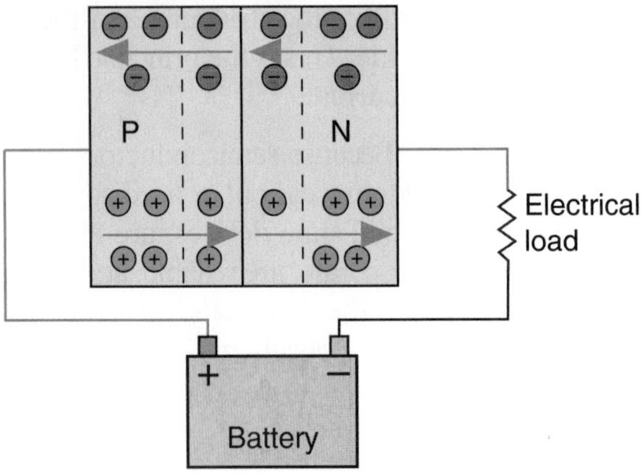

Figure 15-4 Forward biased voltage allows current flow through the diode.

are arranged so that current can leave the AC generator windings in one direction only (as direct current).

Inside a diode are small positive and negative areas, which are separated by a thin boundary area. The boundary area is called the **PN junction.** When a diode is placed in a circuit with the positive side of the circuit connected to the positive side of the diode and the negative side of the circuit connected to the negative side of the circuit, the diode is said to have **forward bias (Figure 15–4).**

Unlike electrical charges are attracted to each other and like charges repel each other. Therefore, the positive charge from the circuit's power supply is attracted to the negative side of the circuit. The voltage in the circuit is much stronger than the charges inside the diode and causes the charges inside the diode to move. The diode's P conductive material is repelled by the positive charge of the circuit and is pushed toward the N material and the N material is pushed toward the P. This causes the PN junction to become a conductor, allowing the circuit's current to flow.

When **reverse bias** is applied to the diode, the P and N areas of the diode are connected to opposite charges. Since opposites attract, the P material moves toward the negative part of the circuit, whereas the N material moves toward the positive part of the circuit. This empties the PN junction and current flow stops.

Light-emitting diodes (LEDs) are much the same as regular diodes **(Figure 15–5)**. However, they emit light when they are forward biased. LEDs are very current sensitive and can be damaged if they are subjected to more than 50 mAmps. LEDs also require higher amounts of voltages to turn on than do regular diodes. Normally 1.5 to 2.5 volts are required to forward bias an LED enough to cause it to light up. LEDs also offer much less resistance to reverse bias voltages. High reverse bias voltages may cause the LED to light or cause it to burn up.

Zener diodes (Figure 15–6) are a more complex type of diode. They are used to regulate voltage in a circuit and are available in many voltage sizes, typically from 2 to 200 volts. Although zener diodes work in the same way as regular diodes when forward biased, they are placed backwards in a circuit. When the zener voltage is reached, the zener diode begins to allow current flow but maintains a voltage drop across itself. This voltage drop is regulated at whatever voltage the diode is rated. A zener diode functions just like a regular diode until a certain voltage is reached. When the voltage level reaches this point, the zener diode allows current to flow in the reverse direction. Zener diodes are often used in electronic voltage regulators.

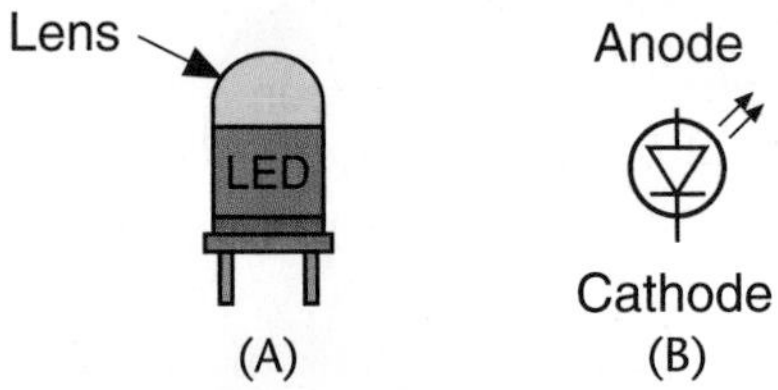

Figure 15-5 (A) An LED uses a lens to emit the light generated by current flow. (B) The schematical symbol for an LED.

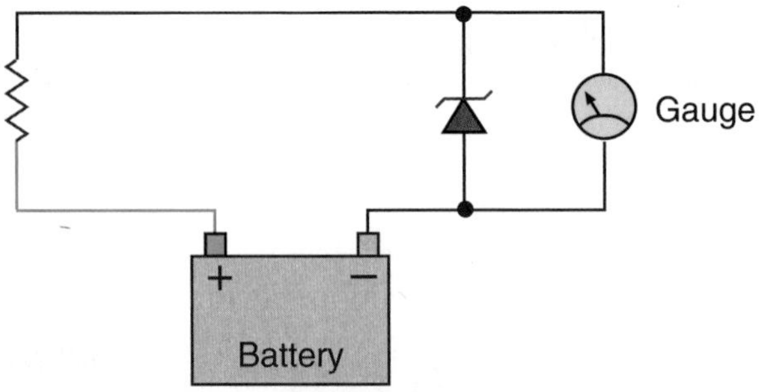

Figure 15-6 A simplified gauge circuit with a zener diode used to maintain a constant supply voltage to the gauge.

Whenever the current flow through a coil of wire (such as used in a solenoid or relay) stops, a voltage surge or spike is produced. This surge results from the collapsing of the magnetic field around the coil. The movement of the field across the winding induces a very high voltage spike that can damage electronic components. In the past, a capacitor was used as a "shock absorber" to prevent component damage from this surge. On today's vehicles, a clamping diode is commonly used to prevent this voltage spike. By installing a **clamping diode** in parallel to the coil, a bypass is provided for the electrons during the time the circuit is opened **(Figure 15–7)**.

An example of the use of clamping diodes is on some air-conditioning compressor clutches. Because the clutch operates by electromagnetism, opening the clutch coil produces a voltage spike. If the spike was left unchecked, it could damage the clutch coil relay contacts or the vehicle's computer. The clamping diode is connected to the circuit in reverse bias.

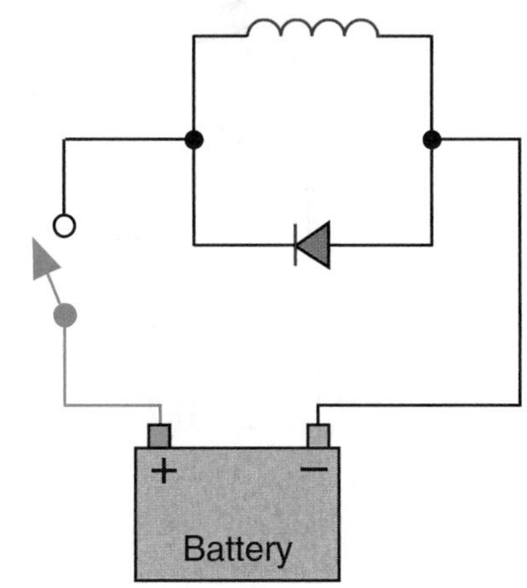

Figure 15-7 A clamping diode in parallel to a coil prevents voltage spikes when the switch is opened.

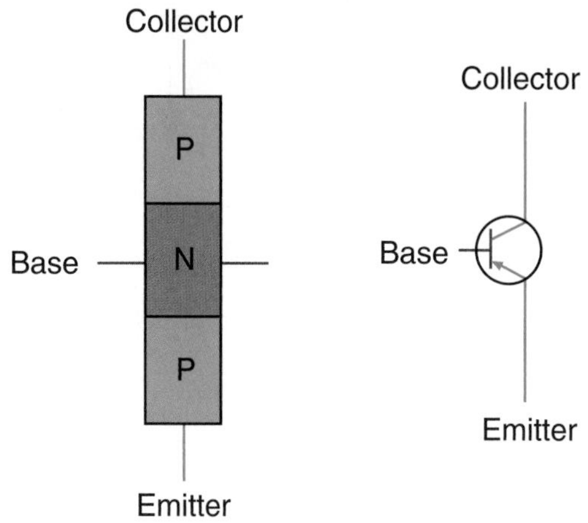

Figure 15-9 A PNP transistor and its schematical symbol.

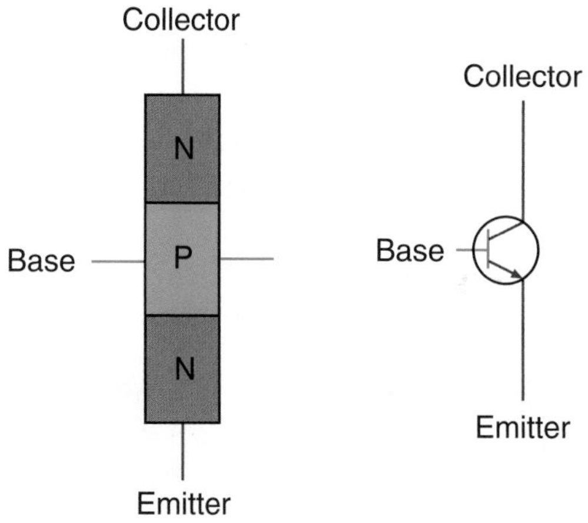

Figure 15-10 A NPN transistor and its schematical symbol.

Transistor. Another semiconductor commonly used in the automotive industry is the **transistor**. These devices are used in place of switches and relays in electronic circuits. They are used to turn an electrical circuit on or off and are controlled by another electrical circuit. Perhaps the easiest way to describe the action of a transistor is to say it is a switch controlled by conditions. The conditions are the voltages present at the parts of the transistor.

A transistor is produced by joining three sections of semiconductor materials. Like the diode, it is very useful as a switching device, functioning as either a conductor or an insulator. **Figure 15–8** shows two designs of transistors, which come in many different sizes and types.

A transistor resembles a diode with an extra side. It can consist of two P-type materials and one N-type material or two N-type materials and one P-type material. These are called **PNP** and **NPN** types. In both types, junctions occur where the materials are joined. **Figure 15–9** shows a PNP junction transistor in a circuit, while **Figure 15–10** shows an NPN transistor. Notice that each of the three sections has a lead connected to it, which allows any of the three sections to be connected to the circuit. The different names for the legs of the transistor are the **emitter**, **base**, and **collector**.

The center section is called the base and is the controlling part of the circuit or where the larger controlled part of the circuit is switched on and off. The path to ground for the base circuit is through the emitter leg of the transistor. A resistor is normally in the base circuit to keep current flow low. This prevents damage to the transistor. The emitter and collector make up the controlled circuit. When a transistor is drawn in electrical schematics, the arrow on the emitter points to the direction of current flow. When positive voltage is applied to the base of an NPN transistor, the collector-to-emitter circuit is turned on **(Figure 15–11)**.

The base of a PNP transistor is controlled by its ground. Current flows in from the emitter through the base, then to ground. A negative voltage or ground must be applied to the base to turn on a PNP transistor. When this transistor is on, the circuit from the emitter to the collector is completed.

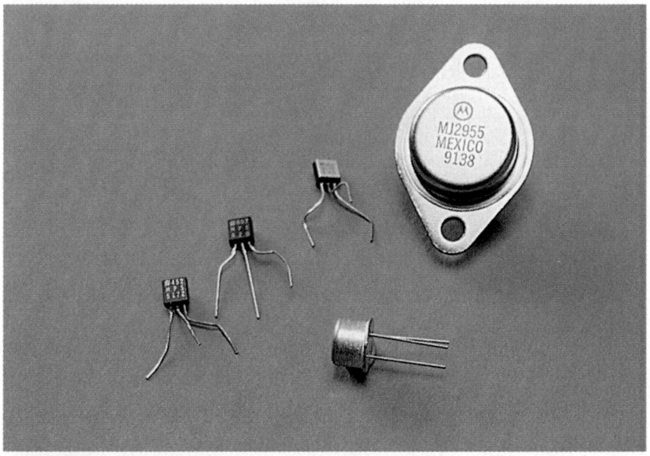

Figure 15-8 Typical transistors.

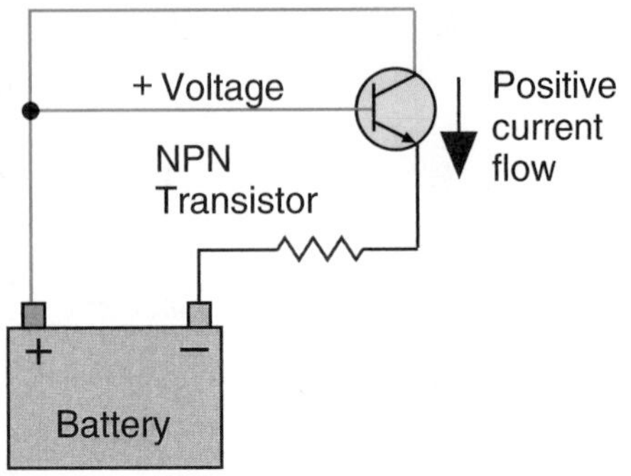

Figure 15-11 When positive voltage is applied to the base of an NPN transistor, current flows through the collector and the emitter.

Transistors can also serve as variable switches. By varying the voltage applied to the base, the completeness of the emitter and collector circuit will also vary. This is done simply by the presence of a variable resistor in the base circuit. This type of arrangement is typically part of a light-dimming circuit.

Semiconductor Circuits

One transistor or diode is limited in its ability to do complex tasks. However, when many semiconductors are combined into a circuit, they can perform complex functions. An example of this is the electronic voltage regulator.

The heart of the electronic regulator is a zener diode. As mentioned, the zener diode has the ability to conduct in the reverse direction without being damaged. The zener diode used in regulators is doped, so it conducts in reverse once the maximum battery voltage has been achieved.

INTEGRATED CIRCUITS

An **integrated circuit** is simply a large number of diodes, transistors, and other electronic components, such as resistors and capacitors, all mounted on a single piece of semiconductor material **(Figure 15–12)**. These circuits are commonly called **ICs** or **chips**. These circuits are extremely small. Circuitry that used to take up entire rooms can now fit into a pocket. The principles of semiconductor operation remain the same in integrated circuits.

The size of chips is constantly becoming smaller, which means electronics is no longer confined to the simple tasks, such as rectifying alternator current. Enough transistors, diodes, and other solid-state components can be installed in a car to make logical decisions and issue commands to automobile systems, such as the ignition, fuel injection, transmission, brake, and suspension systems.

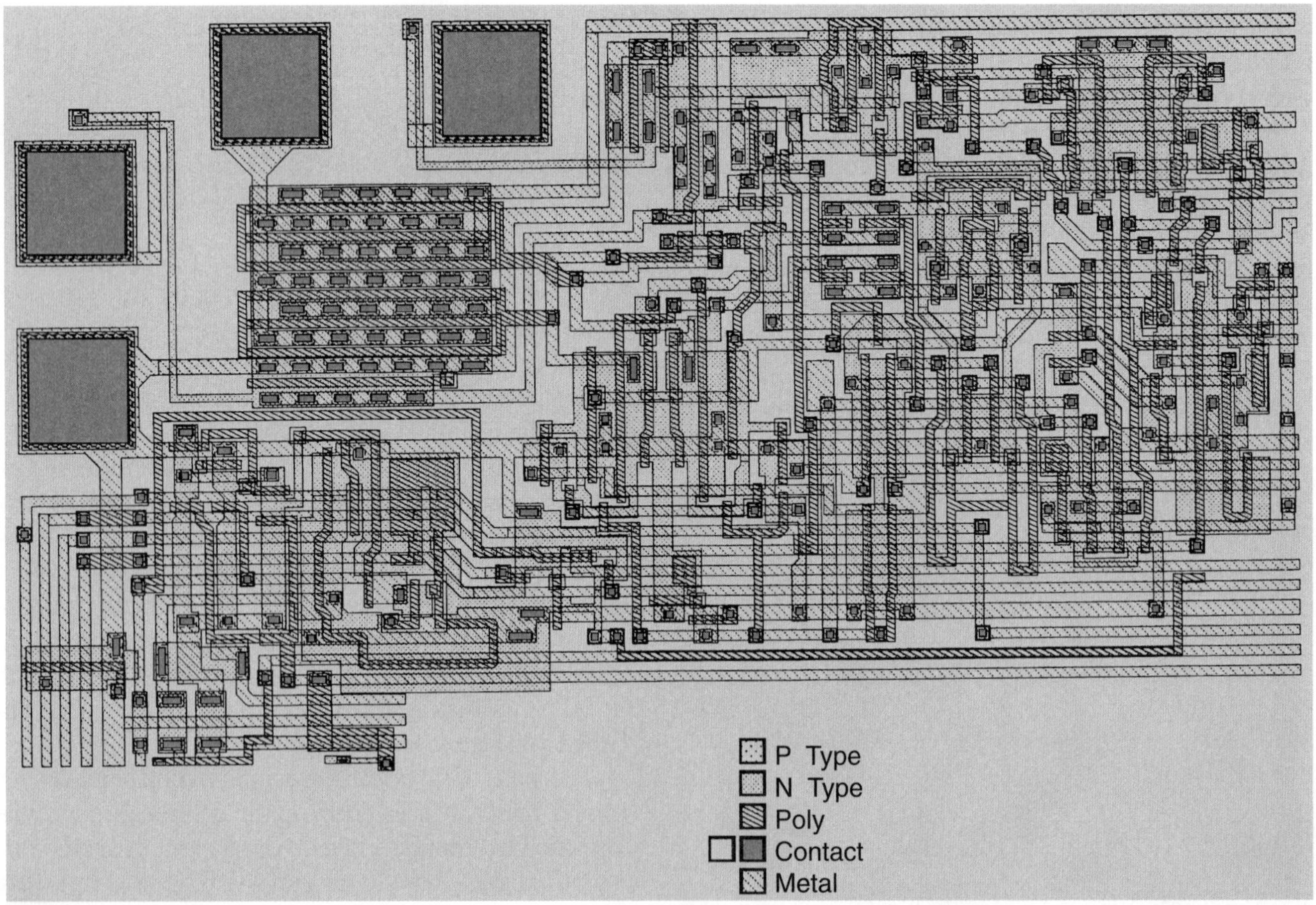

Figure 15-12 An enlarged illustration of an IC with thousands of transistors, diodes, resistors, and capacitors. The actual size of this chip can be less than ¼ of an inch square.

MICROPROCESSOR OPERATION

The microprocessor has taken over many of the tasks in cars and trucks that were formerly performed by vacuum, mechanical, or electromechanical devices. When properly programmed, they can carry out explicit instructions with blinding speed and almost flawless consistency.

A typical electronic control system is made up of sensors, actuators, and related wiring that is tied into a central processor called a **microprocessor (Figure 15–13)** or microcomputer (a smaller version of a computer).

The **central processing unit (CPU)** is the brain of a computer and is constructed of thousands of transistors placed on a small chip. The CPU brings information into and out of the computer's memory. The input information is processed in the CPU and checked against the program in the computer's memory. The CPU also checks the memory for any other information regarding programmed parameters. The information obtained by the CPU can be altered according to the instructions of the program. The CPU may be ordered to make logic decisions on the information received. Once these decisions, or calculations, are made, the CPU sends out commands to make the required corrections or adjustments to the system being controlled.

Sensors

The CPU receives inputs that it checks against programmed values. Depending on the input, the computer controls the actuator(s) until the programmed results are obtained. The inputs can come from other computers, the driver, the technician, or through a variety of sensors.

Driver input signals are usually provided by momentarily applying a ground through a switch. The computer receives this signal and performs the desired functions. For example, if the driver wishes to reset the trip odometer on a digital instrument panel, a reset button is depressed. This switch provides a momentary ground that the computer receives as an input and sets the trip odometer to zero.

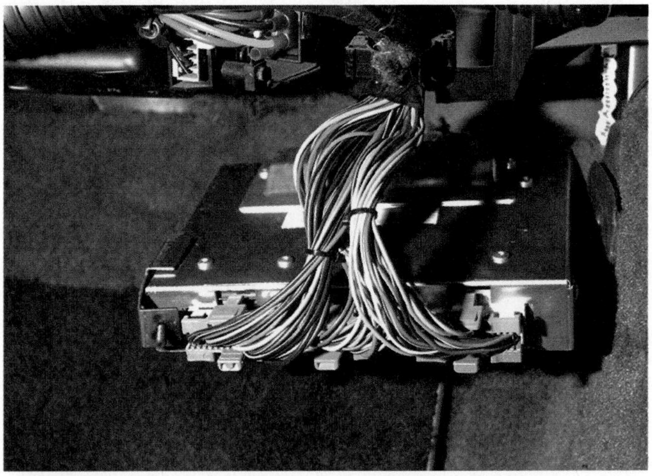

Figure 15–13 A typical automotive computer.

Switches can be used as inputs for any operation that only requires a yes-no, or on-off, condition. Other inputs include those supplied by means of a sensor and those signals returned to the computer in the form of **feedback**. Feedback means that data concerning the effects of the computer's commands are fed back to the computer as an input signal.

If the computer sends a command signal to actuate an output device, a feedback signal may be sent back from the actuator to inform the computer that the task was performed. The feedback signal confirms both the position of the output device and the operation of the actuator. Another form of feedback is for the computer to monitor voltage as a switch, relay, or other actuator is activated. Changing positions of an actuator should result in predictable changes in the computer's voltage sensing circuit. The computer may set a diagnostic code if it does not receive the correct feedback signal.

All sensors perform the same basic function. They detect a mechanical condition (movement or position), chemical state, or temperature condition and change it into an electrical signal that can be used by the computer to make decisions. The CPU makes decisions based on information it receives from the sensors. Each sensor used in a particular system has a specific job to do (for example, monitor throttle position, vehicle speed, manifold pressure). Together these sensors provide enough information to help the computer form a complete picture of vehicle operation. Even though there are a variety of different sensor designs, they all fall under one of two operating categories: reference voltage sensors or voltage-generating sensors.

Reference voltage (Vref) sensors provide input to the computer by modifying or controlling a constant, predetermined voltage signal. This signal, which can have a reference value from 5 to 9 volts, is generated and sent out to each sensor by a reference voltage regulator located inside the CPU. Because the computer knows a certain voltage value has been sent out, it can indirectly interpret things like motion, temperature, and component position, based on what comes back. For example, consider the operation of the throttle position sensor (TPS). During acceleration (from idle to wide-open throttle), the computer monitors throttle plate movement based on the changing reference voltage signal returned by the TPS. (The TPS is a type of variable resistor known as a rotary potentiometer that changes circuit resistance based on throttle shaft rotation.) As TPS resistance varies, the computer is programmed to respond in a specific manner (for example, increase fuel delivery or alter spark timing) to each corresponding voltage change.

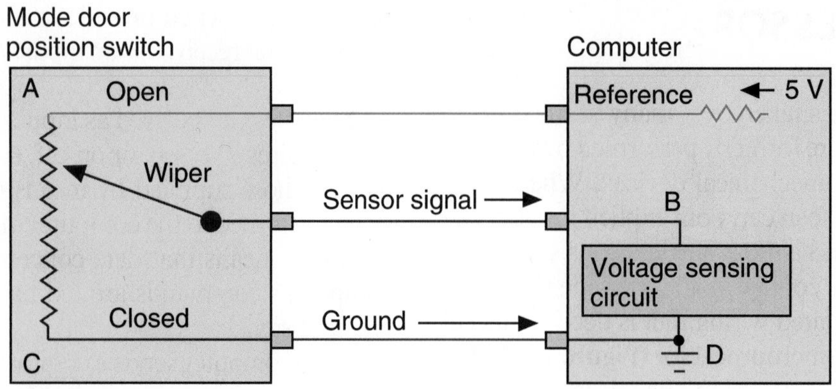

Figure 15-14 A potentiometer sensor circuit measures the amount of voltage drop to determine the position of the A/C mode door.

Most sensors presently in use are variable resistors or potentiometers **(Figure 15–14)**. They modify a voltage to or from the computer, indicating a constantly changing status that can be calculated, compensated for, and modified. That is, most sensors simply control a voltage signal from the computer. When varying internal resistance of the sensor allows more or less voltage to ground, the computer senses a voltage change on a monitored signal line. The monitored signal line may be the output signal from the computer to the sensor (one- and two-wire sensors), or the computer may use a separate return line from the sensor to monitor voltage changes (three-wire sensors).

Another commonly used variable resistor is a thermistor. A thermistor is a solid-state variable resistor made from a semiconductor material that changes resistance in relation to temperature changes. Thermistors are used to monitor engine coolant, intake air, and ambient temperatures. By monitoring the thermistor's resistance value, the CPU is capable of observing small changes in temperature. The CPU sends a reference voltage to the thermistor, normally 5 volts, through a fixed resistor. As the current flows through the thermistor to ground, voltage is dropped by the thermistor. A voltage-sensing circuit compares the voltage sent out by the CPU to the voltage returned to the CPU and determines the voltage drop. Using its programmed values, the computer is able to translate the voltage drop into a temperature reading.

There are two basic types of thermistors: NTC and PTC. A **negative temperature coefficient (NTC)** thermistor reduces its resistance as temperature increases. This type is the most commonly used thermistor. A **positive temperature coefficient (PTC)** thermistor increases its resistance with an increase in temperature. To diagnose thermistors accurately, you must be able to identify the type by looking at the values versus temperature table given in the service manual.

Wheatstone bridges (Figure 15–15) are also used as variable resistance sensors. These are typically constructed of four resistors, connected in series-parallel between an input terminal and a ground terminal. Three of the resistors are kept at the same value. The fourth resistor is a sensing resistor. When all four of the resistors have the same value, the bridge is balanced and the voltage sensor will have a value of 0 volts. If the sensing resistor changes value, a change occurs in the circuit's balance. The sensing circuit will receive a voltage reading proportional to the amount of resistance change. If the

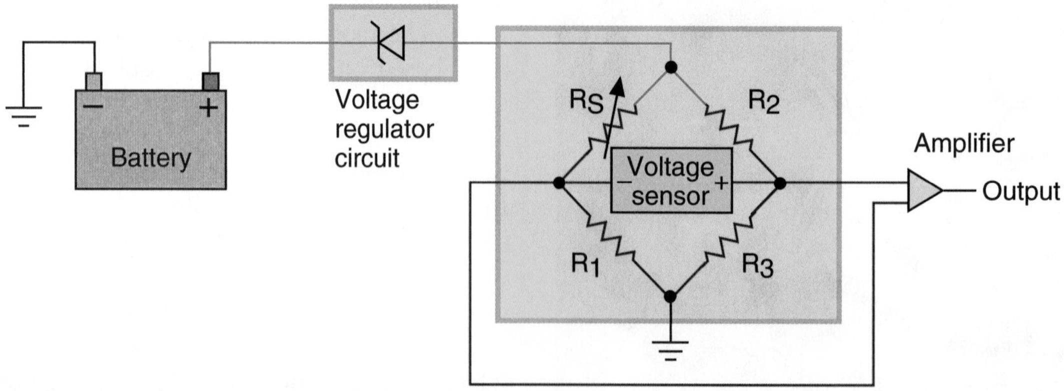

Figure 15-15 Wheatstone bridge.

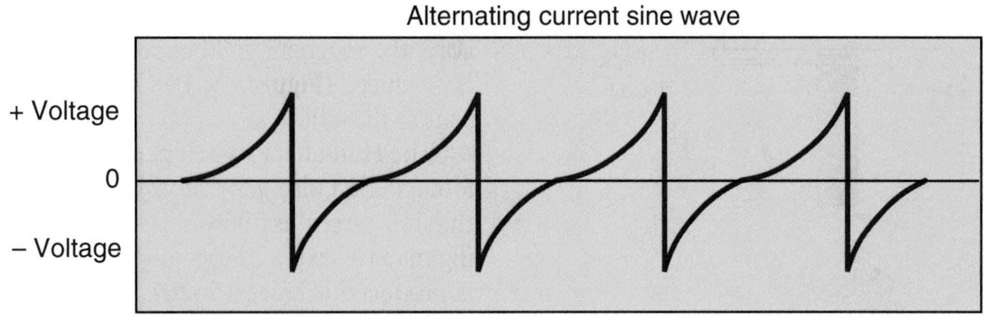

Figure 15-16 Pulse signal sine wave.

Wheatstone bridge is used to measure temperature, temperature changes are indicated as a change in voltage by the sensing circuit. Wheatstone bridges are also used to measure pressure **(piezoresistive)** and mechanical strain.

In addition to variable resistors, another commonly used reference voltage sensor is a switch. By opening and closing a circuit, switches provide the necessary voltage information to the computer so vehicles can maintain the proper performance and driveability.

Computers are digital devices, which means they look at things and react in a series of on-off or high-low voltage signals. When a sensor sends a variable or constant voltage, an analog signal, it must be converted to a digital signal. This is done by the analog/digital (A/D) converter.

Although most sensors are variable resistors, there is another category of sensors—the **voltage-generating devices**. These sensors include components like the magnetic pulse generators, Hall-effect switches, oxygen sensors (zirconium dioxide), and knock sensors (piezoelectric), which are capable of producing their own input voltage signal. This varying voltage signal, when received by the computer, enables the computer to monitor and adjust for changes in the computerized engine control system.

Magnetic pulse generators use the principle of magnetic induction to produce a voltage signal. They are also called permanent magnet (PM) generators. These sensors are often used to send data to the computer about the speed of the monitored component. This data provides information about vehicle speed, shaft speed, and wheel speed. The signals from speed sensors are used for instrumentation, cruise control systems, antilock brake systems, ignition systems, speed-sensitive steering systems, and automatic ride control systems. A magnetic pulse generator is also used to inform the computer about the position of a monitored device. This is common in engine controls where the CPU needs to know the position of the crankshaft in relation to rotational degrees.

The major components of a pulse generator are a timing disc and a pickup coil. The **timing disc** is attached to a rotating shaft or cable. The number of teeth on the timing disc is determined by the manufacturer and the application. If only the number of revolutions is required, the timing disc may have only one tooth. Whereas, if it is important to track quarter revolutions, the timing disc needs at least four teeth. The teeth will cause a voltage generation that is constant per revolution of the shaft. For example, a vehicle speed sensor may be designed to deliver 4,000 pulses per mile. The number of pulses per mile remains constant regardless of speed. The computer calculates how fast the vehicle is going based on the frequency of the signal. The timing disc is also known as an armature, reluctor, trigger wheel, pulse wheel, or timing core.

The **pickup coil** is also known as a stator, sensor, or pole piece. It remains stationary while the timing disc rotates in front of it. The changes of magnetic lines of force generate a small voltage signal in the coil. A pickup coil consists of a permanent magnet with fine wire wound around it.

An air gap is maintained between the timing disc and the pickup coil. As the timing disc rotates in front of the pickup coil, the generator sends a pulse signal **(Figure 15–16)**. As a tooth on the timing disc aligns with the core of the pickup coil, it repels the magnetic field. The magnetic field is forced to flow through the coil and pickup core **(Figure 15–17)**. When the tooth passes the core, the

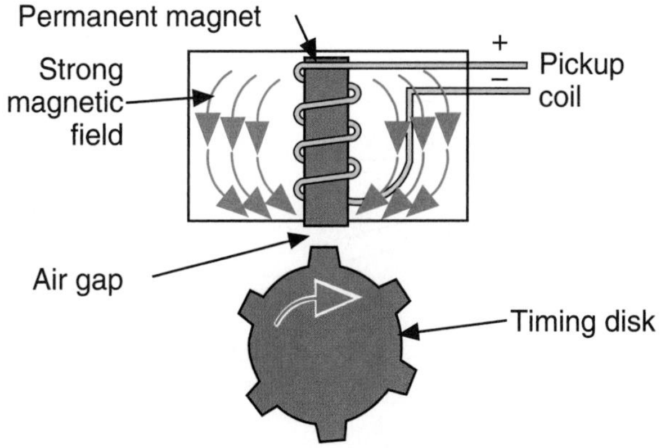

Figure 15-17 A strong magnetic field is produced in the pickup coil as the teeth align with the core.

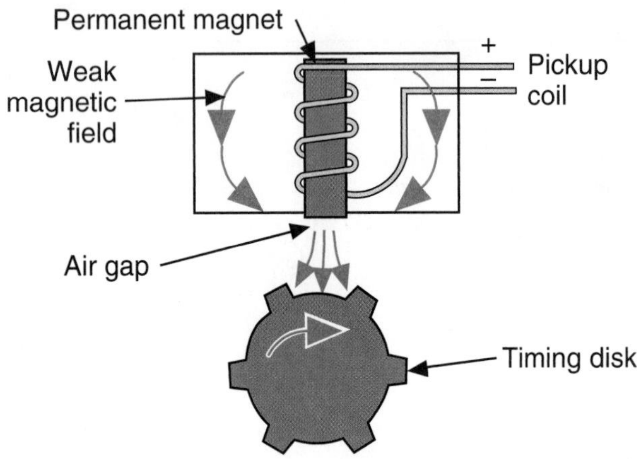

Figure 15-18 The magnetic field expands and weakens as the teeth pass the core.

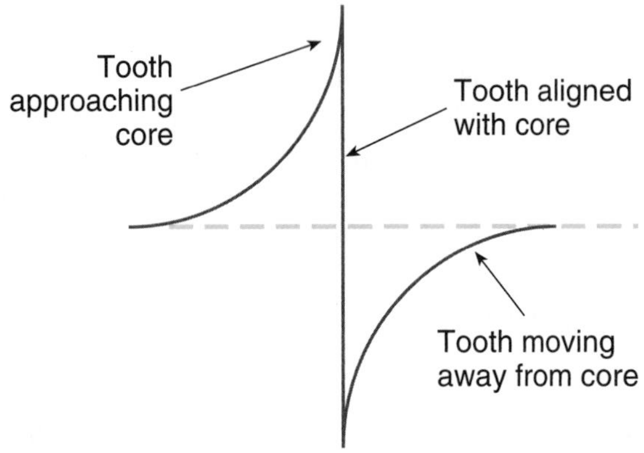

Figure 15-19 The waveform produced by a magnetic pulse generator.

magnetic field is able to expand **(Figure 15–18)**. This action is repeated every time a tooth passes the core. The moving lines of magnetic force cut across the coil windings and induce a voltage signal.

When a tooth approaches the core, a positive current is produced as the magnetic field begins to concentrate around the coil. When the tooth and core align, there is no more expansion or contraction of the magnetic field and the voltage drops to zero. When the tooth passes the core, the magnetic field expands and a negative current is produced **(Figure 15–19)**. The resulting pulse signal is sent to the CPU.

The **Hall-effect switch** performs the same function as a magnetic pulse generator. It operates on the principle that if a current is allowed to flow through thin conducting material exposed to a magnetic field, another voltage is produced **(Figure 15–20)**.

A Hall-effect switch contains a permanent magnet and a thin semiconductor layer made of gallium arsenate crystal (Hall layer) and a shutter wheel **(Figure 15–21)**. The Hall layer has a negative and a positive terminal connected to it. Two additional terminals located on either side of the Hall layer are used for the output circuit.

The permanent magnet is located directly across from the Hall layer so its lines of flux bisect the layer at right angles to the current flow. The permanent magnet is stationary; and a small air gap is between it and the Hall layer.

A steady current is applied to the crystal of the Hall layer. This produces a signal voltage perpendicular to the direction of current flow and magnetic flux. The signal voltage produced is a result of the effect the magnetic field has on the electrons. When the magnetic field bisects the supply current flow, the electrons are deflected toward the Hall layer negative terminal, which results in a weak voltage potential being produced in the Hall switch.

The shutter wheel consists of a series of alternating windows and vanes. It creates a magnetic shunt that changes the strength of the magnetic field from the permanent magnet. The shutter wheel is attached to a rotating component. As the wheel rotates, the vanes pass through the air gap. When a shutter vane enters the gap, it intercepts the magnetic field and shields the Hall layer from its lines of force. The electrons in the supply current are no longer disrupted and return to a normal state. This results in low voltage potential in the signal circuit of the Hall switch.

The signal voltage leaves the Hall layer as a weak analog signal. To be used by the CPU, the signal must be con-

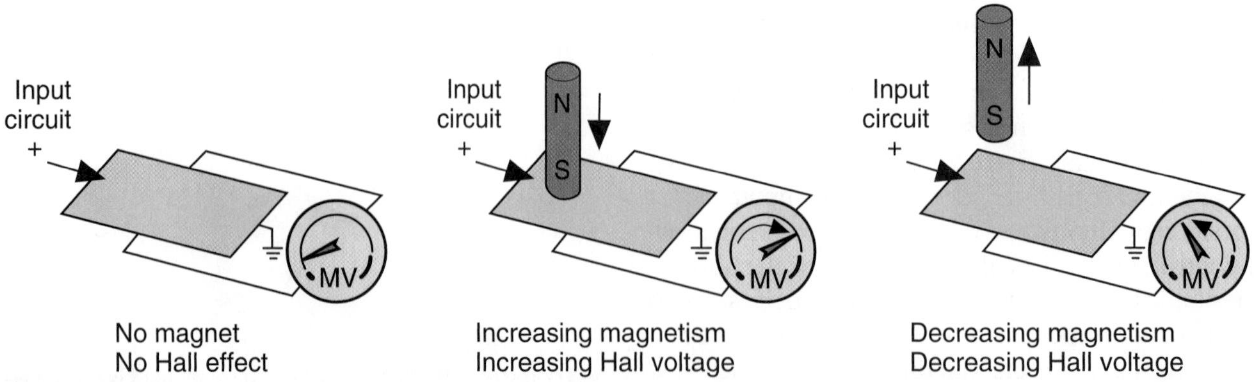

Figure 15-20 Hall-effect principles of voltage induction.

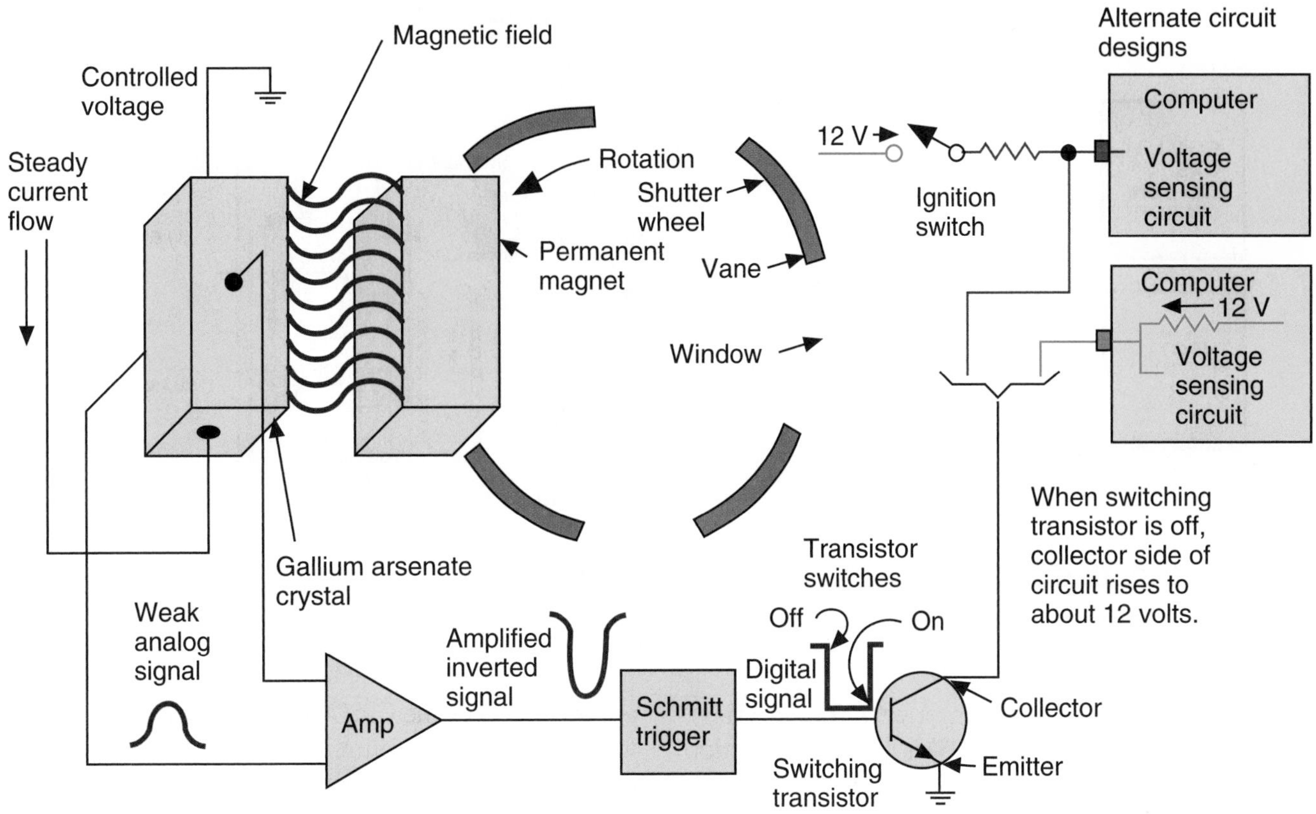

Figure 15-21 Typical circuit of a Hall-effect switch.

ditioned. It is first amplified because it is too weak to produce a desirable result. The signal is also inverted; a low input signal is converted into a high output signal. It is then sent through a **Schmitt trigger**, which is a type of A/D converter, where it is digitized and conditioned into a clean square wave signal. The signal is finally sent to a switching transistor. The computer senses the turning on and off of the switching transistor to determine the frequency of the signals and calculates speed.

Regardless of the types of sensors used in electronic control systems, the computer is incapable of functioning properly without input signal voltage from sensors.

Communication Signals

Voltage does not flow through a conductor; current flows while voltage is the pressure that pushes the current. However, voltage can be used as a **signal**; for example, difference in voltage levels, frequency of change, or switching from positive to negative values can be used as a signal.

A computer is capable of reading voltage signals. The programs used by the CPU are "burned" into IC chips using a series of numbers. These numbers represent various combinations of voltages that the computer can understand. The voltage signals to the computer can be either analog or digital. **Analog** means a voltage signal is infinitely variable, or can be changed, within a given range. **Digital** means a voltage signal that is in one of three states—either on-off, yes-no, or high-low. Most input sensors are designed to produce a voltage signal that varies within a given range, which is an analog signal. For example, ambient temperature sensors do not change abruptly. The resistance varies in very small steps from low to high. The same is true for several other outputs, such as engine speed, vehicle speed, and fuel flow.

Compared to an analog voltage signal, digital voltage patterns are square-waved because the transition from one voltage to another is very abrupt **(Figure 15–22)**. A digital signal is produced by an on-off or high-low voltage. In a digital signal, the voltage is represented by a series of digits, which create a **binary code**.

A computer can only read a digital binary signal. To overcome this communication problem, all analog voltage signals are converted to a digital format by a device

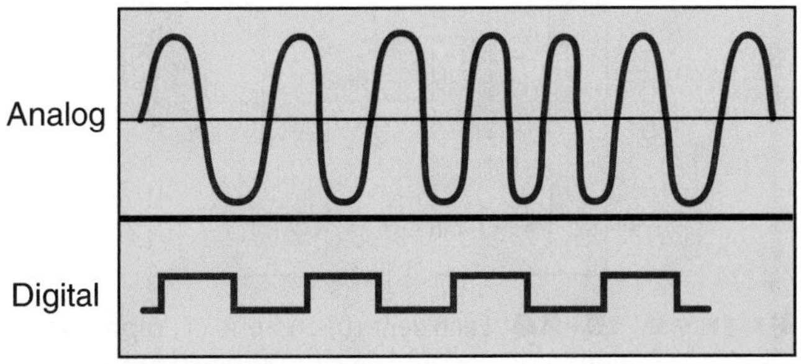

Figure 15-22 Analog signals can be constantly variable. Digital signals are either on/off or low/high.

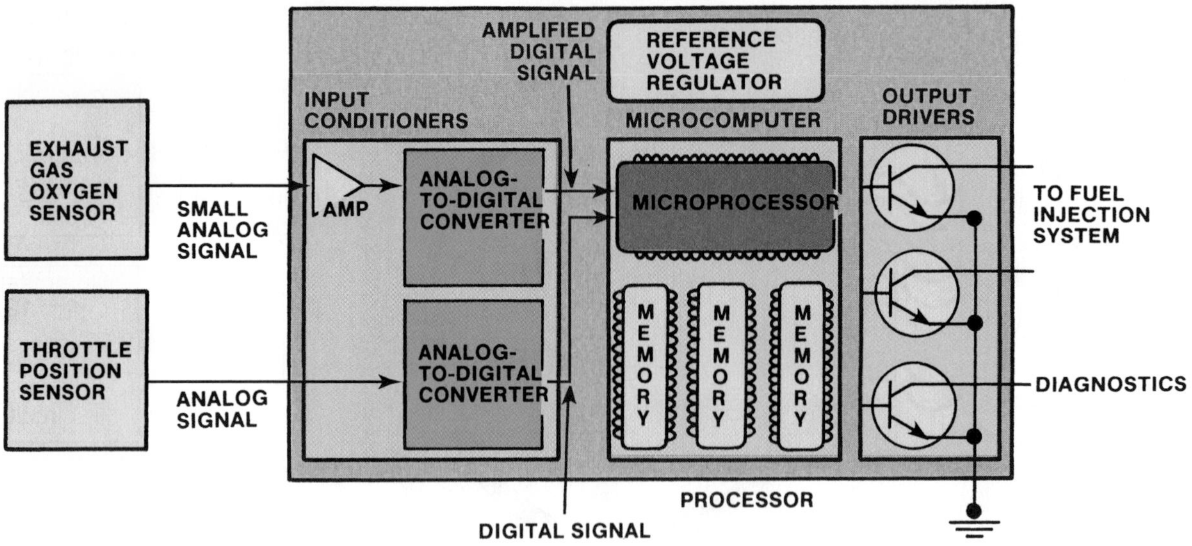

Figure 15-23 The A/D converter prepares input signals for the CPU.

known as an analog-to-digital converter **(A/D converter)**. The A/D converter **(Figure 15–23)** is located in a section of the processor called the input signals. However, some sensors, such as the Hall-effect switch, produce a digital or square-wave signal that can go directly to the CPU as an input. The term **square wave** is used to describe the appearance of a digital circuit after it has been plotted on a graph. The abrupt changes in circuit condition (on and off) result in a series of horizontal and vertical lines that connect to form a square-shaped pattern.

The A/D converter changes a series of signals to a binary number made up of 1s and 0s. Voltage above a given value converts to 1, and zero voltage converts to 0 **(Figure 15–24)**. Each 1 or 0 represents a **bit** of information. Eight bits equal a **byte** (sometimes referred to as a *word*). All communication between the CPU, the memories, and the interfaces is in binary code, with each information exchange in the form of a byte.

To get an idea of how binary coding works, let us see how signals from the coolant temperature sensor (CTS)

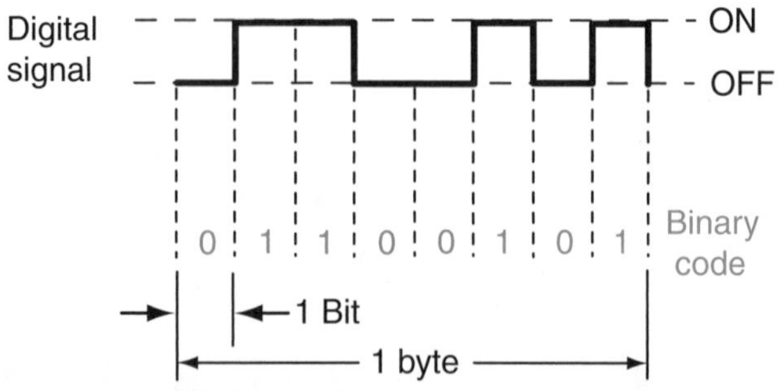

Figure 15-24 Each zero (0) and one (1) represents a bit of information. When eight bits are combined in specific sequence, they form a byte or word that makes up the basis of a computer's language.

are processed by the CPU. The CTS is a type of thermistor (negative temperature coefficient to be exact) that controls a reference signal based on temperature changes. Upon receiving the CTS's analog signals, the input conditioner immediately groups each signal value into a predetermined voltage range and assigns a numeric value to each range. In our example, use the following ranges and numeric values: 0 to 2 volts = 1, 2 to 4 volts = 2, and 4 to 5 volts = 3 (assuming a Vref of 5 volts). If you are wondering where these ranges and numeric values come from, they are written into the computer's memory by a human programmer at the time of the computer's development.

When the CTS is hot, its resistance is low and the modified voltage signal it sends back falls into the high range (4 to 5 volts). Upon entering the A/D converter, the voltage value is assigned a numeric value of 3 (based on the ranges previously cited) and is ready for further translation into a binary code format. Binary numbers are represented by the numbers 0 and 1. Any number and word can be translated into a combination of binary 1s and 0s.

Without going into the finer points of binary numbering, the number 3 in binary is expressed as 11. To the thousands of tiny transistors and diodes that act as the on-off switches inside digitally oriented microprocessors, 11 instructs the computer to turn on or apply voltage to a specific circuit for a predetermined length of time (based on its program). **Table 15–1** illustrates how binary numbers can be converted into decimal or base ten numbers. Note how the right-hand binary number equals one and the left number equals eight.

In addition to A/D conversion, some voltage signals require amplification before they can be relayed to the

CPU. To perform this task, an input conditioner known as an **amplifier** is used to strengthen weak voltage signals.

After input has been generated, conditioned, and passed along to the computer, it is ready to be processed for the purposes of performing work and displaying information. The computer contains a crystal oscillator or clock that delivers a constant time pulse. The clock is a crystal that electrically vibrates when subjected to current at certain voltage levels. As a result, the chip produces very regular series of voltage pulses. The clock maintains an orderly flow of information through the computer circuits by transmitting one bit of binary code for each pulse. In this manner, the computer is capable of distinguishing between the binary codes such as 101 and 1001. For the CPU to make the most informed decisions regarding system operation, sensor input is sent through different logic gates constructed within the CPU.

Communication Rates The amount of information processed by a computer is dependent on its speed or **baud rate**. Baud rate is the speed of communication and is roughly equal to the number of bits per second a computer can process.

Logic Gates

Logic gates are the thousands of **field-effect transistors (FETs)** incorporated into the computer's circuitry. The FETs use the incoming voltage patterns to determine the pattern of pulses leaving the gate. These circuits are called logic gates because they act as gates to output voltage signals depending on different combinations of input signals. The following are the most common logic gates and their operation. The symbols in the figures represent functions and not electronic construction.

TABLE 15–1 BINARY NUMBER CODE

Decimal Number	Binary Number Code 8421	Binary to Decimal Conversion
0	0000	= 0 + 0 = 0
1	0001	= 0 + 1 = 1
2	0010	= 2 + 0 = 2
3	0011	= 2 + 1 = 3
4	0100	= 4 + 0 = 4
5	0101	= 4 + 1 = 5
6	0110	= 4 + 2 = 6
7	0111	= 4 + 2 + 1 = 7
8	1000	= 8 + 0 = 8

1. **NOT gate:** A NOT gate simply reverses binary 1s and 0s and vice versa. A high input results in a low output, and a low input results in a high output.
2. **AND gate:** The AND gate has at least two inputs and one output. The operation of the AND gate is similar to two switches in series with a load. The only way the load will turn on is if both switches are closed. Before current can be present at the output of the gate, current must be present at the base of both transistors.
3. **OR gate:** The OR gate operates similarly to two switches that are wired in parallel to a light. If one switch is closed, the light will turn on. A high signal to either input will result in a high output.
4. **NAND** and **NOR gates:** A NOT gate placed behind an OR or AND gate inverts the output signal.
5. **Exclusive-OR (XOR) gate:** A combination of gates that will produce a high output signal only if the inputs are different.

These different gates are combined to perform the processing function. The following are some of the most common combinations.

1. **Decoder circuits:** A combination of AND gates used to provide a certain output based on a given combination of inputs **(Figure 15–25)**. When the correct bit pattern is received by the decoder, it will produce the high-voltage signal to activate the relay coil.
2. **Multiplexer (MUX):** The basic computer is not capable of looking at all of the inputs at the same time. A multiplexer is used to examine one of many inputs depending on a programmed priority rating **(Figure 15–26)**.
3. **Demultiplexer (DEMUX):** Operates similarly to the MUX except that it controls the order of the outputs **(Figure 15–27)**. The process that the MUX and DEMUX operate on is called **sequential sampling**. This means the computer deals with all the sensors and actuators one at a time.
4. **RS and clocked RS flip-flop circuits:** Flip-flop circuits remember previous inputs and do not change their outputs until they receive new input signals. The clocked flip-flop circuit has an inverted clock signal as an input so circuit operations occur in the proper order. Flip-flop circuits are called **sequential logic circuits** because the output is determined by the sequence of inputs.
5. **Registers:** Used in the computer to temporarily store information. A register is a combination of flip-flops that transfers bits from one to another every time a clock pulse occurs **(Figure 15–28)**.

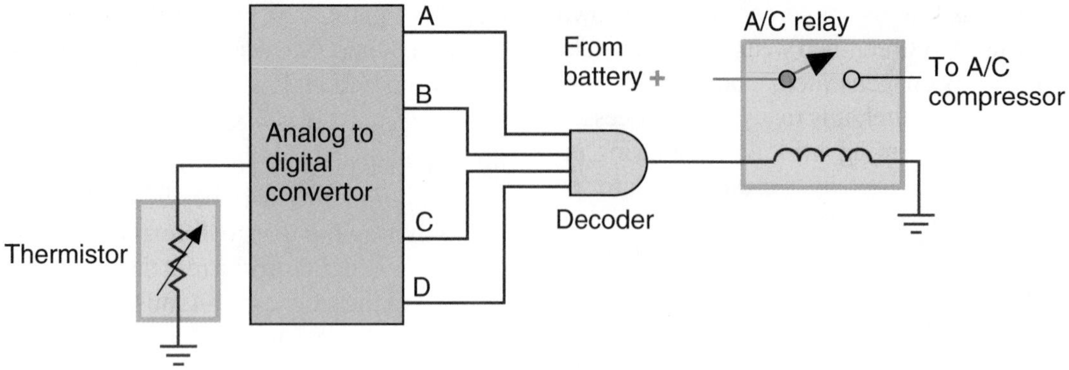

Figure 15-25 A simplified temperature-sensing circuit that turns on the A/C compressor when the inside temperature reaches a predetermined level.

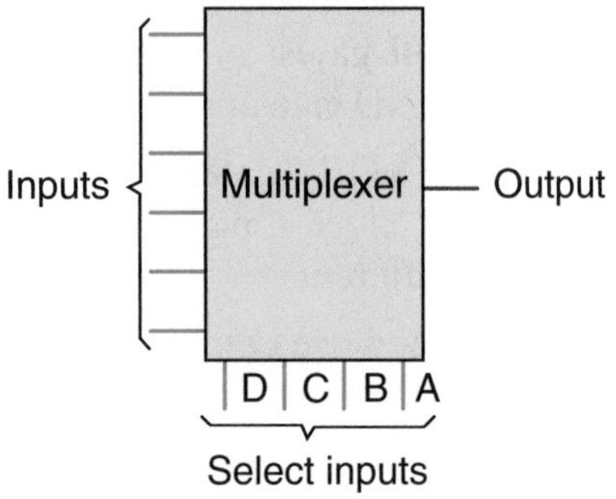

Figure 15-26 The selection at inputs DCBA determine which data input will be processed.

6. **Accumulators:** Registers designed to store the results of logic operations that can become inputs to other computers or modules.

Logic gates process input information to command output devices. The order of this logic or the instructions to the computer are held in the computer's memory.

Memories

A computer's memory holds the programs and other data, such as vehicle calibrations, which the microprocessor refers to in performing calculations. To the CPU, the program is a set of instructions or procedures that it must follow. Included in the program is information that tells the microprocessor when to retrieve input (based on temperature, time, and so forth), how to process the input, and what to do with it once it has been processed.

The microprocessor works with memory in two ways: it can read information from memory or change information in memory by writing in or storing new information. To write information in memory, each memory location is assigned a number (written in binary code also) called an address. These addresses are sequentially numbered, starting with zero, and are used by the microprocessor to retrieve data and write new information into memory. During processing, the CPU often receives more data than it can immediately handle. In these in-

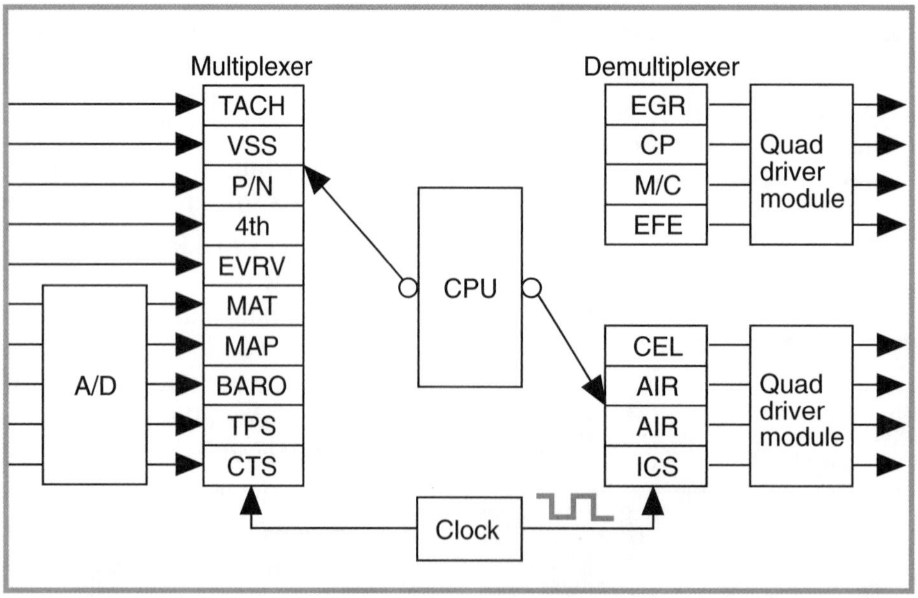

Figure 15-27 A block diagram representation of the MUX and DEMUX circuit.

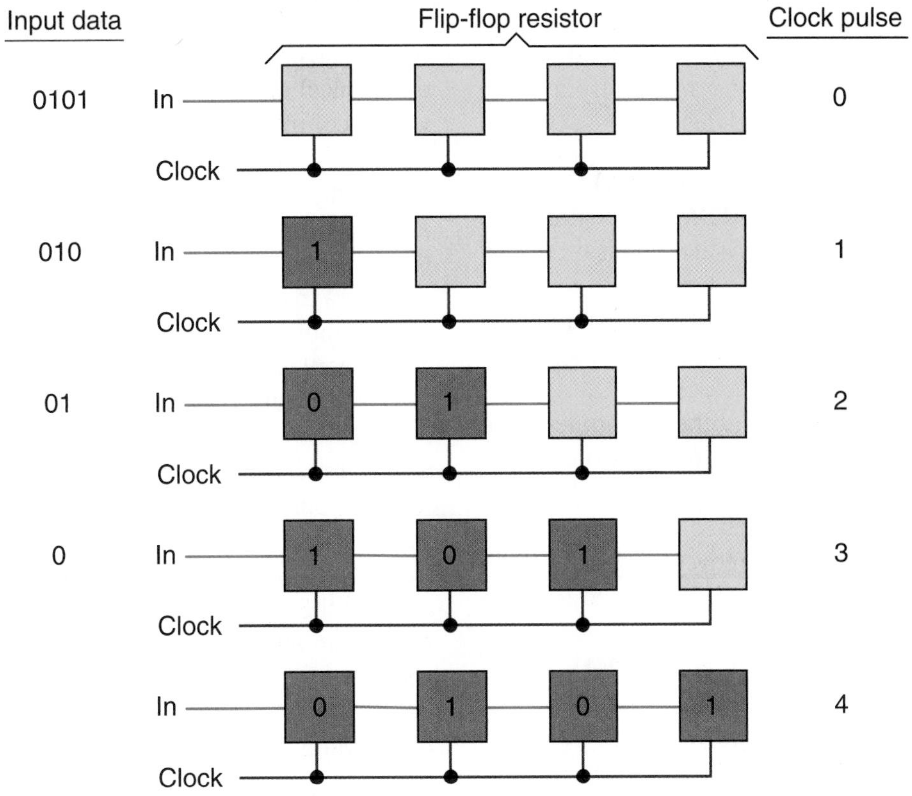

Figure 15-28 It takes four clock pulses to load four bits into the register.

stances, some information has to be temporarily stored or written into memory until the microprocessor needs it.

When ready, the microprocessor accesses the appropriate memory location (address) and is sent a copy of what is stored. By sending a copy, the memory retains the original information for future use.

Basically, three types of memory are used in automotive CPUs today **(Figure 15–29)**: read-only memory, programmable read-only memory, and random-access memory.

Read-Only Memory (ROM) Permanent information is stored in **read-only memory (ROM)**. Information in ROM cannot be erased, even if the system is turned off or the CPU is disconnected from the battery. As the name implies, information can only be read from ROM.

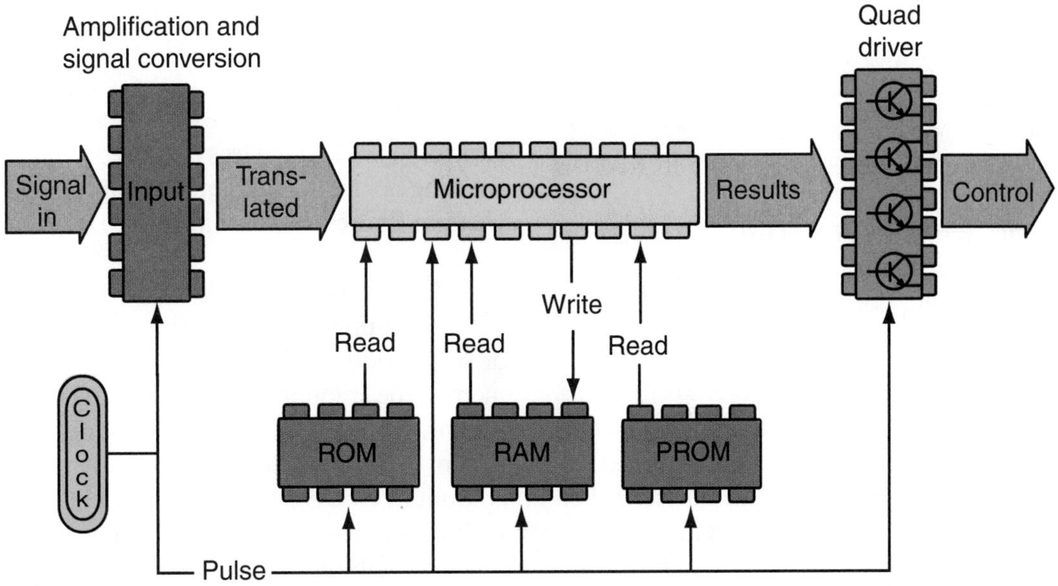

Figure 15-29 The three memories within a computer.

When making decisions, the microprocessor is constantly referring to the stored information and the input from sensors. By comparing information from these sources, the CPU makes informed decisions.

Programmable Read-Only Memory (PROM) The programmable read-only memory **(PROM)** differs from the ROM in that it plugs into the computer and is more easily removed and reprogrammed or replaced with one containing a revised program. It contains program information specific to different vehicle model calibrations. The PROM in some computers is replaceable and can serve as a way to upgrade the system. In other computers, the PROM is soldered to the computer's memory board.

Erasable PROM (EPROM) is similar to PROM except that its contents can be erased to allow new data to be installed. A piece of Mylar tape covers a window. If the tape is removed, the memory circuit is exposed to ultraviolet light that erases its memory.

Electrically erasable PROM (EEPROM) allows changing the information electrically one bit at a time. Some manufacturers use this type of memory to store information concerning mileage, vehicle identification number, and options.

Random-Access Memory (RAM) The random-access memory **(RAM)** is used during computer operation to store temporary information. The CPU can write, read, and erase information from RAM in any order, which is why it is called random. One characteristic of RAM is that when the ignition key is turned off and the engine is stopped, information in RAM is erased. RAM is used to store information from the sensors, the results of calculations, and other data that are subject to constant change.

There are currently two other versions of RAM in use: volatile and nonvolatile. A volatile RAM, usually called **keep-alive memory (KAM)** has most of the features of RAM. Information can be written into KAM and can be read and erased from KAM. Unlike RAM, information in KAM is not erased when the ignition key is turned off and the engine is stopped. However, if battery power to the processor is disconnected, information in KAM is erased.

A **nonvolatile RAM (NVRAM)** does not lose its stored information if its power source is disconnected. Vehicles with digital display odometers usually store mileage information in nonvolatile RAM.

Actuators

Once the computer's programming determines that a correction or adjustment must be made in the controlled system, an output signal is sent to control devices called **actuators**. These actuators, which are solenoids, switches, relays, or motors, physically act on or carry out the command sent by the computer.

Actually, actuators are electromechanical devices that convert an electrical current into mechanical action. This mechanical action can then be used to open and close valves, control vacuum to other components, or open and close switches. When the CPU receives an input signal indicating a change in one or more of the operating conditions, the CPU determines the best strategy for handling the conditions. The CPU then controls a set of actuators to achieve a desired effect or strategy goal. For the computer to control an actuator, it must rely on a component called an **output driver**.

The output driver usually applies the ground circuit of the actuator. The ground can be applied steadily if the actuator must be activated for a selected amount of time, or the ground can be pulsed to activate the actuator in pulses.

Output drivers operate by the digital commands issued by the CPU. Basically, the output driver is nothing more than an electronic on-off switch used to control a specific actuator.

To illustrate this relationship, let us suppose the computer wants to turn on the engine's cooling fan. Once it makes a decision, it sends a signal to the output driver that controls the cooling fan relay (actuator). In supplying the relay's ground, the output driver completes the power circuit between the battery and cooling fan motor and the fan operates. When the fan has run long enough, the computer signals the output driver to open the relay's control circuit (by removing its ground), thus opening the power circuit to the fan.

For actuators that cannot be controlled by a digital signal, the CPU must turn its digitally coded instructions back into an analog signal. This conversion is completed by the A/D converter.

Displays can be controlled directly by the CPU. They do not require digital-to-analog conversion or output drivers because they contain circuitry that decodes the microprocessor's digital signal. The decoded information is then used to indicate such things as vehicle speed, engine rpm, fuel level, or scan tool values. Common types of electronic readout devices used as displays include light-emitting diodes (LED), **liquid crystal display (LCD)**, and **vacuum fluorescent display (VFD)**.

Duty Cycle versus Pulse Width Often the computer controls the results of the output by controlling the duty cycle or pulse width of the actuator. **Duty cycle** is a measurement of the amount of time something is on compared to the time of one cycle and is measured in a percentage. When measuring duty cycle, you are looking at the amount of time something is on during one cycle. **Pulse width** is similar to duty cycle except that it is the exact time something is turned on and is measured in milliseconds.

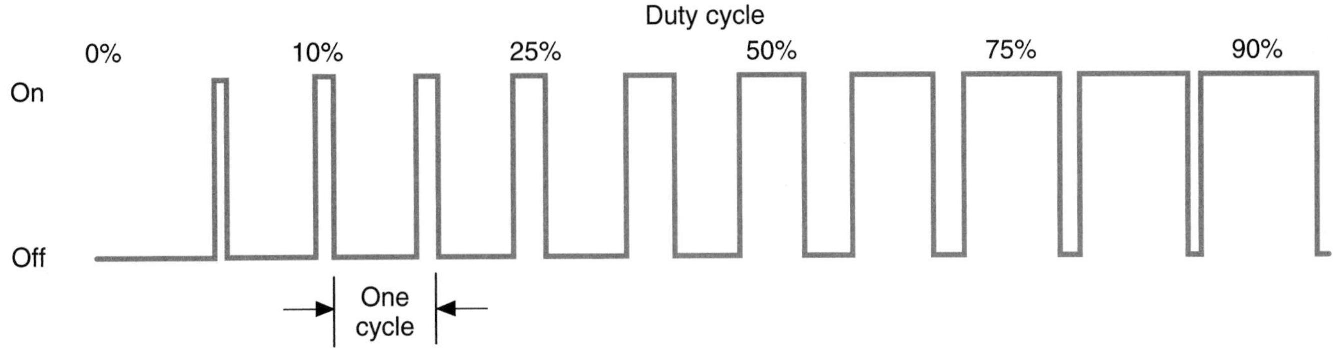

Figure 15–30 Duty cycle is the percentage of on-time per cycle. Duty cycle can be changed; however, total cycle time remains constant.

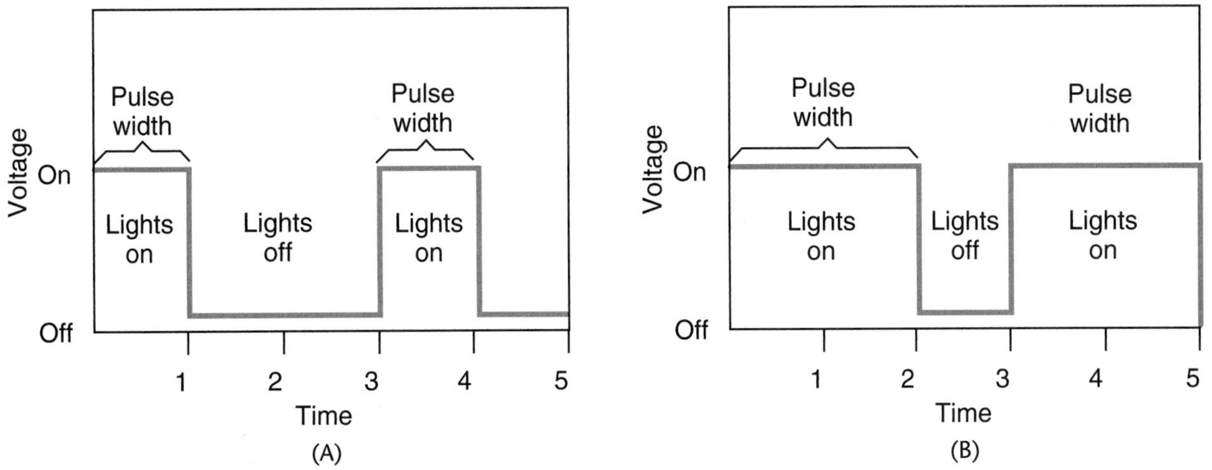

Figure 15–31 Pulse width is the duration of on-time. (A) Pulse width modulation to achieve dimmer dash lights, and (B) pulse width modulation to achieve brighter dash illumination.

Most duty-cycled actuators cycle ten times per second. To complete a cycle, it must go from off to on to off again. If the cycle rate is ten times per second, one actuator cycle is completed in 1/10th of a second. If the actuator is turned on for 30% of each tenth of a second and off for 70%, it is referred to as a 30% duty cycle **(Figure 15–30)**.

If the actuator is cycled on and off very rapidly, the pulse width can be varied to provide the desired results. For example, the computer program will select an illumination level of a digital instrument panel based on the intensity of the ambient light in the vehicle. The illumination level is achieved through pulse width modulation of the lights. If the lights need to be bright, the pulse width is increased, which increases the length of on-time. As light intensity needs to be reduced, the pulse width is decreased **(Figure 15–31)**.

Power Supply

The CPU also contains a power supply that provides the various voltages required by the microprocessor and internal clock that provides the clock pulse, which in turn controls the rate at which sensor readings and output changes are made. Also contained are protection circuits that safeguard the microprocessor from interference caused by other systems in the vehicle and diagnostic circuits that monitor all inputs and outputs and signal a warning light if any values go outside the specified parameters. This warning light is called the malfunction indicator lamp (MIL).

Complete specific information concerning the use of computers (microprocessors) in various automotive electronic systems is presented in later chapters.

MULTIPLEXING

Today's vehicles have hundreds of circuits. In order for these systems to work correctly there must be some communication between the various circuits. Communication can take place through dedicated wires connecting each sensor and circuit to the appropriate control module. If more than one control module is involved, additional pairs of wires must connect the sensor or circuit to the other modules. The result of this communication network is miles of wires and hundreds of connectors. Whenever something electronic or electrical is added to a vehicle, more wires and connectors are needed. To eliminate the need to add wires and their subsequent weight as more electronics are added, manufacturers are using multiplexing **(Figure 15–32)**.

Multiplexing, also called in-vehicle networking, provides an efficient way for different systems to communicate. Multiplexing relies on one wire for communication instead of many wires. A multiplex wiring system uses a serial data bus that connects different computers or control modules. Each module can transmit and receive digital codes over the serial data bus, allowing one module to share the same information with other modules. For example, the signal relating to engine speed may be required by the engine control, transmission control, electronic brake control, and suspension control modules. Instead of having separate engine speed inputs into each module, the serial data bus carries the information, as well as other information, to all of the control modules.

Each sensor is wired directly to the control module that relies heavily on the sensor's signal. That control module sends the information to the serial data bus. Each control module has a code-reading device or chip that reads and sends messages to the serial data bus. Some chips can only send or receive, depending on their purpose. All information (called serial data) on the serial data bus is available for all control modules. However, the chip of each device compares the coded message to its memory list to see whether the information is relevant to its own function.

The chip is also used to prevent the signals from overlapping by allowing only one signal to be transmitted at a time. Each digital signal is preceded by an identification code that establishes its priority. If two modules attempt to send a message at the same time, the signal with the higher priority code is transmitted first.

Keep in mind that data is conveyed in binary numbers and therefore must be interpreted by some type of data processing before it becomes information. On many vehicles, serial data can be monitored with a scan tool connected to the vehicle's data link connector (DLC). Monitoring serial data allows technicians to diagnose the various control modules.

The serial data bus is typically made of two wires, a ground wire and a transmission wire, twisted together to reduce magnetic interference, which can cause false information.

Advantages

Multiplexing offers many advantages over traditional wiring, including the following:

- Fewer wires are required for each system function, which means smaller wiring harnesses, lower cost and weight, and improved serviceability, reliability, and installation.
- The need for redundant sensors is eliminated because sensor data, such as vehicle speed and engine tem-

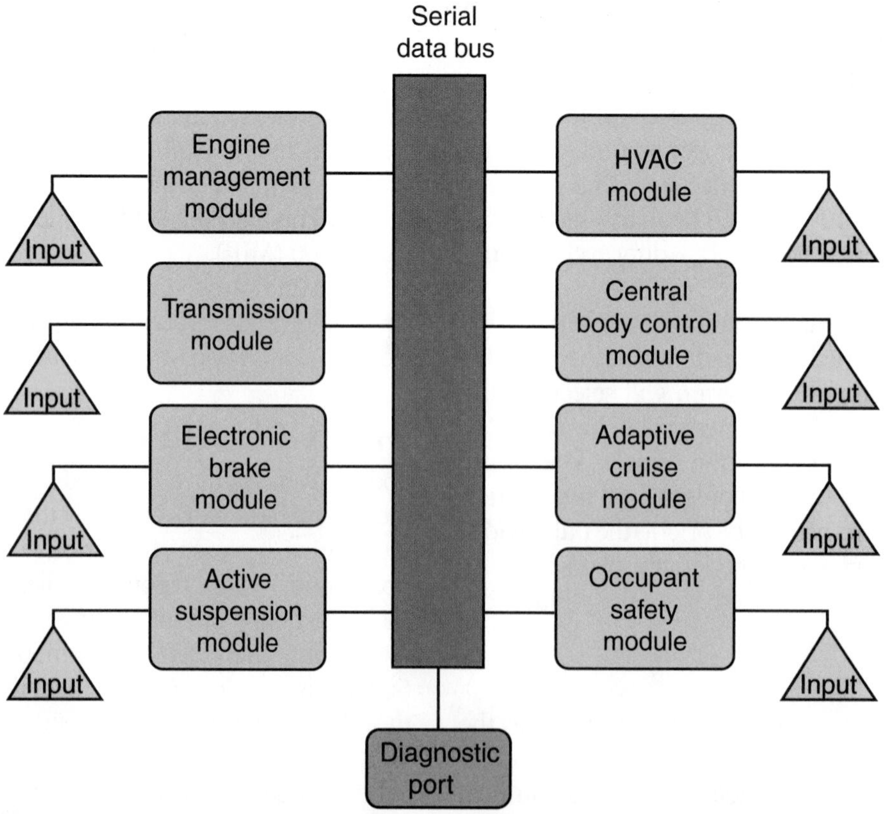

Figure 15-32 A multiplexed system uses a serial data bus to allow communications between the various control modules.

perature, are available on the serial data bus where it can be used by several control modules.

- Accessories and vehicle features can be easily added to the vehicle through software changes.
- Improved communications between control modules allows for more accurate recording and reporting of faults, which in turn helps in locating and solving problems.

As the electrical content of today's vehicles continues to increase, the need for networking is even more evident.

Types of Multiplexing

Four basic techniques are used for multiplexing: frequency division multiplexing (FDM), time division multiplexing (TDM), statistical time division multiplexing (STDM), and wavelength division multiplexing (WDM).

FDM is an analog setup where each communication channel is assigned a carrier frequency. To make sure the different channels do not interfere with each other, a small range of frequencies separates each channel. FDM depends on the bandwidth of the network. The bandwidth is the difference between the highest and lowest frequencies of a communication channel. FDM is prone to noise problems and was used in early multiplexing systems.

TDM is a digital technique where a sample of each channel is inserted into the multiplexed data stream. TDM works like a very fast mechanical switch, selecting each channel for a very short time, then going on to the next channel. TDM is the technique used in nearly all automobiles.

STDM uses chips to allocate time to channels only when it is needed. This allocation means more channels can be connected to the bus because the chips statistically compensate for times when a particular system is not being used. This technique increases the time allocated to the active communication channels.

WDM is used in fiber-optic networks in which multiple signals are transmitted as light split into different wavelengths. The use of fiber-optic material in automobiles will increase in the future. Optical fiber is a small plastic or glass fiber used to transmit information using infrared or visible light as a carrier. The light beams do not escape from the cable because the material used provides total internal reflection. Noise has little effect on the information carried by fiber optics, allowing for the transmission of cleaner information. Fiber-optic material is inexpensive but requires special connection and service techniques.

SAE Classifications

Early multiplexed systems were often based on proprietary serial buses using generic universal asynchronous receiver/transmitter (UART) or custom devices. This use called for dedicated and specific scan tools, each with the ability to work only on specific systems. OBD II called for standardized diagnostic tools, which meant standard protocols had to be implemented. A protocol is a special set of rules for transmitting data between two devices. The protocol determines the type of error checking to be used, the data compression method, how the sending device will indicate that it is finished sending a message, and how the receiving device will indicate that it has received the message.

Automobile manufacturers and various automotive organizations have been working to develop standards for in-vehicle networking. To classify the various standards, SAE has defined three basic categories of in-vehicle networks based on speed and functions:

- Class A (low-speed communication): Class A multiplexing is used for convenience systems, load-control features such as entertainment systems, audio, trip computer, seat controls, windows, and lighting. Most Class A functions require inexpensive, low-speed communication and use generic UARTs. These functions are proprietary and have not been standardized by the international organizations.
- Class B (medium-speed communication): Class B multiplexing is used primarily with the instrument cluster, vehicle speed, and emissions data recording. The SAE J1850 standard can be used for the class B protocol and is used in many production vehicles for data sharing and diagnostic purposes. The SAE J1850 standard has two basic versions. The first is a variable pulse width (VPW) type that uses a single bus wire. The second is a pulse width modulation (PWM) type that uses a two-wire bus. OBD II specifies that stored fault codes be available through a diagnostic port via a standard protocol. Currently OBD II specifies J1850 and the European standard, ISO 9141-2.
- Class C (high-speed communication): Class C multiplexing is used for real-time control of the powertrain, vehicle dynamics, and brake-by-wire. This level of communication can use a twisted pair, but a shielded coaxial cable or fiber optics may be used. The predominant class C protocol is CAN 2.0 (Controller Area Network version 2.0). CAN assigns a unique identifier to every message. The identifier classifies the content of the message and the priority of the message being sent. Each module processes only those messages whose identifiers are stored in the module's acceptance list.

It is common to find a variety of the different classes of multiplexing in a single vehicle **(Figure 15–33)**. Some systems such as powertrain control and vehicle dynamics require high-speed communications, whereas other systems do not.

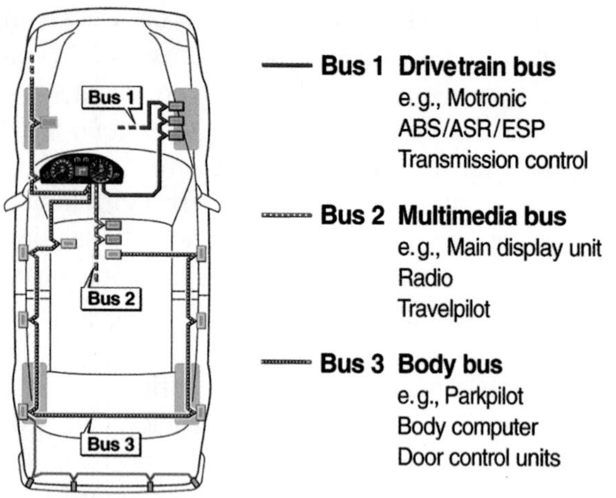

Figure 15-33 It is common to find a variety of the different classes of multiplexing in a single vehicle. *Courtesy of Robert Bosch Corp.*

KEY TERMS

Accumulator
Actuators
A/D converter
Amplifier
Analog
AND gate
Base
Baud rate
Binary code
Bit
Byte
Central processing unit (CPU)
Chip
Clamping diode
Collector
Decoder circuit
Demultiplexer (DEMUX)
Digital
Diode
Duty cycle
Electrically erasable PROM (EEPROM)
Electrically inert
Emitter
Erasable PROM (EPROM)
Exclusive-OR (XOR) gate
Feedback
Field-effect transistor (FET)
Forward bias
Hall-effect switch
IC
Impurities
Input
Integrated circuit
Keep alive memory (KAM)
Liquid crystal display (LCD)
Light-emitting diode (LED)
Logic gate
Magnetic pulse generator
Microprocessor
Multiplexer (MUX)
Multiplexing
NAND gate
Negative temperature coefficient (NTC)
Nonvolatile RAM (NVRAM)
NOR gate
NOT gate
NPN
OR gate
Output
Output driver
Pickup coil
Piezoresistive
PN junction
PNP
Positive temperature coefficient (PTC)
Processing
Programmable read-only memory (PROM)
Pulse width
Random-access memory (RAM)
Read-only memory (ROM)
Reference voltage (Vref)
Registers
Reverse bias
RS and clocked RS flip-flop circuit
Schmitt trigger
Sequential logic circuits
Sequential sampling
Signal
Square wave
Storage
Timing disc
Transistor
Vacuum fluorescent display (VFD)
Voltage-generating devices
Wheatstone bridge
Zener diode

SUMMARY

- A computer is an electronic device that stores and processes data.
- A semiconductor is a material or device that can function as either a conductor or an insulator.
- Some common semiconductor materials include silicon (Si) and germanium (Ge).
- N-type semiconductors have loose, or excess, electrons. They have a negative charge.
- P-type semiconductors are positively charged materials.
- A diode allows current to flow in one direction.
- When a diode is placed in a circuit with the positive side of the circuit connected to the positive side of the diode and the negative side of the circuit connected to the negative side of the circuit, the diode is said to have forward bias.
- Light-emitting diodes (LEDs) are much the same as regular diodes. They emit light when they are forward biased.
- Zener diodes are used to regulate voltage in a circuit.
- Transistors are used in place of switches and relays in electronic circuits.
- A transistor resembles a diode with an extra side. It can consist of two P-type materials and one N-type material or two N-type materials and one P-type material. These are called PNP and NPN types.
- The names for the three legs of a transistor are the emitter, base, and collector.
- A very small current applied to the base of the transistor controls a much larger current flowing through the entire transistor.

- An integrated circuit is simply a large number of diodes, transistors, and other electronic components, such as resistors and capacitors, all mounted on a single piece of semiconductor material.
- Input devices called sensors feed information to the computer. The computer processes this information and sends signals to controlling devices.
- A typical electronic control system is made of sensors, actuators, CPU, and related wiring.
- Most input sensors are variable resistance/reference types, switches, and thermistors.
- The central processing unit (CPU) is the brain of a computer. It brings information into and out of the computer's memory. The input information is processed in the CPU and checked against the program in the computer's memory.
- Feedback means that data concerning the effects of the computer's commands are fed back to the computer as an input signal.
- All sensors perform the same basic function. They detect a mechanical condition (movement or position), chemical state, or temperature condition and change it into an electrical signal that can be used by the computer to make decisions.
- Reference voltage (Vref) sensors provide input to the computer by modifying or controlling a constant, predetermined voltage signal.
- Most sensors presently in use are variable resistors or potentiometers. They modify a voltage to or from the computer, indicating a constantly changing status that can be calculated, compensated for, and modified.
- A thermistor is a solid-state variable resistor made from a semiconductor material that changes resistance in relation to temperature changes.
- There are two basic types of thermistors: NTC and PTC. A negative temperature coefficient (NTC) thermistor reduces its resistance as temperature increases. This type is the most commonly used thermistor. A positive temperature coefficient (PTC) thermistor increases its resistance with an increase in temperature.
- A Wheatstone bridge is also used as a variable resistance sensor. These are typically constructed of four resistors, connected in series-parallel between an input terminal and a ground terminal. Three of the resistors are kept at the same value. The fourth resistor is a sensing resistor.
- Voltage-generating sensors include components like the magnetic pulse generators, Hall-effect switches, oxygen sensors (zirconium dioxide), and knock sensors (piezoelectric), which are capable of producing their own input voltage signal.
- Magnetic pulse generators use the principle of magnetic induction to produce a voltage signal. These sensors are commonly used to send data to the computer about the speed of the monitored component.
- The major components of a pulse generator are a timing disc and a pickup coil.
- The Hall-effect switch performs the same function as a magnetic pulse generator. It operates on the principle that if a current is allowed to flow through thin conducting material exposed to a magnetic field, another voltage is produced.
- A Hall-effect switch contains a permanent magnet, a thin semiconductor layer made of gallium arsenate crystal (Hall layer), and a shutter wheel.
- Outputs or actuators are electromechanical devices that convert current into mechanical action.
- A computer is capable of reading voltage signals. The programs used by the CPU are "burned" into IC chips using a series of numbers. These numbers represent various combinations of voltages that the computer can understand. The voltage signals to the computer can be either analog or digital. Analog means a voltage signal is infinitely variable, or can be changed, within a given range. Digital means a voltage signal that is in one of two states, either on-off, yes-no, or high-low.
- A computer can only read a digital binary signal. To overcome this communication problem, all analog voltage signals are converted to a digital format by a device known as an analog-to-digital converter (A/D converter).
- Logic gates are the thousands of field effect transistors (FET) incorporated into the computer's circuitry. The FETs use the incoming voltage patterns to determine the pattern of pulses leaving the gate. These circuits are called logic gates because they act as gates to output voltage signals depending on different combinations of input signals.
- The most common logic gates are NOT gates, AND gates, OR gates, NAND and NOR gates, and XOR gates.
- A computer's memory holds the programs and other data, such as vehicle calibrations, which the microprocessor refers to in performing calculations.
- The microprocessor works with memory in two ways: it can read information from memory or change information in memory by writing in or storing new information.
- Basically, three types of memory are used in automotive CPUs today: read-only memory, programmable read-only memory, and random-access memory.

- Permanent information is stored in read-only memory (ROM).
- The PROM differs from the ROM in that it plugs into the computer and is more easily removed and reprogrammed or replaced with one containing a revised program. It contains program information specific to different vehicle model calibrations.
- The RAM is used during computer operation to store temporary information. The CPU can write, read, and erase information from RAM in any order, which is why it is called random.
- Once the computer's programming instructs that a correction or adjustment must be made in the controlled system, an output signal is sent to control devices called actuators. These actuators, which are solenoids, switchers, relays, or motors, physically act or carry out the command sent by the computer.
- For the computer to control an actuator, it must rely on a component called an output driver. These drivers usually apply the ground circuit of the actuator.
- Some systems require the actuator to either be turned on and off very rapidly or for a set amount of cycles per second. It is duty cycled if it is turned on and off a set amount of cycles per second. Duty cycle is the percentage of on-time to total cycle time.
- If the actuator is cycled on and off very rapidly, the pulse width can be varied to provide the desired results. Pulse width is the length of time in milliseconds an actuator is energized.
- A multiplex wiring system uses bus data links that connect different computers or control modules together. Each module can transmit and receive digital codes over the bus data links. This allows the module to share the same information with other modules.

TECH MANUAL

The following procedure is included in Chapter 15 of the *Tech Manual* that accompanies this book:

1. Test for continuity across electronic and electrical components and circuits with an ohmmeter.

REVIEW QUESTIONS

1. The three main types of memory in a computer are called ______, ______, and ______.
2. A device that stores and processes data is a ______.
3. ______ signals show any change in voltage.
4. Digital signals are typically called ______ ______ patterns.
5. ______ means that data concerning the effects of the computer's commands are fed back to the computer as an input signal.
6. The type of memory that contains specific information about the vehicle and can be removed and reprogrammed or replaced is called ______.
7. What are the major components of an electronic control system?
8. A ______ is the simplest type of semiconductor.
9. What is meant by the term *pulse width*?
10. What is the major difference between ROM and RAM memory in a microprocessor?
11. Most semiconductors are made of which of the following materials?
    - **a.** gallium
    - **b.** silicon
    - **c.** copper
    - **d.** none of the above
12. Technician A says when positive voltage is present at the base of an NPN transistor, the transistor is turned on. Technician B says when an NPN transistor is turned on, current flows through the collector and emitter of the transistor. Who is correct?
    - **a.** Technician A
    - **b.** Technician B
    - **c.** Both A and B
    - **d.** Neither A nor B
13. Which of the following is *not* a type of information stored in ROM?
    - **a.** strategy
    - **b.** look-up tables
    - **c.** sensor input
    - **d.** none of the above
14. *True or False?* In a multiplex wiring system, a serial data bus is used to allow communication between the various control modules in a vehicle.
15. Which of the following is *not* made of semiconductor material?

a. zener diodes
b. thermistors
c. PM generators
d. Hall-effect switches

16. Technician A says knock sensors are basically Wheatstone bridges. Technician B says oxygen sensors are basically thermistors. Who is correct?
    a. Technician A
    b. Technician B
    c. Both A and B
    d. Neither A nor B

17. Which of the following is a basic function of a computer?
    a. to store information
    b. to process information
    c. to send out commands
    d. all of the above

18. Technician A says some types of voltage sensors provide input to the computer by modifying or controlling a constant, predetermined voltage signal. Technician B says another type of voltage sensor is a voltage generating sensor. Who is correct?
    a. Technician A
    b. Technician B
    c. Both A and B
    d. Neither A nor B

19. Technician A says multiplexing is a way to add wires without increasing the weight of the vehicle. Technician B says multiplexing uses bus data links. Who is correct?
    a. Technician A
    b. Technician B
    c. Both A and B
    d. Neither A nor B

20. Technician A says when a diode is placed in a circuit with the positive side of the circuit connected to the positive side of the diode and the negative side of the circuit connected to the negative side of the diode, the diode is said to have reverse bias. Technician B says the diode's P material is repelled by the positive charge of the circuit and is pushed toward the N material and the N material is pushed toward the P. Who is correct?
    a. Technician A
    b. Technician B
    c. Both A and B
    d. Neither A nor B

16

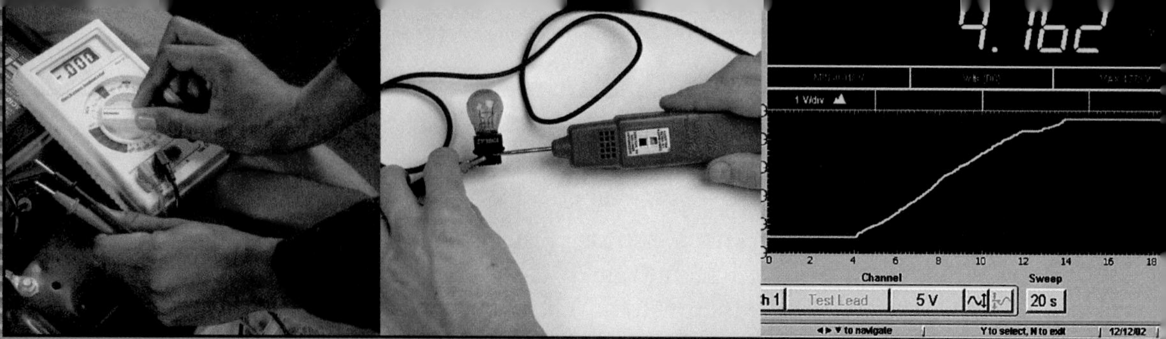

GENERAL ELECTRICAL SYSTEM DIAGNOSTICS AND SERVICE

OBJECTIVES

■ Describe the different possible types of electrical problems. ■ Read electrical automotive diagrams. ■ Perform troubleshooting procedures using meters, test lights, and jumper wires. ■ Describe how each of the major types of electrical test equipment is connected and interpreted. ■ Explain how to use a DMM for diagnosing electrical and electronic systems. ■ Explain how to use an oscilloscope for diagnosing electrical and electronic systems. ■ Test common electrical components. ■ Identify the proper procedures to safeguard electronic systems. ■ Test common electronic components. ■ Properly repair wiring and connectors.

Diagnosing nearly every system of a vehicle involves troubleshooting electrical and electronic systems. Understanding how electrical/electronic systems work and knowing how to use the various types of test equipment are keys to efficient diagnosis.

ELECTRICAL PROBLEMS

All electrical problems can be classified into one of three categories: opens, shorts, or high-resistance problems. Identifying the type of problem allows you to determine the correct tests to perform when diagnosing an electrical problem. The different classifications of electrical problems follow.

Open Circuits

An **open** occurs when a circuit has a break in the wire. Without a completed path, current cannot flow and the load or component cannot work. An open circuit can be caused by a disconnected wire, a broken wire, or a switch in the off position. When a circuit is off, it is open. When the circuit is on, it is closed. Although voltage will be present up to the open point, there is no current flow. Without current flow, there are no voltage drops across the various loads.

Short Circuits

A shorted circuit is one with an unwanted current path. **Short** circuits cause an increase in current flow. This increased current flow can burn wires or components. Sometimes two circuits become shorted together. When this happens, one circuit powers another, which may result in strange happenings, such as the horn sounding every time the brake pedal is depressed **(Figure 16–1)**. In this case, the brake light circuit is shorted to the horn circuit. Improper wiring and damaged insulation are the two major causes of short circuits.

A short can also be an unwanted short to ground **(Figure 16–2)**. This type of short provides a low resistance path to ground and causes extremely high current flow, which will damage wires and electrical components **(Figure 16–3)**. Some call this problem a grounded circuit.

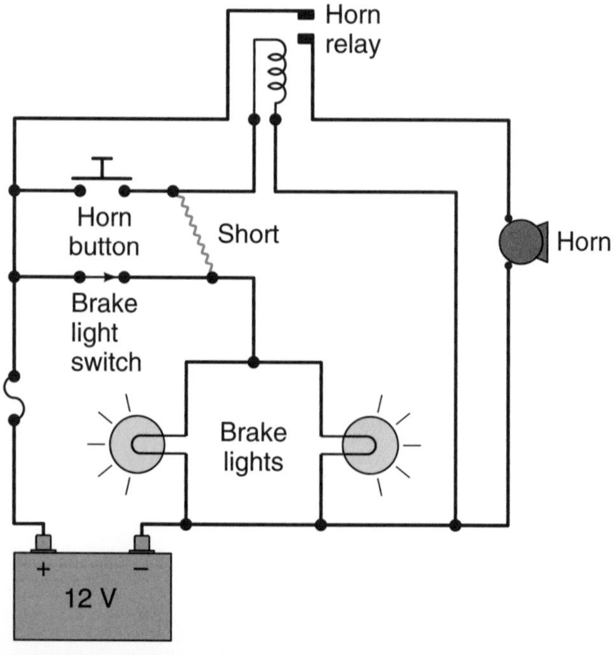

Figure 16–1 A wire-to-wire short.

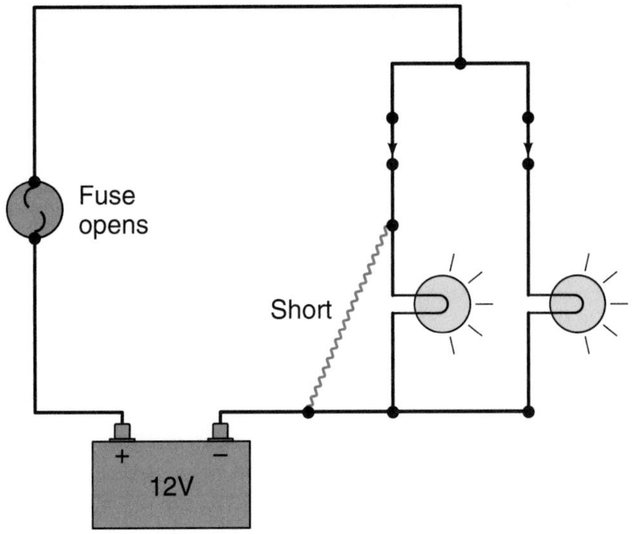

Figure 16-2 A short to ground.

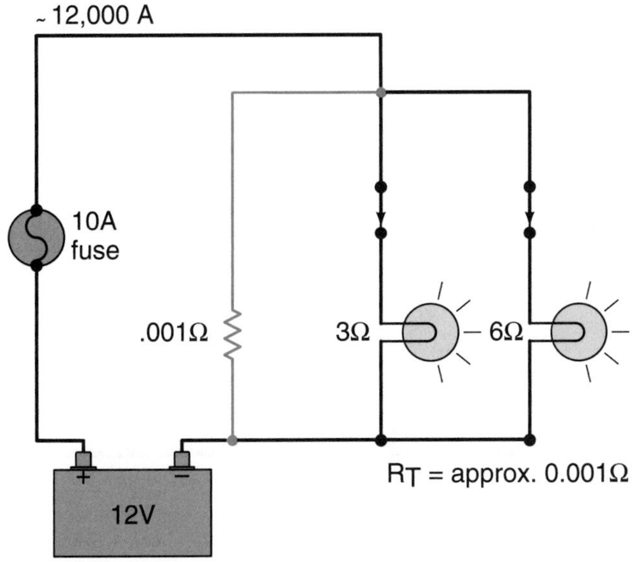

Figure 16-3 Ohm's law applied to Figure 16-2. Notice the rise in circuit amperage.

High-Resistance Circuits

High-resistance problems occur when there is unwanted resistance in the circuit. The higher than normal resistance causes the current flow to be lower than normal and the components in the circuit are unable to operate properly. A common cause of this type of problem is corrosion at a connector. The corrosion becomes an additional load in the circuit **(Figure 16–4)**. This load uses some of the circuit's voltage, which prevents full voltage to the normal loads in the circuit.

Many sensors on today's vehicles are fed a 5-volt reference signal. The signal or voltage from the sensor can be less than 5 volts, depending on the condition it is measuring. A poor ground in the reference voltage circuit can cause higher than normal readings back to the computer.

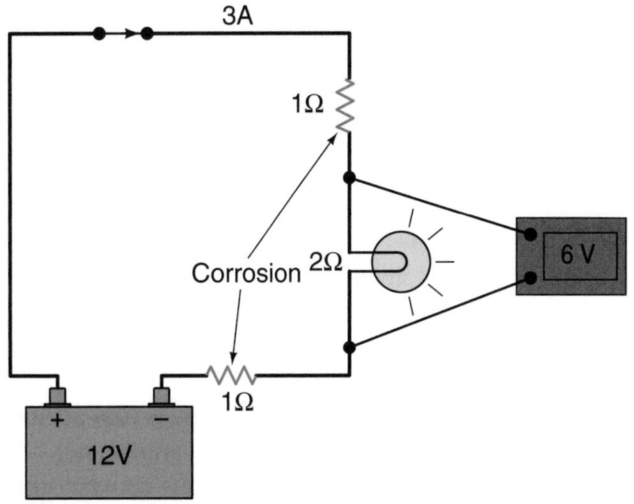

Figure 16-4 A simple light circuit with unwanted resistance. Notice the reduced voltage drop across the lamp and the reduced circuit current.

This seems to be contradictory to other high-resistance problems. However, if you look at a typical voltage divider circuit used to supply the reference voltage, you will understand what is happening. Look at **Figure 16–5**. There are two resistors in series with the voltage reference tap between them. Because the total resistance in the circuit is 12 ohms, the circuit current is 1 amp. Therefore, the voltage drop across the 7-ohm resistor is 7 volts, leaving 5 volts at the tap.

Figure 16–6 is the same circuit, but a bad ground of 1 ohm was added. This low of resistance could be caused by corrosion at the connection. With the bad ground, the total resistance is now 13 ohms. This decreases our circuit current to approximately 0.92 amps. With this lower amperage, the voltage drop across the 7-ohm resistor is now about 6.46 volts, leaving 5.54 volts at the voltage tap. This means the reference voltage would be more than

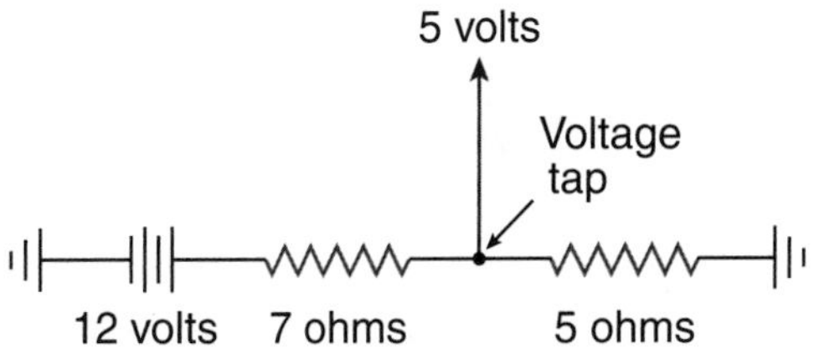

Figure 16-5 A voltage divider circuit.

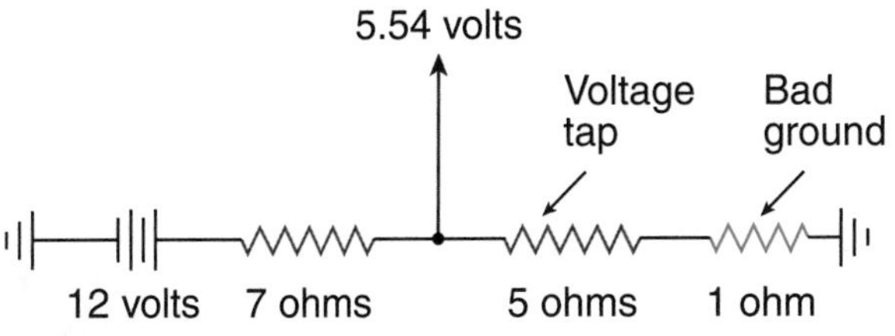

Figure 16-6 A voltage divider circuit with a bad ground.

one-half volt higher than it should be. As a result, the computer will be receiving a return signal of at least one-half volt higher than it should. Depending on the sensor and the operating conditions of the vehicle, this increase in voltage could be critical.

ELECTRICAL WIRING DIAGRAMS

Wiring diagrams, sometimes called **schematics**, are used to show how circuits are constructed. A typical service manual contains dozens of wiring diagrams vital to the diagnosis and repair of the vehicle.

A wiring diagram does not show the actual position of the parts on the vehicle or their appearance, nor does it indicate the length of the wire that runs between components. It usually indicates the color of the wire's insulation **(Figure 16–7)**, and sometimes the wire gauge size. Typically the primary wire insulation is color coded as shown in **Table 16–1**. The first letter in a combination of letters usually indicates the base color. The second letter usually refers to the strip color (if any). Tracing a circuit through a vehicle is basically a matter of following the colored wires.

Many different symbols are also used to represent components such as resistors, batteries, switches, transistors, and many other items. Some of these symbols have already been shown in earlier discussions. Other common symbols are shown in **Figure 16–8**. Part of a typical wiring diagram is shown in **Figure 16–9**; notice that the components are also labeled.

Wiring diagrams can become quite complex. To avoid this, most diagrams usually illustrate only one distinct system, such as the backup light circuit, oil pressure indica-

TABLE 16–1 COMMON WIRE COLOR CODES

Color	Abbreviations		
Aluminum	AL		
Black	BLK	BK	B
Blue (Dark)	BLU DK	DB	DK BLU
Blue (Light)	BLU LT	LB	LT BLU
Brown	BRN	BR	BN
Glazed	GLZ	GL	
Gray	GRA	GR	G
Green (Dark)	GRN DK	DG	DK GRN
Green (Light)	GRN LT	LG	LT GRN
Maroon	MAR	M	
Natural	NAT	N	
Orange	ORN	O	ORG
Pink	PNK	PK	P
Purple	PPL	PR	
Red	RED	R	RD
Tan	TAN	T	TN
Violet	VLT	V	
White	WHT	W	WH
Yellow	YEL	Y	YL

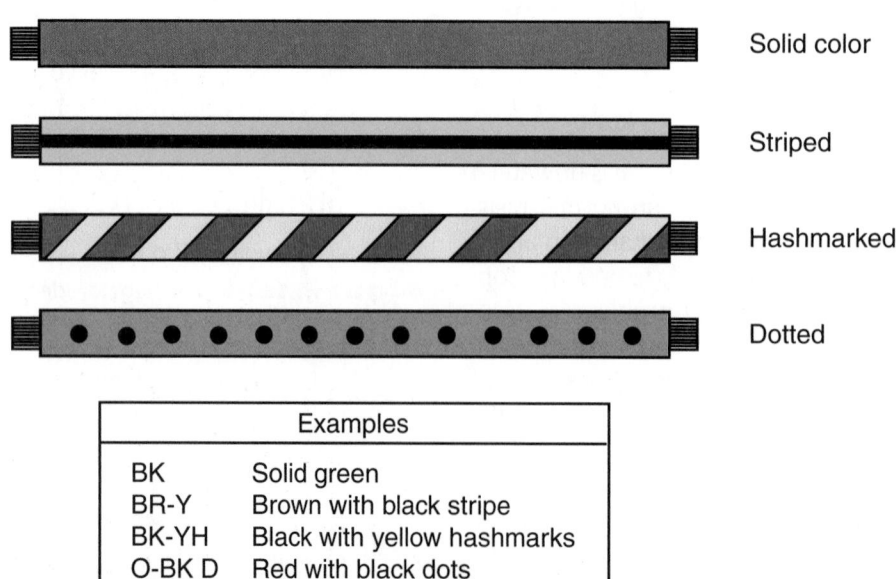

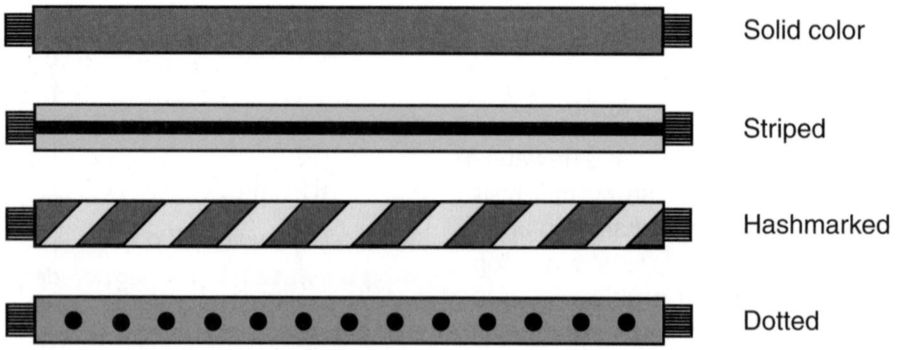

Examples	
BK	Solid green
BR-Y	Brown with black stripe
BK-YH	Black with yellow hashmarks
O-BK D	Red with black dots

Figure 16–7 The different multicolor schemes of wires.

SYMBOLS USED IN WIRING DIAGRAMS			
+	Positive		Temperature switch
—	Negative		Diode
	Ground		Zener diode
	Fuse		Motor
	Circuit breaker	C101	Connector 101
	Condenser	→	Male connector
Ω	Ohms	⤚	Female connector
	Fixed value resistor		Splice
	Variable resistor	S101	Splice number
	Series resistors		Thermal element
	Coil		Multiple connectors
	Open contacts		Digital readout
	Closed contacts		Single filament bulb
	Closed switch		Dual filament bulb
	Open switch		Light-emitting diode
	Ganged switch (N.O.)		Thermistor
	Single pole double throw switch		PNP bipolar transistor
	Momentary contact switch		NPN bipolar transistor
	Pressure switch		Gauge

Figure 16-8 Common electrical symbols used on wiring diagrams.

tor light circuit, or wiper motor circuit. In more complex ignition, electronic fuel injection and computer control systems, a diagram may be used to illustrate only part of the entire circuit.

USING SERVICE MANUALS

Keep in mind that electrical symbols are not standardized throughout the automotive industry. Different manufacturers may have different methods of representing certain components, particularly the less common ones. Always refer to the symbol reference charts, wire color code charts, and abbreviation tables listed in the vehicle's service manual to avoid confusion when reading wiring diagrams. ■

ELECTRICAL TESTING TOOLS

With a basic understanding of electricity and simple circuits it is easier to understand the operation and purpose of the various types of electrical test equipment described in the following sections. Several meters, including the voltmeter, ohmmeter, ammeter, and volt/amp meter are used to test and diagnose electrical systems. These meters should be used along with jumper wires, test lights, and variable resistors (piles).

Circuit Testers

Circuit testers are used to identify shorted and open circuits in any electrical circuit. Low-voltage testers are used to troubleshoot 6- to 12-volt circuits. High-voltage circuit testers diagnose higher voltage systems, such as the secondary ignition circuit.

WARNING!

Do not use a conventional 12-V test light to diagnose components and wires in electronic systems. The current draw of these test lights may damage computers and system components. High-impedance test lights are available for diagnosing electronic systems. Always be sure the test light you are using is recommended by the manufacturer before testing electronic or computer systems.

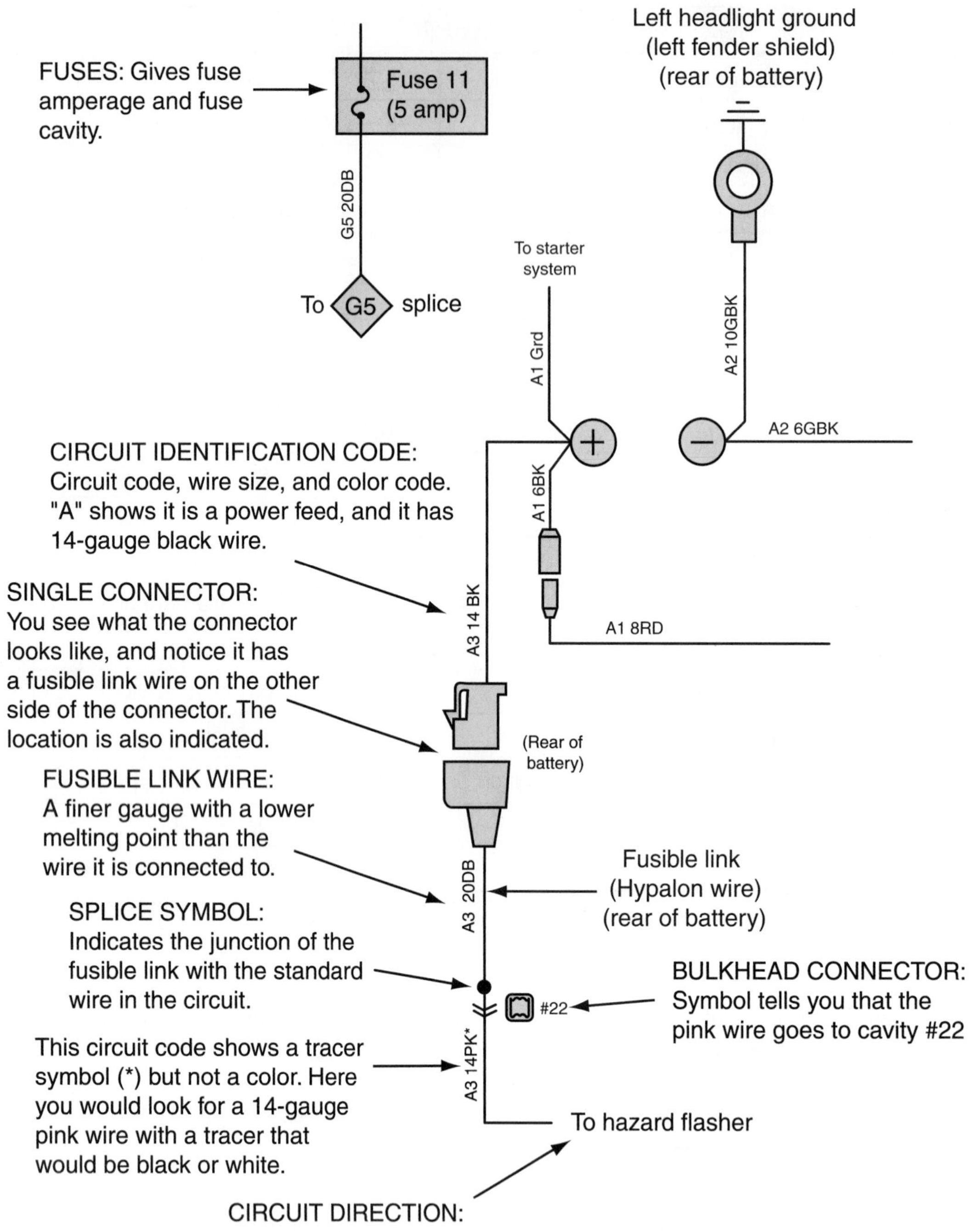

FUSES: Gives fuse amperage and fuse cavity.
Fuse 11 (5 amp)
G5 20DB
To G5 splice
Left headlight ground (left fender shield) (rear of battery)
To starter system
A1 Grd
A2 10GBK
A2 6GBK
CIRCUIT IDENTIFICATION CODE: Circuit code, wire size, and color code. "A" shows it is a power feed, and it has 14-gauge black wire.
A1 6BK
SINGLE CONNECTOR: You see what the connector looks like, and notice it has a fusible link wire on the other side of the connector. The location is also indicated.
A3 14 BK
A1 8RD
(Rear of battery)
FUSIBLE LINK WIRE: A finer gauge with a lower melting point than the wire it is connected to.
A3 20DB
Fusible link (Hypalon wire) (rear of battery)
SPLICE SYMBOL: Indicates the junction of the fusible link with the standard wire in the circuit.
BULKHEAD CONNECTOR: Symbol tells you that the pink wire goes to cavity #22
#22
This circuit code shows a tracer symbol (*) but not a color. Here you would look for a 14-gauge pink wire with a tracer that would be black or white.
A3 14PK*
To hazard flasher
CIRCUIT DIRECTION:

Figure 16-9 Information contained in a typical wiring diagram (schematic).

A circuit tester, commonly called a test light, looks like a stubby ice pick. Its handle is transparent and contains a light bulb. A probe extends from one end of the handle and a ground clip and wire extend from the other end. When the ground clip is attached to a good ground and the probe is touched to a live connector, the bulb in the handle will light up. If the bulb does not light, voltage is not available at the connector.

Two types of test lights are usually used: the nonpowered and the powered test light. Nonpowered test lights are used to check for available voltage. With the wire lead connected to a good ground and the tester's probe at a point of voltage, the light turns on with the presence of voltage **(Figure 16–10)**. The amount of voltage determines the brightness of the light.

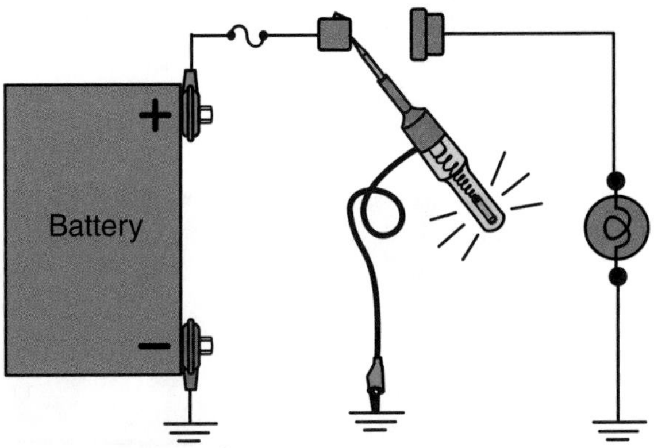

Figure 16-10 A test light will illuminate when the probe is touched to part of the circuit that is powered.

WARNING!

Do not use any type of test light or circuit tester to diagnose automotive air bag systems. Use only the vehicle manufacturer's recommended equipment on these systems.

A self-powered test light is used to check for continuity. Hooked across a circuit or component, the light turns on if the circuit is complete. A powered test light should only be used if the power for the circuit or component has been disconnected.

A self-powered test light is called a continuity tester **(Figure 16–11)**. It is used on open circuits. It is like a regular test light, except that it has a small internal battery. When the ground clip is attached to the negative side of a component and the probe is touched to the positive side, the lamp will light if there is continuity in the circuit. If an open circuit exists, the lamp will not light. Typically, the positive side of a component is called the power or feed side.

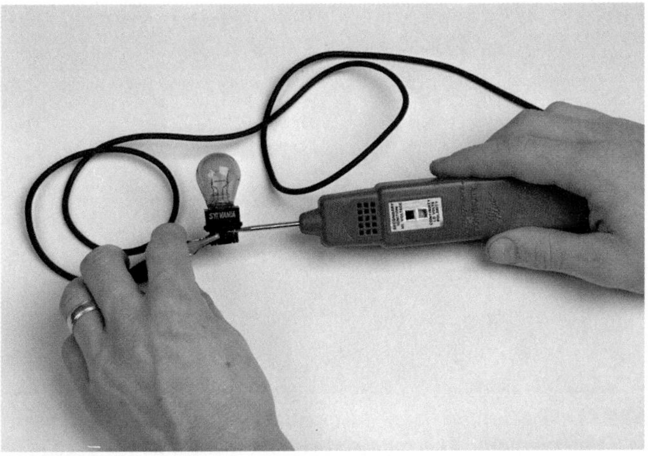

Figure 16-11 A typical self-powered circuit tester, commonly called a continuity tester.

Voltmeters

A voltmeter measures the voltage available at any point in an electrical system. For example, it can be used to measure the voltage available at the battery or at the terminals of any component or connector. A voltmeter can also be used to test voltage drop across an electrical circuit, component, switch, or connector.

The loss of voltage due to resistance in wires, connectors, and loads is called **voltage drop (Figure 16–12)**. Voltage drop is checked using a voltmeter. The procedure for checking voltage is shown in Photo Sequence 10.

A voltmeter has two leads: a red positive lead and a black negative lead. The red lead should be connected to the positive side of the circuit or component. The black lead should be connected to the negative side of the component or a good ground.

Consider a simple circuit **(Figure 16–13)**. If there are 12 volts available at the battery and the switch is closed, there should also be 12 volts available at each light. For example, if less than 12 volts is indicated, that would mean that some additional resistance exists somewhere else in the circuit. The lights may light, but not as brightly as they should.

Figure 16–14 illustrates two headlights connected to a 12-volt battery using two wires. Each wire has a resistance of 0.05 ohm. Each headlight has a resistance of 2

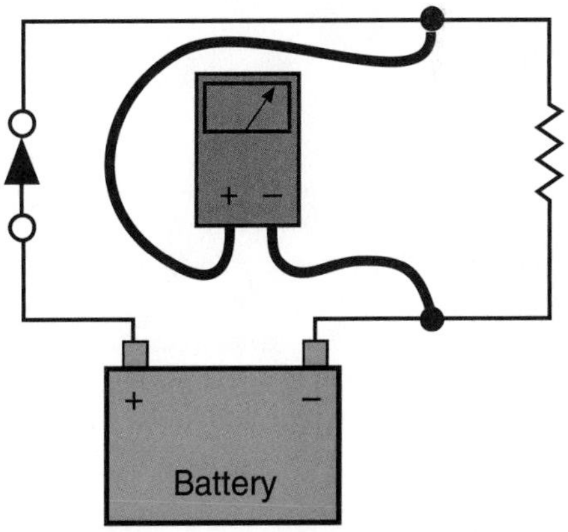

Figure 16-12 A voltmeter is connected in parallel or across a component or part of a circuit to measure voltage drop.

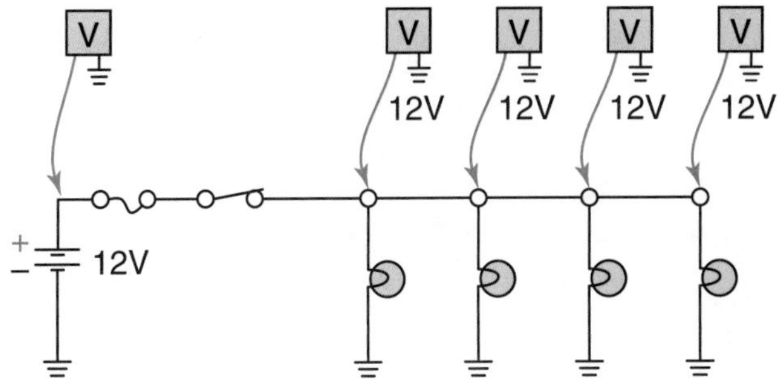

Figure 16-13 Checking a circuit using a voltmeter.

PHOTO SEQUENCE 10

Performing a Voltage Drop Test

P10-1 *Tools required to perform this task: voltmeter and fender covers.*

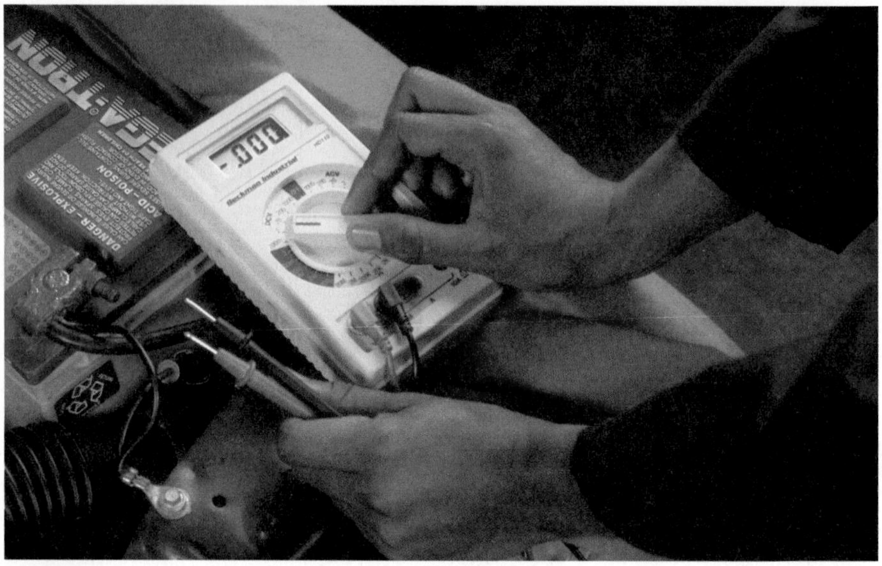

P10-2 *Set the voltmeter on its lowest DC volt scale.*

P10-3 *To test the voltage drop of the entire headlamp system, connect the positive (red) lead to the battery positive terminal.*

P10-4 *Connect the negative (black) lead to the low beam terminal of the headlight socket. Make sure you are connected to the battery "+" or power feed wire of the headlight.*

P10-5 *Turn on the headlights (low beam) and look at the voltmeter reading. The voltmeter will show the amount of voltage dropped between the battery and the headlight. This reading should be very low.*

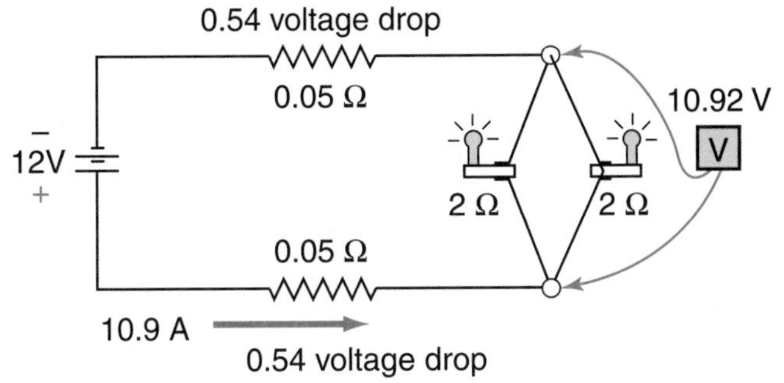

Figure 16–14 Wire resistance results in a slight voltage drop in the circuit.

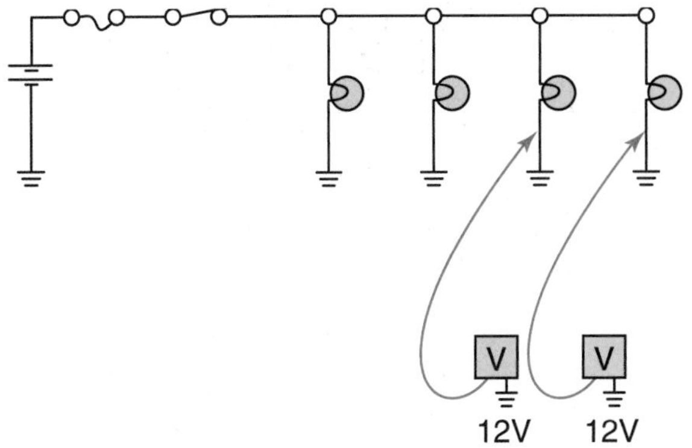

Figure 16–15 Using a voltmeter to check for open grounds.

ohms. As you can see, the two headlights are wired parallel so their effective resistance is

$$\frac{2 \text{ ohms} \times 2 \text{ ohms}}{2 \text{ ohms} + 2 \text{ ohms}} = 1 \text{ ohm}$$

The total circuit resistance is

$$1 \text{ ohm} + 0.05 \text{ ohm} + 0.05 \text{ ohm} = 1.1 \text{ ohms}$$

Therefore, the current in the circuit is

$$I = \frac{E}{R} = \frac{12}{1.1} = 10.9 \text{ amperes}$$

The voltage drop across each wire is

$$E = I \times R$$
$$E = 10.9 \times 0.05 = 0.54 \text{ V}$$

This means there is a total of 1.08 volts dropped across the wires. When the voltage drop of the wires is subtracted from the 12-volt source voltage, 10.92 volts remain for the headlights.

Without the resistance in the wires, the headlights receive 12 amperes. With the resistance, the current flow was reduced to 10.9 amperes.

All wiring must have resistance values low enough to allow enough voltage to the load for proper operation. Many manufacturers suggest that no more than 0.2 volt be dropped across the wires, connectors, and other conductors in a circuit.

A voltmeter can also be used to check for proper circuit grounding. For example, if the voltmeter reading indicates battery voltage at a lamp, but no lighting is seen, the bulb or socket could be bad or the ground connection is faulty.

An easy way to check for a defective bulb is to replace it with one known to be good. You can also use an ohmmeter to check for electrical continuity through the bulb.

If the bulbs are not defective, the problem lies in either the light sockets or ground wires. Connect the voltmeter to the ground wire and a good ground as shown in **Figure 16–15**. If the light socket were defective, there would be no voltage to the light and the voltmeter would read 0 volt. If the socket were not defective but the ground wire was broken or disconnected, the voltmeter would read very close to 12 volts. Any voltage reading would indicate a bad or poor ground circuit. The higher the voltage, the greater the problem.

Ohmmeters

An ohmmeter measures resistance to current flow in a circuit. The two leads of the ohmmeter are placed across or in parallel with the circuit or component being tested **(Figure 16–16)**. The red lead is placed on the positive side of the circuit and the black lead is placed on the negative side of the circuit. The meter sends current through the component and determines the amount of resistance based on the voltage dropped across the load. The scale of an ohmmeter reads from 0 to infinity (∞). A 0 reading means there is no resistance in the circuit and may indicate a short in a component that should show a specific resistance. An infinity reading indicates a number higher

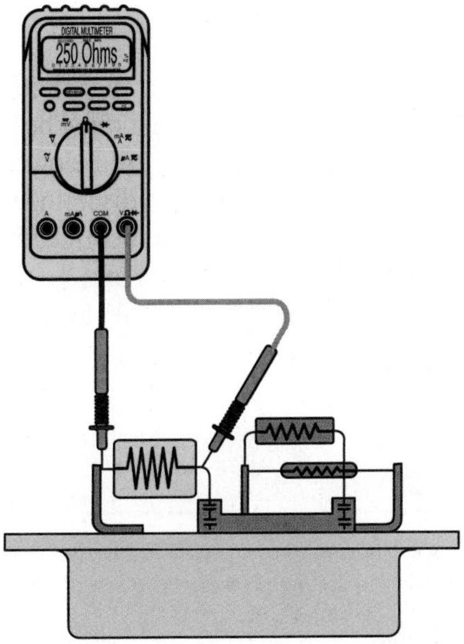

Figure 16–16 An ohmmeter can test the resistance of a component after it has been removed from the circuit.

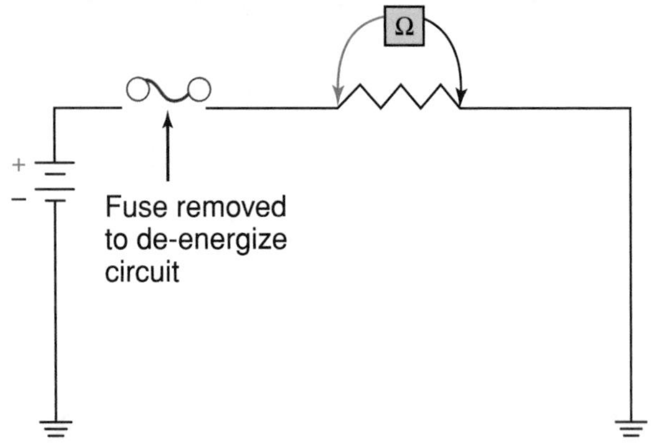

Figure 16–17 Measuring resistance using an ohmmeter. Note that the circuit fuse is removed to de-energize the circuit.

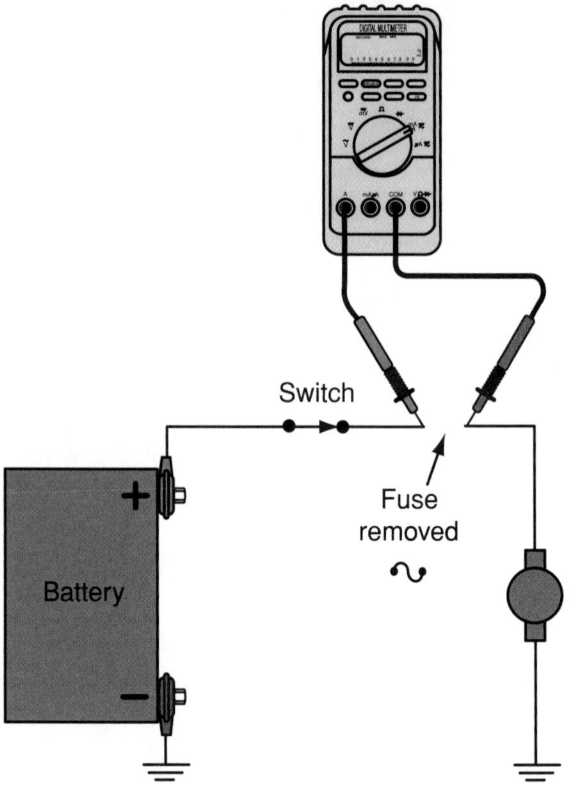

Figure 16–18 An ammeter should be connected in series with the circuit; the best way to do this is to remove the fuse and insert the meter.

than the meter can measure. This reading usually is an indication of an open circuit.

Ohmmeters are used to test circuit continuity and resistance with no power applied. In other words, the circuit or component must first be disconnected from the power source **(Figure 16–17)**. Connecting an ohmmeter into a live circuit usually results in damage to the meter.

Ohmmeters are also used to trace and check wires or cables. Connect one probe of the ohmmeter to the known wire at one end of the cable and touch the other probe to each wire at the other end of the cable. Any evidence of resistance indicates the correct wire. Using this same method, you can check for a defective wire. If low resistance is shown on the meter, the wire is sound. If no resistance is measured, the wire is open. If the wire is okay, continue checking by connecting the probe to other leads. Any indication of resistance indicates that the wire is shorted to one of the other wires and the harness is defective.

Ammeters

An ammeter measures current flow in a circuit. Current is measured in amperes. An ammeter must be placed into the circuit or in series with the circuit being tested **(Figure 16–18)**. Normally, this requires disconnecting a wire or connector from a component and connecting the ammeter between the wire or connector and the component. The red lead of the ammeter should always be connected to the positive side of the connector and the black lead to the other side.

WARNING!

Never place the leads of an ammeter across the battery or a load. Doing so puts the meter in parallel with the circuit and will blow the fuse in the ammeter or possibly destroy the meter.

It is much easier to test current using an ammeter with an inductive pickup or a current probe **(Figure 16–19)**. Both use an inductive pickup to measure the current in a wire. The amount of current is displayed on the meter or the handle of the current probe. The pickup clamps around the circuit's power wire so there is no need to separate the circuit. The amperage reading is based on the strength of the magnetic field created by the current flowing through the wire.

Let us look at a circuit that normally draws 5 amps and is protected by a 6-amp fuse. If the circuit constantly blows the fuse, a short exists somewhere in the circuit

Figure 16–19 A low-amp current probe. *Courtesy of OTC Tool and Equipment, Division of SPX Corporation*

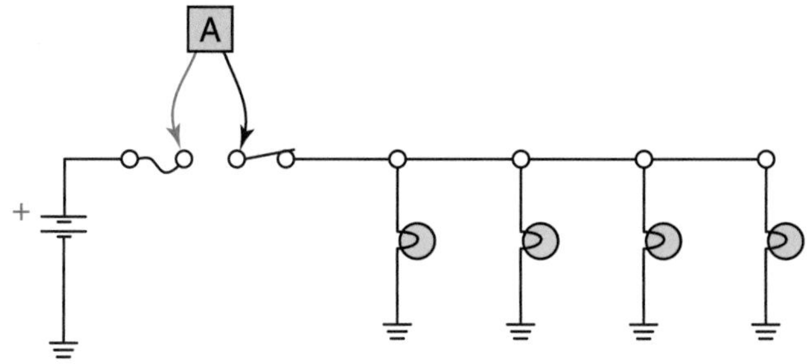

Figure 16-20 Checking a circuit using an ammeter.

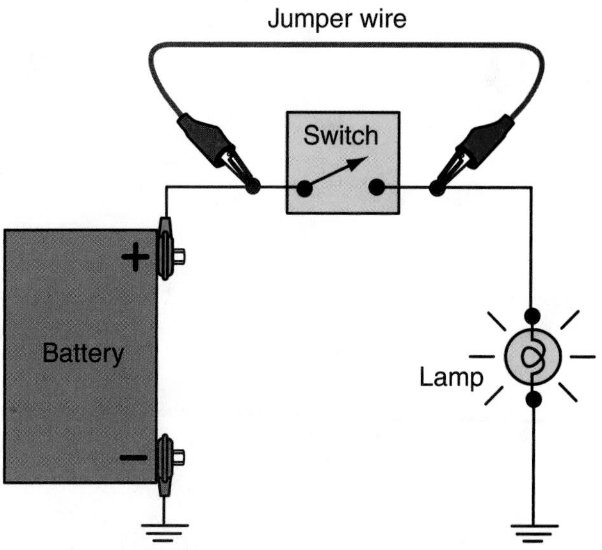

Figure 16-21 A jumper wire can be used to bypass a switch.

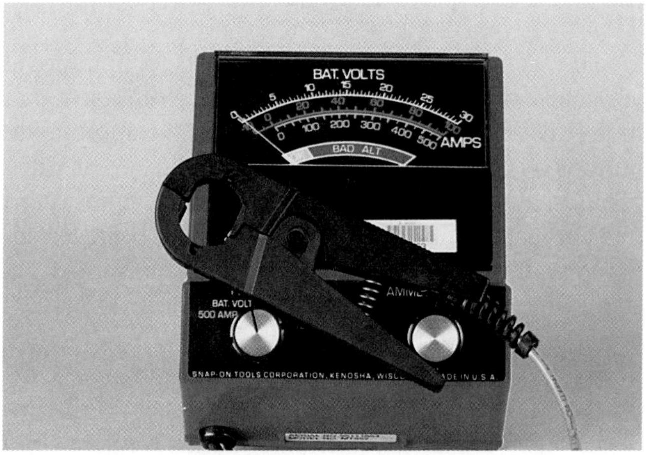

Figure 16-22 A VAT with an adjustable resistor pile.

(Figure 16–20). Mathematically, each light should draw 1.25 amperes (5 ÷ 4 = 1.25). To find the short, disconnect all lights by removing them from their sockets. Then, close the switch and read the ammeter. With the loads disconnected, the meter should read 0 ampere. If there is any reading, the wire between the fuse and the sockets is shorted to ground.

If 0 amp was measured, reconnect each light in sequence. The reading should increase 1.25 amperes with each bulb. If, when making any connection, the reading is higher than expected, the problem is in that part of the light circuit.

WARNING!

When testing for a short, always use a fuse. Never bypass the fuse with a wire. The fuse should be rated at no more than 50% higher capacity than specifications. This rating offers circuit protection and provides enough amperage for testing. After the problem is found and corrected, be sure to install a fuse with the specified rating.

Other Test Equipment

Other electrical test equipment may be needed to accurately diagnose an electrical circuit.

Jumper Wires Jumper wires are used to bypass individual wires, connectors, or components. Bypassing a component or wire helps to determine if that part is faulty **(Figure 16–21)**. If the symptom is no longer evident after the jumper wire is installed, the part bypassed is faulty. Technicians typically have jumper wires of various lengths; usually some of the wires have a fuse or circuit breaker in them to protect the circuits being tested.

Volt/Ampere Tester A volt/ampere tester (VAT), shown in **Figure 16–22**, is used to test batteries, starting systems, and charging systems. The tester contains a voltmeter, ammeter, and carbon pile. The carbon pile is a variable resistor. When the tester is attached to the battery, the carbon pile draws current out of the battery. The ammeter reads the amount of current draw. When testing a battery, the resistance of the carbon pile must be adjusted to match the ratings of the battery. The resistance of the carbon pile is either adjusted by the technician or is automatically adjusted by the tester.

Figure 16–23 shows a small battery tester. This battery tester does not have a carbon pile. However, it can be used to test circuit voltage and to perform battery load tests on 6- and 12-volt batteries. It can also be used to check the condition of battery cables and connectors and to check running voltage.

Scan Tools When plugged into the vehicle's diagnostic connector, a scan tool can retrieve fault codes from a computer's memory and digitally display these codes. A scan tool may also perform other diagnostic functions depending on the year and make of the vehicle. Many scan tools have removable modules that are updated each year. These modules are designed to test the computer systems on various makes of vehicles.

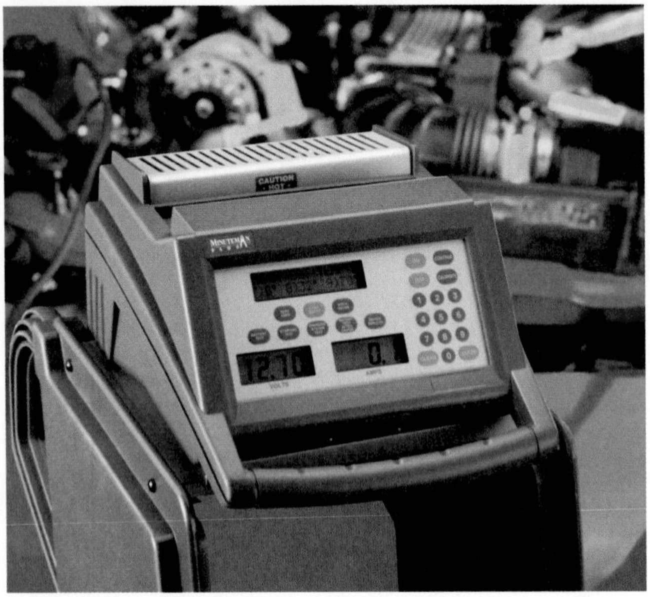

Figure 16-23 A typical automatic VAT. *Courtesy of OTC Tool and Equipment, Division of SPX Corporation*

The scan tool must be programmed for the model year, make of vehicle, and type of engine. With some scan tools, this selection is made by pressing the appropriate buttons on the tester, as directed by the digital tester display. On other scan tools, the appropriate memory card must be installed in the tester for the vehicle being tested. Some scan tools have a built-in printer to print test results, while other scan tools may be connected to an external printer.

Some scan tools display diagnostic information based on the trouble code. Service bulletins may also be indexed on the tool after vehicle information is entered into the tester. Other scan tools display sensor specifications for the vehicle being tested.

Trouble codes are only set by the vehicle's computer when a voltage signal is entirely out of its normal range. The codes help technicians identify the cause of the problem. If a signal is within its normal range but is still not correct, the vehicle's computer will not display a trouble code. However, a problem will still exist. As an aid to identify this type of problem, most manufacturers recommend that the signals to and from the computer be looked at carefully. This is done by observing the serial data stream **(Figure 16–24)** or through the use of a breakout box.

Any scan tool designed for OBD II will work on all OBD-II systems; therefore, the need to have designated scan tools or cartridges is eliminated. The OBD-II scan tool has the ability to run diagnostic tests on all systems and has freeze frame capabilities.

Computer Memory Saver Whenever the vehicle's battery needs to be disconnected, connect a memory saver to the vehicle. This saver may preserve the memory in the radio and electronic accessories and in the engine, transmission, and body computers thus saving you and the vehicle's owner much time and frustration after the services are complete.

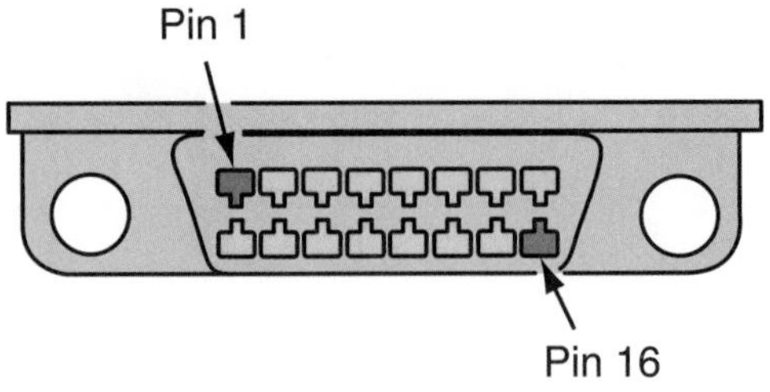

Pin 1	Secondary UART 8192 baud serial data Class B 160 baud serial data
Pin 2	J1850 Bus + line on 2 wire systems, or single wire (class 2)
Pin 3	Ride control diagnostic enable
Pin 4	Chassis ground pin
Pin 5	Signal ground pin
Pin 6	PCM/VCM diagnostic enable
Pin 7	K line for International Standards Organization (ISO) application
Pin 8	Keyless entry enable, or MRD theft diagnostic enable
Pin 9	Primary UART
Pin 10	J1850 Bus-line for J1850-2 wire applications
Pin 11	Electronic variable orifice (EVO) steering
Pin 12	ABS diagnostic, or CCM diagnostic enable
Pin 13	SIR diagnostic enable
Pin 14	E&C bus
Pin 15	L line for International Standards Organization (ISO) application
Pin 16	Battery power from vehicle unswitched (4 amps max)

Figure 16-24 Pin identification of an OBD-II DLC. Serial data is observed through pins 1 and 16.

Two types of memory savers are available or can be made for satisfactory use on late-model cars. For a power source, you can use a 12-volt automotive battery or a 12-volt dry-cell lantern battery. If the cigarette lighter on the car is powered continuously, a cigarette lighter adapter plug with suitable wire leads and large alligator clips can be attached to the auxiliary battery. If the cigarette lighter is controlled by the ignition switch, a set of jumper wires with suitable alligator clips can be connected to the vehicle's electrical system under the hood.

To make either kind of adapter or jumper leads, use number 14 or 16 wire. Install a 5-ampere in-line fuse in the positive lead, along with a diode. The fuse protects the memory saver and the electrical system from an accidental short. The diode prevents current feedback from the vehicle's electrical system to the auxiliary battery.

If you connect the memory saver to the vehicle under the hood, do not connect it to the vehicle's battery cable clamps. Removing and reinstalling the battery will likely dislodge the memory saver alligator clips and render it useless. Instead, connect the memory saver's negative (–) lead to a good engine ground and the positive (+) lead to a point that is hot at all times, such as the battery connection at the alternator or the starter relay. Check a wiring diagram if you are unsure about connection points.

If the memory saver must stay connected while the vehicle is raised and lowered on a hoist, locate the battery securely and out of the way of moving components. If you place the battery inside the car, set it in a large plastic tray or tub to protect the vehicle from electrolyte or battery corrosion.

USING MULTIMETERS

A multimeter is one of the most versatile tools used to diagnose electrical and electronic systems. Multimeters can test DC and AC volts, ohms, and amperes. Usually several test ranges are provided for each of these functions. In addition to these basic electrical tests, multimeters may also test engine speed (rpm), duty cycle, pulse width, diode condition, frequency, and even temperature. The desired test is selected by turning a control knob or depressing keys on the meter.

Multimeters are available with either analog or digital displays. Analog meters **(Figure 16–25)** enjoyed wide popularity prior to electronic control systems. Today, the most commonly used multimeter is the digital volt/ohmmeter (DVOM), which is often referred to as a digital multimeter (DMM). These meters do not use a sweeping needle and scales to display the measurement. Rather, they display the measurement digitally on the meter **(Figure 16–26)**.

There are several drawbacks to using analog-type meters when testing electronic systems. Many electronic components require very precise test results. Digital meters can measure volts, ohms, or amperes in tenths and hundredths. Another problem with analog meters is their

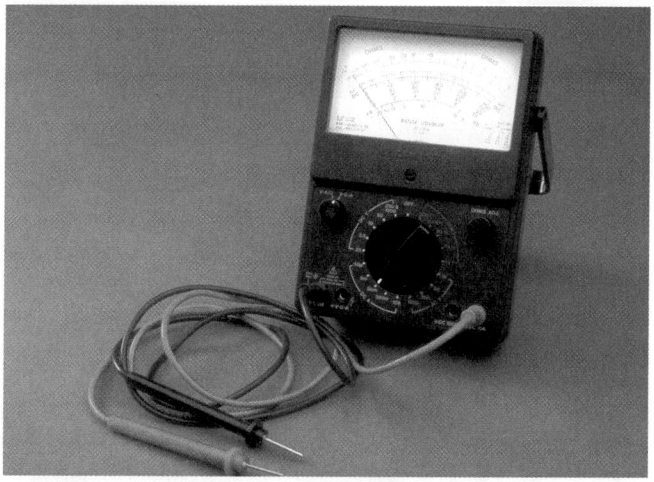

Figure 16–25 An analog meter.

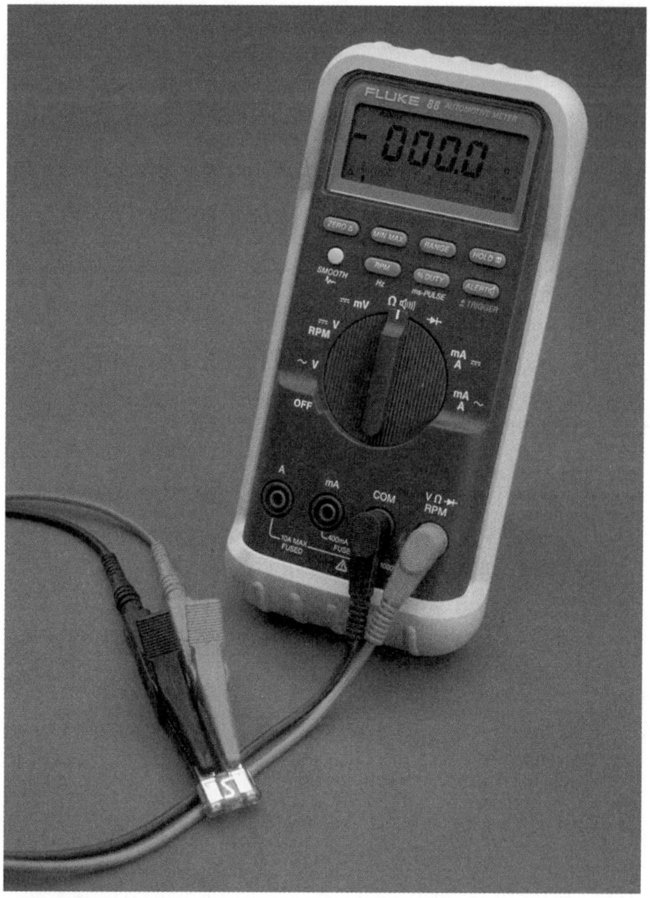

Figure 16–26 A typical DMM.

low internal resistance (input impedance). The low input impedance allows too much current to flow through circuits and should not be used on delicate electronic devices.

Digital meters have high input impedance, usually at least 10 megohms (10 million ohms). Metered voltage for resistance tests is well below 5 volts, reducing the risk of damage to sensitive components and delicate computer circuits. A DMM also gives quite an exact reading and one that is not dependent on the view of a needle against a scale.

Multimeters have either an auto range feature, in which the appropriate scale is automatically selected by the meter, or they must be set to a particular range. In either case, you should be familiar with the ranges and the different settings available on the meter you are using. To designate particular ranges and readings, meters display a prefix before the reading or range. If the meter has a setting for mAmps, this means the readings will be given in milliamps or thousandths of an amp. Ohmmeter scales are expressed as a multiple of ten or use the prefix K or M. K stands for kilo or 1000. A reading of 10K ohms equals 10,000 ohms. An M stands for mega or 1,000,000. A reading of 10M ohms equals 10,000,000 ohms. When using a meter with auto range, make sure you note the

PREFIX	SYMBOL	RELATION TO BASIC UNIT
Mega	M	1,000,000
Kilo	k	1,000
Milli	m	.001 or $\frac{1}{1000}$
Micro	μ	.000001 or $\frac{1}{1000000}$
Nano	n	0.000000001
Pico	p	0.000000000001

Figure 16-27 Common prefixes used on meters.

range being used by the meter. There is a big difference between 10 ohms and 10,000,000 ohms. The common abbreviations and symbols used on multimeters are shown in **Figure 16–27**.

WARNING!

Many DMMs with auto range display the measurement with a decimal point. Make sure you observe the decimal and the range being used by the meter **(Figure 16–28)**. *A reading of .972 K ohm equals 972 ohms. If you ignore the decimal point you will read 972,000 ohms.*

After the test range has been selected, the meter is connected to the circuit in the same way as if it were an individual meter. Be careful; because multimeters do not have heavy leads, the highest ammeter scale on this type of meter is often 10 amperes.

WARNING!

Always be sure the proper scale is selected on the multimeter and the correct connections are made for the component or system being tested. Improper connections or scale selections may blow the internal fuse in the meter or cause meter damage.

0.345 KΩ = 345 Ω

1025 mAmps = 1.025 Amps

Figure 16-28 The placement of the decimal point and the scale should be noticed when measuring with a meter with auto range.

There are two ways multimeters display AC voltage: **root mean square (RMS)** and **average responding**. When an AC voltage signal is a true sine wave, both methods will display the same reading. Since most automotive sensors do not produce pure sine wave signals, it is important to know how the meter will display the AC voltage reading when comparing measured voltage to specifications. RMS meters convert the AC signal to a comparable DC voltage signal. Average responding meters display the average voltage peak. Always check the voltage specification to see if the specification is for RMS voltage. If it is, use an RMS meter.

When using the ohmmeter function, the DMM will show a zero or close to zero when there is good continuity. If there is no continuity, the meter will display an infinite reading. This reading is usually shown as a blinking "1.000," a blinking "1," or an "OL" **(Figure 16–29)**.

Before taking a resistance measurement, calibrate the meter by holding the two leads together and adjusting the meter reading to zero. Not all meters need to be calibrated; some digital meters automatically calibrate when a scale is selected. On meters that require calibration, it is recommended that the meter be zeroed after changing scales.

Some multimeters also feature a MIN MAX function. This function displays the maximum, minimum, and average voltage recorded during the time of the test. This feature is valuable when checking sensors or when looking for electrical noise. Noise is primarily caused by radio frequency interference (RFI), which may come from the ignition system. RFI is an unwanted voltage signal that rides on a signal. This noise can cause intermittent problems with unpredictable results. The noise causes slight increases and decreases in the voltage. When a computer receives a voltage signal with noise, it tries to react to the minute changes. As a result, the computer responds to the noise rather than the voltage signal.

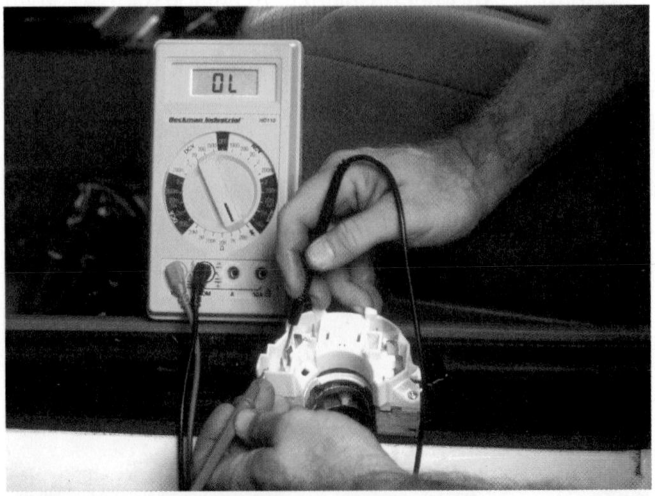

Figure 16-29 On some DMMs, an infinite reading is displayed as "OL".

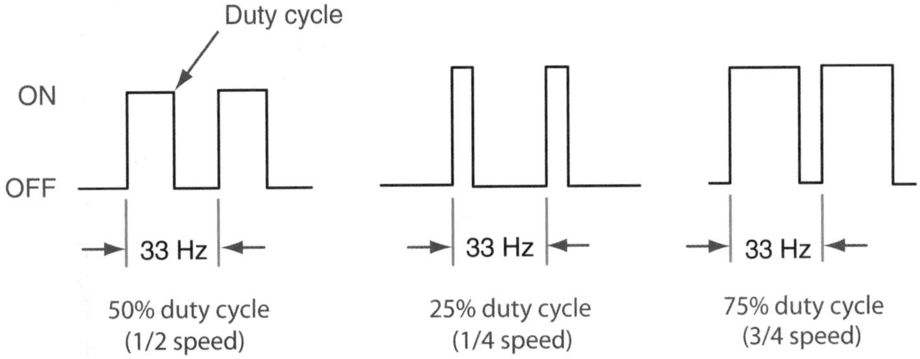

Figure 16-30 Duty cycle is expressed in a percentage.

Multimeters may also have the ability to measure duty cycle, pulse width, and frequency. Duty cycle **(Figure 16–30)** is measured in a percentage. A 60% duty cycle means that a device is on 60% of the time and off 40% of one cycle.

Pulse width is normally measured in milliseconds. When measuring duty cycle, you are measuring the amount of time something is on during one cycle. When measuring pulse width, you are looking at the amount of time something is on.

To accurately measure duty cycle, pulse width, and frequency, the meter's trigger level must be set. The trigger level tells the meter when to start counting. Trigger levels can be set at certain voltage levels or at a rise or fall in the voltage. Normally, meters have a built-in trigger level that corresponds with the voltage range setting. If the voltage does not reach the trigger level, the meter does not recognize a cycle. On some meters you can select between a rise or fall in voltage to trigger the cycle count. Most technicians refer to this as a positive or negative slope trigger. A rise in voltage is a positive increase in voltage. This setting is used to monitor the activity of devices whose power feed is controlled by a computer. A decrease in voltage is negative voltage. This setting is used to monitor ground-controlled devices.

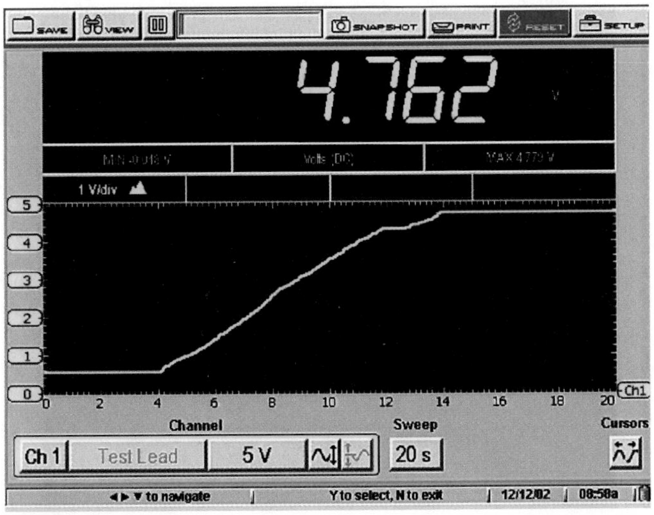

Figure 16-31 A screen on a digital graphing multimeter. *Courtesy of Snap-on Tools Company*

Graphing Multimeters

One of the latest trends in diagnostic tools is a graphing digital multimeter. These meters display readings over time, similar to a lab scope. The graph displays the minimum and maximum readings on a graph, as well as displaying the current reading **(Figure 16–31)**. By observing the graph, a technician can note any undesirable changes during the transition from a low reading to a high reading, or vice versa. These glitches are some of the more difficult problems to identify without a graphing meter or a lab scope.

USING LAB SCOPES

In recent years, an electronic scope, referred to as a *lab scope*, has become the diagnostic tool of choice for many good technicians. A scope may be considered a very fast reacting voltmeter that reads and displays voltages within a specific time. The voltage readings appear as a waveform or trace on the scope's screen. An upward movement of the trace indicates an increase in voltage, and a decrease in voltage results in a downward movement.

When measuring voltage with an analog voltmeter, the meter only displays the average values at the point being probed. Digital voltmeters simply sample the voltage several times each second and update the meter's reading at a particular rate. If the voltage is constant, good measurements can be made with both types of voltmeters. The scope displays any change in voltage as it occurs, which is especially important for diagnosing intermittent problems.

The screen of a lab scope is divided into small divisions of time and voltage **(Figure 16–32)**. These divisions set up a grid pattern on the screen. Time is represented by the horizontal movement of the trace. Voltage is measured with the vertical position of the trace. Because the scope displays voltage over time, the trace moves from

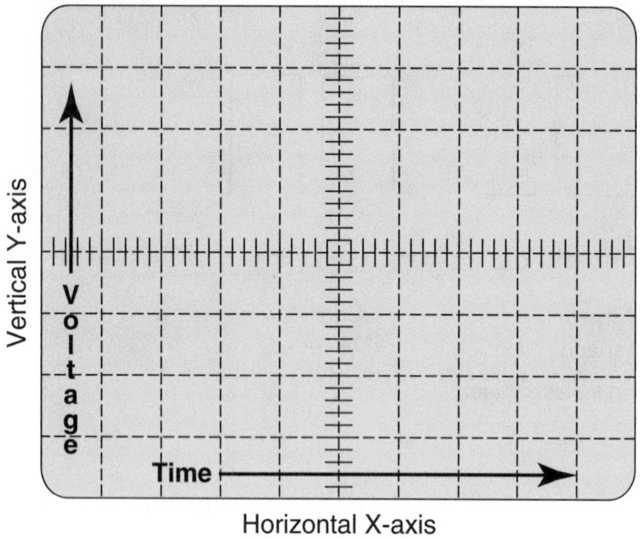

Figure 16-32 Grids on a scope screen that serve as a time and voltage reference.

the left (the beginning of measured time) to the right (the end of measured time). The value of the divisions can be adjusted to improve the view of the voltage trace. For example, the vertical scale can be adjusted so that each division represents 0.5 volt and the horizontal scale can be adjusted so that each division equals 0.005 second (5 milliseconds). This allows for viewing small changes in voltage that occur during a very short period of time. The grid serves as a reference for measurements.

Because a scope displays actual voltage, it displays any electrical noise or disturbances that accompany the voltage signal **(Figure 16–33)**. Electrical disturbances or **glitches** are momentary changes in the signal, which can be caused by intermittent shorts to ground, shorts to power, or opens in the circuit. These problems can occur for only a moment or may last for some time. A lab scope is handy for finding these and other causes of intermittent problems. By observing a voltage signal and wiggling or pulling a wiring harness, any looseness can be detected by a change in the voltage signal.

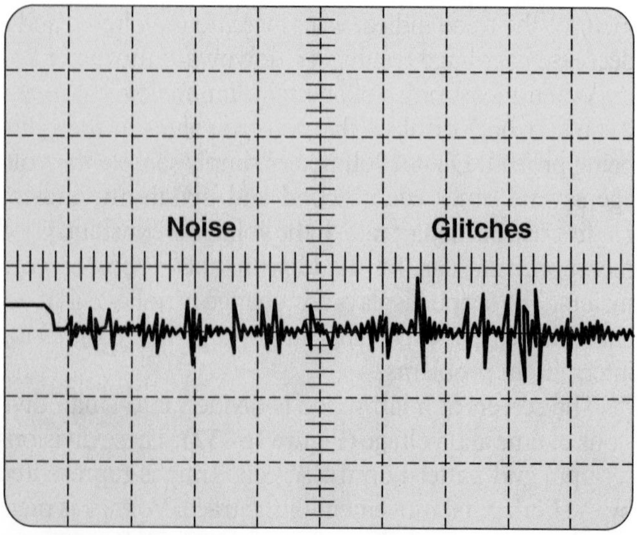

Figure 16-33 RFI noise and glitches may appear on a voltage signal.

Analog versus Digital Scopes

Analog scopes, called *real-time scopes,* show the actual activity of a circuit. This simply means that what is taking place at that time is what you see on the screen. Analog scopes have a fast update rate that allows for the display of activity without delay.

A digital scope, commonly called a **digital storage oscilloscope** or **DSO**, converts the voltage signal into digital information and stores it in its memory. Some DSOs send the signal directly to a computer or a printer or save it to a disk. To help in diagnostics, a technician can "freeze" the captured signal for close analysis. DSOs also have the ability to capture low-frequency signals. Low-frequency signals tend to flicker when displayed on an analog screen. To have a clean waveform on an analog scope, the signal must be repetitive and occurring in real-time. The signal on a DSO is not quite real-time. Rather it displays the signal as it occurred a short time before.

This delay is actually very slight. Most DSOs have a sampling rate of one million samples per second. This rate is quick enough to serve as an excellent diagnostic tool because slight changes in voltage can be observed. Slight and quick voltage changes cannot be observed on an analog scope.

Since digital signals are based on binary numbers, the trace appears slightly choppy when compared to an analog trace. However, the voltage signal is sampled more often, which results in a more accurate waveform. The waveform is constantly being refreshed as the signal is pulled from the scope's memory.

Both an analog and a digital scope can be dual trace **(Figure 16–34)** or multiple trace **(Figure 16–35)** scopes. This means they both have the capability of displaying

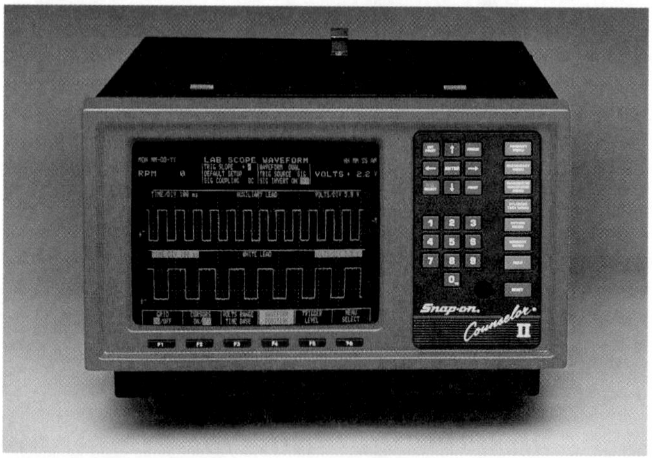

Figure 16-34 A dual trace lab scope. *Courtesy of Snap-on Tools Company*

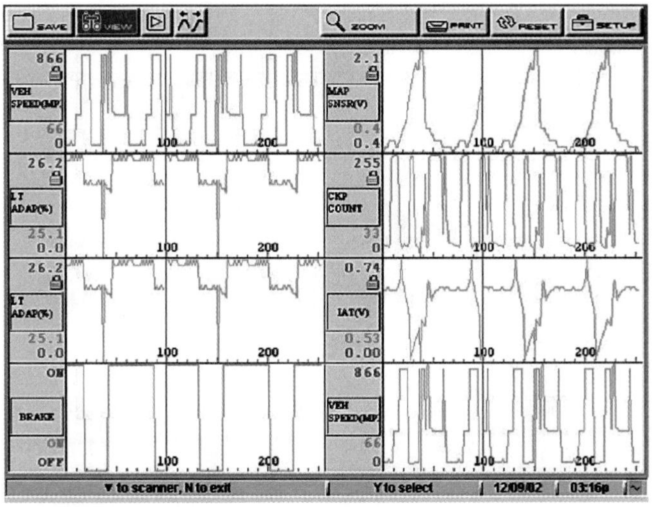

Figure 16-35 Some scopes and graphing scan tools can display many channels. *Courtesy of Snap-on Tools Company*

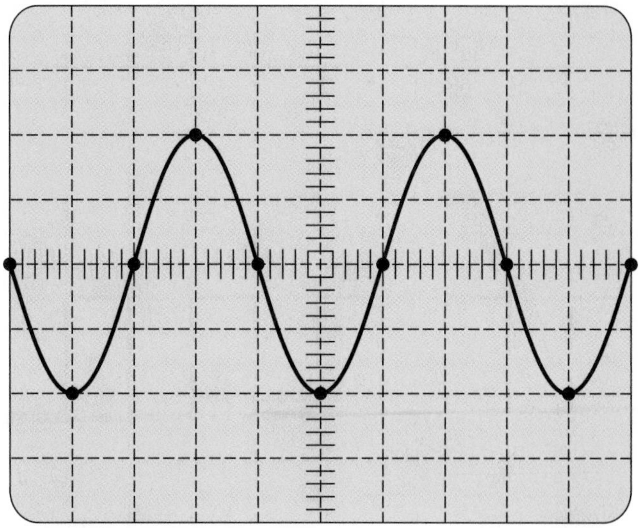

Figure 16-36 An AC voltage sine wave.

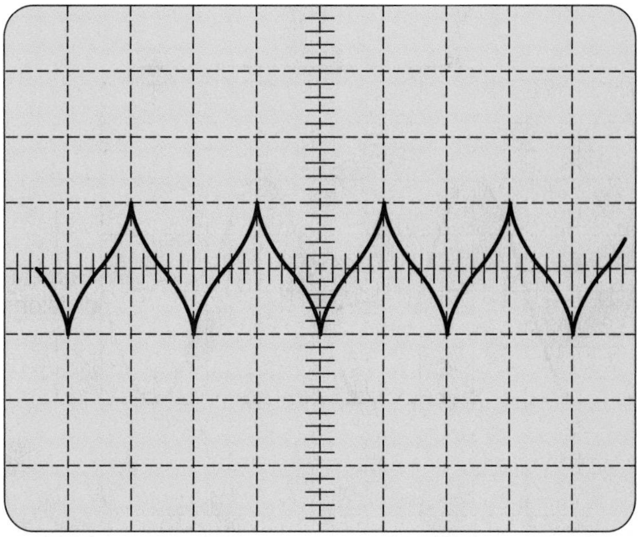

Figure 16-37 An AC voltage trace from a typical permanent magnet generator-type pickup or sensor.

two traces at one time. By watching two traces simultaneously, you can watch the cause and effect of a sensor and compare a good or normal waveform to the one being displayed.

Waveforms

A waveform represents voltage over time. Any change in the amplitude of the trace indicates a change in the voltage. When the trace is a straight horizontal line, the voltage is constant. A diagonal line up or down represents a gradual increase or decrease in voltage. A sudden rise or fall in the trace indicates a sudden change in voltage.

Scopes can display AC and DC voltage, either one at a time or both at the same time, as in the case of noise caused by RFI. Noise results from AC voltage riding on a DC voltage signal. The consistent change of polarity and amplitude of the AC signal causes slight changes in the DC voltage signal. A normal AC signal changes its polarity and amplitude over a period of time. The waveform created by AC voltage is typically called a sine wave **(Figure 16–36)**. One complete sine wave shows the voltage moving from zero to its positive peak then moving down through zero to its negative peak and returning to zero.

One complete sine wave is a cycle. The number of cycles that occur per second is the frequency of the signal. Checking frequency or cycle time is one way of checking the operation of some electrical components. Input sensors are the most common components that produce AC voltage. Permanent magnet voltage generators produce an AC voltage that can be checked on a scope **(Figure 16–37)**. AC voltage waveforms should also be checked for noise and glitches that may send false information to the computer.

DC voltage waveforms may appear as a straight line or line showing a change in voltage. Sometimes a DC voltage waveform will appear as a square wave, which shows voltage making an immediate change **(Figure 16–38)**. Square waves are identified by having straight vertical sides and a flat top. This type of wave represents voltage being applied (circuit being turned on), voltage being maintained (circuit remaining on), and no voltage applied (circuit is turned off). Of course, a DC voltage waveform may also show gradual voltage changes.

Scope Controls

Depending on manufacturer and model of the scope, the type and number of its controls will vary. However nearly all scopes have intensity, vertical (Y-axis), horizontal (X-axis), and trigger adjustments. The intensity control is used to adjust the brightness of the trace, which allows for clear viewing regardless of the light around the scope screen.

The vertical adjustment actually controls the voltage that is shown per division **(Figure 16–39)**. If the scope is

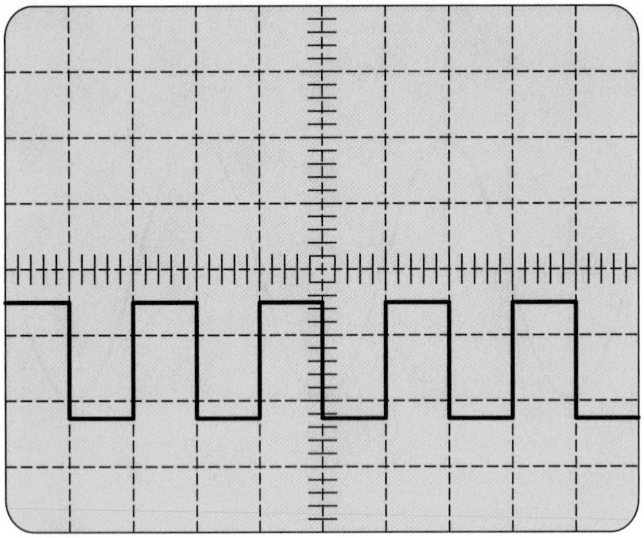

Figure 16–38 A typical square (on-off or high-low) wave.

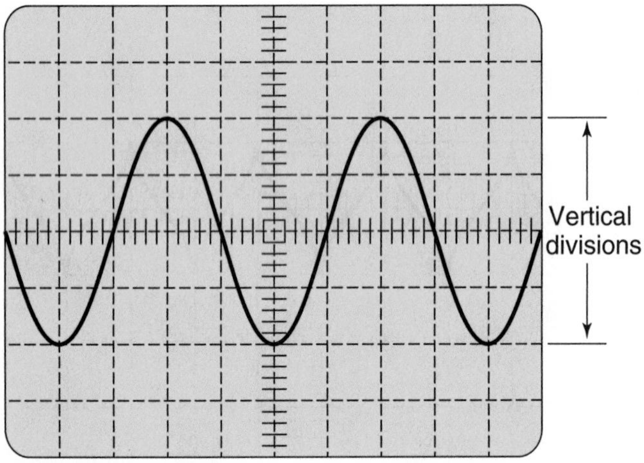

Figure 16–39 Vertical divisions represent voltage.

set at 0.5 (500 milli) volt, a 5-volt signal will need 10 divisions. Likewise if the scope is set to one volt, 5 volts will need only 5 divisions. While using a scope, it is important to set the vertical so that voltage can be read accurately. Setting the voltage too low may cause the waveform to move off the screen, while setting it too high may cause the trace to be flat and unreadable. The vertical position control allows the vertical position of the trace to be moved anywhere on the screen.

The horizontal position control allows the horizontal position of the trace to be set. The horizontal control is actually the time control of the trace **(Figure 16–40)**. If the time per division is set too low, the complete trace may not show across the screen. Also if the time per division is set too high, the trace may be too crowded for detailed observation. The time per division (TIME/DIV) can be set from very short periods of time (millionths of a second) to full seconds.

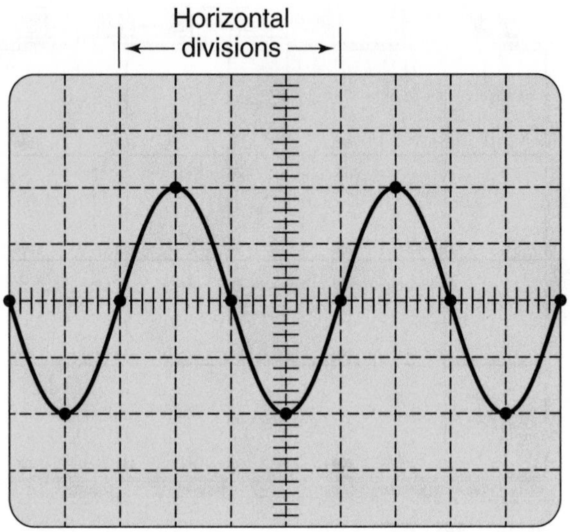

Figure 16–40 Horizontal divisions represent time.

Trigger controls tell the scope when to begin a trace across the screen. Setting the trigger is important when trying to observe the timing of something. Proper triggering allows the trace to repeatedly begin and end at the same points on the screen. There are typically numerous trigger controls on a scope. The trigger mode selector has a NORM and AUTO position. In the NORM setting, no trace appears on the screen until a voltage signal occurs within the set time base. The AUTO setting will display a trace regardless of the time base.

Slope and level controls are used to define the actual trigger voltage. The slope switch determines whether the trace will begin on a rising or falling of the signal **(Figure 16–41)**. The level control sets when the time base is triggered according to a certain point on the slope.

A trigger source switch tells the scope which input signal to trigger on. This can be Channel 1, Channel 2, line

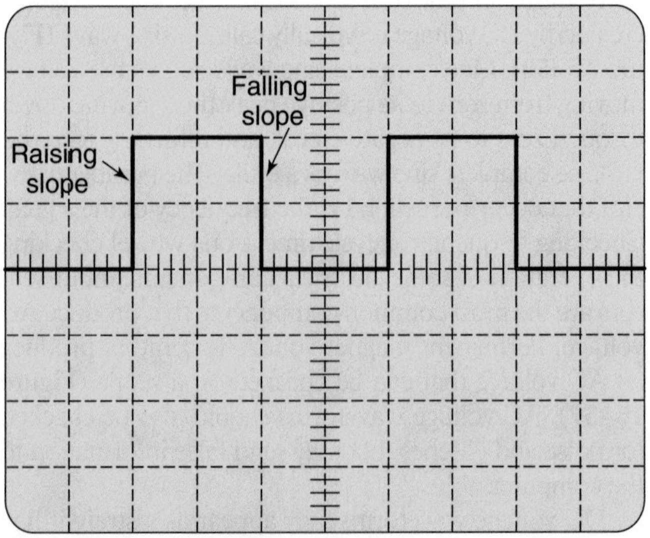

Figure 16–41 The trigger can be set to start the trace with a rise or fall of voltage.

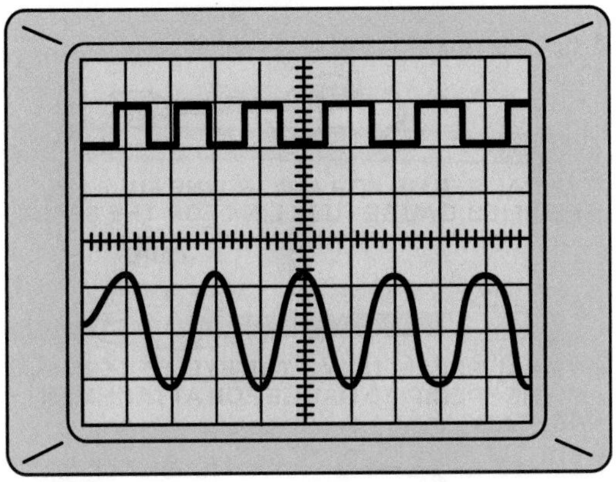

Figure 16-42 Two or more different signals can be observed on a multiple-channel scope, which is invaluable for diagnosis.

voltage, or an external signal. External signal triggering is very useful when observing the trace of a component that may be affected by the operation of another component. An example of this would be observing fuel injector activity when changes in throttle position are made. The external trigger would be voltage changes at the Throttle Position Sensor. The displayed trace would be the cycling of a fuel injector. Channel 1 and Channel 2 inputs are determined by the points of the circuit being probed **(Figure 16–42)**. Some scopes have a switch that allows inputs from both channels to be observed at the same time or alternately.

TESTING BASIC ELECTRICAL COMPONENTS

All electrical components can fail. Testing them is the best way of determining if they are good or bad. For the most part, the proper way of checking electrical components is determined by what the component is supposed to do. If we think about what something is supposed to do, we can figure out how to test it. Often, removing the component and testing it on a bench is the best way to check it.

Protection Devices

When overloads or shorts in a circuit cause too much current, the wiring in the circuit heats up, the insulation melts, and a fire can result unless the circuit has some kind of protective device. Fuses, fuse links, maxi-fuses, and circuit breakers are designed to provide protection from high current. They may be used singularly or in combination.

WARNING!

Fuses and other protection devices normally do not wear out. They go bad because something went wrong. Never replace a fuse or fusible link, or reset a circuit breaker, without finding out why it went bad.

All types of circuit protection devices can be checked with an ohmmeter or test light. If the fuse is good, there will be continuity through it. To test a circuit protection device with a voltmeter, check for available voltage at both terminals of the unit **(Figure 16–43)**. If the device is good, voltage will be present on both sides. A test light can be used in place of a voltmeter.

Measuring voltage drop across a fuse or other circuit protection device will tell you more about its condition than just whether it is open. If a fuse, a fuse link, or a circuit breaker is in good condition, a voltage drop of zero will be measured. If 12 volts is read, the fuse is open. Any reading between zero and 12 volts indicates it has resistance and should be replaced. Make sure you check the fuse holder for resistance as well.

Fuses A cartridge fuse can be visually checked by looking for a break in the internal metal strip. Discoloration of the glass cover or glue bubbling around the metal end caps is an indication of overheating. To visually check a blade or spade fuse, pull it from the fuse panel and look at the fuse element through the transparent plastic

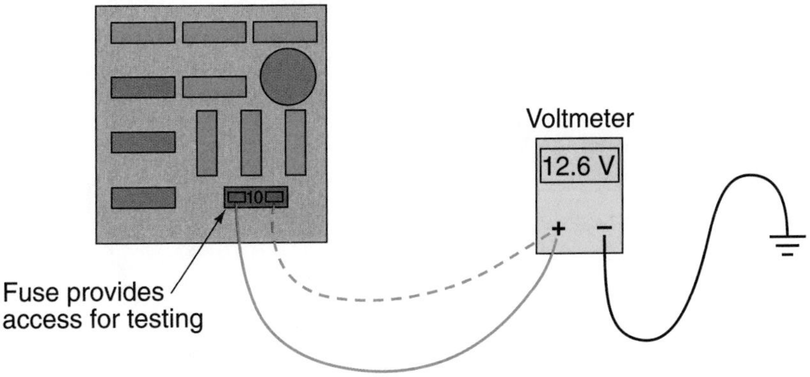

Figure 16-43 Circuit protection devices can be tested with a voltmeter. Make sure there is voltage present on both sides of the device.

housing. Look for internal breaks and discoloration. The ceramic fuse is checked by looking for a break in the contact strip on the outside of the fuse.

Sometimes it is necessary to protect a device in a portion of a circuit even though the entire circuit is protected by a fuse in the fuse panel. This is done by installing an **in-line fuse** in the power wire for the device. In-line fuses are primarily used on accessories that are very sensitive to power surges, such as radios and compact disc players. They are also used with driving lights and power antennas. Normally these fuses are close to the units they protect.

SHOP TALK

To calculate the correct fuse rating, use Watt's law: watts divided by volts equals amperes. For example, if you are installing a 55-watt pair of fog lights, divide 55 by battery voltage to find how much current the circuit will draw. In this case, the current is approximately 5 amperes. To allow for current surges, the correct in-line fuse should be rated slightly higher than the normal current flow. In this case, an 8- or 10-ampere fuse would do the job. ■

Fuse Links Fuse link wire is covered with a special insulation that bubbles when it overheats, indicating that the link has melted. If the insulation appears good, pull lightly on the wire. If the link stretches, the wire has melted. Of course, when it is hard to determine if the fuse link is burned out, check for continuity through the link with a test light or ohmmeter.

WARNING!

Do not mistake a resistor wire for a fuse link. A resistor wire is generally longer and is clearly marked "Resistor—do not cut or splice."

To replace a fuse link, cut the protected wire where it is connected to the fuse link. Then, tightly crimp or solder a new fusible link of the same rating as the original link **(Figure 16–44)**.

WARNING!

Always disconnect the battery ground cable prior to servicing any fuse link.

Maxi-Fuses Maxi-fuses are easier to inspect and replace than fuse links. To check a maxi-fuse, look at the fuse element through the transparent plastic housing. If there is a break in the element, it has blown. To replace it, pull it from its fuse box or panel. Always replace a blown maxi-fuse with one that has the same ampere rating.

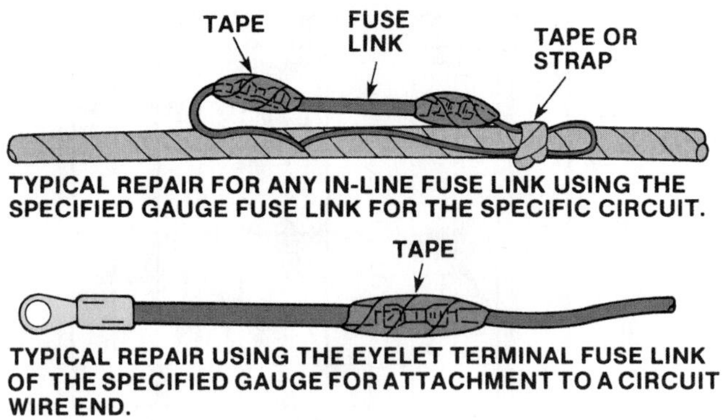
TAPE
FUSE LINK
TAPE OR STRAP
TYPICAL REPAIR FOR ANY IN-LINE FUSE LINK USING THE SPECIFIED GAUGE FUSE LINK FOR THE SPECIFIC CIRCUIT.
TAPE
TYPICAL REPAIR USING THE EYELET TERMINAL FUSE LINK OF THE SPECIFIED GAUGE FOR ATTACHMENT TO A CIRCUIT WIRE END.

Figure 16-44 Typical fusible link repair.

Circuit Breakers Two types of noncycling or resettable circuit breakers are used. One is reset by removing the power from the circuit. The other type is reset by depressing a reset button. If a circuit breaker cannot be reset and remains open, replace it after you make sure there is not excessive current in the circuit.

Thermistors Some systems use a positive temperature coefficient (PTC) thermistor as a protection device. When there is high current, the resistance of the thermistor increases and causes a decrease in current flow. These can be checked with an ohmmeter. If an infinite reading is displayed, the thermistor is open. Another way of checking a thermistor is to change its temperature and see if its resistance changes.

Switches

To check a switch, disconnect the connector at the switch. Check for continuity between the terminals of the switch **(Figure 16–45)** with the switch in the on and off positions. While in the off position, there should be no continuity between the terminals. With the switch on, there should be good continuity between the terminals. If the switch is activated by something mechanical and does not complete the circuit when it should, check the adjustment

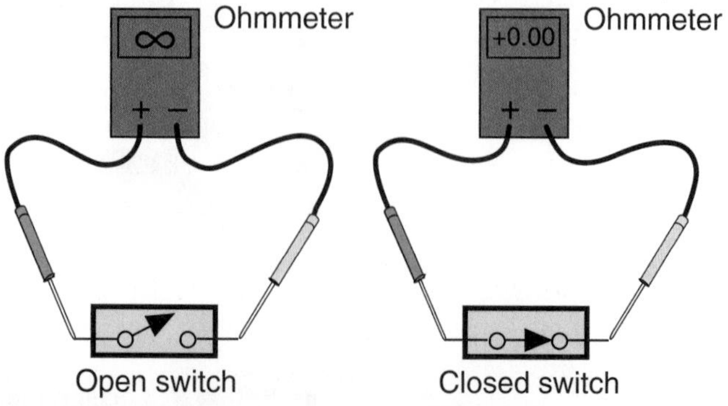
Ohmmeter
∞
+ –
Ohmmeter
+0.00
+ –
Open switch
Closed switch

Figure 16-45 Checking a switch with an ohmmeter.

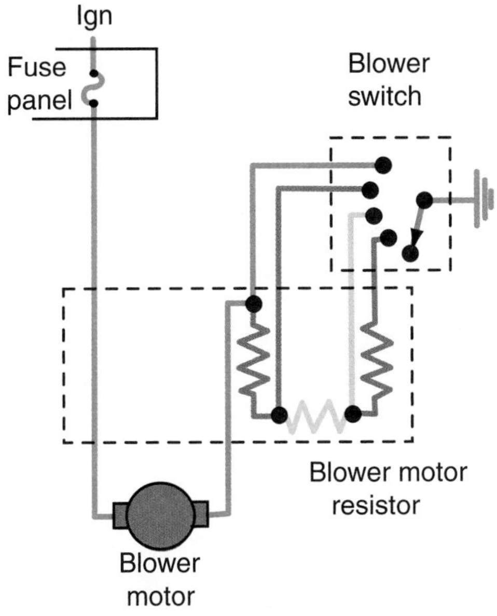

Figure 16–46 An MPMT switch should be checked in all of its possible positions. Use a wiring diagram to guide your tests.

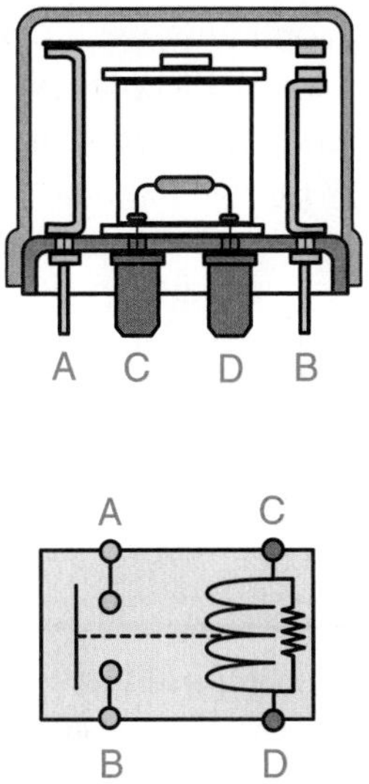

Figure 16–47 Use a wiring diagram to identify the terminals of a relay so it can be tested properly.

of the switch. (Some switches are not adjustable.) If the adjustment is correct, replace the switch. Another way to check a switch is to simply bypass it with a jumper wire. If the component works when the switch is jumped, the switch is bad. An MPMT switch should be checked in each of its possible positions **(Figure 16–46)**. Use a wiring diagram to identify which terminals of the switch should have continuity during each switch position.

Voltage drop across switches should also be checked. Ideally when the switch is closed there should be no voltage drop. Any voltage drop indicates resistance, and the switch should be replaced.

Relays

A relay can be checked with a jumper wire, voltmeter, ohmmeter, or test light. If the terminals are accessible and the relay is *not* controlled by a computer, a jumper wire and test light will be the quickest method. Check the wiring diagram for the relay being tested to identify the terminals **(Figure 16–47)** and to determine if the control is through an insulated or ground switch.

If the relay is controlled on the ground side, follow this procedure to test the relay.

1. Use the test light to check for voltage at the battery side of the relay. If voltage is not present, the fault is in the battery feed to the relay. If there is voltage, continue testing.
2. Probe for voltage at the control terminal. If voltage is not present, the relay coil is faulty. If voltage is present, continue testing.
3. Use a jumper wire to connect the control terminal to a good ground. If the relay works, the fault is in the control circuit. If the relay does not work, continue testing.
4. Connect the jumper wire from the battery to the output terminal of the relay. If the device operated by the relay works, the relay is bad. If the device does not work, the circuit between the relay and the device or the device's ground is faulty.

If the relay is controlled by a computer, do not use a test light; rather, use a high impedance voltmeter set to the 20-V DC scale, then:

1. Connect the negative lead of the voltmeter to a good ground.
2. Connect the positive lead to the output wire. If no voltage is present, continue testing. If there is voltage, disconnect the relay's ground circuit. The voltmeter should now read 0 volt. If it does, the relay is good. If voltage is still present, the relay is faulty.
3. Connect the positive voltmeter lead to the power input terminal. If near battery voltage is not measured there, the relay is faulty. If it is, continue testing.
4. Connect the positive lead to the control terminal. If near battery voltage is not measured there, check the circuit from the battery to the relay. If it is, continue testing.
5. Connect the positive lead to the relay ground terminal. If more than 1 volt is present, the circuit has a poor ground.

If the relay terminals are not accessible, remove the relay from its mounting and test it on a bench. Use an ohmmeter to test for continuity between the relay coil terminals. If the meter indicates an infinite reading, replace the coil. If there is continuity, use a pair of jumper wires to energize the coil. Check for continuity through the relay contacts. If there is an infinite reading, the relay is faulty. If there is continuity, the relay is good and the circuits need to be tested.

Check the service manual for resistance specifications and compare your readings to them. Low resistance across a coil would indicate that it is shorted. If the resistance is too low, the transistors and/or driver circuits could be damaged due to the excessive current flow that would result.

Stepped Resistors

The best way to test a stepped resistor is to use an ohmmeter. To do this, remove the resistor from its mounting. Connect the ohmmeter leads to the two ends of the resistor. Compare the results against specifications. Make sure the ohmmeter is set to the correct scale for the anticipated amount of resistance.

A stepped resistor can also be checked with a voltmeter. Measure the voltage after each part of the resistor block and compare the readings to specifications.

Variable Resistors

A common way to test a variable resistor is with an ohmmeter; however, one can also be checked by observing its output voltage. To test a rheostat, identify the input and output terminals and then connect the ohmmeter across them. Rotate the control while observing the meter. If the resistance values do not match their specifications, or if there is a sudden change in resistance as the control is moved, the unit is faulty. When using a voltmeter, the readings should be smooth and consistent as the control is moved.

To test a potentiometer, connect an ohmmeter across the resistor. The reading should be within the range listed in the specifications. Then move the leads to the input and output of the switch. Manually change the resistance. The readings should sweep evenly and consistently within the specified resistance values. The condition of a potentiometer can also be checked with a voltmeter or a lab scope. Compare the voltage changes with the specifications.

Wiring

Wire insulation should always be in good condition. Broken, frayed, or damaged insulation that exposes live wires can cause shorts. These conditions can also create a safety hazard. Replace all wires that have damaged insulation.

When checking a circuit, make sure to check the ground connection as well, including the ground strap from the engine or other component to the chassis. An engine ground is typically a braided, flat cable. A bad ground cable can cause problems in many different circuits.

The best way to check a wire is to check the voltage drop across it. If the wire is in good shape, there should be very little or no voltage drop.

The current-carrying capacity and the resistance of an electrical wire are determined by its length and gauge (size). The wire sizes are established by the SAE, using the American wire gauge (AWG) system. Sizes are identified by a numbering system ranging from 0 to 20, with 0 being the largest and 20 the smallest in a cross-sectional area. Metric wiring sizes are expressed in a measurement of the cross-sectional area of the wire in square millimeters (mm^2). Metric wire gauge size increases with an increase in the actual size of the wire.

Selection of the correct gauge wire is very important. The wire should be large enough to ensure safe and reliable performance. However, overly large wires add weight and expense to the vehicle. If too small a wire is used, a voltage drop can occur due to electrical resistance. The two factors that should always be considered when determining the correct size of a wire are the total amperage in the circuit and the total length of wire used in each circuit, including the ground. Allowance for the return circuits, including grounded returns, has been computed and is shown in **Table 16–2.**

Wires are usually grouped together in harnesses. A single-plug harness connector may form the connections for many circuits. Harnesses and harness connectors help organize the vehicle's electrical system and provide a convenient starting point for testing many circuits. Most major wiring harness connectors are located in the vehicle's dash or firewall area.

Printed Circuits

Late-model vehicles use flexible printed circuit boards. Printed circuit boards are not serviceable and in some cases not visible. When these boards fail, the entire unit is replaced.

The following precautions should be observed when working with a printed circuit:

- Never touch the surface of the board. Dirt, salts, and acids on your fingers can etch the surface and set up a resistive condition.
- The copper conductors can be cleaned with a commercial cleaner or by lightly rubbing a pencil eraser across the surface.
- A printed circuit board is easily damaged because it is very thin. Be careful not to tear the surface, especially when plugging in connectors or bulbs.

TROUBLESHOOTING CIRCUITS

To troubleshoot a problem, always begin by verifying the customer's complaint. Operate the system and others, to get a complete understanding of the problem. Often there

TABLE 16–2 WIRE SIZE AND LENGTH

Total Approximate Circuit Amperes	**Wire Gauge (for Length in Feet)**								
12V	**3**	**5**	**7**	**10**	**15**	**20**	**25**	**30**	**40**
1.0	18	18	18	18	18	18	18	18	18
1.5	18	18	18	18	18	18	18	18	18
2	18	18	18	18	18	18	18	18	18
3	18	18	18	18	18	18	18	18	18
4	18	18	18	18	18	18	18	16	16
5	18	18	18	18	18	18	18	16	16
6	18	18	18	18	18	18	16	16	16
7	18	18	18	18	18	18	16	16	14
8	18	18	18	18	18	16	16	16	14
10	18	18	18	18	16	16	16	14	12
11	18	18	18	18	16	16	14	14	12
12	18	18	18	18	16	16	14	14	12
15	18	18	18	18	14	14	12	12	12
18	18	18	16	16	14	14	12	12	10
20	18	18	16	16	14	12	10	10	10
22	18	18	16	16	12	12	10	10	10
24	18	18	16	16	12	12	10	10	10
30	18	16	16	14	10	10	10	10	10
40	18	16	14	12	10	10	8	8	6
50	16	14	12	12	10	10	8	8	6
100	12	12	10	10	6	6	4	4	4
150	10	10	8	8	4	4	2	2	2
200	10	8	8	6	4	4	2	2	1

Note: 18 AWG as indicated above this line could be 20 AWG electrically. 18 AWG is recommended for mechanical strength.

are other problems that are not as evident or bothersome to the customer that will provide helpful information for diagnostics. Obtain the correct wiring diagram for the car and study the affected circuit. From the diagram, you should be able to identify testing points and likely problem areas. Test and use logic to identify the cause of the problem.

The logic that you use should be based on the wiring diagram of the vehicle and the type of problem you suspect. If there is an open anywhere in the circuit, the ammeter will read zero current. Without current flow there will be no voltage drops. Voltage drops across a load that are less than battery voltage indicate other resistances in the circuit. Battery voltage after a load indicates an open circuit. A short is indicated by excessive current and/or abnormal voltage drops.

According to many manuals, the maximum allowable voltage drop for an entire circuit, except for the drop across the load, is 10% of the source voltage. Although 1.2 volts is the maximum acceptable amount, it is still too much. Many good technicians use 0.5 volt as the maximum allowable drop. However, there should be no more than 0.1 volt dropped across any one wire or connector. This is the most important specification to consider and remember.

An ohmmeter may be used to measure the values of a component and compare them to specifications. If there is no continuity, it is open. If there is more resistance than called for, there is high internal resistance. If there is less resistance than specified, the part is shorted.

Before you try to use a wiring diagram, you must first have an idea of what you are looking for. The problem may be a particular component, circuit, or connector. The best way to start the process is by identifying the component or one of the components that does not work correctly. Then look in the index for the wiring diagram and find where that component is shown.

The electrical section in most service manuals divides the automobile into individual circuits. This approach makes it easier to find a particular component. Of course you still need to use the index to find the page the component and its circuit are on.

If the service manual uses a total vehicle wiring diagram, finding the component may be a little trickier. Wiring diagrams are usually indexed by grids. The diagram is marked into equal sections like a street map. The wiring diagram's index will list a letter and number for each major component and many different connection points. If the wiring diagram is not indexed, you can locate the component by relating its general location in the vehicle to a general location on the wiring diagram. Most system diagrams are drawn so the front of the car is on the left of the diagram.

Once you have found the component or part of the circuit you are looking for, identify all of the components, connectors, and wires related to that component by tracing through the circuit, starting at the component. Tracing does not mean taking a pencil and marking on the wiring diagram. Tracing means drawing out the circuit on another piece of paper. It does not need to be pretty, it just needs to be accurate. Doing this will prevent you from being distracted by wires that probably are not the cause of the problem. You should identify the malfunctioning component's source of power, its control circuit (such as the switch), and the ground. If the power lead feeds more than one component or the ground is shared by other components, check the operation of those components. If they operate normally, you know the common power and ground circuits are good. Therefore the problem must be between the common points. Likewise if the other components do not work correctly, you know the problem is within the common part of the circuit.

After you have traced the circuit, study it and make sure you know how the circuit is supposed to work. Then describe the problem and ask yourself what could cause it? Limit your answers to what is included in your traced wiring diagram and to the description of the problem. It is wise to make a list of all the most likely causes of the problem, then number them according to probability. For example, if no dashlights come on, it is possible that all of the bulbs are burned out. However, it is not as probable as a blown fuse. After you have listed the probable causes, in order of probability, then look at the wiring diagram to identify how you can quickly test to find out if each is the cause.

Testing for Opens

An open is evident by an inoperative component or circuit. Begin your testing at the most accessible place in the circuit and work from there. Check for voltage at the positive side of the load. If there are zero volts, move to the output of the control **(Figure 16–48)**. If there are at least 10.5 volts, the open is between the control and the load. If the reading is 10.5 volts or higher, check the ground side of the load. If the voltage there is 1 volt or lower and the load does not work, the load is bad. If the voltage at ground is greater than 1 volt, there is excessive resistance or an open in the ground circuit. If the voltage at the positive side of the load was less than 10.5 volts, move the positive lead of the voltmeter toward the battery, testing all connections along the way. If 10.5 volts or more are present at any connector, there is an open between that point and the point previously checked. If battery voltage was present at the ground of the load, there is an open in the ground circuit. Use a jumper wire to verify the location of the open.

Testing for Shorts

Use an ohmmeter to check for an internal short in a component. If the component is good, the meter will read the specified resistance or at least some resistance. If it is shorted, it will read lower than normal or zero resistance.

If the short is between circuits, check the wiring of the affected circuits for signs of burned insulation and melted conductors. Also, check the connectors shared by the two

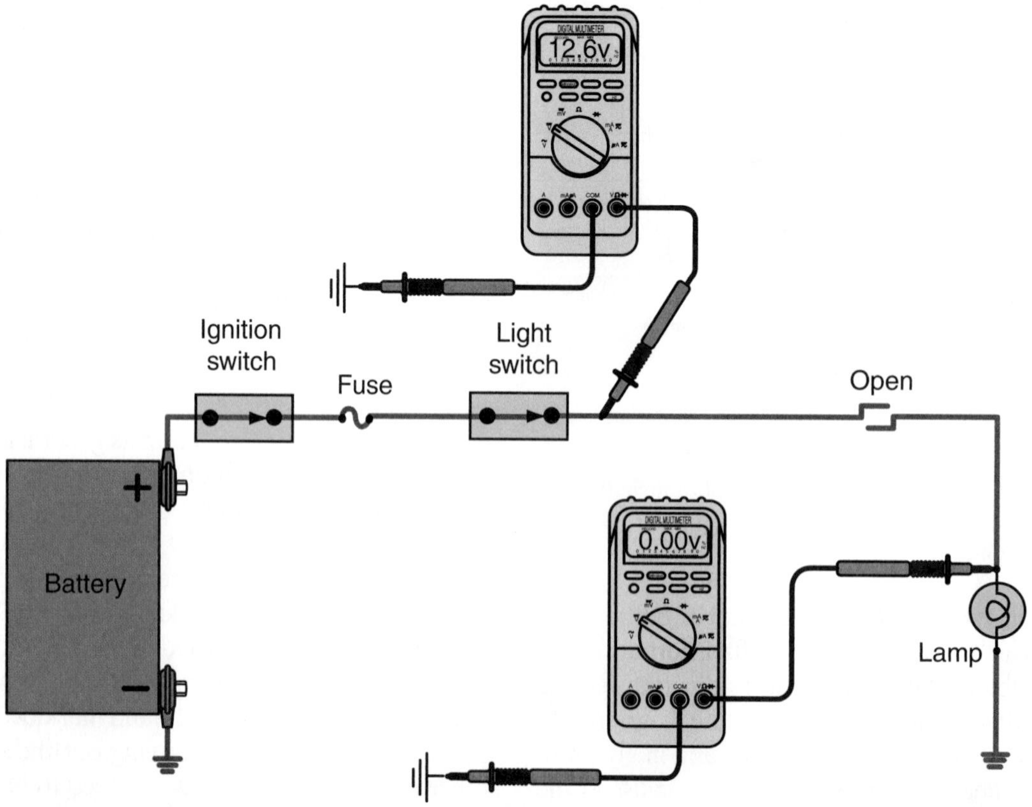

Figure 16–48 If you have zero volts at the load, test the output of the switch. If there is power there, the open is between the switch and the load.

affected circuits. If a visual inspection does not identify the cause of a wire-to-wire short, remove one of the fuses for the affected circuits. Install a buzzer across the fuse holder terminals. Activate the circuit that the buzzer is connected to. Disconnect the loads that are supposed to be activated by the switch. Then disconnect the wire connectors in the circuit that connects the load to the switch. If the buzzer stops when a connector is disconnected, the short is in that circuit.

If the problem is a short to ground, the circuit's fuse or other protection device will be open. If the circuit is not protected, the wire, connector, or component will be burned or melted. To keep current flowing in the circuit so you can test it, connect a cycling circuit breaker across the fuse holder. The circuit breaker will continue to cycle open and closed, allowing you to test for voltage in the circuit. Connect a test light in series with the cycling circuit breaker. While observing the test light, disconnect individual circuits and components one at a time until the light stays out. The short is in the circuit that was disconnected when the light went out.

Testing for Unwanted Resistance

High-resistance problems are typically caused by corrosion on terminal ends, loose or poor connections, or frayed and damaged wires. Carefully inspect the affected circuit.

Whenever excessive resistance is suspected, both sides of the circuit should be checked. Begin by checking the voltage at the positive side of the load. This should be close to battery voltage unless the circuit contains a resistor located before the load. If the voltage is less than desired, check the voltage drop across the circuit from the switch to the load. If the voltage drop is excessive, that part of the circuit contains the unwanted resistance. If the voltage drop was normal, the high resistance is in the switch or in the circuit feeding the switch.

Check the voltage drop across the switch. If the voltage drop is excessive, the problem is the switch. If the voltage drop is normal, the high resistance is in the circuit feeding the switch. If battery voltage is present at the load, the ground circuit for the load should be checked. Connect the red voltmeter lead to the ground side of the load and the black lead to the grounding point for the circuit. If the voltage drop was normal, the problem is the grounding point. If the voltage drop is excessive, move the black meter lead toward the red. Check voltage drop at each step. Eventually you will read high voltage drop at one connector and then a low voltage drop at the next. The point of high resistance is between those two test points. If the voltage drop is normal, the high resistance is in the switch or in the circuit feeding the switch.

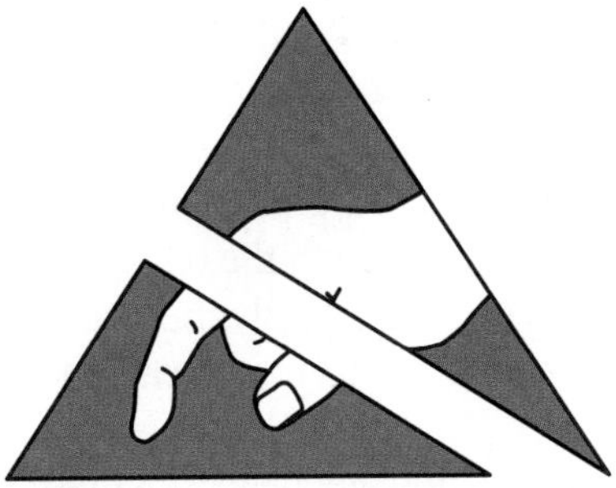

Figure 16-49 General Motors's electrostatic discharge (ESD) symbol warns technicians that a part or circuit is sensitive to static electricity.

PROTECTING ELECTRONIC SYSTEMS

The last thing a technician wants to do when a vehicle comes into the shop for repair is create problems, especially when it comes to electronic components. It is a must to be aware of the proper ways to protect automotive electrical systems and electronic components during storage and repair.

Never touch the electrical contacts on any electrical or electronic part. Avoid touching bare metal contacts. Oils from the skin can cause corrosion and poor contacts.

Many manufacturers mark certain components and circuits with a code or symbol to warn technicians that they are sensitive to electrostatic discharge **(Figure 16–49)**.

Before touching a computer, whether removing or replacing it, always touch a good ground first. This safely discharges any static electricity. Static electricity can generate up to 25,000 volts and can easily damage a computer or other electronic part.

Many tool companies offer static-proof work mats that allow a technician to work inside the vehicle without the fear of creating static electricity. Tool companies also have grounding wrist straps that have a wire to connect the wrist strap to a good ground. When choosing one, make sure it is designed to separate in case of a sudden movement. This feature could prevent a serious wrist injury if you fall or if the vehicle moves while the wrist strap is attached to it.

Accidentally touching two terminals at the same time with a test probe can cause a short circuit. Expensive computer modules or sensors can be destroyed instantly, without warning, by incorrect testing procedures.

If diagnosis indicates a problem with the oxygen sensor, extra caution is required. The oxygen sensor wire carries a very low voltage and must be isolated from other wires. If it is not, nearby wires could add more induced voltage. This gives false data to the computer and can result in a driveability problem. Some vehicle manufacturers use a foam sleeve around the oxygen sensor wire for this purpose. Do not allow grease, lubricants, or cleaning solvents of any kind to touch the end of the sensor or its electrical connector plug. Apply the manufacturer's special antiseize compound to the threads before installing the sensor.

Avoid jump-starting whenever possible. You may create voltage spikes that could damage electronic components. This is true for both the car you are trying to start and the car providing the jump. However, if you must jump-start a car, observe these safety precautions. Connect positive terminal to positive terminal on the batteries, connect the negative terminal of the good battery to a ground other than the ground terminal of the bad battery. Make sure the two vehicles are not touching each other and that every electrical device in the car, including the dome light, is turned off before connecting the batteries. Only after the hookups are properly made should you turn the key in the dead car to get it started. Once the dead car is running, remove all jumper connections before turning on any electrical devices.

Computer diagnostic tests can be performed using a scan tool. Connect the tool to the diagnostic connector. Follow the sequence given in the service manual or the scanning tool's instruction manual. The trouble codes are displayed on the scan tool readout.

Be careful not to damage connectors and terminals when removing electronic components. Some may require special tools to remove them.

When procedures call for connecting test leads or wires to electrical connections, use extreme care and follow the manufacturer's instructions. Identify the correct test terminals before attempting to connect test leads.

The charging system, including the battery, must be in good operating condition. If the electrical system falls below 12 volts, the computer may not work properly but will not be damaged. However, if system voltage exceeds 15 volts, the computer may be damaged.

TESTING ELECTRONIC CIRCUITS AND SYSTEMS

Most electronic circuits can be checked in the same way as other electrical circuits; however, only high-impedance meters should be used.

A DMM may have AC selection modes for voltage and amperage. These modes are used to measure voltages and amperages that change polarity or levels very quickly. Most meters display the average voltage or current in an AC circuit. Some meters display RMS readings, which are very close to being average readings; however, there may be slight differences as this scale compensates for extreme fluctuations in voltage and current flow.

USING SERVICE MANUALS

Specifications given in service manuals are normally not RMS voltage readings. If you compare an RMS reading against specs, you may end up with the wrong conclusion. ■

RMS refers to the effective or useful value of an AC signal. To determine the RMS value, peak voltage readings are multiplied by the RMS constant (0.707). The effective RMS value is calculated by squaring the instantaneous values of all the points on the sine wave, taking the average of these values and extracting the square root. Therefore, the effective value is the root of the mean (average) square of the values along the sine wave. The resulting answer from this mathematical formula is the root mean square **(Figure 16–50)**.

Multimeters can be used to check diodes. Regardless of the bias of the diode, it should allow current flow in one direction only. To test a diode, use an analog ohmmeter. Connect the meter's leads across the diode. Observe the reading on the meter. Then reverse the meter's leads and again observe the reading. The resistance in one direction should be very high or infinite and close to zero in the other direction. If any other readings were ob-

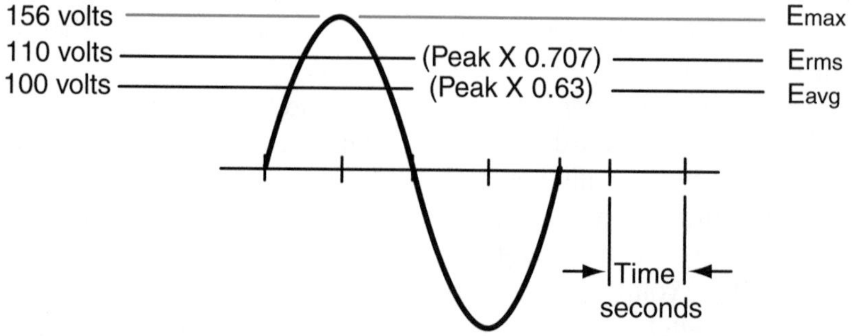

Figure 16-50 AC voltage: RMS.

served, the diode is bad. A diode that has low resistance in both directions is shorted. A diode that has high resistance or an infinite reading in both directions is open.

You may run into problems when checking a diode with a high impedance DMM. Since many diodes do not allow current flow through them unless the voltage is at least 0.6 volt, a digital meter may not be able to forward bias the diode, resulting in readings that indicate the diode is open when in fact it may not be. Because of this problem, many multimeters are equipped with a diode testing feature. This test allows for increased voltage at the test leads. Again, continuity should be present in one direction and not the other. Some meters display the voltage required to forward bias the diode. If the diode is open, the meter will display OL or another reading to indicate infinity or out of range. Some meters make a beeping noise during a diode check when there is continuity.

Diodes may also be tested with a voltmeter. Using the same logic as when testing with an ohmmeter, test the voltage drop across the diode. The meter should read low voltage in one direction and higher voltage in the other direction. Most automotive diodes will drop 500 to 650 mVolts.

A lab scope is a valuable tool for understanding and diagnosing electronic circuits. The scope is primarily used to measure voltages. Scopes are relatively easy to use. Most have two leads that are connected to the circuit in the same way as a voltmeter.

The screen on the scope displays voltage over time, and the increments for voltage and time can be changed to provide a good look at the activity of a component or circuit. The time periods show the frequency of a voltage signal. Frequency is a term that describes how often a signal performs a complete cycle. Frequency is measured in hertz. To determine the frequency of something, divide the length of time it takes to complete one cycle into 1.

Whenever using a scope on a circuit, always follow the meter's instruction manual for hookup and proper settings. Most service manuals illustrate the patterns expected from the different electronic components. If the patterns do not match those in the manual, a problem is indicated. The problem may be in the component or in the circuit. Further testing is required to locate the exact cause of the problem.

SHOP TALK

It is very helpful to know if the component being tested with a scope is digital or analog based. With this knowledge, it is easier to recognize an incorrect pattern. Digital patterns should show a clear on/off signal, whereas analog signals should show steady voltage changes as conditions change. ■

CONNECTOR AND WIRE REPAIRS

Many automotive electrical problems can be traced to faulty wiring. Loose or corroded terminals, frayed, broken, or oil-soaked wires, and faulty insulation are the most common causes. Wires, fuses, and connections should be checked carefully during troubleshooting. Keep in mind that the insulation does not always appear to be damaged when the wire inside is broken. Also, a terminal may be tight but still may be corroded.

WARNING!

Always follow the vehicle manufacturer's wiring and terminal repair procedure given in the service manual. On some components and circuits, manufacturers recommend component replacement rather than wiring repairs. For example, some manufacturers recommend replacing air bag system components such as sensors if the wiring or terminals are damaged on these components.

Wire end terminals are connecting devices. They are generally made of tin-plated copper and come in many shapes and sizes. They may be either soldered or crimped in place. When installing a terminal, select the appropriate size and type of terminal **(Figure 16–51)**. Be sure it fits the unit's connecting post or prongs and has enough current-carrying capacity for the circuit. Also, make sure it is heavy enough to endure normal wire flexing and vibration.

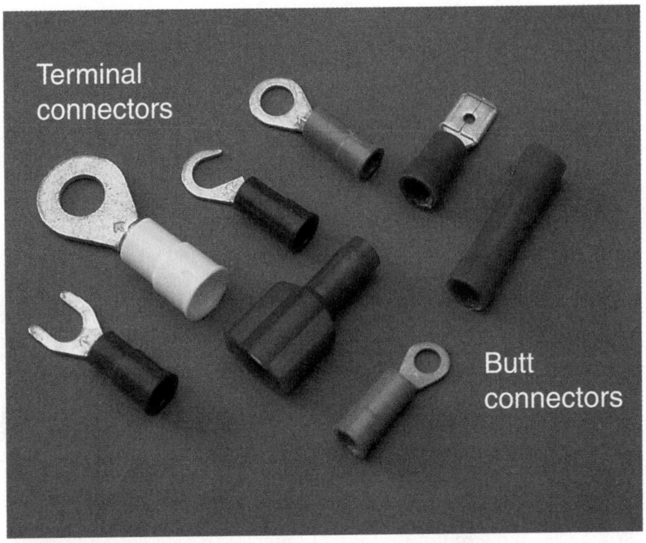

Figure 16–51 Various types of solderless connectors.

PHOTO SEQUENCE 11

Soldering Two Copper Wires Together

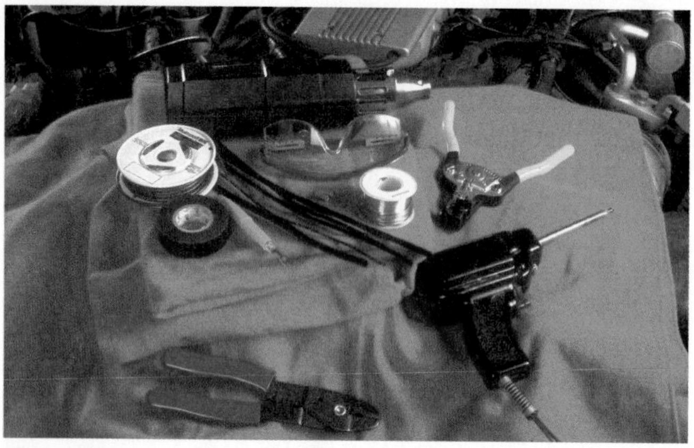

P11-1 *Tools required to solder copper wire: 100-watt soldering iron, 60/40 rosin core solder, crimping tool, splice clip, heat shrink tube, heating gun, and safety glasses.*

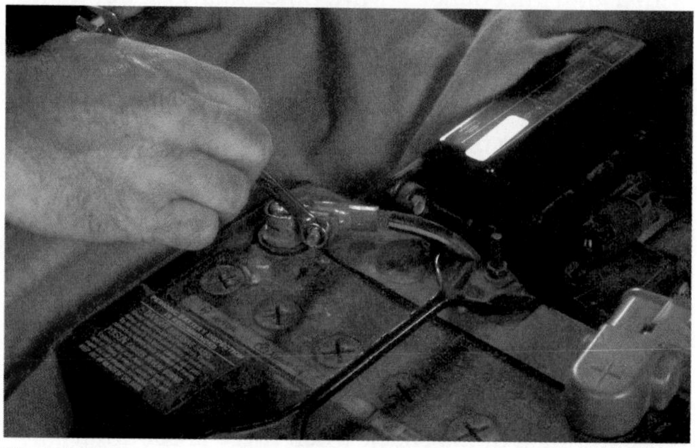

P11-2 *Disconnect the fuse that powers the circuit being repaired. Note: If the circuit is not protected by a fuse, disconnect the ground lead of the battery.*

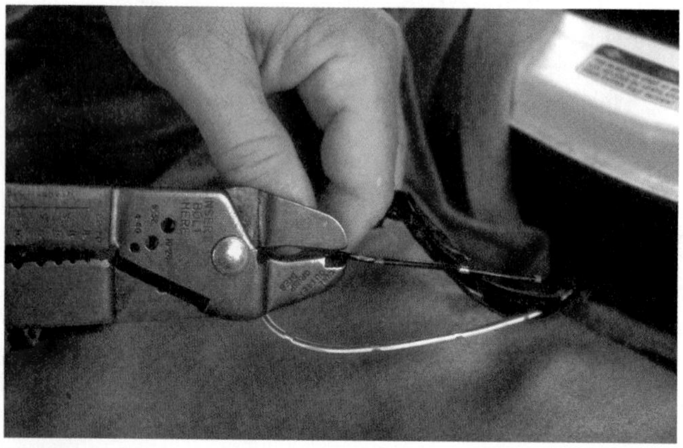

P11-3 *Cut out the damaged wire.*

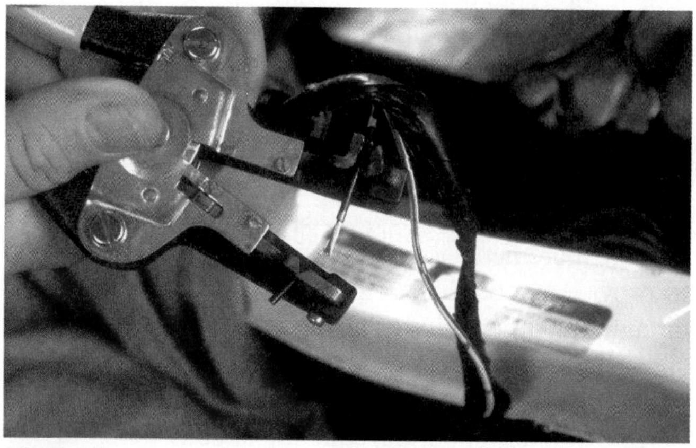

P11-4 *Using the correct size stripper, remove about 1/2 inch of the insulation from both wires.*

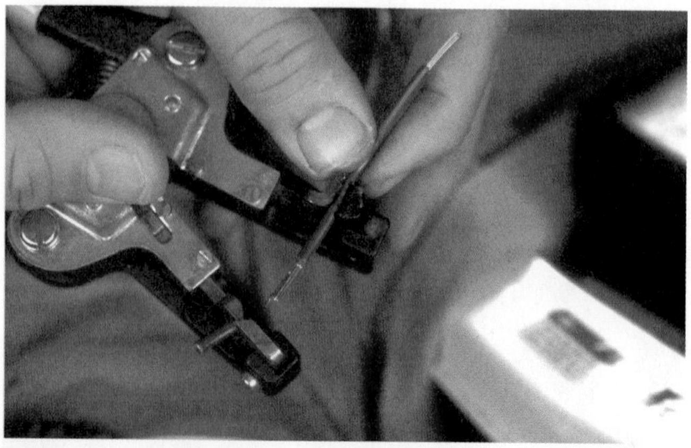

P11-5 *Now remove about 1/2 inch of the insulation from both ends of the replacement wire. The length of the replacement wire should be slightly longer than the length of the wire removed.*

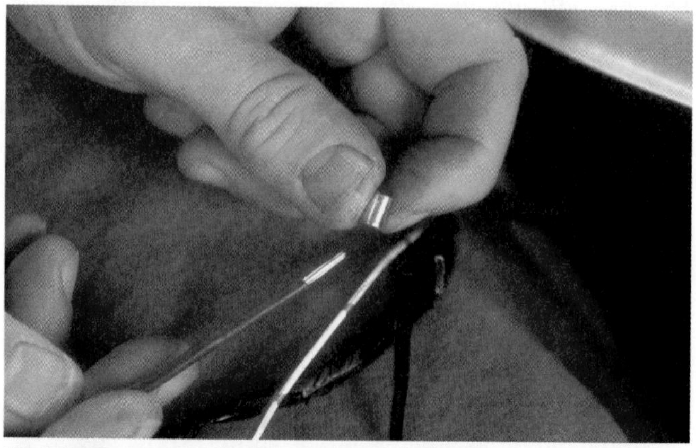

P11-6 *Select the proper size splice clip to hold the splice.*

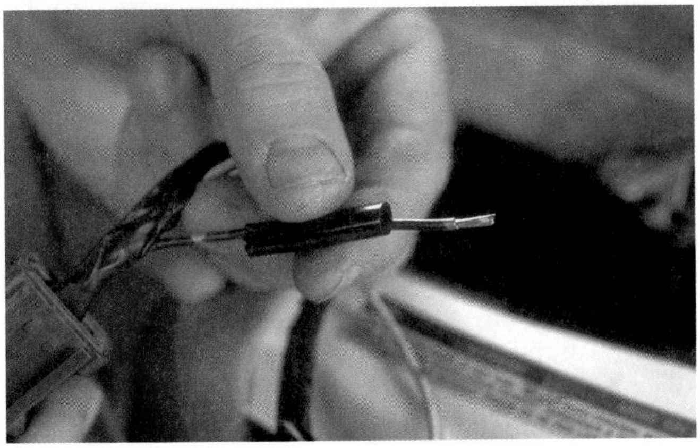

P11-7 *Place the correct size and length of heat shrink tube over the two ends of the wire.*

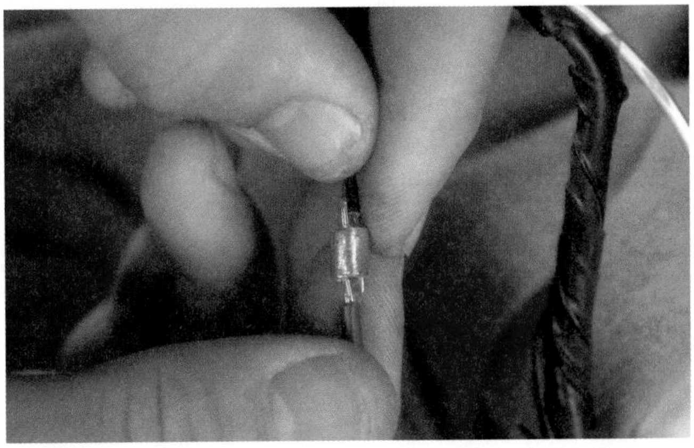

P11-8 *Overlap the two splice ends and center the splice clip around the wires, making sure the wires extend beyond the splice clip in both directions.*

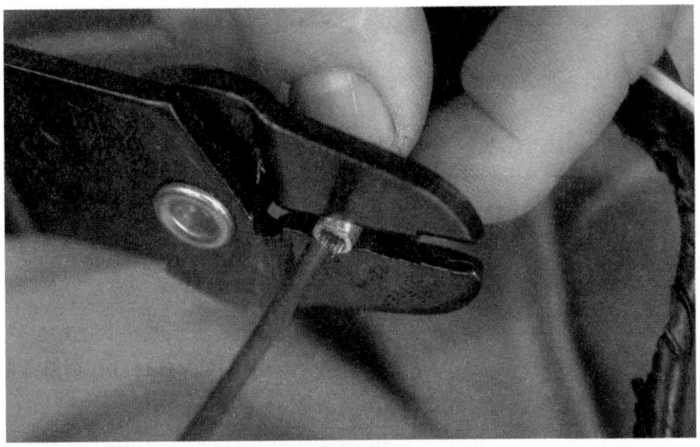

P11-9 *Crimp the splice clip firmly in place.*

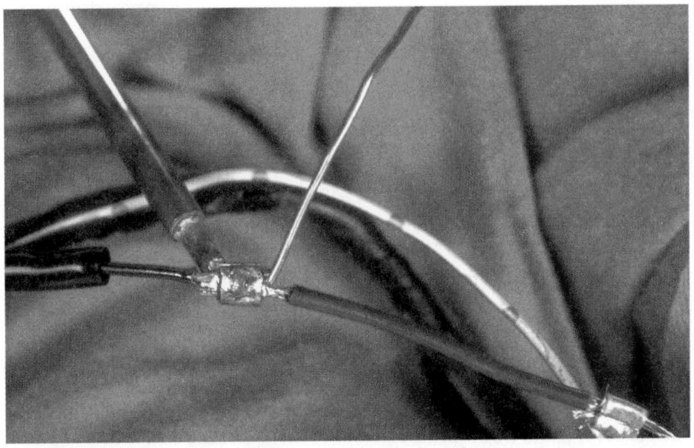

P11-10 *Heat the splice clip with the soldering iron while applying solder to the opening of the clip. Do not apply solder to the iron. The iron should be 180 degrees away from the opening of the clip.*

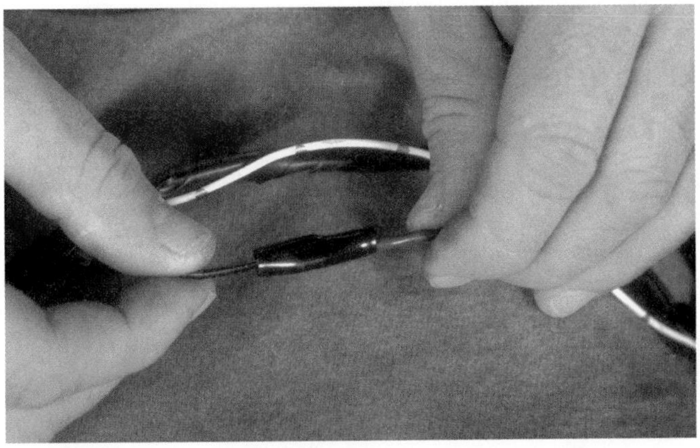

P11-11 *After the solder cools, slide the heat shrink tube over the splice.*

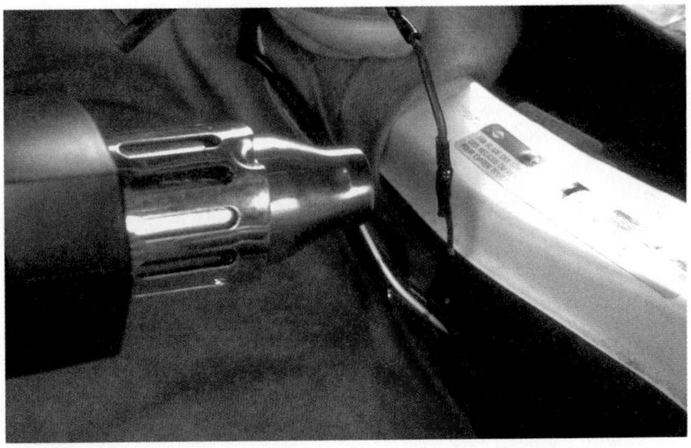

P11-12 *Heat the tube with the hot air gun until it shrinks around the splice. Do not overheat the heat shrink tube.*

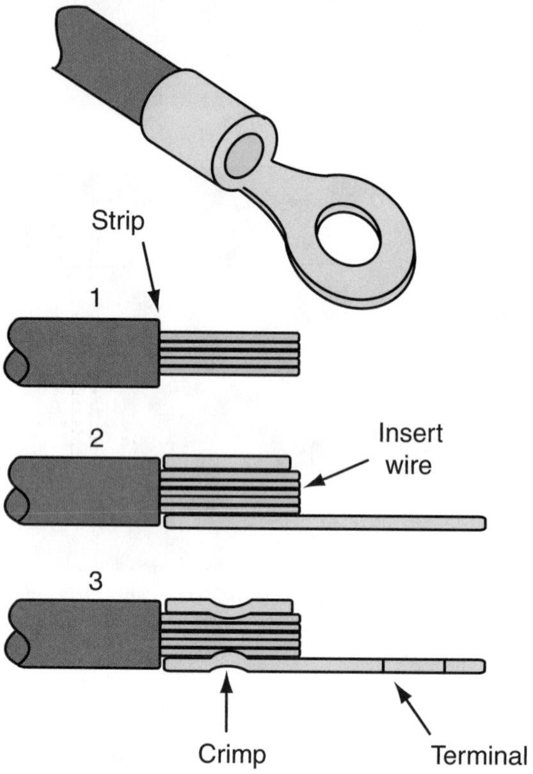

Figure 16–52 The proper way to install a solderless connector.

WARNING!

Do not crimp a terminal with the cutting edge of a pair of pliers. While this method may crimp the terminal, it also weakens it.

General procedures for crimping on a connector **(Figure 16–52)** are as follows:

1. Use the correct size stripping opening on the crimping tool **(Figure 16–53)** and remove enough insulation to allow the wire to completely penetrate the connector.
2. Place the wire into the connector and crimp the connector. To get a proper crimp, place the open area of the connector facing toward the anvil. Make sure the wire is compressed under the crimp.
3. Insert the stripped end of the other wire into the connector.
4. Use electrical tape or heat shrink tubing to tightly seal the connection to provide good protection for the wire and connector.

The preferred way to connect wires or to install a connector is by soldering. Photo Sequence 11 demonstrates the proper procedure for soldering two copper wires together. Some car manufacturers use aluminum in their wiring. Aluminum cannot be soldered. Follow the manufacturer's guidelines and use the proper repair kits when repairing aluminum wiring.

WARNING!

Never use acid core solder. It creates corrosion and can damage electronic components.

When working with wiring and connectors, never pull on the wires to separate the connectors. This can create an intermittent contact and an intermittent problem that may be very difficult to find later. When required, always use the special tools designed for separating connectors.

Check all connectors for corrosion, dirt, and looseness. Nearly all connectors have pushdown release-type locks **(Figure 16–54)**. Make sure these are not damaged when disconnecting the connectors. Many connectors have covers over them to protect them from dirt and moisture. Make sure these are properly installed to provide for that protection.

Never reroute wires when making repairs. Rerouting wires can result in induced voltages in nearby components. These stray voltages can interfere with the function of electronic circuits.

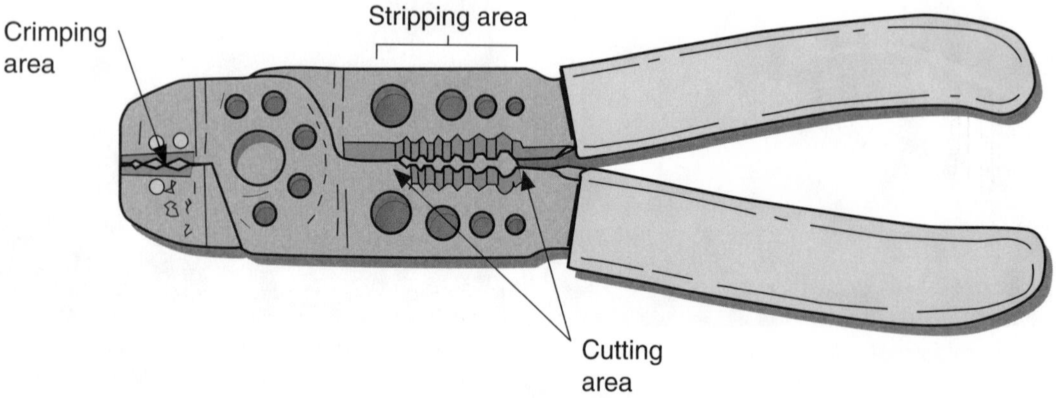

Figure 16–53 A typical crimping tool used for making electrical repairs.

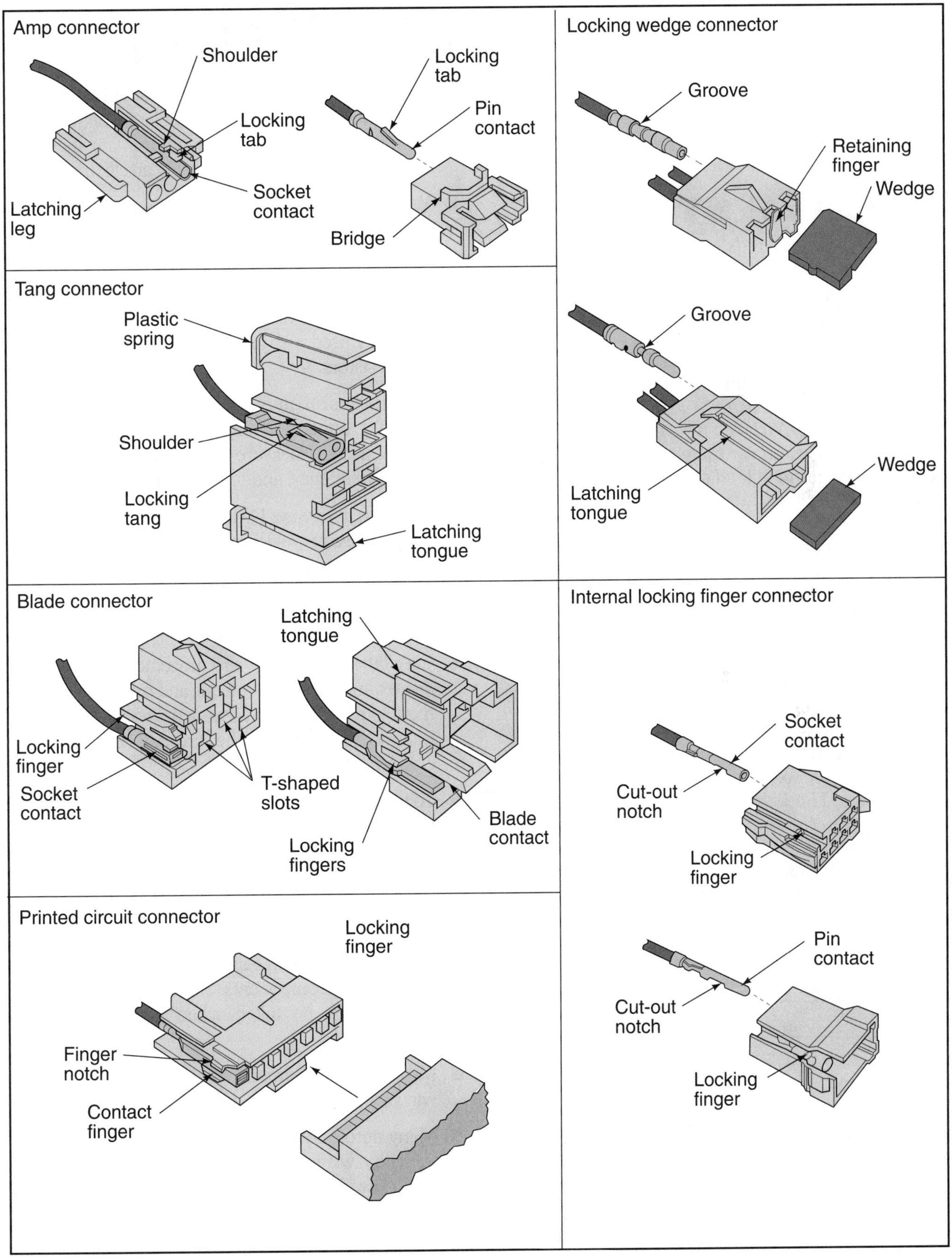

Figure 16-54 Different multiple-wire shell connectors and their locking mechanisms.

Dielectric grease is used to moistureproof and to protect connections from corrosion. Some car manufacturers suggest using petroleum jelly to protect connection points.

CASE STUDY

A customer tells the technician that the fuse for the windshield wiper blows as soon as he replaces it and turns on the wipers.

The technician removes the fuse and observes that it is black inside. The technician knows that these symptoms indicate a short to ground and so substitutes a 12-volt test light for the blown fuse. With the test light in place, the technician disconnects the wiper motor (a mechanically grounded load component), turns the ignition switch on, and notes the status of the test light. Because the motor is disconnected, the circuit should be open and the test light is lit. The technician knows that the problem is not in the motor. The problem is a short to ground in the wiring that leads to the motor. The technician continues the search for the short by separating circuit connectors one at a time. He starts at the connector that is farthest from the test light and works toward the fuse panel. The light remains on when the technician unplugs the first connector so he knows that the short exists in some part of the circuit that is still intact. The light goes off when the technician unplugs the next connector one step closer to the fuse panel. The technician then realizes that the short is somewhere between the last two connectors unplugged because the test light indicates that there is no longer a path to ground. The technician visually inspects the wiring between the two connectors and discovers and repairs a bare spot in the wire that was contacting ground.

KEY TERMS

Average responding
Digital storage oscilloscope (DSO)
Glitches
In-line fuse
Open
Root mean square (RMS)
Schematics
Short
Voltage drop

SUMMARY

- All electrical problems can be classified as an open, a short, or a high-resistance problem. Identifying the type of problem allows for determining the correct tests to conduct when diagnosing an electrical circuit.
- Wiring diagrams show where wires are connected, the circuit's components, the color of the wire's insulation, and sometimes the wire gauge size.
- Voltmeters, ohmmeters, ammeters, and volt/amp meters are used to test and diagnose electrical systems. These are used with jumper wires, test lights, and variable resistors.
- Multimeters are multifunctional and can test DC and AC volts, ohms, and amperes. Some multimeters can also be used to measure engine rpm, duty cycle, pulse width, frequency, and temperature.
- There are two ways DMMs display AC voltage: RMS and average responding.
- Some DMMs also feature a MIN MAX function, which displays the maximum, minimum, and average voltage the meter recorded during the time of the test.
- On a lab scope, an upward movement of the trace indicates an increase in voltage, and a downward movement of this trace represents a decrease in voltage. As the trace moves across the screen, it represents a specific length of time.
- To troubleshoot a problem, begin by verifying the customer's complaint. Then operate the system and others to get a complete understanding of the problem. Use the correct wiring diagram and identify testing points and likely problem areas. Test and use logic to identify the cause of the problem.
- Static electricity can generate up to 25,000 volts and can damage components. Precautions for static discharge must be taken when handling electronic components.
- Most electronic circuits can be checked in the same way as other electrical circuits.
- Many automotive electrical problems can be traced to faulty wiring, such as loose or corroded terminals, frayed, broken, or oil-soaked wires, and faulty insulation.
- The preferred way to connect wires or to install a connector is by soldering. Never use acid core solder. It creates corrosion and can damage electronic components.

TECH MANUAL

The following procedures are included in Chapter 16 of the *Tech Manual* that accompanies this book:

1. Checking continuity in a circuit with a test light.
2. Testing for voltage with a DMM.
3. Testing voltage drop across connectors.
4. Inspecting and testing fuses, fusible links, and circuit breakers.
5. Part identification on a wiring diagram.
6. Using a DSO on sensors and switches.

REVIEW QUESTIONS

1. How will an electrical circuit behave if there is an open in the circuit?

2. What happens to an electrical circuit when there is unwanted resistance in it?

3. How should you test a diode with a multimeter?

4. An ammeter is always connected __________ with the circuit, while a voltmeter is connected in __________ with the circuit.

5. Which of the following is *not* a typical cause of unwanted or high resistance in a circuit?

a. corrosion on terminal ends
b. a power wire contacting the chassis
c. loose or poor connections
d. frayed and damaged wires

6. *True or False?* A zero reading on an ohmmeter means the circuit or component is open.

7. *True or False?* The maximum allowable voltage loss due to voltage drops across wires, connectors, and other conductors in a 12-volt circuit is 1.2 volts.

8. Technician A uses an ohmmeter to test circuit protection devices. Technician B uses a voltmeter to test circuit protection devices. Who is correct?

a. Technician A **c.** Both A and B
b. Technician B **d.** Neither A nor B

9. What type of solder should be used to repair electrical wiring?

10. Which of the following lab scope controls must be set when trying to observe the timing of something?

a. intensity **c.** horizontal
b. vertical **d.** trigger

11. Technician A uses a test light to detect resistance. Technician B uses a jumper wire to test circuit breakers, relays, and lights. Who is correct?

a. Technician A **c.** Both A and B
b. Technician B **d.** Neither A nor B

12. While diagnosing the location of a wire-to-wire short, Technician A checks the wiring of the affected circuits for signs of burned insulation and melted conductors. Technician B checks common connectors shared by the two affected circuits. Who is correct?

a. Technician A **c.** Both A and B
b. Technician B **d.** Neither A nor B

13. Which of the following statements is true?

a. A short causes decreased current flow.
b. An open causes unwanted voltage drops.
c. High-resistance problems cause increased current flow.
d. Both opens and high-resistance problems may cause a load not to work.

14. While measuring resistance, Technician A uses an ohmmeter to measure resistance of a component before disconnecting it from the circuit. Technician B uses a voltmeter to measure voltage drop. Who is correct?

a. Technician A **c.** Both A and B
b. Technician B **d.** Neither A nor B

15. While discussing how to test a switch, Technician A says the action of the switch can be monitored by a voltmeter. Technician B says continuity across the switch can be checked by measuring the resistance across the switch's terminals when the switch is in its different positions. Who is correct?

a. Technician A **c.** Both A and B
b. Technician B **d.** Neither A nor B

16. While discussing electricity, Technician A says an open causes unwanted voltage drops. Technician B says high-resistance problems cause increased current flow. Who is correct?

 a. Technician A
 b. Technician B
 c. Both A and B
 d. Neither A nor B

17. *True or False?* Oils from your skin can cause corrosion on electrical connectors.

18. While measuring the resistance of a wire with an ohmmeter, Technician A says if low resistance is shown on the meter, the wire is basically good. Technician B says if no resistance is measured, the wire is shorted. Who is correct?

 a. Technician A
 b. Technician B
 c. Both A and B
 d. Neither A nor B

19. While testing variable resistors, Technician A says while checking a rheostat with a voltmeter, the voltage should change smoothly with a change in the control. Technician B says a potentiometer should be checked with a test light. Who is correct?

 a. Technician A
 b. Technician B
 c. Both A and B
 d. Neither A nor B

20. While using an ohmmeter to measure the resistance values of a component, Technician A says if the component has less than the specified resistance, the part is open. Technician B says if there is more resistance than called for, the part is shorted. Who is correct?

 a. Technician A
 b. Technician B
 c. Both A and B
 d. Neither A nor B

17

BATTERIES: THEORY, DIAGNOSIS, AND SERVICE

OBJECTIVES

■ Explain the purpose of a battery. ■ Describe the basic parts of an automotive battery. ■ Compare conventional and maintenance-free batteries. ■ Explain the chemical reaction that occurs to produce current in a battery. ■ Describe the differences, advantages, and disadvantages of the different types of batteries. ■ Describe the different types of battery terminals used. ■ Describe the different types of ratings used with batteries. ■ Explain the effects of temperature on battery output. ■ Demonstrate all safety precautions and rules associated with servicing batteries. ■ Perform a visual inspection of a battery. ■ Test a conventional battery's specific gravity. ■ Perform open circuit tests. ■ Test the capacity and conductance of a battery. ■ Correctly slow- and fast-charge a battery. ■ Jump-start a vehicle by using a booster battery and jumper cables. ■ Remove, clean, and reinstall a battery properly.

An automotive battery is an **electrochemical** device that stores and provides electrical energy. When the battery is connected to an external load, an energy conversion occurs that results in current flow through the circuit to operate the load. Electrical energy is produced in the battery by the chemical reaction that occurs between two dissimilar metal plates that are immersed in an electrolyte solution.

When the battery is discharging, it changes chemical energy into electrical energy. It is through this change that the battery releases stored energy. During charging, electrical energy is converted into chemical energy. As a result, the battery can store energy until it is needed.

The storage battery is the heart of a vehicle's electrical and electronic systems. It plays an important role in the operation of the starting, charging **(Figure 17–1)**, ignition, and accessory circuits. The largest demand placed on the battery occurs when it must supply current to operate the starter motor. The amperage requirements of a starter motor may be over several hundred amps.

After the engine has started, the vehicle's charging system recharges the battery. It also provides the power to run the electrical accessories.

If the vehicle's charging system fails, the battery must supply all the current needed to run the vehicle. Most batteries can supply 25 amperes for two hours before they become so low that they are unable to keep the engine running. The amount of time a battery can be discharged at a certain rate before the voltage drops to a specified level is referred to as the **reserve capacity** of the battery.

An automotive storage battery has three main functions. It provides voltage and serves as a source of current for starting, lighting, and ignition. It acts as a voltage stabilizer for the entire electrical system of the vehicle. And, it provides current whenever the vehicle's electrical demands exceed the output of the charging system.

The battery must be able to maintain a good charge when the engine is off. The condition of the battery

Figure 17-1 The starter motor is the largest drain on a battery, and the AC generator replenishes that drain. *Courtesy of Delco-Remy Co.*

determines this capability, as do the electrical demands of the vehicle. When the engine is off, electrical power is still needed to maintain the memory in the various computers used in the vehicle and to keep clocks going. The electrical loads present when the ignition switch is off are called **parasitic loads**. At times the parasitic loads are so great that a battery will go dead if the vehicle has not been driven for a while.

The battery in a fuel cell vehicle serves a different purpose than a battery in a typical vehicle. In most vehicles, the battery's primary purpose is to provide a short powerful burst of power to start the engine. This type of battery is typically called a starting battery. This design has a high number of thin plates. The battery in a fuel cell vehicle and in many RVs is a **deep cycle battery**. This design has less instant energy but greater long-term energy delivery. Deep-cycle batteries have thicker plates and can survive a great number of discharge cycles. The battery in a fuel cell vehicle absorbs the energy generated by the fuel cells and uses it to power electronic features, such as power steering. It also is constantly **cycling**, which means it constantly is discharging and being recharged.

Forty-two-volt systems are based on a single 36-volt battery but are dual voltage systems. Part of the vehicle will be powered by 12 to 14 volts and the rest by 36 to 42 volts. The battery may have two positive connectors, one for each voltage or the voltage will be divided by a converter.

The most common automotive batteries are lead-acid designs. The wet cell, gel cell, and absorbed glass mat are versions of a lead-acid battery.

Other Battery Designs With the many different types of automotive power sources being tried by the automotive industry come new battery designs. In fact, the cost and performance of electrical vehicles depends on battery technology. There are several types of batteries available and under development, from advanced lead-acid batteries to nickel-metal hydride (NiMH) to lithium polymer batteries.

Electric vehicles typically run on 200 to 300 volts. The batteries used as the power source are 6- or 12-volt batteries connected in series. If the electric motor that powers the vehicle requires 240 volts, the vehicle would need forty 6-volt batteries or twenty 12-volt batteries.

BATTERY CONSTRUCTION

A storage battery consists of grids, positive plates, negative plates, separators, elements, electrolyte, a container, cell covers, vent plugs, and cell containers **(Figure 17–2)**. The **grids** form the basic framework of the battery **plates**. Grids are the lead alloy framework that supports the ac-

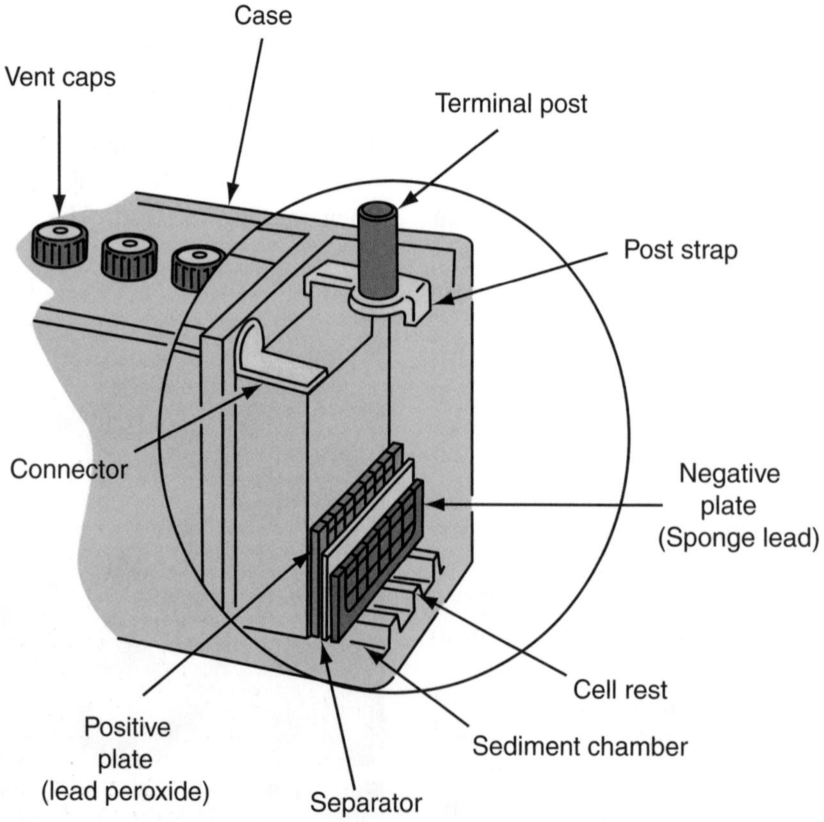

Figure 17–2 Components of a typical conventional storage battery.

tive material of a plate and conducts current. Plates are typically flat, rectangular components that are either positive or negative, depending on the active material they hold.

A positive plate consists of a grid filled with **lead peroxide** as its active material. Lead peroxide (PbO_2) is a dark brown, crystalline material. Its high degree of porosity allows the liquid electrolyte to penetrate freely. The material pasted onto the grids of the negative plates is **sponge lead** (Pb), a porous gray lead that allows the electrolyte to penetrate freely.

Elements and Cells

Each battery contains a number of elements. An **element** is a group of positive and negative plates. The plates are formed into a plate group, which holds a number of plates of the same polarity. The like-charged plates are welded to a lead alloy post or **plate strap**. The plate groups are then alternated within the battery—positive, negative, positive, negative. There is usually one extra set of negative plates to balance the charge. To ensure that the different plate groups do not touch each other, nonconductive sheets called separators are inserted between them. **Separators** are porous plastic sheets that allow the transfer of ions between plates but prevent physical contact between them, which would cause the plates to lose their stored energy.

When the element is placed inside the battery case and immersed in **electrolyte**, it becomes a cell. The lead peroxide and sponge lead that make up the element plates cannot become active until they are immersed in electrolyte. A 12-volt battery has six cells that are connected in series with each other. Each cell has an open circuit voltage of approximately 2.1 volts; therefore, a 12-volt storage battery has an actual open circuit voltage of 12.6 volts.

Electrolyte is a solution of sulfuric acid and water. The sulfuric acid of the electrolyte supplies sulfate, which chemically reacts with both the lead and lead peroxide to release electrical energy. In addition, the sulfuric acid is the carrier for the electrons inside the battery between the positive and negative plates.

To achieve the chemical reaction that creates voltage in a battery, the electrolyte solution must be the correct mixture of water and sulfuric acid. At 12.6 volts, the electrolyte solution is 65% water and 35% sulfuric acid. Sometimes the electrolyte breaks down and the acid moves onto the plates, so there is less acid in the water. Whenever the percentage of acid in the solution decreases, the charge drops.

Discharging and Charging

Remember, a chemical reaction between active materials on the positive and negative plates and the acid in the electrolyte produces electrical energy. When a battery discharges **(Figure 17–3)**, lead in the lead peroxide of the positive plate combines with the sulfate radical (SO_4) to form lead sulfate ($PbSO_4$).

A similar reaction takes place at the negative plate. In this plate, lead (Pb) combines with sulfate (SO_4) to also form lead sulfate ($PbSO_4$), a neutral and inactive material. Thus, lead sulfate forms at both plates as the battery discharges.

As the chemical reaction occurs, the oxygen from the lead peroxide and the hydrogen from the sulfuric acid combine to form water (H_2O). During discharge, the electrolyte becomes weaker and the positive and negative plates become like one another. Since the charge of a battery depends on the difference between the two plate materials and the concentration of the electrolyte and this difference decreases during discharging, the battery loses power.

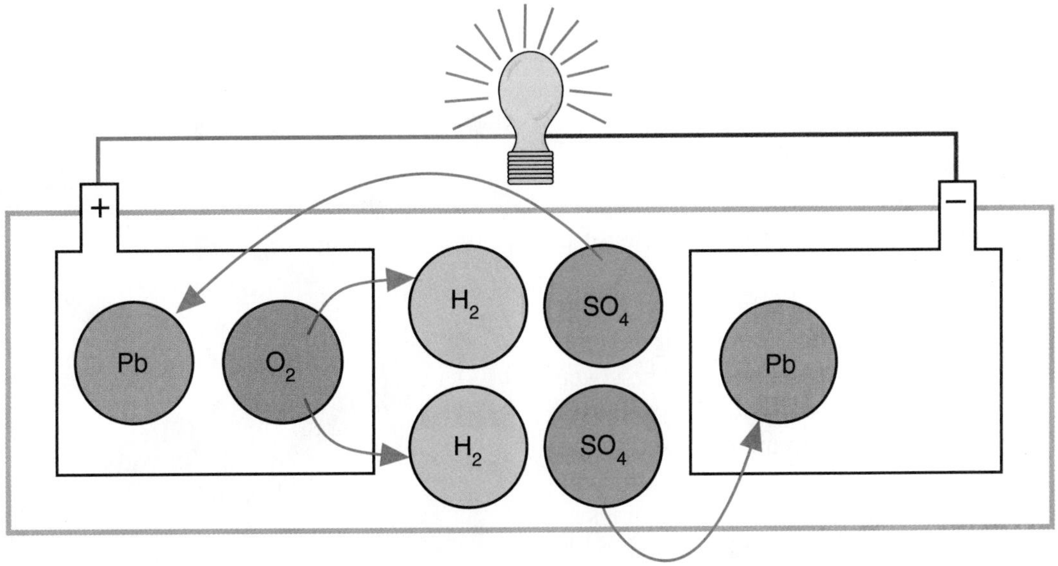

Figure 17–3 Chemical action that occurs inside a battery during a discharge cycle.

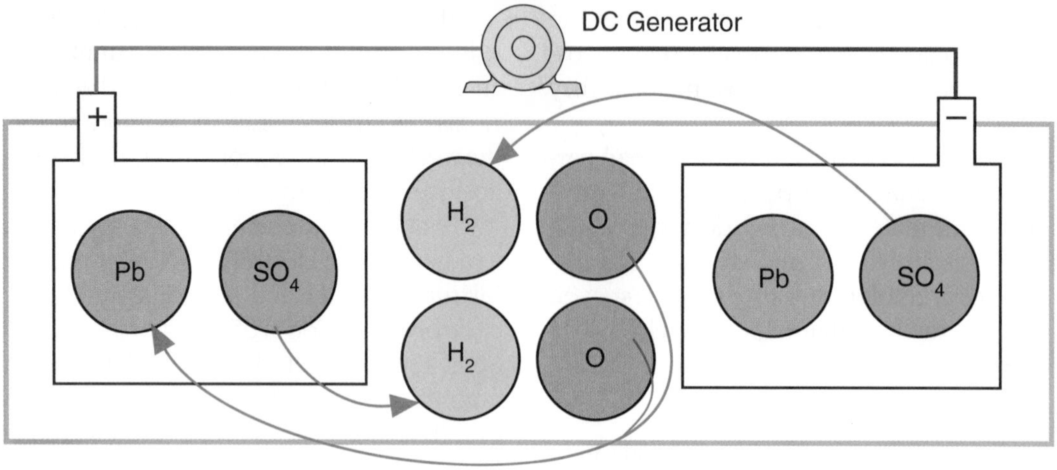

Figure 17-4 Chemical action inside a battery during the charge cycle.

The recharging process **(Figure 17–4)** is the reverse of the discharging process. Electricity from an outside source such as the vehicle's generator or a battery recharger is forced into the battery. The lead sulfate ($PbSO_4$) on both plates separates into lead (Pb) and sulfate (SO_4). As the sulfate (SO_4) leaves both plates, it combines with hydrogen in the electrolyte to form sulfuric acid (H_2SO_4). At the same time, the oxygen (O_2) in the electrolyte combines with the lead (Pb) at the positive plate to form lead peroxide (PbO_2). As a result, the negative plate returns to its original form of lead (Pb), and the positive plate reverts to lead peroxide (PbO_2).

An unsealed battery gradually loses water due to its conversion into hydrogen and oxygen. These gases escape into the atmosphere through the **vent caps**. If the lost water is not replaced, the level of the electrolyte falls below the tops of the plates. This results in a high concentration of sulfuric acid in the electrolyte and permits the exposed material of the plates to dry and harden. In this situation, premature failure of the battery is certain. The electrolyte level in the battery must be checked frequently.

Casing Design

The container or shell of the battery is usually a one-piece, molded assembly of polypropylene, hard rubber, or plastic. The case has a number of individual cell compartments. Cell connectors are used to join the cells of a battery in series.

The top of the battery is encased by a **cell cover (Figure 17–5)**. The cover may be a one-piece design or the cells might have their own individual covers. The cover has vent holes to permit the escape of hydrogen and oxygen gases. Battery vents can be permanently fixed to the cover or be removable, depending on battery design. Vent caps are used on some batteries to close the openings in the cell cover and to allow for topping off the cells with water.

CAUTION!

When lifting a battery, excessive pressure on the end walls could cause acid to spew through the vent caps, resulting in personal injury. Lift with a battery carrier or with your hands on opposite corners.

Terminals

The battery has two external terminals: a positive (+) and a negative (–). These terminals are either two tapered posts, L terminals, threaded studs on top of the case, or two internally threaded connectors on the side **(Figure 17–6)**. The terminals have either a positive (+) or a negative (–) marking, depending on which end of the series they represent.

Tapered terminals have a given dimension in accordance with standards agreed upon by the **Battery Council International (BCI)** and the Society of Automotive Engineers (SAE). This ensures that all positive and negative cable clamps would fit any corresponding battery terminal, regardless of the battery's manufacturer. The

Figure 17-5 A battery with removable cell caps.

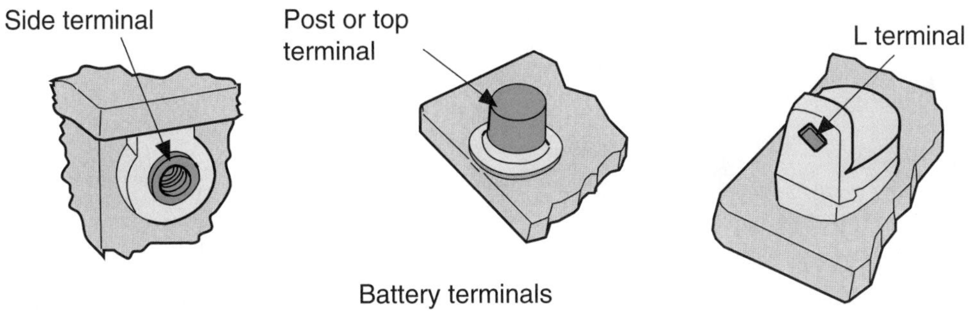

Figure 17-6 The most common types of automotive battery terminals. *Courtesy of Ford Motor Company*

positive terminal is slightly larger, usually around $\frac{11}{16}$ inch in diameter at the top, while the negative terminal usually has a $\frac{5}{8}$-inch diameter. This design minimizes the danger of installing the battery cables in reverse polarity.

Side terminals are positioned near the top of the battery case. These terminals are threaded and require a special bolt to connect the cables. Some batteries are fitted with both top and side terminals to allow them to be used in many different vehicles.

BATTERY DESIGNS

In many lead-acid batteries, the grids are made of lead alloyed with approximately 6% antimony for strength. Antimony added to the lead grids acts as a catalyst but makes outgassing (the loss of hydrogen and oxygen during use) worse, and frequent water replenishing is required.

Maintenance-Free and Low-Maintenance Batteries

Most batteries sold and installed today are either low-maintenance or maintenance-free designs **(Figure 17–7)**. A maintenance-free battery is similar in design to a conventional battery, but many of the components have thicker construction. Different, more durable materials are used in low-maintenance batteries and the amount of antimony is reduced to about 3%. In maintenance-free batteries, the antimony is eliminated and replaced by calcium or strontium **(Figure 17–8)**. Reducing the amount of antimony or replacing it with calcium or strontium alloy reduces both the battery's internal heat and the amount of **gassing** that occurs during charging. Since heat and gassing are the principal reasons for battery water loss, these changes reduce or eliminate the need to periodically add water.

Additionally, non-antimony-lead alloys have better conductivity, so a maintenance-free battery has about a 20% higher cold-cranking power rating than a comparably sized conventional battery. Calcium lead alloy grids are somewhat prone to grid growth and cracking. These problems can be controlled by modifying the alloy and roll hardening the positive grid, or through the use of calcium/silver alloys.

Maintenance-free batteries are equipped with small gas vents that prevent gas-pressure buildup in the case **(Figure 17–9)**. Water is never added to maintenance-free batteries.

Figure 17-7 A maintenance-free battery. *Courtesy of Delco-Remy Co.*

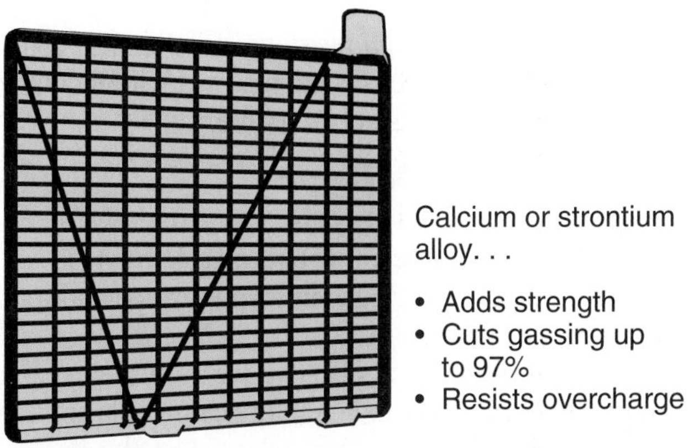

Figure 17-8 Maintenance-free battery grids with support bars give increased strength and faster electrical delivery. *Courtesy of Ford Motor Company*

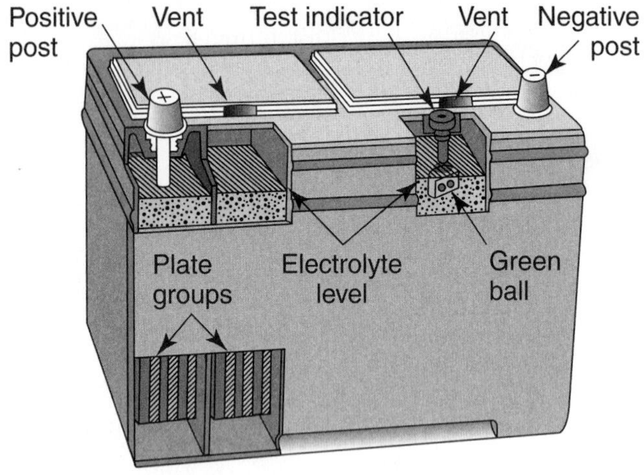

Figure 17–9 Construction of a maintenance-free battery showing the location of the gas vents.

Low-maintenance batteries are still equipped with vent holes and caps, which allow water to be added to the cells. A low-maintenance battery requires additional water substantially less often than a conventional battery.

Hybrid Batteries

A **hybrid battery** can withstand six deep cycles and still retain 100% of its original reserve capacity. The grid construction of the hybrid battery consists of approximately 2.75% antimony alloy on the positive plates and a calcium alloy on the negative plates. This allows the battery to withstand deep cycling while retaining reserve capacity for improved cranking performance. Also, the use of antimony alloys reduces grid growth, corrosion, and water loss.

Grid construction differs from other batteries in that the plates have a lug located near the center of the grid. In addition, the vertical and horizontal grid bars are arranged in a radial design **(Figure 17–10)**. With this grid design, current has less resistance and a shorter path to follow, which means the battery is capable of providing more current at a faster rate.

The separators used are constructed of glass covered with a resin or fiberglass. The separators offer low electrical resistance with high resistance to chemical contamination. This type of construction provides for increased cranking performance and battery life.

Recombination Batteries

A **recombination battery** is a completely sealed maintenance-free battery that uses an electrolyte in a gel form. In a gel-cell battery, gassing is minimized and vents are not needed.

During charging, the negative plates never reach a fully charged condition and therefore cause little or no release of hydrogen. Oxygen is released at the positive plates, but it passes through the separators and recombines with the negative plates. Because the oxygen released by the electrolyte is forced to recombine with the negative plate, these batteries are called recombination batteries.

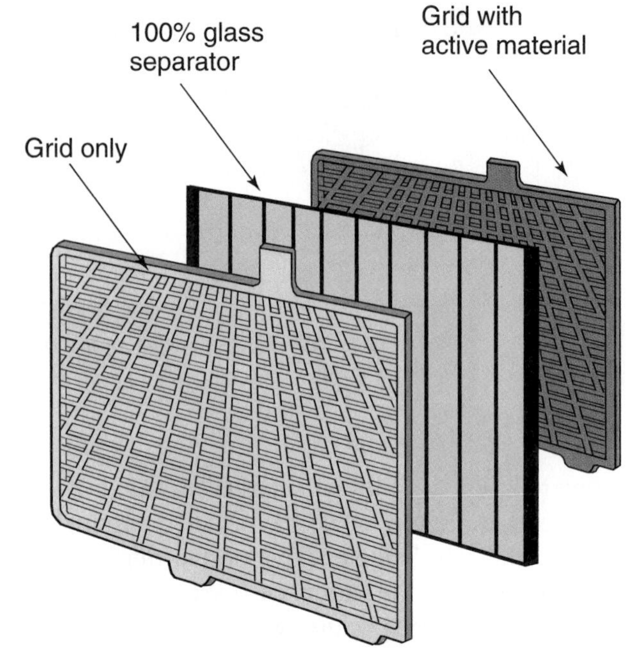

Figure 17–10 Hybrid battery grid and separator construction. *Courtesy of Ford Motor Company*

Absorbed Glass Mat Batteries

The electrolyte in **absorbed glass mat (AGM)** batteries is held in moistened fiberglass matting instead of existing as a liquid or gel. The matting is sandwiched between the battery's lead plates, where it also serves as a vibration dampener **(Figure 17–11)**.

Figure 17–11 The construction of an AGM battery. *Courtesy of Exide Batteries*

Rolls of high-purity lead plates are tightly compressed into six spiral-wound cells. The plates are separated by acid-permeated vitreous separators. Vitreous separators absorb acid the same way a paper towel absorbs water. A small amount of silver is added to the plates and some sodium sulfate is added to the electrolyte.

Each of the six spiral cells is enclosed in its own cylinder within the battery case, forming a sealed, closed system that resembles a six-pack of soda. During normal use, hydrogen and oxygen within the battery are captured and recombined to form the water supply within the bound electrolyte, eliminating the need to ever add water to the battery.

Even if cracked, broken, or punctured, AGM batteries will never leak. They also have short recharging times and low internal resistance, which provides increased output. AGM batteries also have exceptional durability in both high-heat and subzero climates.

BATTERY HARDWARE

In order to connect the battery to the vehicle's electrical system, battery cables are used. Battery hold-downs are used to prevent damage to the battery, and heat shields are sometimes used to keep battery temperatures low.

Battery Cables

Battery cables must be of sufficient capacity to carry the current required to meet all demands **(Figure 17–12)**. The normal 12-volt cable size is 4 or 6 gauge. Various forms of clamps are used to ensure a good electrical connection at each end of the cable. Connections must be clean and tight to prevent arcing between the terminal and clamp, corrosion, and high-voltage drops.

The positive cable is normally red and the negative cable is black. The positive cable fastens to the battery and the starter solenoid or relay. The negative cable fastens to a ground on the engine block or chassis.

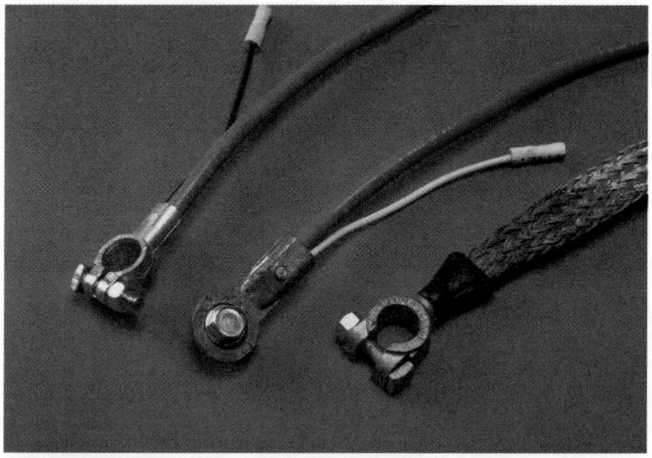

Figure 17-12 The battery cable is designed to carry the high current required to start the engine and supply the vehicle's electrical systems.

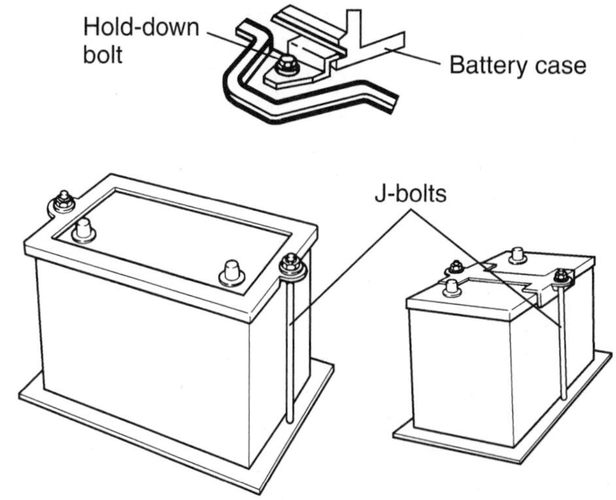

Figure 17-13 Examples of the different types of hold-downs used with batteries. *Courtesy of DaimlerChrysler Corporation*

Battery Hold-Downs

All batteries must be held securely in the vehicle to prevent the possibility of shorting across the terminals if they move or fall. Normal vibrations cause the plates to shed their active materials. Hold-downs reduce the amount of vibration and help increase the life of the battery. Battery hold-downs are made of metal or plastic **(Figure 17–13)**.

Heat Shields

Some vehicles have a heat shield made of plastic or another material to protect the battery from high underhood temperatures. While heat shields do not need to be removed for most testing and inspection procedures, they must be removed and then correctly installed during battery replacement.

Vehicles equipped for cold climates may have a battery blanket or heater to keep the battery warm during extremely cold weather.

BATTERY VOLTAGE AND CAPACITY

Cell size, state of charge, rate of discharge, battery condition and design, and electrolyte temperature all strongly influence the voltage of a cell during discharge. When cranking an engine, the voltage of an average battery at 80°F may be about 11.5 to 12 volts. At 0°F (–17.7°C), the voltage is significantly lower.

At low temperatures, the viscosity of the electrolyte increases, making it more difficult for the acid to move freely into the plate pores and around the separators. This slows the rate of the chemical reaction and lowers battery voltage, limiting the output of the battery, especially at cranking rates.

Battery capacity is the ability to deliver a given amount of current over a period of time. It depends on the number and size of the plates used in the cells and the amount of acid used in the electrolyte.

BATTERY RATING METHODS

The BCI rates batteries according to reserve capacity and cold-cranking power. When replacing a battery, always refer to an application chart to select a battery with the correct BCI group number. Vehicle options, such as air conditioning and a number of major electrical accessories, may indicate the need for an optional heavy-duty battery with a higher rating. Remember, to handle cranking power and the vehicle's other electrical needs, the replacement unit should never have a lower rating than the original battery.

Reserve Capacity

A reserve capacity (RC) rating represents the approximate time in minutes it is possible to travel at night with battery ignition and minimum electrical load, but without a charging system in operation. The time in minutes is based on a current draw of 25 amperes while maintaining a minimum battery terminal voltage of 10.5 volts (12-volt batteries) at 80°F (26.7°C). A battery with a reserve capacity of 100 would be able to deliver 25 amps for 100 minutes before the voltage would drop below 10.5 volts. This rating represents the electrical load that must be supplied by the battery in the event of a charging system failure.

Ampere-Hour Rating

The **ampere-hour** rating is the amount of steady current that a fully charged battery can supply for 20 hours at 80°F (26.7°C) without the cell voltage falling below 1.75 volts or 10.5 volts at the terminals. For example, if a battery can be discharged for 20 hours at a rate of 4.0 amperes before its terminal voltage reads 10.5 volts, it would have a rating of 80 ampere-hours.

Watt-Hour Rating Some battery manufacturers rate their batteries in watt-hours. The watt-hour rating is determined at 0°F (–17.7°C) because the battery's capacity changes with temperature. The rating is determined by multiplying a battery's amp-hour rating by the battery's voltage.

Cold Cranking

A **cold-cranking amperes (CCA)** rating specifies the minimum amperes available at 0°F (–17.7°C) and at –20°F (–28.8°C). CCA is the common standard for low-maintenance batteries. The previous standard, amp-hour rating, is no longer used except with some imported vehicles. This rating allows cranking capability to be related to such significant variables as engine displacement, compression ratio, temperature, cranking time, condition of engine and electrical system, and lowest practical voltage for cranking and ignition. This rating indicates the amperes that a fully charged battery will maintain for 30 seconds without the terminal voltage falling below 7.2 volts for a 12-volt battery. The cold-cranking rating is given in total amperage and is identified as 300 CCA, 400 CCA, and so on. The usual range for passenger cars and light trucks is between 300 and 600 CCA; some batteries have a rating as high as 1,100 CCA.

Cranking AMP Rating This rating is similar to CCA and is a measure of the number of amperes a battery can deliver at 32°F for 30 seconds and maintain at least 1.2 volts per cell (7.2 volts in a 12-volt battery). This rating is more commonly used in climates that rarely see close to 0°F temperatures.

CUSTOMER CARE

An excellent service you can offer your customer is to help select the right type of battery based on the make of the vehicle and the customer's driving habits. In cold climates, putting in a larger battery with higher CCAs is a good idea, but do not get carried away. Reserve capacity and proper mounting are important, too. ■

BATTERY SIZE SELECTION

Besides selecting a battery based on its capacity and rating, the proper battery is also one that fits. The battery should fit the battery holding fixture and the hold-down must be able to be installed. It is also important that the height of the battery not allow the terminals to short across the vehicle's hood when it is closed. BCI group numbers are normally given on the battery **(Figure 17–14)** and are used to indicate the physical size and

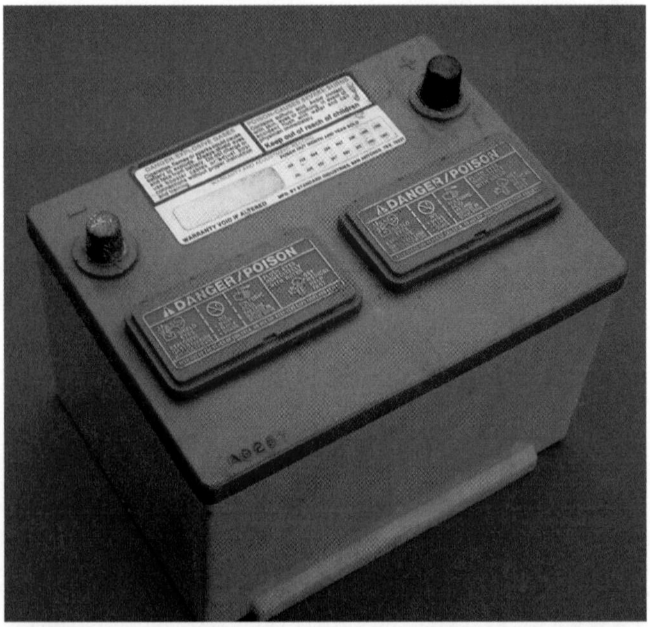

Figure 17–14 Battery sticker with identification and warnings.

BCI Group Number	Maximum Overall Dimensions						Assembly Figure No.	PERFORMANCE RANGES	
	Millimeters			Inches				Cold Cranking Performance Amps. @ 0°F (−18°C)	Reserve Capacity Minutes @ 80°F (27°C)
	L	W	H	L	W	H			
PASSENGER CAR AND LIGHT COMMERCIAL BATTERIES 12-VOLT (6 CELLS)									
21	208	173	222	8 3/16	6 13/16	8 3/4	10	310-400	50-70
21R	208	173	222	8 3/16	6 13/16	8 3/4	11	310-500	50-70
22F	241	175	211	9 1/2	6 7/8	8 5/16	11F	220-425	45-90
22HF	241	175	229	9 1/2	6 7/8	9	11F	400	69
22NF	240	140	227	9 7/16	5 1/2	8 15/16	11F	210-325	50-60
22R	229	175	211	9	6 7/8	8 5/16	11	290-350	45-90
24	260	173	225	10 1/4	6 13/16	8 7/8	10	165-625	50-95
24F	273	173	229	10 3/4	6 13/16	9	11F	250-585	50-95
24H	260	173	238	10 1/4	6 13/16	9 3/8	10	305-365	70-95
24R	260	173	229	10 1/4	6 13/16	9	11	440-475	70-95
24T	260	173	248	10 1/4	6 13/16	9 3/4	10	370-385	110
25	230	175	225	9 1/16	[illegible]	8 7/8	[illegible]	[illegible]	[illegible]

Figure 17–15 BCI battery group numbers indicate the size and features of the battery.

other features of the battery **(Figure 17–15)**. The size of the battery does not always indicate the current capacity of a battery.

The label may also include the date the battery was shipped from the manufacturer. This information may also appear on a label on the side of the battery. The letter on the label corresponds with the month, starting with A for January, B for February, and so on. The letter I is skipped, so September is represented by the letter J. The number represents the year, with 8 standing for 1998, 1 for 2001, and so on.

FACTORS AFFECTING BATTERY LIFE

All storage batteries have a limited service life, but many conditions can decrease service life.

Improper Electrolyte Levels

With nonsealed batteries, water should be the only portion of the electrolyte lost due to evaporation during hot weather and gassing during charging. Maintaining an adequate electrolyte level is the basic step in extending battery life for these designs.

Temperature

Batteries do not work well when they are cold. At 0°F a battery is only capable of working at 40% of its capacity. Like everything else, the electrons find it hard to move when they are cold. Also in the cold, the engine's oil is thicker and it is harder to crank the engine; therefore, the demands on the starter and battery are much higher. There is a possibility that the battery will freeze when it is low on charge and subjected to very cold weather. When the weather is extremely hot, the electrons get hyperactive and there is a possibility of boiling over as the electrons move rapidly. Plus, at high temperatures the water tends to evaporate and heat causes the positive plate grids to corrode more rapidly. Batteries used in hot climates need to have their electrolyte level checked often and distilled water added if necessary.

Corrosion

Battery corrosion is commonly caused by spilled electrolyte or electrolyte condensation from gassing. In either case, the sulfuric acid from the electrolyte corrodes, attacks, and can destroy not only connectors and terminals but hold-down straps and the battery tray as well.

Corroded connections increase resistance at the battery terminals, which reduces the applied voltage to the vehicle's electrical system. Corrosion on the battery cover can also create a path for current, which can allow the battery to slowly discharge. Finally, corrosion can destroy the hold-down straps and battery tray, which can result in physical damage to the battery.

Overcharging

Batteries can be overcharged by either the vehicle's charging system or a battery charger. In either case, the result is a violent chemical reaction within the battery that causes a loss of water in the cells. This can permanently reduce the capacity of the battery. Overcharging can also cause excessive heat, which can oxidize the positive plate grid material and even buckle the plates, resulting in a loss of cell capacity and early battery failure.

Undercharge/Sulfation

The vehicle's charging system might not fully recharge the battery due to stop-and-go driving or a fault in the charging system. In these cases, the battery operates in a partially discharged condition. A battery in this condition will become sulfated when the sulfate normally formed on the plates becomes dense, hard, and chemically irreversible. This happens because the sulfate has been allowed to remain in the plates for a long period.

Sulfation of the plates causes two problems. First, it lowers the specific gravity levels and increases the danger of freezing at low temperatures. Second, in cold weather, a sulfated battery often fails to crank the engine because of its lack of reserve power.

Poor Mounting

Loose hold-down straps allow the battery to vibrate or bounce during vehicle operation. This vibration can shake the active materials off the grid plates, severely shortening battery life. It can also loosen the plate connections to the plate strap, loosen cable connections, or even crack the battery case.

Cycling

Heavy and repeated cycling can cause the positive plate material to break away from its grids and fall into the sediment chambers at the base of the case. This problem reduces battery capacity and can lead to short circuiting between the plates.

SAFETY PROCEDURES

The potential dangers caused by the sulfuric acid in the electrolyte and the explosive gases generated during battery charging require that battery service and troubleshooting be conducted under absolute safe working conditions. According to the National Society to Prevent Blindness, over 14,000 Americans suffered serious eye damage from acid in wet-cell batteries in a recent year.

CAUTION!

Always wear safety glasses or goggles when working with batteries, no matter how small the job.

Sulfuric acid can also cause severe skin burns. If electrolyte contacts your skin or eyes, flush the area with water for several minutes. When eye contact occurs, force your eyelid open. Always have a bottle of neutralizing eyewash on hand and flush the affected areas with it. Do not rub your eyes or skin.

CAUTION!

Receive prompt medical attention if electrolyte contacts your skin or eyes. Call a doctor immediately.

When a battery is charging or discharging, it gives off quantities of highly explosive hydrogen gas. Some hydrogen gas is present in the battery at all times. Any flame or spark can ignite this gas, causing the battery to explode violently **(Figure 17–16)**, propelling the vent caps at a high velocity and spraying acid in a wide area. To prevent this dangerous situation, take these precautions.

- Do not smoke near the top of a battery, and never use a lighter or match as a flashlight.
- Remove wristwatches and rings before servicing any part of the electrical system. This helps to prevent the possibility of electrical arcing and burns.
- Even sealed, maintenance-free batteries have vents and can produce dangerous quantities of hydrogen if severely overcharged.
- Always disconnect the battery's ground cable when working on the electrical system or engine.
- A battery that has been overworked should be allowed to cool down and let air circulate around it before attempting to jump-start the vehicle.
- Never connect or disconnect charger leads when the charger is turned on. Doing so generates a dangerous spark.
- Never lay metal tools or other objects on the battery, because a short circuit across the terminals can result.

Other battery and electrical system safety precautions follow:

- Improper connection of charger cables to the battery can reverse the current flow and damage the generator.

Figure 17–16 Careless use of a battery charger caused this battery to explode.

- When removing a battery from a vehicle, always disconnect the battery ground cable first. When installing a battery, connect the ground cable last.
- Never reverse the polarity of the battery connections. Generally, all vehicles use a negative ground. Reversing this polarity damages the generator and circuit wiring.
- Never attempt to use a fast charger as a boost to start the engine.
- As a battery gets closer to being fully discharged, the acidity of the electrolyte is reduced, and the electrolyte starts to behave more like pure water. A dead battery may freeze at temperatures near 0°F. Never try to charge a battery that has ice in the cells. Passing current through a frozen battery can cause it to rupture or explode. If ice or slush is visible or the electrolyte level cannot be seen, allow the battery to thaw at room temperature before servicing. Do not take chances with sealed batteries. If there is any doubt, allow them to warm to room temperature before servicing.
- As batteries get old, especially in warm climates and especially with lead-calcium cells, the grids start to grow. The chemistry is rather involved, but the point is that plates can grow to the point where they touch, producing a shorted cell. If you see a battery with normal fluid levels in five cells and one nearly dry cell, you are probably looking at a battery that has shorted one cell and turned its electrolyte into hydrogen gas.
- Acid from the battery damages a vehicle's paint and metal surfaces and harms shop equipment. Neutralize any electrolyte spills during servicing.

ROUTINE INSPECTIONS

As part of any electrical system work, always check the battery.

1. Visually inspect the battery cover and case for dirt and grease.
2. Check the electrolyte level. When adding water to the cells, use distilled or clean, soft water. Fill each cell to just above the top of the plates.
3. Inspect the case for cracks, loose terminal posts, and other signs of physical damage.
4. Check for missing cell plug covers and caps.
5. Inspect all cables for broken or corroded wires **(Figure 17–17)**, frayed insulation, or loose or damaged connectors.
6. Visually check battery terminals, cable connectors, metal parts, hold-downs, and trays for corrosion damage or buildup.
7. Check the heat shield for proper installation on vehicles so equipped.

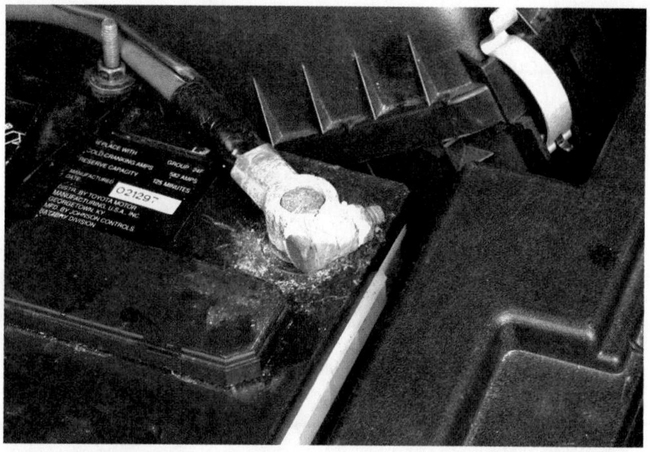

Figure 17-17 Corroded battery terminals reduce the efficiency of the battery and cause voltage drops.

ROUTINE CLEANING

Before removing battery connectors or the battery itself for cleaning or other service, always neutralize accumulated corrosion on terminals, connectors, and other metal parts. Apply a solution of baking soda and water or ammonia and water. Photo Sequence 12 shows the correct procedure for cleaning a battery, battery tray, and battery cables.

Spring-type cable connectors are removed by squeezing the ends of their prongs together with wide-jaw, vise-gripping, channel lock, or battery pliers. This pressure expands the connector so it can be lifted off the terminal post.

For connectors tightened with nuts and bolts, loosen the nut using a box wrench or cable-clamp pliers. Ordinary pliers or an open-end wrench might slip off under pressure with enough force to break the cell cover or damage the casing.

Always grip the cable while loosening the nut. This eliminates unnecessary pressure on the terminal post that could break it or loosen its mounting in the battery. If the connector does not lift easily off the terminal when loosened, use a clamp puller to free it. Prying with a screwdriver or bar strains the terminal post and the plates attached to it. This can break the cell cover or pop the plates loose from the terminal post.

Once the connectors have been removed, open the connector using a connector-spreading tool. Neutralize any remaining corrosion by dipping it in a baking soda or ammonia solution. Next, clean the inside of the connectors and the posts using a wire brush with external and internal bristles **(Figure 17–18)**.

When removing or installing a battery, always use the built-in battery strap or a battery lifting tool to lift the battery in or out of its tray.

PHOTO SEQUENCE 12

Typical Procedure for Cleaning a Battery Case, Tray, and Cables

P12-1 *Loosen the battery negative terminal clamp.*

P12-2 *Use a terminal clamp puller to remove the negative cable.*

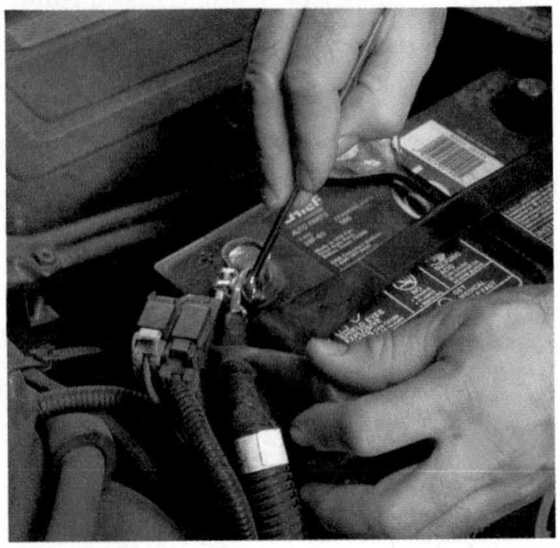

P12-3 *Loosen the battery positive terminal clamp.*

P12-4 *Use a terminal clamp puller to remove the positive clamp.*

P12-5 *Remove the battery hold-down hardware and any heat shields.*

P12-6 *Remove the battery from the tray.*

P12-7 *Mix a solution of baking soda and water.*

P12-8 *Brush the baking soda solution over the battery case, but don't allow the solution to enter the cells of the battery.*

P12-9 *Flush the baking soda off with water.*

P12-10 *Use a scraper and wire brush to remove corrosion from the hold-down hardware.*

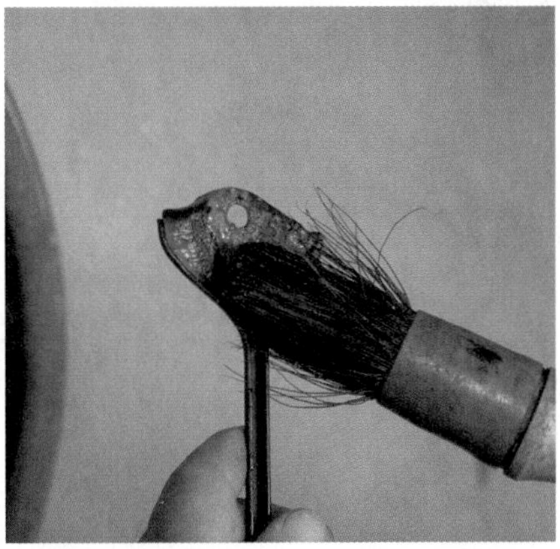

P12-11 *Brush the baking soda solution over the hold-down hardware and then flush with water.*

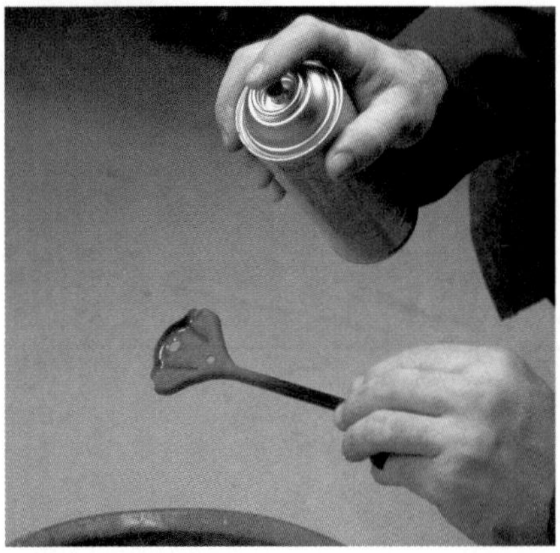

P12-12 *Allow the hardware to dry, then paint it with corrosion-proof paint.*

P12-13 *Use a terminal cleaner brush to clean the battery cables.*

P12-14 *Use a terminal cleaner brush to clean the battery posts.*

P12-15 *Install the battery back into the tray. Also install the hold-down hardware.*

P12-16 *Install the positive battery cable. Then install the negative cable.*

Figure 17–18 Combination external/internal wire brushes clean both terminals and inside cable connector surfaces. *Courtesy of Mac Tools*

Begin reinstallation by correctly positioning the battery in its tray and then installing the hold-down assembly. Now connect the positive connector on its post. Do not overtighten any nuts or bolts because this could damage the post or connector. Finally, coat the connectors with petroleum jelly or battery anticorrosion paste or paint.

BATTERY TESTING

Testing batteries is an important part of electrical system service. Depending on the design of the battery, state of charge and capacity can be determined in several ways: specific gravity tests, visual inspection of batteries with a built-in hydrometer, open circuit voltage tests, capacity test, and capacitance test.

Specific Gravity Tests

On unsealed batteries, the specific gravity of the electrolyte can be measured to give a fairly good indication of the battery's state of charge. A **hydrometer (Figure 17–19)** consists of a glass tube or barrel, rubber bulb, rubber tube, and a glass float or hydrometer with a scale built

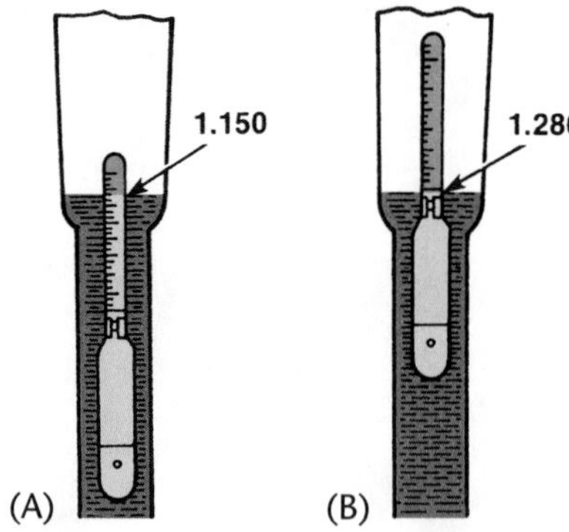

Figure 17–20 (A) When the scale sinks in the electrolyte, the specific gravity is low; (B) when it floats high, the specific gravity is high.

into its upper stem. The glass tube encases the float and forms a reservoir for the test electrolyte. Squeezing the bulb pulls electrolyte into the reservoir.

When filled with test electrolyte, the sealed hydrometer float bobs in the electrolyte. The depth to which the glass float sinks in the test electrolyte indicates its relative weight compared to water **(Figure 17–20)**. The reading is taken off the scale by sighting along the level of the electrolyte.

The electrolyte of a fully charged battery is usually about 64% water and 36% sulfuric acid, which corresponds to a specific gravity of 1.270. Specific gravity is the weight of a given volume of any liquid divided by the weight of an equal volume of water. Pure water has a specific gravity of 1.000, while battery electrolyte should have a specific gravity of 1.260 to 1.280 at 80°F (26.7°C). In other words, the electrolyte should be 1.260 to 1.280 times heavier than water.

The specific gravity of the electrolyte decreases as the battery discharges. This is why measuring the specific gravity of the electrolyte with a hydrometer can be a good indicator of how much charge a battery has lost. **Table 17–1** lists specific gravity readings in various stages of charge with respect to a battery's ability to crank an engine at a temperature of 80°F (26.7°C).

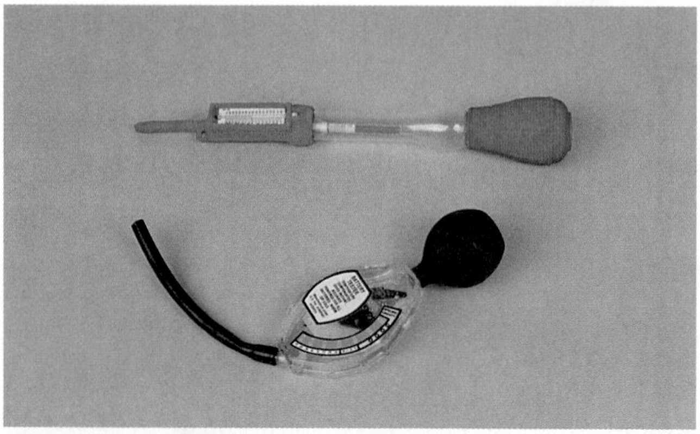

Figure 17–19 Two types of battery hydrometers.

TABLE 17–1 ELECTROLYTE SPECIFIC GRAVITY AS RELATED TO CHARGE

Specific Gravity	Percent of Charge
1.265	100%
1.225	75%
1.190	50%
1.155	25%
1.120 or lower	discharged

CAUTION!

Electrolyte is very corrosive. It can cause severe injuries if it comes in contact with your skin or eye. If electrolyte gets on you, immediately wash with baking soda and water. If the acid gets in your eyes, immediately flush them with cool water. Then get medical help.

Temperature Correction It is necessary to correct the reading by adding or subtracting 4 points (0.004) for each 10°F (–12°C) above or below the standard of 80°F (26.7°C). Most hydrometers have a built-in thermometer to measure the temperature of the electrolyte **(Figure 17–21)**. The hydrometer reading can be misleading if the hydrometer is not adjusted properly. For example, a reading of 1.260 taken at 20°F (–6.6°C) would be 1.260 – (6 × 0.004 or 0.024) = 1.236. This lower reading means the cell has less charge than indicated.

It is important to make these adjustments at high and low temperatures to determine the battery's true state of charge.

Interpreting Results The specific gravity of the cells of a fully charged battery should be near 1.265 when adjusted for electrolyte temperature.

Recharge any battery if the specific gravity drops below an average of 1.230. A specific gravity difference of more than 50 points between cells is a good indication of a defective battery in need of replacement.

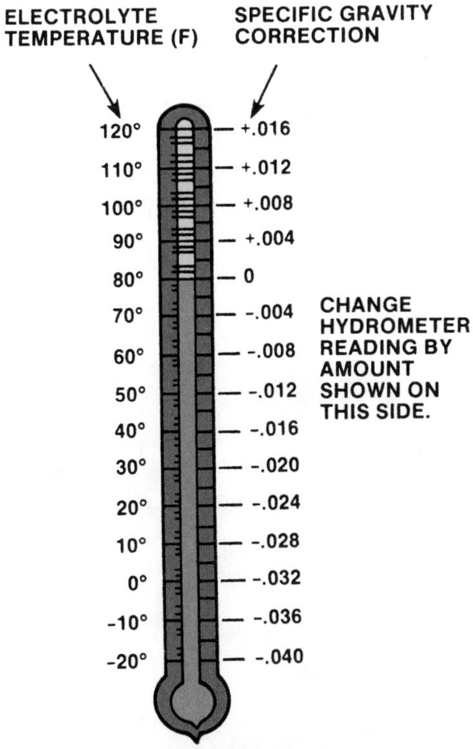

Figure 17-21 Hydrometers with thermometer correction scales make adjusting for electrolyte temperature easy.

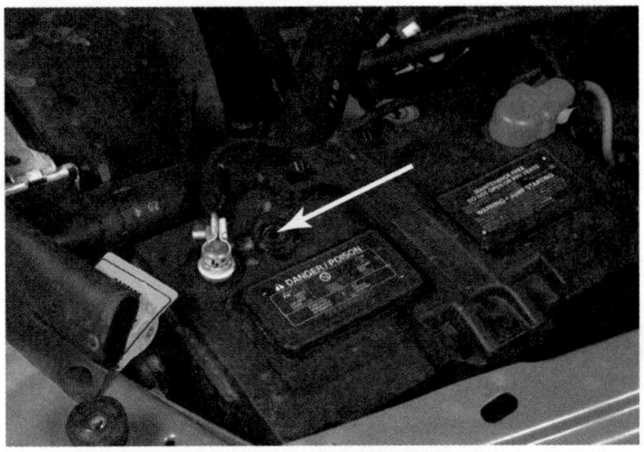

Figure 17-22 Sight glass in a maintenance-free battery.

Built-In Hydrometer

On some sealed maintenance-free batteries, a special temperature-compensated hydrometer is built into the battery cover **(Figure 17–22)**. A quick visual check indicates the battery's state of charge **(Figure 17–23)**. It is important when observing the hydrometer that the battery has a clean top to see the correct indication. A flashlight may be required in poorly lit areas. Always look straight down when viewing the hydrometer.

Many maintenance-free batteries do not have a built-in hydrometer. A voltage check is the only way to check this type of battery's state of charge. The specific gravity of these batteries cannot be checked because they are sealed. *Never* pry off the cell caps to check the electrolyte levels or condition of a sealed battery.

A few battery designs incorporate a charge indicator into the top of the battery **(Figure 17–24)**. Rather than a built-in hydrometer, these batteries use a color display to

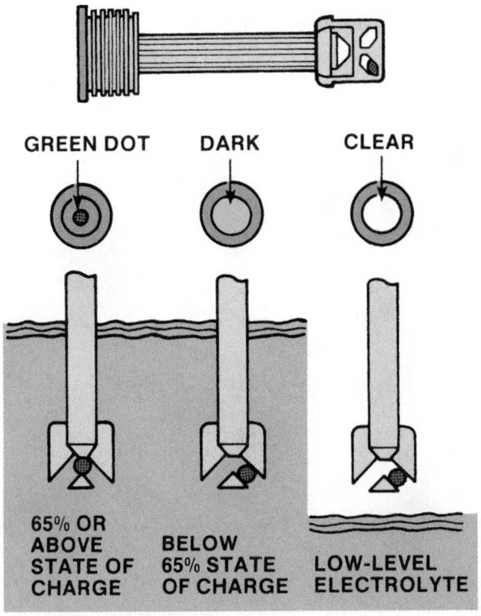

Figure 17-23 Design and operation of built-in hydrometers on maintenance-free sealed batteries.

Figure 17-24 A battery with a built-in charge indicator. *Courtesy of Robert Bosch Corp.*

note the battery's state of charge. The color green stands for "OK," gray for "check or recharge," and white for "change or replace."

Open Circuit Voltage Test

An open circuit voltage check can be used as a substitute for the hydrometer specific gravity test. As the battery is charged or discharged, slight changes occur in the battery's voltage. So battery voltage with no load applied can give some indication of the state of charge.

The battery's temperature should be between 60° and 100°F (15.5° and 37.7°C). The voltage must be allowed to stabilize for at least 10 minutes with no load applied. On vehicles with high drains (computer controls, clocks, and accessories that always draw a small amount of current), it may be necessary to disconnect the battery ground cable. On batteries that have just been recharged, apply a heavy load for 15 seconds to remove the surface charge. Then allow the battery to stabilize. Once voltage has stabilized, use a voltmeter to measure the battery voltage to the nearest one-tenth of a volt **(Figure 17–25)**. Use **Table 17–2** to interpret the results. As you can see, minor changes in battery open circuit voltage can indicate major changes in state of charge.

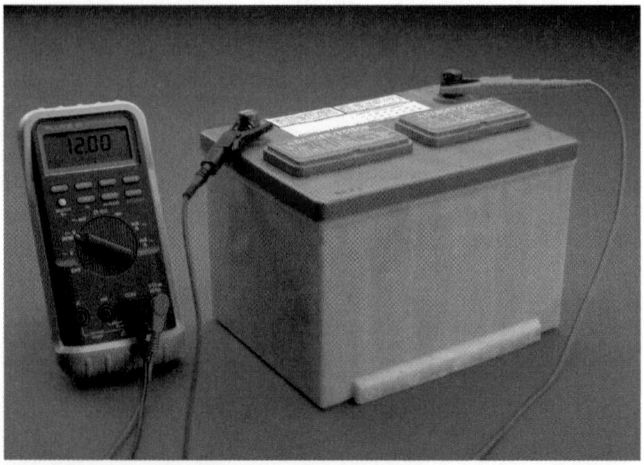

Figure 17-25 Measuring open circuit voltage across battery terminals using a voltmeter.

TABLE 17–2 BATTERY OPEN CIRCUIT VOLTAGE AS AN INDICATOR OF STATE OF CHARGE

Open Circuit Voltage	State of Charge
12.6 or greater	100%
12.4 to 12.6	75–100%
12.2 to 12.4	50–75%
12.0 to 12.2	25–50%
11.7 to 12.0	0–25%
11.7 or less	0%

If the open circuit voltage test indicates a charge of below 75% of full charge, recharge the battery and perform the capacity test to determine battery condition.

Battery Leakage Test

To perform a battery leakage test, set a voltmeter on a low DC volt range. Connect the negative test lead to the battery negative terminal. Then move the meter's positive lead across the top and sides of the battery case **(Figure 17–26)**. If some voltage is read on the voltmeter, current is leaking out of the battery cells. The battery should be cleaned, then rechecked. If the battery again has some leakage, it should be replaced because the case is excessively porous or is cracked.

Battery Drain Test

If a vehicle's battery is dead after it has not been used for a short while, the problem may be a current drain caused by one of the electrical systems. The most common cause for this type of drain is a light that is not turning off—such as a light in the glove box, trunk, or engine compartment.

Constant drains on the battery due to accessories that draw small amounts of current are called parasitic drains.

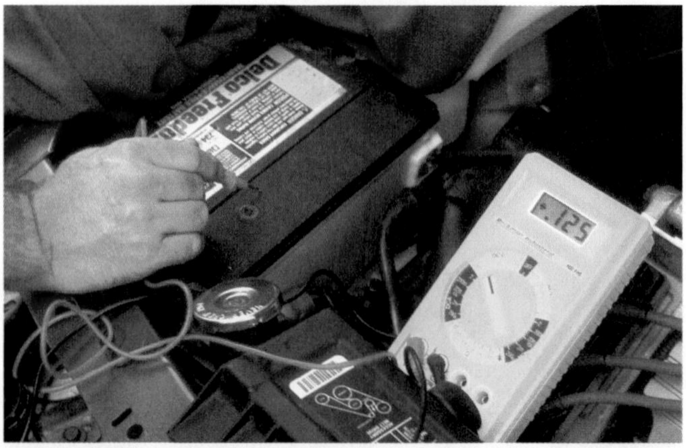

Figure 17-26 Performing a battery leakage test.

These drains on the battery can cause various driveability problems. With low battery voltage several problems can result; the computer may go into the backup mode of operation, the computer may set false codes, or the computer may raise idle speeds to compensate for the low battery voltage.

The procedure for performing a battery drain test varies according to the manufacturer. However, battery drain can often be observed by connecting an ammeter in series with the negative cable or by placing the inductive ammeter lead around the negative cable. If the meter reads 0.25 or more amps, there is excessive drain. Visually check the trunk, glove box, and under hood lights to see if they are on. If they are, remove the bulb and watch the battery drain. If the drain is now within specifications, find out why the circuit is staying on and repair the problem. If the cause of the drain is not the lights, go to the fuse panel or distribution center and remove one fuse at a time while watching the ammeter. When the drain decreases, the circuit protected by the fuse you removed last is the source of the problem.

To test for battery drains using a high-current tester, follow these steps:

PROCEDURE

Battery Drain Test

STEP 1 *Turn off all accessories and close the doors.*
STEP 2 *Remove the underhood lamp, if equipped.*
STEP 3 *Disconnect the negative battery cable.*
STEP 4 *Attach the multiplying coil between the negative battery cable and the battery terminal post* ***(Figure 17–27)****.*
STEP 5 *Zero the ammeter.*
STEP 6 *Connect the inductive pickup probe around the multiplying coil.*
STEP 7 *Read the ammeter.*

The maximum permissible current drain is 0.05 ampere. If the current drain exceeds this limit, remove the fuses, one at a time, as discussed earlier.

Capacity Test

The load or **capacity test** determines how well any type of battery, sealed or unsealed, functions under a load. In other words, it determines the battery's ability to furnish starting current and still maintain sufficient voltage to operate the ignition system.

The load or capacity test can be performed with the battery either in or out of the vehicle. The battery must be at or very near a full state of charge. For best results, the electrolyte should be as close to 80°F (26.7°C) as possible. Cold batteries show considerably lower capacity.

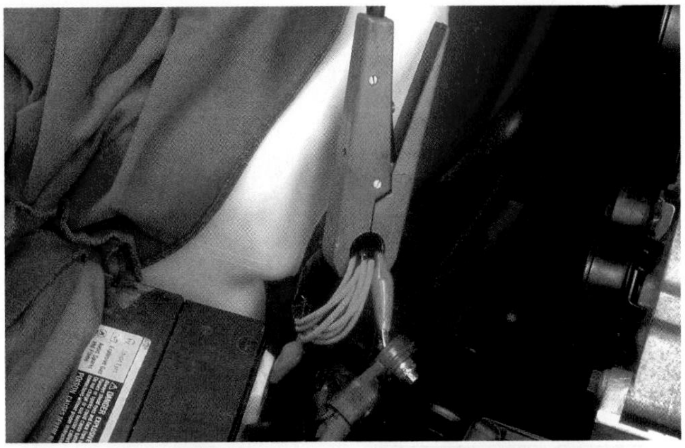

Figure 17-27 Using a multiplying coil to obtain accurate readings when measuring parasitic drains with a VAT-40 or similar tester.

Never load test a sealed battery if its temperature is below 60°F (15.5°C).

On batteries with side terminals, obtaining a sound connection can be a problem. The best solution is to screw in the appropriate manufacturer's adapter **(Figure 17–28)**. If an adapter is not available, use a ⅜-inch (9.54-mm) coarse bolt with a nut on it. Bottom out the bolt. Back it off a turn. Then tighten the nut against the contact. Now, attach the lead to the nut.

A battery tester with a carbon pile should be used to check the capacity of a battery. The typical connections for this type of meter are shown in **Figure 17–29**.

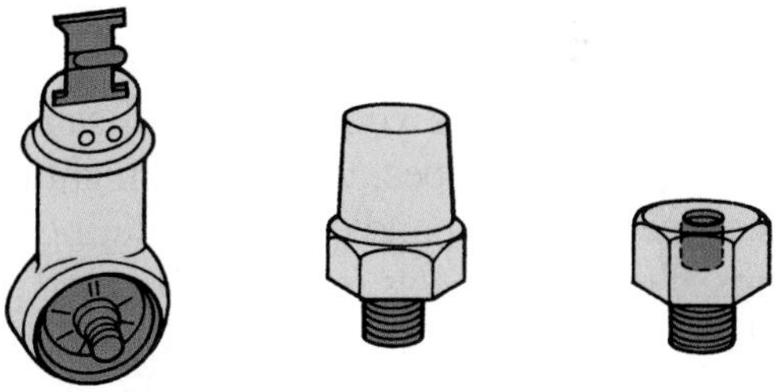

Figure 17-28 Adapters may be needed to test and charge batteries with side-mount terminals.

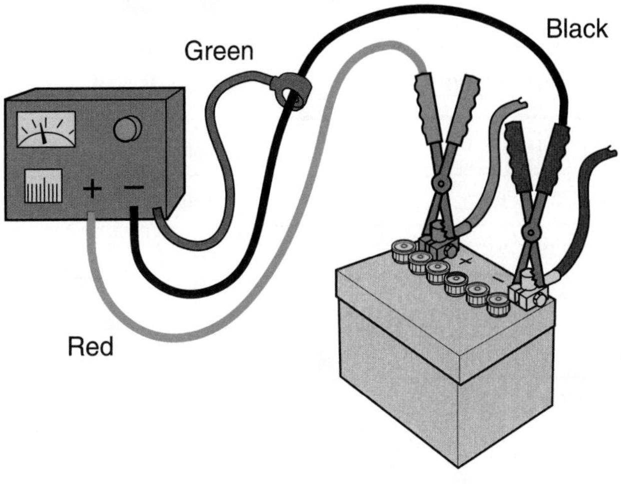

Figure 17-29 Typical tester hook-up for conducting a battery load test.

When performing a battery load test, follow these guidelines:

PROCEDURE

Battery Load Test

STEP 1 *The inductive pickup must surround the negative cable of the tester.*

STEP 2 *Observe the correct polarity and make sure the test leads are making good contact with the battery posts.*

STEP 3 *Turn the load control knob (if the tester is so equipped) to draw current at the rate of three times the battery's ampere-hour rating or one-half of its CCA rating.*

STEP 4 *Maintain the load for 15 seconds. Observe the tester's voltmeter.*

STEP 5 *Discontinue the load after 15 seconds of current draw.*

STEP 6 *At 70°F (21°C) or above or on testers that are temperature compensated, voltage at the end of 15 seconds should not fall below 9.6 volts. If the tester is not temperature compensated, use* ***Table 17–3*** *to determine the adjusted minimum voltage reading for a particular temperature.*

Interpreting Results If the voltage reading exceeds the specification by a volt or more, the battery is supplying sufficient current with a good margin of safety. If the reading is right on the spec, the battery might not have the reserve necessary to handle cranking during low temperatures. If the battery was at 75% charge and fell right on the load specifications, it is probably in good shape.

TABLE 17–3 MINIMUM LOAD TEST VOLTAGES AS AFFECTED BY TEMPERATURE

Battery Temperature (F)	Minimum Test Voltage
70° (21°C)	9.6 volts
60° (15.5°C)	9.5 volts
50° (10°C)	9.4 volts
40° (4.4°C)	9.3 volts
30° (–1°C)	9.1 volts
20° (–6.6°C)	8.9 volts
10° (–12.2°C)	8.7 volts
0° (–17.7°C)	8.5 volts

If the voltage reads below the temperature-corrected minimum, continue to observe the voltmeter of the tester after removing the load. If it rises above 12.4 volts, the battery is bad. It can hold a charge but has insufficient cold-cranking amperes. The battery can be recharged and retested, but the results are likely to be the same.

If the voltage tests below the minimum and the voltmeter does not rise above 12.4 volts when the load is removed, the problem may only be a low state of charge. Recharge the battery and **load test** again.

If a volt-ampere tester is not available, the starter motor can be used as a loading device to conduct a capacity test. By observing a voltage reading before and after the starting motor has run, the condition of the battery can be determined. Connect the voltmeter across the battery. Make sure the ignition is disabled to prevent engine starting.

Battery Capacitance Test

Many manufacturers recommend that a capacitance or **conductance test** be performed on batteries. Conductance describes a battery's ability to conduct current. It is a measurement of the plate surface available in a battery for chemical reaction. Measuring conductance provides a reliable indication of a battery's condition and is correlated to battery capacity. Conductance can be used to detect cell defects, shorts, normal aging, and open circuits, which can cause the battery to fail.

A fully charged new battery will have a high conductance reading, anywhere from 110% to 140% of its CCA rating. As a battery ages, the plate surface can sulfate or shed active material, which will lower its capacity and conductance.

When a battery has lost a significant percentage of its cranking ability, the conductance reading will fall well below its rating and the test decision will be to replace the battery. Because conductance measurements can track the life of the battery, they are also effective for predicting end of life before the battery fails.

To measure conductance, the tester **(Figure 17–30)** creates a small signal that is sent through the battery and then measures a portion of the AC current response. The tester displays the service condition of the battery. The tester indicates that the battery is good, needs to be recharged and tested again, has failed, or will fail shortly.

BATTERY CHARGING

Both fast- and slow-charging units are used to recharge batteries. Each has its advantages. Fast chargers are the most popular. They charge batteries at a higher rate or charge—usually 40 amperes for 12-volt batteries and 70 amperes for 6-volt batteries. At this rate, fast chargers can recharge most batteries in about one hour. However, batteries must be in good condition to accept a fast charge.

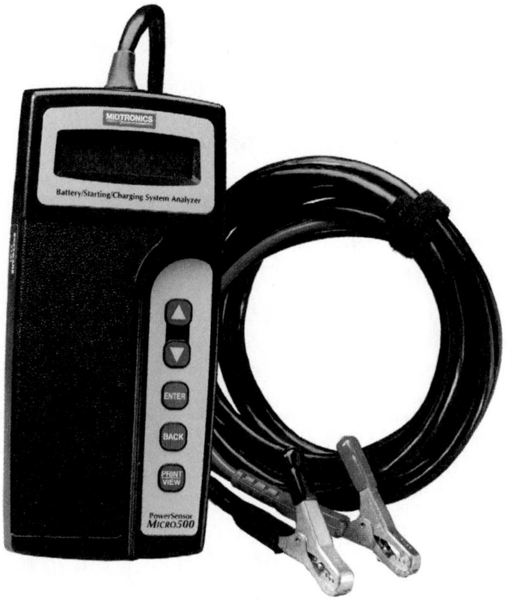

Figure 17-30 A conductance (capacitance) battery tester. *Courtesy of Midtronics Inc.*

Sulfation on the plates of the battery can lead to excessive gassing, boiling, and heat buildup during fast charging. Never fast-charge a battery that shows evidence of sulfation buildup or separator damage.

Trickle chargers provide low charging currents of about 5 to 15 amperes. Slow charging may require 12 to 24 hours but is the only safe way of charging sulfated batteries. In general, almost any battery can be charged at any current rate as long as excessive electrolyte gassing does not occur and the electrolyte temperature does not exceed 125°F (51.6°C). However, when time is available, slow charging is the safest and easiest method to use. In fact, many fast chargers can be adjusted to provide slow charging. The basic rule of thumb for slow charging a battery is 1 ampere for each positive plate in one cell. Use the chart shown in **Figure 17–31** to determine the rate of charge according to the reserve capacity of the battery.

Battery Charging Guide

(6-Volt and 12-Volt Batteries)

Recommended rate and time for fully discharged condition ("flat discharged")

CAUTION – For low water loss batteries, see precautions on page 19 of chapter VI

Rated Battery Capacity *(Reserve Minutes)*	Slow Charge	Fast Charge
80 minutes or less	15 hours @ 3 amperes	2.5 hours @ 20 amperes 1.5 hours @ 30 amperes
Above 80 to 125 minutes	21 hours @ 4 amperes	3.75 hours @ 20 amperes 1.5 hours @ 50 amperes
Above 125 to 170 minutes	22 hours @ 5 amperes	5 hours @ 20 amperes 2 hours @ 50 amperes
Above 170 to 250 minutes	23 hours @ 6 amperes	7.5 hours @ 20 amperes 3 hours @ 50 amperes
Above 250 minutes	24 hours @ 10 amperes	6 hours @ 40 amperes 4 hours @ 60 amperes

Figure 17-31 Table showing the rate and time of slow charging a battery according to reserve capacity.

Perform charging in a well-ventilated area away from sparks and open flames. Always be sure the charger is off before connecting or disconnecting the leads to the battery. Remember to wear eye protection, and never attempt to charge a frozen battery.

All battery chargers have manufacturer-specific characteristics and operating instructions that must be followed. When charging a battery in the vehicle, always disconnect the battery cables to avoid damaging the generator or other electrical components.

WARNING!

Do not exceed the manufacturer's battery-charging limits. Also, never charge the battery if the built-in hydrometer registers clear or light yellow. Replace the battery.

RECYCLING BATTERIES

When a battery is replaced, the old battery should be recycled. Batteries are recycled more often than any other item. Most battery shops and parts stores will accept old batteries for recycling; some will even pay for an old battery.

Battery acid is recycled by neutralizing it into water or converting it to sodium sulfate for laundry detergent, glass, and textile manufacturing. Plastic is recycled by cleaning the battery case, melting the plastic and reforming it into uniform pellets. Lead, which makes up 50% of every battery, is melted, poured into slabs, and purified.

In many places in Canada a five-dollar fee is added to the price of a new battery. When the old battery is brought back to the store, the five dollars is given back.

JUMP-STARTING

When it is necessary to jump-start a car with a discharged battery using a booster battery and jumper cables, follow the instructions shown in **Figure 17–32** to avoid damaging the charging system or creating a hazardous situation.

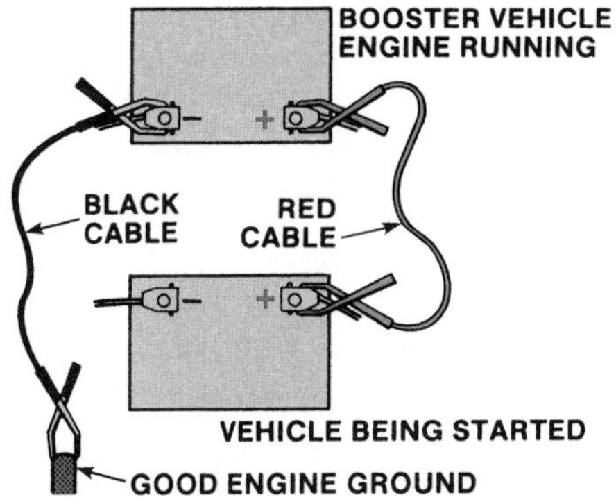

Figure 17-32 Proper setup and connections for jump-starting a vehicle with a low battery.

Always wear eye protection when making or breaking jumper cable connections.

The following steps should be followed to safely jump-start most vehicles:

PROCEDURE

Jump-Starting a Battery

STEP 1 *Make sure the two vehicles are not touching each other. The excessive current flow through the vehicle's bodies can damage the small ground straps that attach the engine block to the frame. These small wires are designed to carry only 30 amperes. If the vehicles are touching, as much as 400 amperes may be carried through them.*

STEP 2 *For each vehicle, engage the parking brake and put the transmission in neutral or park.*

STEP 3 *Turn off the ignition switch and all accessories, on both vehicles.*

STEP 4 *Attach one end of the positive jumper cable to the disabled battery's positive terminal.*

STEP 5 *Connect the other end of the positive jumper cable to the booster battery's positive terminal.*

STEP 6 *Attach one end of the negative jumper cable to the booster battery's negative terminal.*

STEP 7 *Attach the other end of the negative jumper cable to an engine ground on the disabled vehicle.*

STEP 8 *Attempt to start the disabled vehicle. If the disabled vehicle does not readily start, start the jumper vehicle and run it at fast idle to prevent excessive current draw.*

STEP 9 *Once the disabled vehicle starts, disconnect the ground connected negative jumper cable from its engine block.*

STEP 10 *Disconnect the negative jumper cable from the booster battery.*

STEP 11 *Disconnect the positive jumper cable from the booster battery, then from the other battery.*

CASE STUDY

***A** customer complains that the vehicle will not start without having to jump the battery. The technician learns this happens every time the customer attempts to start the vehicle. The customer also says the voltmeter in the dash has been reading higher than normal.*

The technician verifies the complaint. The engine turns over very slowly for a few seconds then does not turn. After jumping the battery to get the vehicle into the shop, the technician makes a visual inspection of the battery and cables. The open circuit voltage test shows a voltage of 12.5 volts across the terminals. When the battery is subjected to the capacity test, the voltage drops to 7.8 volts at 80°F (26.7°C). After ten minutes, the open circuit voltage is back up to 12.5 volts. The technician determines that the battery is sulfated and calls the customer with an estimate of repair costs. The customer agrees to have the battery replaced, which cures the problem.

WARNING!

Consult the manufacturer's service manual for procedures and precautions when jump-starting late-model vehicles with electronic control systems. Excessive battery voltages can damage sensitive electronic components.

KEY TERMS

Absorbed glass mat (AGM)
Ampere-hour
Battery Council International (BCI)
Capacity test
Cell cover
Cold-cranking amperes (CCA)
Conductance test
Cycling
Deep-cycle battery
Electrochemical
Electrolyte
Element
Gassing
Grid
Hybrid battery
Hydrometer
Lead peroxide
Load test
Parasitic load
Plate
Plate strap
Recombination battery
Reserve capacity
Separator
Sponge lead
Trickle charger
Vent cap

SUMMARY

- A battery consists of grids, positive plates, negative plates, separators, elements, electrolyte, a container, cell covers, and vent plugs. Maintenance-free batteries do not have removable cell covers or caps, but they do have gas vents.
- A battery has three main functions: provides voltage and serves as a source of current, acts as a voltage stabilizer for the entire electrical system of the vehicle, and provides current whenever the electrical demands exceed the output of the charging system.
- The electrical loads present when the ignition switch is off are called parasitic loads.
- A battery's positive plate consists of a grid filled with lead peroxide as its active material. The material pasted onto the grids of the negative plates is sponge lead.
- Each battery contains a number of elements. An element is a group of positive and negative plates. All of the positive plates are connected together, as are the negative plates.
- A separator is placed between groups to prevent contact between the positive and negative plates.
- A 12-volt battery has six cells connected in series with each other. Each cell has an open circuit voltage of approximately 2.1 volts; therefore, a 12-volt battery has an actual open circuit voltage of 12.6 volts.
- The container or shell of the battery is usually a one-piece, molded assembly of polypropylene, hard rubber, or plastic.
- The battery has two external terminals, either two tapered posts, L terminals, threaded studs on top of the case, or two internally threaded connectors on the side.
- The positive battery cable fastens to the battery and the starter solenoid or relay. The negative cable fastens to ground on the engine block.
- In low-maintenance batteries, the amount of antimony is reduced to about 3%. In maintenance-free batteries, the antimony is eliminated and replaced by calcium or strontium. Reducing the amount of antimony reduces both the battery's internal heat generation and the amount of gassing that occurs during charging.
- The open-circuit voltage of a fully charged cell is roughly 2.1 volts.
- Cell size, state of charge, rate of discharge, battery condition and design, and electrolyte temperature all strongly influence the voltage of a cell during discharge.
- A reserve capacity (RC) rating represents the approximate time in minutes that it is possible to travel at night with battery ignition and minimum electrical load, without a charging system in operation.
- The ampere-hour rating is the amount of steady current a fully charged battery can supply for twenty hours at 80°F (26.7°C) without the cell voltage falling below 1.75 volts.
- A cold-cranking amperes (CCA) rating specifies the minimum amperes available at 0°F (–17.7°C) and at –20°F (–28.8°C).
- Battery selection should be based on capacity and size.
- Improper electrolyte levels, corrosion, overcharging, undercharge/sulfation, poor mounting, and cycling will affect the life of a battery.
- Sulfuric acid can cause severe skin burns. If electrolyte contacts your skin or eyes, flush the area with water for several minutes. When eye contact occurs, force your eyelid open. Do not rub your eyes or skin.
- When a battery is charging or discharging, it gives off quantities of highly explosive hydrogen gas. Some hydrogen gas is present in the battery at all times. Any flame or spark can ignite this gas, causing the battery to explode.
- On unsealed battery designs, the specific gravity of the electrolyte can be measured to give a fairly good indication of the battery's state of charge.
- On some sealed maintenance-free batteries, a special temperature-compensated hydrometer is built into the battery cover.
- An open circuit voltage check can be used as a substitute for the hydrometer specific gravity test on maintenance-free sealed batteries with no built-in hydrometer.
- The load or capacity test determines how well any type of battery, sealed or unsealed, functions under a load.
- A capacitance or conductance test measures a battery's ability to produce current and to supply power.
- To charge a battery, charging current is passed through the battery for a period of time. Fast-chargers are more popular, but slow charging is the only safe way to charge a sulfated battery.

TECH MANUAL

The following procedures are included in Chapter 17 of the *Tech Manual* that accompanies this book:

1. Removing, cleaning, and replacing a battery.
2. Inspecting a battery for condition, state of charge, and capacity.
3. Charging a maintenance-free battery.

REVIEW QUESTIONS

1. What is a parasitic load?
2. What design characteristics determine the current capacity of a battery?
3. How is a battery leakage test conducted?
4. *True or False?* A fully charged new battery will have a low conductance reading.
5. Battery corrosion is commonly caused by spilled ______ and electrolyte condensation from ______.
6. A ______ battery is also called a deep-cycling battery.
7. What causes gassing in a battery?
8. List five things that shorten the life of a battery.
9. How many volts are present in a fully charged 12-volt battery?
10. What is meant by the term *specific gravity*?
11. Which of the following purposes does an automotive battery have?
    a. It provides voltage and serves as a source for current flow.
    b. It serves as a voltage stabilizer.
    c. It supplies the power needed when the vehicle's charging system fails.
    d. All of these statements are true.
12. Technician A says larger batteries have higher current capacities. Technician B says BCI group numbers identify the features of the battery. Who is correct?
    a. Technician A c. Both A and B
    b. Technician B d. Neither A nor B
13. *True or False?* Batteries are recycled more than any other item.
14. Technician A uses a voltmeter to test the open circuit voltage of a battery. Technician B uses the results of an open circuit voltage test to determine the battery's state of charge. Who is correct?
    a. Technician A c. Both A and B
    b. Technician B d. Neither A nor B
15. What should happen if a battery fails a load test but its voltage increases above 12.4 volts when the load is released?
16. Technician A says baking soda can be used to neutralize battery corrosion. Technician B says baking soda should be used to clean batteries. Who is correct?
    a. Technician A c. Both A and B
    b. Technician B d. Neither A nor B
17. What is the first step when removing an old battery?
    a. Disconnect the positive cable.
    b. Remove the battery hold-down straps and cover.
    c. Inspect and clean the area.
    d. Install a memory saver.
18. Technician A says the reserve rating of a battery is the amount of steady current that a fully charged battery can supply for twenty hours without the voltage dropping below 10.5 volts. Technician B says ampere-hour ratings state how many hours the battery is capable of supplying 25 amperes. Who is correct?
    a. Technician A c. Both A and B
    b. Technician B d. Neither A nor B
19. Technician A says a battery converts chemical energy into electrical energy. Technician B says a battery supplies the power for some accessories in a fuel cell vehicle. Who is correct?
    a. Technician A c. Both A and B
    b. Technician B d. Neither A nor B
20. Technician A says battery straps should be used whenever a battery is lifted. Technician B says when a battery is removed, the positive cable should be disconnected first. Who is correct?
    a. Technician A c. Both A and B
    b. Technician B d. Neither A nor B

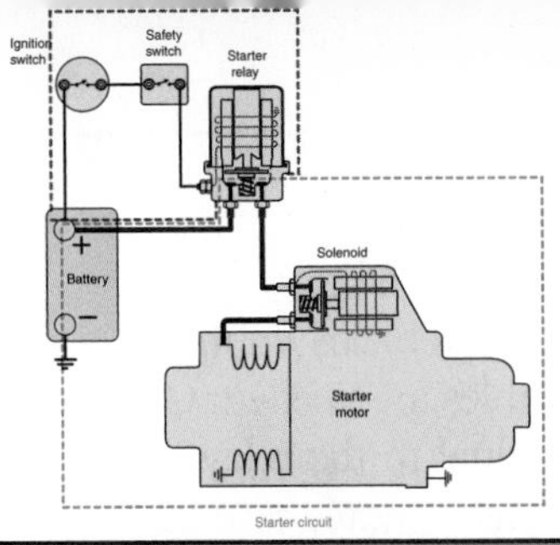

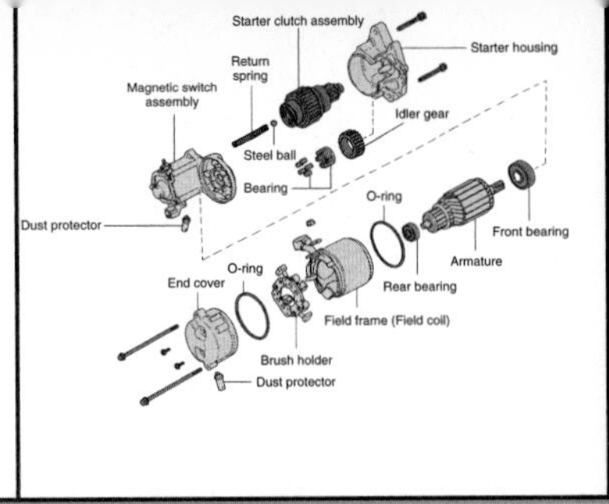

18

STARTING SYSTEMS

OBJECTIVES

■ Explain the purpose of the starting system. ■ List the components of the starting system, starter circuit, and control circuit. ■ Explain the different types of magnetic switches and starter drive mechanisms. ■ Explain how a starter motor operates. ■ Describe the operation of the different types of starter motors. ■ Perform basic tests to determine the problem areas in a starting system. ■ Perform and accurately interpret the results of a current draw test. ■ Disassemble, clean, inspect, repair, and reassemble a starter motor.

The vehicle's starting system is designed to turn or crank the engine over until it can operate under its own power. To do this, the starter motor receives electrical power from the battery. The starter motor then converts this energy into mechanical energy, which it transmits through the drive mechanism to the engine's flywheel **(Figure 18–1)**.

The only function of the starting system is to crank the engine fast enough to run. The vehicle's ignition and fuel systems provide the spark and fuel for engine operation, but they are not considered components of the basic starting system.

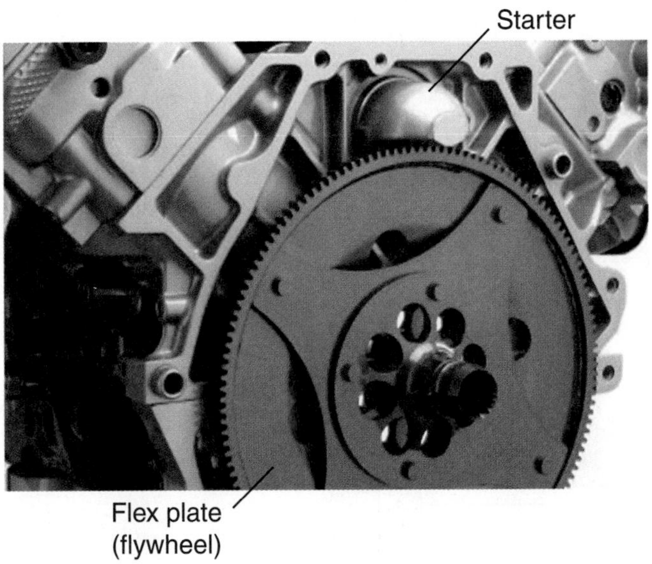

Figure 18–1 A starter motor meshed with the engine's flywheel.

STARTING SYSTEM—DESIGN AND COMPONENTS

A typical starting system has six basic components and two distinct electrical circuits. The components are the battery, ignition switch, battery cables, magnetic switch (either an electrical relay or a solenoid), starter motor, and the starter safety switch.

The starter motor **(Figure 18–2)** draws a great deal of current from the battery. A large starter motor might require 250 or more amperes of current. This current flows through the large cables that connect the battery to the starter and ground.

The driver controls the flow of this current using the ignition switch normally mounted on the steering column. The battery cables are not connected to the switch.

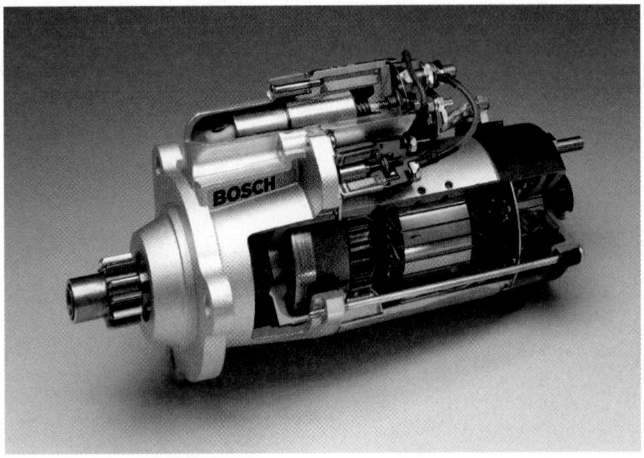

Figure 18–2 A cutaway of a heavy-duty starter motor. *Courtesy of Robert Bosch Corp.*

Rather, the system has two separate circuits: the starter circuit and the control circuit **(Figure 18–3)**. The starter circuit carries the heavy current from the battery to the starter motor through a magnetic switch in a relay or solenoid. The control circuit connects battery power at the ignition switch to the magnetic switch, which controls the high current to the starter motor.

Starter Circuit

The starter circuit carries the high current flow within the system and supplies power for the actual engine cranking. Components of the starter circuit are the battery, battery cables, magnetic switch or solenoid, and the starter motor.

Battery and Cables Many of the problems associated with the starting system can be solved by troubleshooting the battery and its related components.

The starting circuit requires two or more heavy-gauge cables. One of these cables connects between the battery's negative terminal and the engine block or transmission case. The other cable connects the battery's positive terminal with the solenoid. On vehicles equipped with a **starter relay**, two positive cables are needed. One runs from the positive battery terminal to the relay and the second from the relay to the starter motor terminal. In any case, these cables carry the required heavy current from the battery to the starter and from the starter back to the battery.

Cables must be heavy enough to comfortably carry the required current load. Cranking problems can be created when undersized cables are installed. With undersized cables, the starter motor does not develop its greatest turning effort and even a fully charged battery might be unable to start the engine.

Magnetic Switches Every starting system contains some type of magnetic switch that enables the control circuit to open and close the starter circuit. This magnetic switch can be one of several designs.

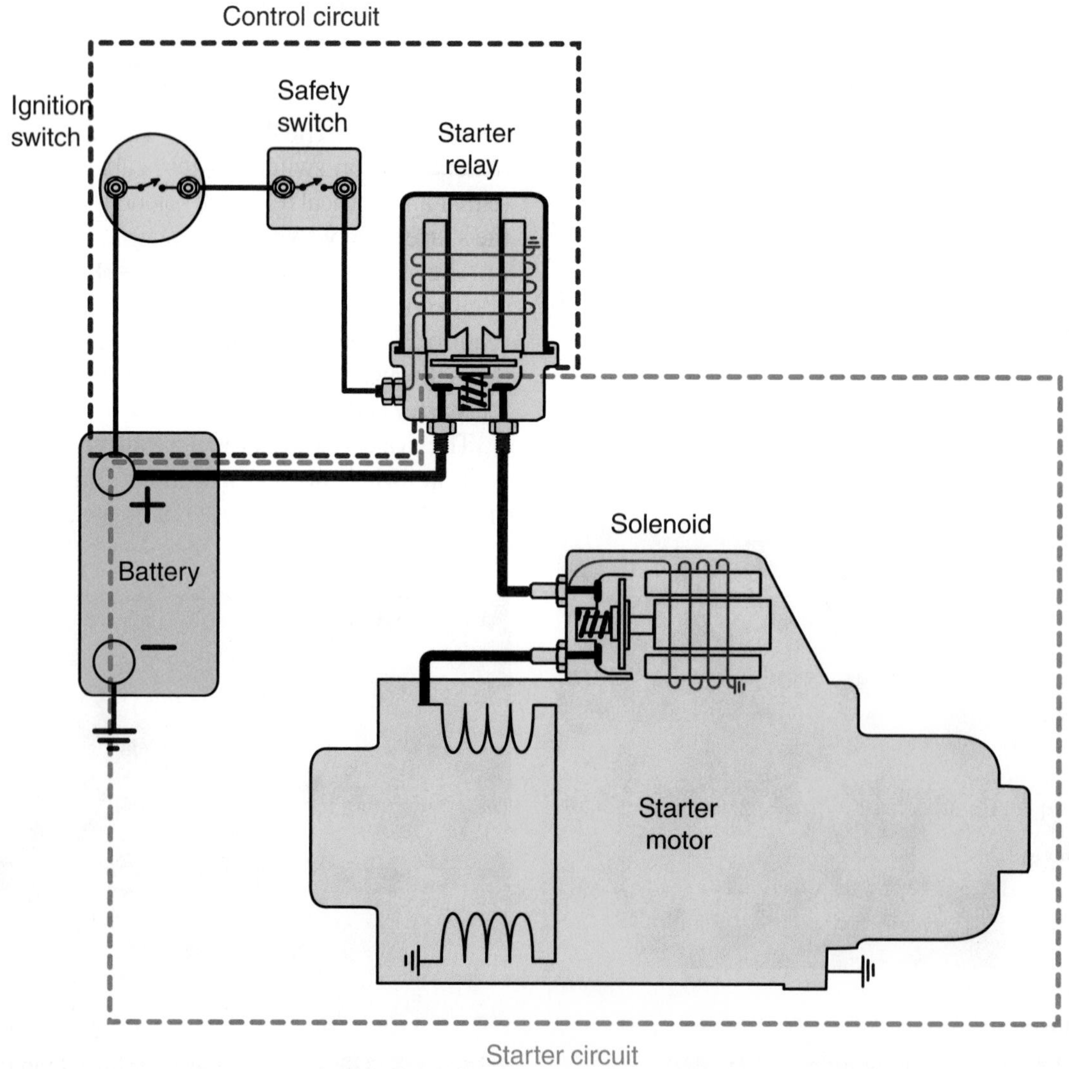

Figure 18–3 A simple diagram showing the starter and starter control circuits.

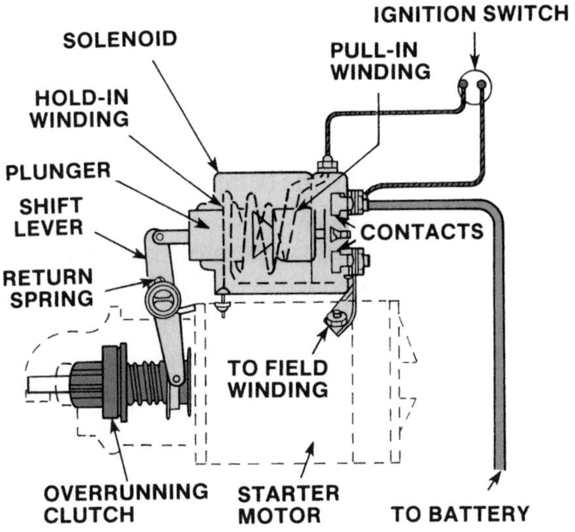

Figure 18-4 An example of a solenoid-actuated starter where the solenoid mounts directly to the starter motor.

Solenoid. The solenoid-actuated starter is by far the most common starter system used. A **solenoid** is an electromechanical device that uses the movement of a plunger to exert a pulling or holding force. As shown in **Figure 18–4**, the solenoid mounts directly on top of the starter motor.

In this type of starting system, the solenoid uses the electromagnetic field generated by its coil to perform two distinct jobs.

The first is to push the drive pinion of the starter motor into mesh with the engine's flywheel. This is the solenoid's mechanical function. The second job is to act as an electrical relay switch to energize the motor once the drive pinion is engaged. Once the contact points of the solenoid are closed, full battery current flows to the starter motor.

The solenoid assembly has two separate windings: a **pull-in winding** and a **hold-in winding**. The two windings have approximately the same number of turns but are wound from different size wire. Together these windings produce the electromagnetic force needed to pull the plunger into the solenoid coil. The heavier pull-in windings draw the plunger into the solenoid, while the lighter-gauge windings produce enough magnetic force to hold the plunger in this position.

Both windings are energized when the ignition switch is turned to the start position. When the plunger disc makes contact with this solenoid terminal, the pull-in winding is deactivated. At the same time, the plunger contact disc makes the motor feed connection between the battery and the starting motor, directing current to the field coils and starter motor armature for cranking power.

As the solenoid plunger moves, the shift fork also pivots on the pivot pin and pushes the starter drive pinion into mesh with the flywheel ring gear. When the starter motor receives current, its armature starts to turn. This motion is transferred through an overrunning clutch and pinion gear to the engine flywheel and the engine is cranked.

With this type of solenoid-actuated direct drive starting system, teeth on the **pinion gear** may not immediately mesh with the flywheel ring gear. If this occurs, a spring located behind the pinion compresses so the solenoid plunger can complete its stroke. When the starter motor armature begins to turn, the pinion teeth quickly line up with the flywheel teeth and the spring pressure forces them to mesh.

Starter Relay. Relays are the second major type of magnetic switch used. All **positive engagement starters** (described later in this chapter) use a relay in series with the battery cables to deliver the high current necessary through the shortest possible battery cables. **Figure 18–5** shows a typical starter relay. It is very similar to the solenoid. However, it is not used to move the drive pinion into mesh. It is strictly an electrical relay or switch. When current from the ignition switch arrives at the ignition switch terminal of the relay, a strong magnetic field is generated in the coil of the relay. This magnetic force pulls the plunger contact disc up against the battery terminal and the starter terminal of the relay, allowing full current flow to the starter motor.

A secondary function of the starter relay is to provide an alternate electrical path to the ignition coil during cranking. This current flow bypasses the resistance wire (or ballast resistor) in the ignition primary circuit. This is done when the plunger disc contacts the ignition bypass terminal on the relay. Not all systems have an ignition bypass setup.

Some vehicles use both a starter relay and a starter motor mounted solenoid. The relay controls current flow to the solenoid, which in turn controls current flow to the starter motor. This reduces the amount of current flowing through the ignition switch.

Basically, all the different starting systems in use today fit into one of three categories: the solenoid shift, solenoid

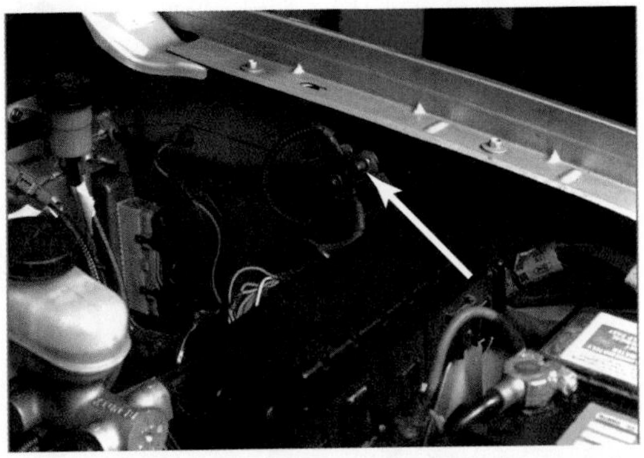

Figure 18-5 A starter relay/solenoid mounted on a vehicle.

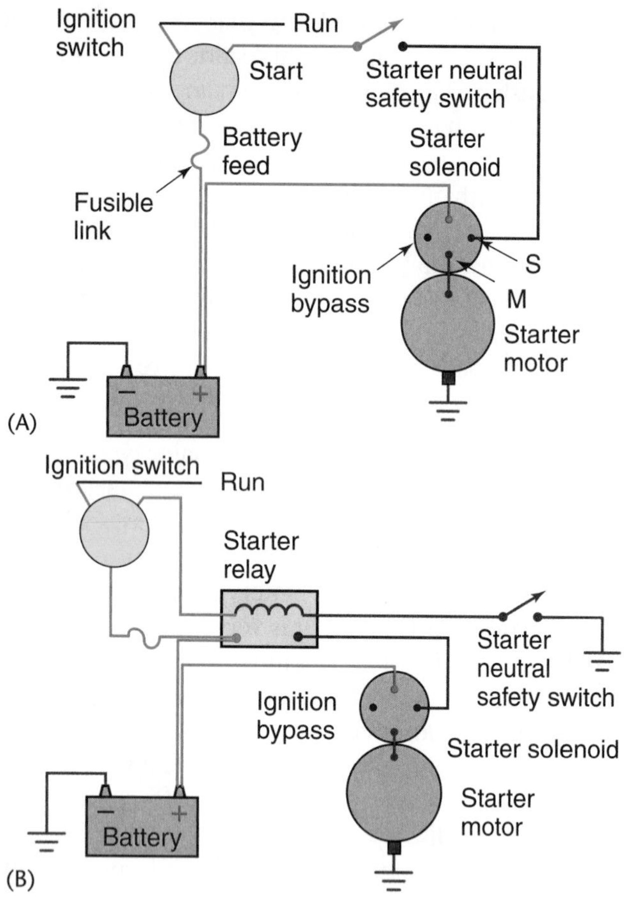

Figure 18-6 (A) A solenoid shift and (B) a solenoid shift with starter relay starting systems.

shift with relay, or positive engagement with relay. Typical wiring diagrams for these systems are shown in **Figures 18–6** and **18–7**.

Starter Motor The starter motor **(Figure 18–8)** converts the electrical energy from the battery into mechanical energy for cranking the engine. The starter is a special type of electric motor designed to operate under great electrical overloads and to produce very high horsepower.

All starting motors are generally the same in design and operation. Basically, the starter motor consists of a housing, field coils, an armature, a commutator and brushes, and end frames.

The **starter housing** or **starter frame** encloses the internal starter components and protects them from damage, moisture, and foreign materials. The housing supports the field coils.

The **field coils** and their **pole shoes (Figure 18–9)** are securely attached to the inside of the iron housing. The field coils are insulated from the housing but are connected to a terminal that protrudes through the outer surface of the housing.

The field coils and pole shoes are designed to produce strong stationary electromagnetic fields within the starter body as current is passed through the starter. These magnetic fields are concentrated at the pole shoe. Fields have an N or S magnetic polarity depending on the direction of current flow. The coils are wound around respective

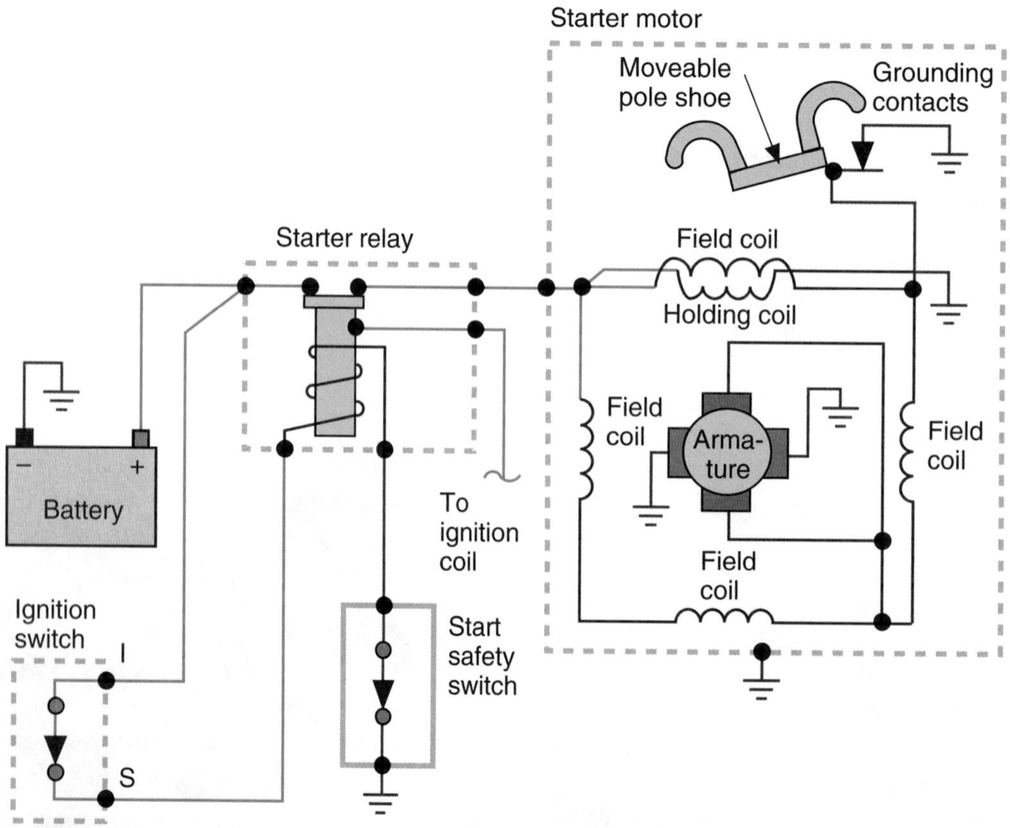

Figure 18-7 A schematic of a positive engagement starter.

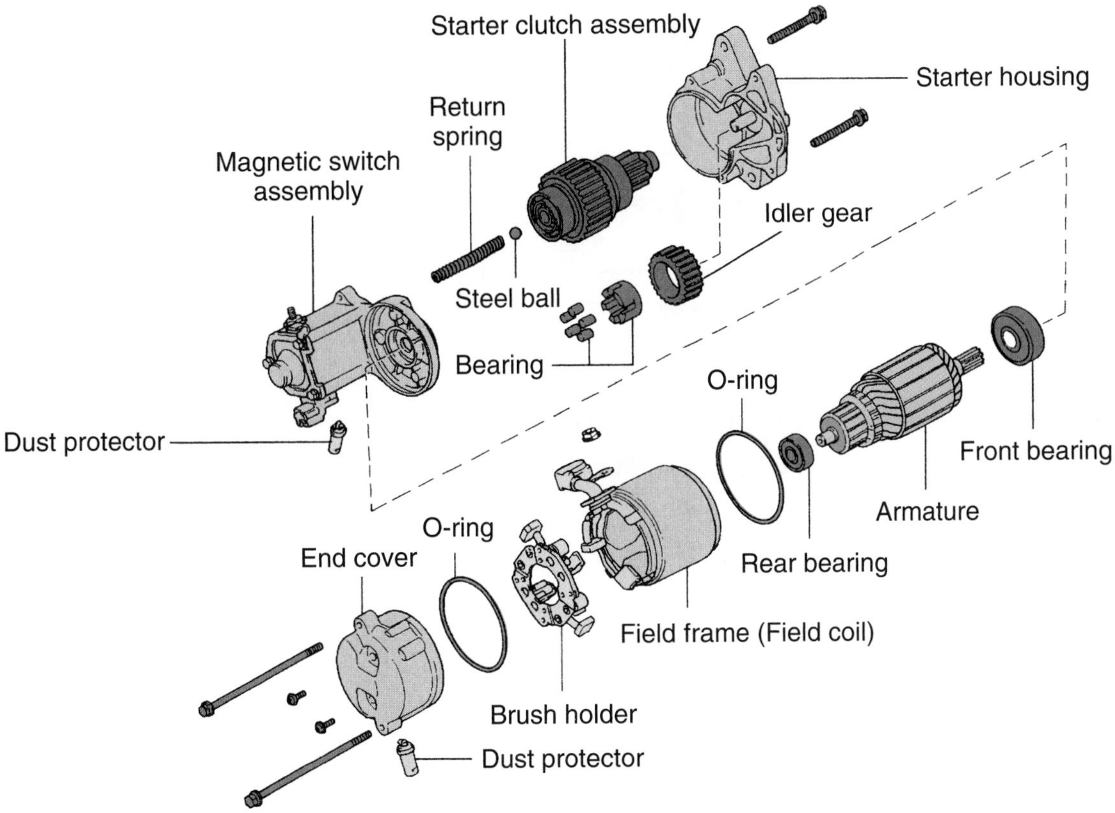

Figure 18-8 A typical starter motor assembly. *Reprinted with permission*

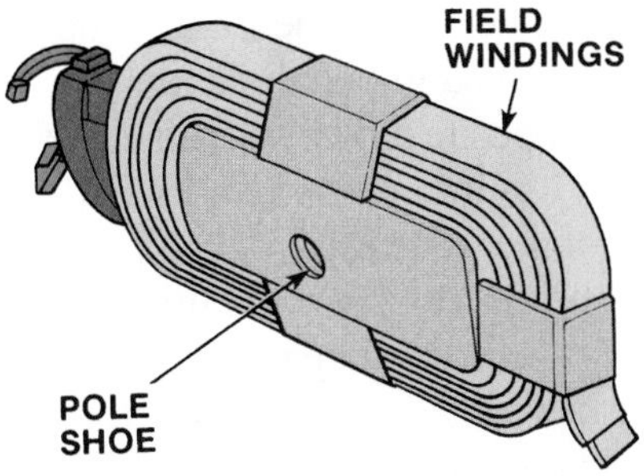

Figure 18-9 An example of a field coil and pole shoe.

pole shoes in opposite directions to generate opposing magnetic fields.

The field coils connect in series with the armature winding through the starter **brushes**. This design permits all current passing through the field coil circuit to also pass through the armature windings.

The **armature** is the only rotating component of the starter. It is located between the drive and commutator end frames and the field windings. When the starter operates, the current passing through the armature produces a magnetic field in each of its conductors. The reaction between the armature's magnetic field and the magnetic fields produced by the field coils causes the armature to rotate. This mechanical energy is then used to crank the engine.

The armature has two main components: the armature windings and the **commutator**. Both mount to the armature shaft. The windings are made of several coils of a single loop each. The sides of these loops fit into slots in the armature core or shaft, but they are insulated from it.

The coils connect to each other and to the commutator so current from the field coils flows through all of the armature windings at the same time. This action generates a magnetic field around each armature winding, resulting in a repulsion force all around the conductor. This repulsion force causes the armature to turn.

The commutator assembly is made up of heavy copper segments separated from each other and the armature shaft by insulation. The commutator segments connect to the ends of the armature windings.

Most starter motors have two to six brushes that ride on the commutator segments and carry the heavy current flow from the stationary field coils to the rotating armature windings via the commutator segments.

The brushes mount on and operate in some type of holder, which may be a pivoting arm design inside the starter housing or frame **(Figure 18–10)**. However, in many starters the brush holders are secured to the starter's end frame. Springs hold the brushes against the commutator with the correct pressure. Finally, alternate brush holders are insulated from the housing or end frame. Those in between the insulated holders are grounded.

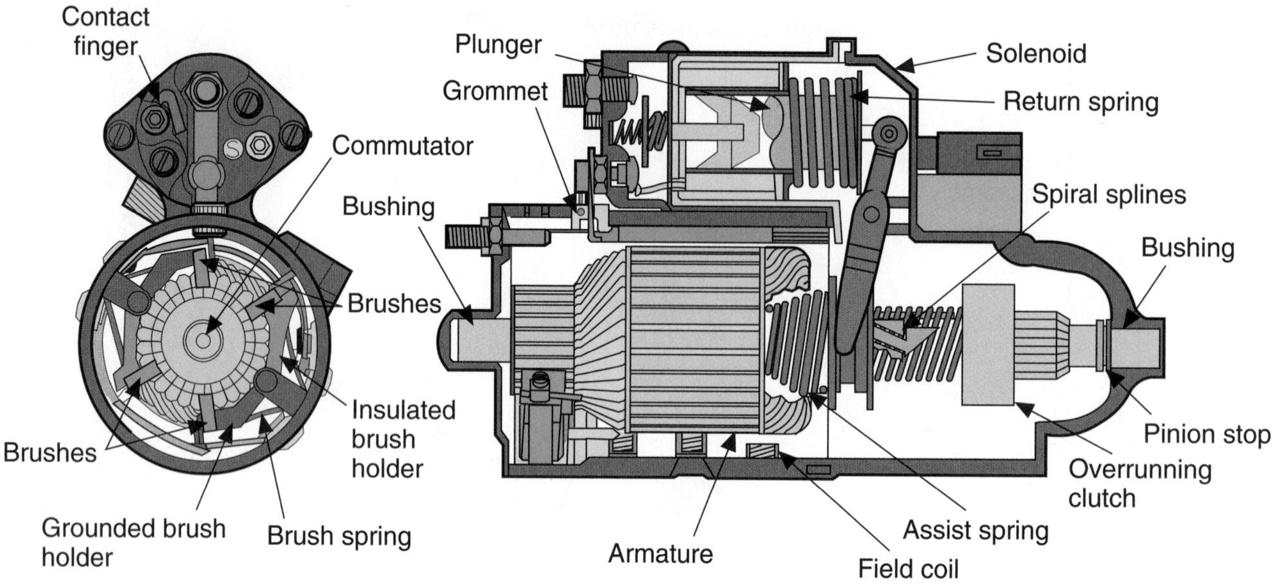

Figure 18-10 The location of the starter motor brushes and commutator.

The end frame is a metal plate that bolts to the commutator end of the starter housing. It supports the commutator end of the armature with a bushing and often contains the brush holders that support the brushes.

Operating Principle. The starter motor converts electric current into torque or twisting force through the interaction of magnetic fields. It has a stationary magnetic field, the field windings, and a current-carrying conductor, the armature windings **(Figure 18–11)**. When the armature windings are placed in this stationary magnetic field and current is passed through the windings, a second magnetic field is generated with its lines of force wrapping around the wire. Since the lines of force in the stationary magnetic field flow in one direction across the winding, they combine on one side of the wire, increasing the field strength, but are opposed on the other side, weakening the field strength. This creates an unbalanced magnetic force, pushing the wire in the direction of the weaker field **(Figure 18–12)**.

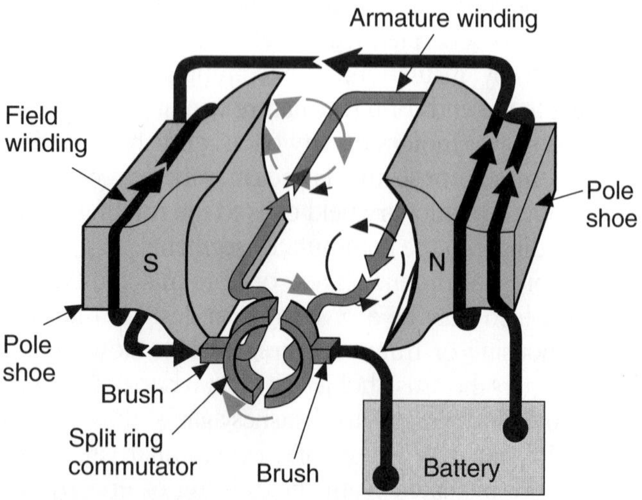

Figure 18-11 A simple DC motor.

Since the armature windings are formed in loops or coils, current flows outward in one direction and returns in the opposite direction. Because of this, the magnetic lines of force are oriented in opposite directions in each of the two segments of the loop. When placed in the stationary magnetic field of the field coils, one part of the armature coil is pushed in one direction. The other part is pushed in the opposite direction, causing the coil and the shaft to which it is mounted to rotate.

Each end of the armature winding is connected to one segment of the commutator **(Figure 18–13)**. Carbon brushes are connected to one terminal of the power supply. The brushes contact the commutator segments conducting current to and from the armature coils.

As the armature coil turns through a half revolution, the contact of the brushes on the commutator causes the current flow to reverse in the coil. The commutator segment attached to each coil end has traveled past one brush and is now in contact with the other. In this way, current flow is maintained constantly in one direction, while al-

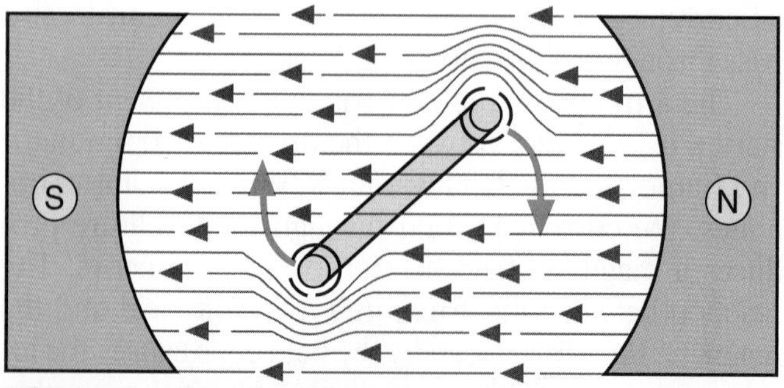

Figure 18-12 Rotation of the conductor is in the direction of the weaker magnetic field.

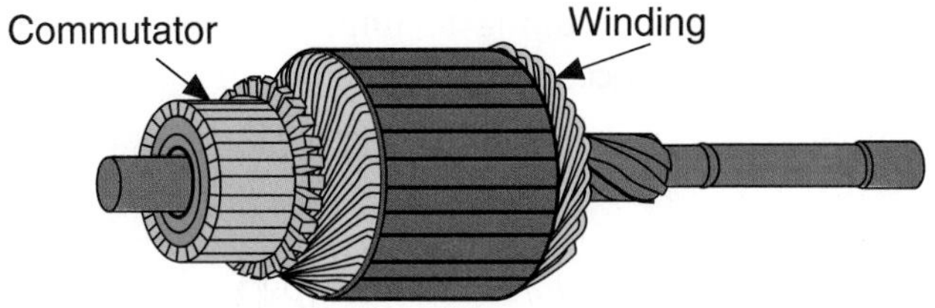

Figure 18–13 The armature of a starter motor.

lowing the segment of the rotating armature coils to reverse polarity as they rotate.

In a starter motor, many armature segments must be used. As one segment rotates past the secondary magnetic field pole, another segment immediately takes its place. The turning motion is made uniform and the torque needed to turn the flywheel is constant rather than fluctuating, as it would be if only a few armature coils were used.

The number of coils and brushes may differ between starter motor models. The armature may be wired in series with the field coils **(series motor)**; the field coils may be wired parallel or shunted across the armature **(shunt motors)**; or a combination of series and shunt wiring may be used **(compound motors) (Figure 18–14)**.

The amount of turning torque from a starter motor depends on a number of factors. One of the most important factors is current draw. The slower the motor turns, the more current it will draw. This is why a starter motor will draw excessive amounts of current when the engine is very difficult to turn over or crank. A motor needs more torque to crank a difficult-to-turn engine. The relationship between current draw and motor speed is explained by the principles of **counter EMF (CEMF)**.

When the armature rotates within the field windings of a motor, conditions exist to induce a voltage in the armature. Voltage is induced any time a wire is passed

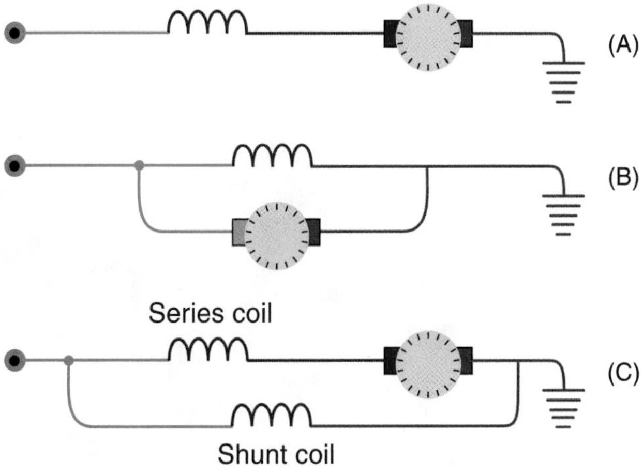

Figure 18–14 Starter motors are grouped according to how they are wired: (A) in series, (B) in parallel (shunt), or (C) as a compound motor using both series and shunt coils.

through a magnetic field. When the armature, which is a structure with many loops of wire, rotates past the field windings, a small amount of voltage is induced. This voltage opposes the voltage supplied by the battery to energize the armature. As a result, less current is able to flow through the armature.

The faster the armature spins, the more induced voltage is present in the armature. The more voltage in the armature, the more opposition there is to normal current flow to the armature. The induced voltage in the armature opposes or is counter to the battery's voltage. This is why the induced voltage is called CEMF.

The effects of CEMF are quite predictable. When the armature of the motor turns slowly, low amounts of voltage are induced and, therefore, low amounts of CEMF are present. The low amount of CEMF allows a high amount of current draw. In fact, the only time a starter motor draws its maximum amount of current is when the armature is not rotating.

A series-wound motor develops its maximum torque at start-up and develops less torque as speed increases. It is ideal for applications involving heavy starting loads.

Shunt or parallel-wound motors develop considerably less start-up torque but maintain a constant speed at all operating loads. Compound motors combine the characteristics of good starting torque with constant speed. The compound design is particularly useful for applications in which heavy loads are suddenly applied. In a starter motor, a shunt coil is frequently used to limit the maximum free speed at which the starter can operate.

Starter Motor Drive Mechanisms The area in which starters differ most is in their drive mechanisms used to crank the engine. The solenoid-actuated direct drive system has been explained earlier in this chapter.

Some starters use a planetary gear set to increase the torque of a starter motor. Planetary gear sets offer the advantage of quiet operation and compactness.

Positive Engagement Movable Pole Shoe Drive. Positive engagement movable pole shoe drive starters are found mostly on older Ford products. In this design, the drive mechanism is an integral part of the motor, and the drive pinion is engaged with the flywheel before the motor is energized.

When the ignition switch is moved to the start position, the system's starter relay closes, and full battery current is delivered to the starter. This current runs through the winding of a movable pole shoe and through a set of contacts to ground. This generates a magnetic force that pulls down the movable pole shoe. It also forces the drive pinion to engage the flywheel ring gear using a lever action and opens the contacts. A small holding coil helps keep the movable shoe and lever assembly engaged during cranking. When the engine starts, an overrunning clutch prevents the engine's flywheel from spinning the armature of the motor.

When the ignition switch is released from the start position, both the pole shoe and lever return to their original positions.

Solenoid-Actuated Gear Reduction Drive. Solenoid-actuated **gear reduction-drive starters** use a solenoid to engage the pinion with the flywheel and to close the motor circuit. The starter armature does not drive the pinion directly. Instead a gear set is used to reduce speed and increase the turning torque of the pinion gear. The gear set may be as simple as a small gear meshed with a larger one, or the reduction may take place through the use of a planetary gear set. This design allows a small, high-speed motor to develop increased turning torque at a satisfactory cranking rpm. The solenoid and starter drive operation is basically the same as in solenoid-actuated direct drive systems.

Permanent Magnet Starter Motors The most recent change in starter motors has been in the use of permanent magnets rather than electromagnets as field coils. Electrically, this starter motor is simpler. It does not require current for field coils. Current is delivered directly to the armature through the commutator and brushes. **Figure 18–15** shows this type of starter motor. This unit functions exactly as the other styles considered. Increased use of this style is expected in the future as production costs are greatly reduced. Maintenance and testing procedures are the same as for other designs. Notice the use of a planetary gear reduction assembly on the front of the armature. This assembly allows the armature to spin with increased torque, resulting in improved starter cold-cranking performance.

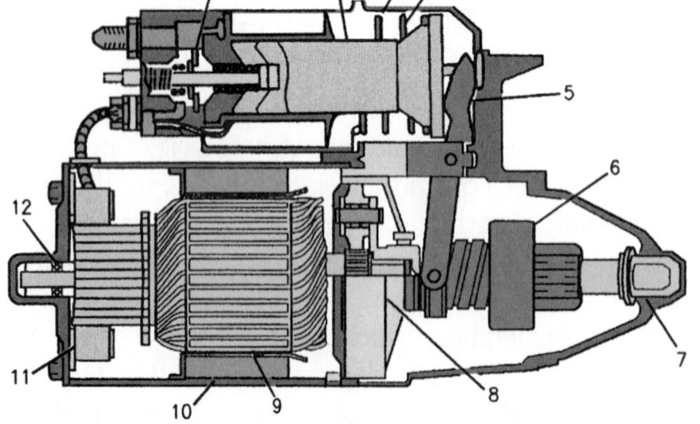

Figure 18-15 A permanent magnet-type starter assembly.

WARNING!

Permanent magnet starters require special handling because the permanent magnet material is quite brittle and can be destroyed with a sharp blow or if the starter is dropped.

Starter Drive A **starter drive** includes a pinion gear set that meshes with the flywheel on the engine's crankshaft **(Figure 18–16)**. To prevent damage to the pinion gear or the flywheel's ring gear, the pinion gear must mesh with the ring gear before the starter motor rotates. To help ensure smooth engagement, the end of the pinion gear is tapered **(Figure 18–17)**. Also, the movement of the armature must always be caused by the action of the motor, not the engine. For this reason, starter drive assemblies include an overrunning clutch.

Overrunning Clutch. The **overrunning clutch** performs a very important job in protecting the starter motor. When

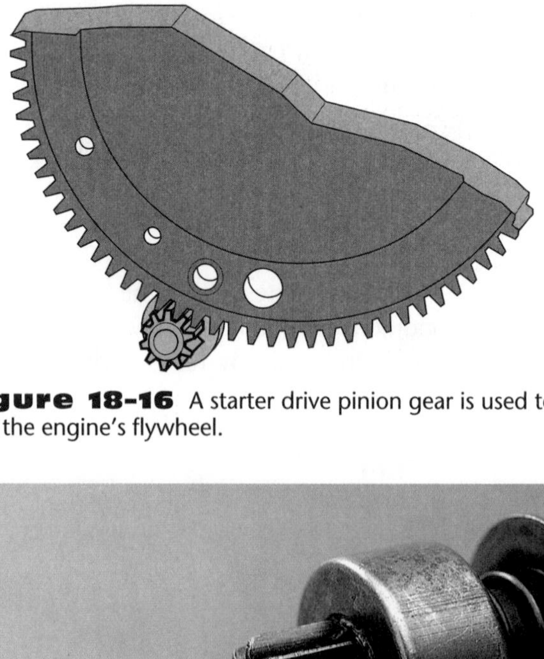

Figure 18-16 A starter drive pinion gear is used to turn the engine's flywheel.

Figure 18-17 The pinion gear teeth are tapered to allow for smooth engagement.

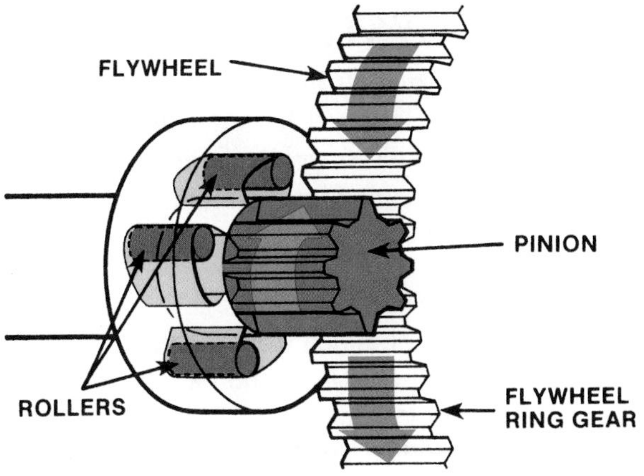

Figure 18-18 When the engine starts, the flywheel spins the pinion gear faster, which releases the rollers from the wedge.

the engine starts and runs, its speed increases. If the starter motor remained connected to the engine through the flywheel, the starter motor would spin at very high speeds, destroying the armature winding.

To prevent this, the armature must disengage from the engine as soon as the engine turns more rapidly than the starter has cranked it. However, with most starter designs the pinion remains engaged until electricity stops flowing to the starter. In these cases, an overrunning clutch is used to disengage the starter.

The clutch housing is internally splined to the starting motor armature shaft. The drive pinion turns freely on the armature shaft within the clutch housing. When the clutch housing is driven by the armature, the spring-loaded rollers are forced into the small ends of their tapered slots and wedged tightly against the pinion barrel. This locks the pinion and clutch housing solidly together, permitting the pinion to turn the flywheel and, thus, crank the engine.

When the engine starts **(Figure 18–18)**, the flywheel spins the pinion faster than the armature. This releases the rollers, unlocking the pinion gear from the armature shaft. The pinion then overruns the armature shaft freely until being pulled out of the mesh without stressing the starter motor. The overrunning clutch is moved in and out of mesh by the starter drive linkage.

CONTROL CIRCUIT

The control circuit allows the driver to use a small amount of battery current to control the flow of a large amount of current in the starting circuit.

The entire circuit usually consists of an ignition switch connected through normal-gauge wire to the battery and the magnetic switch (solenoid or relay). When the ignition switch is turned to the start position, a small amount of current flows through the coil of the magnetic switch, closing it and allowing full current to flow directly to the starter motor. The ignition switch performs other jobs besides controlling the starting circuit. It normally has at least four separate positions: accessory, off, on (run), and start.

Starting Safety Switch

The **starting safety switch**, often called the **neutral safety switch**, is a normally open switch that prevents the starting system from operating when the transmission is in gear. This eliminates the possibility of a situation that could make the vehicle lurch unexpectedly forward or backward. Safety switches are more commonly used with automatic transmissions but are also used on manual transmissions.

Starting safety switches can be located in either of two places within the control circuit. One location is between the ignition switch and the relay or solenoid. In this position, the safety switch must be closed before current can flow to the relay or solenoid. A second location for the safety switch is between the relay and ground. The switch must be closed before current can flow from the relay to ground.

The safety switch used with an automatic transmission can be either an electrical switch or a mechanical device. Contact points on the electrical switch are closed only when the shift selector is in park or neutral. The switch can be mounted near the shift selector or on the transmission housing **(Figure 18–19)**. The switch

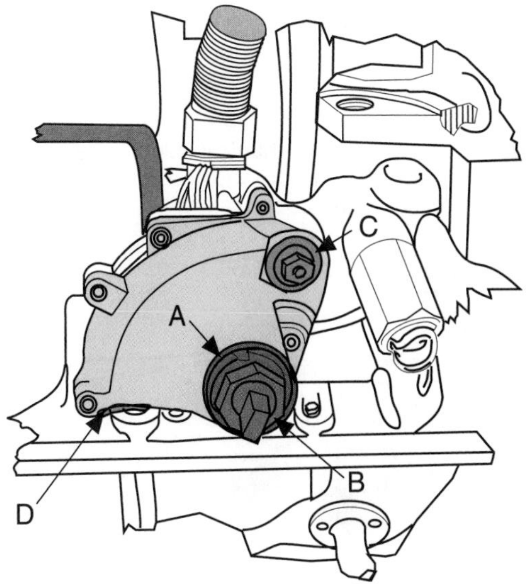

A—Locking washer

B—Switch-attaching nut

C—Switch-adjusting bolt

D—Neutral safety switch

Figure 18-19 A neutral safety switch attached to a transmission.

contacts are wired in series with the control circuit so that no current can flow through the relay or solenoid unless the transmission is in neutral or park.

Mechanical safety switches for automatic transmissions are simply devices that physically block the movement of the ignition key when the transmission is in a gear **(Figure 18–20)**. The ignition key can only be turned when the shift selector is in park or neutral.

The safety switches used with manual transmissions are usually electrical switches mounted near the gear-shift lever or on the transmission housing. A clutch switch is a second type of safety switch used with manual transmissions. This electrical switch mounts on the floor or fire wall. Its contacts are closed only when the clutch pedal is fully depressed **(Figure 18–21)**.

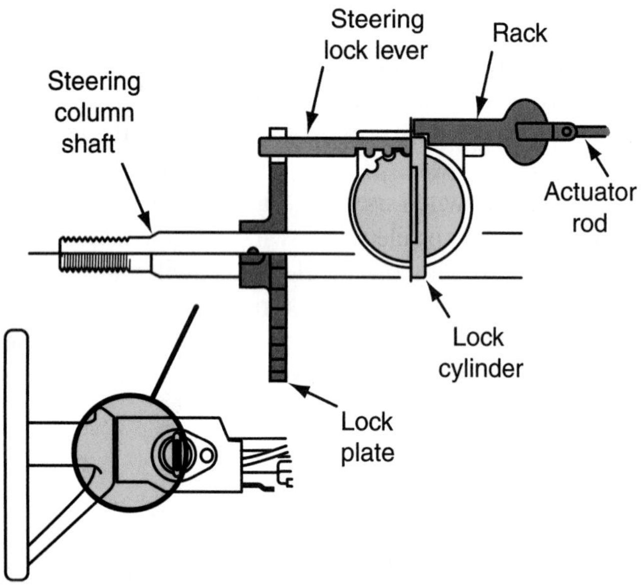

Figure 18-20 A mechanical linkage used to prevent starting the engine while the transmission is in gear. *Courtesy of General Motors Corporation*

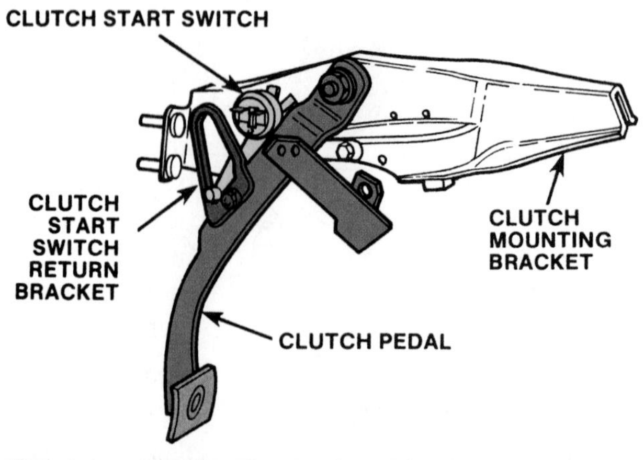

Figure 18-21 The clutch pedal must be fully depressed to close the clutch switch and complete the control circuit.

STARTING SYSTEM TESTING

As mentioned earlier, the starter motor is a special type of electrical motor designed for intermittent use only. During testing, it should never be operated for more than 15 seconds without resting for 2 minutes in between operation cycles to allow it to cool.

Preliminary Checks

The cranking output obtained from the motor is affected by the condition and charge of the battery, the circuit's wiring, and the engine's cranking requirement.

The battery should be checked and charged as needed before testing the starting system.

Check the wiring and cables for clean, tight connections. Loose or dirty connections will cause excessive voltage drops. Cables can be corroded by battery acid, and contact with engine parts and other metal surfaces can fray the cable insulation. Frayed insulation can cause a dead short that can seriously damage some of the electrical units of the vehicle.

Cables should also be checked to make sure they are not undersized (too small a gauge) or too long. Both conditions can limit the amount of current delivered to the starter motor.

When checking cables and wiring, always check any fusible links in the wiring. Most late-model vehicles are equipped with maxi-fuses in place of the fusible links. Both should be checked during any routine starting system inspection. When a maxi-fuse or fusible link has failed, always troubleshoot the system and locate the cause before replacing the fuse or link.

Make certain the engine is filled with proper weight oil as recommended by the vehicle manufacturer. Heavier-than-specified oil when coupled with low operating temperatures can drastically lower cranking speed to the point where the engine does not start and excessively high current is drawn by the starter.

Check the ignition switch for loose mounting, damaged wiring, sticking contacts, and loose connections. Check the wiring and mounting of the safety switch, if so equipped, and make certain the switch is properly adjusted. Check the mounting, wiring, and connections of the magnetic switch and starter motor. Also, be sure the starter pinion is properly adjusted.

Safety Precautions

Almost all starting system tests must be performed while the starter motor is cranking the engine. However, the engine must not start and run during the test or the readings will be inaccurate.

To prevent the engine from starting, the ignition switch can be bypassed with a remote starter switch that allows current to flow to the starting system but not to the ignition system.

During testing, be sure the transmission is out of gear during cranking and the parking brake is set. When servicing the battery, always follow safety precautions. Always disconnect the battery ground cable before making or breaking connections at the system's relay, solenoid, or starter motor.

Troubleshooting Procedures

A systematic troubleshooting procedure is essential when servicing the starting system. Consider the fact that nearly 80% of starters returned as defective on warranty claims work perfectly when tested. This is often the result of poor or incomplete diagnosis of the starting and related charging systems. Testing the starting system can be divided into area tests, which check voltage and current in the entire system, and more detailed pinpoint tests, which target one particular component or segment of the wiring circuit.

Starter Solenoid Problems

A typical symptom of solenoid problems is the presence of a clicking noise when the ignition switch is turned to the start position. The clicking noise is caused by the solenoid's plunger moving back and forth. Normally the plunger moves to the battery contacts and is held there by a magnetic field until the ignition switch is moved from the start position.

In order for the solenoid's plunger to move enough to complete the starter motor circuit and remain in that position, a strong magnetic field must be present around the solenoid's windings. The strength of the magnetic field depends on the current flowing through the windings. Therefore anything that would reduce current flow would affect the operation of the solenoid. Common causes of the clicking are low battery voltage, low voltage available to the solenoid, or an open in the hold-in winding.

Checking voltage at the battery and to the solenoid will help you identify the cause of the problem. If the solenoid is bad, it can be replaced as a unit on some starter motors or replaced with the starter motor on other designs.

Starting Safety Switches

Safety switches can be checked with a voltmeter or an ohmmeter. When the transmission is placed in park or neutral or when the clutch pedal is depressed, the switch should be closed. In other gear positions and when the clutch pedal is released, the switch should be open. Often these switches just need to be properly adjusted to correct their action. This is not possible on all vehicles. If adjustment does not correct the problem, the switch should be replaced.

Battery Load Test

A slow cranking engine is often caused by insufficient current from the battery or other problems such as incorrect ignition timing. The battery must be able to crank the engine under all load conditions while maintaining enough voltage to supply ignition current for starting. Perform a battery load test before checking the starting systems.

Cranking Voltage Test

The **cranking voltage test** measures the available voltage to the starter during cranking. To perform the test, disable the ignition or use a remote starter switch to bypass the ignition switch. Normally, the remote starter switch leads are connected to the positive terminal of the battery and the starter terminal of the solenoid or relay **(Figure 18–22)**. Refer to the service manual for specific instructions on the model car being tested. Connect the voltmeter's negative lead to a good chassis ground. Connect the positive lead to the starter motor feed at the starter relay or solenoid. Activate the starter motor and observe the voltage reading. Compare the reading to the specifications given in the service manual. Normally, 9.6 volts is the minimum required.

Test Conclusions If the reading is above specifications but the starter motor still cranks poorly, the starter motor is faulty. If the voltage reading is lower than specifications, a cranking current test and circuit resistance test should be performed to determine if the problem is caused by high resistance in the starter circuit or an engine problem.

Cranking Current Test

The **cranking current test** measures the amount of current the starter circuit draws to crank the engine. Knowing the amount of current draw helps to identify the cause of starter system problems.

Nearly all starter current testers use an inductive pickup **(Figure 18–23)** to measure the current draw. However, some earlier models were equipped with an ammeter that needed to be connected in series with the battery.

To conduct the cranking current test, connect a remote starter switch or disable the ignition prior to testing. Follow the instructions given with the tester when connecting the test leads. Crank the engine for no more than 15 seconds. Observe the voltmeter. If the voltage drops below 9.6 volts, a problem is indicated. Also, watch the ammeter and compare the reading to specifications.

Table 18–1 summarizes the most probable causes of too low or high starter motor current draw. If the problem appears to be caused by excessive resistance in the system, conduct an insulated circuit resistance test.

Insulated Circuit Resistance Test

The complete starter circuit is made up of the insulated circuit and the ground circuit. The insulated circuit includes all of the high current cables and connections from the battery to the starter motor.

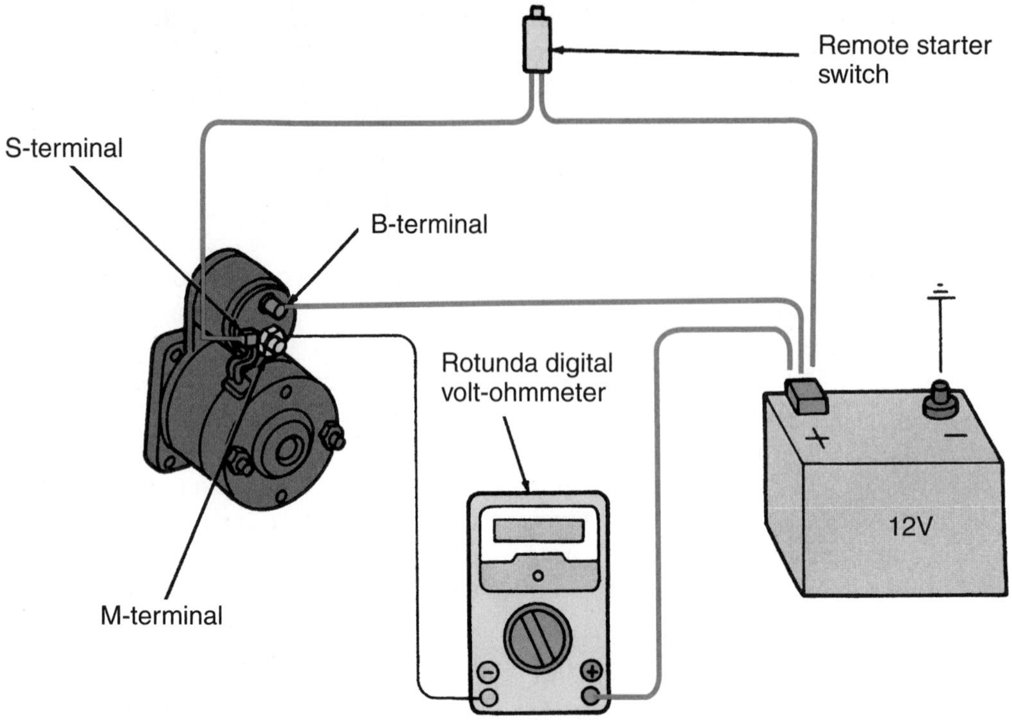

Figure 18-22 Using a remote starter switch to bypass the control circuit and ignition system. *Courtesy of Ford Motor Company*

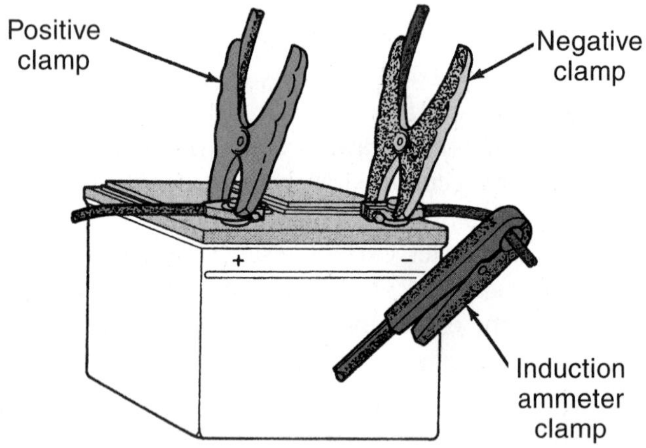

Figure 18-23 Connecting the test leads of a typical charging/starting/battery tester. *Courtesy of DaimlerChrysler Corporation*

TABLE 18–1 RESULT OF CRANKING CURRENT TESTING	
Problem	**Possible Cause**
Low current draw	Undercharged or defective battery. Excessive resistance in circuit due to faulty components or connections.
High current draw	Short in starter motor. Mechanical resistance due to binding engine or starter system component failure or misalignment.

To test the insulated circuit for high resistance, disable the ignition or bypass the ignition switch with a remote starter switch. Connect the positive (+) lead of the voltmeter to the battery's positive (+) terminal post or nut. By connecting the lead to the cable, the point of high resistance (cable-to-post connection) may be bypassed. Connect the negative (–) lead of the voltmeter to the starter terminal at the solenoid or relay. Crank the engine and record the voltmeter reading. If the reading is within specifications (usually 0.2 to 0.6 voltage drop), the insulated circuit does not have excessive resistance. Proceed to the ground circuit resistance test outlined in the next section. If the reading indicates a voltage loss above specifications, move the negative lead of the tester progressively closer to the battery, cranking the engine at each test point. Normally, a voltage drop of 0.1 volt is the maximum allowed across a length of cable.

Photo Sequence 13 goes through the correct procedure for conducting a voltage drop test on a typical starter circuit.

Test Conclusions When excessive voltage drop is observed, the trouble is located between that point and the preceding point tested. It is either a damaged cable or poor connection, an undersized wire, or possibly a bad

TABLE 18–2 MAXIMUM VOLTAGE DROPS	
Each large cable	0.1 volt
Each connection	0.1 volt
Each small wire	0.2 volt
Starter relay	0.3 volt

contact assembly within the solenoid. Repair or replace any damaged wiring or faulty connections. Refer to **Table 18–2** to find the maximum allowable voltage drops for the starter circuit.

Starter Relay Bypass Test

The starter relay bypass test is a simple way to determine if the relay is operational. First, disable the ignition. Connect a heavy jumper cable between the battery's positive (+) terminal and the starter relay's starter terminal. This bypasses the relay. When the connection is made, the engine should crank.

CAUTION!

Make sure the vehicle's transmission is in park or neutral before doing this test. The cranking starter can move the vehicle, which could injure you and others around you.

Test Conclusions If the engine cranks with the jumper installed and did not before the relay was bypassed, the starter relay is defective and should be replaced.

Ground Circuit Resistance Test

The ground circuit provides the return path to the battery for the current supplied to the starter by the insulated circuit. This circuit includes the starter-to-engine, engine-to-chassis, and chassis-to-battery ground terminal connections.

To test the ground circuit for high resistance, disable the ignition, or bypass the ignition switch with a remote starter switch. Refer to **Figure 18–24** for the proper test connection. Crank the engine and record the voltmeter reading.

Test Conclusions Good results would be less than a 0.2 volt drop for a 12-volt system. A voltage drop in excess of this indicates the presence of a poor ground circuit connection, resulting from a loose starter motor bolt, a poor battery ground terminal post connector, or a damaged or undersized ground system wire from the battery to the engine block. Isolate the cause of excessive voltage drop in the same manner as recommended in the insulated circuit resistance test by moving the positive (+) voltmeter lead progressively closer to the battery. If the ground circuit tests out satisfactorily and a starter problem exists, move on to the control circuit test.

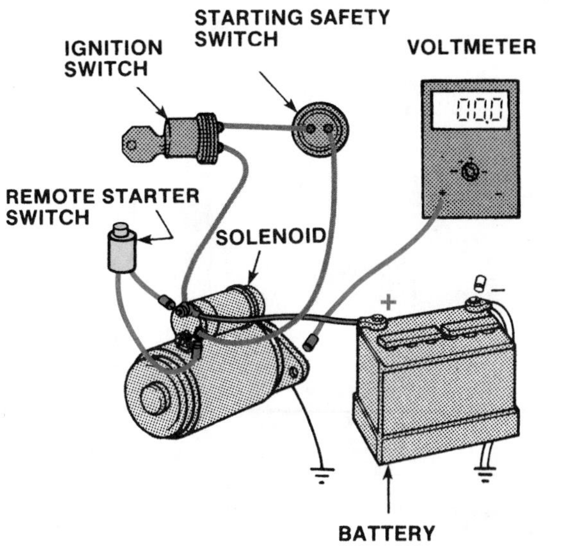

Figure 18–24 The setup for checking voltage drop across the ground circuit.

Voltage Drop Test of the Control Circuit

The control circuit test examines all the wiring and components used to control the magnetic switch, whether it is a relay, a solenoid acting as a relay, or a starter motor-mounted solenoid.

High resistance in the solenoid switch circuit reduces current flow through the solenoid windings, which can cause improper functioning of the solenoid. In some cases of high resistance, it may not function at all. Improper functioning of the solenoid switch generally results in the burning of the solenoid switch contacts, causing high resistance in the starter motor circuit.

Check the vehicle wiring diagram, if possible, to identify all control circuit components. These normally include the ignition switch, safety switch, the starter solenoid winding, or a separate relay.

To perform the test, disable the ignition system. Connect the positive meter lead to the battery's positive terminal and the negative meter lead to the starter switch terminal on the solenoid or relay. Crank the engine and record the voltmeter reading.

Test Conclusions Generally, good results would be less than 0.5 volt, indicating that the circuit condition is good. If the voltage reading exceeds 0.5 volt, it is usually an indication of excessive resistance. However, on certain vehicles, a slightly higher voltage loss may be normal.

Identify the point of high resistance by moving the negative test lead back toward the battery's positive terminal, eliminating one wire or component at a time.

PHOTO SEQUENCE 13

Voltage Drop Testing of a Starter Circuit

P13-1 *The tools required to measure the voltage drop at various points within the starter circuit are fender covers, a DMM, and a remote starter switch.*

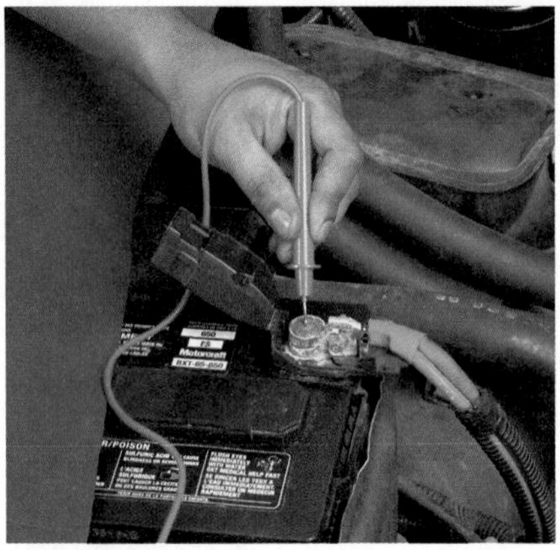

P13-2 *Connect the positive lead of the meter to the positive battery post. If at all possible, do not connect it to the battery clamp.*

P13-3 *Connect the negative lead to the main battery connection at the starter.*

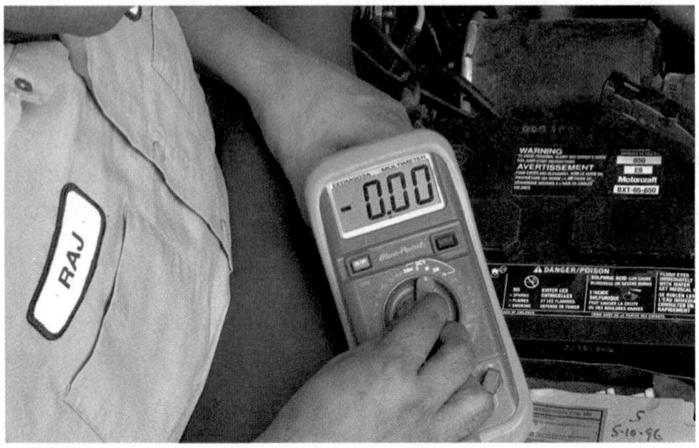

P13-4 *Set the voltmeter to the scale that is close to, but greater than, battery voltage.*

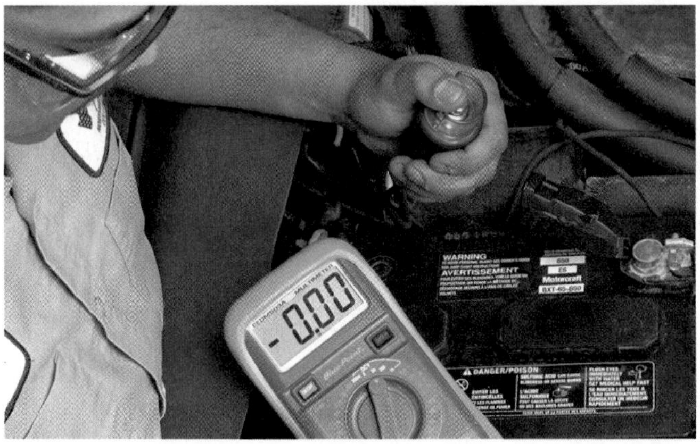

P13-5 *Disable the ignition and/or connect a remote starter switch.*

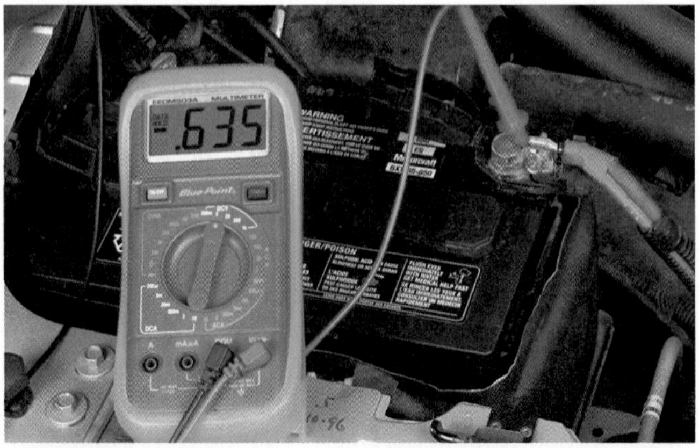

P13-6 *Crank the engine and read the voltmeter. This reading shows the voltage drop on the positive side of the starter circuit.*

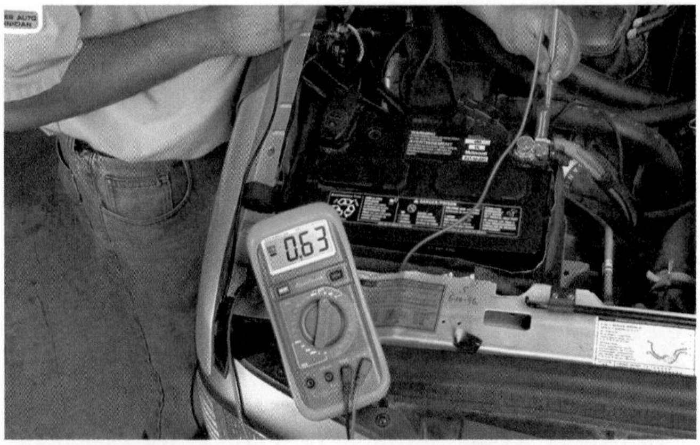

P13-7 *This reading showed excessive resistance in the circuit. To locate the resistance, move the meter's negative lead to the next location toward the battery. In this case it is the starter side of the starter relay.*

P13-8 *Crank the engine and observe the reading on the meter. This is the voltage drop across the positive circuit from the battery to the output of the relay.*

P13-9 *There is still too much voltage drop, so we continue our test by moving the negative lead to the battery side of the relay.*

P13-10 *Crank the engine and observe the reading on the meter. This is the voltage drop across the cable from the battery to the relay. Notice that hardly any voltage was dropped. This cable is okay.*

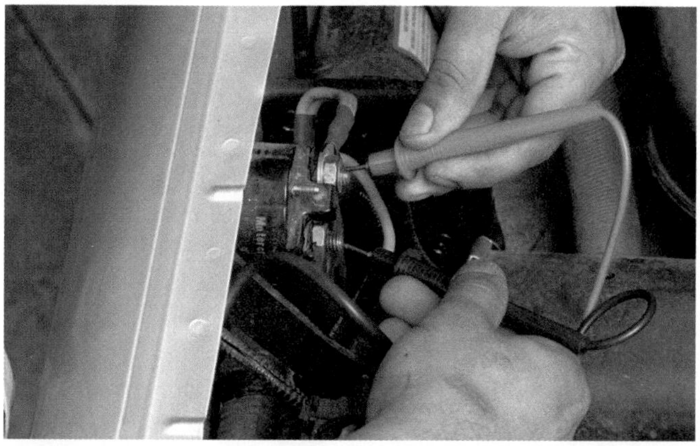

P13-11 *Now connect the meter across the relay with the red lead on the battery side and black lead on the starter side.*

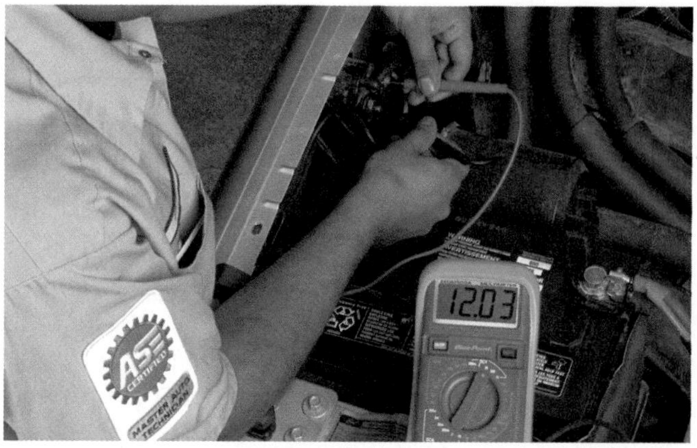

P13-12 *Ignore any voltage reading you may have at this point.*

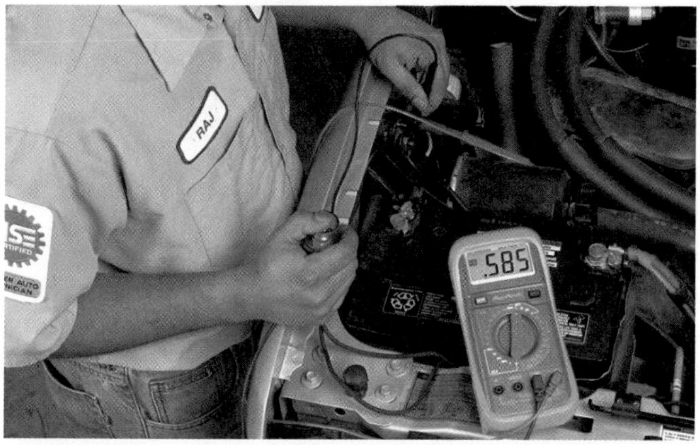

P13-13 *Crank the engine and observe the reading on the meter. This is the voltage drop across the contacts inside the relay.*

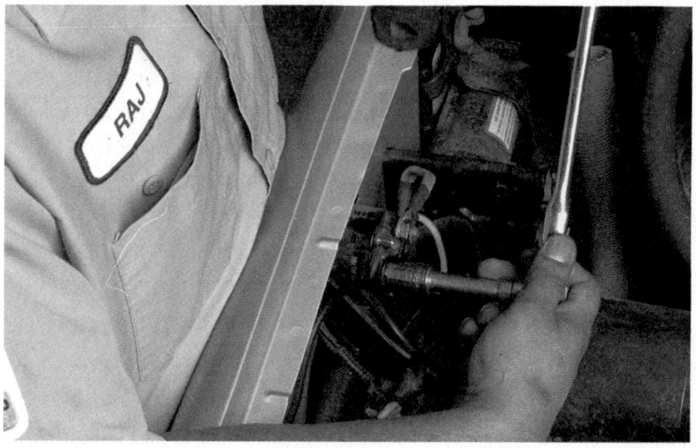

P13-14 *The reading was higher than normal; therefore the starter relay has high resistance and needs to be replaced.*

A reading of more than 0.1 volt across any one wire or switch is usually an indication of trouble. If a high reading is obtained across the safety switch used on an automatic transmission, check the adjustment of the switch according to the manufacturer's service manual. Clutch-operated safety switches cannot be adjusted. They must be replaced.

Test Starter Drive Components

This test detects a slipping starter drive without removing the starter from the vehicle. First, disable the ignition system or bypass the ignition switch with a remote starter switch. Turn the ignition switch to start and hold it in this position for several seconds. Repeat the procedure at least three times to detect an intermittent condition.

Test Conclusions If the starter cranks the engine smoothly, that is an indication that the starter drive is functioning properly. If the engine stops cranking and the starter spins noisily at high speed, the drive is slipping and should be replaced.

If the drive is not slipping, but the engine is not being cranked, inspect the flywheel for missing or damaged teeth. Remove the starter from the vehicle and check its drive components. Inspect the pinion gear teeth for wear and damage. Test the overrunning clutch mechanism. If good, the overrunning clutch should turn freely in one direction, but not in the other. A bad clutch will turn freely in the overrun direction or not at all. If a drive locks up, it can destroy the starter by allowing the starter to spin at more than 15 times engine speed.

The weak point in the movable pole starter is the pole shoe that pulls in toward the armature to engage the starter. This starter requires a minimum of 10.5 volts and high amperes to operate. Otherwise, it simply clicks and does not engage.

As a movable pole starter wears, the pivot bushing sometimes hangs up and prevents the movable pole shoe from being pulled down. When this happens, the starter motor will not spin and the drive will not engage with the flywheel.

A similar problem can occur on solenoid-actuated starters. If the solenoid is too weak to overcome the force of the return springs, the starter does not operate.

Removing the Starter Motor

If your testing indicates that the starter must be removed, the first step is to disconnect the negative cable at the battery and wrap the clamp with electrical tape. It may be necessary to place the vehicle on a lift to gain access to the starter. Before lifting the vehicle, disconnect all wires, fasteners, and so on that can be reached from under the hood.

Disconnect the wires leading to the solenoid terminals. To avoid confusion when reinstalling the starter, it is wise to mark the wires so they can be reinstalled on their correct terminals.

On some vehicles you may need to disconnect the exhaust system to be able to remove the starter. Loosen the starter mounting bolts and remove all but one. Support the starter while removing the remaining bolt. Then pull the starter out and away from the flywheel. Once the starter is free, remove the last bolt and the starter.

Once the starter is out, inspect the starter drive pinion gear and the flywheel ring gear **(Figure 18–25)**. When the teeth of the starter drive are abnormally worn, make sure you inspect the entire circumference of the flywheel. If the starter drive or the flywheel ring gear show signs of wear or damage, they must be replaced.

Reverse the procedure to install the starter. Make sure all electrical connections are tight. If you are installing a new or remanufactured starter, sand away the paint at the mounting point before installing it. Also, make sure you have a good hold on the starter while installing it.

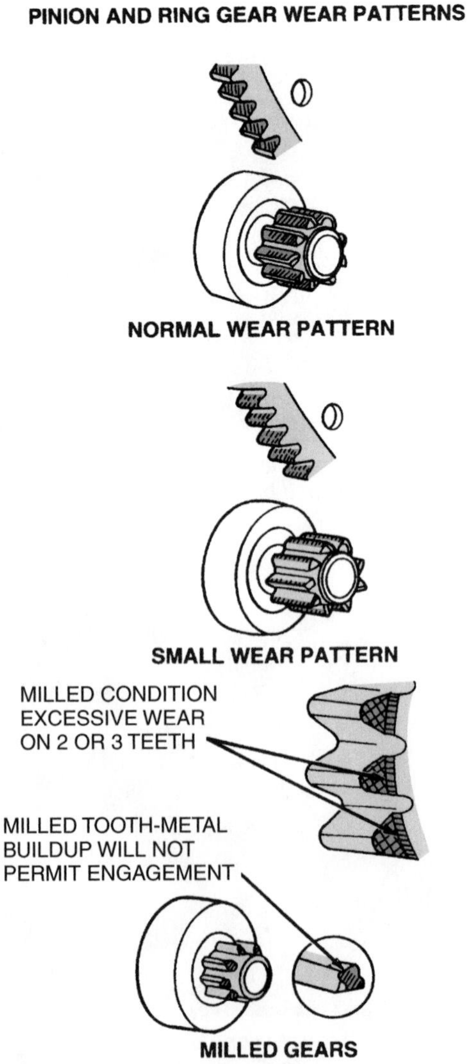

Figure 18–25 Starter drive and flywheel ring gear wear patterns. *Courtesy of Ford Motor Company*

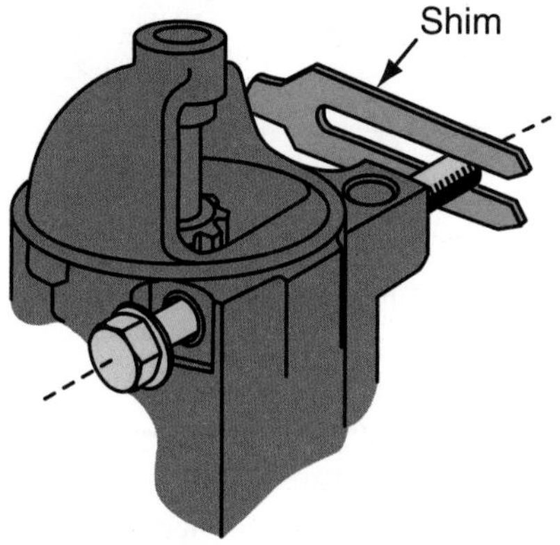

Figure 18-26 Shimming the starter to obtain proper pinion-to-ring gear clearance.

Many General Motors starters use shims between the starter and the mounting pad **(Figure 18–26)**. To check this clearance, install the starter and insert a flat blade screwdriver into the access slot on the side of the drive housing. Pry the drive pinion gear into the engaged position. Use a wire feeler gauge or a piece of 0.020-inch (.5080-mm) diameter wire to check the clearance between the gears **(Figure 18–27)**.

If the clearance between the two gears is incorrect, shims will need to be added or subtracted to bring the clearance within specs. If the clearance is excessive, the starter will produce a high-pitched whine while it is cranking the engine. If the clearance is too small, the starter will make a high-pitched whine after the engine starts and the ignition switch is returned to the RUN position.

Free Speed (No-Load) Test

Every starter should be bench tested after it is removed and before it is installed. To conduct a free speed or no-load test on a starter, follow these steps:

PROCEDURE

Free Speed or No-Load Test

STEP 1 *Clamp the starter firmly in a bench vise.*
STEP 2 *Connect an ammeter to the battery cable and the starter to a battery. This should cause the motor to run.*
STEP 3 *Check current draw and motor speed and compare them to specifications. If they meet specs when the battery has at least 11.5 volts, the starter is working properly.*

If the current draw was excessive or the motor speed too low, there is excessive physical resistance, which can be caused by worn bushings or bearings, a shorted armature, shorted field windings, or a bent armature.

If there was no current draw and the starter did not rotate, the problem could be caused by open field windings, open armature coils, broken brushes, or broken brush springs.

Low armature speed with low current draw indicates excessive resistance. There may be a poor connection

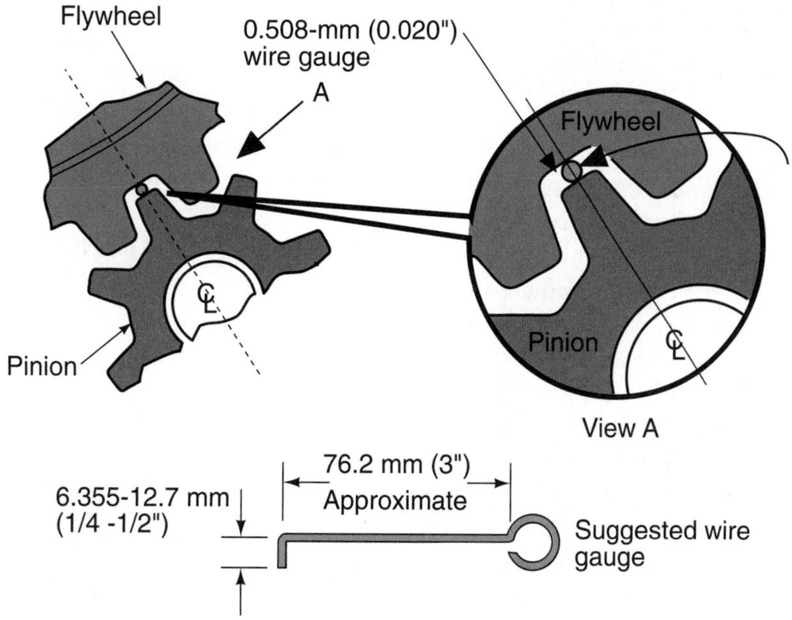

Figure 18-27 Checking the clearance between the pinion gear and the ring gear.

between the commutator and the brushes, or the connections to the starter are bad. If the speed and current draw are both high, check for a shorted field winding.

STARTER MOTOR SERVICE

If the starter is bad, it should be replaced or rebuilt. Often technicians opt for replacing it rather than spending the time to repair or rebuild it. This decision, however, depends on a number of things, including the customer's desire, shop's policies, availability of repair parts, cost, and time.

Photo Sequence 14 covers the typical procedure for disassembling a Delco-Remy starter. Always refer to the manufacturer's procedures when repairing a starter.

WARNING!

Do not clean the starter motor in solvent. The residue left on the parts can ignite and cause a fire and/or destroy the starter. Use denatured alcohol, compressed air regulated to 25 psi (172.3 Kpa), and/or clean rags to clean the unit and its parts.

The starter should be cleaned and inspected as it is disassembled. Inspect the end frame and drive housing for cracks or broken ends. Check the frame assembly for loose pole shoes and broken or frayed wires. Inspect the drive gear for worn teeth and proper overrunning clutch operation. The commutator should be free of flat spots and should not be excessively burned. Check the brushes for wear. Replace them if worn past specifications.

Starter Motor Component Tests

With the starter motor disassembled, tests can be conducted to determine the reason for failure. The armature and field coils should be checked for shorts and opens first. Normally, if the armature or coils are bad, the entire starter is replaced.

Field Coil Tests The field coil and frame assembly can be wired in a number of different ways. Accurate testing of the coils can only be done if you follow the specific guidelines of the manufacturer or if you know how the coils are wired. To do this, look at the wiring diagram and figure out where the coils get their power and where they ground. When you have this information, you will know if the coils are wired in series or parallel.

The usual way to check the field coils for opens is to connect an ohmmeter between the coils' power feed wire and the field coil brush lead **(Figure 18–28)**. If there is no continuity, the field is open. To check the field coil for a short to ground, connect the ohmmeter from the field coil brush lead and the starter (field frame) housing. If there is continuity, the field coil is shorted to the housing.

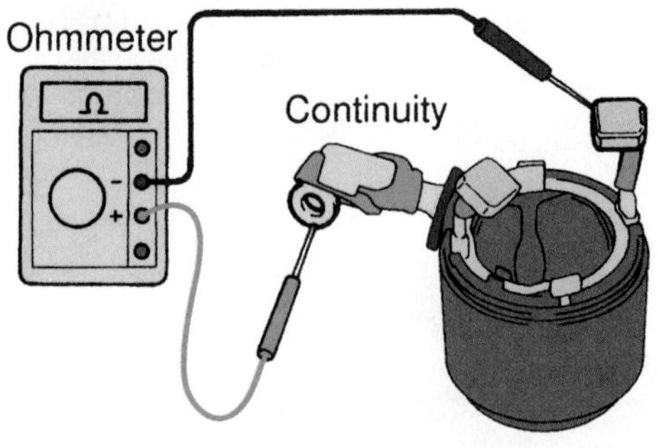

Figure 18-28 Checking a field coil for an open. *Reprinted with permission*

Armature Tests The armature should be inspected for wear or damage caused by contact with the permanent magnets or field windings. If there is wear or damage, check the pole shoes for looseness and repair as necessary. A damaged armature must be replaced.

Next, check the commutator of the armature. If the surface is dirty or burnt, clean it with emery cloth or cut it down with an armature lathe. Measure the diameter of the commutator with an outside micrometer or vernier caliper. If the diameter is less than specifications require, replace the armature.

Measure commutator runout by mounting the armature in V-blocks. Position a dial indicator so that it rides on the center of the commutator. If the runout is within specs, check the commutator for carbon dust or brass chips between the segments. If the commutator runout is beyond specs, replace the armature.

Check the depth of the insulating material (mica) between the commutator segments **(Figure 18–29)**. Check each one and compare the depth with specifications. If necessary, undercut the mica with the proper tool or a hacksaw to achieve the proper depth. If the proper depth cannot be achieved, replace the armature.

Check for continuity between the segments of the commutator **(Figure 18–30)**. If an open circuit exists between any segments, replace the armature.

Place the armature in an armature tester, commonly called a **growler**. Hold a hacksaw blade on the armature core **(Figure 18–31)**. If the blade is attracted to the ar-

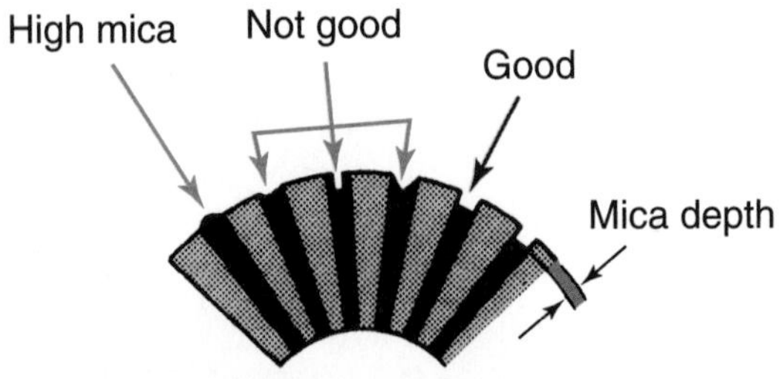

Figure 18-29 Check the depth of the mica between the commutator segments. *Courtesy of American Honda Motor Co., Inc.*

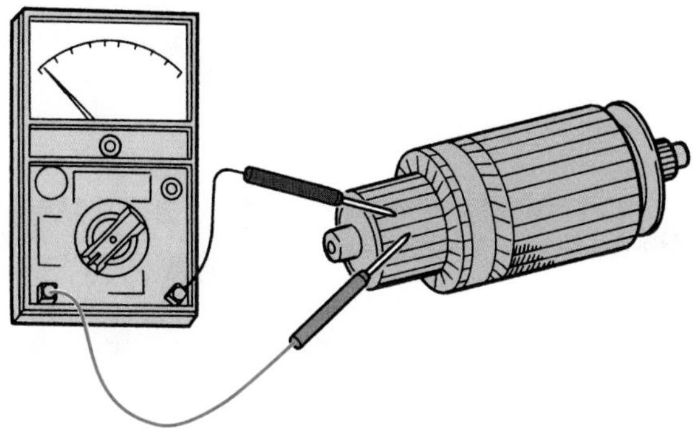

Figure 18-30 Checking the armature for an open.
Courtesy of American Isuzu Motors, Inc.

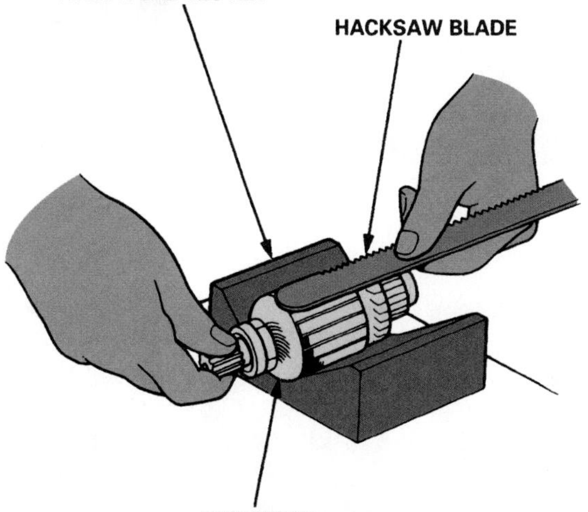

Figure 18-31 Testing an armature on a growler.
Courtesy of American Honda Motor Co., Inc.

mature's core or vibrates while the core is turned, the armature is shorted and must be replaced.

With an ohmmeter, check the armature windings for a short to ground. Hold one meter lead to a commutator segment and the other on the armature core. Also check between the armature shaft and the commutator. If there is continuity at either of these two test points, the armature needs to be replaced.

Brush Inspection Brush inspection begins with an ohmmeter check of the brush holder. Connect one meter lead to a positive brush and the other lead to a negative brush. There should be no continuity between them. If there is, replace the brush holder. Install the brushes into the brush holder and slip the unit over the commutator. Using a spring scale, measure the spring tension of the holders at the moment the spring lifts off the brush. Compare the tension with specs. If the tension is incorrect, replace the spring or the brush holder assembly.

Measure the length of the brushes **(Figure 18–32)**. If the brushes are not within specs, replace the brush or the brush holder assembly. To seat new brushes after in-

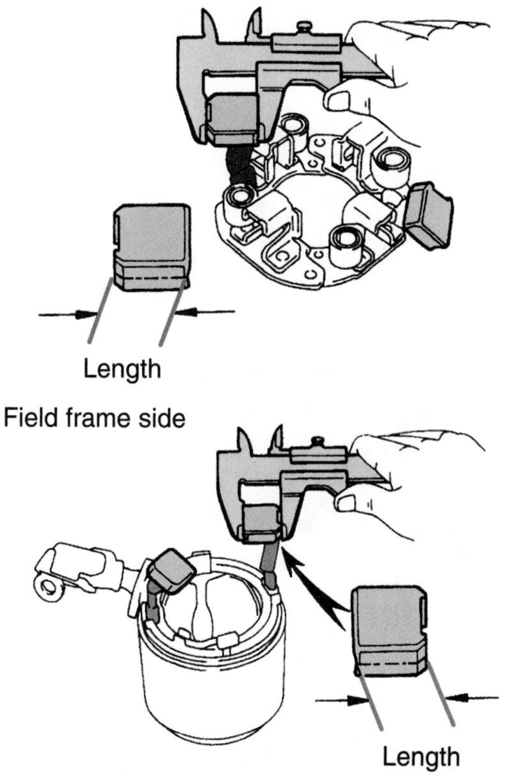

Figure 18-32 Measure the length of the brushes.
Reprinted with permission

stalling them in the brush holder, slip a piece of fine sandpaper between the brush and the commutator. Then rotate the armature. This will put the contour of the commutator on the face of the brushes.

Bearings and Bushings Check each bearing and bushing by placing the armature into the bushing and paying attention to the fit and feel as the armature is rotated in the bushing. If the bushing or bearing feels too loose, tight, or rough, it should be replaced. Bushings can often be visually inspected for uneven and excessive wear. If the bushing is bad, replace it. Many bearings are held in the case by a retainer, while bushings are typically pressed out and into their bore.

Starter Drives and Clutches Carefully inspect the teeth on the starter drive. If the teeth are chipped, excessively worn, or damaged in any way, replace the drive assembly. Also check the teeth on the starter ring gear on the engine's flywheel. Often the same thing that caused damage to the starter drive will damage the teeth on the flywheel. If either or both are damaged, they should be replaced.

To check the operation of the overrunning clutch, slide the drive and clutch assembly onto the armature shaft. Rotate the clutch in both directions. Check its movement.

PHOTO SEQUENCE 14

Typical Procedure for Disassembling a Delco-Remy Starter

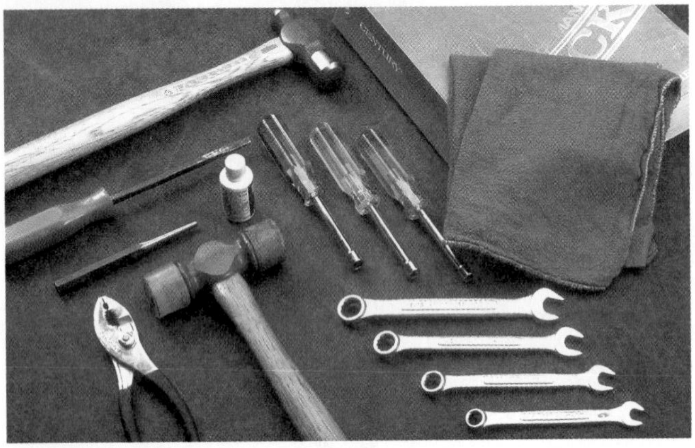

P14-1 *Always have a clean and organized work area. The tools required for disassembling this starter are: rags, assorted wrenches, snap-ring pliers, flat blade screwdriver, ball-peen hammer, plastic head hammer, punch, scribe, safety glasses, and small press.*

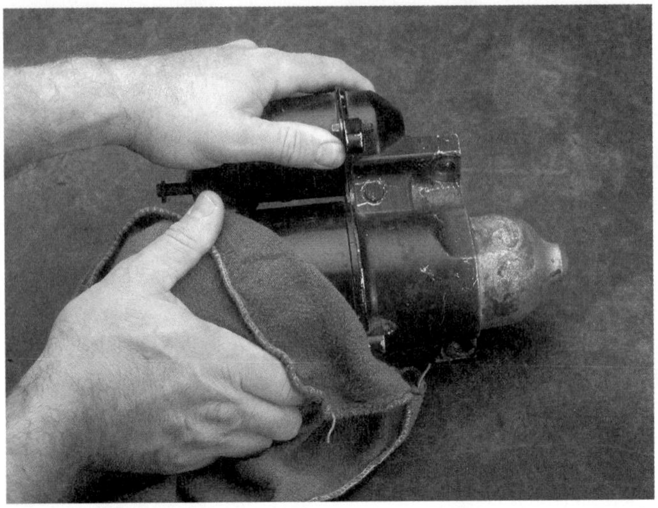

P14-2 *Clean the case.*

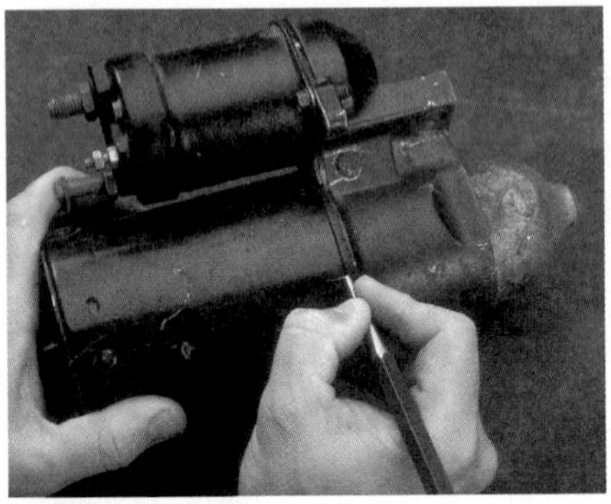

P14-3 *Scribe reference marks at each end of the starter end housings and the frame.*

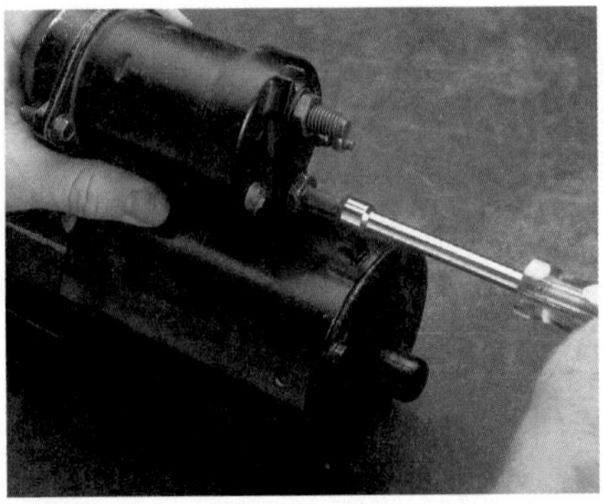

P14-4 *Disconnect the field coil connection at the solenoid's M terminal.*

P14-5 *Remove the two screws that attach the solenoid to the starter drive housing.*

P14-6 *Rotate the solenoid until the locking flange of the solenoid is free. Then remove the solenoid.*

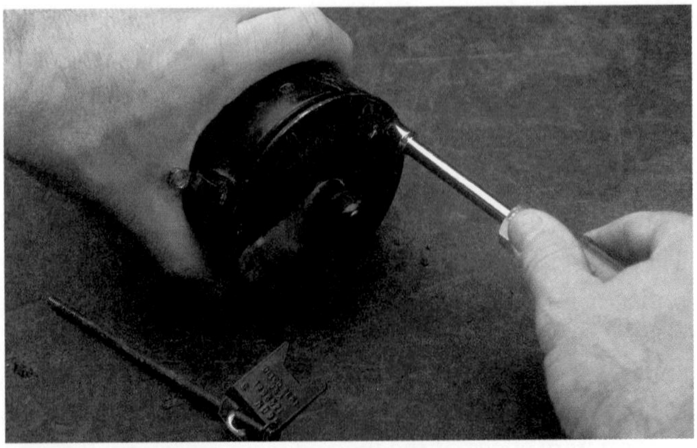

P14-7 *Remove the through-bolts from the end frame.*

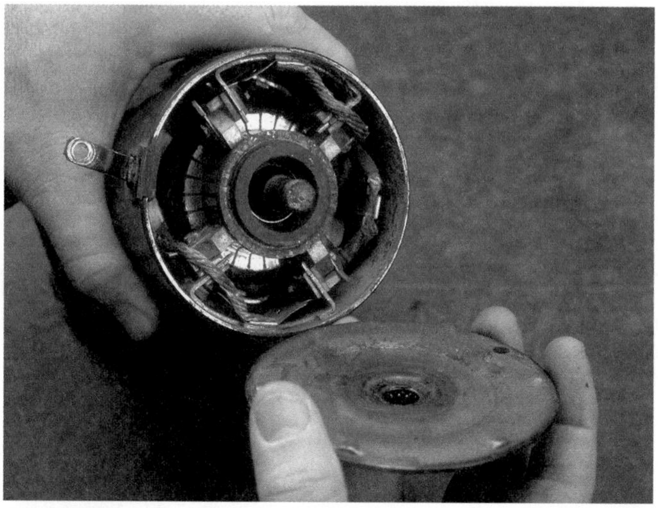

P14-8 *Remove the end frame.*

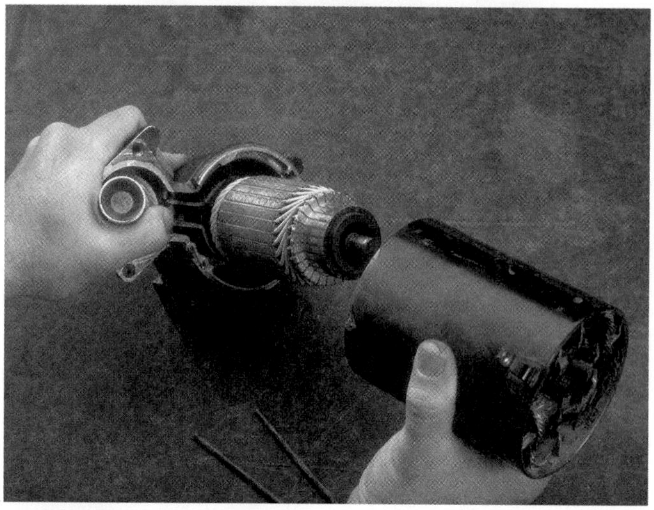

P14-9 *Remove the frame.*

P14-10 *Remove the armature. Note: On some units it may be necessary to remove the shift lever from the drive housing before removing the armature.*

P14-11 *Place a 5/8-inch deep socket over the armature shaft until it contacts the retaining ring of the starter drive.*

P14-12 *Tap the end of the socket with a plastic-faced hammer to drive the retainer toward the armature. Move it only far enough to access the snap ring.*

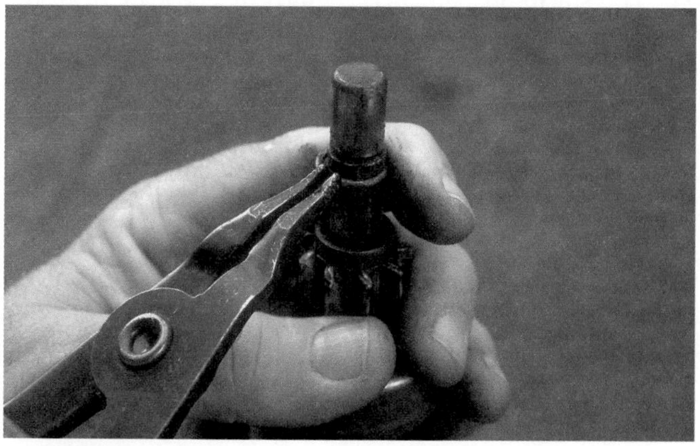

P14-13 *Remove the snap ring.*

P14-14 *Remove the retainer from the shaft and remove the clutch and spring from the shaft. Press out the drive housing bushing and the end-frame bushing.*

Figure 18-33 Check the overrunning clutch by attempting to rotate it in both directions.

It should rotate smoothly in one direction and lock in the other **(Figure 18–33)**. If it does not lock in either direction or if it locks or barely moves in both directions, the assembly must be replaced.

CASE STUDY

A vehicle equipped with a solenoid-actuated direct-drive starting system is towed into the shop. The owner complains that the starter does not crank the engine when the ignition switch is turned to the start position.

The technician performs a battery load test to confirm that the battery is in good working order. It is. The technician then tests for voltage to the M or motor terminal on the solenoid using a voltmeter. Voltage reading at the M terminal is 12.5 volts.

Because there is voltage at the M terminal of the solenoid, inspection of individual connections and components, such as the starter safety switch, are not needed at this time. They are obviously allowing current to pass through the insulated circuit.

The technician then performs a ground circuit check to verify that the ground return path is okay. It is. Since the insulated and ground circuits have checked out okay, the only other source of an open circuit is the starter motor. The technician can now confidently pull the starter motor from the vehicle for rebuilding or replacement.

Starter Motor Reassembly

To reassemble the starter, basically reverse the disassembly procedures. Additional guidelines for reassembly follow:

- Lubricate the splines on the armature shaft that the drive gear rides on with a high-temperature grease.
- Lubricate the bearings and/or bushings with a high-temperature grease.
- Apply sealing compound to the solenoid flange before installing the solenoid to the starter motor.
- Check the pinion depth clearance.
- Perform a no-load test on the starter before installing it.

KEY TERMS

Armature
Brushes
Commutator
Compound motor
Counter EMF (CEMF)
Cranking current test
Cranking voltage test
Field coil
Gear reduction-drive starter
Growler
Hold-in winding
Neutral safety switch
Overrunning clutch
Pinion gear
Pole shoes
Positive engagement starter
Pull-in winding
Series motor
Shunt motor
Solenoid
Starter drive
Starter frame
Starter housing
Starter relay
Starting safety switch

SUMMARY

- The starting system has two distinct electrical circuits: the starter circuit and the control circuit.
- The starter circuit carries high current flow from the battery, through heavy cables, to the starter motor.
- The control circuit uses a small amount of current to operate a magnetic switch that opens and closes the starter circuit.
- The ignition switch is used to control current flow in the control circuit.
- Solenoids and relays are the two types of magnetic switches used in starting systems. Solenoids use electromagnetic force to pull a plunger into a coil to close the contact points. Relays use a hinged armature to open and close the circuit.
- The starter motor is an electric motor capable of producing very high horsepower for very short periods.
- The drive mechanism of the starter motor engages and turns the flywheel to crank the engine for starting.

- An override clutch protects the starter motor from spinning too fast once the vehicle engine starts.
- Starting safety switches prevent the starting system from operating when the transmission is engaged.
- During starter system testing, the ignition system must be bypassed or disabled so the engine can not start.
- Battery load, cranking voltage, cranking current, insulated circuit resistance, starter relay bypass, ground circuit resistance, control circuit, and drive component tests are all used to troubleshoot the starting system.
- With the starter removed, inspect the starter drive pinion gear and the flywheel ring gear.
- If the starter is bad, it should be replaced or rebuilt.
- To check the field coils for opens, connect an ohmmeter between the coils' power feed wire and the field coil brush lead. If there is no continuity, the field is open.
- To check the field coil for a short to ground, connect the ohmmeter from the field coil brush lead and the starter (field frame) housing. If there is continuity, the field coil is shorted to the housing.
- The armature should be inspected for wear or damage caused by contact with the permanent magnets or field windings. A damaged armature must be replaced.
- With an ohmmeter, check across the brushes to see if they are shorted. If there is a short, replace the brush holder.
- Measure the length of the brushes. If the brushes are not within specs, replace the brush or the brush holder assembly.
- Inspect the overrunning clutch by sliding it over the armature shaft and checking its movement.

TECH MANUAL

The following procedures are included in Chapter 18 of the *Tech Manual* that accompanies this book:

1. Removing, inspecting, and replacing a starter motor.
2. Testing the starter and starter circuit.
3. Control circuit test.

REVIEW QUESTIONS

1. Why does proper starter operation depend on good battery cables and connectors?
2. Which of the following is *not* part of the high-current starter circuit?
   a. battery
   b. starting safety switch
   c. starter motor
   d. relay/solenoid
3. Which of the following could result in a hard starting condition?
   a. corroded battery cables
   b. excessive CCA capacity
   c. heavy-gauge battery cables
   d. all of the above
4. Which of the following is *not* a part of the starter control circuit?
   a. ignition switch
   b. starter relay
   c. ballast resistor
   d. starting safety switch
5. Which of the following tests would be performed to check for high resistance in the battery cables?
   a. cranking voltage test
   b. insulated circuit resistance test
   c. starter relay bypass test
   d. ground circuit resistance test
6. If the solenoid clicks while trying to crank the engine with the starter, which of the following is *not* a probable cause?
   a. a faulty neutral safety switch
   b. low battery voltage
   c. low voltage available to the solenoid
   d. an open in the hold-in winding
7. The usual minimum cranking voltage specification is approximately ______ volts.
   a. 9.6
   b. 10.5
   c. 11.0
   d. 12.65

8. A cranking current test is performed, and the amperage is found to be less than specification. Technician A says the starter is bad and it should be replaced. Technician B insists on testing the resistance of the cables, grounds, and connections. Who is correct?

   a. Technician A **c.** Both A and B
   b. Technician B **d.** Neither A nor B

9. A control circuit test may uncover which of the following conditions?

   a. high resistance in the solenoid switch circuit
   b. worn starter motor brushes
   c. loose battery cable connection
   d. short in the starter armature windings

10. An engine cranks slowly. Technician A says a possible cause of the problem is poor starter circuit connections. Technician B says a possible cause is incorrectly set ignition timing. Who is correct?

    a. Technician A **c.** Both A and B
    b. Technician B **d.** Neither A nor B

11. If a ground circuit test reveals a voltage drop of more than 0.2 volt, the problem may be ______.

    a. a loose starter motor mounting bolt
    b. a poor battery ground terminal post connector
    c. a damaged battery ground cable
    d. all of the above

12. While discussing armature testing, Technician A says that to test for shorts, place the armature in a growler and hold a thin metal blade parallel to the armature. If the blade vibrates while the armature is turned, there is a short. Technician B says an ohmmeter can be used to test for shorts. Who is correct?

    a. Technician A **c.** Both A and B
    b. Technician B **d.** Neither A nor B

13. Pinion gear to flywheel ring gear clearance is being discussed. Technician A says if there is too much clearance there will be a high-pitched whine after the engine starts. Technician B says if there is too little clearance, there will be a high-pitched whine while the starter is cranking the engine. Who is correct?

    a. Technician A **c.** Both A and B
    b. Technician B **d.** Neither A nor B

14. *True or False?* Many armature bearings are held in their cases by retainers, while bushings are typically pressed out and into their bores.

15. Which of the following would *not* cause excessive current draw by the starter motor?

    a. a short in the motor
    b. high resistance in the circuit
    c. using oil that is too heavy in the engine
    d. very cold weather

16. Technician A checks a starter's field coils for opens by connecting an ohmmeter between the coils' power feed wire and the field coil brush lead. Technician B checks a starter's field coil for a short to ground by connecting an ohmmeter from the field coil brush lead and the starter housing. Who is correct?

    a. Technician A **c.** Both A and B
    b. Technician B **d.** Neither A nor B

17. The device that prevents the engine from turning the armature of the starter motor is the ______.

    a. overrunning clutch **c.** flywheel
    b. pinion gear **d.** pole shoe

18. The part of the armature that the brushes ride on is called the ______.

19. Technician A says the purpose of the starter relay is to complete the circuit from the battery to the starter motor. Technician B says the purpose of the solenoid is to complete the circuit from the battery to the starter motor. Who is correct?

    a. Technician A **c.** Both A and B
    b. Technician B **d.** Neither A nor B

20. When checking for excessive resistance in the starting circuit, Technician A connects a voltmeter across the ground cable from the battery to the ground connection of the cable. Technician B connects an ohmmeter across the positive battery cable after it has been removed from the vehicle. Who is correct?

    a. Technician A **c.** Both A and B
    b. Technician B **d.** Neither A nor B

19

CHARGING SYSTEMS

OBJECTIVES

■ Explain the purpose of the charging system. ■ Identify the major components of the charging system. ■ Explain the purposes of the major parts of an AC generator. ■ Explain half- and full-wave rectification and how they relate to AC generator operation. ■ Identify the different types of AC voltage regulators. ■ Describe the two types of stator windings. ■ Explain the features enabled by the use of a starter/generator unit. ■ Perform charging system inspection and testing procedures using electrical test equipment.

The primary purpose of a charging system is to recharge the battery. After the battery has supplied the high current needed to start the engine, the battery, even a good battery, has a low charge. The charging system recharges the battery by supplying a constant and relatively low charge to the battery. Charging systems work on the principles of magnetism to change mechanical energy into electrical energy. This is done by inducing voltage.

Voltage is induced in a wire when it moves through a magnetic field. The wire or conductor becomes a source of electricity and has a polarity or distinct positive and negative ends. However, this polarity can be switched depending on the relative direction of movement between the wire and magnetic field. This is why an AC generator produces alternating current **(Figure 19–1)**.

Figure 19–1 An AC generator. *Courtesy of Robert Bosch Corp.*

ALTERNATING CURRENT CHARGING SYSTEMS

During cranking, the battery supplies all of the vehicle's electrical power. However, once the engine is running, the charging system is responsible for producing enough energy to meet the demands of all of the loads in the electrical system, while also recharging the battery. With all of the electrical and electronic devices found on today's vehicles, the charging system has a difficult job.

Several decades ago the charging system depended on a **DC generator**. The DC generator provided direct current (DC) and was similar to an electric motor in construction. The biggest difference between a generator and a motor is the wiring to the armature. In a motor, the armature receives current from the battery. This creates the magnetic field that opposes the magnetic fields in the motor's coils, which causes the armature to rotate. The armature in a DC generator is driven by the engine. It is not magnetized and the windings simply rotate through the stationary magnetic field of the field windings, inducing a voltage in the conductors inside the armature. A motor can become a generator by allowing current to flow from the armature instead of to it. In a DC generator, the placement of the brushes on the commutator changes the induced AC voltage to a DC voltage output.

DC generators had a very limited current output, especially at low speeds. They could not keep up with demands of the modern automobile and were replaced by AC generators. AC generators are capable of providing high current output even at low engine speeds.

SHOP TALK

With the implementation of OBD II, new terminology was given to many parts of an automobile. Prior to that law, an **AC generator** was called an **alternator**. In fact, in many cases and by many manufacturers an AC generator is still referred to as an alternator. To avoid confusion, just remember that an alternator is an AC generator and vice versa. ■

AC generators **(Figure 19–2)** use a design that is basically the reverse of a DC generator. In an AC generator **(Figure 19–3)**, a spinning magnetic field (called the rotor) rotates inside an assembly of stationary conductors (called the stator). As the spinning north and south poles of the magnetic field pass the conductors, they induce a voltage that first flows in one direction and then in the opposite direction (AC voltage). Because automobiles use DC voltage, the AC must be changed or rectified into DC. This is done through an arrangement of diodes that are placed between the output of the windings and the output of the AC generator.

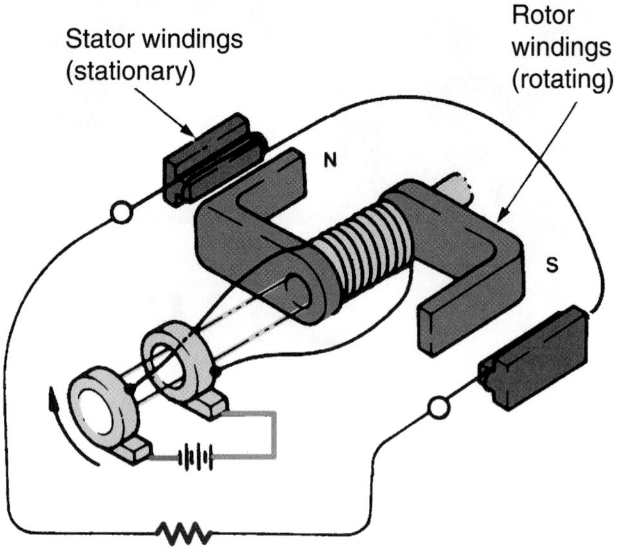

Figure 19-3 A simplified AC generator.

AC Generator Construction

Rotor The rotor assembly consists of a drive shaft, coil, and two pole pieces **(Figure 19–4)**. A pulley mounted on one end of the shaft allows the rotor to be spun by a belt driven by the crankshaft pulley.

The **rotor** is a rotating magnetic field inside the alternator. The coil is simply a long length of insulated wire wrapped around an iron core. The core is located be-

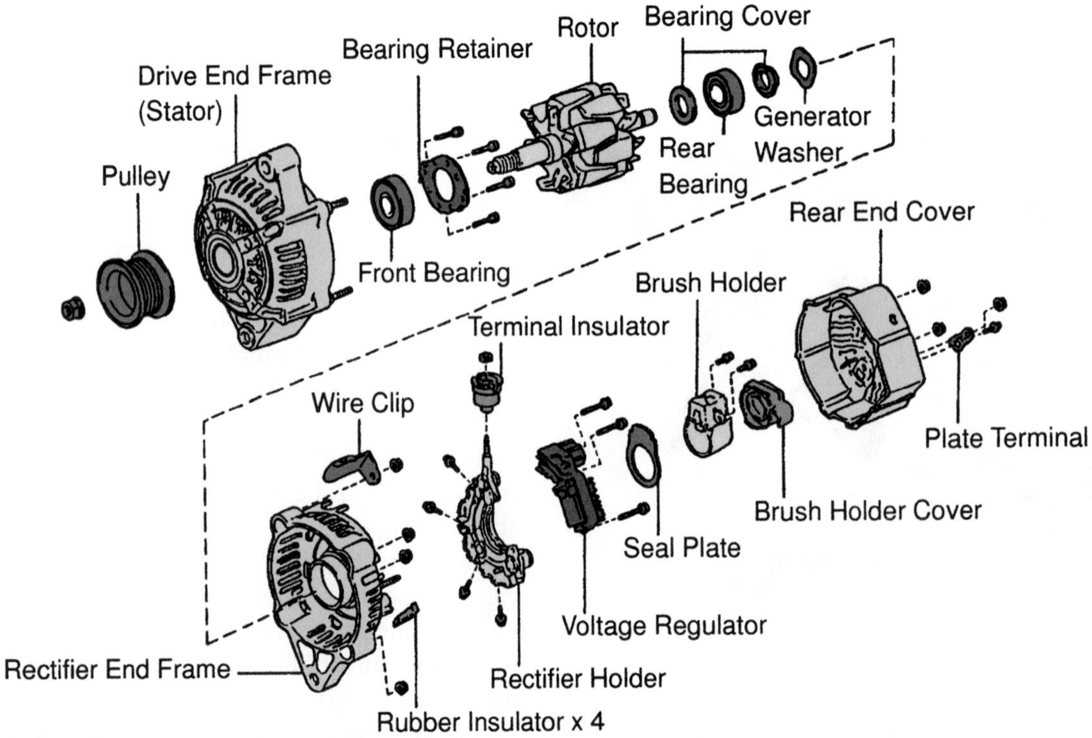

Figure 19-2 An exploded view of an AC generator. *Reprinted with permission*

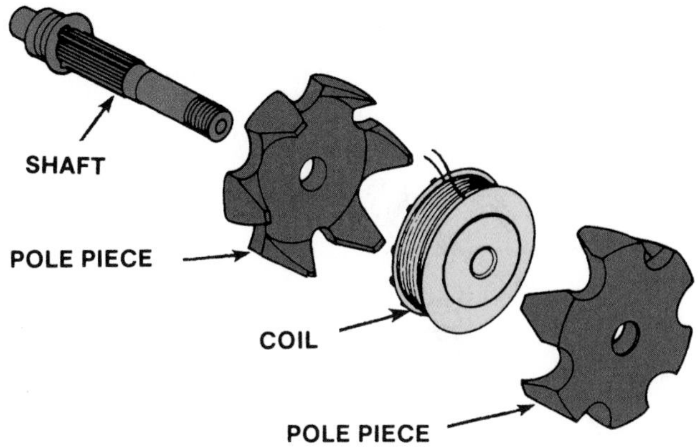

Figure 19-4 The rotor is made up of a coil, pole pieces, and a shaft.

tween the two sets of **pole pieces**. A magnetic field is formed by a small amount (4.0 to 6.5 amperes) of current passing through the coil winding. As current flows through the coil, the core is magnetized and the pole pieces assume the magnetic polarity of the end of the core that they touch. Thus, one pole piece has a north polarity and the other has a south polarity. The extensions of the pole pieces, known as **fingers**, form the actual magnetic poles. A typical rotor has fourteen poles, seven north and seven south, with the magnetic field between the pole pieces moving from the N poles to the adjacent S poles **(Figure 19–5)**.

Slip Rings and Brushes Current to create the magnetic field is supplied to the coil from one of two sources, the battery or the AC generator itself. In either case, the current is passed through the AC generator's voltage regulator before it is applied to the coil. The voltage regulator varies the amount of current supplied. Increasing field current through the coil increases the strength of the magnetic field. This, in turn, increases AC generator voltage output. Decreasing the field voltage to the coil has the opposite effect. Output voltage decreases.

Slip rings and brushes conduct current to the spinning rotor. Most AC generators have two slip rings mounted directly on the rotor shaft. They are insulated from the shaft and each other. Each end of the **field coil** connects to one of the slip rings. A carbon brush located on each slip ring carries the current to and from the field coil. Current is transmitted from the field terminal of the voltage regulator through the first brush and slip ring to the field coil. Current passes through the field coil and the second slip ring and brush before returning to ground **(Figure 19–6)**.

Stator The **stator** is the stationary member of the alternator. It is made up of a number of conductors, or wires, into which the voltage is induced by the rotating magnetic field. Most AC generators use three windings to generate the required amperage output. They can be arranged in either a **delta** configuration or a **wye** configuration **(Figure 19–7)**. The delta winding **(Figure 19–8)** received its name because its shape resembles the Greek letter delta, Δ. The wye winding resembles the letter *Y*. Alternators use one or the other. Usually, a wye winding is used in applications in which high charging voltage at low engine speeds is required. AC generators with delta windings are capable of putting out higher amperages at high speeds but low engine speed output is poor.

The rotor rotates inside the stator. A small air gap between the two allows the rotor to turn without making contact with the stator. The magnetic field of the rotor is able to energize all three of the stator windings at the same time. Therefore, the generation of AC can be quite high if needed.

Alternating current produces a positive pulse and then a negative pulse. The resultant waveform is known as a sine wave. This **sine wave** can be seen on an oscilloscope. The complete waveform starts at zero, goes positive, then drops back to zero before turning negative. The angle and polarity of the field coil fingers are what cause this sine

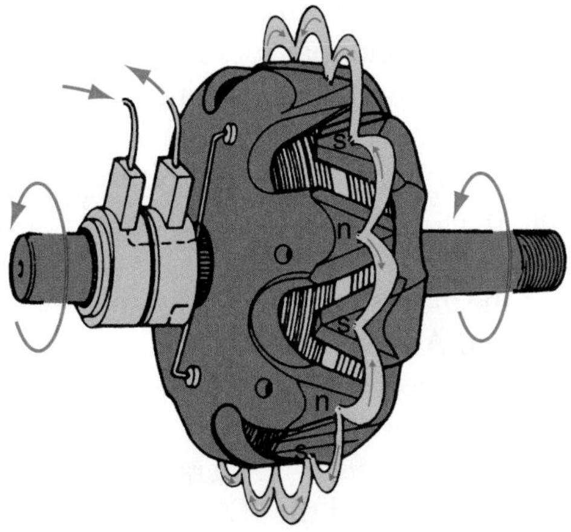

Figure 19-5 The magnetic field moves from the N poles, or fingers, to the S poles.

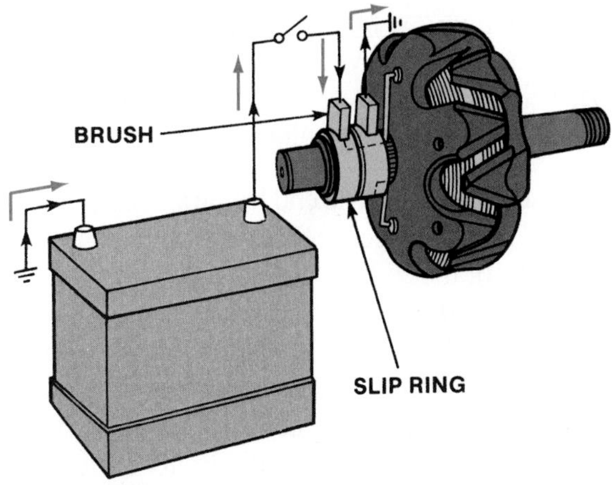

Figure 19-6 Current is carried by the brushes to the rotor windings via the slip rings.

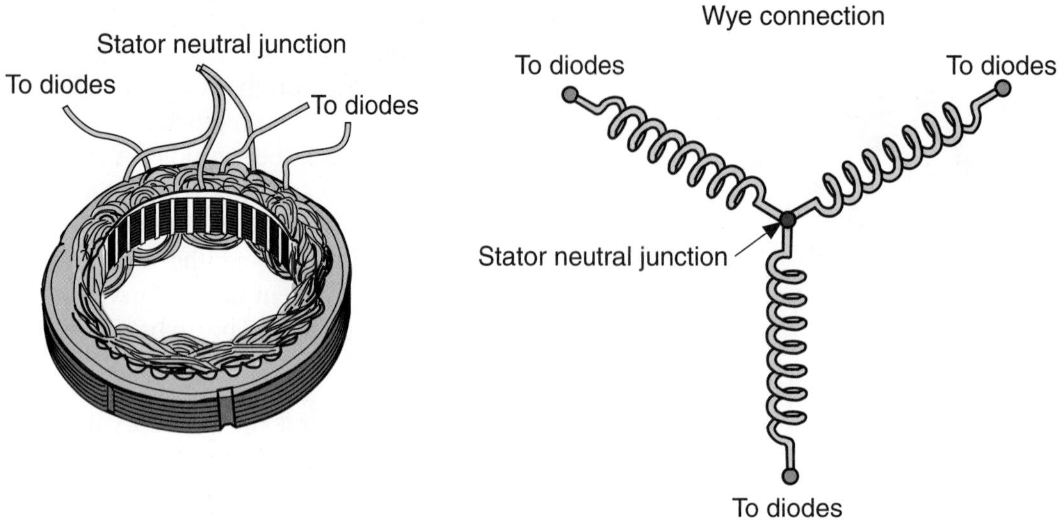

Figure 19-7 A wye-connected stator winding.

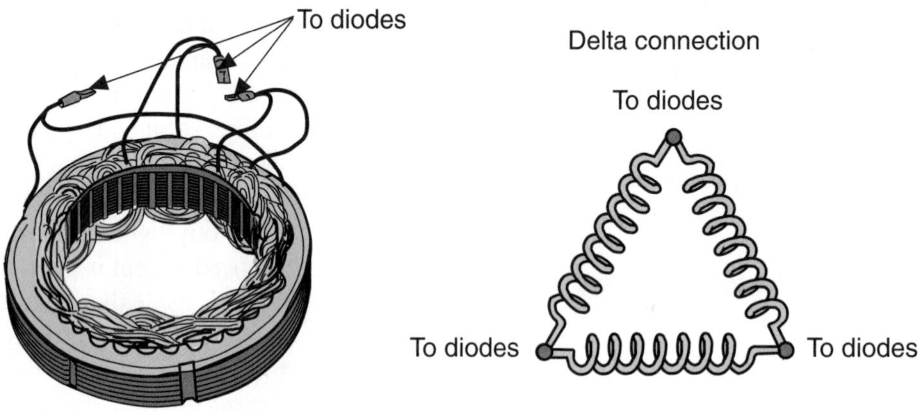

Figure 19-8 A delta-connected stator winding.

wave in the stator. When the north pole magnetic field cuts across the stator wire, it generates a positive voltage within the wire. When the south polarity magnetic field cuts across the stator wire, a negative voltage is induced in the wire. A single loop of wire energized by a single north then a south results in a single-phase voltage. Remember that there are three overlapping stator windings. This produces three overlapping sine waves **(Figure 19–9)**. This voltage, since it was produced by three windings, is called **three-phase voltage**.

End Frame Assembly The end frame assembly, or housing, is made of two pieces of cast aluminum. It contains the bearings for the end of the rotor shaft where the drive pulley is mounted. Each end frame also has built-in ducts so the air from the rotor shaft fan can pass through the AC generator. Normally, a heat sink containing three positive rectifier diodes is attached to the rear end frame. Heat can pass easily from these diodes to the moving air **(Figure 19–10)**. Three negative rectifier diodes are contained in the end frame itself. Because the end frames are bolted together and then bolted directly to the engine, the end frame assembly is part of the electrical ground path. This means that anything connected to the housing that is not insulated from the housing is grounded.

Cooling Fans Behind the drive pulley on most AC generators is a cooling fan that rotates with the rotor. This cooling fan draws air into the housing through the openings at the rear of the housing. The air leaves through openings behind the cooling fan **(Figure 19–11)**. The moving air pulls heat from the diodes and their heat decreases.

Cooling the diodes is important for the efficiency and durability of an AC generator. Several different generator designs have been introduced recently that increase the cooling efficiency of a generator. One of these is the AD-series generator from General Motors. The "A" stands for air-cooled and the "D" means dual fans. This series is lighter than most other generators but capable of

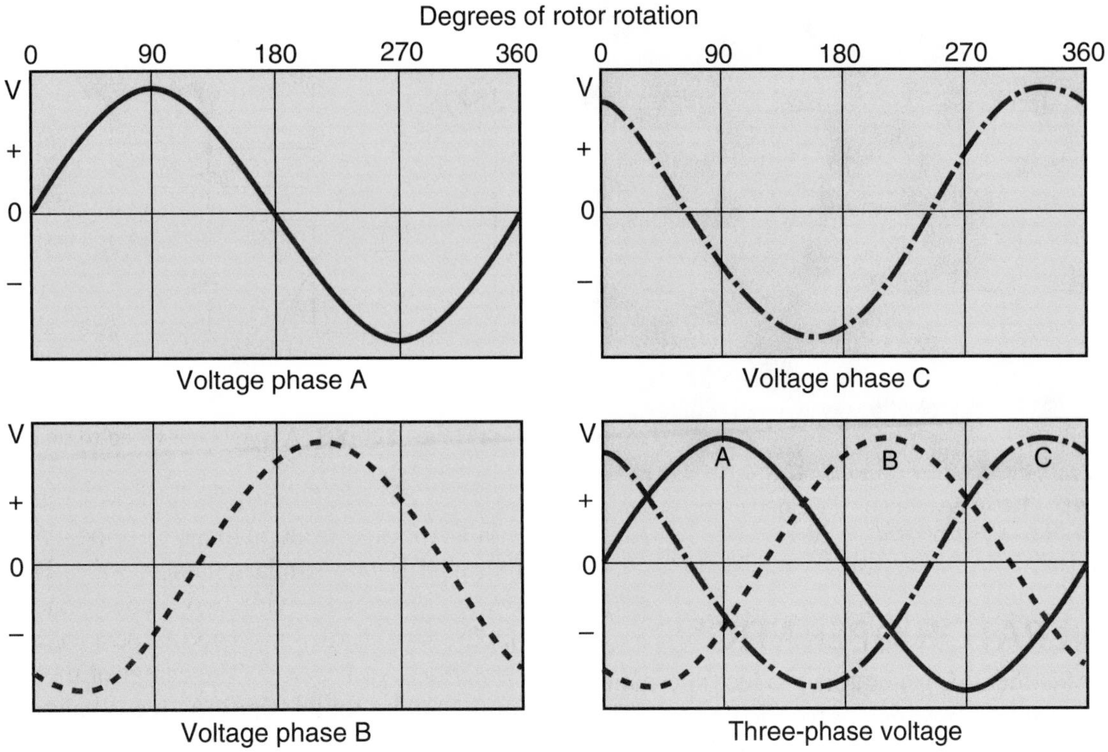

Figure 19-9 The voltage produced in each stator winding is added together to create a three-phase voltage.

Figure 19-10 A bridge rectifier.

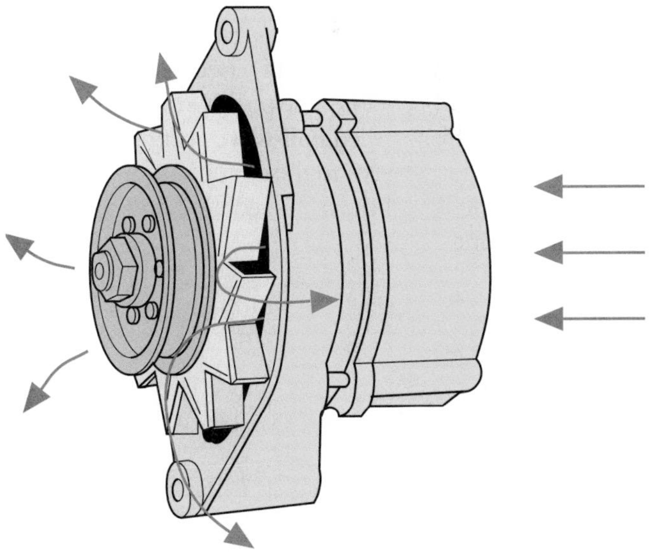

Figure 19-11 The cooling fan draws air in from the rear of the generator to keep the diodes cool. *Courtesy of Robert Bosch Corp.*

very high outputs. This type generator does not have an external fan; instead, it has two internal fans.

Liquid Cooled Generators Another recent design uses liquid cooling **(Figure 19–12)**. Using water or coolant to cool a generator is a very efficient way to keep diode temperatures down. But the real reason for eliminating the fan and using liquid to cool the generator is to reduce noise. The rotating fan is a source of underhood noise that some automobile manufacturers want to eliminate. These new generators have water jackets cast into the housing. Hoses connect the housing to the engine's cooling system. Not only do these generators make less noise, they have higher power output and should last longer in the high-temperature environment of the engine compartment.

Figure 19–12 A water-cooled AC generator. *Courtesy of BMW of North America, Incorporated*

AC GENERATOR OPERATION

As mentioned earlier, AC generators produce alternating current that must be converted, or rectified, to DC. This is accomplished by passing the AC through diodes.

DC Rectification

Figure 19–13 shows that when AC passes through a diode, the negative pulses are blocked off to produce the scope pattern shown. If the diode is reversed, it blocks off current during the positive pulse and allows the negative pulse to flow **(Figure 19–14)**. Because only half of the AC current pulses (either the positive or the negative) is able to pass, this is called **half-wave rectification**.

By adding more diodes to the circuit, more of the AC is rectified. When all of the AC is rectified, **full-wave rectification** occurs.

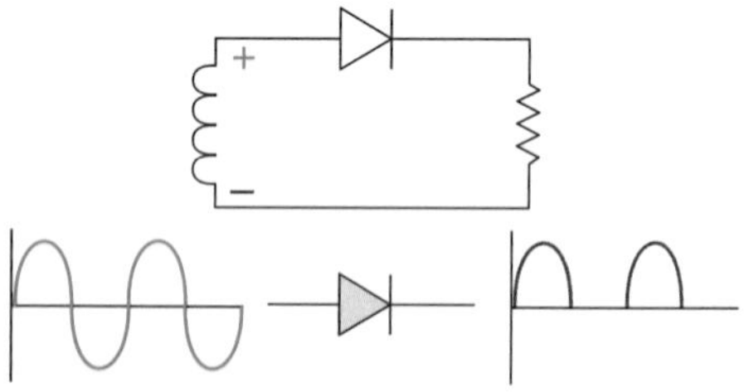

Figure 19–13 Half-wave rectification, diode positively biased.

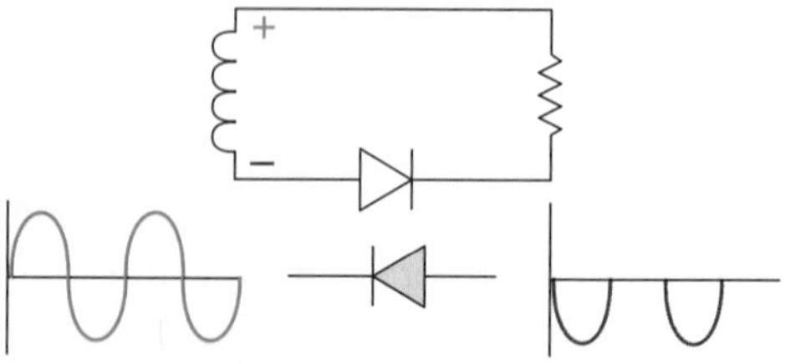

Figure 19–14 Half-wave rectification, diode negatively biased.

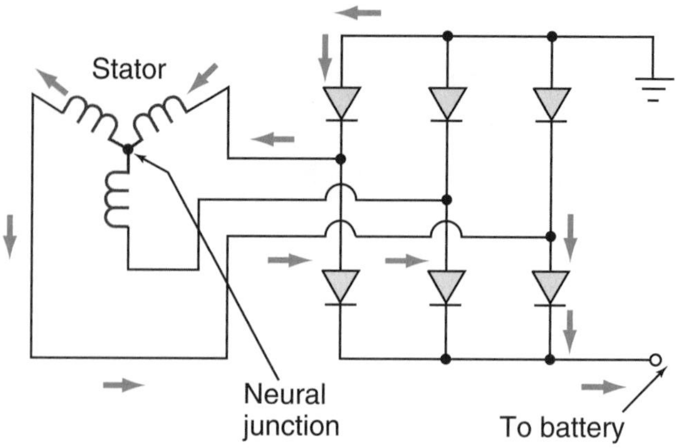

Figure 19–15 A wye stator wired to six diodes.

Full-wave rectification for stator windings requires another circuit with similar characteristics. **Figure 19–15** shows a wye stator with two diodes attached to each winding. One diode is insulated, or positive, and the other is grounded, or negative. The center of the Y contains a common point for all windings. It can have a connection attached to it. It is called the stator neutral junction. At any time during the rotor movement, two windings are in series and the third coil is neutral and inactive. As the rotor revolves, it energizes the different sets of windings in different directions. However, the uniform result is that current in any direction through two windings in series produces the required DC for the battery.

The diode action does not change when the stator and diodes are wired into a delta pattern. **Figure 19–16** shows the major difference. Instead of having two windings in series, the windings are in parallel. Thus, more current is available from a delta-wound AC generator because the parallel paths allow more current to flow through the diodes. Nevertheless, the action of the diodes remains the same.

Many AC generators have an additional set of three diodes called the **diode trio**. The diode trio is used to rectify current from the stator so that it can be used to create the magnetic field in the field coil of the rotor. Using

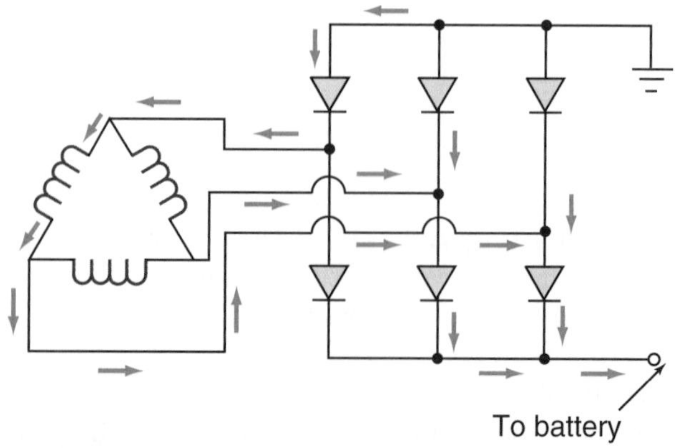

Figure 19–16 A delta stator wired to six diodes.

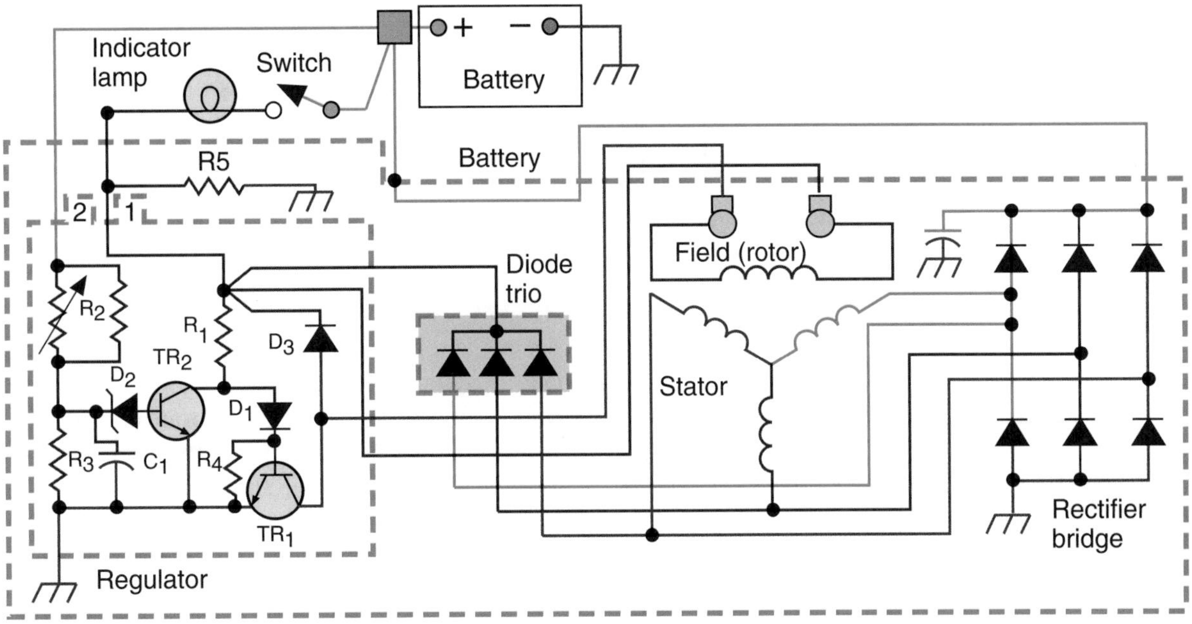

Figure 19–17 A wiring diagram of a charging circuit with a diode trio.

the diode trio eliminates extra wiring. To control generator output, a voltage regulator regulates the current from the diode trio and to the rotor **(Figure 19–17)**.

Voltage Regulation

Voltage output of an AC generator can reach as high as 250 volts if it is not controlled. The battery and the rest of the electrical system must be protected from this excessive voltage. Therefore, the voltage output from a charging system must be controlled. Current output does not need to be controlled because an AC generator naturally limits the current output. The **voltage regulator** controls the voltage output of an AC generator.

Regulation of voltage is accomplished by varying the amount of field current flowing through the rotor. The higher the field current, the higher the voltage output. By controlling the amount of resistance in series with the field coil, control of the field current and voltage output is obtained. To ensure that the battery stays fully charged, most regulators are set for a system voltage between 14.5 and 15.5 volts.

The regulator must receive system voltage as an input in order to regulate the voltage output. This input voltage to an AC generator is called the **sensing voltage**. If the sensing voltage is below the regulator setting, an increase in field current is allowed which causes an increase in charging voltage output. Higher sensing voltage will result in a decrease in field current and voltage output. The regulator will reduce the charging voltage until it is at a level to run the ignition system while putting a low charge (trickle charge) on the battery. If a heavy load is turned on, such as the headlights, the additional draw causes a decrease in battery voltage. The regulator senses the low system voltage and increases current to the rotor. This increases the strength of the magnetic field around the rotor and increases the generator's output voltage. When the load is turned off, the regulator senses the rise in system voltage and reduces the field current.

Another input that affects voltage regulation is temperature. Because ambient temperature influences the rate of charge that a battery can accept, regulators are temperature compensated. Temperature compensation is required because the battery is more reluctant to accept a charge at lower ambient temperatures. The regulator will increase the system voltage until it is at a higher level so the battery will accept it and can become fully charged.

Field Circuits

To properly test and service a charging system, it is important to identify the type of field circuit in that system's generator. There are basically three types of field circuits. The first type is called the A-circuit. It has the regulator on the ground side of the field coil. The battery feed (B+) for the field coil is picked up inside the AC generator **(Figure 19–18)**. By placing the regulator on the ground side

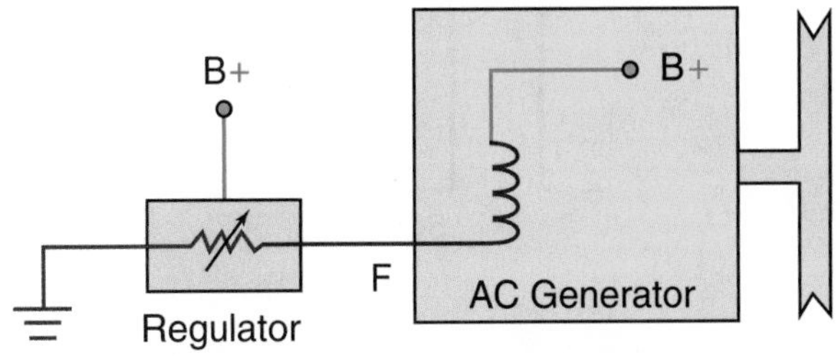

Figure 19–18 An A-circuit.

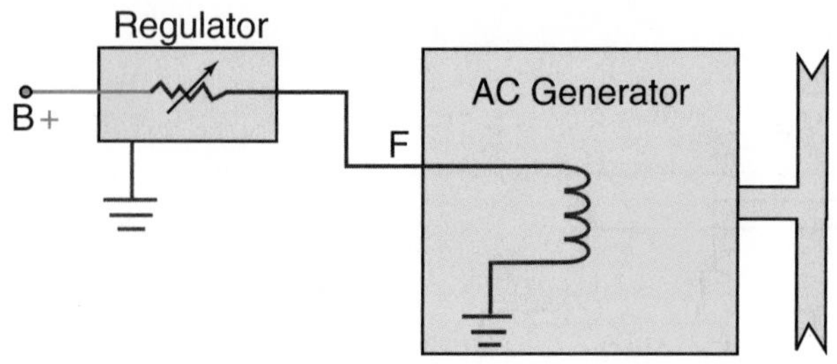

Figure 19-19 A B-circuit.

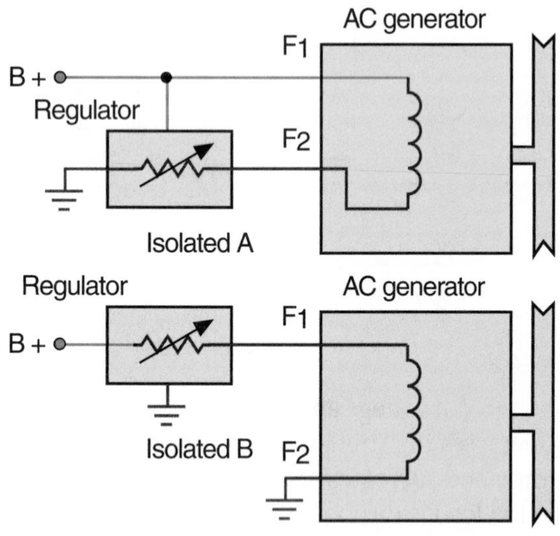

Figure 19-20 In the isolated field circuit AC generator, the regulator can be installed on either side of the field.

of the field coil, the regulator allows the control of field current by varying the current flow to ground.

The second type of field circuit is the B-circuit. In this case the voltage regulator controls the power side of the field circuit. The field coil is grounded inside the AC generator **(Figure 19–19)**. Normally the B-circuit regulator is mounted outside of the generator.

The third type of field circuit is called the isolated field. The AC generator has two field wires attached to the outside of the case. One is the ground, the other is the B+. The voltage regulator can be located on either the ground or the B+ side of the circuit **(Figure 19–20)**.

There are two basic types of regulators: electronic and electromechanical. Older vehicles were equipped with electromechanical regulators, while newer vehicles have electronic regulators. Also, many newer vehicles do not have separate voltage regulators; instead, they control the output of the charging system through the PCM.

Electronic Regulators

Electronic regulators can be mounted externally or internally in relation to the AC generator. The use of electronics allows for quick and accurate control of the field current. Electronic regulation is through the ground side of the field current (A-circuit control).

Pulse width modulation controls the generator's output by varying the amount of time the field coil is energized. For example, assume that a vehicle is equipped with a 100-ampere generator. If the electrical demand placed on the charging system requires 50 amps, the regulator would energize the field coil for 50% of the time **(Figure 19–21)**. If the electrical system's demands were increased to 75 amps, the regulator would energize the field coil 75% of the cycle time.

The electronic regulator uses a zener diode that blocks current flow until a specific voltage is obtained, at which point it allows the current to flow. The schematic for an electronic voltage regulator with a zener diode is shown in **Figure 19–22**.

Integrated circuit voltage regulators are used on most late-model vehicles. This is the most compact regulator design. All of the control circuitry and components are located on a single silicone chip. The chip is sealed in a plastic module and mounted either inside or on the back of the AC generator. Integrated circuit regulators are nonserviceable and must be replaced if defective.

Figure 19–23 illustrates a solid-state integrated regulator. It mounts inside the AC generator slip ring end frame along with the brush holder assembly. All voltage regulator parts are enclosed in a solid mold. The rectifier

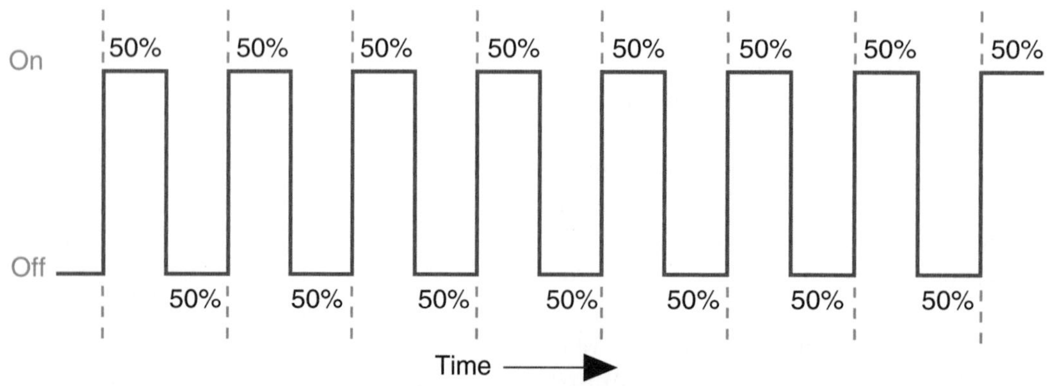

Figure 19-21 Pulse width modulation with 50% on-time.

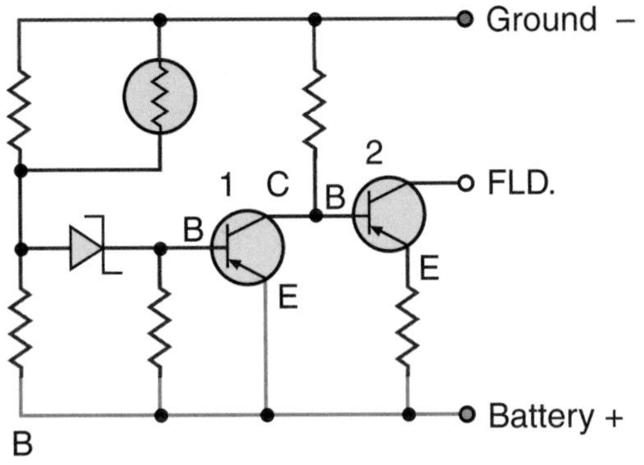

Figure 19-22 A simplified circuit of an electronic regulator with a zener diode.

Figure 19-23 Component locations of an AC generator with an internally mounted voltage regulator. *Courtesy of Robert Bosch Corp.*

bridge contains the six diodes needed to change AC to DC, which is then available at the output battery terminal. Field current is supplied through a diode trio, which is connected to the stator windings.

Fail-Safe Circuits To prevent simple electrical problems from causing high-voltage outputs that can damage delicate electronic components, many voltage regulators contain **fail-safe circuits**.

A detailed explanation of how these circuits operate can be quite confusing. All you need to know is what a fail-safe circuit does, not how it does it. If wire connections to the AC generator become corroded or are accidently disconnected, the regulator's fail-safe circuits may limit voltage output that might otherwise rise to dangerous levels. Under certain conditions, the fail-safe circuits may prevent the AC generator from charging at all. A fusible line in the fail-safe circuitry confines damage to the AC generator. Delicate electronic components in other vehicle systems are not damaged.

Computer Regulation

On a growing number of late-model vehicles, separate voltage regulators are no longer used. Instead, the voltage regulation circuitry is located in the vehicle's powertrain control module or another control module **(Figure 19–24)**. Regardless of where the circuitry is located, it is still used to control current to the field windings in the rotor.

This type of system does not control rotor field current by acting like a variable resistor. Instead, the computer switches or pulses field current on and off at a fixed frequency of about 400 cycles per second. By varying on-off times, a correct average field current is produced to provide correct AC generator output. At high engine speeds with little electrical system load, field circuit on time may be as low as 10%. At low engine speeds with high loads, the computer may energize the field circuit 75% or more of the time to generate the higher average field current needed to meet output demands.

A significant feature of this system is its ability to vary the amount of voltage according to vehicle requirements and ambient temperatures. This precise control allows the use of smaller, lighter, storage batteries. It also reduces the magnetic drag of the AC generator, increasing engine output by several horsepower. Precise manage-

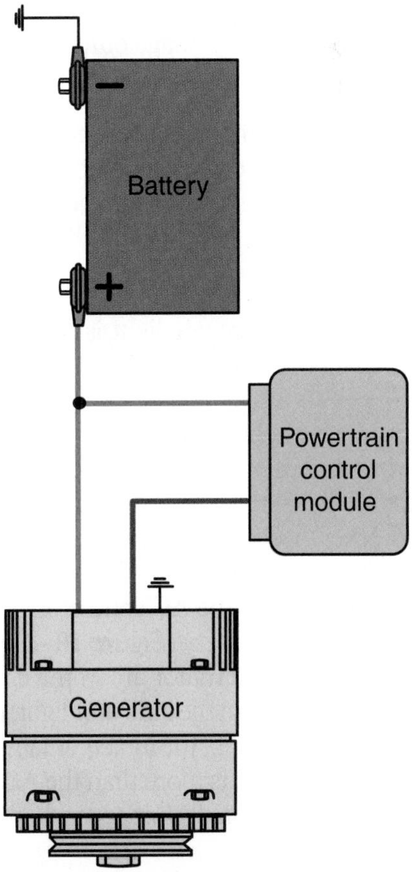

Figure 19-24 The basic circuit for a generator with its regulator as part of the PCM.

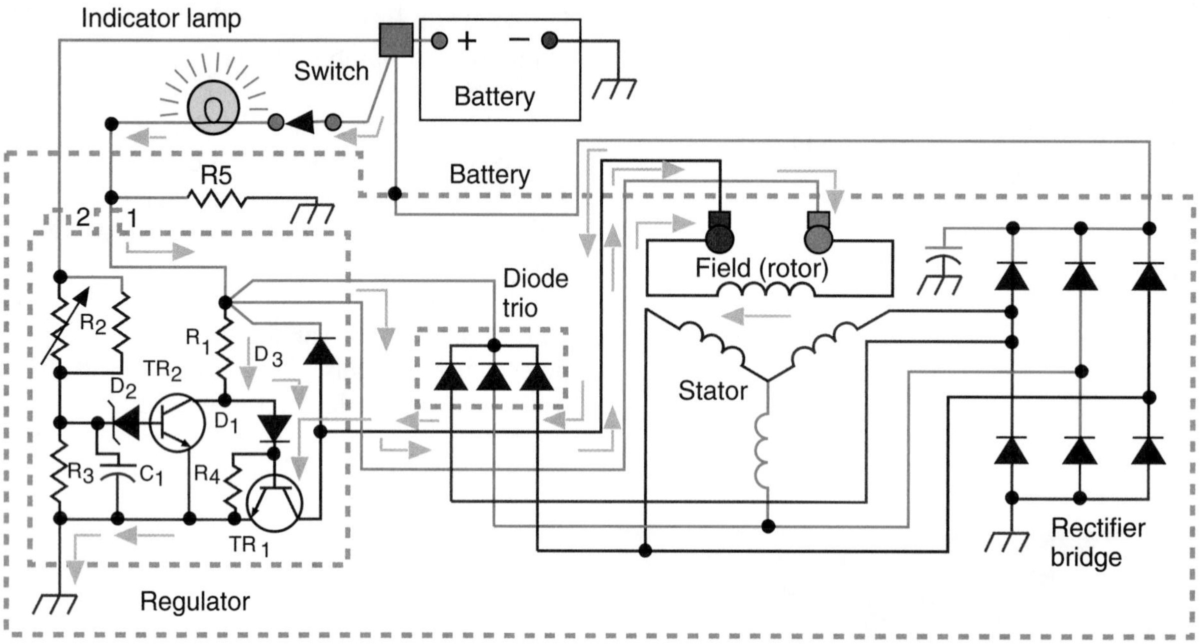

Figure 19–25 An electronic regulator with an indicator light on due to no AC generator output.

ment of the charging rate can result in increased gas mileage and eliminate potential rough idle problems caused by parasitic voltage loss at low idling speeds. Most importantly, it allows the computer's diagnostic capabilities to be used in troubleshooting charging system problems, such as low- or high-voltage outputs.

Indicators

It is very important to monitor charging system performance during the course of vehicle operation. Vehicles are equipped with an ammeter, voltmeter, or indicator light. These allow the driver to monitor the charging system.

Indicator Light The indicator light is the simplest and most common method of monitoring AC generator performance. When the charging system fails to supply sufficient current, the light turns on. However, when the ignition switch is first activated, the light also comes on because the AC generator is not providing power to the battery and other electrical circuits. Thus, the only current path is through the ignition switch, indicator light, voltage regulator, part of the AC generator, and ground, then back through the battery **(Figure 19–25)**. Only the battery, regulator, and alternator are in the circuit. With no current flowing through the indicator light, it goes out.

With the engine running, the indicator light comes on again if the electrical load is more than the AC generator can supply, which occasionally happens when the engine is idling. If there are no problems, the light should go out as the engine speed is increased. If it does not, either the AC generator or regulator is not working properly.

SHOP TALK

On late-model cars, the indicator light can be combined with the oil pressure warning light; it usually is labeled "engine." If this light turns on while the engine is running, either the AC generator is not charging or the oil pressure is low—or both. ■

Meters Some vehicles have an ammeter or voltmeter in their instrument cluster. The voltmeter displays the voltage at the battery. If the charging system is working fine, the voltmeter will read more than 12 volts.

The ammeter monitors current flow in and out of the battery. When the AC generator is delivering current to the battery, the ammeter shows a positive (+) indication. When not enough current (or none at all) is being supplied, the result is a negative (–) indication.

NEW DEVELOPMENTS

In the quest to improve fuel economy, decrease emission levels, and make vehicles more reliable, engineers have applied advanced electronics to starters and generators.

42-Volt Generators

Vehicles with a 42-volt electrical system will have an air- or liquid-cooled generator capable of producing 42 volts and 5 to 10 kilowatts. Currently, a conventional 12-volt generator puts out about 1.5 kilowatts. Depending on the design of the system **(Figure 19–26)**, the vehicle may be fitted with a DC to DC converter that changes some of

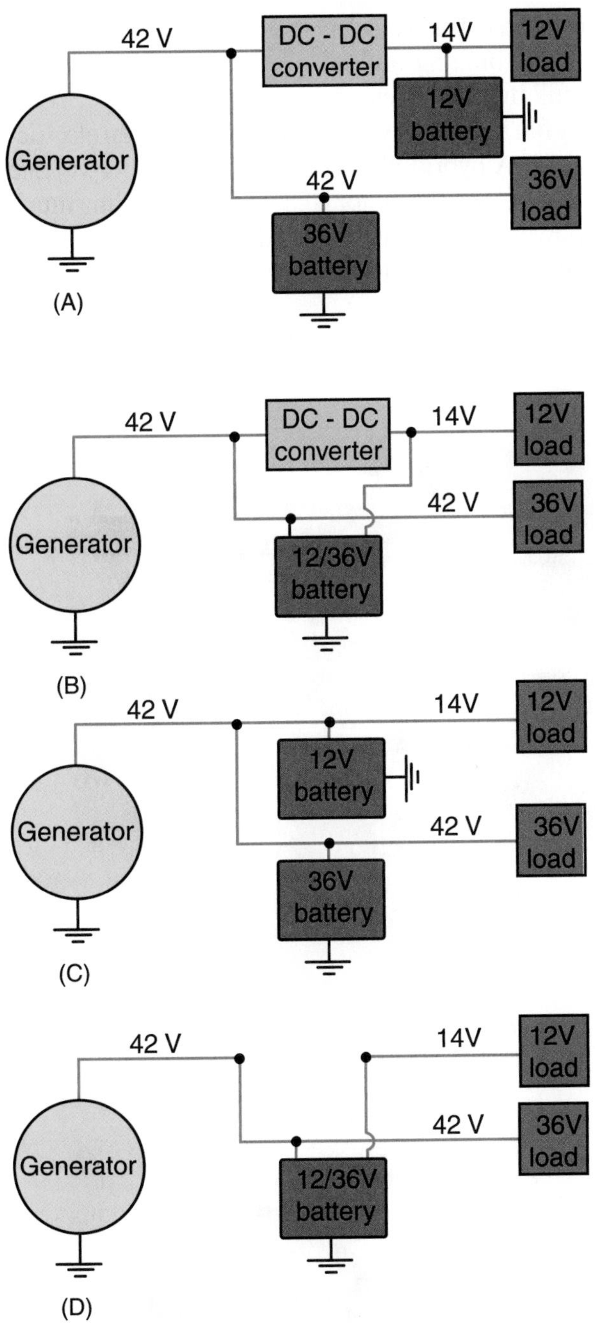

Figure 19-26 The different system layouts for 42-volt electrical systems: (A) a system with two batteries and a converter, (B) a system with one battery and a converter, (C) a two-battery system, and (D) a one-battery system.

the generator's output to charge a 12-volt battery or power the 12-volt loads of the vehicle. Some new generators are built without magnets. Switched reluctance techniques are used to generate the power needed for these high-voltage systems. Plus, switched reluctance systems are very efficient generators at low speeds. This design has a toothed stator and rotor and does not use windings or magnets in its rotor.

Starter/Generators

The main difference between a generator and a motor is that a motor has two magnetic fields that oppose each other, whereas a generator has one magnetic field and wires are moved through the field. Using electronics to control the current to and from the battery, engineers have developed a generator that can also work as a starter motor. These units are commonly called starter/generators.

A starter/generator may be based on two sets of windings and brushes, a brushless design with a permanent magnet, or switched reluctance **(Figure 19–27)**.

A starter/generator can be mounted externally to the engine and connected to the crankshaft with a drive belt. Starter/generators can also be mounted directly on the crankshaft between the engine and the transmission or integrated into the flywheel **(Figure 19–28)**. This unit works as a starter by spinning the crank at starting and serves as a generator, charging both directly from the engine and during braking (regenerative braking).

Figure 19-27 A switched reluctance starter/generator. Note the design of the rotor. *Courtesy of Dana Corporation*

Figure 19-28 An integrated starter/generator assembly built into the flywheel. *Courtesy of BMW of North America, Incorporated*

Starter/generators are capable of high charging outputs and can crank the engine at high speeds. They also allow for other features that make the vehicle more efficient:

- Stop-start. When the engine is not needed, such as at a stoplight, it is automatically turned off. It restarts smoothly and instantly when any demand for power is detected by the control module.
- Regenerative braking. This feature collects energy created from braking and uses it to recharge the vehicle's batteries. As the vehicle decelerates and the brakes are applied, the power flow reverses; then the wheels drive the engine. As the flywheel turns, it produces a charge that develops driveline drag and contributes to stopping. Regenerative braking allows items such as the headlights, stereo, and climate control system to continue to operate when the engine shuts off.
- Electrical assist. The starter/generator helps the engine at startup and during hard acceleration, providing short bursts of added power.

Some starter/generators are belt driven **(Figure 19–29)** and use all of the techniques designed for regenerative braking, torque assistance, and high efficiency. The belt tensioner is mechanically or electrically controlled to allow the starter/generator to drive or to be driven by the belt. A system used by Toyota has an electromagnetic clutch fitted to the crankshaft pulley. During normal operation, the clutch is engaged and the belt is driven by the crankshaft. When the engine is stopped, the crank pulley clutch disengages and the motor/generator works as a motor to keep the accessories going. It also restarts the engine when needed.

Hybrid Vehicles

A vehicle equipped with a starter/generator can be considered a mild hybrid because it is capable of most of the functions of a hybrid vehicle. Functions such as stop-start, regenerative braking, and electrical assist are common to both a full hybrid vehicle and a mild hybrid.

Only full hybrids have the ability to drive in an electric-only mode. A hybrid vehicle has a much higher voltage system; therefore, the motor is capable of providing much more power, more frequently, and for longer periods of time. The motor is located between the transmission and the engine **(Figure 19–30)**.

Honda's integrated motor assist (IMA) is a thin brushless electric motor **(Figure 19–31)** that assists the gaso-

Figure 19-29 A belt-drive starter/generator. *Reprinted with permission*

Figure 19-30 Honda's integrated motor assist unit for hybrid vehicles fits between the engine and the transmission. *Courtesy of American Honda Motor Co., Inc.*

Figure 19-31 Honda's integrated motor assist unit is a brushless motor. *Courtesy of American Honda Motor Co., Inc.*

line engine during acceleration, functions as a generator to recharge the battery pack during deceleration, and serves as the gasoline engine's starter motor. Power for the motor comes mainly from regenerative braking, rather than from the gasoline engine. If the charge of the IMA battery is low, the motor/generator will also recharge while the vehicle is cruising. If the IMA system battery is low, a separate 12-volt battery and starter motor will start the engine.

PRELIMINARY CHECKS

The key to solving charging system problems is getting to the root of the trouble the first time. Once a customer drives away with the assurance that the problem is solved, another case of a dead battery is very costly—both in terms of a free service call and a damaged reputation. Add to this the many possible hours of labor trying to figure out why the initial repair failed, and the importance of a correct initial diagnosis becomes all too clear.

Safety Precautions

- Disconnect the battery ground cable before removing any leads from the system. Do not reconnect the battery ground cable until all wiring connections have been made.
- Avoid contact with the AC generator output terminal. This terminal is hot (has voltage present) at all times when the battery cables are connected.
- The AC generator is not made to withstand a lot of force. Only the front housing is relatively strong. When adjusting belt tension, apply pressure only to the front housing to avoid damaging the stator and rectifier.
- When installing a battery, be careful to observe the correct polarity. Reversing the cables destroys the diodes. Proper polarity must also be observed when connecting a booster battery, positive to positive and negative to ground.
- Keep the tester's carbon pile off at all times, except during actual test procedures.
- Make sure all hair, clothing, and jewelry are kept away from moving parts.

Inspection

In addition to observing the ammeter, voltmeter, or indicator light, there are some common warning signs of charging system trouble. For example, a low state of battery charge often signals a charging problem, as does a noisy AC generator.

Figure 19-32 Start your diagnosis with an inspection of the generator and its drive belt and wires.

Many charging system complaints stem from easily repairable problems that reveal themselves during a visual inspection of the system. Remember to always look for the simple solution before **(Figure 19–32)** performing more involved diagnostic procedures. Use the following inspection procedure when a problem is suspected.

PROCEDURE

Inspections

STEP 1 *Before adjusting belt tension, check for proper pulley alignment, especially critical in serpentine belts.*

STEP 2 *Inspect the generator drive belt. Loose drive belts are a major source of charging problems. The correct procedure for inspecting, removing, replacing, and adjusting a drive belt is shown in Photo Sequence 15.*

STEP 3 *Inspect the battery. It might be necessary to charge the battery to restore it to a fully charged state. If the battery cannot be charged, it must be replaced. Also, make sure the posts and cable clamps are clean and tight, because a bad connection can cause reduced current flow.*

STEP 4 *Inspect all system wiring and connections. Many automotive electrical systems contain fusible links to protect against overloads. Fusible links can blow like a fuse without being noticed. Also, look for a short circuit, an open ground, or high resistance in any of the circuits that could cause a problem that would appear to be in the charging system.*

PHOTO SEQUENCE 15

Typical Procedure for Inspecting, Removing, Replacing, and Adjusting a Drive Belt

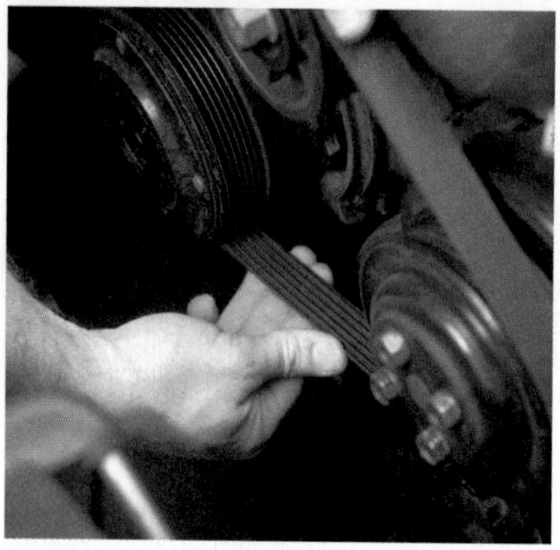

P15-1 *Inspect the belt by looking at both sides.*

P15-2 *Look for signs of glazing.*

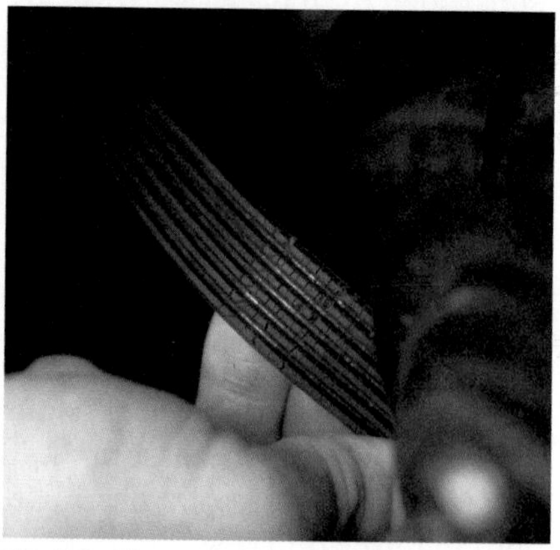

P15-3 *Look for signs of tearing or cracking.*

P15-4 *To replace a worn belt, locate the idler or generator pulley.*

P15-5 *Loosen the hold-down fastener for the idler or generator pulley.*

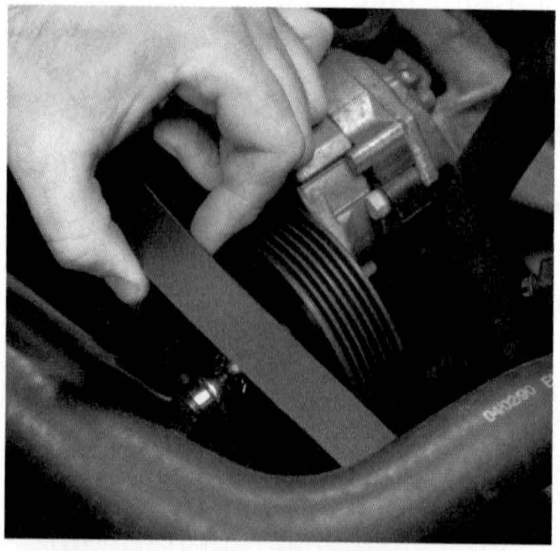

P15-6 *Pry the idler or generator pulley inward to release the belt tension and remove the belt.*

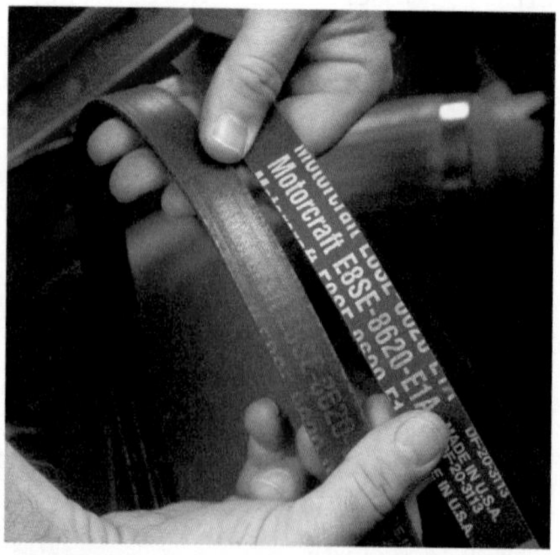

P15-7 *Match the old belt up for size with the new replacement belt.*

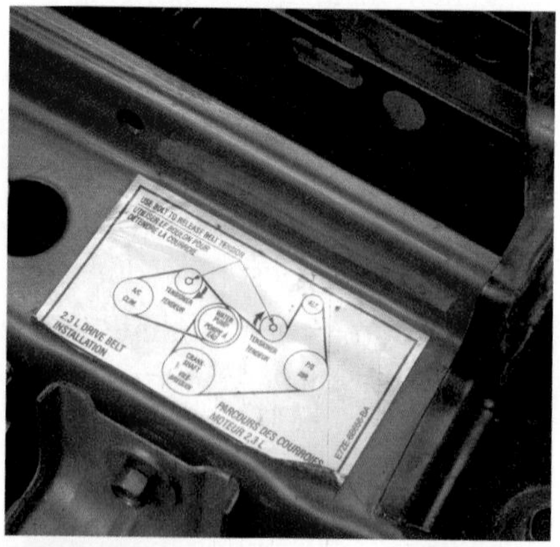

P15-8 *Observe the belt routing diagram in the engine compartment.*

P15-9 *Install the new belt over each of the drive pulleys. Often the manufacturer recommends a sequence for feeding the belt around the pulleys.*

P15-10 *Pry out the idler or generator pulley to put tension on the belt.*

P15-11 *Install the belt squarely in the grooves of each pulley.*

P15-12 *Measure the belt deflection in its longest span. If a belt tension gauge is available, use it and compare the tension to specifications.*

P15-13 *Pry the idler or generator pulley to adjust the belt to specifications.*

P15-14 *Tighten the idler or generator pulley fastener.*

P15-15 *Start the engine and check the belt for proper operation.*

STEP 5 *Inspect the AC generator and regulator mountings for loose or missing bolts. Replace or tighten as needed. Remember that the circuit completes itself through the ground of the AC generator and regulator. Most AC generators and regulators complete their ground through their mountings. If the mountings are not clean and tight, a high resistance ground will result.*

SHOP TALK

Some late-model Chrysler engines require a torque reading to be taken when tension is applied to the generator drive belt. This is especially important on the longer, multiribbed V-belts. ■

If the vehicle passes all preliminary visual checks, listen for noisy belts, bad bearings, or the whining sound of a bad diode. If no unusual sounds are heard, it is time to test the charging system.

GENERAL TESTING PROCEDURES

Diagnosing a charging system is a straightforward task. Tests can be conducted with a VAT, current probe, DMM, or a lab scope. Charging system tests for all cars are basically the same; however, it is very important to refer to the manufacturer's specifications. Even the most accurate test results are no good if they are not matched against the correct specs.

Regulator Tests

Begin your diagnosis by determining if a no-charge problem is caused by the generator or the regulator. To do this, you must first determine if the system has an integral regulator, then whether it has a type-A or type-B field circuit. A type A has one brush connected to the battery terminal and the other brush grounded through the regulator. Type-B circuits have one brush directly grounded and the other connected to the regulator.

With this knowledge you can isolate the problem to the generator or regulator by bypassing the regulator or **full-fielding** the generator. To do this on a type A, ground the wire going from the brush to the regulator. On a type B, apply battery voltage to the wire going from the brush to the regulator. Turn the vehicle's high beam headlights on and start the engine. If the charging system now has an output, the problem is the regulator. It is important that the lights be turned on. The drain on the electrical system protects the vehicle's computers from excessive voltage and helps absorb any damaging voltage spikes. Because the computer systems are so sensitive to voltage, many manufacturers recommend that a full-field test not be done.

A poorly performing charging system is rarely caused by a faulty regulator; therefore, it is not necessary to check it. A bad regulator can cause excessively high voltage outputs.

Voltage Output Test

To check the charging system's voltage output, begin by measuring the battery's open circuit voltage. Connect the voltmeter across the battery and note the reading on the meter. Next, start the engine and run it at the suggested rpm for this test (usually 1,500 rpm). With no electrical load, the voltage reading should be about 2 volts higher than the open circuit voltage.

A reading of less than 13.0 volts immediately after starting the engine indicates a charging problem. No increase in voltage means the system is not producing voltage. A reading of 16 or more volts indicates overcharging. A faulty voltage regulator or control voltage circuit are the most likely causes of overcharging.

If the unloaded charging system voltage is within specifications, test the output under a load. To do this, increase engine rpm to about 2,000 rpm and turn on the headlights and other high-current accessories. Under these conditions, the output should be about 0.5 volt above battery open circuit voltage.

Current Output Test

Using a VAT is an easy way to check the amperage output of a charging system. With the tester connected to the system, the engine is run at a moderate speed and the carbon pile is adjusted to obtain maximum current output. This reading is compared against the rated output. Normally, readings that are more than 10 amperes out of specifications indicate a problem.

Field Current Check

Low generator output can be caused by worn brushes, which limit field current. To measure field current, place a current probe or the VAT's inductive pickup over the field wire at the generator. Now load the charging system with the carbon pile to bring the generator to full output. Observe the ammeter reading on the tester. The procedure for measuring field current is different for generators with an integral regulator and those procedures vary with the model of generator, so follow the instructions given by the manufacturer.

Diode Checks

The output of a generator is highly dependent on the condition of the diodes. Not only do the diodes rectify AC voltage to DC, they also prevent AC voltage from being present in the output. Bad diodes are indicated by the presence of more than 0.5 AC volt in the output wire. To check this, set the DMM to measure AC volts. Then connect the black meter lead to a good ground and the red lead to the generator's battery terminal.

Another check of the diodes while they are still in the generator is done with the engine off and with a low-amperage current probe. Measure the current on the generator's output wire. Any measurement greater than 0.5 milliamp indicates one or more diodes are leaking and the generator or diodes need to be replaced.

Oscilloscope Checks

AC generator output can also be checked using an oscilloscope. **Figure 19–33** illustrates common AC generator voltage patterns for good and faulty generators. The correct pattern looks like the rounded top of a picket fence. A regular dip in the pattern indicates that one or more of the coil windings is grounded or open, or that a diode in

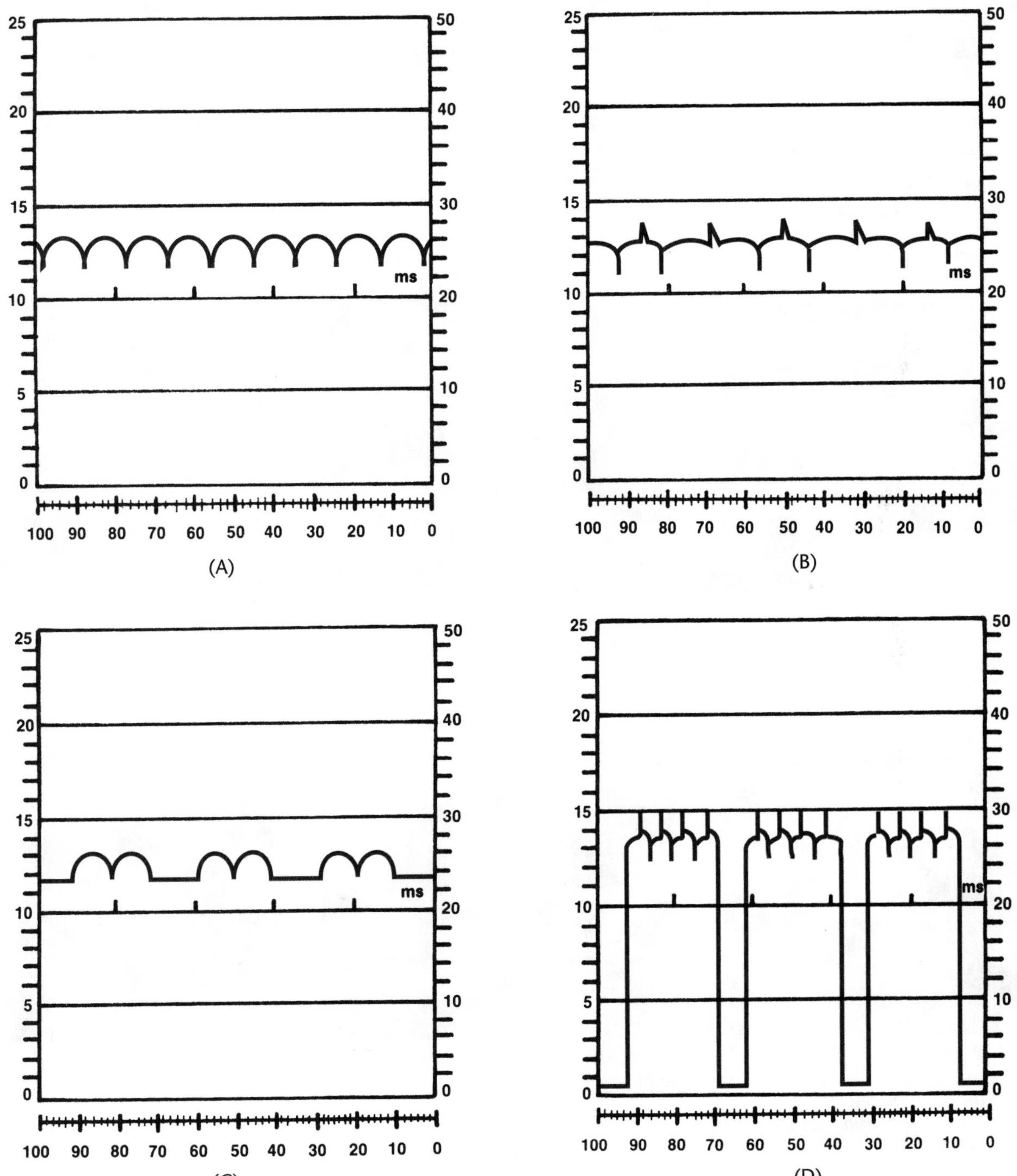

Figure 19-33 AC generator oscilloscope patterns: (A) a good AC generator under full load, (B) a good AC generator under no load, (C) a shorted diode and/or stator winding under full load, and (D) an open diode in diode trio.

PHOTO SEQUENCE 16

Typical Procedure for Disassembling a Ford IAR AC Generator

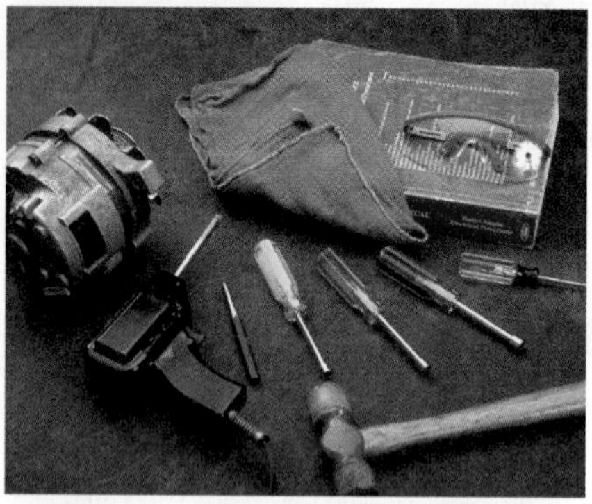

P16-1 *Always have a clean and organized work area. The tools required to disassemble a Ford IAR AC generator are rags, T20 torx wrench, plastic hammer, arbor press, 100-watt soldering iron, soft-jawed vise, safety glasses, and an assortment of sockets or nut drivers.*

P16-2 *Using the torx wrench, remove the four attaching screws that hold the regulator to the AC generator housing.*

P16-3 *Remove the regulator and brush assembly as a unit.*

P16-4 *Using the torx wrench, remove the two screws that attach the regulator to the brush holder. Then separate the regulator from the brush holder.*

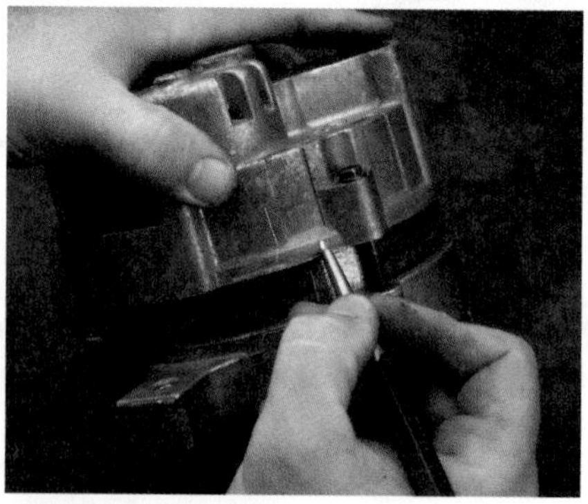

P16-5 *Scribe or mark the two end housings and the stator core for reference during reassembly.*

P16-6 *Remove the three through bolts that secure the two housings.*

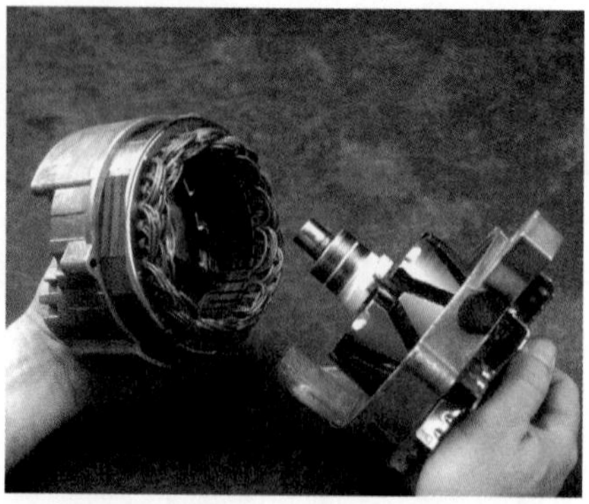

P16-7 *Separate the front housing from the rear housing. The rotor will come out with the front housing, and the stator will stay in the rear housing. It may be necessary to tap the front housing with the plastic hammer to get the two halves to separate.*

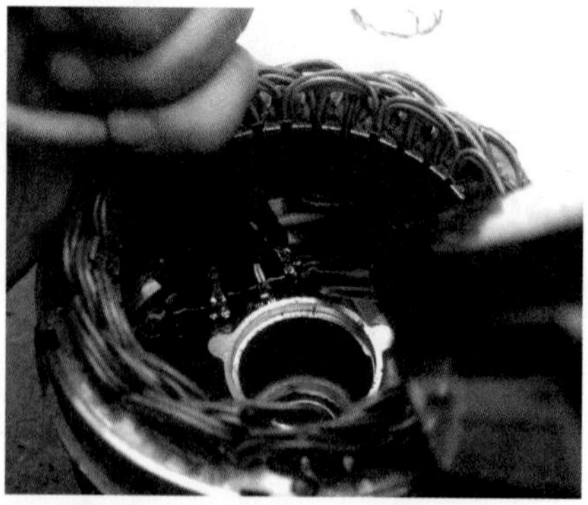

P16-8 *Separate the three stator lead terminals from the rectifier bridge.*

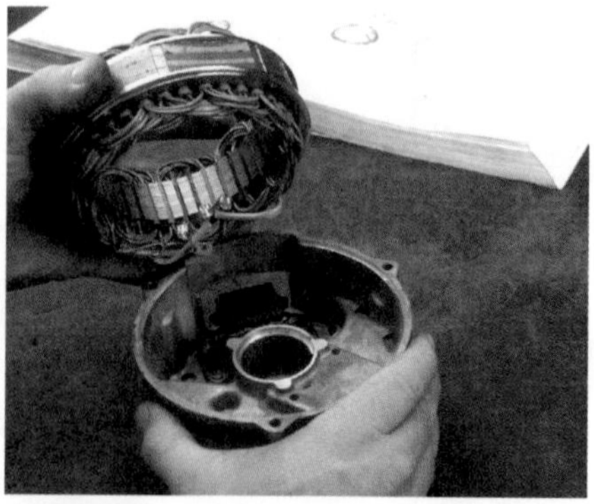

P16-9 *Remove the stator coil from the housing.*

P16-10 *Using the torx wrench, remove the four attaching bolts that hold the rectifier bridge.*

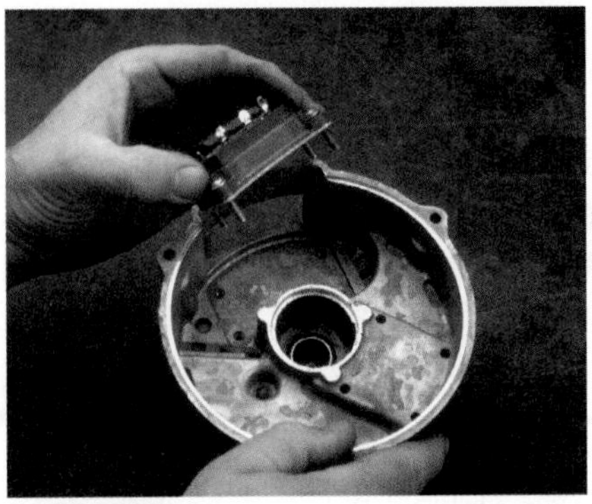

P16-11 *Remove the rectifier bridge from the housing.*

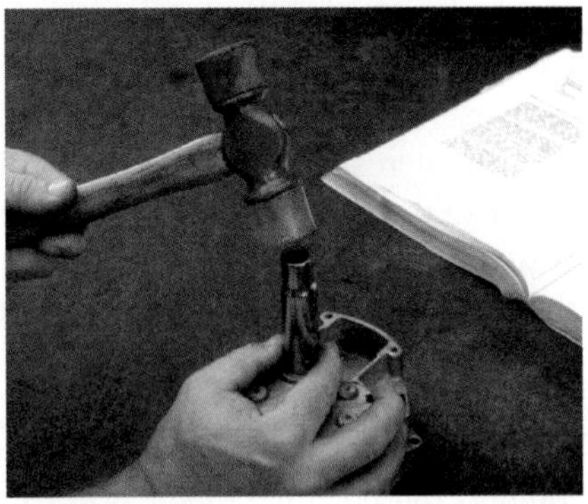

P16-12 *Use a socket to tap out the bearing from the housing.*

P16-13 *Clamp the rotor in the vise.*

P16-14 *Remove the pulley-attaching nut, flat washer, drive pulley, fan, and fan spacer from the rotor shaft.*

P16-15 *Separate the front housing from the rotor. If the stop ring is damaged, remove it from the rotor. If not, leave it on the shaft.*

P16-16 *Remove the three screws that hold the bearing retainer to the front housing.*

P16-17 *Remove the bearing retainer.*

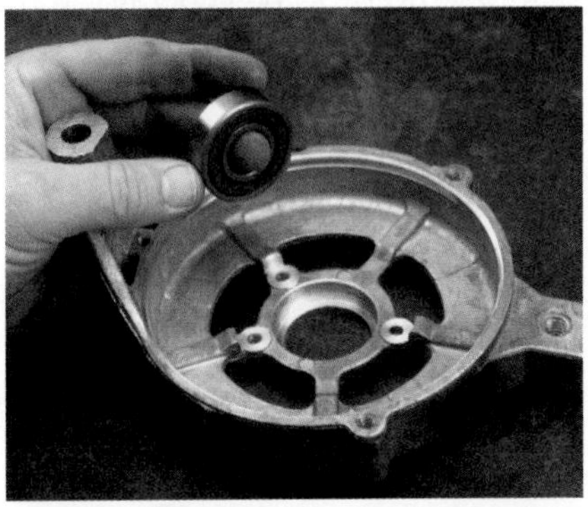

P16-18 *Remove the front bearing from the housing. Test and inspect all parts. Replace any defective ones. Reassembly is the reverse of this procedure.*

the rectifier circuit of a diode trio circuit has failed. One or more bad or leaking diodes will decrease the output of a generator.

Circuit and Ground Resistance

These tests measure voltage drop within the system wiring. They help pinpoint corroded connections or loose or damaged wiring.

Circuit resistance is checked by connecting a voltmeter to the positive battery terminal and the output, or battery terminal of the AC generator. The positive lead of the meter should be connected to the AC generator output terminal and the negative lead to the positive battery terminal. To check the voltage drops across the ground circuit, connect the positive lead to the generator housing and the negative meter lead to the battery negative terminal. When measuring the voltage drop in these circuits, a sufficient amount of current must be flowing through the circuit. Therefore, turn on the headlights and other accessories to ensure that the AC generator is putting out at least 20 amps. If a voltage drop of more than 0.5 volt is measured in either circuit, there is a high resistance problem in that circuit.

AC GENERATOR SERVICE

When the cause of charging system failure is the AC generator, it should be removed and replaced or rebuilt. Whether it is rebuilt or replaced depends on the type of generator it is, the time and cost required to rebuild it, your shop's policy, and your customer's desires. Many late-model AC generators are not rebuilt. They are traded in as a core toward the purchase of a new or remanufactured unit. Just in case you do need to rebuild one, Photo Sequence 16 is given as an example of what it takes to do so. This procedure is for a specific type and model of generator. Make sure you follow the procedures given by the manufacturer for the generator you are working on.

To test the components of an AC generator it must be removed and disassembled.

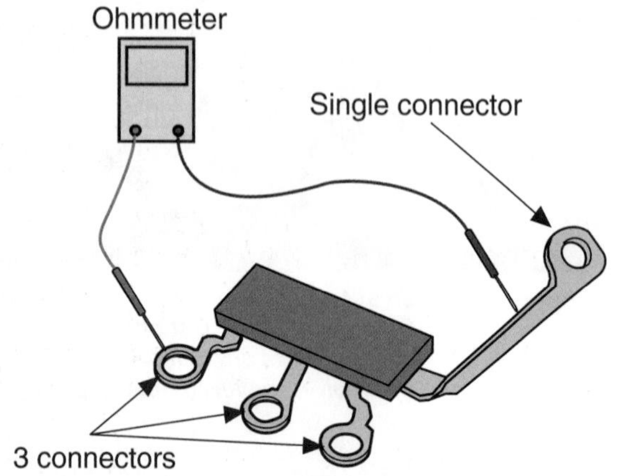

Figure 19-34 Using an ohmmeter to test a diode trio.

A faulty AC generator can be the result of many different types of internal problems. Diodes **(Figure 19–34)**, stator windings **(Figure 19–35)**, and field circuits **(Figure 19–36)** may be open, shorted **(Figure 19–37** and

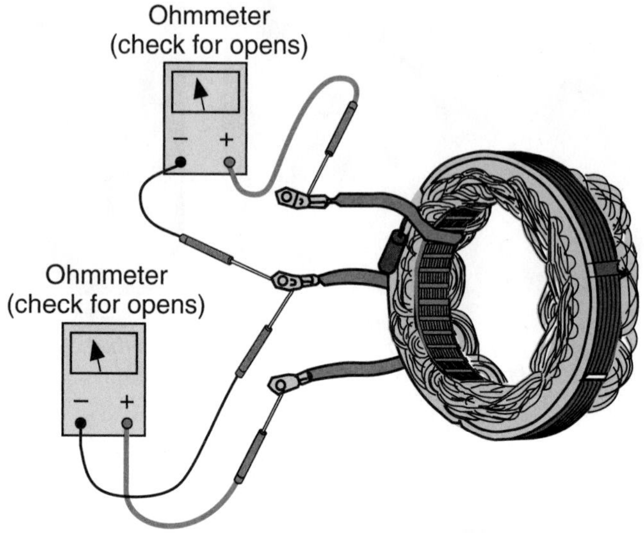

Figure 19-35 Testing a stator for opens.

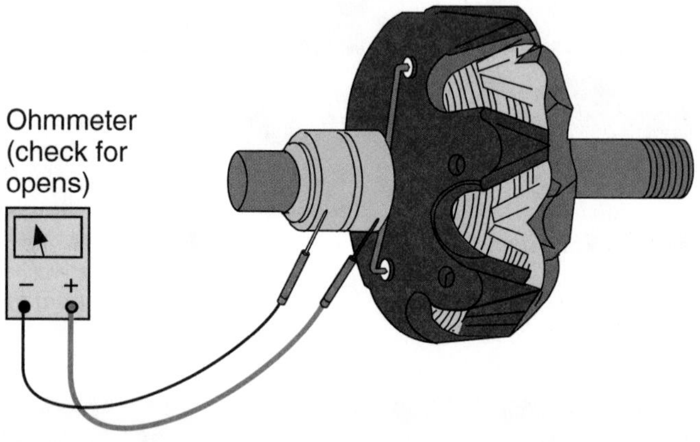

Figure 19-36 Testing a rotor for opens.

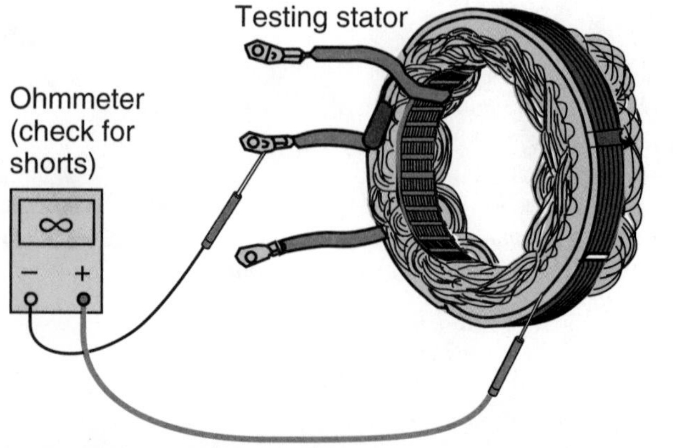

Figure 19-37 Testing a stator for a short to ground.

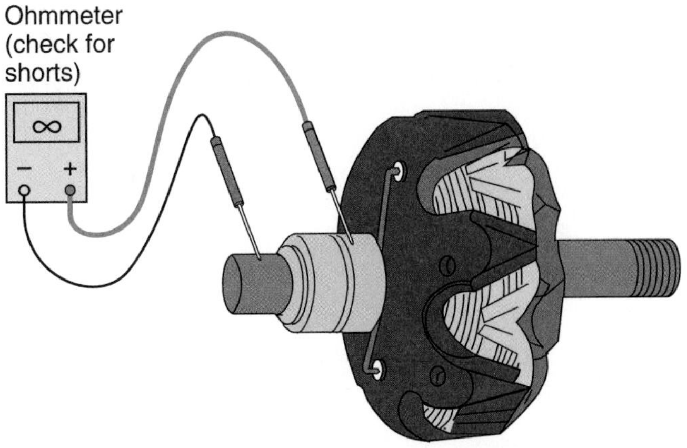

Figure 19-38 Testing a rotor for a short to ground.

Figure 19–38), or improperly grounded. The brushes or slip rings can become worn. The rotor shaft can become bent and the pulley can work loose or bend out of proper alignment.

Follow service manual procedures when removing and installing an AC generator. Remember, improper connections to an AC generator can destroy it.

Follow service manual procedures for disassembling, inspecting, testing, and rebuilding AC generators.

CASE STUDY

A customer brought his pickup truck into the dealership with a complaint that the truck must be jumped to get it started and once it is started, it runs for about 10 minutes then dies.

The technician verified the complaint, then did a visual inspection of the charging system. Based on the complaint, he knew that the battery was not being kept in charge by the AC generator or that the battery was unable to hold a charge. He found the drive belt extremely frayed and glazed. Knowing this could cause the problem, he replaced the belt. However, before releasing the truck back to the customer he tested the battery to make sure it was able to hold a charge if the AC generator was capable of charging it. The battery checked out fine. He then checked the AC generator output and found it to be within specifications. This was a capable technician; he knew that he should not assume that the obvious problem was the only problem. As a result, the customer was happy and should be for quite some time.

KEY TERMS

AC generator
Alternator
DC generator
Delta
Diode trio
Fail-safe circuit
Field coils
Fingers
Full-fielding
Full-wave rectification
Half-wave rectification
Integrated circuit voltage regulator
Pole piece
Rotor
Sensing voltage
Sine wave
Slip rings
Stator
Three-phase voltage
Voltage regulator
Wye

SUMMARY

- Inducing a voltage requires a magnetic field producing lines of force, conductors that can be moved, and movement between the conductors and the magnetic field so the lines of force are cut.
- Modern vehicles use an AC generator to produce electrical current in the charging system. Diodes in the generator change or rectify the alternating current to direct current.
- A voltage regulator keeps charging system voltage above battery voltage. Keeping the AC generator charging voltage above the 12.6 volts of the battery ensures current flows into, not out of, the battery.
- Modern voltage regulators are completely solid-state devices that can be an integral part of the AC generator or mounted to the back of the generator housing. In some vehicles, voltage regulation is the job of the computer control module.
- Voltage regulators work by controlling current flow to the AC generator field circuit. This varies the strength of the magnetic field, which in turn varies current output.
- The driver can monitor charging system operation with indicator lights, a voltmeter, or an ammeter.
- Problems in the charging system can be as simple as worn or loose belts, faulty connections, or battery problems.
- Circuit resistance, current-output, voltage-output, field-current draw, and voltage regulator tests are all used to troubleshoot AC charging systems.

TECH MANUAL

The following procedures are included in Chapter 19 of the *Tech Manual* that accompanies this book:

1. Visually inspecting the charging system.
2. Inspecting, replacing, and adjusting drive belts and pulleys.
3. Removing and replacing an AC generator.
4. Testing the charging system.
5. Disassembling and assembling an AC generator.
6. Inspecting, cleaning, and testing AC generator components.

REVIEW QUESTIONS

1. *True or False?* A faulty voltage regulator can cause a number of problems such as no-charge, undercharging, and overcharging.
2. A voltage regulator controls the voltage output of an AC generator by controlling the current to the ______.
3. To protect electronic circuits, some voltage regulators have a ______ ______ circuit built into them.
4. What is the purpose of a diode trio?
5. How does a voltage regulator regulate the voltage output of an AC generator?
6. Technician A says the waveform produced by an AC generator is called a sine wave. Technician B says the waveform produced by the AC generator after the output moves through the diodes is a straight line because it is a constant DC voltage. Who is correct?
   a. Technician A
   b. Technician B
   c. Both A and B
   d. Neither A nor B
7. What would happen to the output of an AC generator if one of the stator windings has an open?
8. While discussing what is indicated by a vehicle's charge indicator, Technician A says any positive reading on an ammeter shows the battery is in good condition. Technician B says if the voltmeter reading does not increase immediately after starting the engine, the battery is bad. Who is correct?
   a. Technician A
   b. Technician B
   c. Both A and B
   d. Neither A nor B
9. A rotating magnetic field inside a set of conducting wires is a simple description of ______.
   a. a DC generator
   b. an AC generator
   c. a voltage regulator
   d. none of the above
10. What part of the AC generator produces the rotating magnetic field?
    a. stator
    b. rotor
    c. brushes
    d. poles
11. Which type of stator winding produces higher AC generator voltage output at low speeds?
    a. wye
    b. delta
    c. trio
    d. series
12. Slip rings and brushes ______.
    a. mount on the rotor shaft
    b. conduct current to the rotor field coils
    c. are insulated from each other and the rotor shaft
    d. all of the above
13. The alternating current produced by the AC generator is rectified into DC, or direct current, through the use of ______.
    a. transistors
    b. electromagnetic relays
    c. diodes
    d. capacitors
14. Integrated voltage regulators ______.
    a. contain all needed circuitry on a single silicone clip sealed in a plastic module
    b. mount to the back of or inside the AC generator housing
    c. are unserviceable and must be replaced if they are defective
    d. all of the above
15. Voltage regulation or voltage regulator circuitry in the computer control module control AC generator output by ______.
    a. using a variable resistor to vary current flow to the rotor field windings

b. pulsing current flow to the rotor field windings on and off to create a correct average field current supply

c. either a or b, depending on the system

d. neither a nor b

16. Technician A uses a current output test to check AC generator output. Technician B uses a voltage output test to check output. Who is correct?

a. Technician A **c.** Both A and B

b. Technician B **d.** Neither A nor B

17. Technician A says many newer charging systems do not use separate voltage regulators. Technician B says most late-model charging systems are wired as A-type circuits and the voltage is regulated through the ground of the field. Who is correct?

a. Technician A **c.** Both A and B

b. Technician B **d.** Neither A nor B

18. Technician A says if there is no output from the charging system until the AC generator is full-fielded, the voltage regulator is faulty. Technician B says that this indicates one or more leaking diodes in the generator. Who is correct?

a. Technician A **c.** Both A and B

b. Technician B **d.** Neither A nor B

19. When checking AC generator output using an oscilloscope, Technician A says the correct pattern resembles the rounded tops of a picket fence. Technician B says the correct pattern is a square sine-wave pattern. Who is correct?

a. Technician A **c.** Both A and B

b. Technician B **d.** Neither A nor B

20. A voltage drop over ______ indicates high resistance in either the circuit or ground wiring of the charging system.

a. 0.1 **c.** 0.5

b. 0.2 **d.** 1.0

20

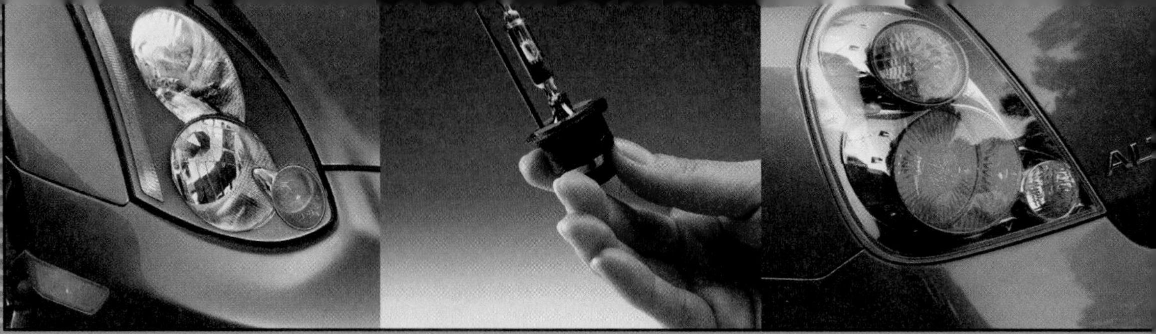

LIGHTING SYSTEMS

OBJECTIVES

■ Explain the operating principles of the various lighting systems. ■ Describe the different types of headlights and how they are controlled. ■ Understand the functions of turn, stop, and hazard warning lights. ■ Know how backup lights operate. ■ Replace headlights and other burned-out bulbs. ■ Explain how to aim headlights. ■ Explain the purpose of auxiliary automotive lighting. ■ Describe the operation and construction of the various automotive lamps. ■ Diagnose lighting problems.

The lighting system provides power to both exterior and interior lights. It consists of the headlights, parking lights, marker lights, taillights, courtesy lights, dome/map lights, instrument illumination or dash lights, coach lights (if so equipped), headlight switch, and various other control switches **(Figure 20–1)**. Other lights, such as vanity mirror lights, the underhood light, the glove box light, and the trunk compartment light, are used on some vehicles and are also part of the lighting system.

Other lights that are not usually in the main lighting system are turn signal, hazard warning, backup, and stop lights. These lights, as well as the pop-up headlights or retractable headlight covers found on some vehicles, are operated by separate control circuits and are covered later in this chapter.

Figure 20–1 Automotive lighting systems.

SHOP TALK

Lighting systems are largely regulated by federal laws, so the systems are similar among the various manufacturers. However, there are many variations. Before attempting to do any repairs on an unfamiliar system, you should always refer to the manufacturer's service manual. ■

LAMPS

A **lamp** generates light as current flows through the filament. This causes it to get very hot. The changing of electrical energy to heat energy in the resistive wire filament is so intense that the filament starts to glow and emits light. The glass envelope that encloses the filament is evacuated so that the filament "burns" in a vacuum. If air enters the envelope, the oxygen would cause the filament to oxidize and burn up.

It is important that any burned-out lamp be replaced with the correct lamp. You can determine what lamp to use by checking the lamp's standard trade number, usually present on the lamp's housing. Lamps are normally one of two types: a single filament **(Figure 20–2)** or a double filament **(Figure 20–3)**. Double-filament bulbs are designed to serve more than one function. They can be used as the lone bulb in the stop light circuit, taillight circuit, and the turn signal circuit.

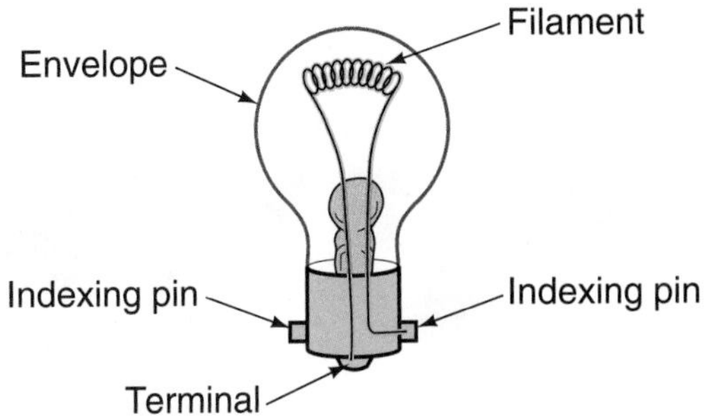

Figure 20–2 A single-filament bulb.

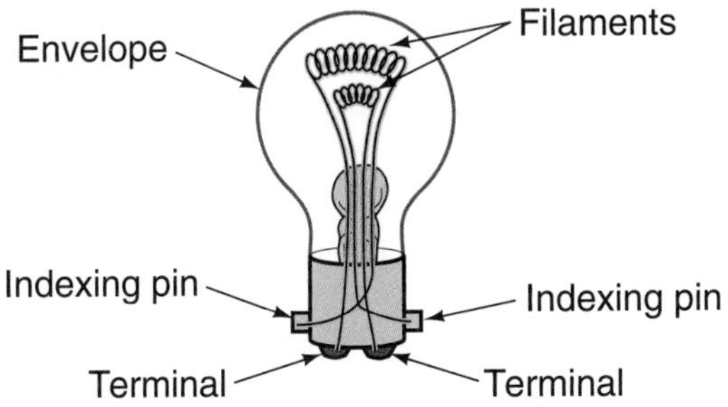

Figure 20–3 A double-filament bulb.

HEADLIGHTS

Headlights light the road ahead during darkness or at other times when normal visibility is poor. Headlight designs and construction have been influenced by the changes in technology. In the past, all cars had two or four round headlights. Now headlights are an integral part of a vehicle's overall design **(Figure 20–4)**. The headlights of today's vehicles are based on sealed-beam, composite, or high intensity discharge lamps.

Sealed-Beam Headlights

The standard **sealed-beam** headlight is an air-tight assembly with a filament, reflector, and lens fused together. The parabolic reflector is sprayed with vaporized aluminum and the inside of the lamp is typically filled with argon gas. The reflector intensifies the light produced by the filament, and the lens directs the light to form the required light beam pattern. The lens is designed to produce a broad flat beam. The light from the reflector is passed through concave prisms in the glass lens.

Figure 20–4 Today's headlights are an integral part of the appearance of vehicles. *Courtesy of Nissan Motor Company*

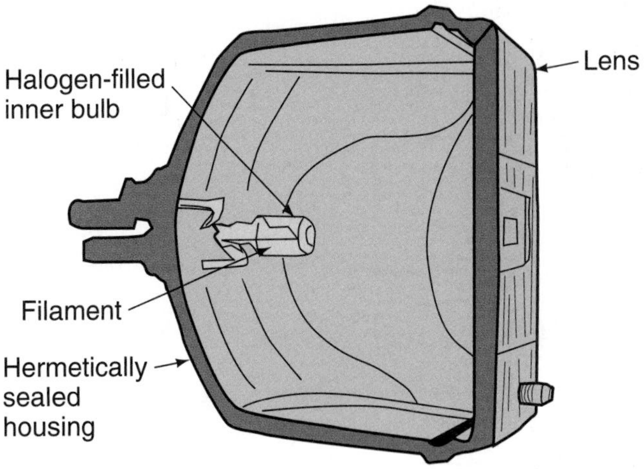

Figure 20–5 A halogen sealed-beam headlight with an iodine vapor bulb.

Today, the most commonly used sealed-beam headlight is the halogen type. A **halogen** lamp typically consists of a small bulb filled with iodine vapor. The bulb is made of high-temperature-resistant glass and it surrounds a tungsten filament. The halogen-filled inner bulb is then installed in a sealed glass or plastic housing **(Figure 20–5)**. With the halogen added to the inner bulb, the tungsten filament is capable of withstanding higher temperatures than that of standard sealed-beam lamps. Because it can withstand higher temperatures, it can burn brighter.

Halogen is the term used to identify a group of chemically related nonmetallic elements. These elements include chlorine, fluorine, and iodine.

SHOP TALK

Because the filament is contained in the inner bulb, cracking or breaking of the housing or lens does not prevent a halogen bulb from working. As long as the filament envelope has not been broken, the filament will continue to operate. However, a broken lens results in poor light quality and the lamp assembly should be replaced. ■

Low- and high-beam filaments are placed at slightly different locations within a sealed-beam bulb. The filament location, relative to the reflector, determines how light passes through the bulb's lens **(Figure 20–6)**, which, in turn, determines the direction in which the light shines.

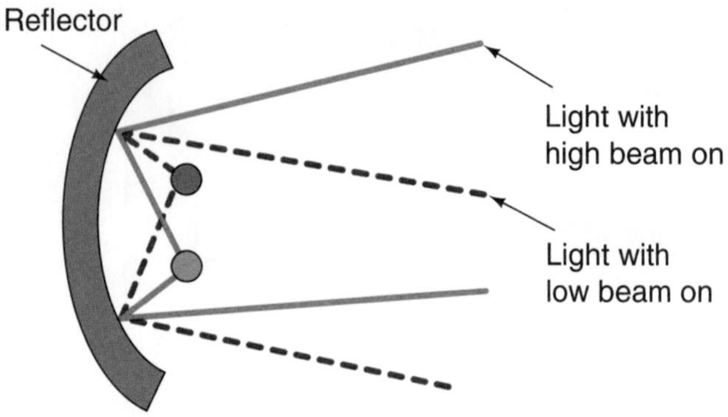

Figure 20-6 Filament placement controls the projection of the light beam.

In a dual filament lamp, the lower filament is used for the high beam and the upper filament is used for the low beam.

Various methods are used to identify sealed-beam headlights, such as 1, 2, and the "halogen" or "H" marking molded on the front of the headlight lens. A type 1 has high beam only and has two electrical terminals on its back. The type 2 has both low and high beam and three terminals. When a type 2 is switched to low beam, only one of its filaments is lit. When the high beam is selected, the second filament lights in addition to the low beam.

If a sealed-beam headlamp has condensation on the lens or inside the assembly or it if is cracked, the headlamp will not work and can only be repaired by replacing it.

Composite Headlights

Many of today's vehicles have halogen headlight systems that use a replaceable bulb **(Figure 20–7)**. These systems are called **composite headlights**. By using the composite headlight system, manufacturers are able to produce any style of headlight lens they desire **(Figure 20–8)**, which improves the aerodynamics, fuel economy, and styling of the vehicle.

Many manufacturers vent the composite headlight housing due to the intense heat developed by these bulbs. Because the housings are vented, condensation may develop inside the lens assembly. This condensation is not

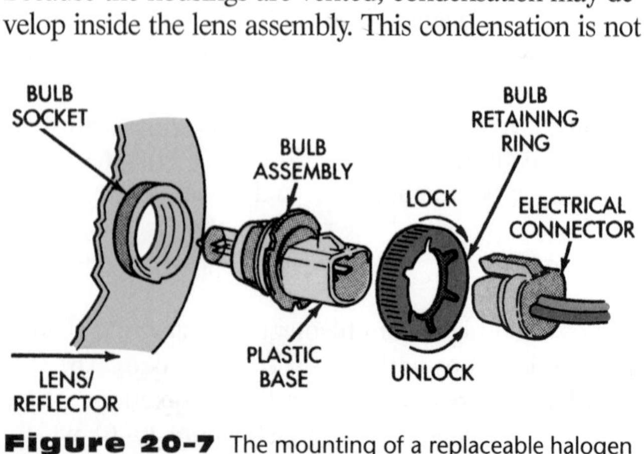

Figure 20-7 The mounting of a replaceable halogen bulb.

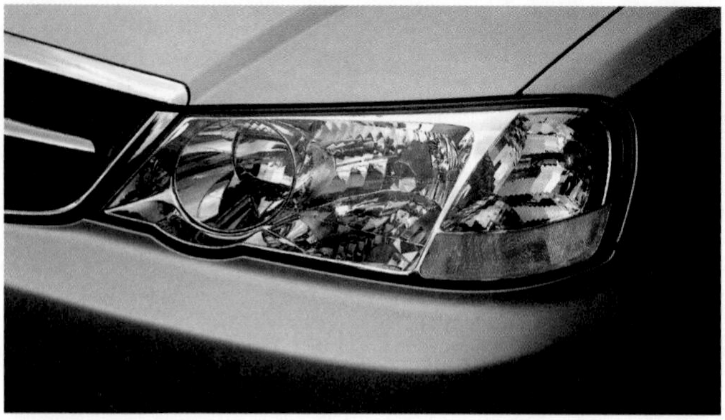

Figure 20-8 Replaceable bulbs and lens technology have allowed manufacturers to integrate headlights to the overall design of their vehicles. *Courtesy of American Honda Motor Co., Inc.*

harmful to the bulb and does not affect headlight operation. When the headlights are turned on, the heat generated by the halogen bulb dissipates the condensation quickly. Ford uses integrated nonvented composite headlights. On these vehicles, condensation is not considered normal. The assembly should be replaced.

WARNING!

Whenever you replace a composite lamp, be careful not to touch the lamp's envelope with your fingers. Staining the bulb with skin oil can substantially shorten the life of the bulb. Handle the bulb only by its base. Also dispose of the bulb properly.

HID Headlamps

High-intensity discharge (HID) or **xenon headlamps** use gas-discharge lamps and are electronically controlled. These lights are recognizable by the blue-white color of their light **(Figure 20–9)**. They have this color because

Figure 20-9 HID (xenon) headlights are readily identifiable by their bluish light. *Courtesy of DaimlerChrysler Corporation*

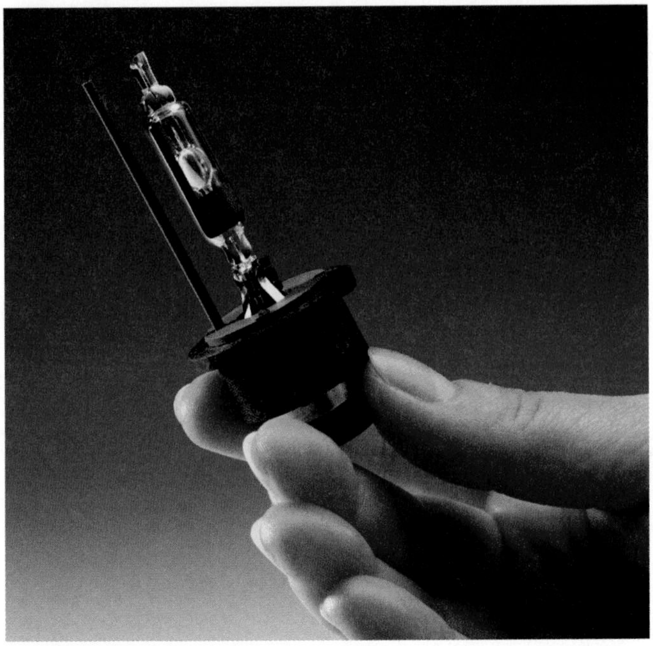

Figure 20-10 A xenon light bulb. *Courtesy of DaimlerChrysler Corporation*

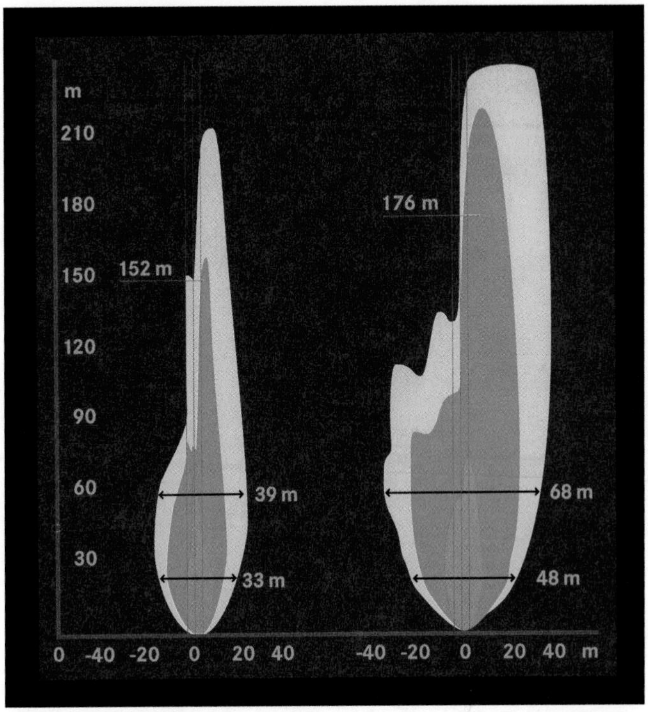

Figure 20-11 A comparison of the light pattern and intensity between a halogen (left) and a xenon (right) headlamp. *Courtesy of DaimlerChrysler Corporation*

the light's spectrum is much closer to daylight than that of a halogen bulb.

Instead of using a filament, an electrical arc is created between two electrodes that excite a gas (usually xenon) inside the headlamp **(Figure 20–10)**, which in turn vaporizes metallic salts that sustain the arc and emit light. The presence of an inert gas amplifies the light given off by this arcing.

More than 15,000 volts are used to jump the gap between the electrodes. To provide this voltage, a voltage booster and controller is required. Once the high voltage bridges the gap, only about 80 volts are needed to keep current flowing across the gap. When the headlights are switched on, it takes approximately 15 seconds for the lamps to reach maximum intensity. However, even during ignition these lamps provide more than adequate light for safe driving.

Xenon headlights illuminate the area to the front and sides of the vehicle with a beam that is both brighter and much more consistent than the light generated by halogen headlamps. The great light output of these lamps allows the headlamp assembly to be smaller and lighter. Xenon lights also produce significantly less heat.

Xenon headlights produce about twice as much light as comparable halogen headlights **(Figure 20–11)** and make night driving safer and less tiring for the driver's eyes. Xenon headlamps also use about two-thirds less power to operate and will last two or three times as long.

Bi-Xenon Lights Some vehicles have bi-xenon headlamps that provide xenon lights for low and high beams. These may also be fitted with halogen lights that are used for the flash-to-pass feature. Bi-xenon lights rely on a mechanical shield plate, or shutter, that physically obstructs a portion of the overall light beam emitted by the arc. When the driver selects high beams, the shutter reacts and allows the headlights to project the complete, unobstructed light beam.

Headlight Switches Headlight switches are either mounted on the dash panel or are part of a multifunction switch on the steering column. The headlight switch controls most of the vehicle's lighting systems. The most common style of headlight switch has three positions: OFF, PARK, and HEADLIGHT. A headlight switch normally receives direct battery voltage at two of its terminals. This allows the light circuits to be operated without having the ignition switch in the RUN or ACC (accessory) position.

When the headlight switch is in the OFF position, the open contacts prevent battery voltage from continuing on to the lamps **(Figure 20–12A)**. When the switch is in the PARK position, battery voltage is applied to the parking lights, side markers, taillights, license plate lights, and instrument panel lamps **(Figure 20–12B)**. This circuit is usually protected by a 15- or 20-amp fuse that is separate from the headlight circuit.

When the switch is in the HEADLIGHT position, battery voltage is applied to the headlights. The lamps lit by the PARK position remain on **(Figure 20–12C)**. Normally, a self-resetting circuit breaker is installed between the battery feed and the headlights. The circuit breaker is

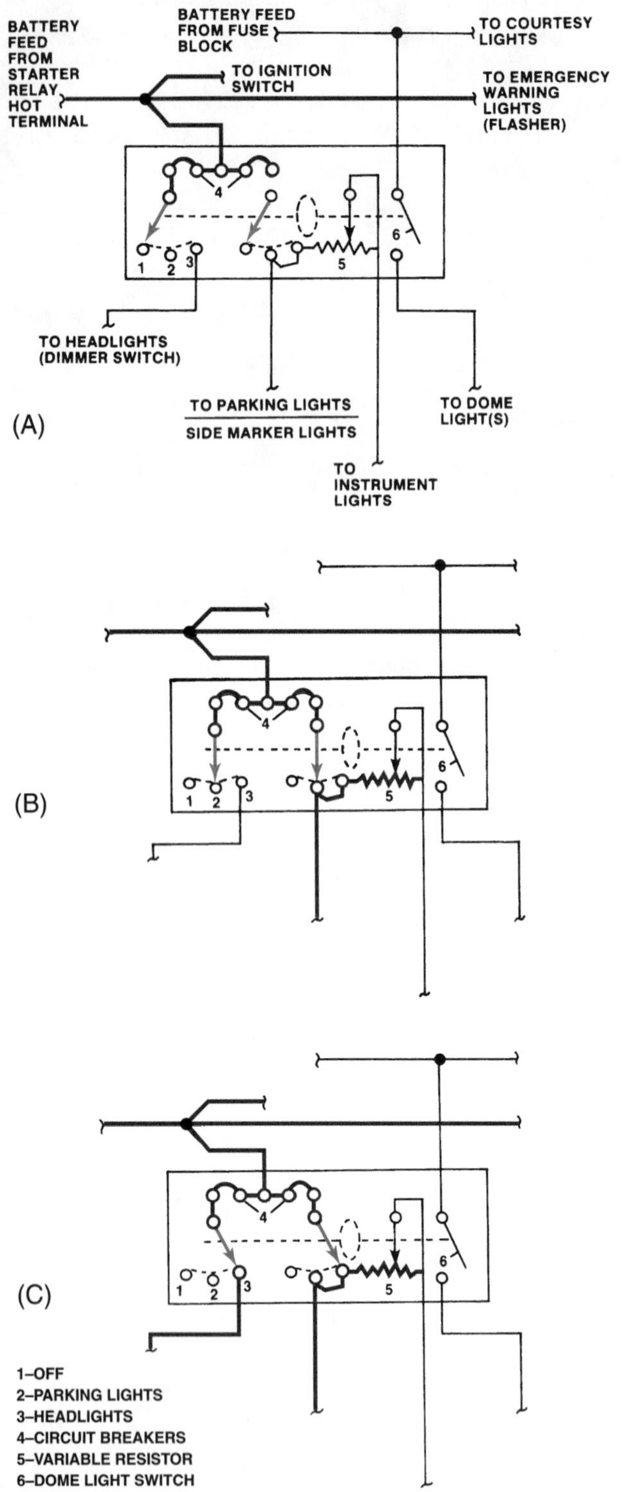

Figure 20–12 A headlight switch (A) in the OFF position; (B) in the PARK position; and (C) in the ON position. *Courtesy of Ford Motor Company*

designed to reset itself. If a problem causes the breaker to open, the lights will go off until the breaker resets. Then the lights will come back on. If there is a serious problem in the circuit, the headlights might flash as the breaker cycles. Some vehicles have a separate fuse for the headlight on each side of the vehicle. This allows one

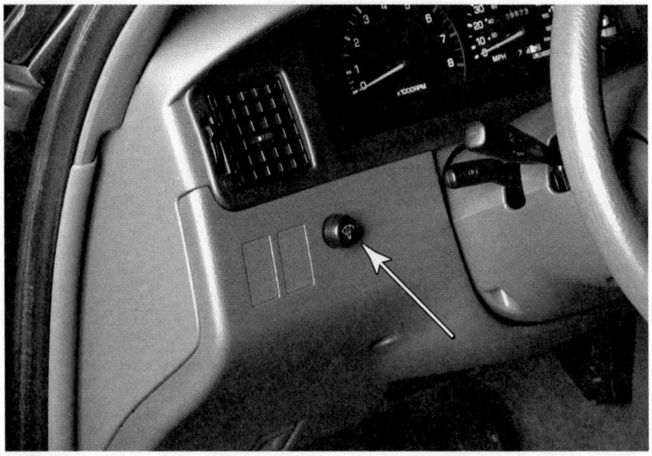

Figure 20–13 An instrument panel light control unit.

headlight to operate if there is a problem in the circuit for one side of the vehicle.

The instrument panel lights come on whenever the headlight switch is in the PARK or HEADLIGHT position. The brightness of these lamps is adjustable. A rheostat is used to allow the driver to control the brightness of the bulbs. This control may be part of the headlight switch, in which case the driver simply rotates the headlight switch knob to adjust the panel lights. Not all headlight switches are designed to control the instrument panel lights. Many vehicles have a separate unit on the dash to control the panel lights **(Figure 20–13)**.

Headlight switches are basically one of three designs. A common switch setup is the rotary switch. Turning the knob of the switch to the PARK or HEADLIGHT position energizes the appropriate lights. The switch's knob also serves as the dimmer control for the instrument panel lights. Some vehicles use a push-button switch. The driver merely pushes a button for the desired set of lights. When this type switch is used, there is a separate instrument panel light control. There is also a separate panel light control when vehicles are equipped with a steering-column-mounted headlight switch **(Figure 20–14)**. To select the desired lighting mode, the driver turns the knob at the end of the switch.

Dimmer Switches

The **dimmer switch** provides a way for the driver to switch between high and low beams. A dimmer switch is connected in series with the headlight circuit and controls the current path to the headlights. The low-beam headlights are wired separately from the high-beam lamps.

Many years ago the dimmer switch was located on the floor. To switch between low and high beams, the driver used his or her foot to press the switch. This type switch worked, but it was also subject to damage because of rust and dirt. Newer vehicles have the dimmer switch on the steering column. This prevents early switch failure and increases driver accessibility.

Figure 20-14 A headlight switch mounted on the steering column.

Headlight Circuits

The complete headlight circuit consists of the headlight switch, dimmer switch, high-beam indicator, and the headlights. When the headlight switch is in the HEADLIGHT position, current flows to the dimmer switch **(Figure 20–15)**. If the dimmer switch is in the LOW position, current flows through the low-beam filament of the headlights. When the dimmer switch is in the HIGH position, current flows to the high beam headlights **(Figure 20–16)**.

The headlight circuits just discussed are designed with switches that control battery voltage and the bulbs have a fixed ground. In this system, battery voltage is present at the headlight switch. The switch must be closed to have voltage present at the headlights. Many manufacturers use a system that relies on a groundside switch to control the headlights. In these systems, voltage is always available at the headlights. A closed headlight switch completes the circuit to ground and the headlights turn on. In this system, the dimmer switch is also a ground control switch.

Daytime Running Lights

Canadian law requires that all new vehicles be equipped with **daytime running lights (DRL)** for added safety. This feature is also standard equipment on all new GM vehicles sold in North America. The system normally uses the vehicle's high-beam lights. The control circuit is connected directly to the vehicle's ignition switch so the lights are turned on whenever the vehicle is running. The circuit is equipped with a module that reduces 12-volt battery voltage to approximately 6 volts. This voltage reduction allows the high beams to burn with less intensity and prolongs the life of the bulbs. When the headlight switch is moved to the HEADLIGHT position, the module is deactivated and the lights work with their normal intensity and brightness. Applying the parking brake

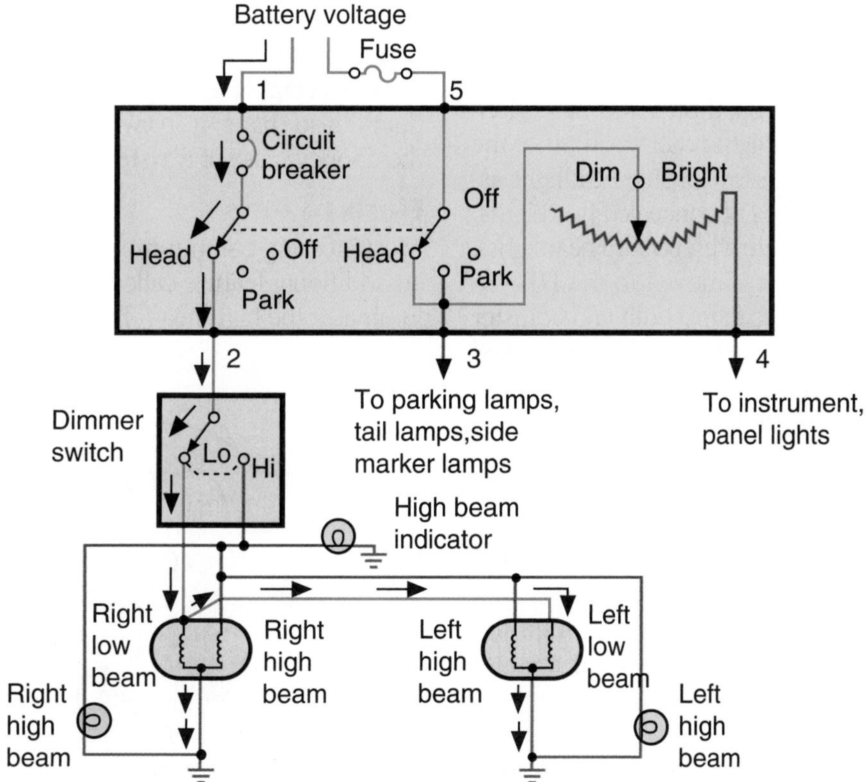

Figure 20-15 A headlight circuit indicating current flow with the dimmer switch in the low-beam position.

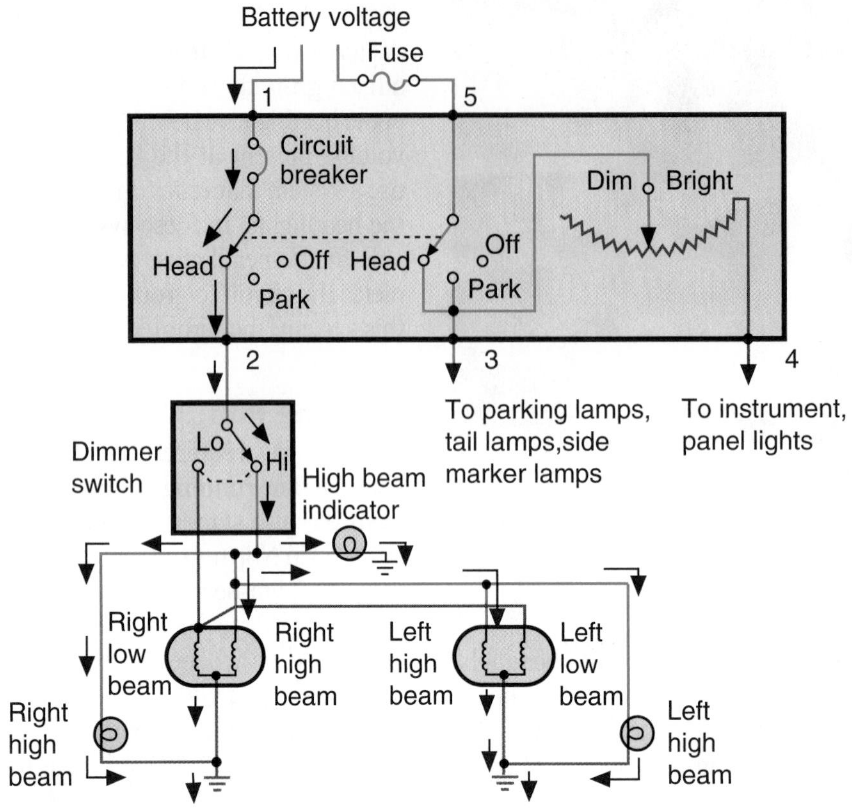

Figure 20-16 A headlight circuit indicating current flow with the dimmer switch in the high-beam position.

also deactivates the DRL system so the lights are not on when the vehicle is parked and the engine is running.

Concealed Headlights

Although not as common as they were a few years ago, concealed headlights are still found on some cars. Manufacturers use a concealed headlight system to improve the vehicle's aerodynamics. Today, low profile headlight assemblies are being used instead of concealed headlights. However there are cars out there with pop-up headlights.

When the headlight switch is moved to the HEADLIGHT position, the entire headlamp bulb and adjuster assembly pivots upward. These headlights are controlled by electric or vacuum motors.

Vacuum systems have a headlight switch and vacuum motors attached to the headlight assembly. With the headlight switch in the OFF position, engine vacuum is supplied to the motors to keep the headlight doors closed. When the headlight switch is moved to the HEADLIGHT position, the vacuum distribution valve vents the vacuum that is held in the vacuum motors, which allows springs at the doors to open the doors. These systems are also equipped with a bypass valve that allows the doors to manually open in case the system fails.

Typically, electrically controlled systems use a torsion bar and a single motor to open both doors or have a separate motor for each headlight door. When the headlight switch is moved to the HEADLIGHT position, current is sent to the motors. This current turns on the motors and causes the doors to open or close. Limit switches stop current flow to the motors when they are completely open or closed **(Figure 20–17)**. Electrically operated headlight doors also have a provision for manually opening the doors in case of a system failure.

Flash to Pass

Most steering-column-mounted dimmer switches have an additional feature called **flash to pass**. This circuit illuminates the high-beam headlights even with the head-

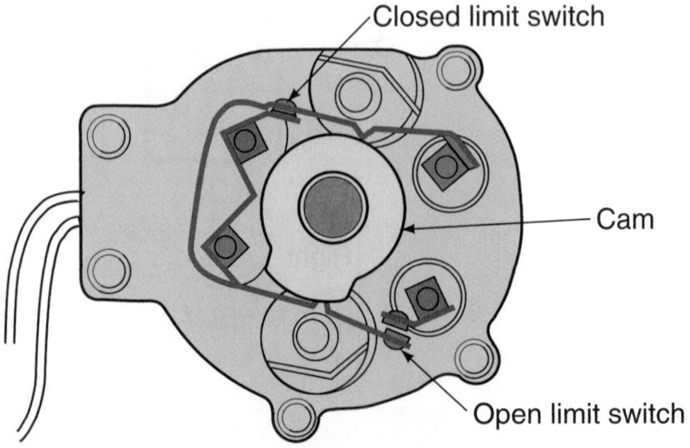

Figure 20-17 Most limit control switches operate off a cam on the motor. *Courtesy of DaimlerChrysler Corporation*

light switch in the OFF or PARK position. When the driver activates the flash to pass feature, the contacts of the dimmer switch complete the circuit to the high-beam filaments.

Automatic Light Systems

These systems provide light-sensitive, automatic on-off control of the light normally controlled by the regular headlight switch. It consists of a light-sensitive **photocell sensor/amplifier** assembly and a headlight control relay. Turning the regular headlight switch on overrides the automatic system. In other words, automatic operation is not possible until the regular headlight switch is turned off.

In normal operation, the photocell sensor/amplifier, which is usually mounted under a group of perforated holes in the upper instrument panel pad or slotted holes in the defroster grille panel, is exposed to ambient light. As the light level decreases, the light sensor's resistance increases. When resistance increases to a preset amount, the amplifier applies power to the headlight relay coil. The headlights, exterior lights, and instrument illumination lights turn on. The lights remain on until the system is turned off or the ambient light level increases.

Some systems have two sensors to monitor the ambient light. The light sensors monitor the intensity of the ambient light at an extended angle above the vehicle and in a narrow angle to the front of the vehicle.

An automatic headlight dimmer system is also available on some vehicles. These systems automatically switch from high beams to low beams when the intensity of light at its photocell increases. The source of the light could be the environment, the headlights of an approaching vehicle, or the taillights of a vehicle. Typically the driver is able to set the sensitivity of the photocell to meet the current driving conditions.

Most automatic light systems have a headlamp delay system as well. This system allows the headlamps to stay on for a period of time after the vehicle is stopped and the ignition switch is turned off. A variable switch **(Figure 20–18)** allows the driver to set the amount of time the headlights should remain on after the ignition is turned off. The system can typically be adjusted to keep the headlights on for up to three minutes after the ignition is turned off. Of course, the driver can turn off the delay system and the headlamps will shut off as soon as the ignition is turned off.

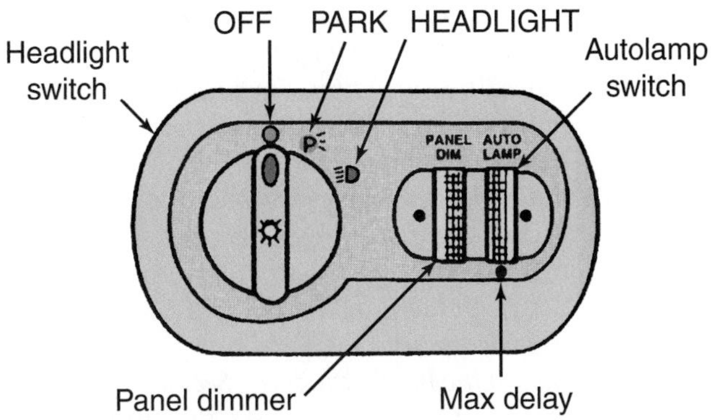

Figure 20-18 A headlight switch, an instrument panel light control, and an auto lamp control. *Courtesy of Ford Motor Company*

CUSTOMER CARE

If the customer's car is equipped with an automatic light control system, point out the location of the perforated holes or slots. Warn the customer not to place any items that may block light from the sensor/amplifier assembly. Blockage causes erratic operation of the system. The photocell must always be exposed to an outside light to function properly. ■

Adaptive Headlights

Adaptive headlight systems swivel the base of the headlamps to illuminate any curves in the road **(Figure 20–19)**. The system responds to signals from a steering wheel angle sensor and swivels the headlamps with small bidirectional motors **(Figure 20–20)**. Adaptive head-

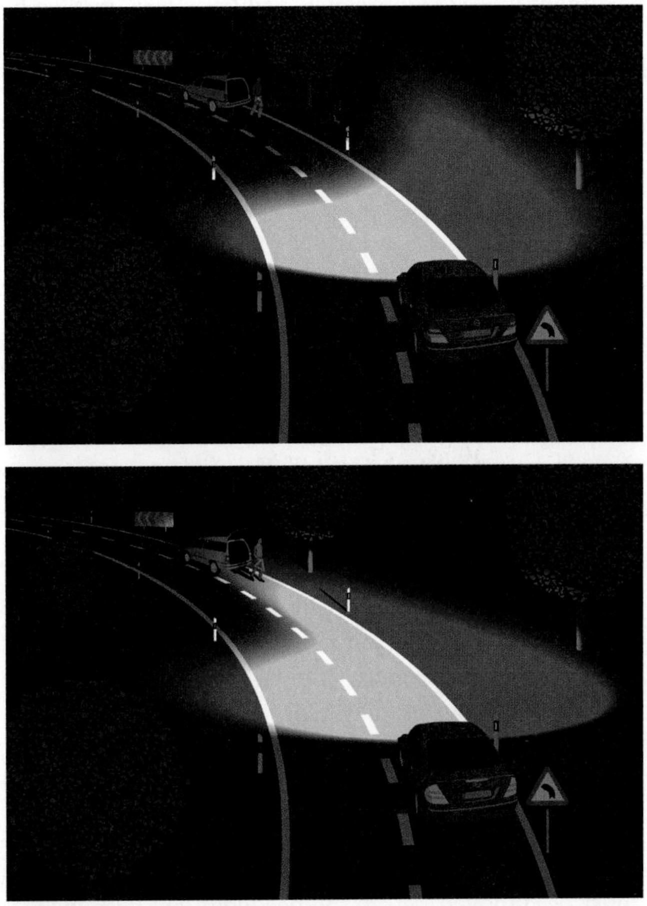

Figure 20-19 A comparison of how the road is lit up with a conventional (top) headlamp system and an adaptive (bottom) system. *Courtesy of DaimlerChrysler Corporation*

Figure 20-20 A headlamp assembly for an adaptive headlight system. *Courtesy of DaimlerChrysler Corporation*

lights are able to rotate up to 15 degrees to the right or to the left. The headlight on the side of the vehicle that is opposite of the direction of the turn swivels about half the distance as the headlamp leading into the curve. The system responds in real time by responding to the car's current steering angle, its yaw rate, and road speed.

Adaptive headlights can be additionally controlled by global positioning system (GPS) satellite navigation and digital road maps. Plotting the road ahead supplies the information needed by the control unit to anticipate road curves and to enable the adaptive headlamps to illuminate curves with optimum brightness and light intensity even before the driver starts to turn the steering wheel.

HEADLIGHT SERVICE

When there is a headlight failure, it is typically caused by a burned-out bulb or lamp, especially if only one lamp fails. However, it possible that the circuit for that one lamp has an open or high resistance. Check for voltage at the bulb before replacing a bulb. If there is no voltage present, the circuit needs work and the original bulb may still be good. If more than one lamp (including the rear lights) is not working, carefully check the circuit. A problem there is much more likely than having a number of burned-out bulbs. Of course, if the charging system is not being regulated properly, the high voltage will cause lamps to burn out prematurely.

Headlight Replacement

There can be slight variations in procedure from one model to another when replacing headlights. For instance, on some models the turn signal light assembly must be removed before the headlight can be replaced. Overall, however, the procedure does not differ much from the following typical instructions.

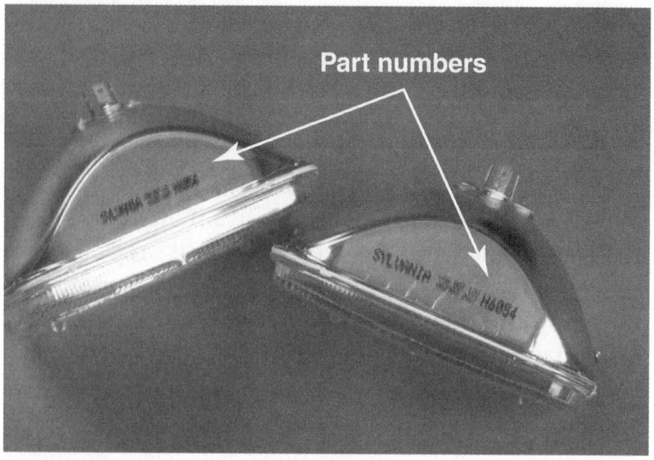

Figure 20-21 Comparing the old and replacement headlamps.

Make sure the replacement bulb is the same type and part number as the one being replaced **(Figure 20–21)**.

SHOP TALK

Because of the extremely high voltages involved, any work on xenon lighting should be done carefully and according to the manufacturer's recommendations. ■

PROCEDURE

Replacing Headlights

STEP 1 *Remove the headlight bezel-retaining screws. Remove the bezel. If necessary, disconnect the turn signal lamp wires.*

STEP 2 *Remove the retaining ring screws from one or both lights.*

STEP 3 *Remove the retaining rings.*

STEP 4 *Remove the light from the housing. Disconnect the wiring connector from the back of the light.*

STEP 5 *Push the wiring connector onto the prongs at the rear of the new light.*

STEP 6 *Place the new light in the headlight housing. Position it so the embossed number in the light lens is on the top.*

STEP 7 *Place the retaining ring over the light and install the retaining ring screws. Tighten them slightly.*

STEP 8 *Check the aim of the headlight and adjust it, if necessary.*

STEP 9 *Install the headlight bezel. Secure it with the retaining screws. Connect the turn signal lamp wiring (if it was disconnected).*

> **SHOP TALK**
>
> Some manufacturers recommend coating the prongs and base of a new sealed beam with dielectric grease for corrosion protection. Use an electrical lubricant approved by the manufacturer. ■

Headlight Adjustments

Headlights must be kept in adjustment to obtain maximum illumination. Properly adjusted sealed beams cover the correct range and afford the driver the proper nighttime view. Headlights that are out of adjustment can cause other drivers discomfort and sometimes create hazardous conditions.

Before adjusting or aiming a vehicle's headlights, however, make the following inspections to ensure that the vehicle is level. Any one of the adverse conditions listed here can result in an incorrect setting.

- If the vehicle is heavily coated with snow, ice, or mud, clean the underside with a high-pressure stream of water. The additional weight can alter the riding height.
- Ensure that the gas tank is half full. Half a tank of gas is the only load that should be present on the vehicle.
- Check the condition of the springs or shock absorbers. Worn or broken suspension components affect the setting.
- Inflate all tires to the recommended air pressure levels. Take into consideration cold or hot tire conditions.
- Make sure the wheel alignment and rear axle tracking path are correct before adjusting the headlights.
- After placing the vehicle in position for the headlight test, bounce the vehicle to settle the suspension.

To properly adjust the headlights, headlight aim must be checked first. Various types of mechanical headlight aiming equipment are available commercially **(Figure 20–22)**. These aimers use mirrors with split images, like split-image finders on some cameras, and spirit levels to determine exact adjustment. When using any mechanical aiming equipment, follow the instructions provided by the equipment manufacturer. Where headlight aiming equipment is not available, headlight alignment can be checked by projecting the upper beam of each light on a screen or chart at a distance of about 25 feet ahead of the headlight **(Figure 20–23)**. The vehicle must be exactly perpendicular to the chart.

The chart should be marked in the following manner. First, measure the distance between the centers of the matching headlights. Use this measurement to draw two vertical lines on the screen with each line corresponding to the center of a headlight. Then, draw a vertical centerline halfway between the two vertical lines. Next, measure the distance from the floor to the centers of the headlights. Subtract 2 inches from this height and then draw a horizontal line on the screen at this new height.

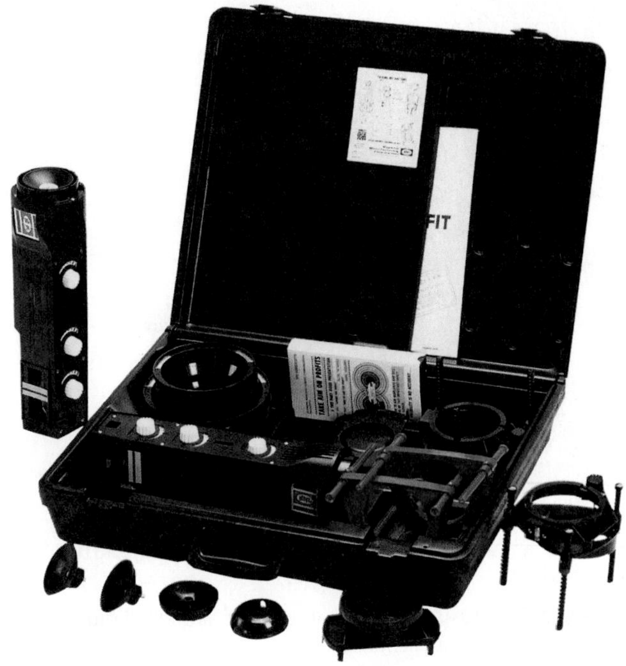

Figure 20–22 A typical headlight aiming kit. *Courtesy of Snap-on Tools Company*

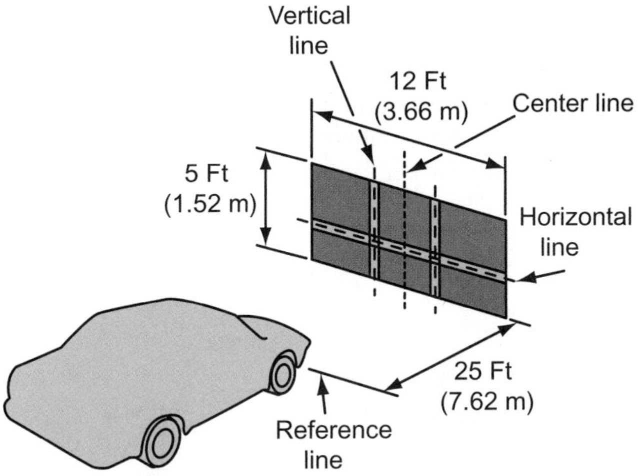

Figure 20–23 Acceptable beam patterns on a wall screen.

With headlights on high beam, the hot spot of each projected beam pattern should be centered on the point of intersection of the vertical and horizontal lines on the chart. If necessary, adjust headlights vertically and laterally to obtain proper aim.

Headlight adjusting screws are provided to move the headlight within its shell assembly to obtain correct headlight aim. Lateral or side-to-side adjustment is accomplished by turning the adjusting screw at the side of the headlight **(Figure 20–24)**. Vertical or up-and-down

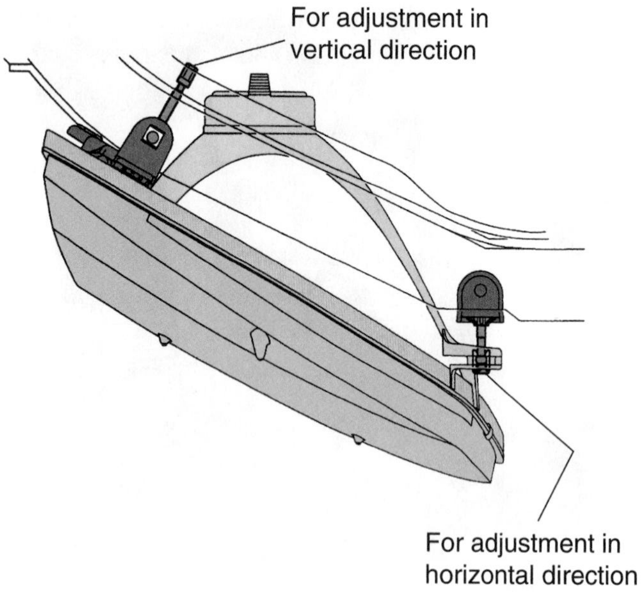

Figure 20-24 An example of headlight adjustment screws. *Reprinted with permission*

adjustment is accomplished by turning the screw at the top of the headlight. Adjustments can be made without removing headlight bezels.

Some vehicles are equipped with indicators to help in the adjustment process. One such setup is used on Hondas. This system uses a horizontal indicator gear **(Figure 20–25)** at each headlamp assembly. Prior to making any adjustments to the headlights, Honda recommends that the horizontal indicator gear be at zero (0). A Phillips screwdriver is used to bring the gear back to zero. After this has been done, the headlamps can be fine adjusted to specifications.

Chrysler's headlamps are equipped with a bubble level to aid vertical headlamp alignment. A horizontal gauge and magnifying window is located next to the bubble level to aid in horizontal alignment **(Figure 20–26)**. The vertical bubble level is used to compensate for vehicle ride height changes due to heavy vehicle loads. The bubble level is calibrated to the earth's surface, therefore the vehicle must be on level ground when the headlights are aimed. If the headlight beam projection appears high to the oncoming traffic, check the headlight alignment using the alignment screen method. If the beam pattern is above or to the left of the specified location on the screen, adjust the headlights and then recalibrate the bubble level and magnifying window. Ideally, if the headlights are aligned, the bubble level and magnifying window will be centered. Never change the calibration of the magnifying window or bubble level if the headlights are out of alignment.

A properly aimed headlight normally does not need to be re-aimed after installation of a new bulb.

Autoleveling Headlamps Some vehicles, primarily those equipped with xenon headlamps, have an automatic headlamp leveling system that keeps headlight range constant regardless of the current conditions. This feature keeps the headlights in position for optimal lighting of the road ahead and prevents glare that could blind oncoming traffic.

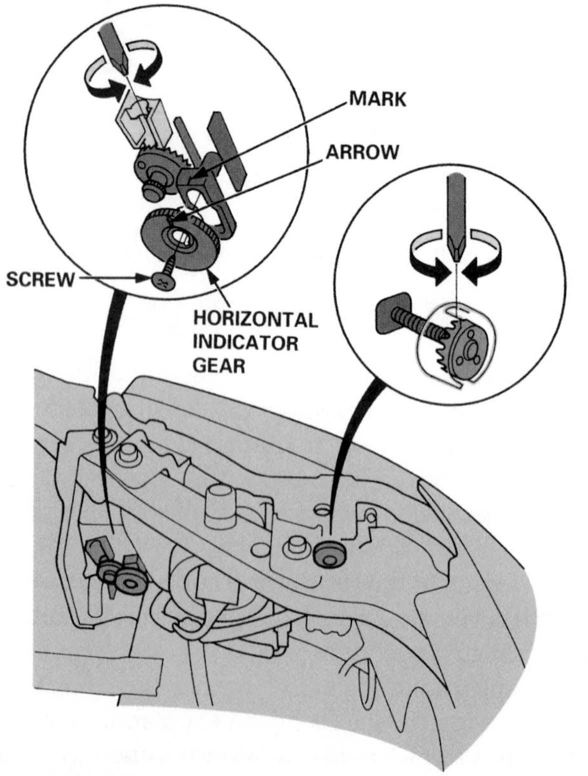

Figure 20-25 Honda's adjustment indicator. *Courtesy of American Honda Motor Co., Inc.*

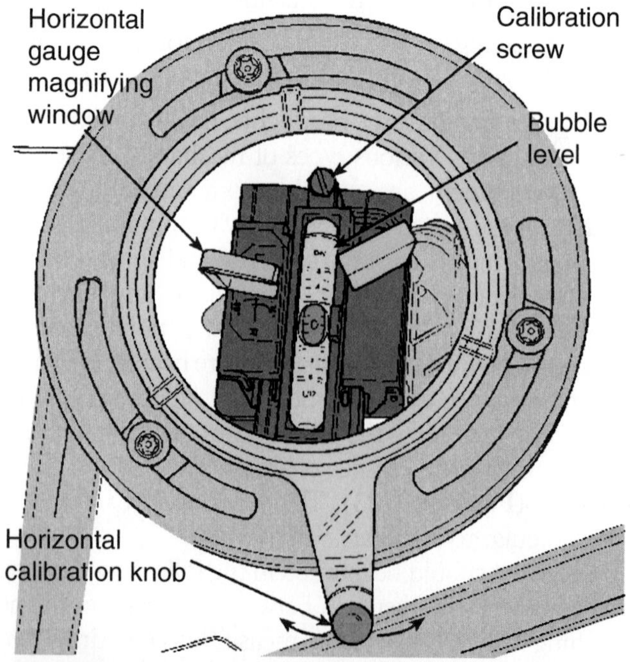

Figure 20-26 Calibrating the bubble level and horizontal gauge. *Courtesy of DaimlerChrysler Corporation*

The aim of headlights changes each time the vehicle is accelerated or the brakes are applied, as well as when weight is added or subtracted from the vehicle. The automatic leveling system continually monitors and re-aims the headlights to compensate for any change due to passenger or cargo weight—even the constantly changing quantity of fuel. The system also adjusts the lamps in response to a change of angle caused by acceleration or braking.

Suspension-mounted sensors monitor variations in the spring compression rates at the front and rear of the vehicle. An electronic control unit processes these two rates to calculate the instantaneous longitudinal pitch of the vehicle. The system also monitors signals from the vehicle's speed sensor, which allows the control unit to recognize braking and acceleration.

Based on these inputs, the control unit determines the optimal headlight angle for the current condition. It then orders electric motors to adjust the inclination of the headlights.

INTERIOR LIGHT ASSEMBLIES

The types and numbers of interior light assemblies used vary significantly from one vehicle to another **(Figure 20–27)**. Following are the more common ones.

Engine Compartment Light Operating the hood causes the engine compartment light mercury switch to close and light the underhood area.

Some pickup trucks and SUVs are equipped with an underhood retractable magnetic base lamp mounted on a reel. The lamp can be used anywhere around the vehicle.

Glove Box Light Opening the glove box door closes the glove box light switch contacts and the light comes on.

Figure 20-27 Full interior illumination is available with this light setup. *Courtesy of DaimlerChrysler Corporation*

Luggage Compartment Light The light is mounted in the underside of the trunk deck lid in the luggage compartment.

Trunk Lid Light Lifting the trunk lid causes the light mercury switch to close and the light comes on.

Vanity Light Pivoting the sun visor downward and opening the vanity mirror cover causes the vanity light switch contacts to close and the light to come on.

Courtesy Lights There are several types of courtesy lights. Some vehicles have courtesy lights that are in the door trim panels, under each side of the instrument panel, and in the center of the headlining. These are illuminated when one of the doors is opened, by rotating the headlight switch to the full counterclockwise position, or by depressing the designated switch. **Figure 20–28** is a wiring diagram of a typical courtesy light circuit. The courtesy lights are also turned on by the illuminated entry or keyless entry systems, if the vehicle is equipped with one or both of these.

Front compartment foot well courtesy lights are mounted on the lower closeout panels at both ends of the instrument panel. The bulbs are accessible from under the instrument panel for replacement without removing other parts.

Some courtesy lights are a combination of map lights located on each side of the dome light housing. The map lights are operated independently of the dome light by two switches located at each map light housing. The dome light is actuated by turning the headlight switch control knob fully counterclockwise.

Power is supplied from the fuse block to the courtesy or dome/map light. The ground for the light is controlled by the position of the door switch. That is, these door switches are held in open position and do not provide for a ground circuit. When the door is opened, a spring pushes the switch closed to ground the circuit, and the dome/map or courtesy lights come on.

Illuminated Entry System This system assists vehicle entry during the hours of darkness by illuminating the door lock cylinder so it may be easily located for key insertion. The vehicle interior is also illuminated by the courtesy lights.

The system consists of four main components: electronic module, illuminated door lock cylinder, door handle switch, and wiring harness.

Activation of the system is accomplished by raising the outside door handle or by pressing a code button on the keyless entry system. This action momentarily closes a switch mounted on the door handle mechanism, which completes the ground circuit of the electronic actuator

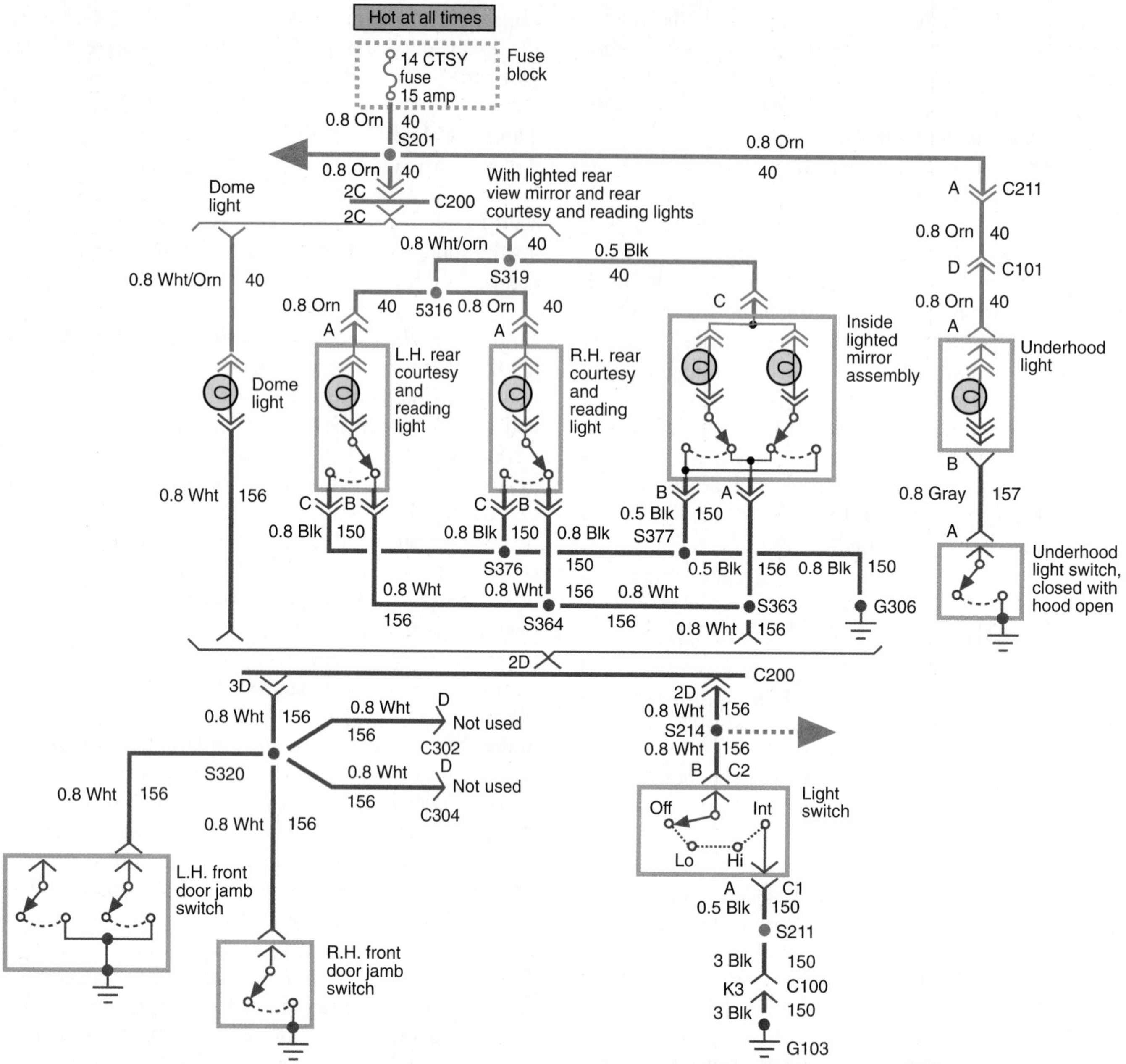

Figure 20-28 A typical courtesy light circuit.

module and switches the system on. The vehicle interior lights turn on, and both front door lock cylinders are illuminated by a ring of light around the area where the key enters. This illumination remains on for approximately 25 seconds, then automatically turns off. During this 25-second period, the system can be manually deactivated by turning the ignition switch to the run position.

The system is activated every time the outside front door handles are operated, whether the vehicle is locked or not. Opening the doors from the inside of the vehicle does not activate the system. If the outside door handle is held up indefinitely so the handle switch is continuously closed, the system operates as normal and turns off after 25 seconds. At the completion of this cycle, if the door handle is still in the raised position, the system remains off. It is impossible to activate the system from the other front door handle until the raised handle is returned to its normal position. This function is built into the logic circuitry of the system to prevent battery discharge should the outside door handle be intentionally propped up or become jammed in any way.

Interior lights all basically operate in the same way. Whether the courtesy lights are on the door, under the seats, under the instrument panel, or on the rear interior quarter panels does not change how they are controlled. Also, whether the illumination lights are just behind the instrument panel or are also used in center consoles or door arm rests does not affect their operation. The only difference is the number of lights and variances in electrical wiring.

Interior and courtesy lights rarely give any trouble. However, if they do not operate, check the fuse, bulb, switch, and wiring.

REAR EXTERIOR LIGHT ASSEMBLIES

The rear light assembly includes the taillights, turn signal/stop/hazard lights/high-mounted stop lights, rear side marker lights, backup lights, and license plate lights **(Figure 20–29)**. Taillights operate when the parking lights or headlights are turned on.

Turn, Stop, and Hazard Warning Light Systems

Power for the turn (directional signal), stop, and hazard warning light systems is provided by the fuse panel **(Figure 20–30)**. Each system has a switch that must close to turn on the lights in the circuit. Hazard lights are commonly referred to as 4-way flashers because the lights at all four corners of the vehicle will flash when the circuit is turned on.

The turn signal and hazard light switches on many current vehicles are part of a **multifunction switch**. When the turn or directional signal switch is activated, only one set of the switch contacts is closed—left or right. However, when the hazard switch is activated, all contacts are closed and all turn signal lights and indicators flash together and at the same time.

The power for the turn signals is provided through the fuse panel, but only when the ignition switch is on. The hazard lights are also powered through the fuse panel; however, they have power at all times regardless of ignition switch position.

Some cars are equipped with cornering side lights, which are generally fed from the multifunction main switch. When the turn signal switch is activated, the cornering light on the appropriate side burns with a steady glow.

Figure 20-29 The rear lights on a late-model car.
Courtesy of Nissan North America, Inc.

Side markers are connected in parallel with the feed circuit (from the headlight switch) that feeds the minor filaments of the front parking lights and rear taillights.

What a multifunction switch controls depends on the make, model, and year of the vehicle. Some control the directional signals and serve as the dimmer switch. Others control the turn and hazard signals and serve as the headlight, dimmer, windshield wiper, and windshield washer switch **(Figure 20–31)**.

This switch is not repairable and must be replaced if defective. Photo Sequence 17 outlines the typical procedure for removing a multifunction switch. Some of the steps shown in this procedure may not apply to all types of vehicles; always refer to the service manual before removing this switch. Also carefully study the procedures beforehand and identify any special warnings that should be adhered to, especially those concerning the air bag.

Flashers **Flashers** are components of both turn and hazard systems. They contain a temperature-sensitive bimetallic strip and a heating element **(Figure 20–32)**. The bimetallic strip is connected to one side of a set of contacts. Voltage from the fuse panel is connected to the other side. When the left turn signal switch is activated, current flows through the flasher unit to the turn signal bulbs. This current causes the heating element to emit heat, which in turn causes the bimetallic strip to bend and open the circuit. The absence of current flow allows the strip to cool and again close the circuit. This intermittent on/off interruption of current flow makes all left turn signal lights flash. Operation of the right turn is the same as the operation of the left turn signals.

Turn signal flashers are installed on the fuse panel on current car models and most current truck models **(Figure 20–33)**. However, on earlier models this is not true. Hazard flashers are also mounted in various locations. Refer to the service manual for locations on the model being serviced.

A testlight can be used to determine which flasher is used for the turn signals and which is used for the hazard warning light. An easier way is to turn on both the directionals and the hazards. This activates both flasher units. By removing one of the flashers, the affected circuit no longer flashes. Therefore, that flash unit controls that particular circuit. If the turn signals fail to operate and the fuse is good, the flasher has probably failed.

Occasionally, the flasher does not flash as fast as it once did, or it flashes faster. This is also cause for replacement. If it flashes too slowly or not at all, check for a burned-out bulb first.

A flasher features two or three prongs that plug into a socket. Just pull the flasher out of the socket and replace it with a new one.

Flashers are designed to operate a specific number of bulbs to give a specific **candlepower** (brightness). If the

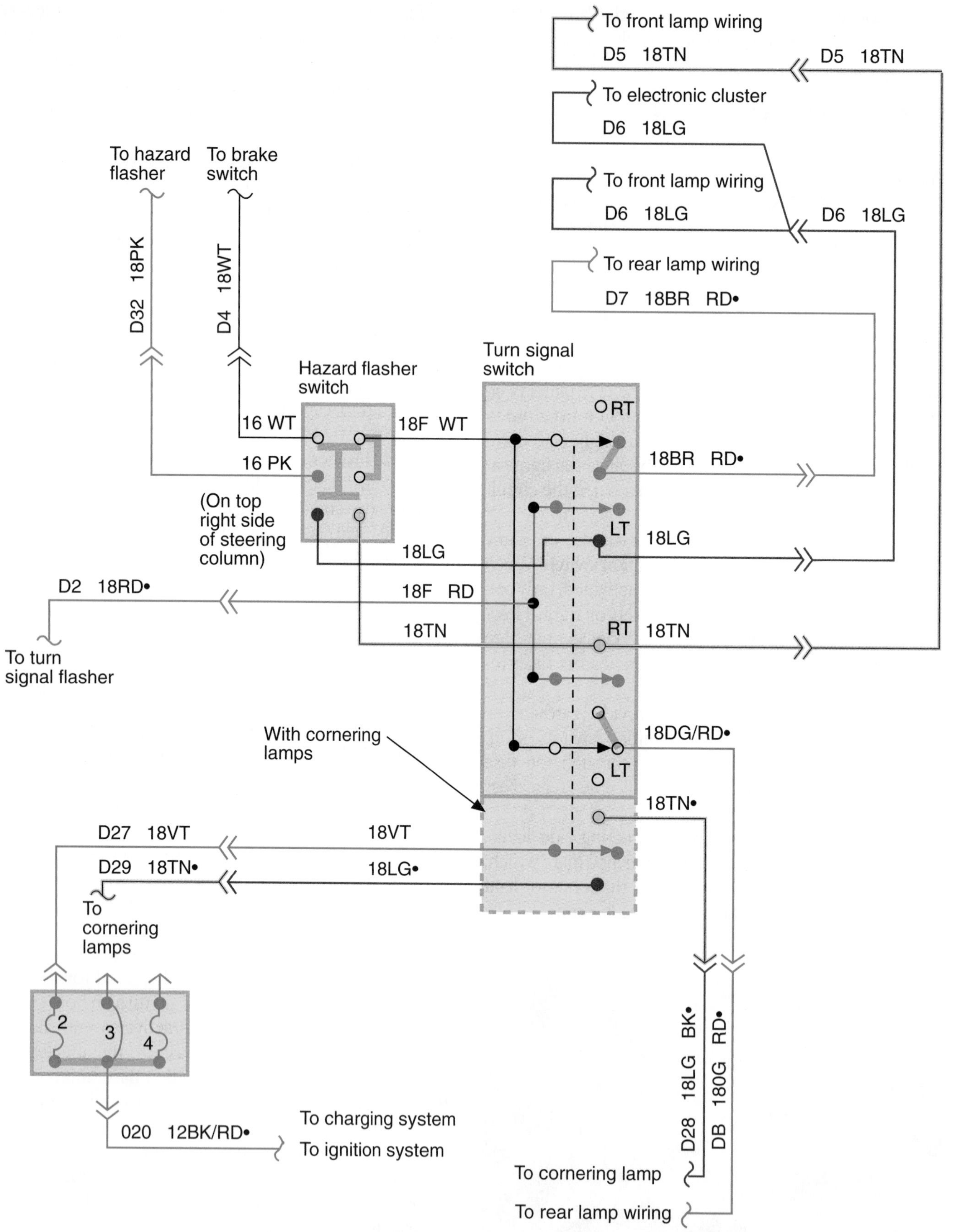

Figure 20-30 The turn signal circuit for a two-bulb system. *Courtesy of DaimlerChrysler Corporation*

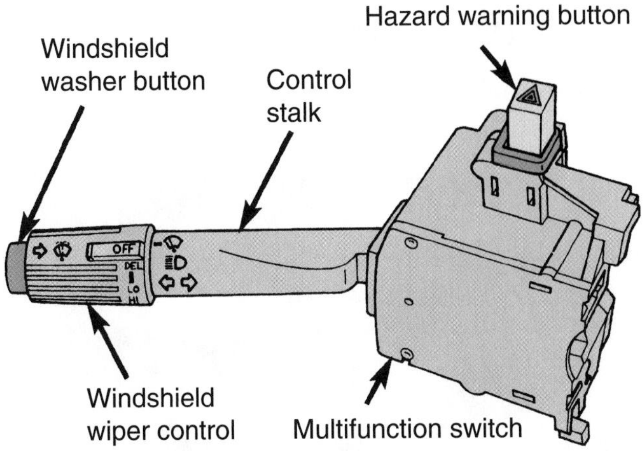

Figure 20-31 A typical multifunction switch. *Courtesy of DaimlerChrysler Corporation*

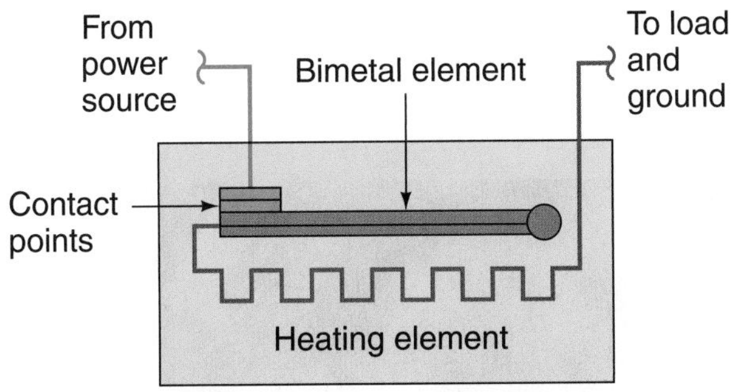

Figure 20-32 A typical turn signal flasher. *Courtesy of Ford Motor Company*

candlepower on the turn signal bulbs is changed, or if additional bulbs are used (if a vehicle is hooked up to a trailer, for instance), a heavy-duty flasher must be used. This usually fits the socket without modifications. Although heavy-duty flashers will operate additional bulbs, they have one big disadvantage and should not be used unless it is necessary. These flashers will not cause the turn signals to flash slower if a bulb burns out. When a turn signal bulb fails, the driver has no idea that it did.

WARNING!

The flasher unit for turn signals should not be switched with a flasher unit for the hazard lights.

Some newer vehicles have a combination flasher unit that controls the flash rate of both the turn signals and the hazard lights. These combination flashers are electronic units **(Figure 20–34)**. The actual turning off and

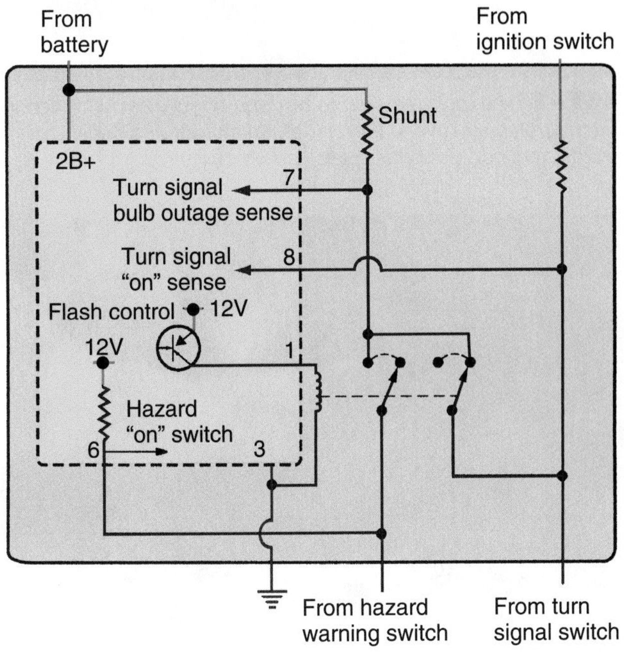

Figure 20-34 A typical electronic combination flasher unit. *Courtesy of DaimlerChrysler Corporation*

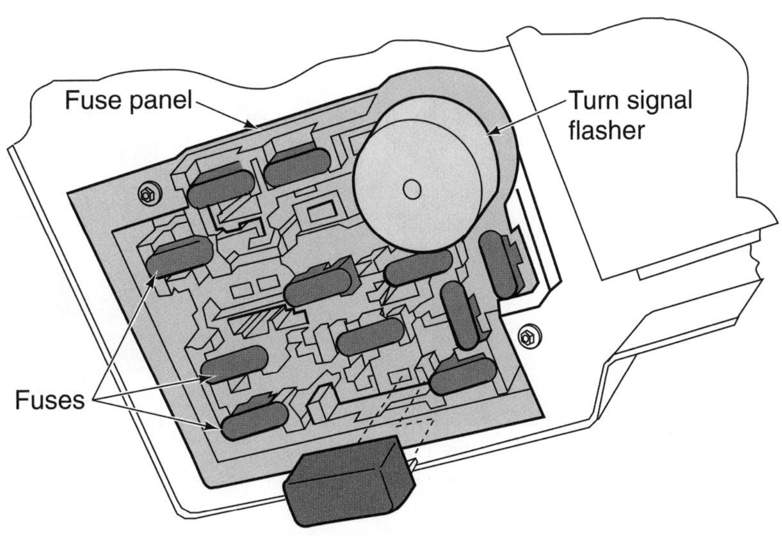

Figure 20-33 A common location for the flashers in the fuse panel. *Courtesy of Ford Motor Company*

PHOTO SEQUENCE 17

Removing a Multifunction Switch

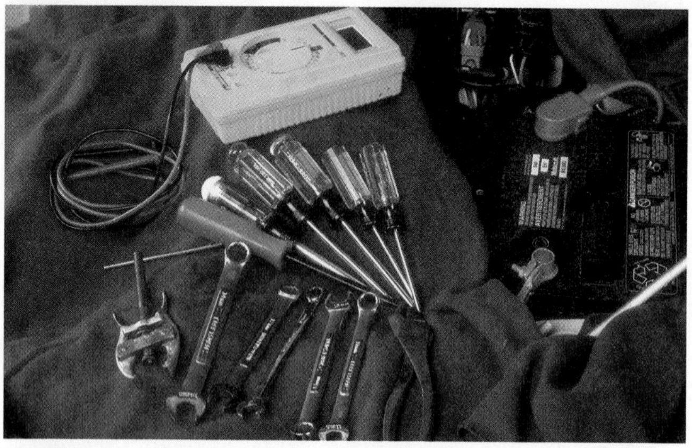

P17-1 *The tools required to test and remove a multifunction switch are fender covers, battery terminal pliers and pullers, assorted wrenches, Torx driver set, and an ohmmeter.*

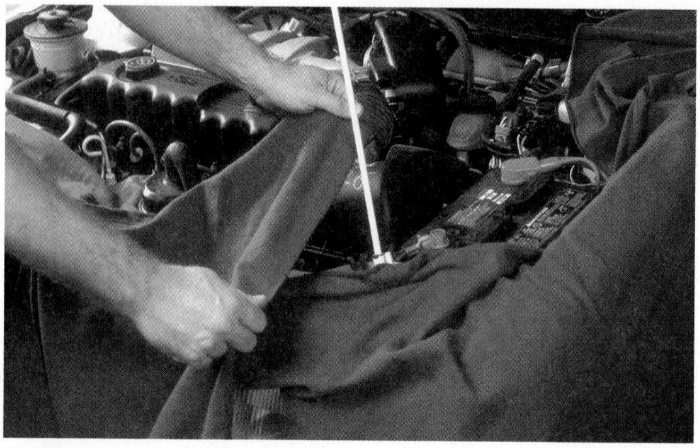

P17-2 *Place the fender covers over the fenders of the vehicle.*

P17-3 *Loosen the negative battery clamp bolt and remove the battery clamp. Place the cable where it cannot contact the battery.*

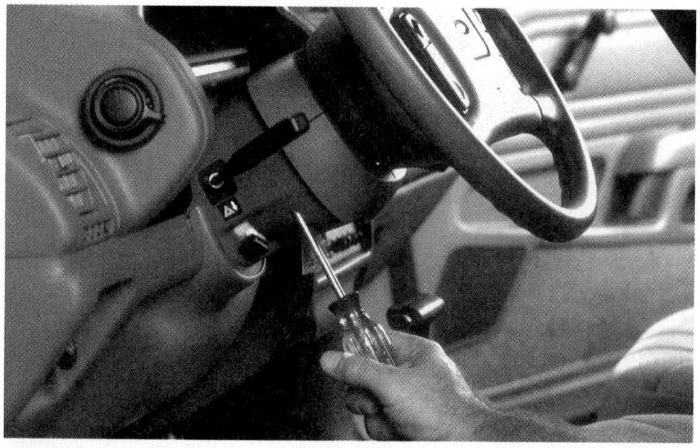

P17-4 *Remove the shroud retaining screws and remove the lower shroud from the steering column.*

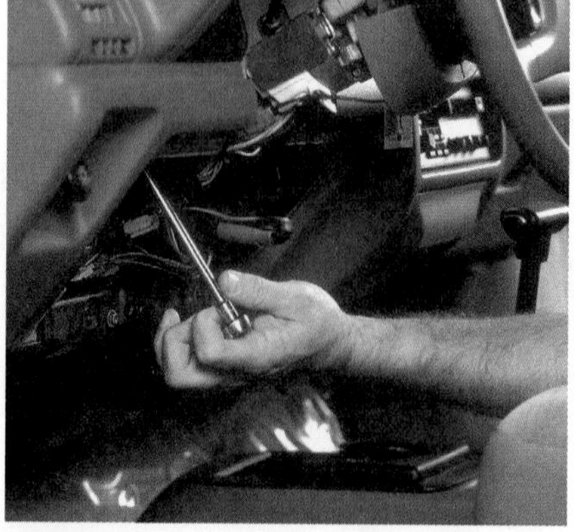

P17-5 *Loosen the steering column attaching nuts. Do not remove the nuts.*

P17-6 *Lower the steering column just enough to remove the upper shroud.*

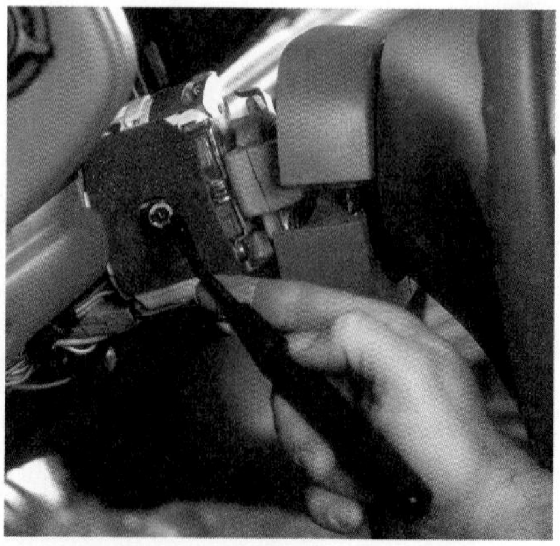

P17-7 *Remove the turn signal lever by simply rotating the outer end of the lever. Then pull it straight out.*

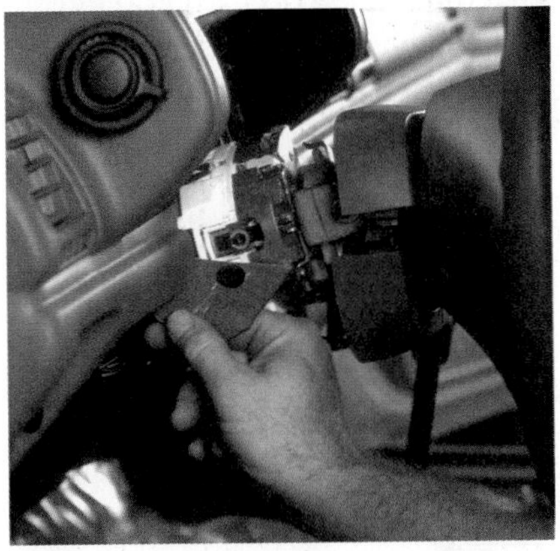

P17-8 *Peel back the foam shield from the turn signal switch.*

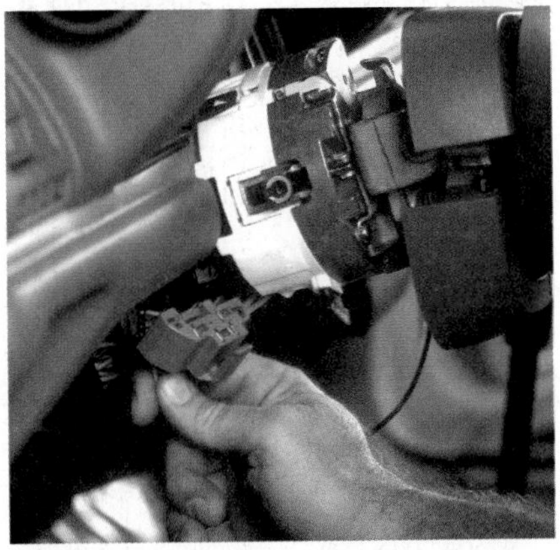

P17-9 *Disconnect the turn signal switch electrical connectors.*

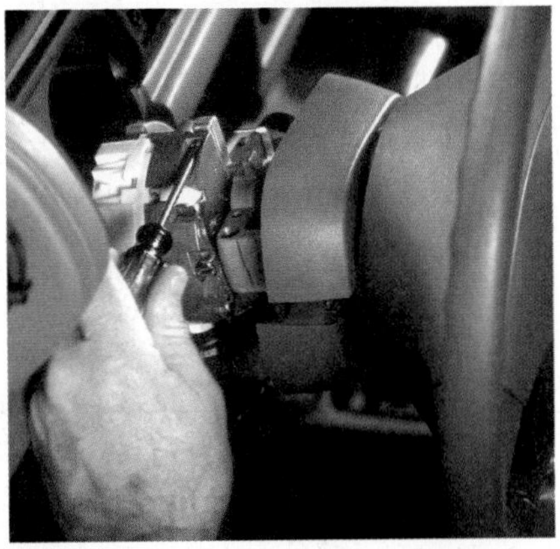

P17-10 *Remove the screws that attach the switch to the lock cylinder assembly.*

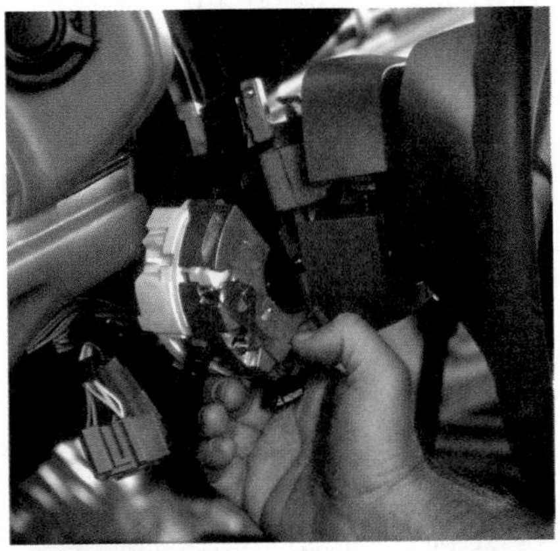

P17-11 *Disengage the switch from the lock assembly.*

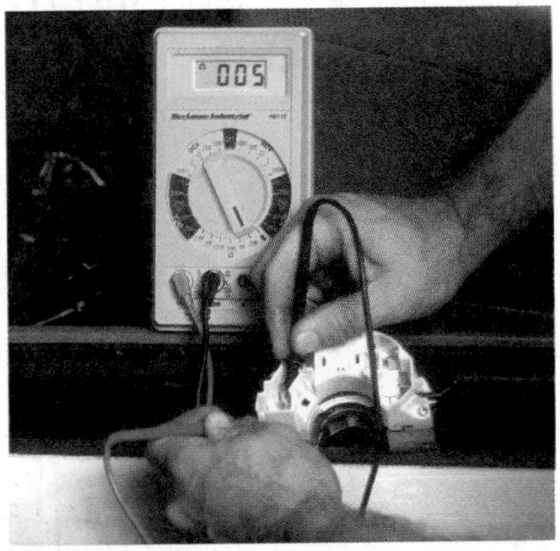

P17-12 *Use an ohmmeter to test the switch. Check for continuity when the dimmer switch is in the low-beam position.*

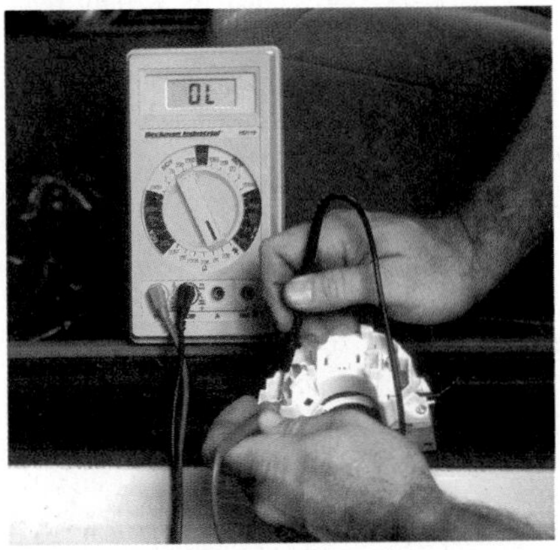

P17-13 *When the switch is in the low-beam position, the circuit should be open between the high-beam terminals.*

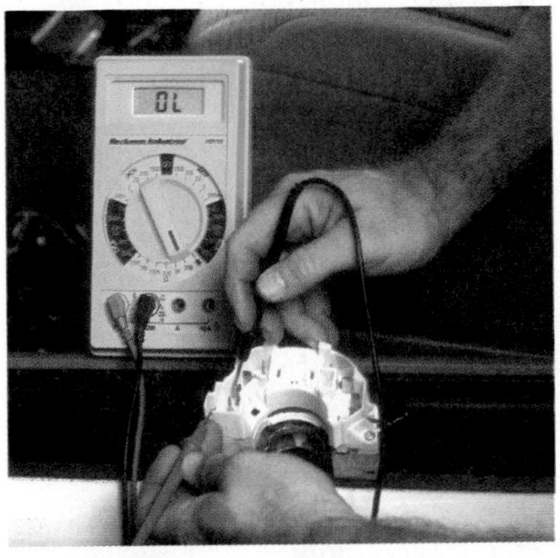

P17-14 *Also check the other terminals and circuits that should be open when the dimmer switch is in the low-beam position.*

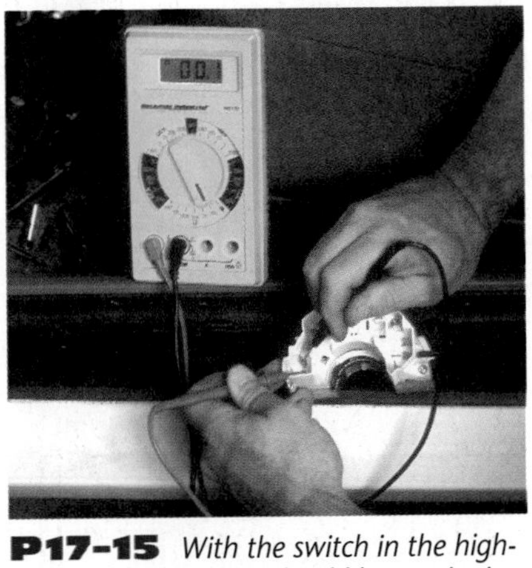

P17-15 *With the switch in the high-beam position, there should be continuity across the high-beam circuit. Also check for continuity across the other circuits that should be open when the switch is in the high-beam position.*

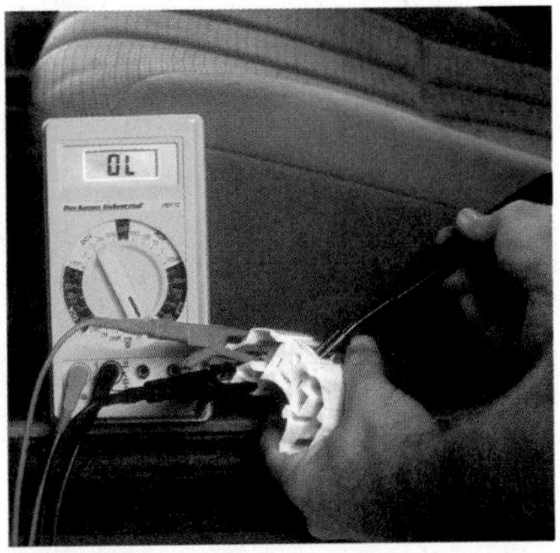

P17-16 *When the dimmer switch is placed in the flash-to-pass position, there should be continuity across those designated terminals and an open across the others.*

on of the lights is caused by the cycling of a transistor. This type flasher also senses when a bulb is burned out and causes the remaining bulbs on that side to flash faster. Because this flasher is an electronic device it cannot be tested with normal test equipment. The only test of the flasher is to substitute it with a known good one. If the lights flash normally, the original flasher unit was bad and needs to be replaced.

Brake Lights

The brake (stop) lights are usually controlled by a stop light switch that is normally mounted on the brake pedal arm **(Figure 20–35)**. Some cars are equipped with a brake or stop light switch mounted on the master cylinder, which closes when hydraulic pressure increases as the brake pedal is depressed. In either case, voltage is present at the stop light switch at all times. Depressing the brake

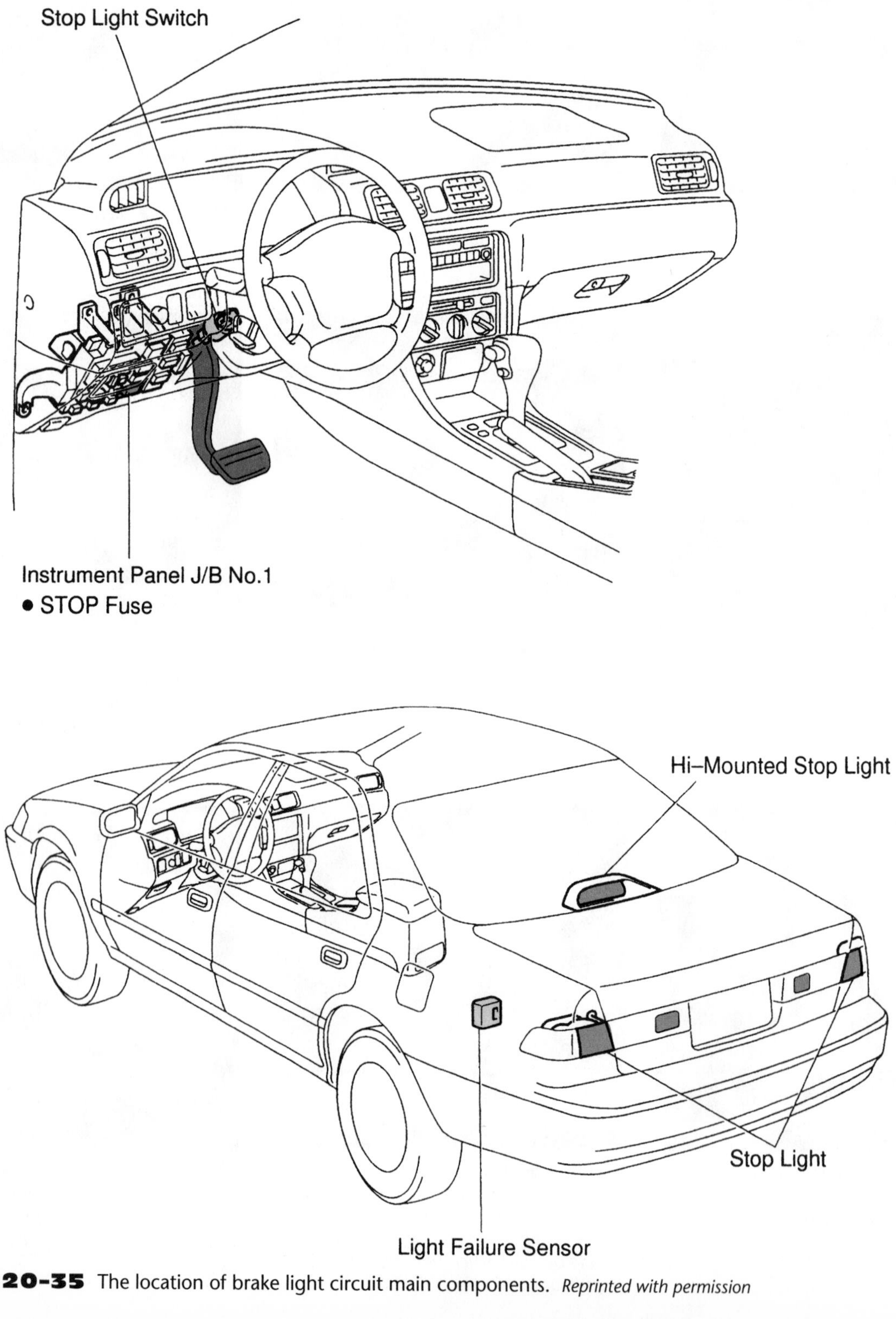

Figure 20-35 The location of brake light circuit main components. *Reprinted with permission*

pedal causes the stop light switch contacts to close. Current can then flow to the stop light filament of the rear light assembly. These stay illuminated until the brake pedal is released.

In addition to the stop lights at the rear of the vehicle, all late-model vehicles have a center high-mounted stop light that provides an additional clear warning signal that the vehicle is braking. Federal studies have shown the additional stop lights to be effective in reducing the number and severity of rear collisions. The high-mounted stop light is activated when current is applied to it from the stop light switch. It stays illuminated until the brake pedal is released. When its contacts are closed, the stop light switch can also provide current to the speed control amplifier, antilock brake control module, and the electric brake controller connector.

LED Lights

Some vehicles use neon lamps and/or LEDs for tail, brake, and turn signal lights. Neon lights are more energy efficient and turn on more quickly than regular lights **(Figure 20–36)**. Because neon bulbs have no filament, the neon bulb will last longer than a conventional light bulb.

Whereas conventional bulbs take around 200 milliseconds to reach their full brightness, neon bulbs turn on within 3 milliseconds. The importance of this time difference is that it gives the driver behind the vehicle an earlier warning to stop. This early warning can give the approaching driver 19 more feet for stopping when driving at 60 miles per hour.

LEDs offer the same advantages as neon bulbs and turn on even quicker because they do not need to heat up to illuminate **(Figure 20–37)**. LEDs achieve their full output in less than one millisecond. Several LEDs are placed behind the lens and are activated at the same time to give a bright illumination of the light assembly. LEDs also require a much smaller space so they are much less intrusive in the trunk. LEDs have a long operating life and provide a more precise contrast and signal pattern, thus attracting attention much more effectively.

Figure 20–37 LEDs are used in this taillight assembly. *Courtesy of Nissan North America, Inc.*

Figure 20–36 A neon lamp used for the third brake light. *Courtesy of BMW of North America, Inc.*

Using the same basic technology as LEDs, laser-lit taillights consume seven times less power than incandescent sources. These savings are extremely important for electric vehicles. The light waves of a laser light beam move in the same direction and the light is all the same color. When used with rear exterior lights, fiber optics carry red light from a diode laser to a series of mirrors, which send the beam across a thin sheet of acrylic material.

Adaptive Brake Lights This system can select one of two available brake light areas for illumination: Moderate braking activates the standard brake lights incorporated within the taillight assemblies as well as the center high-mount brake light. Under intense braking and during all braking maneuvers with active ABS intervention, additional lamps are lit, thereby changing the size of the brake lights and their intensity **(Figure 20–38)**. By increasing the brake lights' illuminated surface area, the system alerts drivers of following vehicles that the vehicle in front has started braking and decelerating at a rapid rate. This warning allows the driver of the following vehicle to react more quickly and reduces the danger of a rear impact.

An electronic control unit processes signals supplied by the speed sensor and the antilock brake system. It then uses these data to calculate the intensity of the braking as reflected by the vehicle's rate of deceleration.

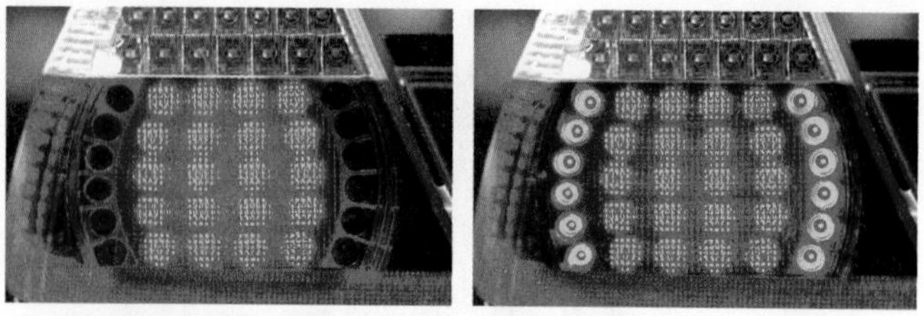

Figure 20–38 (Left) Normal illumination of the brake lights; (right) illumination during hard stops. *Courtesy of BMW of North America, Inc.*

USING SERVICE MANUALS

In addition to the taillight system, the rear of vehicles have many other lighting circuits. Most cars have brake lights, run lights, turn signals, and backup lights. Let us look at these circuits **(Figure 20–39)** and see that their diagnosis is very simple once you are aware of how they appear in the service manual. Start with the brake lights. The easiest circuit to look at first is the three-bulb circuit found on many vehicles. The drawing shows a typical taillight circuit, which contains three separate filaments for each side of the rear of the vehicle. There is a separate filament for each function: brake, turn, and run. A constant source of fused B+ is made available to the brake switch. The brake switch is usually located on the brake pedal and is closed by pushing down on the brake pedal. B+ is now available to the bulbs, wired in parallel, at the rear of the vehicle. Releasing the brake pedal allows the spring-loaded normally open (NO) switch to open and turn the brake lights off. This is a simple circuit that requires only a 12-volt testlight or a voltmeter for diagnosis. The most common cause of failure is bulbs that burn out. Testing for B+ and ground at the bulb socket should verify the circuit. If B+ is not available at the socket, test for power at each connector, moving back toward the switch until it is found. Repair the open. Do not forget that the circuit is only hot if the brake pedal is depressed. ■

Backup Lights When the transmission is placed in reverse gear, backup lights are turned on to illuminate the area behind the vehicle and to let other drivers know that the vehicle is in reverse. The major components in the system are the backup light switch and the lights.

Power for the backup light system is provided by the fuse panel. When the transmission is shifted to reverse, the backup light switch closes and power flows to the backup lights. That is, anytime the transmission is in reverse, current flows from the fuse panel through the backup light switch to the backup lights. On many vehicles, the fuse that protects the backup light system also protects the turn signal system.

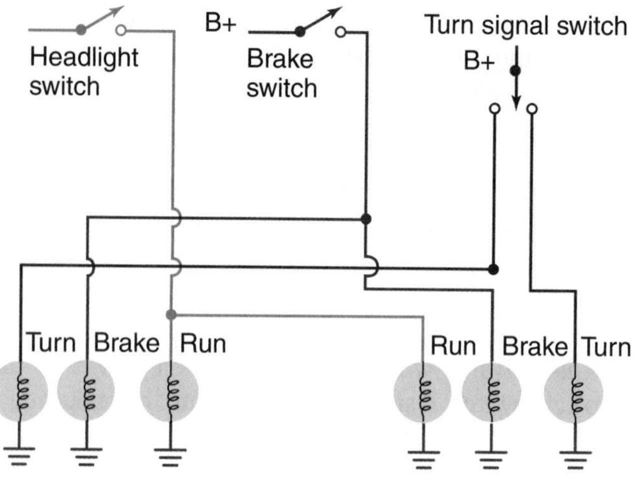

Figure 20–39 A typical three-bulb taillight circuit.

In general, vehicles with a manual transmission have a separate switch. Those with an automatic transmission use a combination neutral start/backup light switch. The combination neutral start/backup light switch used with automatic transmissions is actually two switches combined in one housing. In park or neutral, current from the ignition switch is applied through the neutral start switch to the starting system. In reverse, current from the fuse panel is applied through the backup light switch to the backup lights.

The backup light system is relatively easy to troubleshoot. On vehicles that use one fuse to protect both the turn signals and the backup lights, the fuse can be checked. If the backup lights are not working, check turn signal operation. If they work, the fuse is good. Check for power at the backup light switch input and outlet with the transmission in reverse. (Make sure the parking brake is set.)

If the switch is okay, or there is no power to the switch, check the wiring—especially the connectors. If the backup lights stay on when the transmission is not in reverse, suspect a short in the backup light switch.

LIGHT BULBS

Besides headlight bulbs, there are several different types of light bulbs used in modern vehicles.

Other Bulbs

Bulbs used in most other lighting fixtures fit into sockets and are held in place by spring tension or mechanical force **(Figure 20–40)**. Bulbs are coded with numbers for replacement purposes. Bulbs with different code numbers might appear physically similar but have different wattage ratings.

Light systems normally use one wire to the light, making use of the car body or frame to provide the ground

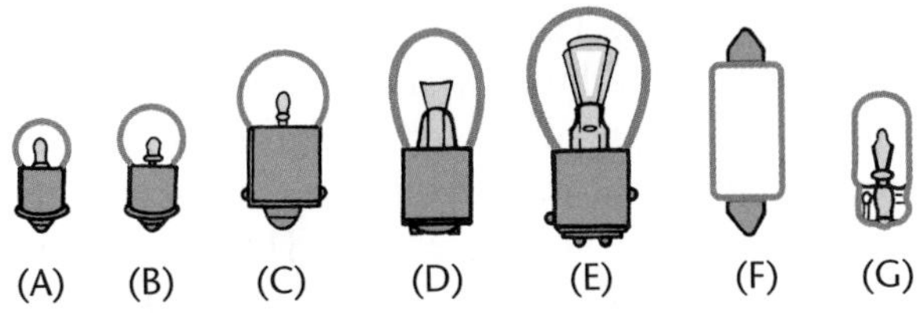

A,B MINIATURE BAYONET FOR INDICATOR AND INSTRUMENT LIGHTS
C—SINGLE CONTACT BAYONET FOR LICENSE AND COURTESY LIGHTS
D—DOUBLE CONTACT BAYONET FOR TRUNK AND UNDERHOOD LIGHTS
E—DOUBLE CONTACT BAYONET WITH STAGGERED INDEXING LUGS FOR STOP, TURN SIGNALS, AND BRAKE LIGHTS
F—CARTRIDGE TYPE FOR DOME LIGHTS
G—WEDGE BASE FOR INSTRUMENT LIGHTS

Figure 20-40 Common types of automotive bulbs.

back to the battery. Since many of the manufacturers have gone to plastic socket and mounting plates (as well as plastic body parts) to reduce weight, many lights must now use two wires to provide the ground connection. Some double-filament lights use two hot wires and a third ground wire. That is, double-filament bulbs have two contacts and two wire connections to them if grounded through the base. If not grounded through the base of the bulb, a two-filament bulb has three contacts and three wires connected to it. Single-filament bulbs may be single- or double-contact types. Single-contact types are grounded through the bulb base, while double-contact, single-filament types have two wires—one live and the other a ground.

When replacing a bulb, inspect the bulb socket. If the socket is rusty or corroded, the socket or light assembly base should be replaced. Also, inspect the lens and gasket for damage while the lens is removed and replace any damaged part.

There are two basic construction designs for exterior lights: those in which the lens is removed and then the bulb removed from the front, and those in which the light assembly must first be removed, then the socket from the back of the assembly, and finally the bulb from the socket. Removing the lens from the latter type assembly could cause serious damage to the reflector due to dust and other contaminants. Wiping the reflector surface to clean it can also seriously reduce the light's brightness. Therefore, do not remove the lens from light assemblies in which the socket and bulb are removed from the back of the assembly.

The bulbs are held in their sockets in a number of ways. Some bulbs are simply pushed into and pulled out of their sockets, and some are screwed in and out. To release a bayonet-style bulb from its socket, the bulb is pressed in and turned counterclockwise. The blade-mount style is removed by pulling the bulb off the mounting tab, then turning the bulb and removing it from the retaining pin.

Auxiliary Lights

While the car's headlights are adequate in normal driving circumstances, some customers desire auxiliary lights for special conditions such as fog or extended night driving. In addition to the standard auxiliary fog, driving, and passing lights, there are off-road lights, worklights, rooflights, decklights, deckbars, and hand-held spot lights.

Driving Lights Driving lights put out more light than the best factory headlights, affording the driver an additional margin of safety. Driving lights typically use an H3 or H4 quartz halogen bulb and a high-quality reflector and lens to project an intense, pencil-thin beam of light far down the road.

Proper aiming of the auxiliary lights is extremely important. Driving lights are used to supplement the high beams for greater distance and width. They should be used only in conjunction with the high beams. That is, driving lights should be wired so they are off when the high beams are off.

SHOP TALK

When adding auxiliary lights, make sure the AC generator and wiring are heavy enough to handle the increased wattage. Installing a higher-output AC generator may be recommended, especially if other electrical accessories are also being installed. The choice of wire size should be based on the load the wire will be powering. ■

Fog Lights Ordinary headlights do not penetrate fog well. Focus a powerful beam of light at the fog and all the driver gets back is a powerful glare. To deal with that problem, fog lights use the same bulbs but, instead of trying to pierce the darkness, they attempt to sneak a flat, wide beam of light under the blanket of fog. This makes it important to mount them low and to aim them low and parallel to the road. Fog lights should only work with the low-beam headlights.

While some vehicles have OEM fog lights, most are auxiliary lighting add-ons. Their circuits, however, are basically the same as driving lights. They involve a relay switch and the lights themselves. A relay is used because the amount of current that fog lights require, especially halogen ones, can be quite high. It is not unusual that they require as much as 25 amperes.

The dash switch controls the current to one side of the relay's coil. A direct ground is supplied to the other side of the relay coil. With both battery voltage and ground applied, current flows through the coil and a magnetic field develops. The field closes the contacts in the relay. One side of the contacts is connected to a fused source of battery voltage. The other side is connected to the fog lights, which are wired in parallel. Each filament has its own remote ground connection.

Driving and fog lights have tremendous output and have correspondingly high electrical requirements. This means the car should have an efficient charging system and a heavy-duty battery.

SHOP TALK

When replacing fog light bulbs, avoid touching the glass part of the new bulb assembly. Skin oil, present on even recently washed hands, will be deposited on the glass. This oil prevents the bulb from dissipating heat. The increased heat inside the bulb causes the filament to burn out prematurely. ■

LIGHTING MAINTENANCE

In addition to replacing all burned-out lights and bulbs, when a vehicle comes in for servicing, periodically check to see that all wiring connections are clean and tight, that light units are tightly mounted to provide a good ground, and that headlights are properly adjusted. Loose or corroded connections can cause a discharged battery, difficult starting, dim lights, and possible damage to the AC generator and regulator. Often moisture gets into a bulb socket and causes corrosion of the electrical contacts and the bulb. Corrosive conditions can be repaired by using sandpaper on the affected areas. For severe cases, replace the socket and bulb. After any repair, always attempt to waterproof the assembly to prevent future problems. Cracked or broken assemblies are easily replaced. They are secured by attaching hardware that is normally readily accessible to the technician.

Another common electrical lighting problem is flickering lights (going on and off). The cause of this is usually a loose electrical connection or a circuit breaker that is kicking out because of a short. If all or several of the lights flicker, the problem is in a section of the circuit common to those lights. Check to see if the lights flicker only with the light switch in one position. For example, if the lights flicker only when the headlights are on high beam, check the components and wiring in the high-beam section of the circuit. If only one light flickers, the problem is in that section of the circuit. Check the bulb socket for corrosion. Also, make sure the bulb terminals are not worn. This could upset the electrical connection. If necessary, replace the bulb socket and bulb.

Look at the turn signal diagram in **Figure 20–41A.** The diagram shows the inside of a turn signal switch for a two-bulb system. The turn signal switch determines whether one of the bulbs is used for turning or brake lighting. The rectangular bars on the diagram are stationary contacts that the circuit wires connect to. Each contact has one wire connected to it. The top connection is from the brake switch and is B+ if the brakes are applied. The middle row connections are for rear combination lights (combination brake/turn signal). The bottom row of connections is for the front lights, including the dash indicators, and the B+ coming from the flasher.

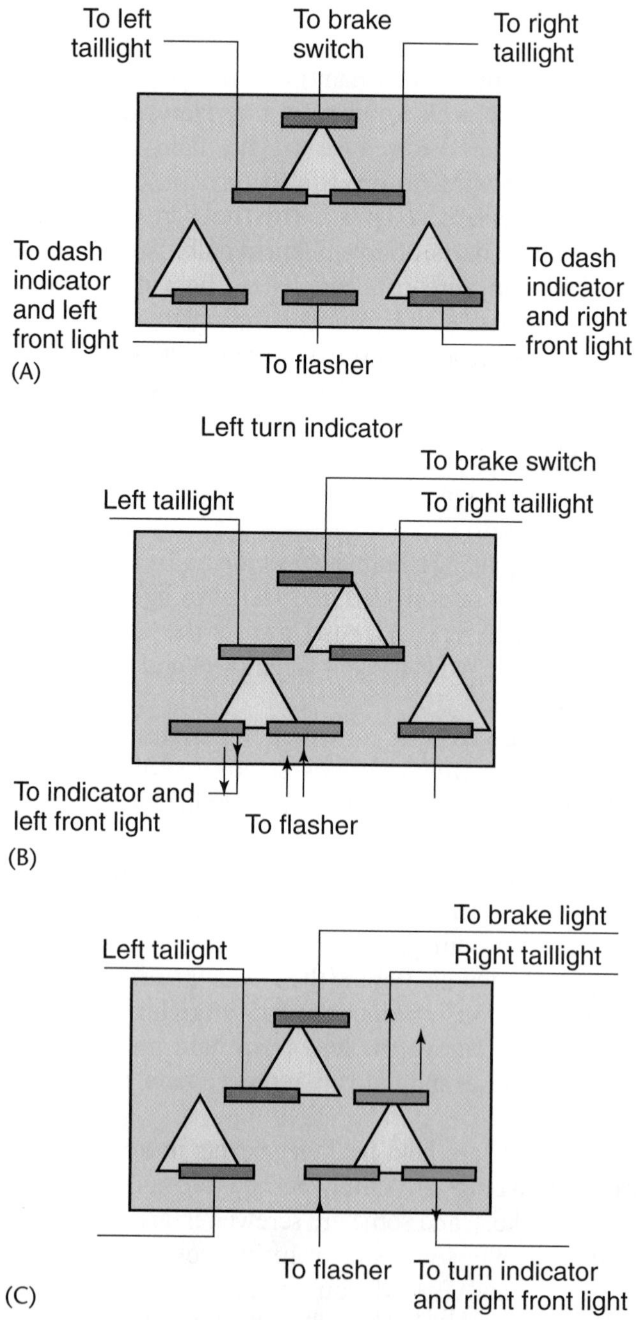

Figure 20–41 A turn signal switch (A) not in use, (B) with a left turn indicated, and (C) with a right turn indicated. *Courtesy of Ford Motor Company*

The triangles drawn over the bars are a set of three movable conductive pads that connect the different bars together depending on the position of the switch. They are drawn in the no turn or neutral position. This allows B+ from the brake switch to activate both rear lights at the same time. **Figure 20–41B** shows the same switch in a left turn. Notice that the conductive pads or triangles have moved to the right. This allows the brake switch to power only the right taillight, while the flasher connection is now in contact with the left rear taillight and the left front/indicator lights. The right taillight is being operated as a brake light, while the left one is in a turn signal operation. **Figure 20–41C** shows the same switch in a right turn mode. Notice that the conductive pads have moved to the left and have connected the brake switch to the left taillight, while the right is now powered off the turn signal flasher. This style of switch is very popular and normally is very durable. The most common problem encountered with the switch is usually mechanical rather than electrical. As the vehicle is driven around the corner, the canceling mechanism must put the switch back into a neutral position so that both taillights can be used for brake warning. When this canceling does not take place, the turn signal switch is normally replaced to correct the problem.

CASE STUDY

The customer complains that the turn signals on his vehicle do not operate as they are supposed to, and they seldom cancel out properly.

Turn signal circuits are frequent sources of difficulties. Their diagnosis, however, is not difficult and can usually be accomplished with just a 12-volt light or a voltmeter. Look at the common circuits, starting first with the flasher. The flasher is actually a type of circuit breaker, which is an overload protection device designed to open the circuit because of heat developed from excessive current. Flashers are usually mounted in the fuse box and made up of a fixed contact and a movable bimetallic contact.

KEY TERMS

Candlepower
Composite headlight
Daytime running lights (DRL)
Dimmer switch
Flash to pass
Flasher
Halogen
High-intensity discharge (HID) headlamp
Lamp
Multifunction switch
Photocell sensor/amplifier
Sealed-beam
Xenon headlamps

SUMMARY

- The headlight switch controls the headlights and all other light systems, with the exception of the turn signals, hazard warning, and stop lights.
- Dimmer switches, located on either the steering wheel or the floor, permit the headlights to change from high to low beam and vice versa.
- An automatic headlight dimmer circuit switches the headlights from high to low beam, in response to either light from an approaching vehicle or light from the taillights of a vehicle being overtaken.
- Depending on the make or model of the vehicle, courtesy lights can be found on the door, under the seats, under the instrument panel, on the rear interior quarter panels, or on the ceiling.
- The rear light assembly includes the taillights, turn signal/stop/hazard lights, high-mounted stop light, rear side marker lights, backup lights, and license plate lights. Taillights operate when the parking lights or headlights are on.
- Headlights must be kept in adjustment to obtain maximum illumination and vehicle occupant safety.
- Flashers are used in turn signal, hazard warning, and side marker light circuits.
- The backup light system illuminates the area behind the vehicle when the transmission is put in reverse gear.
- Headlight systems consist of two or four sealed-beam tungsten or halogen headlight bulbs.

TECH MANUAL

The following procedures are included in Chapter 20 of the *Tech Manual* that accompanies this book:

1. Inspecting and testing a headlight switch.
2. Aiming headlights.
3. Adjusting or replacing a stop light switch.

REVIEW QUESTIONS

1. Name the three types of headlight switches.
2. When is the taillight assembly activated?
3. *True or False?* HID lamps produce more heat and white light than halogen bulbs.
4. Most headlamps either have a replaceable light bulb or they are ______ ______, which must be replaced as a unit.
5. *True or False?* Many turn signal flasher units contain a temperature-sensitive bimetallic strip and a heating element.
6. While troubleshooting a headlight problem, Technician A says when one headlamp does not work the problem is probably a burned-out bulb or lamp. Technician B says you should check for voltage at the bulb before replacing a bulb. Who is correct?

   a. Technician A **c.** Both A and B
   b. Technician B **d.** Neither A nor B
7. An underhood lamp is typically controlled by____.

   a. the headlight switch
   b. a momentary contact switch
   c. the courtesy light switch
   d. a mercury switch
8. The rear light assembly includes the ______.

   a. rear side marker lights **c.** license plate light
   b. taillights **d.** all of the above
9. Circuits that can energize both the high and low beams even if the headlight switch is off are known as ______ circuits.

   a. flash to pass **c.** dimmer
   b. mercury **d.** retractable
10. Technician A says the condition of the vehicle's springs and shocks should be checked before aligning the headight. Technician B says the vehicle should have a full tank of fuel when its headlights are being adjusted. Who is correct?

    a. Technician A **c.** Both A and B
    b. Technician B **d.** Neither A nor B
11. The ______ ______ provides a means for the driver to select between high- or low-beam headlight operation.
12. Composite headlights are being discussed. Technician A says they have replaceable bulbs. Technician B says a cracked or broken lens will prevent the operation of a composite headlight. Who is correct?

    a. Technician A **c.** Both A and B
    b. Technician B **d.** Neither A nor B
13. While troubleshooting a brake light problem, Technician A says when the pedal is depressed the brake lights should come on. Technician B says on some systems only part of the brake light should illuminate when there is slight pressure on the brake pedal. Who is correct?

    a. Technician A **c.** Both A and B
    b. Technician B **d.** Neither A nor B
14. The stop light switch is normally mounted on the ______.

    a. instrument panel **c.** brake pedal arm
    b. transmission **d.** none of the above
15. Which of the following is *not* a true statement about LED-based lights?

    a. LEDs achieve their full output in about than 200 milliseconds.
    b. LEDs require a much smaller space so they are much less intrusive in the trunk.
    c. LEDs have a long operating life.
    d. LEDs provide a more precise contrast and signal pattern, thus attracting attention much more effectively.
16. Technician A says a sealed-beam headlight that is cracked should be replaced. Technician B says condensation in the housing or lens does not prevent a composite headlamp from working. Who is correct?

    a. Technician A **c.** Both A and B
    b. Technician B **d.** Neither A nor B
17. Technician A says if the turn signals blink faster than normal, the problem may be a burned-out bulb on that side of the vehicle. Technician B says if the turn signals blink slower, the problem may be a burned-out light bulb on that side of the vehicle. Who is correct?

    a. Technician A **c.** Both A and B
    b. Technician B **d.** Neither A nor B
18. Technician A says blinking vehicle lights can be caused by a circuit breaker that is kicking out due to a short. Technician B says flickering vehicle lights can be caused by a loose electrical connection. Who is correct?

    a. Technician A **c.** Both A and B
    b. Technician B **d.** Neither A nor B
19. Why do some manufacturers protect the headlignt circuit with a circuit breaker instead of a fuse?
20. What kind of headlight does not have a filament?

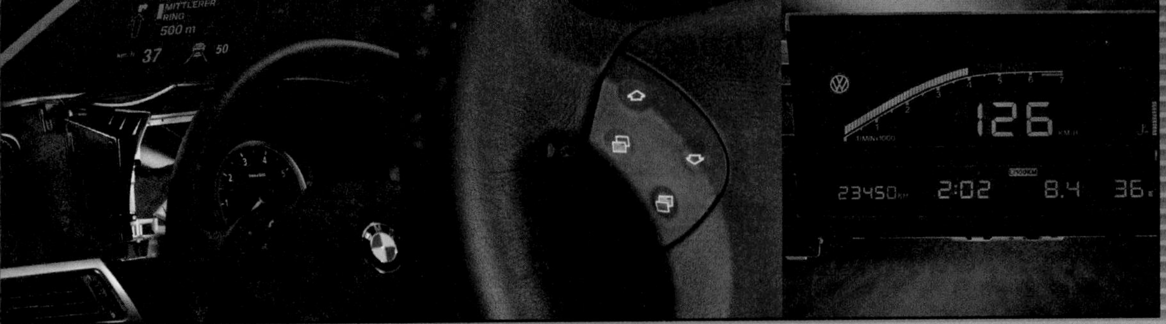

21

ELECTRICAL INSTRUMENTATION

OBJECTIVES

■ Describe the two types of instrument panel displays. ■ Know the purpose of the various gauges used in today's vehicles and how they function. ■ Describe the operation of the common types of gauges found in an instrument cluster. ■ List and explain the function of the various indicators found on today's vehicles. ■ List and explain the function of the various warning devices found on today's vehicles. ■ Explain the basics for diagnosing a gauge or warning circuit.

Every automobile is equipped with a number of electrical instruments. The number and types of these systems and components vary significantly from vehicle model to vehicle model and year to year. The appearance of the gauge layout also varies from the quite simple **(Figure 21–1)** to the elaborate **(Figure 21–2)**. No matter what it looks like, a vehicle's instrumentation must be easy to read and give accurate information. Instrument gauges, lights, and warning indicators provide valuable information to the driver concerning a vehicle's various systems.

Figure 21–2 An elaborate arrangement of the essential gauges in an instrument panel. *Reprinted with permission*

INSTRUMENT PANELS

Today's dashboard is more properly called the **instrument panel**. It contains an array of electrical gauges, switches, and controls connected to mazes of wiring, printed circuitry, and vacuum hoses beneath stylishly finished sheets of plastic or metal.

Figure 21–1 A simple but functional approach to the layout of the essential gauges in an instrument panel. *Courtesy of BMW of North America, Incorporated*

Displays

There are many different instrument panel designs and layouts. The two basic types of instrument panel displays are analog and digital. In a traditional analog display **(Figure 21–3)**, an indicator moves in front of a fixed scale to indicate a condition. The indicator is often a needle but it can also be a liquid crystal or graphic display. A digital display uses numbers instead of a needle or graphic symbol. Analog displays show relative change better than digital displays **(Figure 21–4)**. They are useful when the driver must see something quickly and the exact reading is not important. For example, an analog tachometer shows the rise and fall of the engine speed better than a

Figure 21-3 An analog instrument panel. *Reprinted with permission*

Figure 21-4 A digital instrument panel. *Courtesy of Siemans VDO Automotive*

digital display. The driver does not need to know the exact engine speed. The most important thing is how fast the engine is reaching the red line on the gauge. A digital display is better for showing exact data such as miles. Many speedometer-odometer combinations are examples of both analog (speed) and digital (distance).

Three types of digital electronic displays are used today.

Light-Emitting Diode (LED) These displays are used as either single indicator lights or they can be grouped to show a set of letters or numbers. LED displays are commonly red, yellow, or green and can be hard to see in bright light.

Liquid Crystal Diode (LCD) These displays are made of sandwiches of special glass and liquid. A separate light source is required to make the display work. When there is no voltage, light cannot pass through the fluid. When voltage is applied, the light passes that point of the display. The action of LCDs slows down in cold weather. These displays are also very delicate and must be handled with care. Any rough handling of the display can damage it.

Vacuum Fluorescent These displays use glass tubes filled with argon or neon gas. The segments of the display are little fluorescent lights. When current is passed through the tubes, they glow very brightly. These displays are both durable and bright.

SHOP TALK

Most warning or instrument controls use **International Standards Organization (ISO)** symbols, which were developed by this organization to provide symbols easily recognizable throughout the world. ■

INSTRUMENT GAUGES

Gauges provide the driver with a scaled indication of the condition of a system, for example, that the fuel tank is half full. Two additional components are required for the operation of an electrical gauge: instrument voltage regulators and sending units. **Instrument voltage regulators (IVR)** are used to stabilize and limit voltage for accurate instrument operation **(Figure 21–5)**. Sending or sensor units change electrical resistance in response to changes or movements made by an external component. Movement may be caused by pressure against a diaphragm by heat, or by the motion of a float as liquid fills a fuel tank.

All gauges, analog or digital, require an input from either a sending unit or a sensor. However, with modern computer-controlled displays, the input from the sensor is used in two ways. The engine control computer needs the same information as the electronic display, so the information passes through the computer first. It then goes on to the gauge.

Two types of electrical analog gauges are commonly used with sensor or sending units—magnetic and thermal.

Magnetic Gauges

There are several types of magnetic gauges. The simplest form is the ammeter type **(Figure 21–6)**. in which a permanent magnet attracts a ferrous indicator needle connected to a pivot point and holds it centered on the gauge. An armature, or coil of wire, is wrapped around the base of the needle near the pivot point. When current flows through the armature, a magnetic field is formed. This magnetic field opposes that of the permanent magnet. Attractive or repulsive magnetic forces cause the needle to swing left or right. The direction the needle swings depends on the direction of current flow in the armature.

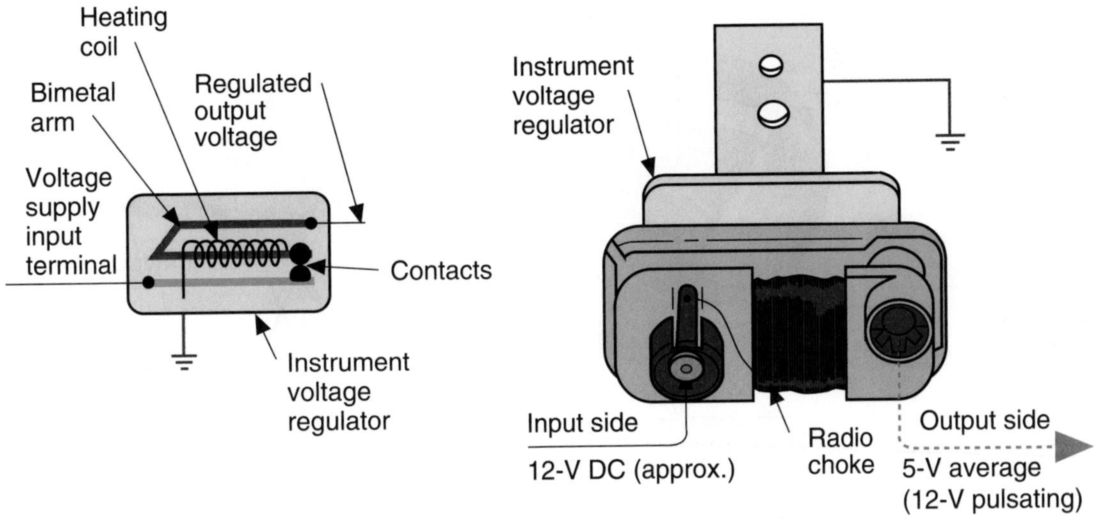

Figure 21-5 An IVR.

The needle of the gauge pivots on the armature, often called a bobbin, which is why this type of gauge is often referred to as a **bobbin gauge**.

When sending unit resistance changes, current flow through the bobbin changes, causing the strength of the magnetic field created around the bobbin to change. As the resistance in the sending unit increases, the circuit's current decreases and a lower indicator position on the gauge results.

A **balancing coil** gauge also operates on principles of magnetic attraction and repulsion. However, a permanent magnet is not used. The base of the indicating arm is pivoted and includes an armature. Two coils are used to create magnetic fields **(Figure 21–7)**.

The two coils are connected so that electricity can flow through either one. When the resistance of the sending unit is low, the right-hand coil receives more current than the left-hand coil, attracting the armature. Thus, the gauge needle moves to the right.

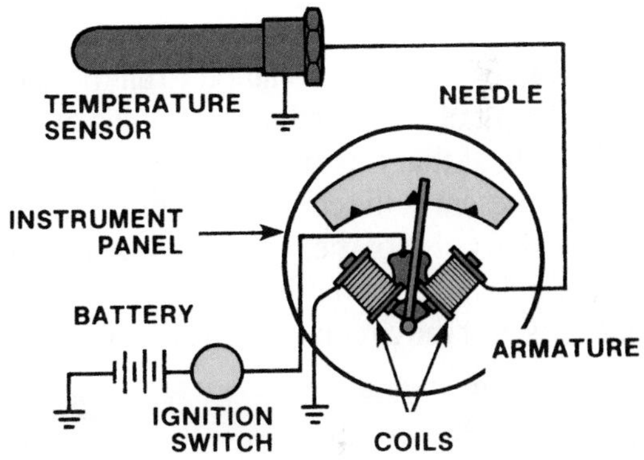

Figure 21-7 A balancing coil gauge. *Courtesy of Ford Motor Company*

When the resistance of the sending unit is high, the left-hand coil receives more current. More magnetic force is created in the left-hand coil and the needle swings to the left.

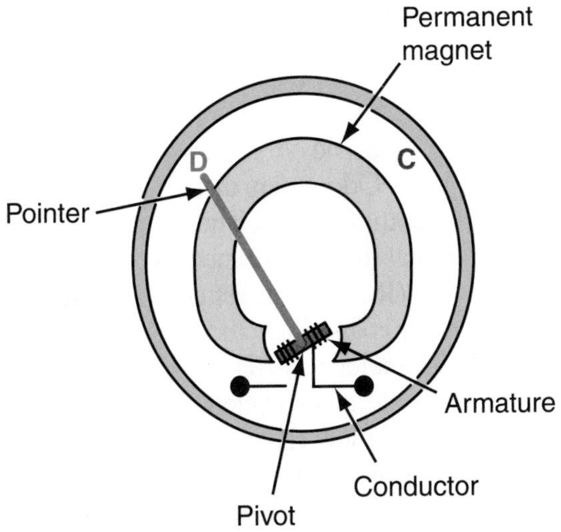

Figure 21-6 A simple ammeter that relies on magnetic principles.

Thermal or Bimetallic Gauge

A bimetallic or **thermal gauge** operates through heat created by current flow **(Figure 21–8)**. A variable-resistance sending unit causes different amounts of current to flow through a heating coil within a gauge. The heat acts on a bimetallic spring attached to a gauge needle. When more heat is created, the needle swings farther up the scale. When less heat is created, the needle moves down the scale.

Diagnosis

If all gauges fail to operate properly, begin by checking the circuit's fuse. Next, test for voltage at the last point common to all the malfunctioning gauges. If voltage is not present, work toward the battery to find the fault.

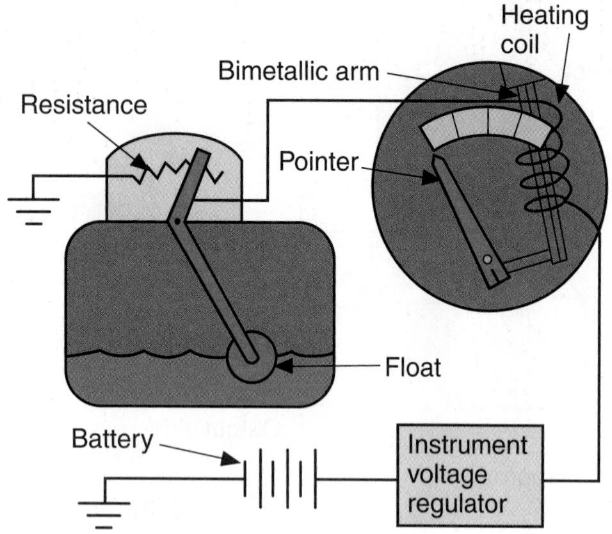

Figure 21-8 A typical bimetallic fuel gauge circuit.
Courtesy of DaimlerChrysler Corporation

If the system uses an IVR, use a voltmeter to test for regulated voltage at a point common to all the gauges. If the voltage is out of specifications, check the ground circuit of the IVR. If that is good, replace the IVR. If there is no voltage present at the common point, check for voltage on the battery side of the IVR. If voltage is present at this point, replace the IVR. If regulated voltage is within specifications, test the circuit from the IVR to the gauges.

WARNING!

Many instruments and warning devices are linked to the vehicle's body control module and multiplexing network. Before troubleshooting a gauge or warning system, check the service manual to identify any special procedures or precautions.

BASIC INFORMATION GAUGES

The following gauges are found on nearly all instrument panels. The detailed operation of some gauges is described in other chapters of this book.

Speedometer

In the past, the speedometer was considered a nonelectrical or mechanical gauge. It had a drive cable attached to a gear in the transmission that turned a magnet inside a cup-shaped metal piece **(Figure 21–9)**. The cup was attached to a needle, which was held at zero by a hairspring (a fine wire spring). As the cable rotated faster with an increase in speed, magnetic forces acted on the cup and forced it to move. As a result, the needle moved up the speed scale.

Electric speedometers are used in nearly all late-model vehicles. While there are several systems in use, one of the most common types receives its speed information from the transmission-mounted vehicle speed sensor **(Figure 21–10)**. This speed signal is also used by other modules in the vehicle, including the speed control module, ride control module, the engine control module, and others.

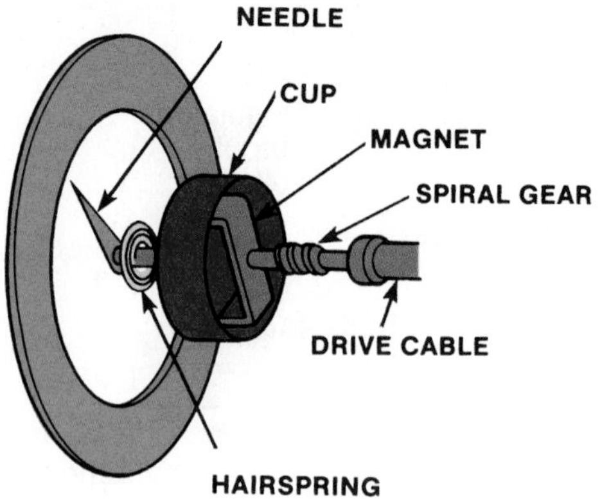

Figure 21-9 A typical conventional speedometer.

For each 40,000 pulses from the vehicle speed sensor (VSS), the trip and total odometers will increase by one mile. Speed is determined by dividing the input pulse frequency (in hertz) by 2.2 hertz/mph. The circuit electronics are calibrated to drive the pointer to a location in proportion to the speed input frequency. That is, as the pulse rate increases, the speedometer records it on an analog display. This display may be limited to a maximum of 85 mph or 199 km/h.

Most digital speedometers have a speed limit. If vehicle speed exceeds these values, the speedometer continues to display the top of its range. Vehicle speed is displayed whether the vehicle is moving forward or backward.

Odometer

The **odometer** is a digital gauge that is usually driven by a spiral gear cut on the speedometer's magnet shaft. The odometer's numbered drums are geared so that when any one drum finishes a complete revolution, the drum to the left is turned one-tenth of a revolution.

Generally, the electric odometer receives its information from the VSS. Odometers display seven digits, with the last digit in tenths of a unit. The accumulated mileage value of the digital display odometer is stored in a nonvolatile memory (ROM) that retains the mileage value even if the battery is disconnected.

Since the odometer records the number of miles or kilometers a vehicle has traveled, federal law requires that the odometer in any replacement speedometer register the same mileage as that registered on the removed speedometer. Therefore, if a speedometer has been replaced, set the odometer of the new one to match the old. The trip odometer may be reset whenever desired.

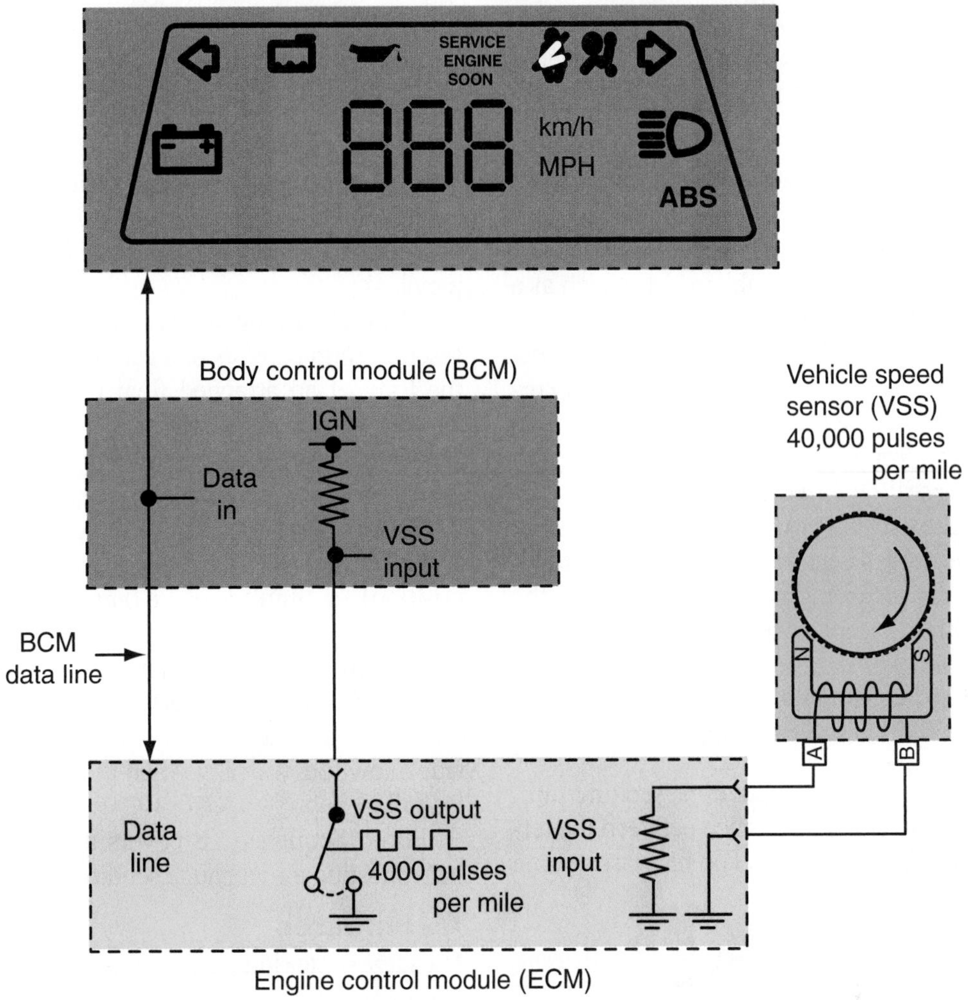

Figure 21-10 The signals from the VSS are shared by the instrument panel and the PCM.

Oil Pressure Gauge

The oil pressure gauge indicates engine oil pressure. The oil pressure typically should be between 45 and 70 psi when the engine is running at a specified engine speed and at operating temperature. A lower pressure is normal at low idle speed.

With low oil pressure (or with the engine shut off), the oil pressure switch is open and no current flows through the gauge winding. The needle points to L. With oil pressure above a specific limit, the switch closes and current flows through the gauge winding to ground. A resistor limits current flow through the winding and ensures that the needle points to about mid-scale with normal oil pressure.

A piezoresistive sensor **(Figure 21–11)** is threaded into the oil delivery passage of the engine. The pressure of the oil causes a flexible diaphragm to move. This movement is transferred to a contact arm that slides down the resistor. The position of the sliding contact arm determines the resistance value and the amount of current flow through the gauge.

Diagnosis To test a piezoresistive-type sending unit, connect an ohmmeter to the sending unit's terminal and to ground. Check the resistance with the engine off and compare it to specifications. Start the engine and allow it to

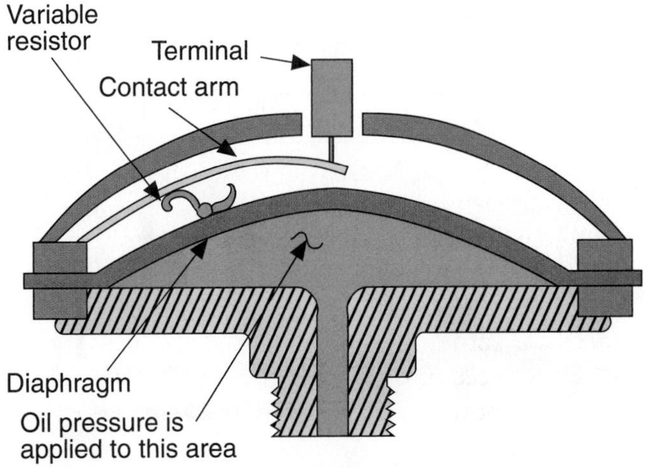

Figure 21-11 A piezoresistive sensor used for measuring engine oil pressure.

idle. Check the resistance value and compare it to specifications. Before replacing the sending unit, connect a shop oil pressure gauge to confirm that the engine is producing adequate oil pressure.

Coolant Temperature Gauge

This gauge indicates engine coolant temperature. It should normally indicate between C (cold) and H (hot). The sending unit is typically a variable resistor such as a thermistor. It regulates the current flow through the temperature gauge winding **(Figure 21–12)**. With low coolant temperature, sender resistance is high and current flow is low. The needle points to C. As coolant temperature increases, sender resistance decreases and current flow increases. The needle moves toward H.

The temperature gauge on a digital panel is of the bar type with a set number of segments. The number of illuminated bars varies according to the current from the gauge sender. With low coolant temperature, sender resistance is high and few segments are turned on. As coolant temperature increases, sender resistance decreases and the number of illuminated segments increases.

Diagnosis To test a coolant temperature sending unit, use an ohmmeter to measure resistance between the terminal and ground **(Figure 21–13)**. The resistance value of the variable resistor should change in proportion to coolant temperature. Check the test results with manufacturer specifications **(Figure 21–14)**.

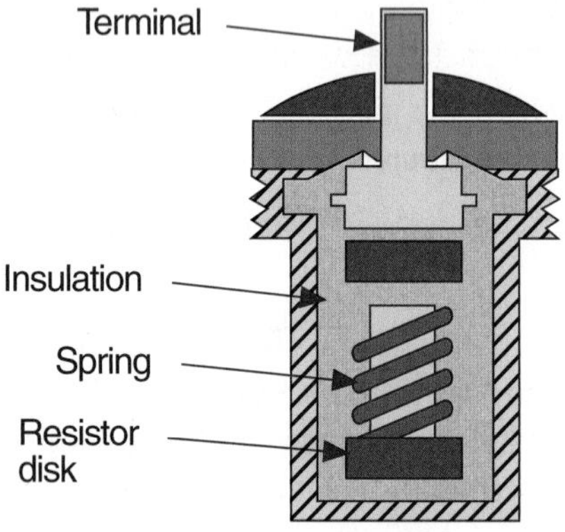

Figure 21-12 A typical temperature sending unit that does not use a thermistor.

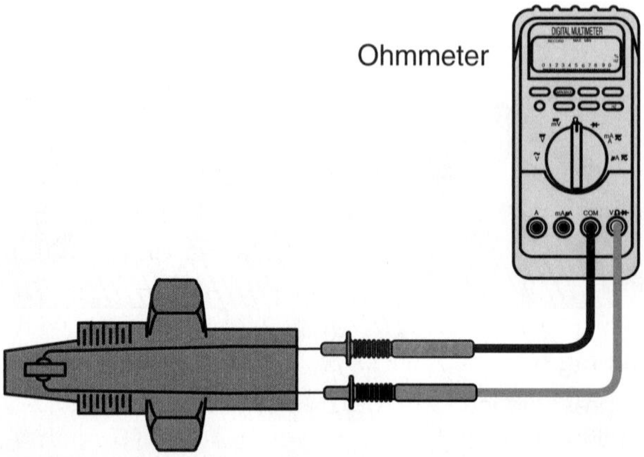

Figure 21-13 Testing a temperature sensor with an ohmmeter.

Fuel Level Gauge

The fuel level gauge indicates the fuel level in the fuel tank. It is a magnetic indicating system that can be found on either an analog (meter) or digital (bars) instrument panel.

The fuel sending unit is combined with the fuel pump assembly and consists of a variable resistor controlled by the level of an attached float in the fuel tank **(Figure 21–15)**. When the fuel level is low, resistance in the sender is low and movement of the gauge needle or number of lit bars is minimal (from empty position). When the fuel level is high, the resistance in the sender is high and movement of the gauge indicator (from the empty position) or number of lit bars on a digital display is greater.

In some fuel gauge systems, an antislosh/**low fuel warning (LFW)** module is used to reduce fuel gauge needle fluctuation caused by fuel motion in the tank and provide a low fuel warning when the fuel tank reaches ⅛ to 1/16 full.

Photo Sequence 18 covers a typical procedure for bench testing a fuel gauge sending unit.

Tachometer

The **tachometer** indicates engine rpm (engine speed). The electrical pulses to the tachometer typically come from the ignition module or PCM. The tachometer, using a balanced coil gauge, converts these impulses to rpm that can be read. The faster the engine rotates, the greater the

COLD 10K-ohm resistor	HOT 909-ohm resistor
-20 °F 4.7V	110° F 4.2V
0° F 4.4V	130° F 3.7V
20° F 4.1V	150° F 3.4V
40° F 3.6V	170° F 3.0V
60° F 3.0V	180° F 2.8V
80° F 2.4V	200° F 2.4V
100° F 1.8V	220° F 2.0V
120° F 1.2V	240° F 1.6V

Figure 21-14 Typical values of a thermistor-type temperature sensor at a variety of temperatures.

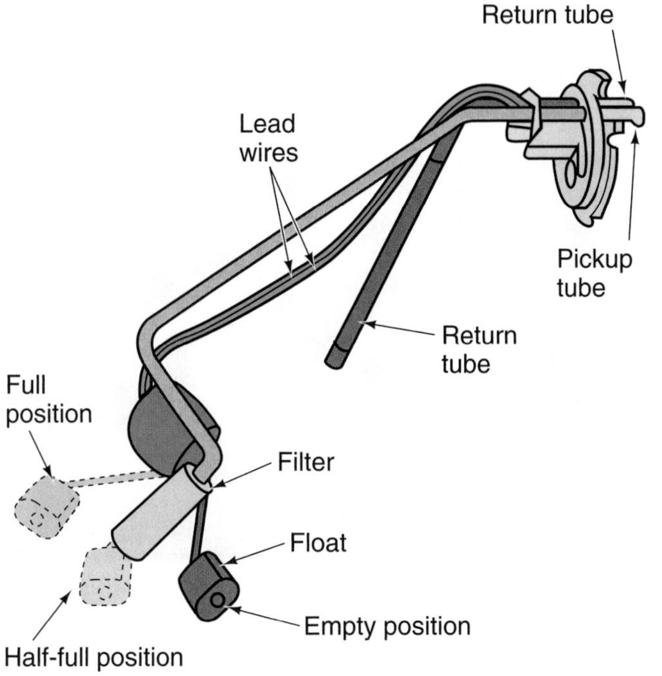

Figure 21-15 A fuel gauge sending unit. *Courtesy of DaimlerChrysler Corporation*

number of impulses from the coil. Consequently, higher engine speeds are indicated.

In vehicles with digital instrumentation, the bar system is used with numbered segments. The numbers represent the engine's rpm times one thousand.

Charging Gauges

Charging gauges allow the driver to monitor the charging system. While a few older cars use a voltmeter, most charging systems employ either an ammeter gauge or an indicator light. The ammeter gauge is placed in series with the battery and generator. When the generator is delivering current to the battery, the gauge displays a positive (+) indication. When the battery is not receiving enough current (or none at all) from the generator, a negative (–) display is obtained.

Other Gauges

A discussion of some of the other electrical gauges commonly used on vehicles follows.

Fuel Consumption Gauge This gauge gives the driver the current fuel consumption in mpg or liters/100 km. The gauge is also called an "energy control" gauge and it informs the driver as to how the current operating conditions are affecting fuel consumption.

Service Interval Display This readout displays the number of miles or kilometers that remain before the vehicle is due for service. The computer may base this display on more than distance traveled. Driving style and previous operating conditions may shorten the mileage interval.

INDICATORS AND WARNING DEVICES

Light Indicators and Warnings

Indicator lights and warning devices are generally activated by the closing of a switch. An indicator light comes on to inform the driver that something has been turned on, such as the rear window defogger. Warning lights notify the driver that something in the system is not functioning properly or that a situation exists that must be corrected. The function of some of the more common warning lights is described here.

WARNING!

Be aware that many of the warning and indicator lamps found on today's vehicles are triggered by a PCM or BCM. Often they are part of the multiplexed system. With this in mind, always refer to the testing methods recommended by the manufacturer before testing these systems. Using conventional testing methods on a computerized system may destroy part or all of the system.

Air Bag Readiness Light The air bag readiness light lets the driver know the air bag system is working and ready to do its job. It lights briefly when the ignition is turned on. A malfunction in the air bag system will cause the light to stay on continuously or to flash.

Fasten Belts Indicator When the ignition is turned to run or start, the warning chime module applies voltage to illuminate the fasten belts indicator for six seconds, whether or not the driver's belt is buckled.

Tire Pressure Monitor When a low inflated or flat tire is found, this warning light is turned on. Some systems illuminate this warning lamp in red or yellow. Red means there is an excessively low or flat tire and yellow means a tire has low pressure. Some systems also emit a sound to alert the driver.

Lamp-Out Warning Light The lamp-out warning module is an electronic unit designed to measure small changes in voltage levels. An electronic switch in the module closes to complete a ground path for the indicator lights in the event of a bulb going out. The key to this system being able to detect one bulb out on a multibulb system is the use of the special resistance wires. With bulbs operating, the resistance wires provide 0.5 volt input to the light-out warning module. If one bulb in a particular system goes out, the input off the resistance wire drops to approximately 0.25 volt. The light-out warning module detects this difference and completes a ground path to the indicator light for the affected circuit.

PHOTO SEQUENCE

18 Bench Testing a Fuel Gauge Sending Unit

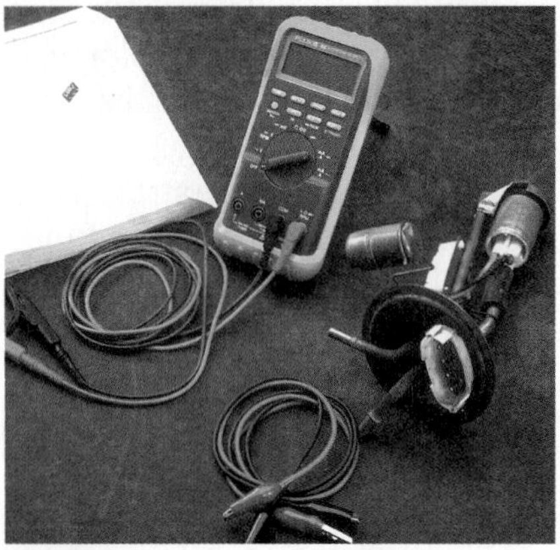

P18-1 *The tools required to perform this task are a DMM, jumper wires, and a service manual.*

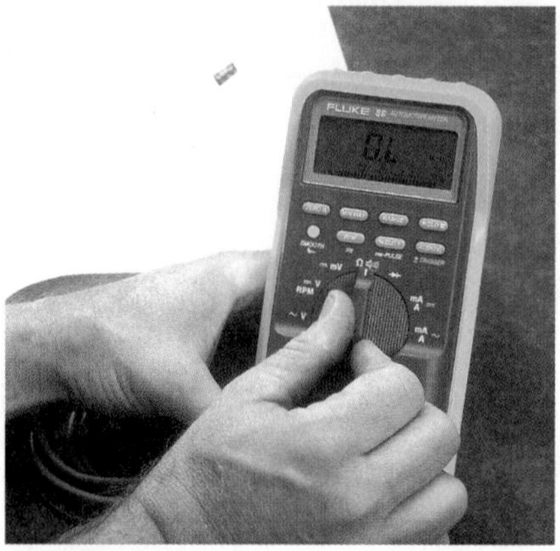

P18-2 *Select the ohmmeter function of the DMM.*

P18-3 *Connect the DMM's negative test lead to the ground terminal of the sending unit.*

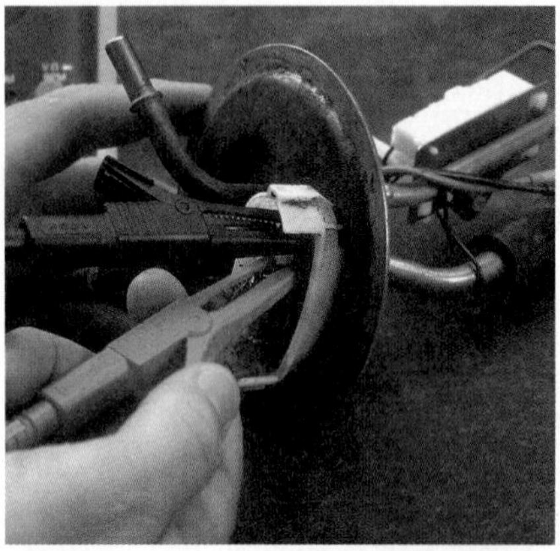

P18-4 *Connect the meter's positive lead to the variable resistor terminal.*

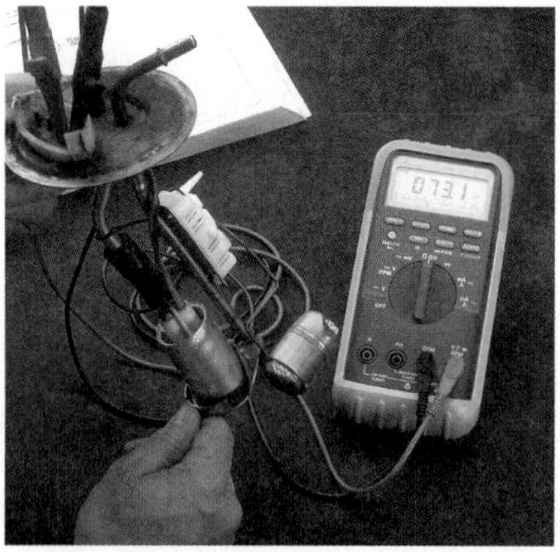

P18-5 *Holding the sender unit in its normal position, place the float rod against the empty stop.*

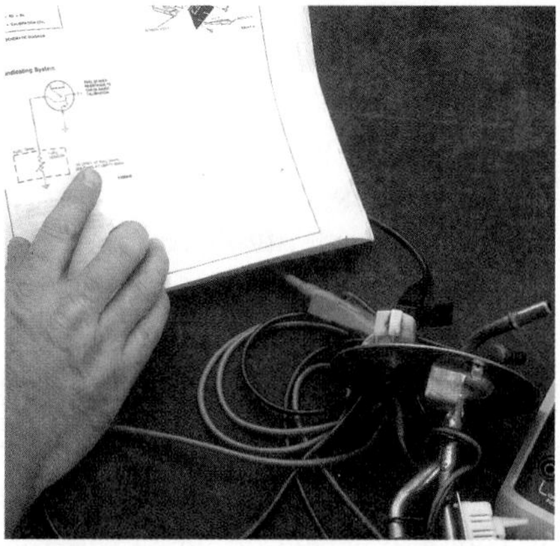

P18-6 *Read the ohmmeter and check the results with the specifications.*

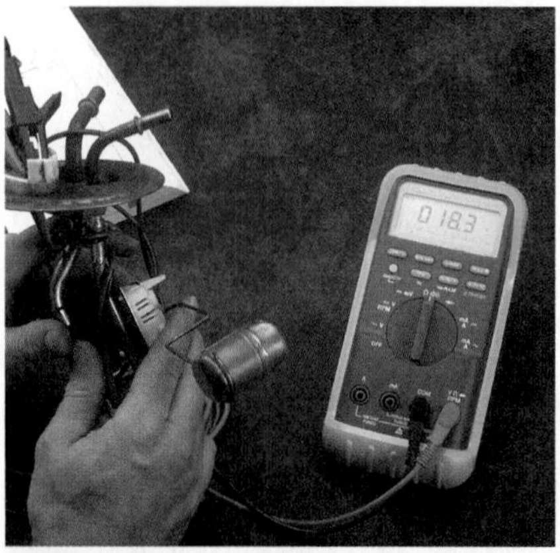

P18-7 *Slowly move the float toward the full stop, while observing the ohmmeter. The resistance change should be smooth and consistent.*

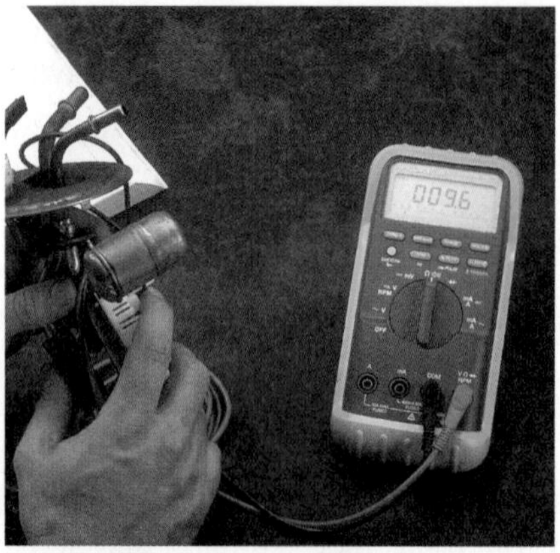

P18-8 *Check the resistance value while holding the float against the full stop. Check the results with the specifications.*

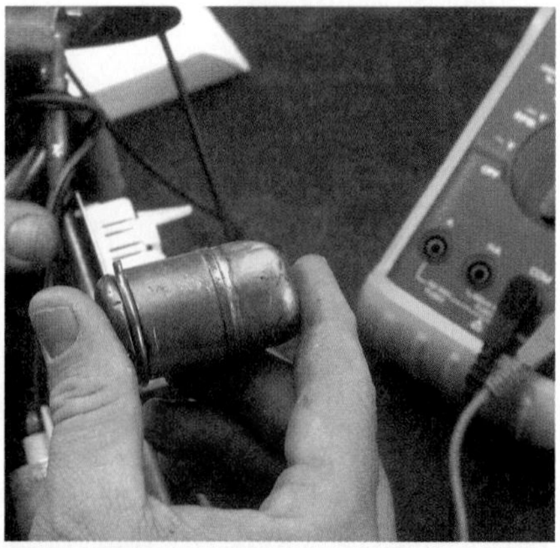

P18-9 *Check the float to be sure it is not filled with fuel, distorted, or loose.*

Brake Warning Light When this light is lit, it is an indication that the parking brake is engaged. Some vehicles use the same warning lamp to indicate hydraulic system failure.

Brake Pad Indicator This lamp illuminates when the sensors at the wheel brake units see that the brake pads are worn too thin. With the lamp lit, the driver should avoid hard braking and have the brakes serviced as soon as possible.

Parking Brake Warning Lamp Vehicles built specifically for Canada may have a separate brake warning lamp that indicates when the parking brake is applied.

Brake Fluid Level Warning Light This light is connected to the brake fluid level sensor in the brake fluid reserve tank. If brake fluid decreases to less than the specified volume in the reservoir, the sensor is actuated and the light comes on while the engine is running.

Low Fuel Warning Light This particular component monitors the fuel level. When it drops below a quarter full, an electronic switch in the module closes and power is applied to illuminate the low fuel indicator light.

Check Engine Warning Light This warning is provided to indicate the condition of the vehicle's engine and its control systems. If there is a fault in the system, the warning light comes on while the engine is running. Check engine lights may be triggered by oil pressure, coolant temperature, or by the engine control computer that monitors several engine systems and illuminates the warning light whenever it senses a fault. It may also illuminate when the computer has stored a fault or diagnostic code in its memory.

Check Filler Cap This lamp will be illuminated when the gas filler cap is not tight or off and when the engine control system senses a problem with the fuel system.

Door Ajar Warning Light When the ignition is turned on, if the doors are left open or are ajar, this light comes on.

Add Washer Fluid Lamp Obviously the purpose of this lamp is to inform the driver of a low level in the windshield washer fluid reservoir.

Add Coolant Lamp The purpose of this lamp is to inform the driver of low coolant levels in the cooling system.

Antilock Light If an antilock brake system fault is present, the antilock brake module grounds the indicator circuit, and the antilock light goes on. Vehicles built for Canada may have a different symbol on their warning lamp.

Traction/Stability Control Lamp This lamp or lamps are illuminated with red lights when there is a problem with the traction control and/or stability control systems. The lamps are lit with yellow lights when the system is actively regulating drive torque and braking force. Vehicles built for Canada may have a different symbol on their warning lamp.

Oil Pressure Indicator Light The light indicates whether the oil pump is feeding oil under normal pressure to various parts of the engine. The indicator light is operated by an oil pressure switch located in the engine's lubricating system. Some vehicles illuminate this lamp in yellow or red to indicate the action the driver should take, red meaning the engine has an oil pressure problem and the engine should be shut down and yellow indicating the oil level is low and should be topped off as soon as possible.

Charge Indicator Light The light indicates the condition of the charging system. If there is something wrong with the charging system, the light comes on while the engine is running.

Transmission Indicator This is part of an automatic transmission's control system. If the system detects a fault, it may operate the transmission in the fail-safe mode and will illuminate the warning light to inform the driver of the problem and to alert him or her that the transmission may not be working normally.

Drive Indicator Light Some four-wheel-drive vehicles have a lamp, which when lit, indicates that the vehicle is in the four-wheel-drive mode of operation.

Air Suspension Light Voltage is present at the air suspension indicator at all times. If an air suspension fault exists, the ground of the light circuit is closed and the indicator illuminates.

Fog Light Indicator This lamp is illuminated when the fog lights are turned on.

High-Beam Light With the headlights turned on and the main light switch dimmer switch in the high-beam position, the indicator illuminates.

Left and Right Turn Indicators With the multifunction switch in the left or right turn position, voltage is applied to the circuit to illuminate the left or right turn indicator. The turn indicator flashes in unison with the exterior turn signal bulbs.

Stop Light Warning Light The light is controlled by the stop light checker. This checker consists of a **reed switch** and magnetic coils. Under normal conditions, magnetic fields form around the coils by the current flowing through each light while the stop light switch is on. These magnetic fields cancel each other because the coils are wound in opposite directions. As a result, the reed switch remains off and the warning light is off. If either the left or right side stop light fails, current flows through only one coil and the resultant magnetic field causes the reed switch to turn on. The warning light remains lighted as long as the brake pedal is depressed.

Cruise Control Light This lamp is lit whenever the cruise control is turned on.

Rear (or Front) Defrost Indicator Light When this light is lit, the defroster or deicer is operating.

Sound Warning Devices

Various types of tone generators, including buzzers, chimes, and voice synthesizers, are used to remind drivers of a number of vehicle conditions. These warnings can include fasten seat belts, air bag operational, key left in ignition, door ajar, and light left on. **Figure 21–16** is a tone generator system schematic.

Park Distance Control (PDC) This feature uses sensors to measure the distance the front and/or rear of the vehicle is from an object **(Figure 21–17)**. An audible signal changes in frequency as the vehicle gets closer to an object. As the distance between the vehicle and object decreases, the intervals between the tones become shorter.

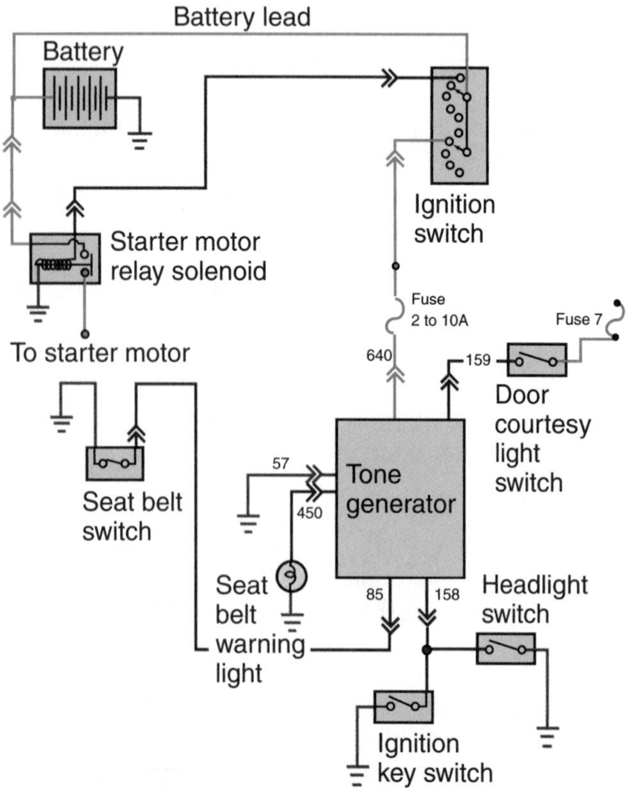

Figure 21-16 A tone generator warning system. *Courtesy of Ford Motor Company*

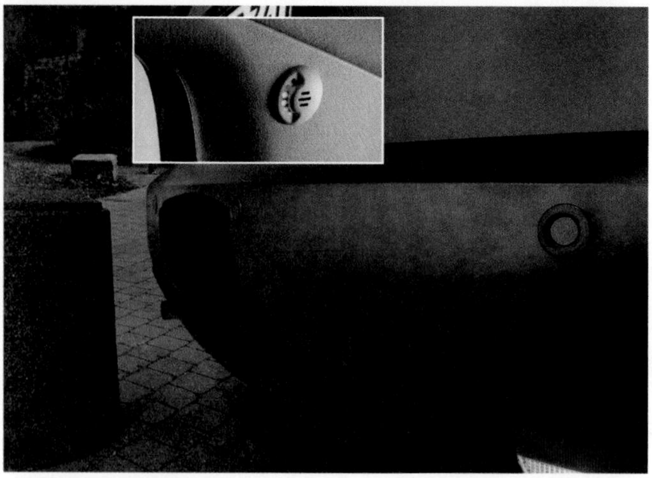

Figure 21-17 The sensors measuring the distance from the rear of the car to obstacles trigger the warning system (insert) inside the vehicle. *Courtesy of Robert Bosch*

When the object is very close, the tone is emitted continuously. The system uses four ultrasonic sensors at the rear and the front of the vehicle. Some systems include a visual indication of the distances to the obstacles, in addition to the audible warning.

Some systems allow the front sensors to be manually turned off in special situations such as stop-and-go traffic. The rear sensors automatically turn on when the transmission is placed into reverse.

Graphic Displays

Graphic displays are translucent drawings or pictures of a vehicle. These displays have lamps located at various spots in the graphic. When a lamp is lit, the area by the lamp has a problem. These indicators can note that the trunk is open or a light bulb is not working **(Figure 21–18)**.

Figure 21-18 Notice the small graphic of the car to see if the trunk or any doors are open. *Courtesy of American Honda Motor Co., Inc.*

DRIVER INFORMATION CENTERS

The various gauges, warning devices, and comfort controls may be grouped together into a driver information center or instrument cluster. This information center may be simple **(Figure 21–19)** or it may be an all-encompassing cluster of information. The purpose of this message center is to keep the driver alert to the information provided by the system. The types and extent of information vary from one system to another.

In addition to standard warning signals, the information center may provide such vital data as fuel range, average or instantaneous fuel economy, fuel used since reset, time, date, estimated time of arrival (ETA), distance to destination, elapsed time since rest, average car speed, percent of oil life remaining, and various engine-operating parameters.

Other electronic displays and controls can be found on today's vehicles.

Heads-Up Display

A **heads-up display (HUD)** projects visual images on the windshield by a vacuum fluorescent light source to complement existing traditional in-dash instrumentation **(Figure 21–20)**. Because these images are projected in the driver's peripheral field of vision, it is not necessary for the driver to refocus attention or remove his or her eyes from the road to obtain certain pertinent information. Among the images HUD may display are vehicle speed, turn-signal indicators, low-fuel warning, and a high-beam indicator.

Steering Wheel Touch Controls

Steering wheel touch controls are standard on many vehicles **(Figure 21–21)**. Large buttons are located conveniently in the steering wheel, providing control over the more frequently used radio and cruise control functions. This provides the driver with fingertip controls where they are easy to use.

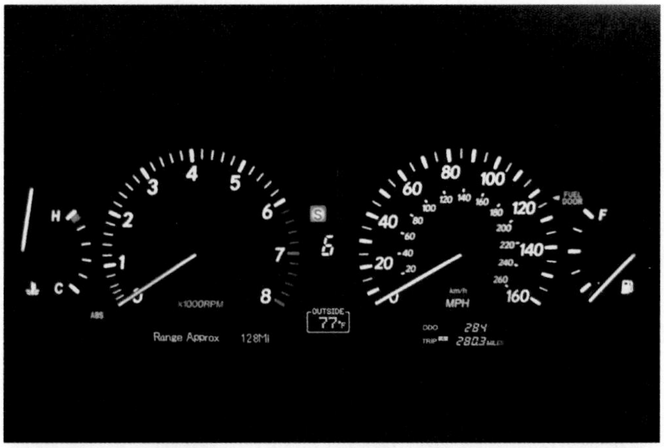

Figure 21-19 Information for the driver appears across the bottom of the instrument cluster. *Reprinted with permission*

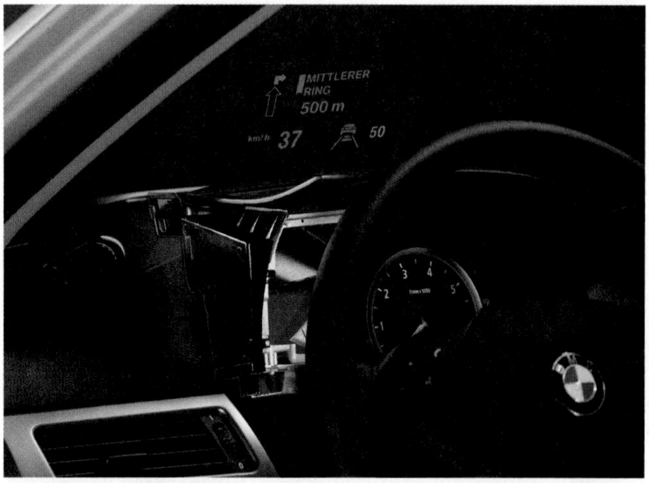

Figure 21-20 The heads-up display (HUD) can project the vehicle's speed onto the windshield, freeing the driver from looking down at the speedometer. *Courtesy of Siemans VDO Automotive*

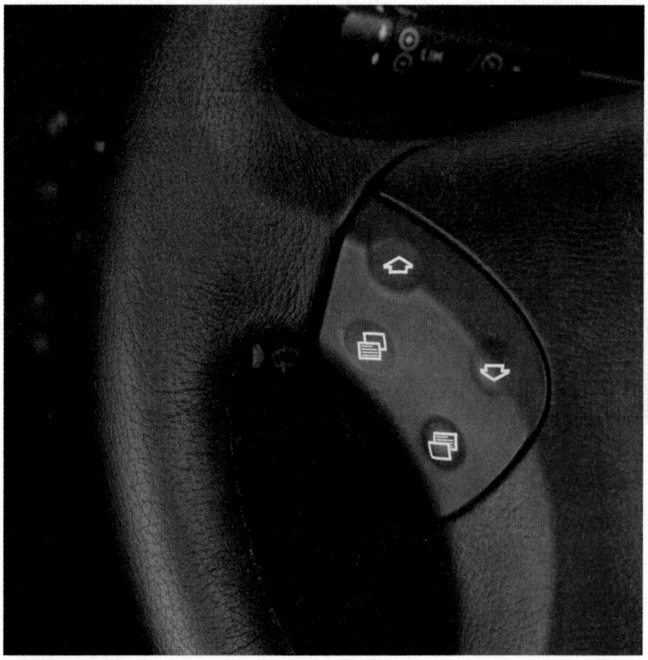

Figure 21-21 Steering wheel touch controls. *Courtesy of DaimlerChrysler Corporation*

iDrive

The iDrive feature offered on some BMWs gives a hint at where the industry, as a whole, may be headed with controls and information displays. The iDrive controller is a modified joystick installed on the center console **(Figure 21–22)**. It controls driving functions, including all active vehicle-control processes, such as those related to information displays. The controller is also used to control the comfort and convenience features of the vehicle. The controller can be pushed, turned, and slid to govern as many as 700 different vehicle functions. The selected functions are then displayed on the control display.

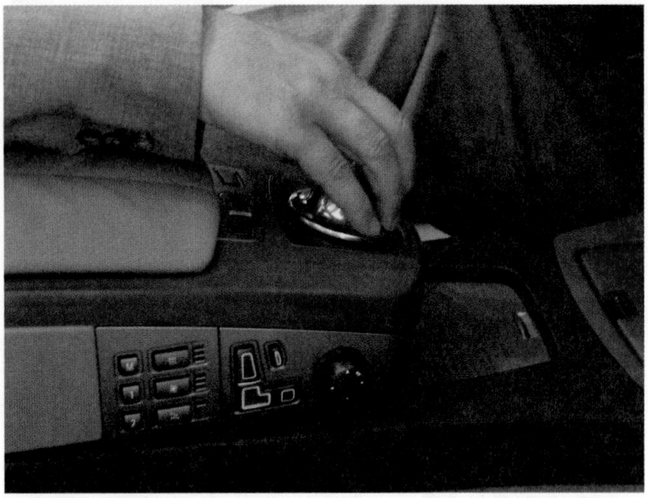

Figure 21-22 The iDrive controller. *Courtesy of BMW of North America, Incorporated*

The controller can be slid in one of eight different directions to access the eight main menus. To move to the submenus, the controller is rotated through electronic detents. The controller is pushed down to select a particular function or setting.

The control display is a color screen with background illumination that adapts to changes in ambient lighting. A photocell monitors the intensity of the ambient light and the illumination of the display responds to that input.

The control display of the iDrive system can also be used to go online to receive and send e-mails, make travel arrangements, get the latest news, and conduct a number of other online tasks. Online access is made through a special Internet portal via the wireless application protocol (WAP).

GENERAL DIAGNOSIS AND TESTING

Diagnosis should begin with a good visual inspection of the circuit. Check all sensors and actuators for physical damage. Check all connections to sensors, actuators, control modules, and ground points. Check wiring for signs of burned or chafed spots, pinched wires, or contact with sharp edges or hot exhaust parts. Also check all vacuum hoses for pinches, cuts, or disconnects.

Service manuals contain detailed information on the diagnosis and testing of the affected system for a particular vehicle. Always refer to it before beginning to diagnose a circuit.

Items to be checked in a malfunctioning gauge or indicator system include the following:

- Fuses
- Indicator bulbs
- Detector switches (indicator systems)
- Sender units (gauge systems)
- IVR (gauge systems)
- Gauges (gauge systems)

Keep in mind intermittent problems are usually caused by:

- Dirty or corroded terminals
- Wire chafing
- Poor wire-to-terminal connections
- Poor mating of connector halves or backed out connector terminals
- Connector body damage

Body Control Modules

Many instruments and warning devices are linked to the vehicle's body control module. In many cases, they are also part of the multiplexing network. Before troubleshooting a gauge or warning system, check the service manual to identify any special procedures or precautions. Often you will need to retrieve codes from the body computer during diagnosis. The typical procedure for doing this involves the use of a scan tool. Some scan tools have the ability to activate the instrument cluster for testing.

CASE STUDY

A customer brings his pickup into the shop because the fuel gauge always shows empty, regardless of how much fuel is in the tank.

The technician verifies the problem and checks to make sure there is gasoline in the tank. Then she checks the fuse for the circuit. Finding all fuses in good shape, she refers to the service manual to determine if the gauge circuit has an IVR. This particular vehicle does not so she disconnects the wire to the fuel tank sending unit and checks it for voltage. Battery voltage is present when the ignition switch is turned on; therefore, she knows the power side of the circuit is good. She then connects a 10-ohm resistor to the wire and grounds the circuit. The fuel gauge now reads FULL. According to the service manual, this is what should happen. She now knows the problem is either a faulty sending unit or a poor circuit ground. She reconnects the wire to the sending unit and then connects a jumper wire from the ground of the sending unit to a known good ground. The fuel gauge now shows a reading that seems to be accurate, so she cleans and corrects the ground circuit for the gauge. Then she verifies the repair by adding a few gallons of fuel and watching the response of the gauge.

The scan tool is plugged into the system's diagnostic connector. Some manufacturers provide a single diagnostic connector and the technician chooses the system to be tested through the scan tool. Always refer to the correct service manual for the vehicle being serviced. Use only the methods recommended by the manufacturer for retrieving diagnostic trouble codes (DTCs). Once the DTCs are retrieved, follow the appropriate diagnostic chart for instructions on isolating the fault. It is also important to check the codes in the order required by the manufacturer.

KEY TERMS

Balancing coil
Bobbin gauge
Heads-up display (HUD)
Instrument panel
Instrument voltage regulator (IVR)
International Standards Organization (ISO)
Low fuel warning (LFW)
Odometer
Reed switch
Tachometer
Thermal gauge

SUMMARY

- The two basic types of instrument panel displays are analog and digital. In an analog display, an indicator moves in front of a fixed scale to indicate a condition. A digital display uses numbers instead of a needle or graphic symbol.
- Three types of digital electronic displays are used today: light-emitting diode, liquid crystal diode, and vacuum fluorescent.
- A gauge circuit is often made of the gauge, a sending unit, and an instrument voltage regulator (IVR).
- Two types of electrical analog gauges—magnetic and thermal—are commonly used with sensors or sending units.
- Indicator lights and warning devices are generally activated by the closing of a switch.
- Various types of tone generators, including buzzers, chimes, and voice synthesizers, are used to remind drivers of a number of vehicle conditions.
- Park distance control uses sensors to measure the distance the front and/or rear of the vehicle is from an object and emits an audible warning as the vehicle gets closer to an object.
- The various gauges, warning devices, and comfort controls may be grouped together into a driver information center or instrument cluster.
- A heads-up display projects visual images on the windshield by a vacuum fluorescent light source to complement existing traditional in-dash instrumentation.
- Diagnosis of gauges, indicators, and warning lights should begin with a good visual inspection of the circuit. Check all sensors and actuators; connections and wires to sensors, actuators, control modules, and ground points; and all vacuum hoses.
- Before troubleshooting a gauge or warning system, check the service manual to identify any special procedures or precautions.

REVIEW QUESTIONS

1. *True or False?* A heads-up display projects indicator and warning lights on the windshield rather than illuminating them in the instrument panel.
2. What is the purpose of an IVR?
3. Explain how a simple magnetic-type ammeter works.
4. What gauge indicates engine speed?
5. Describe the two types of instrument panel displays.
6. What is the device found in some fuel tanks to prevent fuel gauge fluctuations due to rough road surfaces?
7. What is the correct way to check a coolant temperature sensor?
8. The indicator needle of a cable-driven speedometer is held to the zero position by __________.
   a. magnetic force
   b. the weight of the needle
   c. the speedometer cable
   d. a hairspring

TECH MANUAL

The following procedure is included in Chapter 21 of the *Tech Manual* that accompanies this book:

1. Removing, checking, and replacing a temperature sending unit.

9. Which of the following is *not* a likely cause for an intermittent gauge problem?

a. poor mating of connector halves or backed out connector terminals

b. an open ground wire

c. connector body damage

d. poor wire-to-terminal connections

10. While discussing what service interval displays base the time for the next service on, Technician A says future service is based on the type of driving the vehicle has seen since the last service. Technician B says the next interval is based on the number of miles or kilometers since the last service. Who is correct?

a. Technician A **c.** Both A and B

b. Technician B **d.** Neither A nor B

11. None of the engine's gauges work, so Technician A checks the power to the IVR and the IVR's ground. Technician B begins by checking the fuse and then checks for voltage at the last point common to all the malfunctioning gauges. Who is correct?

a. Technician A **c.** Both A and B

b. Technician B **d.** Neither A nor B

12. Which of the following would *not* cause the check engine warning light to come on?

a. a bad generator

b. low oil pressure

c. high coolant temperature

d. A fault code recognized and stored by the computer

13. What type of sending unit is typically used to monitor oil pressure?

14. What is the major difference between an indicator lamp and a warning light?

15. *True or False?* The traction/stability control lamp turns on whenever the system is actively regulating drive torque and braking force.

16. What type of memory is used to store the accumulated mileage in an electronic odometer?

a. RAM **c.** PROM

b. ROM **d.** EPROM

17. The oil pressure light stays on whenever the engine is running. The oil pressure has been checked and it meets specifications. Technician A says that a ground in the circuit between the indicator light and the pressure switch could be the cause. Technician B says that an open in the pressure switch could be the cause. Who is correct?

a. Technician A **c.** Both A and B

b. Technician B **d.** Neither A nor B

18. A digital speedometer constantly reads zero mph. Technician A says the problem may be the vehicle speed sensor. Technician B says the problem may be the throttle position sensor. Who is correct?

a. Technician A **c.** Both A and B

b. Technician B **d.** Neither A nor B

19. Which of the following is *not* a true statement about lamp-out warning lights?

a. The lamp-out warning module measures small changes in voltage levels.

b. An electronic switch in the module completes a ground path for the indicator lights in the event of a bulb going out.

c. A special resistance wire is used in multibulb systems.

d. When a bulb burns out, the module senses the increased voltage drop and turns on the indicator lamp.

20. All gauges read low. Technician A says the power connection to the instrument cluster may be corroded. Technician B says the cluster's IVR may be open. Who is correct?

a. Technician A **c.** Both A and B

b. Technician B **d.** Neither A nor B

22

ELECTRICAL ACCESSORIES

OBJECTIVES

■ Know the basic operation of electric windshield wiper and washer systems. ■ Explain the operation of power door locks, power windows, and power seats. ■ Determine how well the defroster system performs. ■ Identify the components of typical radio and audio systems. ■ Understand how cruise or speed control operates and the differences of various systems. ■ Describe the operation of keyless entry systems. ■ Identify the various security disabling devices. ■ Understand the operation of the various security alarms.

Most electrical accessories make driving safer, easier, and more pleasant for the driver and passengers. This chapter covers many of the common accessories. Other automotive electric and electronic equipment, such as passive seat belts and air bags is described elsewhere in this book.

BODY CONTROL MODULES

Many of the accessories found on today's vehicles are controlled and operated by a body computer or control module (BCM) system. **(Figure 22–1)**. It is important to understand body computer systems and how to diagnose system problems before learning the details of these accessories.

The basic operation of a computer was explained in Chapter 15. Remember that the operation of a computer is divided into four basic functions: input, processing, storage, and output. Understanding these functions will help you organize the troubleshooting process. When a system is tested, you are basically trying to isolate a problem to one of these functions.

In the process of controlling the various systems, the BCM continuously monitors operating conditions for possible system malfunctions. The computer compares system conditions against programmed parameters. If the conditions fall outside of these limits, the BCM will detect the malfunction. A trouble code stored in the BCM memory indicates the portion of the system at fault. A technician can access this code to aid in diagnosis.

If a malfunction results in improper system operation, the computer may minimize the effects of the malfunction by using fail-safe action. During this mode of operation, the computer controls a system based on programmed values instead of the input signals, which allows the system to operate on a limited basis instead of shutting down completely.

Trouble Codes

The method used to retrieve codes from the BCM's memory varies greatly; always refer to the service manual for the correct procedure. Depending on the system, the computer may store codes for long periods of time or lose the code when the ignition is turned off.

On systems that do not retain codes after the ignition is switched off, operate the vehicle until the problem occurs again. Then retrieve the fault code before switching the ignition off. Remember the trouble code does not necessarily indicate the faulty component; it only indicates the circuit of the system that is not operating properly. To locate the problem, follow the diagnostic procedure in the service manual for the code received.

Diagnosis should begin with a good visual inspection of the circuit involved with the code. Check all sensors and actuators for physical damage. Check all connections to sensors, actuators, control modules, and ground points. Check wiring for signs of burned or chafed spots, pinched wires, or contact with sharp edges or hot exhaust parts. Also check all vacuum hoses for pinches, cuts, or disconnects.

Entering Diagnostics

There are as many ways to perform BCM diagnostics as there are automobile manufacturers. Nearly all vehicles need a scan tool. The scan tool is plugged into the diagnostic connector for the system being tested. Some manufacturers provide a single diagnostic connector

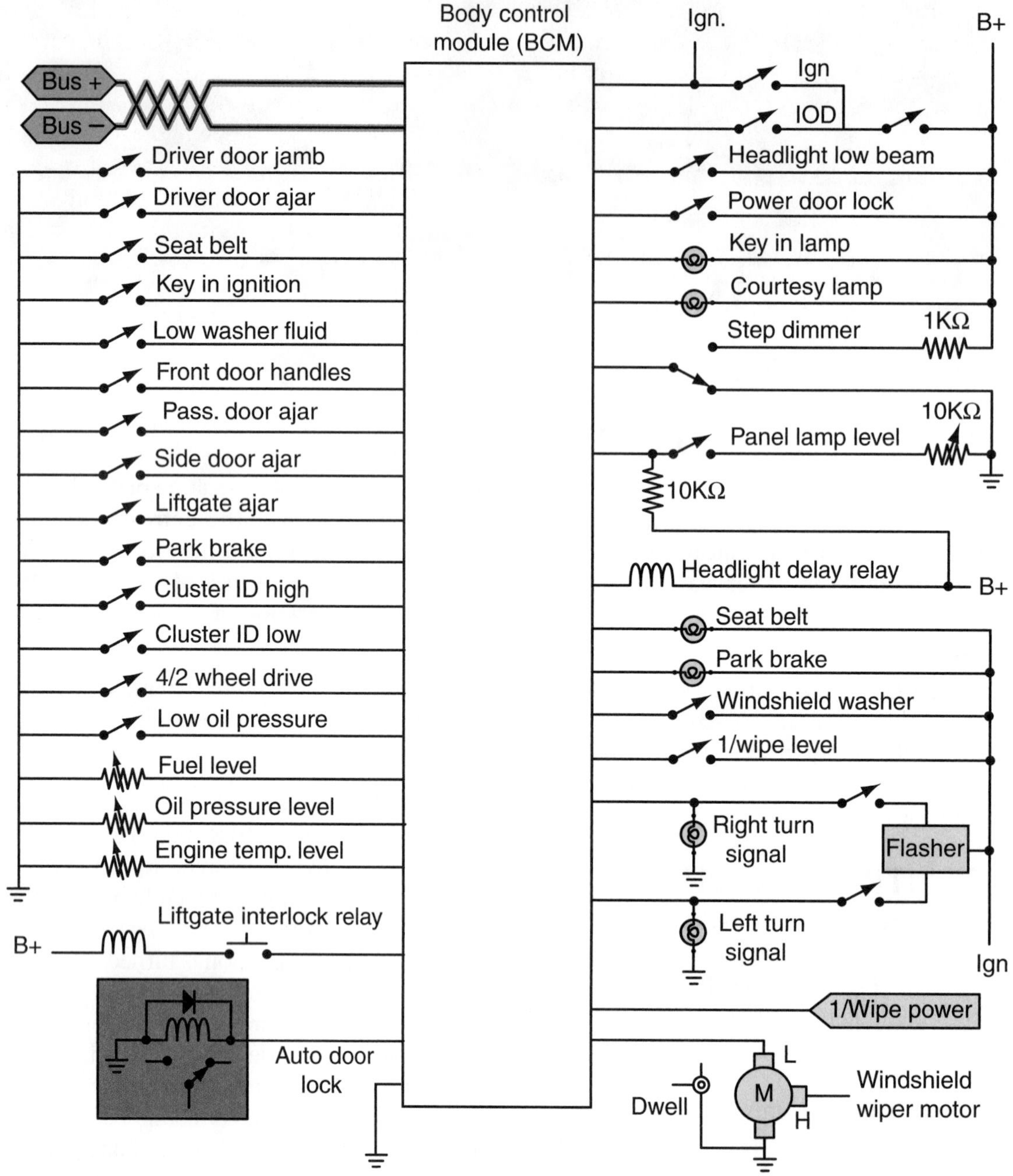

Figure 22-1 The body computer regulates many of the vehicle's electrical systems. *Courtesy of DaimlerChrysler Corporation*

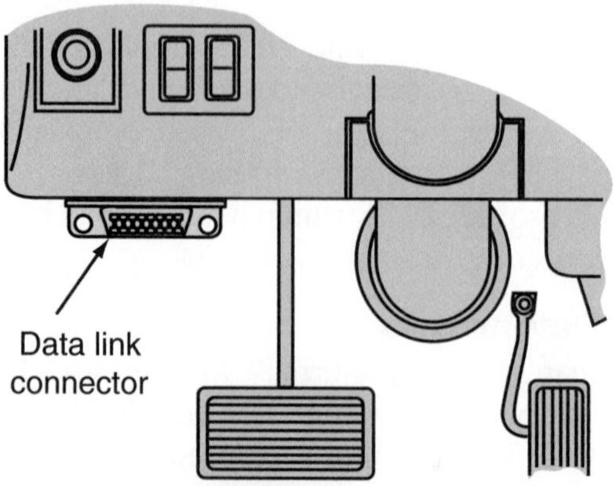

Figure 22-2 The DLC for accessing diagnostic trouble codes.

(Figure 22–2) and the technician chooses the system to be tested through the scan tool. Use only the methods recommended by the manufacturer for retrieving diagnostic trouble codes (DTCs). Once the DTCs are retrieved, follow the appropriate diagnostic chart to isolate the fault.

WINDSHIELD WIPER/WASHER SYSTEMS

There are several types of windshield wiper systems. Both rear and front systems can be found on a vehicle. Headlight wipers and washers are also available that work in unison with the windshield wipers. The two basic designs of windshield wiper systems used on today's vehicles are a standard two- or three-speed system or a two- or three-speed system with an intermittent feature. Windshield wiper motors can have electromagnetic fields or permanent magnetic fields.

Permanent Magnet Motor Circuits

With permanent magnetic fields, motor speed is controlled by the placement of the brushes on the commutator. Three brushes are used: common, high speed, and low speed. The common brush carries current whenever the motor is operating. The low- and high-speed brushes are placed in different locations based on motor design. The high-speed and common brushes can oppose each other, while the low-speed brush can be offset **(Figure 22–3)**. Other designs have opposing low-speed and common brushes with the high-speed brush offset or centered between them. This arrangement is the most common.

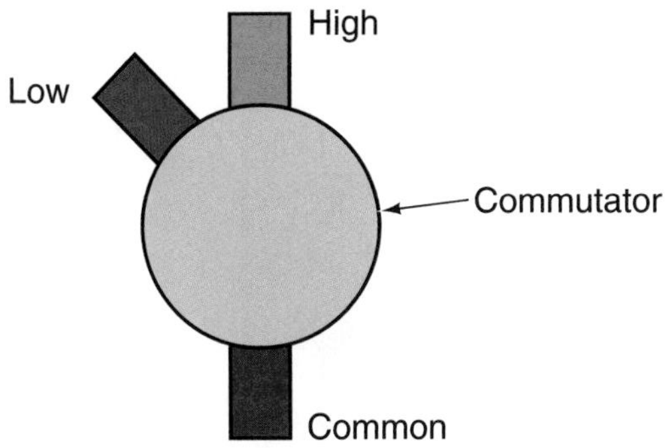

Figure 22-3 One style of brush arrangement has the high-speed brush opposite the common brush.

The placement of the brushes determines how many armature windings are connected in the circuit. When battery voltage is applied to fewer windings, there is less magnetism in the armature and less counter-EMF (CEMF). With less CEMF in the armature, armature current is higher. This high current results in higher motor speeds. When more windings are energized, the magnetic field around the armature is greater and there is more CEMF, resulting in lower current flow and slower motor speeds.

A park switch is incorporated into the motor assembly and operates off a cam or latch arm on the motor's gear **(Figure 22–4)**. The switch supplies voltage to the motor after the wiper control switch has been turned off. This allows the motor to continue running until the wipers have reached their park position. The park switch changes position with each revolution of the motor. The switch remains in the run position for approximately 90% of the revolution. It is in the park position for the remaining 10% of the revolution. This does not affect the operation of the motor until the wiper control switch is placed in the park position.

When the wiper control is in the high-speed position, voltage is applied through the switch to the high-speed brush **(Figure 22–5)**. Wiper 2 moves with wiper 1 but does not complete any circuits. When the switch is moved to the low-speed position, voltage is applied through

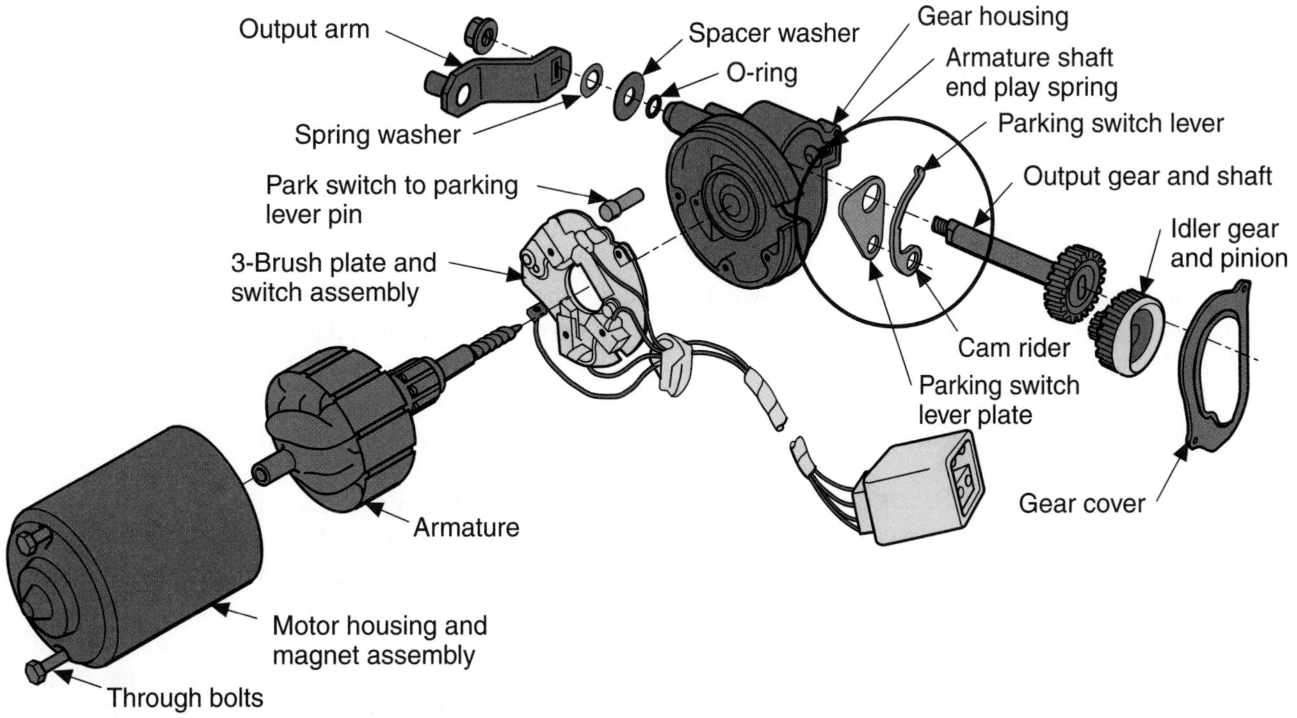

Figure 22-4 An exploded view of a wiper motor with a park mode.

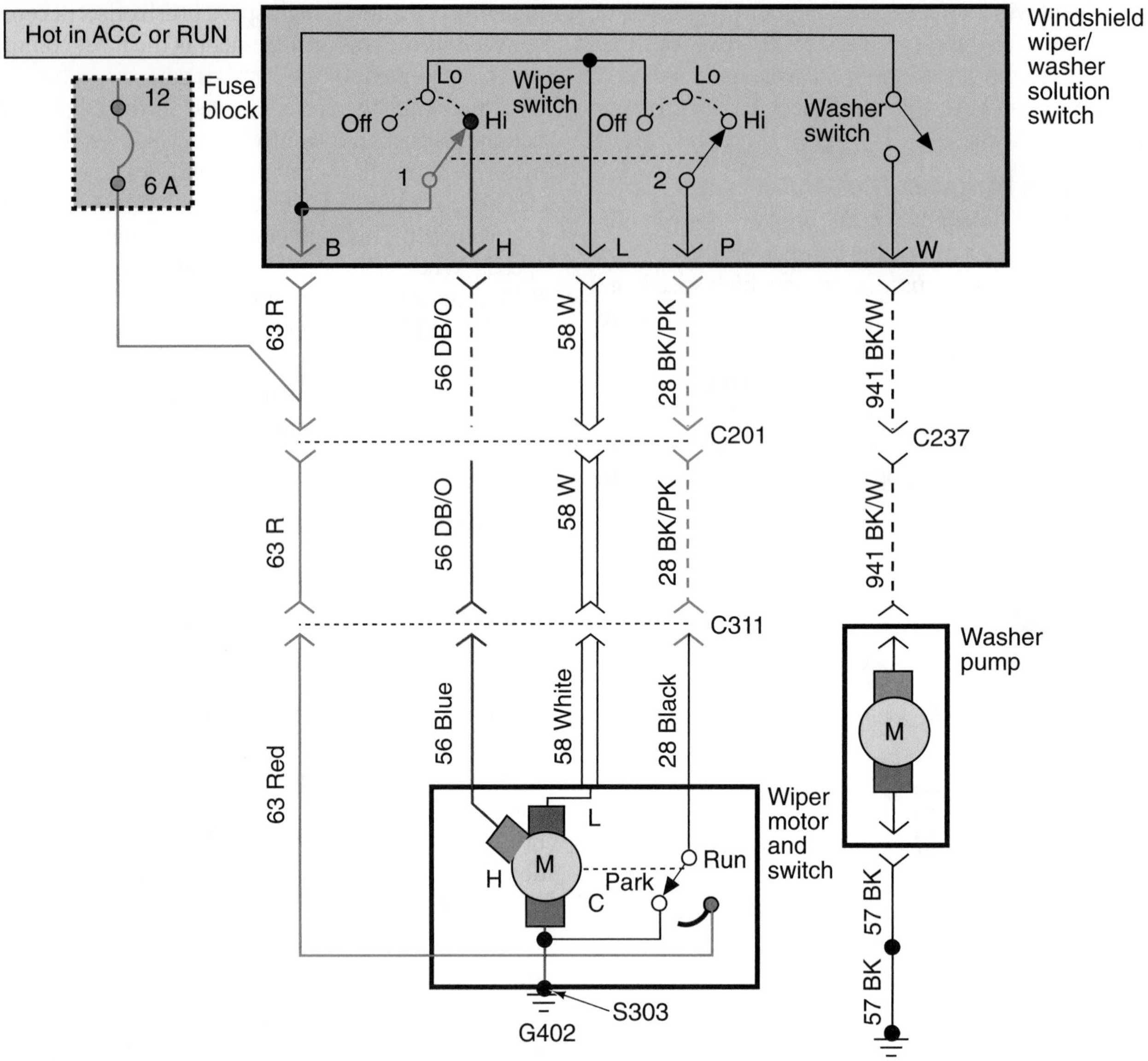

Figure 22-5 Current flow in the high-speed mode of a permanent magnet motor. *Courtesy of Ford Motor Company*

wiper 1 to the low-speed brush. Wiper 2 also moves, but does not complete any circuits.

When the switch is returned to the off position, wiper 1 opens. Voltage is applied to the park switch and wiper 2 allows current to flow to the low-speed brush. When the wiper blades are in their lowest position, the park switch is moved to the park position. This opens the circuit to the low-speed brush and the motor shuts off.

Electromagnetic Field Motor Circuits

Motors with electromagnetic fields have two brushes riding on the armature, a positive and a negative. The speed of the motor depends on the strength of the magnetic fields. Some two-speed and all three-speed wiper motors use two electromagnetic field windings **(Figure 22–6)**. The two field coils are wound in opposite directions so that their magnetic fields oppose each other. The series field is wired in series with the brushes and commutator. The shunt field forms a separate circuit off the series circuit to ground. The strength of the total magnetic field determines the speed of the motor.

A ground side switch determines the path of current and the speed of the motor. One current path is directly to ground after the field coil and the other is to ground through a resistor. With the switch in the off position, voltage is not supplied to the motor. When the switch is placed in the low-speed position, the relay's contacts close and voltage is applied to the motor. The second wiper of the switch provides the path to ground for the shunt field. With no resistance in the shunt field coil, the shunt field is very strong and bucks the magnetic field of the series field. This results in slow motor operation.

When the switch is in the high-speed position, the shunt field finds its ground through the resistor. This re-

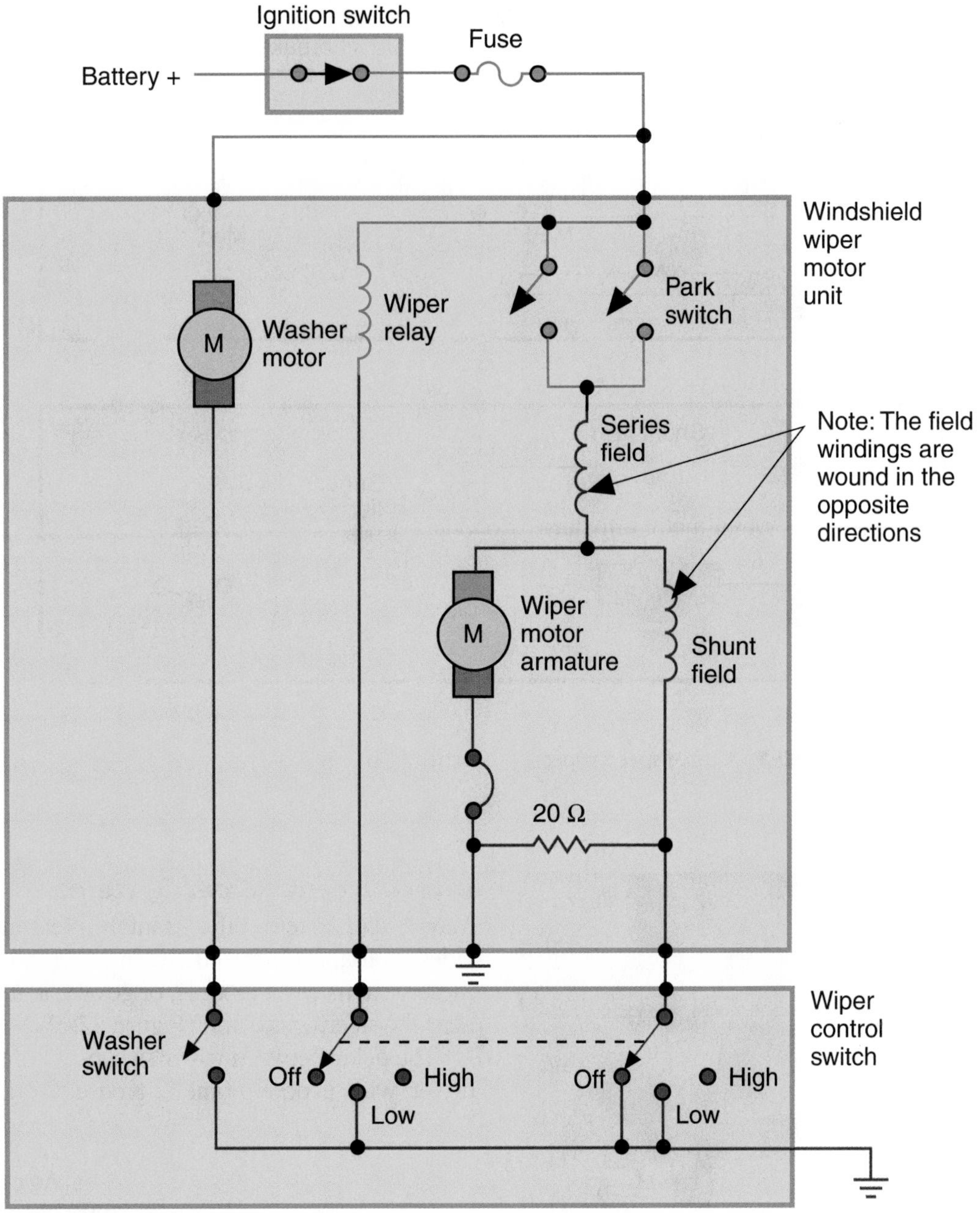

Figure 22-6 The schematic of a two-speed wiper circuit.

sults in low current and a weak magnetic field in the shunt coil; therefore, the armature turns at a higher speed.

Three-Speed Motors The control switch for a three-speed motor determines what resistors, if any, will be connected to the circuit of one of the fields **(Figure 22–7)**. When the switch is in the low-speed position, both field coils have the same amount of current flow. Therefore the total magnetic field is weak and the motor runs slowly.

When the switch is in the medium-speed position, current flows through a resistor before going to the shunt field. This connection weakens the shunt coil and the motor's speed increases.

With the switch in the high-speed position, a resistor of greater value is connected to the shunt field. This connection weakens the magnetic strength of the coil and the motor runs faster.

Windshield Wiper Linkage and Blades Several arms and pivot shafts make up the linkage used to transmit the rotation of the motor to oscillate the windshield wipers. As the wiper motor runs, the linkage rotates the arms from left to right. The arrangement of the linkage causes the wipers' pivot points to oscillate. The wiper arms and blades are attached directly to the two pivot points.

A few wiper systems have two wiper motors that operate in opposite directions, thus creating the oscillation motion of the wipers **(Figure 22–8)**. These systems also occupy less space in the engine's cowl area.

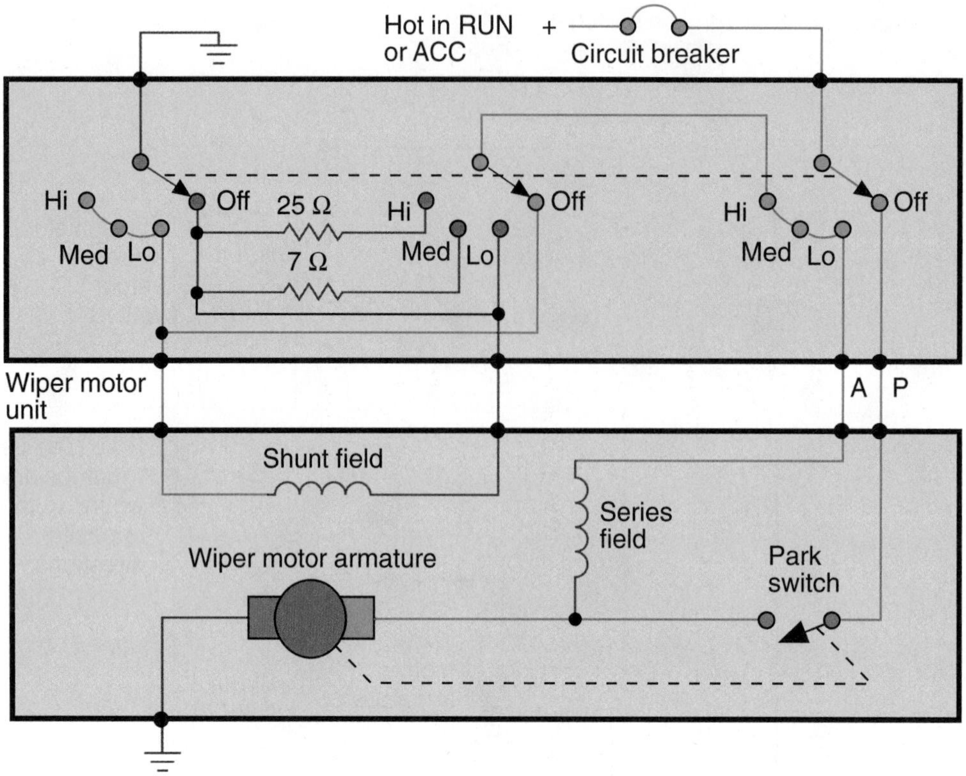

Figure 22-7 A three-speed wiper motor schematic.

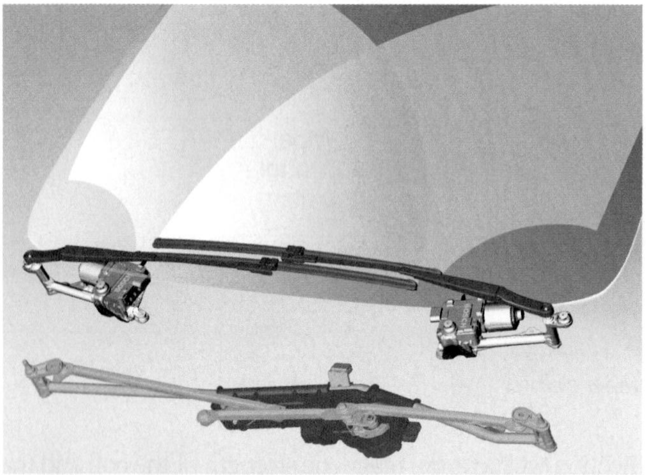

Figure 22-8 (Top) This system uses two motors that operate in the opposite direction. (Bottom) A typical single motor linkage. Note the differences in required space and complexity. *Courtesy of Robert Bosch*

Rear-Window Wiper/Washer System This system is typically found on hatchbacks, vans, and SUVs and has a separate switch to control power to the wiper motor. The parking function is completed within the rear-window wiper motor and switch. Check the fuse or circuit breaker if any wiper/washer system is not working. If the wiper still does not work, trace the power flow through the system following the electrical schematic in the service manual.

Intermittent Wiper Systems

Many wiper systems offer an intermittent mode that provides a variable interval between wiper sweeps. Many of these systems use a module, or governor, mounted on or near the steering column **(Figure 22–9)**.

The delay between wiper sweeps is controlled by the driver with a potentiometer-type control. By rotating

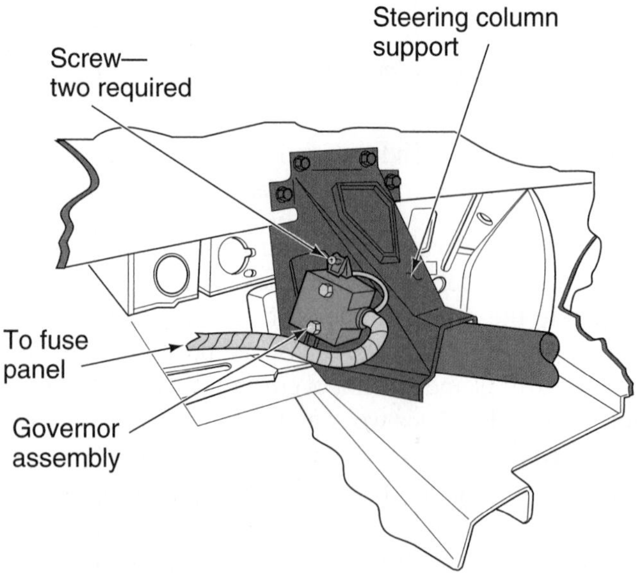

Figure 22-9 An intermittent wiper module. *Courtesy of Ford Motor Company*

the intermittent control, the resistance value changes. The module contains a capacitor that is charged through the potentiometer. Once the capacitor is saturated, the electronic switch is triggered and current flows to the wiper motor. The capacitor discharge is long enough to start the wiper operation and the park switch is returned to the run position. The wiper will continue to run until one sweep is completed and the park switch opens. The amount of time between sweeps is based on the length of time required to saturate the capacitor. As more resistance is added to the potentiometer, it takes longer to saturate the capacitor.

Rain-Sensing Wipers Some vehicles have a setting for windshield wiper operation that responds to water on the windshield. The sensor for these wipers is usually located in the center and at the top of the windshield behind the rear view mirror. The sensor transmits an infrared light onto the windshield's surface through a special optical element **(Figure 22–10)**. When the windshield is dry, all of the light is reflected back to the sensor. The windshield's ability to reflect light starts to change as soon as moisture begins to accumulate on the glass. This allows the infrared beam to penetrate through the windshield, thus reducing the amount of reflected light. This lower level of reflected light serves as an index indicating higher levels of moisture on the windshield's surface. The rain sensor uses all changes in reflected light as the basis for determining the intensity of the rain. In response, the number of sweeps made by the windshield wipers increases or decreases. The sensitivity level of this system can be adjusted by the driver.

Speed-Sensitive Wipers Some vehicles have speed-sensitive wipers that vary the speed or intermittent intervals according to vehicle speed. These systems are typically controlled by the BCM in response to inputs from the VSS.

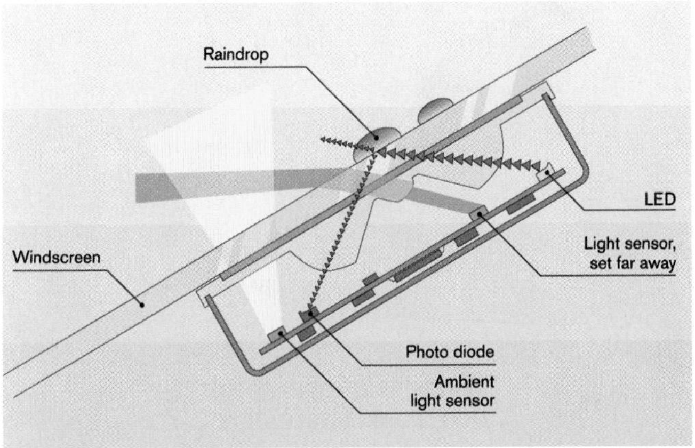

Figure 22–10 The basic operation of a rain sensor. *Courtesy of Robert Bosch Corp.*

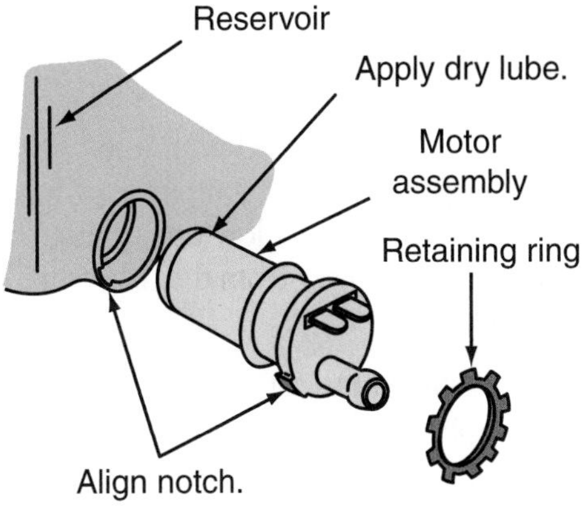

Figure 22–11 Installation of a washer pump and motor in a fluid reservoir.

Windshield Washers

Windshield washers spray a fluid onto the windshield and work in conjunction with the wiper blades to clean the windshield of dirt. Most systems have the washer pump installed in the fluid reservoir **(Figure 22–11)**. A few systems, such as some from GM, use a pulse-type pump that operates off the wiper motor.

Washer systems are activated by holding the washer switch. If the wiper/washer system also has an intermittent control module, a signal is sent to the module when the washer switch is activated. An override circuit in the module operates the wipers at low speed for a programmed length of time. The wipers either return to the parked position or operate in intermittent mode, depending on the design of the system.

Some vehicles have washers that clean the headlights **(Figure 22–12)** and fog lights for maximum visibility.

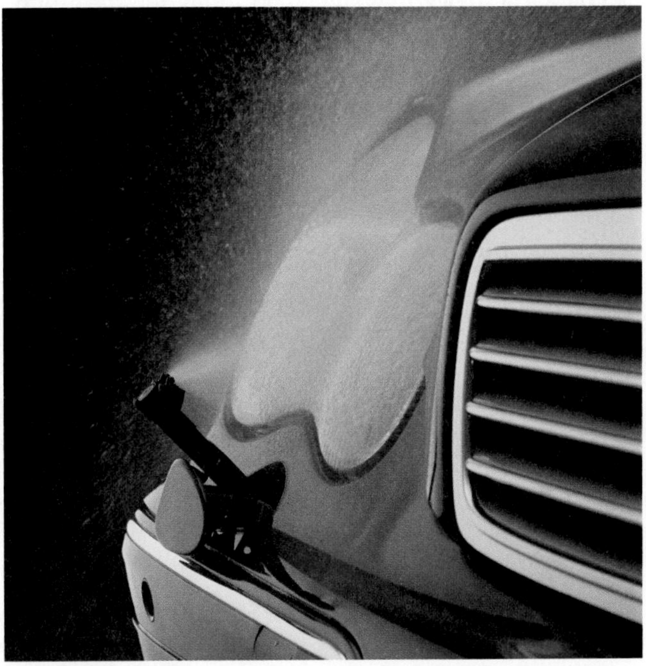

Figure 22–12 Headlight washers. *Courtesy of DaimlerChrysler Corporation*

Headlight washer systems may operate from their own switch and pump or work along with the windshield washer system.

Some vehicles are equipped with a low fluid indicator. The washer fluid level switch closes when the fluid level in the reservoir drops below one-quarter full. Closing the switch allows power from the fuse panel to be applied to the indicator.

Wiper System Service

Customer complaints about windshield wiper operation can include poor wiping, no operation, intermittent operation, continuous operation, and wipers that do not park. Other complaints will be related to wiper arm adjustments, such as slapping the molding or one blade parks lower than the other.

When the wipers move as they should but do not wipe the glass surface they way they should or make noise while moving across the glass, the blades and/or arms should be replaced.

If the wipers work slower than expected, disconnect the wiper linkage at the motor (Figure 22-13). Turn on the wiper system. If the motor runs properly, the problem is the linkage and is not electrical. If the motor runs slower than normal, check for resistance in the circuit.

If the motor does not run at a particular speed or at all, the problem is electrical. Carefully inspect the motor, wires, connectors, and switch. Your diagnosis should continue according to the guidelines given in the service manual. Pay attention to the circuits that could cause the problem, not the entire wiper circuit. Test for voltage at the motor in the various switch positions. Also check the ground circuit. If the motor is receiving the right amount of voltage at the various switch positions and the ground circuits are good, the problem must be the motor. Wiper motors are replaced, not repaired or rebuilt.

WARNING!

Most wiper motors are the permanent magnet type, which can be quite delicate. Do not throw the motor around or hammer on the case. Both of these actions can destroy the magnetic fields.

Washer System Service

Many washer problems are caused by restrictions in the fluid lines or nozzles. To check for restrictions, remove the hose from the pump and operate the system. If the pump ejects a stream of fluid, then the fault is in the delivery system. The exact location of the restriction can be found by reconnecting the fluid line to the pump and disconnecting the line at another location. If the fluid still streams out, the problem is after that new disconnect. If the fluid does not flow out, the problem is before where the hose was disconnected. Repeat this process until the problem is found.

If the pump does not spray out a steady stream of fluid, the problem is in the pump circuit. It should be tested in the same way as any other electrical circuit. Make sure it gets power from the switch when it should, then check the ground. If the power to the pump is good and there is a good ground, the problem is the pump. These are not rebuilt or repaired; they must be replaced.

HORNS/CLOCKS/CIGARETTE LIGHTER SYSTEMS

The purpose and operation of these systems are obvious and their circuits may vary from one model and year to another **(Figure 22–14)**. However, the overall operation remains the same.

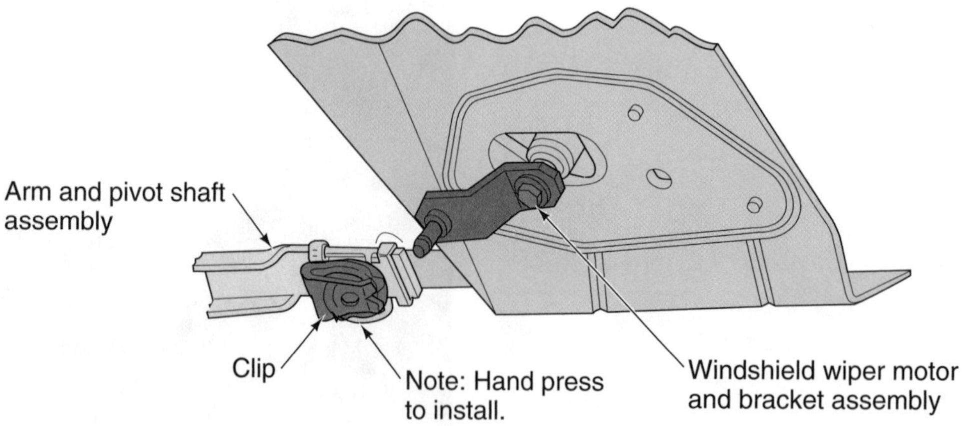

Figure 22-13 Disconnecting the wiper linkage arms at the wiper motor. *Courtesy of Ford Motor Company*

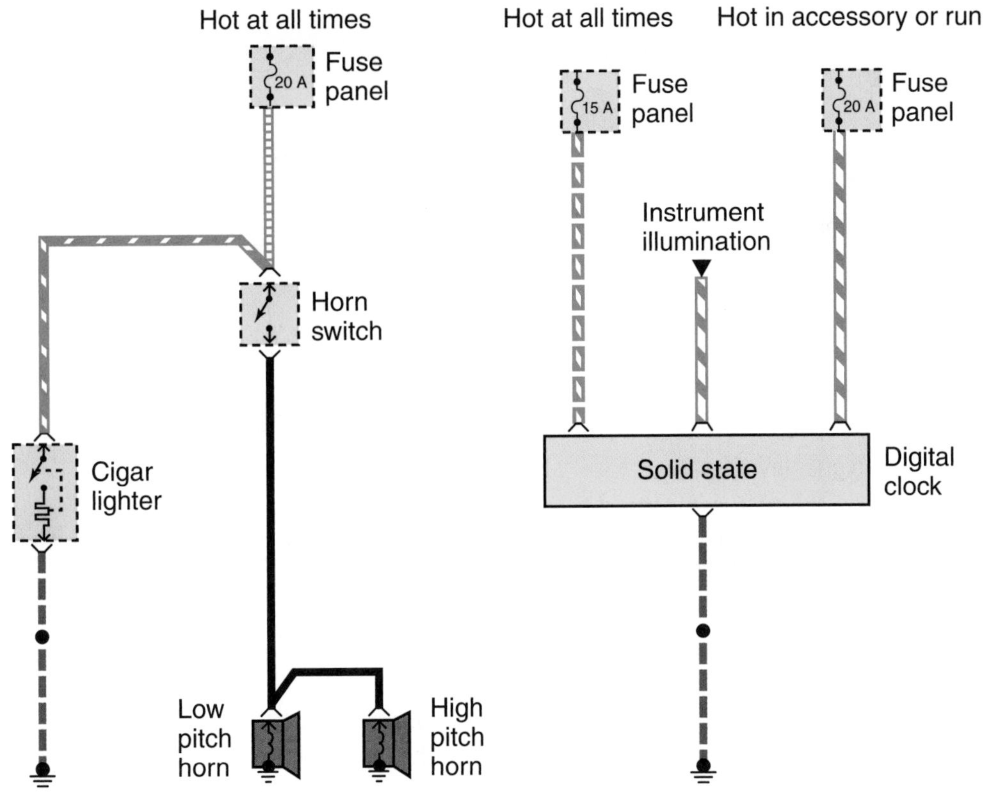

Figure 22-14 A typical horn/clock/cigarette lighter circuit. *Courtesy of Ford Motor Company*

Horns

Most horn systems are controlled by relays. When the horn button, ring, or padded unit is depressed, electricity flows from the battery through a horn lead, into an electromagnetic coil in the horn relay then to the ground. Low current through the coil creates a magnetic field that pulls on the movable arm. This action brings the relay's contacts together, causing the horn to sound.

Most vehicles have the horn switch mounted in the steering wheel assembly. Some horn switches are part of the multifunction switch. Some vehicles are equipped with two horns, each designed to emit a different tone. The two horns provide a fuller sound than one horn can.

Clock

The clock receives power directly from the fuse panel. Some clocks have additional functions. These are explained in the owner's guide for the particular vehicle.

Cigarette Lighter

The cigarette (cigar) lighter is a heating element that automatically releases itself from the pushed-in position when the appropriate heat level is reached.

For an inoperative system, first check the fuse(s). If the fuse is good, make certain that power is present at the lighter receptacle. If power is present, the lighter unit is probably bad. Refer to the service manual for additional troubleshooting information.

Many vehicles also have an additional power outlet that looks like an additional cigarette lighter receptacle. These can be used to power or recharge 12-volt appliances, such as cell phones. Some vehicles also have 110- to 155-volt receptacles to run normal household appliances. Of course you must have the correct plug to put into the receptacle.

CRUISE (SPEED) CONTROL SYSTEMS

Cruise control systems are designed to allow the driver to maintain a constant speed (usually above 30 mph or 48 km/h) without having to apply continual pressure on the accelerator pedal. Selected cruise speeds are easily maintained and can be easily changed. When engaged, the system sets the throttle position to the desired speed. The speed is maintained unless heavy loads and steep hills interfere. The cruise control switch is located on the end of the turn signal or near the center or sides of the steering wheel **(Figure 22–15)**. There are usually several functions on the switch, including off-on, resume, and engage buttons. Cruise control is disengaged whenever the brake pedal or clutch pedal is depressed.

Vacuum Systems

Until recently, the most common type of cruise control system relied on engine vacuum and mechanical linkages.

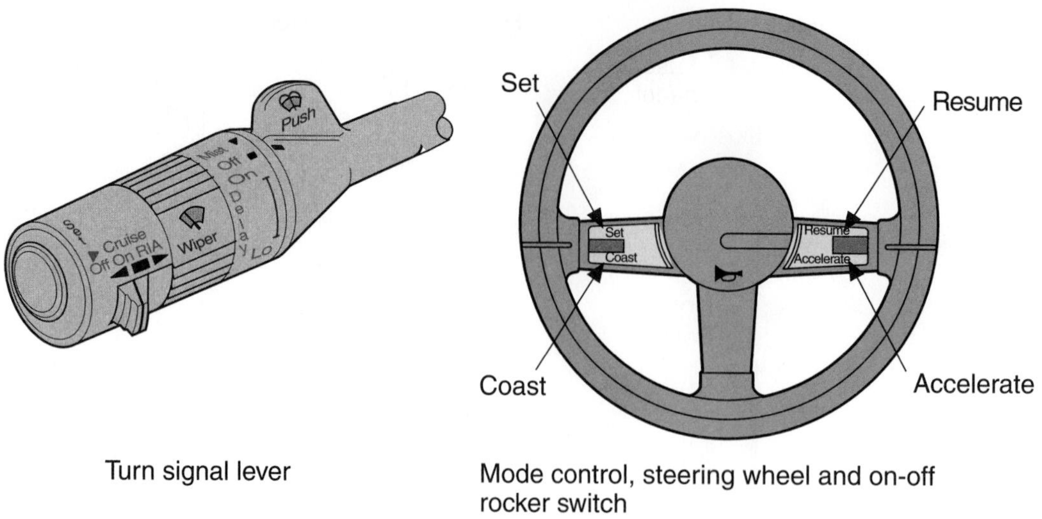

Figure 22-15 The cruise control switch is used to set or increase speed, resume speed, or turn the system off and on. *Courtesy of Ford Motor Company*

The following are the common components in these systems **(Figure 22–16)**:

- When the system is turned on, the **transducer** senses vehicle speed and controls the amount of vacuum applied to a vacuum servo. The amount of vacuum sets the servo position. Vehicle speed is sensed from the lower speedometer cable.
- The **servo unit** is connected to the throttle by a rod or linkage, chain, or cable. The servo unit maintains the desired speed by receiving a controlled amount of vacuum from the transducer. Variations in vacuum change the position of the throttle. When a vacuum is applied, the servo spring is compressed and the throttle is positioned correctly. When the vacuum is released, the servo spring is relaxed and the system is not operating.
- There are two brake-activated switches operated by the position of the brake pedal. When the pedal is depressed, the brake release switch disengages the system. A vacuum release valve is also used to disengage the system when the brake pedal is depressed.

Electronic Cruise Control Systems

Cruise control can use electronic components rather than mechanical components. An electronic control module can be used to move a servo unit **(Figure 22–17)** or it can control an electric stepper motor. Many of today's vehicles are fitted with a stepper motor. The motor moves a strap attached to the cruise control cable, which moves the throttle linkage. The motor may be a separate unit or built into the cruise control module. Vehicles with electronic throttle control (throttle-by-wire) do not need a separate cruise control module, stepper motor, or cable to control engine speed. The PCM has full control of the throttle and therefore the circuitry of the PCM operates the cruise control system.

The vehicle speed sensor (VSS) is used to monitor or sense vehicle speed. The computer has several other inputs to help determine the operation of the servo. These inputs include a brake release switch (clutch release switch); a speedometer, buffer amplifier, or generator speed sensor; and a lever-mounted mode switch or speed control on the steering wheel (signal to control the cruise control). Most cruise control systems are connected to the vehicle's multiplexing network and should be serviced and tested according the manufacturer's recommendations.

Adaptive Cruise Control

Like other cruise control systems, adaptive cruise control automatically maintains the desired speed of the vehicle, but it also maintains a safe distance between vehicles. The desired distance between vehicles is set by the driver. The system also adjusts the speed of the vehicle to mirror that of a slower vehicle in front of it, then maintains that speed. A laser or radar sensor **(Figure 22–18)** mounted near the front bumper serves as the eyes for the system. Other vehicles traveling within the sensor's range reflect the radar waves **(Figure 22–19)**, and the sensor picks up the returning signals. The control unit uses this information to determine the position and speed of the preceding vehicle. When the system detects a slower moving vehicle in the same lane, it reduces the throttle or gently applies the brakes to reduce speed. The vehicle then follows behind the preceding vehicle at the speed required to maintain a predefined distance. As soon as the vehicle in front has moved or increased speed, the system will accelerate the vehicle back up to the set and desired speed.

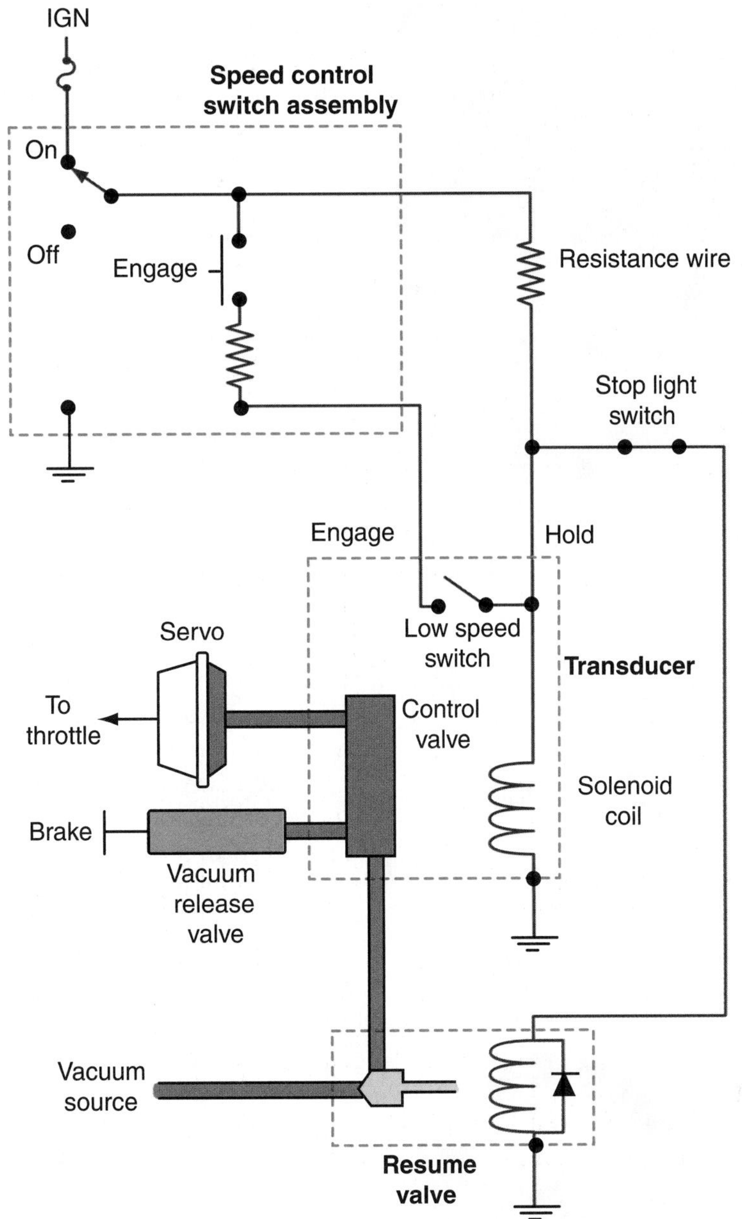

Figure 22-16 A cruise control circuit with vacuum and electrical systems.

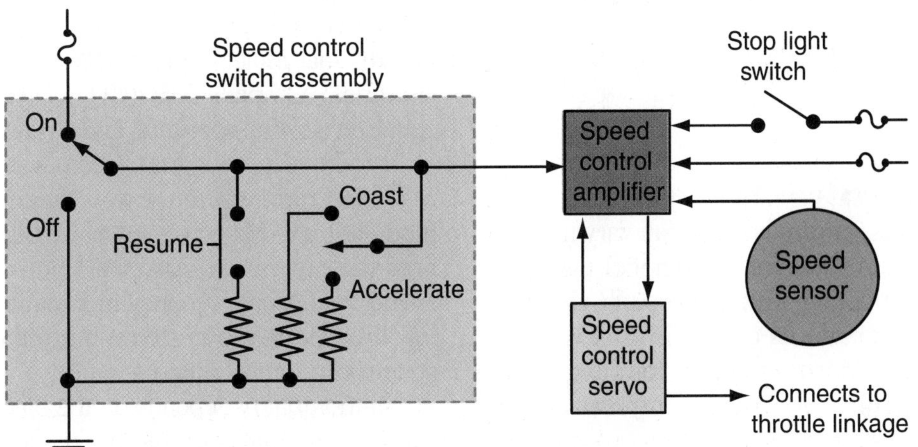

Figure 22-17 Electronic cruise control uses an electronic control module (controller) to operate a servo that controls the position of the throttle.

Figure 22-18 The distance sensor for an intelligent or adaptive cruise control system. *Courtesy of DaimlerChrysler Corporation*

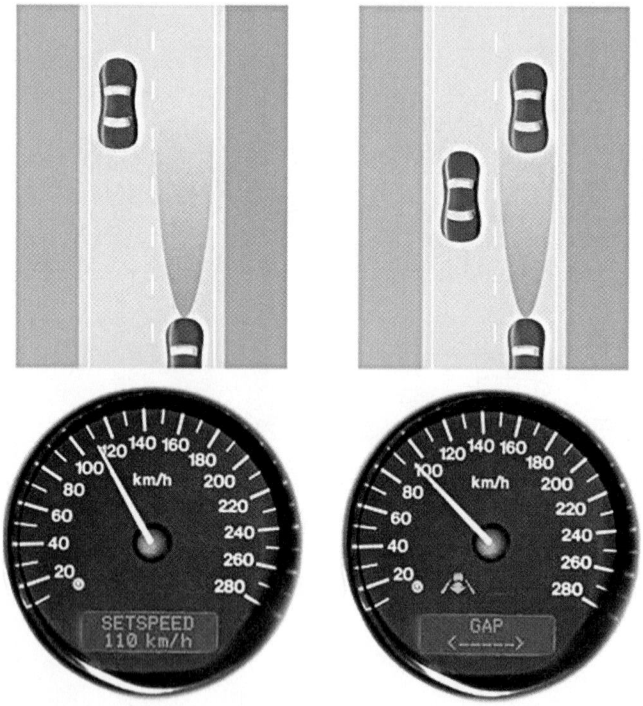

Figure 22-19 The response of an adaptive cruise control system. *Courtesy of Delphi Corporation*

Cruise Control System Service

Problems with a cruise control system can vary from no operation and intermittent operation to not disengaging. Begin your troubleshooting with a thorough visual inspection of the system and its components. Check the system's fuses. Check all vacuum hoses for disconnects, pinches, loose connections, and so forth. Inspect all wiring for tight and clean connections, condition of insulation, and wire routing. Check and adjust the linkage according to specifications.

Diagnosis of PCM-controlled and electronic cruise control systems is the same as for any other electronic system of the automobile. Diagnostic work is done with a scan tool and service manual. Typically, on most late-model vehicles, cruise control problems are caused by faulty circuits, sensors, and/or switches.

If the cruise control does not work, apply the brake pedal and verify brake light operation. If the brake lights do not work, test and repair them. Then road test the vehicle to see if this was the cause of the cruise control problem. If the vehicle has a manual transmission, check the clutch switch with an ohmmeter or voltmeter. When the cruise does not disengage, check the operation of the brake switch. Also, check the wires leading from the switch to the cruise control unit.

Vacuum systems and electronic systems that do not have the capability of self-diagnostics require additional diagnostic steps. Check for proper operation of the actuator lever and throttle linkage. Then disconnect the vacuum hose between the check valve and the servo. Apply 18 inches of vacuum to the check valve. The valve should hold the vacuum. If it does not, replace it.

Check the vacuum dump valve, servo, and speed sensor according to the procedures outlined in the service manual. If everything checks out fine, replace the controller (amplifier).

If the vehicle surges while cruise is activated, check the actuator linkage. It should move smoothly. Then, check the speedometer cable for proper routing. Check the servo and vacuum dump valve with a handheld vacuum pump. If everything is okay, replace the controller. If the servo is found to be faulty, it must be replaced. Photo Sequence 19 covers the procedure for replacing a servo unit on a typical vehicle.

SOUND SYSTEMS

Sound systems are available in a wide variety of models. The complexity of the system varies significantly from the basic AM radio to more complex stereo systems **(Figure 22–20)** that include an AM/FM radio receiver, a stereo amplifier, compact disc (CD) player, cassette player, equalizer, several speakers, and a power antenna system.

A radio receives signals (radio waves) that are broadcast from radio station towers or antennas. Amplitude modulation (AM) waves travel far but cannot be used to broadcast in stereo. Also, AM does not have as good sound quality as frequency modulation (FM). Nearly all FM broadcasts are in stereo but the distance range for good reception is limited.

Sound quality depends on the basic system but especially on the quality of the speakers and their placement. Many sound systems are equipped with several speakers, each designed to produce a different range of sound.

PHOTO SEQUENCE 19

Typical Procedure for Replacing a Servo Unit

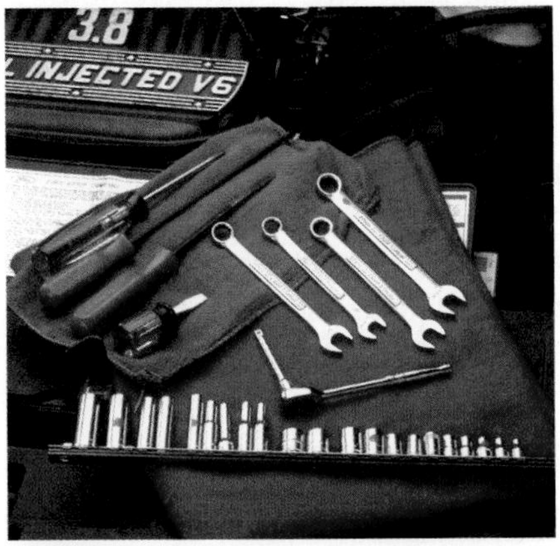

P19-1 *The tools required to replace a servo assembly are fender covers, a screwdriver set, combination wrenches, a ratchet, and a socket set.*

P19-2 *Remove the retaining screws attaching the speed control actuator cable to the accelerator cable bracket and intake manifold support bracket.*

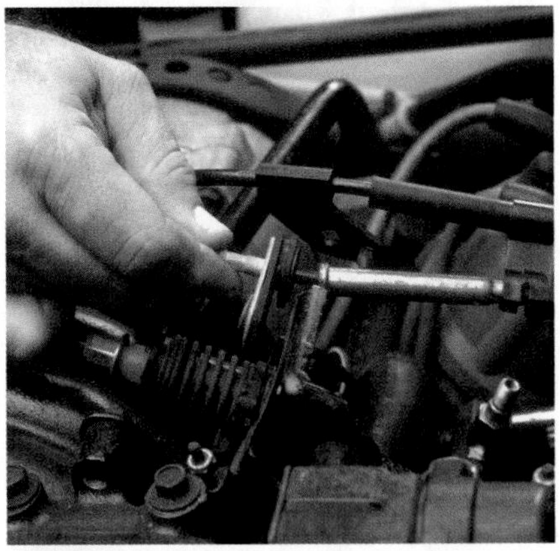

P19-3 *Disconnect the cable from the brackets.*

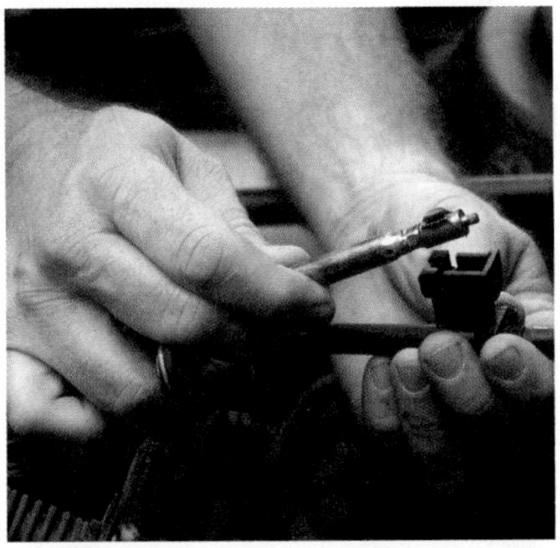

P19-4 *Disconnect the speed control cable from the accelerator cable.*

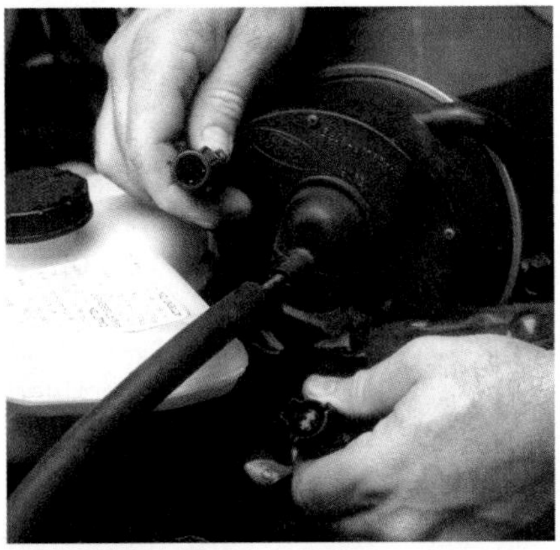

P19-5 *Disconnect the electrical connection to the servo assembly.*

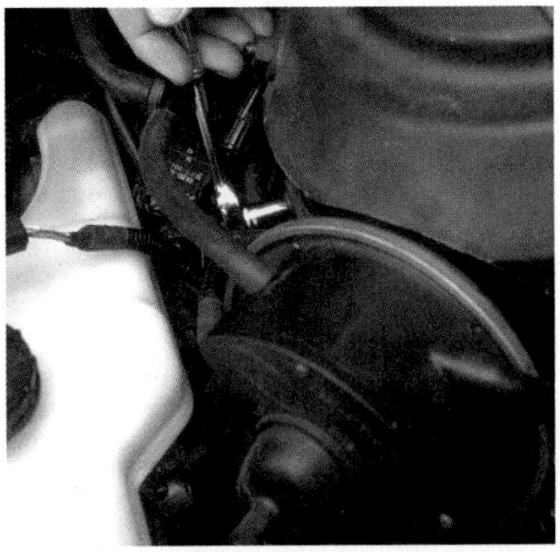

P19-6 *Remove the retaining nuts that attach the servo assembly bracket to the shock tower.*

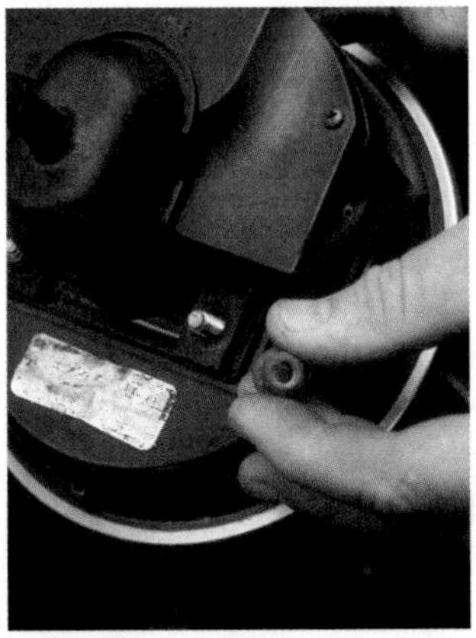

P19-7 *Remove the nuts that attach the servo assembly to the bracket.*

P19-8 *Remove the servo and cable assembly.*

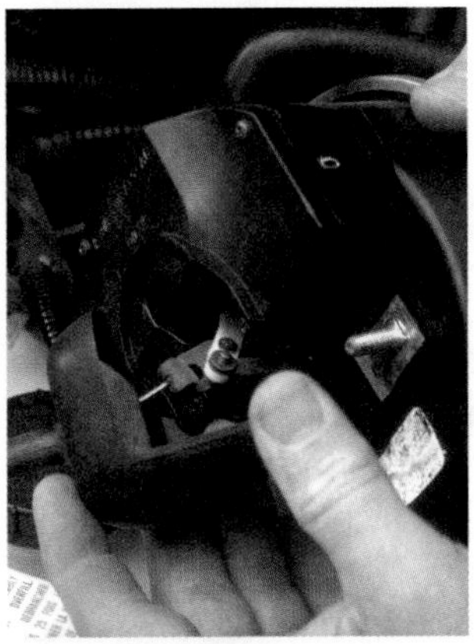

P19-9 *Remove the two cable cover retaining bolts on the servo assembly and pull off the cover.*

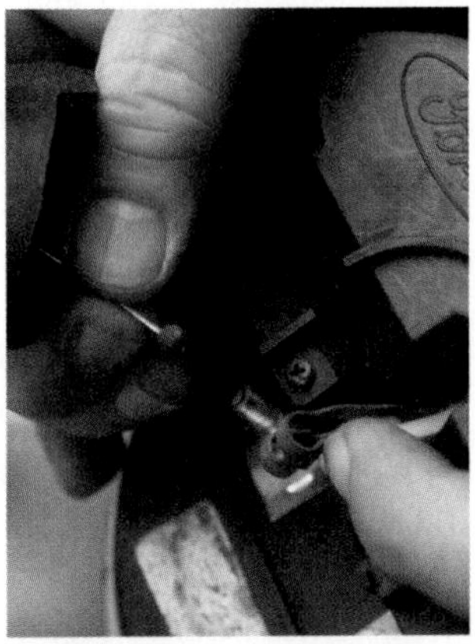

P19-10 *Remove the cable from the servo assembly. Replacement of the servo unit follows the reverse of the removal procedure.*

MACH Audio System With Digital Audio Compact Disc Player

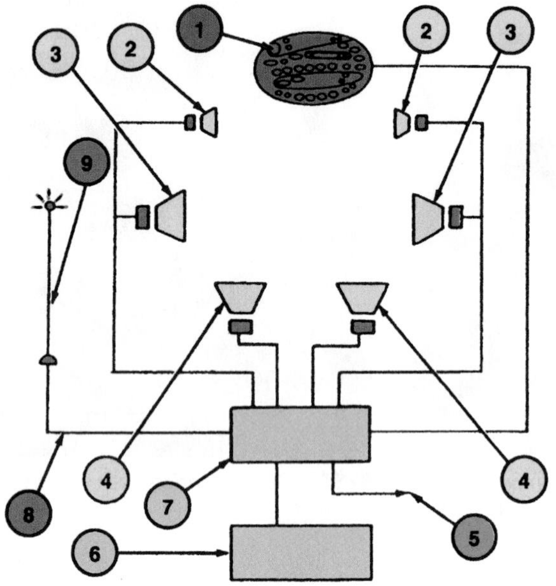

Item	Description
1	Integrated Control Panel
2	Speaker — 5 cm (2 Inch) Round Jensen® (Mounted in Sail)
3	Radio Speaker — 14 x 19 cm (5.5 x 7.5 inch) Jensen® (Front Door)
4	Radio Speaker — 14 x 19 cm (5.5 x 7.5 inch) Jensen® (Package Tray on Sedan, Liftgate on Wagon)
5	To Optional Mobile Telephone
6	Digital Audio Compact Disc Player
7	Rear Chassis Unit
8	Radio Antenna Lead In Cable
9	Electric Antenna

Figure 22-20 The layout of the components in a premium sound system. *Courtesy of Ford Motor Company*

Matching speakers to the system is done by selecting and wiring speakers so they have the same impedance as the rest of the sound system. Impedance requirements are typically noted on the rear of the sound unit and/or in the installation instructions. The placement of the speakers is critical to good clean sound. Sound waves from the speaker will bounce off anything they hit, including other sound waves. This bouncing of sound can cause noise or distortion. To achieve a high-quality sound system, the speakers must be placed so that all bouncing is anticipated.

Many optional sound systems have very a high wattage output and use several amplifiers. Some even have automatic sound level systems that discreetly adjust the volume to compensate for changes in ambient noise and vehicle speed. An amplifier increases the volume of a sound without distorting it. Amplifiers are typically rated by the maximum power (watts) they can put out. In order to take advantage of the power output, speakers must be chosen that match the output.

CD players vary from being able to insert one or more CDs into the main unit to having an auxiliary unit where many CDs can be installed. The control unit allows the operator to select the CD of choice. A few vehicles are fitted with MP3 systems.

Antenna

An antenna is designed to receive radio sound waves for both AM and FM stations. The design of the antenna must satisfy two basic requirements: for good AM reception the antenna should be as high as possible and for FM stations it should be 31 inches high. Therefore most antennas are 31 inches high when they reach full height. Shorter antennas will have poor reception in both AM and FM. Many late-model vehicles are equipped with shorter antennas designed to enhance reception and therefore have no need to be long to provide good reception. The placement of these antennas is also selected to achieve optimal reception.

Power Antennas Many vehicles are equipped with electrically operated antennas that extend when the radio is turned on and lower when it is turned off. These antennas are powered by a small reversible electric motor. The motors are turned off by limit switches that open when the mast has extended or lowered to its desired height. Power antennas (even black-colored antennas) need to be cleaned with chrome polish on a regular basis to keep them working properly. Often when there is a problem with a power antenna, it is caused by dirt or a lack of lubricant on the telescoping mast. When there is a problem with the power unit, it is normally replaced as a unit.

Satellite Radio

To provide high quality radio that is not interrupted by distance, some vehicles can be purchased with satellite radio. Satellite radios are also available as add-on items. These radios pick up sound waves from satellites many miles above the earth. Since the radio waves are transmitted by more than one satellite at all times, and each in their own orbit or place within the orbit, the same radio station can be heard from coast-to-coast **(Figure 22–21)**. Although distance does not hamper the reception, the radio waves cannot penetrate buildings, tunnels, or large groupings of trees. Therefore, to enjoy satellite radio, stay on the open road.

Figure 22-21 The reception of satellite radio systems is not affected by distance because the radio waves are transmitted by more than one satellite. *Courtesy of Delphi Corporation*

Diagnosis

Internal inspection and service to the radio should be left to an authorized radio service center. However, technicians should be able to determine the cause of poor sound quality and radio reception. Most sound system problems are caused by the unit itself, the wiring in the circuit, the antenna, or the speakers.

If the radio system is not working, check the fuse. If the fuses are okay, refer to the service manual. If you determine the radio itself is the problem, remove it, and send it to a qualified shop.

CUSTOMER CARE

Many customers do not understand the limitations of FM reception. Refer your customer to the owner's guide for information about the limitations of FM radio performance. ■

If reception is bad and antenna height is correct, use an ohmmeter and check the ground of the antenna. Also connect one lead of the ohmmeter to the antenna's mast and its case; there should be no continuity between the two.

Poor speaker or sound quality is usually caused by one of the following:

- Damaged speaker cones, internal mountings, or wiring.
- Interference from the ignition system, neon signs, or electrical power lines.
- Distortion caused by the speaker, radio chassis, or wiring. If the concern is in the radio chassis, both speakers on the same side of the vehicle will have poor quality. Distortion caused by damaged wiring is most often accompanied by lower-than-normal sound output.
- Bent package tray sheet metal around the speaker opening, lack of mounting brackets, or missing or loose attaching hardware or speaker covers. Be careful not to overtighten hardware as this may bend or deform the speaker baskets, causing buzzes or distorted sound.

SHOP TALK

Antitheft audio systems have built-in devices that make the audio system soundless if stolen. If the power source for the audio system is cut, even if it is later reconnected, the audio system will not work unless its ID number is put in. When performing repairs on vehicles equipped with this system, before disconnecting the battery terminals or removing the audio system, ask the customer for the ID number so you can input the ID number afterwards, or request that the customer input the ID number after the repairs are completed. With antitheft radio installation, there is a procedure in the service manual to obtain the factory backup code using a touch tone phone if the owner's code is not available. A memory saver or backup battery can be used to maintain the radio's code and settings during service. ■

POWER LOCK SYSTEMS

Although systems for automatically locking doors vary from one vehicle to another, the overall purpose is the same—to lock all outside doors. As a safety precaution against being locked in a car due to an electrical failure, power locks can be manually operated. Many late-model systems include automatic locking when the transmission selector is moved out of park or when the vehicle reaches a particular speed.

When either the driver's or passenger's control switch is activated (either locked or unlocked), power from the fuse panel is applied through the switch to a reversible motor. A rod that is part of the lock assembly moves up or down to lock or unlock the door. On some models the signal from the switch is applied to a relay that, when energized, applies an activating voltage to the door lock actuator. The door lock actuator consists of a motor and a built-in circuit breaker. Since the motors are reversible, each does not have its own ground. The ground for the lock circuits is at the master or door circuits. On station wagon models with power door locks, the tailgate lock actuator is also controlled by the door lock switches. Station wagon models without power door locks can be equipped with a separate tailgate power lock system.

Most models use a control switch mounted in the door arm rest or in the door trim panel. However, some models

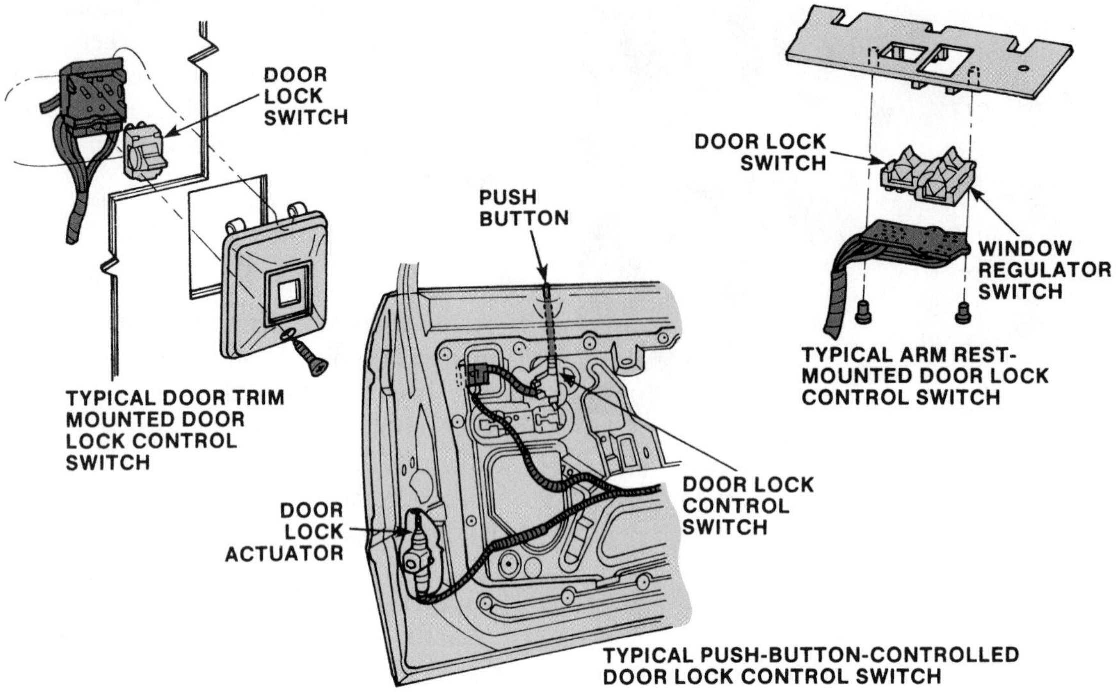

Figure 22-22 Typical door lock control switches. *Courtesy of Ford Motor Company*

use switches controlled by the front door push button locks **(Figure 22–22)**.

Power Trunk Release

The power trunk release system is a relatively simple electrical circuit that consists of a switch and a solenoid. When the trunk release switch is pressed, voltage is applied through the switch to the solenoid. With battery voltage on one side and ground on the other, the trunk release solenoid energizes and the trunk latch releases to open the trunk lid.

Diagnosis

Power door lock or trunk release systems rarely give trouble. Basic diagnosis includes:

- If none of the locks work, check the circuit protection devices. If they are okay, check the wiring.
- If the doors lock or unlock but not both, check the system relays. On vehicles without relays, check the wiring to and from the control switch on the nonworking side.
- If only one door lock does not operate, check the actuator for that door.
- If only one door lock switch does not work, check that switch.
- If the trunk release does not work, check the fuse first. Then check the switch. If the switch is okay, check for continuity through the trunk release solenoid to ground. If the solenoid is okay, check the wiring.

POWER WINDOWS

Obviously, the purpose of any power window system **(Figure 22–23)** is to raise and lower windows. The systems do not vary significantly from one model to another. The major components of a typical system are the master control switch, individual window control switches, and the window drive motors **(Figure 22–24)**.

The master control switch provides overall system control. Power for the system comes directly from the fuse panel on two-door models and from an in-line circuit breaker on four-door models. The window safety relay used on four-door models prevents operation of the system if the ignition switch is not in the run or accessory position. Power for the individual window control switches comes through the master control switch.

Four-door model master control switches usually have four segments while two-door models have two. Each segment operates as a separate independent switch and controls power to a separate window motor. A lock switch included on four-door model master control switches is a safety device to prevent children from opening windows without the drivers knowledge.

Circuit Operation

Typically a permanent magnet motor operates each power window. Each motor raises or lowers the glass when voltage is applied to it. The direction that the motor moves the glass is determined by the polarity of the supply voltage.

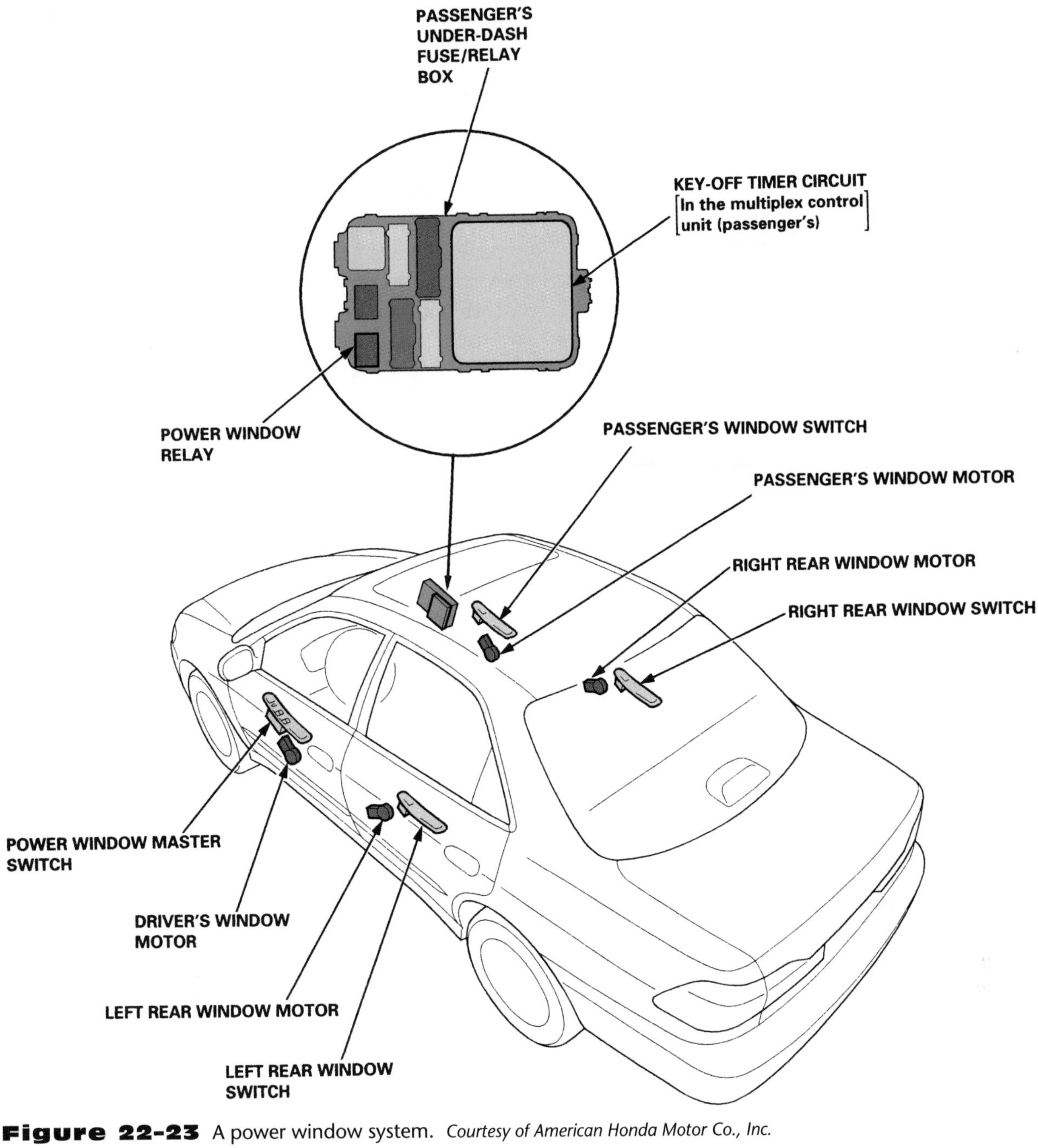

Figure 22-23 A power window system. *Courtesy of American Honda Motor Co., Inc.*

Voltage is applied to a window motor when any UP switch in the master switch assembly is activated. The motor is grounded through the DOWN contact. Battery voltage is applied to the motor in the opposite direction when any DOWN switch in the master switch assembly is activated. The motor is then grounded through the master switch's UP contact.

The operation of the individual window switches is much the same. When the UP switch is activated, voltage is applied to the window's motor. The motor is grounded through the DOWN contact at the switch and the DOWN contact at the master switch. When the DOWN switch in the window switch is activated, voltage is applied to the motor in the opposite direction. The motor is grounded through the UP contact at the window switch and the UP contact in the master switch. This runs the motor in the opposite direction.

Each motor is protected by an internal circuit breaker. If the window switch is held too long with the window obstructed or after the window is fully up or down, the circuit breaker opens the circuit.

Some vehicles are equipped with an express down window feature. This feature fully opens the window by holding the window switch down for more than 0.3

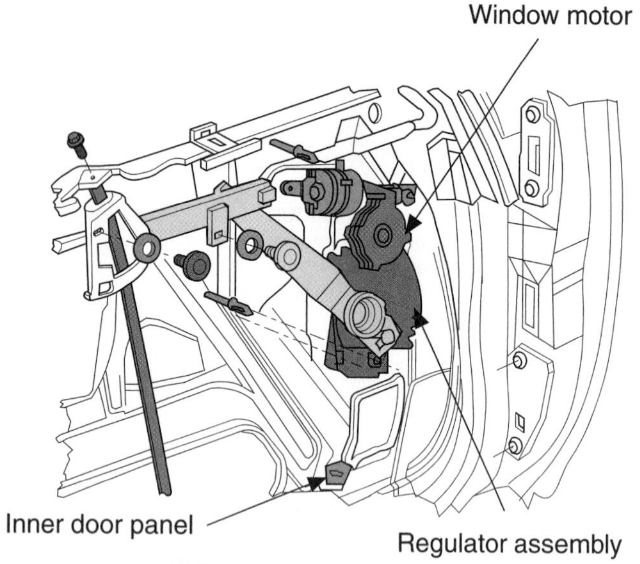

Figure 22-24 An electric window motor and regulator. *Courtesy of DaimlerChrysler Corporation*

seconds then releasing it. The window may be stopped at any time by depressing the UP or DOWN switch. The express window option relies on an electronic module and a relay. When signaled, the module energizes the relay, which completes the motor's circuit. When the window is fully down, the module opens the relay control switch, which stops power to the motor. The motor will also stop 10 to 30 seconds after the DOWN switch is depressed.

Obstacle-Sensing Windows This feature is available on some vehicles and prevents the window from closing if something is in the way, such as fingers. On some systems, if the window is closing and contacts something, the window will reverse and open fully. Other systems do not require direct contact with an obstacle. Rather, they rely on infrared sensors **(Figure 22–25)**. When the light beams are broken by the presence of an obstacle, the window will reverse and go down.

Figure 22-25 The infrared sensor for a smart window system. *Courtesy of Delphi Corporation*

Diagnosis

The first step in diagnosing a power window system is to determine whether the whole system or just one or two windows are not working correctly. If it is the whole system, the problem can be isolated to fuses, circuit breakers, or the master control switch. If only a portion of the system does not work, check the components used in the portion that is not working. Removing the door trim panels allows access to the motor and window linkage. A special tool is often required to release the retaining clips without damaging the trim or clips.

POWER SEATS

Power seats allow the driver or passenger to adjust the seat to the most comfortable position **(Figure 22–26)**. The major components of the system are the seat control and the motors.

Power seats generally come in two configurations: four-way and six-way. However, some vehicles allow the seats to be adjusted in up to twelve directions. In a four-way system, the whole seat moves up or down, or forward and rearward. A six-way system **(Figure 22–27)** has the same adjustments, plus the capability to adjust the height of the front and rear of the seat. Generally, a four-way system is used on bench seats and a six-way system is used on split-bench and bucket seats. Some units also control the tilt, rear/forward movement, height, and angle of the seat back. The adjuster for the seat back may also control the height of the head rest or restraint.

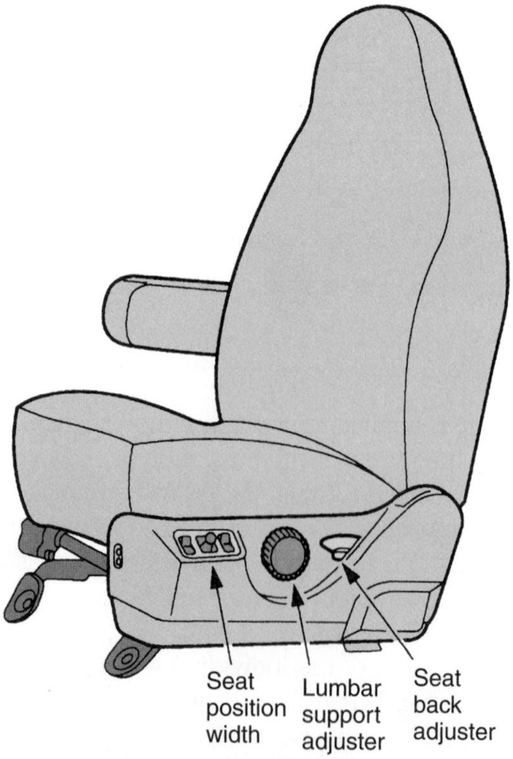

Figure 22-26 A typical power seat. *Courtesy of Ford Motor Company*

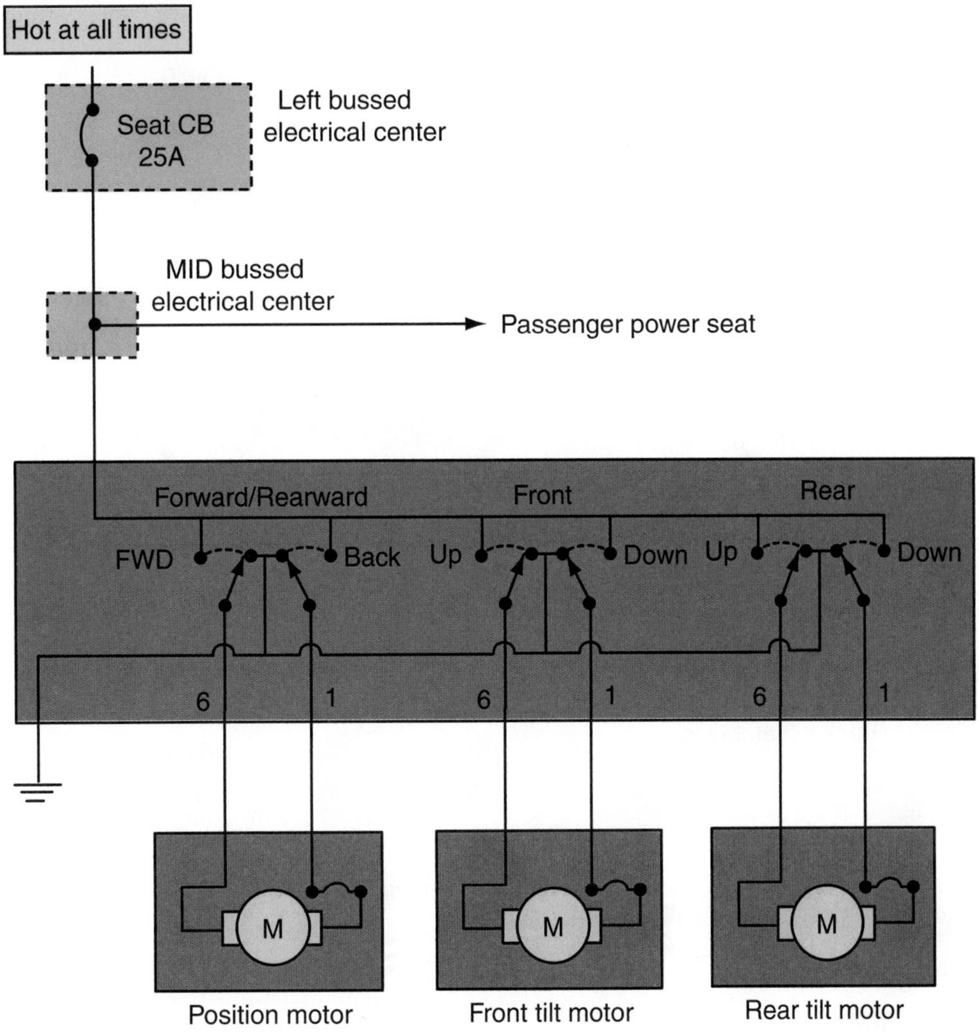

Figure 22-27 A power seat circuit.

Two motors are typically used on four-way systems, while three are used on a six-way system. The names of the motors identify their function. To raise or lower the entire seat on a six-way system, both the front height and the rear height motors are operated together. The motors are generally two-directional motor assemblies that include a circuit breaker to protect against circuit overload if the control switch is held in the actuate position for long periods. A typical six-way motor rack and pinion system is shown in **Figure 22–28**. The four-way system and six-way screwdrive motor arrangements are similar.

Diagnosis

Before testing a power seat system, conduct a visual inspection of the wires and connectors in the system. Two types of problems affect power seats: One is a tripped or constantly tripping circuit breaker and the other is the inability of the seat to move in a direction.

The resettable circuit breaker should protect the system from a short circuit or from high current due to an obstructed or stuck seat adjuster. The circuit breaker must be replaced if it is faulty. Before condemning or testing the circuit breaker, make sure the seat tracks are not damaged and that nothing is physically preventing the seat from moving.

When a seat does not move in a particular direction, turn on the dome light, then move the power seat switch in the problem direction. If the dome light dims, the seat may be binding or have some physical resistance on it. Check under the seat for binding or obstructions. If the dome light does not dim, test the system.

Disconnect the negative terminal at the battery. Remove the power seat switch from the seat or door armrest. Check for battery voltage to the switch. If there is no voltage and the circuit breaker is okay, check for an open in the power feed circuit. If voltage is present, check for continuity between the ground connection at the switch and a good ground. If there is no continuity, repair the ground circuit. If there is continuity, the switch must be tested. Use an ohmmeter to test the continuity of the switch in each position. Check the service manual for the expected continuity between the various terminals of

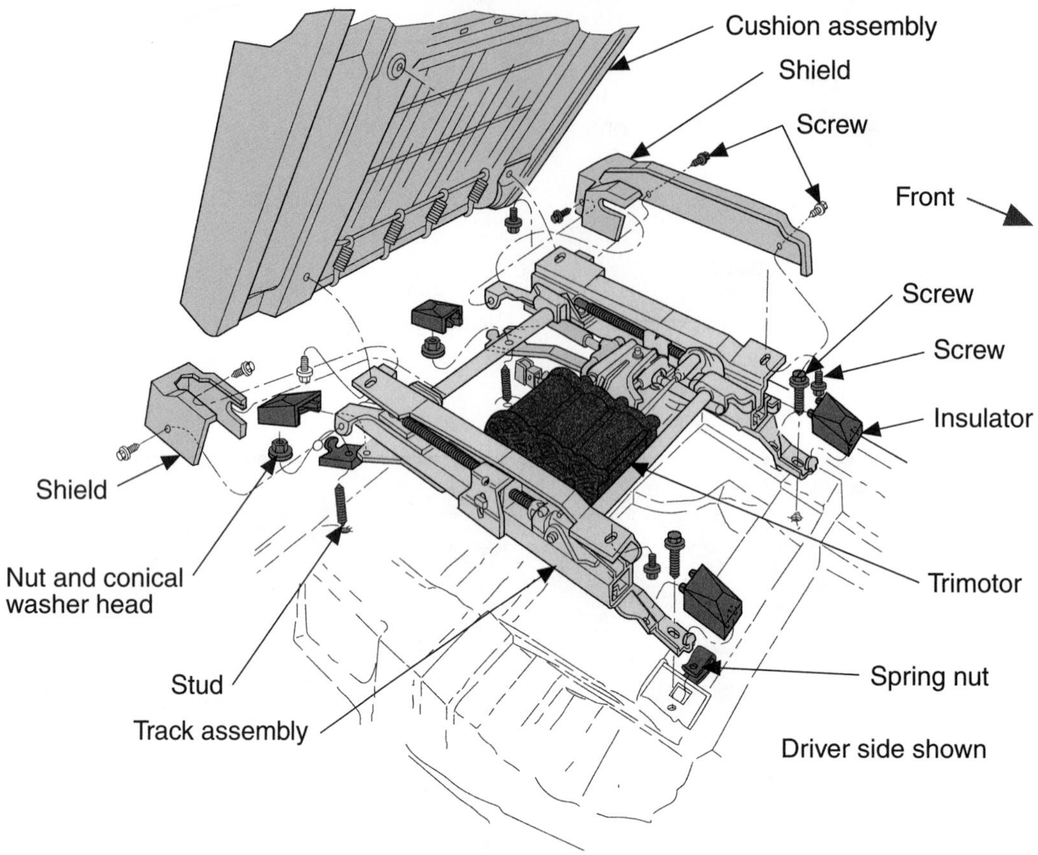

Figure 22-28 A three-motor (trimotor) unit installed in the base of the seat. *Courtesy of Ford Motor Company*

the switch **(Figure 22–29)**. If the switch checks out, test the motor. If the switch is bad, replace it.

Other Seat Options

Many different options are available for the seats of a vehicle. Some of the following features are available on vehicles with manual seats; others are only available with the power seat option.

Temperature-Controlled Seats Some vehicles have an option that warms up the seats, an especially nice feature in cold climates. The system relies on heating coils in the seat cushion and back controlled by relays and switches. In addition to warming the seats in cold weather, some vehicles allow the seats to cool by passing air through them during warm weather **(Figure 22–30)**. The heating grids and switches are normally tested with an ohmmeter. Refer to the appropriate service manual for specifications and testing instructions.

Power Lumbar Supports A power lumbar support allows the driver to inflate or deflate a bladder located in the lower seat back. Adjusting this support improves the driver's comfort and gives support at the lower lumbar region of the spinal column.

Memory Seats The memory seat option allows for automatic positioning of the driver's seat to different, often up to three, programmable positions **(Figure 22–31)**. This feature allows different drivers to have their desired seating position automatically adjusted by the system. It also allows an individual driver to set different positions for different driving situations.

Some systems with a remote key fob can be programmed to move the seat to its memory position whenever the unlock button of the key fob is depressed. Each driver can have his or her own key fob and their desired seat position will be selected when the door is unlocked. Also, other systems automatically adjust the power mirrors to a setting for each driver.

Adaptive and Active Seating Adaptive seating, a feature that moves the seat slightly as the driver shifts positions, is offered in some luxury vehicles **(Figure 22–32)**. Moving the seat improves the driver's comfort and support while driving for a long time. Active seating stimulates the spine and surrounding muscles with continuous yet virtually imperceptible motion. This type of seating is designed to prevent the driver from getting saddle sore while sitting without moving for a long time. The right and left halves of the seat cushion move up and down at cyclical

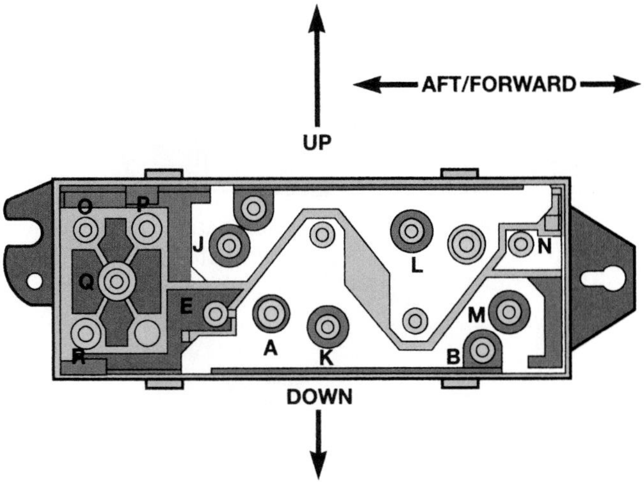

TERMINALS SHOWN AS VIEWED FROM REAR OF SWITCH

POWER SEAT SWITCH	
SWITCH POSITION	CONTINUITY BETWEEN
OFF	B-N, B-J, B-M, B-E, B-L, B-K
VERTICAL UP	A-E, A-M, B-N, B-J
VERTICAL DOWN	A-J, A-N, B-M, B-E
HORIZONTAL FORWARD	A-L, B-K
HORIZONTAL AFT	A-K, B-L
FRONT TILT UP	A-M, B-N
FRONT TILT DOWN	A-N, B-M
REAR TILT UP	A-E, B-J
REAR TILT DOWN	A-J, B-E
LUMBAR OFF	O-P, P-R
LUMBAR UP (INFLATE)	O-P, Q-R
LUMBAR DOWN (DEFLATE)	O-R, P-Q

Figure 22-29 The terminals of a power seat switch identified for continuity checks. *Courtesy of DaimlerChrysler Corporation*

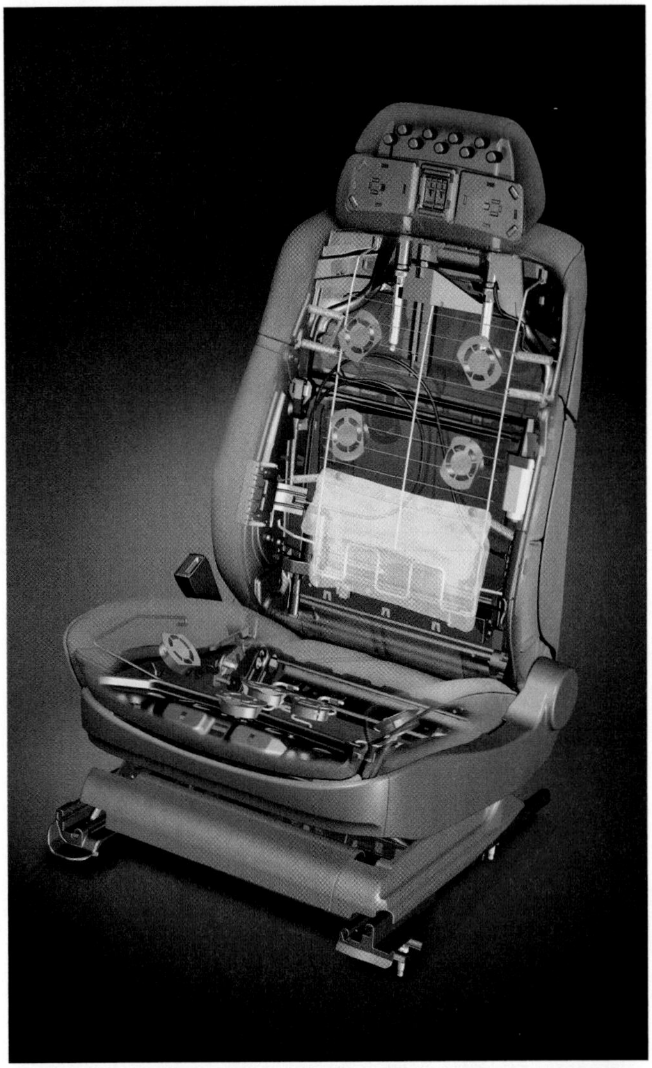

Figure 22-30 This seat features ventilation, heating, and multiple position adjustments including the headrest and lumbar support. The small round features are fans. *Courtesy of BMW of North America, Inc.*

intervals. To do this, two pillows are integrated within the seat's upholstery. A hydraulic pump alternately inflates the two cavities with a mixture of water and glysantine.

Massaging Seats To help reduce driver fatigue, rows of rollers in the seat back move up and down when the driver depresses the control button. This massaging motion continues for about ten seconds at a time.

POWER MIRROR SYSTEM

The power mirror system consists of a joystick-type control switch and a dual motor drive assembly located in each mirror assembly.

Rotating the power mirror switch to the left or right position selects one of the mirrors for adjustment. Moving the joystick control up and down or right and left moves the mirror to the desired position. The dual motor drive assembly is located behind the mirror glass **(Figure 22–33)**. The position of the mirrors may be tied to the

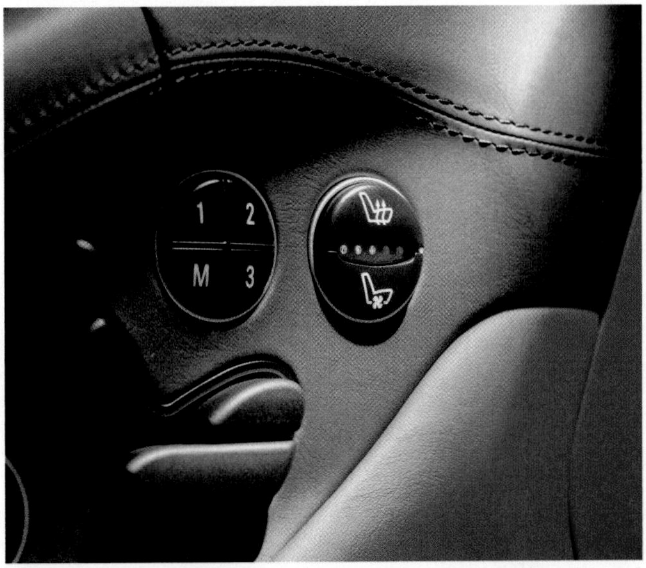

Figure 22-31 Seat adjustment controls and seat memory buttons. *Courtesy of DaimlerChrysler Corporation*

Figure 22-32 A cutaway of an adaptive seat. *Courtesy of DaimlerChrysler Corporation*

memory power seats and will automatically adjust when a seating position is selected.

A few vehicles automatically tilt the passenger-side outside mirror downward whenever the transmission is placed into reverse. This allows the driver to see the area directly next to the vehicle during parking.

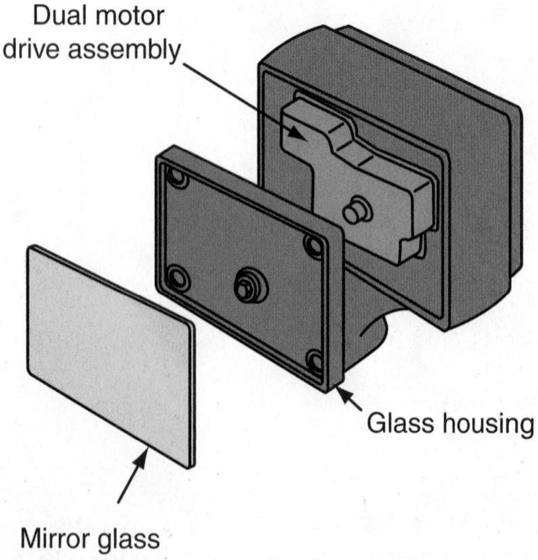

Figure 22-33 A power mirror motor assembly.

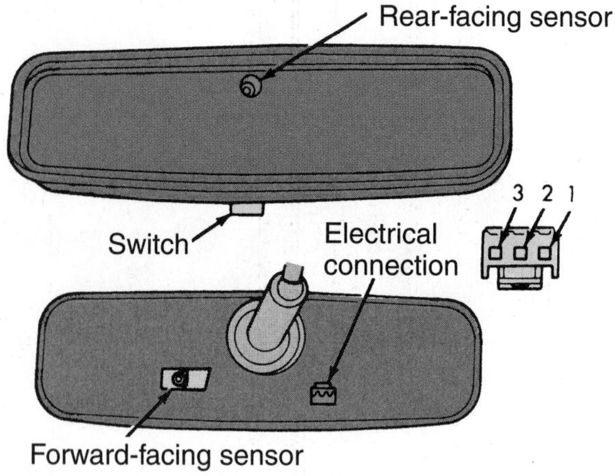

Figure 22-34 An automatic day/night mirror. *Courtesy of DaimlerChrysler Corporation*

Many new vehicles are being equipped with electrochromic side view mirrors. Requiring no electrical connections, these mirrors operate on the intensity of the glare in much the same manner as a pair of eyeglasses with photochromatic lenses. When glare is heavy, the mirrors darken fully (down to 6% reflectivity). When glare is mild, the mirrors provide 20% to 30% reflectivity. When glare subsides, the mirrors change to the clear daytime state.

Inside Mirrors

Some automatic day/night inside rearview mirrors use a thin layer of electrochromic material between two pieces of conductive glass. A switch located on the mirror allows the driver to turn the feature on or off. When it is turned on, the mirror switch is lighted by an LED. The self-dimming feature is disabled whenever the transmission is placed into reverse.

When the mirror is turned on, two photocell sensors monitor external light levels and adjust the reflection of the mirror **(Figure 22–34)**. The ambient photocell sensor detects the light levels outside and in front of the vehicle. The headlamp photocell faces rearward to detect the level of light coming in from the rear of the vehicle. When there is a difference in light levels between the two photocells, the mirror begins to darken.

The automatic day/night mirror cannot be repaired. If it is faulty, it must be replaced.

Some rearview mirrors are fitted with a directional compass display. The readings on the compass appear on the mirror's reflective surface, normally in the lower left corner.

REAR-WINDOW DEFROSTERS AND HEATED MIRROR SYSTEMS

The rear-window defroster (also called a **defogger** or **deicer**) heats the rear-window surface to remove mois-

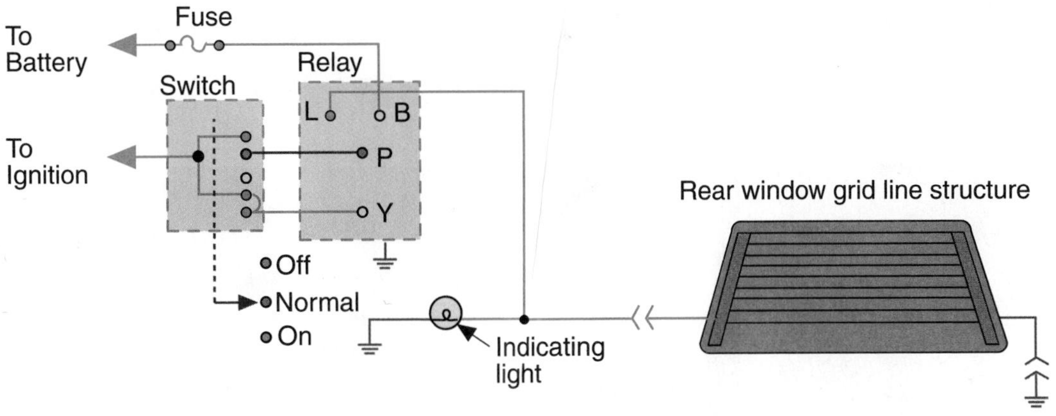

Figure 22-35 A rear defogger circuit schematic.

ture and ice from the window. On some vehicles, the same control heats the outside mirrors. The major components of a rear-window defroster include a switch, relay assembly, and the heating elements on the glass surface **(Figure 22–35)**.

Pressing the rear-window defroster switch momentarily energizes the relay. Battery voltage is then applied through closed contacts of the relay to the rear-window defroster grid. On models with a heated mirror, current also flows through a separate fuse to the mirror's heated grid.

After about 10 minutes, a time-delay circuit opens the ground path to the relay's coil and the coil de-energizes, shutting off power to the grids. The time-delay circuit prevents the system from remaining on during periods of extended driving. The system can also be manually turned off.

Diagnosis

One of the most common problems with rear-window defrosters is damage to the grids on the window. Damage can be caused by hard objects rubbing across the inside surface of the glass or by using harsh chemicals to clean the window. When a segment of the grid breaks, it opens the circuit. Often the customer's complaint is that the unit does not defrost the entire window. Normally one or two lines of the grid are open. The open can be found by using a test light. With the defroster turned on, voltage should be present at all points of the grid. If part of the grid does not have voltage, move the probe toward the positive side of the grid. Once voltage is present, you know the open is between those two points. Opens can be repaired by painting a special compound over the open. The correct procedure for doing this is shown in Photo Sequence 20.

If none of the grids heat up, check for voltage at the connection to the grids, and then check the ground circuit. If no voltage is present at the grids, check the circuit for the switch.

OTHER ELECTRONIC EQUIPMENT

Vehicles are being equipped with many electrical and electronic features. Examples of some of the more common newer accessories are discussed here.

Adjustable Pedals

Shorter drivers normally must move their seat very close, sometimes uncomfortably and unsafely close, to the steering wheel. By moving the pedals toward the driver, drivers may be able to adjust their seat position away from the steering wheel and still comfortably reach and use the pedals. An electric motor at the brake pedal **(Figure 22–36)**, with a cable connection to the accelerator pedal moves both pedals back and forth (up to three inches). A switch on the dash controls the motor. This feature may also be part of the seat memory system so the driver can quickly bring both the seat and pedals to the most comfortable position.

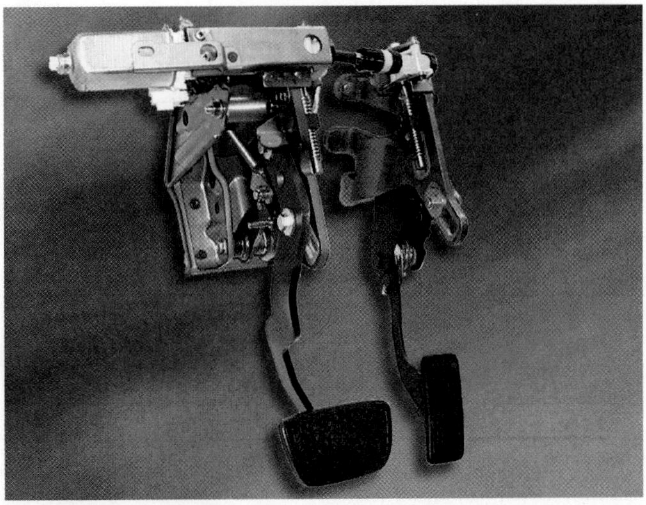

Figure 22-36 An adjustable pedal assembly.
Reprinted with permission

PHOTO SEQUENCE 20

Typical Procedure for Grid Wire Repair

P20-1 *The tools required to perform this task include masking tape, repair kit, 500°F heat gun, test light, steel wool, alcohol, and a clean cloth.*

P20-2 *Clean the grid line area to be repaired. Buff with fine steel wool. Wipe clean with a cloth dampened with alcohol. Clean an area about 1/4 inch (6 mm) on each side of the break.*

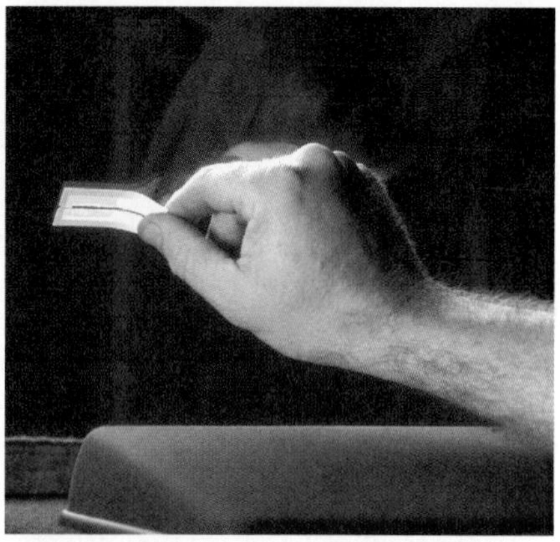

P20-3 *Position a piece of tape above and below the grid. The tape is used to control the width of the repair, so try to match the width with the original grid.*

P20-4 *Mix the hardener and silver plastic thoroughly. If the hardener has crystallized, immerse the packet in hot water.*

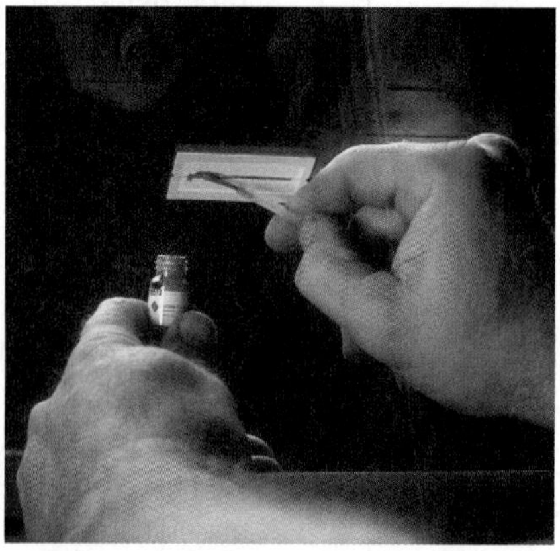

P20-5 *Apply the grid repair material to the repair area using a small stick.*

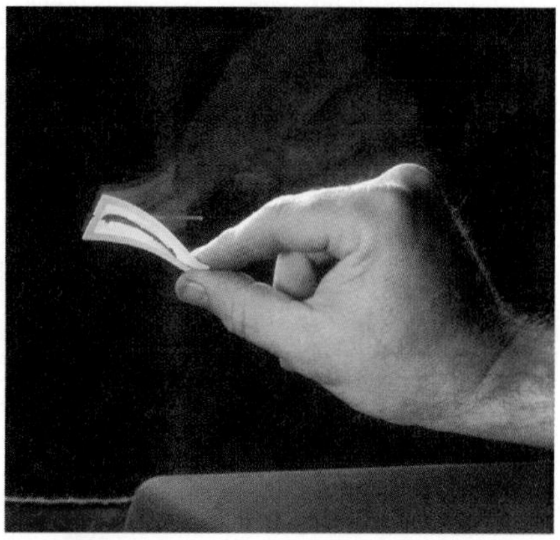

P20-6 *Carefully remove the tape.*

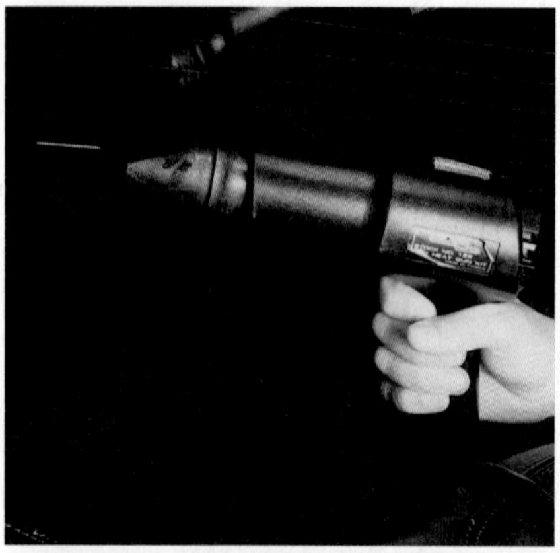

P20-7 *Apply heat to the repair area for 2 minutes. Hold the heat gun one inch (25 mm) away from the repair.*

P20-8 *Inspect the repair. If it is discolored, apply a coat of tincture of iodine to the repair. Allow to dry for 30 seconds, then wipe off the excess with a cloth.*

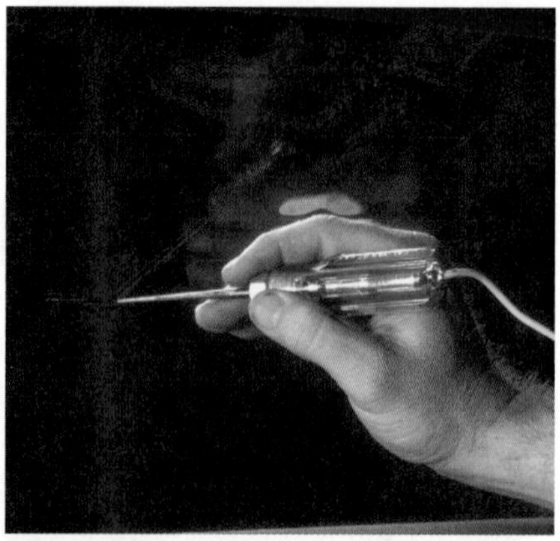

P20-9 *Test the repair with a test light. Note: it takes 24 hours for the repair to fully cure.*

Heated Windshields

Heated, or self-defrosting, front windshield systems work like rear-window defrosters, heating the glass directly. However, instead of using a wire grid that could hinder the driver's vision, a microthin metallic coating inside the windshield is used.

Two basic designs are used for heating a windshield. One system uses glass that contains three layers of material. A plastic laminate is sandwiched between two layers of glass. A silver and zinc oxide coating is fused onto the back of the outer glass layer to carry electrical current. This coating gives the windshield a slight gold tint. Silver bus bars fused to the coating at the top and bottom of the windshield connect the coating to power and ground circuits.

The other design also has three layers. However, the inner layer is not plastic; rather, it is a thin film of resistive coating sprayed between the inner and outer windshield glass. This coating is transparent, so the windshield does not appear tinted.

In a typical windshield defroster system, the generator's output is redirected from the normal electrical system to the windshield circuit. This leaves all other electrical circuits to operate on power from the battery. When the voltage regulator senses a drop in battery voltage, it full fields the generator so it can put out between 30 and 90 volts. If battery voltage drops below 11 volts, the system is turned off and the AC generator's output is again directed to the battery. Check the service manual for specific voltage details.

The defrost cycle generally lasts for about four minutes. After that, the generator is switched to the normal charging operation controlled by the voltage regulator. If the windshield is not clear, the system can be selected again.

Moon (Sun) Roof System

A power **moon roof** or **sun roof** can slide the roof panel open or closed. It can also tilt the panel up in the back to allow fresh air and natural light into the passenger compartment. The major components of any sun roof panel system are a relay, control switch, sliding roof panel, and motor. This circuit is normally protected by the in-line circuit breaker.

When the two-position switch is moved to the open position, the roof panel moves into a storage area between the headliner and the roof. The panel stops moving any time the switch is released. Moving the switch to the closed position reverses the power flow through the motor.

If the system is not operating, check the fuse or circuit breaker. If these are okay, check for power at the switch with the ignition switch. If the voltage is present, the relay is okay. Check for power to the motor with the switch held in the open position. Refer to the service manual for additional diagnosis and testing information.

Solar Sunroof This feature uses light-sensitive elements under the glass sunroof to produce electricity to power a ventilator inside the vehicle. Even with the ignition off, the interior is supplied with a continuous flow of fresh filtered air. The solar sunroof puts no additional demands on the battery or charging system and reduces the energy used by the climate control system.

Cellular Phones

Many auto manufacturers are installing cellular telephones as optional equipment in their vehicles. These cellular phones offer hands-free operation, which means the phone can respond to verbal commands so the driver does not need to fumble and look at the phone while driving. Some states have passed laws that prohibit the use of a cell phone while driving, unless it is hands-free. Problems with a cell phone can be caused by problems with the communication link provider or problems with the phone itself. The service manual for vehicles with a cell phone option contains all the necessary servicing information, except for problems with the communication provider.

Night Vision

A few vehicles have a feature that uses military-style thermal imaging to allow drivers to see things they normally may not see until it is too late **(Figure 22–37)**. The thermal-imaging system uses a camera with a fixed lens mounted behind the front grille and projects the image onto the bottom of the driver's side of the windshield by a HUD. The lens, which is an infrared sensor, is designed to operate at room temperature; therefore, the lens has its own heating and cooling system to keep it at the desired temperature.

The images seen on the HUD display are from the area in front of the light beams from the car's normal high-

Figure 22–37 With night vision, driving is much safer. *Reprinted with permission*

beam headlamps. The system allows the driver to see up to five times more of the road than with just headlights.

The system works by registering small differences in temperature and displaying them in 16 different shades of gray on the HUD screen. It has the ability to display animals, or people, behind bushes or trees and can see through rain, fog, and smoke. Cold objects appear as dark images, whereas warm objects are white or a light color.

Navigation Systems

Navigation systems use global positioning satellites to help drivers make travel decisions. These global positioning systems (GPSs) set up a mathematical grid between the satellites and radio stations on the ground. The exact position of a vehicle can be plotted on the grid; therefore, the system knows exactly where the vehicle is **(Figure 22–38)**. GPS can display traffic and travel information on a display screen **(Figure 22–39)**. It can display a road map marking the exact location of the vehicle. It can plot out the best way of getting to a destination and can also tell the driver how many miles have been traveled and how many remain before reaching a destination. It can also display traffic information regarding traffic backups due to congestion, roadwork, and/or accidents, then display alternative routes so travel is not delayed.

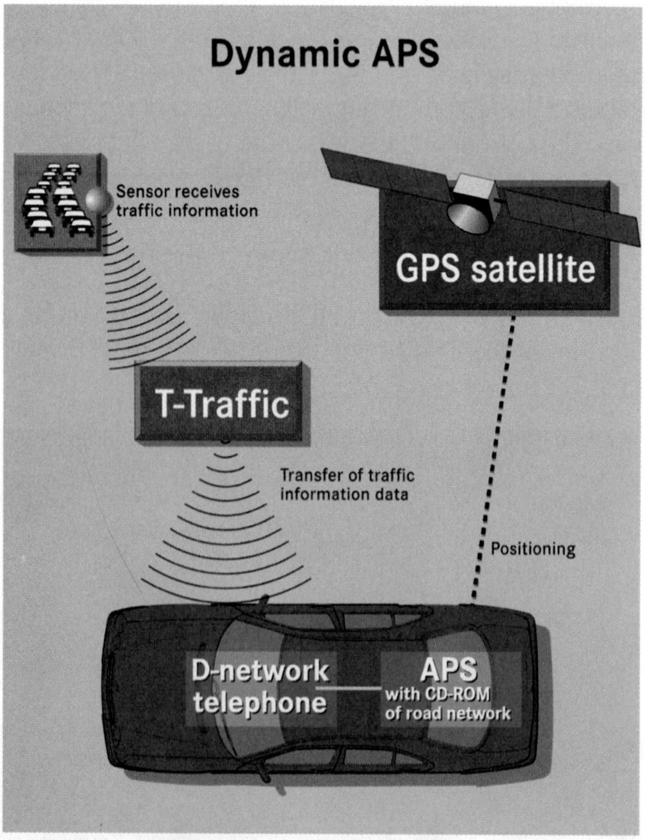

Figure 22-38 The communication network for GPS traffic and travel information. *Courtesy of DaimlerChrysler Corporation*

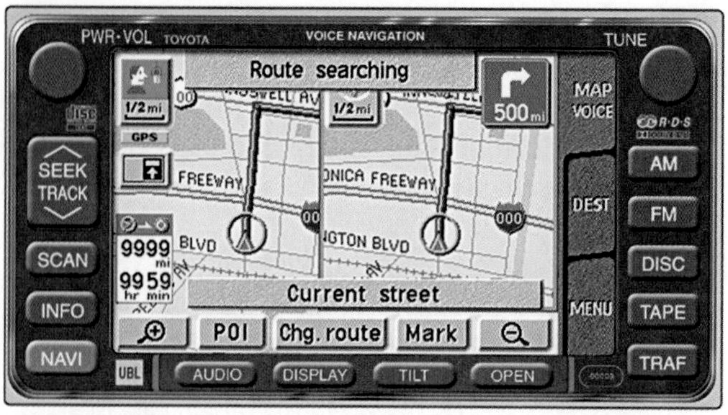

Figure 22-39 The screen of a typical navigational center. *Reprinted with permission*

A computer inside the vehicle compares the data it receives from the global positioning satellites with the information it has in its memory or what it reads from a CD or digital video disk (DVD). Periodically new disks are required to maintain the latest navigational information.

Many navigational systems give turn-by-turn guidance either by voice, on-screen displays, or both. Most of these systems feature touch-screen technology and can display and control other systems, such as air-conditioning, heating, and sound systems. Other systems allow you watch a movie on DVD while the vehicle is parked.

Vehicle Tracking Systems

Vehicle tracking systems can monitor the location of a vehicle if it has been stolen or lost. The system is based on the vehicle's navigational system or the cellular phone inside the vehicle. When the cell phone is the identifier, the tracking system is triggered when the thief attempts to place a phone call. If the correct code is not entered into the phone, the satellite begins to track the vehicle. This tracking signal is then monitored by an operator who can call the police in the area where the vehicle is being tracked. When the system relies on the GPS, a security or police officer can watch the movement of the vehicle on a remote computer screen.

Some systems automatically send a signal to the vehicle tracking operator if an air bag was deployed. This signal, in addition to the global positioning satellite network, lets the authorities immediately know when a serious accident has occurred and the emergency squad can respond without waiting to be called on the phone.

Systems, such as OnStar, can also provide the driver with emergency assistance and advice, roadside assistance, remote diagnostics (while the vehicle is driven), remote activation of the vehicle's horns and lights, remote door unlocking, news, and e-mail.

Voice Activation System

Voice activation or control systems allow for the control and operation of some accessories with voice commands. The voice commands are in addition to normal manual

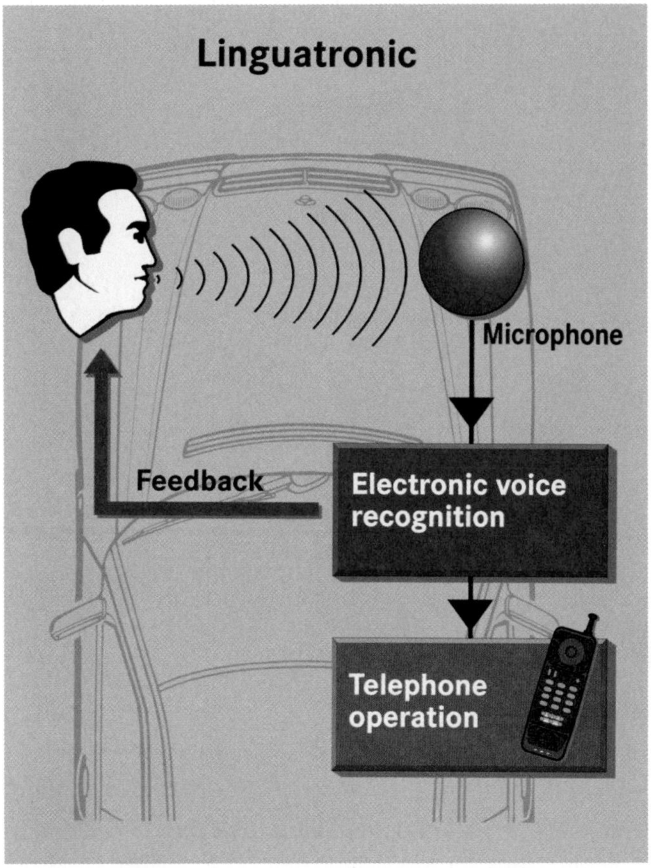

Figure 22-40 A simplified look at how voice activation systems work. *Courtesy of DaimlerChrysler Corporation*

controls. Voice activation is commonly used for cell phone operation but can be used on other controls **(Figure 22–40)**. Voice activation systems recognize the driver's voice and can respond with answers in response to the driver's questions. Once the system has understood the driver's request, it responds by carrying out the desired function, such as changing the stations on the radio. The system works through the microphones located near the driver. The voice activation system can recognize up to 2,000 commands and numeric sequences.

SECURITY AND ANTITHEFT DEVICES

Three basic types of antitheft devices are available: locking devices, disabling devices, and alarm systems. Many of the devices are available as optional equipment from the manufacturers; others are aftermarket installed.

Locks and Keys

Locks are designed to deny entry to the engine, passenger, and trunk compartments of the car as well as to prevent a thief from driving the car away. Most locks deny entry by moving a mechanical block between the vehicle's body and the door. Latches and keys simply move those blocks.

To prevent theft, manufacturers use specially cut keys that can not be easily duplicated, and the lock mechanisms for these keys are extremely difficult to pick. The master key is often built into a remote control key handle. The master key can lock and unlock the doors, trunk, fuel filler door, and glove compartment, all at the same time. These locks can be operated with the key or with the remote. The battery for the remote is charged each time the key is inserted into the lock. These systems also automatically unlock the doors in case of an accident and turn on the hazard and interior lights.

A special key, often called a valet key, only works in the doors and ignition, thereby preventing entry to the trunk and glove compartment.

Both internal and external hood locks are available to prevent theft of anything under the hood. These locks are especially useful on a vehicle with a battery-powered alarm system because they prevent a thief from disconnecting the power source or disabling the alarm.

Many cars are equipped with special fuel filler doors that help to prevent the theft of gas from the fuel tank. Voltage is present at the fuel filler door release switch at all times. When the switch is closed, the door release solenoid is energized and the fuel door opens.

Passkey Systems

The passkey is a specially designed key **(Figure 22–41)**, or transponder, that is selected and programmed just for the vehicle for which it was intended. Although another key may fit into the ignition switch or door lock, the system does not allow the engine to start without the correct electrical signal from the key.

A **resistance key** appears to be a normal key but has a small resistor bonded to it. When the key is inserted into the ignition switch, the circuit must recognize that resistance as being the correct amount for the vehicle before the engine will start.

Transponder key systems are based on a communication scheme between the vehicle's PCM and the transponder in the key **(Figure 22–42)**. Each time the key is inserted into the ignition switch, the PCM sends out a different radio signal. If the key's transponder is not capable of returning the same signal, the engine will not start.

Some late-model vehicles use an ignition-kill system that prevents the vehicle from being hot-wired. In fact, when the ignition kill is activated, not even the ignition key starts the car. An ignition-kill system has a definite advantage as a theft deterrent. There is nothing to transport, no codes to remember, and nothing visible to mar the exterior of the car or to alert a thief.

Passwords On some passkey systems, a new password must be learned by the PCM when the BCM or PCM has been replaced. When a computer is replaced, the EEP-

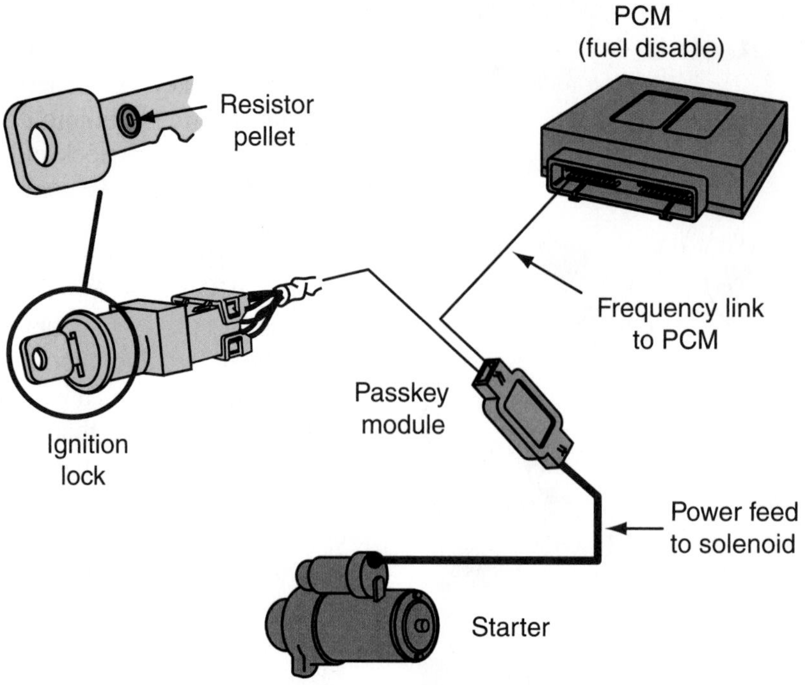

Figure 22–41 A passkey with a resistor.

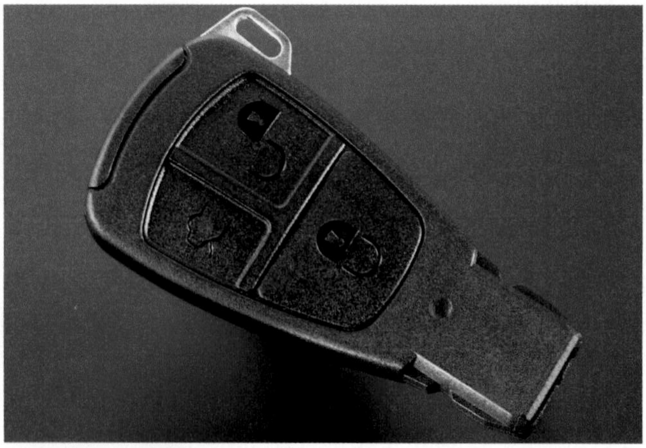

Figure 22–42 A totally electronic ignition key. *Courtesy of DaimlerChrysler Corporation*

ROM calibration is flashed into the new PCM. After the calibration is complete, attempt to start the vehicle. It will not, but leave the ignition on until the theft system warning lamp turns off. Then turn the ignition off and attempt to start the engine. The engine should start this time.

Keyless Entry Systems

A keyless entry system allows the driver to unlock the doors or trunk lid from outside of the vehicle without using a key. It has two main components: an electronic control module and a coded-button keypad on the driver's door or a key fob **(Figure 22–43)**. Some keyless systems also have an illuminated entry system.

The electronic control module typically can unlock all doors, unlock the trunk, lock all doors, lock the trunk, turn on courtesy lamps, and illuminate the keypad or keyhole after any button on the keypad is pushed or either front door handle is pulled.

Remote keyless entry systems rely on a handheld transmitter, frequently part of the key fob. With a press of the unlock button on the transmitter from 25 to 50 feet away (depending on the type of transmitter) in any direction range, the interior lights turn on, the driver's door unlocks, and the theft security system is disarmed. The trunk can also be unlocked. Pressing the lock button locks

Figure 22–43 An assortment of electronic keys and key fobs. *Courtesy of DaimlerChrysler Corporation*

all doors and arms the security system. For maximum security, some remote units and their receiver change access codes each time the remote is used.

Some remote units can also open and close all of the vehicle's windows, including the sun roof. They may also be capable of setting off the alarm system in the case of panic.

Alarm Systems

The two methods for activating alarm systems are passive and active. Passive systems switch on automatically when the ignition key is removed or the doors are locked. They are often more effective than active systems. Active systems are activated manually with a key fob transmitter, keypad, key, or toggle switch.

Switches similar to those used to turn on the courtesy lights as the doors are opened are often used to arm the alarm. When a door, hood, or trunk is opened, the switch closes and the alarm sounds. It turns itself off automatically (provided the intruder has stopped trying to enter the car) to prevent the battery from being drained. It then automatically rearms itself.

For most systems, the driver's door lock is the master switch for the unit. If the alarm is activated, it can be turned off by inserting and turning the door lock or the ignition switch. The driver's door lock may be equipped with a pair of magnetic **reed-type switches (Figure 22–44)**. The reed switches move with the door lock to arm and disarm the system.

Ultrasonic sensors are used to detect motion and will trigger the alarm if there is movement inside the vehicle. Current-sensitive sensors activate the alarm if there is a change in current within the electrical system, such as when a courtesy light goes on or the ignition starts. Motion detectors monitor changes in the vehicle's tilt, such as when someone is attempting to steal the tires.

Many alarm systems are designed to sound an alarm, turn on the hazard lights, and cause the high beams to flash along with the hazard lamps. Indicator lamps on the inside of the vehicle alert others that the alarm is set and also remind the driver to turn the alarm off before entering. To avoid false alarms, some systems allow for the disabling of particular sensors, such as the motion detector inside the vehicle that could be set off by a pet inside the vehicle.

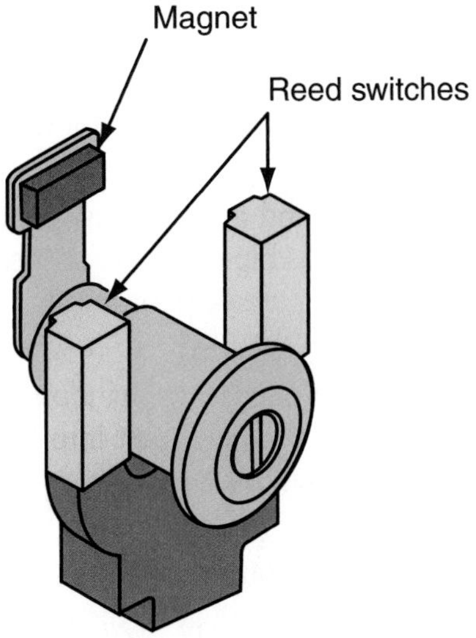

Figure 22-44 A driver's door lock assembly for an antitheft system.

CASE STUDY

***A** customer complains that the horn does not operate.*

*To correct this situation, the technician must determine whether the horn system has a relay. **Figure 22–45** shows a two-horn circuit without a relay. Fused power goes up the steering column to the horn button, which is normally open (NO). Closing the switch closes the circuit, bringing power down the column to a common point where the two horns are connected in parallel. On the other side of the horn is a ground to complete the circuit. The circuit is easily diagnosed with a 12-volt test light, following the usual procedure of looking for power with the horn button pushed in. The ground circuit should also be checked. Power in at the switch and no power out with the button depressed would be the only reason to replace the switch unit.*

***Figure 22–46** shows an adaptation of the simple horn circuit. Notice that fused power is applied to the horn relay rather than to the switch. An NO relay is used. Notice that the relay coil receives power from the same fused source as the contacts do. The other end of the relay coil runs up into the steering column where an NO switch is located. Closing the switch grounds the relay coil, which energizes the coil and causes the contacts to close. The closed contacts allow power to the common point for the two horns, which are grounded. Remember, any relay circuit must be diagnosed as two separate circuits. The relay coil is a separate circuit that needs a path for it to develop the magnetic field necessary to close the contacts. Closing the contacts allows current flow through the other circuit to the horns. Diagnose these circuits separately, as you would any relay circuit, and you are less likely to make mistakes. Use a DMM with a wiring diagram to trace the power through the circuit.*

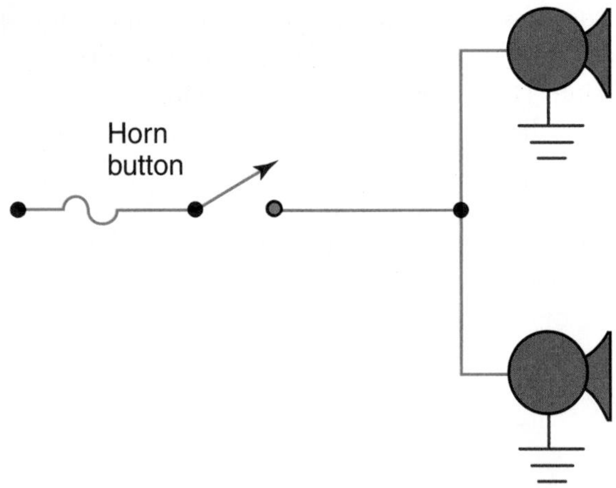

Figure 22-45 A two-horn circuit without relay.

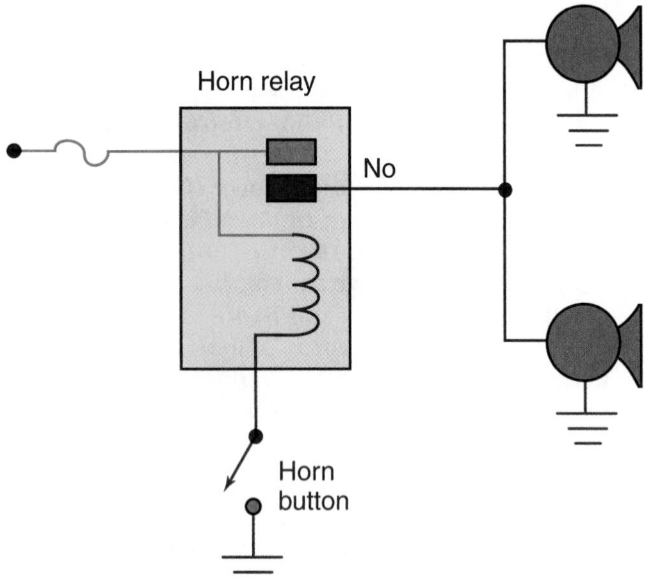

Figure 22-46 A horn circuit utilizing a relay.

Most law enforcement officials recommend audible alarms over silent page alarms. An audible alarm calls attention to the car and startles the intruder. With a silent page alarm, the owner carries a pager that beeps when someone tampers with the car. However, this can be dangerous because many people recklessly confront the thief alone rather than call the police.

Although increased numbers of vehicles are being equipped with security and antitheft devices, most installation and repair work is done by a general auto technician. It is important for you to keep abreast of both original equipment and aftermarket systems for the latest information on security and antitheft equipment.

KEY TERMS

Defogger
Deicer
Moon roof
Reed-type switches
Resistance key
Servo unit
Sun roof
Transducer
Transponder key

SUMMARY

- Many accessories may be controlled and operated by a body computer, and therefore diagnosis may involve retrieving trouble codes from the computer.
- The basic designs of windshield wiper systems used on today's vehicles include a standard two- or three-speed system with or without an intermittent or rain-sensing feature.
- The motors can have electromagnetic fields or permanent magnetic fields. With permanent magnetic fields, motor speed is regulated by the placement of the brushes on the commutator. The speed of electromagnetic motors is controlled by the strength of the magnetic fields.
- Diagnosis of a wiper/washer system should begin by determining if the problem is mechanical or electrical.
- Most current vehicles have electronically regulated cruise control systems that are controlled with a separate control module or by the vehicle's PCM.
- Diagnosis of PCM-controlled cruise control systems is aided by a scan tool.
- Complex sound systems may include an AM/FM radio receiver, stereo amplifier, CD player, cassette player, MP3 player, equalizer, several speakers, and a power antenna system.
- Poor speaker or sound quality is usually caused by a bad antenna, damaged speakers, loose speaker mountings or surrounding areas, poor wiring, or damaged speaker housings.
- Diagnosis of power door locks, windows, and seats is best done by dividing the circuit into individual circuits and the total circuit and basing your testing on the symptoms.
- One of the most common problems with rear-window defrosters is damage to the grids on the window, which can usually be repaired.

- Navigation systems use global positioning satellites to help drivers make travel decisions while they are on the road.
- Three basic types of antitheft devices are available: locking devices, disabling devices, and alarm systems.
- Most ignition keys for late-model vehicles are either resistance or transponder passkeys that only work on the vehicle for which they were intended.
- Passive alarm systems switch on automatically when the ignition key is removed or the doors are locked. Active systems are activated manually.

TECH MANUAL

The following procedures are included in Chapter 22 of the *Tech Manual* that accompanies this book:

1. Testing and repairing a rear-window defogger grid.
2. Performing a speed or cruise control simulated road test.
3. Identifying the source of static on a radio.

REVIEW QUESTIONS

1. What is the primary purpose of adaptive cruise control?
2. Why are amplifiers added to sound systems?
3. What should be checked if none of the grids in a rear-window defogger work?
4. What component in a wiper system makes it possible to adjust the wiper interval period? How does it work?
5. Name the two most common problems that occur with power seats.
6. While discussing power door locks, Technician A says if none of the locks work, the actuators at each door should be checked first. Technician B says if the power locks can lock or unlock but not both, the relays and wiring to the master switch should be checked. Who is correct?
   a. Technician A **c.** Both A and B
   b. Technician B **d.** Neither A nor B
7. *True or False?* On some passkey systems, when the PCM or BCM is replaced, the key must be reprogrammed with a new password.
8. Technician A says horn buttons in circuits with relays are power switches. Technician B says horn buttons in circuits without relays are grounding switches. Who is correct?
   a. Technician A **c.** Both A and B
   b. Technician B **d.** Neither A nor B
9. Rear defrosters generally have a relay with a timer. This allows __________.
   a. the defogger to operate for only a specific amount of time
   b. the defogger to function just until the rear window is clear
   c. the defogger to be independent of the ignition switch
   d. none of the above
10. The right front power window does not work. Technician A says the motor might be insulated from ground. Technician B says either the master or right door switch could be the problem. Who is correct?
    a. Technician A **c.** Both A and B
    b. Technician B **d.** Neither A nor B
11. Windshield wiper systems generally park the blades by __________.
    a. turning off power to the motor
    b. applying power to the motor
    c. ungrounding the switch
    d. reversing the motor direction
12. The reasons for slower-than-normal wiper operation are being discussed. Technician A says the problem may be in the mechanical linkage. Technician B says there may be excessive resistance in the electrical circuit. Who is correct?
    a. Technician A **c.** Both A and B
    b. Technician B **d.** Neither A nor B

13. A six-way power seat does not work in any switch position. Technician A says to check the circuit breaker. Technician B says to use a continuity chart to test the switch. Who is correct?

a. Technician A
b. Technician B
c. Both A and B
d. Neither A nor B

14. Technician A says some antitheft systems only sound an alarm if someone enters the vehicle. Technician B says some antitheft devices prevent the engine from starting. Who is correct?

a. Technician A
b. Technician B
c. Both A and B
d. Neither A nor B

15. The circuit breaker at a power seat motor continuously trips. Technician A says this could be caused by a seat track problem that is causing mechanical resistance. Technician B says this could be caused by corrosion at the motor that is causing electrical resistance. Who is correct?

a. Technician A
b. Technician B
c. Both A and B
d. Neither A nor B

16. Technician A says a resistance key has a small thermistor bonded to it. Technician B says a transponder key sends a radio signal to the PCM. Who is correct?

a. Technician A
b. Technician B
c. Both A and B
d. Neither A nor B

17. *True or False?* Vehicles with electronic throttle control do not need a separate cruise control module, stepper motor, or cable to control engine speed. The PCM has full control of the throttle and therefore the circuitry of the PCM operates the cruise control system.

18. Technician A says that most law enforcement officials recommend silent page alarms over audible alarms. Technician B says that devices used to trigger an alarm system include mechanical switches, ultrasonic sensors, and current-sensitive sensors. Who is correct?

a. Technician A
b. Technician B
c. Both A and B
d. Neither A nor B

19. Technician A says navigational systems rely on global positioning satellites. Technician B says navigational systems rely on information programmed in their memory and information stored on a CD or DVD. Who is correct?

a. Technician A
b. Technician B
c. Both A and B
d. Neither A nor B

20. Which of the following can be sources of radio interference?

a. ignition system wiring
b. electrical power lines
c. neon signs
d. all of the above

23

RESTRAINT SYSTEMS: THEORY, DIAGNOSIS, AND SERVICE

OBJECTIVES

■ Identify and describe devices that contribute to automotive safety. ■ Explain the difference between active and passive restraint systems. ■ Know how to service and repair passive belt systems. ■ Describe the function and operation of air bags. ■ Identify the major parts of a typical air bag system. ■ Safely disarm and inspect an air bag assembly. ■ Know how to diagnose and service an air bag system.

Safety is foremost in the minds of automobile manufacturers. According to a survey by the Insurance Institute for Highway Safety, occupant protection has emerged as a leading factor in determining which car people will buy. According to the institute, 68% of the households surveyed ranked the "degree to which the car protects people" as a very important purchase-decision factor.

Many safety features are now available as standard equipment or as options. Some of these include side impact barriers, crumple zone in the body, seat belts, antilock brakes, traction control, stability control, and air bags **(Figure 23–1)**. There are many safety items that have been around for many years, such as laminated and tempered glass.

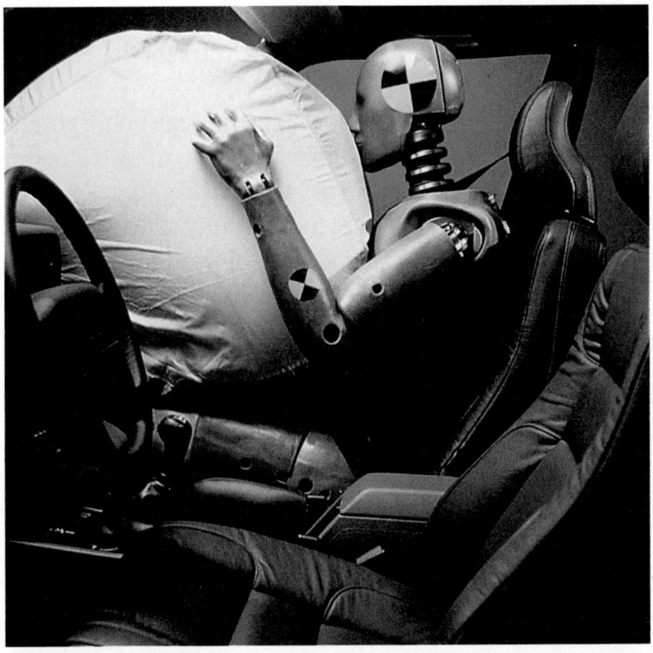

Figure 23-1 A passenger-side air bag. *Courtesy of Volvo Car Corporation*

Common restraint systems—seat belts and air bags—are covered in this chapter. It is important for a technician to understand how these systems work and how to diagnose and service them.

An **active restraint system** is one that a vehicle's occupant must make a manual effort to use **(Figure 23–2)**. For example, in most vehicles the passenger must fasten the seat belts for crash protection. A **passive restraint system** operates automatically **(Figure 23–3)**. No action is required of the occupant to make it functional.

Passive safety equipment includes the safety belt system, the air bags, the rigid occupant cell, and the crumple zones at the front of the vehicle. The rear and sides of the body are among the most important safety features of today's cars and are designed to dissipate most of the impact energy for the protection of vehicle occupants.

All vehicles built or sold in the United States must have one or both of these passive restraints: automatic seat belts or air bags. Both have made a tremendous impact on driver and passenger safety over the last few years. It is very important that technicians be able to service these systems.

SEAT BELTS

A passive seat belt system uses electric motors to automatically move shoulder belts across the driver and front seat passenger. The upper end of the belt is attached to a carrier that moves in a track at the top of the doorframe. The other end is secured to an **inertia lock retractor (Figure 23–4)**. The retractors are mounted in the center console. When the door is opened, the outer end of the

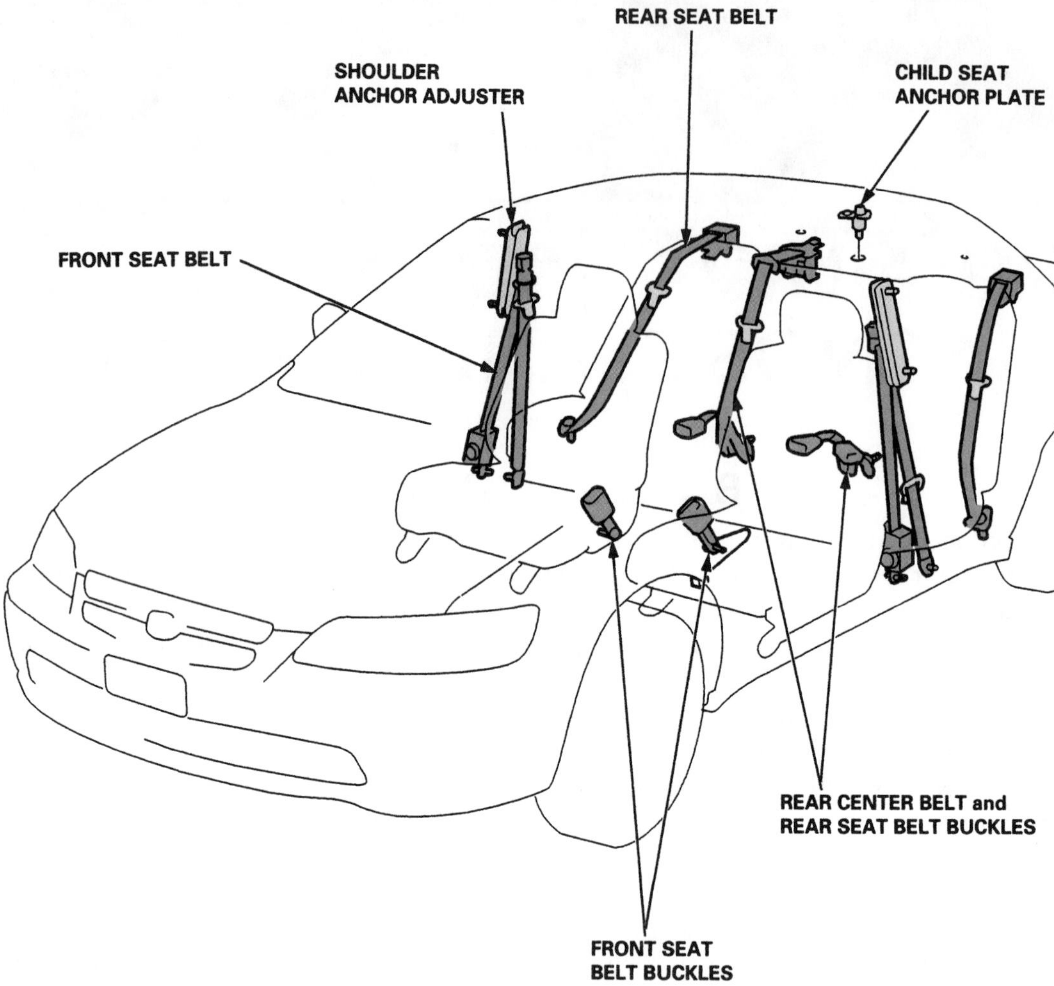

Figure 23-2 Active restraint systems. *Courtesy of American Honda Motor Co., Inc.*

shoulder belt moves forward to allow for easy entry or exit. When the doors are closed and the ignition turned on, the belts move rearward and secure the occupants. The active lap belt is manually fastened by the occupant and should be worn with the passive belt.

Most vehicles, especially those with air bags, have two active belts. One is a lap belt that goes across the occupant's lap; the other is a shoulder belt that goes across the occupant's shoulder and chest. The two belts join together at a single point where they can be inserted into buckle that is anchored to the vehicle's floor.

Seat Belt Retractors

When unbuckled, seat belts are stowed away by the seat belt retractors **(Figure 23–5)**. The retractors may also work as pretensioners to take up the slack in the belt during an accident to limit the forward movement of the occupant's body. Besides inertia lock retractors, vehicles may be equipped with electric or pyrotechnic-type **pretensioners**. Electric pretensioners rely on a motor that quickly tightens the belt. Pyrotechnic pretensioners rely on an explosive charge that quickly retracts the belt and locks it in place.

The speed and force at which the seat belt tightens as a result of a collision is variable on some vehicles. These systems rely on sensors to measure the weight of the person in the seat and the amount of force on the seat belt as that person is moving forward during an impact. Some vehicles are equipped with two-stage belt force limiters.

A mechanized pretensioning seat belt retractor is capable of removing seat belt slack and pulling the occupant into position using various inputs from the brake and chassis control systems or from a precrash sensing system **(Figure 23–6)**.

Warning Lights

All modern seat belt systems have a warning lamp and a buzzer or chime that is turned on when the vehicle is started to remind the occupants to buckle up.

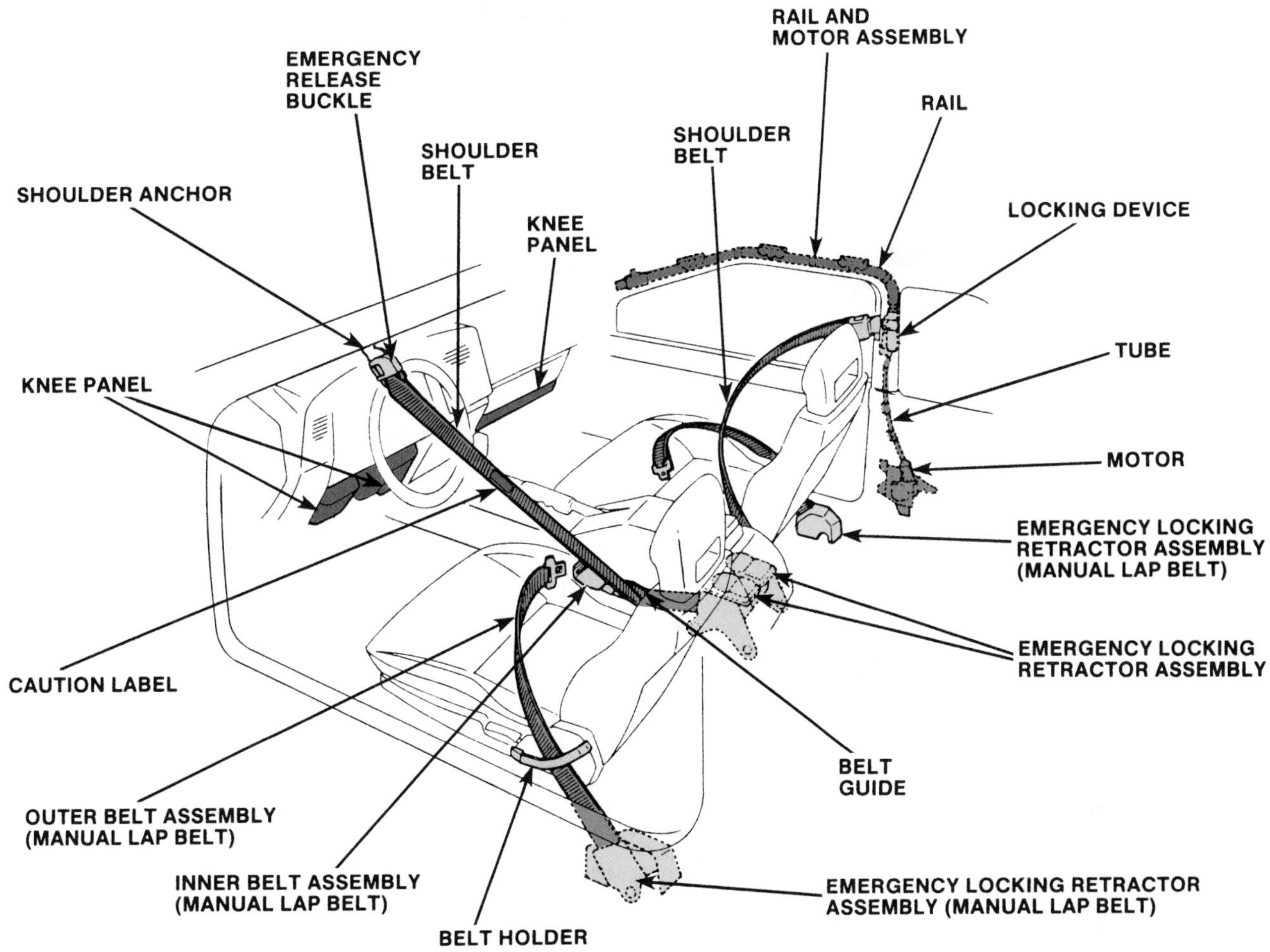

Figure 23-3 Passive seat belt restraint system. *Courtesy of Ford Motor Company*

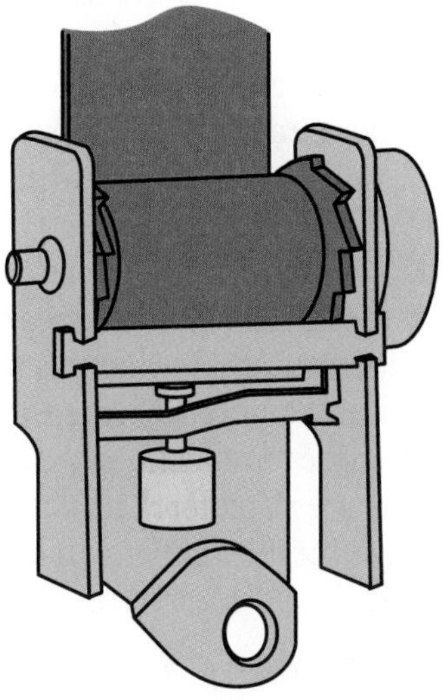

Figure 23-4 An inertia lock seat belt retractor.

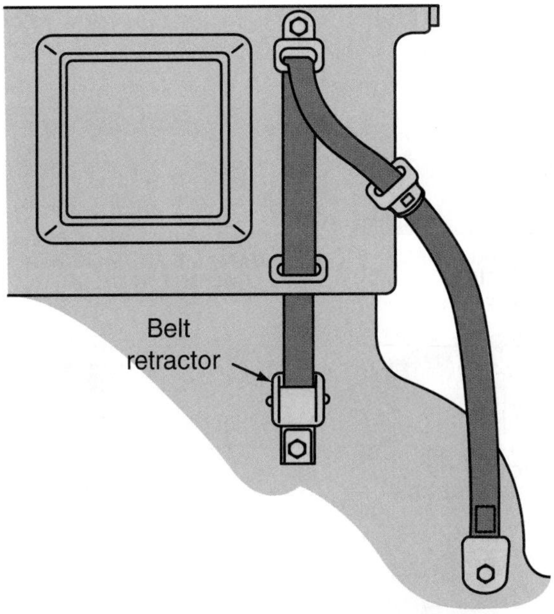

Figure 23-5 A typical seat belt retractor.

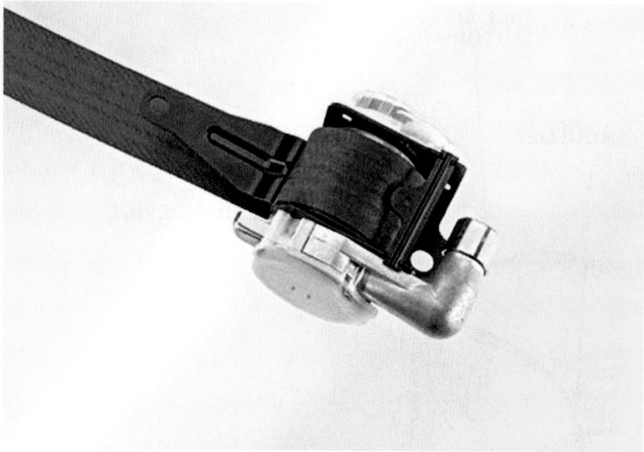

Figure 23-6 A mechanized pretensioning retractor/variable load-limiting seat belt retractor. *Courtesy of Delphi Corporation*

SEAT BELT SERVICE

Inspecting seat belt systems should follow a systematic approach. Always take as much time as necessary to do your inspection; remember that they are designed to protect people.

Webbing Inspection

Pay special attention to where the webbing contacts maximum stress points, such as the buckle, D-ring, and retractor. Collision forces center on these locations and can weaken the belt. Signs of damage at these points require belt replacement. Check for twisted webbing due to improper alignment when connecting the buckle. Fully extend the webbing from the retractor. Inspect the webbing and replace it with a new assembly if the following conditions are noted **(Figure 23–7)**: cut or damaged webbing, broken or pulled threads, cut loops at the belt edge, color fading as a result of exposure to sun or chemical agents, or bowed webbing.

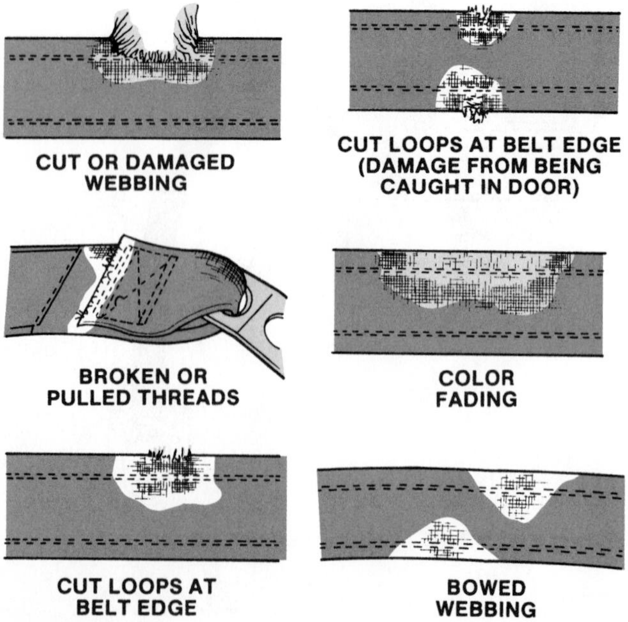

Figure 23-7 Examples of webbing defects.

If the webbing cannot be pulled out of the retractor or will not retract to the stowed position, check for the following conditions and clean or correct as necessary: webbing soiled with gum, syrup, grease, or other material; twisted webbing; or the retractor or loop on the B-pillar out of position.

SHOP TALK

Never bleach or dye the belt webbing. Clean it with a mild soap solution and water. ■

Buckle Inspection

To determine if the buckle works or if the buckle housing has been damaged, insert the seat belt into the buckle until a click is heard. Pull back on the webbing quickly to ensure that the buckle is latched properly. Replace the seat belt assembly if the buckle does not latch. Depress the button on the buckle to release the belt. The belt should release with a pressure of approximately 2 pounds. Replace the seat belt assembly if the buckle cover is cracked, the push button is loose, or the pressure required to release the button is too high.

Retractor Inspection

Retractors for lap belts should lock automatically once the belt is fully out. Either webbing-sensitive or vehicle-sensitive seat belt retractors are used with passive seat belt systems. Webbing-sensitive retractors can be tested by grasping the seat belt and jerking it. The retractor should lock up; if it does not, replace the seat belt retractor.

Vehicle-sensitive belt retractors will not lock up using the same procedure. To test these belts, a braking test is required. Perform this test in a safe place. A helper is required to check the retractors on the passenger side and in the back if the vehicle is equipped with rear lap/shoulder belts.

Test each belt by driving the car at 5 to 8 mph and quickly applying the brakes. If a belt does not lock up, replace the seat belt assembly. During this test, it is important for the driver and helper to brace themselves in the event the retractor does not lock up.

Most retractors are not interchangeable. That is, an R marked on the retractor tab indicates that it is for the right side only, and an L should be used on the left side only.

Drive Track Assembly and Anchor Inspection

Seat belt anchors are found where the retractors attach to the car body. High-impact forces occur here during a collision; therefore, carefully inspect the anchor areas and attaching bolts. Loose bolts should be replaced. Look for cracks and distortion in the metal where the retractors and D-rings anchor. Upper body damage must be properly repaired and brought back to exact dimensional specifications before any repairs are performed on the seat belt system. If there is damage to the metal in the mounting area, proper repairs, such as welding in reinforcement metal, must be completed before reattaching the anchor. Be sure to restore corrosion protection to the area. When spraying anticorrosion materials, make sure they do not enter the retractor. They can keep it from operating properly. Finally, look for dirt and corrosion around the anchor area.

The drive motor is usually located at the base of the track assembly behind the rear seat side trim panel. Its purpose is to pull the tape that positions the belt. If a check of the drive motor reveals that it is faulty, replace it. Like any motorized system, the seat belt parts need periodic lubrication. To service the motorized seat belt system, follow the instruction given in the service manual.

Rear Seat Restraint System

Rear seat belts are inspected in the same way as the front. However, some vehicles have a center seat belt. These belts do not have a retractor. Check the webbing, anchors, and the adjustable locking slide for the belt. Fasten the tongue to the buckle and adjust by pulling the webbing end at a right angle to the connector and buckle. Release the webbing and pull upward on the connector and buckle. If the slide lock does not hold, remove and replace that seat belt assembly.

Warning Light and Sound Systems

When the ignition is turned to the on or run position, the Fasten Seat Belt light should come on. There should also be a buzzer or chime. If these warning light and sound systems do not come on, check for a blown fuse or circuit breaker. If that checks out fine, and there is sound but no light, check for a damaged or burned out bulb. If the bulb lights but there is no sound, check for damaged or loose wiring, switches, or buzzer (voice module).

Servicing Seat Belts

If the seat belt is damaged in any way, it should be replaced. Some guidelines for servicing lap and shoulder belts follow:

- Keep sharp edges from damaging any portion of the belt buckle or latch plate.
- Avoid bending or damaging any portion of the belt buckle or latch plate.
- Do not attempt repairs on lap or shoulder belt retractor mechanisms or lap belt retractor covers. Replace them with new replacement parts.
- Tighten all seat and shoulder belt anchor bolts as specified in the service manual.

AIR BAGS

An air bag is much like a nylon balloon that quickly inflates to stop the forward movement of the occupant's upper body. Air bags are passive systems designed to be used with an active system—seat belts. Air bags were not designed to replace seat belts. An air bag's job during a crash is normally over in less than one second after it begins **(Figure 23–8)**. Consider this sequence:

- Time zero—Impact begins and the air bag system is doing nothing.
- Twenty milliseconds later—the sensors are sending an impact signal to the air bag module and the air bag begins to inflate.
- Three milliseconds later (total time from impact is now 23 milliseconds)—The air bag is inflated and is up against the occupant's chest. The occupant's body has not yet begun to move as a result of the impact.
- Seventeen milliseconds later (total time from impact is now 40 milliseconds)—The air bag is almost fully

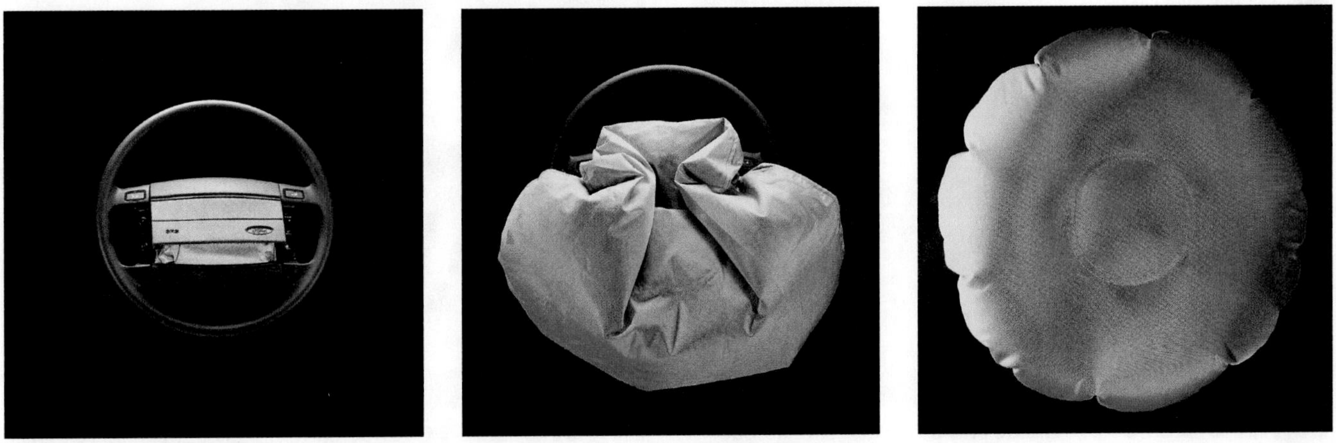

Figure 23-8 Various stages of air bag inflation. *Courtesy of TRW, Incorporated*

deployed and the occupant's body begins to move forward because of the impact.

- Thirty milliseconds later (total time from impact is just 70 milliseconds)—The air bag begins to absorb the forward movement of the occupant and the air bag begins to deflate through its vents. Once the air bag deflates, its job is over.

The systems and parts used to deploy an air bag vary with the year and manufacturer of the vehicle, as well as the location of the air bag. An air bag is inflated or deployed by rapid expansion (explosion) of a gas. The gas is fired by an igniter commonly called a **squib**.

Different manufacturers also call their air bag systems by different names, such as **supplemental inflatable restraint (SIR)** and **supplemental restraint system (SRS)**. All late-model vehicles have a driver-side and a passenger-side air bag **(Figure 23–9)**.

A driver-side air bag system may include a knee diverter, which is also called a knee bolster. This unit is designed to cushion the driver's knee from impact and help prevent the driver from sliding under the air bag during a collision. It is located underneath the steering column and behind the steering column trim.

Passenger-side air bag modules are located in the vehicle's dash. These air bags are very similar in design and operation to those on the driver's side. However, many manufacturers use a different set of sensors. The actual capacity of gas required to inflate the passenger-side air bag is much greater because the bag must span the extra distance between the occupant and the dashboard. The steering wheel and column make up this difference on the driver's side.

Figure 23-9 Driver-side and passenger-side air bags. *Courtesy of DaimlerChrysler Corporation*

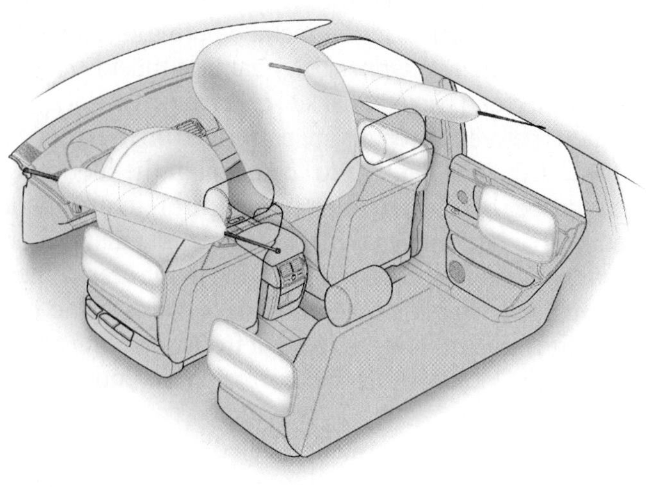

Figure 23-10 This vehicle has a total of eight air bags. *Courtesy of BMW of North America, Incorporated*

Because of the concern for babies and small children, pickups and other two-seat vehicles either do not have a passenger side SIR or have a switch that prevents it from deploying. The switch is typically moved with a key and allows the driver to activate or deactivate the SIR. Deactivation is recommended whenever a child is in a rear-facing child seat. An indicator light in the instrument panel lets the driver know the current status of the passenger side SIR.

Some vehicles are equipped with more than these two air bags **(Figure 23–10)**. The occupants in the front seat may be further protected by side air bags and/or side curtain air bags. The rear passengers may be protected by air bags in the rear of the front seat backs, side air bags, and/or side curtain air bags.

Side air bags can take on many different shapes and be deployed from various locations. Some manufacturers offer side air bags, called side curtains **(Figure 23–11)**, that blanket the entire side of the car. Side air

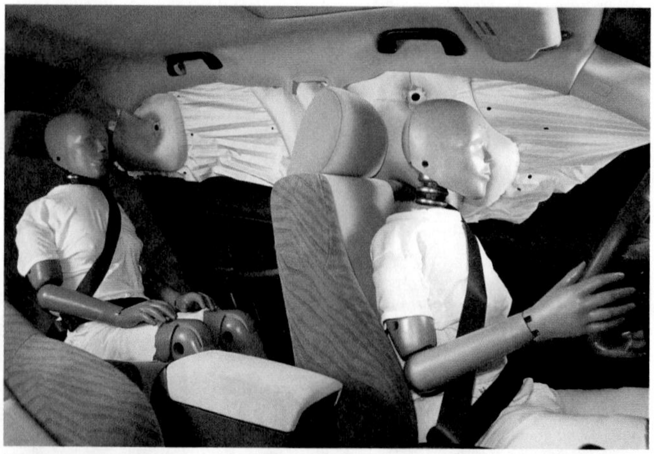

Figure 23-11 A side curtain. *Courtesy of American Honda Motor Co., Inc.*

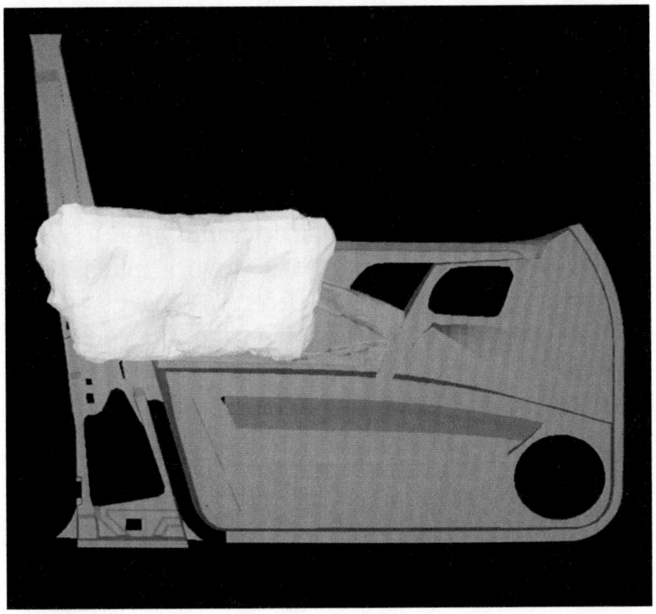

Figure 23-12 A side air bag. *Courtesy of DaimlerChrysler Corporation*

bags **(Figure 23–12)** are available for the front and rear doors on some cars. These air bags are deployed from the interior trim on the door or are deployed from the outside of the seat. Side impact head protection systems inflate a long narrow air bag that extends from the windshield area to behind the front seat back **(Figure 23–13)**.

Door-mounted side air bags must begin deploying in five to six milliseconds. This requirement is based on the simple fact that only a few inches separate the occupant from the other vehicle during a side impact. Seat-back-mounted side air bags do not need to operate at these great speeds. The head air bag, or inflatable tubular structure, is designed to stay inflated for about five seconds to offer protection against a second or third impact.

Figure 23-13 A side impact head air bag. *Courtesy of BMW of North America, Incorporated*

Current air bag systems work in conjunction with the seat belt pretensioners and retractors. When the air bag circuit is turned on, so is the pretensioner circuit. These actions limit the movement of the occupants.

A current trend with SIR systems is to make the impact of the bag's deployment less powerful, thereby accommodating occupants of different sizes and reducing the number of injuries caused by the air bag itself. These systems are referred to as *second-generation* air bags.

Second-generation air bags have been depowered to inflate with less force than earlier air bags. The air bags are depowered by reducing the peak inflation pressure and/or rise rate. Rise rate is the force and speed at which an air bag inflates. Rise rate is controlled by the type and amount of inflator gas, the size of the air bag, and the air bag's vent design.

Depending on the specific model vehicle, air bag size, and seat belt system, a second-generation air bag inflates with an average of 20% to 35% less energy.

Smart air bags are also used by some manufacturers. Many smart systems have two possible stages of air bag deployment. The force of the expanding air bag is controlled to match the severity of the impact and/or by the size and weight of the seat's occupant **(Figure 23–14)**. Other things considered to alter air bag deployment are seat-track position and seat belt use. All of these factors require different deployment rates.

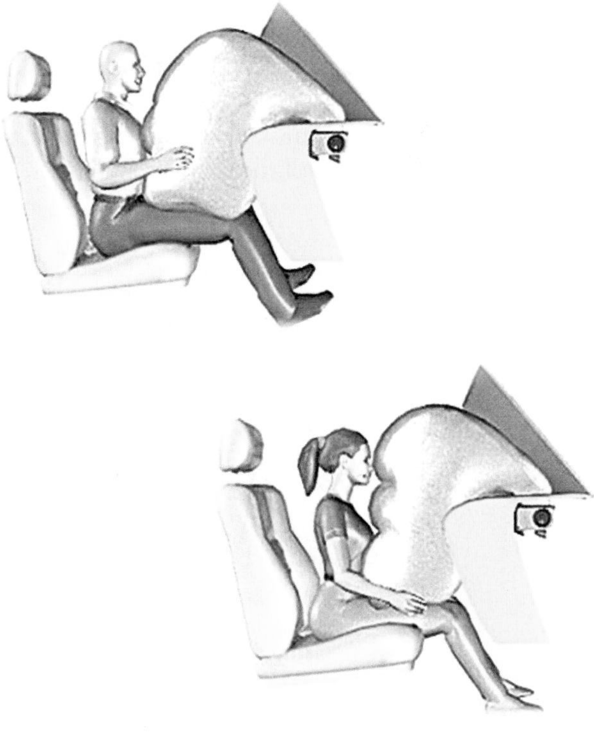

Figure 23-14 (top) Full strength air bag deployment because of the weight of the occupant. (bottom) Reduced air bag deployment because of the weight of the occupant. *Courtesy of Delphi Corporation*

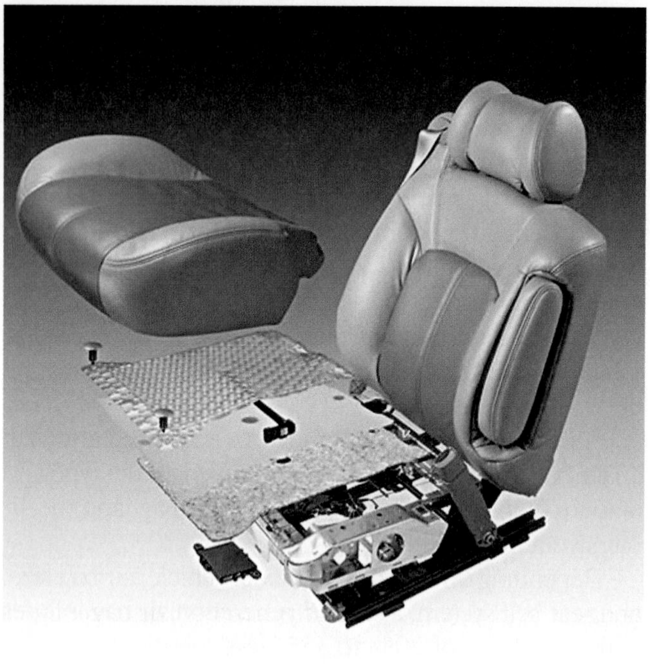

Figure 23-15 An occupant detection system with a seat belt tension sensor. *Courtesy of Delphi Corporation*

Two-stage air bags have two containers of gas and two squibs in their air bag module. When low pressure deployment is desired, only one squib is fired. During a severe collision, the occupants need maximum protection and both squibs fire. The speed of deployment can also be controlled by the firing of the squibs. For rapid deployment, both squibs fire at the same time. To phase in full deployment, one squib is fired a few milliseconds later than the other.

A few late-model systems offer rollover protection. Sensors monitor the speed of the rollover and inflate the air bags according to the sensed severity of the rollover. During a rollover, the air bags stay inflated longer than normal to keep the occupants safe until the vehicle comes to a rest. These systems also take the slack out of the seat belts, shut off the fuel pump, and disconnect the battery when a rollover is sensed.

All year 2006 and newer vehicles must have a system that allows for air bag suppression when infants, children, or small adults are in the front passenger seat. This system is based on a load sensor, seat belt tension sensor, and an electronic control unit **(Figure 23–15)**. The load sensor measures the loading force on the seat, classifies the occupant as an adult, infant, or child, or the seat as empty, and provides the classification to the air bag controller, which enables or suppresses passenger air bag deployment. The belt tension sensor identifies cinched child seats.

Electrical System Components

The electrical circuit of an air bag system includes impact sensors and an electronic control module **(Figure 23–16)**. The electrical system conducts a system self-check to let the driver know that it is functioning properly, detects an impact, and sends a signal that inflates the air bag.

Sensors To prevent accidental deployment of the air bag, most systems require that at least two sensor switches be closed to deploy the air bag **(Figure 23–17)**. The number of sensors used in a system depends on the design of the system. Some systems use only a single sensor and others use up to five. The name used to identify the sensors also varies among manufacturers. Normally sensors are located in the engine and passenger compartments.

Typically, ignition of the air bag only occurs when an outside (impact or crash) sensor and an inside (safing or

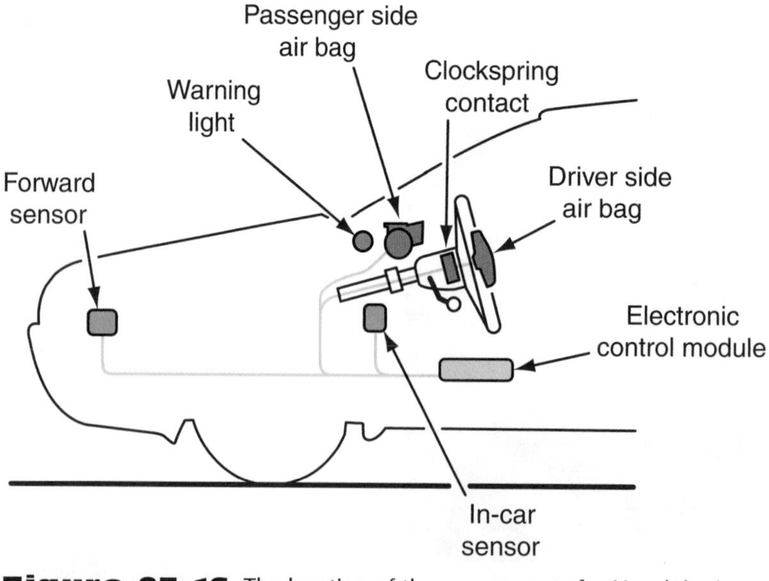

Figure 23-16 The location of the components for Honda's air bag system.

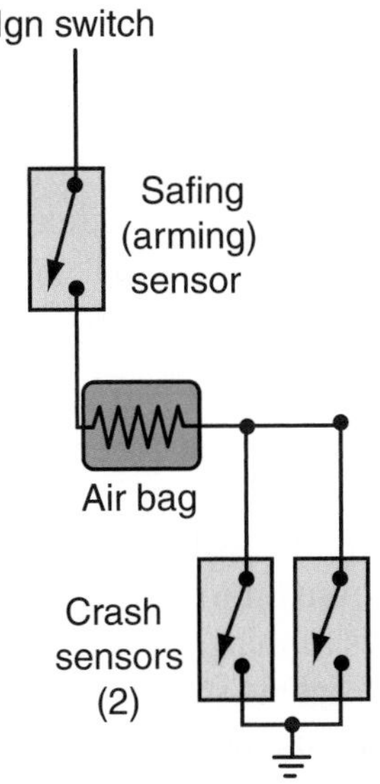

Figure 23–17 A simple air bag circuit with sensors.

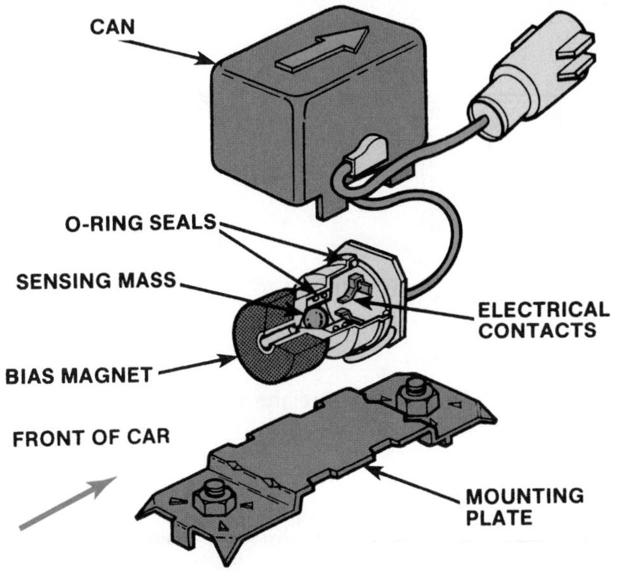

Figure 23–19 A typical ball and magnet sensor for an air bag system. *Courtesy of DaimlerChrysler Corporation*

arming) sensor are closed. Once the two sensors are closed, the electrical circuit to the igniter is complete. The igniter starts the chemical chain reaction that produces heat. The heat causes the generant to produce nitrogen gas, which fills the air bag.

Roller-Type Sensors. This type of sensor has a roller located on a ramp **(Figure 23–18)**. One terminal of the sensor is connected to the ramp and the second terminal is connected to a spring contact extending through an opening in the ramp but not touching the ramp. The roller is held against a stop by small springs. If the vehicle is involved in a collision at a high enough speed, the roller moves up the ramp and strikes the spring contact. This movement completes the circuit between the ramp and the spring contact through the roller and the air bag deploys.

Mass-Type Sensor. This type of sensor contains a normally open set of gold-plated switch contacts and a gold-plated ball, the sensing mass, which is held in place by a magnet **(Figure 23–19)**. At the point of sufficient force, the ball will break loose from the magnet and make contact with the electrical contacts to complete the circuit.

Accelerometer. This sensor contains a piezoelectric element that is distorted during a collision **(Figure 23–20)**. This element generates an analog voltage in relation to the severity of deceleration forces. The accelerometer also senses the direction of an impact force. Typically, this sensor is located within the computer for the air bag system.

Diagnostic Monitor Assembly The **air bag sensing diagnostic monitor (ASDM)** constantly monitors the readiness of the SIR electrical system. If the module determines there is a fault, it will illuminate the warning lamp. Depending on the fault, the SIR system may be disarmed until the fault is corrected.

The diagnostic module also supplies backup power to the air bag module in the event that the battery or cables are damaged during the accident. The stored charge can last up to 30 minutes after the battery has been disconnected.

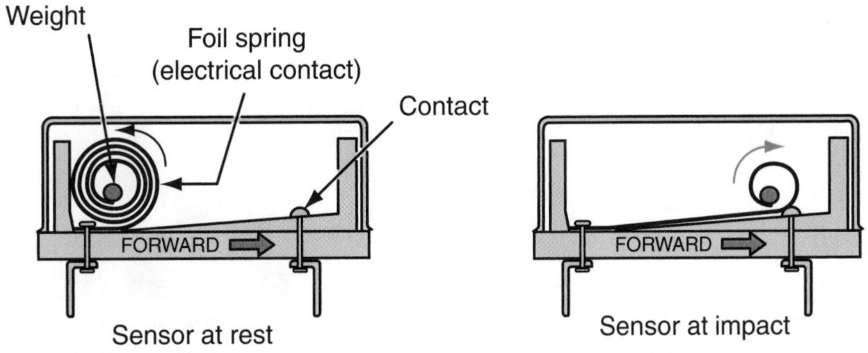

Figure 23–18 A roller-type air bag sensor.

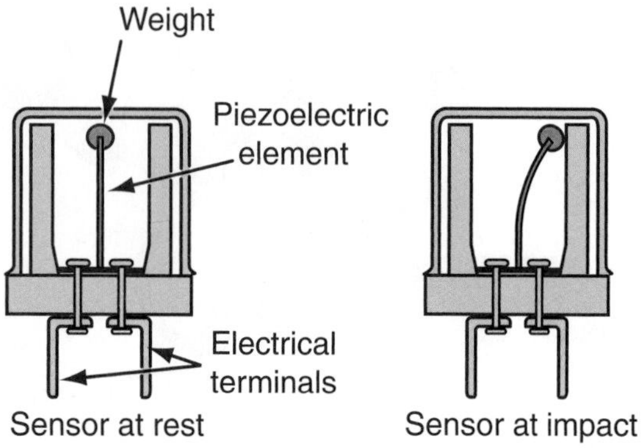

Figure 23-20 An accelerometer air bag sensor. *Courtesy of DaimlerChrysler Corporation*

WARNING!

The backup power supply must be depleted before any air bag service is performed. To deplete this backup power, disconnect the positive battery cable and wait at least 30 minutes. Refer to the information in the service manual to determine exactly how long you should wait.

Wiring Harness For identification and safety purposes, the electrical harnesses of the SIR system typically have yellow connectors. Single-stage air bags have one inflator and one pair of wires that connects to the air bag module. Two-stage air bags have two inflators and two pairs of wires that connect to the air bag module.

Clockspring The **clockspring** allows for electrical contact to the air bag module at all times. Since the air bag module sits in the center of the steering wheel, the clockspring is designed to provide voltage to the module regardless of steering wheel position. The clockspring is located between the steering wheel and the steering column **(Figure 23–21)**.

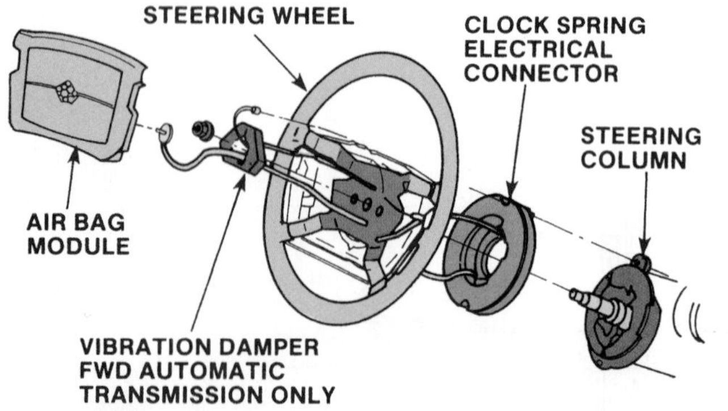

Figure 23-21 The clockspring is located between the steering wheel and the steering column. *Courtesy of DaimlerChrysler Corporation*

The clockspring's electrical connector contains a long conductive ribbon. The wires from the air bag's electrical system are connected from the underside of the clockspring to the conductive ribbon. The other end of the ribbon is connected to the air bag module. When the steering wheel is turned, the ribbon coils and uncoils without breaking the electrical connection.

SIR or Air Bag Readiness Light This light lets the driver know the air bag system is ready to do its job. The warning lamp is operated by the diagnostic module. The readiness lamp lights briefly when the driver turns the ignition key from off to run. The lamp should go out once the engine is running. A malfunction in the air bag system causes the light to stay on continuously or to flash. Some systems have a tone generator that sounds if there is a problem in the system or if the readiness light is not functioning.

Air Bag Module

The air bag module is the air bag and inflator assembly packaged into a single unit or module. The module is located in the steering wheel for the driver and in the dash panel for the front-seat passenger **(Figure 23–22)**. The various types of side protection air bags have the module located at the point where the bag is deployed.

The inflation of the air bag is typically accomplished through an explosive release of nitrogen gas. The igniter **(Figure 23–23)** is an integral part of the inflator assembly. It starts a chemical reaction to inflate the air bag. At the center of the igniter assembly is the squib, which con-

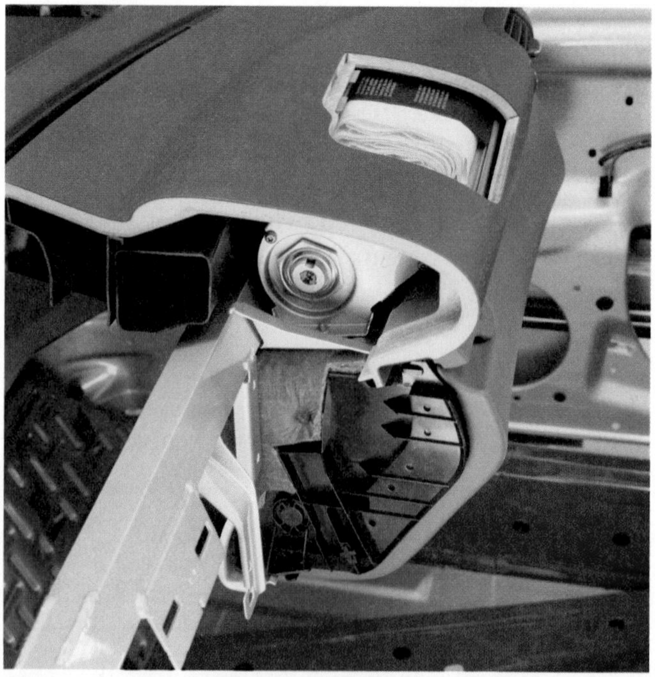

Figure 23-22 A cutaway view showing the complete passenger-side air bag module in the dash. *Courtesy of DaimlerChrysler Corporation*

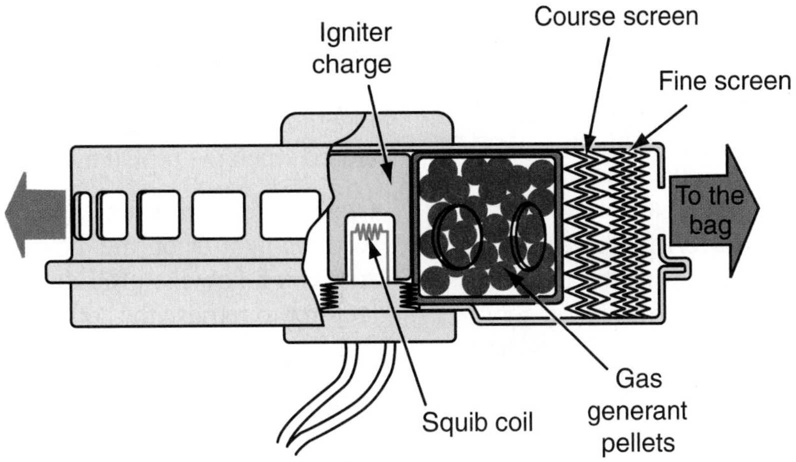

Figure 23-23 The action of an air bag module releasing nitrogen gas to an air bag.

tains **zeronic potassium perchlorate (ZPP)**. When voltage is supplied through the squib, an electrical arc is formed between two pins. The spark ignites a canister of gas and causes a rapid expansion of the gas, which deploys, or inflates, the air bag.

The inflation assembly is composed of a gas generator (called a generant) containing sodium azide and copper oxide or potassium nitrate propellant. The ZPP ignites the propellant charge. During ignition, large quantities of hot, expanding nitrogen gas are produced very quickly and quickly inflate the air bag. As the nitrogen moves into the air bag, it is filtered to remove sodium hydroxide dust formed during the chemical reaction.

CAUTION!

Wear gloves and eye protection when handling a deployed air bag module. Sodium hydroxide residue may remain on the bag and cause a skin irritation.

Not all air bags use nitrogen gas to inflate the bag; some use a solid propellant and compressed argon gas **(Figure 23–24)**. Argon has a stable structure, cools more quickly, and is inert as well as nontoxic. Argon is commonly used for passenger-side and side-protection air bags.

A mounting plate and retainer ring attach the air bag assembly to the inflator. They also keep the entire air bag module connected to the steering wheel.

The bag itself is made of a thin, nylon fabric that is folded into the steering wheel, dash, seat, or door. The powdery substance released from the air bag when it is deployed is regular cornstarch or talcum powder. These powders are used by the air bag manufacturers to keep air bags lubricated and pliable while they are in storage.

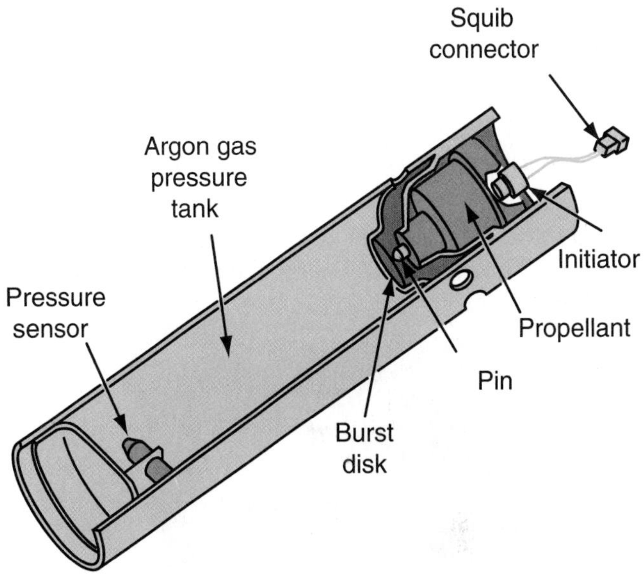

Figure 23-24 An inflator module that uses argon gas.

The entire module must be serviced as one unit when repair of the air bag system is required.

Diagnosis

Before diagnosing the system, perform a system check by observing the air bag warning light and comparing your findings with those described in the service manual for the vehicle. Typically, an air bag system problem is indicated by any of the following warning lamp conditions:

- If the lamp does not come on when the ignition is initially turned on.
- If the lamp does not steadily light while the engine is cranking.
- If the lamp remains on but does flash when the ignition is turned on.

- If the lamp flashes 7 to 9 times and then remains on when the ignition is turned on.
- If the lamp comes on while the engine is running.

If any of these conditions are present, the air bag system needs to be checked. A thorough visual inspection of sensor integrity is the best place to start when diagnosing a system that is disarmed because of a fault. Damage from a collision or mishandling during a nonrelated repair can set up a fault area, which will disarm the air bag system.

When a technician places the system into its diagnostic mode, the warning lamp may flash trouble codes or the codes can be retrieved with a scan tool.

Retrieving Trouble Codes The control module of most air bag systems will store trouble codes that can be retrieved by either a scan tool **(Figure 23–25)** or flash codes. Since 1996, most systems require the use of a scan tool.

Scan Tool DTC Retrieval. Normally an air bag system stores two types of faults. Active DTCs will turn the air bag warning lamp on whereas stored codes are intermittent problems and probably will not turn on the warning lamp.

To retrieve codes, connect the scan tool to the DLC and turn the ignition on. Follow the instructions for the scan tool to retrieve air bag information. Record all stored and active codes. Diagnose the cause of the codes in order, from the lowest number to the highest. Stored codes can be erased with the scan tool but active codes are only erased when the problem is corrected.

Figure 23-25 An OBD-II scan tool that checks the air bag, antilock brake, and engine control systems. *Courtesy of Service Solutions, SPX Corporation*

Flash Codes. On vehicles that display codes with the warning light or on the digital instrument panel, make sure you follow the procedure prescribed by the manufacturer to retrieve the codes. Flash code systems do not display stored codes; therefore, the cause of any intermittent problem must be found through normal diagnostics and reasoning.

SERVICING THE AIR BAG SYSTEM

Whenever working on or around air bag systems, it is important to follow all safety warnings **(Figure 23–26)**. Examples of these warnings follow:

- Wear safety glasses when servicing the air bag system.
- Wear safety glasses when handling an air bag module.
- Wait at least 30 minutes after disconnecting the battery before beginning any service on or around the air bag system. The reserve energy module is capable of storing enough power to deploy the air bag for up to 30 minutes after battery voltage is lost.
- Handle all air bag sensors with care. Do not strike or jar a sensor in such a manner that deployment may occur.
- When carrying a live air bag module, face the trim and bag away from your body.
- Do not carry the module by its wires or connector.
- When placing a live module on a bench, face the trim and air bag up.
- Deployed air bags may have a powdery residue on them. Sodium hydroxide is produced by the deployment reaction and is converted to sodium carbonate when it comes into contact with atmospheric moisture. It is unlikely that sodium hydroxide will still be present. However, wear safety glasses and gloves when handling a deployed air bag. Wash your hands immediately after handling the bag.
- A live air bag module must be deployed before disposal. Because the deployment of an air bag is through an explosive process, improper disposal may result in injury and fines. A deployed air bag should be disposed of according to EPA and manufacturer procedures.
- Do not use a battery- or AC-powered voltmeter, ohmmeter, or any other type of test equipment not specified in the service manual. Never use a test light to probe for voltage.

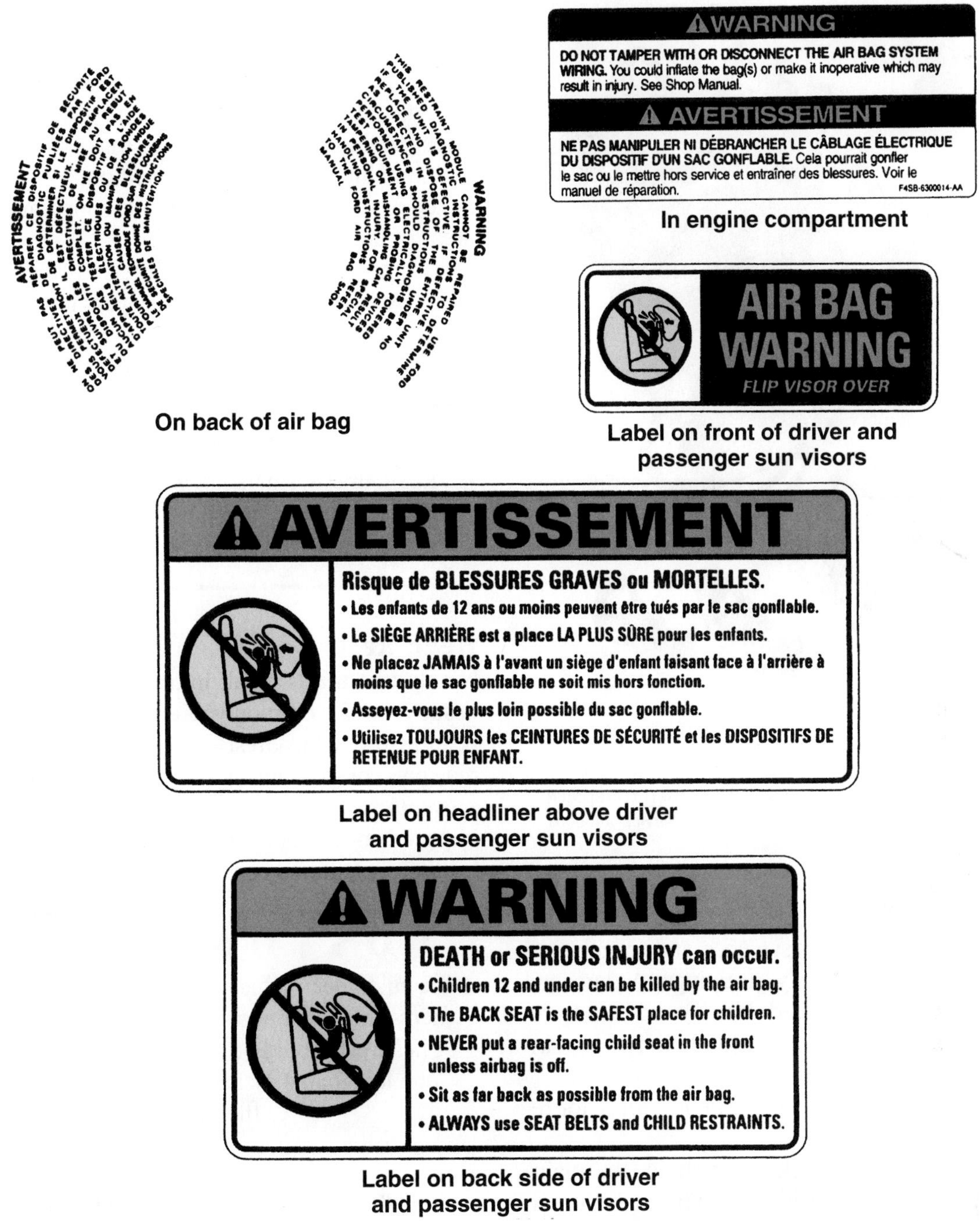
AVERTISSEMENT
WARNING
On back of air bag
⚠WARNING
DO NOT TAMPER WITH OR DISCONNECT THE AIR BAG SYSTEM WIRING. You could inflate the bag(s) or make it inoperative which may result in injury. See Shop Manual.
⚠ AVERTISSEMENT
NE PAS MANIPULER NI DÉBRANCHER LE CÂBLAGE ÉLECTRIQUE DU DISPOSITIF D'UN SAC GONFLABLE. Cela pourrait gonfler le sac ou le mettre hors service et entraîner des blessures. Voir le manuel de réparation.
F4S8-6300014-AA
In engine compartment
AIR BAG WARNING
FLIP VISOR OVER
Label on front of driver and passenger sun visors
⚠AVERTISSEMENT
Risque de BLESSURES GRAVES ou MORTELLES.
• Les enfants de 12 ans ou moins peuvent être tués par le sac gonflable.
• Le SIÈGE ARRIÈRE est a place LA PLUS SÛRE pour les enfants.
• Ne placez JAMAIS à l'avant un siège d'enfant faisant face à l'arrière à moins que le sac gonflable ne soit mis hors fonction.
• Asseyez-vous le plus loin possible du sac gonflable.
• Utilisez TOUJOURS les CEINTURES DE SÉCURITÉ et les DISPOSITIFS DE RETENUE POUR ENFANT.
Label on headliner above driver and passenger sun visors
⚠WARNING
DEATH or SERIOUS INJURY can occur.
• Children 12 and under can be killed by the air bag.
• The BACK SEAT is the SAFEST place for children.
• NEVER put a rear-facing child seat in the front unless airbag is off.
• Sit as far back as possible from the air bag.
• ALWAYS use SEAT BELTS and CHILD RESTRAINTS.
Label on back side of driver and passenger sun visors

Figure 23-26 Various air bag warning labels found on today's vehicles. *Courtesy of Ford Motor Co.*

CAUTION!

A two-stage air bag may appear to be fully deployed when only its first stage has deployed. Care must be taken to make sure that two-stage air bags have been fully deployed before handling them. Always assume that any deployed two-stage air bag has an active stage two. Improper handling or servicing can activate the inflator module and cause personal injury. Always follow the manufacturer's recommended handling procedures.

Service Guidelines

Keep in mind that wiring difficulties of any kind call for careful removal of the air bag module and putting a simulator in its place. The simulator **(Figure 23–27)** replaces the air bag to eliminate any chance of air bag deployment during diagnosis. This tool allows for a safe test sequence that should identify any continuity problems or short circuits. If a condition is present in the system that would cause deployment of the air bag, the simulator will indicate it and will remain in that mode until the problem is fixed.

An air bag module is serviced as a complete assembly. Technicians repairing these systems are also advised to

Figure 23-27 An air bag simulator used to replace the air bag module during testing. *Courtesy of Service Solutions, SPX Corporation*

service crash sensors, mercury switches, and any other related components in assembly groupings. A damaged crash sensor should be replaced. It is a good idea to replace the entire set if a failure or degradation of any single sensor is found.

Photo Sequence 21 covers a typical procedure for replacing an air bag module.

The steering column clockspring should be maintained in its correct index position at all times. Failure to so can cause damage to the enclosure, wiring, or module. Any of these situations can cause the air bag system to default into a nonoperative mode. The clockspring should be replaced any time it has been removed.

Before returning the vehicle to the customer after service, make sure the sensors are firmly fastened to their mounting fixtures, with their arrows facing forward. Be certain all the fuses are correctly rated and replaced. Make sure a final check is made for codes using the approved scan tool. Carefully recheck the wire and harness routing before releasing the car.

OTHER PROTECTION SYSTEMS

To make vehicles safe and to protect the occupants inside the vehicle, manufacturers include many different systems and options. What follows is a quick look at a few of these. By no means is this discussion conclusive; there are many things about how a vehicle is made that influence the protection and safety it offers. Basically, cars that offer good protection are those that are constructed to maintain integrity when impacted on. This construction includes side door beams, crumple zones, and reinforced areas of the frame.

Crumple zones are areas of the body that will bend or break away to protect the passengers inside the vehicle. These crumple zones absorb or take on the impact while keeping the passenger compartment undisturbed.

CUSTOMER CARE

Let your customers know that a well-maintained vehicle is a prerequisite for a safe vehicle. Some problems will lead to accidents and/or injury. ■

Head Rests

Nearly two-thirds of all injuries in collisions are soft-tissue related, commonly referred to as whiplash. Good head restraint (headrest) design and proper adjustment may prevent these injuries. A properly adjusted headrest prevents the head and neck from extending backward on impact. Nine out of ten people do not adjust the headrest for their height and comfort.

CUSTOMER CARE

If a customer's vehicle has adjustable headrests, do them a big favor and tell them how to adjust them. A headrest should be at least to the top of their ear and less than 4 inches (10 cm) from the back of their head. ■

In high-speed rear crashes, the backs of seats must be strong enough to transfer the energy of the impact from the occupant while keeping him or her in the seat. In low-to-high speed crashes, the headrest and upper part of the seat must reduce the relative motion between the head and neck to prevent whiplash. The reclining function of the seats must also be strong enough to withstand the force of the occupant being thrown back into the seat.

New systems have been developed that automatically adjust the headrest. This self-aligning headrest system moves the headrest up and forward when the vehicle is hit from behind.

Rollover Protection

Some convertibles have a built-in roll bar to protect the passengers in the vehicle in the case of a rollover. These units are permanent structures of the vehicle. Others have

PHOTO SEQUENCE 21

Removing an Air Bag Module

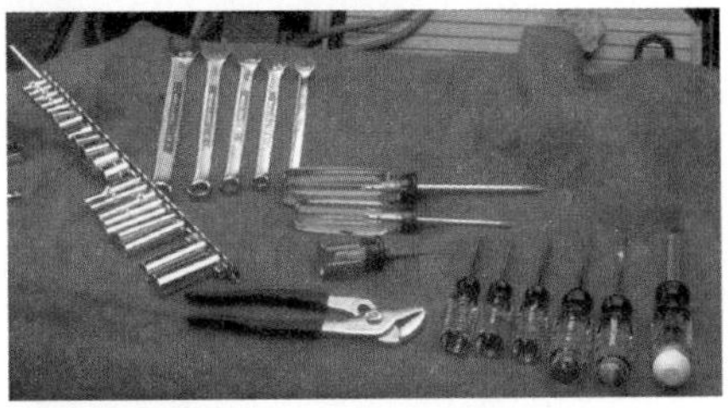

P21-1 *Tools required to remove the air bag module: safety glasses, seat covers, screwdriver set, Torx driver set, battery terminal pullers, battery pliers, assorted wrenches, ratchet and socket set, and service manual.*

P21-2 *Place the seat and fender covers on the vehicle.*

P21-3 *Place the front wheels in the straight ahead position and turn the ignition switch to the LOCK position.*

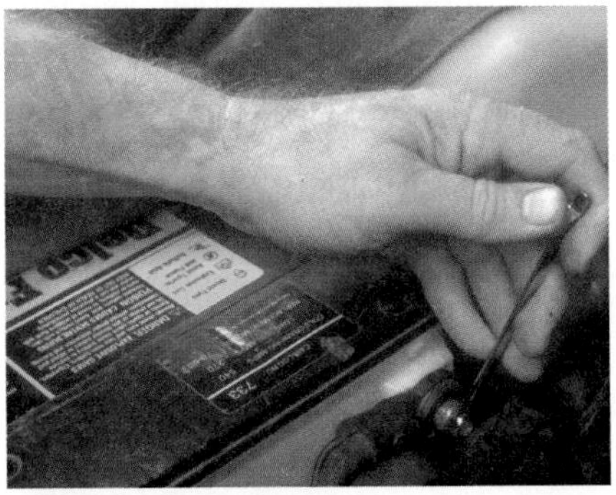

P21-4 *Disconnect the negative battery cable.*

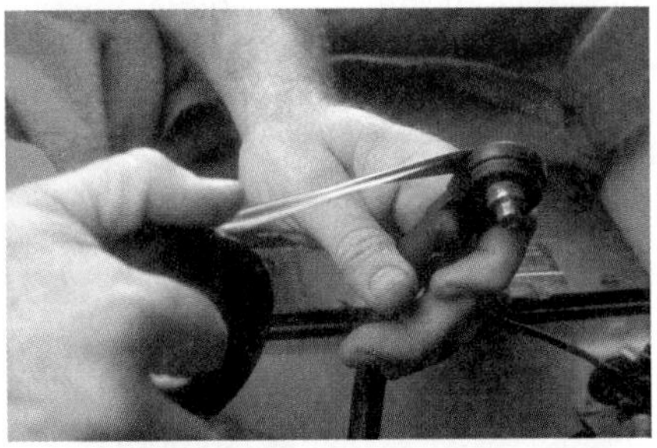

P21-5 *Tape the cable terminal to prevent accidental connection with the battery post. Note: A piece of rubber hose can be substituted for the tape.*

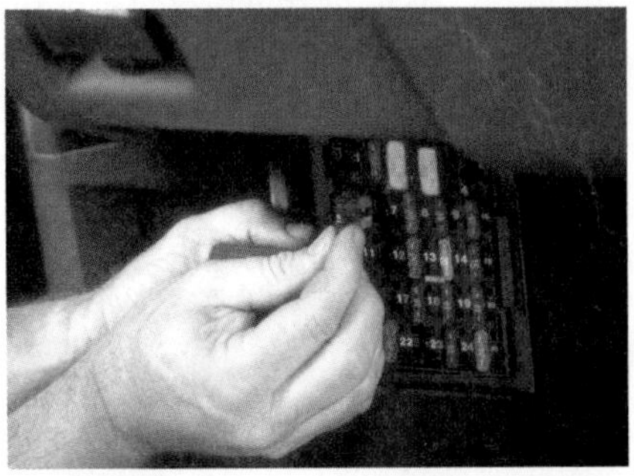

P21-6 *Remove the SIR fuse from the fuse box. Wait 30 minutes to allow the reserve energy to dissipate.*

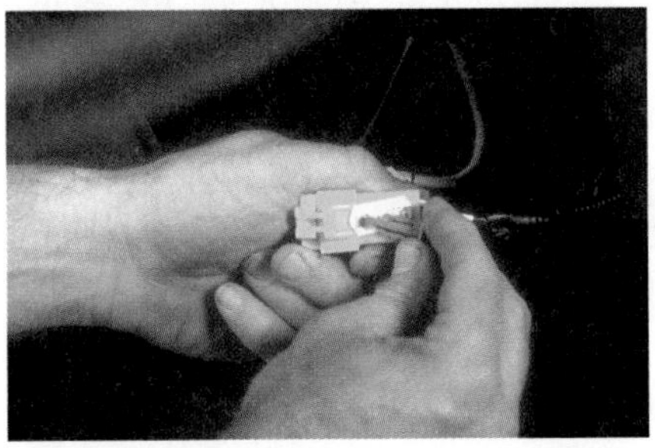

P21-7 *Remove the connector position assurance (CPA) from the yellow electrical connector at the base of the steering column.*

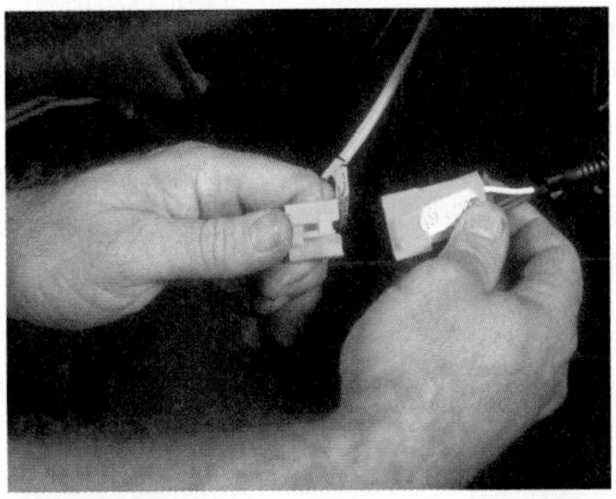

P21-8 *Disconnect the yellow two-way electrical connector.*

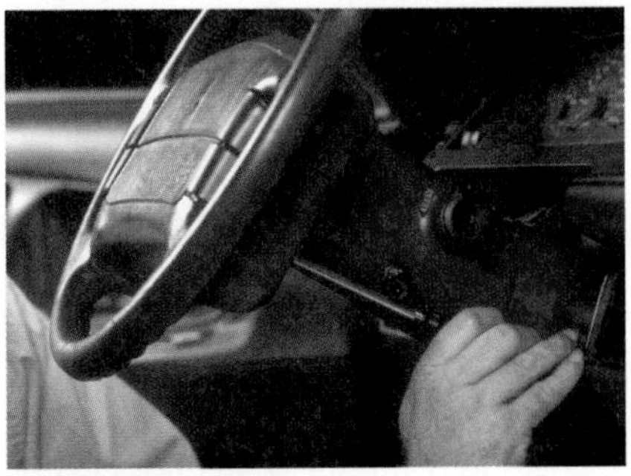

P21-9 *Remove the four bolts that secure the module from the rear of the steering wheel.*

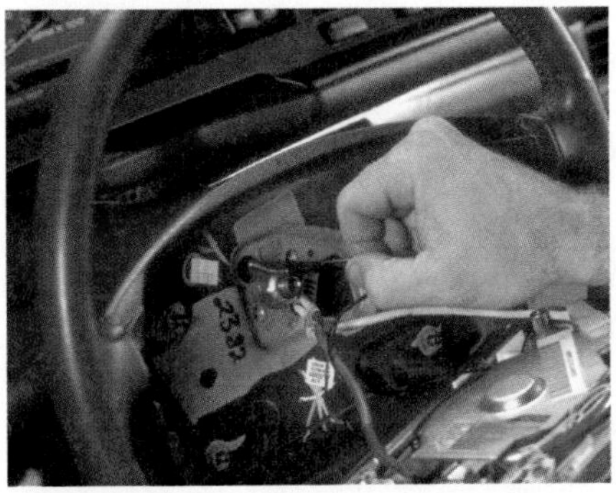

P21-10 *Rotate the horn lead ¼ turn and disconnect.*

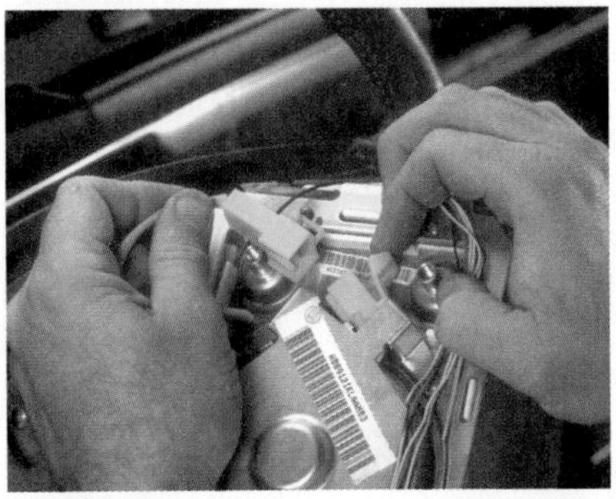

P21-11 *Disconnect the electrical connectors.*

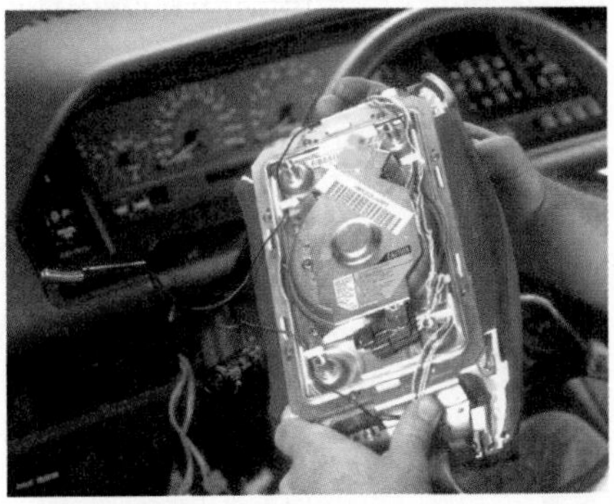

P21-12 *Remove the module.*

Figure 23-28 This roll bar pops out when the system anticipates a potential rollover. *Courtesy of DaimlerChrysler Corporation*

automatic systems that provide for this. These systems deploy a roll bar from behind the headrests when the vehicle experiences extreme tilting, the wheels lose contact with the ground, or the vehicle is involved in a serious accident **(Figure 23–28)**.

CASE STUDY

A late-model car with air bags was brought to the shop because its air bag warning lamp was always on. The customer was afraid the air bags would deploy while he was driving. The service writer explained what was required for the air bag to inflate. The customer appreciated the explanation but was still concerned about the warning light. The car was given to a technician for diagnostics. She verified the complaint and then conducted a visual inspection of all of the air bag circuits. Finding nothing noticeably wrong, she connected a scan tool to retrieve trouble codes. The DTC indicated an open driver's side air bag squib circuit. Following the procedure outlined in the vehicle's service manual, she removed the air bag module and tested the clockspring circuit. This circuit was fine so she replaced the air bag module. She knew this would take care of the problem, simply because everything before the module was good. The service manual had firm cautions about not testing the resistance of the squib because that could cause accidental deployment. Therefore, she did not attempt to test it. After the new module was installed, she ran a diagnostic check to make sure the problem was solved. The warning lamp was off and the customer was satisfied.

KEY TERMS

Active restraint system
Air bag sensing diagnostic monitor (ASDM)
Clockspring
Inertia lock retractors
Passive restraint system
Pretensioner
Second-generation air bags
SIR
Smart air bags
SRS
Squib
ZPP

SUMMARY

- All new vehicles built or sold in the United States must have one or both types of passive restraints: seat belts or air bags.
- Restraint systems are either active or passive.
- When servicing seat belts, inspect the webbing, buckles, retractors, and anchorage.
- An air bag is inflated or deployed by rapid expansion of a gas fired by igniter or squib.
- Second-generation air bags have been depowered to inflate with less force than earlier air bags.
- Smart air bags have two possible stages of deployment in an attempt to match the severity of the impact and/or by the size and weight of the seat's occupant.
- The electrical circuit of an air bag system includes impact sensors and an electronic control module.
- The air bag module is the air bag and inflator assembly packaged into a single unit.
- A system check of an air bag system consists of observing the air bag warning light.
- The control module will store trouble codes that can be retrieved by either a scan tool or flash codes.
- Before doing any work on an air bag system, disconnect the battery.
- Care must be taken when removing a live (not deployed) air bag. Be sure the bag and trim cover are pointed away from you.

REVIEW QUESTIONS

1. What is the difference between a passive and an active restraint system?
2. *True or False?* Second generation air bags rely on compressed air to deploy the bag.

TECH MANUAL

The following procedures are included in Chapter 23 of the *Tech Manual* that accompanies this book:

1. Inspecting seat belts.
2. Working safely around air bags.

3. What air bag system device contains the ZPP?

4. *True or False?* The air bag diagnostic monitor supplies backup power to the air bag module in the event that the battery or cables are damaged during the accident.

5. Technician A says that compressed argon gas is often used to deploy passenger-side and side impact air bags. Technician B says compressed argon gas is used to deploy some driver's-side air bags. Who is correct?

a. Technician A **c.** Both A and B
b. Technician B **d.** Neither A nor B

6. Technician A says seat belt webbing should be replaced if it is bowed. Technician B says webbing does not need to be replaced if the color has merely faded due to exposure to the sun. Who is correct?

a. Technician A **c.** Both A and B
b. Technician B **d.** Neither A nor B

7. *True or False?* The powdery substance released from the air bag when it is deployed is sodium hydroxide, which can cause skin irritation.

8. Technician A says that the module assembly will disarm the air bag system if certain faults occur. Technician B says that at least two sensors must signal the air bag in order to trigger it. Who is correct?

a. Technician A **c.** Both A and B
b. Technician B **d.** Neither A nor B

9. Which of the following statements about two-stage air bags is *not* true?

a. When low-pressure air bag deployment is desired, only one squib is fired.
b. When the vehicle is in a severe collision and the occupant needs maximum protection the squib related to the larger container of gas fires.
c. For rapid deployment, both squibs fire at the same time.
d. To phase in full deployment, one squib is fired a few milliseconds later than the other is.

10. While testing set belt retractors, Technician A grasps a vehicle-sensitive-type belt and jerks it. Technician B grasps a webbing-sensitive-type belt and jerks it. Who is correct?

a. Technician A **c.** Both A and B
b. Technician B **d.** Neither A nor B

11. Which of the following is false?

a. The air bag igniter assembly is a spark plug-type device with two pins that current must jump across.
b. The air bag igniter assembly creates a spark that ignites a canister of gas, generating zerconic potassium perchlorate.
c. The air bag igniter assembly is housed in the module assembly.
d. None of the above.

12. Technician A checks the clockspring electrical connections for signs of damage when replacing a deployed air bag module. Technician B back probes an air bag system with a multimeter to determine if the system is in good working order. Who is correct?

a. Technician A **c.** Both A and B
b. Technician B **d.** Neither A nor B

13. Technician A removes the ground battery cable before servicing any component in the air bag system. Technician B wears safety glasses and protective gloves when handling inflated air bags. Who is correct?

a. Technician A **c.** Both A and B
b. Technician B **d.** Neither A nor B

14. What type of seat belt operates automatically, with no action required by the vehicle's occupant?

a. passive restraint **c.** active restraint
b. retractor **d.** anchorage

15. *True or False?* Wait 15 minutes after disconnecting the battery before beginning any service on or around the air bag system.

16. Technician A says ignition of the air bag only occurs when an outside sensor and an inside sensor are closed. Technician B says the safing sensor determines if the collision is severe enough to inflate the air bag. Who is correct?
    - **a.** Technician A
    - **b.** Technician B
    - **c.** Both A and B
    - **d.** Neither A nor B

17. Technician A says the steering column clockspring should be maintained in its correct index position at all times. Technician B says failure to keep the clockspring straight ahead or in the neutral position, relative to the steering wheel, can cause damage to the enclosure, wiring, or module. Who is correct?
    - **a.** Technician A
    - **b.** Technician B
    - **c.** Both A and B
    - **d.** Neither A nor B

18. Which of these statements about air bag sensors is *not* true?
    - **a.** Roller-type sensors rely on a roller held in place with a magnet. The circuit is closed when the roller moves and strikes the spring contact.
    - **b.** Mass-type sensors rely on a ball and a magnet. The circuit is complete when the ball makes contacts with the electrical contacts in the sensor.
    - **c.** An accelerometer contains a piezoelectric element that generates an analog voltage in relation to the severity of the deceleration forces.
    - **d.** An accelerometer also senses the direction of an impact force.

19. Which of the following statements about seat belt retractors is *not* true?
    - **a.** They stow away the seat belts when they are not being used.
    - **b.** They allow freedom of movement for the occupant of the seat.
    - **c.** They can tighten up and pull the occupant back during a crash.
    - **d.** On some vehicles, they work in concert with the air bag system.

20. What is a crumple zone?

SECTION 4

Engine Performance

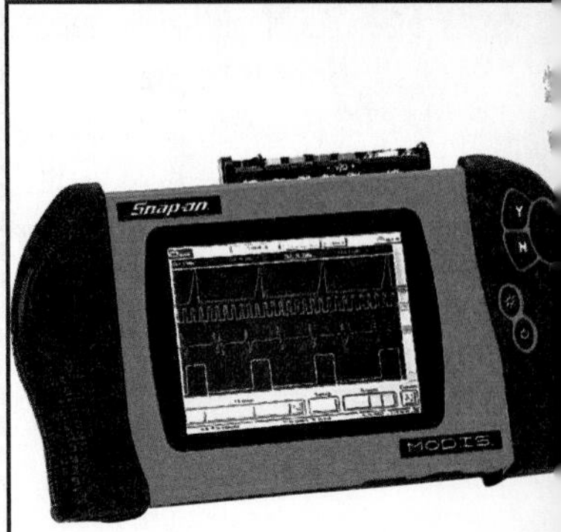

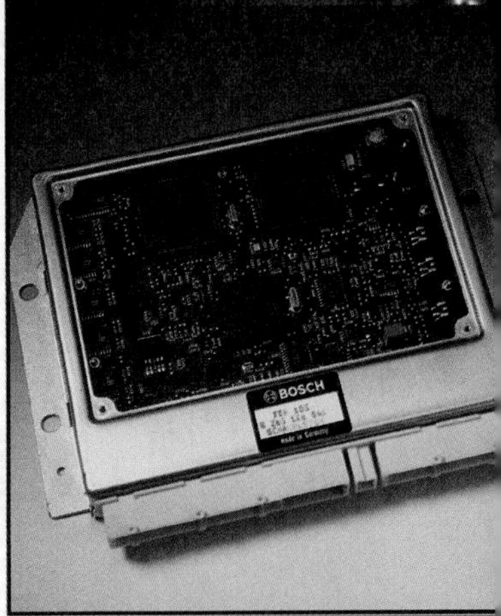

To have an efficient running engine, there must be the correct amount of fuel mixed with the correct amount of air. These must be present in a sealed container and shocked by the right amount of heat at the correct time. With total efficiency, the engine would burn all the fuel it receives and would release extremely low amounts of pollutants in its exhaust.

Today's vehicles use a variety of ignition, fuel, and emissions system designs, but they all operate in much the same way and improve engine efficiency. Vehicles are also available with alternate sources of power—electric, electric and gasoline, alternative fuels, and hydrogen—designed to decrease emission levels and fuel consumption. Engine performance systems are some of the most technologically advanced systems found on modern vehicles. They all rely on a network of input sensors and output devices and tie directly into the electronic engine control system. Mastery of these systems is necessary for an automotive technician.

The material in Section 4 matches the content areas of the ASE certification test on engine performance and sections of the certification test on engine repair.

WE ENCOURAGE PROFESSIONALISM

ASE

THROUGH TECHNICIAN CERTIFICATION

Delmar's Automotive Engine Performance Video Series and CD-ROM Courseware

24

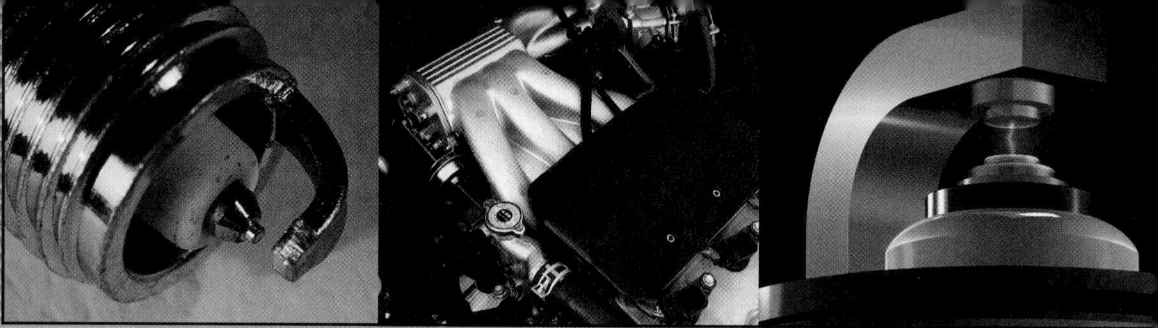

IGNITION SYSTEMS

OBJECTIVES

■ Describe the three major functions of an ignition system. ■ Name the operating conditions of an engine that affect ignition timing. ■ Name the two major electrical circuits used in ignition systems and their common components. ■ Describe the operation of ignition coils, spark plugs, and ignition cables. ■ Explain how high voltage is induced in the coil secondary winding. ■ Describe the various types of spark timing systems, including electronic switching systems and their related engine position sensors. ■ Explain the basic operation of a computer-controlled ignition system. ■ Explain how the fuel injection system may rely on components of the ignition system. ■ Describe the operation of distributor-based ignition systems. ■ Describe the operation of distributorless ignition systems.

One of the requirements for an efficient engine is the correct amount of heat shock, delivered at the right time. This requirement is the responsibility of the ignition system. The ignition system supplies properly timed, high-voltage surges to the spark plugs. These voltage surges cause combustion inside the cylinder.

The ignition system must create a spark, or current flow, across each pair of spark plug electrodes **(Figure 24–1)** at the proper instant, under all engine operating conditions. This may sound relatively simple, but when one considers the number of spark plug firings required and the extreme variation in engine operating conditions, it is easy to understand why ignition systems are so complex.

Figure 24-1 An ignition system has the sole purpose of providing the spark to start combustion. *Courtesy of Autolite*

If a six-cylinder engine is running at 4,000 revolutions per minute (rpm), the ignition system must supply 12,000 sparks per minute because the ignition system must fire three spark plugs per revolution. These plug firings must also occur at the correct time and generate the correct amount of heat. If the ignition system fails to do these things, fuel economy, engine performance, and emission levels will be adversely affected.

PURPOSE OF THE IGNITION SYSTEM

For each cylinder in an engine, the ignition system has three main jobs. First, it must generate an electrical spark that has enough heat to ignite the air/fuel mixture in the combustion chamber. Second, it must maintain that spark long enough to allow for the combustion of all the air and fuel in the cylinders. Last, it must deliver the spark to each cylinder so combustion can begin at the right time during the compression stroke of each cylinder.

When the combustion process is completed, a very high pressure is exerted against the top of the piston. This pressure pushes the piston down on its power stroke and is the force that gives the engine power. For an engine to

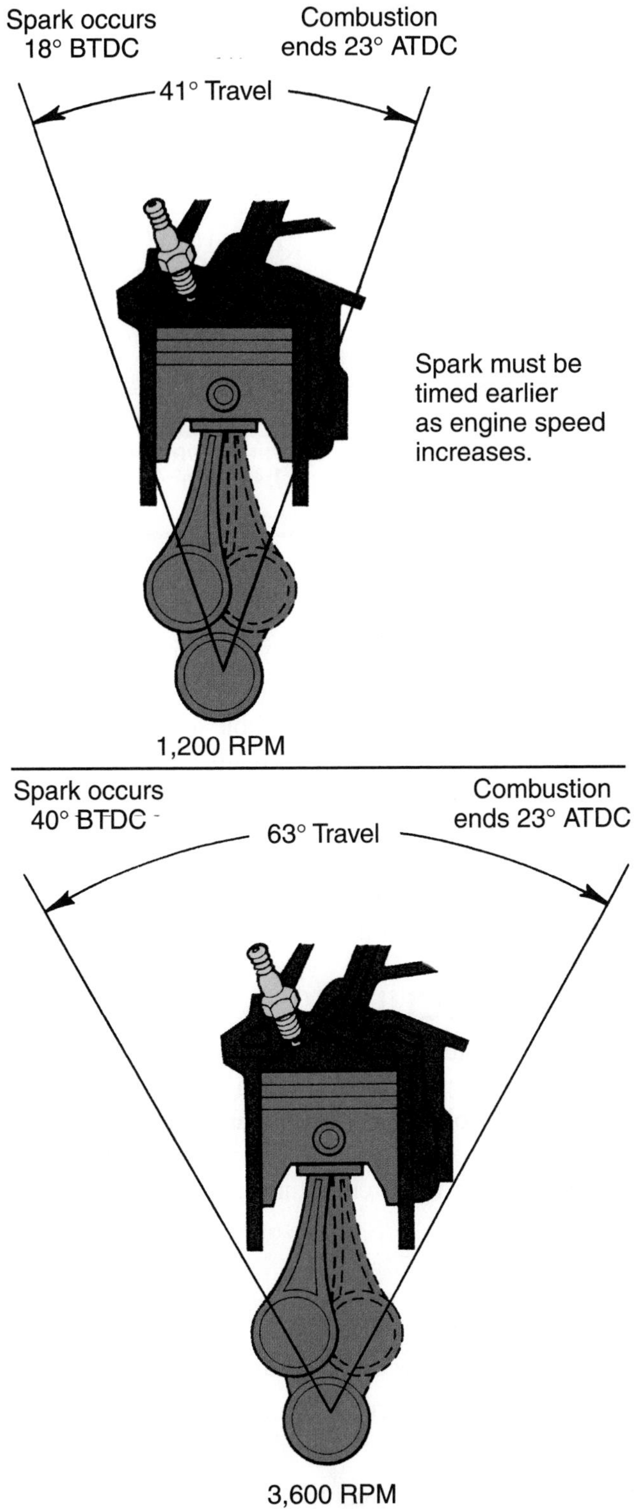

Figure 24-2 Ignition must begin earlier as engine speed increases. *Courtesy of Ford Motor Company*

produce the maximum amount of power it can, the maximum pressure from combustion should be present when the piston is at 10 to 23 degrees after top dead center (ATDC). Because combustion of the air/fuel mixture within a cylinder takes a short period of time, usually measured in thousandths of a second (milliseconds), the combustion process must begin before the piston is on its power stroke. Therefore, the delivery of the spark must be timed to arrive at some point before the piston reaches top dead center.

Determining how much before TDC the spark should begin gets complicated because of the fact that as the speed of the piston moving from its compression stroke to its power stroke increases, the time needed for combustion stays about the same. This means the spark should be delivered earlier as the engine's speed increases **(Figure 24–2)**. However, as the engine has to provide more power to do more work, the load on the crankshaft tends to slow down the acceleration of the piston and the spark should be somewhat delayed.

Figuring out when the spark should begin gets more complicated due to the fact that the rate of combustion varies according to certain factors. Higher compression pressures tend to speed up combustion. Higher octane gasolines ignite less easily and require more burning time. Increased vaporization and turbulence tend to decrease combustion times. Other factors, including intake air temperature, humidity, and barometric pressure, also affect combustion. Because of all of these complications, delivering the spark at the right time is a difficult task.

Two types of ignition systems are found on today's vehicles: **distributor ignition (DI)** and **electronic ignition (EI)**, which is also called a **distributorless** or **direct ignition system (DIS)**. The various designs and operation of these types are discussed in this chapter.

IGNITION TIMING

Ignition timing refers to the precise time spark occurs and is specified by referring to the position of the number 1 piston in relation to crankshaft rotation. Ignition timing reference marks can be located on engine parts and on a pulley or flywheel to indicate the position of the number 1 piston. Vehicle manufacturers specify initial or **base ignition timing**.

When the marks are aligned at TDC, or 0, the piston in cylinder 1 is at TDC of its compression stroke. Additional numbers on a scale indicate the number of degrees of crankshaft rotation before TDC **(BTDC)** or after TDC **(ATDC)**. In a majority of engines, the initial timing is specified at a point between TDC and 20 degrees BTDC.

If optimum engine performance is to be maintained, the ignition timing of the engine must change as the operating conditions of the engine change. These conditions affect the speed of the engine and the load on the engine. All ignition timing changes are made in response to these primary factors.

Engine RPM

At higher rpms, the crankshaft turns through more degrees in a given period of time. If combustion is to be completed by 10 degrees ATDC, ignition timing must occur sooner or be advanced.

However, air/fuel mixture turbulence **(swirling)** increases with rpm. This causes the mixture inside the cylinder to turn faster. Increased turbulence requires that ignition must occur slightly later or be slightly retarded.

These two factors must be balanced for best engine performance. Therefore, while the ignition timing must be advanced as engine speed increases, the amount of advance must be decreased some to compensate for the increased turbulence.

Engine Load

The load on an engine is related to the work it must do. Driving up hills or pulling extra weight increases engine load. Under load there is resistance on the crankshaft, therefore the pistons have a harder time moving through their strokes. This is evident by the low measured vacuum during heavy loads.

Under light loads and with the throttle plate(s) partially opened, a high vacuum exists in the intake manifold. The amount of air/fuel mixture drawn into the manifold and cylinders is small. On compression, this thin mixture produces less combustion pressure and combustion time is slow. To complete combustion by 10 degrees ATDC, ignition timing must be advanced.

Under heavy loads, when the throttle is opened fully, a larger mass of air/fuel mixture can be drawn in, and the vacuum in the manifold is low. High combustion pressure and rapid burning results. In such a case, the ignition timing must be retarded to prevent complete burning from occurring before 10 degrees ATDC.

Firing Order

Up to this point, the primary focus of discussion has been ignition timing as it relates to any one cylinder. However, the function of the ignition system extends beyond timing the arrival of a spark to a single cylinder. It must perform this task for each cylinder of the engine in a specific sequence.

Each cylinder of an engine must produce power once in every 720 degrees of crankshaft rotation. Each cylinder must have a power stroke at its own appropriate time during the rotation. To make this possible, the pistons and rods are arranged in a precise fashion called the engine's firing order. The firing order is arranged to reduce rocking and imbalance problems. Because the potential for this rocking is determined by the design and construction of the engine, the firing order varies from engine to engine. Vehicle manufacturers simplify cylinder identification by numbering each cylinder **(Figure 24–3)**. Regardless of the particular firing order used, the number 1 cylinder always starts the firing order, with the rest of the cylinders following in a fixed sequence.

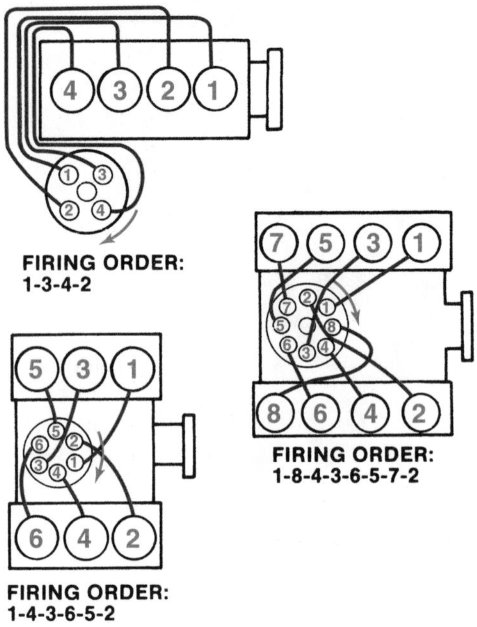

Figure 24–3 Examples of typical firing orders.

The ignition system must be able to monitor the rotation of the crankshaft and the relative position of each piston to determine which piston is on its compression stroke. It must also be able to deliver a high-voltage surge to each cylinder at the proper time during its compression stroke. How the ignition system does these things depends on the design of the system.

BASIC CIRCUITRY

All ignition systems consist of two interconnected electrical circuits: a **primary** (low-voltage) **circuit** and a **secondary** (high-voltage) **circuit (Figure 24–4)**.

Depending on the exact type of ignition system, components in the primary circuit include the following:

- Battery
- Ignition switch
- Ballast resistor or resistance wire (some systems)
- Starting bypass (some systems)
- Ignition coil primary winding
- Triggering device
- Switching device or control module

The secondary circuit includes these components.

- Ignition coil secondary winding (some systems)
- Distributor cap and rotor (some systems)
- Ignition, or spark plug, cables (some systems)
- Spark plugs

Primary Circuit Operation

When the ignition switch is on, current from the battery flows through the ignition switch and primary circuit re-

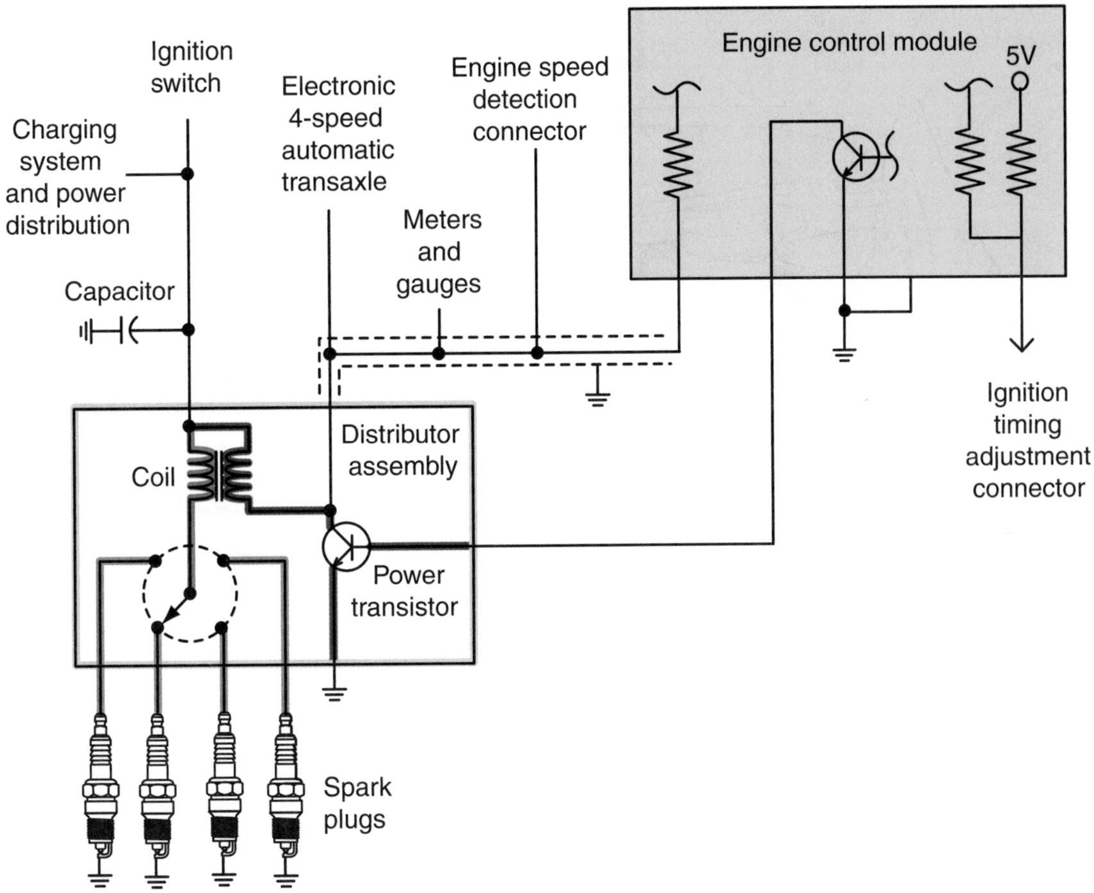

Figure 24-4 Ignition systems have a primary and a secondary (high-voltage) circuit.

sistor to the primary winding of the ignition coil. From there it passes through some type of switching device and back to ground. The switching device can be electronically or mechanically controlled by the triggering device. The current flow in the ignition coil's primary winding creates a magnetic field. The switching device or control module interrupts this current flow at predetermined times. When it does, the magnetic field in the primary winding collapses. This collapse generates a high-voltage surge in the secondary winding of the ignition coil. The secondary circuit of the system begins at this point.

Some older ignition systems had a **ballast resistor** or resistance wire connected between the ignition switch and the positive coil terminal. This resistor controlled the amount of voltage and current to the coil. Today, most ignition systems are not equipped with the resistor and battery voltage is at the coil at all times.

Secondary Circuit Operation

The secondary circuit carries high voltage to the spark plugs. The exact manner in which the secondary circuit delivers these high-voltage surges depends on the system design. Until 1984 all ignition systems used some type of distributor to accomplish this job. However, in an effort to reduce emissions, improve fuel economy, and boost component reliability, most auto manufacturers are now using distributorless or electronic ignition (EI) systems.

In a distributor ignition (DI) system, high-voltage from the secondary winding passes through an ignition cable running from the coil to the distributor. The **distributor** then distributes the high voltage to the individual spark plugs through a set of ignition cables **(Figure 24–5)**. The cables are arranged in the distributor cap according to the firing order of the engine. A **rotor** driven by the distributor shaft rotates and completes the electrical path from the secondary winding of the coil to the individual spark plugs. The distributor delivers the spark to match the compression stroke of the piston. The distributor assembly may also have the capability of advancing or retarding ignition timing.

The distributor cap is mounted on top of the distributor assembly and an alignment notch in the cap fits over a matching lug on the housing. Therefore the cap can only be installed in one position, which ensures the correct firing sequence.

The rotor is positioned on top of the distributor shaft, and a projection inside the rotor fits into a slot in the shaft. This allows the rotor to be installed in only one position. A metal strip on the top of the rotor makes contact with the center distributor cap terminal, and the outer

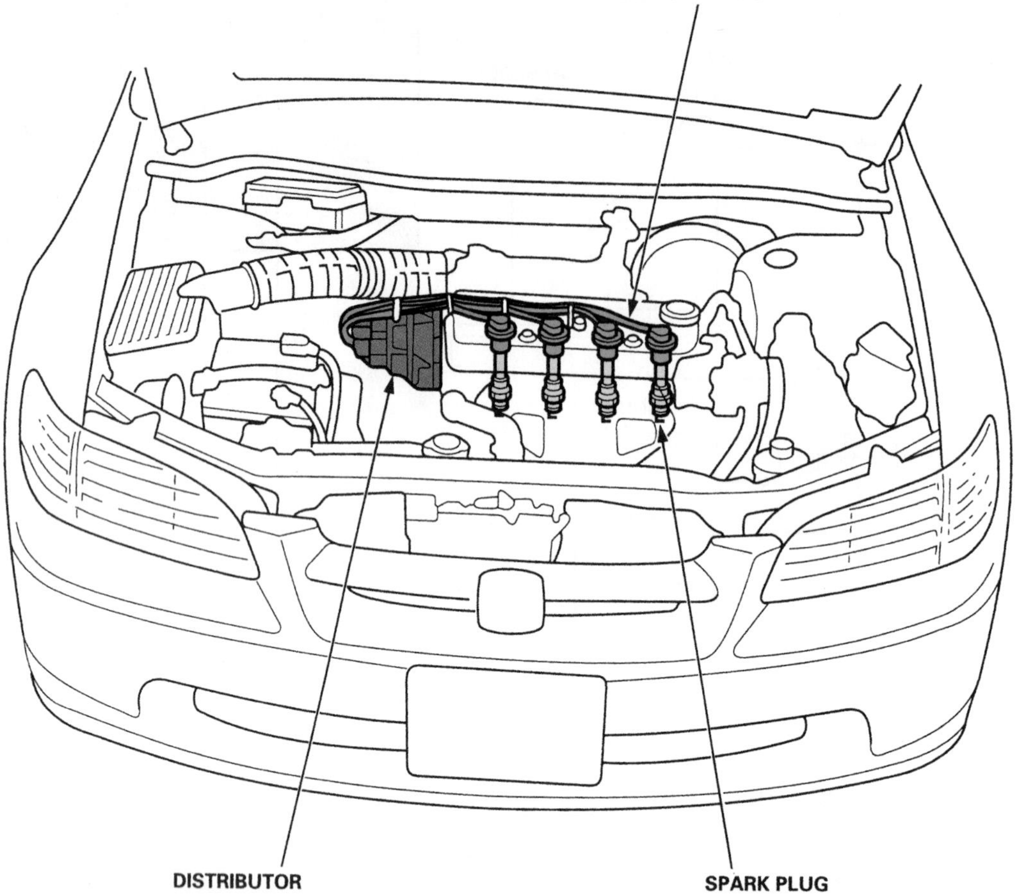

Figure 24-5 A typical spark distribution system. *Courtesy of American Honda Motor Co., Inc.*

end of the strip rotates past the cap terminals as it rotates **(Figure 24–6)**. This action completes the circuit between the ignition coil and the individual spark plugs according to the firing order.

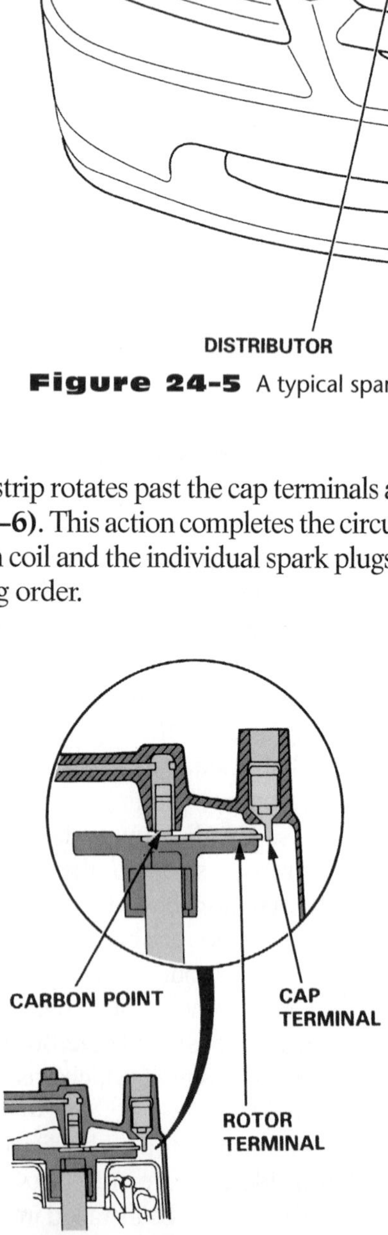

Figure 24-6 The relationship of a rotor and distributor cap. *Courtesy of American Honda Motor Co., Inc.*

EI systems have no distributor; spark distribution is controlled by an electronic control unit and/or the vehicle's computer **(Figure 24–7)**. Instead of a single ignition coil for all cylinders, each cylinder may have its own ignition coil, or two cylinders may share one coil. The coils are wired directly to the spark plug they control. An ignition control module, tied into the vehicle's computer control system, controls the firing order and the spark timing and advance. This module is typically located under the coil assembly.

A specific amount of energy is available in a secondary ignition circuit. In a secondary ignition circuit, the energy is normally produced in the form of voltage required to start firing the spark plug and then a certain amount of current flow across the spark plug electrodes to maintain the spark. Distributorless ignition systems are capable of producing much higher energy than conventional ignition systems.

Since DI and EI systems are both firing spark plugs with approximately the same air gap across the electrodes, the voltage required to start firing the spark plugs in both systems is similar. If the additional energy in the EI systems is not released in the form of voltage, it will be released in the form of current flow. This results in

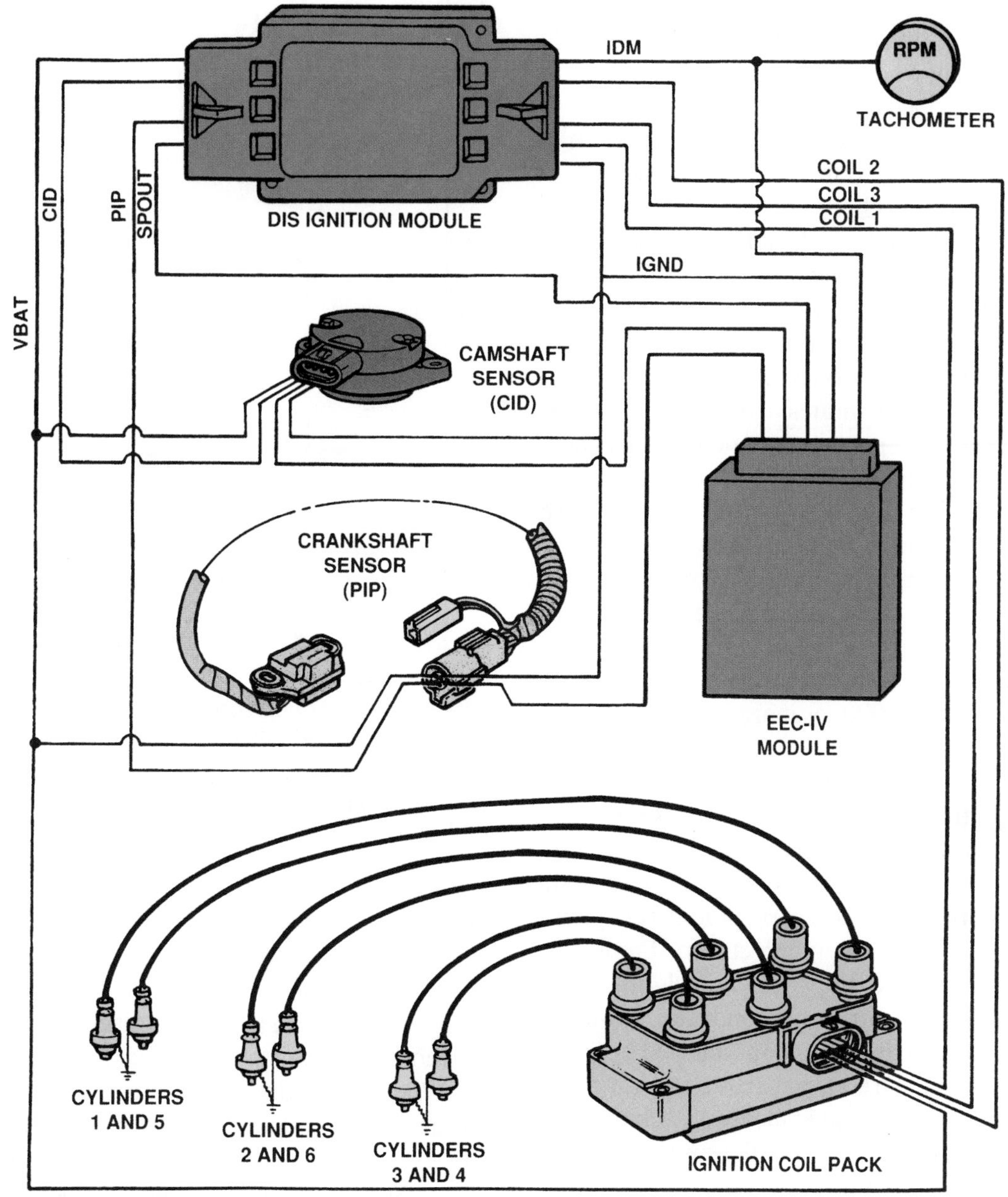

Figure 24-7 An electronic ignition system for a six-cylinder engine. *Courtesy of Ford Motor Company*

higher firing current and longer spark plug firing times. The average firing time across the spark plug electrodes in an EI system is 1.5 milliseconds compared to approximately 1 millisecond in a DI system. This extra time may seem insignificant, but it is very important. Current emission standards demand leaner air/fuel ratios, and this additional spark duration on EI systems helps to prevent cylinder misfiring with leaner air/fuel ratios. For this reason most car manufacturers have equipped their engines with EI systems.

IGNITION COMPONENTS

All ignition systems share a number of common components. Some, such as the battery and ignition switch, perform simple functions. The battery supplies low-voltage current to the ignition primary circuit. The current flows when the ignition switch is in the start or run position. Full-battery voltage is always present at the ignition switch, as if it were directly connected to the battery.

Ignition Coils

To generate a spark to begin combustion, the ignition system must deliver high voltage to the spark plugs. Because the amount of voltage required to bridge the gap of the spark plug varies with the operating conditions, most late-model vehicles can easily supply 30,000 to 60,000 volts to force a spark across the air gap. Since the battery delivers 12 volts, a method of stepping up the voltage must be used. Multiplying battery voltage is the job of a coil.

The ignition coil is a **pulse transformer** that transforms battery voltage into short bursts of high voltage. As explained previously, when a magnetic field moves across a wire, voltage is induced in the wire.

If a wire is bent into loops forming a coil and a magnetic field is passed through the coil, an equal amount of voltage is generated in each loop of wire. The more loops of wire in the coil, the greater the total voltage induced. If the speed of the magnetic field is doubled, the voltage output doubles.

An ignition coil uses these principles and has two coils of wire wrapped around an iron core. An iron or steel core is used because it has low **inductive reluctance**. In other words, iron freely expands or strengthens the magnetic field around the windings. The first, or primary, coil is normally composed of 100 to 200 turns of 20-gauge wire. This coil of wire conducts battery current. When a current is passing through the primary coil, it magnetizes the iron core. The strength of the magnet depends directly on the number of wire loops and the amount of current flowing through those loops. The secondary coil of wires may consist of 15,000 to 25,000, or more, turns of very fine copper wire.

Because of the effects of counter EMF on the current flowing through the primary winding, it takes some time for the coil to become fully magnetized or saturated. Therefore current flows in the primary winding for some time between firings of the spark plugs. The period of time during which there is primary current flow is often called **dwell**. The length of the dwell period is important.

When current flows through a conductor, it will immediately reach its maximum value as allowed by the resistance in the circuit. If a conductor is wound into a coil, maximum current will not be immediately achieved. As the magnetic field begins to form as the current begins to flow, the magnetic lines of force of one part of the winding pass over another part of the winding. This tends to cause an opposition to current flow. This occurrence is called **reactance**. Reactance causes a temporary resistance to current flow and delays the flow of current from reaching its maximum value. When maximum current flow is present in a winding, the winding is said to be saturated and the strength of its magnetic field will also be at a maximum.

Saturation can only occur if the dwell period is long enough to allow for maximum current flow through the primary windings. A less-than-saturated coil will not be able to produce the voltage it was designed to produce. If the energy from the coil is too low, the spark plugs may not fire long enough or may not fire at all. If the current is applied longer than needed to fully saturate the winding, energy is wasted.

A typical coil requires 2 to 6 milliseconds to become saturated. The actual required time depends on the resistance of the coil's primary winding and the voltage applied to it. Some early systems electronically limit the primary current flow at low speeds to prevent the coil from overheating. When the engine reaches higher speeds, the current limitation feature is disabled.

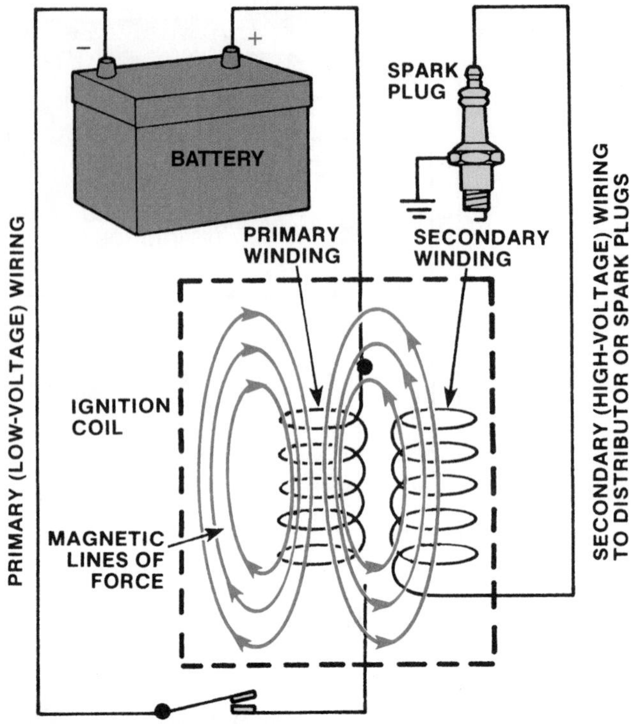

Figure 24-8 Current passing through the coil's primary winding creates magnetic lines of force that cut across and induce voltage in the secondary windings.

When the primary coil circuit is suddenly opened, the magnetic field instantly collapses. The sudden collapsing of the magnetic field produces a very high voltage in the secondary windings. This high voltage is used to push current across the gap of the spark plug. **Figure 24–8** shows the coil's primary and secondary circuits in basic terms.

Ignition Coil ConstructionA laminated soft iron core is positioned at the center of the ignition coil and an insulated primary winding is wound around the core. The two ends of the primary winding are connected to the primary terminals on top of the coil **(Figure 24–9)**. These terminals are usually identified with positive and negative symbols. Enamel-type insulation prevents the primary windings from touching each other. Paper insulation is also placed between the layers of windings.

Secondary coil windings are on the inside of the primary winding. A similar insulation method is used on the secondary and primary windings. The ends of the secondary winding are usually connected to one of the primary terminals and to the high-tension terminal in the coil

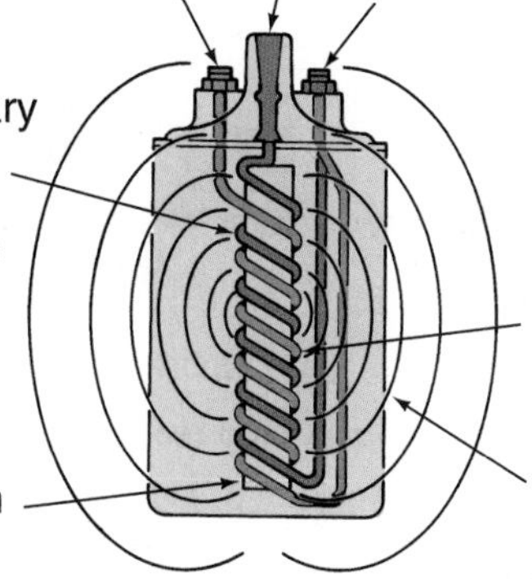

Figure 24–9 An ignition coil design. *Courtesy of DaimlerChrysler Corporation*

tower. Metal sheathing is placed around the outside of the coil windings. This sheathing concentrates the magnetic field on the outside of the windings. The coil assembly is filled with oil. The unit is sealed to protect the windings and to keep the oil inside. The oil helps to cool the coil.

Most newer coils are not oil-cooled; they are air-cooled instead. These coils are constructed in much the same way, except that the core is constructed from laminated sheets of iron. These sheets are shaped like the letter "E," and the primary and secondary windings are wound around the center of the E-core **(Figure 24–10)**.

Secondary Voltage The typical amount of secondary coil voltage required to jump the spark plug gap is 10,000 volts. Most coils have at least 25,000 volts available from the secondary. The difference between the required voltage and the maximum available voltage is referred to as secondary reserve voltage. This reserve voltage is necessary to compensate for high cylinder pressures and increased secondary resistances as the spark plug gap increases through use. The maximum available voltage must always exceed the required firing voltage or ignition misfire will occur. If there is an insufficient amount of voltage available to push current across the gap, the spark plug will not fire.

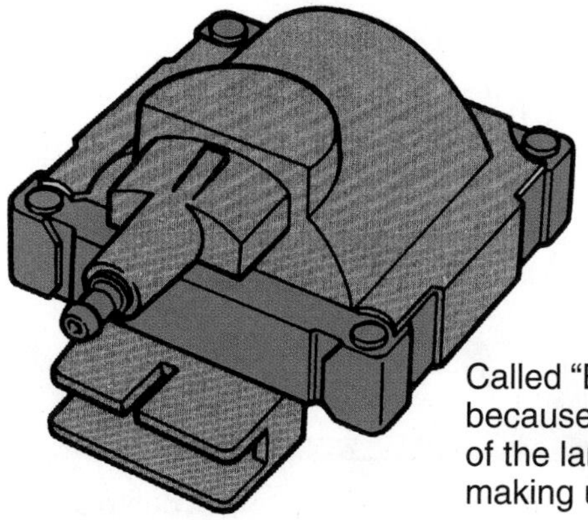

"E" core coil has a higher energy transfer due to the laminations providing closed magnetic path.
Called "E" core because of the shape of the laminations making up the core.

Figure 24–10 A typical E-core ignition coil. *Courtesy of Ford Motor Company*

Since DI and EI systems are both firing spark plugs with approximately the same air gaps, the amount of voltage required to fire the spark is nearly the same in both systems. However, EI systems have higher voltage reserves. If the additional voltage in an EI system is not used to start the spark across the plug's gap, it is used to maintain the spark for a longer period of time.

The number of ignition coils used in an ignition system varies with the type of ignition system found on a vehicle. In most ignition systems with a distributor, only one ignition coil is used. The high voltage of the secondary winding is directed, by the distributor, to the various spark plugs in the system. Therefore, there is one secondary circuit with a continually changing path.

While distributor systems have a single secondary circuit with a continually changing path, distributorless systems have several secondary circuits, each with an unchanging path.

Capacitive Discharge Ignition A typical distributor ignition system is incapable of providing a strong enough spark for 8-, 10-, and 12-cylinder engines when they are running at high engine speeds. To overcome this problem, these engines are often fitted with a capacitive discharge (CD) ignition system. Some CD systems may have a multispark feature that provides additional sparks after the first one to ensure ignition of the air/fuel mixture.

CD systems charge capacitors to store high voltage (30 to 500 volts), which is sent to the primary of the coil. The high-voltage surge greatly reduces the time required for coil saturation; therefore, the coil can have high secondary voltage output even at high engine speeds.

SPARK PLUGS

Spark plugs provide the crucial **air gap** across which the high voltage from the coil causes an arc or spark. The main parts of a spark plug are a steel shell; a ceramic core or insulator, which acts as a heat conductor; and a pair of electrodes, one insulated in the core and the other grounded on the shell. The shell holds the ceramic core and electrodes in a gas-tight assembly and has threads for plug installation in the engine **(Figure 24–11)**. The insulator material may be alumina silicate or a black-glazed, zirconia-enhanced ceramic insulator to provide for increased durability and strength. The shell may be coated with corrosion resistance material and/or materials that prevent the threads from seizing to the cylinder head.

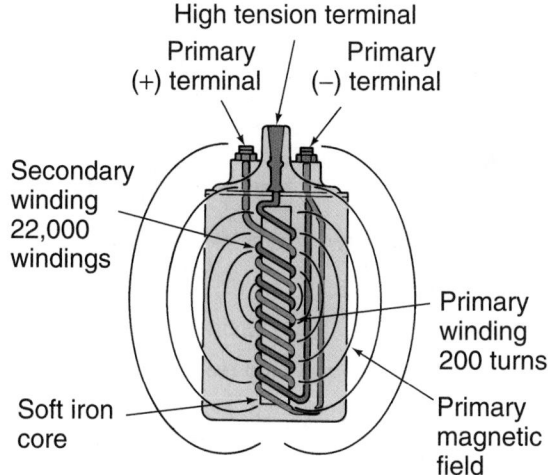

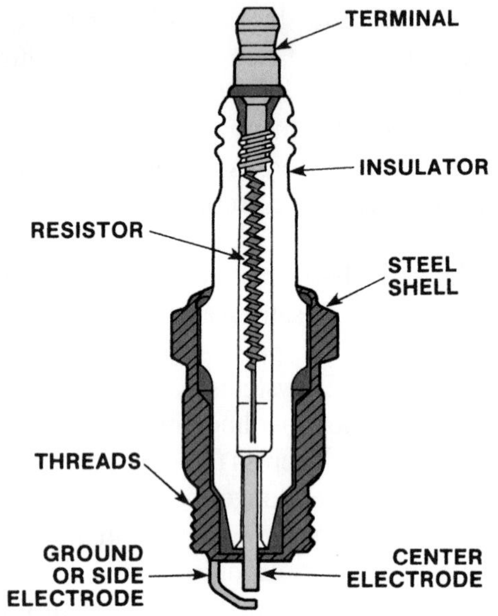

Figure 24-11 Components of a typical spark plug.

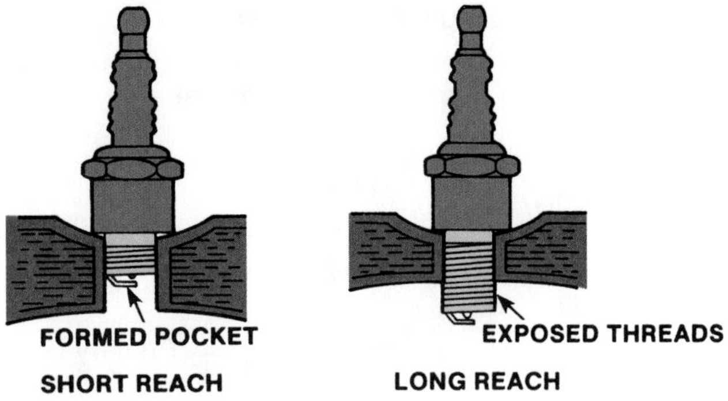

Figure 24-12 Spark plug reach: long versus short.

A terminal post on top of the center electrode is the connecting point for the spark plug cable. Current flows through the center of the plug and arcs from the tip of the center electrode to the ground electrode. The center electrode is surrounded by the ceramic insulator and is sealed to the insulator with copper and glass seals. These seals prevent combustion gases from leaking out of the cylinder. Ribs on the insulator increase the distance between the terminal and the shell to help prevent electric arcing on the outside of the insulator. The steel spark plug shell is crimped over the insulation, and a ground electrode, on the lower end of the shell, is positioned directly below the center electrode. There is an air gap between these two electrodes.

Spark plugs come in many different sizes and designs to accommodate different engines.

Size Automotive spark plugs are available in either 14-mm or 18-mm diameters. The 14-mm variety can have either a flat seat that requires a gasket or a tapered seat that does not. The latter is the most commonly used. All 18-mm plugs feature tapered seats that match similar seats in the cylinder head and need no gasket. All spark plugs have a hex-shaped shell that accommodates a socket wrench for installation and removal. The 14-mm, tapered seat plugs have shells with a ⅝-inch (47.7-mm) hex; 14-mm gasketed and 18-mm tapered seat plugs have shells with a 13⁄16-inch (20.67-mm) hex.

Reach One important design characteristic of spark plugs is the **reach (Figure 24–12)**. This refers to the length of the shell from the contact surface at the seat to the bottom of the shell, including both threaded and nonthreaded sections. Reach is crucial because the plug's air gap must be properly placed in the combustion chamber to produce the correct amount of heat. When a plug's reach is too short, its electrodes are in a pocket and the arc is not able to adequately ignite the mixture. If the reach is too long, the exposed plug threads can get so hot they will ignite the air/fuel mixture at the wrong time and cause **preignition**. Preignition is a term used to describe abnormal combustion, which is caused by something other than the heat of the spark.

Heat Range When the engine is running, most of the plug's heat is concentrated on the center electrode. Heat is quickly dissipated from the ground electrode because it is attached to the shell, which is threaded into the cylinder head. Coolant circulating in the head absorbs the heat and moves it through the cooling system. The heat path for the center electrode is through the insulator into the shell and then to the cylinder head. The **heat range** of a spark plug is determined by the length of the insulator before it contacts the shell. In a cold-range spark plug, there is a short distance for the heat to travel up the insulator to the shell. The short heat path means the electrode and insulator will maintain little heat between firings **(Figure 24–13)**.

In a hot spark plug, the heat travels farther up the insulator before it reaches the shell. This provides a longer

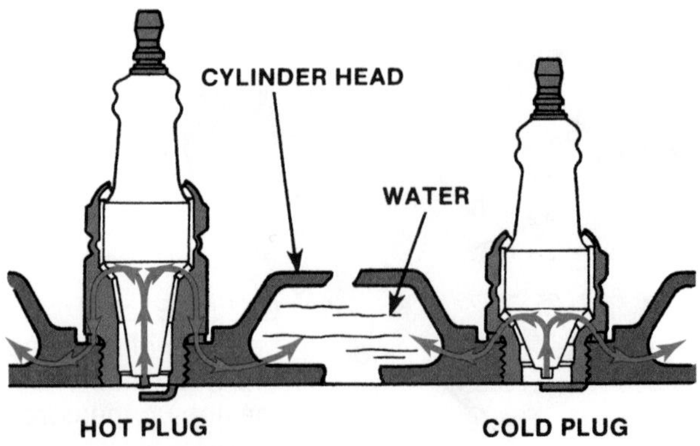

Figure 24-13 Spark plug heat range: hot versus cold.

heat path and the plug retains more heat. A spark plug needs to retain enough heat to clean itself between firings, but not so much that it damages itself or causes premature ignition of the air/fuel mixture in the cylinder.

The heat range is indicated by a code imprinted on the side of the spark plug, usually on the porcelain insulator.

Resistor Plugs Most automotive spark plugs have a resistor (normally about 5 K ohms) between the top terminal and the center electrode. Some spark plugs use a semiconductor material to provide for this resistance. The resistor reduces RFI, which can interfere with, or damage, radios, computers, and other electronic accessories, such as GPS systems. If an engine was originally equipped with resistor plugs, resistor plugs should be installed when the originals are replaced.

WARNING!

Using a nonresistor plug on some engines may cause erratic idle, high-speed misfire, engine run-on, power loss, and abnormal combustion.

Spark Plug Gaps The correct spark plug air gap is essential for achieving optimum engine performance and long plug life. A gap that is too wide requires higher voltage to jump the gap. If the required voltage is greater than what is available, the result is **misfiring**. Misfiring results from the inability of the ignition to jump the gap or maintain the spark. A gap that is too narrow requires lower voltages and can lead to rough idle and prematurely burned electrodes, due to higher current flow.

Electrodes The materials used in the construction of a spark plug's electrodes determine the longevity, power, and efficiency of the plug. The construction and shape of the tips of the electrodes are also important.

The electrodes of a standard spark plug are made with copper and some use a copper-nickel alloy. Copper is a good electrical conductor and offers some resistance to corrosion. Copper melts at 1981°F so it is more than suitable for use in an internal combustion engine.

Platinum electrodes are used to extend the life of a plug **(Figure 24–14)**. Platinum melts at 3,200°F (1,760°C) and is highly resistant to corrosion. Although platinum is an extremely durable material, it is an expensive precious metal; therefore, platinum spark plugs cost more than copper plugs. Also, platinum is not as good a conductor as copper. Spark plugs are available with only the center electrode made of platinum (called single-platinum) and with the center and ground electrodes made of platinum (called double-platinum). Some platinum plugs have a very small center electrode combined with a sharp pointed ground electrode designed for better performance.

Figure 24-14 A platinum-tipped spark plug.

Until recently, platinum was considered the best material to use for electrodes because of its durability. However, iridium is six times harder, eight times stronger, and has a melting point 1200 degrees higher than platinum. Iridium is a precious, silver-white metal and one of the densest materials found on earth. A few spark plugs use an iridium alloy as the primary metal complimented by rhodium to increase oxidation wear resistance. This iridium alloy is so durable that it allows for an extremely small center electrode. A typical copper/nickel plug has a 2.5 mm diameter center electrode and a platinum plug has a 1.1 mm diameter. An iridium plug can have a diameter as small as 0.4 mm **(Figure 24–15)**, which means

Figure 24-15 The spark plug has a small diameter iridium center electrode and a grooved ground electrode.
Courtesy of Denso Sales California, Inc.

the firing voltage requirements are decreased. Iridium is also used as an alloying material for platinum.

Another rare and hard material used to make electrodes is yttrium. Yttrium has a silvery-metallic luster and has a melting point of 2773.4°F (1523°C). Yttrium is fairly stable in air but oxidizes readily when heated. (Moon rocks contain yttrium.) Yttrium produces a highly adhesive oxide layer that makes the spark plug very durable and reliable, thereby extending its service life.

Electrode Designs. Spark plugs are available with many different shapes and numbers of electrodes. When trying to ascertain the advantages of each design, remember the spark is caused by electrons moving across an air gap. The electrons will always jump in the direction of the least electrical resistance. Therefore, if there are four ground electrodes to choose from, the electrons will jump to the closest. Also, keep in mind that the quality and pressure of the air in the air gap influences the resistance of the air gap. Again, the electrons will jump across the path of least resistance. Therefore, spark plugs with four ground electrodes do not typically supply a spark to all four electrodes **(Figure 24–16)**.

The shape of the ground electrode may also be altered. A flat, conventional electrode tends to crush the spark, and the overall volume of the flame front is smaller. A tapered ground electrode increases flame front expansion and reduces the heat lost to the electrode.

Some ground electrodes have a U-groove machined into the side that faces the center electrode. The U-groove allows the flame front to fill the gap formed by the U. This ball of fire develops a larger and hotter flame front, leading to a more complete combustion.

One brand of spark plug has a V-shaped ground electrode. This style of electrode does not block the flame front and allows it to travel upward through the V notch into the combustion chamber. These spark plugs may be equipped with three separate points of platinum, one at each end of the V and the other at the center electrode.

There are also different center electrode designs. Often these variations are based on their diameter and shape. A small diameter center electrode requires less firing voltage and tends to have a longer service life. Some center electrodes are machined with a fine point on their end.

Some center electrodes have a V groove machined in them to force the spark to the outer edge of the ground electrode, placing it closer to the air/fuel mixture. This allows for quicker ignition of the air/fuel mixture. V-grooved center electrodes also require lower firing voltages.

On some spark plugs, the center electrode does not extend from the insulator and the spark is generated across the end of the plug. With this design, the ground electrode does not block the flame front. This arrangement is called a surface gap and is intended to prevent carbon fouling, timing drift, and misfiring.

Ignition Cables

Spark plug cables, or ignition cables, make up the secondary wiring. These cables carry the high voltage from the distributor or the multiple coils to the spark plugs. The cables are not solid wire; instead they contain carbon fiber cores that act as resistors in the secondary circuit **(Figure 24–17)**. They cut down on radio and television interference, increase firing voltages, and reduce spark plug wear by decreasing current. Insulated boots on the ends of the cables strengthen the connections as well as prevent dust and water infiltration and voltage loss.

Some ignition cables are called *variable pitch* resistor cables. These cables rely on tightly wound and loosely wound copper wire around a layer of ferrite magnetic material wrapped over a fiberglass strand core. This construction creates the necessary resistance with a fraction of the impedance found in solid carbon core-type wire sets.

Some engines have spark plug cable heat shields **(Figure 24–18)** pressed into the cylinder head. These shields surround each spark plug boot and spark plug. They protect the spark plug boot from damage due to the extreme heat generated by the nearby exhaust manifold.

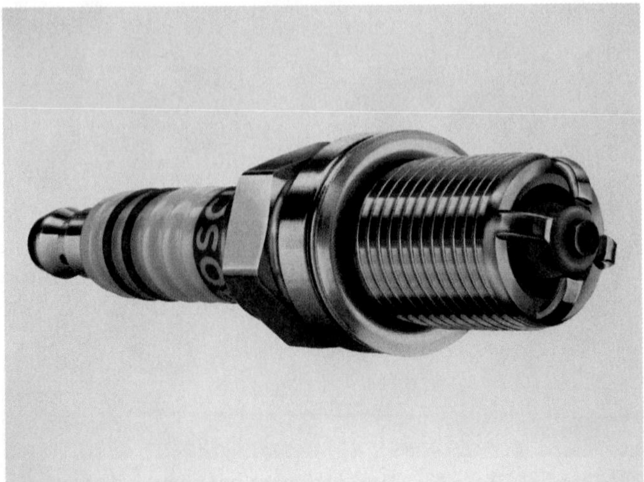

Figure 24–16 A spark plug with four ground electrodes. *Courtesy of Robert Bosch Corp.*

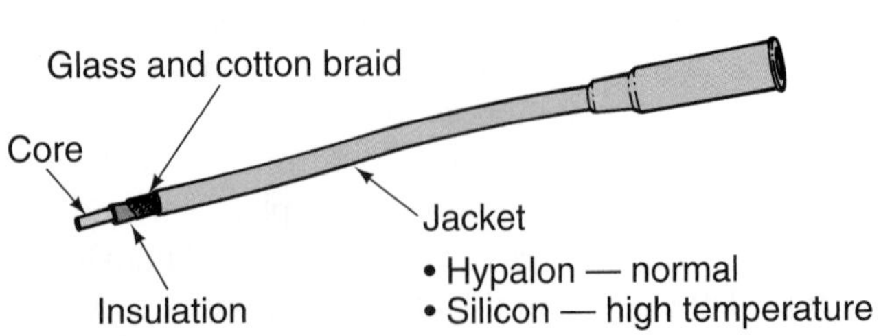

Figure 24–17 Spark plug cable construction. *Courtesy of DaimlerChrysler Corporation*

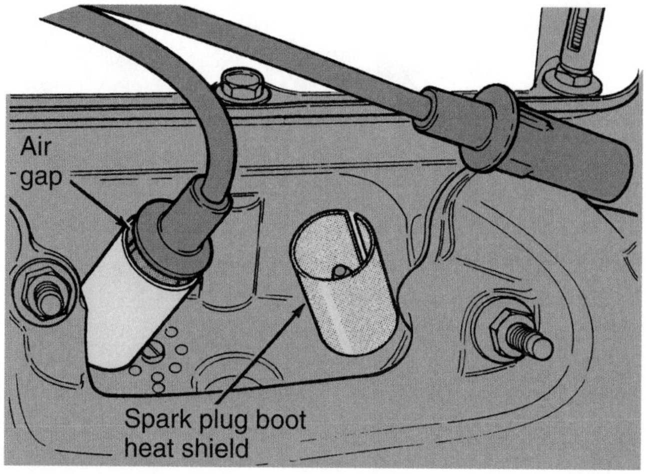

Figure 24-18 Spark plug boot heat shields. *Courtesy of DaimlerChrysler Corporation*

TRIGGERING AND SWITCHING DEVICES

Triggering and switching devices are used to ensure the spark occurs at the correct time. A triggering device is simply a device that monitors the movement of the engine's pistons. In older engines, the triggering device was a set of contact points. The points opened and closed to turn the primary circuit on or off. When the circuit was turned off, a spark plug fired. All newer ignition systems use an electronic triggering device.

Electronic switching components are normally located in an ignition control module, which may be part of the vehicle's PCM. On older vehicles, the control module may be built into the distributor or mounted in the engine compartment.

The ignition module advances or retards the ignition timing in response to engine conditions. Early systems had little control of timing and used mechanical or vacuum devices to alter timing. Today's computer-controlled systems have full control and can adjust ignition timing in response to the input signals from a variety of sensors and the programs in the computer.

Most electronically controlled systems use an NPN transistor to control the primary ignition circuit, which ultimately controls the firing of the spark plugs. The transistor's emitter is connected to ground. The collector is connected to the negative terminal of the coil. When the triggering device supplies a small current to the base of the transistor, current flows through the primary winding of the coil. When the current to the base is interrupted, the current to the coil is also interrupted. An example of how this works is shown in **Figure 24–19**, which is a simplified diagram of an electronic ignition system.

ENGINE POSITION SENSORS

The time when the primary circuit must be opened and closed is related to the position of the pistons and the crankshaft. Therefore, the position of the crankshaft is used to control the flow of current to the base of the switching transistor.

A number of different types of sensors are used to monitor the position of the crankshaft and control the flow of current to the base of the transistor. These engine position sensors and generators serve as triggering devices and include magnetic pulse generators, metal detection sensors, Hall-effect sensors, and photoelectric (optical) sensors.

The mounting location of these sensors depends on the design of the ignition system. All four types of sensors can be mounted in the distributor, which is turned by the camshaft.

Magnetic pulse generators and Hall-effect sensors can also be located on the crankshaft. These sensors are also commonly used on EI systems.

Magnetic Pulse Generator

Basically, a magnetic pulse generator or inductance sensor consists of two parts: a timing disc and a pickup coil. The timing disc may also be called a reluctor, trigger wheel, pulse ring, armature, or timing core. The pickup coil, which consists of a length of wire wound around a weak permanent magnet, may also be called a stator,

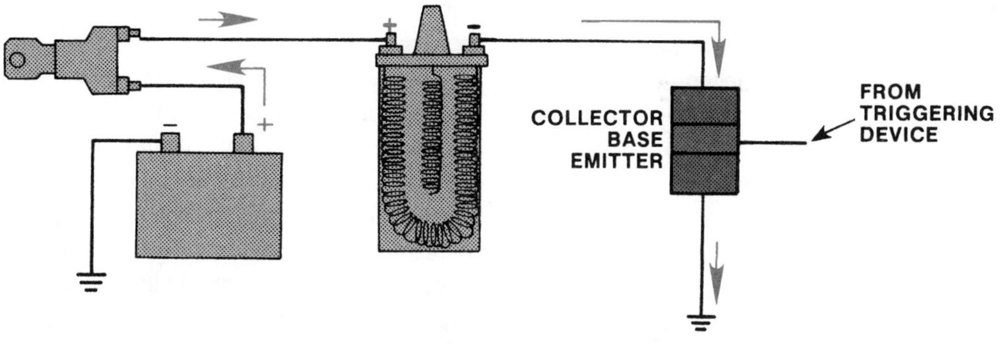

Figure 24-19 When the triggering device supplies a small amount of current to the transistor's base, the primary coil circuit is closed and current flows.

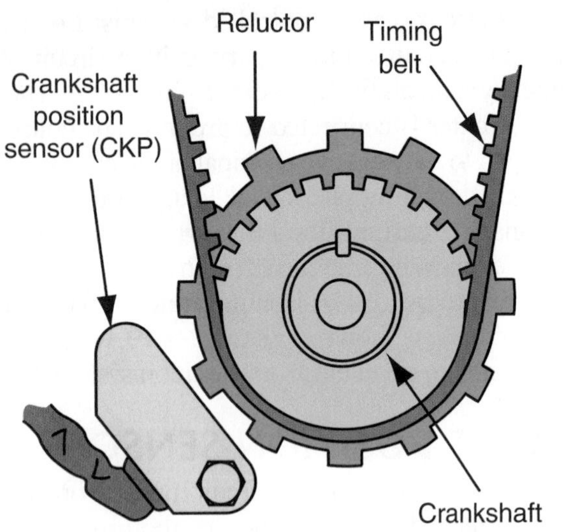

Figure 24-20 The location of a typical crankshaft position sensor.

sensor, or pole piece. Depending on the type of ignition system used, the timing disc may be mounted on the distributor shaft, at the rear of the crankshaft, or on the crankshaft vibration damper **(Figure 24–20)**.

The magnetic pulse or PM generator operates on basic electromagnetic principles. Remember that a voltage can only be induced when a conductor moves through a magnetic field. The magnetic field is provided by the pickup unit and the rotating timing disc provides the movement through the magnetic field needed to induce voltage.

As the disc teeth approach the pickup coil, they repel the magnetic field, forcing it to concentrate around the pickup coil **(Figure 24–21A)**. Once the tooth passes by the pickup coil, the magnetic field is free to expand or unconcentrate **(Figure 24–21B)**, until the next tooth on the disc approaches. Approaching teeth concentrate the magnetic lines of force, while passing teeth allow them to expand. This pulsation of the magnetic field causes the lines of magnetic force to cut across the winding in the pickup coil, inducing a small amount of AC voltage that is sent to the switching device in the primary circuit.

When a disc tooth is directly in line with the pickup coil, the magnetic field is not expanding or contracting. Since there is no movement or change in the field, voltage at this precise moment drops to zero. At this point, the switching device inside the ignition module reacts to the zero voltage signal by turning the ignition's primary circuit current off. As explained earlier, this forces the magnetic field in the primary coil to collapse, discharging a secondary voltage to the distributor or directly to the spark plug.

As soon as the tooth rotates past the pickup coil, the magnetic field expands again and another voltage signal is induced. The only difference is that the polarity of the charge is reversed. Negative becomes positive or positive becomes negative. On sensing this change in voltage, the switching device turns the primary circuit back on and the process begins all over.

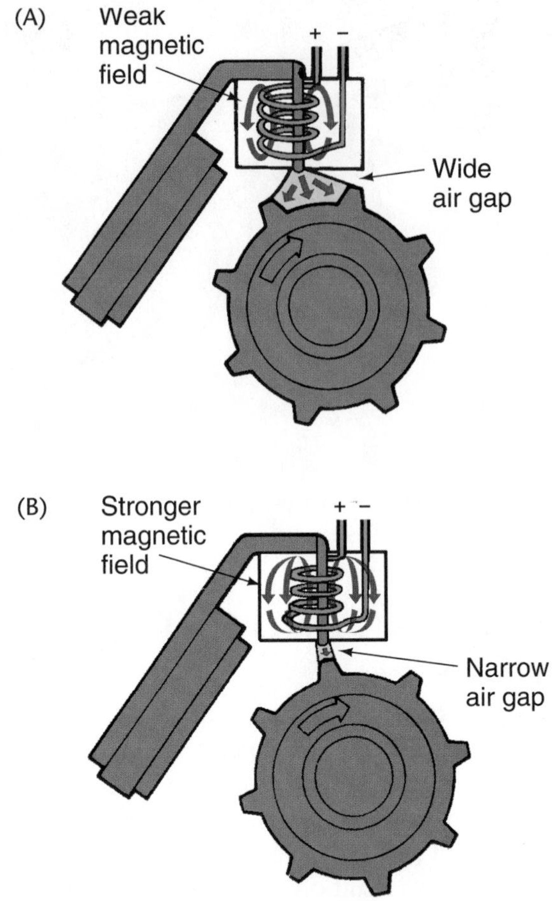

Figure 24-21 A (A) wide gap produces a weak magnetic signal; (B) a narrow gap produces a strong magnetic field. *Courtesy of DaimlerChrysler Corporation*

The slotted disc is mounted on the crankshaft, vibration damper, or distributor shaft in a very precise manner. When the disc teeth align with the pickup coil, this corresponds to the exact time certain pistons are nearing TDC. This means the zero voltage signal needed to trigger the secondary circuit occurs at precisely the correct time.

Metal Detection Sensors

Metal detection sensors are found on early electronic ignition systems. They work much like a magnetic pulse generator with one major difference.

A trigger wheel is pressed over the distributor shaft and a pickup coil detects the passing of the trigger teeth as the distributor shaft rotates. However, unlike a magnetic pulse generator, the pickup coil of a metal detection sensor does not have a permanent magnet. Instead, the pickup coil is an electromagnet. A low level of current is supplied to the coil by an electronic control unit, inducing a weak magnetic field around the coil. As the reluctor on the distributor shaft rotates, the trigger teeth pass very close to the coil **(Figure 24–22)**. As the teeth pass

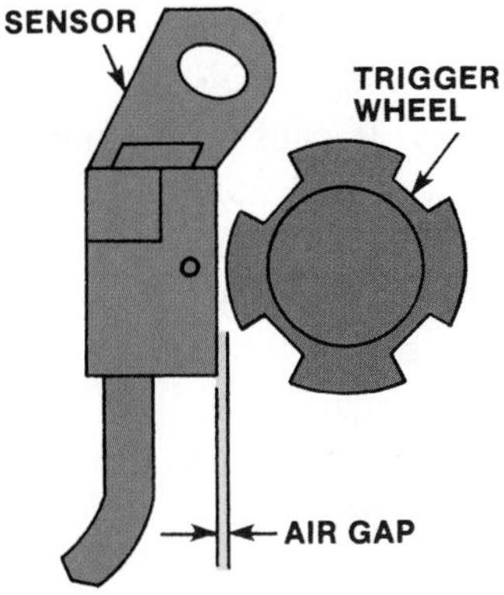

Figure 24–22 In a metal detecting sensor, the revolving trigger wheel teeth alter the magnetic field produced by the electromagnet in the pickup coil.

in and out of the coil's magnetic field, the magnetic field builds and collapses, producing a corresponding change in the coil's voltage. The voltage changes are monitored by the control unit to determine crankshaft position.

Hall-Effect Sensor

The Hall-effect sensor or switch is the most commonly used engine position sensor. There are several good reasons for this. Unlike a magnetic pulse generator, the Hall-effect sensor produces an accurate voltage signal throughout the entire rpm range of the engine. Furthermore, a Hall-effect switch produces a square wave signal that is more compatible with the digital signals required by on-board computers.

Functionally, a Hall switch performs the same tasks as a magnetic pulse generator. But the Hall switch's method of generating voltage is quite unique. It is based on the Hall-effect principle, which states: If a current is allowed to flow through a thin conducting material, and that material is exposed to a magnetic field, voltage is produced.

The heart of the Hall generator is a thin semiconductor layer (Hall layer) derived from a gallium arsenate crystal. Attached to it are two terminals—one positive and the other negative—that are used to provide the source current for the Hall transformation.

Directly across from this semiconductor element is a permanent magnet. It is positioned so that its lines of flux bisect the Hall layer at right angles to the direction of current flow. Two additional terminals, located on either side of the Hall layer, form the signal output circuit.

When a moving metallic shutter blocks the magnetic field from reaching the Hall layer or element, the Hall-effect switch produces a voltage signal. When the shutter blade moves and allows the magnetic field to expand and reach the Hall element, the Hall-effect switch does not generate a voltage signal.

The Hall switch is described as being "on" any time the Hall layer is exposed to a magnetic field and a Hall voltage is being produced **(Figure 24–23)**. However, before this signal voltage can be of any use, it has to be modified. After leaving the Hall layer, the signal is routed to an amplifier where it is strengthened and inverted so the signal reads high when it is actually coming in low and vice versa. Once it has been inverted, the signal goes through a pulse-shaping device called the Schmitt trigger where it is turned into a clean square wave signal. After conditioning, the signal is sent to the base of a switching transistor that is designed to turn on and off in response to the signals generated by the Hall switch assembly.

The shutter wheel is the last major component of the Hall switch. The shutter wheel consists of a series of alternating windows and vanes that pass between the Hall layer and magnet. The shutter wheel may be part of the distributor rotor or be separate from the rotor.

The points where the shutter vane begins to enter and begins to leave the air gap are directly related to primary

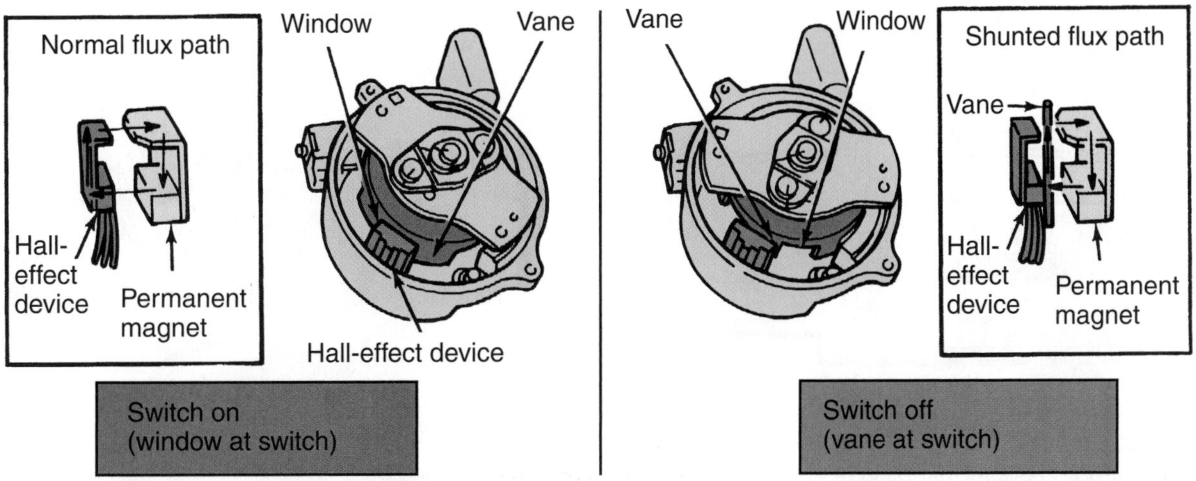

Figure 24–23 The operation of a Hall-effect switch. *Courtesy of Ford Motor Company*

circuit control. As the leading edge of a vane enters the air gap, the magnetic field is deflected away from the Hall layer; Hall voltage decreases. When that happens, the modified Hall output signal increases abruptly and turns on the switching transistor. Once the transistor is turned on, the primary circuit closes and the coil's energy storage cycle begins.

Primary current continues to flow as long as the vane is in the air gap. As the vane starts to leave the gap, however, the reforming Hall voltage signal prompts a parallel decline in the modified output signal. When the output signal goes low, the bias of the transistor changes. Primary current flow stops.

In summary, the ignition module supplies current to the coil's primary winding as long as the shutter wheel's vane is in the air gap. As soon as the shutter wheel moves away and the Hall voltage is produced, the control unit stops primary circuit current, high secondary voltage is induced, and ignition occurs.

In addition to ignition control, a Hall switch can also be used to generate precise rpm signals (by determining the frequency at which the voltage rises and falls) and provide the sync pulse for sequential fuel ignition operation.

Photoelectric Sensor

A fourth type of crankshaft position sensor is the **photoelectric sensor**. The parts of this sensor include a light-emitting diode (LED), a light-sensitive phototransistor (photo cell), and a slotted disc called a light beam interrupter **(Figure 24–24)**.

The slotted disc is attached to the distributor shaft. The LED and the photo cell are situated over and under the disc opposite each other. As the slotted disc rotates between the LED and photo cell, light from the LED shines through the slots. The intermittent flashes of light are translated into voltage pulses by the photo cell. When the voltage signal occurs, the control unit turns on the primary system. When the disc interrupts the light and the voltage signal ceases, the control unit turns the primary system off, causing the magnetic field in the coil to collapse and sending a surge of voltage to a spark plug.

The photoelectric sensor sends a very reliable signal to the control unit, especially at low engine speeds. These units have been primarily used on Chrysler and Mitsubishi engines. Some Nissan and General Motors' products have used them as well.

Timing Advance

Older ignition systems were equipped with a centrifugal advance mechanism that advanced ignition timing in response to engine speed. This setup used a set of pivoted weights and springs connected to the distributor shaft. The advance assembly was typically mounted below the triggering unit's mounting plate, although some had the plate above it. When engine speed increased, the weights moved out, shifting the plate where the triggering device was mounted. The shifting of the plate caused the triggering device to send its signal earlier, causing an advance in ignition timing.

Vacuum Advance Units that changed ignition timing in response to engine load were also fitted to distributors **(Figure 24–25)**. A vacuum advance unit, comprised of a spring-loaded diaphragm, was attached to the triggering device plate. Vacuum was applied to one side of the diaphragm and atmospheric pressure was applied to the other side. Any increase in vacuum allowed atmospheric pressure to move the diaphragm, which caused a movement of the triggering device's mounting plate. The more vacuum that was present on one side, the more the atmospheric pressure could move the diaphragm and advance the timing. A spring was used to return the plate toward its rest position when vacuum decreased.

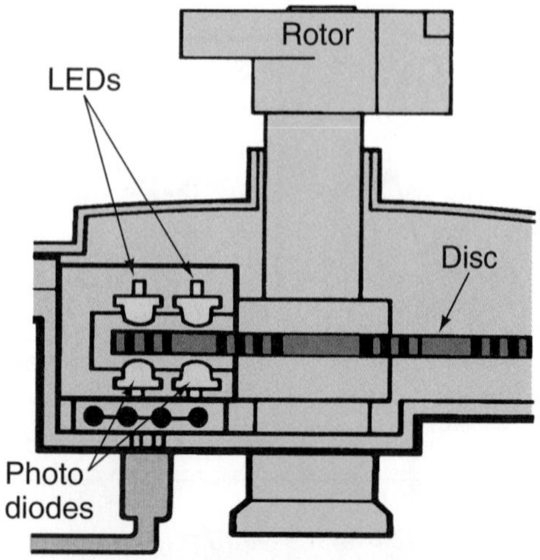

Figure 24-24 A distributor with optical-type pickups. *Courtesy of DaimlerChrysler Corporation*

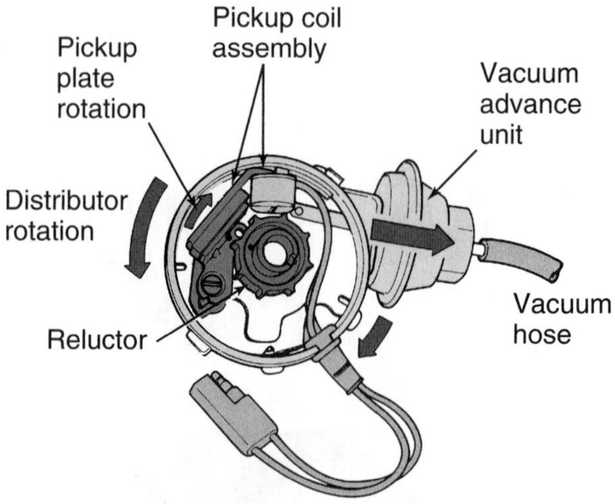

Figure 24-25 Typical vacuum advance unit operation. *Courtesy of DaimlerChrysler Corporation*

DISTRIBUTOR IGNITION SYSTEM OPERATION

The primary circuit of a DI system is controlled electronically by one of the sensors just described and an electronic control unit (module) that contains some type of switching device.

Distributor

The reluctor, or trigger wheel, and distributor shaft assembly rotate on bushings in the aluminum distributor housing. A roll pin extends through a retainer and the distributor shaft to hold the shaft in place in the distributor. Another roll pin is used to fasten the drive gear to the lower end of the shaft. This drive gear typically meshes with a drive gear on the engine's camshaft. The gear size is designed to drive the distributor shaft at the same speed as the camshaft, which rotates at one-half the speed of the crankshaft.

Primary Circuit

When the ignition switch is in the on position, current from the battery flows through the ignition switch and primary circuit resistor to the primary winding of the ignition coil. From there it passes through some type of switching device and back to ground. The switching device is controlled by the triggering device. The current flow in the ignition coil's primary winding creates a magnetic field. The switching device or control module interrupts this current flow at predetermined times. When it does, the magnetic field in the primary winding collapses. This collapse generates a high-voltage surge in the secondary winding of the ignition coil. The secondary circuit of the system begins at this point and as a result the spark plug fires.

Once the plug stops firing, the transistor closes the primary coil circuit. The length of time the transistor allows current flow in the primary ignition circuit is determined by the electronic circuitry in the control module.

DI System Design

Through the years there have been many different designs of DI systems. All operate in the basically the same way but are configured differently. The systems described in this section represent the different designs used by manufacturers. These designs are based on the location of the electronic control module (unit) (ECU) and/or the type of triggering device used.

DI Systems with External Ignition Module Ford Motor Company used two generations of Dura-Spark ignition systems. The second design (Dura-Spark II) was based on the first (Dura-Spark I). The Dura-Spark II had the ECU mounted away from the distributor, typically on a fender wall. The negative primary coil terminal is referred to as a distributor electronic control (dec) or tachometer (tach) terminal. This terminal is connected to the ignition module.

The distributor pickup coil is connected through two wires to the module. A wire is also connected from the distributor housing to the module. This wire supplies a ground connection from the module to the pickup plate; therefore, the module does not need to be grounded at its mounting.

DI Systems with Module Mounted on the Distributor A **thick film integrated (TFI)** ignition system has the ECU mounted on the distributor. This ignition system from Ford uses an epoxy (E) core ignition coil. The windings of the coil are set in epoxy and are surrounded by an iron core. The negative primary coil terminal is connected to the module. A wire is connected from the ignition switch start terminal to the module. The module-to-pickup terminals extend through the distributor housing, and three pickup lead wires are connected to these terminals **(Figure 24–26)**. Heat-dissipating grease must be placed on the back of the module to prevent overheating of the module.

DI Systems with Internal Ignition Module Perhaps the best example of a DI system with the ignition module inside the distributor is the General Motors' **high energy ignition (HEI)** system. Some HEI units also contain the ignition coil, others have the coil remotely mounted away from the distributor.

The pickup coil surrounds the distributor shaft, and a flat magnetic plate is bolted between the pickup coil

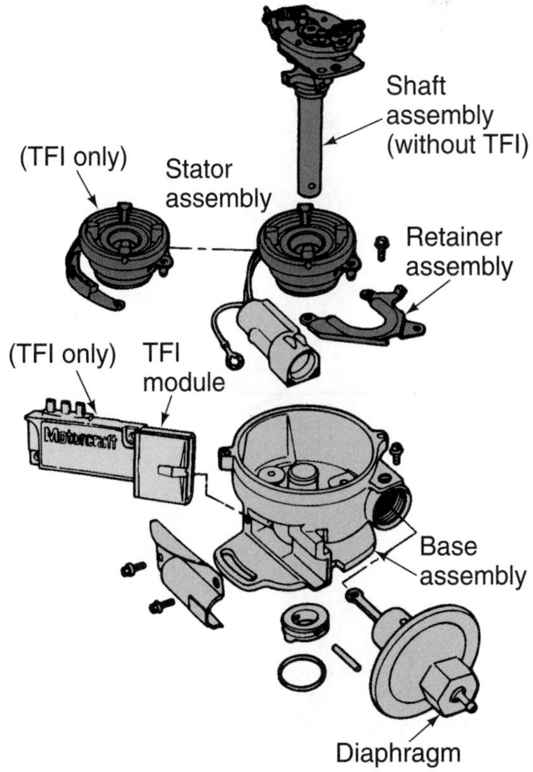

Figure 24-26 A TFI distributor assembly. *Courtesy of Ford Motor Company*

and the pole piece. A **timer core** that has one high point for each engine cylinder is attached to the distributor shaft. The number of timer core high points matches the number of teeth on the pole piece. The module is mounted to the breaker plate and is set in heat-dissipating grease. A capacitor is connected from the module voltage supply terminal to ground on the distributor housing.

The coil battery terminal is connected directly to the ignition switch, and the coil tachometer (tach) terminal is connected to the module **(Figure 24–27)**. HEI coils are basically E core coils and rely on the surrounding air to dissipate the coil's heat.

A wire also extends from the coil's battery terminal to the module. In systems with an internal ignition coil, a ground wire is connected between the frame of the coil and the distributor housing. This lead is used to dissipate any voltage induced in the coil's frame.

When the ignition switch is on and the distributor shaft is not turning, the module opens the primary ignition circuit. As the engine is cranked and the timer core high points approach alignment with the pole piece teeth, a positive voltage is induced in the pickup coil. This voltage signal causes the module to close the primary circuit, and current begins to flow through the primary windings, causing a magnetic field to form around them.

At the instant of alignment between the timer core high points and pole piece teeth, the pickup coil voltage drops to zero. As these high points move out of alignment, a negative voltage is induced in the pickup coil. This voltage signal to the module causes the module to open the primary circuit. When this action occurs, the magnetic field collapses across the ignition coil windings, and the induced secondary voltage forces current through the secondary circuit and across the spark plug gap.

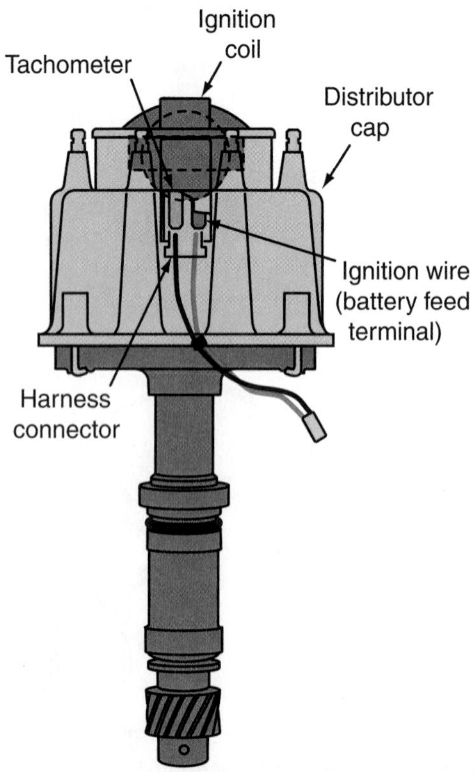

Figure 24-27 HEI distributor terminal identification.

Computer-Controlled DI Systems Spark timing on these systems is controlled by a computer that continuously adjusts ignition timing to obtain optimum combustion. The computer monitors the engine operating parameters with various sensors. Based on this input, the computer signals an ignition module to collapse the primary circuit, allowing the secondary circuit to fire the spark plugs.

Timing control is selected by the computer's program. During engine starting, computer control is bypassed and the mechanical setting of the distributor controls spark timing. Once the engine is started and running, spark timing is controlled by the computer. This scheme or **strategy** allows the engine to start regardless of whether the electronic control system is functioning properly.

The goal of computerized spark timing is to produce maximum engine power, top fuel efficiency, and minimum emissions levels during all types of operating conditions. The computer does this by continuously adjusting ignition timing. The computer determines the best spark timing based on certain engine operating conditions such as crankshaft position, engine speed, throttle position, engine coolant temperature, and initial and operating manifold or barometric pressure. Once the computer receives input from these and other sensors, it compares the existing operating conditions to information permanently stored or programmed into its memory. The computer matches the existing conditions to a set of conditions stored in its memory, determines proper timing setting, and sends a signal to the ignition module to fire the plugs.

The computer continuously monitors existing conditions, adjusting timing to match what its memory tells it is the ideal setting for those conditions. It can do this very quickly, making thousands of decisions in a single second. The control computer typically has the following types of information permanently programmed into it.

- Speed-related spark advance. As engine speed increases to a particular point, there is a need for more advanced timing. As the engine slows, the timing should be retarded or have less advance. The computer bases speed-related spark advance decisions on engine speed and signals from the TP sensor.
- Load-related spark advance. This is used to improve power and fuel economy during acceleration and heavy load conditions. The computer defines the load and the ideal spark advance by processing information from the TP sensor, MAP, and engine speed sensors. Typically, the more load that is on an engine, the less spark advance is needed.

- Warm-up spark advance. This is used when the engine is cold, because a greater amount of advance is required while the engine warms up.
- Special spark advance. This is used to improve fuel economy during steady driving conditions. During constant speed and load conditions, the engine will be more efficient with much advance timing.
- Spark advance due to barometric pressure. This is used when barometric pressure exceeds a preset calibrated value.

All of this information is looked at by the computer to determine the ideal spark timing for all conditions. The calibrated or programmed information in the computer is contained in what is called software **lookup tables**.

Ignition timing can also work in conjunction with the electronic fuel control system to provide emission control, optimum fuel economy, and improved driveability. They are all dependent on spark advance. Some examples of computer-controlled DI systems follow.

Chrysler's Dual Pickup System. This system has two Hall-effect switches in the distributor when the engine is equipped with port fuel injection. In some units, the pickup unit used for ignition triggering is located above the pickup plate in the distributor and is referred to as the reference pickup. The second pickup unit is positioned below the plate. A ring with two notches is attached to the distributor shaft and rotates through the lower pickup unit. This lower pickup is called the synchronizer (SYNC) pickup.

In other designs, the two pickup units are mounted below the pickup plate and one set of blades rotates through both Hall-effect units **(Figure 24–28)**. The shutter blade representing the number one cylinder has a large opening in the center of the blade. When this blade rotates through the SYNC pickup, a different signal is produced compared to the other blades. This number one blade signal informs the PCM when to activate the injectors.

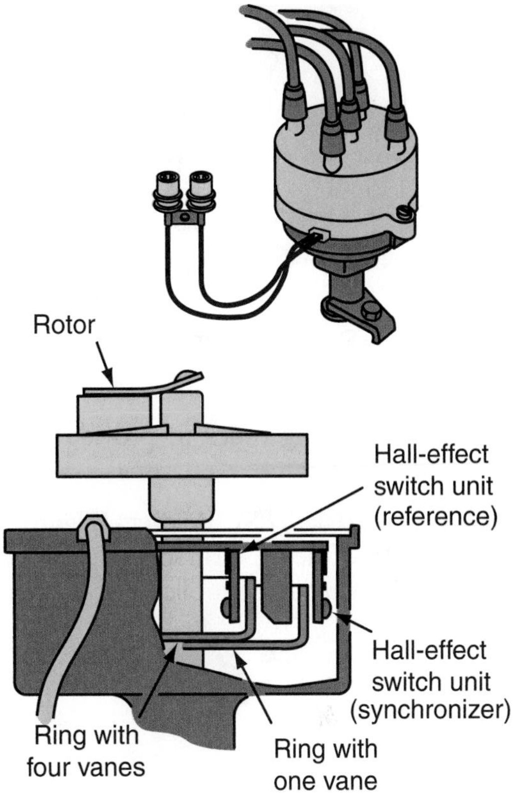

Figure 24–28 A distributor assembly with two Hall-effect switches below the pickup plate.

Distributors with Optical-Type Pickups. The 3.0 L V-6 engine available in some Chrysler products has a distributor fitted with an optical pickup assembly with two LEDs and two photo diodes. A thin plate attached to the distributor shaft rotates between the LEDs above the plate and the photo diodes below the plate. This plate contains six equally spaced slots, which rotate directly below the inner LED and photo diode.

The inner LED and photo diode act as the reference pickup. As in Hall-effect pickup systems, the reference pickup in the optical distributor provides a crankshaft position and speed signal to the PCM. When the ignition switch is on, the PCM supplies voltage to the optical pickup, which causes the LEDs to emit light. If a solid part of the plate is under the reference LED, this light does not shine on the photo diode. Under this condition, the photo diode does not conduct current and the reference voltage signal to the PCM is 5 volts. As a reference slot moves under the LED, the light shines on the photo diode. The diode then conducts current and the reference voltage signal to the PCM is 0 volt.

The outer LED, photo diode, and row of slots perform a function similar to that of the SYNC pickup in a distributor with Hall-effect pickups. The outer row of slots is closely spaced, and the width between each slot represents 2 degrees of crankshaft rotation. On the outer row there is one area where the slots are missing. When this blank area rotates under the LED, a different SYNC voltage signal is produced, which informs the PCM regarding the number one piston position. The PCM uses this signal for injector control. As the outer row of slots rotates under the outer LED, the SYNC voltage signal to the PCM cycles from 0 volt to 5 volts. The reference pickup signal informs the PCM when each piston is a specific number of degrees before TDC on the compression stroke.

When this signal is received, the PCM scans the inputs and calculates the spark advance required by the engine. The SYNC sensor signals always keep the PCM informed of the exact position of the crankshaft. The PCM opens the primary ignition circuit and fires the next

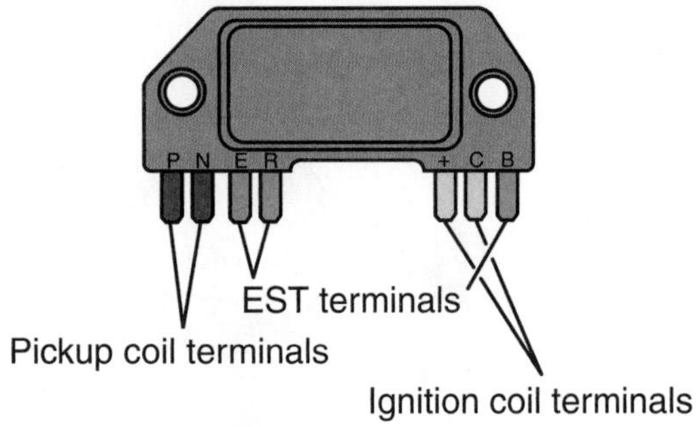

Figure 24-29 A seven-terminal GM HEI/EST ignition module.

spark plug in the firing order to provide the calculated spark advance.

General Motors' HEI with EST. A seven- **(Figure 24–29)** or eight-terminal module is used in some General Motors' distributors with computer-controlled spark advance and fuel injection. Two of the module terminals are connected to the coil primary terminals, and two other module terminals are connected to the pickup coil. The other four module terminals are connected through a four-wire harness to the PCM. These four wires are identified as bypass, **electronic spark timing (EST)**, ground, and reference wires.

Crankshaft position and speed information are obtained from the pickup signal to the PCM. The PCM scans the input sensors and then sends a signal on the EST wire to the module. This signal commands the module to open the primary circuit and fire the next spark plug at the right instant to provide the precise spark advance indicated by the input signals.

Honda's DI System with EST. Honda, as well as other manufacturers, also fits the ignition module inside the distributor **(Figure 24–30)**. The distributor is also fitted

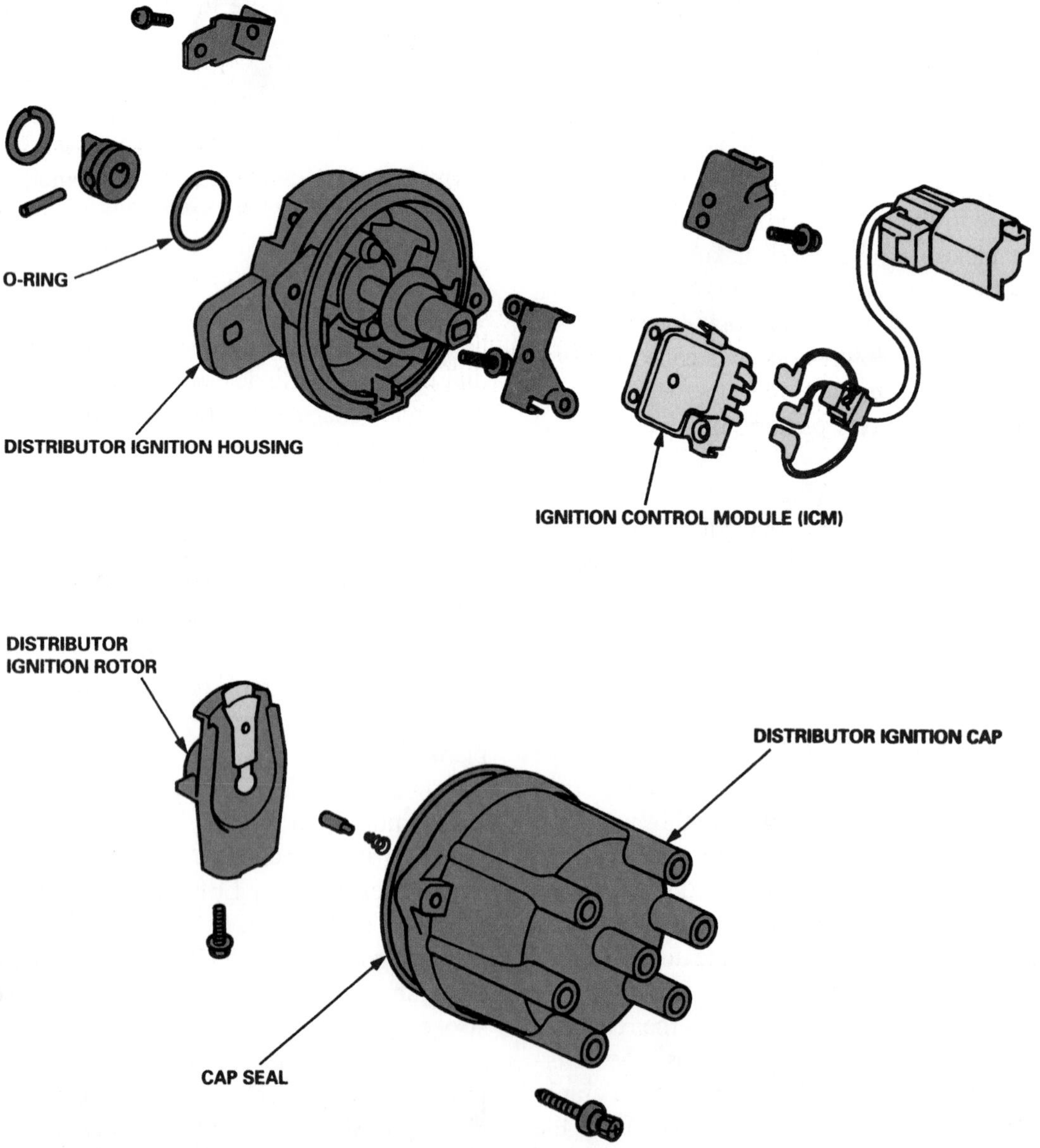

Figure 24-30 Honda's distributor with a built-in control module. *Courtesy of American Honda Motor Co., Inc.*

with a Hall-effect switch that is directly connected to the control module. The only external electrical connections for the module are from the PCM and to the ignition coil. The PCM controls the activity of the module, which in turn controls the dwell of the primary and therefore the ignition timing.

ELECTRONIC IGNITION SYSTEM OPERATION

Distributors are seldom found on today's engines. Nearly all of today's engines are equipped with an EI system. There are primarily two different designs of EI systems: double-ended coil and coil-per-cylinder systems.

In both cases, coil longevity is expected to be significantly longer than that of a coil used with a distributor because individual coils are directly connected to the spark plugs they control. An ignition module, controlled by the PCM, regulates and controls the firing order and ignition timing.

In many EI systems, a crank sensor located at the front of the crankshaft is used to trigger the ignition system. When a distributor is used in the ignition system, the distributor drive gear, shaft, and bushings are subject to wear. Worn distributor components cause erratic ignition timing and spark advance, which results in reduced economy and performance plus increased exhaust emissions. Since the distributor is eliminated in EI systems, ignition timing remains more stable over the life of the engine, which means improved economy and performance with reduced emissions.

There are many advantages of a distributorless ignition system over one that uses a distributor. Here are some of the more important ones:

- No moving parts and therefore requires little maintenance.
- It is possible to control the ignition of individual cylinders to meet specific needs.
- The parts have a longer service life and run cooler because they do not need to fire as often.
- Flexibility in mounting location. This is important because of today's smaller engine compartments.
- Reduced radio frequency interference because there is no rotor to cap gap.
- Elimination of a common cause of ignition misfire, the buildup of water and ozone/nitric acid in the distributor cap.
- Elimination of mechanical timing adjustments.
- Places no mechanical load on the engine in order to operate.
- Increased available time for coil saturation.
- Increased time between firings, which allows the coil to cool more.

Double-Ended Coil or Waste Spark Systems

Double-ended or waste spark ignition systems use one ignition coil for two spark plugs **(Figure 24–31)**. Both ends of the coil's secondary side are directly connected to a spark plug, which means that two plugs are ignited at the same time; one is fired on the compression stroke of one cylinder and the other is fired on the exhaust stroke of another.

A four-cylinder engine has two ignition coils, a six-cylinder has three, and an eight-cylinder has four. The computer, ignition module, and various sensors combine to control spark timing.

The computer collects and processes information to determine the ideal amount of spark advance for the operating conditions. The ignition module uses crank/cam sensor data to control the timing of the primary circuit in the coils **(Figure 24–32)**. Remember that there is more than one coil in a distributorless ignition system. The ignition module synchronizes the coils' firing sequence in relation to crankshaft position and firing order of the engine. Therefore, the ignition module takes the place of the distributor.

Primary current is controlled by transistors in the control module. There is one switching transistor for each ignition coil in the system. The transistors complete the ground circuit for the primary, thereby allowing for a dwell period. When primary current flow is interrupted,

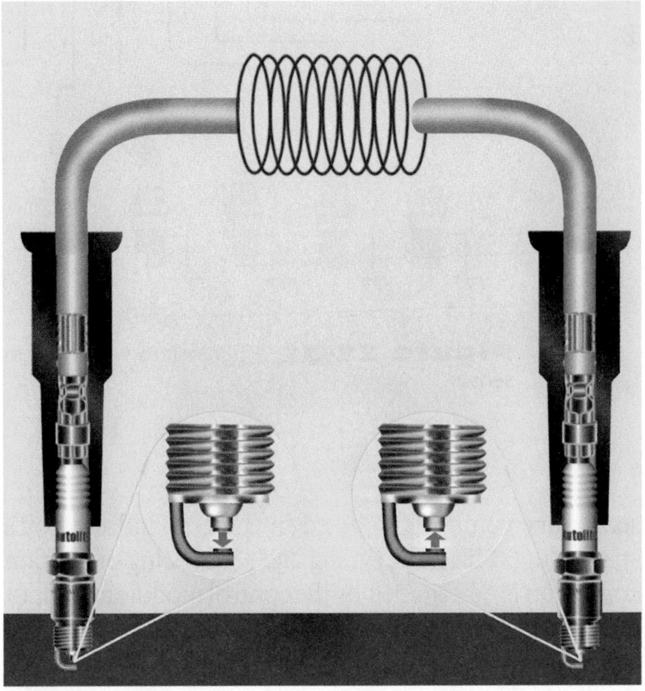

Figure 24–31 An EI system with a double-ended coil.
Courtesy of Autolite

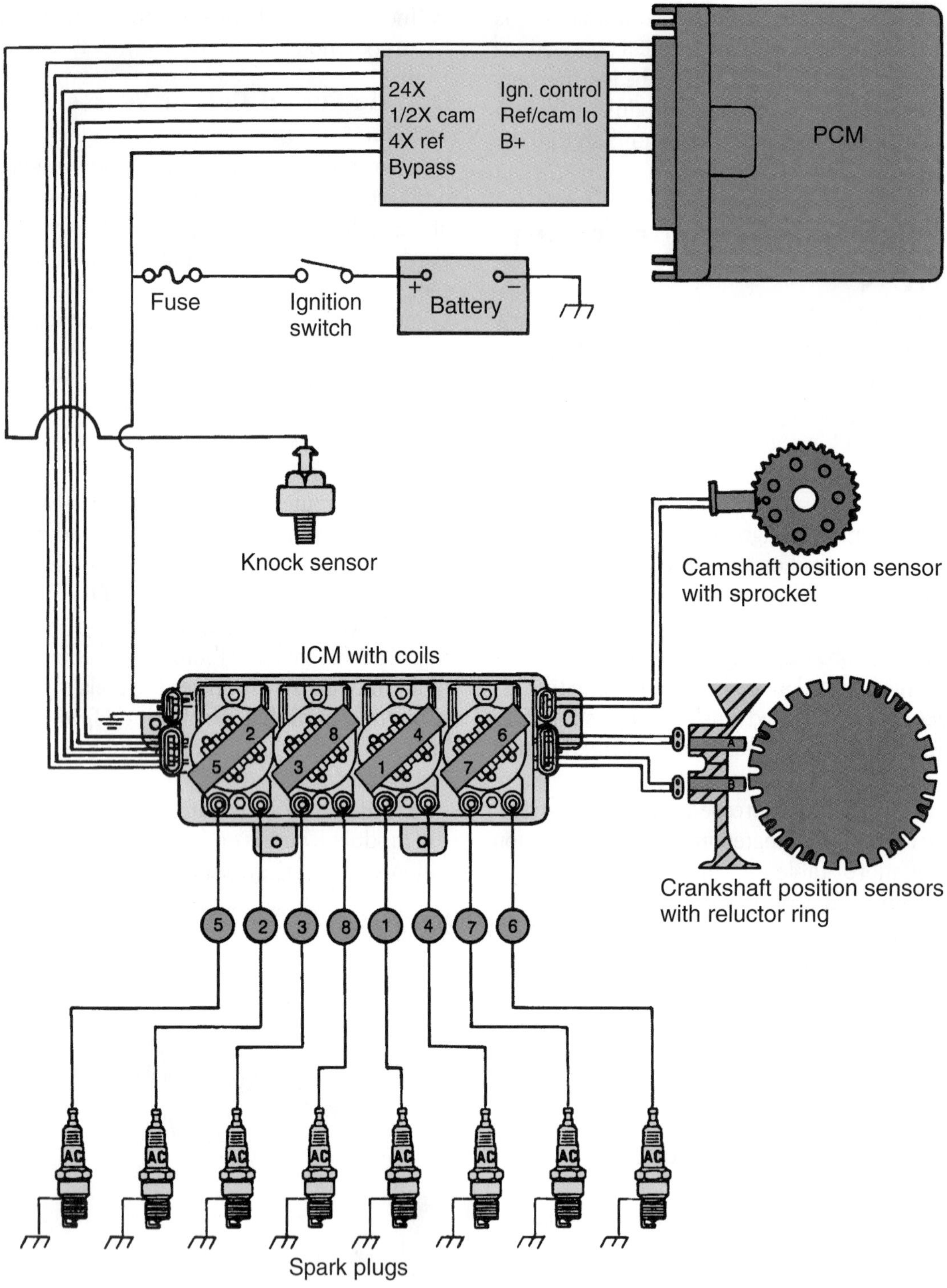

Figure 24-32 An EI system with two crankshaft position sensors and one camshaft position sensor.

secondary voltage is induced in the coil and the coil's spark plug(s) fire. The timing and sequencing of ignition coil action is determined by the control module and input from a triggering device.

The control module is also responsible for limiting the dwell time. In EI systems there is time between plug firings to saturate the coil. Achieving maximum current flow through the coil is great if the system needs the high voltage that may be available. However, if the high voltage is not needed, the high current is not needed and the heat it produces is not desired. Therefore, the control module is programmed to only allow total coil saturation when the very high voltage is needed or the need for it is anticipated.

The ignition module also adjusts spark timing below 400 rpm (for starting) and when the vehicle's control computer bypass circuit becomes open or grounded. Depending on the exact EI system, the ignition coils can be serviced as a complete unit or separately. The coil assembly is typically called a **coil pack** and is comprised of two or more individual coils.

Double-ended coil systems are based on the **waste spark** method of spark distribution. Each end of the coil's secondary winding is attached to a spark plug. Each coil is connected to a pair of spark plugs in cylinders whose pistons rise and fall together. When the field collapses in the coil, voltage is sent to both spark plugs attached to the coil. In all V6s, the paired cylinders are 1 and 4, 2 and 5, and 3 and 6 (or 4 and 1 and 3 and 2 on 4-cylinder engines). With this arrangement, one cylinder of each pair is on its compression stroke while the other is on the exhaust stroke. Both cylinders get spark simultaneously, but only one spark generates power, while the other is wasted out the exhaust. During the next revolution, the roles are reversed.

Due to the way the secondary coils are wired, when the induced voltage cuts across the primary and secondary windings of the coil, one plug fires in the normal direction—positive center electrode to negative side electrode—and the other plug fires just the reverse side to center electrode **(Figure 24–33)**. Both plugs fire simultaneously, completing the series circuit. Each plug always fires the same way on both the exhaust and compression strokes.

The coil is able to overcome the increased voltage requirements caused by reversed polarity and still fire two plugs simultaneously because each coil is capable of producing up to 100,000 volts. There is very little resistance across the plug gap on exhaust, so the plug requires very little voltage to fire, thereby providing its mate (the plug that is on compression) with plenty of available voltage.

Figure 24–34 shows a waste spark system in which the coils are mounted directly over the spark plugs so no wiring between the coils and plugs is necessary. On other systems, the coil packs are mounted remote from the spark plugs. High-tension secondary wires carry high-voltage current from the coils to the plugs.

Some EI systems use the waste spark method of firing but only have one secondary wire coming off each ignition coil. In these systems **(Figure 24–35)**, one spark plug is connected directly to the ignition coil and the companion spark plug is connected to the coil by a high-tension cable.

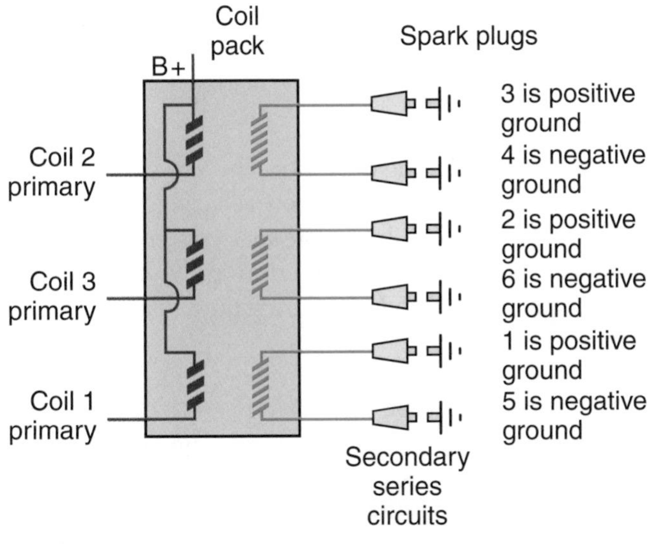

Figure 24-33 The polarity of spark plugs in an EI system. *Courtesy of Ford Motor Company*

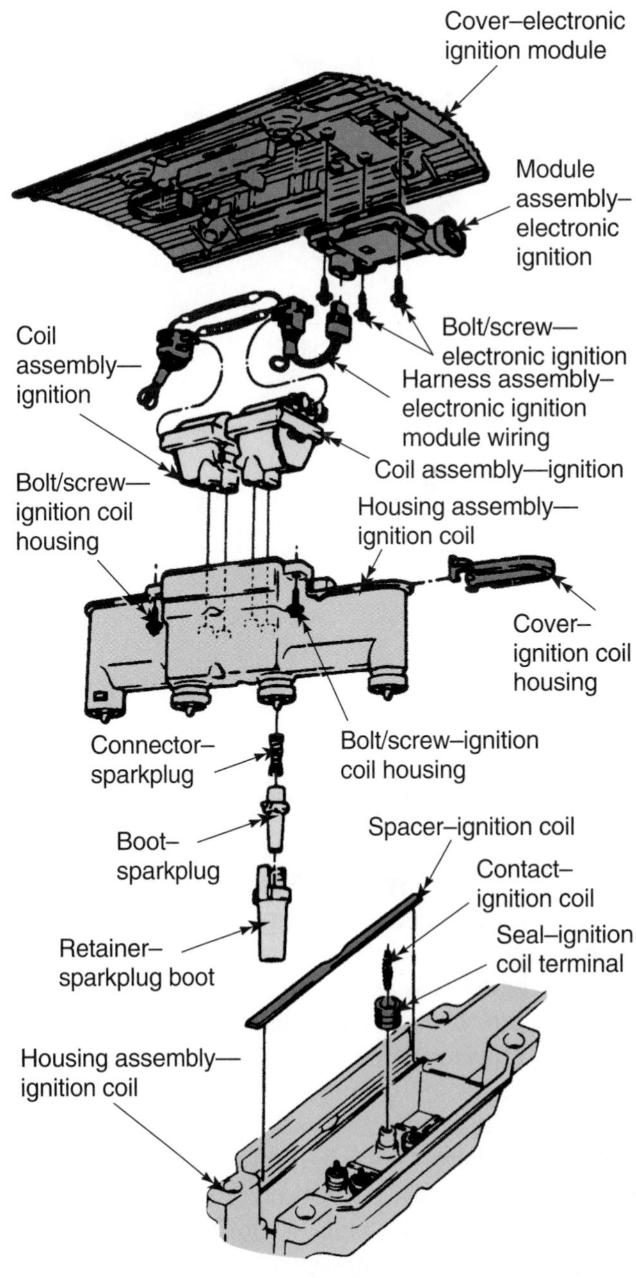

Figure 24-34 A cableless EI system.

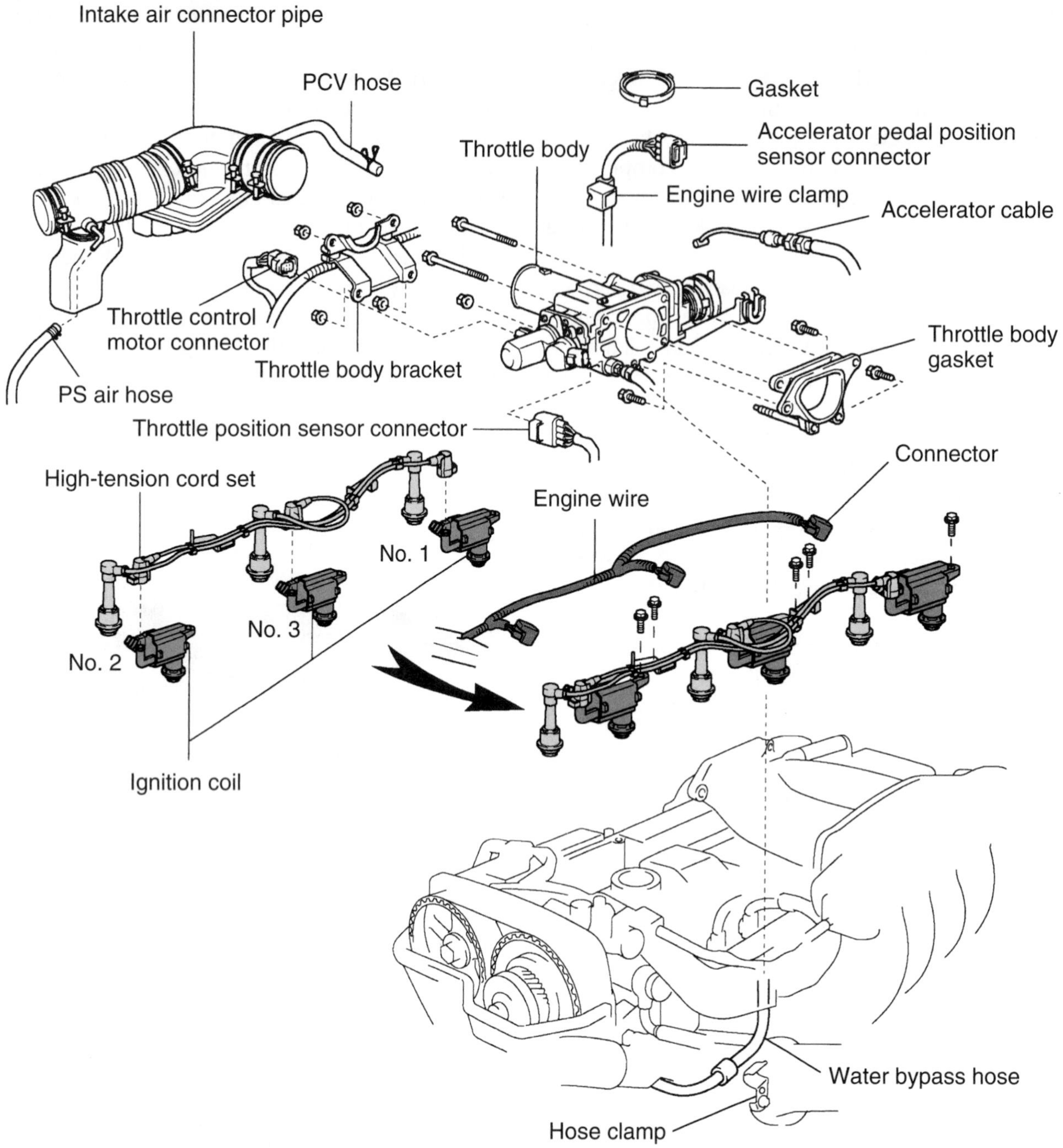

Figure 24-35 A six-cylinder engine with three coils and three spark plug wires. *Reprinted with permission*

Twin Spark Plug Systems Normal engines have one spark plug per cylinder, but some have two. One spark plug is located on the intake side of the combustion chamber and the other is at the exhaust side. When ignition takes place in two locations within the combustion chamber, more efficient combustion and cleaner emissions are possible. The result is improved combustion efficiency, which leads to more power and cleaner emissions.

Two coil packs are used, one for the intake side and the other for the plugs on the exhaust side. These systems are called **dual** or **twin plug** systems **(Figure 24–36)**.

Some engines fire only one plug per cylinder during starting. The additional plug fires once the engine is running. During dual plug operation, the two coil packs are synchronized so the two plugs of each cylinder fire at the same time. Therefore, in a waste spark system, four spark plugs are fired at a time: two during the compression stroke of a cylinder and two during the exhaust stroke of another cylinder.

Coil-Per-Cylinder Ignition

The operation of a coil-per-cylinder ignition system is basically the same as that of any other ignition system. By definition, these systems have an individual coil for each spark plug. There are two different designs of coil-per-cylinder systems used today: the **coil-on-plug (COP)** and

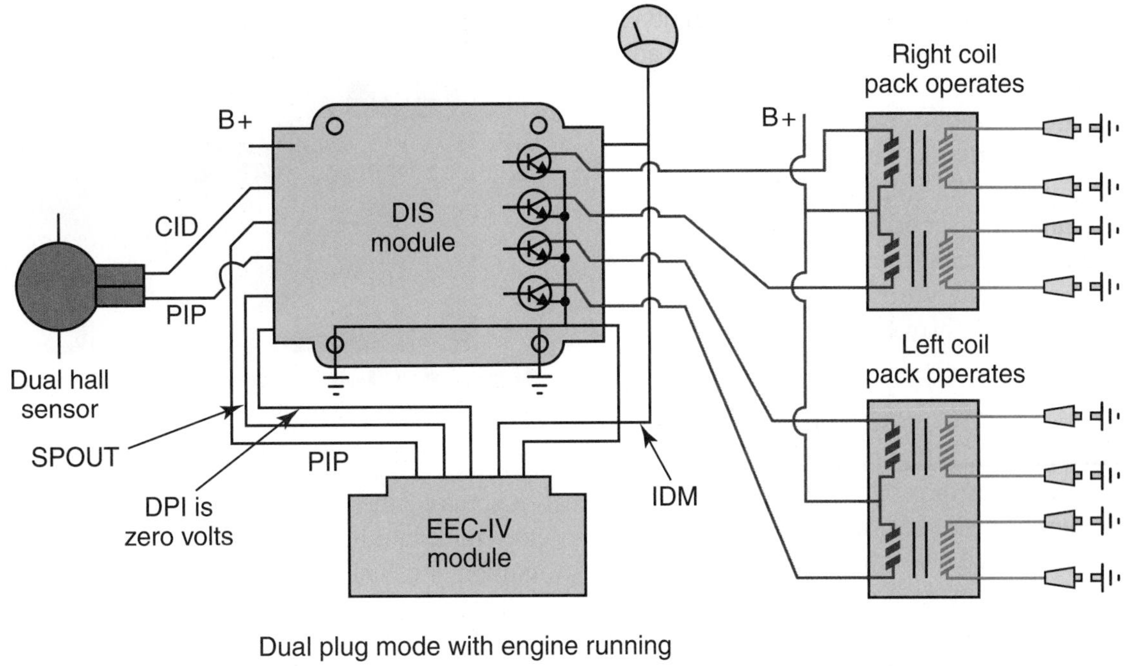

Figure 24–36 A dual plug system for a four-cylinder engine. *Courtesy of Ford Motor Company*

the separate coil. COP systems rely on a single assembly of an ignition coil and spark plug **(Figure 24–37)**. In these systems, the spark plug is directly attached to the coil and there is no spark plug wire.

The separate coil system may also be called the coil-by-plug or coil-near-plug ignition system **(Figure 24–38)**. These systems have individual coils mounted near the plugs and use a short secondary plug wire to connect the coil to the plug. These engines do not use COP ignitions because the location of the spark plug does not allow enough room to mount individual coils over the plugs, or the plugs are too close to the exhaust manifold.

Having one coil for each spark plug allows for more time between each firing, which increases the life of the coil by allowing it to cool. In addition, it also allows for more saturation time, which increases the coil's voltage output at high engine speeds. The increased output makes the coils more effective with lean fuel mixtures, which require higher firing voltages.

On some systems, there is also a coil capacitor for each bank of coils for additional radio noise suppression.

COP systems are controlled in the same way as a double-ended coil system, except the coils are not connected to two spark plugs and therefore there is no wasted spark. In a typical coil-per-cylinder system, a crankshaft position sensor provides a basic timing signal. This signal is sent to the PCM. The PCM is programmed with the fir-

Figure 24–37 Note the three COPs that are visible in this photo of a V-6 engine. *Reprinted with permission*

Figure 24–38 This engine has separate coils for each cylinder. Notice the short spark plug wires. *Courtesy of General Motors Corporation*

ing order for the engine and determines which ignition coil should be turned on or off. Some engines require an additional timing signal from the camshaft position sensor.

The advantage of a coil-per-cylinder system is that the cylinders can be controlled individually and can be changed on a cylinder-by-cylinder basis for maximum performance and to respond to knock sensor signals. Other advantages of a coil-per-cylinder system are that all of the engine's spark plugs fire in the same direction and coil failure will affect only one cylinder.

Coil-on-Plug Ignition The true difference between COP and other ignition systems is that each coil is mounted directly atop the spark plug **(Figure 24–39)** so that the voltage from the coil goes directly to the plug's electrodes without passing through a distributor or plug wire. It is a direct connection that delivers the hottest spark possible. COP ignition systems do not use spark plug wires, so there are no wires to come loose, burn, leak current, break down, or be replaced. Eliminating plug wires also reduces RFI and electromagnetic interference (EMI), which can interfere with computer systems. However, the absence of plug wires also means the coils need to be removed and reconnected with adapters or plug wires to test for spark, to connect a pickup for an ignition scope, or to perform a manual cylinder power balance test.

EI System Operation

From a general operating standpoint, most distributorless ignition systems are similar. However, there are variations in the way different distributorless systems obtain a timing reference in regard to crankshaft and camshaft position.

Figure 24-39 A coil-on-plug assembly. *Courtesy of Visteon Corporation*

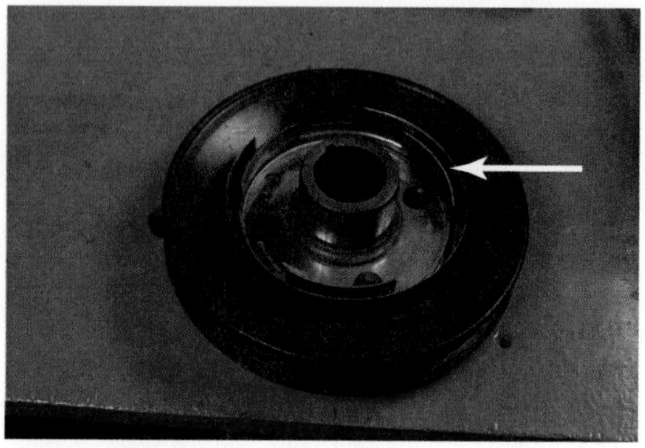

Figure 24-40 A crankshaft pulley for a six-cylinder engine has three interrupter rings.

Some engines use separate Hall-effect sensors to monitor crankshaft and camshaft position for the control of ignition and fuel injection firing orders. The crankshaft pulley has interrupter rings that are equal in number to half of the cylinders of the engine **(Figure 24–40)**. The resultant signal informs the PCM as to when to fire the plugs. The camshaft sensor helps the computer determine when the number 1 piston is at TDC on the compression stroke.

Defining the different types of EI systems used by manufacturers focuses on the location and type of sensors used. There are other differences, such as the construction of the coil pack, wherein some are a sealed assembly and others have individually mounted ignition coils. Some EI systems have a camshaft sensor mounted in the opening where the distributor was mounted. The camshaft sensor ring has one notch and produces a leading edge and trailing edge signal once per camshaft revolution. These systems also use a crankshaft sensor. Both the camshaft and crankshaft sensors are Hall-effect sensors.

Locating the cam sensor in the opening previously occupied by the distributor merely takes advantage of the bore and gear that was already present. Seeing that the distributor was driven at camshaft speed, driving a camshaft position sensor by the same mechanism just made sense **(Figure 24–41)**. This modification really made sense when older engine designs were modified for distributorless ignition.

As the crankshaft rotates and the interrupter passes in and out of the Hall-effect switch, the switch turns the module reference voltage on and off. The three signals are identical, and the control module cannot distinguish which of these signals to assign to a particular coil. The signal from the cam sensor gives the module the information it needs to assign the signals from the crankshaft sensors to the appropriate coils **(Figure 24–42)**. The camshaft sensor synchronizes the crankshaft sensor signals with the position of the number 1 cylinder. From there the module can energize the coils according to the

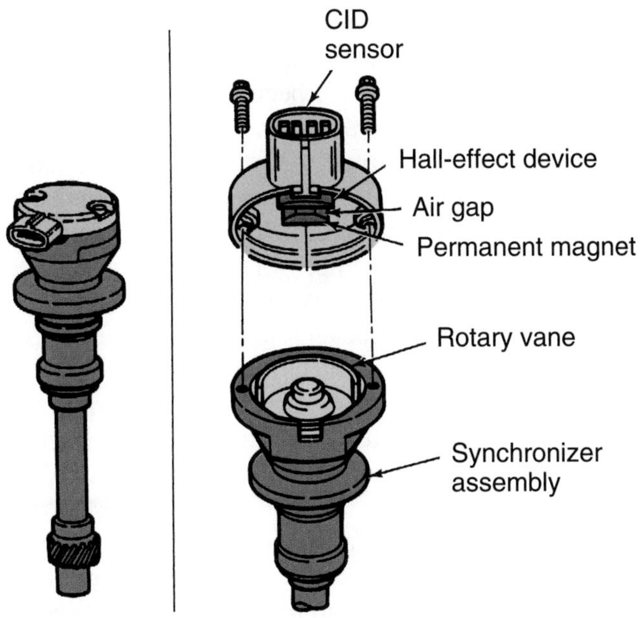

Figure 24-41 A camshaft sensor assembly designed to fit into the distributor bore. *Courtesy of Ford Motor Company*

firing order of the engine. Once the engine has started, the camshaft signal serves no purpose.

Some systems have the camshaft sensor mounted in the front of the timing chain cover. A magnet on the camshaft gear rotates past the inner end of the camshaft sensor and produces a signal for each camshaft revolution. General Motors' 3.8 L nonturbocharged sequential fuel ignition (SFI) V6 engine has a firing order of 1-6-5-4-3-2. Spark plugs 1-4, 6-3, and 5-2 are paired together on the coil assembly. When a trailing edge camshaft sensor signal is received during initial starting, the coil module prepares to fire the coil connected to spark plugs 5-2. After the camshaft sensor signal is received, the next trailing edge crankshaft sensor signal turns on the primary circuit of the 5-2 coil, and the next leading edge crankshaft sensor signal informs the coil module to open the primary circuit of the 5-2 coil **(Figure 24–43)**. When this coil fires, one of these cylinders is always on the compression stroke and the other cylinder is on the exhaust stroke. After the 5-2 coil firing, the coil module fires the 1-4 coil and the 6-3 coil in sequence. This firing sequence provides the correct firing order.

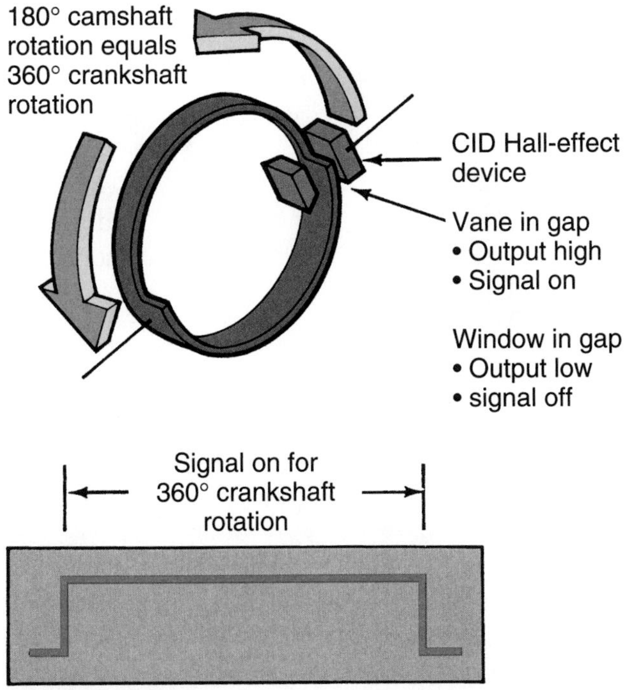

Figure 24-42 Camshaft sensor operation and the resulting signal. *Courtesy of Ford Motor Company*

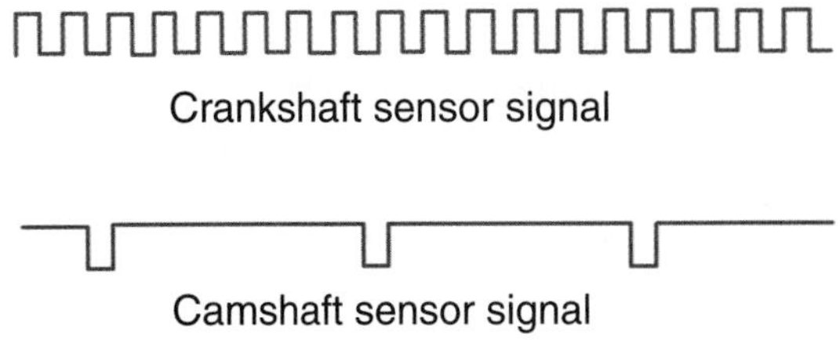

Figure 24-43 The relationship between crankshaft and camshaft signals on a GM 3.8L SFI engine.

On an SFI engine, the PCM grounds each injector individually. The cam sensor signal is also used for injector sequencing. This cam sensor signal is sent from the cam sensor through the coil module to the PCM. The PCM grounds each injector in the intake port when the piston for that cylinder is at 70 degrees before TDC on the intake stroke.

When a crankshaft sensor failure occurs, the engine does not start. If the camshaft sensor signal becomes defective with the engine running, the engine continues to run, but the PCM reverts to multiport fuel injection without the camshaft signal information. Under this condition, engine performance and economy decrease and emission levels may increase. When an engine with a defective cam sensor is shut off, it will not restart.

Other systems use a dual crankshaft sensor located behind the crankshaft pulley. When this type of sensor is used, there are two interrupter rings on the back of the pulley that rotate through the Hall-effect switches at the dual crankshaft sensor. The inner ring with three equally spaced blades rotates through the inner Hall-effect switch, whereas the outer ring with one opening rotates through the outer Hall-effect switch.

In this dual sensor, the inner sensor provides three leading-edge signals and the outer sensor produces one leading-edge signal during one complete revolution of the crankshaft. The outer sensor is referred to as a synchronizer (sync) sensor. The signal from this sensor informs the coil module regarding crankshaft position. The sync sensor signal occurs once per crankshaft revolution, and this signal is synchronized with the inner crankshaft sensor signal to fire the 6-3 coil.

The examples given so far depend on two revolutions of the crankshaft to inform the PCM as to which number cylinder is ready. These systems are referred to as slow-start systems because the engine must crank through two crankshaft revolutions before ignition begins.

The **Fast-Start** EI system used in General Motors' Northstar system uses two crankshaft position sensors **(Figure 24–44)**. A reluctor ring with twenty-four evenly spaced notches and eight unevenly spaced notches is cast onto the center of the crankshaft. When the reluctor ring rotates past the magnetic-type sensors, each sensor produces thirty-two high- and low-voltage signals per crankshaft revolution. The A sensor is positioned in the upper crankcase, and the B sensor is positioned in the lower crankcase. Since the A sensor is above the B sensor, the signal from the A sensor occurs 27 degrees before the B sensor signal.

The signals from the two sensors are sent to the ignition control module. This module counts the number of signals from one of the sensors that are between the other sensor signals to sequence the ignition coils properly. This allows the ignition system to begin firing the spark plugs within 180 degrees of crankshaft rotation while starting the engine. This system allows for much quicker starting than other EI systems which require the crankshaft to rotate one or two times before the coils are sequenced.

The camshaft position sensor is located in the rear cylinder bank in front of the exhaust camshaft sprocket. A reluctor pin in the sprocket rotates past the sensor, and this sensor produces one high- and one low-voltage signal every camshaft revolution, or every two crankshaft revolutions. The PCM uses the camshaft position sensor signal to sequence the injectors properly.

Another example of a fast-start system also uses a dual crankshaft sensor at the front of the crankshaft. The cam sensor is mounted in the timing gear cover. Two Hall-effect switches are located in the dual crankshaft sensor, and two matching interrupter rings are attached to the

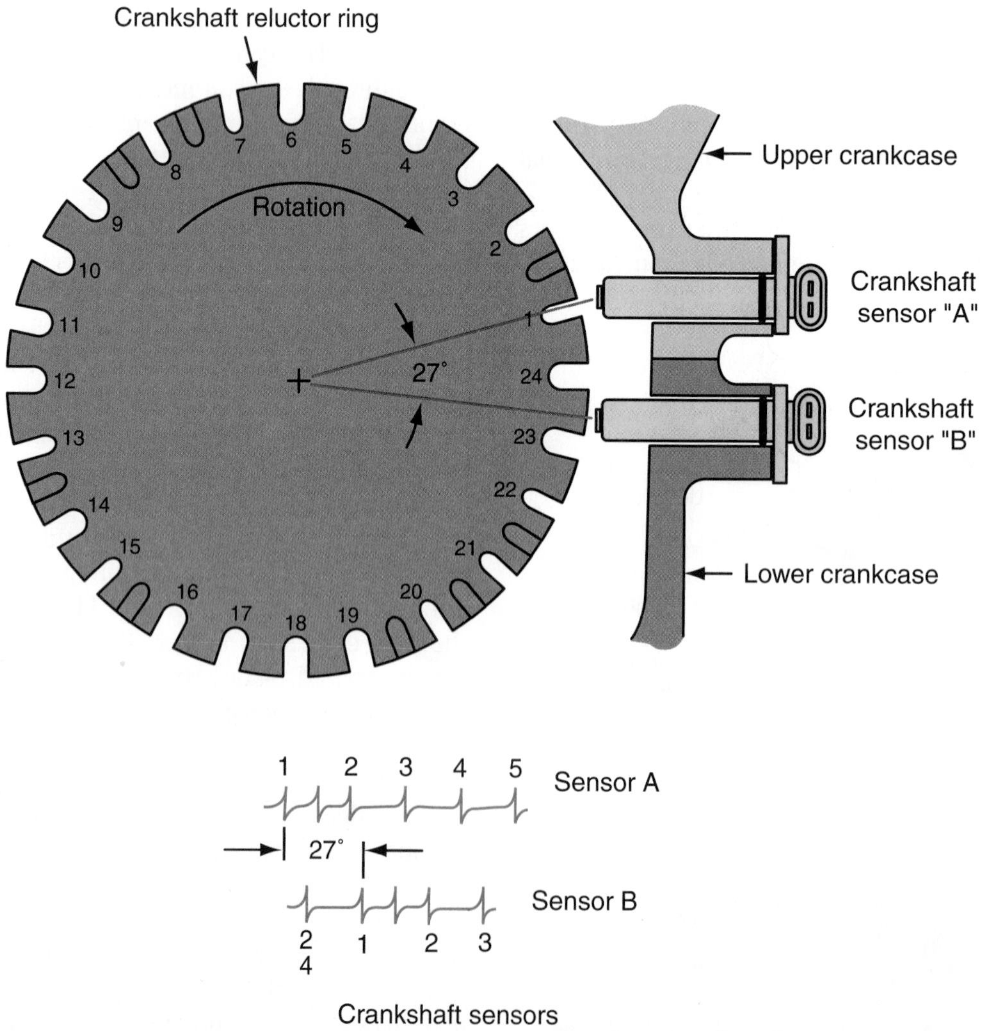

Figure 24-44 A and B crankshaft sensors in a Northstar engine.

back of the crankshaft pulley. The inner ring on the crankshaft pulley has three blades of unequal lengths with unequal spaces between the blades. On the outer ring, there are eighteen blades of equal length with equal spaces between the blades. The signal from the inner Hall-effect switch is referred to as the 3X signal, while the signal from the outer Hall-effect switch is called the 18X signal. These signals are sent from the dual crankshaft sensor to the coil module.

The coil module knows which coil to fire from the number of 18X signals received during each 3X window rotation **(Figure 24–45)**. For example, when two 18X signals are received, the coil module is programmed to sequence coil 3-6 next in the firing sequence. Within 120 degrees of crankshaft rotation, the coil module can identify which coil to sequence and thus start firing the spark plugs. Therefore, the system fires the spark plugs with less crankshaft rotation during initial starting than the previous slow-start systems.

Once the engine is running, the system switches to the EST mode. The PCM uses the 18X signal for crankshaft position and speed information. The 18X signal may be referred to as a high-resolution signal. If the 18X signal is not present, the engine will not start. When the 3X signal fails with the engine running, the engine continues to run but refuses to restart.

In this system, the cam sensor signal is used for injector sequencing, but it is not required for coil sequencing. If the cam sensor signal fails, the PCM logic begins sequencing the injectors after two cranking revolutions. There is a one-in-six chance that the PCM logic will ground the injectors in the normal sequence. When the PCM logic does not ground the injectors in the normal sequence, the engine hesitates on acceleration.

Finally, some engines use a magnetic pulse generator. The timing wheel is cast on the crankshaft and has machined slots on it. If the engine is a six-cylinder, there will be seven slots, six of which are spaced exactly 60 degrees apart. The seventh notch is located 10 degrees from the number-six notch and is used to synchronize the coil firing sequence in relation to crankshaft position **(Figure 24–46)**. The same triggering wheel can be and is used on four-cylinder engines. The computer only needs to be programmed to interpret the signals differently than on a six-cylinder.

The magnetic sensor, which protrudes into the side of the block, generates a small AC voltage each time one of the machined slots passes by. By counting the time between pulses, the ignition module picks out the unevenly spaced seventh slot, which starts the calculation of the ignition coil sequencing. Once its counting is synchronized with the crankshaft, the module is programmed to accept the AC voltage signals of the select notches for firing purposes.

Ford uses a similar system; however, the reluctor ring has many more slots. The crankshaft sensor for their 4.6 L V8 engine is a variable reluctance sensor that is triggered by a 36-minus-1 (or 35-) tooth trigger wheel located inside the front cover of the engine **(Figure 24–47)**. The sensor provides two types of information: crankshaft position and engine speed.

The trigger wheel has a tooth every 10 degrees, with one tooth missing. When the part of the wheel that is missing a tooth passes by the sensor, there is a longer-

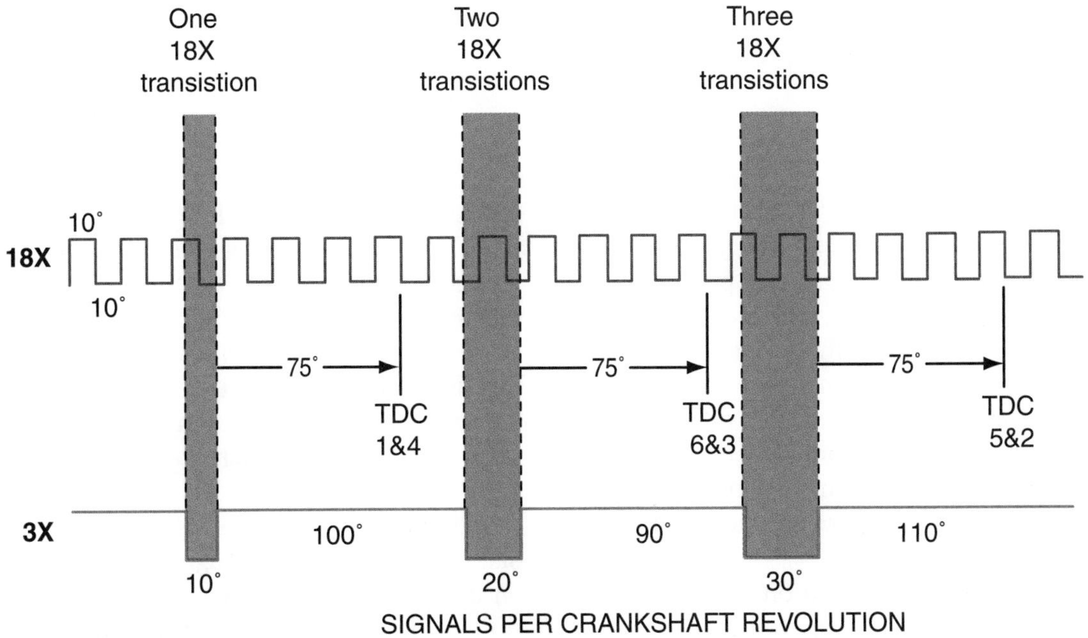

Figure 24-45 3X and 18X crankshaft signals.

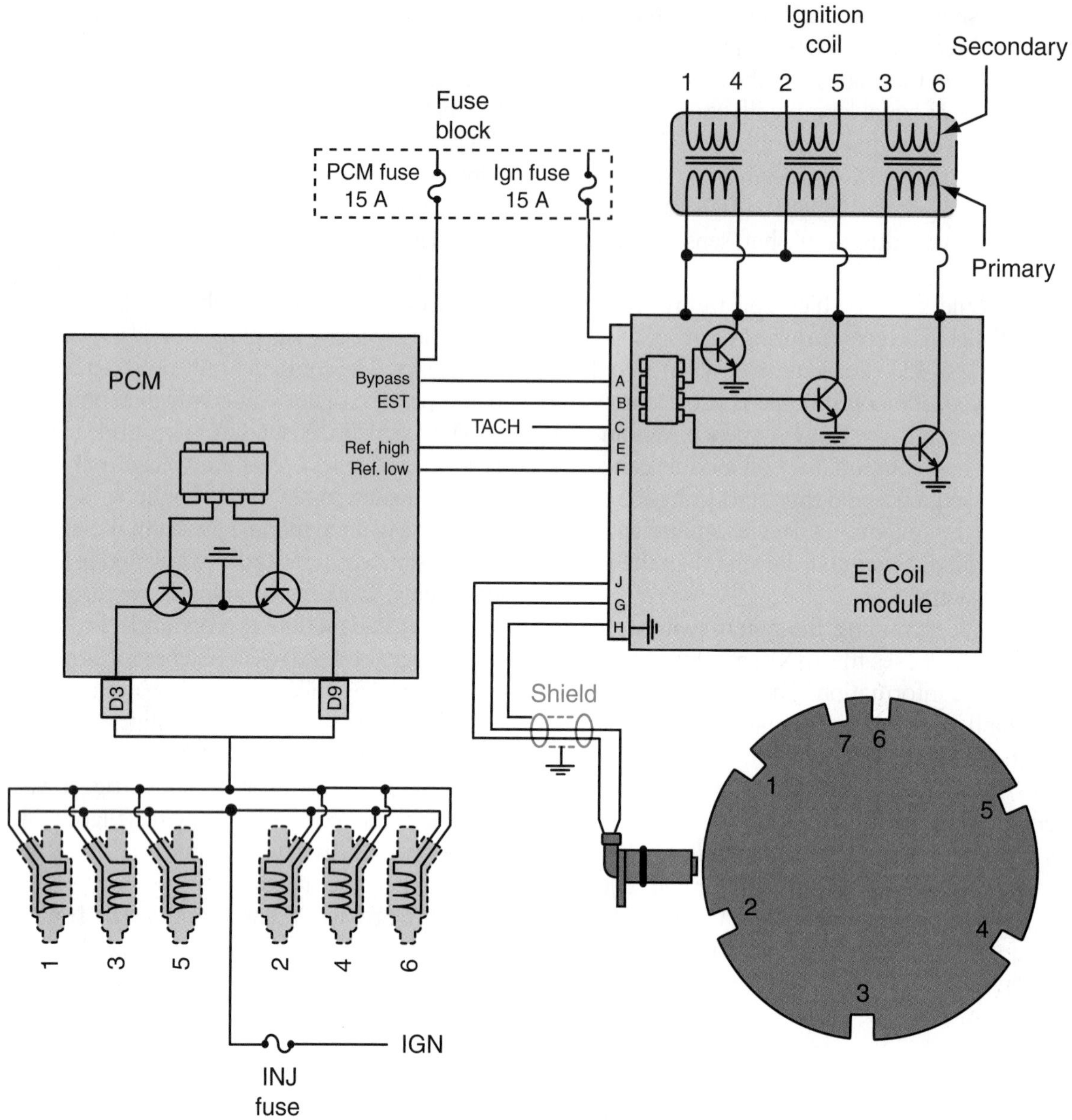

Figure 24-46 Schematic of an EI system with a magnetic pulse generator-type crankshaft sensor. Note the notches on the crankshaft-timing wheel.

than-normal pause between signals from the sensor. The ignition control module recognizes this and is able to identify this long pause as the location of piston 1.

Chrysler also uses a similar system. However, it uses a different number of teeth on the reluctor, a camshaft sensor, and a camshaft reluctor; therefore, the signals received by the control module are also different. The crankshaft timing sensor is mounted in an opening in the transaxle bell housing. The inner end of this sensor is positioned near a series of notches and slots that are integral with the engine's flywheel.

A group of four slots is located on the flywheel for each pair of engine cylinders. Thus, a total of twelve slots are positioned around the flywheel. When the slots rotate past the crankshaft timing sensor, the voltage signal from the sensor changes from 0 volt to 5 volts. This varying voltage signal informs the PCM regarding crankshaft position and speed. The PCM calculates spark advance from this signal. The PCM also uses the crankshaft timing sensor signal along with other inputs to determine air/fuel ratio. Base timing is determined by the signal from the last slot in each group of slots.

The camshaft reference sensor is mounted in the top of the timing-gear cover **(Figure 24–48)**. A notched ring on the camshaft gear has two single slots, two double slots, and a triple slot. When a notch rotates past the

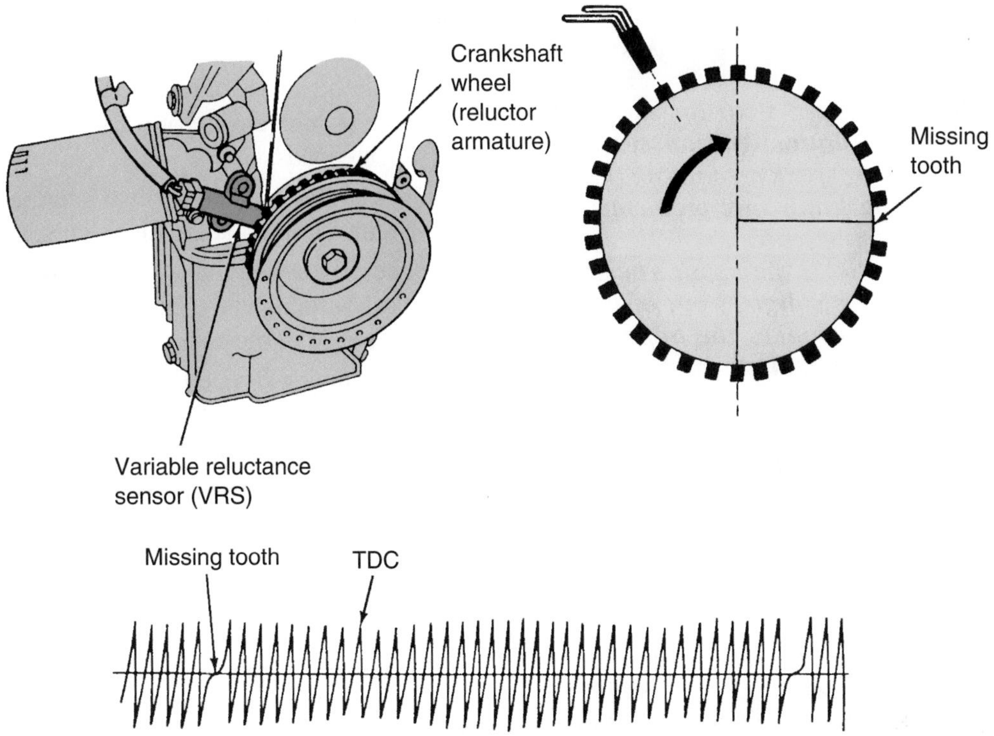

Figure 24-47 Sensor activity to monitor engine speed and crankshaft position, as well as the location of the number 1 piston. *Courtesy of Ford Motor Company*

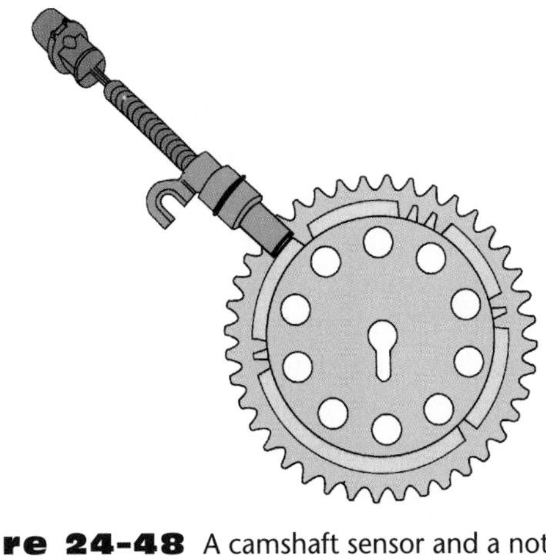

Figure 24-48 A camshaft sensor and a notched cam gear. *Courtesy of DaimlerChrysler Corporation*

camshaft reference sensor, the signal from the sensor changes from 0 volt to 5 volts. The single, double, and triple notches provide different voltage signals. These signals are sent to the PCM. The PCM determines the exact camshaft and crankshaft position from the camshaft reference sensor signals, and the PCM uses these signals to sequence the coil primary windings and each pair of injectors at the correct instant.

The development and spreading popularity of EI is the result of reduced emissions, improved fuel economy, and increased component reliability brought about by these systems.

EI also offers advantages in production costs and maintenance considerations. By removing the distributor, the manufacturers realize substantial savings in ignition parts and related machining costs. Also, by eliminating the distributor, they do away with cracked caps, eroded carbon buttons, burned-through rotors, moisture misfiring, base timing adjustments, and the like.

CAUTION!

Since EI systems have considerably higher maximum secondary voltage compared to distributor-type ignition systems, greater electrical shocks are obtained from EI systems. Although such shocks may not be directly harmful to the human body, they may cause you to jump or react suddenly, which could result in personal injury. For example, when you jump suddenly as a result of an EI electrical shock, you may hit your head on the vehicle hood or push your hand into a rotating cooling fan.

CASE STUDY

***A** vehicle equipped with an early electronic distributor ignition system is experiencing spark detonation (knocking) and erratic spark advance problems. The vehicle has 82,000 miles on it.*

The technician checks the engine's base timing and finds it to be 5 degrees out of adjustment. The technician makes the adjustment, but it does not seem to hold steady. In fact, the problem still occurs on the test drive made immediately after the timing adjustment is made.

The technician then removes the distributor for closer inspection. The centrifugal advance mechanism appears to be in good order, but the technician notices shiny worn areas on the tangs of the distributor shaft's drive coupling. Wear on the tangs could mean the distributor shaft is not in proper mesh with the camshaft. The technician replaces the worn drive coupling and reinstalls the distributor. After resetting initial timing, the problem of erratic advance disappears.

KEY TERMS

Air gap
ATDC
Ballast resistor
Base ignition timing
BTDC
Coil-on-plug (COP)
Coil pack
Direct ignition system (DIS)
Distributor ignition (DI)
Distributorless ignition system (DIS)
Distributor
Dual plug
Dwell
Electronic ignition (EI)
Electronic spark timing (EST)
Fast-Start
Heat range
High energy ignition (HEI)
Ignition timing
Inductive reluctance
Lookup tables
Misfiring
Photoelectric sensor
Preignition
Primary circuit
Pulse transformer
Reach
Reactance
Rotor
Secondary circuit
Strategy
Swirling
Thick film integrated (TFI)
Timer core
Twin plug
Waste spark

SUMMARY

- The ignition system supplies high voltage to the spark plugs to ignite the air/fuel mixture in the combustion chambers.
- The arrival of the spark is timed to coincide with the compression stroke of the piston. This basic timing can be advanced or retarded under certain conditions, such as high engine rpm or extremely light or heavy engine loads.
- The ignition system has two interconnected electrical circuits: a primary circuit and a secondary circuit.
- The primary circuit supplies low voltage to the primary winding of the ignition coil. This creates a magnetic field in the coil.
- A switching device interrupts primary current flow, collapsing the magnetic field and creating a high-voltage surge in the ignition coil secondary winding.
- The switching device used in electronically controlled systems is an NPN transistor.
- The secondary circuit carries high-voltage surges to the spark plugs. On some systems, the circuit runs from the ignition coil, through a distributor, to the spark plugs.
- The distributor may house the switching device plus centrifugal or vacuum timing advance mechanisms. Some systems locate the switching device outside the distributor housing.
- Ignition timing is directly related to the position of the crankshaft. Magnetic pulse generators and Hall-effect sensors are the most widely used engine position sensors. They generate an electrical signal at certain times during crankshaft rotation. This signal triggers the electronic switching device to control ignition timing.
- Distributors are seldom found on today's engines. Nearly all of today's engines are equipped with an EI system for which there are primarily two different designs: double-ended coil and coil-per-cylinder.
- Computer-controlled ignition eliminates centrifugal and vacuum timing mechanisms. The computer receives input from numerous sensors. Based on this data, the computer determines the optimum firing time and signals an ignition module to activate the secondary circuit at the precise time needed.
- In some EI systems, the camshaft sensor signal informs the computer when to sequence the coils and fuel injectors.
- In some EI systems, the crankshaft sensor signal provides engine speed and crankshaft position information to the computer.
- Some EI systems are called fast-start systems because the spark plugs begin firing within 120 degrees of crankshaft rotation.
- Some EI systems have a combined crankshaft and SYNC sensor at the front of the crankshaft.
- Some EI systems may be called slow-start systems because as many as two crankshaft revolutions are required before the ignition system begins firing.

TECH MANUAL

The following procedures are included in Chapters 24/25 of the *Tech Manual* that accompanies this book:

1. Scope testing an ignition system.
2. Testing an ignition coil.
3. Testing individual components.
4. Setting ignition timing.
5. Visually inspecting an EI system.

REVIEW QUESTIONS

1. Explain how the voltage is induced in the distributor pickup coil as the reluctor high point approaches alignment with the pickup coil.
2. Explain why dwell time is important to ignition system operation.
3. Name the engine operating conditions that affect ignition timing requirements.
4. Explain how the plugs fire in a two-plugs-per-coil EI system.
5. Explain the components and operation of a magnetic pulse generator.
6. What happens when the low-voltage current flow in the coil primary winding is interrupted by the switching device?
   a. The magnetic field collapses.
   b. A high-voltage surge is induced in the coil secondary winding.
   c. Both a and b.
   d. Neither a nor b.
7. Which of the following is a function of all ignition systems?
   a. to generate sufficient voltage to force a spark across the spark plug gap
   b. to time the arrival of the spark to coincide with the movement of the engine's pistons
   c. to vary the spark arrival time based on varying operating conditions
   d. all of the above
8. Reach, heat range, and air gap are all characteristics that affect the performance of which ignition system component?
   a. ignition coils **c.** spark plugs
   b. ignition cables **d.** breaker points
9. Technician A says a magnetic pulse generator is equipped with a permanent magnet. Technician B says a Hall-effect switch is equipped with a permanent magnet. Who is correct?
   a. Technician A **c.** Both A and B
   b. Technician B **d.** Neither A nor B
10. While discussing ignition systems, Technician A says an ignition system must supply high voltage surges to the spark plugs. Technician B says the system must maintain the spark long enough to burn all of the air/fuel mixture in the cylinder. Who is correct?
    a. Technician A **c.** Both A and B
    b. Technician B **d.** Neither A nor B
11. While discussing ignition timing requirements, Technician A says more advanced timing is desired when the engine is under a heavy load. Technician B says more advanced timing is desired when the engine is running at high engine speeds. Who is correct?
    a. Technician A **c.** Both A and B
    b. Technician B **d.** Neither A nor B
12. While discussing secondary voltage, Technician A says the normal required secondary voltage is higher at idle speed than at wide-open throttle conditions. Technician B says the maximum available secondary voltage must always exceed the normally required secondary voltage. Who is correct?
    a. Technician A **c.** Both A and B
    b. Technician B **d.** Neither A nor B
13. Technician A says an ignition system must generate sufficient voltage to force a spark across the spark plug gap. Technician B says the ignition system must time the arrival of the spark to coincide with the movement of the engine's pistons and vary it according to the operating conditions of the engine. Who is correct?

a. Technician A
b. Technician B
c. Both A and B
d. Neither A nor B

14. *True or False?* The spark plug wire is eliminated in all coil-per-cylinder ignition systems to reduce maintenance and the chances of EMI and RFI.

15. Modern ignition cables contain fiber cores that act as a _____ in the secondary circuit to cut down on radio and television interference and reduce spark plug wear.

a. conductor
b. resistor
c. semiconductor
d. heat shield

16. In EI systems using one ignition coil for every two cylinders, Technician A says two plugs fire at the same time, with one wasting the spark on the exhaust stroke. Technician B says one plug fires in the normal direction (center to side electrode) and the other in reversed polarity (side-to-center electrode). Who is correct?

a. Technician A
b. Technician B
c. Both A and B
d. Neither A nor B

17. The magnetic field surrounding the pickup coil in a magnetic pulse generator moves when the ______.

a. reluctor tooth approaches the coil
b. reluctor tooth begins to move away from the pickup coil pole
c. reluctor is aligned with the pickup coil pole
d. both a and b

18. The pickup coil in a magnetic pulse generator does not produce a voltage signal when ______.

a. a reluctor tooth approaches the coil
b. a reluctor tooth is aligned with the coil
c. a reluctor tooth begins to move away from the coil
d. the coil is midway between two reluctor teeth

19. Which type of engine position sensor requires its voltage signal be amplified, inverted, and shaped into a clean square wave signal?

a. magnetic pulse generator
b. metal detection sensor
c. Hall-effect sensor
d. photoelectric sensor

20. Which of the following electronic switching devices has a reluctor with wide shutters rather than teeth?

a. magnetic pulse generator
b. metal detection sensor
c. Hall-effect sensor
d. all of the above

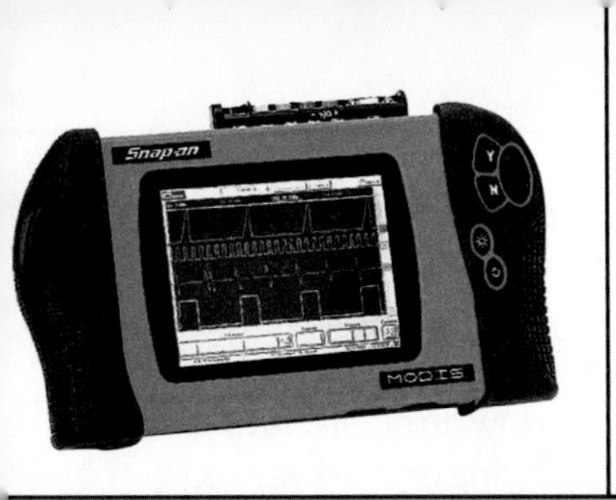

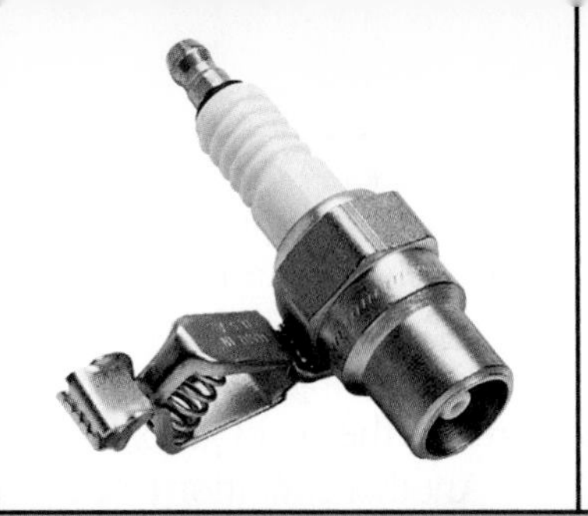

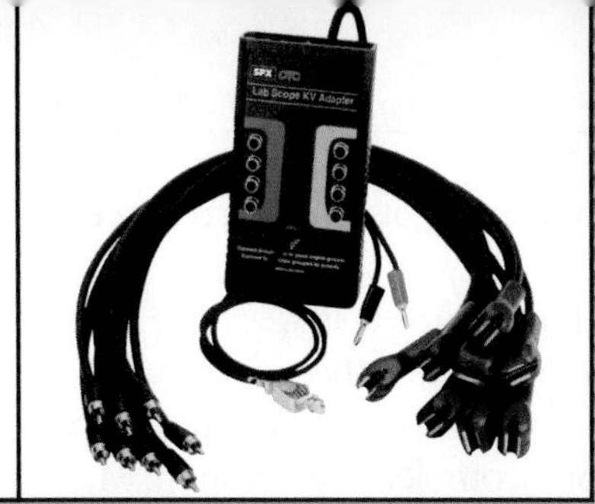

25

IGNITION SYSTEM DIAGNOSIS AND SERVICE

OBJECTIVES

■ Perform a no-start diagnosis and determine the cause of the condition. ■ Determine the cause of an engine misfire. ■ Perform a visual inspection of ignition system components, primary wiring, and secondary wiring to locate obvious trouble areas. ■ Describe what an oscilloscope is, its scales and operating modes, and how it is used in ignition system troubleshooting. ■ Test the components of the primary and secondary ignition circuits. ■ Test individual ignition components using test equipment such as a voltmeter, ohmmeter, and test light. ■ Service and install spark plugs. ■ Describe the effects of incorrect ignition timing. ■ Check and set (when possible) ignition timing. ■ Diagnose engine misfiring on EI-equipped engines.

This chapter concentrates on testing ignition systems and their individual components. It must be stressed, however, that there are many variations in the ignition systems used by auto manufacturers. The tests covered in this chapter are those generally used as basic troubleshooting procedures. Exact test procedures and the ideal troubleshooting sequence will vary among vehicle makers and individual models. Always consult the vehicle's service manual when performing ignition system service.

Two important precautions should be taken during all ignition system tests:

1. Turn the ignition switch off before disconnecting any system wiring.
2. Do not touch any exposed connections while the engine is cranking or running.

COMBUSTION

Although many different things and events can affect combustion in the engine's cylinders, the ignition system has the responsibility for beginning and maintaining the combustion process. Obviously, when combustion does not occur in any of the cylinders, the engine will not run. If combustion occurs in all but one or two cylinders, the engine may start and run but will do neither well. The lack of combustion is not always caused by a problem in the ignition system. Remember that in order to have combustion there must be air, fuel, spark, and compression.

When normal combustion occurs, the burning process moves from the gap of the spark plug across the compressed air/fuel mixture. The movement of this flame front should be rapid and steady, and should end when all of the air/fuel mixture has been involved **(Figure 25–1)**. When normal combustion occurs, the rapidly

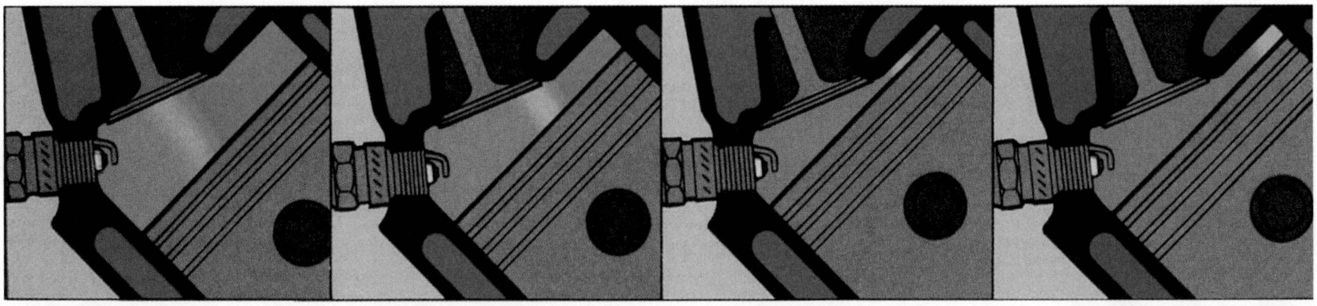

1. Spark occurs 2. Combustion begins 3. Continues rapidly 4. And is completed

Figure 25–1 Normal combustion. *Courtesy of Champion Spark Plug Company*

expanding gases push down on the piston with a powerful but constant force.

When all of the air and fuel in the cylinder have been involved in the combustion process, there is complete combustion. When something prevents complete combustion, there is a misfire or incomplete combustion. Misfires cause a variety of driveability problems, such as a lack of power, poor gas mileage, excessive exhaust emissions, and a rough-running engine. Misfires are not always caused by the ignition system. In order to have complete combustion, the correct amount of air must be mixed with the correct amount of fuel. This mixture must be compressed in a sealed container (cylinder), then shocked by the correct amount of heat by the spark plug.

A spark plug misfires when it has a weak spark or does not fire at all. Misfires can be caused by a fouled spark plug, a bad coil, problems in the primary or secondary ignition circuit, or an incorrect plug gap.

Incomplete combustion is not the only abnormal condition engines may experience; they may also experience detonation. **Detonation** occurs when part of the air/fuel mixture begins to ignite on its own. This results in the collision of two flame fronts **(Figure 25–2)**. One flame front is the normal front moving from the spark plug tip. The other front begins at another point in the combustion chamber. The air/fuel mixture at that point was ignited by heat, not by the spark. The colliding flame fronts cause high-frequency shock waves (heard as a knocking or pinging sound) that could cause physical damage to the pistons, valves, bearings, and spark plugs.

Detonation is usually caused by excessively advanced ignition timing, engine overheating, excessively lean mixtures, or the use of gasoline with too low an octane rating.

Another condition also causes pinging or spark knocking. This condition, called **preignition**, occurs when combustion begins before the spark plug fires **(Figure 25–3)**. Any hot spot within the combustion chamber can cause preignition. Common causes of preignition are incandescent carbon deposits in the combustion chamber, a faulty cooling system, a spark plug that is too hot, poor engine lubrication, and cross firing. Preignition usually leads to detonation; preignition and detonation are two separate events.

Cleaning Carbon Deposits

A buildup of carbon on the top of the piston or in the combustion chamber can cause a number of driveability concerns, including preignition. There are several ways to remove or reduce this carbon. One way is to disassemble the engine and remove the carbon with a scraper or wire wheel. Two other methods are commonly used. One is simply to add chemicals to the fuel. These chemicals work slowly, so do not expect quick results.

The other method uses a carbon blaster, which uses compressed air to force crushed walnut shells into the cylinders. The shells beat on the piston top and combus-

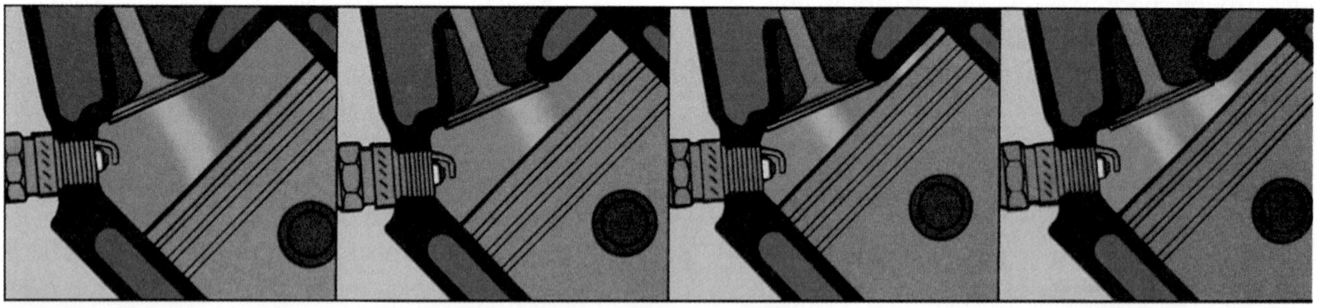

Figure 25–2 Detonation. *Courtesy of Champion Spark Plug Company*

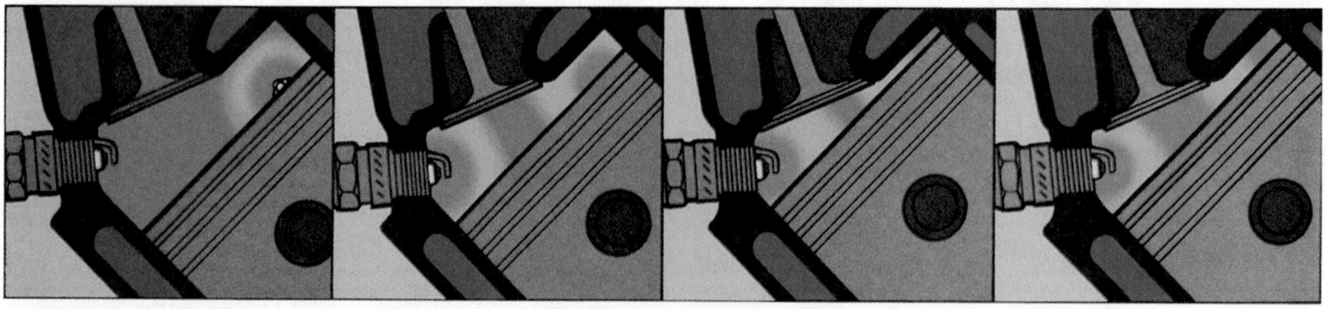

Figure 25–3 Preignition. *Courtesy of Champion Spark Plug Company*

tion chamber walls to loosen and remove the carbon. To use a carbon blaster, the intake manifold and spark plugs are removed. The output hose of the blaster is attached to a cylinder's intake port or inserted into the bore for the fuel injector. Another hose is inserted into the spark plug bore to serve as the exiting point for the shells and the carbon. The blaster forces a small amount of shells in and out of the cylinder. To help remove any remaining shell bits, compressed air is applied to the cylinder. This operation is done at each cylinder. Any remaining shell bits will be burned once the engine is run again.

GENERAL IGNITION SYSTEM DIAGNOSIS

The ignition system should be tested whenever you know or suspect there is no spark, not enough spark, or when the spark is not being delivered at the correct time to the cylinders. Typically, all ignition problems can be divided into two types: common and noncommon. Common problems are those that affect all cylinders, and noncommon problems are those that affect one or more, but not all, cylinders. Common ignition components include the parts of the primary circuit and the secondary circuit up to the distributor's rotor. Noncommon parts are the individual spark plug terminals inside the distributor cap, spark plug wires, and the spark plugs.

The best indicator of a noncommon problem is the reading on a vacuum gauge. For example, suppose that a vacuum gauge is connected to the intake manifold of a 4-cylinder engine at idle and the needle of the gauge is within the normal vacuum range for three-fourths of the time and drops one-fourth of the time. This indicates that three of the cylinders are working normally while the fourth is not. The cause of the problem is noncommon to the rest of the cylinders. If the cylinder is sealed and all cylinders are receiving the correct amount of air and fuel, the problem must be in the ignition system and must involve the distributor cap, spark plug wires, or spark plugs.

Determining whether the ignition problem is common or noncommon is a good way to start troubleshooting the ignition system. There are many parts in the system, and many tests can be conducted on them. By dividing the system into common and noncommon components, you can decide to test only those parts that could cause the problem.

Generally when an engine runs unevenly, the problem is a noncommon problem. Most other times, the problem is a common one. Since the primary circuit of nearly all ignition systems is common to all cylinders, deciding whether to test the primary or secondary circuits comes with deciding if the problem is common or noncommon.

EI systems, especially coil-per-cylinder systems, make troubleshooting a little easier. The PCM or control module may be the only part that is common to all cylinders. The coils and all of the secondary circuit are common to only one or two cylinders. For example, if a coil in a waste fire system is bad, two cylinders will be affected and not the entire engine.

VISUAL INSPECTION OF IGNITION SYSTEMS

All ignition system diagnosis should begin with a visual inspection. The system should be checked for obvious problems. Although no-start problems and incorrect ignition timing are caused by the primary circuit, the secondary circuit should also be checked.

Nearly all ignition systems today are tied into the vehicle's PCM and a fault may generate a DTC that can be retrieved with a scan tool **(Figure 25–4)**, especially in vehicles with OBD-II control systems. Part of a thorough visual inspection is the retrieval of codes. Often the codes will lead you to the exact cause of the problem; at other times you will need to do further diagnostics.

If DTCs are retrieved, correct the problems causing the trouble codes before moving on. Fixing the cause of the codes may correct the ignition problem.

Primary Circuit

Primary ignition system wiring should be checked for tight connections, especially on vehicles with electronic- or computer-controlled ignitions. Electronic circuits operate on very low voltage. Voltage drops caused by corrosion or dirt can cause running problems. Missing tab locks on wire connectors are often the cause of intermittent ignition problems due to vibration or thermal related failure.

Test the integrity of a suspect connection by tapping, tugging, and wiggling the wires while the engine is running. Be gentle. The object is to recreate an ignition

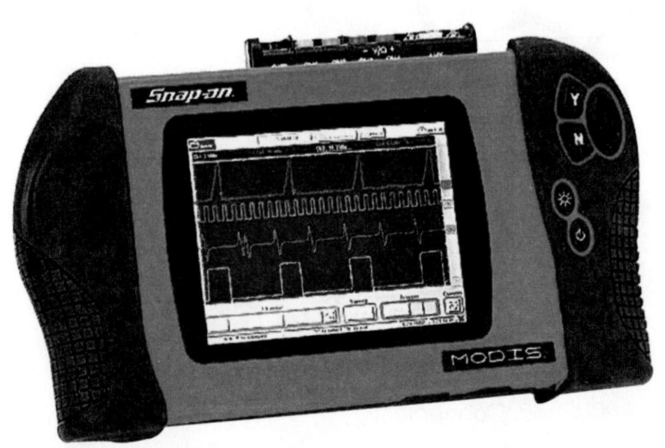

Figure 25-4 A diagnostic tool that performs the functions of many different testers, including a scan tool. *Courtesy of Snap-on Tools Company*

interruption, not to cause permanent circuit damage. With the engine off, separate the suspect connectors and check them for dirt and corrosion. Clean the connectors according to the manufacturer's recommendations.

Do not overlook the ignition switch as a source of intermittent ignition problems. A loose mounting rivet or poor connection can result in erratic spark output. To check the switch, gently wiggle the ignition key and connecting wires with the engine running. If the ignition cuts out or dies, the problem is located.

Check the battery connection to the starter solenoid. Remember, some vehicles use this connection as a voltage source for the coil. A bad connection can result in ignition interruption.

Carefully inspect the wires and belts for the charging system. Also, check the charging voltage at the battery. The efficiency of an ignition system depends on the voltage it receives. If battery or charging system voltage is low, the input to the primary side of the coil will also be low.

Moisture can cause a short to ground or reduce the amount of voltage available to the spark plugs. This can cause poor performance or a no-start condition. Carefully check the ignition system for signs of moisture. **Figure 25–5** shows the common places where moisture may be present in an ignition circuit.

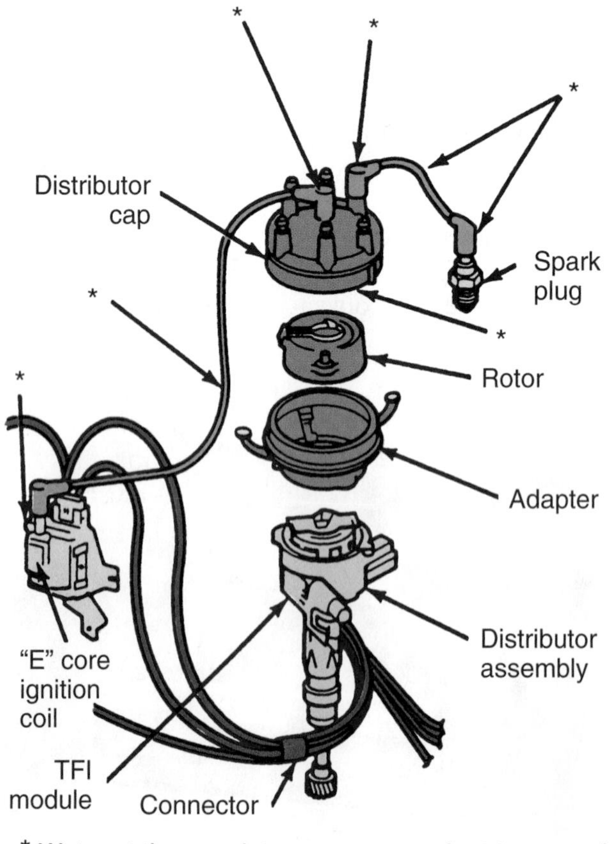

Figure 25–5 Places to check for moisture in an ignition system. *Courtesy of Ford Motor Company*

Ground Circuits

Keep in mind that to simplify a vehicle's electrical system, auto makers use body panels, frame members, and the engine block as current return paths to the battery.

Unfortunately, ground straps are often neglected, or worse, left disconnected after routine service. With the increased use of plastics in today's vehicles, ground straps may mistakenly be reconnected to a nonmetallic surface. The result of any of these problems is that the current that was to flow through the disconnected or improperly grounded strap is forced to find an alternate path to ground. Sometimes the current attempts to back up through another circuit. This may cause the circuit to operate erratically or fail altogether. The current may also be forced through other components, such as wheel bearings or shift and clutch cables that are not meant to handle current flow, causing them to wear prematurely or become seized in their housing.

Examples of bad ground-circuit-induced ignition failures include burned ignition modules resulting from missing or loose coil ground straps and intermittent ignition operation resulting from a poor ground at the control module. Poor ground can be identified by conducting voltage drop tests and by monitoring the circuit with a lab scope.

When conducting a voltage drop test, remember that the circuit must be turned on and have current flowing through it. If the circuit is tested without current flow, the circuit will show zero voltage drop, which would indicate that it is good regardless of the amount of resistance present.

The same is also true when checking a ground with the lab scope. Make sure the circuit is on. If the ground is good, the trace on the scope should be at zero volts and be flat. If the ground is bad, some voltage will be indicated and the trace will not be flat **(Figure 25–6)**.

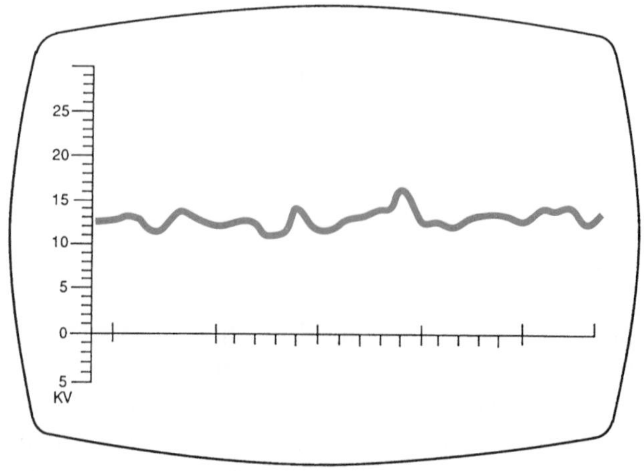

Figure 25–6 A voltage signal caused by a poor ignition module ground. *Courtesy of SPX Corporation, Aftermarket Tool and Equipment Group*

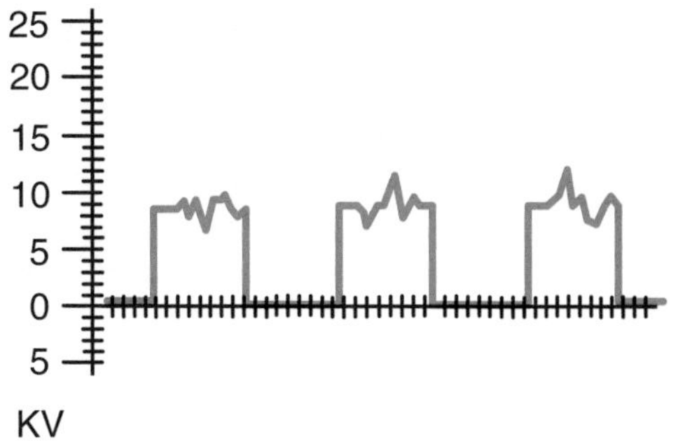

Figure 25–7 A voltage trace with ignition system noise.

Often a bad sensor ground will cause the same symptoms as a faulty sensor. Before condemning a sensor, check its ground with a lab scope. **Figure 25–7** shows the output of a good Hall-effect switch with a bad ground.

> **SERVICE TIP**
>
> When checking the ignition system with a lab scope, gently tap and wiggle the components while observing the trace. This may indicate the source of an intermittent problem. ■

Electromagnetic Interference

Electromagnetic interference (EMI) can cause problems with the vehicle's on-board computer. EMI is produced when electromagnetic radio waves of sufficient amplitude escape from a wire or conductor. Unfortunately, an automobile's spark plug wires, ignition coil, and AC generator coils all possess the ability to generate these radio waves. Under the right conditions, EMI can alter signals from sensors and to actuators. The result may be an intermittent driveability problem that may appear to be caused by many different systems.

To minimize the effects of EMI, check to make sure that sensor wires running to the computer are routed away from potential EMI sources. Rerouting a wire by no more than an inch or two may keep EMI from falsely triggering or interfering with computer operation.

Connecting a lab scope to voltage and ground wires can identify EMI problems. Common problems such as poor spark plug wire insulation will allow EMI.

Sensors

A voltage pulse from a crankshaft position sensor activates the transistor in the control module. In most ignition systems, this sensor is either a magnetic pulse generator or Hall-effect sensor. These sensors are mounted either on the distributor shaft or the crankshaft.

The reluctor or pole piece of a magnetic pulse generator is replaced only if it is broken or cracked. The pickup coil wire leads can become grounded if their insulation wears off as the breaker plate moves with the vacuum advance unit **(Figure 25–8)**. Inspect these leads carefully. Position these wires so that they do not rub the breaker plate as it moves.

Under unusual circumstances, the nonmagnetic reluctor can become magnetized and upset the pickup coil's voltage signal to the control module. Use a steel feeler gauge to check for signs of magnetic attraction and replace the reluctor if the test is positive. On some systems, the gap between the pickup and the reluctor must be checked and adjusted to manufacturer's specifications. To do this, use a properly sized nonmagnetic feeler gauge to check the air gap between the coil and reluctor. Adjust the gap if it is out of specification.

Hall-effect sensor problems are similar to those of magnetic pulse generators. This sensor produces a voltage when it is exposed to a magnetic field. The Hall-effect assembly is made up of a permanent magnet located a short distance away from the sensor. Attached to the distributor shaft is a shutter wheel. When the shutter is between the sensor and the magnet, the magnetic field is interrupted and voltage immediately drops to zero. This drop in voltage is the signal to the ignition module. When the shutter leaves the gap between the magnet and the sensor, the sensor produces voltage again.

Control Modules

Electronic and computer controlled ignitions use transistors as switches. These transistors are contained inside a control module housing that can be mounted on or in the distributor or remotely mounted on the vehicle's fire wall or another engine compartment surface. Control modules should be tightly mounted to clean surfaces. A loose mounting can cause a heat buildup that can damage and destroy transistors and other electronic components contained in the module. Some manufacturers recommend the use of special heat-conductive silicone grease between the control unit and its mounting. This helps conduct heat away from the module, reducing the chance of heat-related failure. During the visual inspection, check all electrical connections to the module. They must be clean and tight.

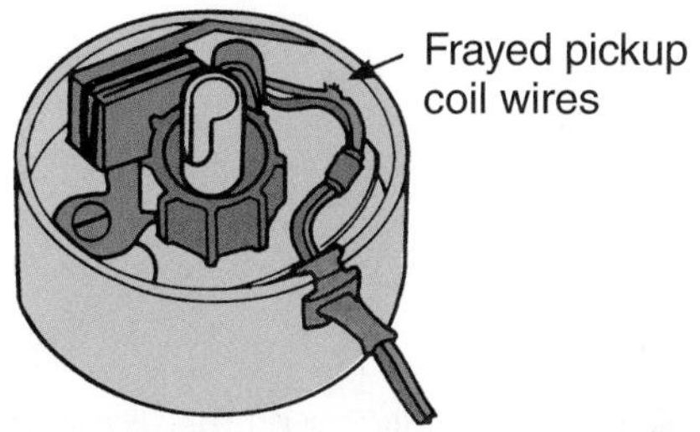

Figure 25–8 Inspect pickup coil wiring for damage.

Secondary Circuit

Spark plug (ignition) and coil cables should be pushed tightly into the distributor cap and coil and onto spark plugs. Inspect all secondary cables for cracks and worn insulation, which cause high-voltage leaks. Inspect all of the boots on the ends of the secondary wires for cracks and hard, brittle conditions. Replace the wires and boots if they show evidence of these conditions. Some vehicle manufacturers recommend spark plug wire replacement only in complete sets.

The secondary coil cable should also be inspected **(Figure 25–9)**. When checking this cable, check the ignition coil. The coil should be inspected for cracks or any evidence of leakage in the coil tower. The coil container should be inspected for oil leaks. If oil is leaking from the coil, air space is present in the coil, which allows condensation to form internally. Condensation in an ignition coil causes high-voltage leaks and engine misfiring.

Secondary cables must be connected according to the firing order. Refer to the manufacturer's service manual to determine the correct firing order and cylinder numbering.

White or grayish powdery deposits on secondary cables at the point where they cross or near metal parts indicate that the cables' insulation is faulty. The deposits occur because the high voltage in the cable has burned the dust collected on the cable. Such faulty insulation may produce a spark that sometimes can be heard and seen in the dark. An occasional glow around the spark plug cables, known as a **corona effect**, is not harmful but indicates that the cable should be replaced.

Spark plug cables from consecutively firing cylinders should cross rather than run parallel to one another. Spark plug cables running parallel to one another can induce firing voltages in one another and cause the spark plugs to fire at the wrong time.

On distributorless or electronic ignition (EI) systems, visually inspect the secondary wiring connections at the individual coil modules. Make sure all of the spark plug wires are securely fastened to the coil and the spark plug. If a plug wire is loose, inspect the terminal for signs of burning. The coils should be inspected for cracks or any evidence of leakage in the coil tower. Check for evidence of terminal resistance. A loose or damaged wire or bad plug can lead to carbon tracking of the coil. If this condition exists, the coil must be replaced.

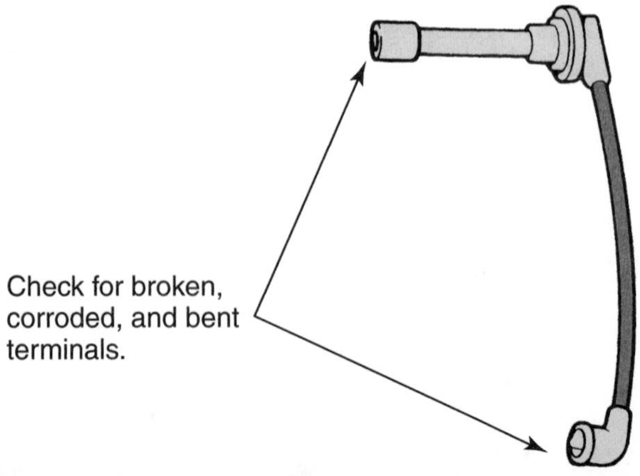
Check for broken, corroded, and bent terminals.

Figure 25-9 Carefully inspect the secondary cables. *Courtesy of American Honda Motor Co., Inc.*

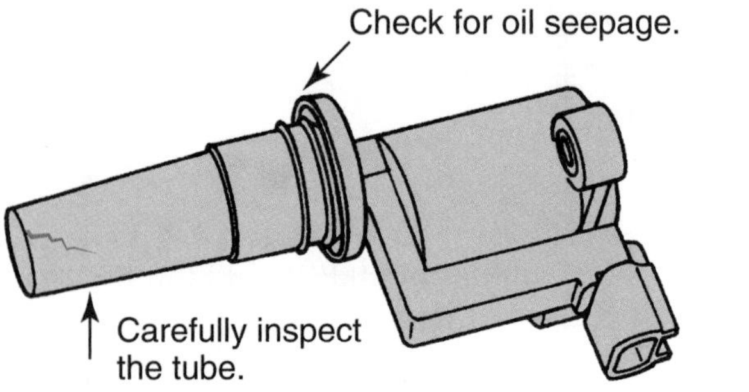
Check for oil seepage.
Carefully inspect the tube.

Figure 25-10 The plastic assembly for COP assembly should be carefully inspected.

On COP systems, carefully check the tubes that fit around the terminal of the spark plugs **(Figure 25–10)**. If the tube is cracked, voltage can leak out, jump to the cylinder head, and cause a misfire. Also, make sure the coil assembly fits snugly over the spark plug.

Distributor Cap and Rotor

The distributor cap should be properly seated on its base. All clips or screws should be tightened securely.

The distributor cap and rotor should be removed for visual inspection **(Figure 25–11)**. Physical or electrical damage is easily recognizable. Electrical damage from high voltage can include corroded or burned metal terminals and **carbon tracking** inside distributor caps. Carbon tracking is the formation of a line of carbonized dust between distributor cap terminals or between a terminal and the distributor housing. Carbon tracking indicates that high-voltage electricity has found a low-resistance conductive path over or through the plastic. The result is a misfire or a cylinder that fires at the wrong time. Check the outer cap towers and metal terminals for defects. Cracked plastic requires replacement of the unit. Damaged or carbon-tracked distributor caps or rotors should be replaced **(Figure 25–12)**.

The rotor should be inspected carefully for discoloration and other damage **(Figure 25–13)**. Inspect the top and bottom of the rotor carefully for grayish, whitish, or rainbow-hued spots. Such discoloration indicates that the rotor has lost its insulating qualities. High voltage is being conducted to ground through the plastic.

If the distributor cap or rotor has a mild buildup of dirt or corrosion, it should be cleaned. If it cannot be cleaned up, it should be replaced. Small round brushes are available to clean cap terminals. Wipe the cap and

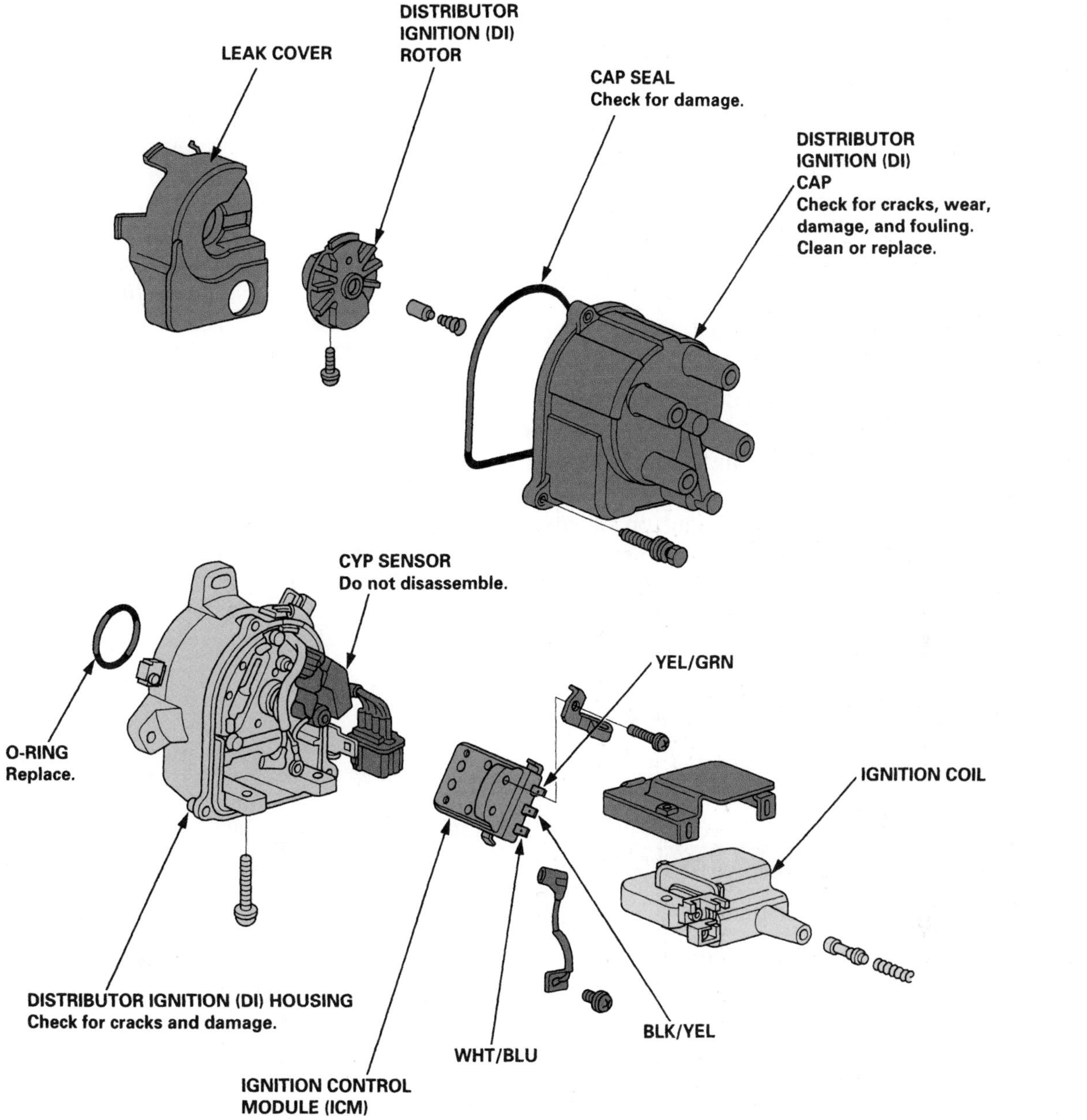

Figure 25-11 Inspect the distributor cap and rotor. *Courtesy of American Honda Motor Co., Inc.*

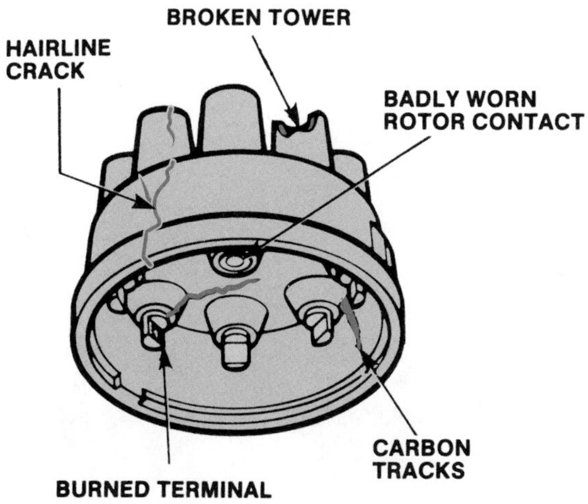

Figure 25-12 Things to look for when inspecting a distributor cap.

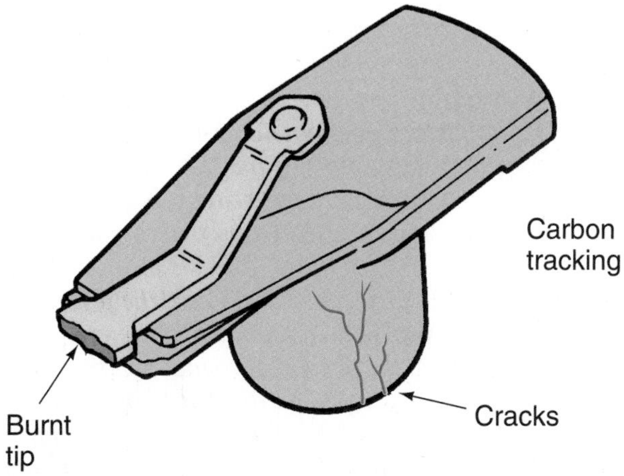

Figure 25-13 Inspect the rotor for damage and signs of high-voltage leaks. *Courtesy of DaimlerChrysler Corporation*

rotor with a clean shop towel, but avoid cleaning these components in solvent or blowing them off with compressed air, which may contain moisture. Cleaning these components with solvent or compressed air may result in high-voltage leaks.

Check the distributor cap and housing vents. Make sure they are not blocked or clogged. If they are, the internal ignition module will overheat. It is good practice to check these vents whenever a module is replaced.

NO-START DIAGNOSIS

When the engine does not run, it is difficult to decide whether the problem is common or noncommon, or whether a fault in the primary or secondary ignition circuit is stopping the engine from running. The cause of a no-start condition can be in the air, fuel, or ignition system. Since the primary purpose of the ignition system is to provide a spark to ignite the air/fuel mixture, if there is no spark there will be no combustion. If the problem is caused by an ignition fault, follow this procedure to determine the exact cause of the problem. Often manufacturers include a detailed troubleshooting tree in their service manuals to help locate the cause of the no-start condition.

PROCEDURE

No-Start Diagnosis

STEP 1 *Connect a 12-volt test lamp from the coil tachometer (tach) terminal to ground. Turn on the ignition switch. The test light normally should be on. If the test light is off, there is an open circuit in the coil primary winding or in the circuit from the ignition switch to the coil battery terminal. On many Chrysler and Ford systems, the test light should be off because the module primary circuit is closed. Since there is primary current flow, most of the voltage is dropped across the primary coil winding. This action results in very low voltage at the tach terminal, which does not illuminate the test light. On these systems, if the test light is illuminated, there is an open circuit in the module or in the wire between the coil and the module.*

STEP 2 *Crank the engine and observe the test light. If the test light flutters while the engine is cranked, the pickup coil signal and the module are okay. When the test lamp does not flutter, the pickup and/or module are bad. A pickup is tested with an ohmmeter. If the pickup coil is satisfactory, the module is defective. Before testing the pickup, check the voltage supply to the positive primary coil terminal with the ignition switch on before the diagnosis is continued.*

STEP 3 *If the test light flutters, connect a test spark plug* **(Figure 25–14)** *to the coil secondary wire and ground the spark plug case.*

STEP 4 *Crank the engine and observe the spark plug. If the test spark plug fires, the ignition coil is satisfactory. If the test spark plug does not fire, the coil is probably bad.*

STEP 5 *Connect the test spark plug to several spark plug wires and crank the engine while observing the spark plug. If the test spark plug fired in step 4 but does not fire at some of the spark plugs, the secondary voltage and current is leaking through a defective distributor cap, rotor, or spark plug wires, or a plug wire is open. If the test spark plug fires at all the spark plugs, the ignition system is working fine.*

SERVICE TIP

On most Chrysler fuel-injected engines, the voltage is supplied through the automatic shutdown (ASD) relay to the coil positive primary terminal and the electric fuel pump. Therefore, a defective ASD relay may cause 0 volt at the positive primary coil terminal. This relay is controlled by the PCM and is designed to shut down the fuel pump and prevent any spark from the ignition system if the vehicle is involved in a collision with the ignition switch on and the engine stalled. A fault code should be present in the computer memory if the ASD relay is defective. ■

Figure 25-14 A test spark plug for high-voltage ignition systems. *Courtesy of Snap-on Tools Corporation*

No-Start Diagnosis of EI Systems

When an EI system has a no-start problem, begin your diagnosis by determining if the problem is caused by problems within the engine, fuel and air system, or ignition system. To determine if the no-start condition is caused by the ignition system, perform a spark intensity check at one spark plug with a test spark plug. A bright, snapping spark indicates that the ignition system is working and the problem is undoubtedly another system. If the spark is weak or if there is no spark, check another spark plug. If the spark is weak or there is no spark, check the wiring to and from the PCM.

When the test spark plug does not fire on any plug, connect a voltmeter from the input (battery) terminal on each coil pack to ground. With the ignition switch on, the voltmeter should read 12 volts. If the voltage is less than that, check the system's wiring diagram to determine what is included in the coil's power feed circuit. Then check the voltage drop across each of the components and wires to identify the location of an open or high resistance.

Included in the circuit is the crankshaft position sensor **(Figure 25–15)**, which will prevent the engine from starting if the sensor is bad. On some engines, a bad camshaft sensor will also prevent starting, as it is the time reference for the fuel injection system. Both of these sensor circuits can be checked with a voltmeter or DSO. If the sensors are receiving the correct amount of voltage and have good ground circuits, their output should be a digital signal or a pulsing voltmeter reading while the engine is cranking. If any of these conditions does not exist, the circuit needs to be repaired or the sensor needs to be replaced.

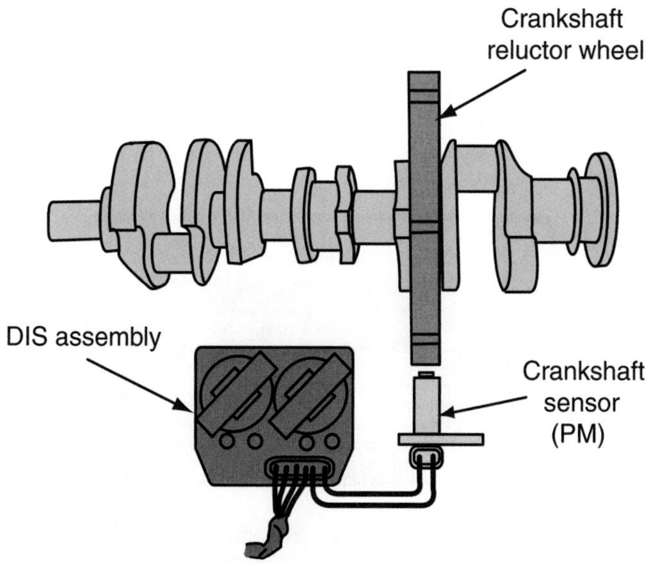

Figure 25-15 A typical electrical connection for a crankshaft position sensor.

> **SERVICE TIP**
>
> When using a digital voltmeter to check a crankshaft or camshaft sensor signal, crank the engine a very small amount at a time and observe the voltmeter. The voltmeter reading should cycle from almost 0 volt to a higher voltage of 9 to 12 volts. Since digital voltmeters do not react instantly, it is difficult to see the change in voltmeter reading if the engine is cranked continually. ■

GENERAL TESTING

It is impossible to accurately troubleshoot any ignition system without performing various electrical tests. An engine analyzer houses all of the necessary test equipment to do a complete engine performance analysis. Using an engine analyzer is a good way to determine whether the driveability problem is caused by the ignition system. Also, the testing instruments available in the analyzer, such as a scope and a digital multimeter, can be used to check individual ignition parts and circuits,

Diagnosing with an Engine Analyzer

On some engine analyzers, a service test screen appears that lists all the tests that may be performed. Any individual test may be selected. For example, you may want to use the multimeter to measure the resistance in the coil windings. To do this, simply move the cursor to and select MULTIMETER to display the multimeter readings on the screen. Now the multimeter leads may be connected to the desired component or circuit to test volts, amperes, or ohms.

If COMPREHENSIVE TESTS is selected, the analyzer automatically performs a complete test sequence. When RUN CUSTOM TESTS is selected, the analyzer allows you to design your own test sequence.

An engine analyzer's test capabilities vary depending on the manufacturer and the model of the analyzer. Always follow the test procedures given in the operating manual. Most engine analyzers perform diagnostic tests in these main areas: cylinder performance; ignition performance; battery, starting, and charging systems; exhaust emission levels; and engine computer systems.

Cylinder Performance Test During the cylinder performance, the analyzer momentarily stops the ignition system from firing one cylinder at a time. During this brief time, the rpm drop is recorded. When a cylinder is not contributing to engine power because it has low compression or some other problem, there is very little rpm drop when the spark plug in that cylinder stops firing. This test is similar to the power balance test.

6.5%	7.5%	7%	3%	6.5%	6.5%	7%	7.5%	rpm drop
1725	1625	1650	850	1725	1750	1650	1700	HC increase
1	5	6	3	4	2	7	8	Firing order

Figure 25-16 Cylinder performance test results showing one bad cylinder.

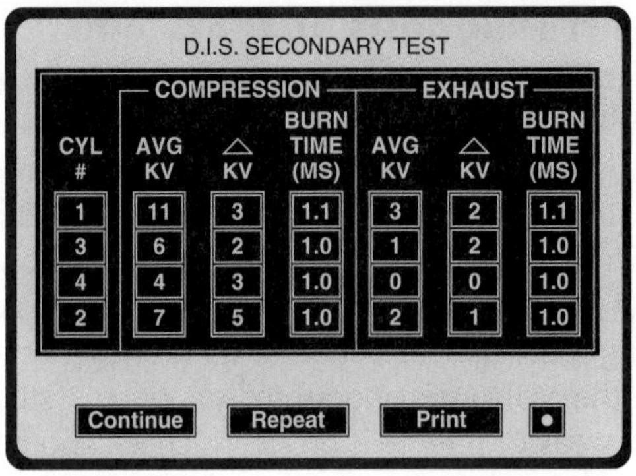

Figure 25-17 A secondary kV display on an EI system. *Courtesy of Bear Automotive Service Equipment Company*

During the test, some analyzers record the actual rpm drop, while others indicate the percentage of rpm drop. Many analyzers also record the amount of hydrocarbon (HC) change in parts per million (ppm) when each cylinder stops firing. If a cylinder is misfiring prior to the cylinder performance test, the cylinder has high HC emissions. Therefore, when this cylinder stops firing during the cylinder performance test, there is not much change in HC emissions. A cylinder with low compression or a problem that causes incomplete combustion will not have much rpm drop or HC change during the cylinder performance test **(Figure 25–16)**.

Ignition Performance Tests On many engine analyzers, ignition performance tests include primary circuit tests, secondary **kilovolt (kV)** tests, acceleration tests, scope patterns, and cylinder miss recall. Primary circuit tests include coil input voltage, coil primary resistance, dwell, curb idle speed, and idle vacuum. Secondary kV tests include the average kV, maximum kV, and minimum kV for each cylinder. The kV is the voltage required to start firing the spark plug. A high-resistance problem in a spark plug circuit causes a higher firing kV, whereas a fouled spark plug, shorted plug circuit, or a cylinder with low compression results in a lower firing kV.

Some secondary kV tests include a snap kV test in which the analyzer directs the technician to accelerate the engine suddenly. When this action is taken, the firing kV should increase evenly on each cylinder.

Some analyzers display the burn time for each cylinder with the secondary kV tests. The burn time is the length of the spark line in milliseconds (ms). When using other analyzers, the burn time is read on the scope screen, preferably using a raster pattern. The average burn time should be 1 ms to 1.5 ms.

The secondary kV display from an EI system includes average kV for each cylinder on the compression stroke and average kV for each matching cylinder that fires at the same time on the exhaust stroke. The burn time is also included on the secondary kV display from an EI system **(Figure 25–17)**.

Many engine analyzers also provide primary and secondary scope patterns. Some analyzers are capable of freezing the scope patterns into the analyzer memory and recalling this information on request. During the recall procedure, individual cylinder firings are played back to easily identify intermittent cylinder misfirings.

Battery, Starting, and Charging System Tests During a battery test, the engine analyzer places a load on the battery and may record the battery's open circuit voltage, load voltage, and available cold-cranking amperes.

While performing the starter, or cranking, test, some analyzers record cranking volts, cranking current, cranking vacuum, cranking speed, cranking dwell, coil input volts, and HC.

On some analyzers, the cranking amperes are displayed for each cylinder. When a cylinder has low compression, the starter draw in amperes will be reduced for that cylinder. Other analyzers assign a value of 100% to the cylinder with the highest cranking amperes, and then provide a percentage reading for each of the other cylinders in relation to the cylinder with the 100% reading **(Figure 25–18)**. Charging circuit tests include regulator voltage, alternator current, and alternator waveforms.

Emission Level Analysis The emission level analysis involves diagnosis using an exhaust gas analyzer. The exhaust gas analysis screen may include HC, CO, O_2, CO_2, NO_x, engine temperature, exhaust temperature, vacuum, and engine rpm.

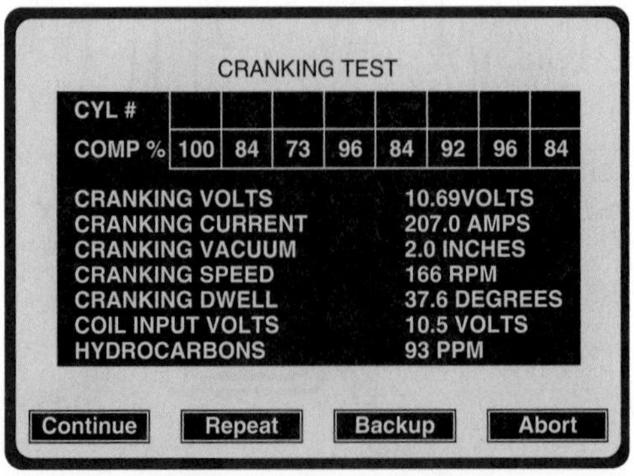

Figure 25-18 Cranking test results. *Courtesy of Bear Automotive Service Equipment Company*

Engine Computer System Diagnosis Many engine analyzers diagnose engine computer systems in much the same way as a scan tester. The analyzer performs tests on common engine control systems.

TESTING WITH A SCOPE

No discussion on ignition troubleshooting would be complete without a discussion of oscilloscope use. An oscilloscope or tune-up scope converts the electrical signals of the ignition system into a visual image showing voltage changes over a given period of time. This information is displayed on a screen in the form of a continuous voltage line called a pattern or trace. By studying the pattern, a technician can determine what is happening inside the ignition system.

Always follow the specific instructions for connecting the test leads and operating the scope you are using. Scopes typically have at least four leads **(Figure 25–19)**: a primary pickup that connects to the negative terminal of the ignition coil; a ground lead that connects to a good engine ground; a secondary pickup, which clamps around the coil's high tension wire; and a trigger pickup, which clamps around the spark plug wire of the number 1 cylinder.

Connecting the scope (or engine analyzer) to an EI system requires adapters or additional test leads **(Figure 25–20)**. On some scopes, the companion cylinders of a waste spark setup are viewed at the same time. Other scopes, with the necessary adapters, display all of the cylinders. Adapters must also be used to gain access to secondary ignition data on COP systems **(Figure 25–21)**.

Most COP systems can be tested with a low-amperage probe through the ignition primary circuit. The inductive pickup on most scopes and engine analyzers will not work on COP systems unless there is a spark plug cable.

Figure 25–20 A DIS adapter assembly. *Courtesy of SPX-OTC/Automotive Electronic Diagnostic Tools*

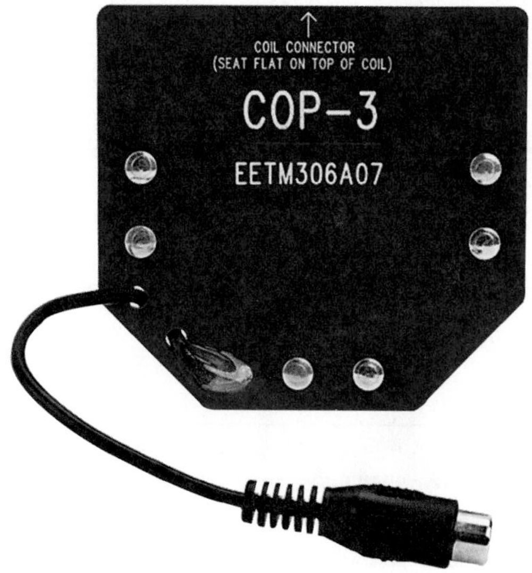

Figure 25–21 A COP test adapter. *Courtesy of Snap-on Tools Company*

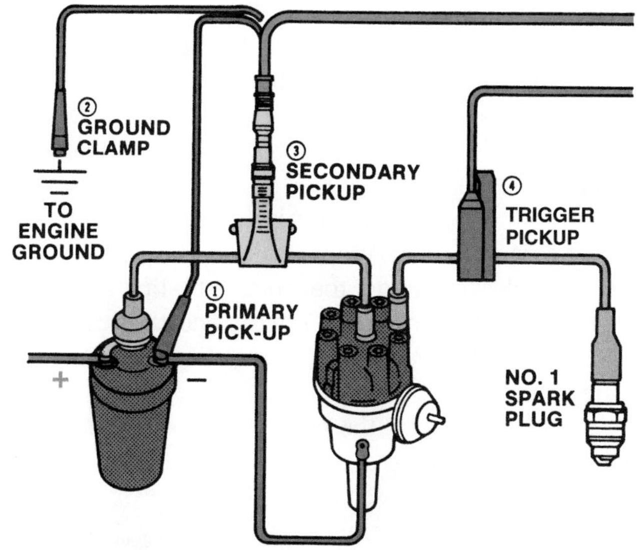

Figure 25–19 Typical oscilloscope pickup connections to an ignition system.

Scales

A typical scope screen has two vertical scales, one on the left and one on the right, and a horizontal time scale. The time may be expressed as percent of dwell or in milliseconds. The vertical scales display voltage. Typically, the scale on the left side of the screen is divided into increments of 1 kilovolt (1000 volts) and ranges from 0 to 25 kilovolts (kV). This scale is useful for testing secondary voltage. It can also be used to measure primary voltage by interpreting the scale in volts rather than kilovolts. The scale on the right side is divided into increments of 2 kilovolts and has a range of 0 to 50 kV. This scale is also used to test secondary voltage and to measure primary voltage in the 0- to 500-volt range.

The horizontal time scale is located at the bottom of the scope screen. The percent of dwell scale is divided into increments of 2 percentage points and ranges from 0 to 100%. This represents one complete ignition cycle. The millisecond scale is typically broken down into units of 0 to 5 milliseconds (ms) or 0 to 25 milliseconds. The 5-ms scale is often used to measure the duration of the spark **(Figure 25–22)**. In the 25-millisecond mode, the complete firing pattern can normally be displayed. A scope displays changes in voltage over time from left to right, similar to reading a book.

Pattern Phases

To monitor various phases of ignition system performance, a typical scope pattern is divided into three main sections: firing, intermediate, and dwell. Information received from specific testing in each one of these areas can be used to piece together a complete picture of the ignition system. Depending on the oscilloscope function selected, the scope can display a pattern for either the secondary or primary circuit. A typical secondary pattern is shown in **Figure 25–23**.

A typical primary circuit pattern is shown in **Figure 25–24**. The primary scope pattern is used when secondary circuit connections are not possible. It is also useful for observing dwell and cylinder timing problems.

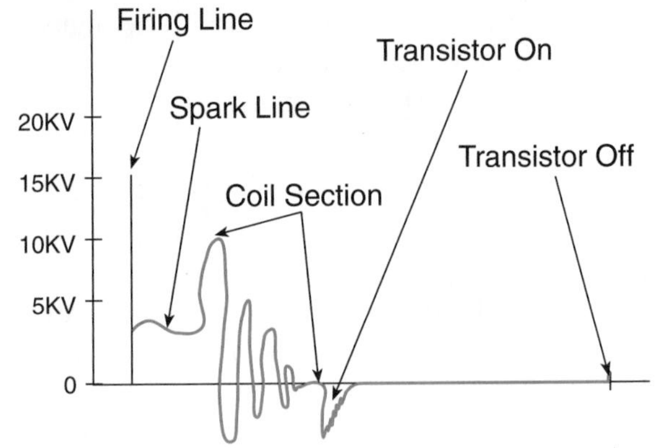

Figure 25-23 A typical secondary pattern.

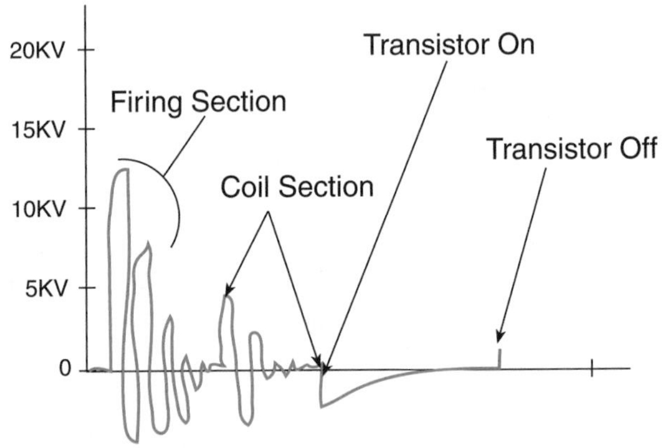

Figure 25-24 A typical primary pattern.

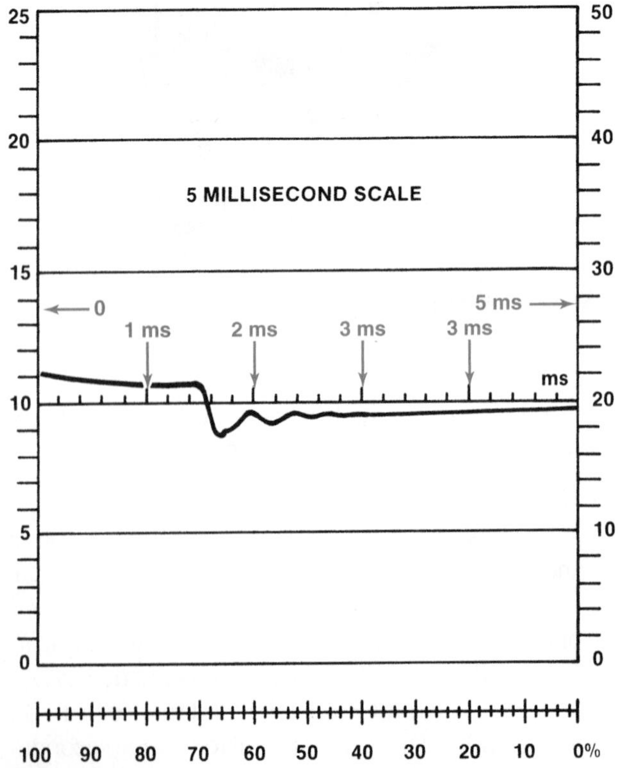

Figure 25-22 A 5-millisecond pattern showing spark duration.

Pattern Display Modes

The scope can display patterns in several ways. When the **parade** or **display pattern** is selected, the oscilloscope displays the patterns of all the cylinders in a row from left to right as shown in **Figure 25–25**. Each cylinder's ignition cycle is displayed according to the engine's firing order. The pattern begins with the spark line of the cylinder 1 and ends with the firing line for cylinder 1. This display pattern is most commonly used to compare the voltage peaks for each cylinder.

A second choice of patterns is the **raster pattern (Figure 25–26)**. A raster pattern stacks the voltage patterns of the cylinders one above the other. The #1 pattern is displayed at the bottom of the screen, and the rest of the cylinders are arranged above it according to the engine's firing order.

In a raster pattern, the pattern for each cylinder begins with the spark line and ends with the firing line. This allows for a much closer inspection of the voltage and time trends than is possible with the display pattern.

A **superimposed pattern** displays all of the patterns one on top of the other. Like the raster pattern, the su-

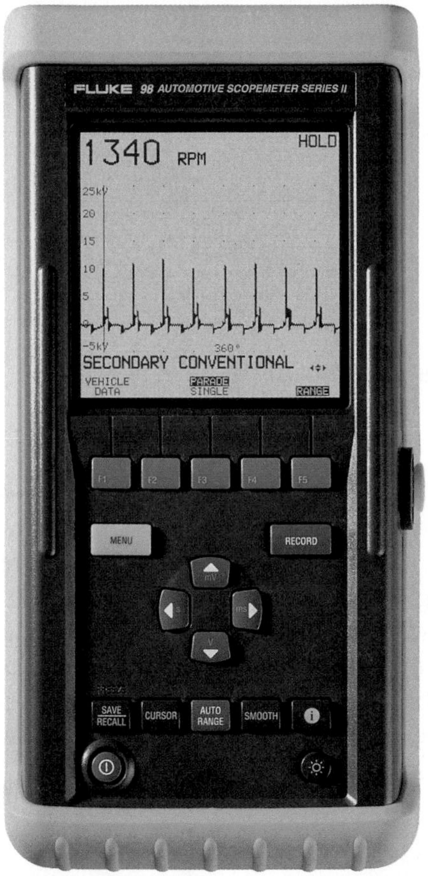

Figure 25–25 A typical parade (display) pattern.
Reprinted with the permission of Fluke Corporation

perimposed voltage patterns are displayed the full width of the screen, beginning with the spark line and ending with the firing line. A superimposed pattern is used to identify variations of one cylinder's pattern from the others. A superimposed pattern for the primary circuits is shown in **Figure 25–27**.

Understanding Single Cylinder Patterns

Firing Line In a secondary pattern, the firing line appears on the left side of the screen. The height of the firing line represents the voltage needed to overcome the resistance in that secondary circuit. Typically, around 10,000 volts are required to overcome this resistance and initiate a spark across the gap of the spark plug. Keep in mind that cylinder conditions have an effect on this resistance. Leaner air/fuel mixtures increase the resistance and increase the required **firing voltage**.

Spark Line Once secondary resistance is overcome, the spark jumps the plug gap, establishing current flow, and ignites the air/fuel mixture in the cylinder. The length of time the spark actually lasts is represented by the **spark line** portion of the pattern. The spark line begins at the firing line and continues until the coil's voltage drops below the level needed to keep current flowing across the gap.

Intermediate Section Once coil voltage drops below the level needed to sustain the spark, the next major section

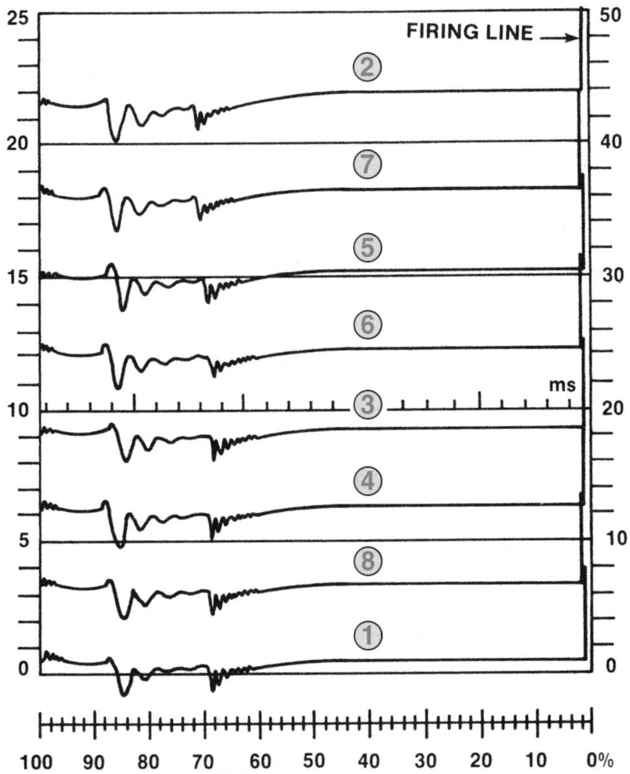

Figure 25–26 A secondary raster pattern.

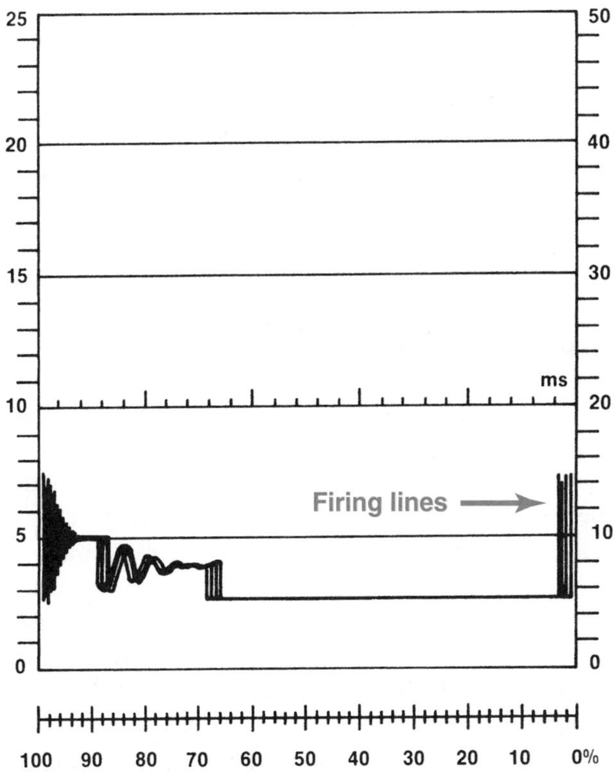

Figure 25–27 A primary superimposed pattern.

of the pattern begins. This section is called the **intermediate section** or coil-condenser zone. It shows the remaining coil voltage as it dissipates or drops to zero. Remember, once the spark has ended, there is quite a bit of voltage stored in the ignition coil. The voltage remaining in the coil then oscillates or alternates back and forth within the primary circuit until it drops to zero. Notice that the lines representing the coil-condenser section steadily drop in height until the coil's voltage is zero.

Dwell Section The next section of the trace pattern begins with the primary circuit current on signal. It appears as a slight downward turn followed by several small oscillations. The slight downward curve occurs just as current begins to flow through the coil's primary winding. The oscillations that follow indicate the beginning of the magnetic field buildup in the coil. This curve marks the beginning of a period known as the dwell section or zone. The end of the dwell zone occurs when the primary current is turned off by the switching device. The trace turns sharply upward at the end of the dwell zone. Turning the primary current off collapses the magnetic field in the coil and generates another high-voltage spark for the next cylinder in the firing order. Remember, the primary current off signal is the same as the firing line for the next cylinder. The length of the dwell section represents the amount of time that current is flowing through the primary.

In general, most scope patterns look more or less like the one just described. The patterns produced by some systems have fewer oscillations in the intermediate section. Patterns may also vary slightly in the dwell section. The length of this section depends on when the control module turns the transistor on and off.

Older breaker point and early electronic DI systems used a fixed dwell period. In a fixed dwell system, the number of dwell degrees remains the same during all engine speeds.

Most control modules provide for a variable dwell. In these systems, dwell changes significantly with engine speed. At idle and low rpm speeds, a short dwell provides enough time for complete ignition coil saturation **(Figure 25–28A)**. The current on signal and the current off signal appear very close to each other, usually less than 20 degrees. As engine speed increases, the control module lengthens the dwell time **(Figure 25–28B)**. This, of course, increases the available time for coil saturation.

Many late-model DI systems are **current limiting**. These systems saturate the ignition coil quickly by passing very high current through the primary winding for a fraction of a second. Once the coil is saturated, the need for high current is eliminated, and a small amount of current is used to keep the coil saturated. This type of system extends coil life.

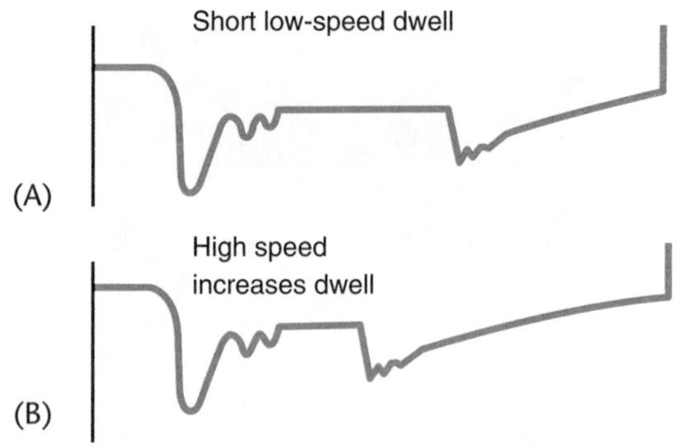

Figure 25-28 An example of variable dwell ignition system.

The point at which the control module cuts back from high to low current appears as a small blip or oscillation during the dwell section of the pattern **(Figure 25–29)**. At very high engine speeds, this telltale blip may be missing because the module keeps high current flow going to keep the coil continually saturated for fast firing.

If the primary winding of the coil has developed excessive resistance or the coil is otherwise faulty, the cutback blip may never occur. Further testing of the coil is needed to pinpoint the cause of the missing blip.

Spark Duration Testing

Spark duration is measured in milliseconds, using the millisecond scale on the scope. Most vehicles have a spark duration of approximately 1.5 milliseconds. A spark duration of approximately 0.8 millisecond is too short to provide complete combustion. A short spark also increases pollution and power loss. If the spark duration is too long (more than approximately 2.0 milliseconds) the spark plug electrodes might wear prematurely. When the oscilloscope shows a long spark duration, it normally follows a short firing line, which may indicate a fouled spark plug, low compression, or a spark plug with a narrow gap.

Spark Plug Firing Voltage

In a secondary pattern, the firing line is seen as the highest line in the pattern. The firing line is affected by anything that adds resistance to the secondary circuit, including the condition of the spark plugs or the secondary circuit, engine temperature, fuel mixture, and compression pressures. When looking at firing lines, make sure you compare all of the cylinders to one another.

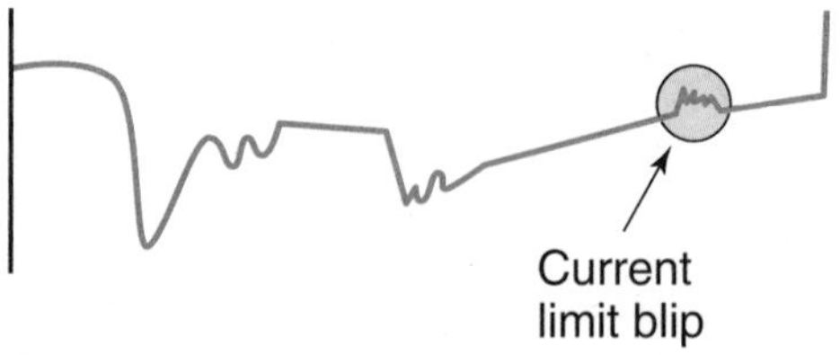

Figure 25-29 A pattern showing an ignition system limiting current during dwell.

The normal height of the firing voltages should be between 7 and 13 kV, with no more than a 3 kV variation between cylinders. If one or more firing lines are too low or too high, the cause is something that is only common to those cylinders. If the firing lines are all too high or too low, the problem is something that is common to all cylinders. **Table 25–1** covers most of the things that could cause abnormal firing lines.

Secondary resistance can be further examined by studying the height and slope of the spark line. Excessive resistance in the secondary circuit causes the spark line to have a steep slope with a shorter firing duration. A good spark line should be relatively flat and measure 2 to 4 kV in height. High resistance in the secondary circuit produces a firing line and spark line that are higher in voltage with shorter firing durations.

To pinpoint the cause of high resistance, use a grounding probe and jumper wire to bypass each component of the secondary circuit on all abnormal cylinders. Connect one end of the jumper lead to ground and the other end to the large portion of a grounding probe. Start the engine and adjust the speed to 1,000 rpm. Touch each secondary connector with the point of the grounding probe and observe the spark lines. If, after grounding, the abnormal spark lines appear normal, the part just bypassed is the cause of the problem.

Spark Plugs Under Load

The voltage required to fire the spark plugs increases when the engine is under load. The voltage increase is moderate and uniform if the spark plugs are in good condition and properly gapped. However, if any unusual characteristics are displayed on the scope patterns when load is applied to the engine, the spark plugs are probably faulty. This condition is most evident in the firing voltages displayed on the scope. To test spark plug operation

TABLE 25–1 FIRING LINE DIAGNOSIS

Condition	Probable Cause	Remedy
Firing voltage lines the same, but abnormally high	1. Retarded ignition timing 2. Fuel mixture too lean 3. High resistance in coil wire 4. Corrosion in coil tower terminal 5. Corrosion in distributor coil terminal	1. Rest ignition timing. 2. Readjust carburetor or check for vacuum leak. 3. Replace coil wire. 4. Clean or replace coil. 5. Clean or replace distributor cap.
Firing voltage lines the same, but abnormally low	1. Fuel mixture too rich 2. Breaks in coil wire causing arcing 3. Cracked coil tower causing arcing 4. Low coil output 5. Low engine compression	1. Readjust carburetor or check for plugged air filter. 2. Replace coil wire. 3. Replace coil. 4. Replace coil. 5. Determine cause and repair.
One or more, but not all firing voltage lines higher than the others	1. Idle mixture not balanced 2. EGR valve stuck open 3. High resistance in spark plug wire 4. Cracked or broken spark plug insulator 5. Intake vacuum leak 6. Defective spark plugs 7. Corroded spark plug terminals	1. Readjust idle mixture. 2. Inspect or replace EGR valve. 3. Replace spark plug wires. 4. Replace spark plugs. 5. Repair leak. 6. Replace spark plugs. 7. Replace spark plugs.
One or more, but not all firing voltage lines lower	1. Curb idle mixture not balanced 2. Breaks in plug wires causing arcing 3. Cracked coil tower causing arcing 4. Low compression 5. Defective or fouled spark plugs	1. Readjust idle mixture. 2. Replace spark plug wires. 3. Replace coil. 4. Determine cause and repair. 5. Replace spark plugs.
Cylinders not firing	1. Cracked distributor cap terminals 2. Shorted spark plug wire 3. Mechanical problem in engine 4. Defective spark plugs 5. Spark plugs fouled	1. Replace distributor cap. 2. Determine cause of short and replace wire. 3. Determine problem and correct. 4. Replace spark plugs. 5. Replace spark plugs.

under load, note the height of the firing lines at idle speed. Then, quickly open and release the throttle (snap accelerate), and note the rise in the firing lines while checking the voltages for uniformity. A normal rise would be between 3 and 4 kV upon snap acceleration.

Coil Condition

The energy remaining in the coil after the spark plugs fire gradually diminishes in a series of oscillations. These oscillations are observable in the intermediate sections of both the primary and secondary patterns. If the scope pattern shows an absence of normal oscillations in the intermediate section, check for a possible short in the coil by testing the resistance of the primary and secondary windings.

Primary Circuit Checks

The secondary pattern is mostly comprised of the secondary circuit but also shows the primary when the secondary is doing nothing. When there is primary current flow, the secondary circuit is idle with no electrical activity. The secondary pattern shows the electrical activity in the primary during this time.

A primary ignition pattern shows the action of the primary circuit. To be able to spot abnormal sections of a primary waveform, you must know what causes each change of voltage and time in a normal primary waveform. Although the true cycle of the primary circuit begins and ends when the switching transistor is turned on, the displayed pattern begins right after the transistor is turned off. At that moment, the magnetic field around the windings collapses and a spark plug is fired.

Looking at the primary pattern shown in **Figure 25–30**, the trace at the left represents the collapsing of the primary winding after the transistor turns off and primary current flow is interrupted. The height of these oscillations depends on the current that was flowing through the winding right before it was stopped. The amount of current flow depends on the time it was able to flow, the voltage or pressure applied to the winding, and the resistance of the winding. High primary circuit resistance reduces the maximum amount of current that can flow through the winding. Reduced current flow through the winding reduces the amount of voltage that can be induced when the field collapses.

During the collapsing of the primary winding, the spark plug is firing. The primary circuit's trace shows sharp oscillations of decreasing voltages. The overall shape of this group of oscillations should be conical and should last until the spark plug stops firing.

After the firing of the plug, some electrical energy remains in the coil. This energy must be released prior to the next dwell cycle. The next set of oscillations shows the dissipation of this voltage. These oscillations should be smooth and become gradually smaller until the zero volt line is reached. At that point there is no voltage left and the coil is ready for the next dwell cycle.

The transistor on signal immediately follows this dissipation of coil energy. This is when current begins to flow through the primary circuit. When the transistor turns on, there should be a clean and sharp turn in the trace. A clean change indicates the circuit was instantly turned on. If there is any sloping or noise at this part of the signal, something is preventing the circuit from being instantly turned on. Or if the scope is showing a superimposed primary pattern, any variation between cylinders will show up as a blurred or noisy transistor on signal **(Figure 25–31)**.

To help define the cause of erratic on signals, change the pattern display to raster, allowing the patterns to be stacked on top of each other. Sometimes the erratic signal is actually transistor on signals that are occurring at different times. Each cylinder has its own and unique time for events. This is not good. It can cause rough engine operation at all speeds, especially idle. It can also cause a

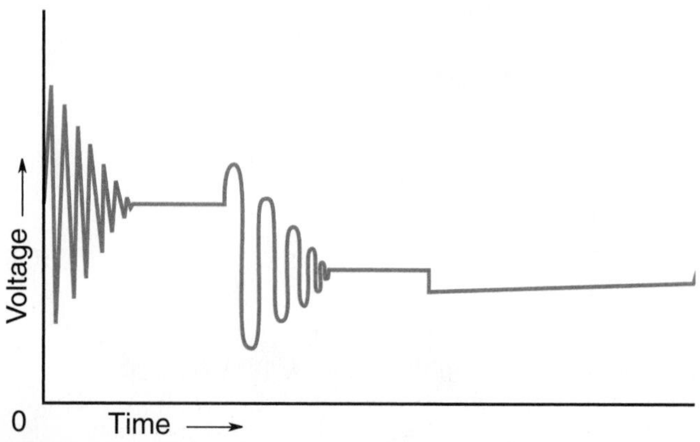

Figure 25-30 A typical primary pattern.

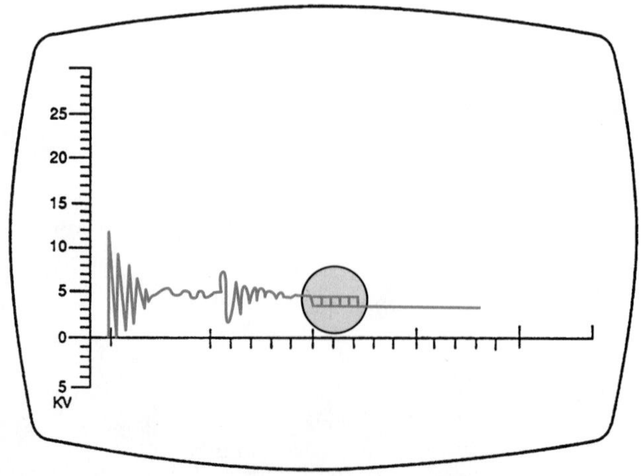

Figure 25-31 A superimposed primary pattern showing cylinders with different transistor ON times. *Courtesy of SPX Corporation, Aftermarket Tool and Equipment Group*

general lack of power, poor acceleration, and backfiring. This can be caused by:

1. Worn distributor shaft bushing or drive gear
2. Bad ignition module
3. Worn camshaft sprocket
4. Stretched timing chain or belt
5. Excessive camshaft end play
6. Loose armature or pickup unit

Stress Testing Components

Often, an intermittent ignition problem only occurs under certain conditions such as extremes in heat or cold, or during rainy or humid weather. Careful questioning of the customer should lead to determining if the problem is stress-condition-related. Does the problem occur on cold mornings? Does it occur when the engine is fully warmed up? Is it a rainy day problem? If the answer to any of these questions is positive, you can reproduce the same conditions in the shop during stress testing.

CAUTION!

When using cool-down sprays, wear eye protection and avoid spraying your skin or clothing. Use extreme caution.

Cold Testing With the scope on raster, cool major ignition components such as the control module, pickup coil, and major connections one at a time using a liquid cool-down agent. After cooling a component, watch the pattern for any signs of malfunction, particularly in the dwell zone. If there is no sign of malfunction, cool down the next component after the first has warmed to normal operating temperature. Cooling (or heating) more than one component at a time provides inconclusive results.

Heat Testing To heat stress components, use a heat gun or hair dryer to direct hot air into the component. Heat guns intended for stripping paint and other household jobs can become extremely hot and melt plastic, wire insulation, and other materials. Use a moderate setting and proceed cautiously. Look for changes in the dwell section of the trace, particularly in the variable dwell or current-limiting areas. If connections appear to be the problem, disconnect them, clean the terminals, and coat them with dielectric compound to seal out dirt and moisture.

Moisture Testing A wet stress test is performed by lightly spraying the components, coil and ignition cables, and connections with water. Do not flood the area; a light mist does the job. A scope set on raster or display helps pinpoint problems, but it is often possible to hear and feel the miss or stutter without the use of a scope. As with heat and cold testing, do not spray down more than one area at a time or results could be misleading. If you suspect a poor connection, clean and seal it, then retest it.

TESTING WITH A GRAPHING METER

A graphing meter functions much like a scope and can also display digital MIN/MAX readings. Many have four-channel capability allowing observation of signals from four different sensors and/or outputs at the same time **(Figure 25–32)**. This is a great help during diagnosis.

Some graphing meters contain specific service information and tips, which make diagnosis simpler. Some even have a database of known-good waveforms in their memory, which again helps in the diagnosis of a problem.

Ignition components can be tested with the multimeter functions of the graphing meter. The display will show any glitches or noise that may affect operation. Accessories are available for the meters that allow them to measure secondary ignition data and monitor COP activity.

Graphing meters can be connected to the engine and its operating systems, and then the vehicle can be taken for a road test. Intermittent problems can be observed and the data around those problems can be stored in the meter's memory for review after the road test.

With the standard hook-up, the meter will display all of the cylinders at once. If it appears that a cylinder is operating at a different time or with a lower or higher voltage than the others, the trigger lead can be moved from one cylinder to another to isolate the problem cylinder(s).

Both the power and waste spark of an EI system can be observed and compared for each coil. This helps in isolating specific problems within the system. For example,

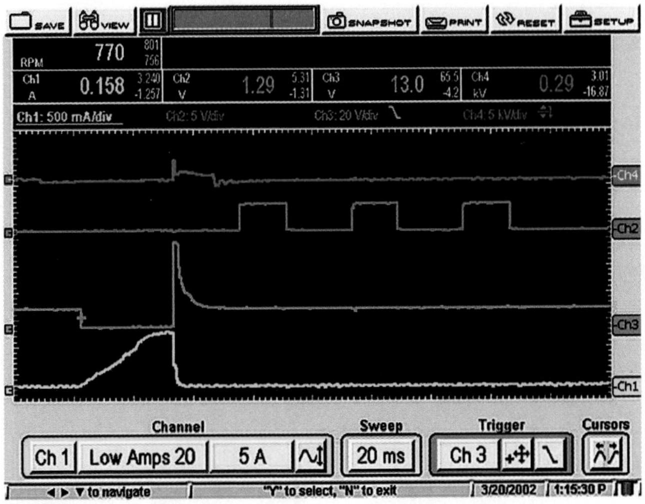

Figure 25–32 A four-channel graphing meter. CH1 = primary current, CH2 = primary voltage, CH3 = SPOUT (spark output), and CH4 = PIP. *Courtesy of Snap-on Tools Company*

one plug in an EI system can be fouled and that will affect the other plug in the circuit. By observing the activity of both, the problem plug can be identified.

EFFECTS OF IMPROPER TIMING

The primary circuit controls the secondary circuit. Therefore, the primary circuit controls ignition timing. Most problems within the primary circuit of computer-controlled ignition systems result in starting problems or poor performance due to incorrect timing.

If engine performance is poor, the cause of the problem can be many things. There can be a problem with the engine: poor compression, incorrect valve timing, overheating, and so on. The air/fuel mixture or the ignition timing may be incorrect. When the ignition timing is not correct, many tests will point to the problem. Incorrect ignition timing causes incomplete combustion at one or all engine speeds. Incomplete combustion causes excessive oxygen in the exhaust and the PCM tries to correct the apparent lean mixture. Incorrect timing is not a lean condition, but the PCM cannot tell that the timing is wrong. It only knows there is too much oxygen in the exhaust. Under this condition the waveform from the O_2 sensor will be lean-biased **(Figure 25–33)**.

Excessive O_2 in the exhaust will also show up on an exhaust gas analyzer. With incorrect timing you should also see higher-than-normal amounts of HC. Remember, it takes approximately seven seconds for the exhaust to be analyzed. If you slowly accelerate the engine and see the HC and O_2 levels on the exhaust gas analyzer rise, the condition that existed seven seconds earlier was the cause of the rise in emissions levels. To make this easier to track, make sure you hold the engine at each test speed for at least seven seconds. This way you will be able to observe the rise (or fall) of the emissions levels at that particular speed.

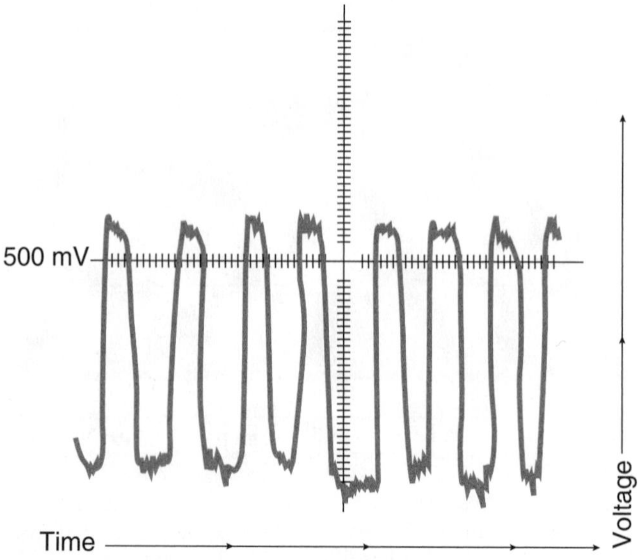

Figure 25-33 A lean-biased O_2 sensor.

Incorrect ignition timing also affects manifold vacuum readings and ignition system waveforms on a scope. When anything indicates a problem with the primary ignition circuit, the suspected parts should be tested. To save time, don't check the components in the secondary circuit until you know all of the primary is working fine. Symptoms of overly advanced timing include pinging or engine knock. Insufficient advance or retarded timing at higher engine rpms could cause hesitation and poor fuel economy.

SETTING IGNITION TIMING

The PCM controls ignition timing on nearly all new engines. Only engines equipped with a distributor may need to have their ignition timing set or adjusted. Although the computer assumes control of ignition timing, having the base timing correct is critical for the proper operation of the engine. Because the computer bases its control or change in ignition timing on the base setting, if the base is wrong all other ignition timing settings will also be wrong. On some systems, the base timing is adjustable. On others, if the base timing is wrong, the ignition module, distributor, or PCM may need to be replaced.

Each ignition system has its own set of procedures to check ignition timing. Always refer to the vehicle's emissions label or service manual before proceeding. These will give the correct procedure for disabling the computer's control of the timing, the correct timing specifications, and the conditions that must be present when checking or adjusting base ignition timing.

To check the ignition timing, the flashing timing light is aimed at the ignition timing marks. The timing marks are usually located on the crankshaft pulley or on the flywheel. A stationary pointer, line, or notch is positioned above the rotating timing marks. The timing marks are lines on the crankshaft pulley or flywheel that represent various positions of the piston as it relates to TDC. When piston 1 is at TDC, the timing line or notch will line up with the zero reference mark on the timing plate. Usually an engine is timed so the number 1 spark plug fires several degrees BTDC. The timing light flashes every time the number 1 spark plug fires. When pointed at the timing marks, the strobe effect of the light will freeze the spinning timing marks as it passes the timing scale. The ignition timing is checked by observing the degree of crankshaft rotation BTDC or ATDC when the spark plug fires.

It is important to remember that most computerized engine control systems use a knock sensor to retard the ignition timing when detonation occurs. The use of the sensor allows the PCM to set ignition timing with the most advance possible. If the knock sensor is faulty, two things can occur: the engine will run flat because of the lack of timing advance or the engine will have a heavy spark knock because the timing is too far advanced. This spark control system should be checked while checking the ignition system's base timing.

Because Ford has used so many different types of engine control and ignition systems, the exact procedure for checking the timing depends entirely on what systems the engine is equipped with. On EEC-IV systems, there is a base timing or SPOUT connector that needs to be disconnected **(Figure 25–34)**. Doing this prevents the PCM from controlling the timing and keeps the ignition at its base setting. The self-test mode of the EEC-IV system also allows timing advance to be monitored. Again, this procedure varies and the service manual should be referred to for the correct procedure.

Typically, General Motors' vehicles with fuel-injected engines require that an underhood test connector be grounded, connecting two terminals in the DLC, or disconnecting a connector at the distributor. Most carbureted engines require the disconnecting of a four-wire terminal at the distributor. The latter disconnects the ignition module from the computer but cannot be disconnected on fuel-injected engines.

Honda's programmed ignition controls the ignition timing according to inputs from several sensors and the PCM's ignition timing program. Ignition timing is checked with the engine at idle and a jumper wire connected across the terminals of a service check connector **(Figure 25–35)** located under the dash.

Spark plug gap and idle speed must be correct before checking or setting ignition timing. Also, the engine must be at operating temperature. After you have a base timing reading, compare it to the specifications. As an example, if the specification calls for 10 degrees BTDC **(Figure 25–36A)** and your reading was 3 degrees BTDC **(Figure 25–36B)**, the timing is retarded 7 degrees. This means the timing must be advanced by 7 degrees. To do this, rotate the distributor until the timing marks align at

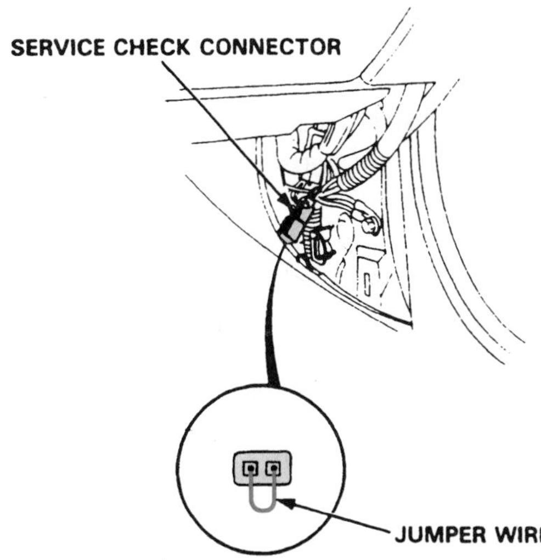

Figure 25–35 A service check connector with a jumper wire to check ignition timing on Honda engines. *Courtesy of American Honda Motor Co., Inc.*

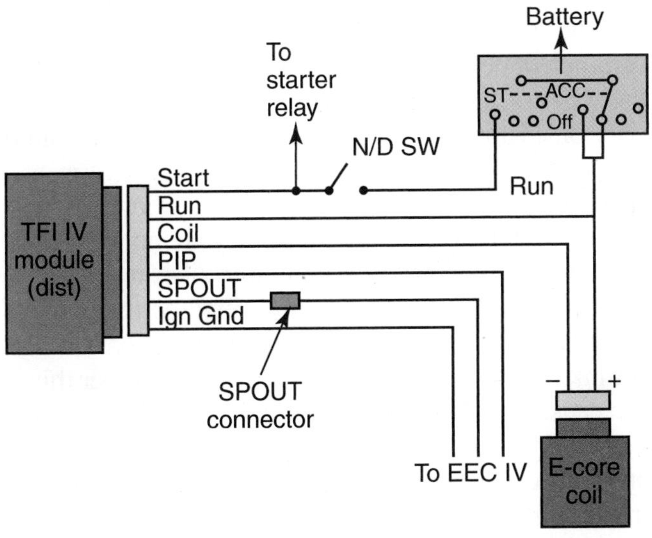

Figure 25–34 A TFI system with SPOUT connector. *Courtesy of Ford Motor Company*

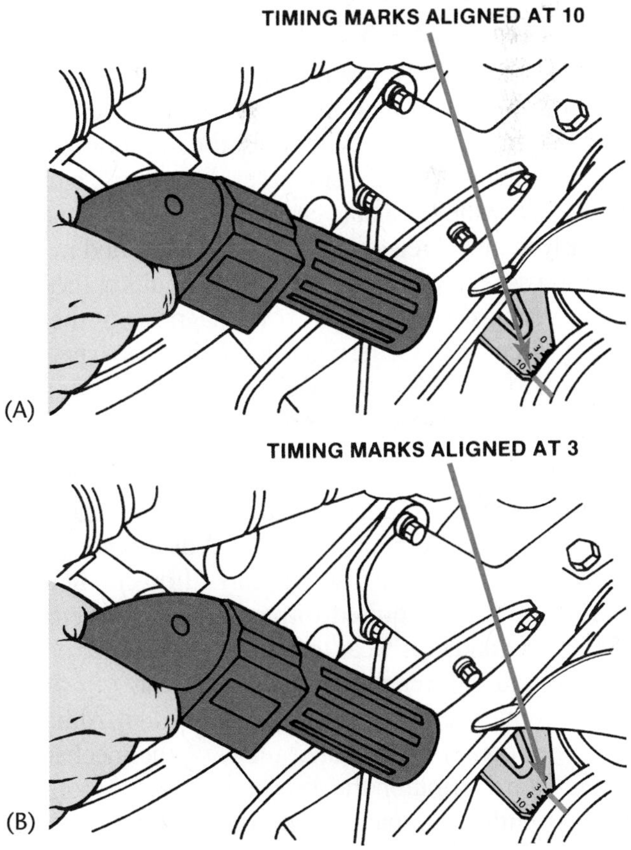

Figure 25–36 (A) Timing marks illuminated by a timing light at 10 degrees BTDC and (B) timing marks at 3 degrees BTDC.

10 degrees. Then retighten the distributor hold-down bolt.

Crank/Cam Sensors

If a crankshaft or camshaft sensor needs to be replaced, always clean the sensor tip and install a new spacer (if so equipped) on the sensor tip. New sensors typically have a new spacer already installed on the sensor. Install the sensor until the spacer lightly touches the sensor ring and tighten the sensor mounting bolt.

The air gap between the sensor and the trigger wheel will affect the operation of the ignition system. On many EI systems, this gap is not adjustable. This does not mean that the gap is not important and should not be checked. The gap should be checked, whenever possible. If there is no provision for adjusting the gap and the gap is incorrect, the sensor should be replaced. If the gap between the blades and the crankshaft sensor is not correct, the engine may fail to start, stall, misfire, or hesitate on acceleration.

On a few engines, the cam sensor must be properly positioned or timed. For example, on General Motors' 3.8-liter engines that are turbocharged, the cam sensor has a dot that must face away from the timing chain toward the passenger's side of the vehicle. Also, the sensor's wiring harness must face toward the driver's side of the vehicle.

DI AND EI SYSTEMS

This section concentrates on computer-controlled ignition systems. The biggest difference between computer-controlled DI and EI systems is the location of the crankshaft sensor and the circuits used to control ignition timing. Since the secondary circuit of all types of ignitions is nearly identical regardless of vehicle make and model, this discussion focuses on the primary circuits of these ignition systems. For the most part, diagnosis of the secondary circuit is the same regardless of ignition type. This holds true for EI systems as well. Secondary circuit diagnosis and service follows this discussion.

DI Systems

Distributor ignition systems are basically mechanical systems with some electronics thrown in. A distributor is driven by the camshaft via a set of gears. If precise ignition timing control is required to have an efficient engine, then engine and gear wear that affects distributor operation will cause the engine to be less efficient.

Many things can go wrong with a DI system. Although computer controls eliminated many of the mechanical parts of the distributor, the distributor is still responsible for distributing the spark to the cylinders, and wear will affect its operation.

Distributor shafts are typically supported, in the distributor housing, at one end and in the middle by a bushing or bearing. On its other end is the rotor. When the bushings keep the shaft centered in the housing, there is an even gap between the rotor and the spark plug terminals inside the distributor cap. There is also an even clearance between the armature mounted onto the shaft and the pickup unit held stationary on the distributor's breaker plate. As the bushings wear, the shaft will spin off center, closing the gap between the rotor and distributor cap and the armature and the pickup unit. If the bushings wear elliptically, the gaps will vary for different cylinders. This causes ignition timing to vary between cylinders.

Because there is an air gap between the rotor and the distributor cap, electrical arcing takes place between the two, causing a deterioration of the rotor tip and the distributor cap terminals. This added resistance could cause cylinder misfires.

EI Systems

Standard test procedures using an oscilloscope, ohmmeter, and timing light can be used to diagnose problems in distributorless ignition systems. Keep in mind, however, that problems involving one cylinder may also occur in its companion cylinder that fires off the same coil. Some scopes require that their pickups be placed on each pair of cylinders to view all patterns. Special adapters are available to make these hookups less troublesome. Many newer engine analyzers use a single adapter that allows viewing of all cylinder patterns at one time.

Follow the testing procedures outlined in the vehicle's service manual for the engine control system. Specific computer-generated trouble codes are designed to help troubleshoot ignition problems in these systems. The diagnostic procedure for EI systems varies depending on the vehicle make and model year. Always follow the recommended procedure.

When diagnosing these systems, keep in mind that there is a separate primary circuit for each coil. If one coil does not work properly, it may be caused by something common or not common to the other coils. Regardless of the system's design, there are common components in all electronic ignitions: an ignition module, crankshaft and/or camshaft sensors, ignition coils, secondary circuit, and spark plugs. Most of these components are common only to the spark plug or spark plugs they are connected to. The components that are common to all cylinders (such as the camshaft sensor) will most often be the cause of no-start problems. The other components that are not common will cause misfire problems. Many other things can cause a misfire; check those before moving on to the ignition system. Test for intake manifold vacuum leaks, and also check the engine's compression and the fuel-injection system.

Although there are many variations in the design of these systems, all operate in much the same way. A sensor sends a signal noting the position of the crankshaft to the PCM. The PCM, based on this signal and input from other sensors, controls the action of an ignition module. The ignition module controls the primary ignition circuit and the firing of the spark plugs. There is either one coil for each cylinder of the engine or one coil for each pair of cylinders. In the latter case, each coil fires two spark plugs at the same time. One firing takes place when one cylinder is on the compression stroke, and the other firing takes place when another cylinder is on its exhaust stroke. The spark during the exhaust stroke is wasted and has no effect on the engine or ignition system.

Coil-On-Plug Systems

Although COP systems are unique compared to other ignition systems, preignition and detonation can still occur and spark plugs can still foul or misfire.

Remember, an individual coil problem will cause misfiring in only one cylinder. Ignition coils are tested with an ohmmeter in much the same way as other ignition coils. If the resistance is out of specification, the coil should be replaced. Intermittent coil problems can be caused by corrosion at the electrical connectors to the coil.

Codes retrieved from the PCM will indicate whether the misfire is a general one or an individual cylinder. Since a COP ignition problem affects only one cylinder, the general misfire code is probably caused by a fuel delivery problem or a vacuum leak.

DTCs that indicate a misfire at an individual cylinder are typically caused by a dirty or defective fuel injector, a fouled spark plug, a bad coil, or an engine mechanical problem. If the misfire is caused by a fuel injector problem, a fuel injector code that identifies the affected cylinder will also be retrieved.

If a crankshaft position sensor is bad, there will be no timing reference, and this can prevent the engine from starting or make it hard to start.

Each ignition coil has a driver circuit in the PCM that controls primary current flow. If there is a bad driver circuit, that spark plug will not fire. Also, keep in mind that an engine may start with a faulty camshaft sensor but it may only run in the fail-safe or limp-in mode because the fuel injectors cannot be synchronized without the camshaft signal.

Coil-near-plug systems can be checked with a scope or graphing meter the same way as other ignition systems. The pickup for secondary signals is installed over the spark plug wire. However, COP systems require special adapters to connect the scope or analyzer to the ignition system. Some low-amperage current probes can also be used to monitor the activity of an individual coil in a COP system.

PRIMARY CIRCUIT COMPONENTS

Ignition systems have many characteristics unique to the system's manufacturer. It would be impossible in any one textbook to explain all the variations. However, the basic goal of any electrical troubleshooting procedure is always to identify electrical activity under a given set of circumstances.

The secret to component testing is to use good troubleshooting practices. Work systematically through a circuit, testing each wire, connector, and component. Do not jump around back and forth between components. The component inadvertently overlooked is probably the one causing the trouble. Checks must be made for available voltage, voltage output, resistance of wires and connectors, and available ground. Always compare the readings with specifications given in the manufacturer's service manual.

The following sections briefly outline common test procedures using different test equipment. For accurate testing always refer to service manual wiring diagrams and testing instructions.

USING SERVICE MANUALS

Service manuals are indispensable when troubleshooting ignition problems. They contain such vital information as base and advance timing specifications, color-coded wiring diagrams and terminal connector descriptions, and illustrations showing test connections for voltage, continuity, and resistance checks. The diagnostic charts found in the service manual are particularly helpful in troubleshooting ignition problems. Ignoring the valuable data contained in the ignition service section of these manuals can lead to many frustrating and costly service mistakes. ■

Ignition Switch

An ignition switch supplies voltage to the ignition control module and/or the ignition coil. Often an ignition system has two wires connected to the run terminal of the ignition switch. One is connected to the module. The other is connected to the primary resistor and coil. The start terminal of the switch is also wired to the module.

You can check for voltage using either a 12-volt test light or a DMM. To use a test light, turn the ignition key off and disconnect the wire connector at the module. Also, disconnect the S terminal of the starter solenoid to prevent the engine from cranking when the ignition is in the run position. Turn the key to the run position and probe the red wire connection to check for voltage. Also check for voltage at the battery terminal of the ignition coil using the test light.

Next, turn the key to the start position and check for voltage at the white wire connector at the module and the battery terminal of the ignition coil. If voltage is present, the switch and its circuit are okay.

To make the same test using a DMM, turn the ignition switch to the off position and back-probe, with the meter's positive lead, the power feed wire at the module. Connect the meter's negative to a good ground at the distributor base. Turn the ignition to the run or start position as needed, and measure the voltage. The reading should be at least 90% of battery voltage.

Primary Resistor

Measure the resistance of the primary (ballast) resistor with an ohmmeter. Remember, the key must be off when this test is performed because power in the circuit damages the meter. In the ignition system shown in **Figure 25–37**, ohmmeter leads are connected at the battery terminal of the coil and the wiring harness connector wire that join the red wire in the ignition module connector. Compare the resistance reading to specifications.

Ignition Coil Resistance

With the key off and the battery lead to the ignition coil disconnected, use an ohmmeter to measure the primary and secondary winding resistance of the ignition coil.

The secondary windings of a waste spark ignition coil are not checked in the same way as other coils because there are two secondary terminals on each coil. However, they can still be checked with an ohmmeter. COP coils are checked in the same way as other coils. All coils should be carefully inspected. When checking the resistance across the windings, pay particular attention to the meter reading. If the reading is out of specifications, even if it is only slightly out, the coil or coil assembly should be replaced.

To check the primary windings, calibrate an ohmmeter on the X1 scale and connect the meter leads to the primary coil terminals to test the winding **(Figure 25–38)**. An infinite ohmmeter reading indicates an open winding. The winding is shorted if the meter reading is below the specified resistance. Most primary windings have a resistance of 0.5 to 2 ohms, but the exact manufacturer's specifications must be compared to the meter readings.

To check the secondary winding, calibrate the meter on the X1,000 scale and connect it from the coil's secondary terminal to one of the primary terminals **(Figure 25–39)**. A meter reading below the specified resistance

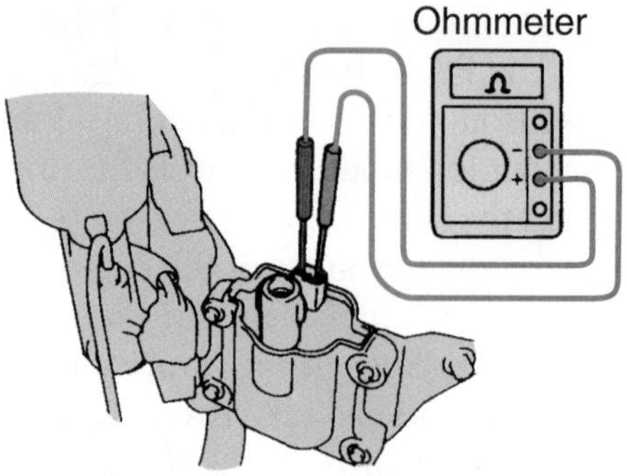

Figure 25-38 An ohmmeter connected to primary coil terminals. *Reprinted with permission*

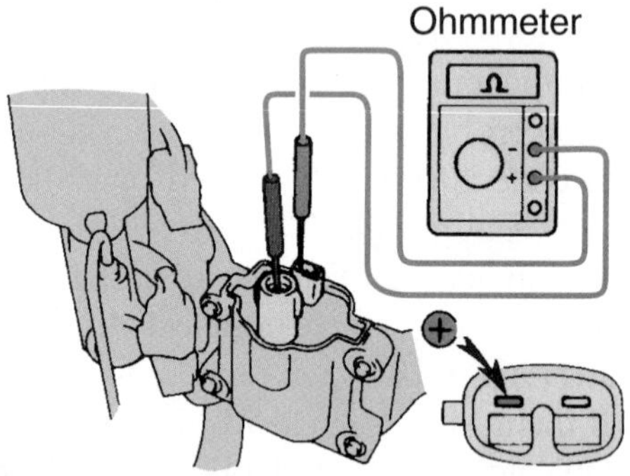

Figure 25-39 An ohmmeter connected from one primary terminal to the coil tower to test secondary winding. *Reprinted with permission*

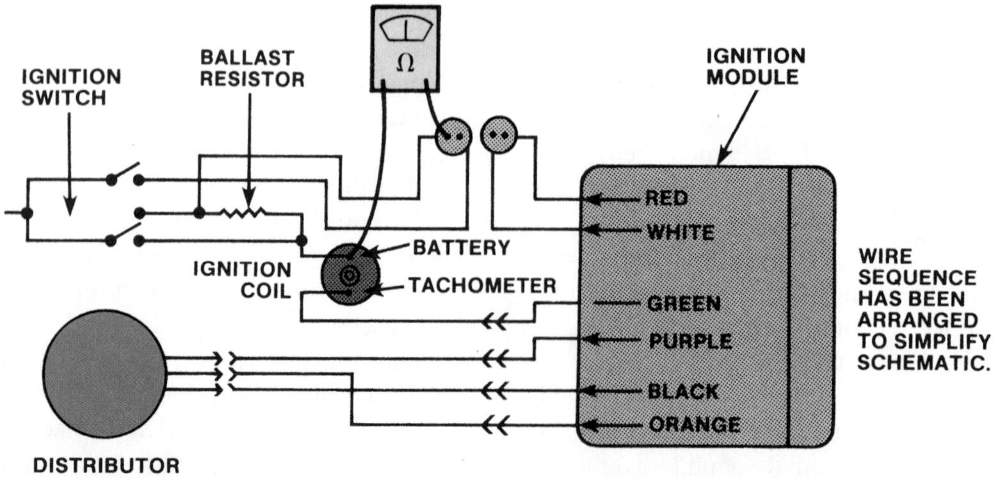

Figure 25-37 Checking primary (ballast) resistance. *Courtesy of Ford Motor Company*

indicates a shorted secondary winding. An infinite meter reading proves that the winding is open.

The secondary resistance of waste spark ignition coils is checked by connecting a lead of the ohmmeter to each of the secondary terminals **(Figure 25–40)**. As with other coils, compare your readings to specifications.

In some coils, the secondary winding is connected from the secondary terminal to the coil frame. When the secondary winding is tested in these coils, the ohmmeter must be connected from the secondary coil terminal to the coil frame or to the ground wire terminal extending from the coil frame.

Many secondary windings have 8,000 to 20,000 ohms resistance, but the meter readings must be compared to the manufacturer's specifications. Ohmmeter tests do not indicate such defects as defective insulation around the coil windings, which causes high-voltage leaks. Therefore, an accurate indication of coil condition is the coil maximum voltage output test with a test spark plug

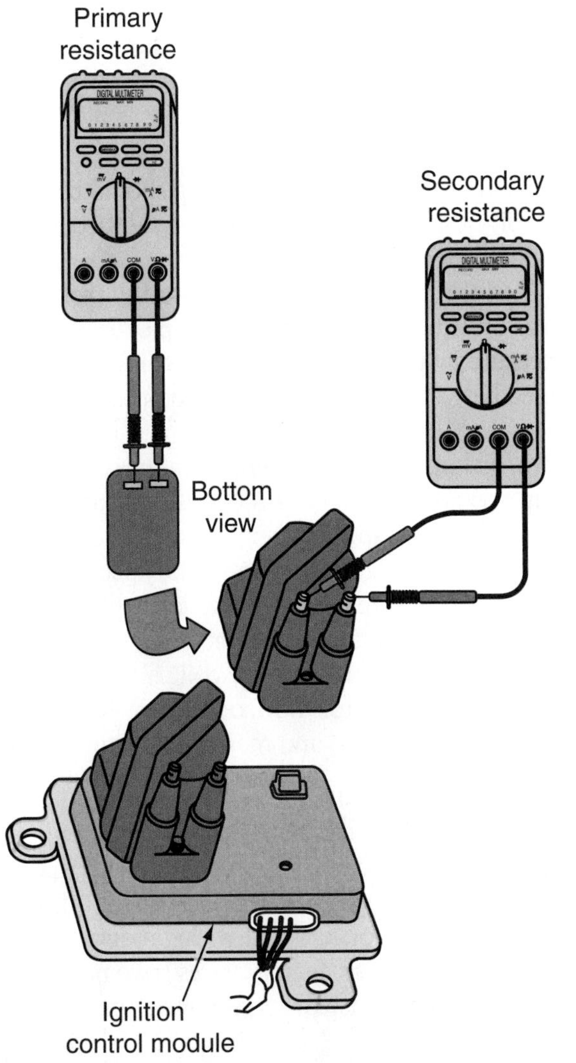

Figure 25–40 Meter connections for testing the resistance of a double-ended ignition coil.

connected from the coil secondary wire to ground as explained in the no-start diagnosis.

Pickup Coil

The pickup coil of a magnetic pulse generator or metal detection sensor is also checked with an ohmmeter. Connect the ohmmeter to the pickup coil terminals to measure the resistance and to test the pickup coil for an open or a shorted condition. While the ohmmeter leads are connected, pull on the pickup leads and watch for an erratic reading, which indicates an intermittent open in the pickup leads. Most pickup coils have 150 to 900 ohms resistance, but always refer to the manufacturer's specifications. If the pickup coil is open, the ohmmeter will display an infinite reading. When the pickup coil is shorted, the ohmmeter will display a reading lower than specifications.

Connect the ohmmeter from one of the pickup leads to ground to test the pickup for a short to ground. If there is no short to ground, the ohmmeter will give an infinite reading. Refer to **Figure 25–41**, which shows that ohmmeter 1 is testing for a short to ground, while ohmmeter 2 is testing the resistance of the pickup coil.

Another method of measuring the pickup coil's AC signal is with a simple voltmeter set on its low-voltage scale. The meter registers AC voltage during cranking. Measure this voltage as close to the control module as possible to account for any resistance in the connecting wire from the pickup coil to the module.

It is important that all wires and connectors between the distributor and module and module to the engine control computer be visually checked as well as checked for excessive resistance with an ohmmeter.

If the resistance checks are within specifications, the circuit should be checked with a digital voltmeter. Turn the ignition on and connect the voltmeter across the voltage input wire and ground. Compare the reading to specifications. The voltmeter should also be used to check the resistance across the ground circuit of the distributor. Do this by measuring the voltage drop across the circuit; a drop of more than 0.1 volt is excessive.

On Chrysler optical distributors, the pickup voltage supply and ground wires may be tested at the four-wire connector near the distributor. With the ignition switch on, connect the voltmeter from the orange voltage supply wire to ground. The reading should be 9.2 to 9.4 volts **(Figure 25–42)**. Because the expected reading may vary depending on the model year, always refer to the manufacturer's service manual before conducting the test.

When the voltmeter is connected between the black/light-blue ground wire and an engine ground, the reading should be less than 0.2 volt. When the voltmeter is connected from the grey/black reference pickup wire or the tan/yellow sync pickup wire to an engine ground, the

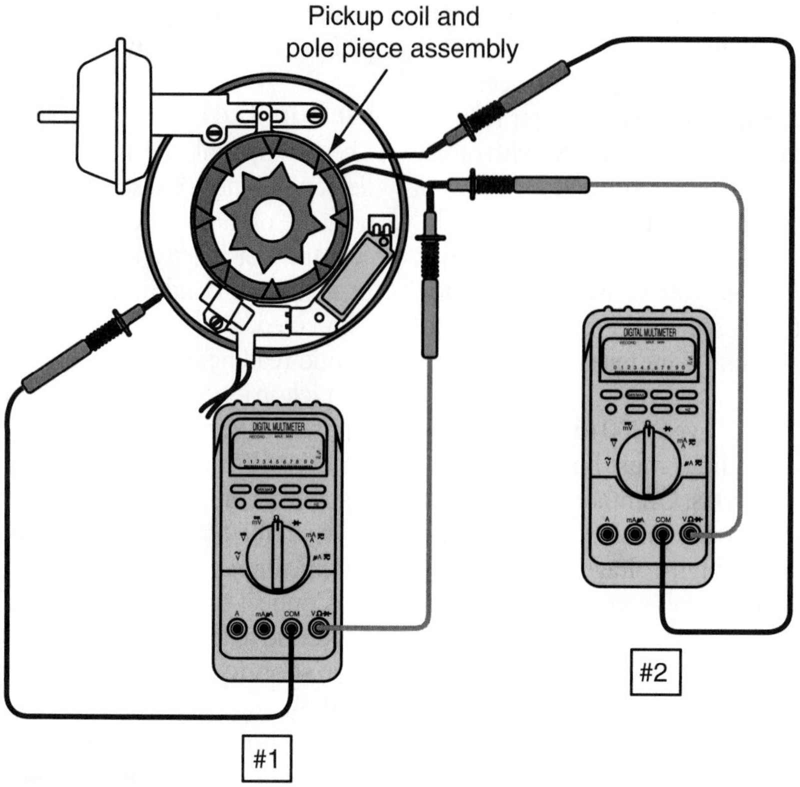

Figure 25-41 Ohmmeter-to-pickup coil test connections.

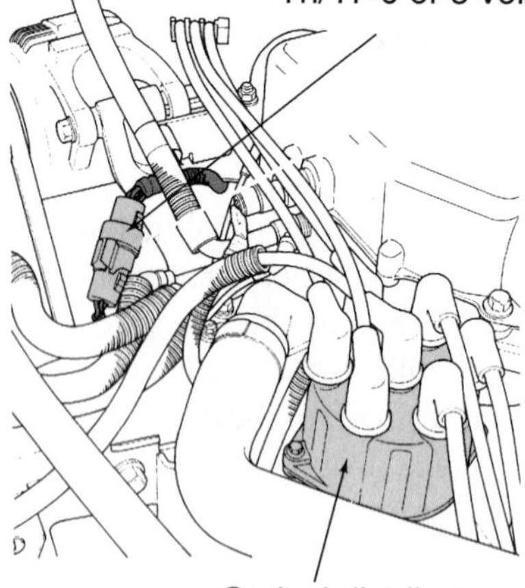

Figure 25-42 A Chrysler optical distributor four-wire connector. *Courtesy of DaimlerChrysler Corporation*

voltmeter should cycle from nearly 0 volt to 5 volts while the engine is cranking. If the pickup signal is not within specifications, the pickup is defective. A defective sync pickup in this distributor should not cause a no-start problem but will affect engine performance.

The ignition module is contained in the PCM on Chrysler products. On Chrysler fuel-injected engines, the reference pickup and the sync pickup should be tested. If either of these pickups is defective, a fault code may be stored in the computer memory.

When the pickup coil is bolted to the pickup plate, such as on Chrysler distributors, the pickup air gap may be measured with a nonmagnetic copper feeler gauge positioned between the reluctor high points and the pickup coil.

If a pickup gap adjustment is required, loosen the pickup mounting bolts and move the pickup coil until the manufacturer's specified air gap is obtained. Retighten the pickup coil retaining bolts to the specified torque. Some pickup coils are riveted to the pickup plate. A pickup gap adjustment is not required for these pickup coils.

When checking the pickup coil, always check the distributor bushing for horizontal movement, which changes the pickup gap and may cause engine misfiring.

Hall-Effect Sensors

Prior to testing a Hall-effect pickup, the ohmmeter should be connected across each of the wires between the pickup and the computer with the ignition switch off. A computer terminal and pickup coil wiring diagram is essential for these tests. Satisfactory wires have nearly 0 ohm resistance, while higher or infinite readings indicate defective wires.

Most Hall-effect sensors can be tested by connecting a 12-volt battery across the plus (+) and minus (–) volt-

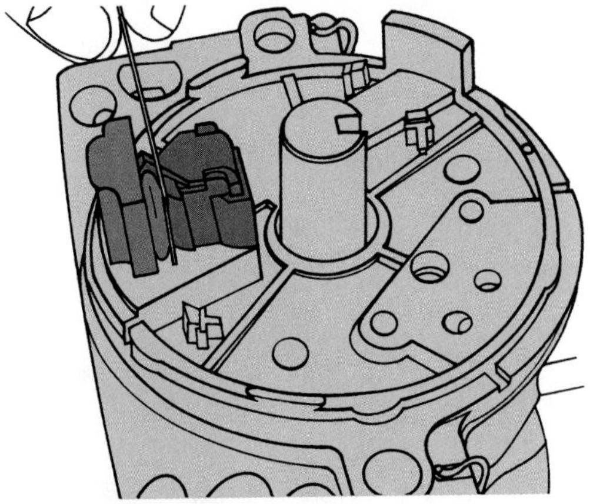

Figure 25-43 Test Hall-effect sensor operation with a steel feeler gauge.

age (supply current) terminals of the Hall layer and a voltmeter across the minus (–) and signal voltage terminals.

With the voltmeter hooked up, insert a steel feeler gauge or knife blade between the Hall layer and magnet **(Figure 25–43)**. If the sensor is good, the voltmeter should read within 0.5 volt of battery voltage when the feeler gauge or knife blade is inserted and touching the magnet. When the feeler gauge or blade is removed, the voltage should read less than 0.5 volt.

In the following tests, the distributor connector is connected to the unit and the connector back-probed. With the ignition switch on, a voltmeter should be connected from the voltage input wire to ground. The specified voltage should appear on the meter. The ground wire should be tested with the ignition switch on and a voltmeter connected from the ground wire to a ground connection near the distributor. With this meter connection, the meter indicates the voltage drop across the ground wire, which should not exceed 0.2 volt.

Connect a digital voltmeter from the pickup signal wire to ground. If the voltmeter reading does not fluctuate while cranking the engine, the pickup is defective. However, if the voltmeter reading fluctuates from nearly 0 volt to between 9 and 12 volts, that indicates a satisfactory pickup. During this test, the voltmeter reading may not be accurate because of the short duration of the voltage signal. If the Hall-effect pickup signal is satisfactory and the 12-volt test lamp did not flutter during the no-start test, the ignition module is probably defective.

Using a Logic Probe

The primary can also be checked with a logic probe. There are three lights on a logic probe. The red light illuminates when the probe senses more than 10 volts. When the monitored signal has less than 4 volts, the green light flashes. The yellow light flashes whenever the voltage changes. This light is used to monitor a pulsing signal, such as one produced by a digital sensor like a Hall-effect switch.

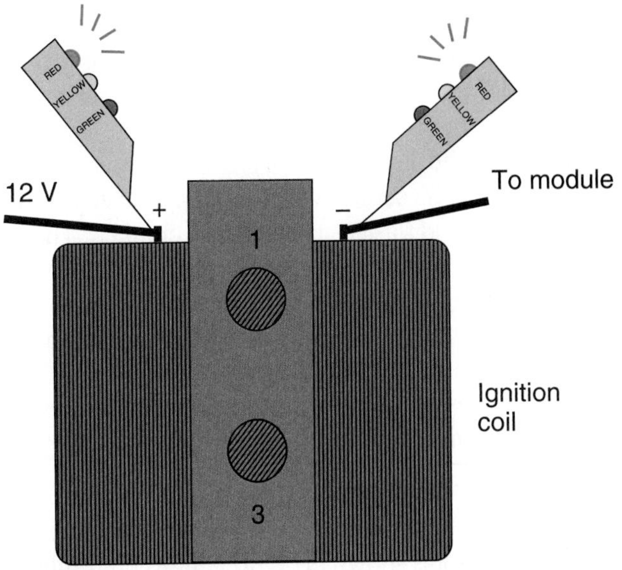

Figure 25-44 The red light on a logic probe should turn on when touched to both sides of the primary winding.

To check the primary circuit with a logic probe, turn the ignition on. Touch the probe to both (positive and negative) primary terminals at the coil. The red light should come on at both terminals **(Figure 25–44)**, indicating that at least 10 volts are available to the coil and that there is continuity through the coil. If the red light does not come on when the positive side of the coil is probed, check the power feed circuit to the coil. If the light comes on at the positive terminal but not at the negative, the coil has excessive resistance or is open.

Now move the probe to the negative terminal of the coil and crank the engine. The red and green lights should alternately flash, indicating that over 10 volts are available to the coil while cranking and that the circuit is switching to ground. If the lights do not come on, check the ignition power feed circuit from the starter. If the red light comes on but the green light does not, check the crankshaft or camshaft sensor. If these are working properly, the ignition module is probably defective.

A Hall-effect switch is also easily checked with a logic probe. If the switch has three wires, probe the outer two wires with the ignition on **(Figure 25–45)**. The red light

Figure 25-45 One end terminal of a Hall-effect sensor connector should cause the red light to come on; the other end terminal should cause the green light to come on.

should come on when one of the wires is probed, and the green light should come on when the other wire is probed. If the red light does not turn on at either wire, check the power feed circuit to the sensor. If the green light does not come on, check the sensor's ground circuit.

Back-probe the center wire and crank the engine. All three lights should flash as the engine is cranked. The red light will come on when the sensor's output is above 10 volts. As this signal drops below 4 volts, the green light should come on. The yellow light will flash each time the voltage changes from high to low. If the logic probe's lights do not respond in this way, check the wiring at the sensor. If the wiring is okay, replace the sensor.

Using a Lab Scope

The waveform for a normally operating inductive-type crankshaft sensor is shown in **Figure 25–46**. Carefully examine the waveform for traces of noise and false pulses.

The pattern of the trace will vary according to the position and number of slots machined into the trigger wheel. The trigger wheel in **Figure 25–47** has nine slots. Eight of them are evenly spaced and one slot is placed close to one of the eight. The trace for this type of sensor will also have eight evenly spaced pulses. One of the pulses will be quickly followed by another pulse **(Figure 25–48)**. Any waveform that does not match the configuration of the trigger wheel indicates a problem with the sensor or its circuit.

If the crankshaft sensor is a Hall-effect switch, digital waves should be seen on the scope. All pulses should be identical in spacing, shape, and amplitude. By using a dual-trace lab scope, the relationship between the crankshaft sensor and the ignition module can be observed. During starting, the module will provide a fixed amount of timing advance according to its program and the cranking speed of the engine. By observing the crankshaft sensor output and the ignition module, this advance can be observed **(Figure 25–49)**. The engine will not start if the ignition module does not provide for a fixed amount of timing advance.

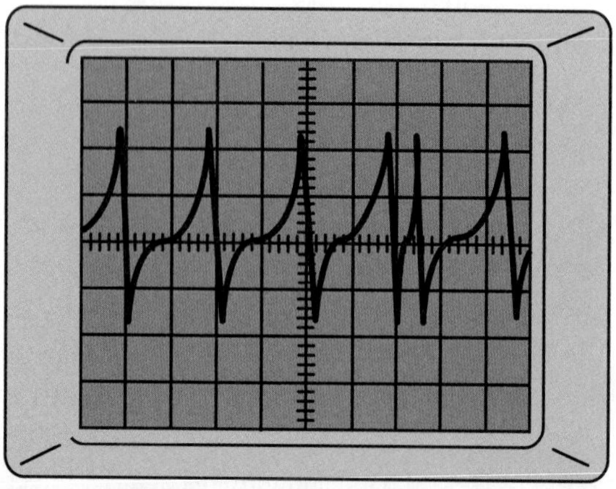

Figure 25-46 A normal waveform for an inductive crankshaft sensor.

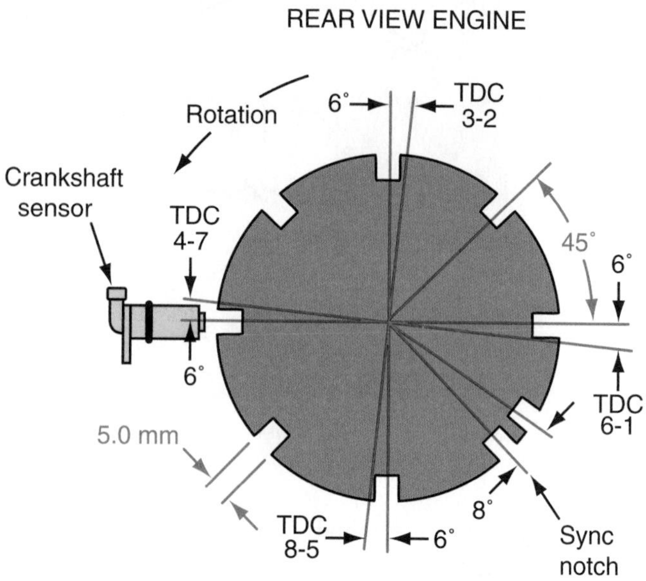

Figure 25-47 A nine-slot trigger wheel for a crankshaft sensor.

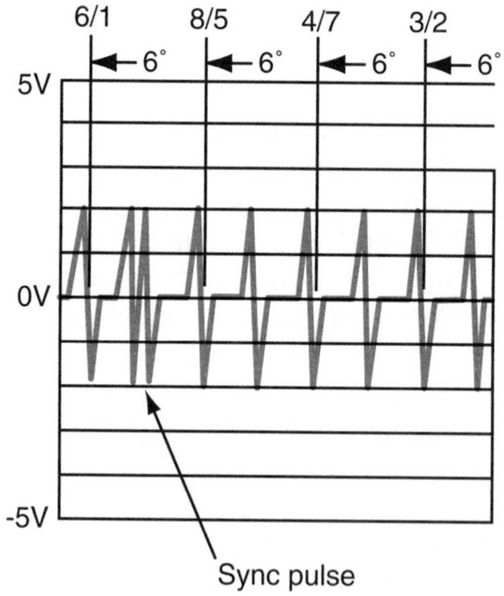

Figure 25-48 A waveform for the sensor shown in Figure 25–47.

Using Special Equipment

Knock Sensors Most computer-controlled systems use knock sensors **(Figure 25–50)** to retard timing when the engine is experiencing pinging or knocking. Obviously, a faulty knock sensor produces problems that appear to be ignition-system-related. A quick check of a knock sensor is to watch the ignition's timing while the engine is running and the engine block is tapped with a hammer. The noise should cause a change in timing.

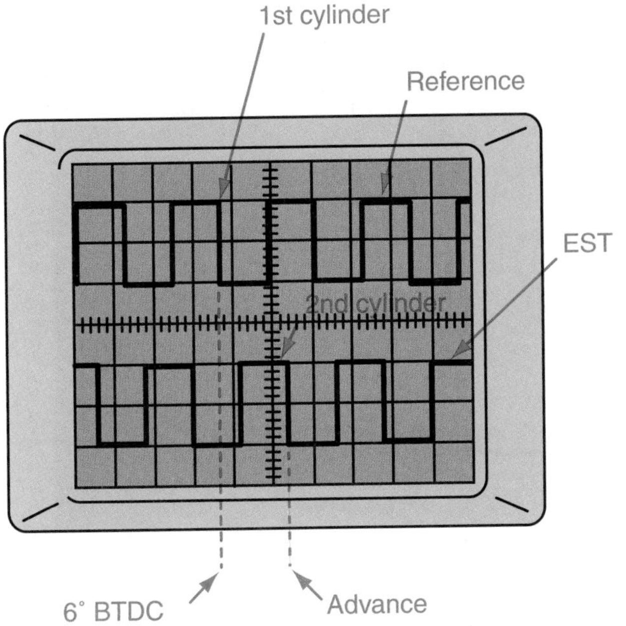

Figure 25-49 EST and crankshaft sensor signals compared on a dual trace scope.

Figure 25-50 A knock (detonation) sensor.

The knock sensor can also be checked by tapping near the sensor and watching for an AC pulse on a DMM or lab scope.

Control Module The most effective method of testing for a defective control module is to use an ignition module tester. This electronic tester evaluates and determines whether the module is operating within a given set of design parameters. It does so by simulating normal operating conditions while looking for faults in key components. Unfortunately, many module testers are designed to troubleshoot specific makes and models of ignitions. Many shops find it impractical to have testers for every type of system they service.

Keep in mind that control modules are very reliable. They are also one of the most expensive ignition system components, or they may be part of the PCM. So, if a module tester is not available, check out all other system components before condemning the control module.

DISTRIBUTOR SERVICE

Often you will need to remove the distributor to service it or to service another part of the engine. Before removing the distributor, check the condition of the distributor's bushing. Do this by grasping the distributor shaft and moving it toward the outside of the distributor. If any movement is detected, remove the distributor and check the bushing on the bench. A typical procedure for removing a distributor follows:

PROCEDURE

Removing a Distributor

STEP 1 *Disconnect the distributor wiring connector and the vacuum advance hose, if the vehicle is so equipped.*

STEP 2 *Remove the distributor cap and note the position of the rotor. On some vehicles, it may be necessary to remove the spark plug wires from the cap prior to cap removal.*

STEP 3 *Note the position of the vacuum advance, then remove the distributor hold-down bolt and clamp.*

STEP 4 *Pull the distributor from the engine. Most distributors need to be twisted as they are pulled out of their bore.*

STEP 5 *Once the distributor is removed, install a shop towel in the distributor opening to keep foreign material out of the engine block.*

Distributor Bushing Check

Before proceeding with the disassembly of the distributor, the shaft's bushings need to be checked. Do this regardless of the type of repair you plan for the distributor. Lightly clamp the distributor housing in a soft-jaw vise. Clamp a dial indicator on the top of the distributor housing. Position the plunger of the indicator so that it rests on the distributor shaft. When the shaft is pushed horizontally, observe the movement of the shaft on the indicator. Compare this movement to the specifications given in the service manual. If the movement exceeds the allowed amount, the distributor bushings and/or shaft are worn. Many manufacturers recommend complete distributor replacement rather than bushing or shaft replacement.

Distributor Inspection

Some guidelines to follow while inspecting a distributor assembly are:

- Inspect all lead wires for worn insulation and loose terminals. Replace these wires as necessary.

- Inspect the centrifugal advance mechanism for wear. In particular, check the weights for wear on the pivot holes. Replace the weights or the complete shaft assembly, if necessary.
- Inspect the pickup plate for wear and rotation. If this plate is loose or seized, replacement is required.
- Connect a vacuum hand pump to the vacuum advance outlet and apply 20 inches of vacuum. The advance diaphragm should hold this vacuum without leaking.
- Check the distributor gear for worn or chipped teeth.
- Inspect the reluctor for damage. If the high points are damaged, the distributor bushing probably is worn, allowing the high points to hit the pickup coil.
- The entire assembly should be cleaned thoroughly. Make sure all solvent residue is completely removed. Do not clean the vacuum advance unit or the pickup unit with solvent.

Installing and Timing the Distributor

Photo Sequence 22 shows the correct way to reinstall a distributor on a typical engine. The following procedure may be followed to install the distributor and time it to the engine, if all reference marks were made during the removal of the distributor:

PROCEDURE

Installing and Timing a Distributor

STEP 1 *Lubricate the O-ring on the distributor shaft.*

STEP 2 *Position the rotor so that it is aligned with the mark made on the distributor housing prior to removal.*

STEP 3 *Align the distributor housing with the mark made on the engine block during removal.*

STEP 4 *Lower the distributor into the engine block, making sure the distributor drive is fully seated. Distributors equipped with a helical drive gear will rotate as the distributor is being installed, causing the distributor to move away from the reference marks. Pay attention to how much the rotor moves, then remove the distributor and move the rotor backward the same amount. This should allow the shaft to rotate while the distributor is being installed and still be aligned with the reference marks.*

STEP 5 *Then make sure the distributor housing is fully seated against the engine block. Sometimes it may be necessary to wiggle or rock the distributor to seat it fully into the drive gear. Distributors with* ***drive lugs (Figure 25–51)*** *must be mated with the drive grooves in the camshaft. Both are offset to eliminate the possibility of installing the distributor 180 degrees out of time.*

STEP 6 *Rotate the distributor a small amount so the timer core teeth and pickup teeth are aligned.*

STEP 7 *Install the distributor hold-down clamp and bolt, leaving the bolt slightly loose.*

STEP 8 *Install the spark plug wires in the direction of distributor shaft rotation and in the cylinder firing order.*

STEP 9 *Connect the distributor wiring connectors. The vacuum advance hose is usually left disconnected until the timing is set with the engine running.*

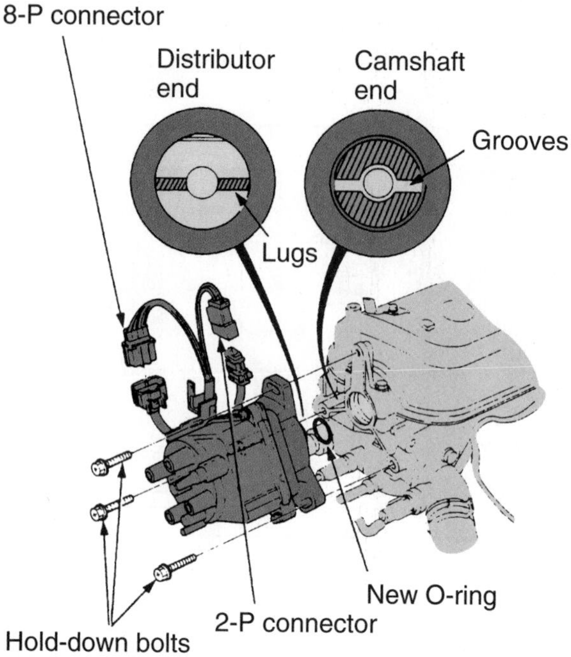

Figure 25–51 The distributor drive must be aligned with the camshaft drive to seat the distributor in the engine block or cylinder head. *Courtesy of American Honda Motor Co., Inc.*

SECONDARY COMPONENT TESTS AND SERVICE

Two major components make up the secondary of nearly all ignition systems: the spark plug and spark plug wire. DI systems also have a distributor cap and rotor in the secondary.

Testing the secondary circuit of an EI system is just like testing the secondary of any other type of ignition system. The spark plug wires and spark plugs should be tested to ensure they have the appropriate amount of resistance. Since the resistance in the secondary dictates the amount of voltage with which the spark plug will fire, it is important that secondary resistance be within the desired range.

A quick way to examine the secondary circuit is with a tune-up scope. On the scope, the firing line and the spark line indicate the resistance of the secondary. While observing the firing line, remember the height of the line increases with an increase in resistance. The length of the spark line decreases as the firing line goes higher. This means that high resistance will cause excessively high firing voltages and reduced spark times.

When checking an EI system with a tune-up scope, remember that on waste spark systems half of the plugs fire with reverse polarity. This means half of the firing lines will be higher than the other firing lines. Normally, reverse firing requires 30% more voltage than normal firing.

Excessive resistance is not the only condition that will affect the firing of a spark plug. Spark plug wires and the spark plug itself can allow the high voltage to leak and establish current through another metal object instead of the electrodes of the spark plug. When this happens, the spark plug does not fire and combustion does not take place within the cylinder.

Another thing to keep in mind while working on waste-fire-type systems is that the secondary circuit is completed through the metal of the engine. If the spark plugs are not properly torqued into the cylinder heads, the threads of the spark plug may not make good contact and the circuit may offer resistance. Always tighten spark plugs to their specified torque. Many of today's engines have aluminum cylinder heads. Not only do these heads require different torque specifications than iron heads, they also require extra care when installing the plugs. Make sure not to cross-thread them.

Most manufacturers recommend the use of an antiseize compound on spark plug threads. This compound must be applied in the correct amount and at the correct place **(Figure 25–52)**. Too little compound will cause gaps in the contact between the spark plug threads and the spark plug bores. Too much may allow the spark to jump to a buildup rather than to the spark plug electrode.

Spark Plugs

All of the ignition system is designed to do no more than supply the voltage necessary to cause a spark across the gap of a spark plug. This simple event is what starts the combustion process. Needless to say, a healthy spark plug is extremely important to the combustion process. Spark plug replacement is part of the preventive maintenance program for all vehicles. The recommended replacement interval depends on a number of factors but ranges from 20,000 to 100,000 miles (32,000 to 160,000 km).

Removal of an engine's spark plugs is pretty straightforward. Remove the cables from each plug, being careful not to pull on the cables. Instead, grasp the boot **(Figure 25–53)** and gently twist it off. (To save time and avoid confusion later, use masking tape to mark each of the cables with the number of the plug it attaches to.)

Using a spark plug socket and ratchet, loosen each plug a couple of turns. Regardless of what other tools may need to be used, a spark plug socket is essential for plug removal and installation. A spark plug socket has an internal rubber bushing to prevent plug insulator breakage. Spark plug sockets are available in two sizes: $^{13}/_{16}$ inch (for 14-millimeter gasketed and 18-millimeter tapered-seat plugs) and $^{5}/_{8}$ inch (for 14-millimeter tapered-seat plugs). They can be either $^{3}/_{8}$- or $^{1}/_{2}$-inch drive, and many feature an external hex so that they can be turned using an open end or box wrench.

Once the spark plugs are loose, use compressed air to blow dirt away from the base of the plugs. Then remove the plugs, making sure to remove the gasket as well (if applicable). When the spark plugs are removed, they should be set in order so the technician can identify the spark plug from each cylinder.

Check the threads in the cylinder head for damage. Normally you can do this by feel as you remove a spark plug. If the plug does not turn out smoothly after it is loose, the threads may be damaged. Often the threads can be cleaned up with a spark plug thread chaser. Also, check the threads on the spark plug. Look for damage or metal embedded in the threads, as these are sure signs of

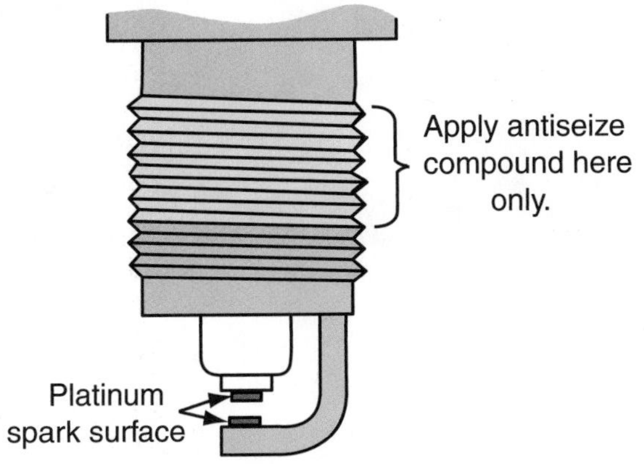

Figure 25-52 Proper placement of antiseize compound on the threads of a spark plug.

Figure 25-53 Grasp the boot and twist while pulling it off a spark plug. *Reprinted with permission*

PHOTO SEQUENCE 22

Installing a Distributor and Setting the Timing of an Engine

P22-1 *Remove the spark plug from cylinder 1. Install a test plug in the end of the spark plug wire and position the test plug away from you and cylinder 1.*

P22-2 *Place your thumb over the bore for the spark plug and crank the engine until compression is felt.*

P22-3 *Lightly crank the engine while observing the timing marks on the crankshaft pulley. Once the timing marks are aligned, stop cranking.*

P22-4 *Identify the position for the number 1 spark plug wire on the distributor cap.*

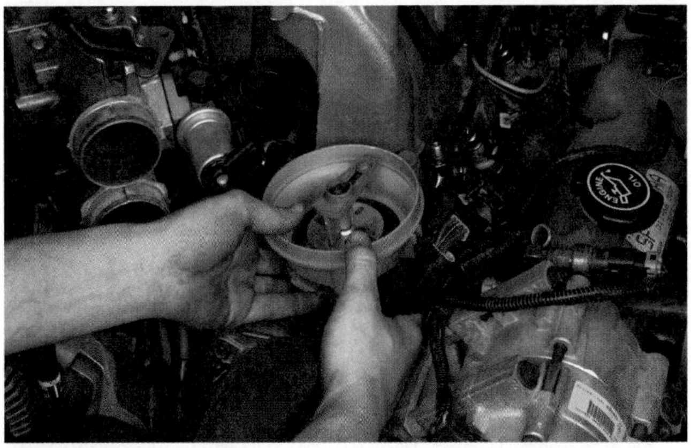

P22-5 *Determine which way the distributor shaft will rotate when it is lowered onto the camshaft gear. Then position the rotor so that it will be under the number 1 terminal on the distributor cap after the distributor gear is meshed with the camshaft gear.*

P22-6 *After the distributor is installed in the engine, turn the distributor housing slightly so the pickup coil is aligned with the reluctor.*

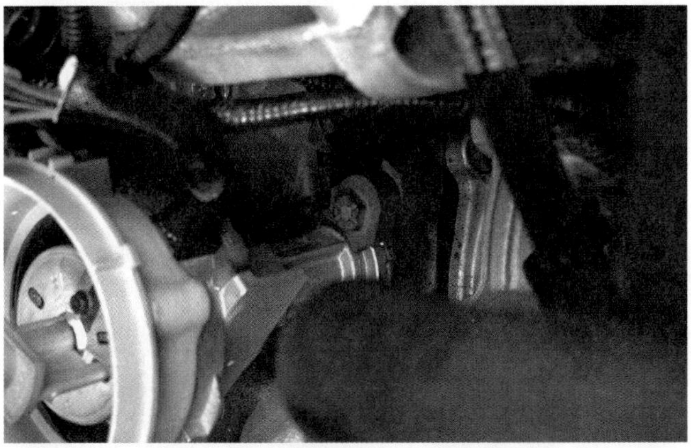

P22-7 *Install the distributor hold-down clamp and bolt. Hand tighten the bolt.*

P22-8 *Connect the pickup leads to the wiring harness.*

P22-9 *Install the spark plug wires onto the distributor cap according to the engine's firing order. Check to make sure everything that was disconnected before the distributor was removed is now connected. Connect a timing light and prepare the engine for an ignition timing check and adjustment.*

problems. If the cylinder head is aluminum, it may be necessary to install a threaded insert into the spark plug bore.

Inspecting Spark Plugs

Once the spark plugs have been removed, it is important to "read" them. In other words, inspect them closely, noting in particular any deposits on the plugs and the degree of electrode erosion. A normal-firing spark plug will have a minimum amount of deposits on it and will be colored light tan or gray **(Figure 25–54)**. However, there should be no evidence of electrode burning, and the increase of the air gap should be no more than 0.001 inch (.0254 mm) for every 10,000 miles (16,000 km) of engine operation. A plug that exceeds this wear should be replaced and the cause of excessive wear corrected **(Figure 25–55)**. Worn or dirty spark plugs may work fine at idle or low speeds, but they frequently fail during heavy loads or higher engine speeds.

It is possible to diagnose a variety of engine conditions by examining the electrodes of the spark plugs. Ideally, all of the plugs from an engine should look alike. Whenever plugs from different cylinders look different, a problem exists in those cylinders. The following are examples of plug problems and how they should be dealt with.

Cold Fouling This condition is the result of an excessively rich air/fuel mixture. It is characterized by a layer of dry, fluffy, black carbon deposits on the tip of the plug **(Figure 25–56)**. **Cold fouling** is caused by a rich air/fuel mixture or an ignition fault that causes the spark plug not to fire. If only one or two of the plugs show evidence of cold fouling, sticking valves are the likely cause. The plug can be used again, provided its electrodes are filed and the air gap is reset. Correct the cause of the problem before reinstalling or replacing the plugs.

Figure 25-54 A normal spark plug. *Courtesy of Champion Spark Plug Company*

Figure 25-55 A worn spark plug. *Courtesy of Champion Spark Plug Company*

SHOP TALK

If cold fouling is present on a vehicle that operates a great deal of the time at idle and low speeds, plug life can be lengthened by using hotter spark plugs. ■

Wet Fouling When the tip of the plug is practically drowned in excess oil, this condition is known as **wet fouling (Figure 25–57)**. In an overhead valve engine, the oil may be entering the combustion chamber by flowing past worn valve guides or valve guide seals. If the vehicle has an automatic transmission, a likely cause of wet-fouled plugs is a defective vacuum modulator that is allowing transmission fluid to enter the chamber. On

Figure 25-56 A cold- or carbon-fouled spark plug. *Courtesy of Champion Spark Plug Company*

Figure 25-57 A wet- or oil-fouled spark plug. *Courtesy of Champion Spark Plug Company*

high-mileage engines, check for worn rings or excessive cylinder wear. The best solution is to correct the problem and replace the plugs with the specified type.

Splash Fouling **Splash fouling** occurs immediately following an overdue tune-up. Deposits in the combustion chamber, accumulated over a period of time due to misfiring, suddenly loosen when the temperature in the chamber returns to normal. During high-speed driving, these deposits can stick to the hot insulator and electrode surfaces of the plug **(Figure 25–58)**. These deposits can actually bridge across the gap, stopping the plug from sparking. Normally splash-fouled plugs can be cleaned and reused.

Gap Bridging A plug with a bridged gap **(Figure 25–59)** is not frequently seen in automobile engines. It occurs when flying carbon deposits within the combustion chamber accumulate over a long period of stop-and-go driving. When the engine is suddenly placed under a hard load, the deposits melt and bridge the gap, causing misfire. This condition is best corrected by replacing the plug.

Glazing Under high-speed conditions, the combustion chamber deposits can form a shiny, yellow glaze over the insulator. When it gets hot enough, the glaze acts as an electrical conductor, causing the current to follow the deposits and short out the plug. **Glazing** can be prevented by avoiding sudden wide-open throttle acceleration after sustained periods of low-speed or idle operation. Because it is virtually impossible to remove glazed deposits, glazed plugs should be replaced.

Overheating This condition is characterized by white or light-gray blistering of the insulator. There may also be considerable electrode gap wear **(Figure 25–60)**. Overheating can result from using too hot a plug, over-advanced ignition timing, detonation, a malfunction in the cooling system, an overly lean air/fuel mixture, using fuel too low in octane, an improperly installed plug, or a heat-riser valve that is stuck closed. Overheated plugs must be replaced.

Turbulence Burning When turbulence burning occurs, the insulator on one side of the plugs wears away as the result of normal turbulence in the combustion chamber. As long as the plug life is normal, this condition is of little consequence. However, if the spark plug shows premature wear, overheating can be the problem.

Preignition Damage is caused by excessive engine temperatures. Preignition damage is characterized by melting of the electrodes or chipping of the electrode tips **(Figure 25–61)**. When this problem occurs, look for the general causes of engine overheating, including over-advanced ignition timing, a burned head gasket, and

Figure 25-58 A splash-fouled spark plug. *Courtesy of Champion Spark Plug Company*

Figure 25-59 There is no longer a gap on this spark plug. *Courtesy of Champion Spark Plug Company*

Figure 25-60 This spark plug shows signs of overheating. *Courtesy of Champion Spark Plug Company*

Figure 25-61 A spark plug with preignition damage. *Courtesy of Champion Spark Plug Company*

using fuel too low in octane. Other possibilities include loose plugs or using plugs of the improper heat range. Do not attempt to reuse plugs with preignition damage.

Regapping Spark Plugs

Both new and used spark plugs should have their air gaps set to manufacturer's specifications. Technicians often use a spark adjusting tool **(Figure 25–62).** This tool should only be used with new plugs because it uses flat gauges for measurement. The gauges are mounted on the tool like spokes on a wheel. Above the gauges is an anvil, which is used to apply pressure to the ground electrode. On the opposite end of the tool is a curved seat. This seat performs two functions: It supports the plug shell during the procedure and it compresses the ground electrode against the gauge, thus setting the air gap.

In many instances, the electrodes of a used spark plug are no longer flat. Therefore, flat feeler gauges should not be used; instead, round wire gauges should be used **(Figure 25–63)**. In fact, it is recommended that new spark plugs be checked with a round gauge.

Always check the air gap of a new spark plug before installing it. Never assume the gap is correct just because the plug is new. Do not try to reduce a plug's air gap by tapping the side electrode on a bench. Use needle-nose pliers or a spark plug gapping tool to bend the ground electrode to its correct height. When doing this, be careful not to contact or put pressure on the center electrode. This is especially critical with fine wire platinum and iridium plugs. Also while bending the ground electrode, try to keep it in alignment with the center electrode.

Some engines are equipped with spark plugs that have more than one ground electrode. The gap between the center electrode and each ground electrode should be checked **(Figure 25–64).** If the gap between the center electrode and one of the ground electrodes is less than that of the others, spark will occur only at the smallest gap. This is also true of V-shaped ground electrodes. If one leg of the vee is closer to the center electrode than the other, the spark will always occur across the shortest distance.

The gap of spark plugs with a surface gap and of some with more than one ground electrode cannot be adjusted with conventional tools, and most manufacturers recommend that the gap be left alone.

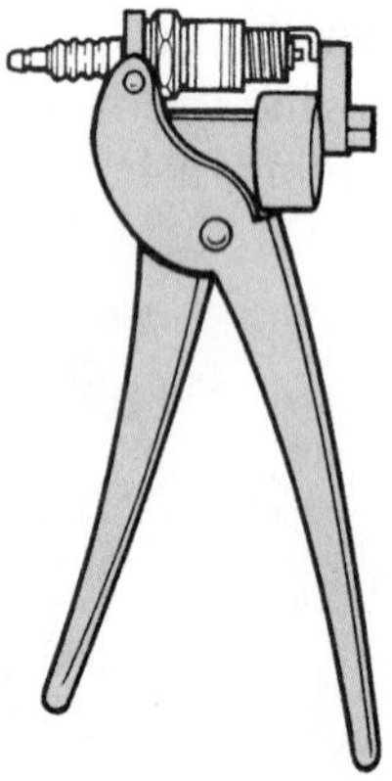

Figure 25-62 A combination spark plug gauge and adjusting tool.

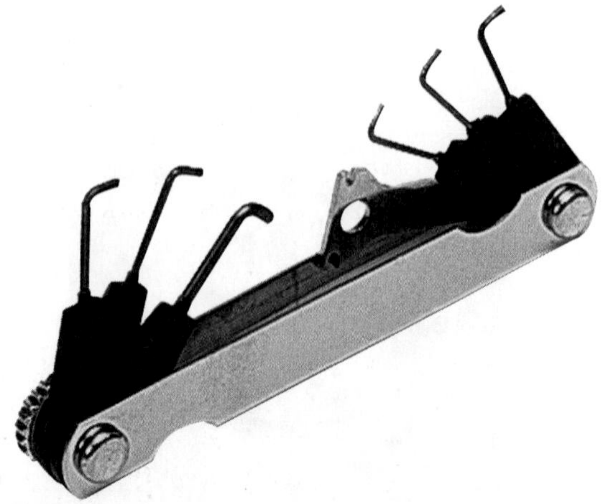

Figure 25-63 A round-wire-type feeler gauge set. *Courtesy of Snap-on Tools Company*

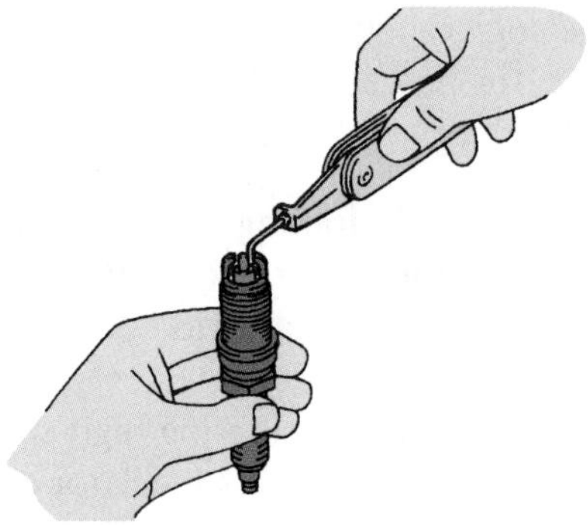

Figure 25-64 The gap between the center electrode and both ground electrodes should be checked and adjusted to specifications. *Reprinted with permission*

PROCEDURE

Spark Plug Installation

STEP 1 *Wipe dirt and grease from the plug seats with a clean cloth.*

STEP 2 *Be sure the gaskets on gasketed plugs are in good condition and properly placed on the plugs. If reusing a spark plug, install a new gasket on it. Be sure that there is only one gasket on each plug.*

STEP 3 *Adjust the air gap as needed.*

STEP 4 *Check the service manual to see if antiseize compound should be applied to the plug's threads. If it should, do so now.*

STEP 5 *Install the plugs and tighten them with your hand. If the plugs cannot be installed easily by hand, the threads in the cylinder head may need to be cleaned with a thread-chasing tap. Be especially careful not to cross-thread the plugs when working with aluminum heads.*

STEP 6 *Tighten the plugs with a torque wrench, following the vehicle manufacturer's specifications.*

Secondary Ignition Wires

Inspect all the spark plug wires and the secondary coil wire for cracks and worn insulation, which cause high-voltage leaks. Inspect all the boots on the ends of the plug wires and coil secondary wire for cracks and hard, brittle conditions. Replace the wires and boots if they show evidence of these conditions. Some vehicle manufacturers recommend spark plug wire replacement only in complete sets.

The spark plug wires may be left in the distributor cap for test purposes, in which case the cap terminal connections are tested with the spark plug wires. Calibrate an ohmmeter on the X1,000 scale, and connect the ohmmeter leads from the end of a spark plug wire to the distributor cap terminal inside the cap to which the plug wire is connected **(Figure 25–65)**.

If the ohmmeter reading is more than specified by the vehicle manufacturer, remove the wire from the cap and check the wire alone. If the wire has more resistance than specified, replace the wire. When the spark plug wire resistance is satisfactory, check the cap terminal for corrosion. Repeat the ohmmeter tests on each spark plug wire and the coil secondary wire.

Replacing Spark Plug Wires

When spark plug wires are being installed, make sure they are routed properly as indicated in the vehicle's service manual. When removing the spark plug wires from a spark plug, grasp the spark plug boot tightly, and twist while pulling the cable from the end of the plug. When installing a spark plug wire, make sure the boot is firmly seated around the top of the plug, then squeeze the boot to expel any air that may be trapped inside.

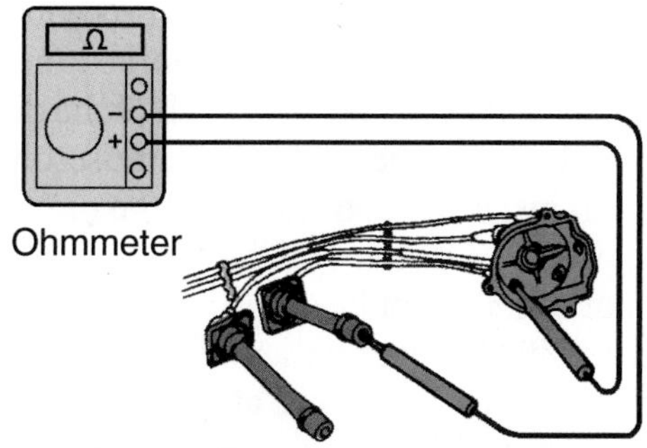

Figure 25-65 An ohmmeter connected to the spark plug wire and the distributor cap terminal to test the secondary circuit. *Reprinted with permission*

Two spark plug wires should not be placed side by side for a long span if these wires fire one after the other in the cylinder firing order. When two spark plug wires that fire one after the other are placed side by side for a long span, the magnetic field from the wire that is firing builds up and collapses across the other wire. This magnetic collapse may induce enough voltage to fire the other spark plug and wire when the piston in this cylinder is approaching TDC on the compression stroke. This action may cause detonation and reduced engine power.

SPECIFIC EI SYSTEM SERVICE

Specific diagnostic and service procedures vary quite a bit from year to year and model to model; always check with the service manual before testing and servicing specific systems.

Chrysler EI Systems

Chrysler's EI systems use the waste spark method for firing the spark plugs. These systems rely on a camshaft and a crankshaft sensor to inform the PCM as to the position of piston 1 and the time at which the other pistons reach TDC. Basic ignition timing is not adjustable in these systems.

If an engine misfire problem is caused by the ignition system, the basic operation of the secondary circuits should be checked. This testing should begin with a complete visual inspection of the circuit. With any waste spark system, the secondary circuit should be looked at as groups of two. Each coil fires two cylinders. Each coil and its associated spark plugs should be checked as a group.

For each group of cylinders, the coil, spark plug wires, and spark plugs should be checked on a scope or with an ohmmeter. Although Chrysler gives a specification for maximum allowable resistance through a spark plug, they add that this test may not be valid. Resistance checks of coils and spark plug wires are recommended. Make sure the plug wires are fully seated onto the spark plug and the ignition coil.

Ford EI Systems

Ford has used several designs of EI through the years. It is important that the design be identified prior to testing the system. Throughout all the designs, certain basic operational features can be applied. The PIP signal is an indication of crankshaft position and engine speed. The PIP signal is sent to both the ignition module and the PCM.

The CID signal is used in conjunction with PIP for the identification of cylinder 1 for the PCM. Most systems utilize the waste-spark method of firing.

Many of Ford's recommended diagnostic procedures for their EI systems involve the use of their special scan tool (STAR tester) and a breakout box. Both of these concentrate on testing the primary circuit. The primary circuit controls the action of the secondary. Problems in the primary can cause no-start and ignition timing problems. Most misfire problems are caused by problems in the secondary.

Make sure the secondary cables are secured tightly to the spark plugs and the ignition coil. Also check the cables for any damage or signs of arcing. The resistance of the cables can be checked with an ohmmeter. Do this by removing the cable and measuring across the cable. Most Ford EI systems use locking tabs to secure the spark plug cable to the ignition coil. To remove the cable from the coil, squeeze the locking tabs at the spark plug terminal and pull the boot straight up. If the cable does not freely release from the coil, squeeze the tabs and twist the boot while pulling the cable up. When reconnecting the cable to the coil, make sure the locking tabs are in place by pressing down on the center of the cable terminal.

Make sure the factory-supplied wire separators are used and are in their original position after the spark plug wires are put in place. The spark plugs and ignition coils can be tested in the same way as other ignition systems.

Although the ignition timing on Ford's EI system is not adjustable, the air gap at the crankshaft and camshaft sensors is critical **(Figure 25–66)**. On some engines, this gap is adjustable and on others it is an indication that the sensor should be replaced. When checking the gap, make sure there are no signs of damage to the rotating vane assembly.

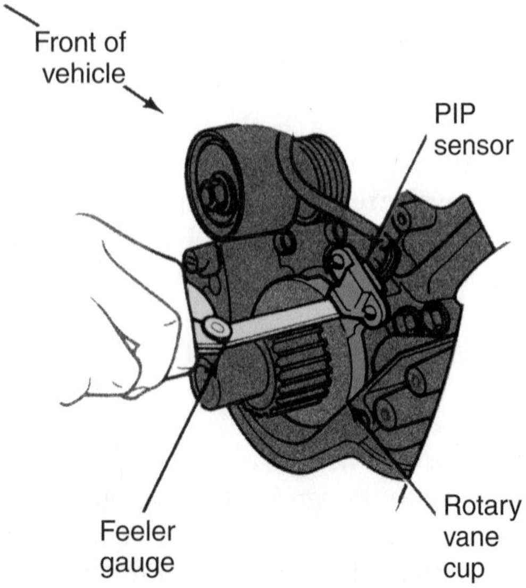

Figure 25-66 Crankshaft position (PIP) sensor gap measurements and alignment. *Courtesy of Ford Motor Company*

Ford's EDIS system is a high-data-rate system. The crankshaft position sensor is a **variable-reluctance**-type **sensor (VRS)** triggered by a 36-minus-1 tooth trigger wheel pressed onto the rear of the crankshaft dampener. The signal generated by this sensor is called a VRS. The VRS signal provides engine position and rpm information to the ignition module. Base timing is 10 degrees BTDC and is not adjustable. The ignition module receives information from the crankshaft position sensor and other sensors to calculate the on and off times for the coils in order to achieve the desired dwell and spark advance. The ignition module also synthesizes a PIP signal for use by the PCM's engine control strategy.

Low-Data-Rate and High-Data-Rate EI Service and Diagnosis The diagnostic procedure for the low-data-rate and high-data-rate systems varies according to the system being tested. It is important that the correct wiring diagram for the system being tested be referred to and the procedure for that system should be followed. Ford provides separate diagnostic harnesses for the two systems. The diagnostic harness has various leads that connect in series with each component in the system **(Figure 25–67)**. A large connector on the harness is connected to Ford's sixty-pin breakout box. Overlays for the ignition system on each engine are available to fit the breakout box terminals. These overlays identify the ignition terminals connected to the breakout box terminals.

The technician must follow the test procedures given in the service manual and measure the resistance or voltage at the specified breakout box terminals connected to the ignition system. If a diagnostic harness, overlay, and breakout box is not available, the voltage and resistance measurements must be taken at the terminals of the individual components.

General Motors' EI Systems

General Motors has used many different varieties of EI systems. Some are waste-spark systems, while others are direct or integrated systems. With each basic design there have been different operational designs as well. Each of these variances has its own diagnostic procedure. Make sure you identify the system being worked on prior to doing any exhaustive diagnostics. Each of the diagnostic procedures is based on the premise that the secondary circuit has been inspected and tested before moving on to testing the primary. Obviously, if the engine does not start, the secondary should not be suspect. Therefore, there is no need to test it. However, if the engine misfires, the secondary circuit should be looked at.

The sparks and plug wires should be visually inspected as well as their resistance checked with an ohmmeter. Check the spark plug wires across the entire length of the

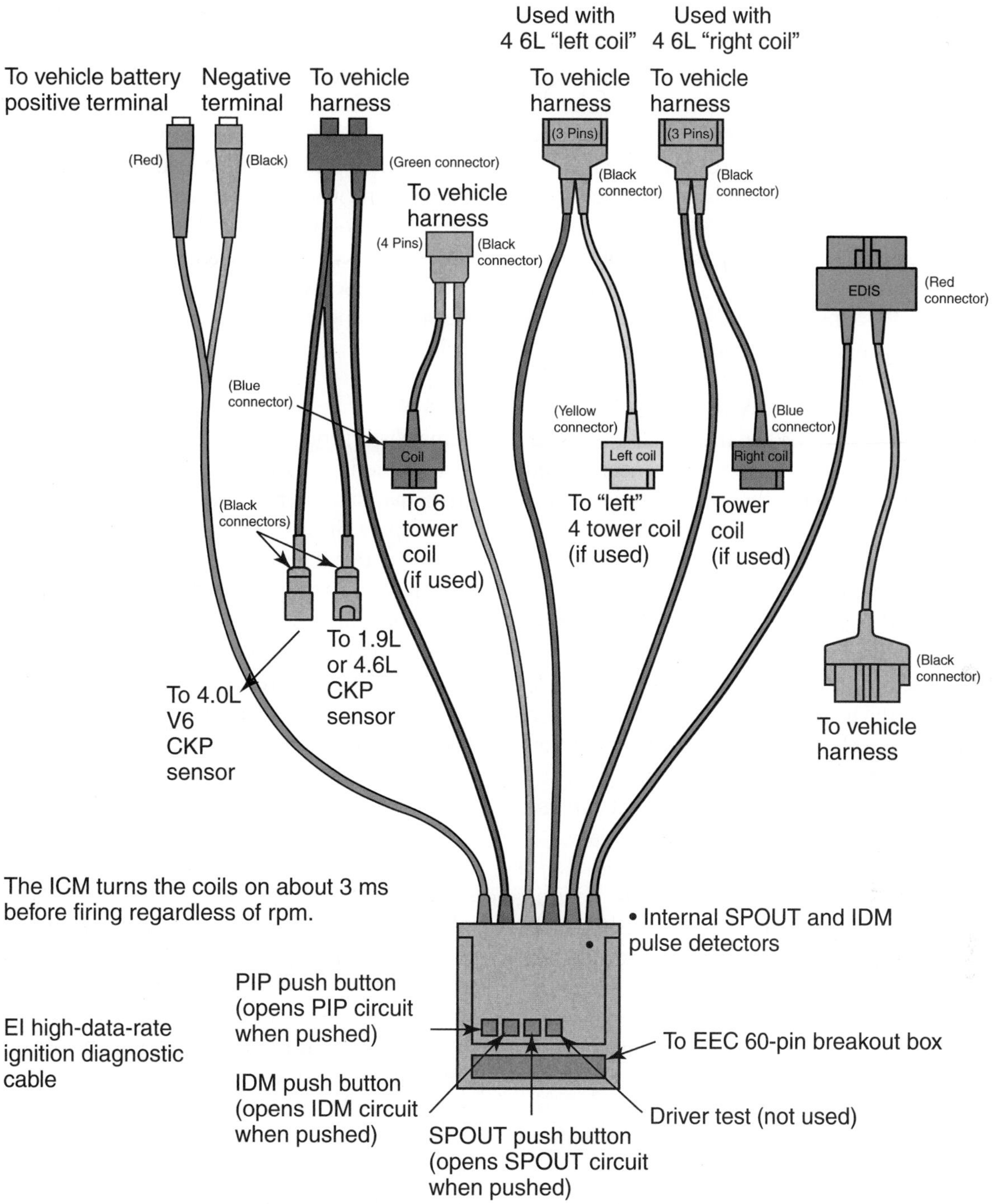

Figure 25-67 Ford's diagnostic wiring harness for a high-data-rate EI system. *Courtesy of Ford Motor Company*

cable, including the cable ends. Some General Motors' engines are equipped with an integrated ignition coil and module assembly. This assembly must be removed to access the spark plugs. Once the assembly is removed, all of its components should be inspected and tested with an ohmmeter. This testing includes the various connectors and wiring harnesses as well as the ignition coils mounted on the cover of the assembly.

General Motors' EI Systems with Magnetic Sensors With the wiring harness connector to the magnetic sensor disconnected and an ohmmeter calibrated on the ×10 scale connected across the sensor terminals, the meter should read 900 to 1,200 ohms (9 to 12K) on 2.0-liter, 2.8-liter, and 3.1-liter engines. The meter should indicate 500 to 900 ohms (5 to 9K) on a Quad 4 engine and 800 to 900 ohms (8 to 9K) on a 2.5-liter engine.

Meter readings below the specified value indicate a shorted sensor winding, whereas infinite meter readings prove that the sensor winding is open. Because these sensors are mounted in the crankcase, they are continually splashed with engine oil. In some sensor failures, the engine oil enters the sensor and causes a shorted sensor winding. If the magnetic sensor is defective, the engine fails to start.

With the magnetic sensor wiring connector disconnected, an alternating current (AC) voltmeter may be connected across the sensor terminals to check the sensor signal while the engine is cranking. On 2.0-liter, 2.8-liter, and 3.1-liter engines, the sensor signal should be 100 millivolts (mV) AC. The sensor voltage on a Quad 4 engine should be 200 mV AC. When the sensor is removed from the engine block, a flat steel tool placed near the sensor should be attracted to the sensor if the sensor magnet is satisfactory.

Mitsubishi EI Systems

Mitsubishi direct EI systems are found on some Chrysler products and all Mitsubishi vehicles. This system uses a crank angle sensor to monitor engine speed and the location of cylinder 1. The signals from the sensor are sent to the PCM, which sends ignition signals to a power transistor to control ignition timing and dwell. There is a power transistor for each ignition coil. This system uses the waste-spark method of firing cylinders.

Like all EI systems, the ignition coils can be tested with an ohmmeter. The resistance of the primary winding should be 0.77 to 0.95 ohms. To measure the resistance of the secondary winding, connect the meter across the adjacent spark plug terminals at the coil—for example, the towers for cylinders 1 and 4 and cylinders 2 and 3. The readings should be 10,300 to 13,900 ohms (10.3 to 13.9K). Readings outside the specifications indicate that the coil should be replaced.

The power transistor can also be checked with an ohmmeter. However, the transistor must be energized with an external power source. Since there are different system designs used on different Mitsubishi models, always refer to the service manual for proper terminal identification. The following procedure refers to the connector shown in **Figure 25–68**. Unplug the transistor's electrical connector. Connect the positive lead of the meter to terminal 3 and the negative lead to terminal 1. The reading should be infinite. Now connect the positive side of a 1.5-volt battery to terminal 2 and the negative side to 3. With the ohmmeter still connected to terminals 1 and 3, it should now read low resistance.

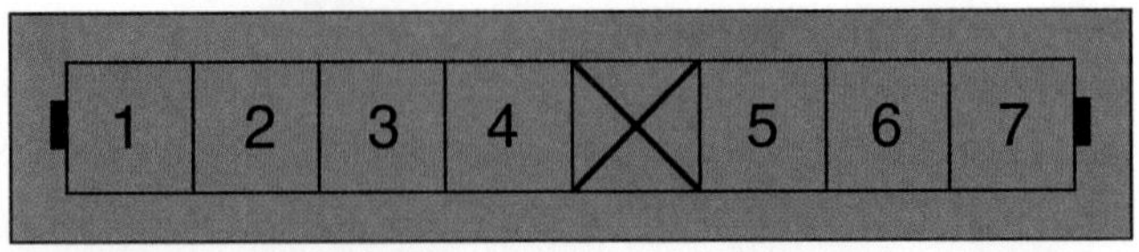

Figure 25-68 A typical Mitsubishi power transistor connector.

Now move the positive lead from the battery to terminal 5 and the negative lead to terminal 3. Connect the positive ohmmeter lead to terminal 3 and the negative lead to terminal 6. There should be good continuity between these two points. If any readings are outside of the specifications, the power transistor assembly should be replaced.

Nissan EI Systems

Some Nissan and Infiniti vehicles are equipped with a direct ignition system. The system is similar to those of other manufacturers; however, Nissan does not rely on waste-spark firing. Instead, individual coils are mounted directly above the spark plugs. The system uses a crankshaft position sensor to monitor engine speed and the location of cylinder 1. Signals from this sensor and other sensors are sent to the PCM, which controls a power transistor. The power transistor controls the action of the coils. A relay, called the power transistor relay, sends 12 volts to the ignition coils when the relay is energized. The power transistor controls the ground circuit of the primary windings in each coil.

The ignition coils are tested with an ohmmeter in the same way as other coils. Primary winding resistance and secondary winding resistance are measured across the windings on each coil. Make sure the coils are removed from the spark plugs and any source of power before testing them with an ohmmeter. Typically, 0.7 ohm is the desired primary resistance, and 7,000 to 8,000 ohms (7 to 8K) are desired for the secondary. If the resistance readings are outside the specifications stated in the service manual, replace the coil.

If the coils check out with an ohmmeter but still do not spark, check the voltage from the relay. There should be 12 volts. If this voltage is not present, check to see if battery voltage is available to the relay. If voltage is present there, check the relay control wires from the PCM. If the circuit is good, the relay should be replaced. If no voltage is available to the relay, check the relay control circuit from the PCM and the ignition switch. If the wires and the ignition switch are good, continue testing by the PCM.

The power transistor can be checked with an ohmmeter. Refer to the service manual for the proper terminal combinations and meter polarity required for testing the transistor. **Figure 25–69** is an example of these testing requirements. If any combination does not give the desired readings, replace the transistor assembly.

Terminal Combinations				Meter Polarity	Continuity?
1	2	3	4	+	Yes
a	**b**	**c**	**d**	**–**	
1	2	3	4	–	No
a	**b**	**c**	**d**	**+**	
1	2	3	4	+	Yes
e	**e**	**e**	**e**	**–**	
1	2	3	4	–	No
e	**e**	**e**	**e**	**+**	
e	e	e	e	+	Yes
a	**b**	**c**	**d**	**–**	
e	e	e	e	–	Yes
a	**b**	**c**	**d**	**+**	

Figure 25-69 Terminal combinations and desired ohmmeter readings at the power transistor connector in Nissan EI systems.

Toyota EI Systems

Some Toyota and Lexus engines are equipped with electronic ignition. One of the systems used by Toyota is very similar to the waste-spark systems used by everyone else. It uses multiple camshaft sensors, one crankshaft sensor, an ignition module, and one coil for each pair of cylinders. There is one camshaft sensor for each coil. A major difference with this system is that the ignition coil is mounted directly over one spark plug. This configuration eliminates the second spark plug wire from the ignition coil. The other systems have an individual ignition coil mounted over each spark plug.

Toyota uses a special locking feature to secure the spark plug wires to the spark plugs and the ignition coils. To remove the wires, begin by using a pair of needle-nose pliers. Disconnect the cable clamp from the engine wire protector **(Figure 25–70)**.

Then disconnect the cables from the spark plugs by firmly holding the boot, and with a twist-pull effort remove the cable from the plug. Using a screwdriver, lift up the locking tab and separate the spark plug wire from the ignition coils. On some engines, it may be necessary to remove a wire protector assembly to remove the spark plug wires. Once the wires are removed, they can be checked with an ohmmeter. The maximum allowable resistance across a secondary cable is 26K ohms. If the resistance is higher than this, check the condition of the terminals. If they are fine, replace the cable.

To reinstall the spark plug cables, reverse the procedure. Make sure all locking mechanisms are secure and that the routing of the cables is correct.

Critical to the proper operation of the system is the air gap at the camshaft sensor. Although the gap is not adjustable, it should be checked whenever diagnostics point to a faulty camshaft signal. To measure the gap, use a nonmagnetic feeler gauge. The gap should be between 0.008 to 0.016 inch (.2032 to .4064 mm). If the gap is not within these specifications, the sensor should be replaced.

The camshaft sensor should also be checked with an ohmmeter **(Figure 25–71)**. Connect the negative ohmmeter lead to the ground terminal at the sensor and measure the resistance between that terminal and the others, one at a time. The resistance should be within the specified range; if not, replace the sensor.

The secondary ignition coil windings may contain a high-voltage diode, which is why Toyota only recommends and gives specifications for measuring the resistance across the primary windings of some of their coils. When the resistance is measurable, the normal secondary winding resistance is 11,000 to 17,500 ohms (11 to 17.5K). The specified primary winding resistance is normally less than 1 ohm. Also check the coils for shorts to ground by connecting one ohmmeter lead to a primary or secondary terminal and the other to ground. There should be no continuity, and the meter should read infinity.

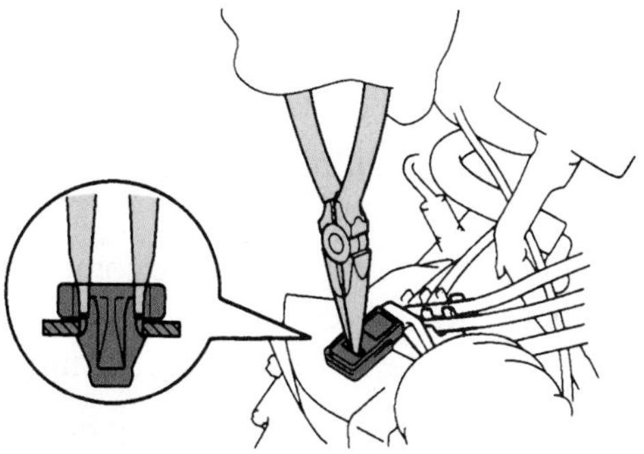

Figure 25-70 Use needle-nose pliers to remove the wires from the spark plug cable protector. *Reprinted with permission*

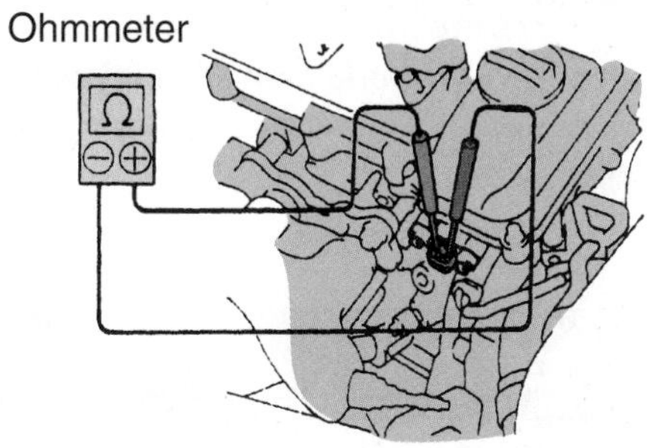

Figure 25-71 Checking a camshaft sensor with an ohmmeter. *Reprinted with permission*

CASE STUDY

An irate customer brought his late-model Chevrolet back to the service department, complaining that the car ran worse than it did before he had it serviced the previous day. The previous complaint was extreme detonation. After interviewing the customer, the service writer found that the engine no longer spark knocked but was very sluggish. According to the repair order, a new knock sensor was installed.

The technician who had done the previous work was assigned this repair. When the engine was started, the MIL was on. The technician immediately connected a scan tool and found a service code of PO325, which indicated a problem with the knock sensor circuit. This was the same code that led him to replace the knock sensor earlier. Since he could not remember erasing the codes after installing the new sensor, he erased the codes. He assumed this was the problem and that the PCM was compensating for a bad knock sensor when in fact it was new and good. Fortunately, the technician took the car for a test drive to verify the repair. Almost immediately after starting the test drive, the MIL was lit again.

Back in the shop, he reconnected the scan tool and code PO325 was present again. A bit bewildered and frustrated, he resorted to looking in the service manual for guidance. He followed the troubleshooting sequence for the code, only to find that the wire from the sensor to the electronic spark control module was disconnected. Apparently he did not connect the wire to the sensor after he replaced the sensor. He connected the wire and visually inspected all hoses and wires around the area where he had been working. He then erased the code and took the car on another test drive. This time the MIL did not come on, and the car ran quite well. The disconnected wire prevented the control module from advancing the timing as needed, and therefore the engine ran poorly. This experience taught the technician many lessons, one of which was that a trouble code identifies a problem circuit, not necessarily a problem component.

KEY TERMS

Carbon tracking
Cold fouling
Corona effect
Current limiting
Detonation
Display pattern
Drive lugs
Electromagnetic interference (EMI)
Firing voltage
Glazing
Intermediate section
Kilovolt (kV)
Parade
Preignition
Raster pattern
Spark duration
Spark line
Splash fouling
Superimposed pattern
Variable reluctance sensor (VRS)
Wet fouling

SUMMARY

- Sound wiring and connections are extremely important in ignition systems. Loose connections, corrosion, and dirt can adversely affect performance.
- Wires, connections, and ignition components can be tested for intermittent failure by wiggling them or stress-testing them by applying heat, cold, or moisture.
- A scope can look inside the ignition system by giving the technician a visual representation of voltage changes over time.
- The trace patterns of the scope can be viewed in several different modes and scales. Both secondary and primary circuit patterns can be viewed.
- Trace patterns can be divided into three main sections or zones: the firing section, the intermediate section, and the dwell section.
- The firing line and spark line of the firing section indicate firing voltage and spark duration.
- The intermediate section shows coil voltage dissipation.
- The dwell section shows the activation of primary coil current flow and primary coil current switch-off. The primary current off signal is also the firing line for the next cylinder in the firing order.
- Precautions must always be taken to avoid open circuits during ignition system testing. A special test plug is used to limit coil output during testing.
- Firing voltages are normally between 7 and 13 kV, with no more than 3 kV variation between cylinders.
- High secondary circuit resistance produces a higher-than-normal firing line.
- Individual ignition components are commonly tested for excessive internal resistance using an ohmmeter. A voltmeter or scope can also be used to monitor their operating voltages.
- Proper spark plug gapping and installation is very important to ignition system operation. Spark plug condition, such as cold fouling, wet fouling, and glazing, is often a good indication of performance problems.

- Standard test procedures using an oscilloscope, ohmmeter, and timing light can be used to diagnose problems in distributorless ignition systems.
- If a crank or cam sensor fails, the engine will not start. Both of these sensor circuits can be checked with a voltmeter. If the sensors are receiving the correct amount of voltage and have good low-resistance circuits, their output should be a digital signal or a pulsing voltmeter reading while the engine is cranking. If any of these conditions do not exist, the circuit needs to be repaired or the sensor needs to be replaced.
- The resistance of coils in a COP system can be checked in the same way as that of conventional coils; however, different meter connections are required for waste spark coils.

TECH MANUAL

The following procedures are included in Chapters 24/25 of the *Tech Manual* that accompanies this book:

1. Scope testing an ignition system.
2. Testing an ignition coil.
3. Testing individual components.
4. Setting ignition timing.
5. Visually inspecting an EI system.

REVIEW QUESTIONS

1. Name the three types of stress testing used to test for intermittent ignition component problems and list the procedures for conducting each type of test.
2. Why is the procedure for checking the resistance of a waste spark ignition coil different from the procedures for checking other types of ignition coils?
3. Name the three types of trace pattern display modes used on an oscilloscope and give examples of when each mode is most useful.
4. List the common types of spark plug fouling and the typical problems each type of fouling indicates.
5. List at least two methods of checking the operation of Hall-effect sensors.
6. What happens if one of the ground electrodes of a spark plug with two or more electrodes is closer to the center electrode than the rest?
7. The firing lines on an oscilloscope pattern are all abnormally low. Technician A says the problem is probably low coil output. Technician B says the problem could be an overly rich air/fuel mixture. Who is correct?
   a. Technician A
   b. Technician B
   c. Both A and B
   d. Neither A nor B
8. Leaner air/fuel mixtures ______.
   a. decrease the electrical resistance inside the cylinder and decrease the required firing voltage
   b. increase the electrical resistance inside the cylinder and increase the required firing voltage
   c. increase the electrical resistance inside the cylinder and decrease the required firing voltage
   d. have no measurable effect on cylinder resistance
9. While testing the coils in a DIS system, Technician A says an infinite reading means that the windings have zero resistance and are shorted. Technician B says the primary windings in each coil should be checked for shorts to ground. Who is correct?
   a. Technician A
   b. Technician B
   c. Both A and B
   d. Neither A nor B
10. While discussing no-start diagnosis with a test spark plug, Technician A says if a test light flutters at the coil's tach terminal but the test spark plug does not fire when connected from the coil secondary wire to ground with the engine cranking, the ignition coil is defective. Technician B says if a test spark plug connected to a terminal of a waste-spark ignition coil fires when the engine is cranking but the engine does not run, a bad PCM is indicated. Who is correct?
    a. Technician A
    b. Technician B
    c. Both A and B
    d. Neither A nor B
11. While discussing a pickup coil test with an ohmmeter connected to the pickup leads, Technician A says an ohmmeter reading higher than the specified resistance indicates the pickup coil is shorted. Technician B says an ohmmeter reading less than

the specified resistance indicates the pickup coil is open. Who is correct?

a. Technician A
b. Technician B
c. Both A and B
d. Neither A nor B

12. While discussing distributor shaft bushing wear, Technician A says worn bushings can cause ignition timing to vary from one cylinder to another. Technician B says worn bushings can allow the rotor to hit the spark plug terminal inside the distributor cap. Who is correct?

a. Technician A
b. Technician B
c. Both A and B
d. Neither A nor B

13. Technician A says EMI can affect sensor signals. Technician B says EMI can cause intermittent driveability problems. Who is correct?

a. Technician A
b. Technician B
c. Both A and B
d. Neither A nor B

14. While diagnosing an ignition problem on a scope, Technician A says a worn distributor drive gear can cause an erratic transistor on signal. Technician B says an erratic transistor off signal can be caused by a bad ignition module. Who is correct?

a. Technician A
b. Technician B
c. Both A and B
d. Neither A nor B

15. While discussing waste-spark EI systems, Technician A says a high-resistance spark plug can affect the firing of its companion spark plug. Technician B says improper spark plug torque can cause an engine misfire. Who is correct?

a. Technician A
b. Technician B
c. Both A and B
d. Neither A nor B

16. While discussing how to test a Hall effect sensor, Technician A says a logic probe can be used. Technician B says a DMM can be used. Who is correct?

a. Technician A
b. Technician B
c. Both A and B
d. Neither A nor B

17. While discussing the possible causes for a no-start condition on an EI-equipped engine, Technician A says a shorted crankshaft sensor may prevent the engine from starting. Technician B says a shorted spark plug may stop the engine from starting. Who is correct?

a. Technician A
b. Technician B
c. Both A and B
d. Neither A nor B

18. Two technicians are discussing the diagnosis of an electronic ignition (EI) system in which the crankshaft and camshaft sensor tests are satisfactory but a test spark plug connected from the spark plug wires to ground does not fire. Technician A says the coil assembly may be defective. Technician B says the voltage supply wire to the coil assembly may be open. Who is correct?

a. Technician A
b. Technician B
c. Both A and B
d. Neither A nor B

19. While discussing EI service and diagnosis, Technician A says the crankshaft sensor may be rotated to adjust the basic ignition timing. Technician B says the crankshaft sensor may be moved to adjust the clearance between the sensor and the rotating blades on some EI systems. Who is correct?

a. Technician A
b. Technician B
c. Both A and B
d. Neither A nor B

20. While discussing engine misfire diagnosis, Technician A says a defective EI coil may cause cylinder misfiring. Technician B says the engine compression should be verified first if the engine is continually misfiring. Who is correct?

a. Technician A
b. Technician B
c. Both A and B
d. Neither A nor B

26

FUELS AND OTHER ENERGY SOURCES

OBJECTIVES

■ Describe the basic composition of gasoline. ■ Explain why materials are added to gasoline to make it more efficient. ■ Name the common substances used as oxygenates in gasoline and explain what they do. ■ Describe how the quality of a fuel can be tested. ■ Explain the advantages and disadvantages of the various alternative fuels. ■ Explain the differences between electric, hybrid, and fuel cell vehicles.

This chapter looks at the fuels and other energy sources used to propel a vehicle. Although there are several different types of fuels for automotive use, gasoline is the most commonly used and most readily available. However, there is much interest in finding suitable alternatives to gasoline, including the use of electricity. Much research is being done on electric, hybrid electric, and fuel cell vehicles. These are discussed, along with the various automotive fuels.

AIR/FUEL MIXTURES

Regardless of the fuel type used during the combustion, combustion efficiency depends on having the correct amount of air mixed with the correct amount of fuel. The amount of air mixed with the fuel is called the air/fuel ratio. The ideal air/fuel ratio for most operating conditions of a gasoline engine is approximately 14.7 pounds of air mixed with one pound of gasoline. This provides a ratio of 14.7:1. Because air is so much lighter than gasoline, it takes nearly 10,000 gallons of air mixed with one gallon of gasoline to achieve an air/fuel ratio of 14.7:1. This is why proper air delivery is as important as fuel delivery.

When the mixture has more air than the ideal ratio calls for, the mixture is said to be **lean**. Ratios of 15 to 16:1 provide the best fuel economy from gasoline engines. Mixtures that have a ratio below 14.7:1 are considered **rich** mixtures. Rich mixtures (12 to 13:1) provide more power production from the engine but greater fuel consumption **(Figure 26–1)**.

If the combustion process is complete, all of the hydrocarbons (HC) will be completely combined with the

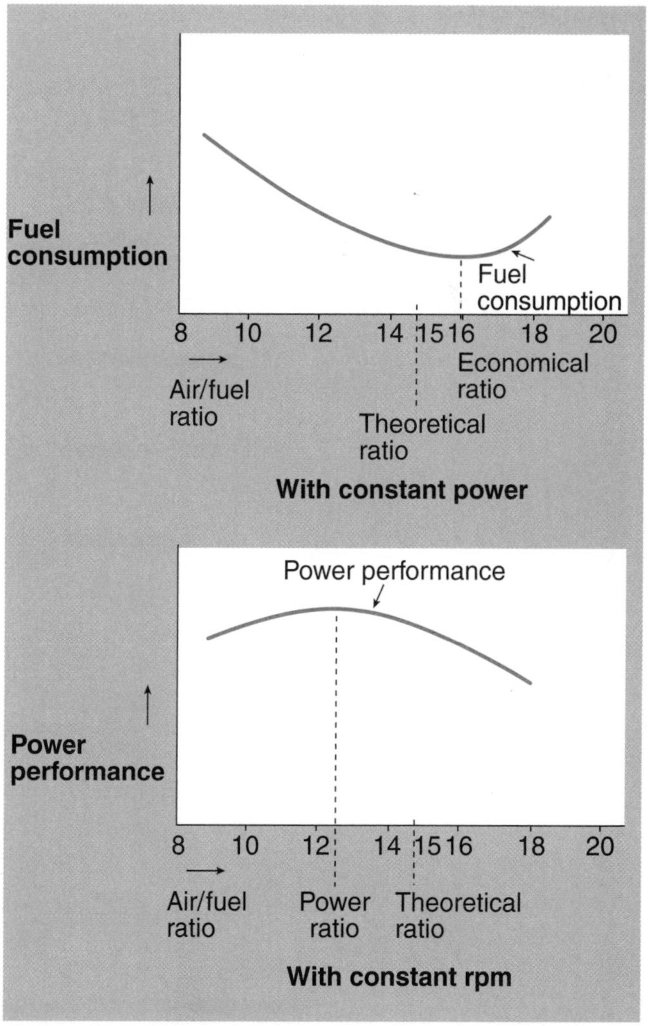

Figure 26-1 Fuel consumption and performance at various air/fuel ratios.

available oxygen. The ratio of air to fuel that accomplishes this is referred to as **stoichiometric**. The stoichiometric quantities for gasoline are 14.7 parts of air per 1 part gasoline by weight. Different fuels have different stoichiometric ratios.

GASOLINE

Gasoline is a complex mixture of approximately 300 various ingredients, mainly hydrocarbons, refined from crude petroleum oil for use as fuel in engines. **Crude oil**, as removed from the earth, is a mixture of hydrocarbon compounds ranging from gases to heavy tars and waxes **(Figure 26–2)**. The crude oil can be refined into products, such as lubricating oils, greases, asphalts, kerosene, diesel fuel, gasoline, and natural gas. The refining process separates the hydrocarbons so they can be used. During refining, the crude oil is heated by pumping it through pipes routed through hot furnaces and into a fractioning column. During the refining, the light hydrocarbon molecules are separated from the heavier ones. Located at different heights in the fractioning tube are draw pipes used to pull the desired petroleum materials out of the tower. The lightest products are taken from the top and so on **(Figure 26–3)**. Before its widespread use in the internal combustion engine, gasoline was an unwanted byproduct of refining for oils and kerosene.

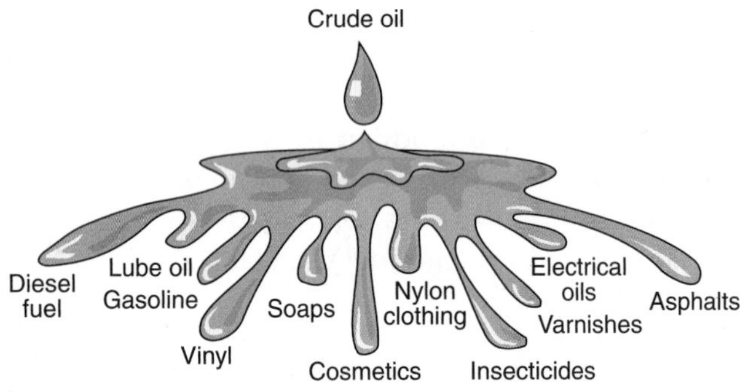

Figure 26-2 Crude oil is the source for many different products.

Gasoline contains hydrogen and carbon molecules. The chemical symbol for this liquid is C8H15, which indicates that each molecule of gasoline contains 8 carbon atoms and 15 hydrogen atoms. Gasoline is a colorless liquid with excellent vaporization capabilities.

Oil refiners must meet gasoline standards set by the American Society for Testing and Materials (ASTM), the EPA, some state requirements, and their own company standards.

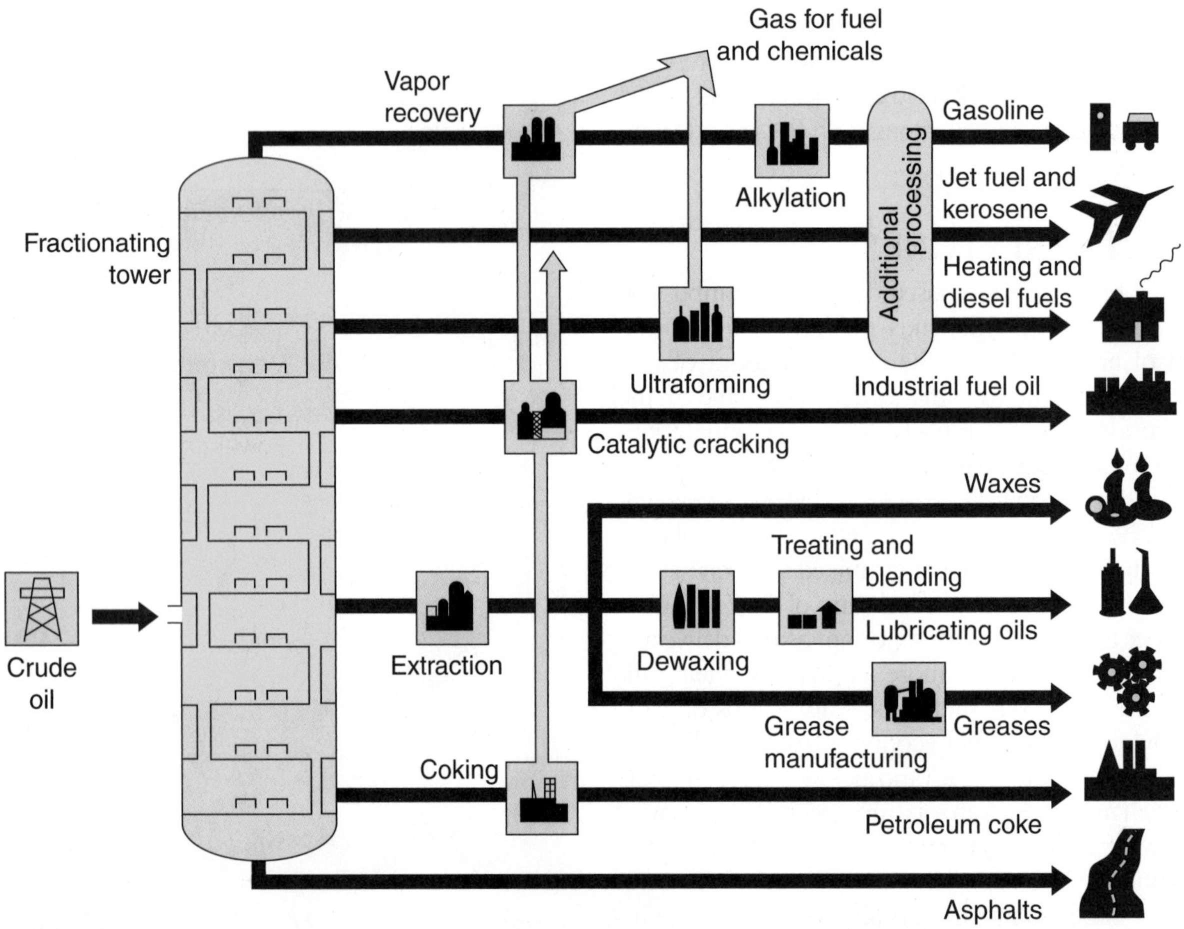

Figure 26-3 The refining process for crude oil. *Courtesy of the American Petroleum Institute*

Two important factors affect the power and efficiency of a gasoline engine—compression ratio and detonation. The higher the compression ratio, the greater the engine's power output and efficiency. The better the efficiency, the less fuel is consumed to produce a given power output. To have a high compression ratio requires an engine of greater structural integrity. Due to the use of unleaded gasoline, compression ratios now generally range from 8:1 to 10:1. High-performance engines may have higher compression ratios.

Normal combustion occurs gradually in each cylinder. The flame front advances smoothly across the combustion chamber until all the air/fuel mixture has been burned. Detonation occurs when the flame front fails to reach a pocket of mixture before the temperature in that area reaches the point of self-ignition. Normal burning at the start of the combustion cycle raises the temperature and pressure of everything inside the cylinder. The last part of the mixture is both heated and pressurized, and the combination of those two factors can raise it to the point of self-ignition. At that moment, the remaining mixture burns almost instantaneously. The two flame fronts create a pressure wave between them that can destroy cylinder head gaskets, break piston rings, and burn pistons and exhaust valves. When detonation occurs, a hammering, pinging, or knocking sound is heard. However, when the engine is operating at high speed, these sounds cannot be heard because of the noise from the engine and the road.

FUEL PERFORMANCE

Many of the performance characteristics of gasoline can be controlled in refining and blending. The major factors affecting fuel performance are antiknock quality, volatility, sulfur content, and deposit control.

Antiknock Quality

An **octane** number or rating was developed by the petroleum industry so the **antiknock** quality of a gasoline could be rated. The octane number is a measure of the fuel's tendency not to experience detonation in the engine. The higher the octane rating, the less of a tendency the engine has to knock. By itself, the antiknock rating has nothing to do with fuel economy or engine efficiency.

Two methods are used for determining the octane number of gasoline: the **motor octane number (MON)** method and the **research octane number (RON)** method. Both use a laboratory single-cylinder engine equipped with a variable head and knock meter to measure knock intensity. A test sample of the fuel is used in the engine as the engine's compression ratio and air/fuel mixture are adjusted to develop a specific knock intensity. There are two primary standard reference fuels: **isooctane** and **heptane.** Isooctane does not knock in an engine but is not used in gasoline because of its expense. Heptane knocks severely in an engine. Isooctane has an octane number of 100. Heptane has an octane number of zero.

A fuel of unknown octane value is run in a special test engine, and the severity of knock is measured. Various proportions of isooctane and heptane are run in the engine to duplicate the severity of the engine knock when the test fuel was run. When the knock caused by the isooctane/heptane mixture matches that caused by the fuel being tested, the octane number is established by the percentage of isooctane in the mixture. For example, if 85% isooctane and 15% heptane produced the same knock severity as the tested fuel, that fuel would be rated as having an octane rating of 85.

The octane rating required by law and the one displayed on gasoline pumps is the **Antiknock Index (AKI)**. It is the average of RON and MON. The antiknock index is stated as $(R+M)/2$.

The following factors affect knock:

- *Lean fuel mixture.* A lean mixture burns slower than a rich mixture. This longer burning time causes higher combustion chamber temperatures, which promotes the tendency for unburned fuel in front of the spark-ignition flame to detonate.
- *Overadvanced ignition timing.* Advancing the ignition timing induces knock. Retarding ignition timing suppresses knock.
- *Compression ratio.* Compression ratio affects knock because cylinder pressures are increased with the increase in compression ratio.
- *Valve timing.* Valve timing that fills the cylinder with more air/fuel mixture promotes higher cylinder pressures, increasing the chances for detonation.
- *Turbocharging and supercharging.* Both turbocharging and supercharging force additional air into the engine's cylinders, which induces higher cylinder pressures and promotes knock.
- *Coolant temperature.* Hotspots in the cylinder or combustion chamber due to inefficient cooling or a damaged cooling system raise combustion chamber temperatures and promote knock.
- *Excessive carbon deposits.* The accumulation of carbon deposits on the pistons, valves, and combustion chamber causes poor heat transfer from the combustion chamber. Carbon accumulation also artificially increases the compression ratio. Both conditions cause knock.
- *Air inlet temperature.* The higher the air temperature when it enters the cylinder, the greater the tendency to knock.
- *Combustion chamber shape.* The optimum combustion chamber shape for reduced knocking is the

hemispherical design with the spark plug located in the center of the combustion chamber. The hemi head allows for faster combustion, allowing less time for detonation to occur ahead of the flame front.

- *Octane number.* Only when an engine is designed to take advantage of the higher octane gasoline can the value of the fuel be obtained. Most modern engines are designed to operate efficiently with regular grade gasoline and do not require high-octane gasoline.

Most electronically controlled ignition systems have a sensor to detect if knock is occurring so the PCM can retard the ignition timing to prevent detonation.

One of the things to remember about high-octane fuel is that it burns slower than low-octane gasoline; therefore, it is less likely to cause detonation.

Volatility

Gasoline is very volatile. It readily evaporates so its vapor adequately mixes with air for combustion. Only vaporized fuel supports combustion. To ensure complete combustion, complete vaporization must occur.

The volatility of gasoline is a significant factor in the following performance conditions:

- *Cold starting and warmup.* A fuel can cause hard starting, hesitation, and stumbling during warmup if it does not readily vaporize. A fuel that vaporizes too easily in hot weather can form vapor bubbles in the fuel delivery system, causing **vapor lock** or a loss of performance. If gasoline vaporizes while it is in a fuel line, it can stop the flow of gasoline through the line. Rather than flow through the line, the pressurized fuel will compress the vapor, not move it. Vapor lock can cause a variety of driveability problems.
- *Altitude.* Gasoline vaporizes more easily at high altitudes, so volatility is controlled in blending according to the elevation of the place where the fuel is sold.
- *Crankcase oil dilution.* A fuel must vaporize well to prevent diluting the crankcase oil with liquid fuel or break down the oil film on the cylinder walls, causing scuffing or scoring. The liquid eventually enters the crankcase oil and results in the formation of sludge, gum, and varnish accumulation as well as the lubrication properties of the oil.

The difference in gasoline blends is the vapor pressure of the finished product. Gasoline blended for use in the summer is less volatile (does not burn as easily) than gasoline for use in the winter. Also, in high-altitude areas, fuels must be blended to have higher volatility because they can boil at lower temperatures. The definition of volatility assumes the vapors will remain in the fuel tank or fuel line and will cause a certain pressure based on the temperature of the fuel.

There are three methods of measuring the volatility of a fuel. The most common is the **Reid vapor pressure (RVP) test.** The RVP test is performed by placing a sample of gasoline into a sealed metal container that has a pressure measuring device attached to it. The container is submerged in heated (100°F or 38°C) water. As the fuel is heated, it vaporizes.

Remember, the more volatile a fuel is, the easier it will vaporize. As the fuel vaporizes, it creates vapor pressure within the container. Fuels that are more volatile will create more pressure. The vapor pressure is measured in psi.

Sulfur Content

Gasoline can contain some of the sulfur present in the crude oil. Sulfur content is reduced at the refinery to limit the amount of corrosion it can cause in the engine and exhaust system.

When the hydrogen in the hydrocarbons of the fuel is burned, one of the by-products of combustion is water. Water leaves the combustion chamber as steam but can condense back to water when passing through a cool exhaust system. When the engine is shut off and cools, steam condenses back to a liquid and forms water droplets. Steam present in crankcase blowby also condenses to water.

When the sulfur in the fuel is burned, it combines with oxygen to form **sulfur dioxide**. This compound can combine with water to form sulfuric acid, a highly corrosive compound. This type of corrosion is the leading cause of exhaust valve pitting and exhaust system deterioration. With catalytic converters, the sulfur dioxide can cause the obnoxious odor of rotten eggs during engine warmup. To reduce corrosion caused by sulfuric acid, the sulfur content in gasoline is limited to less than 0.01%.

Deposit Control

Several additives are put into gasoline to control harmful deposits, including gum or oxidation inhibitors, detergents, metal deactivators, and rust inhibitors.

BASIC FUEL ADDITIVES

At one time, all a gasoline-producing company had to do to produce its product was pump the crude from the ground, run it through the refinery to separate it, dump in a couple of grams of lead per gallon, and deliver the finished product to the service station. Of course, automobiles were much simpler then and what they burned was not very critical. As long as gasoline vaporized easily and did not cause the low-compression engines to knock, everything was fine.

For many years, lead compounds, such as **tetraethyl lead (TEL)** and **tetramethyl lead (TML)** were added to gasoline to increase its octane rating. However, since the mid-1970s, vehicles have been designed to run on unleaded gasoline only. Leaded fuels are no longer available as automotive fuels.

Because of the deactivating or poisoning effect that lead has on a catalytic converter, gasolines are limited to a lead content of 0.06 gram per gallon. To achieve the desired octane rating, **methylcylopentienyl manganese tricarbonyl (MMT)** is added to gasoline.

Not all additives improve the performance of gasoline. Some, such as olefins, have been identified as a cause of deposits on port fuel injectors.

Gasoline additives have different properties and a variety of purposes.

Anti-Icing or Deicer

Isopropyl alcohol is added seasonally to gasoline as an anti-icing agent to prevent fuel line freeze-up in cold weather.

Metal Deactivators and Rust Inhibitors

Metal deactivators and rust inhibitors are used to inhibit reactions between the fuel and the metals in the fuel system that can form abrasive and filter-plugging substances.

Gum or Oxidation Inhibitors

Some gasolines contain aromatic amines and phenols to prevent the formation of gum and varnish. During storage, harmful gum deposits can form due to the reaction of some gasoline components with each other and with oxygen. Oxidation inhibitors are added to promote gasoline stability. They help control gum, deposit formation, and staleness.

Gum content is influenced by the age of the gasoline and its exposure to oxygen and certain metals such as copper. If gasoline is allowed to evaporate, the residue left can form gum and varnish.

Detergents

The use of detergent additives in gasoline has been the subject of some public confusion. Detergent additives are designed to do only what their name implies—clean certain critical parts inside the engine. They do not affect octane.

Nitrous Oxide

Adding nitrous oxide to the air/fuel mixture is not done by oil refineries; rather, it is commonly done by those seeking more instantaneous power from their engines. A brief discussion of what nitrous oxide does and how it improves engine performance follows.

Nitrous oxide is injected as a dense liquid. When nitrous oxide is heated, it breaks down into nitrogen and oxygen, which provides more oxygen atoms inside the cylinder when the fuel ignites. Because there is more oxygen, more fuel can be injected into the cylinder. The engine therefore produces more power. Nitrous oxide also improves engine performance by cooling the gases in the cylinder, thereby making the air denser.

Nitrous oxide is injected into the engine's intake when the driver pushes a button to activate the system. Nitrous kits, which include nearly all that is needed to add the system to an engine, are available for many engines. The nitrous tanks typically store enough nitrous for 3 to 5 minutes of operation.

OXYGENATES

Oxygenates are compounds such as alcohols and ethers that contain oxygen in their molecular structure. Oxygenates improve combustion efficiency, thereby reducing polluting emissions. Many oxygenates also serve as excellent octane enhancers when blended with gasoline **(Figure 26–4)**. Oxygenated fuels tend to have lower carbon monoxide emissions.

Ethanol

By far the most widely used gasoline additive today is **ethanol** (ethyl alcohol), or grain alcohol. Ethanol is a noncorrosive and relatively nontoxic alcohol made from renewable biological sources. Blending 10% ethanol into gasoline results in an increase of 2.5 to 3 octane points. With ethanol-blended gasoline, air toxics are about 50% less.

In addition to octane enhancement, ethanol blending keeps the fuel injectors cleaner and less subject to corrosion due to the detergent additives found in most ethanol. Ethanol can loosen contaminants and residues that may have gathered in the vehicle's fuel system.

All alcohols have the ability to absorb water. Water in the fuel system, originating from condensation, is absorbed by the alcohol. This reduces the chances of fuel line freeze-up during cold weather. Ethanol also decreases carbon monoxide emissions at the tailpipe due to the higher oxygen content of the fuel.

Ethanol blends are approved by all auto manufacturers because of their clean air benefits. Older engines with nonhardened valve seats may need a lead substitute added to gasoline or ethanol blends to prevent premature valve seat wear. The chance of valve burning is decreased when ethanol is used because ethanol burns cooler than gasoline.

The biggest concern with using ethanol or methanol is they have low volatility and therefore can cause cold-start problems or misfiring during warmup.

Methanol

Methanol is the lightest and simplest of the alcohols and is also known as wood alcohol. It can be distilled from coal or renewable sources, but most of what is used today is derived from natural gas.

Many automakers continue to warn motorists about using a fuel that contains more than 10% methanol and

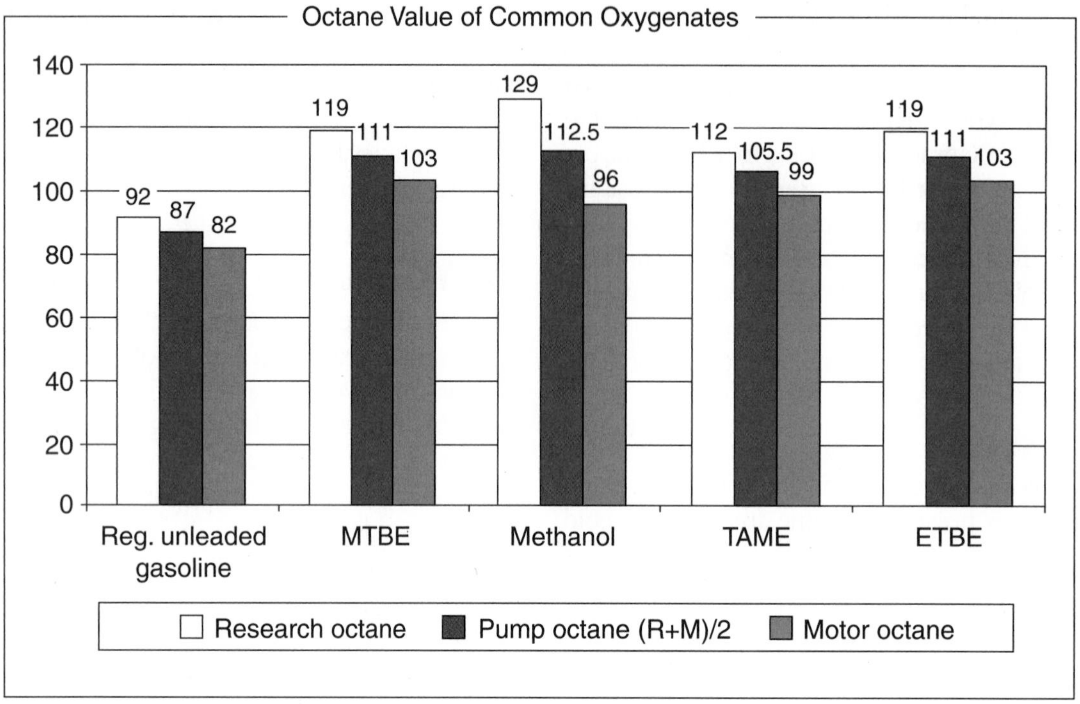

Figure 26-4 Octane values of gasoline and common oxygenates.

co-solvents by volume. Methanol is recognized as being far more corrosive to fuel system components than ethanol, and this corrosion concerns automakers.

Methanol is also highly toxic and there are safety concerns with ingestion, eye or skin contact, and inhalation.

Methanol can be used directly as an automotive fuel but the engine must be modified for its use. It can also be used in flexible-fuel vehicles as M85, which is 85% methanol. However, this is not very common because car manufacturers are no longer supplying methanol-powered vehicles.

In the future, methanol could be the fuel of choice for providing hydrogen to power fuel cell vehicles.

MTBE

Methyl tertiary butyl ether (MTBE) has been used as an octane enhancer and supply extender because of its excellent compatibility with gasoline. Current U.S. EPA restrictions on oxygenates limit MTBE in unleaded gasoline to 11% of volume. At that level, it increases pump octane (R+M/2) by 2.5 points. However, it is usually found in concentrations of 7% to 8% of volume. MTBE increases octane while reducing carbon monoxide emissions at the tailpipe and does it at a cost that makes it very attractive to gasoline marketers across the country.

Methanol can be used to make MTBE. However, MTBE production and use have declined because it has been found to contaminate groundwater. As of 2004, MTBE is no longer used in gasoline and has been replaced by ethanol.

Other oxygenates are being tested as a replacement for MTBE. These include TAME (tertiary amyl methyl ether) and ETBE (ethyl tertiary butyl ether). Both have a slightly higher octane rating than MTBE and are manufactured from ethanol.

Aromatic Hydrocarbons

Aromatic hydrocarbons are petroleum-derived compounds including benzene, xylene, and toluene that are being used in some gasolines as octane boosters.

Reformulated Gasoline

MTBEs and ethanol are the most commonly used oxygenates for producing **reformulated gasoline (RFG)**. By blending oxygen into the gasoline, the fuel requires less ambient oxygen for complete burning. Therefore, for the same carburetor or fuel injector settings, oxygenated gasoline produces a leaner air/fuel mixture and generates less carbon monoxide. Reformulated gasoline is also called cleaner-burning gasoline and costs slightly more than normal gasoline.

RFG can be used in existing engines with no modifications or special refueling facility requirements.

FUEL QUALITY TESTING

Two tests can be done to test the quality of gasoline: the Reid vapor pressure test and the alcohol content test.

Testing the RVP of Gasoline

RVP is a measure of the volatility of gasoline. Fuels that are more volatile vaporize more easily, creating more pressure. Increasing the RVP of a gasoline permits the engine to start easier in cold weather. The RVP of winter-blend gasoline is about 9.0 psi. Summer-blend gasoline is typically around 7.0 psi.

A special fuel vapor pressure tester is needed to test the RVP of gasoline. Make sure the gasoline being tested is cool. Then, put a sample in the tester's container and securely seal the container as soon as the gasoline is in it. Put hot water in another container and put the container holding the fuel into it. Make sure that most of the container holding the fuel is covered by water. Connect the pressure gauge assembly to the container holding the gasoline. Put a thermometer into the water. When the water temperature is 105°F (40°C) for at least 2 minutes, take your pressure reading and compare it to specifications.

Alcohol in Fuel Test

Pump gasoline may contain a small amount of alcohol, normally up to 10%. However if the amount is greater than that, problems may result such as fuel system corrosion, fuel filter plugging, deterioration of rubber fuel system components, and a lean air/fuel ratio. These fuel system problems caused by excessive alcohol in the fuel may cause driveability complaints such as lack of power, acceleration stumbles, engine stalling, and no-start. If the correct amount of fuel is being delivered to the engine and there is evidence of a lean mixture, check for air leaks in the intake, then check the gasoline's alcohol content.

There are many different ways to check the percentage of alcohol in gasoline. Some are more exact than others are and some require complex instruments. The following alcohol-in-fuel test procedure requires only the use of a calibrated cylinder and is accurate enough for diagnosis:

PROCEDURE

STEP 1 *Obtain a 100-milliliter (mL) cylinder graduated in 1-mL divisions.*

STEP 2 *Fill the cylinder to the 90-mL mark with gasoline.*

STEP 3 *Add 10 mL of water to the cylinder so it is filled to the 100-mL mark.*

STEP 4 *Install a stopper in the cylinder and shake it vigorously for 10 to 15 seconds.*

STEP 5 *Carefully loosen the stopper to relieve any pressure.*

STEP 6 *Install the stopper and shake vigorously for another 10 to 15 seconds.*

STEP 7 *Carefully loosen the stopper to relieve any pressure.*

STEP 8 *Place the cylinder on a level surface for 5 minutes to allow liquid separation.*

STEP 9 *Any alcohol in the fuel is absorbed by the water and settles to the bottom. If the water content in the bottom of the cylinder exceeds 10 mL, there is alcohol in the fuel. For example, if the water content is now 15 mL, there was 5% alcohol in the fuel. Note: Since this procedure does not extract 100% of the alcohol from the fuel, the percentage of alcohol in the fuel may be higher than indicated.*

ALTERNATIVE FUELS

Concerns of fossil fuel dilution and emission levels have led to a search for an alternative fuel. Many things are considered when determining the viability of an alternative fuel, including emissions, cost, fuel availability, fuel consumption, safety, engine life, fueling facilities, weight and space requirements of fuel tanks, and the range of a fully fueled vehicle. Currently, the major competing alternative fuels include ethanol, methanol, propane, natural gas, P-series fuel, and electricity. Although diesel fuel has been in use for many years, its properties are included in this section. Diesel fuel has not proved, yet, to be a successful alternate fuel for automobiles.

Ethanol and methanol were presented earlier as oxygenates for blending with gasoline. Engines can be designed to run on either of these alcohols. However, since ethanol is made from renewable sources it is used more often. **Liquefied petroleum gas** or **LP gas** is commonly called propane. Propane is a constituent of natural gas. Natural gas comes in two forms: compressed natural gas (CNG) and liquefied natural gas (LNG). Both of these are based on fossil hydrocarbons and therefore their combustion contributes to increased levels of atmospheric carbon dioxide.

Many vehicles on the road are designed to use something other than gasoline as fuel. These are commonly referred to as dedicated vehicles or bi- or multiple-fuel vehicles. Dedicated vehicles are those designed to use one particular type of alternative fuel, such as diesel fuel. Bi- and multiple-fuel vehicles are designed to use more than one fuel. Bifuel vehicles have two separate tanks and can operate on one fuel or the other, such as natural gas or unleaded gasoline. A bifuel vehicle has two separate fuel systems; both fuels are mixed inside the engine.

Flexible-fuel vehicles (FFV) can operate solely on an alcohol-based fuel, unleaded gasoline, or a mixture of the two **(Figure 26–5)**. This property gives the driver flexibility and convenience when refilling the fuel tank.

Diesel Fuel

Diesel fuel is a petroleum product. Diesel fuel has properties and characteristics different from gasoline. Diesel

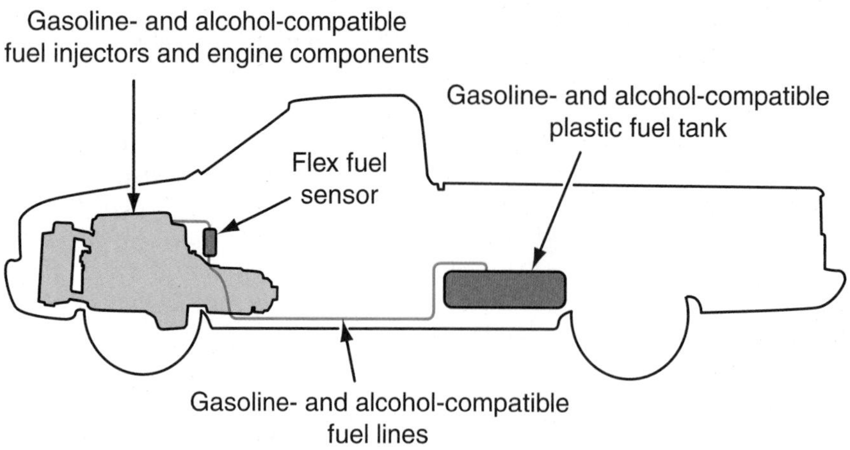

Figure 26-5 A flexible fuel vehicle.

fuel contains more energy per gallon than gasoline and diesel engines are more efficient. They also produce more torque than gasoline engines of the same size. Because diesel engines burn less fuel, they also emit less carbon dioxide. However, they have other emission-related concerns **(Figure 26–6)**:

- Diesel fuel is a high sulfur content fuel.
- Diesel engines emit particulates, commonly called soot.
- Diesel engines have high combustion temperatures, which result in excessive nitrogen oxide emissions.

Diesel engines are built to run on diesel fuel and therefore diesel fuel is not really an alternative fuel for gasoline engines.

The shape of the fuel spray, turbulence in the combustion chamber, the duration of injection, and the chemical properties of the diesel fuel all affect the power output of a diesel engine. Since diesel fuel ignites by the heat of com-

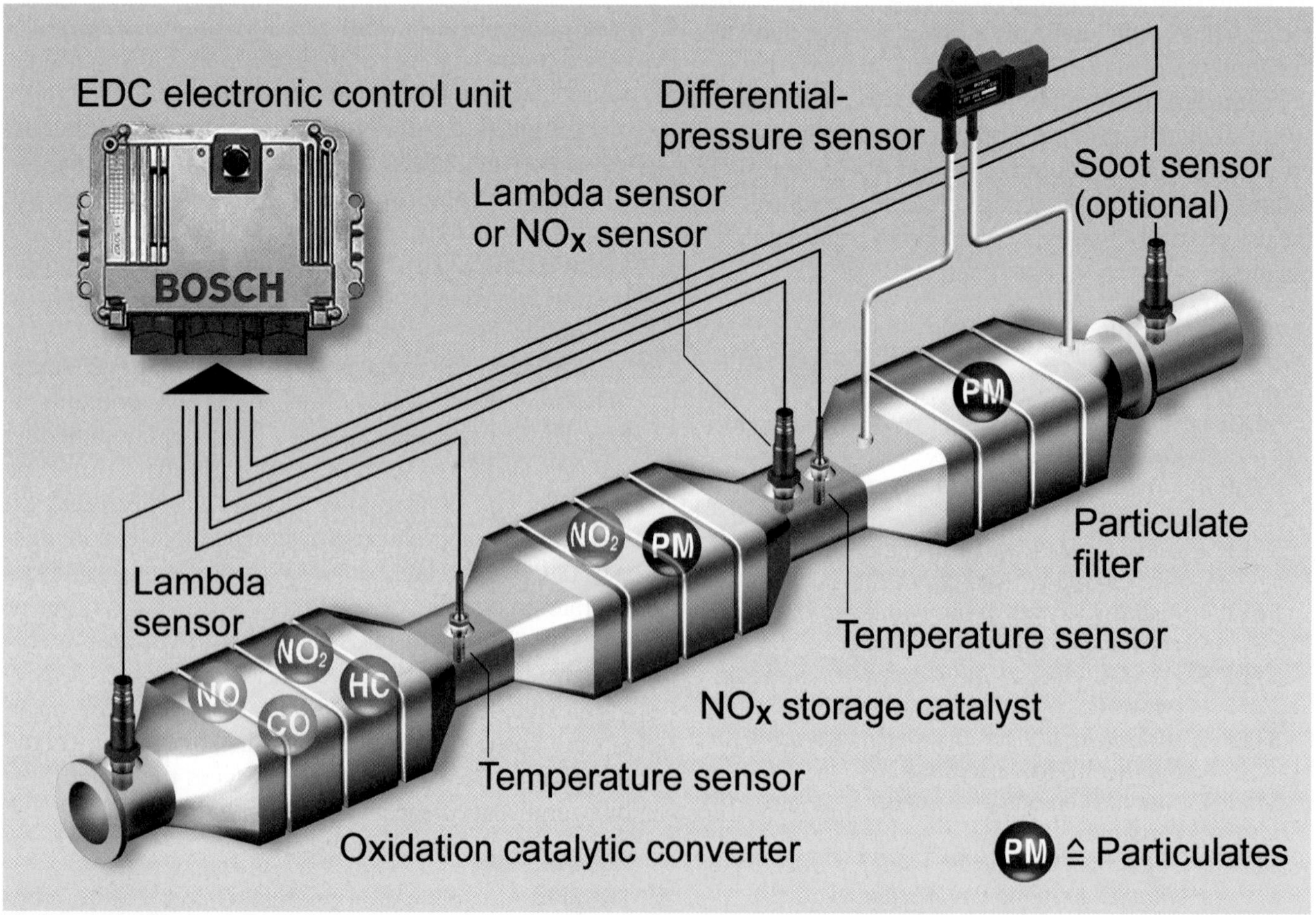

Figure 26-6 The components required to clean up the exhaust from a diesel engine. *Courtesy of Robert Bosch Corp.*

pression rather than by the heat of a spark, the characteristics of the fuel must be different from those of gasoline.

Gasoline is extremely volatile compared to diesel fuel. The amount of carbon residue left by diesel fuel depends on the quality and the volatility of that fuel. Fuel that has a low volatility is much more prone to leaving carbon residue. The small, high-speed diesels found in automobiles require a high quality and highly volatile fuel because they cannot tolerate excessive carbon deposits. Large, low-speed industrial diesels are relatively unaffected by carbon deposits and can run on low-quality fuel.

Diesel fuel's ignition quality is measured by the **cetane** rating. Much like the octane number, the cetane number is measured in a single-cylinder test engine with a variable compression ratio. The diesel fuel to be tested is compared to cetane, a colorless, liquid hydrocarbon that has excellent ignition qualities. Cetane is rated at 100. The higher the cetane number, the shorter the ignition lag time (delay time) from the point the fuel enters the combustion chamber until it ignites.

In fuels that are readily available, the cetane number ranges from 40 to 55 with values of 40 to 50 being most common. These cetane values are satisfactory for medium-speed engines whose rated speeds are from 500 to 1,200 rpm and for high-speed engines rated over 1,200 rpm. Low-speed engines rated below 500 rpm can use fuels in the above-30 cetane number range. The cetane number improves with the addition of certain compounds such as ethyl nitrate, acetone peroxide, and amyl nitrate. Amyl nitrate is commercially available for this purpose.

Minimum quality standards for diesel fuel grades have been set by the ASTM. Two grades of diesel fuels, number 1 and Number 2, are available. number 2 diesel fuel is the most popular and widely distributed. Number 1 diesel fuel is less dense than number 2, with a lower heat content. Number 1 diesel fuel is blended with number 2 to improve starting in cold weather. In the winter, diesel fuel is likely to be a mixture of number 1 and 2 fuels. In moderately cold climates, the blend may be 90% number 2 to 10% number 1. In very cold climates, the ratio may be as high as 50:50. Diesel fuel economy can be expected to drop off during the winter months due to the use of number 1 diesel in the fuel blend.

Biodiesel Fuels derived from renewable biological sources for use in diesel engines are known as **biodiesel fuels**. Animal fats, recycled restaurant grease, and vegetable oils derived from crops such as soybeans, canola, corn, and sunflowers are used in the production of biodiesel fuel. Biodiesel fuel can be used directly or be blended with diesel fuel. Pure biodiesel is biodegradable, nontoxic, and free of sulfur and aromatics.

To use biodiesel, no engine modifications are needed and its use does not affect the vehicle's payload capacity or driving range. The emissions from biodiesel are much lower compared to diesel fuel.

LP Gas

Fuel grade LP gas is almost pure propane with a little butane and propylene, which is why many people simply refer to LP gas as propane. LP gas is chemically similar to gasoline. It is called liquid petroleum because it is stored as a liquid in a pressurized bottle. The pressure increases the boiling point of the liquid and prevents it from vaporizing.

LP gas burns clean because it vaporizes at atmospheric temperatures and pressures, which means it emits less hydrocarbons and carbon monoxide.

Propane allows for quick starting, even in the coldest of climates. It also has a higher octane rating than gasoline. However, there is a reduction of engine power output (about 5%) because it is difficult to fill the cylinders with the gas. Propane is a dry fuel that enters the engine as vapor. Gasoline, however, enters the engine as tiny droplets of liquid, whether it flows through a carburetor or is sprayed in through a fuel injector. The propane fuel system is a completely closed system **(Figure 26–7)**.

Figure 26-7 The layout of a propane car.

LP gas is a good alternative to gasoline but it is a fossil fuel and therefore is not a favored alternative fuel for the future.

P-Series

P-series is a new fuel classified as an alternative fuel. It is a blend of natural gas liquids, ethanol, and biomass-derived co-solvents. **P-series fuels** are clear, colorless, 89–93 octane, liquid blends that are formulated to be used alone or freely mixed, in any proportion, with gasoline. Like gasoline, low vapor pressure formulations are produced to prevent excessive evaporation during summer, and high vapor pressure formulations are used for easy starting in cold weather.

Each gallon of P-series fuel emits approximately 50% less carbon dioxide, 35% less hydrocarbons, and 15% less carbon monoxide than gasoline. It also has 40% less ozone-forming potential. Other benefits of P-series fuels include:

- They could be 96% derived from domestic sources.
- More than 60% of the energy content in the fuel is derived from renewable sources.
- They could reduce fossil fuel energy use by 49% to 57% and petroleum use by 80% compared to gasoline.
- Greenhouse emissions are 45% to 50% below those of reformulated gasoline.

Compressed Natural Gas

Natural gas comes in two forms: **compressed natural gas (CNG)** and liquefied natural gas (LNG). Vehicles have been designed with gasoline/CNG, diesel/CNG, and dedicated CNG engine applications. CNG vehicles offer the following advantages over gasoline.

- The fuel costs less (because there is no road tax on the fuel).
- It is the cleanest alternative fuel, generating up to 99% less carbon monoxide than gasoline, no particulates, almost no sulfur dioxide, and 85% less reactive hydrocarbons than gasoline.
- Although CNG is highly pressurized (about 3,000 psi), natural gas vehicles are safer. The fuel tanks used for CNG are composite, plastic bladders wrapped in fiber. They are designed to withstand severe crash tests, direct gunfire, dynamite explosions, and burning beyond the capabilities of any standard sheet metal gasoline tank.
- Because it is lighter than air, natural gas dissipates quickly. It also has a higher ignition temperature.
- It generally reduces vehicle maintenance because it burns cleanly. Oil changes may not be needed before 12,000 miles.
- Natural gas is abundant and readily available in the United States.

The chief disadvantage of CNG at present is its limited distribution network. Fuel facilities are needed in greater numbers than are currently in existence due to the relatively shorter range of CNG vehicles. The space taken by the CNG cylinders and their weight, about 300 pounds, also would be considered disadvantages in most applications. The basic components of a natural gas vehicle (NGV) are shown in **Figure 26–8**.

CNG is injected into the cylinders by a high-pressure injector at the instant ignition occurs. The high pressures and temperature in the cylinder speeds the preflame reaction to start the ignition of the fuel injected into the cylinder. The compressed, hot air in the cylinder causes the fuel to quickly vaporize and burn.

Hydrogen Hydrogen will play an important role in the transportation of tomorrow. It can be produced in virtually unlimited quantities using renewable sources. Hydrogen has been used effectively in a number of internal combustion engines as pure hydrogen or mixed with natural gas. Pure hydrogen or hydrogen produced by reforming another fuel is used in fuel cell vehicles. When hydrogen is combusted, the only end product is water vapor.

BMW has produced a car with a V12 engine that runs on supercooled hydrogen **(Figure 26–9)**. It is a bifuel vehicle that also uses gasoline: A pair of fuel injection systems allows the driver to switch from hydrogen to gasoline power.

Today the two most common methods used to produce hydrogen are steam reforming of natural gas and the electrolysis of water. Electrolysis uses electrical energy to split water molecules into hydrogen and oxygen. The electrical energy can come from any electricity production source, including renewable fuels. Hydrogen can be stored as a compressed gas, as liquid hydrogen, or by bonding it with a substrate (for example, metal hydrides).

Hydrogen is nontoxic. Storage of hydrogen is a safety, as well as a practical, concern. Hydrogen is highly flammable and there is an explosion risk similar to that of gasoline. Also, hydrogen displaces air, so any release in an enclosed space could cause asphyxiation. Because of these safety concerns, the fuel systems are engineered with redundant safety systems.

RENEWABLE FUELS

Renewable fuels are those derived from nonfossil sources. They are produced from biomass. During combustion, renewable fuels release the same amount of carbon dioxide that is absorbed from the atmosphere by the growing of plants. Carbon dioxide, a by-product of com-

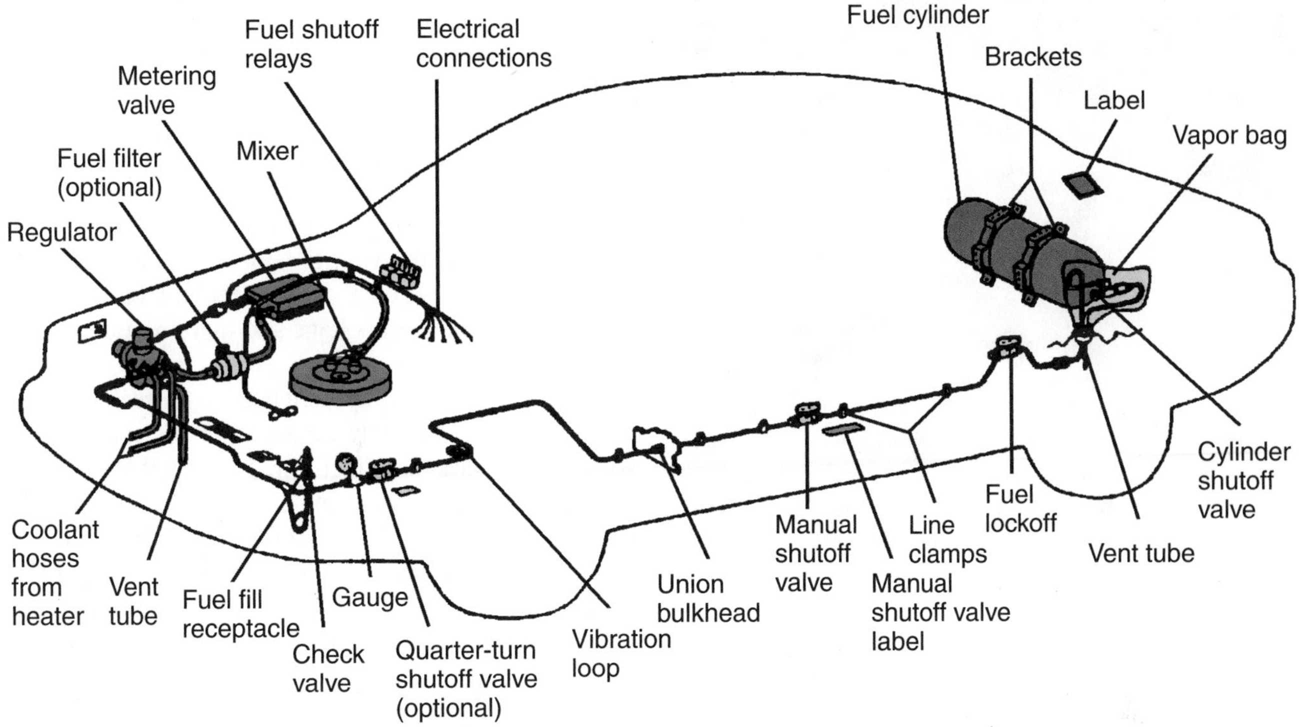

Figure 26-8 NCV system components. *Courtesy of GFI Control Systems, Inc.*

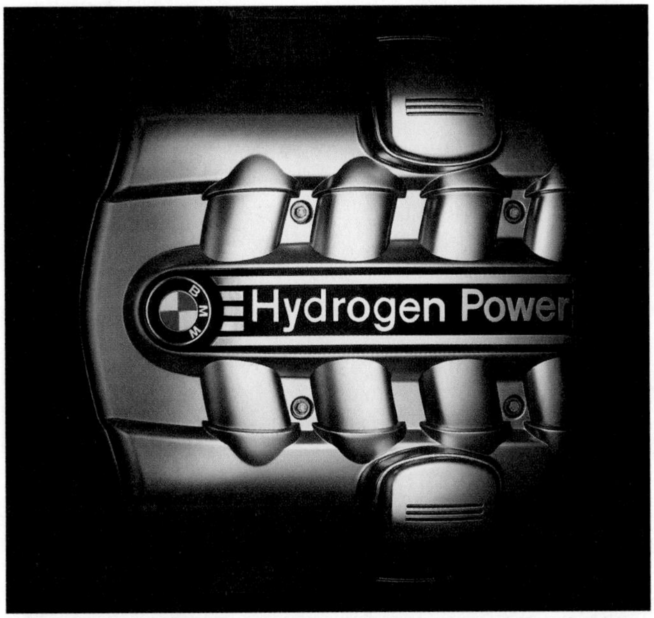

Figure 26-9 A hydrogen-powered internal combustion engine. *Courtesy of BMW of North America, Incorporated*

bustion, is suspected to be a contributing factor to global warning. Biomass fuels are also free of sulfur and aromatic compounds.

Ethanol

Biomass alcohol fuel, or ethanol, is derived almost exclusively from corn, although barley and wheat are also used. It is formed through fermenting and distilling starch crops that have been converted to simple sugars.

Ethanol provides high-quality, low-cost octane fuel and is capable of reducing air pollution because it burns cleaner. With an octane rating of 113, ethanol is the highest performance fuel on the market.

Currently, ethanol is primarily used as an oxygenate in gasoline. Ethanol-blended fuels account for approximately 18% of all automotive fuels sold in the United States. Ethanol-blended fuels are approved under the warranties of all major auto manufacturers. Common blends are the E10 blend, which is 10% ethanol and 90% gasoline and E85, which is used by flexible-fuel vehicles and is 85% ethanol.

Ethanol-blended fuel keeps the fuel system clean because it does not leave varnish or gummy deposits. Being an alcohol, ethanol can absorb moisture in a fuel system and carry it out in suspension as it is consumed. If water contamination becomes too high, it will separate out and fall to the bottom of the vehicle's fuel tank. If this happens, it is best to remove the water-contaminated fuel and refill the tank with ethanol-blended fuel, which will absorb any trace amounts of water that remain.

The downside of using ethanol-blended fuel is a slight increase in fuel consumption (approximately 2%) when compared to gasoline.

ELECTRIC VEHICLES

Nearly all of the major automobile manufacturers have developed an electric vehicle. Electric vehicles (EVs) use electric motors powered by electricity stored in batteries

Figure 26-10 The large battery assembly is being installed in a hybrid electric car during its manufacture. *Reprinted with permission*

(Figure 26–10). Electricity is able to directly produce mechanical energy. Other alternative fuels release stored chemical energy through combustion to provide mechanical energy.

The primary advantage of an EV is a drastic reduction in noise and emission levels. However, the driving range of an EV averages between 50 and 70 miles before the battery must be recharged. Temperature, vehicle load, and speed affect this range. The basic components of an electric vehicle are shown in **Figure 26–11**.

General Motors was the first major automobile manufacturer to offer an electric vehicle to the public. These vehicles were not sold but were leased to their customers. GM has ended the leases on these vehicles and has stopped production of them. There are many reasons for GM doing this, one of which is that the vehicle never gained much popularity.

The driving range of electric vehicles is their biggest disadvantage. Much research is being done to extend the range and to decrease the required recharging times. (Recharging the batteries can take up to ten hours.) Currently the use of nickel-metal hydride or lead-acid batteries and permanent magnet motors has extended the operating range. Other features, such as regenerative braking and highly efficient accessories (such as a heat pump for passenger heating and cooling) are also being installed on electric vehicles. During regenerative braking, the electric motor acts like a generator and recharges the battery when the vehicle is decelerating. Other disadvantages are the cost of replacement batteries and the danger of the high voltage and high frequency of the motors.

The process and equipment needed for recharging the batteries (approximately 26 separate batteries!) varies with the vehicle's manufacturer. Toyota's RAV4-EV carries all of its charging equipment in the vehicle: All that is needed is a 220-volt, 30-amp outlet **(Figure 26–12)**. Other EVs are recharged with an inductive charger **(Figure 26–13)**. This weatherproof plastic paddle is inserted into the charging port at the front of the vehicle. Power is transferred by magnetic fields. Normally it takes four hours for an EV to be recharged.

Although battery-powered electric vehicles are zero-emission vehicles, there are emissions associated with the generation of electricity at the power plant.

Figure 26-12 Recharging the batteries on Toyota's RAV4-EV. *Reprinted with permission*

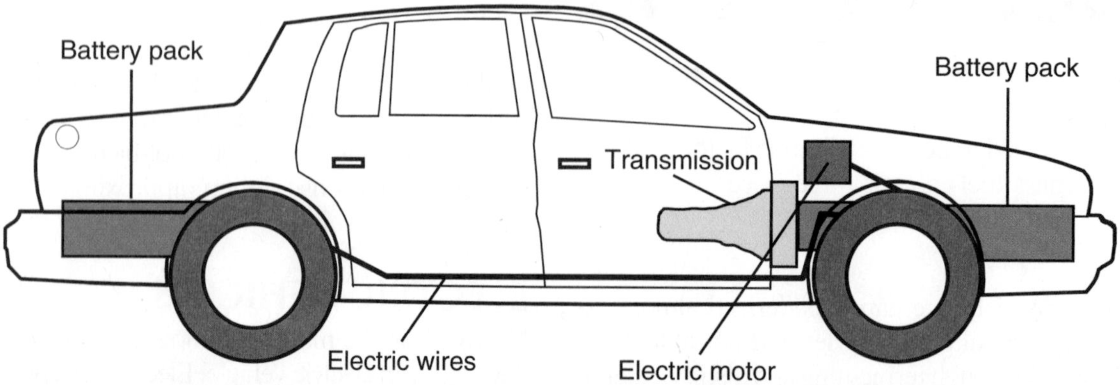

Figure 26-11 An electric car.

Figure 26–13 Recharging an electric vehicle with an inductive charger. *Used with permission from Nissan North America, Inc.*

Electric cars may be more dominant in the future, depending on the development of batteries that can extend the range of operation. Some of the technologies being used or evaluated include lead-acid, nickel cadmium, nickel iron, nickel zinc, nickel metal hydride, sodium nickel chloride, zinc bromide, sodium sulfide, lithium, zinc air, and aluminum air batteries.

Solar Energy Solar energy technologies use sunlight to produce heat and electricity. Electricity produced by solar energy through photovoltaic technologies can be used in conventional electric vehicles.

Solar energy is unlikely to have an impact on transportation in the future.

HYBRID ELECTRIC VEHICLES

Hybrid electric vehicles (HEVs) are or will soon be available from all of the major automobile manufacturers. Any vehicle that combines two or more sources of power is called a hybrid. Current HEVs have an internal combustion engine and an electric motor. How these are used depends on the design of the system.

Series Hybrids

Some HEV designs are close to being an electric vehicle in that the engine is used to drive a generator that charges the battery or powers the electric motor. The electric motor powers the vehicle. The engine is there only to extend the vehicle's driving range. The engine may be a gasoline or diesel engine. A computer controls the operation of the engine depending on the power needs of the battery. The generator also works as a starter motor. When the computer senses that system voltage is low, the generator spins up and quickly starts the engine. An electric motor can only operate at maximum power when it receives full voltage. The batteries, generator, and engine provide for this voltage.

Parallel Hybrids

The most common design of hybrid vehicle relies on power from the electric motor or engine, and in some cases power from both. When the vehicle moves from a stop and has a light load, the motor moves the vehicle. Power for the electric motor comes from stored electricity in the battery pack. During normal driving conditions, the engine is the main power source. Engine power is also used to rotate a generator that recharges the storage batteries. The motor is run to add power to the powertrain. A computer controls the operation of the motor depending on the power needs of the vehicle. During full throttle or heavy load operation, the motor is turned on to increase the output of the powertrain. The electric motor can also act as a generator. During deceleration, the motor works as a generator to charge the batteries and to help slow down the vehicle.

SHOP TALK

Do not be confused between the motor and the engine in a hybrid vehicle. A vehicle's engine has been, and still is, often called a motor. It is not a motor in spite of the fact that we put motor oil in it. The real name for motor oil should be engine oil. By definition, a motor is a machine that converts electrical energy into mechanical energy. An engine converts chemical energy into mechanical energy. ■

The Toyota Prius was the world's first mass-produced hybrid vehicle. This car is fitted with a 44-hp electric motor and an Atkinson cycle four-cylinder engine **(Figure 26–14)**. The Atkinson cycle engine is similar to a Miller cycle engine. The powertrain is called the Toyota Hybrid System (THS). The THS has four main components: the engine, battery pack, hybrid transaxle, and inverter **(Figure 26–15)**.

The hybrid transaxle has a power-splitting device that controls the amount of torque applied to the drive wheels by the engine and/or electric motor. The power-split device uses a planetary gear to vary the amount of power supplied from the engine to either the wheels or the generator **(Figure 26–16)**. The ring gear of the gear set is connected to the electric motor and the differential. The engine is connected to the planetary carrier and the generator is connected to the sun gear. The entire

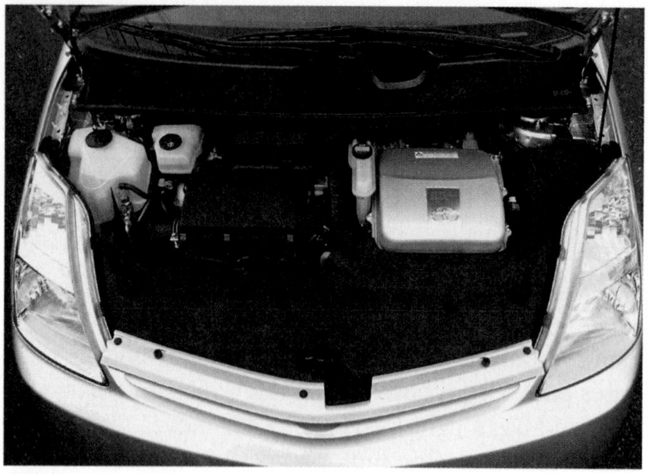

Figure 26–14 The engine compartment of a Toyota Prius. *Reprinted with permission*

Figure 26–15 The Toyota Hybrid System assembly that contains the battery pack, hybrid transaxle, and inverter. *Reprinted with permission*

transmission system functions like an electronically controlled continuously variable transmission. It adjusts the rates of revolution of the engine, motor, and generator.

The inverter is an electric power converter that changes DC voltage of the battery to AC voltage for the electric motor. The inverter also has a voltage converter that drops some of the battery's voltage (273.6 volts) to 12 volts so that normal lights, radios, and other accessories can be used.

The engine is kept within its most efficient speed and torque range to maximize fuel economy and to minimize emission levels. The engine only starts after the car has reached a certain speed. For example, during acceleration the electric motor gets the car up to 15 mph and then the engine starts and helps the motor get the car up to speed **(Figure 26–17)**. Also, when the engine is not needed, fuel to it is cut off and the vehicle relies totally on electric power.

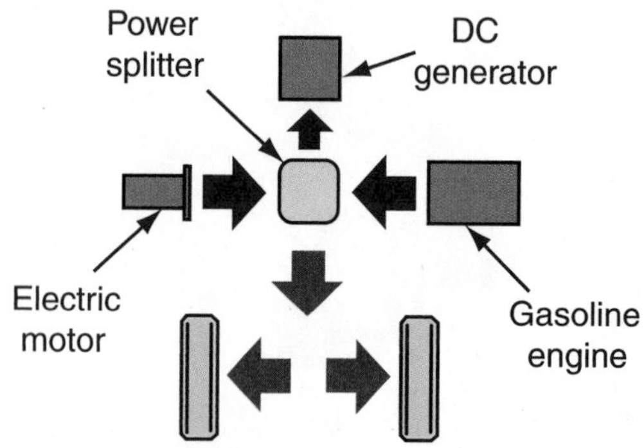

Figure 26–16 The layout of the main components for Toyota's hybrid vehicles.

The Prius is also fitted with a regenerative braking system. When the vehicle is coasting or the brakes are applied, the system causes the electric motor to operate as a generator, recovering energy and transferring it to charge the batteries.

A different approach was taken by Honda. Honda uses a 13-hp motor **(Figure 26–18)** to aid the 85-hp engine, not to replace it, under certain low-load conditions. Placed between the engine and transaxle is the integrated motor assist (IMA) unit **(Figure 26–19)**. The IMA motor is a brushless DC unit that is extremely compact. It serves as both a power booster and a generator.

During acceleration and other times of heavy engine load, the electric motor assists the gasoline engine by providing additional torque resulting in improved acceleration without increased fuel use. At cruising speeds when the load on the engine is low, the IMA lets the engine maintain the speed. During deceleration, the electric motor becomes a generator and converts energy into electricity (regenerative braking).

The engine has an automatic idle stop feature. At a stoplight, the engine turns itself off and is restarted when the brake pedal is released.

Figure 26–17 During full acceleration, in addition to the power from the engine and electric motor, additional current is fed to the motor by the battery. *Reprinted with permission*

Figure 26-18 The electric motor used in Honda's hybrid vehicles. *Courtesy of American Honda Motor Co., Inc.*

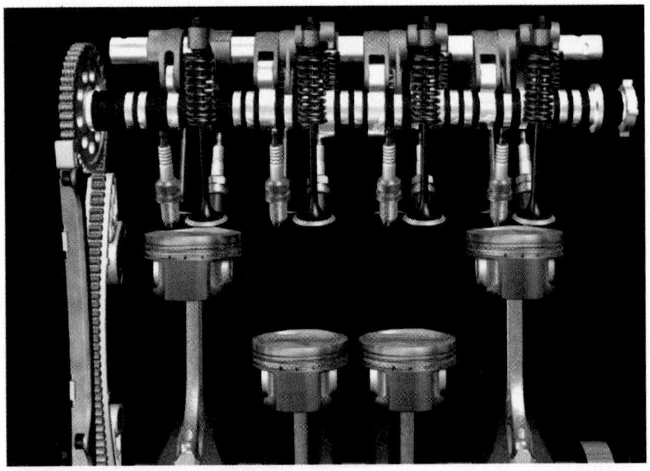

Figure 26-20 The engine used in Honda's hybrid has two spark plugs in each cylinder and has a cylinder activation mechanism built into the valve train. *Courtesy of American Honda Motor Co., Inc.*

The engine has also been modified to achieve maximum fuel mileage with minimal emissions. Each of the cylinders is fitted with dual spark plugs, which are individually controlled by a PCM. Depending on the conditions, an individual cylinder's plugs can be ignited sequentially or simultaneously. The result is more effective combustion and improved emissions. The engine also has electronic valve lift controls that can close the valves on three of the four cylinders during deceleration **(Figure 26–20)**. This action reduces the effect of engine braking, which means there is more energy for the IMA system to regenerate into electricity.

Hybrid technology can also be used to improve the performance of a vehicle. An example of this is the Acura RL, which has a 200-hp V6 engine and a 160-hp electric motor. The engine drives the front wheels and the motor drives the rear wheels. The motor is switched on by demand and its battery is recharged through regenerative braking.

Figure 26-19 The integrated motor assist (IMA) unit is placed between the engine and transaxle. *Courtesy of American Honda Motor Co., Inc.*

FUEL CELL VEHICLES

Fuel cell vehicles (FCVs) are electric vehicles **(Figure 26–21)** that use fuel cells to convert chemical energy into electrical energy.

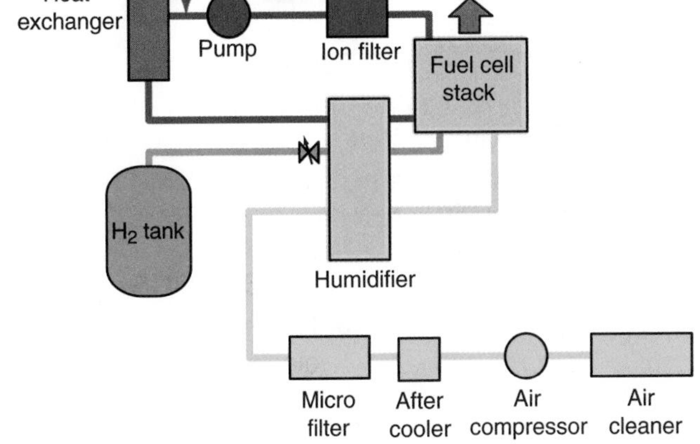

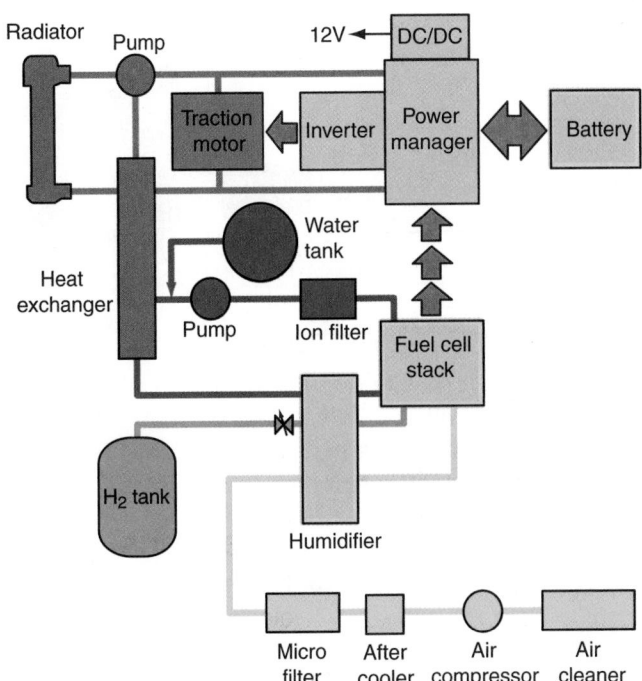

Figure 26-21 The basic layout for a fuel cell vehicle.

General Motors has built a concept car, called the Hy-wire, that has all of the fuel cell powertrain in a skateboard-shaped chassis. The assembly has no mechanical or hydraulic linkages; rather, it relies on drive-by-wire technology for steering, brakes, and other systems. This allows for great flexibility in the design and features of a car.

Fuel Cells

A **fuel cell (Figure 26–22)** is a device that converts fuel directly into electricity. Fuel cells are seen as the solution to the problems inherent to an EV. Fuel cells do not store electricity; rather, electricity is converted, as it is needed, to spin the electric motors that move the vehicle. Fuel cells release energy derived from the reaction between hydrogen and oxygen. They have high efficiency and, depending on the fuel used, produce little or no emissions. Fuel cells operating on hydrogen emit nothing but pure water. There are various options for storing hydrogen and feeding it into a fuel cell, the most common of which are compressed gaseous or liquid hydrogen, methanol, sodium borohydride, metal hydrides, and gasoline.

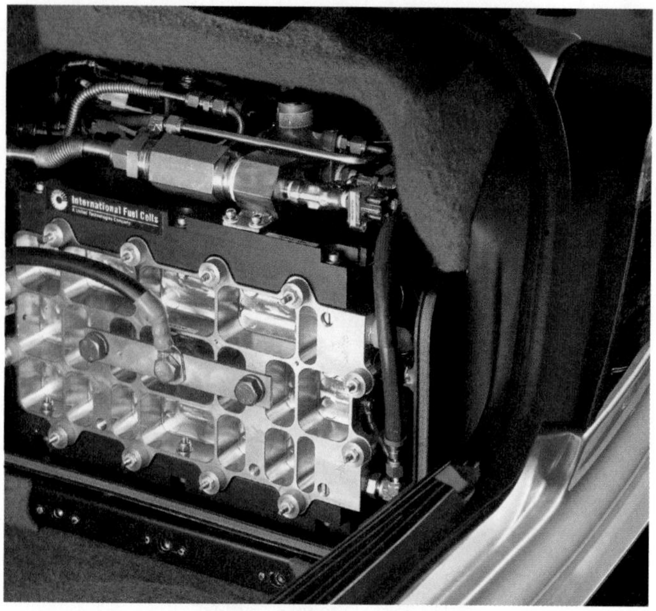

Figure 26-22 A trunk-mounted fuel cell. *Courtesy of BMW of North America, Incorporated*

A fuel cell has no moving parts or chemicals and requires no maintenance.

Like a battery, a fuel cell has two electrodes (a cathode and an anode), but these are not consumed during operation **(Figure 26–23)**. Instead, they are fed fuel and air through channels cut in the plates. Between the electrodes is a thin coat of platinum catalyst, which acts as the electrolyte. Hydrogen fuel is pumped into one side of the fuel cell and air is pumped into the other side. When the hydrogen flows across the catalyst material, it dissociates into free electrons and protons (hydrogen ions) and electrical current flows between the negatively charged anode and the positively charged cathode. While the free electrons become current, the protons move to the cathode and combine with the oxygen in the incoming air to produce water vapor and heat. This is the fuel cell's exhaust.

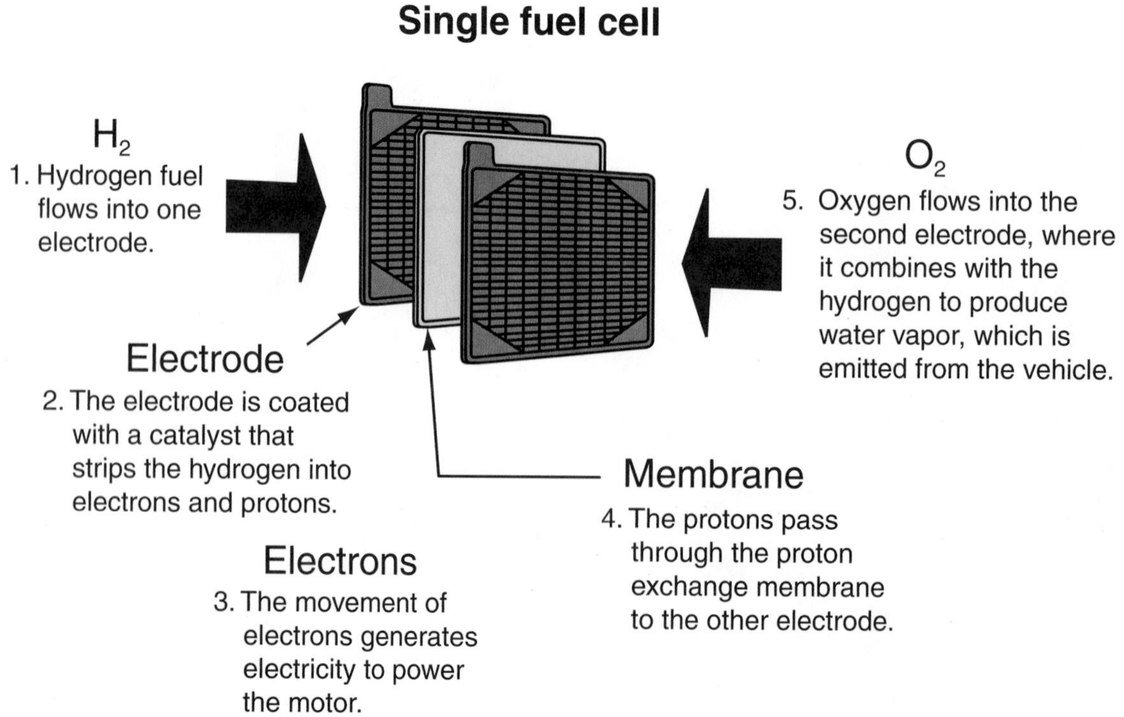
Single fuel cell
H_2
1. Hydrogen fuel flows into one electrode.
Electrode
2. The electrode is coated with a catalyst that strips the hydrogen into electrons and protons.
Electrons
3. The movement of electrons generates electricity to power the motor.
Membrane
4. The protons pass through the proton exchange membrane to the other electrode.
O_2
5. Oxygen flows into the second electrode, where it combines with the hydrogen to produce water vapor, which is emitted from the vehicle.

Figure 26-23 The basic operation of a fuel cell.

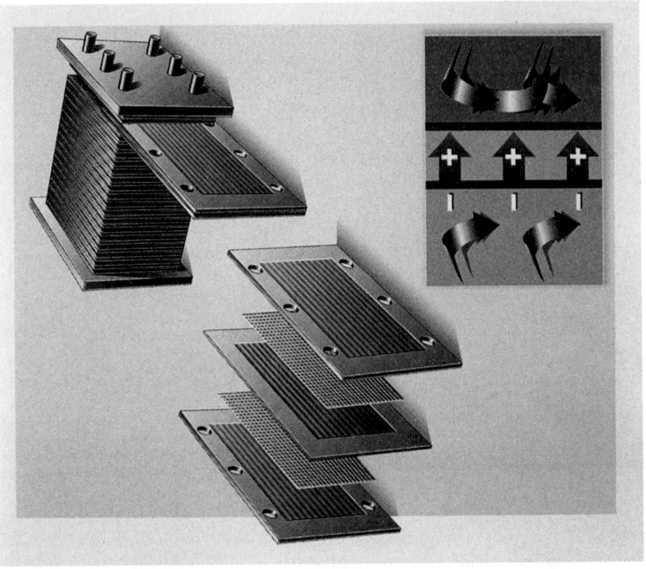

Figure 26-24 A fuel cell stack. *Courtesy of DaimlerChrysler Corporation*

Figure 26-25 The engine compartment of a fuel cell car. *Courtesy of DaimlerChrysler Corporation*

The electricity generated by the fuel cell is used power the vehicle. However, an individual fuel cell produces a relatively small amount of voltage (a little less than 1 volt). To provide the higher voltages required to run the electric motor and to recharge the battery, fuel cells are stacked or connected in series **(Figure 26–24)**. Through the series connection, the voltage output is the sum of the individual cells. When more power than what the fuel cell can generate is needed, additional current is drawn from the storage batteries.

An advantage of a fuel cell vehicle **(Figure 26–25)** is that the batteries are recharged by the fuel cell and do not require plug-in recharging like other electric vehicles. The vehicles also require fewer storage batteries and therefore weigh less. The DC voltage from the batteries passes through an inverter where it is changed to three-phase AC voltage. A converter is used to supply normal accessories and lights with 12 volts.

To control vehicle speed, most FCVs use a throttled fuel cell. Controlling the amount of hydrogen flowing through the fuel cell controls the amount of electricity that flows to the motor.

Reformers

Gaseous hydrogen is difficult to store. Uncompressed, it takes 3,000 times as much volume to store the same amount of energy as gasoline. High-pressure storage reduces the required volume for storage but presents many safety concerns. Methanol is as easy to store and contain as regular gasoline and can be used as a source for hydrogen. The hydrogen is then used in the fuel cell to generate electricity. To begin the chemical process of generating electricity in the fuel cell, the methanol fuel is mixed with water and is passed through a fuel reformer. The output of the reformer (hydrogen and carbon dioxide) is released to the negative side of the fuel cell.

A **reformer** changes the molecular structure of hydrocarbons into hydrogen-rich gas to power fuel cells. Reformers extract hydrogen from normal fuels, such as gasoline **(Figure 26–26)** and methanol. Reforming can be accomplished through the following four-step process:

1. The gasoline is vaporized.
2. The gasoline vapors and air are burned to produce hydrogen gas and carbon monoxide.
3. The hydrogen and CO are combined with steam in the presence of a copper-oxide catalyst that reacts with the CO to form CO_2 and additional hydrogen.
4. Compressed air is injected in the presence of a platinum catalyst to convert the remaining CO to CO_2. The hydrogen, CO_2, and other inert elements from the gasoline are moved into the fuel cell to generate electricity.

Regardless of the fuel used in the fuel cell, it takes time for the cell to produce enough electricity to get the electric motors working. Therefore, these vehicles are equipped with storage batteries to operate the vehicle until the cell is warmed up. The batteries also store electricity regenerated during coasting and braking. **Figure 27–27** shows the layout of a typical fuel cell vehicle.

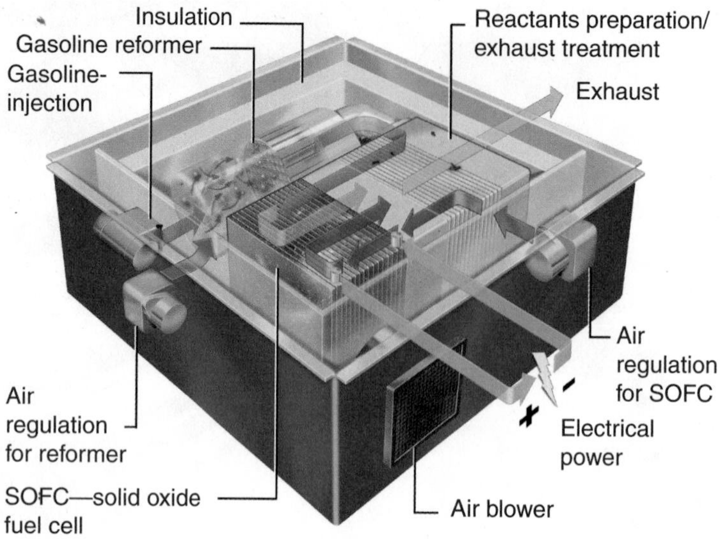

Figure 26-26 A fuel cell with a gasoline reformer. *Courtesy of BMW of North America, Incorporated*

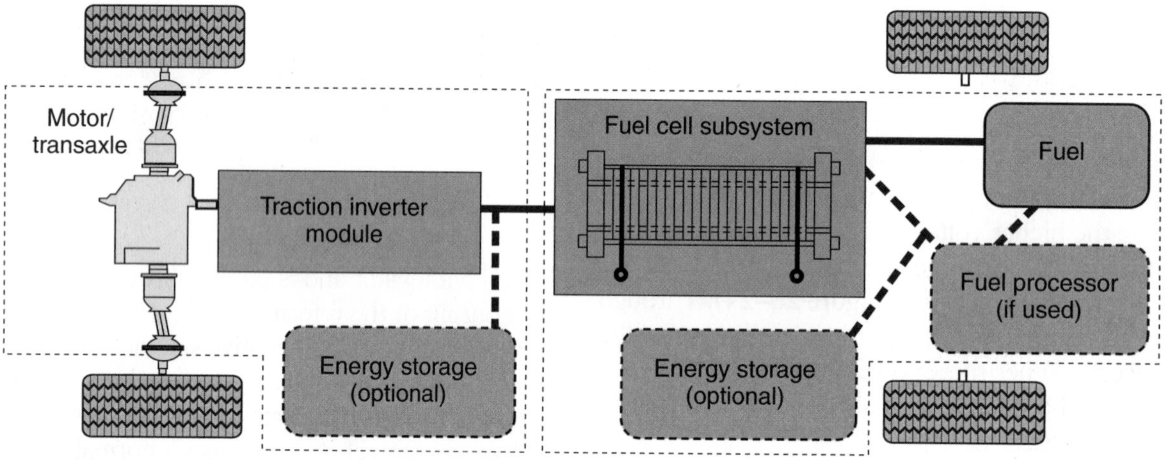

Figure 26-27 Layout of a typical fuel cell vehicle.

CASE STUDY

***A** customer who happens to be able to do a lot with cars had his 1927 Dodge towed into the shop because it would not start. The car had not been driven for two years, but it had been started a few times during those two years. Also, during that time, the engine had been rebuilt and the car was totally restored.*

The owner assumed the gasoline in the tank was stale and drained it after the engine would not start. Even with fresh gasoline, the engine acted as if it did not even want to start. This was aggravating, especially because the engine ran so well after it was rebuilt. The owner was sure it had to be something that could occur because the engine was not run for so long. The owner checked out everything he could think of, but still the engine would not run, so the car was brought to the repair shop.

The technician assigned to the car was thrilled to able to work on the car. He went through the basic checks and found there was no fuel being sprayed into the intake manifold when the throttle was opened. He checked and found fuel was being delivered to the carburetor. He knew, then, that the problem had to be in the carburetor.

He had never worked on a carburetor before so he spent some time searching for information about this carburetor; then he removed it. As he was taking it apart, he found a substantial buildup of gum and varnish in the carburetor. He assumed this was the problem, so he cleaned and rebuilt the carburetor. Fortunately, his assumptions were correct. Once he installed the carburetor back on the engine, the engine started and ran fine.

KEY TERMS

Antiknock
Antiknock Index (AKI)
Biodiesel fuels
Cetane
Compressed natural gas (CNG)
Crude oil
Ethanol
Flexible-fuel vehicles (FFV)
Fuel cell
Fuel cell vehicles (FCVs)
Heptane
Hybrid electric vehicles (HEVs)
Isooctane
Lean mixtures
Liquefied petroleum gas (LP gas)
Methanol
Methyl tertiary butyl ether (MTBE)
Methylcylopentienyl manganese tricarbonyl (MMT)
Motor octane number (MON)
Octane
Oxygenates
P-series fuels
Reformer
Reformulated gasoline (RFG)
Reid vapor pressure (RVP) test
Renewable fuels
Research octane number (RON)
Rich mixtures
Stoichiometric
Sulfur dioxide
Tetraethyl lead (TEL)
Tetramethyl lead (TML)
Vapor lock

SUMMARY

- The ideal air/fuel ratio is referred to as stoichiometric. When the mixture has more air than the ideal ratio calls for, the mixture is said to be lean. Ratios that have more fuel than the ideal are considered rich mixtures.
- Gasoline is a complex mixture of approximately 300 various ingredients, mainly hydrocarbons, refined from crude oil for use as fuel in engines.
- An octane number or rating is assigned to a fuel to indicate its antiknock quality. The octane rating required by law and the one displayed on gasoline pumps is the Antiknock Index (AKI) and is stated as (R+M)/2.
- Oxygenates are compounds such as alcohols and ethers that contain oxygen in their molecular structure and are added to fuels to improve combustion efficiency, increase fuel octane ratings, and reduce emissions.
- The most commonly used additive or oxygenate is ethanol.
- Methanol can be used directly as an automotive fuel but is not recommended by auto manufacturers except for a few flexible-fuel vehicles. In the future, methanol could be the fuel of choice for providing hydrogen to power fuel cell vehicles.
- MTBEs and ethanol are the most commonly used oxygenates for producing reformulated gasoline (RFG), which produces a leaner air/fuel mixture and generates less carbon monoxide.
- Two tests can be done to test the quality of gasoline: the Reid vapor pressure test and the alcohol content test.
- Diesel fuel contains more energy per gallon than gasoline and diesel engines are more efficient.
- Diesel fuel's ignition quality is measured by the cetane rating.
- LP gas is a good alternative to gasoline, but it is a fossil fuel and therefore is not a favored alternative fuel for the future.
- P-series fuels are natural gas liquids, ethanol, and biomass-derived co-solvents that can be used alone or freely mixed, in any proportion, with gasoline.
- CNG offers several advantages over gasoline but is also a fossil fuel.
- Hydrogen can be used directly or mixed with natural gas in internal combustion engines and is the fuel for fuel cells.
- Renewable fuels are produced from biomass and release fewer emissions and lower amounts of carbon dioxide than gasoline.
- Ethanol is typically derived from corn and has a very high octane rating.
- Ethanol is primarily used as an oxygenate in gasoline but is also blended with large amounts for use in flexible fuel vehicles.
- Electric vehicles use electric motors powered by electricity stored in batteries.
- Hybrid electric vehicles (HEVs) have an internal combustion engine and an electric motor.
- Some HEV designs have an engine that is only used to drive a generator that charges the battery or powers the electric motor.
- The most common design of hybrid vehicle relies on power from the electric motor and/or an engine.
- Fuel cell vehicles (FCVs) are electric vehicles that use fuel cells to convert chemical energy into electrical energy.
- A fuel cell is a device that converts fuel directly into electricity, has no moving parts or chemicals, and requires no maintenance.
- A reformer extracts hydrogen from normal fuels, such as gasoline and methanol, for use in a fuel cell.

TECH MANUAL

The following procedure is included in Chapter 26 of the *Tech Manual* that accompanies this book:

1. Testing the RVP of gasoline.

REVIEW QUESTIONS

1. What type of hybrid vehicle only uses the engine to run a generator, charge the battery, and/or supply power for the motor?
2. What is P-series fuel?
3. *True or False?* Diesel fuel's antiknock quality is measured by the cetane rating.
4. Briefly explain how a fuel cell works.
5. The air/fuel ratio that allows all of the hydrocarbons to completely combine with the available oxygen and provide for complete combustion is called the ________________ ratio.
6. What is a hybrid vehicle?
7. Which of the following is the least likely to affect the performance of a fuel?
   - **a.** antiknock quality
   - **b.** volatility
   - **c.** conductivity
   - **d.** deposit control
8. What does the Reid vapor pressure (RVP) test measure?
9. *True or False?* All hybrid vehicles can be powered by the electric motor, a gasoline engine, or both at the same time.
10. *True or False?* Hydrogen-powered vehicles are fuel cell vehicles.
11. The component that changes the molecular structure of hydrocarbons into hydrogen-rich gas is called a:
    - **a.** pressure container
    - **b.** fuel cell
    - **c.** fuel injector
    - **d.** reformer
12. Technician A says the use of methanol in internal combustion engines has declined over the years. Technician B says the use of MTBE as a gasoline additive has declined over the years. Who is correct?
    - **a.** Technician A
    - **b.** Technician B
    - **c.** Both A and B
    - **d.** Neither A nor B
13. Which of the following is not a commonly used oxygenate?
    - **a.** aromatic hydrocarbons
    - **b.** nitrous oxide
    - **c.** methanol
    - **d.** MTBE
14. Which of the following does not affect engine knock?
    - **a.** fuel detergents
    - **b.** a lean fuel mixture
    - **c.** overadvanced ignition timing
    - **d.** the octane number
15. Technician A says reformulated gasoline produces a leaner air/fuel mixture. Technician B says RFG generates more carbon dioxide. Who is correct?
    - **a.** Technician A
    - **b.** Technician B
    - **c.** Both A and B
    - **d.** Neither A nor B
16. *True or False?* During initial acceleration, a Toyota Prius relies solely on electric power.
17. Which of the following statements about hydrogen is *not* true?
    - **a.** Hydrogen displaces air, so any release in an enclosed space could cause asphyxiation.
    - **b.** Hydrogen must be stored as a compressed gas.
    - **c.** Hydrogen is nontoxic.
    - **d.** Hydrogen is highly flammable and there is an explosion risk.
18. Which of the following chemicals is commonly added to gasoline to increase its octane rating?
    - **a.** isooctane
    - **b.** heptane
    - **c.** sulfur
    - **d.** ethanol
19. Technician A says dedicated vehicles are those designed to use one particular type of fuel. Technician B says bifuel vehicles can operate solely on an alcohol-based fuel, unleaded gasoline, or a mixture of the two, which gives the driver flexibility and convenience when refilling the fuel tank. Who is correct?
    - **a.** Technician A
    - **b.** Technician B
    - **c.** Both A and B
    - **d.** Neither A nor B
20. *True or False?* Fuel cell vehicles use a battery to power the electric motor while the fuel cell is warming up.

27

FUEL DELIVERY SYSTEMS

OBJECTIVES

■ Describe the components of a fuel delivery system and the purpose of each. ■ Conduct a visual inspection of a fuel system. ■ Relieve fuel system pressure. ■ Inspect and service fuel tanks. ■ Inspect and service fuel lines and tubing. ■ Describe the different fuel filter designs and mountings. ■ Remove and replace fuel filters. ■ Explain how common electric fuel pump circuits work. ■ Conduct a pressure and volume output test on a mechanical and electric fuel pump. ■ Service and test electric fuel pumps.

The fuel delivery system has the important role of delivering fuel to the injectors. The fuel must also be delivered in the right quantities and at the right pressure. The fuel must also be clean when it is delivered to the engine.

A typical fuel delivery system includes a fuel tank, fuel lines, fuel filters, and a pump. The system works by using a pump to draw fuel from the fuel tank and passing it through fuel lines and filters to the fuel injection system. The filter removes dirt and other harmful impurities from the fuel. A fuel line pressure regulator maintains a constant high fuel pressure. This pressure generates the spraying force needed to inject the fuel. Excess fuel not required by the engine returns to the fuel tank through a fuel return line.

GENERAL FUEL SYSTEM DIAGNOSIS

One of the key requirements for an efficient running engine is the correct amount of fuel. To accomplish this, fuel must be stored, pumped out of storage, piped to the engine, and filtered **(Figure 27–1)**. All of this must be accomplished in an efficient and safe manner.

The fuel system should be checked whenever there is evidence of a fuel leak or smell. Leaks are not only costly to the customer but are also very dangerous. The fuel system should also be checked whenever basic tests suggest there is too little or too much fuel being delivered to the cylinders. Lean mixtures are often caused by insufficient amounts of fuel being drawn out of the fuel tank. Lean mixtures can cause bad results in many different diagnostic tests, including high HC, O_2, and NO_x readings on an exhaust analyzer and high firing lines on a scope.

When no fuel is delivered to the engine, the engine will not start or run. Remember, there must be four things to have combustion–air, fuel, compression, and spark. On carburetor and throttle-body-injected engines, it is easy to determine if fuel is being delivered. Simply look down the throttle body. If the surfaces are wet or you see fuel being sprayed while cranking the engine, fuel is there. With port injection, it is a little more difficult. Connect a fuel-pressure gauge to the fuel line or rail and observe the fuel pressure while cranking the engine. Testing fuel pressure is described later in this chapter. However, if there is no fuel pressure while cranking, no fuel is being delivered to the engine.

CAUTION!

Extreme care should be taken when working with any of the components of a fuel system. Gasoline is very volatile and flammable. Never expose it to open flame or extreme heat. Disconnect the negative battery cable before doing anything that may release gas vapors. Use containers to catch gasoline and rags to wipe up any spills. Use a flashlight or an enclosed fluorescent tube light designed for safe use around fuels. When working with a fuel system, always have a Class B fire extinguisher nearby.

There are many components in the fuel system, which can be grouped into two categories: fuel delivery and the fuel injection system. Diagnosis and basic service to the fuel delivery system is covered in this chapter. All of the tests in this chapter assume that the fuel is good

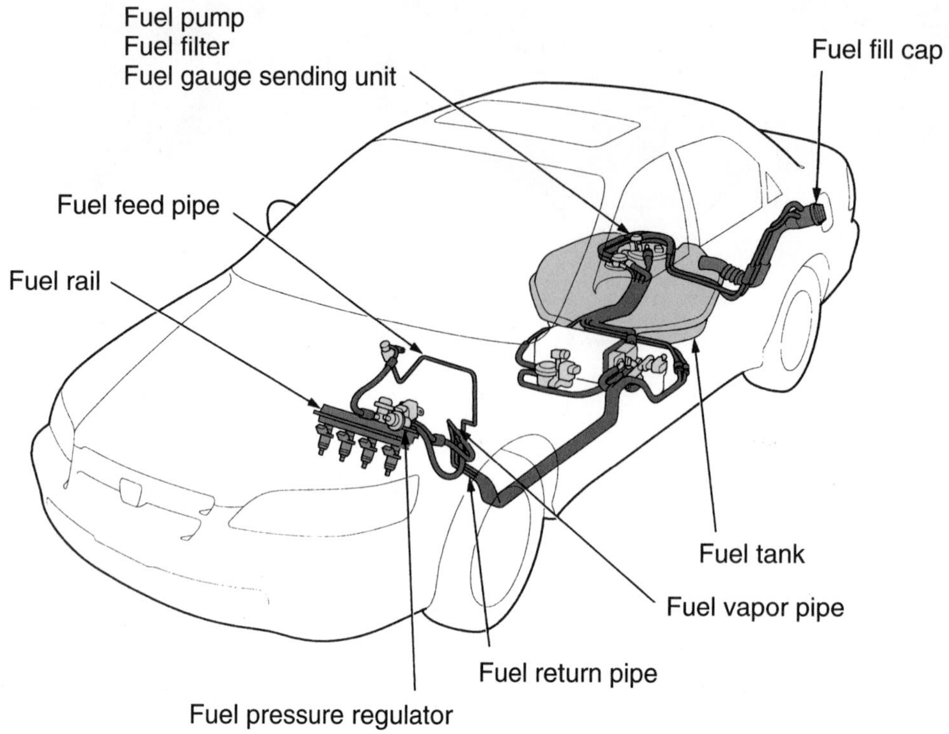

Figure 27-1 A fuel supply (delivery) system for a late-model car. *Courtesy of American Honda Motor Co., Inc.*

and not severely contaminated. Obviously, water does not burn as well as gasoline. Therefore, water in the fuel tank can cause a driveability problem. If water is mixed with the gasoline, drain the tank and refill it with fresh clean gasoline. Also, keep in mind that after gasoline has sat for a while, it becomes less volatile or stale. Additives are available to partially revitalize the fuel; however, if the fuel is so stale that an engine will not run on it, drain it and refill the tank with fresh gasoline.

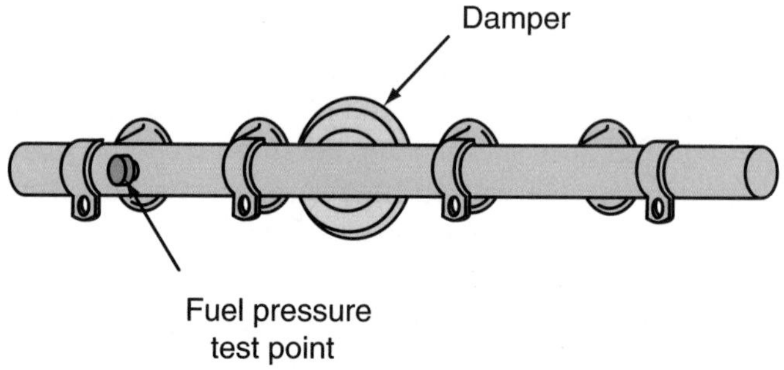

Figure 27-2 A typical fuel pressure test point (port) location.

FUEL SYSTEM PRESSURE RELIEF

CAUTION!

Failure to relieve the fuel pressure on electronic fuel injection systems prior to fuel system service may result in serious personal injury and expensive property damage.

Because electronic fuel injection (EFI) systems have residual fuel pressure, this pressure must be relieved before disconnecting any components in the fuel system. Most port fuel injection (PFI) systems have a fuel pressure test port on the fuel rail. If the fuel rail is fitted with a Schrader valve or test port **(Figure 27–2)**, you can relieve the system's pressure there. Begin by disconnecting the negative battery cable to avoid fuel discharge if an accidental attempt is made to start the engine. Loosen the fuel tank filler cap to relieve any vapor pressure built up in the tank. Wrap a shop towel around the fuel pressure test port on the fuel rail and remove the dust cap from this valve. Connect a fuel pressure gauge to the fuel pressure test port on the fuel rail. Install a bleed hose onto the gauge and put the free end into an approved gasoline container. Then open the gauge bleed valve to relieve fuel pressure from the system into the gasoline container. Be sure all the fuel in the bleed hose is drained into the gasoline container.

On EFI systems that do not have a fuel pressure test port, such as most throttle-body injection (TBI) systems, loosening the fuel tank filler cap will relieve any tank vapor pressure. Then remove the fuel pump fuse. Start and run the engine until the fuel in the lines is used up and the engine stops. Crank the engine with the starter for about three seconds to relieve any remaining fuel pressure.

CAUTION!

Always wear eye protection and observe all other safety rules to avoid personal injury when servicing fuel system components.

FUEL TANKS

Fuel tanks include devices that prevent vapors from leaving the tank. For example, to contain vapors and allow for expansion, contraction, and overflow that result from changes in the temperature, the fuel tank has a separate air chamber dome at the top. All fuel tank designs provide some control of fuel height when the tank is filled. Frequently, this control is achieved by using vent lines with the filler tube or tank **(Figure 27–3)**. These fuel height controls allow only 90% of the tank to be filled. The remaining 10% is for expansion during hot weather. Some fuel tanks have an overfill limiting valve to prevent overfilling of the tank.

Fuel tanks are constructed of pressed corrosion-resistant steel, aluminum, or molded polyethylene plastic. Aluminum and molded plastic tanks are the most commonly used.

Most tanks have slosh baffles or surge plates to prevent fuel from splashing around inside the tank. In addition to slowing down fuel movement, the plates tend to keep the fuel pickup and sending unit for the fuel gauge immersed in the fuel during hard braking and acceleration. The plates or baffles also have holes or slots in them to permit the fuel to move from one end of the tank to the other. With few exceptions, the fuel tank is located in the rear of the vehicle.

A fuel tank has an inlet filler tube and cap. The location of the filler tube depends on the tank design. All current filler tubes have a built-in restrictor that prevents the entry of the larger leaded fuel delivery nozzle at gas pumps. The filler tube can be a rigid one-piece tube soldered to the tank or can be made of multiple pieces.

Filler tube caps (commonly called gas caps) are nonventing and may have some type of pressure-vacuum relief valve arrangement **(Figure 27–4)**. Under normal

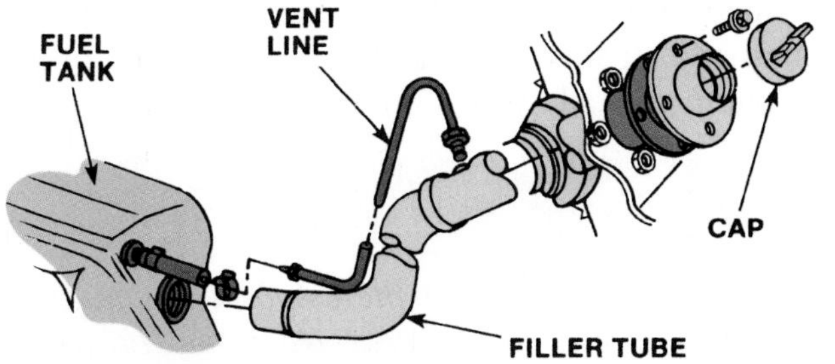

Figure 27–3 Vent lines within the fuel tank filler tube control the fuel level.

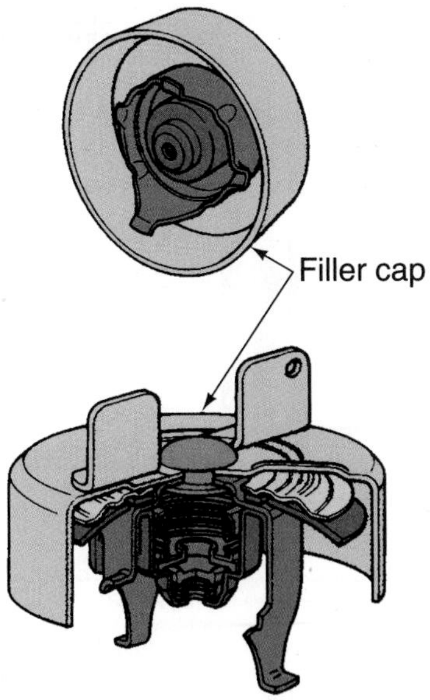

Figure 27–4 A pressure-vacuum gasoline filler cap. *Courtesy of DaimlerChrysler Corporation*

conditions, the valve is closed. When extreme pressure or vacuum is present, the relief valve opens. Once the pressure or vacuum is relieved, the valve closes. Most pressure caps have four antisurge tangs that lock onto the filler neck to prevent the delivery system's pressure from pushing out of the fuel tank.

The filler cap may be threaded into the upper end of the filler pipe. These caps require several turns counterclockwise to remove. The long threaded area on the cap is designed to allow any remaining fuel tank pressure to escape during cap removal. The cap and filler neck are designed to prevent overtightening. When the cap is installed, tighten the cap until a clicking noise is heard. This noise indicates the cap is properly tightened and fully seated.

WARNING!

When a gasoline tank filler cap is replaced, the replacement cap must be exactly the same as the original cap or a malfunction may occur in the filling and venting system, resulting in higher emission levels and the escape of dangerous hydrocarbons.

A check valve might also be fitted in the fuel tank filler cap and the pressure-vacuum relief valve settings increased to prevent fuel leakage in case the vehicle rolls over.

Some form of liquid vapor separator is incorporated into nearly every fuel tank. This separator stops liquid fuel or bubbles from reaching the vapor storage canister or the engine's crankcase **(Figure 27–5)**. It can be located inside the tank, on the tank, in the fuel vent lines, or near

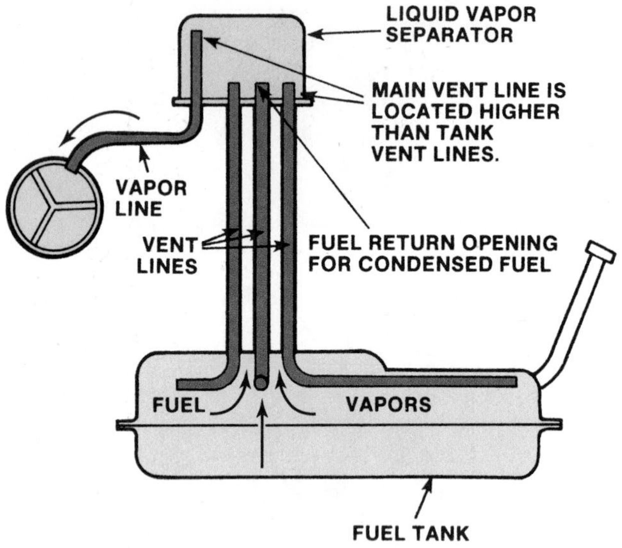

Figure 27-5 A fuel tank vapor separator allows some of the fuel vapors to condense back into liquid and return to the tank.

the fuel pump. Check the service manual for the exact location of the liquid vapor separator and the routing of the hoses to it.

Inside the fuel tank, there is also a sending unit that includes a pickup tube and float-operated fuel gauge sender unit. Most current fuel pumps are installed inside the tank and the pickup is part of that assembly **(Figure 27–6)**. On some systems, the fuel tank pickup tube is connected to the fuel pump by a hose or tube. A pickup tube extends nearly, but not completely, all the way to the bottom of the tank so that rust, dirt, sediment, and water cannot be drawn up into the fuel tank filter, which can cause it to clog.

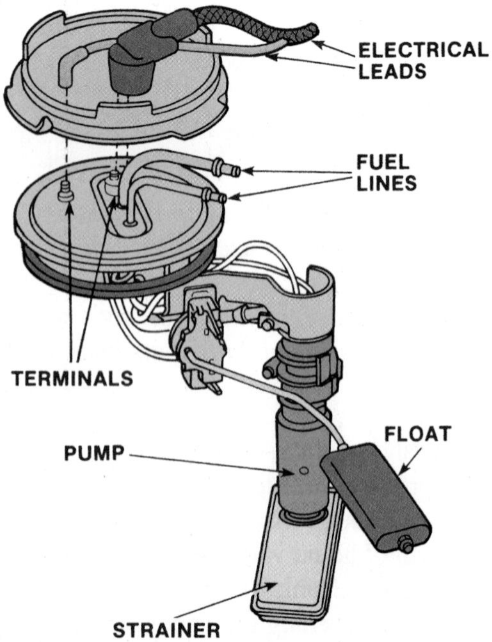

Figure 27-6 A combination electric fuel pump and sending unit.

Inspection

Fuel tanks should be inspected for leaks; road damage; corrosion and rust on metal tanks; loose, damaged, or defective seams; loose mounting bolts; and damaged mounting straps. Leaks in the fuel tank, lines, or filter may cause a gasoline odor in and around the vehicle, especially during low-speed driving and idling.

A weak seam, rust, or road damage can cause leaks in the metal fuel tank. The best method of permanently solving this problem is to replace the tank. Another method is to remove the tank and steam clean or boil it in a caustic solution to remove the gasoline residue. After this has been done, the leak can be soldered or brazed by an appropriately equipped specialty shop.

Holes in a plastic tank can sometimes be repaired by using a special tank repair kit. Be sure to follow manufacturer's instructions when doing the repair.

When a fuel tank is leaking dirty water or has water in it, the tank must be cleaned, repaired, or replaced.

SERVICE TIP

When a fuel tank must be removed, if possible, ask the customer to bring the vehicle to the shop with a minimal amount of fuel in the tank. ■

Fuel Tank Draining

WARNING!

Always drain gasoline into an approved container, and use a funnel to avoid gasoline spills.

The fuel tank must be drained prior to tank removal. Begin by removing the negative cable from the battery. Then raise the vehicle on a hoist. Make sure you have an approved gasoline container and are prepared to catch all of the fuel before proceeding. If the tank has a drain bolt, remove it to drain the fuel. If the fuel tank does not have a drain bolt, locate the fuel tank drainpipe or filler pipe. Using the proper adapter, connect the intake hose from a hand-operated or air-operated pump to the pipe. Insert the discharge hose from the hand-operated or air-operated pump into an approved gasoline container, and operate the pump until all the fuel is removed from the tank.

CAUTION!

Abide by local laws for the disposal of contaminated fuels. Be sure to wear eye protection when working under the vehicle.

Fuel Tank Service

In most cases, the fuel tank must be removed for servicing. The procedure for removing a fuel tank varies depending on the vehicle make and year. Always follow the procedure in the vehicle manufacturer's service manual. What follows is a typical procedure:

PROCEDURE

STEP 1 *Disconnect the negative terminal from the battery.*

STEP 2 *Relieve the fuel system pressure and drain the fuel tank.*

STEP 3 *Raise the vehicle on a hoist or lift the vehicle with a floor jack and lower the chassis onto jack stands.*

STEP 4 *Use compressed air to blow dirt from the fuel line fittings and wiring connectors.*

STEP 5 *Remove the fuel tank wiring harness connector from the body harness connector.*

STEP 6 *Remove the ground wire retaining screw from the chassis if used.*

STEP 7 *Disconnect the fuel lines from the fuel tank. If these lines have quick-disconnect fittings, follow the manufacturer's recommended removal procedure in the service manual. Some quick-disconnect fittings are hand releasable, and others require the use of a special tool* ***(Figure 27–7)****.*

STEP 8 *Wipe the filler pipe and vent pipe hose connections with a shop towel, and then disconnect the hoses from the filler pipe and vent pipe to the fuel tank.*

STEP 9 *Unfasten the filler from the tank. If it is a rigid one-piece tube, remove the screws around the outside of the filler neck near the filler cap. If it is a three-piece unit, remove the neoprene hoses after the clamp has been loosened.*

STEP 10 *Loosen the bolts holding the fuel tank straps to the vehicle* ***(Figure 27–8)*** *until they are about two threads from the end.*

WARNING!

Do not heat the bolts on the fuel tank straps in order to loosen them. The heat could ignite the fumes.

STEP 11 *Holding the tank securely against the underchassis with one hand, remove the strap bolts and lower the tank to the ground. When lowering the tank, make sure all wires and tubes are unhooked. Be careful as small amounts of fuel might still be in the tank.*

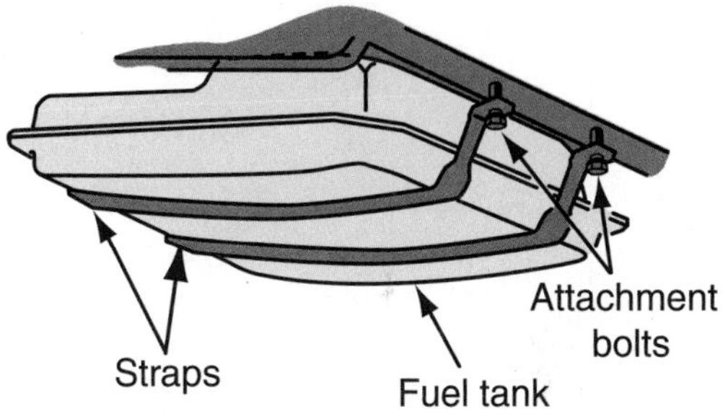

Figure 27-8 Front and rear fuel tank strap mounting bolts.

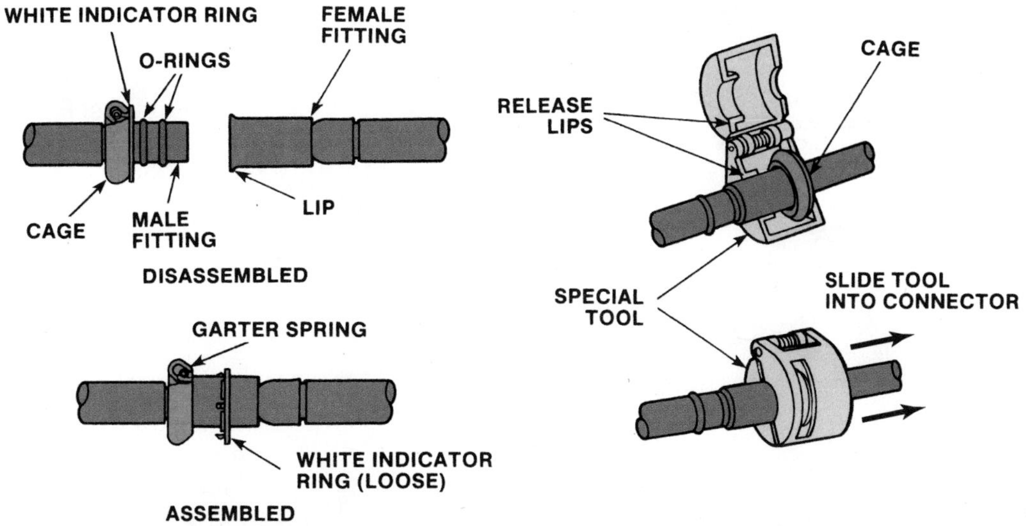

Figure 27-7 Some quick-connect fittings require the use of a special tool to separate them.

To reinstall a repaired or new fuel tank, reverse the removal procedure. Be sure that all the rubber or felt tank insulators are in place. Then, with the tank straps in place, position the tank. Loosely fit the tank straps around the tank, but do not tighten them. Make sure that the hoses, wires, and vent tubes are connected properly **(Figure 27–9)**. Check the filler neck for alignment and for insertion into the tank. Tighten the strap bolts and secure the tank to the car. Install all of the tank accessories (vent line, sending unit wires, ground wire, and filler tube). Fill the tank with fuel and check it for leaks, especially around the filler neck and the pickup assembly. Reconnect the battery and check the fuel gauge for proper operation.

FUEL LINES AND FITTINGS

The fuel lines **(Figure 27–10)** carry fuel from the tank to the fuel filter and fuel injection assembly. Fuel lines can be made of either metal tubing or flexible nylon or synthetic rubber hose. The latter must be able to resist gasoline. The hoses must also be nonpermeable, so gas and gas vapors cannot evaporate through the hose. Ordinary rubber hose, such as that used for vacuum lines, deteriorates when exposed to gasoline. Only hoses made for fuel systems should be used. Similarly, vapor vent lines must be made of material that resists attack by fuel vapors.

Fuel supply lines from the tank to the injectors are routed to follow the frame along the underchassis of the

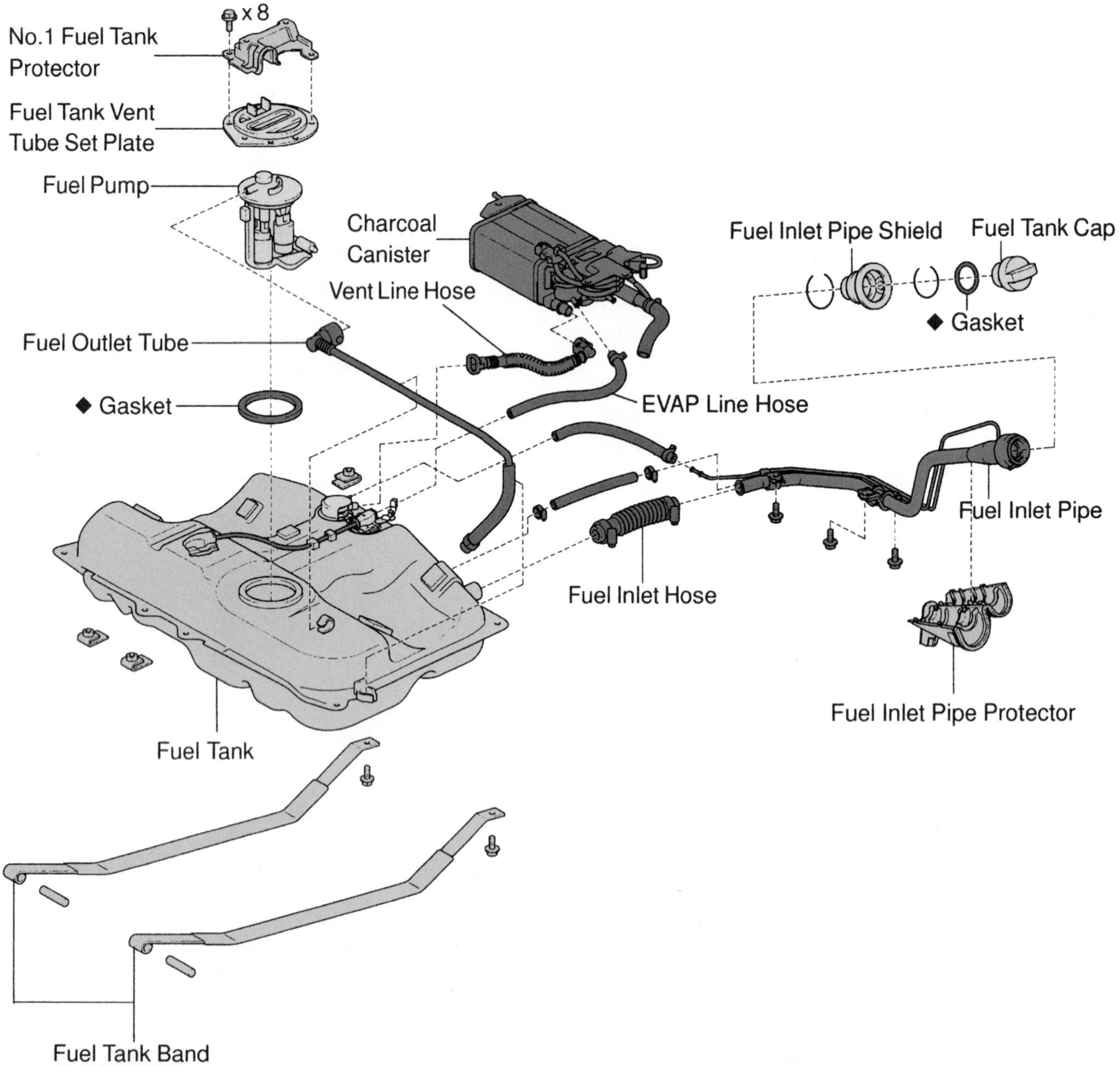

Figure 27-9 The hoses, wires, and tubes normally connected to a fuel tank. *Reprinted with permission*

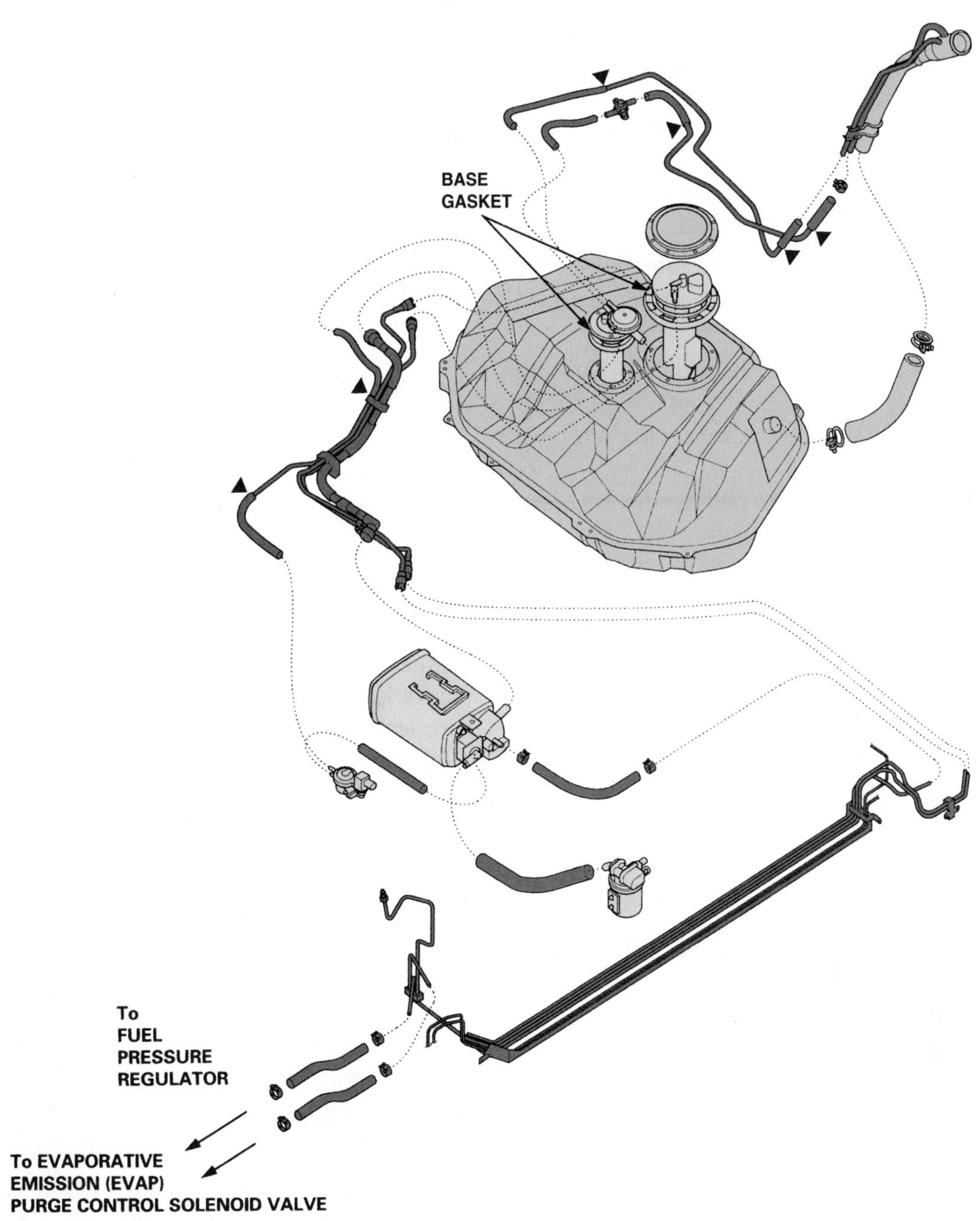

Figure 27-10 A typical layout of the fuel lines on a late-model car. *Courtesy of American Honda Motor Co., Inc.*

vehicle. Generally, rigid lines are used extending from near the tank to a point near the fuel pump and fuel filter. To absorb engine vibrations, the gaps between the frame and tank or fuel pump are joined by short lengths of flexible hose.

Many fuel tanks have vent hoses to allow air in the fuel tank to escape when the tank is being filled with fuel. Vent hoses are usually installed alongside the filler neck. Replacement vent hoses are usually marked with the designation **EVAP** to indicate their intended use. The inside diameter of a fuel delivery hose is generally larger (5⁄16 to 3⁄8 inch [7.94 to 9.35 mm]) than that of a fuel return hose (1⁄4 inch [6.35 mm]).

To control the rate of vapor flow from the fuel tank to the vapor storage tank, a plastic or metal restrictor may be placed in either the end of the vent pipe or in the vapor-vent hose itself. When the latter hose must be replaced, the restrictor must be removed from the old vent hose and installed in the new one.

Fittings

Sections of fuel line are assembled together by fittings. Some of these fittings are a threaded-type fitting, while others are a quick-release design. Many fuel lines have quick-disconnect fittings with a unique female socket and a compatible male connector. These quick-disconnect fittings are sealed by an O-ring inside the female connector. Some of these quick-disconnect fittings have hand-releasable locking tabs **(Figure 27–11)**, while others require a special tool to release the fitting **(Figure 27–12)**.

WARNING!

Other types of O-rings should not be substituted for a Viton O-ring.

The interior components, such as the O-rings and spacers, of quick-connect fittings are not serviceable. If the fitting is damaged, the complete fuel tube or line must be replaced.

When quick-connect fittings are not used, other styles of special fittings are used to join lines together or connect a line to a component. The two most-used threaded fittings are **compression** and **double-flare fittings**. The double-flare, which is the most common, is made with a special tool that has an anvil and a cone **(Figure 27–13)**. The double-flaring process is performed in two steps. First, the anvil begins to fold over the end of the tubing. Then, the cone is used to finish the flare by folding the tubing back on itself, doubling the thickness, and creating two sealing surfaces. The angle and size of the flare are determined by the tool. Careful use of the double flaring helps to produce strong, leakproof connections.

Some fuel lines have threaded fittings with an O-ring seal to prevent fuel leaks. These O-ring seals are usually made from Viton, which resists deterioration from gasoline. On some other fuel lines, the fuel hose is clamped to the steel line and the hose and clamp must be properly positioned on the steel line **(Figure 27–14)**.

A variety of clamps is used on fuel system lines, including the spring and screw types. The crimp clamps shown in **Figure 27–15** are the most commonly used.

Inspection

All fuel lines should occasionally be inspected for holes, cracks, leaks, kinks, or dents. Since the fuel is under pressure, leaks in the line between the pump and injection assembly are relatively easy to recognize.

Rubber fuel hose should be inspected for leaks, cracks, cuts, kinks, oil soaking, and soft spots or deterioration. If any of these conditions is found, the fuel hose should be replaced. When rubber fuel hose is installed,

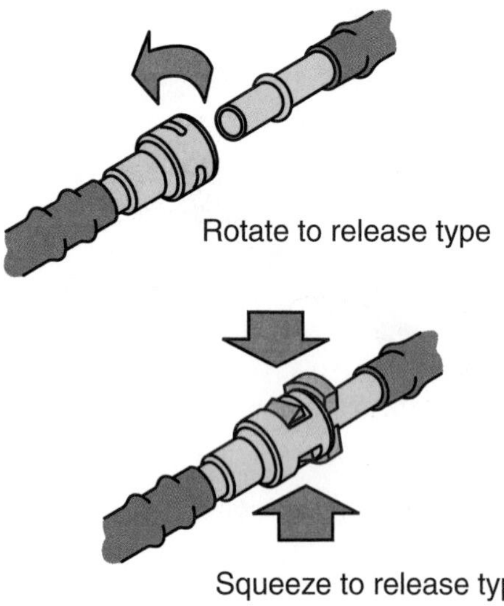

Figure 27-11 Quick-disconnect hand-releasable fuel line fittings.

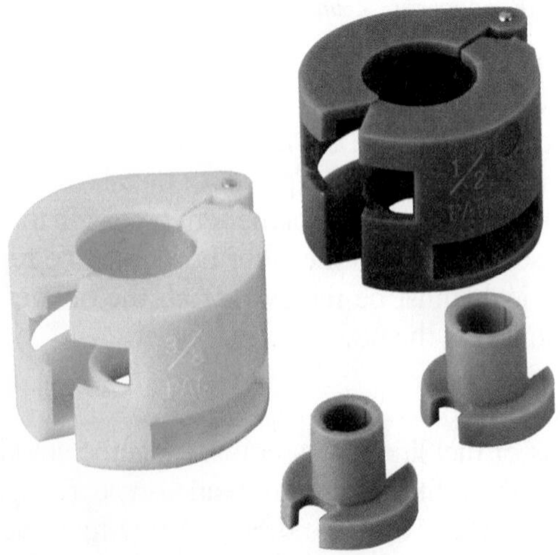

Figure 27-12 An assortment of quick-disconnect tools. *Courtesy of Snap-on Tools Company*

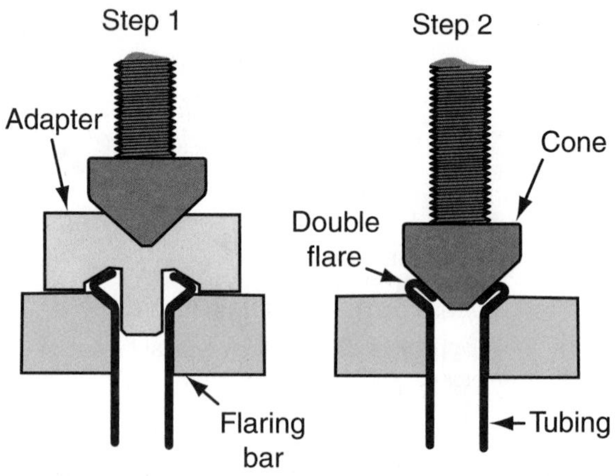

Figure 27-13 The steps to create a flare on a fuel line.

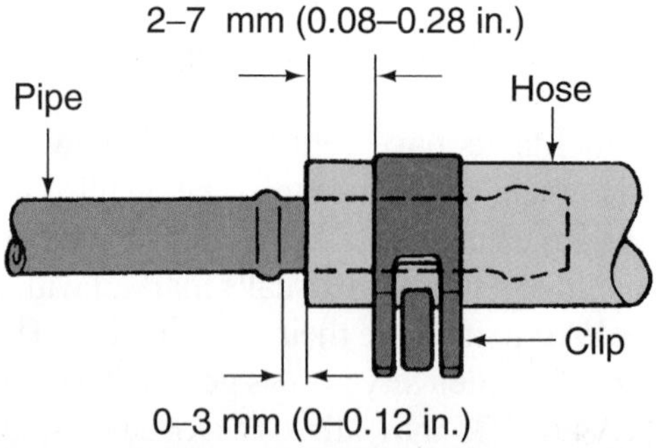

Figure 27-14 A fuel hose clamped to steel tubing.

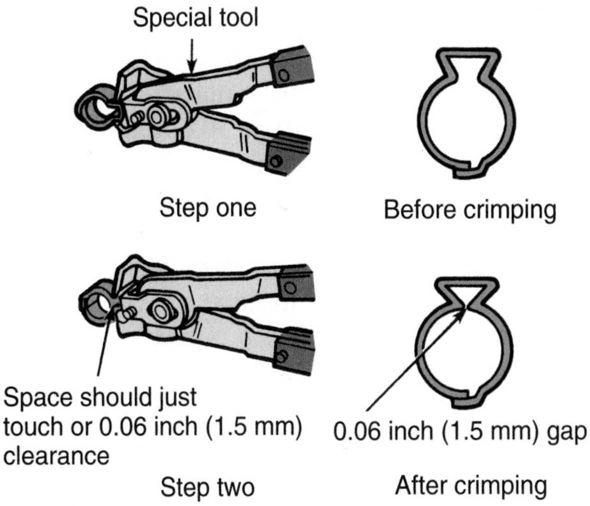

Figure 27-15 A special tool is required to tighten crimp clamps.

the hose should be installed to the proper depth on the metal fitting or line.

Steel tubing should be inspected for leaks, kinks, and deformation **(Figure 27–16)**. Tubing should also be checked for loose connections and proper clamping to the chassis. If the tubing's threaded connections are loose, they must be tightened to the specified torque. Some threaded fuel line fittings contain an O-ring. If the fitting is removed, the O-ring should be replaced.

Nylon fuel pipes should be inspected for leaks, nicks, scratches and cuts, kinks, melting, and loose fittings. If these fuel pipes are damaged in any way, they must be replaced. Nylon fuel pipes must be secured to the chassis at regular intervals to prevent fuel pipe wear and vibration.

WARNING!

Always cover a nylon fuel pipe with a wet shop towel before using a torch or other source of heat near the line. Failure to observe this precaution may result in fuel leaks, personal injury, and property damage.

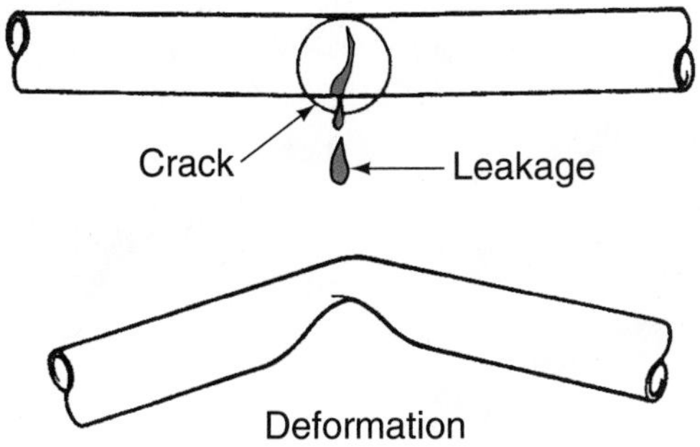

Figure 27-16 Steel tubing should be inspected for leaks, kinks, and deformation. *Reprinted with permission*

WARNING!

If a vehicle has nylon fuel pipes, do not expose the vehicle to temperatures above 194°F (90°C) for any extended period to avoid damage to the pipes.

Line Replacement

When a damaged fuel line is found, replace it with one of similar construction—steel tubing with steel, and flexible tubing with nylon or synthetic rubber. When installing flexible tubing, always use new clamps. The old ones lose some of their tension when they are removed and do not provide an effective seal when used on the new line.

CAUTION!

Do not substitute aluminum or copper tubing for steel tubing. Never use a hose within 4 inches of any hot engine or exhaust system component. A metal line must be installed.

Any damaged or leaking fuel line must be replaced. To fabricate a new fuel line, select the correct tube and fitting dimension and start with a length that is slightly longer than the old line. With the old line as a reference, use a tubing bender to form the same bends in the new line as those that exist in the old. Although steel tubing can be bent by hand to obtain a gentle curve, any attempt to bend a tight curve by hand usually kinks the tubing. To avoid kinking, always use a bending tool like those shown in **Figure 27–17**.

Nylon fuel pipes provide a certain amount of flexibility and can be formed around gradual curves under the

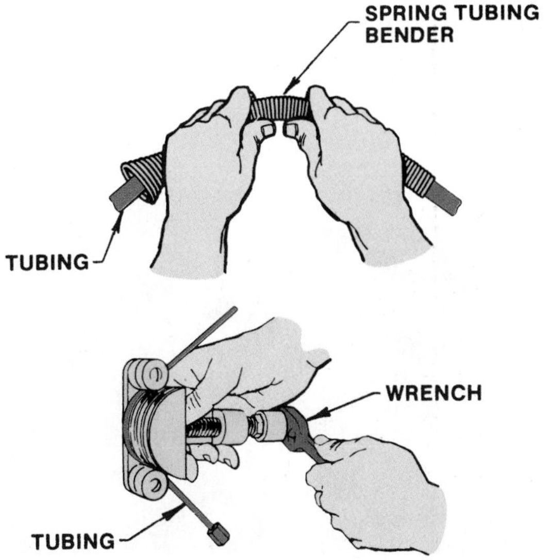

Figure 27-17 Two types of bending tools for steel tubing.

vehicle. Do not force a nylon fuel pipe into a sharp bend, because doing so may kink the pipe and restrict the flow of fuel. When nylon fuel pipes are exposed to gasoline, they may become stiffer, making them more susceptible to kinking. Be careful not to nick or scratch nylon fuel pipes.

WARNING!

Fuel line fittings must be tightened to the specified torque.

FUEL FILTERS

Automobiles and light trucks usually have an in-tank strainer and a gasoline filter. The strainer, located in the gasoline tank, is made of a finely woven fabric. The purpose of this strainer is to prevent large contaminate particles from entering the fuel system where they could cause excessive fuel pump wear or plug fuel metering devices. It also helps to prevent passage of any water that might be present in the tank. Servicing of the fuel tank strainer is seldom required; however, if the gasoline usually used contains large amounts of alcohol, the strainer will need to be replaced often.

A fuel filter is connected in the fuel line between the fuel tank and the engine. Many of these filters are mounted under the vehicle **(Figure 27–18)**, and others are mounted in the engine compartment. Most fuel filters contain a pleated paper element mounted in the filter housing, which may be made from metal or plastic. Paper filter elements are efficient at removing and trapping small particles, as well as large-size contaminates. Fuel filters are typically contained in a metal case, but some have a plastic housing. On many fuel filters, the inlet and outlet fittings are identified, and the filter must be installed properly. An arrow on some filter housings indicates the direction of fuel flow through the filter.

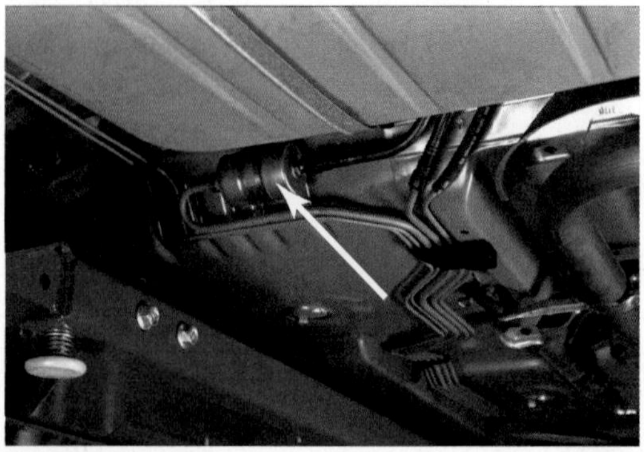

Figure 27-18 An in-line fuel filter mounted under a vehicle.

Servicing Filters

Fuel filters **(Figure 27–19)** and elements are serviced by replacement only. Some vehicle manufacturers recommend fuel filter replacement at 30,000 miles (48,000 km). Always replace the fuel filter at the vehicle manufacturer's recommended mileage. If dirty or contaminated fuel is placed in the fuel tank, the filter may require replacing before the recommended mileage. A plugged fuel filter may cause the engine to surge and cut out at high speed or hesitate on acceleration. A restricted fuel filter causes low fuel pump pressure.

The fuel filter replacement procedure varies depending on the make and year of the vehicle and the type of fuel system. Always follow the filter replacement procedure in the appropriate service manual. Photo Sequence 23 shows a typical procedure for relieving fuel pressure and removing a fuel filter.

CAUTION!

Be sure to properly start the inlet nut threads into the carburetor. If the threads are crossed, a fuel leak inevitably results.

To install a new filter, begin by wiping the male tube ends of the new filter with a clean shop towel. Apply a few drops of clean engine oil to the male tube ends on the filter. Check the quick connectors to be sure the large collar on each connector has rotated back to the original position. The springs must be visible on the inside diameter of each quick connector. Then install the filter, in the proper direction, and leave the mounting bolt slightly loose. Install the outlet connector onto the filter outlet tube and press the connector firmly in place until the spring snaps into position. Grasp the fuel line and try to pull this line from the filter to be sure the quick connec-

Figure 27-19 An assortment of fuel filters. *Courtesy of Robert Bosch Corp.*

PHOTO SEQUENCE 23

Removing a Fuel Filter on an EFI Vehicle

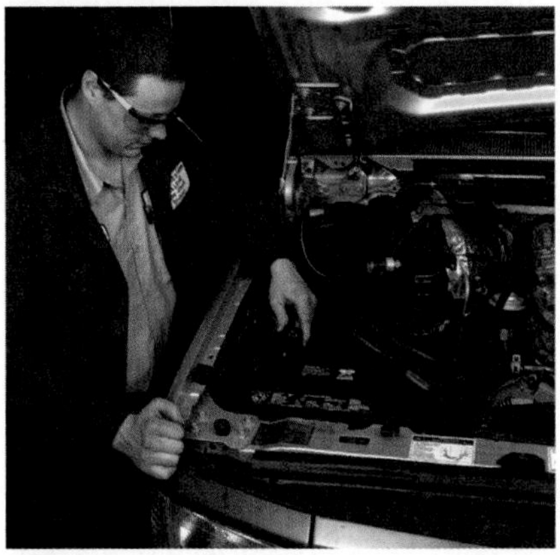

P23-1 *Disconnect the negative cable at the battery.*

P23-2 *Loosen the fuel tank filler cap to relieve any fuel tank vapor pressure.*

P23-3 *Wrap a shop towel around the Schrader valve on the fuel rail and remove the dust cap from the valve.*

P23-4 *Connect the fuel pressure gauge to the Schrader valve.*

P23-5 *Install the free end of the gauge bleed hose into an approved gasoline container, and open the gauge bleed valve to relieve the fuel pressure.*

P23-6 *Place the vehicle on the hoist and position the lift arms according to manufacturer's recommendations. Then raise the vehicle.*

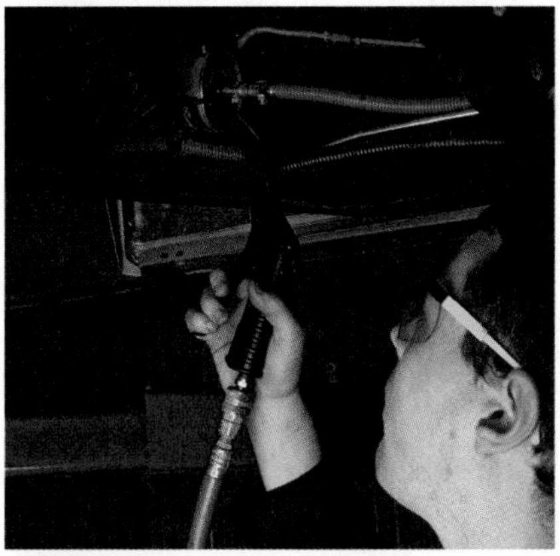

P23-7 *Flush the fuel filter line connectors with water, and use compressed air to blow debris off and away from the connectors.*

P23-8 *Follow the recommended procedures for disconnecting the fuel inlet connector.*

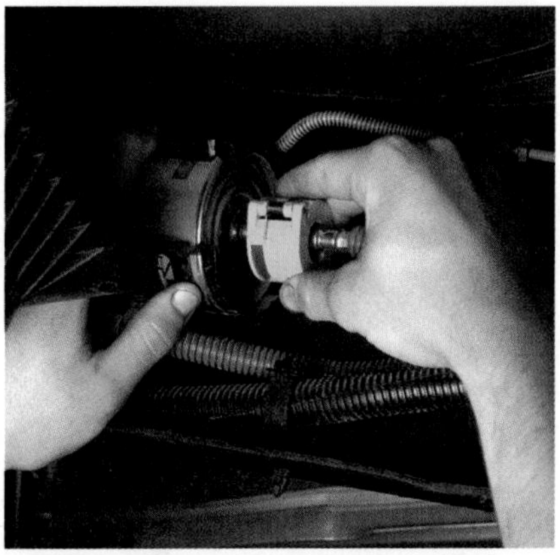

P23-9 *Follow the recommended procedures for disconnecting the fuel outlet connector. Then remove the fuel filter.*

tor is locked in place. Then do the same with the inlet connector. Now tighten the filter-retaining bolt to the specified torque. Once everything is connected, lower the vehicle, start the engine, and check for fuel leaks at the filter.

FUEL PUMPS

A fuel pump is the device that draws the fuel from the fuel tank through the fuel lines to the engine's injection system. All late-model vehicles use an electric fuel pump. Older vehicles with carburetors had mechanical pumps, but these are not discussed here.

An electric fuel pump can be located inside or outside the fuel tank. There are four basic types of electric fuel pumps: the diaphragm, plunger, bellows, and impeller or rotary pump. The in-tank electric pump is usually a rotary type. The diaphragm, plunger, and bellows types are usually of the demand style. That is, when the ignition is turned on, the pump starts to run and shuts off automatically when the fuel line is pressurized. When there is a demand for more fuel, the pump turns on again. **Figure 27–20** shows a typical wiring diagram for an electric fuel pump.

Most vehicles have the fuel pump mounted in the fuel tank. Some vehicles have an in-tank electric pump and a second electric fuel pump mounted under the vehicle. A fuel pump in the fuel tank contains a small direct current (DC) electric motor with an impeller mounted on the end of the motor shaft. A pump cover is mounted over the impeller, and this cover contains inlet and discharge ports. When the armature and impeller rotate, fuel is moved from the tank to the inlet port, and the impeller grooves pick up the fuel and force it around the impeller cover and out the discharge port **(Figure 27–21)**.

Fuel moves from the discharge port through the inside of the motor and out the check valve and outlet connection, which is connected via the fuel line to the fuel filter and underhood fuel system components. A pressure relief valve near the check valve opens if the fuel supply line is restricted and pump pressure becomes very high. When the relief valve opens, fuel is returned through this valve to the pump inlet. This action protects fuel system components from high fuel pressure. Each time the engine is shut off, the check valve prevents fuel from draining out of the underhood fuel system components into the fuel tank. A fuel filter is attached to the pump inlet. This filter prevents dirt or water from entering the pump.

Fuel pumps are typically located in the fuel tank because this helps keep the fuel pump cool while it is operating and keeps the entire fuel line pressurized to prevent premature fuel evaporation. Although it is dangerous to have a spark near gasoline and there is a great potential for sparks between an electric motor's armature and brushes, the in-tank fuel pump is safe because there is no oxygen to support combustion in the tank.

Nearly all electric fuel pump circuits include some sort of rollover protection. Typically this protection includes the installation of an **inertia switch** that shuts off the fuel pump if the vehicle is involved in a collision or rolls over. A typical inertia switch **(Figure 27–22)** consists of a permanent magnet, a steel ball inside a conical ramp, a target plate, and a set of electrical contacts. The magnet holds the steel ball at the bottom of the ramp. In the event of a collision, the inertia of the ball causes it to break away from the magnetic field and roll up the ramp. When it strikes the target plate, the electrical contacts open and the fuel pump shuts off. The switch has a rest button that

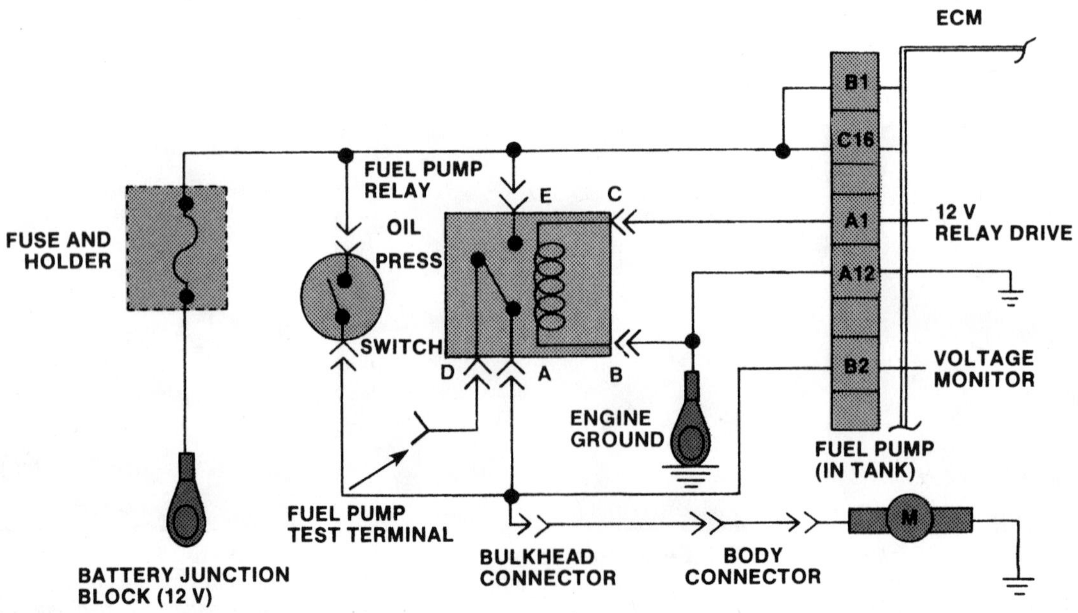

Figure 27–20 A typical wiring diagram for an electric fuel pump.

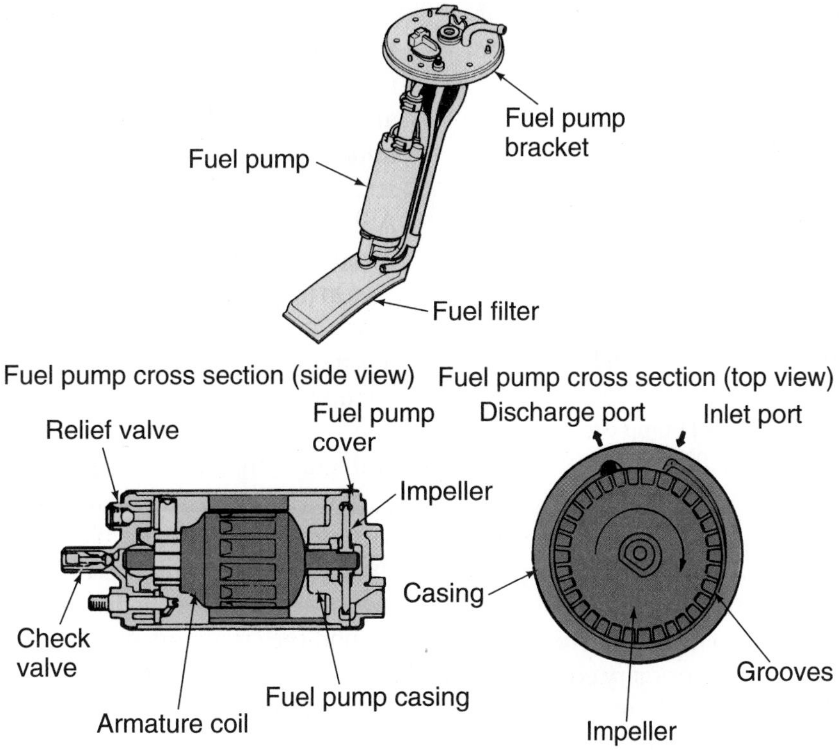

Figure 27-21 An electric fuel pump. *Courtesy of American Honda Motor Co., Inc.*

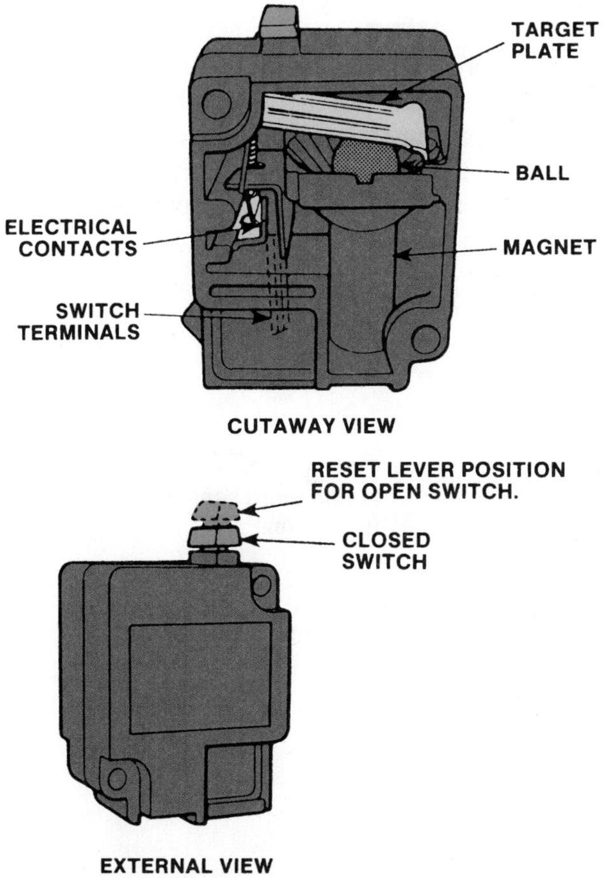

Figure 27-22 Details of a Ford inertia switch.

must be depressed to close the contacts before the pump will operate again.

Fuel Pump Circuits

Electric fuel pump circuits vary depending on the vehicle make and year. Because electric fuel pumps on late-model vehicles are computer controlled, these circuits are explained in great detail in later chapters. However, as an example, let us look at some typical circuits.

In a late-model General Motors fuel pump circuit, the PCM supplies voltage to the winding of the fuel pump relay when the ignition switch is turned on. This action closes the relay points and voltage is supplied through the points to the in-tank fuel pump. The fuel pump remains on while the engine is cranking or running. If the ignition switch is on for two seconds and the engine is not cranked, the PCM shuts off the voltage to the fuel pump relay and the relay points open to stop the pump.

If the ignition switch is on and the fuel line is broken during an accident, PCM and fuel pump relay action is a safety feature that prevents the fuel pump from pumping gasoline from the ruptured fuel line. An oil pressure switch is connected parallel to the fuel pump relay points. If the relay becomes defective, voltage is supplied through the oil pressure switch points to the fuel pump. This action keeps the fuel pump operating and the engine running, even though the fuel pump relay is defective. When the engine is cold, oil pressure is not available immediately, and the engine may be slow to start if the fuel pump relay is defective.

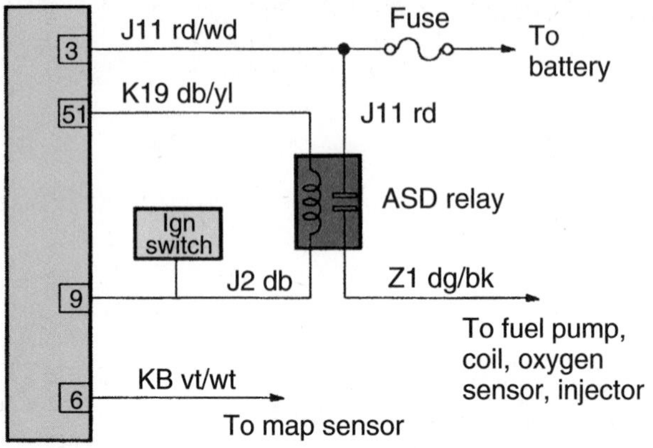

Figure 27-23 A Chrysler fuel pump circuit with an ASD relay. *Courtesy of DaimlerChrysler Corporation*

The fuel pump relay in Chrysler EFI systems is referred to as an automatic shutdown (ASD) relay. With the ignition switch turned on, the PCM grounds the windings of the relay and the relay points close. These supply voltage to the fuel pump, positive primary coil terminal, oxygen sensor heater, and the fuel injectors in some systems **(Figure 27–23)**.

On Chrysler products with a power module and logic module, the engine must be cranked before the power module grounds the ASD relay winding. The later model PCM grounds the ASD relay winding when the ignition switch is turned on, and the relay remains closed while the engine is cranking or running. If the ignition switch is on for one-half second and the engine is not cranked, the PCM opens the circuit from the ASD relay winding to ground. Under this condition, the ASD relay points open and voltage is no longer supplied to the fuel pump, positive primary coil terminal, injectors, and oxygen sensor heater.

Later model Chrysler fuel pump circuits have a separate ASD relay and a fuel pump relay. In these circuits, the fuel pump relay supplies voltage to the fuel pump, and the ASD relay powers the positive primary coil terminal, injectors, and oxygen sensor heater. The ASD relay and the fuel pump relay operate the same as the previous ASD relay. The PCM grounds both relay windings through the same wire.

On many Toyota vehicles, the fuel pump relay is called a circuit opening relay. This relay has dual windings and it is mounted on the firewall **(Figure 27–24)**. One of the windings of the circuit opening relay is connected between the starter relay points and ground, and the second relay winding is connected from the battery positive terminal to the PCM. When the engine is cranking and the starter relay points are closed, current flows through the starter relay points and the circuit opening relay winding to ground. This current flow creates a magnetic field around the circuit opening relay winding that closes the relay points. When these points close, current flows through the points to the fuel pump **(Figure 27–25)**.

Once the engine starts, the starter relay is no longer energized, and current stops flowing through these relay points and the winding of the circuit opening relay. However, the PCM grounds the other winding of the circuit opening relay when the engine starts. This action keeps the relay points closed while the engine is running.

Troubleshooting

Fuel pump problems are usually indicated by improper fuel system pressure or a dead or inoperative fuel pump. Fuel pressure is read with a fuel pressure gauge. The proper procedure for testing fuel pump pressure is shown in Photo Sequence 24. These photos outline the

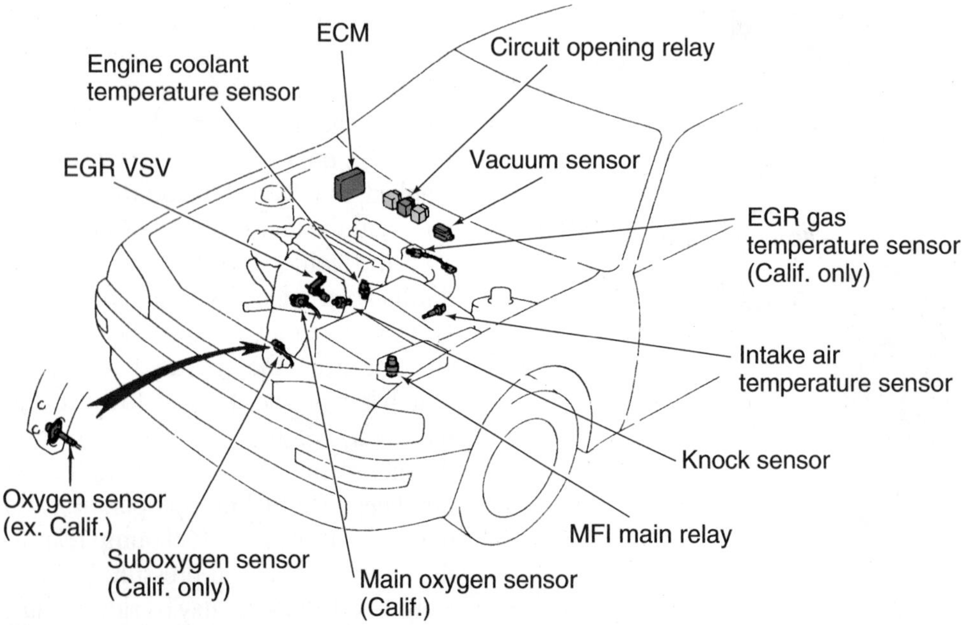

Figure 27-24 The location of a circuit opening relay and other EFI components.

PHOTO SEQUENCE 24

Checking Fuel Pressure on a PFI System

P24-1 *Many problems on today's cars can be caused by incorrect fuel pressure. Therefore, checking fuel pressure is an important step in diagnosing driveability problems.*

P24-2 *Prior to testing the fuel pump, a careful visual inspection of the injectors, fuel rail, and fuel lines and hoses is necessary. Any sign of a fuel leak should be noted and the cause corrected immediately.*

P24-3 *The supply line into the fuel rail is a likely point of leakage. Check the area around the fitting to make sure no leaks have occurred.*

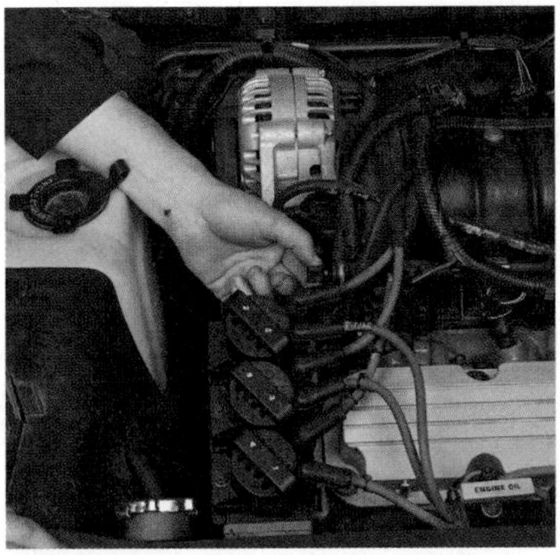

P24-4 *Most fuel rails are equipped with a test fitting that can be used to relieve pressure and to test pressure.*

P24-5 *To test fuel pressure, connect the appropriate pressure gauge to the fuel rail test fitting (Schrader valve).*

P24-6 *Connect a hand-held vacuum pump to the fuel pressure regulator.*

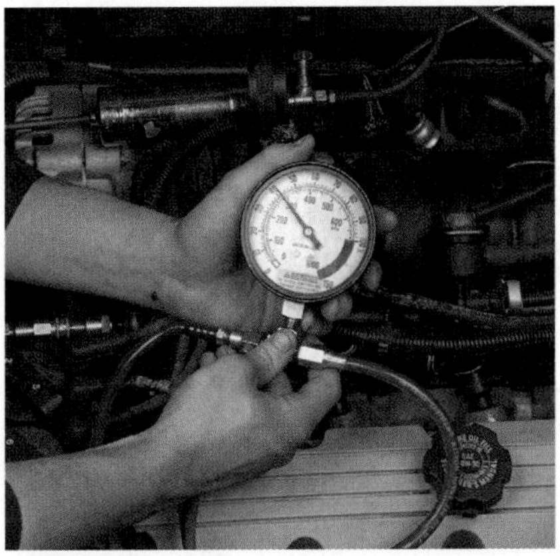

P24-7 *Turn the ignition switch to the RUN position and observe the fuel pressure gauge. Compare the reading to specifications. A reading lower than normal indicates a faulty fuel pump or fuel delivery system.*

P24-8 *To test the fuel pressure regulator, create a vacuum at the regulator with the vacuum pump. Fuel pressure should decrease as vacuum increases. If pressure remains the same, the regulator is faulty.*

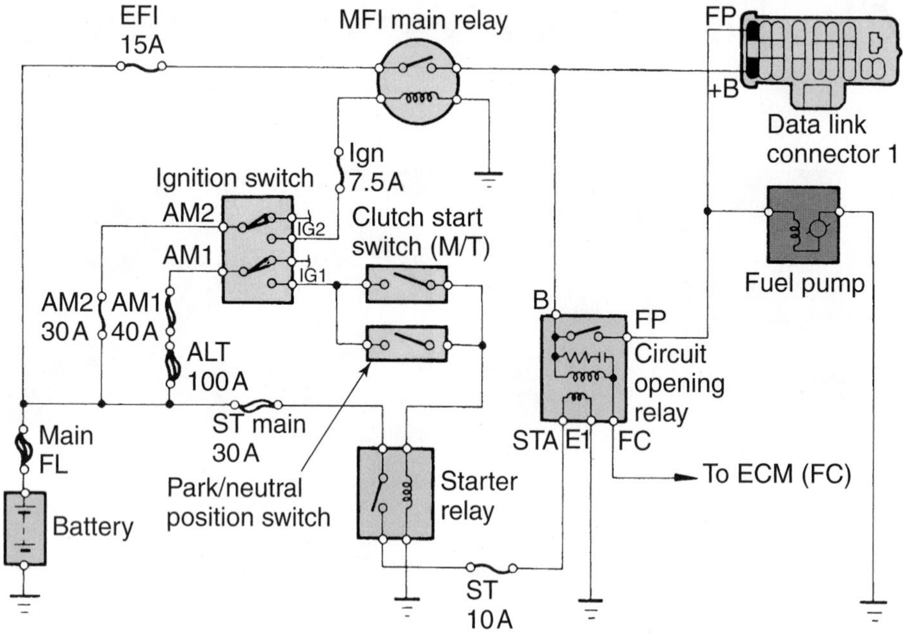

Figure 27-25 A wiring diagram for the circuit opening relay.

steps to follow while performing the test on an engine with fuel injection. To conduct this test on specific fuel-injection systems, refer to the service manual for instructions. However, most systems have a **Schrader valve** on the fuel rail that can be used to connect the fuel pressure gauge.

If the system does not have a Schrader valve, a tee should be installed in the fuel supply line to connect the gauge **(Figure 27–26)**. This same method is also used to test the fuel pump in a TBI system or a carbureted engine with an electric fuel pump. Always follow the manufacturer's recommendations for the placement of the tee. Make sure you relieve the system of any pressure before loosening a fitting to install the pressure gauge.

On some PFI engines, the fuel rail is fitted with a fuel pulsation damper. The point where the damper attaches to the fuel rail is the recommended place for connecting the pressure gauge. To connect the gauge, the damper is removed. To do this, place a rag over the damper unit and loosen it with a wrench. Loosen it only one turn. After all pressure is released, remove the damper unit and connect the fuel pressure gauge into the damper's fitting **(Figure 27–27)**.

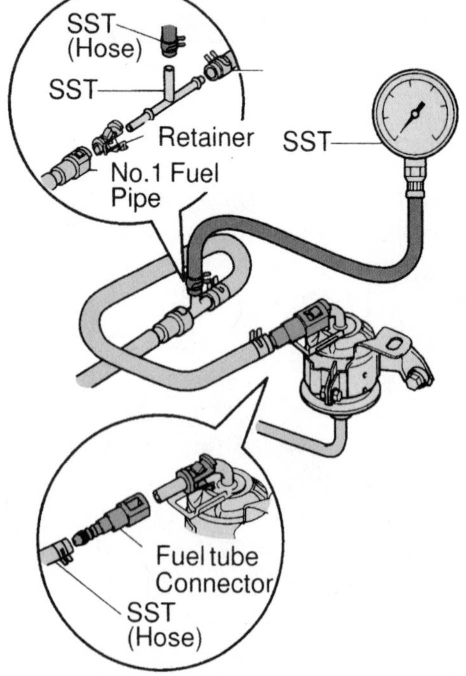

Figure 27-26 On fuel systems that do not have a Schrader valve, it may be necessary to fit a tee in the fuel line to connect the fuel pressure gauge. *Reprinted with permission*

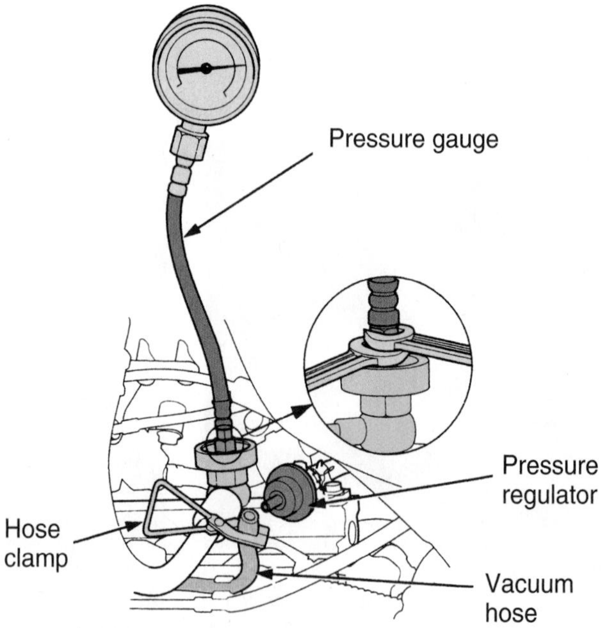

Figure 27-27 Connecting a fuel pressure gauge to the fuel pulsation damper fitting. *Courtesy of American Honda Motor Co., Inc.*

WARNING!

When testing an electric fuel pump, do not let fuel contact any electrical wiring. The smallest spark (electric arc) could ignite the fuel.

A high fuel-pressure reading usually indicates a faulty pressure regulator or an obstructed return line. To identify the cause, disconnect the fuel return line at the tank. Use a length of hose to route the returning fuel into an appropriate container. Start the engine and note the pressure reading at the engine. If fuel pressure is now within specifications, check for an obstruction in the in-tank return plumbing. The fuel reservoir check valve or aspirator jet might be clogged.

If the fuel pressure still reads high with the return line disconnected from the tank, note the volume of fuel flowing through the line. Little or no fuel flow can indicate a plugged return line. Shut off the engine and connect a length of hose directly to the fuel pressure regulator return port to bypass the return hose. Restart the engine and again check the pressure. If bypassing the return line brings the readings back within specifications, a plugged return line is the culprit.

If pressure is still high, apply vacuum to the pressure regulator **(Figure 27–28)** to see if that makes a difference. It should. If there is still no change, replace the faulty pressure regulator. If applying vacuum directly to the regulator lowers fuel pressure, the vacuum hose that controls the operation of the regulator might be plugged, leaking, or misrouted.

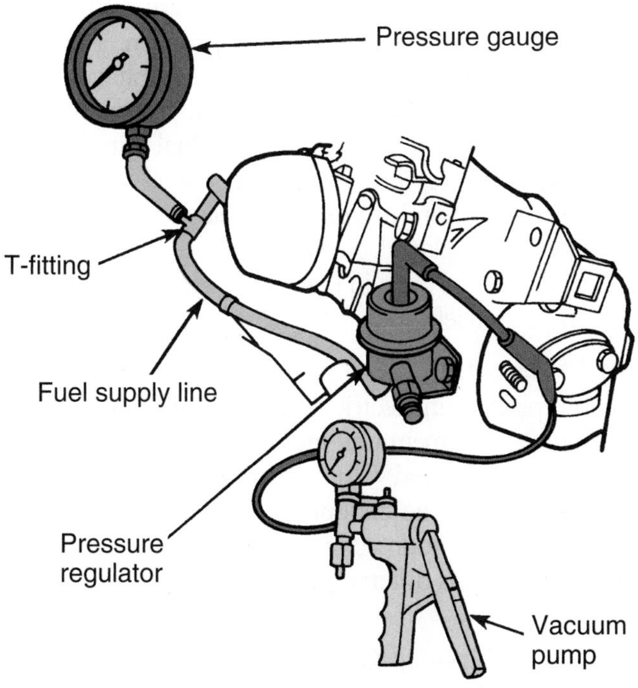

Figure 27-28 Applying vacuum to the fuel pressure regulator should change the fuel pressure in the system.

Low fuel pressure can be due to a clogged fuel filter, restricted fuel line, weak pump, leaky pump check valve, defective fuel pressure regulator, or dirty filter sock in the tank. It is possible to rule out filter and line restrictions as a cause of the problem by making a pressure check at the pump outlet. A higher reading at the pump outlet (at least 5 psi) means there is a restriction in the filter or line. If the reading at the pump outlet is unchanged, then the pump either is weak or is having trouble picking up fuel (clogged filter sock in the tank). Either way, it is necessary to get inside the fuel tank. If the filter sock is gummed up with dirt or debris, it is also wise to clean out the tank when the filter sock is cleaned or replaced.

Another possible source of trouble is the pump's check valve. Some pumps have one; others have two (positive displacement roller vane pumps). The check valve prevents fuel movement through the pump when the pump is off so residual pressure remains at the injectors. This can be checked by watching the fuel pressure gauge after the engine is shut off.

Remember, if the fuel pressure is outside specifications, driveability problems can result. Excessive pressure causes a rich air/fuel mixture, and insufficient pressure results in a leaner than normal mixture. If the fuel pressure is within specifications, you cannot conclude the fuel delivery system is fine. Fuel volume, or the pump's capacity to cause fuel flow, is also important and should be tested according to the procedures outlined in the service manual.

No-Start Diagnosis

When an engine fails to start because there is no fuel delivery, the first check is the fuel gauge. A gauge that reads higher than a half tank probably means there is fuel in the tank, but not always. A defective sending unit or miscalibrated gauge might be giving a false indication. Sticking a wire or dowel rod down the fuel tank filler pipe tells whether there really is fuel in the tank. If the gauge is faulty, repair or replace it.

Listen for pump noise. When the key is turned on, the pump should buzz for a couple of seconds to build system pressure. The pump is usually energized through an oil-pressure switch (the purpose of which is to shut off the flow of fuel in case of an accident that stalls the engine). On most late-model cars with computerized engine controls, the computer energizes a pump relay **(Figure 27–29)** when it receives a cranking signal from the distributor pickup or crankshaft sensor. An oil-pressure switch might still be included in the circuitry for safety purposes and to serve as a backup in case the relay or computer signal fails. Failure of the pump relay or computer driver signal can cause slow starting because the

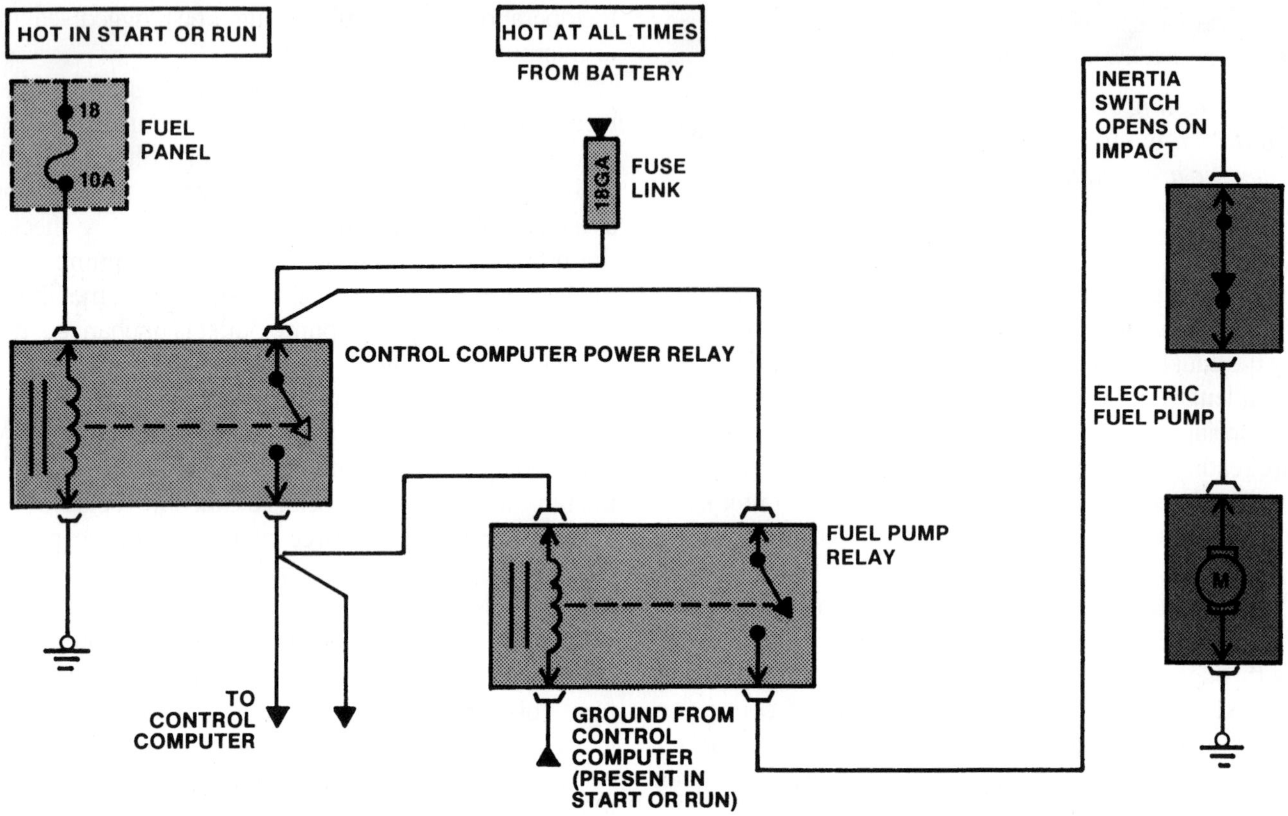

Figure 27-29 An electrical fuel delivery system wiring diagram.

fuel pump does not come on until the engine cranks long enough to build up sufficient oil pressure to trip the oil-pressure switch.

If a buzzing sound is not heard when the key is on or while the engine is being cranked, check for the presence of voltage at the pump electrical connectors. The pump might be good, but if it does not receive voltage and have a good ground, it does not run. To check the ground, connect a test light across the ground and feed wires at the pump to check for voltage, or use a voltmeter to read actual voltage and an ohmmeter to check ground resistance. The latter is the better test technique because a poor ground connection or low voltage can reduce pump operating speed and output. If the electrical circuit checks out but the pump does not run, the pump is probably bad and should be replaced.

No voltage at the pump terminal when the key is on and the engine is cranking indicates a faulty oil-pressure switch, pump relay, relay drive circuit in the computer, or a wiring problem. Check the pump fuse to see if it is blown. Replacing the fuse might restore power to the pump, but until you have found out what caused the fuse to blow, the problem is not solved. The most likely cause of a blown fuse would be a short in the wiring between the relay and pump, or a short inside the oil-pressure switch or relay, or a bad fuel pump.

A faulty oil-pressure switch can be checked by bypassing it with a jumper wire. If doing this restores power to the pump and the engine starts, replace the switch. If an oil-pressure switch or relay sticks in the closed position, the pump can run continuously whether the key is on or off, depending on how the circuit is wired.

To check a pump relay, use a test light to check across the relays and ground terminals to tell if the relay is getting battery voltage and ground. Next, turn off the ignition, wait about ten seconds, then turn it on. The relay should click and you should see battery voltage at the relay's pump terminal. If nothing happens, repeat the test, checking for voltage at the relay terminal wired to the computer. The presence of a voltage signal here means the computer is doing its job but the relay is failing to close and should be replaced. No voltage signal from the computer indicates an open in that wiring circuit or a fault in the computer itself.

Replacement

When replacing an electric pump, be sure that the new or rebuilt replacement unit meets the minimum requirements of pressure and volume for that particular vehicle. This information can be found in the service manual. If the fuel pump is mounted in the fuel tank, the procedure for replacement is different than if the unit is external to the tank.

External Fuel Pump Before removing the fuel pump, disconnect the negative battery cable. Then disconnect the electrical connectors on the fuel pump. Label the wires to aid in connecting it to the new pump. Reversing polarity on most pumps destroys the unit.

Now disconnect the fuel and vapor lines at the pump. These lines should also be labeled so they are installed correctly on the new pump.

Loosen and remove the bolts holding the pump in place. Remove the pump by pushing the pump up until the bottom is clear of the bracket. Swing the pump out to the side and pull it down to free it from the rubber fuel line coupler. The rubber sound insulator between the bottom of the pump and bracket and the rubber coupler on the fuel line are normally discarded because new ones are included with the replacement pump. Some pumps have a rubber jacket around them to quiet the pump. If this is the case, slip off the jacket and put it on the new pump.

CAUTION!

Avoid the temptation to test the new pump before reinstalling the fuel tank by energizing it with a couple of jumper wires. Running the pump dry can damage it because the pump relies on fuel for lubrication and cooling.

Compare the replacement pump with the old one. If necessary, transfer any fuel line fittings from the old pump to the new one. When inserting the new pump back into its bracket, be careful not to bend the bracket. Make sure the rubber sound insulator under the bottom of the pump is in place. Install a new filter sock (if so equipped) on the pump inlet and reconnect the pump wires. Be absolutely certain you have correct polarity.

If the fuel was removed from the tank, replace it. Make sure all electrical connections are reconnected and that all fuel lines and hoses are properly fastened and tightened. Then reconnect the ground terminal at the battery. Start the engine and check all connections for fuel leaks.

CAUTION!

Never turn on the ignition switch or crank the engine with a fuel line disconnected. This action will result in gasoline discharge from the disconnected line, which may result in a fire, causing personal injury and/or property damage.

Internal Fuel Pump On many vehicles, the fuel tank must be removed to replace the fuel pump and/or fuel gauge sending unit. On other vehicles, the unit can be serviced through an opening in the vehicle's trunk or floor **(Figure 27–30)**. Some vehicles have a separate fuel pump and gauge sending unit, while others have both contained

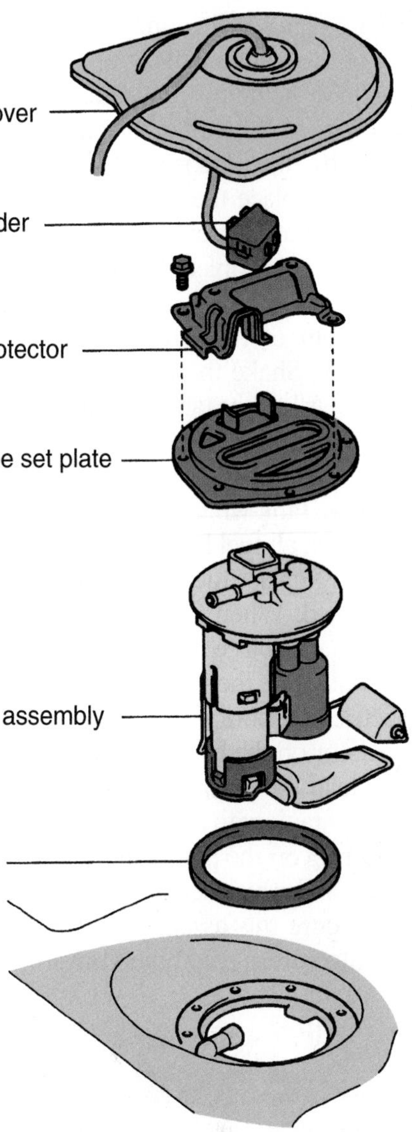

Figure 27–30 Some fuel pumps can be serviced through an access hole in the car's floor. *Reprinted with permission*

in a single unit. Once the fuel tank is out of the vehicle, if necessary, remove the unit from the tank.

These units are often held in the tank by either a retaining ring or screws. The easiest way to remove a retaining ring is to use a special tool designed for this purpose. This tool fits over the metal tabs on the retaining ring, and after about a quarter turn, the ring comes loose and the unit can be removed. If the special tool is not available, a brass drift punch and ball-peen hammer usually can do the job.

When removing the unit from the tank, be very careful not to damage the float arm, the float, or the fuel gauge sender. Check the unit carefully for any damaged components. Shake the float and if fuel can be heard inside, replace it. Make sure the float arm is not bent. It is usually wise to replace the filter and O-ring before replacing the unit. Check the fuel gauge and sender unit as described in the service manual. When reinstalling the pickup pipe-sending unit, be very careful not to damage any of the components.

Once the unit is removed, check the filter on the fuel pump inlet. If the filter is contaminated or damaged, replace the filter. Inspect the fuel pump inlet for dirt and debris. Replace the fuel pump if these foreign particles are found in the pump inlet.

If the pump inlet filter is contaminated, flush the tank with hot water for at least five minutes. Dump all the water out of the tank through the pump opening in the tank. Shake the tank to be sure all the water is removed. Allow the tank to sit and air dry before reinstalling it or adding fuel to it. Remember gasoline fumes are extremely ignitable so keep all open flames and sparks away from the tank while it is drying.

Check all fuel hoses and tubing on the fuel pump assembly. Replace fuel hoses that are cracked, deteriorated, or kinked. When fuel tubing on the pump assembly is damaged, replace the tubing or the pump.

Make sure the sound insulator sleeve is in place on the electric fuel pump, and check the position of the sound insulator on the bottom of the pump.

Clean the pump and sending unit mounting area in the fuel tank with a shop towel, and install a new gasket or O-ring on the pump and sending unit. Install the fuel pump and gauge sending unit assembly in the fuel tank and secure this assembly in the tank using the vehicle manufacturer's recommended procedure.

CASE STUDY

***A** customer complained about severe engine surging in a late-model Ford. When the technician lifted the hood, he noticed many fuel system and ignition system components had been replaced recently. The customer was asked about previous work done on the vehicle, and he indicated this problem had existed for some time. Several shops had worked on the vehicle, but the problem still persisted. The fuel pump and filter were replaced, as were some ignition coils and spark plugs.*

The technician road tested the vehicle and found it did have a severe surging problem at freeway cruising speeds. From past experience, the technician thought the surging was caused by lack of fuel supply. The technician decided to connect a fuel pressure gauge at the fuel rail of the injection assembly. The gauge was securely taped to one of the windshield wiper blades so the gauge could be observed from the passenger compartment. The technician drove the vehicle on a second road test and found when the surging problem occurred, the fuel pump pressure dropped well below the vehicle manufacturer's specifications. Since the fuel pump and filter had been replaced, the technician concluded the problem must be in the fuel line or tank.

The technician returned to the shop and raised the vehicle on a hoist. The steel fuel tubing appeared to be in satisfactory condition. However, a short piece of rubber fuel hose between the steel fuel tubing and the fuel line entering the fuel tank was flattened and soft in the center. This fuel hose was replaced and rerouted to avoid kinking. Another road test proved the surging problem was eliminated.

The flattened fuel hose restricted fuel flow, and at higher speeds the increased vacuum from the fuel pump made the flattened condition worse, which restricted the fuel flow and caused the severe surging problem.

KEY TERMS

Compression fitting
Double-flare fitting
EVAP
Inertia switch
Schrader valve

SUMMARY

- A fuel delivery system consists of a fuel tank, fuel lines, fuel filter, and a fuel pump.
- One of the key requirements for an efficient running engine is the correct amount of fuel.
- The fuel system should be checked whenever there is evidence of a fuel leak or smell.
- The fuel system should also be checked whenever basic tests suggest there is too little or too much fuel being delivered to the cylinders.
- Since electronic fuel injection systems have a residual fuel pressure, this pressure must be relieved before disconnecting any fuel system component.
- Fuel tanks have devices that prevent fuel vapors from leaving the tank. They also have surge plates to prevent the fuel from splashing around inside. Each tank has an inlet filler tube and a nonvented cap.
- Fuel tanks also have a liquid vapor separator to stop liquid fuel and bubbles from reaching the vapor storage canister.
- The fuel tank should be inspected for leaks; road damage; corrosion; rust; loose, damaged, or defective seams; loose mounting bolts; and damaged mounting straps.
- Leaks in the metal fuel tank can be caused by a weak seam, rust, or road damage. The best method of permanently solving this problem is to replace the tank.

- In-tank fuel pumps and fuel level gauge sending units are held in the tank by either a retaining ring or screws.
- Fuel lines can be made of either metal tubing or flexible nylon or synthetic rubber hose. The latter must be able to resist gasoline. It must also be nonpermeable, so gas and gas vapors cannot evaporate through the hose. Ordinary rubber hose, such as that used for vacuum lines, deteriorates when exposed to gasoline. Only hoses made for fuel systems should be used for replacement. Similarly, vapor vent lines must be made of material that resists attack by fuel vapors. Replacement vent hoses are usually marked with the designation EVAP to indicate their intended use.
- All fuel lines should occasionally be inspected for holes, cracks, leaks, kinks, or dents.
- Automobiles and light trucks usually have an in-tank strainer and a gasoline filter. The strainer, located in the gasoline tank, is made of a finely woven fabric. The purpose of this strainer is to prevent large contaminate particles from entering the fuel system where they could cause excessive fuel pump wear or plug fuel-metering devices.
- To determine that the fuel pump is in satisfactory operating condition, tests for both fuel pump pressure and fuel pump capacity should be performed.
- Problems in fuel systems using electric fuel pumps are usually indicated by improper fuel system pressure or a dead or inoperative fuel pump. A high-pressure reading usually indicates either a faulty pressure regulator or an obstructed return line.
- Low pressure can be caused by a clogged fuel filter, restricted fuel line, weak pump, leaky pump check valve, defective fuel pressure regulator, or dirty filter sock in the tank.
- An inertia switch in the fuel pump circuit opens the fuel pump circuit immediately if the vehicle is involved in a collision.
- The oil-pressure switch connected in parallel with the fuel pump relay operates the fuel pump if the fuel pump relay is defective.

REVIEW QUESTIONS

1. Fuel pump ____________ is a statement of the capacity of the flow of the pump.
2. Explain the purpose of the relief valve and one-way check valve in an electric fuel pump.
3. Some fuel tank filler caps contain a pressure valve and a ________________.
   - **a.** vapor separator
   - **b.** vacuum relief valve
   - **c.** one-way check valve
   - **d.** surge plate
4. What type of fire extinguisher should you have close by when you are working on fuel system components?
5. What is the first thing that should be disconnected when removing a fuel tank?
6. Why is a plastic or metal restrictor placed in either the end of the vent pipe or in the vapor-vent hose on some vehicles?
7. Low fuel pump pressure causes a ____________ mixture and excessive pressure causes a ________ mixture.
8. If fuel pump pressure or volume is less than specified, which of the following is *not* a likely cause of the problem?
   - **a.** restricted fuel filter
   - **b.** a faulty fuel pump
   - **c.** restricted fuel return lines
   - **d.** restricted fuel supply lines
9. Technician A says excessively high pressure from an electric fuel pump can be caused by a faulty pressure regulator. Technician B says the problem may be an obstructed return line. Who is correct?
   - **a.** Technician A
   - **b.** Technician B
   - **c.** Both A and B
   - **d.** Neither A nor B

TECH MANUAL

The following procedures are included in Chapter 27 of the *Tech Manual* that accompanies this book:

1. Relieving fuel system pressure in an EFI system.
2. Testing fuel pump pressure on an EFI system.
3. Replacing an in-line fuel filter.

10. Technician A replaces a damaged steel fuel line with one made of synthetic rubber. Technician B replaces a damaged steel fuel line with one made of steel. Who is correct?

a. Technician A **c.** Both A and B
b. Technician B **d.** Neither A nor B

11. While discussing electric fuel pumps, Technician A says some electric fuel pumps are combined in one unit with the gauge sending unit. Technician B says on an engine with an electric fuel pump, low engine oil pressure may cause the engine to stop running. Who is correct?

a. A only **c.** Both A and B
b. B only **d.** Neither A nor B

12. *True or False?* A liquid vapor separator stops liquid fuel or bubbles from reaching the vapor storage canister or the engine's crankcase.

13. To relieve fuel pressure on an EFI car, Technician A connects a pressure gauge to the fuel rail. Technician B disables the fuel pump and runs the car until it dies. Who is correct?

a. Technician A **c.** Both A and B
b. Technician B **d.** Neither A nor B

14. *True or False?* Fuel pressure typically is at its highest level when vacuum is applied to the pressure regulator.

15. While discussing quick-disconnect fuel line fittings, Technician A says some quick-disconnect fittings may be disconnected with a pair of snapring pliers. Technician B says some quick-disconnect fittings are hand releasable. Who is correct?

a. Technician A **c.** Both A and B
b. Technician B **d.** Neither A nor B

16. While discussing quick-disconnect fuel line fittings, Technician A says on some hand-releasable, quick-disconnect fittings, the fitting may be removed by pulling on the fuel line. Technician B says some hand-releasable, quick-disconnect fittings may be disconnected by twisting the large connector collar in both directions and pulling on the connector. Who is correct?

a. Technician A **c.** Both A and B
b. Technician B **d.** Neither A nor B

17. What component, located in the fuel tank, prevents large contaminates from entering the fuel system?

a. fuel filter **c.** vapor separator
b. liquid separator **d.** strainer

18. While discussing fuel tank filler pipes and caps, Technician A says the threaded filler cap should be tightened until it clicks. Technician B says the vent pipe is connected from the top of the filler pipe to the bottom of the fuel tank. Who is correct?

a. Technician A **c.** Both A and B
b. Technician B **d.** Neither A nor B

19. While discussing electric fuel pumps, Technician A says the one-way check valve prevents fuel flow from the underhood fuel system components into the fuel pump and tank when the engine is shut off. Technician B says the one-way check valve prevents fuel flow from the pump to the fuel filter and fuel system if the engine stalls and the ignition switch is on. Who is correct?

a. Technician A **c.** Both A and B
b. Technician B **d.** Neither A nor B

20. While discussing electric fuel pumps, Technician A says some electric fuel pumps in EFI systems are computer controlled. Technician B says fuel pump pressure is determined by measuring the amount of fuel the pump will deliver in a specific length of time. Who is correct?

a. Technician A **c.** Both A and B
b. Technician B **d.** Neither A nor B

28

ELECTRONIC FUEL INJECTION

OBJECTIVES

■ Explain the differences in point of injection in throttle body or port injection systems. ■ Describe the difference between a sequential fuel injection (SFI) system and a multiport fuel injection (MFI) system. ■ Explain the design and function of major EFI components. ■ Describe the inputs used by the computer to control the idle air control and idle air control bypass air motors. ■ Describe how the computer supplies the correct air/fuel ratio on a throttle-body injection (TBI) system. ■ Explain how the clear flood mode operates on a TBI system. ■ Explain why manifold vacuum is connected to the pressure regulator in an MFI system. ■ Describe the operation of the pressure regulator in a returnless EFI system. ■ Describe the operation of the central injector and poppet nozzles in a central port injection (CPI) system. ■ Describe the operation of gasoline direct injection (GDI) systems.

This chapter discusses the common components found in most electronic fuel injection (EFI) systems and explains how the various designs of EFI work. Details on some of the various types of EFI are included, but we could never include all the EFI systems used on domestic and imported vehicles. If you are familiar with EFI components and principles of operation, you will be able to understand the many different EFI systems when you encounter them.

Electronic fuel injection has proven to be the most precise, reliable, and cost-effective method of delivering fuel to the combustion chambers of today's engines. EFI systems are computer controlled and designed to provide the correct air/fuel ratio for all engine loads, speeds, and temperature conditions.

TYPES OF FUEL INJECTION

Although fuel injection technology has been around since the 1920s, it was not until the 1980s that manufacturers began to replace carburetors with **electronic fuel injection (EFI)** systems. Many of the early EFI systems were **throttle-body injection (TBI)** systems in which the fuel was injected above the throttle plates. Recently a similar system, **central port injection (CPI)**, was introduced. In these systems the injector assembly is located in the lower half of the intake manifold. Engines equipped with TBI have gradually become equipped with **port fuel injection (PFI)**, which has injectors located in the intake ports of the cylinders **(Figure 28–1)**. Since the 1995 model year, all new cars are equipped with an EFI system. Recently, some engines have been equipped with **gasoline direct-injection (GDI)** systems. In these systems, the fuel is injected directly into the cylinders. Direct injection has been used for years with diesel fuels but has not been successfully used on gasoline engines until lately.

Throttle-body injection systems have a throttle body assembly mounted on the intake manifold in the position previously occupied by a carburetor. The throttle body assembly usually contains one or two injectors.

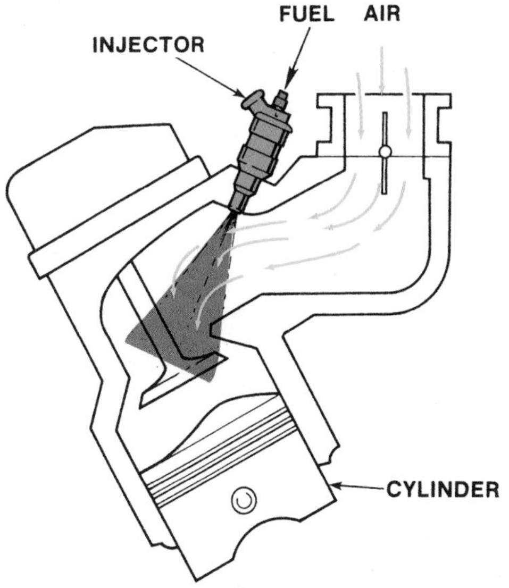

Figure 28-1 In a port fuel injection system, air and fuel are mixed right outside the combustion chamber.

On port fuel injection systems, fuel injectors are mounted at the back of each intake valve. Aside from the differences in injector location and number of injectors, operation of throttle body and port systems is quite similar with regard to fuel and air metering, sensors, and computer operation.

Most electronic fuel injection systems inject fuel only during part of the engine's combustion cycle. The engine's fuel needs are measured by intake airflow past a sensor or by intake manifold pressure (vacuum). The airflow or manifold vacuum sensor converts its reading to an electrical signal and sends it to the engine control computer. The computer processes this signal (and others) and calculates the fuel needs of the engine. The computer then sends an electrical signal to the fuel injector or injectors. This signal determines the amount of time the injector opens and sprays fuel. This interval is known as the injector pulse width.

BASIC EFI

In an EFI system **(Figure 28–2)**, the computer must know the amount of air entering the engine so it can supply the stoichiometric air/fuel ratio. In EFI systems with a MAP sensor, the computer program is designed to calculate the amount of air entering the engine from the MAP and rpm input signals. The ignition pickup or crankshaft position sensor supplies an rpm signal to the computer. The MAP sensor sends a signal relating to the pressure inside the intake manifold to the computer. This type of EFI system is referred to as a **speed density** system, because the computer calculates the air intake flow from the engine rpm, or speed, input, and the density of intake manifold vacuum input. Therefore, the computer must have accurate signals from these inputs to maintain the stoichiometric air/fuel ratio. The other inputs are used by the computer to fine tune the air/fuel ratio.

Powertrain Control Module

The heart of the fuel injection system is the computer or powertrain control module (PCM). The PCM receives signals from all the system sensors, processes them, and transmits programmed electrical pulses to the fuel injectors. Both incoming and outgoing signals are sent through a wiring harness and a multiple-pin connector.

System Operation

Electronic **feedback** means the system is self-regulating and the PCM is controlling the injectors based on operating conditions rather than on preprogrammed instruc-

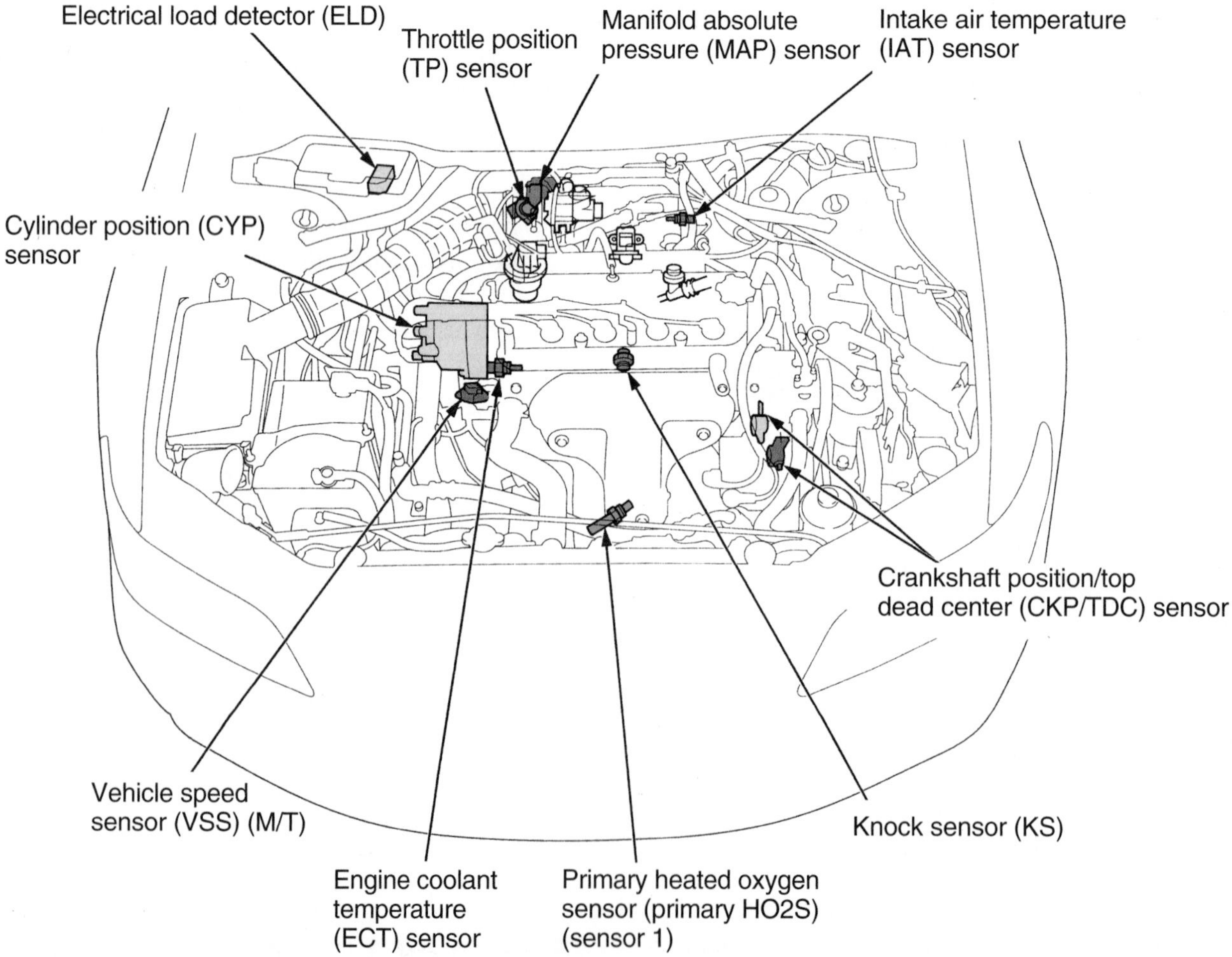

Figure 28-2 A typical electronic fuel injection system. *Courtesy of American Honda Motor Co., Inc.*

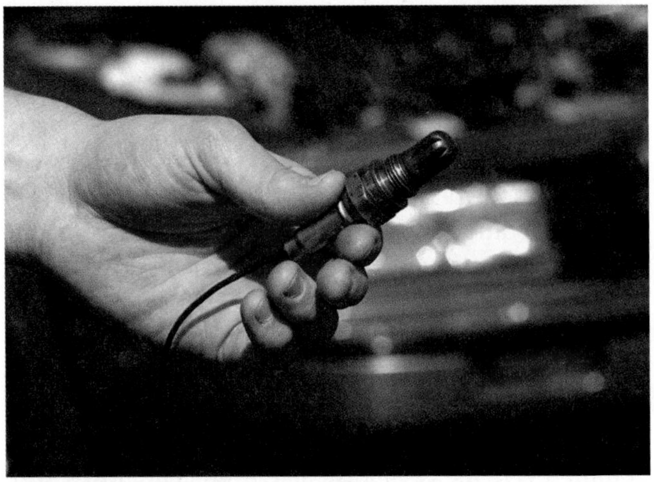

Figure 28-3 An oxygen sensor.

tions. As an example of a feedback loop, the PCM reads signals from the oxygen sensor **(Figure 28–3)**, varies the pulse width of the injectors, and again reads the signals from the oxygen sensor. This cycle is repeated until the injectors are pulsed for just the amount of time needed to get the proper amount of oxygen into the exhaust stream. While this interaction is occurring, the system is operating in **closed loop**. During the closed loop mode, sensor inputs are sent to the PCM; the PCM compares the values to its programs, and then reacts to the information to adjust the air/fuel ratio and other engine systems.

When conditions, such as starting or wide-open throttle, demand that the signals from the oxygen sensor be ignored, the system operates in **open loop**. During open loop, injector pulse length is controlled by set parameters contained in the PCM's memory. Systems with oxygen sensors may also go into the open loop mode while idling, or at any time that the oxygen sensor cools off enough to stop sending a good signal, and at wide-open throttle.

The basic purpose of these control loops is to create an ideal air/fuel ratio, which allows the catalytic converters to operate at maximum efficiency while giving the best mileage and performance possible.

Fuel Injectors

Fuel injectors are electromechanical devices that meter and atomize fuel so it can be sprayed into the intake manifold. Fuel injectors resemble a spark plug in size and shape. O-rings are used to seal the injector at the intake manifold, throttle body, and/or fuel rail mounting positions. These O-rings provide thermal insulation to prevent the formation of vapor bubbles and promote good hot-start characteristics. They also dampen potentially damaging vibration.

When the injector is electrically energized, a fine mist of fuel sprays from the injector tip. Two different valve designs are commonly used.

The first consists of a valve body and a nozzle or needle valve **(Figure 28–4)**. A movable armature is attached to the nozzle valve, which is pressed against the nozzle body sealing seat by a helical spring. The solenoid winding is located at the back of the valve body.

When the solenoid winding is energized, it creates a magnetic field that draws the armature back and pulls the nozzle valve from its seat. When the solenoid is de-energized, the magnetic field collapses and the helical spring forces the nozzle valve back on its seat.

The second popular valve design uses a ball valve and valve seat. In this case, the magnetic field created by the solenoid coil pulls a plunger upward, lifting the ball valve from its seat. Once again, a spring is used to return the valve to its seated or closed position.

Fuel injectors can be either top fuel feeding or bottom fuel feeding **(Figure 28–5)**. Top feed injectors are primarily used in port injection systems that operate using high fuel system pressures. Bottom feed injectors are used in throttle body systems. Bottom feed injectors are able to use fuel pressures as low as 10 psi (68.95 kPa).

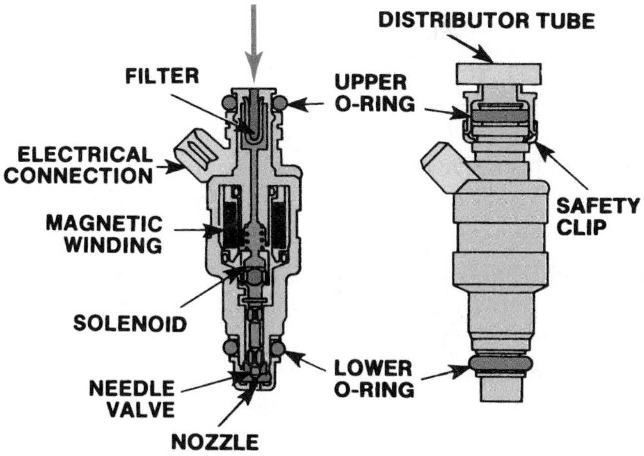

Figure 28-4 This fuel injector design is equipped with a needle valve that has a specially ground pintle for precise fuel control.

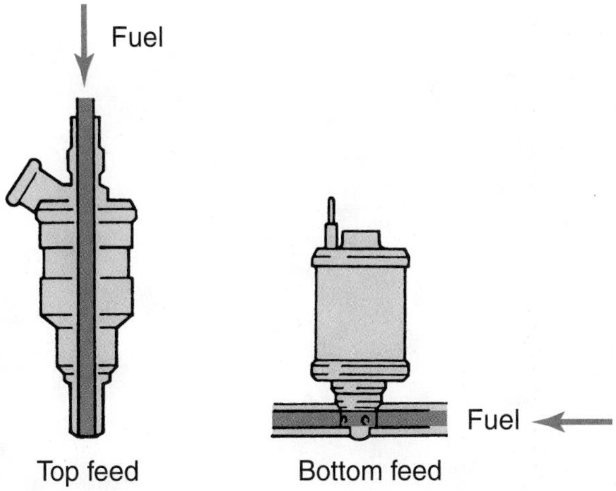

Figure 28-5 Examples of top feed and bottom feed injectors.

There have been some problems with deposits on injector tips. Since small quantities of gum are present in gasoline, injector deposits usually occur when this gum bakes onto the injector tips after a hot engine is shut off. Most oil companies have added a detergent to their gasoline to help prevent injector tip deposits. Car manufacturers and auto parts stores sell detergents to place in the fuel tank to clean injector tips.

Some manufacturers and auto parts suppliers have designed deposit-resistant injectors. These injectors have several different pintle tip and orifice designs to help prevent deposits. On one type of deposit-resistant injector, the pintle seat opens outward away from the injector body and more clearance is provided between the pintle and the body. Another type of deposit-resistant injector has four orifices in a metering plate rather than a single orifice. Some deposit-resistant injectors may be recognized by the color of the injector body. For example, regular injectors supplied by Ford Motor Company are painted black, whereas their deposit-resistant injectors have tan or yellow bodies.

Each fuel injector **(Figure 28–6)** is equipped with a two-wire connector. The connector is often equipped with a spring clip that must be unlocked before the connector can be removed from the injector.

One wire of the connector supplies voltage to the injector. This voltage supply wire may connect directly to the fuse panel or it may connect to the PCM, which in turn connects to the fuse panel. In some systems, a resistor at the fuse panel or PCM is used to reduce the 12-volt battery supply voltage to 3 volts or less. Most other injectors are fed battery voltage.

The second wire of the connector is a ground wire. This ground wire is routed to the PCM. The PCM energizes the injector by grounding its electrical circuit. The pulse width of the injector equals the length of time the injector circuit is grounded. Typical pulse widths range from 1 millisecond to 10 milliseconds at full load. Port fuel injection systems having four, six, or eight injectors use a special wiring harness to simplify and organize injector wiring.

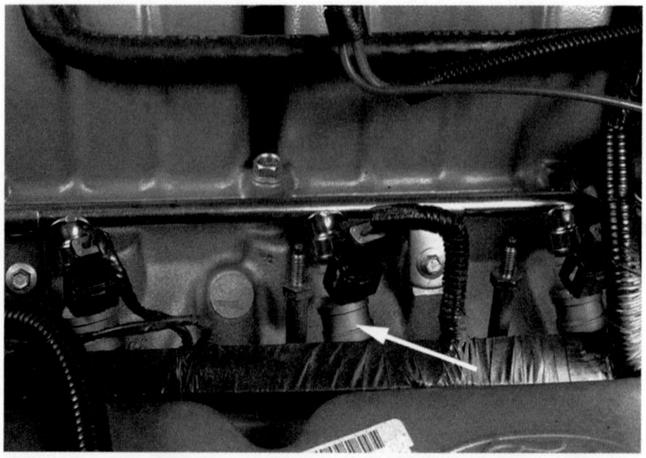

Figure 28-6 Typical injector with its electrical connector attached to it.

Idle Speed Control

Idle speed control is a function of the PCM. Based on operating conditions and inputs from various sensors, the PCM regulates the idle speed to control emissions. In throttle body and port EFI systems, engine idle speed is controlled by bypassing a certain amount of airflow past the throttle valve in the throttle body housing. Two types of air bypass systems are used; auxiliary air valves and **idle air control (IAC)** valves. IAC valve systems are more common **(Figure 28–7)**. Most TBI units are fitted with an idle speed motor.

The IAC system consists of an electrically controlled stepper motor or actuator that positions the IAC valve in the air bypass channel around the throttle valve. The IAC valve is part of the throttle body casting. The PCM calculates the amount of air needed for smooth idling based on input data such as coolant temperature, engine load, engine speed, and battery voltage. It then signals the actuator to extend or retract the idle air control valve in the air bypass channel.

If the engine speed is lower than desired, the PCM activates the motor to retract the IAC valve. This opens the channel and diverts more air around the throttle valve. If engine speed is higher than desired, the valve is extended and the bypass channel is made smaller. Air supply to the engine is reduced, and engine speed falls.

During cold starts, idle speed can be as high as 2,100 rpm to quickly raise the temperature of the catalytic converter for proper control of exhaust emissions. Idle speed that is attained after a cold start is controlled by the PCM. The PCM maintains idle speed for approximately 40 to 50 seconds even if the driver attempts to alter it by kicking the accelerator. After this preprogrammed time interval, depressing the accelerator pedal rotates the throttle position (TP) sensor and signals the PCM to reduce idle speed.

Some engines are equipped with an auxiliary air valve to aid in the control of engine idle speed. The major difference between an IAC valve and an auxiliary air valve is that the auxiliary air valve is not controlled by the PCM. But like the IAC system, the auxiliary air valve provides additional air during cold-engine starts and warm-up.

The auxiliary air valve allows air to bypass the throttle plate, thereby increasing the idle speed. The opening and closing of the auxiliary air valve is controlled by a bimetallic strip. As the strip heats up, it bends to rotate the movable plate, gradually blocking the opening. When the device is closed, there is no auxiliary airflow.

The bimetal strip is warmed by an electric heating element powered from the run circuit of the ignition switch.

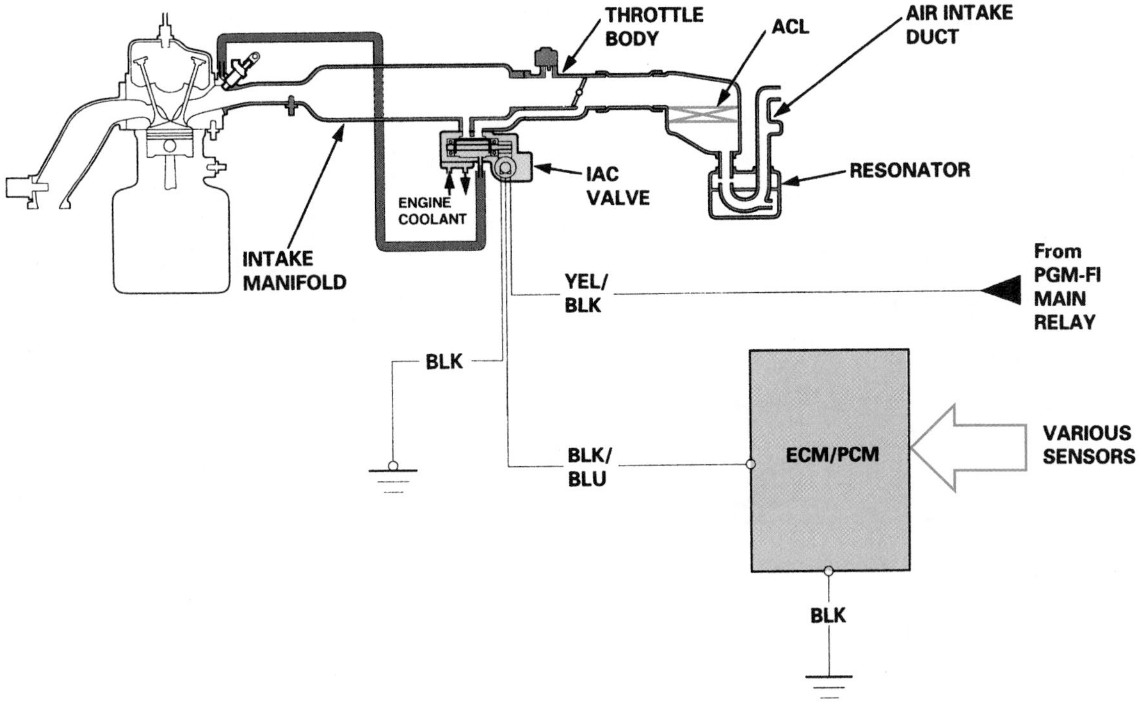

Figure 28-7 An idle air control system. *Courtesy of American Honda Motor Co., Inc.*

The auxiliary air device is independent of the cold start injector. It is not controlled by the PCM but is continuously powered when the ignition key is set to the run position.

THROTTLE BODY FUEL INJECTION

For some auto manufacturers, TBI served as a stepping stone from carburetors to more advanced port fuel injection systems. TBI units were used on many engines during the 1980s and are still used on some engines. The throttle body unit is similar in size and shape to a carburetor, and, like a carburetor, it is mounted on the intake manifold **(Figure 28–8)**. The injector(s) spray fuel down into a throttle body chamber leading to the intake manifold. The intake manifold feeds the air/fuel mixture to all cylinders.

The throttle body assembly is mounted on top of the intake manifold, where the carburetor was mounted on carbureted engines. Four-cylinder engines have a single throttle body assembly with one injector and throttle **(Figure 28–9)**, whereas V6 and V8 engines are equipped with dual injectors and two throttles on a common throttle shaft **(Figure 28–10)**.

Figure 28-8 A throttle body injection assembly.

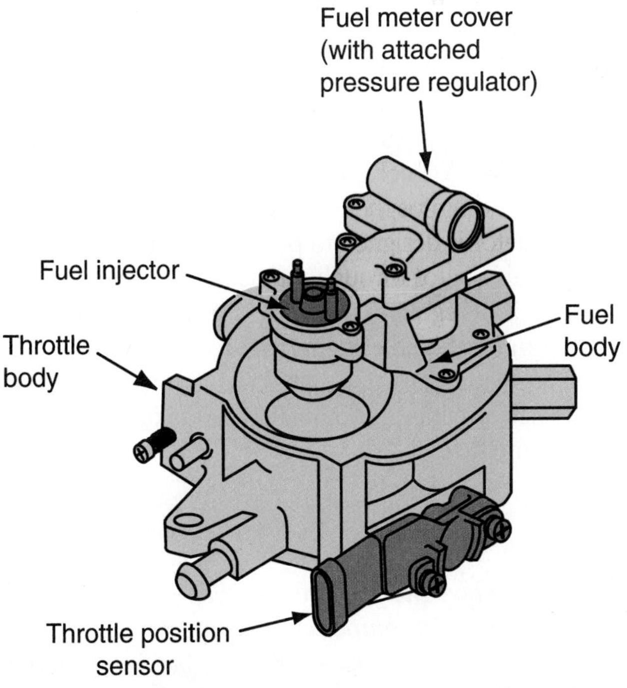

Figure 28-9 Single throttle body assembly used on a four-cylinder engine.

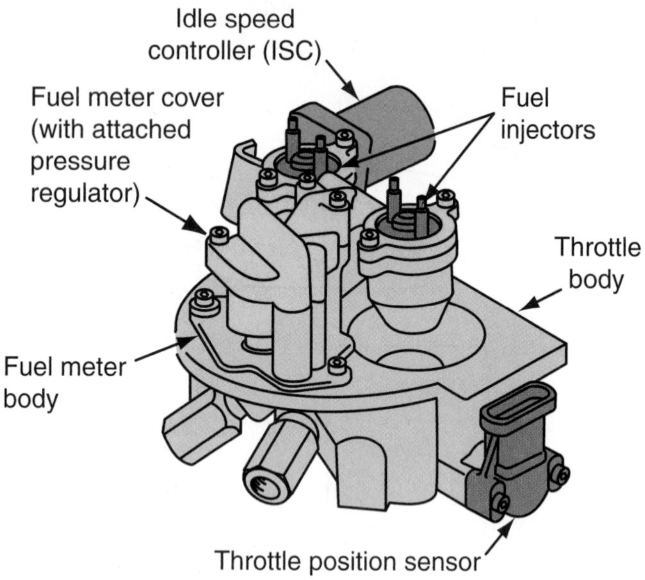

Figure 28–10 Dual throttle body assembly used on an eight-cylinder engine.

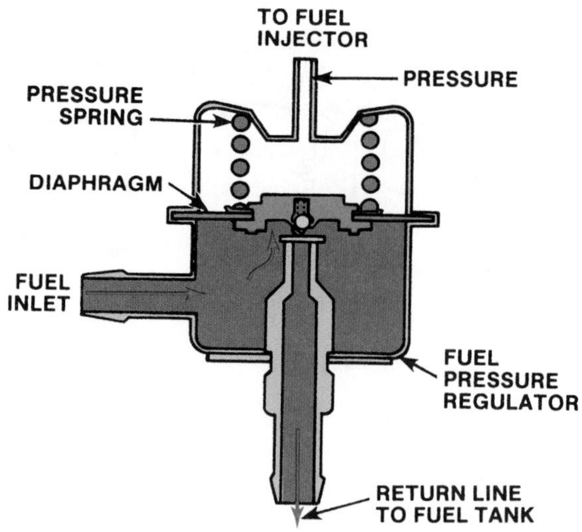

Figure 28–11 Operation of a diaphragm-operated fuel pressure regulator.

The throttle body assembly contains a pressure regulator, injector or injectors, TP sensor, idle speed control motor, and throttle shaft and linkage assembly. A fuel filter is located in the fuel line under the vehicle or in the engine compartment. When the engine is cranking or running, fuel is supplied from the fuel pump through the lines and filter to the throttle body assembly. A fuel return line connected from the throttle body to the fuel tank returns excess fuel to the fuel tank.

The throttle body casting has ports that can be located above, below, or at the throttle valve depending on the manufacturer's design. These ports generate vacuum signals for the manifold absolute pressure sensor and for devices in the emission control system, such as the EGR valve, the canister purge system, and so on.

The fuel pressure regulator used on the throttle body assembly is similar to a diaphragm-operated relief valve **(Figure 28–11)**. Fuel pressure is on one side of the diaphragm and atmospheric pressure is on the other side. The regulator is designed to provide a constant pressure on the fuel injector throughout the range of engine loads and speeds. If regulator pressure is too high, a strong fuel odor is emitted and the engine runs too rich. On the other hand, regulator pressure that is too low results in poor engine performance or detonation can take place, due to the lean mixture.

CAUTION!

Always relieve the fuel pressure before disconnecting a fuel system component to avoid gasoline spills that may cause a fire, resulting in personal injury and/or property damage.

TBI Advantages

Throttle body systems provide improved fuel metering when compared to that of carburetors. They are also less expensive and simpler to service. TBI units also have some advantages over port injection. They are less expensive to manufacture, simpler to diagnose and service, and do not have injector balance problems to the extent that port injection systems do when the injectors begin to clog.

However, throttle body units are not as efficient as port systems. The disadvantages are primarily manifold related. Like a carburetor system, fuel is still not distributed equally to all cylinders, and a cold manifold may cause fuel to condense and puddle in the manifold. Like a carburetor, throttle-body injection systems must be mounted above the combustion chamber level, which eliminates the possibility of tuning the manifold design for more efficient operation.

Injectors

The fuel injector is solenoid operated and pulsed on and off by the vehicle's engine control computer. Surrounding the injector inlet is a fine screen filter, toward which the incoming fuel is directed. When the injector's solenoid is energized, a normally closed ball valve is lifted. Fuel under pressure is then injected at the walls of the throttle body bore just above the throttle plate.

A fuel injector has a movable armature in the center of the injector, and a pintle with a tapered tip is positioned at the lower end of the armature. A spring pushes the armature and pintle downward so that the pintle tip seats in the discharge orifice. The injector coil surrounds the armature, and the two ends of the winding are connected to the terminals on the side of the injector. An integral filter is located inside the top of the injector. When the ignition switch is turned on, voltage is supplied to one injector terminal and the other terminal is connected through the computer. Each time the control unit completes the circuit from the injector winding to ground, current flows through the injector coil, and the coil magnetism moves the plunger and pintle upward. When this occurs, the pintle tip is unseated from the injector orifice, and fuel sprays from this orifice.

Throttle Body Internal Design and Operation

When fuel enters the throttle body fuel inlet, the fuel surrounds the injector or injectors at all times. Each injector is sealed into the throttle body with O-ring seals, which prevent fuel leakage around the injector at the top or bottom. Fuel is supplied from the injector through a passage to the pressure regulator. A diaphragm and valve assembly is mounted in this regulator, and a diaphragm spring holds the valve closed. At a specific fuel pressure, the regulator diaphragm is forced upward to open the valve, and some excess fuel is returned to the fuel tank.

When the pressure regulator valve opens, fuel pressure decreases slightly, and the spring closes the regulator valve. This action causes the fuel pressure to increase and reopen the pressure regulator valve. In most TBI systems, the fuel pressure regulator controls fuel pressure at 10 to 25 psi (70 to 172 kPa). The fuel pressure must be high enough to prevent fuel from boiling in the TBI assembly. When the pressure on a liquid is increased, the boiling point is raised proportionally. If fuel boiling occurs in the TBI assembly, vapor and fuel are discharged from the injectors. The computer program assumes that the injectors are discharging liquid fuel. Vapor discharge from the injectors creates a lean air/fuel ratio, which results in lack of engine power and acceleration stumbles.

Injector Internal Design and Electrical Connections

The plunger and valve seat are held down by a spring. In this position, the seat closes the metering orifices in the end of the injector. Openings in the sides of the injector allow fuel to enter the cavity surrounding the injector tip. A mesh screen filter inside the injector openings removes dirt particles from the fuel. In some injectors, a diaphragm is located between the valve seat and the housing. The tip of the injector may contain up to six metering orifices, but some injectors have a single metering orifice. Injector design varies depending on the manufacturer.

Each injector contains two terminals, across which an internal coil is connected **(Figure 28–12)**. A movable plunger is positioned in the center of the coil, and the lower end of the plunger has a tapered valve seat. When the ignition switch is turned on, 12 volts are supplied to one of the injector terminals, and the other injector terminal is connected to the computer. When the computer

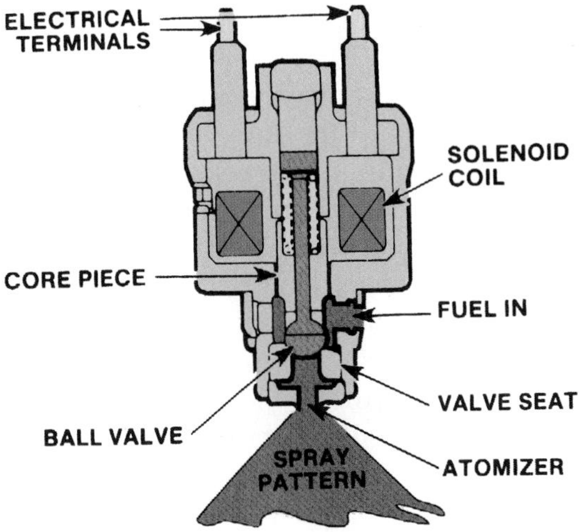

Figure 28–12 Solenoid operated ball-valve-type fuel injector used in a TBI system. When electronically energized, the ball valve lifts off the valve seat, allowing fuel to spray into the throttle body housing.

grounds this terminal, current flows through the injector coil to ground in the computer. When this action occurs, the injector coil magnetism moves the plunger and valve seat upward, and fuel sprays from the injector orifices into the air stream above the throttle.

Pulse Width The length of time that the computer grounds the injector is referred to as pulse width. Under most operating conditions, the computer provides the correct injector pulse width to maintain the stoichiometric air/fuel ratio. For example, the computer might ground the injector for 2 milliseconds at idle speed and 7 milliseconds at wide-open throttle to provide the stoichiometric air/fuel ratio. In many TBI systems, the computer grounds an injector each time a signal is received from the distributor pickup. This type of TBI system may be referred to as a synchronized system, because the injector pulses are synchronized with the pickup signals. In a dual injector throttle body assembly, the computer grounds the injectors alternately under most operating conditions.

Air/Fuel Ratio Enrichment

When the coolant temperature sensor signal to the computer indicates that the engine coolant is cold, the computer increases the injector pulse width to provide a richer air/fuel ratio. This action eliminates the need for a conventional choke on a TBI assembly. The PCM supplies the proper air/fuel ratio and engine rpm when starting a cold engine, eliminating the need for the driver to depress the accelerator pedal while starting the engine.

When a TBI-equipped engine is cold, the computer provides a very rich air/fuel ratio for faster starting. However, if the engine does not start because of an ignition defect, the engine becomes flooded quickly. Under this condition, excessive fuel may run past the piston rings into the crankcase. Therefore, when a cold TBI engine does not start, periods of long cranking should be avoided.

If the driver suspects that the air/fuel ratio is extremely rich, he or she may push the accelerator pedal to the wide-open position when starting a cold engine. Under these conditions, the computer program provides a very lean air/fuel ratio of approximately 18:1. This may be referred to as a **clear flood mode**. However, under normal conditions, the driver should not push on the accelerator pedal at any time when starting an engine with TBI. When the engine is decelerated, the computer reduces injector pulse width in many TBI systems to provide a lean air/fuel ratio, which reduces emissions and improves fuel economy.

PORT FUEL INJECTION

PFI systems use one injector at each cylinder. They are mounted in the intake manifold near the cylinder head, where they can inject a fine, atomized fuel mist as close as possible to the intake valve **(Figure 28–13)**. Fuel lines run to each cylinder from a fuel manifold, usually referred to as a fuel rail **(Figure 28–14)**. The fuel rail assembly on a PFI system of V6 and V8 engines usually consists of a left- and right-hand rail assembly. The two rails can be connected either by crossover and return fuel tubes or by a mechanical bracket arrangement. A typical fuel rail arrangement is shown in **Figure 28–15**. Fuel tubes crisscross between the two rails. Since each cylinder has its own injector, fuel distribution is exactly equal. With little or no fuel to wet the manifold walls, there is no need for manifold heat or any early fuel evaporation system. Fuel does not collect in puddles at the base of the manifold. This means that the intake manifold passages can be tuned or designed for better low-speed power availability. The port type systems provide a more accurate and efficient delivery of fuel. Some engines are now equipped with variable induction intake manifolds that have separate runners for low and high speeds. This technology is only possible with port injection.

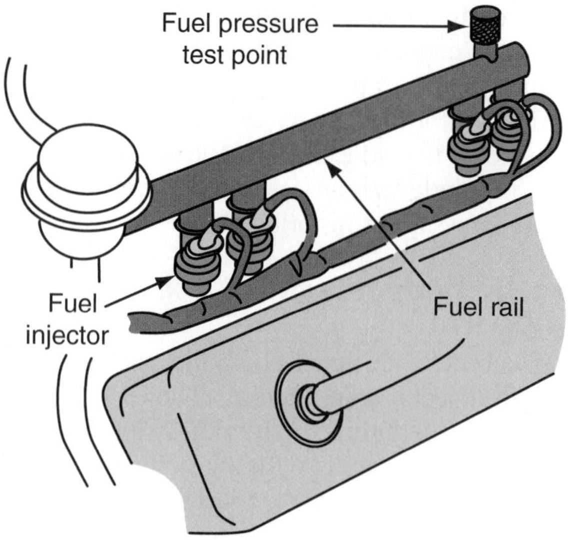

Figure 28–14 A typical fuel rail.

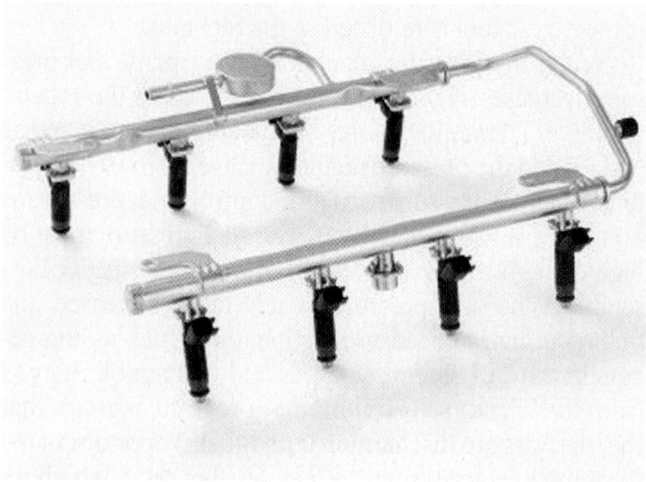

Figure 28–15 A one-piece fuel rail for a V8 engine.
Courtesy of General Motors Corporation

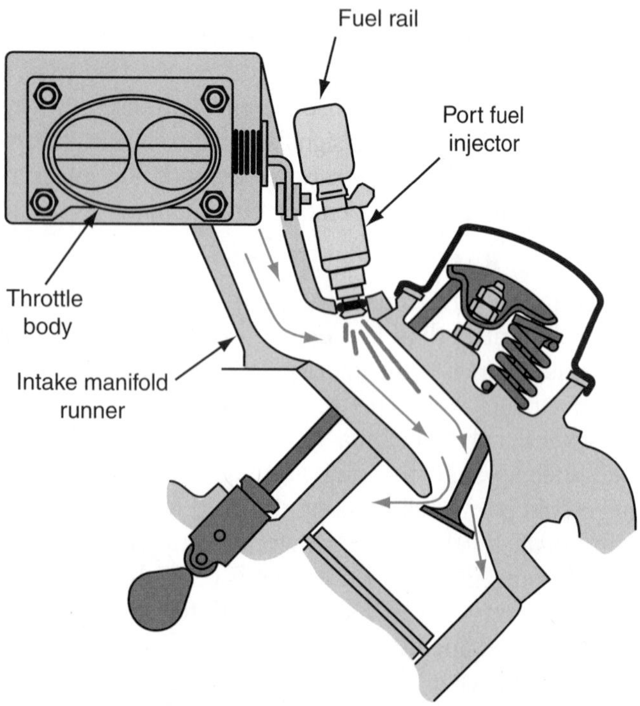

Figure 28–13 Port injection sprays fuel into the intake port and fills the port with fuel vapor before the valve opens.

The throttle body **(Figure 28–16)** in a port fuel injection system controls the amount of air that enters the engine as well as the amount of vacuum in the manifold. It also may house the MAP sensor, idle air control motor, and the throttle position (TP) sensor **(Figure 28–17)**. The TP sensor enables the PCM to know where the throttle is positioned at all times.

The throttle body is a single cast-aluminum housing with a single throttle blade attached to the throttle shaft. The throttle shaft is controlled by the accelerator pedal

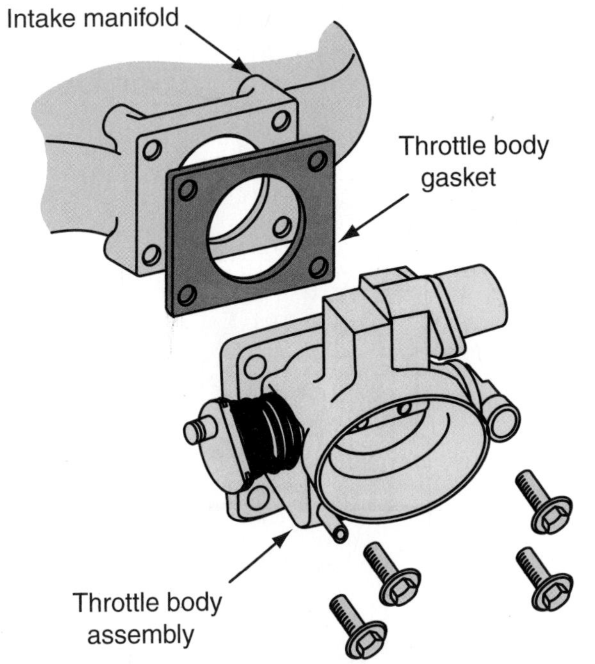

Figure 28-16 A throttle body assembly for a port-type injection system.

Figure 28-17 A TP sensor.

and extends the full length of the housing. The throttle bore controls the amount of incoming air that enters the air-induction system. A small amount of coolant is also routed through a passage in the throttle body to prevent icing during cold weather.

Port systems require an additional control system that throttle body injection units do not require. Throttle body injectors are mounted above the throttle plates and are not affected by fluctuations in manifold vacuum, but port system injectors have their tips located in the manifold, where constant changes in vacuum would affect the amount of fuel injected. To compensate for these fluctuations, port injection systems are equipped with fuel-pressure regulators that sense manifold vacuum and continually adjust the fuel pressure to maintain a constant pressure drop across the injector tips at all times.

Port Firing Control

While all port injection systems operate using an injector at each cylinder, they do not fire the injectors in the same manner. This one statement best defines the difference between typical **multiport injection (MPI)** systems and **sequential fuel injection (SFI)** systems.

SFI systems control each injector individually so that it is opened just before the intake valve opens. This means that the mixture is never static in the intake manifold and that adjustments to the mixture can be made almost instantaneously between the firing of one injector and the next. Sequential firing is the most accurate and desirable method of regulating port injection.

In MPI systems, the injectors are arranged together in pairs or groups **(Figure 28–18)**, and these pairs or groups of injectors are turned on at the same time. When the injectors are split into two equal groups, the groups are fired alternately, with one group firing during each engine revolution.

Because only two injectors can be fired relatively close to the time when the intake valve is about to open, the fuel charge for the remaining cylinders must stand in the intake manifold for varying periods of time. These periods of time are very short, therefore the standing of fuel in the intake manifold is not that great a disadvantage of MPI systems. At idle speeds this wait is about 150 milliseconds and, at higher speeds, the time is much less. The primary advantage of SFI is the ability to make instantaneous changes to the mixture.

In SFI systems, each injector is connected individually into the computer, and the computer completes the ground for each injector, one at a time. In MPI systems, the injectors are grouped and all injectors within the group share the same common ground wire.

Some injection systems fire all of the injectors at the same time for every engine revolution. This type of system offers easy programming and relatively fast adjustments to the air/fuel mixture. The injectors are connected in parallel, so the PCM sends out just one signal for all

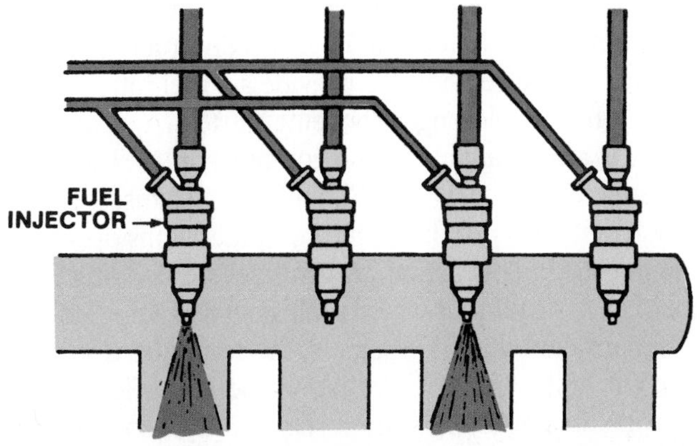

Figure 28-18 Grouped single-fire port injection.

injectors. They all open and close at the same time. It simplifies the electronics without compromising injection efficiency. The amount of fuel required for each four-stroke cycle is divided in half and delivered in two injections, one for every 360 degrees of crankshaft rotation. The fact that the intake charge must still wait in the manifold for varying periods of time is the system's major drawback.

Port Fuel Injection System Design

Basically, the same electric in-tank fuel pumps and fuel pump circuits are found on TBI, MFI, and SFI systems. Some MFI and SFI systems, such as those on Ford trucks, have a booster fuel pump on the frame rail in addition to the in-tank pump. A fuel filter is connected in the fuel line from the tank to the injectors. This filter may be under the vehicle or in the engine compartment. The fuel line from the filter is connected to a hollow fuel rail that is bolted to the intake manifold. The lower end of each port injector is sealed in the intake manifold with an O-ring seal, and a similar seal near the top of the injector seals the injector to the fuel rail.

Each injector has a movable armature in the center of the injector, and a pintle with a tapered tip is positioned at the lower end of the armature. A spring pushes the armature and pintle downward so the pintle tip seats in the discharge orifice. The injector coil surrounds the armature, and the two ends of the winding are connected to the terminals on the side of the injector. An integral filter is located inside the top of the injector. When the ignition switch is turned on, voltage is supplied to one injector terminal and the other terminal is connected through the computer. Each time the computer completes the circuit from the injector winding to ground, current flows through the injector coil, and the coil magnetism moves the plunger and pintle upward. Under this condition, the pintle tip is unseated from the injector orifice, and fuel sprays from this orifice into the intake port **(Figure 28–19)**.

The computer is programmed to ground the injectors well ahead of the actual intake valve openings, so the intake ports are filled with fuel vapor before the intake valves open. In both SFI and MFI systems, the computer supplies the correct injector pulse width to provide the stoichiometric air/fuel ratio. The computer increases the injector pulse width to provide air/fuel ratio enrichment when a cold engine is being started. A clear flood mode is also available in the computer in MFI and SFI systems. On some TBI, MFI, and SFI systems, if the ignition system is not firing, the computer stops operating the injectors. This action prevents severe flooding from long cranking periods when a cold engine is being started. On many MFI and SFI systems, the computer decreases injector pulse width while the engine is decelerating to provide improved emission levels and fuel economy. On some of these systems, the computer stops operating the injectors while the engine is decelerating in a certain rpm range.

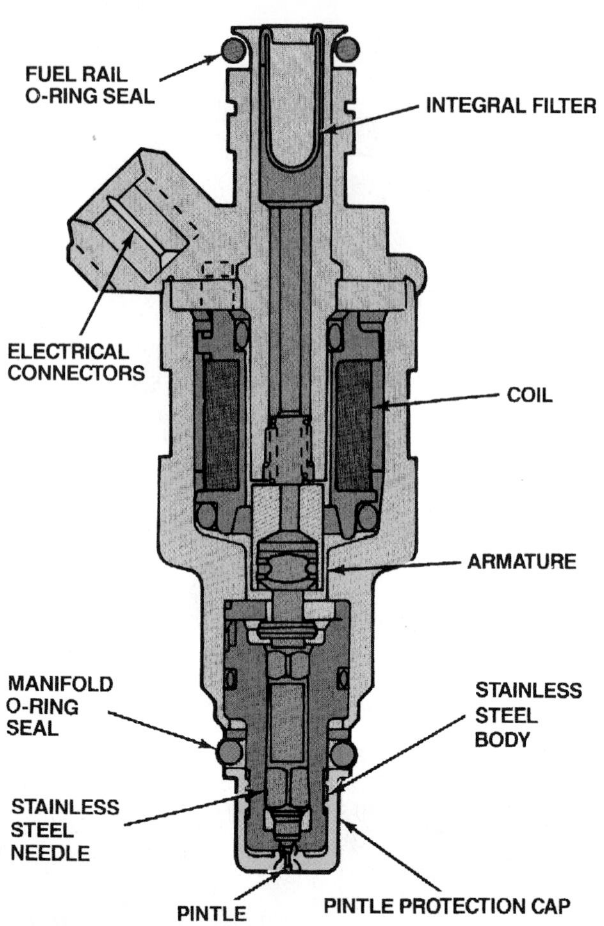

Figure 28-19 A typical fuel injector used in multiport fuel injection systems. *Courtesy of Ford Motor Company*

Cold-Start Injector

Some older EFI engines were fitted with a **cold-start injector**. A pickup pipe was connected from the fuel rail to the cold-start injector, and the end of this injector was mounted in the intake manifold **(Figure 28–20)**.

Unlike the intake port injectors operated by the PCM, the cold-start injector is operated by a thermo-time switch that senses coolant temperature. When the engine is cranked, voltage is supplied from the starter solenoid to one terminal on the cold-start injector. If the coolant temperature is below 95°F (35°C), the thermo-time switch grounds the other cold-start injector terminal. Under this condition, the cold-start injector is energized while cranking the engine, and the injector pintle opens to spray fuel into the intake manifold in addition to the fuel injected by the injectors in the intake ports.

A bimetal switch in the thermo-time switch is heated as current flows through the injector coil. The bimetal switch action opens the circuit through the thermo-time switch in a maximum of 8 seconds. The actual time that the thermo-time switch remains closed is determined by the coolant temperature. In this MFI system, the pulse

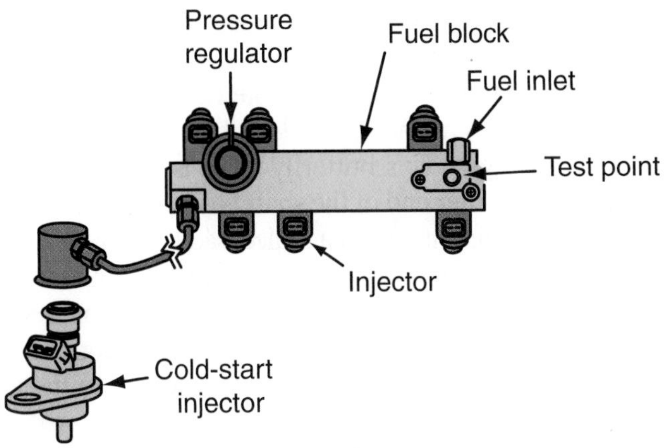

Figure 28-20 A cold-start injector and fuel rail for a V6 engine.

width supplied by the PCM to the intake port injectors is programmed to operate with the cold-start injector and supply the correct air/fuel ratio while cranking a cold engine.

Pressure Regulators

The pressure regulator **(Figure 28–21)** on MFI and SFI systems is similar to the regulator on TBI systems. A diaphragm and valve assembly is positioned in the center of the regulator, and a diaphragm spring seats the valve on the fuel outlet.

When fuel pressure reaches the regulator setting, the diaphragm moves against the spring tension and the valve opens. This action allows fuel to flow through the return line to the fuel tank. The fuel pressure drops slightly when the pressure regulator valve opens, and the spring closes the regulator valve. Under this condition, the fuel pressure increases and reopens the regulator valve.

A vacuum hose is connected from the intake manifold to the vacuum inlet on the pressure regulator. This hose supplies vacuum to the area in which the diaphragm spring is located. This vacuum works with the fuel pressure to move the diaphragm and open the valve. When the engine is running at idle speed, high manifold vacuum is supplied to the pressure regulator. Under this condition, fuel pressure opens the regulator valve. If the engine is operating at wide-open throttle, a very low manifold vacuum is supplied to the pressure regulator.

When this condition is present, the vacuum does not help to open the regulator valve, and a higher fuel pressure is required to open the valve.

When the engine is idling, higher manifold vacuum is supplied to the injector tips, and the injectors are discharging fuel into this vacuum. Under wide-open throttle conditions, the very low manifold vacuum is supplied to the injector tips. When this condition is present, the injectors are actually discharging fuel into a higher pressure compared to idle speed conditions, because the very low manifold vacuum is closer to a positive pressure. If the fuel pressure remained constant at idle and wide-open throttle conditions, the injectors would discharge less fuel into the higher pressure in the intake manifold at wide-open throttle. The increase in fuel pressure supplied by the pressure regulator at wide-open throttle maintains the same pressure drop across the injectors at idle speed and wide-open throttle. When this same pressure drop is maintained, the change in pressure at the injector tips does not affect the amount of fuel discharged by the injectors.

Figure 28-21 A typical fuel pressure regulator for a port injection system.

SEQUENTIAL FUEL INJECTION SYSTEMS

SFI systems control each injector individually so that it is opened just before the intake valve opens. This means that the mixture is never static in the intake manifold and that adjustments to the mixture can be made almost instantaneously between the firing of one injector and the next. Sequential firing is the most accurate and desirable method of regulating port injection.

In SFI systems, each injector is connected individually into the computer, and the computer completes the ground for each injector, one at a time. In MPI systems, the injectors are grouped and all injectors within the group share the same common ground wire.

A Typical Sequential Fuel Injection System

In a DaimlerChrysler SFI system on a 3.5 L engine, each injector has a separate ground wire connected into the PCM. Many Chrysler engines made before 1992 have multiport fuel injection (MPI) systems with the injectors connected in pairs on the ground side. Each pair of injectors shares a common ground wire into the PCM. Voltage is supplied through the ASD relay points to the injectors when the ignition switch is turned on, and a separate fuel pump relay supplies voltage to the fuel pump. This engine is equipped with an electronic ignition system, and the crank and cam sensors are inputs for this system. Since these inputs are connected to the PCM, the ignition module is contained in the PCM.

Returnless Fuel System Pressure Regulators

Some later-model Chrysler SFI systems are referred to as returnless systems. In these systems, the fuel pressure regulator and filter are mounted in the top of the assembly containing the fuel pump and fuel gauge sending unit in the fuel tank **(Figure 28–22)**. The fuel line from the fuel rail under the hood is connected to the filter with a quick-disconnect fitting.

Fuel enters the filter through the fuel supply tube in the center of the regulator and filter assembly. Fuel pressure is applied against the regulator seat washer, which is seated by the seat control spring. At the specified regulator pressure, the seat is forced downward against the spring, and fuel flows past the seat into the cavity around the seat control spring. Fuel returns from this cavity to the fuel tank. When the pressure drops slightly, the seat closes again. With the returnless fuel system, only the fuel needed by the engine is filtered, thus allowing the use of a smaller fuel filter.

The Chrysler SFI system has many similarities to the Chrysler TBI system. For example, the voltage regulator and the cruise control module are contained in the PCM board. The SFI system on the 3.5 L engine has a low-speed and a high-speed cooling fan relay. At a specific coolant temperature, the PCM grounds the low-speed relay winding, which closes the relay points and supplies voltage to the fan motor. If the engine coolant temperature continues to increase, the PCM grounds the high-speed cooling fan relay winding, which closes the fan relay points and supplies voltage to the high-speed fan motor.

The manifold solenoid controls the vacuum supplied to the intake manifold tuning valve. This solenoid is mounted on the right shock tower, and the manifold tuning valve is positioned near the center of the intake manifold. The manifold contains a pivoted butterfly valve that opens and closes to change the length of the intake manifold air passages. This butterfly valve is mounted on a shaft, and the outer end of the shaft is connected through a linkage to a diaphragm in a sealed vacuum chamber. A vacuum hose is connected from the outlet fitting on the manifold solenoid to the vacuum chamber in the manifold tuning valve. Another vacuum hose is connected from the inlet fitting on the manifold solenoid to the intake manifold.

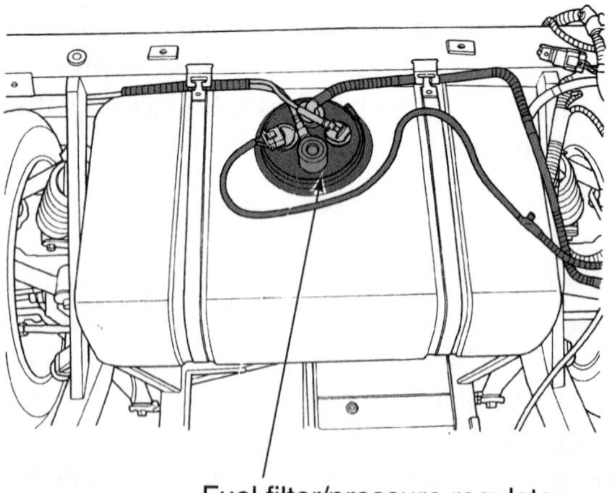

Figure 28-22 A returnless fuel system with the pressure regulator and filter mounted in the fuel tank with the fuel pump and fuel gauge sending unit. *Courtesy of DaimlerChrysler Corporation*

One terminal on the manifold solenoid winding is connected to the ignition switch, and the other terminal on this winding is connected to the PCM. While the engine is running at lower rpm, the PCM opens the manifold solenoid circuit. Under this condition, the solenoid shuts off the manifold vacuum to the intake manifold tuning valve, and the butterfly valve closes some of the air passages inside the intake manifold.

At higher engine rpm, the PCM grounds the manifold solenoid winding and energizes the solenoid. This action opens the vacuum passage through the solenoid and supplies vacuum to the intake manifold tuning valve. Under this condition, the butterfly valve is moved so that it opens additional air passages inside the intake manifold to improve airflow and increase engine horsepower and torque.

Typical Import Sequential Fuel Injection System

The Nissan electronic concentrated engine control system (ECCS) is an SFI system that has many of the same inputs and outputs as the other systems. The system also has some things not found in others, such as the dropping resistor assembly that contains a resistor connected in series with each injector. These resistors protect the injectors from sudden voltage changes and provide constant injector operation. The system uses a vane-type mass airflow sensor. The throttle valve switch is mounted in the throttle chamber, and the switch contacts are closed when the throttle is in the idle position **(Figure 28–23)**.

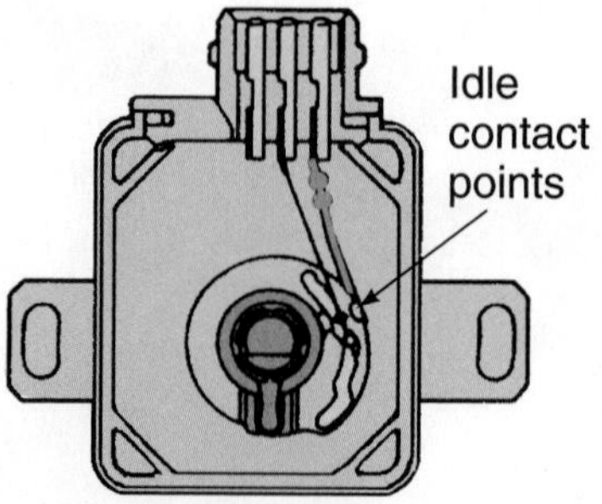

Figure 28-23 A typical idle switch. *Used with permission from Nissan North America, Inc.*

When the throttle is opened from the idle position, the switch contacts open. The throttle valve switch signal informs the PCM when the throttle is in the idle position.

An idle control valve (ICV) solenoid and fast idle control device (FICD) solenoid are mounted on the intake manifold. When the idle speed drops below a specific rpm, the PCM energizes the ICV solenoid, and additional air is bypassed through this solenoid into the intake manifold to increase the idle speed. The FICD solenoid is energized when the A/C is on, and air flows past this solenoid into the intake manifold to maintain idle speed and compensate for the compressor load on the engine.

CENTRAL MULTIPORT FUEL INJECTION (CMFI)

In a central port (CPI) or **central multiport fuel injection (CMFI)** system, a central injector assembly is mounted in the lower half of the intake manifold. The CMFI system uses one injector to control the fuel flow to six (on six-cylinder engines) individual **poppet nozzles (Figure 28–24)**. The CMFI injector assembly consists of a fuel metering body, pressure regulator, one fuel injector, six poppet nozzles with nylon fuel tubes, and a gasket seal. The injector distributes metered fuel through a six-hole distribution gasket. The gasket seals the injector to the six lines connected to the nozzles.

Each nozzle contains a check ball and extension spring that regulates fuel flow. The poppet nozzle opens when high pressure is exerted on the check ball. This action allows the nozzles to feed individual cylinders with atomized fuel.

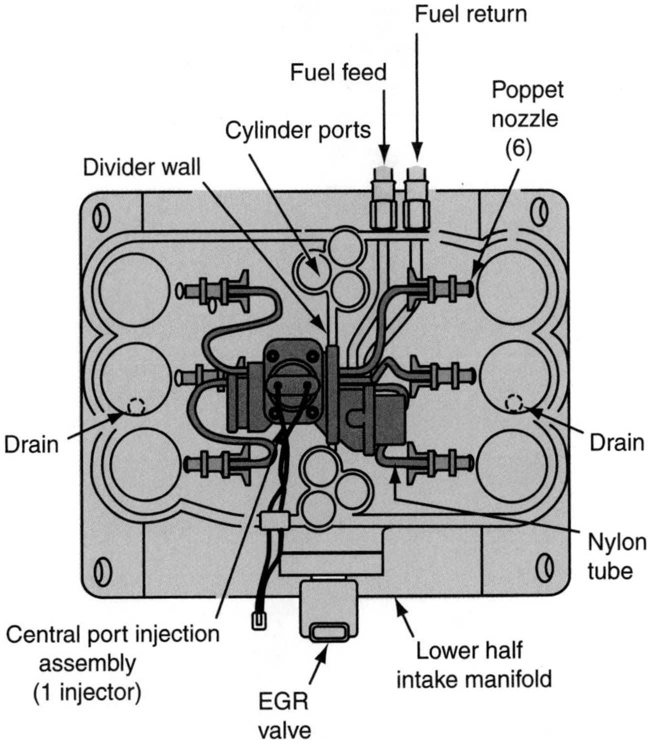

Figure 28-24 Central multiport fuel injection components in the lower half of the intake manifold.

Pressure Regulator

The pressure regulator is mounted with the central injector. Because this regulator is mounted inside the intake manifold, vacuum from the intake is supplied through an opening in the regulator cover to the regulator diaphragm. The regulator spring pushes downward on the diaphragm and closes the valve. Fuel pressure from the in-tank fuel pump pushes the diaphragm upward and opens the valve, which allows fuel to flow through this valve and the return line to the fuel tank **(Figure 28–25)**.

The pressure regulator is designed to regulate fuel pressure to 54 to 64 psi (370 to 440 kPa), which is higher than many port fuel injection systems. Higher pressure is required in the CMFI system to prevent fuel vaporization from the extra heat encountered with the CMFI assembly, poppet nozzles, and lines mounted inside the intake manifold. The pressure regulator operates the same as the regulators explained previously in this chapter.

Injector Design and Operation

A pivoted armature is mounted under the injector winding in the central injector. The lower side of this armature acts as a valve that covers the six outlet ports to the nylon tubes and poppet nozzles. A supply of fuel at a constant pressure surrounds the injector armature while the ignition switch is on. Each time the PCM grounds the injector winding, the armature is lifted upward, which opens the injector ports. Under this condition, fuel is forced from the nylon tubes to the poppet nozzles **(Figure 28–26)**.

The amount of fuel delivered by the central injector is determined by the length of time that the PCM keeps the injector winding grounded. This time is referred to as pulse width. When the PCM opens the injector ground circuit, the injector spring pushes the armature downward and closes the injector ports. The injector winding has low resistance, and the PCM operates the injector with a peak-and-hold current. When the PCM grounds the injector winding, the current flow in this circuit increases rapidly to 4 amperes. When the current flow reaches this value, a current-limiting circuit in the PCM limits the current flow to 1 ampere for the remainder of the injector pulse width. The peak-and-hold function provides faster injector armature opening and closing.

Poppet Nozzles

The poppet nozzles are snapped into openings in the lower half of the intake manifold, and the tip of each nozzle directs fuel into an intake port. Each poppet nozzle contains a valve with a check ball seat in the tip of the nozzle. A spring holds the valve and check ball seat in the closed position. When fuel pressure is applied from the central injector through the nylon lines to the poppet nozzles, this

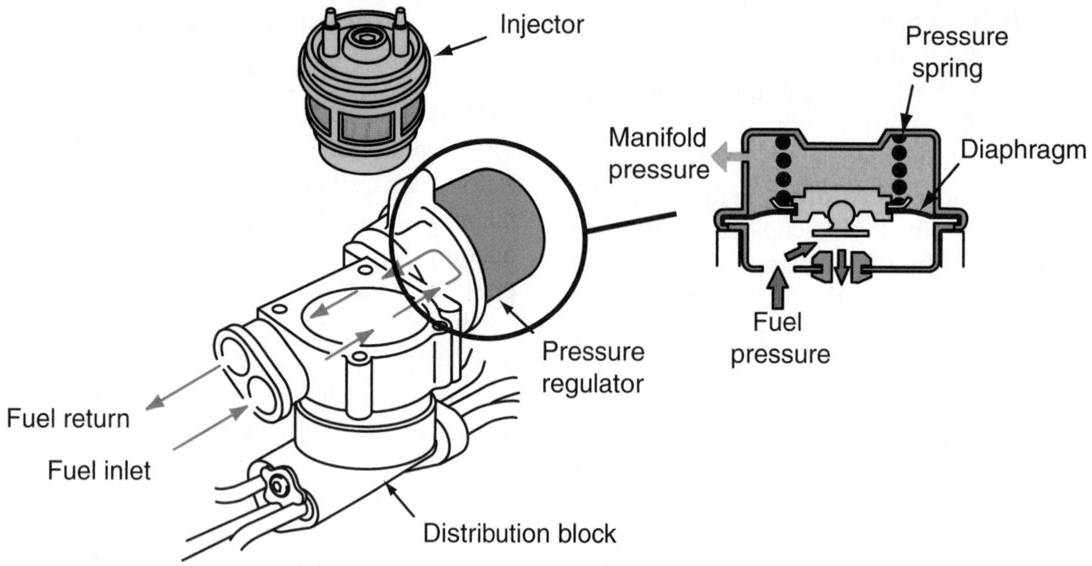

Figure 28-25 A pressure regulator for a CMFI system.

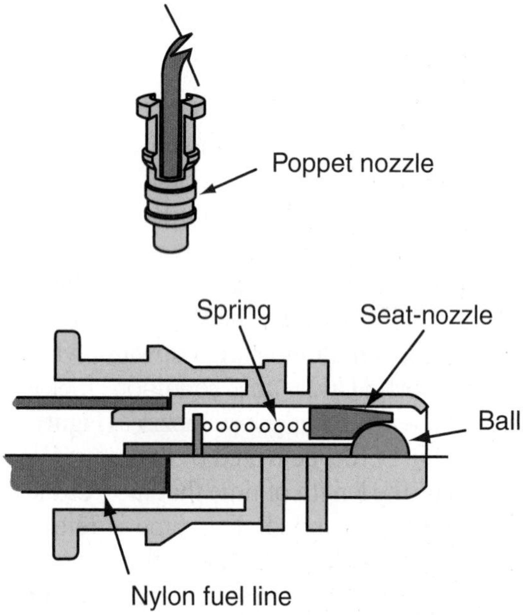

Figure 28-26 The internal design of a poppet nozzle.

pressure forces the valve and check ball seat open against spring pressure. The poppet nozzles open when the fuel pressure exceeds 37 to 43 psi (254 to 296 kPa), and the fuel sprays from these nozzles into the intake ports.

When fuel pressure drops below this value, the poppet nozzles close. Under this condition, approximately 40 psi (276 kPa), fuel pressure remains in the nylon lines and poppet nozzles. This pressure prevents fuel vaporization in the nylon lines and nozzles during hot engine operation or hot soak periods. If a leak occurs in a nylon line or other CMFI component, fuel drains from the bottom of the intake manifold through two drain holes to the center cylinder intake ports. The in-tank fuel pump, fuel filter, lines, and fuel pump circuit used with the CMFI system are similar to those used with SFI and MFI systems.

GASOLINE DIRECT-INJECTION SYSTEMS (GDI)

Direct injection has been around for many years on diesel engines. Until recently, this type of injection system was seldom used with gasoline. With gasoline direct-injection (GDI), the gasoline is injected directly into the combustion chamber. To do this, specially designed injectors deliver the fuel into the high pressures and temperatures in the cylinders **(Figure 28–27)**. To prevent the heat from

Figure 28-27 An injector for gasoline direct injection. *Courtesy of Delphi Corporation*

igniting the fuel in the injector, the injectors are designed to completely seal after the fuel is sprayed. The injectors must also be able to spray the fuel at a much higher pressure than what is in the cylinder. If this did not happen, the fuel would not enter the cylinder; instead the cylinder's pressure would enter the injector. Remember, a high pressure always moves to a point of lower pressure.

GDI allows for very lean operation (as much as 35:1) during cruising. When the engine is operating under heavy loads, the system provides near-stoichiometric air/fuel ratios. By being able to run at such lean ratios, the engine's fuel economy is increased by nearly 30% and the emissions levels are substantially decreased.

Spraying the fuel directly into the cylinder also increases volumetric efficiency because the intake manifold and port only deliver air to the cylinders. GDI decreases an engine's tendency to knock; therefore, it allows for higher compression ratios without the need for higher octane gasoline. Both factors provide for increased horsepower and torque outputs from the engine without consuming more fuel.

With GDI, fuel can be injected at any time, not only when the intake valve is open. Also the injectors can pulse twice during the transition from the compression stroke to combustion. The two pulses promote complete combustion when the PCM senses operating conditions may prevent a complete burning of the fuel.

During most driving conditions **(Figure 28–28)**, a GDI engine operates in a very lean mode and has a stratified charge, resulting in decreased fuel consumption. While in this mode, fuel is injected late in the compression stroke right before ignition. When the engine is under a heavy load or operating at high speed, fuel is injected during the intake stroke. As a result of this action, the cylinder has a homogeneous, cooler air/fuel mixture, which reduces the chances of engine knock.

The fuel is highly pressurized (typically between 400 and 1,500 psi) when it is sprayed into the cylinder **(Figure 28–29)**. Under this pressure, the fuel arrives as a vapor. The injectors release a relatively small, precisely shaped spray of fuel around the spark plug just before ignition. This means only the area around the spark plug has air and fuel to begin combustion; the rest of the combustion chamber is filled with air or recirculated exhaust gas.

The fuel pump that delivers this high pressure to the fuel injectors is driven by the engine. That pump is fed by an in-tank electric fuel pump. A PCM controls the timing of injection and ignition for each cylinder.

Although GDI offers many advantages over port injection, there is concern over the NO_x levels produced by the engine. Because of the very lean air/fuel ratios and the high compression ratios, combustion temperatures are very high during highway cruising. Manufacturers are working on new catalytic converters and exhaust gas recirculation systems to reduce the NO_x levels so the use of GDI can become more common.

Some engines are equipped with a GDI system with moderate compression ratios and normal air/fuel mixtures (not lean). These engines realize an increase in power without the consequence of high NO_x levels.

INPUT SENSORS

The ability of the fuel injection system to control the air/fuel ratio depends on its ability to properly time the injector pulses with the compression stroke of each cylinder and its ability to vary the injector "on" time according to changing engine demands. Both tasks require the use of

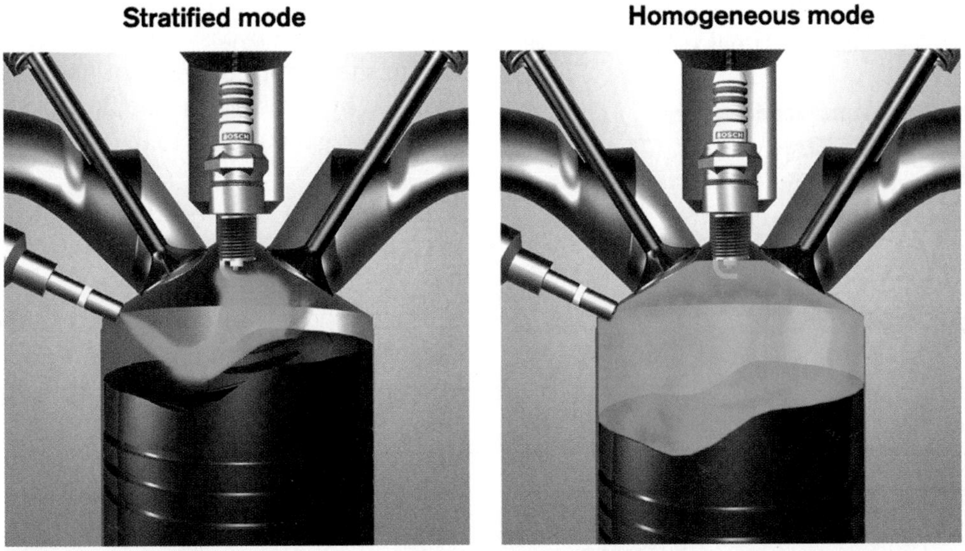

Figure 28-28 The two basic modes of operation for a GDI system. *Courtesy of Robert Bosch Corp.*

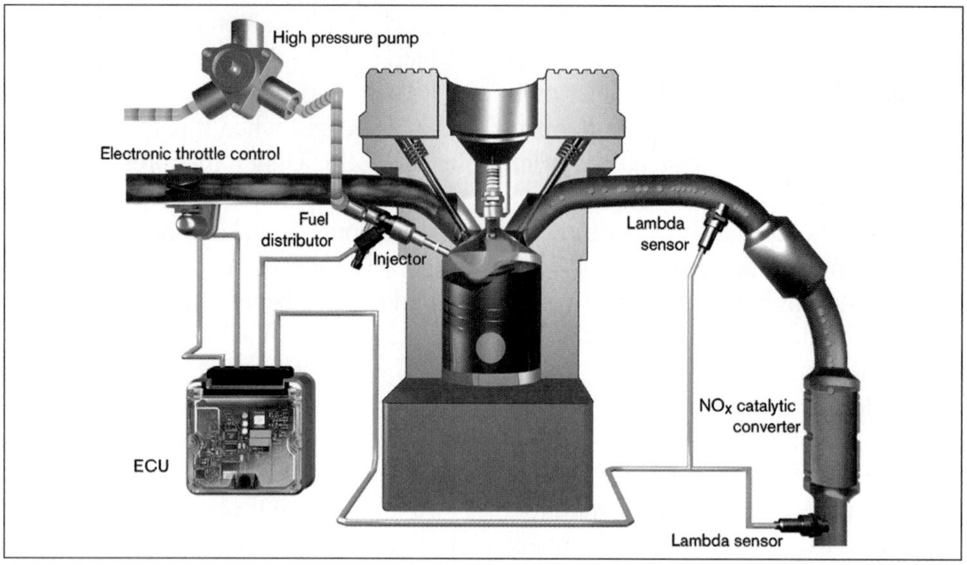

Figure 28-29 The component layout for a GDI system. *Courtesy of Robert Bosch Corp.*

electronic sensors that monitor the operating conditions of the engine. Many of the same sensors used with computer-controlled carburetors are used with EFI systems.

Airflow Sensors

In order to control the proportion of fuel to air in the air/fuel charge, the fuel system must be able to measure the amount of air entering the engine. Several sensors have been developed to do just that.

Volume Airflow Sensor The airflow sensor (commonly called an airflow meter or vane airflow sensor), shown in **Figure 28–30**, measures airflow, or air volume. The sensor consists of a spring-loaded flap, potentiometer, damping chamber, backfire protection valve, and idle bypass channel. As air is drawn into the engine, the flap is deflected against the spring. A potentiometer attached to the flap shaft monitors the flap movement and produces a corresponding voltage signal. The strength of the signal increases as the flap opens. The signal voltage is relayed to the electronic control module and may be used to control the fuel pump.

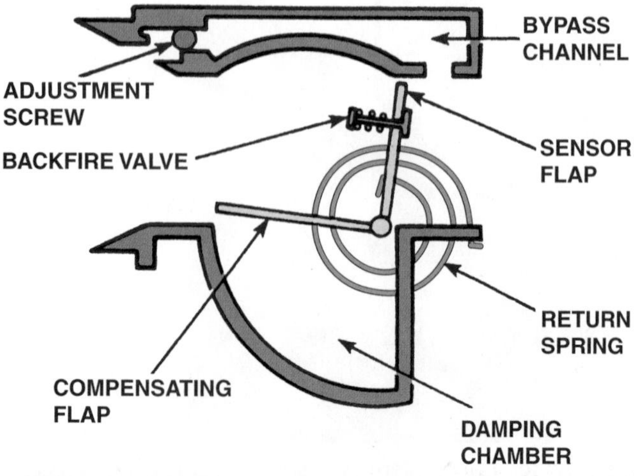

Figure 28-30 A typical airflow sensor.

The curved shape of the airflow sensor is the damping chamber. The damping flap in this chamber is on the same shaft as the airflow sensing flap and is also about the same area. As a result, the damping flap smooths out any possible pulsation caused by the opening and closing of the intake valves. Airflow measurement can be a steady signal, closely related to airflow as controlled by the movement of the flap.

The airflow sensor flap provides for backfire protection with a spring-loaded valve. If the intake manifold pressure suddenly rises because of a backfire, this valve releases the pressure and prevents damage to the system.

The airflow sensor assembly includes an extra air passage for idle, bypassing the airflow sensor plate. When the throttle is closed at idle, the opening and closing of intake valves can cause pulsation in the intake manifold. Without the idle bypass, such pulsations could cause the flap to shudder, resulting in an uneven air/fuel mixture. The idle bypass smooths the flow of the idle intake air, ensuring regular signals to the electronic control.

Karman Vortex Another design of airflow sensor, called a Karman Vortex sensor, works on a different operating principle. Air entering the airflow sensor assembly passes through vanes arranged around the inside of a tube. As the air flows through the vanes, it begins to swirl. The outer part of the swirling air exerts high pressure against the outside of the housing. There is a low-pressure area in the center that moves in a circular motion as the air swirls through the intake tube. Two pressure-sensing tubes near the end of the tube sense the low-pressure area as it moves around. An electronic sensor counts how many times the low-pressure area is sensed.

Figure 28–31 An air charge temperature sensor.

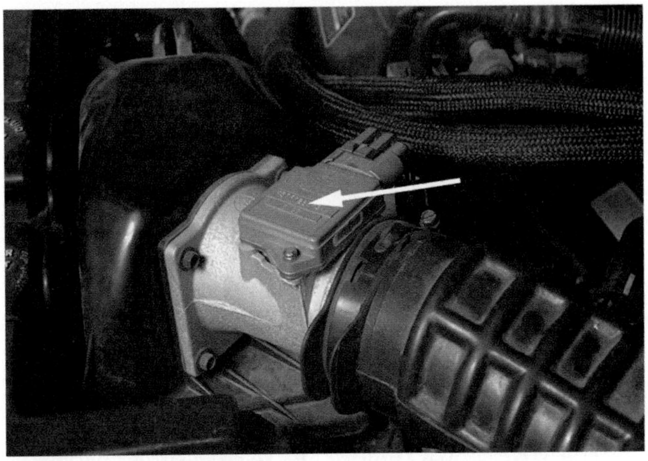

Figure 28–32 A mass airflow sensor.

The faster the airflow, the more times the low-pressure area is sensed. This is translated into a signal that indicates to the combustion control computer how much air is flowing into the intake manifold.

Air Temperature Sensor

Cold air is denser than warm air. Cold, dense air can burn more fuel than the same volume of warm air because it contains more oxygen. That is why airflow sensors that only measure air volume must have their readings adjusted to account for differences in air temperature.

Most systems do this by using an air temperature sensor mounted in the induction system **(Figure 28–31)**. The air sensor measures air temperature and sends an electronic signal to the control computer. The computer uses this input along with the air volume input in determining the amount of oxygen entering the engine.

In some early EFI systems, the incoming air is heated to a set temperature. In these systems an air temperature sensor is used to ensure that this predetermined operating temperature is maintained.

Mass Airflow Sensor

A **mass airflow sensor (Figure 28–32)** does the job of a volume airflow sensor and an air temperature sensor. It measures air mass. The mass of a given amount of air is calculated by multiplying its volume by its density. As explained previously, the denser the air, the more oxygen it contains. Monitoring the oxygen in a given volume of air is important, because oxygen is a prime catalyst in the combustion process. From a measurement of mass, the electronic control unit adjusts the fuel delivery for the oxygen content in a given volume of air. The accuracy of air/fuel ratios is greatly enhanced when matching fuel to air mass instead of fuel to air volume.

The mass airflow sensor converts air flowing past a heated sensing element into an electronic signal. The strength of this signal is determined by the energy needed to keep the element at a constant temperature above the incoming ambient air temperature. As the volume and density (mass) of airflow across the heated element changes, the temperature of the element is affected and the current flow to the element is adjusted to maintain the desired temperature of the heating element. The varying current flow parallels the particular characteristics of the incoming air (hot, dry, cold, humid, high/low pressure). The electronic control unit monitors the changes in current to determine air mass and to calculate precise fuel requirements.

There are two basic types of mass airflow sensors: hot wire and hot film. In the first type, a very thin wire (about 0.2 mm thick) is used as the heated element **(Figure 28–33)**. The element temperature is set at 100° to 200°C above incoming air temperature. Each time the ignition switch is turned to the off position, the wire is heated to

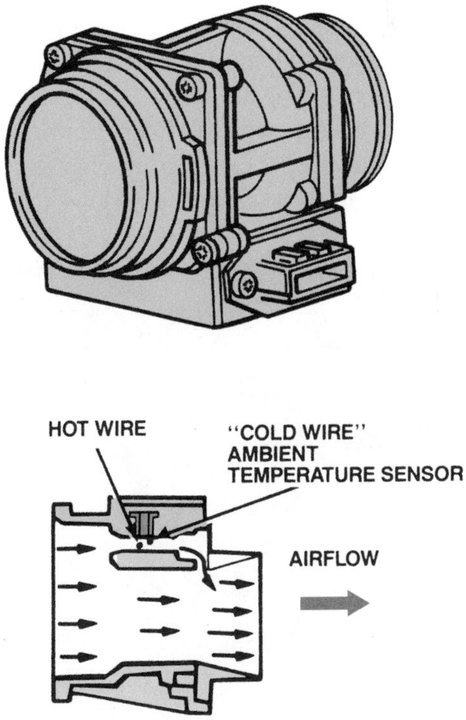

Figure 28–33 Components of a hot wire-type mass airflow sensor. *Courtesy of Ford Motor Company*

approximately 1,000°C for 1 second to burn off any accumulated dust and contaminants.

The second type uses a nickel foil sensor, which is kept 75°C above ambient air temperatures. It does not require a burn-off period and therefore is potentially longer lasting than the hot wire type.

Manifold Absolute Pressure Sensor

Some EFI systems do not use airflow or air mass to determine the base pulse of the injector(s). Instead, the base pulse is calculated on manifold absolute pressure (MAP). The MAP signal may also be used to regulate the EGR.

The MAP sensor **(Figure 28–34)** measures changes in the intake manifold pressure that result from changes in engine load and speed. The pressure measured by the MAP sensor is the difference between barometric pressure and manifold pressure. At closed throttle, the engine produces a low MAP value. A wide-open throttle produces a high value that is produced when the pressure inside the manifold is the same as pressure outside the manifold, and 100% of the outside air is being measured. This MAP output is the opposite of what is measured on a vacuum gauge. The use of this sensor also allows the control computer to adjust automatically for different altitudes.

The control computer sends a voltage reference signal to the MAP sensor. As the MAP changes, the electrical resistance of the sensor also changes. The control computer can determine the manifold pressure by monitoring the sensor output voltage. A high pressure, low vacuum requires more fuel. A low pressure, high vacuum requires less fuel. Like an airflow sensor, an MAP sensor relies on an air temperature sensor to adjust its base pulse signal to match incoming air density.

Many EFI systems with MAF sensors do not have MAP sensors. However, there are a few engines with both of these sensors. In these cases, the MAP is used mainly as a backup if the MAF fails. When the EFI system has an MAF, the computer calculates the intake air flow from the MAF and rpm inputs.

Figure 28-34 A MAP sensor.

Oxygen Sensors (O2S)

The signals from the exhaust gas oxygen sensor (O2S), or lambda sensor **(Figure 28–35)**, are used by the PCM to monitor the air/fuel mixture. The O2S is threaded into the exhaust manifold or into the exhaust pipe near the engine.

The signal from an oxygen sensor is based on the amount of oxygen in the exhaust gas. When the sensor's signal indicates a lean mixture, the computer enriches the air/fuel mixture to the engine. When the sensor reading is rich, the computer leans the air/fuel mixture.

A lean mixture produces a high level of oxygen in the exhaust and a rich mixture produces very little. Typically, if the oxygen level is high, the voltage signal from the oxygen sensor will be low. Likewise, if the oxygen level is low, the sensor's output will be high.

Because the oxygen sensor must be hot to operate properly, most late-model engines use heated oxygen sensors (HO2S). These sensors have an internal heating element that allows the sensor to reach operating temperature more quickly and to maintain its temperature during periods of idling or low engine load.

Late-model vehicles with OBD-II systems have two oxygen sensors in each exhaust system, one before the catalytic converter and one after it. The readings from the two sensors let the PCM know how efficiently the catalytic converter is working.

Other EFI System Sensors

In addition to airflow, air mass, or MAP readings, the control computer relies on input from a number of other system sensors. This input further adjusts the injector pulse width to match engine operating conditions. Operating conditions are communicated to the control computer by the following types of sensors.

Coolant Temperature The coolant temperature sensor signals the PCM when the engine needs cold enrichment, as it does during warm-up. This adds to the base pulse, but decreases to zero as the engine warms up.

Throttle Position The switches on the throttle shaft signal the PCM for idle enrichment when the throttle is closed. These same throttle switches signal the PCM when the throttle is near the wide-open position to provide full load enrichment.

Engine Speed The ignition system sends a tachometer signal reference pulse corresponding to engine speed to the electronic control unit. This signal advises the electronic control unit to adjust the pulse width of the injec-

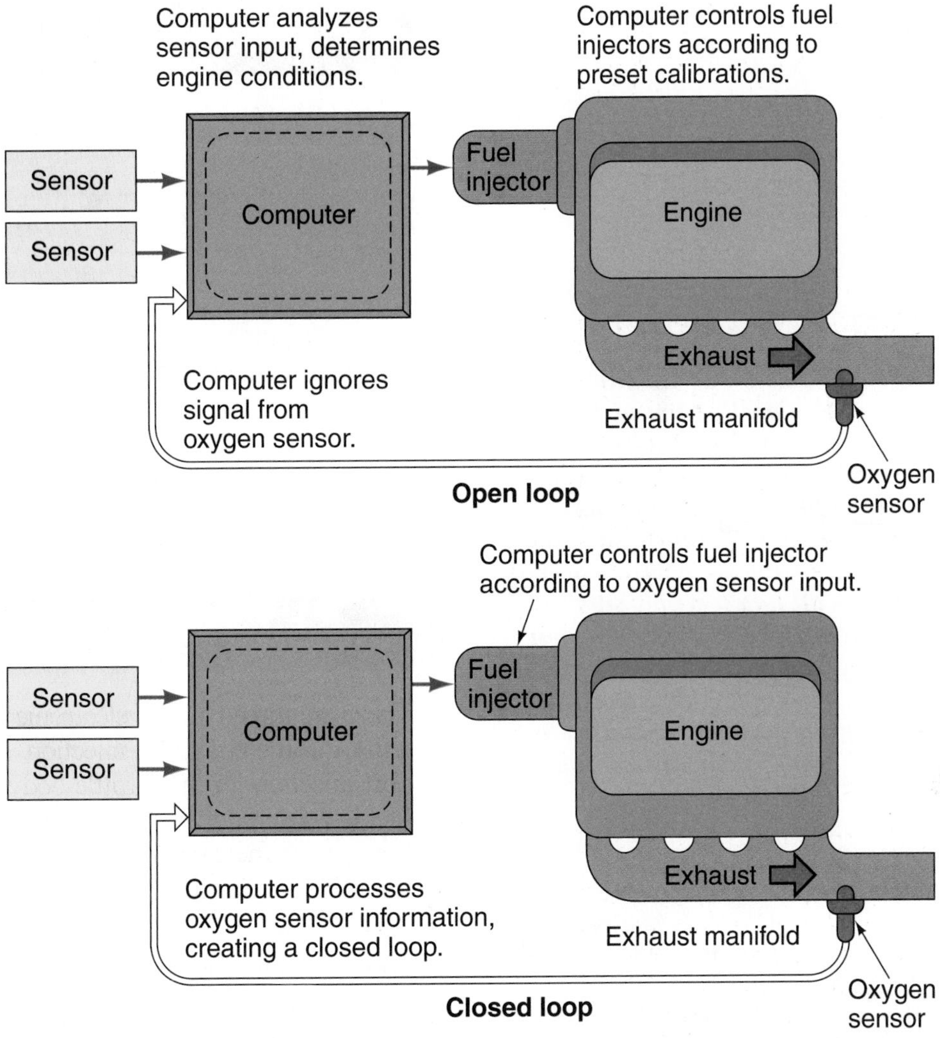

Figure 28-35 The O2S signal is disregarded when the control system is in open loop. *Courtesy of Ford Motor Company*

tors for engine speed. This also times the start of the injection according to the intake stroke cycle.

Cranking Enrichment The starter circuit sends a signal for fuel enrichment during cranking operations even when the engine is warm. This is independent of any cold-start fuel enrichment demands.

Altitude Compensation As the car operates at higher altitudes, the thinner air needs less fuel. Altitude compensation in a fuel injection system is accomplished by installing a sensor to monitor barometric pressure. Signals from the barometric pressure sensor are sent to the PCM to reduce the injector pulse width (or reduce the amount of fuel injected).

Coasting Shutoff Coasting shutoff can be found on a number of control systems. It can improve fuel economy as well as reduce emissions of hydrocarbons and carbon monoxide. Fuel shutoff is controlled in different ways, depending on the type of transmission (manual or automatic). The PCM makes a coasting shutoff decision based on a closed throttle, as indicated by the throttle position or idle switch, or based on engine speed, as indicated by the signal from the ignition coil. When the PCM detects that power is not needed to maintain vehicle speed, the injectors are turned off until the need for power exists again.

Additional Input Information Sensors Additional sensors are also used to provide the following information on engine conditions. *Note:* This list does not attempt to cover all of the sensors that are used by all manufacturers; it contains only the most common:

- Detonation
- Crankshaft position
- Camshaft position
- Air charge temperature
- Air conditioner operation

- Gearshift lever position
- Battery voltage
- Vehicle speed
- EGR valve position

CASE STUDY

A customer phoned to say that he was having his SFI Cadillac towed to the shop because the engine had stopped and would not restart. Before the technician started working on the car, he routinely checked the oil and coolant condition. The engine oil dipstick indicated the crankcase was severely overfilled with oil, and the oil had a strong smell of gasoline. The technician checked the ignition with a test spark plug and found the system to be firing normally. Of course, the technician thought the no-start problem must be caused by the fuel system and the most likely causes would be the fuel pump or filter.

The technician removed the air cleaner hose from the throttle body and removed the air filter to perform a routine check of the filter and the throttle body. The throttle body showed evidence of gasoline lying at the lower edge of the throttle bore. The technician asked a coworker to crank the engine while he looked in the throttle body. While cranking the engine, gasoline was flowing into the throttle body below the throttle plate. The technician thought this situation was impossible. An SFI system cannot inject fuel into the throttle body.

He began thinking about how fuel could be getting into the throttle body and reasoned that the fuel had to be coming through one of the vacuum hoses. Next, he thought about which vacuum hose could be the source of the fuel. He remembered that the pressure regulator vacuum hose is connected to the intake manifold. He then removed the vacuum hose from the fuel pressure regulator and placed the end of the hose into a container. When the engine was cranked, fuel squirted out of the hose. This meant the fuel pressure regulator diaphragm was allowing leaking and allowing fuel to be drawn into the manifold.

A new fuel pressure regulator was installed and the spark plugs cleaned. After this, the engine oil was changed, along with the filter. With all of that done, the engine started and ran normally.

KEY TERMS

Central multiport fuel injection (CMFI)
Central port injection (CPI)
Clear flood mode
Closed loop
Cold-start injector
Electronic fuel injection (EFI)
Feedback
Gasoline direct-injection (GDI)
Idle air control (IAC)
Mass airflow sensor
Multiport injection (MPI)
Open loop
Poppet nozzles
Port fuel injection (PFI)
Sequential fuel injection (SFI)
Speed density
Throttle body injection (TBI)

SUMMARY

- There are three types of electronic fuel injection systems: throttle body, port injection, and central multiport injection. In the throttle-body injection system, fuel is delivered to a central point. In the port injection system, there is one injector at each cylinder. Central multiport is a mixture of both throttle body and port injection.
- Port injection systems use one of four firing systems: grounded single fire, grouped double fire, simultaneous double fire, or sequential fire.
- The volume airflow sensor and mass airflow sensors determine the amount of air entering the engine. The MAP sensor measures changes in the intake manifold pressure that results from changes in engine load and speed.
- The heart of the fuel injection system is the electronic control unit. The PCM receives signals from all the system sensors, processes them, and transmits programmed electrical pulses to the fuel injectors.
- Two types of fuel injectors are currently in use: top feed and bottom feed. Top-feed injectors are used in port injection systems. Bottom-feed injectors are used in throttle-body injection systems.
- In a speed density EFI system, the computer uses the MAP or MAF and engine rpm inputs to calculate the amount of air entering the engine. The computer then calculates the required amount of fuel to go with the air entering the engine.
- In any EFI system, the fuel pressure must be high enough to prevent fuel from boiling.
- In an EFI system, the computer supplies the proper air/fuel ratio by controlling injector pulse width.

- Most computers provide a clear flood mode if a cold engine becomes flooded. Pressing the gas pedal to the floor while cranking the cold engine activates this mode.
- In an SFI system, each injector has an individual ground wire connected to the computer.
- The pressure regulator maintains the specified fuel system pressure and returns excess fuel to the fuel tank.
- In a returnless fuel system, the pressure regulator and filter assembly is mounted with the fuel pump and gauge sending unit assembly on top of the fuel tank. This pressure regulator returns fuel directly into the fuel tank.
- A central multiport injection system has one central injector and a poppet nozzle in each intake port. The central injector is operated by the PCM, and the poppet nozzles are operated by fuel pressure.
- GDI systems inject gasoline directly into the combustion chamber and allow for very lean operation.
- GDI systems use special injectors that spray the gasoline at high pressures and seal extremely well when they are not open.
- EFI systems rely on inputs from various sensors, these include airflow, air temperature, mass airflow, manifold absolute pressure, oxygen, coolant temperature, and throttle position sensors.

TECH MANUAL

The following procedures are included in Chapters 28/29 of the *Tech Manual* that accompanies this book:

1. Visually inspecting an EFI system.
2. Checking the operation of the fuel injectors on an engine.
3. Conducting an injector balance test.

REVIEW QUESTIONS

1. Explain the major differences between throttle-body fuel injection and port fuel injection systems.
2. What is meant by sequential firing of the fuel injectors?
3. Explain the purpose of the TP sensor input in a speed-density fuel injection system.
4. Describe the purpose of an ECT signal on an EFI system.
5. Explain how the computer controls the air/fuel ratio in an EFI system.
6. Describe the purpose of a manifold absolute pressure (MAP) sensor.
7. Explain the basic operation of a CMFI system.
8. Compared to TBI systems, MFI and SFI systems require ______ fuel pressure.
9. In TBI, MFI, and SFI systems, the fuel pressure must be high enough to prevent ______ ______.
10. If the injector pulse width is increased on TBI, MFI, or SFI systems, the air/fuel ratio becomes ______.
11. The length of time that an injector is energized is called ______ ______.
12. The computer determines the air entering the engine from the ______ and ______ input signals in a speed-density EFI system.
13. In EFI systems, the fuel pressure must be high enough to prevent ______ ______.
14. *True or False?* GDI systems allow engines to run at very lean air/fuel ratios and at high compression ratios and eliminate the need to have an EGR valve on the engine.
15. When the engine is idling, the pressure regulator provides ______ fuel pressure compared to the fuel pressure at wide-open throttle.
16. While discussing EFI systems, Technician A says the PCM provides the proper air/fuel ratio by controlling the fuel pressure. Technician B says the PCM provides the proper air/fuel ratio by controlling injector pulse width. Who is correct?

    a. Technician A
    b. Technician B
    c. Both A and B
    d. Neither A nor B
17. While discussing electronic fuel injection principles, Technician A says the PCM always adjusts the air/fuel ratio in response to the O2S signals. Technician B says the cold air is dense, which is why the PCM enriches the mixture when the engine is cold. Who is correct?

a. Technician A **c.** Both A and B
b. Technician B **d.** Neither A nor B

18. While discussing returnless fuel systems, Technician A says in a returnless fuel system the pressure regulator is mounted on the fuel rail. Technician B says in this type of fuel system the filter and pressure regulator are combined in one unit. Who is correct?

a. Technician A **c.** Both A and B
b. Technician B **d.** Neither A nor B

19. While discussing fuel boiling in the fuel rail, Technician A says fuel boiling in the fuel rail causes a lean air/fuel ratio. Technician B says the computer will compensate for the improper air/fuel ratio caused by fuel boiling in the fuel rail. Who is correct?

a. Technician A **c.** Both A and B
b. Technician B **d.** Neither A nor B

20. Why are special fuel injectors needed for GDI systems?

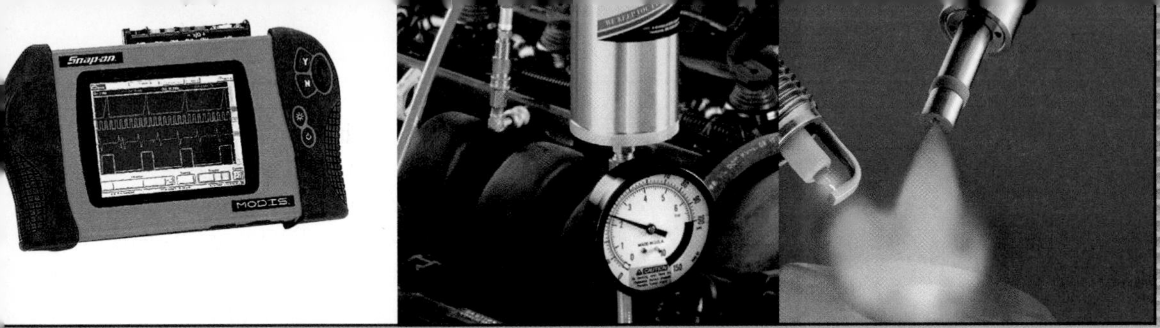

29

FUEL INJECTION SYSTEM DIAGNOSIS AND SERVICE

OBJECTIVES

■ Perform a preliminary diagnostic procedure on a fuel injection system. ■ Remove, clean, inspect, and install throttle body assemblies. ■ Explain the results of incorrect fuel pressure in a TBI, MFI, or SFI system. ■ Perform an injector balance test and determine the injector condition. ■ Clean injectors on an MFI or SFI system. ■ Perform an injector sound, ohmmeter, noid light, and scope test. ■ Perform an injector flow test and determine injector condition. ■ Perform an injector leakage test. ■ Remove and replace the fuel rail, injectors, and pressure regulator. ■ Diagnose causes of improper idle speed on vehicles with fuel injection.

Troubleshooting fuel injection systems requires systematic step-by-step test procedures. With so many interrelated components and sensors controlling fuel injection performance **(Figure 29–1)**, a hit-or-miss approach to diagnosing problems can quickly become frustrating, time-consuming, and costly.

Most fuel injection systems are integrated into engine control systems **(Figure 29–2)**. The self-test modes of these systems are designed to help in engine diagnosis. Unfortunately, when a problem upsets the smooth operation of the engine, many service technicians automatically assume that the computer (PCM) is at fault. But in the vast majority of cases, complaints about driveability, performance, fuel mileage, roughness, or hard starting or no-starting are due to something other than the computer itself (although many problems are caused by sensor malfunctions that can be traced using the self-test mode).

Before condemning sensors as defective, remember that weak or poorly operating engine components can often affect sensor readings and result in poor performance. For example, a sloppy timing chain or bad rings or valves reduce vacuum and cylinder pressure, resulting in a lower exhaust temperature. This can affect the operation of a perfectly good oxygen or lambda sensor, which must heat up to approximately 600°F (315°C) before functioning in its closed loop mode.

A problem such as an intake manifold leak can cause a sensor, the MAP sensor in this case, to adjust engine operation to less than ideal conditions.

PRELIMINARY CHECKS

The best way to approach a problem on a vehicle with electronic fuel injection is to treat it as though it had no electronic controls at all.

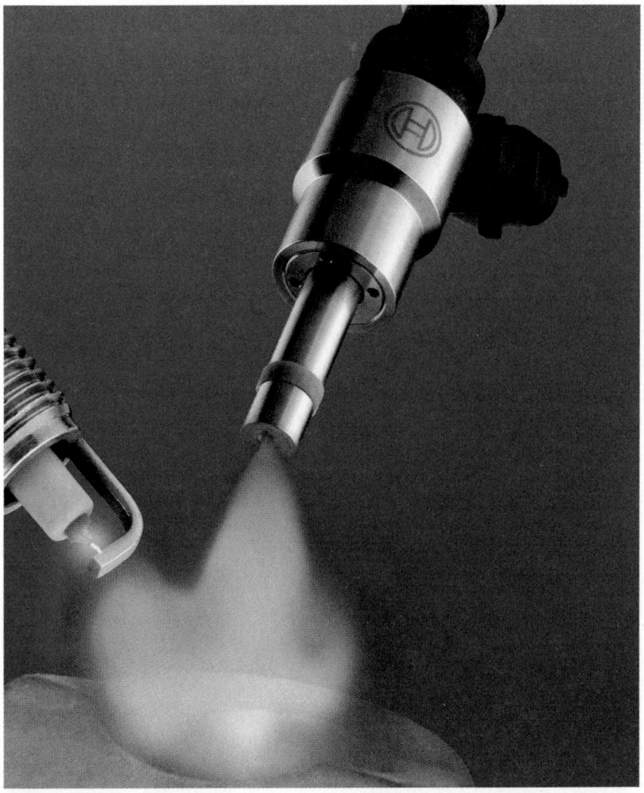

Figure 29–1 The action of a fuel injector and spark plug to start combustion. *Courtesy of Robert Bosch Corp.*

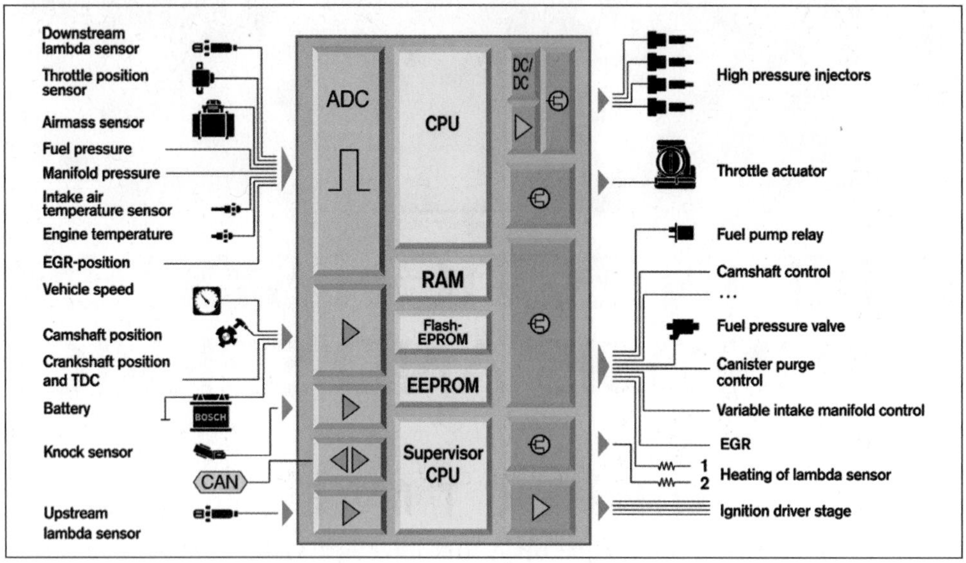

Figure 29-2 The layout of a late-model EFI system. *Courtesy of Robert Bosch Corp.*

Before proceeding with specific fuel injection checks and electronic control testing, be certain of the following:

- That the battery is in good condition, fully charged, with clean terminals and connections.
- That the charging and starting systems are operating properly.
- That all fuses and fusible links are intact.
- That all wiring harnesses are properly routed with connections free of corrosion and tightly attached **(Figure 29–3)**.

Figure 29-3 All underhood wiring should be carefully inspected.

- That all vacuum lines are in sound condition, properly routed, and tightly attached.
- That the PCV system is working properly and maintaining a sealed crankcase.
- That all emission control systems are in place, hooked up and operating properly.
- That the level and condition of the coolant/antifreeze is good and the thermostat is opening at the proper temperature.
- That the secondary spark delivery components are in good shape with no signs of crossfiring, carbon tracking, corrosion, or wear.
- That the engine is in good mechanical condition.
- That the gasoline in the tank is of good quality and has not been substantially cut with alcohol or contaminated with water.

EFI System Component Checks

In any electronic throttle body or port injection system three things must occur for the system to operate:

1. An adequate amount of air must be supplied for the air/fuel mixture.
2. A pressurized fuel supply must be delivered to properly operating injectors.
3. The injectors must receive a trigger signal from the control computer, which must first receive an rpm signal from the ignition.

SERVICE PRECAUTIONS

These precautions must be observed when electronic fuel injection systems are diagnosed and serviced:

- Always relieve the fuel pressure before disconnecting any component in the fuel system.

- Never turn on the ignition switch when any fuel system component is disconnected.
- Use only the test equipment recommended by the vehicle manufacturer.
- Always turn off the ignition switch before connecting or disconnecting any system component or test equipment.
- When arc welding is necessary on a computer-equipped vehicle, disconnect both battery cables before welding is started. Always disconnect the negative cable first.
- Never allow electrical system voltage to exceed 16 volts. This could be done by disconnecting the circuit between the alternator and the battery with the engine running.
- Avoid static electricity discharges when handling computers, modules, and computer chips.

BASIC EFI SYSTEM CHECKS

If all of the preliminary checks do not reveal a problem, proceed to test the electronic control system and fuel injection components. Some older control systems require involved test procedures and special test equipment, but most newer designs have a self-test program designed to help diagnose the problem. These self-tests perform a number of checks on components within the system. Input sensors, output devices, wiring harnesses, and even the computer itself may be among the items tested.

The results of the testing are converted into trouble codes that the technician may read using a scan tool or the **malfunction indicator lamp (MIL)**. The meanings of trouble codes vary from manufacturer to manufacturer, year to year, and model to model, so it is important to have the appropriate service manuals.

Always remember that trouble codes only indicate the particular circuit in which a problem has been detected. They do not pinpoint individual components. So if a code indicates a defective lambda or oxygen sensor, the problem could be the sensor itself, the wiring to it, or its connector. Trouble codes are not a signal to replace components. They indicate that a more thorough diagnosis is needed in that area.

The following sections outline general troubleshooting procedures for the most popular EFI designs in use today.

Oxygen Sensor Diagnosis

Oxygen sensors **(Figure 29–4)**produce a voltage based on the amount of oxygen in the exhaust. Large amounts of oxygen result from lean mixtures and result in low voltage output from the O_2 sensor. Rich mixtures release lower amounts of oxygen into the exhaust; therefore, the O_2 sensor voltage is high.

Carefully check the wiring and connectors to the oxygen sensors for damage and evidence of unwanted resistance. Also, check for intake and exhaust system leaks. These conditions would tend to create more oxygen in the exhaust and the PCM will respond by adding fuel to the mixture. A misfiring spark plug allows unburned fuel and oxygen in the exhaust, which also causes the sensor to give a false lean reading.

Before testing an O_2 sensor, refer to the correct wiring diagram to identify the terminals at the sensor. Most late-model engines use heated oxygen sensors (HO2S). These

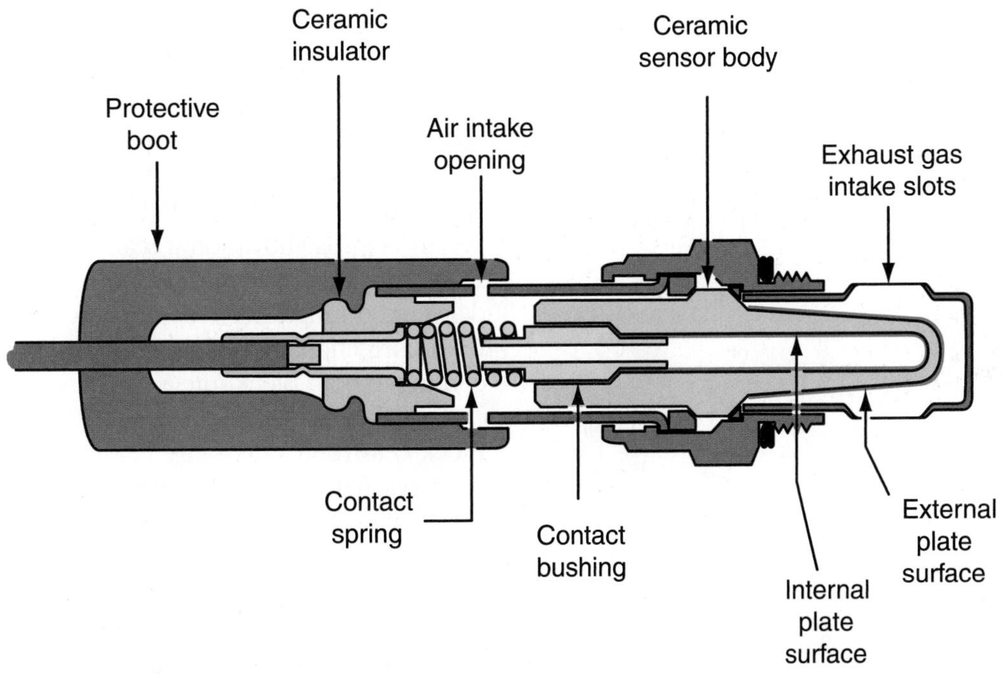

Figure 29-4 The components of an oxygen sensor.

sensors have an internal heater that helps to stabilize the output signals. Most heated oxygen sensors have four wires connected to them. Two are for the heater and the other two are for the sensor **(Figure 29–5)**.

SHOP TALK

On some engines, O2S heater problems (and faulty O2S signals) can result from a poorly grounded exhaust system. A poor ground adds resistance to the circuit, which prevents the heater from heating the sensor sufficiently. This can cause faulty sensor output signals. Often the problem can be corrected by tightening the exhaust manifold-to-engine bolts. ■

An O_2 sensor can be checked with a voltmeter. Connect it between the O_2 sensor wire and ground. The sensor's voltage should be cycling from low voltage to high voltage. The signal from most O_2 sensors varies between 0 and 1 volt. If the voltage is continually high, the air/fuel ratio may be rich or the sensor may be contaminated. When the O_2 sensor voltage is continually low, the air/fuel ratio may be lean, the sensor may be defective, or the wire between the sensor and the computer may have a high-resistance problem. If the O_2 sensor voltage signal remains in a midrange position, the computer may be in open loop or the sensor may be defective.

If the O2S voltage signal sits at, or close to, 0, unplug the sensor. If the voltage signal increases while the sensor is being unplugged, the sensor is probably shorted to ground. If the O2S voltage signal sits at, or close to, 1 volt, check the wiring at the sensor to see if the heater power feed wire or connector is shorted to the sensor's output signal wire.

Some DaimlerChrysler engines are equipped with oxygen sensors that use a 5-volt reference signal. This means the reading on a scan tool, voltmeter, or lab scope is 5 to 6 volts instead of the typical 0 to 1 volt.

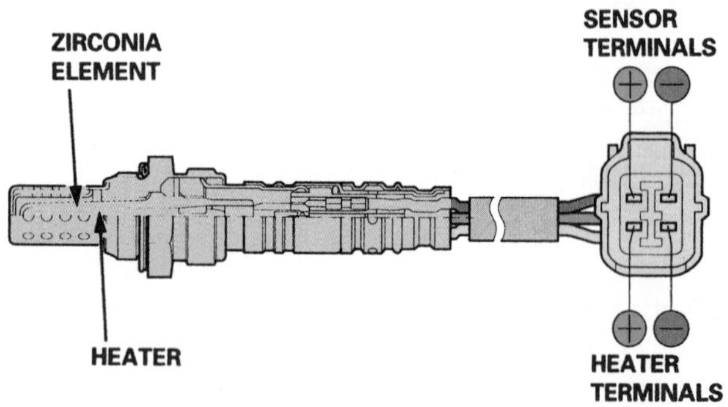

Figure 29-5 Use a wiring diagram to identify the terminals on a heated oxygen sensor. *Courtesy of American Honda Motor Co., Inc.*

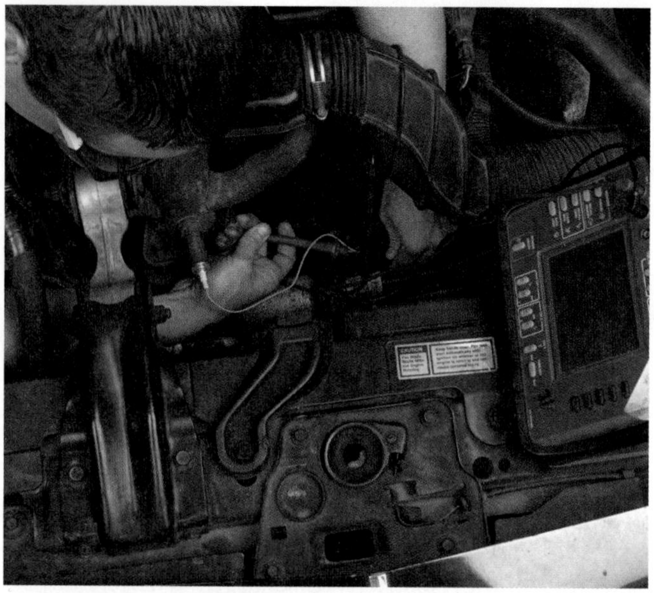

Figure 29-6 The correct way to connect a lab scope to an oxygen sensor.

The activity of an O_2 sensor is best monitored with a lab scope. The scope is connected to the sensor in the same way as a voltmeter **(Figure 29–6)**. The switching of the sensor should be seen as the sensor signal goes to lean to rich to lean continuously **(Figure 29–7)**.

The activity of the sensor can also be monitored on a scanner. While the engine is running, the scanner should show that the O_2 sensor voltage moves to nearly 1 volt then drops back to close to 0 volt. Immediately after it drops, the voltage signal should move back up. This immediate cycling is an important function of an O_2 sensor. If the response is slow, the sensor is lazy and should be replaced. With the engine at about 2,500 rpm, the O_2 sensor should cycle from high to low ten to forty times in ten seconds. When testing the O_2 sensor, make sure the sensor is heated and the system is in closed loop.

A scan tool is also an excellent way to monitor the fuel control of a fuel-injected engine. Many factors determine the pulse width of the injectors, but it should always respond to the O_2 readings. Fuel correction for fuel injection systems is shown on a scan tool. **Block Integrator** represents a short-term correction to the amount of fuel delivered during closed loop. **Block Learn** makes long-term corrections. Injector pulse width is adjusted according to both Block Integrator and Block Learn.

A scale of 0 to 255 is used for Block Integrator and Block Learn and a midrange reading of 128 is preferred. The oxygen (O_2) sensor signal is sent to the integrator chip and then to the pulse-width calculation chip in the PCM, and the block learn chip is connected parallel to the integrator chip. If the O_2 sensor voltage changes once, the integrator chip and the pulse-width calculation chip change the injector pulse width. If the O_2 sensor provides four continually high or low voltage signals, the block

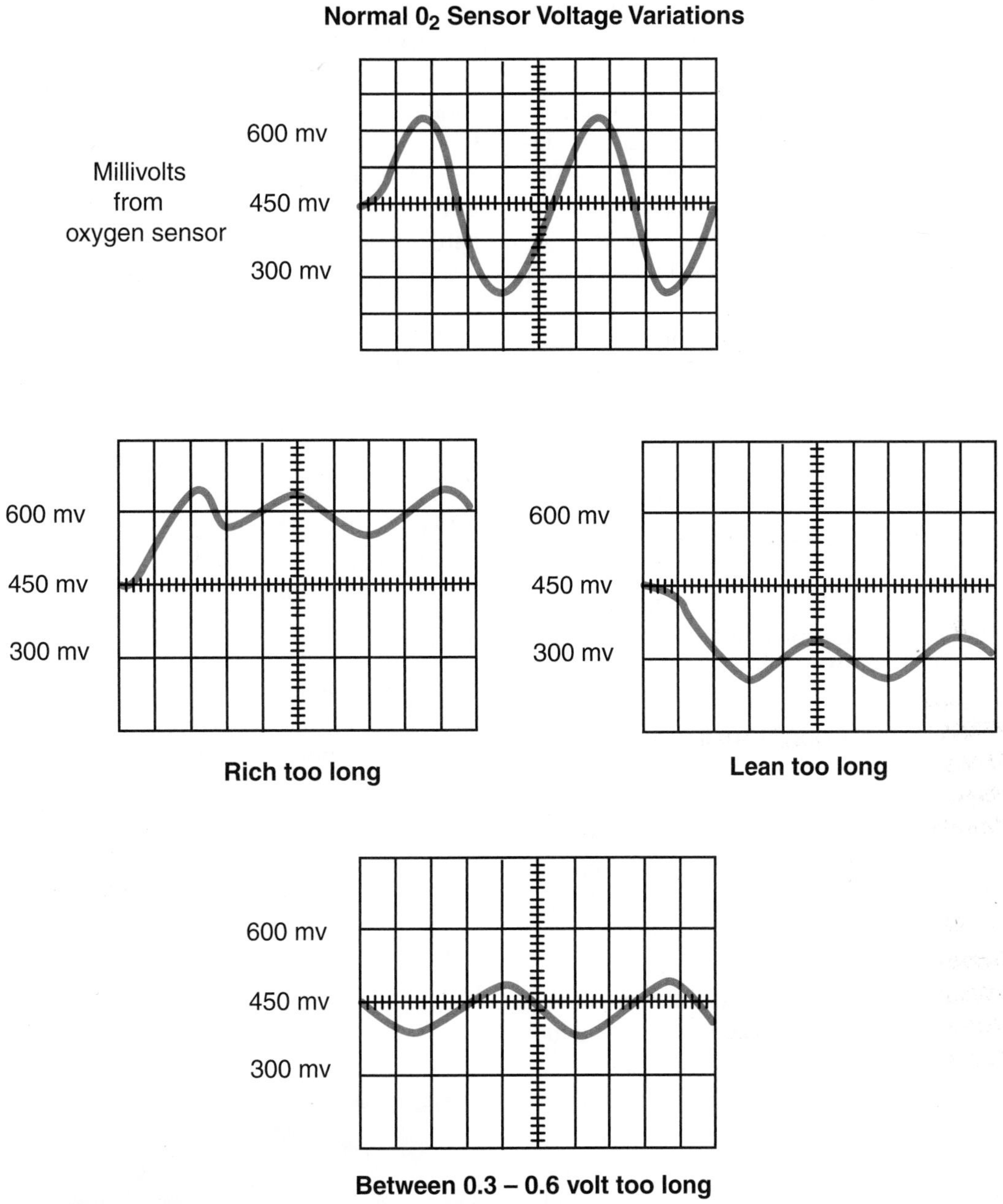

Figure 29-7 Normal and abnormal O_2 sensor waveforms.

learn chip makes a further injector pulse width change. When the integrator, or block learn, numbers are considerably above 128, the PCM is continually attempting to increase fuel; therefore, the O_2 sensor voltage signal must be continually low, or lean. If the integrator, or block learn, numbers are considerably below 128, the PCM is continually decreasing fuel, which indicates that the O_2 sensor voltage must be always high, or rich.

OBD-II Adaptive Fuel Control Strategy

OBD-II systems continuously check the fuel system with its comprehensive and misfire monitors. OBD-II's **short-term fuel trim (STFT)** and **long-term fuel trim (LTFT)** strategies monitor the oxygen sensor signals. The information gathered by the PCM is used to make adjustments to the fuel control calculations. The adaptive fuel control strategy allows for changes in the amount of fuel delivered to the cylinders according to operating conditions. The STFT and LTFT are similar in operation to the block learn and integrator.

During open loop, the PCM changes pulse width without any feedback from the O2S and the short-term adaptive memory value is 1. The number 1 represents a 0% change. Once the engine warms up, the PCM moves into closed loop and begins to recognize the signals from the O2S. The system remains in closed loop until the engine stops, unless the throttle is fully opened or the engine's temperature drops below a specified limit, when it then goes into open loop.

When the system is in open loop, the injectors operate at a fixed base pulse width. During closed loop operation, the pulse width is either lengthened or shortened. As the voltage from the oxygen sensor increases in response to a rich mixture, the short-term fuel trim decreases, which means the pulse width is shortened. Decreases in STFT are indicated on a scan tool as a number below 1. For example, the short-term adaptive value of 0.75 means the pulse width was shortened by 25% and the percent of change on the scan tool will be –25. A short-term adaptive value of 1.25 means the pulse width was lengthened by 25%. The latter will be displayed as +25.

Once the engine reaches a specified temperature (normally 180°F), the PCM begins to update the LTFT. The adaptive setting is based on engine speed and the STFT. If the STFT moves 3% and stays there for a period of time, the PCM adjusts the LTFT. The LTFT becomes a new but temporary base. In other words, the LTFT changes the length of the pulse width that is being changed by the STFT. STFT works to bring LTFT close to 0% correction.

If a lean condition exists because of a vacuum leak or restricted fuel injectors, the LTFT will have a plus number on the scan tool. If the injectors leak or the fuel pressure regulator is faulty, there will be a rich condition. The rich condition will be evident by minus LTFT numbers on the scan tool. If the engine's condition is too far toward either the lean or rich side, the LTFT will not compensate and a DTC will be set.

High LTFT readings indicate there may be a dirty MAF, clogged injectors, a clogged fuel filter, or a bad fuel pump.

On an OBD-II vehicle, zero is the midpoint of the fuel strategy during closed loop. Ford's fuel cells are illustrated as a percentage: numbers without a minus sign indicate fuel is being added and numbers with a minus sign indicate fuel is being subtracted. The constant change or crossing above and below the zero line indicates proper system operation. If the STFT readings are constantly on either side of the zero line, the engine is not operating efficiently.

Some intermittent fuel-related driveability problems may be diagnosed by recording the computer's data stream. Before you jump into the data stream, look at the STFT and LTFT.

To diagnose the fuel control system, connect the scan tool and start the engine. Then pull up the PCM's data stream display on the scan tool. Observe the STFT and note the value. With the engine running, pull off a large vacuum hose and watch for the STFT value to rise above zero to as high as 33%. Now look at the LTFT value; it should have risen above zero as it learned the engine's condition. A smaller vacuum leak must be made to cause the STFT to return to zero as the result of LTFT compensation.

Since vee-type engines have two sets of oxygen sensors, a pair of STFT and LTFT will appear on the scan tool. Use them to isolate the side of the engine that is running rich or lean by simply observing the codes. A high LTFT on one bank indicates there is a problem in that bank, possibly a vacuum leak or EGR problem.

Air Induction System Checks

In a fuel injection system (particularly designs that rely on airflow meters or mass airflow sensors), all the air entering the engine must be accounted for by the air measuring device. If it is not, the air/fuel ratio becomes overly lean. For this reason, cracks or tears in the plumbing between the airflow sensor and throttle body are potential air leak sources that can affect the air/fuel ratio.

During a visual inspection of the air control system, pay close attention to these areas, looking for cracked or deteriorated ductwork **(Figure 29–8)**. Also make sure all induction hose clamps are tight and properly sealed. Look for possible air leaks in the crankcase, for example, around the dipstick tube and oil filter cap. Any extra air entering the intake manifold through the PCV system is not measured either and can upset the delicately balanced air/fuel mixture at idle.

It is important to note that vacuum leaks may not affect the operation of engines fitted with a speed density fuel injection system. This does not mean that vacuum leaks are okay, it just means that the operating system may be capable of adjustments that allow the engine to run well in spite of the vacuum leak. This is true for vacuum

Figure 29-8 Carefully inspect the ductwork and hoses of the air induction system.

leaks that are common to all cylinders. If the vacuum leak affects one or two cylinders, the computer cannot compensate for the unmetered air and those two cylinders will not operate efficiently.

Airflow Sensors

When looking for the cause of a performance complaint that relates to poor fuel economy, erratic performance, hesitation, or hard starting, make the following checks to determine if the airflow sensor is at fault. Detailed testing of these and other sensors is included in Chapter 34.

Mass Airflow Sensors Both mass airflow (MAF) and volume airflow sensors measure intake air before it reaches the throttle plate. If any air bypasses these sensors and enters the combustion chambers without being measured, the engine will run lean. An intake manifold vacuum leak or a leak between the sensors and the throttle plate also reduces fuel delivery. The resulting lean mixture causes severe driveability problems and the PCM will store DTCs, indicating excessive fuel corrections and/or lean misfires.

If a bad MAF is suspected as causing a lean condition, remove the MAF and inspect the hot wire for signs of debris. This wire can be cleaned by gently wiping it with a dry cotton swab. Never soak or immerse the MAF sensor in any cleaner. Some manufacturers recommend that the sensor be replaced if the wire is dirty.

SHOP TALK

To quickly diagnose an intermittent failure of GM's hot film MAF meter, start the engine, let it idle, and lightly tap on the sensor with a plastic mallet or screwdriver handle. If the engine stalls or runs worse, or if the idle quality improves while tapping on the sensor, the MAF meter is probably defective and should be replaced. Similarly, if the engine does not start or idles poorly, unplug the MAF. If the engine starts or runs better with the sensor unplugged, the sensor is defective and should be replaced. ■

Volume Airflow Sensors To check a volume airflow (VAF) sensor, remove the air intake duct from the sensor to gain access to the sensor's flap. Check it for binding, sticking, or scraping by rotating the sensor flap (evenly and carefully) through its operating range. It should move freely, make no noise, and feel smooth. If it does not, a quick spray of carburetor cleaner may loosen it up. If that does not help, replace the sensor.

On some systems, the movement of the sensor flap turns on the fuel pump. If the system is so equipped, turn the ignition on. Do not start the engine. Move the flap toward the open position. The electric fuel pump should come on as the flap is opened. If it does not, turn off the ignition, remove the sensor harness, and check for specific resistance values with an ohmmeter at each of the sensor's terminals.

To check a VAF, connect a voltmeter or scope across its terminals. Watch the meter as you move the flap to its full open position, and then slowly release it. A good VAF should produce a smooth change on the meter or scope screen. If the voltage sweeps are not smooth, the VAF should be replaced.

On most VAFs, it is possible to check the resistance values of the potentiometer by moving the air flap, but in either case a service manual is needed to identify the various terminals and to look up resistance specifications. Readings that are out of range indicate the VAF should be replaced.

Karman Vortex Sensors Karman Vortex sensor problems typically result in surging, hesitation, stalling, and increased emissions levels. These sensors should be carefully inspected as most problems are caused by loose or corroded electrical connectors.

The best way to check a vortex-type airflow sensor is by observing its frequency signal on a scope or a digital meter. The output of Karman Vortex sensors is typically two separate signals: an airflow frequency signal and an air temperature or barometric pressure voltage signal. The airflow signal should be a digital signal that cycles between 0 and 5 volts. The frequency of the signal should increase as airflow increases. Its frequency should increase smoothly and steadily with an increase in engine speed.

Speed Density (MAP) Systems Speed density systems rely heavily on manifold pressure readings when they are making their fuel calculations. Increased manifold pressure increases pulse width, regardless of the cause of the pressure increase (open throttle, vacuum leak, recirculated exhaust gas, or exhaust pressure).

MAP systems measure manifold pressure after the throttle plate; therefore, their response is opposite to that of MAFs and VAFs. Air leaks in speed density systems increase injector pulse width if there is a vacuum leak. This can be verified by watching the IAC counts on a scan tool. Low IAC counts typically indicate a vacuum leak in MAP systems because the vacuum leak is causing a lower pressure in the manifold. The PCM, in turn, increases the amount of fuel to the air, which causes the idle speed to increase. Observing this, the PCM attempts to reduce the idle speed by closing the IAC.

Throttle Body

The throttle body **(Figure 29–9)** allows the driver to control the amount of air that enters the engine, thereby controlling the speed of the engine. Each type of throttle

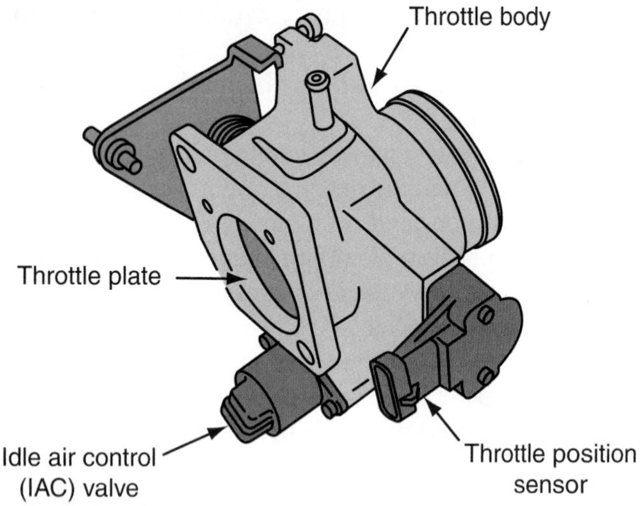

Figure 29-9 A throttle body assembly.

body assembly is designed to allow a certain amount of air to pass through it at a particular amount of throttle opening. If anything accumulates on the throttle plates or in the throttle bore, the amount of air that can pass through is reduced. This normally causes an idle problem.

These deposits can be cleaned off the throttle assembly and the airflow through them restored. Begin by removing the air duct from the throttle assembly; that will give access to the plate and bore. The deposits can be cleaned with a spray cleaner or wiped off with a cloth. If either of these cleaning methods does not remove the deposits, the throttle body should be removed, disassembled, and placed in an approved cleaning solution.

A pressurized can of throttle body cleaner may be used to spray around the throttle area without removing and disassembling the throttle body. The throttle assembly can also be cleaned by soaking a cloth in carburetor solvent and wiping the bore and throttle plate to remove light to moderate amounts of carbon residue. Also, clean the backside of the throttle plate. Then, remove the idle air control valve from the throttle body (if so equipped) and clean any carbon deposits from the pintle tip and the IAC air passage.

If on the scan tool you observed IAC counts around or above 40, these should drop down to about 15 after cleaning, which explains why you may need to adjust the minimum idle air setting. When doing this adjustment, always follow the specified procedures.

If the intake manifold setup has diaphragms or solenoids that control the selection of manifold runners, make sure you allow some of the cleaning solution to be drawn in by the engine while it is running in order to clean the valves, sometimes called *butterflies*, controlled by the solenoids or diaphragms. Dirty switchover valves can cause hard starting, poor performance, increased oil consumption, and DTCs.

Throttle Body Inspection Throttle body inspection and service procedures vary widely depending on the year and make of the vehicle. However, some components such as the TP sensor are found on nearly all throttle bodies. Since throttle bodies have some common components, inspection procedures often involve checking common components.

PROCEDURE

Throttle Body Inspection

STEP 1 *Check for smooth movement of the throttle linkage from idle position to the wide-open position.*

STEP 2 *Check the throttle linkage and cable for wear and looseness.*

STEP 3 *Check the vacuum at each vacuum port on the throttle body while the engine is idling and while it is running at a higher speed.*

STEP 4 *Apply vacuum to the throttle opener. Then, disconnect the TP sensor connector, and test the TP sensor with an ohmmeter connected across the appropriate terminals.*

STEP 5 *Loosen the two TP sensor mounting screws* ***(Figure 29–10)*** *and rotate the TP sensor as required to obtain the specified ohmmeter readings. Then retighten the mounting screws. If the TP sensor cannot be adjusted to obtain the proper ohmmeter readings, replace the TP sensor.*

STEP 6 *Operate the engine until it reaches normal operating temperature, and check the idle speed on a tachometer. The idle speed should be 700 to 800 rpm.*

STEP 7 *Disconnect and plug the vacuum hose from the throttle opener. Maintain 2,500 engine rpm.*

STEP 8 *Release the throttle valve and observe the tachometer reading. When the throttle linkage strikes the throttle opener stem, the engine speed should be between 1,300 and 1,500 rpm. Adjust the throttle opener, as necessary, and reconnect the throttle opener vacuum hose.*

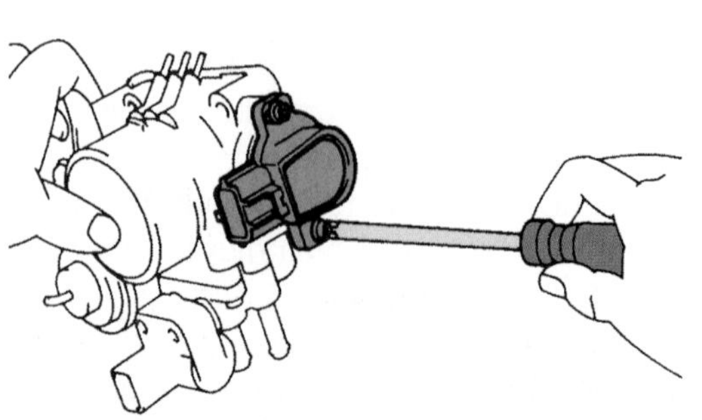

Figure 29-10 Loosen the TP sensor's mounting screws to adjust the sensor. *Reprinted with permission*

Throttle Body Removal and Cleaning Whenever it is necessary to remove the throttle body assembly for replacement or cleaning, make sure you follow the procedures outlined by the manufacturer. Also begin by connecting a 12-volt power supply (if available) to the cigarette lighter socket and disconnect the negative battery cable. If the vehicle is equipped with an air bag, wait one minute. Then, remove the throttle body according to recommendations. Once the assembly has been removed, remove all nonmetallic parts such as the TP sensor, IAC valve, throttle opener, and the throttle body gasket from the throttle body. Now it is safe to clean the throttle body assembly in the recommended throttle body cleaner and blow dry with compressed air. Blow out all passages in the throttle body assembly.

Before reinstalling the throttle body assembly, make sure all metal mating surfaces are clean and free from metal burrs and scratches. Make sure you have new gaskets and seals for all sealing surfaces before you begin to reinstall the assembly. After everything that was disconnected is reconnected, reconnect the negative battery cable and disconnect the 12-volt power supply.

Fuel System Checks

If the air control system is in working order, move on to the fuel delivery system. It is important to always remember that fuel injection systems operate at high fuel pressure levels. This pressure must be relieved before any fuel line connections can be broken. Spraying gasoline (under a pressure of 35 psi [241 kPa] or more) on a hot engine creates a real hazard when dealing with a liquid that has a flash point of –45°F (–7°C).

Follow the specific procedures given in the service manual when relieving the pressure in the fuel lines.

CAUTION!

Dispose of the fuel-soaked towel or rag by placing it in a fireproof container.

Fuel Delivery When dealing with an alleged fuel complaint that is preventing the vehicle from starting, the first step (after spark, compression, and so forth have been verified) is to determine if fuel is reaching the cylinders (assuming there is gasoline in the tank). Checking for fuel delivery is a simple operation on throttle body systems. Remove the air cleaner, crank the engine, and watch the injector for signs of a spray pattern. If a better view of the injector's operation is required, an ordinary strobe light does a great job of highlighting the spray pattern.

It is impossible to visually inspect the spray pattern and volume of port system injectors. However, an accurate indication of their performance can be obtained by performing simple fuel pressure and fuel volume delivery tests. Keep in mind that fuel pressure affects the output of a fuel injector. If an injector has the same pulse rate but receives low pressure, there is less fuel; if the pressure is high, the amount of fuel is increased.

Low fuel pressure can cause a no-start or poor-run problem. It can be caused by a clogged fuel filter, a faulty pressure regulator, or a restricted fuel line anywhere from the fuel tank to the fuel filter connection.

If a fuel volume test shows low fuel volume, it can indicate a bad fuel pump or blocked or restricted fuel line. When performing the test, visually inspect the fuel for signs of dirt or moisture. These indicate that the fuel filter needs replacement.

High fuel pressure readings will result in a rich-running engine. A restricted fuel return line to the tank or a bad fuel regulator may be the problem. To isolate the cause of high pressure, relieve system pressure and connect a tap hose to the fuel return line. Direct the hose into a container and energize the fuel pump. If fuel pressure is now within specifications, the fuel return line is blocked. If pressure is still high, the pressure regulator is faulty.

Ford and other manufacturers are using returnless fuel delivery systems. Rather than a pressure regulator, they use a pressure sensor on the fuel rail **(Figure 29–11)** and

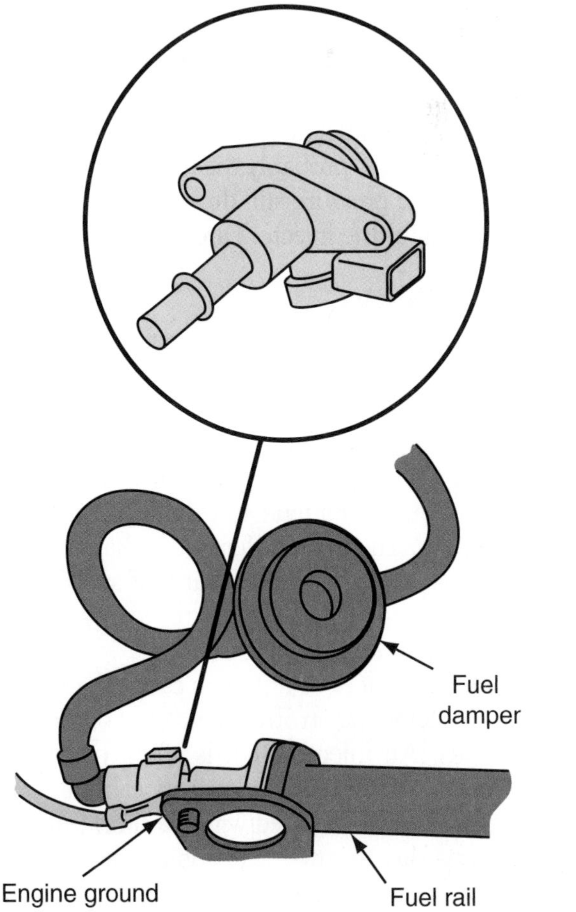

Figure 29-11 The location of a fuel pressure sensor in a fuel rail.

modulate the fuel pump's pulse width to control pressure. In these systems, overly high pressures are caused by the pressure sensor, fuel pump, PCM, or the wiring of the circuit.

If the first fuel pressure reading is within specs but the pressure slowly bleeds down, there may be a leak in the fuel pressure regulator, the fuel pump check valve, or the injectors themselves. Remember, hard starting is a common symptom of system leaks. Also, fuel starvation and lean conditions can occur when the injectors drain the fuel rail faster than the fuel pump can fill it. This could be caused by low fuel pressure or delivery volume.

One of the best ways to check for injector leaks is to conduct the injector rest test. Connect the fuel pressure gauge and run the engine until it reaches normal operating temperature. Then turn off the engine and look at the pressure gauge. The fuel pressure may increase initially after the engine is shut down because of the expansion of the fuel. If the pressure drops faster than specifications, there is a leak in the fuel system.

To find the leak, operate the fuel pump just long enough to restore normal pressure, then stop it and immediately pinch off the fuel return line between the fuel regulator and the tank. If pressure holds with the return line pinched (or plugged if recommended by the manufacturer), the leak is in the regulator or return line. If the pressure still drops quickly, remove the clamp from the line. Run the fuel pump to restore normal pressure, then immediately pinch off the supply hose between the fuel pump and the inlet on the fuel rail. If the system now maintains pressure, probably the fuel pump check valve is leaking. If the pressure still drops, there is an external leak in the rail or the injectors are leaking.

Injector Checks

A fuel injector is nothing more than a solenoid-actuated fuel valve. Its operation is quite basic in that as long as it is held open and the fuel pressure remains steady, it delivers fuel until it is told to stop.

Because all fuel injectors operate in a similar manner, fuel injector problems tend to exhibit the same failure characteristics. The main difference is that, in a TBI design, generally all cylinders will suffer if an injector malfunctions, whereas in port systems the loss of one injector will only affect one cylinder.

An injector that does not open causes hard starts on port-type systems and an obvious no-start on single-point TBI designs. An injector that is stuck partially open causes loss of fuel pressure (most noticeably after the engine is stopped and restarted within a short time period) and flooding due to raw fuel dribbling into the engine. In addition to a rich-running engine, a leaking injector also causes the engine to diesel or run on when the ignition is turned off. Buildups of gum and other deposits on the tip of an injector can reduce the amount of fuel sprayed by the injector or they can prevent the injector from totally sealing, allowing it to leak. Since injectors on MFI and SFI systems are subject to more heat than TBI injectors, port injectors have more problems with tip deposits.

Because an injector adds the fuel part to the air/fuel mixture, it is obvious that any defect in the fuel injection system will cause the mixture to go rich or lean. If the mixture is too rich and the PCM is in control of the air/fuel ratio, a common cause is that one or more injectors are leaking. An easy way to verify this on port-injected engines is to use an exhaust gas analyzer.

With the engine warmed up, but not running, remove the air duct from the airflow sensor. Then, insert the gas analyzer's probe into the intake plenum area. Be careful not to damage the airflow sensor or throttle plates while doing this. Look at the HC readings on the analyzer. They should be low and drop as time passes. If an injector is leaking, the HC reading will be high and will not drop. This test does not locate the bad injector, but does verify that one or more are leaking.

Another cause of a rich mixture is a leaking fuel pressure regulator. If the diaphragm of the regulator is ruptured, fuel will move into the intake manifold through the diaphragm, causing a rich mixture. The regulator can be checked by using two simple tests. After the engine has been run, disconnect the vacuum line to the fuel pressure regulator **(Figure 29–12)**. If there are signs of fuel inside the hose or if fuel comes out of the hose, the regulator's diaphragm is leaking. The regulator can also be tested with a hand-operated vacuum pump. Apply 5 in. Hg (127 mm Hg) to the regulator. A good regulator diaphragm will hold that vacuum.

Checking Voltage Signals. When an injector is suspected as the cause of a lean problem, the first step is to determine if the injector is receiving a signal (from the PCM) to fire. Fortunately, determining if the injector is receiving a voltage signal is easy and requires simple test equipment. Unfortunately, the location of the injector's electrical connector can make this simple voltage check somewhat difficult.

WARNING!

When performing this test, make sure to keep off the accelerator pedal. On some models, fully depressing the accelerator pedal activates the clear flood mode, in which the voltage signal to the injectors is automatically cut off. Technicians unaware of this waste time tracing a phantom problem.

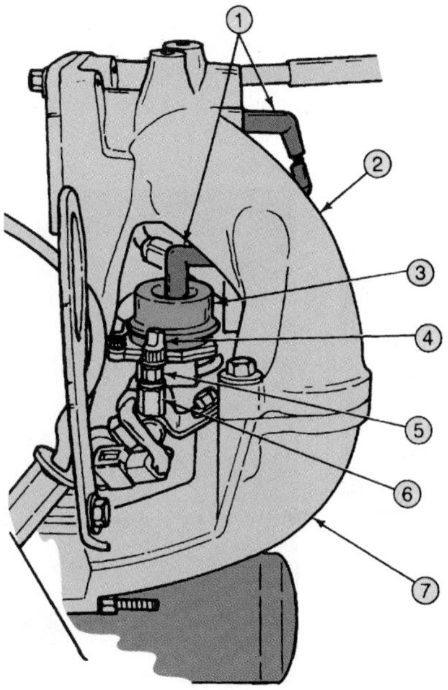

Item	Description
1	Fuel pressure regulator vacuum tube
2	Upper intake manifold
3	Fuel pressure regulator
4	Fuel pressure relief valve cap
5	Fuel pressure relief valve
6	Fuel injection supply manifold
7	Lower intake manifold

Figure 29-12 The location of a fuel pressure regulator on an in-line EFI engine. *Courtesy of Ford Motor Company*

Once the injector's electrical connector has been removed, check for voltage at the injector using a high impedance test light or a convenient **noid light** that plugs into the connector. After making the test connections, crank the engine. The noid light flashes if the computer is cycling the injector on and off. If the light is not flashing, the computer or connecting wires are defective. If sufficient voltage is present after checking each injector, check the electrical integrity of the injectors themselves.

An ohmmeter can be used to test the electrical soundness of an injector. Connect the ohmmeter across the injector terminals **(Figure 29–13)** after the wires to the injector have been disconnected. If the meter reading is infinite, the injector winding is open. If the meter shows more resistance than the specifications call for, there is high resistance in the winding. A reading that is lower than the specifications indicates that the winding is shorted. If the injector is even a little bit out of specifications, it must be replaced.

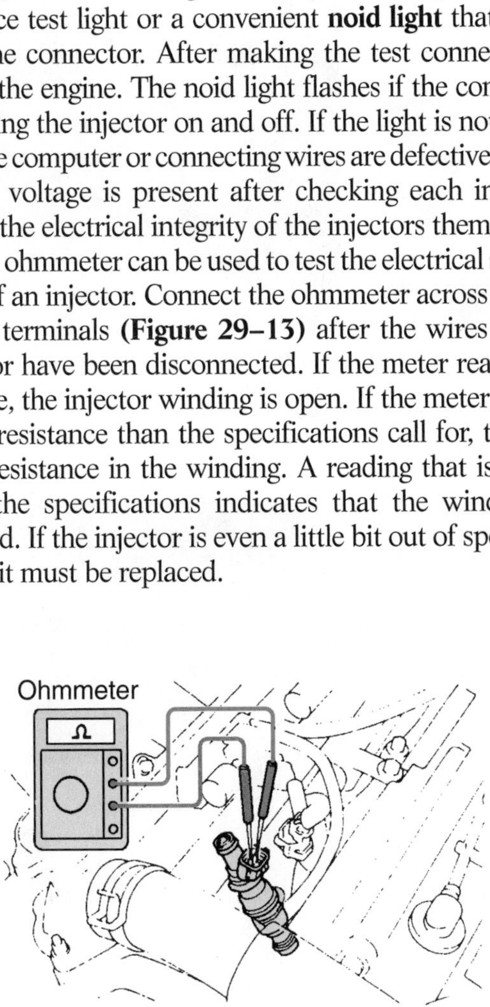

Figure 29-13 Checking an injector with an ohmmeter. *Reprinted with permission*

Injector Balance Test. If the injectors are electrically sound, perform an injector pressure balance test. This test will help isolate a clogged or dirty injector. Photo Sequence 25 shows a typical procedure for testing injector balance. An electronic injector pulse tester is used for this test. As each injector is energized, a fuel pressure gauge is observed to monitor the drop in fuel pressure. The tester is designed to safely pulse each injector for a controlled length of time. The tester is connected to one injector at a time. To prevent oil dilution, the electrical connectors to the other injectors are removed. The ignition is turned on until a maximum reading is on the pressure gauge. That reading is recorded and the ignition turned off. With the tester, activate the injector and record the pressure reading, after the needle has stopped pulsing. This same test is performed on each injector.

The difference between the maximum and minimum reading is the pressure drop. Ideally, each injector should drop the same amount when opened. Typically a variation of 1.5 to 3 psi (20 kPa) after each injector is energized is the cause for concern. If there is no pressure drop or a low pressure drop, suspect a restricted injector orifice or tip. A higher-than-average pressure drop indicates a rich condition. When an injector plunger is sticking in the open position, the fuel pressure drop is excessive. If there are inconsistent readings, the nonconforming injectors either have to be cleaned or replaced.

If the injector's orifice is dirty or otherwise restricted, there will not be much pressure decrease when the injector is energized. Stumbles during acceleration, engine stalling, and erratic idle are all caused by restricted injector orifices.

PHOTO SEQUENCE 25

Typical Procedure for Testing Injector Balance

P25-1 *Connect the fuel pressure gauge to the Schrader valve on the fuel rail, and then relieve the pressure in the system.*

P25-2 *Disconnect the number 1 injector and connect the injector pulse tester to the injector's terminals.*

P25-3 *Connect the injector pulse tester's power supply leads to the battery.*

P25-4 *Cycle the ignition switch several times until the system pressure is at the specified level.*

P25-5 *Push the injector pulse tester switch and record the pressure on the pressure gauge. Subtract this reading from the measured system pressure. The answer is the pressure drop across that injector.*

P25-6 *Move the injector tester to the number 2 injector and cycle the ignition switch several times to restore system fuel pressure.*

P25-7 *Depress the injector pulse tester's switch and observe the fuel pressure. Again the difference between the system pressure and the pressure when an injector is activated is the pressure drop across the injector.*

P25-8 *Move the injector tester's leads to the number 3 injector and cycle the ignition switch to restore system pressure.*

P25-9 *Depress the switch on the tester to activate that injector and record the pressure drop. Continue the procedure for all injectors, then compare the results of each to specifications and to each other.*

If an excessive amount of pressure drop is observed, it is likely that an injector's plunger is sticking open. A sticking injector may result in a rich air/fuel mixture.

Injector Sound Test. If the injector's electrical leads are difficult to access, an injector power balance test is hard to perform. As an alternative, start the engine and use a technician's stethoscope to listen for correct injector operation. A good injector makes a rhythmic clicking sound as the solenoid is energized and de-energized several times each second. If a clunk-clunk instead of a steady click-click is heard, chances are the problem injector has been found. Cleaning or replacement is in order. If an injector does not produce any clicking noise, the injector, connecting wires, or PCM may be defective. When the injector clicking noise is erratic, the injector plunger may be sticking. If there is no injector clicking noise, proceed with the injector resistance test and noid light test to locate the cause of the problem. If a stethoscope is not handy, use a thin steel rod, wooden dowel, or fingers to feel for a steady on/off pulsing of the injector solenoid.

Injector Flow Testing. Some vehicle manufacturers recommend an injector flow test rather than the balance test. To conduct this test, remove the injectors and fuel rail from the engine and place the tip of the injector to be tested in a calibrated container. Leave all of the injectors in the fuel rail. Then connect a jumper wire across the specified terminals in the DLC for fuel pump testing. Turn on the ignition switch and connect a jumper wire from the terminals of the injector to the battery terminals **(Figure 29–14)**. Disconnect the jumper wire from the negative battery cable after 15 seconds. Record the amount of fuel in the calibrated container.

Repeat the procedure on each injector. If the volume of fuel discharged from any injector varies more than 0.3 cu in (5 cc) from the specifications or the others, the injector should be replaced. When you have completed your testing, reconnect the negative battery cable and disconnect the 12-volt power supply.

Connect the fuel pressure gauge to the fuel system. While the fuel system is pressurized with the injectors removed from the fuel rail after the flow test, observe each

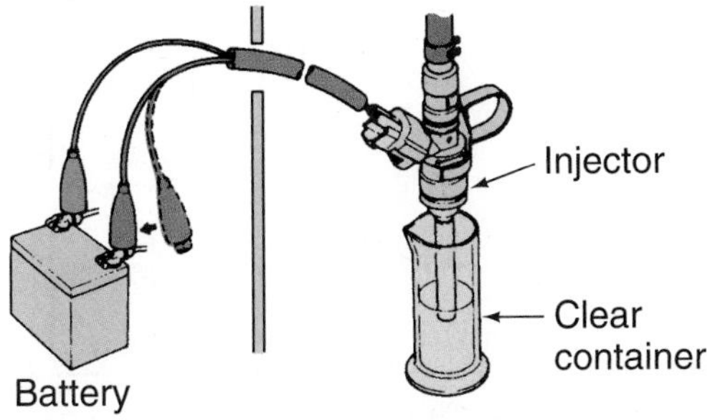

Figure 29-14 The setup for testing the flow of an injector. *Reprinted with permission*

injector for leakage from the injector tip. Injector leakage must not exceed the manufacturer's specifications. If the injectors leak into the intake ports on a hot engine, the air/fuel ratio may be too rich when a restart is attempted a short time after the engine is shut off. When the injectors leak, they drain all the fuel out of the rail after the engine is shut off for several hours. This may result in slow starting after the engine has been shut off for a longer period of time.

While checking leakage at the injector tips, observe the fuel pressure in the pressure gauge. If the fuel pressure drops off and the injectors are not leaking, the fuel may be leaking back through the check valve in the fuel pump. Repeat the test with the fuel line plugged. If the fuel pressure no longer drops, the fuel pump check valve is leaking. If the fuel pressure drops off and the injectors are not leaking, the fuel pressure may be leaking through the pressure regulator and the return fuel line. Repeat the test with the return line plugged. If the fuel pressure no longer drops off, the pressure regulator valve is leaking.

Oscilloscope Checks. An oscilloscope can be used to monitor the injector's pulse width and duty cycle when an injector-related problem is suspected. The pulse width is the time in milliseconds that the injector is energized. The duty cycle is the percentage of on-time to total cycle time.

To check the injector's firing voltage on the scope, a typical hookup involves connecting the scope's positive lead to the injector supply wire and the scope's negative lead to an engine ground.

Fuel injection signals vary in frequency and pulse width. The pulse width is controlled by the PCM, which varies it to control the air/fuel ratio. The frequency varies depending on engine speed. The higher the speed, the more pulses per second there are. Most often the injector's ground circuit is completed by a driver circuit in the PCM. All of these factors are important to remember when setting a lab scope to look at fuel injector activity. Set the scope to read 12 volts, then set the sweep and trigger to allow you to clearly see the on signal on the left and the off signal on the right. Make sure the entire waveform is clearly seen. Also remember the setting may need to be changed as engine speed increases or decreases.

Fuel injectors are either fired individually or in groups. When the injectors are fired in groups, a driver circuit controls two or more injectors. On some V-type engines, one driver fires the injectors on one side of the engine, while another fires the other side. Each fuel injector has its own driver transistor in sequential and throttle body injection. It is extremely important, while troubleshooting that you recognize how the injectors are fired. When the injectors are fired in groups, there can be a common or noncommon cause of the problem. For example, a defective driver circuit in the PCM would affect all of the injectors in a group, not just one. Conversely, if one injector in the group is not firing, the problem cannot be the driver.

To read the injector waveform on group fuel injection systems, the scope must be connected to one injector harness for each group. Since all of the injectors in the group share the same circuit, a problem in one will affect the entire waveform for the group. The only way to isolate an injector electrical problem is to disconnect the injectors, one at a time. If the waveform improves when an injector is disconnected, that injector has a problem. If the waveform never cleans up, the problem is in the driver circuit or the wiring harness.

In sequential fuel injection systems each injector has its own driver circuit and wiring. To check an individual injector, the scope must be connected to that injector. This is great for locating a faulty injector. If the scope has a memory feature, a good injector waveform can be stored and recalled for comparison to the suspected bad fuel injector pattern. To determine if a problem is the injector itself or the PCM and/or wiring, simply swap the injector wires from an injector that had a good waveform to the suspect injector. If the waveform cleans up, the wiring harness or the PCM is the cause of the problem. If the waveform is still not normal, the injector is to blame.

There are three different types of fuel injector circuits. In the conventional circuit, the driver constantly applies voltage to the injector. The circuit is turned on when a ground is provided. To control current flow through the circuit, most injectors have a built-in resistor, while others use a resistor in the circuit. The waveform for this type of injector circuit is shown in **Figure 29–15**. Notice that there is a single voltage spike at the point where the injector is turned off. The total on-time of the injector is measured from the point where the trace drops (on the left) to the point where it rises up (next to the voltage spike).

Peak and hold injector circuits use two driver circuits to control injector action. Both driver circuits complete the circuit to open the injector. This allows for high current at the injector, which forces the injector to open quickly. After the injector is open, one of the circuits turns off. The second circuit remains on to hold the injector open. This is the circuit that controls the pulse width of the injector. This circuit also contains a resistor to limit current flow during on-time. When this circuit turns off, the injector closes. When looking at the waveform for this type of circuit **(Figure 29–16)** there will be two voltage spikes. One is produced when each circuit opens. To measure the on-time of this type injector, measure from the drop on the left to the point where the second voltage spike is starting to move upward.

A **pulse-modulated injector** circuit uses high current to open the injector. Again this allows for quick injector firing. Once the injector is open, the circuit ground is pulsed on and off to allow for a long on-time without allowing high current flow through the circuit. To measure the pulse width of this type of injector **(Figure 29–17)**,

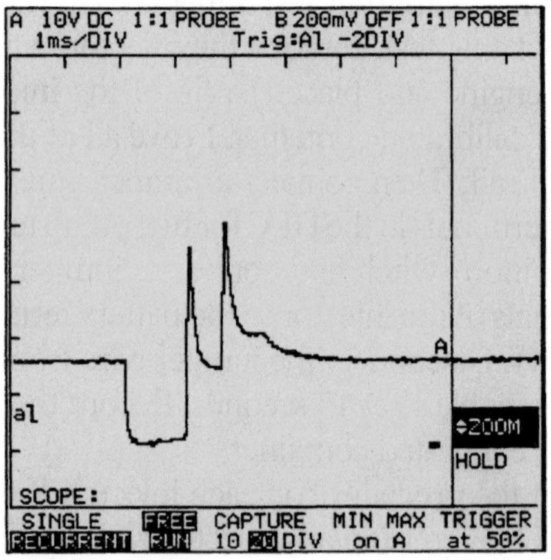

Figure 29–16 The waveform for a peak and hold fuel injector driver circuit. *Reproduced with permission from Fluke Corporation*

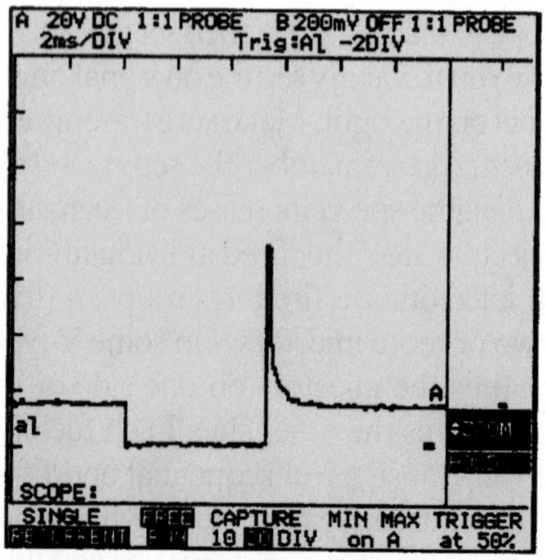

Figure 29–15 The waveform for a conventional fuel injector driver circuit. *Reproduced with permission from Fluke Corporation*

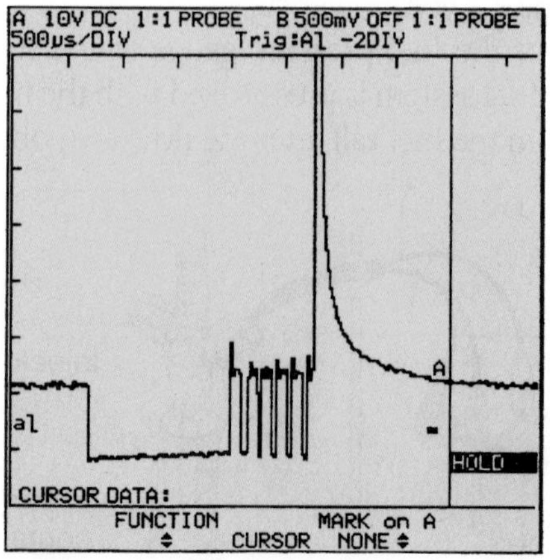

Figure 29–17 The waveform for a pulse-modulated fuel injector driver circuit. *Reproduced with permission from Fluke Corporation*

measure from the drop on the left to the beginning of the large voltage spike, which should be at the end of the pulses.

For all types of injectors, the waveform should have a clean, sudden drop in voltage when it is turned on. This drop should be close to zero volts. Typically, the maximum allowable voltage during the injector's on-time is 600 millivolts. If the drop is not perfectly vertical, either the injector is shorted or the driver circuit in the PCM is bad. If the voltage does not drop to below 600 millivolts, there is resistance in the ground circuit or the injector is shorted. When comparing one injector's waveform to another, check the height of the voltage spikes. The voltage spike of all injectors in the same engine should have approximately the same height. If there is a variance, the power feed wire to the injector with the variance or the PCM's driver for that injector is faulty.

While checking the injectors with a lab scope, make sure the injectors are firing at the correct time. To do this, use a dual trace scope and monitor the ignition reference signal and a fuel injector signal at the same time. The two signals should have some sort of rhythm between them. For example, there can be one injector firing for every four ignition reference signals **(Figure 29–18)**. This rhythm is dependent upon several things; however, it does not matter what the rhythm is, it only matters that the rhythm is constant. If the injector's waveform is fine but the rhythm varies, the ignition reference sensor circuit is faulty and not allowing the injector to fire at the correct time. If the ignition signal is lost because of a faulty sensor, the injection system will also shut down. If the injector circuit and the ignition reference circuit shut down at the same time, the cause of the problem is probably the ignition reference sensor. If the injector circuit shuts off before the ignition circuit, the problem is the injector circuit or the PCM.

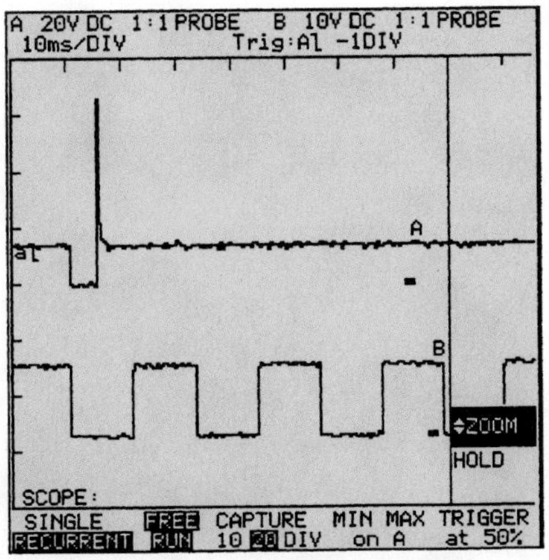

Figure 29–18 On a dual trace scope, you can compare the ignition reference signal to the injector's ON signal. *Reproduced with permission from Fluke Corporation*

INJECTOR SERVICE

Because a single injector can cost up to several hundred dollars, arbitrarily replacing injectors when they are not functioning properly, especially on multiport systems, can be an expensive proposition. If injectors are electrically defective, replacement is the only alternative. However, if the injector balance test indicated that some injectors were restricted or if the vehicle is exhibiting rough idle, stalling, or slow or uneven acceleration, the injectors may just be dirty and require a good cleaning.

Injector Cleaning

Before discussing the typical cleaning systems available and how they are used, several cleaning precautions are in order. First, never soak an injector in cleaning solvent. Not only is this an ineffective way to clean injectors, but it most likely will destroy the injector in the process. Also, never use a wire brush, pipe cleaner, toothpick, or other cleaning utensil to unblock a plugged injector. The metering holes in injectors are drilled to precise tolerances. Scraping or reaming the opening may result in a clean injector but it may also be one that is no longer an accurate fuel-metering device.

The basic premise of all injection cleaning systems is similar in that some type of cleaning chemical is run through the injector in an attempt to dissolve deposits that have formed on the injector's tip. The methods of applying the cleaner can range from single-shot, premixed, pressurized spray cans to self-mix, self-pressurized chemical tanks resembling bug sprayers. The premixed, pressurized spray can systems are fairly simple and straightforward to use since the technician does not need to mix, measure, or otherwise handle the cleaning agent.

Automotive parts stores usually sell pressurized containers of injector cleaner with a hose for Schrader valve attachment. During the cleaning process, the engine runs on the pressurized container of unleaded fuel and injector cleaner. Fuel pump operation must be stopped to prevent the pump from forcing fuel up to the fuel rail. The fuel return line is be plugged to prevent the solution in the cleaning container from flowing through the return line into the fuel tank. Disconnect the wires from the in-tank fuel pump or the fuel pump relay to disable the fuel pump. If you disconnect the fuel pump relay on General Motors products, the oil pressure switch in the fuel pump circuit must also be disconnected to prevent current flow through this switch to the fuel pump. Plug the fuel return line from the fuel rail to the tank. Connect a can of injector cleaner to the Schrader valve on the fuel rail and run the engine for about 20 minutes on the injector solution.

Other systems require the technician to assume the role of chemist and mix up a desired batch of cleaning solution for each application. The chemical solution then is

Figure 29–19 An injector cleaner connected to a fuel rail. *Courtesy of OTC Tool and Equipment, Division of SPX Corporation*

placed in a holding container and pressurized by hand pump or shop air to a specified operating pressure. The injector cleaning solution is poured into a canister on some injector cleaners and shop air supply is used to pressurize the canister to the specified pressure. The injector cleaning solution contains unleaded fuel mixed with injector cleaner.

The container hose is connected to the Schrader valve on the fuel rail **(Figure 29–19)**. Disable the fuel pump according to the car manufacturer's instructions (for example, pull the fuel pump fuse, disconnect a lead at pump). Clamp off the fuel pump return line at the flex connection to prevent the cleaner from seeping into the fuel tank. Set and connect the cleaning system so it can circulate the cleaning solution through the fuel rail with the engine off. To do this, adjust the machine's delivery pressure to a pressure higher than normal delivery pressure. To flush the entire fuel rail, including the injector inlet screens and regulator, adjust the machine's delivery pressure higher than the fuel pressure regulator's normal pressure setting.

Readjust the machine to a pressure slightly lower than the normal regulated pressure and then open the cleaner's control valve one-half turn or so to prime the injectors, and then start the engine. If available, set and adjust the cleaner's pressure gauge to approximately 5 psi (34.47 kPa) below the operating pressure of the injection system and let the engine run at 1,000 rpm for 10 to 15 minutes or until the cleaning mix has run out. If the engine stalls during cleaning, simply restart it. Run the engine until the recommended amount of fluid is exhausted and the engine stalls. Shut off the ignition, remove the cleaning setup, and reconnect the fuel pump.

After removing the clamping devices from around the fuel lines, start the car. Let it idle for 5 minutes or so to remove any leftover cleaner from the fuel lines. In the more severely clogged cases, the idle improvement should be noticeable almost immediately. With more subtle performance improvements, an injector balance test verifies the cleaning results. Once the injectors are clean, recommend the use of an in-tank cleaning additive or a detergent-laced fuel.

The more advanced units feature electrically operated pumps neatly packaged in roll-around cabinets that are quite similar in design to an A/C charging station **(Figure 29–20)**.

After the injectors are cleaned or replaced, rough engine idle may still be present. This problem occurs because the adaptive memory in the computer has learned previously about the restricted injectors. If the injectors were supplying a lean air/fuel ratio, the computer increased the pulse width to try to bring the air/fuel ratio back to stoichiometric. With the cleaned or replaced injectors, the adaptive computer memory is still supplying the increased pulse width. This action makes the air/fuel ratio too rich now that the restricted injector problem does not exist. With the engine at normal operating temperature, drive the vehicle for at least 5 minutes to allow the adaptive computer memory to learn about the cleaned or replaced injectors. Afterward, the computer should supply the correct injector pulse width and the engine should run smoothly. This same problem may occur when any defective computer system component is replaced.

If the fuel delivery system is equipped with a cold-start injector, make sure it gets cleaned along with the pri-

Figure 29–20 A fuel injector cleaner cart.

mary injectors. This step is especially critical if the owner complains of cold-starting problems.

To effectively clean the cold-start injector, hook up the cleaning system and remove the cold-start injector from the engine. Open the control valve on the solvent containers and use an electronic triggering device to manually pulse the injector. Direct the spray into a suitable container. Pulse the injector until the spray pattern looks healthy.

On models where the cold-start injector is not readily accessible, check the fuel flow through the injector before and after the cleaning procedure. Pulse as necessary until the maximum flow through the injector is obtained.

FUEL RAIL, INJECTOR, AND REGULATOR SERVICE

There are service operations that will require removing the fuel injection fuel rail, pressure regulator, and/or injectors. Most of these are not related to fuel system repair. However, when it is necessary to remove and refit them, it is important that it be done carefully and according to the manufacturer's recommended procedures.

Injector Replacement

Photo Sequence 26 outlines a typical procedure for removing and installing an injector. Consult the vehicle's service manual for instructions on removing and installing injectors. Before installing the new one, always check to make sure the sealing O-ring is in place. Also, prior to installation, lightly lubricate the sealing ring with engine oil or automatic transmission fluid (avoid using silicone grease, which tends to clog the injectors) to prevent seal distortion or damage.

WARNING!

Cap injector openings in the intake manifold to prevent the entry of dirt and other particles. Also, after the injectors and pressure regulator are removed from the fuel rail, cap all fuel rail openings to keep dirt out of the fuel rail.

Fuel Rail, Injector, and Pressure Regulator Removal

The procedure for removing and replacing the fuel rail, injectors, and pressure regulator varies depending on the vehicle. On some applications, certain components must be removed to gain access to these components. The system must be relieved of any and all pressure before the fuel lines are opened to remove any of the components. Before removing the fuel rail, wipe off any dirt from the fuel rail with a shop towel. Then, loosen the fuel line clamps on the fuel rail, if so equipped. If these lines have quick-disconnect fittings, grasp the larger collar on

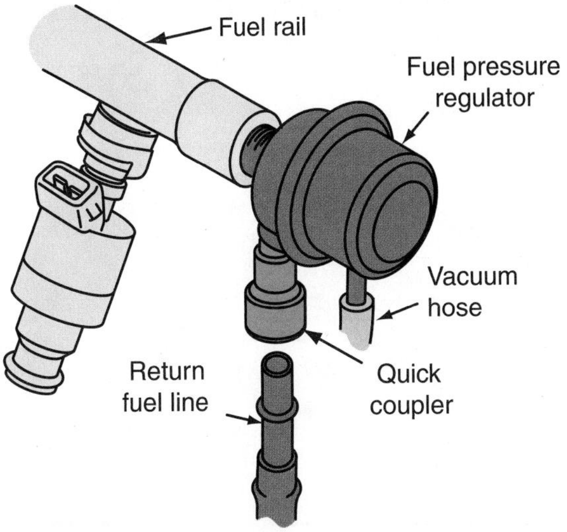

Figure 29-21 Fuel supply and return lines connected to the fuel rail.

the connector and twist in either direction while pulling on the line to remove the fuel supply and return lines **(Figure 29–21)**. Now, remove the vacuum line from the pressure regulator and disconnect the electrical connectors from the injectors. The fuel rail is now ready to be removed. On some engines, the fuel rail is held in place by bolts; they need to be removed before pulling the fuel rail free. When pulling the fuel rail away from the engine, pull with equal force on each side of the fuel rail to remove the rail and injectors.

WARNING!

Do not use compressed air to flush or clean the fuel rail. Compressed air contains water, which may contaminate the fuel rail.

Prior to removing the injectors and pressure regulator, the fuel rail should be cleaned with a spray-type engine cleaner. Normally the approved cleaners are listed in the service manual. After the rail is cleaned the injectors can be pulled from the fuel rail **(Figure 29–22)**. Use snap ring pliers to remove the snap ring from the pressure regulator cavity. Note the original direction of the vacuum fitting on the pressure regulator and pull the pressure regulator from the fuel rail. Clean all components with a clean shop towel. Be careful not to damage fuel rail openings and injector tips. Check all injector and pressure regulator openings in the fuel rail for metal burrs and damage.

WARNING!

Do not immerse the fuel rail, injectors, or pressure regulator in any type of cleaning solvent. This action may damage and contaminate these components.

PHOTO SEQUENCE 26

Removing and Replacing a Fuel Injector on a PFI System

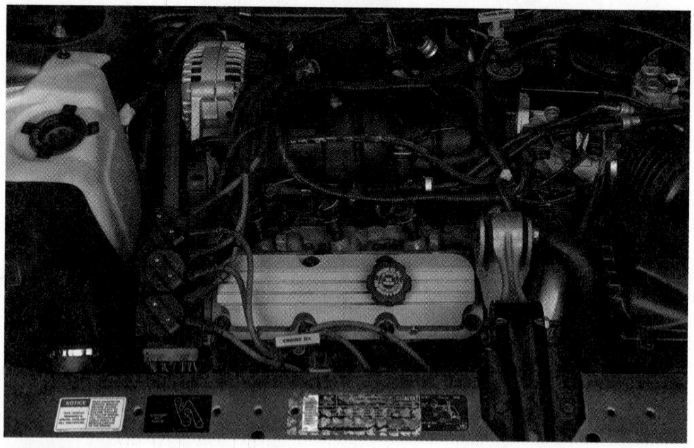

P26-1 *Often an individual injector needs to be replaced. Random disassembly of the components and improper procedures can result in damage to one of the various systems located near the injectors.*

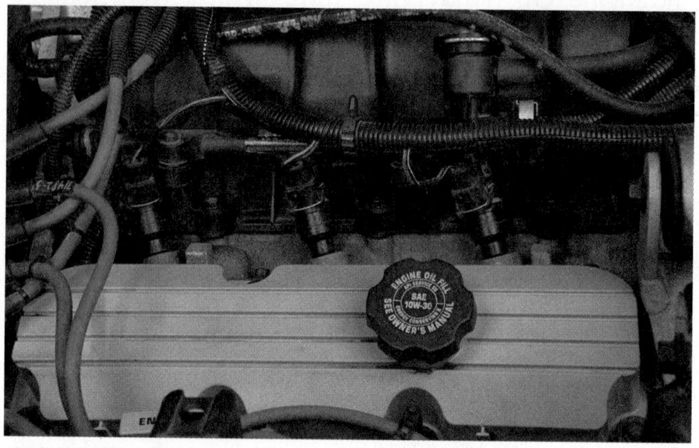

P26-2 *The injectors are normally attached directly to a fuel rail and inserted into the intake manifold or cylinder head. They must be positively sealed because high-pressure fuel leaks can cause a serious safety hazard.*

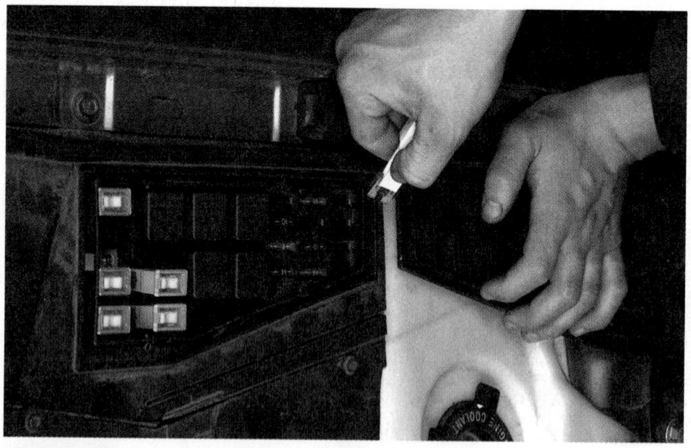

P26-3 *Prior to loosening any fitting in the fuel system, the fuel pump fuse should be removed.*

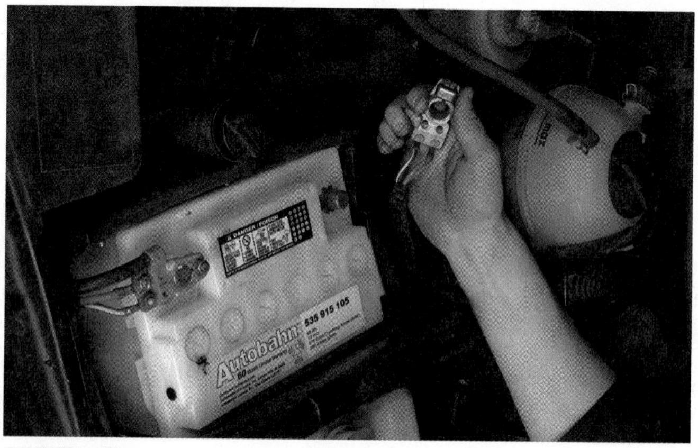

P26-4 *As an extra precaution, many technicians disconnect the negative cable at the battery.*

P26-5 *To remove an injector, the fuel rail must be able to move away from the engine. The rail-holding brackets should be unbolted and the vacuum line to the pressure regulator disconnected.*

P26-6 *Disconnect the wiring harness to the injectors by depressing the center of the attaching wire clip.*

P26-7 *The injectors are held to the fuel rail by a clip that fits over the top of the injector. O-rings at the top and at the bottom of the injector seal the injector.*

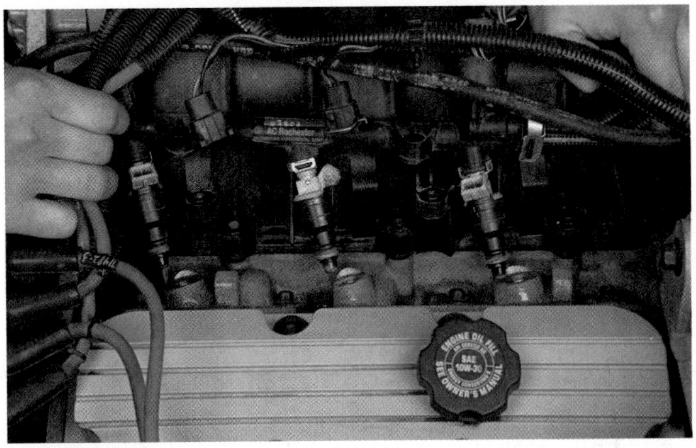

P26-8 *Pull up on the fuel rail assembly. The bottoms of the injectors will pull out of the manifold while the tops are secured to the rail by clips.*

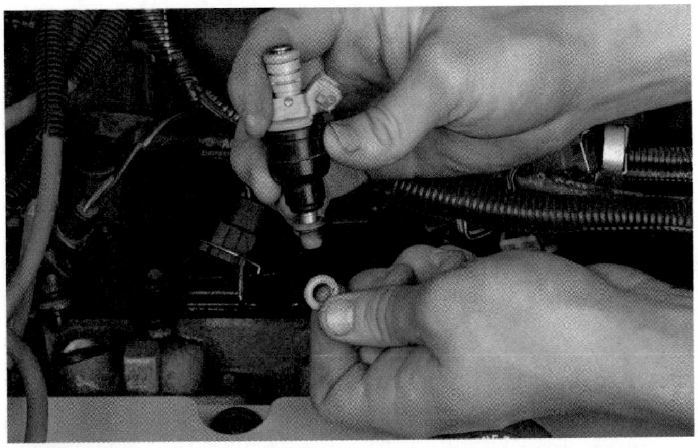

P26-9 *Remove the clip from the top of the injector and remove the injector unit. Install new O-rings onto the new injector. Be careful not to damage the seals while installing them, and make sure they are in their proper locations.*

P26-10 *Install the injector into the fuel rail and set the rail assembly into place.*

P26-11 *Tighten the fuel rail hold-down bolts according to manufacturer's specifications.*

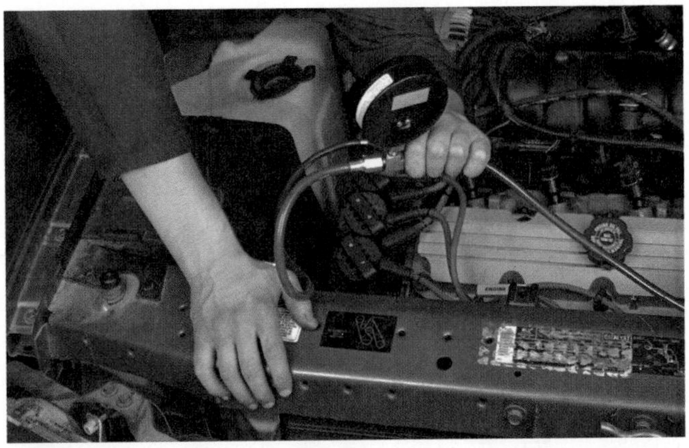

P26-12 *Reconnect all parts that were disconnected. Install the fuel pump fuse and reconnect the battery. Turn the ignition switch to the run position and check the entire system for leaks. After a visual inspection has been completed, conduct a fuel pressure test on the system.*

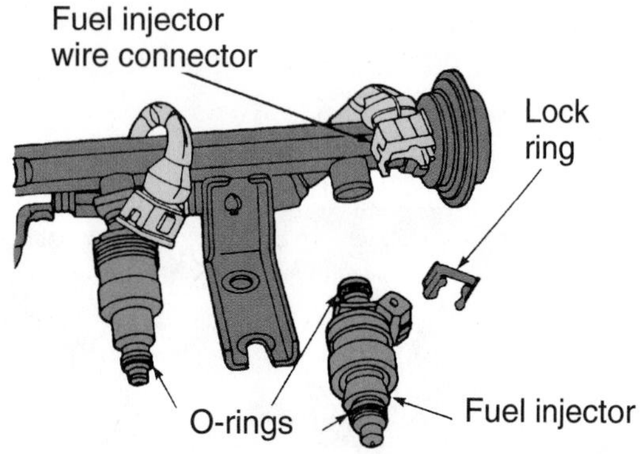

Figure 29-22 To remove the injectors from the fuel rail, remove the lock ring and pull the injector out.

When reassembling the fuel rail with the injectors and pressure regulator, make sure all O-rings are replaced and lightly coated with engine oil. Assemble the fuel rail in the reverse order as that used for disassembly. After the rail and injectors are in place and everything connected to them, reconnect the negative battery terminal and disconnect the 12-volt power supply from the cigarette lighter. Then, start the engine and check for fuel leaks at the rail and be sure the engine operation is normal.

Cold-Start Injector Removal and Testing

To test a typical cold-start injector **(Figure 29–23)**, it should be removed from the engine. Begin the procedure by bleeding the fuel system. Then disconnect the negative battery cable. Wipe excess dirt from the cold-start injector with a shop towel. Remove the electrical connector from the cold-start injector, union bolt, cold-start injector fuel line, cold-start injector retaining bolts, and then the cold-start injector.

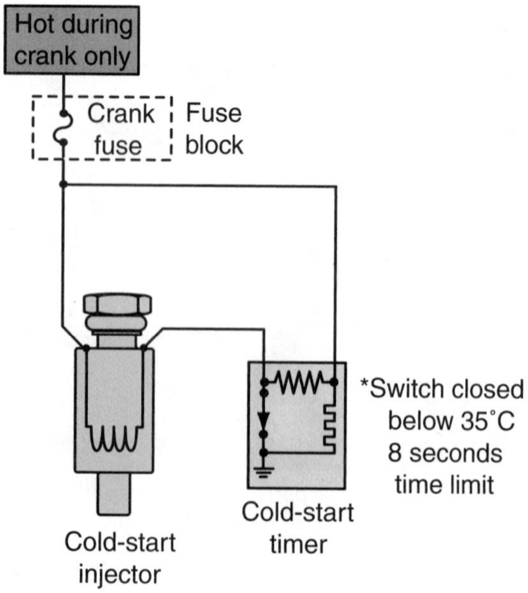

Figure 29-23 The electrical circuit for a cold-start injector.

Connect an ohmmeter across the cold-start injector terminals. If the resistance is more or less than specified, replace the injector.

When installing the cold-start injector, make sure all cold-start injector gaskets are replaced. Also check all mounting surfaces for metal burrs, scratches, and warping. After the cold-start injector gasket and injector is installed, tighten the mounting bolts to the specified torque. Then make sure all connections have been made and start the engine and check for fuel leaks at the cold-start injector.

IDLE SPEED CHECKS

In a fuel injection system, idle speed is regulated by controlling the amount of air that is allowed to bypass the airflow sensor or throttle plates. When presented with a car that tends to stall, especially when coming to a stop, or idles too fast, look for obvious problems like binding linkage and vacuum leaks first. If no problems are found, go through the minimum idle checking/setting procedure described on the underhood decal. The instructions listed on the decal spell out the necessary conditions that must be met prior to attempting an idle adjustment. These adjustment procedures can range from a simple twist of a throttle stop screw to more involved procedures requiring circumvention of idle air control devices, removal of casting plugs, or recalibration of the throttle position sensor.

Some IAC motors, such as those on Chrysler TBIs, have a hex bolt on the end of the motor plunger. However, this hex bolt is not for idle speed adjustment. If this hex bolt is turned, it will not affect idle speed, because the PCM will correct the idle speed. If the hex bolt is turned and the length of the IAC motor plunger changed in an attempt to adjust idle speed, the throttle may not be in the proper position for starting, and hard starting at certain temperatures may occur. On some applications, an idle rpm specification is provided with the plunger fully extended. The plunger can be adjusted under this condition. On DaimlerChrysler products, the plunger may be fully extended by turning the ignition switch off and then disconnecting the IAC motor connector. The IAC motor hex bolt should only require adjustment if it has been improperly adjusted or if a new motor is installed.

Idle Contact Switch Test

The procedure for diagnosing the idle contact switch varies depending on the vehicle make and model year. Basic tests include checking the voltage supply to the switch and checking the switch output when the switch is open and closed. The amount of supply voltage will vary with the system, so make sure you refer to the service manual before testing. The switch can also be checked with an ohmmeter. To do this, the switch must be electrically separated from all electrical power. Once isolated, continuity through the switch is checked in various switch positions. The switch should also be checked for a short to ground.

Scan Tester Diagnosis

If the idle speed is not within specifications, the input sensors and switches should be checked carefully with the scan tester. Photo Sequence 27 shows a typical procedure for performing a scan tester diagnosis of an idle air control motor. If the throttle position sensor voltage is lower than specified at idle speed, the PCM interprets this condition as the throttle being closed too much. Under this condition, the PCM opens the IAC or IAC BPA motor to increase idle speed.

If the engine coolant temperature sensor's resistance is higher than normal, it sends a higher-than-normal voltage signal to the PCM. The PCM thinks the coolant is colder than it actually is, and under this condition, the PCM operates the IAC or IAC BPA motor to increase idle speed. Input sensor defects and low system voltage can cause other problems in addition to improper idle rpm.

Defective input switches result in improper idle rpm. For example, if the A/C switch is always closed, the PCM thinks the A/C is on continually. This action results in the PCM operating the IAC or IAC BPA motor to provide a higher idle rpm. On many vehicles, the scan tester indicates the status of the input switches as closed or open, or high or low. Most input switches provide a high-voltage signal to the PCM when they are open and a low-voltage signal if they are closed.

On some vehicles, a fault code is set in the PCM memory if the IAC or IAC BPA motor or connecting wires are defective. On other systems, a fault code is set in the PCM memory if the idle rpm is out of range. On DaimlerChrysler products, the actuation test mode (ATM), or actuate outputs mode, may be entered with the ignition switch on. The IAC or IAC BPA motor may be selected in the actuate outputs mode, and the PCM is forced to extend and retract the IAC or IAC BPA motor plunger every 2.8 seconds. When this plunger extends and retracts properly, the motor, connecting wires, and PCM are in normal condition. If the plunger does not extend and retract, further diagnosis is necessary to locate the cause of the problem.

WARNING!

When performing a set engine rpm mode test on an IAC motor, always be sure the transmission selector is in the park or neutral position and the parking brake is applied.

On some IAC or IAC BPA motors, a set engine rpm mode may be entered on the scan tester. In this mode, each time a specified scan tester button is touched, the rpm should increase 100 rpm to a maximum of 2,000 rpm. Another specified scan tester button may be touched to decrease the speed in 100 rpm steps. On some scan testers, the up and down arrows are used to increase and decrease the engine rpm during this test. If the IAC or IAC BPA motor responds properly during this diagnosis, the PCM, motor, and connecting wires are in satisfactory condition, and further diagnosis of the inputs is required.

SHOP TALK

If the IAC or IAC BPA motor counts are zero on the scan tool, look for a vacuum leak. The PCM will run the valve all the way down to compensate for the leak. If no leak is found, then check the circuit between the PCM and the motor. The circuit is likely open between the PCM and the motor. Wiggle the wires on the IAC or IAC BPA motor and observe the scan tester reading. If the count reading changes while wiggling the wires, you have found the problem. ■

On some systems, the scan tester reads the IAC or IAC BPA motor counts, and the count range is provided in the scan tester instruction manual. Some of the input switches, such as the A/C, may be operated and the scan tester counts should change. Most technicians look for consistent idle counts on a warm engine to verify that the IAC is not sticking or malfunctioning. If the scan tester counts change when the A/C is turned on and off, the motor, connecting wires, and PCM are operational. When the scan tester counts do not change under this condition, further diagnosis is required.

IAC BPA Motor and Valve Diagnosis

On some systems, such as Toyota, a jumper wire may be connected to terminals E1 and TE1 in the data link connector (DLC) to diagnose the IAC BPA valve with the engine at normal operating temperature. When the engine is started with this jumper wire connection, the engine speed should increase to 1,000 to 1,300 rpm for 5 seconds, and then return to idle speed. If the IAC BPA valve does not respond as specified, further diagnosis of the IAC BPA valve, wires, and PCM is required.

On General Motors' vehicles, the IAC BPA motor extends fully when terminals A and B are connected in the DLC and the ignition switch is turned on. With the IAC BPA motor removed from the throttle body, this jumper connection may be completed while observing the IAC BPA motor. If the motor does not extend, further diagnosis of the motor, connecting wires, and PCM is required.

IAC BPA Motor Removal and Cleaning Carbon deposits in the IAC BPA motor air passage in the throttle

PHOTO SEQUENCE 27

Performing a Scan Tester Diagnosis of an Idle Air Control Motor

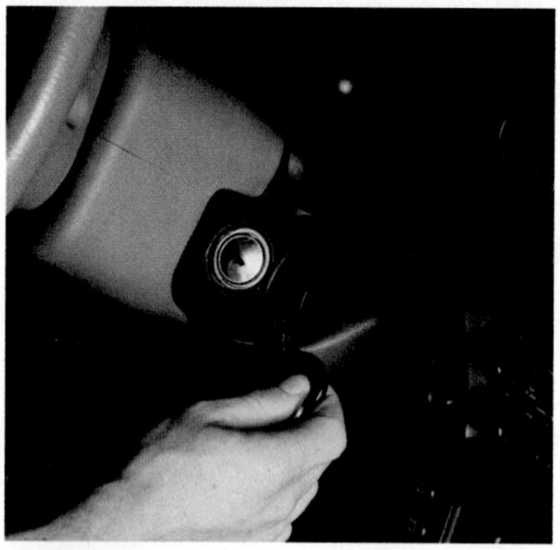

P27-1 *Make sure the ignition switch is turned off.*

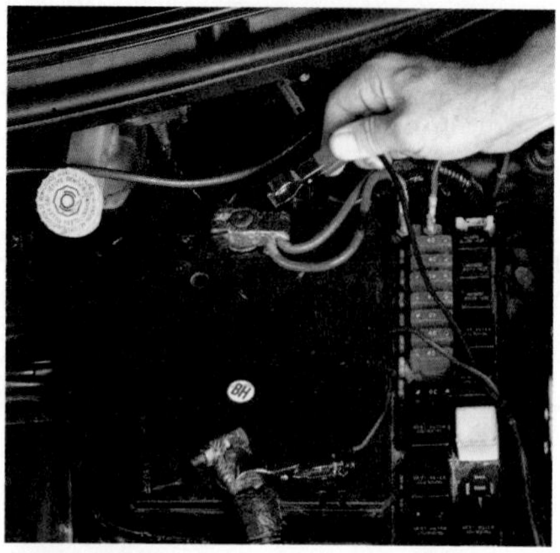

P27-2 *Connect the scan tool leads to the battery terminals or appropriate power source.*

P27-3 *Turn the ignition ON, but do not start the engine.*

P27-4 *Program the scan tool as required for the vehicle being tested.*

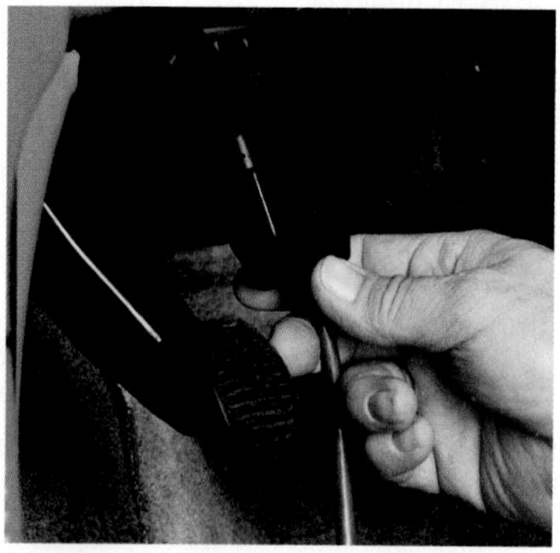

P27-5 *Use the proper adapter to connect the scan tool to the vehicle's DLC.*

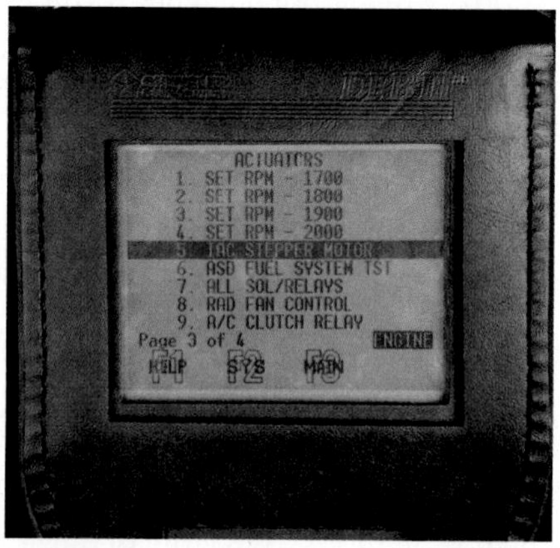

P27-6 *Select idle air control (IAC) motor test on the scan tool.*

P27-7 *Press the appropriate scan tool button to increase engine speed and observe the engine's rpm on the scan tool. The tool will automatically perform a non-running actuation test of the IAC hardware and software.*

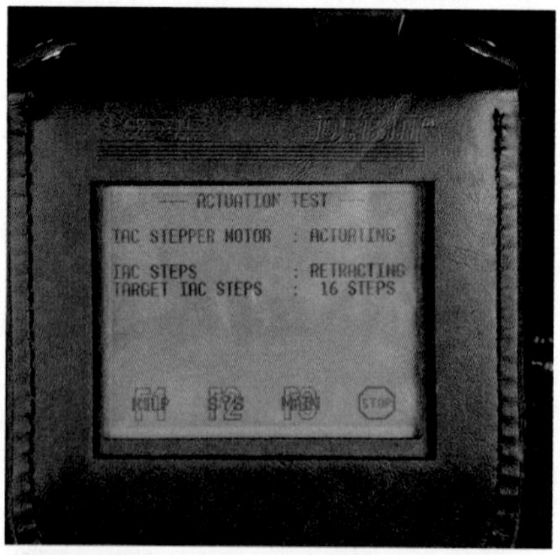

P27-8 *At the low range, the IAC motor has a target range of 16 steps. The scan tool steps the motor through and monitors these steps.*

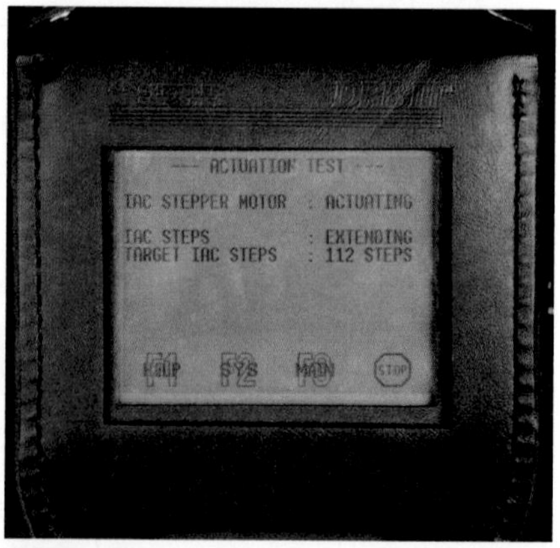

P27-9 *At the high range, the scan tool steps the motor through and monitors 112 steps.*

body or on the IAC BPA motor pintle result in erratic idle operation and engine stalling. Remove the motor from the throttle body, and inspect the throttle body air passage for carbon deposits. If heavy carbon deposits are present, remove the complete throttle body for cleaning. Clean the throttle body IAC BPA air passage, motor sealing surface, and pintle seat with throttle body cleaner. Clean the motor pintle with throttle body cleaner.

WARNING!

IAC BPA motor damage may result if throttle body cleaner is allowed to enter the motor.

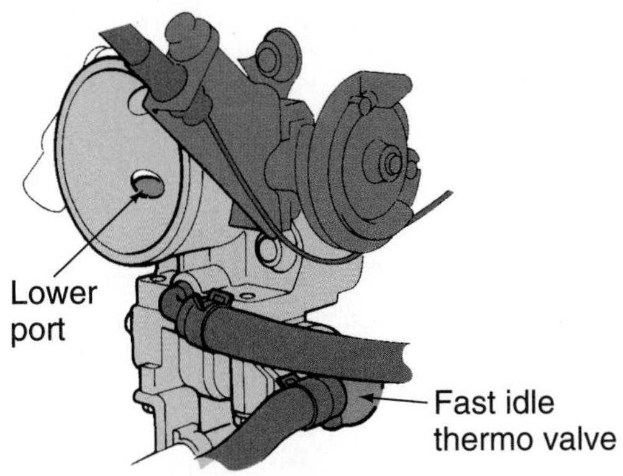

Figure 29-25 The lower port in the throttle body is the fast idle thermo valve air intake. *Courtesy of American Honda Motor Co., Inc.*

IAC BPA Motor Diagnosis and Installation

The motor can be checked with an ohmmeter. If the ohmmeter reading is not within specifications, replace the IAC BPA motor. If a new IAC BPA motor is installed, make sure the part number, pintle shape, and diameter are the same as those on the original motor. Measure the distance from the end of the pintle to the shoulder of the motor casting **(Figure 29–24)**. Move the pintle until it is at the specified distance. If the pintle is extended too far, the motor can be damaged during installation.

Install a new gasket or O-ring on the motor. If the motor is sealed with an O-ring, lubricate the ring with transmission fluid. If the motor is threaded into the throttle body, tighten the motor to the specified torque. When the motor is bolted to the throttle body, tighten the mounting bolts to the specified torque.

Diagnosis of Fast Idle Thermo Valve

The fast idle thermo valve is factory adjusted and should not be disassembled. Remove the air duct from the throttle body, and be sure the engine temperature is below 86°F (30°C). Start the engine and place your finger over the lower port in the throttle body **(Figure 29–25)**. Under this condition, there should be airflow through this lower port and the fast idle thermo valve. If there is no airflow through the lower port, replace the fast idle thermo valve.

As the engine temperature increases, the airflow through the lower throttle body port should decrease. When the engine approaches normal operating temperature, the air should stop flowing through the lower port and the fast idle thermo valve. If there is airflow through the lower throttle body port with the engine at normal operating temperature, check the cooling system for proper operation and temperature. The fast idle thermo valve is heated by engine coolant. If the cooling system operation is normal, replace the fast idle thermo valve.

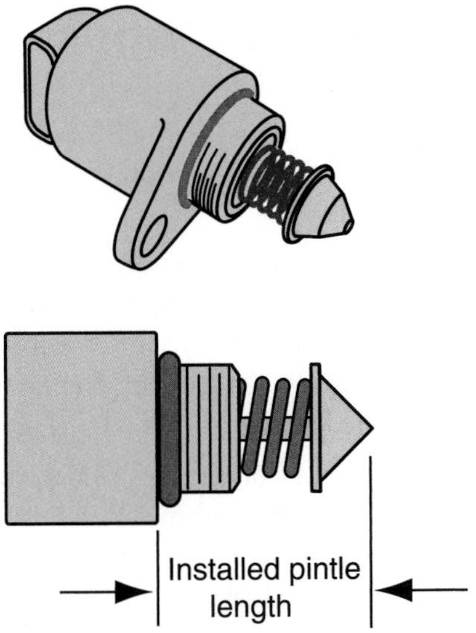

Figure 29-24 Measuring the distance of extension on an IAC BPA motor.

Diagnosis of Starting Air Valve

If the starting air valve is open with the engine running, the idle rpm may be higher than specified. Disconnect the vacuum signal hose from the starting air valve, and connect a vacuum gauge to this hose. With the engine idling, this vacuum should be above 16 in. Hg (406 mm Hg). If the vacuum is lower than specified, check for leaks or restrictions in the signal hose. When the hose is satisfactory, check for late ignition timing or engine conditions (for example, low compression) that may result in low vacuum.

If the signal vacuum is satisfactory, check the hoses from the starting air valve to the intake manifold and the air cleaner for restrictions and leaks. When the hoses are satisfactory, remove the hose from the starting air valve to the air cleaner. With the engine idling, there should be no air flow through this hose. When airflow is present, replace the starting air valve.

With the engine cranking, there should be air flow through the hose from the starting air valve to the air cleaner. If no air flow is present, replace the starting air valve.

CASE STUDY

A customer brings in a late-model General Motors' car, with a 2.0 L port fuel-injected engine and complains of poor performance and fuel economy. The customer also confesses that the MIL had been on for quite some time.

Diagnosis of the problem begins with a visual inspection of the engine, including all hoses and wiring connectors. No obvious problems are found.

Next, a scan tool is connected to the DLC and a code 45 is retrieved from the PCM's memory. Using a service manual, the technician interprets the code as a rich exhaust signal from the oxygen sensor.

At first the technician suspects a faulty oxygen sensor. However, from the scan tool, she knows the system is in closed loop. By referring to the explanation of the code in the service manual, she finds that the code is set when the oxygen sensor voltage is greater than 0.75 volt for 50 seconds or more. Therefore, she knows the sensor or sensor circuit is not open. An open would keep the system in open loop. She also knows a shorted sensor or sensor circuit would cause wire or component damage. None was found. Problems of excessive resistance would cause lower-than-normal voltage readings. The problem here is higher-than-expected voltage readings. Therefore it is unlikely that the problem is caused by a faulty sensor or sensor circuit.

Following the diagnostic procedures given for this specific code, further testing with the scan tool reveals that the problem could be caused by an open in the ignition wiring, a saturated charcoal canister, a faulty MAP sensor or fuel pressure regulator, or by the improper use of RTV sealer. A thorough inspection of testing of these reveals the cause of the problem: a faulty MAP sensor. The sensor was incorrectly signaling low pressures (high vacuum) that caused the computer to run richer than normal.

The sensor was replaced and the codes cleared. The problem was corrected. The technician verified the repair by watching the scan tool while the engine was running. She observed the oxygen sensor switching quickly from rich to lean to rich, just as it should. She also found that the PCM did not set a new code, so the problem was solved.

KEY TERMS

Block Integrator
Block Learn
Long-term fuel trim (LTFT)
Malfunction indicator lamp (MIL)
Noid light
Peak and hold injector
Pulse-modulated injector
Short-term fuel trim (STFT)

SUMMARY

- Always relieve the fuel pressure before disconnecting any component in the fuel system.
- Always turn off the ignition switch before connecting or disconnecting any system component or test equipment.
- An O_2 sensor can be checked with a voltmeter connected between the sensor wire and ground. The sensor's voltage should be cycling from low voltage to high voltage.
- The signal from most O_2 sensors varies between 0 and 1 volt. If the voltage is continually high, the air/fuel ratio may be rich or the sensor may be contaminated. When the O_2 sensor voltage is continually low, the air/fuel ratio may be lean, the sensor may be defective, or the wire between the sensor and the computer may have a high-resistance problem. If the O_2 sensor voltage signal remains in a midrange position, the computer may be in open loop or the sensor may be defective.
- The activity of the sensor can be monitored on a scanner.
- The activity of an O_2 sensor is best monitored with a lab scope.
- Block Integrator represents a short-term correction to the amount of fuel delivered during closed loop. Block Learn makes long-term corrections. Injector pulse width is adjusted according to both Block Integrator and Block Learn.
- The sensor's flap in an airflow sensor should be checked for free, smooth, and quiet movement.
- Some airflow sensors can be checked with an ohmmeter, comparing the resistance values of the potentiometer by moving the air flap to the specifications for that particular sensor.
- An injector that does not open causes hard starts on port-type systems and an obvious no-start on single-point TBI designs.
- An injector that is stuck partially open causes loss of fuel pressure and flooding due to raw fuel dribbling into the engine. In addition to a rich-running engine, a leaking injector also causes the engine to diesel or run on when the ignition is turned off.
- Buildups of gum and other deposits on the tip of an injector can reduce the amount of fuel sprayed by the injector or they can prevent the injector from totally sealing, allowing it to leak.
- Another cause of a rich mixture is a leaking fuel pressure regulator. If the diaphragm of the regulator is rup-

tured, fuel will move into the intake manifold through the diaphragm, causing a rich mixture.

- When an injector is suspected as the cause of a lean problem, the first step is to determine if the injector is receiving a signal to fire. Check for voltage at the injector using a high impedance test light or a convenient noid light that plugs into the connector.
- An ohmmeter can be used to test the electrical soundness of an injector.
- An injector pressure balance test will help isolate a clogged or dirty injector.
- An oscilloscope can be used to monitor the injector's pulse width and duty cycle when an injector-related problem is suspected.
- The pulse width is the time in milliseconds that the injector is energized. The duty cycle is the percentage of on-time to total cycle time.
- For all types of injectors, the waveform on the scope should have a clean, sudden drop in voltage when it is turned on.
- Never soak an injector in cleaning solvent or use a wire brush, pipe cleaner, toothpick, or other cleaning utensil to unblock a plugged injector.
- In a fuel injection system, idle speed is regulated by controlling the amount of air that is allowed to bypass the airflow sensor or throttle plates. When a car tends to stall or idles too fast, look for obvious problems like binding linkage and vacuum leaks first. If no problems are found, go through the minimum idle checking/setting procedure described on the underhood decal.
- If the idle speed is not within specifications, the input sensors and switches should be checked carefully with the scan tester.
- If the engine coolant temperature sensor's resistance is higher than normal, it sends a higher-than-normal voltage signal to the PCM. The PCM thinks the coolant is colder than it actually is, and under this condition, the PCM operates the IAC or IAC BPA motor to increase idle speed.

REVIEW QUESTIONS

1. List the three things that must occur for an EFI to operate properly.
2. What is indicated by trouble codes?
3. What is the correct procedure for checking an oxygen sensor with a DMM?
4. What is the difference between STFT and LTFT?
5. What is indicated by block integrator, or block learn, numbers that are constantly below 128?
6. Which of the following is a likely cause of a no-start condition, if the engine starts when the electrical connector to the MAF is disconnected?
   a. a defective oxygen sensor
   b. a defective PCM
   c. a defective MAP sensor
   d. a defective MAF sensor
7. What problem may result from dirt buildup on an engine's throttle plates?
8. Which of the following would not cause a hard-to-start problem on a port-injected engine?
   a. an electrically open fuel injector
   b. a defective oxygen sensor
   c. a leaking fuel pressure regulator
   d. dirty injectors
9. What is the correct way to test an injector with an ohmmeter?
10. What is the difference between the pulse width of an injector and the duty cycle of an injector?
11. How can you use a dual trace scope to make sure the injectors are firing at the correct time?
12. While discussing the causes of higher-than-specified idle speeds, Technician A says an intake manifold vacuum leak may cause a high idle speed. Technician B says if the TP sensor voltage signal is higher than specified, the idle speed may be higher than normal. Who is correct?

TECH MANUAL

The following procedures are included in Chapters 28/29 of the *Tech Manual* that accompanies this book:

1. Visually inspecting an EFI system.
2. Checking the operation of the fuel injectors on an engine.
3. Conducting an injector balance test.

a. Technician A **c.** Both A and B
b. Technician B **d.** Neither A nor B

13. While discussing IAC BPA valve diagnosis, Technician A says that, on some vehicles, a jumper wire may be connected to specific DLC terminals to check the IAC BPA valve operation. Technician B says that if the scan tester indicates zero IAC BPA valve counts, there may be an open circuit between the PCM and the IAC valve. Who is correct?

a. Technician A **c.** Both A and B
b. Technician B **d.** Neither A nor B

14. While discussing IAC BPA motor removal, service, and replacement, Technician A says throttle body cleaner may be used to clean the IAC BPA motor internal components. Technician B says that, on some vehicles, IAC BPA motor damage occurs if the pintle is extended more than specified during installation. Who is correct?

a. Technician A **c.** Both A and B
b. Technician B **d.** Neither A nor B

15. While discussing injector testing, Technician A says a defective injector may cause cylinder misfiring at idle speed. Technician B says restricted injector tips may result in acceleration stumbles. Who is correct?

a. Technician A **c.** Both A and B
b. Technician B **d.** Neither A nor B

16. While discussing airflow sensors, Technician A says with mass airflow and volume airflow sensor systems, if any air bypasses the sensors the engine will run lean. Technician B says vacuum leaks in a speed density system will decrease injector pulse width. Who is correct?

a. Technician A **c.** Both A and B
b. Technician B **d.** Neither A nor B

17. While discussing scan tool diagnosis of TBI, MFI, and SFI systems, Technician A says the scan tester will erase fault codes quickly on many systems. Technician B says many scan testers will store sensor readings during a road test and then play back the results in a snapshot test mode. Who is correct?

a. Technician A **c.** Both A and B
b. Technician B **d.** Neither A nor B

18. While discussing Block Learn and Block Integrator, when the integrator number is 180 and the Block Learn number is 185, Technician A says these numbers indicate that the PCM is in control of the air/fuel ratio in spite of the lean mixture sensed by the O_2 sensor. Technician B says these numbers indicate the PCM is trying to increase fuel delivery; therefore, the oxygen (O_2) sensor signal must be continually lean. Who is correct?

a. Technician A **c.** Both A and B
b. Technician B **d.** Neither A nor B

19. While discussing a high idle speed problem, Technician A says higher-than-normal idle speed may be caused by low electrical system voltage. Technician B says higher-than-normal idle speed may be caused by a defective coolant temperature sensor. Who is correct?

a. Technician A **c.** Both A and B
b. Technician B **d.** Neither A nor B

20. While discussing the causes of a rich air/fuel ratio, Technician A says a rich air/fuel ratio may be caused by low fuel pump pressure. Technician B says a rich air/fuel ratio may be caused by a defective coolant temperature sensor. Who is correct?

a. Technician A **c.** Both A and B
b. Technician B **d.** Neither A nor B

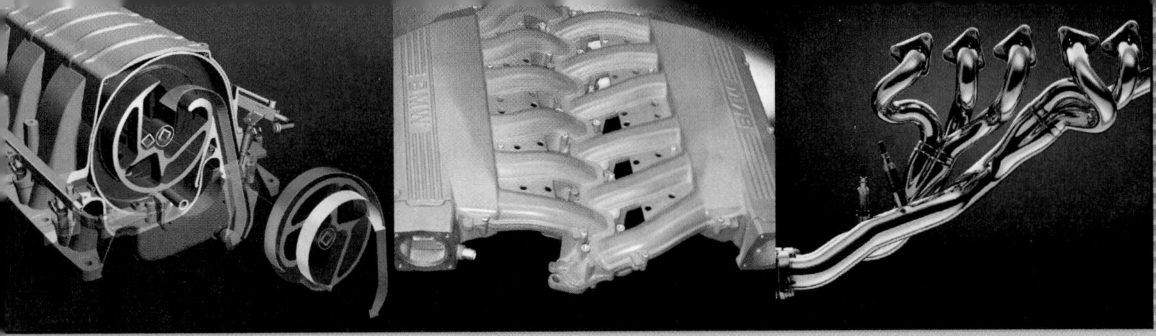

30

INTAKE AND EXHAUST SYSTEMS

OBJECTIVES

■ Explain the operation of the components in the air induction system, including ductwork, air cleaners/filters, and intake manifolds. ■ Describe how the engine creates vacuum and how vacuum is used to operate and control many automotive devices. ■ Inspect and troubleshoot vacuum and air induction systems. ■ Explain the operation of exhaust system components, including exhaust manifold, gaskets, exhaust pipe and seal, catalytic converter, muffler, resonator, tailpipe, and clamps, brackets, and hangers. ■ Properly perform an exhaust system inspection, and service and replace exhaust system components. ■ Explain the purpose and operation of a turbocharger. ■ Inspect a turbocharger, and describe some common turbocharger problems. ■ Explain supercharger operation, and identify common supercharger problems.

An internal combustion engine requires air to operate. This air supply is drawn into the engine by the vacuum created during the intake stroke of the pistons. The air is mixed with fuel and delivered to the combustion chambers. Controlling the flow of air and the air/fuel mixture is the job of the induction system.

THE AIR INDUCTION SYSTEM

Prior to the introduction of emission control devices, the induction system was quite simple. It consisted of an air cleaner housing mounted on top of the engine with a filter inside the housing. Its function was to filter dust and grit from the air being drawn into the carburetor.

The air intake system on a modern fuel-injected engine is rather complicated **(Figure 30–1)**. Ducts channel cool air from outside the engine compartment to the throttle plate assembly. The air filter is placed below the top of the engine to allow for aerodynamic body designs. Electronic meters measure airflow, temperature, and density. Pulse air systems provide fresh air to the exhaust stream to oxidize unburned hydrocarbons in the exhaust. These components allow the air induction system to perform the following functions:

- Silence air intake noise
- Heat or cool the air as required
- Provide the air the engine needs to operate
- Filter the air to protect the engine from wear
- Monitor airflow temperature and density for more efficient combustion and a reduction of hydrocarbon (HC) and carbon monoxide (CO) emissions
- Operate with the PCV system to burn the crankcase fumes in the engine
- Provide air for some air injection systems

Air Intake Ductwork

Ductwork is used to direct the air into the throttle body. Cool outside air is drawn into the air cleaner assembly and, on some engines, warm air from around the exhaust is also brought in for cold engine operation **(Figure 30–2)**.

The most recent designs have remote air cleaner assemblies with an airflow meter installed in the ductwork **(Figure 30–3)**. Other sensors may also be installed in the air cleaner assembly or in the ductwork leading to the throttle body assembly. The air cleaner assembly also provides filtered air to the PCV system.

Be sure that the intake ductwork is properly installed and all connections are airtight—especially those between an airflow sensor or remote air cleaner and the throttle plate assembly. Generally, metal or plastic air ducts are used when engine heat is not a problem. Special paper-metal ducts are used when they will be exposed to high engine temperatures.

Air Cleaner/Filter

The primary function of the air filter is to prevent airborne contaminants and abrasives from entering into the engine. Without proper filtration, these contaminants can

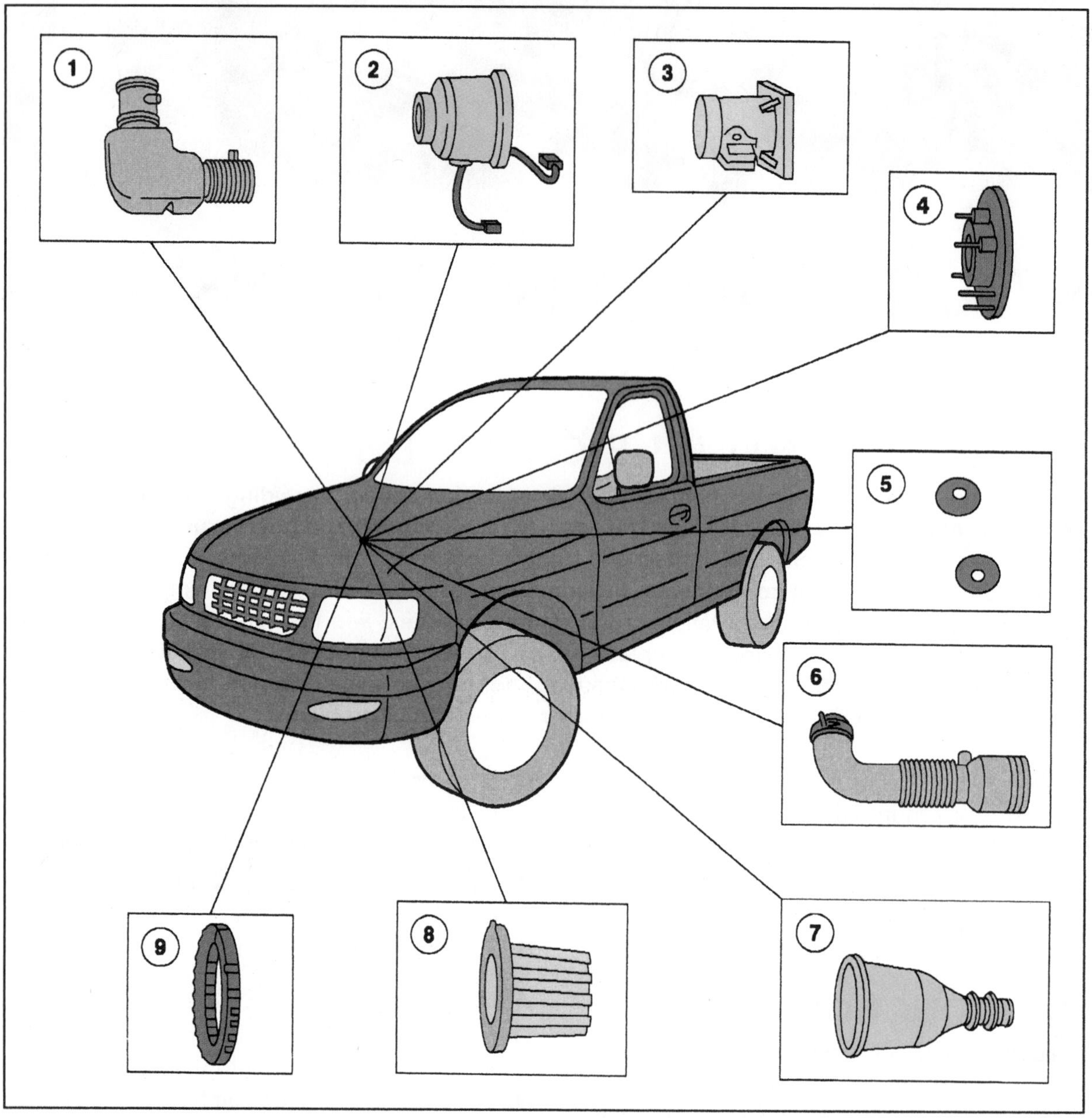

Item	Description
1	Air cleaner outlet tube—(4.2L)
2	Air cleaner, outlet
3	Mass airflow sensor
4	Mass airflow sensor plate
5	Air cleaner support—insulator
6	Air cleaner—outlet tube (4.6L, 5.4L) and resonator
7	Air cleaner, inlet
8	Air cleaner, inlet
9	Air cleaner clamp

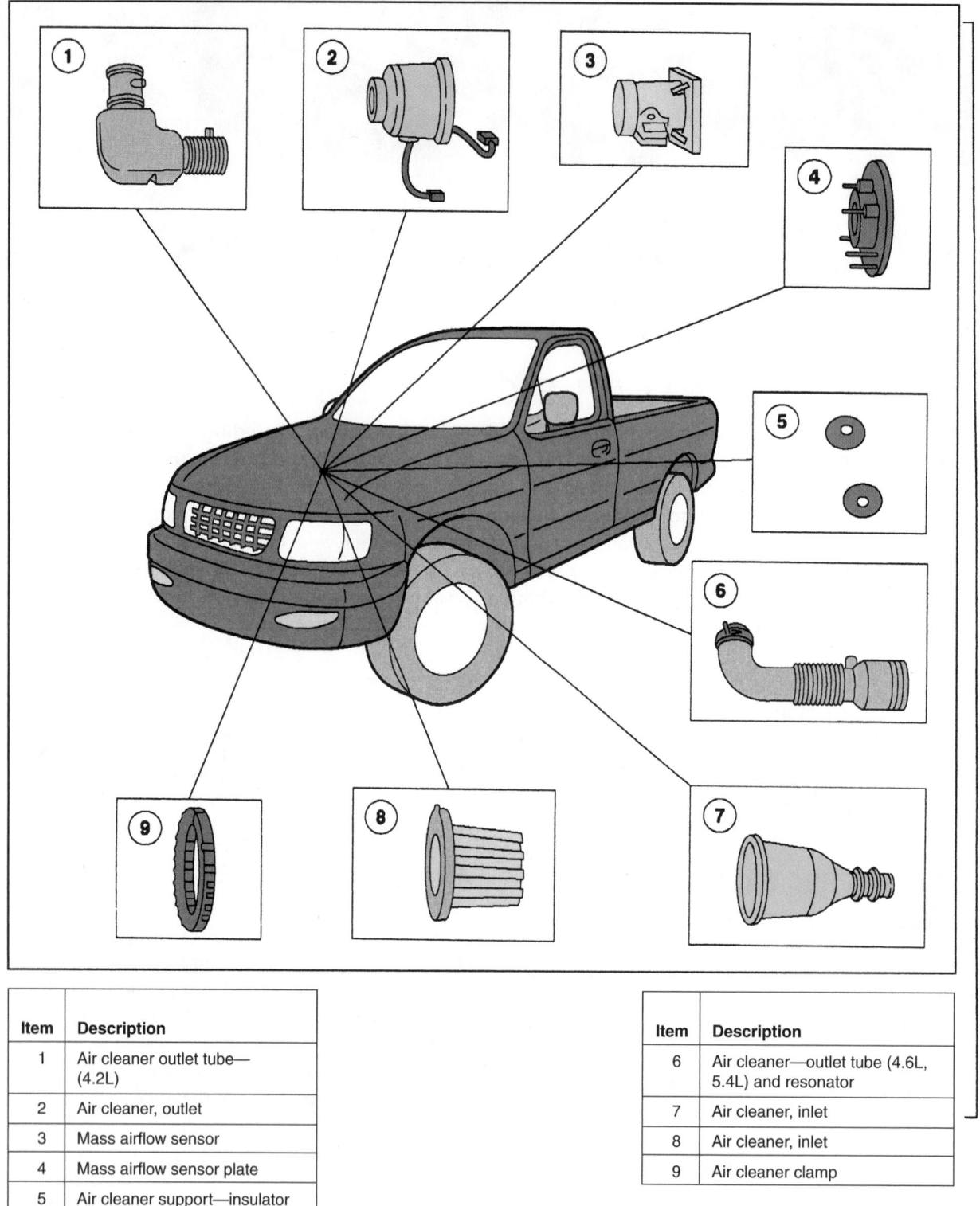

Figure 30-1 A late-model intake air distribution system. *Courtesy of Ford Motor Company*

cause serious damage and appreciably shorten engine life. All incoming air should pass through the filter element before entering the engine.

Air Filter Design Air filters are basically assemblies of pleated paper supported by a layer of fine mesh wire screen. The screen gives the paper some strength and also filters out large particles of dirt. A thick plasticlike gasket material normally surrounds the ends of the filter. This gasket adds strength to the filter and seals the filter in its housing. If the filter does not seal well in the housing, dirt and dust can be pulled into the airstream to the cylinders.

The shape and size of the air filter element depends on its housing; the filter must be the correct size for the

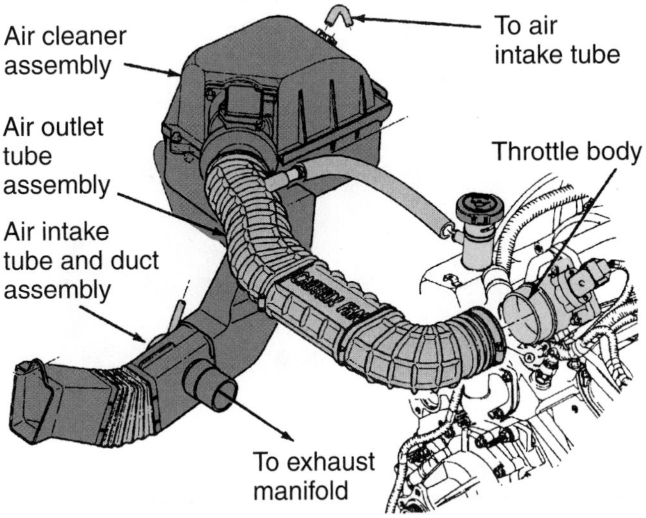

Figure 30–2 A typical air induction system for engines equipped with fuel injection. *Courtesy of Ford Motor Company*

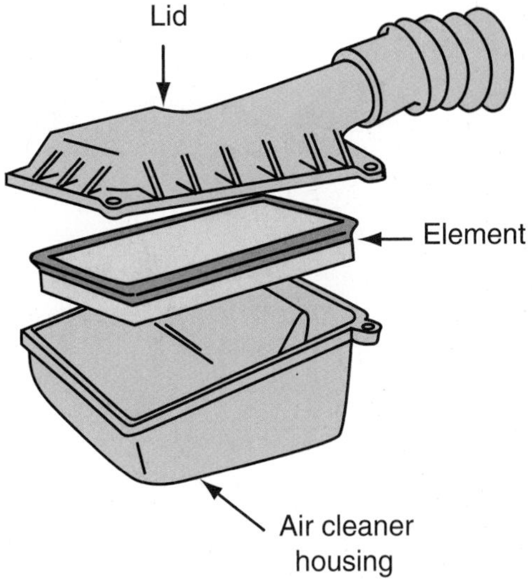

Figure 30–4 A typical flat air filter element.

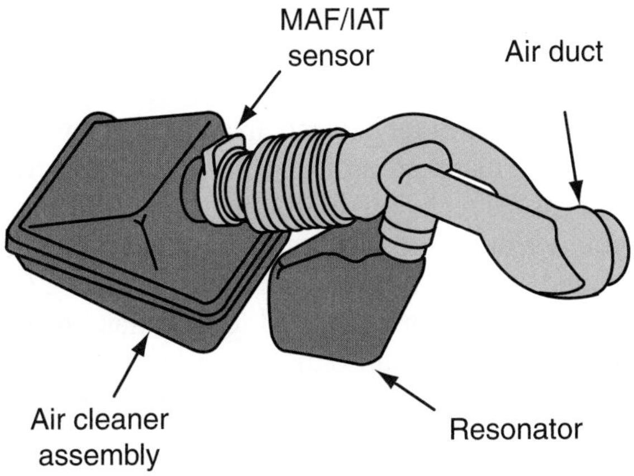

Figure 30–3 The ductwork may include sensors for the engine control system.

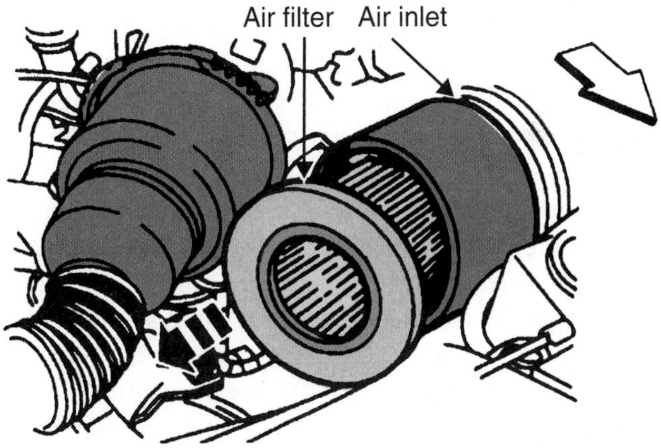

Figure 30–5 A typical round air filter for a late-model vehicle. *Courtesy of Ford Motor Company*

housing or dirt will be drawn into the engine. On today's engines, air filters are either flat **(Figure 30–4)** or round **(Figure 30–5)**.

Air Filter Service If an air filter is doing its job, it will get dirty. That is why filters are made of pleated paper. The paper is the actual filter. It is pleated to increase the filtering area. By increasing the area, the amount of time it will take for dirt to plug the filter becomes longer. As a filter gets dirty, the amount of air that can flow through it is reduced. This is not a problem until less air than the engine needs gets through the filter. Without the proper amount of air, the engine will not be able to produce the power it should, nor will it be as fuel efficient as it should be.

Included in the preventative maintenance plan for all vehicles is the periodic replacement of the air filter. This mileage or time interval is based on normal vehicle operation. If the vehicle is used or has been used in heavy dust, the life of the filter is shorter. Always use a replacement filter that is the same size and shape as the original. When replacing the filter element, carefully remove all dirt from inside of the housing. Large pieces of dirt and stones may accumulate there. It would be disastrous if that dirt got into the cylinders. Also make sure the air cleaner housing is properly aligned and closed around the filter to ensure good airflow of clean air.

Intake Manifold

The intake manifold distributes the clean air or air/fuel mixture as evenly as possible to each cylinder of the engine.

Most engines with throttle-body injection have cast-iron intake manifolds. With this type of engine, the intake manifold delivers air and fuel to the cylinders. Most early intake manifold designs had short runners **(Figure 30–6)**. These manifolds were either wet or dry. Wet manifolds had coolant passages cast directly in them. Dry manifolds did not have these coolant passages but some had exhaust passages. Exhaust gases or coolant was used to

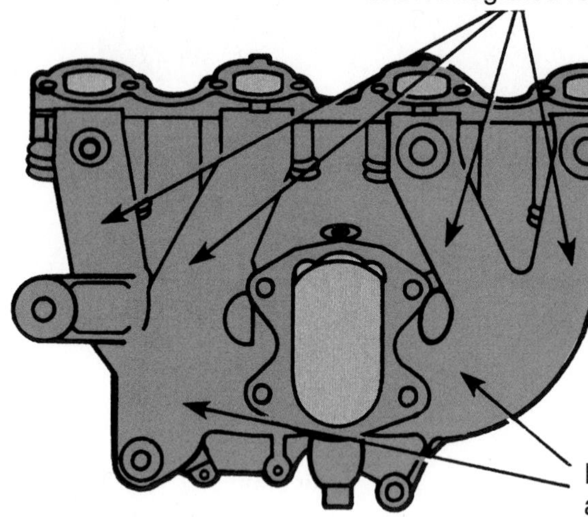

Figure 30–6 The intake manifold on an in-line four-cylinder engine.

Figure 30–8 A plastic intake manifold for an in-line five-cylinder engine. *Courtesy of BMW of North America, Incorporated*

heat up the floor of the manifold. This helped to vaporize the fuel before it arrived in the cylinders. Other dry manifold designs used some sort of electric heater unit or grid to warm up the bottom of the manifold. Heating the floor of the manifold also stopped the fuel from condensing in the manifold's plenum area. Good fuel vaporization and the prevention of condensation allowed for delivery of a more uniform air/fuel mixture to the individual cylinders.

Modern intake manifolds for engines with port fuel injection are typically made of die-cast aluminum **(Figure 30–7)** or plastic **(Figure 30–8)**. These materials are used to reduce engine weight. Because intake manifolds for port-injected engines only deliver air to the cylinders, fuel vaporization and condensation are not design considerations. The primary consideration of these manifolds is the delivery of equal amounts of air to each cylinder. This style manifold is often called a "tuned" intake manifold.

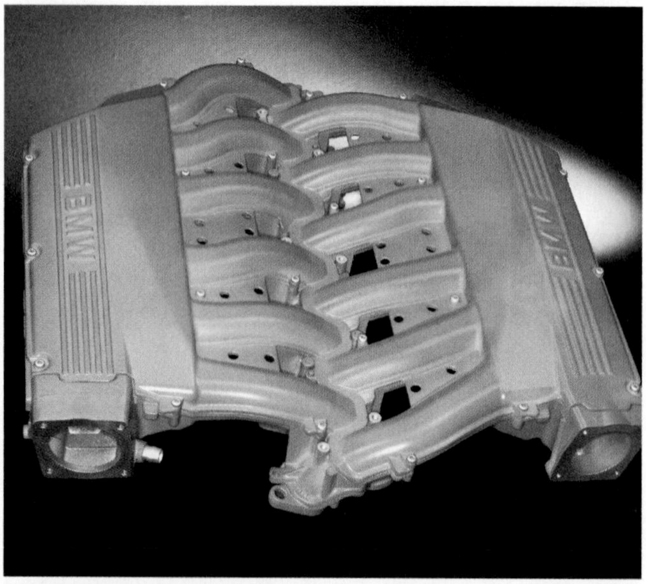

Figure 30–7 A die-cast aluminum intake manifold. *Courtesy of BMW of North America, Incorporated*

Intake manifolds also serve as the mounting point for many intake-related accessories and sensors **(Figure 30–9)**. Some include a provision for mounting the thermostat and thermostat housing. In addition, connections to the intake manifold provide a vacuum source for the exhaust gas recirculation (EGR) system, automatic transmission vacuum modulators, power brakes, and/or heater and air-conditioning airflow control doors. Other devices located on or connected to the intake manifold include the manifold absolute pressure (MAP) sensor, knock sensor, various temperature sensors, and EGR passages.

Most engines cannot produce the amount of power they should at high speeds because they do not receive enough air. This is the reason why many race cars have hood scoops. With today's body styles, hood scoops are not desirable because they increase air drag. However, to get high performance out of high-performance engines, more air must be delivered to the cylinders at high engine speeds. There are a number of ways to do this; increasing the air delivered by the intake manifold is one of them. This can be a little tricky though. Too much airflow at low engine speeds hurt's the engine's efficiency. Therefore, manufacturers have developed manifolds that deliver more air only at high engine speeds.

Variable Intake Manifolds To meet the demands for air at a variety of engine speeds, many engines are equipped with a variable intake manifold. A variable intake manifold is one that has different routes for the air to travel when it is operating at different speeds **(Figure 30–10)**. At low engine speeds, the air travels a longer distance and results in an increase of low-speed torque and responsiveness. At higher speeds, the air travels a short distance because the time allowed to fill the cylinders has been

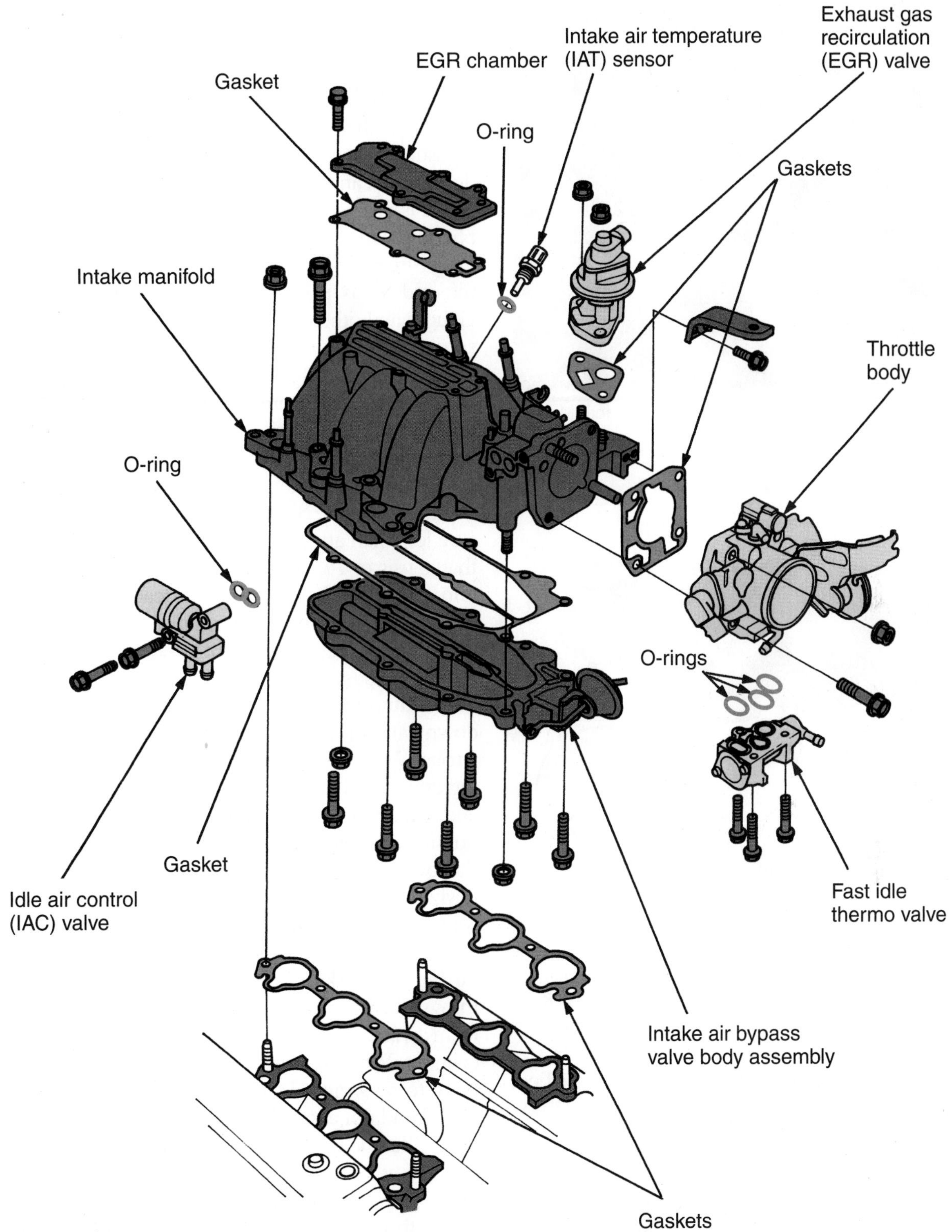

Figure 30-9 A late-model intake manifold for a V6 engine. *Courtesy of American Honda Motor Co., Inc.*

shortened **(Figure 30–11)**, thus increasing high-speed horsepower. Intake manifold variability is the basis for an optimum torque curve over the entire engine-speed range, as well as for good acceleration and responsiveness. Variable intake manifolds provide more power at high speeds without decreasing low-speed torque and fuel economy and without increasing exhaust emissions.

Servicing an Intake Manifold There are few reasons why an intake manifold would need to be replaced. Obviously, if the manifold is cracked or the sealing surfaces severely damaged, it should be replaced. The sealing surfaces should also be checked for flatness. Minor imperfections on the surface can be filed away; however, do not attempt to clean up any serious damage.

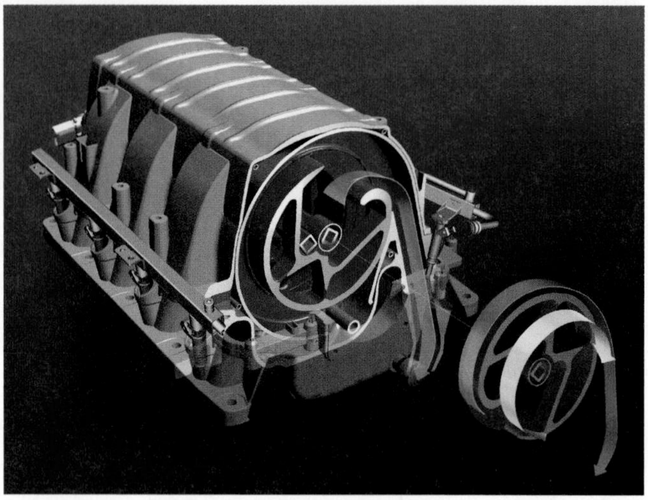

Figure 30-10 A variable-length intake manifold.
Courtesy of BMW of North America, Incorporated

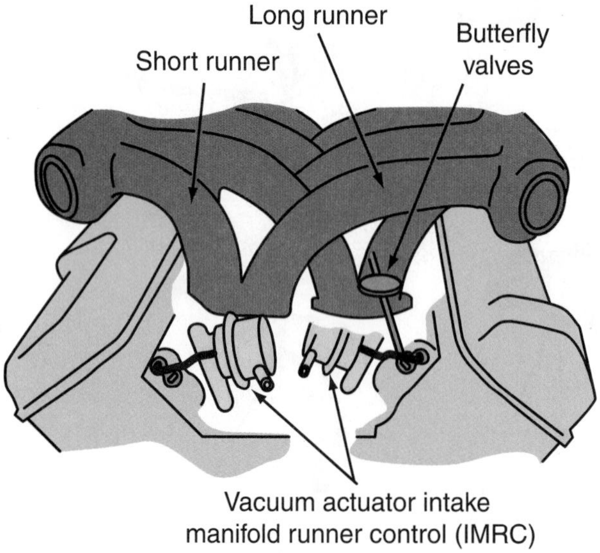

Figure 30-11 The vacuum controls and butterflies used to switch between the long and short intake runners.

When replacing an intake manifold, use new gaskets and seals. Use guide bolts to ensure proper alignment of the manifold. Make sure all of the attaching bolts are tightened to manufacturer's specs and in the specified order. Also make sure that all accessories are properly positioned on and tightened to the manifold.

Vacuum System

The vacuum in the intake manifold is used to operate many systems, such as emission controls, brake boosters, parking brake releases, headlight doors, heater/air conditioners, and cruise controls. Vacuum is applied to these systems through a system of hoses and tubes that can become quite elaborate.

Vacuum Basics Vacuum refers to any pressure that is lower than the earth's atmospheric pressure at any given altitude. The higher the altitude, the lower the atmospheric pressure.

Vacuum is measured in relation to atmospheric pressure. Atmospheric pressure is the pressure exerted on every object on earth and is caused by the weight of the surrounding air. At sea level, the pressure exerted by the atmosphere is 14.7 psi (101.3 kPa). Atmospheric pressure appears as zero on most pressure gauges. This does not mean there is no pressure; rather, it means the gauge is designed to read pressures greater than atmospheric pressure. All measurements taken on this type of gauge are given in pounds per square inch and should be referred to as psig (pounds per square inch gauge). Gauges and other measuring devices that include atmospheric pressure in their readings also display their measurements in psi. However, these should be referred to as psia (pounds per square inch absolute). There is a big difference between 12 psia and 12 psig. A reading of 12 psia is less than atmospheric pressure and therefore would represent a vacuum, whereas 12 psig would be approximately 26.7 psia. Since vacuum is defined as any pressure less than atmospheric, vacuum is any pressure less than 0 psig or 14.7 psia. The normal measure of vacuum is in inches of mercury (in. Hg) instead of psi. Other units of measurement for vacuum are kilopascals and bars. Normal atmospheric pressure at sea level is about 1 bar or 100 kilopascals.

Vacuum in any four-stroke engine is created by the downward movement of the piston during its intake stroke. With the intake valve open and the piston moving downward, a vacuum is created within the cylinder and intake manifold. The air passing the intake valve does not move fast enough to fill the cylinder, thereby causing the lower pressure. This vacuum is continuous in a multicylinder engine, since at least one cylinder is always at some stage of its intake stroke.

The amount of low pressure produced by the piston during its intake stroke depends on a number of a things. Basically it depends on the cylinder's ability to form a vacuum and the intake system's ability to fill the cylinder. When there is high vacuum (15 to 22 inches [381 to 559 mm Hg]), we know the cylinder is well sealed and not enough air is entering the cylinder to fill it. At idle, the throttle plate is almost closed and nearly all airflow to the cylinders is stopped. This is why vacuum is high during idle. Since there is a correlation between throttle position and engine load, it can be said that load directly affects engine manifold vacuum. Therefore, vacuum will be high whenever there is no, or low, load on the engine.

Vacuum Controls Engine manifold vacuum is used to operate and/or control several devices on an engine. Prior to the mid-1970s, vacuum was only used to operate the windshield wipers and/or a distributor vacuum advance unit. Since then the use of vacuum has become extensive.

Today vacuum is typically used to control the following systems:

- Fuel Induction System. Certain vacuum-operated devices are added to carburetors and some fuel-injection throttle bodies to ease engine start-up, warm-and-cold engine driveaway, and to compensate for air conditioner load on the engine.
- Emission Control System. While some emission control output devices are solenoid or linkage controlled, many operate on vacuum. This vacuum is usually controlled by solenoids that are opened or closed, depending on electrical signals received from the **powertrain control module (PCM)**. Other systems use switches controlled by engine coolant temperature, such as a **ported vacuum switch (PVS)** or by ambient air such as a **temperature vacuum switch (TVS)**.
- Accessory Controls. Engine vacuum is used to control operation of certain accessories, such as air conditioner/heater systems, power brake boosters, speed-control components, automatic transmission vacuum modulators, and so on.

Diagnosis and Troubleshooting Vacuum system problems can produce or contribute to the following driveability symptoms:

- Stalls
- No start (cold)
- Hard start (hot soak)
- Backfire (deceleration)
- Rough idle
- Poor acceleration
- Rich or lean stumble
- Overheating
- Detonation, or knock or pinging
- Rotten eggs exhaust odor
- Poor fuel economy

As a routine part of problem diagnosis, a technician who suspects a vacuum problem should first

- Inspect vacuum hoses for improper routing or disconnections (engine decal identifies hose routing).
- Look for kinks, tears, or cuts in vacuum lines.
- Check for vacuum hose routing and wear near hot spots, such as exhaust manifold or the EGR tubes.
- Make sure there is no evidence of oil or transmission fluid in vacuum hose connections. (Valves can become contaminated by oil getting inside.)
- Inspect vacuum system devices for damage (dents in cans; bypass valves; broken nipples on vacuum control valves; broken "tees" in vacuum lines, and so on).

Broken or disconnected hoses allow vacuum leaks that admit more air into the intake manifold than the engine is calibrated for. The most common result is a rough-running engine due to the leaner air/fuel mixture created by the excess air.

Kinked hoses can cut off vacuum to a component, thereby disabling it. For example, if the vacuum hose to the EGR valve is kinked, vacuum cannot be used to move the diaphragm. Therefore, the valve will not open.

To check vacuum controls, refer to the service manual for the correct location and identification of the components. Typical locations of vacuum-controlled components are shown in **Figure 30–12**.

Tears and kinks in any vacuum line can affect engine operation. Any defective hoses should be replaced one at a time to avoid misrouting. OEM vacuum lines are installed in a harness consisting of ⅛-inch (3.18 mm) or larger outer diameter and 1⁄16-inch (1.59 mm) inner diameter nylon hose with bonded nylon or rubber connectors. Occasionally, a rubber hose might be connected to the harness. The nylon connectors have rubber inserts to provide a seal between the nylon connector and the component connection (nipple). In recent years, many domestic car manufacturers have been using ganged steel vacuum lines.

Vacuum Test Equipment. The vacuum gauge is one of the most important engine diagnostic tools used by technicians. With the gauge connected to the intake manifold and the engine warm and idling, watch the action of the gauge's needle. A healthy engine will give a steady, constant vacuum reading between 17 and 22 in. Hg (432 and 559 mm Hg). On some four- and six-cylinder engines, however, a reading of 15 inches (381 mm Hg) is considered acceptable. With high-performance engines, a slight flicker of the needle can also be expected. Keep in mind that the gauge reading will drop about 1 inch (2.54 cm) for each 1,000 feet (305 M) above sea level.

If the amount of vacuum produced by each cylinder is the same, the vacuum gauge will show a steady reading. If one or more cylinders are producing different amounts of vacuum, the gauge will show a fluctuating reading. The amount of vacuum read on the gauge, as well as the movement of the gauge's needle, can tell you quite a bit about the engine. If the gauge reading is low but steady, there is a problem that is common to all cylinders. The severity of the problem is indicated by how low it is. For example, if a vacuum reading is a steady 10 in. Hg (254 mm Hg), the problem is something common to all cylinders, such as a fairly good-sized intake manifold vacuum leak. If the gauge reads a steady 15 in. Hg (381 mm Hg), the problem is less severe but still common to all cylinders, such as retarded ignition timing.

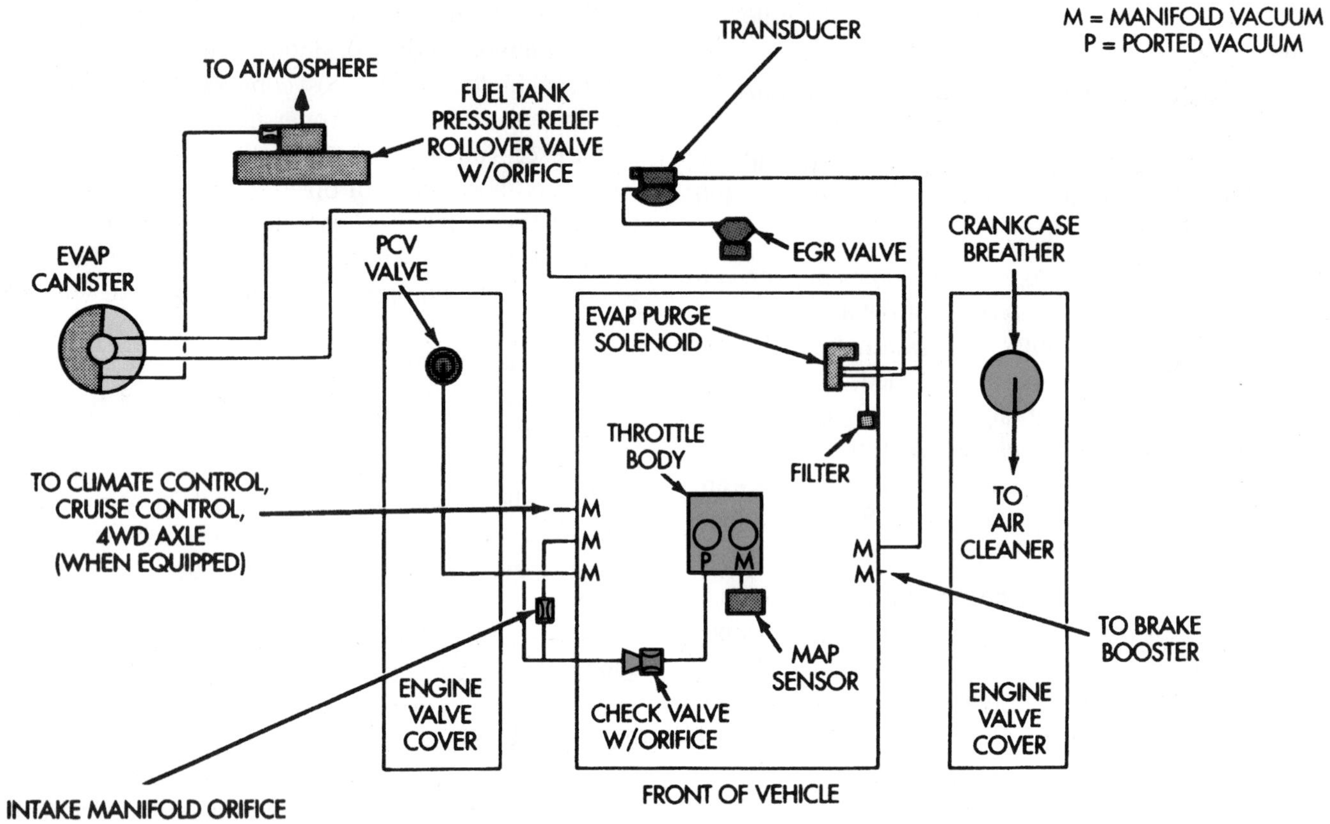

Figure 30-12 Typical vacuum devices and controls. *Courtesy of DaimlerChrysler Corporation*

A fluctuating needle indicates a problem that is not common to all cylinders. If the gauge is connected to a four-cylinder engine and the gauge's needle bounces at an even pace between 10 and 17 in. Hg (254 and 432 mm Hg), we can assume that two of the four cylinders are producing less vacuum than the other two. If the needle spends more time at the 17 reading, we can assume one cylinder is producing less vacuum. Again, how low the needle dips is an indication of the severity of the problem. If the needle dips to zero, we could suspect a hole in the piston or a severely damaged valve in one cylinder. If the needle dips to 15 in. Hg (381 mm Hg), the problem might be worn piston rings on one cylinder.

As shown in **Figure 30–13**, a hand-held vacuum pump/gauge is used to test vacuum-actuated valves and motors. If the component does not operate when the proper amount of vacuum is applied, it should be serviced or replaced.

EXHAUST SYSTEM COMPONENTS

The various components of the typical exhaust system include the following:

- Exhaust manifold
- Exhaust pipe and seal
- Catalytic converter
- Muffler
- Resonator
- Tailpipe
- Heat shields
- Clamps, brackets, and hangers
- Exhaust gas oxygen sensors

All the parts of the system are designed to conform to the available space of the vehicle's undercarriage and yet be a safe distance above the road.

CAUTION!

When inspecting or working on the exhaust system, remember that its components get very hot when the engine is running. Contact with them could cause a severe burn. Also, always wear safety glasses or goggles when working under a vehicle.

Exhaust Manifold

The exhaust manifold **(Figure 30–14)** collects the burnt gases as they are expelled from the cylinders and directs them to the exhaust pipe. Exhaust manifolds for most vehicles are made of cast or nodular iron. Many newer vehicles have stamped, heavy-gauge sheet metal or stainless steel units.

Figure 30-13 A hand-operated vacuum pump being used to test an air cleaner vacuum motor. *Courtesy of Stant Manufacturing Inc.*

In-line engines have one exhaust manifold. V-type engines have an exhaust manifold on each side of the engine. An exhaust manifold will have either three, four, or six passages, depending on the type of engine. These passages blend into a single passage at the other end, which connects to an exhaust pipe. From that point, the flow of exhaust gases continues to the catalytic converter, muffler, and tail pipe, then exits at the rear of the car.

V-type engines may be equipped with a dual exhaust system that consists of two almost identical, but individual systems in the same vehicle.

Exhaust systems are designed for particular engine-chassis combinations. Exhaust system length, pipe size, and silencer size are used to tune the flow of gases within the exhaust system. Proper tuning of the exhaust manifold tubes can actually create a partial vacuum that helps draw exhaust gases out of the cylinder, improving volumetric efficiency. Separate, tuned exhaust headers **(Figure 30–15)** can also improve efficiency by preventing the exhaust flow of one cylinder from interfering with the exhaust flow of another cylinder. Cylinders next to one another may release exhaust gas at about the same time. When this happens, the pressure of the

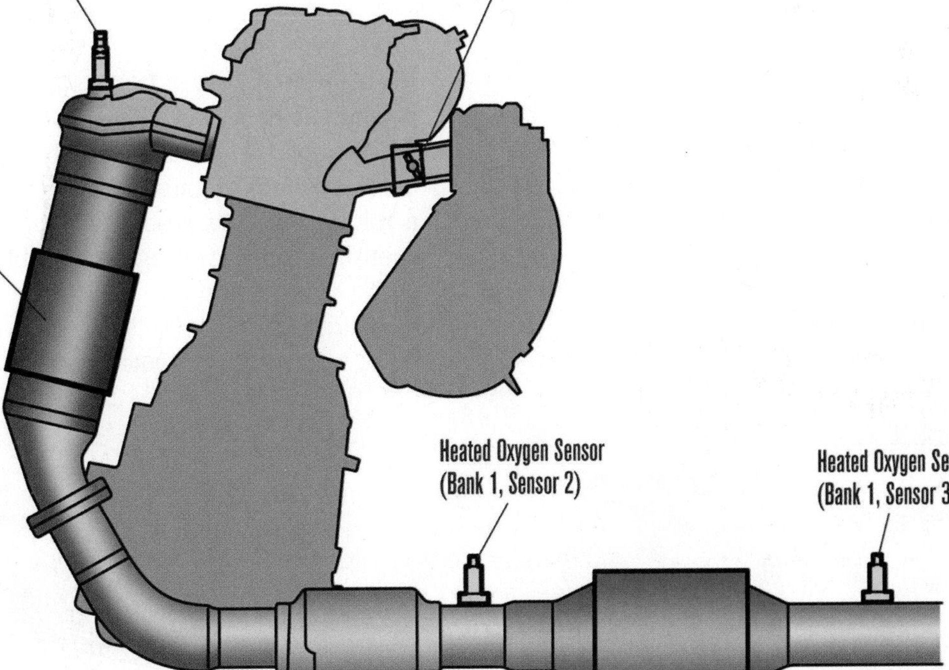

Figure 30-14 The basic configuration of a late-model exhaust manifold and pipe with its oxygen sensors and catalytic converter. *Reprinted with permission*

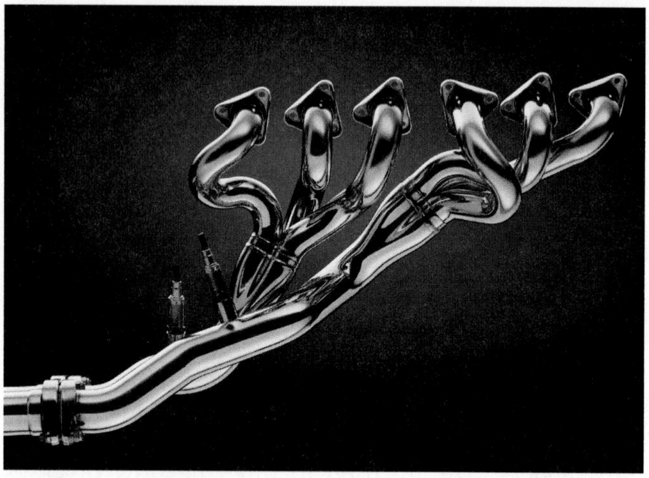

Figure 30-15 A tuned exhaust manifold, called a header. *Courtesy of BMW of North America, Incorporation*

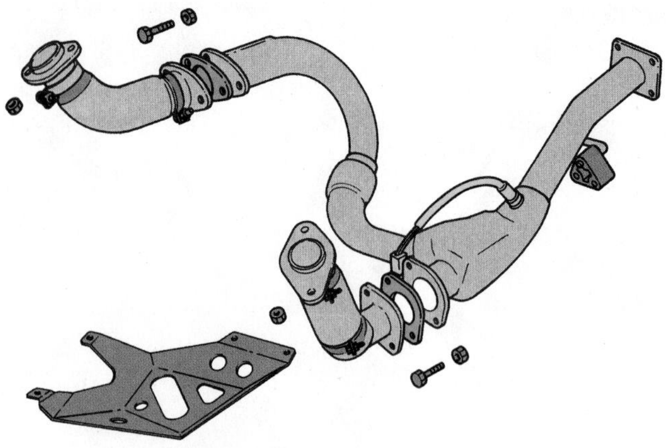

Figure 30-17 The front exhaust pipe assembly for a V6 engine. *Reprinted with permission by Isuzu Motors, Inc.*

exhaust gas from one cylinder can interfere with the flow from the other cylinder. With separate headers, the cylinders are isolated from one another, interference is eliminated, and the engine breathes better. The problem of interference is especially common with V8 engines. However, exhaust headers tend to improve the performance of all engines.

Exhaust manifolds may also be the attaching point for the air injection reaction (AIR) pipe **(Figure 30–16)**. This pipe introduces cool air from the AIR system into the exhaust stream. Some exhaust manifolds have provisions for the EGR pipe. This pipe takes a sample of the exhaust gases and delivers it to the EGR valve. Also, exhaust manifolds have a tapped bore that retains the oxygen sensor.

Exhaust Pipe and Seal

The exhaust pipe is metal pipe—either aluminized steel, stainless steel, or zinc-plated heavy-gauge steel—that runs under the vehicle between the exhaust manifold and the catalytic converter **(Figure 30–17)**.

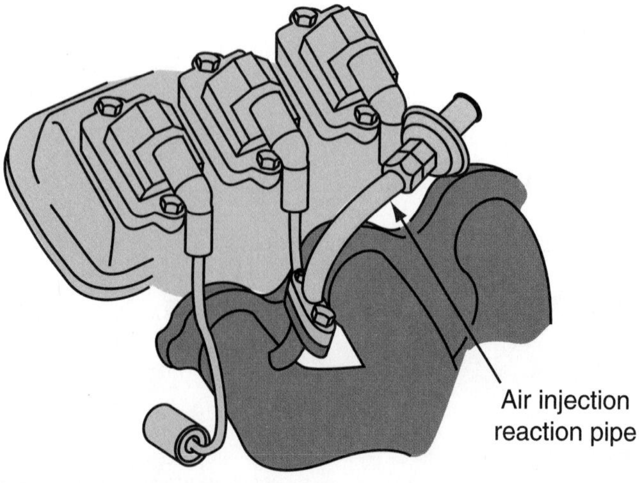

Figure 30-16 An AIR pipe mounting on an exhaust manifold.

SHOP TALK

The exhaust manifold gasket seals the joint between the head and exhaust manifold. Many new engines are assembled without exhaust manifold gaskets. This is possible because new manifolds are flat and fit tightly against the head without leaks. Exhaust manifolds go through many heating/cooling cycles. This causes stress and some corrosion in the exhaust manifold. Removing the manifold usually distorts the manifold slightly so that it is no longer flat enough to seal without a gasket. Exhaust manifold gaskets are normally used to eliminate leaks when exhaust manifolds are reinstalled. ■

Catalytic Converters

A **catalytic converter (Figure 30–18)** is part of the exhaust system and a very important part of the emission control system. Because it is part of both systems, it has a role in both. As an emission control device, it is responsible for converting undesirable exhaust gases into

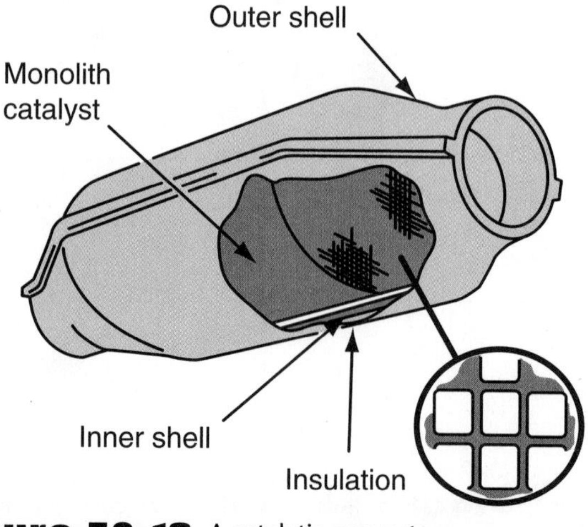

Figure 30-18 A catalytic converter.

harmless gases. As part of the exhaust system, it helps reduce the noise level of the exhaust. A catalytic converter contains a ceramic element coated with a catalyst. A catalyst is a substance that causes a chemical reaction in other elements without actually becoming part of the chemical change and without being used up or consumed in the process.

Catalytic converters may be pellet type or monolithic type. A pellet-type converter contains a bed made from hundreds of small beads. Exhaust gases pass over this bed. In a monolithic-type converter, the exhaust gases pass through a honeycomb ceramic block. The converter beads or ceramic block are coated with a thin coating of cerium, platinum, palladium, and/or rhodium, and are held in a stainless steel container.

Most vehicles are equipped with a minicatalytic converter that is either built into the exhaust manifold or is located next to it **(Figure 30–19)**. These converters are used to clean the exhaust during engine warm-up and are commonly called warm-up converters.

Many catalytic converters have an air hose connected from the AIR system to the oxidizing catalyst. This air helps the converter work by making extra oxygen available. The air from the AIR system is not always forced into the converter; rather, it is controlled by the vehicle's PCM. Fresh air added to the exhaust at the wrong time could overheat the converter and produce NO_x, something the converter is trying to destroy.

OBD-II regulations call for a way to inform the driver that the vehicle's converter has a problem and may be ineffective. The PCM monitors the activity of the converter by comparing the signals of an HO2S located at the front of the converter with the signals from an HO2S located at the rear **(Figure 30–20)**. If the sensors' outputs are the same, the converter is not working properly and the MIL on the dash will light.

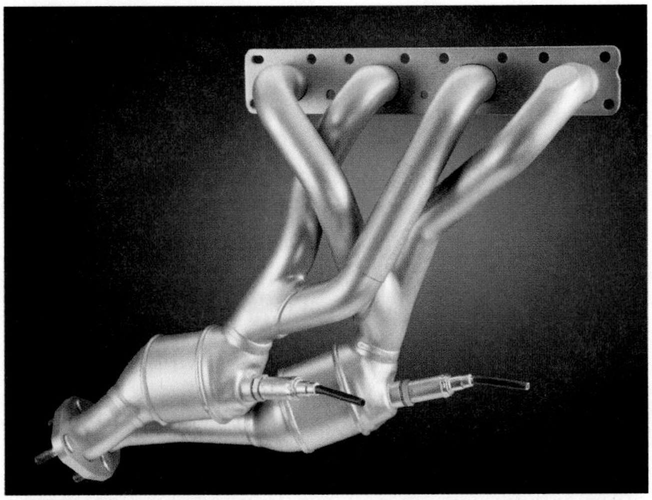

Figure 30–19 This exhaust manifold has two separate mini- or warm-up catalytic converters. *Courtesy of BMW of North America, Incorporated*

Converter Problems The converter is normally a trouble-free emission control device; however, it can go bad or become plugged. Often such problems are caused by overheating the converter. When raw fuel enters the exhaust because of an engine misfiring, the temperature of the converter quickly increases. The heat can melt the catalyst materials inside the converter, causing a major restriction to the flow of exhaust.

A plugged converter or any exhaust restriction can cause: damage to the exhaust valves due to excess heat, loss of power at high speeds, stalling after starting (if totally blocked), a drop in engine vacuum as engine rpm increases, or sometimes popping or backfiring at the carburetor.

The best way to determine if a catalytic converter is working is to check the quality of the exhaust. This is done with a four-gas exhaust analyzer. The results of this test should show low emission levels if the converter is working properly.

Another way to test a converter is to use a hand-held digital **pyrometer**, an electronic device that measures heat. By touching the pyrometer probe to the exhaust pipe just ahead of and just behind the converter, it is possible to read an increase of at least 100°F (37.7°C) as the exhaust gases pass through the converter. If the outlet temperature is the same or lower, nothing is happening inside the converter. This means the converter should be replaced. If there is only a slight difference in temperature, check the activity of the oxygen sensor before condemning the converter. The efficiency of today's converters depends on the normal swings of rich and lean mixtures. A biased O2S can affect converter activity. If the O2S is working fine, the converter should be replaced. Further testing of a catalytic converter is included in Chapter 32.

Mufflers

The **muffler** is a cylindrical or oval-shaped component, generally about 2 feet (.6 meters) long, mounted in the exhaust system about midway or toward the rear of the car. Inside the muffler is a series of baffles, chambers, tubes, and holes to break up, cancel out, or silence the pressure pulsations that occur each time an exhaust valve opens.

Two types of mufflers are frequently used on passenger vehicles **(Figure 30–21)**. Reverse-flow mufflers change the direction of the exhaust gas flow through the inside of the unit. This is the most common type of automotive muffler. Straight-through mufflers permit exhaust gases to pass through a single tube. The tube has perforations that tend to break up pressure pulsations. They are not as quiet as the reverse-flow type.

In recent years there have been several important changes in the design of mufflers. Most of these changes

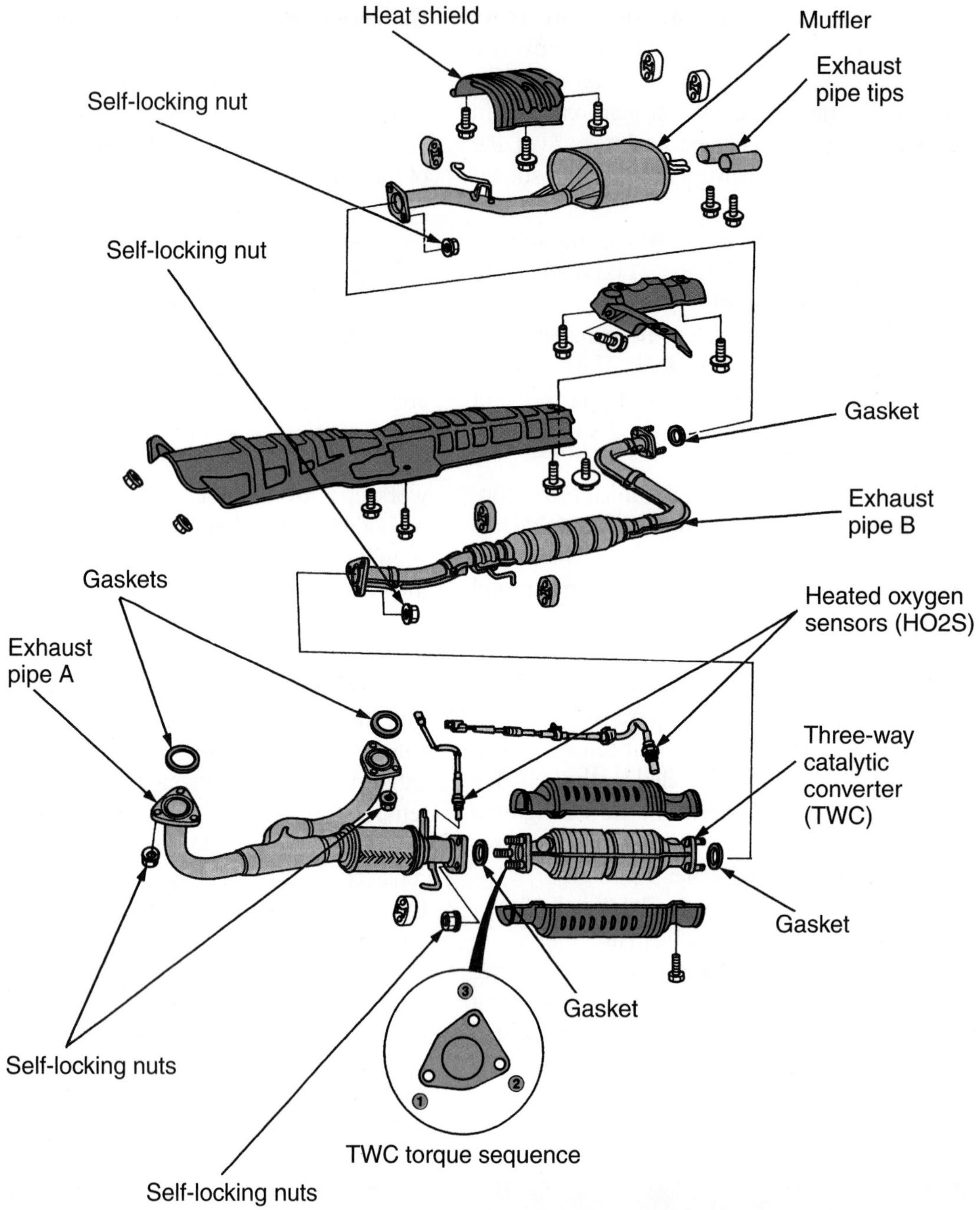

Figure 30-20 An exhaust system for an OBD-II vehicle. *Courtesy of American Honda Motor Co., Inc.*

have been centered at reducing weight and emissions, improving fuel economy, and simplifying assembly. These changes include the following:

New Materials. More and more mufflers are being made of aluminized and stainless steel. Using these materials reduces the weight of the units and extends their lives.

Double-Wall Design. Retarded engine ignition timing that is used on many small cars tends to make the exhaust pulses sharper. Many cars use a double-wall exhaust pipe to better contain the sound and reduce pipe ring.

Rear-Mounted Muffler. More and more often, the only space left under the car for the muffler is at the very rear. This means the muffler runs cooler than before and is more easily damaged by condensation in the exhaust system. This moisture, combined with nitrogen and sulfur oxides in the exhaust gas, forms acids that rot the muffler from the inside out. Many mufflers are being produced with drain holes drilled into them.

Back Pressure. Even a well-designed muffler produces some **back pressure** in the system. Back pressure reduces an engine's volumetric efficiency, or ability to "breathe." Excessive back pressure caused by defects in a muffler or other exhaust system part can slow or stop the engine. However, a small amount of back pressure can be used intentionally to allow a slower passage of exhaust gases

(A)

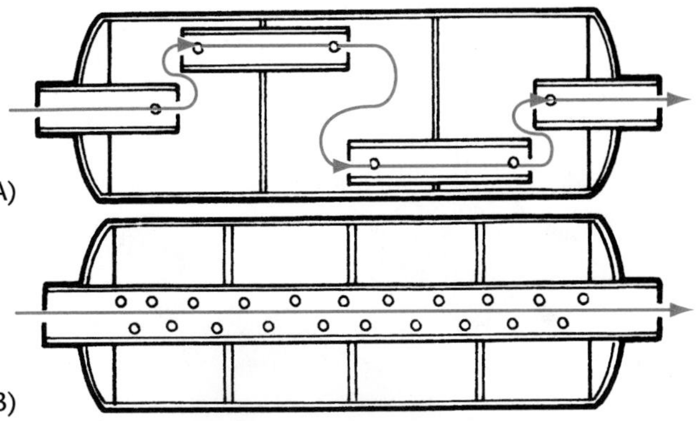

(B)

Figure 30-21 (A) A reverse-flow muffler, and (B) a straight-through muffler.

through the catalytic converter. This slower passage results in more complete conversion to less harmful gases. Also, no back pressure may allow intake gases to enter the exhaust.

Resonator

On some older vehicles, there is an additional muffler, known as a **resonator** or silencer. This unit is designed to further reduce or change the sound level of the exhaust. It is located toward the end of the system and generally looks like a smaller, rounder version of a muffler.

Tailpipe

The **tailpipe** is the last pipe in the exhaust system. It releases the exhaust fumes into the atmosphere beyond the back end of the car.

Heat Shields

Heat shields are used to protect other parts from the heat of the exhaust system and the catalytic converter **(Figure 30–22)**. They are usually made of pressed or perforated sheet metal. Heat shields trap the heat in the exhaust system, which has a direct effect on maintaining exhaust gas velocity.

Clamps, Brackets, and Hangers

Clamps, brackets, and hangers are used to properly join and support the various parts of the exhaust system. These parts also help to isolate exhaust noise by preventing its transfer through the frame **(Figure 30–23)** or body to the passenger compartment. Clamps help to secure exhaust system parts to one another. The pipes are formed in such a way that one slips inside the other. This design makes a close fit. A U-type clamp usually holds this connection tight **(Figure 30–24)**. Another important job of clamps and brackets is to hold pipes to the bottom of the vehicle. Clamps and brackets must be designed to allow the exhaust system to vibrate without transferring the vibrations through the car.

There are many different types of flexible hangers available, each designed for a particular application. Some exhaust systems are supported by doughnut-shaped rubber rings between hooks on the exhaust component and on the frame or car body. Others are supported at the exhaust pipe and tailpipe connections by a combination of metal and reinforced fabric hanger. Both the doughnuts and the reinforced fabric allow the exhaust system to vibrate without breakage that could be caused by direct physical connection to the vehicle's frame.

Some exhaust systems are a single unit in which the pieces are welded together by the factory. By welding instead of clamping the assembly together, car makers save the weight of overlapping joints as well as that of clamps.

EXHAUST SYSTEM SERVICE

Exhaust system components are subject to both physical and chemical damage. Any physical damage to an exhaust system part that causes a partially restricted or blocked exhaust system usually results in loss of power or backfire up through the throttle plate(s). In addition to improper engine operation, a blocked or restricted exhaust system causes increased noise and air pollution. Leaks in the exhaust system caused by either physical or chemical (rust) damage could result in illness, asphyxiation, or even death. Remember that vehicle exhaust fumes can be very dangerous to one's health.

Exhaust System Inspection

Most parts of the exhaust system, particularly the exhaust pipe, muffler, and tailpipe, are subject to rust, corrosion, and cracking. Broken or loose clamps and hangers can allow parts to separate or hit the road as the car moves.

CAUTION!

During all exhaust inspection and repair work, wear safety glasses or equivalent eye protection.

Complete exhaust system inspection and testing procedures are given in the *Tech Manual* that accompanies this textbook. Any inspection should include listening for hissing or rumbling that would result from a leak in the system. An on-lift inspection should pinpoint any of the following types of damage:

- Holes, road damage, separated connections, and bulging muffler seams
- Kinks and dents
- Discoloration, rust, soft corroded metal, and so forth.
- Torn, broken, or missing hangers and clamps
- Loose tailpipes or other components
- Bluish or brownish catalytic converter shell, which indicates overheating

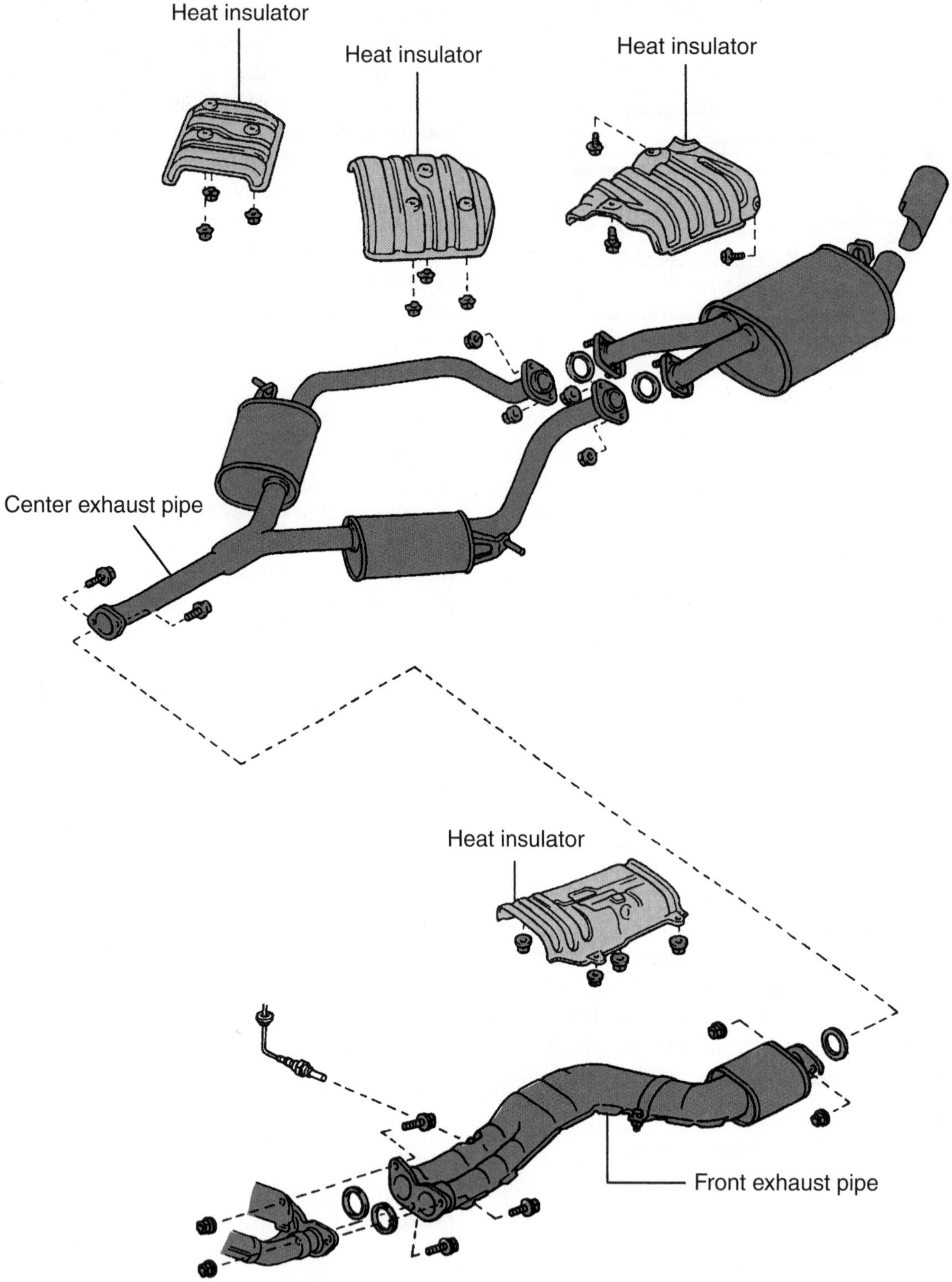

Figure 30-22 The typical location of heat shields in an exhaust system. *Reprinted with permission*

Exhaust Restriction Test Often leaks and rattles are the only things looked for in an exhaust system. The exhaust system should also be tested for blockage and restrictions. Collapsed pipes or clogged converters and/or mufflers can cause these blockages.

There are many ways to check for a restricted exhaust, the most common of which is the use of a vacuum gauge. Connect a vacuum gauge to an intake manifold vacuum source. Bring the engine to a moderate speed and hold it there. Watch the vacuum gauge. If everything is right, the vacuum reading will be high and will either stay at that reading or increase slightly as the engine runs at this speed. If the exhaust is restricted, the vacuum will begin to decrease after a period of time. This is caused by the cylinder's inability to purge itself of all of its exhaust gases during the exhaust stroke. The presence of exhaust in the cylinder when the intake stroke begins will decrease the amount of vacuum that can be formed on that stroke.

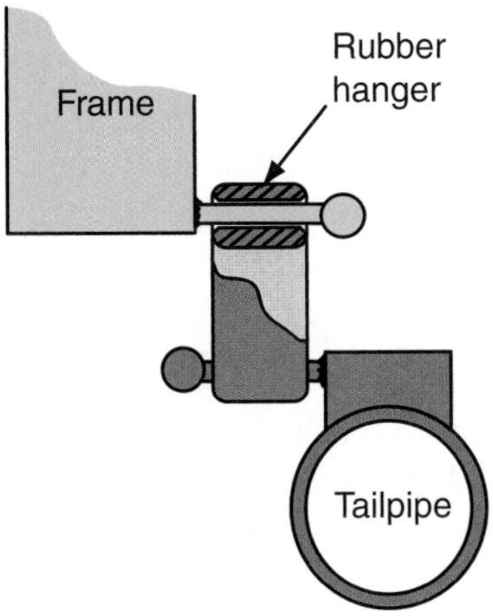

Figure 30-23 Rubber hangers are used to keep the exhaust system in place without allowing it to contact this pickup's frame.

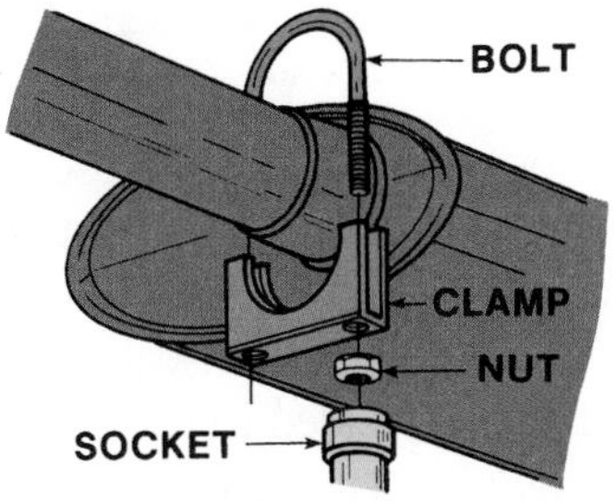

Figure 30-24 A U-type clamp is often used to secure two pipes that slip together.

Replacing Exhaust System Components

Before beginning work on an exhaust system, make sure it is cool to the touch. Some technicians disconnect the battery's negative cable before starting to work to avoid short-circuiting the electrical system. Soak all rusted bolts, nuts, and other removable parts with a good penetrating oil. Finally, check the system for critical clearance points so they can be maintained when new components are installed.

Most exhaust work involves the replacement of parts. When replacing exhaust parts, make sure the new parts are exact replacements for the original parts. Doing this will ensure proper fit and alignment, as well as ensure acceptable noise levels.

Exhaust system component replacement might require the use of special tools **(Figure 30–25)** and welding equipment.

Exhaust Manifold and Exhaust Pipe Servicing As mentioned, the manifold itself rarely causes any problems. On occasion, an exhaust manifold will warp because of excess heat. A straightedge and feeler gauge can be used to check the machined surface of the manifold.

Another problem—also the result of high temperatures generated by the engine—is a cracked manifold. This usually occurs after the car passes through a large puddle and cold water splashes on the manifold's hot surface. If the manifold is warped beyond manufacturer's specifications or is cracked, it must be replaced. Also, check the exhaust pipe for signs of collapse. If there is damage, repair it. These repairs should be done as directed in the vehicle's service manual.

Replacing Leaking Gaskets and Seals The most likely spot to find leaking gaskets and seals is between the exhaust manifold and the exhaust pipe **(Figure 30–26)**.

When installing exhaust gaskets, carefully follow the recommendations on the gasket package label and instruction forms. Read through all installation steps before beginning. Take note of any of the original equipment manufacturer's recommendations in service manuals that

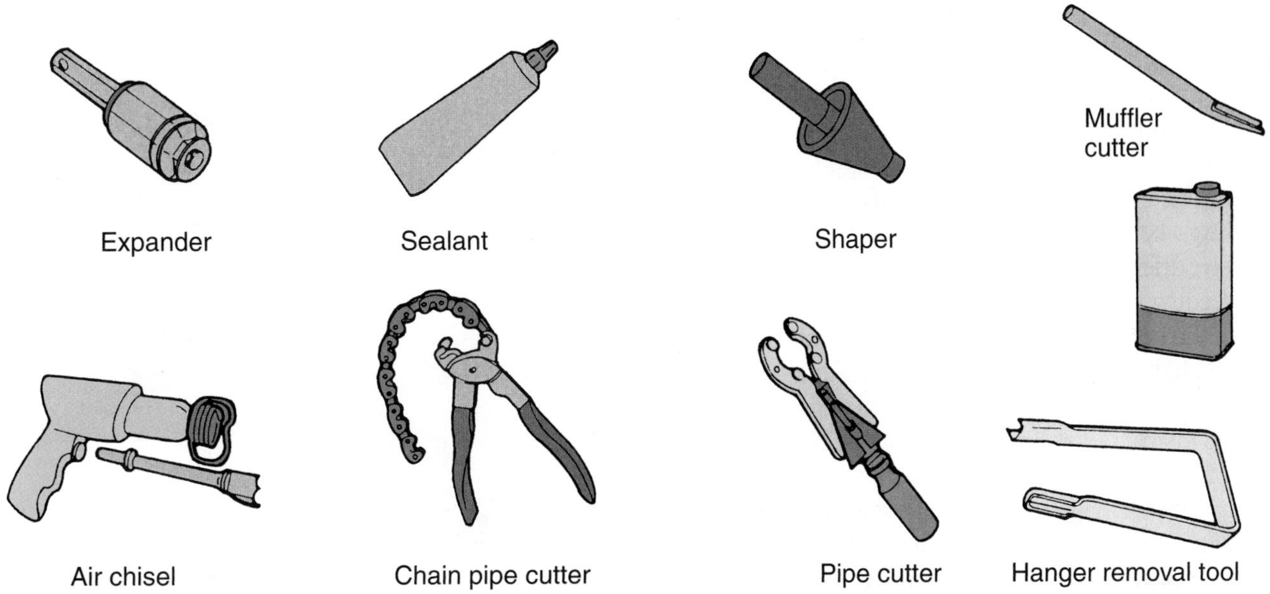

Figure 30-25 Special tools required for exhaust work.

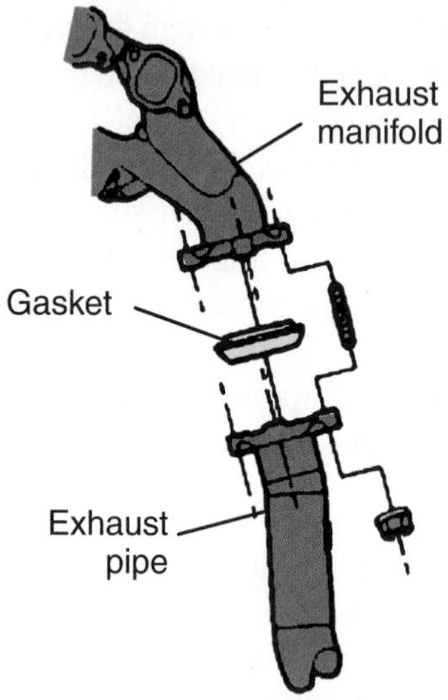

Figure 30-26 Leaking gaskets and seals are often found between the exhaust manifold and pipe.

could affect engine sealing. Manifolds warp more easily if an attempt is made to remove them while they are still hot. Remember, heat expands metal, making assembly bolts more difficult to remove and easier to break.

To replace an exhaust manifold gasket, follow the torque sequence in reverse to loosen each bolt. Repeat the process to remove the bolts. Doing this minimizes the chance that components will warp.

Any debris left on the sealing surfaces increases the chance of leaks. A good gasket remover will quickly soften the old gasket debris and adhesive for quick removal. Carefully remove the softened pieces with a scraper and a wire brush. Be sure to use a nonmetallic scraper when attempting to remove gasket material from aluminum surfaces.

Inspect the manifold for irregularities that might cause leaks, such as gouges, scratches, or cracks. Replace it if it is cracked or badly warped. File down any imperfections to ensure proper sealing of the manifold.

Due to high heat conditions, it is important to retap and redie all threaded bolt holes, studs, and mounting bolts. This procedure ensures tight, balanced clamping forces on the gasket. Lubricate the threads with a good high-temperature antiseize lubricant. Use a small amount of contact adhesive to hold the gasket in place. Align the gasket properly before the adhesive dries. Allow the adhesive to dry completely before proceeding with manifold installation.

Install the bolts finger-tight. Tighten the bolts in three steps—one-half, three-quarters, and full torque—following the torque tables in the service manual or gasket manufacturer's instructions. Torquing is usually begun in the center of the manifold, working outward in an X pattern.

To replace a damaged exhaust pipe, begin by supporting the converter to keep it from falling. Carefully remove the oxygen sensor if there is one. Remove any hangers or clamps holding the exhaust pipe to the frame. Unbolt the flange holding the exhaust pipe to the exhaust manifold. When removing the exhaust pipe, check to see if there is a gasket. If so, discard it and replace it with a new one. Once the joint has been taken apart, the gasket loses its effectiveness. Disconnect the pipe from the converter and pull the front exhaust pipe loose and remove it.

SHOP TALK

An easy way to break off rusted nuts is to tighten them instead of loosening them. Sometimes a badly rusted clamp or hanger strap will snap off with ease. Sometimes the old exhaust system will not drop free of the body because a large part is in the way, such as the rear end or the transmission support. Use a large cold chisel, pipe cutter, hacksaw, muffler cutter, or chain cutter to cut the old system at convenient points to make the exhaust assembly smaller. ■

Although most exhaust systems use flanges or a slip joint and clamps to fasten the pipe to the muffler, a few use a welded connection. If the vehicle's system is welded, cut the pipe at the joint with a hacksaw or pipe cutter. The new pipe need not be welded to the muffler. An adapter, available with the pipe, can be used instead. When measuring the length for the new pipe, allow at least 2 inches (50.8 mm) for the adapter to enter the muffler.

CAUTION!

Be sure to wear safety goggles to protect your eyes and work gloves to prevent cutting your hands on the rusted parts.

When trying to replace a part in the exhaust system, you may run into parts that are rusted together. This is especially a problem when a pipe slips into another pipe or the muffler. If you are trying to reuse one of the parts, you should carefully use a cold chisel or slitting tool **(Figure 30–27)** on the outer pipe of the rusted union. You

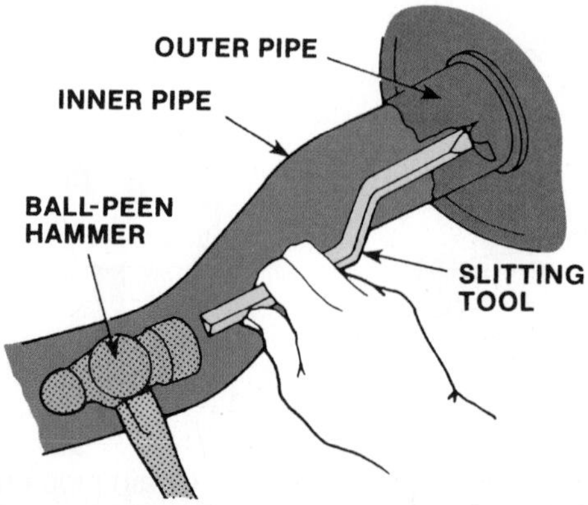

Figure 30-27 Removing a rusted-on muffler.

must be careful when doing this, because you can easily damage the inner pipe. It must be perfectly round to form a seal with a new pipe.

Slide the new pipe over the old. Position the rest of the exhaust system so that all clearances are evident and the parts aligned, then put a U-clamp over the new outer pipe to secure the connection.

CAUTION!

Be sure no exhaust part comes into direct contact with any section of the body, fuel lines, fuel tank, or brake lines.

TURBOCHARGERS AND SUPERCHARGERS

The power generated by the internal combustion engine is directly related to the amount of air compressed in the cylinders. In other words, the greater the compression (within reason), the greater the output of the engine.

The two processes of artificially increasing the amount of airflow into the engine are known as turbocharging and supercharging.

Turbocharger Operation

Turbochargers (Figure 30–28) do not require a mechanical connection between the engine and the pressurizing pump to compress the intake gases. Instead, they rely on the rapid expansion of hot exhaust gases exiting the cylinders. These gases spin the turbine blades (hence the name turbocharger) of the pump. Because exhaust gas is a waste product, the energy developed by the turbine is said to be free since it theoretically does not use any of the engine's power that it helps to produce.

A typical turbocharger, usually called a turbo, consists of the following components **(Figure 30–29)**.

- Turbine or hot wheel
- Shaft
- Compressor or cold wheel
- Wastegate valve
- Actuator
- Center housing and rotating assembly (CHRA). This component contains the bearings, shaft, turbine seal assembly, and compressor seal assembly.

The turbocharger is normally located close to the exhaust manifold. An exhaust pipe runs between the exhaust manifold and the turbine housing to carry the exhaust flow to the turbine wheel **(Figure 30–30)**. Another pipe connects the compressor housing intake to the throttle body assembly.

Inside the turbocharger, the turbine wheel (hot wheel) is attached via a shaft to the intake compressor wheel (cold wheel). Each wheel is encased in its own spiral-shaped housing that controls and directs the flow of exhaust and intake gases. The shaft that joins the two wheels rides on bearings.

The air compressing process typically starts when the engine's speed is above 2,000 rpm. The force of the

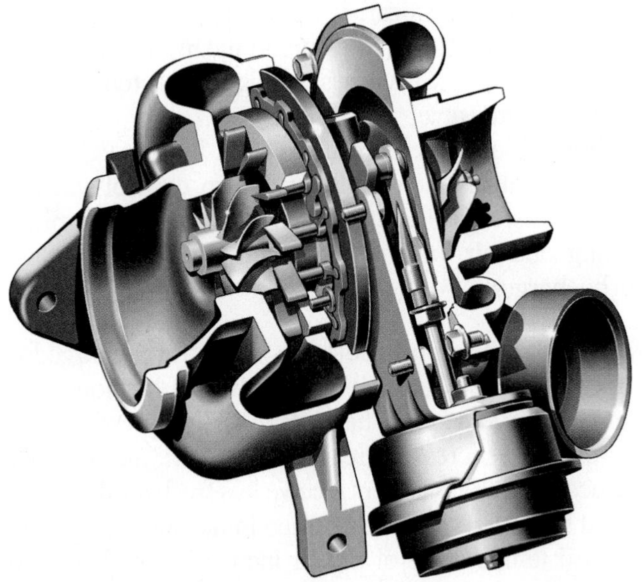

Figure 30–28 A turbocharger. *Courtesy of DaimlerChrysler Corporation*

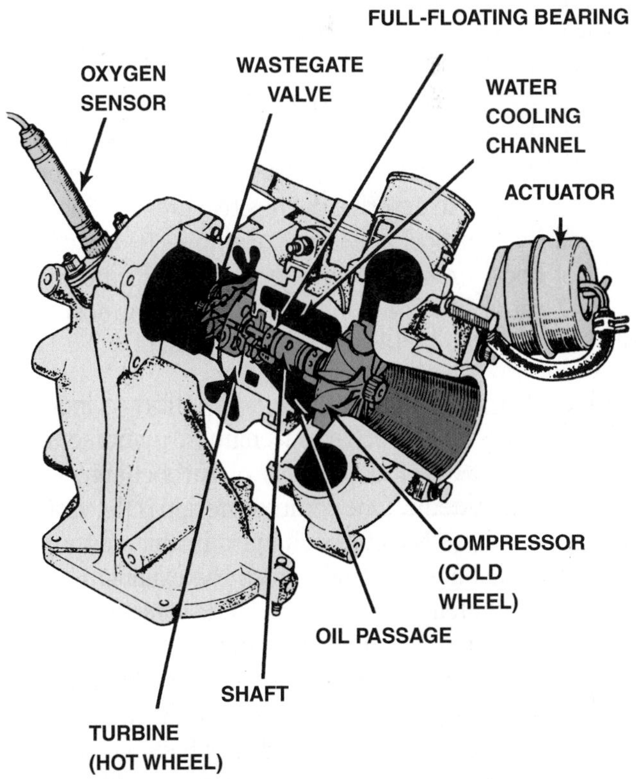

Figure 30–29 A cross section of a turbocharger shows the turbine wheel, the compressor wheel, and their connecting shaft. *Reprinted with permission*

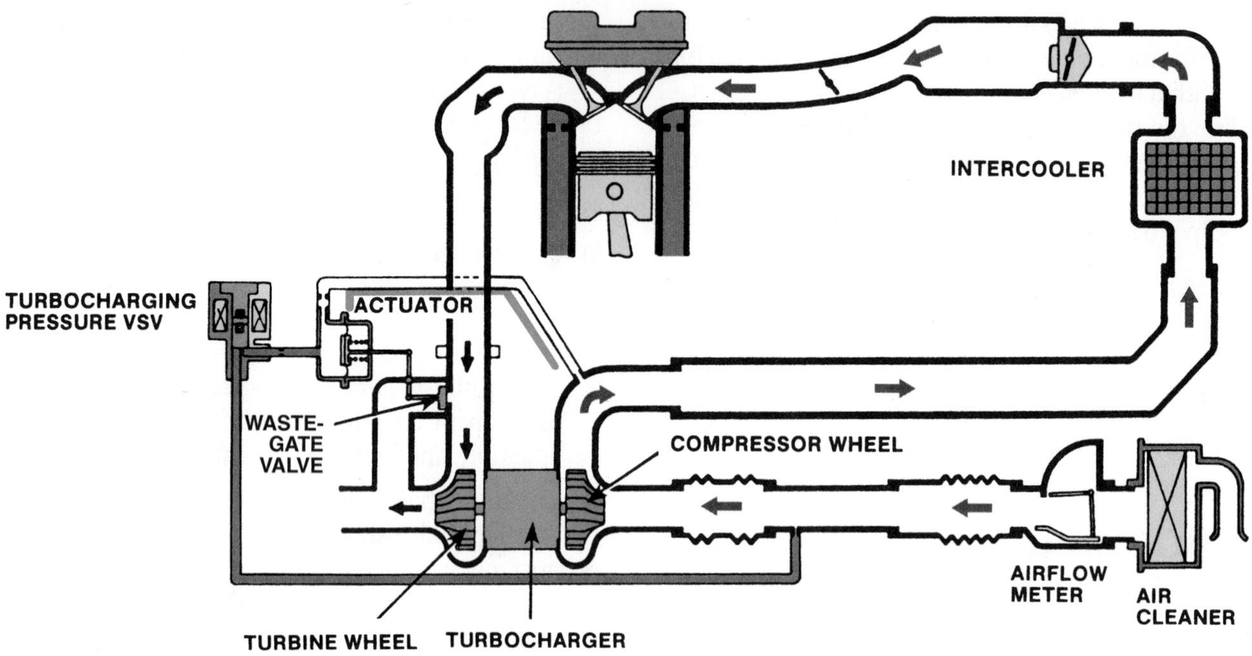

Figure 30-30 Exhaust gas and airflow in a typical turbocharger system. *Reprinted with permission*

exhaust flow is directed through a nozzle against the side of the turbine wheel. As the hot gases hit the turbine wheel causing it to spin, the specially curved turbine fins direct the air toward the center of the housing where it exits. Once the turbine starts to spin, the compressor wheel (shaped like a turbine wheel in reverse) also starts to spin, causing air to be drawn into the center, where it is caught by the whirling blades of the compressor and thrown outward by centrifugal force. From there, the air exits under pressure through the remainder of the induction system on its way to the cylinder.

A turbocharger is capable of pressurizing the intake charge above normal atmospheric pressure. *Turbo boost* is the term used to describe the positive pressure increase created by a turbocharger. For example, 10 psi (68.95 kPa) of boost means the air is being fed into the engine at 24.7 psi (170 kPa) (14.7 psi [101 kPa] atmospheric plus 10 pounds [44.48 Newtons] of boost).

Various Turbocharger Designs In an effort to increase the efficiency of turbocharged engines, manufacturers have developed various designs of turbochargers and their control systems. One common design is the variable nozzle turbine (VNT) turbocharger. In this design, the cross-sectional area through which the exhaust flows is variable. This area is adjusted, via guide vanes, according to engine speed. At lower engine speeds, the vanes restrict exhaust flow, thereby increasing boost pressure. At higher engine speeds, the vanes open wider and exhaust back pressure decreases. VNT turbos do not have a wastegate, provide a higher boost at lower engine speeds, and are more responsive to changes in engine load. They also help reduce the effects of turbo lag.

Some engines have two turbochargers, each of a different size. The smaller of the two spins up to speed very quickly. This reduces turbo lag. The larger one is slower to get up to speed but adds the boost at higher engine speeds. This design is two-stage: one for lower engine speeds and immediate increase of speed and one for sustained power. Other engines with two turbos have one for each bank of a V-type engine.

Wastegate Valve If the turbocharged air pressure becomes too high, knocking occurs and engine output actually decreases. To prevent this, the turbocharger uses a **wastegate** valve. This valve allows a certain amount of exhaust gas to bypass the turbine once the ideal boost is exceeded. The wastegate is usually operated by an actuator that senses the air pressure in the induction system. When the pressure becomes too high, the actuator opens the wastegate valve.

Some wastegates are controlled by the PCM that directly controls a solenoid that, in turn, controls vacuum to the waste gas. The PCM also coordinates ignition timing and air/fuel mixtures with the output of the turbocharger.

Intercooler The **intercooler** cools the turbocharged or supercharged air before it reaches the combustion chamber **(Figure 30–31)**. When intake air is compressed, its temperature increases greatly. As a result, air density is reduced and the cylinders receive less fresh air than what could be gotten from the boosted intake air. Also, the increased temperature of the air increases the chances of engine knock. To offset these consequences, many turbocharger and supercharger systems are fitted with an in-

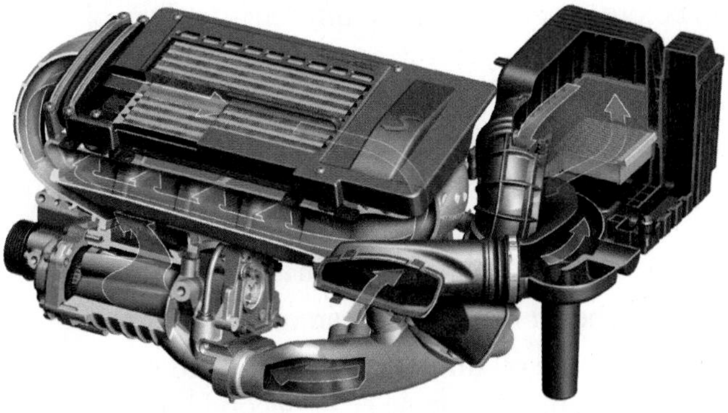

Figure 30-31 Routing of the boosted air in and out of an intercooler. *Courtesy of BMW of North America, Incorporated*

tercooler or charge air cooler. The air leaving the turbo- or supercharger passes through a cooler. Cooling the air makes it denser and lowers the temperature produced in the combustion chamber. These factors help reduce engine knock and increase engine output. Intercoolers are like radiators in that heat from the air passing through them is removed and dissipated to the atmosphere. Intercoolers can be air or water cooled or cooled by the air-conditioning system.

Lubricating System Most turbochargers are lubricated by engine oil that is line-fed to the unit's oil inlet. The oil drains back to the engine through a separate line. A turbocharger should never be operated if the engine has less than 30-psi oil pressure.

WARNING!

Lack of lubrication and oil lag are major causes of turbocharger failure.

SHOP TALK

When oil leakage is noted at the turbine end of the turbocharger, always check the turbocharger oil drain tube and the engine crankcase breathers for restrictions. When sludged engine oil is found, the engine's oil and oil filter must be changed. ■

CUSTOMER CARE

Explain to the customers the oiling needs of a turbocharger and the need for special attention during starting and turning off the engine. ■

Spark-Retard System Retarding spark timing is an often-used method of controlling detonation on turbocharged engine systems. Most systems use knock-sensing devices to retard timing only when detonation is detected.

Computer-Controlled Systems Control devices limit the amount of boost to prevent detonation and engine damage. In fact, some turbocharging systems on computer-controlled vehicles use an electronic control unit to operate the wastegate control valve through sensor signals.

USING SERVICE MANUALS

General service procedures for turbocharger systems are normally in a separate section of service manuals. Individual inputs and outputs that are part of the electronic control system are covered in the engine control or performance section of the manual. ■

Turbocharger Inspection. To inspect a turbocharger, start the engine and listen to the sound the turbo system makes. As a technician becomes more familiar with this characteristic sound, it will be easier to identify an air leak between the compressor outlet and engine or an exhaust leak between engine and turbo by the presence of a higher pitched sound. If the turbo sound cycles or changes in intensity, the likely causes are a plugged air cleaner or loose material in the compressor inlet ducts or dirt buildup on the compressor wheel and housing.

After listening, check the air cleaner and remove the ducting from the air cleaner to turbo and look for dirt buildup or damage from foreign objects. Check for loose clamps on the compressor outlet connections and check the engine intake system for loose bolts or leaking gaskets. Then, disconnect the exhaust pipe and look for restrictions or loose material. Examine the exhaust system for cracks, loose nuts, or blown gaskets. Rotate the turbo shaft assembly. Does it rotate freely? Are there signs of rubbing or wheel impact damage?

Visually inspect all hoses, gaskets, and tubing for proper fit, damage, and wear. Check the low pressure, or air cleaner, side of the intake system for vacuum leaks.

On the pressure side of the system you can check for leaks using soapy water. After applying the soap mixture, look for bubbles to pinpoint the source of the leak.

Leakage in the exhaust system upstream from the turbine housing will also affect turbo operation. If exhaust gases are allowed to escape prior to entering the turbine housing, the reduced temperature and pressure will cause a proportionate reduction in boost and an accompanying loss of power. If the wastegate does not appear to be operating properly (too much or too little boost), check to make sure the connecting linkage is operating smoothly and not binding. Also, check to make sure the pressure sensing hose is clear and properly connected.

SHOP TALK

Specialty shops normally rebuild turbochargers. Because of the needed balance and sealing of the units, automotive technicians do not rebuild the units. Most repair shops do not have the equipment needed. ■

Wastegate Service Wastegate malfunctions can usually be traced to carbon buildup, which keeps the unit from closing or causes it to bind. A defective diaphragm or leaking vacuum hose can result in an inoperative wastegate. But, before condemning the wastegate, check the ignition timing, the spark-retard system, vacuum hoses, knock sensor, oxygen sensor, and computer to be sure that each is operating properly.

Common Turbocharger Problems. The turbocharger, with proper care and servicing, will provide years of reliable service. Most turbocharger failures are caused by lack of lubricant, ingestion of foreign objects, or contamination of lubricant.

Replacing a Turbocharger. If the turbocharger is faulty, it can be replaced with a new or rebuilt unit. Always follow the procedure given in the service manual.

Once the new or rebuilt unit is installed, the turbo should be started up as described in the following section.

Turbo Start-up and Shutdown After replacement of a turbocharger, or after an engine has been unused or stored, there can be a considerable lag after engine start-up before the oil pressure is sufficient to deliver oil to the turbo's bearings. To prevent this problem, follow these simple steps:

1. When installing a new or remanufactured turbocharger, make certain the oil inlet and drain lines are clean before connecting them.
2. Be sure the engine oil is clean and at the proper level.
3. Fill the oil filter with clean oil.
4. Leave the oil drain line disconnected at the turbo and crank the engine at 15-second intervals without starting it until oil flows out of the turbo drain port.
5. Connect the drain line, start the engine, and operate it at low idle for a few minutes before running it at higher speeds.

CUSTOMER CARE

Remind your customer not to immediately stop a turbocharged engine after pulling a trailer or driving at high speed or uphill. Idle the engine for 20 to 120 seconds. The turbocharger will continue to rotate after the engine oil pressure has dropped to zero, which can cause bearing damage. Also, remind the owner of the proper starting procedures and the importance of proper maintenance. ■

Figure 30–32 A drawing showing the drive setup for a supercharger. *Courtesy of DaimlerChrysler Corporation*

Superchargers

Supercharging fascinated auto engineers even before they decided to steer with a wheel instead of a tiller. The 1906 American Chadwick had a supercharger. Since then, many manufacturers have equipped engines with superchargers **(Figure 30–32)**. Supercharged Dusenbergs, Hispano-Suizas, and Mercedes-Benzes were giants among luxury-car marques, as well as winners on the racetracks in the 1920s and 1930s. Then, after World War II, supercharging started to fade, although both Ford and American Motors sold supercharged passenger cars into the late 1950s. However, after being displaced first by larger V8 engines, then by turbochargers, the supercharger started to make a comeback with 1989 models.

Supercharger Operation **Superchargers** are air pumps directly driven by the engine's crankshaft by a belt. They improve horsepower and torque by pumping extra air into the engine in direct relationship to crankshaft speed. This positive connection provides for instant power response.

Figure 30–33 illustrates the flow of the air through the supercharger. The air is inducted into the bottom of the supercharger, pressurized by the spinning rotors, and exits through the top of the supercharger by way of the air outlet adapter. As the air is compressed by the su-

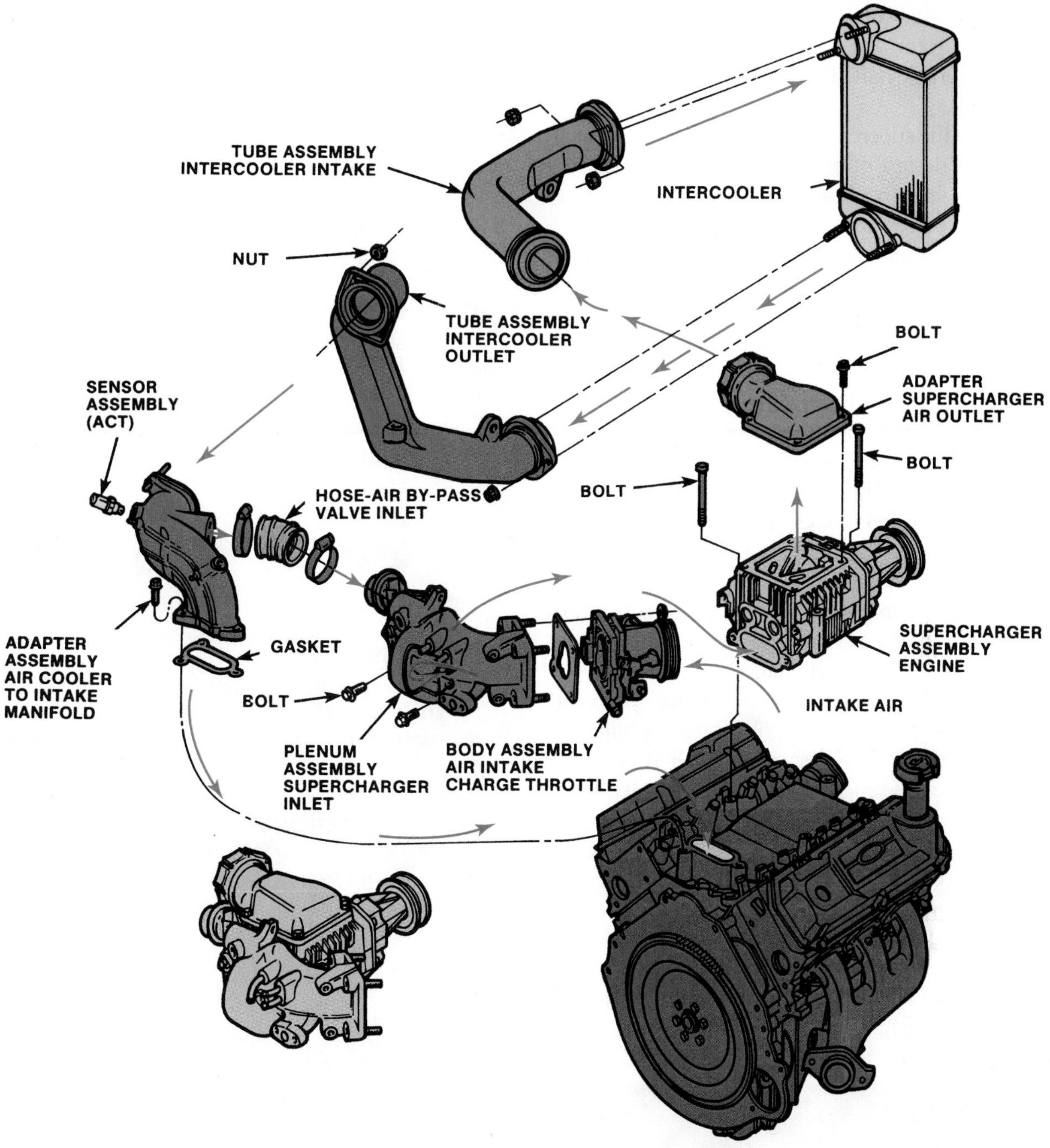

Figure 30-33 Airflow through a supercharger assembly into the engine.

percharger, its temperature increases. Since cooler, denser air is desired for increased power, the heated air is routed through the air-to-air intercooler.

The cooled air then passes through to the intake manifold adapter assembly. When the intake valves open, the air is forced into the combustion chambers, where it is mixed with fuel delivered by the fuel injectors.

Notice that the system also incorporates a bypass, which branches off from the upper portion of the air cooler to the intake manifold adapter assembly. This bypass is designed to allow the supercharger to idle when the extra power is not needed. The bypass routes any excess air in the intake manifold back through the supercharger inlet plenum assembly, allowing the engine to run, in effect, normally aspirated.

Unlike a turbocharger, the supercharger does not require a wastegate to limit boost and prevent a potentially damaging overboost condition. Since the speed of the supercharger is directly linked to the engine speed, its pumping power is limited by the rpm of the engine itself rather than revolutions produced by exhaust gases. Supercharger boost is, therefore, directly controlled through the opening and closing of the throttle.

Supercharger Designs While there have been a number of supercharger designs on the market over the years, the most popular is the **Roots** type **(Figure 30–34)**. The pair of three lobed rotor vanes in the Roots supercharger is driven by the crankshaft. The lobes force air into the intake manifold.

The key to the supercharger's operation, of course, is primarily the design of the rotors. Some Roots-type superchargers use straight-lobe rotors that result in uneven pressure pulses and, consequently, relatively high noise levels. Therefore, the supercharger used with most of today's engines uses a helical design for the two rotors. The helical design evens out the pressure pulses in the blower and reduces noise. It was found that a 60-degree helical twist works best for equalizing the inlet and outlet volumes. However, boost efficiency of this design drops at a faster rate as boost pressure increases.

The rotors are held in a proper relationship to each other by timing gears. They are normally supported by ball bearings in the front and needle roller bearings in the back.

The intercooler extracts excess heat from the air. By cooling the air in this manner, its density and, consequently, its oxygen content, result in increased power output for a given boost pressure. The air-to-air intercooler is located next to the conventional cooling radiator.

To handle the higher operating temperatures imposed by supercharging, an engine oil cooler is usually built into the engine system. This water-to-oil cooler is generally mounted between the engine front cover and oil filter.

Superchargers can be enhanced with electrically operated clutches and bypass valves. These allow the same computer that controls fuel and ignition to kick the boost on and off precisely as needed, resulting in far greater efficiency than a full-time supercharger.

Another popular supercharger design, especially in Europe, is the **G-Lader** supercharger, which is based on a 1905 French design. Spiral ramps in both sides of the rotor intermesh with similar ramps in the housing. Unlike most superchargers, the rotor of the G-Lader does not spin on its axis; instead it moves around an eccentric shaft. This motion draws in air, squeezes it inward through the spiral, which compresses it, then forces it through ducts in the center into the engine. Airflow is essentially constant, so intake noise is lower than that of a Roots blower. Because there is only a slight wiping motion between the spiral and housing, wear is minimal.

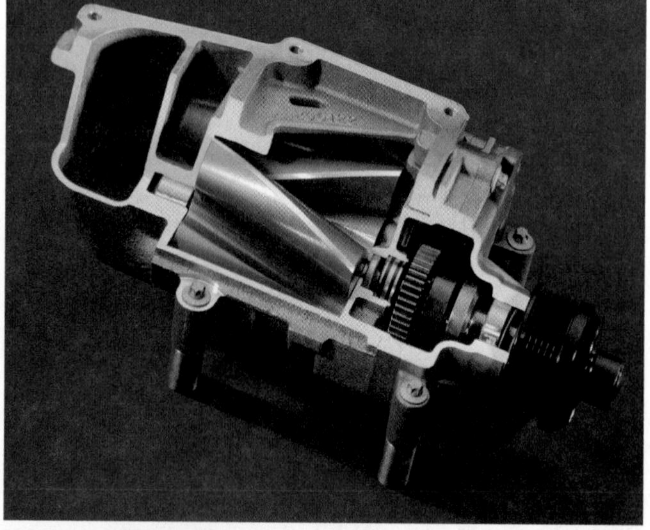

Figure 30–34 A cutaway of a supercharger. *Courtesy of DaimlerChrysler Corporation*

Supercharger Problems Many of the problems and their remedies given for turbochargers hold good for superchargers. There are also problems associated specifically with the supercharger. Refer to the service manual for the symptoms of supercharger failure and a summary of the causes and the recommended repairs.

CASE STUDY

A customer's high-mileage, mid-1980s car has a tendency to run hotter than it used to. It seems to be using more gas and has less power. There are no service records for the car and, based on its overall condition, it is determined that the car was not well maintained. A visual inspection is performed and nothing abnormal is found. However, when the technician removes the air-filter assembly to visually check the action of the carburetor, the smell of raw fuel is evident. Careful inspection reveals no fuel leaks. The engine is started so the carburetor and fuel lines can be observed. A small amount of fuel sprays out the top of the carburetor each time the throttle is opened. This is probably the cause for the gas smell. But what is causing the fuel to spray out? Is this related to the original complaint?

Diagnosis continues with a vacuum test. At idle, the engine has 17 in. Hg (432 mm Hg). When the engine speed is raised to 1,500 rpm and held there, the vacuum gauge climbs to 19 in. Hg (483 mm Hg), then starts to drop. This normally indicates excessive exhaust back pressure. A collapsed exhaust pipe or plugged converter could be the problem. A plugged exhaust will cause fuel to spray out of the carburetor, and it will cause poor gas mileage and poor overall performance.

To determine the location of the exhaust restriction, the catalytic converter is removed and the same vacuum test performed. This time the gauge reacts normally. The restriction in the exhaust is the converter. It is plugged. Replacing the converter will correct the problem. However, the cause of the plugging is still unknown.

A look at the fuel filler assembly reveals the cause of the problem. The owner has enlarged the nozzle opening to accommodate leaded fuel nozzles. The lead in this type fuel will clog a converter. This is explained to the customer and a new fuel filler neck and converter is installed.

KEY TERMS

Back pressure
Catalytic converter
G-Lader
Intercooler
Muffler
Powertrain control module (PCM)
Ported vacuum switch (PVS)
Pyrometer
Resonator
Roots
Supercharger
Tailpipe
Temperature vacuum switch (TVS)
Turbocharger
Wastegate

SUMMARY

- The air induction system allows a controlled amount of clean, filtered air to enter the engine. Cool air is drawn in through a fresh air tube. It passes through a preheater and air cleaner before entering the fuel injection throttle body.
- The intake manifold distributes the air or air/fuel mixture as evenly as possible to each cylinder, helps to prevent condensation, and assists in the vaporization of the air/fuel mixture. Intake manifolds are made of cast iron, plastic, or die-cast aluminum.
- The vacuum in the intake manifold operates many systems such as emission controls, brake boosters, heater/air conditioners, cruise controls, and more. Vacuum is applied through an elaborate system of hoses, tubes, and relays. A diagram of emission system vacuum hose routing is located on the underhood decal. Loss of vacuum can create many driveability problems.
- A vehicle's exhaust system carries away gases from the passenger compartment, cleans the exhaust emissions, and muffles the sound of the engine. Its components include the exhaust manifold, exhaust pipe, catalytic converter, muffler, resonator, tailpipe, heat shields, clamps, brackets, and hangers.
- The exhaust manifold is a bank of pipes that collects the burned gases as they are expelled from the cylinders and directs them to the exhaust pipe. Engines with all the cylinders in a row have one exhaust manifold. V-type engines have an exhaust manifold on each side of the engine. The exhaust pipe runs between the exhaust manifold and the catalytic converter.
- The muffler consists of a series of baffles, chambers, tubes, and holes to break up, cancel out, and silence pressure pulsations. Two types commonly used are the reverse-flow and the straight-through mufflers.
- The tailpipe is the end of the pipeline carrying exhaust fumes to the atmosphere beyond the back end of the car. Heat shields protect vehicle parts from exhaust system heat. Clamps, brackets, and hangers join and support exhaust system components.
- Exhaust system components are subject to both physical and chemical damage. The exhaust can be checked by listening for leaks and by visual inspection. Most exhaust system servicing involves the replacement of parts.
- The turbocharger relies on the rapid expansion of hot exhaust gases exiting the cylinder to spin turbine blades, which compresses the intake air.
- A typical turbocharger consists of a turbine (or hot wheel), shaft, compressor (or cold wheel), turbine housing, compressor housing, and center housing and rotating assembly. A wastegate manages turbo output by controlling the amount of exhaust gas that is allowed to enter the turbine housing. Turbo boost is the positive pressure increase created by a turbocharger.
- Most turbochargers are lubricated by pressurized and filtered engine oil that is line-fed to the unit's oil inlet. Some turbocharged engines are equipped with an intercooler, which is designed to cool the compressed air from the turbocharger.
- To control detonation, most turbocharger systems use knock-sensing devices and an electronic control unit to operate the wastegate control valve through sensor signals.
- If the turbo sound cycles or changes in intensity, the likely causes are a plugged air cleaner or loose material in the compressor inlet ducts or dirt buildup on the compressor wheel and housing. Most turbocharger failures are caused by one of the following reasons: lack of lubricant, ingestion of foreign objects, or contamination of lubricant. Turbo lag occurs when the turbocharger is unable to meet the immediate demands of the engine.
- Superchargers are air pumps connected directly to the crankshaft by a belt. The positive connection yields instant response and pumps air into the engine in direct relationship to crankshaft speed.
- The most popular supercharger design is the Roots type. Another supercharger design is the G-Lader supercharger.

TECH MANUAL

The following procedure is included in Chapter 30 of the *Tech Manual* that accompanies this book:

1. Inspecting an exhaust system.

REVIEW QUESTIONS

1. What can be used to check for leaks on the pressure side of a turbocharger system?
2. What advantages are there to preheating intake air?
3. Name three purposes of the intake manifold.
4. How can the effectiveness of a catalytic converter be checked?
5. A late-model vehicle has at least __________ catalytic converters in its exhaust system.
6. A miniconverter is used
   - **a.** on small engines where a normal converter will not fit properly.
   - **b.** on engines that used leaded fuels.
   - **c.** in conjunction with EGR systems to supply clean exhaust for the cylinders.
   - **d.** to reduce emissions during engine warm-up.
7. Technician A makes sure the exhaust system is cool to the touch before working on it. Technician B disconnects the battery's negative cable before starting to work. Who is correct?
   - **a.** Technician A
   - **b.** Technician B
   - **c.** Both A and B
   - **d.** Neither A nor B
8. Technician A says a vacuum leak results in less air entering the engine, which causes a richer air/fuel mixture. Technician B says a vacuum leak anywhere in the system can cause the engine to run poorly. Who is correct?
   - **a.** Technician A
   - **b.** Technician B
   - **c.** Both A and B
   - **d.** Neither A nor B
9. Technician A says a vacuum leak will cause an engine to run richer than normal. Technician B says a vacuum leak can cause an engine to detonate. Who is correct?
   - **a.** Technician A
   - **b.** Technician B
   - **c.** Both A and B
   - **d.** Neither A nor B
10. Before replacing any exhaust system component, Technician A soaks all old connections with a penetrating oil. Technician B checks the old system's routing for critical clearance points. Who is correct?
    - **a.** Technician A
    - **b.** Technician B
    - **c.** Both A and B
    - **d.** Neither A nor B
11. A restricted exhaust system can cause________.
    - **a.** stalling
    - **b.** loss of power
    - **c.** backfiring
    - **d.** all of the above
12. Technician A says a low vacuum reading can be caused by incorrect ignition timing. Technician B says an engine with low compression will have a low vacuum reading. Who is correct?
    - **a.** Technician A
    - **b.** Technician B
    - **c.** Both A and B
    - **d.** Neither A nor B
13. *True or False?* Engine misfires can cause a catalytic converter to overheat.
14. Which of the following is *not* characteristic of a turbocharger?
    - **a.** used to increase engine power by compressing the air that goes into the combustion chambers
    - **b.** usually located close to the exhaust manifold
    - **c.** utilizes an exhaust-driven turbine wheel
    - **d.** requires a mechanical connection between the engine and the pressurizing pump to compress the intake gases
15. Ten psi of turbo boost means air is being fed into the engine at ______ when the engine is operating at sea level.
    - **a.** 4.7 psi (32.40 kPa)
    - **b.** 10 psi (68.95 kPa)
    - **c.** 14.7 psi (101.3 kPa)
    - **d.** 24.7 psi (170.3 kPa)
16. What manages turbo output by controlling the amount of exhaust gas entering the turbine housing?
    - **a.** wastegate
    - **b.** turbine seal assembly
    - **c.** hot wheel
    - **d.** cold wheel
17. Technician A says a turbocharger has its own self-contained lubrication system. Technician B says a turbocharger should not be operated at an engine oil pressure any lower than 30 psi (6205 kPa). Who is correct?
    - **a.** Technician A
    - **b.** Technician B
    - **c.** Both A and B
    - **d.** Neither A nor B

18. What is the first step in turbocharger inspection?
    - **a.** Check the air cleaner for a dirty element.
    - **b.** Open the turbine housing at both ends.
    - **c.** Start the engine and listen to the system.
    - **d.** Remove the ducting from the air cleaner to turbo and examine the area.
19. Which of the following statements concerning superchargers is incorrect?
    - **a.** Superchargers must overcome inertia and spin up to speed as the flow of exhaust gas increases.
    - **b.** Superchargers do not require a wastegate to limit boost.
    - **c.** A bypass is designed into the system to allow the supercharger to idle along when extra power is not needed.
    - **d.** Superchargers improve horsepower and torque.
20. Technician A says disconnected vacuum hoses admit more air into the intake manifold than the engine is calibrated for. Technician B says the most common result of a vacuum leak is a rough running engine due to a richer air/fuel mixture. Who is correct?
    - **a.** Technician A
    - **b.** Technician B
    - **c.** Both A and B
    - **d.** Neither A nor B

31

EMISSION CONTROL SYSTEMS

OBJECTIVES

■ Explain why hydrocarbon (HC) emissions are released from an engine's exhaust. ■ Explain how carbon monoxide (CO) emissions are formed in the combustion chamber. ■ Describe oxygen (O_2) emissions in relation to air/fuel ratio. ■ Describe how carbon dioxide (CO_2) is formed in the combustion chamber. ■ Describe how oxides of nitrogen (NO_x) are formed in the combustion chamber. ■ Describe the operation of an evaporative control system during the canister purge and nonpurge modes. ■ Explain the purpose of the positive crankcase ventilation system. ■ Describe the operation of the detonation sensor and electronic spark control module. ■ Describe the operation of an exhaust gas recirculation valve. ■ Explain the design and operation of a positive and negative back pressure EGR valve. ■ Explain the operation of a digital EGR valve. ■ Explain the operation of a linear EGR valve. ■ Define the purpose of a catalytic converter. ■ Describe the operation of a secondary air injection system.

Emission controls on cars and trucks have one purpose: to reduce the amount of pollutants and environmentally damaging substances released by the vehicles. The consequences of the pollutants are grievous **(Figure 31–1)**. The air we breathe and the water we drink have become contaminated with chemicals that adversely affect our health. It took many years for the public and the automotive industry to address the problem of these pollutants. Not until smog became an issue did anyone in power really care and do something about these pollutants.

Figure 31–1 Dirty exhaust is bad for everyone.

Smog not only appears as dirty air, it is also an irritant to a person's eyes, nose, and throat. The things necessary to form photochemical smog are HC and NO_x exposed to sunlight in stagnant air. When there is enough HC in the air, it reacts with the NO_x. The energy of sunlight causes these two chemicals to react and form photochemical smog.

There are three main automotive pollutants: HC, CO, and NO_x. Particulate emissions are also present in diesel engine exhaust. HC emissions are caused largely by unburned fuel from the combustion chambers. HC emissions can also originate from evaporative sources such as the gasoline tank. CO emissions are a by-product of the combustion process and result from incorrect air/fuel mixtures. NO_x emissions are caused by nitrogen and oxygen uniting at cylinder temperatures above 2,300°F (1,261°C).

LEGISLATIVE HISTORY

The first Clean Air Act prompted Californians to create the California Air Research Board (ARB). California ARB's purpose was to implement strict air standards; these became the standard for federal mandates. One of the approaches taken by the ARB to clean the air was to start **periodic motor vehicle inspection (PMVI)**. The purpose of the PMVI is to inspect a vehicle's emission controls once a year. This inspection includes a tailpipe emissions test and an underhood inspection. The tailpipe test certifies that the vehicle's exhaust emissions are within the limits set by law. The underhood and/or vehicle inspection verifies that the pollution control equipment has not been tampered with or disconnected.

Today, California is not the only state that requires annual emissions testing. Many states have incorporated an emissions test into their annual vehicle registration procedures. Most states have or are planning to implement an **I/M 240** or similar program. The I/M 240 tests the emissions of a vehicle while it is operating under a variety of load conditions and speeds. This is an improvement over exhaust testing during idle and high speed with no load.

The I/M 240 test requires the use of a chassis (road) dynamometer, commonly called a **dyno**. While on the dyno, the vehicle is operated for 240 seconds and under different load conditions. The test drive on the dyno simulates both in-traffic and highway driving and stopping. The emissions tester tracks the exhaust quality through these conditions **(Figure 31–2)**.

The I/M 240 program also includes a functional test of the evaporative emission control devices and a visual inspection of the total emission control system. If the vehicle fails the test, it must be repaired and certified before it can be registered.

According to a document based on a study by the EPA, passenger cars are responsible for 17.8% of the total hydrocarbon emissions, 30.9% of the total carbon monoxide emissions, and 11.1% of the oxides of nitrogen emissions. After more than thirty years of emission regulations, these figures remain staggering! Imagine what these figures would be if automotive and industrial emissions had remained unregulated during the past thirty years!

Figure 31–2 Emission levels are often checked with the vehicle running on a dyno. *Courtesy of Robert Bosch Corp.*

Emission standards have been one of the driving forces behind many of the technological changes in the automotive industry. Catalytic converters and other emission systems were installed to meet emission standards. Computer-controlled carburetors and fuel injection systems were installed to provide more accurate control of the air/fuel ratio to reduce emission levels and allow the catalytic converter to operate efficiently.

In the 1990s, emission standards in the United States became increasingly stringent. In 1994, an ambitious emission program began in California. Since then the program has initiated stricter emission standards and created several emission categories for vehicles:

- TLEV (transitional low-emission vehicle)
- LEV (low-emission vehicle)
- ULEV (ultralow-emission vehicle)
- SULEV (super ultralow-emission vehicle)
- CFV (clean-fueled vehicle)
- PZEV (partial zero-emission vehicle)
- ZEV (zero-emission vehicle)

These categories are often used to define vehicles and are used for tax credits. Some other states have adopted, or are considering the adoption of, the California emission standards.

Automobile manufacturers have been working toward reduction of automotive air pollutants since the early 1950s, when auto emissions were found to be part of the cause of smog in Los Angeles. Governmental interest in controlling emissions developed around the same time.

Development of Emission Control Devices

In late 1959, California established the first standards for automotive emissions. In 1967, the Federal Clean Air Act was amended to provide for federal standards that would apply to motor vehicles.

The first source of emissions to be brought under control was the crankcase. Positive crankcase ventilation systems designed to route crankcase vapors back to the engine's intake manifold were developed and incorporated into 1961 cars and light trucks sold in California. These systems were installed on all cars nationwide beginning with the 1963 models.

Control of unburned hydrocarbons and carbon monoxide in the engine's exhaust was the next major development. An **air injection reactor (AIR)** system was built into cars and light trucks sold in California in 1966. Other systems, including the controlled combustion system, were developed and used nationwide in 1968. Further progress in the following years improved combustion to reduce hydrocarbon and carbon monoxide emissions.

Fuel vapors from the gasoline tank and the carburetor float bowl were brought under control with the introduction of evaporation control systems. These systems were first installed in 1970 model cars sold in California and in most cars made domestically, beginning with 1971 models.

Most vehicle manufacturers started to provide emission control systems that reduced NO_x as early as 1970. The exhaust gas recirculation system used on some 1972 models was used extensively for 1973 models when federal standards for oxides of nitrogen took effect.

One of the most important developments for lowering emission levels has been the availability and use of unleaded gasolines. Since 1971, engines have been designed to operate on unleaded fuels.

Removing lead from gasoline brings some immediate benefits. It eliminates the emission of lead particles from an automobile's exhaust. It increases spark plug life, which is also important from an emission standpoint. It avoids formation of lead deposits in the combustion chambers that tend to increase hydrocarbon emissions.

The catalytic converter, a later development, provided a means for oxidizing the CO and HC emissions in the engine exhaust. Beginning with the 1975 model year, passenger cars and light trucks have been equipped with converters.

Three basic types of emission control systems are used in modern vehicles: evaporative, precombustion, and postcombustion control systems.

The **evaporative control** system is a sealed system. It traps the fuel vapors (HC) that would normally escape from the fuel tank and carburetor into the air.

Most of the pollution control systems used today prevent emissions from being created in the engine, either during or before the combustion cycle. The common **precombustion control** systems are the positive crankcase ventilation (PCV), engine modification systems, and exhaust gas recirculating (EGR) systems.

Postcombustion control systems clean up the exhaust gases after the fuel has been burned. Secondary air or air injector systems put fresh air into the exhaust to reduce HC and CO to harmless water vapor and carbon dioxide by chemical (thermal) reaction with oxygen in the air. Catalytic converters help this process. Most catalysts reduce NO_x as well as HC and CO.

POLLUTANTS

The gases of most concern to environmentalists, engineers, and technicians are HC, CO, NO_x, CO_2, and O_2. The latter two are not really pollutants but are monitored because they are indicators of combustion efficiency.

Hydrocarbons

Hydrocarbon emissions are caused by incomplete combustion. These emissions are actually molecules of unburned gasoline. Even an engine in good condition with satisfactory ignition and fuel systems releases some HC. When the flame front in the combustion chamber approaches the cooler cylinder wall, the flame front quenches, leaving some unburned HC.

An excessively lean air/fuel ratio also results in cylinder misfiring and high HC emissions. A very rich air/fuel ratio also causes higher-than-normal HC emissions. At the stoichiometric air/fuel ratio, HC emissions are low. Evaporative emissions from fuel tanks, carburetor float bowls, and evaporative systems are also a source of HC emissions.

On cars with catalytic converters, the HC reading at the tail pipe is very low if the engine, ignition, fuel, and emission systems are in normal working condition. When a cylinder misfires, all the unburned gas in the cylinder is delivered to the exhaust system, resulting in a high HC reading. Cylinder misfiring and high HC readings may be caused by ignition defects, such as defective spark plugs, spark plug wires, coil, and distributor cap or rotor. Low cylinder compression also causes high HC emissions.

Carbon Monoxide

Carbon monoxide is a by-product of combustion. CO is a poisonous chemical compound of carbon and oxygen. It forms in the engine when there is not enough oxygen to combine with the carbon during combustion. When there is enough oxygen in the mixture, carbon dioxide (CO_2) is formed. CO_2 is not a pollutant and is the gas used by plants to manufacture oxygen. CO is primarily found in the exhaust, but can also be in the crankcase. CO is odorless and tasteless, but, in concentrated form, it is toxic.

CO emissions are caused by a lack of air or too much fuel in the air/fuel mixture. CO will not occur if combustion does not take place in the cylinders; therefore, the presence of CO means combustion is taking place. As the air/fuel ratio becomes richer, the CO levels increase. At the stoichiometric air/fuel ratio, the CO emissions are very low. If the air/fuel ratio is leaner than stoichiometric, the CO emissions remain very low. Therefore, CO emissions are a good indicator of a rich air/fuel ratio, but they are not an accurate indication of a lean air/fuel ratio. When cylinder misfiring occurs, there is less total combustion in the engine cylinders. Since CO is a by-product of combustion, cylinder misfiring does not increase CO emissions. If a cylinder is misfiring, CO emissions may decrease slightly.

Oxides of Nitrogen

This pollutant is actually various compounds of nitrogen and oxygen gas, both of which are present in the air used for combustion. The formation of oxides of nitrogen is the result of high combustion temperatures. When combustion temperature reaches more than 2,300°F (1,261°C), the N and the O_2 in the air combine to form these oxides of nitrogen. NO_x emissions are a major contributor to the problem of smog. Since outside air is 78% N, the gas cannot be prevented from entering the combustion chamber. The key to controlling NO_x is to prevent N from joining with oxygen during the combustion process.

Preventing the formation of NO_x can be very tricky. When the mixture is slightly rich, there is less chance that NO_x will be produced but a greater chance that CO and HC emissions will occur. Likewise, when the mixture is slightly lean, there is less chance that CO and HC emissions will occur but a greater chance that NO_x will be produced.

The *x* in NO_x stands for the proportion of oxygen that is mixed with a nitrogen atom. The *x* is a variable, which means it could be the number 1, 2, 3, and so forth; therefore, the term NO_x refers to many different oxides of nitrogen (NO, NO_2, NO_3, and so forth). Most of the NO_x emissions from an engine are NO or nitrous oxide. When NO is released in the air, it seeks, finds, and combines with an oxygen atom to form nitrous dioxide (NO_2), which is a very toxic gas that contributes to the formation of smog, ozone, and acid rain.

Oxygen

O_2 is not a pollutant; therefore, its presence in the exhaust does not pose any threat to our environment. However, too much oxygen in the exhaust does indicate that an improper mixture was in the cylinders or poor combustion has occurred in the engine. An improper air/fuel mixture will also cause a high reading of oxygen.

If the air/fuel ratio is rich, all the oxygen in the air is mixed with fuel, and the O_2 levels in the exhaust are very low. When the air/fuel ratio is lean, there is not enough fuel to mix with all the air entering the engine, and O_2 levels in the exhaust are higher. Therefore, O_2 levels are a good indicator of a lean air/fuel ratio and they are not affected by catalytic converter operation.

Carbon Dioxide

CO_2 is also not a pollutant; however, it has been linked to another environmental concern—the greenhouse effect. CO_2 may be one of the causes of global warming. From an efficiency standpoint, carbon dioxide is an ideal by-product of combustion. Therefore, a large amount of CO_2 in the exhaust is desired. If the air/fuel ratio goes from 9:1 to 14.7:1, the CO_2 levels gradually increase from approximately 6% to 13.5%. CO_2 levels are highest when the air/fuel ratio is slightly leaner than stoichiometric. At the stoichiometric air/fuel ratio, CO_2 levels begin to decrease. To reduce CO_2 levels in the exhaust, engineers are working hard to maintain overall engine efficiency while keeping CO_2 low. This is one of the primary factors in the exploration of alternative fuel and power sources for automobiles.

EVAPORATIVE EMISSION CONTROL SYSTEMS

The fuel evaporative emission control system **(Figure 31–3)** reduces the amount of raw fuel vapors that are emitted into the air from the fuel tank. These vapors must not be allowed to escape from the fuel system to the atmosphere. Since the first systems were used nationwide in the early 1970s, several refinements have been added. Current systems include the following components:

- A special filler design to limit the amount of fuel that can be put in the tank.
- A pressure/vacuum relief fuel tank cap instead of a plain vented cap.
- A vapor separator in the top of the fuel tank **(Figure 31–4)**. This device collects droplets of liquid fuel and directs them back into the tank.

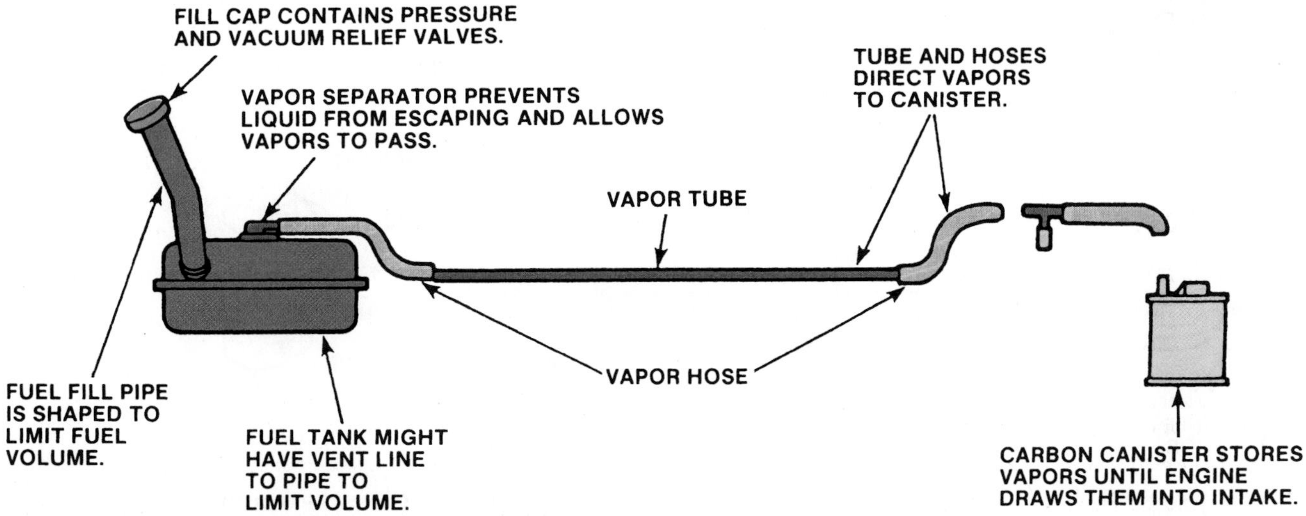
FILL CAP CONTAINS PRESSURE AND VACUUM RELIEF VALVES.
VAPOR SEPARATOR PREVENTS LIQUID FROM ESCAPING AND ALLOWS VAPORS TO PASS.
TUBE AND HOSES DIRECT VAPORS TO CANISTER.
VAPOR TUBE
VAPOR HOSE
FUEL FILL PIPE IS SHAPED TO LIMIT FUEL VOLUME.
FUEL TANK MIGHT HAVE VENT LINE TO PIPE TO LIMIT VOLUME.
CARBON CANISTER STORES VAPORS UNTIL ENGINE DRAWS THEM INTO INTAKE.

Figure 31-3 A typical fuel evaporative emission control system.

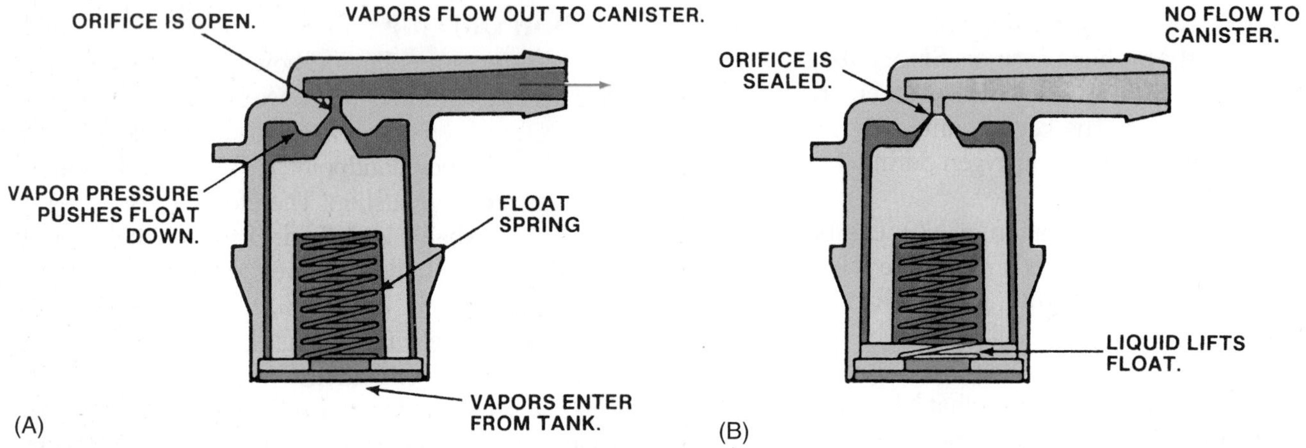

Figure 31-4 (A) Normal operation of a vapor separator, (B) with liquid in the separator.

- A domed fuel tank in which the upper portion is raised. Fuel vapors rise to this upper portion and collect.
- Check, or one-way, valves keep vapors confined. When an engine runs, vacuum- or electrically operated valves open.
- Hoses and tubes connect parts of a vapor-recovery system. Special fuel/vapor rubber tubing must be used.

Figure 31-5 A charcoal canister.

The most obvious part of the evaporative emission control system is the **charcoal (carbon) canister (Figure 31–5)** in the engine compartment. This canister is located in the fuel tank's vapor line, usually in the engine compartment. Fuel vapors from the gas tank are routed to and absorbed by the surfaces of the canister's charcoal granules. When the vehicle is restarted, vapors are drawn by the vacuum into the intake manifold to be burned in the engine. Canister purging varies widely with make and model. On most new vehicles, the PCM controls when the canister will be purged. In some instances a fixed restriction allows constant purging whenever there is manifold vacuum. In others, a staged valve provides purging only at speeds above idle. Generally, the **canister purge valve** is normally closed and opens the inlet to the purge outlet when vacuum is applied. Some units incorporate a thermal-delay valve so the canister is not purged until the engine reaches operating temperature. Purging at idle or with a cold engine creates other problems, such as rough running and increased emissions because of the additional vapor added to the intake manifold. Typical purge valve mountings are shown in **Figure 31–6**.

The canister contains a liquid fuel trap that collects any liquid fuel entering it **(Figure 31–7)**. Condensed fuel vapor forms liquid fuel that is returned from the canister

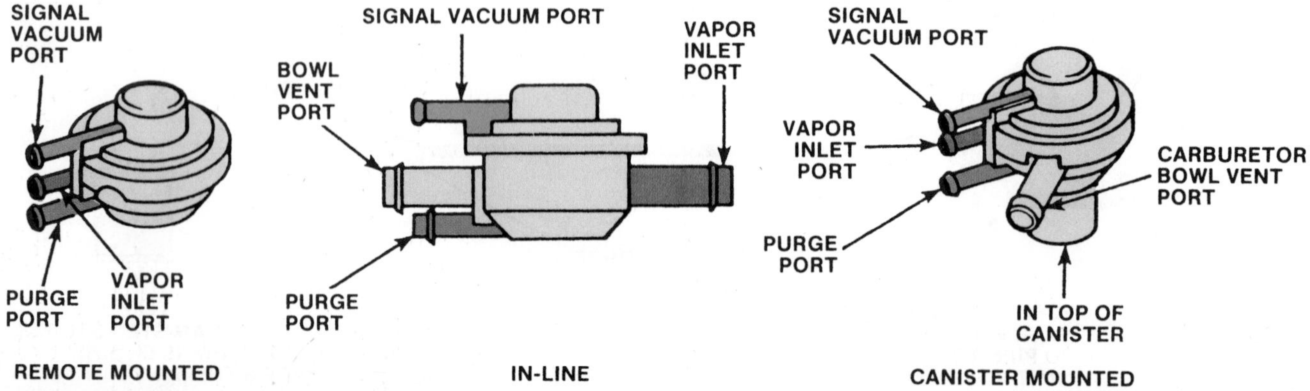

Figure 31-6 Typical purge valve mountings.

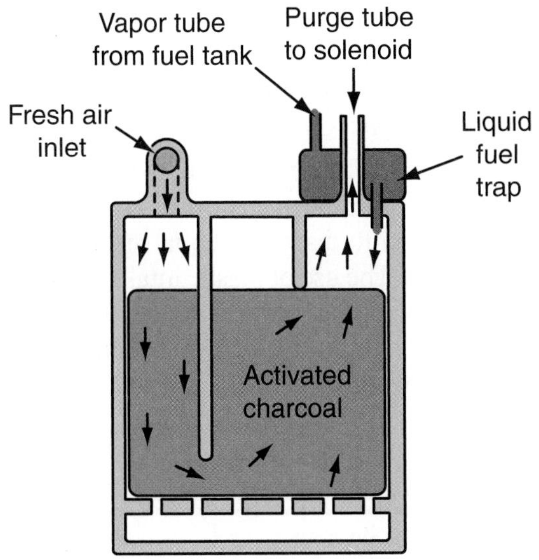

Figure 31-7 Typical charcoal canister operation.

to the tank when a vacuum is present in the tank. This liquid fuel trap prevents liquid fuel from contaminating the charcoal in the canister. The EVAP system reduces the escape of HC evaporative emissions from the gasoline tank to the atmosphere.

WARNING!

Gasoline vapors are extremely explosive! Do not smoke or allow sources of ignition near any component in the EVAP system. Explosion of gasoline vapors may result in property damage or personal injury.

Canister purging may also be electronically controlled **(Figure 31–8)**. The PCM enables a purge solenoid to initiate the purge cycle. A canister purge solenoid is connected in the purge hose from the canister to the intake port near the edge of the throttle. The PCM provides a ground for the canister purge solenoid winding to operate the solenoid. When the solenoid is energized by the computer, the purge valve opens and allows the intake manifold vacuum to draw the trapped fuel vapors from the canister.

The PCM energizes the canister purge solenoid and allows vacuum to purge vapors from the canister after the system has been in closed loop for a period of time and engine temperature and speed are above particular levels. When the correct conditions are not present, the PCM does not energize the canister purge solenoid, and the gasoline vapors from the fuel tank are stored in the canister.

OBD-II EVAP Systems

OBD-II monitors the evaporative system by testing the ability of the fuel tank to hold pressure and the purge system to vent gas fumes from the charcoal canister. Manufacturers often fit their vehicles with a pump to detect leaks in the EVAP system. When the PCM checks the EVAP system, it turns on the pump to pressurize the system. As pressure builds, the cycling rate of the pump decreases. If there is no leak in the system, the pressure builds until the pump shuts off. If there is a leak, pressure does not build up and does not shut down the pump. The pump continues to run until the PCM determines it has run a complete test cycle. If the PCM senses there is no leak in the system, it will run the purge monitor. If no leaks are present and the purge cycle is high, the system passed the test.

Some EVAP systems have a purge flow sensor between the canister purge solenoid and the intake manifold. The PCM monitors the signal from the sensor once per drive cycle to determine if there is vapor flow through the solenoid to the intake manifold.

Other EVAP systems have a vapor management valve between the canister and intake manifold. This valve is

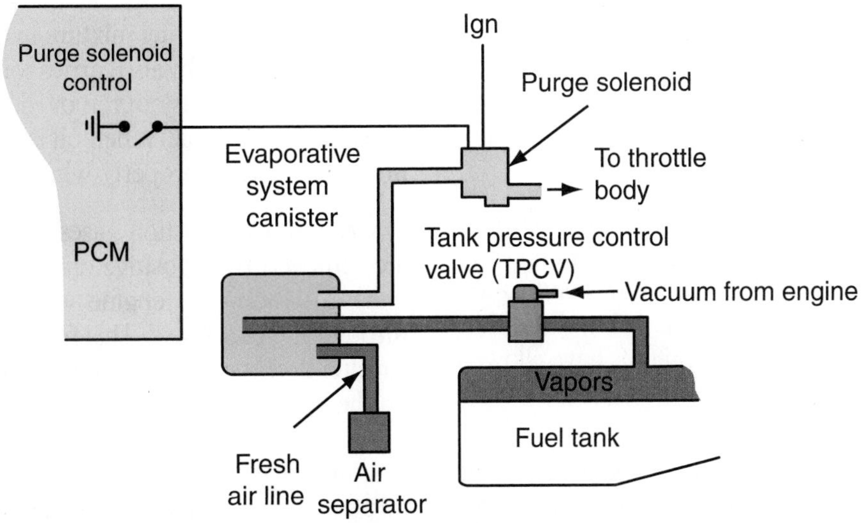

Figure 31-8 The PCM provides a ground to operate the purge solenoid.

normally closed. The PCM operates the valve to control vapor flow from the canister to the intake manifold and monitors the valve's operation to determine if the EVAP system is purging vapors properly.

Some GM vehicles have an enhanced evaporative system monitor, which detects leaks and restrictions in the system. In these enhanced systems, an evaporative system leak or a missing fuel tank cap causes the MIL to turn on.

PRECOMBUSTION SYSTEMS

Systems designed to prevent or limit the amount of pollutants produced by an engine are called precombustion emission control devices. Although there are specific systems and engine designs classified this way, anything that makes an engine more efficient can be categorized as a precombustion emission control.

Engine Design Changes

In recent years, the basic engine has seen many technological changes. For the most part, the basics of operation have not been changed. Engineers have worked overtime in an attempt to squeeze as much out of small engines as they can. Many of these changes have not only increased the efficiency of the engines, but have also decreased the pollutants released by the engines. Also, some of the changes were only necessary or were only brought about to accommodate changes in the engine's fuel and/or ignition systems.

- *Better-Sealing Pistons.* Blowby gases can be and are being reduced through the use of better-sealing piston rings and improved cylinder wall surfaces. Many of these new piston rings also have frictional qualities that make them less of a drag, which increases fuel economy and engine power.
- *Combustion Chamber Designs.* The primary goal in designing combustion chambers is the reduction or elimination of the quench area. Another trend in combustion chamber design is locating the spark plug closer to the center of the chamber. Manufacturers have also worked with designs that cause controlled turbulence in the chamber. This turbulence improves the mixing of the fuel with the air, which results in improved combustion.
- *Lower Compression.* Combustion chamber designs have also affected the compression-ratios of engines. By keeping the compression ratio low, combustion temperatures can be kept below the point where NO_x is formed. However, new developments have allowed the use of higher compression ratios on some high-performance engines.
- *Decreased Friction.* Overcoming the friction of all the engine's moving parts results in a large loss of usable power and energy. By reducing the friction at key points within the engine, engineers have reduced the amount of power lost. Improved engine oils, new component materials, and weight reductions have had the biggest impact on reducing friction.
- *Intake Manifold Designs.* Thanks to the wide and successful use of port fuel injection, intake manifolds are being designed to distribute equal amounts of air to each cylinder. The use of plastic intake manifolds has allowed for smoother runners and better heat control of the air. Intake manifolds can be designed to be more efficient at low or high engine speeds, or both.
- *Improved Cooling Systems.* High engine temperatures reduce HC and CO emissions. However, they also make the formation of NO_x harder to control. Most engine cooling systems have been designed to run at high temperatures but are prevented from getting too hot, thereby limiting the production of NO_x. Today's engine control systems incorporate many features that change air/fuel mixture, ignition timing, and idle speed to control the temperature of the engine.

PCV Systems

During the last stages of combustion in an engine's cylinders, some unburned fuel and products of combustion leak past the piston rings and move into the crankcase. This leakage is called **blowby**. Blowby must be removed from the engine before it condenses in the crankcase and reacts with the oil to form sludge. Sludge, if allowed to circulate with engine oil, corrodes and accelerates wear of pistons, piston rings, valves, bearings, and other internal working parts of the engine.

Blowby gases must also be removed from the crankcase to prevent premature oil leaks. Because these gases enter the crankcase by the pressure formed during combustion, they pressurize the crankcase. The gases exert pressure on the oil pan gasket and crankshaft seals. If the pressure is not relieved, oil is eventually forced out of these seals.

Because the air/fuel mixture in an engine never completely burns, blowby also carries some unburned fuel into the crankcase. If it is not removed, the unburned fuel dilutes the engine's oil. When oil is diluted, it does not lubricate the engine properly, which causes excessive wear.

Operation Combustion gases that enter the crankcase are removed by a positive crankcase ventilation (PCV) system, which uses engine vacuum to draw fresh air through the crankcase. This fresh air enters through the air filter or through a separate PCV breather filter located on the inside of the air filter housing.

When the engine is running, intake manifold vacuum is supplied to the PCV valve. This vacuum moves air through the clean air hose into the rocker arm cover. From this location, air flows through cylinder head openings into the crankcase, where it mixes with blowby gases. The mixture of blowby gases and air flows up through cylinder head openings to the PCV valve. Intake manifold vacuum moves the blowby gas mixture through the PCV valve into the intake manifold **(Figure 31–9)**. The blowby gases mix with the intake charge and enter the combustion chambers, where they are burned.

The PCV system prevents the emission of blowby gases from the engine crankcase to the atmosphere and scavenges the crankcase for vapors that could dilute the oil and cause it to deteriorate or that could build undesirable pressure in the crankcase. An inoperative PCV system could shorten the life of the engine by allowing harmful blowby gases to remain in the engine, causing corrosion and accelerating wear.

PCV Valve The PCV valve **(Figure 31–10)** is usually mounted in a rubber grommet in one of the valve covers. A hose is connected from the PCV valve to the intake manifold. A clean air hose is connected from the air cleaner to the opposite rocker arm cover. A filter is positioned in the air cleaner end of the clean air hose. On some systems, the PCV valve is mounted in a vent module, and the clean air filter is located in this module.

A PCV valve contains a tapered valve. When the engine is not running, a spring keeps the tapered valve seated against the valve housing **(Figure 31–11)**. During idle or deceleration, the high intake manifold vacuum moves the tapered valve upward against the spring tension. Under this condition, the blowby gases flow through a small opening in the valve. Since the engine is not under heavy load, the amount of blowby gas is minimal and the small PCV valve opening is all that is needed to move the blowby gases out of the crankcase.

Manifold vacuum drops off during part-throttle operation. As the vacuum signal to the PCV valve decreases, a spring moves the tapered valve downward to increase the opening **(Figure 31–12)**. Since engine load is higher at part-throttle operation than at idle, blowby gases are increased. The larger opening allows all the blowby gases to be drawn into the intake manifold.

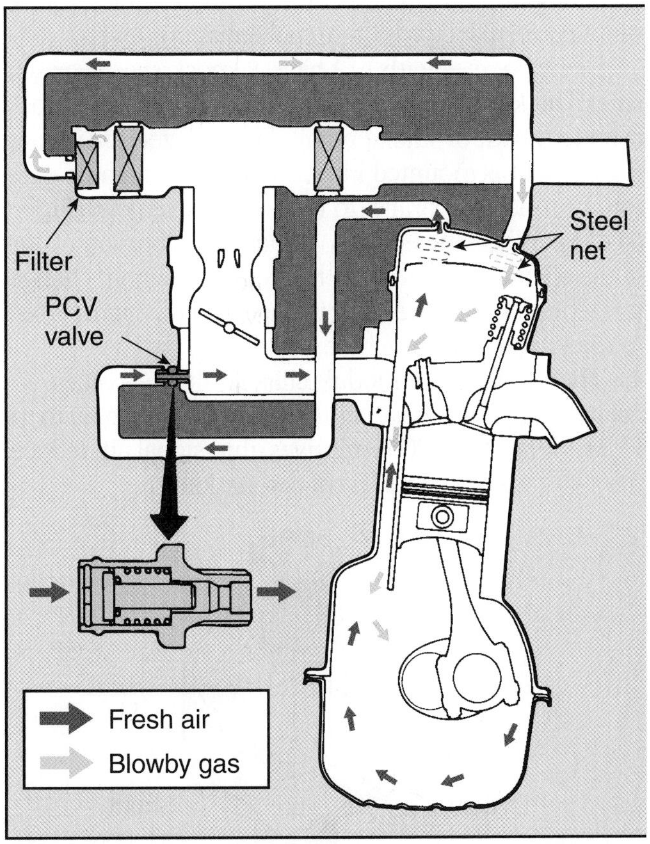

Figure 31-9 A typical PCV system. *Used with permission from Nissan North America, Inc.*

Figure 31-10 A PCV valve.

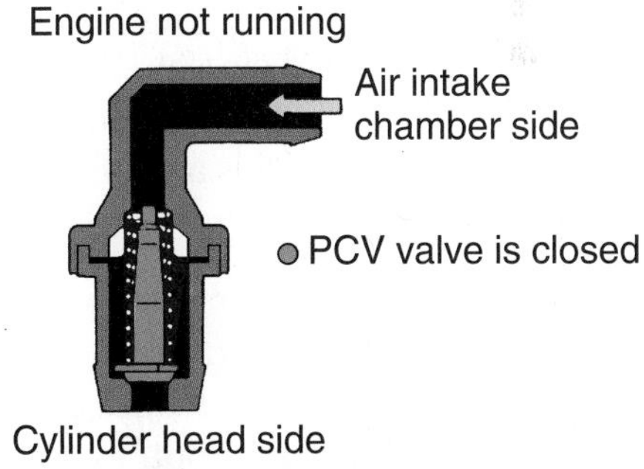

Figure 31-11 The PCV valve position with the engine not running. *Reprinted with permission*

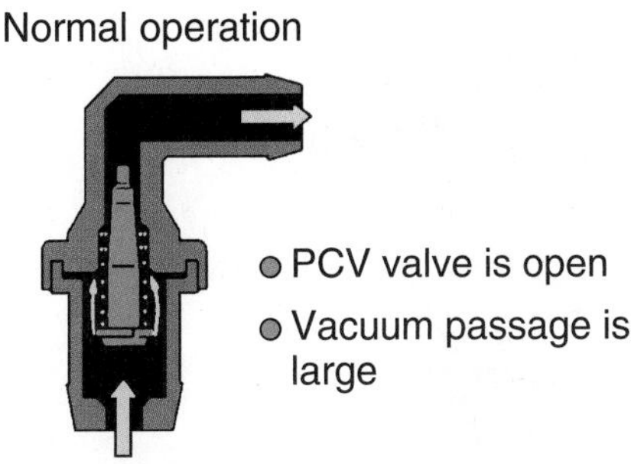

Figure 31-12 The PCV valve position during part-throttle operation. *Reprinted with permission*

When the engine is operating under heavy load with a wide throttle opening, the decrease in intake manifold vacuum allows the spring to move the tapered valve further downward in the PCV valve **(Figure 31–13)**, providing a larger opening through the valve. Since higher engine load results in more blowby gases, the larger PCV valve opening is necessary to allow these gases to flow through the valve into the intake manifold.

When worn rings or scored cylinders allow excessive blowby gases to enter the crankcase, the PCV valve opening may not be large enough to allow these gases to flow into the intake manifold. Under this condition, the blowby gases create a pressure in the crankcase, and some of these gases are forced through the clean air hose and filter into the air cleaner. When this action occurs, there is oil in the PCV filter and air cleaner. This same action occurs if the PCV valve is restricted or plugged.

If the PCV valve sticks in the wide-open position, excessive airflow through the valve causes rough idle operation. If a backfire occurs in the intake manifold, the tapered valve is seated in the PCV valve as if the engine were not running. This action prevents the backfire from entering the engine, where it could cause an explosion.

Fixed Orifice Tube PCV System Some engines are equipped with a PCV system that does not use a PCV valve. Instead, the blowby gases are routed into the intake manifold through a fixed orifice tube **(Figure 31–14)**. The system works the same as if it had a valve, except that the system is regulated only by the vacuum on the orifice. The size of the orifice limits the amount of blowby flow into the intake. The engine's air/fuel system is calibrated for this calibrated air leak. Since the action of the PCV allows unmetered air into the intake, the air/fuel system must be set for this amount of extra air.

Spark Control Systems

Spark control systems have been in use since the earliest gasoline engines. It was discovered that the proper timing of the ignition spark helped to reduce exhaust emissions and develop more power output. Incorrect timing affects the combustion process. Incomplete combustion results in HC emissions. High CO emissions can also result from incorrect ignition timing. Advanced timing can also increase the production of NO_x. When timing is too far advanced, combustion temperatures rise. For every one degree of overadvance, the temperature increases by 125°F (51.67°C). Throughout the years, each car manufacturer developed slightly different spark timing controls according to engine requirements and emission standards for each model year, but the systems and devices all operate on the same principles. Spark control on today's engines is handled by the PCM. Through input signals from various sensors, the PCM adjusts ignition timing for optimal performance with minimal emissions levels.

Many engines with EFI have a knock sensor or sensors. The knock sensors may be mounted in the block, cylinder head, or intake manifold. A piezoelectric sensing element is mounted in the knock sensor, and a resistor is connected parallel to this sensing element **(Figure 31–15)**. When the engine detonates, a vibration occurs in the engine. The piezoelectric sensing element changes this vibration into an analog voltage, and this signal is sent to the knock sensor module.

The knock sensor module changes the AC voltage signal into a digital voltage signal and sends this signal to the PCM. When the PCM receives this signal, it reduces the spark advance to prevent detonation.

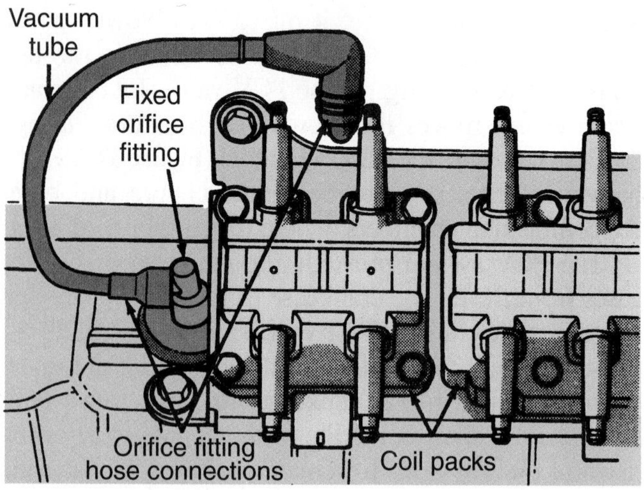

Figure 31–14 A fixed orifice tube-type PCV system. *Courtesy of DaimlerChrysler Corporation*

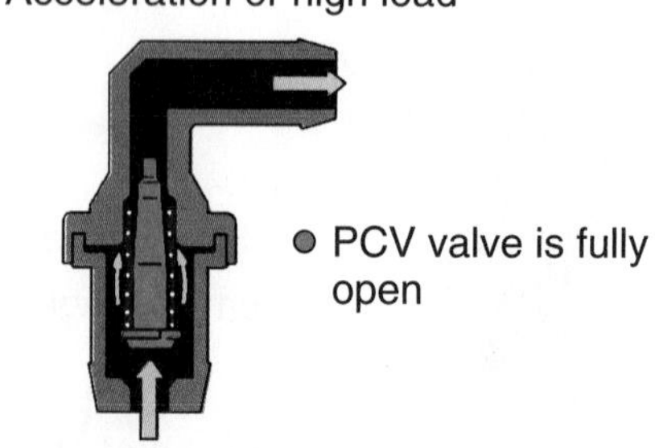

Figure 31–13 The PCV valve position during hard acceleration or heavy load. *Reprinted with permission*

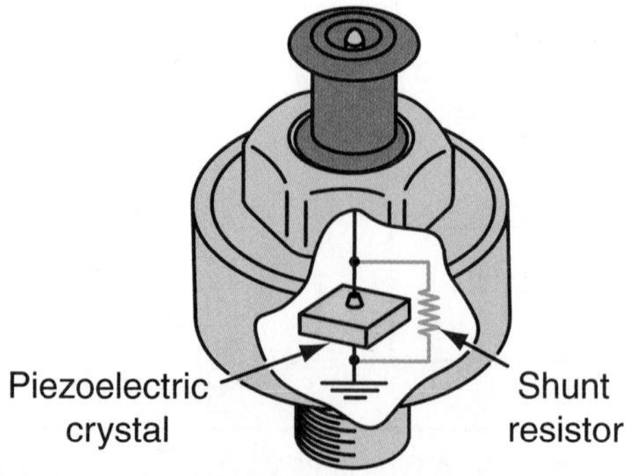

Figure 31–15 A detonation sensor.

EGR Systems

Exhaust gas recirculation (EGR) is the dilution of the air/fuel mixture with controlled amounts of exhaust gas. Since exhaust gas does not burn, this lowers the combustion temperature and reduces NO_x emissions from the engine. At lower combustion temperatures, very little of the nitrogen in the air combines with oxygen to form NO_x. Most of the nitrogen is simply carried out with the exhaust gases.

For driveability, it is desirable for the amount of exhaust gas flow to be proportional to the throttle opening. Driveability is also improved by stopping EGR when the engine is started up cold, at idle, and at full throttle.

There are two methods of exhaust gas recirculation: internally through overlap of valve opening times and externally with recirculating valves and manifolds. EGR systems, including EGR valves, are discussed here. EGR through valve overlap times is not a system; rather, it is a consequence of very careful engineering and precise valve control. The only part of the following discussion that applies to EGR through valve overlap is the purpose of EGR.

EGR Valve Many engines are equipped with a vacuum-operated EGR valve **(Figure 31–16)** to regulate the flow of exhaust gas into the intake manifold. Exhaust crossover passages under the intake manifold channel the exhaust gas to the valve. (Some in-line engines route the exhaust gas to the valve through an external tube.) Typically, the EGR valve is mounted to the intake manifold.

Figure 31–17 illustrates the basic valve design. The EGR valve is a vacuum-operated, flow control valve. A small exhaust crossover passage in the intake manifold admits exhaust gases to the inlet port of the EGR valve. Opening the EGR valve by control vacuum at the diaphragm allows exhaust gases to flow through the valve **(Figure 31–18)**. Here, the exhaust gas mixes with the air/fuel mixture in the intake manifold. The effect is to dilute or lean-out the mixture so that it still burns completely but with a reduction in combustion chamber temperatures.

Figure 31-16 An EGR valve.

On some engines, such as the General Motors' Northstar 4.6 L V8 (which is an SEFI engine), the exhaust gas from the EGR system is distributed through passages in the cylinder heads and distribution plates to each intake port. The distribution plates are positioned between the cylinder heads and the intake manifold. Since the exhaust gas from the EGR system is distributed equally to each cylinder, smoother engine operation results.

Valve Controls. Vacuum is used to control the operation of most EGR valves **(Figure 31–19)**. Many different controls have been used to ensure that the system works when it should. Ideally, the EGR system should operate when the engine reaches operating temperature and/or when the engine is operating under conditions other

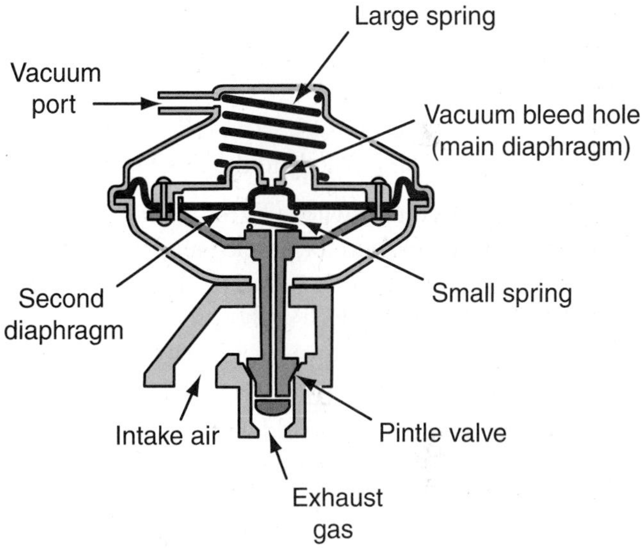

Figure 31-17 The typical design of an EGR valve.

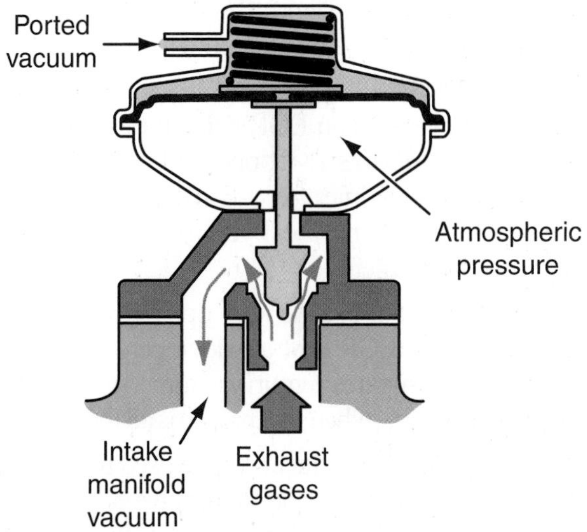

Figure 31-18 When the EGR is open a small amount of exhaust gas recirculates from the exhaust manifold to the intake manifold.

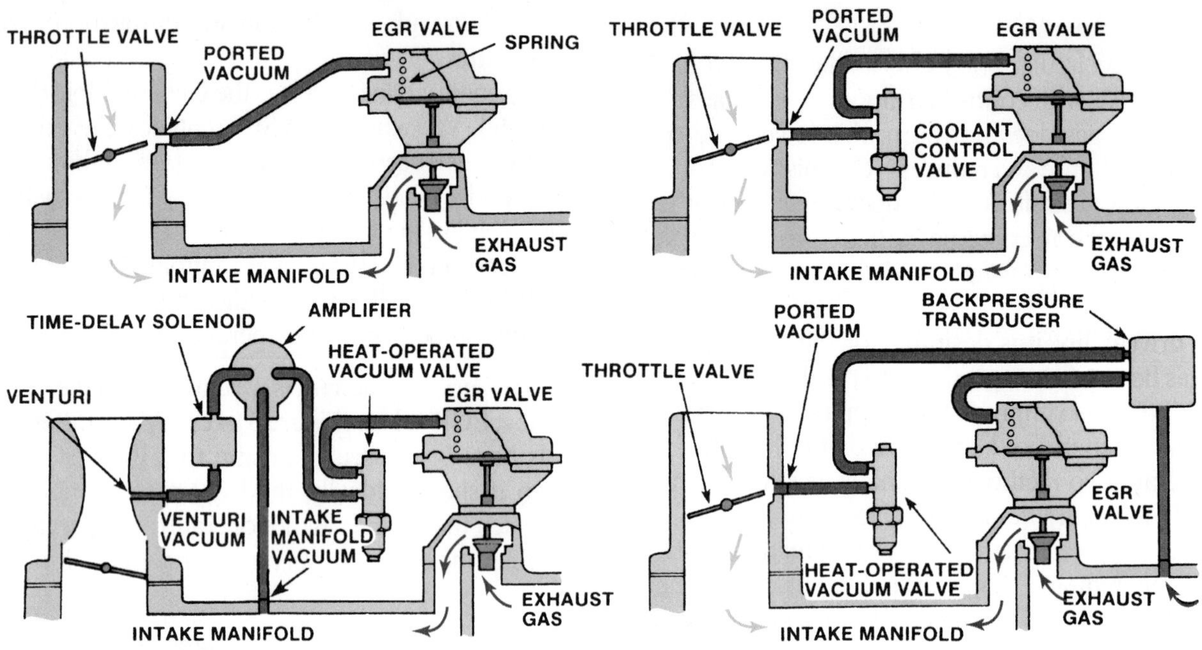

Figure 31-19 Different ways of controlling an EGR valve with vacuum.

than idle or wide-open throttle. The following are various controls that relate directly to the EGR system:

- The thermal vacuum switch (TVS) senses the air temperature in the carburetor air cleaner. When the engine reaches operating temperature, the TVS opens to supply vacuum to the EGR valve.
- The ported vacuum switch (PVS) senses the coolant temperature. The PVS cuts off vacuum to the EGR valve when the engine is cold and connects the vacuum to the EGR valve when the engine is warm.
- Some systems use a venturi vacuum amplifier (VVA) that makes it possible for venturi vacuum to control the EGR valve. Venturi vacuum is proportional to the airflow through the carburetor; however, it is a relatively weak vacuum signal. The VVA uses manifold vacuum to convert venturi vacuum to a strong enough signal to operate the EGR valve.
- Some engines have an EGR delay timer control system, which prevents EGR operation for a predetermined amount of time after warm engine start-up. On cold engine start-ups, the TVS and PVS valves override the delay timer.
- Some applications use the wide-open-throttle (WOT) valve to cut off EGR flow at wide-open throttle. This dump valve compares venturi vacuum to manifold vacuum to determine when the throttle is all the way open.
- A back pressure transducer can be used to modulate, or change, the amount the EGR valve opens. It controls the amount of air bleed in the EGR vacuum line according to the level of exhaust gas pressure, which is dependent on engine speed and load. The EGR valve can be closed or partially opened at different engine speeds and loads. Air bleed is stopped completely when the exhaust back pressure is high. Thus, maximum EGR occurs during acceleration when back pressure is high. When back pressure decreases, the vacuum line bleed is reopened. This decreases the vacuum at the EGR valve, which then reduces the amount of exhaust gas recirculated. In the past, the back pressure **transducer** was a separate unit. Now it is incorporated into the design of the EGR valve itself.

Types of EGR Valves. Depending on the system it is used in, the design of the EGR valve may change. Often these design changes incorporate some of the system controls.

The **positive back pressure** EGR valve has a bleed port and valve positioned in the center of the diaphragm. A light spring holds this bleed valve open, and an exhaust passage is connected from the lower end of the tapered valve through the stem to the bleed valve. When the engine is running, exhaust pressure is applied to the bleed valve. At low engine speeds, exhaust pressure is not high enough to close the bleed valve. If control vacuum is supplied to the diaphragm chamber, the vacuum is bled off through the bleed port, and the valve remains closed.

As engine and vehicle speed increase, the exhaust pressure also increases. At a preset throttle opening, the exhaust pressure closes the EGR valve bleed port. When control vacuum is supplied to the diaphragm, the diaphragm and valve are lifted upward, and the valve is open. If vacuum from an external source is supplied to a positive back pressure EGR valve with the engine not running, the valve will not open because the vacuum is bled off through the bleed port.

In a **negative back pressure** EGR valve, a normally closed bleed port is positioned in the center of the diaphragm. An exhaust passage is connected from the lower end of the tapered valve through the stem to the bleed valve. When the engine is running at lower speeds, there is a high-pressure pulse in the exhaust system. However, between these high-pressure pulses, there are low-pressure pulses. As the engine speed increases, more cylinder firings occur in a given time, and the high-pressure pulses become closer together in the exhaust system.

At lower engine and vehicle speeds, the negative pulses in the exhaust system hold the bleed valve open. When the engine and vehicle speed increase to a preset value, the negative exhaust pressure pulses decrease, and the bleed valve closes. Under this condition, the EGR valve is opened. When vacuum from an external source is supplied to a negative back pressure EGR valve with the engine not running, the bleed port is closed, and the vacuum should open the valve.

A digital EGR valve contains up to three electric solenoids that are operated directly by the PCM **(Figure 31–20)**. Each solenoid contains a movable plunger with a tapered tip that seats in an orifice. When any solenoid is energized, the plunger is lifted and exhaust gas is allowed to recirculate through the orifice into the intake manifold. The solenoids and orifices are different sizes. The PCM can operate one, two, or three solenoids to supply the amount of exhaust recirculation required to provide optimum control of NO_x emissions.

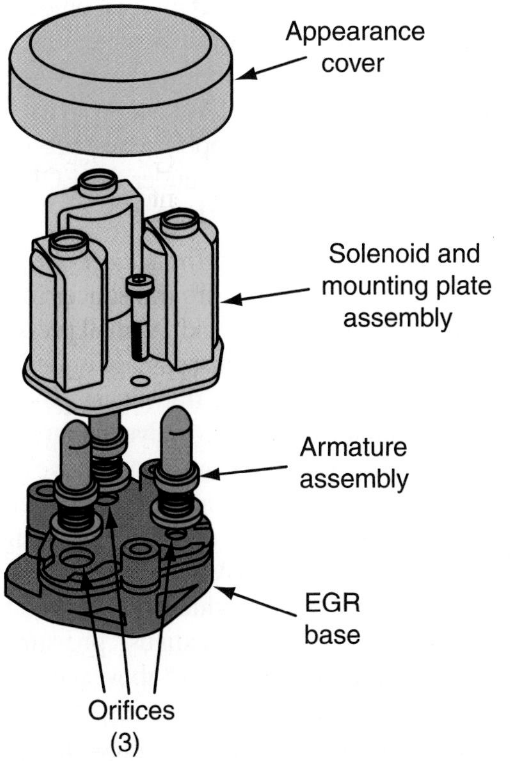

Figure 31-20 A digital EGR with three solenoids.

The linear EGR valve contains a single electric solenoid that is operated by the PCM. A tapered pintle is positioned on the end of the solenoid plunger. When the solenoid is energized, the plunger and tapered valve are lifted, and exhaust gas is allowed to recirculate into the intake manifold **(Figure 31–21)**. The EGR valve contains an EGR valve position or lift sensor, which is a linear potentiometer. The signal from this sensor varies from approximately 1 volt with the EGR valve closed to 4.5 volts with the valve wide open.

The PCM pulses the EGR solenoid winding on and off with a pulse width modulation principle to provide accurate control of the plunger and EGR flow. The EVP sensor acts as a feedback signal to the PCM to inform the PCM if the commanded valve position was achieved.

Some EGR valves, particularly on vehicles sold in California, contain an exhaust gas temperature sensor. This sensor contains a thermistor that changes resistance in relation to temperature. An increase in exhaust temperature decreases the sensor resistance. Two wires are connected from the exhaust gas temperature sensor to the PCM. The PCM senses the voltage drop across this sensor. Cool exhaust temperature and higher sensor resistance cause a high-voltage signal to the PCM, whereas hot exhaust temperature and low sensor resistance result in a low-voltage signal to the PCM.

OBD-II systems monitor the EGR system to determine if the system is operating properly. These monitors can use a variety of sensors and methods. Some monitor the temperature within the EGR passages. Other systems look at the MAP and/or delta pressure feedback EGR (DPFE) signal **(Figure 31–22)**. If a fault is detected in any of the EGR monitor tests, a DTC is set in the PCM memory. If the fault occurs during two drive cycles, the MIL light is illuminated. The EGR monitor operates once per OBD-II trip.

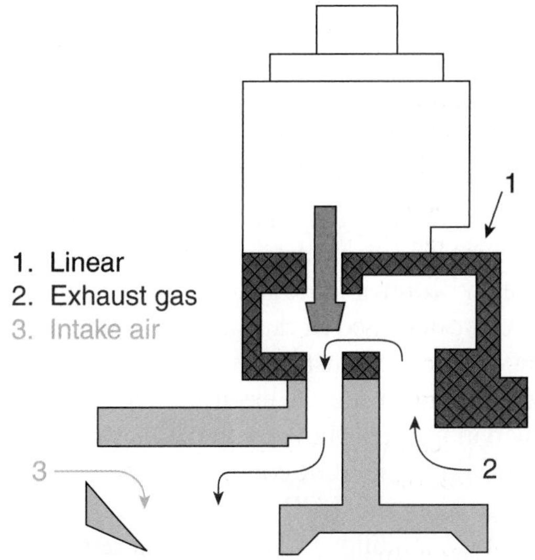

Figure 31-21 Linear EGR valve operation.

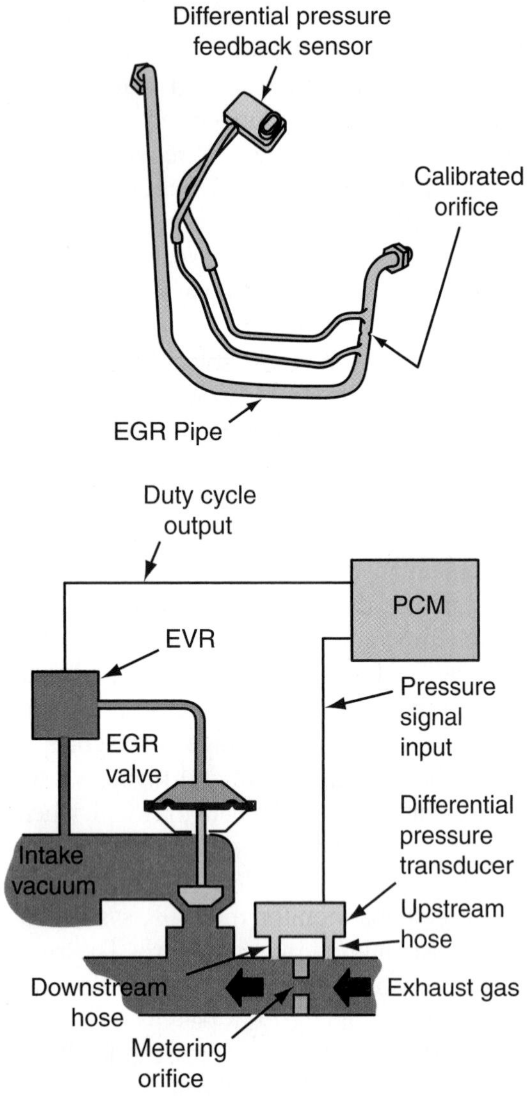

Figure 31-22 An EGR system with a DPFE sensor.

Electronic EGR Controls These various EGR system controls represent some of the common controls currently used by automobile manufacturers. Control devices used in the various systems might have different labels but actually complete the same function within the EGR system. Engines with electronic engine control systems control EGR action in many different ways. The most common of these follow:

- *Twin Solenoid EGR System.* This system uses sensors, solenoids, and an electronic control assembly to modulate and control EGR system components. A pintle valve is often used in the valve to better control the flow rate of exhaust gases. A sensor mounted on the valve stem sends an electronic signal to the PCM, which in turn tells how far the EGR valve is opened. At this time, the EGR control solenoids either maintain or alter the EGR flow, depending on engine operating conditions. The solenoids respond to voltage signals from the PCM. An EGR vent solenoid closes when it is energized. The EGR control solenoid opens when it is energized. Voltage signals from the PCM can trigger the solenoids to increase EGR flow by applying vacuum to the EGR valve, maintain EGR flow by trapping vacuum in the system, or decrease EGR flow by venting EGR vacuum. In actual operation, both solenoids constantly shift between the three operating conditions mentioned as engine operating conditions change.
- *EGR Vacuum Regulator (EVR).* In many EGR systems, the PCM operates a normally closed EGR vacuum regulator solenoid **(Figure 31–23)**, which supplies vacuum to the EGR valve. If the EVR solenoid is not energized, vacuum to the EGR valve is shut off. When the PCM determines the EGR valve should be open, it provides a ground for the EVR solenoid winding. This opens the vacuum passage through the solenoid to the EGR valve. In some systems, the PCM pulses the EVR on and off to supply the precise vacuum and EGR valve opening required by the engine.
- *EGR System with Pressure Feedback Electronic (PFE) Sensor.* Some EGR systems have a pressure feedback electronic sensor. These systems have an orifice located in the exhaust passage below the EGR valve. A small pipe connected from this orifice chamber supplies exhaust pressure to the PFE sensor **(Figure 31–24)**. The PFE sensor changes the exhaust pressure signal into a voltage signal that is sent to the PCM. The PFE signal informs the PCM regarding the amount of EGR flow, and the PCM compares this signal to the EGR flow requested by the input signals. If there is some difference between the actual EGR flow indicated by the PFE signal and the requested EGR flow, the PCM makes the necessary correction to the EVR output signal. In some EGR systems, two pipes supply exhaust pressure from above and below the orifice under the EGR valve to the differential PFE (DPFE) sensor.
- *EGR System with Pressure Transducer (EPT).* Some EGR systems have a pressure transducer in the vacuum hose to the EVR solenoid. A small pressure pipe is connected from the exhaust passage under the EGR valve to the EGR pressure transducer. The exhaust pressure is supplied to a bleed valve diaphragm in the EPT, and ported vacuum from a port above the throttle is connected through a hose to the upper side of this diaphragm. When the engine is operating at low rpm, the vacuum to the EVR is bled off through the open bleed valve in the modulator. At a preset engine rpm, the combination of exhaust pressure below the EPT diaphragm and vacuum above the diaphragm moves the diaphragm upward and closes the bleed valve. When this action occurs, vacuum is supplied through the EPT to the EVR solenoid **(Figure 31–25)**.

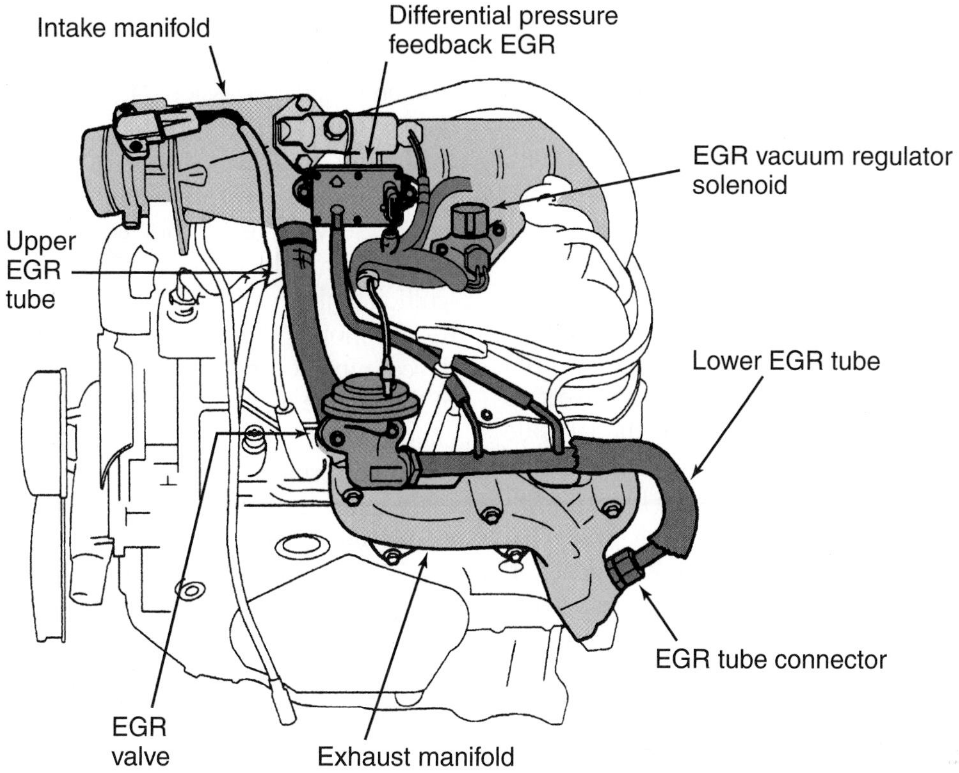

Figure 31-23 An EGR system with an EVR solenoid. *Courtesy of Ford Motor Company*

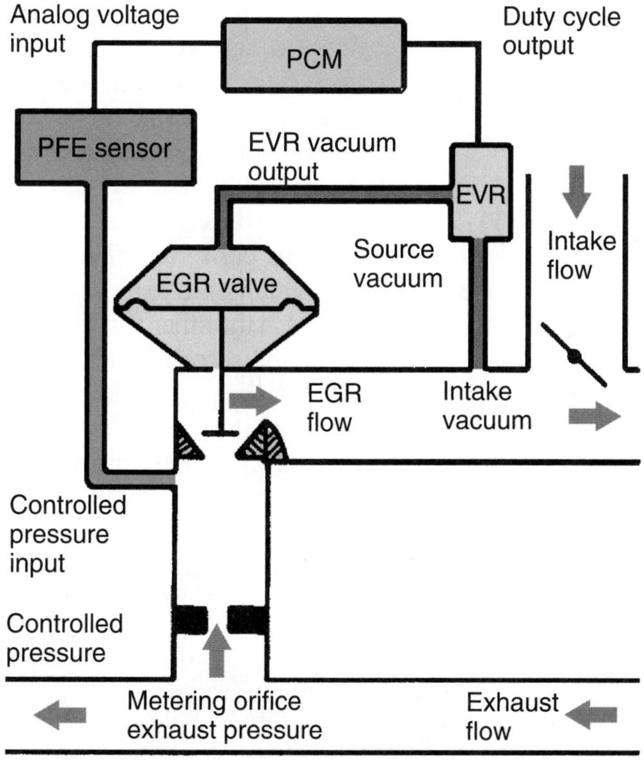

Figure 31-24 A PFE sensor and related circuit. *Courtesy of Ford Motor Company*

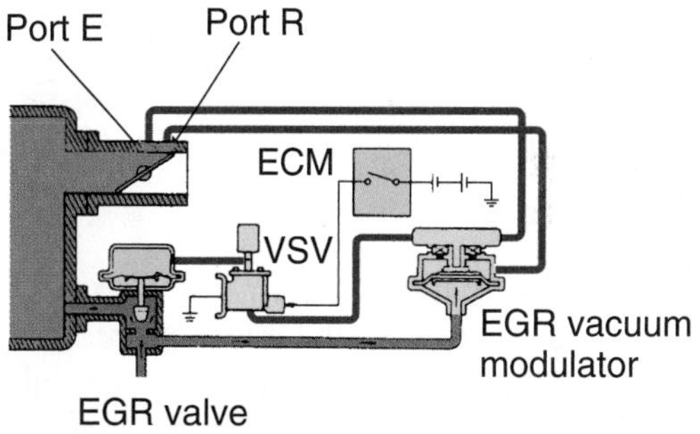

Figure 31-25 An EGR system with an exhaust pressure transducer. *Reprinted with permission*

Intake Heat Control Systems

Hydrocarbon and carbon monoxide exhaust emissions are highest when the engine is cold. The introduction of warm combustion air improves the vaporization of the fuel in the fuel injector throttle body or intake manifold. On older engines equipped with a TBI unit, the fuel is delivered above the throttles, and the intake manifold is filled with a mixture of air and gasoline vapor. Some intake manifold heating is required to prevent fuel condensation, especially when the intake manifold is cool or cold. Therefore, these engines have intake manifold heat control devices such as heated air inlet systems, manifold heat control valves, and **early fuel evaporation (EFE)** heaters.

A heated air inlet control is used on engines with throttle body injection. This system controls the temperature of the air on its way to the throttle body. Another system uses an exhaust manifold heat control valve that routes exhaust gases to warm the intake manifold when the engine is cold. This heats the air/fuel mixture in the intake manifold. These control valves can be either vacuum or thermostatically operated.

Some V8 engines use a more complicated manifold heat control valve, designed to work with a minicatalyst and to preheat the air/fuel mixture for improved cold engine driveability. All right-side exhaust gas travels up through the intake manifold crossover to the left side of the engine. Then, all exhaust gas from the engine passes through a miniconverter just down from the left manifold. As the engine warms up, a coolant-controlled vacuum switch closes. This cuts vacuum to the actuator and allows the valve to open. Exhaust gas flows through both manifolds into the exhaust system and main converter.

Some intake heat systems are computer controlled. These systems use an EFE heater. The engine coolant temperature sensor sends a signal to the PCM in relation to coolant temperature. At a preset temperature, the PCM grounds the mixture heater relay winding, which closes the relay's contacts. Voltage is then supplied through the contacts to the EFE heater. When the coolant temperature reaches a preset point, the ground circuit opens and the relay contacts open and shut off the current flow to the mixture heater. The EFE heater contains a resistance grid that heats the mixture as it passes from the throttle body to the manifold.

Port Fuel Injection In a modern port-injected engine, the fuel is discharged from the injectors into the intake ports near the intake valves. Therefore, the intake ports are filled with fuel vapor, and the rest of the intake manifold passages are filled with air. Because the intake manifold passages are filled with air, the need for intake manifold heating is greatly reduced. Port-injected engines are not equipped with intake manifold heating devices such as heat riser valves or mixture heaters.

POSTCOMBUSTION SYSTEMS

Postcombustion emission control devices clean up the exhaust after the fuel has been burned but before the gases exit the vehicle's tailpipe. An excellent example of this is the catalytic converter. A converter is one of the most effective emission control devices on a vehicle for reducing HC, CO, and NO_x.

Another postcombustion system is the secondary air or air injection system. This system forces fresh air into the exhaust stream to reduce HC and CO emissions.

Catalytic Converters

Catalytic converters are the most effective devices for controlling exhaust emissions. Until 1975, car makers had done a somewhat effective job of controlling emissions by the use of other systems—auxiliary air injection systems, exhaust gas recirculation systems, and positive crankcase ventilation. But controlling emissions with these systems alone also meant lean mixtures and exotic ignition timing, which often severely penalized power and fuel economy. When catalytic converters were introduced, much of the emission control could be taken out of the engine and moved into the exhaust system. This change allowed manufacturers to retune the engine for better performance and improved fuel economy.

A catalytic converter contains a ceramic element coated with a catalyst. A catalyst is something that causes a chemical reaction without being part of the reaction. A catalytic converter causes a chemical change to take place in the exhaust gases as they pass through the converter. Most of the harmful gases are changed to harmless gases.

Three different materials are used as the catalyst in automotive converters: platinum, palladium, and rhodium. Platinum and palladium are the oxidizing elements of a converter **(Figure 31–26)**. When HC and CO are exposed to heated surfaces covered with platinum and palladium, a chemical reaction takes place. The HC and CO are combined with oxygen to become H_2O and CO_2. Rhodium is a reducing catalyst. When NO_x is exposed to hot rhodium, oxygen is removed and NO_x becomes just N. The removal of oxygen is called reduction, which is why rhodium is called a reducing catalyst.

Catalytic converters that contain all three catalysts and reduce HC, CO, and NO_x are called three-way converters **(Figure 31–27)**. Catalytic converters that affect only HC and CO are called oxidizing converters. Three-way converters have the oxidizing catalysts in part of the container and the reducing catalyst in the other. On some engines, fresh air is injected by the secondary air system between the two catalysts. This air helps the oxidizing catalyst work by making extra oxygen available. The air from the secondary air system is not always forced into the converter; rather, it is controlled by the secondary air system. Fresh air added to the exhaust at the wrong time could produce NO_x, something the catalytic converter is trying to destroy.

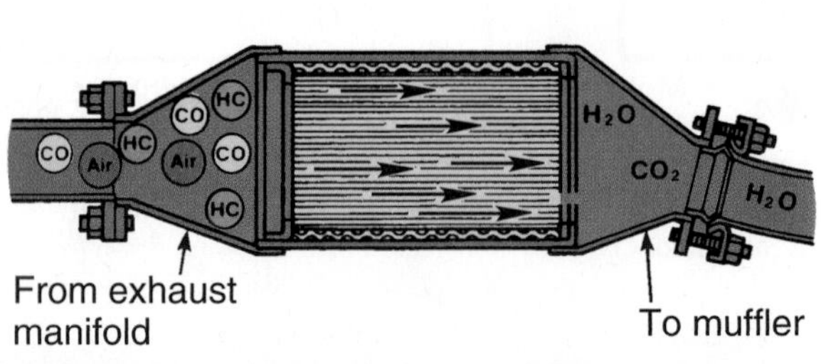

Figure 31-26 An oxidizing catalytic converter.

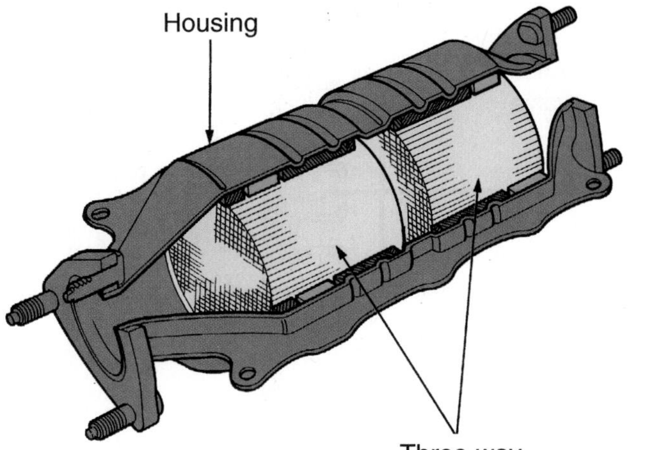

Figure 31-27 A typical three-way catalytic converter. *Courtesy of American Honda Motor Co., Inc.*

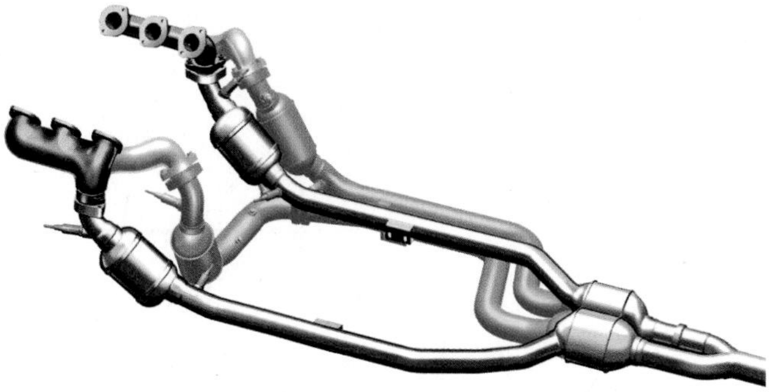

Figure 31-28 Two variations of the same exhaust system, each with a pre-cat right after the exhaust manifold. *Courtesy of DaimlerChrysler Corporation*

Newer converters contain cerium, which is an element that can store oxygen. Oxidation of rich fuel mixtures is accomplished with the oxygen stored in the cerium rather than oxygen supplied by a secondary air system. This need for oxygen explains why the efficiency of a converter is dependent on the constant swings of rich and lean mixtures. When the mixture is lean, oxygen is extracted from the NO_x in the exhaust and stored by the cerium. When the mixture is rich, the stored oxygen is used to oxidize the HC and CO in the exhaust.

GDI engines work with very lean air/fuel mixtures and high compression ratios. Therefore, their NO_x emissions can be very high. To reduce these emissions, some vehicles have an additional catalytic converter, called a storage or adsorber converter. After the exhaust gases leave a three-way converter, they flow through a special NO_x storage converter. This converter is coated with barium and extracts the nitrogen oxides from the exhaust and stores them until its nitrogen oxide sensor senses that the storage converter is filled. At that time, the sensor sends a signal to the PCM and the GDI system starts to deliver a richer air/fuel mixture. As this richer exhaust flows through the storage converter it regenerates the converter, and the nitrogen oxides are converted into harmless nitrogen. When the converter is free of nitrogen oxide, the sensor signals the GDI system to run lean mixtures again.

All late-model vehicles have a pre-cat or warm-up converter **(Figure 31–28)** located on or very near the exhaust manifold(s). These converters warm up rapidly because they are small and close to the engine. Pre-cats are designed to reduce emissions while the main converters are warming up.

Air Injection Systems

One of the earliest methods used to reduce the amount of HCs and CO in the exhaust was by forcing fresh air into the exhaust system after combustion. This additional fresh air causes further oxidation and burning of the unburned HCs and CO. The process is much like blowing on a dwindling fire. Oxygen in the air combines with the HC and CO to continue the burning that reduces the HC and CO concentrations. This allows them to oxidize and produce harmless water vapor and carbon dioxide.

A typical system with an air pump is shown in **Figure 31–29**. On some engines, air from the air injection system is used to pressure purge the charcoal canister, in addition to reacting with the exhaust gases. An air pump **(Figure 31–30)** produces pressurized air that is sent to the exhaust manifold and to the catalytic converter. The air pump is driven by a belt from the crankshaft. An air control valve (or air-switching valve), which is a vacuum-operated valve, routes the air from the pump either to the exhaust manifold or to the catalytic converter. During engine warm-up, the valve directs the air into the exhaust manifold. Once the engine is warm, the extra air in the manifold would affect EGR operation, so the air control valve directs the air to the converter, where it aids the converter in oxidizing emissions. A thermal vacuum switch controls the vacuum to the air control valve. When the coolant is cold, it signals the valve to direct air to the exhaust manifold. Then, when the engine warms to normal operating temperature, the thermal vacuum switch signals the air control valve to reroute the air to the converter.

An air bypass valve (or **diverter valve**) located between the air pump and the air control valve diverts, or detours, air during deceleration. Excess air in an exhaust rich with fuel can produce a backfire or explosion in a muffler. A vacuum signal operates the bypass valve during deceleration. Compressed air is diverted to the atmosphere. The diverter valve prevents backfiring in the exhaust system during sudden acceleration. A very rich mixture is present the moment the throttle closes. The HCs in the exhaust can burn with the air from the air pump, causing a backfire. To prevent this, the air from the air pump is quickly diverted to the atmosphere. One-way check valves allow air into the exhaust but prevent

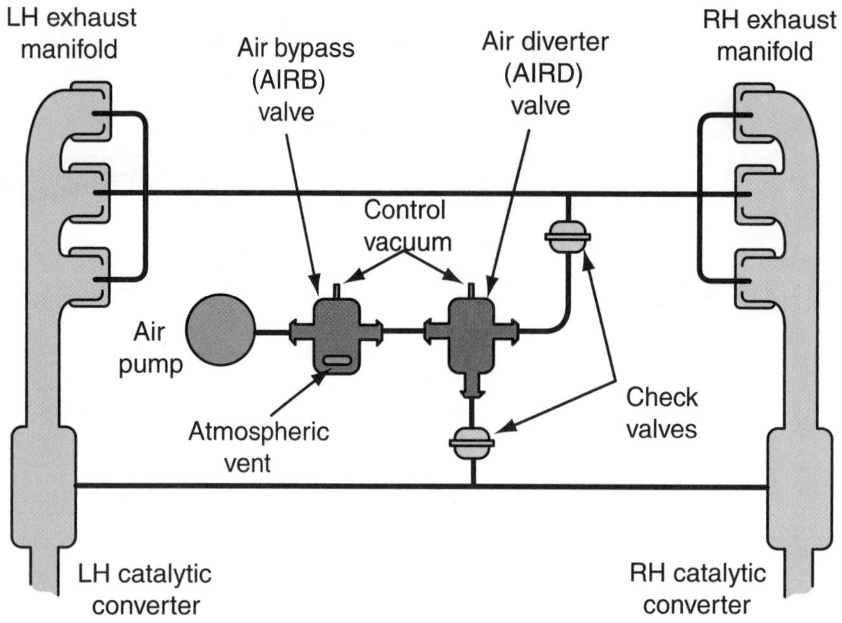

Figure 31–29 A typical pump-type air injection system.

exhaust from entering the pump in the event the drive belt breaks. Their location in the system is behind the air control valve and before the exhaust manifold and catalytic converter.

Electronic Secondary Air System The typical electronic secondary air system, like the conventional air injection system, consists of an air pump to a secondary air bypass valve, which directs the air either to the atmosphere or to the catalytic converter.

The air pump is driven by a belt on the front of the engine and supplies air to the system. Some cars, such as the Corvette, have an air pump driven by an electric motor **(Figure 31–31)**. Intake air passes through a centrifugal filter fan at the front of the pump where foreign materials are separated from the air by centrifugal force. In a commonly used system, air flows from the pump to a secondary **air bypass (AIRB)** valve that directs the air either to the atmosphere or to the secondary **air diverter (AIRD)** valve **(Figure 31–32)**. The AIRD valve directs the air either to the exhaust manifold or to the catalytic converter.

Both the AIRB and AIRD valves have solenoids that are controlled by the PCM. When either solenoid is en-

Figure 31–30 An air injection pump.

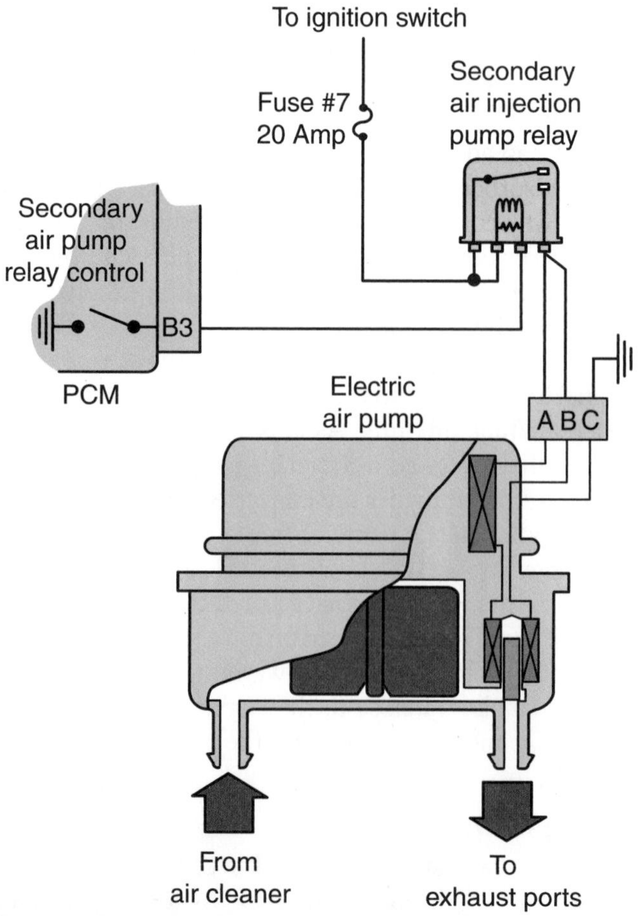

Figure 31–31 An electric air pump circuit.

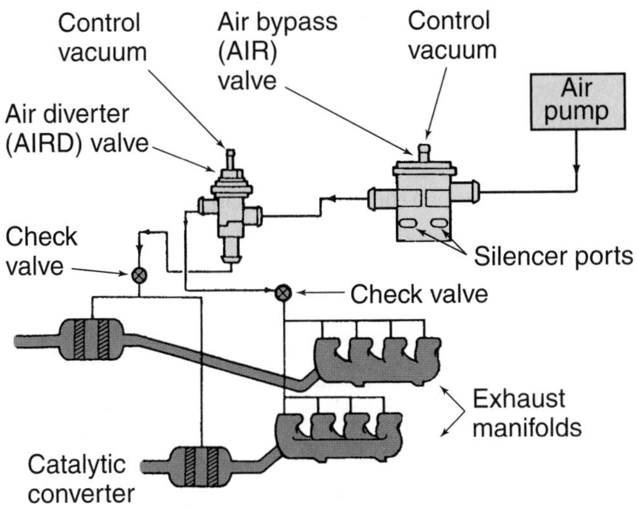

Figure 31-32 A secondary air injection system with AIRB and AIRD valves. *Courtesy of Ford Motor Company*

ergized by the computer, vacuum is applied to the AIRB valve and secondary air is vented to the atmosphere. When no vacuum is applied to the AIRD valve, secondary air (if present) is directed to the catalytic converter.

There are two check valves in the secondary air system. Secondary air must flow through a check valve before it reaches either the exhaust manifold or the catalytic converter. These check valves prevent the backflow of exhaust gases into the pump in the event of an exhaust backfire or if the pump drive belt fails.

Bypass Mode. In the bypass mode, vacuum is not applied either to the AIRB or the AIRD valve and secondary air is vented to the atmosphere. Secondary air may be vented or bypassed due to a fuel-rich condition or during deceleration. Secondary air is also typically bypassed during cold engine cranking and cold idle conditions.

Upstream Mode. During the downstream mode, secondary air is routed through the AIRB valve and the AIRD valve to the exhaust manifold. The upstream mode is actuated when the computer senses a warm crank/start-up condition. The secondary airflow remains upstream for 1 to 3 minutes after start-up to help control emissions.

The air/fuel mixture at start-up is typically very rich. This rich mixture results in unburned HC and CO in the exhaust after combustion. By switching to the upstream mode, the hot HC and CO mix with the incoming secondary air and are burned up.

This reburning of HC and CO compounds causes the exhaust gases to get hotter, which in turn heats up the oxygen sensor. Therefore, switching to the upstream mode allows the electronic engine control system to switch to the closed loop operation sooner because the oxygen sensor is ready to function sooner.

The warm oxygen sensor sends exhaust gas oxygen information to the computer. When the electronic engine control system is in the closed loop operation, the computer uses this information to adjust the air/fuel mixture.

It should be noted that the upstream mode increases the oxygen level in the exhaust gases. As a result, the voltage signal from the oxygen sensor is at a continuous low level. The computer interprets this signal as a continuous lean condition. It can readily be seen, then, that the upstream mode results in inaccurate exhaust gas oxygen measurements. To solve this dilemma, the computer automatically switches to the open loop fuel control whenever the upstream mode is activated. It ignores the oxygen sensor input.

Downstream Mode. During the downstream mode, secondary air is routed through the AIRB valve and the AIRD valve to the catalytic converter. The secondary air system operates in the downstream mode during a majority of engine conditions.

The catalytic converter is most efficient at reducing NO_x when the air/fuel mixture is near stoichiometric. This explains why producing that optimum air/fuel mixture in the closed loop fuel control mode is so important. When secondary air is diverted downstream, the oxygen sensor can provide accurate information about the level of oxygen in the exhaust gases to the computer. Therefore, the PCM can operate in the closed loop only when the secondary air system is diverted downstream.

After the engine has warmed up sufficiently, it is not necessary to run the secondary air system in the upstream mode of operation. The computer automatically switches the system to the downstream mode to allow the secondary air to mix with the exhaust gases inside the catalytic converter. The fresh secondary air allows the converter to reduce the NO_x emissions.

CASE STUDY

A customer complains of severe detonation. The technician checks all the EGR vacuum lines and finds them intact and properly routed.

When the technician pushes the EGR valve stem against spring pressure, it moves freely and returns fully. With the engine at normal operating temperature, a helper opens the throttle enough to reach 2,500 rpm while the technician watches the EGR valve stem. When the throttle is moved, the valve stem does not retract. The technician removes the hose to the EGR valve and feels no vacuum when the engine is revved. Next, the technician pulls off the source vacuum line from the thermostatic vacuum switch and

feels for vacuum. Because vacuum is present, but none is getting to the EGR valve with the engine warm, the technician replaces the thermostatic vacuum switch. The problem should be solved.

KEY TERMS

Air bypass (AIRB)
Air diverter (AIRD)
Air injection reactor (AIR)
Blowby
Canister purge valve
Charcoal (carbon) canister
Diverter valve
Dyno
Early fuel evaporation (EFE)
Evaporative control
I/M 240
Negative back pressure
Periodic motor vehicle inspection (PMVI)
Positive back pressure
Postcombustion control
Precombustion control
Transducer

SUMMARY

- Unburned hydrocarbons, carbon monoxide, and oxides of nitrogen are three types of emissions being controlled in gasoline engines. HC emissions are unburned gasoline released by the engine because of incomplete combustion. CO emissions are a byproduct of combustion and are caused by a rich air/fuel ratio. Oxides of nitrogen (NO_x) are formed when combustion temperatures reach more than 2,300°F (1,261°C).
- Precombustion control systems prevent emissions from being created in the engine, either during or before the combustion cycle. Postcombustion control systems clean up exhaust gases after the fuel has been burned. The evaporative control system traps fuel vapors that would normally escape from the fuel tank and carburetor into the air.
- The PCV system removes blowby gases from the crankcase and recirculates them to the engine intake.
- With the engine running at idle speed, the high intake manifold vacuum moves the PCV valve toward the closed position.
- During part-throttle operation, the intake manifold vacuum decreases and the PCV valve spring moves the valve toward the open position. As the throttle approaches the wide-open position, intake manifold vacuum decreases and the spring moves the PCV valve farther toward the open position.
- A detonation sensor changes a vibration caused by engine detonation into an analog voltage signal. The electronic spark control module changes the analog detonation sensor signal into a digital signal and sends this signal to the PCM.
- An evaporative (EVAP) emission system stores vapors from the fuel tank in a charcoal canister until certain engine operating conditions are present. When the proper conditions are present, fuel vapors are purged from the charcoal canister into the intake manifold.
- A port EGR valve is opened when vacuum is supplied to the chamber above the diaphragm.
- A positive back pressure EGR valve has a normally open bleed valve in the center of the diaphragm. This bleed valve is closed by exhaust pressure at a specific throttle opening.
- A negative back pressure EGR valve contains a normally closed bleed valve in the center of the diaphragm. This bleed valve is opened by negative pressure pulses in the exhaust at low engine speed.
- A digital EGR valve has up to three electric solenoids operated by the PCM.
- A linear EGR valve contains an electric solenoid that is operated by the PCM with a pulse width modulation (PWM) signal.
- A pressure feedback electronic (PFE) sensor sends a voltage signal to the PCM in relation to the exhaust pressure under the EGR valve.
- Many secondary air injection systems pump air into the exhaust ports during engine warm-up, and these systems deliver air to the catalytic converters with the engine at normal operating temperature.

TECH MANUAL

The following procedures are included in Chapters 31/32 of the *Tech Manual* that accompanies this book:

1. Checking the emission levels on an engine.
2. Testing the operation of a PCV system.
3. Testing the operation of an EGR valve.
4. Testing a catalytic converter for efficiency.

REVIEW QUESTIONS

1. Explain why a small PCV valve opening is adequate at idle speed.
2. At what temperature do nitrogen atoms combine with oxygen atoms to form NO_x?
3. Describe the operation of a digital EGR valve.
4. Explain why a secondary air injection system pumps air into the exhaust ports during engine warm-up.
5. Name the three types of emissions being controlled in gasoline engines.
6. The PCV system prevents ______ ______ from escaping to the atmosphere.
7. On some systems, the detonation sensor signal is sent through the ______ ______ ______ to the PCM.
8. In a negative back pressure EGR valve, if the exhaust pressure passage in the stem is plugged, the bleed valve remains ______ .
9. In a secondary air injection system, the one-way check valves prevent ______ from entering the air pump and air pipe to the air cleaner.
10. HC emissions may come from the tail pipe or ______ sources.
11. NO_x, HC, and CO are changed into harmless gases by the ______ in the catalytic converter.
12. While discussing PCV valve operation, Technician A says the PCV valve opening is decreased at part-throttle operation compared to idle operation. Technician B says the PCV valve opening is decreased at wide-open throttle compared to part throttle. Who is correct?
    - **a.** Technician A
    - **b.** Technician B
    - **c.** Both A and B
    - **d.** Neither A nor B
13. *True or False?* All late-model vehicles have a pre-cat or warm-up converter located on or very near the exhaust manifold(s). These converters warm up rapidly because they have electrical heaters built in them and they store oxides of nitrogen.
14. While discussing EGR systems with a pressure feedback electronic (PFE) sensor, Technician A says the PFE sensor sends a signal to the PCM in relation to intake manifold pressure. Technician B says the PCM corrects the EGR flow if the actual flow does not match the requested flow. Who is correct?
    - **a.** Technician A
    - **b.** Technician B
    - **c.** Both A and B
    - **d.** Neither A nor B
15. While discussing exhaust pressure transducers, Technician A says the exhaust pressure transducer is connected in the vacuum hose between the intake manifold and the EGR solenoid. Technician B says the exhaust pressure transducer bleeds off vacuum to the EGR valve when the engine is operating at low rpm. Who is correct?
    - **a.** Technician A
    - **b.** Technician B
    - **c.** Both A and B
    - **d.** Neither A nor B
16. While discussing evaporative (EVAP) systems, Technician A says the coolant temperature has to be above a preset value before the PCM will operate the canister purge solenoid. Technician B says the vehicle speed has to be above a preset value before the PCM will operate the canister purge solenoid. Who is correct?
    - **a.** Technician A
    - **b.** Technician B
    - **c.** Both A and B
    - **d.** Neither A nor B
17. Technician A says that the EGR vent solenoid is normally open. Technician B says that the EGR control solenoid is normally open. Who is correct?
    - **a.** Technician A
    - **b.** Technician B
    - **c.** Both A and B
    - **d.** Neither A nor B
18. Technician A says that the AIRB valve directs secondary air either to the exhaust manifold or to the catalytic converter. Technician B says that secondary air may be vented during deceleration. Who is correct?
    - **a.** Technician A
    - **b.** Technician B
    - **c.** Both A and B
    - **d.** Neither A nor B
19. Explain why a PCV system is critical to an engine's durability?
20. Which of the following systems is designed to reduce NO_x and has little or no effect on overall engine performance?
    - **a.** PCV
    - **b.** EGR
    - **c.** AIR
    - **d.** EVAP

32

EMISSION CONTROL DIAGNOSIS AND SERVICE

OBJECTIVES

■ Describe oxygen (O_2) emissions in relation to air/fuel ratio. ■ Describe how carbon dioxide (CO_2) is formed in the combustion chamber. ■ Describe how oxides of nitrogen (NO_x) are formed in the combustion chamber. ■ Describe the inspection and replacement of PCV system parts. ■ Diagnose engine performance problems caused by improper EGR operation. ■ Diagnose and service the various types of EGR valves. ■ Diagnose EGR vacuum regulator (EVR) solenoids. ■ Diagnose and service the various intake heat control systems. ■ Check the efficiency of a catalytic converter. ■ Diagnose and service secondary air injection systems. ■ Diagnose and service evaporative (EVAP) systems.

The quality of an engine's exhaust depends on two things: the effectiveness of the emission control devices and the efficiency of the engine. A totally efficient engine changes all of the energy in the fuel, in this case gasoline, into heat energy. This heat energy is the power produced by the engine.

In order for an engine to run efficiently, it must have fuel mixed with the correct amount of air. This mixture must be shocked by the correct amount of heat (spark) at the correct time. All of this must happen in a sealed container or cylinder. When these conditions are met, a great amount of heat energy is produced and the fuel and air combine to form water and carbon dioxide. The nitrogen in the incoming air leaves the engine unchanged.

Since it is nearly impossible for an engine to receive the correct amounts of everything, a good-running engine will emit some amounts of pollutants. It is the job of the emission control devices to clean them up.

The three kinds of emissions being controlled in gasoline engines are unburned hydrocarbons, carbon monoxide, and oxides of nitrogen. Federal laws require new cars and light trucks to meet specific emissions levels. State governments also have laws requiring car owners to maintain their vehicles so that the emissions remain below an acceptable level. Most states require an annual emissions inspection to meet that goal. Many shops have an exhaust analyzer for inspection purposes.

TESTING EMISSIONS

Testing the quality of the exhaust is both a procedure for testing emission levels and a diagnostic routine. One of the most valuable diagnostic tools is the exhaust analyzer **(Figure 32–1)**. By looking at the quality of an engine's exhaust, a technician is able to see the effects of the combustion process. Any engine problem can cause a change in exhaust quality. The amount and type of change serves as the basis of diagnostic work.

Exhaust Analyzer

Early exhaust analyzers measured the hydrocarbons and carbon monoxides in the exhaust. Exhaust hydrocarbons are raw, unburned fuel. HC emissions indicate that complete combustion is not occurring in the engine. HC is measured in **parts per million (ppm)** or grams per mile (g/mi). Carbon monoxide is an odorless, toxic gas that is the product of combustion and is typically caused by a lack of air or excessive fuel. CO is typically measured as a percent of the total exhaust.

An exhaust analyzer has a long sample hose with a probe in the end of the hose. The probe is inserted in the vehicle's tailpipe. When the analyzer is turned on, an internal pump moves an exhaust sample from the tailpipe through the sample hose and the analyzer. A water trap and filter in the hose remove moisture and carbon particles.

The pump forces the exhaust sample through a sample cell in the analyzer. The exhaust sample is then vented

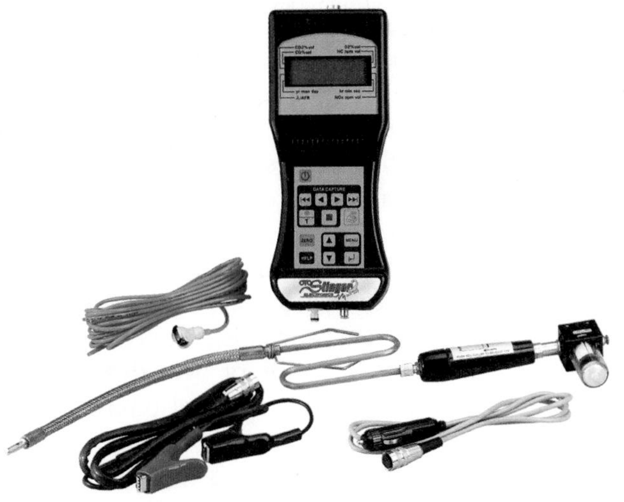

Figure 32-1 A hand-held exhaust gas analyzer. *Courtesy of SPX Corporation Service Solutions*

to the atmosphere **(Figure 32–2)**. In the sample cell, a beam of infrared light passes through the exhaust sample. The analyzer then determines the quantities of HC and CO if the analyzer is a two-gas analyzer, or HC, CO, CO_2, and O_2 if it is a four-gas analyzer. Some analyzers can also measure NO_x and are called five-gas analyzers **(Figure 32–3)**. Nearly all currently used analyzers are either four- or five-gas machines.

Maximum limits for the measured gases are set by regulations according to the model year of the vehicle. These limits also vary by state or locale. It is always desirable to have low amounts of four of the five measured gases at all engine speeds. The only gas that is desired in high percentages is carbon dioxide. Normally the desired amount of CO_2 is 13.4% or more. The other gases should be kept to the lowest levels possible.

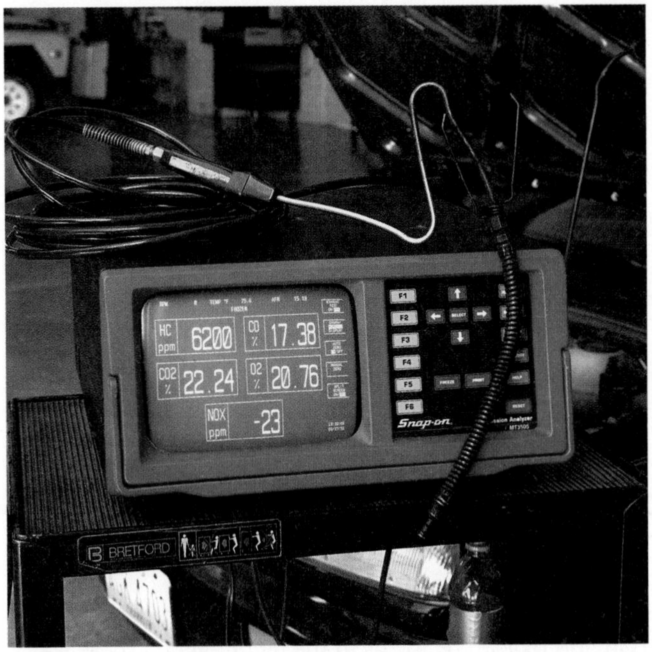

Figure 32-3 A five-gas exhaust analyzer.

Engine Diagnosis

When using the exhaust analyzer as a diagnostic tool, it is important to realize that the severity of the problem dictates how much higher than normal the readings will be. When attempting to identify the exact cause of the abnormal readings, disable the secondary air system. Then rerun the tests. Without the secondary air, the readings will give a more accurate look at the air/fuel mixture. This

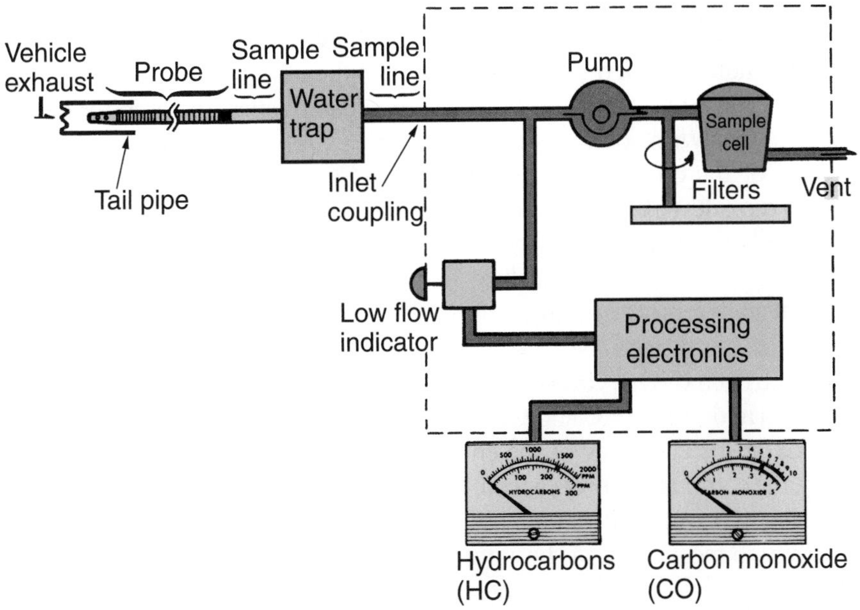

Figure 32-2 The basic circuit of an exhaust gas analyzer. *Courtesy of DaimlerChrysler Corporation*

should help to find the cause of the problem. Also, if you are using a two-gas analyzer it is best to take your readings in front of the catalytic converter.

The levels of HC and CO in the exhaust are a direct indication of engine performance. Unburned hydrocarbons (HC) are particles, usually vapors, of gasoline that have not been burned during combustion, or raw gas that evaporates out of the fuel system. HC in the exhaust is a sign of incomplete combustion; the better the combustion, the less HC in the exhaust. Today's engines are designed to be efficient; therefore, they release lower amounts of HC in the exhaust. A high level of hydrocarbons could indicate a low compression, overly lean mixture, fouled spark plug, defective spark plug wire, or burned valve. All of these problems would decrease the effectiveness of combustion and allow fuel to leave the cylinder unburned.

Carbon monoxide is a poisonous gas comprised of carbon and oxygen. It forms in the engine when there is not enough oxygen to combine with the carbon during combustion. CO is a by-product of combustion. If combustion does not take place, CO will be low. When there is enough oxygen in the air/fuel mixture, carbon dioxide (CO_2) is formed. CO_2 is not a pollutant and is used by plants to manufacture oxygen. Lower levels of CO are to be expected when HC readings are high. High CO indicates an excessively rich air/fuel mixture caused by a restriction in the air intake system or too much fuel being delivered to the cylinders. Low CO does not indicate a lean mixture; rather, as the mixture becomes more lean, HC increases.

Many of the emission control devices that have been added to vehicles over the past thirty years, especially catalytic converters, have decreased the amount of HC and CO in the exhaust. These devices alter the contents of the exhaust. Therefore, checking the HC and CO contents in the exhaust may not be a true indication of the operation of an engine.

Four- and five-gas exhaust analyzers look at the efficiency of an engine in spite of the effectiveness of the emission controls. Carbon dioxide (CO_2) and oxygen (O_2) levels in the exhaust are changed only slightly by the emission controls and therefore can be used to check engine efficiency. Many exhaust analyzers are now available that measure a fifth gas, oxides of nitrogen (NO_x).

By measuring oxides of nitrogen (NO_x), carbon dioxide (CO_2) and oxygen (O_2), in addition to HC and CO, a technician gets a better look at the efficiency of the engine **(Figure 32–4)**. Ideally, the combustion process combines fuel (HC) and oxygen (O_2) to form water and carbon dioxide (CO_2). Although most of the air brought into the engine is nitrogen, this gas should not become part of the combustion process and should pass out of the exhaust as nitrogen.

Oxides of nitrogen (NO_x) are various compounds of nitrogen and oxygen; both gases are present in the air used for combustion. They are formed in the cylinders during combustion and are part of exhaust gas. NO_x is the main chemical that causes smog and results from high combustion temperatures. NO_x is measured in ppm and is typically within 200 to 1,000 ppm. Readings higher than that indicate high combustion temperatures. This can be caused by carbon deposits in the cylinders, cooling system problems, a faulty EGR valve, an overly lean air/fuel mixture, or overadvanced ignition timing. High NO_x readings may also be caused by an ineffective reduction catalytic converter. The efficiency of the reduction converter is related to the fuel control system.

SHOP TALK

Increases in CO tend to lower NO_x emissions, so it is possible that NO_x readings will increase after you have corrected the cause of the high CO readings. ■

CO_2 is a desired element in the exhaust stream and is only present when there is complete combustion. Therefore, the more CO_2 in the exhaust stream, the better.

O_2 is used to oxidize CO and HC into water and CO_2. As the mixture goes lean, O_2 levels increase. As the mixture goes rich, O_2 moves and stays low. Ideally, we would like to see very low amounts of O_2 in the exhaust stream. CO and O_2 are inversely related; when one goes up, the other goes down. If there is a high O_2 reading without a low CO reading, an air leak in the exhaust system may be indicated.

If there is high HC and low CO in the exhaust because of a lean misfire, the amount of O_2 will be high. If there is an extremely rich air/fuel mixture, there will be high HC, CO, and O_2 readings.

Excessive HC emissions may be caused by:

- Ignition system misfiring
- Improper ignition timing
- Excessively lean or rich air/fuel ratio
- Low cylinder compression
- Defective valves, guides, or lifters
- Defective rings, pistons, or cylinders
- Vacuum leaks
- Dirty fuel injector
- Excessive EGR action
- Defective system input sensor

	IDLE	2500 RPM	PROBABLE CAUSE
HC ppm	0–150	0–75	Normal reading
CO%	1–15	0–0.8	
CO_2%	10–12	11–13	
O_2%	0.5–2.0	0.5–1.25	
NO_X ppm	100–300	200–1,000	
HC ppm	0–150	0–75	Rich mixture
CO%	3.0+	3.0+	
CO_2%	8–10	9–11	
O_2%	0–0.5	0–0.5	
NO_X ppm	0–200	100–500	
HC ppm	0–150	0–75	Lean mixture
CO%	0–1.0	0–0,25	
CO_2%	8–10	11	
O_2%	1.5–3.0	1.0–2.0	
NO_X ppm	300–1,000	1,000+	
HC ppm	50–850	50–750	Lean misfire
CO%	0–0.3	0–0.3	
CO_2%	5–9	6–10	
O_2%	4–9	2–7	
NO_X ppm	300–1,000	1,000+	
HC ppm	50–850	50–750	Misfire
CO%	0.1–1.5	0–0.8	
CO_2%	6–8	8–10	
O_2%	4–12	4–12	
NO_X ppm	0–200	100–500	

Figure 32-4 The readings of the chemicals in the exhaust can lead you to the cause of a driveability problem.

Excessive CO emissions may be caused by:

- Rich air/fuel mixtures
- Plugged PCV valve or hose
- Dirty air filter
- Leaking fuel injectors
- Higher than normal fuel pressures
- Ruptured diaphragm in the fuel pressure regulator
- Defective system input sensor

Excessive HC and CO emissions may be caused by:

- Plugged PCV system
- Excessively rich air/fuel ratio
- Stuck open heat riser valve
- AIR pump inoperative or disconnected
- Engine oil diluted with gasoline

Lower-than-normal O_2 emissions may be caused by:

- Rich air/fuel mixtures
- Dirty air filter
- Faulty injectors
- Higher-than-normal fuel pressures
- Defective system input sensor
- Restricted PCV system
- Charcoal canister purging at idle and low speeds

Lower-than-normal CO_2 emissions may be caused by:

- Leaking exhaust system
- Rich air/fuel mixture

Higher-than-normal O_2 emissions may be caused by:

- An engine misfire
- Lean air/fuel mixtures
- Vacuum leaks
- Lower-than-specified fuel pressures
- Defective fuel injectors
- Defective system input sensor

Higher-than-normal NO_x emissions may be caused by:

- An overheated engine

- Lean air/fuel mixtures
- Vacuum leaks
- Overadvanced ignition timing
- Defective EGR system
- Carbon deposits on intake valves

IM240 Test

An IM240 test checks the emission levels of a vehicle while it is operating under a variety of conditions and loads. This test gives a true look at the exhaust quality of a vehicle as it is working. The testing sequence normally begins with a leakage test **(Figure 32–5)** and a functional test **(Figure 32–6)** of the evaporative emission control devices and a complete visual inspection of the emission control system.

The IM240 test is a 240-second test of a vehicle's emissions. During this time the vehicle is loaded to simulate a short drive on city streets, then a longer drive on a highway. The complete test cycle includes acceleration, deceleration, and cruising. During the test, the vehicle's exhaust is collected by a **constant volume sampling (CVS)** system that makes sure a constant volume of ambient air and exhaust pass through the exhaust analyzer. The CVS exhaust hose covers the entire exhaust pipe, therefore it collects all of the exhaust, not just a sample as regular gas analyzers do. The hose contains a mixing tee that draws in outside air to maintain a constant volume to the gas analyzer. This is important for the calculation of mass exhaust emissions. During an IM240 test, the exhaust gases are measured in grams per mile.

The test is conducted on a chassis dynamometer **(Figure 32–7)** that loads the drive wheels to simulate real-world conditions. (Testing a vehicle while it is driven on a chassis dynamometer is called transient testing. Testing a vehicle at a constant load on a dyno is called steady-state testing.)

The rollers of the dyno are loaded by a computer after an inspector inputs basic information about the vehicle to be tested. The computer's program will adjust the resistance on the rollers according to the vehicle's weight, aerodynamic drag, and road friction.

At most I/M testing stations when a vehicle enters an inspection lane, a lane inspector greets the driver. The inspector receives the needed information from the driver, then directs the driver to the customer waiting area. The inspector moves the vehicle to where it will be visually inspected. Upon completion of the inspection, the inspector enters all pertinent information into the computer. After this information has been entered, the inspector moves the car onto the inertia simulation dynamometer.

Once on the dyno, the hood of the vehicle is raised and a large cooling fan is moved in front of the car. The engine is allowed to idle for at least 10 seconds, then test-

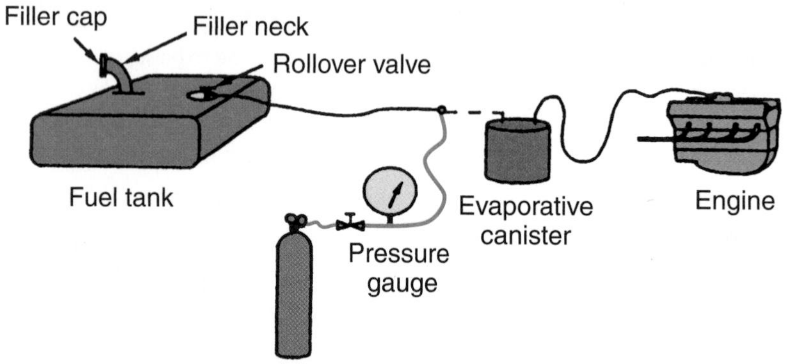

Figure 32–5 A typical setup for conducting a leakage test on an evaporative emission control system. *Courtesy of U.S. EPA*

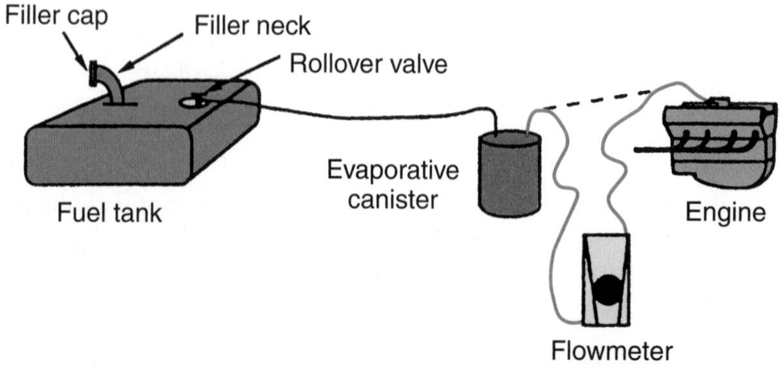

Figure 32–6 A typical setup for conducting a functional test of the evaporative emission control system. *Courtesy of U.S. EPA*

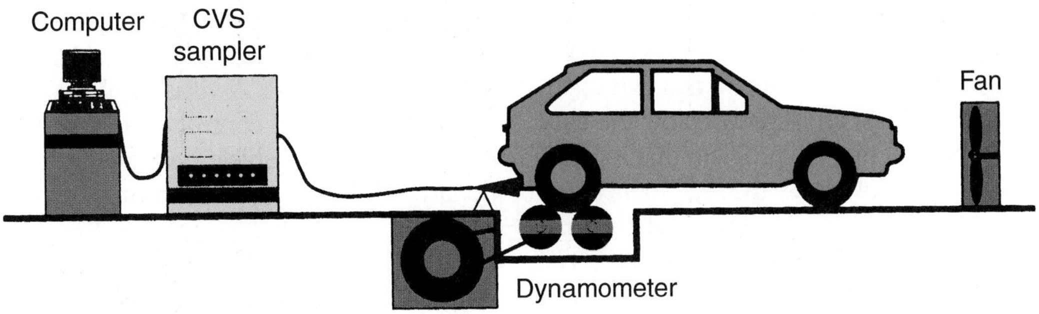

Figure 32-7 Components of a typical IM240 test station.

ing begins. The transient driving cycle is 240 seconds and includes periods of idle, cruise, and varying accelerations and decelerations. During the drive cycle, the car's exhaust is collected by the CVS system. The inspector driving the car must follow an electronic, visual depiction of the speed, time, acceleration, and load relationship of the transient driving cycle. For cars equipped with a manual transmission, the inspector is prompted by the computer to shift gears according to a predetermined schedule.

The IM240 drive cycle is displayed by a trace. The trace is based on road speed versus time. The test is comprised of two phases—Phase 1 and Phase 2. Combining the results of these two phases results in the test composite **(Figure 32–8)**. Phase 1 is the first 95 seconds of the drive cycle. The car travels 56 hundredths of a mile.

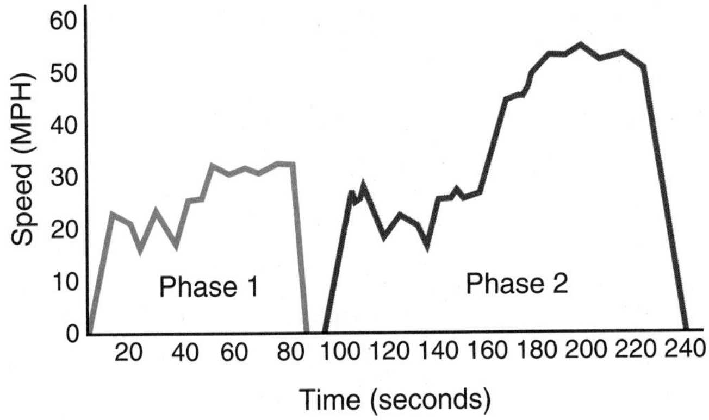

Figure 32-8 The drive trace for an IM240 test.

This phase represents driving on a flat highway at a maximum speed of 32.4 miles per hour (52 km/h) under light to moderate loads. Phase 2 is from the 96th second of the test to the end, or the 240th second. The vehicle travels 1.397 miles (2.2 km) on a flat highway at a maximum speed of 56.7 miles per hour (91 km/h), including a hard acceleration to highway speeds.

The end result of the IM240 test is the measurement of the pollutants emitted by the vehicle during normal, on-the-road driving. After the test, the customer receives an inspection report. If the car failed the emissions test, it must be fixed. This inspection report is a valuable diagnostic tool. To correct the problems that caused the vehicle to fail, the inspection report must be studied.

The report shows the amount of gases emitted during the different speeds of the test. It also shows the cutpoint for the various pollutants. The **cutpoint** represents the maximum amount or limit of each gas that the vehicle is allowed to emit. Nearly all vehicles have some speeds and conditions in which the emissions levels are above the cutpoint. A vehicle that fails the test will have many abnormal speeds or will have an area of very high pollutant output.

To use the report as a diagnostic tool, pay attention to all of the gases and to the loads and speeds at which the vehicle went over the cutpoint **(Figure 32–9)**. Think about what system or systems are responding to the load or speed. Would a malfunction of these systems cause this

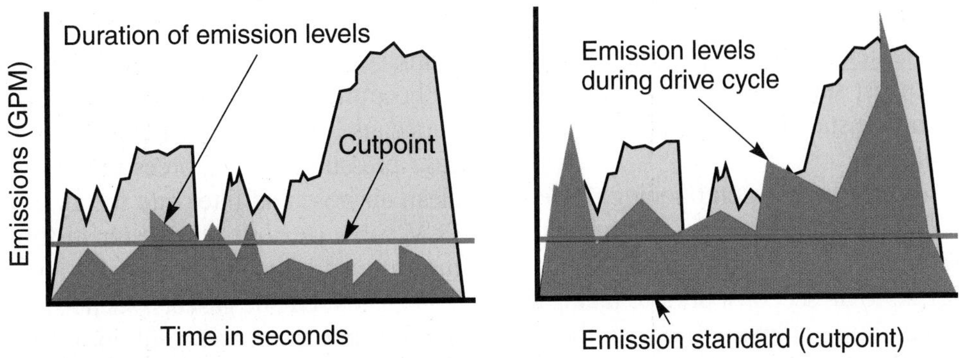

Figure 32-9 (Left) The emissions level is acceptable. (Right) The emissions level is not acceptable.

gas to increase? Pay attention not only to the gases that are above the cutpoint, but also to those that are below. If the HC readings are above the cutpoint at a particular speed and the NO_x readings are slightly below the cutpoint at the same speed, fixing the problem that is causing the high HC will probably cause the NO_x to increase above the cutpoint. As combustion is improved, the chance of forming NO_x also improves. Consider all of the correlations between gases when diagnosing a problem.

The IM240 measures HC, CO, NO_x, and CO_2. The job of a technician is to determine at what loads and engine speeds the vehicle failed. Then test those systems that could affect engine operation and exhaust output at the failed speeds and loads. These tests could include looking at an O_2 waveform, watching a four-gas analyzer, checking data with a scan tool, reading injector pulse width, or doing other tests

Most of the IM240 drive cycle can be simulated either in the diagnostic bay or on the road. It is hard to duplicate the entire drive cycle on a road test, but you can come close. Doing this may be important for diagnostics and is very important for the verification of the repair. The drive cycle is actually six operating modes. These modes need to be duplicated to identify why the vehicle failed that part of the trace.

- Mode one—idle, no load at zero miles per hour.
- Mode two—acceleration from zero to 35 miles per hour (56 km/h).
- Mode three—acceleration from 35 to 55 miles per hour (56 to 88 km/h).
- Mode four—a steady cruise at 35 miles per hour (56 km/h).
- Mode five—a steady high cruise at 55 miles per hour (88 km/h).
- Mode six—decelerations from 35 miles per hour (56 km/h) to zero and from 55 miles per hour (88 km/h) to zero.

Other I/M Testing Programs

Many state I/M programs measure only the emissions output from a vehicle while it is idling. The test is conducted with a certified exhaust gas analyzer. The measurements of the exhaust sample are then compared to standards dictated by the state according to the production year of the vehicle.

Some states also include a preconditioning mode at which the engine is run at a high idle (approximately 2,500 rpm) or the vehicle is run at 30 mph (50 km/h) on a dynamometer for 20 to 30 seconds, prior to taking the idle tests. This preconditioning mode heats up the catalytic converter, allowing it to work at its best. Some programs include the measurement of the exhaust gases during a low constant load on the dyno or during a constant high idle. These measurements are taken in addition to the idle tests. A visual inspection and functional test of the emission control devices is part of some I/M programs. If the vehicle has tampered-with, nonfunctional, or missing emission control devices, the vehicle will fail the inspection.

The California ARB developed a test, called the **Acceleration Simulation Mode (ASM)** test, that incorporates steady-state and transient testing. The ASM includes a high-load steady-state phase and a 90-second transient test. This test is an economical alternative to the IM240 test, which requires a dyno with a computer-controlled power absorption unit. The ASM can be conducted with a normal chassis dyno and a five-gas analyzer.

Another alternative to the IM240 test is the **Repair Grade (RG240)** test. This program uses a chassis dynamometer, constant volume sampling, and a five-gas analyzer. It is very similar to the IM240 test and is conducted in the same way, but it is more economical. The primary difference between the IM240 and the RG240 is the chassis dyno. Although they accomplish basically the same thing, the RG240 dyno is less complicated but still nearly matches the load simulation of the IM240 dyno.

PCV SYSTEM DIAGNOSIS AND SERVICE

No adjustments can be made to the PCV system. Service of the system involves a careful inspection, operation, and replacement of faulty parts. Some engines use a fixed orifice tube in place of a valve. These should be cleaned periodically with a pipe cleaner soaked in carburetor cleaner. Although there is no PCV valve, this type of system is diagnosed in the same way as those systems with a valve. When replacing a PCV valve, match the part number on the valve with the vehicle maker's specifications for the proper valve. If the valve cannot be identified, refer to the part number listed in the manufacturer's service manual.

If the PCV valve is stuck in the open position, excessive air flow through the valve causes a lean air/fuel ratio and possible rough idle operation or engine stalling. When the PCV valve or hose is restricted, excessive crankcase pressure forces blowby gases through the clean air hose and filter into the air cleaner. Worn rings or cylinders cause excessive blowby gases and increased crankcase pressure, which forces blowby gases through the clean air hose and filter into the air cleaner. A restricted PCV valve or hose may result in the accumulation of moisture and sludge in the engine and engine oil.

Leaks at engine gaskets, such as rocker arm cover or crankcase gaskets, will result in oil leaks and the escape of blowby gases into the atmosphere. However, the PCV system also draws unfiltered air through these leaks into the engine. This action could result in wear of engine components, especially when the vehicle is operating in dusty conditions. Check all the engine gaskets for signs of oil leaks **(Figure 32–10)**. Be sure the oil filler cap fits and seals properly.

The first step of PCV servicing is a visual inspection. The PCV valve can be located in several places. The most common location is in a rubber grommet in the valve cover. It can be installed in the middle of the hose connections, as well as installed directly in the intake manifold.

Once the PCV valve is located, make sure all the PCV system hoses are properly connected and that they have no breaks or cracks. Remove the air cleaner and inspect the air and crankcase filters. Crankcase blowby can clog these with oil. Clean or replace such filters. Oil in the air cleaner assembly indicates that the PCV valve or hoses are plugged. Make sure you check these and replace the valve and clean the hoses and air cleaner assembly. When the PCV valve and hose are in satisfactory condition and there is oil in the air cleaner assembly, perform a cylinder compression test to check for worn cylinders and piston rings.

Functional Checks of the PCV System

A rough-idling engine can signal a number of PCV problems, such as a clogged valve or a plugged hose. But before beginning the functional checks, double check the PCV valve part number to make certain the correct valve is installed. If the correct valve is being used, continue by disconnecting the PCV valve from the valve cover, intake manifold, or hose. Start the engine and let it run at idle. If the PCV valve is not clogged, a hissing is heard as air passes through the valve. Place a finger over the end of the valve to check for vacuum **(Figure 32–11)**. If there is little or no vacuum at the valve, check for a plugged or restricted hose. Replace any plugged or deteriorated hoses. Turn off the engine and remove the PCV valve. Shake the valve and listen for the rattle of the check needle inside the valve. If the valve does not rattle, replace it.

Some vehicle manufacturers recommend that the valve be checked by removing it from the valve cover and hose. Connect a hose to the inlet side of the PCV valve, and blow air through the valve with your mouth while holding your finger near the valve outlet **(Figure 32–12)**. Air should pass freely through the valve. If air does not pass freely through the valve, replace the valve. Move the hose to the outlet side of the PCV valve and try to blow back through the valve **(Figure 32–13)**. It should be difficult to blow air through the PCV valve in this direction. When air passes easily through the valve, replace the valve.

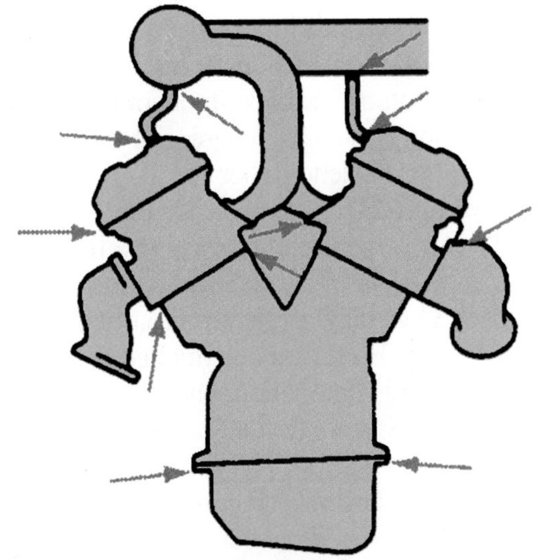

Figure 32–10 Places to check for oil leaks on a typical engine. *Reprinted with permission*

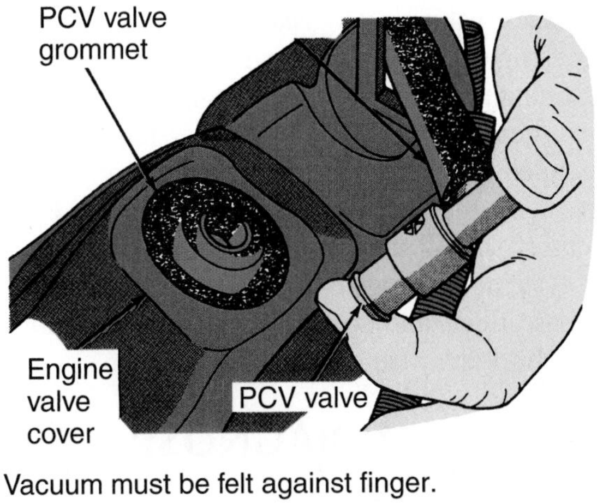

Figure 32–11 With the engine at idle, vacuum should be felt at the PCV valve. *Courtesy of DaimlerChrysler Corporation*

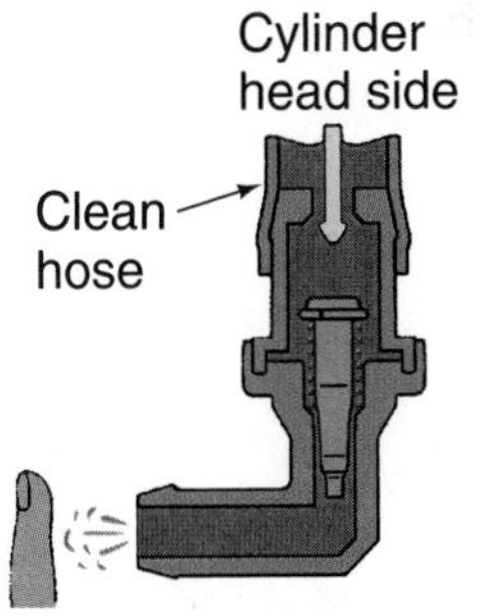

Figure 32–12 Blowing through the inlet of a PCV valve, the air should flow freely through the valve. *Reprinted with permission*

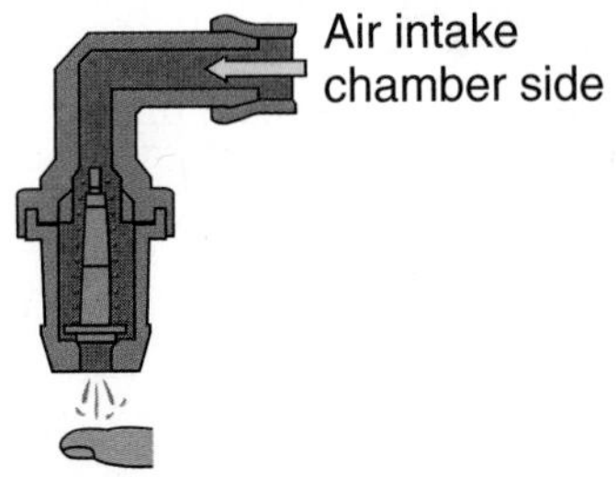

Figure 32–13 Blowing through the outlet of a PCV valve, the air should barely flow through the valve. *Reprinted with permission*

CAUTION!

Do not attempt to suck through a PCV valve with your mouth. Sludge and other deposits inside the valve are harmful to the human body.

Another simple check of the PCV valve can be made by pinching the hose between the valve and the intake manifold **(Figure 32–14)** with the engine at idle. You should hear a clicking sound from the valve when the hose is pinched and unpinched. If no clicking sound is heard, check the PCV valve grommet for cracks or damage. If the grommet is all right, replace the PCV valve.

Remember that proper operation of the PCV system depends on a sealed engine. The crankcase is sealed by the dipstick, valve cover, gaskets, and sealed filler cap. If oil sludging or dilution is found and the PCV system is functioning properly, check the engine for oil leaks and correct them to ensure that the PCV system can function as intended. Also, be aware of the fact that an excessively worn engine may have more blowby than the PCV system can handle. If there are symptoms that indicate the PCV system is plugged (oil in air cleaner, saturated crankcase filter, and so forth) but no restrictions are found, check the wear of the engine.

EGR SYSTEM DIAGNOSIS AND SERVICE

Manufacturers calibrate the amount of EGR gas flow for each engine. If there is too much or too little, it can cause performance problems by changing the engine breathing characteristics. Also, with too little EGR flow, the engine can overheat, detonate, and emit excessive amounts of NO_x. When any of these problems exist and it seems likely that the EGR system is at fault, check the system. Typical problems that show up in ported EGR systems follow:

- Rough idle possibly caused by a stuck-open EGR valve, a PVS that fails to open, dirt on the valve seat, or loose mounting bolts (this also causes a vacuum leak and a hissing noise).
- No-start, surging, or stalling can be caused by an open EGR valve.
- Detonation (spark knock) can be caused by any condition that prevents proper EGR gas flow, such as a valve stuck closed, leaking valve diaphragm, restrictions in flow passages, EGR disconnected, or a problem in the vacuum source.
- Excessive NO_x emissions can be caused by any condition that prevents the EGR from allowing the correct amount of exhaust gases into the cylinder or anything that allows combustion temperatures.
- Poor fuel economy is typically caused by the EGR system if it relates to detonation or other symptoms of restricted or zero EGR flow.

EGR System Troubleshooting

Before attempting to troubleshoot or repair a suspected EGR system on a vehicle, make sure the engine is mechanically sound, the injection system is operating properly, and the spark control system is working properly.

Most often, an electronically controlled EGR valve functions in the same way as a vacuum-operated valve. Apart from the electronic control, the system can have all the problems of any EGR system. Those that are totally electronic and do not use a vacuum signal can have the same problems as others, with the exception of vacuum leaks and other vacuum-related problems. Sticking valves, obstructions, and loss of vacuum produces the same symptoms as those on nonelectronic-controlled systems. If an electronic control component is not functioning, the condition is usually recognized by the PCM. The EGRV and EGRC solenoids, or the EVR, should normally cycle on and off frequently when EGR flow is being controlled (warm engine and cruise rpm). If they do not, it indicates a problem in the electronic control system or the solenoids. Generally, an electronic control failure results in low or zero EGR flow and might cause symptoms such as overheating, detonation, and power loss.

Before attempting any testing of the EGR system, visually inspect the condition of all vacuum hoses for kinks, bends, cracks, and flexibility **(Figure 32–15)**. Replace defective hoses as required. Check vacuum hose routing against the underhood decal **(Figure 32–16)** or the manufacturer's service manual, and correct any misrouted

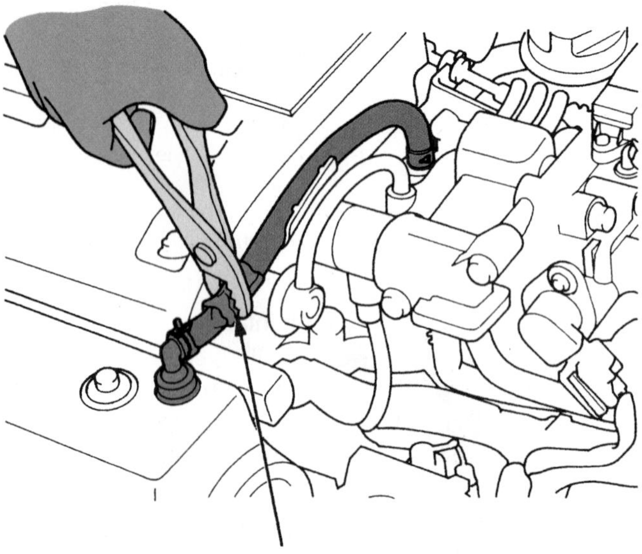

Figure 32–14 When the PCV hose is pinched, the valve should click. *Courtesy of American Honda Motor Co., Inc.*

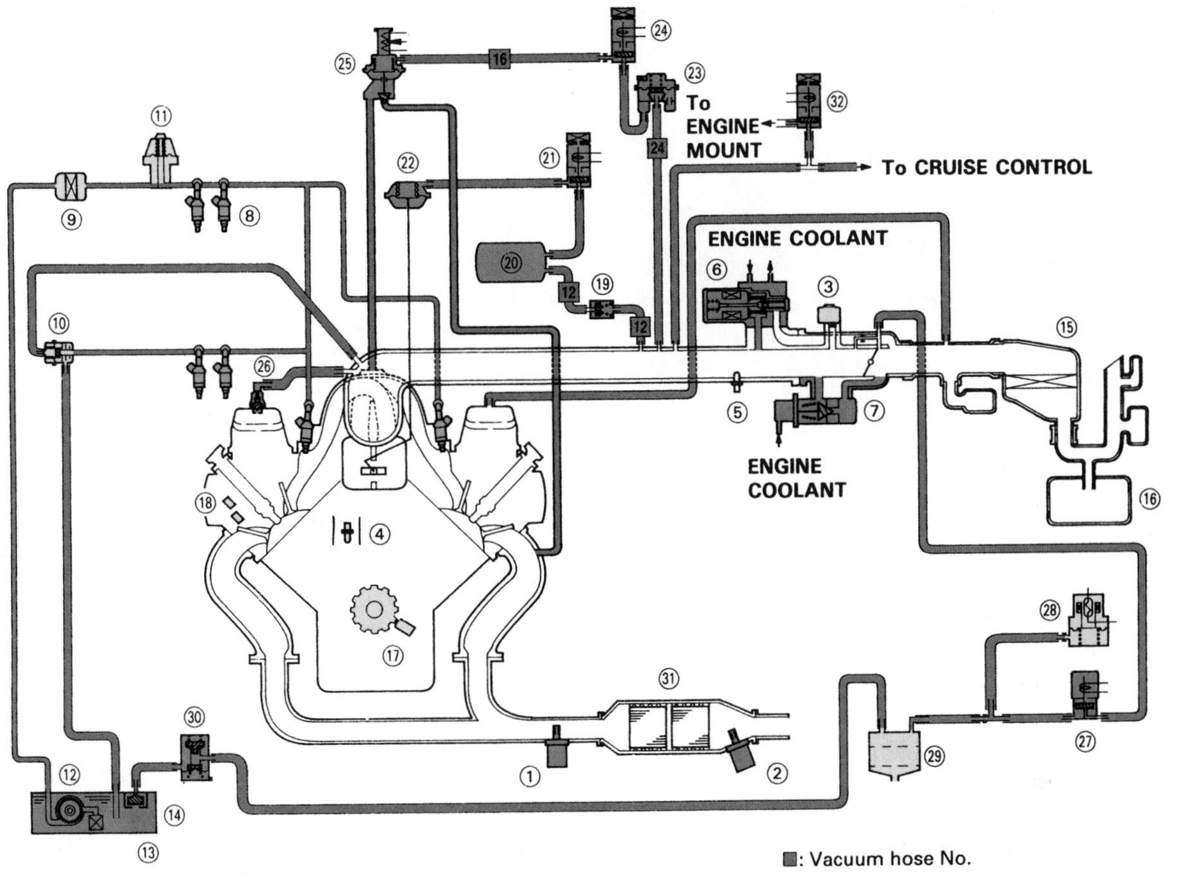

① PRIMARY HEATED OXYGEN SENSOR (HO2S) (SENSOR 1)
② SECONDARY HEATED OXYGEN SENSOR (HO2S) (SENSOR 2)
③ MANIFOLD ABSOLUTE PRESSURE (MAP) SENSOR
④ ENGINE COOLANT TEMPERATURE (ECT) SENSOR
⑤ INTAKE AIR TEMPERATURE (IAT) SENSOR
⑥ IDLE AIR CONTROL (IAC) VALVE
⑦ FAST IDLE THERMO VALVE
⑧ FUEL INJECTOR
⑨ FUEL FILTER
⑩ FUEL PRESSURE REGULATOR
⑪ FUEL PULSATION DAMPER
⑫ FUEL PUMP (FP)
⑬ FUEL TANK
⑭ FUEL TANK EVAPORATIVE EMISSION (EVAP) VALVE
⑮ AIR CLEANER
⑯ RESONATOR
⑰ CRANK POSITION (CKP) SENSOR
⑱ TOP DEAD CENTER/CYLINDER POSITION (TDC/CYP) SENSOR
⑲ INTAKE AIR BYPASS (IAB) CHECK VALVE
⑳ INTAKE AIR BYPASS (IAB) VACUUM TANK
㉑ INTAKE AIR BYPASS (IAB) CONTROL SOLENOID VALVE
㉒ INTAKE AIR BYPASS (IAB) CONTROL DIAPHRAGM
㉓ EXHAUST GAS RECIRCULATION (EGR) VACUUM CONTROL VALVE
㉔ EXHAUST GAS RECIRCULATION (EGR) CONTROL SOLENOID VALVE
㉕ EXHAUST GAS RECIRCULATION (EGR) VALVE
㉖ POSITIVE CRANKCASE VENTILATION (PCV) VALVE
㉗ EVAPORATIVE EMISSION (EVAP) PURGE CONTROL SOLENOID VALVE
㉘ EVAPORATIVE EMISSION (EVAP) PURGE FLOW SWITCH
㉙ EVAPORATIVE EMISSION (EVAP) CONTROL CANISTER
㉚ EVAPORATIVE EMISSION (EVAP) TWO WAY VALVE
㉛ THREE WAY CATALYTIC CONVERTER (TWC)
㉜ ENGINE MOUNT CONTROL SOLENOID VALVE

Figure 32-15 Vacuum to the EGR valve is just one of the vacuum circuits on some late-model vehicles. *Courtesy of American Honda Motor Co., Inc.*

hoses. If the system is fitted with an EVP sensor, the wires routed to it should also be checked.

If the EGR valve remains open at idle and low engine speed, the idle operation is rough and surging occurs at low speed. When this problem is present, the engine may hesitate on low-speed acceleration or stall after deceleration or after a cold start. If the EGR valve does not open, engine detonation occurs. When a defect occurs in the EGR system, a DTC is usually set in the PCM memory.

In many EGR systems, the PCM uses inputs from the ECT, TPS, and MAP sensors to operate the EGR valve. Improper EGR operation may be caused by a defect in one of these sensors. Use a scan tool to retrieve the DTCs. If any are present, correct that problem before continuing your diagnosis of the EGR valve.

EGR Valves and Systems Testing On many engines, the EGR valve can be checked with a hand-operated vacuum pump. Before proceeding with this test, make sure the engine produces enough vacuum to properly operate the valve. This is done by connecting a vacuum gauge to the engine. Make sure the connection is made to a manifold vacuum source. Then start the engine and gradually increase speed to 2,000 rpm with the transmission in

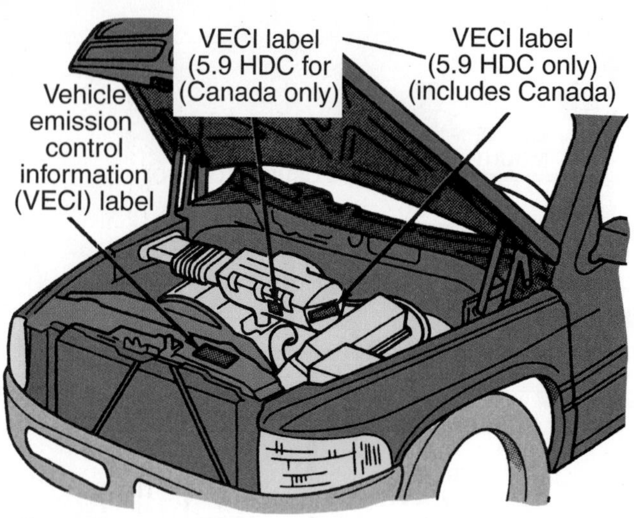

Figure 32–16 Possible locations of a Vehicle Emission Control Information (VECI) label. *Courtesy of DaimlerChrysler Corporation*

neutral. The reading should be above 16 in. Hg (406 mm Hg). If it is not, there could be a vacuum leak or exhaust restriction. Before continuing to test the EGR, check the MAP and/or correct the problem of low vacuum.

To check an EGR valve with a vacuum pump, remove the vacuum supply hose from the EGR valve port. Connect the vacuum pump to the port and supply 18 inches (457 mm Hg) of vacuum. Observe the EGR diaphragm movement. When the vacuum is applied, the diaphragm should move **(Figure 32–17)**. If the valve diaphragm did not move or did not hold the vacuum, replace the valve.

With the engine at normal operating temperature, check the vacuum supply hose to make sure there is no vacuum to the EGR valve at idle. Then plug the hose. On

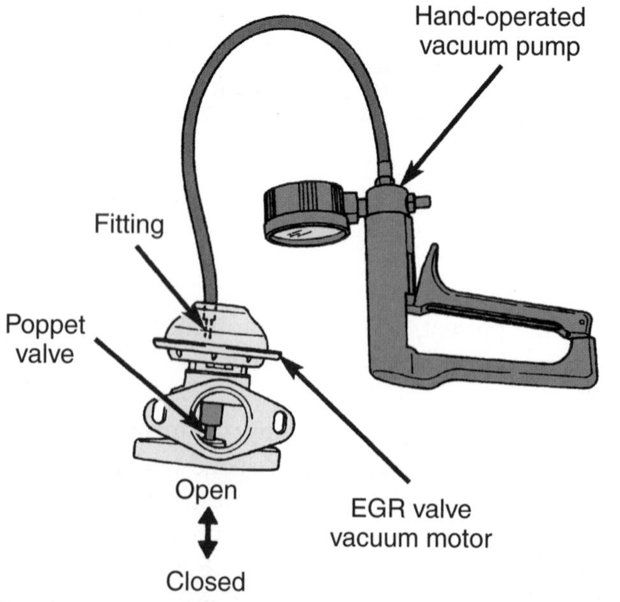

Figure 32–17 Watch the action of the valve when vacuum is applied to it and released. *Courtesy of DaimlerChrysler Corporation*

EFI engines, disconnect the throttle air bypass valve solenoid. Then, observe the engine's idle speed. If necessary, adjust idle speed to the emission decal specification. Slowly apply 5 to 10 inches (127 to 254 mm Hg) of vacuum to the EGR valve. The idle speed should drop more than 100 rpm (the engine may stall), and then return to normal again when the vacuum is removed. If the idle speed does not respond in this manner, remove the valve and check for carbon in the passages under the valve. Clean the passages as required or replace the EGR valve. Carbon may be cleaned from the lower end of the EGR valve with a wire brush, but do not immerse the valve in solvent, and do not sandblast the valve.

Diagnosis of Negative Back Pressure EGR Valve. With the engine at normal operating temperature and the ignition switch off, disconnect the vacuum hose from the EGR valve, and connect a hand vacuum pump to the vacuum fitting on the valve. Supply 18 inches (457 mm Hg) of vacuum to the EGR valve. The EGR valve should open and hold the vacuum for 20 seconds. If the valve does not open or cannot hold the vacuum, it must be replaced.

If the valve had acceptable results in the first test, continue by applying 18 inches (457 mm Hg) of vacuum to the valve and start the engine. The vacuum should drop to zero, and the valve should close. If the valve does not react this way, replace it.

Diagnosis of Positive Back Pressure EGR Valve. Diagnosis of a positive back pressure EGR valve requires the same basic tests as a negative-type but should have the opposite results. Photo Sequence 28 covers the typical procedure for testing a positive back pressure exhaust gas recirculation valve. If the valve fails any part of the test, it should be replaced.

Digital EGR Valve Diagnosis. A digital EGR valve may be diagnosed with a scan tool. With the engine at normal operating temperature and the ignition switch off, connect the scan tool to the DLC. Start the engine and allow it to run at idle speed. Select the EGR control on the scan tool, then energize the solenoids, one at a time. Engine speed should drop slightly as each EGR solenoid is energized. If the EGR valve does not operate properly, basic checks should be made before replacing it. Make sure 12 volts is applied to the power supply terminal at the EGR valve. The resistance of the valve should also be checked. By connecting an ohmmeter across the electrical terminals on the valve **(Figure 32–18)**, the windings can be checked for opens, shorts, and excessive resistance. If any resistance reading is not within specs, the valve should be replaced. Visually check all of the wires between the EGR valve and the PCM. Also make sure the EGR passages are not restricted or plugged. To do this, you will need to remove the valve.

PHOTO SEQUENCE 28

Diagnosing a Positive Back Pressure EGR Valve

P28-1 *With the engine at normal operating temperature, disconnect the vacuum hose from the EGR valve.*

P28-2 *Connect the vacuum hose from a vacuum pump to the EGR valve's vacuum port.*

P28-3 *While the engine is idling, apply vacuum to the EGR valve. You should not be able to build up a vacuum in the valve, and the EGR valve should not open.*

P28-4 *Turn off the engine.*

P28-5 *Disconnect the EGR vacuum supply hose from the throttle body.*

P28-6 *Connect a long vacuum hose from the EGR vacuum port on the throttle body directly to the EGR valve vacuum port.*

P28-7 *Use a tee fitting to connect a vacuum gauge in the vacuum hose to the EGR valve.*

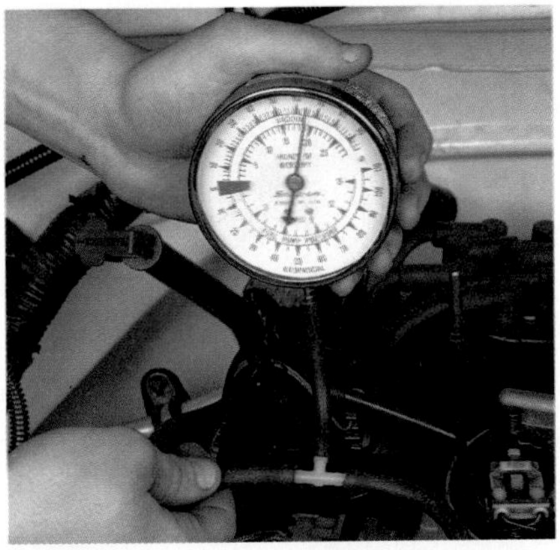

P28-8 *Start the engine and increase its speed to 2,000 rpm. Watch the vacuum gauge. Vacuum should be present, and the EGR should open.*

P28-9 *Turn off the engine and disconnect the vacuum gauge and hose. Reconnect the original vacuum lines.*

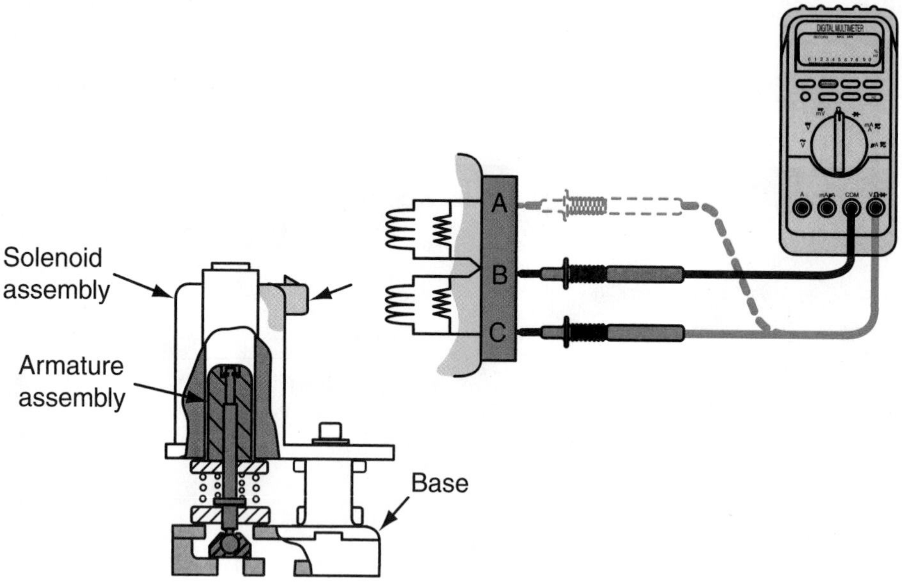

Figure 32-18 Ohmmeter connections for checking a digital EGR valve.

Linear EGR Valve Diagnosis. The correct procedure for diagnosing a linear EGR valve varies, depending on the vehicle make and model year. Always follow the recommended procedure in the vehicle manufacturer's service manual. A scan tool may be used to diagnose a linear EGR valve. The engine should be at normal operating temperature. Since the linear EGR valve has an EVP sensor, the actual pintle position may be checked on the scan tool and should not exceed 3% at idle speed. The scan tool may be operated to command a specific pintle position, such as 75%, and this commanded position should be achieved within 2 seconds. With the engine idling, select various pintle positions and check the actual pintle position. The pintle position should always be within 10% of the commanded position.

If a linear EGR valve does not operate properly, check the fuse in the supply wire to the EGR valve. Also check for open circuits, grounds, and shorts in the wires connected from the EGR valve to the PCM. Verify that the EVP sensor is receiving a 5-volt reference signal and verify that the ground circuit is good. If these are okay, remove the valve with the wiring harness still connected to it. Then connect a DMM across the pintle position wire at the EGR valve to ground, and manually push the pintle upward. The voltmeter reading should change from approximately 1 to 4.5 volts. If the EGR valve did not operate properly, it should be replaced.

Checking EGR Efficiency

Although most testing of EGR valves involves the valve's ability to open and close at the correct time, we are not really testing what the valve was designed to do—control NO_x emissions. EGR systems should be tested to see if they are doing what they were designed to do.

Many technicians incorrectly conclude that an EGR valve is working properly if the engine stalls or idles very rough when the EGR valve is opened. Actually this test just shows that the valve was closed and it will open. A good EGR valve opens and closes, but it also allows the correct amount of exhaust gas to enter the cylinders. EGR valves are normally closed at idle and open at approximately 2,000 rpm. This is where the EGR system should be checked.

To check an EGR system, use a five-gas exhaust analyzer. Allow the engine to warm up, then raise the engine speed to approximately 2,000 rpm. Watch the NO_x readings on the analyzer. The meter measures NO_x in parts per million. In most cases, NO_x should be below 1,000 ppm. It is normal to have some temporary increases over 1,000 ppm; however, the reading should be generally less than 1,000. If the NO_x is above 1,000, the EGR system is not doing its job. The exhaust passage in the valve is probably clogged with carbon.

If only a small amount of exhaust gas is entering the cylinder, NO_x will still be formed. A restricted exhaust passage of only ⅛-inch (3.18 mm) will still cause the engine to run rough or stall at idle, but it is not enough to control combustion chamber temperatures at higher engine speeds. Keep in mind that it should never be assumed the EGR passages are all right just because the engine stalls at idle when the EGR is fully opened.

Electronic EGR Controls

When the EGR valve checks out and visually everything looks fine, and yet a problem with the EGR system is evident, the EGR controls should be tested. Often a malfunctioning electronic control will trigger a DTC. Service manuals give the specific directions for testing these controls. Always follow them. The following tests are given as examples of those test procedures.

EGR Vacuum Regulator (EVR) Tests Connect a pair of ohmmeter leads to the EVR terminals to check the winding for open circuits and shorts. An infinite ohmmeter reading indicates an open circuit, whereas a lower-than-specified reading means the winding is shorted. Then, connect the ohmmeter leads from one of the EVR solenoid terminals to the solenoid case. You should get an infinite reading; a low ohmmeter reading means the winding is shorted to the case.

SHOP TALK

The same quad driver in a PCM may operate several outputs. On General Motors' computers, quad drivers sense high current flow. If a solenoid winding is shorted and the quad driver senses high current flow, the quad driver shuts down all the outputs it controls to prevent damage caused by the high current flow. When the PCM does not operate an output or outputs, always check the resistance of the output's solenoid windings before replacing the PCM. A lower-than-specified resistance in a solenoid winding indicates a shorted condition, and this problem may explain why the PCM quad driver is not operating the outputs. ■

A scan tool may be used to diagnose the operation of an EVR solenoid. The procedure shown in Photo Sequence 29 is given as an example. Always follow the procedures given in the service manual.

SHOP TALK

In some EGR systems, the PCM energizes the EVR solenoid at idle and low speeds. Under this condition, the solenoid shuts off vacuum to the EGR valve. When the proper input signals are available, the PCM deenergizes the EVR solenoid and allows vacuum to the EGR valve. ■

Exhaust Gas Temperature Sensor Diagnosis To test an exhaust gas temperature sensor, remove it and place it in a container of oil. Place a thermometer in the oil and heat the container. Connect the ohmmeter leads to the exhaust gas temperature sensor terminals. The exhaust gas temperature sensor should have the specified resistance at various temperatures.

SPARK CONTROL SYSTEMS

Through the years, many devices and combinations of devices have been used to control ignition timing. Today, most systems rely on the PCM for timing control based on inputs from various sensors. One of these sensors, the knock sensor, has one purpose—spark control.

Diagnosis of Knock Sensor and Knock Sensor Module

WARNING!

Operating an engine with a detonation problem may result in piston, ring, and cylinder wall damage.

If the knock sensor system does not provide an engine detonation signal to the PCM, the engine detonates, especially on acceleration. The first step in diagnosing the knock sensor and knock sensor module is to check all the wires and connections in the system for loose connections, corroded terminals, and damage. With the ignition switch on, be sure that 12 volts are being supplied through the fuse to the knock sensor module.

Connect a scan tool to the DLC and check for DTCs related to the knock sensor system. If DTCs are present, diagnose the cause of these codes. When no DTCs related to the knock sensor system are present, the system needs to be checked. Some guidelines for testing the knock sensor and sensor circuit follow:

- Make sure the engine is at its normal operating temperature.
- If a knock sensor signal is present when the scan tool is first turned on, disconnect the wire from the knock sensor and observe the tester. If the knock sensor signal is no longer present, the engine has an internal knock or the knock sensor is defective. When the knock sensor signal is still present on the scan tool, check the wire from the knock sensor to the knock sensor module for picking up false signals from an adjacent wire. Reroute the knock sensor wire as necessary.
- Run the engine at 1,500 rpm and tap the engine block near the knock sensor. If the knock sensor does not respond, turn the ignition off and disconnect the connector at the sensor's module. Check the ground circuit with an ohmmeter.
- If the ground circuit was found to be all right, reconnect the knock sensor module wiring connector, and disconnect the knock sensor wire. Operate the engine at idle speed and momentarily connect a 12-volt test

PHOTO SEQUENCE 29

Diagnosing an EGR Vacuum Regulator Solenoid

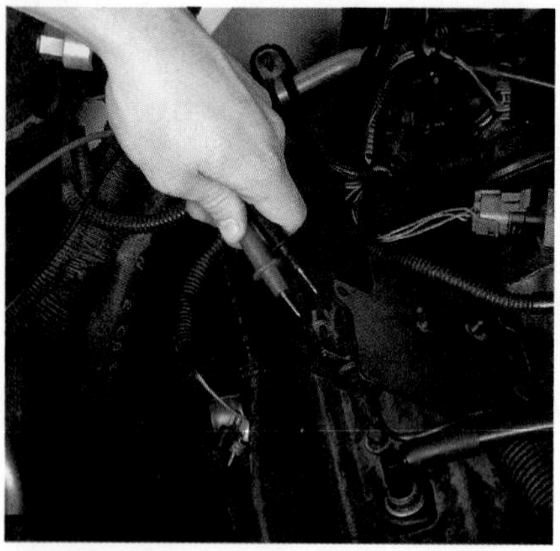

P29-1 *Disconnect the connector to the EGR solenoid and connect the leads of an ohmmeter to the solenoid's terminals.*

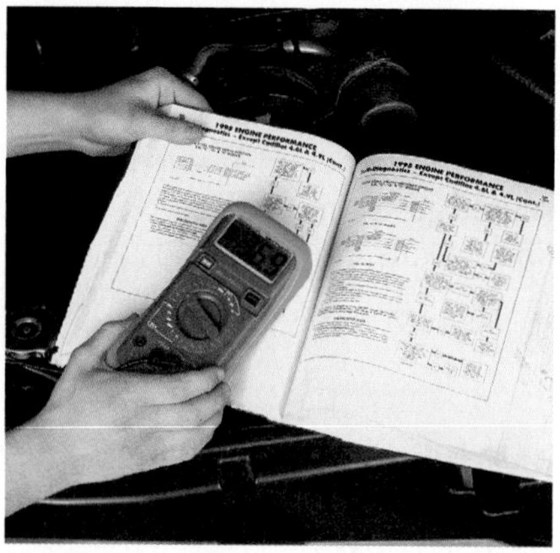

P29-2 *Compare your readings to the specifications for the solenoid.*

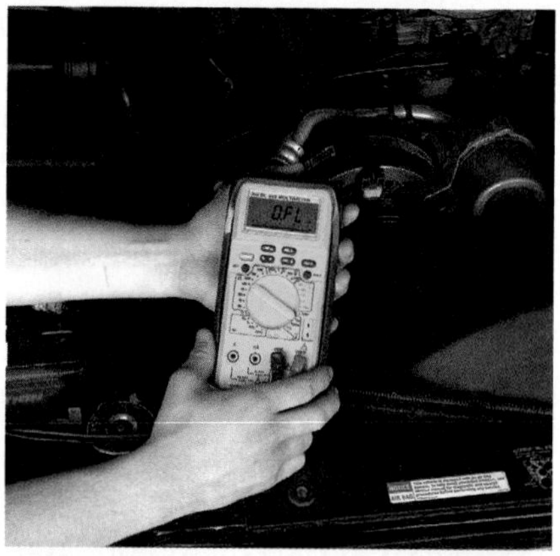

P29-3 *Connect the ohmmeter leads to one of the solenoid's terminals and to ground. An infinite reading means the solenoid is not shorted to ground.*

P29-4 *Reconnect the wiring to the connector and run the engine to bring it to normal operating temperature. While the engine is running, prepare the scan tool for the vehicle.*

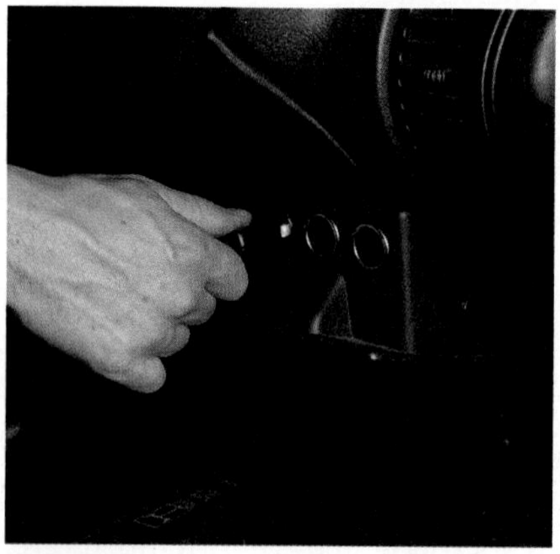

P29-5 *Turn off the engine and connect the power cable of the scan tool to the vehicle.*

P29-6 *Enter the necessary information into the scan tool.*

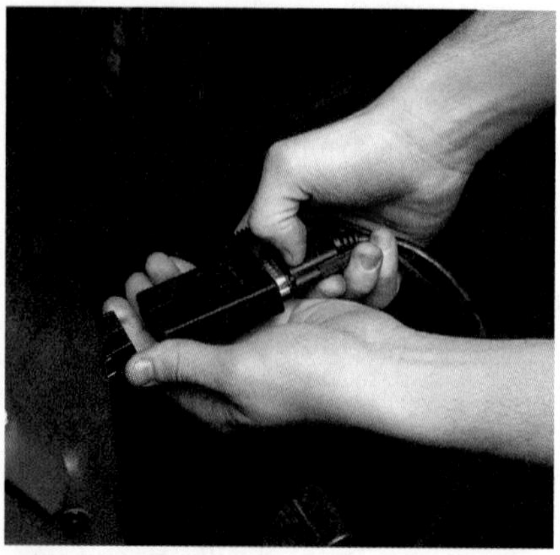

P29-7 *Connect the scan tool to the DLC.*

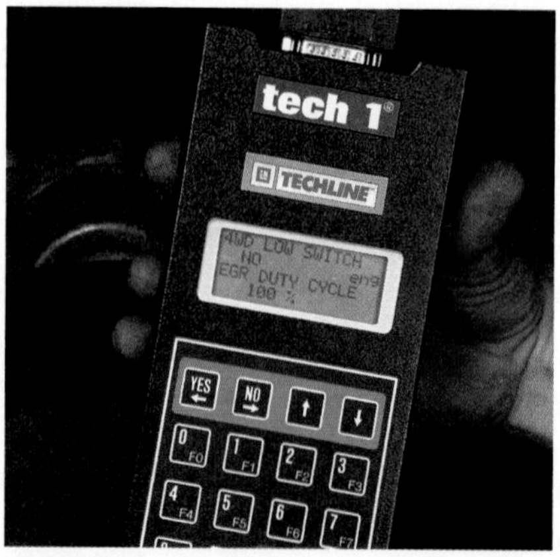

P29-8 *Start the engine and obtain the EGR data on the scanner. The EGR valve should be off and remain off while the engine is idling.*

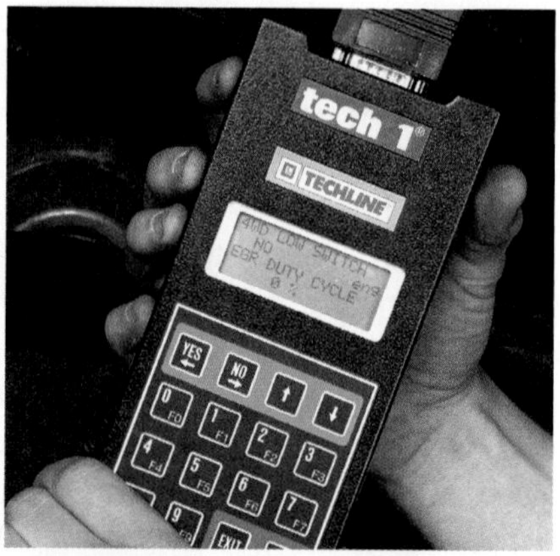

P29-9 *Take the vehicle for a test drive with the scan tool still connected. The EGR solenoid should cycle to ON once the vehicle is at a cruising speed. If it does not, check the solenoid and associated circuits.*

light from a 12-volt source to the knock sensor wire. If a knock sensor AC signal is now generated on the scan tool, there is a faulty connection at the knock sensor or the knock sensor is defective. When a knock sensor signal is not generated, check for faulty wires from the knock sensor to the module or from the module to the PCM. If the wires and connections are satisfactory, the knock sensor module is likely defective.

SHOP TALK

When installing a knock sensor, make sure it is tightened to the proper amount torque. If the knock sensor is installed tighter than it should be, it may become too sensitive and provide an excessively high voltage signal, resulting in more spark retard than required. When the knock sensor is installed looser than it should be, the knock sensor signal will be lower than normal, resulting in engine detonation. ■

INTAKE HEAT CONTROL DIAGNOSIS AND SERVICE

If the mixture heater system is inoperative, the engine may hesitate on acceleration during engine warm-up, and fuel consumption may be higher than normal. The first step in diagnosing a computer-controlled system is to check all the wires and electrical terminals in the system. Check the system for DTCs. If a code is present, diagnose the cause of that code.

Use a DMM to check for 12 volts at the mixture heater relay contact terminal connected to the fuse link. If 12 volts are not available at this terminal, check the fuse link. Turn on the ignition switch, and use a DMM to check for 12 volts at the mixture heater relay terminal connected to the fuse and ignition switch. If 12 volts are not available at this terminal, check the fuse.

The PCM should ground the mixture heater relay winding and close the relay contacts when the coolant temperature is below 122°F (50°C). At this temperature, the PCM grounds the mixture heater relay winding while cranking the engine and when the engine is running. When the coolant temperature is above 122°F (50°C), the PCM does not provide a ground for the mixture heater relay winding and the relay contacts open.

When the mixture heater and the relay are satisfactory, make sure the coolant temperature is below 122°F (50°C) and check for 12 volts at the mixture heater terminal connected to the relay contacts with the engine idling. If 12 volts are not present at this terminal, shut off the ignition switch and disconnect the PCM connector. Identify the terminal that carries the voltage from the PCM to the relay winding, then check for battery voltage. If 12 volts are not available, repair the open circuit in the wire from the relay winding to the PCM. When 12 volts are available, use a jumper wire to connect this terminal to the battery ground. If 12 volts are now available at the mixture heater terminal connected to the relay contacts, the PCM is most likely defective.

CATALYTIC CONVERTER DIAGNOSIS

A plugged converter or any exhaust restriction can cause loss of power at high speeds, stalling after starting (if totally blocked), a drop in engine vacuum as engine rpm increases, or sometimes popping or backfiring at the throttle plate.

In order for a catalytic converter to be effective, the fuel control system must cycle from rich to lean. The lean cycle allows the converter to store oxygen, and the rich cycle uses the stored oxygen. The fuel control system should adjust the air/fuel mixture so it toggles back and forth from a stoichiometric ratio **(Figure 32–19)**.

There are many ways to test a catalytic converter; one of them is to simply smack the converter with a rubber mallet. If the converter rattles, it needs to be replaced and there is no need to do other testing. A rattle indicates loose catalyst substrate, which will soon rattle into small pieces. This could be called a test, but it is generally not used to determine if a catalyst is good.

A vacuum gauge can be used to watch engine vacuum while the engine is accelerated. Another way to check for a restricted exhaust or catalyst is to insert a pressure gauge in the exhaust manifold's bore for the O_2 sensor **(Figure 32–20)**. With the gauge in place, hold the engine's speed at 2,000 rpm and watch the gauge. The desired pressure

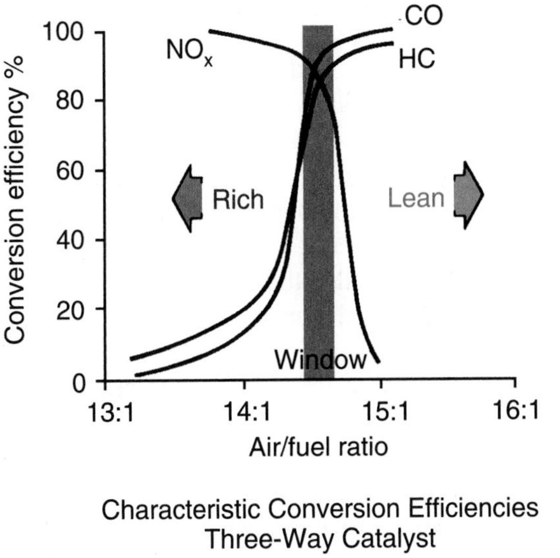

Figure 32–19 In order for a catalytic converter to be efficient, the air/fuel mixture must toggle around stoichiometric. *Courtesy of DaimlerChrysler Corporation*

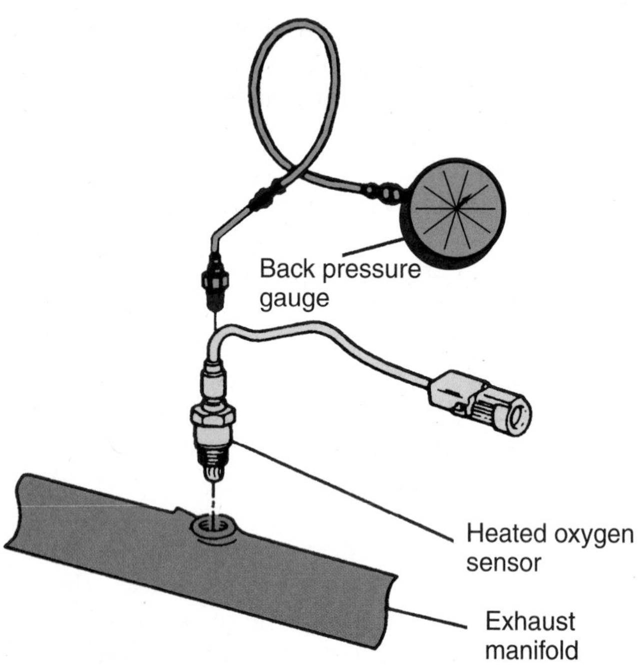

Figure 32-20 To measure exhaust system back pressure, insert a pressure gauge into the oxygen sensor's bore in the exhaust.

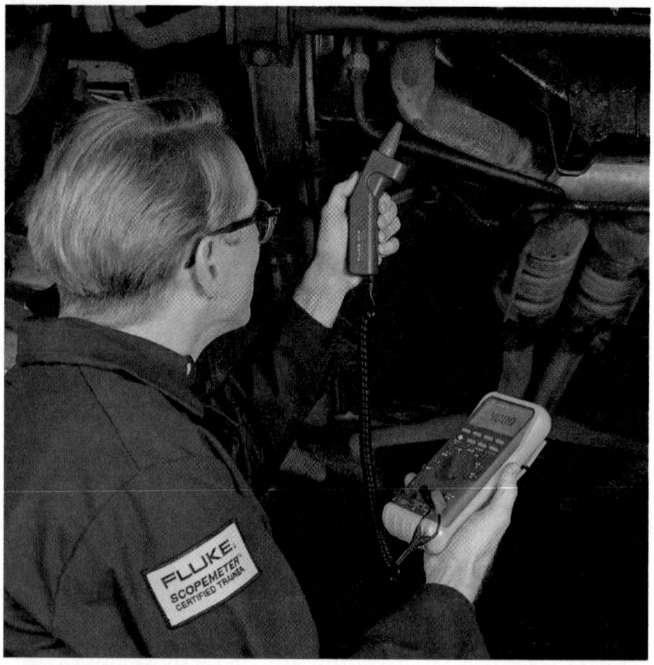

Figure 32-21 Checking the efficiency of a catalytic converter with a pyrometer. *Reproduced with permission from Fluke Corporation*

reading will be less than 1.25 psi (8.61 kPa). A very bad restriction will give a reading of over 2.75 psi (18.96 kPa).

The converter should be checked for its ability to convert CO and HC into CO_2 and water. There are three separate tests for doing this. The first method is the delta temperature test. To conduct this test, use a hand-held digital pyrometer. By touching the pyrometer probe to the exhaust pipe just ahead of and just behind the converter **(Figure 32–21)**, there should be an increase of at least 100°F (38°C) or 8% above the inlet temperature reading as the exhaust gases pass through the converter. If the outlet temperature is the same or lower, nothing is happening inside the converter. To do its job efficiently, the converter needs a steady supply of oxygen from the air pump. A bad pump, faulty diverter valve or control valve, leaky air connections, or faulty computer control over the air injection system could be preventing the needed oxygen from reaching the converter. If the converter fails this test, check those systems.

The next test is called the O_2 storage test and is based on the fact that a good converter stores oxygen. Begin by disabling the air injection system. Once the analyzer and converter are warmed up, hold the engine at 2,000 rpm. Watch the readings on the exhaust analyzer. Once the numbers stop dropping, check the oxygen level on the gas analyzer. The O_2 readings should be about 0.5 to 1%. This shows the converter is using most of the available oxygen. It is important to observe the O_2 reading as soon as the CO begins to drop. If the converter fails the tests, chances are that it is working poorly or not at all.

This final converter test uses a principle that checks the converter's efficiency. Before beginning this test, make sure the converter is warmed up. Calibrate a four- or five-gas analyzer and insert its probe into the tail pipe. Disable the ignition. Then crank the engine for 9 seconds while pumping the throttle. Watch the readings on the analyzer. The CO_2 on fuel-injected vehicles should be over 11% and carbureted vehicles should have a reading of over 10%. As soon as you get your readings, reconnect the ignition and start the engine. Do this as quickly as possible to cool off the catalytic converter. If, while the engine is cranking, the HC goes above 1,500 ppm, stop cranking; the converter is not working. Also, stop cranking once the CO_2 readings reach 10 or 11%; the converter is good. If the catalytic converter is bad, there will be high HC and, of course, low CO_2 at the tailpipe. Do not repeat this test more than *one* time without running the engine in between. If a catalytic converter is found to be bad, it is replaced.

AIR INJECTION SYSTEM DIAGNOSIS AND SERVICE

Not all engines are equipped with an air injection system, only those that need them to meet emissions standards have them. Therefore, air injection systems are vital to proper emissions on engines equipped with them. Each system has its own test procedure; always follow the manufacturer's recommendations for testing.

A typical air injection system has an air pump, distribution manifold, check valve, diverter valve, hoses to the intake and exhaust, and vacuum signal line. The diverter valve sends the air to the exhaust manifold, to the converter, or to the atmosphere **(Figure 32–22)**.

Most AIR systems are computer controlled and rely on solenoids to control the direction of airflow to the exhaust manifold or to the converter. When the system is in closed loop, the air from the air injection system must be directed away from the O_2 sensor. Some systems have switching valves that allow a small amount of air to flow past the O_2 sensor. The computer knows how much and adjusts the O_2 input accordingly. Sometimes the amount of air that can move through a closed switching valve is marked on its housing.

Check Valve Testing

All of the types of air injection systems have at least one thing in common, a one-way check valve. The valve opens to let air in but closes to keep exhaust from leaking out. The check valve can be checked with an exhaust gas analyzer. Start the engine and hold the probe of the exhaust gas analyzer near the check valve port. If any amount of CO or CO_2 is read, the valve leaks. If this valve is leaking, hot exhaust is also leaking, which could ruin the other components in the air injection system.

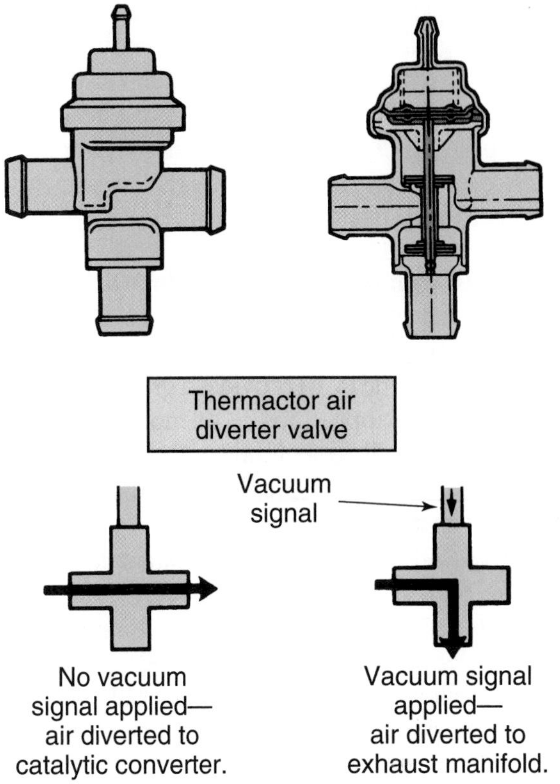

Figure 32–22 The action of a diverter valve in an AIR system. *Courtesy of Ford Motor Company*

Secondary Air Injection Reaction System Service and Diagnosis

The first step in diagnosing a secondary air injection system is to check all vacuum hoses and electrical connections in the system. Many AIR system pumps have a centrifugal filter behind the pulley. Air flows through this filter into the pump, and the filter keeps dirt out of the pump. The pulley and filter are bolted to the pump shaft, and these two components are serviced separately **(Figure 32–23)**. If the pulley or filter is bent, worn, or damaged, it should be replaced. Also check the air pump's belt for condition and tension and correct it as necessary. The pump assembly is usually not serviced.

In some AIR systems, pressure relief valves are mounted in the air injection reactor bypass and air injection reactor divider valves. Other AIR systems have a pressure relief valve in the pump. If the pressure relief valve is stuck open, airflow from the pump is continually exhausted through this valve, which causes high tailpipe emissions.

If the hoses in the AIR system show evidence of burning, the one-way check valves are leaking, which allows exhaust to enter the system. Leaking air manifolds and pipes result in exhaust leaks and excessive noise.

Some AIR systems will set DTCs in the PCM if there is a fault in the AIRB or AIRD solenoids and related wiring. In some AIR systems, DTCs are set in the PCM memory if the airflow from the pump is continually upstream or downstream. Always use a scan tool to check for any DTCs related to the AIR system, and correct the causes of these codes before proceeding with further system diagnosis.

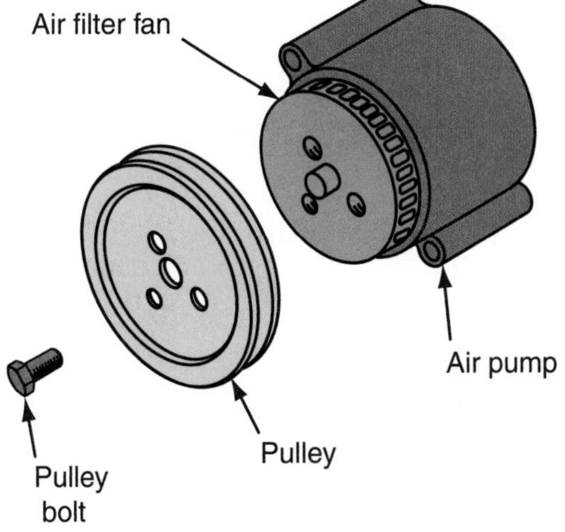

Figure 32–23 An AIR pump pulley and filter assembly.

If the AIR system does not pump air into the exhaust ports during engine warm-up, HC emissions are high during this mode, and the O_2 sensor (or sensors) takes longer to reach normal operating temperature. Under this condition, the PCM remains in open loop longer. Since the air/fuel ratio is richer in open loop, fuel economy is reduced.

When the AIR system pumps air into the exhaust ports with the engine at normal operating temperature, the additional air in the exhaust stream causes lean signals from the O_2 sensor, or sensors. The PCM responds to these lean signals by providing a rich air/fuel ratio from the injectors. This action increases fuel consumption. A vehicle can definitely fail an emission test because of air flowing past the O_2 sensor when it should not be. If the O_2 sensor is always sending a lean signal back to the computer, check the air injection system.

Electric AIR Pumps Some late-model AIR systems use an electric air pump controlled by the PCM **(Figure 32–24)**. These systems have an AIR solenoid and solenoid relay. When the PCM provides a ground for the relay, battery voltage is applied to the solenoid and the pump. Typically, DTCs will be set if one of the components fails or if the hoses or check valves leak. A quick check of the system can be made with a scan tool.

Set the scan tool to watch the voltage at the oxygen sensor(s). Start the engine and allow it to idle. Once the engine has reached normal operating temperature, enable the AIR system and check the HO2S voltages. If the voltages are low, the AIR pump, solenoid, and shut off valve are working properly. If the voltages are not low, each component of the system needs to be checked and tested.

AIR System Component Diagnosis

AIRB Solenoid and Valve Diagnosis When the engine is started, listen for air being exhausted from the AIRB valve for a short period of time. If this air is not exhausted, remove the vacuum hose from the AIRB and start the engine. If air is now exhausted from the AIRB valve, check the AIRB solenoid and connecting wires. When air is still not exhausted from the AIRB valve, check the air supply from the pump to the valve. If the air supply is available, replace the AIRB valve.

During engine warm-up, remove the hose from the AIRD valve to the exhaust ports and check for airflow from this hose. If airflow is present, the system is operating normally in this mode. When air is not flowing from this hose, remove the vacuum hose from the AIRD valve and connect a vacuum gauge to this hose. If vacuum is above 12 in. Hg (305 mm Hg), replace the AIRD valve. When the vacuum is zero, check vacuum hoses, the AIRD solenoid, and connecting wires.

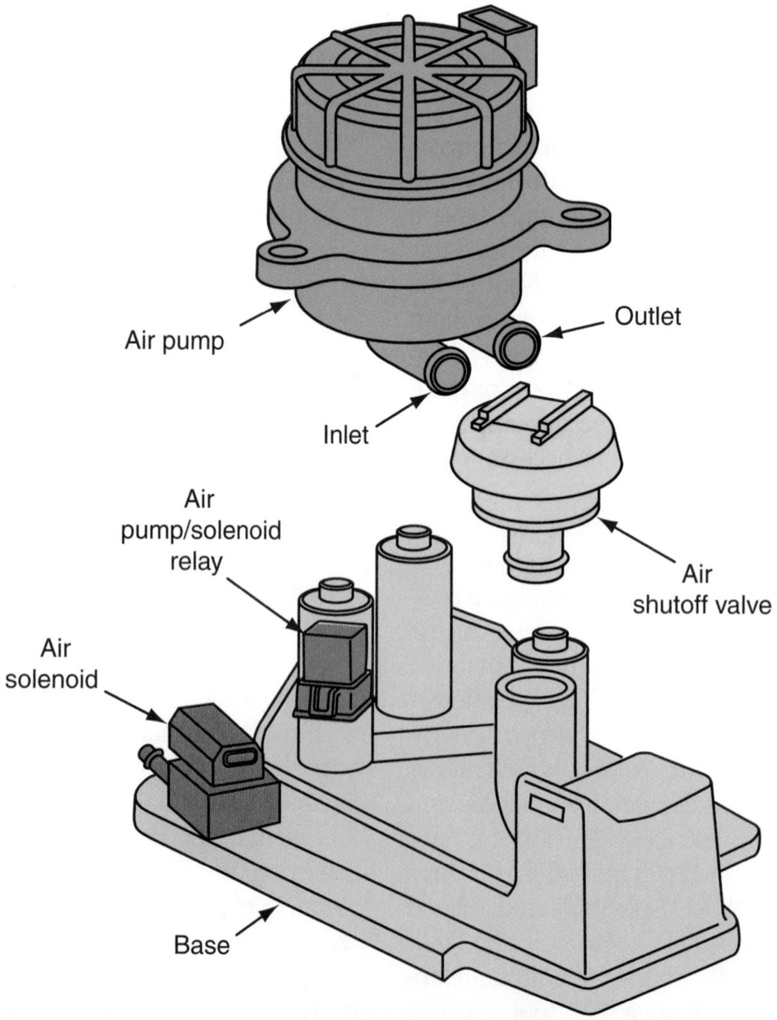

Figure 32-24 An electric air pump and AIR assembly.

SHOP TALK

With the engine at normal operating temperature, the AIR system sometimes goes back into the upstream mode with the engine idling. It may be necessary to increase the engine speed to maintain the downstream mode. ■

With the engine at normal operating temperature, disconnect the air hose between the AIRD valve and the catalytic converters and check for airflow from this hose. When airflow is present, system operation in the downstream mode is normal. If there is no airflow from this hose, disconnect the vacuum hose from the AIRD valve and connect a vacuum gauge to the hose. When the vacuum gauge indicates zero vacuum, replace the AIRD valve. If some vacuum is indicated on the gauge, check the hose, the AIRD solenoid, and connecting wires.

System Efficiency Test

Run the engine at idle with the secondary air system on (enabled). Using an exhaust gas analyzer, measure and record the O_2 levels. Next disable the secondary air system and continue to allow the engine to idle. Again, measure and record the oxygen level in the exhaust gases. The secondary air system should be supplying 2 to 5% more oxygen when it is operational (enabled).

EVAPORATIVE EMISSION CONTROL SYSTEM DIAGNOSIS AND SERVICE

EVAP system diagnosis varies depending on the vehicle make and model year. Always follow the service and diagnostic procedure in the vehicle manufacturer's service manual. If the EVAP system is purging vapors from the charcoal canister when the engine is idling or operating at very low speed, rough engine operation will occur, especially at higher atmospheric temperatures. Cracked hoses or a canister saturated with gasoline may allow gasoline vapors to escape into the atmosphere, resulting in a gasoline odor in and around the vehicle.

CAUTION!

If a gasoline odor is present in or around a vehicle, check the EVAP system for cracked or disconnected hoses and check the fuel system for leaks. Gasoline leaks or escaping vapors may result in an explosion, causing personal injury and/or property damage. The cause of fuel leaks or fuel vapor leaks should be repaired immediately.

All of the hoses in the EVAP system should be checked for leaks, restrictions, and loose connections. The electrical connections in the EVAP system should be checked for looseness, corroded terminals, and worn insulation **(Figure 32–25)**. When a defect occurs in the canister purge solenoid and related circuit, a DTC is usually set in the PCM memory. If a DTC related to the EVAP system is set in the PCM memory, always correct the cause of this code before further EVAP system diagnosis.

A scan tool may be used to diagnose the EVAP system. In the appropriate tester mode, the tester indicates whether the purge solenoid is on or off. Connect the scan tool to the DLC and start the engine. With the engine idling, the purge solenoid should be off. Leave the scan

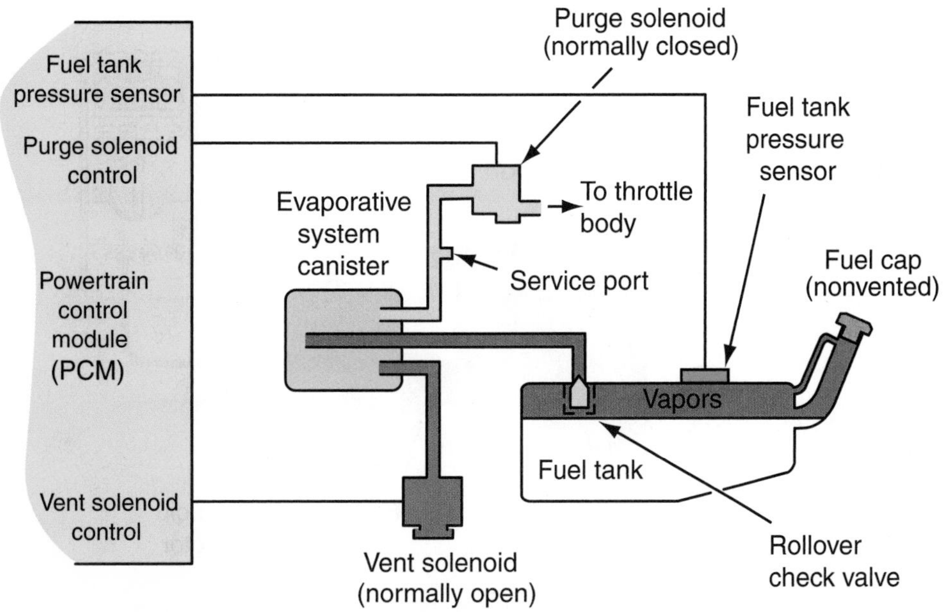

Figure 32-25 Check all hoses and electrical connectors and wires in the EVAP system.

tool connected and road test the vehicle. Be sure all the conditions required to energize the purge solenoid are present, and observe the status of this solenoid on the scan tool. The tester should indicate that the purge solenoid is on when all the conditions are present for canister purge operation. If the purge solenoid is not on under the necessary conditions, check the power supply wire to the solenoid, solenoid winding, and the wire from the solenoid to the PCM.

EVAP System Component Diagnosis

Check the canister to make sure that it is not cracked or otherwise damaged. Also make certain that the canister filter is not completely saturated. Remember that a saturated charcoal filter can cause symptoms that can be mistaken for fuel system problems. Rough idle, flooding, and other conditions can indicate a canister problem. A canister filled with liquid or water causes back pressure in the fuel tank. It can also cause richness and flooding symptoms during purge or start-up. (Some trucks have intentionally pressurized fuel tank systems. Check the calibration and engine decal before diagnosing.)

To test for saturation, unplug the canister momentarily during the diagnosis procedure and observe the engine's operation. If the canister is saturated, either it or the filter must be replaced, depending on its design. That is, some models have a replaceable filter, others do not.

A vacuum leak in any of the evaporative emission components or hoses can cause starting and performance problems, as can any engine vacuum leak. It can also cause complaints of fuel odor. Incorrect connection of the components can cause rich stumble or lack of purging (resulting in fuel odor).

The canister purge solenoid **(Figure 32–26)** winding may be checked with an ohmmeter. With the tank pressure control valve removed, try to blow air through the valve with your mouth from the tank side of the valve. Some restriction to airflow should be felt until the air pressure opens the valve. Connect a vacuum hand pump to the vacuum fitting on the valve and apply 10 in. Hg (254 mm Hg) to the valve. Now try to blow air through the valve from the tank side. Under this condition, there should be no restriction to airflow. If the tank pressure control valve does not operate properly, replace the valve.

If the fuel tank has a pressure and vacuum valve in the filler cap, check these valves for dirt contamination and damage. The cap may be washed in clean solvent. When the valves are sticking or damaged, replace the cap.

CASE STUDY

***A** customer complained about a hesitation on acceleration on a Chevrolet truck with a 5.7 L engine. It was not necessary to road test the vehicle to experience the symptoms. Each time the engine was accelerated with the transmission in park, there was a very noticeable hesitation.*

The technician removed the air cleaner and observed the injector spray pattern, which appeared normal. A fuel pressure test indicated the specified fuel pressure. The technician connected the scan tool to the DLC and checked the PCM for DTCs. There were no DTCs in the PCM memory.

The technician checked all the sensor readings with the scan tool, and all readings were within specifications. Next, the technician visually checked the operation of the EGR valve while accelerating the engine. The EGR valve re-

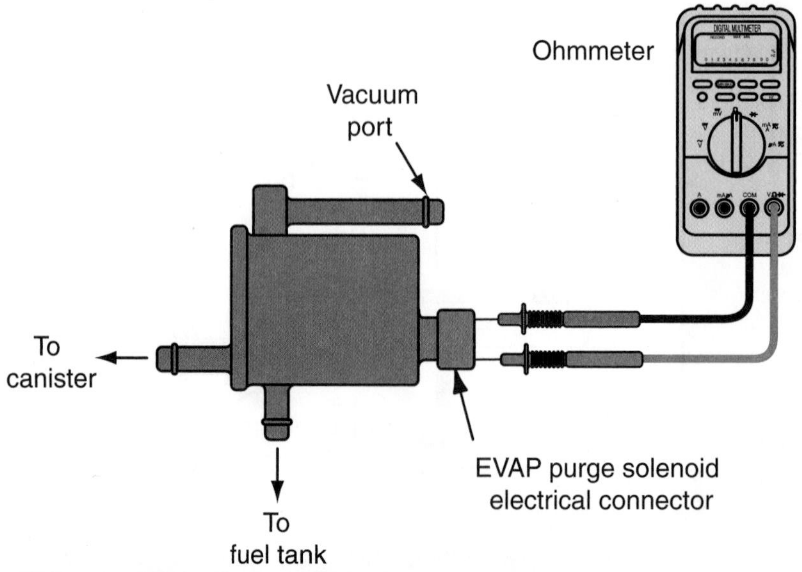

Figure 32-26 An EVAP purge solenoid.

mained closed at idle speed, but each time the engine was accelerated, the EGR valve moved to the wide-open position and remained there. With the hose removed from the EGR valve, the engine accelerated normally.

The technician checked the letters on top of the EGR valve and found that it was a negative back pressure valve. A vacuum hand pump was connected to the EGR valve, and 18 in. Hg (457 mm Hg) of vacuum were supplied to the valve with the ignition switch off. The valve opened and held the vacuum for 20 seconds. With 18 in. Hg (457 mm Hg) of vacuum supplied to the EGR valve, the engine was started. The vacuum dropped slightly, but the valve remained open, indicating the exhaust pressure was not keeping the bleed valve open and the passages in the valve stem or under the valve were restricted.

The EGR valve was removed, and since there was no carbon in the passages under the valve, the valve was replaced. When the replacement EGR valve was installed, the engine accelerated normally.

KEY TERMS

Acceleration Simulation Mode (ASM)
Constant volume sampling (CVS)
Cutpoint
Parts per million (ppm)
Repair Grade (RG240)

SUMMARY

- If the engine has excessive blowby or the PCV valve is restricted, crankcase pressure forces crankcase gases through the clean air hose into the air cleaner.
- Testing the quality of the exhaust is both a procedure for testing emission levels and a diagnostic routine. One of the most valuable diagnostic aids is the exhaust analyzer.
- The IM240 test is a 240-second test of a vehicle's emissions. During this time the vehicle is loaded to simulate a short drive on city streets, then a longer drive on a highway. The complete test cycle includes acceleration, deceleration, and cruising.
- During an IM240 test, the vehicle's exhaust is collected by a constant volume sampling (CVS) system that makes sure a constant volume of ambient air and exhaust pass through the exhaust analyzer.
- An alternative to the IM240 test is the RG240 test, which is very similar to the IM240 test.
- A scan tool may be used to diagnose the EVAP system.
- A saturated charcoal filter can cause symptoms that can be mistaken for fuel system problems. Rough idle, flooding, and other conditions can indicate a canister problem.
- If the PCV valve is stuck in the open position, excessive airflow through the valve causes a lean air/fuel ratio and the possibility of rough idle operation or engine stalling.
- When the PCV valve or hose is restricted, excessive crankcase pressure forces blowby gases through the clean air hose and filter into the air cleaner and may result in the accumulation of moisture and sludge in the engine and engine oil.
- To diagnose a faulty knock sensor, use a scan tool. If no DTCs are present, the system needs to be checked according to the manufacturer's recommendations.
- With too little EGR flow, the engine can overheat, detonate, and emit excessive amounts of NO_x.
- Rough idle can be caused by a stuck-open EGR valve, a PVS that fails to open, dirt on the valve seat, or loose mounting bolts (this also causes a vacuum leak and a hissing noise).
- Detonation (spark knock) can be caused by any condition that prevents proper EGR gas flow, such as a valve stuck closed, leaking valve diaphragm, restrictions in flow passages, EGR disconnected, or a problem in the vacuum source.
- Excessive NO_x emissions can be caused by any condition that prevents the EGR from allowing the correct amount of exhaust gases into the cylinder or anything that allows combustion temperatures.
- On many engines, the EGR valve can be checked by connecting a vacuum pump to the valve and supplying 18 inches of vacuum. The valve should open and should be able to hold the vacuum.
- A true test of how an EGR valve works is testing its efficiency.
- A scan tool may be used to diagnose the operation of an EVR solenoid.
- A plugged converter or any exhaust restriction can cause loss of power at high speeds, stalling after starting (if totally blocked), a drop in engine vacuum as engine rpm increases, or sometimes popping or backfiring at the carburetor.
- A catalytic converter can be tested by smacking it with a rubber mallet. If the converter rattles, it needs to be replaced and there is no need to do other testing.
- A converter can be tested with a vacuum and/or a pressure gauge.

- A converter can also be tested by measuring the temperature in and out of the converter.
- All types of air injection systems have a one-way check valve that can be checked with an exhaust gas analyzer.
- Some AIR systems will set DTCs in the PCM if there is a fault in the AIRB or AIRD solenoids and related wiring.
- An AIR system should be checked for overall efficiency by looking at the changes in exhaust quality when the system is turned on and off.

TECH MANUAL

The following procedures are included in Chapters 31/32 of the *Tech Manual* that accompanies this book:

1. Checking the emission levels on an engine.
2. Testing the operation of a PCV system.
3. Testing the operation of an EGR valve.
4. Testing a catalytic converter for efficiency.

REVIEW QUESTIONS

1. A dirty catalytic converter can cause all of the following *except*
   a. stalling after the engine starts.
   b. decreased HC emissions from the tailpipe.
   c. a drop in engine vacuum.
   d. decreased production of NO_x in the cylinders.
2. What will result from too little EGR flow? And what can cause a reduction in the flow?
3. What can result from a charcoal canister that is filled with liquid or water?
4. What happens if a PCV valve is stuck in the open position?
5. When testing a negative and a positive back pressure EGR valve with a vacuum gauge, what are the differences in the expected results?
6. While discussing the proper way to test a catalytic converter, Technician A says a pressure gauge can be inserted into the oxygen sensor bore and back pressure caused by the converter measured. Technician B says restrictions in the converter can be checked with a vacuum gauge while the engine is being quickly accelerated. Who is correct?
   a. Technician A **c.** Both A and B
   b. Technician B **d.** Neither A nor B
7. Explain how a port EGR valve is opened.
8. Describe the causes of high HC emissions.
9. Describe carbon monoxide (CO) emissions in relation to air/fuel ratio.
10. A lean air/fuel ratio causes HC emissions to ______.
11. While discussing tailpipe emissions, Technician A says CO emissions increase as the air/fuel ratio becomes richer. Technician B says CO emissions increase as the air/fuel ratio becomes leaner. Who is correct?
    a. Technician A **c.** Both A and B
    b. Technician B **d.** Neither A nor B
12. While discussing tailpipe emissions and cylinder misfiring, Technician A says cylinder misfiring causes a significant increase in HC emissions. Technician B says cylinder misfiring results in a large increase in CO emissions. Who is correct?
    a. Technician A **c.** Both A and B
    b. Technician B **d.** Neither A nor B
13. While discussing catalytic converter diagnosis, Technician A says a delta temperature test should be conducted. Technician B says a good converter is evident by low amounts of CO_2 in the exhaust. Who is correct?
    a. Technician A **c.** Both A and B
    b. Technician B **d.** Neither A nor B
14. While discussing EGR valve diagnosis, Technician A says that if the EGR valve does not open, the engine may hesitate on acceleration. Technician B says that if the EGR valve does not open, the engine may detonate on acceleration. Who is correct?
    a. Technician A **c.** Both A and B
    b. Technician B **d.** Neither A nor B
15. While discussing EGR valve diagnosis, Technician A says a defective throttle position sensor may affect the EGR valve operation. Technician B says a defective engine coolant temperature (ECT) sen-

sor may affect the EGR valve operation. Who is correct?

a. Technician A **c.** Both A and B

b. Technician B **d.** Neither A nor B

16. When discussing the diagnosis of a positive back pressure EGR valve, Technician A says that with the engine running at idle speed, if a hand pump is used to supply vacuum to the EGR valve, the valve should open at 12 in. Hg (305 mm Hg) of vacuum. Technician B says that with the engine not running, any vacuum supplied to the EGR valve should be bled off and the valve's diaphragm should not move. Who is correct?

a. Technician A **c.** Both A and B

b. Technician B **d.** Neither A nor B

17. While diagnosing a PCV problem, Technician A says a PCV valve that is stuck open will cause a richer than normal air/fuel mixture. Technician B says oil in the air cleaner assembly can be caused by worn piston rings. Who is correct?

a. Technician A **c.** Both A and B

b. Technician B **d.** Neither A nor B

18. Technician A says that the AIRB valve directs secondary air either to the exhaust manifold or to the catalytic converter. Technician B says that secondary air may be vented during deceleration. Who is correct?

a. Technician A **c.** Both A and B

b. Technician B **d.** Neither A nor B

19. While discussing PCV system diagnosis, Technician A says a defective PCV valve may cause rough idle operation. Technician B says satisfactory PCV system operation depends on a properly sealed engine. Who is correct?

a. Technician A **c.** Both A and B

b. Technician B **d.** Neither A nor B

20. While discussing knock sensor and knock sensor module diagnosis, Technician A says that with the engine running at 1,500 rpm, if the engine block is tapped near the knock sensor, a knock sensor signal should appear on the scan tool. Technician B says that with the engine running at 1,500 rpm, if a 12-V test light is connected from a 12-V source to the disconnected knock sensor wire, a knock sensor signal should appear on the scan tool. Who is correct?

a. Technician A **c.** Both A and B

b. Technician B **d.** Neither A nor B

33

ON-BOARD DIAGNOSTIC SYSTEMS

OBJECTIVES

■ Understand how a typical computerized engine control system operates. ■ Explain the operation of the various input and output sensors. ■ Explain what is meant by open loop and closed loop. ■ Explain the reasons for OBD-II. ■ Describe the primary provisions of OBD-II. ■ Explain the requirements to illuminate the malfunction indicator light in an OBD-II system. ■ Briefly describe the monitored systems in an OBD-II system. ■ Describe the main hardware differences between an OBD-II system and other systems. ■ Describe an OBD-II warm-up cycle. ■ Explain trip and drive cycle in an OBD-II system. ■ Describe how engine misfire is detected in an OBD-II system. ■ Describe the differences between an A misfire and a B misfire. ■ Describe the purpose of having two oxygen sensors in an exhaust system. ■ Briefly describe what the comprehensive component monitor looks at.

Computerized engine control systems present technicians with a totally new way of troubleshooting performance problems. An on-board diagnostic capability allows the computer to aid the technician in pinpointing the source of many performance problems. Basic operation of a computer is discussed in Chapter 15 of this book. The focus of this chapter is computerized engine control systems, specifically those with on-board diagnostics.

Because all manufacturers have continually updated, expanded, and improved their computerized control systems, there are hundreds of different domestic and import systems on the road. Fortunately for technicians, laws have been passed that have mandated standardized test procedures for all electronic control systems. As a result of this mandate, known as *OBD-II,* vehicles use the same terms, acronyms, and definitions to describe their components (commonly referred to as the SAE J1930 standards). They also have the same type of diagnostic connector, the same basic test sequences, and display the same trouble codes. OBD-II was gradually phased into production. It began in 1994 and has been on all vehicles sold in North America since 1997.

SYSTEM FUNCTIONS

In a computerized engine control system, emission levels, fuel consumption, driveability, and durability are carefully balanced to achieve maximum results with minimum waste. The following are some of the things engine control systems are designed to do:

- Air/fuel ratios are held as closely to 14.7 to 1 as possible, allowing maximum catalytic converter efficiency and minimizing fuel consumption.
- Emission control devices, such as EGR valve, carbon canister, and air pump, are operated at predetermined times to increase efficiency.
- The engine is operated as efficiently as possible when it is cold and is warmed up rapidly, reducing unburned hydrocarbon emissions and engine wear due to raw gas washing oil from the piston rings and getting into the crankcase to form sludge and varnish.
- Ignition timing is advanced as much as possible under all conditions.
- Timing and air/fuel ratios are precisely controlled under all operating conditions.
- Control loop operation enables the engine to make rapid changes to match changes in engine temperature, load, and speed.

SYSTEM COMPONENTS

The three basic subsystems of a computer control system **(Figure 33–1)** are the sensors, computer (most commonly referred to as the PCM), and actuators. Sensors

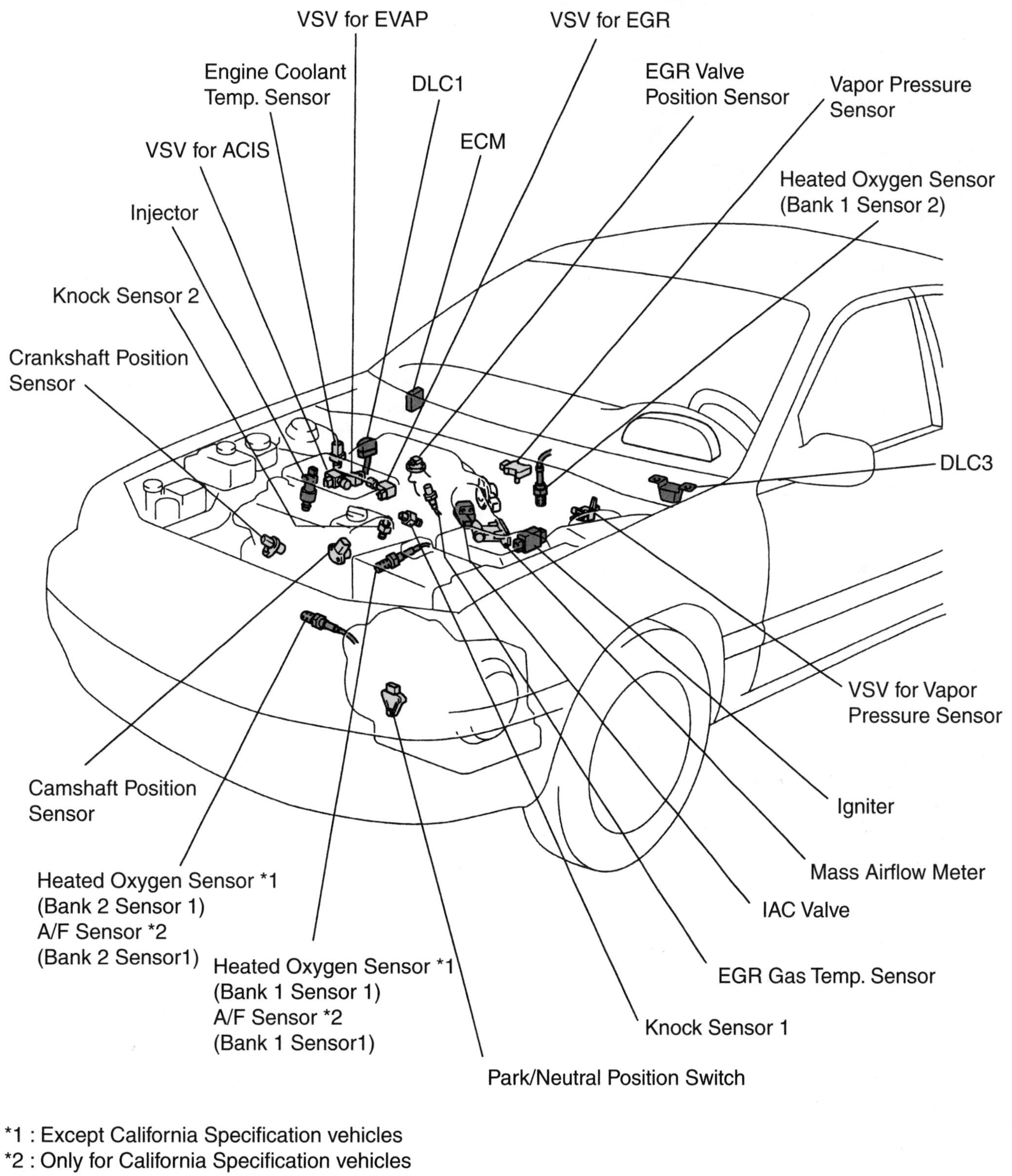

Figure 33-1 Components of a typical computerized engine control system. *Reprinted with permission*

supply the computer with input on engine conditions. The computer analyzes this data and calculates a response to these conditions. It then signals an output or actuator, such as a relay or solenoid, to adjust engine operation.

The sensors, actuators, and computer communicate through the use of electronic circuits. For example, when the incoming voltage signal from the coolant sensor tells the PCM that the engine is getting hot, the PCM sends out a command to turn on the electric cooling fan. The PCM does this by grounding the relay circuit that controls the electric cooling fan. When the relay clicks on, the electric cooling fan starts to spin and cools the engine.

Figure 33-2 A PCM. *Courtesy of DaimlerChrysler Corporation*

In a PCM **(Figure 33–2)**, RAM is used to store data collected by the sensors, the results of calculations, and other information that is constantly changing during engine operation. Information in volatile RAM is erased when the ignition is turned off or when the power to the computer is disconnected. Nonvolatile RAM does not lose its data if its power source is disconnected.

The computer's permanent memory is stored in read-only memory (ROM) or programmable read-only memory (PROM). Like nonvolatile RAM, ROM and PROM are not erased when the power source is disconnected. ROM and PROM are used to store computer control system strategy and look-up tables. PROM normally contains the specific information about the vehicle it is installed in.

System adaptive strategy is a plan, created by engine designers and calibration engineers, for the timing and control of computer-controlled systems. In designing the best strategies, it is necessary to look at all the possible conditions an engine may encounter. It is then determined how the system should respond to these conditions.

The look-up tables (sometimes called maps) contain calibrations and specifications. Look-up tables indicate how an engine should perform. For example, information (a reading of 20 in. Hg [508 mm Hg]) is received from the MAP sensor. This information, plus information from the engine speed sensor, is compared to a table for spark advance. This table tells the computer what the spark advance should be for that throttle position and engine speed **(Figure 33–3)**. The computer then modifies the spark advance.

When making decisions, the PCM is constantly referring to three sources of information: the look-up tables, system strategy, and the input from sensors. By comparing information from these sources, the computer makes informed decisions.

Adaptive Strategy

If a computer has adaptive strategy capabilities, it can actually learn from past experience. For example, the normal voltage signals from the TP sensor to the PCM range from 0.6 to 4.5 volts. If a 0.2 volt signal is received, the PCM may regard this signal as the result of a worn TP sensor and assign this lower voltage to the normal low voltage signal. In other words, the PCM will add 0.4 volt to the 0.2 volt it received. All future signals from the various throttle positions will also have 0.4 volt added to the signal. Doing this calculation adjusts for the worn TP sensor and ensures that the engine will operate normally. If the input from a sensor is erratic or considerably out of range, the PCM may totally ignore the input.

When a computer has adaptive learning, a short learning period is necessary after the battery has been disconnected, when a computer has been disconnected or replaced, or when the vehicle is new. During this learning period, the engine may surge, idle fast, or have a loss of power. The average learning period lasts for five miles of driving.

Most adaptive strategies have two parts: short-term fuel trim and long-term fuel trim. Short-term strategies are those immediately enacted by the computer to overcome a change in operation. These changes are temporary. Long-term strategies are based on the feedback about the short-term strategies. These changes are more permanent.

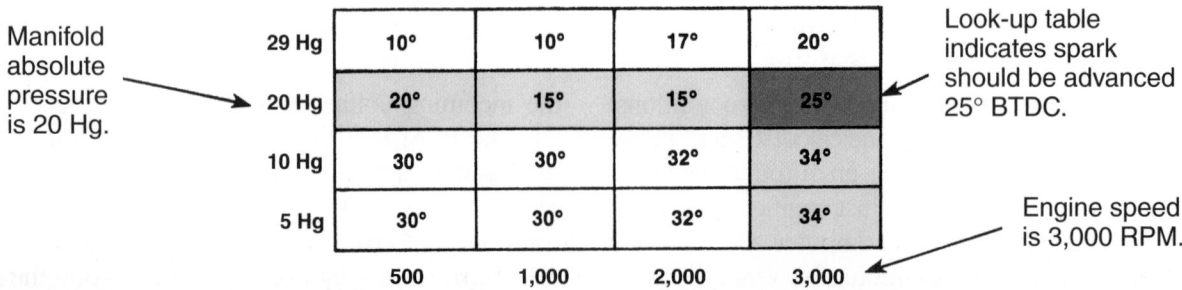

Manifold absolute pressure is 20 Hg.

29 Hg	10°	10°	17°	20°
20 Hg	20°	15°	15°	25°
10 Hg	30°	30°	32°	34°
5 Hg	30°	30°	32°	34°
	500	1,000	2,000	3,000

Look-up table indicates spark should be advanced 25° BTDC.

Engine speed is 3,000 RPM.

Figure 33-3 Example of base spark advance look-up table. *Courtesy of Ford Motor Company*

PRIMARY SENSORS

To monitor engine conditions, the computer uses a variety of sensors. All sensors perform the same basic function. They detect a mechanical condition (movement or position), chemical state, or temperature condition and change it into an electrical signal that can be used by the PCM to make decisions. The following represent the most commonly used sensors. These sensors are shared between the systems. For example, MAP sensor input is used to control fuel, ignition, EGR, emission system airflow, air intake, and idle speed systems. It is also important to remember that not all sensors are used on all vehicles.

Air-Conditioning (A/C) Demand Sensor

This is a pressure switch that signals to the PCM that the A/C clutch is cycling. When the switch is closed, a voltage signal is sent to the computer. The computer uses this data to help determine engine load and control engine idle speed. The timing on some engines can also be altered to prevent hesitation.

Brake Switch

This switch simply lets the PCM know when the brakes are applied. This signal may be used to disengage the torque converter clutch or modify ignition timing or idle speed.

Barometric Pressure (BARO) Sensor

This sensor alters the air/fuel mixture and timing controls, depending upon the altitude at which the car is being operated.

Engine Coolant Temperature (ECT) Sensor

The ECT sensor is very important **(Figure 33–4)**. Its input is used to regulate many engine functions, such as activating and deactivating the early fuel evaporation, EGR, canister purge, and TCC systems and controlling the open- and closed-loop feedback modes of the system.

The ECT sensor is usually located on the cylinder head or intake manifold. The sensor screws into a water jacket. On most systems, the sensor is a variable-resistance thermistor. On older systems, a switch coolant sensor may be used. This type of sensor may be designed to remain closed within a certain temperature range or to open only when the engine is warm.

A faulty ECT sensor or sensor circuit can cause a variety of problems. The most common is the failure to switch to the closed-loop mode once the engine is warm. ECT sensor problems are often caused by wiring faults or loose or corroded connections rather than failure of the sensor itself.

Engine Position Sensors

Engine position sensors tell the computer the speed of the engine and when the piston in each cylinder reaches TDC. This input is used to set ignition timing and fuel injection delivery. Several distinct types of engine position sensors are used, but all communicate with the computer by generating a voltage signal. The sensor does this using a Hall-effect switch of magnetic pulse generator. The engine position sensor may be called a distributor pick-up coil, crankshaft or camshaft position sensor **(Figure 33–5)** or a profile ignition pickup sensor.

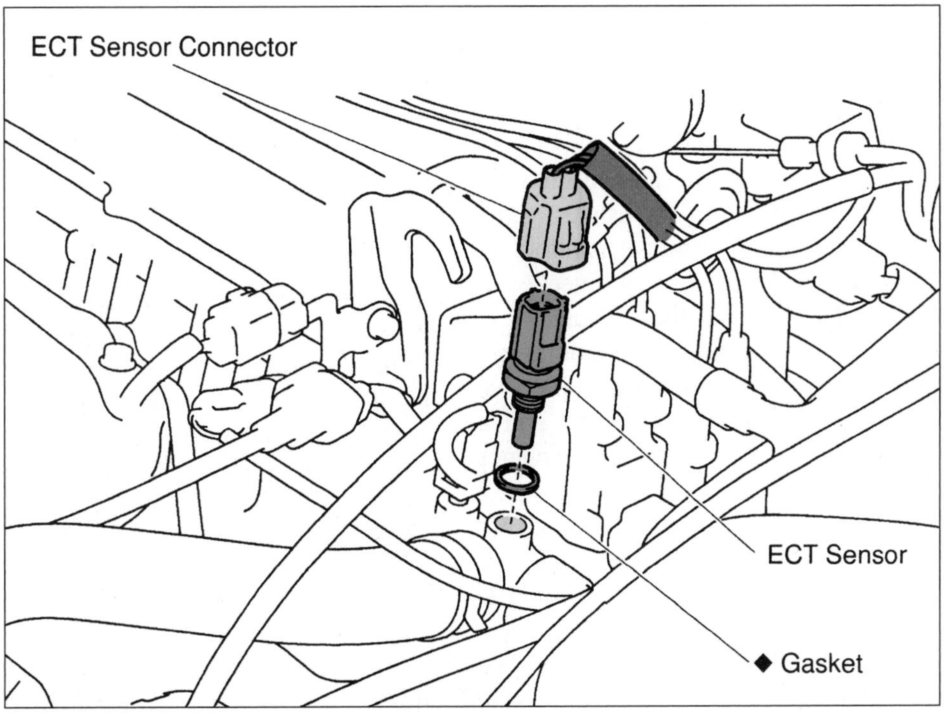

Figure 33–4 An engine coolant temperature (ECT) sensor. *Reprinted with permission*

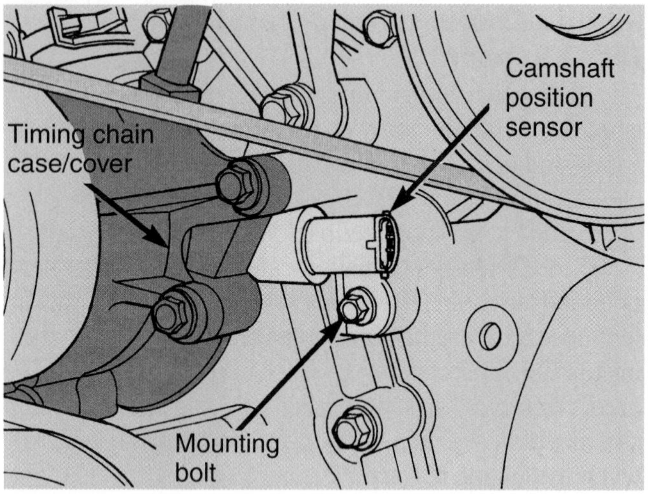

Figure 33-5 A camshaft sensor mounted to the timing chain cover. *Courtesy of DaimlerChrysler Corporation*

EGR Diagnostic Switch

On some vehicles with port fuel injection systems that control EGR operation, an EGR diagnostic switch is used to tell the computer if the EGR is actually being applied when it is commanded to be applied.

Some engines have a vacuum-operated switch tied into the vacuum hose between the EGR valve and the EGR control solenoid. When vacuum is applied to the EGR valve, the switch closes. If the computer detects a closed EGR diagnostic switch during starting, idle, or any other time it has not commanded the EGR to be applied, the PCM will turn on the MIL and set a fault code in its diagnostic memory. If it sees an open switch during any time it has commanded the EGR to be applied, it turns on the MIL and again sets a code.

EGR Valve Position Sensor

Car manufacturers use a variety of sensors or switches to determine when the EGR valve is open. This information is used to adjust the air/fuel mixture. The exhaust gases introduced by the EGR valve into the intake manifold reduce the available oxygen and thus less fuel is needed in order to maintain low HC levels in the exhaust. Most EGR valve position sensors are linear potentiometers mounted on top of the EGR valve. When the EGR valve opens, the potentiometer stem moves upward and a higher voltage signal is sent to the PCM. OBD-II engines are not equipped with these sensors.

Engine Speed Sensor

Similar to the engine position sensor, the information from this sensor may be used by the computer for determining timing (advance based on speed), fuel delivery, emission control, converter clutch operation, and idle speed.

Feedback Pressure EGR Sensor

The feedback pressure EGR sensor used in some vehicles is a pressure-sensing voltage divider (functions as a potentiometer) similar to the ones used as MAP and BARO sensors. It senses exhaust pressure in a chamber just under the EGR valve. This pressure causes the sensor to vary its output voltage signal to the computer. When the EGR valve is closed, the pressure in the sensing chamber is equal to exhaust pressure. When the EGR valve opens, pressure in this chamber drops because of the restricting orifice that lets exhaust into the sensing chamber from the exhaust system. The more the valve opens, the more the pressure drops. The feedback pressure EGR sensor's voltage signal tells the computer how far the EGR valve is open. The computer uses this information to fine tune its control of the electronic vacuum regulator, which controls vacuum to the EGR valve. This information also allows the computer to more accurately control air/fuel ratios and ignition timing.

Heated Windshield Module

This input tells the computer the heated windshield system is operating. This information helps the computer accurately determine engine load and idle speed.

High Gear Switch

This input tells the computer when the car's automatic transmission is in high gear and allows the torque converter clutch to lock up.

Intake Air Temperature (IAT) Sensor

Also referred to as the air change temperature sensor, this sensor is a thermistor. Its resistance decreases as manifold air temperature increases and increases as manifold air temperature decreases **(Figure 33–6)**. The PCM measures the voltage drop across the sensor and uses this input to help calculate fuel delivery. Cold intake air is denser; therefore, a richer air/fuel ratio is required. When the IAT signal indicates colder intake air temperature, the PCM provides a richer air/fuel ratio. The input from this sensor may also be used to control the preheated air and early fuel evaporation systems.

Knock Sensor (KS)

The knock sensor tells the PCM that detonation is occurring in the cylinders. In turn, the computer retards the timing. The knock sensor is a piezoelectric device that converts engine knock vibrations into a voltage signal. Some of the latest systems retard timing on an individual cylinder basis.

Manifold Absolute Pressure (MAP) Sensor

The function of a MAP sensor is to sense air pressure or vacuum in the intake manifold. The computer uses this input as an indication of engine load to adjust the air/fuel mixture and spark timing. The MAP sensor reads vacuum and pressure through a hose connected to the intake man-

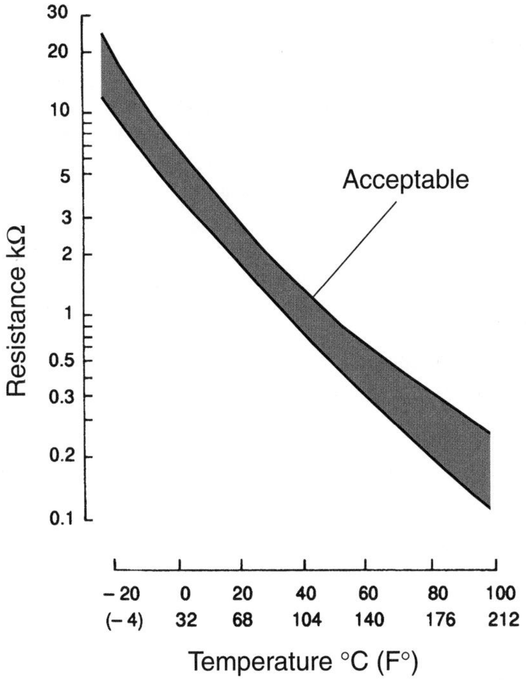

Figure 33–6 A graph showing the change in an IAT's resistance as the temperature changes. *Reprinted with permission*

ifold. A pressure-sensitive ceramic or silicon element and electronic circuit in the sensor generates a voltage signal that changes in direct proportion to pressure.

MAP sensors should not be confused with vacuum sensors or barometric pressure sensors. Whereas a vacuum sensor reads the difference between manifold vacuum and atmospheric pressure, a MAP sensor measures manifold air pressure against a precalibrated absolute pressure. Because it bases its readings on preset absolute pressure, MAP sensor readings are not adversely altered by changes in operating altitudes or barometric pressure.

Mass Airflow (MAF) Sensor

This sensor measures the flow of air entering the engine. This measurement of airflow is a reflection of engine load (throttle opening and air volume). It is similar to the relationship of engine load to MAP or vacuum sensor signal. Since there are several types of MAFs, check the service manual for the one used.

In a vane-type MAF, a pivoted air measuring plate is lightly spring-loaded in the closed position **(Figure 33–7)**. As intake air flows through the sensor, the air measuring plate moves toward the open position. The movement of the plate is proportional to intake airflow. A pointer is attached to the measuring plate shaft. This pointer contacts a resistor to form a potentiometer that sends a voltage signal to the PCM in relationship to intake airflow **(Figure 33–8)**.

In the heated-resistor type, a heated resistor is mounted in the center of the air passage. When the ignition switch is turned on, voltage is applied to the sensor's module and the module allows enough current flow

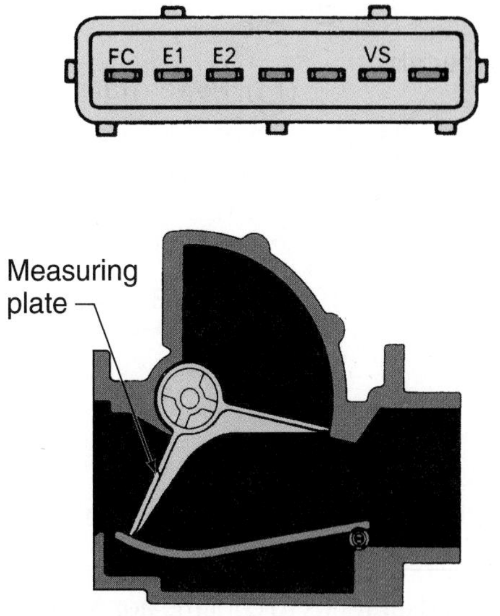

Figure 33–7 The measuring plate movement in a vane-type air flow sensor is proportional to intake airflow. *Reprinted with permission*

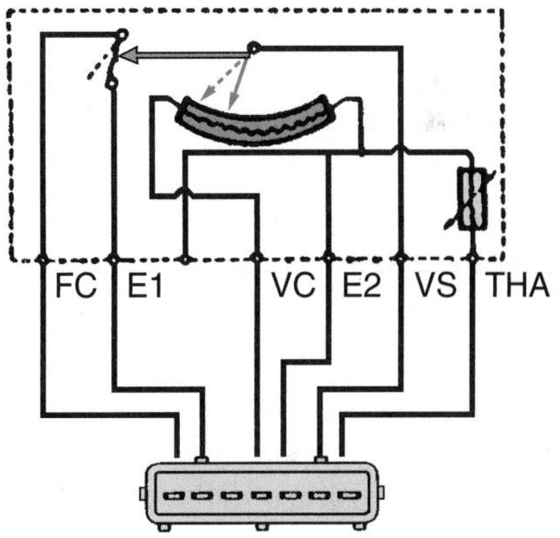

Figure 33–8 A potentiometer and a thermistor in a MAF sensor. *Reprinted with permission*

through the resistor to maintain a specific resistor temperature. If a cold engine is accelerated suddenly, the rush of cool air tries to cool the resistor. Under this condition, more current is needed to maintain the temperature of the resistor. The sensor's module sends a signal to the PCM indicating the amount of current needed to maintain the temperature. When the PCM receives a signal indicating more airflow or higher current, it allows for more fuel to be delivered to the cylinders. Some heated-resistor type MAFs use a heated grid instead of a resistor.

In a hot-wire type MAF **(Figure 33–9)**, a hot wire is positioned in the airstream through the sensor, and an ambient temperature sensor wire is located beside the hot wire. This ambient sensor wire, sometimes called the cold wire, senses intake air temperature. When the ignition switch is turned on, the module in the MAF sensor allows enough current flow through the hot wire to allow it to maintain a specific number of degrees above the ambient temperature measured by the cold wire. Like the heated-resistor type MAF, if the engine is suddenly accelerated,

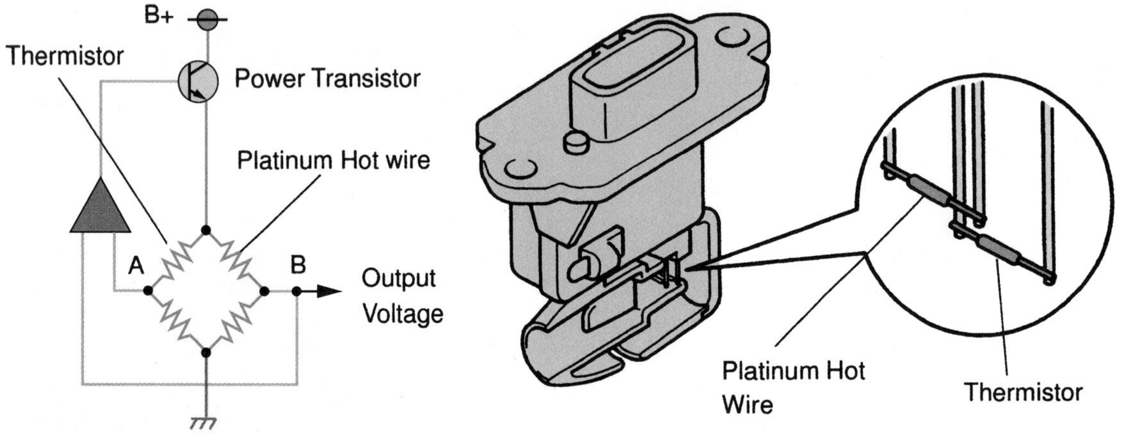

Figure 33-9 A hot-wire MAF. *Reprinted with permission*

the rush of cold air tends to cool the hot wire. As a result, more current is needed to maintain the desired temperature and a signal is sent to the PCM indicating how much current is required. Based on this input, the PCM adds or subtracts fuel in an attempt to achieve the correct air/fuel ratio for the conditions.

Neutral Drive/Neutral Gear Switch (NDS)

A neutral drive switch is used with automatic transmission vehicles to adjust idle speed due to the increased loading of an engaged transmission/transaxle. Vehicles with a manual transmission or transaxle use a neutral gear switch to inform the computer that the vehicle is out of gear. The NDS is operated by the linkage of the transmission. If the transmission is in Park or Neutral, the NDS is closed. A closed NDS sends a voltage signal below 1 volt to the PCM, whereas an open NDS sends a signal above 5 volts to the PCM.

Oxygen Sensors (O2S)

The exhaust gas oxygen sensor, of lambda sensor as it is referred to on many import vehicles, is the key sensor in the closed-loop mode. Its input is used by the computer to maintain a balanced air/fuel mixture. The O2S is threaded into the exhaust manifold or into the exhaust pipe near the engine.

> **SHOP TALK**
>
> Often oxygen sensors are called lambda sensors. The term *lambda* is used to refer to "air/fuel ratio" and "normal." Technically, it refers to normal and is represented by the Greek letter λ. It is best to think of lambda as meaning a reference point for normal or ideal air/fuel mixture. We call this mixture stoichiometric. A lambda sensor measures the variance from stoichiometric, which is about the same thing an oxygen sensor does. ■

One type of oxygen sensor, made with a zirconium dioxide element, generates a voltage signal proportional to the amount of oxygen in the exhaust gas. It compares the oxygen content in the exhaust gas with the oxygen content of the outside air. As the amount of unburned oxygen in the exhaust gas increases, the voltage output of the sensor drops. Sensor output ranges from 0.1 volt (lean) to 0.9 volt (rich). A perfectly balanced air/fuel mixture of 14.7:1 produces an output of around 0.5 volt. When the sensor reading is lean, the computer enriches the air/fuel mixture to the engine. When the sensor reading is rich, the computer leans the air/fuel mixture.

When the sensor is operating at about 600°F, a voltage is generated on two platinum strips inside the sensor. One of these strips is exposed to outside air and the other to the exhaust stream **(Figure 33–10)**. When the amount of oxygen in the exhaust is lower than the amount of oxygen in the outside air, oxygen ions (negatively charged atoms) move from the plate exposed to the outside air to the exhaust plate. This causes a negative voltage at the exhaust plate that results in the generation of 0.45 volt or higher. When the mixture is lean, the opposite occurs.

A second type of oxygen sensor, a titanium type, does not generate a voltage signal. Instead, it acts like a variable resistor, altering a base voltage supplied by the control module. When the air/fuel mixture is rich, sensor resistance is low. When the mixture is lean, resistance increases. Because variable-resistance oxygen sensors do not need an outside air reference, the need for internal venting to the outside is eliminated. They feature very fast warmup, and they operate at lower exhaust temperatures.

A titanium-type oxygen sensor typically is fed a 1-volt reference signal. Most titanium sensors send a low voltage signal with low oxygen content and a high voltage signal with high oxygen.

The accuracy of the oxygen sensor reading can be affected by air leaks in the intake or exhaust manifold. A misfiring spark plug that allows unburned oxygen to pass

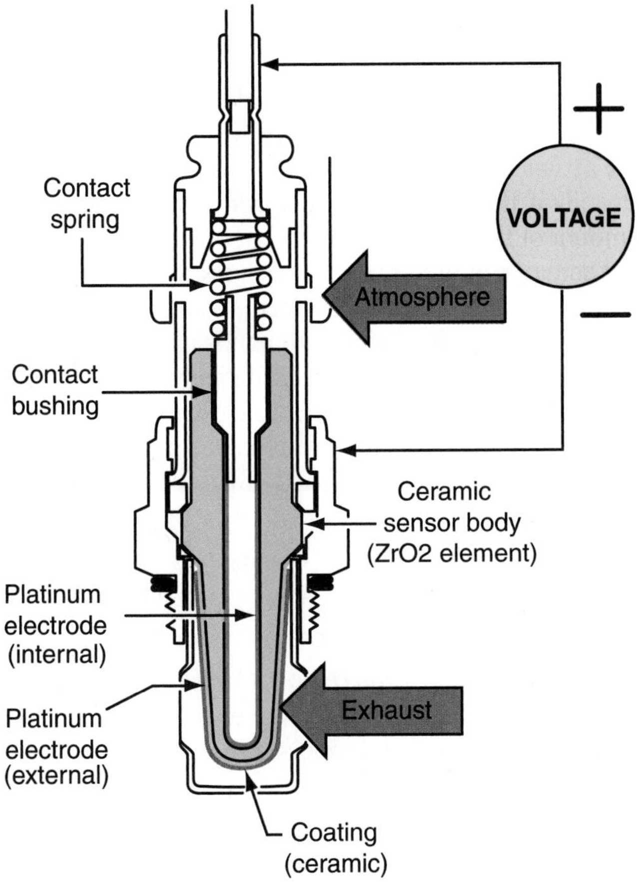

Figure 33–10 The voltage signal from an O2S results from the difference in voltage at the two platinum plates inside the sensor.

into the exhaust also causes the sensor to give a false lean reading.

Heated Oxygen Sensors Because the oxygen sensor must be hot to operate, most late-model engines use heated oxygen sensors. These oxygen sensors have three or four wires connected to them. The additional wires provide voltage for the internal heater in the sensors. Heated oxygen sensors (HO2S) are used to bring the control system into closed loop quickly. O2S signals are not accurate until the sensor is operating at about 600°F. Earlier systems disregarded O2S input during engine warm-up and the control system operated in open loop during that time. HO2Ss are sometimes referred to as **heated exhaust gas sensors (HEGOs)**. The internal heating element allows the sensor to reach operating temperature more quickly and to maintain its temperature during periods of idling or low engine load.

Power Steering (PS) Switch

This switch causes the computer to raise the idle speed when the steering wheel is turned fully in one direction. This prevents stalling when the automobile is engaged in tight turns or is being parked.

System Battery Voltage

This voltage provides power for the system. Poor grounds can cause all or any part of the computer-controlled system to malfunction.

Throttle Position (TP) Sensor

Engines with electronic fuel injection or feedback carburetors use a TP sensor to inform the computer about the rate of throttle opening and the relative throttle position **(Figure 33–11)**. The TP sensor contains a potentiometer with a pointer that is rotated by the throttle shaft **(Figure 33–12)**. As the throttle shaft moves, the pointer moves to a new location on the resistor in the potentiometer. The return voltage signal tells the PCM how much the throttle plates are open. As resistance readings tell the computer that the throttle is opening, it enriches the air/fuel mixture to maintain the proper air/fuel ratio. A separate idle switch or wide-open throttle (WOT) switch may also be used to signal the computer when these throttle positions exist.

The initial setting of the sensor is critical. The voltage signal the computer receives is referenced to this setting. Many service manuals list the initial TP sensor setting to

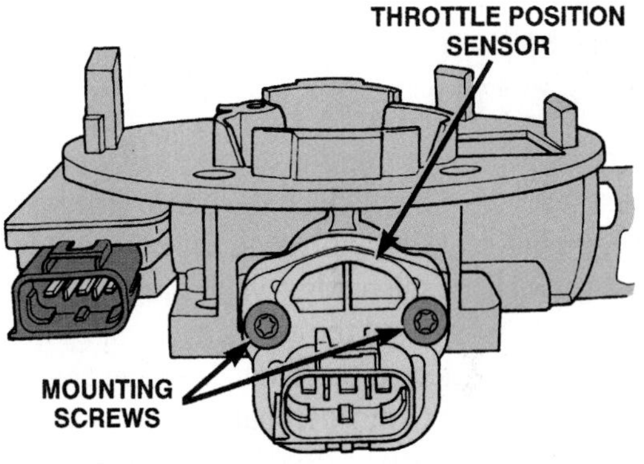

Figure 33–11 A typical TP sensor mounted to a throttle body assembly. *Courtesy of DaimlerChrysler Corporation*

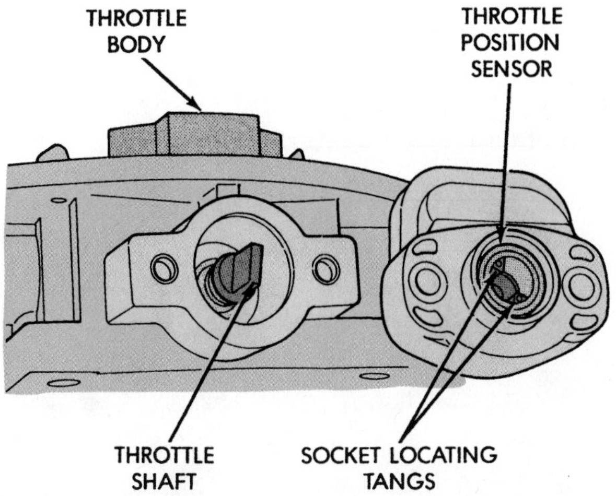

Figure 33–12 The pointer of a TP sensor moves directly with the throttle shaft. Note that the protrusion at the end of the throttle shaft fits into the recess of the TP sensor. *Courtesy of DaimlerChrysler Corporation*

the nearest one-hundredth volt, a clear indication of the importance of this setting.

The most common symptom of a bad or misadjusted TP sensor is hesitation or stumble during acceleration. The fuel mixture leans out because the computer does not receive the right signal telling it to add fuel as the throttle opens. Eventually, the oxygen sensor senses the problem and adjusts the mixture, but not before the engine stumbles.

Vacuum Sensor

Vehicles that do not have MAP or BARO sensors may use a vacuum sensor to provide engine load and speed change information. The vacuum sensor measures the difference between atmospheric pressure (outside air) and engine vacuum.

Vehicle Speed Sensor (VSS)

This sensor tells the computer the vehicle's speed in miles per hour. This input controls when the torque converter clutch locks up and also can be used to control cruise control, EGR flow, and canister purge. The VSS is connected in the speedometer cable or mounted in the transmission/transaxle opening where the speedometer drive was located previously **(Figure 33–13)**.

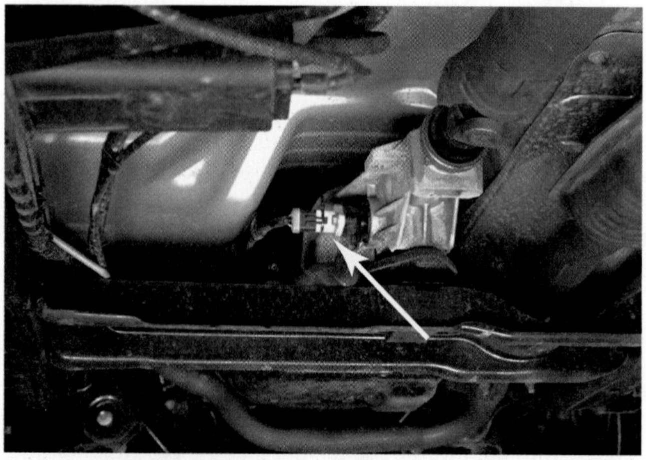

Figure 33–13 A VSS mounted in the tail shaft housing of a transmission.

COMPUTER OUTPUTS AND ACTUATORS

Once the PCM's programming instructs that a correction or adjustment must be made in the controlled system, an output signal is sent to a control device or actuator. These actuators, which are solenoids, switches, relays, or motors, physically act or carry out the command sent by the PCM.

Actuators are electromechanical devices that convert an electrical current into mechanical action. This mechanical action can then be used to open and close valves, control vacuum to other components, or open and close switches. When the PCM receives an input signal indicating a change in one or more of the operating conditions, the PCM determines the best strategy for handling the conditions. The PCM then controls a set of actuators to achieve a desired effect or strategy goal. In order for the computer to control an actuator, it must rely on a component called an output driver.

The circuit driver usually applies the ground circuit of the actuator **(Figure 33–14)**. The ground can be applied steadily if the actuator must be activated for a selected amount of time. Or the ground can be pulsed to activate the actuator in pulses. Output drivers are transistors or groups of transistors that control the actuators. These drivers operate by the digital commands from the PCM. If an actuator cannot be controlled digitally, the output signal must pass through an A/D converter before flowing to the actuator. The major actuators in a computer-controlled engine include the following components:

- *Air Management Solenoids*—Secondary air bypass and diverter solenoids control the flow of air from the air pump to either the exhaust manifold (open loop) or the catalytic converter (closed loop).
- *Evaporative Emission (EVAP) Canister Purge Valve*—This valve is controlled by a solenoid. The valve controls when stored fuel vapors in the canister are drawn into the engine and burned. The computer only activates this solenoid valve when the engine is warm and above idle speed.
- *EGR Flow Solenoids*—EGR flow may be controlled by electronically controlled vacuum solenoids. The solenoid valves supply manifold vacuum to the EGR valve when EGR is required or may vent vacuum when EGR is not required.
- *Fuel Injectors*—These solenoid valves deliver the fuel spray in fuel-injected systems.
- *Idle Speed Controls*—These actuators are small electric motors. On carbureted engines, this idle speed motor is mounted on the throttle linkage. On fuel-injected systems, a stepper motor may be used to control the amount of air bypassing the throttle plate.
- *Ignition Module*—This module is actually an electronic switching device triggered by a signal from the control computer. The ignition module may be a separate unit or may be part of the PCM.
- *Mixture Control (MC) Solenoids*—On some carbureted engines, an electrical solenoid is used to operate the metering rods and idle air bleed valve in the carburetor.
- *Motors and Lights*—Using electrical relays, the computer is used to trigger the operation of electric motors such as the fuel pump, or various warning light or display circuits.
- *Other Solenoids*—Computer-controlled solenoids may also be used in the operation of cruise control sys-

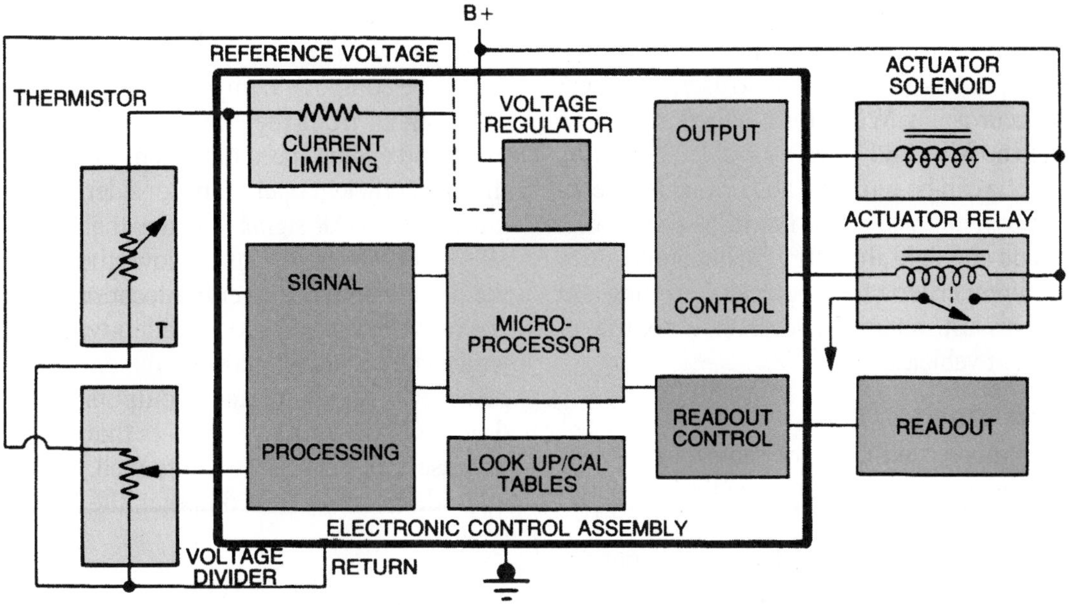

Figure 33–14 Output drivers in the computer usually supply a ground for the actuator solenoids and relays. *Courtesy of Ford Motor Company*

tems, torque converter lockup clutches, automatic transmission shift mechanisms, and many other systems where mechanical action is needed.

Electronic Throttle Control

Like modern aircraft, the acceleration on some late-model vehicles works on the drive-by-wire principle, typically called **electronic throttle control (ETC)**. ETC interprets gas pedal movement and allows for precise throttle control, which helps to improve fuel economy and performance while reducing emissions **(Figure 33–15)**.

Instead of a throttle cable and mechanical linkage to the throttle body, the throttle is moved electrically. Although these systems are electronically operated, some have a mechanical backup system or resort to partial throttle if something goes wrong with the electronic system.

Figure 33–15 A cutaway of an electronic throttle control assembly. *Courtesy of General Motors Corporation*

One or two position sensors are attached to the accelerator pedal assembly and send position and rate of change information to the PCM. The PCM controls an electric motor connected to the throttle plates. A coiled spring in the pedal assembly gives the gas pedal a normal feel. The position or rate of change in the position of the pedal is merely a request to the PCM for throttle opening. The PCM processes this request and sends commands to a driver unit that powers the electric motor attached to the throttle. Signals from a throttle position sensor allow the PCM to track the position of the throttle plate.

Electronic throttles are easily adaptable to support cruise control and traction control systems. In the latter, if the wheels spin, the system can close the throttle until wheel spin is no longer detected. Throttle control is also integrated into automatic shifting. With electronic control, the throttle can be closed slightly to reduce engine output during a shift, providing smoother gear changes.

SYSTEM OPERATION

Control loops are the cycles by which a process can be controlled by information received from input sensors, ROM, computer processing, and output of specific commands to control actuator devices.

The basic purpose of all computerized engine control loops is the same: to create an ideal air/fuel ratio, which allows the catalytic converter to operate at maximum efficiency, while giving the best mileage and performance possible and protecting the engine.

Diagnostics

When the engine is running, the PCM is receiving inputs from its sensors. The PCM knows the normal range of signals from each sensor in its circuit. If an abnormal signal is received, the PCM notices it. In most cases, the PCM will simply put this abnormality into its memory and wait for it to occur again. When it does recur, the PCM will illuminate the MIL and store a trouble code in its memory. The code can be retrieved and erased by a technician. Some problems will cause the PCM to immediately store a code and light the MIL. Each manufacturer has slightly different criteria for trouble code setting and MIL lighting. This last statement, however, is not true of OBD-II compliant vehicles.

Closed-Loop Mode

The closed-loop mode is basically the same for any automotive system. Sensor inputs are sent to the computer; the computer compares the values to its programs, then sends commands to the output devices. The output devices adjust timing, air/fuel ratio, and emission control operation. The resulting engine operation affects the sensors, which send new messages to the computer, completing the cycle of operation. The complete cycle is called a closed loop.

Closed control loops are often referred to as a feedback systems. This means that the sensors provide constant information, or feedback, on what is taking place in the engine. This allows the engine to make constant decisions and changes to output commands.

Open-Loop Mode

When the engine is cold, most electronic engine controls go into open-loop mode. In this mode, the control loop is not a complete cycle because the computer does not react to feedback information. Instead, the computer makes decisions based on preprogrammed information that allows it to make basic ignition or air/fuel settings and to disregard sensor inputs. The open-loop mode is activated when a signal from the temperature sensor indicates that the engine temperature is too low for gasoline to properly vaporize and burn in the cylinders. Systems with oxygen sensors may also go into the open-loop mode while idling, or at any time that the oxygen sensor cools off enough to stop sending a signal, and at wide-open throttle.

Fail Safe or Limp-In Mode

Most computer systems also have what is known as the fail safe or limp-in mode. The limp-in mode is nothing more than the computer's attempt to take control of vehicle operation when input from one of its critical sensors (the MAP, throttle position, coolant temperature, and temperature sensors) has been lost. To be more specific, if the computer sees a problem with the signal from a sensor, it either works with fixed values in place of the failed sensor input, or, depending on which input was lost, it can also generate a modified value by combining two or more related sensor inputs.

To illustrate this last point, let us assume the MAP sensor stops working. Instead of an actual MAP measurement, the computer compensates by creating an artificial MAP signal from the combination of throttle position input and engine speed data. While this may not result in the most efficient operation, considering the alternatives, a modified MAP signal is better than no MAP signal at all. A modified MAP signal allows the engine to run until the driver can reach a service location.

Some systems have an adaptive learning feature, which makes adjustments to the entire system to compensate for faulty inputs or outputs. In this mode, the driver will have little awareness that there is a problem because the engine runs quite well. Depending on the fault identified by the computer, the MIL may not even be lit.

Spark Control Systems

In spark control systems, input to the computer usually consists of engine temperature, engine speed, and manifold vacuum. There may be other sensors for throttle position, incoming air temperature, or engine knocking (detonation).

The computer processes these inputs and advances or retards spark timing as required. This causes changes in engine operation, which sends new messages to the computer, continuing the control loop cycle. The computer can constantly adjust timing for maximum efficiency.

When the engine is cold, the spark control system is in the open loop and runs the engine on preprogrammed settings, disregarding sensors until the engine reaches normal operating temperature.

The advantage of the electronic spark control system is threefold. It compensates for changes in engine (and sometimes outside air) temperature. It makes changes at a rate many times faster than older systems. And, finally, it has a feedback mechanism in which sensor readings allow it to complete its control loop and constantly compensate for changing conditions.

Fuel Control System

In fuel-injected systems, typical inputs to the computer are manifold vacuum, throttle position, engine speed, position, and engine temperature, inlet air temperature, and oxygen in the exhaust.

The computer processes these inputs and sends commands to the fuel injectors, telling them how long to stay open. The constant sensor inputs and precise fuel control provide a very accurate air/fuel mixture.

Emission Control System

The EGR valve, the air pump (thermactor), and the evaporative emissions canister are controlled by the PCM. The computer keeps the level of the three major pollutants (CO, HC, and NO_x) at acceptably low levels. Other emission control devices may also be wholly or partly controlled by the computer.

Cylinder Deactivation

A few engines now, and perhaps more in the future, are equipped with a feature that should increase the fuel economy of larger engines without sacrificing an engine's ability to do work when operating under a light load. This system is called *Displacement on Demand* by General Motors and has various other names by other manufacturers.

This advanced valve train system selectively turns off half the engine's cylinders when there is light load on the engine. The engine continues to fire evenly on the remaining cylinders. The system provides for smooth transition as the cylinders are deactivated and activated, since the deactivation occurs within one camshaft revolution, which is about 120 ms at 1,000 rpm or 40 ms at 3,000 rpm.

The PCM processes data from twenty plus sensors to govern the system. It calculates the current torque from all of the cylinders and projects how hard half of them would need to work to match that output. It then deactivates half the intake and exhaust lifters **(Figure 33–16)**. When deactivated, these hydraulic lifters shorten to keep the valves closed. The control of the lifters is done by sending hydraulic pressure to the central part of the lifter, where it works normally, or to the top, where it is diverted and the lifter does not pump up.

At the same time the lifters are deactivated, the PCM turns off the injectors and adjusts the mixture to the other cylinders. It also advances the ignition timing for the cylinders still working and opens the throttle to allow more to reach those cylinders. These actions allow the engine to produce more torque.

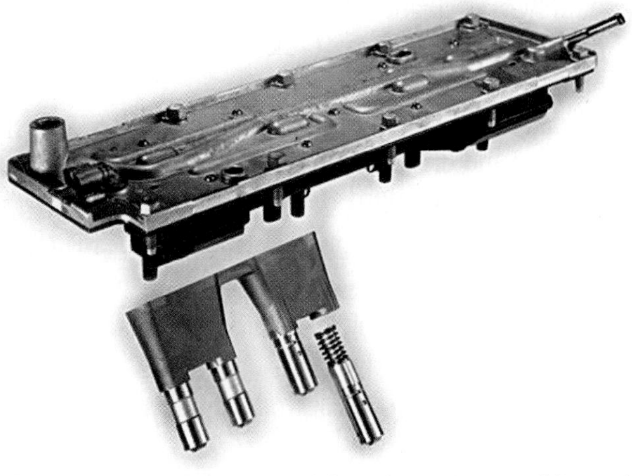

Figure 33-16 Parts of the valve train assembly used for cylinder deactivation. *Courtesy of Delphi Corporation*

Control of Nonengine Functions

Some devices that are not directly connected to the engine are also controlled by the PCM to ensure maximum efficiency.

For example, air conditioner compressor clutches can be turned on or off, depending on various conditions. One common control procedure turns off the compressor when the throttle is fully opened. Doing so allows maximum engine acceleration by eliminating the load of the compressor.

On some vehicles, the torque converter lock-up clutch is applied and released by a signal from the computer. The clutch is applied by transmission hydraulic pressure, which is controlled by electrical solenoids that are in turn controlled by the computer.

Computer Logic

In order to control an engine system, the computer makes a series of decisions. Decisions are made in a step-by-step fashion until a conclusion is reached. Generally, the first decision is to determine the engine mode. For example, to control air/fuel mixture, the computer first determines whether the engine is cranking, idling, cruising, or accelerating. Then, the computer can choose the best system strategy for the present engine mode. In a typical example, sensor input indicates that the engine is warm, rpm is high, manifold absolute pressure is high, and the throttle plate is wide open. The computer determines that the vehicle is under heavy acceleration, or wide-open throttle. Next, the computer determines the goal to be reached. For example, with heavy acceleration, the goal is to create a rich air/fuel mixture. At wide-open throttle, with high manifold absolute pressure and coolant temperature of 170°F (77°C), the table indicates that the air/fuel ratio should be 13:1. That is, 13 pounds (5.85 kg) of air for every 1 pound (.45 kg) of fuel. An air/fuel ratio of 13:1 creates the rich air/fuel mixture needed for heavy acceleration.

In a final series of decisions, the computer determines how the goal can be achieved. In our example, a rich air/fuel mixture is achieved by increasing fuel injector pulse width. The injector nozzle remains open longer and more fuel is drawn into the cylinder, providing the additional power needed.

OBD-II STANDARDS

The On-Board Diagnostics-II (OBD-II) systems were developed in response to the federal government's and the state of California's emission control system monitoring standards for all automotive manufacturers. The main goal of OBD-II was to detect when engine or system wear or when component failure caused exhaust emissions to increase by 50% or more. OBD-II also called for standard service procedures without the use of dedicated special tools. To accomplish these goals, manufacturers needed to change many aspects of their electronic engine control systems. According to the guidelines of OBD-II, all vehicles should have:

- A universal diagnostic test connector, known as the data link connector (DLC), with dedicated pin assignments.
- A standard location for the DLC. It must be under the dash on the driver's side of the vehicle and must be visible (SAE standard J1962).
- A standard list of diagnostic trouble codes (DTCs); this is SAE's standard J2012.
- A standard communication protocol (SAE standard J1850).
- The use of common scan tools on all vehicle makes and modes (SAE standard J1979).
- Common diagnostic test modes (SAE standard J2190).
- Vehicle identification must be automatically transmitted to the scan tool.
- Stored trouble codes must be able to be cleared from the computer's memory with the scan tool.
- The ability to record, and store in memory, a snapshot of the operating conditions that existed when a fault occurred.
- The ability to store a code whenever something goes wrong and affects exhaust quality.
- A standard glossary of terms, acronyms, and definitions must be used for all components in the electronic control systems (SAE standard J1930).

The OBD-II systems must illuminate the MIL **(Figure 33–17)** if the vehicle conditions would allow emissions to exceed 1.5 times the allowable standard for that model year based on a Federal Test Procedure (FTP). When a component or strategy failure allows emissions to exceed that level, the MIL is illuminated to inform the driver of a problem and a diagnostic trouble code is stored in the PCM.

Besides enhancements to the computer's capacities, some additional hardware is required to monitor the emissions performance closely enough to fulfill the tighter constraints and beyond merely keeping track of component failures. In most cases, this hardware consists of an

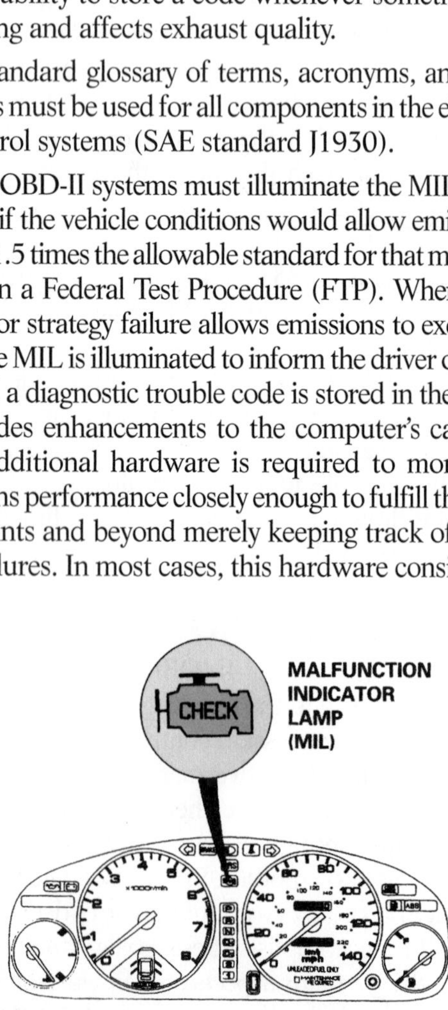

Figure 33–17 A standard MIL. *Courtesy of American Honda Motor Co., Inc.*

additional heated oxygen sensor down the exhaust stream from the catalytic converter, upgrading specific connectors and components to last the mandated 100,000 miles (160,000 km) or 10 years, in some cases a more precise crankshaft or camshaft position sensor (to detect misfires), and a new standardized 16-pin DLC.

On-board diagnostic capabilities are incorporated into a vehicle's on-board computer to monitor virtually every component that can affect emission performance. Each component is checked by a diagnostic routine to verify that it is functioning properly.

OBD-I was the first generation of on-board diagnostic systems and was designed to monitor some of a vehicle's emission control components. Required on all 1991 and newer vehicles, OBD-I systems were not as effective as possible because they monitored only a few of the components and were not calibrated to a specific level of emission performance. OBD-II was developed to address these issues and to allow more accurate diagnosis by technicians.

OBD-II Implementation

Studies estimate that approximately 50% of the total emissions from late-model vehicles are the result of emission-related problems. OBD-II systems are designed to ensure that vehicles remain as clean as possible over their entire life. During an emissions or "smog" check, a scan tool is plugged into the DLC to read the data from the vehicle's computer. If DTCs are present, the vehicle will fail the test.

OBD-II can be found on selected 1994–1996 and all newer vehicles. Because OBD-II systems were not mandated until 1996, some of the 1994 and 1995 OBD-II systems are partial systems that do not have all of the monitors required for a complete OBD-II system. OBD-II systems provide near total control and monitoring of the engine's systems and monitor parts of the chassis, body, and accessory devices as well as the diagnostic control network of the vehicle.

Instead of a fixed, unalterable PROM, the PCMs are equipped with an electronically erasable PROM (EEPROM) to store a large amount of information. The EEPROM is soldered into the PCM and is not replaceable. The EEPROM stores data without the need for a continuing source of electrical power.

The EEPROM is an integrated circuit that contains the program used by the PCM to provide power train control. It is possible to erase and reprogram the EEPROM without removing this chip from the computer. When a modification to the PCM operating strategy is required, it is no longer necessary to replace the PCM. The EEPROM may be reprogrammed through the DLC using computer software.

For example, if vehicle calibrations are updated for a specific car model sold in California, a computer may be used to erase the EEPROM. After the erasing procedure, the EEPROM is reprogrammed with the updated information from a CD ROM. Manufacturers periodically send authorized service facilities the disks required for current updating of the EEPROMs. PCM recalibrations must be directed by a service bulletin or recall letter.

OBD-II systems also use a different type of digital data stream than non-OBD-II vehicles; it is called CLASS 2 communications. Previously used, UART communications toggle from 0–5 volts at 8,192 bits per second. CLASS 2 communications toggle from 0–7 volts at a much faster rate of 10.4 kilobits per second. In CLASS 2 communications, data is pulse-width modulated, having either a short- or long-length digital pulse. UART data had only a short digital pulse; therefore, scan tools for OBD-I systems may not work on OBD-II systems.

SHOP TALK

Diagnosing some 1994 and 1995 OBD systems can be frustrating. Although they have the standard OBD-II diagnostic connector, an OBD-II scan tool will not work because these early systems still use UART communications. ■

OBD-III

Although not implemented at the printing of this book, there is a possibility that soon OBD-III will be used in California and elsewhere. The third generation of on-board diagnostics is aimed at minimizing the delay between the detection of an emissions failure by the OBD-II system and the actual repair of the vehicle. It has been said that the check engine light is a poor motivator for prompt repair and many repairs to emissions-related parts are being delayed until the mandatory emissions inspection is approaching. Vehicles with problems that increase emissions are being driven, and some owners are doing nothing about it.

OBD-III includes a way to automatically read the OBD-II codes from a vehicle. The systems may use cell phone technology and/or remote stations that retrieve information from a vehicle's computer, without physical hookup, via radio or satellite.

A cell phonelike device would be plugged into a car's computer. If the computer found a problem, it would automatically call the Division of Motor Vehicles and report it. The driver would receive notification that the car needed to be fixed.

OBD-III may have communications capabilities enabling information to be passed from the vehicle to remote stations in real time at random, perhaps without the driver's knowledge. If the system's sensors detect a problem, not only would the MIL illuminate, but the associated data would also be received by roadside receivers or satellites that are randomly monitoring vehicles. These stations may automatically receive reports on an emissions problem as soon as it occurs.

Currently, roadside readers are capable of reading eight lanes of bumper-to-bumper traffic at 100 mph and can be used from a fixed location with portable units or a mobile unit. If a fault is detected by the reader unit, it has the capability of sending the vehicle's identification number plus the fault codes to a central location.

A satellite system can be used with a cellular phone hookup or with location monitoring technology. The vehicle would receive an alert through either of these and would send the vehicle's location, VIN, the date and time, and all OBD data.

Another possibility is to have owners drive through specially designed stations where a vehicle's transponder would transmit information about the vehicle, including OBD-II information. The computers in the station would record the data. If the data showed a problem with the vehicle's emission system, the owner of the vehicle would get a notice to have it fixed.

Another possibility, and a very likely one, is to equip each vehicle with a button that is depressed by the driver once a year to establish communications with a central point and send data to it. Owners with unsatisfactory reporting data would be contacted to bring their vehicles into a shop for more detailed inspection. If the OBD system reports no emissions problems, the owner is allowed to renew the vehicle's registration and the vehicle will not need to be tested prior to that. However, if the owner does not depress the button or if the central computer receives signals indicating emissions problems, the owner will receive a notice informing him or her that the vehicle must be brought in for an emissions test.

OBD-III programs can be incorporated into current inspection and maintenance programs. OBD-III might also be used to generate out-of-cycle inspections. Once a fault is detected, a notice could be mailed to the owner requiring an out-of-cycle inspection within a certain number of days, at the next registration, or at resale, or the owner will be fined.

This type of vehicle monitoring has raised some fear in car owners. They feel the government will be able to know too much about driving habits and driving routes, and they want their privacy protected. For this reason, the final design and method for OBD-III is not decided.

Government officials answer these concerns by looking at the positives of OBD-III. Monitoring would not be continuous under the program; instead, there would be a specific number of random checks of a vehicle per year. There would be more accurate detection of high emitters and a more comprehensive inspection of these vehicles. Also, there would be quite a cost-savings and increased convenience for consumers because their vehicles would not need to go to an inspection testing station on a regular basis if the central computer received nothing but good news.

Currently, GM's OnStar systems can determine if the MIL is illuminated in the vehicle. The system can also remotely read DTCs. However, the communications specialists at the OnStar center are not allowed to tell the owner what is specifically wrong with the vehicle. They are only authorized to describe the urgency of the problem for the driver.

MONITORING CAPABILITIES

The California Air Resources Board (CARB) found that by the time a computer or emission system component failure occurs and the MIL is illuminated, the vehicle emissions have been excessive for some time. The CARB developed requirements to monitor the performance of emission systems and indicate component failure. These requirements were accepted by the EPA. The monitoring results must be available to service personnel without special test equipment marketed by the vehicle manufacturer. The monitoring system for engine, computer system, and emission system equipment is called OBD-II.

Computer systems without OBD-II have the ability to detect component and system failure. Computer systems with OBD-II are capable of monitoring the ability of systems and components to maintain low emission levels.

Computer systems with OBD-II capabilities are similar to previous systems except for the monitoring systems and the monitoring strategies in the PCM, which are extensive. New refinements are frequently incorporated into the PCM and other system components as improved technology is developed. Monitors included in OBD-II are:

1. **Catalyst efficiency monitor**
2. **Engine misfire monitor**
3. **Fuel system monitor**
4. Heated exhaust gas oxygen sensor monitor
5. Exhaust gas recirculation monitor
6. Evaporative system monitor
7. Secondary air injection monitor
8. **Comprehensive component monitor**

OBD-II systems will perform certain tests on various subsystems of the engine management system. OBD-II is designed to turn on the MIL when the vehicle has any failure that could potentially cause the vehicle to exceed its designed emission standard by a factor of 1.5. The system does that by the use of a monitor. If one or more monitored systems are out limit, then the MIL turns on to indicate a problem. The various monitors and their functions are discussed in the following sections.

Catalyst Efficiency Monitor

OBD-II vehicles use a minimum of two oxygen sensors: One is used for feedback to the PCM for fuel control and the other, located at the rear of the catalytic converter, gives an indication of the efficiency of the converter. The downstream O_2 sensor is sometimes called the **catalyst monitor sensor (CMS)**.

The conventional HO2S sensor is mounted in the exhaust manifold and the additional HO2S is mounted downstream from the catalytic converter The HO2S sensors are identified by their position in relation to the number 1 cylinder and their location relative to the converters. Sensors on the same side of the vehicle as the number 1 cylinder have a prefix of 1, and sensors mounted downstream from the converter have a prefix of 2 **(Figure 33–18)**.

If the converter is operating properly, the signal from the pre-catalyst O_2 sensor will have oscillations, while the post-catalyst O_2 sensor will be relatively flat. Once the signal from the rear sensor approaches that of the front sensor, the MIL comes on and a DTC is set.

The downstream HO2S sensors have additional protection to prevent the collection of condensation on the ceramic. The internal heater is not turned on until the ECT sensor signal indicates a warmed up engine. This action prevents cracking of the ceramic. Goldplated pins and sockets are used in the HO2S sensors, and the downstream and upstream sensors have different wiring harness connectors.

A catalytic converter stores oxygen during lean engine operation, and gives up this stored oxygen during rich operation to burn up excessive hydrocarbons. Catalytic converter efficiency is measured by monitoring the oxygen storage capacity of the converter during closed loop operation.

When the catalytic converter is storing oxygen properly, the downstream HO2S sensors provide low-frequency voltage signals. If the catalytic converter is not storing oxygen properly, the voltage signal frequency increases on the downstream HO2S sensors until the

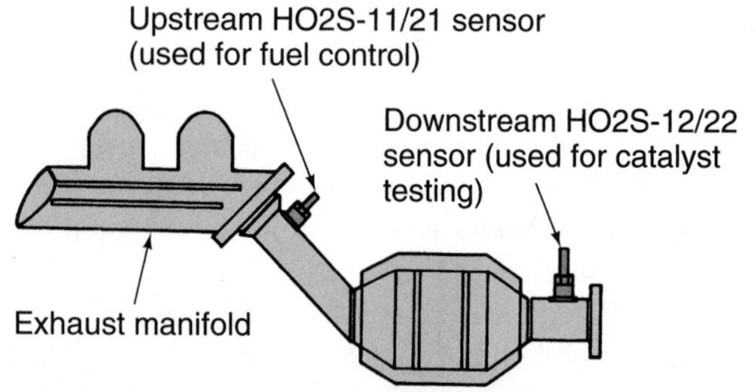

Figure 33-18 An OBD-II system with heated oxygen sensors in the exhaust manifold and after the catalytic converter. *Courtesy of Ford Motor Company*

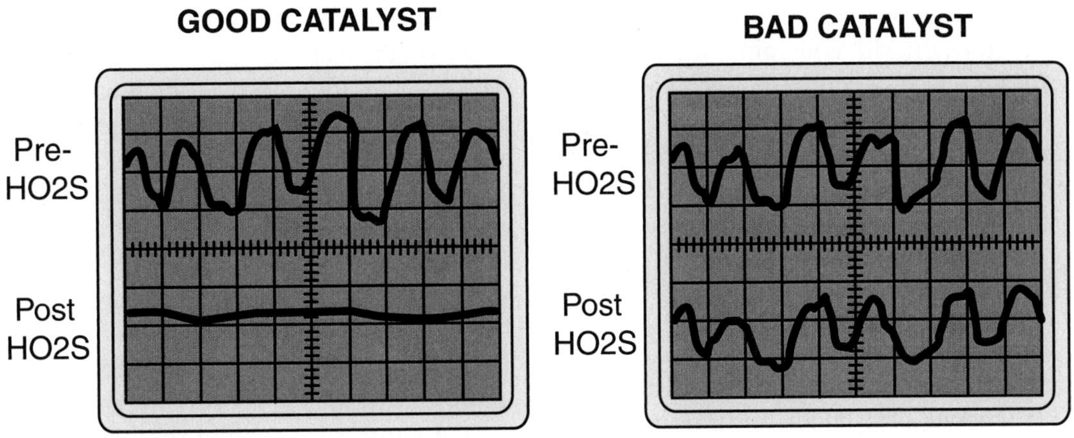

Figure 33-19 An oxygen sensor signal for good and bad catalytic converters.

frequency of its sensors approaches the frequency of the upstream HO2S sensors **(Figure 33–19)**. When the downstream HO2S sensors voltage signals reach a certain frequency, a DTC is set in the PCM memory. If the fault occurs on three drive cycles, the MIL is illuminated.

Misfire Monitor

If a cylinder misfires, unburned HC are exhausted from the cylinder, and these excessive HC emissions enter the catalytic converter. When the catalytic converter changes these excessive HC emissions into carbon dioxide and water, the catalytic converter is overheated. The honeycomb in the converter may melt into a solid mass. If this action occurs, the converter is no longer efficient in reducing emissions.

Cylinder misfire monitoring requires measuring the contribution of each cylinder to engine power. The misfire monitoring system uses a highly accurate crankshaft angle measurement to measure the crankshaft acceleration each time a cylinder fires. A high data rate crankshaft sensor is required for this function **(Figure 33–20)**. The PCM monitors the crankshaft acceleration time for each cylinder firing. If a cylinder is contributing normal power, a specific crankshaft acceleration time occurs. When a cylinder misfires, the cylinder does not contribute to engine power, and crankshaft acceleration for that cylinder is slowed.

Most OBD-II systems allow a random misfire rate of about 2% before a misfire is flagged as a fault. It is important to note that this monitor only looks at the speed of acceleration of the crankshaft during a cylinder's firing stroke. It cannot determine if the problem is fuel-, ignition-, or mechanical-related. Misfire is categorized as Type A, B, or C. Type A could cause immediate catalyst damage. Type B could cause emissions of 1.5 times the design standard and Type C could cause an I/M failure. When there is a Type A misfire, the MIL will flash. If there is a Type B or C misfire, the MIL will turn on but will not flash.

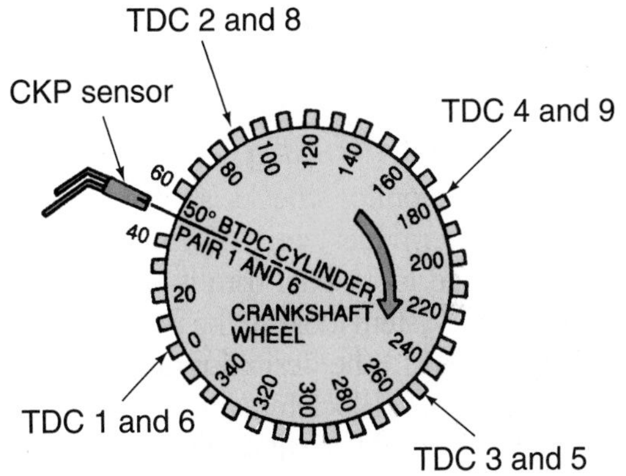

Figure 33-20 A high data rate crankshaft sensor used for misfire detection. *Courtesy of Ford Motor Company*

The misfire monitoring sequence includes an adaptive feature compensating for variations in engine characteristics caused by manufacturing tolerances and component wear. It also has the adaptive capability to allow vibration at different engine speeds and loads. When an individual cylinder's contribution to engine speed falls below a certain threshold, the misfire monitoring sequence calculates the vibration, tolerance, and load factors before setting a misfire code.

Type A Misfires Type A and Type B engine misfires are detected by the misfire monitor. When detecting a **Type A misfire**, the monitor checks cylinder misfiring over a 200-rpm period. If cylinder misfires are between 2% and 20%, the monitor considers the misfiring to be excessive. Under this condition, the PCM may shut off the fuel to the misfiring cylinder or cylinders to limit catalytic converter heat. The PCM may turn off two injectors at the same time on misfiring cylinders. When the engine is operating under heavy load, the PCM will not turn off the injectors on misfiring cylinders.

If the misfire monitor detects a Type A cylinder misfire and the PCM does not shut off the injector or injectors, the MIL begins flashing. When the misfire monitor detects a Type A cylinder misfire, and the PCM shuts off an injector or injectors, the MIL is illuminated continually.

A misfire means a lack of combustion in at least one cylinder for at least one combustion event. A misfire pumps unburned fuels through the exhaust. Although the converter can handle an occasional sample of raw fuel, too much fuel to the converter can overheat and destroy it.

Type B Misfires To detect a Type B cylinder misfire, the misfire monitor checks cylinder misfiring over a 1,000-rpm period. If cylinder misfiring exceeds 2% to 3% during this period, the monitor considers the misfiring to be excessive. This amount of cylinder misfiring may not overheat the catalytic converter, but it may cause excessive emission levels. When a **Type B misfire** is detected, a pending DTC is set in the PCM memory. If this fault is detected on a second consecutive drive cycle, the MIL light is illuminated.

Fuel System Monitoring

The fuel system monitor checks short-term fuel trim (STFT) and long-term fuel trim (LTFT) while the PCM is operating in closed loop. Fuel trim is similar to idle air control in that the system looks at the present state of this indicator compared to the desired range. On GM vehicles, this was known as Block Integrator and Block Learn. When a fuel system problem causes the PCM to make fuel trim corrections for an excessive time period, the fuel system monitor sets a DTC and illuminates the MIL if the fault occurs on two consecutive drive cycles. The fuel system monitor operates continually when the PCM is in closed loop. The fuel system monitor does not involve any new hardware.

Heated Oxygen Sensor Monitor

The system also monitors lean to rich and rich to lean time responses. This test can pick up a lazy O_2 sensor that cannot switch fast enough to keep proper control of the air/fuel mixture in the system. These sensors are the heated type and the amount of time before activity of the sensor signal is present is an indication of whether it is functional. Some systems use current flow to indicate whether the heater is working.

All of the system's HO2S sensors are monitored once per drive cycle, but the **heated oxygen sensor monitor** provides separate tests for the upstream and downstream sensors. The heated oxygen sensor monitor checks the voltage signal frequency of the upstream HO2S sensors. Excessive time between signal voltage frequency indicates a faulty sensor. At certain times, the heated oxygen sensor monitor varies the fuel delivery and checks for HO2S sensor response. A slow response in the sensor voltage signal frequency indicates a faulty sensor. The sensor signal is also monitored for excessive voltage.

The heated oxygen sensor monitor also checks the frequency of the rear HO2S sensor signals, and checks these sensor signals for excessively high voltage. If the monitor does not detect signal voltage frequency within a specific range, the rear HO2S sensors are considered faulty. The heated oxygen sensor monitor will command the PCM to vary the air/fuel ratio to check the rear HO2S sensor response.

EGR System Monitoring

The EGR monitors use several different strategies to determine if the system is operating properly. Some monitor the temperature within the EGR passages. A high temperature indicates that exhaust gas is present. Other systems look at the MAP signal, energize the EGR valve, and look for corresponding change in vacuum levels.

The EGR system may contain a delta pressure feedback EGR (DPFE) sensor. An orifice is located under the EGR valve, and small exhaust pressure hoses are connected from each side of this orifice to the DPFE sensor. During the EGR monitor, the PCM first checks the DPFE signal. If this sensor signal is within the normal range, the monitor proceeds with the tests.

With the engine idling and the EGR valve closed, the PCM checks for pressure difference at the two pressure hoses connected to the DPFE sensor. When the EGR valve is closed and there is no EGR flow, the pressure should be the same at both pipes. If the pressure is different at these two hoses, the EGR valve is stuck open.

The PCM commands the EGR valve to open and then checks the pressure at the two exhaust hoses connected to the DPFE sensor. With the EGR valve open and EGR flow through the orifice, there should be higher pressure at the upstream hose than at the downstream hose **(Figure 33–21)**.

The PCM checks the EGR flow by checking the DPFE signal value against an expected DPFE value for the engine operating conditions at steady throttle within a specific rpm range. If a fault is detected in any of the EGR monitor tests, a DTC is set in the PCM memory. If the fault occurs during two drive cycles, the MIL is illuminated. The EGR monitor operates once per OBD-II trip.

Evaporative (EVAP) Emission System Monitor

In addition to the various components and failures that could affect the tailpipe emissions on a vehicle. OBD-II monitors the evaporative systems as well. The system tests the ability of the fuel tank to hold pressure and of the purge system to vent the gas fumes from the charcoal canister.

Chrysler and others often fit their vehicles with a Leak Detection Pump (LDP) to detect leaks in the EVAP sys-

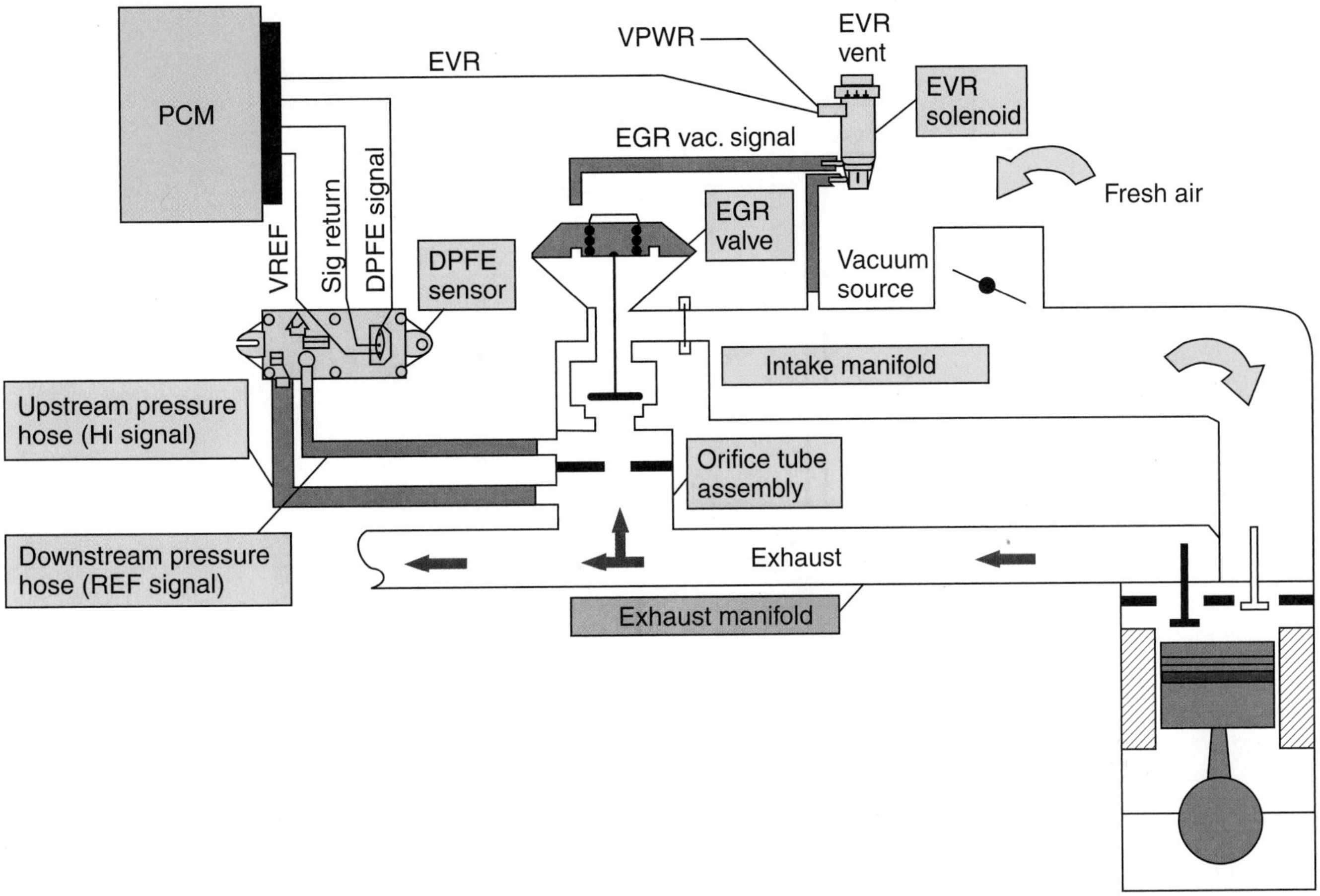

Figure 33–21 An EGR system with a delta pressure feedback sensor. *Courtesy of Ford Motor Company*

tem **(Figure 33–22)**. When the predetermined operating conditions are met, the PCM powers the LDP to test the EVAP system for leaks. The pump pressurizes the system. As pressure builds, the cycling rate of the pump decreases. If there is no leak in the system, the pressure will build until the pump shuts off. If there is a leak, pressure

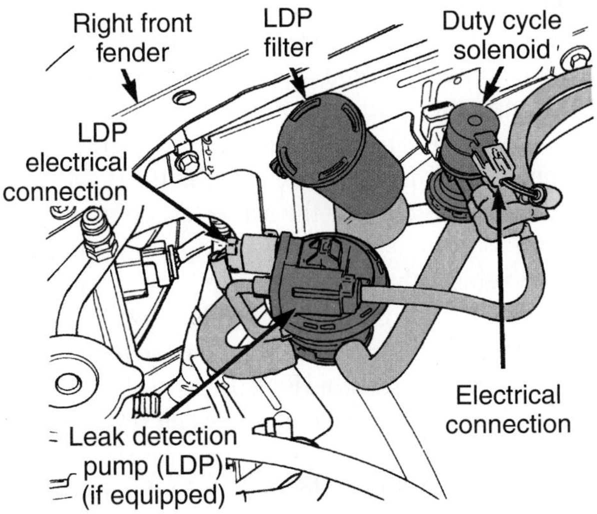

Figure 33–22 The typical location of the LDP and EVAP canister purge solenoid. *Courtesy of DaimlerChrysler Corporation*

will not build up and will not shut down the pump. The pump continues to run until the PCM determines it has run a complete test cycle. If the PCM senses that there is no leak in the system, it will run the purge monitor. Because of the pressure in the system from the LDP, the cycle rate should be high. If no leaks are present and the purge cycle is high, the system passed the test.

Some EVAP systems have a purge flow sensor (PFS) connected to the vacuum hose between the canister purge solenoid and the intake manifold **(Figure 33–23)**. The PCM monitors the PFS signal once per drive cycle to determine if there is vapor flow or no vapor flow through the solenoid to the intake manifold.

Other EVAP systems have a vapor management valve connected in the vacuum hose between the canister and the intake manifold. The vapor management valve is a normally closed valve. The PCM operates the valve to control vapor flow from the canister to the intake manifold. The PCM also monitors the valve's operation to determine if the EVAP system is purging vapors properly.

Fuel Filler Caps Some vehicles have an enhanced evaporative system monitor. This system detects leaks and restrictions in the EVAP system. A specially designed fuel

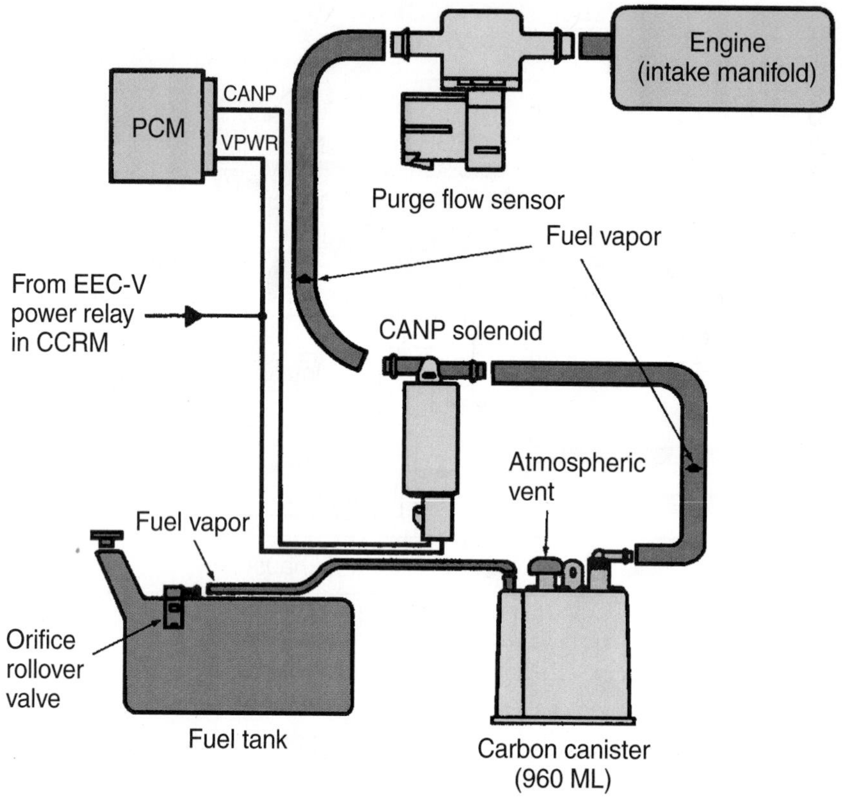

Figure 33-23 An EVAP system with a purge flow sensor. *Courtesy of Ford Motor Company*

tank filler cap is used on these systems. In these enhanced systems, an evaporative system leak or a missing fuel tank cap causes the MIL to turn on. Also, if the fuel filler cap is not on tight enough, the OBD system can detect leaking vapor and the MIL will light up. (This malfunction is listed as Gross Leak Detected.) If the filler cap is then tightened, the indicator will generally go out after a short period of time. On most 2001 and later model year vehicles, once the MIL is set by the detection of a vapor leak, the light will not go out automatically.

Secondary Air Injection Reaction (AIR) System Monitor

The AIR system operation can be verified by turning the AIR system on to inject air upstream of the oxygen sensor while monitoring its signal. Many designs inject air into the exhaust manifold when the engine is in open loop and switch the air to the converter when it is in closed loop. If the air is diverted to the exhaust manifold during closed loop, the O_2 sensor thinks the mixture is lean and the signal should drop.

On some vehicles, the AIR system is monitored with passive and active tests. During the passive test, the voltage of the precatalyst HO2S is monitored from start-up to closed loop operation. The AIR pump is normally on during this time. Once the HO2S is warm enough to produce a voltage signal, the voltage should be low if the AIR pump is delivering air to the exhaust manifold. The secondary AIR monitor will indicate a pass if the HO2S voltage is low at this time. The passive test also looks for a higher HO2S voltage when the AIR flow to the exhaust manifold is turned off by the PCM. When the AIR system passes the passive test, no further testing is done. If the AIR system fails the passive test or if the test is inconclusive, the AIR monitor in the PCM proceeds with the active test.

During the active test, the PCM cycles the AIR flow to the exhaust manifold on and off during closed loop operation and monitors the precatalyst HO2S voltage and the short-term fuel trim value. When the AIR flow to the exhaust manifold is turned on, the sensor's voltage should decrease and the short-term fuel trim should indicate a richer condition. The secondary AIR system monitor illuminates the MIL and stores a DTC in the PCM's memory if the AIR system fails the active test on two consecutive trips.

Some vehicles have an electric air pump system. In this system, the air pump is controlled by a solid-state relay. The relay is operated by a signal from the PCM. An air-injection bypass solenoid is also operated by the PCM. This solenoid supplies vacuum to dual air diverter valves **(Figure 33–24)**.

When the engine is started, the PCM signals the relay to start the air pump. This module supplies the high current required for air pump operation. The air pump may provide a 10-second delay in pump operation after the

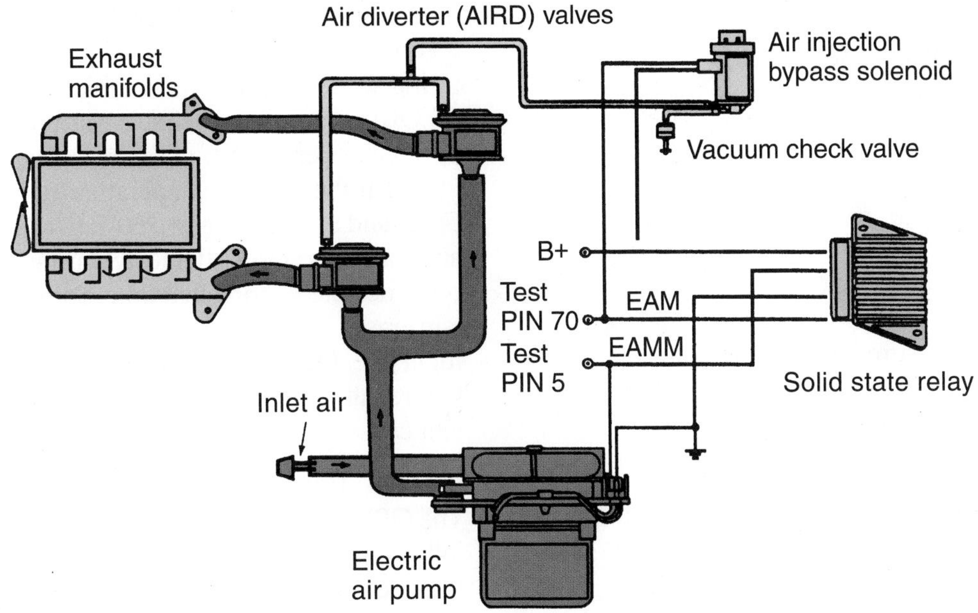

Figure 33-24 An electric air pump system. *Courtesy of Ford Motor Company*

engine is started. The PCM also energizes the air-injection bypass solenoid. When this solenoid is energized, it supplies vacuum to the dual air diverter valves. This action opens the normally closed air diverter valves. Air from the pump is now delivered to the exhaust manifold. The purpose of the air pump is to oxidize HC and CO in the exhaust manifolds for 20 to 120 seconds after the engine has started and until the catalytic converter is working properly. The length of time the air pump is operating depends on the temperature of the engine. Once the catalyst is warmed up, the PCM signals the relay to shut down the air pump. The PCM also deenergizes the air injection bypass solenoid that allows the air diverter valves to close.

The PCM monitors the relay and the air pump to determine if secondary air is present. This PCM monitor for the air pump system functions once per drive cycle. When a malfunction occurs in the air pump system on two consecutive drive cycles, a DTC is stored and the MIL is turned on. If the malfunction corrects itself, the MIL is turned off after three consecutive drive cycles in which the fault is not present.

Comprehensive Component Monitor

The system looks at any electronic input that could affect emissions. The strategy is to look for opens and shorts or input signal values that are out of the normal range. It also looks to see if the actuators have their intended effect on the system and to monitor other abnormalities.

The comprehensive component monitor uses two strategies to monitor inputs and two strategies to monitor outputs. One strategy for monitoring inputs involves checking certain inputs for electrical defects and out-of-range values by checking the input signals at the analog digital converter. The input signals monitored in this way are:

1. Rear HO2S inputs
2. HO2S inputs
3. Mass air flow sensor
4. Manual lever position sensor
5. Throttle position sensor
6. Engine coolant temperature sensor
7. Intake air temperature sensor

The comprehensive component monitor (CCM) checks frequency signal inputs by performing rationality checks. During a rationality check, the monitor uses other sensor readings and calculations to determine if a sensor reading is proper for the present conditions. The CCM checks these inputs with the following rationality checks:

1. Crankshaft position sensor
2. Output shaft speed sensor
3. Ignition diagnostic monitor
4. Camshaft position sensor
5. Vehicle speed sensor

The PCM output that controls the idle air control motor is monitored by checking the idle speed demanded by the inputs against the closed loop idle speed correction supplied by the PCM to the IAC motor.

The output state monitor in the CCM checks most of the outputs by monitoring the voltage of each output solenoid, relay, or actuator at the output driver in the PCM. If the output is off, this voltage should be high, but it is pulled low when the output is on.

Monitored outputs include:

1. Wide-open throttle A/C cutoff
2. Shift solenoid 1
3. Shift solenoid 2
4. Torque converter clutch solenoid
5. HO2S heaters
6. High fan control
7. Fan control
8. Electronic pressure control solenoid

OBD-II DIAGNOSTICS

System Readiness Mode

All OBD-II scan tools include a readiness function showing all of the monitoring sequences on the vehicle and the status of each, which is either complete or incomplete. If vehicle travel time, operating conditions, or other parameters were insufficient for a monitoring sequence to complete a test, the scanner will indicate which monitoring sequence is not yet complete.

OBD-II standards define a warm-up cycle as a period of vehicle operation, after the engine had been turned off, in which the coolant temperature rises by at least 40°F (4°C) and reaches at least 160°F (71°C). Most DTCs are automatically erased after forty warm-up cycles if the fault is not detected during that time. Some manufacturers retain erased DTCs in a flagged condition. This can be useful if the technician notices a pattern of component failure, all of which may be related to a single intermittent cause like low fuel pressure.

OBD-II Trip

The OBD-II **trip** or drive cycle consists of an engine start following an engine off period, with enough vehicle travel **(Figure 33–25)** to allow the following monitoring sequences to complete their tests:

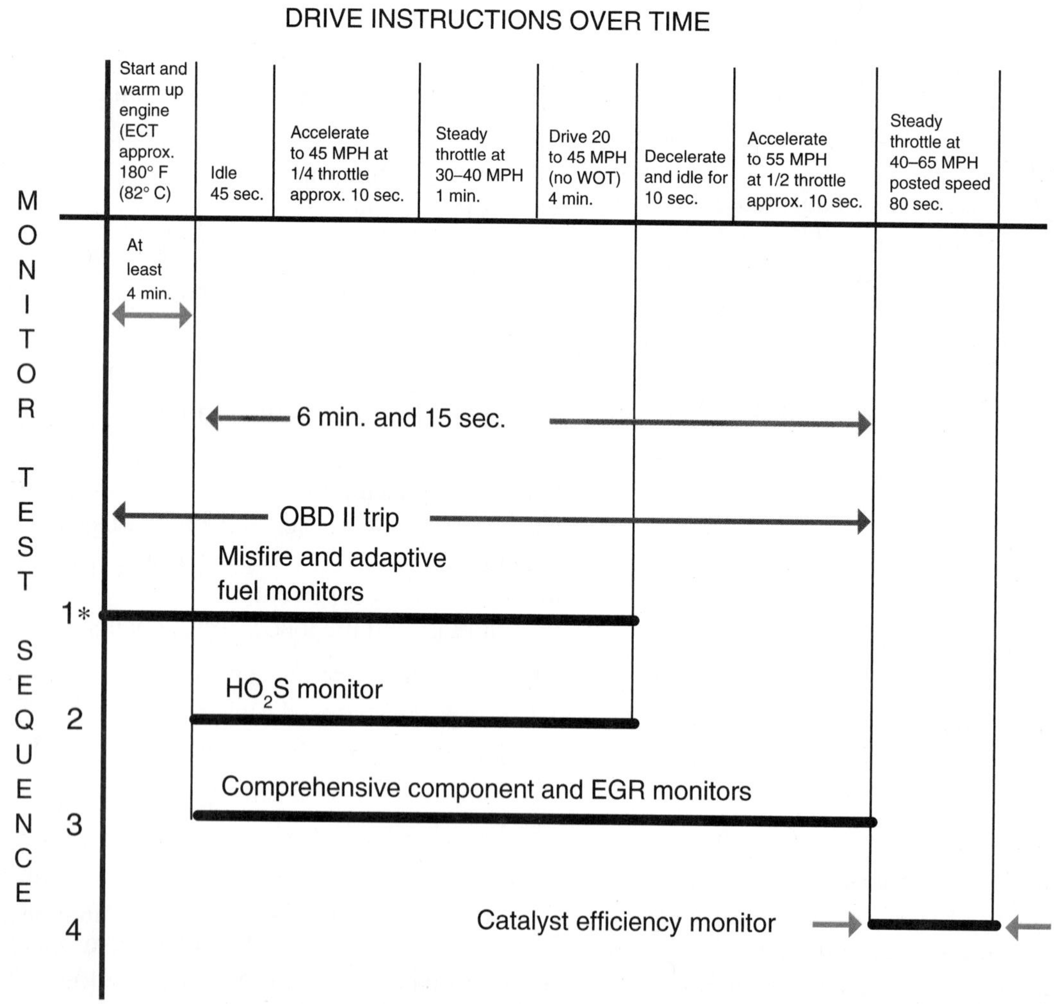

* Since the misfire, adaptive fuel, EGR (requiring idles and accelerations), and comprehensive monitors are continuously checked by the OBD-II system, the test sequence may vary on each vehicle due to outside ambient temperature, engine/vehicle performance temperature, and driving conditions.

Figure 33-25 An OBD trip cycle. *Courtesy of Ford Motor Company*

- Misfire, fuel system, and comprehensive system components—These are checked continuously throughout the trip.
- EGR—This test requires a series of idle speed operations, acceleration, and deceleration to satisfy the conditions needed for completion.
- HO2S—This test requires a steady speed of between 20 and 45 mph (32 and 72 km/h) for about 20 seconds after warm-up to be complete.

The trip display is provided on a scan tester. Some scan testers display YES when the five monitors are completed in a trip. When the five monitors are not completed during a trip, the scan tester displays NO in the trip display.

Depending on the monitor, the system tests the component or system once per trip. Trips are based on the driving cycle that the vehicle experiences during the FTP. Basically the cycle includes operation during warm-up, cruise, acceleration, and deceleration with certain time requirements for each mode of operation.

Test Connector

OBD-II and the Society of Automotive Engineers (SAE) standards require the DLC to be mounted in the passenger compartment out of sight of vehicle passengers. The standard DLC **(Figure 33–26)** is a sixteen-pin connector. The same pins are used for the same information, regardless of the vehicle's make, model, and year. The connector is *D*-shaped and has guide keys that allow the scan tool to be installed only one way. Using a standard connector design and designating the pins allow data retrieval with any scan tool designed for OBD-II. Some European vehicles meet OBD-II standards by providing the designated DLC along with their own connector for their own scan tool.

SHOP TALK

When a vehicle has a sixteen-pin DLC, it does not necessarily mean that the vehicle is equipped with OBD-II. ■

The DLC is designed for scan tool use only. You cannot jump across any of the terminals to display codes on an instrument panel or other indicator lamp. The MIL is used only to inform the driver that the vehicle should be serviced soon. It also informs a technician that the computer has set a trouble code.

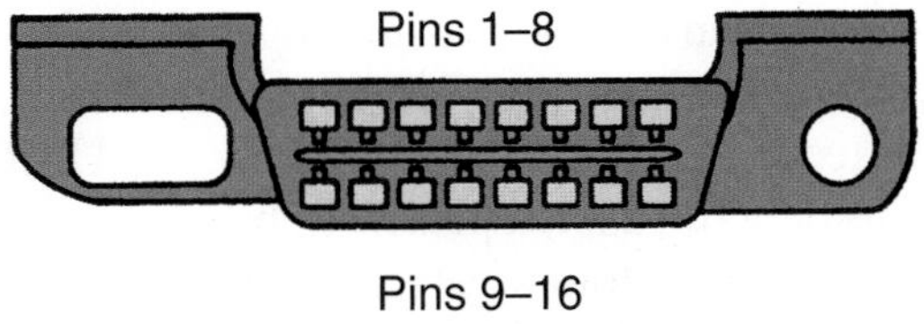

Figure 33-26 A sixteen-pin DLC. *Courtesy of Ford Motor Company*

The DLC must be easily accessible while sitting in the driver's seat. All DLCs must be located somewhere between the left end of the instrument panel and a position 300 millimeters to the right of the center. The DLC cannot be hidden behind panels and must be accessible without tools. Any generic scan tool can be connected to the DLC and can access the diagnostic data stream.

The connector pins are arranged in two rows and are numbered consecutively. Seven of the sixteen pins have been assigned by the OBD-II standard. The remaining nine pins can be used by the individual manufacturers to meet their needs and desires.

Malfunction Indicator Lamp Operation

An OBD-II system continuously monitors the entire emissions system, switches on a MIL if something goes wrong, and stores a fault code in the PCM when it detects a problem. The codes are well defined and can lead a technician to the problem. A scan tool must be used to access and interpret emission-related DTCs regardless of the make and model of the vehicle.

Many of the trouble codes of an OBD-II system mean the same thing regardless of manufacturer. However, some of the codes pertain only to a particular system or mean something different within each system. The DTC is a five-character code with both letters and numbers **(Figure 33–27)**, called the alphanumeric system. The first character of the code is a letter. This defines the system where the code was set. Currently there are four possible first-character codes: B for body, C for chassis, P for power train, and U for undefined. The U-codes are designated for future use.

The second character is a number. This defines the code as being a mandated code or a special manufacturer code. This number will either be a 0, 1, 2, or 3. A 0 code means that the fault is defined or mandated by OBD-II. A 1 code means that the code is manufacturer specific. Codes of 2 or 3 are designated for future use.

The third through fifth characters are numbers. These describe the fault. The third character of a power train code tells you where the fault occurred. The remaining two characters describe the exact condition that set the code.

When the same fault has been detected during two drive cycles, a DTC is stored in the PCM memory. If a misfire occurs, the misfire monitor will store a DTC immediately in the PCM memory, depending on the type of misfire.

If a misfire that threatens engine or catalyst damage occurs, the misfire monitor flashes the MIL on the first

The SAE J2012 standards specify that all DTCs will have a five-digit alphanumeric numbering and lettering system. The following prefixes indicate the general area to which the DTC belongs:

1. P — power train
2. B — body
3. C — chassis

The first number in the DTC indicates who is responsible for the DTC definition.

1. 0 — SAE
2. 1 — manufacturer

The third digit in the DTC indicates the subgroup to which the DTC belongs. The possible subgroups are:

0 — Total system
1 — Fuel-air control
2 — Fuel-air control
3 — Ignition system misfire
4 — Auxiliary emission controls
5 — Idle speed control
6 — PCM and I/O
7 — Transmission
8 — Non-EEC power train

The fourth and fifth digits indicate the specific area where the trouble exists. Code P1711 has this interpretation:

P — Power train DTC
1 — Manufacturer-defined code
7 — Transmission subgroup
11 — Transmission oil temperature (TOT) sensor and related circuit

Figure 33-27 OBD-II trouble codes and their meanings.

occurrence of the misfire. A fault detected by the catalyst monitor must occur on three drive cycles before the MIL is illuminated.

For the misfire and fuel system monitors, if the fault does not occur on three consecutive drive cycles under similar conditions, the MIL is turned off. The system defines similar conditions as:

1. Engine speed within 375 rpm compared to when the fault was detected.
2. Engine load within 10% compared to when the fault was detected.
3. Engine warm-up state or coolant temperature must match the temperature when the fault was detected.

For the catalyst efficiency, HO2S, EGR, and comprehensive component monitors, the MIL is turned off if the same fault does not reappear for three consecutive drive cycles. When the fault is no longer present and the MIL is turned off, the DTC is erased after forty engine warm-up cycles. A technician may use a scan tester to erase DTCs immediately.

A pending DTC is a code representing a fault that has occurred, but that has not occurred enough times to illuminate the MIL. Some scan testers are capable of reading pending DTCs with the continuous DTCs.

Data Links

Each manufacturer must use the same protocol between the PCM and its sensors and actuators. The same protocol must be used to send diagnostic information to the scan tool through the DLC. A **protocol** is an agreed-on digital code that the PCM uses to communicate with a scan tool.

A variety of protocol data links (data links may be referred to as bus wires) are connected to the DLC on OBD-II systems. For example, some General Motors vehicles have universal asynchronous receive and transmit (UART) data links connected to the DLC. When computers or the scan tool are not communicating on the UART data links, the at-rest voltage is 5 volts. When a computer or scan tester is transmitting data on the UART bus wires, the voltage is pulled to ground level. Data is transmitted at 8,192 bits per second on the UART system. This data has a fixed pulse width.

Other General Motors' vehicles have class 2 data links connected to the DLC **(Figure 33–28)**. Data transmission speed is 10.4 kilobits per second on class 2 data links. Most General Motors' OBD-II systems have class 2 data

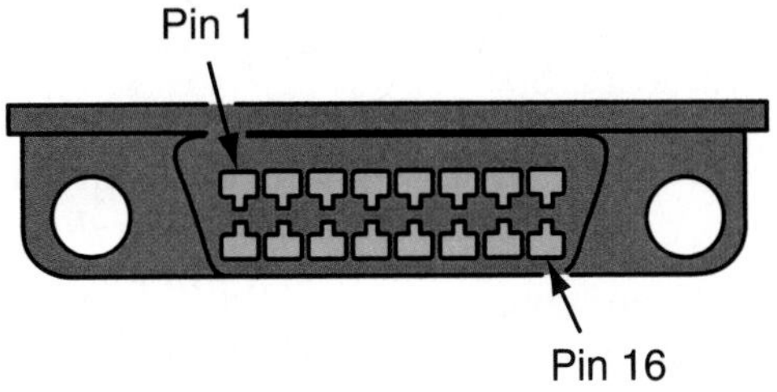

Pin 1	Secondary UART 8192 baud serial data Class B 160 baud serial data
Pin 2	J1850 Bus + line on 2 wire systems, or single wire (class 2)
Pin 3	Ride control diagnostic enable
Pin 4	Chassis ground pin
Pin 5	Signal ground pin
Pin 6	PCM/VCM Diagnostic enable
Pin 7	K line for International Standards Organization (ISO) application
Pin 8	Keyless entry enable, or MRD theft diagnostic enable
Pin 9	Primary UART
Pin 10	J1850 Bus-line for J1850-2 wire applications
Pin 11	Electronic Variable Orifice (EVO) steering
Pin 12	ABS diagnostic, or CCM diagnostic enable
Pin 13	SIR diagnostic enable
Pin 14	E&C bus
Pin 15	L line for International Standards Organization (ISO) application
Pin 16	Battery power from vehicle unswitched (4 amps max)

Figure 33-28 Class 2 data links connected to the sixteen-pin DLC.

links. This data link system meets the SAE J1850 standard for serial data transmission. The at-rest voltage is 0 volts on class 2 data links, while the transmission voltage is 7 volts. A variable pulse width is used to transmit data in class 2 systems. When two devices are attempting to transmit data at the same time, the data is automatically prioritized so that only the high-priority data is allowed to continue. The low-priority data must wait until the high-priority transmission is completed. When a General Motors' OBD-II system has a VCM and class 2 data links, the VCM is referred to a VCM-A. On those systems without class 2 data links, the computer is called a VCM. Some VCMs are used on OBD-II systems.

Some Ford vehicles have international standards organization (ISO) data links connected to the sixteen-pin DLC. These data links transmit data from the antilock brake system module, driver seat module, lighting control module, and driver door module to the DLC.

Standard corporate protocol (SCP) data links are also connected to the sixteen-pin DLC on some Ford vehicles. These data links meet the SAE J1850 standard. Data from the PCM and electronic automatic temperature control is transmitted on the SCP data links. The computers connected to ISO or SCP data links may vary depending on the year, make, and model of the vehicle.

Some Ford cars, such as the Town Car, Grand Marquis, and Crown Victoria, have audio corporate protocol (ACP) data links, which transmit data between the LUX audio system components. Since this system has internal self-diagnostic capabilities and does not require the use of a scan tool, the ACP data links are not connected to the sixteen-pin DLC.

Test Modes

All OBD-II systems have the same basic test modes. These test modes must be accessible with an OBD-II scan tool. Mode 1 is the parameter identification (PID) mode. It allows access to certain data values, analog and digital inputs and outputs, calculated values, and system status information. Some of the PID values are manufacturer specific, others are common to all vehicles.

Mode 2 is the freeze frame data access mode. This mode permits access to emission-related data values from specific generic PIDs. These values represent the operating conditions at the time the fault was recognized and logged into memory as a DTC. Once a DTC and a set of **freeze frame** data are stored in memory, they will stay in memory even if other emission-related DTCs are stored. The number of these sets of freeze frames that can be stored are limited. On 1996 General Motors' vehicles, the possible number of stored sets is five.

One type of failure is an exception to this rule—misfire. Fuel system misfires overwrite any other type of data except for other fuel system misfire data. This data can only be removed with a scan tool. When a scan tool is used to erase a DTC, it automatically erases all freeze frame data associated with the events that led to that DTC.

Mode 3 permits scan tools to obtain stored DTCs. The information is transmitted from the PCM to the scan tool following an OBD-II Mode 3 request. Either the DTC, its descriptive text, or both will be displayed on the scan tool.

The PCM reset mode (Mode 4) allows the scan tool to clear all emission-related diagnostic information from its memory. Once the PCM has been reset, the PCM stores an inspection maintenance readiness code until all OBD-II system monitors or components have been tested to satisfy an OBD trip cycle without any other faults occurring. Specific conditions must be met before the requirements for a trip are satisfied.

Mode 5, the oxygen sensor monitoring test, gives the oxygen sensor fault limits and the actual oxygen sensor outputs during the test cycle. The test cycle includes specific operating conditions that must be met to complete the test. This information helps determine the effectiveness of the catalytic converter.

Mode 6 is the output state mode (OTM) which allows a technician to activate or deactivate the system's actuators through the scan tool. When the OTM is engaged, the actuators can be controlled without affecting the radiator fans. The fans are controlled separately. This gives a pure look at the effectiveness and action of the outputs.

Snapshots

The primary purpose of OBD-II is to make the diagnosis of emissions and driveability problems simple and uniform in the future. No longer is it necessary to learn entirely new systems from each manufacturer. The basic plan is to allow technicians to diagnose any vehicle with the same diagnostic tools. Manufacturers can introduce special diagnostic tools or capacities for their own systems, providing that standard scan tools, along with DMMs and lab scopes, can analyze the system. These special tools can have additional capabilities beyond those given in the standards. One of the mandated capabilities is the freeze frame or **snapshot** feature. This is the ability of the system to record data from all of its sensors and actuators at a time when the system turns on the MIL. General Motors expanded this capability to include "failure reports," which does the same thing as the snapshot, but also includes any fault stored in memory, not just those related to the emissions-related circuits.

The basic advantage of the snapshot feature is the ability to look at the existing conditions when a code was set, which is especially valuable for diagnosing intermittent problems. Whenever a code is set, a record of all related activities is stored in memory. This record allows the technician to look at the action of sensors and actuators when the code was set, which helps identify the cause of the problem.

OBD-II TERMS

All vehicle manufacturers must use the same names and acronyms for all electric and electronic systems related to the engine and emission control systems. Previously, there were many names for the same component. Now all similar components are referred to using the same name. Beginning with the 1993 model year, all service information was required to use the new terms. This new terminology is commonly called J1930 terminology because it conforms to the SAE standard J1930.

CAUTION!

Static electricity can damage the calibration units. Do not open the package until you have grounded it and are ready to install it. Never touch the terminals on the part.

CASE STUDY

Trouble code indicates that a non-OBD-II computer is faulty (Chrysler code 53, Ford code 15, and General Motors code 54.)

Computer modules seldom fail unless something on the vehicle puts an excess load on the unit. Therefore, you should never install a replacement computer without finding what caused it to fail. Otherwise, the replacement module could also be destroyed.

Once it has been determined that the computer should be replaced, follow these precautions:

1. ***Make sure that the ignition key is turned off.***
2. ***Disconnect the battery ground cable and remove the fuse. It is necessary to eliminate all possibilities of static or stray voltage when you connect the new unit.***
3. ***Disconnect the harness and remove the attaching bolts. Then, remove the computer unit.***
4. ***Separate the multipin connector carefully from the computer itself, then remove the computer.***
5. ***Some computers have removable calibration units that must be transferred to the new computer before installation.***

Before installing the new computer, follow the step-by-step instructions for measuring the resistance between particular pins on the wiring harness. Computers rarely fail from internal problems. If any of the resistances do not match, test the subcircuits.

KEY TERMS

Catalyst efficiency monitor
Catalyst monitor sensor (CMS)
Comprehensive component monitor
Electronic throttle control (ETC)
Engine misfire monitor
Freeze frame
Fuel system monitor
Heated oxygen sensor monitor
Heated exhaust gas sensor (HEGO)
Protocol
Snapshot
Trip
Type A misfire
Type B misfire

SUMMARY

- The four major advantages of using electronic engine controls in a vehicle are better emissions, mileage, performance (power and driveability), and protection (durability).
- A typical electronic control system is made up of sensors, actuators, computer, and related wiring.
- Control loops are the cycles by which a process can be controlled by information received from input sensors, data computer processing, and out of specific commands to the output actuators.
- Most computerized engine control systems have self-diagnostic capabilities. By entering a self-test mode, the computer is able to evaluate the condition of the entire electronic engine control system, including itself.
- An OBD-II system has many monitors to check system operation, and the MIL is illuminated if vehicle emissions exceed 1.5 times the allowable standard for that model year.
- According to the guidelines of OBD-II, all vehicles should have a universal diagnostic test connector; a standard location for the DLC; a standard list of diagnostic trouble codes; a standard communication protocol; common use of scan tools on all vehicle makes and models; common diagnostic test modes; the ability to record, and store in memory, a snapshot of the operating conditions that existed when a fault occurred; and a standard glossary of terms, acronyms, and definitions must be used for all components in the electronic control systems.
- Monitors included in OBD-II are: catalyst efficiency, engine misfire, fuel system, heated exhaust gas oxygen sensor, EGR, EVAP, secondary air injection, and comprehensive component monitors.
- OBD-II vehicles use a minimum of two oxygen sensors: One is used for feedback to the PCM for fuel control and the other gives an indication of the efficiency of the converter.
- Cylinder misfire monitoring requires measuring the contribution of each cylinder to engine power.
- Fuel system monitoring checks short-term fuel trim and long-term fuel trim while the PCM is operating in closed loop.
- Heated oxygen sensor monitor checks lean to rich and rich to lean time responses.
- The EGR monitors use several different strategies to determine if the system is operating properly.
- OBD-II monitors the evaporative system: the ability of the fuel tank to hold pressure and of the purge system to vent the gas fumes from the charcoal canister.
- The AIR system operation can be verified by turning the AIR system on to inject air upstream of the oxygen sensor while monitoring its signal.
- The comprehensive monitor looks at any electronic input that could affect emissions.
- An OBD-II drive cycle includes whatever specific operating conditions are necessary either to initiate and complete a specific monitoring sequence or to verify a symptom or verify a repair.
- The OBD-II DLC must be a sixteen-terminal connector with twelve terminals defined by the SAE.
- An OBD-II system continuously monitors the entire emissions system, switches on a MIL if something goes wrong, and stores a fault code in the PCM when it detects a problem.
- OBD-III adds a telecommunication system to OBD-II.
- Each manufacturer must use the same protocol between the PCM and its sensors and actuators. The same protocol must be used to send diagnostic information to the scan tool through the DLC.
- All OBD-II systems have the same basic test modes.
- One of the mandated capabilities is the freeze frame or snapshot feature which gives the ability to record

TECH MANUAL

The following procedures are included in Chapter 33/34 of the *Tech Manual* that accompanies this book:

1. Retrieving diagnostic trouble codes (DTCs) form the computer of an engine control system.
2. Testing an ECT sensor.
3. Testing the operation of a TP sensor.
4. Testing an O_2 sensor.
5. Testing a MAP sensor.
6. Conducting a diagnostic check on an engine equipped with OBD-II.
7. Monitoring the adaptive fuel strategy on an OBD-II-equipped engine.

data from all of its sensors and actuators at a time when the system turns on the MIL.

- Compared to previous systems, the main difference in an OBD-II system is in the software contained in the PCM.
- An EEPROM may be erased and reprogrammed with the proper equipment without removing the chip.

REVIEW QUESTIONS

1. Describe the difference between an open- and closed-loop operation.
2. Explain the use and importance of system strategy and look-up tables in the computerized control system.
3. Describe an OBD-II warm-up cycle.
4. Explain trip and drive cycle in an OBD-II system.
5. Describe how engine misfire is detected in an OBD-II system.
6. Describe the purpose of having two oxygen sensors in an exhaust system.
7. Describe briefly five of the monitors in an OBD-II system.
8. Type B engine misfires are excessive if the misfiring exceeds ______% to ______% in a(n) ______ rpm period.
9. The ______ monitor system checks the action of the canister purge system.
10. The ______ monitor system has a(n) ______ and ______ test to check the efficiency of the air injection system.
11. The fuel monitor checks ______ ______ fuel trim and ______ ______ fuel trim.
12. Technician A says an oxygen sensor can be a voltage-producing sensor. Technician B says an oxygen sensor is a thermistor sensor. Who is correct?
    - **a.** Technician A
    - **b.** Technician B
    - **c.** Both A and B
    - **d.** Neither A nor B
13. Technician A says that the oxygen sensor provides the major input during the open-loop mode. Technician B says that the coolant temperature sensor controls open-loop mode operation. Who is correct?
    - **a.** Technician A
    - **b.** Technician B
    - **c.** Both A and B
    - **d.** Neither A nor B
14. A computer is capable of doing all of the following *except* ______.
    - **a.** receive input data
    - **b.** process input data according to a program and monitor output action
    - **c.** control the vehicle's operating conditions
    - **d.** store data and information
15. Which of the following memory circuits is used to store trouble codes and other temporary information?
    - **a.** read only memory
    - **b.** programmed read only memory
    - **c.** random access memory
    - **d.** all of the above
16. While discussing OBD-II systems, Technician A says the PCM illuminates the MIL is a defect causes emissions levels to exceed 2.5 times the emission standards for that model year vehicle. Technician B says if a misfire condition threatens engine or catalyst damage, the PCM flashes the MIL. Who is correct?
    - **a.** Technician A
    - **b.** Technician B
    - **c.** Both A and B
    - **d.** Neither A nor B
17. While discussing the catalyst efficiency monitor, Technician A says that if the catalytic converter is not reducing emissions properly, the voltage frequency increases on the downstream HO2S. Technician B says that if a fault occurs in the catalyst monitor system on three drive cycles, the MIL will be illuminated. Who is correct?
    - **a.** Technician A
    - **b.** Technician B
    - **c.** Both A and B
    - **d.** Neither A nor B
18. While discussing the misfire monitor, Technician A says that while detecting Type A misfires, the monitor checks cylinder misfiring over a 500-rpm period. Technician B says that while detecting Type B misfires, the monitor checks cylinder misfires over a 1,000-rpm period. Who is correct?
    - **a.** Technician A
    - **b.** Technician B
    - **c.** Both A and B
    - **d.** Neither A nor B
19. While discussing monitoring systems, Technician A says the fuel system monitor checks the short-term and long-term fuel trim. Technician B says the heated oxygen sensor monitoring system checks lean to rich and rich to lean response times. Who is correct?
    - **a.** Technician A
    - **b.** Technician B
    - **c.** Both A and B
    - **d.** Neither A nor B
20. While discussing the comprehensive monitoring system, Technician A says it tests various input circuits. Technician B says it tests various output circuits. Who is correct?
    - **a.** Technician A
    - **b.** Technician B
    - **c.** Both A and B
    - **d.** Neither A nor B

34

ON-BOARD DIAGNOSTIC SYSTEM DIAGNOSIS AND SERVICE

OBJECTIVES

■ Perform flash code diagnosis on various vehicles. ■ Perform a scan tester diagnosis on various vehicles. ■ Conduct preliminary checks on an OBD-II system. ■ Use a symptom chart to set up a strategic approach to troubleshooting a problem. ■ Define the terms associated with OBD-II diagnostics. ■ Identify the cause of an illuminated MIL. ■ Explain the basic format of OBD-II DTCs. ■ Monitor the activity of OBD-II system components. ■ Explain how to diagnose intermittent problems. ■ Diagnose computer voltage supply and ground wires. ■ Test and diagnose switch-type input sensors. ■ Test and diagnose variable resistance-type input sensors. ■ Test and diagnose generating-type input sensors. ■ Test and diagnose output devices (actuators).

For more than 20 years, a computer has played an important role in the way an engine runs. The role of the computer has evolved from the control of a single system to the control of nearly all of the engine's systems. In the process of controlling the various engine systems, the power train control module (PCM) continuously monitors operating conditions for possible system malfunctions. The computer compares system conditions against programmed parameters. If conditions fall outside the limits of these parameters, the computer detects a malfunction. A trouble code (DTC) is set to indicate the portion of the system that is at fault. A technician can access the code as an aid in troubleshooting.

If a malfunction results in improper system operation, the computer may minimize the effects by using fail-safe action. In other words, the computer may substitute a fixed value in place of the real value from a sensor to avoid shutting down the entire system. This fixed value can be programmed into the computer's memory or it can be the last received signal from the sensor prior to failure. This allows the system to operate on a limited basis instead of shutting down completely.

There are several things you need to know about before learning how to access trouble codes in a computer's memory. You need to become familiar with what you are looking at, and you must follow proper precautions when servicing these systems.

ELECTRONIC SERVICE PRECAUTIONS

A technician must take some precautions before servicing a computer or its circuit. The PCM is designed to withstand normal current draws associated with normal operation. However, overloading of the system will destroy the computer. To prevent damage to the PCM and its related components, follow these service precautions:

1. Never ground or apply voltage to any controlled circuit unless the service manual instructs you to do so.
2. Use only a high-impedance multimeter (10 megaohms or higher) to test the circuits. Never use a test light unless instructed to do so in the manufacturer's procedures.
3. Make sure the ignition switch is turned off before disconnecting or connecting electrical terminals at the PCM.
4. Unless instructed otherwise, turn off the ignition switch before disconnecting or connecting any electrical connections to sensors or actuators.
5. Turn the ignition switch off whenever disconnecting or connecting the battery terminals. Also turn it off when replacing a fuse.

6. Do not connect any electrical accessories to the insulated or ground circuits of computer-controlled systems.
7. Use only the manufacturer's specific test and replacement procedures for the year and model of the vehicle being serviced.

WARNING!

Remember all computer system components are sensitive to electrostatic discharge. Some manufacturers mark certain components and circuits with a code or symbol to warn technicians that they are sensitive to electrostatic discharge ***(Figure 34–1)****. Static electricity can destroy or render a component useless.*

When handling any electronic part, especially those that are static sensitive, follow the precautions explained earlier to reduce the possibility of electrostatic buildup on your body and the inadvertent discharge to the electronic part. If you are not sure if a part is sensitive to static, treat it as if it is.

BASIC DIAGNOSIS OF ELECTRONIC ENGINE CONTROL SYSTEMS

Diagnosing a computer-controlled system involves much more than accessing the DTCs in the computer's memory. As is true during diagnosis of any system, you need to know what to test, when to test it, and how to test it. Because the capabilities of the engine control computer have evolved from simple to complex, it is important to know the capabilities of the system you are working with before attempting to diagnose a problem. Refer to the service manual for this information. After you understand the system and its capabilities, begin your diagnosis, using your knowledge and logic.

The importance of logical troubleshooting cannot be overemphasized. The ability to diagnose a problem (to find its cause and its solution) is what separates an automotive technician from a parts changer.

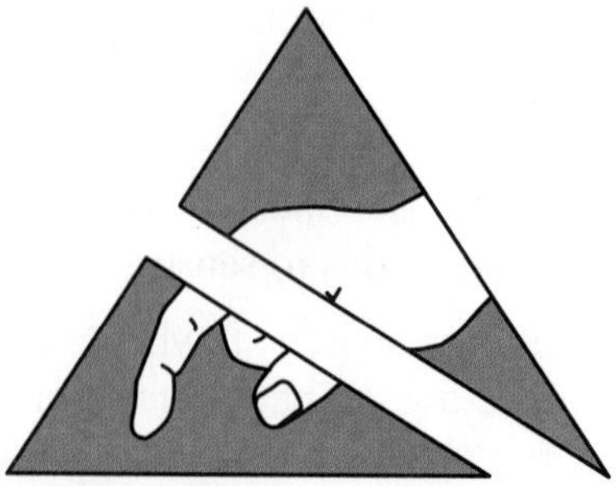

Figure 34–1 GM's Electrostatic Discharge (ESD) symbol that warns a technician that the component or circuit is sensitive to static.

Logical Diagnosis

When faced with an abnormal engine condition, the best automotive technicians compare clues (such as meter readings, oscilloscope readings, visible problems) with their knowledge of proper conditions and discover a logical reason for the way the engine is performing. Logical diagnosis means following a simple basic procedure. Start with the most likely cause and work to the most unlikely. In other words, check out the easiest, most obvious solutions first before proceeding to the less likely, and more difficult, solutions. Do not guess at the problem or jump to a conclusion before considering all of the factors.

The logical approach has a special application to troubleshooting electronic engine controls. Always check all traditional nonelectronic engine control possibilities before attempting to diagnose the electronic engine control itself. For example, something as simple as low battery voltage might be the cause of faulty sensor readings.

Isolating Computerized Engine Control Problems

Determining which part or area of a computerized engine control system is defective requires having a thorough knowledge of how the system works and following a logical troubleshooting process.

Electronic engine control problems are usually caused by defective sensors and, to a lesser extent, output devices. The logical procedure in most cases is, therefore, to check the input sensors and wiring first, then the output devices and their wiring, and, finally the computer.

Most late-model computerized engine controls have self-diagnosis capabilities. A malfunction in any sensor, output device, or in the computer itself is stored in the computer's memory as a trouble code. Stored codes can be retrieved and the indicated problem areas checked further.

Systems without self-diagnostics must be checked out by the process of elimination. All possible causes of the problem are tested. If the fault is not found, the components that cannot be tested may be malfunctioning or faulty.

These methods can be used to check individual system components:

- *Visual Checks*—Look for obvious problems. Any part that is burned, broken, cracked, corroded, or has any other visible problem must be replaced before continuing the diagnosis. Examples of visible problems include disconnected sensor vacuum hoses and broken or disconnected wiring.
- *Ohmmeter Checks*—Most sensors and output devices can be checked with an ohmmeter. For example, an ohmmeter can be used to check a temperature sensor.

Normally the ohmmeter reading is low on a cold engine and high or infinity on a hot engine, if the sensor is a positive temperature coefficient (PTC). If the sensor is a negative temperature coefficient (NTC), the opposite readings would be expected. Output devices such as coils or motors can also be checked with an ohmmeter.

- *Voltmeter Checks*—Many sensors, output devices, and their wiring can be diagnosed by checking the voltage to them, and in some cases, from them. Even some oxygen sensors can be checked in this manner.
- *Lab Scope Checks*—The activity of sensors and actuators can be monitored with a lab scope. By watching their activity, you are doing more than testing them. Often problems elsewhere in the system cause a device to behave abnormally. These situations are identified by the trace on a scope and by the technician's understanding of a scope and the device being monitored.

In some cases, a final check on the computer can be made only by substitution. Also, some vacuum, MAP, and barometric pressure sensors can only be checked by substitution. Substitution is not an allowable diagnostic method under the mandates of OBD-II. Nor is it the most desirable way to diagnose problems. However, to substitute, replace the suspected part with a known good unit and recheck the system. If the system now operates normally, the original part must have been defective.

Service Bulletin Information

When diagnosing engine control system problems, technical service bulletin (TSB) information is absolutely essential. If a technician does not have service bulletin information, many hours of diagnostic time may be wasted. This information is available from different suppliers of CD-ROM and paper TSBs. The following are some difficult problems that are only quickly discovered by using a TSB or by having much experience with similar problems and their causes.

Example 1—Many fuel-injected General Motors' engines are equipped with Multec injectors. Some of these injectors experienced shorting problems in the windings, especially if the fuel contained some alcohol. If the injectors become shorted, they draw excessive current. General Motors' P4 PCMs have a sense line connected to the quad driver that operates the injectors. When this sense line experiences excessive current flow from the shorted injectors, the quad driver shuts off and stops operating the injectors. This action protects the quad driver, but also causes the engine to stall. After a few minutes, the engine will usually restart. If a technician does not have this information available in a service bulletin, he or she may waste a great deal of time locating the problem.

Example 2—On some General Motors' computers, the pins on the internal components extended through the circuit board tracks, and soldering was done on the opposite side of the board from where the components were located. For these PCMs, surface mount technology was developed in which the pins were bent at a 90-degree angle and then soldered on top of the tracks on the circuit boards. In some cases, loose connections developed in these surface-mounted computers. These loose connections usually caused the engine to quit. If a technician suspects this problem, the PCM is removed with the wiring harness connected. The engine is started and the PCM is given a slap with the palm of a hand. If the engine stalls or engine operation changes, a loose connection is present on the circuit board. When a technician does not have this information available in service bulletins, much diagnostic time may be wasted.

Example 3—In 1991, Chrysler experienced some low-speed surging during engine warm-up on 3.3L and 3.8L engines. On these engines, the port fuel injectors sprayed against a hump in the intake port. As a result, fuel puddled behind this hump, especially while the engine was cold. When the engine temperature increased, this fuel evaporated and caused a rich air/fuel ratio and engine surging. Chrysler corrected this problem by introducing angled injectors with the orifices positioned at an angle so the fuel sprayed over the hump in the intake. When angled injectors are installed, the wiring connector must be positioned vertically. Angled injectors have beige exterior bodies. Technicians must have service bulletin information regarding problems like this.

SELF-DIAGNOSTIC SYSTEMS

Most computerized engine controls have self-diagnostic capabilities. By entering a self-test mode, the computer is able to evaluate the condition of the entire electronic engine control system, including itself. If problems are found, they are identified as either hard faults (on-demand) or intermittent failures. Each type of fault or failure is assigned a numerical trouble code that is stored in the computer's memory.

A hard fault means a problem has been found somewhere in the system at the time of the self-test. An intermittent problem indicates a malfunction occurred (for example, a poor connection causing an intermittent open or short), but is not present at the time of the self-test. Nonvolatile RAM allows intermittent faults to be stored for up to a specific number of ignition key on/off cycles. If the trouble does not reappear during that period, it is erased from the computer's memory.

There are various methods of assessing the trouble codes generated by the computer. Most manufacturers have diagnostic equipment designed to monitor and test the electronic components of their vehicles. Aftermarket companies also manufacture scan tools that have the capability to read and record the input and output signals passing to and from the computer.

Visual Inspection

Before looking at trouble codes, do a visual check of the engine and its systems. This quick inspection can save much time during diagnosis. While visually inspecting the vehicle, make sure to include the following:

1. Inspect the condition of the air filter and related hardware around the filter.
2. Inspect the entire PCV system.
3. Check to make sure the EVAP canister is not saturated or flooded.
4. Check the battery and its cables, the vehicle's wiring harnesses, connectors, and the charging system for loose or damaged connections. Also, check the connectors for signs of corrosion.
5. Check the condition of the battery and its terminals and cables.
6. Make sure all vacuum hoses are connected and are not pinched or cut.
7. Check all sensors and actuators for signs of physical damage.

Unlocking Trouble Codes

Although the parts in any computerized system are amazingly reliable, they do occasionally fail. Diagnostic charts in service manuals **(Figure 34–2)** help you through troubleshooting procedures in the proper order. Start at the top and follow the sequence down. There will be branches of the tree—yes or no, on or off, ok or not ok—to follow after making the check required in each step.

Two things would tell the driver there is a problem with some part of the computer system: the MIL (check engine, power loss, or service engine soon light) or the car simply does not start.

The malfunction indicator lamp (MIL) comes on normally for a few seconds when the key is turned on and during cranking. It should go off when the engine starts. The check engine light should not come back on unless the computer finds a problem. If it does come on, there is probably a trouble code stored in memory.

Self-diagnostic procedures and diagnostic trouble codes differ depending on vehicle make and year. Each time the key is turned to the on position, the system does a self-check. The self-check makes sure that all of the bulbs, fuses, and electronic modules are working. If the self-test finds a problem, it might store a code for later servicing. It may also instruct the computer to turn on the MIL to show that service is needed.

Using a Scan Tool

Scan tools are available to diagnose nearly all engine control systems. When test procedures are performed with a scan tester, these precautions must be observed:

1. Always follow the directions in the manual supplied by the scan tester manufacturer.
2. Do not connect or disconnect any connectors or components with the ignition switch on, including the scan tester power wires and the connection from the tester to the vehicle diagnostic connector.
3. Never short across or ground any terminals in the electronic system except those recommended by the vehicle manufacturer.
4. If the computer terminals must be removed, disconnect the scan tester diagnostic connector first.

Photo Sequence 30 shows a typical procedure for using a scanner in diagnostics. Scan tester operation varies depending on the make of the tester, but a typical example of initial entries includes the vehicle year and VIN code.

After the scanner has been programmed with the initial entries, some entry options appear on the screen. The technician presses the number beside the desired selection to proceed with the test procedure. After the test has been selected, the scanner moves on to the actual test selections. These selections vary depending on the scan tester and the vehicle being tested.

Snapshot Testing Many scan testers have snapshot capabilities for some vehicles that allow the technician to operate the vehicle and store the captured sensor voltage readings in the tester's memory when a problem occurs. When the vehicle has been driven back to the shop, the technician can play back the recorded sensor readings. During the play back, the technician watches closely for a momentary change in any sensor reading, which indicates a defective sensor or wire. This action is similar to taking a series of sensor reading "snapshots" and then reviewing the pictures later. Snapshots may be taken on most vehicles with a data line from the computer to the data link connector (DLC).

RETRIEVING TROUBLE CODES

Prior to OBD-II, each car manufacturer required a different method for retrieving stored DTCs. In fact, there were different procedures for the different models and engines made by the same manufacturer. The diagnostics were located in different places and the diagnostic codes represented different things. Although all of the computer systems did basically the same thing, diagnostic methods were very different. What follows are typical procedures

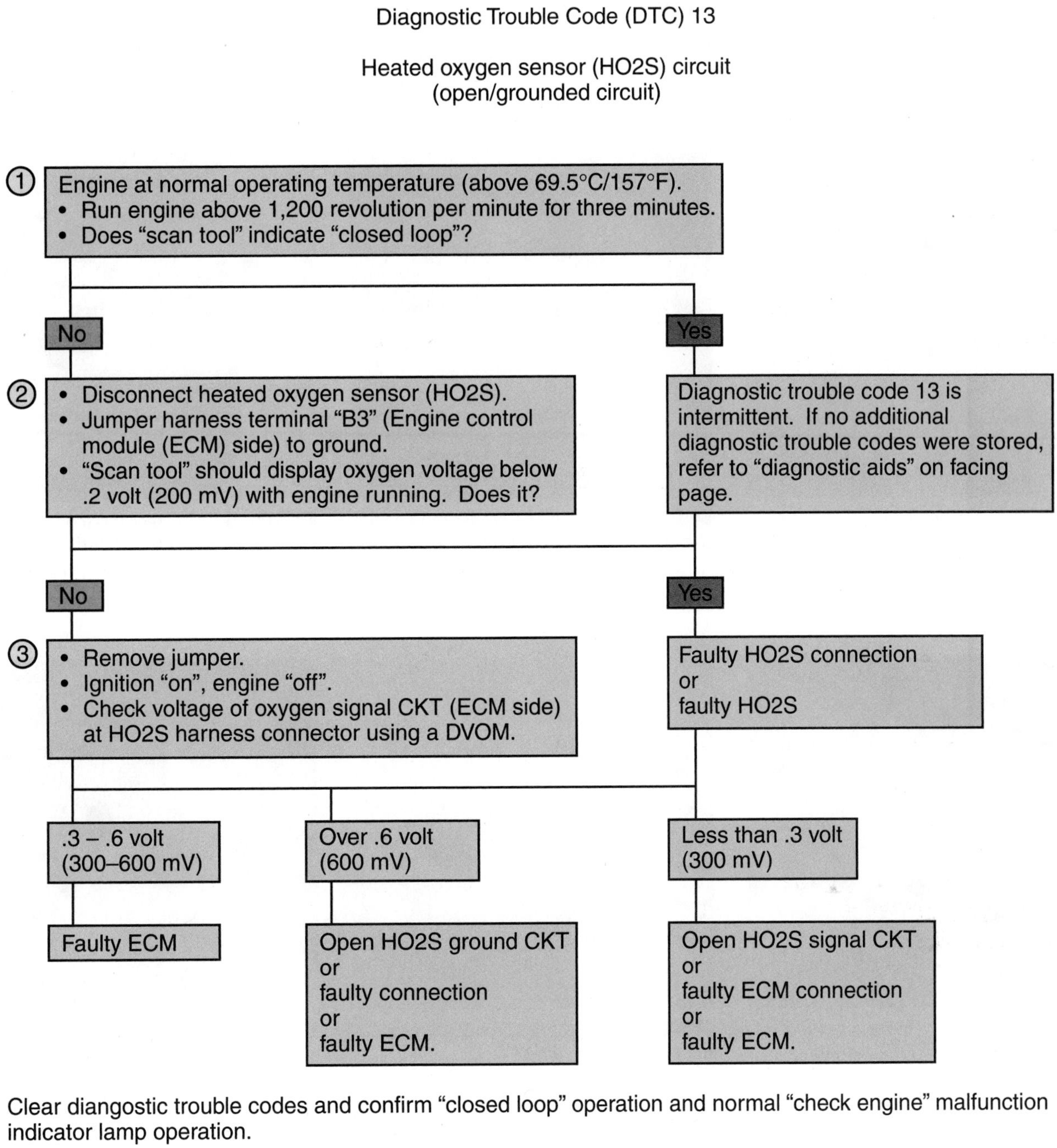
Diagnostic Trouble Code (DTC) 13
Heated oxygen sensor (HO2S) circuit (open/grounded circuit)
1
Engine at normal operating temperature (above 69.5°C/157°F).
• Run engine above 1,200 revolution per minute for three minutes.
• Does "scan tool" indicate "closed loop"?
No
Yes
2
• Disconnect heated oxygen sensor (HO2S).
• Jumper harness terminal "B3" (Engine control module (ECM) side) to ground.
• "Scan tool" should display oxygen voltage below .2 volt (200 mV) with engine running. Does it?
Diagnostic trouble code 13 is intermittent. If no additional diagnostic trouble codes were stored, refer to "diagnostic aids" on facing page.
No
Yes
3
• Remove jumper.
• Ignition "on", engine "off".
• Check voltage of oxygen signal CKT (ECM side) at HO2S harness connector using a DVOM.
Faulty HO2S connection or faulty HO2S
.3 – .6 volt (300–600 mV)
Over .6 volt (600 mV)
Less than .3 volt (300 mV)
Faulty ECM
Open HO2S ground CKT or faulty connection or faulty ECM.
Open HO2S signal CKT or faulty ECM connection or faulty ECM.
Clear diangostic trouble codes and confirm "closed loop" operation and normal "check engine" malfunction indicator lamp operation.

Figure 34-2 A diagnostic chart to be followed for a specific trouble code. *Courtesy of American Isuzu Motors, Inc.*

for retrieving DTCs from some popular non-OBD-II vehicles.

General Motors Vehicles

The main components of General Motors' Computer Command Control (CCC) diagnostic system include an electronic control module or power train control module (PCM), a MIL (Check Engine or Service Engine Soon warning light), and an assembly line communications link (ALCL) or DLC. The MIL was a signal to the driver that a detectable CCC failure had occurred. The MIL was also used to display DTCs.

If a problem was detected by the CCC system, a DTC was set in the PCM's memory and the MIL was illuminated. If the fault disappears, the MIL turns off, but a DTC may be stored in memory. Some problems may cause the system to go into the limp-in mode. In this mode, the PCM provides a rich air/fuel ratio and a fixed spark advance. Therefore, engine performance and economy decrease and emission levels increase, but the vehicle may be driven to a service center.

To retrieve DTCs, begin by performing a diagnostic circuit check to make sure the diagnostic system is working. Turn the ignition key on, but do not start the engine.

PHOTO SEQUENCE 30 Diagnosing with a Scan Tool

P30-1 *Be sure the engine is at normal operating temperature and the ignition switch is off.*

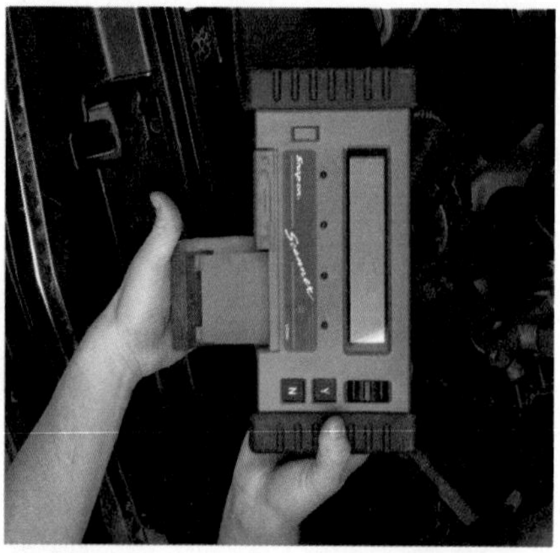

P30-2 *Install the proper module for the vehicle and system into the scan tool.*

P30-3 *Connect the scan tool power leads to the battery or cigar lighter (depending on the design of the scan tool).*

P30-4 *Enter the vehicle's model year and VIN code into the scan tool.*

P30-5 *Select the proper scan tool adapter for the vehicle's DLC.*

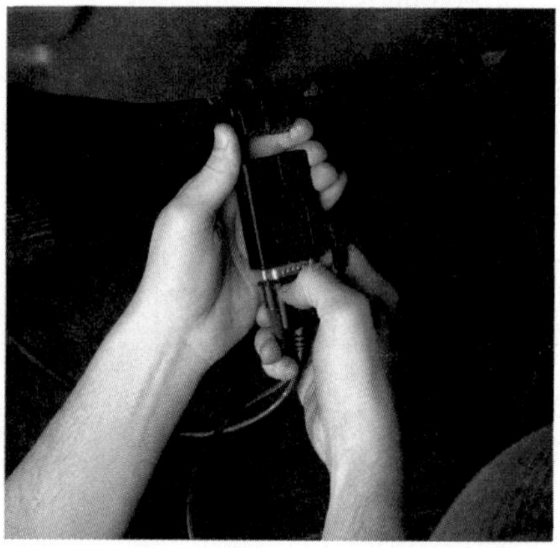

P30-6 *Connect the scan tool to the DLC.*

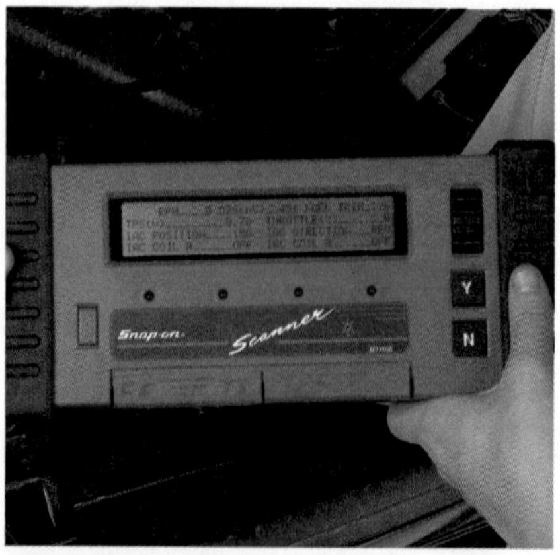

P30-7 *Retrieve the DTCs with the scan tool. Interpret the codes by using the service manual.*

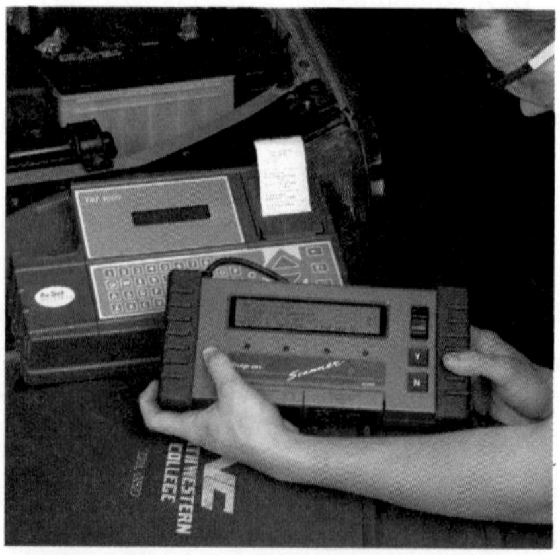

P30-8 *Start the engine and obtain the input sensor and output actuator data on the scan tool. If a printer for the tool is available, print out the data report.*

P30-9 *Compare the input sensor and out actuator data to the specifications given in the service manual. Mark all data that is not within specifications.*

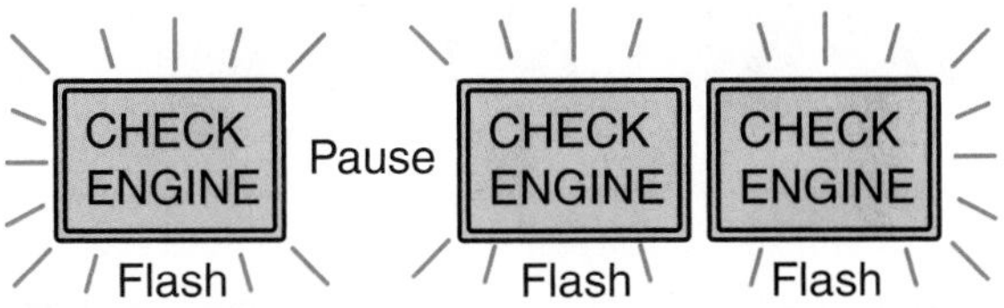

Figure 34-3 The MIL signals the codes.

If the check engine light does not come on, follow the diagnostic chart in the service manual. If the light comes on, connect a jumper wire across terminals A (the ground) and B (the test terminal) in the DLC. Watch the MIL to see if it displays code 12. A code 12 is one flash, a short pause, and two flashes **(Figure 34–3)**. If the computer's diagnostic program is working properly, it flashes code 12 three times. The system then displays any fault codes in the same manner. If code 12 does not appear, there is something wrong in the DLC or related circuits.

Each code is repeated three times. When more than one code is present, the codes are given in numerical order. Code 12, which indicates that the PCM is capable of diagnosis, is given first. The DTC sequence continues to repeat until the ignition switch is turned off.

If the A and B terminals are connected at the DLC and the engine is started, the PCM enters the field service mode. In this mode, the speed at which the MIL flashes indicates whether the system is in open loop or closed loop. If the system is in open loop, the MIL flashes quickly. When the system enters closed loop, the MIL flashes at half the speed of the open-loop flashes.

Once the codes have been identified, check the service manual. Each code directs you to a specific troubleshooting tree. If you have a code, follow the tree exactly. Never try to skip a step, or you will be certain to miss the problem.

A hand-held scan tool can be used by itself or in conjunction with the MIL. When using a scan tool, always follow the manufacturer's instructions.

After correcting all faults, clear the PCM memory of any current codes by pulling the PCM fuse at the fuse panel for ten seconds. Then, to make sure the codes have cleared, remove the test terminal ground and set the parking brake. Put the transmission in park, and run the engine for two to five minutes. Watch the check engine light. If the light comes on again, ground the test lead and note the flashing trouble code. It is a good idea to take the vehicle on a road test. Some parts such as the vehicle speed sensor do not show any problems with the engine just idling.

Ford Motor Company Vehicles

The PCM of Ford's EEC-IV system has self-diagnostic capabilities. By entering the self-test mode, the computer evaluates the condition of the entire system. If problems are found, they are recorded as being either hard faults (on demand) or intermittent failures. On non-OBD-II vehicles, the DTCs are typically displayed by the MIL.

The Ford self-test diagnostic procedure can be divided into four parts.

1. **Key on/engine off (KOEO)**: checks system inputs for hard faults (malfunctions that occur during the self-test) and intermittent faults (malfunctions that occurred sometime prior to the self-test and were stored in memory).
2. Computed ignition timing check: determines the PCM's ability to advance or retard ignition timing.
3. Engine running segment—**key on, engine running (KOER)**: checks system output for hard faults only.
4. Continuous monitoring test (wiggle test): allows the technician to look for and set intermittent faults while the engine is running.

Photo Sequence 31 goes through a flash code procedure for a Ford product with a MIL. A jumper wire must be connected from the self-test input wire to the appropriate DLC terminal to enter the self-test mode. When the ignition switch is turned on after this jumper wire connection, the MIL begins to flash any DTCs in the PCM memory.

Within these four tests, there are six types of service codes: on-demand codes, keep-alive memory codes, separator codes, dynamic response codes, fast codes, and engine I.D. codes. On-demand codes are used to identify hard faults. A hard fault means that a system failure has been detected and is still present at the time of testing.

Keep-alive memory codes mean that a malfunction was noted sometime during the last twenty vehicle warm-ups but is not present now (if it were, it would be recorded as a hard fault). The continuous memory code comes on after an approximate ten-second delay. Make the on-demand code repairs first. Once you have completed repairs, repeat the KOEO test. If all the parts are repaired correctly, a pass code of 11 should be received.

A separator code (10) indicates that the on-demand codes are over and the memory codes are about to begin. The separator code only occurs as part of the KOEO self-test.

When a code 10 appears during the engine running segment of the self-test, it is referred to as a dynamic response code. The dynamic response code is a signal to the technician to goose the throttle momentarily so that the PCM can verify the operation of the TP and MAP sensors. On some models, the technician must goose the throttle, step on the brake pedal, and turn the steering wheel 180 degrees. Failure to respond to the dynamic code within 15 seconds after it appears sets a code 77 (operator did not respond).

Fast codes are of no value to the service technician. They are designed for factory use and are transmitted about

PHOTO SEQUENCE 31

Performing KOEO and KOER Tests

P31-1 *Make sure the engine is at normal operating temperature and the ignition switch is off.*

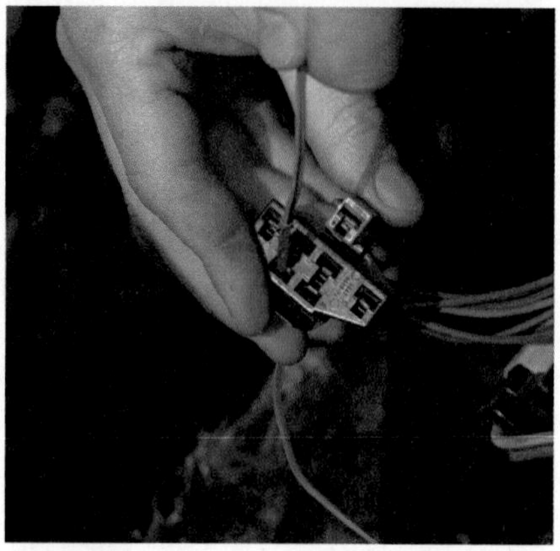

P31-2 *Connect a jumper wire to the proper terminals in the DLC.*

P31-3 *Turn the ignition switch on.*

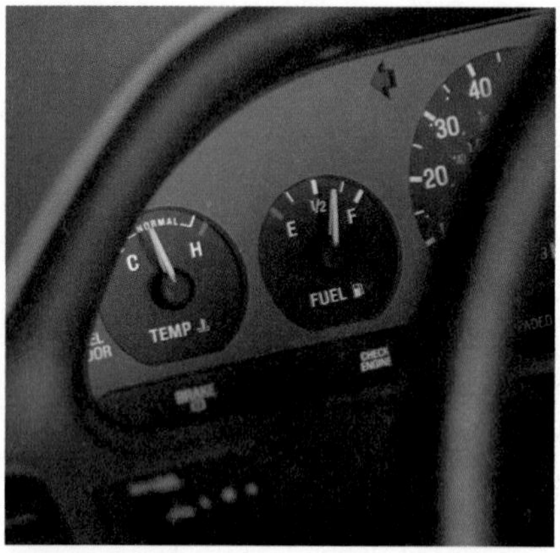

P31-4 *Observe the MIL flashes. The sequence of codes in the KOEO test is hard faults, separator code (10), and intermittent fault codes. Record all codes.*

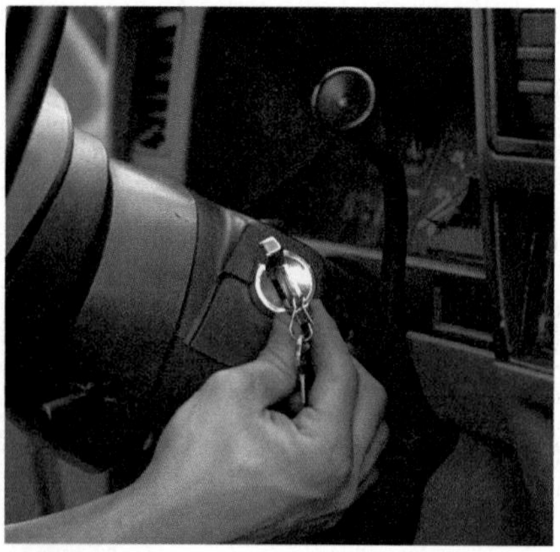

P31-5 *Turn the ignition off and wait ten seconds.*

P31-6 *Start the engine and observe the MIL flashes. The sequence of codes in the KOER test is engine ID code, separator code (10), and hard faults. Record all codes.*

P31-7 *Compare all codes to the fault interpretation chart in the service manual. Determine what tests need to be conducted to find the cause of the problem(s).*

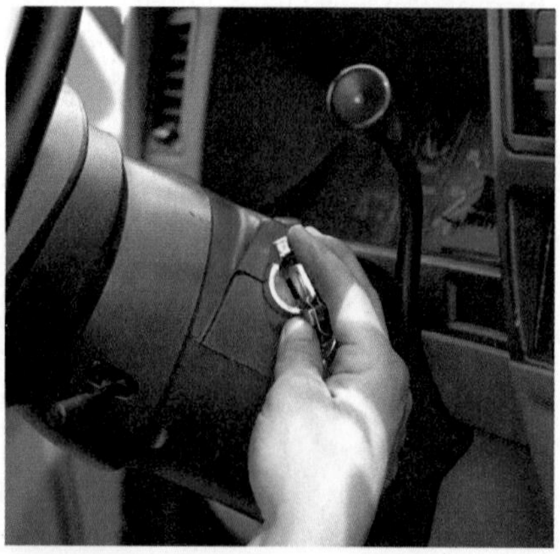

P31-8 *Turn the ignition off and wait ten seconds.*

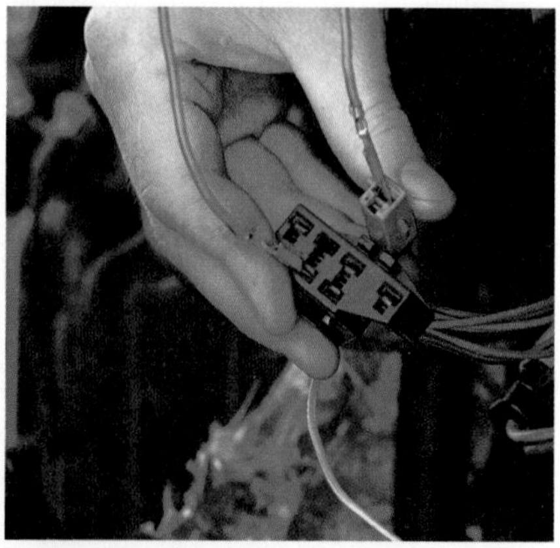

P31-9 *Erase the faults by turning the ignition on and removing the jumper wire at the DLC while the MIL is flashing in the KOEO mode.*

100 times faster than a scan tester can read. Although fast codes have no practical use in the service bay, pay attention to when they occur so you know what is coming next. Fast codes appear twice during the entire self-test: once at the very beginning of the KOEO test (right before the on-demand codes) and again after the dynamic response code.

Engine identification codes appear at the beginning of the engine-running segment only and are used to tell automated assembly line equipment how many cylinders the engine has.

Chrysler Corporation Vehicles

The PCM of early non-OBD-II Chrysler systems is called the logic module (LM). The logic module operates in conjunction with a subordinate control unit called the power module (PM). The power module, located inside the engine compartment, controls the injector's and ignition coil's ground circuit (based on the LM's commands). It also supplies the ground to the automatic shutdown (ASD) relay, which controls the voltage supply to the fuel pump, logic module, injector, and coil drive circuits.

Later non-OBD-II systems have a single unit called the Single Board Electronic Controller (SBEC) that actually consists of the logic and power modules. On these systems, the LM and PM still have the same primary purposes as they did when they were in separate units.

Chrysler has a simple method for checking trouble codes. Without starting the engine, turn the ignition key on and off three times within five seconds, ending with it on. The MIL glows a short time to test the bulb, then starts flashing. Count the first set of flashes as tens. There is a half-second pause before the light starts flashing again. This time, count by ones. Add the two sets of flashes together to obtain the trouble code. For example, three flashes, a half-second pause, followed by five flashes would be read as code 35. Watch carefully, because each trouble code is displayed only once. Look the code up in the service manual to determine which circuit to check. Once the trouble codes are flashed, the computer signals a code 55. If the light does not flash at all, there are no trouble codes stored.

A scan tool **(Figure 34–4)** can also be used to pull the codes. Connect the tool to the diagnostic connector on the left fender apron. Follow the same sequence with the ignition key. The trouble codes are displayed on the scan tool readout. You can also check the circuits, switches, and relays with the scan tool. The service manual contains the troubleshooting trees to continue your diagnosis of the problem.

Some DaimlerChrysler O_2 feedback systems do not have a MIL. These systems must be diagnosed with a scan tool.

To erase codes on early systems, disconnect the quick-disconnect connector at the positive battery cable for ten seconds with the ignition switch off. On later model systems, this connector must be disconnected for thirty minutes to erase fault codes.

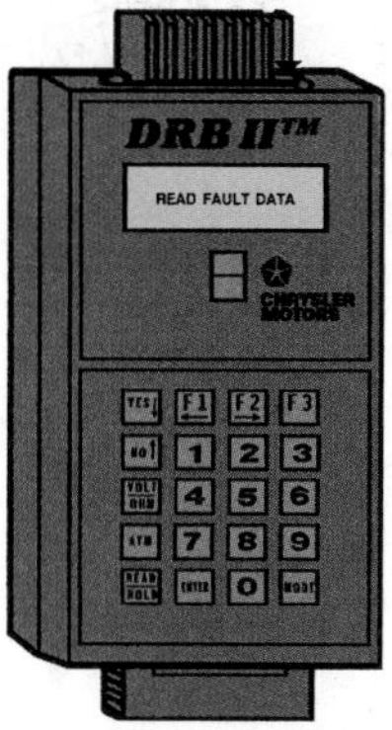

Figure 34–4 Chrysler's DRB-II scan tool. *Courtesy of DaimlerChrysler Corporation*

Toyota Vehicles

To retrieve flash out codes from most non-OBD-II compliant Toyota vehicles, turn the ignition switch on and connect a jumper wire between terminals E1 and TE1 in the DLC. Some round DLCs are located under the instrument panel **(Figure 34–5)**, while the rectangular-shaped DLCs are positioned in the engine compartment. Now, observe the MIL flashes. If the light flashes on and off at 0.26-second intervals, there are no DTCs in the computer memory. If there are DTCs in memory, the MIL flashes out the DTCs in numerical order. The codes will be repeated as long as terminals E1 and TE1 are connected and the ignition switch is on. After all codes have been retrieved, remove the jumper wire from the DLC to end testing.

To identify the cause of intermittent faults, this system can be set into a drive test mode. This mode should only be used after any problems causing a hard code have been identified and repaired. Turn on the ignition switch and connect terminals E1 and TE2 in the DLC. Then, start the engine and drive the vehicle at speeds above 6 mph (10 km/h). Simulate the conditions when the problem occurs. Now, stop the engine, connect a jumper wire between terminals E1 and TE1 on the DLC, and observe the flashes of the MIL to read the DTCs. Remove the jumper wire from the DLC to end the testing.

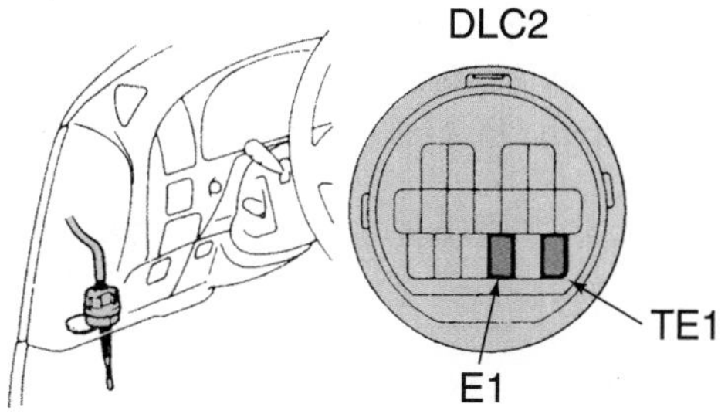

Figure 34–5 Toyota's round DLC. *Reprinted with permission*

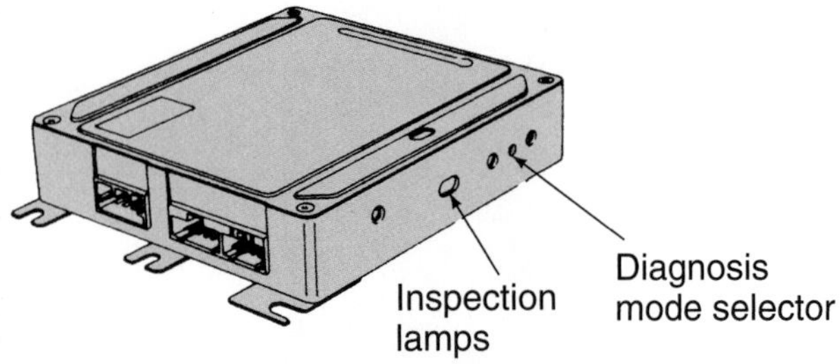

Figure 34-6 The diagnosis mode selector on a Nissan PCM. *Used with permission from Nissan North America, Inc.*

Nissan Vehicles

On some Nissan Electronic Concentrated Engine Control Systems (ECCS), the PCM has two LEDs that flash DTCs. One LED is red and the other is green. The flashing pattern of the two LEDs indicates the DTC. If there are no DTCs, the LEDs flash a system pass code.

To observe the codes, locate and remove the PCM with the wiring harness connected to it. Turn the diagnosis mode selector on the PCM clockwise to enter into the diagnostic mode **(Figure 34–6)**.

The diagnostic modes available on some Nissan products are:

- Mode 1 checks the oxygen sensor signal. With the system in closed loop and the engine idling, the green light should flash on each time the oxygen sensor detects a lean condition. This light goes out when the oxygen sensor detects a rich condition. After five to ten seconds, the PCM "clamps" on the ideal air/fuel ratio and pulse width, and the green light may be on or off. This PCM clamping of the pulse width only occurs at idle speed.
- In Mode 2, the green light comes on each time the oxygen sensor detects a lean mixture, and the red light comes on when the PCM receives this signal and makes the necessary correction in pulse width.
- Mode 3 provides DTCs representing various defects in the system.
- Mode 4 tests switch inputs to the PCM. This mode cancels codes available in Mode 3.
- Mode 5 increases the diagnostic sensitivity of the PCM for diagnosing intermittent faults while the vehicle is driven on the road.

After the defect is corrected, turn off the ignition switch, rotate the diagnosis mode selector counterclockwise, and install the PCM securely in the original position.

OBD-II SYSTEM DIAGNOSIS AND SERVICE

OBD-II regulations require that the PCM monitor and perform some continuous (active) tests on the engine control system and components. Some OBD-II tests are completed at random, at specific intervals, or in response to a detected fault.

To perform the new strategies and tests on the control system, OBD-II PCMs have diagnostic management software. The many diagnostic steps and tests required of OBD-II systems must be performed under specific operating conditions. The PCM's software organizes and prioritizes the diagnostic routines. The software determines if the conditions for running a test are present. Then it monitors the system for each test and records the results of these tests.

The PCM supplies a buffered low voltage to various sensors and switches. The input and output devices in the PCM include analog to digital converters, signal buffers, counters, and special drivers. The PCM controls most components with electronic switches that complete a ground circuit when turned on. These switches are arranged in groups of four and seven and are called one of the following: quad driver module or output driver module. The quad driver module can independently control up to four output terminals. The output driver module can independently control up to seven outputs.

The PCM has a learning ability that allows the module to make corrections for minor variations in the fuel system in order to improve driveability. Whenever the battery cable is disconnected, the learning process resets. The driver may note a change in the vehicle's performance. In order to allow the PCM to relearn, drive the vehicle at part throttle with moderate acceleration.

EEPROM modules are soldered into the PCM. EEPROMs allow the manufacturer to update what is held in the PROM without replacing it. The PCM checks its internal circuits continuously for integrity. Likewise, it checks its EEPROM for accuracy of its data. It checks its files against what they are supposed to be and sets a code if they are different. Besides the hard-wired memory chip, the PCM also monitors its volatile keep-alive memory. If that has been improperly changed or deleted, the PCM sets a code. This type of code will also be set if the vehicle's battery has been disconnected.

WARNING!

The PCM is designed to withstand normal current draws associated with vehicle operation. Avoid overloading any circuit. When testing for opens or shorts, do not ground any of the PCM circuits unless instructed to do so. When testing for opens or shorts, do not apply voltage to any of the control module circuits unless instructed. Only test these circuits with a lab scope or digital voltmeter ***(Figure 34–7)****, while the PCM remains connected.*

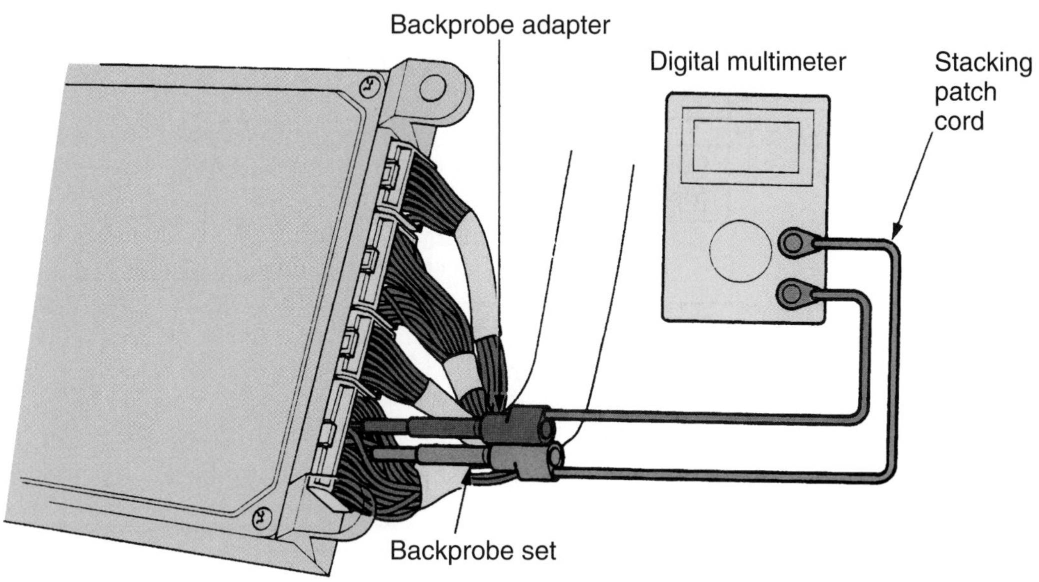

Figure 34-7 Using a digital voltmeter to check the PCM's circuit. *Courtesy of American Honda Motor Co., Inc.*

There is continuous self-diagnosis on certain control functions. This diagnostic capability is complemented by the diagnostic procedures contained in the service manual. The system monitoring diagnostic sequence is a unique segment of the software designed to coordinate and prioritize the diagnostic procedures as well as define the protocol for recording and displaying their results.

The diagnostic tables **(Figure 34–8)** and functional checks given in service manuals are designed to locate a faulty circuit or component through a process of logical decisions. The tables are prepared with the assumption that the vehicle functioned correctly at the time of assembly and that there are not multiple faults present.

OBD-II DIAGNOSTICS

Diagnosing and servicing OBD-II systems is much the same as diagnosing and servicing any other electronic engine control system except OBD-II systems give more useful information than the previous designs and, with OBD-II, each make and model vehicle does not have its own basic testing and servicing procedures. General testing of the engine and its major systems are done in the same way on OBD-II systems as is done on non-OBD systems. Testing of individual components is important in diagnosing OBD-II systems. Although the DTCs in these systems give much more detail on the problems, the PCM does not know the exact cause of the problem. That is the technician's job.

OBD-II systems note the deterioration of certain components before they fail, which allows owners to bring in their vehicles at their convenience and before it is too late. OBD-II monitors the vehicle's exhaust gas recirculation and fuel systems, oxygen sensors, catalytic converter performance, and miscellaneous other components. OBD-II diagnosis is best done with a strategy-based procedure that is based on the flow charts and other information given in the service manuals. Before beginning to diagnose a problem, verify the customer's complaint. In order to verify the complaint, a technician must know the normal operation of the system. Compare what the vehicle is doing to what it should be doing. Sometimes there is another problem or symptom that the customer is not aware of or has disregarded. This symptom may be the key to properly diagnosing the system.

Visual Inspection

The next step in diagnosis is a careful visual inspection. Make sure all of the grounds, including the battery and computer ground, have clean and tight connections. If any connection is suspect, do a voltage drop test across all related ground circuits. The PCM and its systems cannot function correctly with a bad ground.

If the visual inspection did not identify the source of the customer's complaint, proceed with gathering as much information as you can about the symptom. This should include a review of the vehicle's service history. Check for any published TSBs relating to the exhibited symptoms, including videos, newsletters, and any electronically transmitted media. Do not depend solely on the diagnostic tests run by the PCM. A particular system may not be supported by one or more DTCs.

WARNING!

Never use a test light to diagnose power train electrical systems unless specifically instructed by the diagnostic procedures.

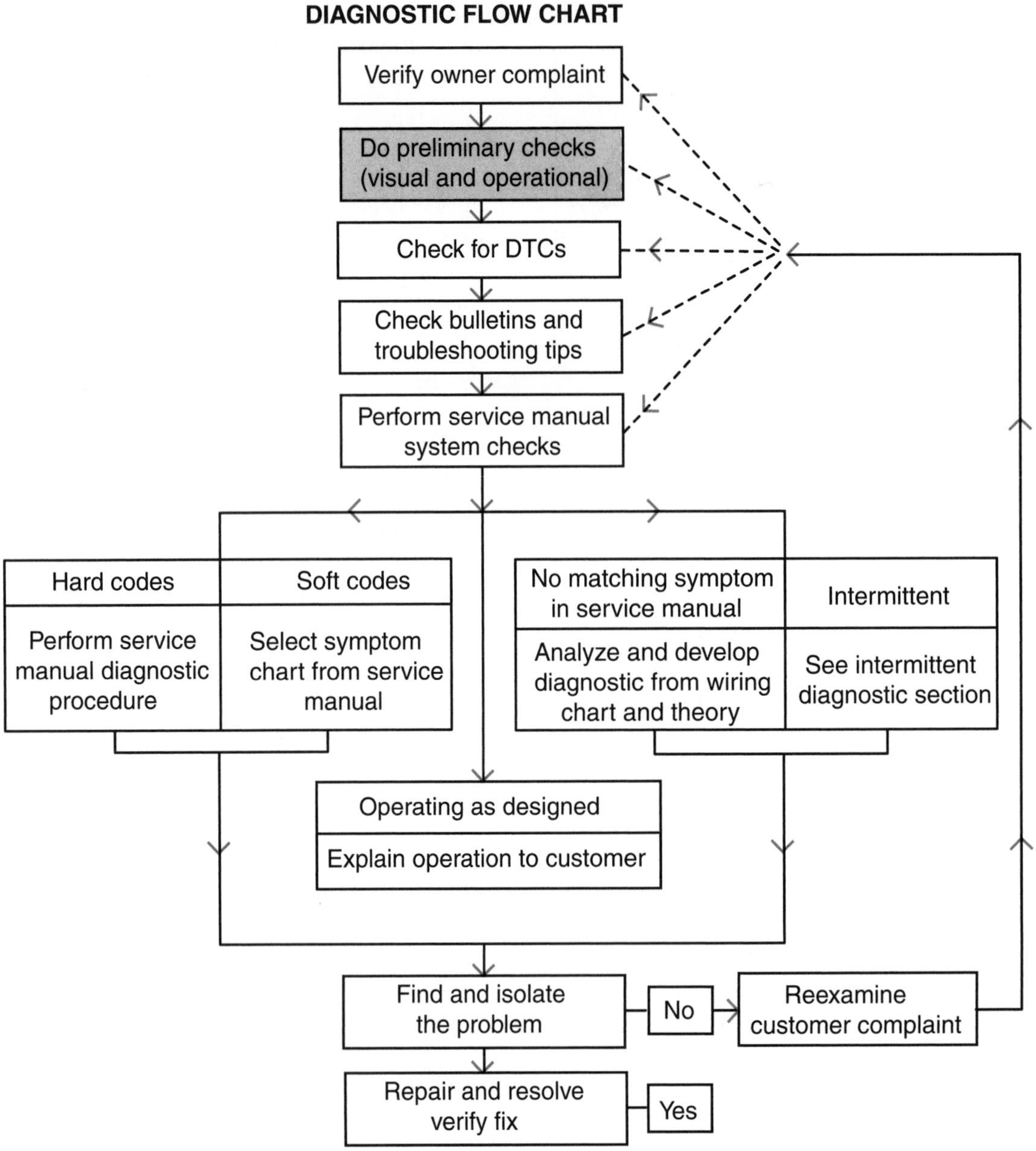
DIAGNOSTIC FLOW CHART
Verify owner complaint
Do preliminary checks (visual and operational)
Check for DTCs
Check bulletins and troubleshooting tips
Perform service manual system checks
Hard codes
Perform service manual diagnostic procedure
Soft codes
Select symptom chart from service manual
No matching symptom in service manual
Analyze and develop diagnostic from wiring chart and theory
Intermittent
See intermittent diagnostic section
Operating as designed
Explain operation to customer
Find and isolate the problem
No
Reexamine customer complaint
Repair and resolve verify fix
Yes

Figure 34-8 A strategy-based diagnostic tree.

Diagnosis should continue with preliminary system checks. The purpose of these checks is to make sure the basic engine and its systems are in good shape. This includes checking for vacuum leaks, checking the engine's compression, ignition system, and air/fuel system.

Scan Tool Diagnosis

When the ignition is initially turned on, the MIL will momentarily flash on then off and remain on until the engine is running or there are no DTCs stored in memory. Now connect the scan tool. Enter the vehicle identification information, then retrieve the DTCs with the scan tool. The scan tool can be used to do more than simply retrieve codes. The information that is available, whether or not a DTC is present, can reduce diagnostic time. Some of the common uses of the scan tool are: identifying stored DTCs, clearing DTCs, performing output control tests, and reading serial data **(Figure 34–9)**.

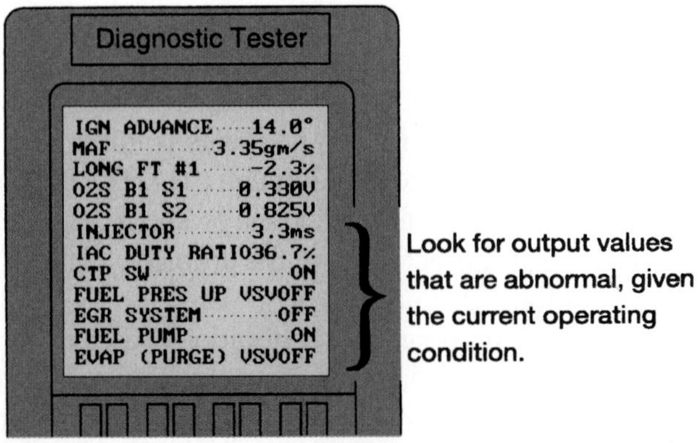
Diagnostic Tester
IGN ADVANCE 14.0°
MAF 3.35gm/s
LONG FT #1 -2.3%
O2S B1 S1 0.330V
O2S B1 S2 0.825V
INJECTOR 3.3ms
IAC DUTY RATIO36.7%
CTP SW ON
FUEL PRES UP VSVOFF
EGR SYSTEM OFF
FUEL PUMP ON
EVAP (PURGE) VSVOFF
Look for output values that are abnormal, given the current operating condition.

Figure 34-9 Use serial data to identify abnormal conditions. *Reprinted with permission*

The scan tool must have the appropriate connector to fit the DLC on an OBD-II system and the proper software for the vehicle being tested. Cable adapters are used to connect a standard scan tool to an OBD-II system. The software inserts make the scan tool compatible with the PCM in the vehicle. Make sure the recommended insert for that make and model of vehicle is used in the scan tool. Always follow the instructions given in the scan tool's manual when testing the system.

If DTCs are stored in memory, the scan tool will display them. Record them. If there are multiple DTCs, they should be interpreted in the following order:

1. PCM error DTCs
2. System voltage DTCs
3. Component level DTCs
4. System level DTCs

WARNING!

Do not clear DTCs unless directed to do so by a diagnostic procedure. Clearing DTCs will also clear valuable freeze frame and failure records data.

After retrieving the codes, use the scan tool to command the MIL to turn off. If it turns off, turn it back on and continue. If it stays on, follow the manufacturer's diagnostic procedure for this problem.

The next step in diagnostics depends on the outcome of the DTC display. If there were DTCs stored in the PCM's memory, proceed by following the designated DTC table to make an effective diagnosis and repair. If no DTCs were displayed, match the symptom to the symptoms listed in the manufacturer's symptom tables and follow the diagnostic paths or suggestions to complete the repair, or refer to the applicable component/system check. If there is not a matching symptom, analyze the complaint. Then develop a plan for diagnostics. Utilize the wiring diagrams and theory of component and system operation. Call technical assistance for similar cases in which repair history may be available. It is possible that the customer's complaint is a normal operating condition for that vehicle. If this is suspected, compare the complaint to the driveability of a known good vehicle.

After the DTCs and the MIL have been checked, the OBD-II data stream can be selected. In this mode, the scan tool displays all of the data from the inputs. This information may include engine data, EGR data, three-way catalyst data, oxygen sensor data, and misfire data. The actual available data will depend on the vehicle and the scan tool; however, OBD-II codes dictate a minimum amount of data. Always refer to the appropriate service manual and the scan tool's operating instructions for proper identification of normal or expected data.

One of the selections on a scan tool is DTC data. This selection contains two important categories of information: the freeze frame that displays the conditions that existed when a DTC was set, and the failure records. The latter is valuable when the MIL was not turned on but a fault was detected. The failure records display the conditions present when the system failed the diagnostic test.

With the scan tool, record the freeze frame and failure records information. By storing the freeze frame and failure records on the scan tool, an electronic copy of the conditions that were present when the fault occurred is made and stored in the scan tool. Using the scan tool with a four- or five-gas exhaust analyzer allows for a comparison of computer data with tailpipe emissions **(Figure 34–10)**. A lab scope is also extremely handy for watching the activity of sensors and actuators. Both of these testers will aid in the diagnosis of a problem.

If the PCM reacts to a misfire by shutting down the injector of a cylinder, that bad cylinder will still have an effect on the engine and emissions. The piston of that cylinder is still moving, the valves are still opening, and the spark plug is still firing. All of this leads to excessive amounts of oxygen in the exhaust stream. Since the O_2 sensor does not know where the oxygen is coming from, it sends a very lean reading to the PCM. The PCM, in turn,

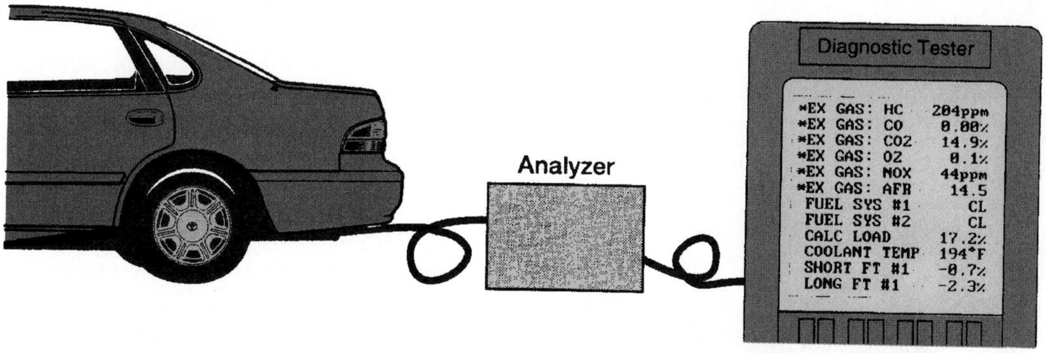

Figure 34–10 Combining the readings of a scan tool with those of an exhaust gas analyzer is very valuable in diagnostics. *Reprinted with permission*

will enrich the mixture. On most engines, that much enriching will be beyond the capacity of the system because no amount of added fuel will burn the pass-through oxygen of the dead cylinder. As a result, the PCM will set a code and turn on the MIL. When a cylinder misfire is present, check to see if two cylinders are involved. If there are, look for something shared by both cylinders, such as the common coil or a common vacuum leak.

PDA Testers Personal data assistants (PDAs) can be used to diagnose OBD-II systems **(Figure 34–11)**. To do this, modules are sold to be installed in the expansion port of the PDA. The PDAs are then equipped to serve as a scan tool and a lab scope. They are capable of recording data during prolonged test drives, which aids in diagnosing intermittent problems.

OBD-II Terminology

The introduction of OBD-II brought forth some new terms to describe the operational modes of the PCM and the operating conditions that must be met before the PCM evaluates its system. The most important of these are:

- *Drive Cycle*—An OBD-II drive cycle is a method of driving that begins with an engine start. The engine is then run until the system goes into closed loop. The drive cycle continues to include whatever specific operating conditions are necessary either to initiate and complete a specific monitoring sequence or to verify a symptom or verify a repair.
- *Monitoring Sequence*—A monitoring sequence is an operational strategy designed to test the operation of a specific system, function, or component.

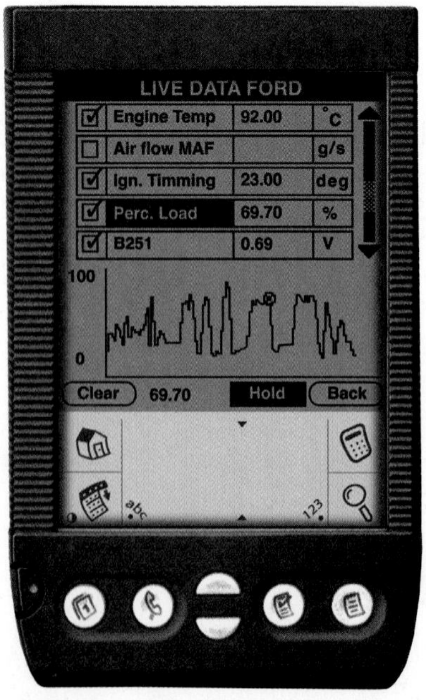

Figure 34–11 A PDA equipped to work as a scan tool and lab scope. *Courtesy of Service Solutions, SPX Corporation*

- *Enable Criteria*—This term defines the criteria that must be met in order for a monitor test to be completed. The **enable criteria** varies with make and model of vehicle; therefore, refer to the service manual to identify the exact requirements. If the enable criteria are not met, that particular monitor test cannot be completed and the condition of the tested system will be unknown.
- *Diagnostic Test*—A diagnostic test is a series of evaluation steps. At the end of the diagnostic test, a pass or fail will be reported to the PCM. When a diagnostic test reports a pass result, the PCM records the following data: "The diagnostic test has been completed since the last ignition cycle, the diagnostic test has passed during the current ignition cycle, and the fault identified by the diagnostic test is not currently active." When a diagnostic test reports a fail result, the PCM records the following data: "The diagnostic test has been completed since the last ignition cycle, the fault identified by the diagnostic test is currently active, the fault has been active during the ignition cycle, and the operating conditions at the time of the failure."
- *Pending Situation*—Under certain circumstances, the PCM will postpone and not run a particular test if the MIL is lit and a DTC is stored. This creates a **pending situation** in that a test will not run or be completed until the fault is corrected or goes undetected for the required number of driving cycles. Pending situations occur when a sensor is malfunctioning but its signal is important to the testing of another component or system. An example of this is the oxygen sensor. If the O_2 sensor signal is out of acceptable range or if it is not responding according to the guidelines set in the PCM, a DTC will be stored. The O_2 signal, however, is needed to measure the efficiency of the catalytic converter. This prevents the catalyst test from being conducted and the test is pending.
- *Serial Data*—**Serial data** is a term that refers to a method of data transmission. Serial data is transmitted, one bit at a time, over a single wire. The data travels through a data bus where it is made available to the PCM and other associated control modules.
- *Similar Conditions*—This term defines the operating conditions that must be present during a second trip to set a code when there is a defect and to erase a code after a repair has been made. For the conditions to be similar, the load conditions must be within 10% of the vehicle load present when the diagnostic test reported the malfunction, the engine speed must be within 375 rpm of the engine speed present when the DTC was set, and the engine coolant temperature must have been in the same range present when the PCM turned on the MIL.

- *Trip*—The ability for a diagnostic test to run depends upon whether a trip has been completed. A trip for a particular diagnostic test is defined as a key on and key off cycle in which all the enabling criteria for a given diagnostic test have been met. The requirements for trips vary because they may involve items of an unrelated nature—driving style, length of trip, ambient temperature, on so on. Some diagnostic tests, such as the catalyst monitor, run only once per trip. Others, such as the misfire and fuel system monitor systems, run continuously. If the proper enabling conditions are not met during that ignition cycle, the tests may not be complete or the test may not have run. Misfire, fuel system, and comprehensive system components are checked continuously throughout the trip. The EGR test requires a series of idle speed operations, acceleration, and deceleration. This test is performed once per trip. An HO2S test requires a steady speed of between 20 and 45 mph (32 and 72 km/h) for about 20 seconds after warm-up to be complete. This test is performed once per trip.
- *Two Trip Monitors*—The first time the OBD-II monitor detects a fault during any drive cycle, it sets a pending code in the memory of the PCM. Before a pending code becomes a DTC and turns on the MIL, the fault must be repeated under similar conditions. A pending code can remain in memory for some time while it waits for the conditions to repeat themselves. When the same conditions exist, the PCM checks the fault. If the fault is still present, a DTC is stored and the MIL is illuminated. Two trip monitors are those monitors that require a second trip to set a code. Some misfire and catalyst faults can cause the PCM to turn on the MIL after only one trip.
- *Warm-Up Cycle*—A warm-up cycle consists of engine start-up and operation that raises the coolant temperature more than 40°F (4°C) from start-up temperature and has reached at least 160°F (71°C). If this condition is not met during the ignition cycle, the diagnostic tests may not run.

Malfunction Indicator Lamp

The MIL is on the instrument panel and informs the driver when a fault that affects the vehicle's emission levels has occurred. The owner should take the vehicle in for service as soon as possible. The MIL will illuminate if a component or system that has an impact on emissions indicates a malfunction or fails to pass an emissions-related diagnostic test **(Figure 34–12)**. The MIL will stay lit until the system or component passes the same test for three consecutive trips with no emissions-related faults. After making the repair, technicians may need to take the vehicle for three trips to ensure the MIL does not illuminate again.

COMPONENTS THAT ILLUMINATE MIL
Automatic transmission temperature sensor
Engine coolant temperature sensor (ECT)
Evaporative emission canister purge (EVAP)
Evaporative emission purge vacuum switch
Idle air control (IAC)
Ignition control (IC)
Ignition sensor (cam sync)
Ignition sensor High resolution (7X)
Intake air temperature sensor (IAT)
Manifold absolute pressure sensor (MAP)
Manifold airflow sensor (MAF)
Spark knock sensor (KS)
Throttle position sensor (TP)
Torque converter control solenoid (TCC)
Transmission range mode pressure switch (TR)
Transmission turbine speed sensor (HI/LO)
Transmission vehicle input speed sensor (HI/LO)
Transmission vehicle output speed sensor (HI/LO)
Transmission TCC enable solenoid
Transmission shift solenoid A
Transmission shift solenoid B
Transmission 3/2 shift solenoid

Figure 34–12 A list of components that will cause the MIL to turn on when the PCM detects they are defective or are malfunctioning.

As a bulb and system check, the MIL comes on with the ignition switch on and the engine off. When the engine is started, the MIL turns off if there are no DTCs set. When the MIL remains on while the engine is running, or when a malfunction is suspected due to a driveability or emissions problem, perform a system check. These checks expose faults a technician may not detect if other diagnostics are performed first.

If the vehicle is experiencing a malfunction that may cause damage to the catalytic converter, the MIL will flash once per second. If the MIL flashes, the driver needs to get the vehicle to the service department right away. If the driver reduces speed or load and the MIL stops flashing, a code is set and the MIL stays on. This means the conditions that presented potential problems to the converter have passed with the changing operating conditions.

On some systems, the MIL may also light if the gas cap is off or is loose because there is a loss of pressure in the fuel vapor system.

MIL History Codes When the PCM turns on the MIL, a history DTC is also recorded. The provision for clearing a history DTC for any diagnostic test requires forty subsequent warm-up cycles during which no diagnostic tests have reported a fail result, a battery disconnect, or a scan tool clear command.

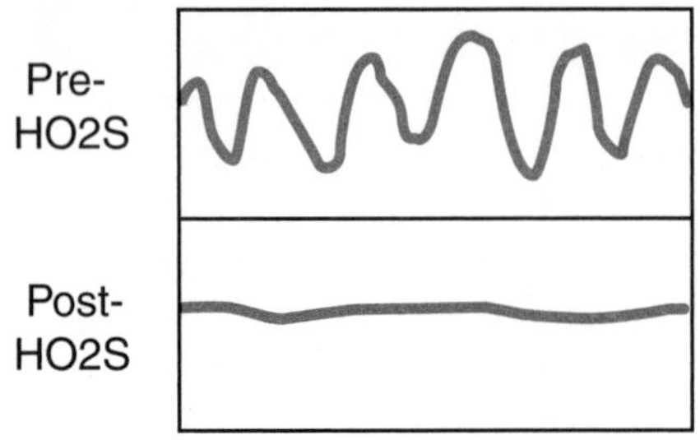

Figure 34-13 O_2 sensor signals from before and after a good catalytic converter.

With OBD-II, there are multiple oxygen sensors. One sensor is located before the catalyst and detects rich-lean swings in the exhaust stream. Another oxygen sensor is located after the catalytic converter. It also looks at the rich-lean swings of the exhaust stream. However, if the catalytic converter is working properly, there should be no swings from the converter's HO2S. If the downstream O_2 sensor signal is the same as the upstream sensor, this means the catalytic converter is not doing anything **(Figure 34–13)**. The PCM responds by setting a fault code.

Each time a fuel trim malfunction is detected, the engine load, engine speed, and the engine's coolant temperature are recorded. When the ignition is turned off, the last reported set of conditions remain stored. During subsequent ignition cycles, the stored conditions are used as a reference for similar conditions. If a fuel trim malfunction occurs during two consecutive trips, the PCM treats the failure as a normal malfunction and does not use the stored data. However, if a fuel trim malfunction occurs on two nonconsecutive trips, the stored conditions are compared with the current conditions. The MIL will light if the present conditions are similar to those present at the time the fault was detected.

In the case of an intermittent fault, the MIL may illuminate and then, after three trips, turn off. However, the corresponding DTC will be stored in memory. When unexpected DTCs appear, check for an intermittent problem.

Freeze Frame

Government regulations require that engine-operating conditions be captured whenever the MIL is illuminated. The data captured is called freeze frame data. Whenever the MIL is lit, the corresponding operating conditions are recorded in the freeze frame buffer. Each time a diagnostic test reports a failure, the current engine operating conditions are recorded in the failure records buffer. A subsequent failure will update the recorded operating conditions. The following information is what is typically stored as freeze frame data:

- Air/fuel ratio
- Air flow rate
- Fuel trim
- Engine speed
- Engine load
- Engine coolant temperature
- Vehicle speed
- TP angle
- MAP/BARO
- Injector base pulse width
- Loop status

Freeze frame data can only be overwritten with data associated with a misfire or fuel trim malfunction. Data from these faults take precedence over data associated with any other fault. The freeze frame data is not erased until the associated history DTC is cleared.

Diagnostic Trouble Codes

OBD-II standards call for standardized diagnostics and DTCs. The standardized diagnostic codes give somewhat detailed descriptions of the faults detected by the PCM **(Figure 34–14)**. In the standardized list, there are gaps in the numbering that will allow codes to be added in the future. Also, the list of codes includes some equipment not included on all engines. This means some of the codes are not available for some engines. The importance of the standardized codes is simply that every car make and model will use the same code to define a fault.

All DTCs are displayed as five-character alphanumeric codes in which each character has a specific meaning. The first character is the prefix letter that indicates the area of the fault. The second character is a number that indicates whether the DTC that follows is a standard or common code or a manufacturer-specific code. The third character indicates the subsystem of the area that the fault lies in. The fourth and fifth characters identify the specific detected fault. These characters not only indicate the component or circuit, but also give a description of the type of fault. As an example, consider DTC P0108. This is a power train DTC. The fault is in the air/fuel system. The last two characters indicate that the condition that triggered the fault is that the input voltage to the MAP sensor is above the acceptable standard. With this sort of information, a technician knows what to check.

OBD-II mandates that all DTCs be stored according to a priority. DTCs with a higher priority take precedence over a DTC with a lower priority. The higher priority codes are set the first time the fault occurs and turn on the MIL immediately. The next level of code priorities are those faults that set a code the first time they are detected but the MIL is not lit until they occur a second time. The lower-level codes are faults that relate to nonemissions systems.

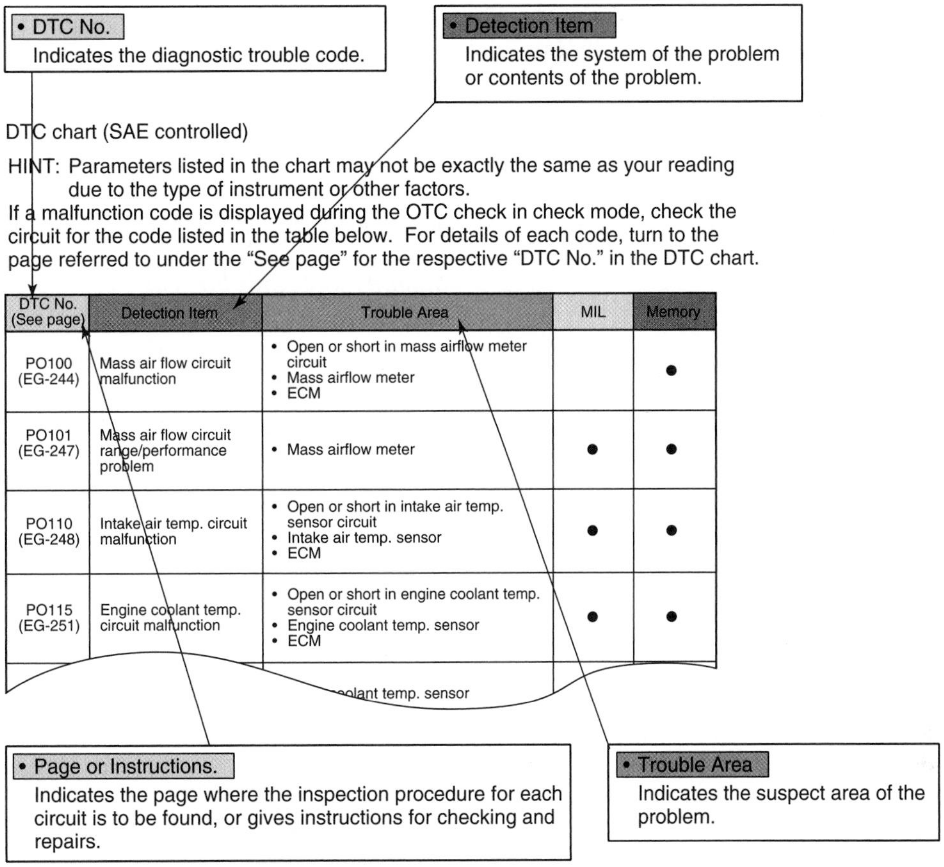

Figure 34-14 An explanation of a typical DTC chart. *Reprinted with permission*

OBD-II Monitor Test Results

All OBD-II scan tools include a readiness function showing all of the monitoring sequences on the vehicle and the status of each, which is either complete or incomplete. If vehicle travel time, operating conditions, or other parameters were insufficient for a monitoring sequence to complete a test, the scanner will indicate which monitoring sequence is not yet complete.

The specific set of driving conditions that will set the requirements for all OBD-II monitoring sequences follows:

1. Start the engine. Do not turn off the engine for the remainder of the drive cycle.
2. Drive the vehicle for at least 4 minutes.
3. Continue to drive until the engine's temperature reaches 180°F (82°C) or more.
4. Idle the engine for 45 seconds.
5. Open the throttle to about 25% to accelerate from a standstill to 45 mph in about 10 seconds.
6. Drive between 30 and 40 mph (50 and 65 km/h) with a steady throttle for at least 1 minute.
7. Drive for at least 4 minutes at speeds between 20 and 45 mph (32 and 72 km/h). If, during this phase, you must slow down below 20 mph, repeat this phase of the drive cycle. During this time, do not fully open the throttle.
8. Decelerate and idle for at least 10 seconds.
9. Accelerate to 55 mph (88 km/h) with about half throttle.
10. Cruise at a constant speed of 40 to 65 mph (65 to 104 km/h) for at least 80 seconds with a steady throttle.
11. Decelerate and allow the engine to idle.
12. Using the scan tool, check for diagnostic system readiness and retrieve any DTCs.

When most monitor tests are run and a system or component fails a test, a pending code is set. When the fault is detected a second time, a DTC is set and the MIL is lit. It is possible that a DTC for a monitored circuit may not be entered into memory even though a malfunction has occurred. This may happen when the monitoring criteria have not been met.

Comprehensive Component Monitor The PCM's output signals are constantly being monitored by the diagnostic management software. These signals are observed to detect opens and shorts in the circuits. Some of the output devices are also monitored for their effectiveness and functionality. The PCM is programmed to expect certain things to happen when it commands a device to do something. Perhaps the simplest example of this is idle speed. The PCM has control over the engine's idle speed. This is accomplished in many different ways, one of which is with a stepper motor. The PCM sends voltage signals to the motor, which changes the idle speed. When the PCM commands the motor to extend, the engine's speed should increase. The PCM watches engine speed and continues to adjust the motor until the desired speed is reached. If the PCM commands the motor to extend and the engine's speed does not increase, the PCM knows the motor is not doing what it is supposed to be doing. As a result, the PCM sets a DTC and illuminates the MIL.

Possible Causes of OBD-II Monitor Failures

Catalyst monitor failure can be caused by:

- Fuel contaminants
- Leaking exhaust
- Engine mechanical problems
- Defective upstream or downstream oxygen sensor circuits
- Defective PCM

EGR monitor failure can be caused by:

- Faulty EGR valve
- Faulty EGR passages or tubes
- Loose or damaged EGR solenoid wiring and/or connectors
- Damaged DPFE or EGR VP sensor
- Disconnected or loose electrical connectors to the DPFE or EGR VP sensors
- Disconnected, damaged, or misrouted EGR vacuum hoses

EVAP monitor failure can be caused by:

- Disconnected, damaged, or loose purge solenoid connectors and/or wiring
- Leaking hoses, tubes, or connectors in the EVAP system
- Vacuum and/or vent hoses to the solenoid and charcoal canister are misrouted
- Plugged hoses from the purge solenoid to the charcoal canister
- Loose or damaged connectors at the purge solenoid
- Fuel tank cap not tightened properly or is missing

Fuel system monitor failure can be caused by:

- A defective fuel pump
- Abnormal signal from the upstream HO2S
- Engine temperature sensor faults
- Malfunctioning catalytic converter
- MAP- or MAF-related faults
- Cooling system faults
- EGR system faults
- Fuel injection system faults
- Ignition system faults
- Vacuum leaks
- Worn engine parts

Misfire monitor failure can be caused by:

- Fuel level too low during drive cycle
- Dirty or defective fuel injectors
- Contaminated fuel
- Defective fuel pump
- Restricted fuel filter
- EGR system faults
- EVAP system faults
- Restricted exhaust system
- Faulty secondary ignition circuit
- Damaged, loose, or resistant PCM power and/or ground circuits

Oxygen sensor monitor failure can be caused by:

- Malfunctioning upstream and/or downstream oxygen sensor
- Malfunctioning heater for the upstream or downstream oxygen sensor
- A faulty PCM
- Defective wiring to and/or from the sensors

AIR system monitor failure can be caused by:

- Faulty secondary AIR solenoid and/or relay
- Damaged, loose, or disconnected wiring in the secondary air solenoid and/or relay circuit
- Defective aspirator valve
- Disconnected or damaged AIR hoses and/or tubes
- A defective electric or mechanical air pump
- Air pump drive belt missing
- Faulty AIR check valve

Intermittent Faults

An intermittent fault is a fault that is not always present. This type of fault may not activate the MIL or cause a DTC to be set. Therefore, intermittent problems can be difficult to diagnose. Prior to testing for the source of an intermittent problem, make sure the preliminary diagnostic checks are completed first.

By studying the system and the relationship of each component to another, you should be able to create a list of possible causes for the intermittent problem. To help identify the cause of an intermittent problem, follow these steps:

PROCEDURE

Identifying the Cause of an Intermittent Problem

STEP 1 *Observe the history DTCs, the DTC modes, and the freeze frame data.*

STEP 2 *Call technical assistance for similar cases where repair history may be available. Combine your knowledge and understanding of the system with the service information available.*

STEP 3 *Evaluate the symptoms and conditions described by the customer.*

STEP 4 *Use a check sheet or other method in order to identify the circuit or electrical system component that may have the problem.*

STEP 5 *Follow the suggestions for intermittent diagnosis found in service material and manuals.*

STEP 6 *Visually inspect the suspected circuit or system.*

STEP 7 *Use the data-capturing capabilities of the scan tool.*

STEP 8 *Test the circuit's wiring for shorts, opens, and high resistance. This should be done with a DMM and in the typical manner, unless instructed differently in the service manual.*

Most intermittent problems are caused by faulty electrical connections or wiring. Refer to a wiring diagram for each of the suspected circuits or components to help identify all of the connections and components in that circuit. The entire electrical system of the suspected circuit should be carefully and thoroughly inspected. Check for burned or damaged wire insulation, damaged terminals at the connectors, corrosion at the connectors, loose connectors, wire terminals loose in the connector, and disconnected or loose ground wire or straps.

To locate the source of the problem, a voltmeter can be connected to the suspected circuit and the wiring harness wiggled **(Figure 34–15)**. As a guideline for what voltage should be expected in the circuit, refer to the reference value table in the service manual. If the voltage reading changes with the wiggles, the problem is in that circuit. The vehicle can also be taken for a test drive with the voltmeter connected. If the voltmeter readings become abnormal with the changing operating conditions, the circuit being observed is probably the circuit with the problem.

The vehicle can also be taken for a test drive with the scan tool connected to it. Scan tools have several features that make intermittent problem identification easier. The scan tool can be used to monitor the activity of a circuit while the vehicle is being driven. This allows you to look at a circuit's response to changing operating conditions. The snapshot or freeze frame feature stores engine conditions and operating parameters at command or when the PCM sets a DTC. If the snapshot can be taken when the intermittent problem is occurring, the problem will be easier to diagnose.

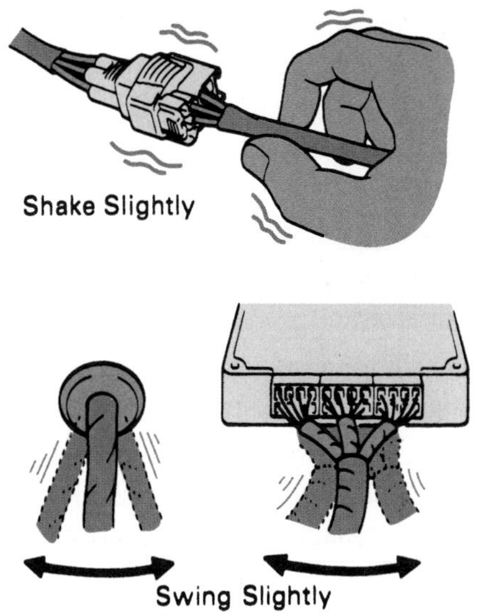

Figure 34–15 The wiggle test can be used to locate intermittent problems. *Reprinted with permission*

With an OBD-II scan tool, actuators can be activated and their functionality tested. The results of the change in output operation can be monitored. Also, the activity of the outputs can be monitored as they respond to changes in sensor signals to the PCM. This can be helpful in locating an intermittent problem.

When an actuator is activated, watch the response on the scan tool. Also, listen for the clicking of the relay that controls that output. If no clicking is heard, measure the voltage at the relay's control circuit. There should be a change of more than 4 volts when the output is activated.

To monitor how the PCM and an output responds to a change in sensor signals, use the scan tool. Select the mode that relates to the suspected circuit and view, then record the scan data for that circuit. Compare the reading to specifications. Then create a condition that would cause the related inputs to change. Observe the scan data to see if the change was appropriate.

Repairing the System

After isolating the source of the problem, the repairs should be made. The system should then be rechecked to verify that the repair took care of the problem. This may involve road testing the vehicle in order to verify that the complaint has been resolved.

When servicing or repairing OBD-II circuits, the following guidelines are important:

- Do not connect aftermarket accessories into an OBD-II circuit.
- Do not move or alter grounds from their original locations.

- Always replace a relay in an OBD-II circuit with an exact replacement. Damaged relays should be thrown away, not repaired.
- Make sure all connector locks are in good condition and are in place.
- After repairing connectors or connector terminals, make sure the terminals are properly retained and the connector is sealed.
- When installing a fastener for an electrical ground, be sure to tighten it to the specified torque.

Verification of repair is more comprehensive for vehicles with OBD-II system diagnostics than earlier vehicles. Following a repair, a technician should perform the following steps:

1. Review the fail records and the freeze frame data for the DTC that was diagnosed. Record the fail records or freeze frame data.
2. Use the scan tool's clear DTCs or clear info functions to erase the DTCs.
3. Operate the vehicle within the conditions noted in the fail records or the freeze frame data.
4. Monitor the status information for the specific DTC until the diagnostic test associated with that DTC runs.

HO2S Repair If the HO2S wiring, connector, or terminal are damaged, the entire oxygen sensor assembly should be replaced. Do not attempt to repair the assembly. In order for this sensor to work properly, it must have a clean air reference. The sensor receives this reference from the air present around the sensor's signal and heater wires. Any attempt to repair the wires, connectors, or terminals could result in the obstruction of the air reference and degrade oxygen sensor performance.

Additional guidelines for servicing a heated oxygen sensor follow:

- Do not apply contact cleaner or other materials to the sensor or wiring harness connectors. These materials may get into the sensor, causing poor performance.
- The sensor pigtail and harness wires must not be damaged in such a way that the wires inside are exposed. This could provide a path for foreign materials to enter the sensor and cause performance problems.
- Neither the sensor or vehicle lead wires should be bent sharply or kinked. Harp bends, kinks, and so forth, could block the reference air path through the lead wire.
- Do not remove or defeat the oxygen sensor ground wire. Vehicles that utilize the ground-wired sensor may rely on this ground as the only ground contact to the sensor. Removal of the ground wire will cause poor engine performance.
- To prevent damage due to water intrusion, be sure the peripheral seal remains intact on the vehicle harness.

DIAGNOSIS OF COMPUTER VOLTAGE SUPPLY AND GROUND WIRES

All PCMs (OBD-II and earlier designs) cannot operate properly unless they have good ground connections and the correct voltage at the required terminals. A wiring diagram for the vehicle being tested must be used for these tests. Backprobe the battery terminal at the PCM and connect a digital voltmeter from this terminal to ground. Always ground the black meter lead.

SHOP TALK

Never replace a computer unless the ground wires and voltage supply wires have proved to be in satisfactory condition. ■

The voltage at this terminal should be 12 volts with the ignition switch off. If 12 volts are not available at this terminal, check the computer fuse and related circuit. Turn on the ignition switch and connect the red voltmeter lead to the other battery terminals at the PCM with the black lead still grounded. The voltage measured at these terminals should also be 12 volts with the ignition switch on. When the specified voltage is not available, test the voltage supply wires to these terminals. These terminals may be connected through fuses, fuse links, or relays.

Computer ground wires usually extend from the computer to a ground connection on the engine or battery. With the ignition switch on, connect a digital voltmeter from the battery ground to the computer ground. The voltage drop across the ground wires should be 30 millivolts or less. If the voltage reading is greater than that or more than that specified by the manufacturer, repair the ground wires or connection.

Not only should the computer ground be checked, but so should the ground (and positive) connection at the battery. Checking the condition of the battery and its cables should always be part of the initial visual inspection before beginning diagnosis of an engine control system.

A voltage drop test is a quick way of checking the condition of any wire. To do this, connect a voltmeter across the wire or device being tested. Place the positive lead on the most positive side of the circuit. Then turn on the circuit. Ideally, there should be a zero volt reading across any wire unless it is a resistance wire that is designed to drop voltage; even then, check the drop against specifications to see if it is dropping too much.

A good ground is especially critical for all reference voltage sensors. The problem here is not obvious until it is thought about. A bad ground will cause the reference voltage (normally 5 volts) to be higher than normal. Normally, the added resistance of a bad ground in a circuit would cause less voltage at a load. Because of the way reference voltage sensors are wired, the opposite is true. If the reference voltage to a sensor is too high, the output signal from the sensor to the computer will also be too high. As a result, the computer will be making decisions based on the wrong information. If the output signal is within the normal range for that sensor, the computer will not notice the wrong information and will not set a DTC.

To explain why the reference voltage increases with a bad ground, let us look at a voltage divider circuit. This circuit is designed to provide a 5-volt reference signal off the tap. A vehicle's computer feeds a regulated 12 volts to a similar circuit to ensure that the reference voltage to the sensors is very close to 5 volts. The voltage divider circuit consists of two resistors connected in series with a total resistance of 12 ohms. The reference voltage tap is between the two resistors. The first resistor drops 7 volts **(Figure 34–16)**, which leaves 5 volts for the second resistor and for the reference voltage tap. This 5-volt reference signal will be always available at the tap, as long as 12 volts are available for the circuit.

If the circuit has a poor ground, one that has resistance, the voltage drop across the first resistor will be decreased, causing the reference voltage to increase. In **Figure 34–17**, to simulate a bad ground, a 4-ohm resistor was added into the circuit at the ground connection

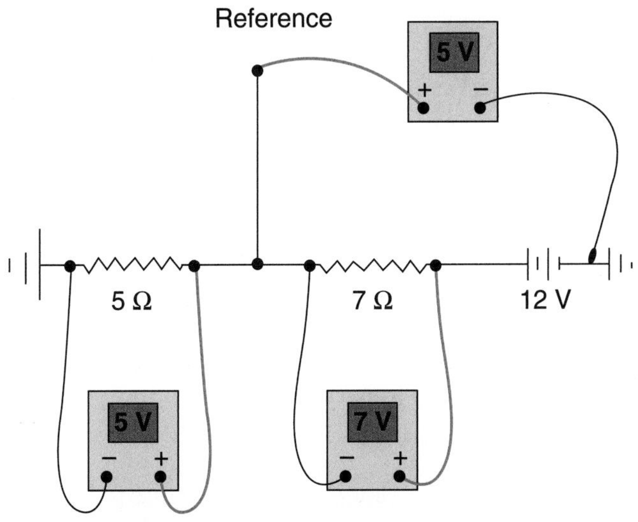

Figure 34–16 A voltage divider circuit with voltage values.

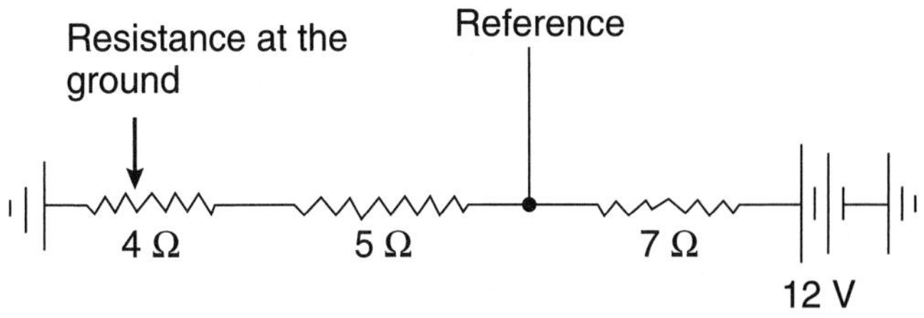

Figure 34–17 Voltage divider circuit with a bad ground.

at the battery. This increases the total resistance of the circuit to 16 ohms and decreases the current flowing throughout the circuit. With less current flow through the circuit, the voltage drop across the first resistor decreases to 5.25 volts **(Figure 34–18)**. This means the voltage available at the tap will be higher than 5 volts; it will be 6.75 volts.

Poor grounds can also allow EMI or noise to be present on the reference voltage signal. This noise causes small changes in the voltage going to the sensor. Therefore, the output signal from the sensor will also have these voltage changes. The computer will try to respond to these small, rapid changes, which can cause a driveability problem. The best way to check for noise is to use a lab scope.

Connect the lab scope between the 5-volt reference signal into the sensor and the ground. The trace on the scope should be flat **(Figure 34–19)**. If noise is present, move the scope's negative probe to a known good ground. If the noise disappears, the sensor's ground circuit is bad or has resistance. If the noise is still present, the voltage feed circuit is bad or there is EMI in the circuit from another source, such as the AC generator. Find and repair the cause of the noise.

Circuit noise may be present in the positive side or the negative side of a circuit. It may also be made evident by a flickering MIL, a popping noise on the radio, or by an

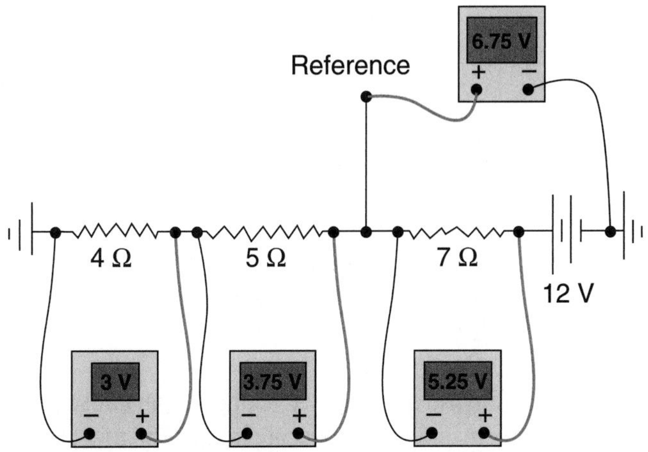

Figure 34–18 Figure 34–17 with voltage readings.

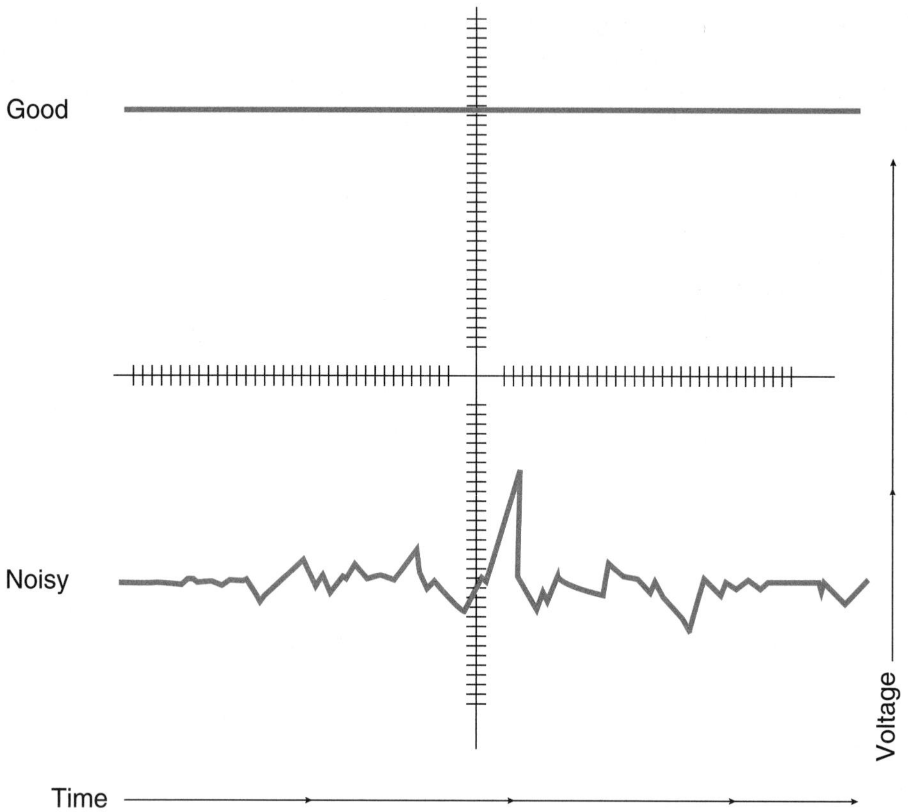

Figure 34-19 (*Top*) A good voltage signal. (*Bottom*) A voltage signal with noise.

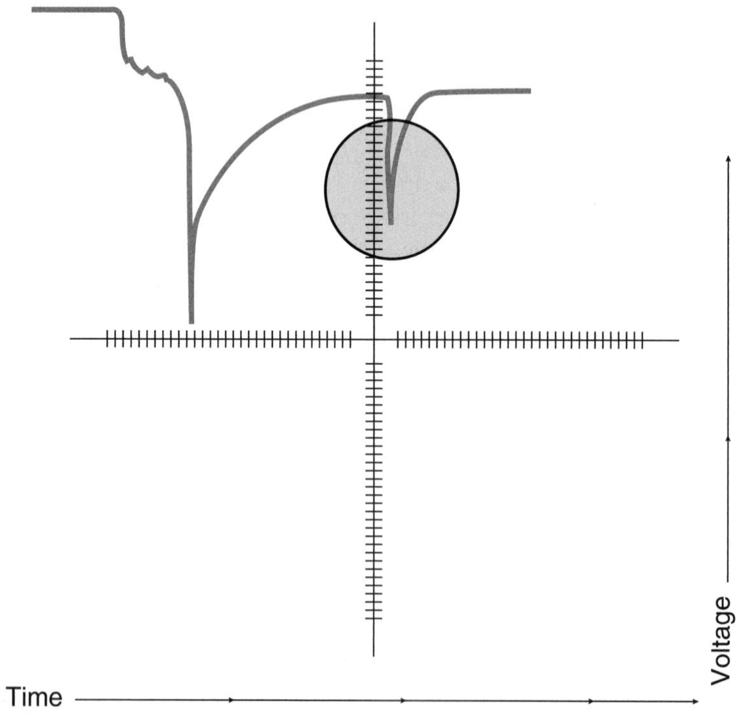

Figure 34-20 A trace of an A/C compressor clutch with a bad clamping diode.

intermittent engine miss. However, noise can cause a variety of problems in any electrical circuit. The most common sources of noise are electric motors, relays and solenoids, AC generators, ignition systems, switches, and A/C compressor clutches. Typically, noise is the result of an electrical device being turned on and off. Sometimes the source of the noise is a defective suppression device. Manufacturers include these devices to minimize or eliminate electrical noise. Some of the commonly used noise suppression devices are resistor-type secondary cables and spark plugs, shielded cables, capacitors, diodes, and resistors. If the source of the noise is not a poor ground or a defective component, check the suppression devices.

Diodes and resistors are the most commonly used noise suppression devices. Resistors do not eliminate the spikes but limit their intensity. If a voltage trace has a large spike and the circuit is fitted with a resistor to limit noise, the resistor may be bad. Clamping diodes are used on devices like A/C compressor clutches to eliminate voltage spikes. If the diode is bad, a negative spike will result **(Figure 34–20)**. Capacitors or chokes are used to control noise from a motor or generator.

TESTING INPUT SENSORS

If a DTC directs you to a faulty sensor or sensor circuit, or if you suspect that a sensor is faulty, it should be tested. Testing sensors is included here to orient you to the basic procedures. The recommended procedures given in a service manual may differ from those described here. Always follow the manufacturer's recommendations. Sensors are tested with a DMM, scanner, and/or lab scope.

Since the controls are different on the various types of lab scopes that can be used for automotive diagnostic work, the connections and settings for a lab scope are loosely defined in this discussion. Make sure you follow the instructions of the scope's manufacturer when using a lab scope. These are not simple test instruments! You will become more familiar and comfortable with a scope as you use it. If the scope is set incorrectly, the scope will not break. It just will not show you what you want to be shown. To help with understanding how to set the controls on a scope, keep the following things in mind. The vertical voltage scale must be adjusted in relation to the voltage expected in the signal being displayed. The horizontal time base or milliseconds per division must be adjusted so that the waveform appears properly on the screen. Many waveforms are clearly displayed when the horizontal time base is adjusted so that three waveforms are displayed on the screen.

The trigger is the signal that tells the lab scope to start drawing a waveform. A marker indicates the trigger line on the screen and minor adjustments of the trigger line may be necessary to position the waveform in the desired vertical position. Trigger slope indicates the direction in which the voltage signal is moving when it crosses the trigger line. A positive trigger slope means the voltage signal is moving upward as it crosses the trigger line, whereas a negative trigger slope indicates the voltage signal is moving downward when it crosses the trigger line.

Personal computer software packages are available to help you properly interpret scope patterns and set up various types of lab scopes. These also contain an extensive waveform library that you can quickly refer to and find what the normal waveform of a particular device should look like. The library also contains the waveforms caused by common problems. You can also add to the library by transferring waveforms from any lab scope. After you have transferred the waveform, you can make notes to the file. The package also includes the theory of operation and diagnostic procedures for common inputs and outputs. Packages such as this can be quite helpful when solving driveability problems.

There are many different types of sensors; their design depends on what they are monitoring. Some sensors are simple on-off switches. Others are some form of variable resistor that changes resistance according to temperature changes. Some sensors are voltage or frequency generators, while others send varying signals according to the rotational speed of another device. Knowing what they are measuring and how they respond to changes are the keys to being able to accurately test an input sensor. What follows are typical testing procedures for common input sensors. The sensors are grouped according to the type of sensor.

Switches

Switches are turned on and off through an action of a device or by the actions of the driver. Some of these switches are grounding switches that complete a circuit when they are closed. Others send a 5- or 12-volt signal to the PCM when they are closed.

Grounding switches always send a digital signal to the PCM; the switch's circuit is either on or off. These circuits contain a fixed resistor to limit the circuit's current and prevent voltage spikes in the circuit as the switch opens and closes. When the switch is closed, the voltage signal to the PCM is low or zero. When the switch is open, there is a high voltage signal.

A typical grounding switch is an idle tracking switch. This switch lets the PCM know when the throttle plates are open. The switch is open when the throttle is closed and closes as soon as the throttle is opened. When the throttle is closed, a low voltage signal is sent to the PCM. This tells the PCM that it can control idle speed, if necessary.

The switch can be easily tested with an ohmmeter. Disconnect the connector to the idle tracking switch. Then connect the ohmmeter across the switch's terminals. When the throttle is opened wide, there should zero ohms of resistance across the terminals. Now slowly allow the throttle to close. The switch should remain closed until the throttle is closed. At that point, the ohmmeter should give an infinite reading. If the switch reacts in any other way, the switch is bad and should be replaced. Some systems have the idle tracking switch as part of the throttle position sensor. The idle tracking function of the TP sensor can be tested in the same way as other idle tracking switches.

Some switches are adjustable and must be set so that they close and open at the correct time. An example of this is the clutch-engaged switch. This switch is used to inform the computer when there is no load (clutch pedal depressed) on the engine. The switch is also connected into the starting circuit. The switch prevents starting of the engine unless the clutch pedal is fully depressed. The switch is normally open when the clutch pedal is released. When the clutch pedal is depressed, the switch closes and completes the circuit between the ignition switch and the starter solenoid. It also sends a signal of no-load to the PCM.

Most grounding switches react to some mechanical action to open or close. There are some, however, that respond to changes of conditions. These may respond to changes in pressure or temperature. An example of this type of switch is the power-steering pressure switch. This switch informs the PCM when power-steering pressure reaches a particular point. When the power-steering pressure exceeds that point, the PCM knows there is an additional load on the engine and will increase idle speed.

To test this type of switch, monitor its activity with a DMM or lab scope. With the engine running at idle speed, turn the steering wheel to its maximum position on one side. The voltage signal should drop as soon as the pressure in the power-steering unit has reached a high level. If the voltage does not drop, either the power-steering assembly is incapable of producing high pressure or the switch is bad.

Temperature-responding switches operate in the same way. When a particular temperature is reached, the switch opens. This type switch is best measured by removing it and submerging it in heated water. Watch the ohmmeter as the temperature increases. A good temperature-responding switch will open (have an infinite reading) when the water temperature reaches the specified amount. If the switch fails this test, it should be replaced.

Voltage input signal switches send a high voltage signal to the PCM when they are closed. An example of this type of switch is the A/C compressor switch. This switch lets the PCM know when the extra load on the engine is caused by the engaging of the air-conditioning compressor's clutch. Some A/C switches close when the driver selects A/C. Always check the service manual to determine what action closes this type of switch.

These switches can be tested with a voltmeter. Connect the meter to the output of the switch and to a good ground. When the A/C is turned on, the meter should read 12 volts. If it does not, check the wiring to the switch and the switch itself. Make sure voltage is available to the input side of the switch before condemning the switch. The switch can also be checked with an ohmmeter. Disconnect the wiring to the switch and connect the meter across the input and output of the switch. The switch should open and close with the cycling of the A/C compressor.

VARIABLE RESISTOR-TYPE SENSORS

Many sensors send a voltage signal to the PCM in direct response to changes in operating conditions. The voltage signal changes as the resistance in the sensor changes. Most often these variable resistor-type sensors are thermistors and potentiometers.

Engine Coolant Temperature (ECT) Sensor

WARNING!

Never apply an open flame to an ECT or IAT sensor for test purposes. This action will damage the sensor.

A defective ECT sensor may cause some of the following problems:

1. Hard engine starting
2. Rich or lean air/fuel ratio
3. Improper operation of emission devices
4. Reduced fuel economy
5. Improper converter clutch lockup
6. Hesitation on acceleration
7. Engine stalling

The ECT sensor may be removed and placed in a container of water with an ohmmeter connected across the sensor terminals. A thermometer is also placed in the water. When the water is heated, the sensor should have the specified resistance at any temperature **(Figure 34–21)**. If the sensor does not have the specified resistance, replace the sensor.

The wiring to the sensor can also be checked with an ohmmeter. With the wiring connectors disconnected from the ECT sensor and the computer, connect an ohmmeter from each sensor terminal to the computer terminal to which the wire is connected. Both sensor wires should indicate less resistance than specified by the vehicle manufacturer. If the wires have higher resistance than specified, the wires or wiring connectors must be repaired.

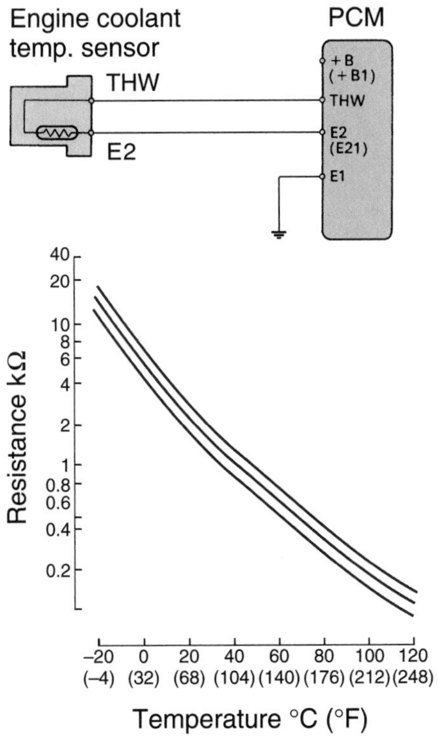

Figure 34-21 Specifications for an ECT sensor. *Reprinted with permission*

WARNING!

Before disconnecting any computer system component, be sure the ignition switch is turned off. Disconnecting components may cause high induced voltages and computer damage.

With the sensor installed in the engine, the sensor terminals may be backprobed to connect a digital voltmeter to the sensor terminals. The sensor should provide the specified voltage drop at any coolant temperature **(Figure 34–22)**.

Some computers have internal resistors connected in series with the ECT sensor. The computer switches these resistors at approximately 120°F (49°C). This resistance change inside the computer causes a significant change in voltage drop across the sensor, as indicated in the specifications. This condition is normal on any computer with this feature. This change in voltage drop is always evident in the vehicle manufacturer's specifications.

Intake Air Temperature Sensors

A defective intake air temperature (IAT) sensor may cause the following problems:

1. Rich or lean air/fuel ratio
2. Hard engine starting
3. Engine stalling or surging

COLD CURVE 10,000-OHM RESISTOR USED		HOT CURVE CALCULATED RESISTANCE OF 909 OHMS USED	
–20°F	4.70 V	110°F	4.20 V
–10°F	4.57 V	120°F	4.00 V
0°F	4.45 V	130°F	3.77 V
10°F	4.30 V	140°F	3.60 V
20°F	4.10 V	150°F	3.40 V
30°F	3.90 V	160°F	3.20 V
40°F	3.60 V	170°F	3.02 V
50°F	3.30 V	180°F	2.80 V
60°F	3.00 V	190°F	2.60 V
70°F	2.75 V	200°F	2.40 V
80°F	2.44 V	210°F	2.20 V
90°F	2.15 V	220°F	2.00 V
100°F	1.83 V	230°F	1.80 V
110°F	1.57 V	240°F	1.62 V
120°F	1.25 V	250°F	1.45 V

Figure 34-22 Voltage drop specifications for an ECT sensor. *Courtesy of DaimlerChrysler Corporation*

4. Acceleration stumbles
5. Excessive fuel consumption

The IAT sensor may be removed from the engine and placed in a container of water with a thermometer. When an ohmmeter is connected to the sensor terminals and the water in the container is heated, the sensor should have the specified resistance at any temperature. If the sensor does not have the specified resistance, sensor replacement is required.

With the IAT sensor installed in the engine, the sensor terminals may be backprobed and a voltmeter connected across the sensor terminals. The sensor should have the specified voltage drop at any temperature. The wires between the air charge temperature sensor and the computer may be tested in the same way as the ECT wires.

Throttle Position Sensor

A defective throttle position sensor may cause acceleration stumbles, engine stalling, and improper idle speed. With the ignition switch on, connect a voltmeter between the reference wire to ground. Normally, the voltage reading should be approximately 5 volts.

If the reference wire is not supplying the specified voltage, check the voltage on this wire at the computer terminal. If the voltage is within specifications at the computer but low at the sensor, repair the reference wire. When this voltage is low at the computer, check the voltage supply wires and ground wires on the computer. If these wires are satisfactory, replace the computer.

With the ignition switch on, connect the voltmeter from the sensor ground wire to the battery ground. If the voltage drop across this circuit exceeds specifications, repair the ground wire from the sensor to the computer.

SHOP TALK

When the throttle is opened gradually to check the throttle position sensor voltage signal, tap the sensor lightly and watch for fluctuations on the voltmeter pointer, indicating a defective sensor. ■

With the ignition switch on, connect a voltmeter from the sensor signal wire to ground. Slowly open the throttle and observe the voltmeter. The voltmeter reading should increase smoothly and gradually. Typical TP sensor voltage readings would be 0.5 volt to 1 volt with the throttle in the idle position, and 4 to 5 volts at wide-open throttle. If the TP sensor does not have the specified voltage or if the voltage signal is erratic, replace the sensor.

Adjustment of the TP sensor can be made on some engines. Incorrect TP sensor adjustment may cause inaccurate idle speed, engine stalling, and acceleration stumbles. Follow these steps to adjust a typical TP sensor:

PROCEDURE

Adjusting a Typical TP Sensor

STEP 1 *Backprobe the TP sensor signal wire and connect a voltmeter from this wire to ground.*

STEP 2 *Turn on the ignition switch and observe the voltmeter reading with the throttle in the idle position.*

STEP 3 *If the TP sensor does not provide the specified voltage, loosen the TP sensor mounting bolts and rotate the sensor housing until the specified voltage is indicated on the voltmeter* ***(Figure 34–23)****.*

STEP 4 *Hold the sensor in this position and tighten the mounting bolts to the specified torque.*

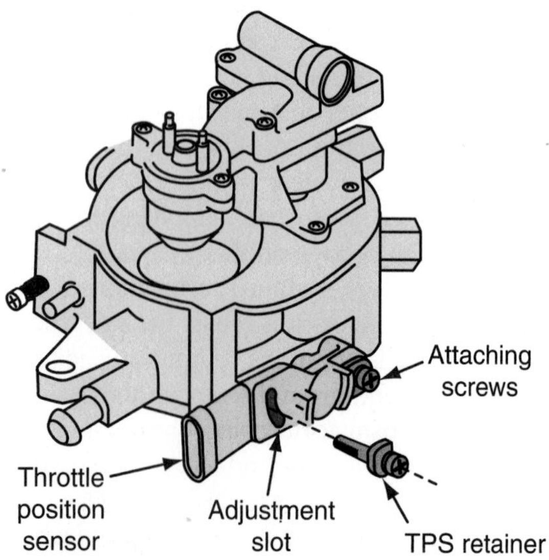

Figure 34-23 A TP sensor with elongated slots for sensor adjustments.

TP sensors can also be tested with a lab scope. Connect the scope to the sensor's output and a good ground and watch the trace as the throttle is opened and closed. The resulting trace should look smooth and clean, without any sharp breaks or spikes in the signal **(Figure 34–24)**. A bad sensor will typically have a glitch (a downward spike) somewhere in the trace **(Figure 34–25)** or will not have a smooth transition from high to low. These glitches are an indication of an open or short in the sensor. Be care-

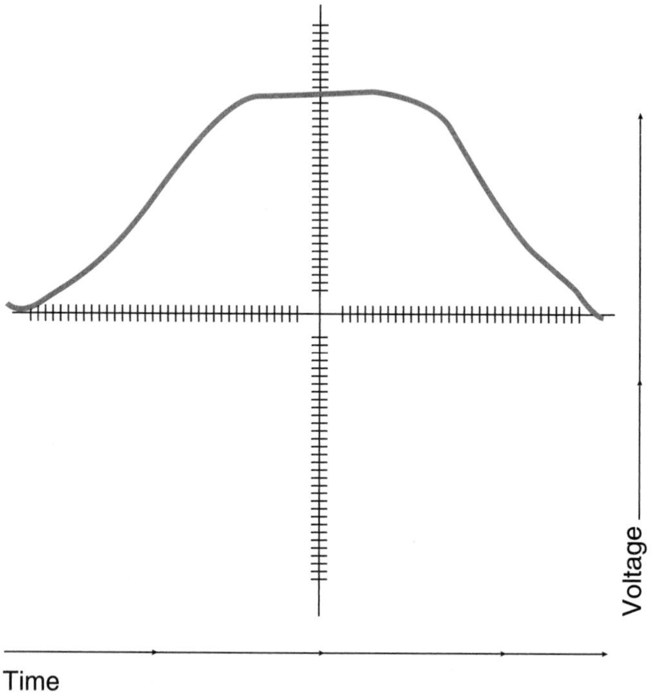

Figure 34-24 A normal TP sensor waveform while it opens and closes.

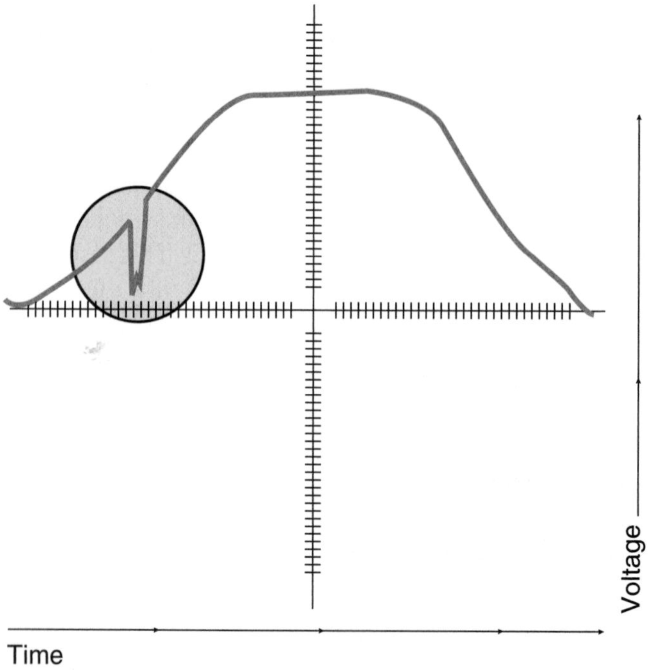

Figure 34-25 The waveform of a defective TP sensor. Notice the glitch while the throttle opens.

ful; some TP sensors have four wires. The additional wire is connected to an idle switch. When this wire is probed, the toggling of the switch will cause spikes. Spikes are expected if the switch is good.

Exhaust Gas Recirculation Valve Position Sensor

Many EGR valve position sensors have a 5-volt reference wire, a voltage signal wire, and a ground wire. This sensor can be tested in the same way as a TP sensor. Connect a voltmeter across the voltage signal wire and ground, and turn on the ignition switch. The voltage signal should be approximately 0.8 volt. Connect a vacuum hand pump to the vacuum fitting on the EGR valve, and slowly increase the vacuum from zero to 20 in. Hg (508 mm Hg). The sensor's voltage signal should gradually increase to 4.5 volts at 20 in. Hg (508 mm Hg). If the sensor does not have the specified voltage, replace the sensor.

It is good practice to check all variable resistor sensors with a lab scope. Any defects in the sensor will show up as glitches in the waveform. Often these are unnoticeable on a DMM. The trace from all variable resistors should be clean and smooth.

GENERATING SENSORS

Some engine sensors generate a voltage or frequency signal in response to changing conditions. The most common voltage generating sensors are the oxygen sensor and crankshaft position sensors. Common frequency generator-type sensors are some MAP sensors, some MAF sensors, and reference signals.

Oxygen Sensors

Oxygen sensors produce a voltage based on the amount of oxygen in the exhaust. Large amounts of oxygen result from lean mixtures and result in low voltage output from the O_2 sensor. Rich mixtures release lower amounts of oxygen into the exhaust, therefore the O_2 sensor voltage is high. The engine must be at normal operating temperature before the oxygen sensor is tested. Always refer to the specifications supplied by the manufacturer.

Keep in mind that bad signals from an oxygen sensor can be caused by problems other than a faulty sensor. Common causes of poor signals include inadequate fuel delivery, a malfunctioning AIR system, a leaking injector, a vacuum leak, and a contaminated MAF sensor.

A bad oxygen sensor can cause a number of problems and set a variety of DTCs. There are several ways an O_2 sensor can be checked. Photo Sequence 32 covers the use of the Min/Max function on a DMM. Other testing procedures are discussed in the following paragraphs.

If the voltage readings from the Min/Max test are not what they should be, watch how the sensor reacts to lean conditions by causing a vacuum leak. Then check its response to a rich mixture by introducing some propane into the intake. If the sensor responds as it should to these changes, the cause for the previous abnormal readings is probably not the sensor.

WARNING!

An oxygen sensor must be tested with a digital voltmeter. If an analog meter is used for this purpose, the sensor may be damaged.

Testing with a DMM Connect the voltmeter between the O_2 sensor wire and ground. Backprobe the connector near the O_2 sensor to connect the voltmeter to the sensor signal wire. If possible, avoid probing through the insulation to connect a meter to the wire. With the engine idling, the sensor voltage should be cycling from low voltage to high voltage. The signal from most O_2 sensors varies between 0 and 1 volt.

If the voltage is continually high, the air/fuel ratio may be rich or the sensor may be contaminated by RTV sealant, antifreeze, or lead from leaded gasoline. When the O_2 sensor voltage is continually low, the air/fuel ratio may be lean, the sensor may be defective, or the wire between the sensor and the computer may have a high-resistance problem. If the O_2 sensor voltage signal remains in a midrange position, the computer may be in open loop or the sensor may be defective.

The sensor can also be tested after it is removed from the exhaust manifold. Connect the voltmeter between the sensor wire and the case of the sensor. Using a propane torch, heat the sensor element. The propane flame keeps the oxygen in the air away from the sensor element, causing the sensor to produce voltage. While the sensor element is in the flame, the voltage should be nearly 1 volt. The voltage should drop to zero immediately when the flame is removed from the sensor. If the sensor does not produce the specified voltage or if the sensor does not quickly respond to the change, it should be replaced.

If a defect in the O_2 sensor signal wire is suspected, backprobe the sensor signal wire at the computer and connect a digital voltmeter from the signal wire to ground with the engine idling. The difference between the voltage readings at the sensor and at the computer should not exceed the vehicle manufacturer's specifications. A typical specification for voltage drop across the average sensor wire is 0.02 volt.

Now check the sensor's ground. With the engine idling, connect the voltmeter from the sensor case to the sensor ground wire on the computer. Typically, the maximum allowable voltage drop across the sensor ground circuit is 0.02 volt. Always use the vehicle manufacturer's specifications. If the voltage drop across the sensor

PHOTO SEQUENCE 32

Testing an Oxygen Sensor

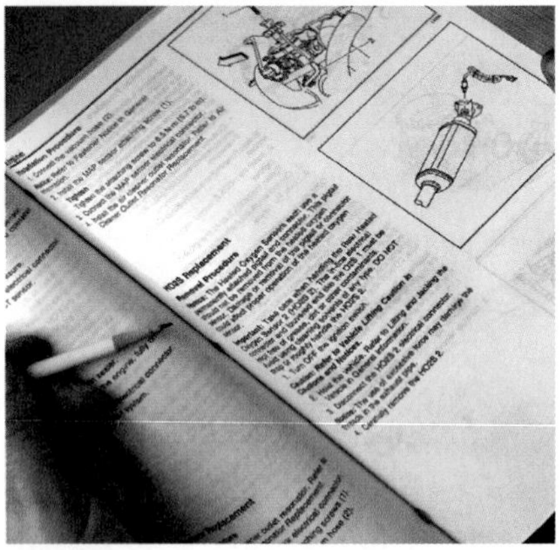

P32-1 *Locate the oxygen sensor in a wiring diagram for the vehicle and identify what part of the sensor each wire is connected to.*

P32-2 *Connect the positive lead of the meter to the power wire for the sensor's heater. Connect the meter's negative lead to a good ground.*

P32-3 *Place the meter where you can see it from the driver's seat.*

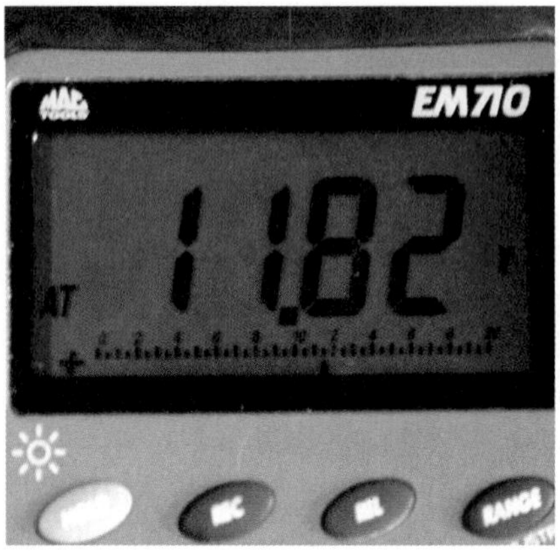

P32-4 *Start the engine and observe the voltage reading as the engine initially starts.*

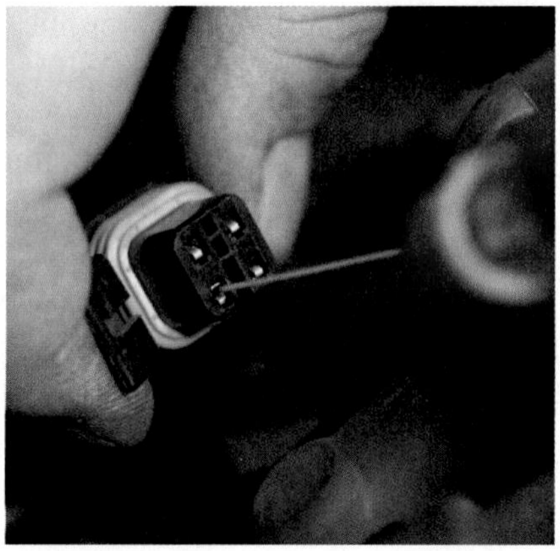

P32-5 *Turn off the engine and move the positive meter lead to the sensor's signal wire. Keep the negative lead grounded.*

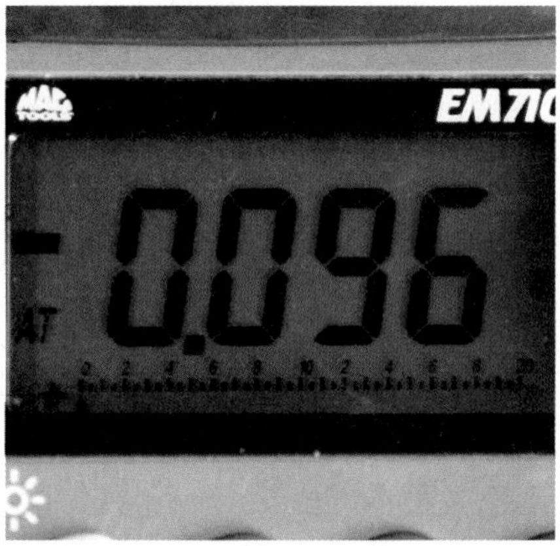

P32-6 *Restart the engine and allow it to reach normal operating temperature. Look at the meter to make sure the sensor's signal is toggling from low to high voltage.*

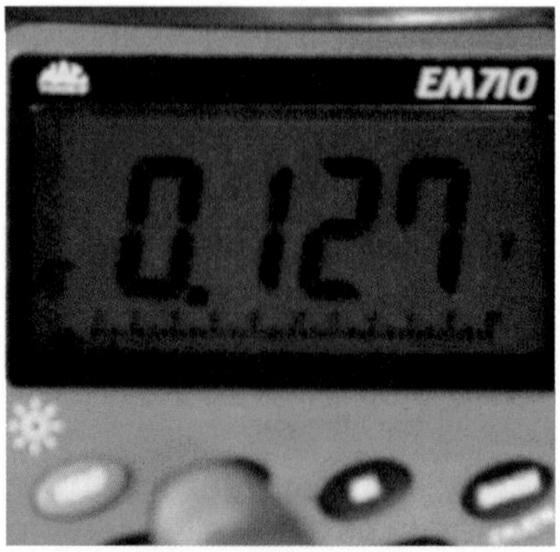

P32-7 *Press the Min/Max button on the meter and observe the voltage. This reading will be the minimum voltage and should be about 0.1 volt.*

P32-8 *Press the Min/Max button again to observe the maximum voltage reading. This should be about 0.9 volt.*

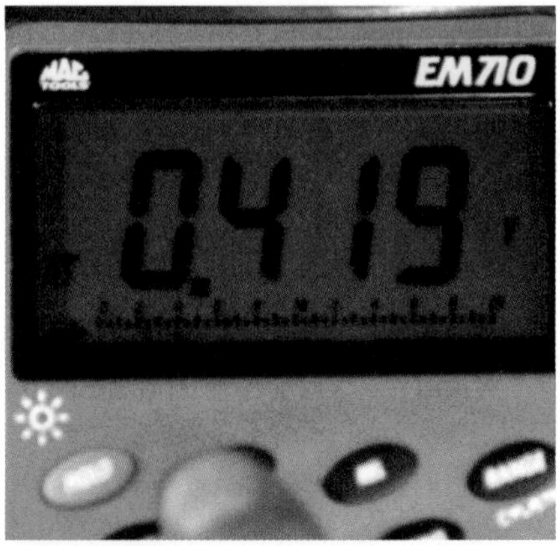

P32-9 *Press the Min/Max button again to read the average voltage. This reading should be about 0.45 volt. Repeat this test at different speeds to get a good look at how well the O_2 sensor responds.*

ground exceeds specifications, repair the ground wire or the sensor ground in the exhaust manifold.

Most late-model engines are fitted with heated O_2 sensors. If the O_2 sensor heater is not working, the sensor warm-up time is extended and the computer stays in open loop longer. In this mode, the computer supplies a richer air/fuel ratio. As a result, the engine's emissions are high and its fuel economy is reduced. To test the heater circuit, disconnect the O_2 sensor connector and connect a voltmeter between the heater voltage supply wire and ground. With the ignition switch on, 12 volts should be supplied on this wire. If the voltage is less than 12 volts, repair the fuse in this voltage supply wire or the wire itself.

With the O_2 sensor wire disconnected, connect an ohmmeter across the heater terminals in the sensor connector. If the heater does not have the specified resistance, replace the sensor.

Testing with a Scanner The output from an O_2 sensor should constantly cycle between high and low voltages as the engine is running in closed loop. This cycling is the result of the computer constantly correcting the air/fuel ratio in response to the feedback from the O_2 sensor. When the O_2 sensor reads lean, the computer will enrich the mixture. When the O_2 sensor reads rich, the computer will lean the mixture. When the computer does this, it is in control of the air/fuel mixture. Many things can occur to take that control away from the computer. One of them is a faulty O_2 sensor.

The activity of the sensor can be monitored on a scanner. Watch the scanner while the engine is running. The O_2 voltage should move to nearly 1 volt then drop back to close to zero volt. Immediately after it drops, the voltage signal should move back up. This immediate cycling is an important function of an O_2 sensor. If the response is slow, the sensor is lazy and should be replaced. With the engine at about 2,500 rpm, the O_2 sensor should cycle from high to low ten to forty times in 10 seconds. The voltage readings shown on the scanner are also an indicator of how well the sensor works. When testing the O_2 sensor, make sure the sensor is heated and the system is in closed loop **(Figure 34–26)**.

Testing with a Lab Scope A faulty O_2 sensor can cause many different types of problems. For example, it can cause excessively high HC and CO emissions and all sorts of driveability problems. Most computer systems monitor the activity of the O_2 sensor and store a code when the sensor's output is not within the desired range. Again, the normal range is between 0 and 1 volt and the sensor should constantly toggle from close to 0.2 to 0.8 volt, then back to 0.2 **(Figure 34–27)**. If the range that the sensor toggles in is within the specifications, the computer will think everything is normal and respond accordingly. This, however, does not mean the sensor is working properly.

Figure 34-26 A graphing scan tool. *Courtesy of Snap-on Tools Company*

The voltage signal from an O_2 sensor should have two to three **cross counts** with the engine without a load at 2,000 rpm. A cross count of an O_2 sensor is the number of times the waveform moves from a high voltage and back to that voltage **(Figure 34–28)**.

If the sensor's voltage toggles between zero volt and 500 millivolts, it is toggling within its normal range but it is not operating normally. It biased low or lean. As a result, the computer will be constantly adding fuel to try to reach the upper limit of the sensor. Something is causing the sensor to be biased lean. If the toggling only occurs at the higher limits of the voltage range, the sensor is biased rich. In either case, the computer does not have true control of the air/fuel mixture because of the faulty O_2 signals.

The O_2 can be biased rich or lean, not work at all, or work too slowly to ensure good emissions and fuel economy. To test the O_2 sensor for all of these concerns, use a lab scope. Begin by allowing the engine and O_2 sensor to warm up. Insert the hose of a propane enrichment tool into the power brake booster vacuum hose or simply install it into the nozzle of the air cleaner assembly. This will drive the mixture rich. Most good O_2 sensors will produce almost 1 volt when driven full rich. The typical specification is at least 800 millivolts.

Connect the lab scope to the sensor and a good ground. Set the scope to display the trace at 200 millivolts per division and 500 milliseconds per division. Inject some propane into the air cleaner assembly. Observe the O_2 signal's trace. The O_2 sensor should show over 800 millivolts. If the voltage does not go high, the O_2 sensor is bad and should be replaced. Now, remove the propane bottle and cause a vacuum leak by pulling off an intake vacuum hose. Watch the scope to see how the O_2 sensor reacts. It should drop to under 175 millivolts. If it does not, replace the sensor. These tests check the O_2 sensor, not the system, therefore they are reliable O_2 sensor checks.

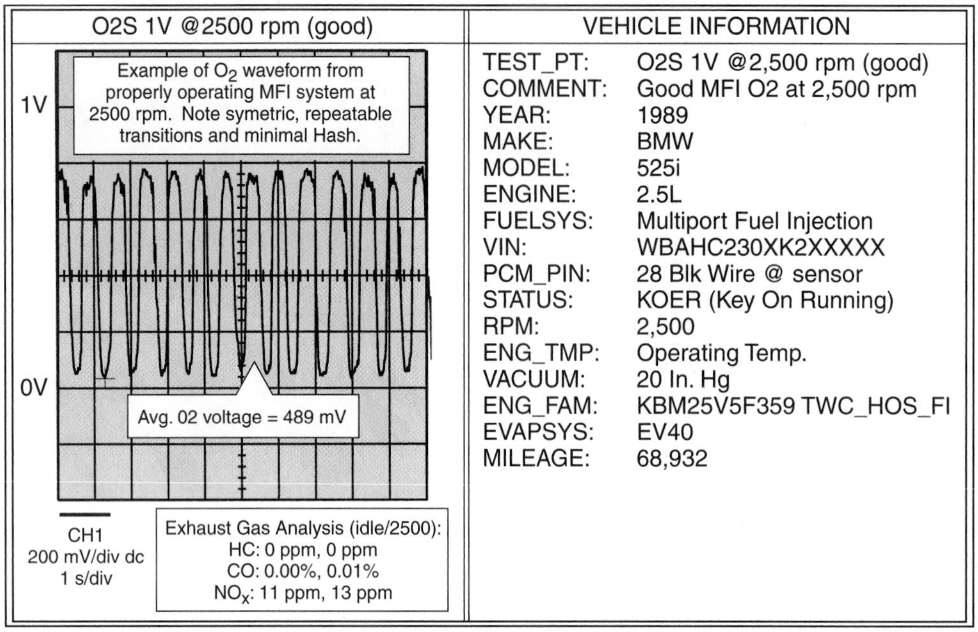

Figure 34-27 A good O_2 sensor trace. *Courtesy of Progressive Diagnostics—WaveFile AutoPro*

Also keep in mind that on an air pump-equipped car, it is a good idea to disable the air pump before doing this test. Unwanted air may bias the results.

Observing the trace of an O_2 sensor can also help in the diagnosis of other engine performance problems. **Figure 34–29** shows how ignition problems affect the signal from the O_2 sensor. Keep in mind that during complete combustion, nearly all of the oxygen in the combustion chamber is combined with the fuel. This means there will be little O_2 in the exhaust of a very efficient engine. As combustion becomes more incomplete, the levels of oxygen increase. Ignition problems cause incomplete combustion and there is much oxygen in the exhaust. This is also true of lean mixtures, overadvanced ignition timing, or anything else that causes incomplete combustion.

When the mixture is rich, combustion has a better chance of being complete. Therefore, the oxygen levels in the exhaust decrease. The O_2 sensor output will respond to the low oxygen with a high voltage signal **(Figure 34–30)**. Remember that the PCM will always try to do the opposite of what it receives from the O_2 sensor. When the O_2 shows lean, the PCM goes rich, and vice versa. When a lean exhaust signal is not caused by an air/fuel problem, the PCM does not know what the true cause is and will enrich the mixture in response to the signal. This may make the engine run worse than it did.

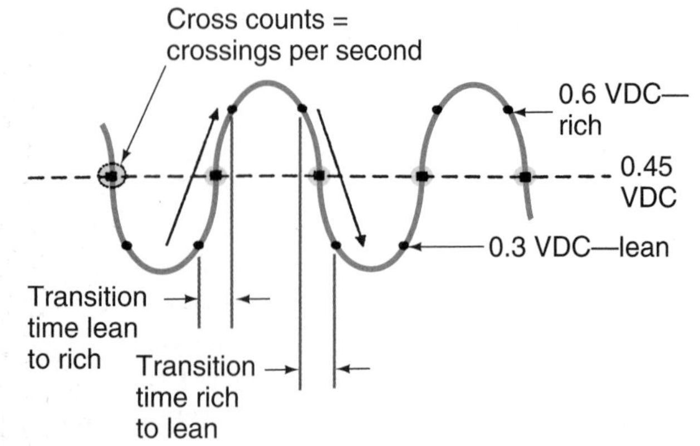

Figure 34-28 O_2 sensor signal cross counts. *Courtesy of OTC Tool and Equipment Division of SPX Corporation*

Identifying the Cause of O2S Contamination Many things can cause an oxygen sensor to become contaminated. Before simply replacing a contaminated sensor, find out why and how it was contaminated. Begin by examining the engine for leaks; oil, coolant, and other liquids that can plug the pores of the sensor and cause it to respond slowly and inaccurately to the amount of oxygen in the outside air or in the exhaust. If no leaks are evident, check the vehicle's service history. It is possible that recent problems that may or may not have been corrected are the cause of the contamination. For example, if the engine had some service done to it, RTV that was not designed for use around oxygen sensors may have been used.

You may also discover the cause by removing the sensor. The color and smell of the sensor may indicate the problem. If the sensor has a sweet smell, it is undoubtedly contaminated by engine coolant. If it smells burnt, there is a good chance that oil has melted onto the sensor. Silicone and engine coolant will leave white deposits on the sensor. Brown coloring may indicate oil contamination, and black means it was contaminated by a rich air/fuel mixture. Contaminated sensors should be replaced and not reinstalled.

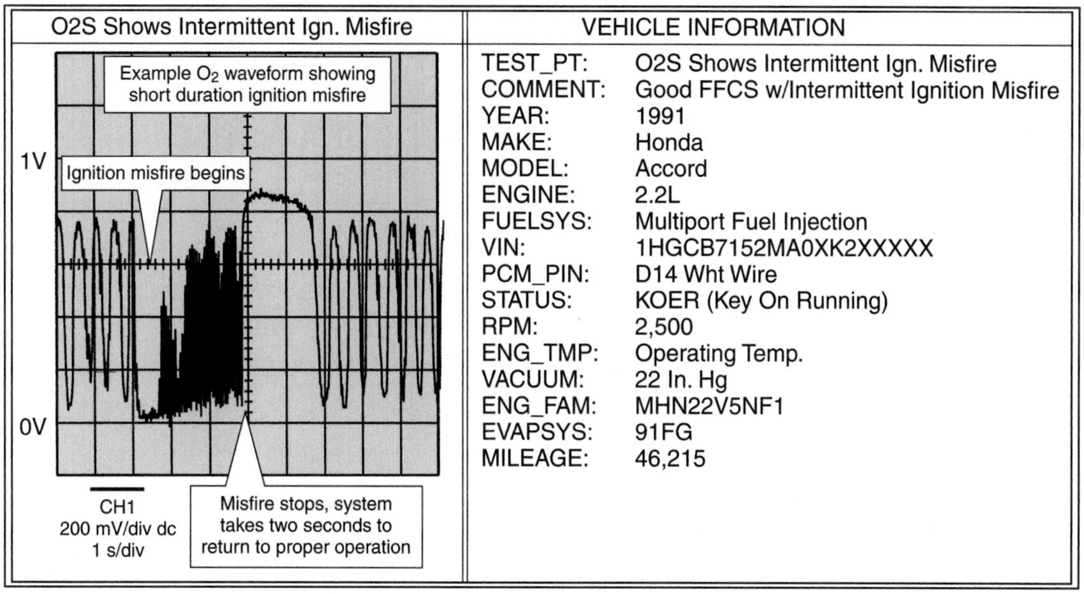

Figure 34-29 An O_2 sensor signal caused by an ignition problem. *Courtesy of Progressive Diagnostics—WaveFile AutoPro*

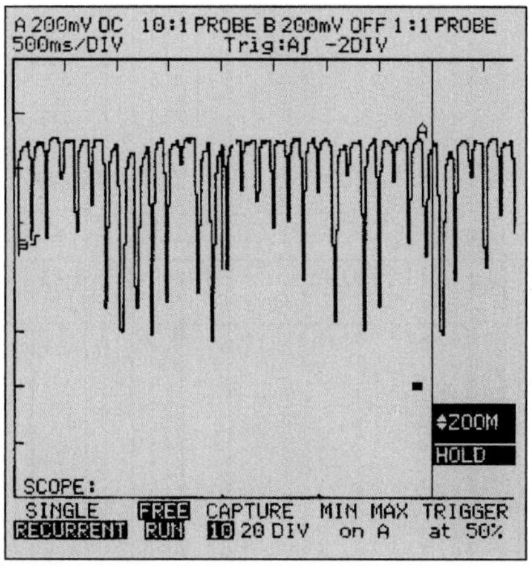

Figure 34-30 An O_2 sensor signal caused by a defective fuel injector. *Reproduced with permission from Fluke Corporation*

Hall-Effect Switches

To test a Hall-effect sensor, disconnect its wiring harness. Connect a voltage source of the correct low voltage level across the positive and negative terminals of the Hall layer. Then connect a voltmeter across the negative and signal voltage terminals.

Insert a metal feeler gauge between the Hall layer and the magnet. Make sure the feeler gauge is touching the Hall element. If the sensor is operating properly, the meter will read close to battery voltage. When the feeler gauge blade is removed, the voltage should decrease. On some units, the voltage will drop to near zero. Check the service manual to see what voltage you should observe when installing and removing the feeler gauge.

PIP and SPOUT are terms specific to Ford vehicles; however, they are similar to General Motors' EST and ignition reference signals. The PIP signal is an ignition reference signal from the crankshaft position sensor, which is a Hall-effect sensor. The PCM uses the signal to determine crankshaft position and engine speed. The computer uses PIP to build the SPOUT, which it sends to the ignition module. The SPOUT signal is a ground-controlled signal. It appears in a trace as a downward pulse that also represents the beginning of dwell **(Figure 34–31)**.

When observing a Hall-effect sensor on a lab scope, pay attention to the downward and upward pulses. These should be straight. If they appear at an angle **(Figure 34–32)**, that indicates the transistor is faulty and is causing the voltage to rise slowly. This can cause a no-start condition. The entire square wave from a Hall-effect unit should be flat. Any change from a normal trace means the sensor should be replaced.

Magnetic Pulse Generators

Ignition pickup coils, wheel speed sensors, and vehicle speed sensors are permanent magnet voltage-producing sensors. The procedures for testing each of these is similar.

A pickup coil is a common AC voltage source. The frequency and amplitude of the output signal from an ignition pickup coil will depend on the numbers of cylinders and the speed of the engine. A good pickup coil will have even peaks that reach at least 300 millivolts when the engine is cranking **(Figure 34–33)**.

The waveforms from most crankshaft position sensors will have a number of equally spaced pulses and one double pulse or sync signal, as shown in **Figure 34–34**. The number of evenly spaced pulses equals the number of cylinders that the engine has. Carefully examine the trace. Any glitches indicate a problem with the sensor or sensor circuit.

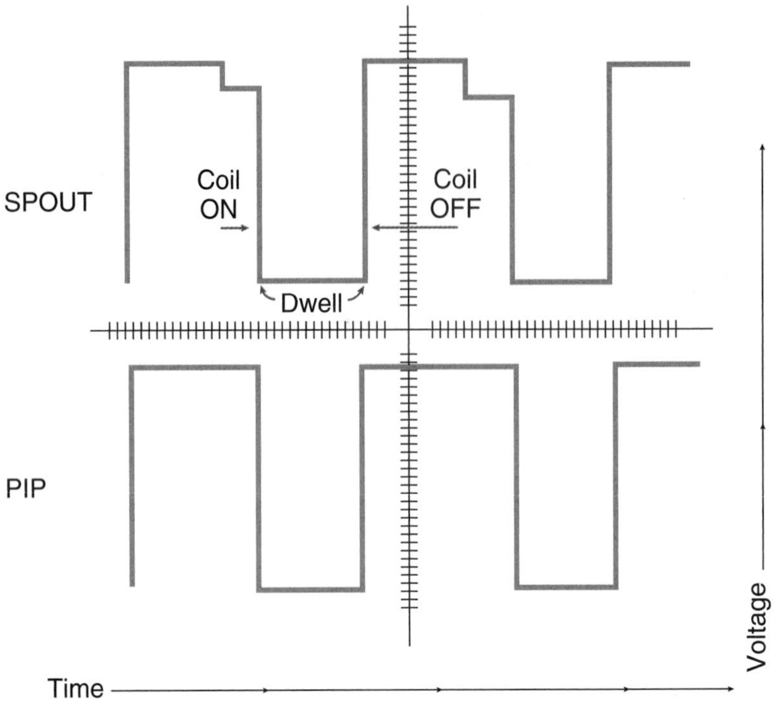

Figure 34-31 PIP and SPOUT signals.

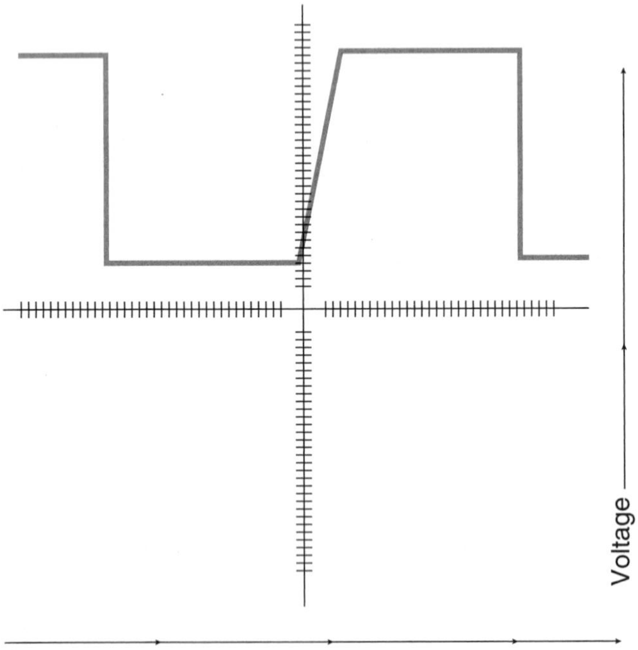

Figure 34-32 A Hall-effect switch with a bad transistor.

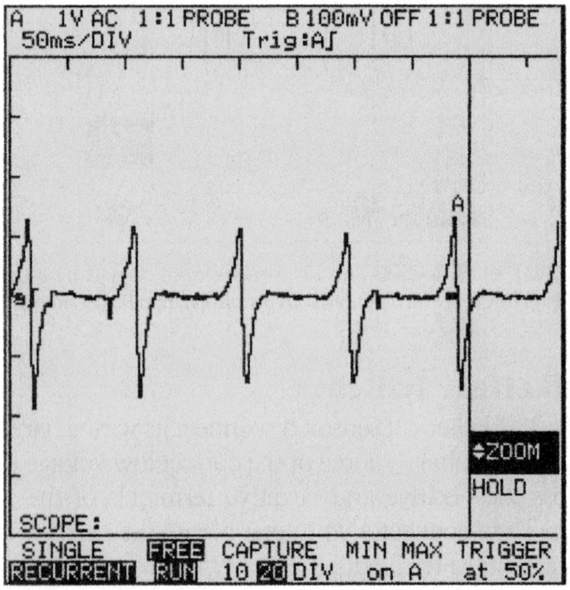

Figure 34-33 The trace of a good PM generator. *Reproduced with permission from Fluke Corporation*

Vehicle Speed Sensor

A defective vehicle speed sensor may cause different problems depending on the computer output control functions. A defective vehicle speed sensor (VSS) may cause improper converter clutch lockup, improper cruise control operation, and inaccurate speedometer operation.

Prior to VSS diagnosis, the vehicle should be lifted on a hoist so the drive wheels are free to rotate. Backprobe the VSS output wire and connect the voltmeter leads from this wire to ground. Select the 20-volt AC scale on the voltmeter. Then start the engine.

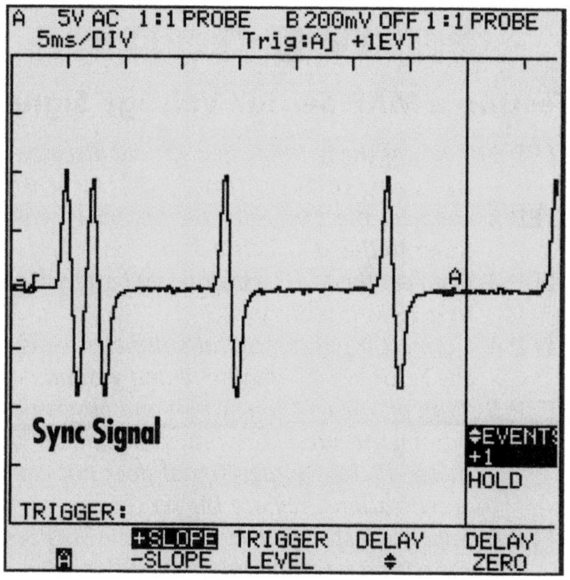

Figure 34-34 The trace of a good crankshaft position sensor. *Reproduced with permission from Fluke Corporation*

Place the transaxle in drive and allow the drive wheels to rotate. If the VSS voltage signal is not 0.5 volt, or more, replace the sensor. When the VSS provides the specified voltage signal, backprobe the VSS terminal at the PCM and repeat the voltage signal test with the drive wheels rotating. If 0.5 volt is available at this terminal, the trouble may be in the PCM.

When 0.5 volt is not available at this terminal, turn the ignition switch off and disconnect the wire from the VSS to the PCM. Connect the ohmmeter leads across the wire. The meter should read zero ohm. Repeat the test with the ohmmeter leads connected to the VSS ground terminal and the PCM ground terminal. This wire should also have zero ohm resistance. If the resistance in these wires is more than specified, repair the wires.

The condition of a speed sensor can be checked by going through the diagnostic routines for the system. If this test indicates that a sensor is faulty, the sensor should be replaced. Speed sensors can also be checked with an ohmmeter. Most manufacturers list a resistance specification. The resistance of the sensor is measured across the sensor's terminals. The typical range for a good sensor is 800 to 1400 ohms of resistance.

Knock Sensors

A defective knock sensor may cause engine detonation or reduced spark advance and fuel economy. When a knock sensor is removed and replaced, the sensor must be tightened to its specified torque. The procedure for checking a knock sensor varies depending on the vehicle make and year. Always follow the vehicle manufacturer's recommended test procedure and specifications. Follow these steps for a typical knock sensor diagnosis:

PROCEDURE

Typical Knock Sensor Diagnosis

STEP 1 *Disconnect the knock sensor wiring connector, and turn on the ignition switch.*

STEP 2 *Connect a voltmeter from the disconnected knock sensor wire to ground. The voltage should be 4 to 6 volts. If the specified voltage is not available at this wire, backprobe the knock sensor wire at the computer and read the voltage at this terminal. If the voltage is satisfactory at this terminal, repair the knock sensor wire. When the voltage is not within specifications at the computer terminal, replace the computer.*

STEP 3 *Connect an ohmmeter from the knock sensor terminal to ground. Some knock sensors should have 3,300 to 4,500 ohms. If the knock sensor does not have the specified resistance, replace the sensor.*

Manifold Absolute Pressure Sensors

A defective manifold absolute pressure sensor may cause a rich or lean air/fuel ratio, excessive fuel consumption, and engine surging. This diagnosis applies to MAP sensors that produce an analog voltage signal. With the ignition switch on, backprobe the 5-volt reference wire and connect a voltmeter from the reference wire to ground.

SHOP TALK

Manifold absolute pressure sensors have a much different calibration on turbocharged engines than on nonturbocharged engines. Be sure you are using the proper specifications for the sensor being tested. ■

If the reference wire is not supplying the specified voltage, check the voltage on this wire at the computer. If the voltage is within specifications at the computer, but low at the sensor, repair the reference wire. When this voltage is low at the computer, check the voltage supply wires and ground wires on the computer. If these wires are satisfactory, replace the computer.

With the ignition switch on, connect the voltmeter from the sensor ground wire to the battery ground. If the voltage drop across this circuit exceeds specifications, repair the ground wire from the sensor to the computer.

Backprobe the MAP sensor signal wire and connect a voltmeter from this wire to ground with the ignition switch on. The voltage reading indicates the barometric pressure signal from the MAP sensor to the computer. Many MAP sensors send a barometric pressure signal to the computer each time the ignition switch is turned on and each time the throttle is in the wide-open position. If the voltage supplied by the barometric pressure signal in the MAP sensor does not equal the vehicle manufacturer's specifications, replace the MAP sensor.

The barometric pressure voltage signal varies depending on altitude and atmospheric conditions. Follow this calculation to obtain an accurate barometric pressure reading:

1. Phone your local weather or TV station and obtain the present barometric pressure reading; for example, 29.85 inches. The pressure they quote is usually corrected to sea level.
2. Multiply your altitude by 0.001; for example, 600 feet × 0.001 = 0.6.
3. Subtract the altitude correction from the present barometric pressure reading: 29.85 – 0.6 = 29.79.
4. Check the vehicle manufacturer's specifications to obtain the proper barometric pressure voltage signal in relation to the present barometric pressure **(Figure 34–35)**.

To check the voltage signal of a MAP, turn the ignition switch on and connect a voltmeter to the MAP sensor signal wire. Connect a vacuum pump to the MAP sensor vacuum connection and apply 5 in. Hg (127 mm Hg) of vacuum. On some MAP sensors, the sensor voltage signal should change 0.7 to 1.0 volt for every 5 in. Hg (127 mm Hg) of vacuum change applied to the sensor. With 5 in. Hg (127 mm Hg) of vacuum applied to the MAP sensor, the voltage should be 3.5 volts to 3.8 volts. When 10 in. Hg (254 mm Hg) of vacuum is applied to the sensor, the voltage signal should be 2.5 volts to 3.1 volts. Check the MAP sensor voltage at 5-inch (127-mm) intervals from 0 to 25 inches (635 mm). If the MAP sensor voltage is not within specifications at any vacuum, replace the sensor.

To check a MAP sensor with a lab scope, connect the scope to the MAP output and a good ground. When the engine is accelerated and returned to idle, the output voltage should increase and decrease **(Figure 34–36)**. If the engine is accelerated and the MAP sensor voltage does not rise and fall, or if the signal is erratic, the sensor or sensor wires are defective.

If the MAP sensor produces a digital voltage signal of varying frequency, check the 5-volt reference wire and the ground wire with the same procedure used on other MAP sensors. Photo Sequence 33 shows a typical procedure for testing a Ford MAP sensor, which has a varying frequency. Follow these steps to test the MAP sensor voltage signal:

PROCEDURE

Testing a MAP Sensor Voltage Signal

STEP 1 *Turn off the ignition switch, and disconnect the wiring connector from the MAP sensor.*

STEP 2 *Connect the connector on the MAP sensor tester to the MAP sensor.*

STEP 3 *Connect the MAP sensor tester battery leads to a 12-volt battery.*

STEP 4 *Connect a pair of digital voltmeter leads to the MAP tester signal wire and ground.*

STEP 5 *Turn on the ignition switch and observe the barometric pressure voltage signal on the meter. If this voltage signal does not equal specifications, replace the sensor.*

STEP 6 *Supply the specified vacuum to the MAP sensor with a vacuum pump.*

STEP 7 *Observe the voltmeter reading at each specified vacuum. If the MAP sensor voltage signal does not equal the specifications at any vacuum, replace the sensor.*

If a special MAP tester is not available, the sensor can be checked with a DMM that measures frequency. Connect the meter to the MAP sensor. Measure the voltage, duty cycle, and frequency at the sensor with no vacuum applied. Then apply about 18 in. Hg (457 mm Hg) of vacuum to the MAP. Observe and record the same readings. A good MAP will have about the same amount of voltage and duty cycle with or without the vacuum. However, the frequency should decrease **(Figure 34–37)**. Normally a frequency of about 155 Hertz is expected at sea level with no vacuum applied to the MAP. When vacuum is applied, the frequency should decrease to around 95 Hz.

A lab scope can be used to check a Ford MAP sensor. The upper horizontal line of the trace should be at 5 volts and the lower horizontal line should be close to zero **(Figure 34–38)**. Check the waveform for unusual movements of the trace. If the waveform is anything but normal, replace the sensor.

Mass Airflow Sensors

The MAF sensor measures the mass of the air being drawn into the engine. The signal from the MAF is used by the PCM to calculate injector pulse width. Two types of MAF sensors are commonly used: vane-type and hot-wire types. A faulty MAF will cause driveability problems resulting from incorrect ignition timing and improper air/fuel ratios.

Vane-Type MAF Sensors Begin checking a vane-type MAF sensor by checking the voltage supply wire and the ground wire to the MAF module before checking the sensor voltage signal. Always follow the recommended test

PHOTO SEQUENCE 33

Testing a Ford MAP Sensor

P33-1 *Remove the MAP sensor's electrical connector and vacuum hose.*

P33-2 *Connect the appropriate connector of the MAP sensor tester to the MAP sensor.*

P33-3 *Connect the remaining tester connector to the MAP sensor's electrical connector.*

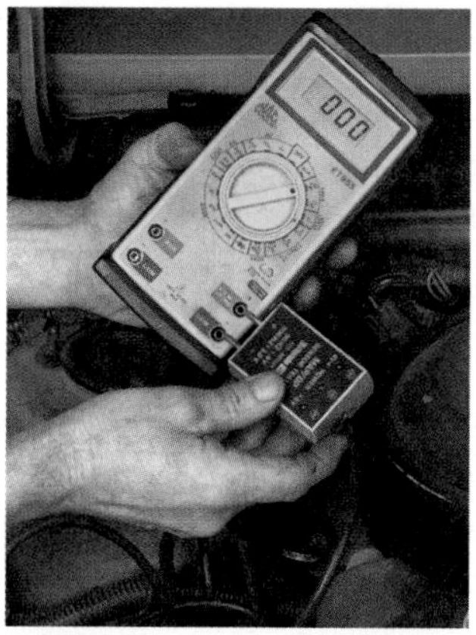

P33-4 *Insert the voltage terminals of the MAP sensor tester into the test lead terminals on a DMM; make sure the polarity is correct.*

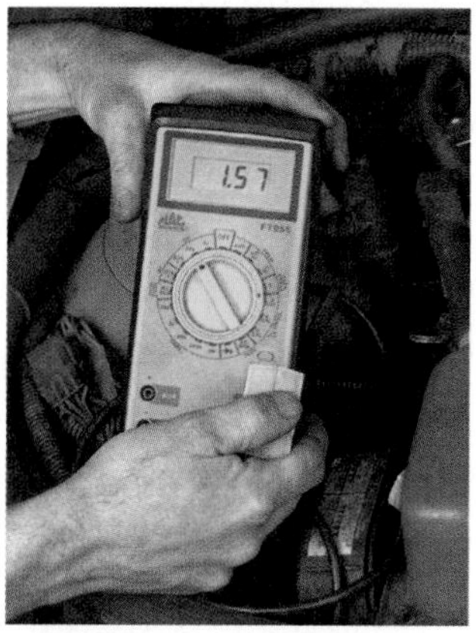

P33-5 *Observe the MAP sensor barometric pressure voltage reading on the voltmeter. Compare this reading to the specifications.*

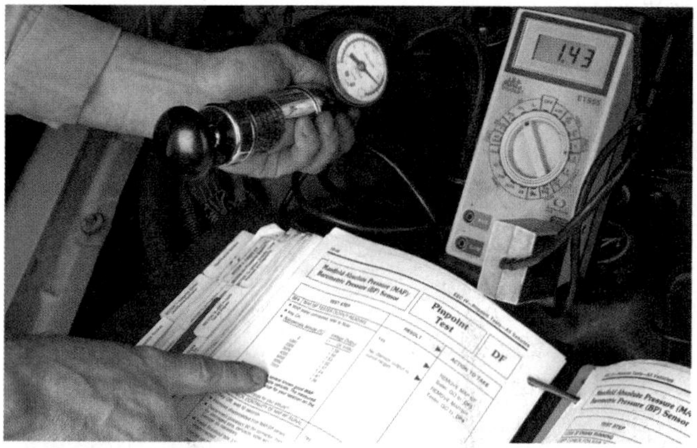

P33-6 *Connect a hand-operated vacuum pump to the MAP sensor and apply 5 in. Hg to the sensor. Observe the sensor's voltage signal on the voltmeter. Compare the reading to specifications.*

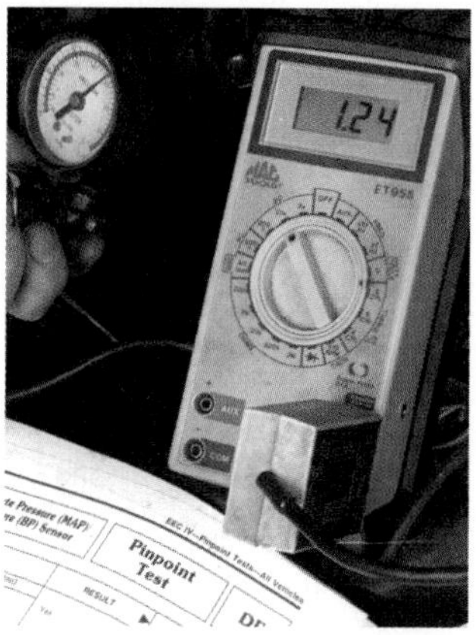

P33-7 *Apply 10 in. Hg to the MAP sensor. Observe the voltage reading now and compare this to specifications.*

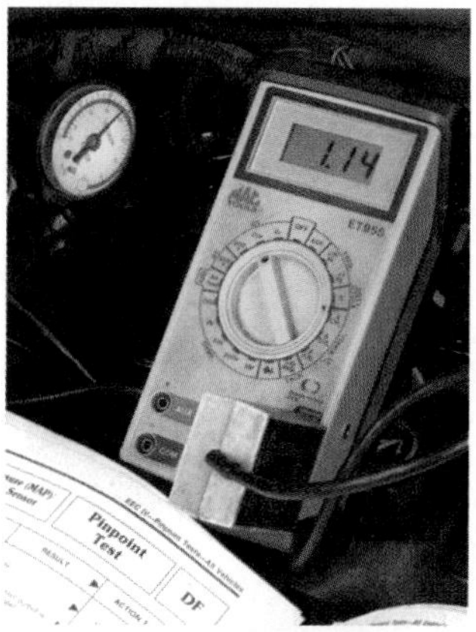

P33-8 *Increase the vacuum to the MAP sensor so that it now has 15 in. Hg. Observe the voltage reading now and compare this to specifications.*

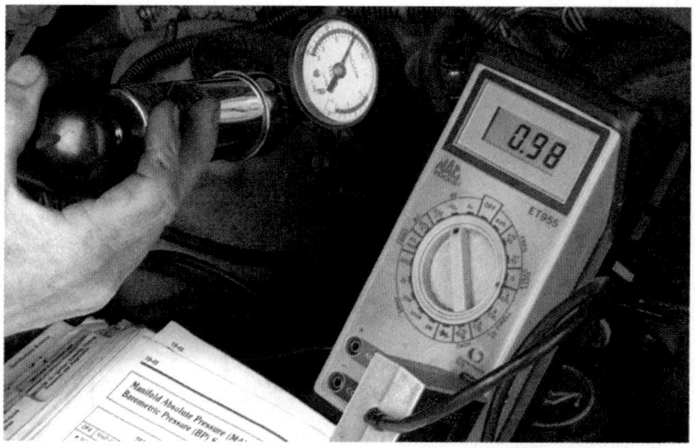

P33-9 *Increase the vacuum to 20 in. Hg. Observe the voltage reading now and compare this to specifications. If the voltage signals from any of the tests do not match specifications, the MAP sensor needs to be replaced. Disconnect the tester and reconnect the electrical connector and vacuum hose to the MAP after you have completed your testing.*

Absolute Baro Reading	Lowest Allowable Voltage at –40°F	Lowest Allowable Voltage at 257°F	Lowest Allowable Voltage at 77°F	TBI MAP Sensor Designed Output Voltage	Highest Allowable Voltage at 77°F	Highest Allowable Voltage at 257°F	Highest Allowable Voltage at –40°F
31.0″	4.548 V	4.632 V	4.716 V	4.800 V	4.884 V	4.968 V	5.052 V
30.9″	4.531 V	4.615 V	4.699 V	4.783 V	4.867 V	4.951 V	5.035 V
30.8″	4.514 V	4.598 V	4.682 V	4.766 V	4.850 V	4.934 V	5.018 V
30.7″	4.497 V	4.581 V	4.665 V	4.749 V	4.833 V	4.917 V	5.001 V
30.6″	4.480 V	4.564 V	4.648 V	4.732 V	4.816 V	4.900 V	4.984 V
30.5″	4.463 V	4.547 V	4.631 V	4.715 V	4.799 V	4.883 V	4.967 V
30.4″	4.446 V	4.530 V	4.614 V	4.698 V	4.782 V	4.866 V	4.950 V
30.3″	4.430 V	4.514 V	4.598 V	4.682 V	4.766 V	4.850 V	4.934 V
30.2″	4.413 V	4.497 V	4.581 V	4.665 V	4.749 V	4.833 V	4.917 V
30.1″	4.396 V	4.480 V	4.564 V	4.648 V	4.732 V	4.816 V	4.900 V
30.0″	4.379 V	4.463 V	4.547 V	4.631 V	4.715 V	4.799 V	4.883 V

Figure 34-35 Barometric pressure voltage signal specifications at different barometric pressures. *Courtesy of DaimlerChrysler Corporation*

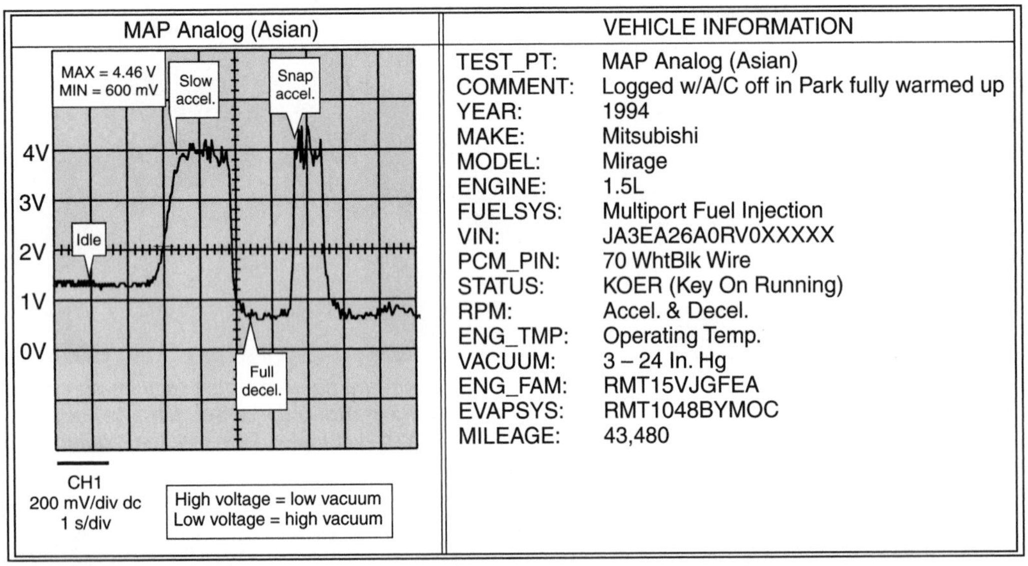

Figure 34-36 A trace of a normal MAP sensor. *Courtesy of Progressive Diagnostics—WaveFile AutoPro*

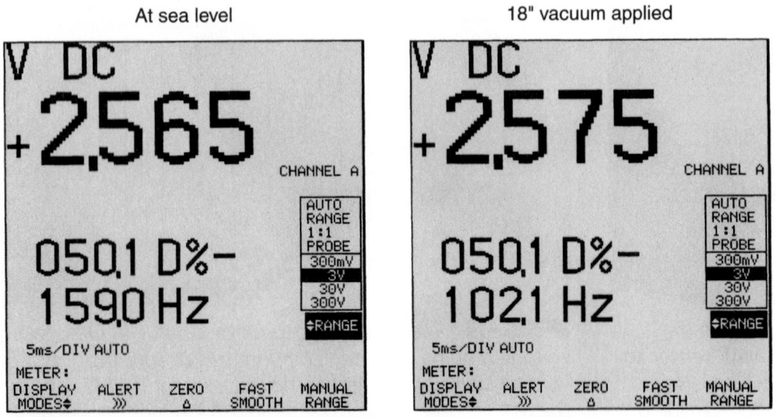

Figure 34-37 The reaction of a good Ford MAP sensor with vacuum applied. *Reproduced with permission from Fluke Corporation*

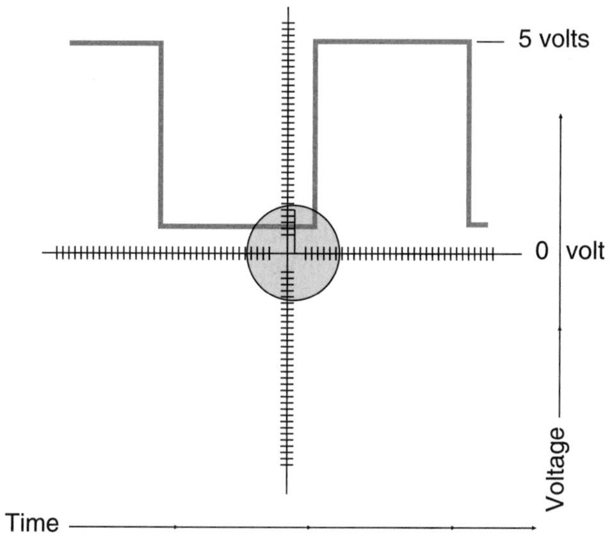

Figure 34-38 A good Ford MAP sensor signal.

procedure in the manufacturer's service manual and use the specifications supplied by the manufacturer.

Typically, to test the sensor, a DMM is used and set on a DC voltage scale. The negative meter lead is connected to ground and the red lead to the MAF signal wire **(Figure 34–39)**. Turn on the ignition switch and press the Min/Max button, if available, on the DMM. Slowly push the MAF vane from the closed to the wide-open position, and allow the vane to slowly return to the closed position. Observe the maximum and minimum voltage readings as the vane was moved. If the minimum voltage signal is zero, there may be an open circuit in the MAF sensor variable resistor. When the voltage signal is not within the manufacturer's specifications, replace the sensor.

WARNING!

While pushing the mass airflow sensor vane open and closed, be careful not to mark or damage the vane or sensor housing.

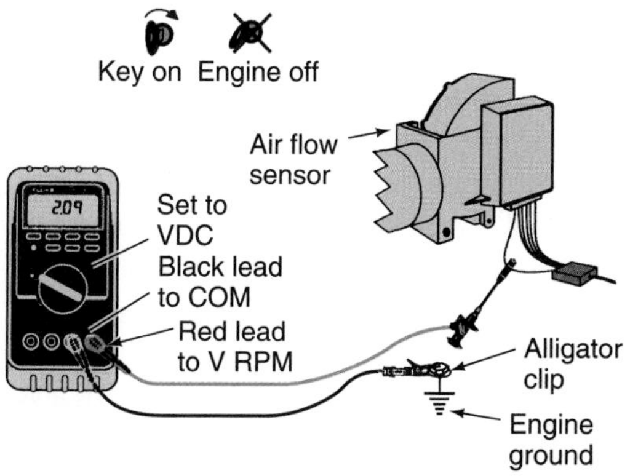

Figure 34-39 A voltmeter connected to measure the signal from a MAF sensor. *Reproduced with permission from Fluke Corporation*

Some vehicle manufacturers specify ohmmeter tests for the MAF sensor. With the MAF sensor removed, connect the ohmmeter across the sensor's output and input terminals. The resistance at these terminals is normally 200 to 600 ohms.

Connect the ohmmeter leads to the specified MAF sensor terminals, and move the vane from the fully closed to the fully open position. With each specified meter connection and vane position, the ohmmeter should indicate the specified resistance **(Figure 34–40)**. When the ohmmeter leads are connected to the sensor's input and output terminals, the ohmmeter reading should increase smoothly as the sensor vane is opened.

To check a vane-type MAF with a lab scope, connect the positive lead to the output signal terminal and the negative scope lead to a good ground. This type MAF should display an analog voltage signal when the engine is accelerated. A defective MAF will have sudden and erratic voltage changes **(Figure 34–41)**.

Hot-Wire–Type MAF Sensors The test procedure for heated resistor and hot-wire MAF sensors varies depending on the vehicle make and year. Always follow the

BETWEEN TERMINALS	RESISTANCE (Ω)	MEASURING PLATE OPENING
FC – E1	Infinity	Fully closed
FC – E1	Zero	Other than closed
VS – E2	200 – 600	Fully closed
VS – E2	20 – 1,200	Fully open

Figure 34-40 Resistance specifications for a typical MAF sensor with the door open and closed. *Reprinted with permission*

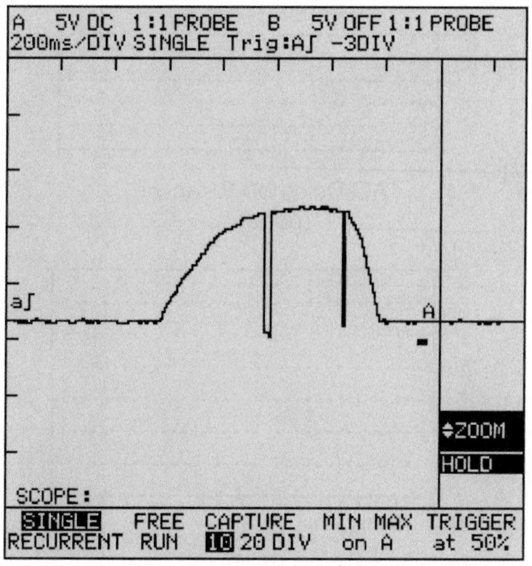

VANEOPEN.PCX

Figure 34-41 The trace of a defective vane-type MAF sensor. *Reproduced with permission from Fluke Corporation*

test procedure in the appropriate service manual. A frequency test may be performed on some MAF sensors, such as the AC Delco MAF on some General Motors' products.

To check the MAF sensor's voltage signal and frequency, connect a voltmeter across the MAF voltage signal wire and ground wire. Start the engine and observe the voltmeter reading. On some MAF sensors, this reading should be 2.5 volts. Lightly tap the MAF sensor housing with a screwdriver handle and watch the voltmeter pointer. If the pointer fluctuates or the engine misfires, replace the MAF sensor. Some MAF sensors have experienced loose internal connections, which cause erratic voltage signals and engine misfiring and surging.

Set the DMM so that it can read the frequency of DC voltage. With it still connected to the signal wire and ground, the meter should read about 30 Hz with the engine idling. Now, increase the engine speed, and record the meter reading at various speeds. Graph the frequency readings. The MAF sensor frequency should increase smoothly and gradually in relation to engine speed. If the MAF sensor frequency reading is erratic, replace the sensor **(Figure 34–42)**.

When a scanner is used to diagnose a General Motors' vehicle, one test mode displays grams per second from the MAF sensor. This mode provides an accurate test of the MAF sensor. The grams per second reading should be 4 to 7 with the engine idling. This reading should gradually increase as the engine speed increases. When the engine speed is constant, the grams-per-second reading should remain constant. If the grams-per-second reading is erratic at a constant engine speed or if this reading varies when the sensor is tapped lightly, the sensor is defective. A MAF sensor fault code may not be present with an erratic grams-per-second reading, but the erratic reading indicates a defective sensor.

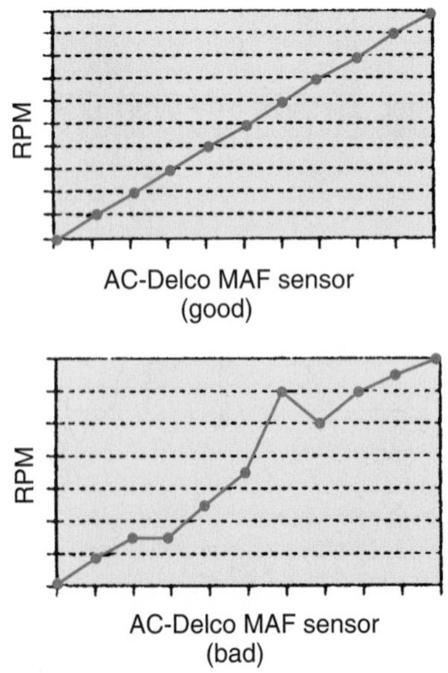

Figure 34–42 Satisfactory and unsatisfactory MAF sensor frequency readings. *Reproduced with permission from Fluke Corporation*

Frequency-varying types of MAF sensors can be tested with a lab scope. The waveform should appear as a series of square waves **(Figure 34–43)**. When the engine speed and intake airflow increase, the frequency of the MAF sensor signals should increase smoothly and proportionately to the change in engine speed. If the MAF or connecting wires is defective, the trace will show an erratic change in frequency **(Figure 34–44)**.

TESTING ACTUATORS

Most computer-controlled actuators are electromechanical devices that convert the output commands from the computer into mechanical action. These actuators are used to open and close switches, control vacuum flow to other components, and operate valves depending on the requirements of the system.

Common computer-controlled outputs are solenoid valves for EGR control and venting, secondary AIR bypass and diversion, and EVAP canister purge. Fuel injectors are also solenoids controlled by the PCM. Many motors and relays that control motors are also activated by the PCM; some examples are the idle speed motor, fuel pump, fuel pump relay, A/C relay, and the mixture heater relay. The PCM also controls the operation of many specific modules, such as the ignition, A/C controller, and cooling fan modules. Information signals from the PCM are also outputs. This information is used for self-diagnosis, ignition timing, shift indicator lamp operation, and the operation of the MIL.

Most systems allow for testing of the actuator through a scan tool. Actuators that are duty cycled by the computer are more accurately diagnosed through this method. Prior to diagnosing an actuator, make sure the engine's compression, ignition system, and intake system are in good condition. Using a scanner, serial data can be used to diagnose outputs. The displayed data should be compared against specifications to determine the condition of any actuator. Also, when an actuator is suspected of being faulty, make sure the inputs related to the control of that actuator are within normal range. Faulty inputs will cause an actuator to appear faulty.

If the actuator is tested by means other than a scanner, always follow the manufacturer's recommended procedures. Because many actuators operate with 5 to 7 volts, never connect a jumper wire from a 12-volt source unless directed to do so by the appropriate service procedure. Some actuators are easily tested with a voltmeter by testing for input voltage to the actuator. If there is the correct amount of input voltage, check the condition of the ground. If both of these are good, then the actuator is faulty. If an ohmmeter needs to be used to measure the

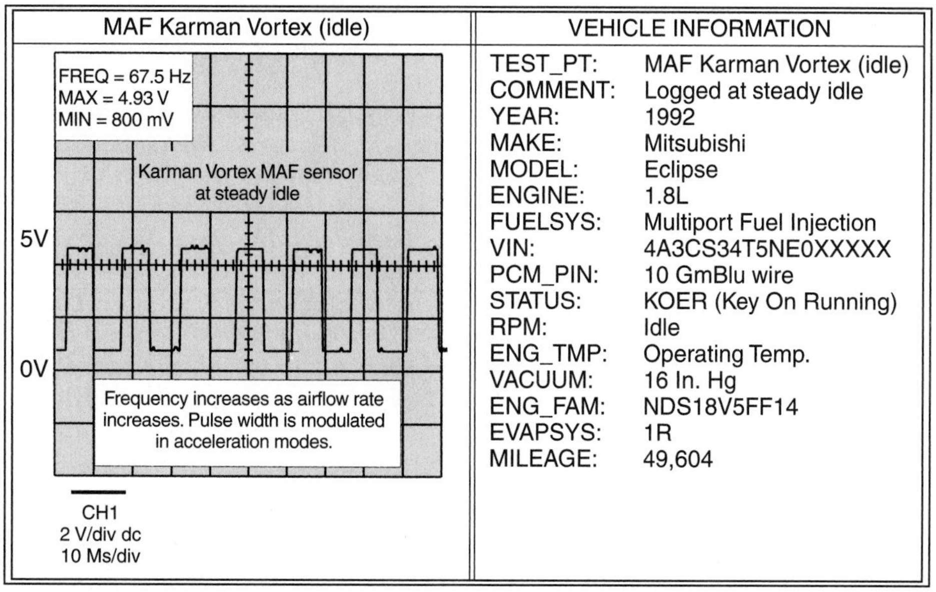

Figure 34-43 A normal trace for a frequency varying MAF sensor. *Courtesy of Progressive Diagnostics—WaveFile AutoPro*

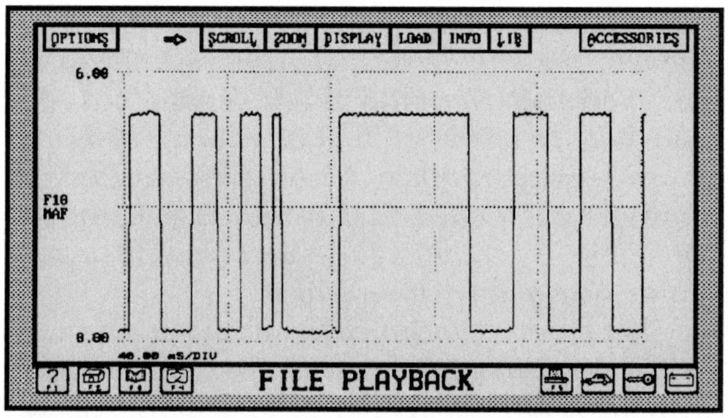

Figure 34-44 The trace of a defective frequency varying MAF sensor. *Courtesy of EDGE Diagnostics Systems*

resistance of an actuator, disconnect it from the circuit first.

When checking anything with an ohmmeter, logic can dictate good and bad readings. If the meter reads infinite, this means there is an open. Based on what you are measuring across, an open could be good or bad. The same is true for very low resistance readings. Across some things, this would indicate a short. For example, you do not want an infinite reading across the windings of a solenoid. You want low resistance. However, you want an infinite reading from one winding terminal to the case of the solenoid. If you have low resistance, the winding is shorted to the case.

Testing Actuators with a Lab Scope

Most computer-controlled circuits are ground-controlled circuits. The PCM energizes the actuator by providing the ground. On a scope trace, the on-time pulse is the downward pulse. On positive-feed circuits, in which the computer is supplying the voltage to turn a circuit on, the on-time pulse is the upward pulse. One complete cycle is measured from one on-time pulse to the beginning of the next on-time pulse.

Actuators are electromechanical devices, meaning they are electrical devices that cause some mechanical action. When actuators are faulty, it is because they are electrically faulty or mechanically faulty. By observing the action of an actuator on a lab scope, you will be able to watch its electrical activity. Normally, if there is a mechanical fault, this will affect its electrical activity as well. Therefore, you get a good sense of the actuator's condition by watching it on a lab scope.

To test an actuator, you need to know what it basically is. Most actuators are solenoids. The computer controls the action of the solenoid by controlling the pulse width of the control signal. By watching the control signal, you can see the turning on and off of the solenoid **(Figure 34–45)**. The voltage spikes are caused by the discharge of the coil in the solenoid

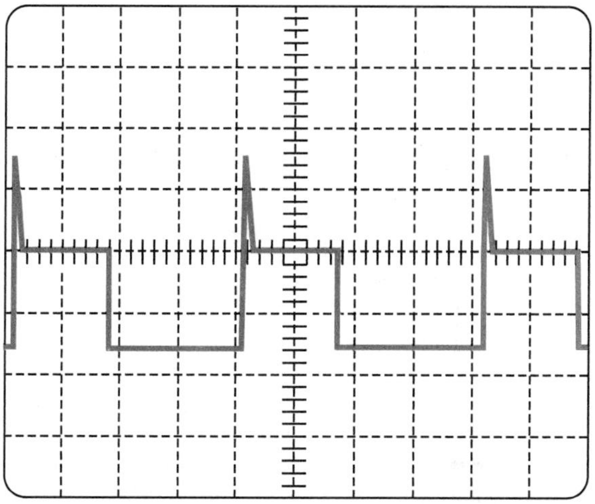

Figure 34-45 A typical solenoid control signal.

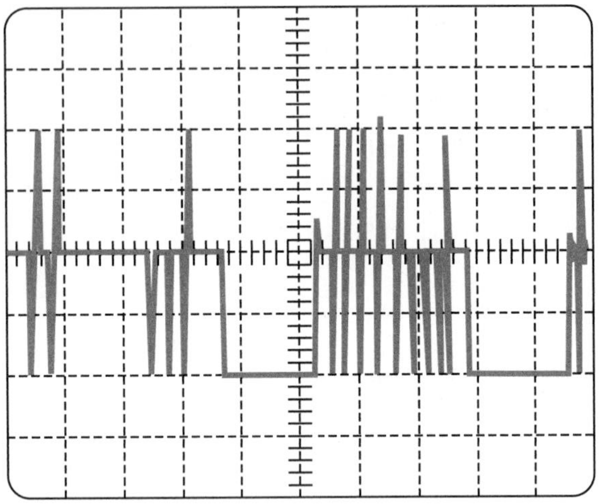

Figure 34-46 A typical pulse-width modulated solenoid control signal.

Some actuators are controlled pulse-width modulated signals **(Figure 34–46)**. These signals show a changing pulse width. These devices are controlled by varying the pulse width, signal frequency, and voltage levels.

Both waveforms should be checked for amplitude, time, and shape. You should also observe changes to the pulse width as operating conditions change. A bad waveform will have noise, glitches, or rounded corners. You should be able to see evidence that the actuator immediately turns off and on according to the commands of the computer.

A fuel injector is actually a solenoid. The PCM's signals to an injector vary in frequency and pulse width. Frequency varies with engine speed and the pulse width varies with fuel control. Increasing an injector's on-time increases the amount of fuel delivered to the cylinders. The trace of a normally operating fuel injector is shown in **Figure 34–47**.

CASE STUDY

***A** customer brought his OBD-II–equipped car to the dealership and complained that the car had poor fuel economy. The customer stated that the gas mileage had been declining since the car was new. This is not normal. Gas mileage normally improves slightly as the engine is broken in. The customer had no other complaints. Because this problem is very difficult to verify, the technician began the diagnosis process with a visual inspection and found nothing out of the ordinary. The car's MIL was not lit.*

He then connected a scan tool to the DLC and reviewed the data. Comparing the input data being displayed to the normal range of values listed in the service manual, he discovered the upstream O_2 sensor was biased rich. That meant the PCM was seeing a lean condition and adding fuel to correct for this problem. To verify this, the technician watched the fuel trim. Sure enough, the long-term fuel trim had moved to add more fuel. Normally this is caused by a vacuum leak or restricted fuel injectors. The latter cause seemed unlikely because the O_2 sensor showed that added fuel was being delivered. Therefore, the technician began to look for a possible cause of a vacuum leak.

The process continued for quite some time as he checked all of the vacuum hoses and components. Nothing appeared to be leaking. As he leaned over to check something in the back of the engine, he heard a slight "pffft." The noise had

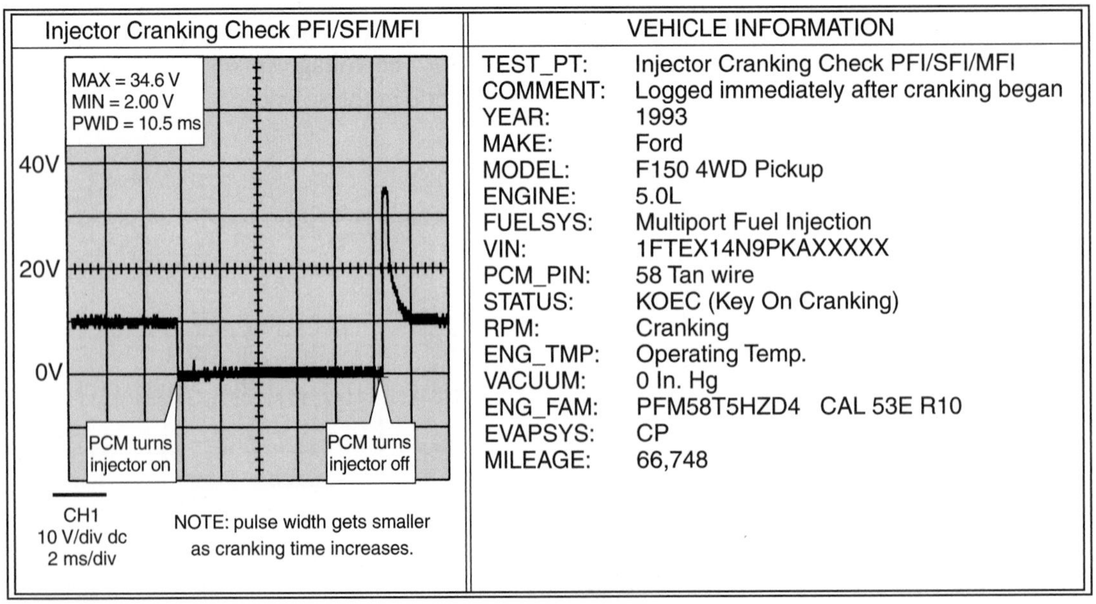

Figure 34-47 The trace of a normally operating fuel injector. *Courtesy of Progressive Diagnostics—WaveFile AutoPro*

somewhat of a rhythm to it and he focused his attention on it. It did not sound like a vacuum leak. As he increased the engine's speed, the noise pulses became closer together and soon became a constant noise. The noise appeared to be coming from the lower part of the engine.

Using a stethoscope, he was able to identify the source of the noise. It appeared that the gasket joining the exhaust manifold to the exhaust pipe was not seated properly. To verify this, he raised the car on a hoist and took a look. He found the retaining bolts were loose. He took a quick look at the gasket and found it to be in reasonable shape. He then tightened the bolts to specifications.

Not sure the noise or the exhaust leak was related to the problem, but suspected that it could be, he connected a lab scope to the oxygen sensor and watched its activity. The sensor's signal no longer showed a bias. It appeared that the leak in the exhaust was pulling air into the exhaust between each pulse. This was adding oxygen to the exhaust stream, causing the computer to think the mixture was lean.

KEY TERMS

Cross counts
Enable criteria
Key on, engine off (KOEO)
Key on, engine running (KOER)
Pending situation
Serial data

SUMMARY

- The role of the computer has evolved from the control of a single system to the control of nearly all of the engine's systems.
- A trouble code (DTC) is set to indicate the portion of the system that is at fault. A technician can access the code as an aid in troubleshooting.
- The logical approach to troubleshooting electronic engine controls includes checking all traditional non-electronic engine control possibilities before attempting to diagnose the electronic engine control itself.
- The following can be used to check individual system components: visual checks, ohmmeter checks, voltmeter checks, and lab scope checks.
- When diagnosing engine control system problems, service bulletin information is absolutely essential.
- Most computerized engine controls have self-diagnostic capabilities. By entering a self-test mode, the computer is able to evaluate the condition of the entire electronic engine control system, including itself. If problems are found, they are identified as either hard faults (on-demand) or intermittent failures. Each type of fault or failure is assigned a numerical trouble code that is stored in computer memory.
- Prior to OBD-II, each car manufacturer required a different method for retrieving stored DTCs. The diagnostic connector looked different and was located in different places. Also, the diagnostic codes represented different things. Although all of the computer systems did basically the same thing, diagnostic methods were as different as day and night.
- OBD-II regulations require that the PCM monitor and perform some continuous tests on the engine control system and components. Some OBD-II tests are completed at random, at specific intervals, or in response to a detected fault.
- Diagnosing and servicing OBD-II systems is much the same as diagnosing and servicing any other electronic engine control system, except that OBD-II systems give more useful information than the previous designs and with OBD-II each make and model vehicle does not have its own basic testing and servicing procedures.
- OBD-II systems note the deterioration of certain components before they fail, which allows owners to bring in their vehicles at their convenience and before it is too late. OBD-II monitors the vehicle's exhaust gas recirculation and fuel systems, oxygen sensors, catalytic converter performance, and miscellaneous other components.
- The MIL informs the driver that a fault that affects the vehicle's emission levels has occurred. The MIL will illuminate if a component or system that has an impact on emissions indicates a malfunction or fails to pass an emissions-related diagnostic test. The MIL will stay lit until the system or component passes the same test for three consecutive trips with no emissions-related faults. After making the repair, technicians may need to take the vehicle for three trips to ensure that the MIL does not illuminate again.
- Government regulations require that engine operating conditions be captured whenever the MIL is illuminated. The data captured is called freeze frame data.
- OBD-II's short-term fuel trim and long-term fuel trim strategies monitor the oxygen sensor signals and use this information to make adjustments to the fuel control calculations. The adaptive fuel control strategy

allows for changes in the amount of fuel delivered to the cylinders according to operating conditions.

- Most intermittent problems are caused by faulty electrical connections or wiring. To locate the source of the problem, a voltmeter can be connected to the suspected circuit and the wiring harness wiggled. The vehicle can also be taken for a test drive with the scan tool connected to it. This allows you to look at a circuit's response to changing operating conditions. The snapshot or freeze frame feature stores engine conditions and operating parameters at command or when the PCM sets a DTC.
- All PCMs (OBD-II and earlier designs) cannot operate properly unless they have good ground connections and the correct voltage at the required terminals.
- A voltage drop test is a quick way of checking the condition of any wire.
- Poor grounds can also allow EMI or noise to be present on the reference voltage signal. The best way to check for noise is to use a lab scope.
- If a DTC directs you to a faulty sensor or sensor circuit, or if you suspect that a sensor is faulty, it should be tested.
- Most computer-controlled actuators are electromechanical devices that convert the output commands from the computer into mechanical action. These actuators are used to open and close switches, control vacuum flow to other components, and operate valves depending on the requirements of the system.
- Most systems allow for testing of the actuator through a scan tool.
- When checking anything with an ohmmeter, logic can dictate good and bad readings. If the meter reads infinite, this means there is an open. Based on what you are measuring across, an open could be good or bad. The same is true for very low resistance readings. Across some things, this would indicate a short. For example, you do not want an infinite reading across the windings of a solenoid. You want low resistance. However, you want an infinite reading from one winding terminal to the case of the solenoid. If you have low resistance, the winding is shorted to the case.
- Actuators can be accurately tested with a lab scope.

TECH MANUAL

The following procedures are included in Chapter 33/34 of the *Tech Manual* that accompanies this book:

1. Retrieving diagnostic trouble codes (DTCs) form the computer of an engine control system.
2. Testing an ECT sensor.
3. Testing the operation of a TP sensor.
4. Testing an O_2 sensor.
5. Testing a MAP sensor.
6. Conducting a diagnostic check on an engine equipped with OBD-II.
7. Monitoring the adaptive fuel strategy on an OBD-II-equipped engine.

REVIEW QUESTIONS

1. Name three ways that a scan tool can help in diagnostics.
2. List the four ways individual components can be checked.
3. *True or False?* A bad ground can cause an increase in the reference voltage to a sensor.
4. State why TSBs are a valuable diagnostic tool.
5. A typical normal oxygen sensor signal will toggle between ______ and ______ volts.
6. Explain how a bad circuit ground can affect a sensor's reference voltage.
7. When an engine is running lean, the voltage signal from the oxygen sensor will be ______ (low or high).
8. Which of the following is the least likely cause for a misfire monitor failure on an OBD-II system?
   a. defective aspirator valve
   b. EVAP system faults
   c. EGR system faults
   d. restricted exhaust

9. While discussing O_2 sensor diagnosis, Technician A says the voltage signal on a satisfactory O_2 sensor should always be cycling between 0.5 volt and 1 volt. Technician B says a contaminated O_2 sensor provides a continually low voltage signal. Who is correct?

a. Technician A
b. Technician B
c. Both A and B
d. Neither A nor B

10. While discussing ECT sensor diagnosis, Technician A says a defective ECT sensor may cause hard cold-engine starting. Technician B says a defective ECT sensor may cause improper operation of emission control devices. Who is correct?

a. Technician A
b. Technician B
c. Both A and B
d. Neither A nor B

11. While discussing vehicle speed sensor tests, Technician A says use an ohmmeter to test the resistance of the coil. Technician B says the voltage generated by the sensor can be measured by connecting a voltmeter across the sensor's terminals. Who is correct?

a. Technician A
b. Technician B
c. Both A and B
d. Neither A nor B

12. While discussing TP sensor diagnosis, Technician A says a four-wire TP sensor contains an idle switch. Technician B says in some applications the TP sensor mounting bolts may be loosened and the TP sensor housing rotated to adjust the voltage signal with the throttle in the idle position. Who is correct?

a. Technician A
b. Technician B
c. Both A and B
d. Neither A nor B

13. While discussing MAF sensor diagnosis, Technician A says that on a vane-type MAF sensor the voltage signal should be checked as the vane is moved from fully closed to fully open. Technician B says on a vane-type MAF sensor, the voltage signal should decrease as the vane is opened. Who is correct?

a. Technician A
b. Technician B
c. Both A and B
d. Neither A nor B

14. While discussing testing OBD-II system components, Technician A says a test light can be used on 12-volt circuits. Technician B says a digital voltmeter can be used on the circuits. Who is correct?

a. Technician A
b. Technician B
c. Both A and B
d. Neither A nor B

15. While discussing the MIL on OBD-II systems, Technician A says the MIL will flash if the PCM detects a fault that would damage the catalytic converter. Technician B says whenever the PCM has detected a fault, it will turn on the MIL. Who is correct?

a. Technician A
b. Technician B
c. Both A and B
d. Neither A nor B

16. While discussing diagnostic procedures, Technician A says after preliminary system checks have been made, the DTCs should be cleared from the memory of the PCM. Technician B says serial data can be helpful when there are no DTCs but there is a fault in the system. Who is correct?

a. Technician A
b. Technician B
c. Both A and B
d. Neither A nor B

17. While discussing freeze frame data, Technician A says this feature displays the conditions that existed when the PCM set a DTC. Technician B says the freeze frame feature stores data even when a fault does not turn on the MIL. Who is correct?

a. Technician A
b. Technician B
c. Both A and B
d. Neither A nor B

18. Technician A says the enable criteria is the criteria that must be met before the PCM completes a monitor test. Technician B says a drive cycle includes operating the vehicle under specific conditions so that a monitor test can be completed. Who is correct?

a. Technician A
b. Technician B
c. Both A and B
d. Neither A nor B

19. While discussing PCM monitor tests, Technician A says the monitor tests are prioritized. Technician B says the monitor tests are run when the scan tool activates them. Who is correct?

a. Technician A
b. Technician B
c. Both A and B
d. Neither A nor B

20. When testing a frequency-varying MAF, Technician A says the frequency should increase with an increase in engine speed. Technician B says the signal from the sensor should be a square wave when observed on a lab scope. Who is correct?

a. Technician A
b. Technician B
c. Both A and B
d. Neither A nor B

SECTION 5

Manual Transmissions and Transaxles

Chapter 35 ■ *Clutches*

Chapter 36 ■ *Manual Transmissions and Transaxles*

Chapter 37 ■ *Manual Transmission/Transaxle Service*

Chapter 38 ■ *Drive Axles and Differentials*

Transmission/transaxles, drive axles, and differentials perform the important task of manipulating the power produced by the engine and routing it to the driving wheels of the vehicle. Precision machined and fitted gearsets and shafts change the ratio of speed and power between the engine and drive axles. The flow of power is controlled through a manually operated clutch and shift lever.

Transmission/transaxle service requires sound diagnostic skills plus the ability to work with intricate gearing and linkages. The material covered in Section 5 corresponds to the ASE certification test on manual transmissions and drive lines.

Delmar's Manual Transmissions and Transaxles Video Series and CD-ROM Courseware

WE ENCOURAGE PROFESSIONALISM

ASE

THROUGH TECHNICIAN CERTIFICATION

35

CLUTCHES

OBJECTIVES

■ Describe the various clutch components and their functions. ■ Name and explain the advantages of the different types of pressure plate assemblies. ■ Name the different types of clutch linkages. ■ List the safety precautions that should be followed during clutch servicing. ■ Explain how to perform basic clutch maintenance. ■ Name the six most common problems that occur with clutches. ■ Explain the basics of servicing a clutch assembly.

The clutch is located between the transmission and engine where it provides a mechanical coupling between the engine's flywheel and the transmission's input shaft. The driver operates the clutch through a linkage that extends from the passenger compartment to the **bell housing** (also called the clutch housing) between the engine and the transmission.

All manual transmissions require a clutch to engage or disengage the transmission. If the vehicle had no clutch and the engine was always connected to the transmission, the engine would stop every time the vehicle was brought to a stop. The clutch allows the engine to idle while the vehicle is stopped. It also allows for easy shifting between gears. (Of course, all of this applies to manual transaxles as well.)

The clutch engages the transmission gradually by allowing a certain amount of slippage between the transmission's input shaft and the flywheel. **Figure 35–1** shows the components needed to do this: the flywheel, clutch disc, pressure plate assembly, clutch release bearing (or throwout bearing), and the clutch fork.

OPERATION

The basic principle of engaging a clutch is demonstrated in **Figure 35–2**. The flywheel and the pressure plate are the drive or driven members of the clutch. The driven member connected to the transmission input shaft is the **clutch disc**, also called the **friction disc**. As long as the clutch is disengaged (clutch pedal depressed), the drive members turn independently of the driven member, and the engine is disconnected from the transmission. However, when the clutch is engaged (clutch pedal released), the pressure plate moves toward the flywheel and the clutch disc is squeezed between the two revolving drive members and forced to turn at the same speed.

WARNING!

Use the appropriate cleaning liquid and equipment before and during disassembling a clutch assembly. Some clutch discs were made with asbestos. Inhalation of asbestos can cause serious illnesses.

Flywheel

The **flywheel**, an important part of the engine, is also the main driving member of the clutch. It is normally made of nodular or grey cast iron, which has a high graphite content to lubricate the engagement of the clutch. Welded to or pressed onto the outside diameter of the flywheel is the starter ring gear. The starter ring gear is replaceable on most flywheels. The large diameter of the flywheel allows for an excellent gear ratio of the starter drive to ring gear, which provides for ample engine rotation during starting. The rear surface of the flywheel is a friction surface machined very flat to ensure smooth clutch engagement. The flywheel also provides some absorption of torsional vibration of the crankshaft. It further provides the inertia to rotate the crankshaft through the four strokes.

The flywheel has two sets of bolt holes drilled into it. The inner set is used to fasten the flywheel to the crankshaft, and the outer set provides a mounting plate for the pressure plate assembly. A bore in the center of the

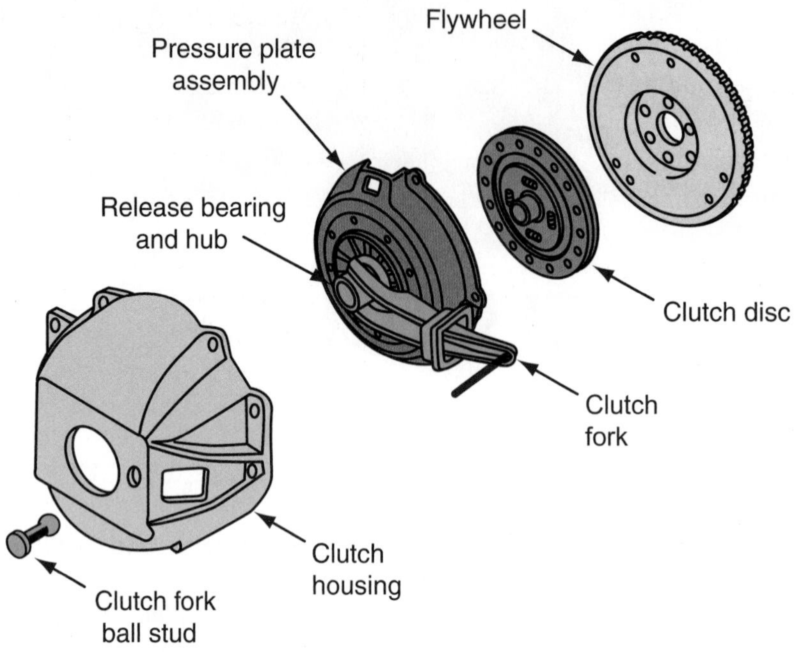

Figure 35-1 Major parts of a clutch assembly.

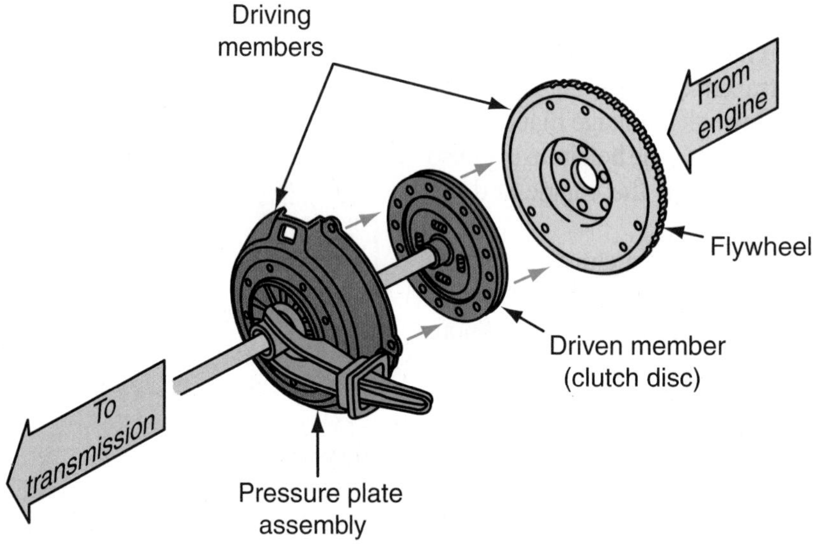

Figure 35-2 When the clutch is engaged, the driven member is squeezed between the two driving members. The transmission is connected to the driven member.

flywheel and crankshaft holds the **pilot bushing** or bearing, which supports the front end of the transmission input shaft and maintains alignment with the engine's crankshaft. Sometimes a ball or roller needle bearing is used instead of a pilot bushing.

Dual-Mass Flywheel A few cars and light trucks use a dual-mass flywheel. These flywheels are used to reduce vibrations transmitted through the transmission, provide for smoother shifting, and reduce gear noise. Dual-mass flywheels can reduce the oscillations of the crankshaft before they move through the transmission **(Figure 35–3)**. The flywheel consists of two rotating plates connected by a spring and damper system. The forwardmost portion of the flywheel is bolted to the end of the crankshaft and smoothes out the crankshaft's oscillations. The pressure plate of the clutch is bolted to the rearward portion of the flywheel. Engine torque moves from the front plate through the damper and spring assembly to the rear plate before it enters the transmission.

Some have a torque-limiting feature that prevents damage to the transmission during peak torque loads. The rotation of the two flywheel plates can differ by as much as 360 degrees. This allows the forward plate to absorb torque spikes and not pass them along through the transmission.

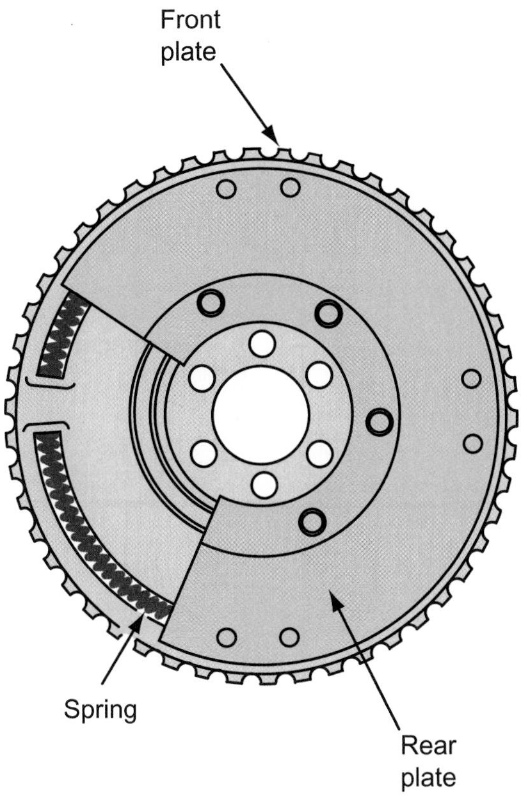

Figure 35-3 A dual-mass flywheel.

Clutch Disc

The clutch disc **(Figure 35–4)** receives the driving motion from the flywheel and pressure plate assembly and transmits that motion to the transmission input shaft. The parts of a clutch disc are shown in **Figure 35–5**.

There are two types of friction facings. Molded friction facings are preferred because they withstand greater pressure plate loading force without damage. Woven friction facings are used when additional cushioning action is needed for clutch engagement. Until recently, the material that was molded or woven into facings was predominantly asbestos. Now, because of the hazards associated with asbestos, other materials such as paper-base and ceramics are being used instead. Particles of cotton, brass, rope, and wire are added to prolong the life of the clutch disc and provide torsional strength.

Figure 35-4 A clutch disc.

Grooves are cut across the face of the friction facings. This promotes clean disengagement of the driven disc from the flywheel and pressure plate; it also promotes better cooling. The facings are riveted to wave springs, also called **cushioning springs**, which cause the contact pressure on the facings to rise gradually as the springs flatten out when the clutch is engaged. These springs eliminate chatter when the clutch is engaged and also reduce the chance of the clutch disc sticking to the flywheel and pressure plate surfaces when the clutch is disengaged. The wave springs and friction facings are fastened to the steel disc.

The clutch disc is designed to absorb such things as crankshaft vibration, abrupt clutch engagement, and driveline shock. Torsional coil springs or rubber grommets allow the disc to rotate slightly in relation to the pressure plate while they absorb the torque forces. The number and tension of these springs is determined by engine torque and vehicle weight. Stop pins limit this torsional movement to approximately 3⁄8 inch.

Pilot Bushing/Bearing

The purpose of the pilot bushing or bearing **(Figure 35–6)** is to support the outer end of the transmission's input shaft. This shaft is splined to the clutch disc and transmits power from the engine (when the clutch is engaged) to the transmission. The transmission end of the input shaft is supported by a large bearing in the transmission case. Because the input shaft extends unsupported from the transmission, a pilot bushing is used to keep it in position. By supporting the shaft, the pilot bushing keeps the clutch disc centered in the pressure plate.

Pressure Plate Assembly

The purpose of the pressure plate assembly **(Figure 35–7)** is two-fold. First, it must squeeze the clutch disc onto the flywheel with sufficient force to transmit engine torque efficiently. Second, it must move away from the clutch disc so the clutch disc can stop rotating, even though the flywheel and pressure plate continue to rotate.

Basically, there are two types of pressure plate assemblies: those with coil springs and those with a **diaphragm spring**. Both types have a stamped steel cover that bolts to the flywheel and acts as a housing to hold the parts together. In both, there is also the pressure plate, which is a heavy, flat ring made of nodular or grey cast iron. The assemblies differ in the manner in which they move the pressure plate toward and away from the clutch disc.

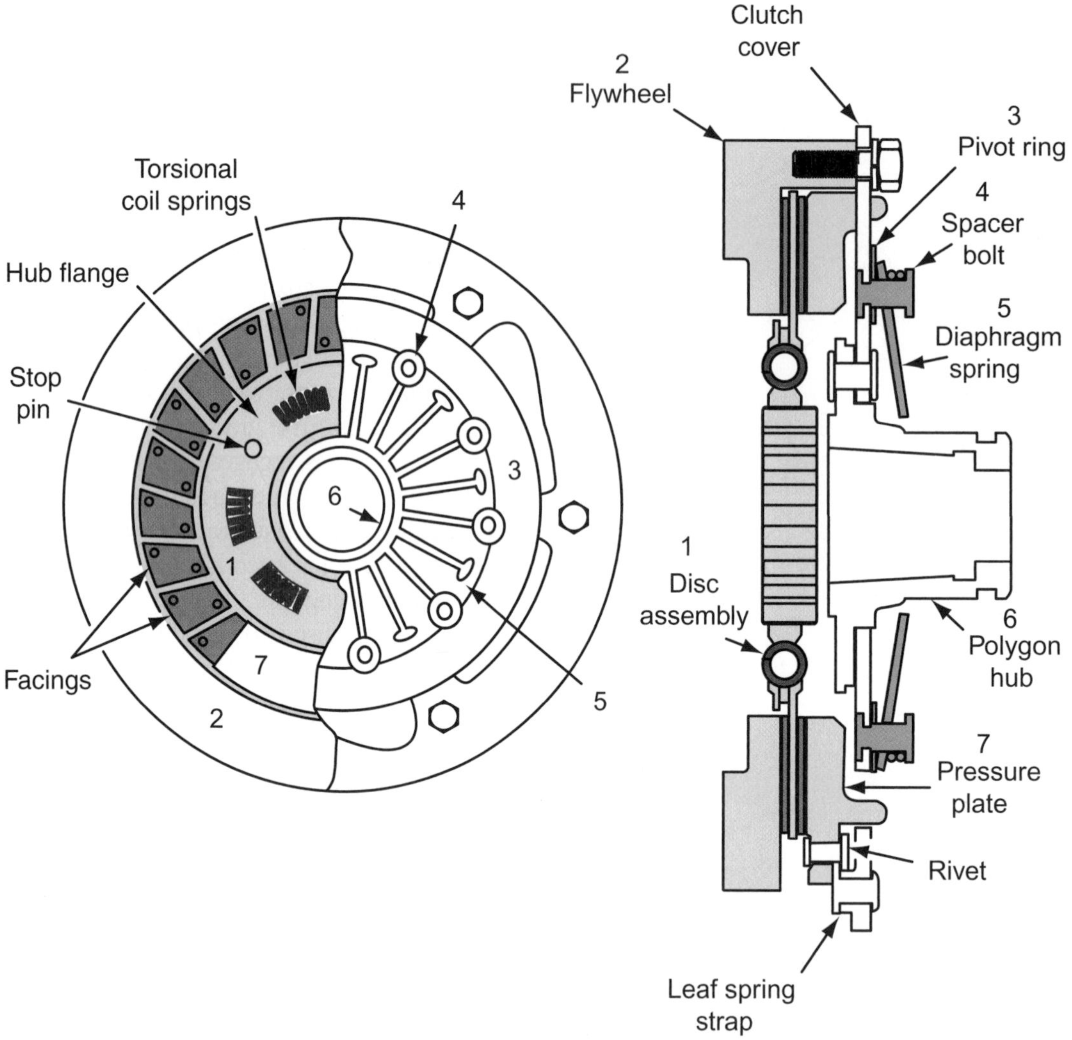

Figure 35-5 Parts of a clutch disc.

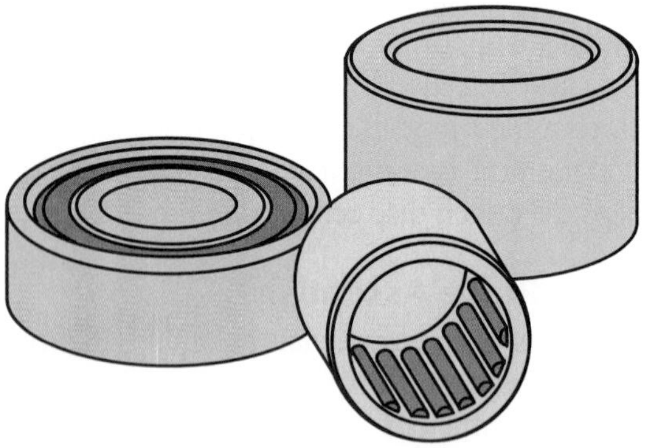

Figure 35-6 Different designs of pilot bushings and bearings used in today's clutch assemblies.

Figure 35-7 A clutch pressure plate.

Coil Spring Pressure Plate Assembly A coil spring pressure plate assembly, shown in **Figure 35–8**, uses coil springs and release levers to move the pressure plate back and forth. The springs exert pressure to hold the pressure plate tightly against the clutch disc and flywheel. This forces the clutch disc against the flywheel. The release levers release the holding force of the springs. There are usually three of them. Each one has two pivot points. One

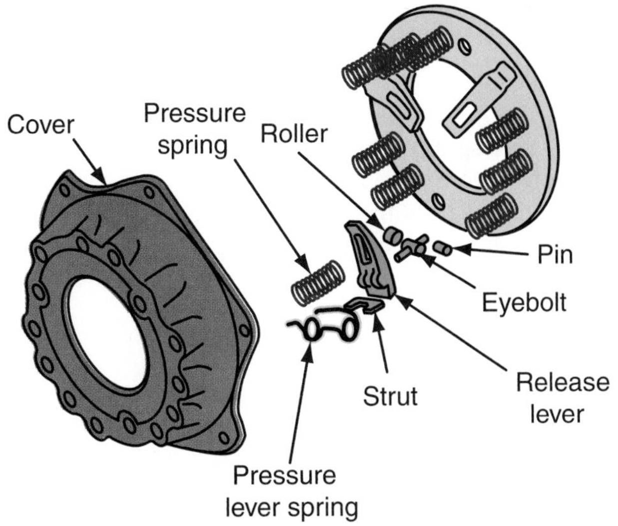

Figure 35-8 Parts of a coil spring pressure plate.

of these pivot points attaches the lever to a pedestal cast into the pressure plate and the other to a release lever yoke/keybolt bolted to the cover. The levers pivot on the pedestals and release lever yokes to move the pressure plate through its engagement and disengagement operations. To disengage the clutch, the release bearing pushes the inner ends of the release levers forward toward the flywheel. The release lever yokes act as fulcrums for the levers and the outer ends of the release levers move backward, pulling the pressure plate away from the clutch disc. This action compresses the coil springs and disengages the clutch disc from the driving members.

When the clutch is engaged, the release bearing moves backward toward the transmission. Without this force against the release levers, the coil springs are able to push the pressure plate and clutch disc against the flywheel with sufficient force to resist slipping.

Diaphragm Spring Pressure Plate Assembly The diaphragm spring pressure plate assembly relies on a cone-shaped diaphragm spring between the pressure plate and the pressure plate cover to move the pressure plate back and forth. The diaphragm spring (sometimes called a Belleville spring) is a single, thin sheet of metal that works in the same manner as the bottom of an oil can. The metal yields when pressure is applied to it. When the pressure is removed, the metal springs back to its original shape. The center portion of the diaphragm spring is slit into numerous fingers that act as release levers **(Figure 35–9)**.

During clutch disengagement, these fingers are moved forward by the release bearing. The diaphragm spring pivots over the fulcrum ring (also called the pivot ring), and its outer rim moves away from the flywheel. The retracting springs pull the pressure plate away from the driven disc and disengage the clutch.

When the clutch is engaged, the release bearing and the fingers of the diaphragm spring move toward the transmission. As the diaphragm pivots over the pivot ring, its outer rim forces the pressure plate against the clutch disc so the clutch is engaged to the flywheel.

Diaphragm spring pressure plate assemblies have the following advantages over other types of assemblies:

- Compactness
- Less weight
- Fewer moving parts to wear out
- Little pedal effort required from the operator
- Provide a balanced force around the pressure plate so rotational unbalance is reduced
- Clutch disc slippage is less likely to occur. Mileage builds because the force holding the clutch disc to the flywheel does not change throughout its service life.

Clutch Release Bearing

The **clutch release bearing**, also called a **throwout bearing**, is usually a sealed, prelubricated ball bearing

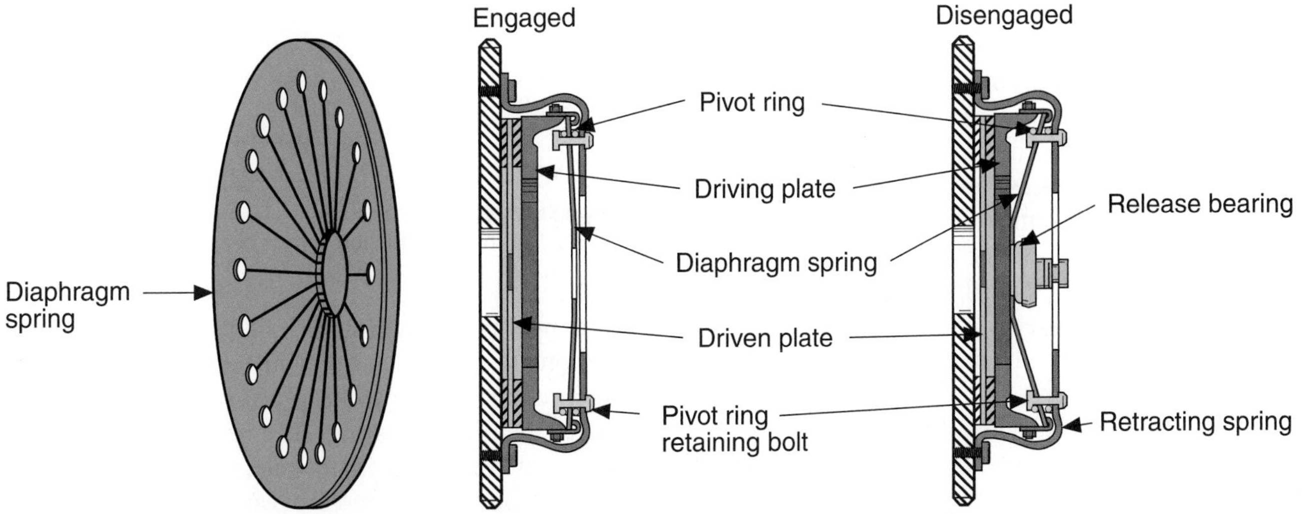

Figure 35-9 A diaphragm spring-type pressure plate assembly.

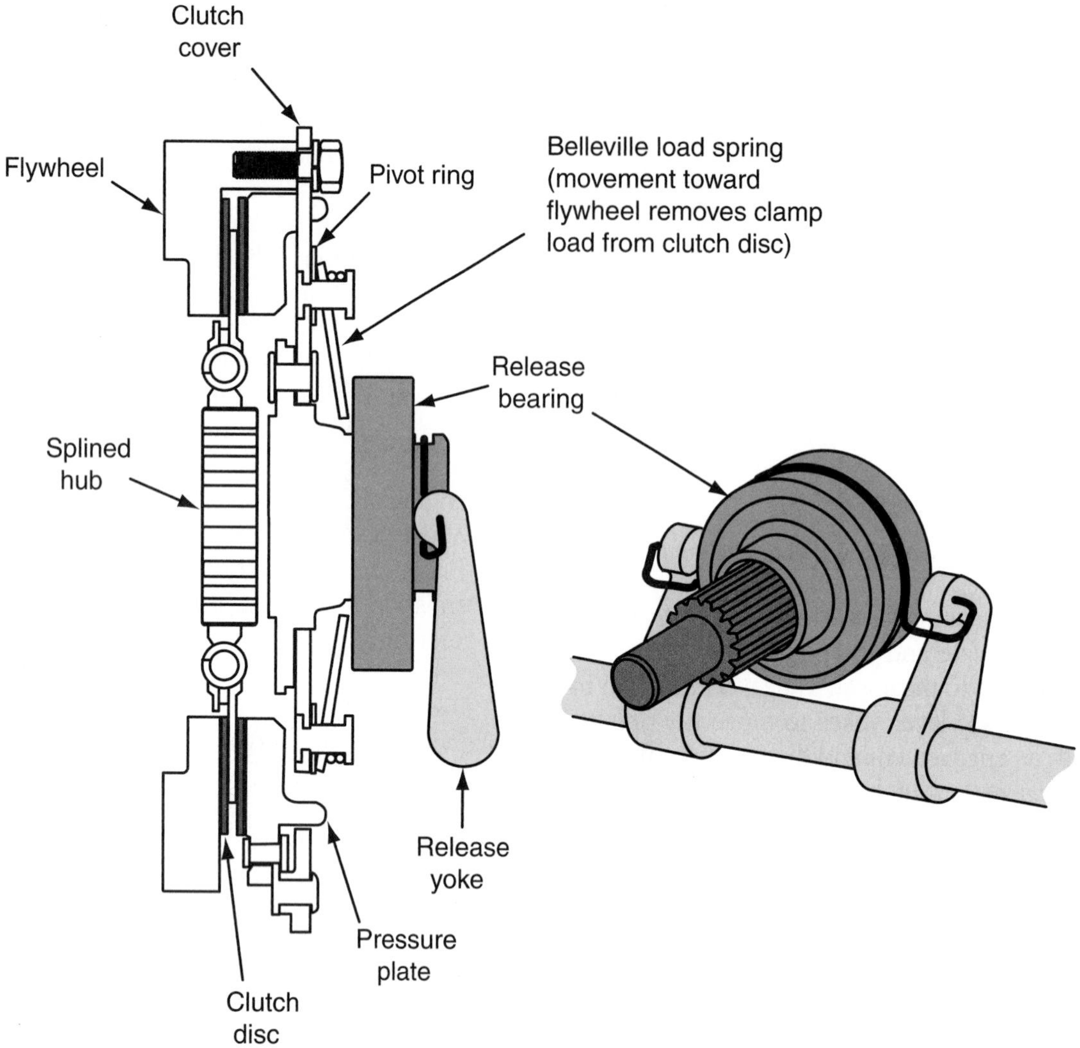

Figure 35-10 The clutch fork and throwout bearing location in the bell housing.

(Figure 35–10). Its function is to smoothly and quietly move the pressure plate release levers or diaphragm spring through the engagement and disengagement process.

The release bearing is mounted on an iron casting called a hub, which slides on a hollow shaft at the front of the transmission housing. This hollow shaft, shown in **Figure 35–11**, is part of the transmission bearing retainer.

To disengage the clutch, the release bearing is moved forward on its shaft by the **clutch fork**. As the release bearing contacts the release levers or diaphragm spring of the pressure plate assembly, it begins to rotate with the rotating pressure plate assembly. As the release bearing continues forward, the clutch disc is disengaged from the pressure plate and flywheel.

To engage the clutch, the release bearing slides to the rear of the shaft. The pressure plate moves forward and traps the clutch disc against the flywheel to transmit engine torque to the transmission input shaft. Once the clutch is fully engaged, the release bearing is normally stationary.

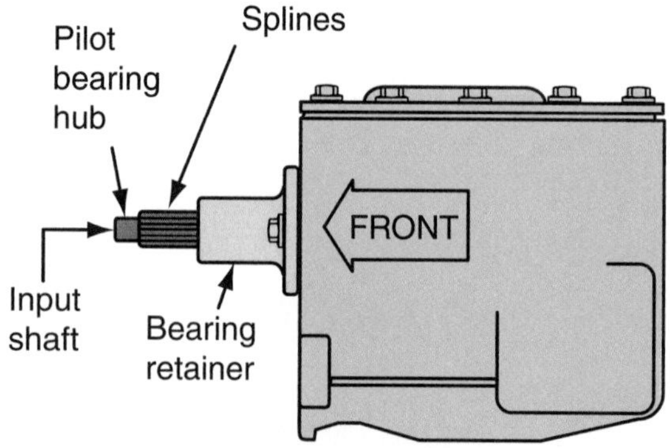

Figure 35-11 The release bearing slides on the hollow shaft of the transmission's front bearing retainer.

Rotating Release Bearing Self-adjusting clutch linkages, used on many vehicles, apply just enough tension to the clutch control cable to keep a constant light pressure against the release bearing. As a result, the release bearing is kept in contact with the release levers or diaphragm spring of the rotating pressure plate assembly. The release bearing rotates with the pressure plate.

Clutch Fork

The clutch fork is a forked lever that pivots on a ball stud located in an opening in the bell housing. The forked end slides over the hub of the release bearing and the small end protrudes from the bell housing and connects to the clutch linkage and clutch pedal. The clutch fork moves the release bearing and hub back and forth during engagement and disengagement.

Clutch Linkage

The clutch linkage is a series of parts that connects the clutch pedal to the clutch fork. It is through the clutch linkage that the operator controls the engagement and disengagement of the clutch assembly smoothly and with little effort.

Cable Linkage A cable linkage can perform the same controlling action as the shaft and lever linkage but with fewer parts. The clutch cable system does not take up much room. It also has the advantage of flexible installation so it can be routed around the power brake and steering units. These advantages help to make it the most commonly used clutch linkage.

The clutch cable **(Figure 35–12)** is made of braided wire. The upper end is connected to the top of the clutch pedal arm, and the lower end is fastened to the clutch fork. It is designed with a flexible outer housing that is fastened at the fire wall and the clutch housing.

When the clutch pedal is pushed to the disengaged position, it pivots on the pedal shaft and pulls the inner cable through the outer housing. This action moves the clutch fork forward to disengage the clutch. The pressure plate springs and springs on the clutch pedal provide the force to move the cable back when the clutch pedal is released.

Self-Adjusting Clutch Self-adjusting clutch mechanisms monitor clutch pedal play and automatically adjust it when necessary.

Usually the self-adjusting clutch mechanism is a ratcheting mechanism located at the top of the clutch pedal behind the dash panel. The ratchet is designed with a pawl and toothed segment, and a pawl tension spring is used to keep the pawl in contact with the toothed segment. The pawl allows the toothed segment to move in only one direction in relation to the pawl.

The clutch cable is guided around and fastened to the toothed segment, which is free to rotate in one direction (backwards) independently of the clutch pedal. The tension spring pulls the toothed segment backwards.

When the clutch cable develops slack due to stretching and clutch disc wear, the cable is adjusted automatically when the clutch is released. The tension spring pulls the toothed segment backwards and allows the pawl to ride over to the next tooth. This effectively shortens the cable. Actually, the cable is not really shortened; but the slack has been reeled in by the repositioning of the

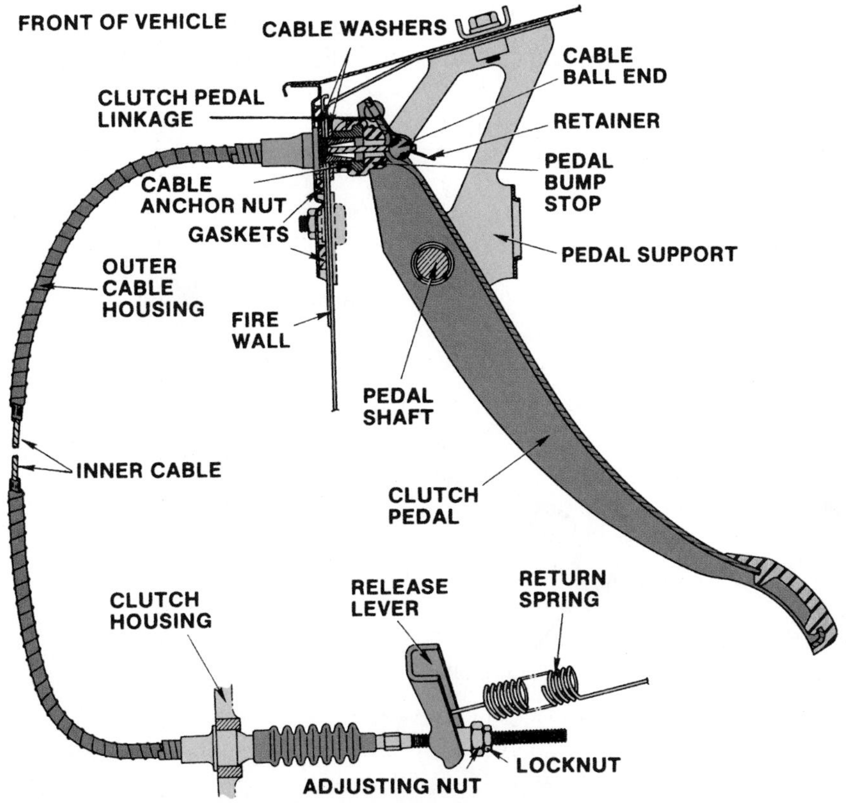

Figure 35-12 A typical clutch cable system.

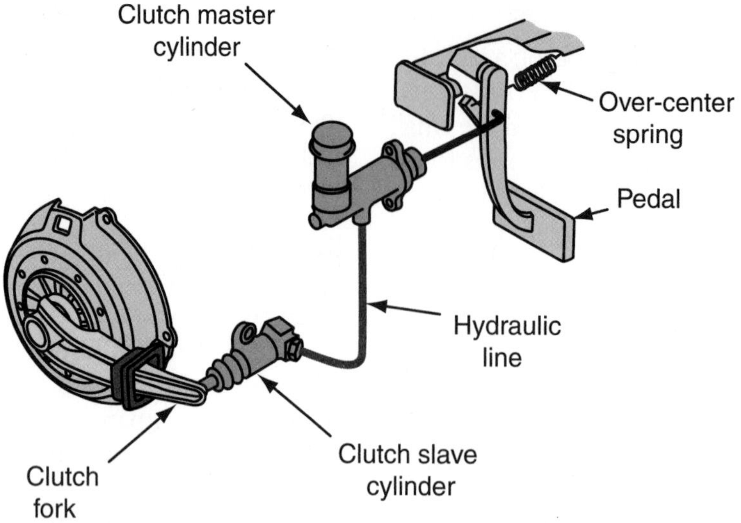

Figure 35–13 A typical hydraulic clutch linkage arrangement.

toothed segment. This self-adjusting action takes place automatically during the clutch's operational life.

Hydraulic Clutch Linkage Frequently, the clutch assembly is controlled by a hydraulic system **(Figure 35–13)**. In the hydraulic clutch linkage system, hydraulic (liquid) pressure transmits motion from one sealed cylinder to another through a hydraulic line. Like the cable linkage assembly, the hydraulic linkage is compact and flexible. Cable linkages also allow engineers to place the release fork anywhere that gives them more flexibility in body design. In addition, the hydraulic pressure developed by the master cylinder decreases required pedal effort and provides a precise method of controlling clutch operation. Brake fluid is commonly used as the hydraulic fluid in hydraulic clutch systems.

A hydraulic clutch master cylinder is shown in **Figure 35–14**. Its pushrod moves the piston and primary cup to create hydraulic pressure. The snap ring restricts the travel of the piston. The secondary cup at the snap ring end of the piston stops hydraulic fluid from dripping into the passenger compartment. The piston return spring holds the primary cup and piston in the fully released position. Hydraulic fluid is stored in the reservoir on top of the master cylinder housing.

The slave cylinder body has a bleeder valve to bleed air from the hydraulic system for efficient clutch linkage operation. The cylinder body is threaded for a tube and

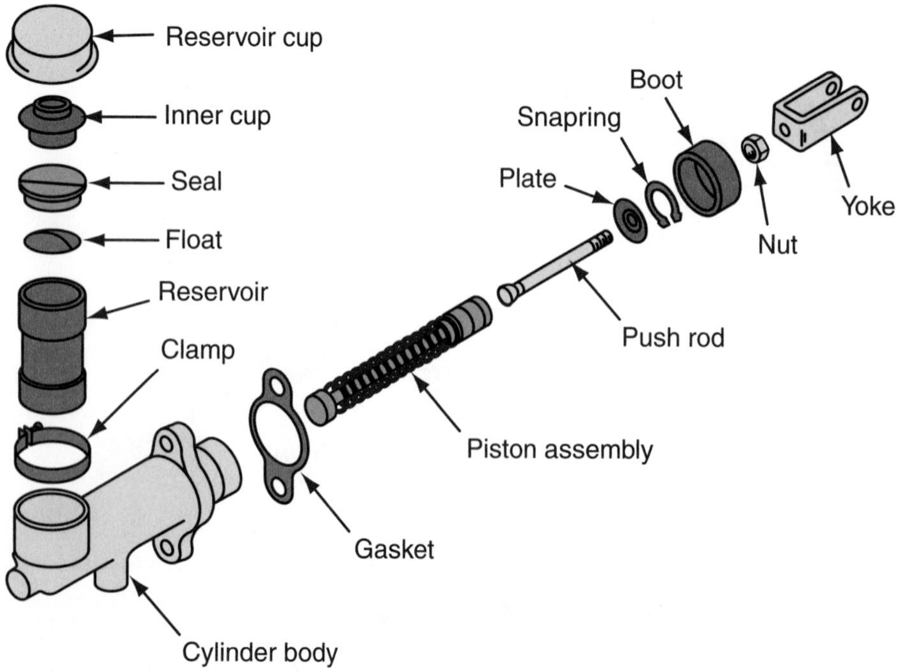

Figure 35–14 Parts of a hydraulic clutch master cylinder.

fitting at the fluid entry port. Rubber seal rings are used to seal the hydraulic pressure between the piston and the slave cylinder walls. The piston retaining ring is used to restrict piston travel to a certain distance. Piston travel is transmitted by a pushrod to the clutch fork. The pushrod boot keeps contaminants out of the slave cylinder.

When the clutch pedal is depressed, the movement of the piston and primary cup develops hydraulic pressure that is displaced from the master cylinder, through a tube, into the slave cylinder. The slave cylinder piston movement is transmitted to the clutch fork, which disengages the clutch.

When the clutch pedal is released, the primary cup and piston are forced back to the disengaged position by the master cylinder piston return spring. External springs move the slave cylinder pushrod and piston back to the engaged position. Fluid pressure returns through the hydraulic tubing to the master cylinder assembly. There is no hydraulic pressure in the system when the clutch assembly is in the engaged position.

Internal Slave Cylinders An internal concentric slave cylinder is found on some cars and light trucks. These units are actually a combination of the slave cylinder and the clutch release bearing **(Figure 35–15)**. By having the slave cylinder directly behind the release bearing, the movement of the release bearing is linear. In other clutch linkage designs, the release bearing moves through an arc as it engages and disengages the clutch.

An internal slave cylinder is a doughnut-shaped unit that mounts to the front of the transmission and the transmission's input shaft passes through it. The slave cylinder is either bolted to the transmission's front bearing cover or is held by a pressed pin. In many cases, the transmission must be removed to properly diagnose and service this type of slave cylinder.

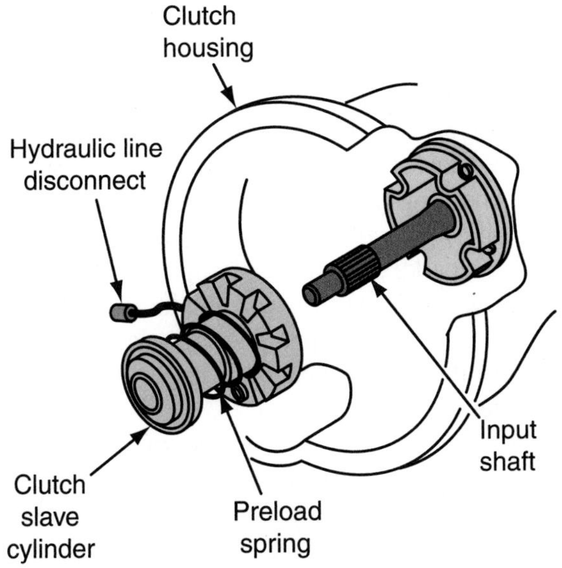

Figure 35-15 A concentric internal clutch slave cylinder.

CLUTCH SERVICE SAFETY PRECAUTIONS

When servicing the clutch, exercise the following precautions.

- Always wear eye protection when working underneath a vehicle.
- Remove asbestos dust only with a special, approved vacuum collection system or an approved liquid cleaning system.
- Never use compressed air or a brush to clean off asbestos dust.
- Follow all federal, state, and local laws when disposing of collected asbestos dust or liquid containing asbestos dust.
- Never work under a vehicle that is not raised on a hoist or supported by safety or jack stands.
- Use jack stands and special jacks to support the engine and transmission.
- Have a helper assist in removing the transmission.
- Be sure the work area is properly ventilated, or attach a ventilating hose to the vehicle's exhaust system when an engine is to be run indoors.
- Do not allow anyone to stand in front of or behind the automobile while the engine is running.
- Set the emergency brake securely and place the gearshift in neutral when running the engine of a stationary vehicle.
- Avoid touching hot engine and exhaust system parts. Whenever possible, let the vehicle cool down before beginning to work on it.

CLUTCH MAINTENANCE

All clutches require checking and adjustment of linkage at regular intervals. Vehicles with external clutch linkage require periodic lubrication. These maintenance procedures are explained in this section.

USING SERVICE MANUALS

Service manuals include adjustment procedures and instructions for clutch removal, inspection, installation, and troubleshooting. They may also offer information to aid in clutch release bearing distress analysis. ■

Clutch Linkage Adjustment

Except for systems with self-adjusting mechanisms, the release bearing should not touch the pressure plate release levers when the clutch is engaged (pedal up). Clearance between these parts prevents premature clutch plate, pressure plate, and release bearing wear. As the clutch disc wears and becomes thinner, this clearance is reduced.

Clearance can be ensured by adjusting the clutch linkage so the pedal has a specified amount of play, or **free travel**. Free travel is the distance a clutch pedal moves when depressed before the release bearing contacts the clutch release lever or diaphragm spring of the pressure plate.

To check pedal play, use a tape measure or ruler. Place the tape measure or ruler beside the clutch pedal and the end against the floor of the vehicle and note the reading **(Figure 35–16)**. Then, depress the clutch pedal just enough to take up the pedal play and note the reading again. The difference between the two readings is the amount of pedal play.

Adjustment should be performed when pedal play is not correct or when the clutch does not engage or disengage properly. To adjust clutch pedal play, refer to the manufacturer's service manual for the correct procedure and adjustment point locations. Often pedal play can be increased or decreased by turning a threaded fastener located either under the dash at the clutch pedal or where the linkage attaches to the clutch fork.

SHOP TALK

Normally clutch condition will dictate the amount of clutch pedal free play there is. However, there will always be free play with hydraulic clutch systems. ■

Clean the linkage with a shop towel and solvent, if necessary, before checking it and replacing any damaged or missing parts or cables. Check hydraulic linkage systems for leaks at the clutch master cylinder, hydraulic hose, and slave cylinder. Then, adjust the linkage to provide the manufacturer's specified clutch pedal play.

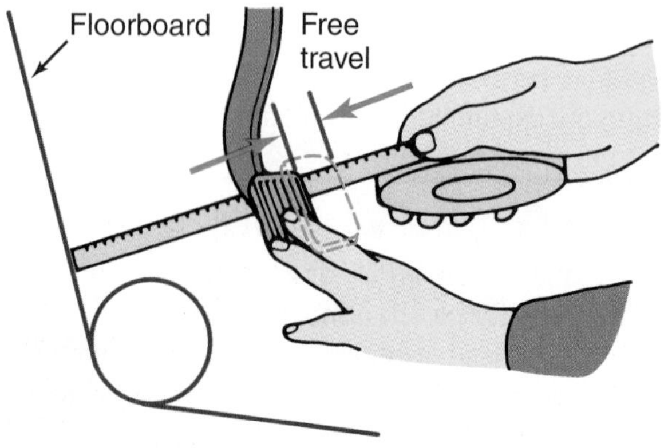

Figure 35-16 Checking clutch pedal play.

External Clutch Linkage Lubrication

External clutch linkage should be lubricated at regular intervals, such as during a chassis lubrication. Refer to the vehicle's service manual to determine the proper lubricant. Many clutch linkages use the same chassis grease used for suspension parts and U-joints. Lubricate all the sliding surfaces and pivot points in the clutch linkage **(Figure 35–17)**. The linkage should move freely after lubrication.

On vehicles with hydraulic clutch linkage, check the clutch master cylinder reservoir fluid level. It should be approximately $\frac{1}{4}$ inch (6.35 mm) from the top of the reservoir. If it must be refilled, use approved brake fluid. Also, because the clutch master cylinder does not consume fluid, check for leaks in the master cylinder, connecting flexible line, and slave cylinder, if the fluid is low.

CLUTCH PROBLEM DIAGNOSIS

Check and attempt to adjust the clutch pedal play before attempting to diagnose any clutch problems. If the friction lining of the clutch is worn too thin **(Figure 35–18)**, the clutch cannot be adjusted successfully. The most common clutch problems are described here.

Slippage

Clutch slippage is a condition in which the engine overspeeds without generating any increase in torque to the driving wheels. It occurs when the clutch disc is not gripped firmly between the flywheel and the pressure plate. Instead, the clutch disc slips between these driving members. Slippage can occur during initial acceleration or subsequent shifts, but it is usually most noticeable in higher gears.

One way to check for slippage is by driving the vehicle. Normal acceleration from a stop and several gear changes indicate whether the clutch is slipping.

Slippage also can be checked in the shop. Check the service manual for correct procedures. A general procedure for checking clutch slippage follows. Be sure to follow the safety precautions stated earlier.

With the parking brake on, disengage the clutch. Shift the transmission into third gear, and increase the engine speed to about 2,000 rpm. Slowly release the clutch pedal until the clutch engages. The engine should stall immediately.

If it does not stall within a few seconds, the clutch is slipping. Safely raise the vehicle and check the clutch linkage for binding and broken or bent parts. If no linkage problems are found, the transmission and the clutch assembly must be removed so the clutch parts can be examined.

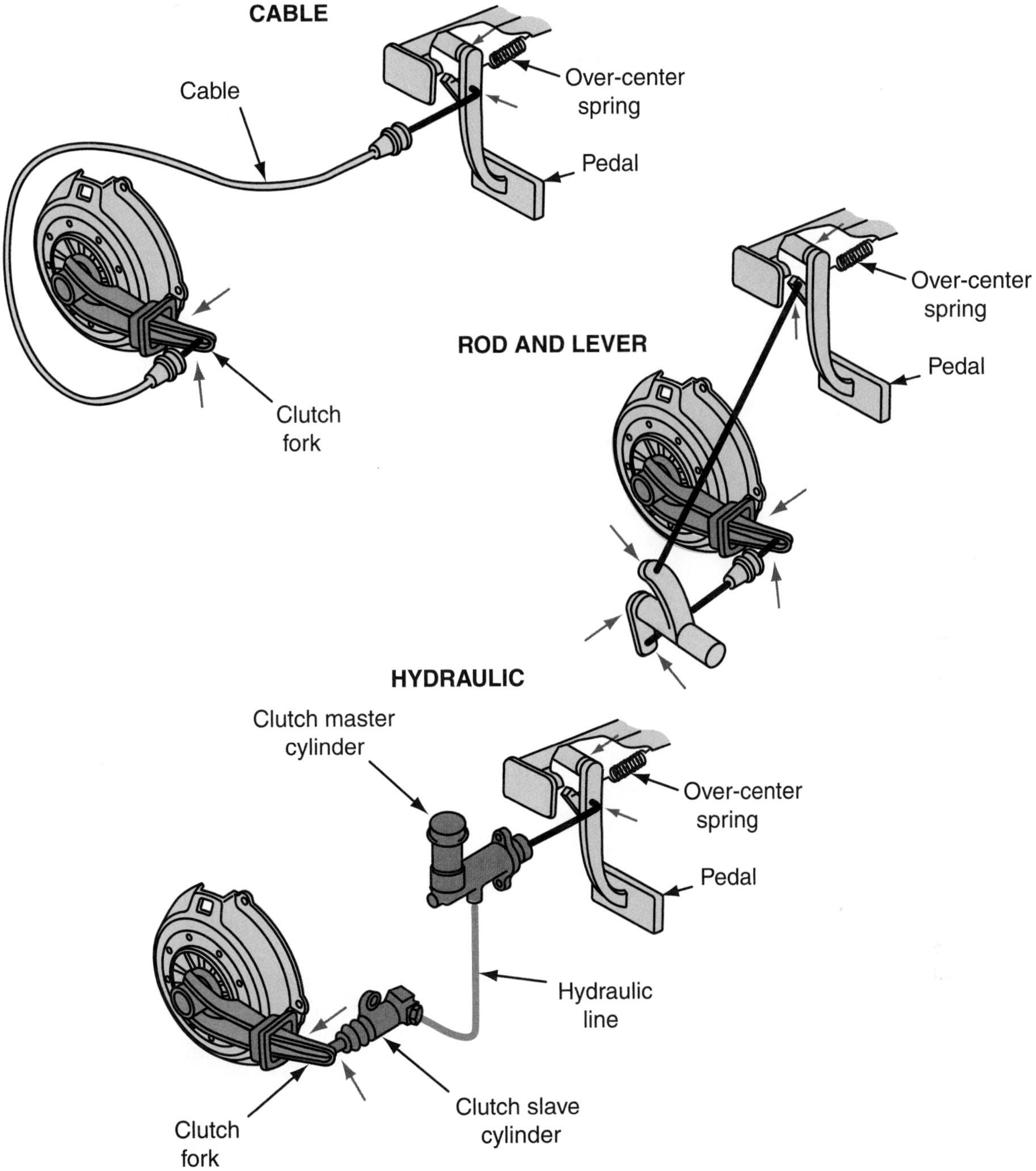

Figure 35-17 Clutch linkage lubrication points. *Used with permission from Nissan North America, Inc.*

SHOP TALK

Severe or prolonged clutch slippage causes grooving or heat damage to the pressure plate and/or flywheel. ■

Clutch slippage can be caused by an oil-soaked **(Figure 35–19)** or worn disc facing, warped pressure plate, weak or broken diaphragm spring, or the release bearing contacting and applying pressure to the release levers.

Drag and Binding

If the clutch disc is not completely released when the clutch pedal is fully depressed, clutch drag occurs. Clutch drag causes gear clash, especially when shifting into reverse. It can also cause hard starting because the engine attempts to turn the transmission input shaft.

To check for clutch drag, start the engine, depress the clutch pedal completely, and shift the transmission into first gear. Do not release the clutch. Then, shift the transmission into neutral and wait five seconds before attempting to shift smoothly into reverse.

Figure 35-18 A severely worn clutch disc. *Courtesy of Luk Automotive Systems*

Figure 35-19 Grease and oil on the hub of this disc is an indication that the disc may be contaminated with oil. *Courtesy of Luk Automotive Systems*

It should take no more than five seconds for the clutch disc, input shaft, and transmission gears to come to a complete stop after disengagement. This period, called the clutch spindown time, is normal and should not be mistaken for clutch drag.

If the shift into reverse causes gear clash, raise the vehicle safely and check the clutch linkage for binding, broken, or bent parts. If no problems are found in the linkage, the transmission and clutch assembly must be removed so that the clutch parts can be examined.

Clutch drag can occur as a result of a warped disc or pressure plate, a loose disc facing, a defective release lever, or incorrect clutch pedal adjustment that results in excessive pedal play. A binding or seized pilot bushing or bearing can also cause clutch drag.

Binding can result when the splines in the clutch disc hub or on the transmission input shaft are damaged or when there are problems with the release levers.

Chatter

A shaking or shuddering that is felt in the vehicle as the clutch is engaged is known as clutch **chatter (Figure 35–20)**. It usually occurs when the pressure plate first contacts the clutch disc and stops when the clutch is fully engaged.

To check for clutch chatter, start the engine, depress the clutch completely, and shift the transmission into first gear. Increase engine speed to about 1,500 rpm, then slowly release the clutch pedal and check for chatter as the pedal begins to engage. Do not release the pedal completely, or the vehicle might jump and cause serious injury. As soon as the clutch is partially engaged, depress the clutch pedal immediately and reduce engine speed to prevent damage to the clutch parts.

Usually clutch chatter is caused by liquid leaking onto the clutch and contaminating its friction surfaces. This results in a mirrorlike shine on the pressure plate or a glazed clutch facing. Oil and clutch hydraulic fluid leaks can occur at the engine rear main bearing seal transmission

Figure 35-20 The marks on the surface of this pressure plate are caused by clutch chatter. *Courtesy of Luk Automotive Systems*

input shaft seal, clutch slave cylinder, and hydraulic line. Other causes of clutch chatter include broken engine mounts, loose bell housing bolts, and damaged clutch linkage.

During disassembly, check for a warped pressure plate or flywheel, hot spots on the flywheel, a burned or glazed disc facing, and worn input shaft splines. If the chattering is caused by an oil-soaked clutch disc and no other parts are damaged, then the disc alone needs to be replaced. However, the cause of the oil leak must also be found and corrected.

Clutch chatter can also be caused by broken or weak torsional coil springs in the clutch disc and by the failure to resurface the flywheel when a new clutch disc and/or pressure plate is installed. It is highly recommended that the flywheel be resurfaced every time a new clutch disc or pressure plate is installed.

Pedal Pulsation

Pedal pulsation is a rapid up-and-down movement of the clutch pedal as the clutch disengages or engages. This pedal movement usually is minor, but it can be felt through the clutch pedal. It is not accompanied by any noise. Pulsation begins when the release bearing makes contact with the release levers.

To check for pedal pulsation, start the engine, depress the clutch pedal slowly until the clutch just begins to disengage, and then stop briefly. Resume depressing the clutch pedal slowly until the pedal is depressed to a full stop.

On many vehicles, minor pulsation is considered normal. If pulsation is excessive, the clutch must be removed and disassembled for inspection.

Pedal pulsations can result from the misalignment of parts. Check for a misaligned bell housing or a bent flywheel. Inspect the clutch disc and pressure plate for warpage. Broken, bent, or warped release levers also create misalignment **(Figure 35–21)**.

Figure 35–21 This pressure plate shows that the release bearing is not evenly contacting the pressure plate.
Courtesy of Luk Automotive Systems

CUSTOMER CARE

If you repair a vehicle with clutch slippage, tactfully inform the customer about the different poor driving habits that can cause this problem. These habits include riding the clutch pedal and holding the vehicle on an incline by using the clutch as a brake. ■

Vibration

Clutch vibrations, unlike pedal pulsations, can be felt throughout the vehicle, and they occur at any clutch pedal position. These vibrations usually occur at normal engine operating speeds (more than 1,500 rpm).

There are several possible sources of vibration that should be checked before disassembling the clutch to inspect it. Check the engine mounts and the crankshaft damper pulley. Look for any indication that engine parts are rubbing against the body or frame.

Accessories can also be a source of vibration. To check them, remove the drive belts one at a time. Set the transmission in neutral, and securely set the emergency brake. Start the engine and check for vibrations. Do not run the engine for more than 1 minute with the belts removed.

If the source of vibration is not discovered through these checks, the clutch parts should be examined. Be sure to check for loose flywheel bolts, excessive flywheel runout, and pressure plate cover balance problems.

Noises

Many clutch noises come from bushings and bearings. Pilot bushing noises are squealing, howling, or trumpeting sounds that are most noticeable in cold weather. These bushing noises usually occur when the pedal is being depressed and the transmission is in neutral. Release bearing noise is a whirring, grating, or grinding sound that occurs when the clutch pedal is depressed and stops when the pedal is fully released. It is most noticeable when the transmission is in neutral, but it also can be heard when the transmission is in gear.

Hydraulic Clutch Diagnosis

Diagnostics of a hydraulic clutch system should begin with an inspection of the fluid. Check the fluid and reservoir for dirt and contamination. Foreign matter in the fluid will destroy the seals and wear grooves in the master and slave cylinders' bores.

A soft clutch pedal, excessive pedal travel, or a clutch that fails to release when the pedal is depressed can be caused by low fluid in the reservoir. To correct this problem, refill the reservoir to the correct level then bleed the system. This problem can also be caused by a faulty or damaged primary or secondary seal in the master cylinder. A leaking secondary seal will be evident by external leaks, whereas a primary seal leak will be internal. To correct either of these problems, replace or rebuild the master cylinder then refill and bleed the system. A leaking slave cylinder should be replaced and the system refilled with fluid and then bled.

If there is an extremely hard pedal, check the pedal mechanism and release fork for binding. If there is evidence of binding, repair and lubricate the assembly to ensure free movement. A hard pedal can also be caused by a blocked compensation port in the master cylinder. The port may be blocked by improper pushrod adjustments or because the piston is binding in the master cylinder bore. If the piston is binding, the master cylinder should be replaced or rebuilt and the hydraulic system flushed, refilled, and bled. This problem may be also caused by swollen cup seals or contamination in the master or slave cylinders. If this is the problem, the master or slave cylinder should be replaced and the system flushed, refilled, and bled. Restricted hydraulic lines can also cause a hard pedal. The restricted lines should be replaced and the system flushed to remove the debris that may have caused the restriction. A worn clutch disc and/or pressure plate may also cause this problem.

If the clutch fails to engage when the clutch pedal is released, check the pedal and release assemblies for binding or improper adjustment and repair them as needed. A swollen primary cup or restricted hydraulic lines can also cause a lack of engagement. Replace the defective parts, then flush and refill the system.

CLUTCH SERVICE

A prerequisite for removing and replacing the clutch in a vehicle is removing the driveline or drive shafts and transmission or transaxle.

Removing the Clutch

After raising the vehicle on a hoist, clean excessive dirt, grease, or debris from around the clutch and transmission. Then, disconnect and remove the clutch linkage. Cable systems need to be disconnected at the transmission.

On rear-wheel-drive automobiles, remove the driveline and the transmission. In some cases the bell housing is removed with the transmission. In others, it is removed after the transmission is removed.

On front-wheel-drive vehicles with transaxles, any parts that interfere with transaxle removal must be removed first. These parts might include drive axles, parts of the engine, brake and suspension system or body parts. Check the service manual for specific instructions.

The clutch assembly is accessible after the bell housing has been removed. Use an approved vacuum collection system or an approved liquid cleaning system to remove asbestos dust and dirt from the clutch assembly.

Photo Sequence 34 outlines the typical procedure for replacing a clutch disc and pressure plate. Always refer to the manufacturer's recommendations for bolt torque specifications prior to reinstalling the assembly.

While working on a clutch assembly, follow these guidelines:

- Check the bell housing, flywheel, and pressure plate for signs of oil leaks.
- Make sure the mating surfaces of the engine block and bell housing are clean. The smallest amount of dirt can cause a misalignment, which can cause premature wear of transmission shafts and bearings.
- Check both mounting surfaces of the bell housing for damage and runout **(Figure 35–22)**.
- Check the flywheel for signs of burning or excessive wear. Check the runout of the flywheel with a dial indicator **(Figure 35–23)**.
- When installing a flywheel, make sure the bolts are tightened to specifications and in the prescribed sequence (normally a star pattern).
- When measuring the lining thickness of a bonded clutch disc, measure the total thickness of the facing

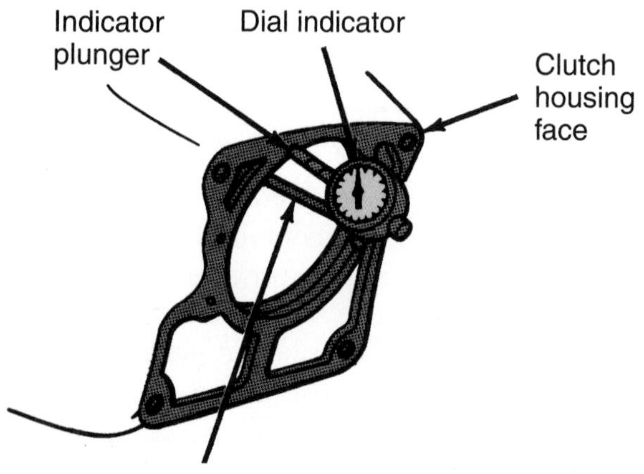

Figure 35-22 Use a dial indicator to check the runout of the clutch housing mounting surfaces. *Courtesy of DaimlerChrysler Corporation*

PHOTO SEQUENCE 34

Installing and Aligning a Clutch Disc

P34-1 *The removal and replacement of a clutch assembly can be completed while the engine is in or out of the car. The clutch assembly is mounted to the flywheel that is mounted to the rear of the crankshaft.*

P34-2 *Before disassembling the clutch, make sure alignment marks are present on the pressure plate and flywheel.*

P34-3 *The attaching bolts should be loosened before removing any of the bolts. With the bolts loosened, support the assembly with one hand while using the other to remove the bolts. The clutch disc will be free to fall as the pressure plate is separated from the flywheel. Keep it intact with the pressure plate.*

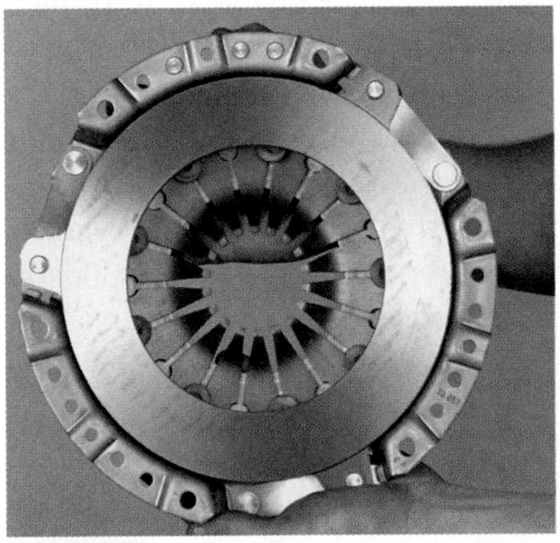

P34-4 *The surface of the pressure plate should be inspected for signs of burning, grooving, warpage, and cracks. Any faults normally indicate that the plate should be replaced.*

P34-5 *The surface of the flywheel should also be carefully inspected. Normally the flywheel surface can be resurfaced to remove any defects. Also check the runout of the flywheel. The pilot bushing or bearing should also be inspected.*

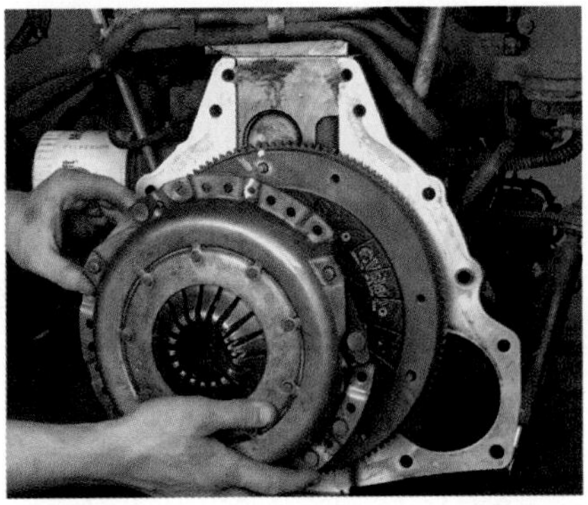

P34-6 *The new clutch disc is placed into the pressure plate as the pressure plate is moved into its proper location. Make sure the disc is facing the correct direction. Most are marked to indicate which side should be seated against the flywheel surface.*

P34-7 *Install the pressure plate according to the alignment marks made during disassembly.*

P34-8 *Install the attaching bolts, but do not tighten. Then install the clutch alignment tool through the hub of the disc and the pilot bearing to center the disc on the flywheel.*

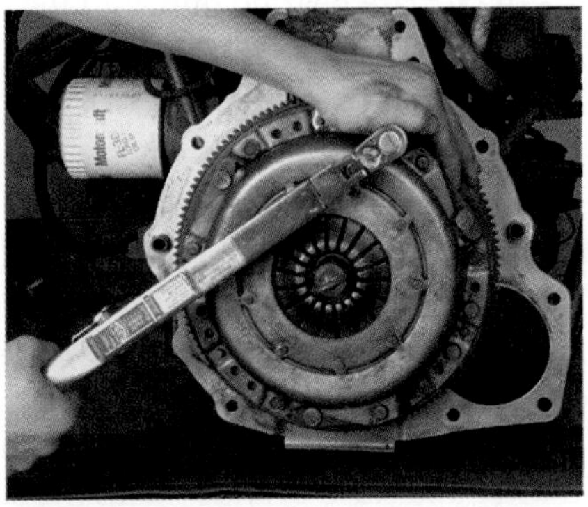

P34-9 *With the disc aligned, tighten the attaching bolts according to the procedures outlined in the service manual and check the release finger/lever height after tightening the bolts.*

Standard (new): 0.05 mm (0.002 in) max.
Service limit: 0.15 mm (0.006 in) max.

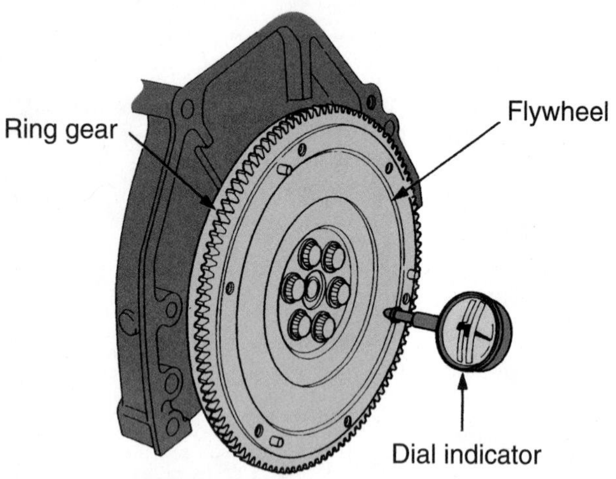
Ring gear
Flywheel
Dial indicator

Figure 35-23 Measure the runout of the flywheel with a dial indicator. *Courtesy of American Honda Motor Co., Inc.*

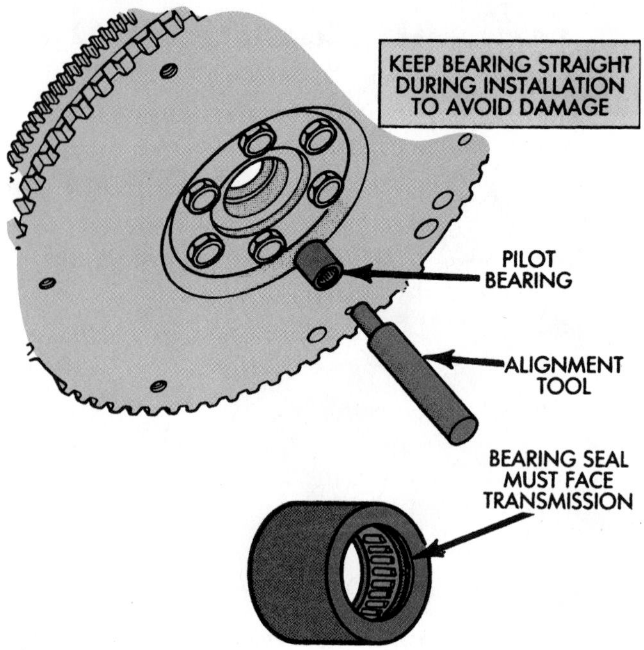
KEEP BEARING STRAIGHT DURING INSTALLATION TO AVOID DAMAGE
PILOT BEARING
ALIGNMENT TOOL
BEARING SEAL MUST FACE TRANSMISSION

Figure 35-24 A typical method for installing a pilot bearing. *Courtesy of DaimlerChrysler Corporation*

or lining. To measure the wear of a riveted lining, measure the material above the rivet heads.

- Keep grease off the frictional surfaces of the clutch disc, flywheel, and pressure plate.
- Check the pressure plate surface for warpage by laying a straightedge across the surface and inserting a feeler gauge between the surface and the straightedge. Compare the measurement against the specifications given in the service manual.
- Check the release levers of the pressure plate for uneven wear or damage.
- Check the release bearing by turning it with your fingers and making sure it rotates freely.
- Check the clutch for damage.
- Check the pilot bushing or bearing for wear. Replace it if necessary **(Figure 35–24)**.
- Lightly lubricate the input shaft and bearing retainer.
- Lubricate the clutch fork pivot points, the inside of the release bearing hub, and the linkages.
- After the clutch assembly has been reinstalled, check the clutch pedal free travel.

Hydraulic Linkage Service

The proper fluid level in the reservoir is usually marked on the reservoir. The reservoir is normally mounted to the top of the master cylinder or is part of the master cylinder assembly. The hydraulic system does not consume fluid; therefore, if the fluid is low, check for leaks at the master and slave cylinders and the connecting hydraulic lines. Fill the reservoir only to the fill line to allow the fluid to rise as the clutch disc wears. Overfilling the system will cause slip and premature failure. Air can enter the system through the compensation and bleed ports if the fluid level in the reservoir is too low. The system must be bled to remove the trapped air.

Master cylinder problems are typically external or internal fluid leaks that require that the unit be replaced or rebuilt. Rebuild kits are available for most master cylinders. If a cast-iron master cylinder is rebuilt, the cylinder bore should be honed to remove any imperfections in the bore and new seals used. The bores of aluminum master cylinders should never be honed.

Internal and external leaks are also typical problems for slave cylinders. Seldom are these cylinders rebuilt; rather, they are replaced. Replacing a slave cylinder is rather straightforward on most vehicles. Simply disconnect the hydraulic lines and unbolt the unit. If it appears that the piston of a slave cylinder is seized in its bore, check the movement of the release fork and lever at the clutch before replacing the slave cylinder. Leaks may also result from damaged or corroded hydraulic lines. These lines should be replaced, if damaged, with the same type of tube as was originally installed.

Bleeding the System Whenever the hydraulic system is opened, the entire system should be bled. **Bleeding** may also be necessary if the system has run low on fluid and air is trapped in the lines or cylinders. Bleeding can be accomplished through the use of a power bleeder (the same device used to bleed a brake system), a vacuum bleeder, or the use of a coworker. On most cars, it is impossible to pressurize the system and bleed the hydraulic lines at the same time; therefore, it is important that you have the proper equipment or someone to assist you. The typical procedure for bleeding the system follows:

PROCEDURE

Bleeding the Hydraulic System

STEP 1 *Check the entire hydraulic circuit to make sure there are no leaks.*

STEP 2 *Check the clutch linkage for wear and repair any defects before continuing.*

STEP 3 *Make sure all mounting points for the master and slave cylinders are solid and do not flex under the pressure of depressing the pedal.*

STEP 4 *Fill the master cylinder with the approved fluid.*

STEP 5 *Attach one end of a hose to the end of the bleeder screw and the other end into a catch can. Loosen the bleed screw at the slave cylinder approximately one-half turn.*

STEP 6 *Fully depress the clutch pedal, and then move the pedal through three quick and short strokes. Allow the fluid and air to exit the system and then immediately close the bleeder screw.*

STEP 7 *Release the pedal rapidly.*

STEP 8 *Recheck the fluid level in the master cylinder.*

STEP 9 *Repeat steps 3 and 4 until no air is evident in the fluid leaving the bleeder screw.*

STEP 10 *Close the bleeder screw immediately after the last downward movement of the pedal.*

KEY TERMS

Bell housing
Bleeding
Chatter
Clutch disc
Clutch fork
Clutch release bearing
Clutch slippage
Cushioning springs
Diaphragm spring
Flywheel
Free travel
Friction disc
Pilot bushing
Throwout bearing

CASE STUDY

A customer brings in a car that has a series of driveability problems. It uses an excess amount of fuel and oil, it lacks acceleration, and seems to lose power while travelling on the highway.

The technician listens to the customer's complaints and asks the right questions. Then he prepares the car for a road test. While doing this, he checks the oil level and finds it to be low. This somewhat verifies that the engine is using excessive amounts of oil, so he removes the spark plugs from the engine to check for signs of oil burning. The spark plugs are clean and show no signs of oil. The technician then checks the engine for leaks hoping to find the cause of the excessive oil loss. The only wet spot is to the rear of the oil pan and the front of the bell housing. A leaking rear main seal is the likely cause of the oil loss. However, this does not explain the other problems of poor fuel mileage and a general lack of power. If the engine had been burning oil, all of the problems could have been related.

He then takes the car on a road test and finds that the clutch is slipping. When the car is in high gear, he steps hard on the accelerator and notices that the engine speed increases while the car's speed stays the same. After some thought, he determines this is the cause of the poor fuel mileage and the lack of power. He further determines that all of the customer's complaints are related to the same problem, a leaking rear main seal. Oil has been leaking out of the engine and onto the clutch disc, causing it to slip. He returns to the shop. He then notifies the customer and gives an estimate for the repairs. The customer authorizes the repair, which includes a new rear main seal, release bearing, pilot bushing, clutch disc, and pressure plate, plus resurfacing the flywheel.

SUMMARY

- The clutch, located between the transmission and the engine, provides a mechanical coupling between the engine flywheel and the transmission's input shaft. All manual transmissions and transaxles require a clutch.
- The flywheel, an important part of the engine, is also the main driving member of the clutch.
- The clutch disc receives the driving motion from the flywheel and pressure plate assembly and transmits that motion to the transmission input shaft.
- The twofold purpose of the pressure plate assembly is to squeeze the clutch disc onto the flywheel and move away from the clutch disc so the disc can stop rotating. There are basically two types of pressure plate assemblies: those with coil springs and those with a diaphragm spring.
- The clutch release bearing, also called a throwout bearing, smoothly and quietly moves the pressure plate release levers or diaphragm spring through the engagement and disengagement processes.

- The clutch fork moves the release bearing and hub back and forth. It is controlled by the clutch pedal and linkage.
- Clutch linkage can be mechanical or hydraulic.
- The self-adjusting clutch is a clutch cable linkage that monitors clutch pedal play and automatically adjusts it when necessary.
- It is important that certain precautions are exercised when servicing the clutch. Clutch maintenance includes linkage adjustment and external clutch linkage lubrication.
- Slippage occurs when the clutch disc is not gripped firmly between the flywheel and the pressure plate. It can be caused by an oil-soaked or worn disc facing, warped pressure plate, weak diaphragm spring, or the release bearing contacting and applying pressure to the release levers.
- Clutch drag occurs if the clutch disc is not completely released when the clutch pedal is fully depressed. It can occur as a result of a warped disc or pressure plate, a loose disc facing, a defective release lever, or incorrect clutch pedal adjustment that results in excessive pedal play.
- Chatter is a shuddering felt in the vehicle when the pressure plate first contacts the clutch disc and it stops when the clutch is fully engaged. Usually chatter is caused when liquid contaminates the friction surfaces.
- Pedal pulsation is a rapid up-and-down movement of the clutch pedal as the clutch disengages or engages. It results from a misalignment of parts.
- Clutch vibrations can be felt throughout the vehicle, and they occur at any clutch pedal position. Sources of clutch vibrations include loose flywheel bolts, excessive flywheel runout, and pressure plate cover balance problems.

TECH MANUAL

The following procedures are included in Chapter 35 of the *Tech Manual* that accompanies this book:

1. Troubleshooting a clutch assembly.
2. Removing the clutch.
3. Inspecting and servicing the clutch.
4. Reassembling and installing a clutch.

REVIEW QUESTIONS

1. Name two types of friction facings.
2. *True or False?* If the fluid reservoir for a hydraulic clutch is low, the entire system should be checked for fluid leaks.
3. What is another name for the diaphragm spring?
4. Name three types of clutch linkages.
5. What is used to measure clutch pedal play?
6. The clutch, or friction, disc always rotates with the ______.
   a. engine crankshaft
   b. transmission input shaft
   c. transmission output shaft
   d. transmission counter shaft
7. Torsional coil springs in the clutch disc ______.
   a. cushion the driven disc engagement rear to front
   b. are the mechanical force holding the pressure plate against the driven disc and flywheel
   c. absorb the torque forces
   d. are located between the friction rings
8. The pressure plate moves away from the flywheel when the clutch pedal is ______.
9. Which of the following is probably *not* the cause of a vibrating clutch?
   a. excessive crankshaft end play
   b. out-of-balance pressure plate assembly
   c. excessive flywheel runout
   d. loose flywheel bolts
10. While discussing the different types of pressure plates, Technician A says coil spring types are commonly used because they have strong springs. Technician B says Belleville-type pressure plates are not commonly used because they require excessive space in the bell housing. Who is correct?
    a. Technician A **c.** Both A and B
    b. Technician B **d.** Neither A nor B
11. When the clutch pedal is released on a hydraulic clutch linkage, the ______.

a. master cylinder piston is released by spring tension
b. master cylinder piston is released by hydraulic pressure
c. slave cylinder is released by hydraulic pressure
d. slave cylinder piston does not move

12. When the clutch is disengaged, the power flow stops at the ______.
a. transmission input shaft
b. driven disc hub
c. pressure plate and flywheel
d. torsion springs

13. While discussing the purpose of a clutch pressure plate, Technician A says the pressure plate assembly squeezes the clutch disc onto the flywheel. Technician B says the pressure plate moves away from the clutch disc so the disc can stop rotating. Who is correct?
a. Technician A
b. Technician B
c. Both A and B
d. Neither A nor B

14. Insufficient clutch pedal clearance results in ______.
a. gear clashing while shifting transmission
b. a noisy front transmission bearing
c. premature release bearing failure
d. premature pilot bearing failure

15. Technician A says an oil-soaked clutch disc can cause clutch chatter. Technician B says clutch chatter can be caused by loose bell housing bolts. Who is correct?
a. Technician A
b. Technician B
c. Both A and B
d. Neither A nor B

16. When making a clutch adjustment, it is necessary to ______.
a. measure clutch pedal free travel
b. lubricate the clutch linkage
c. check hydraulic fluid level
d. place the transmission in reverse

17. Technician A says clutch slippage is most noticeable in higher gears. Technician B says clutch slippage is not noticeable in lower gears. Who is correct?
a. Technician A
b. Technician B
c. Both A and B
d. Neither A nor B

18. While discussing ways to determine whether a pilot bushing is faulty, Technician A says that if it is operating quietly, it does not need to be replaced. Technician B says a careful inspection of the transmission's input shaft can determine the condition of the pilot bushing. Who is correct?
a. Technician A
b. Technician B
c. Both A and B
d. Neither A nor B

19. The surface of the pressure plate contacts the ______.
a. transmission main shaft
b. throwout bearing
c. clutch disc
d. flywheel

20. While discussing different abnormal clutch noises, Technician A says pilot bushing noises are most noticeable in cold weather and usually occur when the pedal is being depressed and the transmission is in neutral. Technician B says release bearing noise is most noticeable when the transmission is in neutral and occurs when the clutch pedal is depressed and stops when the pedal is fully released. Who is correct?
a. Technician A
b. Technician B
c. Both A and B
d. Neither A nor B

36

MANUAL TRANSMISSIONS AND TRANSAXLES

OBJECTIVES

■ Explain the design characteristics of the gears used in manual transmissions and transaxles. ■ Explain the fundamentals of torque multiplication and overdrive. ■ Describe the purpose, design, and operation of synchronizer assemblies. ■ Describe the purpose, design, and operation of internal and remote gearshift linkages. ■ Explain the operation and power flows produced in typical manual transmissions and transaxles.

The transmission or transaxle **(Figure 36–1)** is a vital link in the power train of any modern vehicle. The purpose of the transmission or transaxle is to use gears of various sizes to give the engine a mechanical advantage over the driving wheels. During normal operating conditions, power from the engine is transferred through the engaged clutch to the input shaft of the transmission or transaxle. Gears in the transmission or transaxle housing alter the torque and speed of this power input before passing it on to other components in the drivetrain. Without the mechanical advantage the gearing provides, an engine can generate only limited torque at low speeds. Without sufficient torque, moving a vehicle from a standing start would be impossible.

In any engine, the crankshaft always rotates in the same direction. If the engine transmitted its power directly to the drive axles, the wheels could be driven only in one direction. Instead, the transmission or transaxle provides the gearing needed to reverse direction so the vehicle can be driven backward. There is also a neutral position that stops engine rotation and power from reaching the drive wheels.

TRANSMISSION VERSUS TRANSAXLE

Vehicles are propelled in one of three ways: by the rear wheels, by the front wheels, or by all four wheels. The type of drive system used determines whether a conventional transmission or a transaxle is used.

Vehicles propelled by the rear wheels normally use a transmission. Transmission gearing is located within an aluminum or iron casting called the transmission **case** assembly. The transmission case assembly is attached to the rear of the engine, which is normally located in the front of the vehicle. A drive shaft links the output shaft of the transmission with the differential and drive axles located in a separate housing at the rear of the vehicle. The differential splits the driveline power and redirects it to the two rear drive axles, which then pass it on to the wheels. For many years, rear-wheel-drive systems were the conventional method of propelling a vehicle.

Front-wheel-drive vehicles are propelled by the front wheels. For this reason, they must use a drive design different from that of a RWD vehicle. The transaxle is the special power transfer unit commonly used on FWD vehicles. A transaxle combines the transmission gearing,

Figure 36–1 A late-model transaxle.

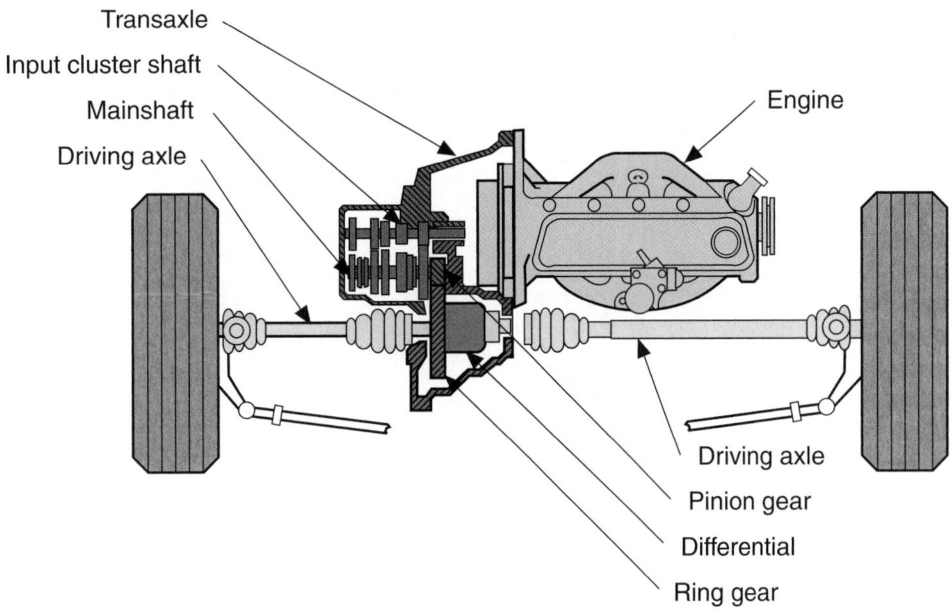

Figure 36–2 The location of typical front-wheel-drive power train components.

differential, and drive axle connections into a single case aluminum housing located in front of the vehicle **(Figure 36–2)**. This design offers many advantages. One major advantage is the good traction on slippery roads due to the weight of the drivetrain components being directly over the driving axles of the vehicle. It is also more compact and lighter than the transmission of a RWD vehicle. Transverse engine and transaxle configurations also allow for lower hood lines, thereby improving the vehicle's aerodynamics.

Four-wheel-drive vehicles typically use a transmission and transfer case. The transfer case mounts on the side or back of the transmission. A chain or gear drive inside the transfer case receives power from the transmission and transfers it to two separate drive shafts. One drive shaft connects to a differential on the front drive axle. The other drive shaft connects to a differential on the rear drive axle.

Most manual transmissions and transaxles are constant mesh, fully synchronized units. Constant mesh means that whether or not the gear is locked to the output shaft, it is in mesh with its counter gear. All gears rotate in the transmission as long as the clutch is engaged. Fully synchronized means the unit uses a mechanism of brass rings and clutches to bring rotating shafts and gears to the same speed before shifts occur. This promotes smooth shifting. In a vehicle equipped with a four-speed manual shift transmission or transaxle, all four forward gears are synchronized. Reverse gearing may or may not be synchronized, depending on the type of transmission/transaxle.

Transmission Designs

All automotive transmissions/transaxles are equipped with a varied number of forward speed gears, a neutral gear, and one reverse speed. Transmissions can be divided into groupings based on the number of forward speed gears they have. In the past, the most commonly used transmission was a three-speed; four-speeds were only found in trucks and high-performance cars. The growing concern for improved gas mileage led to smaller engines with four-speed transmissions. The additional gear allowed the smaller engines to perform better by matching the engine's torque curve with vehicle speeds and loads.

Five-speed transmissions and transaxles are now the commonly used units. Some of the early five-speed units were actually four-speeds with an add-on fifth or overdrive gear. Overdrive reduces engine speed at a given vehicle speed, which increases top speed, improves fuel economy, and lowers engine noise. Most late-model five-speed units incorporate a fifth gear in their main assemblies. This is also true of six-speed transmissions and transaxles. The fifth and sixth gears are included in the main assembly and typically provide two overdrive gears. The addition of the two overdrive gears allows the manufacturers to use lower final drive gears for acceleration. The fifth and sixth gears reduce the overall gear ratio and allow for slower engine speeds during highway operation.

GEARS

The purpose of the gears in a manual transmission or transaxle is to transmit rotating motion. Gears are normally mounted on a shaft, and they transmit rotating motion from one parallel shaft to another **(Figure 36–3)**.

Gears and shafts can interact in one of three ways: the shaft can drive the gear; the gear can drive the shaft; or the gear can be free to turn on the shaft. In this last case, the gear acts as an idler gear.

Sets of gears can be used to multiply torque and decrease speed, increase speed and decrease torque, or transfer torque and leave speed unchanged.

Figure 36-3 The gears in a transmission transmit the rotating power from the engine. *Courtesy of DaimlerChrysler Corporation*

Gear Design

Gear pitch is a very important factor in gear design and operation. Gear pitch refers to the number of teeth per given unit of pitch diameter. A simple way of determining gear pitch is to divide the number of teeth by the pitch diameter of the gear. For example, if a gear has thirty-six teeth and a 6-inch pitch diameter, it has a gear pitch of six **(Figure 36–4)**. The important fact to remember is that gears must have the same pitch to operate together. A five-pitch gear meshes only with another five-pitch gear, a six-pitch only with a six-pitch, and so on.

Spur Gears The **spur gear** is the simplest gear design used in manual transmissions and transaxles. As shown in **Figure 36–5**, spur gear teeth are cut straight across the edge parallel to the gear's shaft. During operation, meshed spur gears have only one tooth in full contact at a time.

Its straight tooth design is the spur gear's main advantage. It minimizes the chances of popping out of gear, an important consideration during acceleration/deceleration and reverse operation. For this reason, spur gears are often used for the reverse gear.

The spur gear's major drawback is the clicking noise that occurs as teeth contact one another. At higher speeds, this clicking becomes a constant whine. Quieter gears, such as the helical design, are often used to eliminate this gear whine problem.

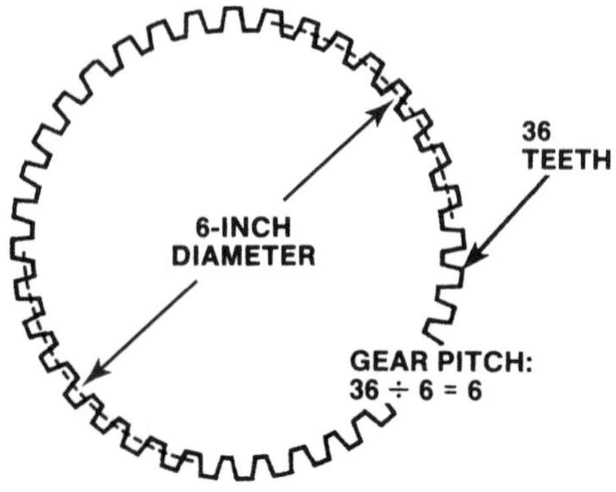

Figure 36-4 Determining gear pitch.

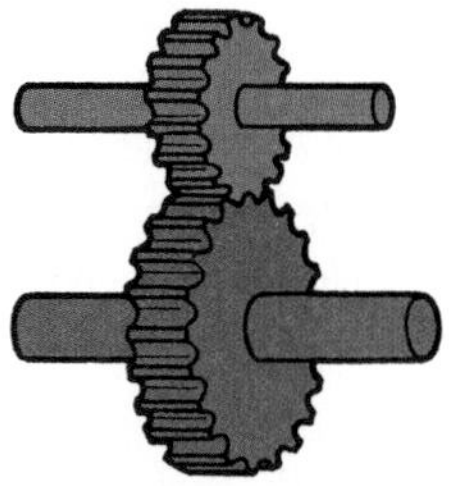

Figure 36-5 Spur gears have teeth cut straight across the gear edge parallel to the shaft.

SHOP TALK

When a small gear is meshed with a much larger gear, the small gear is often called a pinion or pinion gear, regardless of its tooth design. ■

Helical Gears A **helical gear** has teeth that are cut at an angle or are spiral to the gear's axis of rotation **(Figure 36–6)**. This configuration allows two or more teeth to mesh at the same time, which distributes tooth load and produces a very strong gear. Helical gears also run more

Figure 36-6 Helical gears have teeth cut at an angle to the gear's axis of rotation.

quietly than spur gears because they create a wiping action as they engage and disengage the teeth on another gear. One disadvantage is that helical teeth on a gear cause the gear to move fore or aft (axial thrust) on a shaft, depending on the direction of the angle of the gear teeth. This axial thrust must be absorbed by thrust washers and other transmission gears, shafts, or the transmission case.

Helical gears can be either righthanded or lefthanded, depending on the direction the spiral appears to go when the gear is viewed face-on. When mounted on parallel shafts, one helical gear must be righthanded and the other lefthanded. Two gears with the same direction spiral do not mesh in a parallel mounted arrangement.

Spur and helical gears that have teeth cut around their outside diameter edge are called **external gears**. When two external gears are meshed together, one rotates in the opposite direction as the other **(Figure 36–7)**. If an external gear is meshed with an internal gear (one that has teeth around its inside diameter), both rotate in the same direction.

Idler Gears

An **idler gear** is a gear that is placed between a drive gear and a driven gear. Its purpose is to transfer motion from the drive gear to the driven gear without changing the direction of rotation. It can do this because all three gears have external teeth **(Figure 36–8)**.

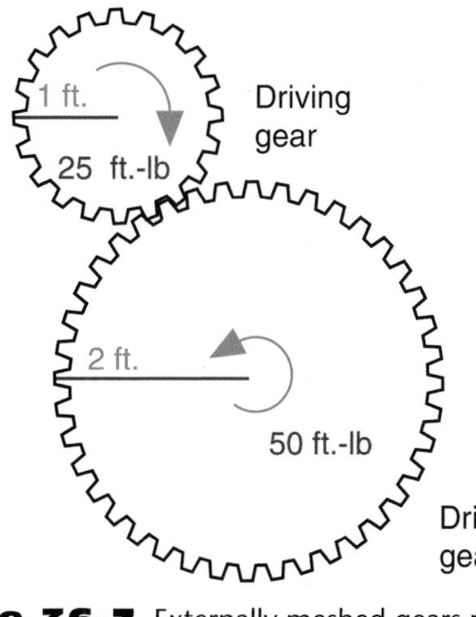

Figure 36-7 Externally meshed gears rotate in opposite directions.

Figure 36-8 An idler gear is used to transfer motion without changing rotational direction.

Idler gears are used in reverse gear trains to change the directional rotation of the output shaft. In all forward gears, the input shaft and the output shaft turn in the same direction. In reverse, the output shaft turns in the opposite direction as the input shaft. This allows the vehicle drive wheels to turn backward.

BASIC GEAR THEORY

Gears apply torque to other rotating parts of the drive train and are used to multiply torque. As gears with different numbers of teeth mesh, each rotates at a different speed and torque. Torque is calculated by multiplying the force by the distance from the center of the shaft to the point where the force is exerted.

A manual transmission is an assembly of gears and shafts that transmits power from the engine to the drive axle. The driver controls the changes in gear ratios. By moving the shift lever, various gear and speed ratios can be selected. The gears in a transmission are selected to give the driver a choice of both speed and torque. Lower gears allow for lower vehicle speeds but more torque. Higher gears provide less torque but higher vehicle speeds. Gear ratios state the ratio of the number of teeth on the driven gear to the number of teeth on the drive gear.

Different gear ratios are necessary because an engine develops relatively little power at low engine speeds. The engine must be turning at a fairly high speed before it can deliver enough power to get the car moving. Through selection of the proper gear ratio, torque applied to the drive wheels can be multiplied.

Transmission Gear Sets

Power is moved through the transmission via four gears (two sets of two gears). Speed and torque are altered in steps. To explain how this works, let us assign numbers to each of the gears. The small gear on the input shaft has twenty teeth. The gear it meshes with has forty. This provides a gear ratio of 2:1. The output of this gear set moves along the shaft of the forty-tooth gear and rotates other gears. The gear involved with first gear has fifteen teeth. This gear rotates with the same speed and with the same torque as the forty-tooth gear. However, the fifteen-tooth gear is meshed with a larger gear with thirty-five teeth. The gear ratio of the fifteen-tooth and the thirty-five-tooth gear set is 2.33:1. However, the ratio of the entire gear set (both sets of two gears) is 4.67:1.

To calculate this gear ratio, divide the driven (output) gear of the first set by the drive (input) gear of the first set. Do the same for the second set of gears, and then multiply the answer from the first by the second. The result is equal to the gear ratio of the entire gear set. The mathematical formula follows:

$$\frac{\text{driven (A)}}{\text{drive (A)}} \times \frac{\text{driven (B)}}{\text{drive (B)}} = \frac{40}{20} \times \frac{35}{15} = 4.67{:}1$$

Most of today's transmissions have at least one overdrive gear. Overdrive gears have ratios of less than 1:1. These ratios are achieved by using a small driving gear meshed with a smaller driven gear. Output speed is increased and torque is reduced. The purpose of overdrive is to promote fuel economy and reduce operating noise while maintaining highway cruising speed.

The driveline's gear ratios are further increased by the gear ratio of the ring and pinion gears in the drive axle assembly. Typical axle ratios are between 2.5 and 4.5:1. The final (overall) drive gear ratio is calculated by multiplying the transmission gear ratio by the final drive ratio. If a transmission is in first gear with a ratio of 3.63:1 and has a final drive ratio of 3.52:1, the overall gear ratio is 12.87:1. If fourth gear has a ratio of 1:1, using the same final drive ratio, the overall gear ratio is 3.52:1. The overall gear ratio is calculated by multiplying the ratio of the first set of gears by the ratio of the second (3.63 times 3.52 equals 12.78).

Reverse Gear Ratios

Reverse gear ratios involve two driving (driver) gears and two driven gears:

- the input gear is driver #1
- the idler gear is driven #1
- the idler gear is also driver #2
- the output gear is driven #2

If the input gear has twenty teeth, the idler gear has twenty-eight teeth and the output gear has forty-eight teeth. However, since a single idler gear is used, the teeth of it are not used in the calculation of gear ratio. The idler gear merely transfers motion from one gear to another. The calculations for determining reverse gear ratio with a single idler gear follow.

$$\text{Reverse gear ratio} = \frac{\text{driven \#2}}{\text{driver \#1}} = \frac{48}{20} = 2.40$$

If the gear set uses two idler gears (one with twenty-eight teeth and the other with forty teeth), the gear ratio involves three driving gears and three driven gears:

- the input gear is driver #1
- the #1 idler gear is driven #1
- the #1 idler gear is also driver #2
- the #2 idler gear is driven #2
- the #2 idler gear is also driver #3
- the output gear is driven #3

The ratio of this gearset would be calculated as follows:

$$\text{Reverse gear ratio} = \frac{\text{driven \#1} \times \text{driven \#2} \times \text{driven \#3}}{\text{driver \#1} \times \text{driver \#2} \times \text{driver \#3}} = \frac{28 \times 40 \times 48}{20 \times 28 \times 40} = \frac{53{,}760}{22{,}400} = 2.40$$

TRANSMISSION/TRANSAXLE DESIGN

The internal components of a transmission or transaxle consist of a parallel set of metal shafts on which meshing gearsets of different ratios are mounted **(Figure 36–9)**. By moving the shift lever, gear ratios can be selected to generate different amounts of output torque and speed.

The gears are mounted or fixed to the shafts in a number of ways. They can be internally splined or keyed to a shaft. Gears can also be manufactured as an integral part of the shaft. Gears that must be able to freewheel around the shaft during certain speed ranges are mounted to the shaft using bushings or bearings.

The shafts and gears are contained in a transmission or transaxle case or housing. The components of this housing include the main case body, side or top cover plates, extension housings, and bearing retainers **(Figure 36–10)**. The metal components are bolted together with gaskets providing a leak-proof seal at all joints. The case is filled with transmission fluid to provide constant lubrication and cooling for the spinning gears and shafts.

Transmission Features

Although they operate in a similar fashion, the layout, components, and terminology used in transmissions and transaxles are not exactly the same.

A transmission has three specific shafts: the input shaft, the **countershaft**, and the mainshaft or output shaft **(Figure 36–11)**. The clutch gear is an integral part of the transmission's input shaft and always rotates with the input shaft.

The countershaft is actually several gears machined out of a single piece of steel. The countershaft may also be called the countergear or **cluster gear**. The counter gear mounts on roller bearings on the countershaft. The countershaft is pinned in place and does not turn. Thrust washers control the amount of end play of the unit in the transmission case.

The main gears on the main shaft or output shaft transfer rotation from the countergears to the output shaft. The main gears are also called speed gears. They

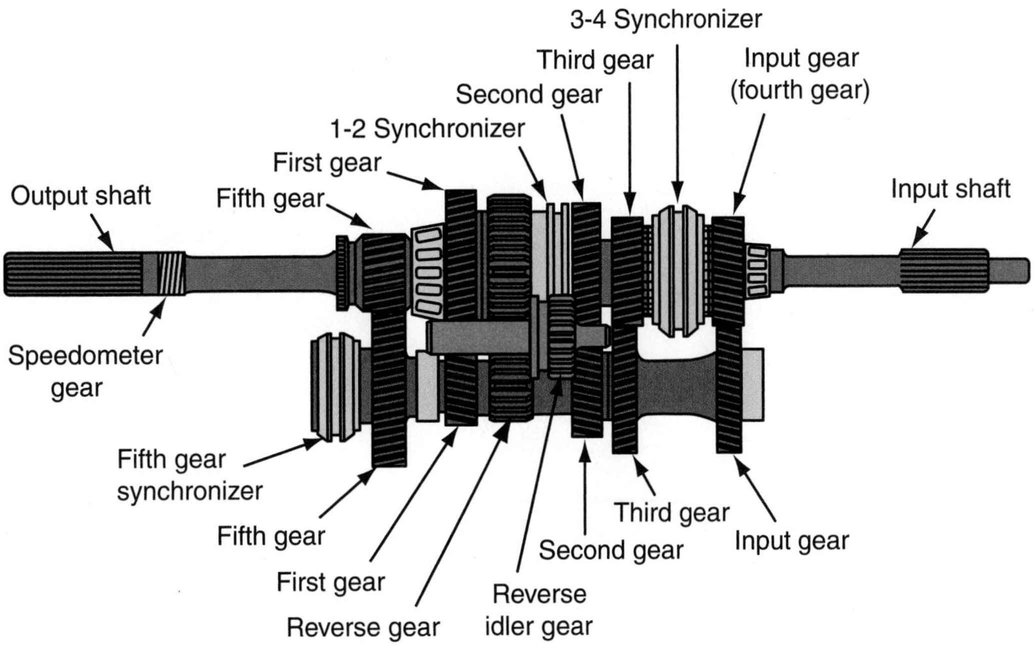

Figure 36–9 The arrangement of the gears and shafts in a typical five-speed transmission.

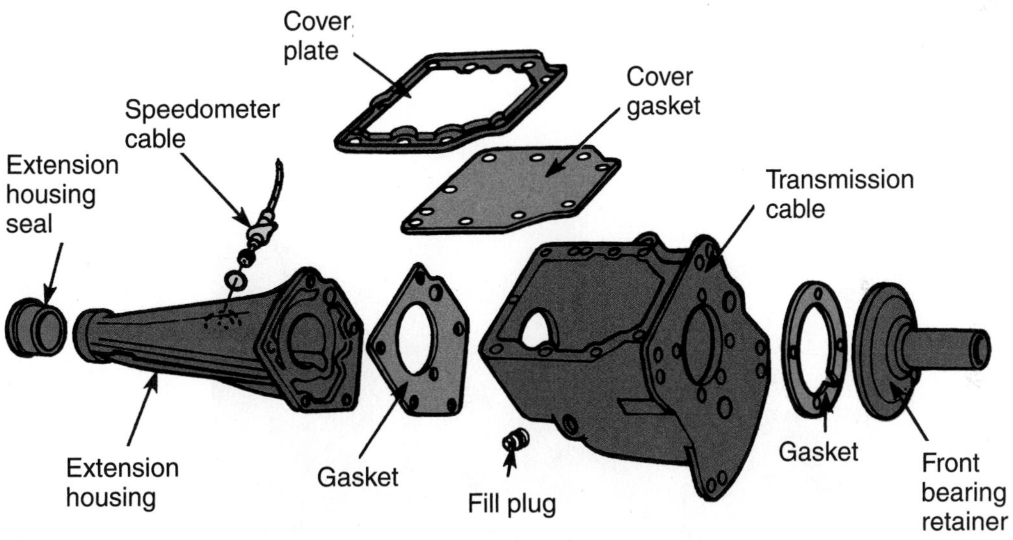

Figure 36–10 Typical four-speed manual transmission case components. *Courtesy of Ford Motor Company*

are mounted on the output shaft using roller bearings. Speed gears freewheel around the output shaft until they are locked to it by the engagement of their shift **synchronizer** unit.

Power flows from the transmission **input shaft** to the **clutch gear**. The clutch gear meshes with the large countergear of the countergear cluster. This cluster gear is now rotating. Since the cluster gear is meshed with the speed gears on the mainshaft, the speed gears are also turning.

There can be no power output until one of the speed gears is locked to the mainshaft. This is done by activating a shift fork, which moves its synchronizer to engage the selected speed gear to the mainshaft. Power travels along the countergear until it reaches this selected speed gear. It then passes through this gear back to the mainshaft and out of the transmission to the driveline.

Transaxle Features

Transaxles **(Figure 36–12)** use many of the design and operating principles found in transmissions. But because the transaxle also contains the differential gearing and drive axle connections, there are major differences in some areas of operation.

A transaxle typically has two separate shafts—an input shaft and an output shaft. The input shaft is the driving shaft. It is normally located above and parallel to the output shaft. The output shaft is the driven shaft. The transaxle's main (speed) gears freewheel around the

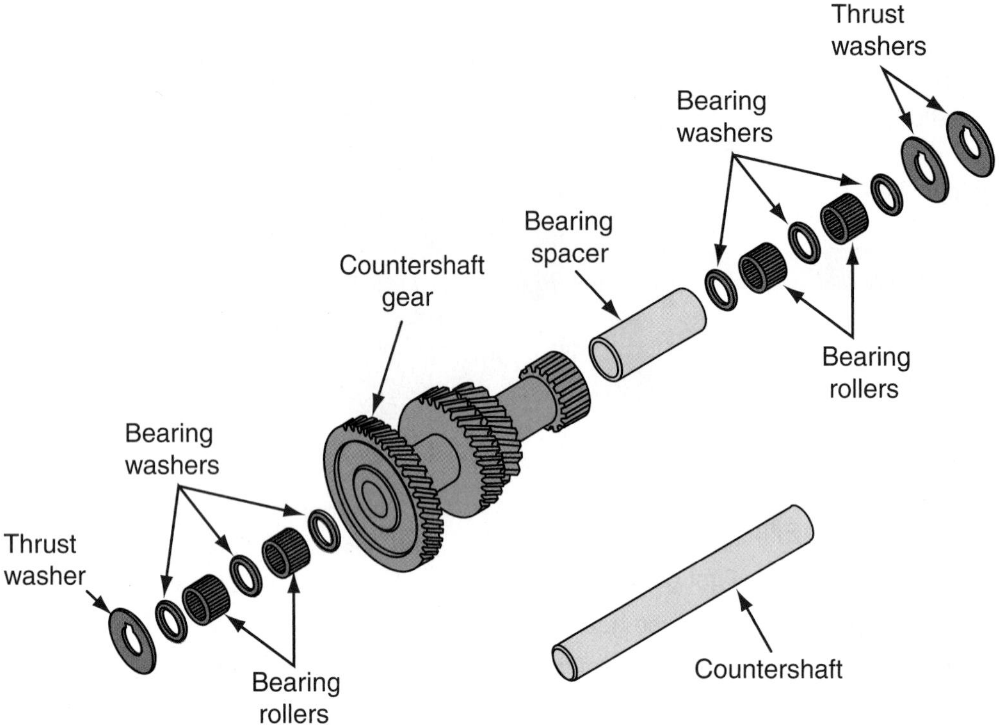

Figure 36-11 A typical countershaft assembly.

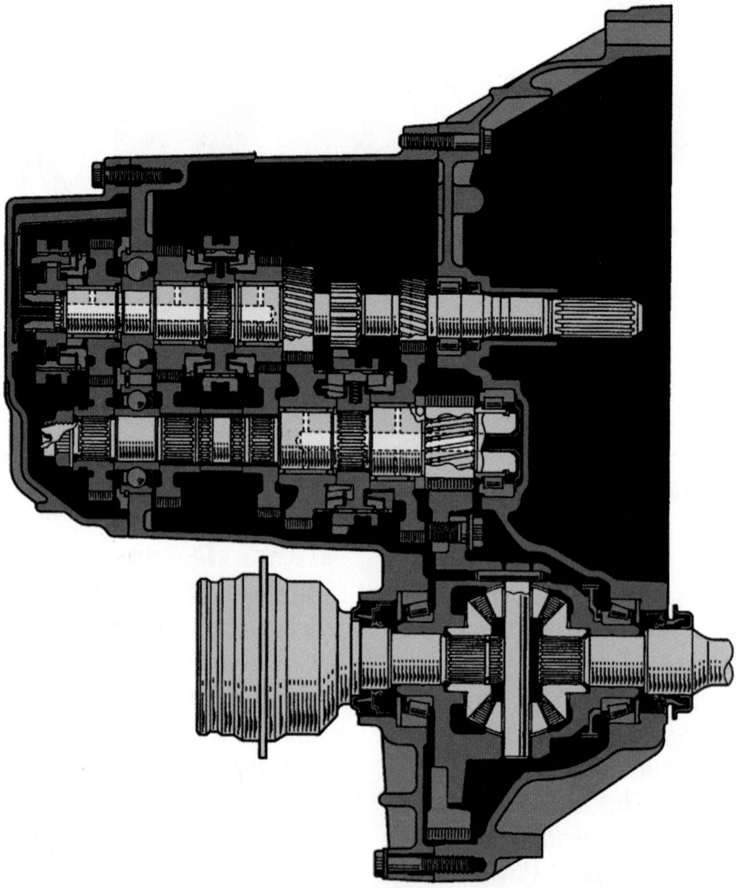

Figure 36-12 A transaxle assembly. *Reprinted with permission*

Item	Description
1.	Mainshaft
2.	Input cluster gear shaft
3.	4th speed gears
4.	3rd speed gears
5.	2nd speed gears
6.	Reverse gears
7.	Reverse idler gears
8.	1st speed gears 5th speed gear
9.	Driveshaft
10.	5th speed gear
11.	5th gear driveshaft pinion gear
12.	Mainshaft pinion gear
13.	Differential oil seals
14.	CV shafts
15.	Differential pinion gears
16.	Differential side gears
17.	Final drive ring gear
18.	1st/2nd synchronizer
19.	3rd/4th synchronizer
20.	5th synchronizer

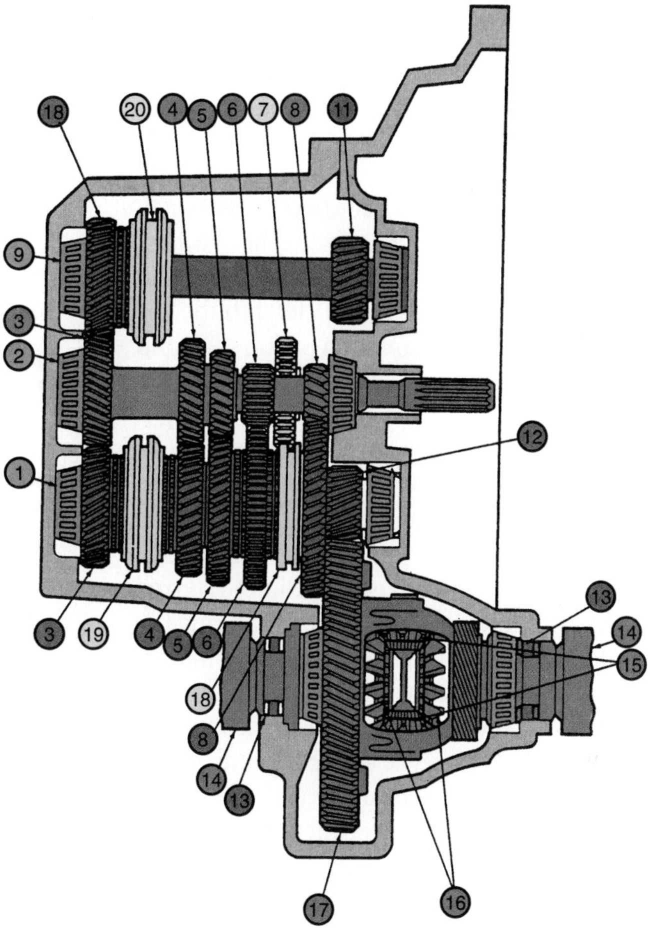

Figure 36–13 A transaxle with three gear shafts. *Courtesy of Ford Motor Company*

output shaft unless they are locked to the shaft by their synchronizer assembly. The main speed gears are in constant mesh with the input shaft drive gears. The drive gears turn whenever the input shaft turns.

The names used to describe transaxle shafts vary between manufacturers. The service manuals of some vehicles refer to the input shaft as the mainshaft and the output as the driven pinion or drive shaft. Others call the input shaft and its gears the input gear cluster and refer to the output shaft as the mainshaft. For clarity, this text uses the terms input gear cluster for the input shaft and its drive gears, and **pinion shaft** for the output shaft.

A pinion gear is machined onto the end of the transaxle's pinion shaft. This pinion gear is in constant mesh with the differential ring gear located in the lower portion of the transaxle housing. Because the pinion gear is part of the pinion shaft, it must rotate whenever the pinion shaft turns. With the pinion rotating, engine torque flows through the ring gear and differential gearing to the drive shafts and driving wheels.

Some transaxles have a third shaft designed to offset the power flow on the output shaft. Power is transferred from the output shaft to the third shaft using helical gears and by placing the third shaft in parallel with the output and input shafts. Other transaxles with a third shaft use an offset input shaft that receives the engine's power and transmits it to a mainshaft, which serves as an input shaft. The third shaft is only added to transaxles when an extremely compact transaxle is required **(Figure 36–13)**.

SYNCHRONIZERS

The synchronizer performs a number of jobs vital to transmission/transaxle operation. Its main job is to bring components that are rotating at different speeds to one synchronized speed. A synchronizer ensures that the pinion shaft and the speed gear are rotating at the same speed. The second major job of the synchronizer is to actually lock these components together. The end result of these two functions is a clash-free shift. In some transmissions, a synchronizer can have another important job. When spur teeth are cut into the outer sleeve of the synchronizer, the sleeve can act as a reverse gear and assist in producing the correct direction of rotation for reverse operation.

In modern transmissions and transaxles, all forward gears are synchronized. One synchronizer is placed between the first and second gears on the pinion shaft. Another is placed between the third and fourth gears on the **mainshaft**. If the transmission has a fifth gear, it is also equipped with a synchronizer. Reverse gear is not normally fitted with a synchronizer. A synchronizer requires gear rotation to do its job and reverse is selected with the vehicle at a stop.

Synchronizer Design

Figure 36–14 illustrates the most commonly used synchronizer—a **block** or **cone synchronizer**. The synchronizer sleeve surrounds the synchronizer assembly and meshes with the external splines of the clutch hub. The clutch hub is splined to the transmission pinion shaft and is held in position by a snap ring. A few transmissions use pin-type synchronizers.

The synchronizer sleeve has a small internal groove and a large external groove in which the shift fork rests. Three slots are equally spaced around the outside of the clutch hub. Inserts fit into these slots and are able to slide freely back and forth. These inserts, sometimes referred to as **shifter plates** or keys, are designed with a ridge in their outer surface. Insert springs hold the ridge in contact with the synchronizer sleeve internal groove.

The synchronizer sleeve is precisely machined to slide onto the clutch hub smoothly. The sleeve and hub sometimes have alignment marks to ensure proper indexing of their splines when assembling to maintain smooth operation.

Brass or bronze synchronizing **blocking rings** are positioned at the front and rear of each synchronizer assembly. Some synchronizer assemblies use frictional material on the blocking rings to reduce slippage. Each blocking ring has three notches equally spaced to correspond with the three insert keys of the hub. Around the outside of each blocking ring is a set of beveled clutching teeth, which is used for alignment during the shift sequence. The inside of the blocking ring is shaped like a cone. This coned surface is lined with many sharp grooves.

The cone of the blocking ring makes up only one-half of the total cone clutch. The second or matching half of the cone clutch is part of the speed gear to be synchronized. As shown in **Figure 36–15**, the shoulder of the speed gear is cone shaped to match the blocking ring. The shoulder also contains a ring of beveled clutching teeth designed to align with the clutching teeth on the blocking ring.

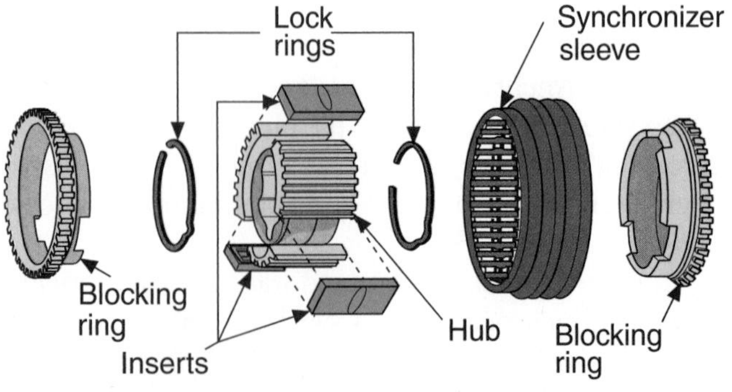

Figure 36–14 An exploded view of a blocking ring-type synchronizer assembly.

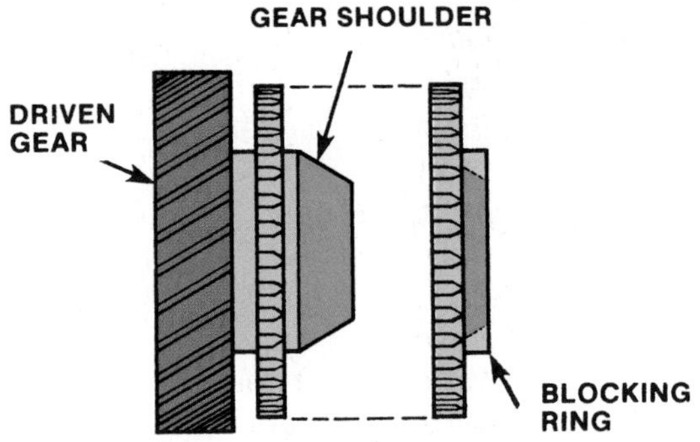

Figure 36–15 Gear shoulder and blocker ring mating surfaces.

Operation

When the transmission is in neutral or reverse, the first/second and third/fourth synchronizers are in their neutral position and are not rotating with the pinion shaft. Gears on the mainshaft are meshed with their countershaft partners and are freewheeling around the pinion shaft at various speeds.

To shift the transmission into first gear, the clutch is disengaged and the gearshift lever is placed in first gear position. This forces the shift fork on the synchronizer sleeve toward the first speed gear on the pinion shaft. As the sleeve moves, the inserts also move because the insert ridges lock the inserts to the internal groove of the sleeve.

The movement of the inserts forces the blocking ring's coned friction surface against the coned surface of the first speed gear shoulder. When the blocking ring and gear shoulder come into contact, the grooves on the blocking ring cone cut through the lubricant film on the first speed gear shoulder and a metal-to-metal contact is made. The contact generates substantial friction and heat. This is one reason bronze or brass blocking rings are used. A nonferrous metal such as bronze or brass minimizes wear on the hardened steel gear shoulder. This frictional coupling is not strong enough to transmit loads for long periods. As the components reach the same speed, the synchronizer sleeve can now slide over the external clutching teeth on the blocking ring and then over the clutching teeth on the first speed gear shoulder. This completes the engagement. Power flow is now from the first speed gear, to the synchronizer sleeve, to the synchronizer clutch hub, to the main output shaft, and out to the driveline.

To disengage the first speed gear from the pinion shaft and shift into second speed gear, the clutch must be disengaged as the shift fork is moved to pull the synchronizer sleeve and disengage it from the first gear. As the transmission is shifted into second gear, the inserts again lock into the internal groove of the sleeve. As the sleeve moves forward, the forward blocking ring is forced by the inserts against the coned friction surface on the second speed gear shoulder. Once again, the grooves on the blocking ring cut through the lubricant on the gear shoulder to generate a frictional coupling that synchronizes the speed gear and shaft speeds. The shift fork can then continue to move the sleeve forward until it slides over the blocking ring and speed gear shoulder clutching teeth, locking them together. Power flow is now from the second speed gear, to the synchronizer sleeve, to the clutch hub, and out through the pinion shaft.

GEARSHIFT MECHANISMS

Figure 36–16 illustrates a typical transmission shift linkage for a five-speed transmission. As you can see, there are three separate shift rails and forks. Each shift rail/shift fork is used to control the movement of a synchronizer, and each synchronizer is capable of engaging and locking two speed gears to the mainshaft. The shift rails transfer motion from the driver-controlled gearshift lever to the shift forks. The **shift forks (Figure 36–17)** are semicircular castings connected to the shift rails with split pins. The shift fork rests in the groove in the synchronizer sleeve and surrounds about one-half of the sleeve circumference.

The gearshift lever is connected to the shift forks by means of a gearshift linkage. Linkage designs vary between manufacturers but can generally be classified as being direct or remote.

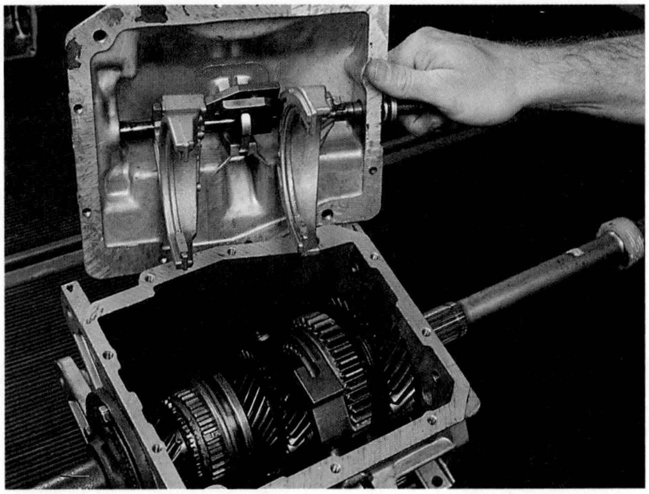

Figure 36–17 The shift forks and rails are assembled in the cover of this transmission with a direct shift linkage.

Gearshift Linkages

There are two basic designs of gearshift linkages: internal and external. Internal linkages are located at the side or top of the transmission. The control end of the shifter is mounted inside the transmission, as are all of the shift controls. Movement of the shifter moves a **shift rail** and shift fork toward the desired gear. This moves the synchronizer sleeve to lock the selected speed gear to the shaft. This type of linkage is often called a direct linkage, because the shifter is in direct contact with the internal gear shifting mechanisms.

Shift rails are machined with interlock and detent notches. The interlock notches prevent the selection of more than one gear during shifting. When a shift rail is moved by the shifter, interlock pins hold the other shift

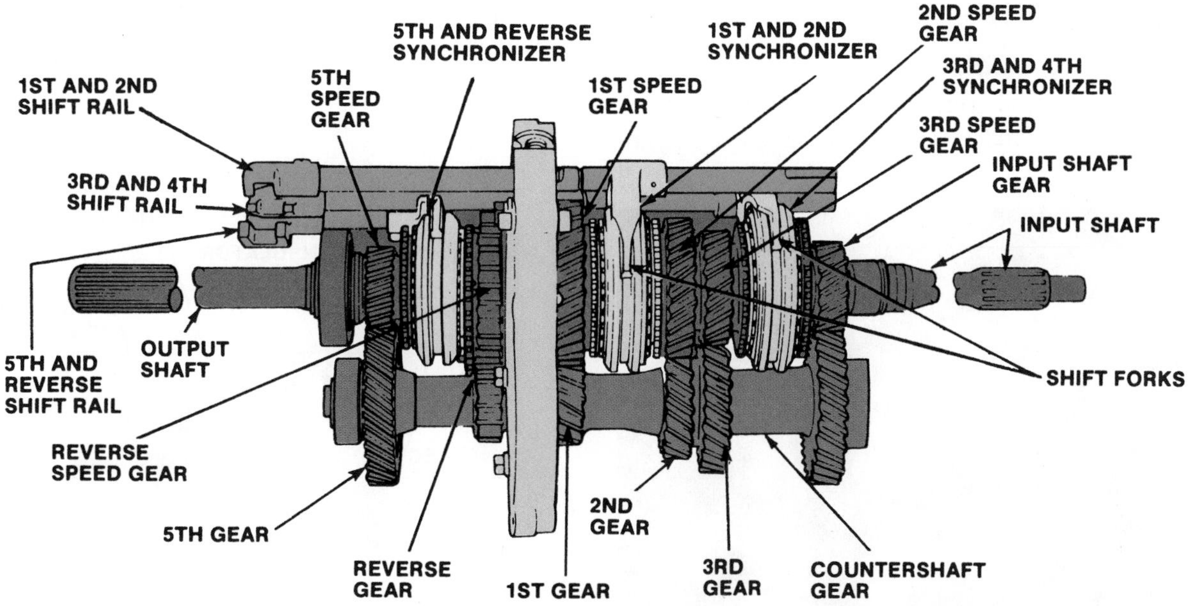

Figure 36–16 An interior view of a five-speed overdrive transmission. Three separate shift rail/shift fork/synchronizer combinations control first/second, third/fourth, and fifth/reverse shifting. *Courtesy of Ford Motor Company*

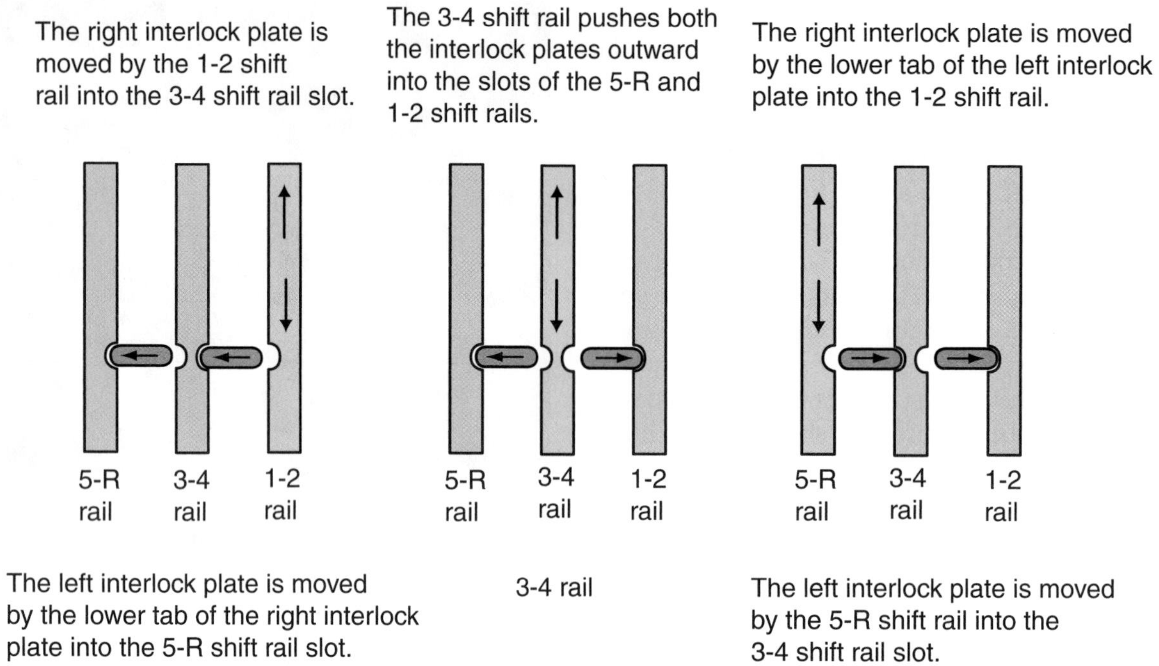
The right interlock plate is moved by the 1-2 shift rail into the 3-4 shift rail slot.
The 3-4 shift rail pushes both the interlock plates outward into the slots of the 5-R and 1-2 shift rails.
The right interlock plate is moved by the lower tab of the left interlock plate into the 1-2 shift rail.
5-R rail
3-4 rail
1-2 rail
5-R rail
3-4 rail
1-2 rail
5-R rail
3-4 rail
1-2 rail
The left interlock plate is moved by the lower tab of the right interlock plate into the 5-R shift rail slot.
3-4 rail
The left interlock plate is moved by the 5-R shift rail into the 3-4 shift rail slot.

Figure 36-18 Interlock pins prevent the selection of one or more gears.

rails in their neutral position **(Figure 36–18)**. The detent notches and matching spring loaded pins or balls give the driver feedback as to when the shift collar is adequately moved.

As the shift rail moves, a detent ball moves out of its detent notch and drops into the notch for the selected gear. At the same time, an interlock pin moves out of its interlock notch and into the other shift rails.

External linkages function in much the same way, except that rods, external to the transmission, are connected to levers that move the internal shift rails of the transmission **(Figure 36–19)**. Some transaxles are shifted by rods **(Figure 36–20)** or by cable **(Figure 36–21)**.

Automatic Manuals

Some manual transmissions are equipped with complex shifting mechanisms that cause them to behave like an automatic transmission. This should not be confused with automatic transmissions that act somewhat like manual transmissions. Although automatic manual transmissions do not have a clutch, their design and operation are like that of other manual transmissions.

BMW's sequential manual gearbox (SMG) is based on a true six-speed manual transmission. The system does not use a clutch pedal but does have a clutch. Gear shifting is done through paddle shifters on the steering column or with a gearshift lever in the center console. The actual shifting, though, is done by computers, solenoids, hydraulics, and linkages **(Figure 36–22)**.

The system has a total of eleven settings—six in manual mode and five in automatic mode—that allow the driver to select the desired style of driving.

Figure 36-19 An external shifter assembly mounted to the transmission.

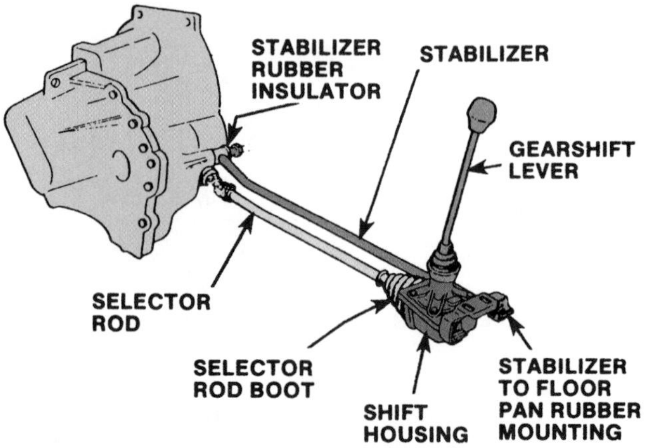
STABILIZER RUBBER INSULATOR
STABILIZER
GEARSHIFT LEVER
SELECTOR ROD
SELECTOR ROD BOOT
SHIFT HOUSING
STABILIZER TO FLOOR PAN RUBBER MOUNTING

Figure 36-20 A remote gearshift showing linkage, selector rod, and stabilizer (stay bars). *Courtesy of Ford Motor Company*

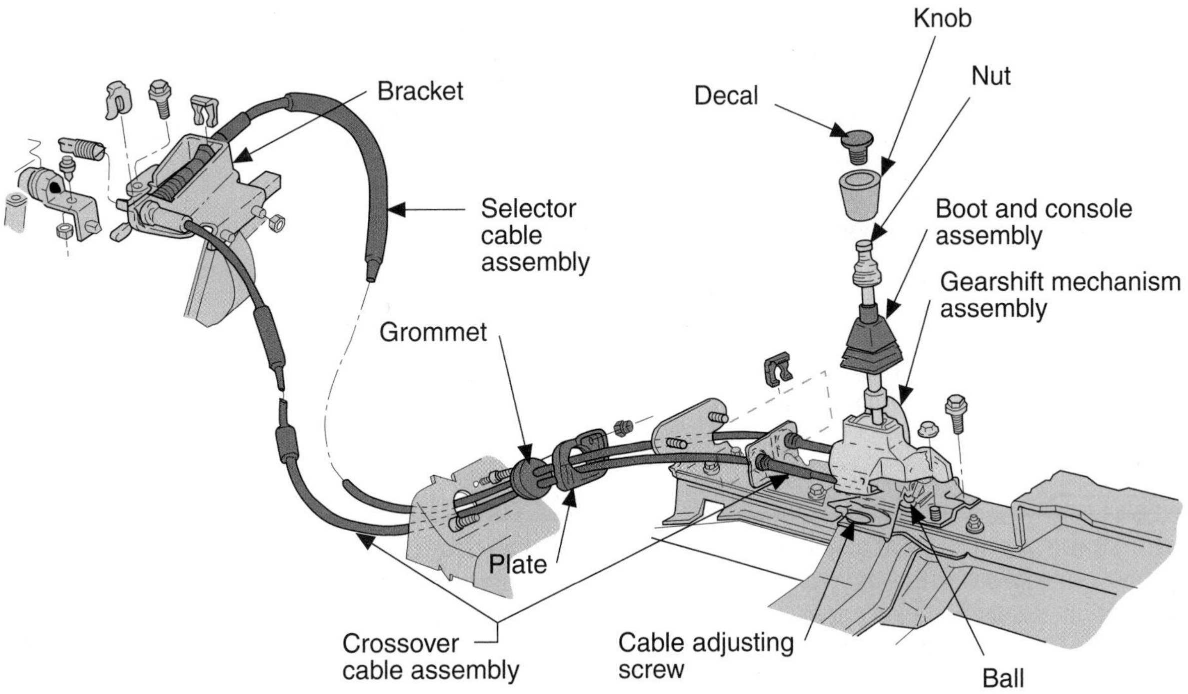

Figure 36-21 A cable-type external gearshift linkage used in a transaxle application. *Courtesy of DaimlerChrysler Corporation*

Direct Shift Gearbox A direct shift gearbox is a manual transmission with high variability in the selection of the transmission's ratio. Using an integrated twin multiplate clutch, two gears can be engaged at the same time. During normal operation, one gear is engaged. When the next gearshift point is approached, the appropriate gear is preselected but its clutch is kept disengaged. The gearshift process opens the clutch of the activated gear and closes the other clutch at the same time. The gear change takes place under load, with the result that a permanent flow of power is maintained.

TRANSMISSION POWER FLOW

The following sections describe the power flow paths in a typical four-speed manual transmission.

Neutral

Neutral power flow is illustrated in **Figure 36–23**. The input shaft rotates at engine speed whenever the clutch is engaged. The clutch gear is mounted on the input shaft and rotates with it. The clutch gear meshes with the countergear, which rotates around the countershaft.

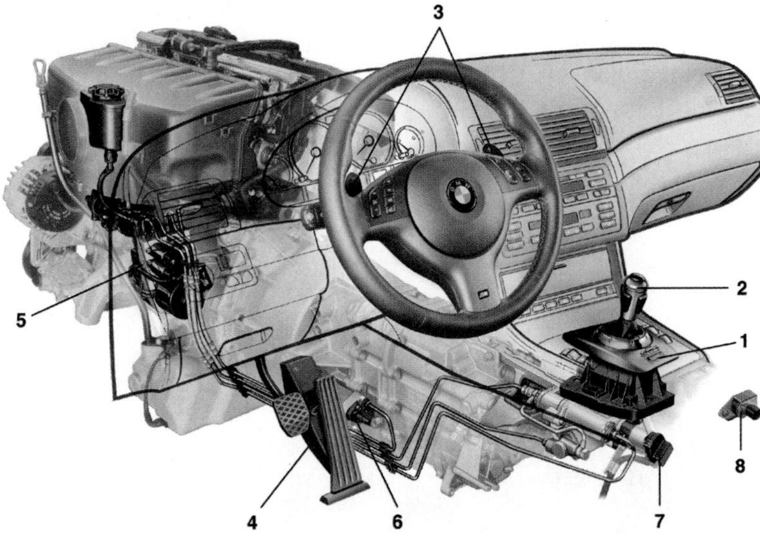

Figure 36-22 BMW's SMG system: 1 = drivelogic control module; 2 = gearshift; 3 = paddles; 4 = accelerator input; 5 = hydraulic unit; 6 = clutch position sensor; 7 = input shaft speed sensor and transmission oil temperature sensor; and 8 = gear selector position sensor. *Courtesy of BMW of North America, Incorporated*

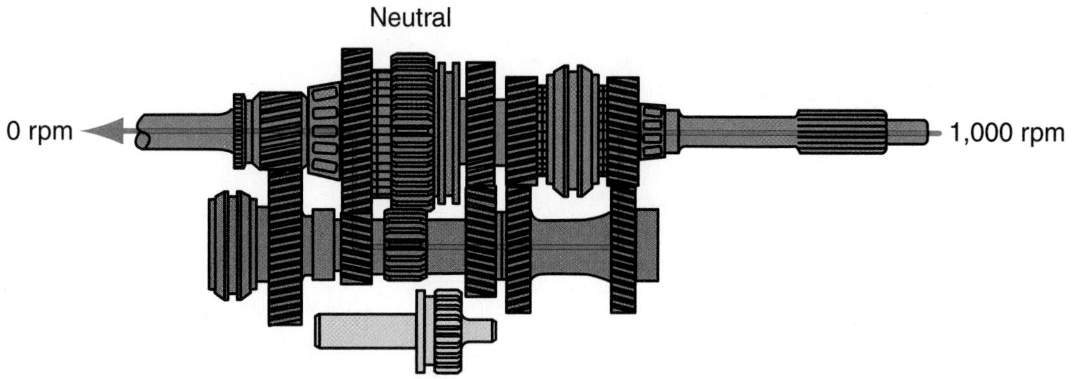

Figure 36-23 Power flow in neutral.

The countergear transfers power to the speed gears on the mainshaft. However, since speed gears one, two, three, and four are not locked to the mainshaft when the transmission is in neutral, they cannot transfer power to the mainshaft. The mainshaft does not turn, and there is no power output to the driveline.

All gear changes pass through the neutral gear position. When changing gears, one speed gear is disengaged, resulting in neutral, before the chosen gear is engaged. This is important to remember when diagnosing hard-to-shift problems.

First Gear

First gear power flow is illustrated in **Figure 36–24**. Power or torque flows through the input shaft and clutch gear to the countergear. The countergear rotates. The first gear on the cluster drives the first speed gear on the mainshaft. When the driver selects first gear, the first/second synchronizer moves to the rear to engage the first speed gear and lock it to the mainshaft. The first speed gear drives the main (output) shaft, which transfers power to the driveline. A typical first speed gear ratio is 3:1 (three full turns of the input shaft to one full turn of the output shaft). So, if the engine torque entering the transmission is 220 ft-lb (298 Nm) it is multiplied three times to 660 ft-lb (895 Nm) by the time it is transferred to the driveline.

Second Gear

When the shift from first to second gear is made, the shift fork disengages the first/second synchronizer from the first speed gear and moves it until it locks the second speed gear to the mainshaft. Power flow is still through the input shaft and clutch gear to the countergear. However, now the second countergear on the cluster transfers power to the second speed gear locked on the mainshaft. Power flows from the second speed gear through the synchronizer to the mainshaft (output shaft) and driveline **(Figure 36–25)**.

In second gear, the need for vehicle speed and acceleration is greater than the need for maximum torque multiplication. To meet these needs, the second speed gear on the mainshaft is designed slightly smaller than the first speed gear. This results in a typical gear ratio of 2.2:1, which reflects a drop in torque and an increase in speed.

Third Gear

When the shift from second to third gear is made, the shift fork returns the first/second synchronizer to its neutral position. A second shift fork slides the third/fourth synchronizer until it locks the third speed gear to the mainshaft. Power flow now goes through the third gear of the countergear to the third speed gear, through the synchronizer to the mainshaft, and driveline **(Figure 36–26)**.

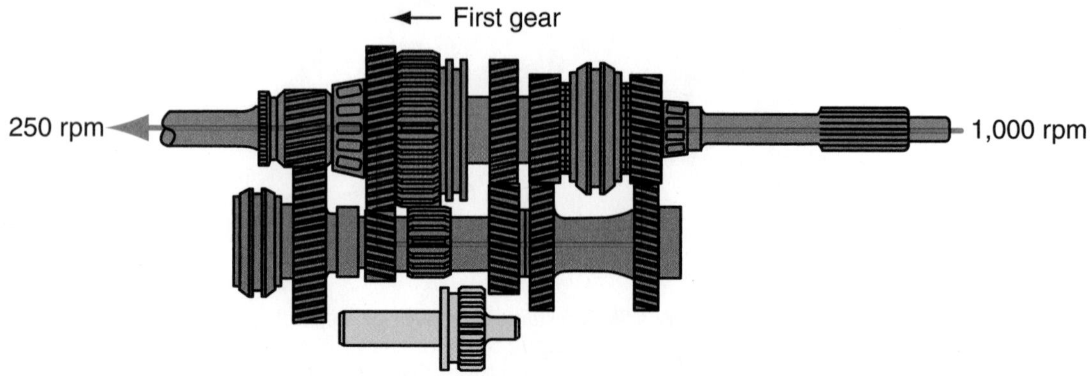

Figure 36-24 Power flow in first gear.

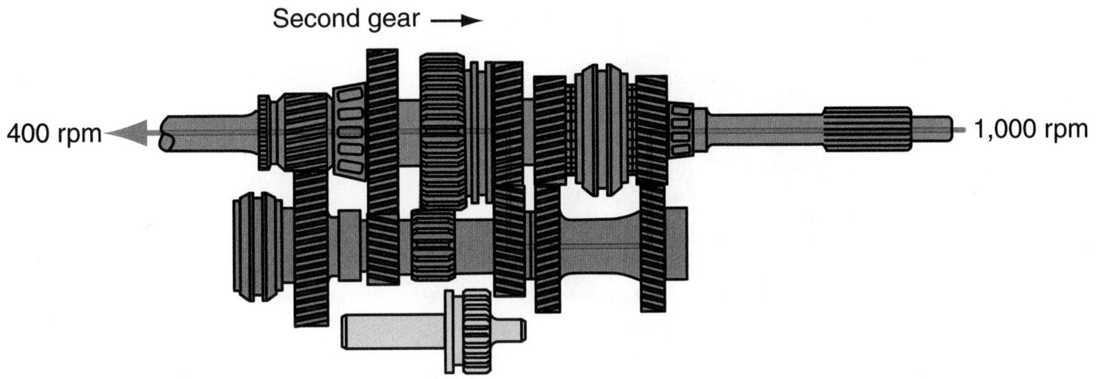

Figure 36-25 Power flow in second gear.

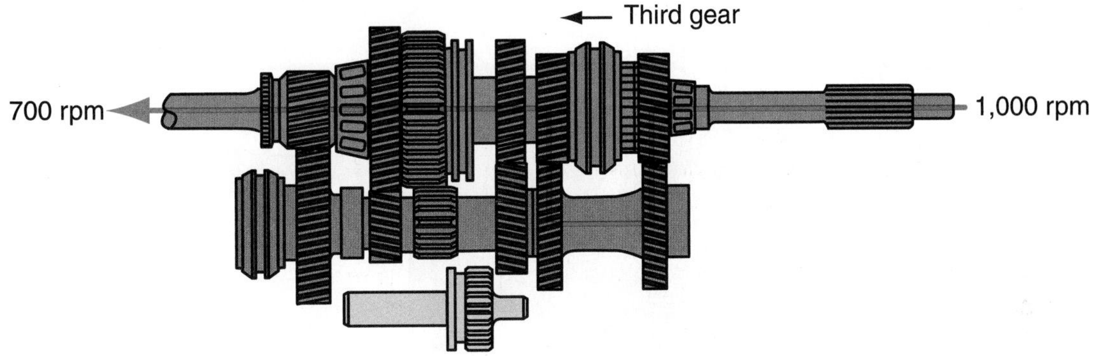

Figure 36-26 Power flow in third gear.

Third gear permits a further decrease in torque and increase in speed. As you can see, the third speed gear is smaller than the second speed gear. This results in a typical gear ratio of 1.7:1.

Fourth Gear

In fourth gear, the third/fourth synchronizer is moved to lock the clutch gear on the input shaft to the mainshaft. This means power flow is directly from the input shaft to the mainshaft (output shaft) at a gear ratio of 1:1 **(Figure 36–27)**. This ratio results in maximum speed output and no torque multiplication. Fourth gear has no torque multiplication because it is used at cruising speeds to promote maximum fuel economy. The vehicle is normally downshifted to lower gears to take advantage of torque multiplication and acceleration when passing slower vehicles or climbing grades.

Fifth Gear

When fifth gear is selected, the fifth gear synchronizer engages fifth gear to the mainshaft **(Figure 36–28)**. This causes a large gear on the countershaft to drive a smaller gear on the mainshaft, which results in an overdrive condition. Overdrive permits an engine speed reduction at higher vehicle speeds.

Reverse

In reverse gear, it is necessary to reverse the direction of the mainshaft (output shaft). This is done by introducing a reverse idler gear into the power flow path. The idler gear is located between the countershaft reverse gear and the reverse speed gear on the mainshaft. The idler assembly is made of a short drive shaft independently mounted in the transmission case parallel to the countershaft. The idler gear may be mounted near the midpoint of the shaft.

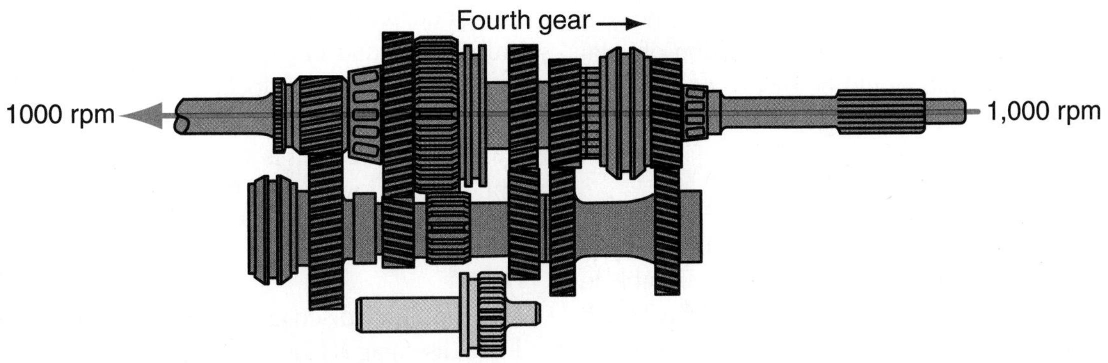

Figure 36-27 Power flow in fourth gear.

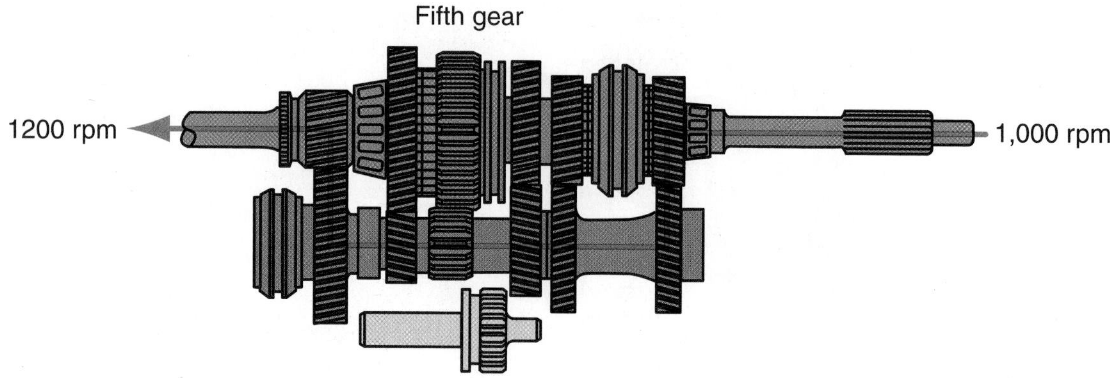

Figure 36-28 Power flow in fifth gear.

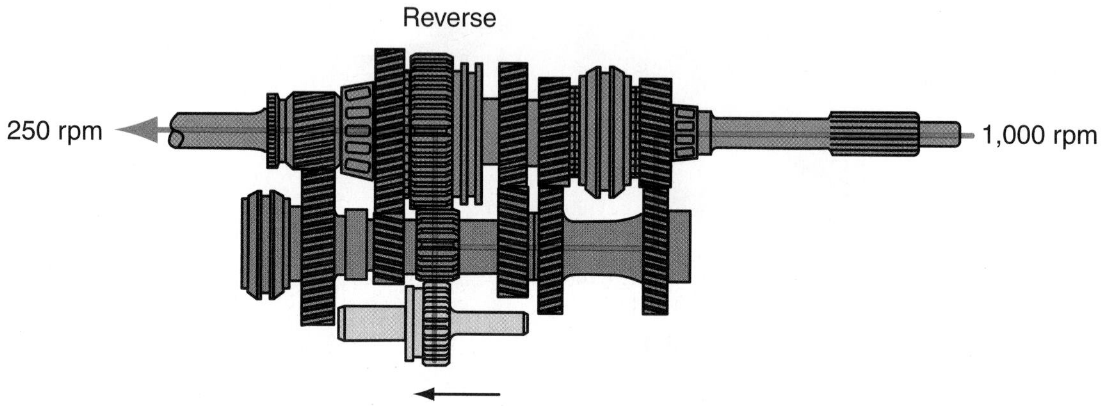

Figure 36-29 Power flow in reverse gear.

The reverse speed gear is actually the external tooth sleeve of the first/second synchronizer.

When reverse gear is selected, both synchronizers are disengaged and in the neutral position. In the transmission shown in **Figure 36–29**, the shifting linkage moves the reverse idler gear into mesh with the first/second synchronizer sleeve. Power flows through the input shaft and clutch gear to the countershaft. From the countershaft, it passes to the reverse idler gear, where it changes rotational direction. It then passes to the first/second synchronizer sleeve. Rotational direction is again reversed. From the sleeve, power passes to the mainshaft and driveline.

TRANSAXLE POWER FLOWS

The following sections describe power flow for the gear ranges of a four-speed transaxle. The direction of rotation used in this example would be as viewed from the vehicle's right front fender looking into the engine compartment.

Neutral

When the transaxle is placed in neutral, the engaged clutch drives the input shaft and gear cluster assembly in a clockwise direction. The first/second and third/fourth synchronizers on the mainshaft are not engaged, so the pinion shaft gears are not locked to the pinion shaft. The pinion shaft and the pinion gear do not turn, so there is no output to the transaxle differential ring gear.

First

In first gear, the first/second synchronizer engages the first speed gear to the mainshaft, locking it to the pinion shaft. The cluster's first gear, rotating clockwise, drives the first speed gear and the pinion shaft in a counterclockwise direction. The counterclockwise turning pinion on the end of the pinion shaft drives the differential ring gear, differential gearing, drive shafts, and wheels in a clockwise direction **(Figure 36–30)**.

Second

As the shift from first to second gear is made, the first/second synchronizer disengages the first speed gear on the pinion shaft and engages the second speed gear. With the second speed gear locked to the pinion shaft, power flow is shown in **Figure 36–31**. As you can see, power flow and direction is similar to first gear, with the exception that flow is now through the second speed gear and synchronizer to the pinion shaft and pinion.

Third

With the clutch disengaged, the first/second synchronizer sleeve disengages from the second speed gear on the pinion shaft and returns to its midway or neutral position be-

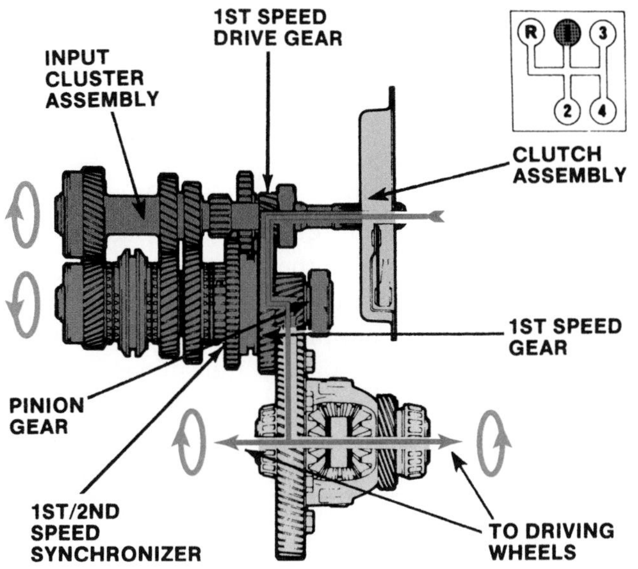

Figure 36-30 Four-speed transaxle power flow in first gear. *Courtesy of Ford Motor Company*

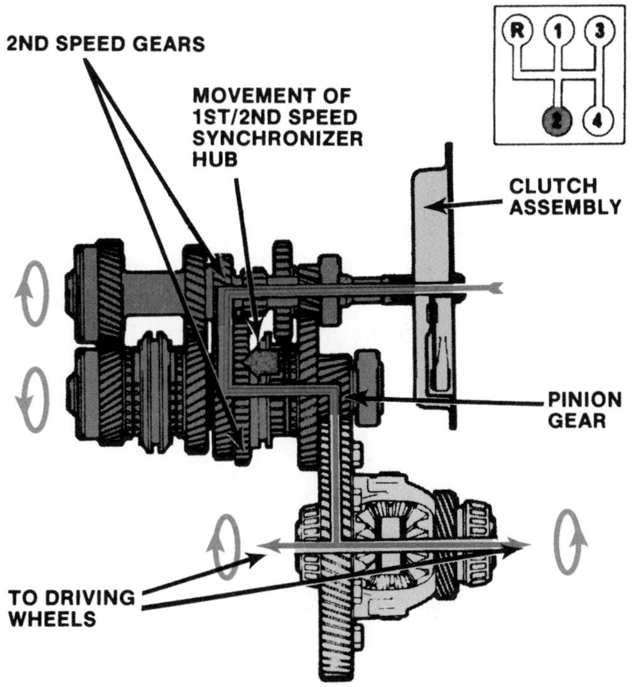

Figure 36-31 Four-speed transaxle power flow in second gear. *Courtesy of Ford Motor Company*

tween the first and second speed gears. As the driver moves the shift lever from its second gear position through neutral to the third gear position, the gear lever inside the transaxle housing moves from the first/second synchronizer position to the third/fourth synchronizer position. It engages the third/fourth synchronizer and locks it to the third speed gear on the pinion shaft. Power flow is then through the third speed gear to the synchronizer and pinion shaft to the pinion gear and differential ring gear **(Figure 36–32)**.

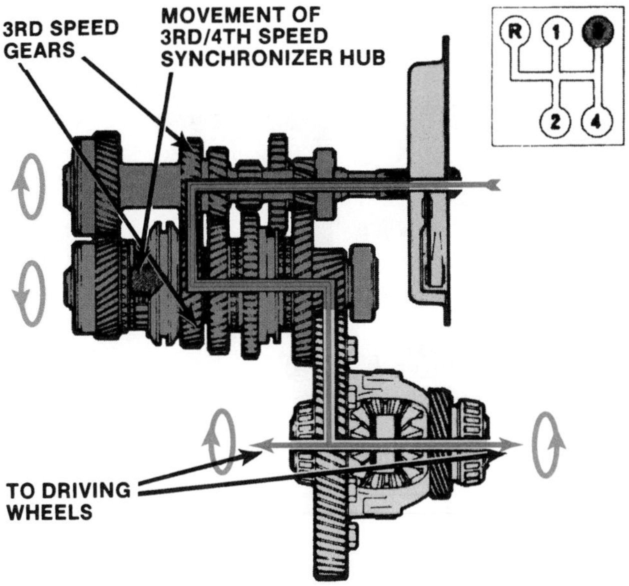

Figure 36-32 Four-speed transaxle power flow in third gear. *Courtesy of Ford Motor Company*

Fourth

The action of the shift lever moves the third/fourth synchronizer sleeve away from the pinion shaft third speed gear and toward the fourth speed gear, locking it to the pinion shaft. Power flow for fourth gear is shown in **Figure 36–33.**

Reverse

When the shift lever is placed in reverse, the reverse idler gear shifts into mesh with the input cluster reverse gear and the reverse speed gear. The reverse speed gear is the sleeve of the first/second synchronizer. To act as the reverse speed gear, the synchronizer sleeve is designed with spur teeth machined around its outside edge.

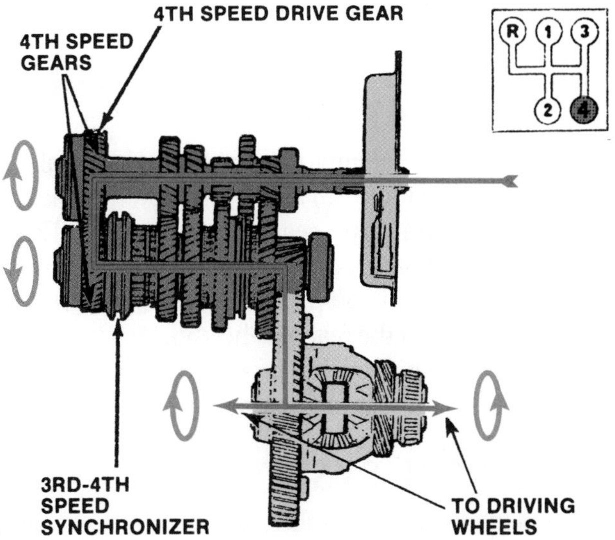

Figure 36-33 Four-speed transaxle power flow in fourth gear. *Courtesy of Ford Motor Company*

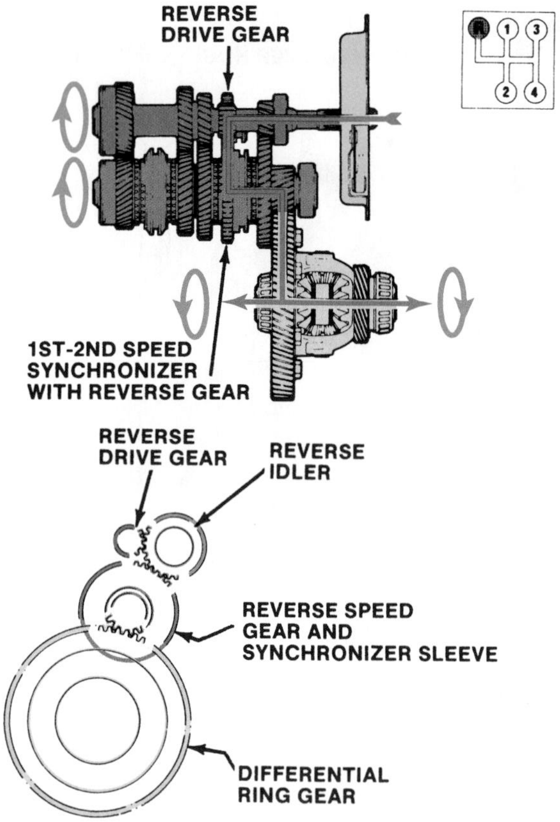

Figure 36–34 Four-speed transaxle power flow in reverse gear. *Courtesy of Ford Motor Company*

The reverse idler gear changes the direction of rotation of the pinion shaft reverse speed gear so that the vehicle backs up. Reverse power flow is illustrated in **Figure 36–34**.

Like transmissions, some transaxles have five forward speeds. Normally, fourth and fifth gears for smaller cars have overdrive ratios. These high gear ratios compensate for very low final drive gear ratios. Low final drive ratios provide great torque multiplication, which is needed to safely accelerate with a small engine.

FINAL DRIVE GEARS AND OVERALL RATIOS

All vehicles use a differential to provide an additional gear reduction (torque increase) above and beyond what the transmission or transaxle gearing can produce. This is known as the **final drive gear**.

In a transmission-equipped vehicle, the differential gearing is located in the rear axle housing. In a transaxle, however, the final reduction is produced by the final drive gears housed in the transaxle case.

ELECTRICAL SYSTEMS

Although manual transmissions are not electrically operated or controlled, a few accessories of the car are controlled or linked to the transmission. The transmission may also be fitted with sensors that give vital information to the computer that controls other car systems. There are a few transmissions that have their shifting controlled or limited by electronics.

Reverse Lamp Switch

All vehicles sold in North America after 1971 have been required to have backup (reverse) lights. Backup lights illuminate the area behind the vehicle and warn other drivers and pedestrians that the vehicle is moving in reverse. Most manual transmissions are equipped with a separate switch located on the transmission **(Figure 36–35)** but can be mounted to the shift linkage away from the transmission. If the switch is mounted in the transmission, the shifting fork closes the switch and completes the electrical circuit whenever the transmission is shifted into reverse gear. If the switch is mounted on the linkage, the switch is closed directly by the linkage.

Vehicle Speed Sensor

Most late-model transmissions and transaxles are fitted with a VSS. This sensor sends an electrical signal to the vehicle's PCM. This signal represents the speed of the transmission's output shaft. The PCM then calculates the speed of the vehicle. This information is used for many systems, such as cruise control, fuel and spark management, and instrumentation.

Upshift Lamp Circuit

Upshift and shift lamps inform the driver when to shift into the highest gear in order to maximize fuel economy. These lights are typically controlled by the PCM, which controls the light according to engine speed, engine load,

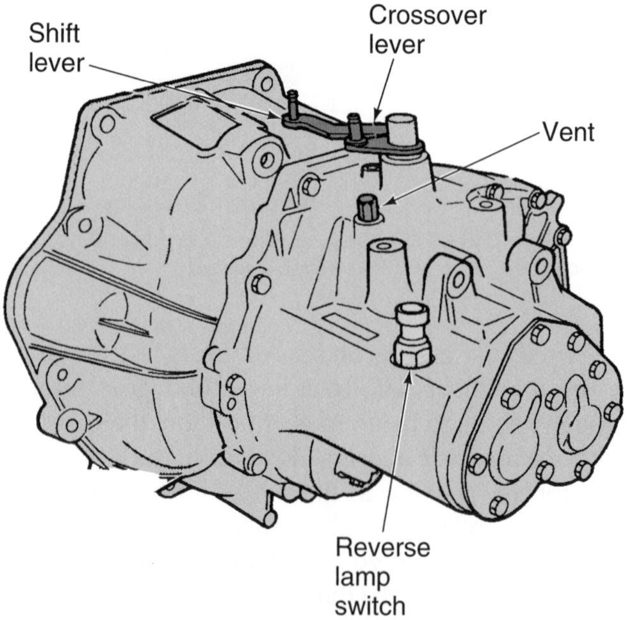

Figure 36–35 The typical location of a backup light switch in a transaxle. *Courtesy of DaimlerChrysler Corporation*

and vehicle speed. Basically, these lights operate much like a vacuum gauge. The shift lamp is lit at those engine speeds and loads when engine vacuum is high and the transmission is in a forward gear. The shift light stays on until the transmission is shifted or the engine's operating conditions change. High gear disables the circuit.

Shift Blocking

Some six-speed transmissions have a feature called shift blocking that prevents the driver from shifting into second or third gears from first gear, when the engine's coolant temperature is below a specified degree, the speed of the car is between 12 and 22 mph (20 and 29 km/h), and the throttle is opened less than 35%. Shift blocking helps improve fuel economy. These transmissions are also equipped with reverse lockout, as are some others. This feature prevents the engagement of reverse whenever the vehicle is moving forward.

Shift blocking is controlled by the PCM. A "skip shift" solenoid is used to block off the shift pattern from first gear to second and third gears. The driver moves the gearshift from its up position to a lower position, as if shifting into second gear, but fourth gear is selected.

CUSTOMER CARE

Just because the technician gets a little dirty in the course of a repair does not mean the vehicle should, too. Treat every car that enters the shop with the utmost care and consideration. Scratches from belt buckles or tools and grease smears on the steering wheel, upholstery, or carpeting are inexcusable, and a sure way of losing business. Always use fender, seat, and floor covers when the job requires them. Check your hands for cleanliness before driving a vehicle or operating the windows and dash controls. ■

CASE STUDY

A customer brought his 1991 Toyota into the shop and said that the transaxle was jumping out of fifth gear. The technician verified the problem, then proceeded to check the adjustment of the shift linkage. He checked the adjustment and inspected the linkage and found nothing wrong. He proceeded to check the alignment of the transaxle to the engine, and again the cause of the problem was not found.

Suspecting internal damage, the technician removed the transaxle from the car and disassembled it, hoping to find the cause of the problem. Again the cause was not found. The technician then carefully reassembled the transaxle and installed it back in the car. During the road test, he again experienced the problem. He returned the car to the customer and told him that the problem could not be fixed.

Taking this approach is a good way to lose customers and your job. Of course the problem could have been fixed! True, he did check all the right things. However, he probably did not totally disassemble the transaxle to check all the internal parts carefully. A slight bit of wear on the dog teeth of the synchronizer can cause this kind of problem. At times, the wear is difficult to see, especially if the wear is even all around the ring. Most technicians, as a matter of course, will replace the blocking rings whenever a transaxle is taken apart because they cannot always see the wear. If the technician in this case study had followed this procedure, he probably would not be delivering pizza today.

KEY TERMS

Block synchronizer
Blocking rings
Case
Cluster gear
Clutch gear
Cone synchronizer
Countershaft
External gear
Final drive gear
Gear pitch
Helical gear
Idler gear
Input shaft
Mainshaft
Pinion shaft
Shift forks
Shift rails
Shifter plate
Spur gear
Synchronizer

SUMMARY

- A transmission or transaxle uses meshed gears of various sizes to give the engine a mechanical advantage over its driving wheels.
- Transaxles contain the gear train plus the differential gearing needed to produce the final drive gear ratios. Transaxles are commonly used on front-wheel-drive vehicles.
- Transmissions are normally used on rear-wheel-drive vehicles.
- Gears in the transmission/transaxle transmit power and motion from an input shaft to an output shaft. These shafts are mounted parallel to one another.

- Spur gears have straight cut teeth, while helical gears have teeth cut at an angle. Helical gears run without creating gear whine.
- When a small gear drives a larger gear, output speed decreases but torque (power) increases.
- When a large gear drives a smaller gear, output speed increases but torque (power) decreases.
- When two external toothed gears are meshed and turning, the driven gear rotates in the opposite direction of the driving gear.
- Synchronizers bring parts rotating at different speeds to the same speed for smooth clash-free shifting. The synchronizer also locks and unlocks the driven (speed) gears to the transmission/transaxle output shaft.
- Idler gears are used to reverse the rotational direction of the output shaft for operating the vehicle in reverse.
- In typical five-speed transmission shift linkage, there are three separate shift rails and forks. Each shift rail/shift fork is used to control the movement of a synchronizer.
- Gear ratios indicate the number of times the input drive gear is turning for every turn of the output driven gear. Ratios are calculated by dividing the number of teeth on the driven gear by the number of teeth on the drive gear. You can also use the rpm speeds of meshed gears to calculate gear ratios.
- A gear ratio of less than one indicates an overdrive condition. This means the driven gear is turning faster than the drive gear. Speed is high, but output torque is low.
- All vehicles use a gearset in the final drive to provide additional gear reduction (torque increase) above and beyond what the transmission or transaxle gearing can produce.

TECH MANUAL

The following procedures are included in Chapter 36/37 of the *Tech Manual* that accompanies this book:

1. Checking the fluid level in a manual transmission and transaxle.
2. Inspecting and adjusting shift linkage.
3. Road testing a vehicle for transmission problems.

REVIEW QUESTIONS

1. What determines whether a conventional transmission or a transaxle is used?
2. Explain the relationship between output speed and torque.
3. *True or False?* A reverse idler gear changes the direction of torque flow to the opposite direction of engine rotation.
4. Define final drive gear.
5. Explain the role of shift rails and shift forks in the operation of transmissions and transaxles.
6. Gears can ______.
   a. transfer speed and torque unchanged
   b. decrease speed and increase torque
   c. increase speed and increase torque
   d. all of the above
7. The number of gear teeth per unit of measurement of the gear's diameter (such as teeth/inch) is known as gear ______.
   a. ratio **c.** size
   b. pitch **d.** load
8. Which of the following gear ratios shows an overdrive condition?
   a. 2.15:1 **c.** 0.85:1
   b. 1:1 **d.** none of the above
9. Which type of gear develops the problem of gear whine at higher speeds?
   a. spur gear **c.** both a and b
   b. helical gear **d.** all of the above
10. Technician A says in a conventional transmission, the speed gears freewheel around the mainshaft until they are locked to it by the appropriate synchronizer. Technician B says speed gears are an integral part of the countershaft. Who is correct?
    a. Technician A **c.** Both A and B
    b. Technician B **d.** Neither A nor B
11. When an idler gear is placed between the driving and driven gear, the driven gear ______.
    a. rotates in the same direction as the driving gear

b. rotates in the opposite direction of the driving gear

c. remains stationary

d. causes the driven gear to rotate faster

12. The component used to ensure that the mainshaft (output shaft) and main (speed) gear to be locked to it are rotating at the same speed is known as a ______.

a. synchronizer **c.** shift fork

b. shift linkage **d.** transfer case

13. Technician A says the countergear or cluster gear is actually several gears machined out of a single piece of steel. Technician B says the countergear is driven by the clutch gear and drives the mainshaft speed gears. Who is correct?

a. Technician A **c.** Both A and B

b. Technician B **d.** Neither A nor B

14. In a transaxle, the pinion gear on the pinion shaft meshes with the ______.

a. reverse idler gear

b. ring gear

c. countershaft drive gear

d. input gear

15. *True or False?* The cone on a synchronizer's blocking ring serves as a cone-clutch assembly during the changing of gears.

16. While discussing the various types of transmissions found on today's vehicles, Technician A says one type is the sliding gear transmission. Technician B says one type is the sliding collar transmission. Who is correct?

a. Technician A **c.** Both A and B

b. Technician B **d.** Neither A nor B

17. Which of the following gear ratios generates the highest torque or power output?

a. 0.85:1 **c.** 5.23:1

b. 2.67:1 **d.** 11.12:1

18. Technician A says most transaxles have a countershaft between the main and output shafts. Technician B says a countershaft is always in motion when the engine is running. Who is correct?

a. Technician A **c.** Both A and B

b. Technician B **d.** Neither A nor B

19. Technician A says reverse lamp switches are activated at the transmission/transaxle by the gearshift lever. Technician B says reverse lamp switches are activated by a sensor at the input shaft. Who is correct?

a. Technician A **c.** Both A and B

b. Technician B **d.** Neither A nor B

20. Technician A says if a single idler gear is used to obtain reverse gear, the size and number of teeth of the idler gear must be used to calculate gear ratios. Technician B says an idler gear is often used for reverse gear because it causes the output shaft to rotate in the reverse direction of the input shaft. Who is correct?

a. Technician A **c.** Both A and B

b. Technician B **d.** Neither A nor B

37

MANUAL TRANSMISSION/ TRANSAXLE SERVICE

OBJECTIVES

■ Perform a visual inspection of transmission/transaxle components for signs of damage or wear. ■ Check transmission oil level correctly, detect signs of contaminated oil, and change oil as needed. ■ Describe the steps taken to remove and install transmissions/transaxles, including the equipment and safety precautions used. ■ Identify common transmission problems and their probable causes and solutions. ■ Describe the basic steps and precautions taken during transmission/transaxle disassembly, cleaning, inspection, and reassembly procedures.

When properly operated and maintained, a manual transmission/transaxle normally lasts the life of the vehicle without a major breakdown. All units are designed so the internal parts operate in a bath of oil circulated by the motion of the gears and shafts. Some units also use a pump to circulate oil to critical wear areas that require more lubrication than the natural circulation provides.

Maintaining good internal lubrication is the key to long transmission/transaxle life. If the amount of oil falls below minimum levels, or if the oil becomes too dirty, problems result.

SHOP TALK

Whenever you are diagnosing or repairing a transaxle or transmission, make sure you refer first to the appropriate service manual before you begin. ■

Prior to beginning any service or repair work, be sure you know exactly which transmission you are working on. This will ensure that you are following the correct procedures and specifications and are installing the correct parts. Proper identification can be difficult because transmissions cannot be accurately identified by the way they look. The only positive way to identify the exact design of the transmission is by its identification numbers.

Transmission identification numbers are found either as numbers stamped on the case or on a metal tag held by a bolt head. Use a service manual to decipher the identification number. Most identification numbers include the model, gear ratios, manufacturer, and assembly date (**Figure 37–1**). Whenever you work with a transmission with a metal ID tag, make sure the tag is put back on the transmission so that the next technician will be able to properly identify the transmission.

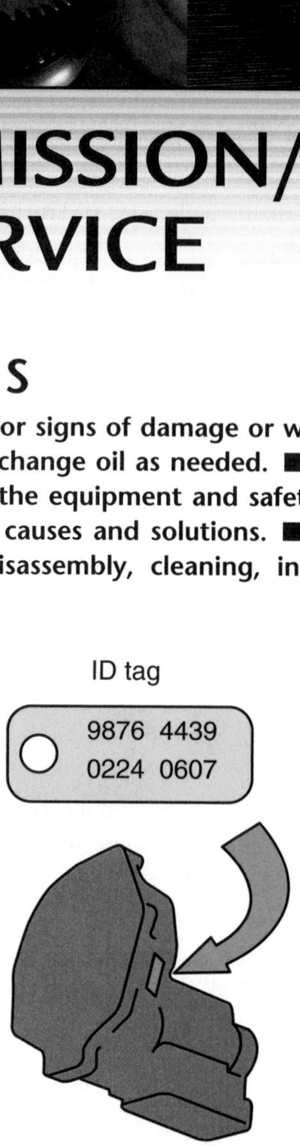

Figure 37–1 A typical transmission ID tag.

If the transmission does not have an ID tag, the transmission must be identified by comparing it with those in the vehicle's service manual.

LUBRICANT CHECK

The transmission/transaxle gear oil level should be checked at the intervals specified in the service manual. Normally, these range from every 7,500 to 30,000 miles (12,000 to 48,000 km). For service convenience, many units are now designed with a dipstick and filler tube accessible from beneath the hood. Check the oil with the engine off and the vehicle resting in a level position. If the engine has been running, wait 2 to 3 minutes before checking the gear oil level.

Some vehicles have no dipstick. Instead, the vehicle must be placed on a lift, and the oil level checked through the **fill plug** opening on the side of the unit. Clean the area around the plug before loosening and removing it. Lubricant should be level with, or not more than ½ inch (12.7 mm) below the fill hole. Add the proper grade lubricant as needed using a filler pump.

Manual transmission/transaxle lubricants in use today include single- and multiple-viscosity gear oils, engine oils, special hydraulic fluids, and automatic transmission fluid. Always refer to the service manual to determine the correct lubricant and viscosity range for the vehicle and operation conditions **(Figure 37–2)**.

Lubricant Leaks

Normally, the location and cause of a transmission fluid leak can be quickly identified by a visual inspection. The following are common causes for fluid leakage:

1. An excessive amount of lubricant in the transmission or transaxle.
2. The use of the wrong type of fluid; it will foam excessively and leave through the vent.
3. A loose or broken input shaft bearing retainer.
4. A damaged input shaft bearing retainer O-ring and/or lip seal.
5. Loose or missing case bolts.
6. Case is cracked or has a porosity problem.
7. A leaking shift lever seal.
8. Gaskets or seals are damaged or missing.
9. The drain plug is loose.

Fluid leaks from the seal of the extension housing can be corrected with the transmission in the car. Often, the cause for the leakage is a worn extension housing bushing, which supports the sliding yoke. When the drive shaft is installed, the clearance between the sliding yoke and the bushing should be minimal. If the clearance is satisfactory, a new oil seal will correct the leak. If the clearance is excessive, the repair requires that a new seal and a new bushing be installed. If the seal is faulty, the transmission vent should be checked for blockage. If the vent is plugged, the oil will be under high pressure when the transmission is hot, and this pressure can cause seal leakage **(Figure 37–3)**.

An oil leak at the speedometer cable can be corrected by replacing the O-ring seal. An oil leak stemming from the mating surfaces of the extension housing and the transmission case may be caused by loose bolts. To correct this problem, tighten the bolts to the specified torque.

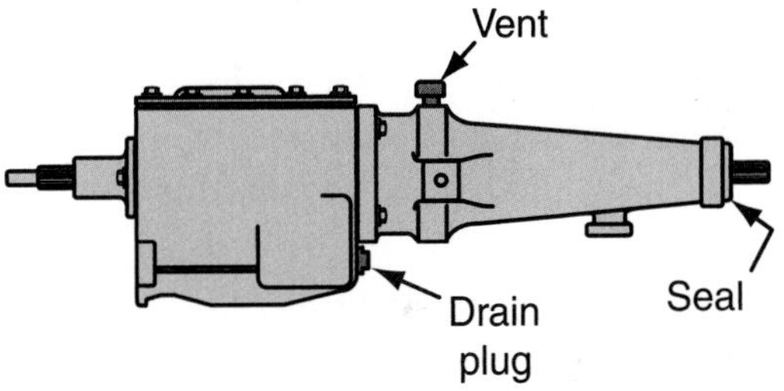

Figure 37–3 Possible sources of fluid leaks.

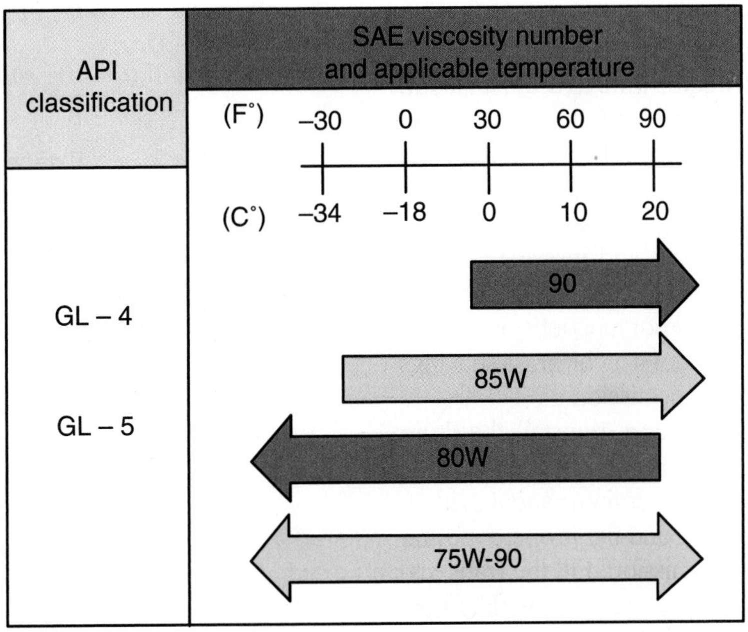

Figure 37–2 Typical transmission/transaxle gear oil classification and viscosity range data.

Lubricant Replacement

Transmission/transaxle lubricant should be changed at the manufacturer's specified intervals. Typical intervals are 24,000 or 30,000 miles (39,000 or 48,000 km) or every two years. Vehicles used for towing trailers, off-road operation, or continuous stop-and-go driving may require shorter change intervals.

Drive the vehicle to warm the lubricant before placing the vehicle on the hoist. Clean and remove the drain plug and allow the lubricant to drain into a clean catch pan. Inspect the lubricant for metal particles, which may appear as a shiny, metallic color in the lubricant. Large amounts of metal particles indicate severe bearing, synchronizer, gear, or housing wear.

Once all lubricant has drained, replace the washer or apply a recommended sealant to the threads on the drain plug and replace it. Tighten the drain plug to the recommended torque. Fill the transmission or transaxle with the proper lubricant.

Fluid Changes

The manufacturers of most transmissions do not recommend that the fluid be changed at any scheduled time. Older transmissions typically had 20,000-mile (32,000-km) fluid change intervals. When a car has been operated under severe conditions, such as in high heat or dusty road conditions, the fluid may need to be periodically changed. Check the service manual for the manufacturer's recommendations.

To change the transmission fluid, raise the car and safely support it on jack stands. Locate the oil drain plug in the bottom of the transmission case or extension housing. Make sure the car is level so that all of the fluid can drain out. Remove the drain plug with a catch pan positioned below the hole, and let the oil drain into the pan. Let the transmission drain completely. The fluid is normally very thick and it takes some time to drain it all out.

Inspect the drained fluid for gold-color metallic and other particles. The gold-color particles come from the brass blocking rings of the synchronizers. Metal shavings are typically from the wearing of gears. After the fluid has drained out, take a small magnet and insert it into the drain hole, then sweep it around the inside to remove all metal particles. Because brass is not magnetic, it will not show on the magnet. An excess of iron or brass shavings indicates severe wear in the transmission.

Before refilling the transmission, reinstall the drain plug. Remove the filler plug, which is normally located above the drain plug. Check your service manual to identify the location of the filler plug and the proper type and quantity of fluid for that transmission. Fill the transmission case until the oil just starts to run out the filler hole or until it is at the bottom of the bore. Reinstall the plug. You should check the case's vent to make sure it is not blocked with dirt. If the case is not properly vented, the fluid can easily break down.

IN-VEHICLE SERVICE

Much service and maintenance work can be done to transmissions while they are in the car. Only when a complete overhaul or clutch service is necessary does the transmission need to be removed from the car. The following are procedures for common service operations: the replacement of a rear oil seal and bushing, linkage adjustments, and replacement of the back-up light switch and the speedometer cable retainer and drive gear.

Rear Oil Seal and Bushing Replacement

Procedures for the replacement of the rear oil seal and bushing on a transmission vary little with each car model. Typically, to replace the rear bushing and seal follow these steps:

PROCEDURE

Replacement of Rear Oil Seal and Bushing

STEP 1 *Remove the drive shaft.*
STEP 2 *Remove the old seal from the extension housing.*
STEP 3 *Insert the appropriate puller tool into the extension housing until it grips the front side of the bushing.*
STEP 4 *Pull the bushing from the housing* ***(Figure 37–4)****.*
STEP 5 *Drive a new bushing into the extension housing.*
STEP 6 *Lubricate the lip of the seal, then install the new seal in the extension housing* ***(Figure 37–5)****.*
STEP 7 *Install the drive shaft.*

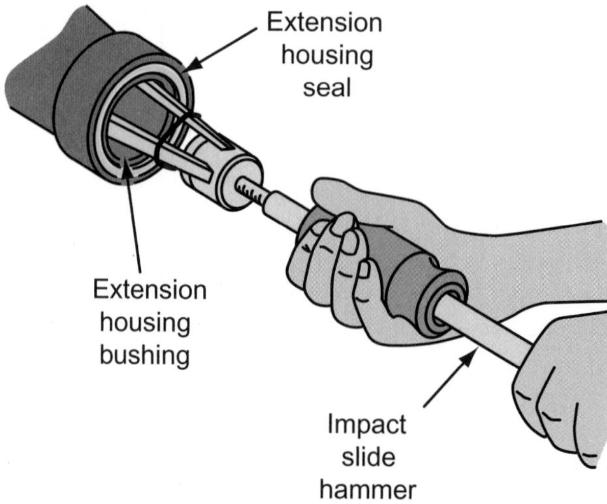

Figure 37-4 Removing the extension housing's seal and bushing.

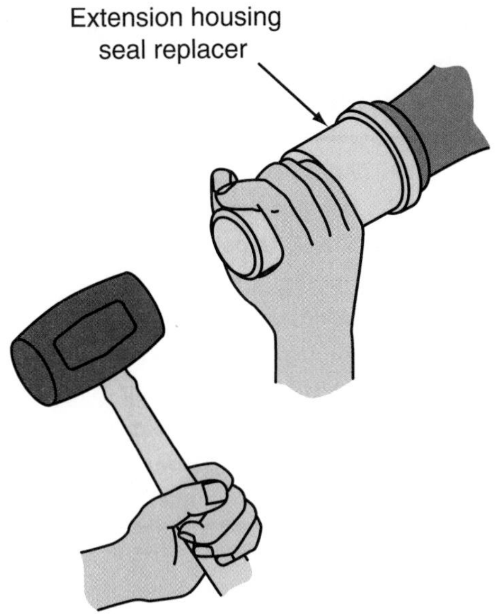

Figure 37-5 Drive the new seal into place with a hammer and seal driver.

Linkage Adjustment

Transmissions with internal linkage have no provision for adjustments. However, external linkages, both floor and column mounted, can be adjusted. Linkages are adjusted at the factory, but worn parts may make adjustments necessary. Also, after a transmission has been disassembled, the shift lever and other controls may need adjustment.

Only externally controlled gearshift levers and linkages can be adjusted. To begin the adjustment procedure, raise the car and support it on jack stands. Then follow the procedure given in your service manual.

Backup Light Switch Service

To replace the back-up light switch, disconnect the electrical lead to the switch. Put the transmission into reverse gear and remove the switch. Never shift the transmission until a new switch has been installed. To prevent fluid leaks, wrap the threads of a new back-up light switch with Teflon tape in a clockwise direction before installing it. Tighten the switch to the correct torque and reconnect the electrical wire to it.

Speedometer Drive Gear Service

Begin to remove the speedometer cable retainer and drive gear by cleaning off the top of the speedometer cable retainer. Then remove the hold-down screw that keeps the retainer in its bore. Carefully pull up on the speedometer cable, pulling the speedometer retainer and drive gear assembly from its bore. Unscrew the speedometer cable from the retainer.

To reinstall the retainer, lightly grease the O-ring on the retainer and gently tap the retainer and gear assembly into its bore while lining the groove in the retainer with the screw hole in the side of the clutch housing case. Install the hold-down screw and tighten it in place.

DIAGNOSING PROBLEMS

Service manuals list the most common problems associated with manual transmissions and transaxles. Proper diagnosis involves locating the exact source of the problem. Many problems that seem transmission/transaxle related may actually be caused by problems in the clutch driveline or differential. Check these areas along with the transmission/transaxle, particularly if you are considering removing the transmission/transaxle for service.

Table 37–1 is a troubleshooting chart for common transmission and transaxle problems.

Visual Inspection

Visually inspect the transmission/transaxle at regular intervals. Perform the following checks:

1. Check for lubricant leaks at all gaskets and seals. The transmission rear seal at the driveline is particularly prone to leakage.
2. Check the case body for signs of porosity that show up as leakage or seepage of lubricant.
3. Push up and down on the unit. Watch the transmission mounts to see if the rubber separates from the metal plate. If the case moves up, but not down, the mounts require replacement.
4. Move the clutch and shift linkages around and check for loose or missing components. Cable linkages should have no kinks or sharp bends, and all movement should be smooth.
5. Transaxle drive axle boots should be checked for cracks, deformation, or damage.
6. The constant velocity joints on transaxle drive axles should be thoroughly inspected.

Transmission Noise

Most manual transmission/transaxle complaints center around noise in the unit. Once again, be certain the noise is not coming from other components in the drivetrain. Unusual noises may also be a sign of trouble in the engine or transmission mounting system. Improperly aligned engines, improperly torqued mounting bolts, damaged or missing rubber mounts, cracked brackets, or even a stone rattling around inside the engine compartment can create noises that appear to be transmission/transaxle related.

SHOP TALK

If during the test drive you hear a noise you suspect is coming from inside the transmission/transaxle, bring the vehicle to a stop. Disengage the clutch. If the noise stops with the engine at idle and the clutch disengaged, the noise is probably inside the unit. ■

TABLE 37–1 TRANSMISSION/TRANSAXLE TROUBLESHOOTING CHART

Problem	Possible Cause	Remedy
Gear clash when shifting from one gear to another	1. Clutch adjustment incorrect 2. Clutch linkage or cable binding 3. Clutch housing misalignment 4. Lubricant level low or incorrect lubricant 5. Gearshift components or synchronizer blocking rings worn or damaged	1. Adjust clutch. 2. Lubricate or repair as necessary. 3. Check runout at rear face of clutch housing. Correct runout. 4. Drain and refill transmission/transaxle and check for lubricant leaks if level was low. Repair as necessary. 5. Remove, disassemble, and inspect transmission/transaxle. Replace worn or damaged components as necessary.
Clicking noise in any one gear range	1. Damaged teeth on input or intermediate shaft gears (transaxles) or damaged teeth on the countergear, cluster gear assembly, or output shaft gears (transmissions)	1. Remove, disassemble, and inspect unit. Replace worn or damaged components as necessary.
Does not shift into one gear	1. Gearshift internal linkage or shift rail assembly worn, damaged, or incorrectly assembled 2. Shift rail detent plunger worn, spring broken, or plug loose 3. Gearshift lever worn or damaged 4. Synchronizer sleeves or hubs damaged or worn	1. Remove, disassemble, and inspect transmission/transaxle cover assembly. Repair or replace components as necessary. 2. Tighten plug or replace worn or damaged components as necessary. 3. Replace gearshift lever. 4. Remove, disassemble, and inspect unit. Replace worn or damaged components.
Locked in one gear—cannot be shifted out of that gear	1. Shift rails worn or broken, shifter fork bent, setscrew loose, center detent plug missing or worn 2. Broken gear teeth on countershaft gear input shaft, or reverse idler gear 3. Gearshift lever broken or worn, shift mechanism in cover incorrectly assembled or broken, worn or damaged gear train components	1. Inspect and replace worn or damaged parts. 2. Inspect and replace damaged part. 3. Disassemble transmission/transaxle. Replace damaged parts of assembly correctly.
Slips out of gear	1. Clutch housing misaligned 2. Gearshift offset lever nylon insert worn or lever attachment nut loose 3. Gearshift mechanisms, shift forks, shift rail, detent plugs, springs, or shift cover worn or damaged 4. Clutch shaft or roller bearings worn or damaged 5. Gear teeth worn or tapered, synchronizer assemblies worn or damaged, excessive end play caused by worn thrust washers or output shaft gears 6. Pilot bushing worn	1. Check runout at rear face of clutch housing. 2. Remove gearshift lever and check for loose offset lever nut or worn insert. Repair or replace as necessary. 3. Remove, disassemble, and inspect transmission cover assembly. Replace worn or damaged components as necessary. 4. Replace clutch shaft or roller bearings as necessary. 5. Remove, disassemble, and inspect transmission/transaxle. Replace worn or damaged components as necessary. 6. Replace pilot bushing.

TABLE 37–1 (CONT.)		
Problem	**Possible Cause**	**Remedy**
Vehicle moving—rough growling noise isolated in transmission/transaxle and heard in all gears	1. Intermediate shaft front or rear bearings worn or damaged (transaxle) or output shaft rear bearing worn or damaged (transmission)	1. Remove, disassemble, and inspect transmission/transaxle. Replace damaged components as necessary.
Rough growling noise when engine operating with transmission/transaxle in neutral	1. Input shaft front or rear bearings worn or damaged (transaxle) or input shaft bearing, countergear, or countershaft bearings worn or damaged (transmission)	1. Remove, disassemble, and inspect transmission/transaxle. Replace damaged components as necessary.
Vehicle moving—rough growling noise in transmission—noise heard in all gears except direct drive	1. Output shaft pilot roller bearings	1. Remove, disassemble, and inspect transmission. Replace damaged components as needed.
Transmission/transaxle shifts hard	1. Clutch adjustment incorrect 2. Clutch linkage binding 3. Shift rail binding	1. Adjust clutch. 2. Lubricate or repair as necessary. 3. Check for mispositioned roll pin, loose cover bolts, worn shift rail bores, worn shift rail, distorted oil seal, or extension housing not aligned with case. Repair as necessary.
	4. Internal bind in transmission/transaxle caused by shift forks, selector plates, or synchronizer assemblies 5. Clutch housing misalignment	4. Remove, disassemble, and inspect unit. Replace worn or damaged components as necessary. 5. Check runout at rear of clutch housing. Correct runout.
	6. Incorrect lubricant	6. Drain and refill.

Once you have eliminated all other possible sources of noise, concentrate on the transmission/transaxle unit. Noises from inside the transmission/transaxle may indicate worn or damaged bearings, gear teeth, or synchronizers. A noise that changes or disappears in different gears can indicate a specific problem area in the transmission.

CAUTION!

When the transmission/transaxle is in gear and the engine is running, the driving wheels and related parts turn Avoid touching moving parts. Severe physical injury can result from contact with spinning drive axles and wheels.

The type of noise detected may also help indicate the problem.

Rough, Growling Noise This noise can be a sign of several problems in a transaxle or transmission depending on when it occurs. If the noise occurs when the transaxle is in neutral and the engine running, the problem may be the input shaft roller bearings. The input shaft is supported on either end by tapered roller bearings, and these are the only bearings in operation when the transaxle is in neutral. In its early stages, the problem should not cause operational difficulties; but left uncorrected, it grows worse until the bearing race or rolling element fractures. Solving the problem involves transaxle disassembly and bearing replacement.

When the vehicle is moving, both the input and mainshaft (output shaft) are turning in the transaxle. If the noise occurs in forward and reverse gears, but not in neutral, the output or mainshaft bearings are the likely failed component. Replacement is the solution.

In transmissions, the problem is also bearing related. If the rough growling noise occurs when the engine is running, the clutch engaged, and the transmission in neutral, the front input shaft bearing is likely at fault. Rough growling when the vehicle is moving in all gears indicates faulty countergear bearings or countershaft-to-cluster assembly needle bearings. If the problem occurs in all gears except direct drive, the bearing at the rear of the transmission input shaft may be at fault. This bearing supports the pilot journal at the front of the transmission output shaft. In all forward gears except direct drive, the input shaft and output shaft turn at two different speeds. In reverse, the two shafts turn in opposite directions. In direct drive, the two shafts are locked together and this bearing does not turn. If the growling noises stop during direct drive operation, the rear input shaft bearing may be at fault. Disassembly, inspection, and replacement of damaged parts is needed.

Clicking or Knocking Noise Normally, the helical gears used in modern transmissions/transaxles are quiet because the gear teeth are constantly in contact. (When spur cut gear teeth are found in the reverse gearing, clicking or a certain amount of **gear whine** is normal, particularly when backing up at faster speeds.)

Clicking or whine in forward gear ranges may indicate worn helical gear teeth. This problem may not require immediate attention.

Chipped or broken teeth are dangerous because the loose parts can cause severe damage in other areas of the transmission/transaxle. Broken parts are usually indicated by a rhythmical knocking sound, even at low speeds. Complete disassembly, inspection, and replacement of damaged parts is the solution to this problem.

Gear Clash

Gear clash is indicated by a grinding noise during shifting. The noise is the result of one gearset remaining partly engaged while another gearset attempts to turn the mainshaft. Gear clash can be caused by incorrect clutch adjustment or binding of clutch or gearshift linkage. Damaged, worn, or defective synchronizer blocking rings can cause gear clash, as can use of an improper gear lubricant.

Hard Shifting

If the shift lever is difficult to move from one gear to another, check the clutch linkage adjustment. Hard shifting may also be caused by damage inside the transmission/transaxle, or by a lubricant that is too thick. Common hard shifting includes badly worn bearings and damaged clutch gears, control rods, shift rails, shift forks, and synchronizers.

Jumping Out of Gear

If the car jumps out of gear into neutral, particularly when decelerating or going down hills, first check the shift lever and internal gearshift linkage. Excessive clearance between gears and the input shaft or badly worn bearings can cause jumping out of gear. Other internal transmission/transaxle parts to inspect are the clutch pilot bearing, gear teeth, shift forks, shift rails, and springs or detents.

Locked in Gear

If a transmission or transaxle locks in one gear and cannot be shifted, check the gearshift lever linkage for misadjustment or damage. Low lubricant level can also cause needle bearings, gears, and synchronizers to seize and lock up the transmission.

If these checks do not resolve the problem, the transmission or transaxle must be removed from the vehicle and disassembled. After disassembly, inspect the internal countergear, clutch shaft, reverse idler, shift rails, shift forks, and springs or detents for damage. Also, check for worn support bearings.

If the problem seems to be in the clutch assembly, make sure the transmission/transaxle is out of gear, set the parking brake, and start the engine. Increase the engine speed to about 1,500–2,000 rpm and gradually apply the clutch until the engine torque causes tension at the drive train mounts. Watch the torque reaction of the engine. If the engine's reaction to the torque appears to be excessive, broken or worn drive train mounts may be the cause and not the clutch.

The engine mounts on FWD cars are important to the operation of the clutch and transaxle **(Figure 37–6)**. Any engine movement may change the effective length of the shift and clutch control cables and therefore may affect the engagement of the clutch and/or gears. A clutch may slip due to clutch linkage changes as the engine pivots on its mounts. To check the condition of the transaxle mounts, pull up and push down on the transaxle case while watching the mount. If the mount's rubber separates from the metal plate or if the case moves up but not down, replace the mount. If there is movement between the metal plate and its attaching point on the frame, tighten the attaching bolts to an appropriate torque.

If it is necessary to replace the transaxle mount, make sure you follow the procedure for maintaining the alignment of the drive line. Some manufacturers recommend that a holding fixture or special bolt be used to keep the unit in its proper location. A broken clutch cable may be caused by worn mounts and improper cable routing. Inspect all clutch and transaxle linkages and cables for kinks or stretching. Often transaxle problems can be corrected by replacing or repairing the clutch or gearshift cables and linkage.

Shift Linkage

Check the shift linkage for smooth gear changes and full travel. If the linkage cannot move enough to fully engage a gear, the transmission/transaxle will jump out of gear while it is under a load. Some FWD cars have experienced the problem of jumping out of second or fourth gear. Two

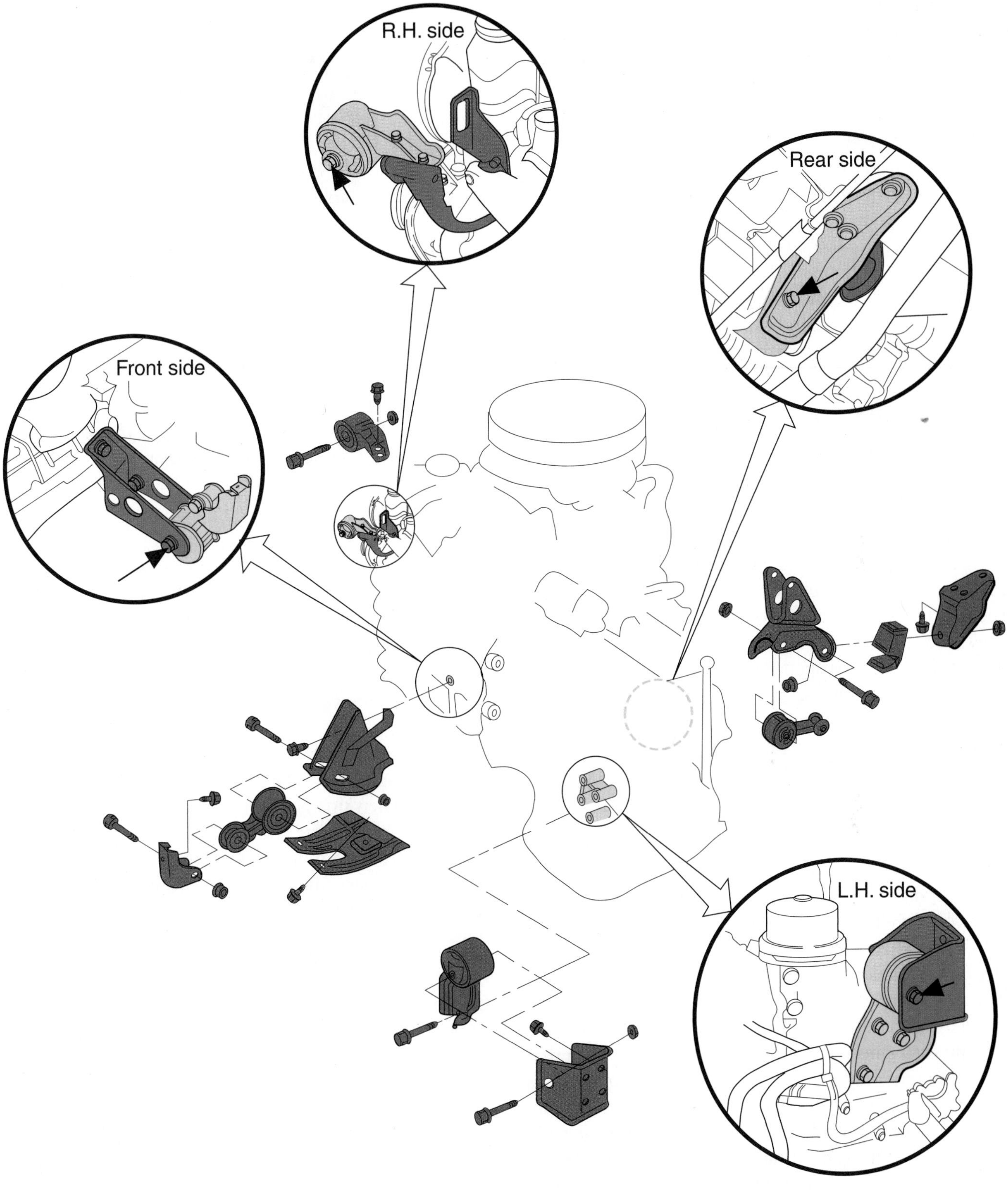

Figure 37-6 Typical engine and transaxle mounts. *Used with permission from Nissan North America, Inc.*

causes have been identified with this problem: the upshift light interferes with the shifter or there are improper shifter-to-shifter boot clearances. Both conditions prevent the transaxle's shift forks from moving enough to fully engage the synchronizer collars to their mating gears. If correcting these problems does not solve the complaint, the cause may be the engine mounts or an internal problem in the transaxle.

USING SERVICE MANUALS

A service manual is absolutely necessary when performing any type of transmission/transaxle disassembly work. Not only does the manual clearly illustrate all components and their disassembly procedure, it also lists many vital specifications, such as shaft and gear thrust (side) clearances, synchronizer ring and cone clearances, and bolt torque values. Special service tools, such as transmission service stands, oil seal presses, bearing replacers, shaft removers, pullers, and installing tools are also illustrated and explained. ■

Figure 37-7 To ensure proper balance and phasing of the drive shaft, make alignment marks on the rear flange and the rear yoke. *Reprinted with permission*

TRANSMISSION/TRANSAXLE REMOVAL

Removing the transmission from a RWD vehicle is generally more straightforward than removing one from a FWD model, as there is typically one cross member, one drive shaft, and easy access to the cables, wiring, and bell housing bolts. Transmissions in FWD cars, because of their limited space, can be more difficult to remove as you may need to disassemble or remove large assemblies such as engine cradles, suspension components, brake components, splash shields, or other pieces that would not usually affect RWD transmission removal. The engine may also need to be supported with fixtures while removing the transmission.

RWD Vehicles

The correct procedure for removing a transmission varies with each year, make, and model of vehicle; always refer to the service manual for the correct procedure. Normally the procedure begins with placing the vehicle on a hoist.

Once the vehicle is in position, disconnect the negative battery cable and place it away from the battery. Carefully check under the hood to identify anything that may interfere with transmission removal. Then raise the vehicle and disconnect the parts of the exhaust system that may get in the way. Disconnect all electrical connections and the speedometer cable at the transmission. Make sure you place these away from the transmission so they are not damaged during transmission removal or installation.

Place a drain pan under the transmission and drain the transmission's fluid. Then move the drain pan to the rear of the transmission. Before removing the drive shaft, use chalk to show the alignment of the rear U-joint and the pinion flange **(Figure 37–7)**. Then remove the drive shaft.

Disconnect and remove the transmission linkage. It is best to do this by disconnecting as little as possible.

Place a transmission jack under the transmission and secure the transmission to it. Then loosen and remove the lower bell housing-to-engine block bolts and the cross member at the transmission. After the mount is free from the transmission, lower the transmission slightly so you can easily access the top transmission-to-engine bolts. Loosen and remove the remaining transmission-to-engine bolts.

Slowly and carefully move the transmission away from the engine until the input shaft is out of the clutch assembly. Then slowly lower the transmission. Once the transmission is out of the vehicle, carefully move it to the work area and mount it to a stand or bench.

On some cars, the engine and transmission must be removed as a unit. The assembly is lifted with an engine hoist or lowered underneath the car.

FWD Vehicles

On some vehicles, the recommended procedure may include removing the engine with the transaxle. Always refer to the service manual before proceeding to remove the transaxle. Identify any special tool needs and precautions recommended by the manufacturer. You will waste much time and energy if you do not check the manual first.

Begin removal by placing the vehicle on a lift. Working under the hood, disconnect the battery before loosening any other components. Then, disconnect all electrical connectors and the speedometer cable at the transaxle.

Now disconnect the shift linkage or cables and the clutch cable. Identify the transaxle-to-engine bolts that cannot be removed from under the vehicle and remove them. Install the engine support fixture to hold the engine in place while removing the transaxle. Disconnect and remove all items that will interfere with the removal of the transaxle.

Loosen the large nut that retains the outer CV joint, which is splined shaft to the hub. It is recommended that this nut be loosened with the vehicle on the floor and the brakes applied. This will make the job easier and reduces the chance of damaging the CV joints and wheel bearings.

Now raise the vehicle and remove the front wheels. Tap the splined CV joint shaft with a soft-faced hammer to see if it is loose. Most will come loose with a few taps. Some vehicles have an interference fit spline at the hub and you will need a special puller for this type of CV joint. The tool pushes the shaft out and on installation pulls the shaft back into the hub.

The lower ball joint must now be separated from the steering knuckle. The ball joint will either be bolted to the lower control arm or held in the knuckle with a pinch bolt. Once the ball joint is loose, the control arm can be pulled down and the knuckle can be pushed outward to allow the splined CV joint shaft to slide out of the hub. The inboard joint can be pried out or will slide out. Some transaxles have retaining clips that must be removed before the inner joint can be removed. Pull the drive axles out of the transaxle **(Figure 37–8)**.

While removing the axles, make sure the brake lines and hoses are not stressed. Suspend them with wire to relieve the weight on the hoses and to keep them out of the way.

On some cars, the inner CV joints have flange-type mountings. These must be unbolted for removal of the shafts. In some cases, the flange-mounted drive shafts may be left attached to the wheel and hub assembly and only unbolted at the transmission flange. The free end of the shafts should be supported and placed out of the way.

Now the remaining shift linkages, electrical connections, and speedometer cables should be disconnected. The exhaust system may also need to be lowered or partially removed.

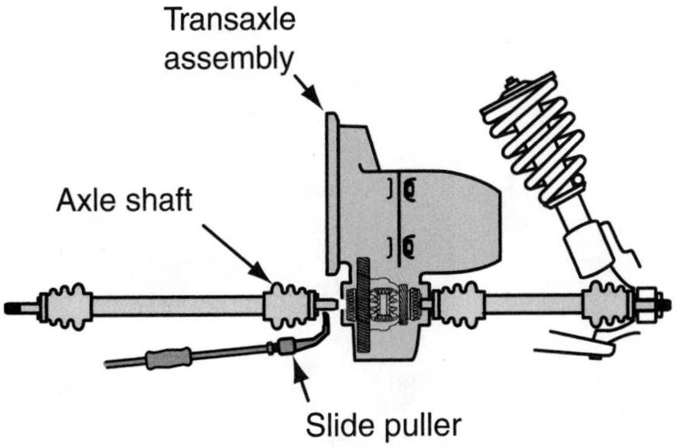

Figure 37–8 The inboard joints are typically pulled out of the transaxle.

Remove the starter. The starter's wiring may be left connected or you may remove the starter from the vehicle to get it totally out of the way.

With the transmission jack supporting the transmission, remove the transaxle mounts. If the car is equipped with an engine cradle that will separate, remove the half of the cradle that allows for transaxle removal. Then remove all remaining transaxle to engine bolts. Slide the transaxle away from the engine. Make sure the input shaft is out of the clutch assembly before lowering the transmission.

CLEANING AND INSPECTION

Disassembly and overhaul procedures can vary greatly between transmission/transaxle models, so always follow the exact steps outlined in the service manual.

Clean the transmission/transaxle with a steam cleaner, degreaser, or cleaning solvent. As you begin to disassemble the unit, pay close attention to the condition of its parts. Using a dial indicator, measure and record the endplay of the input and main shafts. This information will be needed during the reassembly of the unit for selecting the appropriate selective shims and washers.

SHOP TALK

Before disassembling a transmission, observe the effort it takes to rotate the input shaft through all the forward gears and reverse. Extreme effort in any or all gears may indicate an endplay or preload problem. ■

Remove the bell housing from the transmission case, extension housing, and the side or top cover. The seal and bushing should be removed from the extension housing (tail shaft) prior to cleaning. With the housing and cover removed, the gears, synchronizers, and shafts are exposed and the shift forks can be removed.

SHOP TALK

It is good practice to lay the parts on a clean rag as you remove them, and to keep them in order to aid you during reassembly. ■

Each transmission design has its own specific service procedures. Photo Sequence 35 guides you through the disassembly of a typical transaxle. Always refer to your service manual prior to overhauling a transmission or transaxle.

PHOTO SEQUENCE 35

Disassembling a Typical Transaxle

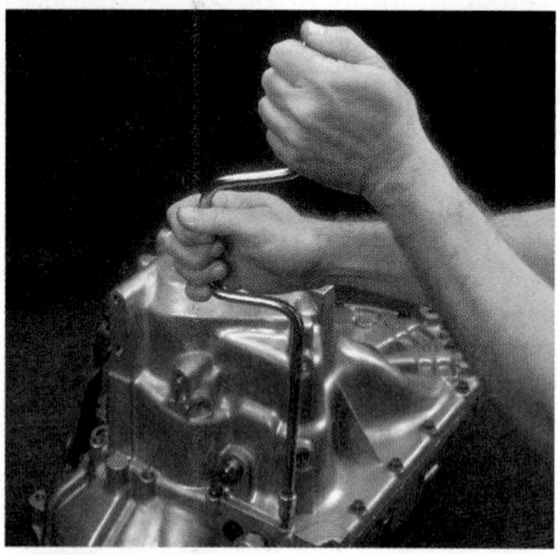

P35-1 *Place the transaxle into a suitable work stand. Remove the reverse idler shaft retaining bolt and detent plunger retaining screw. Then loosen and remove all transaxle case-to-clutch housing attaching bolts.*

P35-2 *Separate the housing from the case. If the housing is difficult to loosen, tap it with a soft mallet.*

P35-3 *Remove the C clip retaining ring from the fifth-gear shift-relay lever pivot pin.*

P35-4 *Remove the fifth-gear shift-relay lever, reverse idler shaft, and reverse idler gear from the case.*

P35-5 *Use a punch to drive the roll pin from the shift-lever shaft.*

P35-6 *Remove the shift-lever shaft by gently pulling on it.*

P35-7 *Remove the reverse kickdown spring assembly.*

P35-8 *Grasp the input and main shafts and lift them as an assembly from the case. Note the position of the shift forks as an aid when reinstalling them.*

P35-9 *Remove the fifth-gear shaft assembly and the fifth-gear shift fork assembly.*

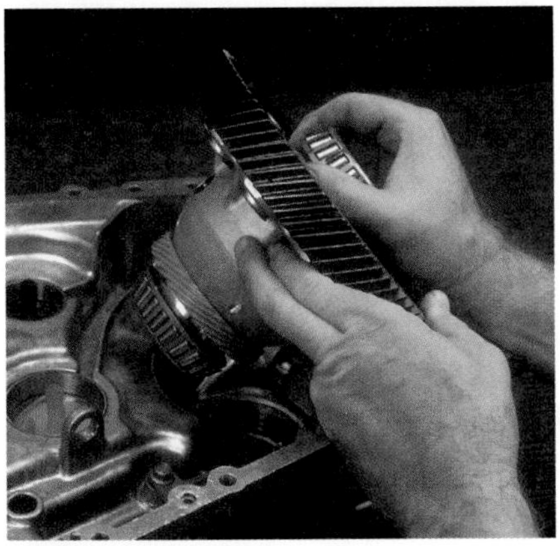

P35-10 *Remove the differential assembly from the case.*

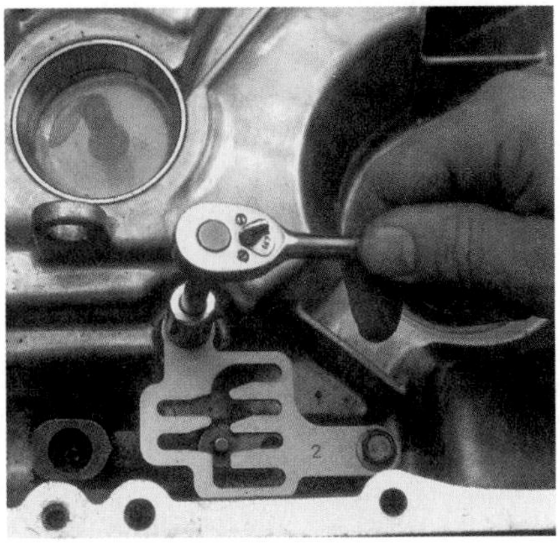

P35-11 *Remove the bolts for the shift-relay lever support bracket, then remove the assembly.*

P35-12 *Carefully separate the shift rail and forks from the gears on the main shaft.*

P35-13 *Remove the bearing at the fourth-gear end of the main shaft.*

P35-14 *Slide the fourth gear from the shaft.*

P35-15 *Remove the synchronizer blocking ring from the assembly.*

P35-16 *Remove the third- and fourth-gear synchronizer retaining ring*

P35-17 *Lift the assembly off the shaft. Then remove the remaining gears and synchronizer assembly as a unit.*

P35-18 *Separate the synchronizer's hub, sleeve, and keys, noting their relative positions and scribing their location on the hub and the sleeve prior to separation.*

In some cases the countershaft must be removed before the input and mainshaft. In other cases, the mainshaft is removed with the extension housing. It may be removed through the shift cover opening. To avoid difficulty in disassembly, follow the recommended sequence. A **gear puller** or hydraulic press is often needed to remove gears and synchronizer assemblies from transmission/transaxle pinion shafts.

Bearing removal and installation procedures require that the force applied to remove or install the bearing should always be placed on the tight bearing race. In some cases, the inner race is tight on the shaft, while in others it is the outer race that is tight in its bore. Removal or installation force should be applied to the tight race. Serious damage to the bearing can result if this practice is not followed.

Use a soft-faced hammer or a brass drift and ball-peen hammer if tapping is required. Never use excessive force or hammering.

During assembly of the transmission, never attempt to force parts into place by tightening the front bearing retainer bolts or extension housing bolts. All parts must be fully in place before tightening any bolts. Check for free rotation and shifting. New gaskets and seals should always be used.

The following are some general cleaning and inspection guidelines that result in quality workmanship and service:

1. Wash all parts, except sealed ball bearings and seals, in solvent. Brush or scrape all dirt from the parts. Remove all traces of old gasket. Wash roller bearings in solvent; dry them with a clean cloth. Never use compressed air to spin the bearings.
2. Inspect the front of the transmission case for nicks or burrs that could affect its alignment with the flywheel housing. Remove all nicks and burrs with a fine stone (cast-iron casing) or fine file (aluminum casing).
3. Replace any cover that is bent or distorted. If there are vent holes in the case, make certain they are open.
4. Inspect ball bearings by holding the outer ring stationary and rotating the inner ring several times. Inspect the raceway of the inner ring from both sides for pits and spalling. Light particle indentation is acceptable wear, but all other types of wear merit replacement of the bearing assembly. Next, hold the inner ring stationary and rotate the outer ring. Examine the outer ring raceway for wear and replace as needed.
5. Examine the external surfaces of all bearings. Replace the bearings if there are radial cracks on the front and rear faces of the outer or inner rings, cracks on the outside diameter or outer ring, or deformation or cracks in the ball cage.
6. Lubricate the cleaned bearing raceways with a light coat of oil. Hold the bearing by the inner ring in a vertical position. Spin the outer ring several times by hand. If roughness or vibration is felt, or the outer ring stops abruptly, replace the bearing.
7. Replace any roller bearings that are broken, worn, or rough. Inspect their respective races. Replace them as needed.
8. Replace the counter (cluster) gear if its gear teeth are chipped, broken, or excessively worn **(Figure 37–9)**. Replace the countershaft if the shaft is bent, scored, or worn. Also, inspect the bore for the countershaft. If the bore is excessively worn or damaged, the needle bearings will not seat properly against the shaft.

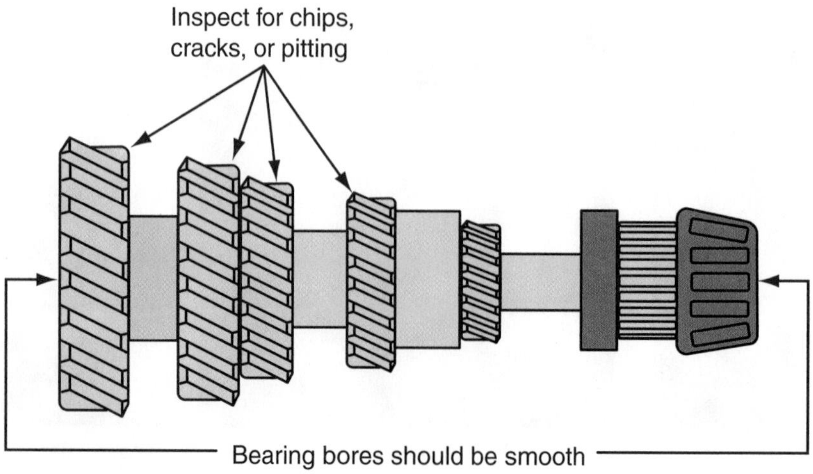
Inspect for chips, cracks, or pitting
Bearing bores should be smooth

Figure 37-9 Carefully inspect the countergear assembly.

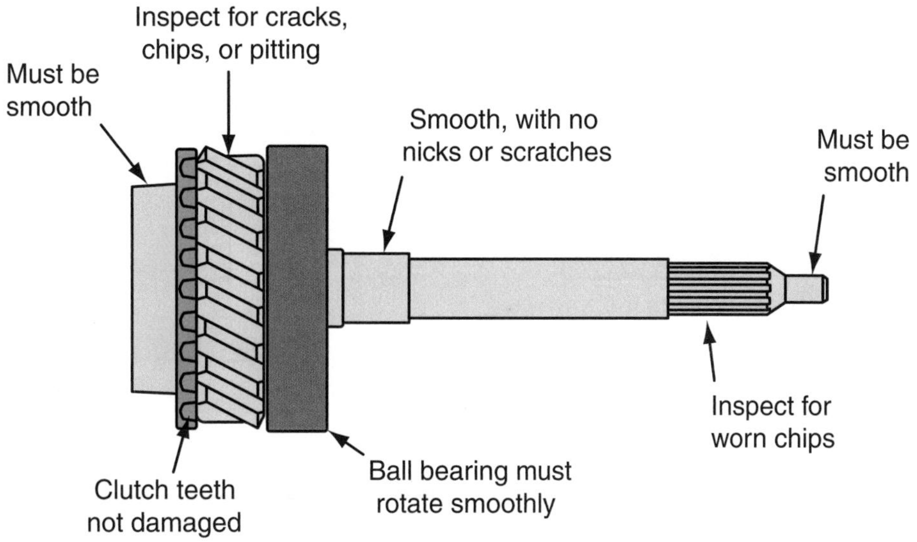

Figure 37–10 The input shaft, including the splines, should also be carefully inspected.

9. Replace the reverse idler gear or sliding gear if its teeth are chipped, worn, or broken. Replace the idler gear shaft if it is bent, worn, or scored.
10. Replace the input shaft and gear if its splines are damaged or if the teeth are chipped, worn, or damaged **(Figure 37–10)**. If the roller bearing surface in the bore of the gear is worn or rough, or if the cone surface is damaged, replace the gear and the gear rollers.
11. Replace all main or speed gears that are chipped, broken, or worn.
12. Check the synchronizer sleeves for free movement on their hubs. Alignment marks (if present) should be properly indexed **(Figure 37–11)**.
13. Inspect the synchronizer blocking rings for widened index slots, rounded clutch teeth, and smooth internal surfaces. Remember, the blocking rings must have machined grooves on their internal surfaces to cut through lubricant **(Figure 37–12)**. Units with worn, flat grooves must be replaced. Also, check the clearance between the block ring and speed gear clog teeth against service manual specifications **(Figure 37–13)**.

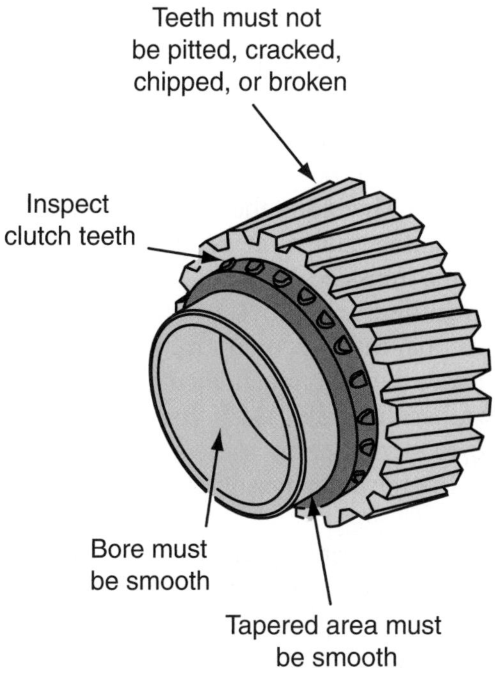

Figure 37–11 Every gear should be checked.

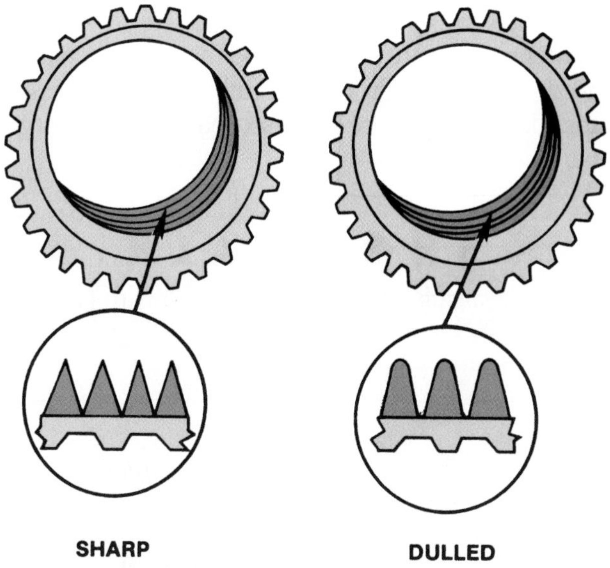

Figure 37–12 Grooves on the internal surface of the synchronizer blocker ring must be sharp.

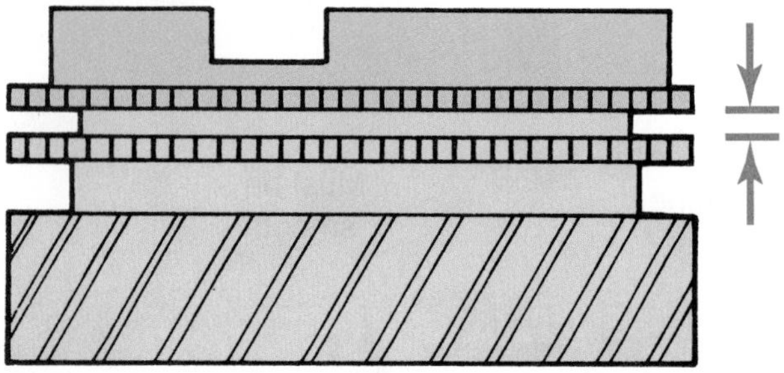

Figure 37-13 The clearance between the synchronizer blocker ring and the gear's clutching teeth must meet specifications.

14. Replace the speedometer drive gear if its teeth are stripped or damaged. Install the correct size replacement gear.
15. Replace the output shaft if there is any sign of wear or runout or if any of the splines are damaged.
16. Inspect the bushings and seal in the extension housing, and replace if worn or damaged. The bushing and seal should be replaced once the extension housing has been reinstalled on the transmission.

Aluminum Case Repair

Normally, the case is replaced if it is cracked or damaged. However some manufacturers recommend the use of an epoxy-based sealer on some types of leaks in some locations on the transmission. Refer to the manufacturer's recommendations before attempting to repair a crack or correct for porosity leaks.

If a threaded area in an aluminum housing is damaged, helicoil-type service kits can be used to insert new threads in the bore. Some threads should never be repaired; check the service manual to identify which ones can be repaired.

After all parts are inspected and the defective parts replaced, you can begin to reassemble the transmission/transaxle. While you are doing so, coat all parts with gear lube.

SHOP TALK

If the transmission/transaxle is fitted with paper-type blocking rings, soak them in ATF prior to installing them. ■

Many late-model transmissions and transaxles have specifications for endplay, backlash, and preload; make sure these specifications are met. Follow the procedures given in the service manual for the particular transmission/transaxle being worked on **(Figure 37–14)**. For most transmissions, there are specifications for the endplay and preload of the input shaft, the countershaft, and the differential. These are usually set by shims under the bearing caps. Reuse the original shims, if possible.

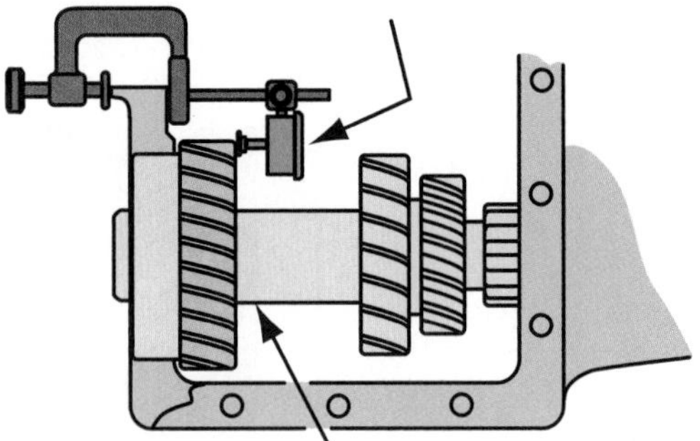

Figure 37-14 A typical setup for checking countershaft endplay.

Specific repair and assembly instructions will vary from transaxle to transaxle and from transmission to transmission. Therefore, before beginning to reassemble the unit, gather the specific information about the unit you are working on. Photo Sequence 36 shows the reassembly of a typical transaxle.

DISASSEMBLY AND REASSEMBLY OF THE DIFFERENTIAL CASE

Although it is a part of the transaxle, the differential is often kept together while making a repair to the transmission part of the transaxle **(Figure 37–15)**. The differential case normally can be removed as soon as the transaxle case has been separated. It may be the source of the problem and be the only part of the transaxle that needs service. Therefore, the disassembly and reassembly of the differential is set aside from the procedures listed for the transaxle.

Begin the disassembly by separating the ring gear from the differential case. Then remove the pinion shaft lock bolt. Remove the pinion shaft, then remove the gears and thrust washers from the case. If the differential side bearings are to be replaced, use a puller to remove the bearings. Use the correct installer for reinstallation of the side bearings.

Clean and inspect all parts. Replace any damaged or worn parts. Install the gears and thrust washers into the case and install the pinion shaft and lock bolt. Tighten the bolt to the specified torque. Attach the ring gear to the differential case and tighten to the specified torque.

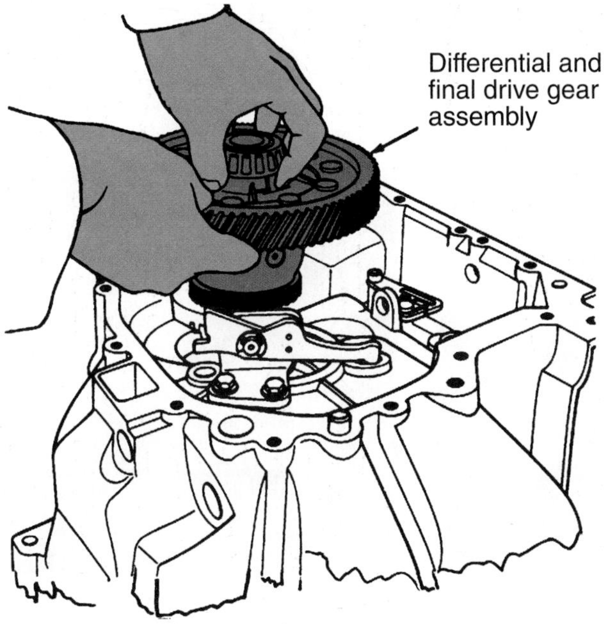

Figure 37-15 The differential assembly is normally pulled out from the transaxle case as a unit. *Courtesy of Ford Motor Company*

Shim Selection

While you are disassembling the differential or transaxle, make sure to keep all shims and bearing races together and identified for reinstallation in their original location. Carefully inspect the bearings for wear and/or damage and determine if a bearing should be replaced. Replacement tapered roller bearings will be available with a nominal thickness service shim. A nominal thickness service shim will handle reshimming the input shaft and output shaft bearings during normal repair.

When it is necessary to replace a bearing, race, or housing, refer to the manufacturer's recommendation for nominal shim thickness. If only other parts of the differential or transaxle are replaced, reuse the original shims. When repairs require the use of a service shim, discard the original shim. Never use the original shim together with the service shim. The shims must be installed only under the bearing cups at the transaxle case end of both the input and output shafts.

REASSEMBLY/REINSTALLATION OF TRANSMISSION/TRANSAXLE

Transmission/transaxle reassembly and reinstallation procedures are basically the reverse of disassembly. Once again, refer to the service manual for any special procedures. New parts are installed as needed, and new gaskets and seals are always used.

Serviceable gears are pressed onto the main shaft using special press equipment. Separate needle bearings should be held in place with petroleum jelly so shafts can be inserted into place. During reassembly, measure shaft end play. Adjust it to specifications with shims, spacers, or snap rings of different thicknesses. All fasteners are tightened to the manufacturer's torque specifications.

Soft-faced mallets can be used to tap shafts and other parts into place. After reassembly, secure the transmission to a transmission jack with safety chains and raise it into place. Before the transmission is reinstalled, inspect and service the clutch as necessary.

Installing the Transmission/Transaxle

SHOP TALK

Always check for free rotation of the transmission in all gears before installing in a vehicle. If the shafts do not rotate freely, identify the cause and correct it. ■

After the unit is together, install the clutch assembly, and a new throw-out bearing, prior to installing the transmission/transaxle. Generally, installation is the reverse procedure as removal. When installing the transmission, never let the transmission hang by its input shaft. Use a transmission jack to hold the transmission while you guide it into place. The input shaft should be lightly coated with grease prior to installation to aid installation and serve as a lubricant for the pilot bearing. Avoid putting too much grease on the shaft, because the excess may fly off and get on the clutch disc, causing it to slip and/or burn.

Most transmissions are doweled to the engine or bell housing. Use the dowels to locate and support the transmission during installation. Tighten the mounting bolts evenly, making sure nothing is caught between the housings. Then, lightly coat the drive shaft's slip joint and carefully insert it into the extension housing to prevent possible damage to the rear oil seal. Reattach and adjust the shift linkage and fill the transmission with the proper fluid.

WARNING!

If the transmission/transaxle does not fit snugly against the engine block or if you cannot move the transmission into place, do not force it. Pull the transmission back and lower it. Inspect the input shaft splines for dirt or damage. Also check the mating surfaces for dirt or obstructions. If you try to force the transmission into place by tightening the bolts, you may break the case.

PHOTO SEQUENCE 36

Reassembling a Typical Transaxle

P36-1 *Lightly oil the parts of the synchronizer.*

P36-2 *Assemble the synchronizer assemblies, being careful to align the index marks made during disassembly.*

P36-3 *Install the synchronizer assemblies onto the main shaft.*

P36-4 *Install fourth gear and its bearing. The bearing may need to be lightly pressed into position.*

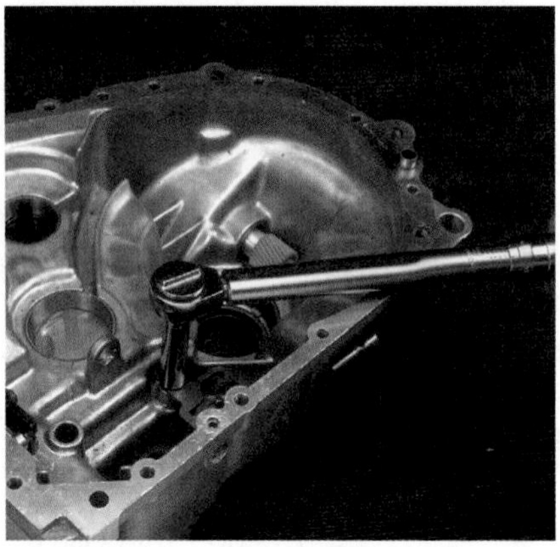

P36-5 *Install and tighten the shift-relay lever support bracket.*

P36-6 *Install the differential assembly into the transaxle case.*

P36-7 *Install the fifth-gear shaft assembly and fork shaft into the case.*

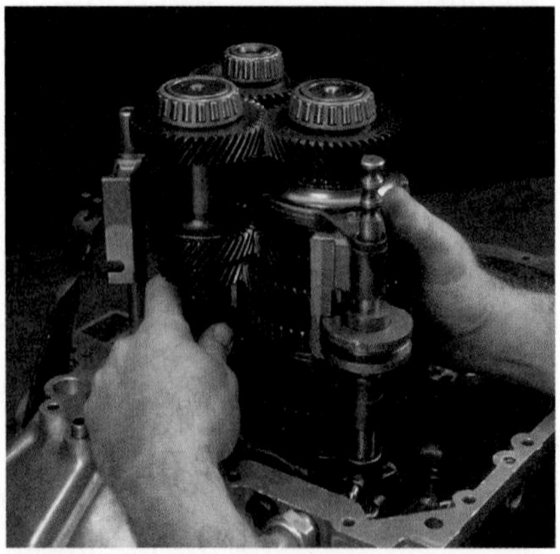

P36-8 *Place the main shaft control-shaft assembly on the main shaft so that the shift forks engage in their respective slots in the synchronizer sleeves. Then install the main shaft assembly.*

P36-9 *Properly position the shift lever, bias and kickdown springs.*

P36-10 *Install the inhibitor spring and ball in the fifth and reverse-inhibitor shaft hole.*

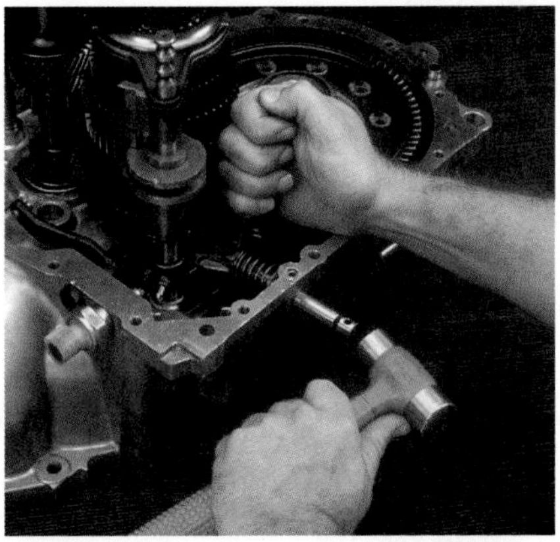

P36-11 *Depress the inhibitor ball and spring using a drift, and slide the shift-lever shaft through the shift lever. Then tap the shaft into its bore in the clutch housing.*

P36-12 *Install the reverse idler gear shaft and gear into the appropriate bore in the case.*

P36-13 *Install the fifth-gear relay shaft and align it with the fifth-gear fork slot and interlock spring.*

P36-14 *Install the retaining clip onto the fifth-gear relay shaft.*

P36-15 *Apply a thin bead of anaerobic sealant on the case's mating surface for the clutch housing.*

P36-16 *Install the clutch housing to the case. Be careful that the main shaft, shift-control shaft, and fifth-gear shaft align with the bores in the case.*

P36-17 *After the housing and case are fit snugly together, tighten the attaching bolts to the specified torque.*

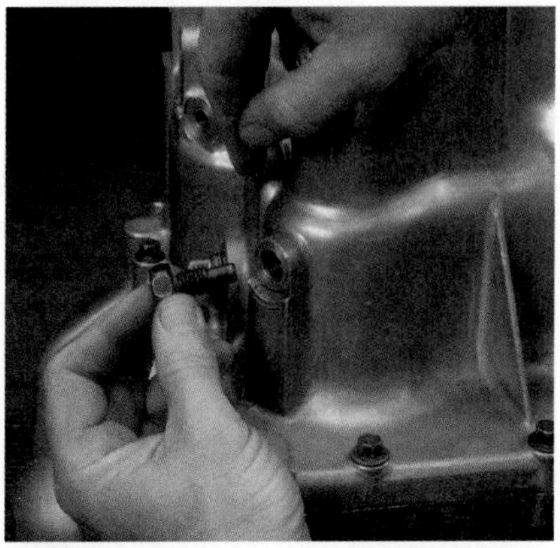

P36-18 *Install and tighten the reverse-idler shaft retaining bolt.*

CASE STUDY

A customer complains that his rear-wheel-drive, manual transmission vehicle is experiencing intermittent operating noise, particularly when the vehicle is "just warming up." On the test drive, the technician notices a low growling noise in the lower gear ranges. The noise disappears at cruising speeds in high (fourth) gear.

The technician returns to the shop and places the vehicle on a lift for further inspection. The driveline, differential, clutch, and wheel bearings all appear to be in good condition. This confirms that the problem is transmission related. The technician suspects a damaged bearing at the rear of the input shaft. This bearing supports the pilot journal at the front of the transmission's output shaft. In all forward gear ranges except direct drive, the input and output shafts turn at different speeds. In reverse gear they turn in opposite directions; but in direct drive (fourth gear in this case), the two shafts are locked together by a synchronizer and turn at the same speed. This relieves the operating pressure placed on the input shaft rear bearing, thus eliminating the growling noise.

The only way this problem can be corrected is to disassemble the transmission and clean, examine, and replace any damaged components. The teardown confirms the technician's diagnosis. The roller bearings are cracked and disintegrating. The pilot journal on the output shaft is also slightly damaged. The shaft is sent to a machine shop where it is undercut and fitted with a press-on steel bushing to return it to the manufacturer's specified diameter. The roller bearings are then replaced and the transmission reassembled. While the transmission is apart, all the components are cleaned and examined closely for damage and wear. The oil-cutting grooves on the first/second synchronizer are dull and flat. Although the customer has not yet experienced jumping out of gear problems in these gear ranges, the technician shows the customer the worn synchronizer and explains the problem it could cause in the near future. He strongly suggests replacement at this time and the customer agrees to this additional work.

KEY TERMS

Fill plug	Gear puller
Gear clash	Gear whine

SUMMARY

- Proper lubrication is vital to long transmission/transaxle life. The transmission gear lubricant must be checked and changed at manufacturer's suggested intervals.
- Metal particles or shavings in the gear lubricant indicate extensive internal wear or damage.
- The first step in diagnosing transmission/transaxle problems is to confirm that the problem exists inside the transmission/transaxle. Clutch and driveline problems may often appear to be transmission/transaxle problems.
- The initial visual inspection should include checks for lubricant leakage at gaskets and seals, transmission mount inspection, clutch and gearshift linkage checks, and drive axle and CV joint inspection.
- Rough growling noise inside the transmission/transaxle housing is an indication of bearing problems.
- A clicking noise may indicate excessive gear tooth wear. Rhythmical knocking is a sign of loose or broken internal components.
- Hard shifting can be caused by shift linkage problems, improper lubricant, or worn internal components, such as bearings, gears, shift forks, or synchronizers.
- Jumping out of gear can be caused by misaligned drivetrain mounts, a worn or poorly adjusted shift linkage, excessive clearance between gears, or badly worn bearings.
- Low lubricant levels, poorly adjusted shift linkages, or damaged internal components can result in transmission/transaxle lockup.
- Always follow service manual recommendations for removing the transmission/transaxle from the vehicle and disassembling it.
- Use recommended bearing pullers, gear pullers, and press equipment to remove and install gears and synchronizers on shafts.
- Clean and inspect all parts carefully, replacing worn or damaged components. Never force components in place during reassembly. Follow all clearance specifications listed in the service manual.
- Always use new snap rings, gaskets, and seals during reassembly.

TECH MANUAL

The following procedures are included in Chapter 36/37 of the *Tech Manual* that accompanies this book:

1. Checking the fluid level in a manual transmission and transaxle.
2. Inspecting and adjusting shift linkage.
3. Road testing a vehicle for transmission problems.

REVIEW QUESTIONS

1. After draining gear oil from a transaxle, the technician notices the oil has shiny, metallic particles in it. What does this indicate?
2. List at least five separate checks that should be made during the visual inspection of transmission/transaxle components.
3. List at least three causes of noise that are not transmission related but may appear to be.
4. What tool is often needed to remove gears and synchronizer assemblies from the transmission/transaxle mainshaft?
5. When removing or installing bearings, where should force be applied?
6. Technician A says broken or worn engine and transaxle mounts can cause a transaxle to have shifting problems. Technician B says poor shift boot alignment can cause a transaxle to jump out of gear. Who is correct?
   a. Technician A **c.** Both A and B
   b. Technician B **d.** Neither A nor B
7. While inspecting a transaxle's gears, Technician A says wear on the back of the gear teeth is normal. Technician B says it is normal for a reverse idler gear to have small chips on its engagement side. Who is correct?
   a. Technician A **c.** Both A and B
   b. Technician B **d.** Neither A nor B
8. A rough, growling noise is heard from a transaxle while it is in neutral with the engine running. If the vehicle is stationary and the clutch is engaged, the noise is a likely indication that there is a problem in the ______.
   a. transaxle input shaft bearings
   b. transaxle main (intermediate) shaft bearings
   c. first/second synchronizer assembly
   d. pinion and ring gear interaction
9. A clicking noise during transmission/transaxle operation may be an indication of ______.
   a. worn mainshaft (input shaft) bearings
   b. faulty synchronizer operation
   c. failed oil seals
   d. worn, broken, or chipped gear teeth
10. Low lubricant levels will most likely cause ______.
   a. gear clash
   b. hard shifting
   c. gear lockup or seizure
   d. gear jump out
11. Using a lubricant that is thicker than service manual specifications can lead to ______.
   a. gear jump out **c.** gear lockup
   b. hard shifting **d.** gear slippage
12. During a test drive, a noise that appears to be transmission related disappears when the driver brings the vehicle to a stop and disengages the clutch with the engine at idle. Technician A says this indicates the noise is most likely coming from inside the transmission. Technician B says the problem is most likely not transmission related. Who is correct?
   a. Technician A **c.** Both A and B
   b. Technician B **d.** Neither A nor B
13. While inspecting the synchronizers of a transaxle, Technician A says that, if the dog teeth of the synchronizer are rounded, the synchronizer assembly must be replaced. Technician B says the movement of the synchronizer sleeve on the shaft should be checked. Who is correct?
   a. Technician A **c.** Both A and B
   b. Technician B **d.** Neither A nor B
14. Technician A says spinning cleaned bearings with compressed air is a fast, convenient way to dry them. Technician B says this can damage the bearing and should never be done. Who is correct?

a. Technician A c. Both A and B
b. Technician B d. Neither A nor B

15. A poorly adjusted shift linkage can cause which of the following problems?
a. gear clash c. gear jump out
b. hard shifting d. all of the above

16. Technician A says the transmission rear seal at the driveline is particularly prone to leakage. When Technician B pushes up and down on the transmission, he says the mounts require replacement because the case moves up and down. Who is correct?
a. Technician A c. Both A and B
b. Technician B d. Neither A nor B

17. Noise occurs in forward and reverse gears, but not in neutral. Technician A says the input shaft bearing is the likely failed component. Technician B says the mainshaft bearing is the likely failed component. Who is correct?
a. Technician A c. Both A and B
b. Technician B d. Neither A nor B

18. A rough, growling noise occurs when a vehicle is moving in any gear. Technician A says the rear input shaft bearing may be at fault. Technician B says this condition indicates the countergear bearings may be faulty. Who is correct?
a. Technician A c. Both A and B
b. Technician B d. Neither A nor B

19. A car jumps out of gear into neutral, particularly when decelerating or going down hills. Technician A checks the shift lever and internal gearshift linkage first. Technician B says the clutch pilot bearing could be the problem. Who is correct?
a. Technician A c. Both A and B
b. Technician B d. Neither A nor B

20. While diagnosing a noise from a transmission, Technician A says the noise is caused by something internal if it is most noticeable during a test drive. Technician B says the noise is caused by the clutch if it disappears when the clutch is disengaged. Who is correct?
a. Technician A c. Both A and B
b. Technician B d. Neither A nor B

38

DRIVE AXLES AND DIFFERENTIALS

OBJECTIVES

■ Name and describe the components of a front-wheel-drive axle. ■ Describe the operation of a front-wheel-drive axle. ■ Diagnose problems in CV joints. ■ Perform preventive maintenance on CV joints. ■ Explain the difference between CV joints and universal joints. ■ Name and describe the components of a rear-wheel-drive axle. ■ Describe the operation of a rear-wheel-drive axle. ■ Explain the function and operation of a differential and drive axles. ■ Describe the various differential designs including complete, integral carrier, removable carrier, and limited slip. ■ Describe the three common types of driving axles. ■ Explain the function of the main driving gears, drive pinion gear, and ring gear. ■ Describe the operation of hunting, nonhunting, and partial nonhunting gears. ■ Describe the different types of axle shafts and axle shaft bearings.

The drive axle assembly transmits torque from the engine and transmission to drive the vehicle's wheels. The drive axle changes the direction of the power flow, multiplies torque, and allows different speeds between the two drive wheels. Drive axles are used for both front-wheel-drive and rear-wheel-drive vehicles.

FRONT-WHEEL-DRIVE (FWD) AXLES

Front-wheel-drive axles, also called axle shafts, typically transfer engine torque from the transaxle's differential to the front wheels. One of the most important components of FWD axles is the constant velocity (CV) joint. These joints are used to transfer uniform torque at a constant speed, while operating through a wide range of angles.

On front- or four-wheel-drive cars, operating angles of as much as 40 degrees are common **(Figure 38–1)**. The drive axles must transmit power from the engine to front wheels that must drive, steer, and cope with the severe angles caused by the up-and-down movement of the vehicle's suspension. To accomplish this, these cars must have a compact joint that ensures the driven shaft is rotated at a constant velocity, regardless of angle. CV joints also allow the length of the axle assembly to change as the wheel travels up and down.

TYPES OF CV JOINTS

Constant velocity joints come in a variety of styles. The different types of joints can be referred to by position (inboard or outboard), by function (fixed or plunge,) or by design (ball-type or tripod).

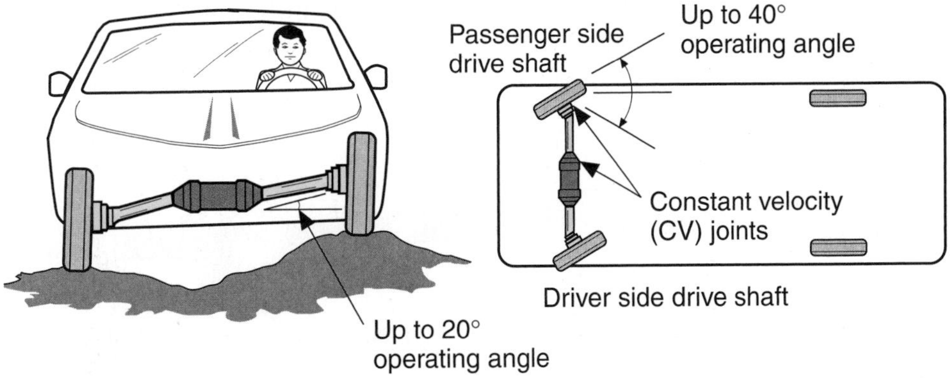

Figure 38–1 FWD drive axle shaft angles.

Inboard and Outboard Joints

On FWD vehicles, two CV joints are used on each half shaft **(Figure 38–2)**. The joint nearer the transaxle is the inner or **inboard joint**, and the one nearer the wheel is the outer or **outboard joint**. In a RWD vehicle with independent rear suspension, the joint nearer the differential can also be referred to as the inboard joint. The one closer to the wheel is the outboard joint.

Fixed and Plunge Joints

CV joints are either a **fixed joint** (meaning it does not plunge in and out to compensate for changes in length) or a **plunge joint** (one that is capable of in-and-out movement).

In FWD applications, the inboard joint is also a plunge joint. The outboard joint is a fixed joint. Both joints do not have to plunge if one can handle the job. Further, the outboard joint must also be able to handle much greater operating angles needed for steering (up to 40 degrees).

In RWD applications with IRS, one joint on each axle shaft can be fixed and the other a plunge or both can be plunge joints. Because the wheels are not used for steering, the operating angles are not as great. Therefore, plunge joints can be used at either or both ends of the axle shafts.

Ball-Type Joints

There are two basic varieties of CV joints: the **ball-type** and **tripod-type** joints. Both types are used as either inboard or outboard joints, and both are available in fixed or plunge designs.

Fixed Ball-Type CV Joints The **Rzeppa** or fixed ball-type joint consists of an inner ball race, six balls, a cage to position the balls, and an outer housing **(Figure 38–3)**. Tracks machined in the inner race and outer housing allow the joint to flex. The inner race and outer housing form a ball-and-socket arrangement. The six balls serve both as bearings between the races and the means of transferring torque from one to the other.

If viewed from the side, the balls within the joint always bisect the angle formed by the shafts on either side of the joint regardless of the operating angle. This reduces the effective operating angle of the joint by half and virtually eliminates all vibration problems. The input speed to the joint is always equal to the output velocity of the joint—thus the description "constant velocity." The cage helps to maintain this alignment by holding the six balls snugly in its windows. If the cage windows become worn or deformed over time, the resulting play between ball and window typically results in a clicking noise when turning. It is important to note that opposing balls in a Rzeppa CV joint always work together as a pair. Heavy wear in the tracks of one ball almost always results in identical wear in the tracks of the opposing ball.

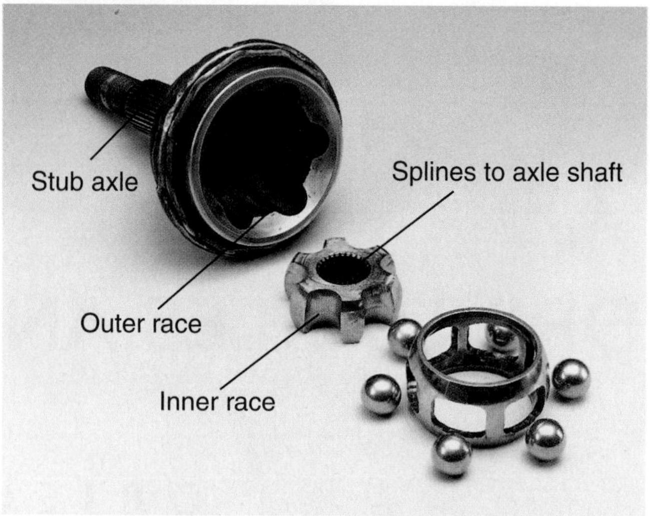

Figure 38–3 A Rzeppa ball-type fixed CV joint. *Courtesy of Moog Automotive, Inc.*

Another ball-type joint is the dish-style CV joint, which is used predominantly on Volkswagen as well as on many German RWD models. Its design is very similar to the Rzeppa joint.

Plunging Ball-Type Joints There are two basic styles of plunging ball-type joints: the **double-offset** and the **cross groove joint**. This is a more compact design with a flat, doughnut-shaped outer housing and angled grooves.

The double-offset joint **(Figure 38–4)** uses a cylindrical outer housing with straight grooves and is typically used in applications that require higher operating angles (up to 25 degrees) and greater plunge depth (up to 2.4 inches [60 mm]). This type of joint can be found at the inboard position on some front-wheel-drive half shafts as

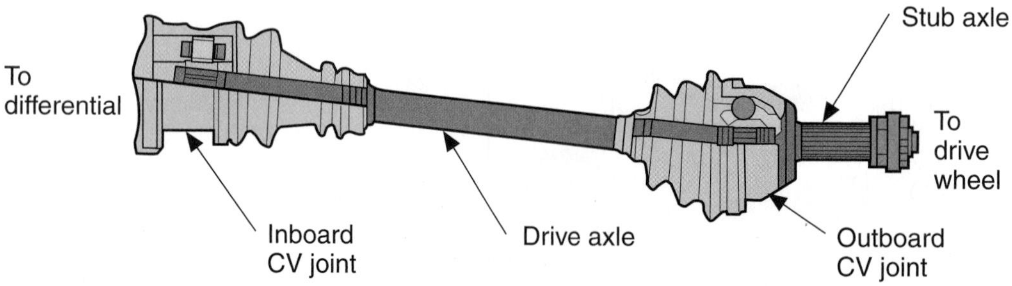

Figure 38–2 A typical FWD drive axle assembly. *Courtesy of Perfect Circle/ Dana Corporation*

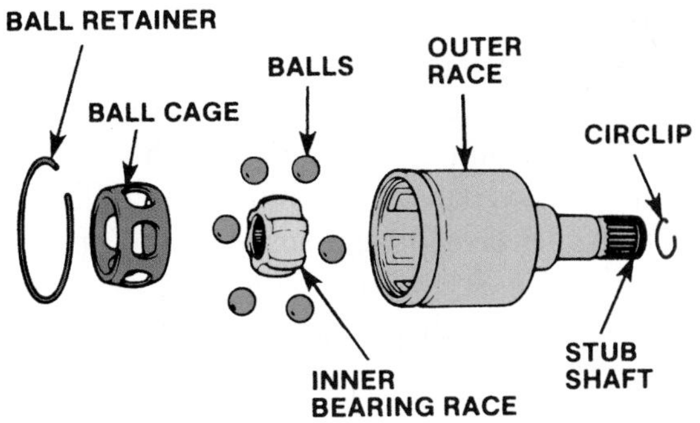

Figure 38-4 A double-offset CV joint.

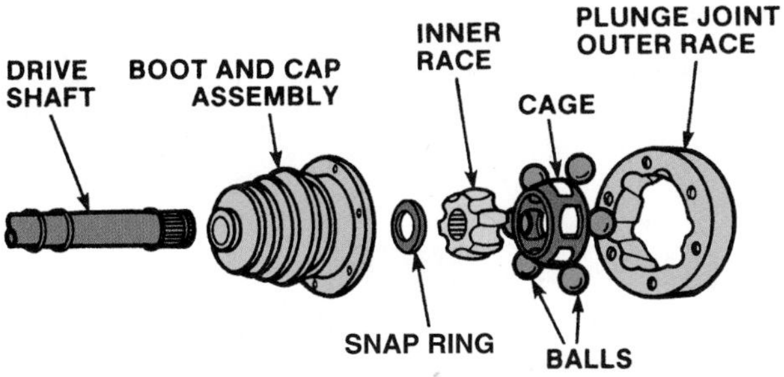

Figure 38-5 A cross-groove joint.

well as on the propeller shaft of some four-wheel-drive shafts.

The cross groove joint **(Figure 38–5)** has a much flatter design than any other plunge joint. It is used as the inboard joint on FWD half shafts or at either end of a RWD independent rear suspension axle shaft.

The feature that makes this joint unique is its ability to handle a fair amount of plunge (up to 1.8 inches [46 mm]) in a relatively short distance. The inner and outer races share the plunging motion equally, so less overall depth is needed for a given amount of plunge. The cross groove can handle operating angles up to 22 degrees.

Tripod CV Joints

As with ball-type CV joints, tripod joints come in two varieties: plunge and fixed.

Tripod Plunging Joints Tripod plunging joints **(Figure 38–6)** consist of a central drive part or tripod (also known as a "spider"). This has three trunnions fitted with spherical rollers on needle bearings and an outer housing (sometimes called a "tulip" because of its three-lobed, flowerlike appearance). On some tripod joints, the outer housing is closed, meaning the roller tracks are totally enclosed within it. On others, the tulip is open and the roller tracks are machined out of the housing. Tripod joints are most commonly used as FWD inboard plunge joints.

Fixed Tripod Joints The fixed tripod joint is sometimes used as the outboard joint in FWD applications. In this design, the trunnion is mounted in the outer housing and

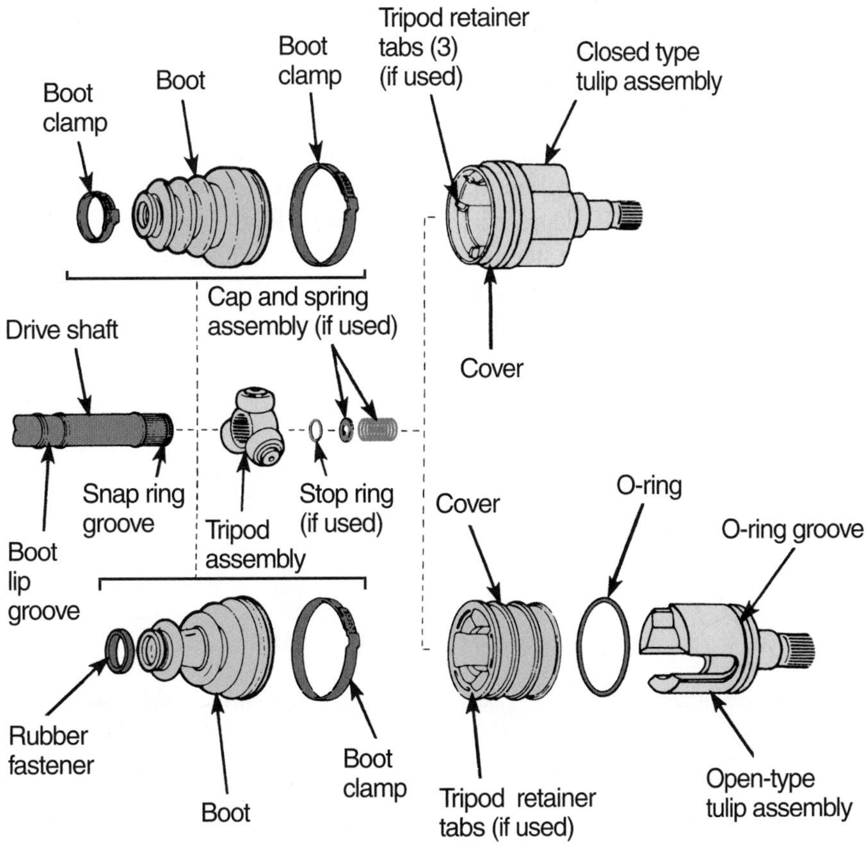

Figure 38-6 Inner tripod plunge-type joints: closed housing and open housing.

the three roller bearings turn against an open tulip on the input shaft. A steel locking spider holds the joint together.

The fixed tripod joint has a much greater angular capability. The only major difference from a service standpoint is that the fixed tripod joint cannot be removed from the drive shaft or disassembled because of the way it is manufactured. The complete joint and shaft assembly must be replaced if the joint goes bad.

FRONT-WHEEL-DRIVE APPLICATIONS

FWD half shafts can be solid or tubular, of equal **(Figure 38–7)** or unequal length **(Figure 38–8)**, and come with or without damper weights. Equal-length shafts are used in some vehicles to help reduce torque steer (the tendency to steer to one side as engine power is applied). In these applications, an intermediate shaft is used as a link from the transaxle to one of the half shafts. This intermediate shaft can use an ordinary Cardan universal joint (described later in this chapter) to a yoke at the transaxle. At the outer end is a support bracket and bearing assembly. Looseness in the bearing or bracket can create vibrations. These items should be included in any inspection of the drivetrain components. The small damper weight, called a **torsional damper**, that is sometimes attached to one half shaft serves to dampen harmonic vibrations in the drivetrain and to stabilize the shaft as it spins, not to balance the shaft.

Regardless of the application, outer joints typically wear faster than inner joints because of the increased range of operating angles to which they are subjected. Inner joint angles may change only 10 to 20 degrees as the suspension travels through jounce and rebound. Outer joints can undergo changes of up to 40 degrees in addition to jounce and rebound as the wheels are steered. That, combined with more flexing of the outer boots, is why outer joints have a higher failure rate. On average, nine outer CV joints are replaced for every inner CV joint. That does not mean the technician should overlook the inner joints. They wear too. Every time the suspension travels through jounces and rebound, the inner joints must plunge in and out to accommodate the different arcs between the drive shafts and suspension. Tripod inner joints tend to develop unique wear patterns on each of the three rollers and their respective tracks in the housing, which can lead to noise and vibration problems.

Other Applications

CV joints are also found on the front axles of many 4WD vehicles and on vehicles with rear independent suspension systems **(Figure 38–9)**. Their use in these designs offers the same benefits as when they are used for front-wheel drive.

CV JOINT SERVICE

With proper service, CV joints can have a long life, despite having to perform extremely difficult jobs in hostile environments. They must endure extreme heat and cold and survive the shock of hitting potholes at high speeds. Fortunately, high-torque loads during low-speed turns and many thousands of high-speed miles normally do not bother the CV joint. It is relatively trouble-free unless damage to the boot or joint goes unnoticed.

All CV joints are encased in a protective rubber (neoprene, natural, or silicone) or thermoplastic (Hycrel) boot. The job of the boot is to retain grease and to keep dirt and water out. The importance of the boot cannot be overemphasized because without its protection the joint does not survive. For all practical purposes, a CV joint is lubed for life. Once packed with grease and installed, it requires no further maintenance. A loose or missing boot clamp, or a slit, tear, or a small puncture in the boot itself allows grease to leak out and water or dirt to enter. Consequently, the joint is destroyed.

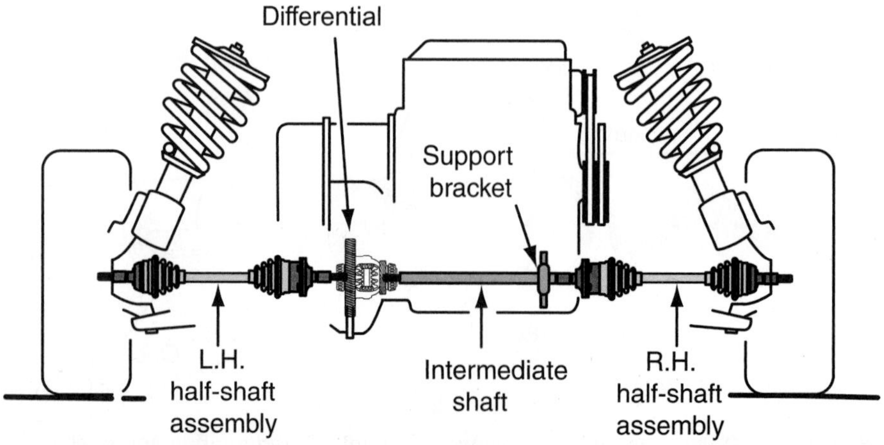

Figure 38-7 Equal length FWD half shafts with an intermediate shaft.

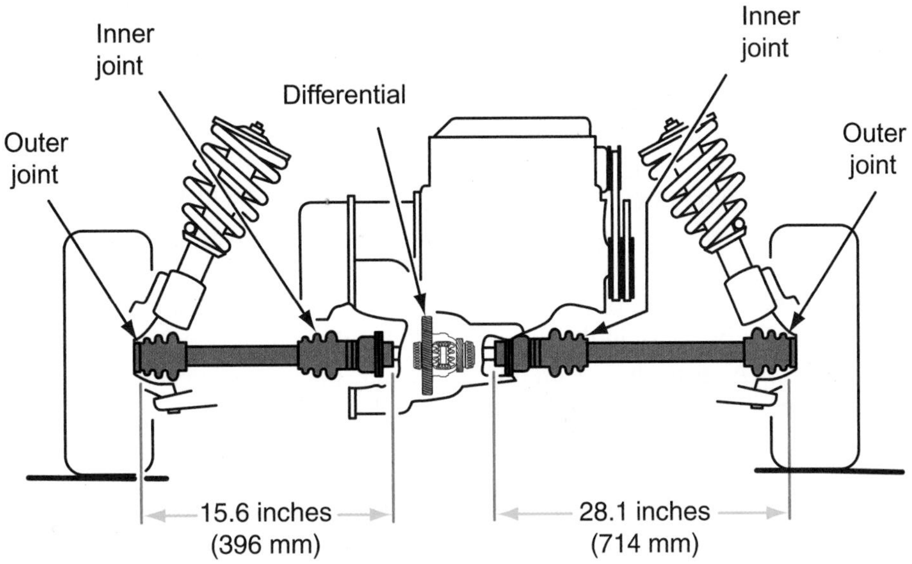

Figure 38-8 Unequal length FWD half shafts.

Although outboard joints tend to wear faster than the inboard ones, the decision as to whether to replace both joints when the half shaft is removed depends on the circumstances. If the vehicle has low miles and joint failure is the result of a defective boot, there is no reason to replace both joints. On a high-mileage vehicle where the bad joint has actually just worn itself out, it might be wise to save the expense and inconvenience of having the half shaft removed twice for CV joint replacement.

Diagnosis and Inspection

Any noise in the engine, drive axle, steering, or suspension is a good reason for a thorough inspection of the vehicle. A road test on a smooth surface is a good place to begin. The test should include driving at average highway speeds, some sharp turns, acceleration, and coasting. Look and listen for the following signs:

- A popping or clicking noise when turning indicates a possible worn or damaged outer joint **(Figure 38–10)**. To help identify the exact cause, put the vehicle in

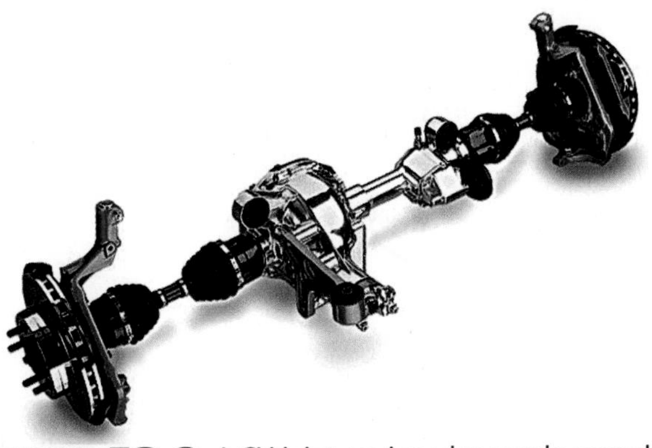

Figure 38-9 A CV-joint-equipped rear axle assembly for a vehicle with independent rear suspension. *Courtesy of Dana Corporation*

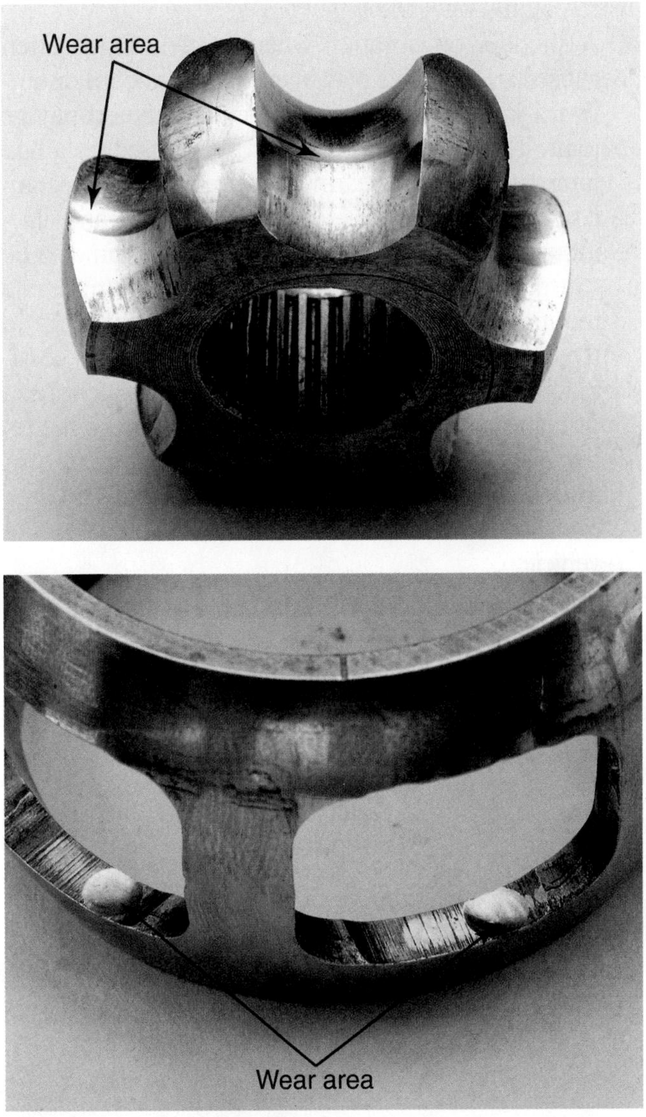

Figure 38-10 A worn cage or race can cause a clicking sound during a turn. *Courtesy of Moog Automotive, Inc.*

reverse and back up in a circle. If the noise gets louder, the outer joints should be replaced.

- A clunk during accelerating, decelerating, or putting an automatic transaxle into drive can be caused by excessive play in the inner joint on FWD vehicles. A clunking noise when putting an automatic transmission into gear or when starting out from a stop usually indicates excessive play in an inner or outer joint. Be warned, though, that the same kind of noise can also be produced by excessive backlash in the differential gears and transmission. Alternately accelerating and decelerating in reverse while driving straight can reveal worn inner plunge joints. A bad joint clunks or shudders.
- A humming or growling noise is sometimes due to inadequate lubrication of either the inner or outer CV joint. It is more often due to worn or damaged wheel bearings, a bad intermediate shaft bearing on equal-length half-shaft transaxles, or worn shaft bearings within the transmission.
- A shudder or vibration when accelerating is often caused by excessive play in either the inboard or outboard joint but more likely it is the inboard plunge joint. These vibrations can also be caused by a bad intermediate shaft bearing on transaxles with equal-length half shafts. On FWD vehicles with transverse-mounted engines, this kind of vibration can also be caused by loose or deteriorated engine/transaxle mounts. Be sure to inspect the rubber bushings in the engine's upper torque strap to rule out this possibility. A vibration or shudder that increases with speed or comes and goes at a certain speed may be the result of excessive play in an inner or outer joint. A bent axle shaft can cause the same problem. Note, however, that some shudder could also be inherent to the vehicle.
- A cyclic vibration that comes and goes between 45 and 60 mph (72 and 100 km) may lead the technician to think there is a wheel that is out of balance. However, as a rule, an out-of-balance wheel produces a continuous vibration. A more likely cause is a bad inner tripod CV joint. The vibration occurs because one of the three roller tracks has become dimpled or rough. Every time the tripod roller on the bad track hits the rough spot, it creates a little jerk in the driveline, which the driver feels as a cyclic vibration.
- If a noise is heard while driving straight ahead but it ceases while turning, the problem is usually not a defective outer CV joint but a bad front wheel bearing. Turning changes the side load on the bearing, which may make it quieter than before.
- A vibration that increases with speed is rarely due to CV joint problems or FWD half-shaft imbalance. An out-of-balance tire or wheel, an out-of-round tire or wheel, or a bent rim are the most likely causes. It is possible that a bent half shaft, as the result of collision or towing damage, could cause the vibration. A missing damper weight could also be the culprit.

Begin CV joint inspection **(Figure 38–11)** by checking the condition of the boots. Splits, cracks, tears, punctures, or thin spots caused by rubbing call for immediate boot replacement. If the boot appears rotted, this indicates improper greasing or excessive heat, and it should be replaced. Squeeze all boots. If any air escapes, replace the boot.

If the inner boot appears to be collapsed or deformed, venting it (allowing air to enter) might solve the problem. Place a round-tipped rod between the boot and drive shaft. This equalizes the outside and inside air and allows the boot to return to its normal shape.

Make sure that all boot clamps are tight. Missing or loose clamps should be replaced. If the boot appears loose, slide it back and inspect the grease inside for possible contamination. A milky or foamy appearance indicates water contamination. A gritty feeling when rubbed between the fingers indicates dirt. In most cases, a water- or dirt-contaminated joint should be replaced.

The drive axles should be checked for signs of contact or rubbing against the chassis. Rubbing can be a symptom of a weak or broken spring or engine mount, as well as chassis misalignment. On FWD transaxles with equal-

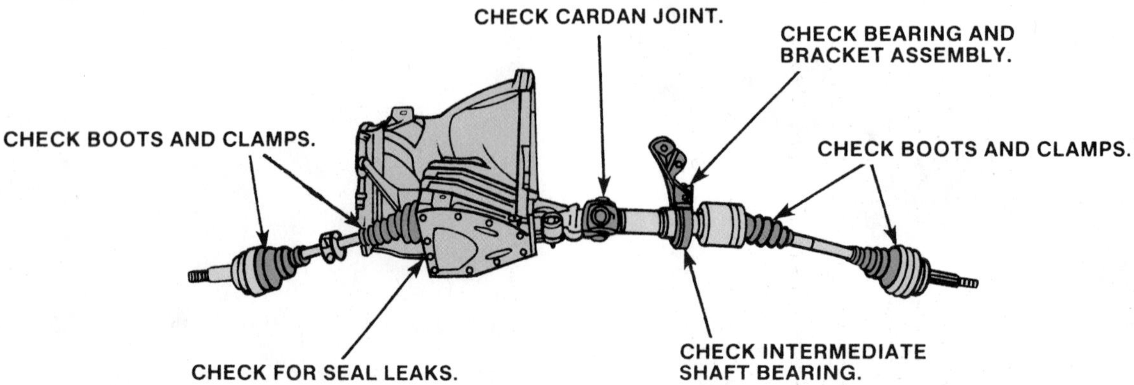
CHECK CARDAN JOINT.
CHECK BEARING AND BRACKET ASSEMBLY.
CHECK BOOTS AND CLAMPS.
CHECK BOOTS AND CLAMPS.
CHECK FOR SEAL LEAKS.
CHECK INTERMEDIATE SHAFT BEARING.

Figure 38-11 Inspection points for a FWD vehicle.

length half shafts, inspect the intermediate shaft U-joint, bearing, and support bracket for looseness by rocking the wheel back and forth and watching for any movement. Oil leakage around the inner CV joints indicates a faulty transaxle shaft seal. To replace the seal, the half shaft must be removed.

Obtaining CV Repair Parts

To repair a drive axle, a complete shaft should be installed. Most aftermarket part suppliers offer a complete line of original equipment drive shafts for FWD vehicles. These shafts come fully assembled and ready for installation. This repair method eliminates the need to tear down and rebuild an old shaft.

If only the CV joints need service, a CV joint service kit should be installed. Joint service kits typically include a CV joint, boot, boot clamps and seals, special grease for lubrication (various joints require different amounts of grease; the correct quantity is packed in each kit), retaining rings, and all other attachment parts.

Part manufacturers also produce a line of complete boot sets for each application, including new clamps and the appropriate type and amount of grease for the joint. CV joints require a special high-temperature, high-pressure grease. Substituting any other type of grease may lead to premature failure of the joint. Be sure to use all the grease supplied in the joint or boot kit. The same rule applies to the clamps. Use only those clamps supplied with the replacement boot. Follow the directions for positioning and securing them.

Old boots should never be reused when replacing a CV joint. In most cases, failure of the old joint is caused by some deterioration of the old boot. Reusing an old boot on a new joint usually leads to the quick destruction of the joint.

Photo Sequence 37 shows the procedure for removing a typical drive axle and replacing a CV joint boot. Always refer to the Service Manual for the exact service procedure. The diagnosis and service chart shown in **Table 38–1** gives an idea of the types of front-wheel drivetrain problems that can occur.

CV Joint Service Guidelines

The following are some guidelines to follow when servicing CV joints:

- Never jerk or pull on the axle shaft when removing it from a vehicle with tripod inner joints. Doing so may pull the joint apart, allowing the needle bearings to fall out of the roller. Pull on the inner housing, and support the outer end of the shaft until the shaft is completely out.
- Always torque the hub nuts to the vehicle manufacturer's specifications. This is absolutely necessary to properly preload the wheel bearings. Do not guess. The specifications can vary from 75 to

TABLE 38–1 PROBLEM DIAGNOSIS AND SERVICE FOR FWD DRIVELINES

Problem	Possible Cause	Corrective Remedy
Vibrations in steering wheel at highway speeds	Front-wheel balance	Front-wheel unbalance is felt in the steering wheel. Front wheels must be balanced.
Vibrations throughout vehicle	Worn inner CV joints	Worn parts of the inner CV joint not operating smoothly.
Vibrations throughout vehicle at low speed	Bent axle shaft	Axle shaft does not operate on center of axis; thus, vibration develops.
Vibrations during acceleration	Worn or damaged outer or inner CV joints	CV joints not operating smoothly due to damage or wear on parts.
	Fatigued front springs	Sagged front springs are causing the inner CV joint to operate at too great an angle, causing vibrations.
Grease dripping on ground or sprayed on chassis parts	Ripped or torn CV joint boots	Front-wheel-drive CV joints are immersed in lubricant. If the CV joint boot has a rip or is torn, lubricant leaks out. Condition must be corrected as soon as possible.
Clicking or snapping noise heard when turning curves and corners	Worn or damaged outer CV joint Bent axle shaft	Worn parts are clicking and noisy as loading and unloading on CV joint takes place. Irregular rotation of the axle shaft causing a snapping, clicking noise.

PHOTO SEQUENCE 37

Removing and Replacing a CV Joint Boot

P37-1 *Removing the axle from the car begins with the removal of the wheel cover and wheel hub cover. The hub nut should be loosened before raising the car and removing the wheel.*

P37-2 *After the car is raised and the wheel is removed, the hub nut can be unscrewed from the axle shaft.*

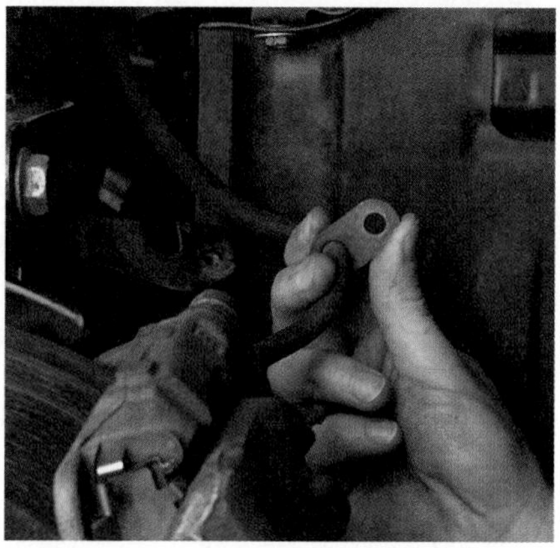

P37-3 *The brake line holding clamp must be loosened from the suspension.*

P37-4 *The ball joint must be separated from the steering knuckle assembly. To do this, first remove the ball joint retaining bolt. Then pry down on the control arm until the ball joint is free.*

P37-5 *The inboard joint can be pulled free from the transaxle.*

P37-6 *A special tool is normally needed to separate the axle shaft from the hub allowing the axle to be removed from the car.*

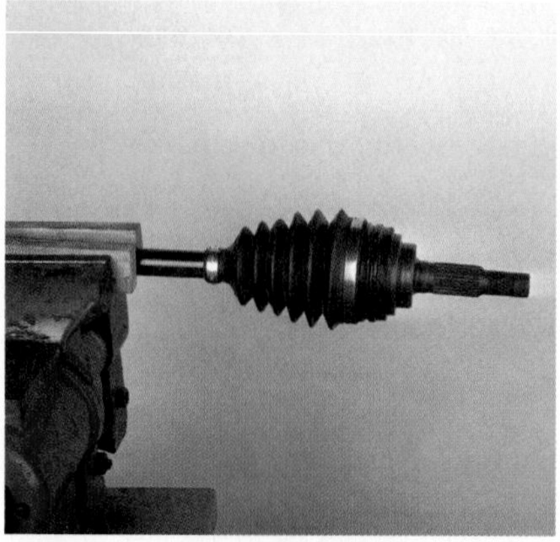

P37-7 *The axle shaft should be mounted in a soft-jawed vise for work on the joint. Pieces of wood on either side of the axle work well to secure the axle without damaging it.*

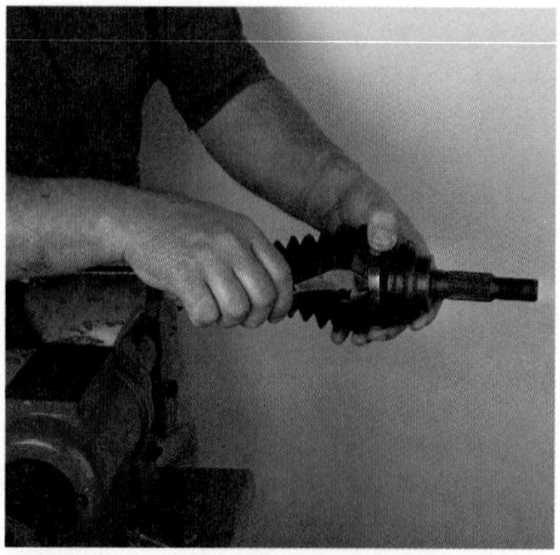

P37-8 *Begin boot removal by cutting and discarding the boot clamps.*

P37-9 *Scribe a mark around the axle to indicate the boot's position on the shaft. Then, move the boot off the joint.*

P37-10 *Remove the circlip and separate the joint from the shaft.*

P37-11 *Slide the old boot off the shaft.*

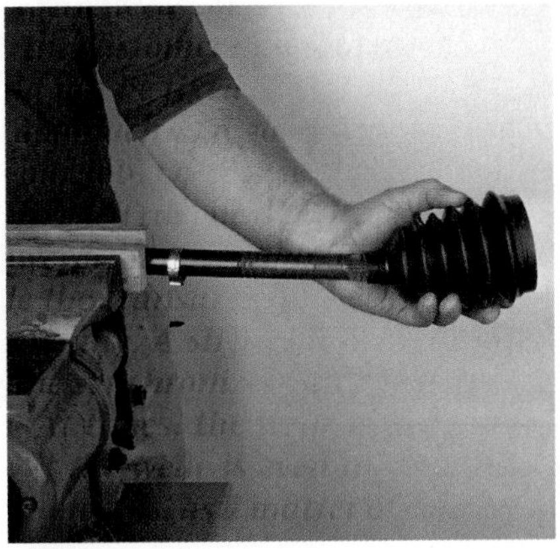

P37-12 *Wipe the axle shaft clean and install the new boot onto the shaft.*

P37-13 *Place the boot into its proper location on the shaft and install a new clamp.*

P37-14 *Using a new circlip, reinstall the joint on the shaft. Pack joint grease into the joint and boot. The entire packet of grease that comes with a new boot needs to be forced into the boot and joint.*

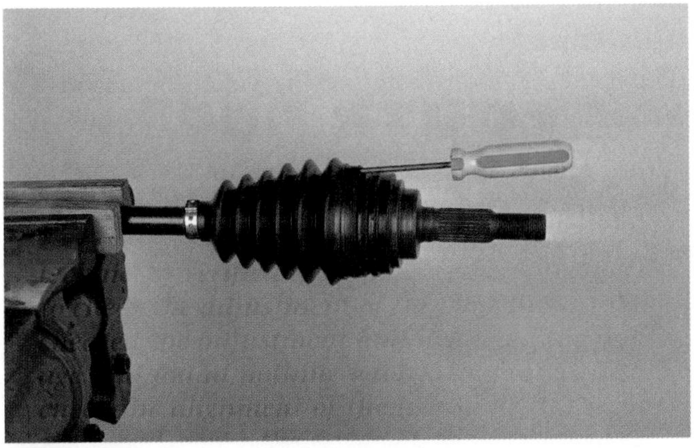

P37-15 *Pull the boot over the joint and into its proper position. Use a dull screwdriver to lift an edge of the boot up to equalize the pressure inside the boot with the outside air.*

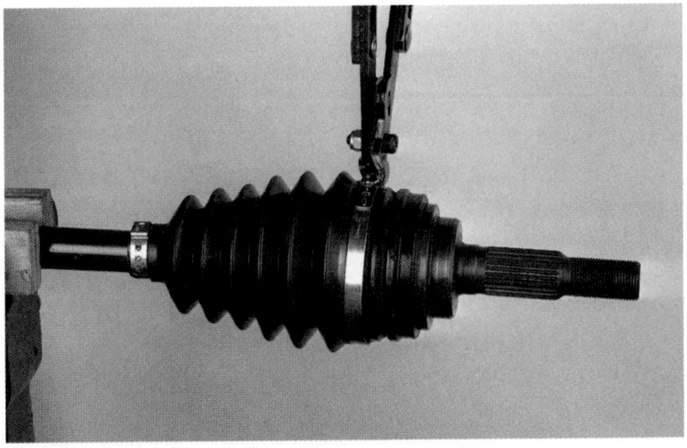

P37-16 *Install the new large boot clamp and reinstall the axle into the car. Torque the hub nut after the wheels have been reinstalled and the car is sitting on the ground.*

Figure 38-12 Most axle hub nuts are staked after they are tightened to lock them in place. *Courtesy of Moog Automotive, Inc.*

235 ft.-lb (101 to 318 Nm). Most axle hub nuts are staked in place after they have been tightened **(Figure 38–12)**.

- Never use an impact wrench to loosen or tighten axle hub nuts. Doing so may damage the wheel bearings as well as the CV joints.
- On vehicles with antilock brakes, use care to protect the wheel speed sensor and tone ring on the outer CV joint housings. If misaligned or damaged during joint replacement, it can cause wheel speed sensor problems.
- Always recheck the alignment after replacing CV joints. Marking the camber bolts is not enough, because camber can be off as much as three-quarters of a degree due to differences between the size of the camber bolts and their holes.

CV Shaft and Rubber Boot Care Tips

The rubber boots need special care when you are servicing the CV joints, engine, or transaxle. The following tips might save you trouble later:

- Always support the control arm when doing on-the-car balancing of the front wheels to avoid high-speed operation at a steep half-shaft angle. Off-the-car balancing might be a wiser choice.
- Do not use half shafts as lift points for raising a car.
- Use a plastic or metal shield over rubber boots to protect them from accidental tool damage when performing other wheel, brake, suspension, or steering system maintenance.
- Clean only with soap and water.
- Avoid boot contact with gasoline, oil, or degreaser compounds.

REAR-WHEEL DRIVE SHAFTS

A drive shaft must smoothly transfer torque while rotating, changing length, and moving up and down. The different designs of drive shafts all attempt to ensure a vibration-free transfer of the engine's power from the transmission to the differential. This goal is complicated by the fact that the engine and transmission are bolted solidly to the frame of the car, whereas the differential is mounted on springs. As the rear wheels go over bumps in the road or changes in the road's surface, the springs compress or expand, changing the angle of the drive shaft between the transmission and the differential, as well as the distance between the two. To allow for these changes, the Hotchkiss-type drive shaft is fitted with one or more U-joints to permit variations in the angle of the drive, and a slip joint that permits the effective length of the drive shaft to change.

SHOP TALK

When a vehicle is intentionally raised or lowered, the length of the drive shaft should be changed to allow for normal travel of the slip yoke on the output shaft. ■

Starting at the front or transmission end of a RWD shaft, there is a slip yoke, universal joint, drive shaft yoke, and drive shaft **(Figure 38–13)**. At the rear or differential end, there is another drive shaft yoke and a second universal joint connected to the differential pinion flange.

In addition to these basic components, some drivetrain systems use a center carrier bearing for added support **(Figure 38–14)**. Large cars with long drive shafts often use a double U-joint arrangement, called a double Cardan joint or a constant velocity U-joint, to help minimize driveline vibrations.

Slip Yoke

The most common sliding or **slip yoke (Figure 38–15)** has an internally splined, externally machined bore that lets the yoke rotate at transmission output shaft speed and slide at the same time (hence the name slip yoke). While the need for rotation is obvious, without the linear flexibility, the drive shaft would bend like a bow the first time the suspension jounced.

Drive Shaft and Yokes

The drive shaft is nothing more than an extension of the transmission output shaft. The drive shaft, which is usually made from seamless steel tubing, transfers engine torque from the transmission to the rear driving axle. The yokes, which are either welded or pressed onto the shaft,

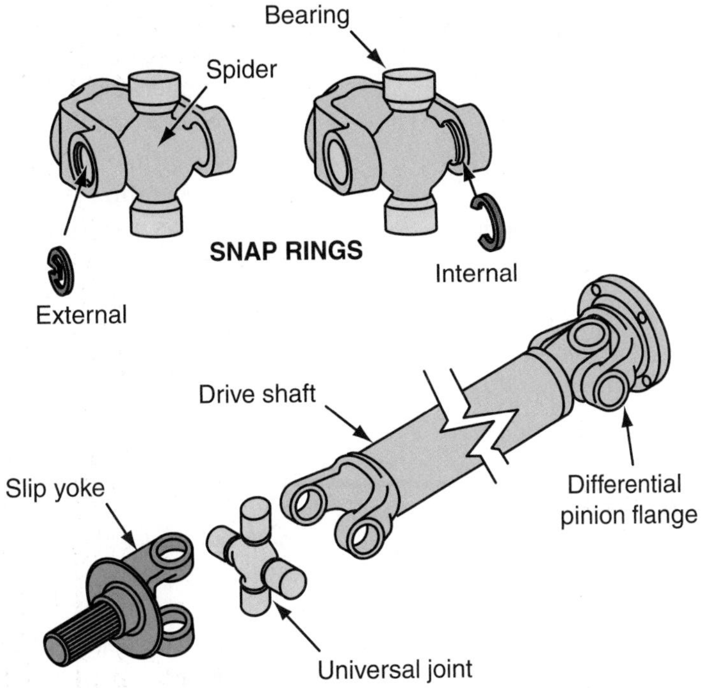

Figure 38-13 A drive shaft assembly with exploded U-joints.

Figure 38-14 A center bearing assembly.

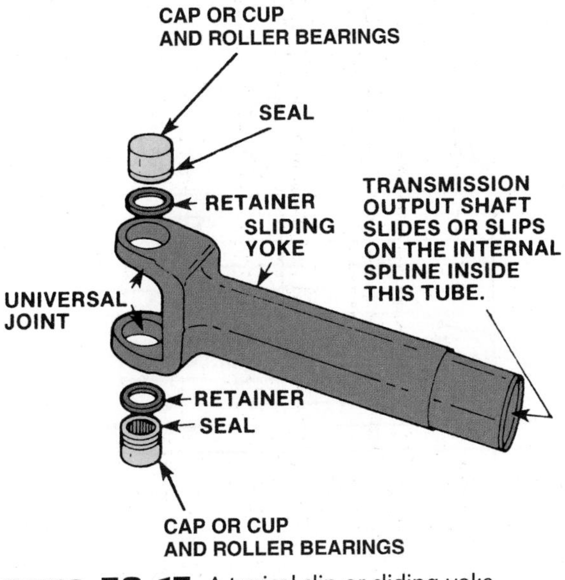

Figure 38-15 A typical slip or sliding yoke.

provide a means of connecting two or more shafts together. At the present time, a limited number of vehicles are equipped with fiber composite—reinforced fiberglass, graphite, and aluminum—drive shafts. The advantages of using these materials are weight reduction, torsional strength, fatigue resistance, easier and better balancing, and reduced interference from shock loading and torsional problems. Some drive shafts are fitted with a torsional damper to reduce torsional vibrations.

The drive shaft, like any other rigid tube, has a natural vibration frequency. If one end were held tightly, it would vibrate at its own frequency when deflected and released. It reaches its natural frequency at its critical speed. Critical drive shaft speed depends on the diameter of the tube and its length. Diameters are as large as possible and shafts as short as possible to keep the critical speed frequency above the driving speed range. It should be remembered that since the drive shaft generally turns three to four times faster than the tires, proper drive shaft balance is required for vibration-free operation.

OPERATION OF U-JOINTS

The U-joint allows two rotating shafts to operate at a slight angle to each other. A French mathematician named Cardan developed the original joint in the sixteenth century. In 1902, Clarence Spicer modified Cardan's invention for the purpose of transmitting engine torque to an automobile's rear wheels.

The universal joint is basically a double-hinged joint consisting of two Y-shaped yokes, one on the driving or

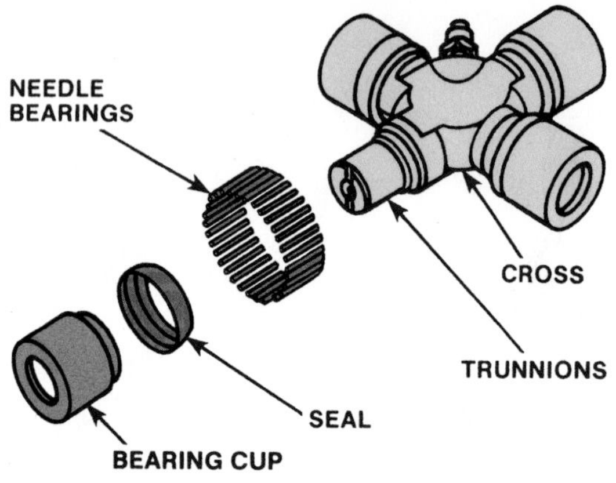

Figure 38–16 A Cardan joint.

input shaft and the other on the driven or output shaft, plus a cross-shaped unit called the cross **(Figure 38–16)**. A yoke is used to connect the U-joints together. The four arms of the cross are fitted with bearings in the ends of the two shaft yokes. The input shaft's yoke causes the cross to rotate, and the two other trunnions of the cross cause the output shaft to rotate. When the two shafts are at an angle to each other, the bearings allow the yokes to swing around on their trunnions with each revolution. This action allows two shafts, at a slight angle to each other, to rotate together.

Universal joints allow the drive shaft to transmit power to the rear axle through varying angles that are controlled by the travel of the rear suspension. Because power is transmitted on an angle, U-joints do not rotate at a constant velocity, nor are they vibration free.

Speed Variations (Fluctuations)

Although simple in appearance, the universal joint is more intricate than it seems because its natural action is to speed up and slow down twice in each revolution while operating at an angle. The amount that the speed changes varies according to the steepness of the U-joint's angle.

U-joint **operating angle** is determined by taking the difference between the transmission installation angle and the drive shaft installation angle. When the universal joint is operating at an angle, the driven yoke speeds up and slows down twice during each drive shaft revolution.

These four speed changes are not normally visible during rotation. But they may be understood more easily after examining the action of a U-joint. A universal joint is a coupling between two shafts not in direct alignment, usually with changing relative positions. It would be logical to assume that the entire unit simply rotates. This is true only for the universal joint's input yoke.

The output yoke's circular path looks like an ellipse because it can be viewed at an angle instead of straight on. This effect can be obtained when a coin is rotated by the fingers. The height of the coin stays the same even though the sides seem to get closer together.

This illusion might seem to be a merely visual effect, but it is more than that. The U-joint rigidly locks the circular action of the input yoke to the elliptical action of the output yoke. The result is similar to what would happen when changing a clock face from a circle to an ellipse.

Like the hands of a clock, the input yoke turns at a constant speed in its true circular path. The output yoke, operating at an angle to the other yoke, completes its path in the same amount of time. However, its speed varies, or is not constant, compared to the input.

Speed fluctuation is more easily visualized when looking at the travel of the yokes by 90-degree quadrants **(Figure 38–17)**. The input yoke rotates at a steady or constant speed through the complete 360-degree turn. The output yoke quadrants alternate between shorter and longer dis-

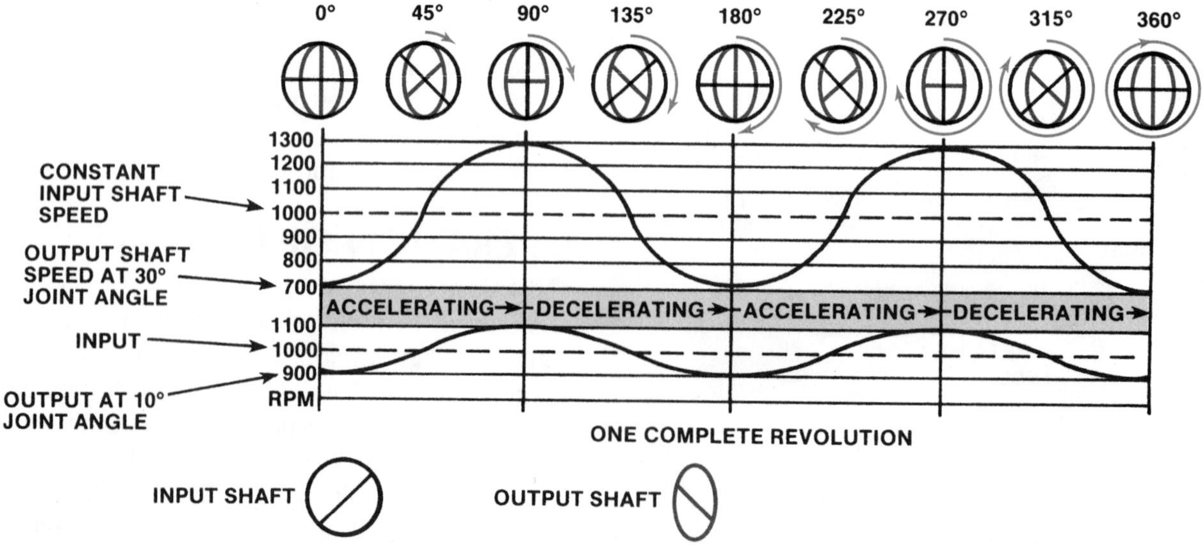

Figure 38–17 A graph showing typical drive shaft yoke speed fluctuations.

tance travel than the input yoke quadrants. When one point of the output yoke covers the shorter distance in the same amount of time, it must travel at a slower rate. Conversely, when traveling the longer distance (but only 90 degrees) in the same amount of time, it must move faster.

Because the average speed of the output yoke through the four 90-degree quadrants (360 degrees) equals the constant speed of the input yoke during the same revolution, it is possible for the two mating yokes to travel at different speeds. The output yoke is falling behind and catching up constantly. The resulting acceleration and deceleration produces a fluctuating torque and torsional vibrations characteristic of all Cardan U-joints. The steeper the U-joint angle, the greater the fluctuations in speed will be. Conversely, the smaller the angle, the speed will change less.

Phasing of Universal Joints

The torsional vibrations set up by the fluctuations in velocity are transferred down the drive shaft to the next U-joint. At this joint similar speed fluctuation occurs. Since these speed variations take place at equal and opposite angles to the first joint, they cancel out each other. To provide for this canceling effect, drive shafts should have at least two U-joints and their operating angles must be equal to each other. Speed fluctuations can be cancelled if the driven yoke has the same point of rotation, or same plane, as the driving yoke. When the yokes are in the same plane, the joints are said to be "in phase."

On a two-piece drive shaft, you may encounter problems if you are not careful. The center U-joint must be disassembled to replace the center support bearing. The center driving yoke is splined to the front drive shaft. If the yoke's position on the drive shaft is not indicated in some manner, the yoke could be installed in a position that is out of phase. Manufacturers use different methods of indexing the yoke to the shaft. Some use aligning arrows. Others machine a master spline that is wider than the others. The yoke and shaft cannot be reassembled until the master spline is aligned properly. When there are no indexing marks, you should index the yoke to the drive shaft before disassembling the U-joint. This saves time and frustration during reassembly.

Canceling Angles

Vibrations can be reduced by using canceling angles **(Figure 38–18)**. Carefully examine the illustration, and note that the operating angle at the front of the drive shaft is offset by the one at the rear of the drive shaft. When the front universal joint accelerates, causing a vibration, the rear universal joint decelerates, causing a vibration. The vibrations created by the two joints oppose and dampen the vibrations from one to the other. The use of canceling angles provides a smoother drive shaft operation.

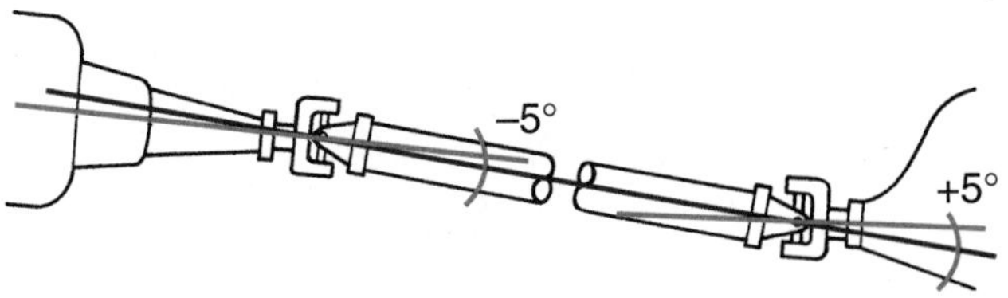

Figure 38–18 When a drive shaft's joints are in phase and have canceling angles, inherent vibrations are reduced.

TYPES OF U-JOINTS

There are three common designs of U-joints: single U-joints retained by either an inside or outside snap ring, coupled U-joints, and U-joints held in the yoke by U-bolts or lock plates.

Single Universal Joints

The single Cardan/Spicer universal joint is also known as the cross or four-point joint. These two names aptly describe the single Cardan, since the joint itself forms a cross, with four machined trunnions or points equally spaced around the center of the axis. Needle bearings used to abate friction and provide smoother operation are set in bearing cups. The trunnions of the cross fit into the cup assemblies and the cup assemblies fit snugly into the driving and driven universal joint yokes. U-joint movement takes place between the trunnions, needle bearings, and bearing cups. There should be no movement between the bearing cup and its bore in the universal joint yoke. The bearings are normally held in place by snap rings that drop into grooves in the yoke's bearing bores. The bearing caps allow free movement between the trunnion and yoke. The needle bearing caps may also be pressed into the yokes, bolted to the yokes, or held in place with U-bolts or metal straps.

There are other styles of single U-joints. The method used to retain the bearing caps is the major difference between these designs. The Spicer style **(Figure 38–19A)** uses an outside snap ring that fits into a groove machined in the outer end of the yoke. The bearing cups for this style are machined to accommodate the snap ring.

The Mechanics or Detroit/Saginaw style **(Figure 38–19B)** uses an inside snap ring or C-clip that fits into a groove machined in the bearing cup on the side closer to the grease seal. When installed, the clip rests against the machined inside portion of the yoke. The snap rings are retained by nylon injected into the retaining ring grooves.

The Cleveland style is an attempt to combine different joint styles to have more applications from one joint. The bearing cups for this U-joint are machined to accommodate either Spicer or Mechanics style snap rings. If a replacement U-joint comes with both style clips, use the clips that pertain to your application.

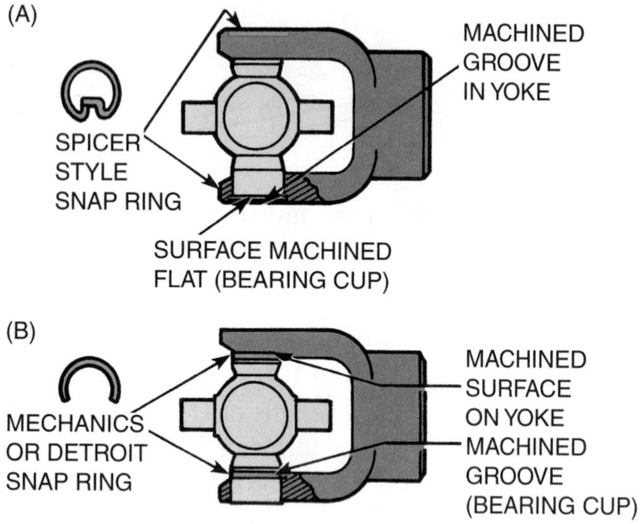

Figure 38-19 (A) A Spicer-style U-joint and (B) a mechanics or Detroit-style U-joint.

Double-Cardan Universal Joint

A **double-Cardan** U-joint is used with split drive shafts and consists of two Cardan universal joints closely connected by a centering ball socket and a center yoke, which functions as a ball-and-socket. The ball-and-socket splits the angle of the two shafts between two U-joints **(Figure 38–20)**. Because of the centering socket yoke, the total operating angle is divided equally between the two joints. Since the two joints operate at the same angle, the normal fluctuations that result from the use of a single U-joint are cancelled out. The acceleration and deceleration of one joint is cancelled by the equal and opposite action of the other.

The double-Cardan joint is classified as a constant velocity universal joint. It is most often found in front-engine RWD luxury-type vehicles.

DIAGNOSIS OF DRIVE-SHAFT AND U-JOINT PROBLEMS

A failed U-joint or damaged drive shaft can exhibit a variety of symptoms. A clunk that is heard when the transmission is shifted into gear is the most obvious. You can also encounter unusual noise, roughness, or vibration.

To help differentiate a potential drivetrain problem from other common sources of noise or vibration, it is important to note the speed and driving conditions at which the problem occurs. As a general guide, a worn U-joint is most noticeable during acceleration or deceleration and is less speed sensitive than an unbalanced tire (commonly occurring in the 30 to 60 mph [50 to 100 km/h] range) or a bad wheel bearing (more noticeable at higher speeds). Unfortunately, it is often very difficult to accurately pinpoint drivetrain problems with only a road

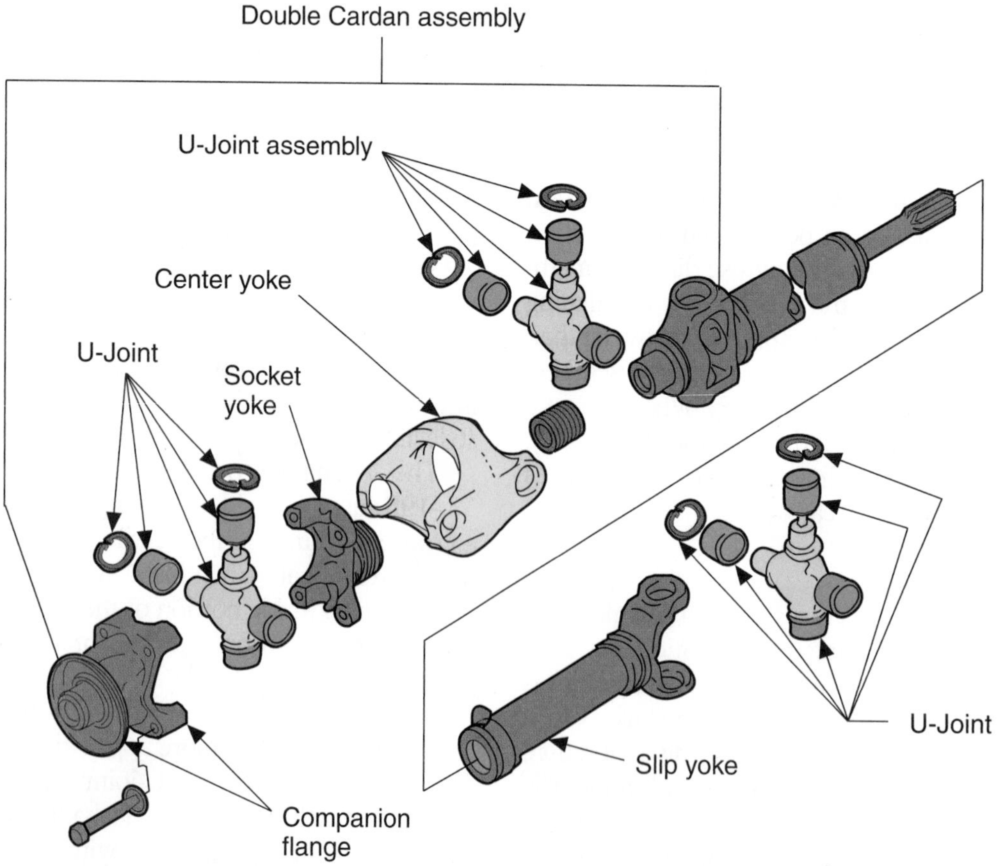

Figure 38-20 A double Cardan joint. *Courtesy of Ford Motor Company*

test. Therefore, expand the undercar investigation by putting the vehicle up on the lift, where it is possible to get a good view of what is going on underneath.

The first problem most likely encountered is an undercar fluid leak. If a lot of lube is escaping from the pinion shaft seal, the drivetrain noise could be caused by a bad pinion bearing. To confirm the problem, start the engine, put the transmission in gear, and listen at the carrier. If the bearing is noisy, it is necessary to make one of those difficult judgment calls. If the bearing sounds fine but the pinion seal is still leaking, suggest an on-the-car seal replacement.

On some vehicles, seal replacement is a simple procedure that involves removing the pinion flange and replacing the seal. However, always refer to the service manual for the correct procedure and note any special precautions to be taken.

At the other end of the driveline, inspect the transmission's extension housing seal the same way. If it is leaking, the seal itself can be easily replaced. Check the extension housing bushing. That is the most likely reason the seal went bad in the first place. Once the yoke is removed, an internal expanding bearing/bushing puller makes short work of bushing replacement. Before pushing the slip yoke back in after the new seal is installed, make sure the machined surface of the bore is free of scratches, nicks, and grooves that could damage the seal. For that added margin of safety, a little transmission lube or petroleum jelly on the lip of the seal helps the parts slide in easily.

If the seals pass the test, continue driveline examination by inspecting the U-joint's grease seals for signs of rust, leakage, or lubrication breakdown. Also, check for excessive joint movement by firmly grasping and attempting to rotate the coupling yokes back and forth in opposite directions. If any perceptible trunnion-to-bearing movement is felt, the joint should be replaced.

The runout of the drive shaft should also be checked. If there is excessive runout, determine the cause and make the necessary repairs. If the runout is fine, check the phasing of the joints and their angle. To check their operating angle, use an inclinometer. This instrument, when attached to the drive shaft, displays the angle of the drive shaft along any point. Your finding from this test should be compared to specifications. Normally, if the angles are wrong, the rear axle has moved in its mounting.

As a final diagnosis inspection point, check the entire length of the drive shaft for excess undercoating, dents, missing weights, or other damage that could cause an imbalance and result in a vibration. If no damage is found, the drive shaft should be removed and its balance checked by a specialty shop.

When a U-joint is damaged or excessively worn, it must be replaced. Photo Sequence 38 covers the typical procedure for removing a U-joint from a drive shaft. After a replacement joint is obtained, it needs to be installed. Photo Sequence 39 covers the reassembly of a common U-joint.

DIFFERENTIALS AND DRIVE AXLES

The differential is a geared mechanism located between the driving axles of a vehicle. It rotates the driving axles at different speeds when the vehicle is turning a corner **(Figure 38–21)**. It also allows both axles to turn at the same speed when the vehicle is moving in a straight line. The drive axle assembly directs driveline torque to the vehicle's drive wheels. The gear ratio of the drive axle's ring and pinion gears is used to increase torque. The differential serves to establish a state of balance between the forces between the drive wheels and allows the drive wheels to turn at different speeds when the vehicle changes direction.

On a FWD car or truck, the differential is normally an integral part of the transaxle assembly located at the front of the vehicle. Transaxle design and operation depends on whether the engine is mounted transversely or longitudinally. With a transversely mounted engine, the crankshaft centerline and drive axle are on the same plane. With a longitudinally mounted power plant, the differential must change the power flow 90 degrees.

On RWD vehicles, the differential is located in the rear axle housing or carrier. The drive shaft connects

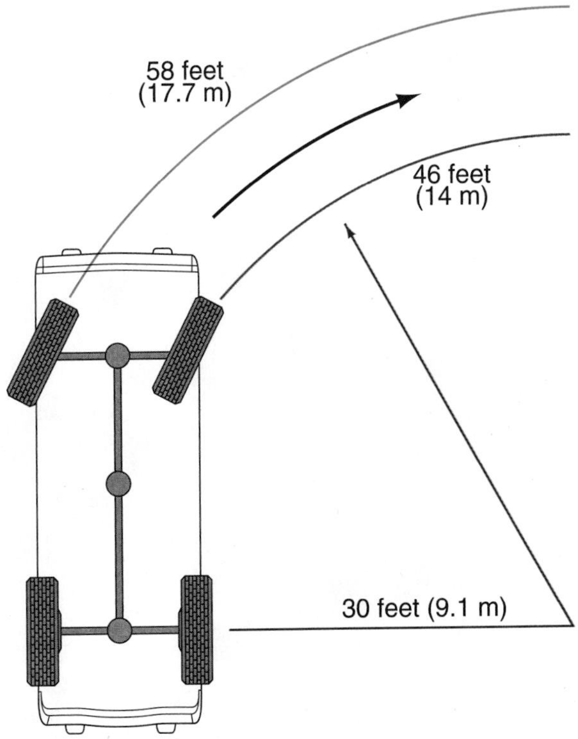

Figure 38-21 Travel of wheels when a vehicle is turning a corner.

PHOTO SEQUENCE 38

Disassembling a Single Universal Joint

P38-1 *Clamp the slip yoke in a vise and support the outer end of the drive shaft.*

P38-2 *Remove the lock rings on the tops of the bearing cups. Make index marks in the yoke so that the joint can be assembled with the correct phasing.*

P38-3 *Select a socket that has an inside diameter large enough for the bearing cup to fit into; usually a 1-1/4 inch socket works.*

P38-4 *Select a second socket that can slide into the shaft's bearing cup bore—usually a 9/16-inch socket.*

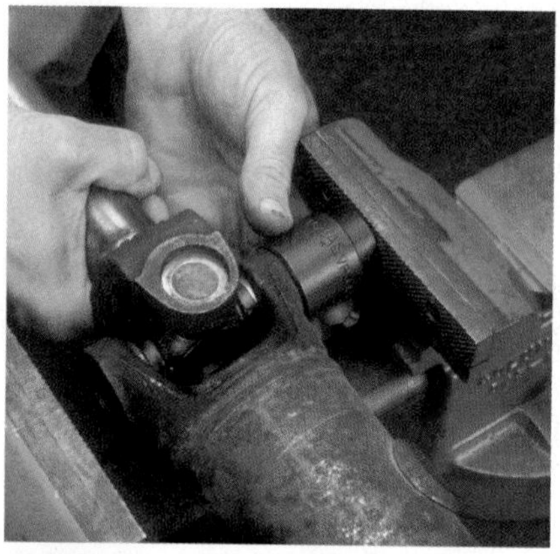

P38-5 *Place the large socket against one vise jaw. Position the drive shaft yoke so that the socket is around a bearing cup.*

P38-6 *Position the other socket to the center of the bearing cup opposite to the one in line with the large socket.*

P38-7 *Carefully tighten the vise to press the bearing cup out of the yoke and into the large socket.*

P38-8 *Separate the joint by turning the shaft over in the vise and driving the spider and remaining bearing cup down through the yoke with a brass drift and hammer.*

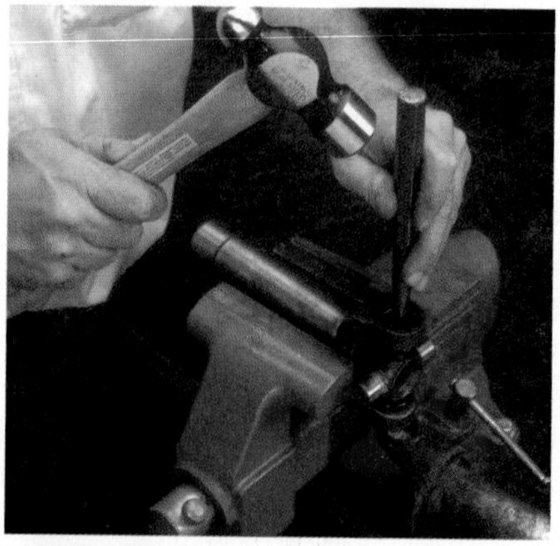

P38-9 *Use a drift and hammer to drive the joint out of the other yokes.*

PHOTO SEQUENCE 39

Reassembling a Single Universal Joint

P39-1 *Clean any dirt from the yoke and the retaining ring grooves.*

P39-2 *Carefully remove the bearing cups from the new U-joint.*

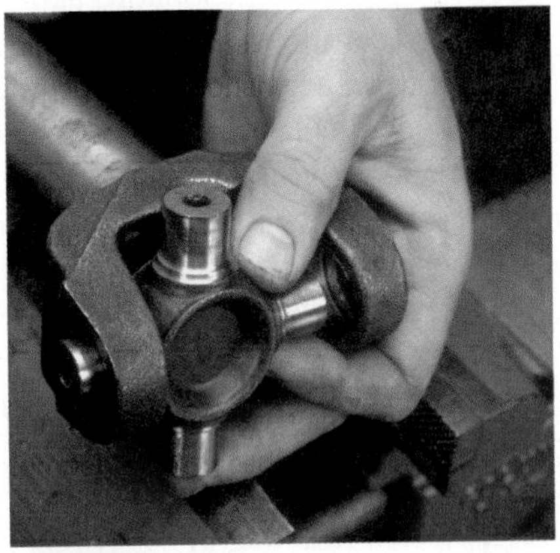

P39-3 *Place the new spider inside the yoke and push it to one side.*

P39-4 *Start one cup into the yoke's ear and over the spider's trunnion.*

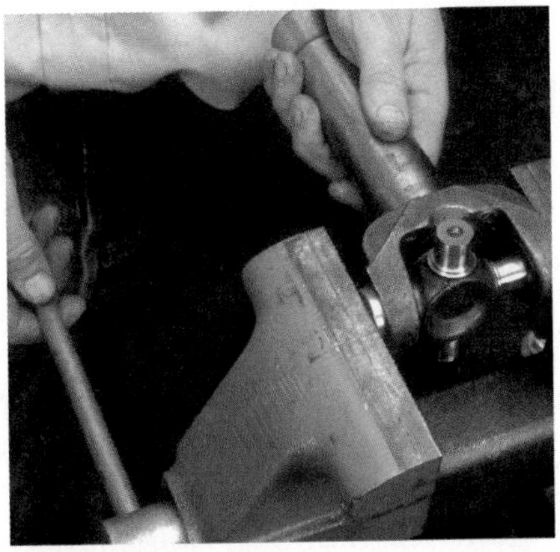

P39-5 *Carefully place the assembly in a vise and press the cup partially through the ear.*

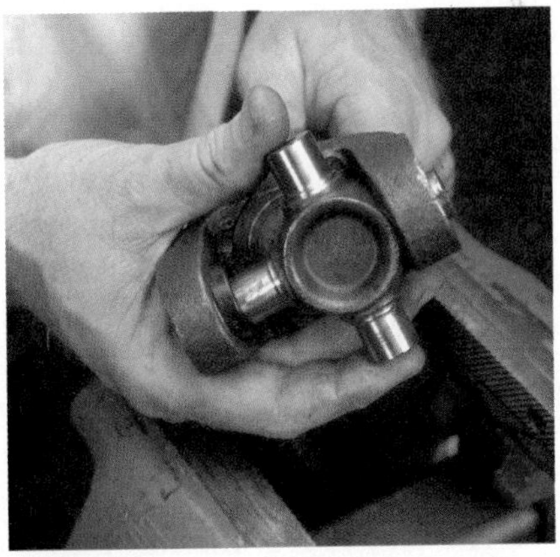

P39-6 *Remove the shaft from the vise and push the spider toward the other side of the yoke.*

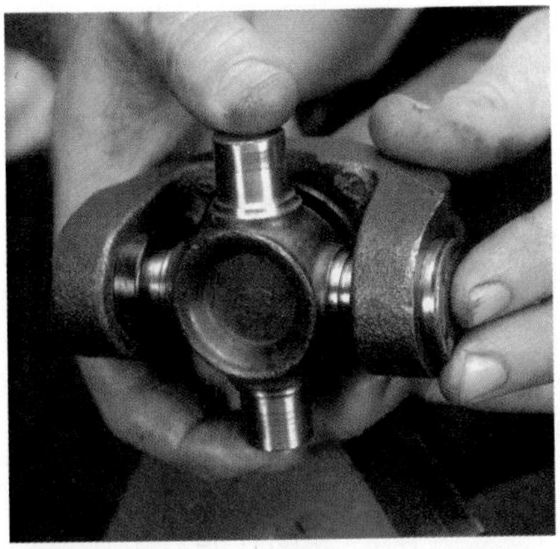

P39-7 *Start a cup into the yoke's ear and over the trunnion.*

P39-8 *Place the shaft in the vise and tighten the jaws to press the bearing cup into the ear and over the trunnion. Then install the snap rings. Make sure they are seated in their grooves.*

P39-9 *Position the joint's spider in the drive shaft yoke and install the two remaining bearing cups.*

the transmission with the rear axle gearing. Four-wheel-drive vehicles have differentials on both their front and rear axles.

The differential allows the drive wheels to rotate at different speeds when negotiating a turn or curve in the road and redirects the engine torque from the drive shaft to the rear drive axle shafts. The drive shaft turns in a motion perpendicular to the rotation of the drive wheels. The final drive gears redirect the torque so that the drive axle shafts turn in a motion parallel to the rotation of the drive wheels.

The final drive gears in the drive axle assembly are also sized to provide a gear reduction, or a torque multiplication. Axles with a low (numerically high) gear ratio allow for fast acceleration and good pulling power. Axles with high gear ratios allow the engine to run slower at any given speed, resulting in better fuel conservation.

Differential Components

The components of commonly used final drive units are shown in **Figure 38–22**. There are several other basic design arrangements. However, the most commonly used design has pinion/ring gears and a **pinion shaft**. The latter is normally a spiral bevel gear mounted on an input (pinion) shaft. The shaft is mounted in the front end of the carrier and supported by two or three bearings. An overhung pinion gear is supported by two tapered bearings spaced far enough apart to provide the needed leverage to rotate the ring gear and drive axles. A straddle-mounted pinion gear rests on three bearings: two tapered bearings on the front support the input shaft and one roller bearing is fitted over a short shaft extending from the rear end of the **pinion gear**.

The pinion gear meshes with a **ring gear**. The ring gear is a ring of hardened steel with curved teeth on one side and threaded holes on the other. The ring gear is bolted to the differential case. When the pinion gear is rotated by the drive shaft, the ring gear is forced to rotate, turning the differential case and axle shafts. In most automotive applications, two pinion gears are mounted on a straight shaft in the differential case. On heavier trucks, the differential contains four pinion gears mounted on a cross-shaped spider in the differential case. The pinion shafts are mounted in holes in the case (or in matching grooves in the case halves) and are secured in place with a lock bolt or retaining rings.

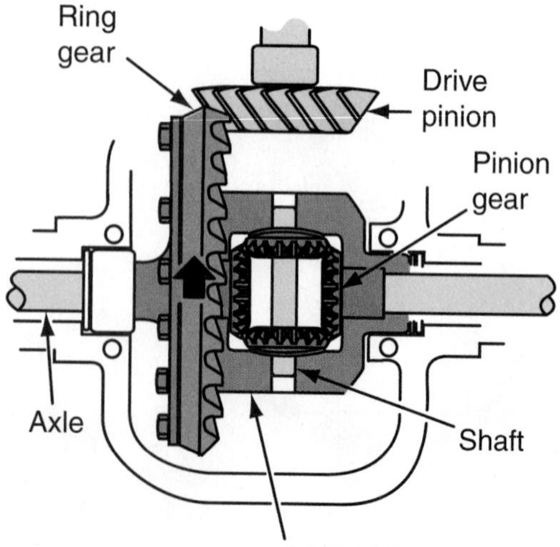

Figure 38-22 The components of a typical final drive unit.

Ring and pinion gears are normally classified as hunting, nonhunting, or partial nonhunting gears. Each type of gearset has its own requirements for a satisfactory gear tooth contact pattern. These classifications are based on the number of teeth on the pinion and ring gears.

- *Hunting Gearset.* When one drive pinion gear tooth contacts every ring gear tooth after several revolutions, the gearset is a **hunting gearset**. In other words, the drive pinion hunts out each ring gear tooth. A typical hunting gearset may have nine drive pinion teeth and thirty-seven ring gear teeth. The rear-axle ratio for this combination would be 4.11:1.
- *Nonhunting Gearset.* When one drive pinion gear tooth contacts only certain ring gear teeth, the gearset is a **nonhunting gearset**. A typical nonhunting gearset may have ten drive pinion teeth and thirty ring gear teeth. The rear-axle ratio for this combination would be 3.00:1. For every revolution of the ring gear, each drive pinion tooth would contact the same three teeth of the ring gear. The drive pinion gear teeth do not hunt out all ring gear teeth.
- *Partial Nonhunting Gearset.* The difference between a nonhunting and a **partial nonhunting gearset** is the amount of ring gear teeth that are contacted. In a partial nonhunting gearset, one drive pinion tooth contacts six ring gear teeth instead of three. During the first revolution of the ring gear, one drive pinion tooth contacts three ring gear teeth. During the second revolution of the ring gear, the drive pinion tooth contacts three different ring gear teeth. During every other ring gear revolution, one drive pinion tooth contacts the same ring gear teeth. A typical partial nonhunting gearset may have ten drive pinion teeth and thirty-five ring gear teeth. The rear-axle ratio for this combination would be 3.50:1.

The number of teeth on the drive pinion and ring gear determine whether the gearset is hunting, nonhunting, or partial nonhunting. Knowing the type of gearset is important in diagnosing ring and pinion problems.

A **hypoid gear** contacts more than one tooth at a time. The hypoid gear also makes contact with a sliding motion. This sliding action, however, is smoother than that of the spiral gear, resulting in quieter operation. The biggest difference is that, in a hypoid gear, the centerlines of the ring and pinion gears do not match. Using hypoid gears, the drive pinion gear is placed lower in the differential. The drive pinion meshes with the ring gear at a point below its centerline.

The sliding effect of two hypoid gears meshing tends to wipe lubricant from the face of the gears **(Figure 38–23)**, resulting in eventual damage. Differentials require the use of extreme pressure-type lubricants. The additives in this type of lubricant allow the lubricant to withstand the wiping action of the gear teeth without separating from the gear face.

The differential also contains two side gears. The inside bore of the side gears is splined and mates with splines on the ends of the axles. The differential pinion gears and side gears are in constant mesh. The pinion gears are mounted on a pinion gear shaft, which is mounted in the differential casing. As the casing turns with the ring gear, the pinion shaft and gears also turn. The pinion gears deliver torque to the side gears.

When the pinion and ring gears are manufactured, the faces of the gear teeth are machined to provide smooth mating surfaces. The pinion gear and ring gears are always matched to provide a good mesh **(Figure 38–24)**. Pinion gears and the ring gears should always be installed as a set. Otherwise, the mismatched gearset might operate noisily. A matched gearset code is etched in each drive pinion and ring gearset.

Rear Axle Housing and Casing

The differential and final drive gears in a rear-drive vehicle are housed in the rear axle housing, or carrier. The axle housing also contains the two drive axle shafts. Two types of axle housings are found on modern automobiles: the removable carrier and the integral carrier. The removable carrier axle housing is open on the front side. Because it resembles a banjo, it is often called a banjo housing. The backside of the housing is closed to seal out dirt and contaminants and keep in the lubricant. The differential is mounted in a carrier assembly that can be removed as a unit from the axle housing **(Figure 38–25)**. Removable carrier axle housings are most commonly used today on trucks and other heavy-duty vehicles.

The **integral housing** is most often found on late-model cars and light trucks **(Figure 38–26)**. A cast-iron carrier forms the center of the axle housing. Steel axle tubes are pressed into both sides of the carrier to form the housing. The housing and carrier have a removable rear cover that allows access to the differential assembly. Because the carrier is not removable, the differential components must be removed and serviced separately. For many operations, a case spreader **(Figure 38–27)** must be used to remove the components. In addition to providing a mounting place for the differential, the axle housing also contains brackets for mounting suspension components such as control arms, leaf springs, and coil springs.

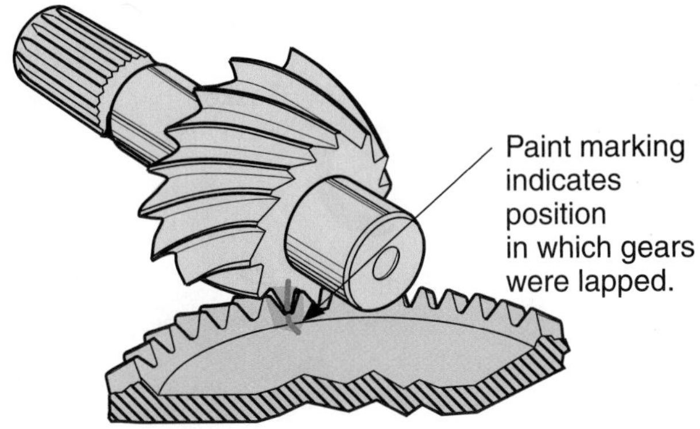

Figure 38-24 Index marks on a ring and pinion gearset. *Courtesy of Ford Motor Company*

Figure 38-23 The flow of oil in a hypoid gear set as it spins. *Courtesy of Dana Corporation*

Figure 38-25 A typical removable-carrier axle housing.

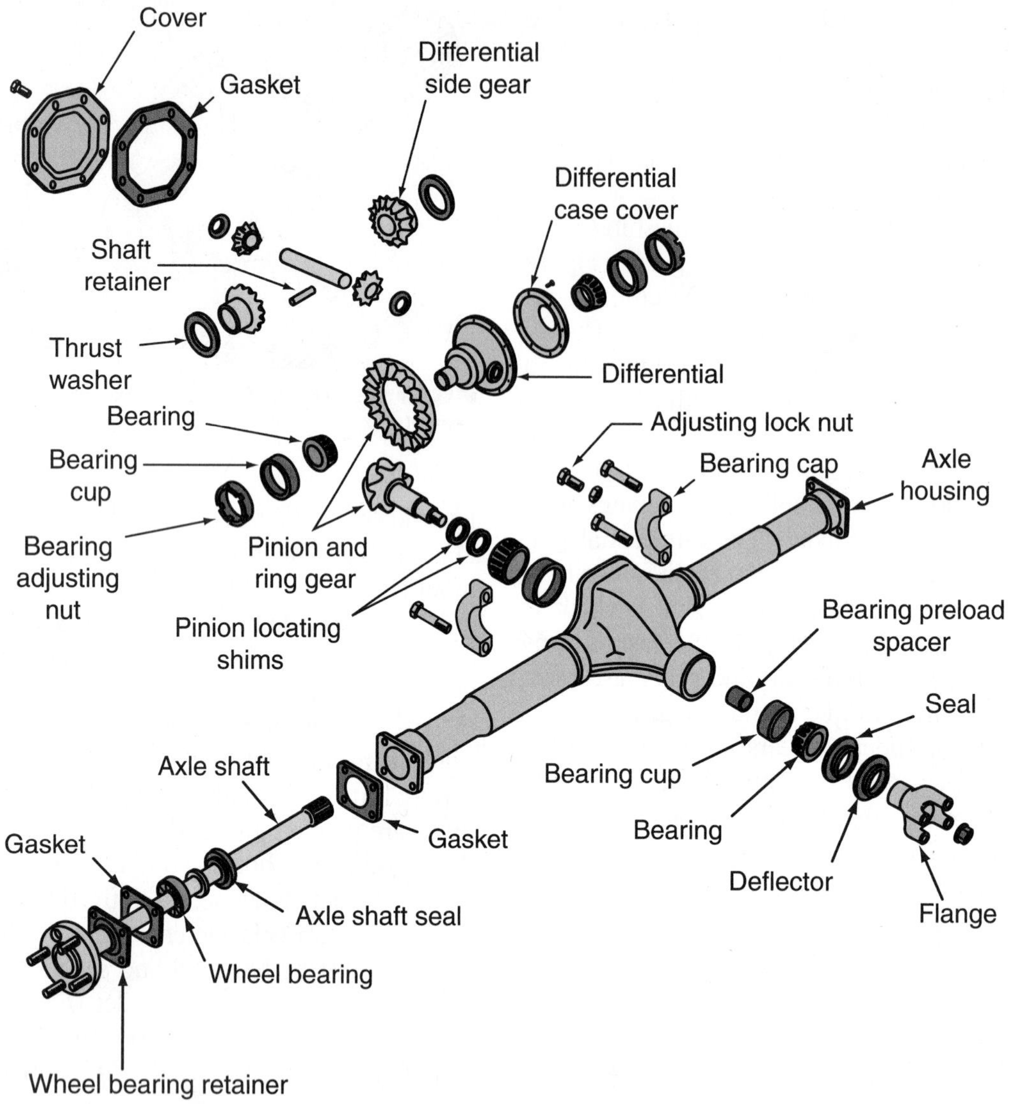

Figure 38-26 An exploded view of integral-carrier axle housing.

Some vehicles have an ABS speed sensor attached to the carrier housing for rear-wheel lockup prevention during braking.

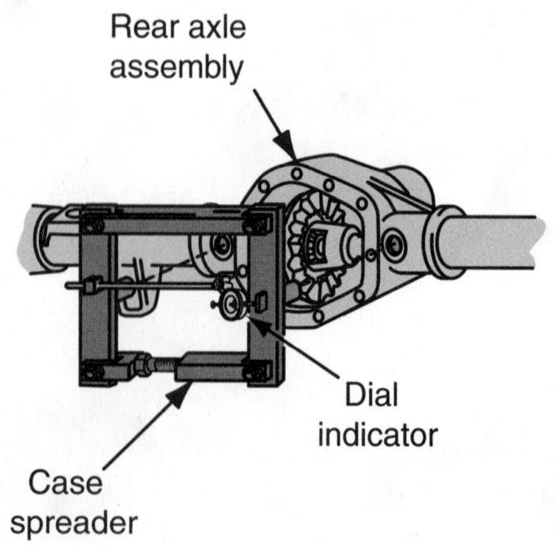

Figure 38-27 A case spreader.

Differential Operation

The amount of power delivered to each driving wheel by differential is expressed as a percentage. When the vehicle moves straight ahead, each driving wheel rotates at 100% of the differential case speed. When the vehicle is turning, the inside wheel might be getting 90% of the differential case speed. At the same time, the outside wheel might be getting 110% of the differential case speed.

Power flow through the axle begins at the drive pinion yoke, or companion flange **(Figure 38–28)**. The companion flange accepts torque from the rear U-joint. The companion flange is attached to the drive pinion gear, which transfers torque to the ring gear. As the ring gear turns, it turns the differential case and the pinion shaft. The differential pinion gears transfer torque to the side gears to turn the driving axle shafts. The differential pinion gears determine how much torque goes to each driving axle, depending on the resistance an axle shaft or

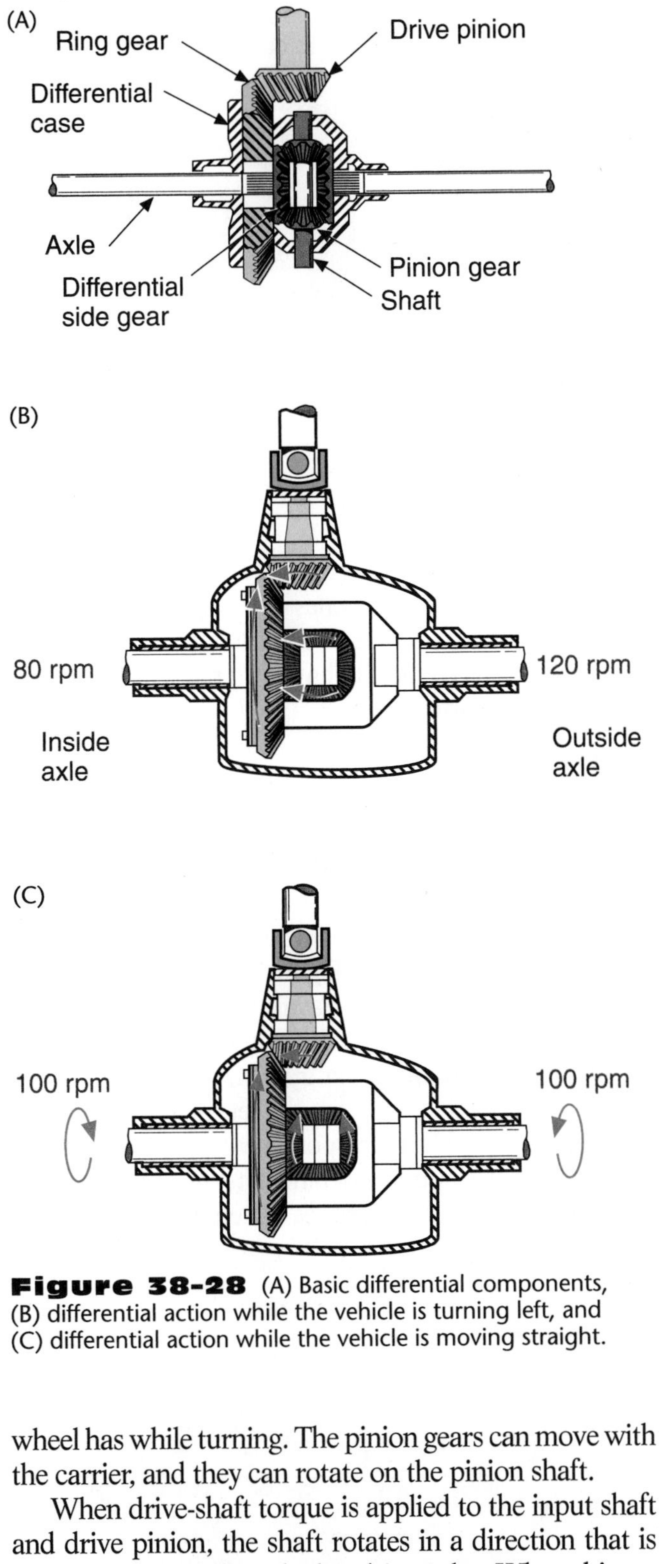

Figure 38-28 (A) Basic differential components, (B) differential action while the vehicle is turning left, and (C) differential action while the vehicle is moving straight.

wheel has while turning. The pinion gears can move with the carrier, and they can rotate on the pinion shaft.

When drive-shaft torque is applied to the input shaft and drive pinion, the shaft rotates in a direction that is perpendicular to the vehicle's drive axles. When this rotary motion is transferred to the ring gear, the torque flow changes direction and becomes parallel to the axle shafts and wheels. Because the ring gear is bolted to the differential case, the case must rotate with the ring gear. The pinion gear shaft mounted in the differential case must also rotate with the case and the ring gear. The pinions turn end over end. Gears do not rotate on the pinion shaft when both driving wheels are turning at the same speed. They rotate end over end as the differential case rotates. Because the pinions are meshed with both side gears, the side gears rotate and turn the axle shafts. The ring gear, differential gears, and axle shafts turn together without variation in speed as long as the vehicle is moving in a straight line.

When a vehicle turns into a curve or negotiates a turn, the wheels on the outside of the curve must travel a greater distance than the wheels on the inside of the curve. The outer wheels must then rotate faster than the inside wheels. This would be impossible if the axle shafts were locked solidly to the ring gear. However, the differential allows the outer wheels and axle shaft to increase in speed and the inner wheels and axle to slow down, thus preventing the skidding and rapid tire wear that would otherwise occur. The differential action also makes the vehicle much easier to control while turning.

For example, when a car makes a sharp right-hand turn, the left-side wheels, axle shaft, and side gear must rotate faster than the right-side wheels, axle shaft, and side gear. The left side of the axle must speed up and the right side must slow down. This is possible because the pinions to which the side gears are meshed are free to rotate on the pinion shaft. The increased speed of the left-side wheels causes the side gear to rotate faster than the differential case. This causes the pinions to rotate and walk around the slowing down side gear. As the pinions turn to allow the left-side gear to increase speed, a reverse action—known as a reverse walking effect—is produced on the right-side gear. It slows down an amount that is inversely proportional to the increase in the left-side gear.

LIMITED-SLIP DIFFERENTIALS

Driveline torque is evenly divided between the two rear drive axle shafts by the differential. As long as the tires grip the road, providing a resistance to turning, the drivetrain forces the vehicle forward. When one tire encounters a slippery spot on the road, it loses traction, resistance to rotation drops, and the wheel begins to spin. Because resistance has dropped, the torque delivered to both drive wheels changes. The wheel with good traction is no longer driven. If the vehicle is stationary in this situation, only the wheel over the slippery spot rotates. When this is occurring, the differential case is driving the differential pinion gears around the stationary side gear.

This situation places stress on the differential gears. When the wheel spins because of traction loss, the speed of some of the differential gears increases greatly, while others remain idle. The amount of heat developed increases rapidly, the lube film breaks down, metal-to-metal contact occurs, and the parts are damaged. If spinout is allowed to continue long enough, the axle could break. The final drive or differential gears can also be damaged from prolonged spinning of one wheel. This is especially true if the spinning wheel suddenly has traction. The shock of the sudden traction can cause severe damage to the drive axle assembly.

To overcome these problems, differential manufacturers have developed the **limited-slip differential (LSD)**. Limited-slip differentials are manufactured under such names as sure-grip, no-spin, positraction, or equal-lock. Some vehicles use a viscous clutch in their limited-slip drive axles. These units are predominantly used in 4WD vehicles and are discussed in Chapter 41.

Clutch-Based Units

Many LSDs use friction material to transfer the torque applied to a slipping wheel to the one with traction. Those that use a clutch pack **(Figure 38–29)** have two sets (one for each side gear) of clutch plates and friction discs to prevent normal differential action. The friction discs are steel plates with an abrasive coating on both sides. These discs fit over the external splines on the side gears' hub. The clutch plates are also made of steel but have no friction material bonded to them. The plates are placed between the friction discs and fit into internal splines in the differential case. Pressure is kept on the clutch packs by either an S-shaped spring or coil springs.

As long as the friction discs maintain their grip on the steel plates, the differential side gears are locked to the differential case **(Figure 38–30)**, allowing the case and drive axles to rotate at the same speed and preventing one wheel from spinning faster than the other.

A common LSD uses two cone-shaped parts to lock the side gears to the differential case. The cones are located between the side gears and the case and are splined to the side gear hubs. The exterior surface of the cones is coated with a friction material that grabs the inside surface of the case. Four to six coil springs mounted in thrust plates between the side gears maintain a preload on the cones. When the cones are forced against the case, the axles rotate with the differential case.

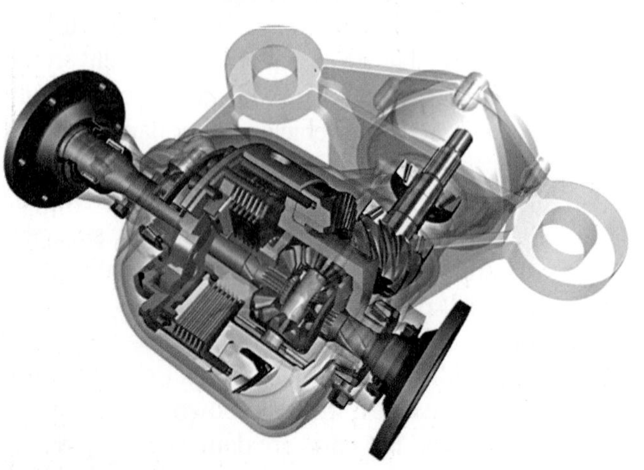

Figure 38-29 A late-model sophisticated LSD with friction clutches. *Courtesy of Dana Corporation*

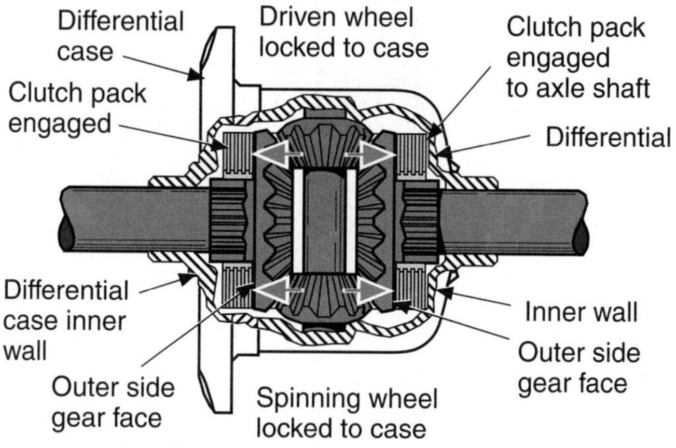

Figure 38-30 Action of the clutches in a limited-slip differential.

The clutch plates and cones are designed to slip when a predetermined amount of torque is applied to them, which allows the vehicle to have differential action when it is turning a corner.

Gear-Based Units

Manufacturers are using a wide range of LSD designs other than the typical clutch type. These designs were born out of the need to improve vehicle stability and tire traction. Many are gear-based and are often called torque-bias or torque-sensing (Torsen) units. The basis of these units is a parallel-axis helical gearset **(Figure 38–31)**. The Torsen differential multiplies the torque available from the wheel that is starting to spin or lose traction and sends it to the slower turning wheel with the better traction. This action is initiated by the resistance between the sets of gears in mesh.

Helical-geared LSDs respond very quickly to changes in traction. They also do not bind in turns and do not lose their effectiveness with wear as clutch-based units can.

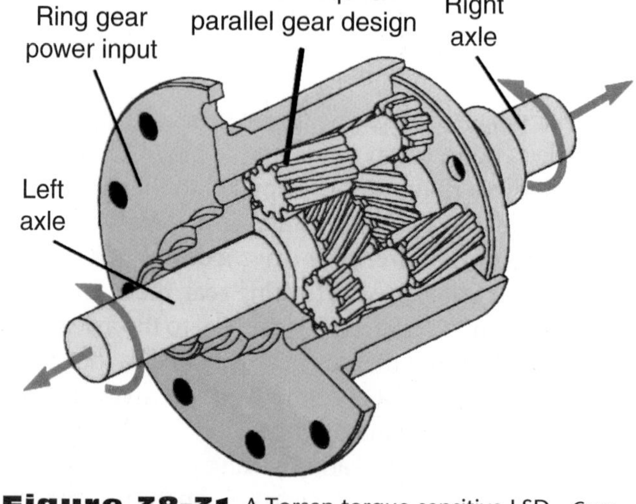

Figure 38-31 A Torsen torque-sensitive LSD. *Courtesy of Zexel Torsen Inc.*

AXLE SHAFTS

The purpose of an axle shaft is to transfer driving torque from the differential assembly to the vehicle's driving wheels. There are two types of axles: dead and live or drive. A **dead axle** does not drive a vehicle. It merely supports the vehicle load and provides a mounting place for the wheels. The rear axle of a FWD vehicle is a dead axle, as are the axles used on trailers.

A **live axle** is one that drives the vehicle. Drive axles transfer torque from the differential to each driving wheel. Depending on the design, rear axles can also help carry the weight of the vehicle or even act as part of the suspension. Three types of driving axles are commonly used **(Figure 38–32)**: semifloating, three-quarter floating, and full-floating.

All three use axle shafts that are splined to the differential side gears. At the wheel ends, the axles can be attached in any one of a number of ways. This attachment defines the type of axle it is and the manner in which the shafts are supported by bearings.

Semifloating Axle Shafts

Semifloating axles help to support the weight of the vehicle. The axles are supported by bearings located in the axle housing. An axle shaft bearing supports the vehicle's weight and reduces rotational friction. The inner ends of the axle shafts are splined to the axle side gears. The axle shafts transmit only driving torque and are not acted upon by other forces. Therefore, the axle shafts are said to be floating.

The driving wheels are bolted to the outer ends of the axle shafts. The outer axle bearings are located between the axle shaft and axle housing. This type of axle has a bearing pressed into the end of the axle housing. This bearing supports the axle shaft. The axle shaft is held in place with either a bearing retainer belted to a flange on the end of the axle housing or by a C-shaped washer that fits into grooves machined in the splined end of the shaft.

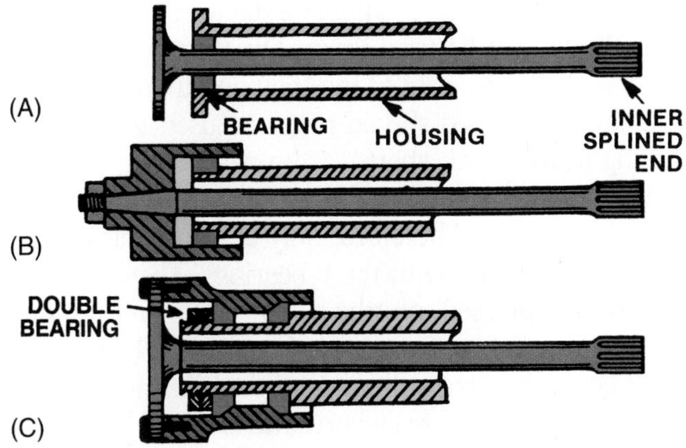

Figure 38–32 The types of rear axle shafts: (A) semifloating; (B) three-quarter floating; and (C) full-floating.

A flange on the wheel end of the shaft is used to attach the wheel.

When semifloating axles are used to drive the vehicle, the axle shafts push on the shaft bearings as they rotate. This places a driving force on the axle housing, springs, and vehicle chassis, moving the vehicle forward. The axle shaft faces the bending stresses associated with turning corners and curves, skidding, and bent or wobbling wheels, as well as the weight of the vehicle. In the semifloating axle arrangement with a C-shaped washer-type retainer, if the axle shaft breaks, the driving wheel comes away from or out of the axle housing.

Three-Quarter Floating Axle

The wheel bearing on a **three-quarter floating axle** is on the outside of the axle housing instead of inside the housing as in the semifloating axle. The wheel hubs are bolted to the end of the axle shaft and are supported by the bearing. In this arrangement, the axle shaft only supports 25% of the vehicle's weight. The weight is transferred through the wheel hub and bearing to the axle housing. Three-quarter floating axles are found on older vehicles and some trucks.

Full-Floating Axle Shafts

Most medium- and heavy-duty vehicles use a **full-floating axle shaft**. This design is similar to the three-quarter floating axle except that two bearings rather than one are used to support the wheel hub. These are slid over the outside of the axle housing and carry all of the stresses caused by torque loading and turning. The wheel hubs are bolted to flanges on the outer end of each axle shaft.

In operation, the axle shaft transmits only the driving torque. The driving torque from the axle shaft rotates the axle flange, wheel hub, and rear driving wheel. The wheel hub forces its bearings against the axle housing to move the vehicle. The stresses caused by turning, skidding, and bent or wobbling wheels are taken by the axle housing through the wheel bearings. If a full-floating axle shaft should break, it can be removed from the axle housing. Because the rear wheels rotate around the rear axle housing the disabled vehicle can be towed to a service area for replacement of the axle shaft.

Independently Suspended Axles

In an independently suspended axle system, the driving axles are usually open instead of being enclosed in an axle housing. The two most common suspended rear driving axles are the DeDion axle system and the swing axle system.

The DeDion axle system resembles a normal driveline. The driving axles look like a drive shaft with U-joints at each end of the axles. A slip joint is attached to the innermost U-joint. The outboard U-joint is connected to the wheel hub, which allows the driving axle to move up and down as it rotates.

On vehicles that use a swing axle, the driving axle shafts can be open or enclosed. An axle fits into the differential by way of a ball-and-socket system. The ball-and-socket system allows the axle to pivot up and down. As the axle pivots, the driving wheel swings up and down. This system best describes the drive axles of a FWD vehicle.

Axle Shaft Bearings

The axle shaft bearing supports the vehicle's weight and reduces rotational friction. In an axle mount, radial and thrust loads are always present on the axle shaft bearing when the vehicle is moving. Radial bearing loads act at 90 degrees to the axle shaft's center of axis. **Radial loading** is always present whether or not the vehicle is moving.

Thrust loading acts on the axle bearing parallel with the center of axis. It is present on the driving wheels, axle shafts, and axle bearings when the vehicle turns corners or curves.

There are three designs of axle shaft bearings used in semifloating axles: ball-type bearing, straight roller bearing, and tapered roller bearing.

The bearing load of primary concern is axle shaft end thrust. When a vehicle moves around a corner, centrifugal force acts on the vehicle body, causing it to lean to the outside of the curve. The vehicle's chassis does not lean because of the tires' contact with the road's surface. As the body leans outward, a thrust load is placed on the axle shaft and axle bearing. Each type of axle shaft bearing handles end thrust differently.

Normally, the way the axles are held in the housing is quite obvious after the rear wheels and brake assemblies have been removed. If the axle shaft is held in by a retainer and three or four bolts, it is not necessary to remove the differential cover to remove the axle. Most ball and tapered roller bearing supported axle shafts are retained in this manner **(Figure 38–33)**. To remove the axle, remove the bolts that hold the retainer to the backing plate, then pull the axle out. Normally, the axle shaft slides out without the aid of a puller. Sometimes a puller is required.

A straight-roller bearing supported axle shaft does not use a retainer to secure it. Rather, a C-shaped washer is used to retain the axle shaft **(Figure 38–34)**. This C-shaped washer is located inside the differential, and the differential cover must be removed to gain access to it. To remove this type of axle, first remove the wheel, brake drum, and differential cover. Then, remove the differential pinion shaft retaining bolt and differential pinion shaft. Now push the axle shaft in and remove the C-shaped washer. The axle can now be pulled out of the housing.

Ball bearings are lubricated with grease packed in the bearing at the factory. An inner seal, designed to keep the gear oil from the bearing, rides on the axle shaft just in front of the retaining ring. This type of bearing also has an outer seal to prevent grease from spraying onto the rear brakes. Ball-type axle bearings are pressed on and off the axle shaft. The retainer ring is made of soft metal and is pressed onto the shaft against the wheel bearing. Never use a torch to remove the ring. Rather, drill into it or notch it in several places with a cold chisel to break the seal **(Figure 38–35)**. The ring can then be slid off the shaft easily. Heat should not be used to remove the ring because it can take the temper out of the shaft and thereby weaken it. Likewise, a torch should never be used to remove a bearing from an axle shaft.

Roller axle bearings are lubricated by the gear oil in the axle housing. Therefore, only a seal to protect the brakes is necessary with these bearings. These bearings are typically pressed into the axle housing and not onto the axle. To remove them, the axle must first be removed and then the bearing pulled out of the housing. With the axle out, inspect the area where it rides on the bearing for pits or scores. If pits or score marks are present, replace the axle.

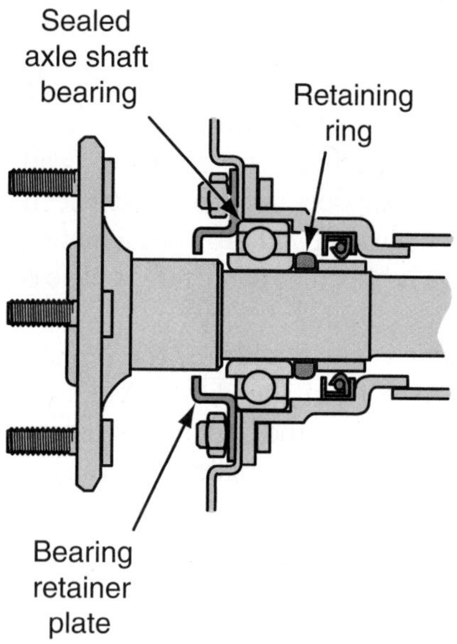

Figure 38–33 The location of an axle bearing retainer.

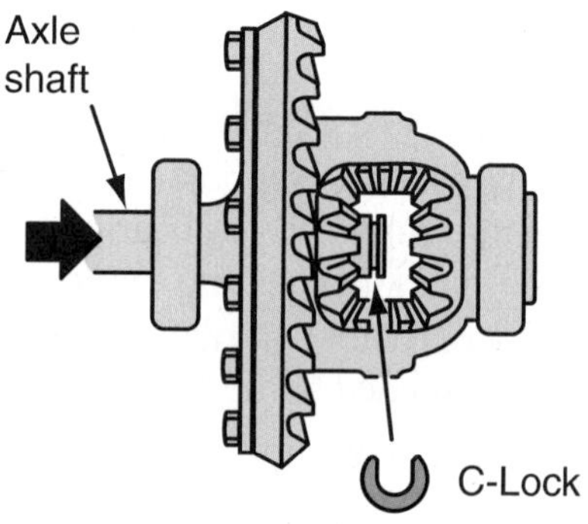

Figure 38–34 The location of C-lock-type axle shaft retainers.

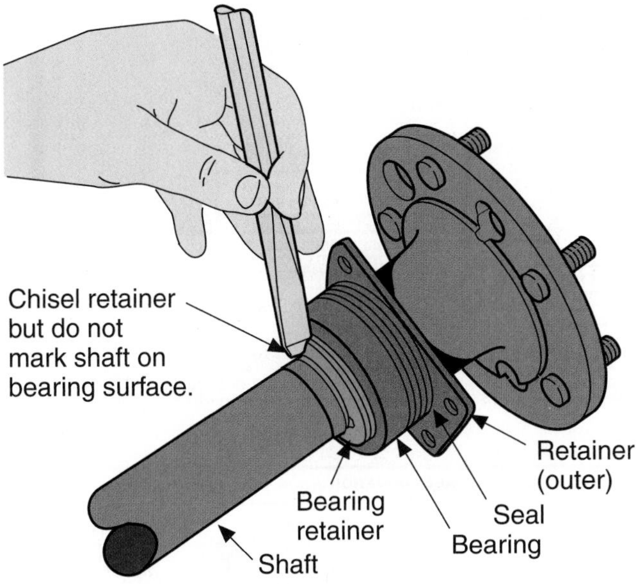
Chisel retainer but do not mark shaft on bearing surface.
Retainer (outer)
Seal
Bearing
Bearing retainer
Shaft

Figure 38-35 Freeing the retainer ring from an axle shaft. *Reprinted with permission*

Tapered-roller axle bearings are not lubricated by gear oil. They are sealed and lubricated with wheel grease. This type of bearing uses two seals and must be pressed on and off the axle shaft using a press. After the bearing is pressed onto the shaft, it must be packed with wheel bearing grease. After packing the bearing, install the axle in the housing. Shaft endplay must be checked. Use a dial indicator and adjust the endplay to the specifications given in the service manual. If the endplay is not within specifications, change the size of the bearing shim.

The installation of new axle shaft seals is recommended whenever the axle shafts have been removed. Some axle seals are identified as being either right or left side. When installing new seals, make sure to install the correct seal in each side. Check the seals or markings of right or left or for color coding.

USING SERVICE MANUALS

The driveline can create some especially difficult diagnostic problems. The driveline easily picks up vibrations and noises from other parts of the vehicle. A test drive is the best way to begin diagnosis. Note what happens during any of these speeds or during speed changes. ■

Diagnosis

Before the test drive, check out the test-drive chart **(Figure 38–36)**. The four modes of driving given in most service manuals should be checked out for driving-axle and differential problems. For the drive mode, accelerate the vehicle. The throttle must be depressed enough to apply sufficient engine torque. In the cruise mode, vehicle speed must be constant, which means that the throttle must be applied at all times. The speed must be held at a predetermined rpm on a level road. For the coast mode, take the foot off the throttle. Let the vehicle coast down from a specific speed. The float mode is a controlled deceleration. Back off the throttle gradually and continually. Do not brake or accelerate during this mode.

Remember that driving safely is always important! Hard cornering or sudden braking should be avoided. Driving conditions other than normal can worsen a problem or create a new one. Carelessness during the test drive can result in an accident. Remember, you are responsible for someone else's vehicle.

SERVICING THE FINAL DRIVE ASSEMBLY

Before removing a final drive unit for service, make sure it needs to be serviced. Typically, problems with the differential and drive axles are first noticed as a leak or noise. As the problem worsens, vibrations or a clunking noise might be felt during certain operating conditions. Diagnosis of the problem should begin with a road test in which the vehicle is taken through the different modes of operation.

Basic Diagnosis

It is common for the source of a noise to be tires, not the final drive unit. To make sure the noises are not caused by tire tread patterns and/or wear, drive the car on various types of road surfaces (asphalt, concrete, and packed dirt). If the noise changes with the road surfaces, it means the tires are the cause of the noise.

Another way to isolate tire noises is to coast at speeds less than 30 mph (48 km/h). If the noise is still heard, the tires are probably the cause. Drive axle and differential noises are less noticeable at these speeds. Accelerate and compare the sounds to those made while coasting. Drive axle and differential noises change. Tire noise remains constant.

Sometimes it is difficult to distinguish between axle-bearing noises and noises coming from the differential. Differential noises often change with the driving mode whereas axle-bearing noises are usually constant. The sound of the bearing noise usually increases in speed and loudness as vehicle speed increases.

Operational noises are generally caused by bearings or gears that are worn, loose, or damaged. Bearing noises might be a whine or a rumble. A whine is a high-pitched, continuous "whee" sound. A rumble sounds like distant thunder.

Gears can also whine or emit a howl—a very loud, continuous sound. Howling is often caused by low lubricant in the drive axle housing. The meshing teeth scrape metal from each other and can be heard in all gear ranges.

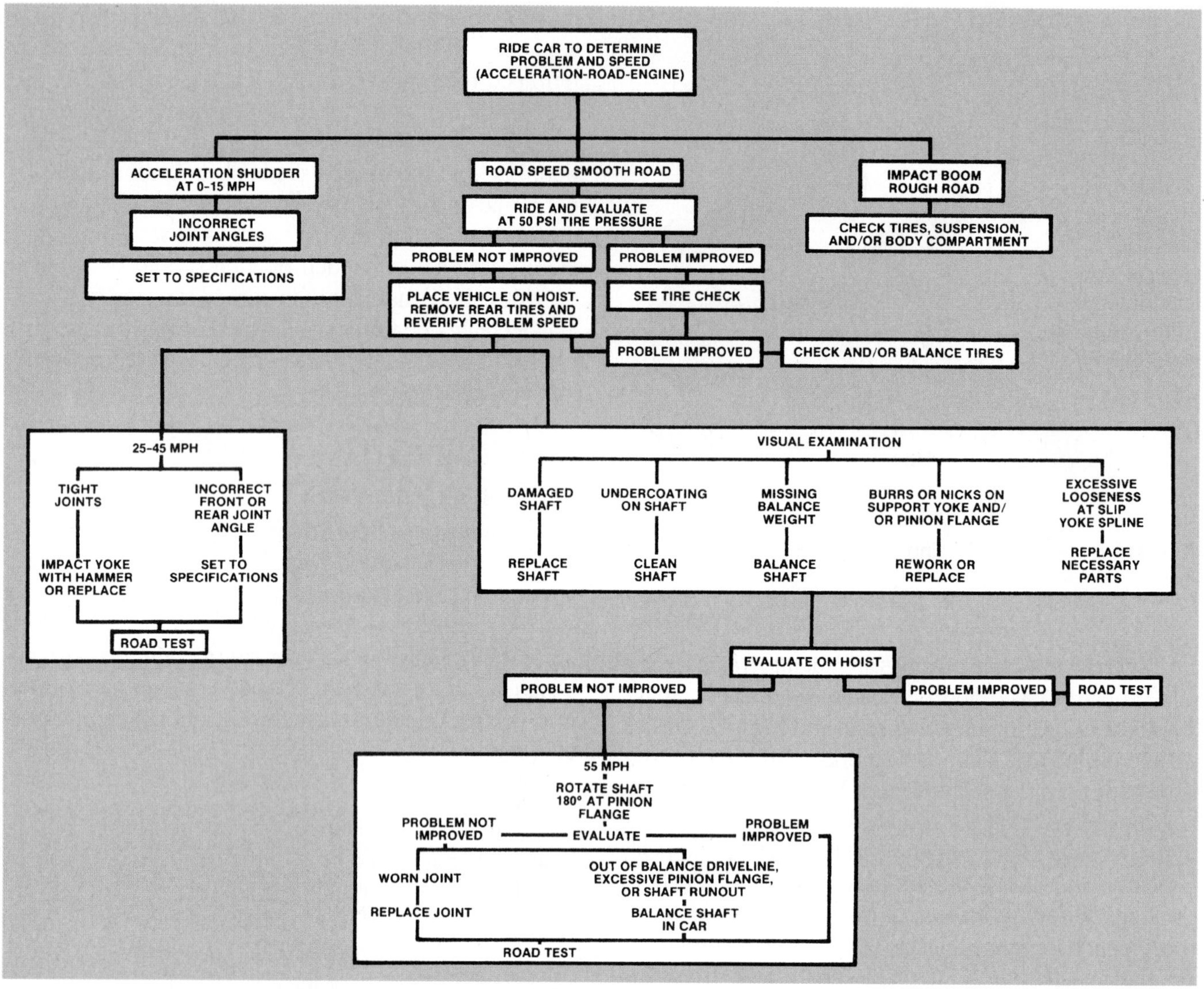

Figure 38-36 A typical test-drive troubleshooting chart.

If topping up the lubrication level does not alleviate the howling noise, then the drive pinion and ring gear must be replaced.

Disassembly

Although FWD axle final drive units are normally an integral part of the transaxle, most of the procedures for servicing RWD units apply to them. To service a final drive assembly in removable carrier housing, the unit must be removed from the housing. Units in integral carriers are serviced in the housing.

A highly important step in the procedure for disassembling any final drive unit is a careful inspection of each part as it is removed. The bearings should be looked at and felt to determine if there are any defects or evidence of damage.

After the ring and pinion gears have been inspected and before they have been removed from the assembly, check the side play. Using a screwdriver, attempt to move the differential case assembly laterally. Any movement is evidence of side play. Side play normally indicates that as the result of loose bearing cones on the differential case hubs, the differential case must be replaced.

Prior to disassembling the assembly, measure the runout of the ring gear. Excessive runout can be caused by a warped gear, worn differential side bearings, warped differential case, or particles trapped between the gear and case. Runout is checked with a dial indicator mounted on the carrier assembly. The plunger on the indicator should be set at a right angle to the gear. With the dial indicator in position and its dial set to zero, rotate the ring gear and note the highest and lowest readings. The difference between these two readings indicates the total runout of the ring gear. Normally, the maximum permissible runout is 0.003 to 0.004 inches (.0762 to .1016 mm).

To determine if the runout is caused by a damaged differential case, remove the ring gear and measure the runout of the ring gear mounting surface on the differential case. Runout should not exceed 0.004 inch (.1016 mm). If runout is greater than that, the case should be replaced. If the runout was within specifications, the ring gear is probably warped and should be replaced. A ring gear is never replaced without replacing its mating pinion gear.

Some ring gear assemblies have an excitor ring, used in antilock brake systems. This ring is normally pressed onto the ring gear hub and can be removed after the ring gear is removed. If the ring gear assembly is equipped with an excitor ring, carefully inspect it and replace it if it is damaged.

Prior to disassembling the unit, the drive shaft must be removed. Before disconnecting it from the pinion's companion flange, locate the shaft-to-pinion alignment marks. If they are not evident, make new ones. This avoids assembling the unit with the wrong index, which can result in driveline vibration.

During disassembly, keep the right and left shims, cups, and caps separated. If any of these parts are reused, they must be installed on the same side as they were originally located.

Assembly

When installing a ring gear onto the differential case, make sure the bolt holes are aligned before pressing the gear in place. While pressing the gear, pressure should be evenly applied to the gear. Likewise, when tightening the bolts, always tighten them in steps and to the specified torque. These steps reduce the chances of distorting the gear.

Examine the gears to locate any timing marks on the gearset that indicate where the gears were lapped by the manufacturer. Normally, one tooth of pinion gear is grooved and painted, while the ring gear has a notch between two painted teeth. If the paint marks are not evident, locate the notches. Proper timing of the gears is set by placing the grooved pinion tooth between the two marked ring gear teeth. Some gearsets have no timing marks. These gears are hunting gears and do not need to be timed. Nonhunting and partial nonhunting gears must be timed.

Whenever the ring and pinion gears or the pinion or differential case bearings are replaced, pinion gear depth, pinion bearing preload, and the ring and pinion gear tooth patterns and backlash must be checked and adjusted. This holds true for all types of differentials except most FWD differentials that use helical-cut gears, and taking tooth patterns is not necessary. Nearly all other final drive units use hypoid gears that must be properly adjusted to ensure a quiet operation.

Pinion gear depth is adjusted with shims placed behind the pinion bearing **(Figure 38–37)** or in the housing. The thickness of the drive pinion rear bearing shim controls the depth of the mesh between the pinion and ring gear. To determine and set pinion depth a special tool is normally used to select the proper pinion shim **(Figure 38–38)**. Always follow the procedures in the service manual when setting up the tool and determining the proper shim.

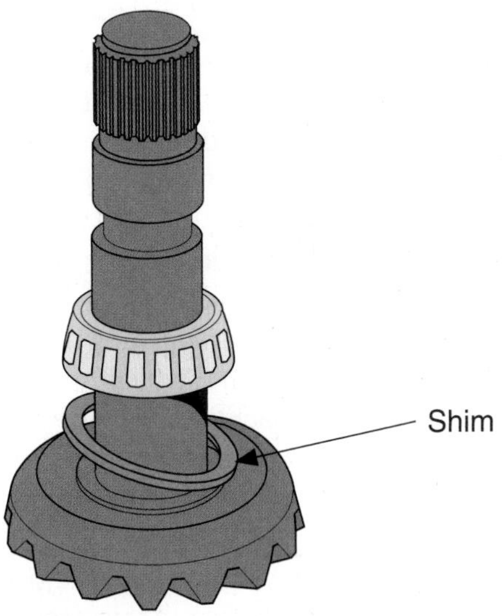

Figure 38–37 The typical placement of a pinion gear depth shim. *Reprinted with permission*

Pinion bearing **preload** is set by tightening the pinion nut until the desired number of inch-pounds is required to turn the shaft. Tightening the nut crushes the collapsible pinion spacer, which maintains the desired preload. Never overtighten and then loosen the pinion nut to reach the desired torque reading. Tightening and loosening the pinion nut damages the collapsible spacer. It must then be replaced. For the exact procedures and specifications for bearing preload, refer to the service manual. Incorrect bearing preload can cause differential noise. Some cases use shims to set pinion bearing preload.

It is recommended that a new pinion seal be installed whenever the pinion shaft is removed from the differen-

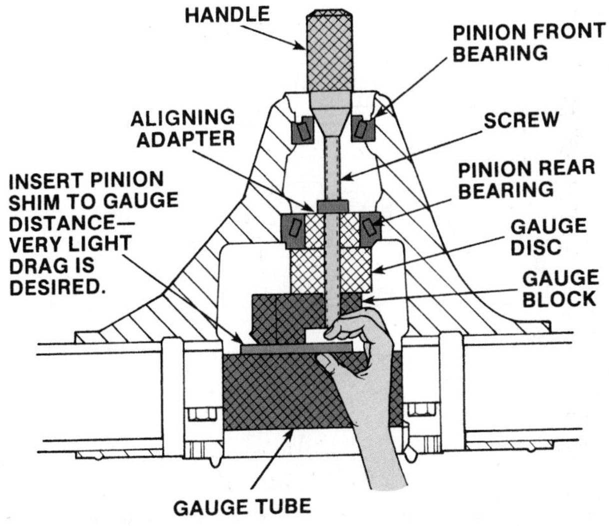

Figure 38–38 A special tool for measuring proper pinion gear depth.

tial. To install a new seal, thoroughly lubricate it and press it in place with an appropriate seal driver.

Backlash of the gearset is adjusted at the same time as the side-bearing preload. Side-bearing preload limits the amount the differential is able to move laterally in the axle housing. Adjusting backlash sets the depth of the mesh between the ring and pinion gear teeth. Both of these are adjusted by shim thickness or by the adjustments made by the side-bearing adjusting nuts. Photo Sequence 40 goes through the typical procedure for measuring and adjusting backlash and side-bearing preload on a gearset that uses shims for adjustment. Photo Sequence 41 covers the same steps for a unit that has adjusting nuts.

A typical procedure for measuring and adjusting backlash and preload involves rocking the ring gear and measuring its movement with a dial indicator. Compare measured backlash with the specifications. Make the necessary adjustments. Then recheck the backlash at four points equally spaced around the ring gear. Normally, backlash should be less than 0.004 inch (.1016 mm).

The pattern of gear teeth determines how quietly two meshed gears run. The pattern also describes where on the faces of the teeth the two gears mesh. The pattern should be checked during teardown for gear noise diagnosis, after adjusting backlash and side-bearing preload, or after replacing the drive pinion and setting up the pinion bearing preload. The terms commonly used to describe the possible patterns on a ring gear and the necessary corrections are shown in **Figure 38–39**.

To check the gear tooth pattern, paint several ring gear teeth with nondrying Prussian blue, ferric oxide, or red or white lead marking compound. White marking compound is preferred by many technicians because it tends to be more visible than the others are. Use the pinion gear

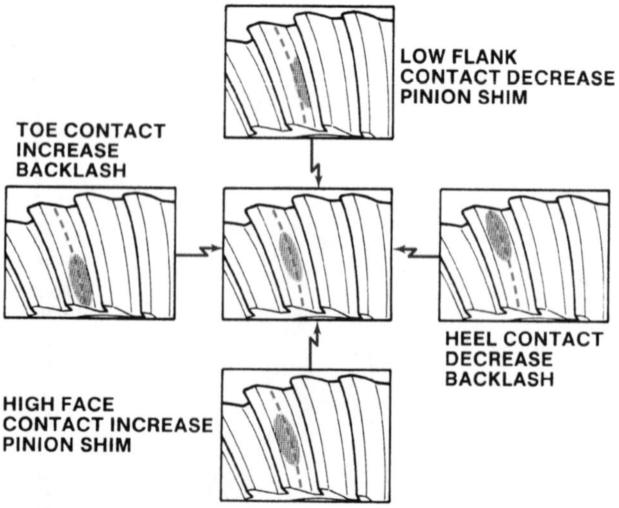

Figure 38–39 Commonly used terms for describing the possible patterns on a ring gear with the recommended corrections. *Courtesy of Ford Motor Company*

yoke or companion flange to rotate the ring gear. This will preload the ring gear while it is rotating and will simulate vehicle load. Rotate the ring gear so the painted teeth contact the pinion gear. Move it in both directions enough to get a clearly defined pattern. Examine the pattern on the ring gear and make the necessary corrections.

Most new gearsets purchased today come with a pattern prerolled on the teeth. This pattern provides the quietest operation for that gearset. Never wipe this pattern off or cover it up. When checking the pattern on a new gearset, only coat half of the ring gear with the marking compound and compare the pattern with the prerolled pattern.

Maintenance

Maintenance includes inspecting the level of and changing the gear lubricant, and lubricating the U-joints if they are equipped with zerk or grease fittings. Most modern U-joints are of the extended life design, meaning they are sealed and require no periodic lubrication. However, it is wise to inspect the joints for hidden grease plugs or fittings.

Proper lubrication is necessary for drive axle durability. Different applications require different gear lubes. The American Petroleum Institute (API) has established a rating system for the various gear lubes available. In general, rear axles use either SAE 80- or 90-weight gear oil for lubrication, meeting API GL-4 or GL-5 specifications. With limited-slip axles, it is very important that the proper gear lube be used. Most often, a special friction modifier fluid should be added to the fluid. If the wrong lubricant is used, damage to the clutch packs and grabbing or chattering on turns will result. If this condition exists, try draining the oil and refilling with the proper gear lube before servicing it.

DIAGNOSING DIFFERENTIAL NOISES

If a whining is heard when turning corners or rounding curves, the problem might be damaged differential pinion gears and pinion shaft. This damage is caused when the inside diameter of the differential pinions and the outside diameter of the differential pinion shaft is scored and damaged. The damage is usually caused by allowing one driving wheel to revolve at high speeds while the opposite wheel remains stationary.

Another gear noise that is common in differentials is the chuckle. A chuckle is a low "heh-heh" sound that occurs when gears are worn to the point where there is excessive clearance between the pinion gear and the ring gear. Chuckle sounds occur most often in the decelerating mode, particularly below 40 mph (65 km/h). As the vehicle decelerates, the chuckle also slows and can be heard all the way to a stop.

PHOTO SEQUENCE 40

Measuring and Adjusting Backlash and Side-Bearing Preload on a Final Drive Assembly with a Shim Pack

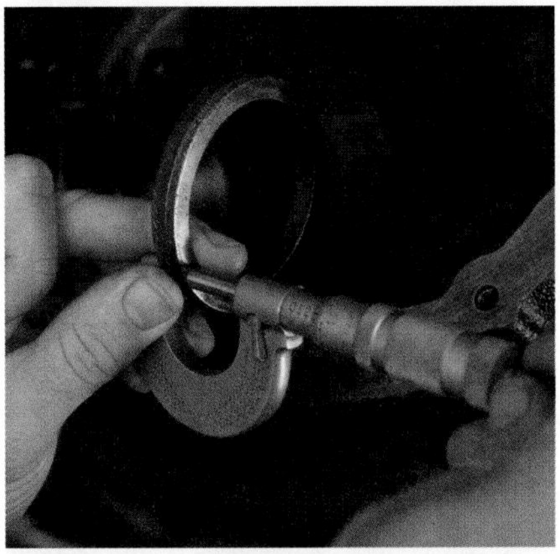

P40-1 *Measure the thickness of the original side bearing preload shims.*

P40-2 *Install the differential case into the housing.*

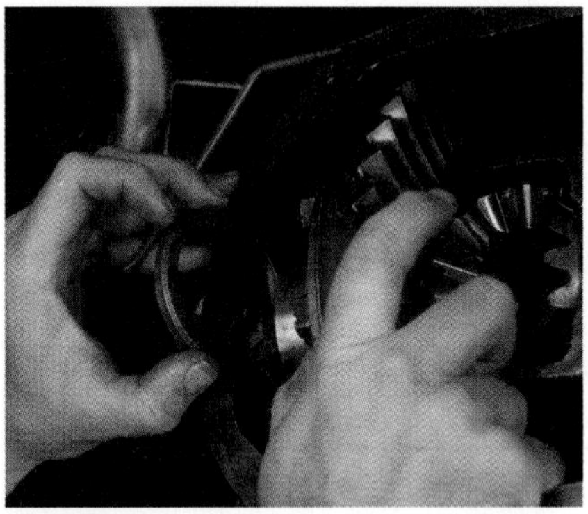

P40-3 *Install service spacers that are the same thickness as the original preload shims between each bearing cup and the housing.*

P40-4 *Install the bearing caps and finger tighten the bolts.*

P40-5 *Mount a dial indicator to the housing so that the button of the indicator touches the face of the ring gear. Using two screwdrivers, pry between the shims and the housing. Pry to one side and set the dial indicator to zero, then pry to the opposite side and record the reading.*

P40-6 *Select two shims with a combined thickness to that of the original shims plus the indicator reading, then install them.*

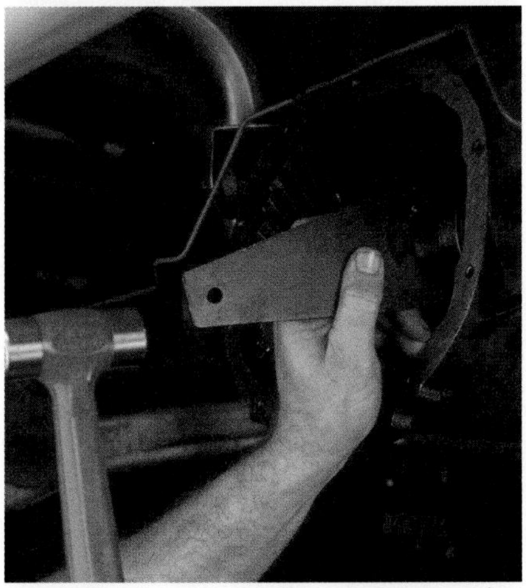

P40-7 *Using the proper tool, drive the shims into position until they are fully seated.*

P40-8 *Install and tighten the bearing caps to specifications.*

P40-9 *Check the backlash and preload of the gearset. Check the backlash by rocking the ring gear and noting the movement on the dial indicator. Adjust the shim pack to allow for the specified backlash. Recheck the backlash at four points equally spaced around the ring gear.*

PHOTO SEQUENCE 41 Measuring and Adjusting Backlash and Side-Bearing Preload on a Final Drive Assembly with Adjusting Nuts

P41-1 *Lubricate the differential bearings, cups, and adjusters.*

P41-2 *Install the differential case into the housing.*

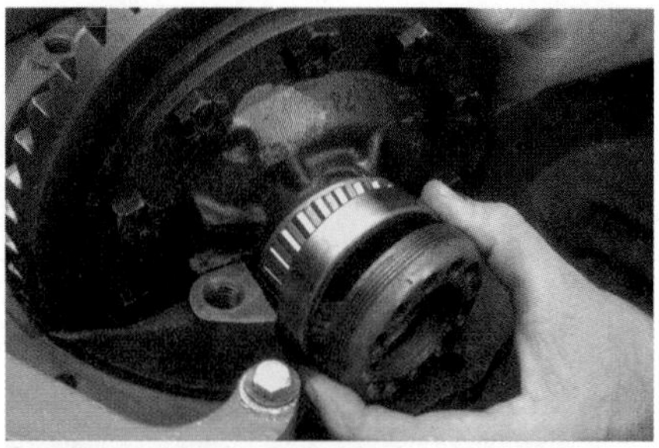

P41-3 *Install the bearing cups and adjusting nuts onto the differential case.*

P41-4 *Snugly tighten the top bearing cup bolts and finger tighten the lower bolts.*

P41-5 *Turn each adjuster until bearing free play is eliminated with little or no backlash present between the ring and pinion gears.*

P41-6 *Seat the bearings by rotating the pinion several times each time the adjusters are moved.*

P41-7 *Install a dial indicator and position the plunger against the drive side of the ring gear. Set the dial to zero. Using two screwdrivers, pry between the differential case and the housing. Observe the reading.*

P41-8 *Determine how much the preload needs to be adjusted and set the preload by turning the right adjusting nut.*

P41-9 *Check the backlash by rocking the ring gear and noting the movement on the dial indicator.*

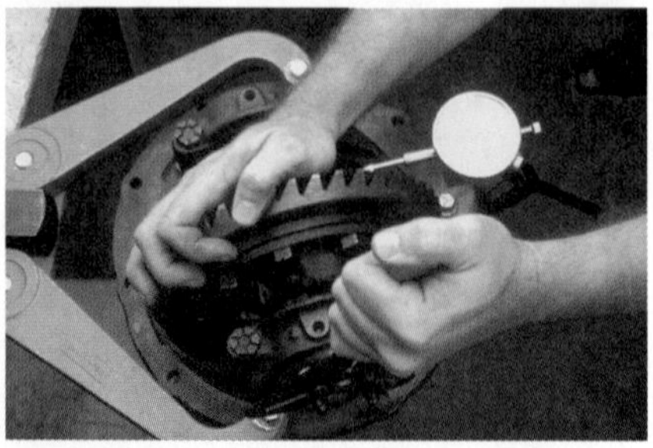

P41-10 *Adjust the backlash by turning both adjusting nuts the same amount so that the preload adjustment remains unchanged.*

P41-11 *Install the locks on the adjusting nuts.*

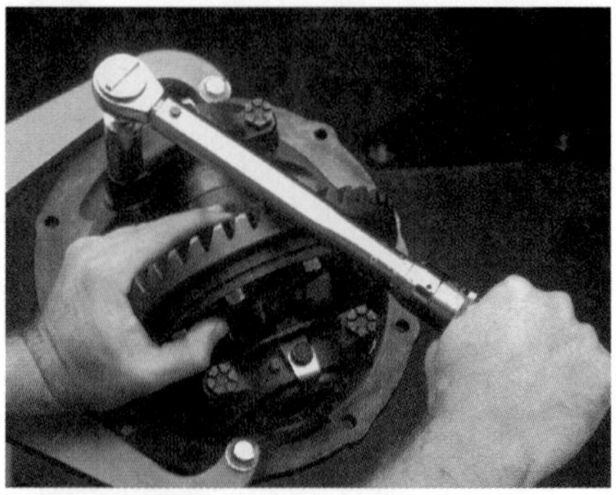

P41-12 *Tighten the bearing cup bolts to the specified torque.*

A knock or clunk is caused by excessive wear or loose or broken parts. A knock is a repetitious rapping sound that occurs during all phases of driving but is most noticeable during acceleration and deceleration when the gears are loaded.

A clunk is a sharp, loud noise caused by one part hitting another. Unlike a knock, a clunk can be felt as well as heard. Clunks are generally caused by loose parts striking each other.

Limited-slip clutch packs or cones that need servicing might be heard as a chatter or a rapid clicking noise that creates a vibration in the vehicle. Chattering is usually noticed when rounding a corner. A change of differential lubricant and adding friction modifier to the fluid sometimes corrects this problem. After draining the oil, replace it with the manufacturer's suggested friction modifier and lubricant. Road test the vehicle again.

To make sure that the noise heard during the test drive is coming from the differential, stop the vehicle and shift the transmission into neutral. Run the engine at various rpm levels. If the noise is heard during this procedure, it is caused by a problem somewhere other than in the differential.

Vibration Problems

Often the source of vibration is a bent axle or axle flange, or improper mounting of the wheel to the flange. To check the runout of the flange, position a dial indicatior against the outer flange surface of the axle. Apply slight pressure to the center of the axle to remove the endplay in the axle, and then zero the indicator. Slowly rotate the axle one complete revolution and observe the readings on the indicator. The total amount of indicator movement is the total amount of axle flange lateral runout. Compare the measured runout with specifications.

Inspect the wheel studs in the axle flange. If they are broken or bent, they should be replaced. Also check the condition of the threads. If they have minor distortions, run a die over the stud. If the threads are severely damaged, the stud should be replaced. Studs are normally pressed in and out of the flange. Make sure you install the correct size.

CASE STUDY

A customer brings his subcompact, front-wheel-drive car into the shop complaining of recurring noise in the front wheels. The noise is most noticeable when the car is making turns. The owner states he had a similar complaint a few months ago and the shop had replaced an outer CV joint. This corrected the problem until recently. The customer suspects that the same joint went bad again and demands that the shop replace it, free of charge, because it is obvious that the replacement joint was defective.

The service writer records the information from the customer and tells him that he will be notified as soon as the problem is diagnosed. As soon as the customer leaves, the service writer looks up the customer's file and finds that a CV joint had been replaced two months ago.

The service writer gives the repair order to the technician, along with the old repair order. The technician begins the diagnostic procedure with a test drive to verify the complaint. From the test drive and a visual inspection, the technician concludes that the same CV joint is faulty. What could cause the joint to fail so soon? Was the replacement joint defective? Was the replacement joint installed incorrectly? Is some other fault causing the joints to fail? No matter what the answer, it seems that the customer will not be charged for this repair. Also, it is likely that the technician will not get paid for the repair.

Upon disassembly of the axle, the technician finds the joint's lubricant to be contaminated with metal shavings and moisture. A thorough inspection of the boot reveals no tears or punctures. While inspecting the boot, it is noticed that the inner end of the boot moves freely on the axle shaft. The technician knows then what had caused the contamination and resulting premature failure of the joint.

When installing the replacement joint and boot, the technician failed to properly tighten the inner boot clamp. This allowed lubricant to leak from and water to enter the boot. A new joint and boot is installed on the axle. The technician verifies the repair by a test drive. The replacement joint took care of the noise. Before releasing the car back to the owner, the technician rechecks the position and tightness of the boot clamps. The customer is called and is told what had happened. Although the customer is not happy about the mistake, he appreciates the honesty of the technician. Two months later he returned to the shop for an oil change. He has been a regular customer ever since.

KEY TERMS

Backlash
Ball-type joint
Cross groove joint
Dead axle
Double-Cardan joint
Double-offset joint
Fixed joint
Full-floating axle shaft

Hunting gearset
Hypoid gear
Inboard joint
Integral housing
Limited-slip differential (LSD)
Live axle
Nonhunting gearset
Operating angle
Outboard joint
Partial nonhunting gearset
Pinion gear
Pinion shaft
Plunge joint
Preload
Radial loading
Ring gear
Rzeppa joint
Semifloating axle
Slip yoke
Three-quarter floating axle
Thrust loading
Torsional damper
Tripod-type joint

SUMMARY

- FWD axles generally transfer engine torque from the transaxle to the front wheels.
- Constant velocity (CV) joints provide the necessary transfer of uniform torque and a constant speed while operating through a wide range of angles.
- In FWD drivetrains, two CV joints are used on each half shaft. The different types of joints can be referred to by position (inboard or outboard) by function (fixed or plunge), or by design (ball-type or tripod).
- Front-wheel-drive half shafts can be solid or tubular, of equal or unequal length, and with or without damper weights.
- Most problems with front-wheel-drive systems are noted by noise and vibration.
- Lubricant is the most important key to a long life for the CV joint.
- A U-joint is a flexible coupling located at each end of the drive shaft between the transmission and the pinion flange on the drive axle assembly.
- A U-joint allows two rotating shafts to operate at a slight angle to each other; this is important to RWD vehicles.
- A failed U-joint or damaged drive shaft can exhibit a variety of symptoms. A clunk that is heard when the transmission is shifted into gear is the most obvious. You can also encounter unusual noise, roughness, or vibrations.
- A differential is a geared mechanism located between the driving axle shafts of a vehicle. Its job is to direct power flow to the driving axle shafts. Differentials are used in all types of power trains.
- The differential performs several functions. It allows the drive wheels to rotate at different speeds when negotiating a turn or curve in the road, and the differential drive gears redirect the engine torque from the drive shaft to the rear-drive axles.
- The final drive and differential of a RWD vehicle is housed in the axle housing, or carrier housing.
- The purpose of the axle shaft is to transfer driving torque from the differential and final drive assembly to the vehicle's driving wheels.
- There are three types of driving axle shafts commonly used: semifloating, three-quarter floating, and full-floating.
- Axle shaft bearings may support the vehicle's weight but always reduce rotational friction.
- Problems with the differential and drive axle shafts are usually first noticed as a leak or noise. As the problem progresses, vibrations or a clunking noise might be felt in various modes of operation.

TECH MANUAL

The following procedures are included in Chapter 38 of the *Tech Manual* that accompanies this book:

1. Inspecting and diagnosing a drive axle.
2. Servicing outer CV joints.
3. Servicing inner CV joints.
4. Inspecting U-joints and the drive shaft.
5. Road testing differential noises.
6. Measuring and adjusting pinion depth, bearing preload, and backlash.

REVIEW QUESTIONS

1. Name the three ways in which CV joints can be classified.
2. What type of axle housing resembles a banjo?
3. What type of axle merely supports the vehicle load and provides a mounting place for the wheels?
4. What type of floating axle has one wheel bearing per wheel on the outside of the axle housing?
5. How are problems with the differential and drive axles usually first noticed?
6. Constant velocity joints are used in ______.
   a. front-wheel-drive vehicles
   b. rear-wheel-drive vehicles
   c. rear-wheel-drive vehicles with independent rear suspension
   d. all of the above
7. In front-wheel drivetrains, the CV joint nearer the transaxle is the ______.
   a. inner joint
   b. inboard joint
   c. outboard joint
   d. both a and b
8. A CV joint that is capable of in and out movement is a ______.
   a. plunge joint
   b. fixed joint
   c. inboard joint
   d. both A and C
9. Technician A says that a gear tooth pattern identifies ring gear runout. Technician B says that gear patterns are not accurate if there is excessive ring gear runout. Who is correct?
   a. Technician A
   b. Technician B
   c. Both A and B
   d. Neither A nor B
10. The double-offset joint is typically used in applications that require ______.
    a. higher operating angles and greater plunge depth
    b. lower operating angles and lower plunge depth
    c. higher operating angles and lower plunge depth
    d. lower operating angles and greater plunge depth
11. Which type joint has a flatter design than any other?
    a. double-offset
    b. disc
    c. cross groove
    d. fixed tripod
12. Which of these is the best way to determine which CV joint is faulty?
    a. a squeeze test
    b. a runout test
    c. a visual inspection
    d. a road test
13. The single Cardan/Spicer universal joint is also known as the ______.
    a. cross joint
    b. four-point joint
    c. both a and b
    d. neither a nor b
14. The drive shaft component that provides a means of connecting two or more shafts together is the ______.
    a. pinion flange
    b. U-joint
    c. yoke
    d. biscuit
15. Large cars with long drive shafts often use a double-U-joint arrangement called a ______.
    a. Spicer style U-joint
    b. constant velocity U-joint
    c. Cleveland style U-joint
    d. none of the above
16. Which type of driving axle supports the weight of the vehicle?
    a. semifloating
    b. three-quarter floating
    c. full-floating
    d. none of the above
17. Technician A says that limited-slip differential clutch packs are designed to slip when the vehicle turns a corner. Technician B says that a special additive is placed in the hypoid gear lubricant to promote clutch pack slippage on corners. Who is correct?
    a. Technician A
    b. Technician B
    c. Both A and B
    d. Neither A nor B
18. Technician A says side bearing preload limits the amount the differential case is able to move laterally in the axle housing. Technician B says that adjusting backlash sets the depth of the mesh between the ring and pinion gears' teeth. Who is correct?
    a. Technician A
    b. Technician B
    c. Both A and B
    d. Neither A nor B
19. Technician A says that a hunting gearset is one in which one drive pinion gear tooth contacts only certain ring gear teeth. Technician B says that a partial nonhunting gearset is one in which one pinion tooth contacts only six ring gear teeth. Who is correct?
    a. Technician A
    b. Technician B
    c. Both A and B
    d. Neither A nor B
20. Which of the following describes the double-Cardan universal joint?
    a. It is most often installed in front-engine rear-wheel-drive luxury automobiles.
    b. It is classified as a constant velocity U-joint.
    c. A centering ball socket is inside the coupling yoke between the two universal joints.
    d. All of the above.

SECTION 6

Automatic Transmissions and Transaxles

Many vehicles are equipped with automatic transmissions or transaxles. These systems perform the same job as standard or manual shift units, but they do so in quite a different manner. A torque converter replaces the clutch as the link between the engine and transmission/transaxle gearing. The planetary gear trains found in automatic units generate speed and torque using a completely different set of operating principles. Finally, the hydraulic and electronic control systems used to initiate shifting are unique to automatic transmissions/transaxles.

The information found in Section 6 corresponds to materials covered on the ASE certification test on automatic transmissions/transaxles.

Delmar's Automatic Transmissions and Transaxles Video Series and CD-ROM Courseware

WE ENCOURAGE PROFESSIONALISM

ASE

THROUGH TECHNICIAN CERTIFICATION

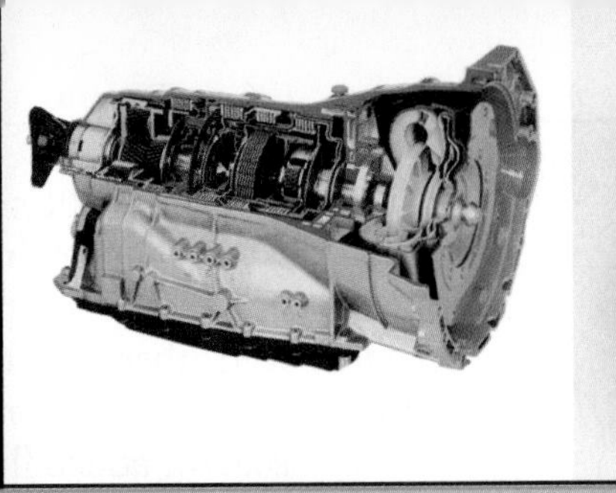

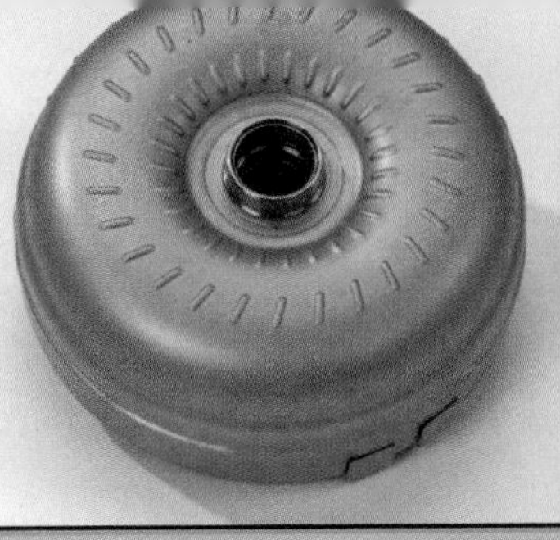

39

AUTOMATIC TRANSMISSIONS AND TRANSAXLES

OBJECTIVES

■ Explain the basic design and operation of standard and lockup torque converters. ■ Describe the design and operation of a simple planetary gearset and Simpson gear train. ■ Name the major types of planetary gear controls used on automatic transmissions and explain their basic operating principles. ■ Describe the construction and operation of common Simpson gear-train-based transmissions and transaxles. ■ Describe the construction and operation of common Ravigneaux gear-train-based transmissions. ■ Describe the construction and operation of transaxles that use planetary gearsets in tandem. ■ Describe the construction and operation of automatic transmissions that use helical gears in constant mesh. ■ Describe the construction and operation of CVTs. ■ Describe the design and operation of the hydraulic controls and valves used in modern transmissions and transaxles. ■ Explain the role of the following components of the transmission control system: pressure regulator valve, throttle valve, governor assembly, manual valve, shift valves, and kickdown valve. ■ Identify the various pressures in the transmission, state their purpose, and tell how they influence the operation of the transmission. ■ Explain the advantages of using electronic controls for transmission shifting. ■ Briefly describe what determines the shift characteristics of each selector lever position. ■ Identify the input and output devices in a typical electronic control system and briefly describe the function of each.

Many rear-wheel-drive and four-wheel-drive vehicles are equipped with automatic transmissions **(Figure 39–1)**. Automatic transaxles, which combine an automatic transmission and final drive assembly in a single unit, are used on FWD, all-wheel-drive, and some RWD vehicles **(Figure 39–2)**.

An automatic transmission or transaxle selects gear ratios according to engine speed, power train load, vehicle speed, and other operating factors. Little effort is needed on the part of the driver, because both upshifts and downshifts occur automatically. A driver-operated clutch is not needed to change gears, and the vehicle can

Figure 39-1 A six-speed automatic transmission. *Courtesy of BMW of North America, Incorporated*

Figure 39–2 A five-speed automatic transaxle. *Reprinted with permission*

be brought to a stop without shifting to neutral. This is a great convenience, particularly in stop-and-go traffic. The driver can also manually select a lower forward gear, reverse, neutral, or park. Depending on the forward range selected, the transmission can provide engine braking during deceleration.

The most widely used automatic transmissions and transaxles are four-speed units with an overdrive fourth gear. Five- and six-speed transmissions are also used. Many older cars had three speeds and a select group of newer cars have five speeds. Most new automatics also feature a lockup torque converter. Until recently, all automatic transmissions were controlled by hydraulics. However, many systems now feature computer-controlled operation of the torque converter and transmission. Based on input data supplied by electronic sensors and switches, the computer sets the torque converter's operating mode, controls the transmission's shifting sequence, and in some cases regulates transmission oil pressure.

TORQUE CONVERTER

Automatic transmissions use a fluid clutch known as a torque converter to transfer engine torque from the engine to the transmission.

The torque converter operates through hydraulic force provided by automatic transmission fluid, often simply called transmission oil. The torque converter changes or multiplies the twisting motion of the engine crankshaft and directs it through the transmission.

The torque converter automatically engages and disengages power from the engine to the transmission in relation to engine rpm. With the engine running at the correct idle speed, there is not enough fluid flow for power transfer through the torque converter. As engine speed is increased, the added fluid flow creates sufficient force to transmit engine power through the torque converter assembly to the transmission.

Design

Nearly all torque converters **(Figure 39–3)**, or T/Cs, are one-piece welded units that cannot be dismantled for repair. The torque converter, located between the engine and transmission, is a sealed, doughnut-shaped unit that is always filled with automatic transmission fluid.

A special flex plate is used to mount the torque converter to the crankshaft. The purpose of the flex plate is to transfer crankshaft rotation to the shell of the torque converter assembly. The flex plate bolts to a flange machined on the rear of the crankshaft and to mounting pads located on the front of the torque converter shell.

The starter motor ring gear is attached to the flex plate on most vehicles; however, some have the ring gear welded to the torque converter. A flywheel is not required because the mass of the torque converter and flex disc acts like a flywheel to smooth out the intermittent power strokes of the engine.

Components

A standard torque converter consists of three elements **(Figure 39–4)**: the pump assembly, often called an impeller, the stator assembly, and the turbine.

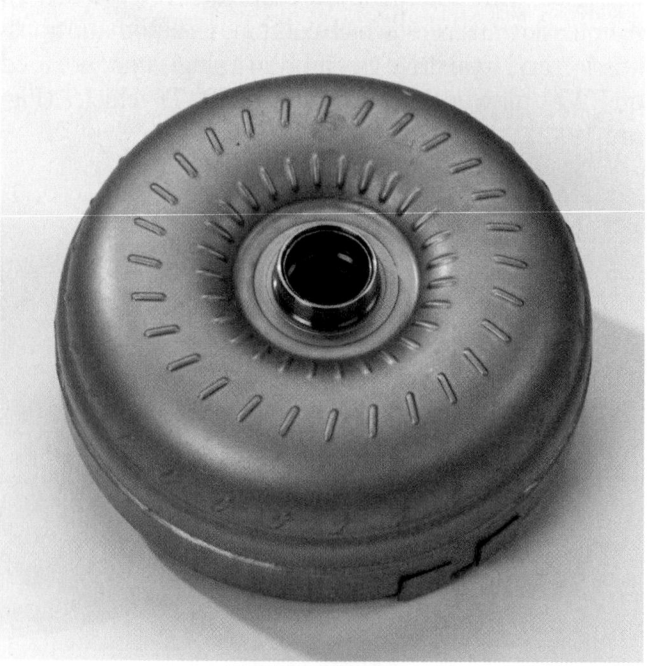

Figure 39–3 A torque converter. *Courtesy of Transtar Industries Inc.*

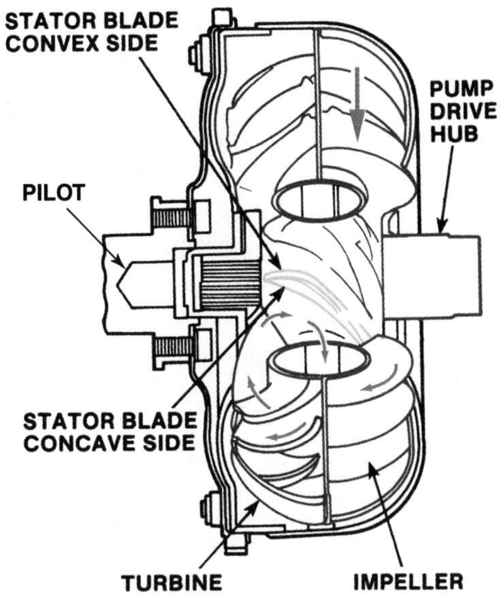

Figure 39-4 A torque converter's major internal parts are its impeller, turbine, and stator. *Courtesy of Ford Motor Company*

The **impeller** assembly is the input (drive) member. It receives power from the engine. The **turbine** is the output (driven) member. It is splined to the transmission's turbine shaft. The **stator** assembly is the reaction member or torque multiplier. The stator is supported on a one-way clutch, which operates as an overrunning clutch and permits the stator to rotate freely in one direction and lock up in the opposite direction.

The exterior of the torque converter shell is shaped like two bowls standing on end, facing each other. To support the weight of the torque converter, a short stubby shaft projects forward from the front of the torque converter shell and fits into a pocket at the rear of the crankshaft. At the rear of many torque converter shells is a hollow hub with notches or flats at one end, ground 180 degrees apart. This hub is called the pump drive hub. The notches or flats drive the transmission pump assembly. At the front of the transmission within the pump housing is a pump bushing that supports the pump drive hub and provides rear support for the torque converter assembly. Some other transaxles have a separate shaft to drive the pump.

The impeller forms one internal section of the torque converter shell and has numerous curved blades that rotate as a unit with the shell. It turns at engine speed, acting like a pump to start the transmission oil circulating within the torque converter shell.

While the impeller is positioned with its back facing the transmission housing, the turbine is positioned with its back to the engine. The curved blades of the turbine face the impeller assembly.

The turbine blades have a greater curve than the impeller blades, which helps eliminate oil turbulence between the turbine and impeller blades that would slow impeller speed and reduce the converter's efficiency.

The stator is located between the impeller and turbine. It redirects the oil flow from the turbine back into the impeller in the direction of impeller rotation with minimal loss of speed or force. The side of the stator blade with the inward curve is the concave side. The side with an outward curve is the convex side.

Basic Operation

Transmission oil is used as the medium to transfer energy in the T/C. **Figure 39–5A** illustrates the T/C impeller or pump at rest. **Figure 39–5B** shows it being driven. As the pump impeller rotates, centrifugal force throws the oil outward and upward due to the curved shape of the impeller housing.

The faster the impeller rotates, the greater the centrifugal force becomes. In **Figure 39–5B**, the oil is simply flying out of the housing and is not producing any work. To harness some of this energy, the turbine assembly is mounted on top of the impeller **(Figure 39–5C)**. Now the oil thrown outward and upward from the impeller strikes the curved vanes of the turbine, causing the turbine to rotate. (There is no direct mechanical link between the impeller and turbine.) An oil pump driven by the converter shell and the engine continually delivers oil under pressure into the T/C through a hollow shaft at the center axis of the rotating torque converter assembly. A seal prevents the loss of fluid from the system.

The turbine shaft is located within this shaft. As mentioned earlier, the turbine shaft is splined to the turbine and transfers power from the torque converter to the transmission's main drive shaft. Oil leaving the turbine is directed out of the torque converter to an external oil cooler and then to the transmission's oil sump or pan.

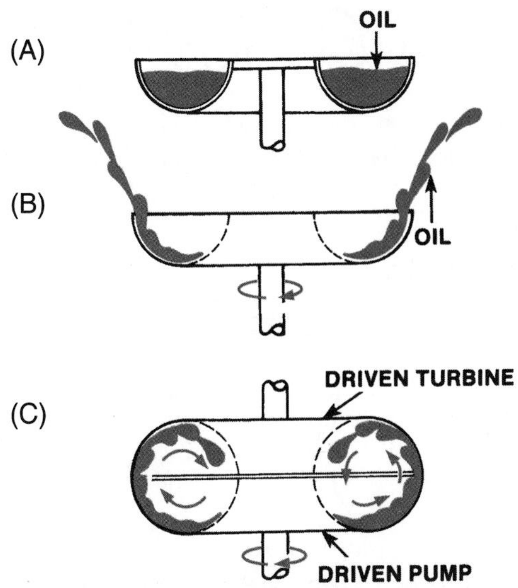

Figure 39-5 Fluid travel inside the torque converter: (A) fluid at rest in the impeller/pump, (B) fluid thrown up and outward by the spinning pump, and (C) fluid flow harnessed by the turbine and redirected back into the pump.

With the transmission in gear and the engine at idle, the vehicle can be held stationary by applying the brakes. At idle, engine speed is slow. Since the impeller is driven by engine speed, it turns slowly creating little centrifugal force within the torque converter. Therefore, little or no power is transferred to the transmission.

When the throttle is opened, engine speed, impeller speed, and the amount of centrifugal force generated in the torque converter increase dramatically. Oil is then directed against the turbine blades, which transfer power to the turbine shaft and transmission.

Types of Oil Flow

Two types of oil flow take place inside the torque converter: rotary and vortex flow **(Figure 39–6)**. **Rotary oil flow** is the oil flow around the circumference of the torque converter caused by the rotation of the torque converter on its axis. **Vortex oil flow** is the oil flow occurring from the impeller to the turbine and back to the impeller, at a 90-degree angle from engine rotation.

Figure 39–7 also shows the oil flow pattern as the speed of the turbine approaches the speed of the impeller. This is known as the **coupling point**. The turbine and the impeller are running at essentially the same speed. They cannot run at exactly the same speed due to slippage between them. The only way they can turn at exactly the same speed is by using a lockup clutch to mechanically tie them together.

The stator mounts through its splined center hub to a mating stator shaft. The stator freewheels when the impeller and turbine reach the coupling stage.

The stator redirects the oil leaving the turbine back to the impeller, which helps the impeller rotate more efficiently **(Figure 39–8)**. Torque converter multiplication can only occur when the impeller is rotating faster than the turbine.

A stator is either a rotating or fixed type. Rotating stators are more efficient at higher speeds because there is less slippage when the impeller and turbine reach the coupling stage.

Overrunning Clutch

An **overrunning clutch** keeps the stator assembly from rotating when driven in one direction and permits overrunning (rotation) when turned in the opposite direction. Rotating stators generally use a roller-type overrunning clutch that allows the stator to freewheel (rotate) when the speed of the turbine and impeller reach the coupling point.

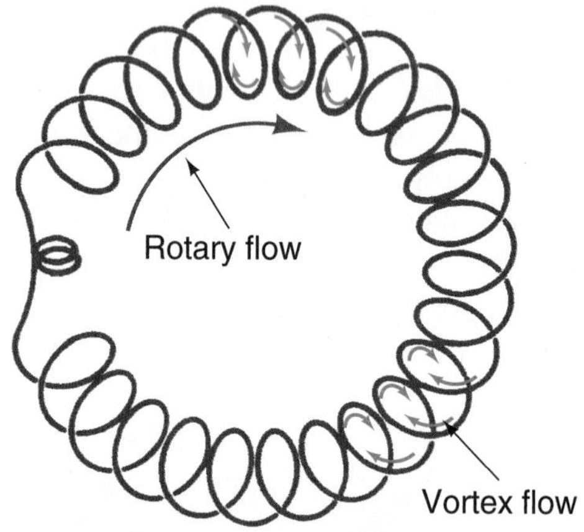

Figure 39-6 The difference between rotary and vortex flow. Note that vortex flow spirals its way around the converter.

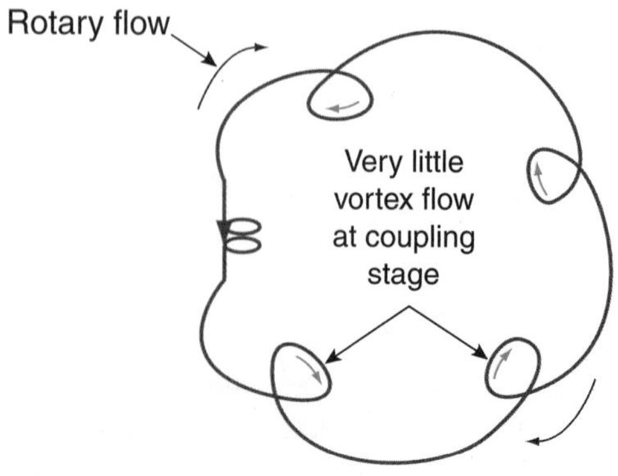

Figure 39-7 Rotary flow is at its greatest at the coupling stage.

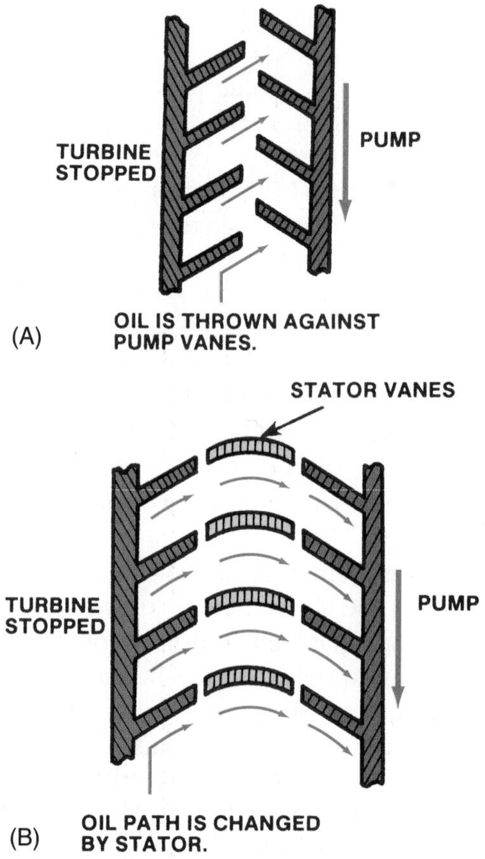

Figure 39-8 (A) Without a stator, fluid leaving the turbine works against the direction in which the impeller or pump is rotating. (B) With a stator in its locked (noncoupling) mode, fluid is directed to help push the impeller in its rotating direction.

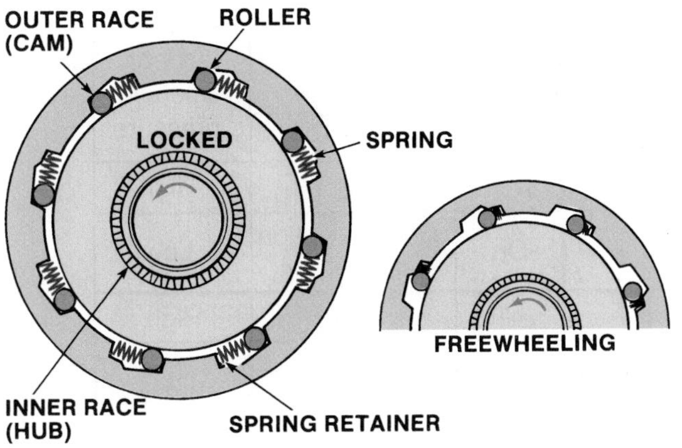

Figure 39-9 A typical roller-type overrunning clutch.

The roller clutch **(Figure 39–9)** is designed with a movable inner race, rollers, accordion (apply) springs, and outer race. Around the inside diameter of the outer race are several cam-shaped pockets. The rollers and accordion springs are located in these pockets.

As the vehicle begins to move, the stator stays in its stationary or locked position because of the difference between the impeller and turbine speeds. This locking mode takes place when the inner race rotates counterclockwise. The accordion springs force the rollers up the ramps of the cam pockets into a wedging contact with the inner and outer races.

As vehicle road speed increases, turbine speed increases until it approaches impeller speed. Oil exiting the turbine vanes strikes the back face of the stator, causing the stator to rotate in the same direction as the turbine and impeller. At this higher speed, clearance exists between the inner stator race and hub. The rollers at each slot of the stator are pulled around the stator hub. The stator freewheels or turns as a unit.

If the vehicle slows, engine speed also slows along with turbine speed. This decrease in turbine speed allows the oil flow to change direction. It now strikes the front face of the stator vanes, halting the turning stator and attempting to rotate it in the opposite direction.

As this happens, the rollers jam between the inner race and hub, locking the stator in position. In a stationary position, the stator now redirects the oil exiting the turbine so torque is again multiplied.

LOCKUP TORQUE CONVERTER

A lockup torque converter eliminates the 10% slip that takes place between the impeller and turbine at the coupling stage. The engagement of a clutch between the impeller and the turbine assembly greatly improves fuel economy and reduces operational heat and engine speed. The assembly of a lockup torque converter is typically called a **torque converter clutch (TCC)**.

Through the years, many different types of TCC systems have been used. The most common design is the electronically controlled lockup piston clutch. Clutch lockup systems can also be fully mechanical, centrifugally controlled, or dependent on a viscous coupling.

Lockup Piston Clutch

The lockup piston clutch has a piston-type clutch located between the front of the turbine and the interior front face of the shell **(Figure 39–10)**. Its main components are a piston plate and damper assembly and a clutch friction plate. The damper assembly is made of several coil springs and is designed to transmit driving torque and absorb shock.

The clutch is controlled by hydraulic valves, which are controlled by the PCM **(Figure 39–11)**. The PCM monitors operating conditions and controls lockup according to those conditions.

To understand how this system works, consider this example. To provide for clutch control, Chrysler adds a three-valve module to its standard transmission valve body. The three valves are the lockup valve, fail-safe valve, and switch valve. The lockup valve actually controls the clutch. The fail-safe valve prevents lockup until the transmission is in third gear. The switch valve directs fluid through the turbine shaft to fill the torque converter.

When the converter is not locked, fluid enters the converter and moves to the front side of the piston, keeping it away from the shell or cover. Fluid flow continues around the piston to the rear side and exits between the neck of the torque converter and the stator support.

During the lockup mode, the switch valve moves and reverses the fluid path. This causes the fluid to move to the rear of the piston, pushing it forward to apply the clutch to the shell and allowing for lockup. Fluid from the front side of the piston exits through the turbine shaft that is now vented at the switch valve.

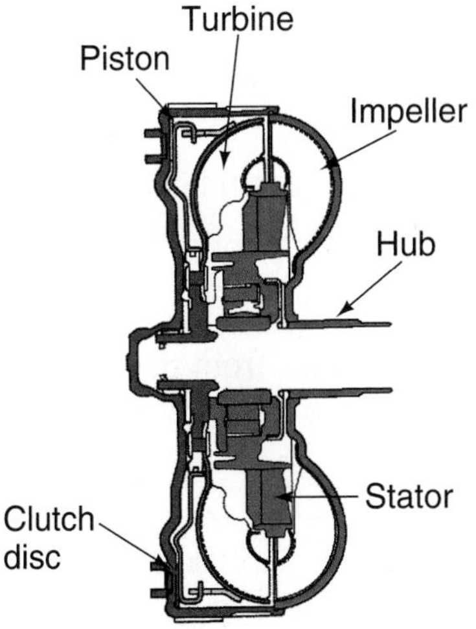

Figure 39-10 A piston-type converter lockup clutch assembly. *Courtesy of DaimlerChrysler Corporation*

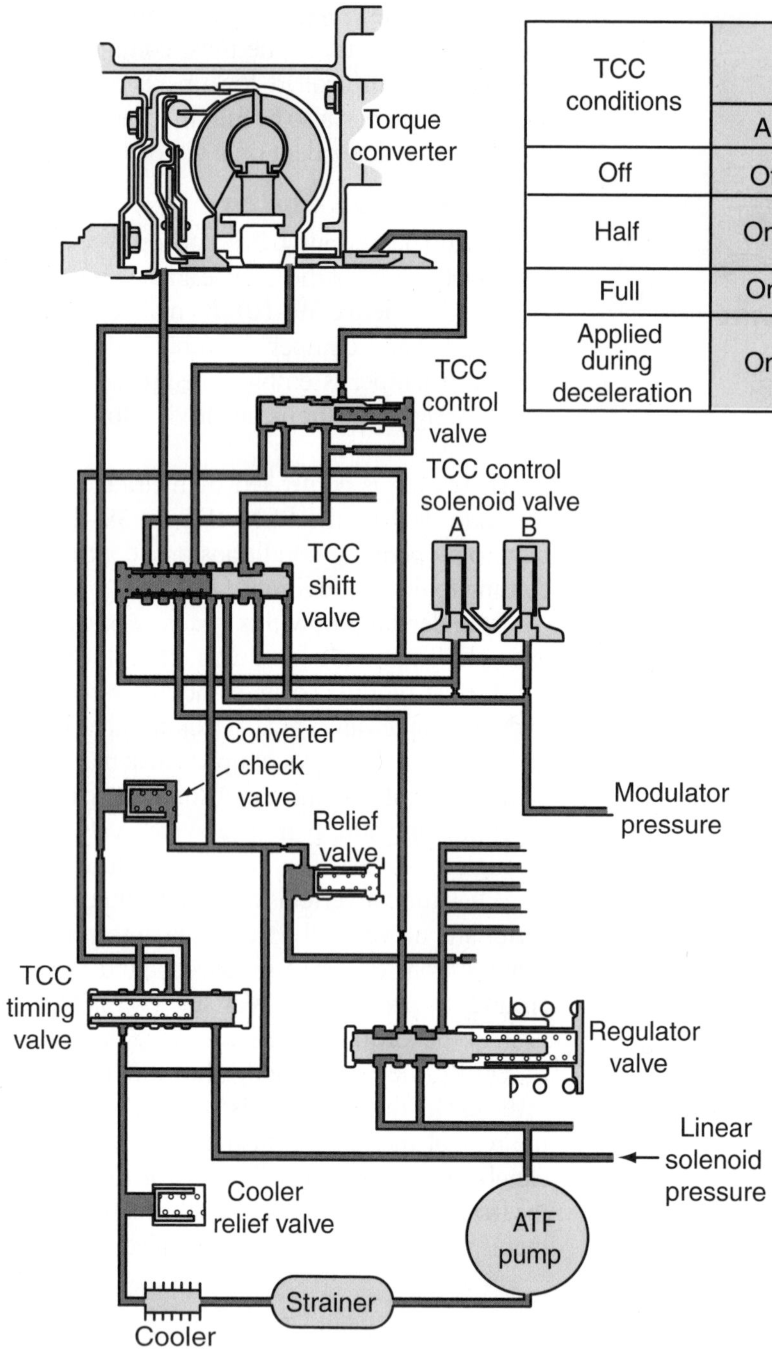

TCC conditions	TCC control solenoid valve		Linear solenoid pressure
	A	B	
Off	Off	Off	High
Half	On	Duty operation Off ↔ On	Low
Full	On	On	High
Applied during deceleration	On	Duty operation Off ↔ On	Low

Figure 39-11 A typical circuit for activating the TCC.

During acceleration, system fluid pressure increases. If the converter is in its lockup mode, the higher pressure moves the fail-safe valve to block fluid pressure to the lockup valve. Spring tension moves the switch valve, directing fluid pressure to the front side of the piston. The torque converter then returns to its nonlockup mode.

PLANETARY GEARS

Nearly all automatic transmissions rely on planetary gearsets **(Figure 39–12)** to transfer power and multiply engine torque to the drive axle. Compound gearsets combine two simple planetary gearsets so load can be spread

Figure 39-12 A single planetary gearset.

over a greater number of teeth for strength and also to obtain the largest number of gear ratios possible in a compact area.

A simple planetary gearset consists of three parts: a sun gear, a carrier with planetary pinions mounted to it, and an internally toothed ring gear or **annulus**. The **sun gear** is located in the center of the assembly **(Figure 39–13)**. It can be either a spur or helical gear design. It meshes with the teeth of the planetary pinion gears. Planetary pinion gears are small gears fitted into a framework called the **planetary carrier**. The planetary carrier can be made of cast iron, aluminum, or steel plate and is designed with a shaft for each of the planetary pinion gears. (For simplicity, planetary pinion gears are called **planetary pinions**.)

Planetary pinions rotate on needle bearings positioned between the planetary carrier shaft and the planetary pinions. The carrier and pinions are considered one unit—the midsize gear member.

The planetary pinions surround the sun gear's center axis and they are surrounded by the annulus or ring gear, which is the largest part of the simple gearset. The ring gear acts like a band to hold the entire gearset together and provide great strength to the unit. To help remember the design of a simple planetary gearset, use the solar system as an example. The sun is the center of the solar system with the planets rotating around it; hence, the name planetary gearset.

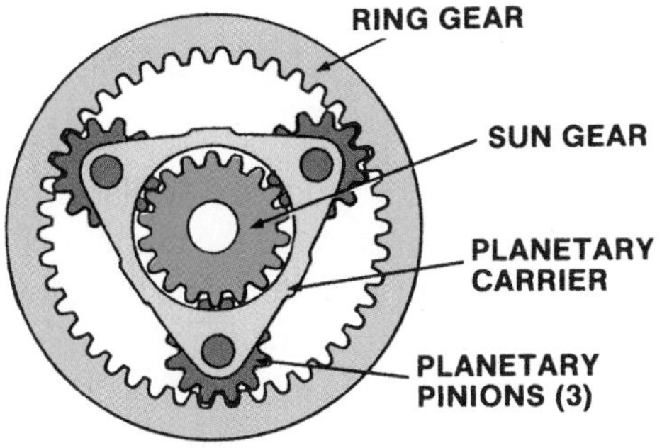

Figure 39–13 Planetary gear configuration is similar to the solar system, with the sun gear surrounded by the planetary pinion gears. The ring gear surrounds the complete gearset.

How Planetary Gears Work

Each member of a planetary gearset can spin (revolve) or be held at rest. Power transfer through a planetary gearset is only possible when one of the members is held at rest, or if two of the members are locked together.

Any one of the three members can be used as the driving or input member. At the same time, another member might be kept from rotating and thus becomes the held or stationary member. The third member then becomes the driven or output member. Depending on which member is the driver, which is held, and which is driven, either a torque increase (underdrive) or a speed increase (overdrive) is produced by the planetary gearset. Output direction can also be reversed through various combinations.

Table 39–1 summarizes the basic laws of simple planetary gears. It indicates the resultant speed, torque, and direction of the various combinations available. Also, remember that when an external-to-external gear tooth set is in mesh, there is a change in the direction of rotation at the output. When an external gear tooth is in mesh with an internal gear, the output rotation for both gears is the same.

Maximum Forward Reduction With the ring gear held and the sun gear (the input) turning clockwise, the sun gear rotates the planetary pinions counterclockwise on their shafts. The small sun gear (driving) rotates several times, driving the midsize planetary carrier (the output) one complete revolution, resulting in the most gear reduction or the maximum torque multiplication that can

TABLE 39–1 LAWS OF SIMPLE PLANETARY GEAR OPERATION

Sun Gear	Carrier	Ring Gear	Speed	Torque	Direction
1. Input	Output	Held	Maximum reduction	Increase	Same as input
2. Held	Output	Input	Minimum reduction	Increase	Same as input
3. Output	Input	Held	Maximum increase	Reduction	Same as input
4. Held	Input	Output	Minimum increase	Reduction	Same as input
5. Input	Held	Output	Reduction	Increase	Reverse of input
6. Output	Held	Input	Increase	Reduction	Reverse of input
7. When any two members are held together, speed and direction are the same as input. Direct 1:1 drive occurs.					
8. When no member is held or locked together, output cannot occur. The result is a neutral condition.					

be achieved in one planetary gearset. Input speed is high, but output speed is low.

Minimum Forward Reduction In this combination the sun gear is held and the ring gear (input) rotates clockwise. The ring gear drives the planetary pinions clockwise and walks around the stationary sun gear (held). The planetary pinions drive the planetary carrier (output) in the same direction as the ring gear—forward. This results in more than one turn of the input as compared to one complete revolution of the output. The result is torque multiplication. The planetary gearset is operating in a forward reduction with the large ring gear driving the small planetary carrier. Therefore, the combination produces minimum forward reduction.

Maximum Overdrive With the ring gear held and the planetary carrier (input) rotating clockwise, the three planetary pinion shafts push against the inside diameter of the planetary pinions. The pinions are forced to walk around the inside of the ring gear, driving the sun gear (output) clockwise. In this combination, the midsize planetary carrier is rotating less than one turn and driving the smaller sun gear at a speed greater than the input speed. The result is overdrive with maximum speed increase.

Slow Overdrive In this combination, the sun gear is held and the carrier rotates (input) clockwise. As the carrier rotates, the pinion shafts push against the inside diameter of the pinions and they are forced to walk around the held sun gear. This drives the ring gear (output) faster and the speed increases. The carrier turning less than one turn causes the pinions to drive the ring gear one complete revolution in the same direction as the planetary carrier and a slow overdrive occurs.

Slow Reverse Here the sun gear (input) is driving the ring gear (output) with the planetary carrier held stationary. The planetary pinions, driven by the sun gear, rotate counterclockwise on their shafts. While the sun gear is driving, the planetary pinions are used as idler gears to drive the ring gear counterclockwise. This means the input and output shafts are operating in the opposite or reverse direction to provide a reverse power flow. Since the driving sun gear is small and the driven ring gear is large, the result is slow reverse.

Fast Reverse For fast reverse, the carrier is held, but the sun gear and ring gear reverse roles, with the ring gear (input) now being the driving member and the sun gear (output) driven. As the ring gear rotates counterclockwise, the pinions rotate counterclockwise as well, while the sun gear turns clockwise. In this combination, the input ring gear uses the planetary pinions to drive the output sun gear. The sun gear rotates in reverse to the input ring gear, providing fast reverse.

Direct Drive In the direct drive combination, both the ring gear and the sun gear are input members. They turn clockwise at the same speed. The internal teeth of the clockwise turning ring gear try to rotate the planetary pinions clockwise as well. But the sun gear, which rotates clockwise, tries to drive the planetary pinions counterclockwise. These opposing forces lock the planetary pinions against rotation so the entire planetary gearset rotates as one complete unit, providing direct drive. Whenever two members of the gearset are locked together, direct drive results.

Neutral Operation When no member is held or locked, a neutral condition exists.

The following are helpful tips for remembering the basics of simple planetary gearset operation:

- When the planetary carrier is the drive (input) member, the gearset produces an overdrive condition. Speed increases, torque decreases.
- When the planetary carrier is the driven (output) member, the gearset produces a forward underdrive direction. Speed decreases, torque increases.
- When the planetary carrier is stationary (held), the gearset produces a reverse.

COMPOUND PLANETARY GEARSETS

A limited number of gear ratios are available from a single planetary gearset. To increase the number of available gear ratios, gearsets can be added. The typical automatic transmission with three or four forward speeds has at least two planetary gearsets.

In automatic transmissions, two or more planetary gearsets **(Figure 39–14)** are connected together to provide the various gear ratios needed to efficiently move a vehicle. There are two common designs of compound gearsets, the Simpson gearset in which two planetary gearsets share a common sun gear and the Ravigneaux gearset which has two sun gears, two sets of planet gears, and a common ring gear. Some transmissions are fitted with an additional single planetary gearset that is used to provide an add-on overdrive gear.

Simpson Gear Train

The **Simpson gear train** is an arrangement of two separate planetary gearsets with a common sun gear, two ring gears, and two planetary pinion carriers. A Simpson gear train is the most commonly used compound planetary gearset and is used to provide three forward gears. One half of the compound set or one planetary unit is referred to as the front planetary and the other planetary unit is the rear planetary **(Figure 39–15)**. The two planetary units do not need to be the same size or have the same number of teeth on their gears. The size and number of

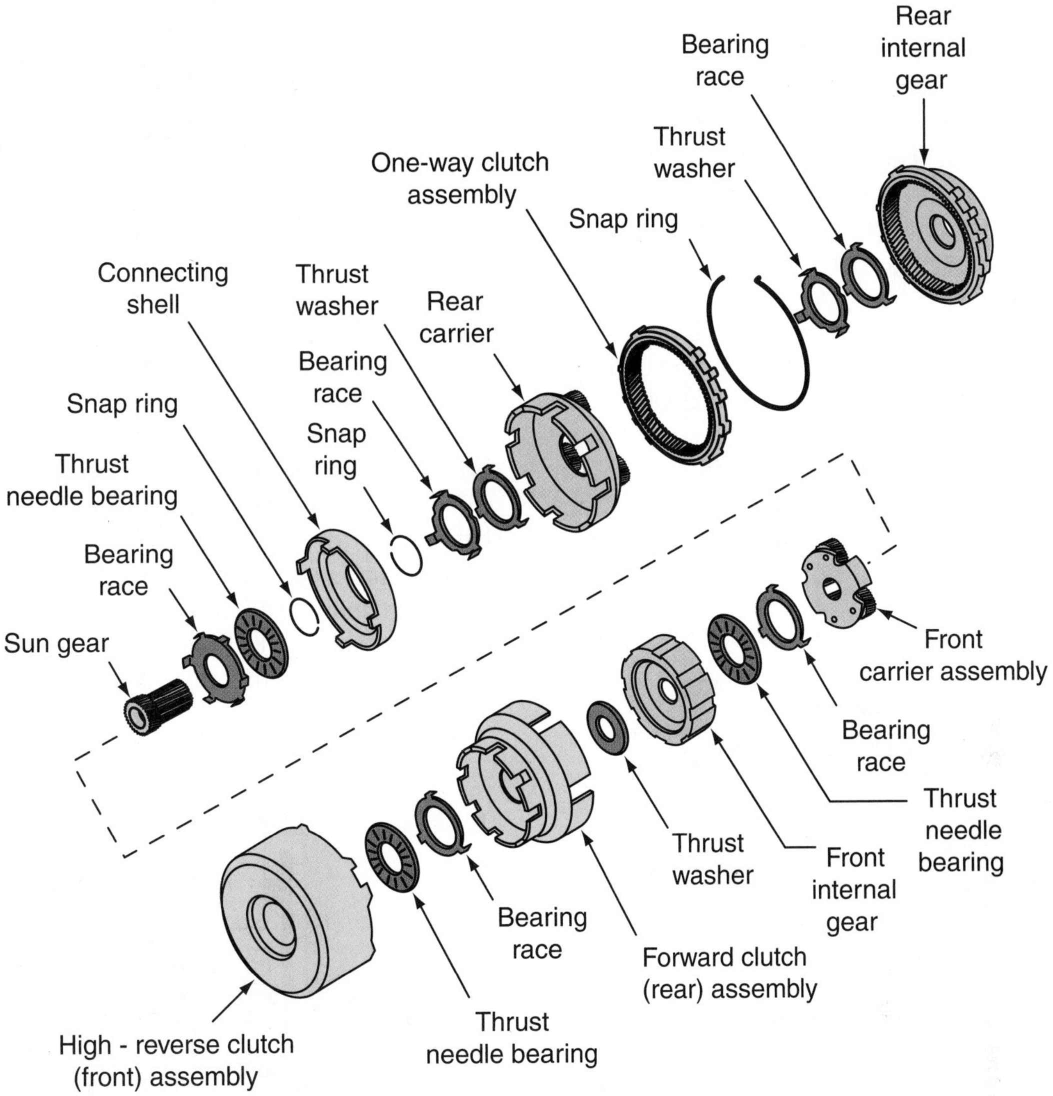

Figure 39-14 A Simpson planetary gearset.

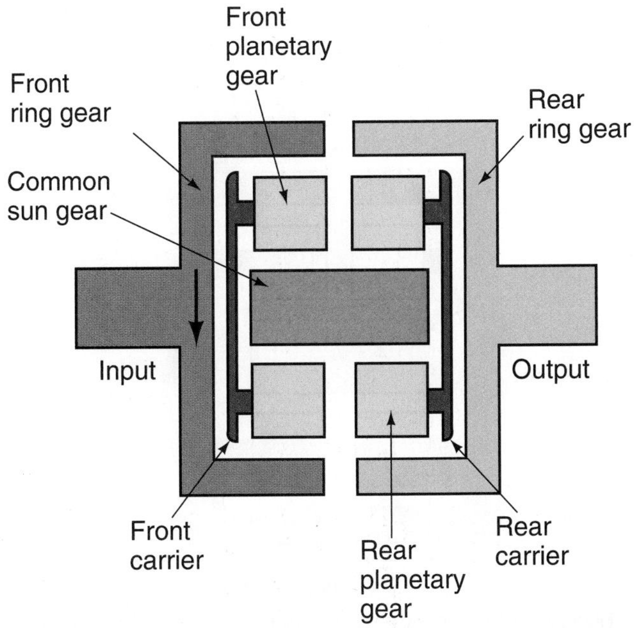

Figure 39-15 Components of a Simpson gearset.

gear teeth determine the actual gear ratios obtained by the compound planetary gear assembly.

Gear ratios and direction of rotation are the result of applying torque to one member of either planetary unit, holding at least one member of the gearset, and using another member as the output. For the most part, each automobile manufacturer uses different parts of the planetary assemblies as inputs, outputs, and reaction members. The role of the planetary members also varies with the different transmission models from the same manufacturer. There are also many different apply devices used in the various transmission designs.

A Simpson gearset can provide the following gear ranges: neutral, first reduction gear, second reduction gear, direct drive, and reverse. The typical power flow through a Simpson gear train when it is in neutral has engine torque being delivered to the transmission's input shaft by the torque converter's turbine. No planetary gearset member is locked to the shaft, therefore engine torque enters the transmission but goes nowhere else.

When the transmission is shifted into first gear **(Figure 39–16)**, engine torque is again delivered into the transmission by the input shaft. The input shaft is now locked to the front planetary ring gear that turns clockwise with the shaft. The front ring gear drives the front planet gears, also in a clockwise direction. The front planet gears drive the sun gear in a counterclockwise direction. The rear planet carrier is locked; therefore, the sun gear spins the rear planet gears in a clockwise direction. These planet gears drive the rear ring gear, which is locked to the output shaft, in a clockwise direction. The result of this power flow is a forward gear reduction, normally with a ratio of 2.5:1 to 3.0:1.

When the transmission is operating in second gear **(Figure 39–17)**, engine torque is again delivered into the transmission by the input shaft. The input shaft is locked to the front planetary ring gear that turns clockwise with the shaft. The front ring gear drives the front planet gears, also in a clockwise direction. The front planet gears walk around the sun gear because it is held. The walking of the planets forces the planet carrier to turn clockwise. Since the carrier is locked to the output shaft, it causes the shaft to rotate in a forward direction with some gear reduction. A typical second gear ratio is 1.5:1.

When operating in third gear **(Figure 39–18)**, the input is received by the front ring gear, as in the other forward positions. However, the input is also received by the sun gear. Since the sun and ring gear are rotating at the same speed and in the same direction, the front planet carrier is locked between the two and is forced to move with them. Since the front carrier is locked to the output shaft, direct drive results.

To obtain a suitable reverse gear in a Simpson gear train, there must be a gear reduction, but in the opposite direction as the input torque **(Figure 39–19)**. The input is received by the sun gear, as in the third gear position,

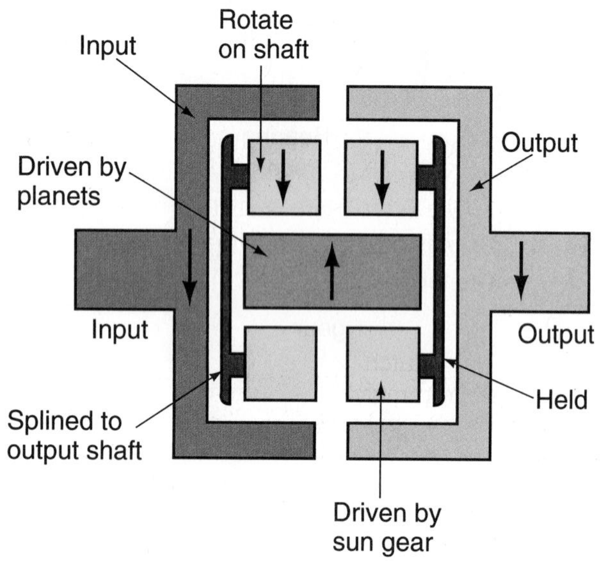

Figure 39-16 Power flow through a Simpson gearset while operating in first gear.

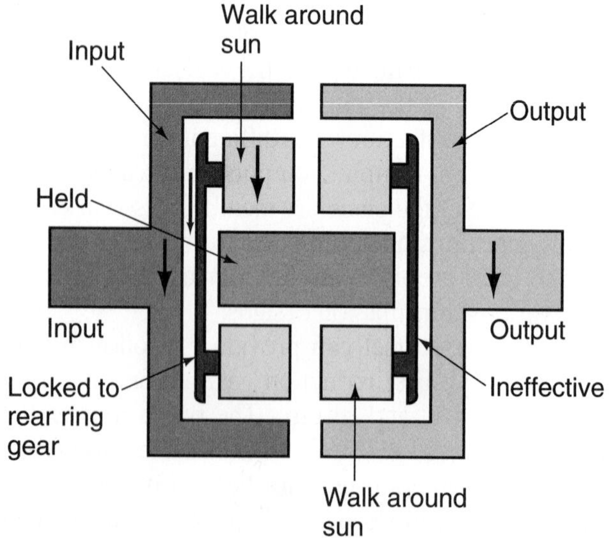

Figure 39-17 Power flow through a Simpson gearset while operating in second gear.

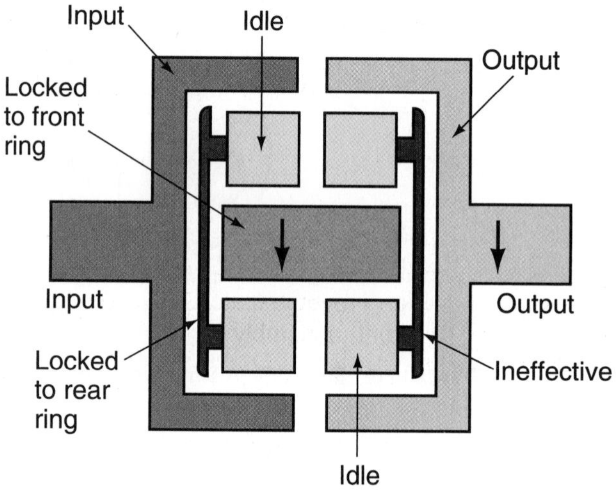

Figure 39-18 Power flow through a Simpson gearset while operating in direct drive.

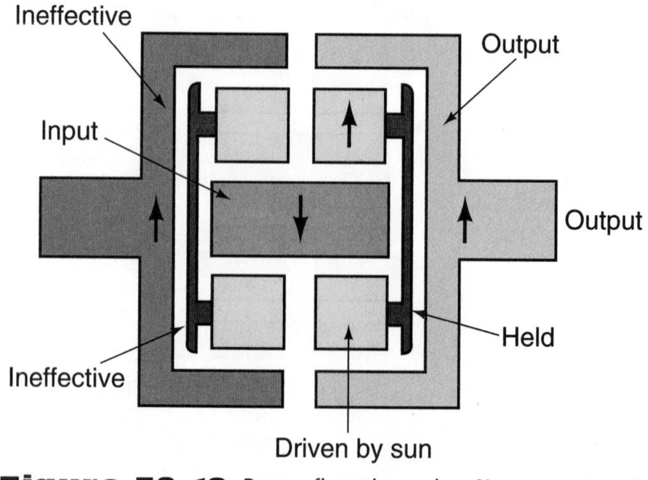

Figure 39-19 Power flow through a Simpson gearset while operating in reverse.

and rotates in a clockwise direction. The sun gear then drives the rear planet gears in a clockwise direction. The rear planet carrier is held; therefore, the planet gears drive the rear ring gear in a counterclockwise direction. The ring gear is locked to the output shaft that turns at the same speed and direction as the rear ring gear. The result is a reverse gear with a ratio of 2.5:1 to 2.0:1.

Typically, when the transmission is in neutral or park, no apply devices are engaged, allowing only the input shaft and the transmission's oil pump to turn with the engine. In park, a pawl is mechanically engaged to a parking gear that is splined to the transmission's output shaft, locking the drive wheels to the transmission's case.

Ravigneaux Gear Train

The **Ravigneaux gear train**, like the Simpson gear train, provides forward gears with a reduction, direct drive, overdrive, and a reverse operating range. The Ravigneaux offers some advantages over a Simpson gear train. It is very compact. It can carry large amounts of torque because of the great amount of tooth contact. It can also have three different output members. However, it is disadvantageous to students and technicians, because it is more complex and therefore its actions are more difficult to understand.

The Ravigneaux gear train is designed to use two sun gears, one small and one large **(Figure 39–20)**. This type gear train also has two sets of planetary pinion gears: three long pinions and three short pinions. The planetary pinion gears rotate on their own shafts that are fastened to a common planetary carrier. A single ring gear surrounds the complete assembly.

The small sun gear is meshed with the short planetary pinion gears. These short pinions act as idler gears to drive the long planetary pinion gears. The long planetary pinion gears mesh with the large sun gear and the ring gear.

Typically, when the gear selector is in neutral position, engine torque, through the converter turbine shaft, drives the forward clutch drum. Since the forward clutch is not applied, the power is not transmitted through the gear train, and there is no power output.

When the transmission is operating in first gear **(Figure 39–21)**, engine torque drives the forward clutch drum. The forward clutch is applied and drives the forward sun gear clockwise. The planetary carrier is prevented from rotating counterclockwise by the one-way clutch, therefore the forward sun gear drives the short planet gears counterclockwise. The direction of rotation is reversed as the short planet gears drive the long planet gears, which drive the ring gear and output shaft in a clockwise direction but at a lower speed than the input. This results in a gear reduction and a ratio of approximately 2.5:1.

In second gear **(Figure 39–22)** operation, the intermediate clutch is applied and locks the outer race of the one-way clutch. This prevents the reverse clutch drum, shell, and reverse sun gear from turning counterclockwise. The forward clutch is also applied and locks the input to the forward sun gear and it rotates in a clockwise direction. The forward sun gear drives the short planet gears counterclockwise. The direction of rotation is reversed as the short planet gears drive the long planet

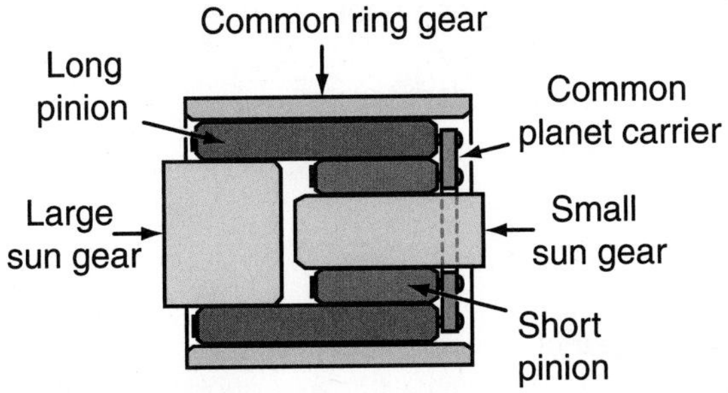

Figure 39–20 The parts of a Ravigneaux gearset.

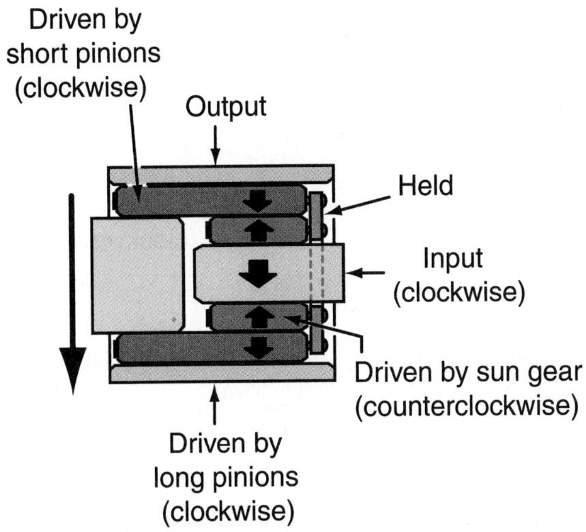

Figure 39–21 Power flow through a Ravigneaux gearset while operating in first gear.

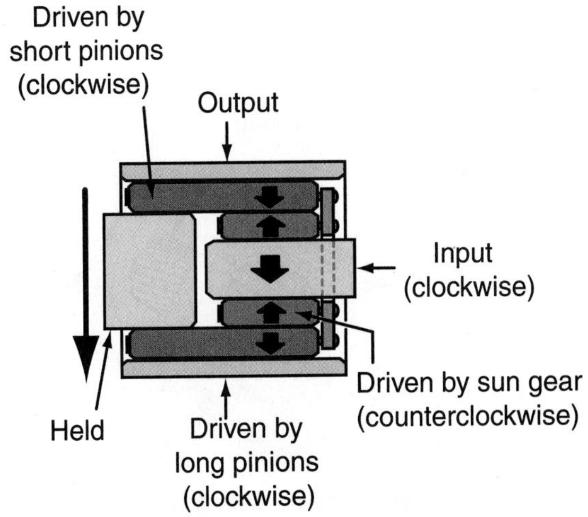

Figure 39–22 Power flow through a Ravigneaux gearset while operating in second gear.

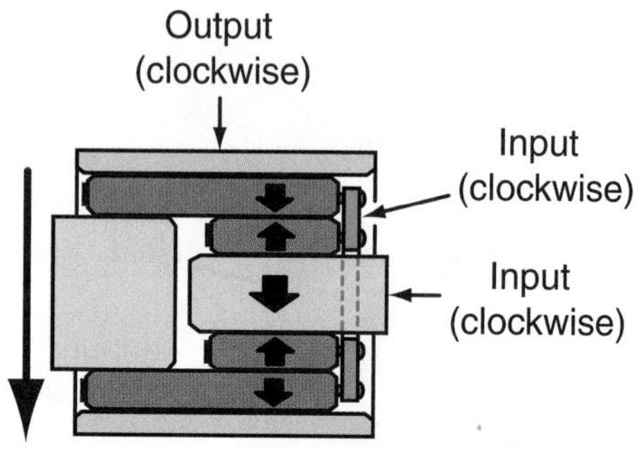

Figure 39-23 Power flow through a Ravigneaux gearset while operating in direct drive.

gears, which walk around the stationary reverse sun gear. This walking drives the ring gear and output shaft in a clockwise direction and at a reduction with a gear ratio of approximately 1.5:1.

During third gear **(Figure 39–23)** operation, there are two inputs into the planetary gear train. As in other forward gears, the turbine shaft of the torque converter drives the forward clutch drum. The forward clutch is applied and drives the forward sun gear in a clockwise direction. Input is also received by the direct clutch that is driven by the torque converter cover. The direct clutch is applied and drives the planetary gear carrier. Since two members of the gear train are being driven at the same time, the planetary gear carrier and the forward sun gear rotate as a unit. The long planet gears transfer the torque, in a clockwise direction, through the gearset to the ring gear and output shaft. This results in direct drive.

To operate in overdrive or fourth gear, input is received only by the direct clutch from the torque converter cover. The direct clutch is applied and drives the planetary carrier in a clockwise direction. The long planet gears walk around the stationary reverse sun gear in a clockwise direction and drive the ring gear and output shaft. This results in an overdrive condition with an approximate ratio of 0.75:1.

During reverse gear operation, input is received through the torque converter's turbine shaft to the reverse clutch. The reverse clutch, when applied, connects the turbine shaft to the reverse sun gear. The low/reverse band is applied and holds the planetary gear carrier. The clockwise rotation of the reverse sun gear drives the long planet gears in a counterclockwise direction. The long planets then drive the ring gear and output shaft in a counterclockwise direction with a speed reduction and a gear ratio of approximately 2.0:1.

Planetary Gearsets in Tandem

Rather than relying on the use of a compound gearset, some automatic transmissions use two simple planetary units in series **(Figure 39–24)**. In this type of arrangement, gearset members are not shared; instead, the holding devices are used to lock different members of the planetary units together.

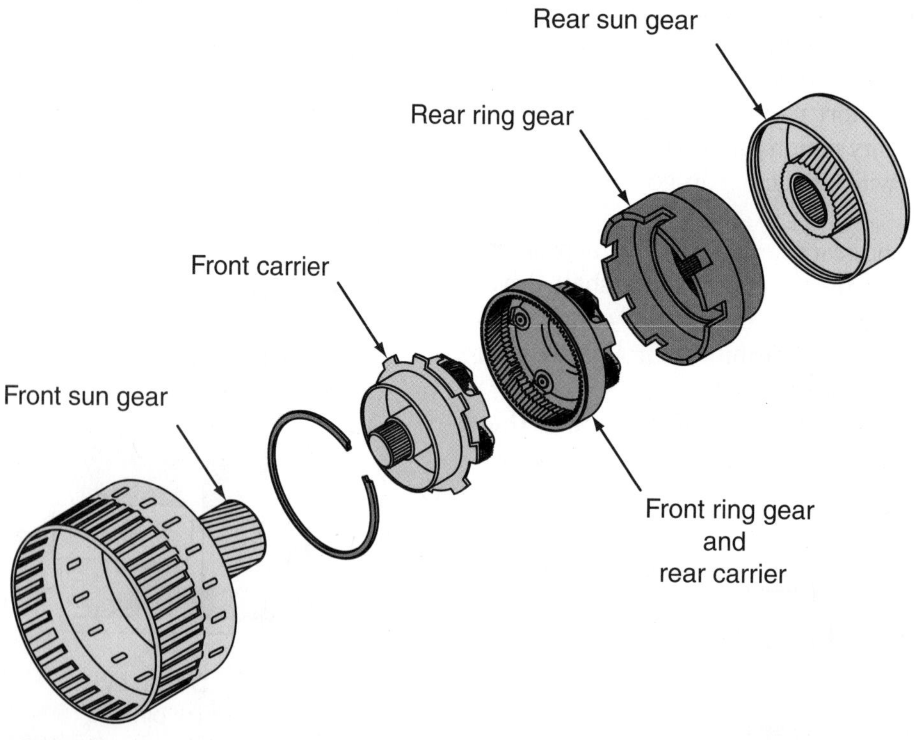

Figure 39-24 Two planetary units with the ring gear of one gearset connected to the planet carrier of the other.

Although the gear train is based on two simple planetary gearsets operating in tandem, the combination of the two planetary units does function much like a compound unit. The two tandem units do not share a common member; rather, certain members are locked together or are integral with each other. The front planetary carrier is locked to the rear ring gear and the front ring gear is locked to the rear planetary carrier. The transaxle houses a third planetary unit used only as the final drive unit and not for overdrive.

HONDA'S NONPLANETARY-BASED TRANSMISSION

The Honda nonplanetary based transaxles are used in many Honda and Acura cars. Saturn automatic transaxles are also based on this design. These transmissions are unique in that they use constant-mesh helical and square-cut gears **(Figure 39–25)** in a manner similar to that of a manual transmission.

These transaxles have a mainshaft and countershaft on which the gears ride. To provide the four forward and one reverse gear, different pairs of gears are locked to the shafts by hydraulically controlled clutches **(Figure 39–26)**. Reverse gear is obtained through the use of a shift fork that slides the reverse gear into position. The power flow through these transaxles is also similar to that of a manual transaxle.

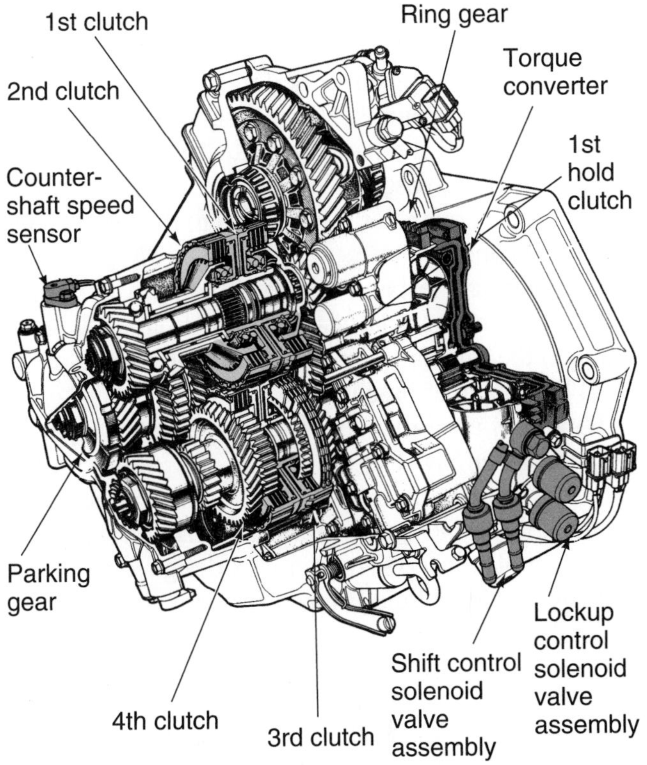

Figure 39-25 Honda automatic transaxles use constant-mesh helical gears instead of planetary gearsets. *Courtesy of Automatic Transmission Rebuilders Association (ATRA)*

Power flow through these transaxles is similar to that through a manual transaxle. The action of the clutches is much the same as the action of the synchronizer assemblies in a manual transaxle. Honda uses four multiple-disc clutches, a sliding reverse gear, and a one-way clutch to control the gears.

CONTINUOUSLY VARIABLE TRANSMISSIONS (CVT)

Another unconventional transmission design, the **continuously variable transmission (CVT)**, is a transmission with no fixed forward speeds. The gear ratio varies with engine speed and temperature. These transmissions are, however, fitted with a one-speed reverse gear. Some CVT transaxles do not have a torque converter; rather, they use a manual transmission-type flywheel with a start clutch. Instead of relying on planetary or helical gearsets to provide drive ratios, a CVT uses belts and pulleys **(Figure 39–27)**.

One pulley is the driven member and the other is the drive. Each pulley has a moveable face and a fixed face. When the moveable face moves, the effective diameter of the pulley changes. The change in effective diameter changes the effective pulley (gear) ratio. A steel belt links the driven and drive pulleys **(Figure 39–28)**.

To achieve a low pulley ratio, high hydraulic pressure works on the moveable face of the driven pulley to make it larger. In response to this high pressure, the pressure on the drive pulley is reduced. Since the belt links the two pulleys and proper belt tension is critical, the drive pulley reduces just enough to keep the proper tension on the belt. The increase of pressure at the driven pulley is proportional to the decrease of pressure at the drive pulley. The opposite is true for high pulley ratios. Low pressure causes the driven pulley to decrease in size, whereas high pressure increases the size of the drive pulley.

Different speed ratios are available any time the vehicle is moving. Because the size of the drive and driven pulleys can vary greatly, vehicle loads and speeds can be changed without changing the engine's speed. With this type of transmission, attempts are made to keep the engine operating at its most efficient speed, thus increasing fuel economy and decreasing emissions.

Many late-model CVTs are equipped with a feature that simulates the activity of a manual shifting automatic transmission. These transmissions have five or six predetermined areas that the pulleys stop in, thereby giving the feel and shift effect of distinct shifts.

Nissan has recently introduced a CVT, called the Extroid CVT, that is based on discs and rollers. This design

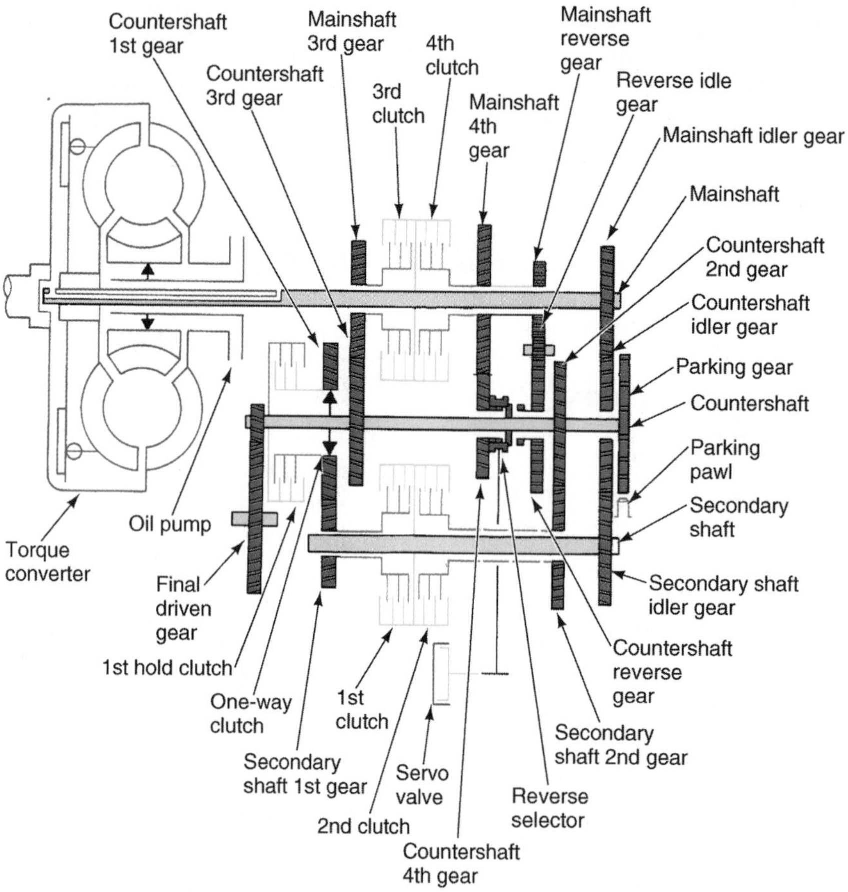

Figure 39-26 The arrangement of gears and reaction devices in a typical nonplanetary gearset transaxle. *Courtesy of American Honda Motor Co., Inc.*

may be more widely used in the future because it has the capability of withstanding large amounts of engine torque.

PLANETARY GEAR CONTROLS

Certain parts of the planetary gear train must be held, while others must be driven to provide the needed torque multiplication and direction for vehicle operation. Planetary gear controls is the general term used to describe transmission bands, servos, and clutches.

Transmission Bands

A **band** is a braking assembly positioned around a stationary or rotating drum or carrier. The band brings a drum to a stop by wrapping itself around the drum and holding it. The band is hydraulically applied by a servo assembly. Connected to the drum is a member of the planetary gear train. The purpose of a band is to hold a member of the planetary gearset by holding the drum and connecting planetary gear member stationary. Bands provide excellent holding characteristics and require a minimum amount of space within the transmission housing.

When a band closes around a rotating drum, a wedging action takes place to stop the drum from rotating. The wedging action is known as self-energizing action. A typical band is designed to be larger in diameter than the drum it surrounds. This design promotes self-disengagement of the band from the drum when servo apply force is decreased to less than servo release spring tension. A friction material is bonded to the inside diameter of the band.

Typically, if the band will be holding a low-speed drum, the lining material of a band is a semimetallic compound. If the band is designed to hold a high-speed drum, it will have a paper-based lining.

Band lugs are either spot welded or cast as a part of the band assembly. The purpose of the lugs is to connect the band with the servo through the actuating (apply) linkage and the band anchor (reaction) at the opposite end. The band's steel strap is designed with slots or holes to release fluid trapped between the drum and the applying band.

SHOP TALK

A holding planetary control unit is also called a brake or reaction unit because it holds a gear train member stationary, reacting to rotation. ■

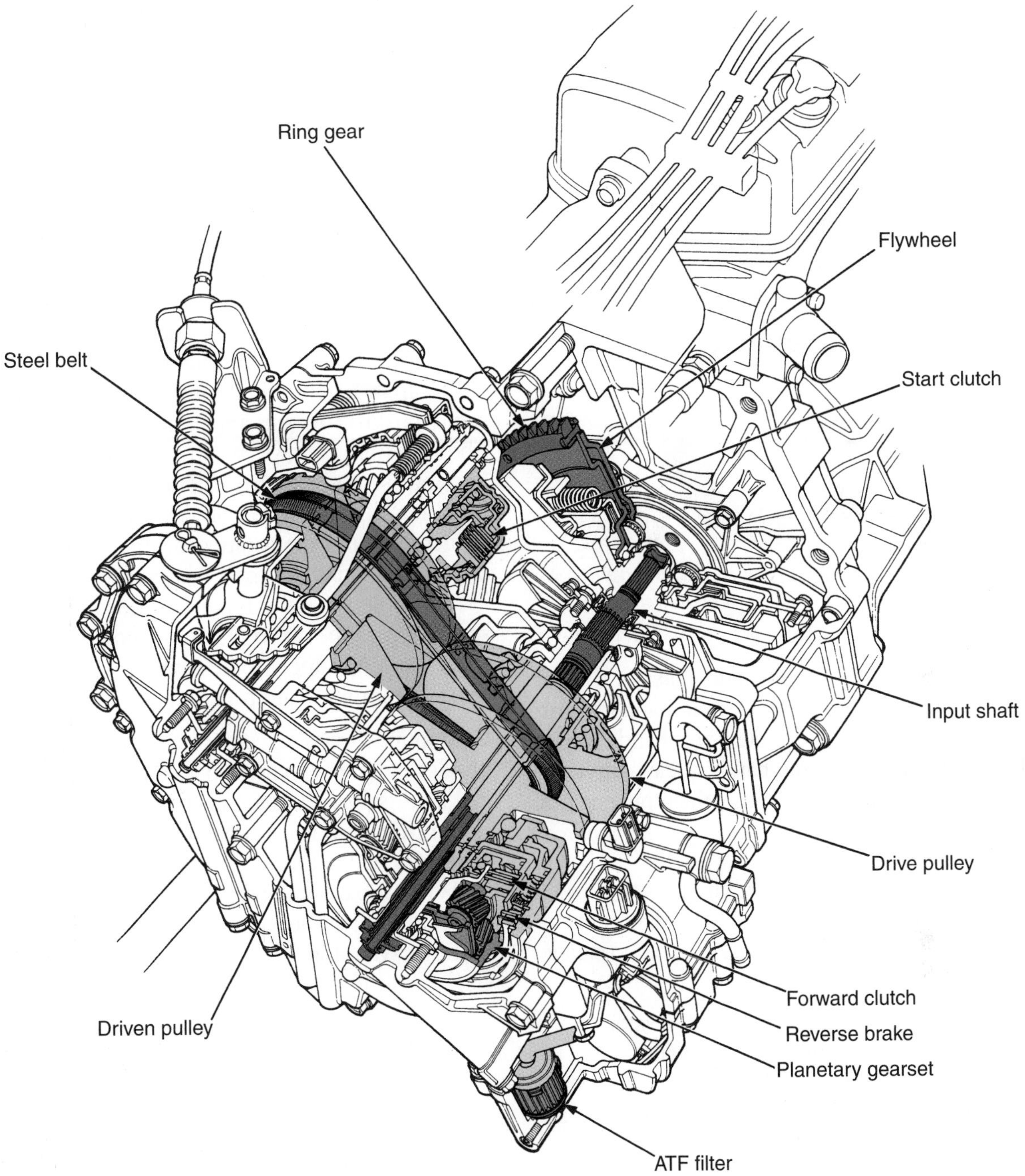

Figure 39–27 Honda's CVT. *Courtesy of American Honda Motor Co., Inc.*

Figure 39–28 CVTs use pulleys that change size and are connected by a belt. *Courtesy of Nissan North America, Inc.*

The bands used in automatic transmissions are rigid, flexible, single wrap, or double wrap types. Steel single wrap bands **(Figure 39–29A)** are used to hold gear train components driven by high-output engines. Self-energizing action is low because of the rigidity of the band's design. Thinner steel bands are not able to provide a high degree of holding power, but because of the flexibility of design, self-energizing action is stronger and provides more apply force.

The double wrap band is a circular external contracting band normally designed with two or three segments **(Figure 39–29B)**. As the band closes, the segments align themselves around the drum and provide a cushion. The

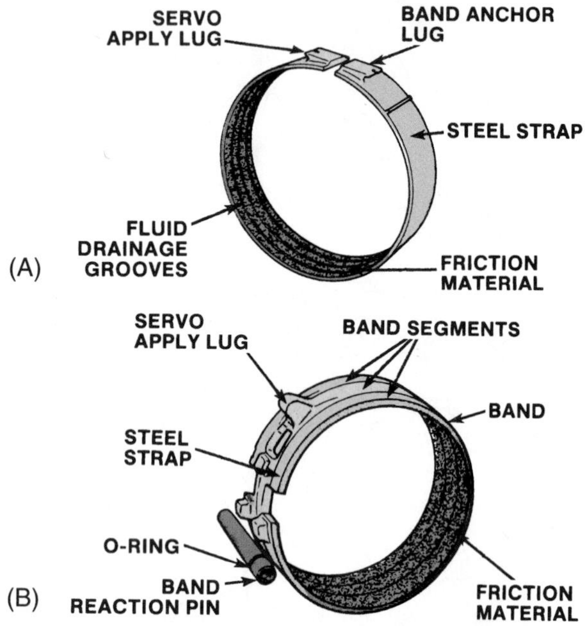

Figure 39-29 (A) Typical single-wrap and (B) double-wrap transmission band designs. *Courtesy of DaimlerChrysler Corporation*

steel body of the double wrap band may be thin or thick steel strapping material. Modern automatic transmissions use thin single or double wrap bands for increased efficiency. Double wrap bands made with heavy thick steel strapping are required for high output engines.

Transmission Servos

The **servo** assembly converts hydraulic pressure into a mechanical force that applies a band to hold a drum stationary. Simple and compound servos are used to engage bands in modern transmissions.

Simple Servo In a simple servo **(Figure 39–30)**, the servo piston fits into the servo cylinder and is held in the released position by a coil spring. The piston is sealed with a rubber ring, which keeps fluid pressure confined to the apply side of the servo piston.

In the illustration (but not on all servo designs), the piston pushrod is drilled through the center, which permits fluid pressure to be directed to the apply side of the servo piston. The pushrod moves the band apply strut, which is seated in the band apply lug, to tighten the band. At the opposite end of the band is the anchor strut and adjustment screw. They receive the engagement force of a band.

To apply a band, fluid pressure is directed down the servo pushrod to the apply side of the servo piston. The servo piston moves against the servo coil spring and develops servo apply force. This force is applied to the band lug through the apply lever and strut. The band tightens around the rotating drum. The rotating drum comes to a stop and is held stationary by the band.

When servo apply force is released, the servo coil spring forces the servo piston to move up the cylinder. With the servo apply force removed, the band springs free and permits drum rotation.

Compound Servo A compound servo **(Figure 39–31)** has a cylinder that is cast as part of the transmission housing. If the servo is located near the front of the transmission, it uses seal rings capable of withstanding the heat generated by the torque converter and engine.

When the compound servo is applied, fluid pressure flows through the hollow piston pushrod to the apply side of the servo piston. The piston compresses the servo coil spring, and forces the pushrod to move one end of the

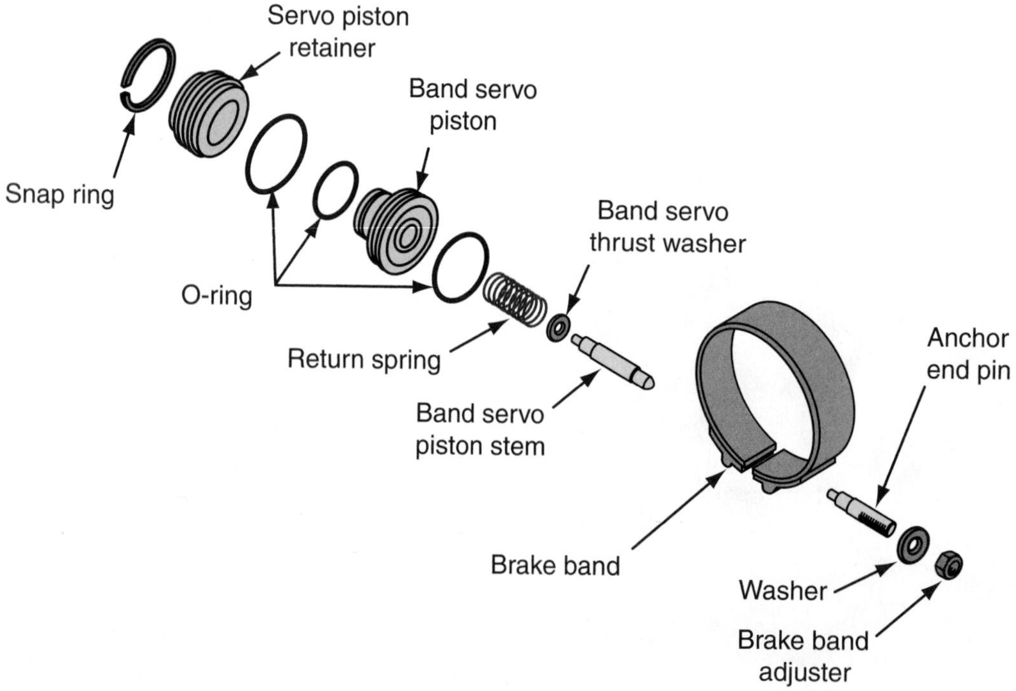

Figure 39-30 A typical band and servo assembly.

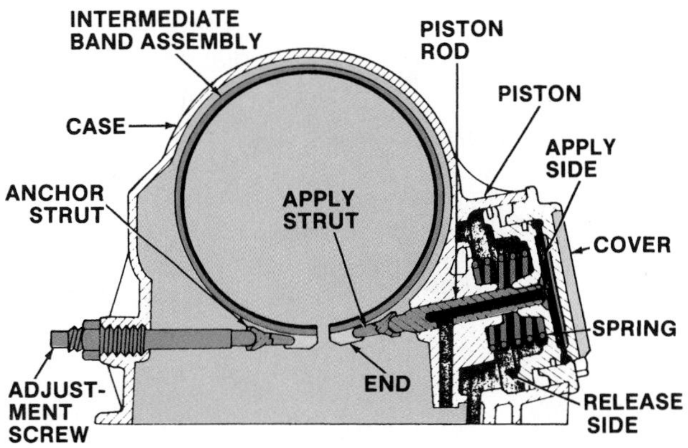

Figure 39-31 A typical compound servo design.
Courtesy of Ford Motor Company

band toward the adjusting screw and anchor. The band tightens around the rotating drum and brings it to a stop. The apply of the compound servo piston is much like the simple servo, but there the similarity ends.

Fluid pressure is applied to the release side of the servo piston when the band is to be released. This provides equal pressure on both sides of the piston and allows the tension of the servo spring to push the piston back up its bore. This action releases the band.

In some transmissions, the servo piston has a larger area on the release side, which causes a more positive release. This design is used to ensure that a band is released before another reaction member is applied.

TRANSMISSION CLUTCHES

In contrast to a band, which can only hold a planetary gear member, transmission clutches are capable of both holding and driving members.

Overrunning Clutches

In an automatic transmission operation, both sprag and roller overrunning clutches are used to hold or drive members of the planetary gearset. These clutches operate mechanically. An overrunning clutch allows rotation in only one direction and operates at all times. One-way overrunning clutches can be either roller-type or sprag-type clutches.

In a roller-type **(Figure 39–32)**, roller bearings are held in place by springs to separate the inner and outer race of the clutch assembly. Around the inside of the outer race are several cam-shaped indentations. The rollers and springs are located in these pockets. Rotation of one race in one direction locks the rollers between the two races, causing both to rotate together and prevent both from moving. When the race is rotated in the opposite direction, the roller bearings move into the pockets and are not locked and the races are free to rotate independently.

A one-way sprag clutch **(Figure 39–33)** consists of a hub and drum separated by figure-eight-shaped metal pieces called sprags. The sprags are shaped so that they lock between the races when a race is turned in one direction only. The sprags are longer than the distance between the two races. Springs hold the sprags at the correct angle and maintain the sprags' contact with the races, thereby allowing for instantaneous engagement. When a race rotates in one direction, the sprags lift and allow the races to move independently. When a race is moved in the opposite direction, the sprags straighten and lock the two races together.

Multiple-Disc Clutches

A **multiple-disc clutch** uses a series of friction discs to transmit torque or apply braking force. The discs have internal teeth that are sized and shaped to mesh with splines

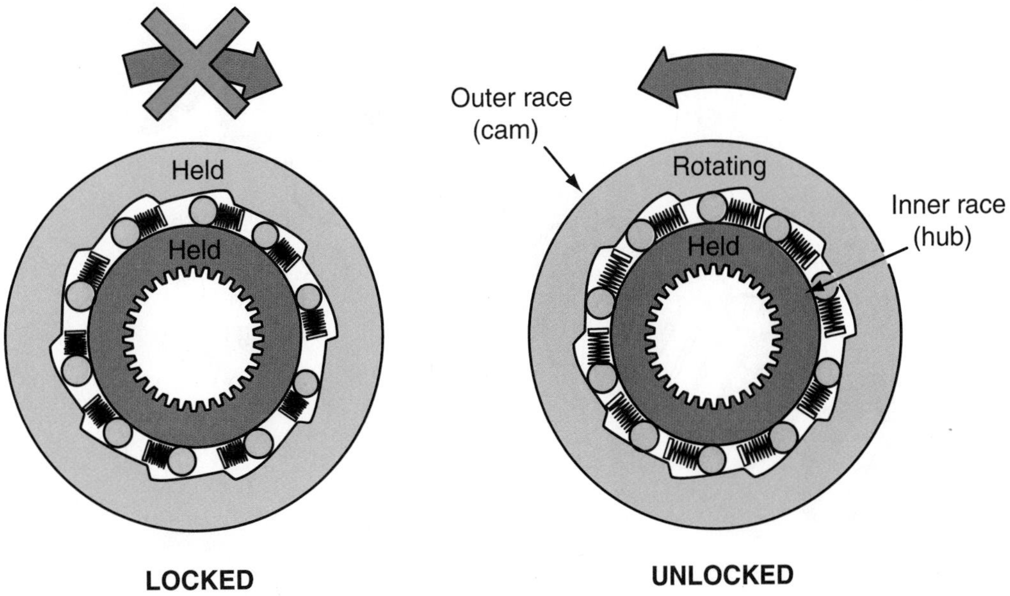

Figure 39-32 The action of a one-way roller clutch.

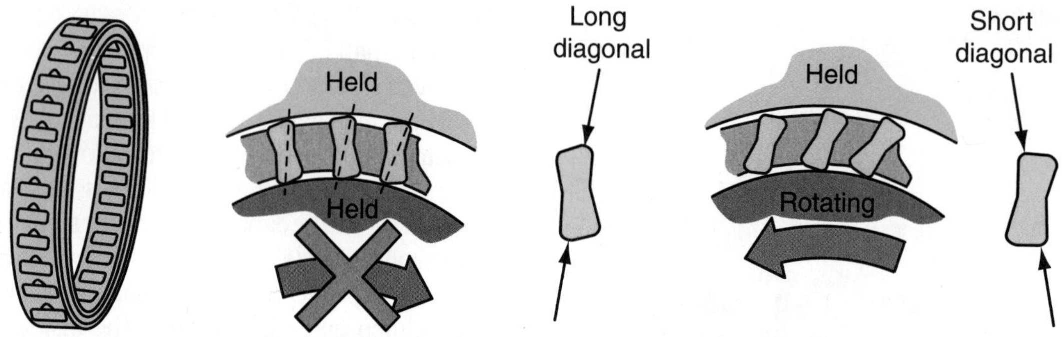

Figure 39-33 The action of a sprag-type one-way clutch.

on the clutch assembly hub. In turn, this hub is connected to a member of the planetary gearset that will receive the desired braking or transfer force when the clutch is applied or released.

Multiple-disc clutches are enclosed in a large drum-shaped housing that can be either a separate casting **(Figure 39–34)** or part of the existing transmission housing. This drum housing also holds the other clutch components: cylinder, hub, piston, piston return springs, seals, pressure plate, steel plates, friction plates, and snap rings.

The piston is made of cast aluminum or stamped steel with a seal ring groove around the outside diameter. A rubber seal ring is installed in this groove to retain the fluid pressure that disengages the clutch. Piston return springs overcome the reduced fluid pressure in the clutch and move the piston to the disengaged position when clutch action is no longer needed.

The clutch pack consists of steel clutch plates, friction discs, and one very thick plate known as the pressure plate. The pressure plate has tabs around the outside diameter to mate with the channels in the clutch drum. It is held in place by a large snap ring. At engagement, the clutch piston forces the clutch pack against the fixed pressure plate. Most steel clutch plates should be flat, and although the surface of the plate might appear smooth, it is specifically machined to promote a coefficient of friction to help transmit engine torque.

The friction discs are sandwiched between the clutch plates and pressure plate. Friction discs are steel core plates with friction material bonded to one or both sides. Asbestos was once the universal friction material used, but because it is hazardous to human health, cellular paper fibers, graphites, and ceramics are now being used as friction materials.

Planetary Control Terminology

Table 39–2 is a crossover chart listing the names that the different manufacturers call the same planetary gear control.

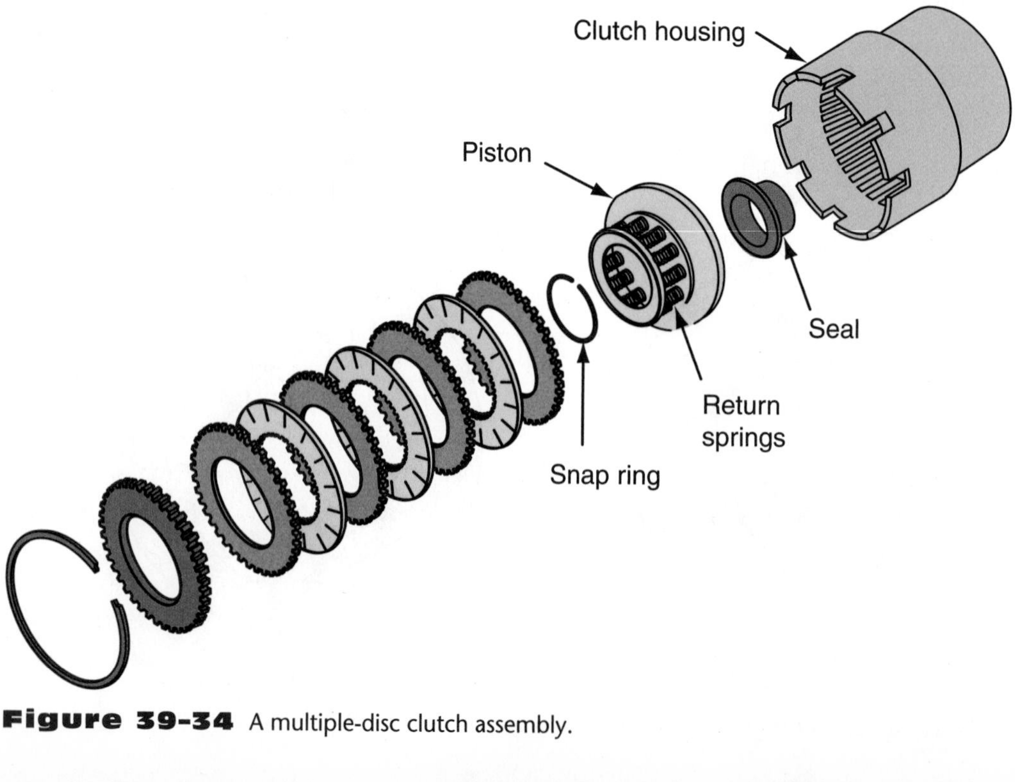

Figure 39-34 A multiple-disc clutch assembly.

TABLE 39–2 PLANETARY CONTROL TERMINOLOGY CROSSOVER CHART

Chrysler	Ford	General Motors
Front clutch	Reverse and high clutch	Direct clutch
Rear clutch	Forward clutch	Forward clutch
Front kickdown band	Intermediate band	Intermediate band
Low and reverse rear band	Low and reverse band or clutch	Low and reverse band or clutch
Overrunning clutch	One-way clutch	Low roller clutch

This terminology is a sure point of confusion while you are first learning about automatic transmissions. Sometimes manufacturers refer to a clutch or band by the speed gear(s) it controls, others use terms that define what the control does, and others, such as DaimlerChrysler, refer to the planetary controls by their location in the transaxle or transmission housing.

To help eliminate confusion, this text follows DaimlerChrysler's logic as much as possible. In doing so there may some discrepancies between what is in this text and what is in a service manual. Just keep in mind the purpose of this text is to give you a working knowledge of automatic transmissions. The purpose of a service manual is to give you the specific information needed to service or diagnose a specific transmission. Once you have a good working knowledge, you will be able to understand and use the information in the service manual.

BEARINGS, BUSHINGS, AND THRUST WASHERS

When a component slides over or rotates around another part, the surfaces that contact each other are called bearing surfaces. A gear rotating on a fixed shaft can have more than one bearing surface; it is supported and held in place by the shaft in a radial direction. Also the gear tends to move along the shaft in an axial direction as it rotates and is therefore held in place by some other components. The surfaces between the sides of the gear and the other parts are bearing surfaces.

A bearing is a device placed between two bearing surfaces to reduce friction and wear. Most bearings have surfaces that either slide or roll against each other. In automatic transmissions, sliding bearings are used where one or more of the following conditions prevail: low rotating speeds, very large bearing surfaces compared to the surfaces present, and low use applications. Rolling bearings are used in high-speed applications, high load with relatively small bearing surfaces, and high use.

Transmissions use sliding bearings composed of a relatively soft bronze alloy. Many are made from steel with the bearing surface bonded or fused to the steel. Those that take radial loads are called bushings and those that take axial loads are called thrust washers **(Figure 39–35)**. The bearing's surface usually runs against a harder surface such as steel to produce minimum friction and heat wear characteristics.

Bushings are cylindrically shaped and usually held in place by press fit. Since bushings are typically made of a soft metal, they act like a bearing and support many of the transmission's rotating parts **(Figure 39–36)**. They are also used to precisely guide the movement of various valves in the transmission's valve body. Bushings can also be used to control fluid flow; some restrict the flow from one part to another, while others are made to direct fluid flow to a particular point or part in the transmission.

Often serving both as a bearing and a spacer, thrust washers are made in various thicknesses. They may have one or more tangs or slots on the inside or outside circumference that mate with the shaft bore to keep them from turning. Some thrust washers are made of nylon or Teflon, which are used when the load is low. Others are fitted with rollers to reduce friction and wear.

Thrust washers normally control free axial movement or endplay. Since some endplay is necessary in all transmissions because of heat expansion, proper endplay is often accomplished through selective thrust washers. These thrust washers are inserted between various parts of the transmission. Whenever endplay is set, it must be set to manufacturer's specifications. Thrust washers work by filling the gap between two objects and become the primary wear item because they are made of softer materials than the parts they protect. Normally, thrust washers are made of copper- or babbit-faced soft steel, bronze, nylon, or plastic.

Torrington bearings (Figure 39–37) are thrust washers fitted with roller bearings. These thrust bearings are primarily used to limit endplay but also to reduce the friction between two rotating parts. Most often Torrington bearings are used in combination with flat thrust washers to control endplay of a shaft or the gap between a gear and its drum.

The bearing surface is greatly reduced through the use of roller bearings. The simplest roller bearing design leaves enough clearance between the bearing surfaces of two sliding or rotating parts to accept some rollers. Each roller's two points of contact between the bearing surfaces are so small that friction is greatly reduced. The bearing surface is more like a line than an area.

If the roller length to diameter is about 5:1, or more, the roller is called a needle and such a bearing is called a

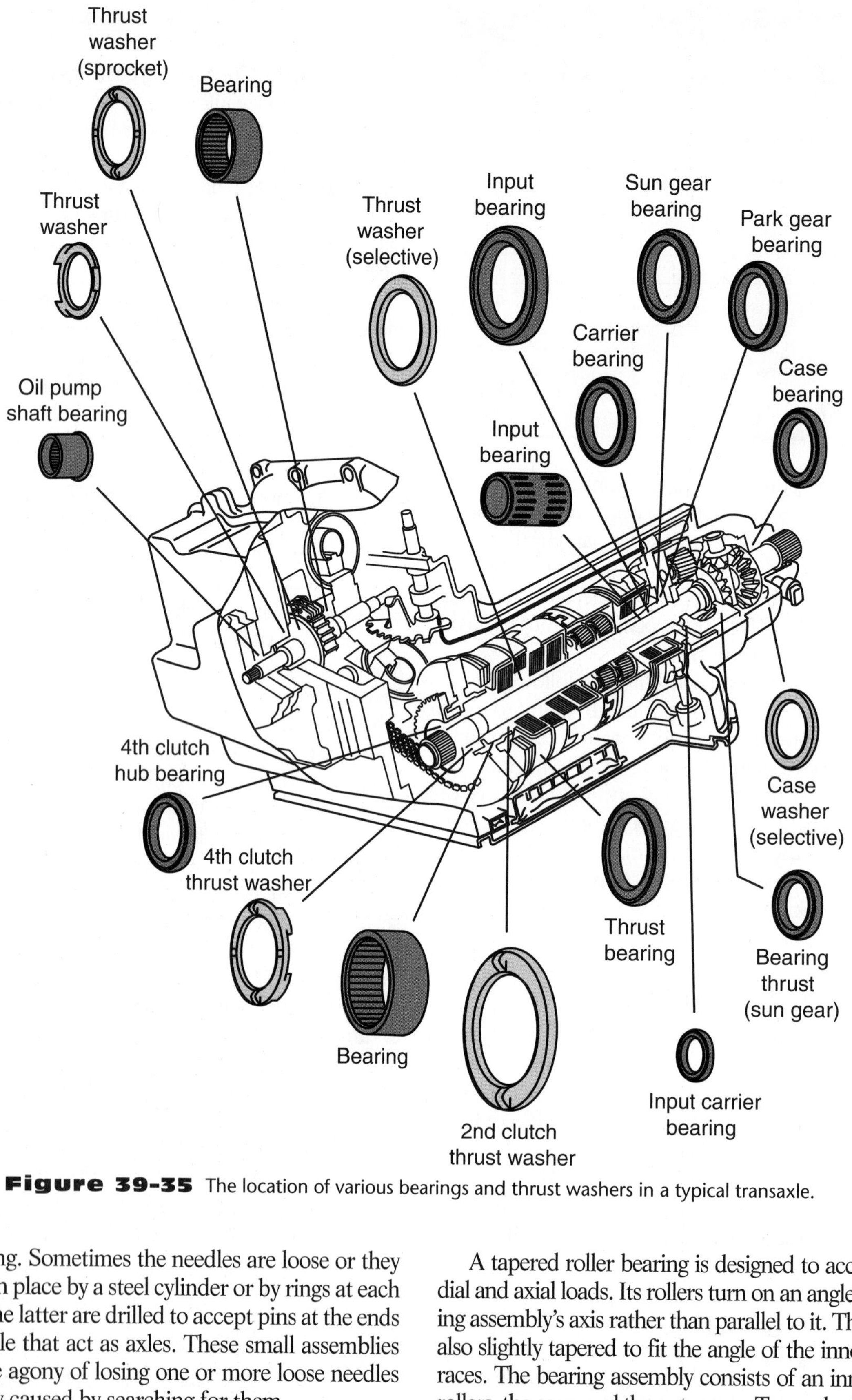

Figure 39-35 The location of various bearings and thrust washers in a typical transaxle.

needle bearing. Sometimes the needles are loose or they can be held in place by a steel cylinder or by rings at each end. Often the latter are drilled to accept pins at the ends of each needle that act as axles. These small assemblies help save the agony of losing one or more loose needles and the delay caused by searching for them.

Many other roller bearings are designed as assemblies. The assemblies consist of an inner and outer race, the rollers, and a cage. There are roller bearings designed for radial loads and others designed for axial loads.

A tapered roller bearing is designed to accept both radial and axial loads. Its rollers turn on an angle to the bearing assembly's axis rather than parallel to it. The rollers are also slightly tapered to fit the angle of the inner and outer races. The bearing assembly consists of an inner race, the rollers, the cage, and the outer race. Tapered roller bearings are normally used in pairs and are rarely used in automatic transmissions. They are commonly used in final drive units.

The heaviest radial loads in automatic transmissions are carried by either roller or ball bearings. Ball bearings

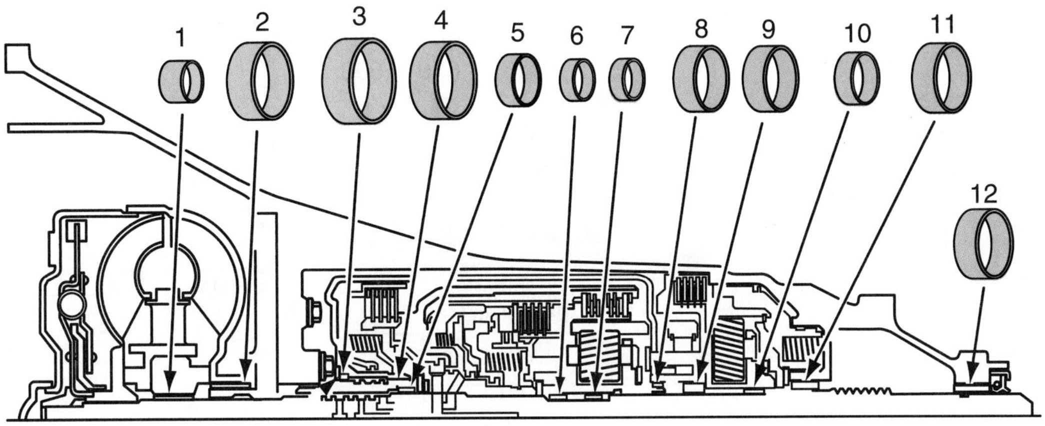

LEGEND

1. Bushing, Stator shaft (front)
2. Bushing, Oil pump body
3. Bushing, Reverse input clutch (front)
4. Bushing, Reverse input clutch (rear)
5. Bushing, Stator shaft (rear)
6. Bushing, Input sun gear (front)
7. Bushing, Input sun gear (rear)
8. Bushing, Reaction carrier shaft (front)
9. Bushing, Reaction gear
10. Bushing, Reaction carrier shaft (rear)
11. Bushing, Case
12. Bushing, Case extension

Figure 39-36 Bushings are used throughout a transmission.

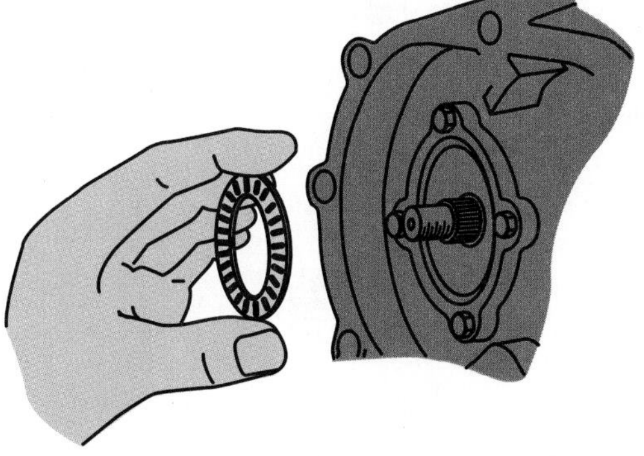

Figure 39-37 A Torrington-type thrust washer.

are constructed similarly to roller bearings, except that the races are grooved to accept the balls. The groove radius is slightly larger than the ball radius, which reduces the bearing surface area more than the roller bearing does. A ball bearing can also withstand light axial loads. Lip seals are sometimes built into ball bearings to retain lubricants.

SNAP RINGS

Many different sizes and types of snap rings are used in today's transmissions. External and internal snap rings are used as retaining devices throughout the transmission. Internal snap rings are used to hold servo assemblies and clutch assemblies together. In fact, snap rings are also available in several thicknesses and may be used to adjust the clearance in multiple-disc clutches. Some snap rings for clutch packs are waved to smooth clutch application. External snap rings are used to hold gear and clutch assemblies to their shafts.

GASKETS AND SEALS

The gaskets and seals of an automatic transmission help to contain the fluid within the transmission and prevent the fluid from leaking out of the various hydraulic circuits. Different types of seals are used in automatic transmissions; they can be made of rubber, metal, or Teflon **(Figure 39–38)**. Transmission gaskets are made of rubber, cork, paper, synthetic materials, metal, or plastic.

Gaskets

Gaskets are used to seal two parts together or to provide a passage for fluid flow from one part of the transmission to another. Gaskets are easily divided into two separate groups, hard and soft, depending on their application. Hard gaskets are used whenever the surfaces to be sealed are smooth. This type of gasket is usually made of paper. A common application of a hard gasket is the gasket used to seal the valve body and oil pump against the transmission case. Hard gaskets are also often used to direct fluid flow or to seal off some passages between the valve body and the separator plate.

Gaskets that are used when the sealing surfaces are irregular or in places where the surface may distort when the component is tightened into place are called soft gaskets. A typical location of a soft gasket is the oil pan gasket that seals the oil pan to the transmission case. Oil pan gaskets are typically a composition-type gasket made with rubber and cork. However, some late-model transmissions use an RTV sealant instead of a gasket to seal the oil pan.

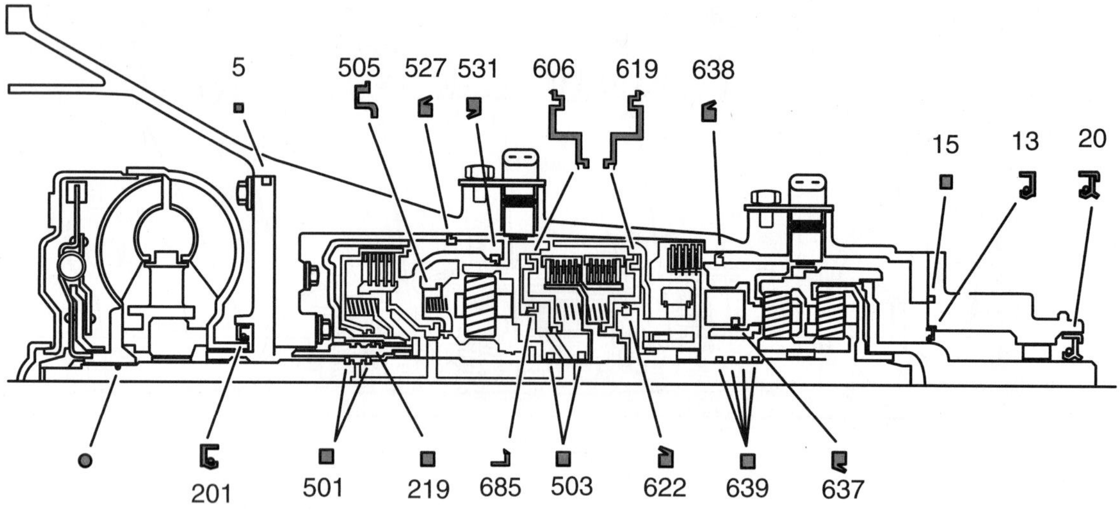

LEGEND

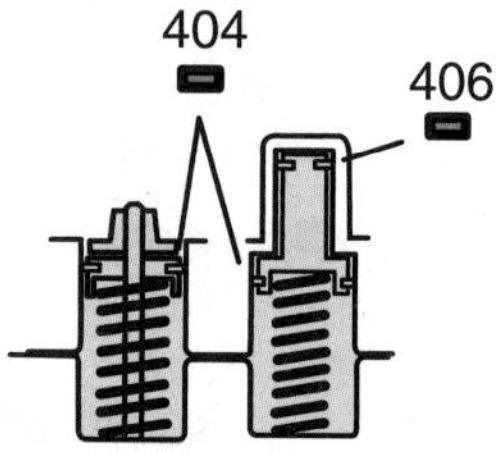

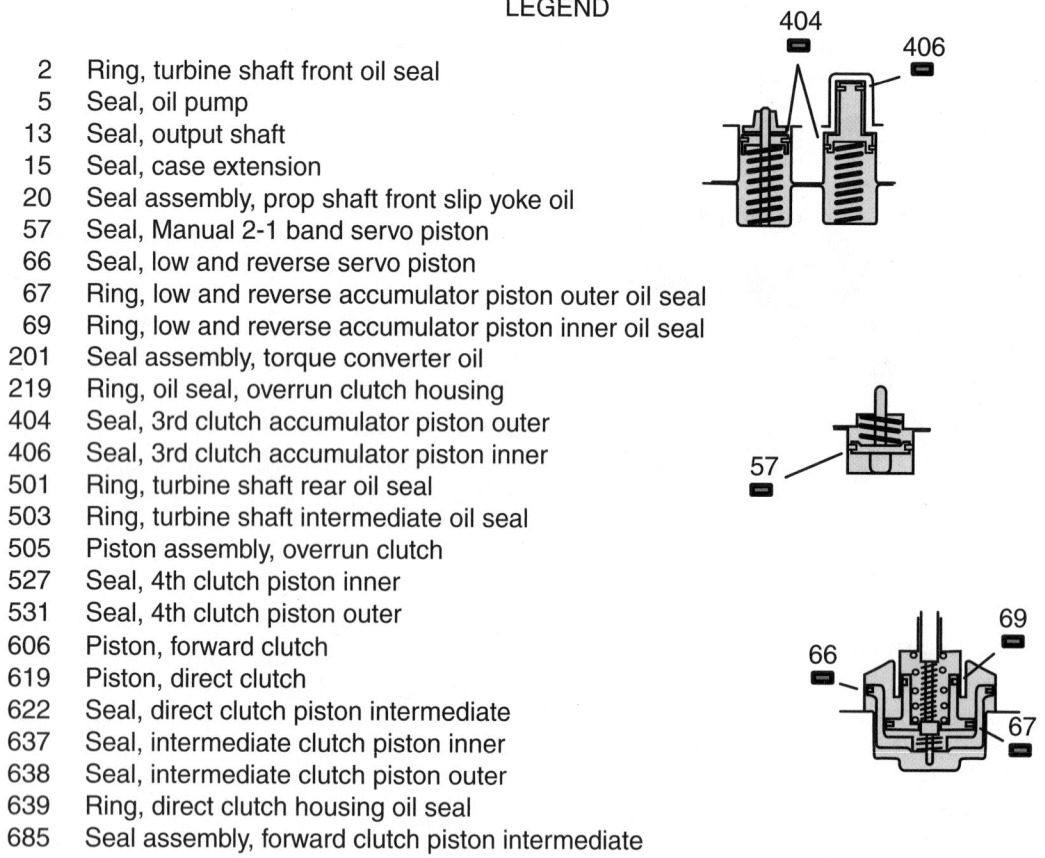

2 Ring, turbine shaft front oil seal
5 Seal, oil pump
13 Seal, output shaft
15 Seal, case extension
20 Seal assembly, prop shaft front slip yoke oil
57 Seal, Manual 2-1 band servo piston
66 Seal, low and reverse servo piston
67 Ring, low and reverse accumulator piston outer oil seal
69 Ring, low and reverse accumulator piston inner oil seal
201 Seal assembly, torque converter oil
219 Ring, oil seal, overrun clutch housing
404 Seal, 3rd clutch accumulator piston outer
406 Seal, 3rd clutch accumulator piston inner
501 Ring, turbine shaft rear oil seal
503 Ring, turbine shaft intermediate oil seal
505 Piston assembly, overrun clutch
527 Seal, 4th clutch piston inner
531 Seal, 4th clutch piston outer
606 Piston, forward clutch
619 Piston, direct clutch
622 Seal, direct clutch piston intermediate
637 Seal, intermediate clutch piston inner
638 Seal, intermediate clutch piston outer
639 Ring, direct clutch housing oil seal
685 Seal assembly, forward clutch piston intermediate

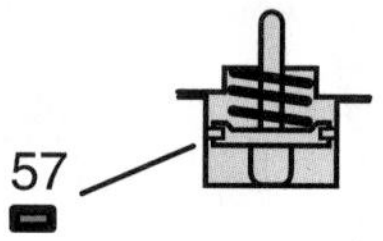

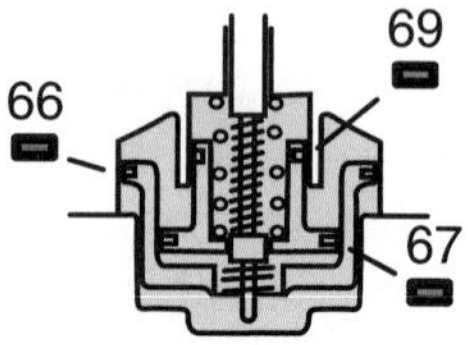

Figure 39-38 The location of various seals and gaskets in a typical transmission.

Seals

As valves and transmission shafts move within the transmission, it is essential that the fluid and pressure be contained within its bore. Any leakage would decrease the pressure and result in poor transmission operation. Seals are used to prevent leakage around valves, shafts, and other moving parts. Rubber, metal, or Teflon materials are used throughout a transmission to provide for static and dynamic sealing. Both static and dynamic seals can provide for positive and nonpositive sealing. A definition of each of the different basic classifications of seals follows:

- *Static.* A seal used between two parts that do not move in relationship to each other.
- *Dynamic.* A seal used between two parts that do move in relationship to each other. This movement is either a rotating or reciprocating (up and down) motion.

- *Positive.* A seal that prevents all fluid leakage between two parts.
- *Nonpositive.* A seal that allows a controlled amount of fluid leakage. This leakage is typically used to lubricate a moving part.

Three major types of rubber seals are used in automatic transmissions: the O-ring, the lip seal, and the square-cut seal. Rubber seals are made from synthetic rubber rather than natural rubber.

O-rings are round seals with a circular cross section. Normally an O-ring is installed in a groove cut into the inside diameter of one of the parts to be sealed. When the other part is inserted into the bore and through the O-ring, the O-ring is compressed between the inner part and the groove. This pressure distorts the O-ring and forms a tight seal between the two parts.

O-rings can be used as dynamic seals but are most commonly used as static seals. An O-ring can be used as a dynamic seal when the parts have relatively low amounts of axial movement. If there is a considerable amount of axial movement, the O-ring will quickly be damaged as it rolls within its groove. O-rings are never used to seal a shaft or part that has rotational movement.

Lip seals are used to seal parts that have axial or rotational movement. They are round to fit around a shaft but the entire seal does not serve as a seal; rather, the sealing part is a flexible lip. The flexible lip is normally made of synthetic rubber and shaped so that it is flexed when it is installed to apply pressure at the sharp edge of the lip. Lip seals are used around input and output shafts to keep fluid in the housing and dirt out. Some seals are double-lipped.

When the lip is around the outside diameter of the seal, it is used as a piston seal **(Figure 39–39)**. Piston seals are designed to seal against high pressures and the seal is positioned so that the lip faces the source of the pressurized fluid. The lip is pressed firmly against the cylinder wall as the fluid pushes against the lip; this forms a tight seal. The lip then relaxes its seal when the pressure on it is reduced or exhausted.

Lip seals are also commonly used as shaft seals. When used to seal a rotating shaft, the lip of the seal is around the inside diameter of the seal and the outer diameter is bonded to the inside of a metal housing. The outer metal housing is pressed into a bore. To help maintain good sealing pressure on the rotating shaft, a garter spring is fitted behind the lip. This toroidal spring pushes on the lip to provide for uniform contact on the shaft. Shaft seals are not designed to contain pressurized fluid; rather, they are designed to prevent fluid from leaking over the shaft and out of the housing. The tension of the spring and of the lip is designed to allow an oil film of about 0.0001 (.00254 mm) of an inch. This oil film serves as a lubricant for the lip. If the tolerances increase, fluid will be able to leak past the shaft and if the tolerances are too small, excessive shaft and seal wear will result.

A **square-cut seal** is similar to an O-ring; however, a square-cut seal can withstand more axial movement than an O-ring can. Square-cut seals have a rectangular or square cross section. They are designed this way to prevent the seal from rolling in its groove when there are large amounts of axial movement. Added sealing comes from the distortion of the seal during axial movement. As the shaft inside the seal moves, the outer edge of the seal moves more than the inner edge causing the diameter of the sealing edge to increase, which creates a tighter seal.

Metal Sealing Rings

There are some parts of the transmission that do not require a positive seal and in which some leakage is acceptable. These components are sealed with ring seals that fit into a groove on a shaft **(Figure 39–40)**. The outside diameter of the ring seals slide against the walls of the bore into which the shaft is inserted. Most ring seals in a transmission are placed near pressurized fluid outlets on rotating shafts to help retain pressure. Ring seals are made of cast iron, nylon, or Teflon.

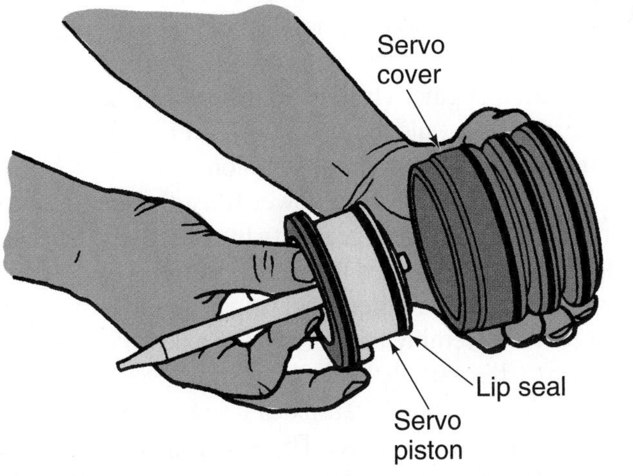

Figure 39-39 Typical application of a lip seal. *Courtesy of Ford Motor Company*

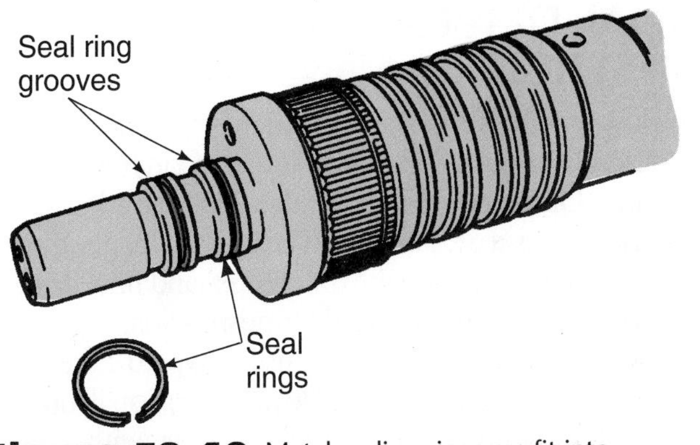

Figure 39-40 Metal sealing rings are fit into grooves on a shaft.

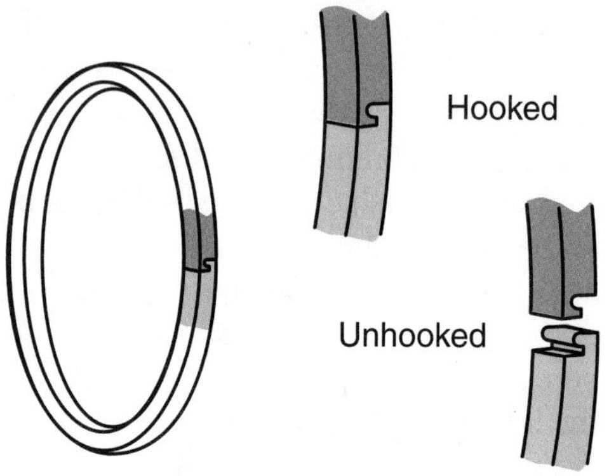

Figure 39-41 Hook-end sealing rings.

Three types of metal seals are used in automatic transmissions: **butt-end seals**, open-end seals, and hook-end seals. In appearance, butt-end and open-end seals are much the same; however, when an open-end seal is installed, there is a gap between the ends of the seal. **Hook-end seals (Figure 39–41)** have small hooks at their ends that are locked together during installation to provide better sealing than the open-end or butt-end seals.

Teflon Seals

Some transmissions use Teflon seals instead of metal seals. Teflon provides for a softer sealing surface, which results in less wear on the surface that it rides on and therefore a longer-lasting seal. Teflon seals are similar in appearance to metal seals except for the hook-end type. The ends of locking-end Teflon seals are cut at an angle **(Figure 39–42)** and the locking hooks are somewhat staggered. These seals are often called scarf-cut seals.

Many late-model transmissions are equipped with solid one-piece Teflon seals. Although the one-piece seal requires some special tools for installation, they provide for a nearly positive seal. These Teflon rings form a better seal than other metal sealing rings.

General Motors uses a different type of synthetic seal on some late-model transmissions. The material used in these seals is Vespel, which is a flexible but highly durable plasticlike material.

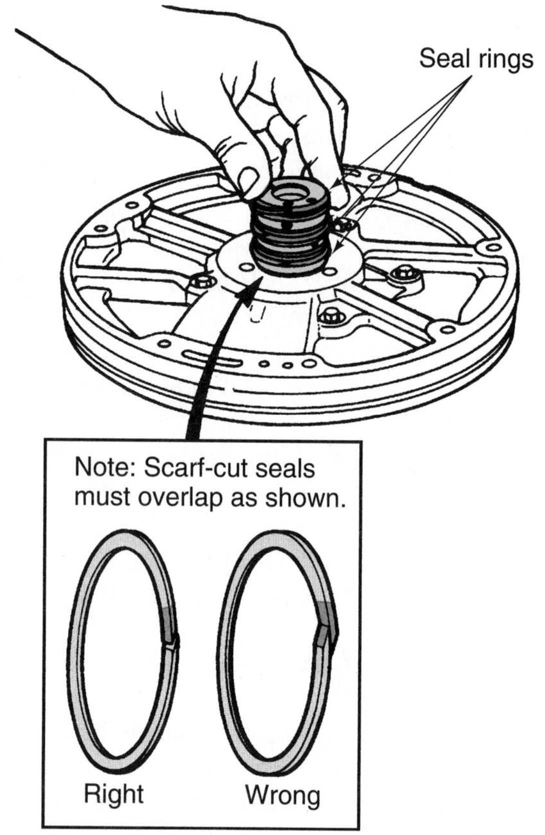

Figure 39-42 Scarf-cut seals. *Courtesy of Ford Motor Company*

FINAL DRIVES AND DIFFERENTIALS

The last set of gears in the drive train is the final drive. In most RWD cars, the final drive is located in the rear axle housing. On most FWD cars, the final drive is located within the transaxle. Some FWD cars with longitudinally mounted engines locate the differential and final drive in a separate case that bolts to the transmission.

RWD final drives normally use a hypoid final drive gearset that turns the power flow 90 degrees from the drive shaft to the drive axles. On FWD cars with a transversely mounted engines, the power flow axis is naturally parallel to that of the drive axles, therefore the power does not need to turn. Simple gear connections can be made to connect the output of the transmission to the final drive.

Final Drive Assemblies

A transaxle's final drive gears provide a way to transmit the transmission's output to the differential section of the transaxle. There are four common configurations used as the final drives on FWD vehicles: helical gear, planetary gear, hypoid gear, and chain drive. The helical, planetary, and chain final drive arrangements are found with transversely mounted engines. Hypoid final drive gear assemblies are normally found in vehicles with a longitudinally placed engine. The hypoid assembly is basically the same unit as would be used on RWD vehicles and is mounted directly to the transmission.

Some transaxles route power from the transmission through two helical-cut gears to a transfer shaft. A helical-cut pinion gear attached to the opposite end of the transfer shaft drives the differential ring gear and carrier. The differential assembly then drives the axles and wheels.

Rather than use helical-cut or spur gears in the final drive assembly, some transaxles use a simple planetary gearset for its final drive. The sun gear of this planetary unit is driven by the final drive sun gear shaft, which is splined to the front carrier and rear ring gear of the transmission's gearset. The final drive sun gear meshes with the final drive planetary pinion gears, which rotate on their shafts in the planetary carrier. The pinion gears mesh with the ring gear, which is splined to the transaxle case. The planetary carrier is part of the differential case, which contains typical differential gearing, two pinion gears, and two side gears.

The ring gear of a planetary final drive assembly has lugs around its outside diameter that fit into grooves machined inside the transaxle housing. These lugs and grooves hold the ring gear stationary. The transmission's output is connected to the planetary gearset's sun gear. In operation, the transmission's output drives the sun gear that, in turn, drives the planetary pinion gears. The pinion gears walk around the inside of the stationary ring gear. The rotating planetary pinion gears drive the planetary carrier and differential case. This combination provides maximum torque multiplication from a simple planetary gearset.

Chain-drive final drive assemblies use a multiple-link chain to connect a drive sprocket, connected to the transmission's output shaft, to a driven sprocket that is connected to the differential's pinion shaft. This design allows for remote positioning of the differential within the transaxle housing. Final drive gear ratios are determined by the size of the driven sprocket compared to the drive sprocket.

HYDRAULIC SYSTEM

A hydraulic system uses a liquid to perform work. In an automatic transmission, this liquid is automatic transmission fluid (ATF). ATF is one of the most complex fluids produced by the petroleum industry for the automobile.

ATF is a special oil designed for transmission operation. Transmissions are equipped with a fluid cooler that prevents the overheating of the fluid, which can result in damage to the transmission. The transmission's pump is the source of all fluid flow and provides a constant supply of fluid under pressure to operate, lubricate, and cool the transmission. **Pressure-regulating valves** change the fluid's pressure to control the shift quality and, in some transmissions, the shift points of the transmission. **Flow-directing valves** direct the pressurized fluid to the appropriate apply device to cause a change in gear ratios. The hydraulic system also keeps the torque converter filled with fluid.

Hydraulic Principles

An automatic transmission uses ATF fluid pressure to change gears automatically through the use of various pressure regulators and control valves.

Basic hydraulic theory is explained in Chapter 7 of this book. Hydraulic circuits are direct applications of Pascal's law, which states, "Pressure exerted on a confined liquid or fluid is transmitted undiminished and equally in all directions and acts with equal force on all areas."

Fluids are perfect conductors of pressure. Therefore, when a piston in a cylinder moves and displaces fluid, that fluid is distributed equally within the circuit.

To form a complete, working hydraulic system, the following elements are needed: a fluid reservoir, a pressure source, control valving, and output devices.

The automatic transmission reservoir is the transmission oil pan. Transmission fluid is drawn from the pan and returned to it. The pressure source in the system is the oil pump. The valve body contains control valving to regulate or restrict the pressure and flow of fluid within the transmission. Output devices are the servos or clutches operated by hydraulic pressure.

APPLICATION OF HYDRAULICS IN TRANSMISSIONS

A common hydraulic system within an automatic transmission is the servo assembly, which is used to control the application of a band. The band must tightly hold the drum or planetary carrier it surrounds when it is applied. The holding capacity of the band is determined by the construction of the band and the pressure applied to it. This pressure or holding force is the result of the action of a servo. The servo multiplies the force through hydraulic action.

If a servo has a diameter of 4 inches (102 mm) and has a pressure of 70 psi (4.9 kg/cm^2) applied to it, the apply force of the servo is 880 pounds (399 kg) **(Figure 39–43)**. The force exerted by the servo is further increased by its lever-type linkage and the self-energizing action of the band. The total force applied by the band stops and holds the rotating drum connected to a planetary gearset member.

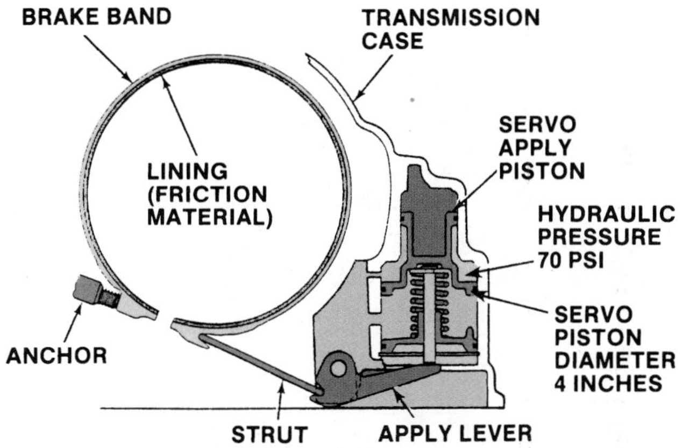

FORCE = $\pi r^2 \times$ psi
= 3.14 × 2 × 2 × 70 = 880 POUNDS

Figure 39–43 Calculating the output force developed by a servo assembly. *Courtesy of DaimlerChrysler Corporation*

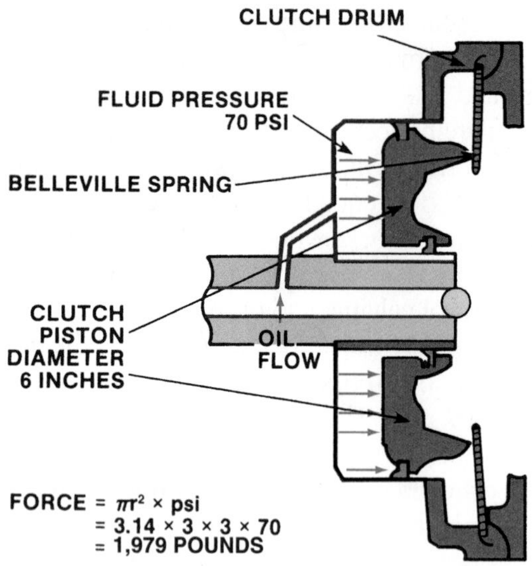

Figure 39-44 Using hydraulics to increase work in a multiple-disc clutch. *Courtesy of DaimlerChrysler Corporation*

A multiple-disc assembly is also used to stop and hold gear set members. This assembly also uses hydraulics to increase its holding force. If the fluid pressure applied to the clutch assembly is 70 psi (4.9 kg/cm^2) and the diameter of the clutch piston is 6 inches (152 mm), the force applying the clutch pack is 1,979 pounds (898 kg). If the clutch assembly uses a **Belleville spring** or piston spring **(Figure 39–44)**, which adds a mechanical advantage of 1.25, the total force available to engage the clutch will be 1,979 pounds (898 kg) multiplied by 1.25, or 2,474 pounds (1,122.5 kg).

Functions of ATF

The ATF circulating through the transmission and torque converter and over the parts of the transmission cools the transmission. The heated fluid typically moves to a transmission fluid cooler, where the heat is removed. As the fluid lubricates and cools the transmission, it also cleans the parts. The dirt is carried by the fluid to a filter, where the dirt is removed.

Another critical job of ATF is its role in shifting gears. ATF moves under pressure throughout the transmission and causes various valves to move. The pressure of the ATF changes with changes in engine speed and load.

ATF is also used to operate the various apply devices (clutches and bands) in the transmission. At the appropriate time, a switching valve opens and sends pressurized fluid to the apply device that engages or disengages a gear. The valving and hydraulic circuits are contained in the valve body.

Reservoir

A fluid reservoir stores fluid and provides a constant source of fluid for the system. In an automatic transmission, the reservoir is the pan, typically located at the bottom of the transmission case. ATF is forced out of the pan by atmospheric pressure and into the pump, and is then returned to it after it has circulated through the selected circuits. A transmission dipstick placed within a filler tube is typically used to check the level of the fluid and to add ATF to the transmission. Other transmissions have a side plug on the pan or the transmission to check and replenish fluid level.

Venting

All reservoirs must have an air vent that allows atmospheric pressure to force the fluid into the pump when the pump creates a low pressure at its inlet port. The pans of many automatic transmissions vent through the handle of the dipstick; others rely on a vent in the transmission case. Transmissions must also be vented to allow for the exhaust of built-up air pressure that results from heat and the moving components inside the transmission. The movement of these parts can force air into the ATF, which would not allow it to increase in pressure, cool, or lubricate the transmission properly.

Transmission Coolers

The removal of heat from ATF is extremely important to the durability of the transmission. Excessive heat causes the fluid to break down. Once broken down, ATF no longer lubricates well and has poor resistance to oxidation. Oxidized ATF may damage transmission seals. When a transmission is operated for some time with overheated ATF, varnish is formed inside the transmission. Varnish buildup on valves can cause them to stick or move slowly. The result is poor shifting and glazed or burned friction surfaces. Continued operation can lead to the need for a complete rebuilding of the transmission.

It is important to note that ATF is designed to operate at 175°F (80°C). At this temperature, the fluid should remain effective for 100,000 miles (160,000 km). However, when the operating temperature increases, the useful life of the fluid quickly decreases. A 20°F increase in operating temperature will decrease the life of ATF by one-half!

Transmission housings are fitted with ATF cooler lines **(Figure 39–45A)** that direct the hot fluid from the torque converter to the transmission cooler, normally located in the vehicle's radiator. The heat of the fluid is reduced by the cooler and the cool ATF returns to the transmission. In some transmissions, the cooled fluid flows directly to the transmission's bushings, bearings, and gears. Then, the fluid is circulated through the rest of the transmission. The cooled fluid in other transmissions is returned to the oil pan, where it is drawn into the pump and circulated throughout the transmission.

Some vehicles, such as those designed for heavy-duty use, are equipped with an auxiliary fluid cooler **(Figure 39–45B)**, in addition to the one in the radiator. This

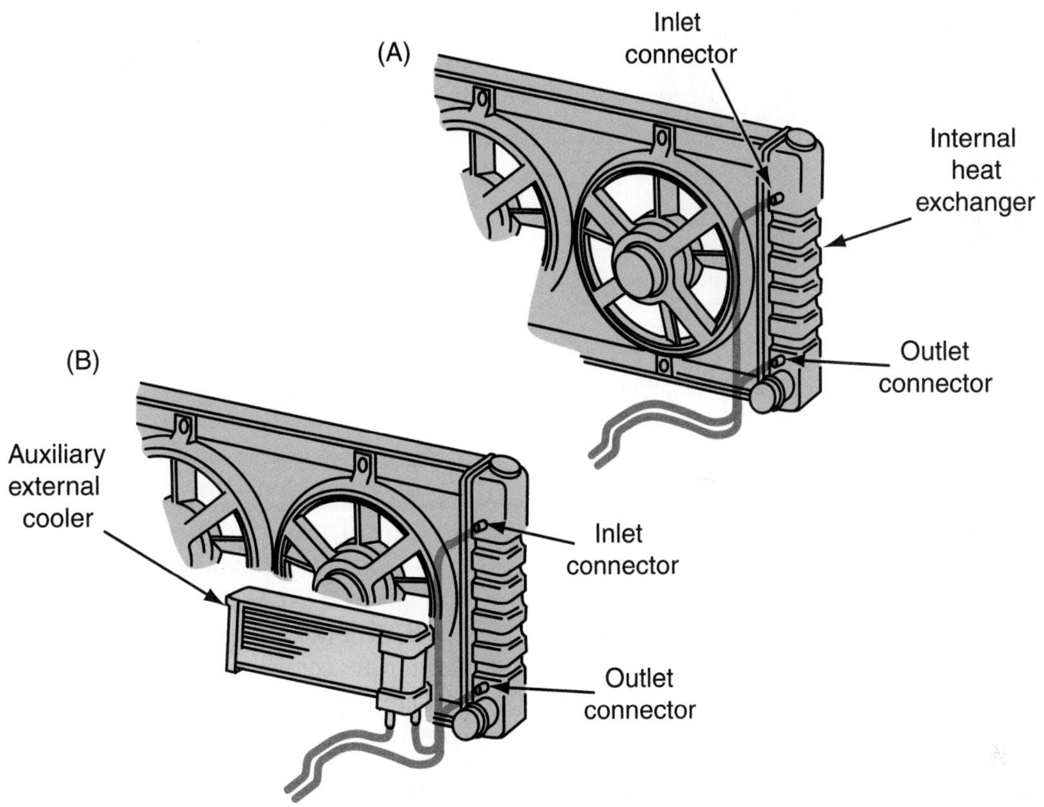

Figure 39–45 (A) A transmission cooler (heat exchanger) located in a radiator. (B) An auxiliary cooler added to the normal cooler circuit.

cooler remove additional amounts of heat from the fluid before it is sent back to the transmission.

Valve Body

For efficient transmission operation, the bands and multiple-disc packs must be released and applied at the proper time. The **valve body** assembly **(Figure 39–46)** is responsible for the control and distribution of pressurized fluid throughout the transmission. This assembly is made of two or three main parts: a valve body, separator plate, and transfer plate. These parts are bolted as a single unit to the transmission housing. The valve body is machined from aluminum or iron and has many precisely machined bores and fluid passages. Various valves are fitted into the bores, and the passages direct fluid to various valves and other parts of the transmission. The separator and transfer plates are designed to seal off some of these passages and to allow fluid to flow through specific passages.

Figure 39–46 A typical valve body.

The purpose of a valve body is to sense and respond to engine and vehicle load as well as to meet the needs of the driver. Valve bodies are normally fitted with three different types of valves: spool valves, check ball valves, and poppet valves. The purpose of these valves is to start, to stop, or to use movable parts to regulate and direct the flow of fluid throughout the transmission.

Check Ball Valve The **check ball valve** is a ball that operates on a seat located on the valve body. The check ball operates by having a fluid pressure or manually operated linkage force it against the ball seat to block fluid flow **(Figure 39–47)**. Pressure on the opposite side unseats the

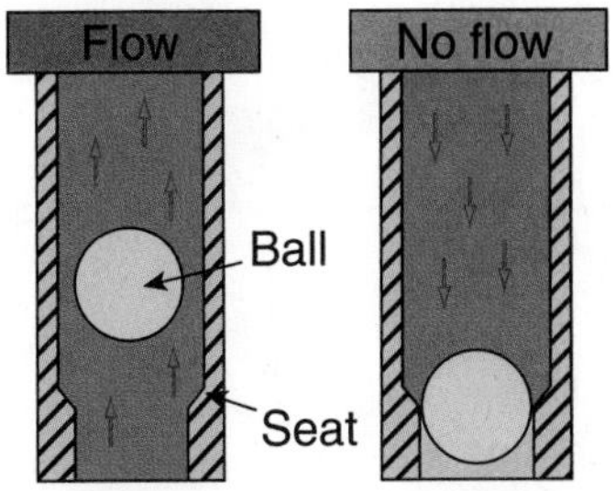

Figure 39–47 The operation of a check ball valve.
Courtesy of DaimlerChrysler Corporation

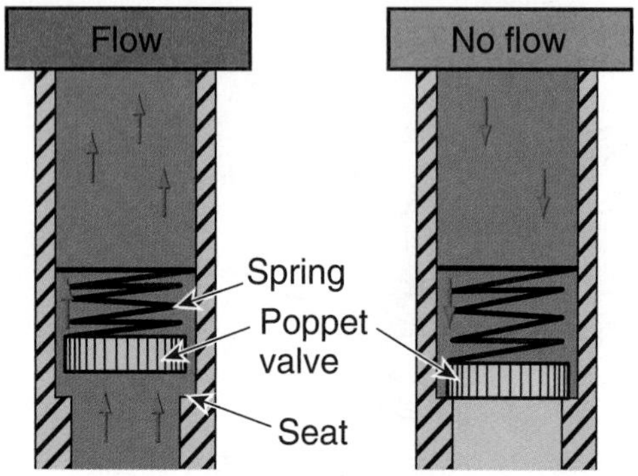

Figure 39-48 Typical poppet valve operation. *Courtesy of DaimlerChrysler Corporation*

check ball. Check balls and poppet valves can be normally open, which allows free flow of fluid pressure, or normally closed, which blocks fluid pressure flow.

At times, the check ball has two seats to check and direct fluid flow from two directions, being seated and unseated by pressures from either source.

Poppet Valve A **poppet valve (Figure 39–48)** can be a ball or a flat disc. In either case, the poppet valve blocks fluid flow. Often the poppet valve has a stem to guide the valve's operation. The stem normally fits into a hole acting as a guide to the valve's opening and closing. Poppet valves tend to pop open and closed, hence their name. Normally poppet valves are held closed by a spring.

Spool Valve The most commonly used valve in a valve body is the **spool valve**. A spool valve **(Figure 39–49)** looks similar to a sewing thread spool. The large circular parts of the valve are called the lands. There is a minimum of two lands per valve. Each land of the assembly is connected by a stem. The space between the lands and stem is called the valley. Valleys form a fluid pressure chamber between the spools and valve body bore. Fluid flow can be directed into other passages depending on the spool valve and valve body design.

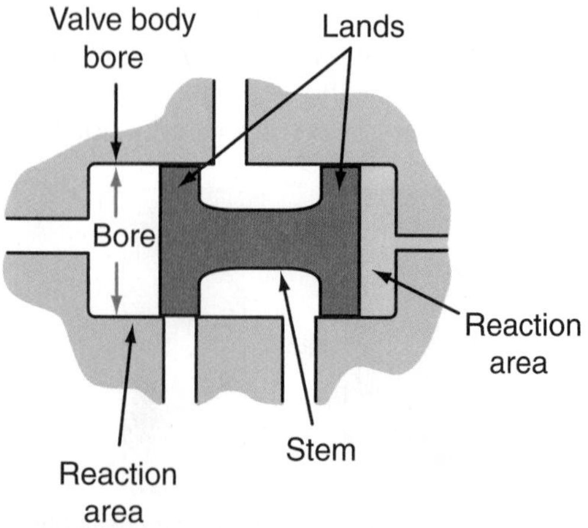

Figure 39-49 Components of a spool valve assembly.

Precisely machined around the periphery of each valve, the land is the part of the valve that rides on a thin film of fluid in a bore of the valve body. The land must be treated very carefully because any damage, even a small score or scratch, can impair smooth valve operation. As the spool valve moves, the land covers (closes) or uncovers (opens) ports in the valve body.

The reaction area, also known as the face, is the space at the outside of the lands at the end of the valve. Forces acting against the reaction area that cause the valve to move include spring tension, fluid pressure, or mechanical linkage.

Oil Pump

The source of fluid flow through the transmission is the oil pump **(Figure 39–50)**. Three types of oil pumps are commonly used in automatic transmissions: the gear type **(Figure 39–51A)**, rotor type, and vane type **(Figure 39–51B)**. Oil pumps are driven by the pump drive hub of the torque converter or oil pump shaft and/or converter cover on transaxles. Therefore, whenever the torque converter cover is rotating, the oil pump is driven. The oil pump creates fluid flow throughout the transmission.

Pressure Regulator Valve

Transmissions are capable of creating excessive amounts of pressure that may cause damage, therefore, the transmission is equipped with a pressure regulator valve, normally located in the valve body. Pressure regulating valves are typically spool valves that toggle back and forth in their bores to open and close an exhaust passage. By opening the exhaust passage, the valve decreases the pressure of the fluid. As soon as the pressure decreases to a predetermined amount, the spool valve moves to close off

Figure 39-50 An oil pump. *Courtesy of Transtar Industries Inc.*

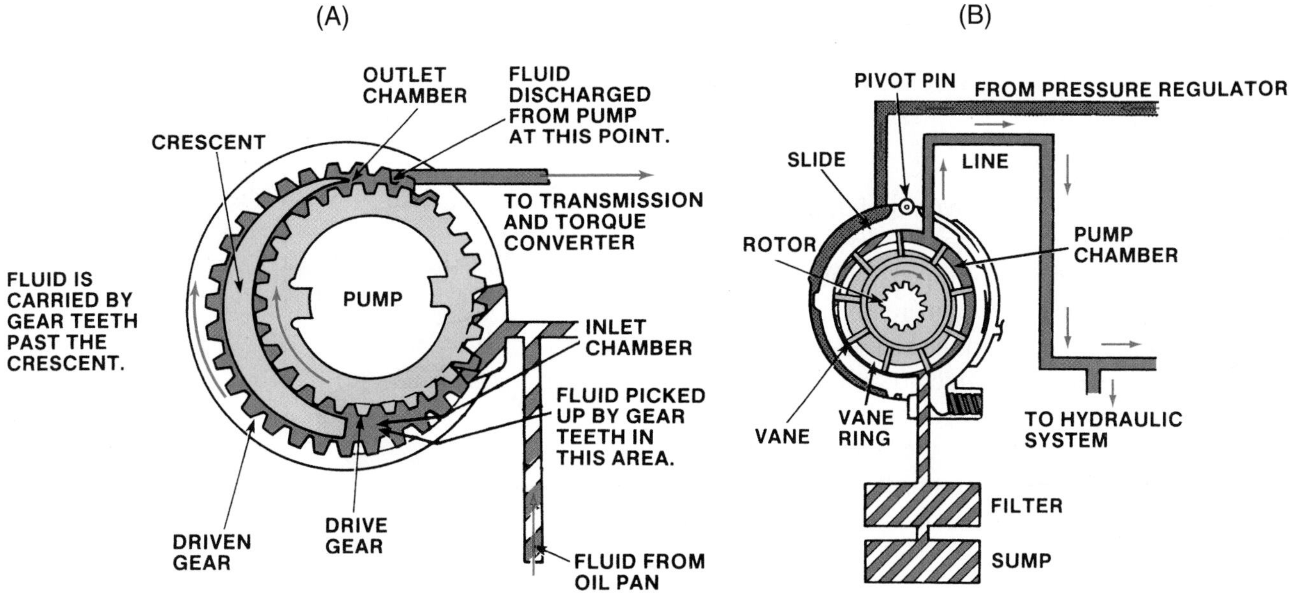

Figure 39-51 Operation of (A) a gear-type pump and (B) a vane-type pump.

the exhaust port and pressure again begins to build. The action of the spool valve regulates the fluid pressure.

Many late-model transmissions use an electronic pressure control (EPC or PC) solenoid to regulate system pressure.

Governor Assembly

The **governor** assembly is driven by the transmission's output shaft, senses road speed, and sends a fluid pressure signal to the valve body to either upshift or downshift. When vehicle speed is increased, the pressure developed by the governor is directed to the shift valve. As the speed (and therefore the pressure) increases, the spring tension and throttle pressure on the shift valve are overcome and the valve moves. This action causes an upshift. Likewise, a decrease in speed results in a decrease in pressure and a downshift.

Although the governor sends a signal that will force an upshift, engine load may cause a delay in the shift. This allows for operation in a lower gear when there is a heavy load and the vehicle needs the gear reduction. During heavy-load operation, the governor pressure must be strong enough to overcome the high throttle pressure plus the spring tension on the shift valve before it can force an upshift. Because of this, the transmission will remain in a particular gear range until a higher-than-normal engine speed is reached.

PRESSURE BOOSTS

When the engine is operating under heavy load conditions, fluid pressure must be increased to increase the holding capacity of a hydraulic member. Increasing the fluid pressure holds the band and clutch control units tighter to reduce the chance of slipping while under heavy load. This is accomplished by sending pressurized fluid to one side of the pressure regulator's spool valve. This pressure works against the spool valve's normal movement to open the exhaust port and allows pressure to build to a higher point than normal.

Engine load can be monitored electronically by various electronic sensors that send information to an electronic control unit, which in turn controls the pressure at the valve body. Load can also be monitored by throttle pressure. Throttle pedal movement moves a **throttle valve** in the valve body via a throttle cable. When the throttle plate is opened, the throttle valve opens and applies pressure to the pressure regulator. This delays the opening of the pressure regulator valve, which allows for an increase in pressure. When the driver lets off the throttle pedal, the pressure regulator valve is free to move and normal pressure is maintained.

Many early transmissions were equipped with a **vacuum modulator**, which uses engine vacuum to change transmission pressure. The vacuum modulator allows for an increase in pressure when vacuum is low and decreases it when vacuum is high.

MAP Sensor

Engine load can be monitored electronically through the use of various electronic sensors that send information to an electronic control unit, which in turn controls the pressure at the valve body. The most commonly used sensor is the MAP sensor. The MAP sensor senses air pressure in the intake manifold. The control unit uses this information as an indication of engine load. A pressure-sensitive ceramic or silicon element and electronic circuit in the sensor generates a voltage signal that changes in direct proportion to pressure. A MAP sensor measures manifold air pressure against a precalibrated absolute pressure; therefore, the readings from these sensors are not adversely affected by changes in operating altitudes or barometric pressures.

Kickdown Valve

The valve body is also fitted with a **kickdown** circuit, which provides a downshift when the driver requires additional power. When the throttle pedal is quickly opened wide, throttle pressure rapidly increases and directs a large amount of pressure onto the kickdown valve. This moves the kickdown valve, which opens a port and allows mainline pressure to flow against the shift valve. The spring tension on the shift valve, the kickdown pressure, and throttle pressure will push on the end of the shift valve, causing it to move to the downshift position and forcing a quick downshift.

SHIFT QUALITY

All transmissions are designed to change gears at the correct time according to engine and vehicle speed, load, and driver intent. However, transmissions are also designed to provide for positive change of gear ratios without jarring the driver or passengers. If a band or clutch is applied too quickly, a harsh shift will occur. **Shift feel** is controlled by the pressure at which each hydraulic member is applied or released, the rate at which each is pressurized or exhausted, and the relative timing of the apply and release of the members.

To improve shift feel during gear changes, a band is often released while a multiple-disc pack is being applied. The timing of these two actions must be just right or both components will be released or applied at the same time, which would cause engine flare-up or clutch and band slippage. Several other methods are used to smooth gear changes and improve shift feel.

Multiple friction disc packs sometimes contain a wavy spring-steel separator plate that helps smooth the application of the clutch. Shift feel can also be smoothed out by using a restricting **orifice** or an **accumulator** piston **(Figure 39–52)** in the band or clutch apply circuit. A restricting orifice or check ball in the passage to the apply piston restricts fluid flow and slows the pressure increase at the piston by limiting the quantity of fluid that can pass in a given time. An accumulator piston slows pressure buildup at the apply piston by diverting a portion of the pressure to a second spring-loaded piston in the same hydraulic circuit. This delays and smooths the application of a clutch or band.

Manufacturers have also applied electronics to get the desired shift feel. One of the most common techniques is the pulsing (turning on and off) of the shift solenoids, which prevents the immediate engagement of a gear by allowing some slippage.

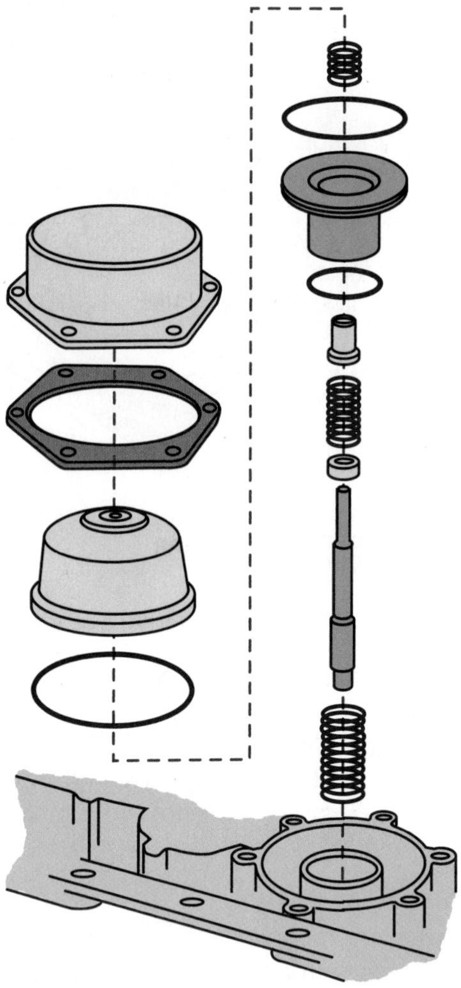

Figure 39-52 An accumulator assembly is used to control shift feel.

Shift Timing

Shift timing is determined by throttle pressure and governor pressure acting on opposite ends of the shift valve. When a vehicle is accelerating from a stop, throttle pressure is high and governor pressure is low. As vehicle speed increases, the throttle pressure decreases and the governor pressure increases. When governor pressure overcomes throttle pressure and the spring tension at the shift valve, the shift valve moves to direct pressure to the appropriate apply device and the transmission upshifts.

HYDRAULIC CIRCUITS

To provide a practical review of the valve body's operation, the following sections examine the three-speed transaxle with computer-controlled lockup torque converter that is installed in Chrysler FWD vehicles. This transaxle and valve body are not highly sophisticated, but are a good example of general, fundamental valve body operation and the control of various fluid pressures. The hydraulic circuits for each gear selector position are explained using flowcharts, accompanied by the torque converter's lockup clutch controls and operation along with schematic diagrams of transaxle power flows. The flowcharts and transaxle diagrams bring together the hydraulic controls and mechanical operation that combined give transaxle operation.

Preceding the explanation for each flowchart, note the gear range, torque converter mode, planetary controls engaged, approximate vehicle speed, and throttle position. Be sure to read the flow indication and pressure chart to establish the parameters for valve body and transaxle operation.

Gear Range: Neutral and park

Gear Selector Position: N and P

Throttle Position: 0 to 10 psi (0 to 70 kPa) (approximately closed)

Pump pressure leaves the transmission pump and is directed to the pressure regulator valve and manual valve **(Figure 39–53)**. At the pressure regulator valve, pump pressure is regulated to become line pressure. Line pressure enters the pressure regulator valve and leaves as converter pressure, flowing to the switch valve. The switch valve allows line pressure to enter the torque converter. Converter pressure circulates from the switch valve to fill the torque converter and returns to the switch valve to become cooling and lubrication pressure.

From the pressure regulator valve, line pressure flows to the manual valve. From the manual valve, line pressure seats check ball 9 and flows around check ball 8 to stop at the land of the closed throttle valve. Throttle pressure is low because the throttle valve is not open. Line pressure flows to the accumulator to cushion the engagement of the planetary controls when the gear selector is moved to D or R ranges. The accumulator is basically a hydraulic shock absorber designed to absorb the shock of engaging planetary controls.

In neutral, line pressure is maintained by the pressure regulator valve and flows to the manual and throttle valves.

Gear Range: D first gear

Selector Position: D

Torque Converter Mode: Unlock

Planetary Controls Engaged: Rear clutch; overrunning clutch

Approximate Speed: 8 mph (13 km/h)

Throttle Position: Half throttle

In **Figure 39–54**, pressure between the manual valve and pressure regulator valve is considered to be line pressure. This line pressure circulates to the switch valve. Since the switch valve is held in the torque converter unlocked position, line pressure flows no further.

Line Pressure Beginning at the first manual valve outlet 1, line pressure seats #9 check ball flows past #8 check ball to enter the throttle valve and establish throttle pressure. Line pressure also flows to the pressure regulator valve to regulate pressure. Line pressure moves the accumulator piston against coil spring tension, cushioning the engagement of the rear clutch. At outlet port 2 of the manual valve, line pressure fills the worm track, which engages the rear clutch and flows to the governor assembly. When the rear clutch is engaged and the governor assembly is filled with line pressure, the forward circuit is ready to drive the vehicle forward.

Throttle Pressure As line pressure passes through the valley of the throttle valve, it becomes throttle pressure. Throttle pressure circulates around the kickdown valve valley. With throttle pressure at the kickdown valve, a very quick downshift response to full-throttle operation is provided. Throttle pressure is directed to the pressure

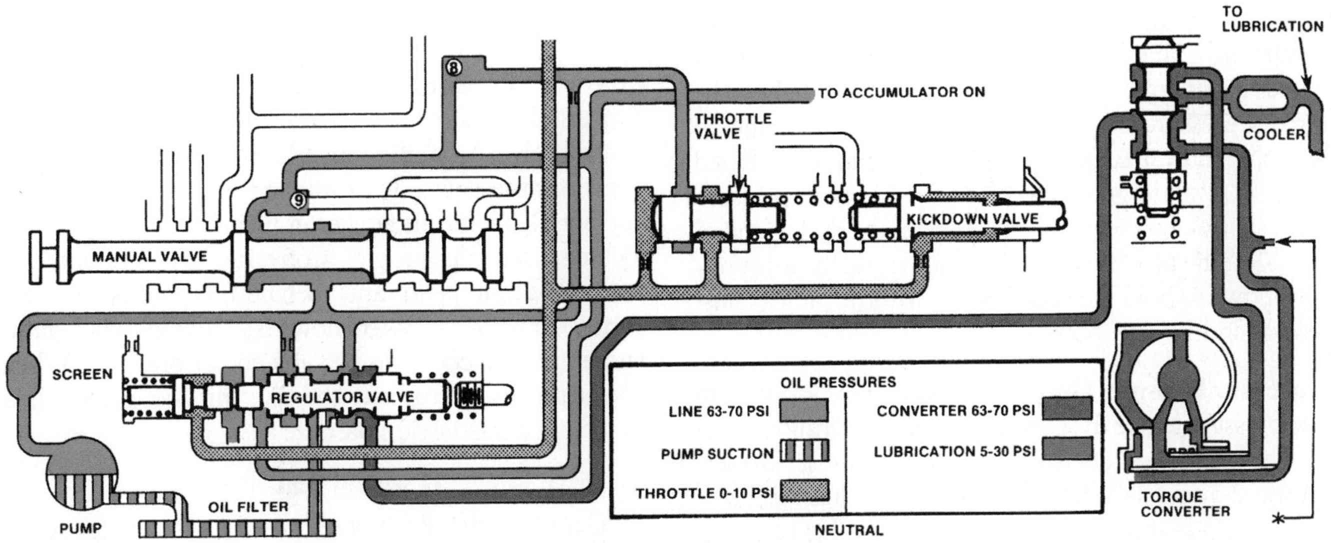

Figure 39–53 Fluid flows in the neutral gear range. *Courtesy of DaimlerChrysler Corporation*

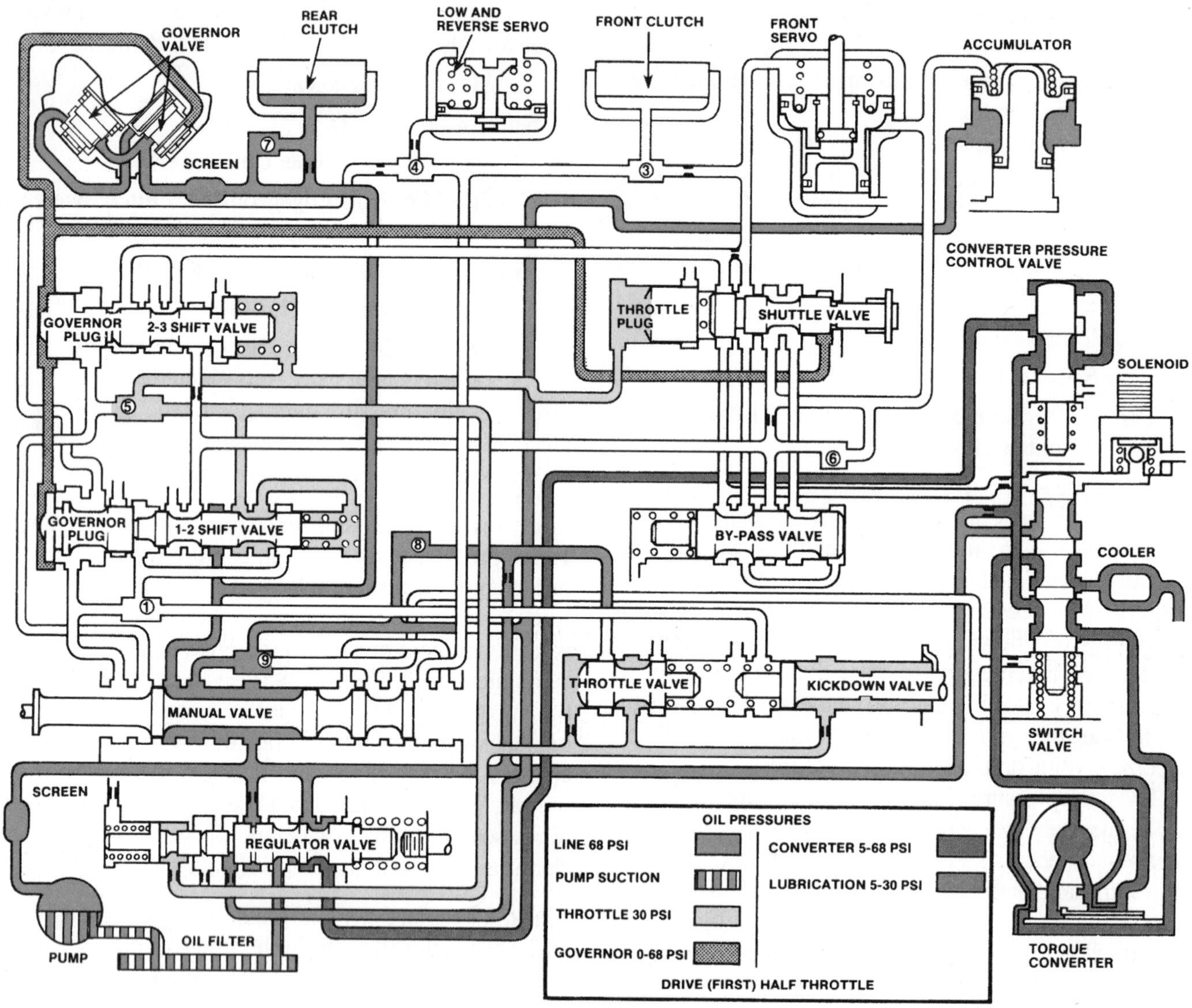

Figure 39-54 In drive range first gear, the rear clutch and overrunning clutch are engaged. *Courtesy of DaimlerChrysler Corporation*

regulator throttle plug. It acts on the throttle plug, compressing the throttle plug spring. The result is that the pressure regulator valve closes the exhaust port, which results in a line pressure increase. Throttle pressure moves to act on the spring end of the 1-2 shift valve. The throttle pressure and coil spring tension work together to hold the shift valve and governor plug in the downshifted position against governor pressure. Throttle pressure passes check ball 5, which is acting on the spring end of the 2-3 shift valve. From the 2-3 shift valve, throttle pressure flows to hold the shuttle valve throttle plug against its stop in the valve body.

Governor Pressure Governor pressure developed from line pressure leaves the governor assembly terminating at the shuttle valve spool land. Governor pressure also acts on the 2-3 and 1-2 shift valve governor plugs. Because the vehicle is traveling at 8 mph (13 km/h), governor pressure is not strong enough to overcome throttle pressure and spring tension at the opposite end of the shift valves. Therefore, the transmission stays downshifted in drive range first gear.

Converter Pressure From the pressure regulator valve, converter pressure is directed to the converter pressure control valve. From the converter pressure control valve, converter pressure flows through the switch valve valley and enters the torque converter turbine shaft to keep the lockup piston disengaged. Converter pressure entering between the impeller and turbine fills the torque converter. Converter pressure flows back to the switch valve and enters the cooler to become cooler pressure. When cooler pressure returns to the transmission, it cools and lubricates transmission parts.

Drive Range: D second gear
Gear Selector Position: D
Torque Converter Mode: Unlock

Planetary Controls Engaged: Rear clutch; front kickdown band

Approximate Vehicle Speed: 15 mph (25 km/h)

Throttle Position: Half throttle

Line pressure leaving the area between the pressure regulator valve and manual valve develops throttle pressure and torque converter control pressure.

Referring to the manual valve **(Figure 39–55)**, line pressure leaves outlet port 1, seating check ball 9. Line pressure passing check ball 8 moves to the throttle valve to become throttle valve pressure. From outlet 1, line pressure also flows to the pressure regulator valve, regulating line pressure. Line pressure from the same circuit flows to the accumulator, opposing line pressure and spring tension. The accumulator cushions the engagement of the intermediate band.

From manual valve outlet 2, fluid moves to the upshifted 1-2 shift valve. Line pressure flows to engage the rear clutch, then around check ball 7, through the governor screen to the governor assembly.

From the 1-2 shift valve, line pressure circulates to the shuttle valve and bypass valve through the restriction, or around check ball 6 to operate the front servo. When the front servo piston strokes in the cylinder, the front kickdown band engages around the front clutch drum.

Throttle Pressure The throttle pedal is in the half-open position, developing throttle pressure, which is directed to the kickdown valve. Throttle pressure is also directed to the throttle plug of the pressure regulator valve. Throttle pressure moving the throttle plug forces it against the throttle plug spring tension. The action of the throttle plug removes some of the opposition to the pressure regulator valve spring, resulting in a boost in line

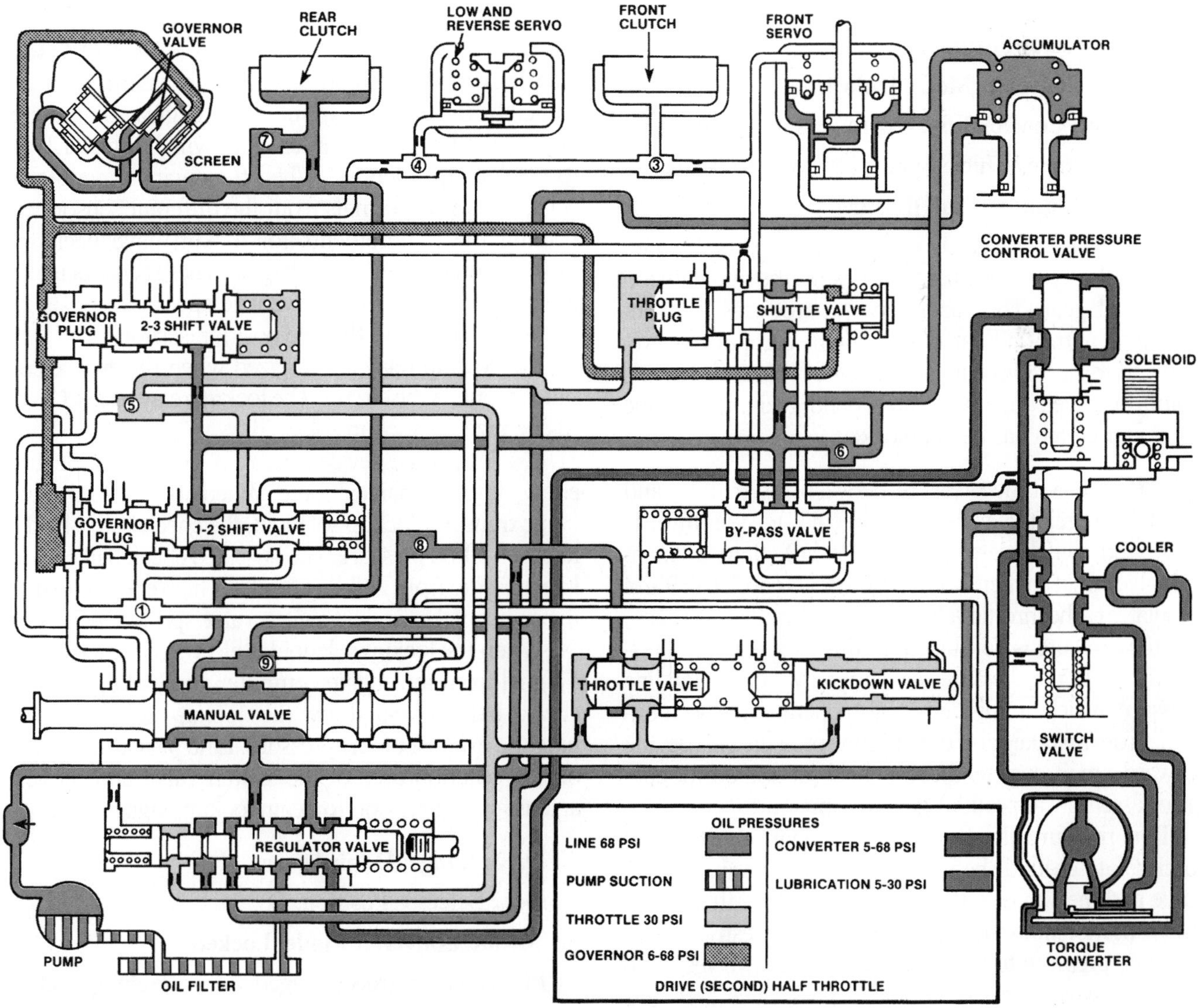

Figure 39-55 In drive range second gear, the rear clutch and front kickdown band are applied. *Courtesy of DaimlerChrysler Corporation*

pressure. Throttle pressure moves to the 1-2 shift valve where it is blocked by the upshifted valve's spool land. After seating check ball 5, throttle pressure and coil spring tension push on the 2-3 shift valve reaction area and move it to the downshifted position against governor pressure. Throttle pressure leaves the 2-3 shift valve to hold the shuttle valve throttle plug against its seat.

Governor Pressure Governor pressure leaves the rotating governor assembly and flows to the shuttle valve throttle plug. Governor pressure also acts on the governor plugs of the two shift valves. Vehicle speed and governor pressure are high enough to overcome throttle valve pressure and coil spring tension at the 1-2 shift valve. Governor pressure forces the 1-2 shift valve to move against throttle pressure. Therefore, throttle pressure is blocked from acting on the shift valve reaction area.

Converter Pressure Converter pressure flow is the same as in first gear.

Drive Range: D third gear

Gear Selector Position: D

Torque Converter Mode: Unlock

Planetary Controls Engaged: Rear clutch; front clutch

Approximate Vehicle Speed: 25 mph (40 km/h)

Throttle Position: Half throttle

Drive range third gear is covered in two parts. The first explains the hydraulic operation to shift the transmission into third gear. The second part introduces the sensors and controls affiliated with engine computer operation and torque converter lockup clutch control. (To this point, in drive range first and second gears, the torque converter clutch has been unlocked.)

Line Pressure Line pressure between the manual and pressure regulator valves is directed to the throttle valve and switch valve **(Figure 39–56)**. Start at manual valve outlet 1 where line pressure seats check ball 9 and circulates to the throttle valve. With the throttle valve open, throttle pressure is developed. In this circuit, pump pressure flows to the pressure regulator valve to develop line pressure, which, in this circuit, also operates the accumulator. From manual valve outlet 2, line pressure charges the forward circuit, engages the rear clutch, and enters the governor assembly to produce governor pressure.

Line pressure also flows from manual valve outlet 2 to circulate around the valley of the upshifted 1-2 shift valve and then around the valley of the 2-3 shift valve to the restriction above the shuttle valve. This restriction biases line pressure to the shuttle and bypass valves. Line pressure leaves the shuttle and bypass valve area and moves to the release side of the front servo and the front clutch. The feed line to the front clutch has a restriction that causes pressure to build on the release side of the front servo, which disengages the front kickdown band. Line pressure flowing through the front clutch feed restriction engages the front clutch for third gear direct drive operation.

Throttle Pressure With throttle valve open, throttle pressure flows to the kickdown valve, the throttle plug of the pressure regulator valve, 1-2 and 2-3 shift valves, and the shuttle valve throttle plug. At the 2-3 shift valve, throttle pressure and coil spring tension are opposed by increasing governor pressure.

Governor Pressure Governor pressure leaves the governor to move the shuttle valve, opening line pressure circuits. The shuttle valve's movement buries the coil spring in the hollow shuttle valve throttle plug. The governor pressure has moved the 1-2 and 2-3 shift valves to the upshifted position, directing line pressure to engage the front clutch and hold the front servo released during third gear operation.

Torque Converter Controls and Pressure

The lockup torque converter clutch assembly is controlled by the engine computer. The engine computer energizes the torque converter relay, which sends 12 volts to the lockup solenoid. When the engine computer receives electronic signals from the different sensors confirming the requirements for lockup have been met, lockup clutch engagement begins. These sensors include an engine coolant sensor, vehicle speed sensor, engine vacuum sensor, and throttle position sensor.

The lockup relay is energized when the engine computer grounds the circuit. The lockup relay sends 12 volts to energize the lockup solenoid.

A solenoid is a device capable of converting electrical energy into mechanical force. When applied to the automatic transaxle lockup solenoid, the check ball is held off its seat by fluid pressure. The unseated check ball prevents line pressure from building until it is high enough to move the switch valve against switch valve spring tension.

When the lockup solenoid is electrically energized by a signal from the engine computer and lockup relay, the check ball is seated by the lockup solenoid, stopping the exhaust of line pressure. Line pressure builds up on the reaction area of the switch valve and moves it against spring tension to begin lockup engagement.

Drive Range: D third gear

Gear Selector Position: D

Torque Converter Mode: Locked

Planetary Controls Engaged: Rear clutch; front clutch

Approximate Vehicle Speed: 40 mph (65 km/h)

Throttle Position: Half throttle

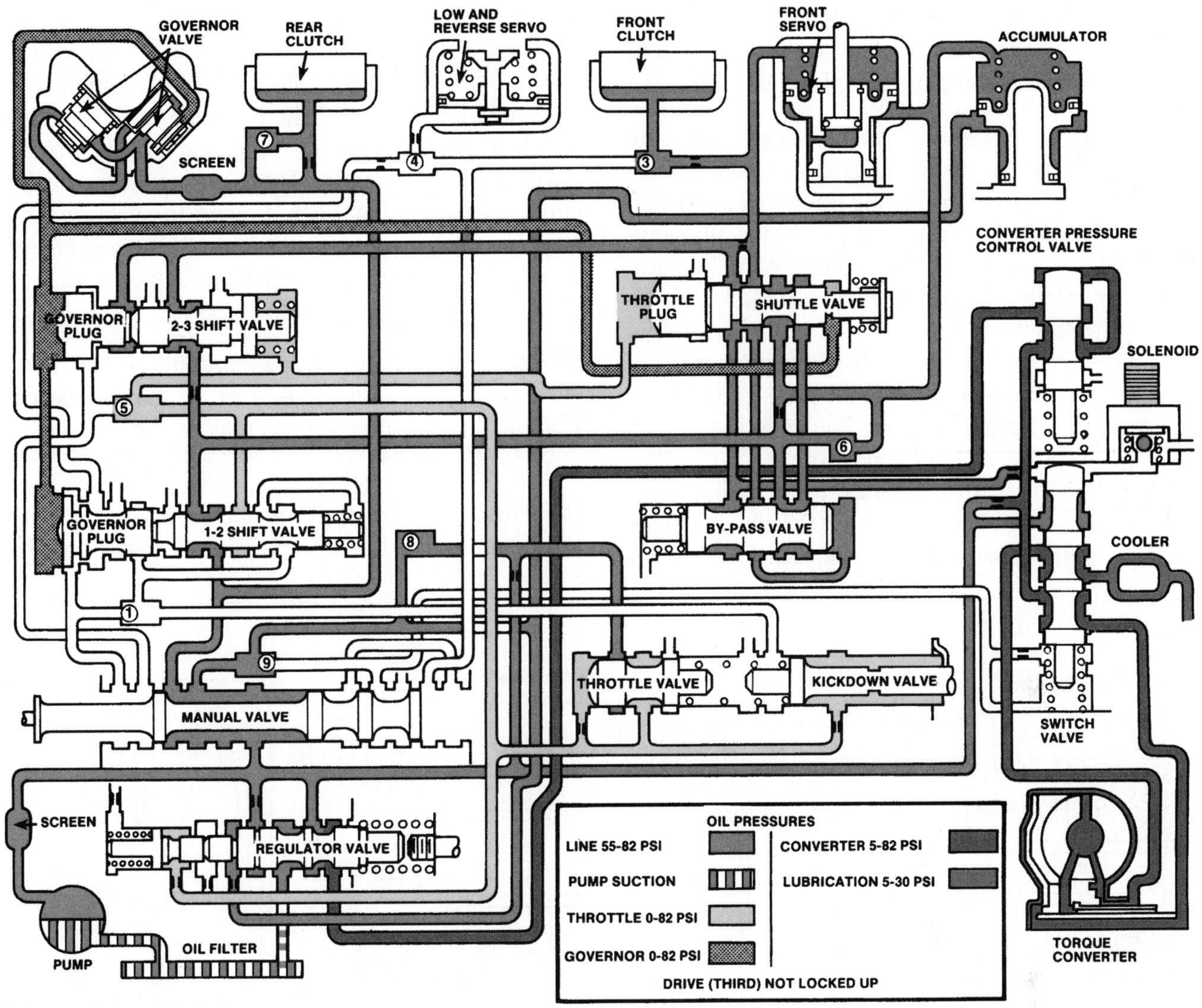

Figure 39-56 Drive range third gear with the lockup torque converter clutch disengaged and front and rear clutches engaged. *Courtesy of DaimlerChrysler Corporation*

In drive range third gear lockup, the transaxle operates in the same manner as third gear unlock. The focus of attention is on the torque converter lockup clutch control system.

Lockup Clutch Engagement

The coolant sensor reports to the computer that the engine has reached a temperature of at least 150°F (66°C). The vehicle speed sensor located on the speedometer cable sends an electronic signal to the computer, which reports that vehicle speed is above 40 mph (65 km/h). Since the vehicle is traveling at engagement speed, the throttle must be open (driver's foot on throttle). The vacuum transducer reports to the computer via electronic signal that engine vacuum is above 4 inches (102 mm Hg) and within 22 inches (560 mm Hg). Based on these inputs, the computer energizes the clutch relay and lockup solenoid to move the solenoid check ball on its seat. The lockup solenoid check ball stops the lockup solenoid from exhausting line pressure. The increasing line pressure forces the switch valve to move against the coil's spring tension. Line pressure from the switch valve is directed to the pump drive hub and stator support to fill the torque converter with fluid. Fluid in the torque converter during lockup operation resides there to become the cooling and lubricating pressure. Line pressure flows from the impeller and turbine to fill the space behind the torque converter clutch piston and force engagement.

Drive Range: Reverse gear

Gear Selector Position: R

Torque Converter Mode: Unlock

Planetary Controls Engaged: Low and reverse rear band; front clutch

Approximate Vehicle Speed: 5 mph (8 km/h)

Throttle Position: Part throttle

Line pressure from a manual valve outlet not used before circulates through a bypass around the manual valve valley. Line pressure circulates to the low and reverse servo and front clutch **(Figure 39–57)**. During the process of engaging the low and reverse servo and front clutch, check balls 4 and 3 are seated by line pressure.

Line Pressure Line pressure from between the pressure regulator and manual valve circulates around the pressure regulator valve. After flowing through a restriction to seat check ball 8, line pressure enters the throttle valve, which produces line-to-throttle pressure. With the throttle valve open, throttle pressure charges the kickdown valve and strokes the throttle plug to its extreme left position at the pressure regulator valve. Line pressure from the manual valve does not flow to the pressure regulator valve to oppose spring tension. The pressure regulator valve coil spring pushes the pressure regulator valve over to close the exhaust port. Line pressure builds to approximately 200 to 300 psi (1380 to 2070 kPa). You may wonder why line pressure must be increased so much in reverse gear. The planetary control units engaged in reverse are the front clutch and the low and reverse rear band. The front clutch, unlike the rear clutch, does not have a Belleville spring. The Belleville spring multiplies clutch piston apply force on the clutch pack, which reduces possible slippage. Therefore, to keep the front clutch from slipping when moving the vehicle from a stop, reverse line pressure is increased. The concept of increasing line pressure in reverse is common in many automatic transaxles and transmissions.

Throttle Pressure Throttle pressure circulates to the shift valve area to keep both the 1-2 and 2-3 shift valves downshifted. Throttle pressure also keeps the shuttle valve throttle plug against its seat in the valve body.

Converter Pressure In reverse, the transaxle needs the torque multiplication of vortex flow to start the vehicle

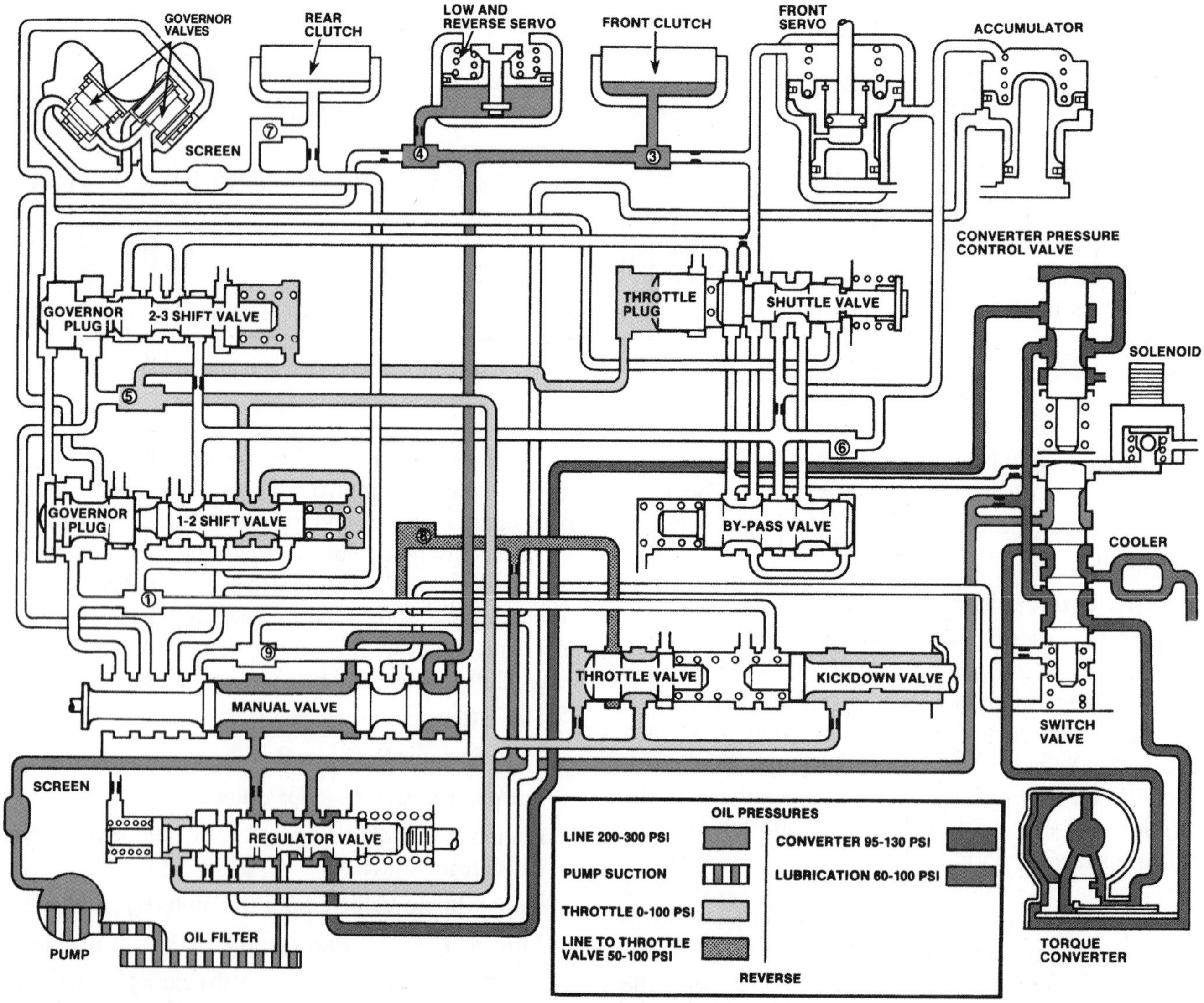

Figure 39–57 Reverse line pressure is increased to keep the front clutch, which has no Belleville spring, from slipping. The front clutch and the low reverse bands are engaged. *Courtesy of DaimlerChrysler Corporation*

moving from a stop. Therefore, the switch valve maintains the position and holds the torque converter piston in the unlocked position.

USING SERVICE MANUALS

Before beginning to service a transaxle or transmission, have the service manual and latest service bulletins on the bench for ready reference. Many service manual publishers have complete volumes dedicated to automatic transmissions and transaxles. These volumes contain detailed information on domestic and imported transmissions and transaxles. ■

ELECTRONIC CONTROLS

Through the use of electronics, today's transmissions have better shift timing and quality. As a result, they contribute to improved fuel economy, lower exhaust emission levels, and improved driver comfort. Electronically controlled transmissions function in the same basic way as hydraulically based transmissions. However, a computer uses inputs from several different sensors and matches this information to a predetermined schedule, and then changes gears.

Hydraulically controlled transmissions rely on signals from a governor and throttle pressure to force a change in gears. Electronically controlled transmissions usually do not have governors or throttle pressure devices. Hydraulically and electronically controlled transmissions rely on pressure differentials at the sides of a shift valve to hold or change a gear. However, the pressure differential in electronically controlled units is caused by the action of shift solenoids **(Figure 39–58)**. The computer controls these solenoids. The solenoids do not directly control the transmission's clutches and bands. These are engaged or disengaged in the same way as hydraulically controlled units. The solenoids simply control the fluid pressures.

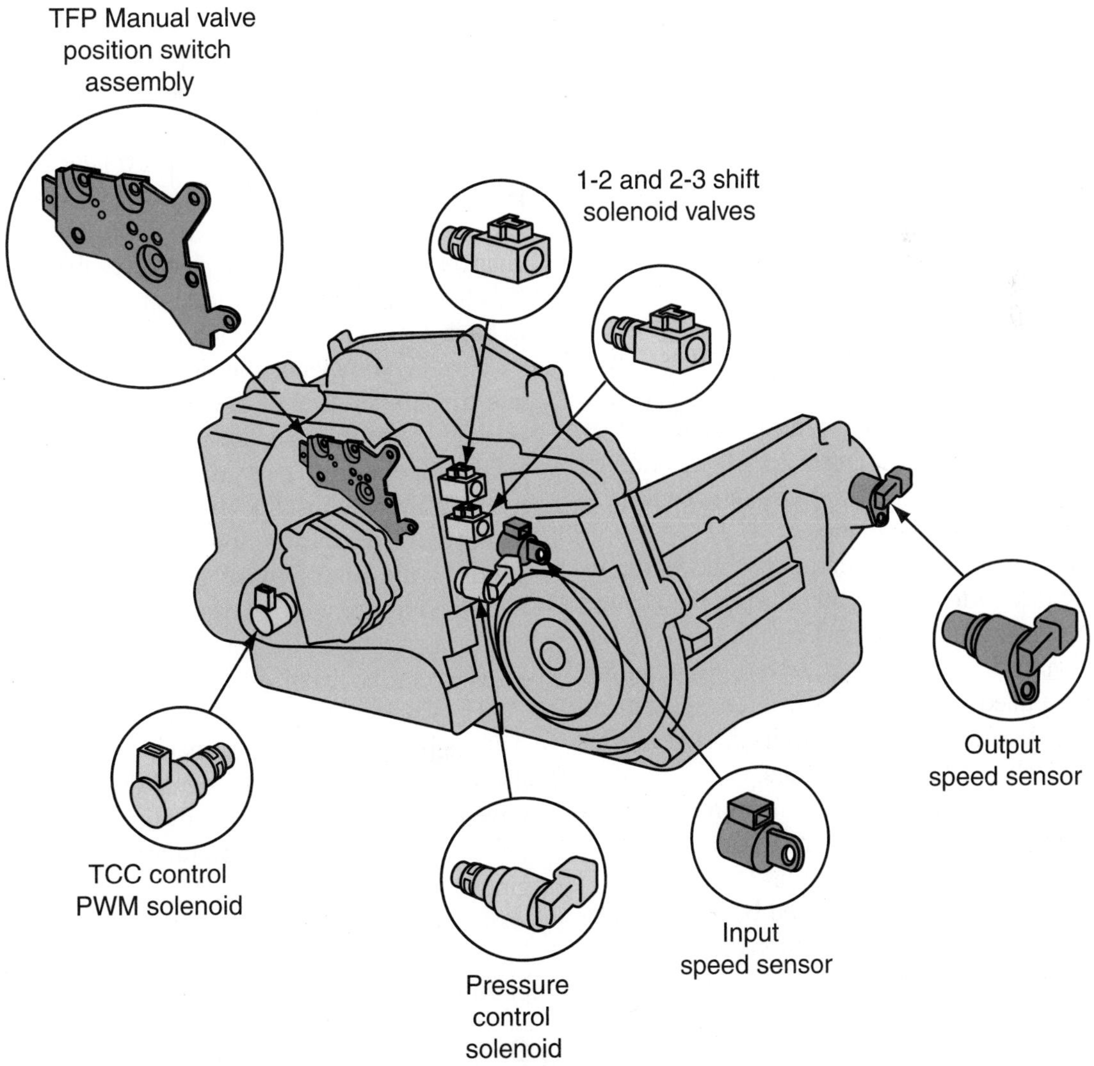

Figure 39–58 An electronically controlled transaxle.

Most electronically controlled systems are complete computer systems. There is a central processing unit with various inputs and outputs. Often the central processing unit is a separate computer designated for transmission control. The computer receives information from the inputs and controls two to five (perhaps more!) solenoids that control hydraulic pressure and fluid flow to the various apply devices and to the clutch of the torque converter. This computer may be the **transmission control module (TCM)**, powertrain control module (PCM), or the body control module (BCM). When transmission control is not handled by the PCM, the controlling unit communicates with the PCM. For simplicity, the transmission's controlling computer will be referred to as the TCM whether it is a separate computer or integrated into another computer.

Transmission Control Module

The TCM relies on programming stored in its memory to provide gear shifting at the optimum time. The decision to shift or not to shift is based on shift schedules and logic programmed into its memory. A **shift schedule** contains the actual shift points to be used by the computer according to the input data it receives from the sensors. Shift schedule logic chooses the proper shift schedule for the current operating conditions. It uses the shift schedule to select the appropriate gear and then determines the correct shift schedule that should be followed.

The first input a computer looks at to determine the correct shift logic is the position of the gearshift lever. All shift schedules are based on the gear selected by the driver. Each possible engine/transmission combination for a vehicle has a different set of shift schedules. These schedules are coded by the position of the gear selector and the current gear range and use throttle angle and vehicle speed as primary determining factors. The computer also looks at different temperature, load, and engine operation inputs for more information.

The basic shift logic of the computer allows the releasing apply device to slip slightly during the engagement of the engaging apply device. Once the apply device has engaged and the next gear is driven, the releasing apply device is pulled totally away from its engaging member and the transmission is fully into its next gear. This allows for smooth shifting into all gears.

The TCM frequently reviews the input information and can make quick adjustments to the shift schedule, if and *as* needed.

The electronic control systems used by the manufacturers differ with the various transmission models and the engines to which they are attached. The components in each system and the overall operation of the system also vary with the different transmissions. However, all operate in a similar fashion and use basically the same parts.

Inputs

The computer may receive information directly from a sensor or through a multiplexed system that connects all of the vehicle computer systems **(Figure 39–59)**. Normal engine-related inputs are used by the computer to determine the best shift points. Many of these inputs are available through and are inputted from the common data bus. Other information, such as engine and body identification, the TCM's target idle speed, and speed control operation are not the result of sensor input; rather, these have been calculated and are made available on the bus.

Typical bus inputs used by the TCM are from the throttle position (TP), manifold absolute pressure (MAP), mass airflow (MAF), intake air temperature (IAT), barometric pressure (BARO), engine coolant temperature (ECT), and crankshaft position (CKP) sensors. These provide the TCM with information about operating conditions. Through these, the TCM is able to control shifting and TCC operation for optimum performance.

The TP sensor sends a voltage signal to the TCM in response to throttle position. This signal is not only used to inform the TCM of the driver's intent; it is also used in place of the throttle pressure linkage. ECT signals are critical to the operation of the transmission. If the engine's coolant temperature is cold, the TCM may delay upshifts to improve driveability. The TCM may also engage the converter clutch in second or third gear if coolant temperature rises.

The MAP sensor keeps the computer informed of changes in engine load. This signal, combined with the signals from the TP and MAF sensors, allows the computer to have a good idea of the load and the driver's intent.

The TCM may use the signal from the IAT to calculate the temperature of the battery. The TCM then uses this temperature calculation to estimate transmission fluid temperature. The signals from the BARO are used by the TCM to adjust line pressures according to changes in altitude. This sensor input may not be used; its use depends on the type of intake air monitoring system the vehicle is equipped with. On those vehicles using the BARO sensor as an input, the sensor may be integrated on the PCM circuit board or mounted externally.

The CKP sensor provides the TCM with engine speed. Although engine speed information is available at the bus, the TCM may receive this signal directly from the CKP sensor. With the direct feed, any time delay at the bus circuit is avoided and the computer is aware of current engine speeds.

Direct inputs are from those sensors that provide information to the TCM and are not available on bus circuit.

On-Off Switches The brake switch is used to disengage the TCC when the brakes are applied. Its input has little

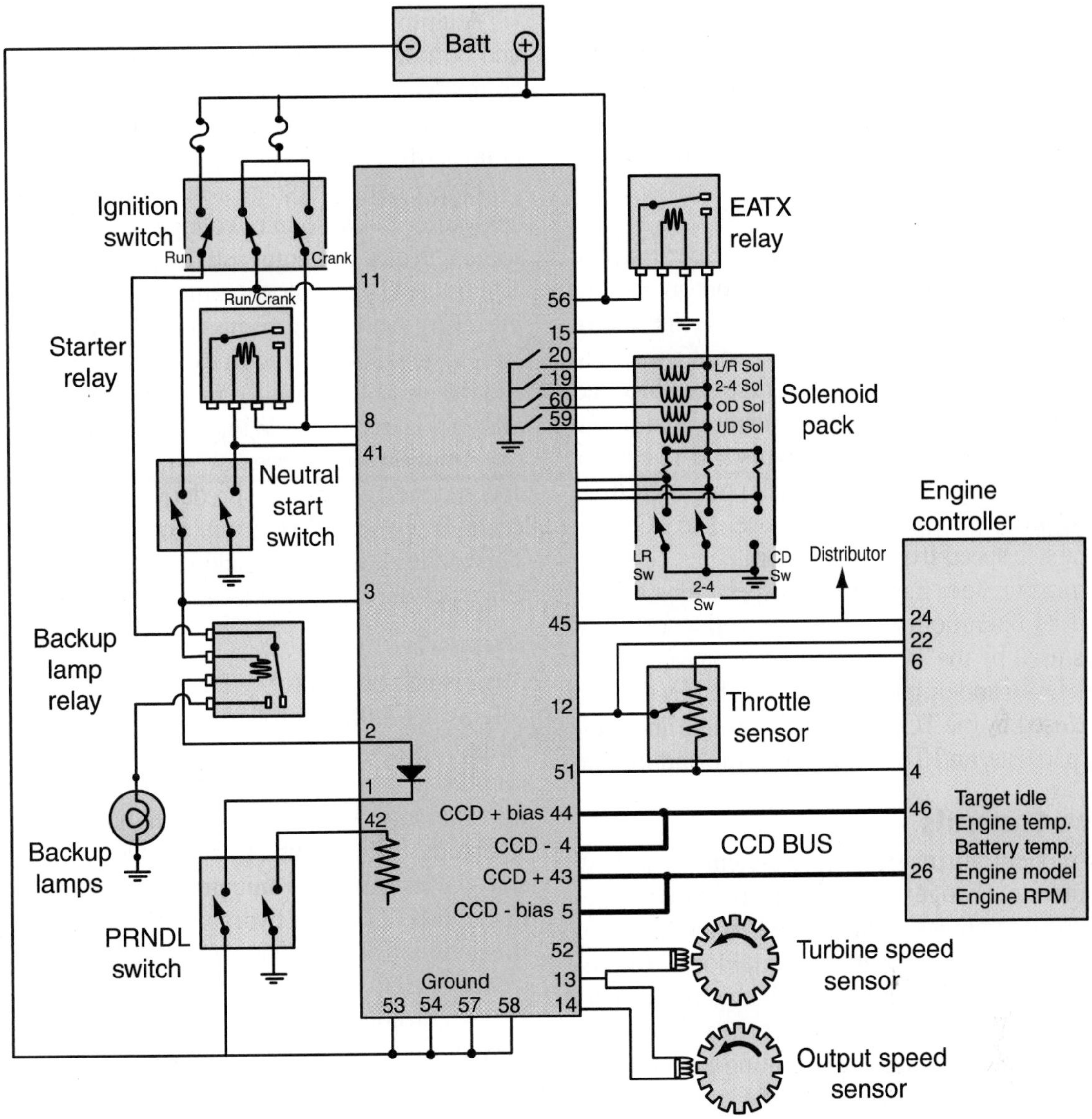

Figure 39-59 The electric circuit for a typical electronically controlled transmission. Note: The CCD BUS is the data source for other inputs in this multiplexed circuit.

to do with the up and down shifting of gears, except that in some systems it signals a need for engine braking. An A/C request switch informs the TCM that the A/C has been turned on. The TCM then changes line pressure and shift timing to accommodate the extra engine load created by A/C compressor operation. The transmission range (TR) sensor informs the TCM of the gear selected by the driver. This sensor normally also contains the neutral safety switch and the reverse light switch. The TR sensor is typically a multiple pole-type on-off switch.

Transmission Fluid Temperature (TFT) Sensor The TFT sensor monitors the temperature of the transmission's fluid. When the signal from this sensor is normal, the transmission will operate within its normal range. However, when the signal indicates that the fluid is overly hot, the TCM allows the transmission to only operate in such a way that will allow the transmission to cool down, thus preventing damage to the transmission. When the TFT signal indicates that the fluid is cooler than normal, the TCM will alter the shift schedule.

The TFT sensor is a thermistor typically located in the solenoid assembly. Its electrical resistance varies with ATF temperature. The TCM integrates this input with others to control TCC operation. Typically, the TCM prevents TCC engagement until fluid temperatures reach about 68°F (20°C). If fluid temperature reaches about 250°F (122°C), the TCM applies the TCC in second, third, or fourth gears. If mechanically connecting the engine to the input shaft of the transmission does not reduce fluid temperature and it reaches 300°F (150°C), the TCM will release the TCC to prevent damage to the converter clutch from excessive temperatures. If the fluid reaches about 310°F (154°C), the TCM sets a fluid temperature trouble code and uses a fixed value as the fluid temperature input signal.

Transmission Pressure Switches Various transmission pressure switches can be used to keep the TCM informed as to which hydraulic circuits are pressurized and which clutches and brakes are applied. These input signals can serve as verification to other inputs and as self-monitoring or feedback signals.

Voltage-Generating Sensors The vehicle speed sensor (VSS) and **output shaft speed (OSS)** sensors are used to monitor transmission output and/or vehicle speeds. In some electronic control systems, only one of these sensors is used. Some transmissions use these speed-related inputs instead of a governor. These signals are used to regulate hydraulic pressure and shift points and to control TCC operation. Four-wheel-drive vehicles may use a third speed sensor installed in the transfer case. The TCM determines vehicle speed from this sensor.

Some transmissions have an input speed sensor. This sensor and its operation are identical to the OSS and its signal is used by the TCM to calculate converter turbine speed. Input and output speeds provided by the two sensors are used by the TCM to help determine line pressure, shift patterns, and TCC apply pressure and timing.

Adaptive Controls

Many late-model transmissions have systems that allow the computer to change transmission behavior in response to the operating conditions and to the habits of the driver. The system monitors the condition of the engine and compensates for any changes in the engine's performance. It also monitors and memorizes the typical driving style of the driver and the operating conditions of the vehicle. With this information, the computer adjusts the timing of shifts and converter clutch engagement to provide good shifting at the appropriate time.

These systems are constantly learning about the vehicle and driver. The computer adapts its normal operating procedures to best meet the needs of the vehicle and the driver. When systems are capable of doing this, they are said to have **adaptive learning** capabilities. To store this information, the computer includes a long-term adaptive memory.

The computer also learns the characteristics of the transmission and changes its programming accordingly. It learns the release and application rates of various components during various operating conditions. Adaptive learning allows the computer to compensate for wear and other events that might occur and cause the normal shift programming to be inefficient. Doing this, the adaptive learning capability of the transmission computer allows for this smooth shifting throughout the life of the transmission. As component wear and shift overlap times increase, the TCM adjusts line pressure to maintain proper shift timing calibrations.

Adaptive learning takes place as the TCM reads input and output speeds more than 140 times per second. The computer responds to each new reading. This learning process allows the TCM to make adjustments to its program so that quality shifting always occurs.

Direct battery voltage is supplied to the TCM. If the computer loses source voltage, the transmission on some vehicles will enter into a default or limp-in mode. The transmission will also enter into the default mode if the TCM senses a transmission failure. At this point, a fault code will be stored in the memory of the computer and the transmission will remain in default until the transmission is repaired. While in the default mode, the transmission will operate only in park, neutral, reverse, and second (or another predetermined forward range) gears. The transmission will not upshift or downshift. This mode allows the vehicle to be operated, although its efficiency and performance is hurt.

Outputs

A transmission control system uses two to five solenoids **(Figure 39–60)**, controlled by the TCM, to regulate shift timing, feel, and TCC application **(Figure 39–61)**. The number and purpose of each depends on the model of transmission. Normally, two solenoids are used as shift solenoids. They control the delivery of fluid to the shift valves. An **electronic pressure control (EPC)** solenoid is used to control hydraulic pressures throughout the transmission. This solenoid operates on a duty cycle controlled by the TCM. Its purpose is to regulate line pressure according to engine running conditions and engine torque. An additional solenoid is used to provide for engine braking during coasting. This solenoid operates when the vehicle is slowing down and the throttle is closed. The other solenoid controls the operation of the converter clutch.

The shift solenoids offer four possible on/off combinations to control fluid to the various shift valves. These

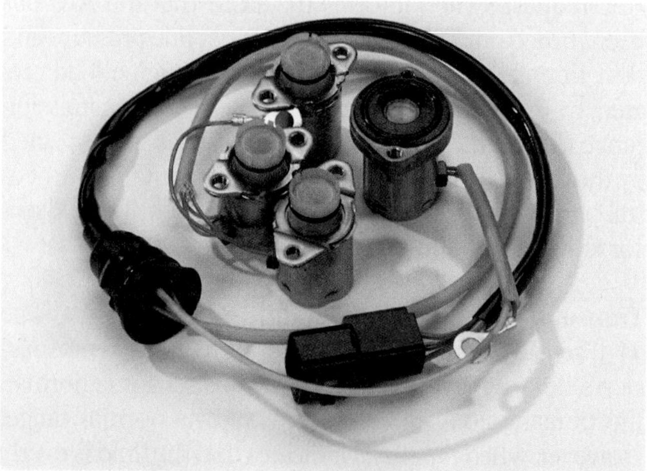

Figure 39–60 The solenoids for a transmission and their wiring harness. *Courtesy of Transtar Industries Inc.*

SHIFT SOLENOID OPERATION CHART					
Transaxle range selector lever position	Powertrain control module gear commanded	Eng braking	AX4N solenoids		
			SS 1	SS 2	SS 3
P / N[a]	P / N	NO	OFF[b]	ON[b]	OFF
R (Reverse)	R	YES	OFF	ON	OFF
Overdrive	1	NO	OFF	ON	OFF
	2	NO	OFF	OFF	OFF
	3	NO	ON	OFF	ON
	4	YES	ON	ON	ON
D (Drive)	1	NO	OFF	ON	OFF
	2	NO	OFF	OFF	OFF
	3	YES	ON	OFF	OFF
Manual 1	2[c]	YES	OFF	ON	OFF
	3[c]	YES	OFF	OFF	OFF
		YES	ON	OFF	OFF

[a] When transmission fluid temperature is below 50° then SS1 = OFF, SS2 = ON, SS3 = ON to prevent cold creep.
[b] Not contributing to powerflow.
[c] When a manual pull-in occurs above calibrated speed, the transaxle will downshift from the higher gear until the vehicle speed drops below this calibrated speed.

Figure 39-61 An example of solenoid activity during different gear ranges.

solenoids are on/off solenoids that are normally off and in the open position. Being open, the solenoid valves allow line pressure to the bore of the shift valve and keep the shift valve closed. When the shift solenoids are energized, they block the line pressure and exhaust line pressure from the valve, allowing the shift valve to open.

The EPC solenoid replaces the conventional TV cable setup to provide changes in pressure in response to engine load. This solenoid is a **variable force solenoid (VFS)** and contains a spool valve and spring. To control fluid pressure, the PCM sends a varying signal to the solenoid that varies the amount the solenoid will cause the spool valve to move. When the solenoid is off, the spring tension keeps the valve in place to maintain maximum pressure. As more current is applied to the solenoid, the solenoid moves the spool valve more, which moves to uncover the exhaust port more, thereby causing a decrease in pressure. The EPC solenoid controls line pressure at all times based on the programming of the system's computer and is able to match shift timing and feel with the current needs of the vehicle.

Typically, the operation of the converter clutch is also totally controlled by the computer. The only exception to this is during first gear and reverse gear operation when the clutch is disabled hydraulically to prevent engagement regardless of the commands by the computer. Normally, the converter clutch is hydraulically applied and electrically controlled through a pulse-width-modulated (PWM) solenoid **(Figure 39–62)**, which is controlled by the TCM. Modulating the pressure to the converter clutch allows for smooth engagement and disengagement and also allows for partial engagement of the clutch.

The PWM solenoid is installed in the valve body. It controls the position of the TCC apply valve. When the solenoid is off, TCC signal fluid exhausts and the converter clutch remains released. Once the solenoid is energized, the plunger moves the metering ball to allow TCC signal fluid to pass to the TCC regulator valve. The

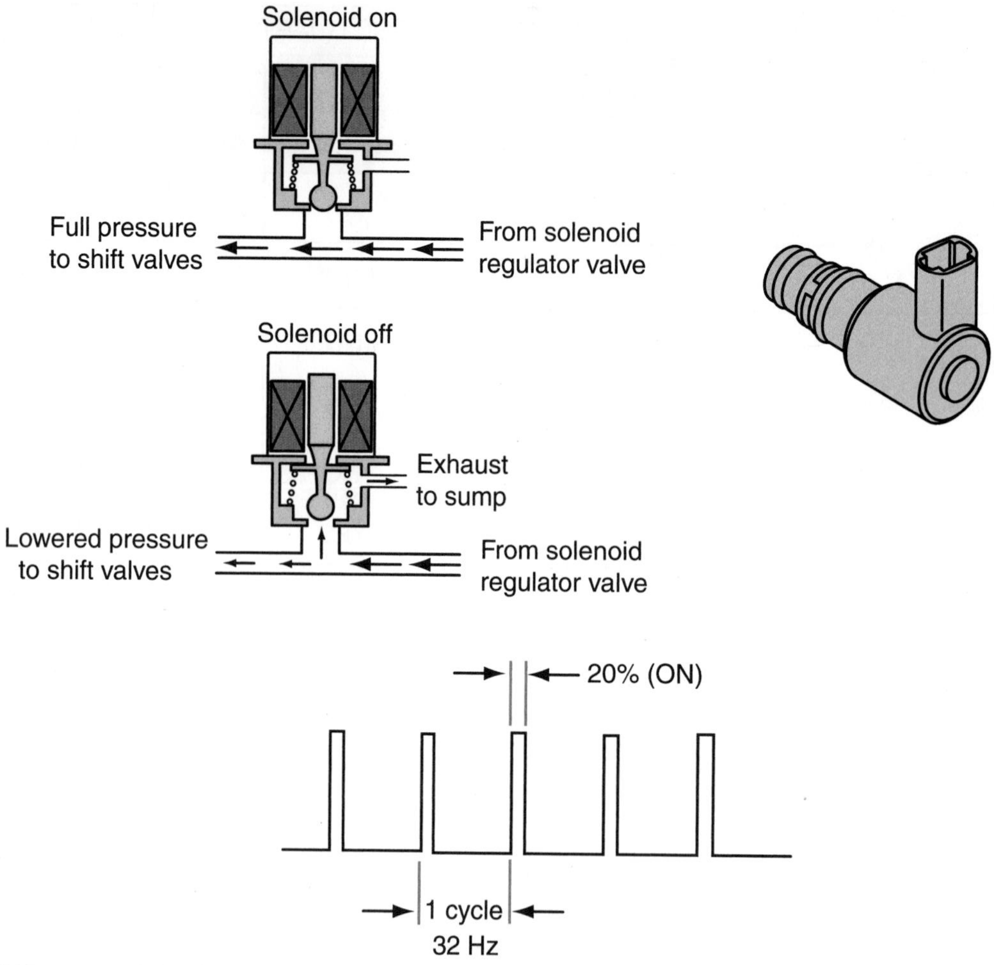

Figure 39-62 (TOP) A typical PWM solenoid. (BOTTOM) The signal representing the control or ordered duty cycle from the computer.

TCM cycles the PWM solenoid on and off thirty-two times per second but varies the length of time it is energized in each cycle.

Operational Modes

With electronic controls, automatic transmissions can be programmed to operate in different operational modes. The desired mode is selected by the driver. The mode selection switch can be located on the center console or the instrument panel. Most transmissions with this feature have two selective modes, usually called "Normal" and "Power." During the normal mode, the transmission operates according to the shift schedule and logic set for normal operation. In the power mode, the TCM uses different logic and shift schedules to provide for better acceleration and performance with heavy loads. Normally this means delaying upshifts.

If three modes are available, the third mode is called the "Auto" mode. The auto mode is a mixture between the normal and power modes. While in this mode, the TCM will control the shifts in a normal way. However, if the throttle is quickly opened, the shift pattern will switch to the power mode.

Manual Shifting

One of the most publicized features of electronically controlled transmissions is the availability of manual shift controls. Although not all electronic transmissions have this feature, they all could. Basically, these systems allow the driver to manually upshift and downshift the transmission at will, much like a manual transmission. Unlike a manual transmission, the driver does not need to depress a clutch pedal nor is there a clutch assembly on the flywheel. The driver simply moves the gear selector or hits a button and the transmission changes gears. If the driver does not change gears and engine speed is high, the transmission shifts on its own. If the driver elects to let the transmission shift automatically, a switch disconnects the manual control and the transmission operates automatically.

Marketed as a sport option and a combination of a manual and an automatic transmission, these transmissions are still based on an automatic transmission with a torque converter. Therefore, performance numbers are not quite as good as if the vehicle were equipped with a manual transmission. In fact, manual control of an automatic transmission often results in slower acceleration times than when the transmission shifts by itself.

Not all manually shifted automatics behave in the same way, nor do they control the same things. Actually, the behavior of the transmission depends on the car it is installed in. Some are pure high-performance cars, while others are moderate-performance family sedans. What follows are basic descriptions of some of the manually shifted automatic transmissions available, which are mentioned only to serve as examples. Many other similar systems are available.

BMW's Steptronic Steptronic systems are based on five- or six-speed transmissions and offer the option of shifting gears manually by using the selector lever or through steering-wheel-mounted buttons. The driver can select to shift the gears manually or let the transmission shift automatically. The system relies on shift-by-wire technology, which replaced mechanical linkages to the transmission. These transmissions also are equipped with adaptive transmission controls (ATC), which responds to the driver's style as well as the operating conditions. The ATC looks at the travel and movement patterns of the accelerator, deceleration rates during braking, and lateral acceleration in curves and then selects the most suitable shift schedule. Upshifts are delayed on uphill stretches to allow better use of the engine's power. On downhill grades, ATC downshifts when you are forced to brake to counteract undesired acceleration.

In the manual mode, ATC works to avoid overspeeding the engine by upshifting just before the engine reaches its automatic cutoff speed. At low speeds, it downshifts automatically without input from the driver. In the kickdown mode, the system downshifts to the lowest gear possible without overrevving the engine.

Chrysler Autostick This is the most familiar system. Manual shifting is performed by moving a control on the console. Moving the selector to the right provides for an upshift. A movement to left allows for a downshift.

The transaxle is fitted with a special gear selector and switch assembly. The driver can either manually shift the gears or allow the transaxle to shift automatically. The selected gear is displayed on the instrument panel to keep the driver informed of the selected gear.

Although the driver has control of the shifting, the TCM will override the controls during some conditions. Regardless of the action taken by the driver, these automatic shifts will occur under these conditions:

4–3 coast downshift at 13 mph

3–2 coast downshift at 9 mph

2–1 coast downshift at 5 mph

1–2 upshift at 6,300 engine rpm

2–3 upshift at 6,300 engine rpm

4–3 kickdown shift at 13–31 mph and with sufficient throttle

Manual shifts are not permitted under the following conditions:

3–4 upshift when the vehicle is traveling at less than 15 mph

3–2 downshift when the vehicle is traveling above 74 mph with a closed throttle or 70 mph when the throttle is open

2–1 downshift when the vehicle is traveling above 41 mph with a closed throttle or 38 mph when the throttle is open

The Autostick feature is deactivated if the TCM senses problems and/or sets a trouble code that relates to the TR sensor or Autostick switch or when there is a high engine and transmission temperature code.

Honda's Sequential Sportshift Manual shifting is performed by moving a control on the console. Moving the selector forward provides for an upshift. A movement down allows for a downshift. This transmission is unique in that it will not automatically upshift if the driver brings the engine's speed too high. All other transmissions of this design will upshift automatically at a predetermined engine speed to prevent damage to the engine.

Tiptronic This is a five- or six-speed transmission available in a few European cars, such as Porsches and Audis. Although the concept of all Tiptronic units is the same, the driver control varies quite a bit. Manual shifting on the Audi is performed by moving a control on the console. Moving the selector forward provides for an upshift. A movement down allows for a downshift. To shift Porsche's transmission, the driver moves the gear selector into the manual gate, next to the automatic ranges, and depresses buttons on the steering wheel. These systems also are typically controlled by a computer that controls the change of gears and tries to mimic the action of a manual transmission. While operating in the manual shift mode, these systems study the driving habits of the driver and select the optimum driving range, reducing the driver's workload particularly in stop-and-go traffic.

Toyota/Lexus Systems Toyota has a series of high-performance cars that feature five-speed transmissions that can operate in either of two modes, providing fully automatic shifting or electronic manual control. The TCM is programmed to allow for rapid shifts in response to the driver's commands. It also prevents shifting during conditions that may cause engine or transmission failure. The transmission may also be manually shifted by its gated console-mounted shift lever. The shift lever allows the driver to select individual gear ranges as well as the full-automatic mode.

Manual shifting may also be controlled by fingertip shifting buttons located on both horizontal spokes of the steering wheel **(Figure 39–63)**. Touching a button on the front of the steering wheel triggers downshifts. Contacting the buttons on the backside of the steering wheel controls upshifts. The buttons are located so that either thumb can be used to downshift and either index finger can be used to upshift.

CVT Controls

The electronic control system for Honda's CVT consists of a TCM, various sensors, three linear solenoids, and an inhibitor solenoid. Pulley ratios are always controlled by the control system. Input from the various sensors determines which linear solenoid the TCM will activate. Activating the shift control solenoid changes the shift control valve pressure, causing the shift valve to move. This changes the pressures applied to the driven and drive pulleys, which change the effective pulley ratio. Activating the start clutch control solenoid moves the start clutch valve. This valve allows or disallows pressure to the start clutch assembly. When pressure is applied to the clutch, power is transmitted from the pulleys to the final drive gearset.

The start clutch allows for smooth starting. Because this transaxle does not have a torque converter, the start clutch is designed to slip just enough to get the car moving without stalling or straining the engine. The slippage is controlled by the hydraulic pressure applied to the start clutch. To compensate for engine loads, the TCM monitors the engine's vacuum and compares it to the measured vacuum of the engine while the transaxle is in park or neutral.

Figure 39–63 Fingertip controls for manually shifting Lexus' automatic/manual transmission. *Reprinted with permission*

The TCM controls the pulley ratios to reduce engine speed and maintain ideal engine temperatures during acceleration. If the car is continuously driven at full throttle acceleration, the TCM causes an increase in pulley ratio. This reduces engine speed and maintains normal engine temperature while not adversely affecting acceleration. After the car has been driven at a lower speed or not accelerated for a while, the TCM lowers the pulley ratio. When the gear selector is placed into reverse, the TCM sends a signal to the PCM. The PCM then turns off the car's air conditioning and causes a slight increase in engine speed.

Audi's stepless Multitronic CVT is based on what it refers to as a variator. The variator is made from vanadium-hardened steel that is encased in oil and offers more durability than belt-driven CVTs. Manual gear selection mode is available and six simulated gear ratios can be selected. In the automatic mode, Multitronic calculates the optimum gear ratio with the aid of a dynamic regulating program, according to engine load, the driver's preferences, and driving conditions.

CASE STUDY

A fast-lube business has recently been faced with numerous complaints from customers who had their transmission fluid and filters changed. All of these customers owned late-model Ford products and have a common complaint of harsh shifting.

Concerned about these complaints, the business owner reviews the files of each of these customers and their vehicles. He finds something else the cars have in common. All of the vehicles were refilled with Type F transmission fluid. The owner wonders if the fluid was faulty. After doing some research, he finds that late-model Ford products call for Dexron-III fluid, which has special friction modifiers in it to provide smoother shifts.

This could be the cause of the complaints. The owner contacts one of the customers and requests that the vehicle be brought in. The fluid was drained and the correct fluid installed. The vehicle was then returned to the customer, because the transmission now shifted normally. All of the other customers were contacted and the same service was performed to their vehicles. The shop owner quickly realized that the money saved by buying Type F fluid from a surplus supplier was more than lost. Not only did he lose what he paid for the fluid; he also risked losing his customers. The correct type of fluid should always be used.

KEY TERMS

Accumulator
Adaptive learning
Annulus
Band
Belleville spring
Butt-end seals
Check ball valve
Continuously variable transmission (CVT)
Coupling point
Electronic pressure control (EPC) solenoid
Flow-directing valves
Governor
Hook-end seals
Impeller
Kickdown
Lip seals
Multiple-disc clutch
Orifice
Output shaft speed sensor (OSS)
Overrunning clutch
Planetary carrier
Planetary pinions
Poppet valve
Pressure regulating valves
Ravigneaux gear train
Rotary oil flow
Servo
Shift feel
Shift schedule
Simpson gear train
Spool valve
Square-cut seal
Stator
Sun gear
Throttle valve
Torque converter clutch (TCC)
Torrington bearing
Transmission control module (TCM)
Turbine
Vacuum modulator
Valve body
Variable force solenoid (VFS)
Vortex oil flow

SUMMARY

- The torque converter is a fluid clutch used to transfer engine torque from the engine to the transmission. It automatically engages and disengages power transfer from the engine to the transmission in relation to engine rpm. It consists of three elements: the impeller (input), turbine (output), and stator (torque multiplier).
- Two types of oil flow take place inside the torque converter: rotary and vortex flow.
- An overrunning clutch keeps the stator assembly from rotating in one direction and permits overrunning when turned in the opposite direction.
- A lockup torque converter eliminates the 10% slip that takes place between the impeller and turbine at the coupling stage of operation.
- Planetary gearsets transfer power and alter the engine's torque. Compound gearsets combine two simple planetary gearsets so that load can be spread over a greater number of teeth for strength and also so the largest number of gear ratios possible can be obtained in a compact area. A simple planetary gearset consists of a sun gear, a carrier with planetary pinions mounted to it, and an internally toothed ring gear.
- Planetary gear controls include transmission bands, servos, and clutches. A band is a braking assembly positioned around a drum. There are two types: single wrap and double wrap. Simple and compound servos are used to engage bands. Transmission clutches, either overrunning or multiple-disc, are capable of both holding and driving members.
- There are two common designs of compound gearsets, the Simpson gearset in which two planetary gearsets share a common sun gear and the Ravigneaux gearset which has two sun gears, two sets of planet gears, and a common ring gear.
- Most Honda transaxles do not use a planetary gearset, rather constant-mesh helical and square-cut gears are used in a manner similar to that of a manual transmission.
- The operation of most CVTs is based on a steel belt linking two variable pulleys.
- An automatic transmission uses ATF pressure to control the action of the planetary gearsets. This fluid pressure is regulated and directed to change gears automatically through the use of various pressure regulators and control valves. To form a complete working hydraulic system, the following elements are needed: fluid reservoir (transmission oil pan), pressure source (oil pump), control valving (valve control body), and output devices (servos and clutches).
- The transmission's pump is driven by the torque converter shell at engine speed. The purpose of the pump is to create fluid flow and pressure in the system. Excessive pump pressure is controlled by the pressure regulator valve. Three common types of oil pumps are installed in automatic transmissions: the gear, rotor, and vane.
- The valve body is the control center of the automatic transmission. It is made of two or three main parts. Internally, the valve body has many fluid passages called worm tracks.
- The purpose of a valve is to start, stop, or to direct and regulate fluid flow. Generally, in most valve bodies used in automatic transmissions, three types of valves are used: check ball, poppet, and most common, the spool.
- To prevent stalling, the automatic transmission pump has a pressure regulator valve normally located in the valve body. It maintains a basic fluid pressure. The valve's movement to the exhaust position is controlled

by calibrated coil spring tension. The three stages of pressure regulation are charging the torque converter, exhausting fluid pressure, and establishing a balanced position. There are times when fluid pressure must be increased above its baseline pressure to accomplish these stages.

- The purpose of the governor assembly is to sense vehicle road speed and send a fluid pressure signal to the transmission valve body to either upshift or permit the transmission to downshift. Throttle pressure delays the transmission upshift and forces the downshift.
- Bearings that take radial loads are called bushings and those that take axial loads are called thrust washers.
- The gaskets and seals of an automatic transmission help to contain the fluid within the transmission and prevent the fluid from leaking out of the various hydraulic circuits. Different types of seals are used in automatic transmissions; they can be made of rubber, metal, or Teflon.
- Three major types of rubber seals are used in automatic transmissions: the O-ring, the lip seal, and the square-cut seal.
- Three types of sealing rings are used in automatic transmissions: butt-end seals, open-end seals, and hook-end seals.
- There are four common configurations used as the final drives on FWD vehicles: helical gear, planetary gear, hypoid gear, and chain drive.
- A shift schedule contains the actual shift points to be used by the computer according to the input data it receives from the sensors. Its logic chooses the proper shift schedule for the current conditions of the transmission.
- In most electronically controlled transmission systems, the hydraulically operated clutches are controlled by the transaxle controller.
- The controller receives information from various inputs and controls a solenoid assembly. The solenoid assembly normally is comprised of four solenoids that control hydraulic pressure to four of the five clutches in the transaxle and to the lockup clutch of the torque converter.
- Adaptive learning allows the controller to compensate for wear and other events that might occur and cause the normal shift programming to be inefficient.

TECH MANUAL

The following procedures are included in Chapter 39/40 of the *Tech Manual* that accompanies this book:

1. Visually inspecting an automatic transmission.
2. Road testing a vehicle to check the operation of a lockup torque converter clutch.
3. Road testing a vehicle to check the operation of an automatic transmission.
4. Conducting a stall test.
5. Pressure testing an automatic transmission.
6. Servicing a valve body.
7. Overhauling a multiple-disc clutch assembly.

REVIEW QUESTIONS

1. Explain the difference between rotary and vortex fluid flow in a torque converter.
2. What component keeps the stator assembly from rotating when driven in one direction and permits rotation when turned in the opposite direction?
3. When a transmission is described as having two planetary gearsets in tandem, what does this mean?
4. The four common configurations used as the final drives on FWD vehicles are the ______ gear, ______ gear, ______ gear, and ______ ______.

5. Three major types of rubber seals are used in automatic transmissions: the ______, the ______, and the ______ seal.
6. Although computers receive different information from a variety of sensors, the decisions for shifting are actually based on more than the inputs. On what are they based?
7. To achieve a slow overdrive in a simple planetary gearset, the ______.
   a. sun gear must be the input member
   b. ring gear must be the input member
   c. planetary carrier must be the input member
   d. ring gear must be held
8. In a simple planetary gearset, when the planetary carrier is held, the gearset produces a ______.
   a. reverse
   b. direct drive
   c. fast overdrive
   d. forward reduction
9. Overrunning clutches are capable of ______.
   a. holding a planetary gear member stationary
   b. driving a planetary gear member
   c. both a and b
   d. neither a nor b
10. While discussing electronically controlled transmissions, Technician A says shifting relies on inputs from a TP sensor for shift timing. Technician B says shift feel is controlled by inputs from the generator. Who is correct?
    a. Technician A
    b. Technician B
    c. Both A and B
    d. Neither A nor B
11. Technician A says some systems use a special modulated shift control solenoid. Technician B says some systems use a special modulated converter clutch control solenoid. Who is correct?
    a. Technician A
    b. Technician B
    c. Both A and B
    d. Neither A nor B
12. Technician A says shift solenoids direct fluid flow to and away from the various apply devices in the transmission. Technician B says shift solenoids are used to mechanically apply a friction brake or multiple-disc clutch assembly. Who is correct?
    a. A only
    b. B only
    c. Both A and B
    d. Neither A nor B
13. While discussing transmission solenoids Technician A says an EPC solenoid replaces the conventional TV cable setup to provide changes in pressure in response to engine load. Technician B says an EPC solenoid is used to control hydraulic pressures throughout the transmission. Who is correct?
    a. A only
    b. B only
    c. Both A and B
    d. Neither A nor B
14. What is necessary for lockup engagement to take place?
    a. The lockup solenoid check ball must be off its seat and the switch valve held up by the spring tension.
    b. The lockup solenoid must be energized with the check ball on its seat and the switch valve held down by line pressure.
    c. The lockup check ball must be off its seat and the switch valve held down by the line pressure.
    d. The solenoid must have the check ball seated and the switch valve held up by spring tension.
15. The three types of sealing rings used in automatic transmissions are the ______, ______, and ______ seals.
16. Technician A says bearings that take radial loads are called Torrington washers. Technician B says bearings that take axial loads are called thrust washers. Who is correct?
    a. Technician A
    b. Technician B
    c. Both A and B
    d. Neither A nor B
17. Technician A says rotary oil flow is the oil flow around the circumference of the torque converter caused by the rotation of the torque converter on its axis. Technician B says the centrifugal lockup clutch is the type installed in most automatic transmissions. Who is correct?
    a. Technician A
    b. Technician B
    c. Both A and B
    d. Neither A nor B

18. *True or False?* The vent in a transmission housing is designed to allow fluid to escape when there is excessive pressure in the system.

19. Technician A says a Simpson gearset is two planetary gearsets that share a common sun gear. Technician B says a Ravigneaux gearset has two sun gears, two sets of planet gears, and a common ring gear. Who is correct?

    a. Technician A

    b. Technician B

    c. Both A and B

    d. Neither A nor B

20. Technician A says throttle position is an important input in most electronic shift control systems. Technician B says vehicle speed is an important input for most electronic shift control systems. Who is correct?

    a. Technician A

    b. Technician B

    c. Both A and B

    d. Neither A nor B

40

AUTOMATIC TRANSMISSION AND TRANSAXLE SERVICE

OBJECTIVES

■ Listen to the driver's complaint, road test the vehicle, and then determine the needed repairs. ■ Diagnose unusual fluid usage, level, and condition problems. ■ Replace automatic transmission fluid and filters. ■ Diagnose electronic control systems and determine needed repairs. ■ Conduct preliminary checks on EAT systems and determine needed repairs or service. ■ Perform converter clutch system tests and determine needed repairs or service. ■ Inspect, test, and replace electrical/electronic sensors. ■ Inspect, test, bypass, and replace actuators. ■ Diagnose noise and vibration problems. ■ Diagnose hydraulic and vacuum control systems. ■ Perform oil pressure tests and determine needed repairs. ■ Inspect and adjust external linkages. ■ Describe the basic steps for overhauling a transmission.

Because of the many similarities between a transmission and a transaxle, most diagnostic and service procedures are similar. Therefore, all references to a transmission apply equally to a transaxle unless otherwise noted. Whenever you are diagnosing or repairing a transaxle or transmission, make sure you refer first to the appropriate service manual before you begin.

Transmissions are strong and typically trouble-free units that require little maintenance. Normal maintenance usually includes fluid checks, scheduled linkage adjustments, and oil and filter changes.

Nearly all current automatic transmissions have electronic controls that work with the hydraulic and mechanical systems of the transmission to provide reliable and efficient operation.

IDENTIFICATION

Prior to beginning any service or repair work, be sure you know exactly which transmission you are working on to ensure that you are following the correct procedures and specifications and are installing the correct parts. Proper identification can be difficult because transmissions cannot be accurately identified by the way they look. The only exception to this is the shape of the oil pan, which can sometimes be used for identification.

The only positive way to identify the exact design of the transmission is by its identification numbers. **Transmission identification numbers** are found on stickers on the transmission **(Figure 40–1)**, as stamped numbers in the case, or on a metal tag held by a bolt head. Use a service manual to decipher the identification number. Most identification numbers include the model, manufacturer, and assembly date. Whenever you work with a transmission with a metal ID tag, make sure the tag is put back on the transmission so the next technician can properly identify the transmission.

DIAGNOSTICS

Automatic transmission problems are usually caused by poor engine performance, problems in the hydraulic system, abuse resulting in overheating, mechanical malfunctions, electronic failures, and/or improper adjustments. Diagnosis of these problems should begin by checking the condition of the fluid and its level, conducting a thorough visual inspection, checking the various linkage adjustments, and scanning the TCM for DTCs.

Engine performance can affect torque converter clutch operation. If the engine is running too poorly to maintain a constant speed, the converter clutch will engage and disengage at higher speeds. The customer complaint may be that the vehicle vibrates.

If the vehicle has an engine performance problem, the cause should be found and corrected before any conclusions on the transmission are made. A quick way to identify if the engine is causing shifting problems is to connect a vacuum gauge to the engine and take a reading while the engine is running. The gauge should be connected to intake manifold vacuum. A normal vacuum gauge

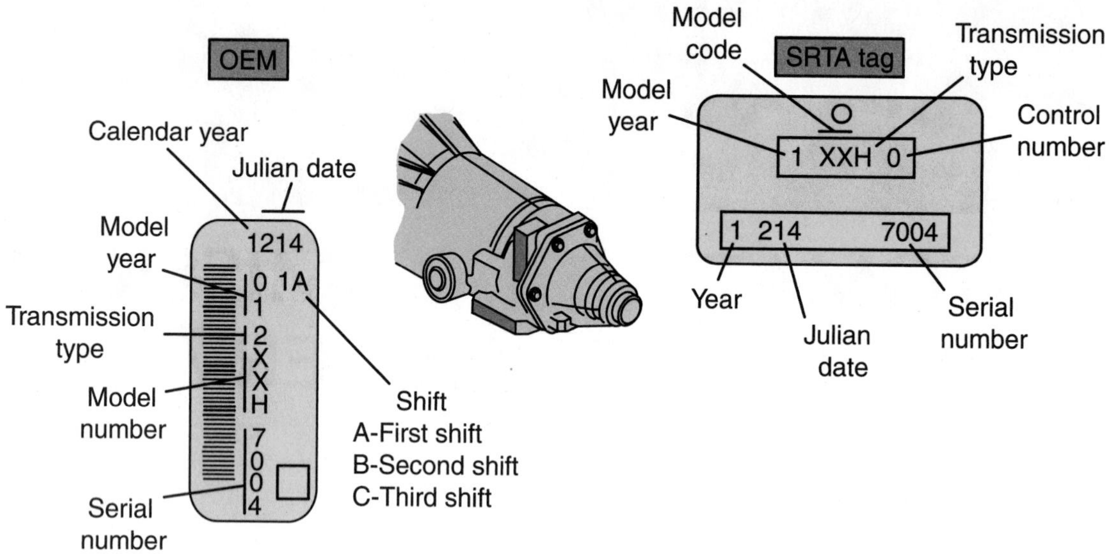

Figure 40-1 The location of and information contained on a GM transaxle ID plate.

reading is steady and at 17 in. Hg (431.8 mm/Hg). The rougher the engine runs, the more the gauge readings will fluctuate.

In order to properly diagnose a problem, you must totally understand the customer's concern or complaint. Make sure you know the conditions that exist when the problem occurs.

Fluid Check

Diagnosis should begin with a fluid level check. Make sure the vehicle is on a level surface when checking the level. Many late-model transmissions do not have a dipstick and the fluid level is checked in the same way as a manual transmission.

If the transmission has a dipstick, wipe all dirt off the protective disc and the dipstick handle. On most automobiles, the ATF level can be checked accurately only when the transmission is at operating temperature and the engine is running. Remove the dipstick and wipe it clean with a lint-free white cloth or paper towel. Reinsert the dipstick, remove it again, and note the reading. Markings on a dipstick indicate ADD levels, and, on some models, FULL levels for cool, warm, or hot fluid.

If the fluid level is low, the problem could be external fluid leaks. Check the transmission case, oil pan, and cooler lines for signs of leaks.

Excessively high fluid levels can also cause **aeration**. As the planetary gears rotate in high fluid levels, air can be forced into the fluid. Aerated fluid can foam, overheat, and oxidize. All of these problems can interfere with normal valve, clutch, and servo operation. Foaming may be evident by fluid leakage from the transmission's vent.

The condition of the fluid should be checked while checking its level. Examine the fluid carefully. The normal color of ATF is red. If the fluid has a dark brownish or blackish color and/or a burned odor, the fluid has been overheated. A milky color indicates that engine coolant has been leaking into the transmission's cooler in the radiator. If there is any question about the condition of the fluid, drain out a sample for closer inspection.

Synthetic ATF is normally a darker red than petroleum-based fluid. Synthetic fluids tend to look and smell burnt after normal use; therefore, the appearance and smell of these fluids is not a good indicator of the fluid's condition.

After checking the ATF level and color, wipe the dipstick on absorbent white paper and look at the stain left by the fluid. Dark particles are normally band and/or clutch material, while silvery metal particles are normally caused by the wearing of the transmission's metal parts. If the dipstick cannot be wiped clean, it is probably covered with varnish, which results from fluid oxidation. Varnish will cause the spool valves to stick, causing shifting malfunction. Varnish or other heavy deposits indicate the need to change the transmission's fluid and filter.

Contaminated fluid can sometimes be felt better than seen. Place a few drops of fluid between two fingers and rub them together. If the fluid feels dirty or gritty, it is contaminated with burned friction material.

Recommended Applications

Several ratings or types of ATF are available **(Figure 40–2)**; each type is designed for a specific application. The different classifications of transmission fluid have resulted from the inclusion of new or different additives that enhance the operation of the different transmission designs. Each automobile manufacturer specifies the proper type of ATF that should be used in its transmissions. Both the design of the transmission and the desired shift characteristics are considered when a specific ATF is chosen.

To reduce wear and friction inside a transmission, most commonly used transmission fluids are mixed with

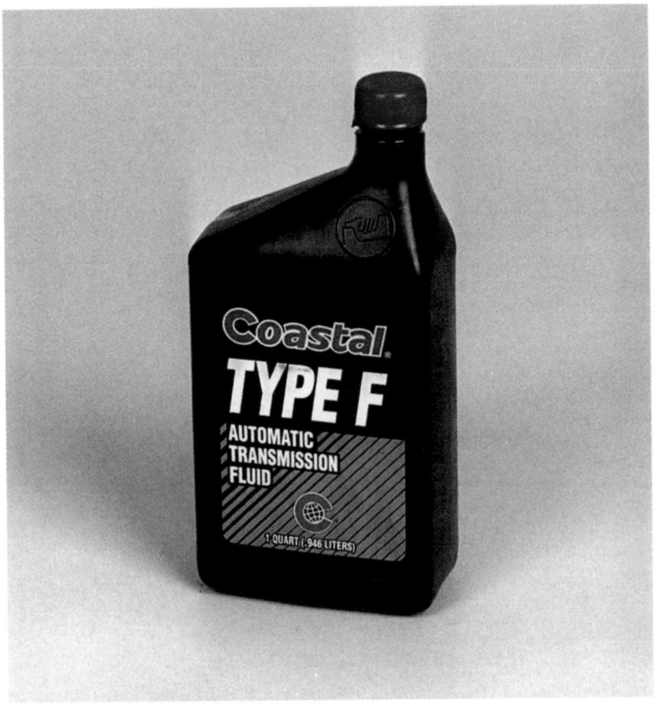

Figure 40-2 The classification of an ATF is clearly marked on its container.

friction modifiers. Transmission fluids with these additives allow for the use of lower clutch and band application pressures, which, in turn, provide for a very smooth feeling shift. Transmission fluids without a friction modifier, such as types F and G, tend to have a firmer shift because higher clutch and band application pressures are required to avoid excessive slippage during gear changes.

Nearly all manufacturers recommend the use of **Dexron**-III ATF in their automatic transmissions. This type of ATF has been developed through the years to meet the needs of the newer transmissions. The series or classification of the fluid is designated by a progressing alphabetical suffix. For example, Dexron-IIE was developed to replace all previous friction-modified fluids in all Hydra-Matic transmissions and transaxles and many other transmissions. It replaced Dexron-IID and was recommended wherever Dexron-II was previously specified. It may be used in all General Motors' automatic transmissions produced since 1949. It may be mixed, in any ratio, in a transmission already filled with Dexron-II without draining or flushing. Both fluids are fully compatible and may be substituted when necessary. Dexron-III has the following advantages over Dexron-II fluids: better oxidation resistance, compatibility, and flow.

Some manufacturers recommend the use of other types of fluid, but normally list Dexron-III as a secondary choice. Always use the exact automatic transmission fluid type specified in the vehicle service manual. The dipsticks of some transmissions indicate the type of fluid that should be used in the transmission.

Fluid Changes

The transmission's fluid and filter should be changed whenever there is an indication of oxidation or contamination. Periodic fluid and filter changes are also part of the preventative program for most vehicles. The mileage interval recommended depends on the type of transmission.

Change the fluid only when the engine and transmission are at normal operating temperatures. On most transmissions, you must remove the oil pan to drain the fluid. Some transmission pans include a drain plug. A filter or screen is normally attached to the bottom of the valve body. Filters are made of paper or fabric and are held in place by screws, clips, bolts, or by the pressure on the pickup tube seal **(Figure 40–3)**. Filters should be replaced, not cleaned.

After draining the fluid, carefully remove the pan. Check the bottom of the pan for deposits and metal particles. Slight contamination—blackish deposits from clutches and bands—is normal. Other contaminants should be of concern. Clean the oil pan and its magnet.

Remove the filter and inspect it. Use a magnet to determine if metal particles are steel or aluminum. Steel particles indicate severe internal transmission wear or damage. If the metal particles are aluminum, they may be part of the torque-converter stator. Some torque converters use phenolic plastic stators; therefore, aluminum particles found in these transmissions must be from the transmission itself.

Remove any traces of the old pan gasket on the case and oil pan. Then, install a new filter and gasket and tighten the retaining bolts to the specified torque. If the filter is sealed with an O-ring, make sure it is properly installed. Then, reinstall the pan using the gasket or sealant recommended by the manufacturer. Tighten the pan retaining bolts to the specified torque. Make sure you check the specifications. The required torque is often given in inch-pounds. You can easily break the bolts or damage something if you tighten the bolts to foot-pounds.

Pour a little less than the required amount of fluid into the transmission through the filler tube or fill point. Always use the recommended type of ATF. The wrong fluid will alter the shifting characteristics of the transmission. Start the engine and allow it to idle for at least one minute. Then, with the parking and service brakes applied, move the gear selector lever momentarily to each position, ending in the park. Recheck the fluid level and add a sufficient amount of fluid to bring the level to about 1/8 inch below the ADD mark.

Run the engine until it reaches normal operating temperature. Then recheck the fluid level; it should be in the HOT region on the dipstick. Make sure the dipstick is fully seated into the dipstick tube opening to prevent dirt from entering into the transmission.

Figure 40–3 Transmission fluid filters are attached to the transmission case by screws, bolts, retaining clips, and/or the pickup tube.

Parking Pawl

Any time you have the oil pan off, you should inspect all of the exposed parts, especially the parking pawl assembly. This component is typically not hydraulically activated; rather, the gearshift linkage moves the pawl into position to lock the output shaft of the transmission. Unless the customer's complaint indicates a problem with the parking mechanism, no test will detect a problem here.

Check the pawl assembly for excessive wear and other damage. Also, check to see how firmly the pawl is in place when the gear selector is shifted into the PARK mode. If the pawl can be moved out easily, it should be repaired or replaced.

FLUID LEAKS

Continue your diagnostics with a quick and careful visual inspection. Check all drive train parts for looseness and leaks. If the transmission fluid was low or there was no fluid, raise the vehicle and carefully inspect the transmission for signs of leakage. Leaks are often caused by defective gaskets or seals. Common sources of leaks are the oil pan seal, rear cover and final drive cover (on transaxles), extension housing, speedometer gear assembly, and electrical switches mounted into the housing **(Figure 40–4)**. The housing may have a porosity problem, allowing fluid to seep through the metal. Case porosity may be repaired with an epoxy sealer.

Oil Pan

A common source of fluid leakage is between the oil pan and the transmission housing. If fluid is present around the rim of the pan, retorquing the pan bolts may correct the problem. If it does not correct the problem, the pan must be removed and a new gasket installed. Make sure the sealing surface of the pan is flat and capable of providing a seal before reinstalling it.

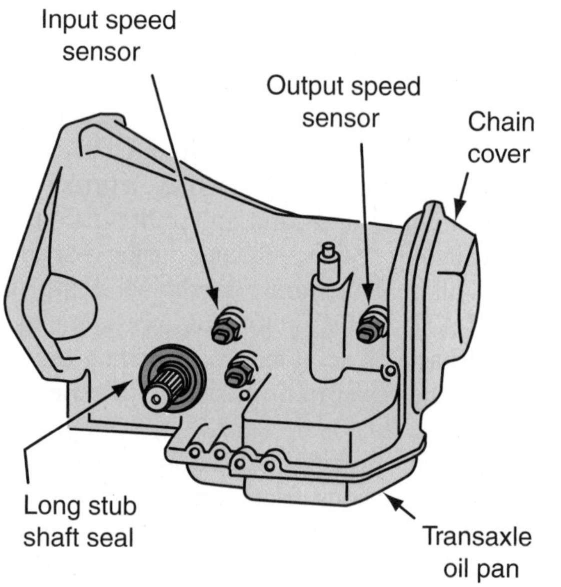

Figure 40–4 Possible sources of fluid leaks on this transaxle.

Torque Converter

Torque converter problems can be caused by a leaking converter **(Figure 40–5)**. This type of problem may be the cause of customer complaints of slippage and a lack of power. To check the converter for leaks, remove the

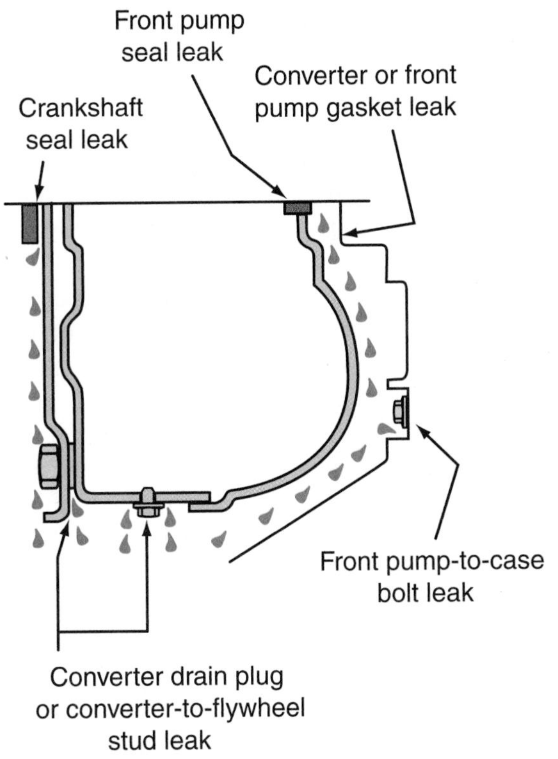

Figure 40-5 By determining the direction of fluid travel, the cause of a fluid leak around the torque converter can be identified.

converter access cover and examine the area around the torque converter shell. An engine oil leak may be falsely diagnosed as a converter leak. The color of engine oil is different from that of transmission fluid and may help identify the true source of the leak. However, if the oil or fluid has absorbed much dirt, both will look the same. An engine leak typically leaves an oil film on the front of the converter shell, whereas a converter leak will cause the entire shell to be wet. If the transmission's oil pump seal is leaking, only the back side of the shell will be wet. If the converter is leaking or damaged, it should be replaced.

Extension Housing

An oil leak stemming from the mating surfaces of the extension housing and the transmission case may be caused by loose bolts. To correct this problem, tighten the bolts to the specified torque. Also, check for signs of leakage at the rear of the extension housing. Fluid leaks from the seal of the extension housing can be corrected with the transmission in the car. Often, the cause for the leakage is a worn extension housing bushing, which supports the sliding yoke of the drive shaft. When the drive shaft is installed, the clearance between the sliding yoke and the bushing should be minimal. If the clearance is satisfactory, a new oil seal will correct the leak. If the clearance is excessive, the repair requires that a new seal and a new bushing be installed. If the seal is faulty, the transmission vent should be checked for blockage.

Speedometer Drive

The vehicle's speedometer can be purely electronic, which requires no mechanical hookup to the transmission, or it can be driven, via a cable, off the output shaft. If the transmission is equipped with a VSS, the bore and sensor can be a source of leaks. The sensor is retained in the bore with a retaining nut or bolt. An oil leak at the speedometer cable or VSS can be corrected by replacing the O-ring seal. While replacing the seal, inspect the drive gear for chips and missing teeth. Always lubricate the O-ring and gear prior to installation.

Electrical Connections

Check all electrical connections to the transmission. Faulty connectors or wires can cause harsh or delayed and missed shifts. On transaxles, the connectors can normally be inspected through the engine compartment, whereas they can only be seen from under the vehicle on longitudinally mounted transmissions. To check the connectors, release the locking tabs and disconnect them, one at a time, from the transmission. Carefully examine them for signs of corrosion, distortion, moisture, and transmission fluid. A connector or wiring harness may deteriorate if ATF reaches it. Also, check the connector at the transmission. Using a small mirror and flashlight may help you get a good look at the inside of the connectors. Inspect the entire transmission wiring harness for tears and other damage. Road debris can damage the wiring and connectors mounted underneath the vehicle.

Because the operation of the engine and transmission are integrated through the control computer, a faulty engine sensor or connector may affect the operation of the engine and the transmission. The various sensors and their locations can be identified by referring to the appropriate service manual. The engine control sensors that are the most likely to cause shifting problems are the TP, MAP, and VSS.

Checking Transmission and Transaxle Mounts

The engine and transmission mounts on FWD cars are important to the operation of the transaxle. Any engine movement may change the effective length of the shift and throttle cables or wiring harnesses and therefore may affect the engagement of the gears. Delayed or missed shifts may result from linkage changes as the engine pivots on its mounts.

Visually inspect the mounts for looseness and cracks. With a pry bar, pull up and push down on the transaxle case while watching the mount. If the mount's rubber separates from the metal plate or if the case moves up but not down, replace the mount. If there is movement between the metal plate and its attaching point on the frame, tighten the attaching bolts.

Then, from the driver's seat, apply the foot brake, set the parking brake, and start the engine. Put the transmission into a gear and gradually increase the engine speed to about 1,500 to 2,000 rpm. Watch the torque reaction of the engine on its mounts. If the engine's reaction to the torque appears to be excessive, broken or worn drive train mounts may be the cause.

If it is necessary to replace the transaxle mount, make sure you follow the manufacturer's recommendations for maintaining the alignment of the driveline. Failure to do so may result in poor gear shifting, vibrations, and/or broken cables. Some manufacturers recommend that a holding fixture or special bolt be used to keep the unit in its proper location.

Transmission Cooler and Line Inspection

Transmission coolers are a possible source of fluid leaks. The efficiency of the coolers is also critical to the operation and longevity of the transmission. Follow these steps when inspecting the transmission cooler and associated lines and fittings:

1. Check the engine's cooling system. The transmission cooler cannot be efficient if the engine's cooling system is defective. Repair all engine cooling system problems before continuing to check the transmission cooler.
2. Inspect the fluid lines and fittings between the cooler and transmission. Check these for looseness, damage, signs of leakage, and wear. Replace any damaged lines and fittings.
3. Inspect the engine's coolant for traces of transmission fluid. If ATF is present in the coolant, the transmission cooler leaks.
4. Check the transmission's fluid for signs of engine coolant. Water or coolant will cause the fluid to appear milky with a pink tint. This milky appearance is also an indication that the transmission cooler leaks and is allowing engine coolant to enter into the transmission fluid.

The cooler can be checked for leaks by disconnecting and plugging the transmission to cooler lines at the radiator. Then remove the radiator cap to relieve any pressure in the system. Tightly plug one of the ATF line fittings at the radiator. Using the shop air supply with a pressure regulator, apply 50 to 70 psi (3.52 to 4.92 kg/cm^2) of air pressure into the cooler at the other cooler line fitting. Look into the radiator. If bubbles are observed, the cooler leaks.

BASIC EAT TESTING

Some electronic transmissions are only partially controlled, that is, only the engagement of the converter clutch and third-to-fourth shifting is electronically controlled. Other models feature electronic shifting into all gears plus electronic control of system pressures and the TCC **(Figure 40–6)**. The techniques for diagnosing electronic transmissions are basically the same techniques used to diagnose TCC systems.

One of the first tasks during diagnosis of an electronic automatic transmission (EAT) is to determine if the problem is caused by the transmission or by electronics. To determine this, the transmission must be observed to see if it responds to commands given by the computer. Identifying whether the problem is the transmission or electrical will determine what steps need to be followed to diagnose the cause of the problem.

An EAT will work only as well as the commands it receives from the computer, even if the hydraulic and mechanical parts of the transmission are fine. All diagnostics should begin with a scan tool to check for trouble codes in the system's computer. After the received codes are addressed, you can begin a more detailed diagnosis of the system and transmission. Your next step may be manually activating the shift solenoids by connecting a jumper wire to them or by using a transmission tester that allows you to manually activate the solenoids. Prior to doing this, the wiring to the solenoids should be studied to determine if the computer activates them by supplying voltage to them or by completing the ground circuit. In addition, you need to know in what gear certain solenoids are activated. This information can be found in the service manual.

The best way to diagnose an electronically controlled transmission is to approach solving the problem in a logical way. To do this, you should follow these seven steps, in order:

1. Verify the customer's complaint.
2. Conduct preliminary inspections and checks.
3. Check all service information for information about the complaint and the system, including service bulletins and recall notices.
4. Follow the diagnostic procedures outlined in the service manual for the specific complaint.
5. Interpret and respond to all diagnostic codes.
6. Define and isolate the cause of the complaint or problem.
7. Fix the problem and verify the repair.

Since many EAT problems are caused by the basics, it is wise to conduct all of the preliminary checks required for a nonelectronically controlled transmission. Also, thoroughly inspect the electronic system. This inspection should include the retrieval of diagnostic codes, which not only allows you to see what the PCM sees as a transmission problem, but also allows you to check for any engine problems. Whenever diagnosing a transmission, remember that an engine problem can and will cause the transmission to act abnormally.

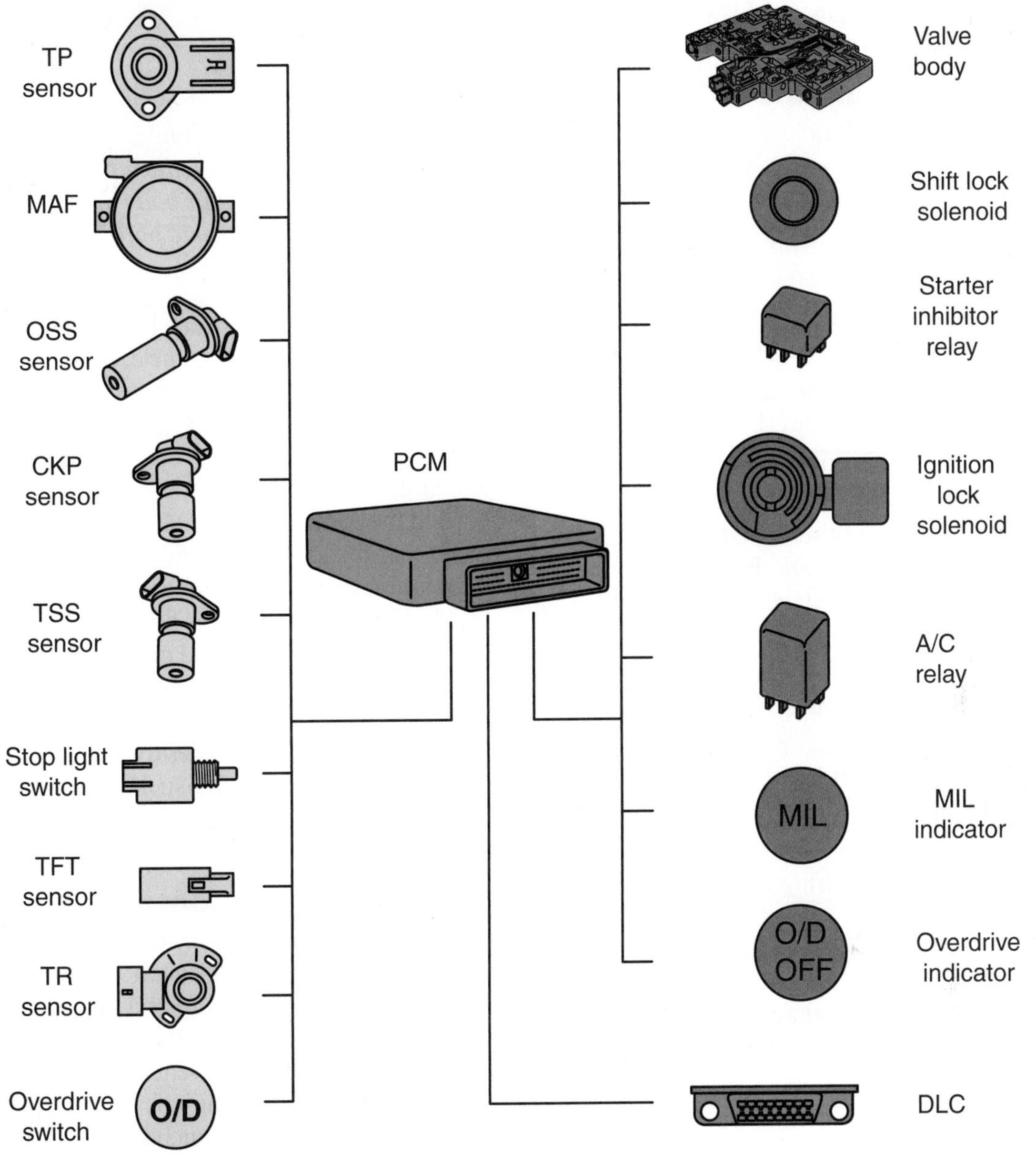

Figure 40-6 The components of a typical late-model EAT system.

Often accurately defining the problem and locating related information in TSBs and other materials can identify the cause of the problem. When a manufacturer recognizes common occurrences of a problem, a bulletin will be issued regarding the fix of the problem. Also for many DTCs and symptoms, service manuals will give a simple diagnostic chart or path for identifying the cause of the problem. These are designed to be followed step by step and will lead to a conclusion if you follow the path matched exactly to the symptom. Check all available information before moving on in your diagnostics.

Sometimes the symptom will not match any of those described in the service manual. This does not mean it is time to guess; rather, it is time to clearly identify what is working right. By eliminating those circuits and components that are working correctly from a list of possible causes, you can identify what may be causing the problem and what should be tested further.

Although the first steps in diagnosis include retrieving DTCs, there are problems that will not be evident by a code. These problems may be solved with the diagnostic charts or pure logic. This logic must be based on an understanding of the transmission and its controls.

DTCs that relate to transmission faults can be caused or detected by engine input or transmission input and/or output devices. Although codes may appear to be caused only by faulty input sensors or outputs, they may actually be caused by internal transmission problems. These problems may cause the inputs or outputs to appear out-of-normal range to the computer. To pinpoint the exact cause of a transmission problem, use basic electrical troubleshooting equipment, such as wiring diagrams, diagnostic charts, DMMs, lab scopes, special transmission testers, and scan tools.

Guidelines for Diagnosing EATs

When working on EATs, observe the following:

1. Make sure the battery has at least 12.6 volts before troubleshooting the transmission.
2. Check all fuses and identify the cause of any blown fuses.
3. Compare the wiring to all suspected components with the wire colors given in the service manual.
4. When testing electronic circuits, always use a high-impedance test light or DMM.
5. If an output device is not working properly, check the power circuit to it.
6. If an input device is not sending the correct signal back to the computer, check the reference voltage it is receiving and the voltage it is sending back to the computer.
7. Compare the voltages in and out of a sensor with the voltages the computer is sending out and receiving.
8. Before replacing a computer, check the solenoid isolation diodes according to the procedures outlined in the service manual.
9. Make sure computer wiring harnesses do not run parallel with any high-current wires or harnesses. The magnetic field created by the high current may induce a voltage in the computer harness. You should also be aware that antenna cables and CB radios could cause interference.
10. Take necessary precautions to prevent the possibility of static discharge while working with electronic systems.
11. While checking individual components, always check the voltage drop of the ground circuits. This becomes more and more important as cars are made of less material that conducts electricity well.
12. Make sure the ignition is off when you disconnect or connect an electronic component.
13. All sensors should be checked in cold and hot conditions.
14. All wire terminals and connections should be checked for tightness and cleanliness.
15. Use TV-tuner cleaning spray to clean all connectors and terminals.
16. Use dielectric grease at all connections to prevent future corrosion.
17. If you must break through the insulation of a wire to take an electrical measurement, make sure you tightly tape over the area after you are finished testing.

WARNING!

Static electricity can destroy or render an electronic part useless. When handling any electronic part, do whatever is possible to reduce the chances of electrostatic buildup on your body and the inadvertent discharge to the electronic part.

Although DTCs are very helpful during diagnosis, they also can become an obstacle, especially if more than one DTC is present. If two or more codes are present, you should look at the relationship of the codes to identify if they could have a common cause. Reacting to the individual codes may move your troubleshooting efforts beyond the common cause. Many manufacturers recommend that the lowest numeric code be considered first.

It is also important to remember that codes can be set by out-of-range signals. This does not mean the sensor or sensor circuit is bad. It could mean the sensor is working properly but there is a mechanical or hydraulic problem causing the abnormal signals. Not only can internal transmission problems cause codes to be set; so can basic electrical problems. Problems such as loose connections, broken wires, corrosion, and poor grounds will affect the signals.

Converter Clutch Control Diagnostics

On early TCC equipped vehicles, clutch engagement was controlled hydraulically. A switch valve was controlled by two other valves, the lockup and fail-safe valves, in the clutch control assembly. The lockup valve responds to governor pressure and prevents lockup at speeds below 40 mph (64 km/h). The fail-safe valve responds to throttle pressure and permits clutch engagement in high gear only. Problems with this system are diagnosed in the same way as other hydraulic circuits.

On early electronically controlled systems with a TCC solenoid, the clutch is typically applied when oil flow through the torque converter is reversed. This change can be observed with a pressure gauge. Connect a pressure gauge to the fluid line from the transmission to the cooler. Position the gauge so it is easily seen from the driver's seat. Then raise the vehicle on a hoist with the drive wheels off the ground and able to spin freely. Operate the vehicle until the transmission shifts into high gear. Then maintain a speed of approximately 55 mph (88 km/h). Once the speed is maintained, watch the pressure gauge.

If the pressure decreases 5 to 10 psi (0.35 to 0.70 kg/cm^2), the converter clutch is applied. With this action, you should feel the engagement of the clutch, as well as a drop in engine speed. If the pressure changes but the clutch does not engage, the problem may be inside the converter or at the end of the input shaft. If the input shaft end is worn or the O-ring at the end is cut or worn, there will be a pressure loss at the converter clutch. This loss in pressure will prevent full engagement of the clutch. If the pressure does not change and the clutch does not engage, suspect a faulty clutch valve or control solenoid or a fault in the solenoid control circuit.

If the clutch does not engage, check for power to the solenoid. If power is available, make sure the ground of the circuit is good. If there is power available and the ground is good, check the voltage drop across the solenoid. The solenoids should drop very close to source voltage. If less than that is measured, check the voltage drop across the power and ground sides of the circuit. If the voltage drop testing gives good results remove the solenoid and test it with an ohmmeter. If the solenoid checks out fine with the ohmmeter, suspect clutch material, dirt, or other material plugging up the solenoid valve passages. If a blockage is found, attempt to flush the valve with clean ATF. If the solenoid has a filter assembly, replace the filter after cleaning the fluid passages. If the blockage cannot be removed, replace the solenoid.

If the clutch engages at the wrong time, a sensor or switch in the circuit is probably the cause. If clutch engagement occurs at the wrong speed, check all speed-related sensors. A faulty temperature sensor may cause the clutch not to engage. If the sensor is not reading the correct temperature, the PCM may never realize the temperature is suitable for engagement. Checking the appropriate sensors can be done with a scan tool, DMM, and/or lab scope. A check of the sensors is normally part of the manufacturer's system check.

PRELIMINARY EAT CHECKS

Common problems that affect shift timing and quality as well as the timing and quality of TCC engagement are incorrect battery voltage, a blown fuse, poor connections, a defective TP sensor or VSS, defective solenoids, wires to the solenoids or sensors crossed, corrosion at an electrical terminal, or faulty installation of an accessory, such as a cellular telephone.

Electrical circuit problems, faulty electrical components, or bad connectors, as well as a defective governor or governor drive gear assembly can cause improper shift points. Most EATs do not rely on the hydraulic signals from a governor; rather, they rely on the electrical signals from electrical sensors to determine shift timing.

Often computer-controlled transmissions will start off in the wrong gear. This can happen because of internal transmission problems or external control system problems. Internal transmission problems can be faulty solenoids or stuck valves. External problems can be the result of a complete loss of power or ground to the control circuit or a fail-safe protection strategy initiated by the computer to protect itself or the transmission from an observed problem. Typically, the default gear is simply the gear that is applied when the shift solenoids are off.

Basic System Checks

The diagnostic procedure for most EAT systems includes checking the system with a scan tool, which will display any trouble codes in the system. More important to the technician, most scan tools will display serial data. The serial data stream allows you to monitor system sensor and actuator activity during operation. Comparing the test values to the manufacturer's specifications will help greatly in diagnostics.

It is possible that the data displayed by a scan tool is not the actual value. Most computer systems will disregard inputs that are well out of range and rely on a default value held in memory. These default values are hard to recognize and do little for diagnostics; this is why detailed testing with a DMM or lab scope is preferred by many technicians. These test instruments are also used to further test the system after a scan tool has identified a problem.

Diagnostic Trouble Codes

Diagnostic trouble codes are designed to help technicians identify and locate problems in the transmission's system. DTCs from pre-OBD-II systems can indicate a variety of things. Each manufacturer (and sometimes each vehicle model) uses different codes to identify detected problems. It is very important that you refer to the service manual when interpreting DTCs.

There are basically two types of DTCs. A hard code is a DTC that represents a problem present at the time of retrieval. These are the codes that should be responded to first during diagnostics. Soft codes are those DTCs that are not currently present. These codes can be retrieved and represent an intermittent problem or a problem that existed but is no longer present.

Basic Diagnostics of Non-OBD-II Systems

Diagnosis on vehicles equipped with non-OBD-II electronic control systems can be done by the self-diagnostics mode of the system's computer and/or by plugging in a manufacturer-provided or aftermarket-designed scan tool into the DLC or ALDL. Always refer to the correct service manual when diagnosing electronic control systems. The manufacturers list the procedures for self-diagnostics in their service manuals. Also listed are the interpretation tables for the trouble codes retrieved from the computer. Keep in mind that by law all 1996 and newer vehicles must have an OBD-II compliant system.

Basic Diagnostics of OBD-II Systems

General testing of the engine and its major systems is done in the same way as is done on non-OBD-II systems.

OBD-II diagnosis is best done with a strategy-based procedure based on the flow charts and other information given in the service manuals. Check for any published TSBs, including videos, newsletters, and any electronically transmitted media, relating to the exhibited symptoms. Do not depend solely on the diagnostic tests run by the PCM. A particular system may not be supported by one or more DTCs.

Connect a scan tool to the vehicle's DLC and retrieve the codes. The DTCs have a prefix indicating whether there is a problem with the engine, transmission, or body. No special procedures are necessary to retrieve transmission-related DTCs.

WARNING!

Do not clear DTCs unless directed by a diagnostic procedure. Clearing DTCs also clears valuable freeze frame and failure records data.

The next step in diagnostics depends on the DTCs. If DTCs were retrieved, proceed by following the designated DTC diagnostic table to make an effective diagnosis and repair. If no DTCs were displayed, match the symptom to the manufacturer's symptom tables and follow the diagnostic paths or suggestions to complete the repair, or refer to the applicable component/system check. If there is not a matching symptom, analyze the complaint. Then develop a plan for diagnostics. Utilize the wiring diagrams and theory of component and system operation. Call technical assistance for similar cases where repair history may be available. It is possible that the customer's complaint is a normal operating condition for that vehicle. If this is suspected, compare the complaint to the driveability of a known good vehicle.

After the DTCs and the MIL have been checked, the OBD-II data stream can be selected. In this mode, the scan tool displays all of the data from the inputs. The actual available data will depend on the vehicle and the scan tool; however, OBD-II codes dictate a minimum amount of data. Always refer to the appropriate service manual and the scan tool's operating instructions for proper identification of normal or expected data.

One of the selections on a scan tool is DTC data. This selection contains two important categories of information: a freeze frame that displays the conditions that existed when a DTC was set and failure records. Failure records are valuable when a fault was detected but the MIL was not lit. Failure records display the conditions present when the system failed the diagnostic test.

At times, it may be difficult to determine if the problem is caused by something inside the transmission or if the cause is electronic. If you face this dilemma, take the vehicle on a road test and note all of the things that do not seem right. Then, remove the fuse for the transmission control unit. Doing this disables electronic control of the transmission and will cause it to operate in its default mode. Road test the vehicle again and observe the operation of the transmission. Keep in mind that during default operation there will be no forward upshifts. If the problem or problems observed in the first road test still exist, the cause must be in the transmission.

Electronic Defaults

Some electronically controlled transmissions will exhibit strange characteristics if the computer senses a problem within its system. These transmissions will go into a default mode and the computer will disregard all input signals. During this period of time, the transmission may lock itself into a single forward gear or will bypass all electronic controls. While diagnosing a problem in an electronically controlled transmission, always refer to the appropriate service manual to identify the normal default operation of the transmission. By not recognizing that the transmission is operating in default, you could spend time tracing the wrong problem.

Whenever the computer sees a potential problem that may increase wear and/or damage to the transmission, the system also defaults to limp-in. Minor slipping can be sensed by the computer through its input and output sensors. This slipping will cause premature wear and may cause the computer to move into its default mode. Some systems may increase fluid pressure to compensate for the problem. A totally burnt clutch assembly will cause limp-in operation, as will some internal pressure leaks that may not be apparent until pressure tests are run.

DETAILED TESTING OF INPUTS

There are many different designs of sensors, depending on the operation of the system they monitor. Some sensors are nothing more than a switch that completes a circuit. Others are complex devices that react to chemical reactions and generate their own voltage during certain conditions. If the self-test sequence points to a problem in an input circuit, the input circuit should be tested to determine the exact malfunction. Often the manufacturers list specific procedures for specific sensors; always follow them.

Testing Switches

Many different switches are used as inputs or control devices for EATs. Most of the switches are either mechanically or hydraulically controlled. The operation of these switches can be easily checked with an ohmmeter. With the meter connected across the switch's leads, there should be continuity or low resistance when the switch is closed and there should be infinite resistance across the switch when it is open. A test light can also be used. When the switch is closed, power should be present at both sides of the switch. When the switch is open, power should be present at only one side.

Pressure switches can be checked by applying air pressure to the part that would normally be exposed to oil pressure **(Figure 40–7)**. When applying air pressure to these switches, check them for leaks. Although a malfunctioning electrical switch will probably not cause a shifting problem, it will if it leaks. If the switch leaks off the applied pressure in a hydraulic circuit to a holding device, the holding member may not be able function properly. When possible, you should check pressure switches when they are installed and controlled by the vehicle.

Throttle Position Sensor

Another type of switch is a potentiometer. Rather than open and close a circuit, a potentiometer controls the circuit by varying its resistance in response to something. A TP sensor is a potentiometer. It sends very low voltage back to the computer when the throttle plates are closed and increases the voltage as the throttle is opened. A bad TP sensor can cause the following problems: no upshifts, quick upshifts, delayed shifts, line-pressure problems with transmissions that have a line pressure control solenoid, and erratic converter clutch engagement.

A TP sensor can be checked with an ohmmeter or a voltmeter. If checked with an ohmmeter, you should be able to watch the resistance across the TP sensor change as the throttle is opened and closed. Often a resistance specification is given in the service manual. Compare your reading to this specification.

With a voltmeter, you will be able to measure the reference voltage and the output voltage, both of which should be within specified amounts. If the reference voltage is lower than normal, check the voltage drop across the reference voltage circuit from the computer to the TP sensor. If the TP sensor is found to be defective, it should be replaced.

Testing with a lab scope is a good way to watch the sweep of the resistor. The waveform is a DC signal that moves up as the voltage increases. Most potentiometers in computer systems are fed a reference voltage of 5 volts. Therefore, the voltage output of these sensors will typically range from 0.5 to 4.5 volts. The change in voltage should be smooth. Look for glitches in the signal. These can be caused by changes in resistance or an intermittent open.

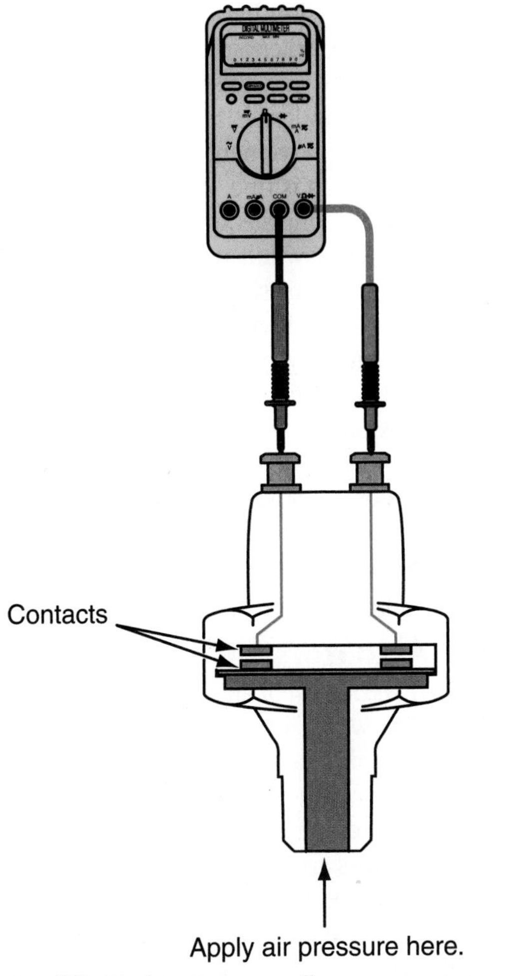

Figure 40–7 A typical normally open pressure switch. To check, apply air to the bottom oil passage and check for continuity across the terminals.

Mass Airflow Sensor

When a mass airflow sensor fails or sends faulty signals, the engine runs roughly and tends to stall as soon as you put the transmission into gear. This sensor can typically be checked with a multimeter set to the Hz frequency range. Check the service manual for specific values. Normally, 30 Hz is measured at idle, and the frequency will increase as the throttle opens. A scan tool may also be used to test this sensor; most have a test mode that monitors MAF sensors. The output of some MAFs can be observed with a DMM, as their output is variable DC voltage. While diagnosing these systems, keep in mind that cold air is denser than warm air.

Temperature Sensors

Temperature sensors can be checked with an ohmmeter. To do so, disconnect the sensor. Then determine the temperature of the sensor and measure the resistance across it. Compare your reading to the chart of normal resistances given in the service manual.

Thermistor activity can be monitored with a lab scope. Connect the scope across to the output of the thermistor or temperature sensor. Run the engine and watch the waveform. As the temperature increases, there should be a smooth increase or decrease in voltage. Look for glitches in the signal. These can be caused by changes in resistance or an intermittent open.

Speed Sensors

Speed sensors negate the need for hydraulic signals from a governor. When this sensor fails or sends faulty readings, it can cause complaints similar to those caused by a bad TP sensor. The most common complaints are no overdrive, no converter-clutch engagement, and no upshifts. Vehicle speed sensors provide road speed information to the computer.

The operation of a PM generator-type speed sensor can be tested with a DMM set to measure AC voltage. Raise the vehicle on a lift. Allow the wheels to be suspended and to rotate freely. Connect the meter to the speed sensor. Start the engine and put the transmission in gear. Slowly increase the engine's speed until the vehicle is at approximately 20 mph (32 km/h), and then measure the voltage at the speed sensor. Slowly increase the engine's speed and observe the voltmeter. The voltage should increase smoothly and precisely with an increase in speed.

Magnetic pulse generator speed sensors can be tested with a lab scope. Connect the lab scope's leads across the sensor's terminals. The expected pattern is an AC signal, which should be a perfect sine wave when the speed is constant. When the speed is changing, the AC signal should change in amplitude and frequency. If the readings are not steady and do not smoothly change with speed, suspect a faulty connector, wiring harness, or sensor.

A speed sensor can also be tested when it is out of the vehicle. Connect an ohmmeter across the sensor's terminals. The desired resistance readings across the sensor will vary with every individual sensor; however, you should expect to have continuity across the leads. If there is no continuity, the sensor is open and should be replaced. Reposition the leads of the meter so that one lead is on the sensor's case and the other is on a terminal. There should be no continuity in this position. If there is any measurable amount of resistance, the sensor is shorted.

DETAILED TESTING OF ACTUATORS

If you were unable to identify the cause of a transmission problem through the previous checks, you should continue your diagnostics by testing the solenoids. This allows you to determine if the shifting problem is the solenoids or their control circuit, or if it is a hydraulic or mechanical problem in the transmission.

Before continuing, however, you must first determine if the solenoids are case grounded and fed voltage by the computer or if they always have power applied to them and the computer merely supplies the ground. While looking in the service manual to find this information, also find the section that tells you which solenoids are on and which are off for each of the different gears.

To begin this test you should collect the tools and/or equipment necessary to manually activate the solenoids. Switch panels are available that connect into the solenoid assembly and allow the technician to switch gears by depressing or flicking a switch.

SHOP TALK

This type of tester is easily made. Get a wiring harness for the transmission you want to test. Connect the harness to simple switches. Follow the solenoid/gear pattern when doing this. To change gears, all you need to do is turn off one switch and turn on the next. ■

With the tester, the solenoids will be energized in the correct pattern. Observe the action of the solenoids. A simple electronically controlled transmission has two shift solenoids and the computer sends voltage to the solenoids to activate them. In first gear, solenoid 1 is on. In second gear, both solenoids are on. In third gear, only solenoid 2 is on and neither of the solenoids is on in fourth gear. One wire in the transmission harness goes to solenoid 1 and another connects to solenoid 2.

To totally test the transmission, you should shift gears under light, half, and full throttle. If the transmission shifts well with the movement of the switches, you know that the transmission is fine. Any shifting problem must be caused by something electrical. If the transmission did not respond to the switch movements, the problem is probably in the transmission.

At times, a solenoid will work well during light throttle operation but may not allow the valve to exhaust enough fluid when pressure increases. To verify that the valve is not exhausting, activate the solenoid, then increase engine speed while pulling on the throttle cable. If the valve cannot exhaust, the transmission will downshift. Restricted solenoids are a common cause of rough shifting under heavy loads or full throttle but of good shifting under light throttle.

Testing Actuators with a Lab Scope

By observing the action of an actuator on a lab scope, you will be able to watch its electrical activity. Normally if there is a mechanical fault, its electrical activity will be affected. Therefore, you get a good sense of the actuator's condition by watching it on a lab scope.

Some actuators are controlled by pulse width–modulated signals **(Figure 40–8)**. These devices are controlled by varying the pulse width, signal frequency, and voltage levels. By watching the control signal, you can see the turning on and off of the solenoid **(Figure 40–9)**. All waveforms should be checked for amplitude, time, and

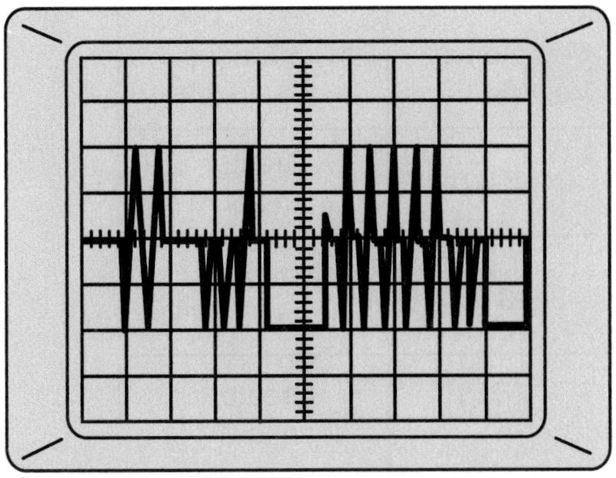

Figure 40-8 A typical control signal for a pulse width–modulated solenoid.

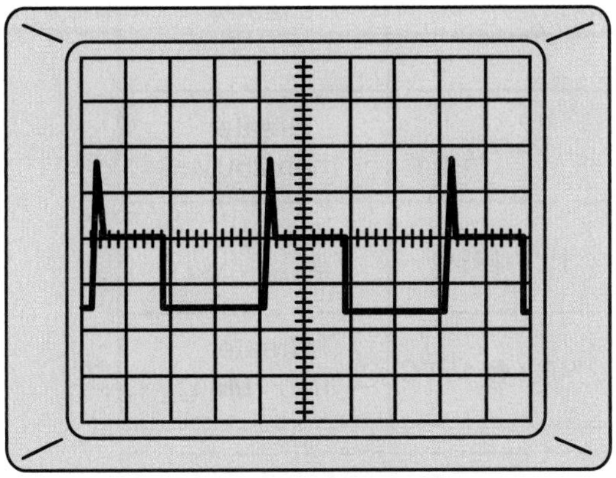

Figure 40-9 A typical control signal for a solenoid.

shape. You should also observe changes to the pulse width as operating conditions change. A bad waveform will have noise, glitches, or rounded corners. You should be able to see evidence that the actuator immediately turns off and on according to the commands of the computer.

Testing Actuators with an Ohmmeter

Solenoids can be checked for circuit resistance and shorts to ground. This procedure can typically be done without removing the oil pan. The test can be conducted at the transmission case connector. By identifying the proper pins in the connector, individual solenoids can be checked with an ohmmeter. Remember, lower-than-normal resistance indicates a short, whereas higher-than-normal resistance indicates a problem of high resistance. If you get an infinite reading across the solenoid, the solenoid windings are open. The ohmmeter can also be used to check for shorts to ground. Simply connect one lead of the ohmmeter to one end of the solenoid windings and the other lead to a good ground. The reading should be infinite. If there is any measurable resistance, the winding is shorted to ground.

Solenoids can also be tested on a bench. Resistance values are typically given in service manuals for each application **(Figure 40–10)**. A solenoid may be electrically fine but still may fail mechanically or hydraulically. A solenoid's check valve may fail to seat or the porting can be plugged. This problem is not electrical; rather, it could be caused by the magnetic field collecting metal particles in the ATF and clogging the port or check valve. These would cause erratic shifting, no shift conditions, wrong gear starts, no or limited passing (kickdown) gear, or binding shifts. When a solenoid affected in this way is activated, it will make a slow dull thud. A good solenoid tends to snap when activated.

ROAD TESTING THE VEHICLE

Critical to proper diagnosis of EAT and TCC control systems is a road test. The road test should be conducted in the same way as one for a nonelectronic transmission except that a scan tool should also be connected to the circuit to monitor engine and transmission operation.

During the road test, the vehicle should be driven in the normal manner. All pressure and gear changes should be noted. Also, the various computer inputs should be monitored and the readings recorded for future reference. Some scan tools have the capability of printing out a report of the test drive. Critical information from the inputs includes engine speed, vehicle speed, manifold vacuum, operating gear, and the time it took to shift gears. If the scanner does not have the ability to give a summary of the road test, you should record this same information after each gear or operating condition change.

All transmission complaints should be verified by road testing the vehicle and attempting to duplicate the customer's complaint. Knowing the exact conditions that cause the symptom will allow you to accurately diagnose problems. Many problems that appear to be transmission problems may be caused by problems in the engine, drive shaft, U- or CV joints, wheel bearings, wheel/tire imbalance, or other conditions.

Make sure these conditions are not the cause of the problem before you begin to diagnose and repair a transmission. Diagnosis becomes easy if you think about what is happening in the transmission when the problem occurs. If there is a shifting problem, think about the parts that are being engaged and what these parts are attempting to do.

SHOP TALK

Always refer to your service manual to identify the particulars of the transmission you are diagnosing. It is also helpful to check for any TSBs that may be related to the customer's complaint.

Components	Pass thru pins	Resistance at 20°C	Resistance at 100°C	Resistance to ground (case)
1-2 shift solenoid valve	A, E	19–24Ω	24–31Ω	Greater than 250M Ω
2-3 shift solenoid valve	B, E	19–24Ω	24–31Ω	Greater than 250M Ω
TCC solenoid valve	T, E	21–24Ω	26–33Ω	Greater than 250M Ω
TCC PWM solenoid valve	U, E	10–11Ω	13–15Ω	Greater than 250M Ω
3-2 shift solenoid valve assembly	S, E	20–24Ω	29–32Ω	Greater than 250M Ω
Pressure control solenoid valve	C, D	3–5Ω	4–7Ω	Greater than 250M Ω
Transmission fluid temp. (TFT) sensor	M, L	3088–3942Ω	159–198Ω	Greater than 10M Ω
Vehicle speed sensor	A, B Vss conn	1420Ω @ 25°C	2140Ω @ 150°C	Greater than 10M Ω

IMPORTANT: The resistance of this device is necessarily temperature dependent and will therefore vary far more than that of any other device. Refer to transmission fluid temp (TFT) sensor specifications.

Figure 40-10 Service manuals list the resistance values and test points for transmission solenoids and other components.

Diagnosis of Noise and Vibration Problems

Often a customer will complain of a transmission noise, which in reality is caused by something in the driveline and not the transmission or torque converter. Bad CV or U-joints, wheel bearings, brake calipers, and dragging brake pads can generate noises that customers and, unfortunately, some technicians, mistakenly blame on the transmission or torque converter. The entire driveline should be checked before assuming the noise is transmission related.

Abnormal transmission noises and vibrations can be caused by faulty bearings, damaged gears, worn or damaged clutches and bands, a bad oil pump, contaminated fluid, or improper fluid levels.

Most vibration problems are caused by an unbalanced torque converter assembly, a poorly mounted torque converter, or a faulty output shaft. The key to determining the cause of the vibration is to pay particular attention to the vibration in relationship to engine and vehicle speed. If the vibration changes with a change in engine speed, the cause of the problem is most probably the torque converter. If the vibration changes with vehicle speed, the cause is probably the output shaft or the driveline connected to it. The latter type of problem can be a bad extension housing bushing or U-joint, which would become worse at higher speeds.

Begin your diagnosis by determining if the cause of the problem is the driveline or the transmission. To do this, put the transmission in gear and apply the foot brakes. If the noise is no longer evident, the problem can be in the driveline or the output of the transmission. If the noise is still present, the problem must be in the transmission or torque converter.

Noise problems are also best diagnosed by paying a great deal of attention to the speed and the conditions at which the noise occurs. If the noise is engine speed related and is present in all gears, including park and neutral, the most probable source of the noise is the oil pump because it rotates whenever the engine is running. However, if the noise is engine related and is present in all gears except park and neutral, the most probable sources of the noise are those parts that rotate in all gears, such as the drive chain, the input shaft, and the torque converter.

Noises that only occur when a particular gear is operating must be related to those components responsible for providing that gear, such as a band or clutch. Often the exact cause of noise and vibration problems can only be identified through a careful inspection of a disassembled transmission. All noises and vibrations that occur during the road test should be noted.

Before beginning your road test, find and duplicate, from a service manual, the chart **(Figure 40–11)** that shows the band and clutch application for different gear selector positions. Using these charts will greatly simplify your diagnosis of automatic transmission problems. It is also wise to have a notebook or piece of paper to jot down notes about the operation of the transmission. If the transmission is electronically controlled, take a scan tool on the road test with you. It is an added convenience if the scan tool has memory or a printer.

Begin the road test with a drive at normal speeds to warm the engine and transmission. If a problem appears only when starting and/or when the engine and transmission are cold, record this symptom on the chart or in your notebook.

During the road test, the transmission should be operated in all possible modes and its operation noted. Check for proper gear engagement as the selector lever is moved to each gear position, including park. There should be no hesitation or roughness as the gears are engaging. Check for proper operation in all forward ranges, especially the upshifts and converter clutch engagement during light throttle operation. These shifts should be smooth and positive and occur at the correct speeds. These same shifts should feel firmer under medium to heavy throttle pressures. Transmissions equipped with a torque converter clutch should be brought to the specified apply speed and their engagement noted. Again, record the operation of the transmission in these different modes in your notebook or on the diagnostic chart.

Force the transmission to kick down and pay attention to the quality of this shift. Manual downshifts should also be made at a variety of speeds. The reaction of the transmission should be noted, as should all abnormal noises and the gears and speeds at which they occur.

After the road test, check the transmission for signs of leakage. Any new leaks and their probable cause should be noted. Then compare your written notes from the road test to the information given in the service manual to identify the cause of the malfunction. The service manual usually has a diagnostic chart to aid you in this process.

Torque Converter

Many transmission problems are related to the operation of the torque converter. Normally torque converter problems will cause abnormal noises, poor acceleration in all gears, normal acceleration but poor high-speed performance, or transmission overheating.

To test the torque converter, many technicians perform a stall test. The stall test checks the holding capacity of the converter's stator overrunning clutch assembly, as well as the clutches and bands in the transmission. Some manufacturers do not recommend the stall test.

Range		Gear ratio	Clutch		Low and reverse brake	Lockup	Band servo		One-way clutch	Parking pawl
			High-reverse clutch (front)	Forward clutch (rear)			Operation	Release		
Park										on
Reverse		2.364	on		on					
Neutral										
Drive	D_1 Low	2.826		on					on	
	D_2 Second	1.543		on			on			
	D_3 Top (3rd)	1.000	on	on		on	(on)	on		
2	2_1 Low	2.856		on					on	
	2_2 Second	1.543		on			on			
1	1_1 Low	2.826		on	on				on	
	1_2 Second	1.543		on			on			

Figure 40-11 A typical band and clutch application chart. This chart should be referred to during a road test and when determining the cause of any shifting problems. *Used with permission from Nissan North America, Inc.*

Torque converter problems can often be identified by the symptoms and therefore the need for a stall test is minimized. If the vehicle lacks power during acceleration, it has a restricted exhaust or the torque converter's one-way stator clutch is slipping. To determine which of these problems is causing the power loss, test for a restricted exhaust first. Other possible causes of this problem include a restricted air or fuel filter and a defective fuel pump.

If there is no evidence of a restricted exhaust, it can be assumed that the torque converter's stator clutch is slipping and not allowing any torque multiplication to take place in the converter. To correct this problem, the torque converter should be replaced.

If the engine's speed flares up during acceleration in drive and does not have normal acceleration, the clutches or bands in the transmission are slipping. This symptom is similar to the slipping of a clutch in a manual transmission. Often this problem is mistakenly blamed on the torque converter.

Technicians often blame the torque converter for problems simply based on the customer's complaint. Complaints of thumping or grinding noises are often blamed on the converter when they are really caused by bad thrust washers or damaged gears and bearings in the transmission. This type of noise can also be caused by nontransmission components, such as bad CV joints and wheel bearings.

Also, many engine problems can cause a vehicle to act as if it has a torque converter problem, especially with converter clutches, which may engage early or not at all. Ignition and fuel injection problems can behave the same as a malfunctioning converter. Vacuum leaks and bad electrical sensors can prevent the converter from engaging at the correct time.

TESTING CONVERTER CLUTCHES

Late-model transmissions are equipped with a torque converter clutch. Most converter clutches are controlled by the power train control module. The computer turns on the converter clutch solenoid, which opens a valve and allows fluid pressure to engage the clutch. When the computer turns the solenoid off, the clutch disengages.

A malfunctioning converter clutch can cause a wide variety of driveability problems. Normally, the application of the clutch should feel like a smooth engagement into another gear. It should not feel harsh, nor should there be any noises related to the application of the clutch.

To properly diagnose converter clutch problems, you must know when they should engage and disengage and understand the function of the various controls involved with the system. Although the actual controls for a converter clutch vary with the different manufacturers and models of transmissions, they all have certain operating conditions that must be met before the clutch can be engaged.

Diagnosis of a converter clutch circuit should be conducted in the same way as that of any other computer system. The computer will recognize problems within the system and store trouble codes that reflect the problem area of the circuit. The codes can be retrieved and displayed by an instrument panel light or a hand-held scanner tool.

Engagement Quality

If the clutch engages prematurely or is not applied with full pressure, a shudder or vibration results from the rapid grabbing and slipping of the clutch. The clutch begins to engage, then slips because it cannot hold the engine's torque and fully engage. The torque capacity of the clutch is determined by the oil pressure applied to it and the condition of the frictional surfaces of the clutch assembly.

If the shudder is only noticeable during the engagement of the clutch, the problem is typically in the converter. When the shudder is only evident after the engagement of the clutch, the cause of the shudder is the engine, the transmission, or another component of the driveline. If the shudder is caused by the clutch, the converter must be replaced to correct the problem.

A faulty clutch solenoid or its return spring may cause low apply pressure. The valve controlled by the solenoid is normally held in position by a coil-type return spring. If the spring loses tension, the clutch will engage too soon. Because there is insufficient pressure to hold the clutch, shudder occurs as the clutch begins to grab and then slip. If the solenoid valve and/or return spring are faulty, they should be replaced, as should the torque converter.

An out-of-round torque converter prevents full clutch engagement, which also causes shudder, as does contaminated clutch frictional material. The frictional material can become contaminated by metal particles circulating through the torque converter and collecting on the clutch. Broken or worn clutch dampener springs can also cause shudder.

TC-Related Cooler Problems

Vehicles equipped with a converter clutch may stall when the transmission is shifted into reverse gear. The cause of this problem may be plugged transmission cooler lines, or the cooler itself may be plugged. Fluid normally flows from the torque converter through the transmission cooler. If the cooler passages are blocked, fluid is unable to exhaust from the torque converter and the converter clutch piston remains engaged. When the clutch is engaged, there is no vortex flow in the converter and therefore little torque multiplication is taking place in the converter.

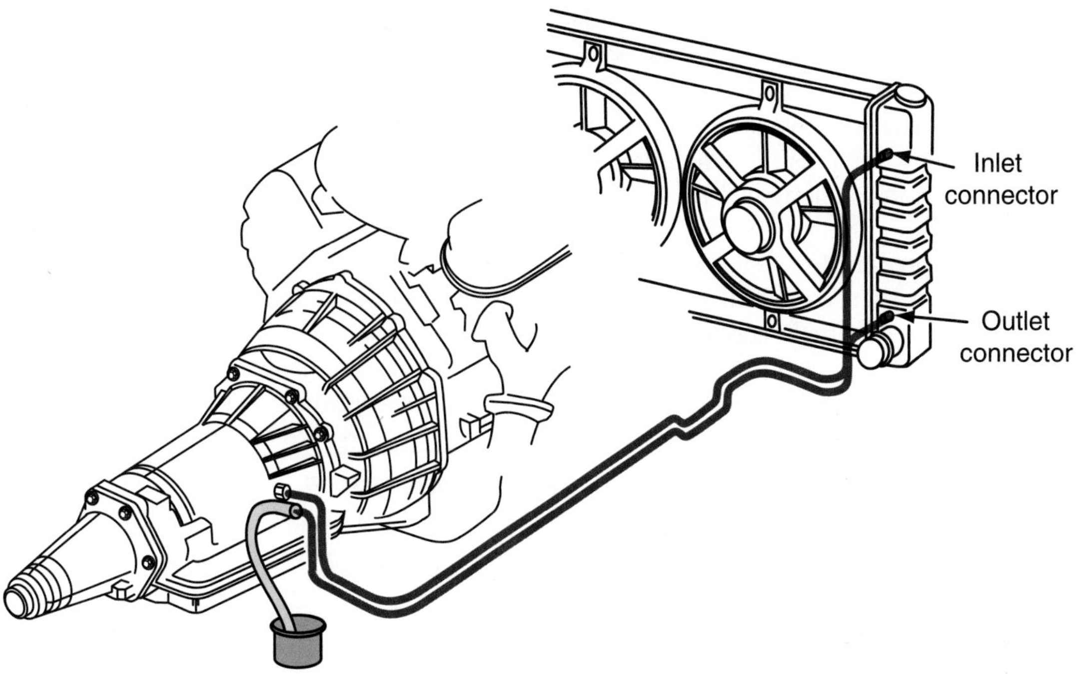

Figure 40–12 To check for a plugged cooler, disconnect the return line at the transmission and measure the flow of fluid out of the cooler.

To verify that the transmission cooler is plugged, disconnect the cooler return line from the radiator or cooler **(Figure 40–12)**. Connect a short piece of hose to the outlet of the cooler and allow the other end of the hose to rest inside an empty container. Start the engine and measure the amount of fluid that flows into the container after twenty seconds. Normally one quart of fluid should flow into the container. If less than that filled the container, a plugged cooler is indicated.

To correct a plugged transmission cooler, disconnect the cooler lines at the transmission and the radiator. Blow air through the cooler, one end at a time, then through the cooler lines. The air will clear large pieces of debris from the transmission cooler. Always use low air pressure, no more than 50 psi (3.5 kg/cm^2). Higher pressures may damage the cooler. If there is little airflow through the cooler, the radiator or external cooler must be removed and flushed or replaced.

DIAGNOSING HYDRAULIC AND VACUUM CONTROL SYSTEMS

The best way to identify the exact cause of the problem is to use the results of the road test, logic, and the oil circuit charts for the transmission being worked on. Before doing this, however, you should always check all sources for information about the symptom first. Also, make sure you check the basics: trouble codes in the computer, fluid level and condition, leaks, and mechanical and electrical connections. Using the oil circuits, you can trace problems to specific valves, servos, clutches, and bands.

The basic oil flow is the same for all transmissions. The oil pump creates the fluid flow used throughout the transmission. Fluid from the pump always goes to the pressure-regulating valve and torque converter. From there, the fluid is directed to the manual shift valve. When the gear selector is moved, the manual valve directs the fluid to other valves and to the apply devices. By following the flow of the fluid on the oil circuit chart, you can identify which valves and apply devices should be operating in each particular gear selector position. Through a process of elimination, you can identify the most probable cause of the problem.

In most cases, the transmission or transaxle is removed to repair or replace the items causing the problem. However, some transmissions allow for a limited amount of service to the apply devices and control valves.

Mechanical and/or vacuum controls can also contribute to shifting problems. The condition and adjustment of the various linkages and cables should be checked whenever there is a shifting problem. If all checks indicate that the problem is either an apply device or in the valving, an air pressure test can help identify the exact problem. Air pressure tests are also performed during disassembly to locate leaking seals and during reassembly to check the operation of the clutches and servos.

Pressure Tests

If you cannot identify the cause of a transmission problem from your inspection or road test, a pressure test should be conducted. This test measures the fluid pressure of the different transmission circuits during the various operating gears and gear selector positions **(Figure 40–13)**. The number of hydraulic circuits that can be tested varies with the different makes and models of transmissions. However, most transmissions are equipped with pressure taps, which allow the pressure test equipment to be connected to the transmission's hydraulic circuits **(Figure 40–14)**.

Before conducting a pressure test on an electronic automatic transmission, check and correct all trouble codes retrieved from the system. Also make sure the transmission fluid level and condition is okay and that the shift linkage is in good order and properly adjusted.

The test is best conducted with three pressure gauges, but two will work. Two of the gauges should read up to 400 psi (28 kg/cm^2) and the other to 100 psi (7 kg/cm^2). The two 400-psi (28 kg/cm^2) gauges are usually used to check mainline and an individual circuit, such as mainline and direct or forward circuits. If a circuit is 15 psi (1 kg/cm^2) lower than mainline pressure when they are both tested at exactly the same time, a leak is indicated. A 100-psi (7 kg/cm^2) gauge may be used be used on TV and governor circuits.

The pressure gauges are connected to the pressure taps in the transmission housing and routed so that the

PRELIMINARY CHECK PROCEDURE

Check transmission oil level.
Check and adjust TV cable.
Check and adjust outside manual linkage.
Check engine tune.
Install oil pressure gauge and tachometer.
Check oil pressures in the following order.

Minimum TV line pressure check
Set TV cable to specifications; with brakes applied, take pressure readings in the ranges and rpm shown below.

Full TV line pressure check
Hold the TV cable to the full extent of its travel; with brakes applied, take pressure readings in the ranges and rpm shown below.

CAUTION Brakes must be applied at all times

NOTICE Total running time must not exceed 2 minutes

RANGE	MODEL	MIN. TV		MAX. TV	
		kPa	PSI	kPa	PSI
Park @ 1,000 rpm	7BPC	459–507	66–74	459–507	66–74
	7HLC	511–581	74–84	511–581	74–84
Reverse @ 1,000 rpm	7BPC	804–887	117–129	1630–1847	236–268
	7HLC	895–1018	130–148	1721–1978	250–287
Neutral @ 1,000 rpm	7BPC	459–507	67–74	931–1055	135–153
	7HLC	511–581	74–84	983–1130	143–164
Intermediate LO @ 1,000 rpm	7BPC	788–869	114–126	788–869	114–126
	7HLC	877–998	127–145	877–998	127–145

Figure 40–13 Typical manufacturer's line pressure chart. This chart should be constantly referred to when checking oil pressure to diagnose a problem.

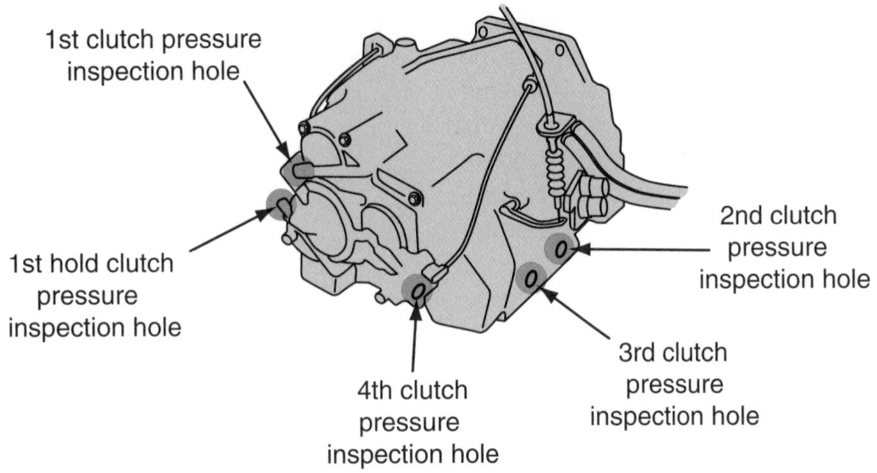

Figure 40–14 Pressure taps on the outside of a typical Honda transaxle case.

gauges can be seen by the driver. The vehicle is then road tested and the gauge readings observed during the following operational modes: slow idle, fast idle, and WOT.

During the road test, observe the starting pressures and the steadiness of the increases that occur with slight increases in load. The pressure drops as the transmission shifts from one gear to another also should be noted. The pressure should not drop more than 15 psi (1 kg/cm^2) between shifts.

Any pressure reading not within the specifications indicates a problem **(Figure 40–15)**. Typically, when the fluid pressures are low, there is an internal leak, clogged

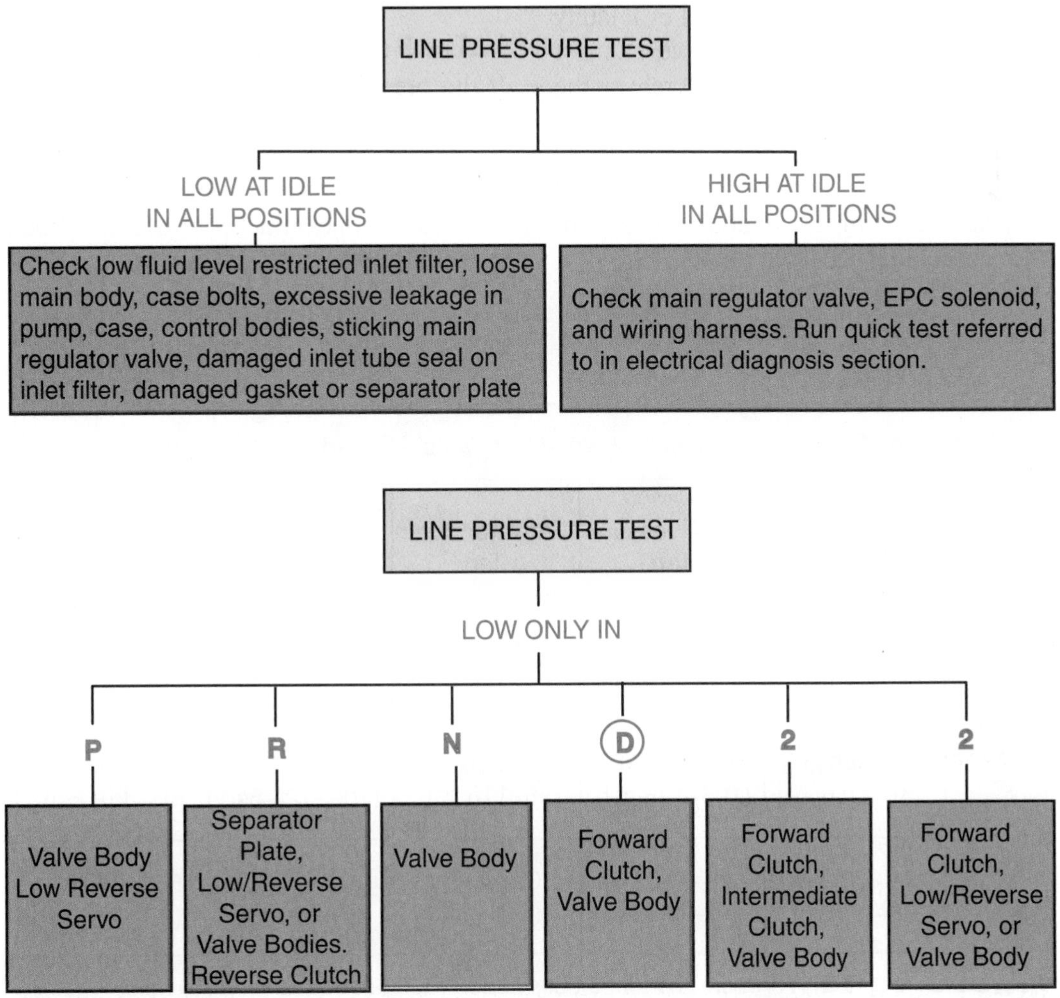

Figure 40–15 A troubleshooting chart for abnormal line pressure.

filter, low oil pump output, or faulty pressure regulator valve. If the pressure increased at the wrong time or the pressure was not high enough, sticking valves or leaking seals are indicated. If the pressure drop between shifts was greater than approximately 15 psi (1 kg/cm^2), an internal leak at a servo or clutch seal is indicated. Always check the manufacturer's specifications for maximum drop off.

On transmissions equipped with an electronic pressure control (EPC) solenoid, if the line pressure is not within specifications, the EPC pressure needs to be checked. To do this, connect the pressure gauge to the EPC tap. Start the engine and check EPC pressure, then compare it to specifications. If the pressure is not within specifications **(Figure 40–16)**, follow the procedures for testing the EPC. If the pressure is okay, there is a mainline pressure problem.

If the pressure tests suggest a governor problem, it should be removed, disassembled, cleaned, and inspected. Some transmissions require that the transmission be removed to service the governor. In other transmissions, it can be serviced by removing the extension housing or oil pan, or by detaching an external retaining clamp and then removing the unit.

Improper shift points are typically caused by a faulty governor or governor drive gear system. However, most electronically controlled transmissions do not rely on the hydraulic signals from a governor; rather, they rely on the electrical signals from sensors. Sensors, such as speed and load sensors, signal to the TCM when gears should be shifted. Faulty electrical components and/or loose connections can also cause improper shift points.

COMMON PROBLEMS

The following problems and their causes are given as examples. The actual causes of these types of problems vary with the different models of transmissions. Refer to the appropriate Band and Clutch Application Chart while diagnosing shifting problems. Doing so allows you to identify the cause of the shifting problems through the process of elimination.

Normally if the shift for all forward gears is delayed, the clutch that is applied in all forward gears may be slipping. Likewise, if the slipping occurs in one or more gears but not all, suspect the clutch that is applied only during those gear ranges.

It is important to remember that delayed shifts or slippage may also be caused by leaking hydraulic circuits or sticking spool valves in the valve body. Since the application of bands and clutches is controlled by the hydraulic system, improper pressures will cause shifting problems. Other components of the transmission can also contribute to shifting problems. For example, on transmissions equipped with a vacuum modulator, if upshifts do not occur at the specified speeds or do not occur at all, the modulator may be faulty or the vacuum supply line may be leaking.

Valve Body

If the pressure problem was associated with the valve body, a thorough disassembly, cleaning in fresh solvent, careful inspection, and the freeing and polishing of the valves may correct the problem. Disconnect the lever and detent assemblies attached to the valve body, and then remove the valve body screws. Before lowering the valve

Transmission Pressure with TP at 1.5 Volts and Vehicle Speed Above 8 Km/h (5 mph)					
Gear	EPC Tap	Line Pressure Tap	Forward Clutch Tap	Intermediate Clutch Tap	Direct Clutch Tap
1	276–345 kPa (40–50 psi)	689–814 kPa (100–118 psi)	620–745 kPa (90–108 psi)	641–779 kPa (93–113 psi)	0–34 kPa (0–5 psi)
2	310–345 kPa (45–50 psi)	731–869 kPa (106–126 psi)	662–800 kPa (96–116 psi)	689–827 kPa (100–120 psi)	655–800 kPa (95–116 psi)
3	341–310 kPa (35–45 psi)	620–758 kPa (90–110 psi)	0–34 kPa (0–5 psi)	586–724 kPa (85–105 psi)	551–689 kPa (80–100 psi)

Figure 40-16 A pressure chart for a transmission equipped with an EPC solenoid.

body and separating the assembly, hold the assembly with the valve body on the bottom and the transfer and separator plates on top. Holding the assembly in this way reduces the chances of dropping the steel balls located in the valve body. Lower the valve body and note where these steel balls are located in the valve body **(Figure 40–17)**, then remove them and set them aside along with the various screws.

After all of the valves and springs have been removed from the valve body, soak the valve body, separator, and transfer plates in mineral spirits for a few minutes. Thoroughly clean all parts and make sure all passages within the valve body are clear and free of debris. Carefully blow-dry each part individually with dry compressed air. Never wipe the parts of a valve body with a wiping rag or paper towel. Lint from either will collect in the valve body passages and cause shifting problems. As the parts of the valve body are dried, place them in a clean container.

Examine each valve for nicks, burrs, and scratches. Check that each valve properly fits into its respective bore. If a valve cannot be cleaned enough to move freely in its bore, the valve body is normally replaced. Individual valve body parts are available, as well as bore reamers. Care must be taken when rebuilding valve bodies.

During reassembly of the valve body, lube the valves with fresh ATF. Check the valve body gasket (if used) to make sure it is the correct one by laying it over the separator plate and holding it up to the light. No oil holes should be blocked. Then install the bolts to hold valve body sections together and the valve body to the case. Tighten the bolts to the torque specifications to prevent valve body warpage and possible leakover.

Servo Assemblies

On some transmissions, the servo assemblies are serviceable with the transmission in the vehicle **(Figure 40–18)**. Others require the complete disassembly of the transmission. Internal leaks at the servo or clutch seal will cause excessive pressure drops during gear changes.

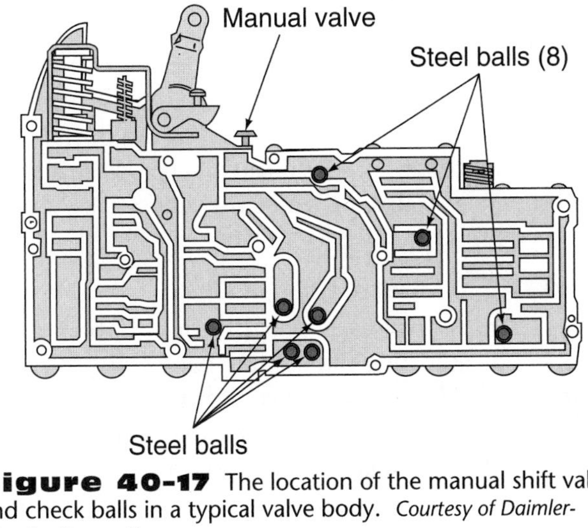

Figure 40–17 The location of the manual shift valve and check balls in a typical valve body. *Courtesy of DaimlerChrysler Corporation*

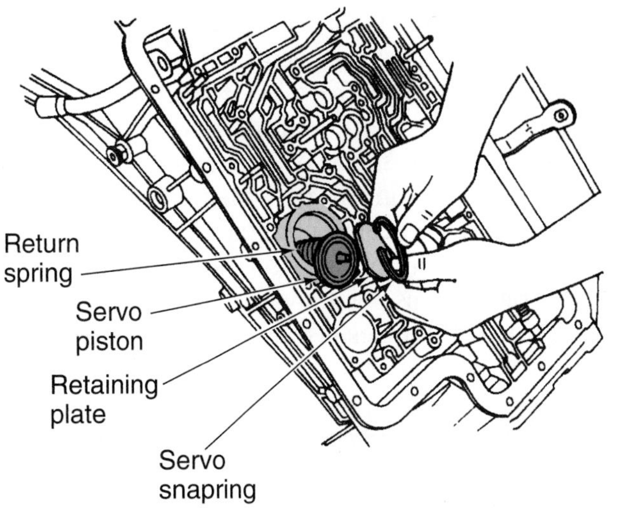

Figure 40–18 The servos in some transmissions are serviceable while the transmission is in the vehicle. The servos are contained in their own bores. *Courtesy of Ford Motor Company*

When removing the servo, check both the inner and outer parts of the seal for wet oil that means leakage. When removing the seal, inspect the sealing surface, or lips, before washing. Look for unusual wear, warping, cuts and gouges, or particles embedded in the seal.

The servo piston, spring, piston rod, and guide should be cleaned and dried. Then, check the sealing rings to make sure they are able to turn freely in the groove of the piston ring. These seal rings are not typically replaced unless they are damaged, so carefully inspect them. Check the servo piston for cracks, burrs, scores, and wear. Inspect the servo cylinder for scores or other damage. Move the piston rod through the piston rod guide and check for freedom of movement. If all of the parts are in good condition, the servo assembly can be reassembled.

Lubricate the sealing ring with ATF and carefully install it on the piston rod. Lubricate and install the piston rod guide with its snap ring into the servo piston. Then, install the servo piston assembly, return spring, and piston guide into the servo cylinder. Some servos are fitted with rubber lip seals that should be replaced. Lubricate and install the new lip seal. On lip seals, make sure the spring is seated around the lip and that the lip is not damaged.

LINKAGES

Many transmission problems are caused by improper adjustment of the linkages. All transmissions have either a cable or a rod-type gear selector linkage. Some transmissions also have a throttle valve linkage, while others use an electric switch connected to the throttle to control forced downshifts.

Normal operation of a neutral safety switch provides a quick check for the adjustment of the gear selector linkage. To do this, move the selector lever slowly until it clicks into the park position. Turn the ignition key to the start position; if the starter operates, the park position is correct. After checking the park position, move the lever slowly toward the neutral position until the lever drops at the end of the N stop in the selector gate. If the starter also operates at this point, the gearshift linkage is properly adjusted. This quick test also tests the adjustment of the neutral safety switch. If the engine does not start in either or both of these positions, the neutral safety switch or the gear selector linkage needs adjustment or repair.

Gear Selector Linkage

A worn or misadjusted gear selection linkage will affect transmission operation. The transmission's manual shift valve must completely engage the selected gear **(Figure 40–19)**. Partial manual shift valve engagement will not allow the proper amount of fluid pressure to reach the rest of the valve body. If the linkage is misadjusted, poor gear engagement, slipping, and excessive wear can result. The gear selector linkage should be adjusted so the manual shift valve detent position in the transmission matches the selector level detent and position indicator.

To check the adjustment of the linkage, move the shift lever from park to the lowest drive gear. Detents should be felt at each of these positions. If the detent cannot be felt, the linkage needs to be adjusted. While moving the shift lever, pay attention to the gear position indicator. Although the indicator will move with an adjustment of the linkage, the pointer may need to be adjusted so that it shows the exact gear after the linkage has been adjusted.

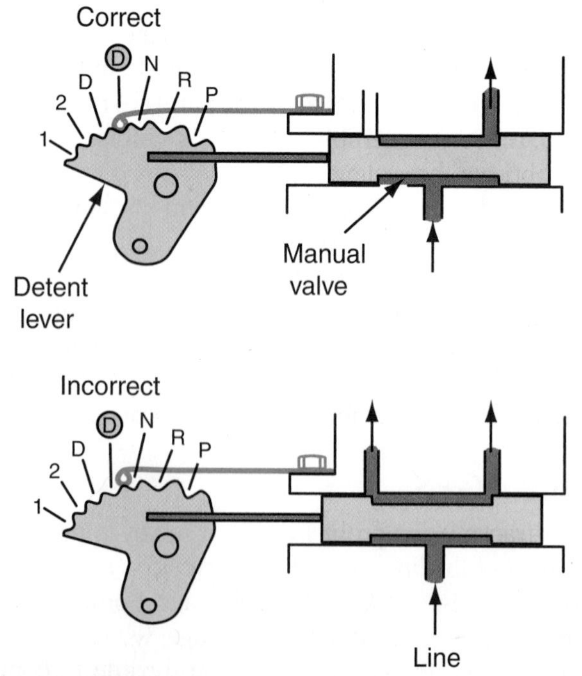

Figure 40-19 Incorrect linkage adjustments may cause the manual shift valve to be positioned improperly in its bore and cause slipping during gear changes.

Throttle Valve Linkages

The throttle valve (TV) cable connects the movement of the throttle pedal movement to the throttle valve in the transmission's valve body. On some transmissions, the TV linkage may control both the downshift valve and the TV. Others use a vacuum modulator to control the TV and a throttle linkage to control the downshift valve. Late-model transmissions may not have a throttle cable. Instead, they rely on electronic sensors and switches to monitor engine load and throttle plate opening. The action of the TV produces throttle pressure. Throttle pressure is used as an indication of engine load and influences the speed at which automatic shifts take place.

A misadjusted TV linkage may also result in throttle pressure that is too low in relation to the amount the throttle plates are open, causing early upshifts. Throttle pressure that is too high can cause harsh and delayed upshifts, and downshifts will occur earlier than normal. When adjusting the TV linkage, always follow the manufacturer's recommended procedures. An adjustment as small as a half turn can make a big difference in shift timing and feel.

Kickdown Switch Adjustment

Some transmissions have a **kickdown switch** located at the upper post of the throttle pedal. On other transmissions, kickdown control is based on signals from the TP sensor. In both designs, the movement of the throttle pedal to the wide-open position signals to the transmission that the driver desires a forced downshift.

To check the operation of the switch, fully depress the throttle pedal and listen for a click that should be heard just before the pedal reaches its travel stop. If the click is not heard, loosen the locknut and extend the switch until the pedal lever makes contact with the switch. If the pedal contacts the switch too early, the transmission may downshift during part-throttle operation.

If you feel and hear the click of the switch but the transmission still does not kick down, check the continuity of the switch when it is depressed. An open switch will prevent forced downshifting, whereas a shorted switch can cause upshift problems. Defective switches should be replaced.

Band Adjustment

On some transmissions, slippage during shifting can be corrected by adjusting the holding bands. To help identify if a band adjustment will correct the problem, refer to the written results of your road test. Compare your results with the Clutch and Band Application Chart in the service manual. If slippage occurs when there is a gear change that requires the holding by a band, the problem may be corrected by tightening the band.

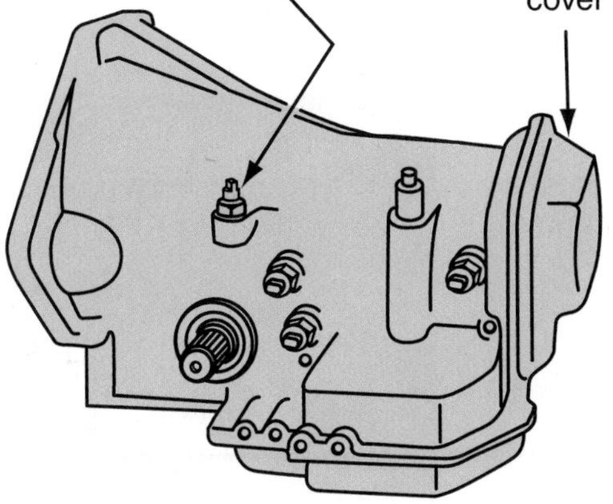

Figure 40–20 An example of the location of an external band adjusting screw.

On some vehicles, the bands can be adjusted externally with a torque wrench. On others, the transmission fluid must be drained and the oil pan removed. Locate the band-adjusting nut **(Figure 40–20)**, then clean off all dirt on and around the nut. Now, loosen the band-adjusting bolt locknut and back it off approximately five turns. Use a calibrated pound-inch torque wrench to tighten the adjusting bolt to the specified torque. Then, back off the adjusting screw the specified number of turns and tighten the adjusting bolt locknut while holding the adjusting stem stationary. Reinstall the oil pan with a new gasket and refill the transmission with fluid. If the transmission problem still exists, an oil pressure test or transmission teardown must be done.

WARNING!

Do not excessively back off the adjusting stem as the anchor block may fall out of place. It will then be necessary to remove and disassemble the transmission to fit it back in place.

REBUILDING A TRANSMISSION

The exact procedures for rebuilding a transmission depend on the specific transmission as well as the problems the transmission may have. Photo Sequence 42 outlines the procedure for a commonly used transaxle. Always refer to the service manual for the specific procedures for a specific transmission/transaxle.

The following guidelines will help you service any transmission:

- For the most part, torque converters are replaced. If you want to have the torque converter rebuilt, a specialist should do this.
- If the transaxle is fitted with a drive chain, it should be inspected for side play and stretch.
- All pumps, valve bodies, and cases should be checked for warpage and should be flat filed to take off any high spots or burrs prior to reassembly.
- Keep every part absolutely clean and air dry all parts. Only use lint-free rags to wipe off parts. Lint can collect and damage the transmission.
- Make sure the correct size thrust washers are used throughout the transmission. Also make sure the thrust washers, bearings, and bushings are well lubricated before you install them.
- Soak all friction materials in clean ATF prior to installing them.
- When tightening any fastener that directly or indirectly involves a rotating shaft, rotate the shaft during and after tightening to ensure the part does not bind.
- Always use new gaskets and seals throughout the transmission.
- Always flush out the transmission cooling system before using a rebuilt or new transmission. The cooling system is a good place for debris to collect. Some manufacturers require the cooler be replaced, rather than flushed.

CASE STUDY

A customer came into the shop with a high-mileage Hyundai Excel with a Mitsubishi KM-175 transmission. A tow truck delivered the car because the transmission would not shift gears. The technician assigned to the car pulled the transmission and overhauled it. After installing an overhaul kit and inspecting all other transmission parts, the transmission was reassembled and installed back into the car. The transmission still did not shift.

Then the technician began to check the electronics involved with the transmission. After a little research in the service manual, he found that the transmission was in the fail-safe mode. With a multimeter, the technician then checked the solenoids, connectors, and wiring only to find nothing wrong. However, after becoming frustrated, he began to check complete circuits and found that the computer was doing nothing.

He located the computer under the passenger's seat and found heavy corrosion on the computer and the wiring harness terminals. The technician then cleaned the area and replaced the computer. The transmission then shifted well.

Typical Procedure for Overhauling a 4T60E Transaxle

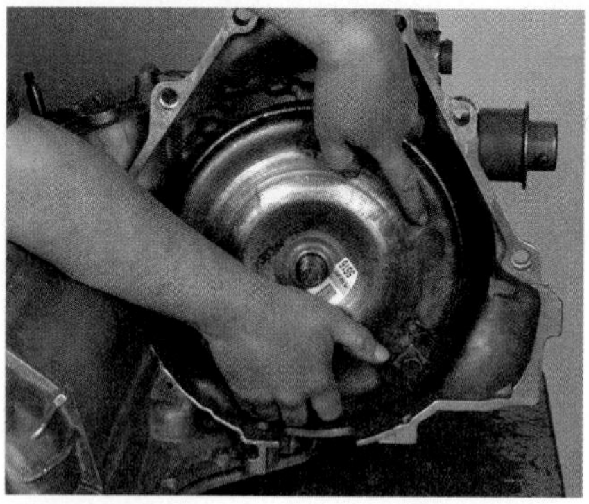

P42-1 *Remove the torque converter. Then place the transaxle in a holding fixture. Remove the speedometer sensor and governor assembly.*

P42-2 *Remove the bottom oil pan, oil filter, and modulator valve.*

P42-3 *Remove the accumulator cover with governor feed and return pipes, accumulator pistons, gaskets, retainers, and pipes from their bores.*

P42-4 *Remove the reverse servo cover by applying pressure to it and removing the retaining ring. Then remove the servo assembly from the case.*

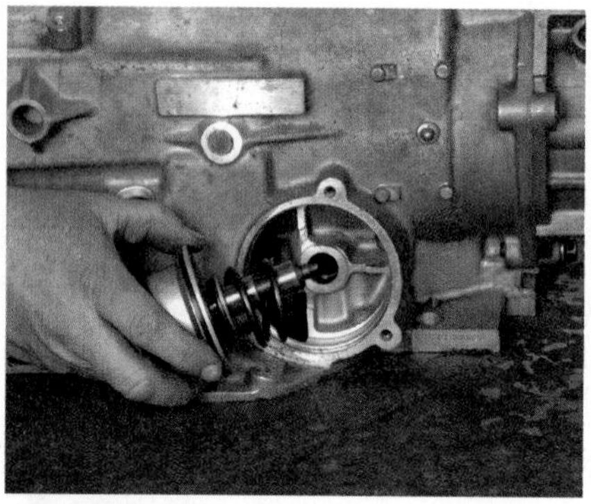

P42-5 *Apply pressure to the 1-2 servo cover and remove the cover's retaining ring. Then remove the cover and servo assembly.*

P42-6 *Remove the side cover and gaskets. Disconnect the wiring harness to the pressure switches, solenoid, and case connector. Unplulg solenoids, and switches.*

P42-7 *Remove the pump assembly cover. Remove the servo pipe retainer bolt, retainer plate, bolts, and valve body.*

P42-8 *Mark the location of and remove all of the check balls between the spacer plate and the valve body and between the channel plate and the spacer plate.*

P42-9 *Disconnect the manual valve link from the manual valve. Place the detent lever in the park position and remove the retaining clip. Then remove the channel plate with its gaskets.*

P42-10 *Remove the oil pump drive shaft. Remove the input clutch accumulator and converter clutch piston assemblies (if so equipped). Remove the plates from the fourth clutch and the thrust bearing, hub, and shaft.*

P42-11 *Rotate the final drive unit until both ends of the output shaft retaining rings are showing.*

P42-12 *With a C-ring removal tool, loosen the ring from the output shaft. Rotate the shaft 180 degrees and remove the ring. Then remove the output shaft.*

P42-13 *Remove and discard the O-ring from the input shaft. Then remove the drive sprocket, driven sprocket, and chain as an assembly with the selective thrust washers.*

P42-14 *Remove the driven sprocket support and thrust washer from between the sprockets and the channel plate. Remove the scavenging scoop and driven sprocket with the second clutch's thrust washer.*

P42-15 *Using the correct tool, remove the second clutch and input shaft clutch housings as an assembly. Remove the reverse band.*

P42-16 *Measure the endplay of the input shaft clutch housing. Select the correct thickness of thrust washer and set it aside for reassembly. Remove the input clutch sprag assembly, third clutch assembly, and the input sun gear. Then remove the reverse band, reverse reaction drum input carrier assembly and thrust washer.*

P42-17 *Remove the reaction carrier, reaction sun gear/drum assembly, and forward band.*

Typical Procedure for Overhauling a 4T60E Transaxle (*Continued*)

P42-18 *Remove the reaction sun gear thrust bearing and final drive sun gear shaft.*

P42-19 *Using the proper tool, remove the final drive assembly and selective thrust washers and bearings.*

P42-20 *Check the final drive endplay and select the correct size thrust washers for the unit.*

P42-21 *Clean and inspect the transaxle case and components. Then position the correct thrust washers and bearings onto the final drive and install the unit. Use petroleum jelly to hold the bearings in place.*

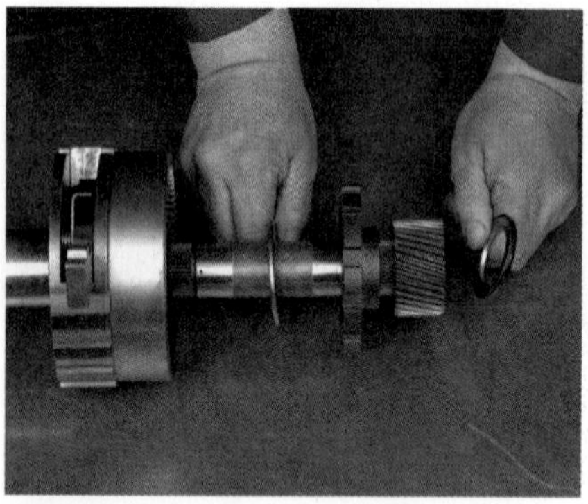

P42-22 *Install the final drive sun gear shaft through the final drive ring gear. The splines must engage with the parking gear and the final drive sun gear.*

P42-23 *Install the forward band into the case, making sure the band is aligned with the anchor pin.*

P42-24 *Install the reaction sun gear to final drive ring gear thrust bearing. Then assemble the reaction sun gear and drum assembly onto the final drive ring gear.*

P42-25 *Check the endplay of the carrier in the reaction gearset. Then install the thrust washer and carrier assembly into the case. Rotate the carrier until the pinions engage with the reaction sun gear.*

P42-26 *Install the input carrier with its thrust bearing in the case. Then install the reverse reaction drum, making sure its spline engages with the input carrier.*

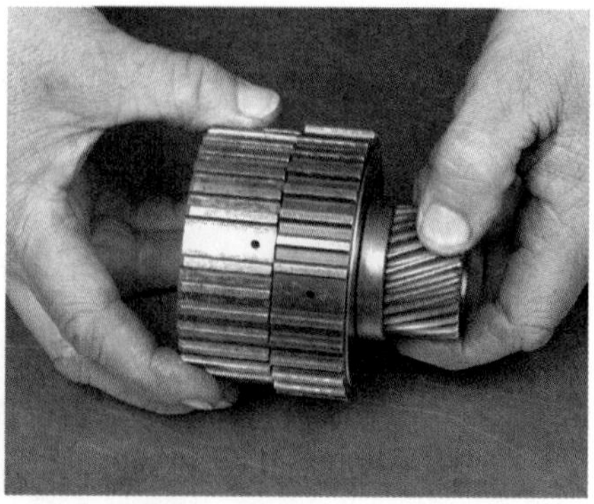

P42-27 *Install the spacer, input sprag retainer, sprag assembly, and roller clutch onto the sun gear. Hold the input sun gear and make sure the sprag and roller clutch hold and freewheel in opposite directions.*

P42-28 *Lubricate the inner seal on the input clutch's piston. Then install the seal and assemble the input piston into the input housing. Install a new O-ring onto the input shaft.*

P42-29 *Install the spring retainer and guide into the piston. Then install the third clutch piston housing into the input housing. Using the proper compressor, install the retaining snap ring.*

P42-30 *Install the inner seal for the third clutch. Then install the third clutch piston into the housing. Compress the spring retainer and install the retaining snap ring.*

P42-31 *Install the wave plate, the correct number of clutch plates (in the correct sequence), backing plate, and retaining snap ring into the input clutch housing. Air check its operation.*

P42-32 *Assemble the second clutch piston in the housing. Install the apply ring, return spring, snap ring, and wave plate. Then install the correct number of plates (in the correct sequence), backing plate, and retaining snap ring. Air check its operation.*

P42-33 *Install the thrust washers. Using the correct tool, install the second clutch and input shaft clutch housings as an assembly. Install the reverse band.*

P42-34 *Install the scavenging scoop and driven sprocket support with the second clutch thrust washer. Install the driven sprocket support and thrust washer between the sprockets and the channel plate.*

P42-35 *Install the drive sprocket, driven sprocket, and chain as an assembly.*

P42-36 *Install the output shaft. Start the C-ring onto the output shaft, then rotate the shaft 180 degrees and fully seat the ring onto the shaft.*

P42-37 *Install the fourth clutch's thrust bearing, hub, and shaft. Then install the fourth clutch's plates and the apply plate. Install the input clutch accumulator and converter clutch piston assemblies. Install the oil pump drive shaft.*

P42-38 *Install the channel plate with new gaskets. Connect the manual valve link to the manual valve. Place the detent lever in the park position and install the retaining clip.*

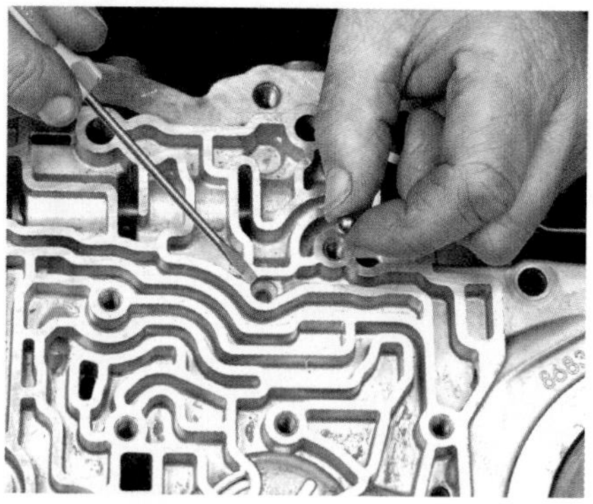

P42-39 *Install the oil reservoir weir and the check balls in their proper location between the spacer plate and the valve body and between the channel plate and the spacer plate.*

P42-40 *Install the servo pipe retainer bolt, retainer plate, mounting bolts, and valve body. Install the pump assembly cover. Tighten the bolts to specifications. Install the TV lever, linkage, and bracket onto the valve body.*

P42-41 *Connect and install the wiring harness to the pressure switches, solenoids, and case connector. Then install the side covers and gaskets. Tighten the bolts to the specified torque.*

P42-42 *Install the 1-2 servo cover, servo assembly, and retaining ring.*

P42-43 *Install the reverse servo cover, servo assembly, and retaining ring.*

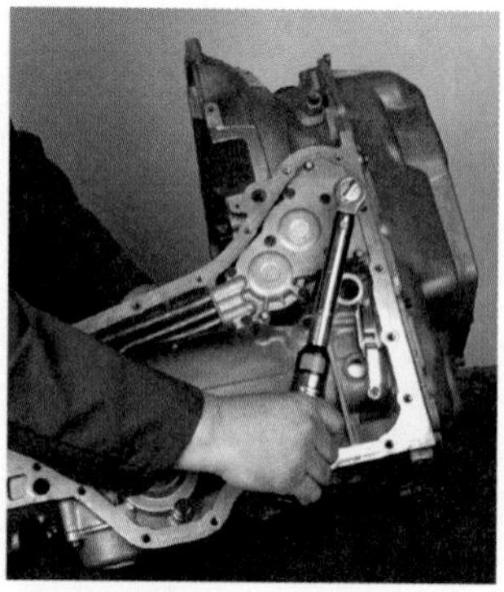

P42-44 *Install the accumulator cover with governor feed and return pipes and the accumulator pistons, gaskets, retainers, and pipes into their bores.*

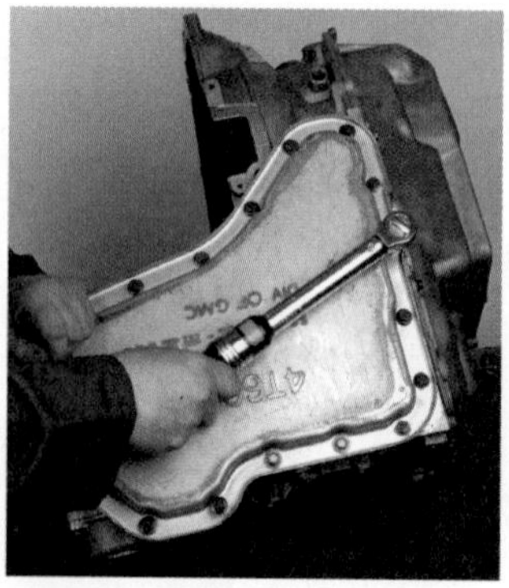

P42-45 *Install the bottom oil pan, oil filter, modulator, and modulator valve.*

P42-46 *Install the torque converter, speedometer sensor, and governor assembly.*

KEY TERMS

Aeration
Dexron
Kickdown switch
Transmission identification number

SUMMARY

- The ATF level should be checked at regular mileage and time intervals. Typically, when the fluid is checked, the vehicle should be level and running and the transmission should be at operating temperature.
- Both low fluid levels and high fluid levels can cause aeration of the fluid, which, in turn, can cause a number of transmission problems.
- Uncontaminated ATF is red in color and has no dark or metallic particles suspended in it.
- The fluid should be changed when the engine and transmission or transaxle are at normal operating temperatures. After draining the fluid, the pan should be inspected and the filter replaced.
- If ATF is leaking from the pump seal, the transmission must be removed from the vehicle so the seal can be replaced. Other worn or defective gaskets or seals can be replaced without removing the transmission. Case porosity may be repaired using an epoxy-type sealer.
- If a computer's input signals are correct and its output signals are incorrect, the computer must be replaced.
- A solenoid valve can be checked by measuring its resistance or by applying a current to it and listening and feeling for its movement. It can also be checked with a lab scope or DMM.
- Slippage during shifting can indicate the need for band adjustment.
- Improper shift points can be caused by a malfunction in the governor or governor drive gear system or a misadjusted throttle linkage.
- The road test gives the technician the opportunity to check the transaxle or transmission operation for slipping, harshness, incorrect upshift speeds, and incorrect downshift.
- Accurate diagnosis depends on knowing what planetary controls are applied in a particular gear range.
- A pressure test checks hydraulic pressures in the transmission by using gauges attached to the transmission. Pressure readings reveal possible problems in the oil pump, governor, and throttle circuits.
- Proper adjustment of the gear selector or manual linkage is important to have the manual valve fluid inlet and outlets properly aligned in the valve body. If the manual valve does not align with the inlet and outlets, line pressure could be lost to an open circuit.

TECH MANUAL

The following procedures are included in Chapter 39/40 of the *Tech Manual* that accompanies this book:

1. Visually inspecting an automatic transmission.
2. Road testing a vehicle to check the operation of a lockup torque converter clutch.
3. Road testing a vehicle to check the operation of an automatic transmission.
4. Conducting a stall test.
5. Pressure testing an automatic transmission.
6. Servicing a valve body.
7. Overhauling a multiple-disc clutch assembly.

REVIEW QUESTIONS

1. What is the most probable cause of a low fluid level?
2. What does milky-colored ATF indicate?
3. What do varnish or gum deposits on the dipstick indicate?
4. Typically during a pressure test, the pressure should not drop more than ________ psi or ________ kPa between shifts.
5. How can air pressure be used to check an electrical switch?
6. What should you do if a valve does not move freely in its bore in the valve body?

7. *True or False?* On most late-model EAT systems, separate procedures must be followed to retrieve the DTCs from the transmission control system and the TCC system.
8. Which of the following is the most likely cause for a shudder during the engagement of a lockup torque converter?
   a. a bad converter
   b. worn or damaged CV or U-joints
   c. a worn front planetary gearset
   d. a loose flex plate
9. When should the ATF level be checked on most vehicles?
   a. when the engine is cool
   b. when the engine is at operating temperature and the engine is off
   c. when the engine is at operating temperature and the engine is on
   d. It does not matter.
10. Technician A says a prerequisite to accurate road testing analysis is knowing what planetary controls are applied in a particular gear range. Technician B says all slipping conditions can be traced to a leaking hydraulic circuit. Who is correct?
   a. Technician A
   b. Technician B
   c. Both A and B
   d. Neither A nor B
11. Pressure readings reveal possible problems in which of the following?
   a. oil pump
   b. governor
   c. apply circuits
   d. all of the above
12. While discussing proper band adjustment procedures, Technician A says on some vehicles the bands can be adjusted externally with a torque wrench. Technician B says a calibrated inch-pound torque wrench is normally used to tighten the band-adjusting bolt to a specified torque. Who is correct?
   a. Technician A
   b. Technician B
   c. Both A and B
   d. Neither A nor B
13. Technician A says if the shift for all forward gears is delayed, a slipping forward clutch is normally indicated. Technician B says a bad forward clutch is indicated when there is a slip when the transmission shifts into any forward gear. Who is correct?
   a. Technician A
   b. Technician B
   c. Both A and B
   d. Neither A nor B
14. Technician A says the only positive way to identify the exact design of the transmission is by the shape of its oil pan. Technician B says identification numbers only identify the manufacturer and assembly date of the transmission. Who is correct?
   a. Technician A
   b. Technician B
   c. Both A and B
   d. Neither A nor B
15. Technician A says a faulty TP sensor can cause delayed shifts. Technician B says delayed shifts can be caused by an open shift solenoid. Who is correct?
   a. Technician A
   b. Technician B
   c. Both A and B
   d. Neither A nor B
16. Technician A says delayed shifting can be caused by worn planetary gearset members. Technician B says delayed shifts or slippage may be caused by leaking hydraulic circuits or sticking spool valves in the valve body. Who is correct?
   a. Technician A
   b. Technician B
   c. Both A and B
   d. Neither A nor B
17. While checking the condition of a car's ATF, Technician A says if the fluid has a dark brownish or blackish color and/or a burned odor, the fluid has been overheated. Technician B says if the fluid has a milky color it indicates that engine coolant has been leaking into the transmission's cooler. Who is correct?
   a. Technician A
   b. Technician B
   c. Both A and B
   d. Neither A nor B
18. While discussing the results of an oil pressure test, Technician A says when the fluid pressures are high, internal leaks, a clogged filter, low oil pump output, or a faulty pressure regulator valve are indicated. Technician B says if the fluid pressure

increased at the wrong time, an internal leak at the servo or clutch seal is indicated. Who is correct?

a. Technician A

b. Technician B

c. Both A and B

d. Neither A nor B

19. Technician A says some shift solenoids can be activated by providing a ground for the solenoid. Technician B says some shift solenoids can be activated by applying hydraulic pressure to their valve. Who is correct?

a. Technician A

b. Technician B

c. Both A and B

d. Neither A nor B

20. An electronically controlled transmission has erratic shifting. Technician A says a poor PCM ground causes this problem. Technician B says a bad AC generator-to-battery circuit causes erratic performance of a transmission or transaxle. Who is correct?

a. Technician A

b. Technician B

c. Both A and B

d. Neither A nor B

41 FOUR- AND ALL-WHEEL DRIVE

OBJECTIVES

■ Identify the advantages of four- and all-wheel drive. ■ Name the major components of a conventional four-wheel-drive system. ■ Name the components of a transfer case. ■ State the difference between the transfer, open, and limited slip differentials. ■ State the major purpose of locking/unlocking hubs. ■ Name the five shift lever positions on a typical 4WD vehicle. ■ Understand the difference between four- and all-wheel drive. ■ Know the purpose of a viscous clutch in all-wheel drive.

With the popularity of SUVs **(Figure 41–1)** and pickup trucks, **(Figure 41–2)** the need for technicians who can diagnose and service four-wheel-drive systems has drastically increased. This type of vehicle has topped the sales charts because it offers utility and comfort. Although all-wheel-drive passenger cars are available, most prospective buyers for all-wheel and four-wheel-drive vehicles are opting for truck-based SUVs and pickups.

Four-wheel-drive (4WD) and **all-wheel-drive (AWD)** systems can dramatically increase a vehicle's traction and handling ability in rain, snow and off-road driving **(Figure 41–3)**. Considering that the vehicle's only contact with the road is the small areas of the tires, driving and handling is vastly improved if the work load is spread out evenly among four wheels rather than two.

Factors such as the side forces created by cornering and wind gusts have less effect on vehicles with four driving wheels. The increased traction also makes it possible to apply greater amounts of energy through the drive system. Vehicles with 4WD and AWD can maintain control while transmitting levels of power that would cause two wheels to spin either on takeoff or while rounding a curve.

Four-wheel-drive vehicles can transfer large amounts of horsepower through the drivelines without fear of drivetrain damage. This is not necessarily true of front-wheel-drive and all-wheel-drive vehicles (both of which use transaxles).

Both 4WD and AWD systems add initial cost and weight. With most passenger cars, the weight problem is minor. A typical 4WD system adds approximately 170 pounds (77 kg) to a passenger car. An AWD system adds

Figure 41-1 A 4WD sport utility vehicle. *Reprinted with permission*

Figure 41-2 A 4WD pickup truck.

Figure 41-3 A 4WD passenger car moving through the dirt. *Courtesy of Subaru of America Inc.*

even less weight. The additional weight in larger 4WD trucks can be as much as 400 pounds (180 kg) or more.

The systems also add initial cost to the vehicle. Vehicles equipped with 4WD and AWD require special service and maintenance not needed on 2WD vehicles. However, the slight disadvantages of 4WD and AWD are heavily outweighed by the traction and performance these systems offer. Their popularity is increasing at a rapid rate, and technicians must be prepared to diagnose and repair these systems.

FOUR-WHEEL DRIVE VERSUS ALL-WHEEL DRIVE

Because of the many names manufacturers give their drive systems, it is often difficult to clearly define the difference between 4WD and AWD.

In this text, 4WD systems are those having a separate **transfer case**. They also give the driver the choice of operating in either 2WD or 4WD through the use of a shift lever or shift button. The **locking hubs** used in 4WD systems can be either manual or automatic locking. Finally, a 4WD system may or may not use an **interaxle differential** or **viscous clutch** to distribute driving power to its front and rear drivelines.

All-wheel-drive systems differ in several major ways. They may not have a separate transfer case. All-wheel-drive systems use a front-wheel-drive transaxle equipped with a viscous clutch, center differential, or transfer clutch that transfers power from the transaxle to a rear driveline and rear axle assembly. All-wheel-drive systems do not give the driver the option of selecting 2WD or 4WD modes. The system operates in continuous 4WD.

FOUR-WHEEL-DRIVE SYSTEMS

The typical truck or utility 4WD vehicle contains the following components **(Figure 41–4)**: a front-mounted, longitudinally positioned engine; the transmission (either manual or automatic); two driveline shafts (front and rear); front and rear axle assemblies; and the transfer case.

The heart of most conventional 4WD vehicles is the transfer case, which is usually mounted to the side or the back of the transmission **(Figure 41–5)**. A chain or gear drive within the case receives the power flow from the transmission and transfers it to two separate drive shafts leading to the front and rear axles.

A selector switch or shifter located in the driving compartment **(Figure 41–6)** controls the transfer case so power is directed to the axles as the driver desires. Power can be directed to all four wheels, two wheels, or no

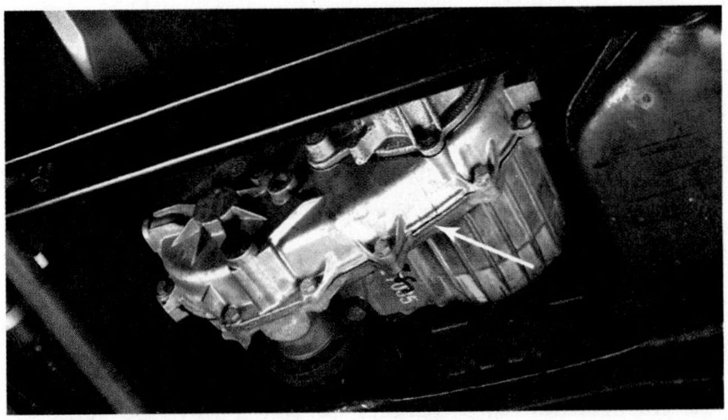

Figure 41-5 A typical transfer case.

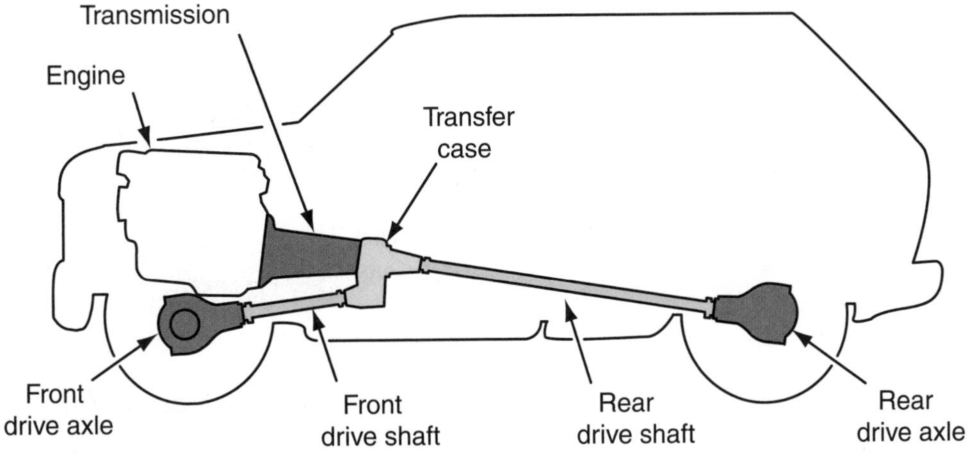

Figure 41-4 A typical arrangement of 4WD components.

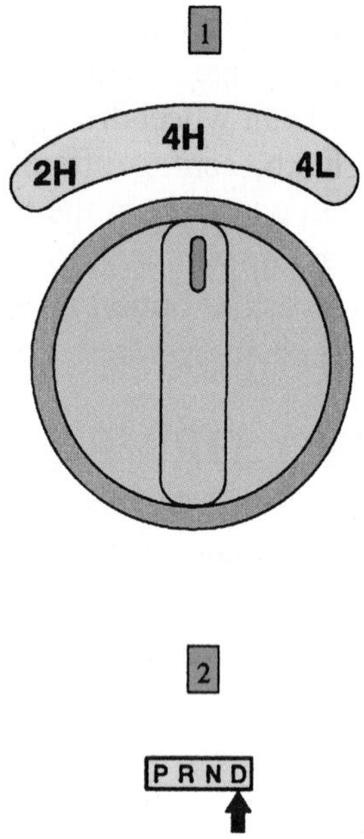

Figure 41–6 A 4WD-mode selector switch. *Courtesy of Ford Motor Company*

wheels (neutral). On many vehicles, the driver is also given the option of low four-wheel-drive range for extra traction in especially rough conditions such as deep snow or mud.

The driveline from the transfer case shafts run to differentials at the front and rear axles. As on two-wheel-drive vehicles, these axle differentials are used to compensate for road and driving conditions by adjusting the rpm to opposing wheels. For example, the outer wheel must roll faster in a turn than the inner wheel during a turn because it has more ground to cover. To permit this action, the differential cuts back the power delivered to the inner wheel and boosts the amount of power delivered to the outer wheel.

U-joints are used to couple the driveline shafts with the differentials and transfer cases on all these vehicles. U-joints can also be used on some vehicles to connect the rear axle and wheels. Normally, however, rear axles are simply bolted to the wheel hubs.

The coupling between front wheels and axles is normally done with U-joints or CV (constant velocity) joints **(Figure 41–7)**. Generally, half axles or half shafts with CV joints are found on 4WD passenger cars. They can also be found on a number of passenger vans and on mini pickups and trucks.

To provide independent front suspension, some vehicles have one half shaft and one solid axle for the front drive axle. The half shaft is able to move independently of the solid axle, thereby giving the vehicle the ride characteristics desired from an independent front suspension.

On 4WD systems adapted from front-wheel-drive systems, a separate front differential and driveline are not needed. The front wheels are driven by the transaxle differential of the base model. A **power takeoff** is added to the transaxle to transmit power to the rear wheels in four-wheel drive. This takeoff gearing is housed in a transfer case mounted to the transaxle housing. The gearing connects to a rear driveline and rear axle assembly that includes the rear differential.

TRANSFER CASE

In a 4WD vehicle, as mentioned earlier, the transfer case delivers power to both the front and rear assemblies. Two drive shafts normally operate from the transfer case, one to each drive axle.

The transfer case itself is constructed similarly to a standard transmission. It uses shift forks to select the operating mode, plus splines, gears, shims, bearings, and other components found in manual and automatic transmissions. The outer case of the unit is made of cast iron, magnesium, or aluminum. It is filled with lubricant (oil) that cuts friction on all moving parts. Seals hold the lubricant in the case and prevent leakage around shafts and yokes. Shims set up the proper clearance between the internal components and the case.

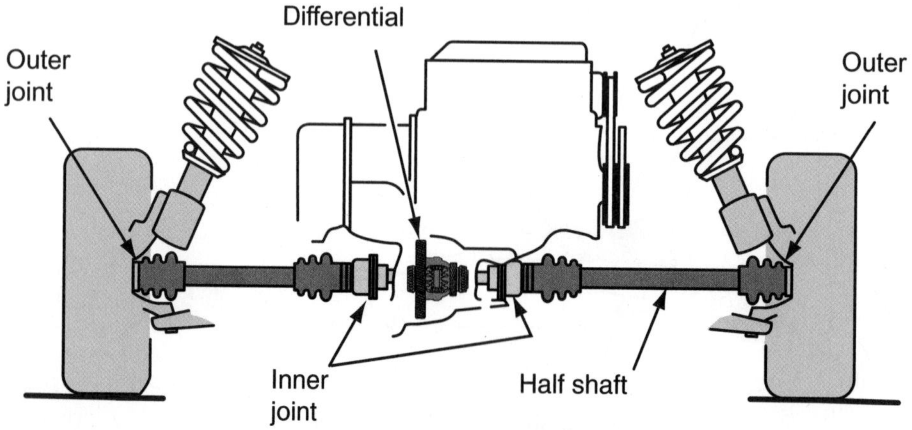

Figure 41–7 A front drive axle with half shafts and CV joints.

Interaxle Differentials

The most common method of dissipating driveline windup is to include a third or interaxle differential in the transfer case gearing.

The front and rear drivelines are connected to the interaxle differential inside the transfer case. Just as a drive axle differential allows for different left and right drive axle shaft speeds, the interaxle differential allows for different front and rear driveline shaft speeds. The driveline windup, developed as a result of different front and rear axle gear ratios, is dissipated by the interaxle differential.

While the interaxle differential solves the problem of driveline windup during turns, it also lowers performance in poor traction conditions because the interaxle differential tends to deliver more power to the wheels with the least traction. The result is increased slippage, the exact opposite of what is desired.

To counteract this problem, some interaxle differentials are designed much like a limited-slip differential. They use a multiple-disc clutch pack to maintain a predetermined amount of torque transfer before the differential action begins to take effect. Other systems, such as the one shown in **Figure 41–8**, use a transfer case or a

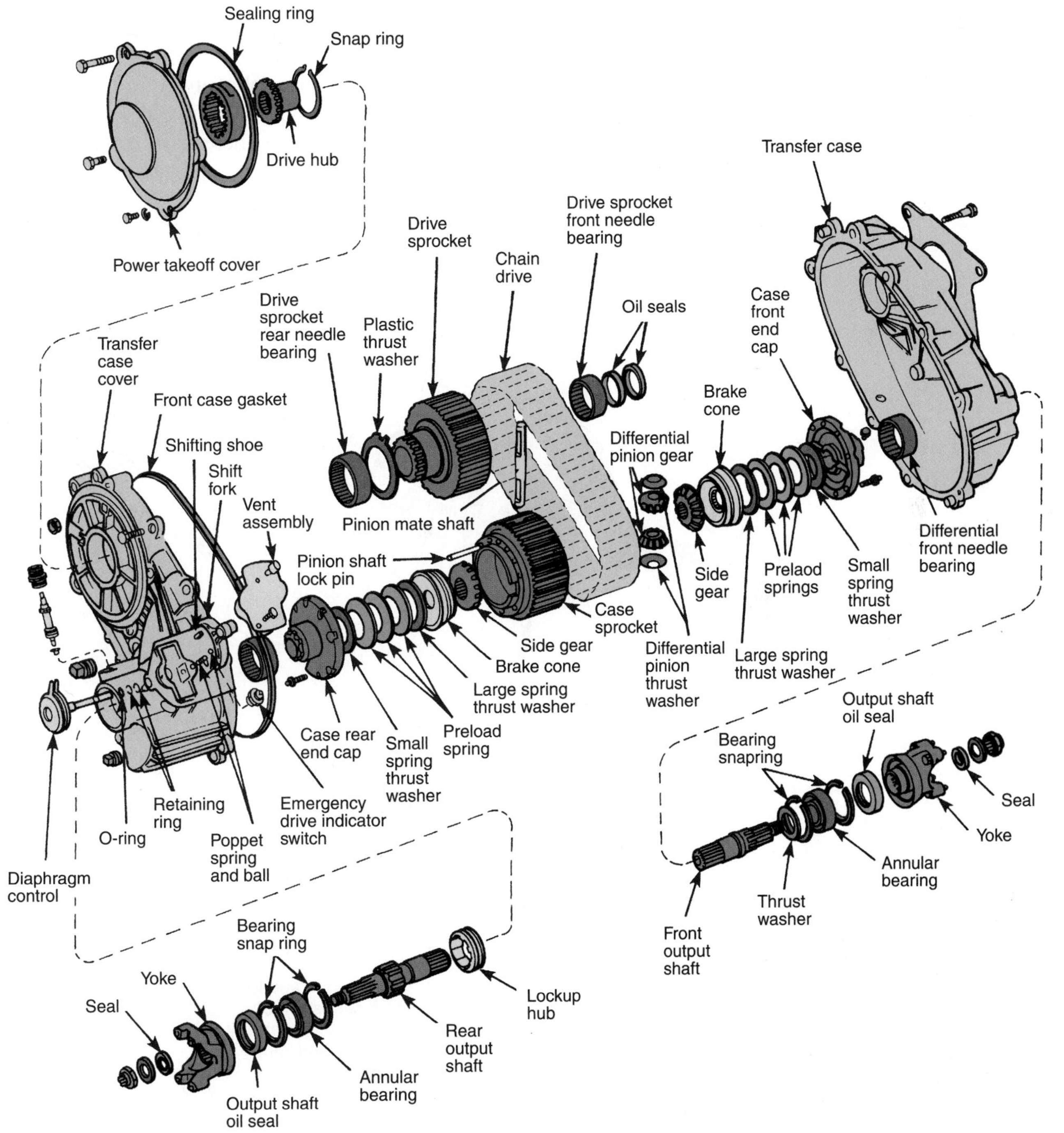

Figure 41–8 Four-wheel-drive transfer case with integral differential and cone brakes for limited slip.

cone braking system rather than a clutch pack. However, the end result is the same. Power is supplied to both axles regardless of the traction encountered.

Most systems also give the driver the option of locking the interaxle differential in certain operating modes. This eliminates the differential action altogether. However, the interaxle differential should only be locked when driving in slippery conditions.

LOCKING/UNLOCKING HUBS

Many older 4WD systems on trucks had front-wheel-drive hubs that disengaged the front drive axles when the vehicle was operating in two-wheel drive. When unlocked, the entire front drivetrain, including the front axles and front differential, stop rotating, which helped reduce wear to these items.

The front hubs had to be locked during 4WD operation. Some front hub designs locked automatically **(Figure 41–9)**. Others required the driver to get out and turn a lever or knob at the center of each front wheel. By locking the hubs, the wheels turned with the axle. The hubs were locked only when the vehicle was driving off the road or in adverse weather conditions. When they were engaged on dry, smooth pavement, the tires scrubbed against the surface when the vehicle turned.

The latest development in hubs is the fully automatic type. Once four-wheel drive is engaged, they lock automatically when power is applied, either forward or reverse. That is, power can be applied in either forward or reverse direction once locked.

Axle Disconnects

Some 4WD vehicles are equipped with ordinary front wheel hubs but have a vacuum-operated axle disconnect shift mechanism **(Figure 41–10)**. The vacuum motor moves a splined collar to connect or disconnect one of the front axle drive shafts from the front differential. The axle shafts and differential continue to turn. However, the ring gear, pinion, and front drive axle remain stationary, which reduces wear to these major components. This system is used on both trucks and passenger cars.

Some axle disconnects are operated electrically instead of by vacuum. An electric motor can be used to con-

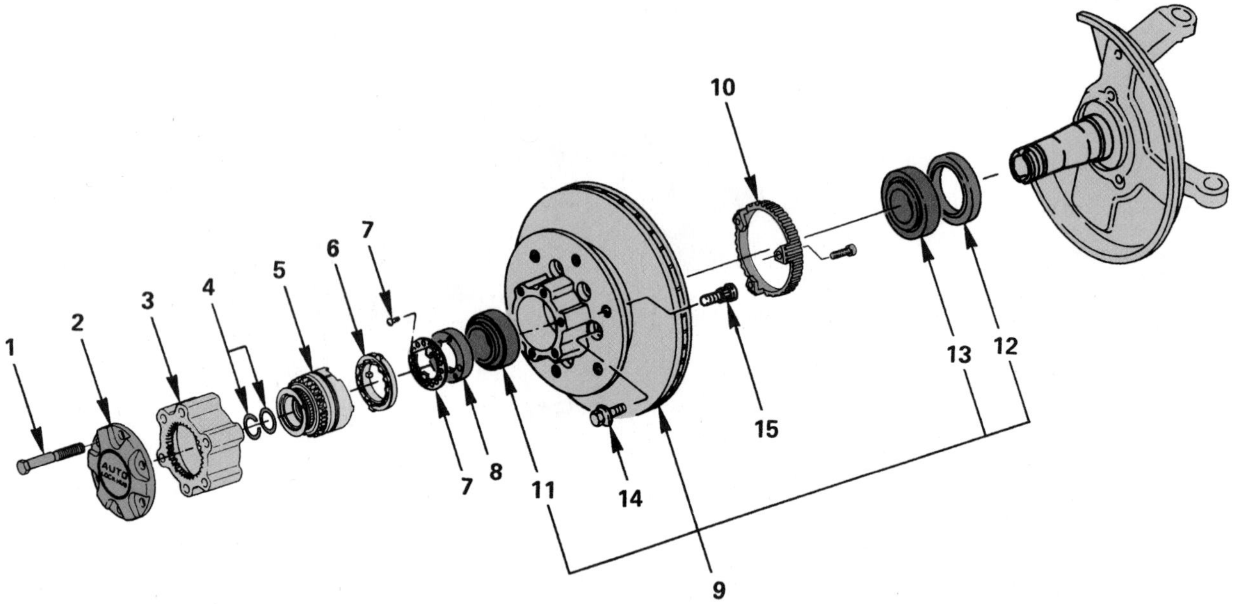

Disassembly steps
1. Bolt
2. Hub cap
3. Housing
4. Snap ring and shim
5. Drive clutch assembly
6. Inner cam assembly
7. Lock washer and lock screw
8. Hub nut
9. Hub and disc assembly
10. ABS sensor ring
11. Outer bearing outer race
12. Oil seal
13. Inner bearing outer race
14. Bolt
15. Wheel pin

Reassembly steps
15. Wheel pin
14. Bolt
13. Inner bearing outer race
12. Oil seal
11. Outer bearing outer race
10. ABS sensor ring
9. Hub and disc assembly
8. Hub nut
7. Lock washer and lock screw
6. Inner cam assembly
5. Drive clutch assembly
4. Snap ring and shim
3. Housing
2. Hub cap
1. Bolt

Figure 41–9 Automatic locking front hub components. *Reprinted with permission by American Isuzu Motors Ltd. Inc.*

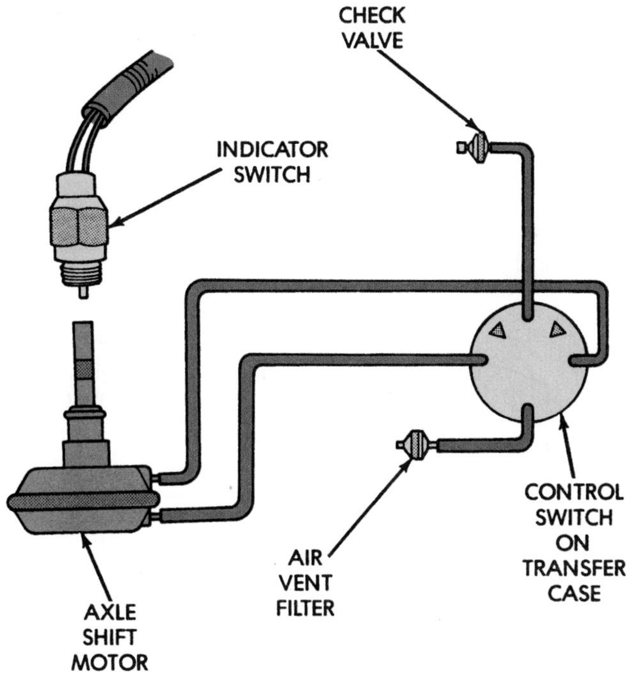

Figure 41-10 An axle disconnect system using a vacuum-operated shifter. *Courtesy of DaimlerChrysler Corporation*

nect and disconnect the axle **(Figure 41–11)**. This system allows for a smooth transition from two- to four-wheel drive. General Motors used a system whereby selecting 4WD on the selector switch energizes a heating element in the axle disconnect. The heating element heats a gas causing the plunger to operate the shift mechanism.

CONVENTIONAL FOUR-WHEEL-DRIVE OPERATING MODES

The typical 4WD vehicle has a conventional gear selector plus a gear selector for 4WD. This selector may have three positions: 2-high (2H), 4-high (4H), and 4-low (4L) **(Figure 41–12)**. In the 2H position, the vehicle operates as a normal two-wheel-drive vehicle. In the 4H position, the front axle is connected to the driveline and operates at the normal final drive ratio for the vehicle. In 4L, the vehicle is in four-wheel drive but the transfer case adds further gear reduction to the driveline. Generally the added ratio is 2:1 but on some vehicles it can be as low as 6:1. This extra low ratio allows the vehicle to move

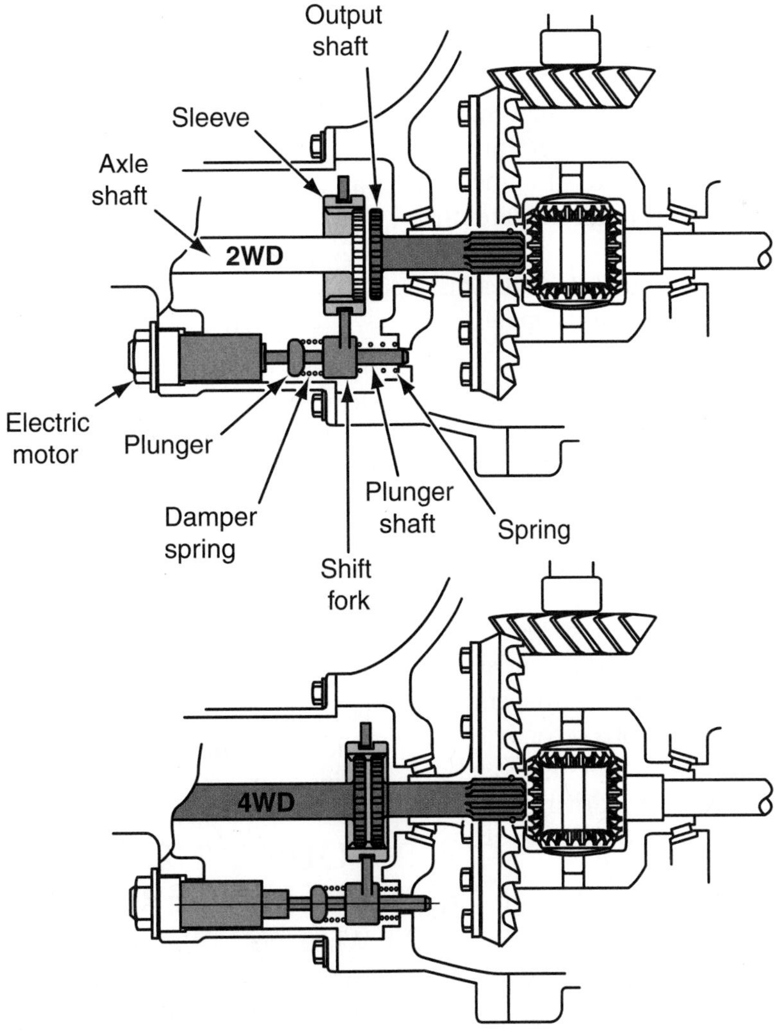

Figure 41-11 An electric motor disconnects the axles in this system.

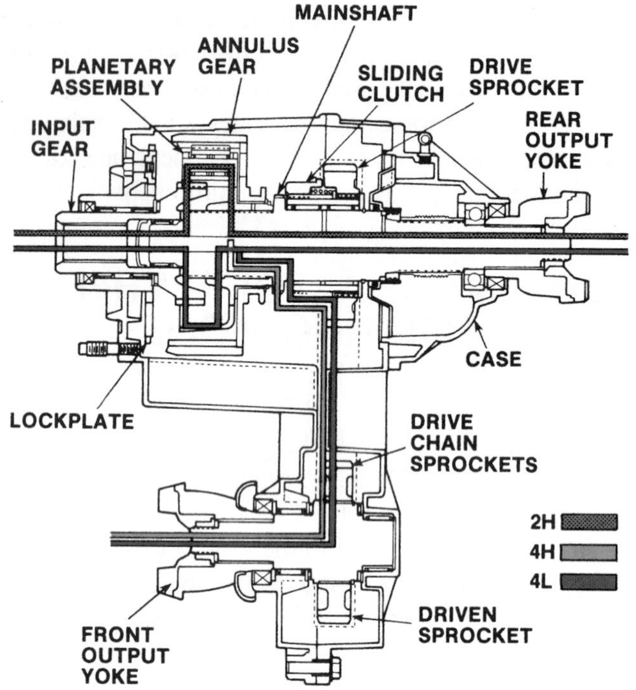

Figure 41–12 The power flow through transfer case in 2H, 4H, and 4L modes. *Courtesy of DaimlerChrysler Corporation*

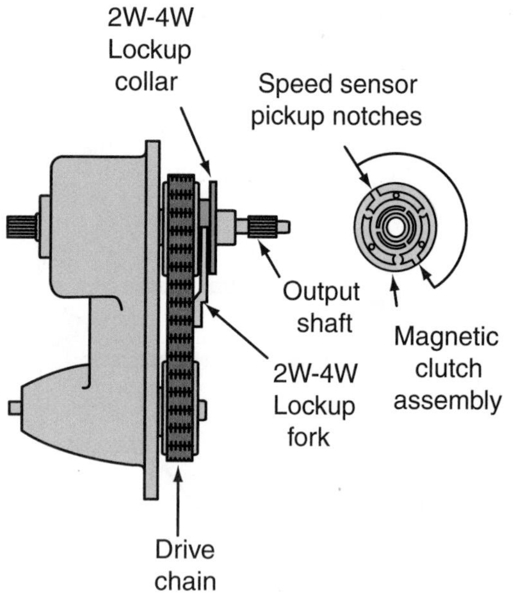

Figure 41–13 A magnetic clutch that synchronizes the shaft speeds before allowing the shift.

through very deep snow or mud, or gives it extra wheel torque to pull very heavy loads.

On some models, particularly more recent ones, the shift to 4WD can be accomplished while the vehicle is moving (shifting on-the-fly). Most vehicles must be stopped before the change can be made. Also, the shift into a four-wheel-drive low mode normally requires that the vehicle is stopped first.

Some of the power on-the-fly systems feature a magnetic clutch on the transfer case that speeds up the front drive shaft, differential, and axles to the same speed as the transmission **(Figure 41–13)**. At the instant the speed is synchronized, an electric motor on the transfer case completes the shaft, smoothly and quietly. The operator cannot damage the system because it does not shift until the speeds are synchronized.

The newest of all 4WD systems is the automatic design. While there are several different systems on the market, one of the more commonly used systems has a microprocessor to control the interaxle differential, thus providing varying amounts of torque to the front and rear driving wheels. That is, in a fraction of a second, a computer device senses the difference in wheel speeds and engages a clutch-type device to begin driving the rear wheels. When the wheel speeds are equal again, the system reverts to front drive.

Many newer RWD-based vehicles have an automatic 4WD feature that switches from 2WD to 4WD when the transfer case shift control module receives wheel rotating slip information from the wheel sensors. The transfer case shift control module then engages the transfer case motor/encoder to go into 4WD. The transfer case has the typical three-gear selection positions. When any of these positions is selected, the transfer case motor is locked and the transfer case stays at the gear selected. However, when the control switch is in the AUTO (or A4WD) position, the transfer case will cycle to 4WD when necessary. This is called the *adaptive mode*.

Some pickups and SUVs are fitted with an all-wheel-drive transfer case. In this system, all-wheel-drive is always activated. A viscous clutch is used as a torque distribution device. Normal torque distribution is 35% to the front axle and 65% to the rear.

A takeoff from electronically operated all-wheel-drive systems for trucks is Chrysler's Quadra-drive. This system uses a gerotor pump to react to variations between front and rear axle speeds. When the front and rear driveshafts are rotating at the same speed, the gerotor pump produces no pressure and everything is normal. If a rear wheel loses traction and begins to spin, the speed difference between the front and rear axles builds hydraulic pressure in the gerotor. This pressure gradually locks the clutch pack inside the gerotor pump, transferring power to the front axle. This system also uses gerotor pumps in each axle, which provides limited-slip operation at both axles. All of the torque splitting takes place without driver intervention.

FOUR-WHEEL-DRIVE PASSENGER CARS

While most 4WD trucks and utility vehicles are design variations of most basic rear-wheel-drive vehicles **(Figure 41–14)**, most passenger cars featuring 4WD were developed from a front-wheel-drive base model. They feature

Figure 41-14 A 4WD chassis based on a normal RWD chassis. *Courtesy of DaimlerChrysler Corporation*

a transaxle and differential that drive the front wheels, plus some type of mechanism for connecting the transaxle to a rear driveline. In many cases this is a simple clutch-type engagement.

4WD passenger cars normally differ from heavier-duty 4WD trucks and vehicles in several other ways. First, there is no separate transfer case; any gearing needed to transfer power to the rear driveline is usually contained in the transaxle housing or small bolt-on extension housing. Four-wheel-drive passenger cars do not use front-wheel locking hubs. Since the cars do not have lockout hubs, traction is not quite as good in the worst conditions and tire wear not so severe should the driver forget to take the car out of four-wheel drive. Interaxle differentials are also very uncommon.

Most passenger cars with 4WD are not designed for the rigors of off-road driving, in which clearance over rocks and debris is needed. For simple road driving there are no clearance or durability problems.

Limited-Slip and Open Differentials

To help compensate for the lack of locking hubs, many 4WD passenger cars use limited slip or open differential designs for the front and rear differentials. These systems allow enough differential slippage to substantially reduce wheel scrubbing and driveline windup.

A limited-slip differential (LSD) uses a friction clutch to join the right and left rear axle shafts to the differential case. This system ensures that most power is supplied where the traction is greatest. It permits the inner and outer wheels to rotate at different speeds when turns are made, just as in general differentials.

In addition, the LSD provides these features. When one front and one rear wheel, which are diagonal to each other, slip on snow-covered roads and driving force cannot be delivered, or when one rear wheel is caught in a ditch and runs idle, the LSD delivers strong torque to the other wheel so the vehicle can run normally. Even if the vehicle bumps with one rear wheel off the road, such as when driving on rough terrain, gravel, or snow-covered roads, the LSD ensures easy straightforward driveability by its limited differential function.

Other systems use an open differential. Here the greatest power is supplied where there is the least resistance. This is an excellent system for fuel economy and gear wear in full-time, four-wheel drive, but it does not provide the true traction advantages of four-wheel drive. For example, the greatest power would be supplied to the slipping wheels (least resistance) on icy road conditions when it is actually needed at the wheels facing the most resistance (for example, where the greatest traction is).

USING SERVICE MANUALS

Because of the wide variety of AWD and 4WD systems on the road today, make sure you locate the specific system in the service manual before beginning to service it. Normally, four-wheel-drive systems are covered in two distinct sections: transfer cases and propeller shafts. ■

Systems using an open differential have a drive mode on the selector that permits the differential to be locked up in such conditions. This causes the drivetrain to function the same as a part-time four-wheel-drive system. The power is distributed equally to all the wheels. Vehicles should not be driven in this mode, except under poor road conditions, for the same reasons that part-time systems should be used only when needed.

CUSTOMER CARE

Advise your customer that wear on 4WD systems is much greater than on a two-wheel-drive transaxle or transmission, especially if drivers leave the four-wheel drive engaged on dry pavement. Some car makers warn that abuse of the system is not covered under the warranty. ■

SERVICING FOUR-WHEEL-DRIVE VEHICLES

Components of 4WD vehicles can be serviced in basically the same manner as the identical components of a 2WD vehicle. Before doing any servicing, however, it is necessary to give the undercar a complete inspection. Pay particular attention to the steering dampers, steering linkage, wheel bearing, ball joints, coil springs, and radius arm bushings.

The U-joints, slip joints, or CV joints on a 4WD drivetrain must be lubed on a regular basis. To service the drive shafts on a four-wheel-drive vehicle, use the general instructions given earlier for a cross universal joint. A four-wheel drive simply uses two drive shafts instead of one.

Servicing the Transfer Case

As with all automobile servicing procedures, be sure to check the manufacturer's service manual for specific transfer case repair and overhaul procedures. It gives details for the particular make and model of transfer case to be worked on.

CAUTION!

When removing and working on a transfer case, be sure to support it on a transmission jack or safety stands. The unit is heavy and if dropped, could cause part damage and personal injury.

When removing the transfer case, disconnect and remove all driveline or propeller shaft assemblies. Be sure to mark the parts and their relative positions on their yokes so the proper driveline balance can be maintained when reassembled. Disconnect the linkage to the transfer case shift lever. Also disconnect wires to switches for 4WD dash indicator lights, if used. Remove all fasteners holding the case and move it away from the transaxle or transmission.

Once the transfer case has been removed from the vehicle and safely supported, take off the case cover and disconnect any electrical connections. Visually inspect for any oil leaks. Then carefully loosen and drive out the pins that hold the shift forks in place. Remove the front output shafts and chain drive or gearsets from the case. Keep in mind that some cases use chain drives while others use spur or helical cut gearsets to transfer torque from the transaxle or transmission to the output shafts. Planetary gearsets provide the necessary gear reductions in some transfer cases. Photo Sequence 43 shows the procedure for disassembling a Warner 13-56 transfer case, which is a commonly used unit.

Clean and carefully inspect all parts for damage and wear. Check the slack in the chain drive by following the procedure given in the service manual **(Figure 41–15)**. Replace any defective parts. It may be necessary to measure the shaft assembly end play. If excessive, new snap rings and shims may be used to correct the situation.

When reassembling the transfer case, the procedure is essentially the reverse of the removal. Be sure to use new gaskets between the covers when reassembling the unit. Photo Sequence 44 shows the procedure for reassembling a Warner 13-56 transfer case.

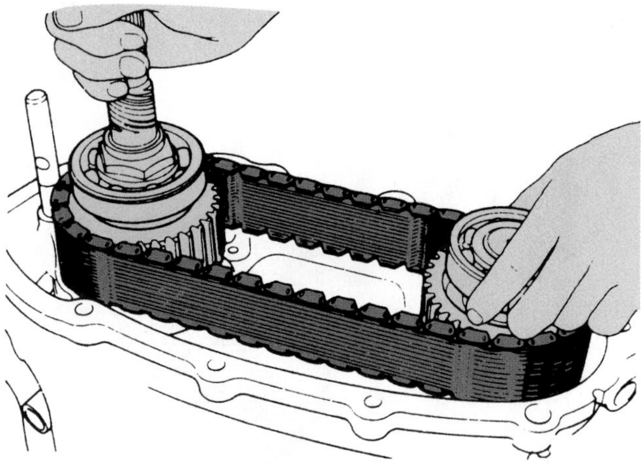

Figure 41–15 Check the chain drive for stretch. *Courtesy of Mitsubishi Motor Sales of America, Inc.*

CUSTOMER CARE

It is very important to remind your customers that the fluid level in a transfer case must be checked at recommended time intervals. (Check the service manual.) It is wise to show your customers where the transfer case fill plug is located and how to remove it. Also tell them the lubricant should be almost even with the fill hole. Always refer to the service manual for recommended transfer case lubricants. Many transfer cases require extreme pressure (EP) lubricants as used in differentials and in some manual transmissions ■

ALL-WHEEL-DRIVE SYSTEMS

All-wheel-drive systems do not give the driver the option of 2WD or 4WD. They always drive all wheels. All-wheel-drive vehicles are usually passenger cars that are not designed for off-road operation. The all-wheel-drive-vehicle is designed to increase vehicle performance in poor traction situations, such as icy or snowy roads, and in emergencies. All-wheel drive gives the vehicle operator maximum control in adverse operating conditions by biasing the driving torque to the axle with driving traction. The advantage of all-wheel drive can be compared to walking on snowshoes. Snowshoes prevent the user from sinking into the snow by spreading the body weight over a large surface. All-wheel-drive vehicles spread the driving force over four wheels when needed rather than two wheels. When a vehicle travels over the road, the driving wheels transmit a tractive force to the road's surface. The ability of each tire to transmit tractive force is a result of vehicle weight pressing the tire into the road's surface and the coefficient of friction between the tire and the road. If the road's surface is dry and the tire is dry, the coefficient of friction is high and four driving wheels are not needed. If the road's surface is wet and slippery, the coefficient of friction between the tire and road is low. The tire loses its coefficient of friction on slippery road surfaces, which could result in loss of control by the operator. Unlike a two-wheel-drive vehicle, an all-wheel-drive vehicle spreads the tractive effort to all four driving wheels. In addition to spreading the driving torque, the all-wheel-drive vehicle biases the driving torque to the axle that has the traction only when it is needed.

Viscous Clutch

The viscous clutch **(Figure 41–16)** is used in the driveline of vehicles to drive the axle with low tractive effort, taking the place of the interaxle differential. In existence for several years, the viscous clutch is installed to improve the mobility factor under difficult driving conditions. It is similar in action to the viscous clutch described for the cooling system fan. The viscous clutch in AWD is a self-contained unit. When it malfunctions, it is simply replaced as an assembly. The viscous clutch assembly is very compact, permitting installation within a front transaxle housing. Viscous clutches operate automatically while constantly transmitting power to the axle assembly as soon as it becomes necessary to improve driving wheel traction. This action is also known as biasing driving torque to the axle with tractive effort. The viscous clutch assembly is designed similarly to a multiple-disc clutch with alternating driving and driven plates **(Figure 41–17)**.

The viscous clutch parts fit inside a drum that is completely sealed. The clutch pack is made up of alternating steel driving and driven plates. One set of steel plates is splined internally to the clutch assembly hub. The second set of clutch plates is splined externally to the clutch drum. The clutch housing is filled with a small quantity of air and special silicone fluid for the purpose of transmitting force from the driving plates to the driven plates.

Based on practical experience, vehicles operating with this clutch transmit power automatically, smoothly, and with the added benefit of the fluid being capable of dampening driveline shocks. When a difference in speed of 8% exists between the input shaft driven by the driving axle with tractive effort, the clutch plates begin shearing (cutting) the special silicone fluid. The shearing action causes heat to build within the housing very rapidly, which results in the silicone fluid stiffening. The stiffening action causes a locking action between the clutch plates to take place within approximately one-tenth second. The locking action results from the stiff silicone fluid becoming very hard for the plates to shear. The stiff silicone fluid transfers power flow from the driving to the driven plates. The driving shaft is then connected to the driven shaft through the clutch plates and stiff silicone fluid.

Figure 41-16 A viscous clutch assembly.

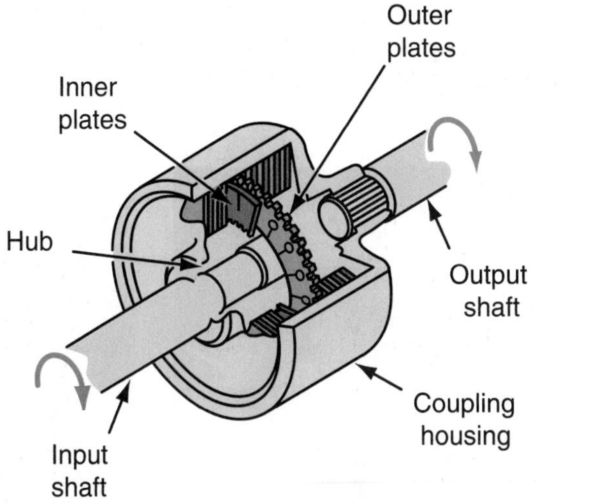

Figure 41-17 Inside a typical viscous clutch assembly.

The viscous clutch has a self-regulating control. When the clutch assembly locks up, there is very little, if any, relative movement between the clutch plates. Because there is little relative movement, silicone fluid temperature drops, which reduces pressure within the clutch housing. But as speed fluctuates between the driving and driven members, heat increases, causing the silicone fluid to stiffen. Speed differences between the driving and driven members regulate the amount of slip in a viscous clutch driveline. The viscous clutch takes the place of the interaxle differential, biasing driving torque to the normally undriven axle during difficult driving conditions.

The viscous clutch is also used in some part-time 4WD vehicles, replacing the transfer case.

Center Differential All-Wheel Drive

One of the more recent AWD designs features a center differential to split the power between the front and rear axles. On the manual transmission model, the driver can lock the center differential with a switch. On the automatic transmission model, the center differential locks automatically, depending on which transaxle range the driver selects and whether there is any slippage between front and rear wheels.

PHOTO SEQUENCE 43

Typical Procedure for Disassembling a Warner 13-56 Transfer Case

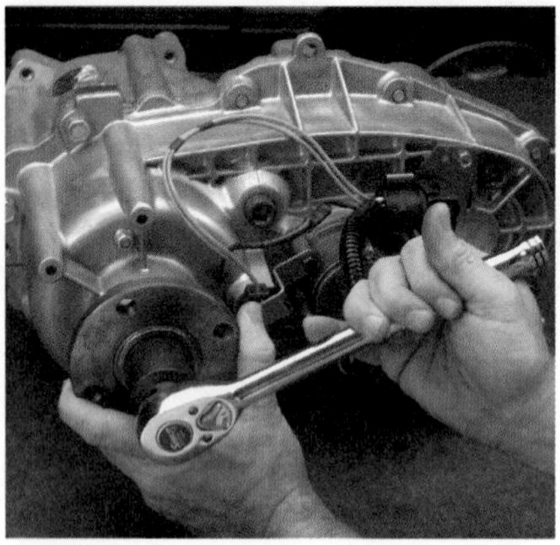

P43-1 *If not previously drained, remove the drain plug and allow the oil to drain, then reinstall the plug. Loosen the flange nuts.*

P43-2 *Remove two output shaft yoke nuts, washers, rubber seals, and output yokes from the case.*

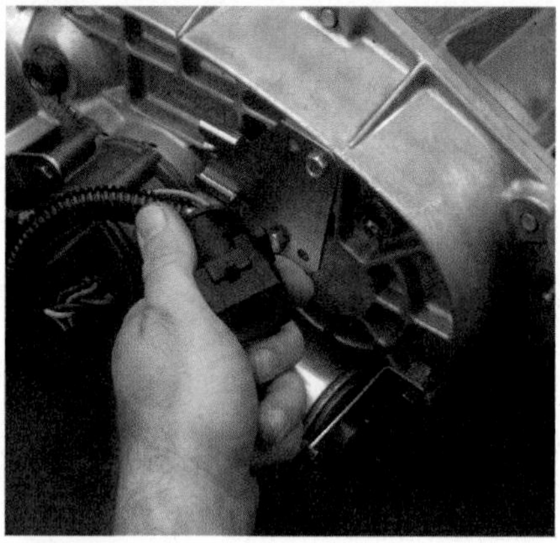

P43-3 *Remove the four-wheel-drive indicator switch from the cover.*

P43-4 *Remove the wires from the electronic shift harness connector.*

P43-5 *Remove the speed sensor retaining bracket screw, bracket, and sensor.*

P43-6 *Remove the bolts securing the electric shift motor and remove the motor.*

P43-7 *Note the location of the triangular shaft in the case and the triangular slot in the electric motor.*

P43-8 *Loosen and remove the front case to rear case retaining bolts.*

P43-9 *Separate the two halves by prying between the pry bosses on the case.*

P43-10 *Remove the shift rail for the electric motor.*

P43-11 *Pull the clutch coil off the main shaft.*

P43-12 *Pull the 2WD/4WD shift fork and lockup assembly off the main shaft.*

P43-13 *Remove the chain, driven sprocket, and drive sprocket as a unit.*

P43-14 *Remove the main shaft with the oil pump assembly.*

P43-15 *Slip the high-low range shift fork out of the inside track of the shift cam.*

P43-16 *Remove the high-low shift collar from the shift fork.*

P43-17 *Unbolt and remove the planetary gear mounting plate from the case.*

P43-18 *Pull the planetary gearset out of the mounting plate.*

PHOTO SEQUENCE 43

Typical Procedure for Reassembling a Warner 13-56 Transfer Case

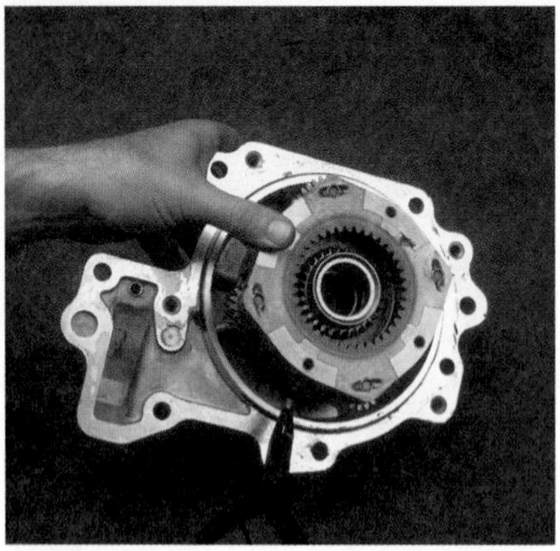

P44-1 *Install the input shaft and front output shaft bearings into the case.*

P44-2 *Apply a thin bead of sealer around the ring gear housing.*

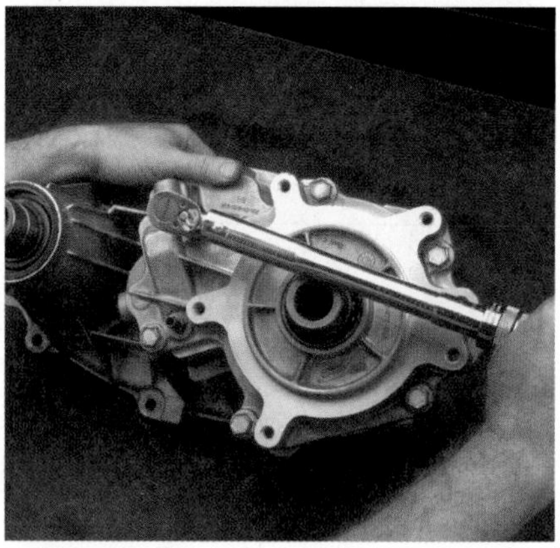

P44-3 *Install the input shaft with planetary gear set and tighten retaining bolts to specifications.*

P44-4 *Install the high-low shift collar into the shift fork.*

P44-5 *Install the high-low shift assembly into the case.*

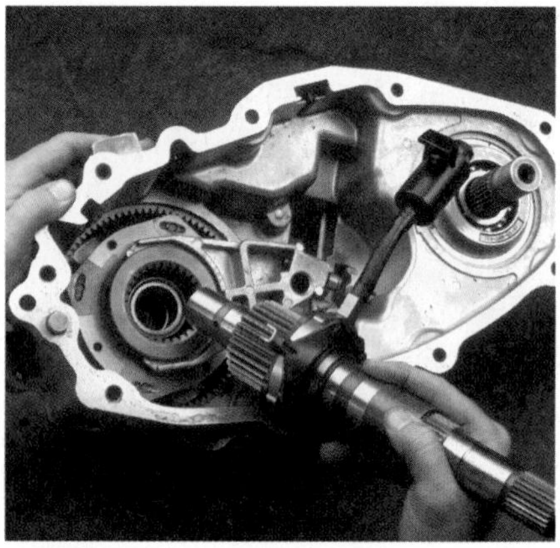

P44-6 *Install the main shaft with oil pump assembly into the case.*

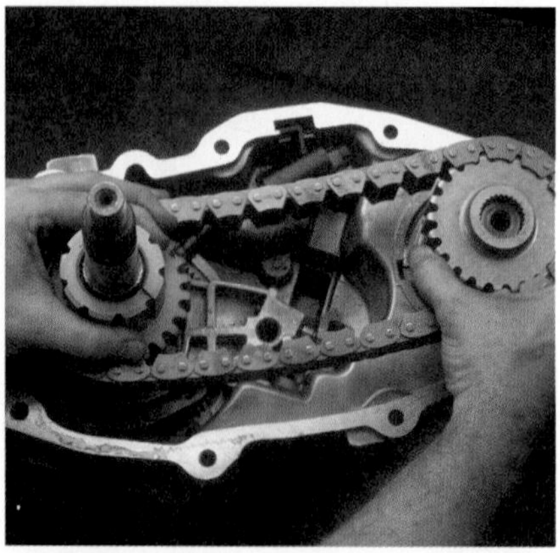

P44-7 *Install the drive and driven sprockets and chain into position in the case.*

P44-8 *Install the shift rails.*

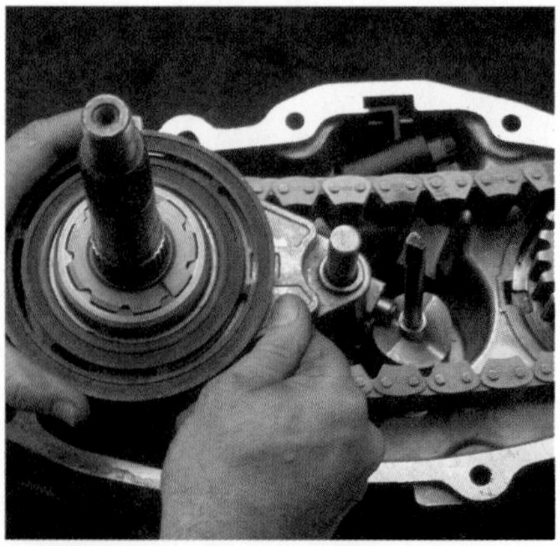

P44-9 *Install the 2WD/4WD shift fork and lockup assembly onto the main shaft.*

P44-10 *Install the clutch coil onto the main shaft.*

P44-11 *Clean the mating surface of the case.*

P44-12 *Position the shafts and tighten the case halves together. Tighten attaching bolts to specifications.*

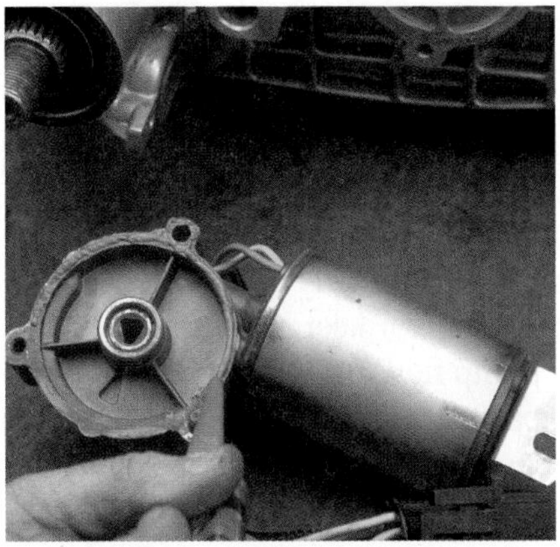

P44-13 *Apply a thin bead of sealer to the mating surface of the electric shift motor.*

P44-14 *Align the triangular shaft with the motor's triangular slot.*

P44-15 *Install the motor over the shaft, and wiggle the motor to make sure it is fully seated on the shaft.*

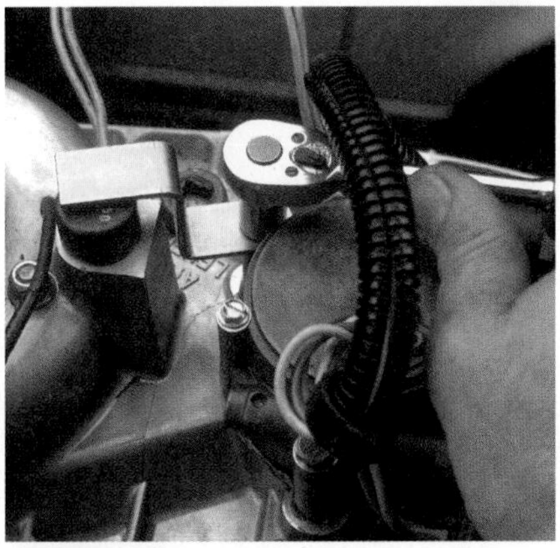

P44-16 *Tighten the motor's retaining bolts to specifications.*

P44-17 *Reinstall the wires into the connector and connect all electric sensors.*

P44-18 *Install the companion flanges' seal, washer, and nut. Then tighten the nut to specifications.*

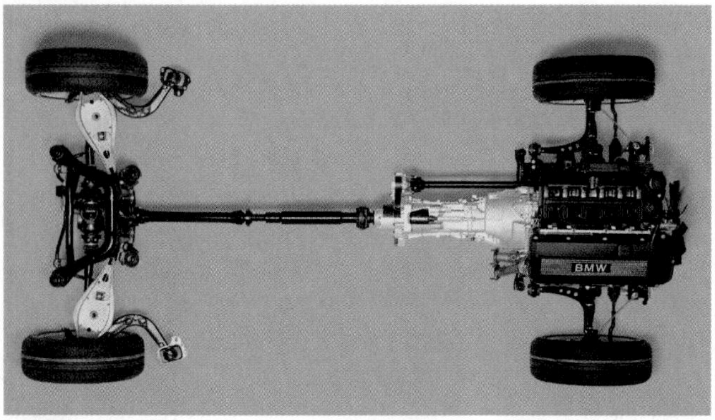

Figure 41-18 An all-wheel-drive system with three differentials based on a RWD platform. *Courtesy of BMW of North America, Incorporated*

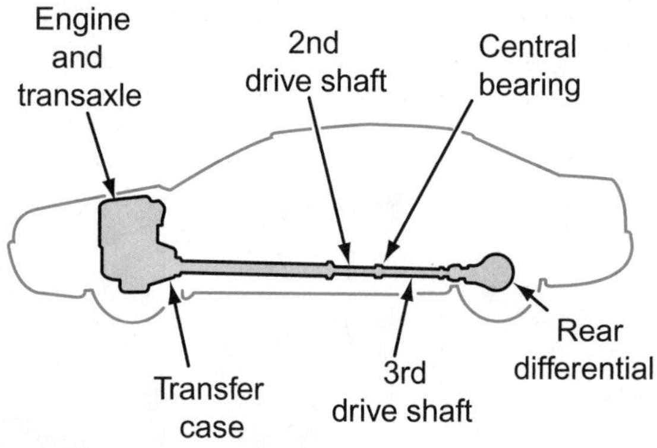

Figure 41-19 The location of 4WD driveline components on a typical FWD platform.

The system shown in **Figure 41–18** is electronically controlled and is based on a RWD setup. While most 4WD trucks and SUVs are design variations of basic RWD vehicles, most passenger cars equipped with 4WD or AWD are based on FWD designs. These modified FWD systems consist of a transaxle and a differential to drive the front wheels, plus some type of mechanism for connecting the output of the transaxle to a rear driveline. In many cases the mechanism is a simple clutch or differential. On some models, a transfer case, which transfers power to the rear axle, is fitted to the transaxle **(Figure 41–19)**.

Haldex Clutch

Some AWD vehicles are equipped with a Haldex clutch **(Figure 41–20)**, which serves as a center differential. This clutch unit distributes the drive force variably between two axles. In a typical application, the Haldex unit mounts in front of the rear differential and receives torque from the front axle.

The Haldex unit has three main parts: the hydraulic pump driven by the slip between the axles or wheels, a wet multidisc clutch, and an electronically controlled valve. The unit is much like a hydraulic pump in which the housing and a piston are connected to one shaft and a piston actuator is connected to the other.

When a front wheel slips, the input shaft to the Haldex unit spins faster than its output shaft, causing the pump to immediately generate oil flow. The oil flow and pressure engages the multidisc clutch to send power to the rear wheels. This happens extremely quickly because an electric pump and accumulator keep the circuit primed.

The oil from the pump flows to the clutch's piston to compress the clutch pack. The oil returns to the reservoir through a controllable valve, which adjusts the oil pres-

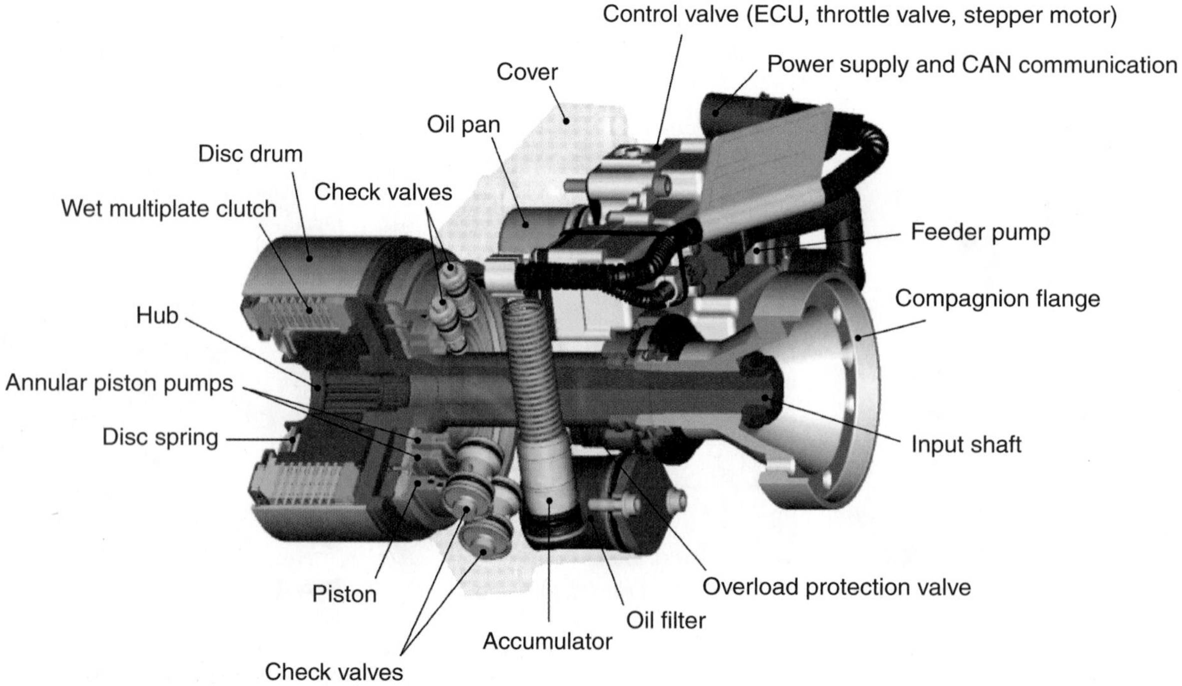

Figure 41-20 A Haldex clutch assembly. *Courtesy of Haldex Traction Systems*

sure and the force on the clutch pack. An electronic control module controls the valve and also determines when to decouple the axles to prevent the rear brakes from braking the front axle.

In high-slip conditions, a high pressure is delivered to the clutch pack; in tight curves or at high speeds, a much lower pressure is provided. When there is no difference in speed between the front and rear axles, the pump does not supply pressure to the clutch pack.

CASE STUDY

A four-wheel-drive pickup truck is brought into the shop. There is a severe vibration through the truck when it is traveling at road speeds. New wheels and tires were recently installed on the truck. The new wheels and tires are drastically oversized. Lift kits were installed by the owner to provide the necessary tire clearance. The owner would like the alignment checked and the wheels balanced.

The technician carefully balances and aligns the wheels. He takes the truck for a test drive to verify the repair. The truck still has a harsh vibration. It seems to get worse as speed increases. A bent wheel or bad wheel bearing or hub is suspected.

After returning to the shop, he measures the runout and checks the play of each wheel. Finding nothing wrong, he carefully checks the roundness of the tires. While doing so on the rear tires, he discovers the cause of the problem. When the lift kit was installed, the operating angle of the rear U-joint was affected. As the drive shaft turns, it hesitates slightly. This hesitation is caused by a binding of the U-joint.

The technician informs the owner of the cause and suggests that the lift kits be removed or the drive shaft operating angle be corrected.

KEY TERMS

All-wheel drive (AWD)
Four-wheel drive (4WD)
Interaxle differential
Locking hubs
Power takeoff
Transfer case
Viscous clutch

SUMMARY

- The importance of excellent traction and the benefits of four-wheel- and all-wheel-drive systems become readily apparent in snow or heavy rain.
- The heart of most conventional 4WD systems is the transfer case.
- The interaxle is placed in the transfer case to operate in the same fashion as the differentials at the axles. The only difference is that the transfer differential controls the speed of the drivelines instead of the axles.
- Many vehicles require that the front hubs be in a locked condition to operate as 4WD vehicles. This lockup may be made either manually or automatically.
- On 4WD vehicles with on-the-fly shifting, it is not necessary to come to a complete stop when changing operational modes.
- Components of 4WD vehicles can be serviced in basically the same manner as the identical components on a 2WD vehicle.
- All-wheel-drive vehicles may use a viscous clutch, rather than a transfer case, to drive the axle with low tractive effort.
- Some AWD vehicles have a third differential, called an interaxle differential, instead of a transfer case.
- All-wheel-drive passenger cars are typically based on FWD platforms.

REVIEW QUESTIONS

1. What is the main advantage of all-wheel drive?

2. Name the three main driveline components that are added to a 2WD vehicle to make it a 4WD vehicle.

3. Describe the purpose of a viscous clutch.

4. What is the purpose of the interaxle differential?

5. *True or False?* All 4WD and AWD vehicles have at least three drive shafts.

6. Tractive effort is defined as a ______.

a. driving force where a driving gear meshes with a driven gear

b. place where the clutch disc makes contact with the flywheel and pressure plate

c. driving force in which the wheel contacts the road's surface

d. driving force at the wheel

7. When the plates of a viscous coupling (clutch) rotate at different speeds, the plates ______ the fluid.

8. Which shift position is used full-time for all surfaces?

a. 4WD

b. hi lock

c. lo lock

d. none of the above

TECH MANUAL

The following procedures are included in Chapter 41 of the *Tech Manual* that accompanies this book:

1. Inspecting the fluid level in a transfer case.
2. Road testing a transfer case.
3. Replacing a transfer case output shaft bushing and seal.

9. How does a viscous clutch assembly work? Why is it used in AWD systems?
10. What results from having different axle ratios on the front and rear of a 4WD vehicle when it is operating in 4WD?
    - **a.** poor handling on dry surfaces
    - **b.** driveline windup
    - **c.** mechanical damage to the driveline
    - **d.** all of the above
11. In full-time, four-wheel-drive vehicles all four wheels are ______.
    - **a.** constantly receiving an equal amount of power
    - **b.** constantly open to receiving power
    - **c.** receiving power when the driver asks for it
    - **d.** all of the above
12. Technician A says viscous couplings are electronically controlled. Technician B says Haldex clutches are electronically controlled. Who is correct?
    - **a.** Technician A
    - **b.** Technician B
    - **c.** Neither A nor B
    - **d.** Both A and B
13. The transfer clutch in the all-wheel-drive automatic transaxle takes the place of the ______.
    - **a.** transmission
    - **b.** reduction gears
    - **c.** torque converter
    - **d.** interaxle differential
14. In a typical all-wheel-drive automatic transaxle, the transfer clutch is controlled by the _____.
    - **a.** transmission control unit operating the pressure regulator valve
    - **b.** engine computer controlling line pressure to the various clutches in the transmission
    - **c.** transmission control unit and the duty solenoid
    - **d.** front axle speed sensor and the vehicle speed sensor
15. Technician A says some AWD systems have a center differential. Technician B says some AWD vehicles have a viscous clutch. Who is correct?
    - **a.** Technician A
    - **b.** Technician B
    - **c.** Neither A nor B
    - **d.** Both A and B
16. In a viscous clutch, when the silicone fluid is heated, it ______.
    - **a.** becomes a very thin fluid
    - **b.** boils to a vapor
    - **c.** thickens to a solid mass
    - **d.** stiffens
17. When servicing transfer cases, all of the following are correct, except ______.
    - **a.** it must be supported on a transmission jack or safety stands
    - **b.** all transfer cases have both a chain drive and a planetary gearset
    - **c.** visual inspections for leaks and damage are necessary
    - **d.** follow the manufacturer's recommendations for lubricants
18. Which of the following is an advantage of four-wheel drive?
    - **a.** It weighs less.
    - **b.** It requires less maintenance.
    - **c.** Tires cost less.
    - **d.** None of the above.
19. The viscous clutch is used to ______.
    - **a.** engage engine torque to the transmission
    - **b.** improve mobility when driving conditions become difficult
    - **c.** drive both rear driving wheels when a vehicle is stuck
    - **d.** drive the rear axle assembly
20. The primary difference(s) between an AWD and a 4WD driveline is (are) that the AWD has ______.
    - **a.** a center differential
    - **b.** a transfer case
    - **c.** a differential unit at both axles
    - **d.** all of the above

SECTION 7

Suspension and Steering Systems

A vehicle's tires, wheels, and suspension and steering systems provide the contact between the driver and the road. They allow the driver to safely maneuver the vehicle in all types of conditions as well as to ride in comfort and security.

Once straightforward, mechanical systems, suspension and steering systems have become increasingly sophisticated. New suspension and steering systems have been developed to provide good handling for lighter vehicles as well as to provide car-like comfort in pickups and SUVs. Computer technology has made possible such features as variable ride qualities and active and adaptive suspensions. These computer systems are integrated with powertrain controls to ensure improved handling and safety. Systems such as stability control are also linked to the brake system through a computer. Four-wheel alignment is required or recommended on all vehicles. All of these new systems require special servicing techniques.

The information found in Section 7 corresponds to materials covered on the ASE certification test on suspension and steering.

Delmar's Automotive Suspension & Steering Systems Video Series and CD-ROM Courseware

WE ENCOURAGE PROFESSIONALISM

ASE

THROUGH TECHNICIAN CERTIFICATION

42

TIRES AND WHEELS

OBJECTIVES

■ Describe basic wheel and hub design. ■ Recognize the basic parts of a tubeless tire. ■ Explain the differences between the three types of tire construction in use today. ■ Explain the tire ratings and designations in use today. ■ Describe why certain factors affect tire performance, including inflation pressure, tire rotation, and tread wear. ■ Remove and install a wheel and tire assembly. ■ Dismount and remount a tire. ■ Repair a damaged tire. ■ Describe the differences between static balance and dynamic balance. ■ Balance wheels both on and off a vehicle. ■ Describe the three popular types of wheel hub bearings.

Tire and wheel assemblies provide the only connection between the road and the vehicle. Tire design has improved dramatically during the past few years. Modern tires require increased attention to achieve their full potential of extended service and correct ride control. Tire wear that is uneven or premature is usually a good indicator of steering and suspension system problems. Tires, therefore, become not only a good diagnostic aid to a technician, but can also be clear evidence to the customer that there is need for service.

WHEELS

Wheels are made of either stamped or pressed steel discs riveted or welded together. They are also available in the form of aluminum or magnesium rims that are die-cast or forged **(Figure 42–1)**. Magnesium wheels are commonly referred to as mag wheels, although they are usually made of an aluminum alloy. Aluminum wheels are lighter in weight when compared with the stamped steel type. This weight savings is important because the wheels and tires on a vehicle are unsprung weight. This means the weight is not supported by the vehicle's springs.

Figure 42–1 An alloy wheel on a late-model car.

Near the center of the wheel are mounting holes that are tapered to fit tapered mounting nuts (**lug nuts**) that center the wheel over the hub. The rim has a hole for the tire's **valve stem** and a **drop center** area designed to allow for easy tire removal and installation. **Wheel offset** is the vertical distance between the centerline of the rim and the mounting face of the wheel. The offset is considered positive if the centerline of the rim is inboard of the mounting face and negative if outboard of the mounting face. The amount and type of offset is critical because changing the wheel offset changes the front suspension loading as well as the scrub radius.

The wheel is bolted to a **hub**, either by lug bolts that pass through the wheel and thread into the hub, or by studs that protrude from the hub. In the case of studs, special lug nuts are required. A few vehicles have left-hand threads (which turn counterclockwise to tighten) on the driver's side and right-hand threads (which turn clockwise to tighten) on the passenger's side. All other vehicles use right-hand threads on both sides.

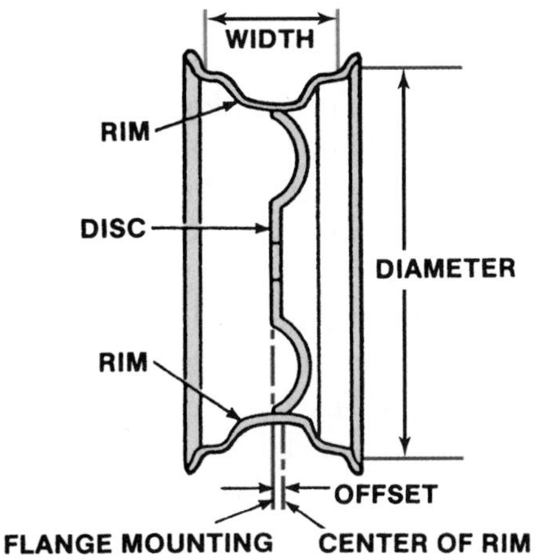

Figure 42-2 Wheel dimensions are important when replacing tires.

Wheel size is designated by rim width and rim diameter **(Figure 42–2)**. Rim width is determined by measuring across the rim between the flanges. Rim diameter is measured across the bead seating areas from the top to the bottom of the wheel. Some rims have safety ridges near their lips. In the event of a tire blowout, these ridges tend to keep the tire from moving into the dropped center and from coming off the wheel.

Replacement wheels must be equal to the original equipment wheels in load capacity, diameter, width, offset, and mounting configuration. An incorrect wheel can affect wheel and bearing life, ground and tire clearance, or speedometer and odometer calibrations. A wrong size wheel can also affect the antilock brake system.

TIRES

The primary purpose of tires is to provide traction. Tires also help the suspension absorb road shocks, but this is a side benefit. They must perform under a variety of conditions. The road might be wet or dry or paved with asphalt, concrete, or gravel, or there might be no road at all. The car might be traveling slowly on a straight road, or moving quickly through curves or over hills. All of these conditions call for special requirements that must be present, at least to some degree, in all tires.

In addition to providing good traction, tires are also designed to carry the weight of the vehicle, to withstand side thrust over varying speeds and conditions, and to transfer braking and driving torque to the road. As a tire rolls on the road, friction is created between the tire and the road. This friction gives the tire its traction. Although good traction is desirable, it must be limited. Too much traction means there is much friction. Too much friction means there is a lot of rolling resistance. Rolling resistance wastes engine power and fuel; therefore, it must be kept to a minimal level. This dilemma is a major concern in the design of today's tires.

Tube and Tubeless Tires

Early vehicle tires were solid rubber. These were replaced with pneumatic tires, which are filled with air.

There are two basic types of pneumatic tires: those that use inner tubes and those that do not. The latter are called tubeless tires and are about the only type used on passenger cars today. A tubeless tire has a soft inner lining that keeps air from leaking between the tire and rim. On some tires, this inner lining can form a seal around a nail or other object that punctures the tread. A self-sealing tire holds in air even after the object is removed. The key to this sealing is a lining of sticky rubber compound on the inside of the tread area that will seal a hole up to 3/16 inch (4.77 mm).

A tubeless tire air valve has a central core that is spring-loaded to allow air to pass inward only, unless the pin is depressed. If the core becomes defective, it can be unscrewed and replaced. The airtight cap on the end of the valve provides extra protection against valve leakage. A tubeless tire is mounted on a special rim that retains air between the rim and the tire casing when the tire is inflated.

Figure 42–3 shows a cutaway view of a typical tubeless tire. The basic parts are shown. The cord body or casing consists of layers of rubber-impregnated cords, called **plies**, that are bonded into a solid unit. Typically tires are made of 2, 4, or 8 plies; thus, the reference to 2-, 4-, and 8-ply tires. The plies determine a tire's strength, handling, ride, amount of road noise, traction, and resistance to fatigue, heat, and bruises. The **bead** is the portion of the tire that helps keep it in contact with the rim of the wheel. It also provides the air seal on tubeless tires.

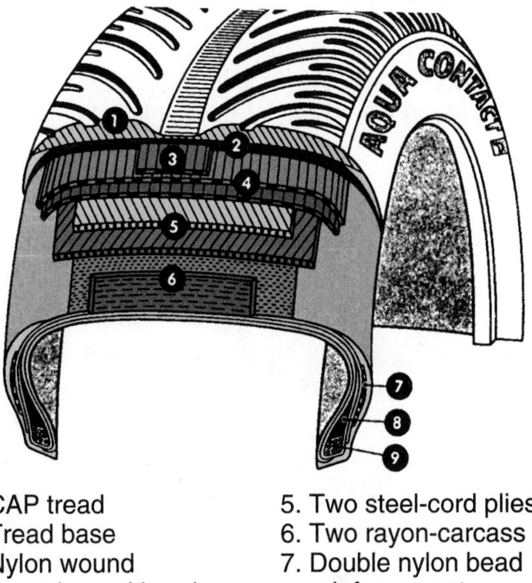

1. CAP tread
2. Tread base
3. Nylon wound aquachannel breaker
4. Two-ply nylon wound breaker
5. Two steel-cord plies
6. Two rayon-carcass plies
7. Double nylon bead reinforcement
8. Bead filler
9. Bead core

Figure 42-3 A typical tubeless tire. *Courtesy of SAE International*

The bead is constructed of a heavy band of steel wire wrapped into the inner circumference of the tire's ply structure. The **tread**, or crown, is the portion of the tire that comes in contact with the road surface. It is a pattern of grooves and ribs that provides traction. The grooves are designed to drain off water, while the ribs grip the road surface. Tread thickness varies with tire quality. On some tires, small cuts, called **sipes**, are molded into the ribs of the tread. These sipes open as the tire flexes on the road, offering additional gripping action, especially on wet road surfaces. The **sidewalls** are the sides of the tire's body. They are constructed of thinner material than the tread to offer greater flexibility.

The tire body and belt material can be made of rayon, nylon, polyester, fiberglass, steel, amarid, or kevlar. Each has its advantages and disadvantages. For instance, rayon and cord tires are low in cost and give a good ride, but do not have the inherent strength needed to cope with long high-speed runs or extended periods of abusive use on rough roads. Nylon-cord tires generally give a slightly harder ride than rayon—especially for the first few miles after the car has been parked—but offer greater toughness and resistance to road damage. Polyester and fiberglass tires offer many of the best qualities of rayon and nylon, but without the disadvantages. They run as smoothly as rayon tires but are much tougher. They are almost as tough as nylon, but give a much smoother ride. Steel is tougher than fiberglass or polyester, but it gives a slightly rougher ride because the steel cord does not give under impact, as do fabric plies. Amarid and kevlar cords are lighter than steel cords and, pound for pound, stronger than steel.

Types of Tire Construction

There are three basic types of tire construction in use today **(Figure 42–4)**. The oldest design currently in use is the **bias ply**. It has a body of fabric plies that run alternately at opposite angles to form a crisscross design. The angle varies from 30 to 38 degrees with the centerline of the tire and has an effect on high-speed stability, ride harshness, and handling. Generally speaking, the lower the cord angle, the better the high-speed stability, but also the harsher the ride. Bias ply tires usually are available in 2- or 4-ply.

Belted bias ply tires are similar to bias ply tires, except that two or more belts run the circumference of the tire under the tread. This construction gives strength to the sidewall and greater stability to the tread. The belts reduce tread motion during contact with the road, thus improving tread life. Plies and belts of various combinations of rayon, nylon, polyester, fiberglass, and steel are used with belted bias construction. Belted bias ply tires generally cost more than conventional bias ply tires, but last up to 40% longer.

Radial ply tires have body cords that extend from bead to bead at an angle of about 90 degrees—"radial" to the circumferential centerline of the tire—plus two or more layers of relatively inflexible belts under the tread. The construction of various combinations of rayon, nylon, fiberglass, and steel gives greater strength to the tread area and flexibility to the sidewall. The belts restrict tread motion during contact with the road, thus improving tread life and traction. Radial ply tires also offer greater fuel economy, increased skid resistance, and more positive braking.

One of the newer tire designs has steel reinforced sidewalls that allow the tire to maintain shape with no air pressure. The so-called run-flat tires give the driver a limited amount of travel after a tire has been punctured. Since they were designed to be filled with air, their ride and handling characteristics are much better when they have air than when they do not. However, the tires will not leave the driver stranded or forced to put on a spare tire when the tire is punctured.

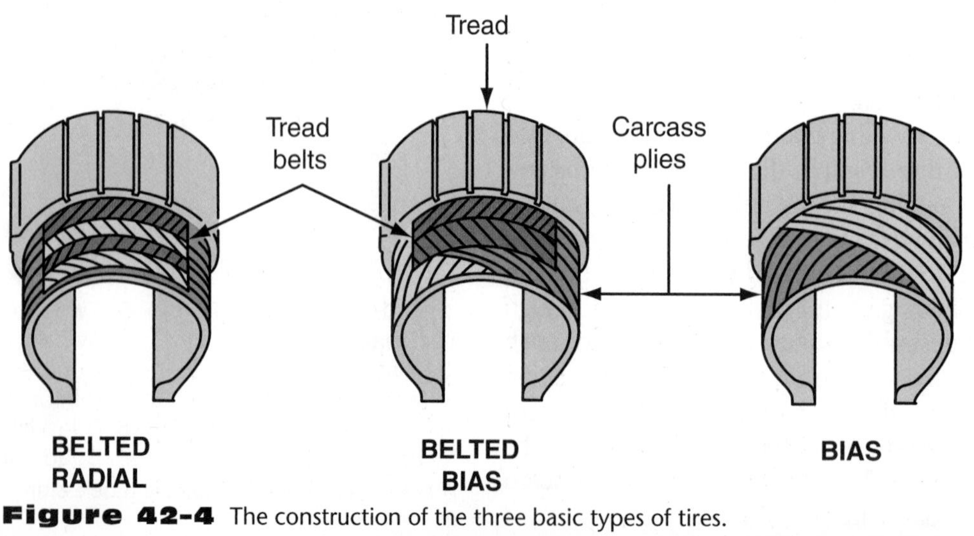

Figure 42–4 The construction of the three basic types of tires.

Specialty Tires

Specialty tires reflect the advances made in the conventional tire field. Special snow and mud tires are available in all three construction types. Studded tires provide superior traction on ice, but are slowly disappearing from the tire market because their performance on dry surfaces is poor. In addition, many states have outlawed their use because they damage the road and can be a safety hazard on dry surfaces. Because the studs have more contact on the road than the rubber, it is easy for the tire to slide on the road during cornering and stopping. The studs offer much less friction than the rubber tire tread.

Most tire manufacturers offer all-season or all-weather tires. These tires are designed to perform well on all types of road conditions, but they will not be excellent performers on every road surface. Tires can be designed to be great on dry surfaces or wet surfaces. However, it is near impossible to have a tire that performs extremely well on both. Tires designed for dry and smooth roads do not really need a tread pattern. They can be "slicks." The smooth surface would give the tires maximum grip on a smooth, dry surface. However when a slick hits a wet spot, there is no traction. The tire simply slides on the water.

Tires designed for wet surfaces have tread designs that move the road's water behind and to the side of the tire. Moving the water is the only way a tire can grip the road. Needless to say, when a tire has many directed and open channels for water in its tread, less rubber meets the road.

Tread Designs

The real purpose of a tire is to get a grip on the road. The ideal tire is one that wears little, holds the road well to provide sure handling and braking, and provides a cushion from road shock. The ideal tire should also provide maximum grip on dry roads, wet roads, and snow and ice, and operate quietly at any speed. This is a tall order, so tire manufacturers compromise on one or two of these qualities for the sake of excelling at another. A tire's tread design dictates what the tire will excel at.

There are basically three categories of tread patterns: directional, nondirectional, and symmetric and asymmetric.

A directional tire is mounted so that it revolves in a particular direction. These tires have an arrow on the sidewalls that show the designed direction of travel. A directional tire offers good performance only when it is rotating in the direction in which it was designed to rotate **(Figure 42–5)**. A nondirectional tire has the same handling qualities in either direction of rotation. A symmetric tire has the same tread pattern on both sides of the tire. An asymmetrical tire has a tread design that is different from one side to the other. Asymmetrical tires are typically designed to provide good grip when travelling straight (the inside half) and good grip in turns (the outside half of the tread). Most asymmetric tires are also directional tires.

Figure 42–5 There are many different tread designs available for today's tires. *Courtesy of the Goodyear Tire Company*

The number and size of the blocks, sipes, and grooves on a tire's tread not only determine how much rubber contacts the road and how much water can be displaced, they also determine how quiet the tire will be during travel. The more aggressive the tread, the more noise it will make. This statement is especially true if the tire's tread is made of a hard compound. Softer tires typically make less noise but wear more quickly. Soft tires also adhere to the road better.

Channels are cut into a tire's tread to allow water to move away from the tire's direction of travel. Obviously the deeper the channel, the more water the tire can move. The disadvantage of these channels is decreased road contact.

SHOP TALK

Whenever a customer wants a better handling tire, make sure he or she knows a better gripping tire may make more noise and not wear as long as other tires. Knowing what design of tire will meet a customer's needs is a science. Always consult with a tire specialist before recommending one tire or another. ■

Spare Tires

Nearly all vehicles are equipped with a spare tire to be used in case one of the vehicle's tires loses air and goes flat. A spare tire can be a tire that matches the tires on the vehicle or can be a compact spare. Compact spares are designed to reduce weight and storage space but still provide the driver with a tire in the case of an emergency. Compact tires are typically one of three types: high-pressure mini spare, space-saver spare, and lightweight skin spare.

A high-pressure mini-spare tire is a temporary tire. It should not be used for extended mileage or for speeds above 50 mph. A space-saver spare must be blown up with a compressor that operates from the cigarette lighter or a built-in air compressor. A skin spare is a normal bias ply type tire with a reduced tread depth.

CUSTOMER CARE

Make sure you warn your customer that a mini, space-saver, or similar type compact spare should be used only as a temporary tire. It should never be used as a regular tire. Any continuous load use of a temporary spare will result in tire failure, loss of vehicle control, and possible injury to the vehicle's occupants. ■

High-Performance (Speed-Rated) Tires

The so-called high-performance tire has made an increasing appearance on many new cars. Most drivers do not care if they actually use the speed capabilities of the tire, just as long as the performance is there. In fact, many of their cars are not capable of the speeds the tires are rated at. The speed rating of a tire is really nothing more than an expression of how well the tire will withstand the temperatures of high speed. This does not necessarily mean a high-speed rated tire will perform better at low speeds than a lower rated tire.

Table 42–1 lists the various letters used to designate the speed rating of a tire and the maximum speed the tire was designed to safely operate at. Driving a vehicle at speeds greater than the speed rating of the tires is risky. The heat generated can cause the tire to come apart. If this happens at high speed, it will be close to impossible for the driver to maintain control of the vehicle.

Although high-performance tires withstand heat better than normal tires, they still wear and must be replaced. In some European countries, the replacement tire must have, by law, the same speed rating as the OE tire. Although it is not a law in the United States, preventing the practice of trading down in speed ratings probably would not be a bad idea.

TABLE 42–1 SPEED RATINGS

Symbol	Maximum Speed
F	50 mph
G	56 mph
J	62 mph
K	68 mph
L	75 mph
M	81 mph
N	87 mph
P	93 mph
Q	100 mph
R	106 mph
S	112 mph
T	118 mph
U	124 mph
H	130 mph
V	149 mph
Z	+149 mph

Tire Ratings and Designations

Tires are also rated by their profile (aspect) ratio, size, and load range. A **tire's profile** is the relation of its cross-sectional height (from tread to bead) compared to the cross-sectional width (from sidewall to sidewall). Today, this ratio is also known as **series**.

For many years, the accepted profile ratio for standard bias ply passenger car tires was approximately 83. This meant the tire was 83% as high as it was wide. Since the introduction of bias-belted and radial-ply construction, lower profile tires and ratios of 78, 70, 60, 50, even 40 have become popular **(Figure 42–6)**. The lower the number, the wider the tire. As an example, a 50-series tire appears to be low and flat because its height is 50% of its width.

In the designation P195/75R14, P identifies passenger car tire. The width in millimeters is 195. The height-to-width ratio is 75. R identifies radial construction. The rim diameter is 14. If this designation is followed by M, S, or MS, the tire's tread is rated for use in mud, snow, or both.

All-metric system measurements are now being given along with a standard translation. A typical metric tire shows its width in millimeters, its inflation pressure in kilopascals (kPa), and its load capacity in kilograms (kg). One kilopascal equals 6.895 psi. A typical all-metric radial size is 190/65R-390. It fits a 390-mm diameter wheel.

Also on the sidewall of a tire are ratings for maximum load, road wear, traction, temperature, maximum inflation, tread plies, and sidewall plies. These ratings are nor-

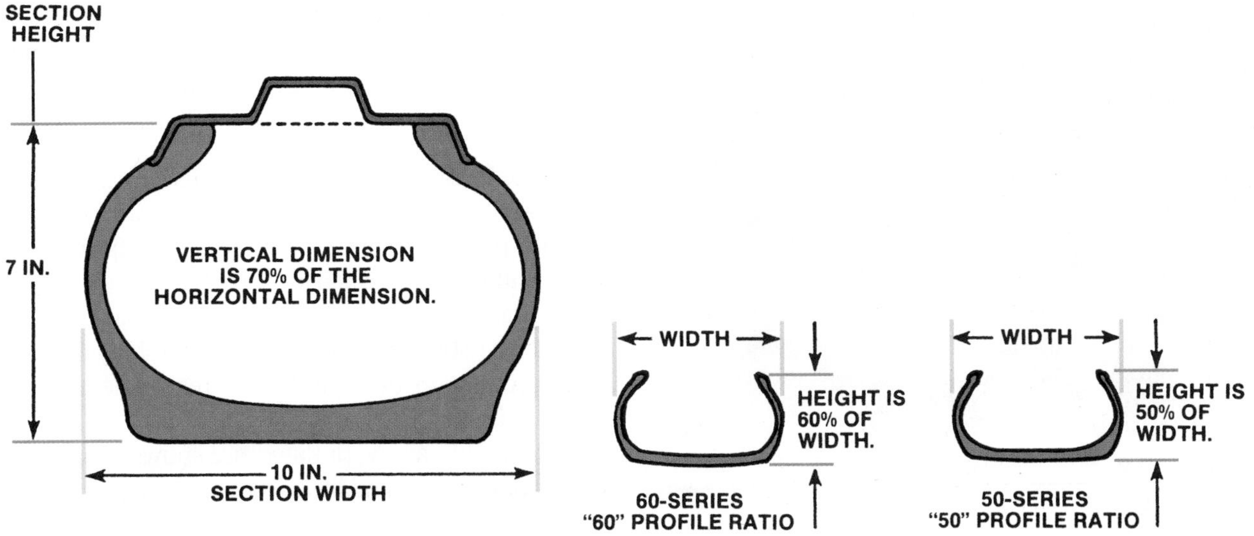

Figure 42-6 The aspect ratio (profile) of a tire is its cross-sectional height compared to its cross-sectional width expressed in a percentage.

mally listed separately from the size of the tire **(Figure 42–7)**.

The maximum load rating lists the maximum amount of weight the tire can carry at the recommended tire pressure. Tread wear is listed as a number that ranges from 100 to 500. A rating of 100 means the tire will wear easily and quickly, whereas a rating of 500 means the tire is hard and will resist wear. The traction rating is given as an A, B, or C. A tire rated as C will provide less traction than one rated with an A. Also given as A, B, or C is the tire temperature rating. A tire with an A temperature rating will be able to withstand high temperatures better than those rated B or C.

It is important to remember that the maximum inflation number on a tire is the maximum inflation and not the recommended inflation. Tires should never be inflated beyond this maximum rating.

Typically, the load rating and the number of tread and sidewall plies are proportional. In most cases, the more plies a tire has, the more weight it can support.

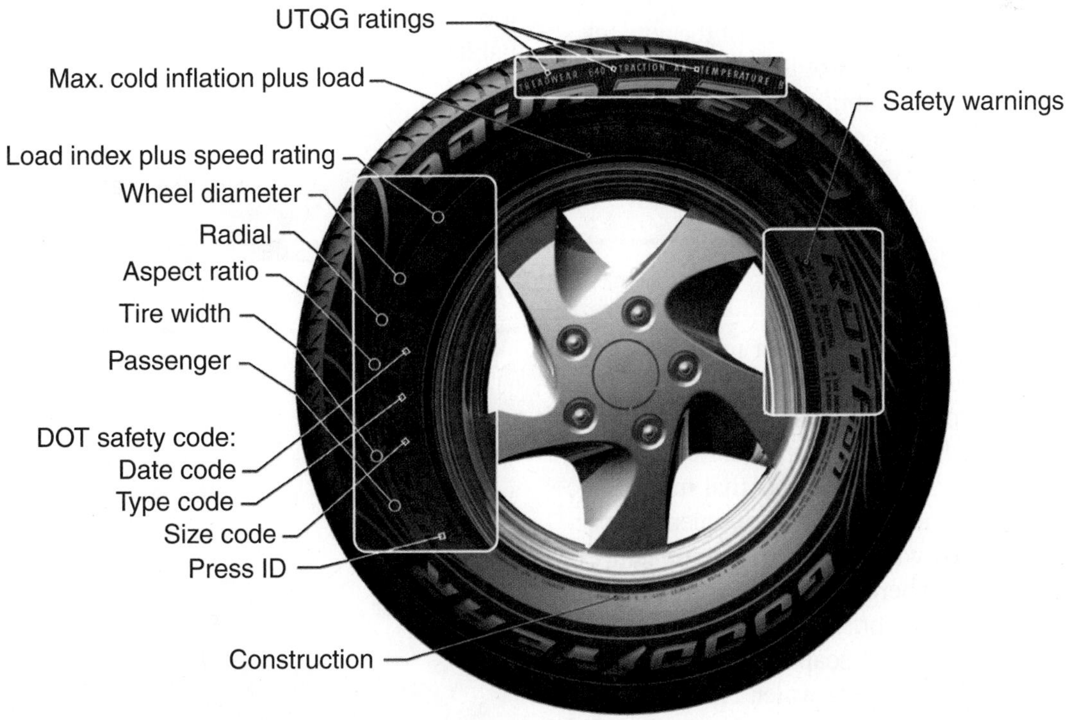

Figure 42-7 There is a lot of information about a tire on its sidewall.
Courtesy of the Goodyear Tire Company

Figure 42-8 A tire placard on a doorjamb.

Tire Placard

The tire placard, or safety compliance certification label, is generally found on the driver's door jamb. It includes recommended maximum vehicle load, tire size (including spare), and the *correct* cold tire inflation for each tire of the vehicle **(Figure 42–8)**. Never use this information for other cars.

As a general rule, tires should be replaced with the same size designation or an approved optional size as recommended by the auto or tire manufacturer. Also, always follow the manufacturer's recommendations for tire type, inflation pressures, and rotation patterns.

SHOP TALK

Tires with a larger or smaller diameter than originally installed will affect the operation of the antilock brake system and the accuracy of the speedometer. It might be necessary to recalibrate the ABS computer and change the speedometer drive gears when tire size has been changed. Check the vehicle's service manual for details. ■

Tire Care

To maximize tire performance, inspect for signs of improper inflation and uneven wear, which can indicate a need for balancing, rotation, or wheel alignment. Tires should also be checked frequently for cuts, stone bruises, abrasions, and blisters, and for objects that might have become imbedded in the tread. More frequent inspections are recommended when rapid or extreme temperature changes occur, or where road surfaces are rough or occasionally littered with debris.

To clean tires, use a mild soap and water solution only. Rinse thoroughly with clear water. Do not use any caustic solutions or abrasive materials. Never use steel wool or wire brushes. Avoid gasoline, paint thinner, and similar materials having a mineral oil base. These materials will cause premature drying of the tire's rubber. As the rubber dries, it gets harder and the tire will lose some of its performance characteristics.

Inflation Pressure A properly inflated tire gives the best tire life, riding comfort, handling stability, and fuel economy for normal driving conditions. Too little air pressure can result in tire squeal, hard steering, excessive tire heat, abnormal tire wear, and increased fuel consumption by as much as 10%. An underinflated tire shows maximum wear on the outside edges of the tread. There is little or no wear in the center.

Conversely, an overinflated tire shows its wear in the center of the tread and little wear on the outside edges. A tire inflated higher than recommended can cause a hard ride, tire bruising, and rapid wear at the center of the tire **(Figure 42–9)**.

Many inflation pressures listed for imported vehicles are given in kilopascals (kPa) rather than psi. **Table 42–2** converts kPa to psi.

A few vehicles are fit with tire inflation monitoring systems. These systems have air pressure sensors strapped around the drop center of each wheel. The sensors monitor the pressure in the tire. When the pressure is below or above a specified range, the vehicle's computer causes a warning light on the dash to illuminate. This alerts the driver to the problem.

Tire Rotation To equalize tire wear, most car and tire manufacturers recommend that the tires be rotated. Remember that front and rear tires perform different jobs and can wear differently, depending on driving habits and the type of vehicle. In a RWD vehicle, for instance, the front tires usually wear along the outer edges,

TABLE 42–2 INFLATION PRESSURE CONVERSION (KILOPASCALS TO psi)

kPa	psi	kPa	psi
140	20	215	31
145	21	220	32
155	22	230	33
160	23	235	34
165	24	240	35
170	25	250	36
180	26	275	40
185	27	310	45
190	28	345	50
200	29	380	55
205	30	415	60

Conversion: 6.9 kPa = 1 psi.

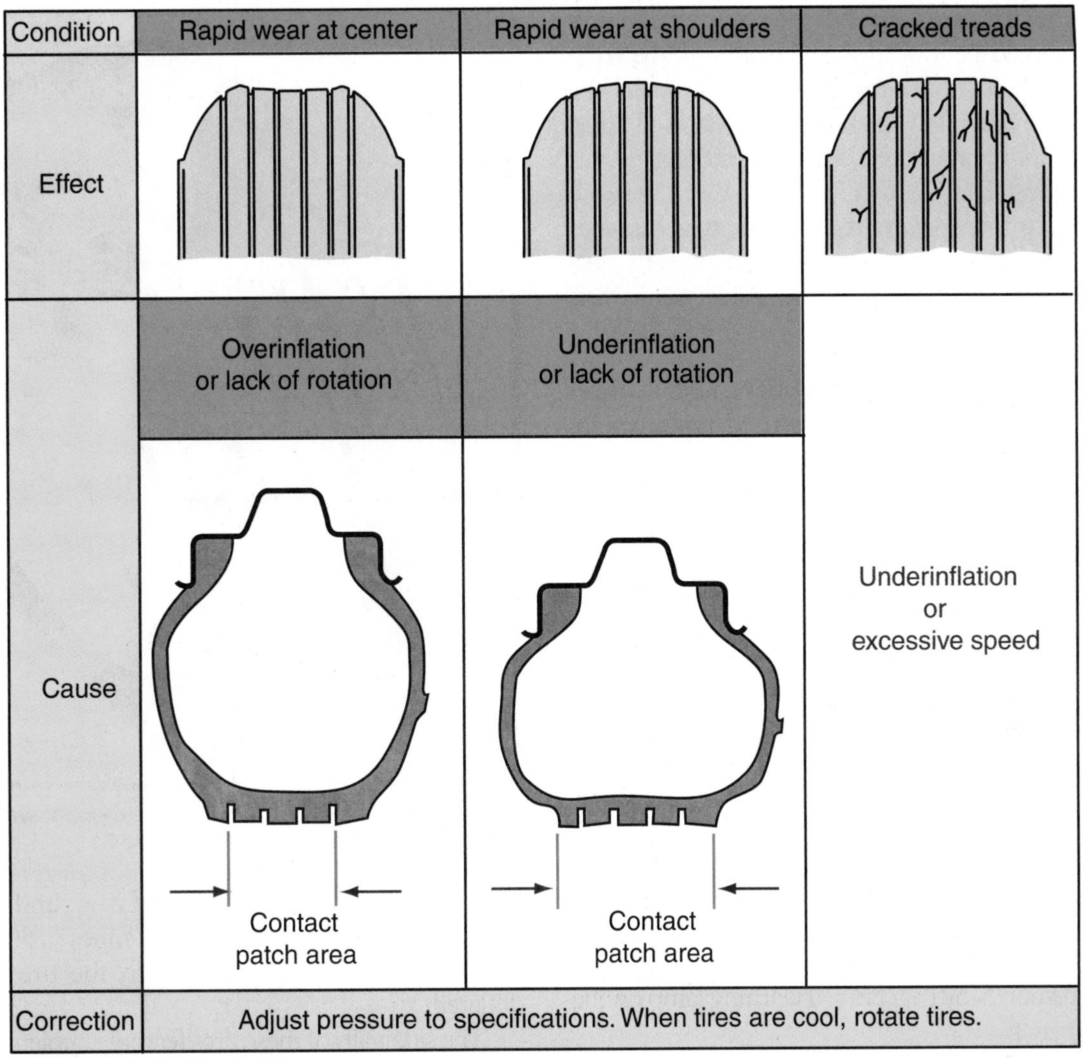
Condition
Rapid wear at center
Rapid wear at shoulders
Cracked treads
Effect
Cause
Overinflation or lack of rotation
Underinflation or lack of rotation
Underinflation or excessive speed
Contact patch area
Contact patch area
Correction
Adjust pressure to specifications. When tires are cool, rotate tires.

Figure 42-9 Effects of inflation on tread contact.

primarily because of the scuffing and slippage encountered in cornering. The rear tires wear in the center because of acceleration thrusts. To equalize wear, it is recommended that tires be rotated as illustrated in **Figure 42–10**. Bias-ply and bias-belted tires should be rotated about every 6,000 miles (10,000 km). Radial tires should be initially rotated at 7,500 miles (12,000 km) and then at least every 15,000 miles (24,000 km) thereafter. It is important that directional tires are kept rotating in the correct direction after rotating them. This means the tires may need to be dismounted from the wheel, flipped, and reinstalled on the rim before being put on the other side of the car. Many auto shops keep a record of tire rotation periods so they can notify their customers when it should be done next.

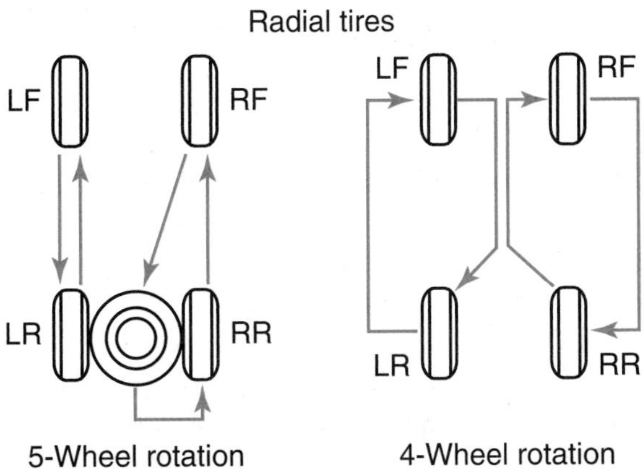
Radial tires
LF
RF
LR
RR
LF
RF
LR
RR
5-Wheel rotation
4-Wheel rotation

Figure 42-10 Rotation for radial tires.

SHOP TALK

Although the rotation patterns shown in **Figure 42–10** can serve as the general rule, always refer to a service manual. Some vehicles equipped with radial tires should have their tires rotated in a crisscross pattern. The correct rotation method depends on the design of the vehicle's suspension and steering systems. ■

When snow tires are installed, the regular tread tires on the rear should be moved to the front and the front tires stored. When snow tires are removed, install the stored tires on the rear. Do not rotate studded tires. Always remount them in their original positions.

When storing tires, lay them flat on a clean, dry, oil-free floor. Keep them away from ozone, which comes from the electrical sparking frequently produced by electric motors. Store them in the dark. Direct sunlight is hard on tires.

Tread Wear Most tires used today have built-in tread wear indicators to show when they need replacement. These indicators appear as ½-inch (12-mm) wide bands when the tire tread depth wears to 1⁄16 inch (1.5 mm). When the indicators appear in two or more adjacent grooves at three locations around the tire, or when cord or fabric is exposed, tire replacement is recommended.

If the tires do not have tread wear indicators, a tread depth indicator **(Figure 42–11)** quickly shows in 32nds of an inch how much tire tread is left. When only 2⁄32 inch (1.5 mm) is left, it is time to replace a tire.

Tire Pressure Monitor (TPM)

The tire pressure monitor (TPM) checks the inflation pressures in all four tires at frequent periodic intervals. The TPM's sensors keep track of the tire pressures both when the vehicle is moving and when it is stationary. When the TPM detects changes in any tire's inflation pressure, it responds by triggering a warning lamp on the instrument panel.

Each wheel is equipped with an electronic sensor assembly located just behind the valve stem **(Figure 42–12)**. This assembly is made up of a pressure sensor, a transmitter, and a battery. The pressure sensor measures the tire's inflation pressure and relays this information to the vehicle via radio waves. These signals are picked up by separate body-mounted antennas for each wheel. A central electronic control unit processes the signals from the four wheels and reports any variations to the system.

Run-Flat Tires

A run-flat tire consists of a self-supporting tire, a specially designed wheel, and a flat-tire monitor that displays a warning when a flat tire is detected. Depending on the load of the vehicle and the type of tire, run-flat tires can typically be driven for about 100 miles (160 km) at up to 50 mph (80 km/h) even with the tire completely deflated.

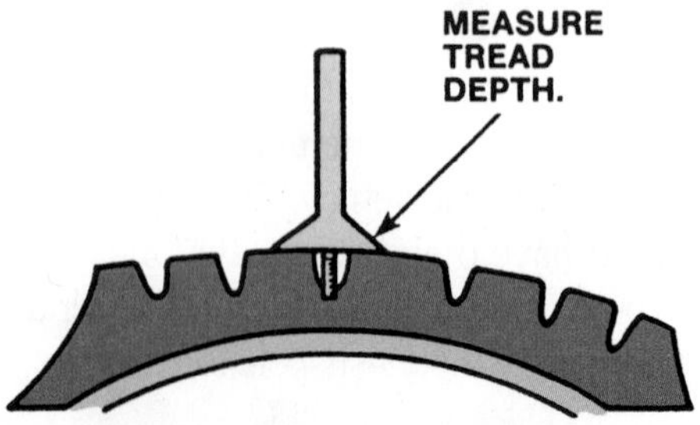

Figure 42-11 Checking tread depth.

Figure 42-12 A tire pressure monitor with its valve stem. *Courtesy of Adaptive Activity Network BV*

The sidewalls of these tires feature supplementary reinforcement belts manufactured in a special, heat-resistant rubber compound. This design allows the deflated tire to cover the specified distance without compressing against the wheel's rim flange. The contribution provided by the specially shaped safety rim ensures the tire remains securely seated **(Figure 42–13)**. Together these two features allow the driver to continue traveling without installing the spare tire.

WARNING!

Special rim clamps must be used for mounting run-flat tires.

The flat-tire monitor uses the rotation rate of each tire as its index for detecting pressure loss; the reduction in rolling radius that accompanies pressure loss produces a change in the tire's rotation speed. The ABS wheel speed sensors monitor rotation rates of each wheel. If there is a pressure loss, the system turns on a warning lamp. On some vehicles, a graphic will display which tire is flat.

Safety tires are labeled on the tire sidewall with a circular symbol containing the letters "RSC." Safety tires

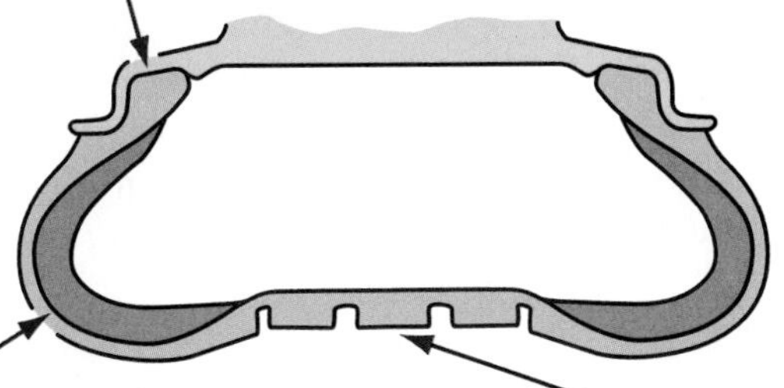

Figure 42-13 Features of a run-flat tire.

consist of self-supporting tires and special rims **(Figure 42–14)**. The tire reinforcement ensures that the tire retains some residual safety in the event of pressure drop and that driving remains possible to a restricted degree. Vehicles with a TPM system or flat-tire monitor use these tires.

Figure 42-14 A cutaway of a run-flat tire with an insert for support in case the tire goes very flat. *Courtesy of the Goodyear Tire Company*

TIRE REPAIR

The most common tire problem besides wear is a puncture. When properly repaired, the tire can be put back in service without the fear of an air leak recurring. Punctures in the tread area are the only ones that should be repaired or even attempted to be repaired. Never attempt to service punctures in the tire's shoulders or sidewalls. In addition, do not service any tire that has sustained the following damage:

- Bulges or blisters
- Ply separation
- Broken or cracked beads
- Fabric cracks or cuts
- Wear to the fabric or visible wear indicators
- Punctures larger than ¼-inch (6-mm) diameter

Some car owners attempt to seal punctures with tire sealants. These sealants are injected into the tire through the valve stem. Sometimes the chemicals in the sealant do a great job sealing the hole, other times they fail. The sealants should never be used and will not work on sidewall punctures. Some of the sealants are very flammable and carry a warning that the tire should be marked so that the next technician knows the sealant has been used.

WARNING!

Tire sealants injected through the valve stem can produce wheel rust and tire imbalance.

To locate a puncture in a tire, inflate it to the maximum inflation pressure indicated on its sidewall. Then submerge the tire/wheel assembly in a tank of water or sponge it with a soapy water solution. Bubbles will identify the location of any air leakage.

Mark the location of the leak with a crayon so it can be easily found once the tire is removed from the wheel. Also use the crayon to mark the location of the valve stem so that original tire and wheel balance can be maintained after the tire is put back on the wheel.

The proper procedure for dismounting and remounting a tire is illustrated in Photo Sequence 45. Do not use hand tools or tire irons alone to change a tire because they might damage the beads or wheel rim. When mounting or dismounting tires on vehicles using aluminum or wire spoke wheels, it is recommended that the tire changer manufacturer be contacted about the accessories that are required to protect the wheel's finish.

Repair Methods

Once the tire is off the wheel and the cause of the puncture is removed, the tire can be permanently serviced from the inside using a combination service plug and vulcanized patch. While the service kit's instructions should always be followed, there are some general guidelines that help make a good, permanent patch of the puncture. The following methods are the most common methods used to repair a tire.

Plug Repair The head-type plug **(Figure 42–15)** is commonly used. A plug that is slightly larger than the size of the puncture is inserted into the hole from the inside of the tire with an insertion tool. Before doing this, insert the plug into the eye of the tool and coat the hole, plug, and tool with vulcanizing fluid.

While holding and stretching the long end of the plug, insert it into the hole. The plug must extend above both the tread and inner liner surface. If the plug pops through, throw it away and insert a new plug. Once the plug is in place, remove the tool and trim off the plug 1⁄32 inch (0.7 mm) above the inner surface. Be careful not to pull on the plug while cutting it.

Cold Patch Repair When using a cold patch, carefully remove the backing from the patch. Spread vulcanizing fluid on the punctured area. Let it dry, then center the patch base over the punctured area. Run a stitching tool over the patch to help bind it to the tire.

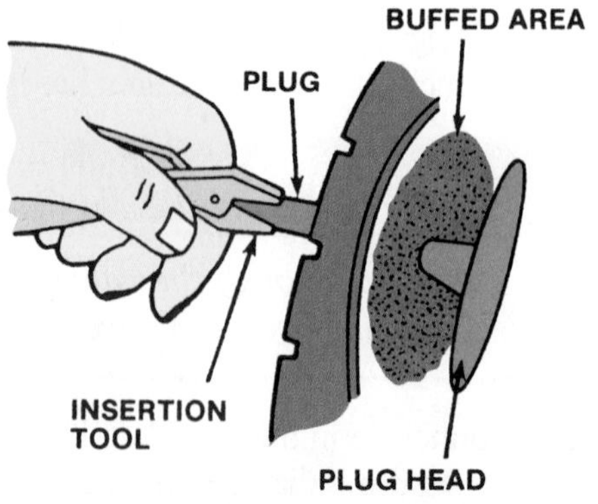

Figure 42-15 A plug for a radial tire.

WARNING!

When repairing radial tires, use only a patch specially approved for that application. These special patches have arrows that must be lined up parallel to the radial plies.

Hot Patch Repair A hot tire patch application is similar to a cold patch. The difference is that the hot patch is clamped over the puncture and heat is applied to the patch to make it adhere.

Installation of Tire/Wheel Assembly on the Vehicle

A wheel should be carefully inspected each time a tire is to be mounted on it. The major causes of wheel failure are improper maintenance, overloading, age, and accidents, including pothole damage. Wheels must be replaced when they are bent, dented, or heavily rusted; have leaks or elongated bolt holes; and have excessive **lateral** or **radial runout**. Wheels with lateral or radial runout greater than specifications can cause high-speed vibrations. Wobble or shimmy caused by a damaged wheel eventually damages the wheel bearings. Stones wedged between the wheel and disc brake rotor or drum can unbalance the wheel. Also, check the lug nuts to be sure that they are set according to the torque given in the vehicle's service manual. Loose lug nuts can cause shimmy and vibration and can also distort the stud holes in the wheels. Another wheel mounting problem is caused by improperly positioning the wheel on the wheel hub or by improperly tightening the lug nuts.

SHOP TALK

When mounting new tires, always install new valve stems. The life of tire rubber is close to the life of the valve stem rubber. Most stems are the snap-in type. These are installed from inside the wheel with a pulling tool. Make sure that the stem is properly seated. Another style of stem has a retaining nut that must be removed when pulling off the old stem. Be sure to completely tighten the new nut. ■

Before reinstalling a tire/wheel assembly on a vehicle, inspect the wheel bearings as described later in this chapter, then clean the axle/rotor flange and wheel bore with a wire brush or steel wool. Coat the axle pilot flange with disc brake caliper slide grease or an equivalent.

PHOTO SEQUENCE 45

Dismounting and Mounting a Tire on a Wheel Assembly

P45-1 *Dismounting the tire from the wheel begins with releasing the air, removing the valve stem core, and unseating the tire from its rim. The machine does the unseating. The technician merely guides the operating lever.*

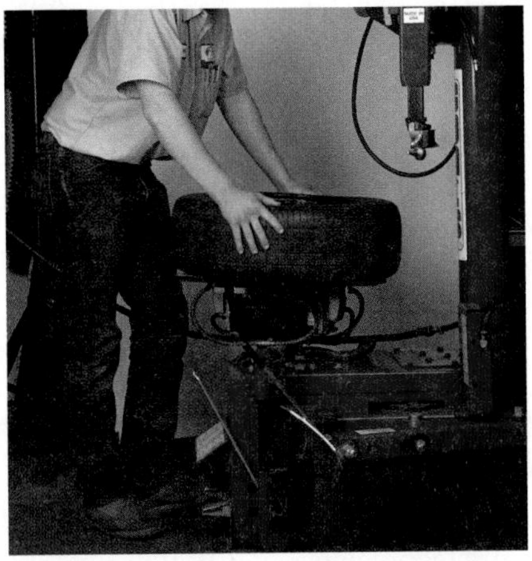

P45-2 *Once both sides of the tire are unseated, place the tire and wheel onto the machine. Then depress the pedal that clamps the wheel to the tire machine.*

P45-3 *Lower the machine's arm into position on the tire and wheel assembly.*

P45-4 *Insert the tire iron between the upper bead of the tire and the wheel. Depress the pedal that causes the wheel to rotate. Do the same with the lower bead.*

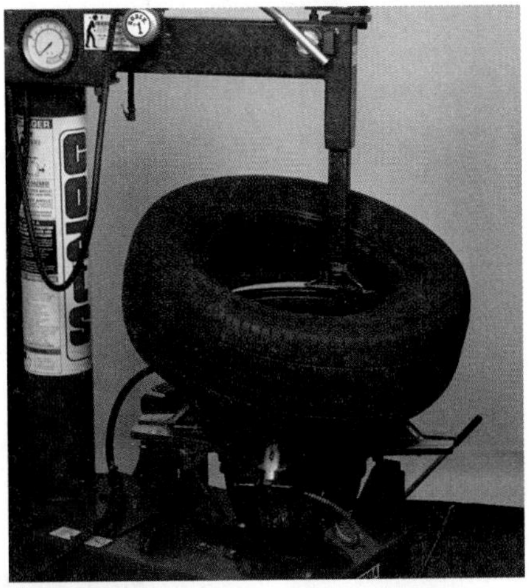

P45-5 *After the tire is totally free from the rim, remove the tire.*

P45-6 *Prepare the wheel for the mounting of the tire by using a wire brush to remove all dirt and rust from the sealing surface. Apply rubber compound to the bead area of the tire.*

P45-7 *Place the tire onto the wheel and lower the arm into place. As the machine rotates the wheel, the arm will force the tire over the rim. After the tire is completely over the rim, install the air ring over the tire. Activate it to seat the tire against the wheel.*

P45-8 *Reinstall the valve stem core and inflate the tire to the recommended inflation.*

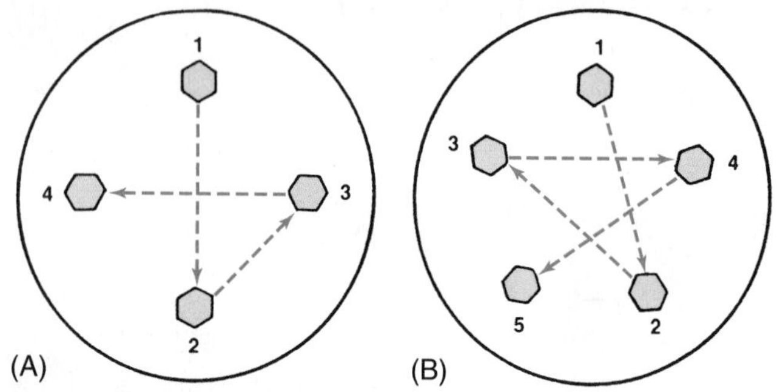

Figure 42-16 The lug nut tightening sequence for a four-lug wheel (A) and a five-lug wheel (B).

Place the wheel on the hub. Install the locking wheel-cover pedestal (if used) and lug nuts, and tighten them alternately to draw the wheel evenly against the hub. They should be tightened to a specified torque and sequence **(Figure 42–16)** to avoid distortion. Many tire technicians snug up the lug nuts. Then when the car is lowered to the floor, they use a torque wrench for the final tightening.

WARNING!

Overtorqueing of the lug nuts is the most common cause of disc brake rotor distortion. Also an overtorqued lug distorts the threads of the lug and could lead to premature failure.

Be sure the wheels and hub are clean. To clean aluminum wheels, use a mild soap and water solution and rinse thoroughly with clear water. Do not use a steel wool abrasive cleaner or strong detergent containing high alkaline or caustic agents because they might damage the protective coating and cause discoloration. Once the vehicle is on the ground, check and adjust the air pressure in all tires.

TIRE/WHEEL RUNOUT

A tire that is off center is said to run out; that is, it has radial runout or eccentricity. One that wobbles side to side is said to have lateral runout. If a tire with some built-in runout is mismatched with a wheel's runout, the resulting total runout can exceed the ability of the balance weights to correct the problem. For this reason, part of a wheel balance check should be a check for excessive runout. Sometimes tires or wheels can be remounted to lessen or correct runout problems.

To avoid false readings caused by temporary flat spots in the tires, check runout only after the vehicle has been driven. Visually inspect the tire for abnormal bulges or distortions. The extent of runout should be measured with a dial indicator. All measurements should be made on the vehicle with the tires inflated to recommended load inflation pressures and with the wheel bearing adjusted to specification.

Measure tire radial runout at the center and outside ribs of the tread face. Measure tire lateral runout just above the buffing rib on the sidewall **(Figure 42–17)**. Mark the high points of lateral and radial runout for future references. On bias or belted bias tires, radial runout must not exceed 0.104 inch (2.64 mm) and lateral runout must not exceed 0.099 inch (2.5 mm). On radial ply tires, radial runout must not exceed 0.081 inch (2.05 mm) and lateral runout must not exceed 0.099 inch (2.5 mm).

If total radial or lateral runout of the tire exceeds specified limits, it is necessary to check wheel runout to determine whether the wheel or tire is at fault. Wheel radial runout is measured at the wheel rim just inside the wheelcover retaining nibs. Wheel lateral runout is measured at the wheel rim bead flange just inside the curved lip of the flange. Wheel radial runout should not exceed 0.035 inch (.88 mm) and wheel lateral runout should not

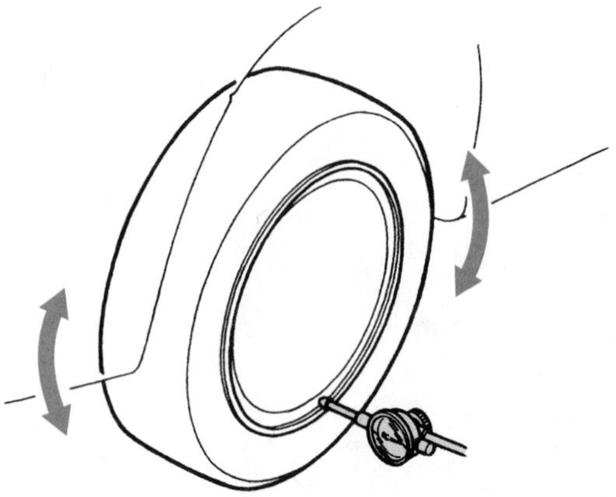

Front and Rear Wheel Radial Runout:
Standard:
Steel Wheel: **0 – 1.0 mm (0 – 0.04 in)**
Aluminum Wheel: **0 – 0.7 mm (0 – 0.03 in)**
Service Limit: **1.5 mm (0.06 in)**

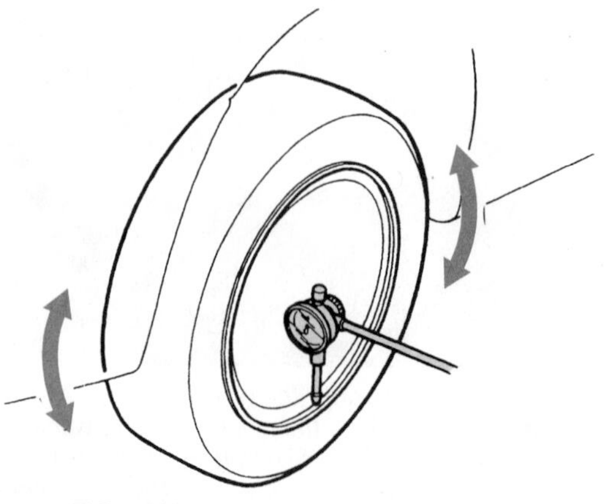

Figure 42-17 Checking wheel runout. *Courtesy of American Honda Motor Co., Inc.*

exceed 0.040 inch (1.0 mm). Mark the high points of radial and lateral runout for future reference.

If total tire runout, either lateral or radial, exceeds the specified limit but wheel runout is within the specified limit, it might be possible to reduce runout to an acceptable level. This is done by changing the position of the tire on the wheel so that the previously marked high points are 180 degrees apart.

TIRE/WHEEL ASSEMBLY SERVICE

For most tire/wheel service, the assembly must first be removed from the vehicle. The wheel and the tire must be separated whenever tires are placed or repaired. The rear-wheel drum or disc brake rotor is usually attached to studs on the rear axle shaft's hub flange. The wheel and tire mount on the same studs and are held against the hub and drum or rotor by the wheel nuts.

Tire/Wheel Balance

Proper wheel alignment allows the tires to roll straight without excessive tread wear. The wheels can go out of alignment from striking raised objects or potholes. Misalignment subjects the tires to uneven and/or irregular wear. An out-of-balance condition can also cause increased wear on the ball joints, as well as deterioration of shock absorbers and other suspension components.

Should an inspection show uneven or irregular tire wear, wheel alignment and balance service is a must. Wheel balancing distributes weights along the wheel rim, which counteract heavy spots in the wheels and tires and allow them to roll smoothly without vibration. The wheel weights are adhered to the wheel or are clipped over the edge of the wheel's rim. There are two types of wheel imbalance: static and dynamic.

Static Balance **Static balance** is the equal distribution of weight around the wheel. Wheels that are statically unbalanced cause a bouncing action called **wheel tramp**. This condition eventually causes uneven tire wear. As the name implies, static balance means balancing a wheel at rest. This is done by adding a compensating weight. A statically unbalanced wheel tends to rotate by itself until the heavy portion is down. A bubble balancer is used to statically balance a tire and wheel. When it is placed on the balancer, any imbalance moves the bubble off center.

Many equipment manufacturers recommend static balancing a wheel at equal distances from the center of the light area. Balance weights are normally hammered on with their holding tabs between the tire bead and rim **(Figure 42–18)**. Wheel weights are not normally hammered onto alloy or mag wheels, rather special tape weights are adhered to the wheels to balance them.

Figure 42-18 A typical wheel weight attached to a wheel.

Dynamic Balance **Dynamic balance** is the equal distribution of weight on each side of the centerline. When the balanced tire spins, there is no tendency for the assembly to move from side to side. Wheels that are dynamically unbalanced can cause **wheel shimmy** and a wear pattern **(Figure 42–19)**. Dynamic balance, simply stated, means balancing a wheel in motion. Once a wheel starts to rotate and is in motion, the static weights try to reach the true plane of rotation of the wheel because of the action of centrifugal force. In an attempt to reach the true plane of rotation when there is an imbalance, the static weights force the spindle to one side.

At 180 degrees of wheel rotation, static weights kick the spindle in the opposite direction. The resultant side thrusts cause the wheel assembly to wobble or wiggle. When the imbalance is severe enough, as already mentioned, it causes vibration and front-wheel shimmy.

To correct dynamic unbalance, equal weights are placed 180 degrees opposite each other, one on the inside of the wheel and one on the outside, at the point of unbalance. This corrects the couple action or wiggle of the wheel assembly. Also, note that dynamic balance is obtained, while static balance remains unaffected.

The most commonly used dynamic wheel balancer requires that the tire/wheel assembly be taken off and mounted on the balancer's spindle **(Figure 42–20)**. The machine spins the entire assembly to indicate the heavy spot with a strobe light or other device. Two tests must be done, one for static and one for dynamic imbalance. One set of weights is placed to correct for static imbalance, and others are placed to correct for dynamic imbalance. Sometimes proper positioning of the static balance weights also corrects dynamic imbalance.

There are several electronic dynamic/static balancer units available that will permit balancing while the wheel

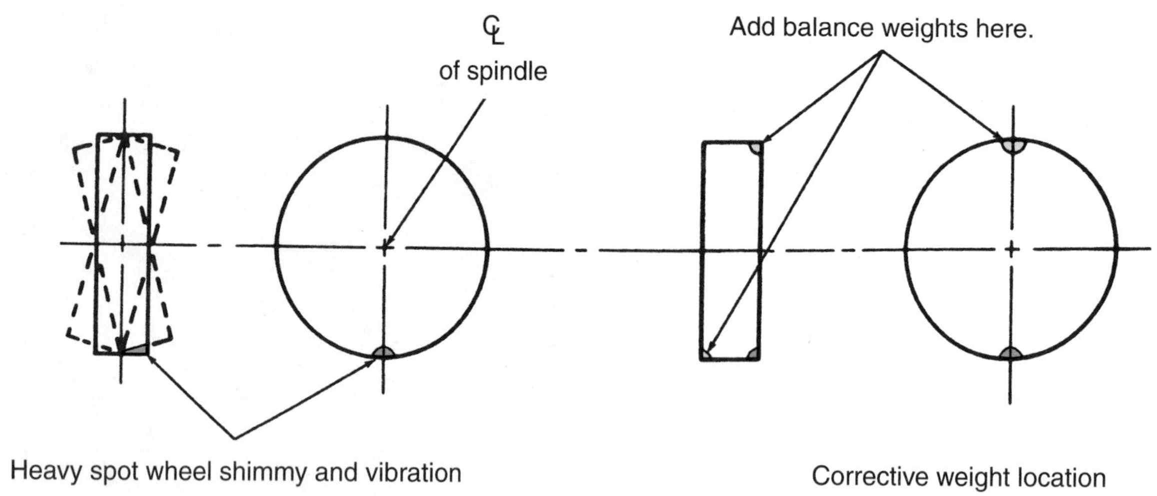

Figure 42-19 Dynamic wheel balancing calls for a weight to be attached to the wheel to compensate for a heavy spot on the wheel. *Courtesy of DaimlerChrysler Corporation*

and tire are on the car. A switch on the console sets the machine for either static or dynamic balancing. When the wheel balancing assembly is mounted for static balancing, it rotates until the heavy spot falls to the bottom. Weights are added to balance the assembly.

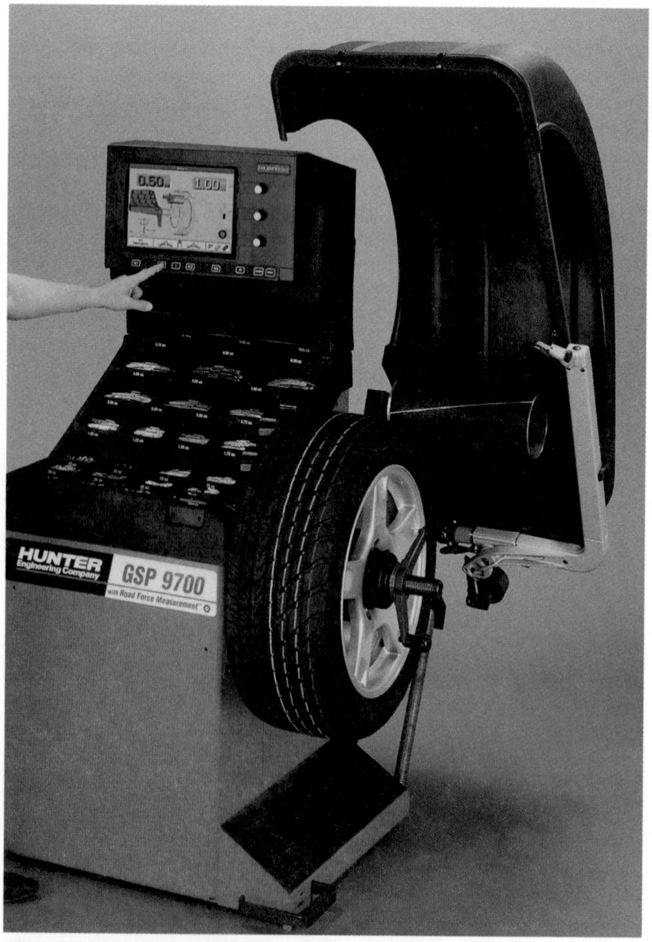

Figure 42-20 A computerized tire balancer. *Courtesy of Hunter Engineering Company*

In the dynamic balance mode, the wheel assembly is rotated at high speed. Observing the balance scale, the operator reads out the amount of weight that has to be added and the location where the weights should be placed.

WHEEL BEARINGS

The purpose of all bearings is to allow a shaft to rotate smoothly in a housing or to allow the housing to rotate smoothly around a shaft. Wheel and axle bearings do this for a vehicle's wheels. Typically, the bearings on the axle that drives the wheels are called *axle bearings*. The wheel is mounted to the hub of an axle shaft and the shaft rotates within a housing. Wheel bearings are used on nondriving axles. The wheel's hub typically rotates on a shaft called the spindle. Axle bearings are typically serviced with the drive axle. Wheel bearings, however, require periodic maintenance service and are often serviced with suspension and brake work. Although there is a distinction between axle and wheel bearings, the bearings for the front wheels on a FWD and 4WD vehicle are commonly called *wheel bearings*. Regardless of what they are called, bad bearings will cause handling and tire wear problems.

Front Wheel Hubs

Often the front wheel hub bearing assembly for driven and nondriven wheels is actually two tapered bearings **(Figure 42–21)** facing each other. Each of the bearings rides in its own race. Some front wheel bearings are sealed units and are lubricated for life. They are replaced and serviced as an assembly. Others are serviceable and require periodic lubrication and adjustment.

Except when making slight adjustments to the bearings, the bearing assembly must be removed for all service work. This is done with the vehicle on lifts and the

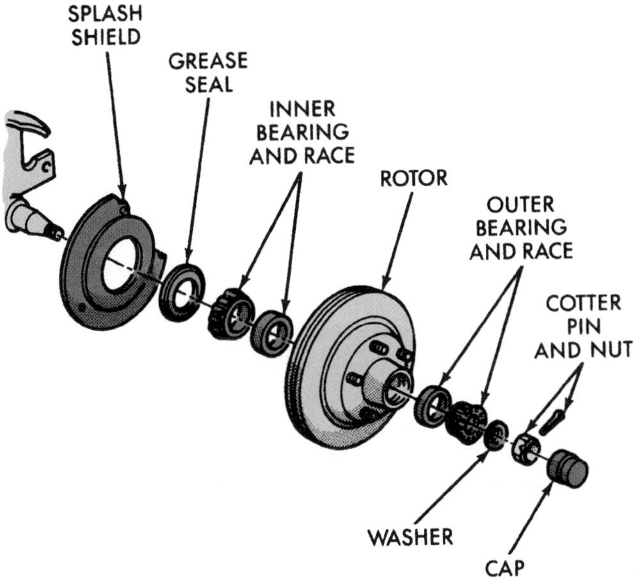

Figure 42–21 An exploded view of a typical front wheel bearing assembly for a RWD vehicle. *Courtesy of DaimlerChrysler Corporation*

wheel assembly removed. In the center of the hub there is a dust (grease) cap. Using slip-joint pliers or a special dust cap removal tool **(Figure 42–22)**, wiggle the cap out of its recess in the hub. Now remove the cotter pin and nut lock from the end of the spindle. Loosen the spindle nut while supporting the brake assembly and hub. On many vehicles you will need to remove the brake caliper to remove the brake disc and hub. Once the hub is free to come off the spindle, remove the spindle nut and the washer behind the nut. Move the hub slightly forward, then push it back. This should free the outer bearing so you can remove it. Now remove the hub assembly.

A grease seal located on the back of the hub normally keeps the inner bearing from falling out when the hub is removed. To remove the bearing assembly, the grease must be removed first. In most cases, all you need to do

Figure 42–22 A special tool for removing a dust cap. If one is not available, use slip-joint pliers.

to remove the seal is pry and pop it out of the hub. The inner bearing should then fall out. Keep the outer bearing and inner bearing separated if you plan on reusing them.

Wipe the bearings and races or use brake parts cleaner to clean them. While doing this, pay close attention to the condition and movement of the bearings. The bearings need to rotate smoothly. Also visually inspect the bearings and races after they have been cleaned. Any noticeable damage means they should be replaced. Also inspect the spindle. If it is damaged or excessively worn, the steering knuckle assembly should be replaced.

Whenever a bearing is replaced, its race must also be replaced. Races are pressed in and out of the hub. Typically the old race can be driven out with a large drift and a hammer. Once the race has been removed, wipe all grease from the inside of the hub. The new race should be installed with the proper driver.

During assembly, the bearings and hub assembly must be thoroughly and carefully lubricated. Care must be taken not to get grease on the brake disc or on any part that will directly contact the disc. Always use the recommended grease on this assembly. The grease must be able to withstand much heat and friction. If the wrong grease is used, it may not offer the correct protection or it may liquefy from the heat and leak out of the seals.

The bearings should be packed with grease. It is important that the grease is forced into and around all of the rollers in the bearing. Merely coating the outside of the bearing with grease will not do the job. A bearing packer does the best job at packing in the grease. If one is not available, force grease into the bearing with your hand.

Install the greased inner bearing into the hub. Install a new grease seal into the hub. To avoid damaging the seal, use the correct size driver to press the seal into the hub. Lubricate the spindle, then slip the hub over the spindle. Install the outer bearing, washer, and lock nut. The lock nut should be adjusted to the exact specifications given in the service manual **(Figure 42–23)**. Often it is tightened until the hub cannot rotate, then it is loosened about one-half turn before it is set to the specified free play. The initial tightening seats the bearings into their races. Once the lock nut is tightened, install the nut lock and use a new cotter pin to retain the lock.

The adjustment of the bearings can be checked with a dial indicator. Mount the base of the indicator as close as possible to the center of the hub. Locate the tip of the indicator's plunger on the tip of the spindle. Set the indicator to zero. Firmly grasp the brake disc and move it in and out. The total movement shown on the indicator is the amount of free play at the bearing. Compare your reading to the specifications and make adjustments as necessary.

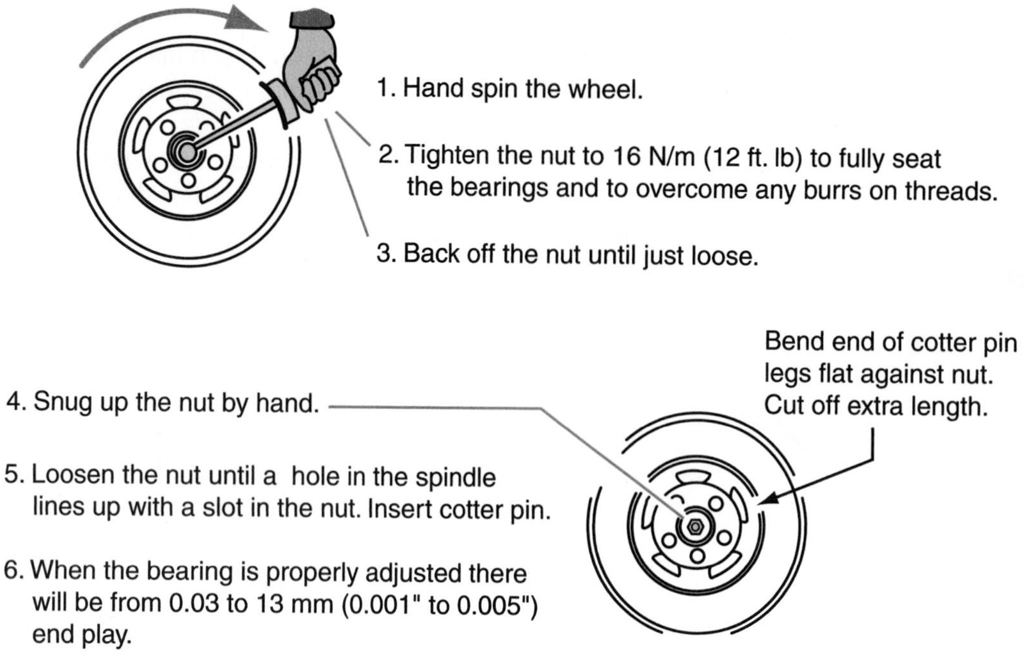
1. Hand spin the wheel.
2. Tighten the nut to 16 N/m (12 ft. lb) to fully seat the bearings and to overcome any burrs on threads.
3. Back off the nut until just loose.
Bend end of cotter pin legs flat against nut. Cut off extra length.
4. Snug up the nut by hand.
5. Loosen the nut until a hole in the spindle lines up with a slot in the nut. Insert cotter pin.
6. When the bearing is properly adjusted there will be from 0.03 to 13 mm (0.001" to 0.005") end play.

Figure 42-23 A typical procedure for adjusting wheel bearings.

WARNING!

Throughout this entire process, your hands will have grease on them. Be very careful not to touch the brake assembly with your greasy hands. Clean them before handling the brake parts or use a clean rag to hold the brake assembly.

The front bearing arrangement often found on FWD and 4WD vehicles is often nonserviceable. These bearings are pressed in and out of the steering knuckle. To do this, the axle or half shaft is removed, as is the steering knuckle and hub assembly. The bearings may be sealed and require no additional lubrication or they may need to be packed with grease when they are reassembled. In most cases, the bearings are not adjusted. A heavily torqued axle nut is used to hold the assembly in place on the axle. This nut is typically replaced after it has been removed and is staked in place after it is tightened.

SHOP TALK

Because an axle nut is heavily torqued, it is wise to loosen the nut before the vehicle is raised for service. The weight of the vehicle on the tires will stop the hub from rotating while the nut is being loosened. The same holds true for final torquing. Tighten the nut as tight as possible with the vehicle raised, then lower the vehicle and torque the nut to specifications. ■

Rear Hubs

The rear bearings on a FWD vehicle are serviced in the same way as the nondriving front wheel bearings. Most RWD axle bearings are of the straight roller bearing design, in which the drive axle tube serves as the bearing race. Some rear wheel axle bearings are of the ball or tapered roller bearing type.

Wheel Bearing Grease Specification

The grease for wheel bearings should be smooth textured, consist of soaps and oils, and be free of filler and abrasives. Recommended are lithium complex (or equivalent) soaps or solvent-refined petroleum oils. Additives could be used to inhibit corrosion and oxidation. The grease should be noncorrosive to bearing parts, with no chance of separating during storage or use.

Always make sure you use the grease recommended by the manufacturer. Greases are classified by the National Lubricating Grease Institute (NLGI) to indicate their application. Failure to maintain proper lubrication might result in bearing damage, causing a wheel to lock **(Figure 42–24)**. To lubricate a bearing, force grease around the outside of the bearing and between the rollers, cone, and cage. Also pack grease into the wheel hub. The depth of the grease should be level with the outside diameter of the cup.

Bearing Troubleshooting

Wheel bearings are designed for longevity. Their life expectancy, based on metal fatigue, can usually be calculated if general operating conditions are known. Bearing failures not caused by normal material fatigue are called

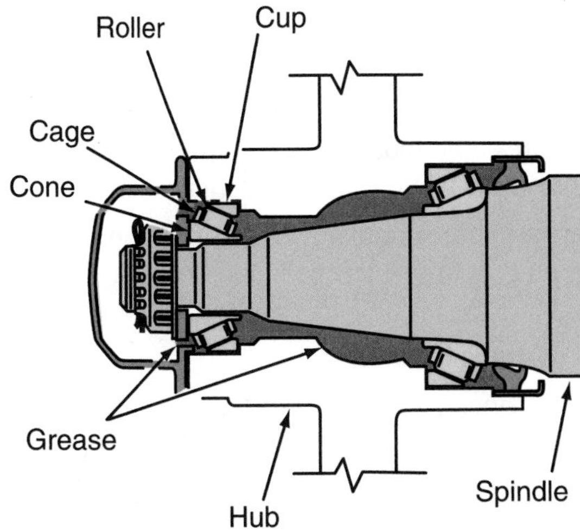

Figure 42-24 Wheel bearing lubrication.

permanent failures. The causes can range from improper lubrication or incorrect mounting, to poor condition of the shaft housing or bearing surfaces.

CASE STUDY

A customer complains of a bearing noise from the right front of his late-model 2WD pickup. The customer says this bearing has already been replaced twice in the last year and he is not very happy about paying for it again. Since not many miles have expired between bearing replacements, something must be causing the premature bearing failure.

The technician removes the right front wheel and hub. The outer bearing and race are heavily scored. After he cleans the bearing for a closer look at the wear, he notices that the bearing has an uneven wear pattern, which indicates poor bearing alignment. The same pattern is visible on the bearing race. Both must be replaced. After he removes the race from the hub, the technician inspects the hub only to find a small metal burr behind the bearing race. This small burr is the cause of the misalignment of the bearing. The burr is removed with a fine file, inner and outer bearing assemblies are installed, and the truck taken for a test drive.

There was no noise from the bearing during the test drive. The truck was returned to the customer. After six months, the technician called the customer to find out if the bearing noise had reappeared and found that it had not. Isn't it funny how the little things, like a small metal burr, can make life difficult? In this case the burr had frustrated the technician and the customer.

When servicing, replacing, or installing wheel bearings, always follow the procedure given in the service manual.

KEY TERMS

Bead
Belted bias ply
Bias ply
Drop center
Dynamic balance
Hub
Lateral runout
Lug nuts
Plies
Radial ply
Radial runout
Series
Sidewall
Sipes
Static balance
Tire profile
Tread
Valve stem
Wheel offset
Wheel shimmy
Wheel tramp

SUMMARY

- Wheels are made of either stamped or pressed steel discs riveted or welded into a circular shape or are die-cast or forged aluminum or magnesium rims.
- The primary purpose of tires is to provide traction. They are also designed to carry the weight of the vehicle, to withstand side thrust over varying speeds and conditions, to transfer braking and driving torque to the road, and to absorb much of the rock shock from surface irregularities.
- Pneumatic tires are of two types: those that use inner tubes and those that do not. The latter are called tubeless tires and are about the only type used on passenger cars today.
- There are three types of tire construction in use today: bias ply, belted bias ply, and radial ply.
- Tires are rated by their profile, ratio, size, and load range.
- Tire construction affects both dimensions and ride characteristics, creating differences that can seriously affect vehicle handling.
- An ideal tire is one that wears little, holds the road well to provide sure handling and braking, and provides a cushion from road shock. It should also provide maximum grip on dry roads, wet roads, and snow and ice, and operate quietly at any speed.
- There are basically three categories of tread patterns: directional, nondirectional, and symmetric and asymmetric. A directional tire is mounted so that it revolves in a particular direction and will perform well only when it is rotating in that direction. A

nondirectional tire has the same handling qualities in either direction of rotation. A symmetric tire has the same tread pattern on both sides of the tire. An asymmetrical tire has a tread design that is different from one side to the other.

- The number and size of the blocks, sipes, and grooves on a tire's tread determines how much rubber contacts the road, how much water can be displaced, and how quiet the tire will be during travel.
- To maximize tire performance, inspect for signs of improper inflation and uneven wear, which can indicate a need for balancing, rotation, or alignment. Tires should also be checked frequently for cuts, bruises, abrasions, and blisters, and for stones or other objects that might have become imbedded in the tread.
- A properly inflated tire gives the best tire life, riding comfort, handling stability, and even gas mileage during normal driving conditions.
- To equalize tire wear, most car and tire manufacturers recommend that the tires be rotated. It must be remembered that front and rear tires perform different jobs and can wear differently, depending on driving habits and the type of vehicle.
- Most tires used today have built-in tread wear indicators to show when tires need replacement.
- There are three popular methods of tire repair: head-type plug, cold patch repair, and hot patch repair.
- There are two types of wheel balancing: static balance and dynamic balance.
- The bearings on an axle that drives the wheels are called axle bearings. Wheel bearings are used on nondriving axles. Axle bearings are typically serviced with the drive axle. Wheel bearings require periodic maintenance service and are often serviced with suspension and brake work.
- Bad axle and wheel bearings will cause handling and tire wear problems.
- The front wheel hubs on ball or tapered roller bearings are lubricated by wheel bearing grease.
- Rear wheels are bolted to integral or detachable hubs.

TECH MANUAL

The following procedures are included in Chapter 42 of the *Tech Manual* that accompanies this book:

1. Inspecting tires for inflation and wear.
2. Removing and installing front wheel bearings on a rear-wheel-drive vehicle.
3. Balancing a tire and wheel off the vehicle.

REVIEW QUESTIONS

1. How is a modern tire wheel constructed? What materials are used?

2. Define dynamic and static wheel balance.

3. What is the section of the wheel called that allows the tire to be easily removed and installed?

4. Describe the procedure for using a plug to seal a puncture in a tire.

5. What type of balance is checked on a bubble balancer?

6. What can cause an increase in a tire's rolling resistance and why is this important?

7. Series is another name for a tire's ______.

a. size
b. load range
c. bead
d. profile

8. The three belting systems available for today's tires are ______.

a. steel, nylon, and fiberglass
b. lateral, cut, and block
c. cut, folded, and woven
d. folded, straight ahead, and turning

9. Technician A says the wheel bearings on a typical FWD car are pressed in and out of the steering knuckle assembly. Technician B says the rear wheel bearings on a typical FWD car are serviced in the same way as the front bearings for a RWD car. Who is correct?

a. Technician A
b. Technician B
c. Both A and B
d. Neither A nor B

10. Where does an overinflated tire show its wear?

a. in the center of the tread
b. on the outside edges

c. both a and b

d. neither a nor b

11. Which of the following is a major cause of wheel failure?
   a. overloading
   b. age
   c. improper maintenance
   d. all of the above

12. Technician A uses tire sealant injected through the valve stem to repair a tire punctured in the tread area. Technician B uses tire sealant injected through the valve stem to repair a tire punctured in the sidewall. Who is correct?
   a. Technician A
   b. Technician B
   c. Both A and B
   d. Neither A nor B

13. Technician A says dynamic wheel unbalance causes wheel shimmy. Technician B says static wheel unbalance causes wheel tramp. Who is correct?
   a. Technician A
   b. Technician B
   c. Both A and B
   d. Neither A nor B

14. Most information about a car's tires can be found on the ______.
   a. engine block
   b. tire placard
   c. undercarriage of the vehicle
   d. steering wheel

15. A tire that wobbles from side to side is said to have ______.
   a. radial runout
   b. lateral runout
   c. eccentric runout
   d. none of the above

16. Technician A says front bearing assembly lock nuts for RWD vehicles are typically heavily torqued to maintain the bearing adjustment. Technician B says the axle nut for a FWD wheel bearing assembly is often staked in place to maintain its specified torque. Who is correct?
   a. Technician A
   b. Technician B
   c. Both A and B
   d. Neither A nor B

17. Which of the following statements is *not* true? ______
   a. A cold patch should be stitched to the inner surface of the carcass.
   b. A tread puncture that is less than ¼ inch (6.35 mm) in diameter is repairable.
   c. Wheel balancing is required only when new tires are installed.
   d. For best results, punctures should be repaired from inside the tire.

18. Which of the following statements is *not* correct? ______
   a. Belts are reinforcing materials that encircle the tire under the tread.
   b. The carcass of a tire is made up of plies, layers of cloth, and rubber.
   c. Most tread patterns are designed to work well on both wet and dry roads.
   d. The bead is the decorative pattern at the outer edge of the tread.

19. Which of the following statements about sidewall markings is correct? ______
   a. Aspect ratio, or profile, indicates a grading for tread life.
   b. Important information and safety warnings are indicated.
   c. The tire's maximum safe inflation pressure and load are indicated.
   d. Section width is the distance between the tread and the bead.

20. Which of the following statements is *not* correct? ______
   a. Tire inflation pressure directly affects traction.
   b. Recommended tire pressure for front and rear tires may be different.
   c. Recommended tire pressures are often lower than maximum pressures.
   d. Recommended tire pressure is molded into the sidewall of the tire.

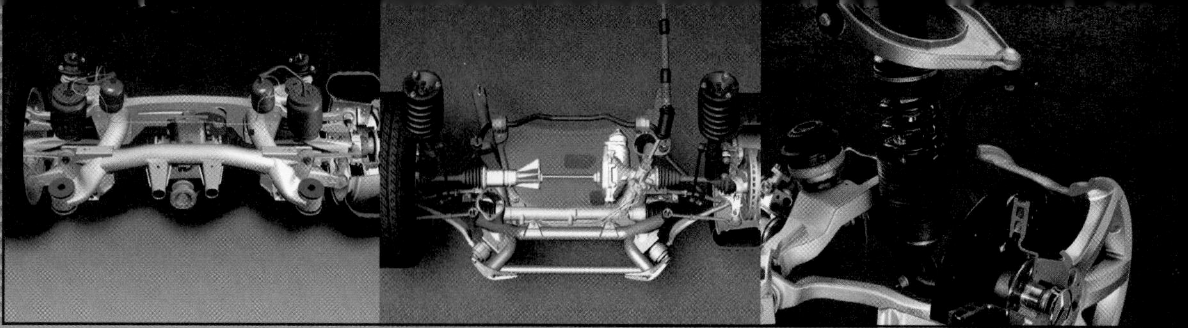

43 SUSPENSION SYSTEMS

OBJECTIVES

■ Name the different types of springs and how they operate. ■ Name the advantages of ball joint suspensions. ■ Explain the important differences between sprung and unsprung weight with regard to suspension control devices. ■ Identify the functions of shock absorbers and struts and describe their basic construction. ■ Identify the components of a MacPherson strut system and describe their functions. ■ Identify the functions of bushings and stabilizers. ■ Perform a general front suspension inspection. ■ Check chassis height measurements to specifications. ■ Identify the three basic types of rear suspensions and know their effects on traction and tire wear. ■ Identify the various types of springs, their functions, and their locations in the rear axle housing. ■ Describe the advantages and operation of the three basic electronically controlled suspension systems: level control, adaptive, and active. ■ Explain the function of electronic suspension components including air compressors, sensors, control modules, air shocks, electronic shock absorbers, and electronic struts. ■ Explain the basic towing, lifting, jacking, and service precautions that must be followed when servicing air springs and other electronic suspension components.

Like the rest of the systems on cars and light trucks, the suspension system has greatly changed through the years. Not only has technology brought about these changes, so has the quest for great handling and comfortable vehicles. Suspension systems for the front and rear of a vehicle can get quite complex **(Figure 43–1)**.

The reason suspension systems are complex is that they perform a very complicated function. These systems must keep the vehicle's wheels lined up with the travel of the vehicle, limit the movement of the vehicle's body during cornering and when going over bumps, and provide a smooth and comfortable ride.

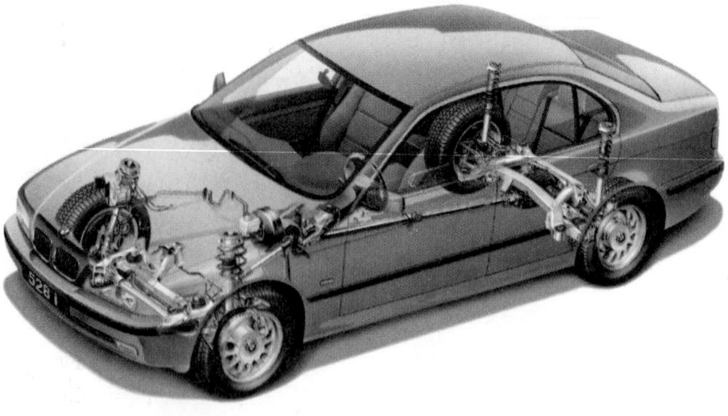

Figure 43–1 The suspension system for a late-model car. The front uses a strut setup, while the rear has a multilink system. *Courtesy of the BMW of North America, Incorporated*

FRAMES

To provide a rigid structural foundation for the vehicle body and a solid anchorage for the suspension system, a frame of some type is essential. There are two basic frames in common use today.

Conventional Frame Construction

In the conventional body-over-frame construction, the frame is the vehicle's foundation. The body and all major parts of a vehicle are attached to the frame. It must provide the support and strength needed by the assemblies and parts attached to it. In other words, the frame is an independent, separate component because it is not welded to any of the major units of the body shell.

Unibody Construction

Unibody construction has no separate frame. The body is constructed in such a manner that the body parts themselves supply the rigidity and strength required to maintain the structural integrity of the car. The unibody design significantly lowers the base weight of the car, and that, in turn, increases gas mileage capabilities.

SUSPENSION SYSTEM COMPONENTS

Most automotive suspension systems have the same basic components and operate similarly. Their differences are found in the method in which the basic components are arranged.

Springs

The spring is the core of nearly all suspension systems. It is the component that absorbs shock forces while maintaining correct riding height. If the spring is worn or damaged, the other suspension elements shift out of their proper positions and are subject to increased wear. The increased effect of shock impairs the vehicle's handling.

Various types of springs are used in suspension systems—coil, torsion bar, leaf (both mono- and multileaf types) **(Figure 43–2)** and air springs. Springs are rubber mounted to reduce road shock and noise.

Automotive springs are generally classified by the amount of deflection exhibited under a specific load. This is referred to as the **spring rate**. According to the law of physics, a force (weight) applied to a spring causes it to compress in direct proportion to the force applied. When that force is removed, the spring returns to its original position if not overloaded. Remember that a heavy vehicle requires stiffer springs than a lightweight car.

The springs take care of two fundamental vertical actions: jounce and rebound. **Jounce** occurs when a wheel hits a bump and moves up **(Figure 43–3A)**. When this happens, the suspension system acts to pull in the top of the wheel, maintaining an equal distance between the two front wheels and preventing a sideways scrubbing action as the wheel moves up and down. **Rebound** occurs when the wheel hits a dip or hole and moves downward **(Figure 43–3B)**. In this case, the suspension system acts to move the wheel in at both the top and bottom equally, while maintaining an equal distance between the wheels.

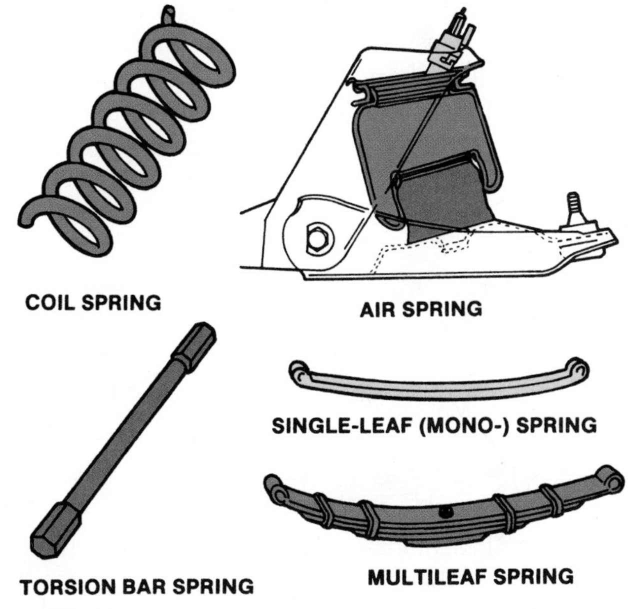

Figure 43–2 Various types of automotive springs.

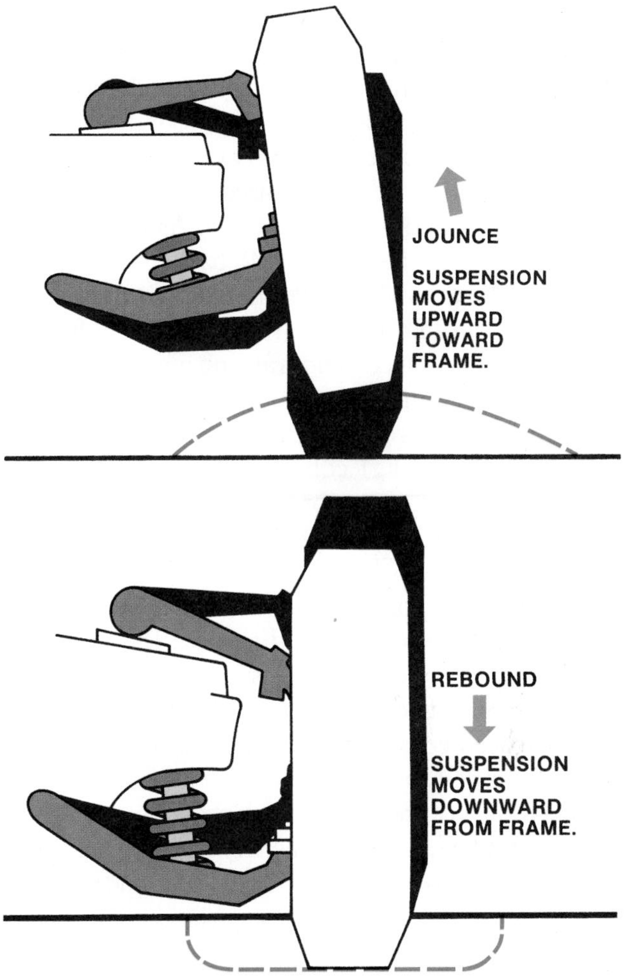

Figure 43–3 (*A*) Upward and (*B*) downward suspension movement.

The spring goes back and forth from jounce to rebound. Each time, jounce and rebound become smaller and smaller. This is caused by friction of the spring's molecular structure and the suspension pivot joints. A **shock absorber** is added to each suspension to dampen and stop the motion of the spring after jounce.

All of the vehicle's weight supported by the suspension system is known as **sprung weight**. The weight of those components not supported by the springs is known as **unsprung weight**. The vehicle's body, frame, engine, transmission, and all of its components are considered sprung weight. Undercar parts classified as unsprung weight include the steering knuckles and rear axle assemblies (but not always the differentials). Keep in mind that, in general, the lower the ratio of unsprung weight to sprung weight, the better the vehicle's ride will be.

Coil Springs Two basic designs of coil springs are used: linear rate and variable rate. **Linear rate** springs characteristically have one basic shape and a consistent wire diameter. All linear springs are wound from a steel rod into a cylindrical shape with even spacing between the coils. As the load is increased, the spring is compressed and the coils twist (deflect). As the load is removed, the coils flex (unwind) back to the normal position. The amount of load necessary to deflect the spring 1 inch (25.4 mm) is the spring rate. On linear rate springs this is a constant rate, no matter how much the spring is compressed. For example, 250 pounds (112 kg) compress the spring 1 inch (25.4 mm) and 750 pounds (340 kg) compress the spring 3 inches (76.2 mm). Spring rates for linear rate springs are normally calculated between 20% and 60% of the total spring deflection.

Variable rate spring designs are characterized by a combination of wire sizes and shapes. The most commonly used variable rate springs have a consistent wire diameter, are wound in a cylindrical shape, and have unequally spaced coils. This type of spring is called a progressive rate coil spring.

The design of the coil spacing gives the spring three functional ranges of coils: inactive, transitional, and active. Inactive coils are usually the end coils and introduce force into the spring. Transitional coils become inactive as they are compressed to their point of maximum load-bearing capacity. Active coils work throughout the entire range of spring loading. Theoretically in this type of design, at stationary loads the inactive coils are supporting all of the vehicle's weight. As the loads are increased, the transitional coils take over until they reach maximum capacity. Finally, the active coils carry the remaining overload. This allows for automatic load adjustment while maintaining vehicle height.

Another common variable rate design uses tapered wire to achieve this same type of progressive rate action. In this design, the active coils have a large wire diameter and the inactive coils have a small wire diameter.

Later designs of variable rate springs deviate from the old cylindrical shape. These include the truncated cone, the double cone, and the barrel spring. The major advantage of these designs is the ability of the coils to nest, or bottom out, within each other without touching, which lessens the amount of space needed to store the springs in the vehicle.

Unlike a linear rate spring, a variable rate spring has no predictable standard spring rate. Instead, it has an average spring rate based on the load of a predetermined spring deflection. This makes it impossible to compare a linear rate spring to a variable rate spring. Variable rate springs, however, handle a load of up to 30% over standard rate springs in some applications.

Servicing Coil Springs. A technician must know how to check and replace coil springs, select the proper replacement, and recommend the proper size and type of spring to the customer.

The first step in coil spring selection is to check for the original equipment part number. This is usually on a tag wrapped around the coil. In many instances, this tag falls off before replacement is necessary. If a set of aftermarket springs has been installed, the part number might be stamped on the end of the coil. Next, check to see what type of ends the coil springs have. There are three types of ends used in automotive applications: full wire open, tapered wire closed, and pigtail. Springs with full wire open ends are cut straight off and sometimes flattened or ground into a D or square shape. Tapered wire closed ends are wound to ensure squareness and ground into a taper at the ends. Pigtail ends are wound into a smaller diameter at the ends.

The final step is to check the application in the catalog. To do this, it is necessary to know the make, year, model, body style, and engine size (number of cylinders), and if the vehicle is equipped with air-conditioning. In some cases, it is also good to know the type of transmission, seating capacity, and other specifics that add extra weight to the vehicle. In most catalogs, springs are listed by vehicle application in two sections: front and rear.

Leaf Springs Although leaf springs were the first type of suspension spring used on automobiles, today they are generally found only on light-duty trucks, vans, and some passenger cars. There are three basic types of leaf springs: multiple leaf, monoleaf, and fiber composite.

Multiple-Leaf Springs. **Multiple-leaf** springs consist of a series of flat steel leaves that are bundled together and held with clips or by a bolt placed slightly ahead of the center of the bundle **(Figure 43–4)**. One leaf, called the **main leaf**, runs the entire length of the spring. The next leaf is a little shorter and attaches to the main leaf. The next leaf is shorter yet and attaches to the second leaf, and so on. This system allows almost any number of

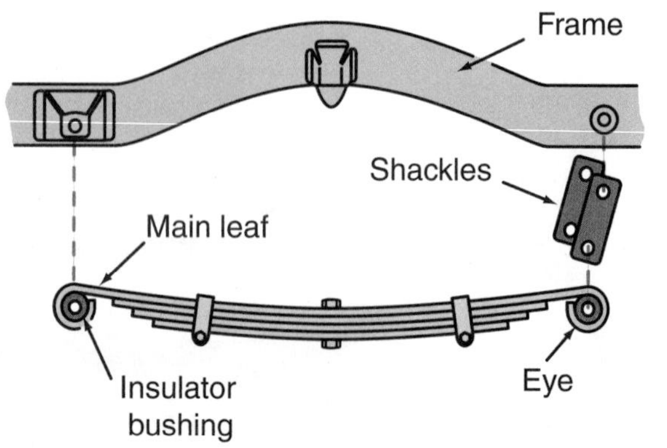

Figure 43–4 A leaf spring and its related hardware.

Figure 43–5 To provide more ground clearance, a lift kit that included ten leaf rear springs was installed on this pickup. *Courtesy of Superlift Suspension Systems*

leaves to be used to support the vehicle's weight **(Figure 43–5)**. It also gives a progressively stiffer spring. The spring easily flexes over small distances for minor bumps. The farther the spring is deflected, the stiffer it gets. The more leaves and the thicker and shorter the leaves, the stronger the spring. It must be remembered that as the spring flexes, the ends of the leaves slide over one another. This sliding could be a source of noise and can also produce friction. These problems are reduced by interleaves of zinc and plastic placed between the spring's leaves. As the multiple leaves slide, friction produces a harsh ride as the spring flexes. This friction also dampens the spring motion.

Multiple-leaf springs have a curve in them. This curve, if doubled, forms an ellipse. Thus, leaf springs are sometimes called semielliptical or quarter-elliptical. The semi or quarter refers to how much of the ellipse the spring actually describes. The vast majority of leaf springs are semielliptical.

Leaf springs are commonly mounted at right angles to the axle. In addition to absorbing blows from road forces, they also serve as suspension locators by fixing the position of the suspension with respect to the front and rear of the vehicle. A centering pin is frequently used to ensure that the axle is correctly located. If a spring is broken or misplaced, the axle might be mislocated and the alignment impaired.

The front eye of the main leaf at either end of the axle is attached to a bracket on the frame of the vehicle with a bolt and bushing connection. The rear eye of the main leaf is secured to the frame with a shackle, which permits some fore and aft movement in response to physical forces of acceleration, deceleration, and braking.

Monoleaf Springs. **Monoleaf** or single-leaf springs are usually the tapered plate type with a heavy or thick center section tapering off at both ends. This provides a variable spring rate for a smooth ride and good load-carrying ability. In addition, single-leaf springs do not have the

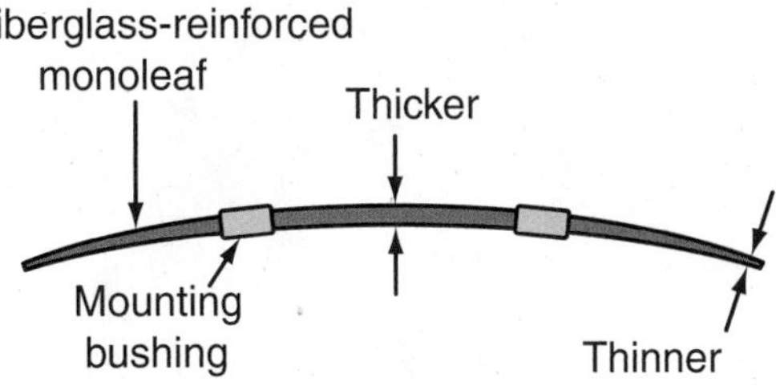

Figure 43–6 The construction of a fiberglass-reinforced monoleaf spring.

noise and static friction characteristic of multiple-leaf springs.

Fiber Composite Springs. While most leaf springs are still made of steel, **fiber composite** types are increasing in popularity **(Figure 43–6)**. Some automotive people call them plastic springs in spite of the fact that the springs contain no plastic at all. They are made of fiberglass, laminated and bonded together by tough polyester resins. The long strands of fiberglass are saturated with resin and bundled together by wrapping (a process called filament winding) or squeezed together under pressure (compression molding).

Fiber composite leaf springs are incredibly lightweight and possess some unique ride control characteristics. Conventional monoleaf steel springs are real heavyweights, tipping the scale at anywhere from 25 to 45 pounds (11 to 20 kg) apiece. Some multiple-leaf springs can weigh almost twice as much. A fiber composite leaf spring is a featherweight by comparison, weighing a mere 8 to 10 pounds (3.6 to 4.5 kg). As every performance enthusiast knows, springs are dead weight. Reducing the weight of the suspension not only reduces the overall weight of the vehicle, but also reduces the sprung mass of the suspension itself. This reduces the spring effort and amount of shock control that is required to keep the wheels in contact with the road. The result is a smoother riding, better handling, and faster responding suspension, which is exactly the sort of thing every performance enthusiast wants.

Air Springs Another type of spring, an air spring, is used in an air-operated microprocessor-controlled system that replaces the conventional coil springs with air springs to provide a comfortable ride and automatic front and rear load leveling. This system, fully described later in this chapter, uses four air springs to carry the vehicle's weight. The air springs are located in the same positions where coil springs are usually found. Each spring consists of a reinforced rubber bag pressurized with air. The bottom of each air bag is attached to an inverted pistonlike mount that reduces the interior volume of the air bag during jounce **(Figure 43–7)**. This has the effect of increasing air pressure inside the spring as it is compressed, making it progressively stiffer. A vehicle equipped

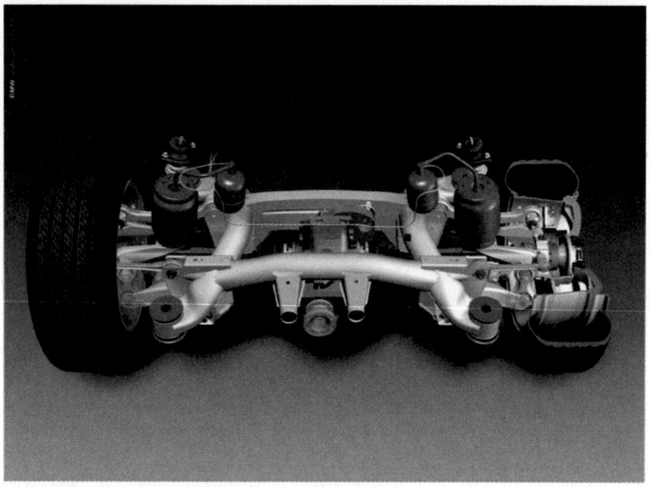

Figure 43–7 A rear suspension setup with air springs.
Courtesy of BMW of North America, Incorporated

with an electronic air suspension system is able to provide a comfortable street ride, about a third softer than conventional coil springs. At the same time, its variable spring rate helps absorb bumps and protect against bottoming.

Torsion Bar Suspension System

Torsion bars serve the same function as coil springs. In fact, they are often described as straightened-out coil springs. Instead of compressing like coil springs, a torsion bar twists and straightens out on the recoil. That is, as the bar twists, it resists up-and-down movement. One end of the bar—made of heat-treated alloy spring steel—is attached to the vehicle frame. The other end is attached to the lower control arm **(Figure 43–8)**. When the wheel moves up and down, the lower control arm is raised and lowered. This twists the torsion bar, which causes it to absorb road shocks. The bar's natural resistance to twisting quickly restores it to its original position, returning the wheel to the road.

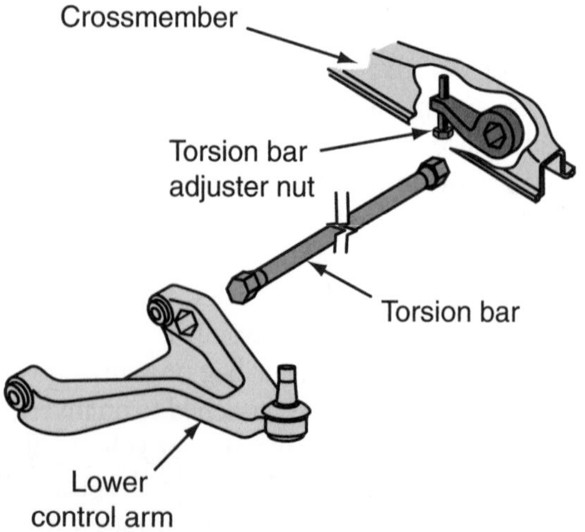

Figure 43–8 A torsion bar setup.

When torsion bars are manufactured, they are prestressed to give them fatigue strength. Because of directional prestressing, torsion bars are directional. The torsion bar is marked either right or left to identify on which side it is to be used.

Because the torsion bar is connected to the lower control arm, the lower ball joint is the load carrier. A shock absorber is connected between the lower control arm and the frame to damp the twisting motion of the torsion bar.

Many late-model pickups and SUVs use torsion bars in their front suspensions. They are primarily used in this type vehicle because they can be mounted low and out of the way of the driveline components.

Shock Absorbers

Shock absorbers damp or control motion in a vehicle. If unrestrained, springs continue expanding and contracting after a blow until all the energy is absorbed. Not only would this lead to a rough and unstable—perhaps uncontrollable—ride after consecutive shocks, it would also create a great deal of wear on the suspension and steering systems. Shock absorbers prevent this. Despite their name, they actually dampen spring movement instead of absorbing shock. As a matter of fact, in England and almost everywhere else but the United States, shock absorbers are referred to as **dampers**.

Today's conventional shock absorber is a velocity-sensitive hydraulic damping device. The faster it moves, the more resistance it has to the movement **(Figure 43–9)**. This allows it to automatically adjust to road conditions. A shock absorber works on the principle of fluid displacement on both its compression (jounce) and extension (rebound) cycles. A typical car shock has more resistance during its extension cycle than its compression cycle. The extension cycle controls motions of the vehicle body spring weight. The compression cycle controls the same motions of the unsprung weight. This motion energy is converted into heat energy and dissipated into the atmosphere.

Shock absorbers can be mounted either vertically or at an angle. Angle mounting of shock absorbers improves vehicle stability and dampens accelerating and braking torque.

Conventional hydraulic shocks are available in two styles: single-tube and double-tube. The vast majority of domestic shocks are double tubed. While they are a little heavier and run hotter than the single-tubed type, they are easier to make. The double-tube shock has an outer tube that completely covers the inner tube. The area between the tubes is the oil reservoir. A compression valve at the bottom of the inner tube allows oil to flow between the two tubes. The piston moves up and down inside the inner tube.

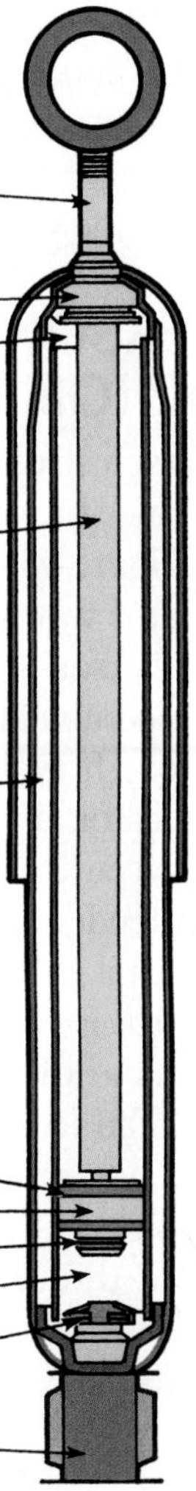

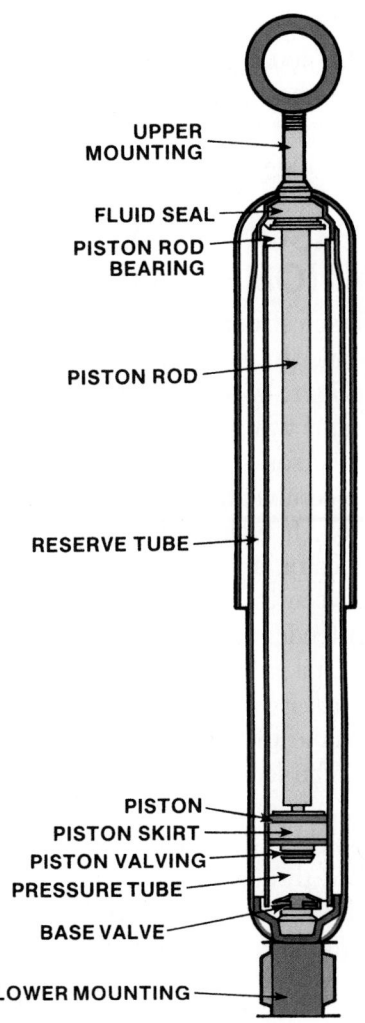

Figure 43-9 A cross section of a conventional shock absorber.

In a single **monoshock,** there is a second floating piston near the bottom of the tube. When the fluid volume increases or decreases, the second piston moves up and down, compressing the reservoir. The fluid does not move back and forth between a reservoir and the main chamber. There are no other valves in a single-tube shock besides those in the main piston. The second piston prevents the oil from splashing around too much and getting air bubbles in it. Air in the shock oil is detrimental. Air, unlike oil, is compressible and slips past the piston easily. When this happens, the result is a shock that offers poor vehicle control on bumpy roads.

In addition to these conventional hydraulic shocks, there are a number of others the technician may encounter.

Gas-Charged Shock Absorbers On rough roads, the passage of fluid from chamber to chamber becomes so rapid that foaming can occur. Foaming is simply the mixing of the fluid with any available air. Since aeration can cause a skip in the shock's action, engineers have sought methods of eliminating it. One is the spiral groove reservoir, the shape of which breaks up bubbles. Another is a gas-filled cell or bag (usually nitrogen) that seals air out of the reservoir so the shock fluid can only contact the gas.

A gas-charged shock absorber **(Figure 43–10)** operates on the same hydraulic fluid principle as conventional shocks. It uses a piston and oil chamber similar to other shock absorbers. Instead of a double tube with a reserve chamber, it has a dividing piston that separates the oil chamber from the gas chamber. The oil chamber contains a special hydraulic oil, and the gas chamber contains either freon or nitrogen gas under pressure equal to approximately 25 times atmospheric pressure.

As the piston rod moves downward in the shock absorber, oil is displaced, just as it is in a double-tube shock. This oil displacement causes the divided piston to press

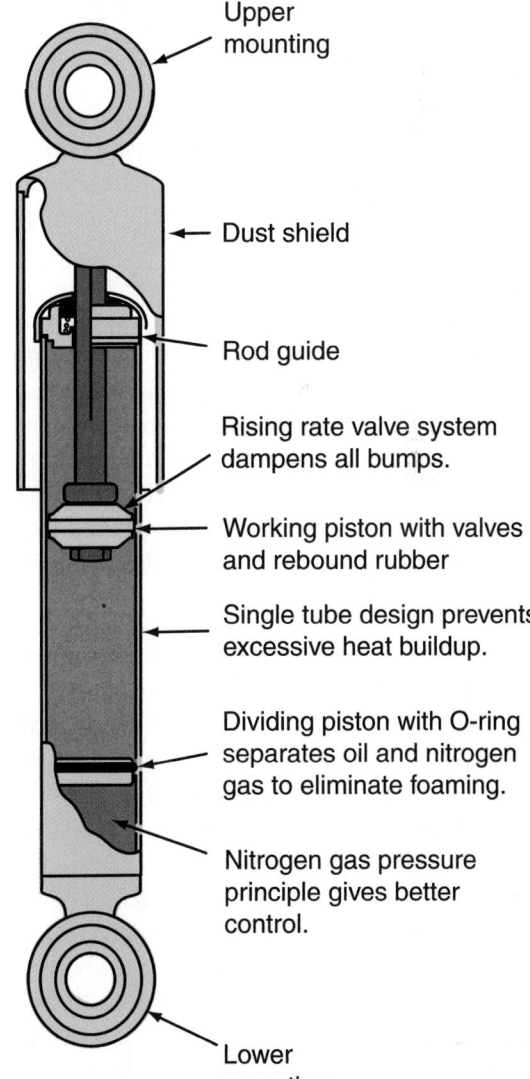

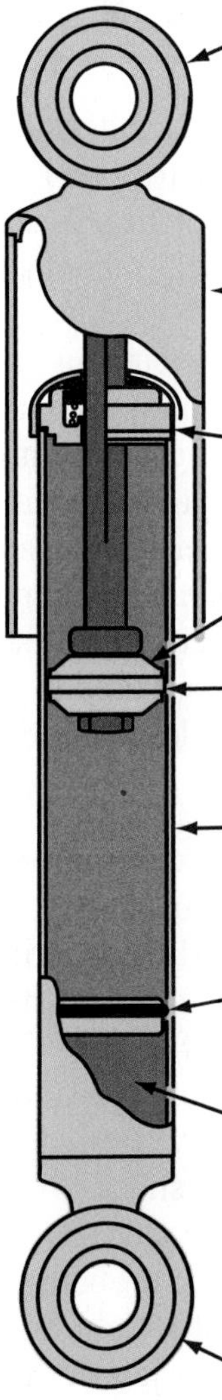

Figure 43-10 Gas-pressure damped shocks operate like conventional oil-filled shocks. Gas is used to keep oil pressurized, which reduces oil foaming and increases efficiency under severe conditions.

on the gas chamber. The gas is compressed and the chamber reduces in size. When the piston rod returns, the gas pressure returns the dividing piston to its starting piston. Whenever the static pressure of the oil column is held at approximately 100 to 360 psi (690 to 2.482 kPa) (depending on the design), the pressure decreases behind the piston and so cannot be high enough for the gas to escape from the oil column. As a result, a gas-filled shock absorber operates without aeration.

SHOP TALK

Some high-pressure gas-charged shocks are monotube shocks with fluid and gas in separate chambers. The gas is charged to 360 psi (2.482 kPa). Its basic design does not allow the valving range needed for a more responsive ride over a broad range of road conditions. The high-pressure gas charge can provide a harsh ride under normal driving conditions and is usually found on small trucks. ■

Air Shock Systems There are two basic adjustable air shock systems: manual fill and automatic load-leveling. The manual fill system can be ordered on new vehicles or can be installed on almost any vehicle manufactured without it.

There are several different types of manual fill air shock systems available. One common manual fill air shock system uses a high-speed, direct current (DC) motor to transfer a command signal that is manually selected from the driver's seat. In another manual air system, the units are inflated through air valves mounted at the rear of the vehicle. Air lines run between the shocks and the valve. A tire air pressure pump is used to fill the shocks to bring the rear of the vehicle to the desired height.

Shock Absorber Ratio Most shock absorbers are valved to offer roughly equal resistance to suspension movement upward (jounce) and downward (rebound). The proportion of a shock absorber's ability to resist these movements is indicated by a numerical formula. The first number indicates jounce resistance. The second indicates rebound resistance. For example, passenger cars with normal suspension requirements use shock absorbers valued at 50/50 (50% jounce/50% rebound). Drag racers, on the other hand, use shocks valued at about 90/10. Small vehicles, because of their light weight and soft springs, require more control in both jounce and rebound in the shock absorbers. Damping rates within the shock absorbers are controlled by the size of the piston, the size of the orifices, and the closing force of the valves.

It is important to keep in mind that the shock absorber ratio only describes what percentage of the shock absorber's total control is compression and what percentage is extension. Two shocks with the same ratio can differ greatly in their control capacity. This is one reason the technician must be sure correct replacement shocks are installed on the vehicle.

MacPHERSON STRUT SUSPENSION COMPONENTS

The **MacPherson strut** suspension is dramatically different in appearance from the traditional independent front suspension **(Figure 43–11)**, but similar components operate in the same way to meet suspension demands.

The MacPherson strut suspension's most distinctive feature is the combination of the main elements into a single assembly. It typically includes the spring, upper suspension locator, and shock absorber. It is mounted vertically between the top arm of the steering knuckle and the inner fender panel.

Struts have taken two forms: a concentric coil spring around the strut itself **(Figure 43–12)** and a spring located between the lower control arm and the frame **(Figure 43–13)**. The location of the spring on the lower control arm, not on the strut as in a conventional MacPherson strut system, allows minor road vibrations to be absorbed through the chassis rather than be fed back to the driver through the steering system. This system is called modified MacPherson suspension.

Struts

The core element of this type of suspension is the strut. With its cylindrical shape and protruding piston rod, it looks quite similar to the conventional shock absorber. In fact, the strut provides the damping function of the shock absorber, in addition to serving to locate the spring and to fix the position of the suspension.

The shock-damping function is accomplished differently on various types of struts. None of them uses a sep-

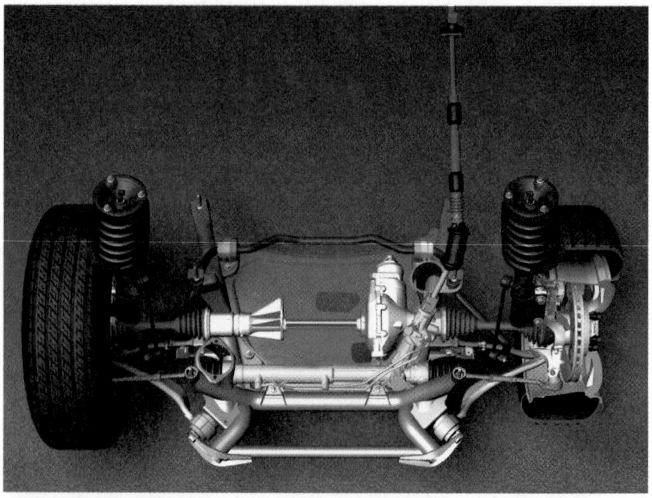

Figure 43–11 A complete MacPherson strut front suspension. *Courtesy of BMW of North America, Incorporated*

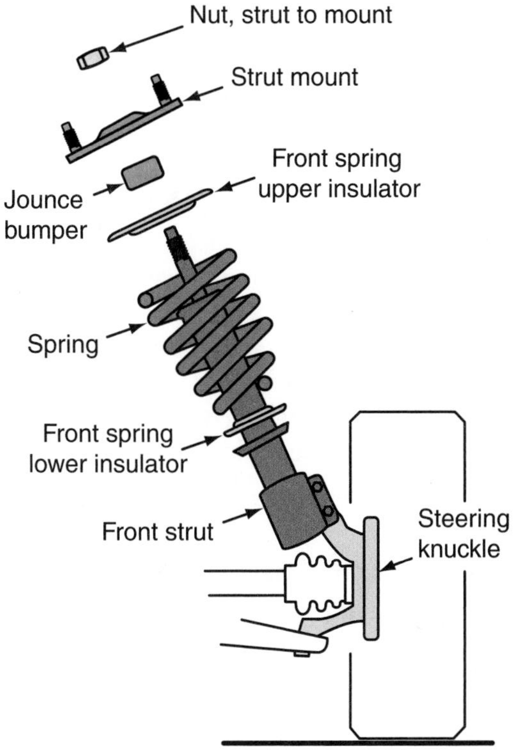

Figure 43-12 A MacPherson strut with a replaceable shock absorber cartridge.

arate shock absorber as the traditional front suspension does. Some versions are designed so the damper can be independently serviced.

Struts fall into two broad categories: sealed and serviceable units. A sealed strut is designed so the top closure of the strut assembly is permanently sealed. There is no access to the shock absorber cartridge inside the strut housing and no means of replacing the cartridge. Therefore, it is necessary to replace the entire strut unit. A serviceable strut is designed so the cartridge inside the housing, which provides the shock-absorbing function, can be replaced with a new cartridge. Serviceable struts use a threaded body nut in place of a sealed cap to retain the cartridge.

The shock absorber device inside a serviceable strut is generally wet. This means the shock absorber contains oil that contacts and lubricates the inner wall of the strut body. The oil is sealed inside the strut by the body nut, O-ring, and piston rod seal. Servicing a wet strut with the equivalent components involves a thorough cleaning of the inside of the strut body, absolute cleanliness, and great care in reassembly (including replenishing the strut with oil).

Cartridge inserts were developed to simplify servicing wet struts. The insert is a factory-sealed replacement for the strut shock absorber. The replacement cartridge is simply substituted for the original shock absorber cartridge and retained with the body nut.

Most OE domestic struts are serviced by replacement of the entire unit. There is no strut cartridge to replace. Sealed OE units can also be serviced by replacement with an aftermarket unit that permits future servicing by cartridge replacement.

The use of the strut reduces suspension space and weight requirements. By mounting the bottom of the

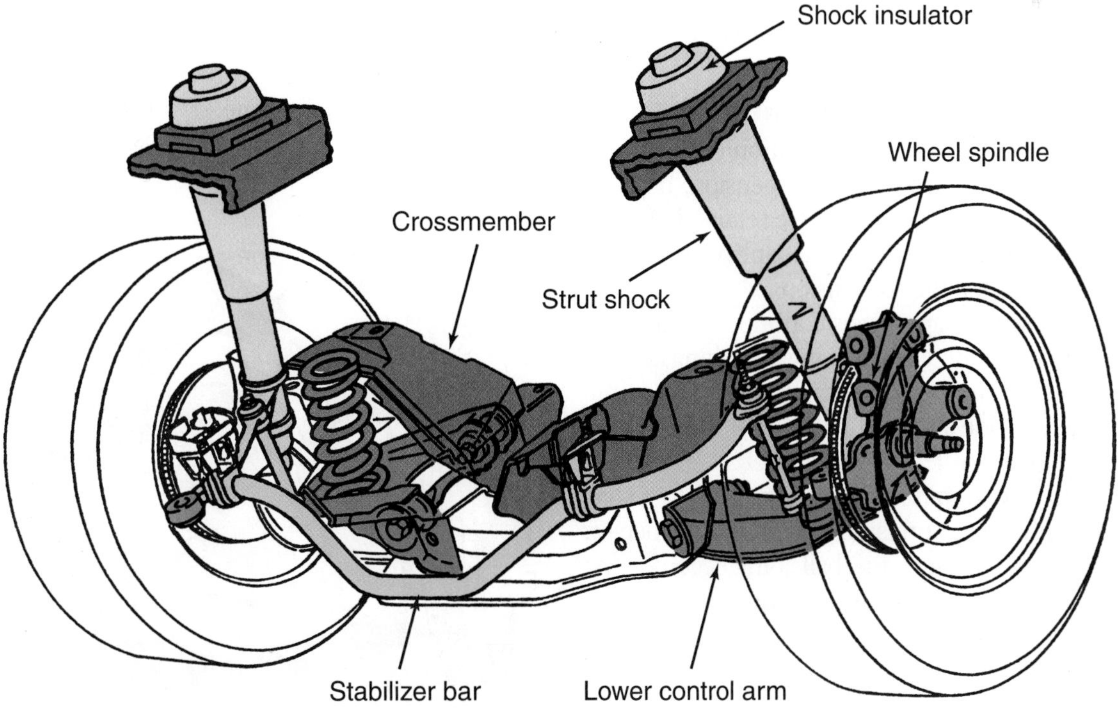

Figure 43-13 A modified MacPherson suspension has the spring mounted separately from the strut.
Courtesy of Ford Motor Company

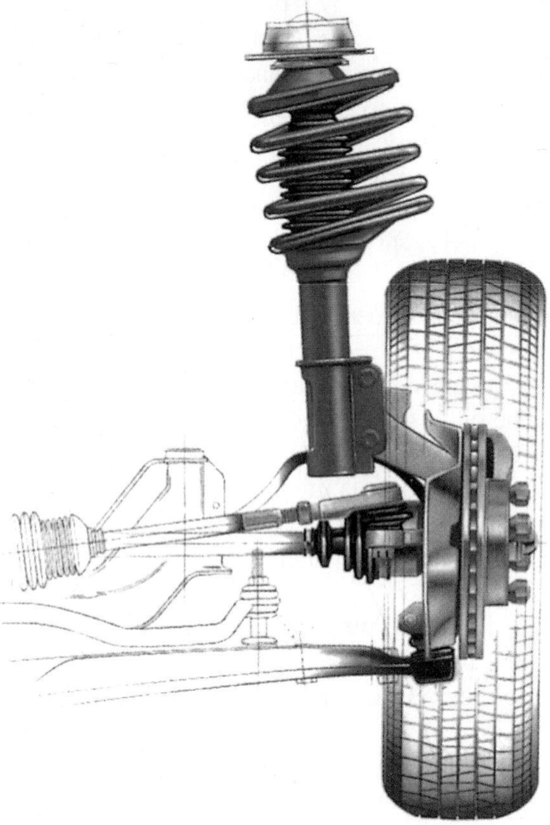

Figure 43-14 A front MacPherson strut assembly. *Courtesy of Ford Motor Company*

strut assembly to the steering knuckle, the upper control arm and ball joint of the traditional suspension are eliminated. In place of the ball joint, the upper mount, which is bolted to the fender panel, is the load-carrying member on MacPherson suspensions.

Lower Suspension Components

The suspension's lower mounting position continues to be the frame, as on the traditional suspension, because the lower control arm and ball joint are retained **(Figure 43–14)**. As on those suspensions, the control arm serves as the lower locator for the suspension.

MacPherson strut suspensions continue to use sway or stabilizer bars. On models with single-bushing control arms, **strut rods** or the sway bar can be fastened to the control arm to provide lateral stability.

The lower **ball joint** is a friction or steering ball joint and is used to stabilize the steering and to retard shimmy. The only exception is on modified MacPherson suspensions. In this design, the ball joint becomes the load bearer; the upper mount becomes the steering component.

Springs

Coil springs are used on all strut suspensions. A mounting plate welded to the strut serves as the lower spring seat. The upper seat is bolted to the strut piston rod. A bearing or rubber bushing in the upper mount permits the spring and strut to turn with the motion of the wheel as it is steered.

INDEPENDENT FRONT SUSPENSION

Front-suspension systems are fairly complex. They have somewhat contradictory jobs. They must keep the wheels rigidly positioned and at the same time allow them to steer right and left. In addition, because of weight transfer during braking, the front suspension system absorbs most of the braking torque. While accomplishing this, it must provide good ride and stability characteristics.

Short-Long Arm Suspension

The unequal length control arm or **short-long arm (SLA)** suspension system has been common on domestic-made vehicles for many years **(Figure 43–15)**. Each wheel is independently connected to the frame by a steering knuckle, ball joint assemblies, and short upper and longer lower control arms. Because the upper arm pivots in a shorter arc, the top of the wheel moves in and out slightly but the tire's road contact remains constant **(Figure 43–16)**.

One design of an SLA uses a narrow lower control arm, shaped like an "I" **(Figure 43–17)**. A strut rod is used to hold the control arm in place. The strut rod is attached to the control arm close to the steering knuckle and to the frame in front of the wheel assembly. Rubber bushings at the frame mounting allow the strut rod to move a little when the tire hits a bump. The bushing dampens the shock and prevents it from transmitting through the vehicle's frame.

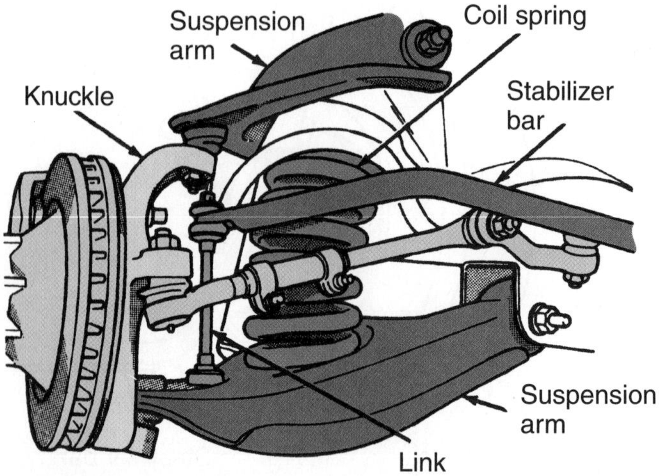

Figure 43-15 A typical SLA front suspension. *Courtesy of DaimlerChrysler Corporation*

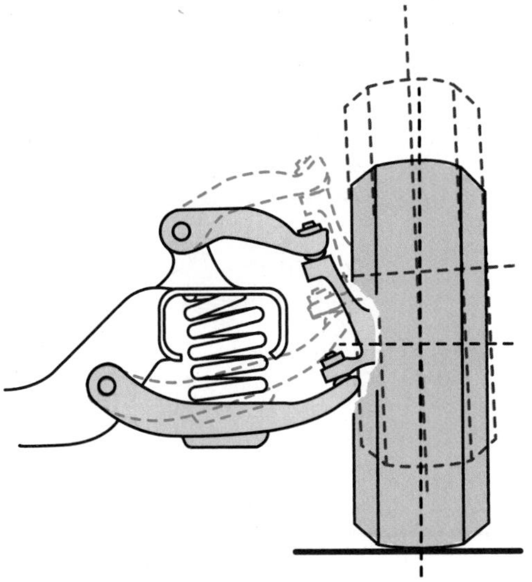

Figure 43-16 The movement of the wheel as a short-long arm suspension system moves up and down.

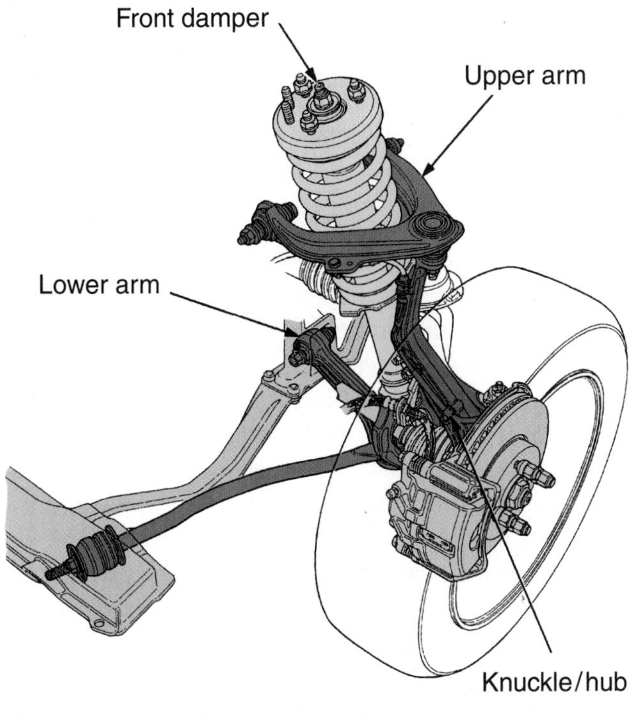

Figure 43-17 A FWD wishbone suspension system with a narrow lower control arm mounted on a single pivot. *Courtesy of American Honda Motor Co., Inc.*

The essential components of SLA systems are the wheel spindle assembly, control arms, ball joints, shock absorbers, and springs, among others.

Wheel Spindle A **wheel spindle** assembly consists of a wheel spindle and a steering knuckle. A wheel spindle is connected to the wheel through wheel bearings and is the point at which the wheel hub and wheel bearings are connected. A **steering knuckle** is connected to control arms. In most cases, a steering knuckle and wheel spindle are forged to form a single piece.

Control Arms The upper and lower **control arms** on the traditional **independent front suspension (IFS)** function primarily as locators. They fix the position of the system and its components relative to the vehicle and are attached to the frame with bushings that permit the wheel assemblies to move up and down separately in response to irregularities in the road surface. The outer ends are connected to the wheel assembly with ball joints **(Figure 43–18)** inserted through each arm into the steering knuckle.

There are two types of control arms: the wishbone, or double-pivot, control arm and the single-pivot, or single-bushing, control arm **(Figure 43–19)**. The wishbone offers greater lateral stability than the single-pivot arm, which is lighter and requires less space than the wishbone but also requires modifications in suspension design to compensate for the reduced lateral stability. Those modifications are discussed further later in this chapter.

Ball Joints A ball joint **(Figure 43–20)** connects the steering knuckle to the control arm, allowing it to pivot on the control arm during steering. Ball joints also permit up-and-down movement of the control arm as the suspension reacts to road conditions. The ball joint stud protrudes from its socket through a rubber seal that keeps lubricating grease in the housing and keeps dirt out. Some ball joints require periodic lubrication, while most do not. These maintenance-free ball joints move in a prelubricated nylon bearing.

Ball joints are either load carrying or are followers. A load-carrying ball joint supports the car's weight and is generally in the control arm that holds or seats the

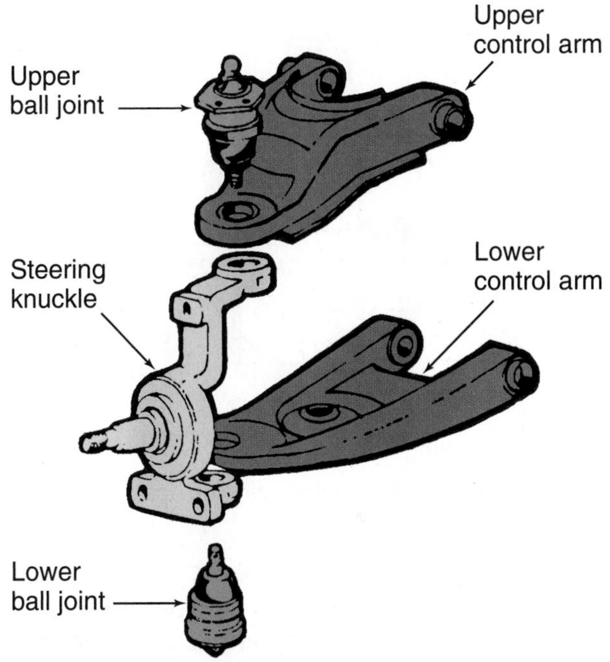

Figure 43-18 Ball joint locations. *Courtesy of Moog Automotive, Inc.*

Figure 43-19 A front suspension system with an upper control arm shaped like a wishbone, which is what this type of suspension is called. *Courtesy of DaimlerChrysler Corporation*

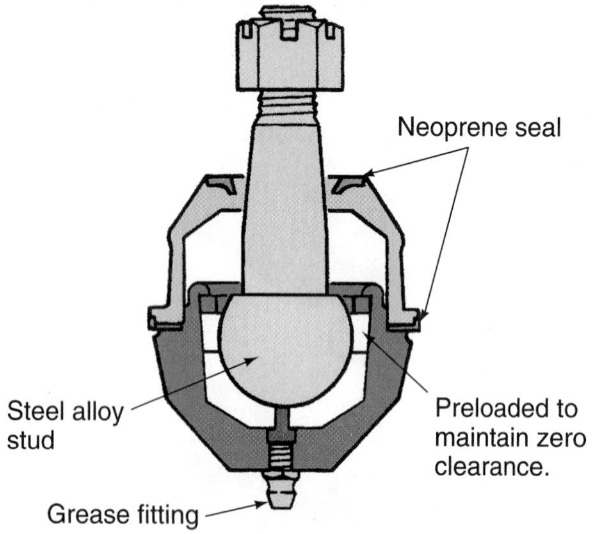

Figure 43-20 A typical ball joint.

spring. Load-carrying joints can be called tension-loaded or compression-loaded ball joints **(Figure 43–21)**. The correct term depends on whether the force of the load tends to push the ball into the socket (compression) or pull it out of the socket (tension).

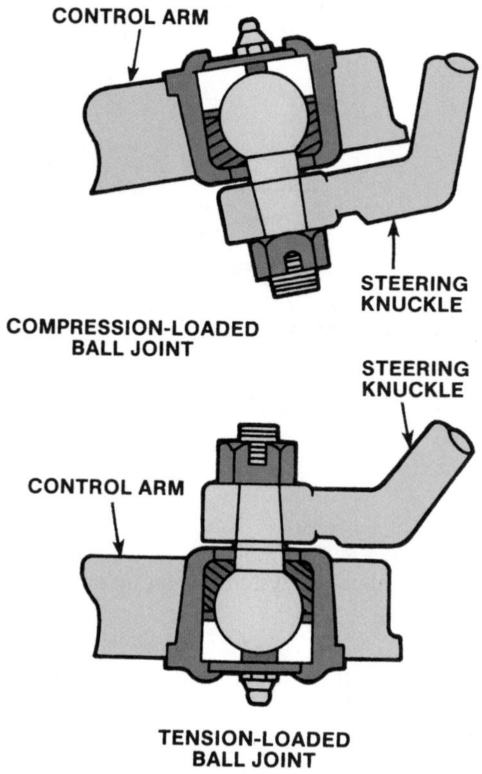

Figure 43-21 Two basic types of load-carrying ball joints.

Follower ball joints are often called friction-loaded ball joints. A follower ball joint mounts on the control arm that does not provide a seat for the spring. The follower does not support vehicle weight and does not get the same stress as the load carrier.

Depending on the location of the suspension system's spring, either the upper or lower ball joint will be the load-carrying joint. In a MacPherson strut suspension, there is usually only one ball joint on each side and it is typically a follower. In modified strut suspensions, the ball joint is a load-carrying joint because the spring is positioned between the frame crossmember and the lower control arm.

Some ball joints have wear indicators. As the joint wears, the grease fitting of the joint recedes into the housing. When the shoulder of the fitting is flush with the housing, the joint needs to be replaced **(Figure 43–22)**.

A ball joint is nothing more than a ball-in-socket joint. As long as the ball is firmly in the socket and the ball and/or socket is not worn, the joint will provide a solid connection. Once the ball or socket is worn, the connection becomes sloppy. How the ball is kept in its socket depends on the type of ball joint it is. Load-carrying ball joints rely on the vehicle's weight to keep the ball in the socket **(Figure 42–23)**. As weight is removed from the joint, the ball relaxes in the socket and will feel loose. Follower ball joints are held in place by friction inside the joint. A spring inside the joint typically

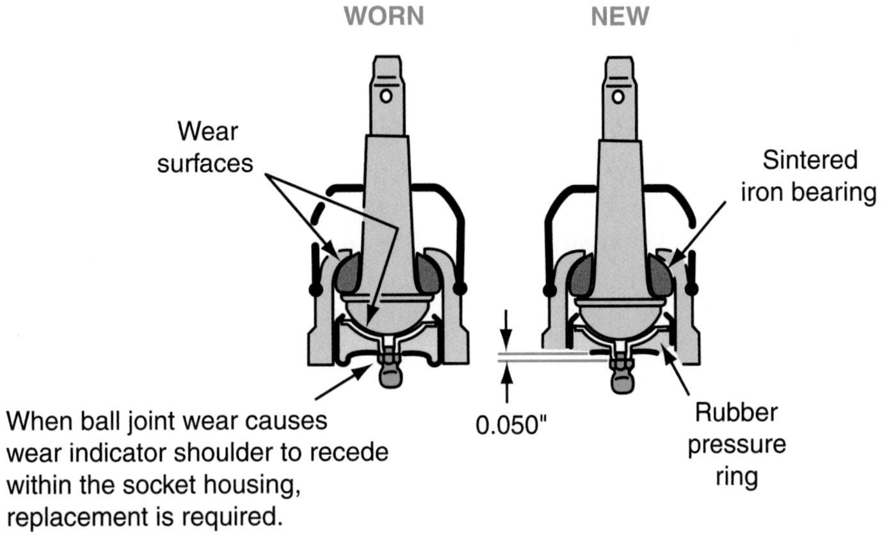

Figure 43-22 A wear indicator on a ball joint.

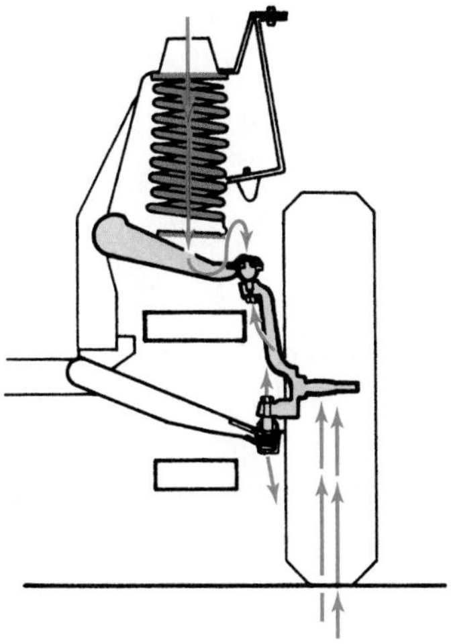

Figure 43-23 The upper ball joint is the load-carrying joint in this system because of the position of the spring.

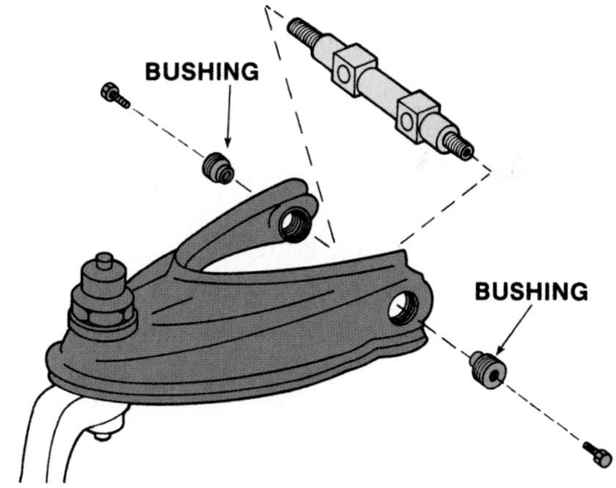

Figure 43-24 Control arm bushings.

keeps the ball tight in the socket but allows for some flexibility. This type of joint should never have any play.

Other Front System Components In addition to shock absorbers, other suspension control devices include bushings, stabilizer or sway bars, and strut rods. In the design of these suspension control devices, the difference between sprung and unsprung weight is important.

Bushings. Rubber or polyurethane bushings are found on many suspension components, such as the control arms **(Figure 43–24)**, radius arms, and strut rods. They make good suspension system pivots, minimize the number of lubrication points, and allow for slight assembly misalignments. Bushings help to absorb road shock, allow some movement, and reduce noise entering the vehicle.

Suspension bushings can deteriorate fairly rapidly, causing tire wear. They are a common cause of misalignment. Replacement bushings come in two basic varieties: stock and performance. The latter are usually made of a high grade polyurethane material and are sold primarily as an upgrade to improve handling response and ride control.

The harder urethane bushings eliminate unwanted compliance or give in the suspension. When hard cornering overtaxes stock bushings and causes them to deflect excessively, undesirable chamber changes occur in the front wheels that result in excessive outer shoulder wear on the tires. Compliance also slows down the action of the sway bar when cornering, which has a significant impact on the vehicle's road holding ability and steering stability (especially in cross winds). Firmer bushings can also help reduce torque steer in front-wheel-drive cars. Torque steer is the tendency to pull to one side (usually to the right) under hard acceleration. It is more of a problem on FWD cars with unequal length drive shafts because the wheel with the shorter shaft (usually the left side) gets more torque than the one with the longer shaft. There is enough give in the suspension that the left wheel tries to pull ahead of the right, throwing wheel alignment off enough to make the car pull to the right.

The procedures for checking bushings and the methods for replacing them are given in service manuals.

Stabilizers. A variety of devices are used with the basic suspension components to provide additional stability. One of the most common is the **sway bar**, which is also known as the **antisway bar** or stabilizer. This is a metal rod running between the opposite lower control arms. As the suspension at one wheel responds to the road surface, the sway bar transfers a similar movement to the suspension at the other wheel. For example, if the right wheel is drawn down by a dip in the road surface, the sway bar is drawn with it creating a downward draw on the left wheel as well. In this way, a more level ride is produced. Sway or lean during cornering is also reduced.

If both wheels go into a jounce, the sway bar simply rotates in its insulator bushings. It is a different matter when only one wheel goes into jounce. The stabilizer bar twists, just like a torsion bar, to lift the frame and the opposite suspension arm. This action reduces body roll.

The sway bar can be a one-piece, U-shaped rod fastened directly into the control arms with rubber bushings, or it can be attached to each control arm by a separate sway bar link. The arm is held to the links with nuts and rubber bushings and is also mounted to the frame in the center with rubber bushings. If the sway bar is too large, it causes the vehicle to wander. If it is too small, it has little effect on stability.

On suspensions that use single-housing lower control arms instead of wishbone types, the sway bar can also be used to add lateral stability to the control arm. Strut rods are used on models that do not use the sway bar like this. They are attached to the arm and frame with bushings, allowing the arm a limited amount of forward and backward movement **(Figure 43–25)**. Strut rods are directly affected by braking forces and road shocks, and their failure can quickly lead to failure of the entire suspension system.

Four-Link Front Suspension

A four-link front suspension fixes the wheel with four rod-type control arms and the tie rod **(Figure 43–26)**. The suspension strut supports the vehicle weight against the body via the load-bearing link. By separating wheel attachment and suspension elements, this suspension op-

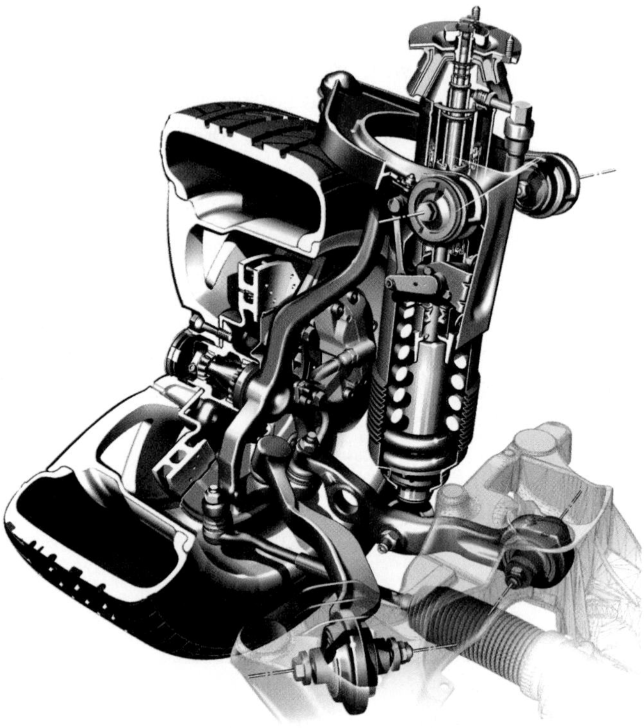

Figure 43-26 A four-link front suspension system.
Courtesy of DaimlerChrysler Corporation

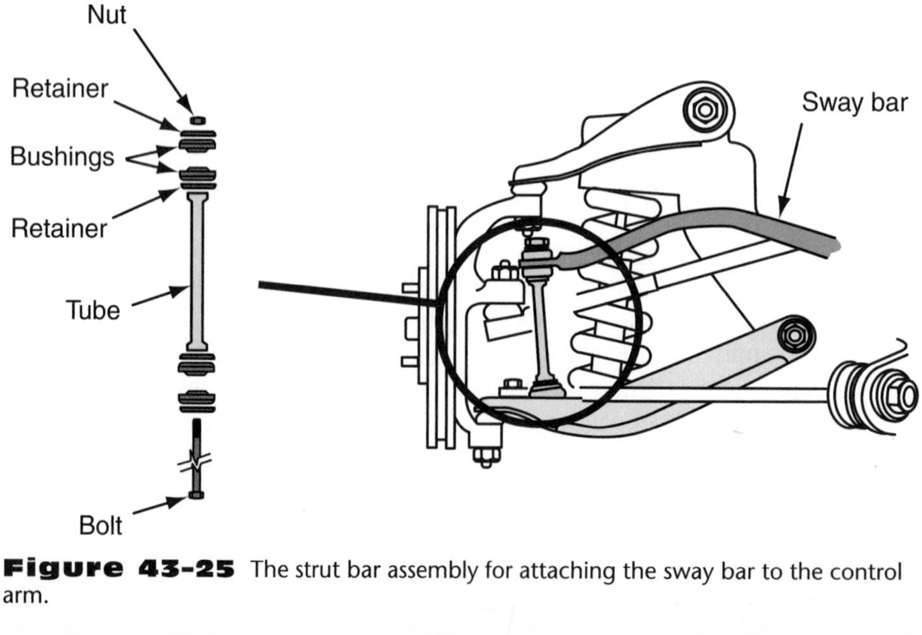

Figure 43-25 The strut bar assembly for attaching the sway bar to the control arm.

timizes ride quality and movement. The influence of drive forces on the steering system is minimal.

GENERAL FRONT-SUSPENSION INSPECTION

To minimize the chance of performing an unnecessary service, the following preliminary or general inspections should be made:

1. Check all tires for proper inflation pressure.
2. Check the tires for telltale indications of improper front-end alignment, wheel and tire imbalance, and physical defects or damage.
3. Check the vehicle for optional suspension equipment, such as that provided for heavy-duty applications or trailer towing packages. They have a firmer ride quality.
4. Check vehicle attitude for evidence of overloading or sagging. Be sure the chassis height is correct.
5. Raise the vehicle off the floor. Grasp the upper and lower surfaces of the tire and shake each front wheel to check for worn wheel bearings.
6. Check the ball joints for looseness and wear.
7. Check the condition of the struts' upper mounts.
8. Check the shock absorbers and struts for signs or fluid leakage **(Figure 43–27)** and damage.
9. Check all of the mountings for the shocks and struts.
10. Check all suspension bushings for looseness, splits, cracks, misplacement, and noises.
11. Check the steering mounts, linkages, and all connections for looseness, binding, or damage.

If any problems or unusual conditions are found during the visual inspection, the parts should be replaced.

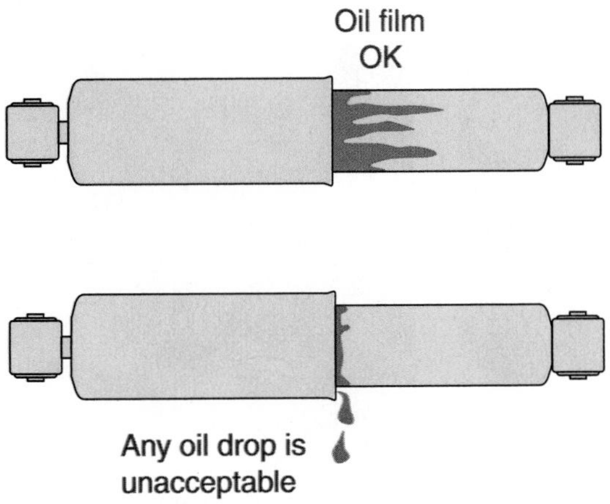

Figure 43-27 Check the shock absorbers for signs of fluid leakage.

SHOP TALK

Before visual inspection or suspension height measurements can be performed, the vehicle must be on a level surface. Tires must be at recommended pressures; gas tank must be full; and there should be no passenger or luggage compartment load. Beginning at the rear bumper, jounce the car up and down several times. Proceed to the front bumper and repeat, releasing during same cycle as rear jounce. ■

Chassis Height Specifications

A quick overall visual inspection detects any obvious sag from front to rear or from side to side. Under the car, at the level of the two ends of the control arms, check for out-of-level, damaged, or worn rubber bumpers, or shiny or worn spring coils. All indicate weak coil springs.

A more accurate inspection reveals less obvious problems by measuring heights at specific points on each side of the suspension system.

USING SERVICE MANUALS

For the most accurate measurement of chassis height, use the service manual to check against the manufacturer's recommendations for the specific model. Photo Sequence 46 shows the typical procedure for checking vehicle ride height. Be careful. The measurement points vary from one model to another even if manufactured by the same company. When coil spring wear is suspected, it might be necessary to load the vehicle to the manufacturer's suggested capacities and measure at the designated points. ■

FRONT-SUSPENSION COMPONENT SERVICING

As is the case in all sections of this text, specific troubleshooting procedures are given in detail in the *Tech Manual*. However, the following is information on servicing the major components or assemblies of front suspension systems.

Coil Springs

The never-ending twisting and untwisting of the coil spring (or the torsion bar) lead to inevitable loss of elasticity and spring sag. Coil springs, then, require replacement because they sag in service. A sagged coil spring upsets vehicle trim height resulting in upset wheel alignment angles, steering angles, headlight aiming, braking distribution, riding quality, tire tread life, shock life, and U-joint life.

PHOTO SEQUENCE 46

Measuring Front and Rear Curb Riding Height

P46-1 *Check the trunk for extra weight.*

P46-2 *Check the tires for normal inflation pressure.*

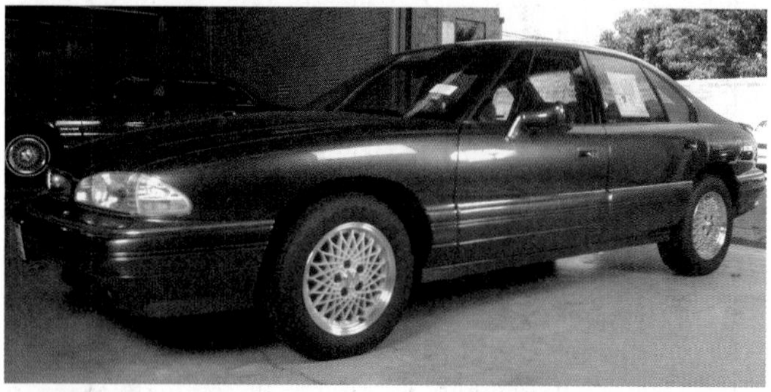

P46-3 *Park the car on a level shop floor or alignment rack.*

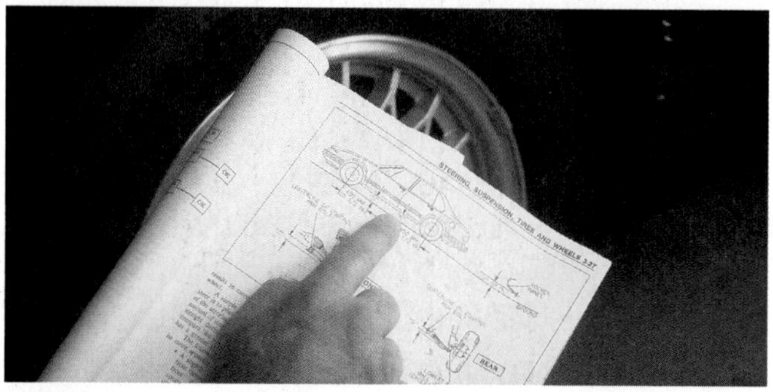

P46-4 *Find the vehicle manufacturer's specified curb riding height measurement locations in the service manual.*

P46-5 *Measure and record the right front curb riding height.*

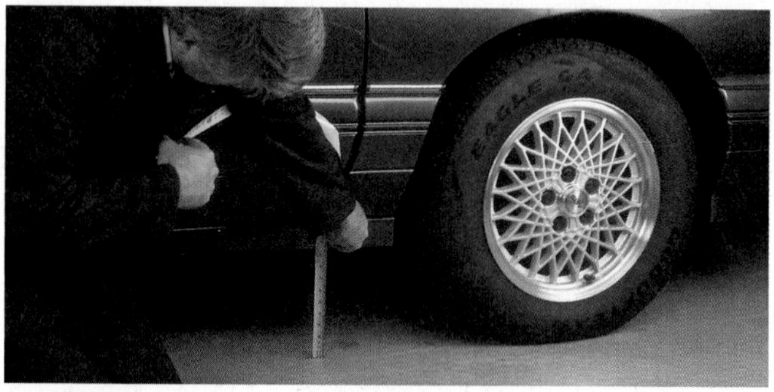

P46-6 *Measure and record the left front curb riding height.*

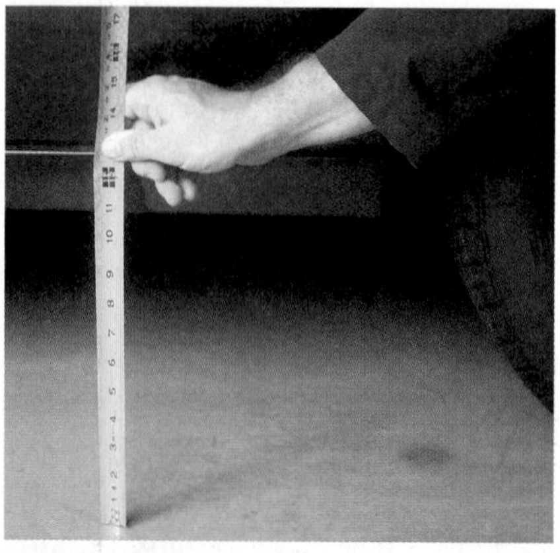

P46-7 *Measure and record the right rear curb riding height.*

P46-8 *Measure and record the left rear curb riding height.*

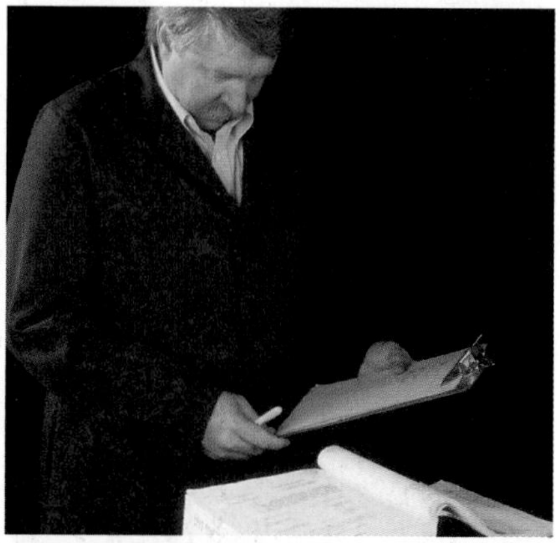

P46-9 *Compare the measurement results to the specified curb riding height in the service manual.*

Coil springs also break. Downsized cars are often forced to carry the same loads as their larger counterparts, mostly because people and their hauling needs did not downsize along with the cars.

CAUTION!

The coil spring exerts a tremendous force on the control arm. Before you disconnect either control arm from the knuckle for any service operation, contain the spring with a spring compressor to prevent it from flying out and causing injury.

Removing a Spring To remove a coil spring, raise and support the vehicle by its frame. Let the control arm hang free. Remove wheels, shock absorbers, and stabilizer links. Disconnect the outer tie-rod ends from their respective arms.

Unload the ball joints with a roll-around floor jack. Jack under the lower control arm from the opposite side of the vehicle. This allows the jack to roll back when the control arm is lowered. Position the jack as close to the lower ball joint as possible for maximum leverage against the spring.

The spring is ready for the installation of the spring compressor **(Figure 43–28)**. There are many different types of spring compressors. One type uses a threaded compression rod that fits through two plates, an upper and lower ball nut, a thrust washer, and a forcing nut. The two plates are positioned at either end of the spring. The compression rod fits through the plates with a ball nut at either end. The upper ball nut is pinned to the rod. The thrust washer and forcing nut are threaded onto the end of the rod. Turning the forcing nut draws the two plates together and compresses the spring.

Figure 43-28 A coil spring compressor is used to compress the spring before disconnecting some suspension parts.

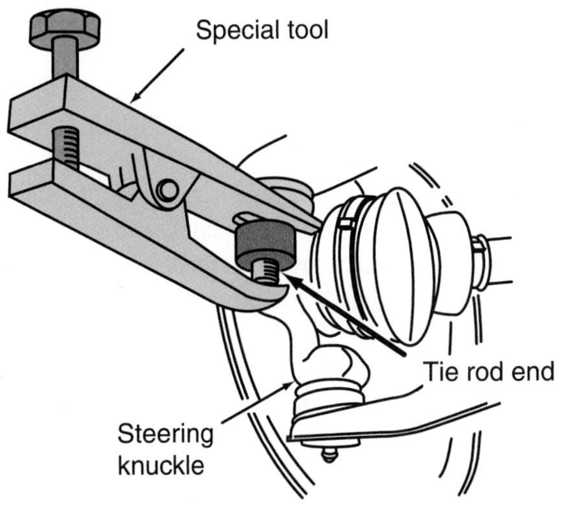

Figure 43-29 When this tool is expanded, it can force the tie rod end or ball joint stud out of the steering knuckle taper.

In some cases, it is necessary to break the tapers of both upper and lower ball joints so the steering knuckle can be moved to one side **(Figure 43–29)**. If the vehicle is equipped with a strut rod, this must be disconnected at the lower control arm. Push the control arm down until the spring can be removed. If necessary, a pry bar can be used to remove the spring from its lower seat. Remove the spring and compressor.

If the same spring is to be reinstalled, leave the compressor in position. If a new spring is to be used, slowly release the pressure on the tool by backing off the forcing nut. Compress the new spring prior to installing it.

Torsion Bars

Torsion bars are subjected to many of the same conditions affecting coil springs. Periodic adjustment of the torsion bars is necessary to maintain the proper height. Replacement is sometimes necessary because of breakage. It should be noted that the bars are not interchangeable from side to side.

Height inspection and measurements for vehicles with torsion bar suspensions are the same for coil springs. However, sagging can usually be corrected by adjusting the bars. Procedures for adjusting torsion bars are given in the service manual.

Ball Joints

Begin your inspection of a ball joint by checking to see if the ball joint has a wear indicator on it. If it does, check the placement of the grease fitting. If it is recessed, the ball joint is worn and should be replaced. On some vehicles it is recommended that you check to see if the grease fitting can wiggle in the ball joint. If it does, the ball joint should be replaced. Always check the service manual when checking ball joints.

Look carefully at the joint's boot. A damaged boot or joint seal will allow lubricant to leak out and allow dirt to enter and contaminate the lubricant. If the boot is damaged, the ball joint should be replaced.

If no boot damage is evident, gently squeeze the boot. If the boot is filled with grease, it will feel somewhat firm. If the joint has a grease fitting and appears not to be filled with grease, use a grease gun and refill the joint. Fill the joint until fresh grease is seen flowing out of the boot's vent. If too much grease is forced into the joint or it is forced in too quickly, the boot can unseat or tear.

Ball joints should be checked for excessive wear. Load-carrying joints will have some slop when the weight of the vehicle is taken off them. Follower joints should never have play. To check a load-carrying joint, it must be unloaded.

When the coil spring is on the lower control arm, raise the vehicle by jacking under the control arm as close to the ball joint as possible. This gives the maximum amount of leverage against the spring. The ball joint is unloaded when the upper strike out bumper is not in contact with the control arm or frame. A quick check for looseness can be made by using a pry bar between the tire and the ground. To find out if the ball joint is loose beyond manufacturer's specifications, use an accurate measuring device. The following checking procedures demonstrate the use of a dial indicator. The dial indicator is a precision instrument and should be handled carefully to prevent damage. The mounting procedure for the checking tool might vary depending on the style of ball joint used on the vehicle. Manufacturer's tolerances can be axial (vertical), radial (horizontal), or both. To conduct these checks, follow these procedures.

Typical Radial Check For a radial check, attach a dial indicator to the control arm of the ball joint being checked. Position and adjust the plunger of the dial indicator against the edge of the wheel rim nearest to the ball joint being checked. Slip the dial ring to the zero marking. Move the wheel in and out and note the amount of ball joint radial looseness registered on the dial **(Figure 43–30)**. The procedure for checking the radial movement of a lower ball joint on a MacPherson strut front suspension is shown in Photo Sequence 47.

Typical Axial Check For an axial check, first fasten the dial indicator to the control arm, then clean off the flat on the spindle next to the ball joint stud nut. Position the dial indicator plunger on the flat of the spindle and depress the plunger approximately 0.350 inch. Turn the lever to tighten the indicator in place. Pry the bar between the floor and tire. Record the reading **(Figure 43–31)**.

If the ball joint looseness reading on the dial indicator exceeds manufacturer's specifications, the ball joint should be replaced.

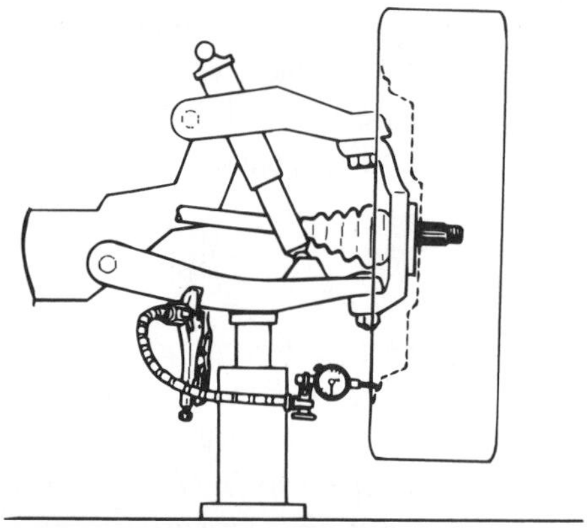

Figure 43-30 A typical mounting of a dial indicator for a radial check. *Courtesy of Moog Automotive, Inc.*

Figure 43-31 A typical mounting of a dial indicator for an axial check. *Courtesy of Moog Automotive, Inc.*

PHOTO SEQUENCE 47

Measuring the Lower Ball Joint Radial Movement on a MacPherson Strut Front Suspension

P47-1 *Raise the front suspension with a floor jack and place jack stands under the chassis at the vehicle's lift points.*

P47-2 *Grasp the front tire at the top and bottom and rock the tire inward and outward while a coworker visually checks for movement in the front wheel bearing. If there is movement, adjust or replace the wheel bearing.*

P47-3 *Position a dial indicator against the inner edge of the rim at the bottom. Preload and zero the dial indicator.*

P47-4 *Grasp the bottom of the tire and push outward.*

P47-5 *With the tire held outward, read the dial indicator.*

P47-6 *Pull the bottom of the tire inward and be sure the dial indicator reading is zero. Adjust the dial indicator as required.*

P47-7 *Grasp the bottom of the tire and push outward.*

P47-8 *With the tire held in this position, read the dial indicator.*

P47-9 *If the dial indicator reading is more than specified, replace the lower ball joint.*

When the load-carrying ball joints are on the upper control arm (spring mounted on the upper arm), raise the vehicle by its frame using support tools to unload the ball joints and hold them in their normal position. To determine the condition of the non-load-carrying (or follower) ball joint, vigorously push and pull on the tire, while watching the ball joint for signs of movement. Refer to the manufacturer's specifications for tolerances.

Inspection of Wear Indicators Wear-indicator-type ball joints must remain loaded to check for wear. The vehicle should be checked with the suspension at curb height. The most common type has a small diameter boss, which protrudes from the center of the lower housing. As wear occurs internally, this boss recedes very gradually into the housing. When it is flush with the housing, the ball joint should be replaced. To remove and install a ball joint, follow the procedure given in the service manual.

Ball joints are mounted to the control arm in one of four basic ways: rivets, bolts, press fit, and threaded. The most common method is press fit. Some manufacturers require you to replace the entire control arm assembly if a ball joint is to be replaced. In these cases, the ball joint and control arm are made as a single assembly and individual parts are not available.

Press-fit ball joints are removed and installed using special tools and a hydraulic press. The control arm must be removed before attempting to remove the ball joint. While pressing the ball joint out of or into the control arm, make sure you do not damage the arm.

WARNING!

Never use heat to remove a ball joint!

Control Arm Bushings

If the bushings, which attach the arms to the frame, are not in good condition, precise wheel alignment settings cannot be maintained.

Visually inspect each rubber bushing for signs of distortion, movement, off-center condition, and presence of heavy cracking. Check metal bushings for noise and loose seals.

To remove the control arm bushings, raise the vehicle and support the frame on safety jack stands. Remove the wheel assembly. Install a spring compressor on the coil spring.

Disconnect the ball joint studs from the steering knuckle as described previously. Remove the bolts attaching the control arm assembly to the frame and remove the control arm.

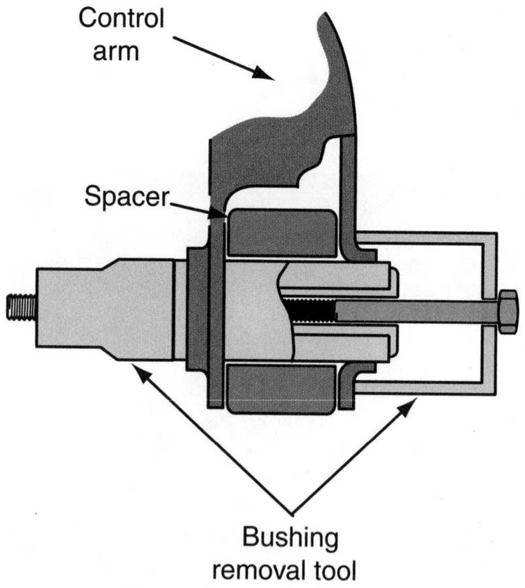

Figure 43–32 Removing a control-arm bushing.

Bushings are pressed in and out of their bores by using special tools. A special tool is installed over the bushing **(Figure 43–32)** after the correct size adapter for the tool has been selected. Tightening the tool pushes the bushing out of the control arm. The same process is used to press a new bushing into the control arm. As the special tool is tightened, the bushing moves into its bore. When installing new bushings, make sure they stay straight while they are being pressed in.

Once the new bushings are started into the control arm, measure and mark the center between mounting holes and center the control arm. Now, alternately press in the bushings on each side, keeping the reference marks aligned. This ensures the shaft is not off center, causing binding. End cap nuts or bolts should not be torqued until the vehicle is at curb height and the suspension has been bounced and allowed to settle out.

Rebolt the control arm and tighten the bolts to specifications, then install the coil spring into position. Install the ball joint studs into the control arms. Remove the coil spring compressor. Install the wheel assembly and lower the car. Road test the car, retighten all bolts, and set wheel alignment.

Strut Rod Bushings

Except in the case of accidental damage, the strut rod itself is rarely replaced. Rather, it is the bushing that wears, deteriorates, and needs replacement.

Sway Bar Bushings

These bushings anchor the sway bar securely to the vehicle frame and the control arms on each side. The condition of the bushings affects the performance of the bar. Visual inspection of mounting bushings indicates if the bushings are worn, have taken a permanent set, or are possibly missing.

Shock Absorbers

A shock absorber that is functioning properly ensures vehicle stability, handling, and rideability. Most motorists fail to notice gradual changes in the operation of their cars as a result of worn shock absorbers. Some common indications of shock absorber failure follow:

- Steering and handling are more difficult.
- Braking is not smooth.
- Bouncing is excessive after stops.
- Tire wear patterns are unusual, especially cupping.
- Springs are bottoming out.

Vibrations set up by a worn shock absorber can cause premature wear in many of the undercar systems. They can cause wear in the front and rear component parts of the suspension system, the linkage component parts of the steering system, and the U-joints and motor or transmission mounts of the driveline. Also, vibrations can cause unnatural wear patterns on the tires.

A shock absorber can be bench tested. First, turn it up in the same direction it occupies in the vehicle. Then extend it fully. Next, turn it upside down and fully compress it. Repeat this operation several times. Install a new shock absorber if a lag or skip occurs near mid-stroke of the shaft's change in travel direction, or if the shaft seizes at any point in its travel, except at the ends. Also, install a new shock absorber if noise, other than a switch or click, is encountered when the stroke is reversed rapidly, if there are any leaks, or if action remains erratic after purging air.

When removing and installing shock absorbers, be sure to follow the aftermarket manufacturer's instructions or those given in the service manual.

MacPherson Strut Suspension

The MacPherson strut suspension system is based on a triangular design. The strut shaft is a structural member that does away with the upper control arm bushings and the upper ball joint. Since this shaft is also the shock absorber shaft, it receives a tremendous amount of force vertically and horizontally. Therefore, this assembly should be inspected very closely for leakage, bent shaft, and poor damping.

To remove and replace the MacPherson strut, proceed as shown in Photo Sequence 48.

During the disassembly of the strut, make sure you check the strut pivot bearing **(Figure 43–33)**. Move the bearing with your hand. If the bearing is hard to move or seems to bind, it must be replaced. When replacing the bearing, make sure the correct side is up. Manufacturers normally mark the up side with paint or some other marking. Make sure you check all rubber insulators for deterioration and other damage and replace them if necessary. Also make sure you mark the eccentric camber bolts before loosening them. Returning the bolts in the same location will help maintain the correct camber angle after reassembly.

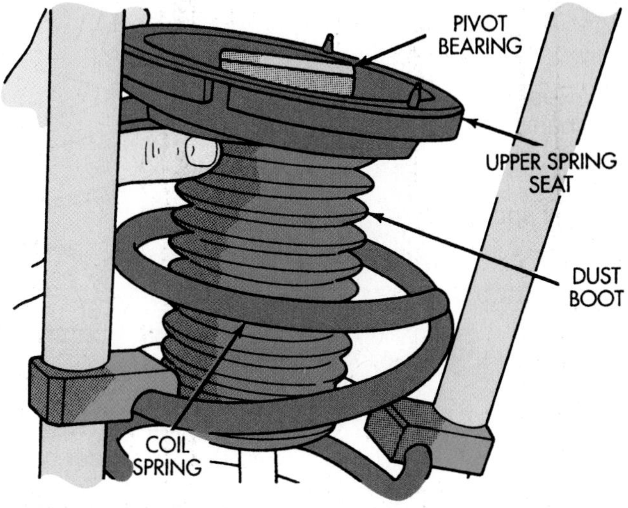

Figure 43–33 Check the strut's pivot bearing for free movement. *Courtesy of DaimlerChrysler Corporation*

REAR-SUSPENSION SYSTEMS

There are three basic types of rear suspensions: live-axle, semi-independent, and independent. There are distinct designs of each, but the types of components and the principles involved are the same as on front suspension systems described earlier in this chapter. Live-axle suspensions are found on rear-wheel-drive (RWD) trucks, vans, and many four-wheel-drive (4WD) passenger cars. Semi-independent systems are used on front-wheel-drive (FWD) vehicles. Independent suspensions can be found on both RWD and FWD vehicles, as well as 4WD cars.

Live-Axle Rear-Suspension Systems

This traditional rear-suspension system consists of springs used in conjunction with a live-axle (one in which the differential axle, wheel bearings, and brakes act as a unit). The springs are either of leaf or coil type.

Leaf Spring Live-Axle System Two springs—either multiple-leaf or monoleaf—are mounted at right angles to the axle and along with the shock absorbers, are positioned below the rear axle housing. The front of the two springs is attached to brackets on the vehicle's frame by a bolt and bushing inserted through the eyes of the springs. While the bushing allows the spring to move, it isolates the rest of the vehicle from noisy road vibrations.

The center of each leaf spring is connected to the rear axle housing with U-bolts. Rubber bumpers are located between the rear axle housing and frame or unit body to dampen severe shocks. The rear eye pivot bushings are

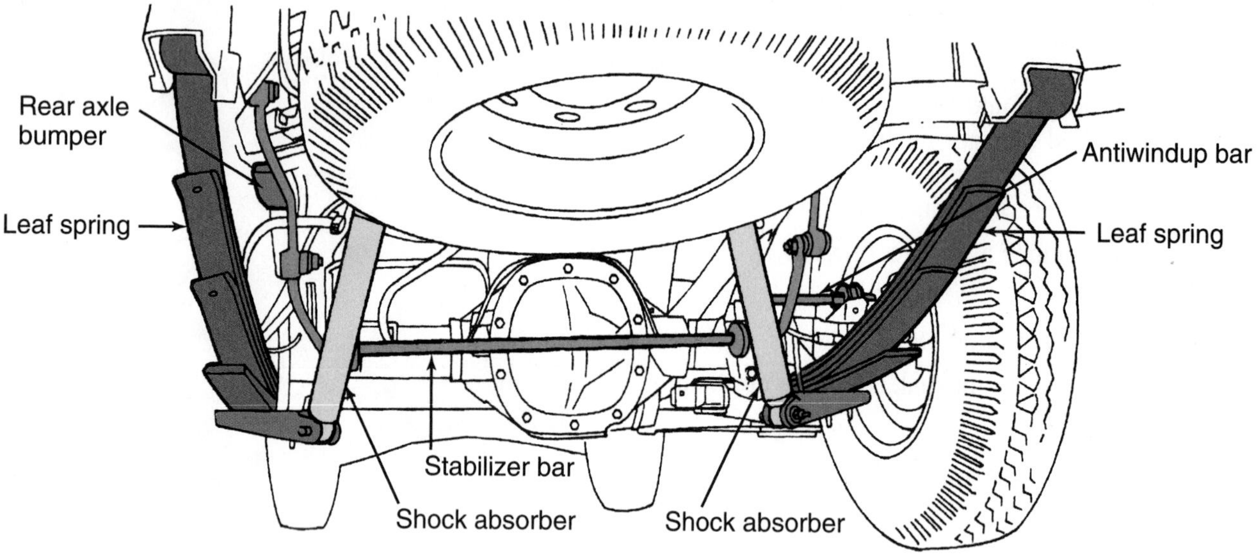

Figure 43-34 A typical rear suspension with a live axle. *Courtesy of Ford Motor Company*

held to the frame with shackles, which attach to the springs by a bolt and bushing **(Figure 43–34)**.

The advantage to this live-axle suspension system is that the leaf springs act as excellent axle locators. The springs allow for the up-and-down movement of the axle and wheels; but at the same time, the springs prevent them from wandering. Therefore, control arms are not required with leaf springs.

There are some disadvantages to the live-axle suspension system. First, this design has a large amount of unsprung weight. Another drawback is the instability caused by the use of a solid axle. Since both rear wheels are connected to the same axle, movement up or down by one wheel affects the other. Consequently, poor traction results because both wheels are pushed out of alignment with the road. Under severe acceleration, this type of suspension is subject to **axle tramp**, a rapid up-and-down jumping of the rear axle due to the torque absorption of the leaf springs. This condition can break spring mounts and shock absorbers and cause premature wear of wheel bearings. Axle tramp is reduced by mounting shock absorbers on the opposing sides (front and back) of the axle. Some heavy-duty vehicles have two-stage springs that allow the vehicle to ride comfortably with both a light or heavy load.

Coil Spring Live-Axle System Some vehicles use two coil springs at the rear with a live rear axle. Because coil springs can only support weight and have little axle-locating capabilities, such vehicles need forward and lateral control arms or links. This type of suspension is called the link-type rigid axle.

The coil springs, located between the brackets on the axle housing and the vehicle body or frame, are held in place by the weight of the vehicle and sometimes by the shock absorbers **(Figure 43–35)**. The control arms are usually made of channeled steel and mounted with rubber bushings. Accelerating, driving, and braking torque are transmitted through three or four control arms, depending on the design. Two forward links are always used, but either one or two lateral links can be found on individual models. **Trailing arms** mount to the underside of the axle, and run forward at a 90-degree angle to the axle to brackets on the car frame. Rubber bushings are used at mounting locations to permit up-and-down movement of the arms and to reduce noise and the effect of shock.

Some rear-axle assemblies are connected to the body by two lower control arms and a tracking bar. A single torque arm is used in place of upper control arms and is rigidly mounted to the rear-axle housing at the rear and through a rubber bushing to the transmission at the front.

Live-Axle Suspension System Servicing Typical service to both coil-spring and leaf-spring systems include the replacement of shock absorbers or springs. Bushings, shackles, or control arms do not need replacement frequently. Always follow the procedures outlined in the service manual whenever servicing the rear suspension.

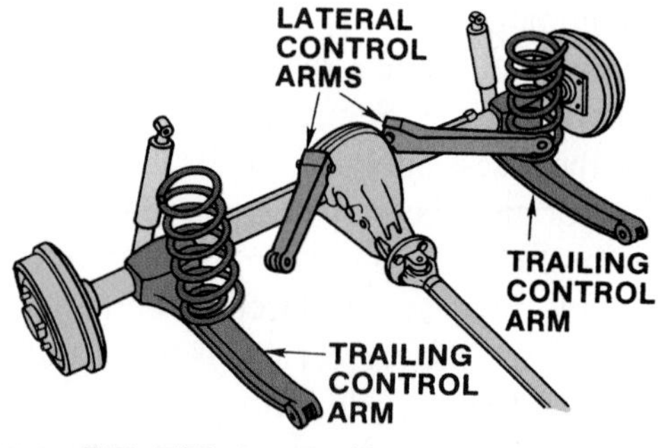

Figure 43-35 A typical live-axle suspension with coil springs.

PHOTO SEQUENCE 48

Removing and Replacing a MacPherson Strut

P48-1 *The top of the strut assembly is mounted directly to the chassis of the car.*

P48-2 *Prior to loosening the strut chassis bolts, scribe alignment marks on the strut bolts and the chassis.*

P48-3 *With the top strut bolts or nuts removed, raise the car to a working height. It is important that the car be supported on its frame and not on its suspension components.*

P48-4 *Remove the wheel assembly. The strut is accessible from the wheel well after the wheel is removed.*

P48-5 *Remove the bolt that fastens the brake line or hose to the strut assembly.*

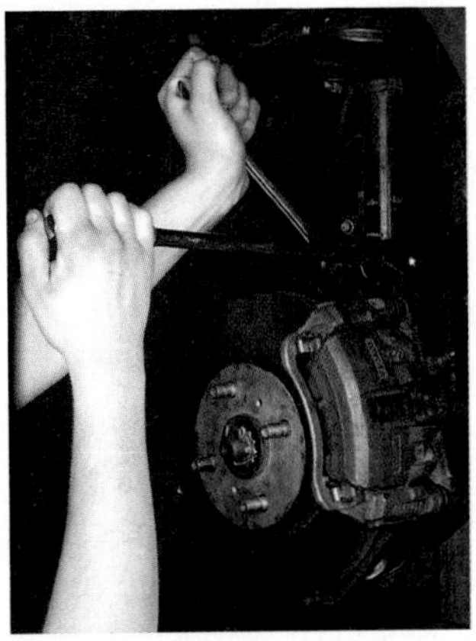

P48-6 *Remove the strut's two steering knuckle bolts.*

P48-7 *Support the steering knuckle with wire and remove the strut assembly from the car.*

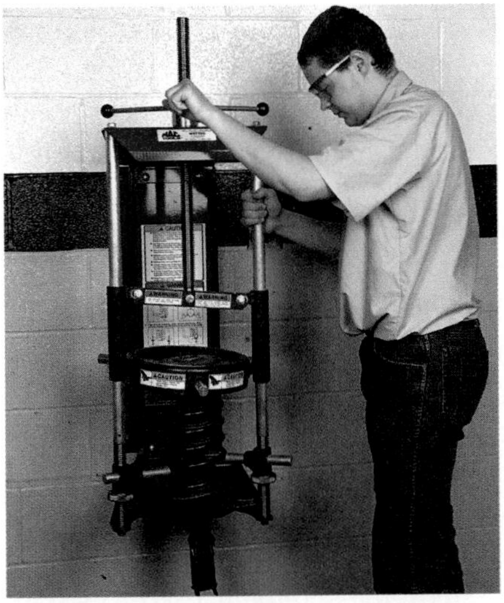

P48-8 *Install the strut assembly into the proper type of spring compressor. Then compress the spring until it is possible to safely loosen the retaining bolts.*

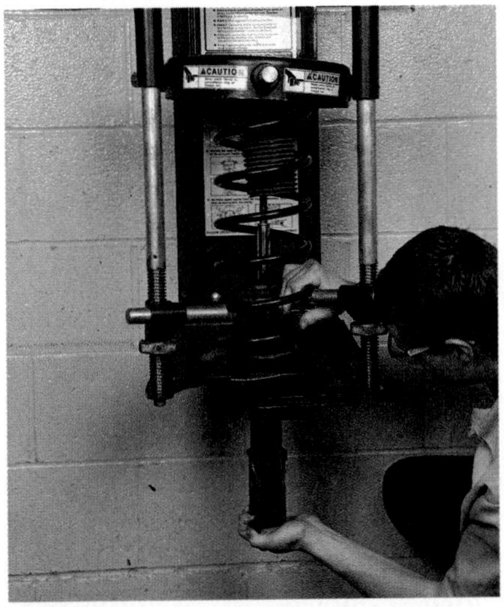

P48-9 *Remove the old strut assembly from the spring and install the new strut. Compress the spring to allow for reassembly and tighten the retaining bolts.*

P48-10 *Reinstall the strut assembly into the car. Make sure all bolts are properly tightened and in the correct locations.*

SEMI-INDEPENDENT SUSPENSION

A semi-independent suspension system is used on many front-wheel-drive models. On some, the suspension position is fixed by an axle beam, or crossmember, running between two trailing arms. Although there is a solid connection between the two halves of the suspension because of the axle beam, the beam twists as the wheel assemblies move up and down. The twisting action not only permits semi-independent suspension movement, but it also acts as a stabilizer. Frequently, a separate shock and spring trailing arm system is also used. In either an integrated or separate shock system, each rear wheel is independently suspended by a coil spring.

A coil spring and shock absorber-strut assembly are ordinarily used with this suspension system. The bottom of the strut is mounted to the rear end of the trailing arm. The top is mounted to the reinforced inner fender panel. Braking torque is transmitted through the trailing control arms and struts. The arms and struts also maintain the fore and aft, and lateral positioning of the wheels. A tracking bar is also used on some trailing arm suspension systems. The tracking bar helps to reduce sideways movement of the axle.

Semi-Independent Suspension System Servicing

As in most rear-system servicing, the first step is to remove the shock absorber. It is important to remember not to remove both shock absorbers at one time. Suspending the rear axle at full length could result in damage to the brake lines and hoses. The servicing of a semi-independent suspension system usually involves the removal and reinstallation of shock absorbers, springs, insulators, and control arm bushings. Follow the procedures given in the vehicle's service manual.

WARNING!

When removing the rear springs, do not use a twin-post hoist. The swing arc tendency of the rear-axle assembly when certain fasteners are removed might cause it to slip from the hoist. Perform this operation on the floor if necessary.

Independent Suspension

Independent suspensions can be found in large numbers on both FWD and RWD vehicles. The introduction of independent rear suspensions was brought about by the same concerns for improved traction ride that prompted the introduction of independent front suspensions. If the wheels can move separately on the road, traction and ride is improved.

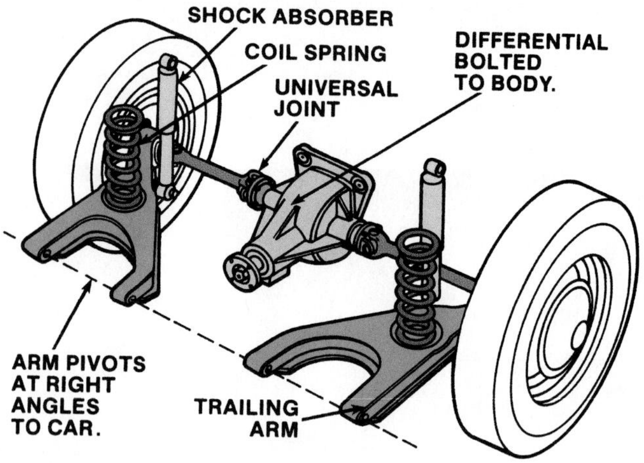

Figure 43–36 Trailing arms are often used with independent rear suspensions.

Independent coil-spring rear suspensions can have several control arm arrangements. For example, A-shaped control arms are sometimes employed. When the wide bottom of a control arm is toward the front of the car and the point turns in to meet the upright, they are called trailing arms **(Figure 43–36)**. When the entire A-shaped control arms are mounted at an angle, they are known as semitrailing control arms or multilink suspensions. Coil springs are used between the control arm and the vehicle body. The control arms pivot on a crossmember and are attached at the other end to a spindle. A shock absorber is attached to the spindle or control arm.

Some vehicles use a rear-suspension system that uses a lower control arm and open driving axles. A crossmember supports the control arms, while the tops of the shock absorbers are mounted to the body. The springs are set in seats at the bottom and top of the crossmember.

A few cars use only lower control arms, but substitute a wishbone-shaped subframe for the upper control arms. Two torque arms transfer the rear-end torque to the subframe. In fact, many cars are now featuring rear **double-wishbone** suspension. Torque loads create bushing and control arm deflection during braking, cornering, acceleration, and deceleration. It is interesting to note this rear-suspension system allows for a small amount of toe-in change to enhance straight line stability. The toe-in change during cornering leads to quicker and more responsive turning. The rear-suspension system can also be tuned to ensure minimal dive under braking and minimal squat under acceleration.

Currently, struts are replacing conventional shock absorbers in rear independent-suspension systems. One of the latest strut rear-suspension designs used by car manufacturers is shown in **Figure 43–37**.

On this type of system, the spindle is used to secure the strut, the outer ends of two of the four control arms, the rear ends of the tie-rods, and a rear wheel. The control arms contain bushings of different sizes at their

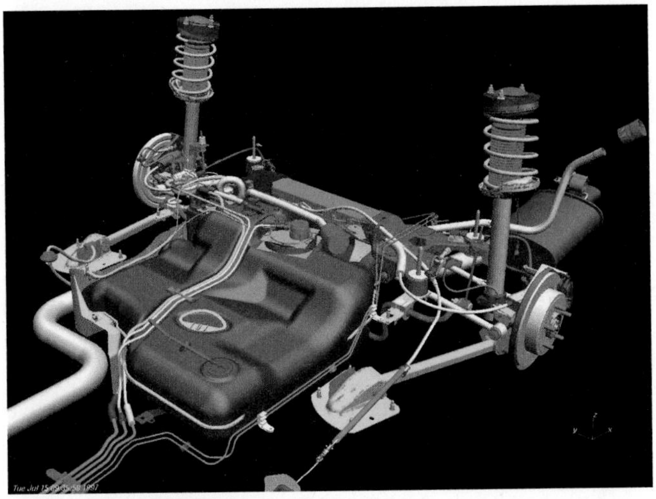

Figure 43-37 A strut-based rear suspension system. *Courtesy of DaimlerChrysler Corporation*

outer ends. The ends with the smaller bushings attach near the body centerline. The ends with the larger bushings attach at the spindle. (When replacing control arms, it is mandatory that offsets at their outer ends and the flanges on the arms face in the direction prescribed in the manufacturer's service manual.) This system is also called the nonmodified MacPherson strut system.

The modified MacPherson strut rear suspension is common for vehicles with front-wheel drive. The major components on each side of the vehicle are a modified MacPherson shock strut, lower control arm, tie-rod, and wheel spindle. A coil spring mounts between the lower control arm and the body crossmember/side rail. The spindle, in addition to supporting the rear wheel, is used as an attaching location for the outer end of the control arm and the rear end of the tie-rod. The inner end of the control arm attaches to the crossmember. The forward end of the tie-rod attaches to the side rail.

Another rear strut design uses a **Chapman strut** that is similar to the modified MacPherson strut. The difference between the two struts is that the MacPherson strut is involved directly in the car's steering system. The Chapman strut is not. In addition, the Chapman strut can be used with conventional springs, often a leaf-type spring. It frees the Chapman strut of load-carrying duties so it can concentrate on providing exact wheel location and shock absorbing functions.

Rear leaf-spring suspension systems are used on many vehicles with conventional rear drives. These leaf springs are generally mounted longitudinally in the same manner as described earlier in this chapter for live-axle systems. A few leaf-spring systems, however, employ springs mounted transversely. Both multiple-leaf and monoleaf, or single-leaf, can be used. The transverse-leaf spring is mounted to the differential housing rather than the vehicle frame as in the longitudinal installation. The transversely mounted spring's eyes are connected to the wheel spindle assemblies.

Rear shock absorbers or shock struts have the same service limitations as those used on the front of the vehicle. They cannot be adjusted, refilled, or repaired. The procedure for inspecting rear shocks or shock struts is similar to previously mentioned front-end parts inspection. Repetition is not necessary.

Multilink Rear Suspension

A multilink rear suspension uses several control arms to guide the wheel **(Figure 43–38)**. Different models feature different types of multilink rear suspensions that satisfy the varying demands of vehicle dynamics, ride comfort, and space requirements. These include the double-wishbone rear suspension, trailing-link double-wishbone rear suspension, and trapezoidal-link rear suspension.

In the double-wishbone suspension, the wheel is guided by two triangulated lateral control arms (the wishbones) and a tie rod. The suspension strut is attached to the lower wishbone to provide vertical support.

The trailing-link double-wishbone suspension has a trailing link that also carries the wheel and upper and lower wish bones. The spring is located on the trailing link ahead of the center of the wheel; the shock absorber is behind it.

The trapezoidal-link rear suspension permits excellent performance, handling, and comfort. The rear wheel is fixed by an upper lateral control arm and a trapezoidal lower link with a tie rod behind it. For reduced weight, the trapezoidal link and upper control are hollow aluminum castings.

Servicing Independent Suspension Systems

Most of the servicing techniques for rear independent suspension systems—except coil, control arm, and strut

Figure 43-38 A multilink rear suspension. *Courtesy of DaimlerChrysler Corporation*

removal and installation—are similar to other front- and rear-suspension parts. They have been covered earlier in this chapter. Of course, check the service manual for all inspection and repair techniques of the vehicle's independent rear system.

Servicing Rear Coil Spring Raise the vehicle on a frame contact hoist or position jack stands under the frame forward of the rear-axle assembly. This allows the shock absorbers to fully extend. Place a floor jack under the center of the rear-axle housing and support the weight of the rear axle, but do not lift the vehicle off the jack stands. Disconnect the lower end of the shock absorber. Then, lower the floor jack until all of the coil spring force is relieved. If a coil spring positioner is used, remove it from the center of the coil spring. The coil spring can usually be removed from the vehicle at this time by lifting it from its spring seat. If the springs are to be used again, mark or tag each one so it can be returned to its original location. When a replacement is needed, always replace coil springs in pairs. This ensures equal height.

To install a spring, place the insulator on top of the coil spring and position the spring on the spring seat. The end of the top coil must be positioned to line up with the recess in the spring seat. Jack up the rear-axle housing so the spring is properly seated at the lower end and the shock absorbers line up. Reconnect the shock absorbers.

There are some definite advantages to working on one spring at a time. First of all, the assembled side of the vehicle helps support the disassembled side. It also keeps the parts aligned and eliminates the possibility of putting the parts on the wrong side of the vehicle.

Servicing Rear Control Arms To remove the upper rear control arms from the vehicle, remove the bolts passing through the control arms at the frame and at the axle ends. Usually the rear coil spring does not have to be removed for this. Service one side of the vehicle at a time. This simplifies realigning the parts during assembly. On a serviceable control arm, replace the control arm bushings by removing the defective bushing with an appropriate puller. Properly position the new bushing and press it into place in the same manner as is done on front suspensions. Position the repaired control arm on the vehicle and loosely install the bolts. Repeat the service on the other control arm if necessary. Properly torque the nuts and bolts once the vehicle's entire weight is on the springs again.

The coil springs must be removed to service the lower rear control arms. Again, one side of the vehicle should be serviced at a time. Once the vehicle is properly supported and the springs are dismantled, remove the nuts and bolts that pass through the control arm. Remove the control arm from the vehicle and service it in the same way as the upper control arm.

Check the service manual to see if there is an adjustment for the driveline working angle. If none is specified, torque the control arm bolts to specification while the full vehicle weight is on the rear axle. This sets neutral bushing tension at normal curb height.

When there is a driveline working angle adjustment, adjust the angle before torquing the control arm bolts. After the rear suspension has been serviced, always check the working angle of the universal joints on the drive shaft. This minimizes the possibility of driveline vibration.

Some independent rear-suspension systems have ball joints that perform a function similar to the front ball joints, and they should be inspected in the same way.

Although very few cars have rear wheels that steer, some independent suspension systems have components that would normally be seen only on vehicles with four-wheel steering. Components such as tie-rod ends may appear to serve the same purpose as if they were on the front suspension. They are used to adjust the angle of the wheels for stable straight ahead performance. These suspension and steering items are covered in greater detail in Chapters 44 and 45 of this book.

ELECTRONICALLY CONTROLLED SUSPENSIONS

All of the suspension systems covered up to this point are known as passive systems. Vehicle height and damping depend on fixed nonadjustable coil springs, shock absorbers, or MacPherson struts. When weight is added, the vehicle lowers as the springs are compressed. Air-adjustable shock absorbers may provide some amount of flexibility in ride height and ride firmness, but there is no way to vary this setting during operation. Passive systems can be set to provide a soft, firm, or compromise ride. Vehicle body motion and tire traction vary due to road conditions and turning and braking forces. Passive systems have no way of adjusting to these changes.

Advances in electronic sensor and computer control technology have led to a new generation of suspension systems. The simplest systems are level control systems that use electronic height sensors to control an air compressor linked to air-adjustable shock absorbers.

More advanced adaptive suspensions are capable of altering shock damping and ride height continuously. Electronic sensors **(Figure 43–39)** provide input data to a computer. The computer adjusts air spring and shock damping settings to match road and driving conditions.

The most advanced computer-controlled suspension systems are true active suspensions. These systems are hydraulically, rather than air, controlled. They use high pressure hydraulic actuators to carry the vehicle's weight rather than conventional springs or air springs.

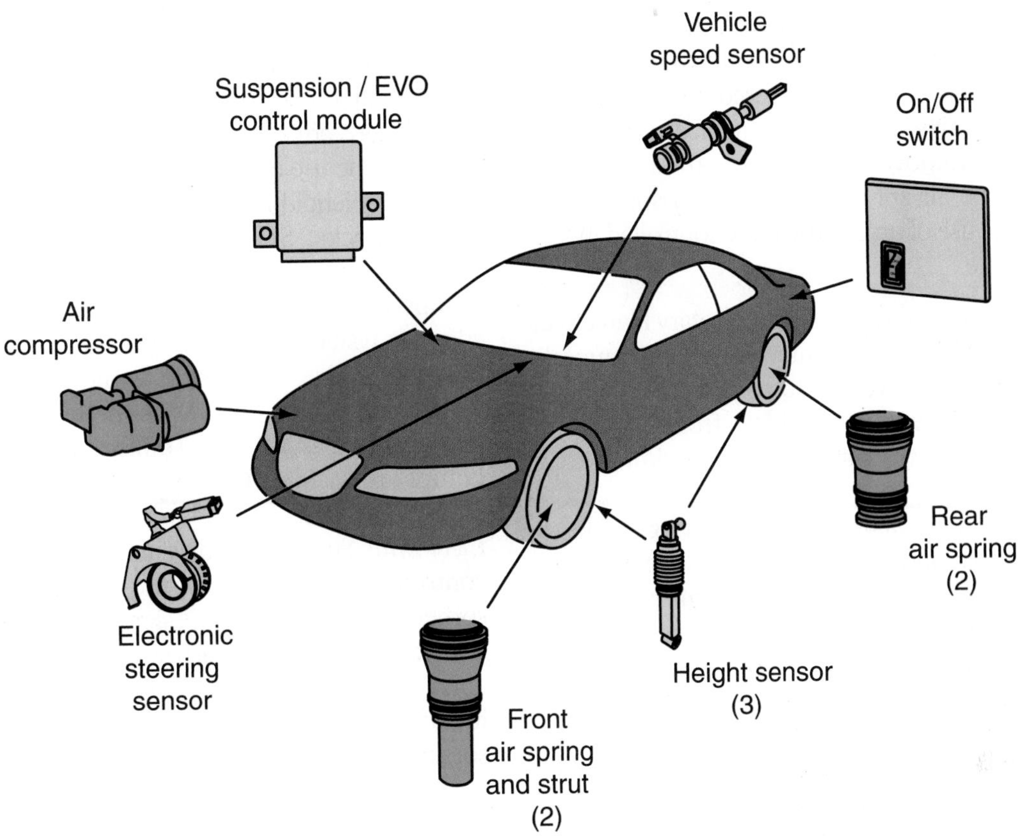

Figure 43-39 The components of a typical electronic air suspension system.

The unique feature of an active suspension is that it can be programmed to respond almost perfectly to various operating conditions. For example, by raising the height of the outside actuators and lowering the inside actuators when going around a curve, the vehicle can be made to lean into a curve, much like a motorcycle. Active systems using hydraulic actuators are presently used on a limited number of high-performance vehicles. Most manufacturers are introducing various adaptive suspension systems that rely on pneumatically actuated air springs and dampers.

Some late-model pickups and SUVs offer air suspension systems. These systems are added to existing leaf-spring suspensions. The air spring is positioned between the center of the leaf spring and the frame of the truck. The air spring serves as an adjustable and additional spring at each end of the axle.

Adaptive Suspensions

Adaptive suspensions use electronic shock absorbers with variable valving. In some cases, variable air spring rates are used to adapt the vehicle's ride characteristics to the prevailing road conditions or driver demands.

Electronic sensors monitor factors such as vehicle height, vehicle speed, steering angle, braking force, door position, shock damping status, engine vacuum, throttle position, and ignition switching. A computer is used to analyze this input and switch the suspension into a preset operating mode that matches existing conditions. Some systems are fully automatic. Others allow the driver to select the ride mode.

At present, adaptive suspensions are less costly and complicated than hydraulically controlled active suspensions. However, they do have some limitations. Although they can reduce body roll, adaptive suspensions cannot eliminate it like true active systems. Adaptive systems also experience a slight delay in their reaction time, although some systems can change shock valving in as little as 150 microseconds.

System Components There are many different designs and components used by the manufacturers to accomplish the same task. Some systems use adjustable shocks, while others use air springs at each side of the axles. The air spring membrane is similar to a tire in construction. A solenoid valve and filter assembly allows clean air to be added or released from the air spring. Adding or removing air changes the ride height of the vehicle.

The airflow to the springs is controlled by the interaction of the air compressor, system sensors, computer control module, and solenoid valves. All of the air-operated parts of the system are connected by nylon tubing.

Compressor. The compressor supplies the air pressure for operating the entire system. It is often a positive displacement single piston pump powered by a 12-volt DC motor. A regenerative air dryer is attached to the compressor output to remove moisture from the air before it is delivered to the air springs. The compressor is operated through the use of an electric relay controlled by the computer module.

Sensors. Vehicle height sensors can be rotary Hall-effect sensors that enable the computer to more accurately measure ride height as well as to compensate for road variations. This prevents the vehicle from bottoming out when crossing over railroad tracks or similar road irregularities.

Advanced systems also read the steering angle by using a photo diode and shutter location inside the steering column. This allows the system to firm up the suspension when the vehicle is turning. The system also reads engine vacuum or throttle position to stiffen the suspension when the vehicle is accelerating. A brake sensor allows the system to compensate for front nose dive during hard braking. Some systems use a special G-sensor to sense sudden acceleration or braking. Other adaptive systems use a yaw sensor to pick up body roll when cornering.

Electronic Shock Absorbers. Many adaptive suspension systems use electronically controlled shock absorbers that feature variable shock damping. The degree of damping is controlled by the computer based on input from the vehicle speed, steering angle, and braking sensors. As explained earlier in this chapter, variable shock damping is accomplished by varying the size of the metering orifices inside the shock. A small actuating motor mounted on top of the shock absorber rotates a control rod that alters the size of the metering orifices.

A recent advancement in adaptive suspension technology is the use of real-time shock damping. These systems use solenoid-actuated shocks rather than the motor driven shocks. Solenoids allow almost instantaneous valving changes. This means the suspension can react to bumps and body motions as they happen. Real-time adaptive systems deliver most of the handling advantages of a full active suspension without increased vehicle weight and power drain. With these systems, changes to shock valving as little as 10 milliseconds is possible when bumps are encountered.

Electronic Struts. Some systems use an electronically controlled strut in place of the air spring and shock absorber **(Figure 43–40)**. Design and operation is similar to electronic shock absorbers. A valve selector or variable orifice located inside the strut controls fluid pressure in the suspension system based on input from many sensors and commands from the system's control module.

Some variable damping suspension systems use air or gas rather than a fluid. At speeds up to 40 mph (65 km/h), the orifice is fully open and provides full flow. From 40 to 60 mph (65 to 100 km/h), the orifice is in the normal position and flow is restricted. At speeds more than 60 mph (100 km/h), or when the vehicle is accelerating or braking, the variable orifice is shifted to the firm position.

The use of a variable orifice in the damper control, coupled with the deflected disc valving, provides optimum fluid flow control for both rebound and jounce strokes. In the comfort mode, the selector is set to allow

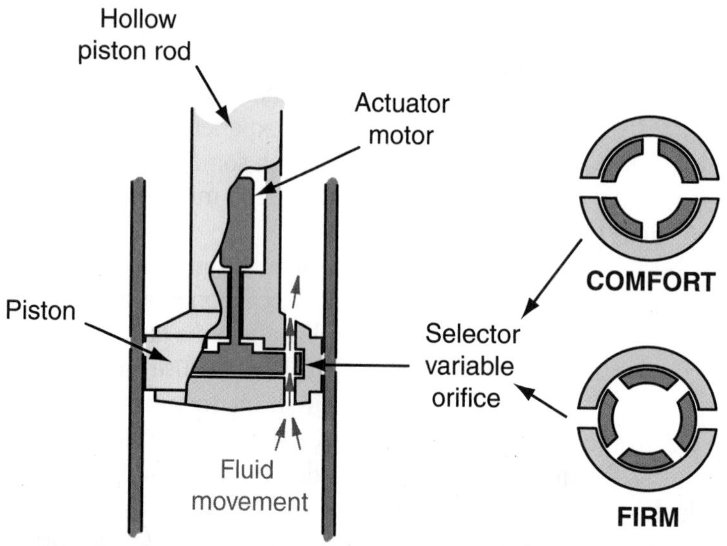

Figure 43–40 A computer command ride strut. Four electronically controlled struts are used on many adaptive suspension systems. Based on input from the computer, the valving selector shifts the variable bypass orifice to a comfort, normal, or firm setting.

fluid flow primarily through the large selector orifice to achieve minimum damping forces. While in the normal mode, the unit is set to balance fluid flows between the small selector orifice and the deflected disc valving to provide moderate damping forces. Under conditions in which the firm mode is needed, the selector is rotated to its firm or blocked position and fluid flows entirely through the deflected disc valving.

The damper control also can raise or lower the vehicle's height. This action also improves the car's aerodynamic characteristics at highway speed. As speed increases, the suspension reduces the vehicle's height and the front end angles downward. This action tends to reduce wind resistance for greater stability and better gas mileage. As the vehicle slows, the suspension brings the body up to its normal height and level position.

Computer Control Module. A microcomputer-based module controls the air compressor motor (through a relay), the compressor vent solenoid, and the four air spring solenoids. The computer module also controls operation of electronic shock absorber actuating motors and electronic strut valving selectors. The control module receives input from all system sensors.

The computer module also has the capability of performing diagnostic tests on the system. It has a preprogrammed routine for properly fitting air springs after servicing. The module also controls the dash-mounted system warning light.

Electrical power to operate the basic air suspension system is distributed by the main body wiring harness. Each wiring harness involved has a special function in the typical air suspension system.

WARNING!

The compressor relay, compressor vent solenoid, and all air spring solenoids have internal diodes for electrical noise suppression and are polarity sensitive. Care must be taken when servicing these components not to switch the battery feed and ground circuits, or component damage results. When charging the battery, the ignition switch must be in the off position if the air suspension switch is on, or damage to the air compressor relay or motor may occur. However, use of a battery charger while performing the diagnostic test or air spring fill option is acceptable. Set to a rate to maintain but not damage the vehicle battery.

Electronic Leveling Control. Adaptive suspension systems are capable of adjusting the suspension system during operation. Less complicated electronic level-control systems are used on many large and mid-size vehicles.

These systems do not use a computer module. In most cases, height sensors are the only types of sensors used. These height sensors sense when passenger weight or cargo is added to or removed from the vehicle **(Figure 43–41)**. The height sensors control two basic circuits. The compressor relay coil grounds circuits that activate the compressor. The exhaust solenoid coil grounds circuits that vent air from the system.

To prevent falsely actuating the compressor relay or exhaust solenoid circuits during normal ride motions, the sensor circuitry provides an 8- to 15-second delay before either circuit can be completed.

In addition, the typical sensor electronically limits compressor run time or exhaust solenoid energized time to a maximum of approximately 3½ minutes. This time limit function is necessary to prevent continuous compressor operation in a case of a solenoid malfunction. Turning the ignition off and on resets the electronic timer circuit to renew the 3½ minute maximum run time. The height sensor is mounted to the frame crossmember in the rear. The sensor actuator arm is attached to the rear upper control arm by a link. The link should be attached to the metal arm when making any trim adjustment.

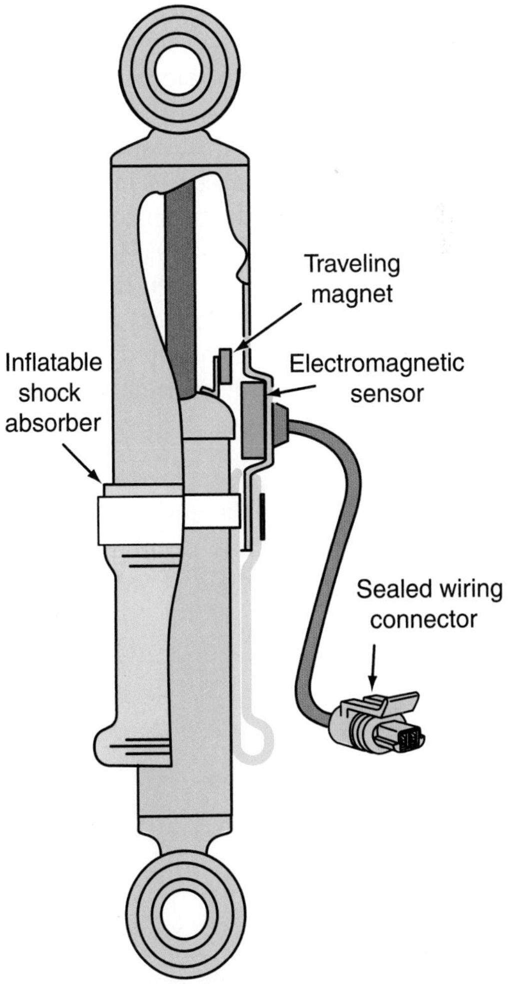

Figure 43–41 A load-sensing shock absorber.

Figure 43-42 An MR fluid-based strut. *Courtesy of the Delphi Corporation*

When the air line is attached to the shock absorber fittings or compressor dryer fitting, the retainer clip snaps into a groove in the fitting, locking the air line in position. To remove the air line, spread the retainer clip, release it from the groove, and pull on the air line.

Adjustable Pneumatic Suspension Adjustable pneumatic suspension at the front and rear wheels is a feature on some AWD vehicles. By varying the vehicle's ground clearance, it can be used off-road but also performs and handles well on the highway. There are four ride-height positions that can be selected either manually or automatically, with a total range of over 8 inches (20 mm) of ground clearance. At highway speeds, the vehicle's ground clearance is 5.6 inches (14 mm). Urban mode raises it a full inch. For moderate off-road and local driving, ground clearance is 7.6 inches (19 mm). For severe off-road conditions at speeds under 25 mph (40 km/h), maximum clearance is 8.2 inches (21 mm). The vehicle will adjust to the desired height based on vehicle speed, or the driver can temporarily override the setting by depressing a button.

MagneRide

MagneRide is a semiactive suspension system, which features shocks or struts with no electromechanical valves or small moving parts. Instead of valve-controlled orifices, MagneRide regulates the flow of fluid by a variable magnetic field produced by a small electric coil mounted in the shock **(Figure 43–42)**. The shocks are filled with magneto-rheological (MR) fluid. MR fluid consists of magnetically soft particles, such as iron, suspended in synthetic hydrocarbon fluid.

The action of the shock forces the MR fluid through a magnetized opening in each shock. When the shock is in its off state, the fluid is not magnetized and flows freely through the orifice. When current is sent to the coil, the fluid becomes magnetized and its viscosity changes instantly **(Figure 43–43)**.

The material changes from a fluid state to a semisolid state that is directly proportional to the magnetic field applied to it. With little or no electrical current, the iron particles are randomly distributed, and the fluid passes freely through the piston orifice. When a strong electrical current is applied to the coil, the resulting magnetic field aligns the iron particles so that the fluid stiffens and the flow is resisted. This condition causes heavy damping. The resulting damping force is proportional to the viscosity, which is proportional to the strength of the magnetic field.

Sensors monitoring wheel position, lateral acceleration, vehicle speed, steering wheel angle, and brake pedal angle are inputs to the control module that sends current to the coil in the shocks. This system provides extremely quick response time, typically about 5 ms, and the fluid is capable of reacting 30,000 times per second.

SERVICING ELECTRONIC SUSPENSION COMPONENTS

Most electronic suspension servicing requires the removal of the component from the system, replacing or repairing it, and then reinstalling the component back into the system. Procedures for individual component replacement are covered in the vehicle's service manual. Serviceable components include the air compressor, charger, mounting brackets, height sensors, air springs, air lines and connections, gas struts, strut mounts, control arm components, shock absorbers, and stabilizer bars.

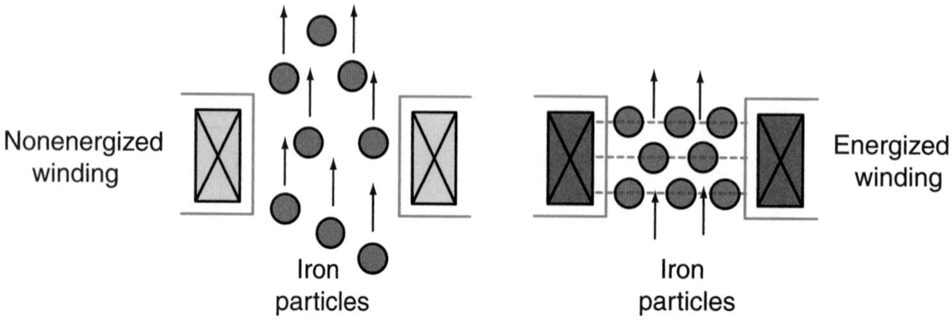

Figure 43-43 The iron particles in the MR fluid align themselves when they pass through the magnetic field, causing the fluid to stiffen.

Failure to keep the following procedures in mind might result in a sudden failure of the air spring or suspension system. Suspension fasteners are important attaching parts. They could affect performance of vital components and systems or could result in major service expenses. They must be replaced with fasteners of the same part number or with an equivalent part, if replacement becomes necessary. Do not use a replacement part of lesser quality or substitute design. Torque values must be used as specified during assembly to ensure proper retention of parts. New fasteners must be used in the place of the old ones whenever they are loosened or removed and when new component parts are installed.

WARNING!

Do not remove an air spring under any circumstances when there is pressure in the air spring. Do not remove any components supporting an air spring without either exhausting the air or providing support for the air spring. Power to the air system must be shut off by turning the air suspension switch (in the luggage compartment) off or by disconnecting the battery when servicing any air suspension components. Most air suspension systems are equipped with a warning light. The light comes on if there is a problem, or when servicing the system.

CUSTOMER CARE

Because the technician is seldom present when a vehicle requires towing, it is important to advise the customer of proper procedures so the tow operator does not damage the electronic suspension system.

You must also know the proper hoist lifting and jacking restrictions. While it is necessary to check the service manual for specific instructions, the following are the basics for electronic suspension.

When towing, it must be remembered that when the ignition is off, the automatic leveling suspension is still on. Before lifting the vehicle, be sure the ignition switch is turned off and the trunk switch deactivated.

When towed from the front, towing should not exceed a speed of 35 mph (60 km/h) or a distance of 50 miles (80 km/h). When the car is towed from the rear, speeds should not exceed 50 mph (80 km/h) (or 35 mph [60 km/h] on bumpy pavement).

A body hoist is usually the only type of life recommended. In most service manuals, manufacturers warn against using a suspension hoist. The proper sequence is to position the car over the lift, shut off the ignition, then deactivate the system.

If a body hoist is not available, a floor jack and jack stands will do. Lift the car by the front cross member and the rear jacking points that are just in front of the rear wheel wells. Jack stands should be used to support the car.

In all situations, the lifting theory is the same. The suspension should be free to hang down while the car is in the air. This allows the wheels to be supported by the struts in the front and the shocks at the rear, both in their full extension (rebound) positions. Thus, the membrane of each air spring retains its proper shape while the car is in the air. ■

Vehicle Alignment

Aligning a vehicle with an electronic suspension system is essentially the same as the aligning procedure described in Chapter 45—with one notable exception—curb height.

Curb height is an important dimension because it affects the other alignment angles. Caster is the most obvious one that is affected, but front camber and toe can also be included. Curb height is especially critical when checking rear camber and toe on independent rear suspensions. With electronic suspension, the ride can vary depending on various circumstances. The only way to guarantee the suspension at curb height is to preset it.

ACTIVE SUSPENSIONS

Some of the advanced adaptive suspension systems may be called **active suspensions**. In this text, active suspensions refer to those controlled by double-acting hydraulic cylinders or solenoids (usually called actuators) that are mounted at each wheel. Each actuator maintains a sort of hydraulic equilibrium with the others to carry the vehicle's weight, while maintaining the desired body attitude. At the same time, each actuator serves as its own shock absorber, eliminating the need for yet another traditional suspension component.

In other words, each hydraulic actuator acts as both a spring (with variable-rate damping characteristics) and a variable-rate shock absorber. This is accomplished in an active suspension system by varying the hydraulic pressure within each cylinder and the rate at which it increases or decreases. By bleeding or adding hydraulic pressure from the individual actuators, each wheel can react independently to changing road conditions.

The components that make such a system possible are the actuator control valves, various sensors, and the chassis computer **(Figure 43–44)**. Feeding information to a computer are a number of specialized sensors. Each actuator has a linear displacement sensor and an acceleration sensor to keep the computer informed about the actuator's relative position. This enables the computer to track the extension and compression of each actuator, and to know when each wheel is undergoing jounce or rebound. There are also load sensors and hub acceleration sensors in each wheel to measure how heavily each wheel is loaded.

A steering angle sensor is used to signal the computer when the vehicle is turning. To monitor body motions, a roll sensor and lateral acceleration and G-sensors are used. The computer also monitors hydraulic pressure within the system and the speed of the pump monitor.

Once it has all the necessary inputs, the computer can then regulate the flow of hydraulic pressure within each individual actuator according to any number of variables and its own built-in program. Another nice feature of a suspension such as this is that it can be programmed to behave in a variety of unique and currently impossible ways: leaning or rolling into turns, for example, or even raising a flat tire on command in order to change the tire without using a separate jack.

When the wheel of an active suspension hits a bump, the sensors detect the sudden upward deflection of the wheel. The computer recognizes the change as a bump, and instantly opens a control valve to bleed pressure from the hydraulic actuator. The rate at which pressure is bled from the actuator determines the cushioning of the bump and the relative harshness or softness of the ride. The rate can be varied at any point during jounce or rebound to produce a variable spring rate effect. In other words, the feel of the suspension can be programmed to respond in an almost indefinite variety of ways. Once the bump has been absorbed by the actuator, pressure is forced back into it to keep the wheel in contact with the road and to maintain the suspension's desired ride height.

During hard braking with a conventional suspension system, there is a tendency for a vehicle to make a dive. The weight of the vehicle seemingly pushes the front of the car downward and the back upward. During hard braking, the active suspension increases air pressure in the front actuators and reduces air pressure in the rear actuators. These actions minimize dive to keep the vehicle level and make it easier for the driver to control. After braking, valves operate to equalize air pressures in front and rear air actuators and level the vehicle again.

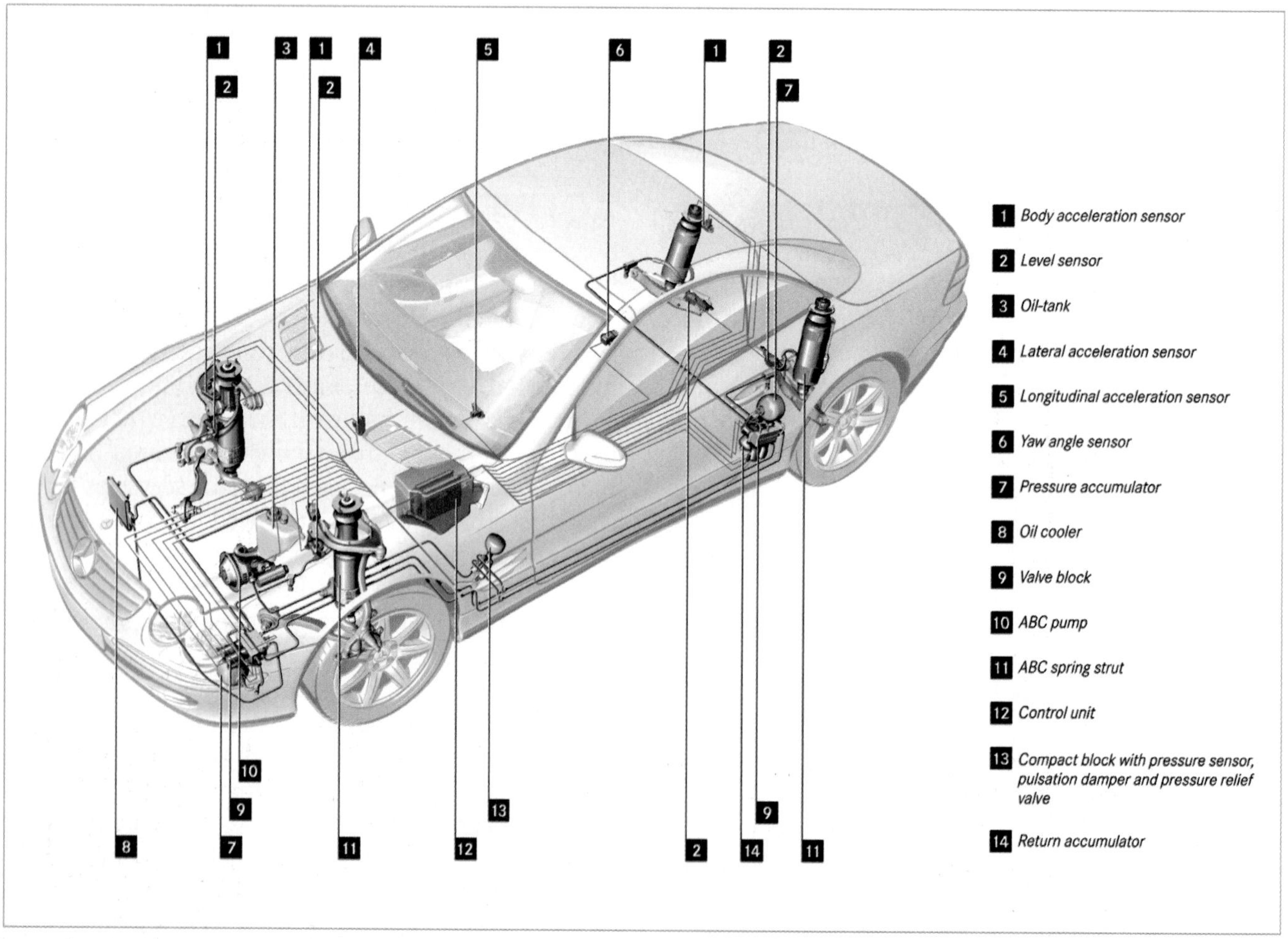

Figure 43-44 An active suspension system. *Courtesy of DaimlerChrysler Corporation*

Frequently, when a driver depresses the accelerator quickly during hard acceleration, the front end of the vehicle tends to lift up, while the rear end lowers. The action is known as **squat**. With an active suspension system, squat is controlled by the operating valve's solenoids, which increase the air pressure in rear wheel actuators and reduce air pressure in front wheel actuators. When the vehicle is no longer accelerating quickly, the control system operates valves to equalize air pressures and level the vehicle. Thus, an active suspension changes the height of the front, rear, or either side of the vehicle to counteract tilting, rolling, and leaning. These active attitude control functions improve vehicle stability and increase tire traction and driver control.

The power required for a totally active system is only 3 to 5 horsepower (about the same as a typical power steering pump). Power consumption is lowest when the system is least active, as when driving on a smooth road. Rough roads and hard maneuvers, on the other hand, put more of a demand on the system. The hydraulic pump works harder and thus requires more power.

Power consumption can be reduced by going with a semiactive suspension that uses small springs with the hydraulic actuators. The springs help to support the vehicle's weight, which reduces the load on the actuators. Smaller actuators that require less hydraulic power can then be used, which reduces the bulk and weight of the system. The addition of springs also adds a certain margin of safety to the system to keep it from going flat should the hydraulics spring a leak.

Although not as widely used as electronic leveling or adaptive suspension systems, hydraulic active suspensions are sure to become more common.

Chassis Lubrication

Grease fittings, called zerk fittings, are threaded into the part that should be lubricated. These fittings have a one-way spring-loaded check valve that allows grease into the joint but prevents it from leaking out. To lubricate the chassis of a vehicle, safely raise the vehicle and lock the lift or set the vehicle on safety stands. Locate all of the lubrication points and wipe the fittings clean with a shop towel. If the lubrication points do not have Zerk fittings but have plugs, remove the plugs and install new Zerk fittings or use a special lubrication adapter.

Carefully look at the joints and determine if the joint boots are lealed. If the boots are good, push the grease gun nozzle straight into the fitting and pump grease slowly into the joint. If the joint has a sealed boot, put just enough grease into the joint to cause the boot to expand slightly. If the boot is not sealed, put in enough grease to push the old grease out. Then wipe off the old grease. Repeat this at all lubrication points and wipe all excess grease off the joints and fittings.

CASE STUDY

***A**n eight-year-old vehicle is brought to the shop. The customer states the vehicle runs well, but the ride has become so bouncy he wants the shock absorbers replaced.*

The work order is written and the technician replaces the shock absorbers that afternoon.

The following day, the customer returns, complaining of only a slight improvement in ride and handling. He states the shocks must be defective or mismatched to his vehicle. Faced with an angry customer, the technician now takes the time to do what he should have done the previous day. He takes the vehicle on a road test and performs a complete inspection. On checking the ride height of the vehicle, he finds the vehicle is riding extremely low. This is a sure sign the coil springs are severely weakened.

Showing the customer the ride height measurement and the factory specification for that measurement convinces the customer of the need for new coil springs.

With new coil springs and shock absorbers, ride and handling improve dramatically.

KEY TERMS

Active suspension
Adaptive suspension
Antisway bar
Axle tramp
Ball joint
Chapman strut
Control arm
Dampers
Double wishbone
Fiber composite
Independent front suspension (IFS)
Jounce
Linear rate
MacPherson strut
Main leaf
Monoleaf
Monoshock
Multiple-leaf
Rebound
Shock absorber
Short-long arm (SLA)
Spring rate
Sprung weight
Squat
Steering knuckle
Strut rod
Sway bar
Torsion bar
Trailing arm
Unsprung weight
Variable rate
Wheel spindle

SUMMARY

- Four types of springs are used in suspension systems: coil, leaf, torsion bar, and air.

- Springs take care of two fundamental wheel actions: jounce and rebound.
- Common coil spring materials include carbon steel, carbon boron, steel, and alloy steels. Alloy steels, such as those containing chromium and silicon, improve the coil's resistance to relaxation. Most coil springs are manufactured by either a cold coiling or a hot coiling process. Hot coiling usually requires hardening and tempering along with short peening to increase the fatigue strength of the base material.
- Two basic designs of coil springs are used in vehicles: linear rate and variable rate.
- Leaf springs are made of steel or a fiber composite.
- In torsion suspension, the bar may either run from front to rear or side to side across the chassis.
- Air springs are generally only used in microprocessor-controlled suspension systems.
- Shock absorbers damp or control motion in a vehicle. A conventional shock absorber is a velocity-sensitive hydraulic damping device. The faster it moves, the more resistance it has to the movement.
- Shock absorbers can be mounted either vertically or at an angle. Angle mounting of shock absorbers improves vehicle stability and dampens accelerating and braking torque.
- There are two basic adjustable air shock systems: the manual fill type and the automatic or electronic load-leveling type.
- MacPherson struts provide the damping function of a shock absorber. In addition, they serve to locate the spring and to fix the position of the suspension.
- Domestic struts have taken two forms: a concentric coil spring around the strut itself and a spring located between the lower control arm and the frame.
- MacPherson suspensions use sway or stabilizer bars. Coil springs are used on all strut suspensions.
- Independent front suspension (IFS) must keep the wheels rigidly positioned and at the same time allow them to steer right and left. In addition, because of weight transfer during braking, the front suspension system absorbs most of the braking torque. When accomplishing this, it must provide good ride and stability characteristics.
- The unequal length arm or short-long arm (SLA) suspension system is most commonly used on domestic vehicles.
- Live-axle is the traditional rear suspension system and consists of springs used in conjunction with a live-axle (one in which the differential axle, wheel bearings, and brakes act as a unit). The springs are either of leaf or coil type.
- Semi-independent suspension is used on many front-wheel-drive models.
- Three strut designs are frequently used in IFS systems; the conventional MacPherson strut, the modified MacPherson strut, and the Chapman strut.
- The two basic types of computer suspension systems are adaptive and active.
- Electronically controlled suspensions can be either simple load-leveling systems, adaptive systems, or fully active systems. Adaptive and active suspension systems are computer controlled. Most load-leveling systems do not use a computer.
- Adaptive suspensions can alter vehicle ride height and shock absorber damping while the vehicle is in motion. Such systems use air springs and electronic shock absorbers or struts.
- Active suspensions are hydraulically operated actuators to control up-and-down and side-to-side movement. They can be programmed to respond to certain road conditions and turning forces.

TECH MANUAL

The following procedures are included in Chapter 43 of the *Tech Manual* that accompanies this book:

1. Inspecting suspension components.
2. Removing and replacing a strut assembly.
3. Replacing the cartridge in a strut assembly.
4. Removing and installing control arm bushings on a rear coil-spring-type suspension system.

REVIEW QUESTIONS

1. How does a stabilizer bar work?
2. Explain the difference between sprung and unsprung weight.
3. What is the principle of the air spring?
4. Explain the action of the conventional shock absorber on both compression (jounce) and rebound strokes.
5. Describe the action of the independent front wheel suspension system.
6. The core of any suspension system is the ______.
   a. wheel spindle assembly
   b. spring
   c. ball joints
   d. control arm
7. What occurs when a wheel hits a dip or hole and moves downward?
   a. jounce
   b. free length
   c. deflection
   d. rebound
8. Which of the following is part of the sprung weight of a vehicle?
   a. steering linkage
   b. tires
   c. engine
   d. all of the above
9. What occurs when a wheel hits a dip or hole and moves upward?
   a. jounce
   b. free length
   c. deflection
   d. rebound
10. Technician A says load-carrying ball joints should always have some play in them. Technician B says follower ball joints should never have any play in them. Who is correct?
    a. Technician A
    b. Technician B
    c. Both A and B
    d. Neither A nor B
11. Technician A says a weak suspension spring can cause a loss of traction during acceleration. Technician B says a weak suspension spring can cause poor braking power. Who is correct?
    a. Technician A
    b. Technician B
    c. Both A and B
    d. Neither A nor B
12. What controls the movement of a vehicle as it moves down a bumpy road?
    a. struts
    b. shock absorbers
    c. both a and b
    d. neither a nor b
13. The coil springs of the vehicle ______.
    a. support the weight of the vehicle
    b. provide axle location
    c. stabilize the up-and-down motion
    d. all of the above
14. Before replacing springs, vehicle height is checked on both sides of the front suspension, and tire pressure is checked. Should you check the amount of fuel in the vehicle?
    a. Yes, the fuel tank should be empty.
    b. Yes, the fuel tank should be full.
    c. No, it does not make any difference.
    d. None of the above.
15. Technician A says leaf-spring-type rear suspensions are subject to wheel tramp. Technician B says an antisway bar is designed to limit wheel tramp. Who is correct?
    a. Technician A
    b. Technician B
    c. Both A and B
    d. Neither A nor B
16. The two SLA systems in common use today are ______.
    a. coil spring and strut suspension
    b. coil spring and torsion bar suspension
    c. coil spring and single control arm suspension
    d. single and double control arm suspension
17. Coil springs are used ______.
    a. on all strut suspensions
    b. on selected strut suspensions
    c. not at all on strut suspensions
    d. only on modified MacPherson suspensions
18. When the wide bottom of a control arm is toward the front of the car and the point turns in to meet the upright, it is called a(n) ______.
    a. trailing arm
    b. semitrailing arm
    c. wishbone arm
    d. A-shaped arm
19. The modified MacPherson strut rear suspension is very common in ______.
    a. front-wheel-drive vehicles
    b. rear-wheel-drive vehicles
    c. pickup trucks
    d. station wagons
20. Technician A says the use of firmer, urethane bushings in the suspension system improves the vehicle's road holding ability and handling. Technician B says firmer bushings also help eliminate torque steer in FWD vehicles. Who is correct?
    a. Technician A
    b. Technician B
    c. Both A and B
    d. Neither A nor B

44

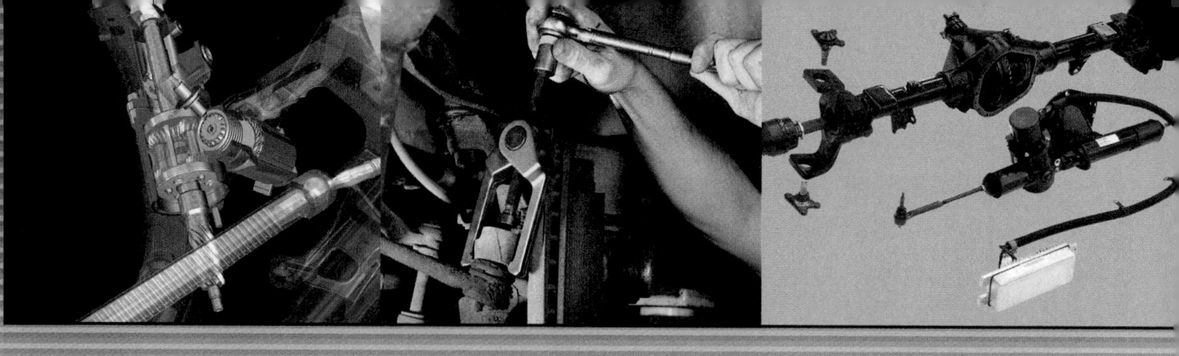

STEERING SYSTEMS

OBJECTIVES

■ Describe the similarities and differences between parallelogram, worm and roller, and rack and pinion steering linkage systems. ■ Identify the typical manual-steering system components and their functions. ■ Name the five basic types of steering linkage systems. ■ Identify the components in a parallelogram steering linkage arrangement and describe the function of each. ■ Identify the components in a manual rack and pinion steering arrangement and describe the function of each. ■ Describe the function and operation of a manual-steering gearbox and the steering column. ■ Explain the various manual-steering service procedures. ■ Describe the service to the various power-steering designs. ■ Perform general power-steering system checks. ■ Describe the common four-wheel steering systems.

The purpose of the steering system is to turn the front wheels. In some cases, it also turns the rear wheels. The wheels constantly change direction, while switching lanes, rounding sharp turns, and when avoiding roadway obstacles.

MANUAL-STEERING SYSTEMS

The steering system is composed of three major subsystems: the steering linkage, steering gear, and steering column and wheel. As the steering wheel is turned by the driver, the steering gear transfers this motion to the steering linkage. The steering linkage turns the wheels to control the vehicle's direction **(Figure 44–1)**. Although there are many variations to this system, these three major assemblies are in all steering systems.

Steering Linkage

The term **steering linkage** is applied to the system of pivots and connecting parts placed between the steering gear and the steering arms that are attached to the front or rear wheels that control the direction of vehicle travel. The steering linkage transfers the motion of the steering gear output shaft to the steering arms, turning the wheels to maneuver the vehicle.

The type of front-wheel suspension (independent wheel suspension as compared with a solid front axle) greatly influences steering geometry. Most passenger cars and many light trucks and recreational vehicles have independent front-wheel suspension systems. Therefore, a steering linkage arrangement that tolerates relatively large wheel movement must be used.

Parallelogram Steering Linkage

A parallelogram type of steering linkage arrangement was at one time the most common type used on passenger cars. Now it is found mostly on larger cars, pickups, and larger SUVs. It is used with the recirculating ball steering gear and can be classified into two distinct configurations: parallelogram steering linkage placed behind the front-wheel suspension **(Figure 44–2A)** and parallelogram steering linkage placed ahead of the front-wheel suspension **(Figure 44–2B)**. This type of steering linkage

Figure 44–1 Allowing a vehicle to do something other than go straight is the job of the steering system. *Reprinted with permission*

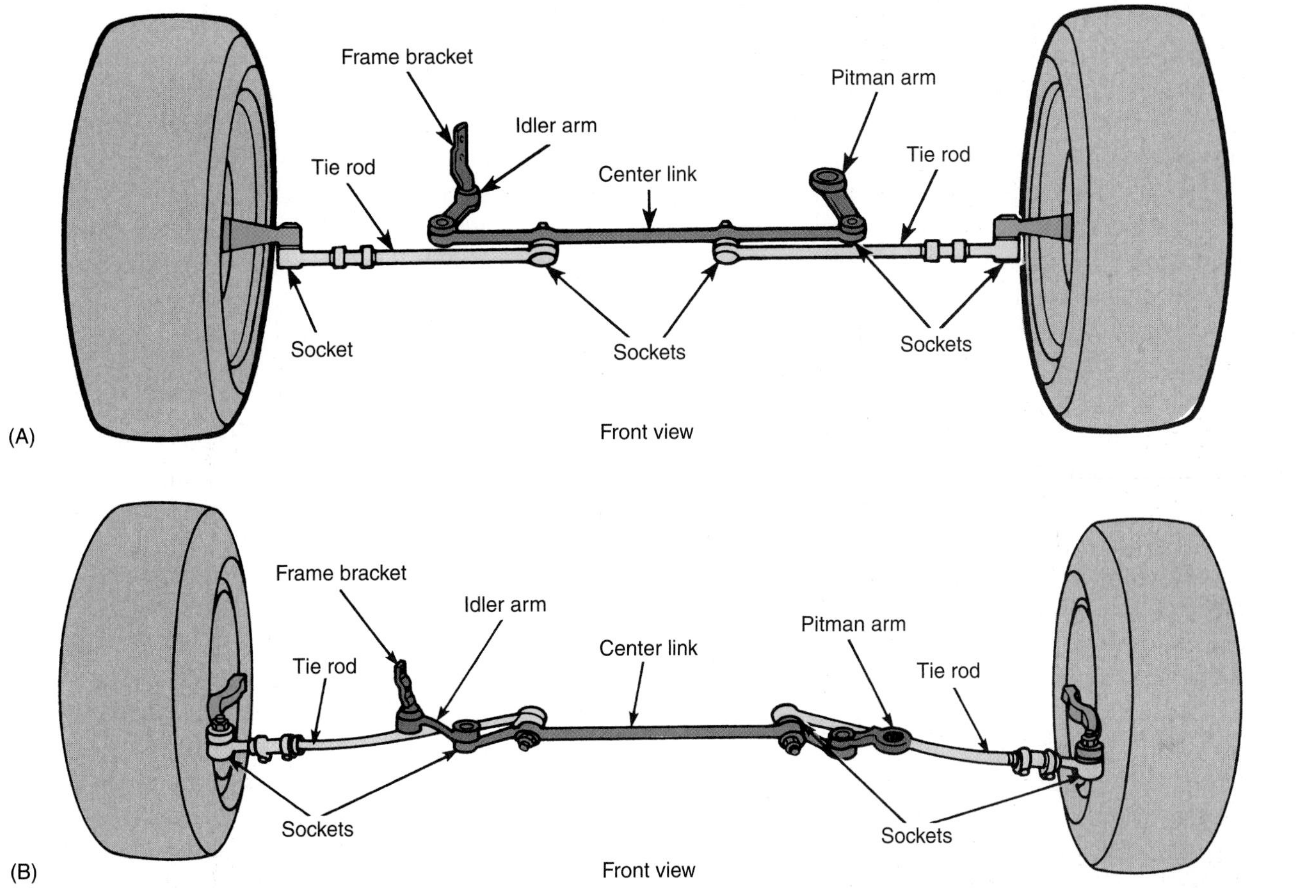

Figure 44-2 A parallelogram steering system mounts (A) behind front suspension and (B) ahead of front suspension.

is most often used where motor and chassis components would interfere with normal operation of the steering linkage.

These designs are the basic steering systems used in conjunction with independent front-wheel suspensions. This type of linkage also provides good steering and suspension geometry. Road vibrations and impact forces are transmitted to the linkage from the tires, causing wear and looseness in the system, which permits intermittent changes in the toe setting of the front wheels, allowing further tire wear.

In a parallelogram steering linkage, the tie rods have ball socket assemblies at each end. One end is attached to the wheel's steering arm and the other end to the center link.

The components in a parallelogram steering linkage arrangement are the pitman arm, idler arm, links, and tie rods.

Pitman Arm The **pitman arm** connects the linkage to the steering column through a steering gear located at the base of the column. It transmits the motion it receives from the gear to the linkage, causing the linkage to move left or right to turn the wheels in the appropriate direction. It also serves to maintain the height of the center link. This ensures that the tie rods are able to be parallel to the control arm movement and avoid unsteady toe settings or **bump steer**. *Toe*, a critical alignment factor, is a term that defines how well the tires point to the direction of the vehicle.

There are two basic types of pitman arms: wear and nonwear. Service needs differ, depending on which type of arm is used. Nonwear arms have tapered holes at their center link ends and normally need to be replaced only if they have been damaged in an accident or have been mounted with excessive tolerance. Wear arms have studs at the center link end and are subject to deterioration from normal operation. These arms must be inspected periodically to determine whether they are still serviceable.

Idler Arm The **idler arm** or idler arm assembly is normally attached, on the opposite side of the center link, from the pitman arm and to the car frame, supporting the center link at the correct height. A pivot built into the arm or assembly permits sideways movement of the linkage. On some linkages, such as those on a few light-duty trucks, two idler arms are used.

Idler arms normally wear more than pitman arms because of this pivot function, with wear usually showing up at the swivel point of the arm or assembly. Worn bushings or stud assemblies on idler arms permit excessive vertical movement in the idler arms.

Links Links, depending on the design application, can be referred to as **center links, drag links**, or steering links. Their purpose is to control sideways linkage movement, which changes the wheel's direction. Because they usually are also mounting locations for tie rods, they are very important for maintaining correct toe settings. If they are not mounted at the correct height, toe is unstable and a condition known as the toe change or bump steer is produced. Center links and drag links can be used either alone or in conjunction with each other, depending on the particular steering design.

There are several common designs of center links. Like pitman arms, they can be broadly characterized as either wear or nonwear. Center links with stud or bushing ends are likely to become unserviceable from the effects of normal operation and should be inspected periodically. Links with open tapers are nonwear and usually need to be replaced only if they have been damaged in an accident or through excessive tolerance at the mounting position of the idler or pitman arms.

SHOP TALK

If a center link is nonwear, the pitman arm is normally a wear arm. If the link is wear, the pitman arm is usually nonwear. Idler arms or assemblies are always subject to wear. ■

Tie Rods **Tie rods** are actually assemblies that make the final connections between the steering linkage and steering knuckles. They consist of inner tie-rod ends, which are connected to the opposite sides of the center link; outer tie-rod ends, which connect to the steering knuckles; and adjusting sleeves or bolts, which join the inner and outer tie-rod ends, permitting the tie-rod length to be adjusted for correct toe settings **(Figure 44–3)**.

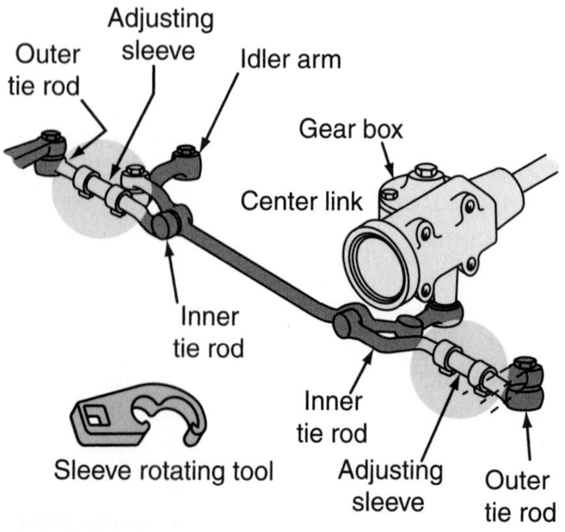

Figure 44–3 A tie-rod assembly.

Tie rods are subject to wear and damage, particularly if the rubber or plastic dust boots covering the ball stud have been damaged or are missing. Contaminants such as dirt and moisture can enter and cause rapid part failure. A special bonded ball stud, in which no boot is used, is available for use on certain light-duty two-wheel-drive and four-wheel-drive trucks. An elastomer bushing bonded to the stud ball provides strong shock absorption and steering return in downsized vehicles.

WARNING!

Never apply heat to any part of the steering linkage while servicing it. If any parts must be heated in order to remove them, they must be replaced and not reinstalled.

Rack and Pinion Steering Linkage

Rack and pinion is lighter in weight and has fewer components than parallelogram steering **(Figure 44–4)**. Tie rods are used in the same fashion on both systems, but the resemblance stops there. Steering input is received from a pinion gear attached to the steering column. This gear moves a toothed rack that is attached to the tie rods.

In the rack and pinion steering arrangement, there is no pitman arm, idler arm assembly, or center link. The **rack** performs the task of the center link. Its movement pushes and pulls the tie rods to change the wheel's direction. The tie rods are the only steering linkage parts used in a rack and pinion system.

Most rack and pinion constructions are composed of a tube in which the steering rack can slide. The rack is a rod with gear teeth cut along one end-spur and helical. The other end is fitted with two balls to which the ends of the divided track rods are attached. The rack meshes with the teeth of a small pinion at the end of the steering column. The two inner tie-rod ends, which are attached to the rack, are covered by rubber bellows boots that protect the rack from contamination. The inner tie rods connect to outer tie-rod ends, which connect to the steer-

Figure 44–4 A rack and pinion steering system.

ing arms. The rack and pinion housing is fastened to the vehicle at two or three points.

In some cases, the rack and pinion steering gear on unibody cars is bolted directly to a body panel, like a cowl. When this is done, the body panel must hold the steering gear in its correct location. The unibody structure must maintain the proper relationship of the steering and suspension parts to each other. Along with other advantages, the rack and pinion steering system combined with the MacPherson strut suspension system is found in most front-wheel-drive unibody vehicles because of their weight- and space-saving feature.

The driver gets a greater feeling of the road with rack and pinion because there are fewer friction points. This means a higher probability of car owners with steering complaints. Fewer friction points can reduce the system's total ability to isolate and dampen vibrations.

Rack The rack is a toothed bar contained in a metal housing. The rack maintains the correct height of the steering components so that the tie-rod movement is able to parallel control arm movement.

The rack is similar to the parallelogram center link in that its sideways movement in the housing is what pulls or pushes the tie rods to change wheel directions.

Pinion The pinion is a toothed or worm gear mounted at the base of the steering column assembly where it is moved by the steering wheel. The pinion gear meshes with the teeth in the rack so that the rack is propelled sideways in response to the turning of the pinion.

Yoke Adjustment The rack-to-pinion lash, or preload, affects steering harshness, feedback, and noise. It is set according to the manufacturer's specifications. An adjustment screw, plug, or shim pack are located on the outside of the housing at the junction of the pinion and rack to correct or set the **yoke lash (Figure 44–5)**.

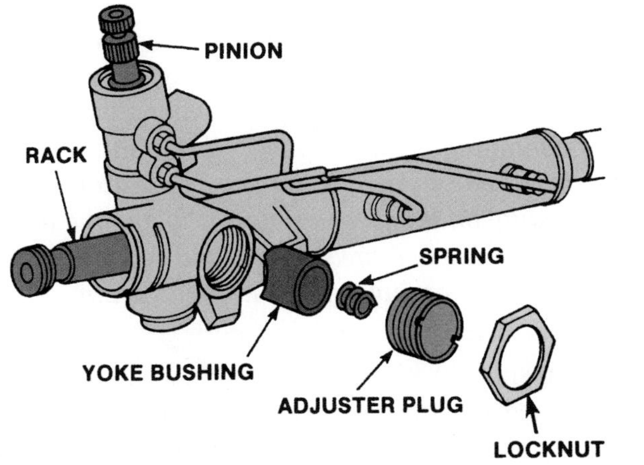

Figure 44–5 The rack preload (yoke lash) is adjusted by a screw, plug, or shim pack.

Tie Rods Tie rods in rack and pinion systems are very similar to those used on parallelogram systems. They consist of inner and outer ends and adjusting sleeves or bolts. The inner tie-rod ends on rack and pinion units are usually spring-loaded ball sockets that screw onto the rack ends **(Figure 44–6)**. They are preloaded and protected against contaminant entry by rubber bellows or boots.

Manual-Steering Gear

The purpose of the steering gear is to change the rotational motion of the steering wheel to a reciprocating motion to move the steering linkage. There are three styles currently in use: the recirculating ball, worm and roller, and the rack and pinion. The latter gear assembly incorporates the already described rack and pinion linkage system and steering gear as a single unit.

The recirculating ball, as shown in **Figure 44–7**, is generally found in larger cars. A sector shaft is supported by needle bearings in the housing and a bushing in the sector cover. A ball nut is used that has threads that mate to the threads of the wormshaft via continuous rows or ball bearings between the two. Ball bearings recirculate through two outside loops, referred to as ball return guide tubes. The ball nut has gear teeth cut on one face that mesh with gear teeth on the sector shaft. As the steering wheel is rotated, the wormshaft rotates, causing the ball nut to move up or down the wormshaft. Since the gear teeth on the ball nut are meshed with the gear

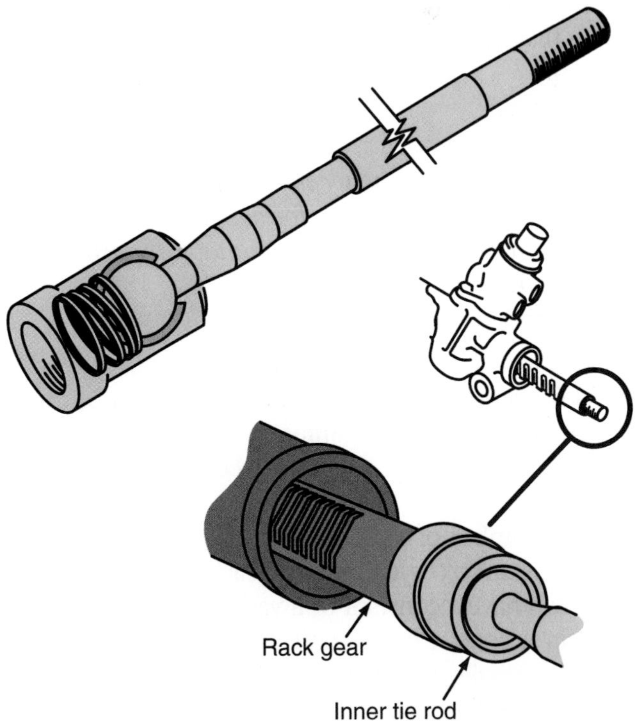

Figure 44–6 The inner tie rod is a spring-loaded ball socket in a rack and pinion steering box.

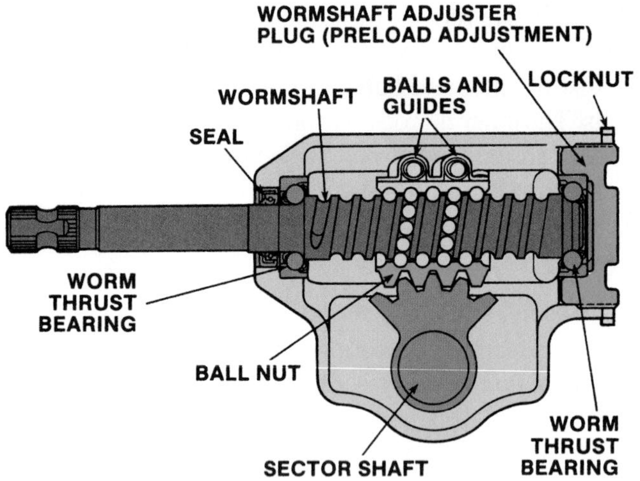

Figure 44-7 A top cutaway of a manual recirculating ball steering gear.

teeth on the sector shaft, the movement of the nut causes the sector shaft to rotate and swing the pitman arm.

The design of two separate circuits of balls results in an almost friction-free operation of the ball nut and the wormshaft. When the steering wheel is turned, the ball bearings roll in the ball thread grooves of the wormshaft and ball nut. When the ball bearings reach the end of their respective circuit, they enter the guide tubes and are returned to the other end of the circuits.

The teeth on the sector shaft and the ball nut are designed so that an interference fit exists between the two when the front wheels are straight ahead. This interference fit eliminates gear tooth lash for a positive feel when driving straight ahead. Proper mesh engagement between the sector and ball nut is obtained by an adjusting screw that moves the sector shaft axially.

CAUTION!

Proper adjustment between the worm gear and the pitman gear is important for adequate steering response. Refer to the vehicle manual for specific adjustment procedures.

The worm thrust bearing adjuster can be turned to provide proper preloading of the worm thrust bearings. Worm bearing preload eliminates worm endplay and is necessary to prevent steering free play and vehicle wander.

The number of input turns per output turn of the steering gearbox is called the gearbox ratio. Steering gears can have a constant or a variable ratio. The sector teeth in a constant ratio unit are identical in size and shape, while the sector of a variable ratio unit has larger center teeth. This makes the steering faster in turns than in a straight direction. Variable ratio is normally used only in power-steering units.

The **worm and roller gearbox** is similar to the recirculating ball except a single roller replaces the balls and ball nut. This reduces internal friction, making it ideal for smaller cars. The steering linkage used with a worm and roller gearbox typically includes a pitman arm, center link, idler arm, and two tie-rod assemblies. The function of these components is the same as in the parallelogram steering linkage described earlier in this chapter.

In operation, the steering shaft rotates the worm gear. It, in turn, engages the roller, causing the roller shaft to turn. The shaft moves the pitman arm left or right to steer the vehicle.

It must be noted that the steering gear does not cause the vehicle to pull to one side nor does it cause road wheel shimmy.

Steering Wheel and Column

The purpose of the steering wheel and column is to produce the necessary force to turn the steering gear. The exact type of steering wheel and column depends on the year and the car manufacturer. The steering column, also called a steering shaft, relays the movement of the steering wheel to the steering gear.

Major parts of the steering wheel and column are shown in **Figure 44–8**. The steering wheel is used to produce the turning effort. The lower and upper covers conceal parts. The universal joints rotate at angles. Support brackets are used to hold the steering column in place. Assorted screws, nuts, bolt pins, and seals are used to make the steering wheel and column perform correctly. Since 1968, all steering columns have a collapsible feature that allows the column to fold into itself, on impact. This feature prevents injury to the driver.

In most vehicles equipped with a driver's side air bag, the air bag assembly is contained in the center portion of the steering wheel. This assembly must be disarmed and removed before the steering wheel can be removed.

WARNING!

Always disconnect the negative battery cable and wait one full minute before beginning to remove an air bag assembly. Failure to do this may result in accidental air bag deployment. (See Chapter 23 of this textbook.) Air bag service precautions vary depending on the vehicle. Always follow the manufacturer's service precautions and recommendations when working with an air bag system.

Differences in steering wheel and column designs include fixed column, telescoping column, tilt column, manual transmission, floor shift, and automatic transmission column shift. The tilt columns **(Figure 44–9)** feature at least five driving positions (two up, two down, and a

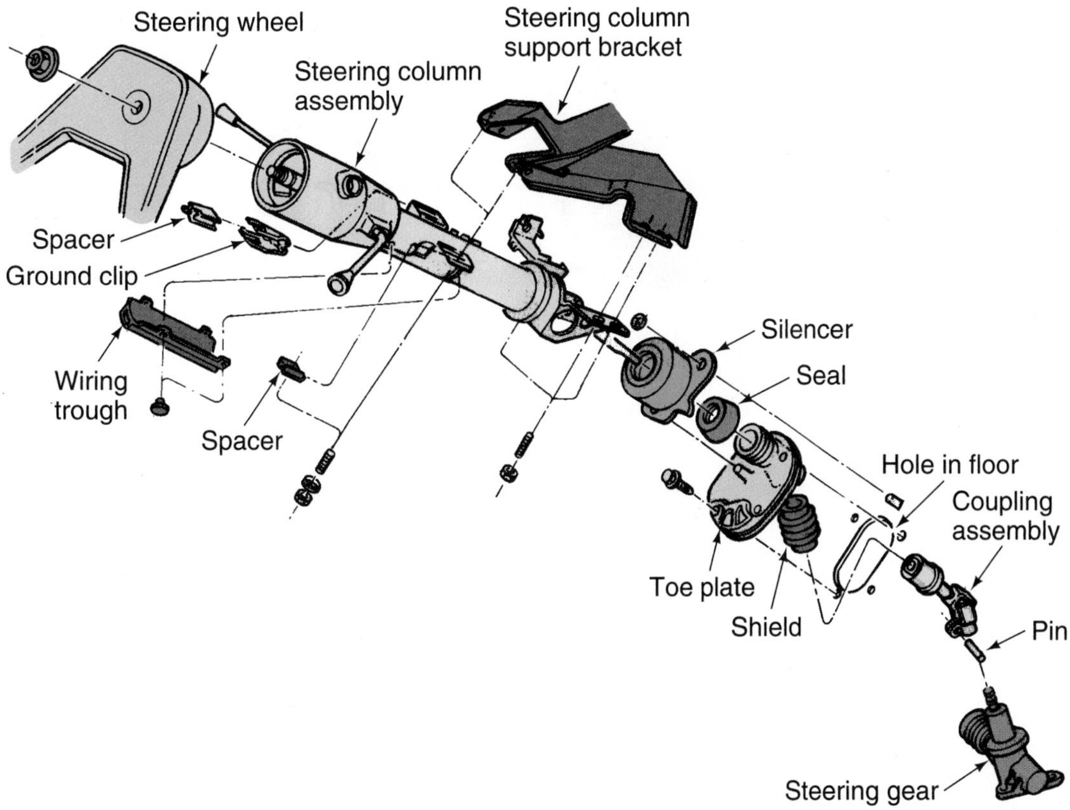

Figure 44-8 Typical steering column components. The steering wheel is splined to the shaft that extends through the column and down to the steering gearbox. *Courtesy of DaimlerChrysler Corporation*

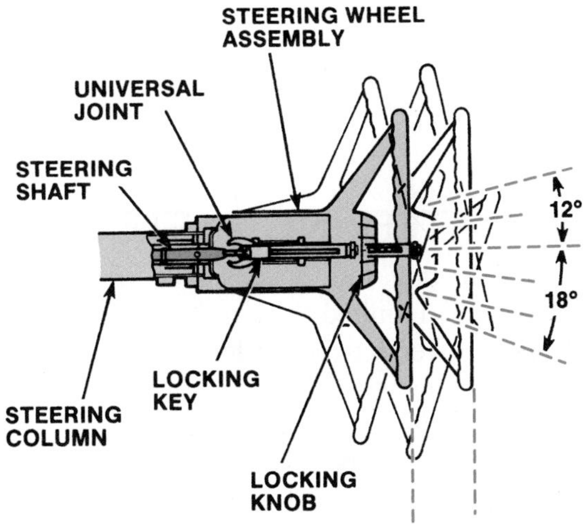

Figure 44-9 Tilt steering column operation.

center position). Both fixed and tilt columns may house an emergency warning flasher control, a turn signal switch, ignition key, lights (high/low beams), horn, windshield wipers and washers, and an antitheft device that locks the steering system. On automatic-transmission-equipped vehicles, the transmission linkage locks also.

Methods used to lock the shaft to the tube include a breakaway plastic capsule or a series of inserts or steel balls held in a plastic retainer that allow the shaft to roll forward inside the tube. There are also collapsible steel mesh **(Figure 44–10)** or accordion-pleated devices that give way under pressure. After the vehicle has been in an accident, the steering column should be checked for evidence of collapse. Although the car can be steered with a collapsed column that has been pulled back, the collapsed portion must be replaced. All service manuals provide explicit instructions for doing this.

The steering wheel is usually held in place on the steering column by either a bolt or nut **(Figure 44–11)**.

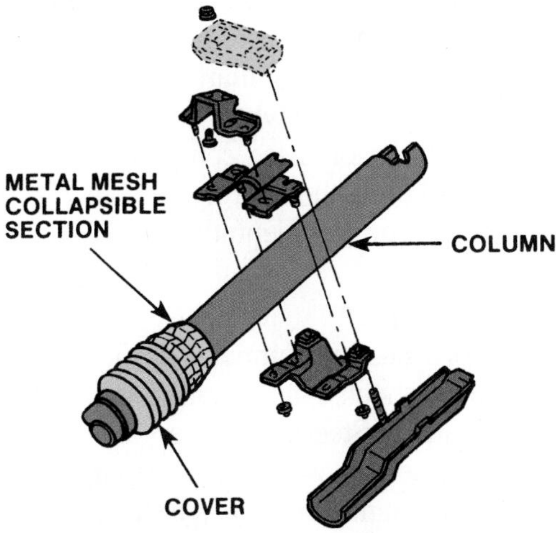

Figure 44-10 A collapsing mesh steering column. *Courtesy of Ford Motor Company*

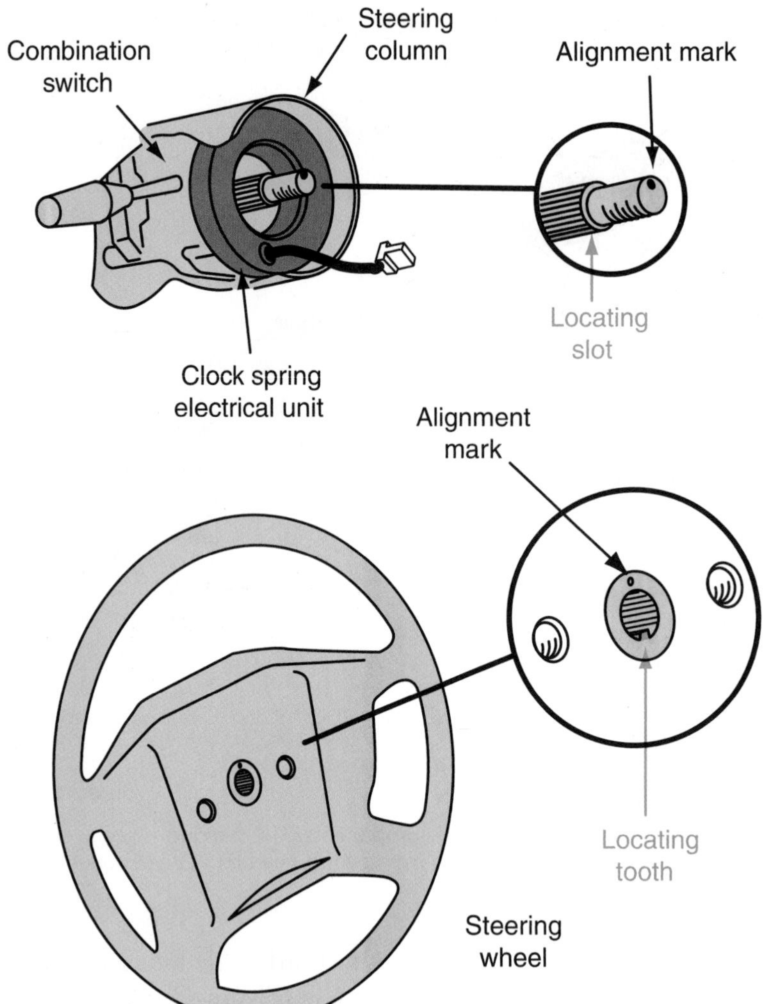

Figure 44-11 The steering wheel is splined to the steering column and held in place by a nut.

When the blocked tooth on the steering gear input shaft is in the 12 o'clock position, the front wheels should be in the straight-ahead position and the steering wheel spokes in their normal position. If the spokes are not in their normal position, they can be adjusted by changing the toe adjustment. This adjustment can be made only when the steering wheel indexing mark is aligned with the steering column indexing marks. As a rule, indexing teeth or mating flats on the wheel hub and steering shaft prevent misindexing of these components. The alignment of the notches on the steering wheel hub and steering shaft confirm correct orientation.

Steering Damper

The purpose of a steering damper is simply to reduce the amount of road shock that is transmitted up through the steering column. Steering dampers are found mostly on 4WD, especially those fitted with large tires. The damper serves the same function as a shock absorber but is mounted horizontally to the steering linkage—one end to the center link and the other to the frame.

POWER-STEERING SYSTEMS

The power-steering unit is designed to reduce the amount of effort required to turn the steering wheel. It also reduces driver fatigue on long drives and makes it easier to steer the vehicle at slow road speeds, particularly during parking.

Power steering can be broken down into two design arrangements: conventional and nonconventional or electronically controlled. In the conventional arrangement, hydraulic power is used to assist the driver. In the nonconventional arrangement, an electric motor and electronic controls provide power assistance in steering.

There are several power-steering systems in use on passenger cars and light-duty trucks. The most common ones are the integral-piston, and power-assisted rack and pinion system **(Figure 44–12)**.

Integral Piston System

The **integral piston** system is the most common conventional power-steering systems in use today. It consists of a power-steering pump and reservoir, power-steering

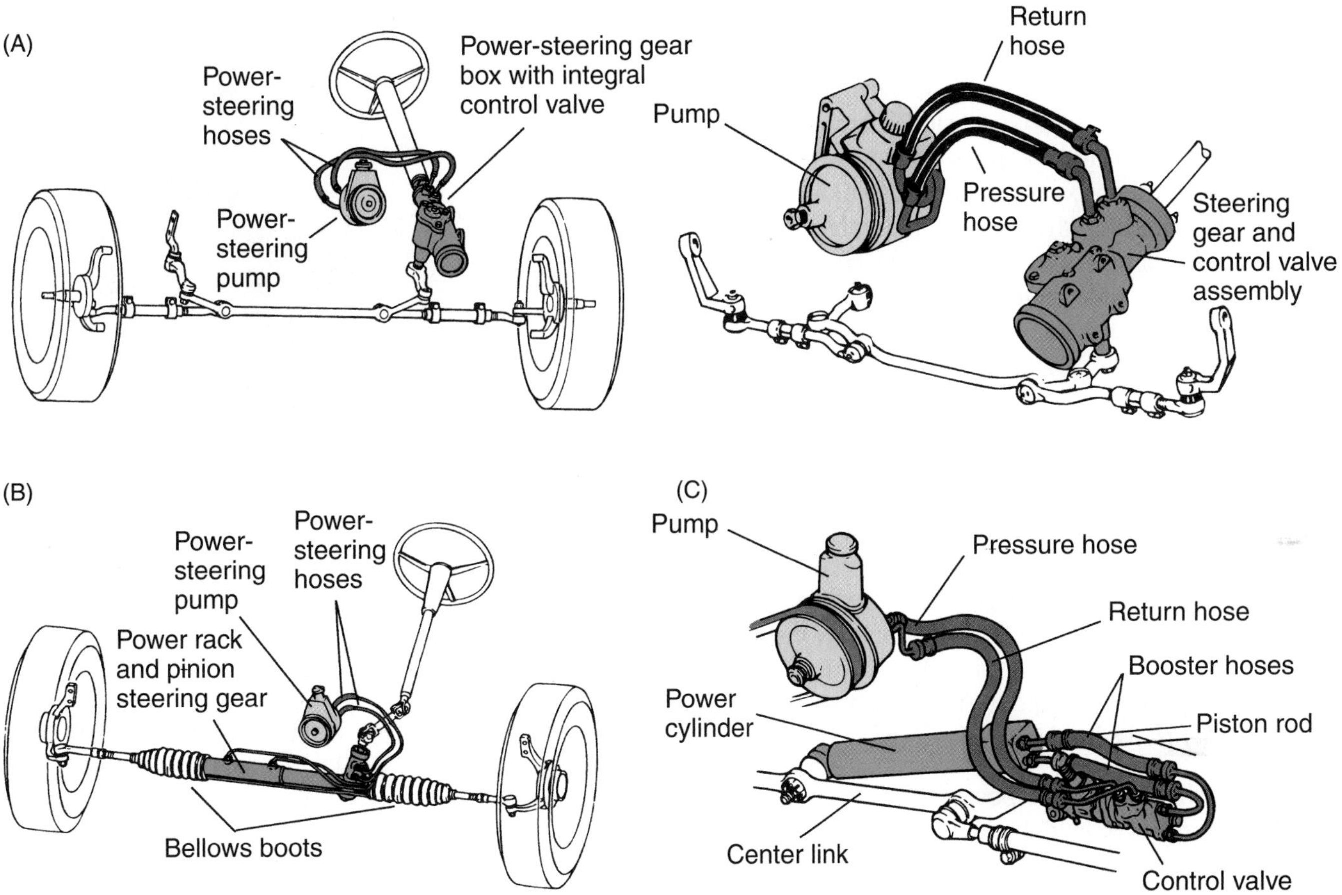

Figure 44-12 The three major power-steering systems: (A) integral-piston linkage, (B) rack and pinion, and (C) external-piston linkage. *Courtesy of Moog Automotive, Inc.*

pressure and return hose, and steering gear. The power cylinder and the control valve are in the same housing as the steering gear.

On some recent model cars and light trucks, instead of the conventional vacuum-assist brake booster, the hydraulic fluid from the power-steering pump is also used to actuate the brake booster. This brake system is called the hydro-boost system **(Figure 44–13)**.

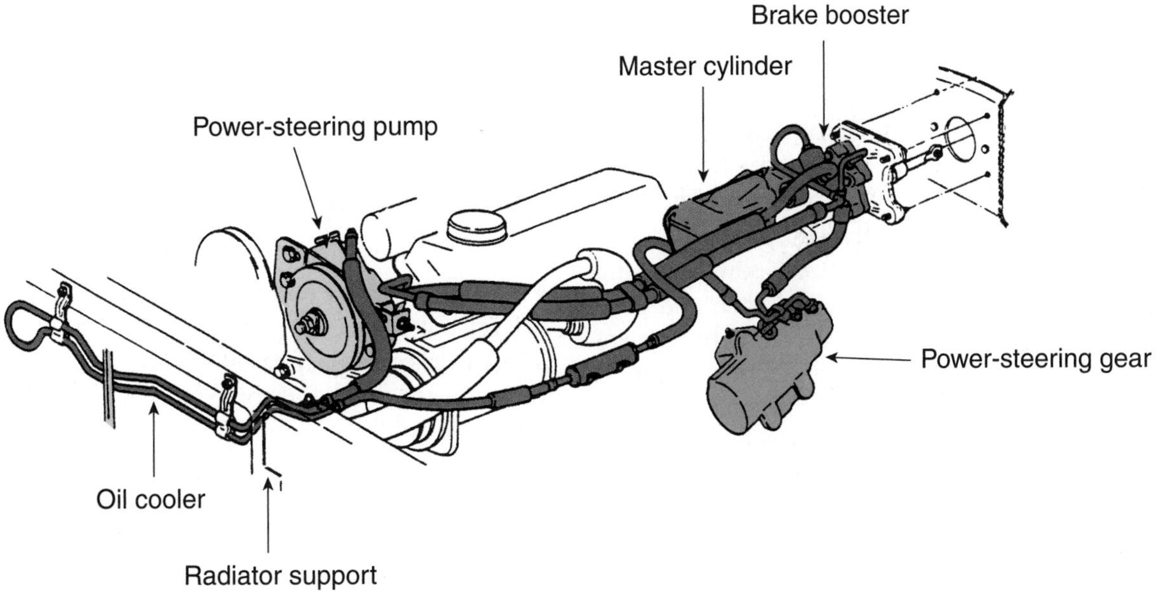

Figure 44-13 A typical hydro-boost system that uses the power-steering pump to power assist brake applications. *Courtesy of Ford Motor Company*

Power-Assisted Rack and Pinion System

The power-assisted rack and pinion system is similar to the integral system because the power cylinder and the control valve are in the same housing. The rack housing acts as the cylinder and the power piston is part of the rack. Control valve location is in the pinion housing. Turning the steering wheel moves the valve, directing pressure to either end of the back piston. The system utilizes a pressure hose from the pump to the control valve housing and a return line to the pump reservoir. This type of steering system is common in front-wheel-drive vehicles.

Components

Several of the manual-steering parts described earlier in this chapter, such as the steering linkage, are used in conventional power-steering systems. The components that have been added for power steering provide the hydraulic power that drives the system. They are the power-steering pump, flow control and pressure relief valves, reservoir, spool valves and power pistons, hydraulic hose lines, and gearbox or assist assembly on the linkage.

Power-Steering Fluid The power-steering fluid can be checked either hot or cold. Fluid level will vary with temperature, however, and a more accurate check is done when the engine is warm. The reservoir cap is usually marked for HOT and COLD fluid levels on opposite sides of the dipstick. Make sure to check the level on the right side of the dipstick. If necessary, add fluid to correct the level. Most manufacturers recommend a specific fluid for use in a power-steering system. Some, however, allow the use of ATF as a substitute. Always check the service or owner's manual for the proper fluid recommendation.

Power-Steering Pump The steering pump is used to develop hydraulic flow, which provides the force needed to operate the steering gear. The pump is belt driven from the engine crankshaft, providing flow any time the engine is running. It is usually mounted near the front of the engine **(Figure 44–14)**. The pump assembly includes a reservoir and an internal flow control valve. The drive pulley is normally pressed onto the pump's shaft.

There are four general types of power-steering pumps: roller, vane **(Figure 44–15)**, slipper, and gear. Functionally, all pumps operate in the same basic manner. Hydraulic fluid for the power-steering pump is stored in a reservoir. Fluid is routed to and from the pump by hoses and lines. Excessive pressure is controlled by a relief valve.

Power-Steering Pump Drive Belts Many power-steering pumps are driven by a belt that connects the crankshaft pulley to the power-steering pump pulley. The belt may also drive other components, such as the water pump. The sides of a V-belt are the friction surfaces that drive the power-steering pump. If the sides of the belt are worn and the lower edge of the belt is contacting the bottom of the pulley, the belt will slip. The power-steering pump pulley, crankshaft pulley, and any other pulleys driven by the V-belt must be properly aligned. If these pulleys are misaligned, excessive belt wear occurs.

Figure 44–14 A power-steering pump. *Courtesy of Visteon Corporation*

A ribbed V-belt is used on many vehicles, and this belt may be used to drive all the belt-driven components. Most ribbed V-belts have a spring-loaded automatic belt tensioner that eliminates periodic belt tension adjustments. Because the ribbed V-belt may be used to drive all of the belt-driven components, these components are placed on the same vertical plane, which saves a considerable amount of underhood space. The smooth backside of the ribbed V-belt may also be used to drive one of the components. Regardless of the type of belt, the belt tension is critical. A power-steering pump will never develop full pressure if the belt is slipping.

Electronic Power-Steering Many new vehicles are using 12-volt motors mounted directly to the steering rack instead of pumps, hoses, and belts. These provide significant weight savings and the parasitic load on the battery is reduced.

Flow Control and Pressure Relief Valves A pressure relief valve controls the pressure output from the pump. This valve is necessary because of the variations in engine rpm and the need for consistent steering ability in all ranges from idle to highway speeds. It is positioned in a chamber that is exposed to pump outlet pressure at one end and supply hose pressure at the other. A spring is used at the supply pressure end to help maintain a balance.

As the fluid leaves the pump rotor, it passes the end of the flow control valve and is forced through an orifice

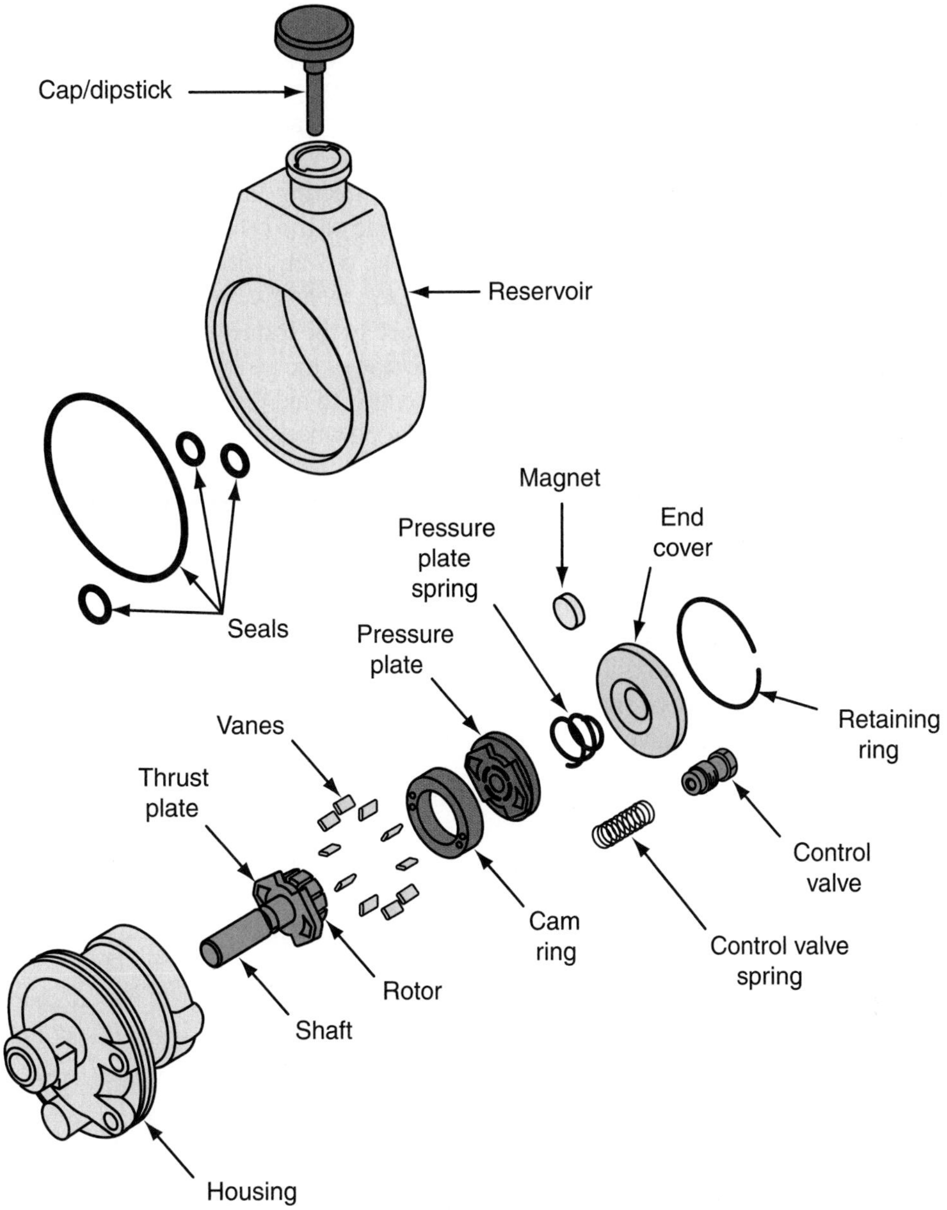

Figure 44-15 A vane-type power-steering pump.

that causes a slight drop in pressure. This reduced pressure, aided by the springs, holds the flow control valve in the closed position. All pump flow is sent to the steering gear.

When engine speed increases, the pump can deliver more flow than is required to operate the system. Since the outlet orifice restricts the amount of fluid leaving the pump, the difference in pressure at the two ends of the valve becomes greater until pump outlet pressure overcomes the combined force of supply line pressure and spring force. The valve is pushed down against the spring, opening a passage that returns the excess flow back to the inlet side of the pump.

A spring and ball contained inside the flow control valve are used to relieve pump outlet pressure. This is done to protect the system from damage due to excessive pressure when the steering wheel is held against the stops. Since flow in the system is severely restricted, the pump would continue to build pressure until a hose ruptured or the pump destroyed itself.

When outlet pressure reaches a preset level, the pressure relief ball is forced off its seat, creating a greater pressure differential at the two ends of the flow control valve. This allows the flow control valve to open wider, permitting more pump pressure to flow back to the pump inlet and pressure is held at a safe level.

Power-Steering Gearbox A power-steering gearbox is basically the same as a manual recirculating ball gearbox with the addition of a hydraulic assist. A power-steering gearbox is filled with hydraulic fluid and uses a control valve.

In a power rack and pinion gear, the movement of the rack is assisted by hydraulic pressure. When the wheel is

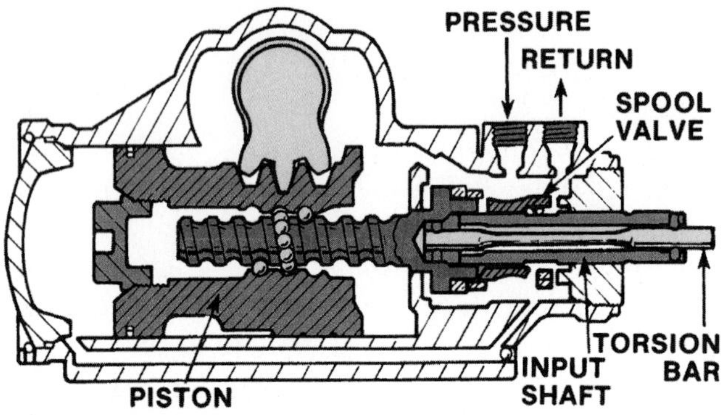

Figure 44–16 A torsion bar moves the spool valve to direct the oil flow to the piston.

turned, the rotary valve changes hydraulic flow to create a pressure differential on either side of the rack. The unequal pressure causes the rack to move toward the lower pressure, reducing the effort required to turn the wheels.

The integral power steering has the spool valve and a power piston integrated with the gearbox. The spool valve directs the oil pressure to the left or right power chamber to steer the vehicle. The spool valve is actuated by a lever or a small torsion bar **(Figure 44–16)**.

In linkage systems, the control valve is connected directly to the steering center link and the pitman arm on the steering gear. Any movement of the steering wheel and the pitman arm compresses the centering spring and moves the valve spool. This opens and closes a series of ports directing fluid under pressure from the pump to one side or the other of the power cylinder piston.

The power cylinder in the linkage system is attached to the steering center link and the piston shaft is attached to the frame. As fluid under pressure is directed to one side of the piston by the control valve, power assist is provided to aid the driver in moving the steering linkage and road wheels. Two lines connect the cylinder to the control valve. Each one functions as both the return line or the supply line, depending on the direction of the turn.

Power-Assisted Rack and Pinion Steering Power-assisted rack and pinion components are basically the same as for manual rack and pinion steering **(Figure 44–17)**, except for the hydraulic control housing. As mentioned earlier, the power rack and pinion steering unit may be classified as integral. The rack functions as the power piston and the spool valve is connected to the pinion gear.

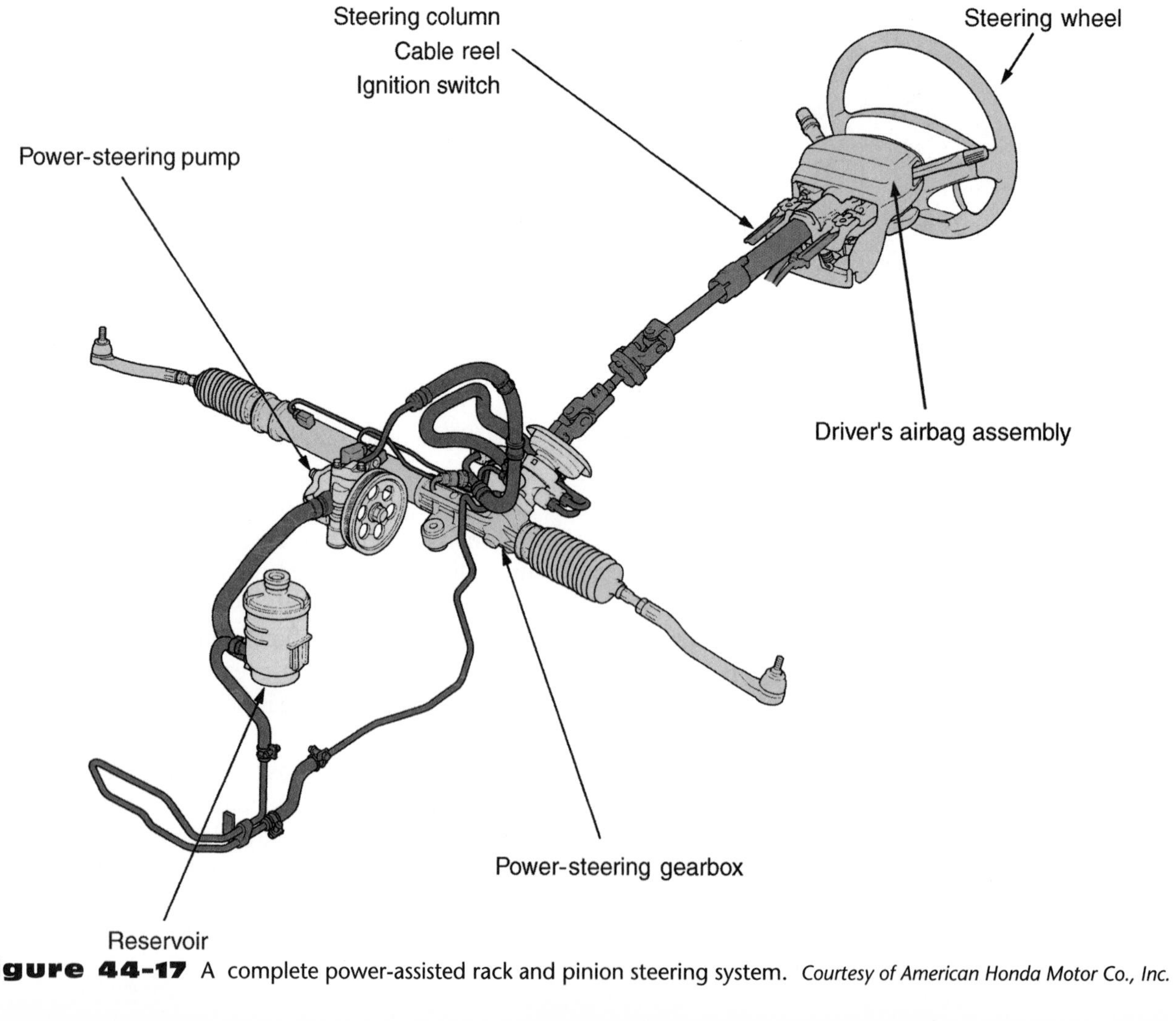

Figure 44–17 A complete power-assisted rack and pinion steering system. *Courtesy of American Honda Motor Co., Inc.*

Figure 44–18 Power-steering hoses may have two internal diameters.

In a power rack and pinion gear, the piston is mounted on the rack, inside the rack housing. The rack housing is sealed on either side of the rack piston to form two separate hydraulic chambers for the left and right turn circuits. When the wheel is turned to the right, the rotary valve creates a pressure differential on either side of the rack piston. This causes the rack to move toward the lower pressure and reduces the total effort required to turn the wheels.

Power-Steering Hoses The primary purpose of power-steering hoses is to transmit power (fluid under pressure) from the pump, to the steering gearbox, and to return the fluid ultimately to the pump reservoir. Hoses also, through material and construction, function as additional reservoirs and act as sound and vibration dampers.

Hoses are generally a reinforced synthetic rubber (neoprene) material coupled to metal tubing at the connecting points. The pressure side must be able to handle pressures up to 1,500 psi (10,342 kPa). For that reason, wherever there is a metal tubing to a rubber connection, the connection is crimped. Pressure hoses are also subject to surges in pressure and pulsations from the pump. The reinforced construction permits the hose to expand slightly and absorb changes in pressure.

Two internal diameters of hose **(Figure 44–18)** may be used on the pressure side; the larger diameter or pressure hose is at the pump end. It acts as a reservoir and as an accumulator absorbing pulsations. The smaller diameter or return hose reduces the effects of kickback from the gear itself. By restricting fluid flow, it also maintains constant backpressure on the pump, which reduces pump noise. If the hose is of one diameter, the gearbox is performing the damping functions internally.

Because of working fluid temperature and adjacent engine temperatures, these hoses must be able to withstand temperatures up to 300°F (150°C). Due to various weather conditions, they must also tolerate subzero temperatures as well. Hose material is specially formulated to resist breakdown or deterioration due to oil or temperature conditions.

CAUTION!

Hoses must be carefully routed away from engine manifolds. Power-steering fluid is very flammable. If it comes in contact with hot engine parts, it could start an underhood fire.

ELECTRONICALLY CONTROLLED POWER-STEERING SYSTEMS

The object of power steering is to make steering easier at low speeds, especially while parking. However, higher steering efforts are desirable at higher speeds in order to provide improved down-the-road feel. The electronically controlled power-steering (EPS) systems **(Figure 44–19)**

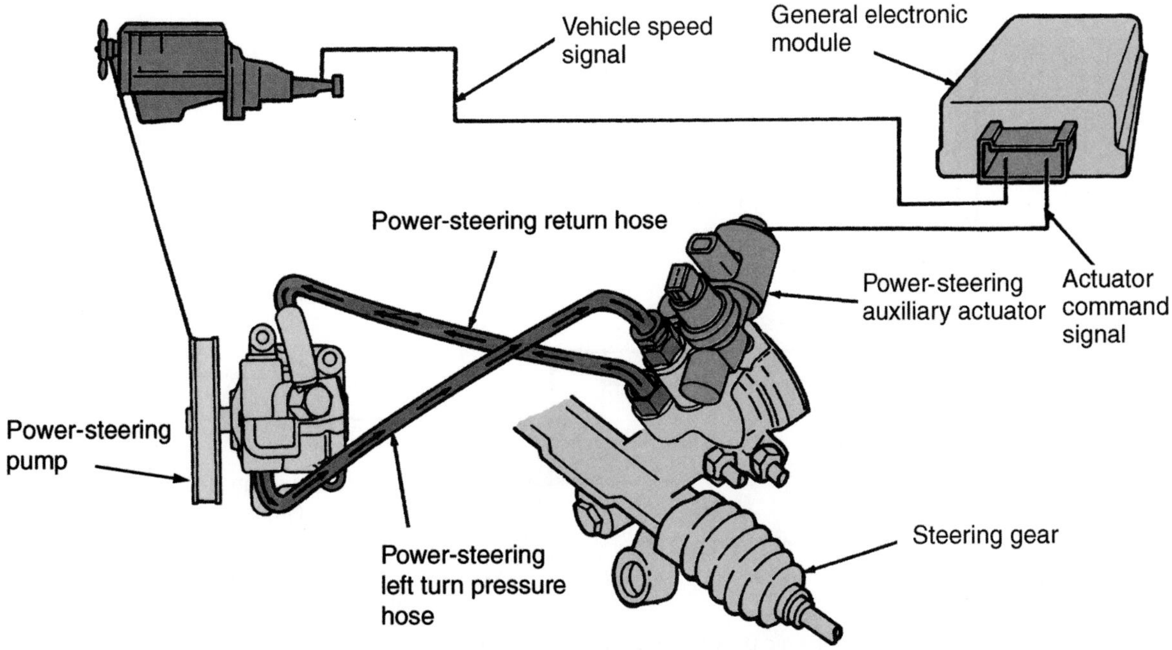

Figure 44–19 A variable-assist power-steering system. *Courtesy of Ford Motor Company*

provide both of these benefits. The hydraulic boost of these systems is tapered off by electronic control as road speed increases. Thus, these systems require well under 1 pound (4.4 N) of steering effort at low road speeds and 3 pounds plus (13.2 N) of steering effort at higher road speeds to enable the driver to maintain control of the steering wheel for improved high-speed handling.

A rotary valve electronic power-steering system consists of the power-steering gearbox, power-steering oil pump, pressure hose, and the return hose. The amount of hydraulic fluid flow (pressure) used to boost steering is controlled by a solenoid valve that is identified as its PCV (pressure control valve).

SHOP TALK

The PCV (pressure control valve) is not to be confused with the PCV (positive crankcase ventilator) used with emission controls systems. ■

The electronic power-steering system's PCV **(Figure 44–20)** is exposed to spring tension on the top and plunger force on the bottom. The plunger slips inside an electromagnet. By varying the electrical current to the electromagnet, the upward force exerted by the plunger can be varied as it works against the opposing spring. Current flow to the electromagnet is variable with vehicle road speed and, therefore, provides steering to match the vehicle's road speed.

General Motors' variable effort steering (VES) system relies on an input signal from the vehicle speed sensor to the VES controller to control the amount of power assist. The controller, in turn, supplies a pulse width modulated voltage to the actuator solenoid in the power-steering pump. The controller also provides a ground connection for the solenoid.

When the vehicle is operating at low speeds, the controller supplies a signal to cycle the solenoid faster so it allows high pump pressure. This provides for maximum power assist during cornering and parking. As the vehicle's speed increases, the solenoid cycles less and the pump provides a lower amount of assist. This gives the driver better road feel during high speeds.

Active Steering

Active steering improves vehicle stability by turning the wheels more or less sharply than commanded by the turn of the steering wheel during some situations. Through inputs and computer programming, this system can adjust the steering to respond quickly to the threat of skidding **(Figure 44–21)**. The system also allows for a variable steering ratio dependent on vehicle speed.

Current active steering systems are not true steer-by-wire systems. There is still a mechanical connection between the steering wheel and vehicle's wheels **(Figure 44–22)**. The systems have an overriding drive built into the steering column. This drive is controlled by an electric motor, which is controlled by the system's computer. The computer determines whether the steering angle

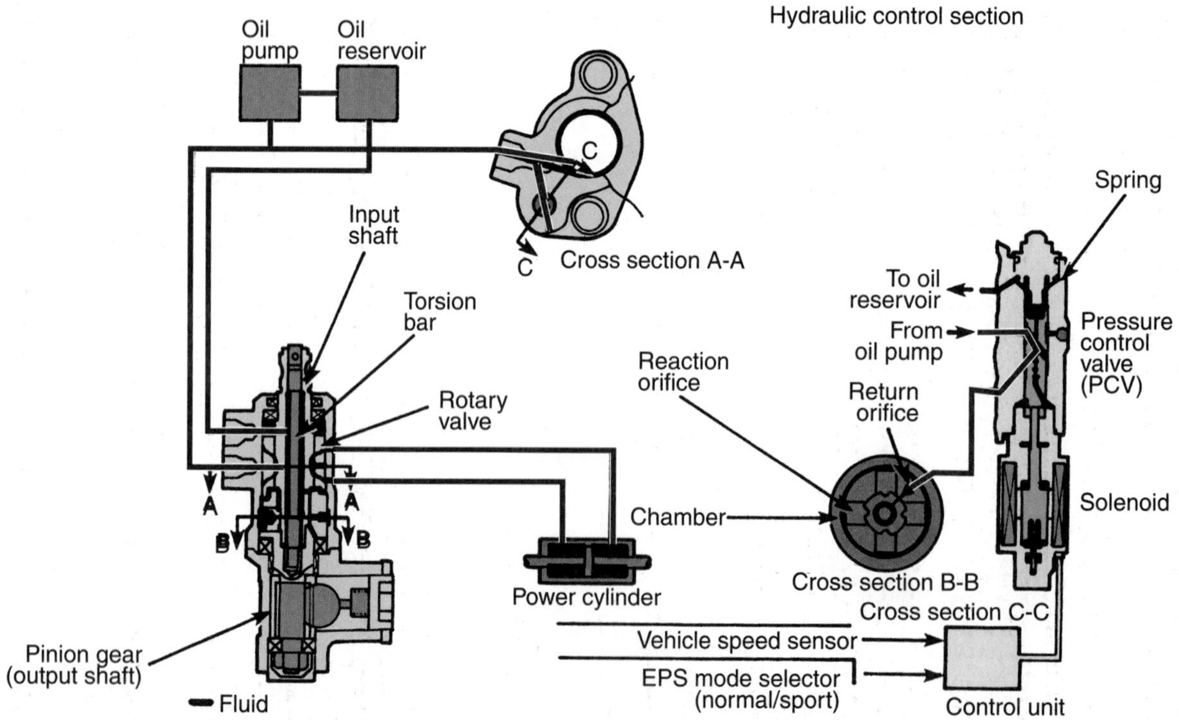

Figure 44–20 An outline of electronic power-steering components. The EPS PCV is exposed to spring tension and plunger force. *Courtesy of Mitsubishi Motor Sales of America, Inc.*

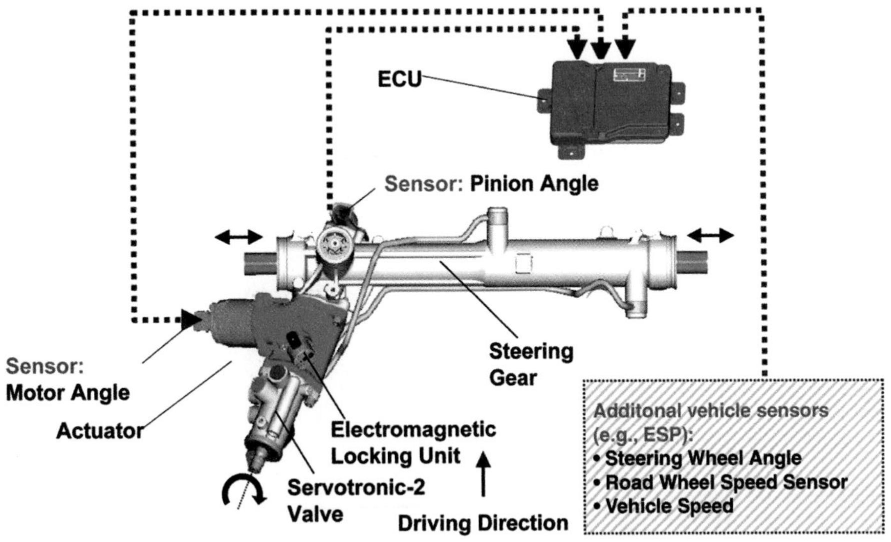

Figure 44-21 The main components and circuits of an active steering system. *Courtesy of Robert Bosch Corp.*

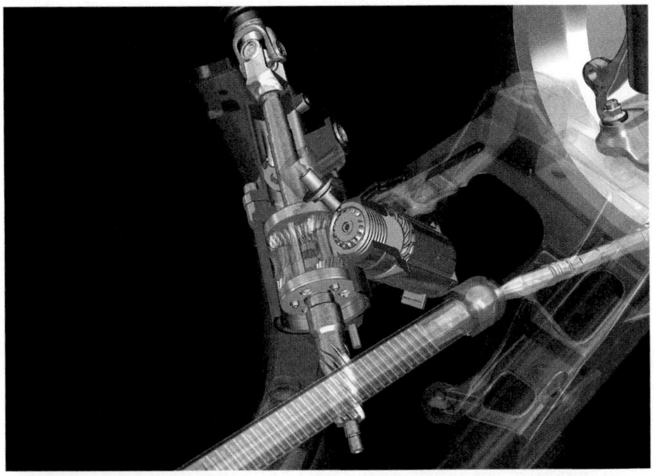

Figure 44-22 The planetary gearset and electric motor that turn the road wheels when the steering wheel is turned. *Courtesy of BMW of North America, Incorporated*

needs to be changed and by how much. If the system fails, the planetary gear unit will rotate directly with the steering wheel.

General Service

When servicing, as with any power-steering system, the first step is to look for fluid leaks, damaged components, a slipping drive belt, and so on. Only after these things have been checked and no problems have been found should the electronics be suspected as the problem. Check the appropriate service manual for the correct troubleshooting procedures of the electronics.

Electric/Electronic Rack and Pinion System

The electric/electronic rack and pinion unit replaces the hydraulic pump, hoses, and fluid associated with conventional power-steering systems with electronic controls and an electric motor located concentric to the rack itself **(Figure 44–23)**. The design features a DC motor armature with a follow shaft to allow passage of the rack through it. The outboard housing and rack are designed so that the rotary motion of the armature can be transferred to linear movement of the rack through a ball nut with thrust bearings. The armature is mechanically connected to the ball nut through an internal/external spline arrangement.

The basis of system operation is its ability to change the rotational direction of the electric motor while being able to deliver the necessary amount of current to meet torque requirements at the same time. The system can deliver up to 75 amperes to the motor. The higher the current, the greater the force exerted on the rack. The direction of the turn is controlled by changing the polarity of the signal to the motor.

The field assembly houses permanent ceramic magnets while providing structural integrity for the gear system. In essence, the electronic/electric rack design allows for a direct power source to the rack and steering linkage. The system monitors steering wheel movement through a sensor mounted on the input shaft of the rack and pinion steering gear. After receiving directional and load information from the sensor, an electronic controller activates the motor to provide power assistance.

These units are readily retrofitted to conventionally equipped vehicles. As for servicing, there are currently no replacement parts available; therefore, if the rack should become faulty, the entire unit should be replaced. Rebuilt kits, with complete installation instructions, are available.

Unlike conventional power steering, electric/electronic units provide power assistance even when the engine stalls, since the power source is the battery rather

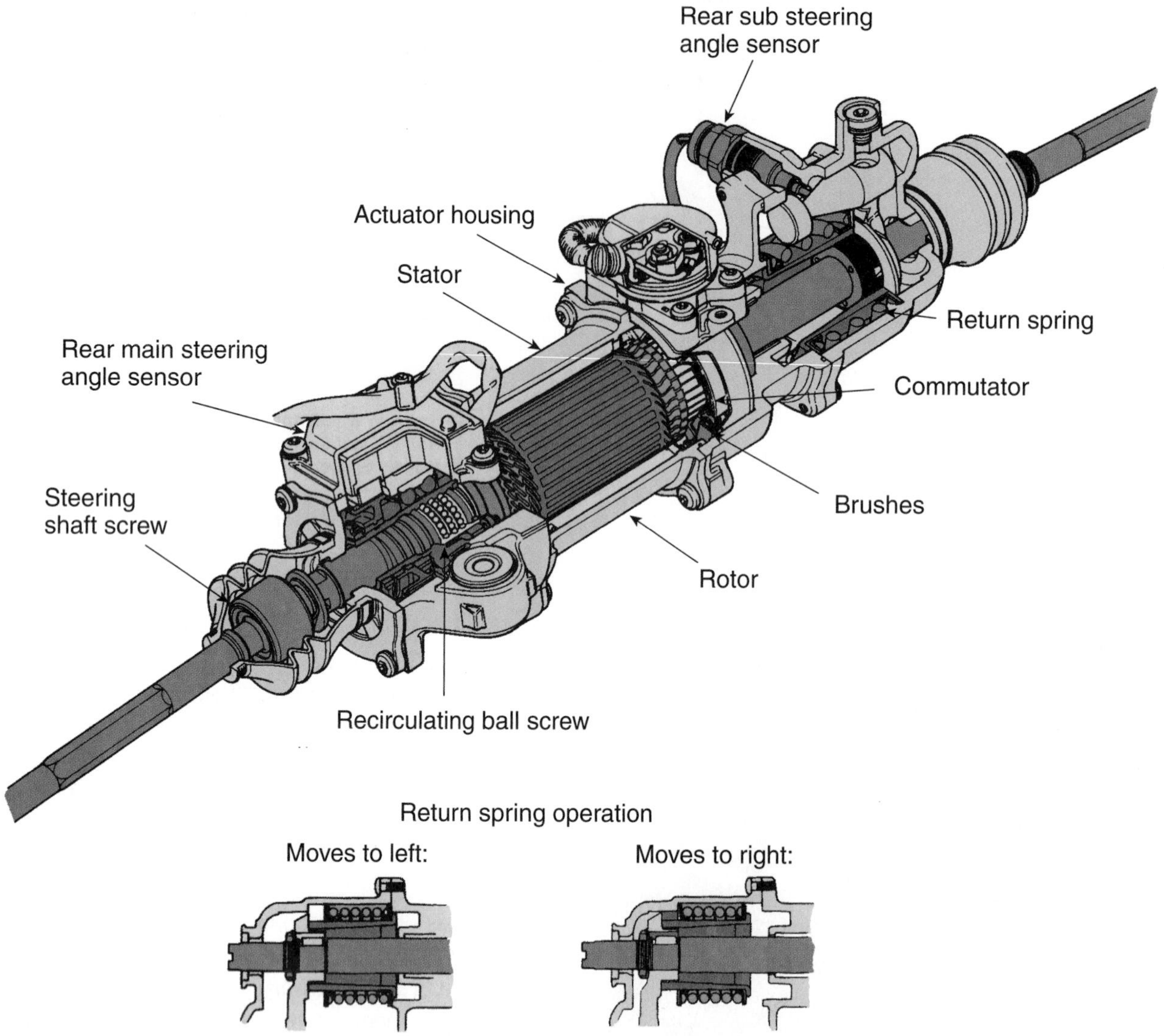

Figure 44-23 An electric/electronic rack and pinion system. *Courtesy of American Honda Motor Co., Inc.*

than the engine-driven pump. The feel of the steering can also be adjusted to match the particular driving characteristics of cars and drivers, from high performance to luxury touring cars. It also eliminates hydraulic oil, which means no leaks.

STEERING SYSTEM DIAGNOSIS

It is important to realize that many steering complaints are caused by problems in areas other than the steering system. A good diagnosis is one that finds the exact cause of the customer's complaint. Although customers may describe the problem in different ways, the most common complaints and their typical causes are discussed next.

Common Complaints

Excessive Steering-Wheel Play Excessive play in the steering wheel is apparent when there is too much steering-wheel movement before the wheels begin to turn. A small amount of play is normal.

This problem can be caused by the following:

- Loose, worn, or damaged steering linkages or tie-rod ends
- Loose, worn, or damaged steering column U-joints
- Loose, worn, or damaged steering column bearings
- Damaged or worn steering gear
- Loose steering gear bolts

Feedback When the driver feels the surface of the road through the steering wheel it is called feedback.

This problem can be caused by the following:

- Loose, worn, or damaged steering linkages or tie-rod ends
- Loose, worn, or damaged steering column U-joints
- Loose or damaged steering gear mounting bolts

- Damaged or worn steering column bearings
- Loose suspension bushings, fasteners, or ball joints

Hard Steering Obviously a complaint of hard steering results when extra effort is needed to turn the steering wheel. This problem may be simply an absence of power assist. Hard steering problems can occur whenever the steering wheel is turned or just when it has been turned close to its limit.

This problem can be caused by the following:

- A faulty power-steering pump
- Damaged or faulty steering column bearings
- Seized steering column U-joints
- Steering gear set too tight or is binding
- Faulty suspension components
- Inadequately inflated tires

Nibble This feeling is similar to a shimmy. Nibble results from the interaction of the tires with the road's surface. The customer's complaint may describe the nibble problem as slight rotational oscillations of the steering wheel.

This problem can be caused by the following:

- Loose, worn, or damaged steering linkages or tie-rod ends
- Loose, worn, or damaged suspension parts

Pulling or Drifting A pull is a tugging sensation felt at the steering wheel. The driver must push the steering wheel in the opposite direction of the pull to keep the vehicle going straight. Drifting is a condition in which the vehicle slowly moves to one side of the road when the driver's hands are taken off the steering wheel.

These problems can be caused by the following:

- Improper frame alignment
- Brake system problem
- Worn or binding suspension components, especially springs
- Poor wheel alignment
- Unevenly loaded or overloaded vehicle
- Loose, worn, or damaged steering linkages or tie-rod ends
- Out of balance steering gear valve
- Torque steer
- Tire problems
- Binding strut bearing

Shimmy When the wheels shimmy, the driver will feel large, consistent, rotational oscillations at the steering wheel. These motions are caused by the lateral movement of the tires.

This problem can be caused by the following:

- Loose, worn, or damaged steering linkages or tie-rod ends
- Loose, worn, or damaged suspension parts
- Out-of-balance tires
- Excessive wheel runout
- A bad tire
- Loose wheel bearings

Sticking Steering or Poor Return Poor returnability and sticky steering describes the steering wheel's resistance to return to center after a turn.

This problem can be caused by the following:

- Binding steering column U-joints
- Loose, worn, or damaged steering linkages or tie-rod ends
- Steering gear set too tight or is binding
- Loose, damaged or worn suspension parts
- Poor wheel alignment
- Binding steering column bearings

Wandering When a vehicle wanders, the driver must constantly turn the steering wheel to the left and right to keep the vehicle going straight on a level road.

This problem can be caused by the following:

- Loose or worn suspension components
- Poor wheel alignment
- Unevenly loaded or overloaded vehicle
- Loose or damaged steering gear bolts
- Loose steering column U-joint bolts
- Loose, worn, or binding steering linkages or tie-rod ends
- Improper steering gear preload adjustment

Noise There may also be abnormal noises that accompany the turning of the steering wheel. The cause of these noises is best identified by paying close attention to where the noise is coming from. Some noises may be caused by tires or interference between the steering wheel and the steering column covers. Others can result from a faulty power-steering pump or system.

- A hissing noise during low speed turning, such as while parking is normal.
- A chirping or squealing normally means the drive belt is loose or slipping.

- Improperly routed hoses, loose mounting bolts on the power steering pump or the steering gear can cause a rattle.
- A groaning or howling can indicate a restriction in the hydraulic system.
- A sticking control valve will cause a buzzing sound.

Diagnosing

As with the diagnosis of any problem, your diagnosis should begin by trying to duplicate the customer's complaint. For steering problems, this is done on a road test. Make sure you drive carefully and cautiously especially since the vehicle has a control problem. It is very important that during the road test the vehicle is driven under conditions similar to the owner's normal driving. While it may be somewhat inconvenient to seek out a particular type of road surface, this may be the only way to verify the condition. Take plenty of time because it could save hours of service time in the final diagnosis. Before going on the road test, do a thorough safety inspection of the vehicle that includes the tires.

Once the road test has been completed and it has been determined that there is an abnormal condition, certain tests are necessary to pinpoint the exact cause. In some cases partial disassembly of the system may be required to complete a test. Do not shortcut test steps to save time. Doing so can alter the results, leading to an inaccurate diagnosis and unnecessary replacement of parts. It is also vital that specifications be confirmed in the service manual. Guessing gives incorrect results. Continue your diagnosis with a thorough visual inspection.

Power-Steering Pressure Checks

A power-steering pressure gauge is used to test the power-steering pump pressure. With the engine off, disconnect the pressure hose at the pump. Install the pressure gauge between the pump and the steering gear and bleed the system.

Run the engine for about 2 minutes, and then stop the engine and add fluid to the power-steering pump if necessary. Restart the engine, allow it to idle, and observe the pressure reading. The readings should be about 30 to 80 psi (200 to 550 kPa). If the pressure is low, the pump may be faulty. If the pressure is too high, the problem may be restricted hoses.

Continue for testing by closing the shut-off valve of the tester and observing the pressure reading **(Figure 44–24)**. Never keep the valve closed for more than 5 seconds. When the valve is closed, the pressure should increase. If the pressure is too high, a faulty pressure relief valve is suggested. If the pressure is too low, the pump may be bad.

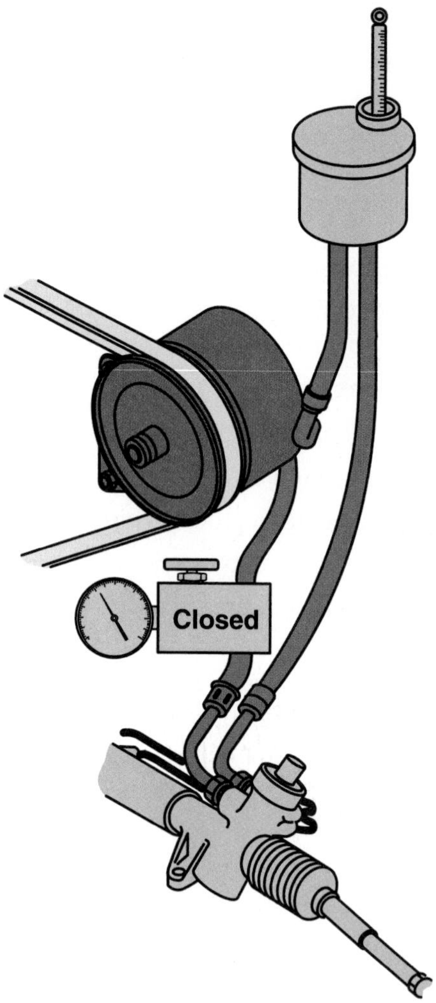

Figure 44–24 The final step when checking power-steering pump pressure is to close the tester's valve and observe the pressure in the system. Never keep the valve closed for more than five seconds.

VISUAL INSPECTION

Often, as you go over the vehicle's systems, you will run across something that appears to be faulty, worn, or damaged. At that point you may wish to check out the component further before continuing your inspection. These checks and some services to the components are given with the details of the visual inspection. Before actually checking or servicing a component, check with the service manual to make sure you are doing it correctly.

Begin your visual inspection of the steering system by inspecting the tires. Check for correct pressure, construction, size, wear, and damage, and for defects that include ply separations, sidewall knots, concentricity problems, and force problems. Keep in mind that tire wear patterns are good indicators of steering and suspension problems **(Figure 44–25)**. Tire wear is also a great indicator of wheel alignment problems that will be covered in the next chapter.

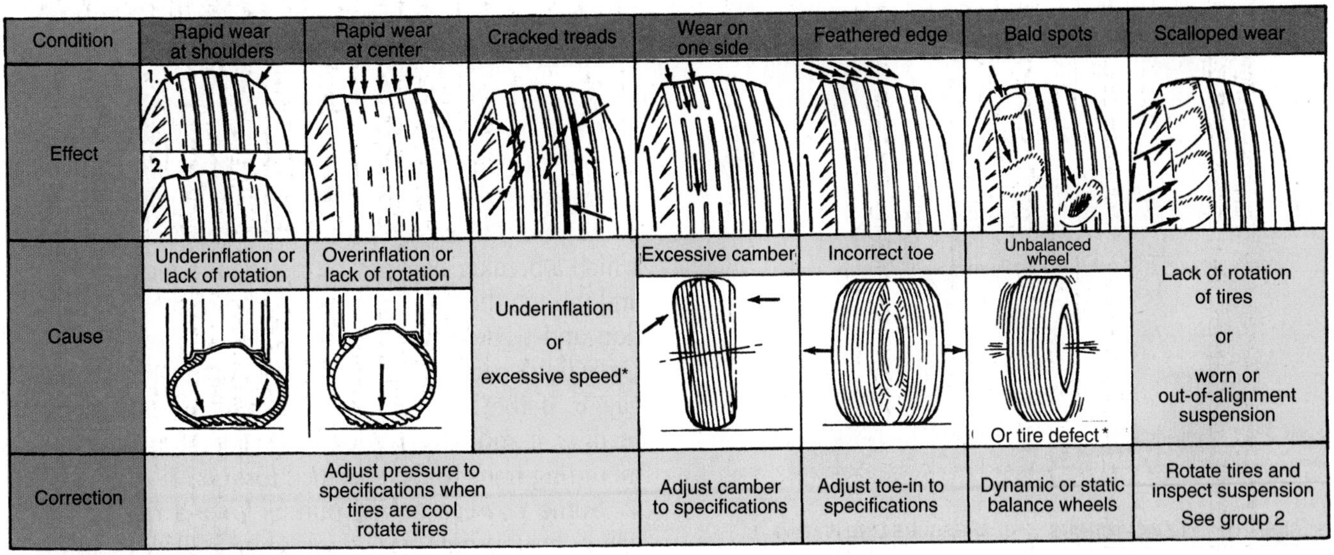
Condition
Rapid wear at shoulders
Rapid wear at center
Cracked treads
Wear on one side
Feathered edge
Bald spots
Scalloped wear
Effect
1.
2.
Cause
Underinflation or lack of rotation
Overinflation or lack of rotation
Underinflation or excessive speed*
Excessive camber
Incorrect toe
Unbalanced wheel
Or tire defect*
Lack of rotation of tires or worn or out-of-alignment suspension
Correction
Adjust pressure to specifications when tires are cool rotate tires
Adjust camber to specifications
Adjust toe-in to specifications
Dynamic or static balance wheels
Rotate tires and inspect suspension See group 2

*Have tire inspected for further use.

Figure 44-25 Tire wear patterns. *Courtesy of DaimlerChrysler Corporation*

Also check the tire and wheel assemblies for radial and lateral runout, and static and dynamic imbalance. Check the adjustment of the wheel bearings.

Check the power-steering fluid level and condition. The fluid is checked at the pump reservoir with a dipstick attached to the reservoir cap. Check the fluid level in the reservoir after the engine has been run at idle for 2 or 3 minutes and the wheel has been cycled from lock to lock several times. This warms the fluid to its normal operating temperature and gives a more accurate reading. Make sure the fluid level is at the full mark.

Examine the condition of the fluid carefully. Check for evidence of contamination such as solid particles or water. If either of these conditions is present or the fluid has a burnt odor, the system should be flushed before returning to service.

Also check the fluid for evidence of air trapped in the system. If the fluid looks foamy, it is likely that air is in the system. To verify this, run the engine until it reaches normal operating temperature. Then turn the steering wheel to the left and to the right several times without hitting the stops. If there is air in the system, bubbles will appear in the fluid reservoir. To remove the air, the system must be bled. The method of bleeding depends on the type of power-steering system. Follow the procedures given in the service manual. The typical procedure involves connecting a vacuum pump to the cap's opening in the reservoir **(Figure 44–26)**. While the engine is running, vacuum is applied to the reservoir and maintained for about 5 minutes. The vacuum is then released and the reservoir refilled with fluid. Vacuum is again applied to the reservoir and the steering wheel is cycled from stop-to-stop every 30 seconds for at least 5 minutes. After this period of time, the vacuum is released and the reservoir cap installed. Then the system is checked for leaks.

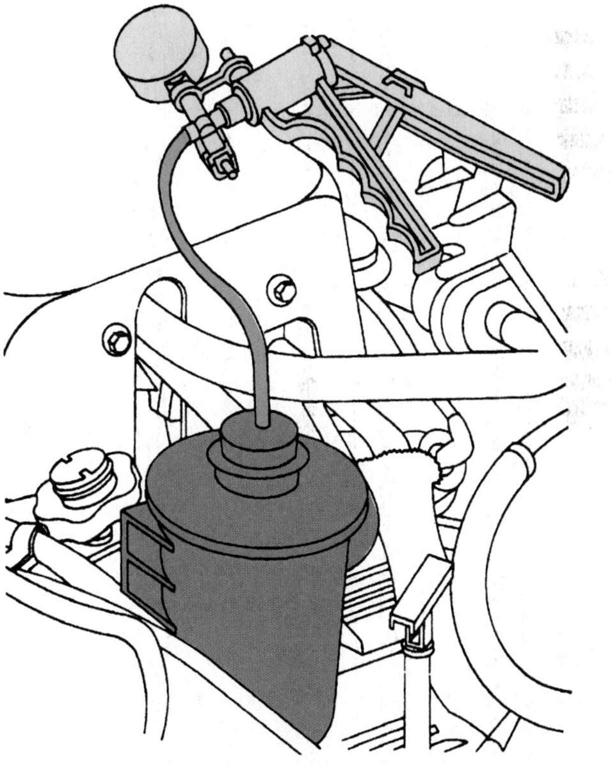

Figure 44-26 To bleed the system, a vacuum pump is connected to the fluid's reservoir. *Courtesy of Ford Motor Company*

With the ignition OFF, wipe off the outside of the power-steering pump, pressure hose, return hose, fluid cooler, and the steering gear **(Figure 44–27)**. Start the engine and turn the steering wheel several times from stop to stop. Check for leaks **(Figure 44–28)**. Fluid leakage will not only cause abnormal noises, but may result in unequal and abnormal steering efforts. If no signs of leakage are apparent, repeat the wheel cycling process and inspection several more times. Hoses should also be

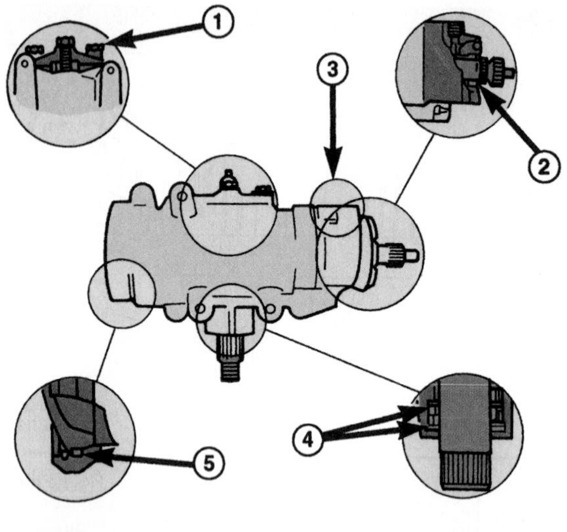

1. SIDE COVER LEAK - TORQUE SIDE COVER BOLTS TO SPECIFICATION. REPLACE THE SIDE COVER SEAL IF THE LEAKAGE PERSISTS.
2. ADJUSTER PLUG SEAL - REPLACE THE ADJUSTER PLUG SEALS.
3. PRESSURE LINE FITTING - TORQUE THE HOSE FITTING NUT TO SPECIFICATIONS. IF LEAKAGE PERSISTS, REPLACE THE SEAL.
4. PITMAN SHAFT SEALS - REPLACE THE SEALS.
5. TOP COVER SEAL - REPLACE THE SEAL.

Figure 44-27 Points for fluid leaks at a steering gear. *Courtesy of DaimlerChrysler Corporation*

carefully inspected for swelling and cracks. Always replace power-steering hoses with an exact replacement hose. Never attempt to patch or seal a leak in a hose or the hose's fittings.

On all systems, carefully check all of the mechanical parts of the steering and suspension system. Many suspension parts affect the operation of the steering system. Worn ball joints can cause erratic steering and premature tire wear. Bad suspension bushings will allow excessive wheel movement that can affect braking, handling, and wheel alignment. If any part is found to be defective it should be replaced.

Power-Steering Pump Belt

Power-steering belt condition and tension are extremely important for satisfactory power-steering pump operation. A loose belt causes low pump pressure and hard steering. A loose, dry, or worn belt may cause squealing and chirping noises, especially during engine acceleration and cornering. The power-steering pump belt should be checked for tension, cracks, oil soaking, worn or glazed edges, tears, and splits. If any of these conditions are present, the belt should be replaced.

Belt tension can be checked by measuring the belt deflection. Press on the belt with the engine stopped to measure the belt deflection, which should be ½ inch (13 mm) per foot of free span. The belt tension may also be checked with a belt tension gauge placed over the belt. The tension on the gauge should equal the vehicle manufacturer's specifications.

To adjust the tension of the belt, loosen the power-steering pump bracket or tension adjusting bolt and the power-steering pump mounting bolts. Pry against the pump ear and hub with a pry bar to tighten the belt. Some pump brackets have a ½-inch square opening in which a breaker bar may be installed to move the pump and tighten the belt. Hold the pump in the desired position and tighten the bracket or tension adjusting bolt. Once tightened, recheck the belt tension with the tension gauge. If the belt does not have the specified tension, readjust it and tighten the tension adjusting bolt and the mounting bolts to the specified torque.

Some power-steering pumps have a ribbed V-belt, which has an automatic tensioning pulley; therefore, a tension adjustment is not required. The belt, however, should be checked to make sure it is installed properly on each pulley in the belt drive system and that it is in good condition.

Pitman Arm

Because of its function, the pitman arm is the most heavily stressed point in the system. To inspect the pitman arm, grasp it and vigorously shake it to detect any looseness. Check the socket to reveal any damage or looseness. Either condition must be corrected by replacing the worn part. Their removal normally requires the use of a special puller **(Figure 44–29)**.

Idler Arm

A worn or damaged idler arm can cause steering instability, uneven tire wear, front-end shimmy, hard steering, excessive play in steering, or poor returnability. Because an idler arm is the weakest link in a parallelogram steering system, it wears more quickly than the rest of the system.

The procedure is simple for checking an idler arm for looseness or wear. The suspension should be normally loaded on the ground or on an alignment rack. When raised by a frame contact hoist, the vehicle's steering linkage is allowed to hang, and proper testing cannot be done. Check the idler arm ends for worn sockets or deteriorated bushings. Grasp the center link firmly with your hand at the idler arm end. Push up with approximately a 25-pound (110-N) load. Pull down with the same load. The allowable movement of the idler arm and support assembly in one direction is ⅛ inch (3 mm), for a total acceptable movement of ¼ inch (6 mm). The load can be accurately measured by using a dial indicator or pull-spring scale located as near the center link end of the idler arm as possible. Keep in mind that the test forces should not exceed 25 pounds (110 N), as even a new idler arm might be forced to show movement due to steel flexing when excessive pressure is applied. It is also nec-

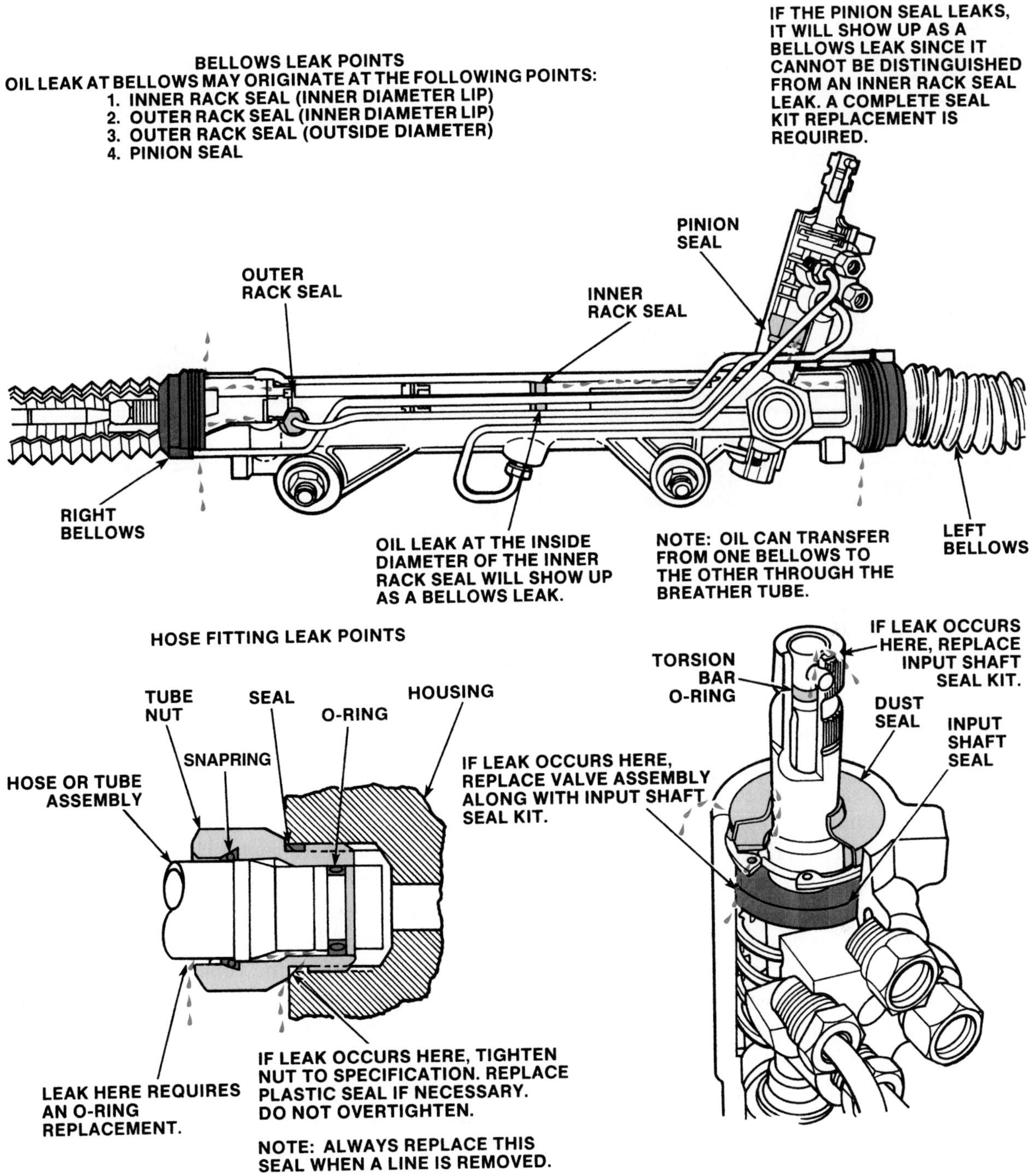

Figure 44-28 Possible leakage points on power-steering systems.

essary that a scale or ruler be rested against the frame and used to determine the amount of movement. Observers tend to overestimate the actual movement when a scale is not used. The idler arm should always be replaced if it fails this test. Jerking the right front wheel and tire assembly back and forth (causing an up-and-down movement in the idler arm) is not an acceptable method of checking, as there is no control on the amount of force being applied.

Center Link

Worn or bent center links can cause front-end shimmy, vehicle pull to one side, or change in the toe setting, causing excessive tire wear.

When inspecting the center link, look closely to insure it has not been bent or damaged. Grasp the center link firmly and try moving it in all directions. Any movement, or sign of damage, is reason for replacement. Tapered openings seldom wear but should be checked for

Figure 44–29 A pitman arm puller. *Courtesy of Moog Automotive, Inc.*

enlargement caused by a loose connection. If necessary, replace the center link.

Tie-Rod Assembly

Worn tie-rod ends result in incorrect toe-in settings, scalloped and scuffed tires, wheel shimmy, understeering, or front-end noise and tire squeal on turns.

Tie-rod end and center link inspections are similar. Grasp the tie-rod end firmly. Push vertically with the stud, and inspect for movement at the joint with the steering knuckle. Any movement over ⅛ inch (3 mm) or observation of damaged or missing parts, such as seals, is sufficient evidence that replacement is necessary.

Adjusting sleeves resemble a piece of internally threaded pipe. They have a slot or separation that runs either their entire length or just part way. Adjusting sleeves also have two crimping or squeezing clamps located at each end to lock the toe adjustment. Badly rusted, worn, or damaged adjusting sleeves should be replaced.

An additional check of the tie rods can be made by rotating each tie-rod end to feel for roughness or binding, which could indicate that the socket has probably rusted internally. A special puller is often required to separate a tie-rod end from the steering knuckle **(Figure 44–30)**.

Steering Damper

The steering dampers found in some steering linkage designs are generally nonadjustable, nonrefillable, and not repairable. At each inspection interval, inspect the mountings and check the assembly for damage (such as being bent) and fluid leaks. A light film of fluid is evidence of fluid leakage. However, a light film of fluid is permissible on the body of the damper near the shaft seal. A dripping damper should be replaced. A bad steering damper may cause wheel shimmy even though the rest of the suspension and steering system is fine.

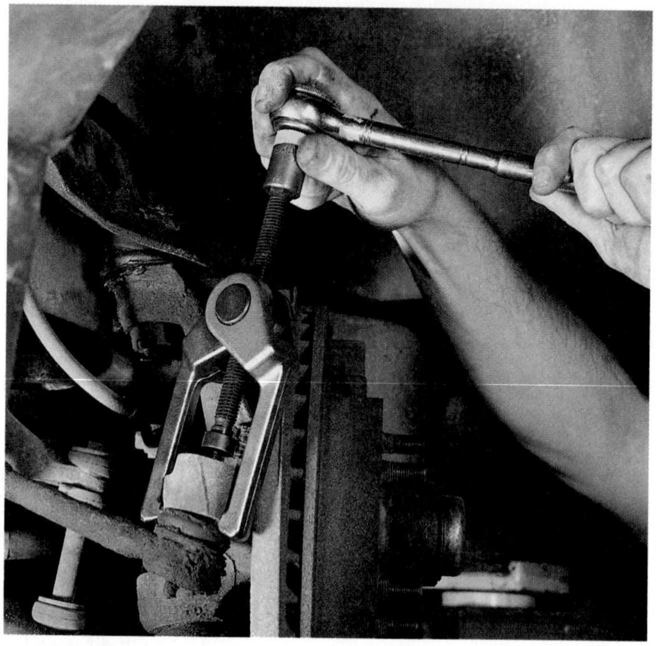

Figure 44–30 A tie-rod end-separating tool. *Courtesy of Moog Automotive, Inc.*

Dry Park Check

An excellent overall check for worn or loose conventional steering components is the **dry park check**. With the full weight of the vehicle on the wheels, have an assistant rock the steering wheel back and forth. Start your inspection from one side to the other side. Note any looseness in tie-rod, center link, idler arm, or pitman arm sockets **(Figure 44–31)**. If a second person is not available, reach up under the vehicle and grasp the flexible coupling on the steering shaft. Rock the linkage.

Before assembling any steering linkage parts, thoroughly check all tapered holes for out-of-roundness and wear. Thoroughly clean all bores that the stud tapers mount in. On new and reused parts, firmly install the tapered stud into its tapered hole. The stud must seat firmly without rocking. Only thread should protrude from the hole. If the parts do not meet these requirements, the mating part is worn and must be replaced, or

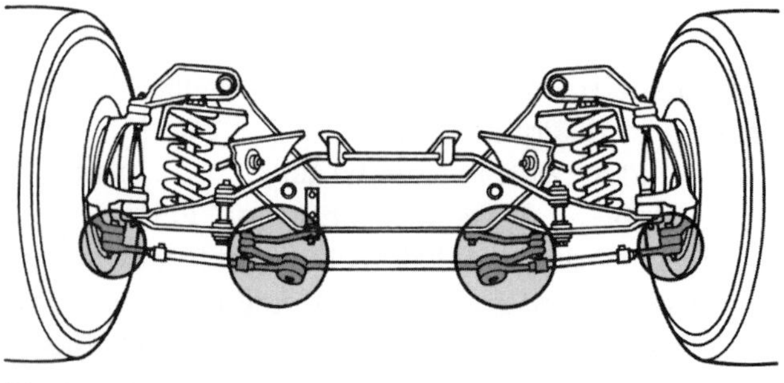

Figure 44–31 Circled areas indicate where a dry park check of steering linkage should be made.

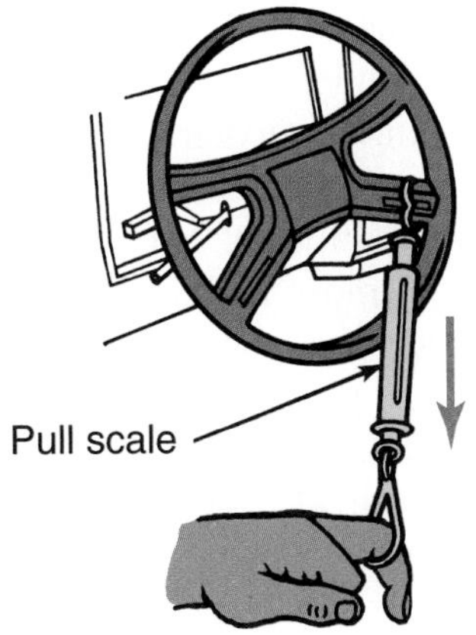

Figure 44–32 A pull scale is used to measure steering effort. *Reprinted with permission*

the correct parts are not being used. Always follow the manufacturer's stud and mounting bolt torque specifications when installing chassis parts.

Turning Effort

If an owner's concern indicates excessive turning effort, a pull scale should be used to read the actual force required to turn the wheel **(Figure 44–32)**. Compare the test results to the specifications in the service manual. If the effort exceeds the maximum, carefully inspect the entire steering system before performing a pressure test.

Tie-Rod Articulation Effort

The effort required to move the tie rod or its inner ball socket should be checked with a pull scale if excessive steering effort or looseness is noted during the road test. If the effort is not within the specified limits, the tie rod must be replaced.

Worm and Roller Steering

Since the worm and roller steering linkage components are almost identical to those of a parallelogram linkage, the same methods of inspection are used.

Rack and Pinion Steering

A rack and pinion system has no idler or pitman arms and no center link. Instead, they are replaced with a rack. Because the rack has no wear points, the number of wear points on rack and pinion systems is reduced to four—each of the tie-rod ends. Tie-rod ends are also wear points on the parallelogram steering system. Power rack and pinion assemblies should be carefully checked for leaks. If leaks cause the pump to run out of fluid, the pump will be damaged.

In order to solve customer complaints, a very thorough inspection of the entire system is needed. Everything, including ball joints, tires, outer tie rods, bellows boots, inner tie rods, rack-mounting bushings, mounting bolts, steering couplings, and gearbox adjustment must be checked. Rack and pinion steering inspection must be very thorough because of the system's sensitivity.

PROCEDURE

Rack and Pinion Steering Inspection

STEP 1 *Check all working components **(Figure 44–33)** of the systems. Inspect the flexible steering coupling or the universal joints for wear or looseness. If any play is found, recommend replacement. Universal joints can also seize or bind. They should be checked closely.*

STEP 2 *Grasp the pinion gear shaft at the flexible steering coupling and try to move it in and out of the gear. If there is movement, the pinion bearing preload might need adjustment. If there is no adjustment, internal components have to be replaced.*

STEP 3 *Carefully inspect the rack housing. In most cases, the rack and pinion steering assemblies are mounted in rubber bushings. As the vehicle gets older, these mounting bushings deteriorate from heat, age, and oil leakage from the engine. When this happens, the housing moves within its mounting and causes loose and erratic steering. Also, be alert for excessive movement of the rack housing. Stiffness in steering can be caused by a bent rack assembly, tight yoke bearing adjustment, loose power-steering belt, weak pump, internal leaks in power-steering system, and damaged CV joints in front-wheel-drive vehicles.*

STEP 4 *Check the inner tie-rod socket assemblies located inside the bellows. The most foolproof way of checking these sockets is to loosen the inner bellows clamp and pull the bellows back, giving a clear view of the socket. During the dry park check, observe any looseness. The inner tie-rod socket can also be checked by squeezing the bellows boot until the inner socket can be felt. Push and pull on the tire. If looseness is found in the tie rod, it should be replaced. On some vehicles the boot might be made of hard plastic. For this type of boot, lock the steering wheel and push and pull on the tire. Watch for in-and-out movement of the tie rod. If movement is observed, replace the inner tie rod.*

One fact to keep in mind is that the condition of the bellows boot determines the life of the inner socket. The bellows boot protects the rack from contamination. It might also contain fluid that helps keep the rack lubricated. If any cracks, splits, or leaks exist, the boot should be replaced. Also, be sure that clamps for the bellows are in their proper place and fastened tightly.

STEP 5 *Inspect the outer tie-rod ends. In addition to the dry park check, grab each end and rotate to feel for any roughness that would indicate internal rusting. Be sure to check for bent or damaged forgings and studs, split or deteriorated seals, and damaged, out-of-round, or loose tapers. If any of these conditions exist, the parts should be replaced.*

STEERING SYSTEM SERVICING

When a steering system component is found to be faulty, it is replaced. Most often part replacement is quite straightforward but you should always refer to the service manual before proceeding. At times, diagnosis will indicate a need to adjust the steering gear or inspect and repair the steering column.

Steering Gear Adjustments

Before any adjustments are made or servicing procedures performed to the steering gear, a careful check should be made of front-end alignment, shock absorbers, wheel balance, and tire pressure for possible steering system problems.

Before adjusting or servicing a manual steering gear, the technician must disconnect the battery ground cable. Raise the vehicle with the front wheels in the straight-ahead position. Remove the pitman arm nut. Mark the relationship of the pitman shaft. Remove the pitman arm with a pitman arm puller. Loosen the steering gear adjuster plug lock nut and back the adjuster plug off one-quarter turn **(Figure 44–34)**. Remove the horn shroud or button cap. Turn the steering wheel gently in one direction until stopped by the gear; then, turn back one-half turn. Measure and record bearing drag by applying a torque wrench with a socket on the steering wheel nut and rotating through a 90-degree arc. Check the service manual for the correct amount of drag.

Once these steps are taken, the steering gear is ready for adjusting or servicing as per instructions in the vehicle's service manual.

Steering Columns

To perform service procedures on the steering column upper end components, it is not necessary to remove the

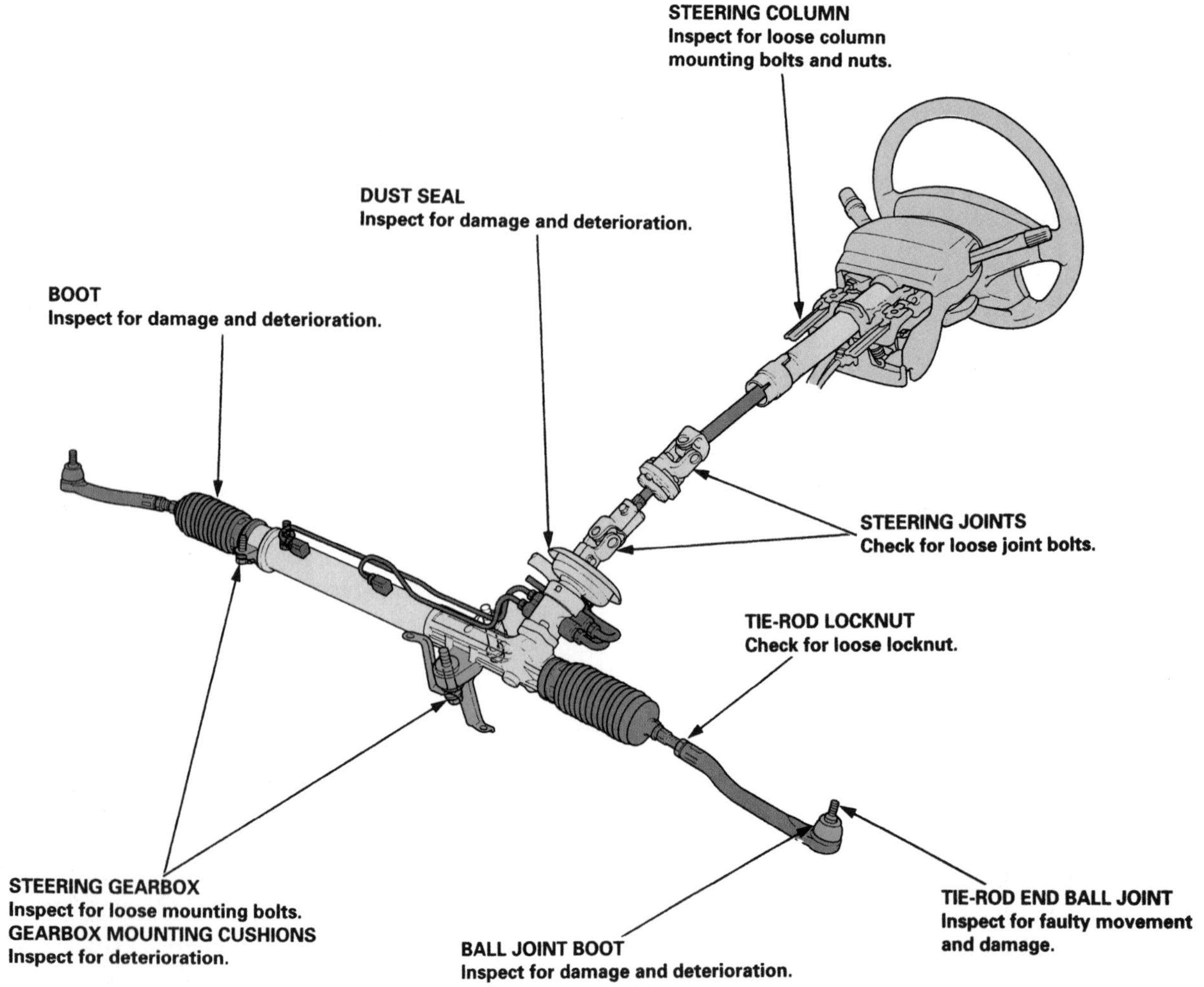
STEERING COLUMN
Inspect for loose column mounting bolts and nuts.
DUST SEAL
Inspect for damage and deterioration.
BOOT
Inspect for damage and deterioration.
STEERING JOINTS
Check for loose joint bolts.
TIE-ROD LOCKNUT
Check for loose locknut.
STEERING GEARBOX
Inspect for loose mounting bolts.
GEARBOX MOUNTING CUSHIONS
Inspect for deterioration.
BALL JOINT BOOT
Inspect for damage and deterioration.
TIE-ROD END BALL JOINT
Inspect for faulty movement and damage.

Figure 44–33 All steering parts should be carefully checked during diagnosis. *Courtesy of American Honda Motor Co., Inc.*

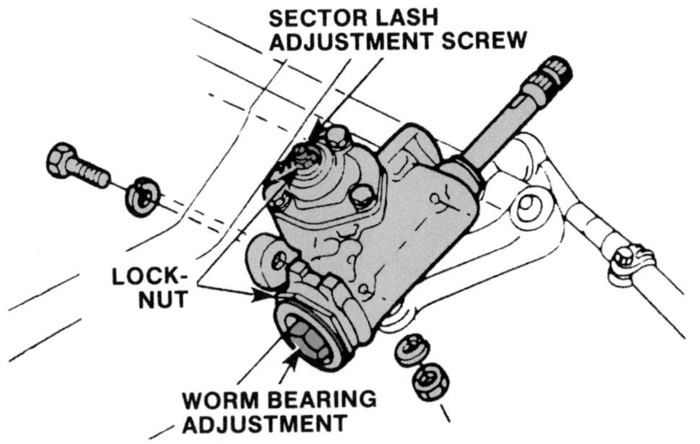

Figure 44-34 Typical steering gear adjustment points.

column from the vehicle. The steering wheel, horn components, directional signal switch, ignition switch, and lock cylinder can be removed with the column remaining in the vehicle.

To determine if the energy-absorbing steering column components are functioning as designed, or if repairs are required, a close inspection should be made. An inspection is called for in all cases where damage is evident or whenever the vehicle is being repaired due to a front-end collision. If damage is evident, the affected parts must be replaced. Because of the differences in the steering column styles and various components, consult the service manual for more explicit inspection and servicing procedures.

WARNING!

Set the parking brake before removing the steering column. Also, remove the battery cable from the negative terminal. Remember that special precautions must be observed before beginning disassembly and during assembly to ensure the correct fitting together of the steering column shaft and steering gear shaft connections.

POWER-STEERING SYSTEM SERVICING

Vehicles with power-steering systems have the same type of steering linkage as manual steering. The power-steering linkage is checked and serviced as previously described. Actually, the only difference is the servicing of the hydraulic components such as the hoses, pump, and power-steering gear that are fully covered in the vehicle's service manual. One of the common procedures that is recommended by manufacturers as part of a preventive maintenance program is flushing the hydraulic system.

WARNING!

Nearly all late-model vehicles are equipped with at least a driver's side air bag. The air bag module is housed in the steering wheel pad that must be removed before the steering wheel can be removed. Plus, it is dangerous to work on or around the steering column without following these precautions:

- *Always wear safety glasses when repairing an air bag system or when handling an air bag module.*
- *To prevent accidental deployment and possible injury, the backup power supply for the air bag must be depleted before replacing an air bag and before servicing, replacing, adjusting, or striking components close to the air bag sensors.*
- *To deplete backup power supply, disconnect the ground cable of the battery and wait at least one minute (some manufacturers recommend a longer waiting period).*
- *Refer to the appropriate service manual to identify the location of the air bag sensors.*
- *Never probe the connectors on the air bag module; this can cause accidental deployment.*
- *Carry a live air bag module with the air bag and trim cover pointed away from your body.*
- *Never set a live air bag module down with the trim cover face down.*
- *If a vehicle with an air bag was involved in an accident, inspect its sensor mounting bracket and wiring harness for damage. Replace any damaged parts.*
- *After deployment of an air bag, the bag and the surrounding surfaces may contain sodium hydroxide, which is a skin irritant. Wash your hands with soap and water after working around the bag.*

Flushing the System

The reason for flushing the system should be obvious. Hydraulic fluid is easily contaminated by moisture and dirt. Flushing removes the old fluid with its contaminants and new fluid is added to the system. It is also wise to flush the system after you have replaced or repaired a part in the system. Before beginning to flush the system, you should disable the engine's ignition. Then disconnect the power-steering return hose and plug the reservoir. Attach an extension hose between the power-steering return hose and an empty container. Raise the vehicle's front wheels off the ground. Fill the reservoir with the correct type of fluid.

WARNING!

Never mix oil types. Any mixture or any unapproved oil can lead to seal deterioration and leaks.

Turn the steering wheel from stop-to-stop while cranking the engine until the fluid leaving the return hose is clean. Never crank the engine for more than 5 seconds at a time. Add fluid to the reservoir to make sure it does not empty. Once the fluid is clear, fill the reservoir to its full mark and lower the vehicle.

Disconnect the extension hose from the power-steering return hose and reconnect the return hose to the reservoir. Check the fluid level again and add fluid as necessary. Now enable the ignition system. Start the engine and turn the steering wheel from stop-to-stop. If the power-steering system is noisy and bubbles are forming in the fluid, the system must be purged of air.

FOUR-WHEEL STEERING SYSTEMS

A few manufacturers have offered four-wheel steering systems in which the rear wheels also help to turn the car by electrical, hydraulic, or mechanical means. Although they certainly are not very common, you should be aware of how they work.

Production-built cars tend to understeer or, in a few instances, oversteer. If a car could automatically compensate for an understeer/oversteer problem, the driver would enjoy nearly neutral steering under varying operating conditions. **Four-wheel steering (4WS)** is a serious effort on the part of automotive design engineers to provide near-neutral steering with the following advantages:

- The vehicle's cornering behavior becomes more stable and controllable at high speeds as well as on wet or slippery road surfaces **(Figure 44–35)**.
- The vehicle's response to steering input becomes quicker and more precise throughout the vehicle's entire speed range.
- The vehicle's straight-line stability at high speeds is improved. Negative effects of road irregularities and crosswinds on the vehicle's stability are minimized.
- Stability in lane changing at high speeds is improved. High-speed slalom-type operations become easier. The vehicle is less likely to go into a spin even in situations in which the driver must make a sudden and relatively large change of direction.
- By steering the rear wheels in the direction opposite the front wheels at low speeds, the vehicle's turning circle is greatly reduced. Therefore, vehicle maneuvering on narrow roads and during parking becomes easier.

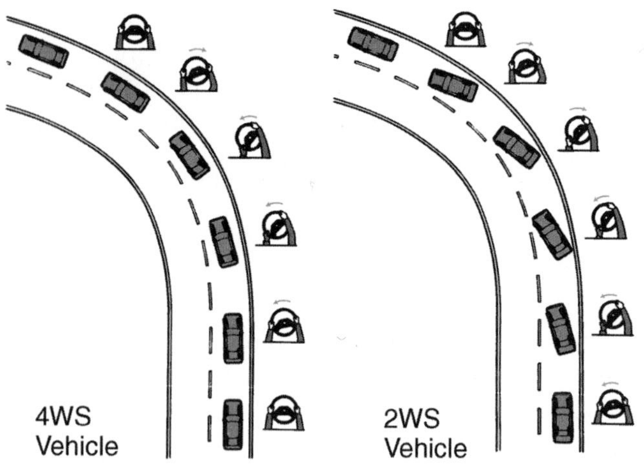

Figure 44–35 A comparison of 2WS and 4WS vehicle behavior during cornering.

To understand the advantages of four-wheel steering, it is wise to review the dynamics of typical steering maneuvers with a conventional front-steered vehicle. The tires are subject to the forces of grip, momentum, and steering input when making a movement other than straight-ahead driving. These forces compete with each other during steering maneuvers. With a front-steered vehicle, the rear end is always trying to catch up to the directional changes of the front wheels. This causes the vehicle to sway. As a normal part of operating a vehicle, the driver learns to adjust to these forces without thinking about them.

When turning, the driver is putting into motion a complex series of forces. Each of these must be balanced against the others. The tires are subjected to road grip and slip angle. Grip holds the car's wheels to the road, and momentum moves the car straight ahead. Steering input causes the front wheels to turn. The car momentarily resists the turning motion, causing a tire slip angle to form. Once the vehicle begins to respond to the steering input, cornering forces are generated. The vehicle sways as the rear wheels attempt to keep up with the cornering forces already generated by the front tires. This is referred to as rear-end lag because there is a time delay between steering input and vehicle reaction. When the front wheels are turned back to a straight-ahead position, the vehicle must again try to adjust by reversing the same forces developed by the turn. As the steering is turned, the vehicle body sways as the rear wheels again try to keep up with the cornering forces generated by the front wheels.

The idea behind four-wheel steering is that a vehicle requires less driver input for any steering maneuver if all four wheels are steering the vehicle. As with two-wheel-steer vehicles, tire grip holds the four wheels on the road. However, when the driver turns the wheel slightly, all four wheels react to the steering input, causing slip angles to form at all four wheels. The entire vehicle moves in one direction rather than the rear half attempting to catch up to the front. There is also less sway when the wheels are turned back to a straight-ahead position. The vehicle responds more quickly to steering input because rear wheel lag is eliminated.

Because each 4WS system is unique in its construction and repair needs, the vehicle's service manual must be followed for proper diagnosis, repair, and alignment of a four-wheel system.

Mechanical 4WS

In a straight-mechanical type of 4WS, two steering gears are used—one for the front and the other for the rear wheels. A steel shaft connects the two steering gearboxes and terminates at an eccentric shaft that is fitted with an offset pin **(Figure 44–36)**. This pin engages a second offset pin that fits into a planetary gear.

The planetary gear meshes with the matching teeth of an internal gear that is secured in a fixed position to the gearbox housing. This means that the planetary gear can rotate but the internal gear cannot. The eccentric pin of the planetary gear fits into a hole in a slider for the steering gear.

A 120-degree turn of the steering wheel rotates the planetary gear to move the slider in the same direction that the front wheels are headed. Proportionately, the rear wheels turn the steering wheel about 1.5 to 10 degrees. Further rotation of the steering wheel, past the 120-degree point, causes the rear wheels to start straightening out due to the double-crank action (two eccentric pins) and rotation of the planetary gear. Turning the steering wheel to a greater angle, about 230 degrees, finds the rear wheels in a neutral position regarding the front wheels. Further rotation of the steering wheel results in the rear wheels going counter phase with regard to the front wheels. About 5.3 degrees maximum counter phase rear steering is possible.

Mechanical 4WS is steering angle sensitive. It is not sensitive to vehicle road speed.

Hydraulic 4WS

The hydraulically operated 4WS system shown in **Figure 44–37** is a simple design, both in components and operation. The rear wheels turn only in the same direction as the front wheels. They also turn no more than 1½ degrees. The system only activates at speeds above 30 mph (50 km/h) and does not operate when the vehicle moves in reverse.

A two-way hydraulic cylinder mounted on the rear stub frame turns the wheels. Fluid for this cylinder is supplied by a rear steering pump that is driven by the differential. The pump only operates when the front wheels are turning. A tank in the engine compartment supplies the rear steering pump with fluid.

When the steering wheel is turned, the front steering pump sends fluid under pressure to the rotary valve in the front rack and pinion unit. This forces fluid into the front power cylinder, and the front wheels turn in the direction steered. The fluid pressure varies with the turning of the steering wheel. The faster and farther the steering wheel is turned, the greater the fluid pressure.

The fluid is also fed under the same pressure to the control valve where it opens a spool valve in the control valve housing. As the spool valve moves, it allows fluid from the rear steering pump to move through and operate the rear power cylinder. The higher the pressure on

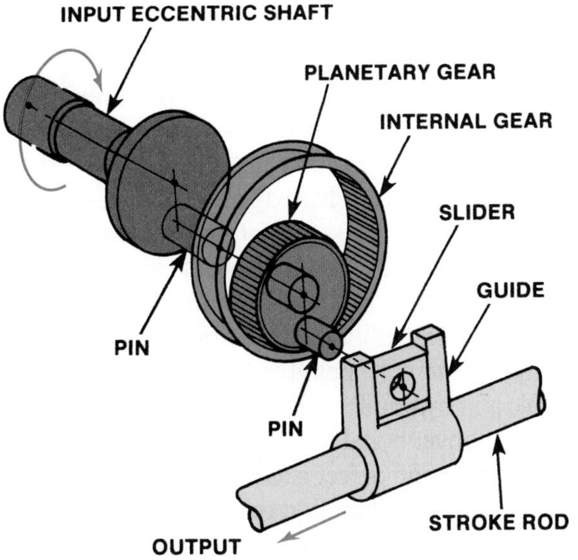

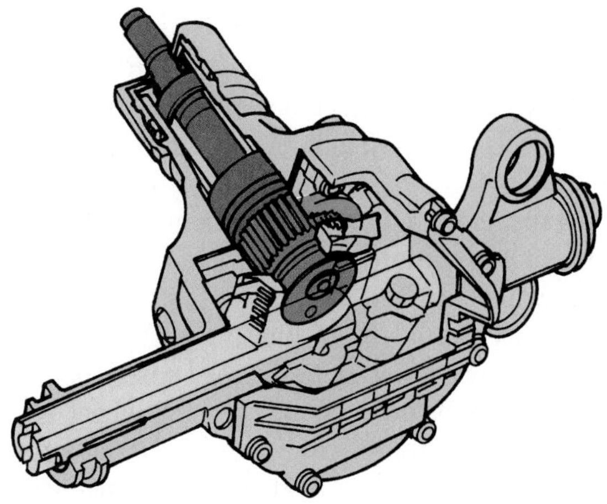

Figure 44-36 Inside a rear-steering gearbox is a simple planetary gear setup. *Courtesy of American Honda Motor Co., Inc.*

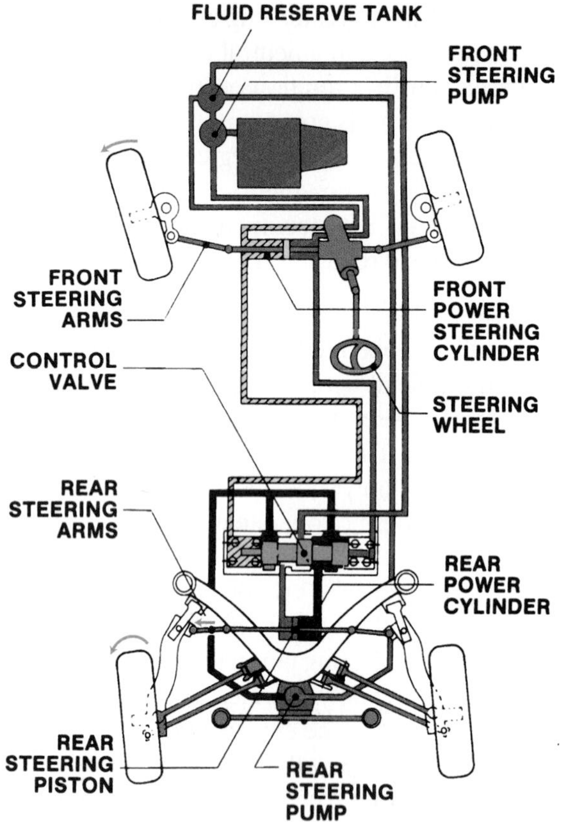

Figure 44–37 A simple hydraulic 4WS system. *Courtesy of Mitsubishi Motor Sales of America, Inc.*

the spool, the farther it moves. The farther it moves, the more fluid it allows through to move the rear wheels. As mentioned earlier, this system limits rear wheel movement to 1½ degrees in either the left or right direction.

Electro/Hydraulic 4WS

Several 4WS systems combine computer electronic controls with hydraulics to make the system sensitive to both steering angle and road speeds. In this design, a speed sensor and steering wheel angle sensor feed information to the electronic control unit (ECU). By processing the information received, the ECU commands the hydraulic system to steer the rear wheels. At low road speed, the rear wheels of this system are not considered a dynamic factor in the steering process.

At moderate road speeds, the rear wheels are steered momentarily counterphase, through neutral, then in phase with the front wheels. At high road speeds, the rear wheels turn only in phase with the front wheels. The ECU must know not only road speed, but also how much and how quickly the steering wheel is turned. These three factors—road speed, amount of steering wheel turn, and the quickness of the steering wheel turn—are interpreted by the ECU to maintain continuous and desired steering angle of the rear wheels.

Another electro/hydraulic 4WS system is shown in **Figure 44–38**. The basic working elements of the design are the control unit, a stepper motor, a swing arm, a set of beveled gears, a control rod, and a control valve with an output rod. Two electronic sensors tell the ECU how fast the car is going.

The yoke is a major mechanical component of this electro/hydraulic design. The position of the control yoke varies with vehicle road speed. For example, at speeds below 33 mph (53 km/h), the yoke is in its downward position, which results in the rear wheels steering in the counterphase (opposite front wheels) direction. As road speeds approach and exceed 22 mph (35 km/h), the control yoke swings up through a neutral (horizontal) position to an up position. In the neutral position, the rear wheels steer in phase with the front wheels.

The stepper motor moves the control yoke. A swing arm is attached to the control yoke. The position of the yoke determines the arc of the swing rod. The arc of the swing arm is transmitted through a control arm that passes through a large bevel gear. Stepper motor action eventually causes a push-or-pull movement of its output shaft to steer the rear wheels up to a maximum of 5 degrees in either direction.

The electronically controlled 4WS system regulates the angle and direction of the rear wheels in response to speed and driver's steering. This speed-sensing system optimizes the vehicle's dynamic characteristics at any speed, thereby producing enhanced stability and, within certain parameters, agility.

The actual 4WS system consists of a rack and pinion front steering that is hydraulically powered by a main twin-tandem pump. The system also has a rear-steering mechanism, hydraulically powered by the main pump. The rear-steering shaft extends from the rack bar of the front-steering assembly to the rear-steering-phase control unit.

The rear steering is comprised of the input end of the rear-steering shaft, vehicle speed sensors, and steering-phase control unit (deciding direction and degree), a power cylinder, and an output rod. A centering lock spring is incorporated that locks the rear system in a neutral (straight-ahead) position in the event of hydraulic failure. Additionally, a solenoid valve that disengages the hydraulic boost (thereby activating the centering lock spring in case of an electrical failure) is included.

All 4WS systems have fail-safe measures. For example, with the electro/hydraulic setup, the system automatically counteracts possible causes of failure, both electronic and hydraulic, and converts the entire steering system to a conventional two-wheel steering type. Specifically, if a hydraulic defect should reduce pressure level (by a movement malfunction or a broken driving belt), the rear-wheel-steering mechanism is automatically locked in a neutral position, activating a low-level warning light.

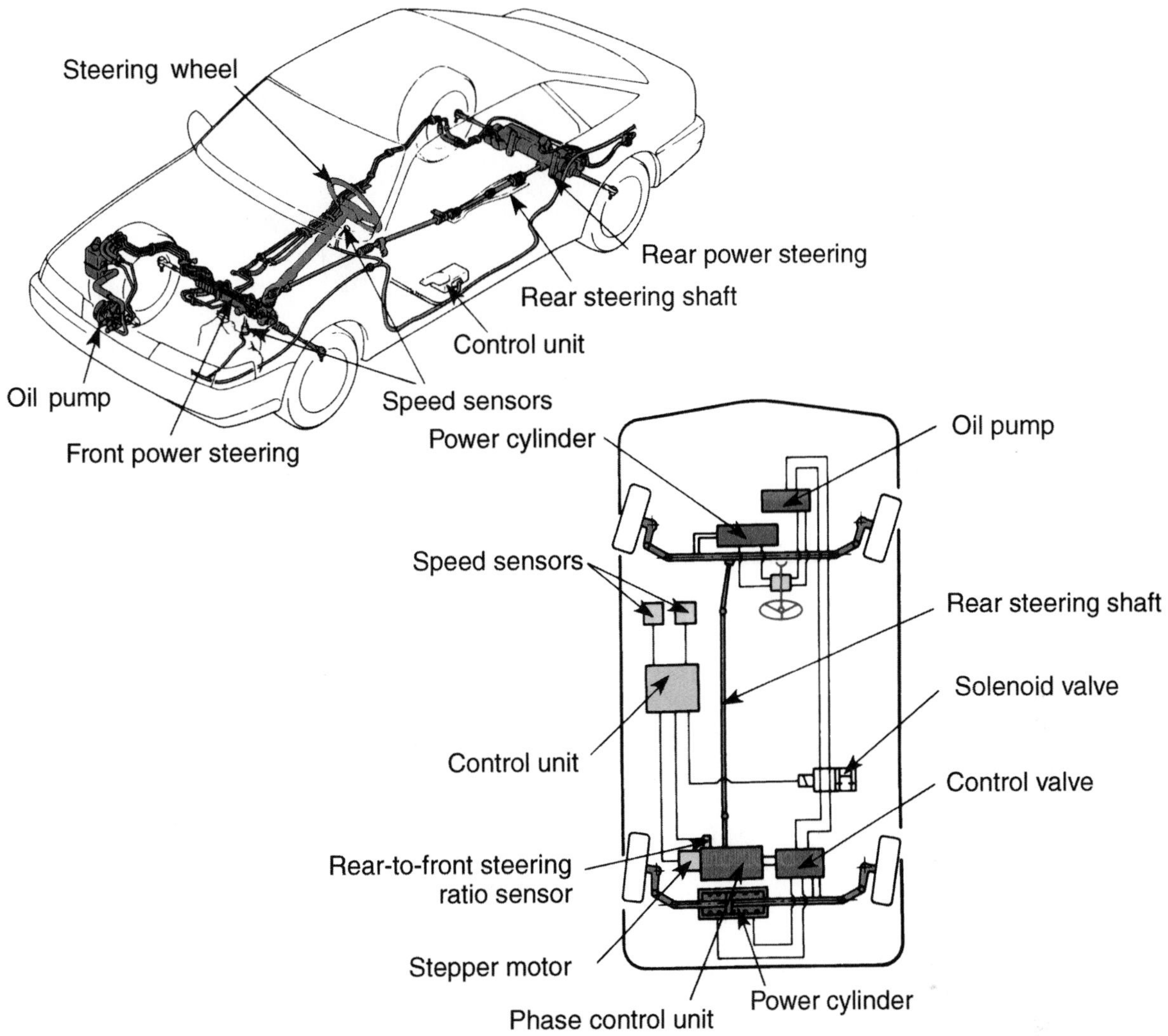

Figure 44–38 An electronically and hydraulically controlled 4WS system using a stepper motor and control yoke. *Courtesy of Mazda Motor of America, Inc.*

An electrical failure would be detected by a self-diagnostic circuit integrated in the four-wheel-steering control unit. The control unit stimulates a solenoid valve, which neutralizes hydraulic pressure, thereby alternating the system to two-wheel steering. The failure would be indicated by the system's warning light in the main instrument display.

On any 4WS system, there must be near-perfect compliance between the position of the steering wheel, the position of the front wheels, and the position of the rear wheels. It is usually recommended that the car be driven about 20 feet (6 meters) in a dead-straight line. Then, the position of the front/rear wheels is checked with respect to steering wheel position. The base reference point is a strip of masking tape on the steering wheel hub and the steering column. When the wheel is positioned dead center, draw a line down the tape. Run the car a short distance straight ahead to see if the reference line holds. If not, corrections are needed, such as repositioning the steering wheel.

Even severe imbalance of a rear wheel on a speed-sensitive 4WS system can cause problems and make basic troubleshooting a bit frustrating.

Quadrasteer

Quadrasteer is a 4WS system that improves low-speed maneuverability (decreasing the turning radius), high-speed stability, and trailering capabilities for full-size pickups, vans, and SUVs. The system combines normal front-wheel steering with an electrically powered and electronically controlled rear-wheel steering system. Besides the mechanical part, the system uses wheel position and vehicle speed sensors and a central control module **(Figure 44–39)**.

At low speeds, the rear wheels turn in the opposite direction of the front wheels **(Figure 44–40)**. At moderate speeds, the rear wheels remain straight. At high speeds, the rear wheels turn in the same direction as the front wheels. If the system were to fail, the truck would default to normal two-wheel steering.

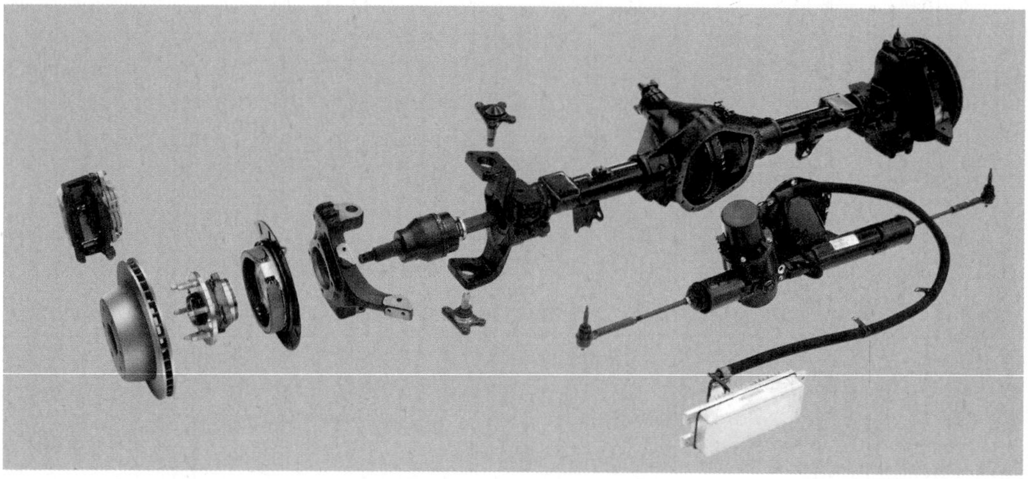

Figure 44-39 An exploded view of the Quadrasteer setup. *Courtesy of the Delphi Corporation*

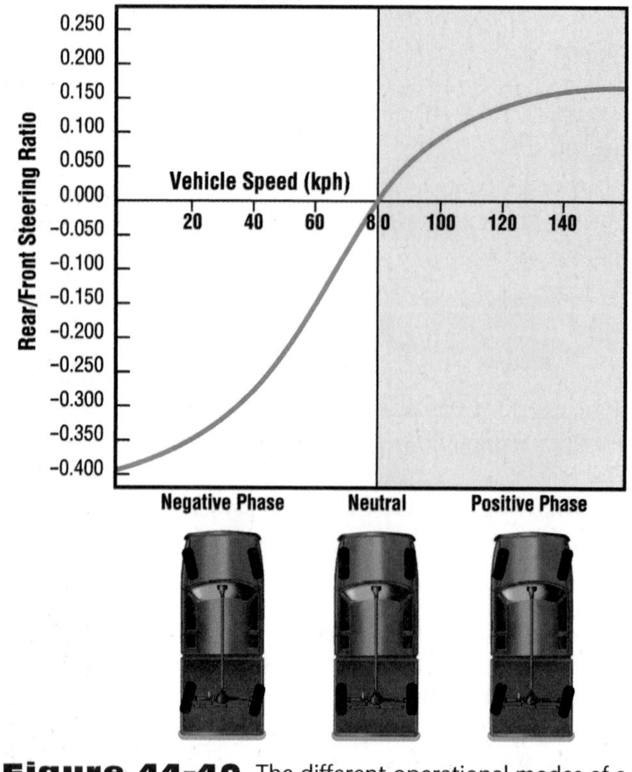

Figure 44-40 The different operational modes of a Quadrasteer system. *Courtesy of the Delphi Corporation*

CASE STUDY

A customer complains of excessive steering effort on his 1998 Oldsmobile Aurora. The technician took the car out for a road test to verify the customer's complaint, but found no evidence of hard steering. Further questioning of the customer revealed that the problem only occurred when the engine was cold. The service writer informed the customer that the car would need to be parked overnight at the shop so the problem could be duplicated. The customer agreed to this.

The next morning prior to starting the engine, the technician connected a power-steering pressure tester in series with the high-pressure hose. When the engine was first started, the technician found the steering effort to be overly hard and it felt as if there was no power assist. But the power-steering pump was working and was putting out a higher than normal pressure. Once the steering wheel was turned in one direction, the steering effort quickly returned to normal. Since the power-steering pump pressure was higher than specified, the technician concluded the pump and drive belt must be in fine shape. He also reasoned that other causes of hard steering, such as binding in the steering column, would be always present. They would not cause the problem to occur only when the system was cold. Therefore, he decided the problem must be in the steering gear.

He then removed and inspected the steering gear. He found that the rack cylinder was severely ridged and scored in the center area, and the Teflon ring on the rack piston was severely worn. A new steering gear assembly was installed, the system flushed, and the car road tested. To verify the repair, the car was left out overnight again and the steering checked the next morning. Everything worked fine.

KEY TERMS

Bump steer
Center link
Drag link
Dry park check
Four-wheel steering (4WS)
Idler arm
Integral piston

Pitman arm
Rack
Rack and pinion
Steering linkage
Tie rod
Worm and roller gearbox
Yoke lash

SUMMARY

- The components of a manual-steering system include the steering linkage, steering gear, and the steering column and wheel.
- The term steering linkage is applied to the system of pivots and connecting parts placed between the steering gear and the steering arms that are attached to the front wheels, controlling the direction of vehicle travel. The steering linkage transfers the motion of the steering gear output shaft to the steering arms, turning the wheels to maneuver the vehicle.
- Basic components of a parallelogram steering linkage system include the pitman arm, idler arm, links, tie rods, and, in some designs, a steering damper.
- The worm and roller steering components are basically the same as found in the parallelogram system.
- In rack and pinion steering linkage, steering input is received from a pinion gear attached to the steering column. This gear moves a toothed rack that is attached to the tie rods that move the wheels.
- There are three types of manual steering gears in use today: recirculating ball, worm and roller, and rack and pinion.
- The steering wheel and column produce the necessary force to turn the steering gear.
- The power-steering unit is designed to reduce the amount of effort required to turn the steering wheel. It also reduces driver fatigue on long drives and makes it easier to steer the vehicle at slow road speeds, particularly during parking.
- There are several power-steering systems in use on passenger cars and light-duty trucks. The most common ones are the integral, linkage, hydro-boost, and power-assisted rack and pinion systems.
- The major components of a conventional power-steering system are the steering linkage, power-steering pump, flow control and pressure relief valves, reservoir, spool valves and power pistons, hydraulic hose lines, and gearbox or assist assembly on the linkage.
- Electronic rack and pinion systems replace the hydraulic pump, hoses, and fluid associated with conventional power-steering systems with electronic controls and an electric motor located concentric to the rack itself.
- Four-wheel steering (4WS) advantages include cornering capability, steering response, straight-line stability, lane changing, and low-speed maneuverability.

TECH MANUAL

The following procedures are given in Chapter 44 of the *Tech Manual* that accompanies this book:

1. Diagnosing, removing, and replacing an idler arm.
2. Removing and replacing an outer tie-rod end on a parallelogram steering linkage system.
3. Flushing a power-steering system.
4. Testing the pressure of a power-steering pump.
5. Inspecting a rack and pinion steering gear.

REVIEW QUESTIONS

1. Describe how rack and pinion steering, parallelogram steering, and worm and roller steering systems operate.
2. A power-steering hose transmits fluid under pressure from the ______ to the ________________.
3. What is an integral power-steering system?
4. Define the term *gearbox ratio.*
5. What are the basic features of all 4WS systems?
6. What is the primary purpose of power-steering hoses?
   a. to lubricate the pump
   b. to relieve pressure
   c. to transmit power through fluid under pressure
   d. none of the above

7. Technician A says electrically operated power-steering systems allow for power assistance even when an engine stalls. Technician B says an electronic rack and pinion system provides power assistance even when the engine stalls. Who is correct?
   a. Technician A
   b. Technician B
   c. Both A and B
   d. Neither A nor B
8. Technician A says that the manual-steering gear is the probable cause of a shimmy. Technician B says it could be loose steering linkage. Who is correct?
   a. Technician A
   b. Technician B
   c. Both A and B
   d. Neither A nor B
9. Technician A says that adjusting the steering gear too tightly can cause hard steering. Technician B says that adjusting the steering gear too tightly can cause poor returnability. Who is correct?
   a. Technician A
   b. Technician B
   c. Both A and B
   d. Neither A nor B
10. Technician A says a worn or damaged idler arm can cause front-end shimmy. Technician B says a worn or damaged idler arm can cause hard steering. Who is correct?
   a. Technician A
   b. Technician B
   c. Both A and B
   d. Neither A nor B
11. Worn tie-rod ends can result in ______.
   a. scalloped and scuffed tires
   b. the vehicle pulling to one side
   c. poor returnability
   d. all of the above
12. In a rack and pinion steering system, what protects the rack from contamination?
   a. the inner tie-rod socket
   b. the outer tie-rod socket
   c. grommets
   d. the bellows boot
13. A variable ratio can be normally found in ______.
   a. manual-steering units
   b. power-steering units
   c. both a and b
   d. neither a nor b
14. Which one of the following is *not* a probable cause of wheel shimmy?
   a. binding steering column U-joints
   b. loose wheel bearings
   c. out-of-balance tires
   d. loose, worn, or damaged suspension parts
15. The main job of the idler arm is to:
   a. support the left side of the center link.
   b. support the right side of the center link.
   c. support the pitman arm.
   d. keep both ends of the steering system level.
16. An owner's concern indicates excessive turning effort. Technician A checks the alignment of the vehicle's frame. Technician B uses a pull scale. Who is correct?
   a. Technician A
   b. Technician B
   c. Both A and B
   d. Neither A nor B
17. Which of the following is true of the steering linkage system?
   a. The power cylinder can be integrated with the steering gear.
   b. The power cylinder can have four hydraulic lines connected to it.
   c. The power cylinder may be attached to the steering center link.
   d. All of the above.
18. In a 4WS steering system, the rear wheels go into ______ when the steering wheel is rotated more than 230 degrees.
   a. parallel
   b. same-phase
   c. counter phase
   d. none of the above
19. Rack and pinion steering ______.
   a. is lighter in weight and has fewer components than parallelogram steering
   b. does not provide as much feel for the road as parallelogram steering
   c. does not use tie rods in the same fashion as parallelogram steering
   d. all of the above
20. What connects the steering column to the wheels?
   a. the steering shaft
   b. the steering linkage
   c. the gearbox
   d. the pitman arm

45

WHEEL ALIGNMENT

OBJECTIVES

■ Explain the benefits of accurate wheel alignment. ■ Explain the importance of correct wheel alignment angles. ■ Describe the different functions of camber and caster with regard to the vehicle's suspension. ■ Identify the purposes of steering axis inclination. ■ Explain why toe is the most critical tire wear factor of all the alignment angles. ■ Identify the purposes of turning radius or toe-out. ■ Explain the condition known as tracking. ■ Perform a prealignment inspection. ■ Describe the various types of equipment that can be used to align the wheels of a vehicle. ■ Describe how alignment angles can be changed on a vehicle. ■ Understand the importance of rear-wheel alignment. ■ Know the difference between two-wheel and four-wheel alignment procedures.

Swift and sure steering responses are needed at today's driving speeds. To accomplish this, the wheels must be in alignment. Wheel alignment allows the wheels to roll without scuffing, dragging, or slipping on different types of road conditions. Proper alignment of both the front and the rear wheels ensures greater safety in driving, easier steering, longer tire life, reduction in fuel consumption, and less strain on the parts that make up the steering and suspension systems of the vehicle **(Figure 45–1)**.

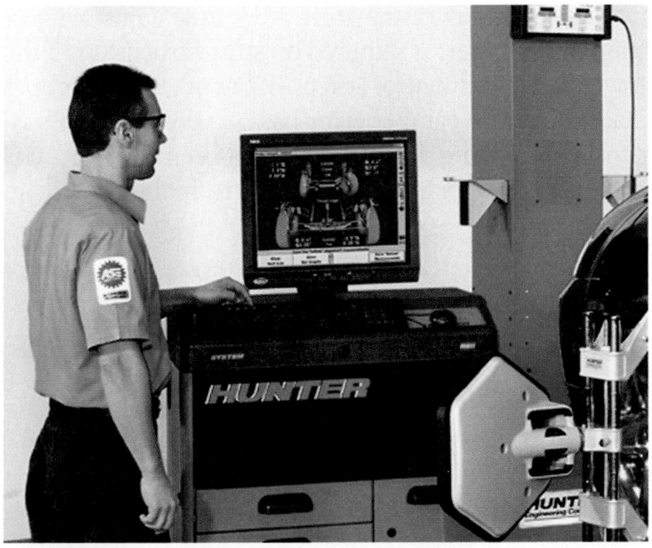

Figure 45-1 Technicians who specialize in wheel alignment are always in demand. *Courtesy of Hunter Engineering Company*

There are a multitude of angles and specifications that the automotive manufacturers must consider when designing a car. The multiple functions of the suspension system complicate things a great deal for design engineers. They must take into account more than basic geometry. Durability, maintenance, tire wear, available space, and production cost are all critical elements. Most elements contain a degree of compromise in order to satisfy the minimum requirements of each.

Most technicians do not need to be concerned with such requirements. All they need to do is restore the vehicle to the condition the design engineer specified. To do this, however, the technician must be totally familiar with the purpose of basic alignment angles.

The alignment angles are designed in the vehicle to properly locate the vehicle's weight on moving parts and to facilitate steering. If these angles are not correct, the vehicle is misaligned. The effects of misalignment are given in **Table 45–1.** It is important to remember that alignment angles, when specified in the text, are those specific angles that should exist when the system is being measured under a given set of conditions. During regular performance, these angles change as the traveling surface and vehicle driving forces change.

An important thing to remember when doing wheel alignment and diagnosis is that roads are not designed to be flat. They are designed to allow rain to flow off the surface rather than to accumulate. To accomplish this drainage, the road's surface is paved at a slight angle.

TABLE 45–1 EFFECTS OF INCORRECT WHEEL ALIGNMENT	
Problem	**Effect**
Incorrect camber setting	Tire wear Ball joint/wheel bearing wear Pull to side of most positive/ least negative camber
Too much positive caster	Hard steering Excessive road shock Wheel shimmy
Too much negative caster	Wander Weave Instability at high speeds
Unequal caster	Pull to side most negative/ least positive caster
Incorrect SAI	Instability Poor return Pull to side of lesser inclination Hard steering
Incorrect toe setting	Tire wear
Incorrect turning radius	Tire wear Squeal in turns

This angle is called road crown. Road crown also can cause a vehicle to pull toward the right. Therefore, alignment angles must compensate for it. Slightly more positive camber on one of the front tires will correct for road crown.

ALIGNMENT GEOMETRY

The proper alignment of a suspension/steering system centers on the accuracy of the following angles.

Caster

Caster is the angle of the steering axis of a wheel from the vertical, as viewed from the side of the vehicle. The forward or rearward tilt from the vertical line **(Figure 45–2)** is caster. Caster is the first angle adjusted during an alignment. Tilting the wheel forward is negative caster. Tilting backward is positive caster.

Caster is designed to provide steering stability. The caster angle for each wheel on an axle should be equal. Unequal caster angles cause the vehicle to steer toward the side with less caster. Too much negative caster can cause the vehicle to have sensitive steering at high speeds. The vehicle might wander as a result of negative caster. Caster is not related to tire wear.

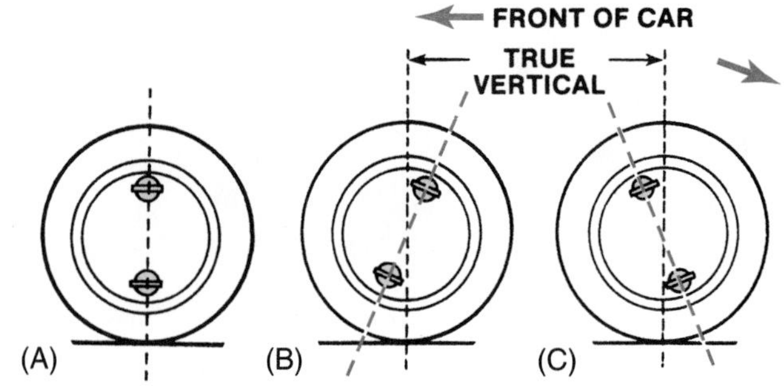

Figure 45-2 Three types of caster are (A) zero, (B) positive, and (C) negative.

Caster is affected by worn or loose strut rod and control arm bushings. Caster adjustments are not possible on some strut suspension systems. Where they are provided, they can be made at the top or bottom mount of the strut assembly.

Camber

Camber is the angle represented by the tilt of either the front or rear wheels inward or outward from the vertical as viewed from the front of the car **(Figure 45–3)**. Camber is designed into the vehicle to compensate for road crown, passenger weight, and vehicle weight. Camber is usually set equally for each wheel. Equal camber means each wheel is tilted outward or inward the same amount. Unequal camber causes tire wear and causes the vehicle to steer toward the side that is more positive.

Camber angle changes, through the travel of the suspension system, are controlled by pivots. Camber is affected by worn or loose ball joints, control arm bushings, and wheel bearings. Anything that changes chassis height also affects camber. Camber is adjustable on most vehicles. Some manufacturers prefer to include a camber adjustment at the spindle assembly. Camber adjustments are also provided on some strut suspension systems at the top mounting of the strut. Very little adjustment of camber (or caster) is required on strut suspensions if the tower and lower control arm positions are in their proper place. If serious camber error has occurred and the suspension mounting positions have not been damaged, it is an indication of bent suspension parts. In this case, diag-

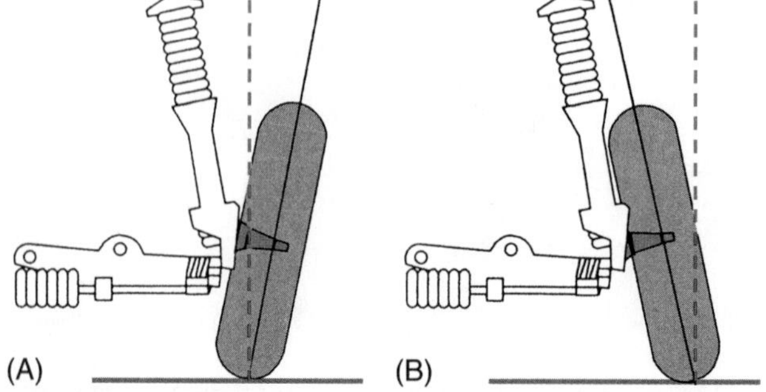

Figure 45-3 (A) Positive and (B) negative camber.

Positive toe (toe-in)

Negative toe (toe-out)

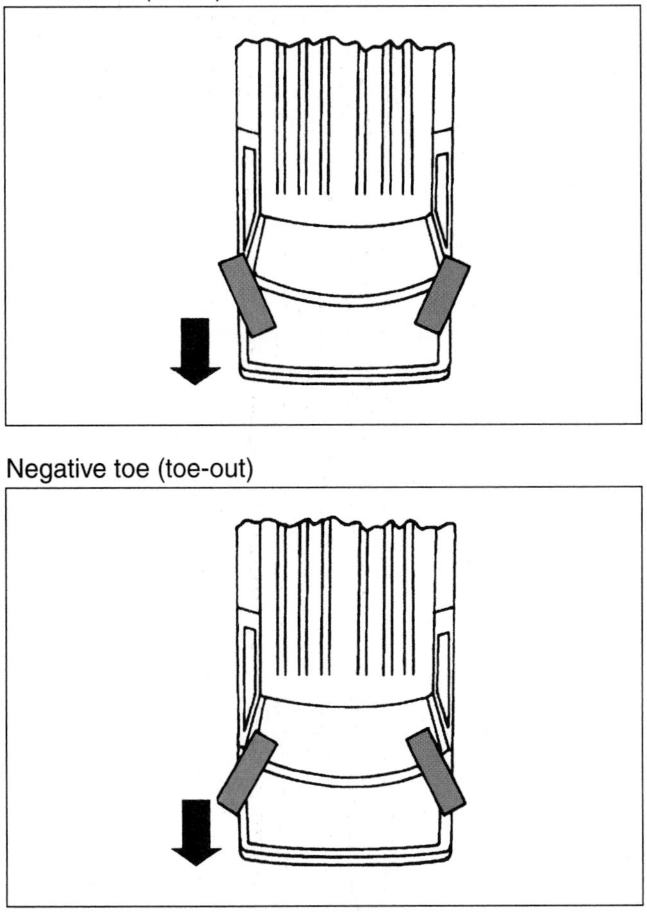

Figure 45-4 The position of the wheels during toe-in and toe-out. *Courtesy of Ford Motor Company*

nostic angle and dimensional checks should be made on the suspension parts. Damaged parts should be replaced.

Toe

Toe is the distance comparison between the leading edge and trailing edge of the front tires. If the leading edge distance is less, then there is toe-in. If it is greater, there is toe-out **(Figure 45–4)**. Actually, toe is critical as a tire-wearing angle. Wheels that do not track straight ahead have to drag as they travel forward. Excessive toe measurements (in or out) cause a sawtooth edge on the tread surface from dragging the tire sideways. Excessive toe-in will cause tire wear on the outside edge of the tire. Toe-out causes wear on the inside edge.

Toe adjustments are made at the tie rod. They must be made evenly on both sides of the car. If the toe settings are not equal, the car may tend to pull due to the steering wheel being off-center. An off-center steering wheel and steering pull should be corrected by making the toe adjustments equal on both sides of the car with the steering wheel centered. Toe is the last adjustment made in an alignment.

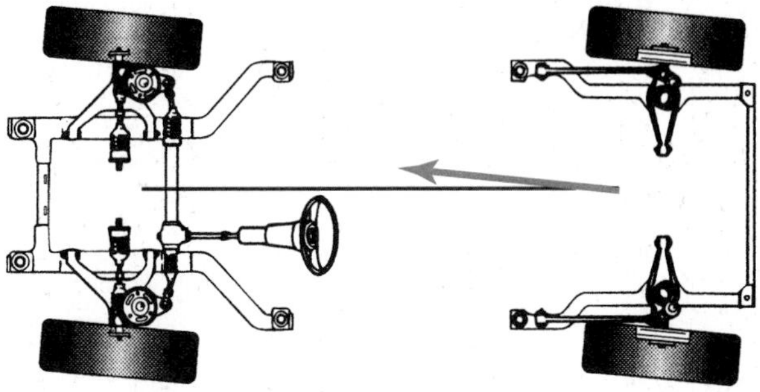

Figure 45-5 The thrust line or driving direction of the rear wheels. *Courtesy of Northstar Manufacturing Company, Inc.*

Thrust Line Alignment

A main consideration in any alignment is to make sure the vehicle runs straight down the road, with the rear tires tracking directly behind the front tires when the steering wheel is in the straight-ahead position. The geometric centerline of the vehicle should parallel the road direction. This is the case when rear toe is parallel to the vehicle's geometric centerline in the straight-ahead position. If rear toe does not parallel the vehicle centerline, a thrust direction to the left or right is created **(Figure 45–5)**. This difference of rear toe from the geometric centerline is called the **thrust angle**. The vehicle tends to travel in the direction of the thrust line, rather than straight ahead.

To correct this problem, begin by setting individual rear-wheel toe equally in reference to the geometric centerline. Four-wheel-alignment machines check individual toe on each wheel. Once the rear wheels are in alignment with the geometric centerline, set the individual front toe in reference to the thrust angle. Following this procedure assures that the steering wheel is straight ahead for straight-ahead travel. If you set the front toe to the vehicle geometric centerline, ignoring the rear toe angle, a cocked steering wheel results.

Steering Axis Inclination (SAI)

Steering axis inclination (SAI) locates the vehicle weight to the inside or outside of the vertical centerline of the tire. The SAI is the angle between true vertical and a line drawn between the steering pivots as viewed from the front of the vehicle **(Figure 45–6)**. It is an engineering angle designed to project the weight of the vehicle to the road surface for stability. The SAI helps the vehicle's steering system return to straight ahead after a turn.

If the vehicle has 0 (zero) SAI, the upper and lower ball joints (or strut pivot points) would be located directly over one another. Problems associated with this simple relationship include tire scrub in turns, lack of control, and increased effort during turn recovery. If the SAI is tilted, a triangle is formed between ball joints and spindle. An arc is then formed when turning. There is a high point at straight-ahead position and a drop

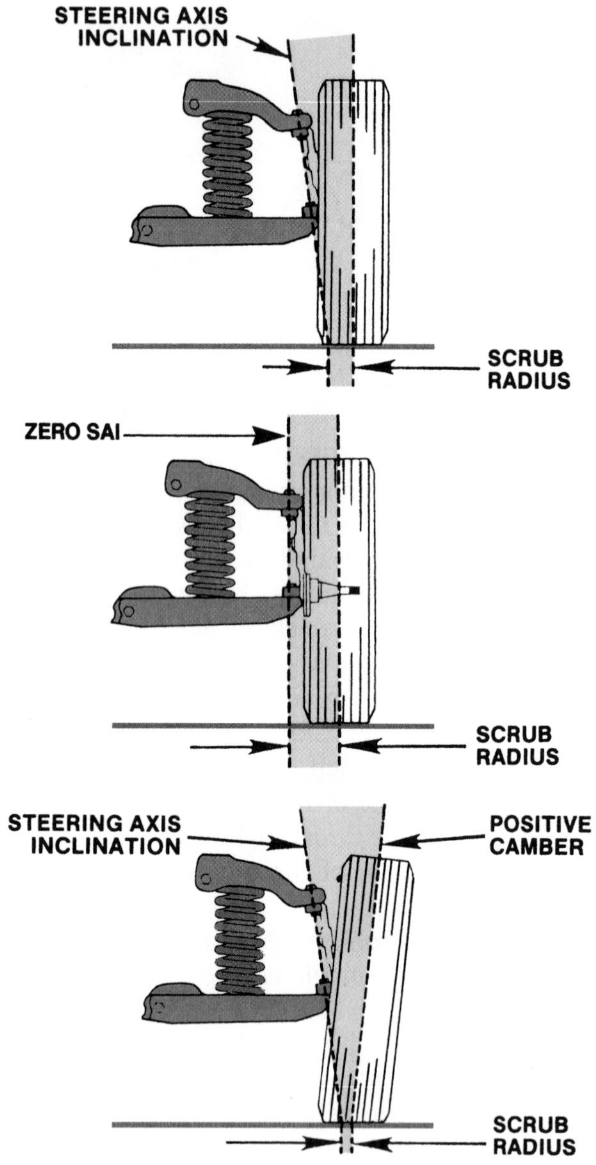

Figure 45-6 The effects of steering axis inclination changes.

downward turning to each side. This motion travels through the control arms to the springs and, finally, to the weight of the vehicle. The forces generated in a turn are actually trying to lift the vehicle. The tilting and loading effect of SAI offsets the lifting forces and helps to pull the tires back to straight ahead when the turn is finished.

Front-wheel-drive vehicles with strut suspensions typically have a higher SAI angle (12 to 18 degrees) than a short-long arm rear-wheel-drive suspension (6 to 8 degrees). This is because the extra leverage provided by a larger angle helps directional stability.

If the SAI angles are unequal side-to-side, torque steer, brake pull, and pump steer (jerking from side to side) can occur even if static camber angles are within specifications.

Checking the SAI angle can help locate various problems that affect wheel alignment. For example, an SAI angle that varies from side to side may indicate an out-of-position upper strut tower, a bowed lower control arm, or a shifted center crossmember.

On a short-long arm suspension, SAI is the angle between true vertical and a line drawn from the upper ball joint through the lower ball joint. In a strut-equipped vehicle this line is drawn through the center of the strut's upper mount down through the center of the lower ball joint.

When the camber angle is added to the SAI angle, the sum of the two is called the **included angle**. Comparing SAI, included, and camber angles can also help identify damaged or worn components. For example, if the SAI reading is correct, but the camber and included angles are less than specifications, the steering knuckle or strut tower may be bent. **Table 45–2** summarizes the various angle combinations used to troubleshoot short-long arm, strut, and twin I-beam suspension system alignment problems.

Turning Radius

Turning radius is the amount of toe-out present in turns **(Figure 45–7)**. Turning radius is also called "toe-out on turns" or "turning angle." As a car goes around a corner, the inside tire must travel in a smaller radius circle than the outside tire. This is accomplished by designing the steering geometry to turn the inside wheel sharper than the outside wheel. The result can be seen as toe-out in turns. This eliminates tire scrubbing on the road surface by keeping the tires pointed in the direction they have to move.

When a car has a steering problem, the first diagnostic check should be a visual inspection of the entire vehicle for anything obvious: bent wheels, misalignment of the cradle, and so on. If there is nothing obviously wrong with the car, make a series of diagnostic checks without disassembling the vehicle. One of the most useful checks that can be made with a minimum of equipment is a jounce-rebound check.

This jounce-rebound check determines if there is misalignment in the rack and pinion gear. For a quick check, unlock the steering wheel and see if it moves during the jounce or rebound. For a more careful check, use a pointer and a piece of chalk. Use the chalk to make a reference mark on the tire tread and place the pointer on the same line as the chalk mark. Jounce and rebound the suspension system a few times while someone watches the chalk mark and the pointer. If the mark on the wheel moves unequally in and out on both sides of the car, chances are there is a steering arm or gear out of alignment. If the mark does not move or moves equally in and out on both sides of the car, the steering arm and gear are probably all right. Each wheel or side should be checked.

TABLE 45–2 ALIGNMENT ANGLE DIAGNOSTIC CHART

Suspension Systems	SAI	Camber	Included Angle	Probable Cause
Short Arm/	Correct	Less	Less	Bent knuckle
Long Arm	Less	Greater	Correct	Bent lower control arm
Suspension	Greater	Less	Correct	Bent upper control arm
	Less	Greater	Greater	Bent knuckle
MacPherson	Correct	Less	Less	Bent knuckle and/or bent strut
Strut	Correct	Greater	Greater	Bent knuckle and/or bent strut
Suspension	Less	Greater	Correct	Bent control arm or strut tower (out at top)
	Greater	Less	Correct	Strut tower (in at top)
	Greater	Greater	Greater	Strut tower (in at top) and spindle and/or bent strut
	Less	Greater	Greater	Bent control arm or strut tower (out at top) plus bent knuckle and/or bent strut
	Less	Less	Less	Strut tower (out at top) and knuckle and/or strut bent or bent control arm
Twin I-Beam	Correct	Greater	Greater	Bent knuckle
Suspension	Greater	Less	Correct	Bent I beam
	Less	Greater	Correct	Bent I beam
	Less	Greater	Greater	Bent knuckle

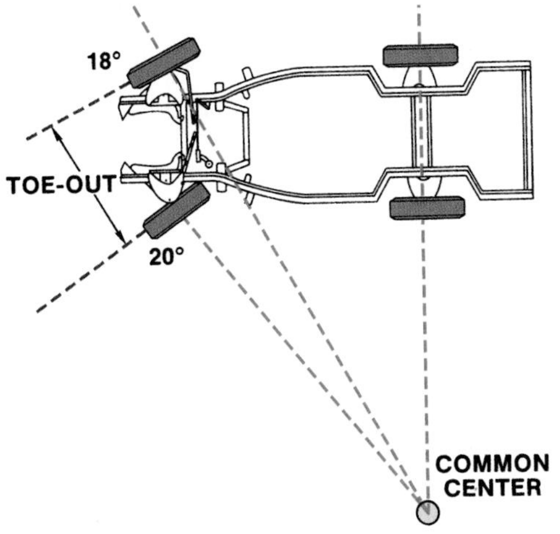

Figure 45–7 Turning radius is affected by toe-out in turns.

Turning radius is not an adjustable angle. If the angle is not correct, steering or suspension parts are damaged and will need to be replaced.

Tracking

All vehicles are built around a geometric centerline that runs through the center of the chassis from the back to the front. The thrust line is the direction the rear axle would travel if unaffected by the front wheels. This condition is also called **tracking**. An ideal alignment has all four wheels parallel to the centerline, making the thrust line parallel to the centerline. However, the rear-wheel thrust line of a vehicle might not always be parallel to the actual centerline of the vehicle, so the angle of the thrust line must be checked first.

Rear-wheel-drive vehicles rarely need thrust line adjustment unless they have been in an accident or experienced severe usage. Independent rear suspension can have offset thrust angles from unequal rear toe adjustments. An offset thrust angle affects handling by pulling in the direction away from the thrust line, and it can cause tire wear similar in pattern to that of incorrect toe settings. As a general rule, minor variations between the thrust line and centerline are not noticeable and do not cause handling problems as long as the front wheels are aligned parallel to the thrust line.

Correct tracking refers to a situation with all suspension and wheels in their correct location and condition and aligned so that the rear wheels follow directly behind the front wheels while moving in a straight line **(Figure 45–8)**. For this to occur, all wheels must be parallel to one another, and axle and spindle lines must be at 90-degree angles to the vehicle centerline. Simply stated, all four wheels should form a perfect rectangle.

Load Distribution

Load distribution refers to the load placed on each wheel. Every vehicle is engineered to operate at a designed curb height (also called **trim height**). At this

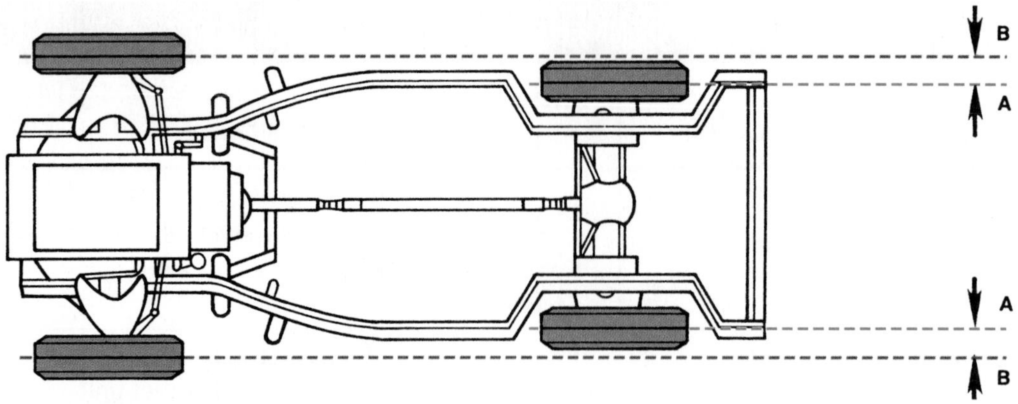

Figure 45-8 When a car is tracking correctly, its rear wheels are the same distance from the front wheels on both sides.

height, each wheel must carry the correct amount of weight. Excessive loading to the front, rear, or one side of the vehicle changes the curb height, upsetting vehicle balance and steering geometry.

In correct alignment, sagging springs and bent suspension parts can also change this condition, upsetting geometry and placing excessive load on only one or two wheels.

All of these elements—springs, shocks, suspension, and geometry—are engineered to work together as a balanced team to provide safe and comfortable riding and handling. Quite naturally, if one wheel is running under a different condition of weight load and steering geometry, the vehicle does not ride and handle as it is capable of doing.

PREALIGNMENT INSPECTION

An extremely important part of a wheel alignment is the preliminary inspection. During this inspection if any parts are found to be defective, they should be replaced before proceeding with the alignment. The following are guidelines for the inspection:

- Begin the alignment with a road test. While driving the car, check to see that the steering wheel is straight. Feel for vibration in the steering wheel as well as in the floor or seats. Notice any pulling or abnormal handling problems, such as hard steering, tire squeal while cornering, or mechanical pops or clunks. The road test helps find problems that must be corrected before proceeding with the alignment.
- Carefully inspect the tire wear patterns and mismatched tire sizes or types. Check the tires' inflation and correct if necessary. Also, look for the results of collision damage and towing damage.
- Check the tires and wheels for radial runout.
- Check the wheel bearings.
- Remove heavy items from the trunk and passenger compartment. However, if these items, such as toolboxes, are normally carried in the vehicle, leave them in.
- Check the vehicle's ride height **(Figure 45–9)**. Every vehicle is designed to ride at a specific curb height. Curb height specifications and the specific measuring

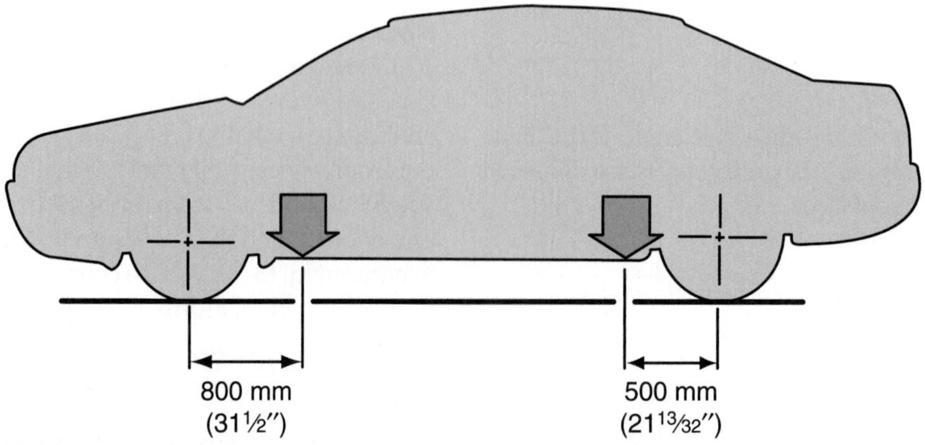

Figure 45-9 Typical measuring points for ride height.

points are given in service manuals. Proper alignment is impossible if the ride height is incorrect. If the height is slightly off, the problem can be corrected with spring spacers.

- Check the play of the steering wheel.
- Jounce the vehicle to check the condition of the shock absorbers.
- With the vehicle raised, inspect all steering components such as control arm bushings, upper strut mounts, pitman arm, idler arm, center link, tie-rod ends, ball joints, and shock absorbers. Check the CV joints (if equipped) for looseness, popping sounds, binding, and broken boots. Damaged components must be repaired before adjusting alignment angles.

WHEEL ALIGNMENT EQUIPMENT

There are many different ways to measure the alignment angles on a vehicle. The most common way is the use of an alignment machine or rack. The equipment used for checking alignment angles has evolved from string and measuring tapes to computerized machines. Today, computerized machines are the most commonly used **(Figure 45–10)**. A typical computerized system gives information on a CRT screen to guide the technician step-by-step through the alignment process.

After vehicle information is keyed into the machine and the wheel units are installed, the machine must be compensated for wheel runout. When compensation is complete, alignment measurements are instantly displayed. Also displayed are the specifications for that vehicle. In addition to the normal alignment specifications, the CRT may display asymmetric tolerances, different left- and right-side specifications, and cross specifications (difference allowed between left and right side). Graphics and text on the screen show the technician where and how to make adjustments **(Figure 45–11)**. As the adjustments are made on the vehicle, the technician can observe the center block slide toward the target. When the block aligns with the target, the adjustment is within half the specified tolerance.

Other alignment equipment often used are turning radius gauges, caster-camber gauges, optical toe gauges, and trammel bar gauges (also known as tram gauges).

Turning Radius Gauges

Turning radius gauges measure how many degrees the front wheels are turned. They are commonly used to measure camber, caster, and toe-out on turns. Turning radius gauges (sometimes called turn tables) may be portable but are commonly found as part of an alignment rack. To use these gauges, the front wheels are centered on the gauge plates. Then the locking pins are removed to allow the plate to turn with the tires. As the tires are turned, a pointer will indicate how many degrees the tires have turned. To check toe-out on turns, turn one of the tires to 20 degrees. Then look at the gauge on the other tire.

Caster-Camber Gauge

A caster-camber gauge is used with the turning radius plate to check caster and camber. This gauge is often

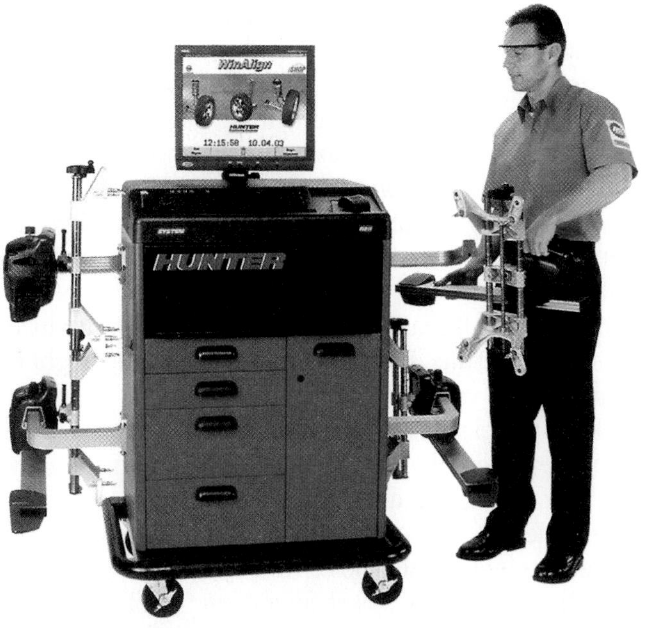

Figure 45–10 A computerized wheel alignment machine. *Courtesy of Hunter Engineering Company*

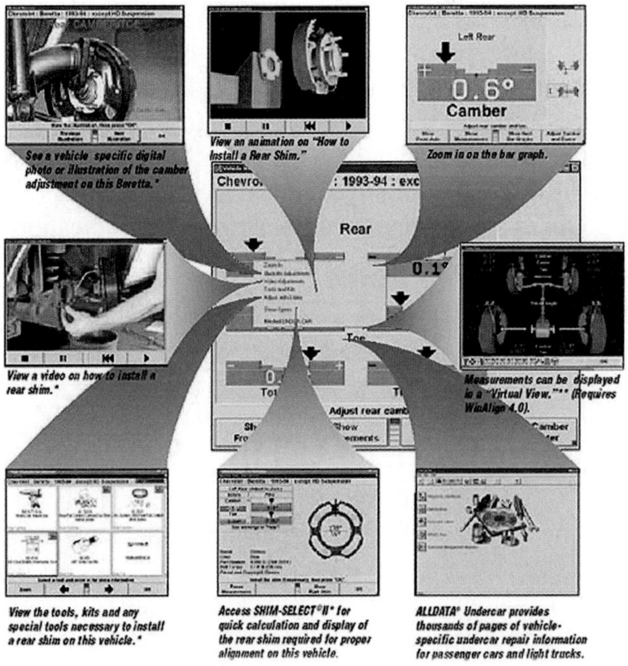

Figure 45–11 Examples of the screens available on the latest alignment machines. *Courtesy of Hunter Engineering Company*

referred to as a bubble gauge. The gauge is normally attached to the wheel hub with a magnet. Make sure the vehicle is on a level surface, and then jounce the front bumper several times to stabilize the suspension. Now look at the bubble gauges to read camber and compare yours with the specifications. Do both front wheels. Apply the brakes and hold the brake pedal down with a brake pedal lock. Turn one front tire 20 degrees out and adjust the bubble gauge to read zero. Now turn the wheel 20 degrees in and take the reading from the bubble gauge. This is the caster reading for that wheel. Compare it to specifications. Now measure caster on the other front tire. If any reading is outside the specifications, the angles need to be adjusted.

Optical Toe Gauges

Checking wheel toe is also done with the tires sitting on the turning radius plates. Make sure the tires are straight ahead and the plates are at zero degrees before removing the locking pins on the plate. Then install a steering wheel clamp to prevent the steering wheel from moving. Then install the brake pedal lock. Remove the hubcaps and grease caps from the front wheels. Jounce the front suspension several times to allow it to stabilize. Note the location of the steering gear. If the steering linkage is at the rear of the front wheels, push outwardly on the rear of the tires. If the linkage is to the front, push on the front of the tires. Mount the optical toe gauges onto the front wheel hubs. Use the level indicators on the gauges to make sure the gauges are level. Then read wheel toe on the gauges and compare them to specifications.

Trammel Bar Gauge

A trammel bar gauge is a purely mechanical way to check toe. It is used to measure the distance between the center of the two front tires. First a distance measurement is taken at the rear of the tires (centerline to centerline), then the same measurement is taken at the front of the tires. The difference in the measurements is the amount of toe. Compare this reading with specifications.

Miscellaneous Tools

Figure 45–12 shows an assortment of the special tools required for wheel alignment and other steering and suspension system work.

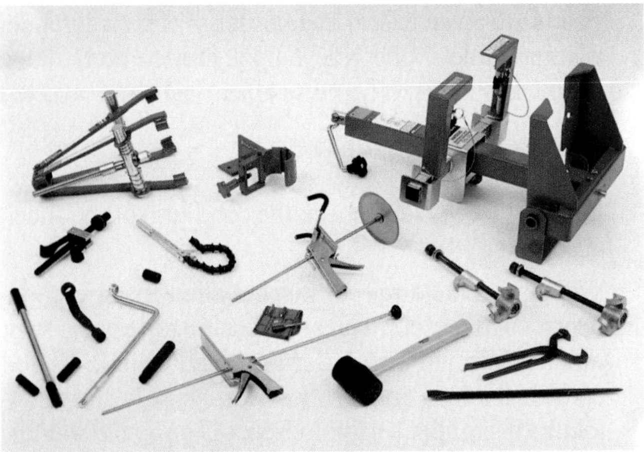

Figure 45-12 An assortment of steering and suspension tools. *Courtesy of Snap-on Tools Company*

ALIGNMENT MACHINES

Most shops use an alignment machine to check all of the alignment angles. Normally an alignment rack is part of the alignment machine's package. The rack is best described as a limited purpose vehicle hoist (lift) equipped with turning radius plates. There are many varieties of alignment machines that have been used through the years. Some are equipped to measure alignment angles at all four wheels of the vehicle, and others measure the angles at only two wheels. Some alignment machines simply display the angle readings, whereas others display the readings plus give advice on how to correct the angles.

Four-wheel alignment (or **total wheel alignment**), whether front or rear drive, solid axle or independent rear suspension, sets the alignment angles on all four wheels so they are positioned straight ahead with the steering wheel centered. The wheels must also be parallel to one another and perpendicular to a common centerline. More than 85% of all new vehicles require that all four wheels are aligned.

To accomplish this, the total for all four wheels must be determined and rear toe adjusted where possible to bring the rear axle or wheels into square with the chassis. The front toe setting can then be adjusted to compensate for any rear alignment deviation that might persist.

Four-wheel alignment also includes checking and adjusting rear-wheel camber as well as toe, and doing all the traditional checks of front camber, toe, caster, toe-out on turns, and steering axis inclination. The most important thing a four-wheel alignment job tells the technician is whether or not the rear axle or rear wheels are square with respect to the front wheels and chassis.

With two-wheel alignment equipment that aligns the front wheels to the geometric centerline, the assumption is made that the rear wheels are square with respect to the centerline. If that is true, the alignment job produces satisfactory results. If not, steering and tracking might be a problem.

Rear-wheel alignment can be checked on a two-wheel machine by simply backing the vehicle onto the rack. The rear wheels can also be checked for square by measuring the wheelbase on both sides with a track bar to make sure it is equal. The obvious drawback to this approach is that four-wheel alignment involves several extra steps. There is also the question of aligning the wheels to the centerline or thrust line.

The best approach in terms of both accuracy and completeness for wheel alignment is to reference the front wheels to the rear wheels. With this approach, individual rear toe is measured so the thrust line can be determined and the front wheels adjusted to the centerline. It also eliminates the need to back the vehicle onto the alignment machine because heads are provided for the rear wheels.

The actual procedure for total four-wheel alignment, as with two-wheel alignment, begins with a thorough inspection of the tires and suspension. Do not forget to check the ride height based on the normal vehicle load.

Once the alignment heads are installed and compensated for wheel runout, read front and rear camber and front and rear toe. The thrust angle created by the rear wheels can then be determined. A typical procedure for checking the alignment of all four wheels with a computerized alignment machine is shown in Photo Sequence 49.

Most alignment machines compensate for wheel setback. Some show how much setback is present and where. Nearly all machines will also show SAI. Steering axis inclination, an angle frequently neglected on quick alignment jobs, is important to include because it can help pinpoint damaged parts. If the SAI does not match specifications, compare both sides. Problems can be caused by bent struts, control arm spindles, or steering knuckles. A mislocated strut tower or control arm anchor can also throw off the reading.

ADJUSTING WHEEL ALIGNMENT

All wheel alignment angles are interrelated. Regardless of the make of a car or the type of suspension, the same adjustment order—caster, camber, toe—should be followed, as far as the car permits such adjustments to be made. Some MacPherson suspensions do not provide for caster or camber adjustments. Additionally, adjustment methods vary from model to model and, occasionally, even in different model years.

Caster/Camber Adjustment

Caster affects steering stability and steering wheel returnability. Zero (0) caster is present when the upper ball joint or top strut bearing and lower ball joint are in the same plane as viewed from the side of the vehicle. Positive caster exists when the upper ball joint or top strut bearing is toward the rear of the vehicle in relationship to the lower ball joint. When the upper ball joint or top strut bearing is toward the front of the vehicle in relationship to the lower ball joint, negative caster is present. If the caster at both wheels is not equal, the vehicle will tend to drift toward the side with the lowest caster.

Camber is the inward or outward tilt of the top of the wheel. Adjusting camber centers the vehicle's weight on the tire. Proper camber adjustment minimizes tire wear. Zero (0) camber is present when the wheel is at a perfectly vertical position. The tires have positive camber when the top of the tire is tilted out, or away from the engine. When the top of the tire is tilted in, there is negative camber. Incorrect camber will cause excessive stress and wear on suspension parts. Too much negative camber will cause wear on the inside tread of the tire, whereas too much positive camber will cause tire wear on the outside tread. If camber is not the same on both wheels, the vehicle will pull toward the side with the most positive camber.

To compensate for road crown, most alignment specifications allow for slightly more negative caster or slightly more positive camber on the left side of the vehicle.

Several methods are used to adjust caster and camber.

Shims Many cars use shims for adjusting caster and camber **(Figure 45–13)**. The shims can be located between the control arm pivot shaft and the inside of the frame. Both caster and camber can be adjusted in one operation requiring the loosening of the shim bolts just once. Caster is changed by adding or subtracting shims from one end of the pivot shaft only. Then, camber is adjusted by adding or subtracting an equal amount of shims from the front and rear bolts. This procedure allows camber to change without affecting the caster setting.

Some cars use shims located between the control arm pivot shaft and the outside of the frame. The adjustment procedure is the same as just described. Always look at the shim arrangements to determine the desired direction of change before loosening the bolts.

Eccentrics and Shims Eccentrics and shims are used on some vehicles to adjust caster and camber. In some designs, an eccentric bolt and cam on the upper control arm adjust both caster and camber. To adjust, the nuts on the upper control arm are loosened first. Then, one eccentric bolt at a time is turned to set caster. Both bolts are turned equally to set camber.

The **eccentric bolt and cam** assembly **(Figure 45–14)** can be located on the inner lower or upper control arm. Unlike other designs, camber is adjusted first. Some car models have a camber eccentric between the steering knuckle and the upper control arm. The camber eccentric is rotated to set camber **(Figure 45–15)**. Caster is set with an adjustable strut rod.

Slotted Frame The slotted frame adjustment has slotted holes under the control arm inner shaft that allow the shaft to be repositioned to the correct caster and camber settings. Caster and camber adjusting tools help in

PHOTO SEQUENCE 49 Typical Procedure for Performing Four-Wheel Alignment with a Computer Wheel Aligner

P49-1 *Position the vehicle on the alignment rack.*

P49-2 *Make sure the front tires are positioned properly on the turn tables.*

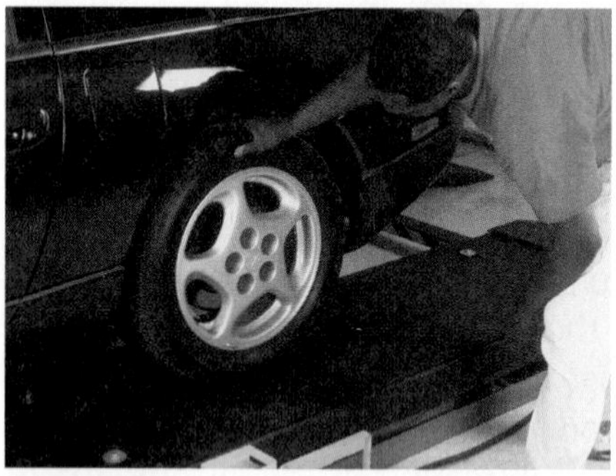

P49-3 *Position the rear wheels on the slip plates.*

P49-4 *Attach the wheel units.*

P49-5 *Select the vehicle make and model year.*

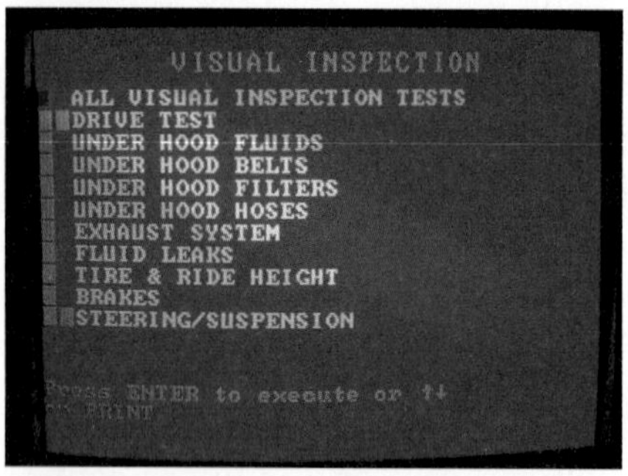

P49-6 *Check the items on the screen during the preliminary inspection.*

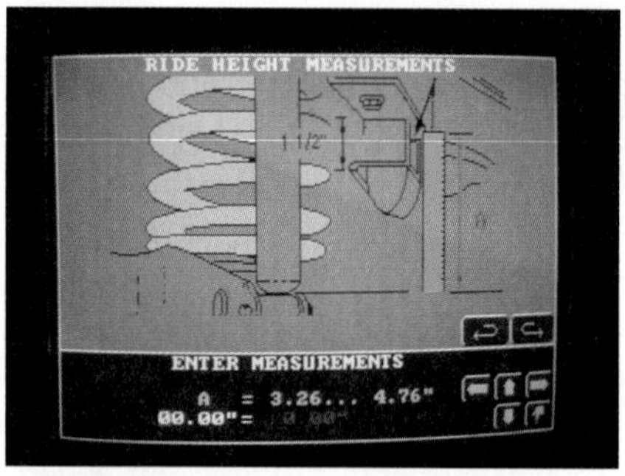

P49-7 *Display the ride height screen. Check the tire condition for each tire on the tire condition screen.*

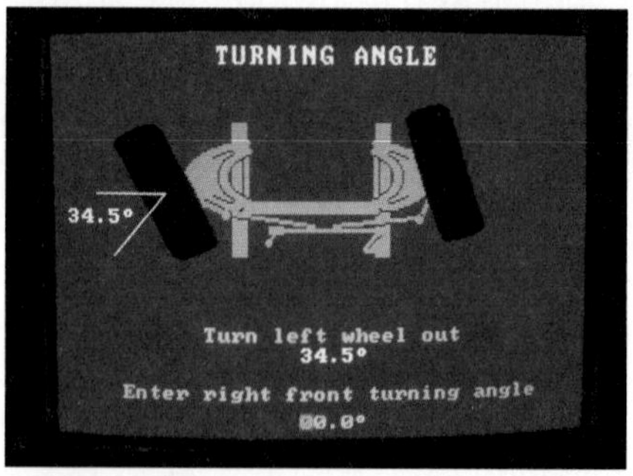

P49-8 *Display the wheel runout compensation screen.*

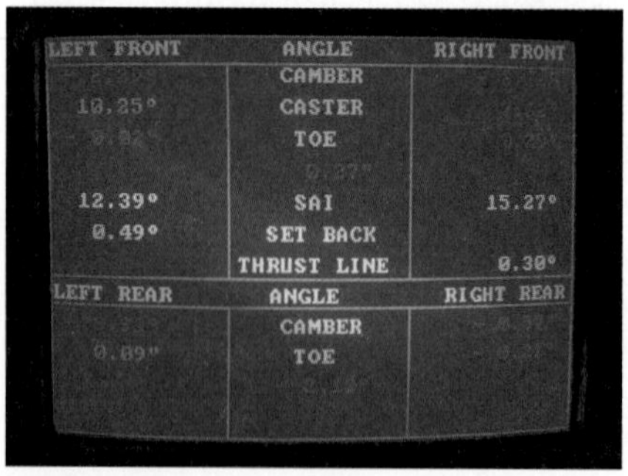

P49-9 *Display the turning angle screen and perform turning angle check.*

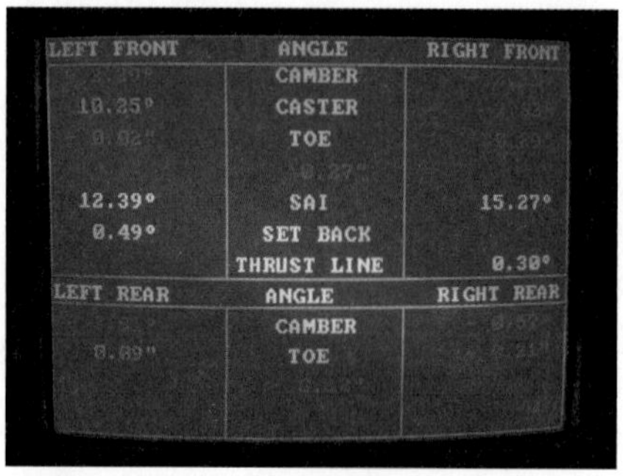

P49-10 *Display the front- and rear-wheel alignment angle screen.*

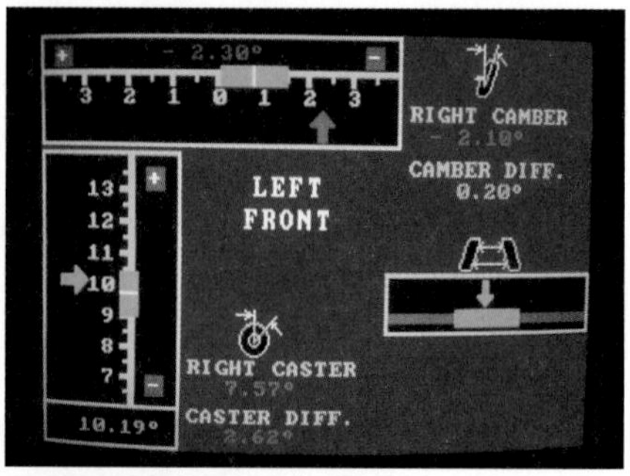

P49-11 *Display the adjustment screen.*

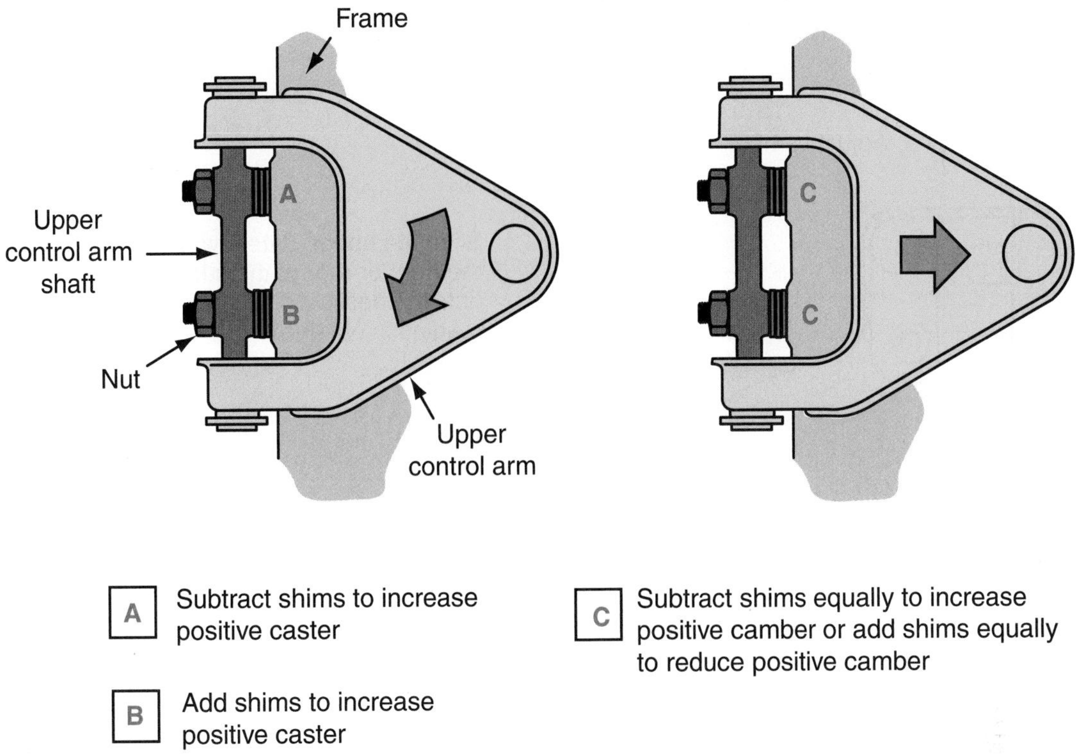
Frame
Upper control arm shaft
Nut
Upper control arm
A
B
C
C
A Subtract shims to increase positive caster
B Add shims to increase positive caster
C Subtract shims equally to increase positive camber or add shims equally to reduce positive camber

Figure 45–13 Adding and subtracting shims between the control arm and the frame will change caster and camber.

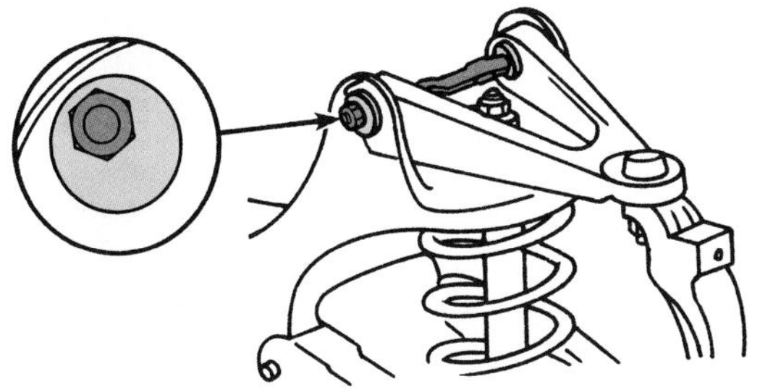

Figure 45–14 Eccentric bolt and cam, shown on an upper control arm.

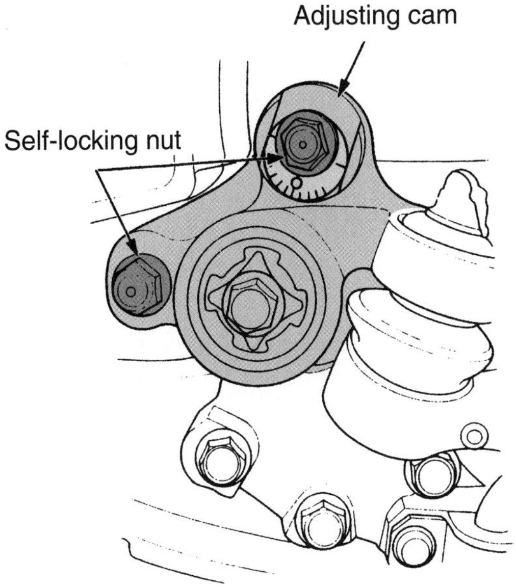
Adjusting cam
Self-locking nut

Figure 45–15 Graduated cam for adjusting camber.
Courtesy of American Honda Motor Co., Inc.

making adjustments. One end of the shaft is moved for caster adjustment. Both ends of the shaft are moved for camber adjustment. Turning a nut on one end of the rod changes its length and adjusts caster. Camber is set by an eccentric at the inner end of the lower control arm, or by a camber eccentric in the steering knuckle of the upper support arm, as described earlier.

Rotating Ball Joint and Washers In this design, camber is increased by disconnecting the upper ball joint, rotating it 180 degrees, and reconnecting. This positions the flat of the ball joint flange inboard and increases camber approximately 1 degree. Caster angle is changed with a kit containing two washers, one 0.12-inch (3-mm) thick and one 0.36-inch (9-mm) thick. The washers are placed at opposite ends of the locating tube between the legs of the upper control arm. Placement of the large washer at the rear leg of the control arm increases caster by 1 degree. Placement of the large washer at the front leg of the control arm decreases caster by 1 degree.

MacPherson Suspension Adjustments

Caster/camber adjustments are made only on certain models with MacPherson suspensions. There are two general OEM procedures for doing this, although aftermarket kit adapters are available for some models. Service information must be consulted for an accurate listing of models on which adjustments can be made.

In one version, a cam bolt at the base of the strut assembly is used to adjust camber **(Figure 45–16)**. On

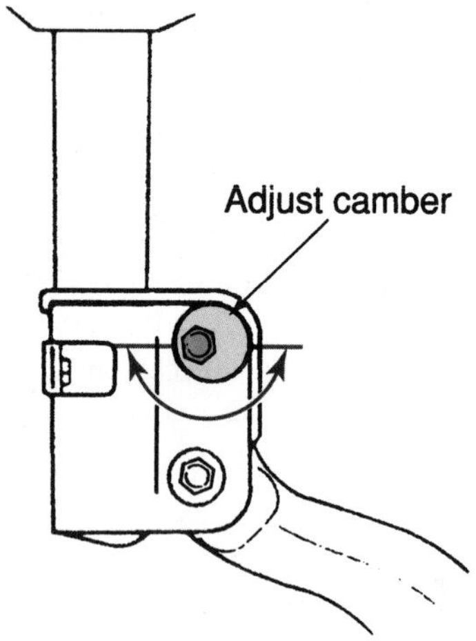

Figure 45-16 Some MacPherson suspensions use cam bolts at the connection to the steering knuckle for camber adjustments. *Courtesy of Hunter Engineering Company*

different models, this bolt can be either the upper or the lower of the two bolts connecting the strut assembly to the steering knuckle. Both bolts must be loosened to make the adjustment, and the wheel assembly must be centered. Turn the cam bolt to reach the correct alignment, then retighten the bolts to the appropriate torque specifications. There is no caster adjustment on this version.

In the other form of kit adapter, both caster and camber are adjustable at the strut upper mount. Slots in the mounting plate permit the strut assembly to be shifted to reach the alignment specifications. To adjust caster, loosen the three locknuts on the mounting studs and relocate the plate. Do not remove the nuts **(Figure 45–17)**. Loosen the center locknut and slide it toward or away from the engine as needed to adjust camber correctly.

While caster cannot be adjusted on many MacPherson strut front suspensions, camber can be adjusted. The camber is such that, although it is locked in place with a pop rivet, it can be adjusted by removing the rivet from the camber plate and loosening the three nuts that hold the plate to the body apron **(Figure 45–18)**. Camber is changed by moving the top of the shock strut to the position in which the desired camber setting is achieved. The nuts are then tightened to specifications. (It is not necessary to install a new pop rivet.)

Rear-Wheel Camber Adjustments

Like front camber, rear camber affects both tire wear and handling. The ideal situation is to have zero running camber on all four wheels to keep the tread in full contact with the road for optimum traction and handling.

Camber is not a static angle. It changes as the suspension moves up and down. Camber also changes as the vehicle is loaded and the suspension sags under the weight.

To compensate for loading, most vehicles with independent rear suspension often call for a slight amount of

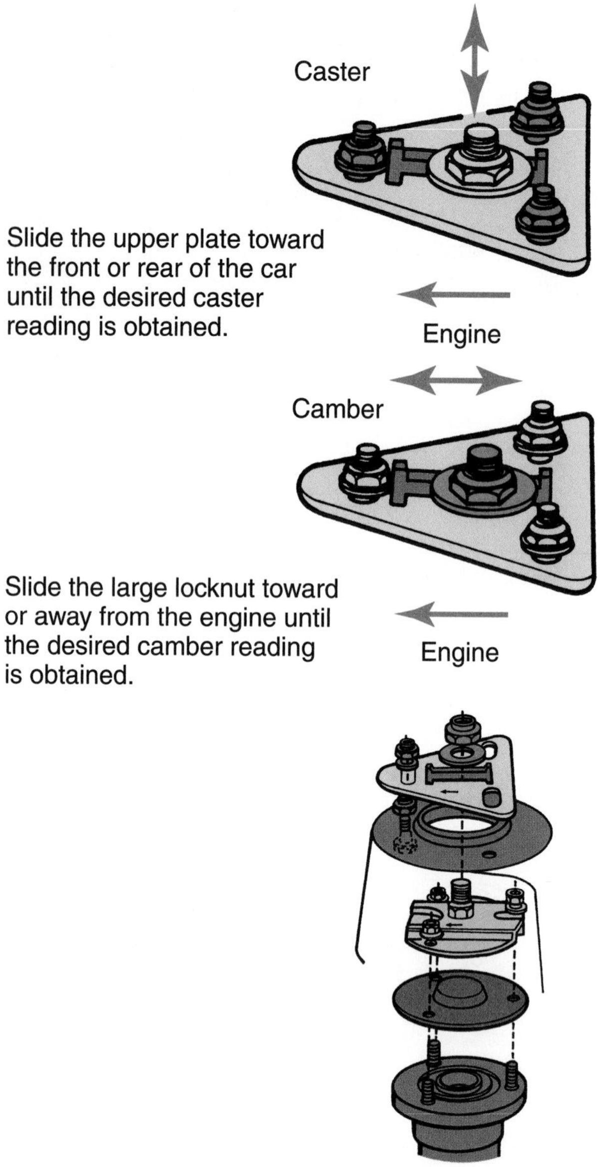

Caster
Slide the upper plate toward the front or rear of the car until the desired caster reading is obtained.
Engine
Camber
Slide the large locknut toward or away from the engine until the desired camber reading is obtained.
Engine

Figure 45-17 Caster and camber adjustment of locknuts.

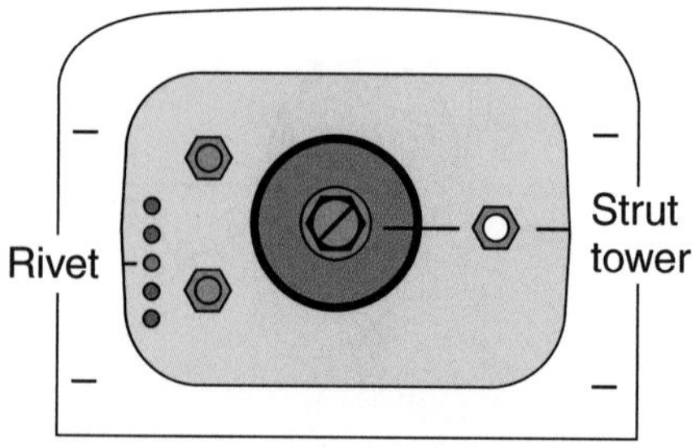

Figure 45-18 An upper strut mount with a camber plate. Note the location of the rivet and the attaching bolts. *Courtesy of Hunter Engineering Company*

positive camber. A collapsed or mislocated strut tower, bent strut, collapsed upper control arm bushing, bent upper control arm, sagging spring, or an overloaded suspension can cause the rear wheels to have negative camber. A bent spindle or strut or bowed lower control arm can cause too much positive camber. Even rigid rear axle housings in rear-wheel-drive vehicles can become bowed by excessive torque, severe overloading, or road damage.

CAUTION!

Never jack up or lift a FWD vehicle on its rear axle. The weight of the vehicle may cause the axle to bend and result in misalignment of the rear wheels. Always lift the vehicle at the recommended lifting points.

Besides wearing the tires unevenly across the tread, uneven side-to-side camber (as when one wheel leans in and the other does not) creates a steering pull just like it does when the camber readings on the front wheels do not match. It is like leaning on a bicycle. A vehicle always pulls toward a wheel with the most positive camber. If the mismatch is at the rear wheels, the rear axle pulls toward the side with the greatest amount of positive camber. If the rear axle pulls to the right, the front of the car drifts to the left, and the result is a steering pull even though the front wheels may be perfectly aligned.

The methods used to adjust rear suspensions vary. On some semi-independent suspensions, camber and toe are adjusted by inserting different sizes of shims between the rear spindle and the spindle mounting **(Figure 45–19)**. The shim thickness is changed between the top or bottom of the spindle to adjust camber. Many shims are now available that are round but have different thicknesses through their diameters.

On others, a camber adjustment can be made by installing a wedge spacer between the top of the knuckle

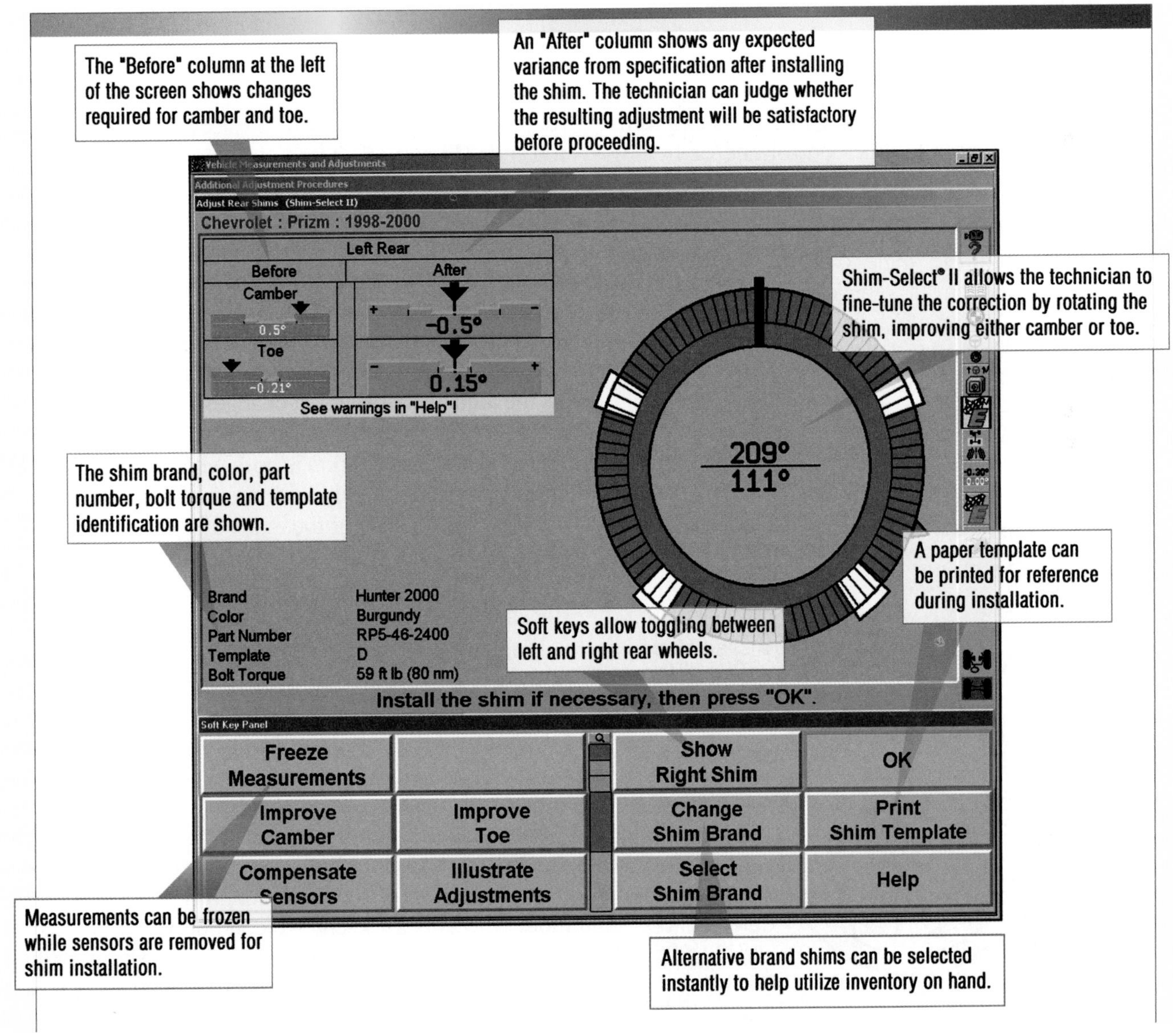

Figure 45–19 A computer screen showing what shim to install in the rear to bring the vehicle into alignment. *Courtesy of Hunter Engineering Company*

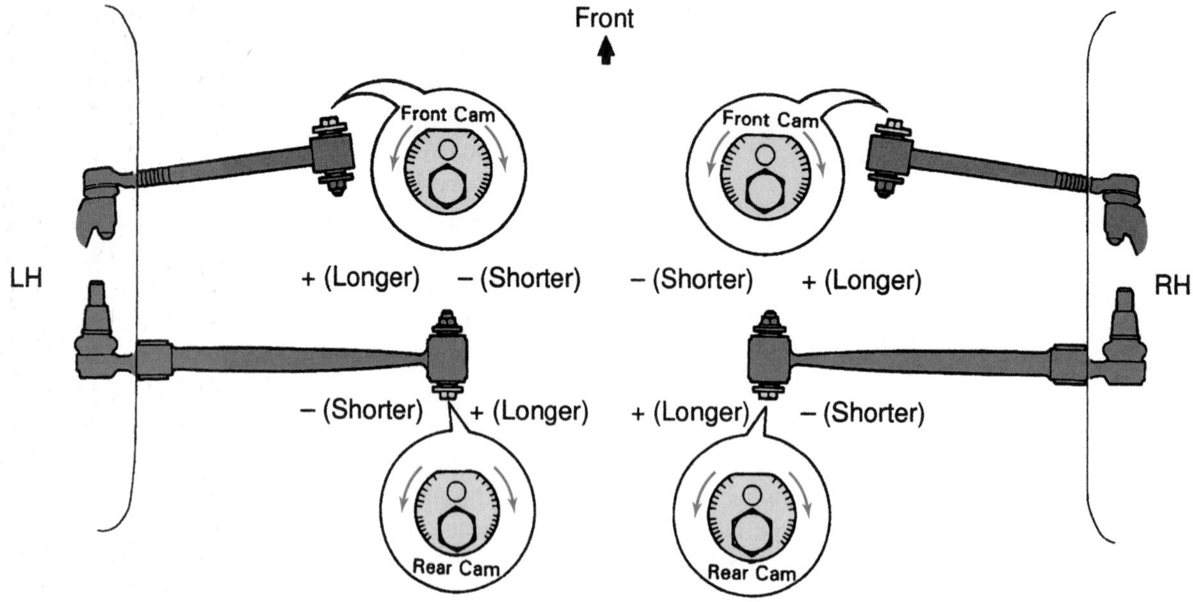

Front
Front Cam
Front Cam
LH
RH
+ (Longer)
– (Shorter)
– (Shorter)
+ (Longer)
– (Shorter)
+ (Longer)
+ (Longer)
– (Shorter)
Rear Cam
Rear Cam

Figure 45-20 An independent rear suspension may have eccentric cams at all of its control arms.
Reprinted with permission

and the strut. Still others have eccentric bolts and cams at the mounting points for the control and/or trailing arms **(Figure 45–20)**.

Toe Adjustment

Toe is the last alignment angle to be set. The same procedure is followed on all vehicles, except those with bonded ball stud sockets. Correct toe will minimize tire wear and rolling friction.

To adjust toe, start by being sure the steering wheel is centered **(Figure 45–21)** when the front wheels point straight ahead. Then loosen the retaining bolts on the tie-rod adjusting sleeves. Turn the sleeves to move the tie-rod ends **(Figure 45–22)**.

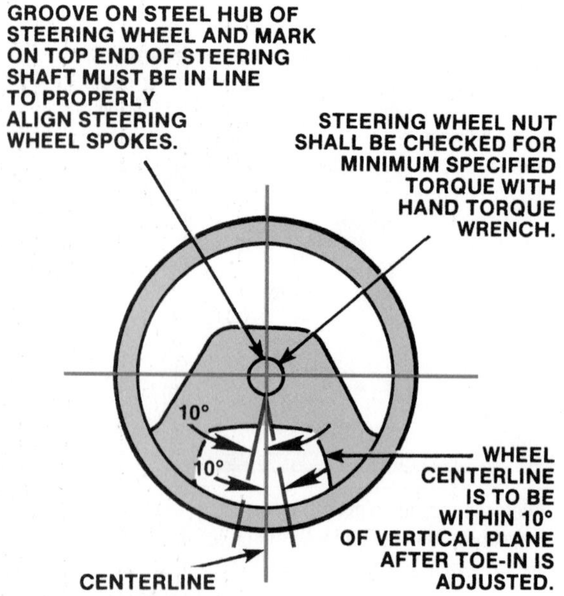

GROOVE ON STEEL HUB OF STEERING WHEEL AND MARK ON TOP END OF STEERING SHAFT MUST BE IN LINE TO PROPERLY ALIGN STEERING WHEEL SPOKES.
STEERING WHEEL NUT SHALL BE CHECKED FOR MINIMUM SPECIFIED TORQUE WITH HAND TORQUE WRENCH.
10°
10°
WHEEL CENTERLINE IS TO BE WITHIN 10° OF VERTICAL PLANE AFTER TOE-IN IS ADJUSTED.
CENTERLINE

Figure 45-21 A typical acceptable steering wheel position—measured from a normal spoke angle.

On many rack and pinion systems, the tie-rod locknut must be loosened and the tie rod rotated to adjust toe at each wheel **(Figure 45–23)**. Before rotating the tie rod, the small outer bellows clamp must be loosened.

Other rack and pinion tie-rod ends have internal threads and a threaded adjuster. One end of the adjuster has right-hand threads, the other has left-hand threads. As the adjuster is turned, it changes the overall length of the tie rod, thereby changing toe.

An ideal toe condition is both wheels exactly straight ahead, which would minimize tire wear. This, however, is not possible because of the many factors affecting alignment. As a result of these numerous conditions dealing with both tire wear and handling, all suspensions are designed with a slight toe-in or toe-out.

Any misalignment of the steering linkage pivot point or control arm pivot point (such as the center link or rack and pinion out of place) causes the condition known as **toe-change**. Toe-change involves turning the wheels from their straight-ahead position as the suspension moves up and down.

The change might be only one wheel, both wheels in the same direction, or both wheels in the opposite direction. Regardless of the condition, any change of one or more wheels is a toe-change condition. The results are tire wear and a hard-to-handle vehicle. The poor handling effects can get to the point that the vehicle is dangerous to drive.

Toe change is not a specification, it is a condition in which the toe setting constantly varies. It must be deter-

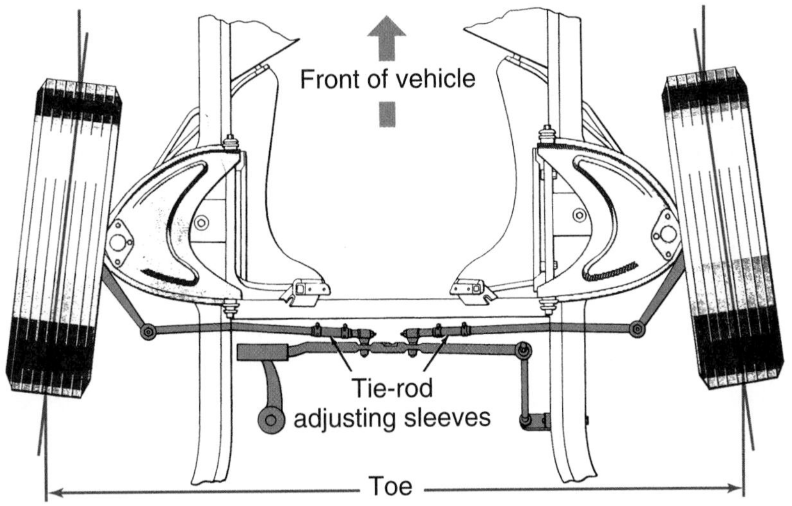

Figure 45-22 Turn the adjusting sleeves to adjust toe. *Courtesy of SPX Corporation, Aftermarket Tool and Equipment Group*

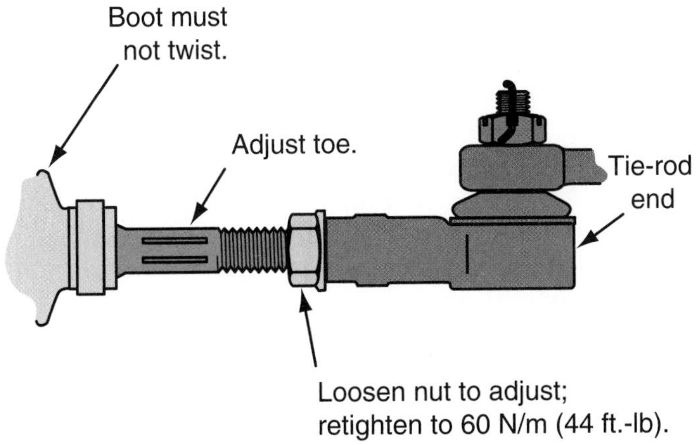

Figure 45-23 Rotating the tie rod to adjust the toe on a rack and pinion steering gear.

mined by equipment or a method that measures individual wheel toe at all suspension heights. There must be a change in suspension heights for any changes to occur.

Lightweight front-wheel-drive vehicles can be affected greatly by toe-change. With these vehicles, the front wheels are no longer being pushed. They actually pull the vehicle forward and as a result, if the wheels are not maintaining a straight-ahead position, they affect directional control. Adverse road conditions, such as wet or icy conditions, can also increase the handling effects created by toe-change in the front-wheel-drive car.

Rear Toe Rear toe, like front toe, is a critical tire wear angle. If toed-in or toed-out, the rear tires scuff just like the front ones. Either condition can also contribute to steering instability as well as reduced braking effectiveness. Keep this in mind with antilock brake systems.

Like camber, rear toe is not a static alignment angle. It changes as the suspension goes through jounce and rebound. It also changes in response to rolling resistance and the application of engine torque. With FWD vehicles, the front wheels tend to toe-in under power while the rear wheels toe-out in response to rolling resistance and suspension compliance. With RWD vehicles, the opposite happens: the front wheels toe-out while the rear wheels on an independent suspension try to toe-in as they push the vehicle ahead.

If rear toe is not within specifications, it affects tire wear and steering stability just as much as front toe. A total toe reading that is within specifications does not necessarily mean the wheels are properly aligned—especially when it comes to rear toe measurements. If one rear wheel is toed-in while the other is toed-out by an equal amount, total toe would be within specifications. However, the vehicle would have a steering pull because the rear wheels would not be parallel to center.

Remember, the ideal situation is to have all four wheels at zero running toe when the car is traveling down the road. This is especially true with antilock brakes where improper toe can affect brakes; such a condition can affect brake balance when braking on slick or wet surfaces, causing the antilock brakes to cycle on and off to prevent a skid. Without antilock brakes, this condition may upset traction enough to cause an uncontrollable skid.

Thrust Line If both rear wheels are square to one another and the rest of the vehicle, the thrust line is perpendicular to the rear axle and coincides with the vehicle's centerline. But if one or both rear wheels are toed-in or toed-out, or one is set back slightly with respect to the other, the thrust line is thrown off-center. This creates a thrust angle that causes a steering pull in the opposite direction. For example, a thrust line that is off-center to the right makes the car pull left.

The presence of a thrust angle can cause poor directional stability on ice, snow, or wet pavement. It can sometimes make a vehicle pull during braking or hard acceleration. It can also increase tire wear as the front wheels fight the rear ones for steering control.

The only way to eliminate the problem is to eliminate the thrust angle. The thrust line can be recentered by realigning rear toe. On most FWD applications, that can be easily done by using the factory-provided toe adjustments, by placing toe/camber shims between the rear spindles and axle, or by using eccentric bushing kits. With RWD vehicles that have a solid rear axle, changing rear toe is not as easy. Sometimes the floorpan or frame rails are misaligned from the factory or from collision damage. Short of pulling the chassis on a collision bench to restore the correct control arm or spring-mount geometry, the only other options are to try some type of offset trailing arm bushing with the coil springs or to reposition the spring shackles or U-bolts with leaf springs.

If rear toe cannot be easily changed, the next best alternative is to align the front wheels to the rear axle thrust line rather than the vehicle centerline. Doing this puts the steering wheel back on center and eliminates the steering pull—but it does not eliminate dog tracking. Dog tracking occurs when the rear wheels do not follow directly behind the front wheels when the vehicle is moving forward.

FOUR-WHEEL-DRIVE VEHICLE ALIGNMENT

With front-wheel-drive and full-time 4WD vehicles, the front wheels are also driving wheels. As the front wheels pull the vehicle, the wheels tend to toe-in when torque is applied. To offset this tendency, the front wheels usually need less static toe-in to produce zero running toe. In fact, the preferred toe alignment specifications in this instance can be zero to slightly toed-out (1/16-inch [1.5-mm] toe-out).

It is important to note that when the front wheels of a part-time 4WD system are freewheeling, they behave the same as the front wheels in a rear-wheel-drive vehicle. That is, they roll rather than pull. The wheels tend to toe-out, so the static toe setting would have to toe-in to achieve zero running toe when driving in the two-wheel mode.

The tires suffer in proportion to toe misalignment. For a tire that is only 1/8 inch (3 mm) off (1/4 degree), the tire is scrubbed sideways 12 feet (3.6 meters) for every mile traveled. That may not sound like much, but 12 feet (3.6 meters) of sideways scrub every mile can cut a tire's life in half.

If rapid tire wear seems to be the problem, look for the telltale feathered wear pattern. If the wheels are running toe-in, the feathered wear pattern leaves sharp edges on the inside edges of the tread. If the wheels are running toe-out, the sharp edges are toward the outside of the tread. It is usually easier to feel the feathered wear pattern than to see it. To tell which way the wear pattern runs, rub your fingers sideways across the tread.

On most 4WD vehicles, caster is not adjustable. Aftermarket companies do provide caster adjustment kits for some pickups. These kits may contain shims or eccentric cam and bolt. For some pickups, the aftermarket cam kit will also provide for camber adjustments.

On other 4WD vehicles, camber is adjusted by installing adjustment shims between the spindle and the steering knuckle or by installing and/or adjusting an eccentric bushing at the upper ball joint. Most aftermarket parts manufacturers have camber adjustment shims available in various thicknesses and diameters. Never stack the shims. Only one shim per side should be used.

CASE STUDY

Two days after having a complete alignment performed on her late-model vehicle, a customer returns to the shop complaining of improper service. Since having the alignment work done, the vehicle now consistently pulls to the right.

The technician who performed the work road tests the vehicle and verifies that the car is drifting to the right. The technician rechecks the alignment against factory specifications. All settings appear to be within range. A check of the suspension system finds nothing wrong. Desperately trying to figure out why the vehicle is drifting, the technician reviews some notes she kept from an alignment class attended several months before. Here, she finds the answer. When setting the right wheel, she forgot to allow for road crown. The technician readjusts the right wheel to account for road crown angle and test drives the vehicle. The drifting has been eliminated.

KEY TERMS

Camber
Caster
Eccentric bolt and cam
Four-wheel alignment
Included angle
Road crown
Steering axle inclination (SAI)
Thrust angle
Toe
Toe-change
Total wheel alignment
Tracking
Trim height

SUMMARY

- Caster is the angle of the steering axis of a wheel from the vertical, as viewed from the side of the vehicle. Tilting the wheel forward is negative caster. Tilting backward is positive.
- Camber is the angle represented by the tilt of either the front or rear wheels inward or outward from the vertical as viewed from the front of the car.
- Toe is the distance comparison between the leading edge and trailing edge of the front tires. If the edge distance is less, then there is toe-in. If it is greater, there is toe-out.
- The difference of rear toe from the geometric centerline of the vehicle is called the thrust angle. The vehicle tends to travel in the direction of the thrust line, rather than straight ahead.
- Steering axis inclination (SAI) angles locate the vehicle weight to the inside or outside of the vertical centerline of the tire. The SAI is the angle between the true vertical and a line drawn between the steering pivots as viewed from the front of the vehicle.
- Turning radius or cornering angle is the amount of toe-out present on turns.
- In correct tracking, all suspensions and wheels are in their correct locations and conditions and are aligned so that the rear wheels follow directly behind the front wheels while moving in a straight line.
- It is important to remember that approximately 85% of today's vehicles not only undergo front-end alignment but require rear-wheel alignment as well.
- The primary objective of four-wheel or total-wheel alignment, whether front or rear drive, solid axle, or independent rear suspension, is to align all four wheels so the vehicle drives and tracks straight with the steering wheel centered. To accomplish this, the wheels must be parallel to one another and perpendicular to a common centerline.

TECH MANUAL

The following procedures are given in Chapter 45 of the *Tech Manual* that accompanies this book:

1. Measuring front and rear wheel alignment angles with a computer-based wheel aligner.
2. Centering a steering wheel.

REVIEW QUESTIONS

1. Define positive camber.
2. What tire wear pattern will result from excessive negative camber?
3. What can be caused by excessive negative caster?
4. Describe the difference between toe-in and toe-out.
5. Describe thrust angle.
6. How will a vehicle handle if it has too much toe-out?
7. In what direction must the bottom of a tire and wheel assembly be moved to add more positive camber?
8. What provides for toe-out on turns?
9. Which of the following is a good definition of SAI?
   a. forward tilt of the top of the steering knuckle
   b. inward tilt of the spindle steering arm
   c. inward tilt of the top of the ball joint, kingpin, or MacPherson strut
   d. outward tilt of the top of the ball joint, kingpin, or MacPherson strut
10. What is the correct sequence for adjusting alignment angles during a four-wheel alignment?
11. Describe the procedure for checking caster with a bubble gauge.
12. Define the term *tracking* as it applies to vehicle handling.
13. What checks should be made before undertaking a wheel alignment?
14. Technician A says positive caster provides directional stability. Technician B says positive caster will result from excessive positive camber. Who is correct?
   a. Technician A
   b. Technician B
   c. Both A and B
   d. Neither A nor B

15. Technician A says the purpose of the steering wheel lock is to make sure the steering wheel is centered after the alignment. Technician B says if the steering wheel is not centered after an alignment, it should be pulled off the column and positioned correctly. Who is correct?

a. Technician A
b. Technician B
c. Both A and B
d. Neither A nor B

16. Technician A says the presence of a thrust angle can cause poor directional stability on ice, snow, or wet pavement. Technician B says it can also increase tire wear as the front wheels fight the rear ones for steering control. Who is correct?

a. Technician A
b. Technician B
c. Both A and B
d. Neither A nor B

17. Technician A says camber changes as the suspension moves up and down. Technician B says camber changes as the vehicle is loaded and the suspension sags under the weight. Who is correct?

a. Technician A
b. Technician B
c. Both A and B
d. Neither A nor B

18. All of the following angles are adjustable during wheel alignment except ______.

a. toe-in
b. toe-out
c. camber
d. steering axis inclination

19. Technician A says on FWD vehicles, the front wheels toe-out while the rear wheels on an independent suspension try to toe-in as the vehicle moves ahead. Technician B says on RWD vehicles, the front wheels tend to toe-in while the rear wheels toe-out in response to rolling resistance and suspension compliance. Who is correct?

a. Technician A
b. Technician B
c. Both A and B
d. Neither A nor B

20. Technician A says if caster at both wheels is not equal, the vehicle will tend to drift toward the side with the most positive caster. Technician B says if camber is not the same on both wheels, the vehicle will pull toward the side with the most positive camber. Who is correct?

a. Technician A
b. Technician B
c. Both A and B
d. Neither A nor B

SECTION 8

Brakes

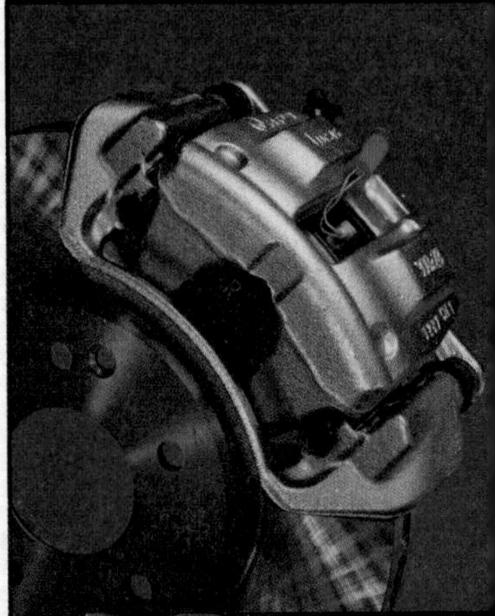

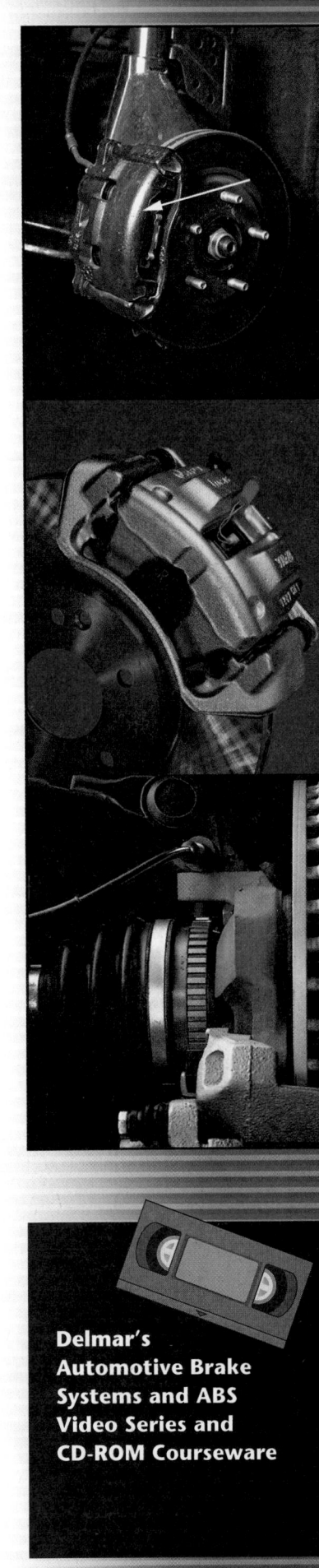

As with the steering and suspension systems, a properly operating brake system is vital to the safe operation and control of any vehicle. A full understanding of hydraulic principles, brake lines and hoses, and disc and drum brake components is needed to perform accurate brake service. Applying the brakes is the normal first response when there is potential danger. A poorly working brake system can be disastrous.

Antilock braking systems (ABS) and traction-assist systems are rapidly gaining popularity. They are standard equipment on most vehicles. Stability control systems have been incorporated into the ABS and traction control systems of many cars. All three of these systems are electronically controlled systems and are linked to the vehicle's PCM.

The information found in Section 8 corresponds to materials covered on the ASE certification test on brake systems.

Delmar's Automotive Brake Systems and ABS Video Series and CD-ROM Courseware

WE ENCOURAGE PROFESSIONALISM

THROUGH TECHNICIAN CERTIFICATION

46 BRAKE SYSTEMS

OBJECTIVES

■ Explain the basic principles of braking, including kinetic and static friction, friction materials, application pressure, and heat dissipation. ■ Describe the components of a hydraulic brake system and their operation, including brake lines and hoses, master cylinders, system control valves, and safety switches. ■ Perform both manual and pressure bleeding of the hydraulic system. ■ Briefly describe the operation of drum and disc brakes. ■ Inspect and service hydraulic system components. ■ Describe the operation and components of both vacuum-assist and hydraulic-assist braking units.

The brake system **(Figure 46–1)** is designed to slow and halt the motion of a vehicle. To do that, various components within a hydraulic brake system must convert the momentum of the vehicle into heat. They do so by using friction.

FRICTION

There are two basic types of friction that explain how brake systems work: **kinetic**, or moving, and **static**, or stationary **(Figure 46–2)**. The amount of friction, or resistance to movement, depends on the type of materials in contact, the smoothness of their rubbing surfaces, and the pressure holding them together (often gravity or weight). Friction always converts moving, or kinetic, energy into heat. The greater the friction between two moving surfaces, the greater the amount of heat produced.

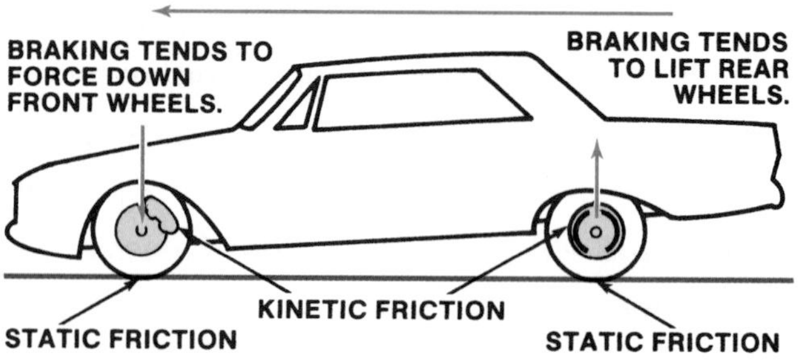

Figure 46–2 Braking action creates kinetic friction in the brakes and static friction between the tire and road to slow the vehicle. When brakes are applied, the vehicle's weight is transferred to the front wheels and is unloaded on the rear wheels.

Figure 46–1 A complex late-model brake system.
Courtesy of DaimlerChrysler Corporation

As the brakes on a moving automobile are applied, rough-textured pads or shoes are pressed against rotating parts of the vehicle—either rotors (discs) or drums. The kinetic energy, or momentum, of the vehicle is then converted into heat energy by the kinetic friction of rubbing surfaces and the car or truck slows down.

When the vehicle comes to a stop, it is held in place by static friction. The friction between the surfaces of the brakes and between the tires and the road resists any movement.

Factors Governing Braking

Four basic factors determine the braking power of a system. The first three factors govern the generation of friction: pressure, coefficient of friction, and frictional contact surface. The fourth factor is a result of friction. It is heat or, more precisely, heat dissipation.

Pressure The amount of friction generated between moving surfaces contacting one another depends in part on the pressure exerted on the surfaces. For example, if you slowly increase the downward pressure on the palm of your hand as you move it across a tabletop, you feel a gradual increase in friction.

In automobiles, hydraulic systems provide application pressure. Hydraulic force is used to move brake pads or brake shoes against spinning rotors or drums mounted to the wheels.

Coefficient of Friction The amount of friction between two surfaces is expressed as a **coefficient of friction (COF)**. The coefficient of friction is determined by dividing the force required to pull an object across a surface by the weight of the object **(Figure 46–3)**. For example, if it requires 100 pounds (445 Newtons) of pull to slide a 100-pound (445 N) block of rubber across a concrete floor, the coefficient of friction is 100 ÷ 100 or 1. To pull a 100-pound (445 N) block of ice across the same floor might require only 2 pounds (9 N) of pull. The coefficient of friction then would be only 0.02. As it applies to automotive brakes, the COF expresses the frictional relationship between pads and rotors or shoes and drums and is carefully engineered to ensure optimum performance. Therefore, when replacing pads or shoes, it is important to use replacement parts with similar COF. If, for example, the COF is too high, the brakes are too sticky to stop the car smoothly. Premature wheel lockup or grabbing would result. If the coefficient is too low, the friction material tends to slide over the machined surface of the drum or rotor rather than slowing it down. Most automotive friction materials are thus engineered with a COF of between 0.25 and 0.55, depending on their intended application.

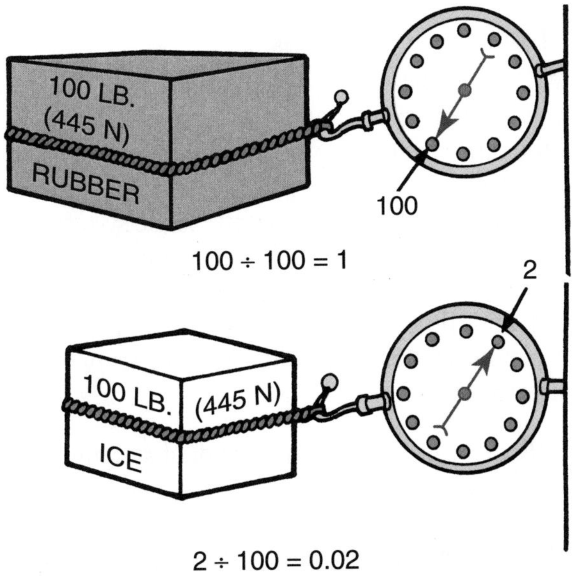

Figure 46-3 Coefficient of friction is equal to the pounds of pull divided by the weight of the object.

Frictional Contact Surface The third factor is the amount of surface, or area, that is in contact. Simply put, bigger brakes stop a car more quickly than smaller brakes used on the same car. Similarly, brakes on all four wheels slow or stop a moving vehicle faster than brakes on only two wheels, assuming the vehicles are equal in size.

Heat Dissipation Any braking system must be able to effectively handle the heat created by friction within the system. The tremendous heat created by the rubbing brake surfaces must be conducted away from the pad and rotor (or shoe and drum) and be absorbed by the air. The greater the surface areas of the brake components, the faster the heat can be dissipated. Thus, the weight and potential speed of the vehicle determine the size of the braking mechanism and the friction surface area of the pad or shoe. Brakes that do not effectively dissipate heat experience brake fade during hard, continuous braking. The linings of the pad and shoe become glazed, and the rotor and drum become hardened. Therefore, the coefficient of friction is reduced and excessive foot pressure must be applied to the brake pedal to produce the desired braking effect.

Brake Lining Friction Materials

The friction materials used on brake pads and shoes are called brake linings. Brake linings are either riveted or bonded to the backing of the pad or shoe. Some newer brake pads are integrally molded. These can be identified by looking at the backing of the pad. Integrally molded pad assemblies will have holes that are partially or totally filled with the lining material.

For many decades, asbestos was the standard brake lining material. It offers good friction qualities, long wear, and low noise. New materials, such as composite/organic, ceramics, and carbon fibers, are being used because of the health hazards of breathing asbestos dust. In fact, the federal government has banned the use of asbestos in new vehicles and in aftermarket replacement parts.

Fully Metallic Fully **metallic** linings of sintered iron have been used for years in heavy-duty and racing applications because they have great fade resistance. However, they require very high pedal pressure and tend to quickly wear out drums and rotors.

Semimetallic Semimetallic lining materials were developed to eliminate the disadvantages of fully metallic linings. Semimetallic linings are made of iron fibers molded with an adhesive matrix. Semimetallic material offers excellent fade resistance that meets the needs of today's vehicles. It has good frictional characteristics so only a moderate amount of application pressure is needed. Finally, semimetallic pads and shoes do not cause excessive wear on rotor or drum surfaces.

Nonasbestos Other **nonasbestos** lining materials made of synthetic substances are now available. The major brake lining manufacturers are constantly experimenting with new materials that meet all established criteria for long life, friction characteristics, drum and rotor wear, and heat dissipation.

PRINCIPLES OF HYDRAULIC BRAKE SYSTEMS

A hydraulic system **(Figure 46–4)** uses a brake fluid to transfer pressure from the brake pedal to the pads or shoes. This transfer of pressure is reliable and consistent because liquids are not compressible. That is, pressure applied to a liquid in a closed system is transmitted by that liquid equally to every other part of that system. Apply a force of 5 pounds (35 kPa) per square inch (psi) through the master cylinder and you can measure 5 psi (35 kPa) anywhere in the lines and at each wheel where the brakes operate.

The force can be increased at output (that is, at the wheel) by increasing the size of the wheel's piston, though piston travel decreases. The force at output can be decreased by decreasing the size of the wheel piston, but the piston travel increases.

Thus, to double the output force of the 5 psi (35 kPa) at the master cylinder to 10 psi (69 kPa) at the wheels, simply use a wheel cylinder piston with a surface area of 2 square inches (13 sq. cm). To triple the force of 100 psi (690 kPa), use a piston with 3 square inches (20 sq. cm) and 300 pounds (2068 kPa) of output result **(Figure 46–5)**. No matter what the fluid pressure is, the output force can be increased with a larger piston, though piston travel decreases proportionately. In actual practice, however, fluid movement in an automotive hydraulic brake system is very slight. In an emergency, when the pedal goes all the way to the floor, the volume of fluid displaced amounts to only about 20 cubic centimeters. About 15 cubic centimeters goes to the front discs and 5 cubic centimeters goes to the rear drums. Even under these conditions, the wheel cylinder and caliper pistons move only slightly.

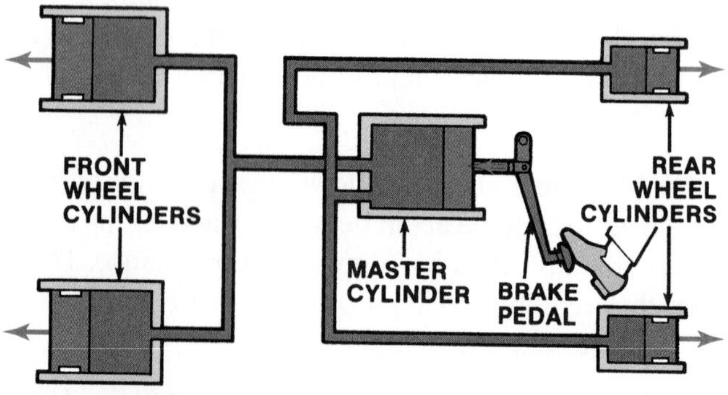

Figure 46–4 A schematic of a basic automotive hydraulic brake system.

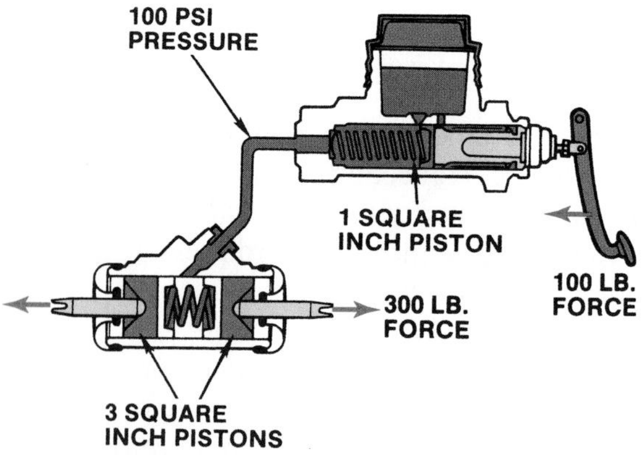

Figure 46–5 Output force increases with piston size.

Of course, the hydraulic system does not stop the car all by itself. In fact, it really just transmits the action of the driver's foot on the brake pedal out to the wheels. In the wheels, sets of friction pads are forced against rotors or drums to slow their turning and bring the car to a stop. Mechanical force (the driver stepping on the brake pedal) is changed into hydraulic pressure, which is changed back into mechanical force (brake shoes and disc pads contacting the drums and rotors). The amount of force acting on the friction pads and shoes is equivalent to the psi applied to the pedal multiplied by the area of the piston affected. A force of 25 psi applied to the pedal times 4 square inches of piston area equals 100 pounds of pressure in the system.

Dual Braking Systems

Since 1967, federal law has required that all cars be equipped with two separate brake systems. If one circuit fails, the other provides enough braking power to safely stop the car.

The dual system differs from the single system by employing a tandem master cylinder, which is essentially two master cylinders formed by installing two separate pistons and fluid reservoirs into one cylinder bore. Each piston applies hydraulic pressure to two wheels **(Figure 46–6)**.

Front/Rear Split System In early dual systems, the hydraulic circuits were separated front and rear. Both front wheels were on one hydraulic circuit and both rear wheels on another. If a failure occurred in one system, the other system was still available to stop the vehicle. However, the front brakes do approximately 70% of the

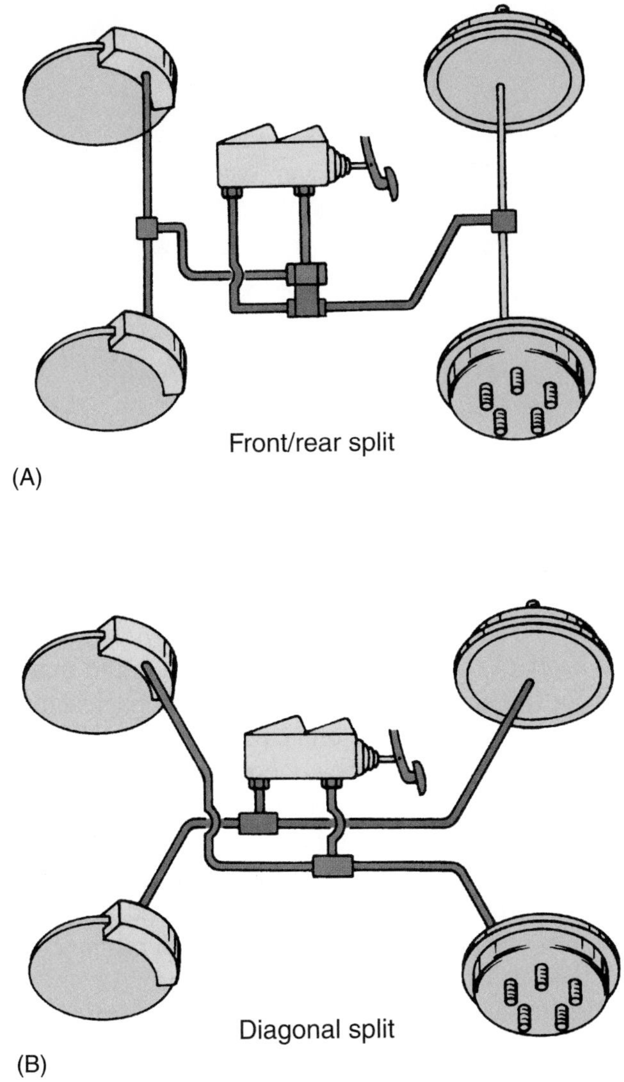

Figure 46-6 (A) A front/rear split of the braking system. (B) A diagonal split. The master cylinder is also split to allow pressure only to its designated wheel units. *Courtesy of Ford Motor Company*

braking work. A failure in the front brake system would only leave 20 to 40% braking power. This problem was somewhat reduced with the development of diagonally split systems.

Diagonally Split System The **diagonally split system** operates on the same principles as the front and rear split system. It uses primary and secondary master cylinders that move simultaneously to exert hydraulic pressure on their respective systems.

The hydraulic brake lines on this system, however, have been diagonally split front to rear (left front to right rear and right front to left rear). The circuit split can occur within the master cylinder or externally at a proportioning valve or pressure differential switch.

In the event of a system failure, the remaining good system would do all the braking on one front wheel and the opposite rear wheel, thus maintaining 50% of the total braking force.

HYDRAULIC BRAKE SYSTEM COMPONENTS

The following sections describe the major components of a hydraulic brake system, including power-assisted systems and antilock braking systems.

Brake Fluid

Brake fluid **(Figure 46–7)** is the lifeblood of any hydraulic brake system and is what makes the system operate properly.

Modern brake fluid is specially blended to enable it to perform a variety of functions. Brake fluid must be able to flow freely at extremely high temperatures (500°F, 260°C) and at very low temperatures (–104°F, –75°C). Brake fluid must also serve as a lubricant to the many parts with which it comes into contact to ensure smooth and even operation. In addition, brake fluid must fight corrosion and rust in the brake lines and various assemblies and components in services. Another important property of brake fluid is that it must resist evaporation. All brake fluids are hydroscopic; that is, they readily absorb water. This is why brake fluid should always be kept in a sealed container and should only be exposed to outside air for limited periods of time.

Some of the earliest brake fluids had chemicals in them that ate away at the rubber components in the brake system (that is, cups and seals). Modern brake fluid must be compatible with rubber to avoid damage to the cups and seals in the system. Brake fluid must provide a controlled amount of swell to the brake system cups and seals. There must be just enough swell to form a good seal. However, the swell cannot be too great. If it is, drag and poor brake response occur. Every can of

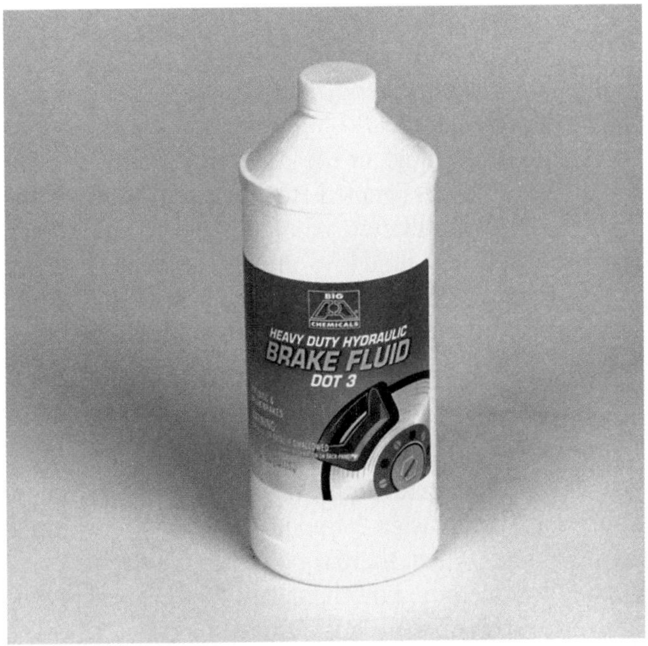

Figure 46-7 A container of brake fluid.

brake fluid carries the identification letters of SAE and **DOT**. These letters (and corresponding numbers) indicate the nature, blend, and performance characteristics of that particular brand of brake fluid.

There are three basic types or classifications of hydraulic brake fluids. DOT 3 is a conventional brake fluid with a minimum dry **equilibrium reflux boiling point (ERBP)** of 401°F (205°C) and a minimum wet ERBP of 284°F (140°C). It is generally recommended for most ABS systems and some power brake setups. DOT 4 is a conventional brake fluid with a minimum dry ERBP of 446°F (230°C) and a minimum wet ERBP of 356°F (180°C). It is the most commonly used brake fluid for conventional brake systems. DOT 5 is a unique silicone-based brake fluid with a minimum dry ERBP of 500°F (260°C) and a minimum wet ERBP of 356°F (180°C). In the last few years, DOT 5 has lost its demand by brake servicing experts. DOT 5.1 is a nonsilicone-based polyglycol fluid and is amber in color. This is a severe duty fluid that has the same boiling point as DOT 5. However, remember that it is best to follow the vehicle manufacturer's recommendations.

WARNING!

Use only approved brake fluid in a brake system. Any other lubricant, such as power steering fluid, automatic transmission fluid, or engine oil, which has a petroleum base, must never be used in the brake system. Petroleum-based fluids attack the rubber components in the brake system, like the piston cups and seals, and cause them to swell and disintegrate.

Some vehicles have brake fluid level sensors that provide the driver with an early warning message that the brake fluid in the master cylinder reservoir has dropped below the normal level.

As the brake fluid in the master cylinder reservoir drops below the designated level, the sensor closes the warning message circuit. About fifteen seconds later, the message "brake fluid low" appears on the instrument panel. At this time, the master cylinder reservoir should be checked and filled to the correct level with the specified brake fluid.

Brake Pedal

The brake pedal is where the brake's hydraulic system gets its start. When the brake pedal is depressed, force is applied to the master cylinder. On a basic hydraulic brake system (where there is no power assist), the force applied is transmitted mechanically. As the pedal pivots, the force applied to it is multiplied mechanically. The force that the pushrod applies to the master cylinder piston is, therefore, much greater than the force applied to the brake pedal.

Figure 46-8 A brake master cylinder.

Master Cylinders

The heart of the hydraulic brake system is the master cylinder **(Figure 46–8)**. It converts the mechanical force of the driver's foot on the brake pedal to hydraulic pressure. The master cylinder has a bore that contains an assembly of two pistons and a return spring. One piston pressurizes one-half of the brake system and one takes care of the other half. Although master cylinders differ in the size of the pistons, reservoir design, and integrated hydraulic components, the operation of all master cylinders is basically the same.

Master cylinders are generally constructed of cast iron **(Figure 46–9A)** with an integral fluid reservoir, or

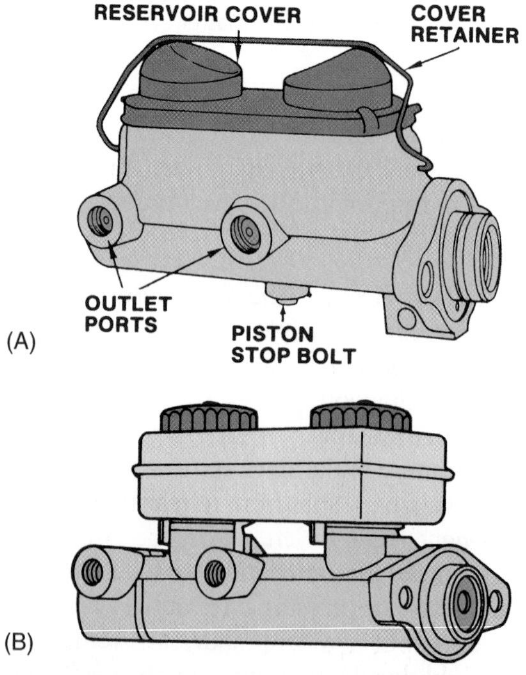

Figure 46-9 (A) A typical cast-iron dual master cylinder and (B) a typical aluminum/composite dual master cylinder.

they are made of aluminum with a separate molded nylon or fiberglass-reinforced plastic reservoir. Aluminum body master cylinders feature an anodized body **(Figure 46–9B)** to protect against corrosion and to extend bore life.

CAUTION!

It is recommended that aluminum master cylinders not be rebuilt if pitting or scoring of the cylinder bore is evident. A new unit should be installed.

MASTER CYLINDER OPERATION

It is important that you understand how a master cylinder works. This understanding will be of great help when diagnosing a brake system. The basic principle of operation is a simple one. The brake pedal, connected to the master cylinder's piston assembly by a pushrod, controls the movement of the pistons. As the pedal is depressed, the piston assembly moves. The piston exerts pressure on the fluid, which flows out under pressure through the fluid outlet ports into the rest of the hydraulic system. When the brake pedal is released, the return spring in the cylinder forces the piston assembly back to its original position.

Because the master cylinder is a hydraulic device, the system must remain sealed and full of fluid. To do this, each piston has a primary and a secondary cup. The primary cup compresses the fluid when the pedal is depressed and keeps the brake fluid ahead of the pistons when they are put under pressure. The movement of the piston and cup shut off the fluid supply from the reservoir and create a sealed system from the primary cup forward. To prevent the fluid in the reservoir from leaking out of the cylinder when the primary cup has moved forward, each piston has a secondary cup. The secondary cup seals nonpressurized fluid.

When the brake pedal is not depressed, fluid from the reservoir keeps the cylinder filled. The fluid enters the cylinder through vent (intake) ports and compensating or replenishing ports that connect the reservoir to the piston chamber.

Each piston has a return spring in front of it. This holds the primary piston cup slightly behind the compensating or replenishing port **(Figure 46–10)** for the reservoir allowing gravity to keep the cylinder filled with fluid. In addition to keeping the replenishing ports uncovered, the return springs also help to return the brake pedal when the force has been removed from it. As the brakes are applied **(Figure 46–11)**, the stiffer primary piston spring pushes the secondary piston and spring slightly. Then the cup at the front end of the secondary piston passes and closes off the primary replenishing port on the secondary side of the master cylinder.

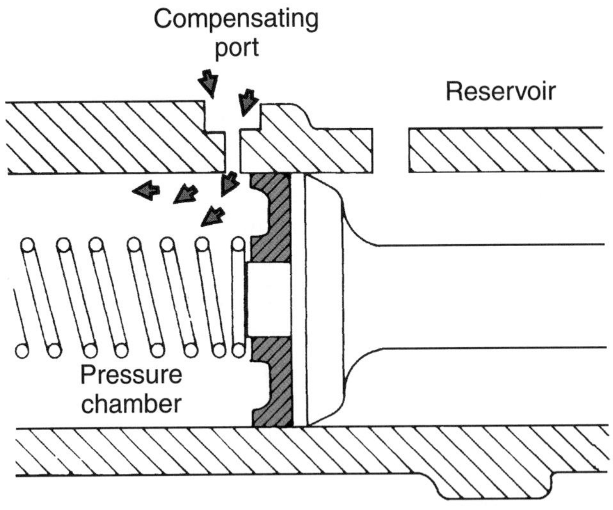

Figure 46–10 Fluid from the reservoir fills the cylinder through the compensating port. *Courtesy of Ford Motor Company*

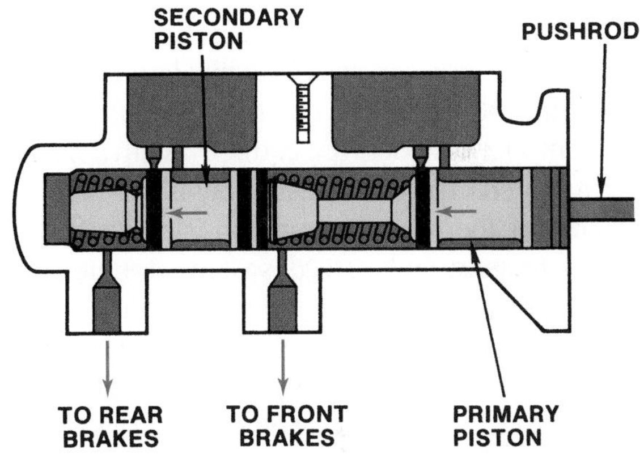

Figure 46–11 The position of the pistons in a master cylinder when the brakes are applied.

When the piston moves quickly back to its resting place, the brake fluid cannot return through the lines fast enough to avoid creating a low-pressure condition ahead of it. The fluid must reach the low-pressure area in time for another stroke of the cylinder. To keep the system filled with fluid and to maintain a sealed system, fluid from the reservoir then enters the cylinder through a vent port to fill the void. Fluid also enters the system by flowing past the primary cups. During the return stroke, the edges of the cup pull away from the bore enough to allow fluid to pass around the piston assembly to the area of low pressure.

Finally, the pads and shoes return to position. That is, once the brake pedal is released and the master cylinder piston has returned to the rest position, shoe return springs (in the drum brakes) and piston seals (in the disc brakes) cause these pistons to retract. When all of the brakes are fully released, any excess fluid is returned to the reservoir through the compensating port to relieve pressure in the system.

Master Cylinder Components

The two pistons in the master cylinder are not rigidly connected. Each piston has a return spring with the primary piston spring between the two pistons. Stepping on the brake pedal moves a pushrod, causing the first or primary piston to move forward. The fluid ahead of it cannot be compressed, so the secondary piston moves. As the pistons progress deeper into the cylinder bore, the brake fluid that is put under pressure transmits this force through both systems to friction pads at the wheels. A retaining ring fits into a groove near the end of the bore and holds the piston inside the cylinder.

Extra brake fluid is stored in two separate reservoirs. The reservoirs are made of either cast iron or specially blended plastic and are designed to function independently as a protection against lost brake fluid **(Figure 46–12)**. The cap or cover for the reservoir keeps an airtight seal on the master cylinder while letting it breathe. It also keeps moisture from entering the system.

Most master cylinders are equipped with a fluid level warning switch. The switch assembly is normally an integral part of the reservoir and consists of a float containing a magnet and a reed switch mounted in the bottom of the reservoir. When the brake fluid level gets to a predetermined level, the floating magnet will activate the reed switch, which causes the red brake warning lamp on the dash to turn on. The switch itself is not serviceable. If there is problem with the switch or float assembly, the entire reservoir needs to be replaced.

The purpose of the **residual check valve, (Figure 46–13)** found only on vehicles with rear drum brakes, is

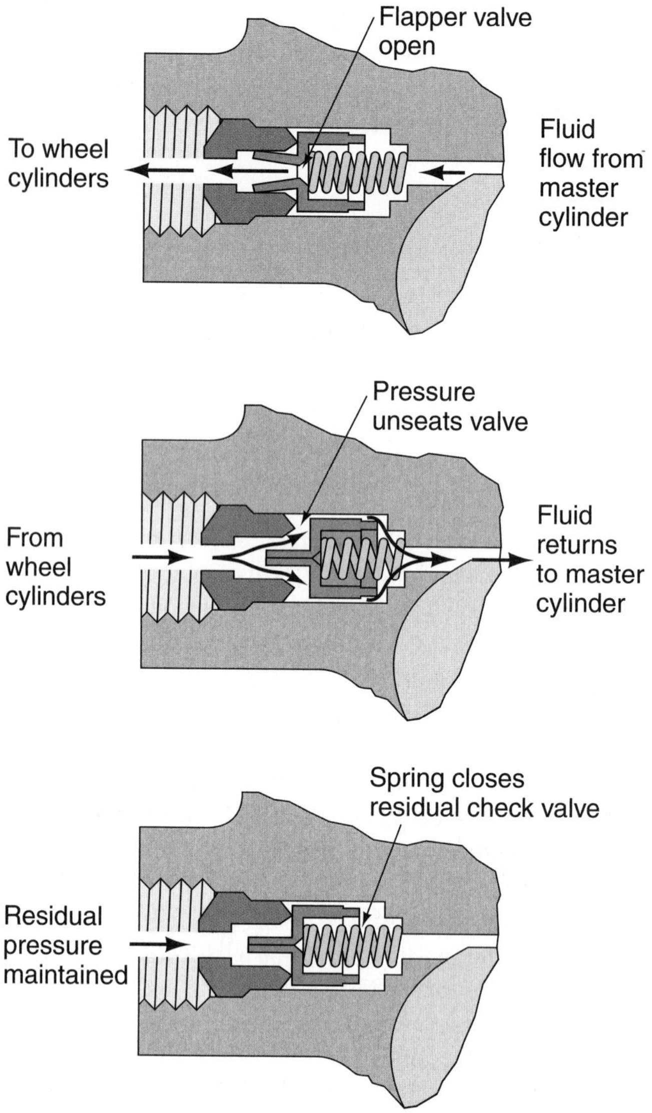

Figure 46–13 The operation of a master cylinder's residual valve.

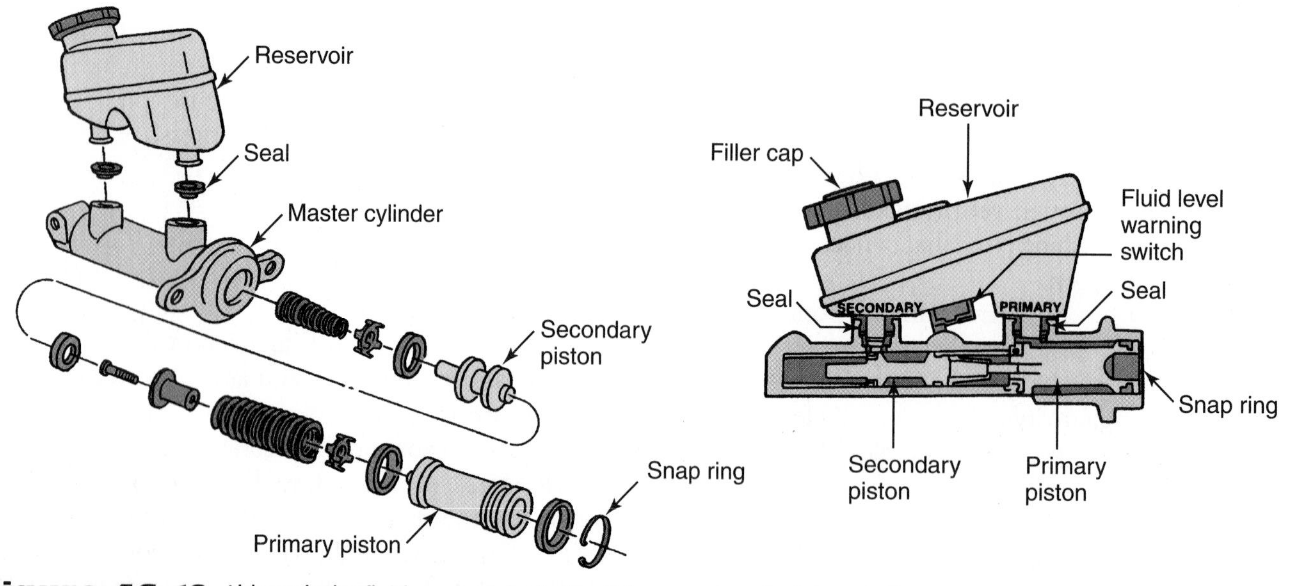

Figure 46–12 Although the fluid reservoir looks like it is one reservoir, it is actually two separate reservoirs, one for each half of the hydraulic system. *Courtesy of Ford Motor Company*

to keep light (or residual) pressure in the brake lines and also at the wheel cylinders. Without residual pressure, air can be sucked past the wheel cylinder cups and into the wheel cylinders. This could occur when the brake pedal is released quickly. Residual check valves are designed to permit hydraulic fluid and pressure to flow from the master cylinder into the hydraulic lines when the pressure in the master cylinder becomes great enough to open the valve. After allowing the fluid to flow back into the master cylinder when the pedal is released, the outlet check valve spring **(Figure 46–14)** takes over and keeps the hydraulic lines and wheel cylinders full of fluid, ready for the next application of the brakes.

On most of the newer car models, automobile manufacturers have eliminated residual pressure check valves from the brake system. Instead, the wheel cylinders are equipped with cup expanders. The cup expanders keep the cup seals tightly pressed against the cylinder walls to prevent air from entering the system.

Light residual pressure is needed for operating the drum brakes. However, when the brakes are used repeatedly, too much residual pressure can build up. Through heavy use the brake drums and wheel cylinders can get very hot. This in turn will cause the brake to heat up and expand. This pressure buildup can become so strong that it keeps the brake shoes from returning after releasing the brake pedal. If the brake shoes do not return properly, they can drag against the drums, which causes more heat to build up and pressure is further increased. The replenishing or compensating port takes care of this problem by allowing the excess pressure to leave the closed system.

Vent ports in the master cylinder help to pump up a low brake pedal by allowing more fluid in the lines. There are several reasons for a brake pedal to be lower than normal. One is that the clearance between the brake lining and the drum is too large. This can happen when the automatic brake adjusters are not working properly. Another reason could be the result of air leaking into the brake system. Vent ports help with this problem because the replenishing ports are not large enough to handle the extra flow of fluid needed when the pedal is pumped.

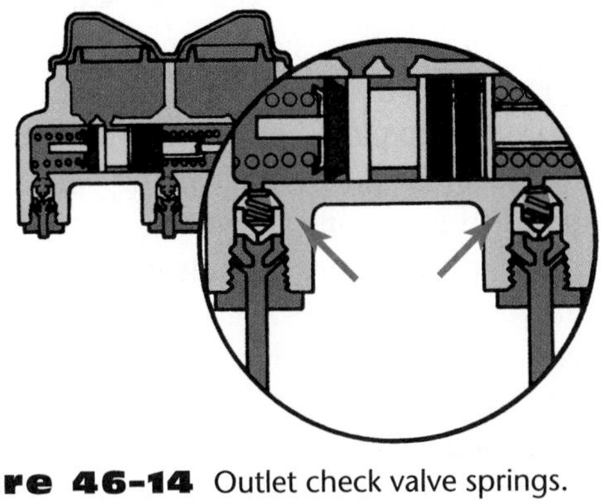

Figure 46–14 Outlet check valve springs.

The design of the master cylinder used on a particular model car depends on the capacity of the reservoir and the displacement of the pistons at the wheel. Disc brakes require larger reservoirs than drum brakes because the caliper cylinder is much larger than that of a wheel cylinder. Some internal parts of the master cylinder may also change with application. For example, some General Motors' products are fitted with wheel caliper pistons that have a low drag. These were designed to retract farther into the caliper's cylinder when the brake pedal was released. This reduction of drag increased fuel economy. However since the piston had to move farther to push the pads against the rotor, more fluid had to be moved by the master cylinder. This caused a delay in pedal response and the feeling of a low pedal. To correct this problem, General Motors uses a **quick-take-up** master cylinder, with a larger rear piston to move more fluid. A quick-take-up valve in the master cylinder controls the flow of fluid from the reservoir to the master cylinder's rear piston.

HYDRAULIC TUBES AND HOSES

Steel tubing and flexible synthetic rubber hosing serve as the arteries and veins of the hydraulic brake system. These brake lines transmit brake fluid pressure (the blood) from the master cylinder (the heart) to the wheel cylinders and calipers (the muscles and working parts) of the drum and disc brakes.

Fluid transfer from the driver-actuated master cylinder is usually routed through one or more valves and then into the steel tubing and hoses **(Figure 46–15)**. The design of the brake lines offers quick fluid transfer response with very little friction loss. Engineering and installing the brake lines so they do not wrap around sharp curves is very important in maintaining this good fluid transfer.

Brake Line Tubing

Most brake line tubing consists of copper-fused double-wall steel tubing in diameters ranging from 1/8 to 3/8 inch (3 mm to 9 mm). Some OEM brake tubing is manufactured with soft steel strips, sheathed with copper. The strips are rolled into a double-wall assembly and then bonded in a furnace at extremely high temperatures. Corrosion protection is often added by tin-plating the tubing.

Fittings

Assorted fittings are used to connect steel tubing to junction blocks or other tubing sections. The most common fitting is the double or inverted flare style. Double flaring is important to maintain the strength and safety of the system. Single flare or sleeve compression fittings may not hold up in the rigorous operating environment of a standard vehicle brake system.

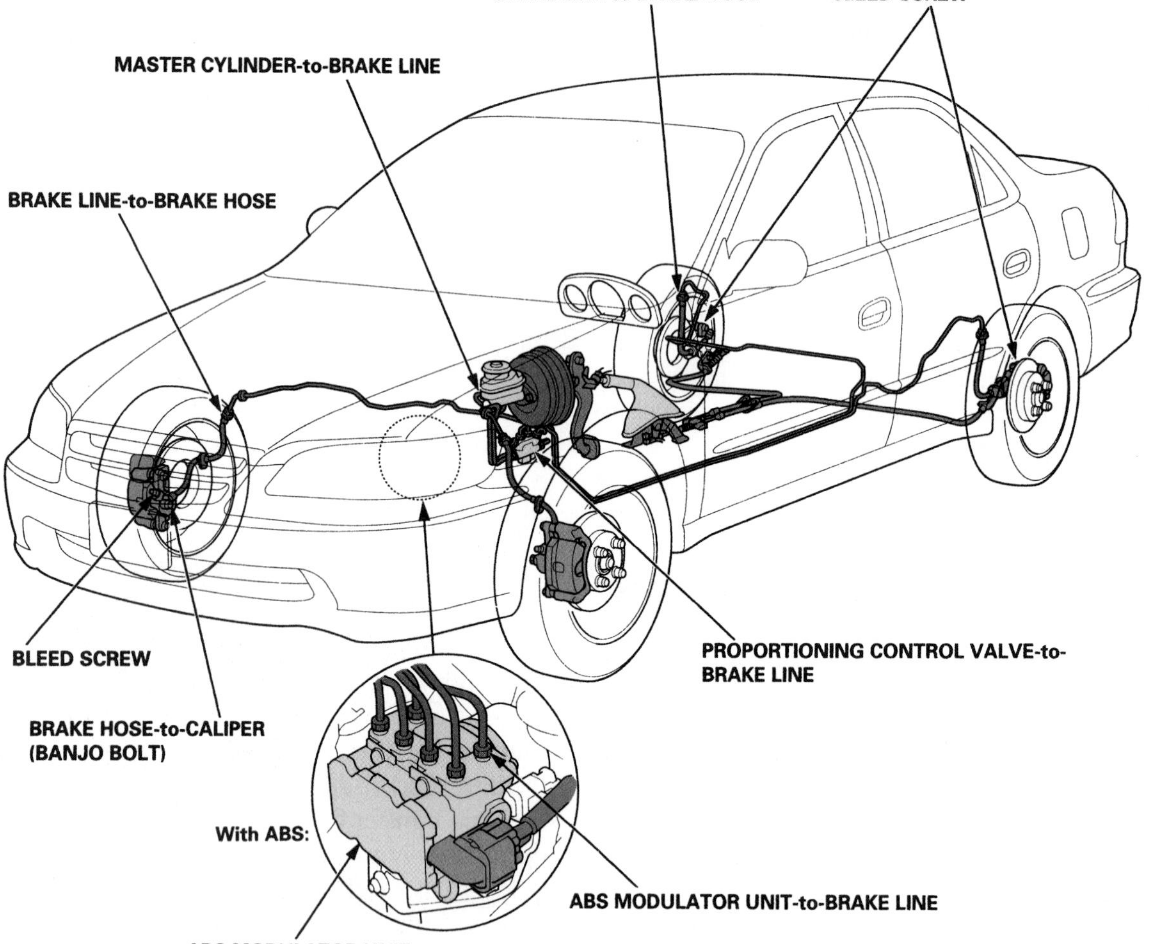

Figure 46-15 The typical layout of the hoses and tubes for the brake system. *Courtesy of American Honda Motor Co., Inc.*

Fittings are constructed of steel or brass. The 37-degree inverted flare or standard flare fitting is the most commonly used coupling. Newer vehicles may use the **ISO** or metric bubble flare fitting.

Never change the style of fitting being used on the vehicle. Replace ISO fittings only with ISO fittings. Replace standard fittings with standard fittings.

The metal composition of the fittings must also match exactly. Using an aluminum-alloy fitting with steel tubing may provide a good initial seal, but the dissimilar metals create a corrosion cell that eats away the metal and reduces the connection's service life.

Brake Line Hoses

Brake line hoses offer flexible connections to wheel units so steering and suspension members can operate without damaging the brake system. Typical brake hoses range from 10 to 30 inches (25 to 76 mm) in length and are constructed of multiple layers of fabric impregnated with a synthetic rubber. Brake hose material must offer high heat resistance and withstand harsh operating conditions.

SHOP TALK

Many brake hose failures can be traced to errors made in the original installation or repair of the hose. Hoses twisted into place become stressed and are prime candidates for leaks and bursting. Most manufacturers now print a natural lay indicator or line on the hose. By making sure this line is not spiralled after fittings are tightened, you can ensure the hose is not overly stressed. Also, always use a hose of the same length and diameter as the original during servicing to maintain brake balance at all wheels. ■

HYDRAULIC SYSTEM SAFETY SWITCHES AND VALVES

Switches and valves are installed in the brake system hydraulic lines to act as warning devices to pressure control devices.

Pressure Differential (Warning Light) Switches

A pressure differential valve is used to operate a warning light switch. Its main purpose is to tell the driver if pressure is lost in either of the two hydraulic systems. Since each brake hydraulic system functions independently, it is possible the driver might not notice immediately that pressure and braking are lost. When a pressure loss occurs, brake pedal travel increases and a more-than-usual effort is needed for braking. Should the driver not notice the extra effort needed, the warning light is actuated by the hydraulic system safety switch.

Under normal conditions, the hydraulic pressure on each side of the pressure differential valve piston is balanced. The piston is located at its center point, so the spring-loaded warning switch plunger fits into the tapered groove of the piston. This leaves the contacts of the warning switch open. The brake warning light stays off.

If there is a leak in the front or rear braking system, the hydraulic pressure in the two systems is unequal. For example, if there is a leak in the system supplying the front brakes, there is lower pressure in the front system when the brake pedal is applied. The hydraulic pressure in the rear system then pushes the piston toward the front side, where the pressure is lower. As the piston moves, the plunger is pushed out **(Figure 46–16)**. This closes the switch and illuminates the brake warning light.

While all brake warning light switches serve the same function, there are three common variations in the design of these switches. These variations include switches with centering springs, without centering springs, and with centering springs and two pistons.

Metering and Proportioning Valves

Metering and proportioning valves are used to balance the braking characteristics of disc and drum brakes.

The braking response of the disc brakes is immediate when the brake pedal is applied. It is directly proportionate to the effort applied at the pedal. Drum brake response is delayed while rear brake hydraulic pressure moves the wheel cylinder pistons to overcome the force of their return springs and force the brake shoes to contact the drum. Their actions are self-energizing and tend to multiply the pedal effort.

Metering Valve A **metering valve** in the front brake line holds off pressure going from the master cylinder to the

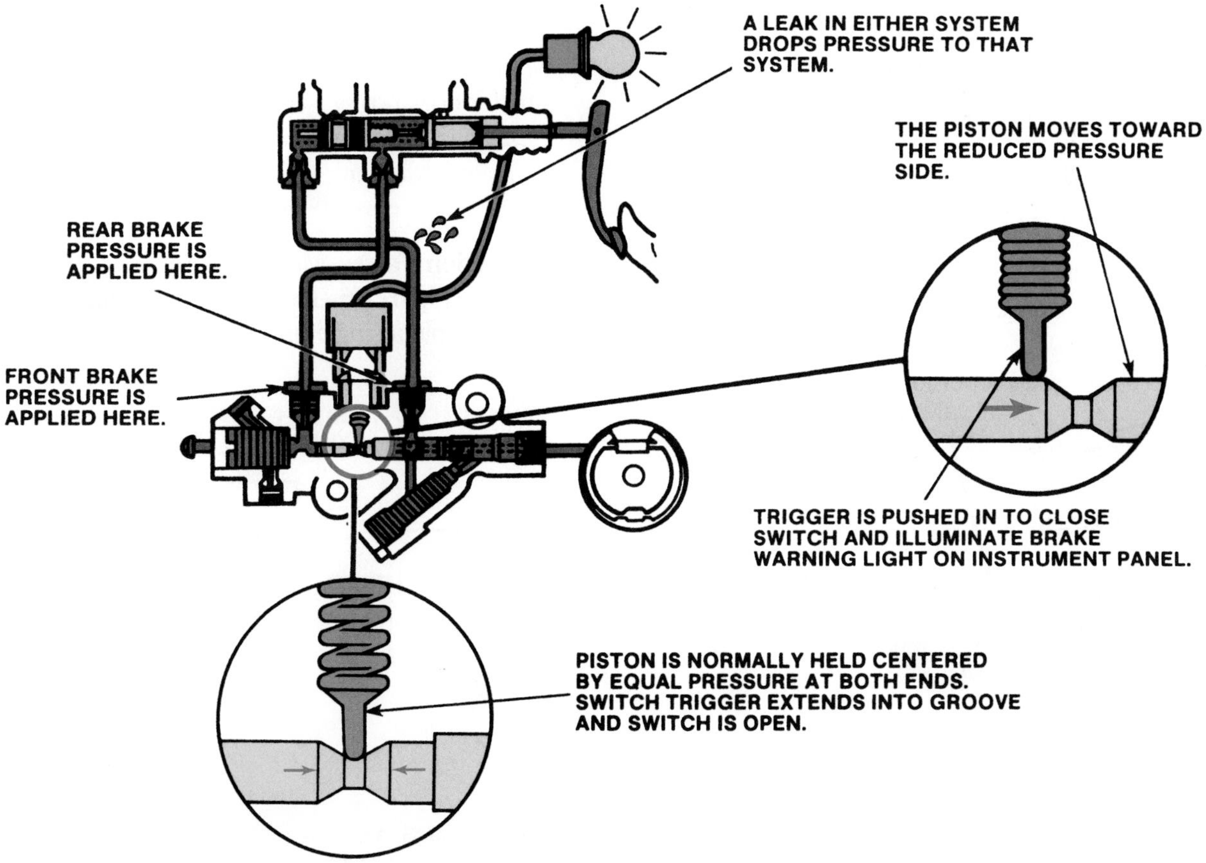

Figure 46–16 A pressure differential valve under normal conditions.

front disc calipers. This delay allows pressure to build up in the rear drums first. When the rear brakes begin to take hold, the hydraulic pressure builds to the level needed to open the metering valve. When the metering valve opens, line pressure is high enough to operate the front discs. This process provides for better balance of the front and rear brakes. It also prevents lockup of the front brakes by keeping pressure from them until the rear brakes have started to operate. The metering valve has the most effect at the start of each brake operation and all during light braking conditions.

Proportioning Valve The self-energizing action of the delayed response rear drum brakes can cause them to lock the rear wheels at a lower hydraulic pressure than the front brakes. The **proportioning valve (Figure 46–17)** is used to control rear brake pressures, particularly during hard stops. When the pressure to the rear brakes reaches a specified level, the proportioning valve overcomes the force of its spring-loaded piston, stopping the flow of fluid to the rear brakes. By doing so, it regulates rear brake system pressure and adjusts for the difference in pressure between front and rear brake systems. This keeps front and rear braking forces in balance.

Height-Sensing Proportional Valve The height-sensing proportional valve provides two different brake balance modes to the rear brakes based on vehicle load. This is accomplished by turning the valve on or off. When the vehicle is not loaded, hydraulic pressure is reduced to the rear brakes. When the vehicle is carrying a full load, the actuator lever moves up to change the valve's setting. The valve now allows full hydraulic pressure to the rear brakes. The valve contains a plunger, cam, torsional clutch spring, and an actuator shaft **(Figure 46–18)**.

The valve is mounted to the frame above the rear axle and has an actuator lever connected by a link to the

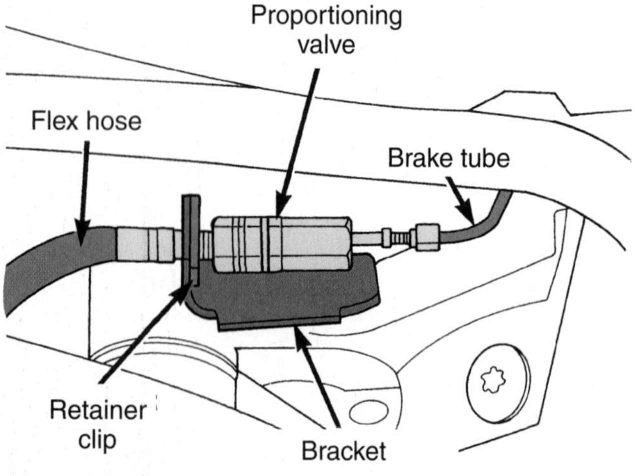

Figure 46–17 A proportioning valve. *Courtesy of DaimlerChrysler Corporation*

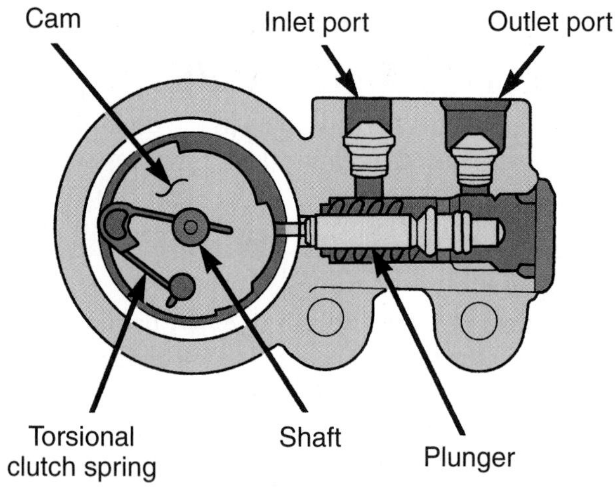

Figure 46–18 A height-sensing proportional valve. *Courtesy of DaimlerChrysler Corporation*

lower shock absorber bracket. The valve is turned on and off as the axle-to-frame height changes due to load in the vehicle. The torsional clutch spring attached to the valve shaft is used as an override. Once the valve is positioned during braking, the spring prevents the valve from changing position if the vehicle goes over a bump or moves off the road.

Height-sensing proportional valves are replaced when defective and are not adjustable.

SHOP TALK

Distribution and weight on some vehicles, such as station wagons, is such that a rear-to-front weight transfer does not present a similar problem. For this reason, standard proportioning valves are not required on all station wagon models. ■

Combination Valves Most newer cars have a **combination valve (Figure 46–19)** in their hydraulic system. This valve is simply a single unit that combines the metering

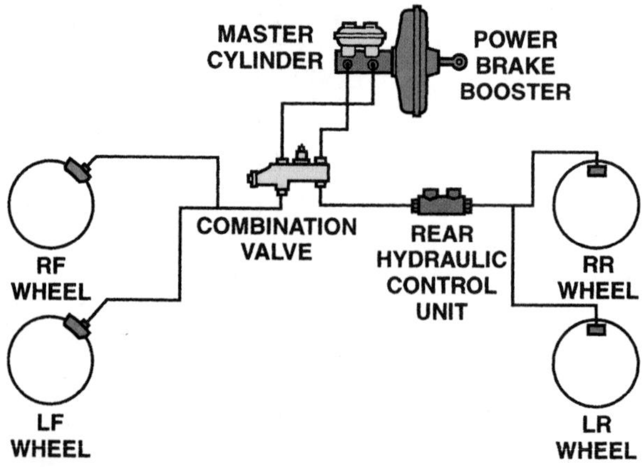

Figure 46–19 The hydraulic circuit for a pickup with rear wheel antilock brakes and a combination valve. *Courtesy of DaimlerChrysler Corporation*

and proportioning valves with the pressure differential valve. Combination valves are described as three-function or two-function valves, depending on the number of functions they perform in the hydraulic system.

Three-Function Valve. This type of valve performs the functions of the metering valve, brake warning light switch, and proportioning valve.

Two-Function Valves. There are two variations of the two-function combination valve. One variation does the proportioning valve and brake warning light switch functions. The other performs the metering valve and brake warning light switch functions.

If any one of its several operations fail, the entire combination valve must be replaced, because these units are not repairable.

Stop Light Switches

The stop light (stop light/speed control) switch and mounting bracket assembly **(Figure 46–20)** is attached to the brake pedal bracket and is activated by pressing the brake pedal.

The mechanical stop light switch is operated by contact with the brake pedal or with a bracket attached to the pedal. The hydraulic switch is operated by hydraulic pressure developed in the master cylinder. In both types, the circuit through the switch is open when the brake pedal is released. When the brakes are applied, the circuit through the switch closes and causes the stop lights to come on. Hydraulic brake light switches are not commonly used today because they provided one more area for possible fluid leaks.

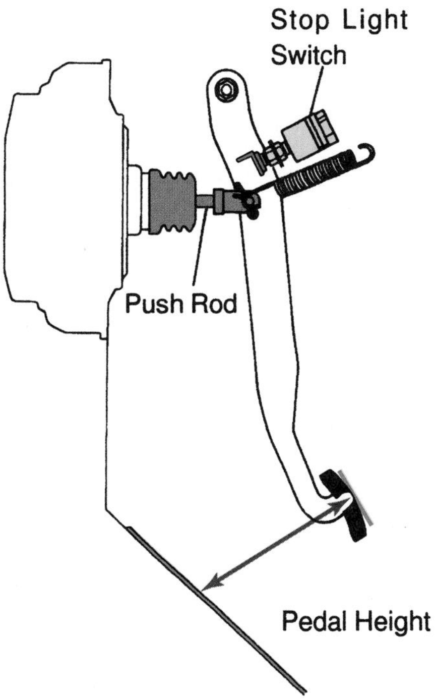

Figure 46-20 A mechanically activated stop light switch. *Reprinted with permission*

DRUM AND DISC BRAKE ASSEMBLIES

Although drum and disc brakes are explained in great detail in later chapters, a brief explanation of their components and operating principles is essential at this point.

Drum Brakes

A drum brake assembly consists of a cast-iron drum, which is bolted to and rotates with the vehicle's wheel, and a fixed backing plate to which the shoes, wheel cylinders, automatic adjusters, and linkages are attached **(Figure 46–21)**. Additionally, there might be some extra hardware for parking brakes. The shoes are surfaced with frictional linings, which contact the inside of the drum when the brakes are applied. The shoes are forced outward by pistons located inside the wheel cylinder. They are actuated by hydraulic pressure. As the drum rubs against the shoes, the energy of the moving drum is transformed into heat. This heat energy is passed into the atmosphere. When the brake pedal is released, hydraulic pressure drops and the pistons are pulled back to their unapplied position by return springs.

Disc Brakes

Disc brakes resemble the brakes on a bicycle: the friction elements are in the form of pads, which are squeezed or clamped about the edge of a rotating wheel. With automotive disc brakes, this wheel is a separate unit, called a **rotor**, inboard of the vehicle wheel **(Figure 46–22)**. The rotor is made of cast iron. Since the pads clamp against both sides of it, both sides are machined smooth. Usually the two surfaces are separated by a finned center section for better cooling (such rotors are called **ventilated rotors**). The pads are attached to metal shoes, which are actuated by pistons, the same as with drum brakes. The pistons are contained within a caliper assembly, a housing that wraps around the edge of the rotor. The caliper is kept from rotating by way of bolts holding it to the car's suspension framework.

The **caliper** is a housing containing the pistons and related seals, springs, and boots as well as the cylinders and fluid passages necessary to force the friction linings or pads against the rotor. The caliper resembles a hand in the way it wraps around the edge of the rotor. It is attached to the steering knuckle. Some models employ light spring pressure to keep the pads close against the rotor. In other caliper designs, this is achieved by a unique type of seal that allows the piston to be pushed out the necessary amount, then retracts it just enough to pull the pad off the rotor.

Unlike shoes in a drum brake, the pads act perpendicular to the rotation of the disc when the brakes are applied. This effect is different from that produced in a brake drum, where frictional drag actually pulls the shoe

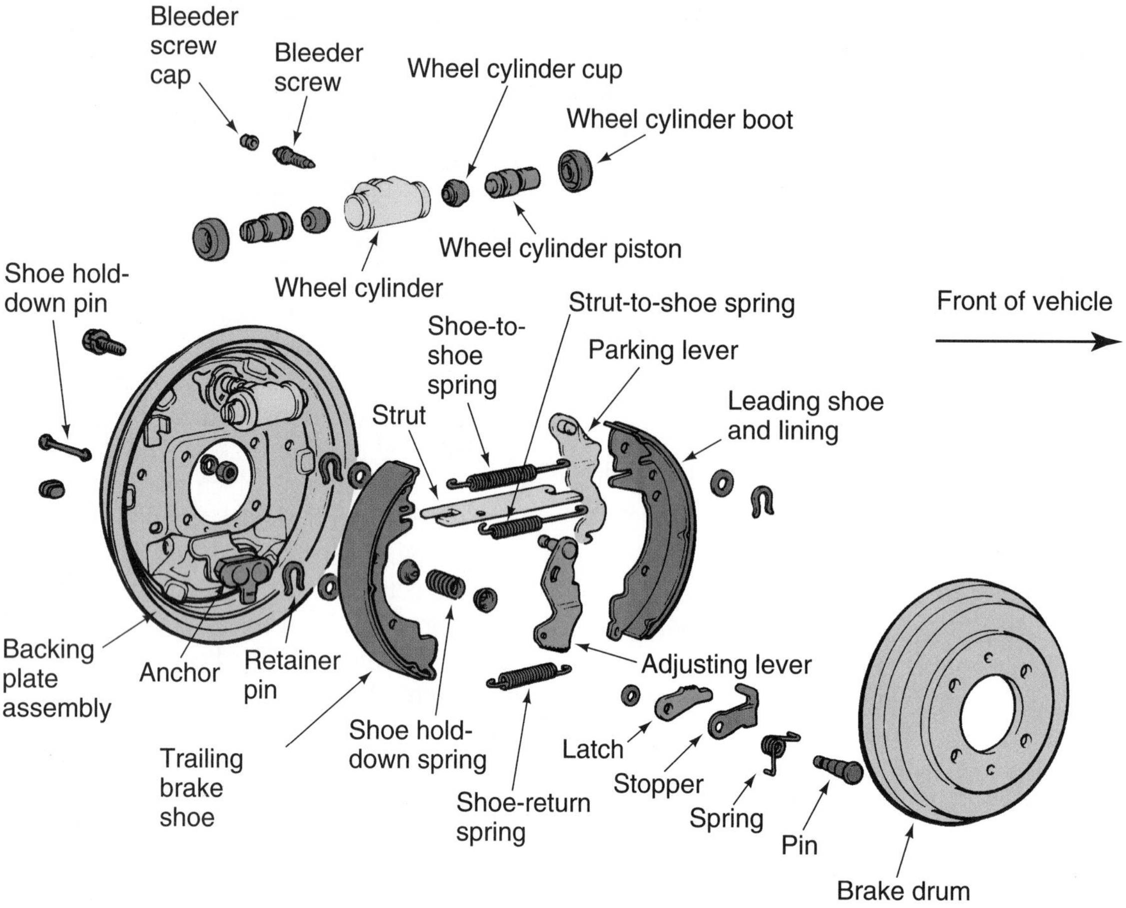

Figure 46-21 A typical drum brake assembly.

Figure 46-22 A disc brake assembly.

into the drum. Disc brakes are said to be nonenergized and so require more force to achieve the same braking effort. For this reason, they are ordinarily used in conjunction with a power brake unit.

HYDRAULIC SYSTEM SERVICE

Hydraulic system service is relatively uncomplicated, but it is vital to the vehicle's safe operation.

Brake Fluid Inspection

The master cylinder is usually located under the hood and near the fire wall on the driver's side.

WARNING!

Clean the cover before removal to avoid dropping dirt into the reservoir.

Remove the cover and check the gasket, or diaphragm **(Figure 46–23)**. Inspect the cover for damage or plugged vent holes. Clean the vent holes, if necessary.

Check the brake fluid level in the master cylinder. A cast-iron reservoir is usually filled to within ¼ inch (6 mm) of the top. A plastic reservoir may have fluid level marks. Do not overfill a reservoir. If fluid must be added, a leak probably has developed or the shoes and/or pads are worn. Check the system carefully to locate the leak.

To check for contaminated fluid, place a small amount of brake fluid in a clear glass jar. If the fluid is dirty or separates into layers, it is contaminated. Contaminated fluid must be replaced.

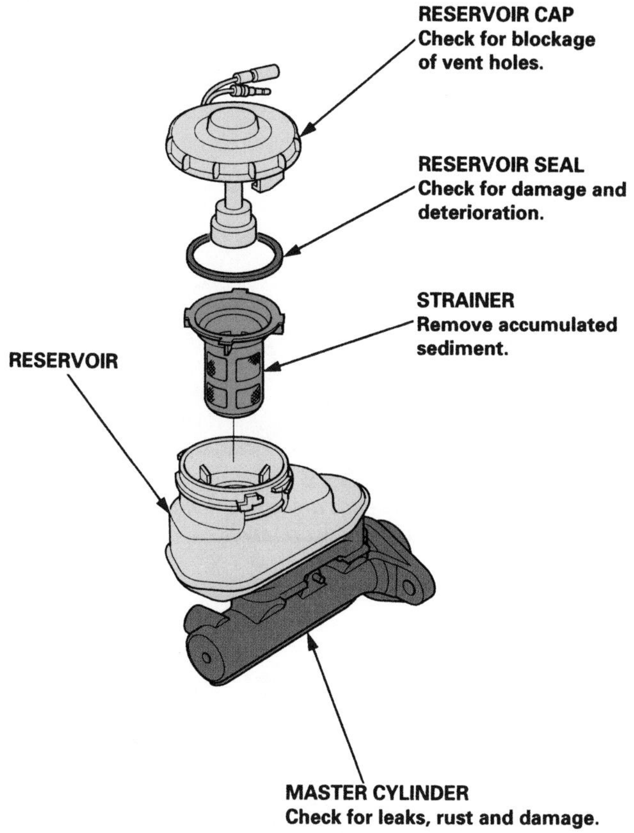
RESERVOIR CAP
Check for blockage of vent holes.
RESERVOIR SEAL
Check for damage and deterioration.
STRAINER
Remove accumulated sediment.
RESERVOIR
MASTER CYLINDER
Check for leaks, rust and damage.

Figure 46-23 When inspecting a master cylinder, make sure to carefully check these items. *Courtesy of American Honda Motor Co., Inc.*

Contaminated brake fluid can damage rubber parts and cause leaks. When replacing contaminated brake fluid, it is necessary to flush and refill the brake system with new fluid. Always use fluid with a DOT rating of 3 or higher. Follow manufacturer's recommendations.

Check the master cylinder for dampness and leaks around the body fittings, especially at the rear. A leak where it is mounted to the fire wall or power brake unit indicates a defective rear piston seal. The master cylinder must be rebuilt or replaced.

System Flushing

Currently, more than a dozen manufacturers specify periodic brake fluid changes for some, or all, of their models built during the past twelve years. Change intervals vary from as often as every twelve months or 15,000 miles to as infrequently as every 60,000 miles. All brake systems accumulate sludge over some period of time. Flushing the system can remove this sludge and any moisture, but once you have disturbed the sludge, you want to be sure you get it *all* out of the system. Stirring up sludge from the master cylinder reservoir may cause it to get into ABS valves and pumps if you do not get it all out of the system.

Brake hoses for disc brakes usually enter the caliper near the top of the caliper body. The bleeder valve is also located at the top of the caliper bore. If sludge accumulates in the caliper bore, it collects at the bottom. A quick, superficial bleeding of the caliper will not flush out the sludge and all of the old fluid. To flush a caliper thoroughly, pump several ounces of fluid through it. On some vehicles during brake pad replacement, you may want to remove the caliper from its mounts and retract the piston to force out all the old fluid. Then reinstall it and thoroughly flush it with fresh fluid.

Flushing should be done at each bleeder screw in the same manner as bleeding. Open the bleeder screw approximately 1 1/2 turns and force fluid through the system until the fluid emerges clear and uncontaminated. Do this at each bleeder screw in the system. After all lines have been flushed, bleed the system using one of the common bleeding procedures. All contaminated fluid should be drawn out of the master cylinder reservoir before bleeding. Make sure you dispose of the old brake fluid in the proper manner.

Brake Line Inspection Check all tubing, hoses, and connections from under the hood to the wheels for leaks and damage. Wheels and tires should also be inspected for signs of brake fluid leaks. Check all hoses for flexibility, bulges, and cracks. Check parking brake linkage, cable, and connections for damage and wear. Replace parts where necessary.

Brake Pedal Inspection Depress and release the brake pedal several times (engine running for power brakes). Check for friction and noise. Pedal movement should be smooth, with no squeaks from the pedal or brakes. The pedal should return quickly when it is released.

When operating the engine, be sure the transmission lever is in neutral or park. Be sure the area is properly ventilated for the exhaust to escape.

Apply heavy foot pressure to the brake pedal (engine running for power brakes). Check for a spongy pedal and pedal reserve. Spongy pedal action is springy. Pedal action should feel firm. **Pedal reserve** is the distance between the brake pedal and the floor after the pedal has been depressed fully. The pedal should not go lower than 1 or 2 inches (25 or 50 mm) above the floor.

With the engine off, hold light foot pressure on the pedal for about fifteen seconds. There should be no pedal movement during this time. Pedal movement indicates a leak. Repeat the procedure using heavy pedal pressure (engine running for power brakes).

If there is pedal movement but the fluid level is not low, the master cylinder has internal leakage. It must be rebuilt or replaced. If the fluid level is low, there is an external leak somewhere in the brake system. The leak must be repaired.

Depress the pedal and check for proper stop light operation.

PHOTO SEQUENCE 50

Typical Procedure for Bench Bleeding a Master Cylinder

P50-1 *Mount the master cylinder firmly in a vise, being careful not to apply excessive pressure to the casting. Position the master cylinder so the bore is horizontal.*

P50-2 *Connect short lengths of tubing to the outlet ports, making sure the connections are tight.*

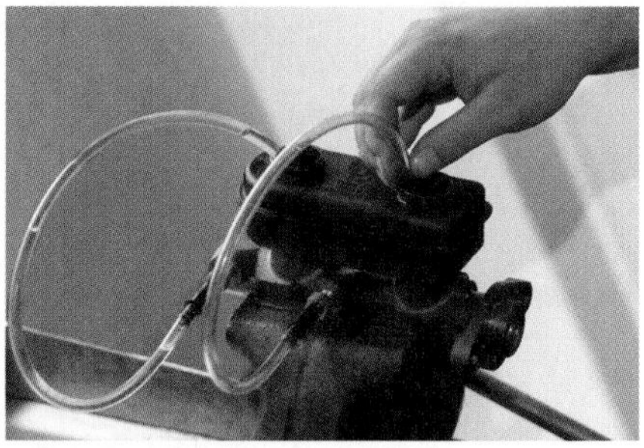

P50-3 *Bend the tubing lines so that the ends are in each chamber of the master cylinder reservoir.*

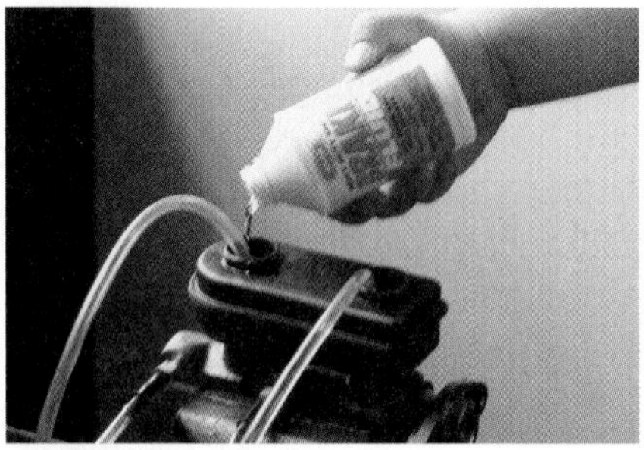

P50-4 *Fill the reservoirs with fresh brake fluid until the level is above the ends of the tubes.*

P50-5 *Using a wooden dowel or the blunt end of a drift or punch, slowly push on the master cylinder pistons until both are completely bottomed out in their bore.*

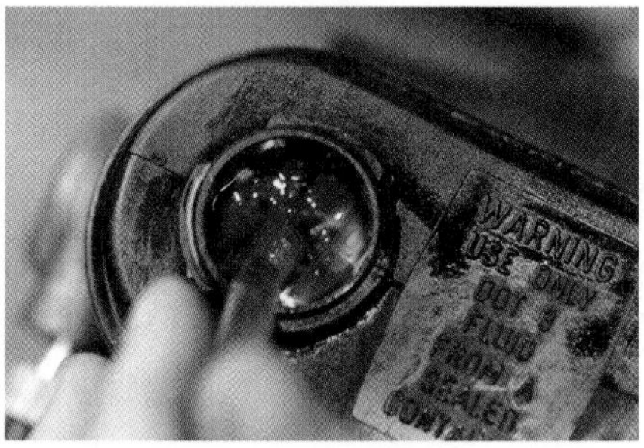

P50-6 *Watch for bubbles to appear at the tube ends immersed in the fluid. Slowly release the cylinder piston and allow it to return to its original position. On quick take-up master cylinders, wait 15 seconds before pushing in the piston again. On other units, repeat the stroke as soon as the piston returns to its original position. Slow piston return is normal for some master cylinders.*

To check power brake operation, depress and release the pedal several times while the engine is stopped. This eliminates vacuum from the system. Hold the brake down with moderate foot pressure and start the engine. If the power unit is operating properly, the brake pedal moves downward when the engine is started.

Master Cylinder Rebuilding A master cylinder is rebuilt to replace leaking seals or gaskets. If a more serious problem exists, the master cylinder should be replaced.

To remove a master cylinder, disconnect the brake lines at the master cylinder. Install plugs in the brake lines and master cylinder to prevent dirt from entering. Remove the nuts that attach the master cylinder to the fire wall power brake unit, and remove the cylinder.

Remove the cover and seal. Drain the master cylinder and carefully mount it in a vise. Remove the piston assembly and seals according to the manufacturer's instructions. New pistons, pushrods, and seals are usually included in rebuilding kits.

Clean master cylinder parts only with brake fluid, brake cleaning solvent, or alcohol. Do not use a solvent containing mineral oil, such as gasoline. Mineral oil is very harmful to rubber seals.

Inspect the master cylinder. Damage, cracks, porous leaks, and worn piston bores mean the master cylinder must be replaced. Check very carefully for pitting or

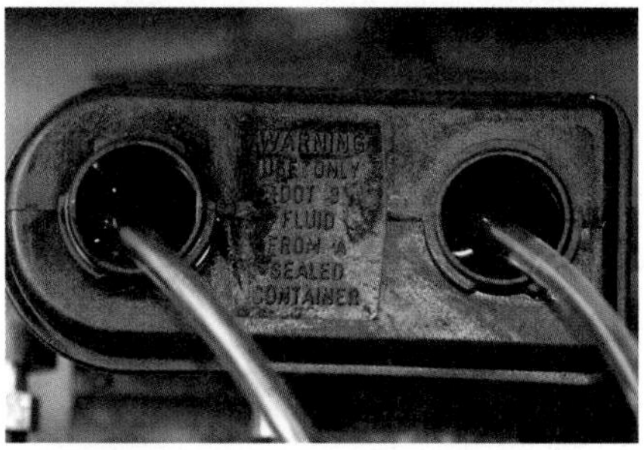

P50-7 *Pump the cylinder piston until no bubbles appear in the fluid.*

P50-8 *Remove the tubes from the outlet ports and plug the openings with temporary plugs or your fingers. Keep the ports covered until you install the master cylinder on the vehicle.*

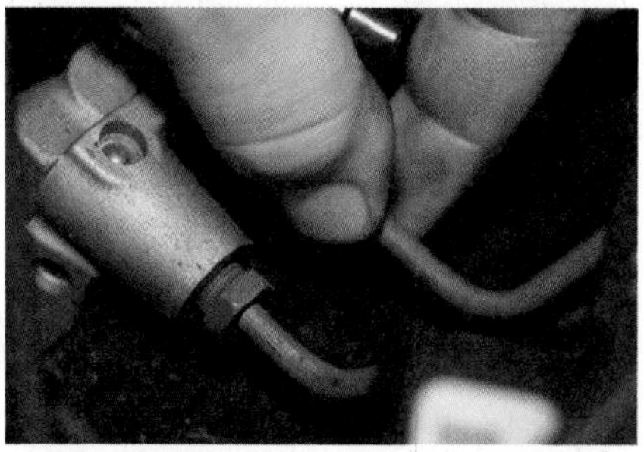

P50-9 *Install the master cylinder on the vehicle. Attach the lines, but do not tighten the tube connections.*

P50-10 *Slowly depress the pedal several times to force out any air that might be trapped in the connections. Before releasing the pedal, tighten the nut slightly and loosen it before depressing the pedal each time. Soak up the fluid with a rag to avoid damaging the car finish.*

P50-11 *When there are no air bubbles in the fluid, tighten the connections to the manufacturer's specifications. Make sure the master cylinder reservoirs are adequately filled with brake fluid.*

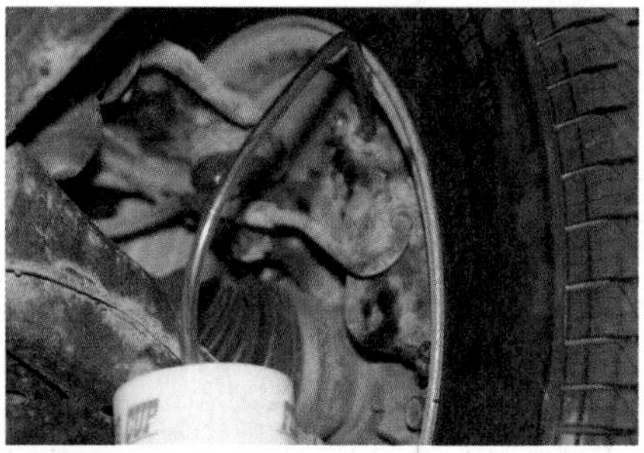

P50-12 *After reinstalling the master cylinder, bleed the entire brake system on the vehicle.*

roughness in the bore. If any are present, the cylinder must be replaced.

Reassemble, install, and bleed the master cylinder according to the manufacturer's directions. Photo Sequence 50 is a typical procedure for bench bleeding a master cylinder.

WARNING!

Brake fluid will remove paint. Always use fender covers to protect the vehicle's finish and take extra care not to spill brake fluid.

Hydraulic System Bleeding

Fluids cannot be compressed, whereas gases are compressible. Any air in the brake hydraulic system is compressed as the pressure increases. This action reduces the amount of force that can be transmitted by the fluid, therefore it is very important to keep all air out of the hydraulic system. To do this, air must be bled from brakes. This procedure is called **bleeding** the brake system.

Bleeding is a process of forcing fluid through the brake lines and out through a bleeder valve or bleeder screw **(Figure 46–24)**. The fluid eliminates any air that might be in the system. Bleeder screws and valves are fastened to the wheel cylinders or calipers. The bleeder

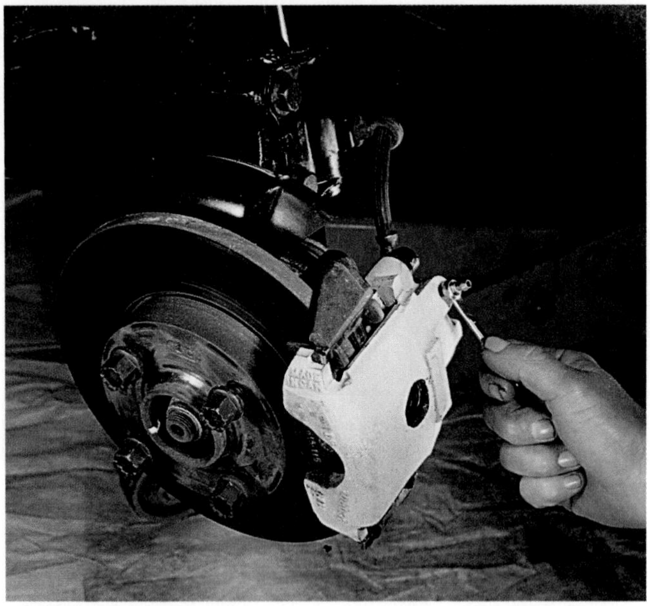

Figure 46–24 The wheel brake units are fitted with bleeder screws.

must be cleaned. A drain hose then is connected from the bleeder to a glass jar **(Figure 46–25)**.

CUSTOMER CARE

Remind your customers to use only approved brake fluid in a brake system. Any other lubricant, such as power steering fluid, automatic transmission fluid, or engine oil, which has a petroleum base, must never be used in the brake system. Petroleum-based fluids attack the rubber components in the brake system, like the piston cups and seals, and cause them to swell and disintegrate. If you open a brake fluid reservoir and find the diaphragm is swelled out of shape and is larger than it should be, you can be sure someone has previously put some kind of petroleum-based fluid in it. Such a discovery is costly to the vehicle's owner as all rubber components in the entire brake system will need to be replaced. ■

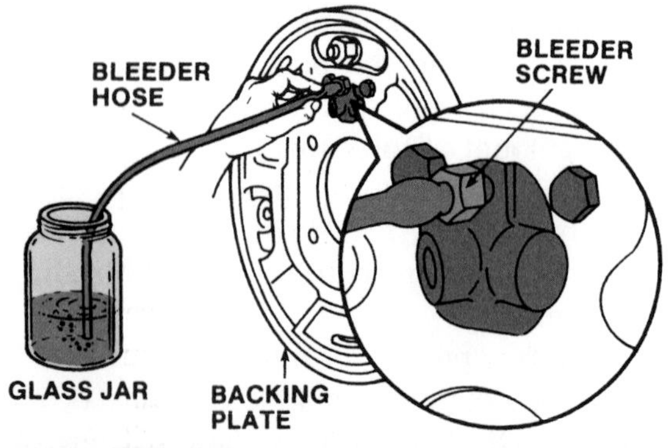

Figure 46–25 While bleeding the brake system, releasing the fluid into a glass jar will help to determine when the system is free of air.

Two types of brake bleeding procedures are used: manual bleeding and pressure bleeding. On some antilock brake systems, a scan tool can be used to help bleed the brakes. Always follow the manufacturer's recommendations when bleeding brakes. The sequence in which bleeding is performed can be critical. When bleeding a power brake system, remove the vacuum line from the power unit and plug the unit. To remove vacuum, the engine must be off. Pump the brake pedal several times.

WARNING!

Always use fresh brake fluid when bleeding the system. Do not use fluid that has been drained. Drained fluid may be contaminated and can damage the system.

Manual Bleeding A manual bleeding procedure requires two people. One person operates the bleeder; the other, the brake pedal. Bleed only one wheel at a time.

WARNING!

Be sure the bleeder hose is below the surface of the liquid in the jar at all times. Do not allow the master cylinder to run out of fluid at any time. If these precautions are not followed, air can enter the system, and it must be bled again. The master cylinder cover must be kept in place.

Place the bleeder hose and jar in position. Have a helper pump the brake pedal several times and then hold it down with moderate pressure. Slowly open the bleeder valve. After fluid/air has stopped flowing, close the bleeder valve. Have the helper slowly release the pedal. Repeat this procedure until fluid that flows from the bleeder is clear and free of bubbles.

Discard all used brake fluid. Fill the master cylinder reservoir. Check the brakes for proper operation.

WARNING!

Clean the master cylinder and cover before adding fluid. This is important for preventing dirt from entering the reservoir.

Pressure Bleeding A pressure bleeding procedure can be done by one person. Pressure bleeding equipment uses pressurized fluid that flows through a special adapter fitted into the master cylinder **(Figure 46–26)**.

The use of pressure bleeding equipment varies with different automobiles and different equipment makers.

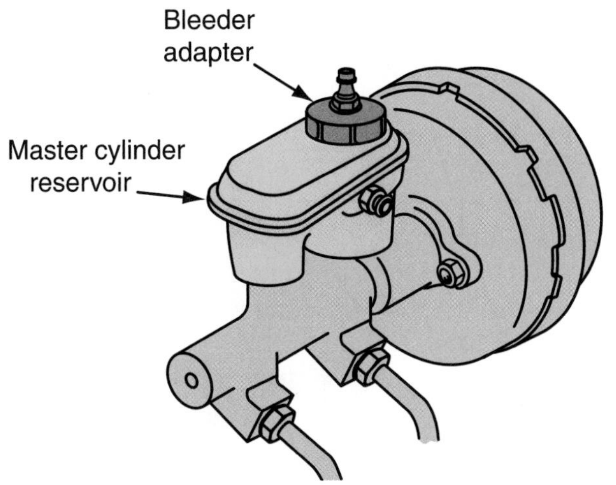

Figure 46-26 An adapter that installs onto the reservoir's fill opening is required to pressure the brake system.

Always follow the automobile manufacturer's recommendations when using pressure bleeding equipment.

On automobiles with metering valves, the valve must be held open during pressure bleeding. A special tool is used to hold open the metering section of a combination valve.

USING SERVICE MANUALS

Consult the service manual to be sure the proper bleeding sequence is followed. If a vehicle requiring a special sequence is bled in the conventional manner, air might be chased throughout the system. ■

Open the bleeder valves one at a time until clear, air-free fluid is flowing. Progress from the wheel cylinder farthest from the master cylinder to the cylinder closest.

Do not exceed recommended pressure while bleeding the brakes. Always release air pressure after bleeding. Clean and fill the master cylinder after pressure bleeding. Check the brakes for proper operation. Be sure to remove the special tool used to hold the metering valve.

Power Brakes

Power brakes are nothing more than a standard hydraulic brake system with a booster unit located between the brake pedal and the master cylinder to help activate the brakes.

Two basic types of power-assist mechanisms are used. The first is **vacuum assist**. These systems use engine vacuum, or sometimes vacuum pressure developed by an external vacuum pump, to help apply the brakes. The second type of power assist is **hydraulic assist**. It is normally found on larger vehicles. This system uses hydraulic pressure developed by the power steering pump or other external pump to help apply the brakes.

Both vacuum and hydraulic assist act to multiply the force exerted on the master cylinder pistons by the driver. This increases the hydraulic pressure delivered to the wheel cylinders or calipers while decreasing driver foot pressure.

Vacuum-Assist Power Brakes

All vacuum-assisted units are similar in design. They generate application energy by opposing engine vacuum to atmospheric pressure. A piston and cylinder, flexible diaphragm, or bellows use this energy to provide braking assistance.

All modern vacuum-assist units are vacuum-suspended systems. This means the diaphragm inside the unit is balanced using engine vacuum until the brake pedal is depressed. Applying the brake allows atmospheric pressure to unbalance the diaphragm and allows it to move generating application pressure.

Atmospheric pressure is normally between 14 and 15 psi (96.5 and 103 kPa). If the diameter of the diaphragm is 12 inches (305 mm), the area of the diaphragm is about 113 square inches (72,907 sq. mm). Since the vacuum to the booster is typically 17 inches of mercury (432 mm Hg) or about 7 psi (48 kPa), the pressure differential on the diaphragm is 7.7 psi (53 kPa). Therefore the resulting force on the diaphragm would be 870 pounds (3870 N) (7.7×113).

Vacuum boosters may be single diaphragm or tandem diaphragm. The unit consists of three basic elements combined into a single power unit **(Figure 46–27)**.

The three basic elements of the single diaphragm follow:

1. A vacuum power section that includes a front and rear shell, a power diaphragm, a return spring, and a pushrod.
2. A control valve built as an integral part of the power diaphragm and connected through a valve rod to the brake pedal. It controls the degree of brake application or release in accordance with the pressure applied to the brake pedal.
3. A hydraulic master cylinder, attached to the vacuum power section that contains all the elements of the conventional brake master cylinder except for the pushrod. It supplies fluid under pressure to the pressure applied by the brake booster.

Operation When the brakes are applied, the valve rod and plunger move to the left in the power diaphragm. This action closes the control valve's vacuum port and opens the atmospheric port to admit air through the valve at the rear diaphragm chamber. With vacuum in the rear chamber, a force develops that moves the power diaphragm, hydraulic pushrod, and hydraulic piston or

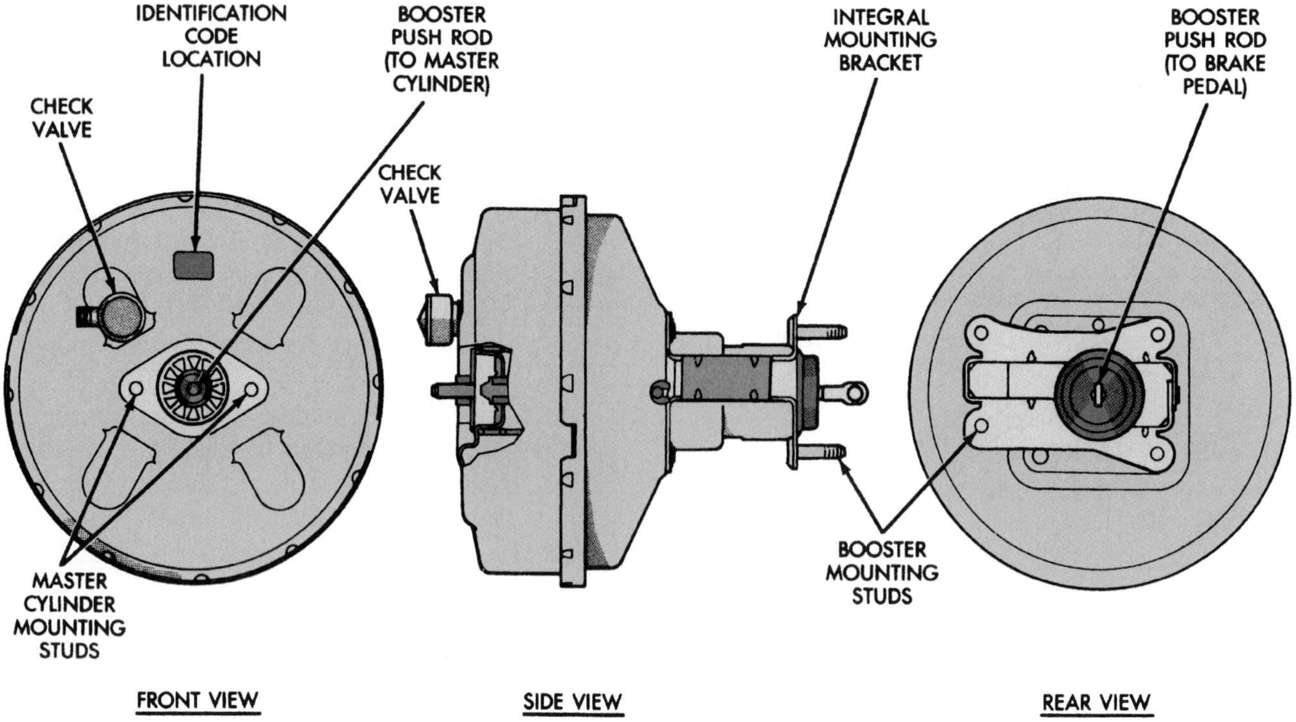

Figure 46-27 A typical vacuum brake booster. *Courtesy of DaimlerChrysler Corporation*

pistons to close the compensating port or ports and force fluid under pressure through the residual check valve or valves and lines into the front and rear brake assemblies.

As pressure develops in the master cylinder, a counterforce acts through the hydraulic pushrod and reaction disc against the vacuum power diaphragm and valve plunger. This force tends to close the atmospheric port and reopen the vacuum port. Since this force is in opposition to the force applied to the brake pedal by the operator, it gives the operator a feel for the amount of brake applied.

Servicing Vacuum-Assist Booster Units The fact that a vehicle's brakes still operate when the vacuum-assist unit fails indicates that the hydraulic brake system and the vacuum-assist system are two separate systems. This means you should always check for faults in the hydraulic system first. If it checks out satisfactorily, start inspecting the vacuum-assist circuit.

For a fast check of vacuum-assist operation, press the brake pedal firmly and then start the engine. The pedal should fall away slightly and less pressure should be needed to maintain the pedal in any position.

Pressure Check. Another simple check can be made by installing a suitable pressure gauge in the brake hydraulic system. Take a reading with the engine off and the power unit not operating. Maintain the same pedal height, start the engine, and read the gauge. There should be a substantial pressure increase if the vacuum-assist booster is operating correctly.

Pedal Travel. **Pedal travel** and total travel are critical on vacuum-assisted vehicles. Pedal travel should be kept strictly to specifications listed in the vehicle's service manual.

Vacuum Reading. If the power unit is not giving sufficient assistance, take a manifold vacuum reading. If manifold vacuum level is below specifications, tune the engine and retest the unit. Loose or damaged vacuum lines and clogged air intake filters reduce braking assistance. Most units have a check valve that retains some vacuum in the system when the engine is off. A vacuum gauge check of this valve indicates if it is restricted or stays open.

Release Problems. Failure of the brakes to release is often caused by a tight or misaligned connection between the power unit and the brake linkage. Broken pistons, diaphragms, bellows, or return springs can also cause this problem.

To help pinpoint the problem, loosen the connection between the master cylinder and the brake booster. If the brakes release, the problem is caused by internal binding in the vacuum unit. If the brakes do not release, look for a crimped or restricted brake line or similar problem in the hydraulic system.

Hard Pedal. Power brakes that have a **hard pedal** may have collapsed or leaking vacuum lines of insufficient manifold vacuum. Punctured diaphragms or bellows and leaky piston seals all lead to weak power unit operation and hard pedal. A steady hiss when the brake is held down indicates a vacuum leak that causes poor operation.

Grabbing Brakes. First, look for all the usual causes of brake grab, such as greasy linings, or scored rotors or drums. If the trouble appears to be in the power unit, check for a damaged reaction control. The reaction control is made up of a diaphragm, spring, and valve that tend to resist pedal action. It is put into the system to give the driver more brake pedal feel.

Check of Internal Binding. Release problems, hard pedal, and dragging (slow releasing) brakes can all be caused by internal binding. To test a vacuum unit for internal binding, place the transmission/ transaxle in neutral and start the engine. Increase engine speed to 1,500 rpm, close the throttle, and completely depress the brake pedal. Slowly release the brake pedal and stop the engine. Remove the vacuum check valve and hose from the vacuum assist unit. Observe for backward movement of the brake pedal. If the brake pedal moves backward, there is internal binding and the unit should be replaced.

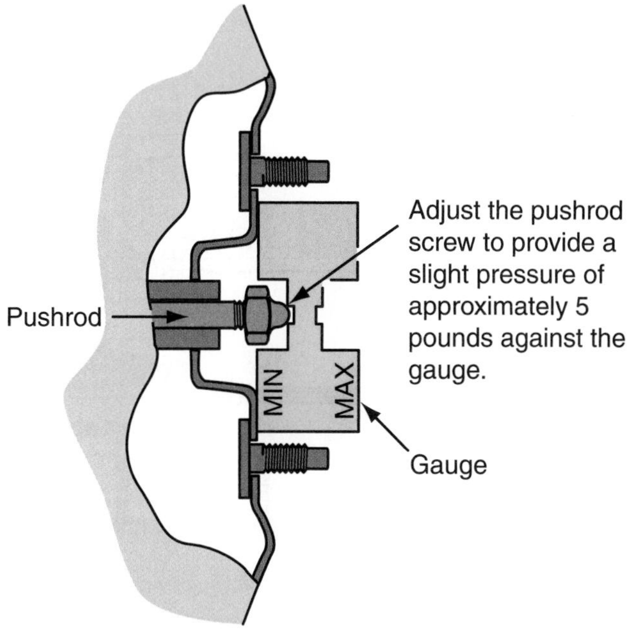

Figure 46-28 A gauge for measuring pushrod length.

PUSHROD ADJUSTMENT

Proper adjustment of the master cylinder pushrod is necessary to ensure proper operation of the power brake system. A pushrod that is too long causes the master cylinder piston to close off the replenishing port, preventing hydraulic pressure from being released and resulting in brake drag. A pushrod that is too short causes excessive brake pedal travel and causes groaning noises to come from the booster when the brakes are applied. A properly adjusted pushrod that remains assembled to the booster with which it was matched during production should not require service adjustment. However, if the booster, master cylinder, or pushrod are replaced, the pushrod might require adjustment.

Two methods can be used to check for proper pushrod length and installation: the gauge method and the air method.

Gauge Method

In most vacuum power units, the master cylinder pushrod length is fixed, and length is usually checked only after the unit has been overhauled or replaced. A typical adjustment using the gauge method is shown in **Figure 46–28**.

Air Method

The air-testing method uses compressed air applied to the hydraulic outlet of the master cylinder. Air pressure is regulated to a value of approximately 5 psi (35 kPa) to prevent brake fluid spraying from the master cylinder.

If air passes through the replenishing port, which is the smaller of the two holes in the bottom of the master cylinder reservoir, the adjustment is satisfactory. If air does not flow through the replenishing port, adjust the pushrod as required, either by means of the adjustment screw (if provided) or by adding shims between the master cylinder and power unit shell until the air flows freely.

HYDRAULIC BRAKE BOOSTERS

Decreases in engine size, plus the continued use of engine vacuum to operate other engine systems, such as emission control devices, led to the development of hydraulic-assist power brakes. These systems use fluid pressure, not vacuum pressure, to help apply the brakes.

Fluid pressure from the power-steering pump provides the power assist to the brakes. The power brake booster is located between the cowl and the master cylinder. Hoses connect the power-steering pump to the booster assembly.

The power-steering pump provides a continuous flow of fluid to the brake booster whenever the engine is running. Three flexible hoses route the power-steering fluid to the booster. One hose supplies pressurized fluid from the pump. Another hose routes the pressurized fluid from the booster to the power-steering gear assembly. The third hose returns fluid from the booster to the power-steering pump.

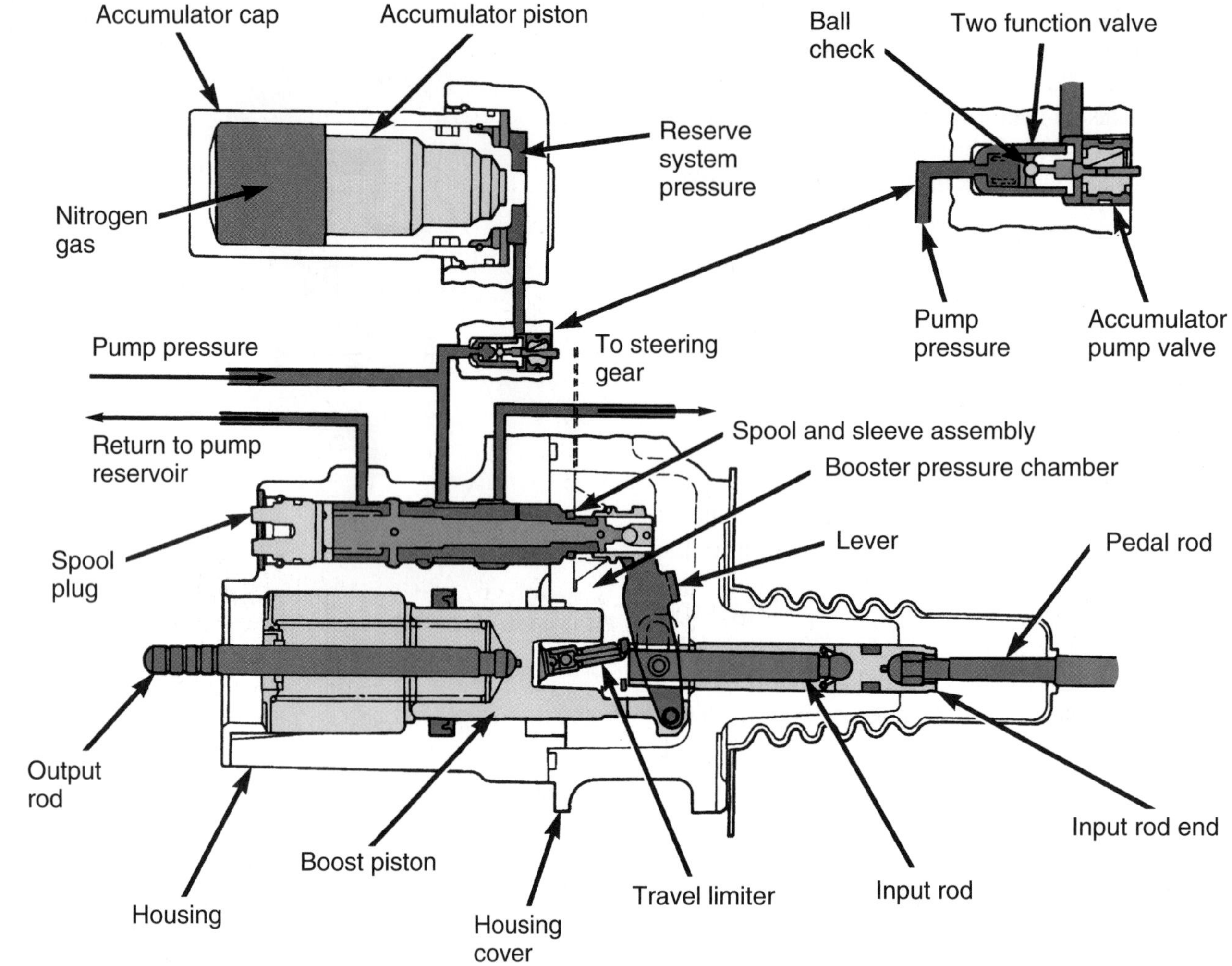

Figure 46-29 A hydraulic brake booster. *Courtesy of DaimlerChrysler Corporation*

The hydraulic pressure in the hydraulic booster should not be confused with the hydraulic pressure in the brake lines. Remember that they are two separate systems and require two different types of fluid: power-steering fluid for the pump and brake fluid for the brake system. Never put power-steering fluid in the brake reservoir. If the brake fluid becomes contaminated by power-steering fluid, the brake system must be flushed.

Some systems have a nitrogen charged pneumatic accumulator on the booster to provide reserve power-assist pressure. If power-steering pump pressure is not available due to belt failure or similar problems, the accumulator pressure is used to provide brake assist.

The booster assembly **(Figure 46–29)** consists of an open center spool valve and sleeve assembly, a lever assembly, an input rod assembly, a power piston, an output pushrod, and the accumulator. The booster assembly is mounted on the vehicle in much the same manner as a vacuum booster. The pedal rod is connected at the booster input rod end.

Power-steering fluid flow in the booster unit is controlled by a hollow center spool valve. The spool valve has lands, annular grooves, and drilled passages. These mate with grooves and lands in the valve bore. The flow pattern of the fluid depends on the alignment of the valve in the bore.

Operation

When the brake pedal is depressed, the pedal's pushrod moves the master cylinder's primary piston forward. This causes the lever assembly of the booster to move a sleeve forward to close off the holes leading to the open center of the spool valve. A small additional lever movement moves the spool valve into the spool valve bore. The spool valve then diverts some hydraulic fluid into the cavity behind the booster piston building up hydraulic pressure that moves the piston and a pushrod forward. The output pushrod moves the primary and secondary master cylinder pistons that apply pressure to the brake system. When the brake pedal is released, the spool and sleeve assemblies return to their normal positions. Excess fluid behind the piston returns to the power-steering pump reservoir through the return line. After the brakes have been released, pressurized fluid from the power-steering pump flows into the booster through the open center of the spool valve and back to the power-steering pump.

There have been many variations of and names given to hydraulic brake booster systems. The most common names are the hydro-boost system produced by Bendix and the Powermaster system produced by General Motors. The Powermaster system is a little different than the rest in that it uses a self-contained hydraulic booster that is built directly onto the master cylinder. Instead of relying on the power-steering pump for hydraulic pressure as is done in the other systems, the Powermaster has its own vane pump and electric motor to provide the hydraulic pressure required for booster operation.

Any investigation of a hydraulic boost complaint should begin with an inspection of the power-steering pump belt, fluid level, and hose condition and connections. Hydraulic boost systems do not work properly if they are not supplied with a continuous supply of clean, bubble-free power-steering fluid at the proper pressure.

WARNING!

Always depressurize the accumulator of any hydraulic boost system before disconnecting any brake lines or hoses. This is usually done by turning the engine off and depressing and releasing the brake pedal up to ten times.

Basic Operational Test

The basic operational test of these systems is as follows. With the engine off, pump the brake pedal numerous times to bleed off the residual hydraulic pressure that is stored in the accumulator. Hold firm pressure on the brake pedal and start the engine. The brake pedal should move downward, then push up against the foot.

Accumulator Test

To be sure the **accumulator** is performing properly, rotate the steering wheel with the engine running until it stops and hold it in that position for no more than 5 seconds. Return the steering wheel to the center position and shut off the engine. Pump the brake pedal. You should feel two to three power-assisted strokes. Now repeat the steps. That pressurizes the accumulator. Wait one hour, then pump the brake pedal. There should be two or three power-assisted strokes. If the system does not perform as just described, the accumulator is leaking and should be replaced.

Noise Troubleshooting

The booster is also part of another major subsystem of the vehicle, the power-steering system. Problems or malfunctions in the steering system may affect brake-assist operation. The following are some common troubleshooting tips.

Moan or low-frequency hum usually accompanied by a vibration in the pedal or steering column might be encountered during parking or other very low-speed maneuvers. This can be caused by a low fluid level in the power-steering pump, or by air in the power-steering fluid due to holding the pump at relief pressure (steering wheel held all the way in one direction) for an excessive amount of time (more than five seconds). Check the fluid level and add fluid if necessary. Allow the system to sit for one hour with the cap removed to eliminate the air. If the condition persists, it might be a sign of excessive pump wear. Check the pump according to the vehicle manufacturer's recommended procedure.

At or near power runout (brake pedal near fully depressed position), a high-speed fluid noise (like a faucet can make) might occur. This is a normal condition and will not be heard, except in emergency braking conditions.

Whenever the accumulator pressure is used, a slight hiss is noticed. It is the sound of the hydraulic fluid escaping through the accumulator valve and is completely normal.

After the accumulator has been emptied and the engine is started again, another hissing sound might be heard during the first brake application or the first steering maneuver. This sound is caused by the fluid rushing through the accumulator charging orifice. It is normal and will only be heard once after the accumulator is emptied. However, if this sound continues even though no apparent accumulator pressure assist was made, it could indicate that the accumulator is not holding pressure. Check for this possibility using the accumulator test discussed previously.

After bleeding, a gulping sound might be present during brake applications, as noted in the bleeding instructions. This sound is normal and should disappear with normal driving and braking.

Diagnosis and testing of the Powermaster unit requires the use of a special adapter and test gauge or aftermarket equivalents. The Powermaster pressure switch is removed and the adapter and test gauge is installed in its port. The unit can then be energized, and the switch's high-pressure cut-off and low-pressure turn-on points observed and checked against specifications. Follow service manual instructions for the connection and operation of the test gauge and all system test procedures.

CASE STUDY

A customer brings her late-model vehicle to a brake specialty shop, complaining of occasional rear brake lockup when the brakes are lightly applied. She is nervous because she feels the vehicle is unsafe to drive. The vehicle is equipped with front disc and rear drum brakes.

A thorough road test by the technician does not verify the complaint. Talking to the customer, the technician learns the problem most often occurs in damp or cold weather or when the vehicle has been sitting for a length of time.

The technician performs a complete visual and operational inspection. He checks the action of the brake pedal, the power brake booster, and the combination valve. All systems are working fine. He then removes the rear drums and inspects the shoes and drums. Nothing appears out of order or broken. The technician turns to his supervisor for a second opinion.

After listening to a summary of the inspection and checks made to date, the shop supervisor checks the file on factory service bulletins for that model vehicle. The answer is found in a bulletin issued about a year after the vehicle was introduced. The OEM brake shoes were found to have a tendency to swell when subjected to cold and wet. The factory recommended another type of shoe should be installed to correct the problem. If the technician had the experience to check the service bulletin file on the vehicle, valuable diagnostic time would have been saved.

The correct shoe is installed, and the vehicle is road tested. The rear brakes no longer lock up.

KEY TERMS

Accumulator
Bleeding
Caliper
Coefficient of friction (COF)
Combination valve
Diagonally split system
DOT
Equilibrium reflux boiling point (ERBP)
Hard pedal
Hydraulic assist
ISO
Kinetic friction
Metallic
Metering valve
Nonasbestos
Pedal reserve
Pedal travel
Proportioning valve
Quick-take-up
Residual check valve
Rotor
Semimetallic
Static friction
Vacuum assist
Ventilated rotor

SUMMARY

- The four factors that determine a vehicle's braking power are pressure, which is provided by the hydraulic system; coefficient of friction, which represents the frictional relationship between pads and rotors or shoes and drums and is engineered to ensure optimum performance; frictional contact surface, meaning that bigger brakes stop a car more quickly than smaller brakes; and heat dissipation, which is necessary to prevent brake fade.
- With asbestos banned as a brake lining material, today full metallics and semimetallics are used, as well as other nonasbestos substances.
- Since 1967, all cars have been required to have two separate brake systems. The dual brake system uses a tandem master cylinder, which is two master cylinders formed by installing two separated pistons and fluid reservoirs in one cylinder bore.
- Brake fluid is the lifeblood of any hydraulic brake system. DOT 4 is the type most commonly used on conventional systems. DOT 3 is recommended for most ABS and some power brake systems.
- The brake lines transmit brake fluid pressure from the master cylinder to the wheel cylinders and calipers of drum and disc brakes. Brake hoses offer flexible connections to wheel units and must offer high heat resistance.
- A pressure differential valve is used in all dual brake systems to operate a warning light switch that alerts the driver if pressure is lost in either hydraulic system.
- The metering valve, located in the front brake line, provides for better balance of the front and rear brakes, while also preventing lockup of the front brakes. The proportioning valve controls rear brake pressure, particularly during hard stops.
- Hydraulic brake system service is vital to safe vehicle operation. It includes brake fluid, brake line, and brake pedal inspection.
- Bleeding removes air from the hydraulic system.
- Power brakes can be either vacuum assist or hydraulic assist. Vacuum-assisted units use engine vacuum or vacuum developed by an external pump to help apply the brakes. Hydraulic-assisted units use fluid pressure.

TECH MANUAL

The following procedures are included in Chapter 46 of the *Tech Manual* that accompanies this book:

1. Bench bleeding a master cylinder.
2. Pressure bleeding a hydraulic brake system.
3. Performing power vacuum brake booster test.

REVIEW QUESTIONS

1. Explain why bleeding air out of a hydraulic system is so important.
2. Explain why modern hydraulic braking systems are dual designs and why this is important.
3. Define the different types of power brake designs and tell how they operate.
4. Describe the purpose of a pressure differential valve.
5. Describe the functions of the hydraulic system combination valve.
6. The friction that is generated between a vehicle's tires and the road when the vehicle is in a non-skidding state is called ______ friction.
   a. static
   b. kinetic
   c. heat
   d. coefficient
7. Bleeding a hydraulic system involves ______ from the system.
   a. removing water
   b. removing contaminated brake fluid
   c. removing air pockets
   d. filtering out dirt
8. The purpose of the master cylinder is to ______.
   a. generate the hydraulic pressure needed to apply the brake mechanisms
   b. automatically pump the brakes during panic stops
   c. apply braking power when wheel slippage occurs
   d. all of the above
9. After bleeding a hydraulic system, Technician A examines the brake fluid and returns it to the master cylinder if the fluid is clean. Technician B always refills the master cylinder with new brake fluid. Who is correct?
   a. Technician A
   b. Technician B
   c. Both A and B
   d. Neither A nor B
10. Which of the following can lead to brake hose failure?
    a. improperly matched fittings
    b. stressing the hose during installations
    c. deterioration from heat and contaminants
    d. all of the above
11. *True or False?* Metering and proportioning valves balance the braking characteristics of disc and drum brakes.
12. Which type of brake requires greater application force and is commonly used with power-boost units?
    a. drum
    b. disc
    c. both a and b
    d. neither a nor b
13. Which of the following is not a factor in determining the effectiveness of a brake system?
    a. heat dissipation
    b. lubricant
    c. pressure
    d. frictional contact area
14. A vehicle's power brakes are grabbing. Technician A says the most likely cause is the power brake booster. Technician B says a likely cause is greasy linings or scored drums. Who is correct?
    a. Technician A
    b. Technician B
    c. Both A and B
    d. Neither A nor B
15. While discussing what affects the amount of pressure exerted by the brakes, Technician A says the shorter the line, the more pressure there will be. Technician B says braking force will increase if the size of the pistons in a master cylinder are increased. Who is correct?
    a. Technician A
    b. Technician B
    c. Both A and B
    d. Neither A nor B
16. Which of the following is *not* a component of a typical dual master cylinder?
    a. primary piston
    b. piston stop bolt
    c. pressure relief valve
    d. secondary cup

17. The metering valve portion of a combination valve fails. Technician A says this means the entire combination valve must be replaced. Technician B says the metering part can be repaired. Who is correct?

 a. Technician A
 b. Technician B
 c. Both A and B
 d. Neither A nor B

18. While discussing quick-take-up master cylinders, Technician A says this design allows for increased braking power. Technician B says this design is only used on drum brake systems. Who is correct?

 a. Technician A
 b. Technician B
 c. Both A and B
 d. Neither A nor B

19. While bleeding a brake system, Technician A loosens the brake line fitting at the master cylinder if a bleeder screw is seized and cannot be loosened. Technician B uses shop air to push the fluid and air from the wheel units to the master cylinder. Who is correct?

 a. Technician A
 b. Technician B
 c. Both A and B
 d. Neither A nor B

20. Which of the following could cause an extremely hard brake pedal?

 a. air in the system
 b. excessively worn brake pads
 c. use of the wrong fluid
 d. a leaking diaphragm in the vacuum power booster

47

DRUM BRAKES

OBJECTIVES

■ Explain how drum brakes operate. ■ Identify the major components of a typical drum brake and describe their functions. ■ Explain the difference between duo-servo and nonservo drum brakes. ■ Perform a cleaning and inspection of a drum brake assembly. ■ Recognize conditions that adversely affect the performance of drums, shoes, linings, and related hardware. ■ Reassemble a drum brake after servicing. ■ Explain how typical drum parking brakes operate. ■ Adjust a typical drum parking brake.

For many years, drum brakes **(Figure 47–1)** were used on all four wheels on virtually every vehicle on the road. Today, disc brakes have replaced drum brakes on the front wheels of most vehicles and some models are equipped with both front and rear disc brakes. One reason for their continued use is that drum brakes can easily handle the 20 to 40% of total braking load placed on the rear wheels. Another is that drum brakes can also be built with a simple parking brake mechanism.

DRUM BRAKE OPERATION

Drum brake operation is fairly simple. The most important feature contributing to the effectiveness of the braking force supplied by the drum brake is the brake shoe pressure or force directed against the drum **(Figure 47–2)**. With the vehicle moving in either the forward or reverse direction with the brakes on, the applied force of the brake shoe pressing against the brake drum increasingly multiplies itself because the brake's anchor pin acts as a brake shoe stop and prohibits the brake shoe from its tendency to follow the movement of the rotating drum. The result is a wedging action between the brake shoe and brake drum. The wedging action combined with the applied brake force creates a self-multiplied brake force.

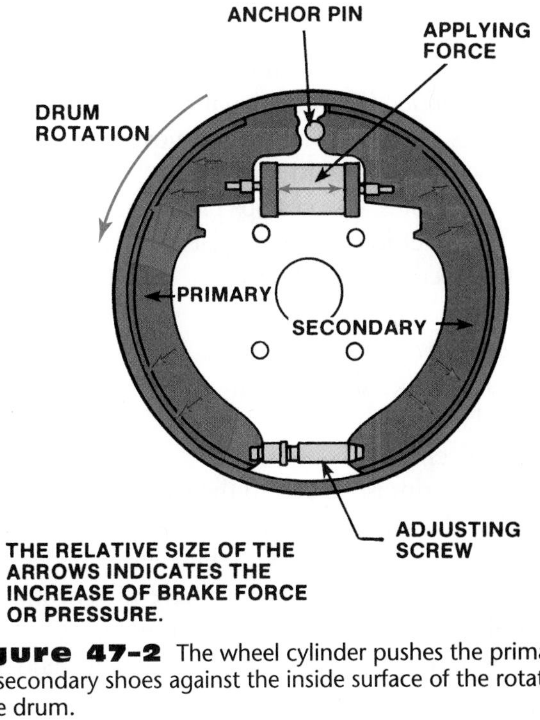

Figure 47–2 The wheel cylinder pushes the primary and secondary shoes against the inside surface of the rotating brake drum.

Figure 47–1 A drum brake assembly.

DRUM BRAKE COMPONENTS

The **backing plate** provides a foundation for the brake shoes and associated hardware. The plate is secured and bolted to the axle flange or spindle. The wheel cylinder, under hydraulic pressure, forces the brake's shoes against the drum. There are also two linked brake shoes attached to the backing plate. Brake shoes are the backbone of a drum brake. They must support the lining and carry it into the drum so the pressure is distributed across the lining surface during brake application. Shoe return springs and shoe hold-down parts maintain the correct shoe position and clearance. Some drum brakes are self-adjusting. Others require manual adjustment mechanisms. Brake drums provide the rubbing surface area for the linings. Drums must withstand high pressures without excessive flexing and must also dissipate large quantities of heat generated during brake application. Finally, the rear drum brakes on most vehicles include the parking brakes.

Wheel Cylinders

Wheel cylinders convert hydraulic pressure from the master cylinder into a mechanical force at the brakes. The wheel cylinder bore is filled with fluid. When the brake pedal is depressed, additional brake fluid is forced into the cylinder. The additional fluid moves the cups and pistons outward. This piston movement forces the brake shoes outward to the contact drum and thus applies the brakes. Piston stops prevent the fluid leakage or air from getting into the system when the pistons move to the end of their bores.

Brake Shoes and Linings

In the same brake shoe sizes, there can be differences in **web** thickness, shape of web cutouts, and positions of any reinforcements.

The shoe rim is welded to the web to provide a stable surface for the lining. The web thickness might differ to provide the stiffness or flexibility needed for a specific application. Many shoes have nibs or indented places along the edge of the rim. These nibs rest against shoe support ledges on the backing plate and keep the shoe from hanging up.

Each drum in the drum braking system contains a set of shoes. The **primary shoe** (or leading shoe) is the one that is toward the front of the vehicle. The friction between the primary shoe and the brake drum forces the primary shoe to shift slightly in the direction that the drum is turning. (An **anchor pin** permits just limited movement.) The shifting of the primary shoe forces it against the bottom of the secondary shoe, which causes the secondary shoe to contact the drum. The **secondary shoe** (or trailing shoe) is the one that is toward the rear of the vehicle. It comes into contact as a result of the movement and pressure from the primary shoe and wheel cylinder piston and increases the braking action.

The brake shoe lining provides friction against the drum to stop the car. It contains heat-resistant fibers. The lining is molded with a high-temperature synthetic bonding agent.

The two general methods of attaching the lining to the shoe are riveting and bonding. Regardless of the method of attachment, brake shoes are usually held in a position by spring tension. They are either held against the anchor by the shoe return springs or against the support plate pads by shoe **hold-down springs**. The shoe webs are linked together at the end opposite the anchor by an adjuster and a spring. The adjuster holds them apart. The spring holds them against the adjuster ends.

Mechanical Components

In the unapplied position, the shoes are held against the anchor pin by the return springs. The shoes are held to the backing plate by hold-down springs or spring clips. Opposite the anchor pin, a star wheel adjuster links the shoe webs and provides a threaded adjustment that permits the shoes to be expanded or contracted. The shoes are held against the star wheel by a spring.

Shoe Return Springs **Return springs** can be separately hooked into a link or a guide **(Figure 47–3)** or strung between the shoes. Springs are normally installed on the anchor in the order shown under each category listed in **Figure 47–4**.

While shoe brake springs look the same, they are usually not interchangeable. Sometimes to help distinguish between them, they are color coded. Pay close attention to the colors and the way they are hooked up.

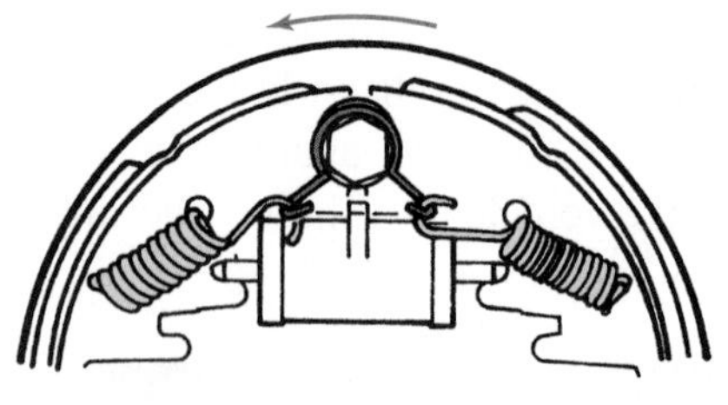

SEPARATE SPRING LINK

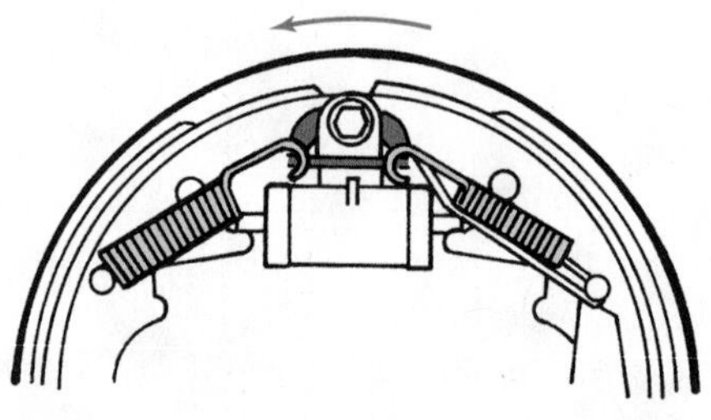

SEPARATE SPRING GUIDE

Figure 47–3 Typical brake shoe return spring alignments.

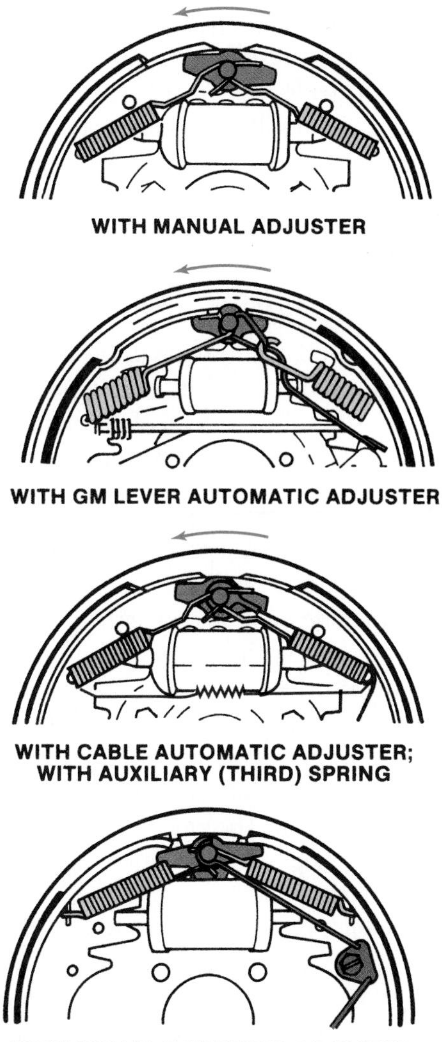

Figure 47-4 Typical brake shoe return spring anchoring points.

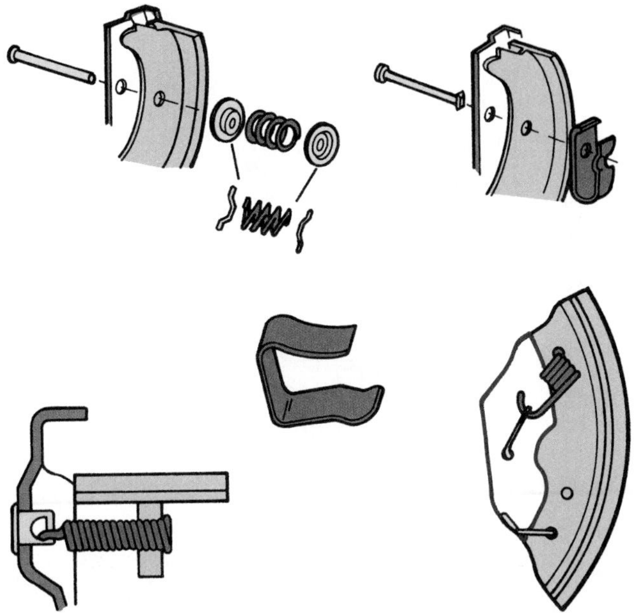

Figure 47-5 Types of brake shoe hold-downs.

Shoe Hold-Downs Various shoe hold-downs are illustrated in **Figure 47–5**. To unlock or lock the straight pin hold-downs, depress the locking cup and coil spring or the spring clip, and rotate the pin or lock 90 degrees. On General Motors' lever adjusters, the inner (bottom) cup has a sleeve that aligns the adjuster lever.

Shoe Anchors There are various types of **shoe anchors** such as the fixed nonadjustable type, self-centering shoe sliding type, or on some earlier models, adjustable fixed-type providing either an eccentric or a slotted adjustment. On some front brakes, fixed anchors are threaded into or are bolted through the steering knuckle and also support the wheel cylinder.

On adjustable anchors, when it is necessary to recenter the shoes in the drum or drum gauge, loosen the locknut enough to permit the anchor to slip but not so much that it can tilt.

Drums

Modern automotive brake drums are made of heavy cast iron (some are aluminum with an iron or steel sleeve or liner) with a machined surface inside against which the linings on the brake shoes generate friction when the brakes are applied. This results in the creation of a great deal of heat. The inability of drums to dissipate as much heat as disc brakes is one of the main reasons discs have replaced drums at the front of all late-model cars and light trucks, and at the rear of some sports and luxury cars.

Sometimes the rear drums of FWD cars are integral with the hub and cannot be removed without disassembling the wheel bearing. The rear drums of other FWD and most RWD cars are held in place by the wheel lugs so they can be removed without tampering with the wheel bearings.

DRUM BRAKE DESIGNS

There are two brake designs in common use. They are **duo-servo** (or self-energizing) **drum brakes** and **nonservo** (or leading-trailing) **drum brakes**.

Most large American cars use the duo-servo design of brake. However, the nonservo type has become popular as the size of cars has become smaller. Because the smaller cars are lighter, this type of brake helps reduce rear brake lockup without reducing braking ability.

Duo-Servo Drum Brakes

The name duo-servo drum brake is derived from the fact that the **self-energizing force** is transferred from one shoe to the other with the wheel rotating in either

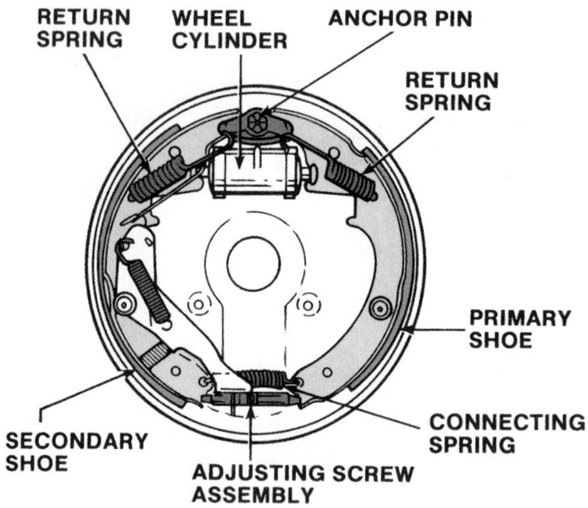

Figure 47-6 A typical duo-servo drum brake.

direction. Both the primary (front) and secondary (rear) brake shoes are actuated by a double-piston wheel cylinder **(Figure 47–6)**. The upper end of each shoe is held against a single anchor by a heavy coil return spring. An adjusting screw assembly and spring connect the lower ends of the shoes.

The wheel cylinder is mounted on the backing plate at the top of the brake. When the brakes are applied, hydraulic pressure behind the wheel cylinder cups forces both pistons outward causing the brakes to be applied.

When the brake shoes contact the rotating drum in either direction of rotation, they tend to move with the drum until one shoe contacts the anchor and the other shoe is stopped by the star wheel adjuster link **(Figure 47–7)**. With forward rotation, frictional forces between the lining and the drum of the primary shoe result in a force acting on the adjuster link to apply the secondary shoe. This adjuster link force into the secondary shoe is many times greater than the wheel cylinder input force acting on the primary shoe. The force of the adjuster link into the secondary shoe is again multiplied by the frictional forces between the secondary lining and rotating drum, and all of the resultant force is taken on the anchor pin. In normal forward braking, the friction developed by the secondary lining is greater than the primary lining. Therefore, the secondary brake lining is usually thicker and has more surface area. The roles of the primary and secondary linings are reversed in braking the vehicle when backing up.

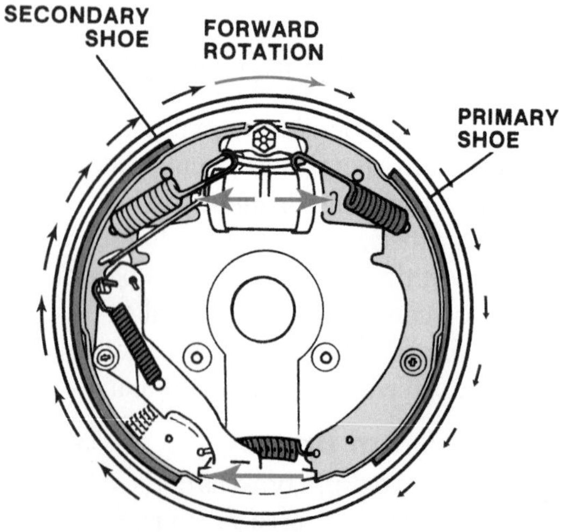

Figure 47-7 Duo-servo braking forces.

Automatically Adjusted Servo Brakes

Since the early 1960s, automatic drum brake adjusters have been used on all American and most import vehicles. There are several variations of automatic adjusters used with servo brakes. The more common types available follow.

Basic Cable **Figure 47–8** shows a typical automatic adjusting system. Adjusters, whether cable, crank, or lever, are installed on one shoe and operated whenever the shoe moves away from its anchor. The upper link, or cable eye, is attached to the anchor. As the shoe moves, the cable pulls over a guide mounted on the shoe web (the crank or lever pivots on the shoe web) and operates a lever (pawl), which is attached to the shoe so it engages a star wheel tooth. The pawl is located on the outer side of the star wheel and, on different styles, slightly above or below the wheel centerline so it serves as a ratchet lock, which prevents the adjustment from backing off. However, whenever lining wears enough to permit sufficient shoe movement, brake application pulls the pawl high enough to engage the next tooth. As the brake is released, the adjuster spring returns the pawl, thus advancing the star wheel one notch.

On most vehicles, the adjuster system is installed on the secondary shoe and operates when the brakes are applied as the vehicle is backing up. On a few models, it is located on the primary shoe and operates when the brakes are applied as the vehicle is moving forward. Left-hand and right-hand threaded star wheels are used on opposite sides of the car, so parts should be kept separated. If the wrong star wheel thread is installed, the system does not adjust at all.

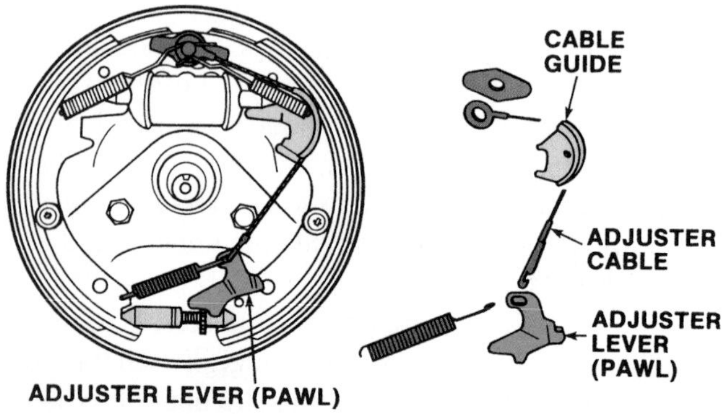

Figure 47-8 Cable self-adjusters.

Another system uses a cable and pawl, with the left brake having right-hand threads and the right brake, left-hand threads. The first cable guide is usually retained on the shoe web by the secondary shoe return spring, and the lever-pawl engages a hole in the shoe web. The adjuster operates in either direction of vehicle movement.

Cable with Overtravel Spring **Figure 47–9** shows a system with an upstroke pawl advance. The left brake has left-hand threads, and the right brake has right-hand threads. The lever (pawl) is installed on a web pin with an additional pawl return mousetrap spring. The cable is hooked to the lever (pawl) by means of an **overtravel spring** installed in the cable hook. The overtravel spring dampens movements and prevents unnecessary adjustment should sudden hard braking cause excessive drum deflection and shoe movement.

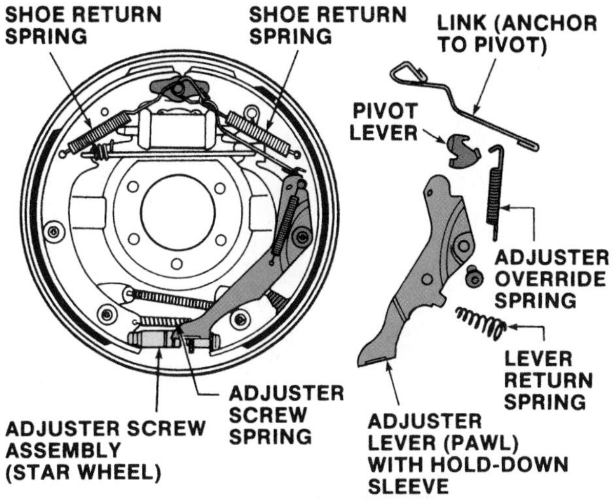

Figure 47-10 An adjusting lever with a pivot and override spring.

Lever with Override The system illustrated in **Figure 47–10** uses a downstroke pawl advance. The left brake has right-hand threads, and the right brake has left-hand threads.

The lever (pawl) is mounted on a shoe hold-down, pivoting on a cup sleeve. It has a separate lever-pawl return spring located between the lever and the shoe table. A pivot lever and an override spring assembled to the upper end of the main lever dampen movement, preventing unnecessary adjustment in the event of excessive drum deflection.

Lever and Pawl The system illustrated in **Figure 47–11** uses a downstroke pawl advance. The left brake has right-hand threads, and the right brake has left-hand threads. The lever is mounted on a shoe hold-down, pivoting on a cup sleeve, and engages the pawl. A separate pawl return spring is located between the pawl and the shoe.

Nonservo Drum Brakes

The nonservo (or as it is better known today as the leading-trailing shoe) drum brake is often used on small cars. The basic difference between this type and the duo-servo brake is that both brake shoes are held against a fixed anchor at the bottom by a retaining spring **(Figure 47–12)**. Nonservo brakes have no servo action.

On a forward brake application, the forward (leading) shoe friction forces are developed by wheel cylinder fluid pressure forcing the lining into contact with the rotating brake drum. The shoe's friction forces work against the anchor pin at the bottom of the shoe. The trailing shoe is also actuated by wheel cylinder pressure

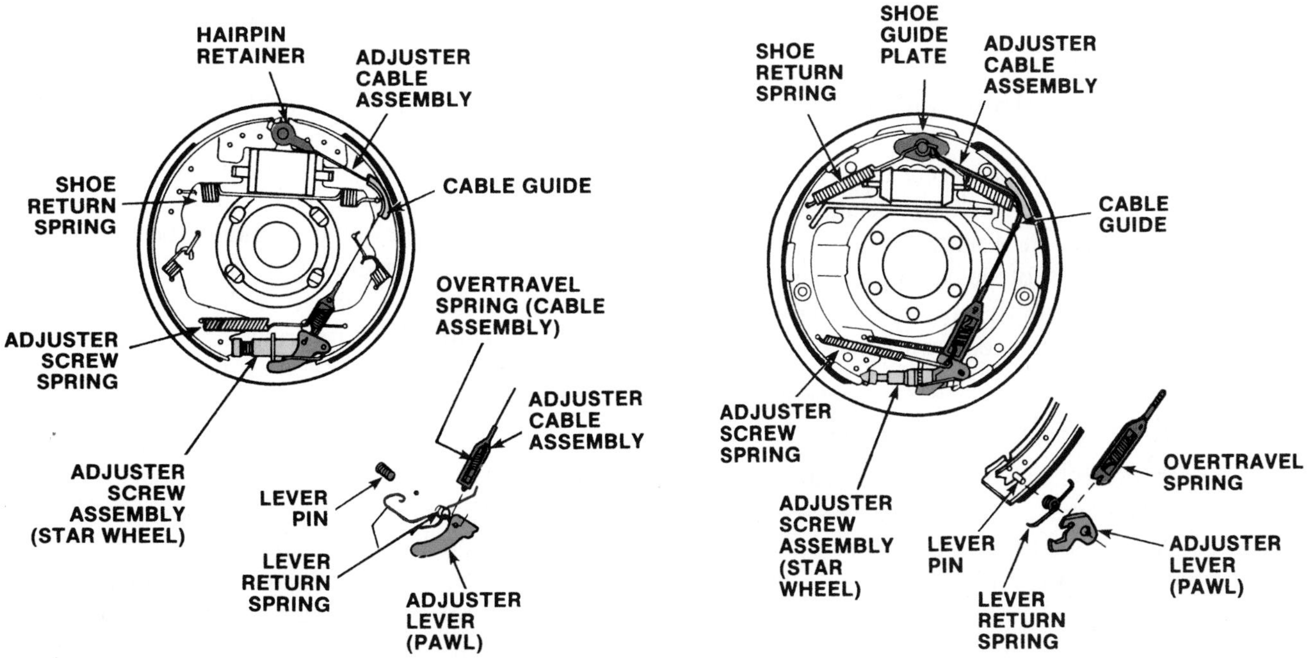

Figure 47-9 Cable automatic adjustment with overtravel springs.

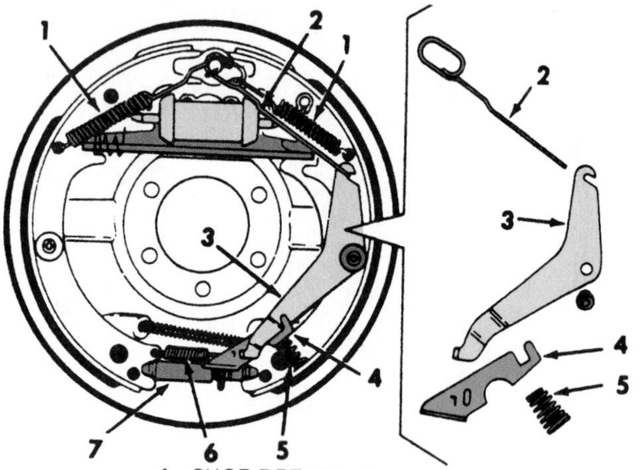

1. SHOE RETURN SPRINGS
2. LINK, ANCHOR TO LEVER
3. ADJUSTER LEVER (w/HOLD-DOWN SLEEVE)
4. ADJUSTER PAWL
5. PAWL RETURN SPRING
6. ADJUSTER SCREW SPRING
7. ADJUSTER SCREW ASSY. (STAR WHEEL)

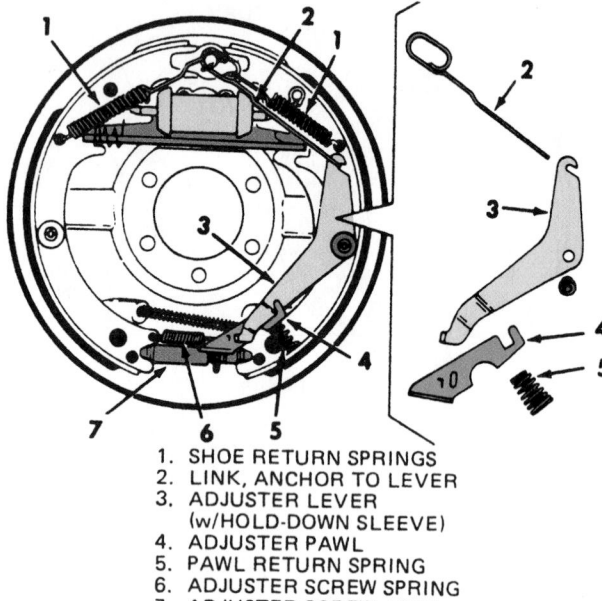

Figure 47-11 Lever and pawl automatic adjustment.
Courtesy of Wagner Brake Products

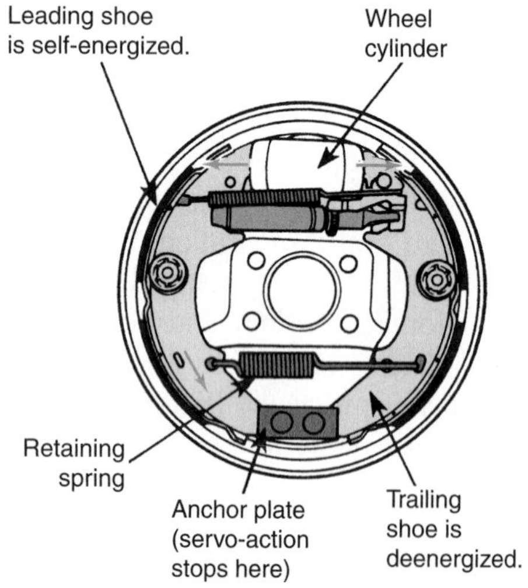

Figure 47-12 A typical nonservo drum brake.

but can only support a friction force equal to the wheel cylinder piston forces. The trailing shoe anchor pin supports no friction load. The leading shoe in this brake is energized and does most of the braking in comparison to the nonenergized trailing shoe. In reverse braking, the leading and trailing brake shoes switch functions.

SHOP TALK

It is important for the technician to remember that on nonservo drum systems the forward shoe is called the leading shoe and the rear one is known as the trailing shoe (when the vehicle is moving in the forward direction). On duo-servo designs, the forward shoe is the primary, and the rear is the secondary. ■

Automatically Adjusted Nonservo Brakes

While some standard automatic adjusters similar to the one already discussed are employed on small cars, some of the automatic adjuster mechanisms are unique and varied, using expanding struts between the shoes, or special ratchet adjusting mechanisms. Among the more common of these designs are automatic cam, ratchet automatic, and semiautomatic adjusters.

Automatic Cam Adjusters This rear nonservo drum brake is for use with front disc brakes and has one forward acting (leading) and one reverse acting (trailing) shoe. Shoes rest against the wheel cylinder pistons at the top and are held against the anchor plate by a shoe-to-shoe pull-back spring. The anchor plate and retaining plate are riveted to the backing plate. Adjustment of the brake shoes takes place automatically as needed when the brakes are applied. The automatic cam adjusters are attached to each shoe by a pin through a slot in the shoe webbing. As the shoes move outward during application, the pin in the slot moves the cam adjuster, rotating it outward. Shoes always return enough to provide proper clearance because the pin diameter is smaller than the width of the slot.

Ratchet Automatic Adjuster These brakes are a leading-trailing shoe design with a ratchet self-adjusting mechanism. The shoes are held to the backing plate by spring and pin hold-downs, and are held against the anchors at the top by a shoe-to-shoe spring. At the bottom, the shoe webs are held against the wheel cylinder piston ends by a return spring **(Figure 47–13)**.

The self-adjusting mechanism consists of a spacer strut and a pair of toothed ratchets attached to the secondary brake shoe. The parking brake actuating lever is pivoted on the spacer strut.

The self-adjusting mechanism automatically senses the correct lining-to-drum clearance. As the linings wear, the clearance is adjusted by increasing the effective length of the spacer strut. This strut has projections to engage the inner edge of the secondary shoe via the hand brake lever and the inner edge of the large ratchet on the secondary shoe. As wear on the linings increases, the movement of the shoes to bring them in contact with the drums becomes greater than the gap. The spacer strut, bearing on the shoe web, is moved together with the primary shoe to close the gap. Further movement causes the large ratchet behind the secondary shoe to rotate inward against the spring-loaded small ratchet, and the serrations on the mating edges maintain this new setting until further wear on the shoe results in another adjustment. On releasing brake pedal pressure, the return springs cause the shoes to move into contact with the shoulders of the spacer strut/hand brake actuating lever.

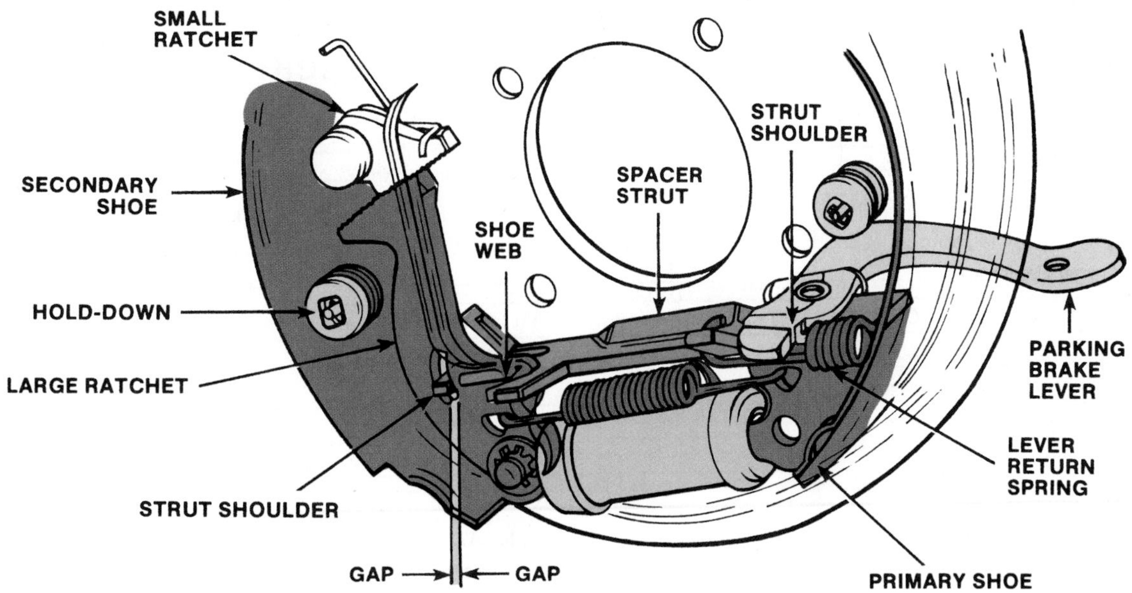

Figure 47–13 A typical nonservo self-adjusting mechanism.

This restores the clearance between the linings and the drum proportionate to the gap.

Inspection and Service

The first rule of quality brake service is to perform a complete job. For example, if new linings are installed without regard to the condition of the hydraulic system, the presence of a leaking wheel cylinder quickly ruins the new linings. Braking power and safety are also compromised.

Problems such as spongy pedal, excessive pedal travel, pedal pulsation, poor braking ability, brake drag, lock, or pulling to one side, and braking noises can be caused by trouble in the hydraulic system or the mechanical components of the brake assembly. To aid in doing a complete inspection and diagnosis, a form like the one shown in **Figure 47–14** is very helpful. Working with such a form helps the technician avoid missing any brake test and components that may cause problems.

Brake Noise

All customer complaints related to brake performance must be carefully considered. The number one customer complaint is brake noise. Noise is often the first indication of wear or problems within the braking system, particularly in the mechanical components. Rattles, clicking, grinding, and hammering from the wheels when the brake is in the unapplied position should be carefully investigated. Be sure the noise is not caused by the bearings or various suspension parts. If the noise is coming from the brake assembly, it is most likely caused by worn, damaged, or missing brake hardware, or the poor fastening or mounting of brake components. Grinding noises usually occur when a stone or other object becomes trapped between the lining material and the rotor or drum.

When the brakes are applied, a clicking noise usually indicates play or hardware failure in the attachment of the pad or shoe. On recent systems, the noise could be caused by the lining tracking cutting tool marks in the rotor or drum. A nondirectional finish on rotors eliminates this and so does a less pointed tip on the cutting tool used to refinish drums.

Grinding noises on application can mean metal-to-metal contact, either from badly worn pads or shoes, or from a serious misalignment of the caliper, rotor, wheel cylinder, or backing plate. Wheel cylinders and calipers that are frozen due to internal corrosion can also cause grinding or squealing noises.

Other noise problems and their solutions are covered later in this chapter.

ROAD TESTING BRAKES

Road testing allows the brake technician to evaluate brake performance under actual driving conditions. Whenever practical, perform the road test before beginning any work on the brake system. In every case, road test the vehicle after any brake work to make sure the brake system is working safely and properly.

WARNING!

Before test driving any car, first check the fluid level in the master cylinder. Depress the brake pedal to be sure there is adequate pedal reserve. Make a series of low-speed stops to be sure the brakes are safe for road testing. Always make a preliminary inspection of the brake system in the shop before taking the vehicle on the road.

PRE-BRAKE-JOB INSPECTION CHECKLIST

Owner ______________ (LAST) ______________ (FIRST) Phone ______________ Date ___/___/___

Address ______________________________ License No. ______________

Make __________ Model __________ Mileage __________ Serial No. __________ Year __________

Special Key for Hubcaps/Wheels Location ______________ Owner Use Parking Brake Yes ☐ No ☐

4 Drum ☐ 4 Disc ☐ Disc/Drum ☐ P/B No ☐ Yes ☐ Vacuum ☐ Hydro ☐ ABS ☐

Owner Comments ____

1. CHECKS BEFORE ROAD TEST	Safe	Unsafe
Stoplight Operation		
Brake Warning Light Operation		
Master Cylinder Checks		
Fluid Level		
Fluid Contamination		
Under Hood Fluid Leaks		
Under Dash Fluid Leaks (No Power)		
Bypassing		

BRAKE PEDAL HEIGHT AND FEEL

Check One		Check One	
Low		Spongy	
Med		Firm	
High			

Power Brake Unit Checks

VACUUM	Safe	Unsafe	HYDRO	Safe	Unsafe
Vacuum Unit			Hydro Unit		
Engine Vacuum			P/S Fluid		
Vacuum Hose			P/S Belt Tension		
Unit Check Valve			P/S Belt Condition		
Reserve Braking			P/S Fluid Leaks		
			Reserve Braking		

Tire Pressure Specs	Front	Rear
Record Pressure Found		

Record Pressure Found: RF ______ LF ______ RR ______ LR ______

Tire Condition: RF ______ LF ______ RR ______ LR ______

2. ROAD TEST	Yes	No	RF	LF	RR	LR
Brake Pull						
Brake Clunk						
Brake Scraping						
Brake Squeal						
Brake Grabby						
Brakes Lock Prematurely						
Wheel Bearing Noise						
Vehicle Vibrates						

STEERING WHEEL MOVEMENT WHEN STOPPING FROM 2-3 MPH YES/NO/RGT/LFT

Does ABS Work	YES	NO
Pedal Pulsation when Braking	YES	NO
Steering Wheel Oscillation when Braking	YES	NO
No Stopping Power	YES	NO
Warning Light Comes on when Braking	YES	NO
Difference in Pedal Height after Cornering	YES	NO
Nose Dive	YES	NO

3. In Shop Checks On Hoist	Yes	No	RF	LF	RR	LR
Brake Drag						
Intermittent Brake Drag						
Brake Pedal Linkage Binding						
Wheel Bearing Looseness						
Missing or Broken Wheel Fasteners						
Suspension Looseness						

	Front					Rear				
	Right			Left		Right			Left	
	Spec	Safe	Unsafe	Safe	Unsafe	Spec	Safe	Unsafe	Safe	Unsafe
Mark Wheels and Remove										
Caliper/Piston Stuck RF LF RR LR										
Mark Drums and Remove										
Measure Rotor Thickness or Drum Diameter.										
Measure Rotor Thickness Variation.										
Measure Rotor Runout.										
Lining Thickness										
Tubes and Hoses										
Fluid Leaks										
Broken Bleeders										
Leaky Seals										
Self-Adjuster Operation										

	Safe	Unsafe
Parking Brake Cables and Linkage		

Figure 47-14 A sample of a pre-brake-job inspection checklist. *Courtesy of Hennessy Industries, Inc.*

Brakes should be road tested on a dry, clean, reasonably smooth, and level roadway. A true test of brake performance cannot be made if the roadway is wet, greasy, or covered with loose dirt. All tires do not grip the road equally. Testing is also adversely affected if the roadway is crowned so as to throw the weight of the vehicle toward the wheels on one side, or if the roadway is so rough that wheels tend to bounce.

Test brakes at different speeds with both light and heavy pedal pressure. Avoid locking the wheels and sliding the tires on the roadway. There are external conditions that affect brake road-test performance. Tires having unequal contact and grip on the road cause unequal braking. Tires must be equally inflated and the tread pattern of right and left tires must be approximately equal. When the vehicle has unequal loading, the most heavily loaded wheels require more braking power than others and a heavily loaded vehicle requires more braking effort. A loose front-wheel bearing permits the drum and wheel to tilt and have spotty contact with the brake linings, causing erratic brake action. Misalignment of the front end causes the brakes to pull to one side. Also, a loose front-wheel bearing could permit the disc to tilt and have spotty contact with brake shoe linings, causing pulsations when the brakes are applied. Faulty shock absorbers that do not prevent the car from bouncing on quick stops can give the erroneous impression that the brakes are too severe.

DRUM BRAKE INSPECTION

Before inspecting drum brakes, place the vehicle in neutral, release the parking brake, and raise the vehicle on the hoist. Once the vehicle is raised, mark the wheel-to-drum and drum-to-axle positions so the components can be accurately reassembled. Relieve all tension from the parking brake cable by loosening or removing the adjusting nut at the equalizer. To access the drum brake assembly, remove the lug nuts and pull the wheel off the hub.

CAUTION!

When servicing wheel brake parts, do not create dust by cleaning with a dry brush or with compressed air. Asbestos fibers can become airborne if dust is created during servicing. Breathing dust containing asbestos fibers can cause serious bodily harm. To clean away asbestos from brake surfaces, use an OSHA-approved washer. Follow manufacturer's instructions when using the washer.

Shoe and Lining Removal

Several different methods are used to mount the drum to the wheel hub flange. It can be fastened with rivets or by swaging the piloting shoulders of the wheel studs, or with speed nut fasteners installed over the threads of the wheel studs.

The most common mounting method is to use the tire rim and lug nuts. The drum is a slip fit over the axle flange and studs. Speed nuts are installed at the vehicle manufacturing plant for temporary retention of the drum on the assembly line until the wheel is installed.

A **floating drum** can be retained by a nut and cotter pin. The drum can also be secured to the axle flange by one or two bolts. Some import applications have two additional holes threaded into the drum face so bolts can be used to press the drum from the hub or flange.

After removing the retaining devices that hold the drum to the axle flange or hub, the drum can be removed for servicing. If the brake drum is rusted or corroded to the axle flange and cannot be removed, lightly tap the axle flange to the drum mounting surface with a plastic mallet. Remember that if the drum is worn, the brake shoe adjustment has to be backed off for the drums to clear the brake shoes. Do not force the drum or distort it. Do not allow the drum to drop.

If the brake shoes have expanded too tightly against the drum or have cut into the friction surface of the drum brake, the drums might be too tight for removal. In such a case, the shoes must be adjusted inward before the brake drum is removed. On most cars with self-adjusting mechanisms, reach through the adjusting slot with a thin screwdriver (or similar tool) and carefully push the self-adjusting lever away from the star wheel a maximum of 1/16 inch (1.5 mm) **(Figure 47–15)**. While holding the lever back, insert a brake adjusting tool into the slot and turn the star wheel in the proper direction until the brake drum can be removed.

Be sure to inspect the rear wheel axle gaskets and wheel seals for leaks. Replace worn components as needed.

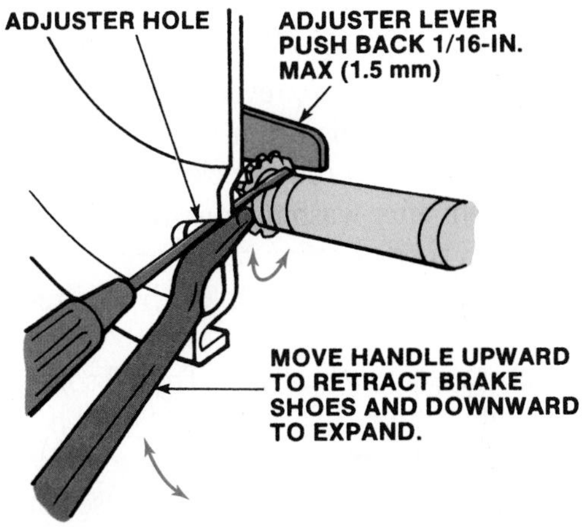
ADJUSTER HOLE
ADJUSTER LEVER PUSH BACK 1/16-IN. MAX (1.5 mm)
MOVE HANDLE UPWARD TO RETRACT BRAKE SHOES AND DOWNWARD TO EXPAND.

Figure 47-15 Backing off the self-adjusters in order to remove the brake drum.

Figure 47-16 Carefully check the inside surface of the brake drum.

Drum Inspection

One of the most important safety inspections to be made is that of the brake drum **(Figure 47–16)**. First, visually inspect the brake shoes, as installed on the car. Often their condition can reveal defects in the drums. If the linings on one wheel are worn more than the others, it might indicate a rough drum. Uneven wear from side to side on any one set of shoes can be caused by a tapered drum. If some linings are worn badly at the toe or heel, it might indicate an out-of-round drum.

Thoroughly clean the drums with a water-dampened cloth or a water-based solution. Equipment for washing brake parts is commercially available. Wet cleaning methods must be used to prevent asbestos fibers from becoming airborne. If the drums have been exposed to leaking oil or grease, thoroughly clean them with a non-oil-based solvent after washing to remove dust and dirt. It is important to determine the source of the oil or grease leak and correct the problem before reinstalling the drums.

Brake drums act as a heat sink. They absorb heat and dissipate it into the air. As drums wear from normal use or are machined, their cooling surface area is reduced and their operating temperatures increase. Structural strength also reduces. This leads to overdistortion, which causes some of the drum conditions covered here.

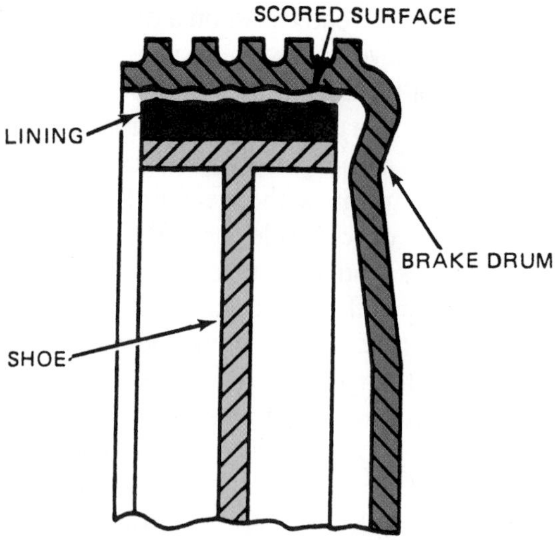

Figure 47-17 An example of a scored brake drum. *Courtesy of Wagner Brake Products*

Scored Drum Surface Figure 47–17 shows a scored drum surface. The most common cause of this condition is buildup of brake dust and dirt between the brake lining and drum. A glazed brake lining, hardened by high heat or in some cases by very hard inferior grade brake lining, can also groove the drum surface. Excessive lining wear that exposes the rivet head or shoe steel will score the drum surface. If the grooves are not too deep, the drum can be turned.

Bell-Mouthed Drum Figure 47–18 shows a distortion due to extreme heat and braking pressure. It occurs mostly on wide drums and is caused by poor support at the outside of the drum. Full drum-to-lining contact cannot be achieved and fading can be expected. Drums must be turned.

Concave Drum Figure 47–19A shows an excessive wear pattern in the center area of the drum brake sur-

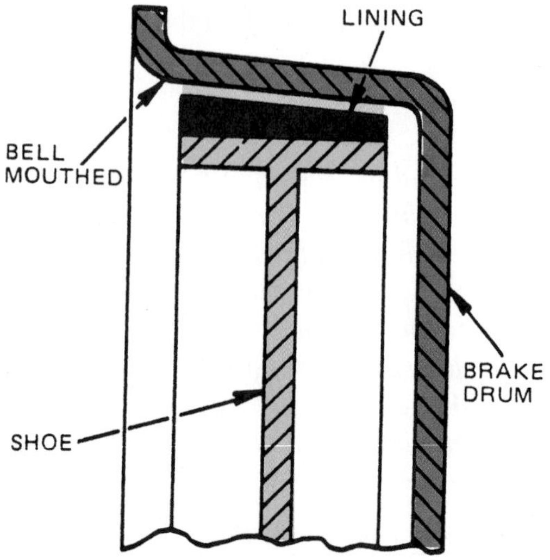

Figure 47-18 An example of a bell-mouthed brake drum. *Courtesy of Wagner Brake Products*

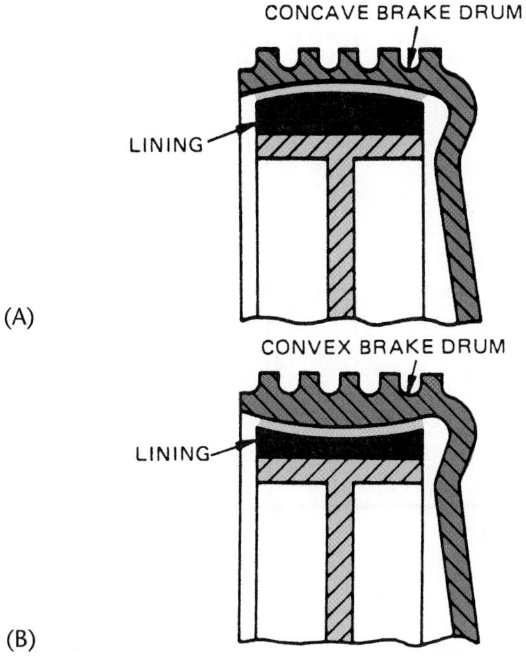

Figure 47–19 An example of concave and convex brake drums. *Courtesy of Wagner Brake Products*

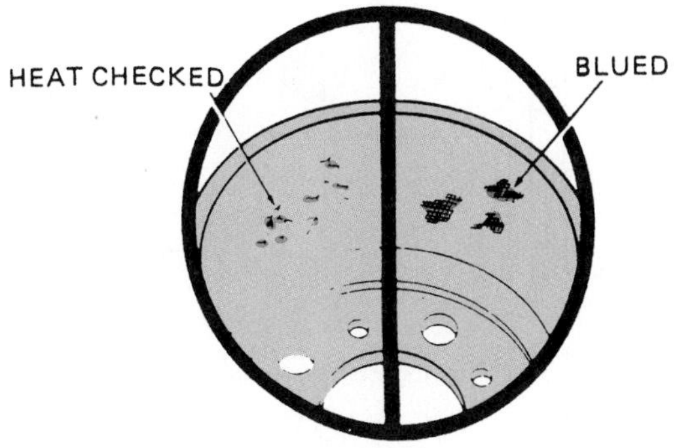

Figure 47–20 An example of a heat-checked and overheated brake drum. *Courtesy of Wagner Brake Products*

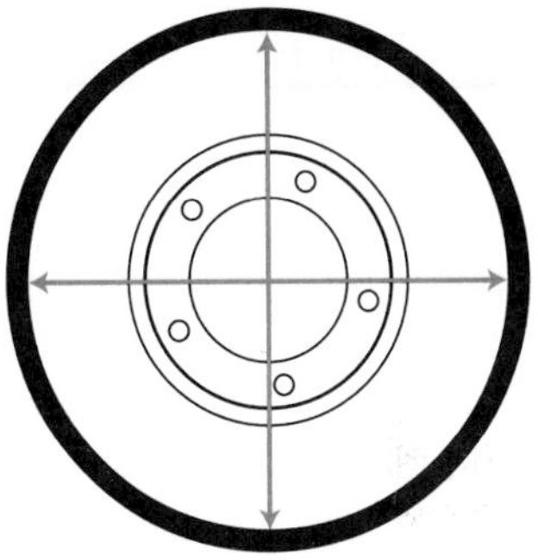

Figure 47–21 Measure the inside diameter of the drum in several spots to determine out-of-roundness.

face. Extreme braking pressure can distort the shoe platform so braking pressure is concentrated at the center of the drum.

Convex Drum This wear pattern is greater at the closed end of the drum **(Figure 47–19B)**. It is the result of excessive heat or an oversized drum, which allows the open end of the drum to distort.

Hard Spots on the Drum This condition in the cast-iron surface, sometimes called chisel spots or islands of steel, results from a change in metallurgy caused by braking heat. Chatter, pulling, rapid wear, hard pedal, and noise can occur. These spots can be removed by grinding. However, only the raised surfaces are removed, and they can reappear when heat is applied. If this condition reappears, the drum must be replaced.

Threaded Drum Surface An extremely sharp or chipped tool bit or a lathe that turns too fast can result in a threaded drum surface. This condition can cause a snapping sound during brake application as the shoes ride outward on the thread, then snap back. To avoid this, recondition drums using a rounded tool and proper lathe speed. Check the edge of the drum surface around the mounting flange side for tool marks indicating a previous machining. If the drum has been machined, it might have worn too thin for use. Check the diameter.

Heat Checks **Heat checks** are visible, unlike hard spots that do not appear until the machining of the drum **(Figure 47–20)**. Extreme operating temperatures are the major cause. The drum might also show a bluish/gold tint, which is a sign of high temperatures. Hardened carbide lathe bits or special grinding attachments are available through lathe manufacturers to service these conditions. Excessive damage by heat checks or hard spots requires drum replacement.

Cracked Drum Cracks in the cast-iron drum are caused by excessive stress. They can be anywhere but usually are in the vicinity of the bolt circle or at the outside of the flange. Fine cracks in the drums are often hard to see and, unfortunately, often do not show up until after machining. Nevertheless, should any cracks appear, no matter how small, the drum must be replaced.

Out-of-Round Drums Drums with eccentric distortion might appear fine to the eye but can cause pulling, grabbing, and pedal vibration or pulsation. An out-of-round or egg-shaped condition **(Figure 47–21)** is often caused by heating and cooling during normal brake operation. Out-of-round drums can be detected before the drum is removed by adjusting the brake to a light drag and feeling the rotation of the drum by hand. After removing the drum, gauge it to determine the amount of eccentric distortion. Drums with this defect should be machined or replaced.

Drum Measurements

Measure every drum with a drum micrometer **(Figure 47–22)**, even if the drum passed a visual inspection, to

Figure 47–22 Measuring the inside diameter with a drum micrometer.

make sure that it is within the safe oversize limits. If the drum is within safe limits, even though the surface appears smooth, it should be turned to ensure a true drum surface and to remove any possible contamination in the surface from previous brake linings, road dust, and so forth. Remember that if too much metal is removed from a drum, unsafe conditions can result.

Take measurements at the open and closed edges of the friction surface and at right angles to each other. Drums with taper or out-of-roundness exceeding 0.006 inch (.152 mm) are unfit for service and should be turned or replaced. If the maximum diameter reading (measured from the bottom of any grooves that might be present) exceeds the new drum diameter by more than 0.060 inch (1.5 mm), the drum cannot be reworked. If the drums are smooth and true but exceed the new diameter by 0.090 inch (2.2 mm) or more, they must be replaced.

If the drums are true, smooth up any slight scores by polishing with fine emery cloth. If deep scores or grooves are present that cannot be removed by this method, the drum must be turned or replaced.

Drum Refinishing

Brake drums can be refinished by either turning or grinding on a **brake lathe (Figure 47–23)**.

Only enough metal should be removed to obtain a true, smooth friction surface. When one drum must be machined to remove defects, the other drum on the same axle set must also be machined in the same manner and to the same diameter so braking is equal.

Brake drums are stamped with a discard dimension **(Figure 47–24)**. This is the allowable wear dimension and not the allowable machining dimension. There must be 0.030 inch (.762 mm) left for wear after turning the drums. Some states have laws about measuring the limits of a brake drum.

Figure 47–23 Brake drums can be resurfaced by grinding or turning them on a brake lathe. *Courtesy of Hennessy Industries, Inc.*

Figure 47–24 The drum's discard diameter is stamped on the drum.

Machining or grinding brake drums increases the inside diameter of the drum and changes the lining-to-drum fit. When remachining a drum, follow the equipment instructions for the specific tool you are using.

Cleaning Newly Refaced Drums

The friction surface of a newly refaced drum contains millions of tiny metal particles. These particles not only remain free on the surface, they always lodge themselves in the open pores of the newly machined surface. If the metal particles are allowed to remain in the drum, they become imbedded in the brake lining. Once the brake lining gets contaminated in this manner, it acts as a fine grinding stone and scores the drum.

PROCEDURE

Mechanical Component Service of Duo-Servo Drum Brakes

STEP 1 *Disconnect the cable from the parking brake lever.*

STEP 2 *If required, install wheel cylinder clamps on the wheel cylinders to prevent fluid leakage or air from getting into the system while the shoes are removed. Some brakes have wheel cylinder stops; therefore, wheel cylinder clamps are not required. Regardless of whether the clamps are needed, do not press down on the brake pedal after shoe return springs have been removed. To prevent this, block up the brake pedal so it cannot be depressed.*

STEP 3 *Remove the brake shoe return springs. Use a brake spring removal and installation tool to unhook the springs from the anchor pin or anchor plate* ***(Figure 47–25)****.*

STEP 4 *Remove the shoe retaining or hold-down cups and springs. Special tools are available, but the hold-down springs can be removed by using pliers to compress the spring and rotating the cup with relation to the pin.*

STEP 5 *Self-adjuster parts can now be removed. Lift off the actuating link, lever and pivot assembly, sleeve (through lever), and return spring. No advantage is gained by disassembling the lever and pivot assembly unless one of the parts is damaged.*

STEP 6 *Spread the shoes slightly to free the parking brake strut and remove the strut with its spring. Disconnect the parking brake lever from the secondary shoe. It can be attached with a retaining clip, bolt, or simply hooked into the shoe.*

STEP 7 *Slip the anchor plate off the pin. No advantage is gained by removing the plate if it is bolted on or riveted. Spread the anchor ends of the shoes and disengage them from the wheel cylinder links, if used. Remove the shoes connected at the bottom by the adjusting screw and spring, as an assembly.*

STEP 8 *Overlap the anchor end of the shoes to relieve spring tension. Unhook the adjusting screw spring, and remove the adjusting screw assembly.*

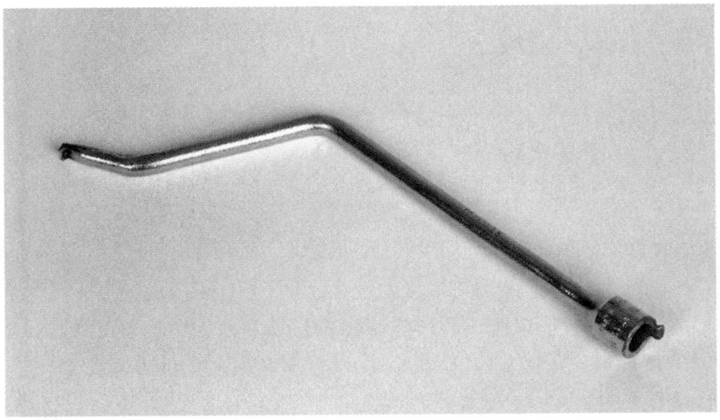

Figure 47-25 A brake spring tool.

SHOP TALK

Keep the adjusting screws and automatic adjuster parts for left and right brakes separate. These parts usually are different. For example, on some automatic adjusters, the adjusting screws on the right brakes have left-hand threads and the adjusting screws on the left brakes have right-hand threads. ■

PROCEDURE

Disassembling Nonservo or Leading-Trailing Brakes

STEP 1 *Install the wheel cylinder clamp. Then unhook the adjuster spring from the parking brake strut and reverse shoe.*

STEP 2 *Unhook the upper shoe-to-shoe spring from the shoes and unhook the antinoise spring from the spring bracket.*

STEP 3 *Remove the parking brake strut and disengage the shoe webs from the flat, clamp shoe hold-down clips.*

STEP 4 *Unhook the lower shoe-to-shoe spring and remove the forward shoe. Disconnect the parking brake cable, then remove the reserve shoe.*

STEP 5 *Remove the shoe hold-down clips from the backing plate.*

STEP 6 *Press off the C-shaped retainers from the pins and remove the parking brake lever, automatic adjuster lever, and adjuster latch.*

STEP 7 *Remove the parking brake lever.*

SHOP TALK

Mark the shoe positions if shoes and linings are to be reused. When disassembling an unfamiliar brake assembly, work on one wheel at a time and use the other wheel as a reference. ■

PROCEDURE

Cleaning and Inspecting Brake Parts

STEP 1 *Clean the backing plates, struts, levers, and other metal parts to be reused using a water-dampened cloth or a water-based solution. Equipment is commercially available to perform washing functions of brake parts. Wet cleaning methods must be used to prevent asbestos fibers from becoming airborne.*

STEP 2 *Carefully examine the raised shoe pads on the backing plate to make sure they are free from corrosion or other surface defects that might prevent the shoes from sliding freely. Use fine emery cloth to remove surface defects, if necessary. Clean them thoroughly.*

STEP 3 *Check to make sure that the backing plates are not cracked or bent. If so, they must be replaced. Make sure backing plate bolts and bolted-on anchor pins are torqued to specifications.*

STEP 4 *If replacement of the wheel cylinder is needed, it should be done at this time. To determine wheel cylinder condition, carefully inspect the boots. If they are cut, torn, heat-cracked, or show evidence of leakage, the wheel cylinders should be replaced. If more than a drop of fluid spills out, leakage is excessive and indicates that replacement is necessary.*

STEP 5 *Disassemble the adjusting screw assembly* ***(Figure 47–26)*** *and clean the parts in a suitable solvent. Make sure the adjusting screw threads into the pivot nut over its complete length without sticking or binding. Check that none of the adjusting screw teeth are damaged. Lubricate the adjusting screw threads with brake lubricant.*

STEP 6 *Examine the shoe anchor, support plate, and small parts for signs of looseness, wear, or damage that could cause faulty shoe alignment. Check springs for spread or collapsed coils, twisted or nicked shanks, and severe discoloration. Operate star wheel automatic adjusters by prying the shoe lightly away from its anchor or by pulling the cable to make sure the adjuster advances easily, one notch at a time. Adjuster cables tend to stretch, and star wheels and pawls become blunted after a long period of use. For rear-axle parking brakes, pull on the cable and shoe linkage to make sure no binding condition is present that could cause the shoes to drag when the parking brake is released.*

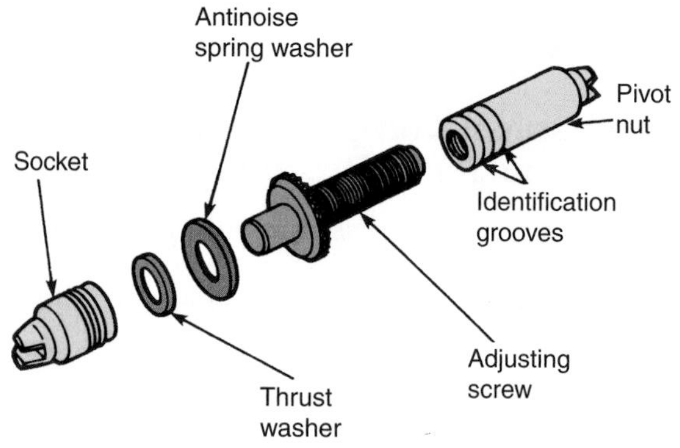

Figure 47–26 An exploded view of a brake adjuster assembly.

WARNING!

Do not use ordinary grease to lubricate drum brake parts. It does not hold up under high temperatures.

These metal particles must be removed by washing or cleaning the drum. Do not blow out the drum with air pressure. Either of the following methods is recommended to clean a newly refaced brake drum. The first method involves washing the brake drum thoroughly with hot water and wiping with a lint-free rag. Then, use the air pressure to thoroughly dry it. If the front hub and drums are being cleaned, be very careful to avoid contaminating the wheel bearing grease. Or, completely remove all the old grease, then regrease and repack the wheel bearing after the drum has been cleaned and dried. The wheel bearings and the grease seals must be removed from the drum before cleaning. The second method involves wiping the inside of the brake drum (especially the newly machined surface) with a lint-free white cloth dipped in one of the many available brake cleaning solvents that do not leave a residue. This operation should be repeated until dirt is no longer apparent on the wiping cloth. Allow the drum to dry before reinstalling it on the vehicle.

Both of these procedures are also good for cleaning disc brake rotors.

Cleaning, Inspecting, and Lubricating Brake Parts

CAUTION!

When servicing wheel brake parts, do not create dust by cleaning with a dry brush or compressed air.

SHOP TALK

Some brake technicians check brake spring tension by the drop method. This method is not overly scientific and the results are not always correct. Drop the brake spring on a clean concrete floor. If it bounces with a chunky sound, it is good. If the bounced spring gives off a tinny sound, it is tired and should be replaced. ■

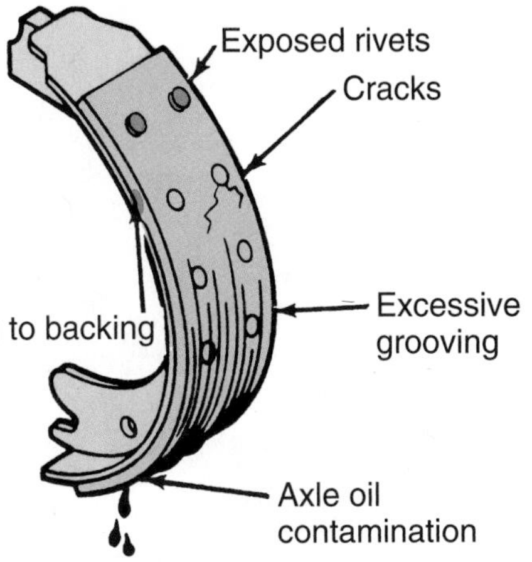

Figure 47–27 Potential brake shoe problems.

To complete the drum brake inspection, examine wheel bearings and hub grease seals for signs of damage. Service or replace, if necessary.

BRAKE SHOES AND LININGS

Lining materials influence braking operation. The use of a lining with a friction value that is too high can result in a severe grabbing condition. A friction value that is too low can make stopping difficult because of a hard pedal.

Overheating a lining accelerates wear and can result in dangerous lining heat fade—a friction-reducing condition that hardens the pedal and lengthens the stopping distance. Continual overheating eventually pushes the lining beyond the point of recovery into a permanent fade condition. In addition to fade, overheating can cause squeal.

Overheating is indicated by a lining that is charred or has a glass-hard glazed surface, or if severe, random cracking of the surface is present.

CAUTION!

Automotive friction materials often contain substantial amounts of asbestos. Studies indicate that exposure to excessive amounts of asbestos dust can be a potential health hazard. It is important that anyone handling brake linings understands this and takes the necessary precautions to avoid injury.

Inspect the linings for uneven wear, imbedded foreign material, loose rivets, and to see if they are oil soaked. If linings are oil soaked, replace them.

If linings are otherwise serviceable, tighten or replace loose rivets, remove imbedded foreign material, and clean the rivet counterbores.

If linings at any wheel show a spotty wear pattern or an uneven contact with the brake drum, it is an indication that the linings are not centered in the drums. Linings should be circle ground to provide better contact with the drum.

Brake Relining

Brake linings that are worn to within $1/32$ inch (.79 mm) of a rivet head or that have been contaminated with brake fluid, grease, or oil must be replaced **(Figure 47–27)**. Failure to replace worn linings results in a scored drum. When it is necessary to replace brake shoes, they must also be replaced on the wheel on the opposite side of the vehicle. Inspect brake shoes for distortion, cracks, or looseness. If these conditions exist, the shoe must be discarded.

Do not let brake fluid, oil, or grease touch the brake lining. If a brake lining kit is used to replace the linings, follow the instructions in the kit and install all the parts provided.

The two general methods of attaching the linings to the brake shoes are bonding and riveting. The **bonded linings** are fastened (glued) with a special adhesive to the shoe, clamped in place, then cured in an oven. Instead of using an adhesive, some linings are riveted to the shoe. Riveted linings allow for better heat transfer than bonded linings.

Sizing New Linings

Modern brake shoes are usually supplied with what is known as cam, offset, contour, or eccentric shape, which is ground in at the factory. That is, the full thickness of the lining is present at the middle of the shoe, but is ground down slightly at the heel and toe. The diameter of the circle the shoes make is slightly smaller than that of the drum. This compensates for the minor tolerance variations of drums and brake mountings and promotes proper wearing-in of the linings to match the drum.

SHOP TALK

On duo-servo shoe designs, the forward shoe is the primary and the rear, the secondary. The secondary shoe lining is longer. ■

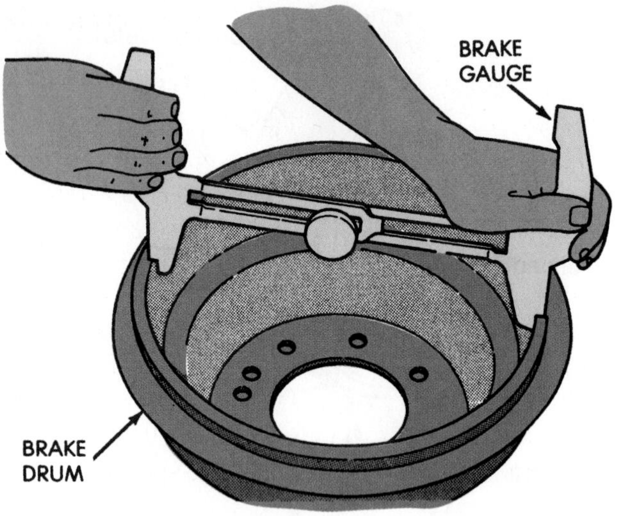

Figure 47–28 Using a brake gauge to measure the diameter of a brake drum. *Courtesy of DaimlerChrysler Corporation*

Lining Adjustment

New **eccentric-ground** linings tolerate a closer new lining clearance adjustment than concentric ground linings. With manual adjusters, the shoes should be expanded into the drums until the linings are at the point of drag but not dragging heavily against the drum. With star wheel automatic adjusters, a drum/shoe gauge **(Figure 47–28)** provides a convenient means of making the preliminary adjustment. This type of gauge, when set at actual drum diameter, automatically provides the working clearance of the shoes **(Figure 47–29)**. If new linings have been concentrically ground, the initial clearance adjustment must be backed off an amount that provides sufficient working clearance.

Many technicians take pride in showing the customer a high pedal, but it should be remembered that with new linings an extremely high pedal indicates tight clearances and can cause seating problems.

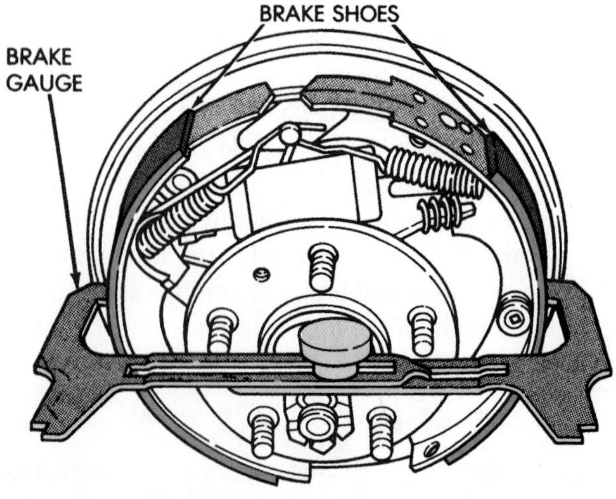

Figure 47–29 Using the other side of the brake gauge to set the brake shoes. *Courtesy of DaimlerChrysler Corporation*

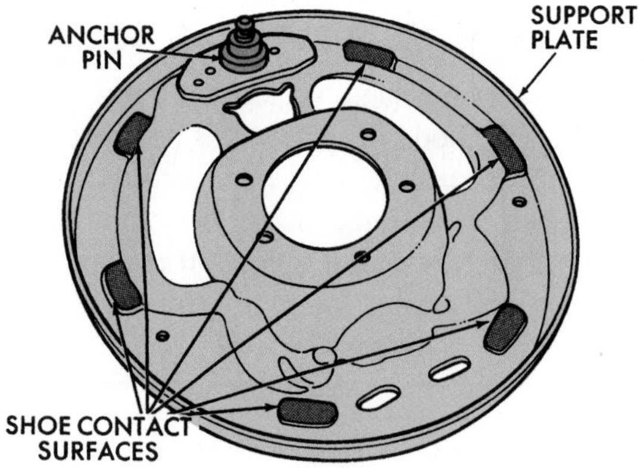

Figure 47–30 The areas or pads where the brake shoe will rub or contact the backing plate. *Courtesy of DaimlerChrysler Corporation*

Drum Shoe and Brake Installation

Before installing the shoes, be sure to sand or stone the inner edge of the shoe to dress down any slight lining or metal nicks and burrs that could interfere with the sliding on the support pads.

A support (backing) plate must be tight on its mount and not bent. Stone the shoe support pads brightly and dress down any burrs or grooves that could cause the shoes to bind or hang up.

Using an approved lubricant, lightly coat the support pads **(Figure 47–30)** and the threads of servo star wheel adjusters. On rear axle parking brakes, lubricate any point of potential binding in the linkage and the cable. Do not lubricate nonservo brake adjusters other than to free a frozen adjuster with penetrating oil.

Reassemble the brakes in the reverse order of disassembly. Make sure all parts are in their proper locations and that both brake shoes are properly positioned in either end of the adjuster. Also, both brake shoes should correctly engage the wheel cylinder pushrods and parking brake links. They should be centered on the backing plate. Parking brake links and levers should be in place on the rear brakes. With all of the parts in place, try the fit of the brake drum over the new shoes. If not slightly snug, pull it off and turn the star wheel until a slight drag is felt when sliding on the drum. A brake preset gauge makes this job easy and final brake adjustment simple. Then install the brake drum, wheel bearings, spindle nuts, cotter pins, dust caps, and wheel/tire assemblies, and make the final brake adjustments as specified in individual instructions in the vehicle's service manual. Torque the spindle and lug nuts to specifications.

WHEEL CYLINDER INSPECTION AND SERVICING

Wheel cylinders might need replacement when the brake shoes are replaced or when they begin to leak.

Inspecting and Cleaning Wheel Cylinders

Wheel cylinder leaks reveal themselves in several ways: (1) fluid can be found when the dust boot is peeled back; (2) the cylinder, linings, and backing plate, or the inside of a tire might be wet; or (3) there might be a drop in the level of fluid in the master cylinder reservoir.

Such leaks can cause the brakes to grab or fail and should be immediately corrected. Note the amount of fluid present when the dust boot is pulled back. A small amount of fluid seepage dampening the interior of the boot is normal. A dripping boot is not.

WARNING!

Hydraulic system parts should not be allowed to come in contact with oil or grease. They should not be handled with greasy hands. Even a trace of any petroleum-based product is sufficient to cause damage to the rubber parts.

Cylinder binding can be caused by rust deposits, swollen cups due to fluid contamination, or by a cup wedged into an excessive piston clearance. If the clearance between the pistons and the bore wall exceeds allowable values, a condition called heel drag might exist. It can result in rapid cup wear and can cause the piston to retract very slowly when the brakes are released.

Evidence of a scored, pitted, or corroded cylinder bore is a ring of hard, crystallike substance. This substance is sometimes noticed in the cylinder bore in which the piston rests after the brakes are released.

Light roughness or deposits can be removed with crocus cloth or an approved cylinder hone. While honing lightly, brake fluid can be used as a lubricant. If the bore cannot be cleaned up readily, the cylinder must be replaced.

Care must be taken when installing new or reconditioned wheel cylinders on cars equipped with wheel cylinder piston stops. The rubber dust boots and the pistons must be squeezed into the cylinder before it is tightened to the backing plate. If this is not done, the pistons jam against the stops causing hydraulic fluid leaks and erratic brake performance.

DRUM PARKING BRAKES

The parking brake keeps a vehicle from rolling while it is parked. It is important to remember that the parking brake is not part of the vehicle's hydraulic braking system. It works mechanically, using a lever assembly connected through a cable system to the rear drum service brakes.

Types of Parking Brake Systems

Parking brakes can be either hand or foot operated. In general, downsized cars and light trucks use hand-operated self-adjusting lever systems **(Figure 47–31)**. Full-size vehicles normally use a foot-operated parking brake pedal **(Figure 47–32A)**. The pedal or lever assembly is designed to latch into an applied position and is re-

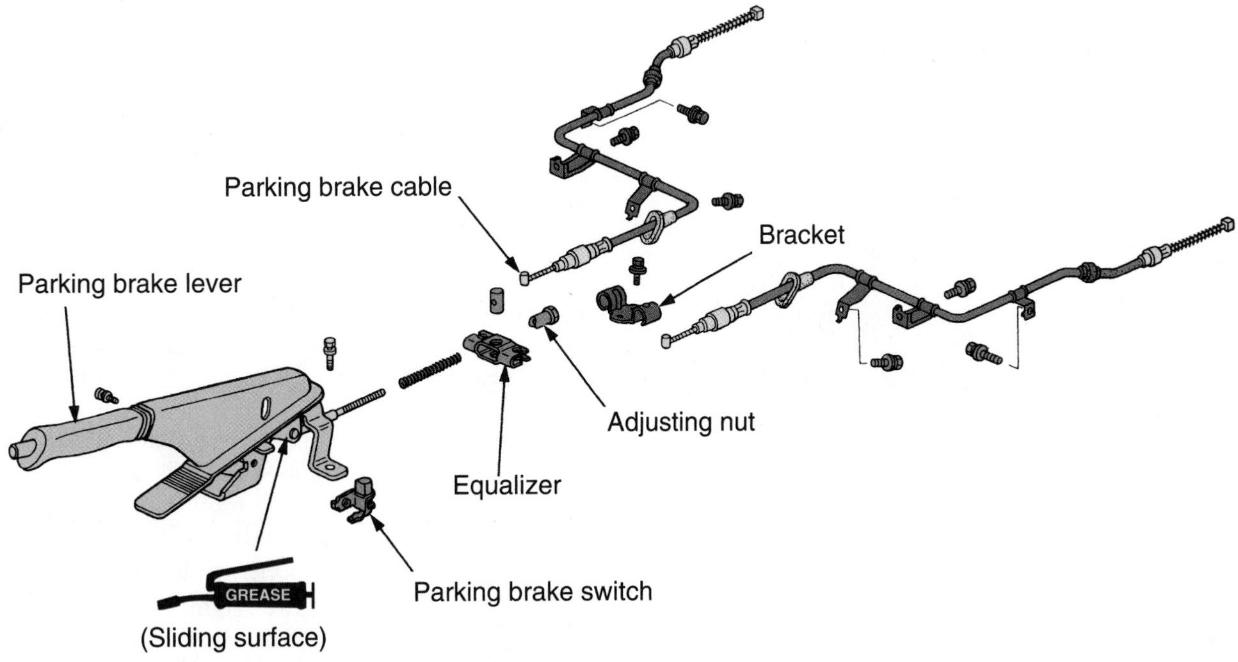

Figure 47–31 A typical setup for a center-mounted hand-operated parking brake. *Courtesy of American Honda Motor Co., Inc.*

leased by pulling a brake release handle or pushing a release button.

On some vehicles, a vacuum power unit **(Figure 47–32B)** is connected by a rod to the upper end of the release lever. The vacuum motor is actuated to release the parking brake whenever the engine is running and the transmission is in forward driving gear. The lower end of the release lever extends down for alternate manual release in the event of vacuum power failure or for optional manual release at any time. Hoses connect the power unit and the engine manifold to a vacuum release valve on the steering column.

The starting point of a typical parking brake cable and lever system is the foot pedal or hand lever. This assembly is a variable ratio lever mechanism that converts input effort of the operator and pedal/lever travel into output force with less travel. Tensile force from the front cable is transmitted through the car's brake cable system to the rear brakes. This tension pulls the flexible steel cables attached to each of the rear brakes. It operates the internal lever and strut mechanism of each rear brake, expanding the brake shoes against the drum. Springs return the shoes to the unapplied position when the parking brake pedal is released and tensile forces in the cable system are relaxed.

An electronic switch, triggered when the brake pedal is applied, lights the brake indicator in the instrument panel when the ignition is turned on. The light goes out when either the pedal or control is released or the ignition is turned off.

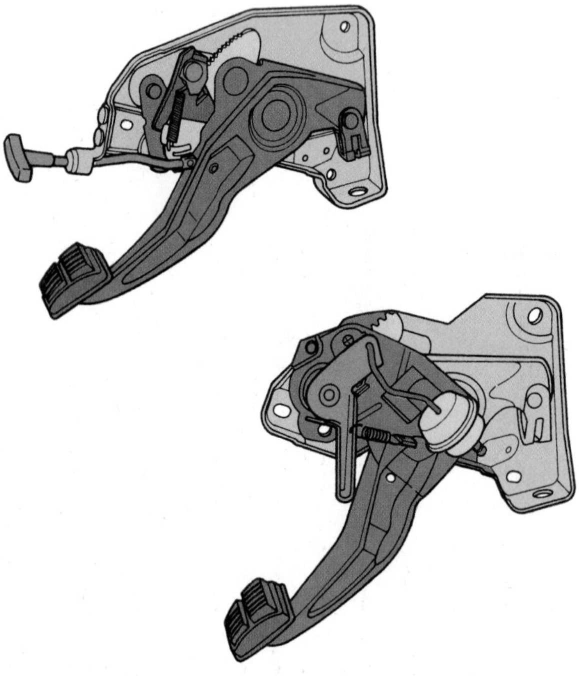

(A)

(B)

Figure 47-32 Typical pedal-operated parking brakes: (A) mechanical release and (B) vacuum release.

PROCEDURE

Replacing a Wheel Cylinder

STEP 1 *Since brake hoses are an important link in the hydraulic system, it is recommended they be replaced when a new cylinder is to be installed or when the old cylinder is to be reconditioned. Remove the brake shoe assemblies from the backing plate before proceeding. The smallest amount of brake fluid contaminates the friction surface of the brake lining.*

STEP 2 *Using two appropriate wrenches, disconnect the hydraulic hose from the steel line located on the chassis. On solid rear axles, use the appropriate tubing wrench and disconnect the hydraulic line where it enters the wheel cylinder. Care must be exercised in removing this steel line. It might be bent at this point and be difficult to install once new wheel cylinders are mounted to the backing plate.*

STEP 3 *Remove the plates, shims, and bolts that hold the wheel cylinder to the backing plate. Some later designed wheel cylinders are held to the backing plate with a retaining ring that can be removed with two small picks.*

STEP 4 *Remove the wheel cylinder from the backing plate and clean the area with a proper cleaning solvent.*

The cable/lever routing system in a typical parking brake arrangement **(Figure 47–33)** employs a three-lever setup to multiply the physical effort of the operator. First is the pedal assembly or hand grip. When moved, it multiplies the operator's effect and pulls the front cable. The front cable, in turn, pulls the equalizer lever.

The **equalizer lever** multiplies the effort of the pedal assembly, or hand grip, and pulls the rear cables. This pulling effort passes through an equalizer, which ensures equal pull on both rear cables. The equalizer functions by allowing the rear brake cables to slip slightly to balance out small differences in cable length or adjustment.

USING SERVICE MANUALS

Service manuals list the standard brake drum inside diameter along with the discard dimension. They also state the standard and minimum lining thickness. Manual illustrations should be used to accurately identify all components plus the disassemble/reassembly procedure. Tightening torques for backing plate nuts and other components should always be followed. ■

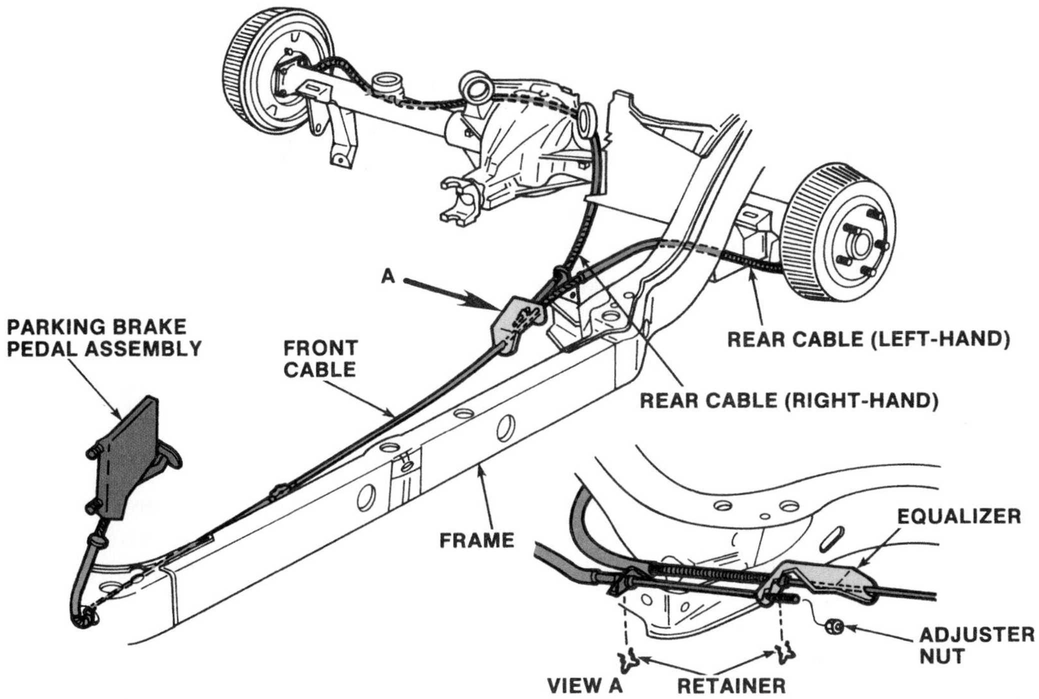

Figure 47–33 Typical parking brake routing to a cable equalizer and the rear drum brakes.

INTEGRAL PARKING BRAKES

Integral parking brakes are for vehicles with rear wheel drum brakes. **Figure 47–34** shows a typical integral parking brake. When the parking brake pedal is applied, the cables and equalizer exert a balanced pull on the parking brake levers of both rear brakes. The levers and the parking brake struts move the shoes outward against the brake drums. The shoes are held in this position until the parking brake pedal is released.

The rear cable enters each rear brake through a conduit **(Figure 47–35)**. The cable end engages the lower end of the parking brake lever. This lever is hinged to the web of the secondary shoe and linked with the primary shoe by means of a strut. The lever and strut expand both shoes away from the anchor and wheel cylinder and into contact with the drum as the cable and lever are drawn forward. The shoe return springs reposition the shoes when the cable is slacked.

To remove and replace the rear brake shoes, it might be necessary to relieve the parking brake cable tension by backing off the adjusting check nuts at the equalizer. Count the turns backed off in order to restore the nuts to their original position.

Adjusting and Replacing Parking Brakes

Regular wheel brake service should be completed before adjusting the parking brake. Then check the parking

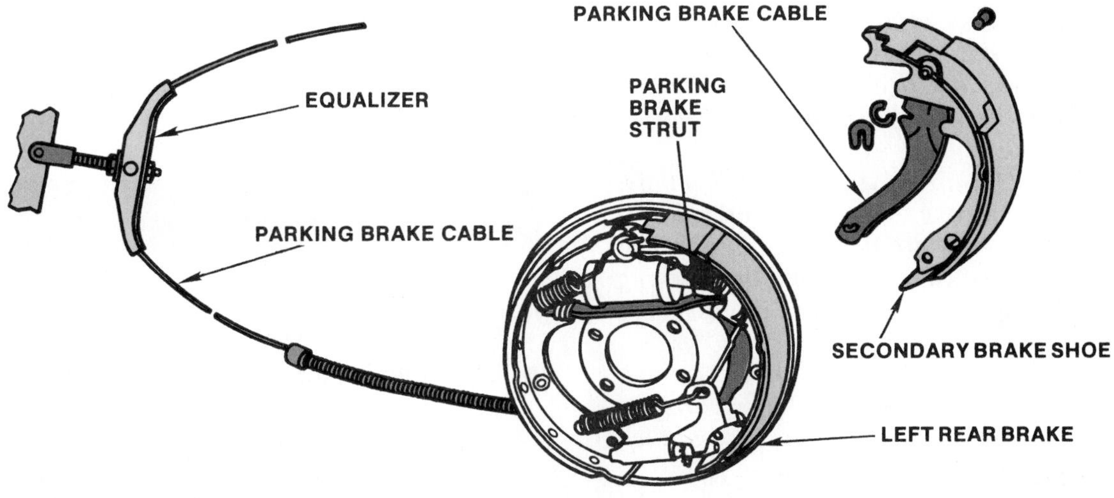

Figure 47–34 Integral parking brake components.

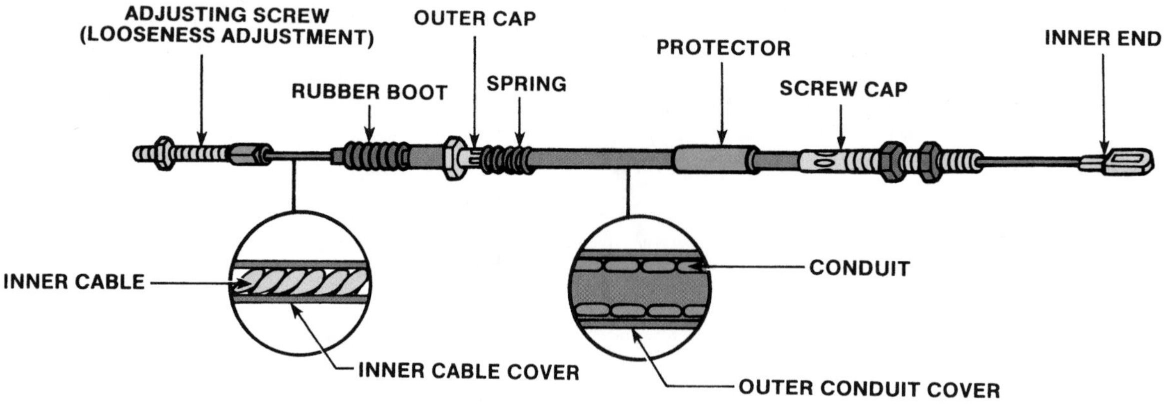

Figure 47-35 Rear cable and conduit details.

brake for free movement of the parking brake cables in the conduits. If necessary, apply a lubricant to free the cables. Check for worn equalizer and linkage parts. Replace any defective parts. Finally, check for broken strands in the cables. Replace any cable that has broken strands or shows signs of wear.

Testing Test the parking brake by parking the vehicle facing up on an incline of 30 degrees or less. Set the parking brake fully and place the transmission in neutral. The vehicle should hold steady. Reverse the vehicle position so it is facing down the incline and repeat the test. If the vehicle creeps or rolls in either case, the parking brake requires adjustment.

CASE STUDY

Shortly after new brake shoes and hardware were installed on a vehicle, it is returned to the shop on the rear of a tow truck. An extremely irate and distressed customer is heard throughout the entire service bay area. It seems when the customer applied the brakes while traveling at a high speed, there was a sudden loss of braking power and a fine metallic grinding noise was heard. Luckily, the customer was able to control the vehicle and avoid an accident, but the situation was potentially disastrous.

The vehicle is immediately placed on a hoist and the drum brakes are disassembled. Although all the parts are new, they had failed. The shoes are pulled away from the backing plate and the hold-down pins and retainers are broken.

The technician replaces all broken components with new parts from the shop's parts department. Everything appears to be installed correctly and in good working order.

The technician is reluctant to release the vehicle to the owner without uncovering the cause of the first failure. Inspecting the assembly, the technician compares the new shoes with the failed shoes. He notices the hold-down pin bores on the defective shoes do not line up perfectly with the pins. More comparisons between the new and failed parts uncover slight differences. The shoes and hardware that had failed were the wrong parts for the vehicle. Although similar in many ways, these slight differences were enough to cause a major failure. The technician who had performed the original replacement should have visually compared the new parts with the old parts.

KEY TERMS

Anchor pin
Backing plate
Bonded linings
Brake lathe
Duo-servo drum brake
Eccentric-ground
Equalizer lever
Floating drum
Heat checks
Hold-down spring
Nonservo drum brake
Overtravel spring
Primary shoe
Return spring
Secondary shoe
Self-energizing force
Shoe anchor
Web

SUMMARY

- Drum brakes are still used on the rear wheels of many cars and light trucks.
- The drum is mounted to the wheel hub. When the brakes are applied, a wheel cylinder uses hydraulic power to press two brake shoes against the inside surface of the drum. The resulting friction between the shoe's lining and drum slows the drum and wheel.

- The brake's anchor pin acts as a brake shoe stop, keeping the shoes from following the rotating drum. This creates a wedging action that multiplies braking force.
- The shoes and wheel cylinder are mounted on a backing plate. Hardware, such as shoe return springs, hold-down parts, and linkages are also mounted on the backing plate.
- The primary or leading shoe is toward the front of the vehicle while the secondary or trailing shoe is toward the rear of the vehicle.
- Brake lining can be attached to the shoes by riveting or a special adhesive bonding process.
- Brake drums act as a heat sink to dissipate the heat of braking friction. Drums can be refinished on a brake lathe provided the inside diameter is not increased above a safe limit (discard dimension).
- Servicing brakes requires performing a complete system inspection. Partial replacement of worn or damaged parts does not solve the braking problems and may ruin the new parts installed.
- When servicing brakes, extreme care must be taken to avoid generating asbestos dust.
- Wheel cylinders should be replaced if they show any signs of hydraulic fluid leakage or component wear.
- Drum brakes allow for the use of a simple parking brake mechanism that can be activated with a hand lever or foot pedal. This is a mechanical system, completely separate from the service brake hydraulic system.

TECH MANUAL

The following procedures are included in Chapter 47 of the *Tech Manual* that accompanies this book:

1. Inspecting and servicing duoservo drum brakes.
2. Inspecting and servicing nonservo drum brakes.
3. Adjusting and replacing a parking brake cable.

REVIEW QUESTIONS

1. Name the two methods of attaching brake lining materials to the brake shoes.
2. Explain how drum brakes create a self-multiplying brake force.
3. List at least five separate types of wear and distortion to look for when inspecting brake drums.
4. What is the job of wheel cylinder stops?
5. Explain the operation of an integral drum brake parking brake.
6. In a typical drum brake, which component provides a foundation for the brake shoes and associated hardware?
   a. wheel cylinder **c.** backing plate
   b. drum **d.** lining
7. *True or False?* The name duo-servo drum brake is derived from the fact that the self-energizing force is transferred from one shoe to another with the wheel rotating forward.
8. Which of the following statements about drum brake shoes is *not* true?
   a. Web thickness varies only with changes in shoe size.
   b. Many shoes have indented places along the edge of the rim.
   c. The shoe rim is welded to the web.
   d. The primary shoe is typically the one that is positioned toward the front of the vehicle.
9. Brake linings should be replaced when ______.
   a. linings are worn to within 1/32 inch (.79 mm) of a rivet head
   b. linings are contaminated with oil or grease
   c. linings are contaminated with brake fluid
   d. all of the above
10. In the unapplied position, drum brake shoes are held against the anchor pin by the ______.
    a. hold-down springs **c.** shoe hold-down
    b. star wheel adjuster **d.** return springs
11. Duo-servo drum brakes are also known as ______.
    a. leading-trailing brakes
    b. self-energizing brakes

c. nonservo brakes

d. none of the above

12. On most vehicles, the automatic adjuster system is ______.

a. installed on the secondary shoe

b. set up to operate when the brakes are applied as the vehicle moves forward

c. installed on the primary shoe

d. set up to operate when the brakes are not applied

13. Technician A says an out-of-round drum can cause a pulsating brake pedal. Technician B says an out-of-round drum can cause the brakes to grab. Who is correct?

a. Technician A
b. Technician B
c. Both A and B
d. Neither A nor B

14. Backing plates, struts, levers, and other metal brake parts should be ______.

a. wet-cleaned using water or a water-based solution

b. wet-cleaned using an alcohol-based solvent

c. wet-cleaned using gasoline

d. dry-cleaned only

15. A buildup of brake dust and dirt between the lining and the drum is the most common cause of ______.

a. concave/barrel-shaped drum

b. convex/tapered drum

c. threaded drum surface

d. scored drum surface

16. Technician A says a grinding noise from a drum brake when it is not applied can be caused by a bad wheel bearing. Technician B says a grinding noise from a drum brake when it is not applied can be caused by worn brake hardware. Who is correct?

a. Technician A
b. Technician B
c. Both A and B
d. Neither A nor B

17. The effort of the parking brake pedal assembly or hand grip is multiplied and passed on to the rear cables by the ______.

a. equalizer lever
b. front cable
c. rear disc
d. drive shaft

18. It has been determined that chatter and brake pull are being caused by hard spots on the brake drum. Technician A says the problem can be solved by grinding off the hard spots. Technician B says the drum must be replaced. Who is correct?

a. Technician A
b. Technician B
c. Both A and B
d. Neither A nor B

19. Technician A says the discard dimension of a brake drum is the drum's allowable machining dimension. Technician B says the discard dimension is the allowable wear dimension. There must be 0.030 inch (.762 mm) left for wear after machining. Who is correct?

a. Technician A
b. Technician B
c. Both A and B
d. Neither A nor B

20. After resurfacing a brake drum, Technician A cleans it using hot water and a lint-free cloth. He then uses compressed air to thoroughly dry it. Technician B cleans the drum using a lint-free cloth dipped in a special brake cleaning solution. She then allows the drum to dry before reinstallation. Who is correct?

a. Technician A
b. Technician B
c. Both A and B
d. Neither A nor B

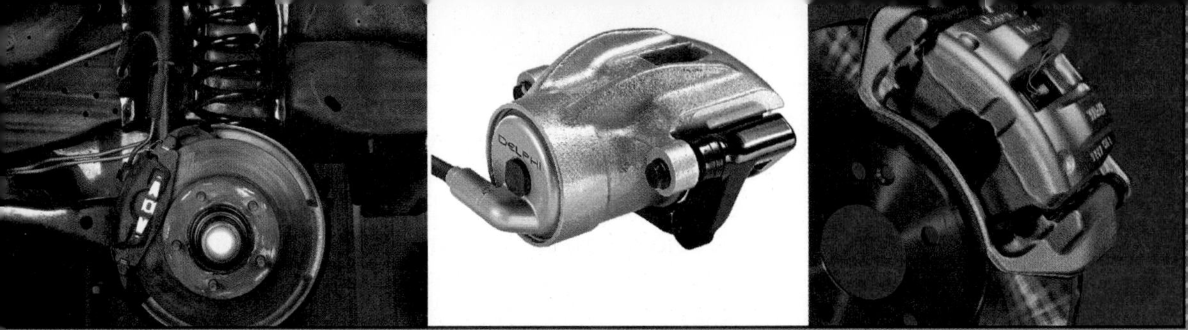

48

DISC BRAKES

OBJECTIVES

■ List the advantages of disc brakes. ■ List disc brake components and describe their functions. ■ Explain the difference between the three types of calipers commonly used on disc brakes. ■ Describe the two types of parking brake systems used with disc brakes. ■ Describe the causes of common disc brake problems. ■ Explain what precautions should be taken when servicing disc brake systems. ■ Describe the general procedure involved in replacing disc brake pads. ■ List and describe five typical disc brake rotor problems.

Disc brakes resemble the brakes on a bicycle. The friction elements are in the form of pads, which are squeezed or clamped about the edge of a rotating wheel. With automotive disc brakes, this wheel is a separate unit mounted to the wheel, called a **rotor (Figure 48–1)**. The rotor is typically made of cast iron. However, much research is being done with composite materials in the design of rotors. Since the pads clamp against both sides of a rotor, both sides are machined smooth. Usually the two surfaces are separated by a finned center section for better cooling. Such rotors are called ventilated rotors. The pads are attached to metal shoes, which are actuated by pistons contained within a **caliper** assembly, a housing that wraps around the edge of the rotor. The caliper is mounted to the steering knuckle to stop it from rotating.

Figure 48-1 A typical disc brake assembly.

The caliper contains the pistons and related seals, springs, and boots as well as the cylinder(s) and fluid passages necessary to force the pads against the rotor. The caliper resembles a hand in the way it wraps around the edge of the rotor.

Unlike shoes in a drum brake, the pads act perpendicularly to the rotation of the disc when the brakes are applied. In a brake drum, frictional drag actually pulls the shoe tighter into the drum. This does not happen with disc brakes, therefore they require more force to achieve the same braking effort. For this reason, they are typically used with a power brake unit.

Disc brakes offer four major advantages over drum brakes. Disc brakes are more resistant to heat fade during high-speed brake stops or repeated stops because the design of the disc brake rotor exposes more surface to the air and thus dissipates heat more efficiently. They are also resistant to water fade because the rotation of the rotor tends to throw off moisture and the squeeze of the sharp edges of the pads clears the surface of water. Disc brakes perform more straight-line stops. Due to their clamping action, disc brakes are less apt to pull. Finally, disc brakes automatically adjust as pads wear.

DISC BRAKE COMPONENTS AND THEIR FUNCTIONS

The disc brakes used today are typically of two basic designs: **fixed caliper** or **floating caliper.** There is also a **sliding caliper,** but its design is very similar to the floating caliper **(Figure 48–2)**. The only difference is that sliding calipers slide on surfaces that have been machined smooth for this purpose, and floating calipers

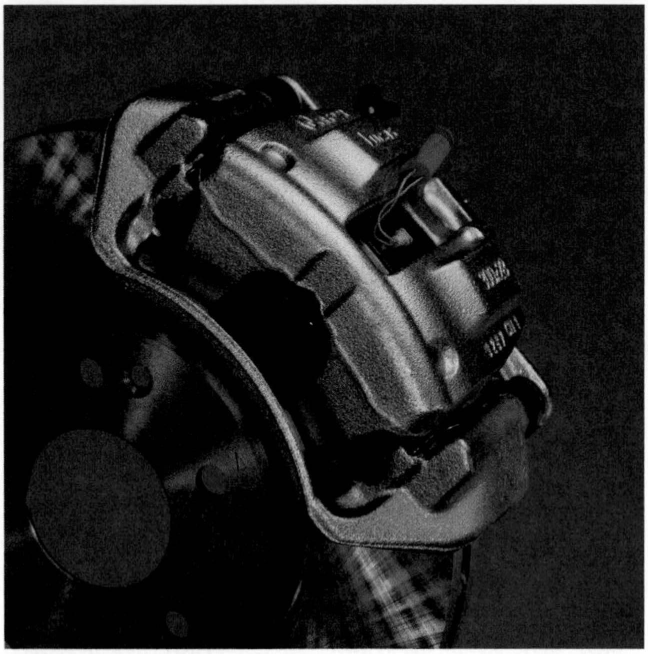

Figure 48-2 A floating caliper assembly. *Courtesy of DaimlerChrysler Corporation*

slide on special pins or bolts. The disc brake, regardless of its design, consists of a hub and rotor assembly, a caliper assembly, and a brake pad assembly.

Hub and Rotor Assembly

Many of the advantages of disc brakes can be attributed to the disc, or rotor, as it is more commonly called. The typical rotor can be solid or ventilated and is made of cast iron that has a high coefficient of friction and withstands wear exceptionally well. Composite style rotors are growing in popularity and are found on many of today's FWD and four-wheel disc vehicles. The rotor is attached to and rotates with the wheel hub assembly **(Figure 48–3)**.

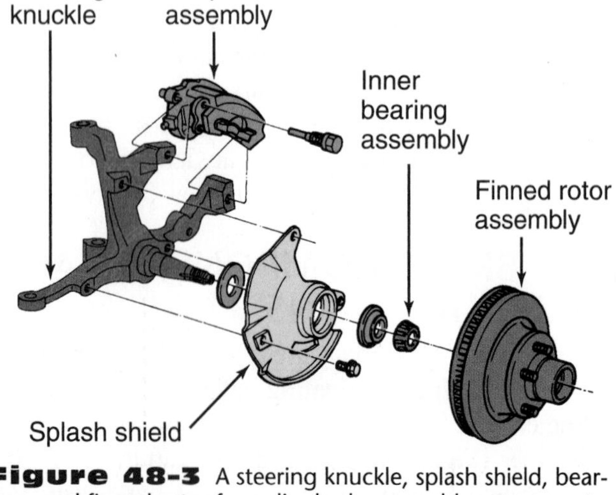

Figure 48-3 A steering knuckle, splash shield, bearings, and finned rotor for a disc brake assembly. *Courtesy of Ford Motor Company*

A ventilated rotor is cast with a weblike construction between two friction surfaces. The webs radiate out from the center of the rotor, much like vanes or fins in a fan. As the rotor turns, air is drawn into the rotor at its center, flows between the friction surfaces, and is discharged along the outer edge. This cools the rotor, which drastically reduces the chances of brake fade, even during multiple hard stops.

A **splash shield** protects the rotor and pads from road splashes and dirt. It is also shaped to channel the flow of air over the exposed rotor surfaces. As long as the car is moving, this flow of air helps to cool the rotor. The splash shield cannot be removed unless the rotor and caliper are first removed. Replacement of the splash shield is necessary only when it has been damaged or when the spindle is replaced.

Caliper Assembly

A brake caliper converts hydraulic pressure into mechanical force. The caliper housing is usually a one-piece construction of cast iron or aluminum and has an inspection hole in the top to allow for lining wear inspection. The housing contains the cylinder bore(s). In the cylinder bore is a groove that seats a square-cut seal. This groove is tapered toward the bottom of the bore to increase the compression on the edge of the seal that is nearest hydraulic pressure. The top of the cylinder bore is also grooved as a seat for the dust boot. A fluid inlet hole is machined into the bottom of the cylinder bore and a bleeder valve is located near the top of the casting.

A caliper can contain one, two, or four cylinder bores and pistons that provide uniform pressure distribution against the brake's friction pads. The pistons are relatively large in diameter and short in stroke to provide high pressure on the friction pad assemblies with a minimum of fluid displacement.

Basically, the hydraulics of disc brakes are the same as for drum brakes, in that the master cylinder piston forces the brake fluid into the wheel cylinders and against the wheel pistons.

The disc brake piston is made of steel, aluminum, or fiberglass-reinforced **phenolic resin**. Steel pistons are usually nickel-chrome plated for improved durability and smoothness. The top of the pistons is grooved to accept the **dust boot** that seats in a groove at the top of the cylinder bore and also in a groove in the piston. The dust boot prevents moisture and road contamination from entering the bore.

A piston hydraulic seal prevents fluid leakage between the cylinder bore wall and the piston. This rubber sealing ring also acts as a retracting mechanism for the piston when hydraulic pressure is released, causing the piston to return in its bore **(Figure 48–4)**. When hy-

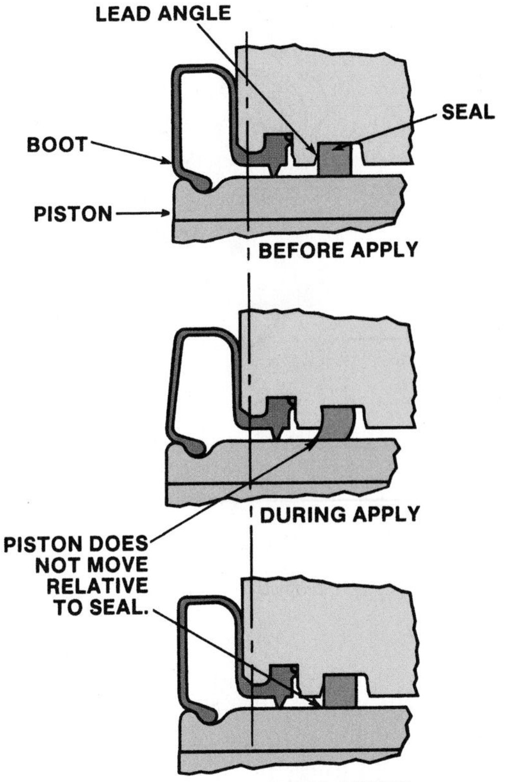

Figure 48-4 Action of the piston's hydraulic seal in the caliper's cylinder.

draulic pressure is diminished, the seal functions as a return spring to retract the piston.

In addition, as the disc brake pads wear, the seal allows the piston to move farther out to adjust automatically for the wear, without allowing fluid to leak. Since the brake pads need to retract only slightly after they have been applied, the piston moves back only slightly into its bore. The additional brake fluid in the caliper bore keeps the piston out and ready to clamp the surface of the rotor.

Fixed Caliper Disc Brakes Fixed caliper disc brakes have a caliper assembly that is bolted in a fixed position and does not move when the brakes are applied. The pistons in both sides of the caliper come inward to force the pads against the rotor **(Figure 48–5)**.

Floating Caliper Disc Brakes A typical floating caliper disc brake is a one-piece casting that has one hydraulic cylinder and a single piston. The caliper is attached to the **spindle anchor plate** with two threaded locating pins. A Teflon sleeve separates the caliper housing from each pin and the caliper slides back and forth on the pins as the brakes are actuated. When the brakes are applied, hydraulic pressure builds in the cylinder behind the piston and seal. Because hydraulic pressure exerts equal force in all directions, the piston moves evenly out of its bore.

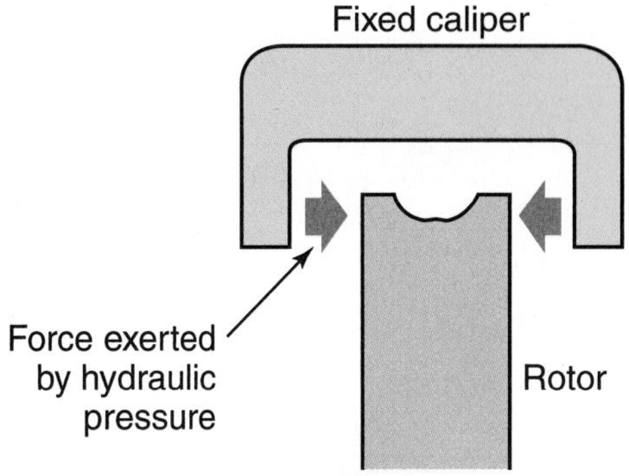

Figure 48-5 Operation of a fixed caliper. *Courtesy of EIS—Standard Motor Products*

The piston presses the inboard pad against the rotor. As the pad contacts the revolving rotor, greater resistance to outward movement is increased, forcing pressure to push the caliper away from the piston. This action forces the outboard pad against the rotor **(Figure 48–6)**. However, both pads are applied with equal pressure.

Sliding Caliper Disc Brakes With a sliding caliper assembly, the caliper slides or moves sideways when the brakes are applied. As mentioned previously, in operation, these brakes are almost identical to the floating type. But unlike the floating caliper, the sliding caliper does not float on pins or bolts attached to the anchor plate. It has angular machined surfaces at each end that slide in mating machined surfaces on the anchor plate. This is where the caliper slides back and forth.

Some sliding calipers use a support key to locate and support the caliper in the anchor plate **(Figure 48–7)**. The caliper support key is inserted between the caliper and the anchor plate. A worn support key may cause tapered brake pad wear. Always inspect the support keys when replacing brake pads. Also make sure they are lubricated when reassembling the unit.

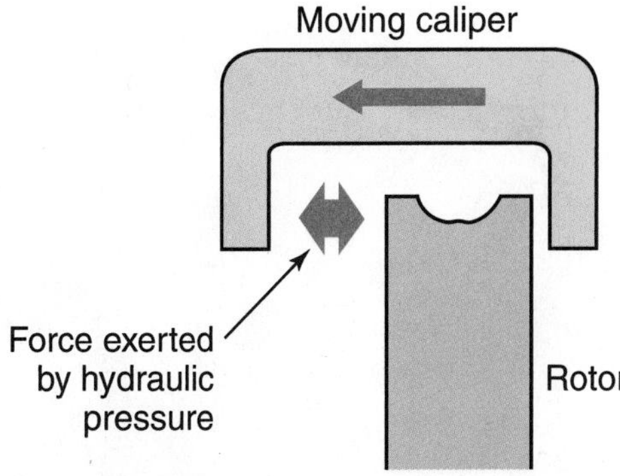

Figure 48-6 Operation of a floating caliper. *Courtesy of EIS—Standard Motor Products*

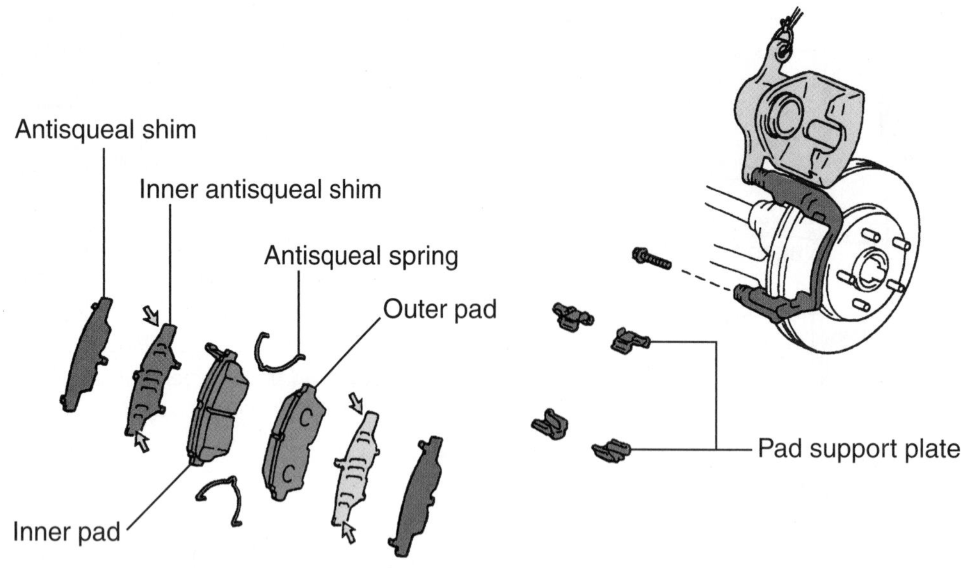

Figure 48-7 A sliding caliper with support plates (keys). *Reprinted with permission*

Electric Calipers A recent development is the use of electrically operated calipers **(Figure 48–8).** Using electric calipers eliminates the need to run hydraulic lines to the wheels and results in better brake control. This system is computer controlled and can evenly distribute braking power at the four wheels. At the rear wheels, this system can also be used as the parking brake, thereby eliminating the need to have cables at the rear wheels.

Brake Pad Assembly

Brake pads are metal plates with the linings either riveted or bonded to them. Pads are placed at each side of the caliper and straddle the rotor. The inner brake pad, which is positioned against the piston, is not interchangeable with the outer brake pad. The linings are made of semimetallic or other nonasbestos material.

Disc Pad Wear Sensors Some brake shoe pads have wear sensing indicators. The three most common design wear sensors are audible, visual, and tactile.

Figure 48-8 An electric brake caliper. *Courtesy of the Delphi Corporation*

Audible sensors are thin, spring steel tabs that are riveted to or installed onto the edge of the pad's backing plate and are bent to contact the rotor when the lining wears down to a point that replacement is necessary. At that point, the sensor causes a high-pitched squeal whenever the wheel is turning, except when the brakes are applied. Then the noise goes away. The noise gives a warning to the driver that brake service is needed and perhaps saves the rotor from destruction **(Figure 48–9).** The tab is generally installed toward the rear of the wheel.

Visual sensors inform the driver of the need for new linings. This method employs electrical contacts recessed

Figure 48-9 Operation of the wear indicator. *Courtesy of Wagner Brake Products*

in the pads that touch the rotor when the linings are worn out, completing a circuit and turning on a dashboard warning light. This system is found mostly on imports.

Tactile sensors create pedal pulsation as the sensor on the rotor face contracts the sensor attached to the lower portion of the disc pad.

CAUTION!

As disc brake pad linings wear thin, more brake fluid is needed in the system.

CUSTOMER CARE

Excessive heat liquefies the resin binder that holds the brake pad material together. Once liquefied, the binder rises to the surface of the pad to form a glaze. A glazed pad may cause squealing because more heat is needed to achieve an amount of friction equal to a pad in good condition. One common cause of pad glazing is the improper break-in of the pads. Remember to inform customers that unnecessary hard braking during the first 200 miles (320 km) of a pad's life can generate enough heat to glaze the pads and ruin the quality of the work just performed. There are pads now available that require no break-in time. ■

Rear Disc/Drum (Auxiliary Drum) Parking Brake

The rear disc/drum or auxiliary drum parking brake arrangement is found on some vehicles. On these brakes, the inside of each rear wheel hub and rotor assembly is used as the parking brake drum **(Figure 48–10)**. The auxiliary drum brake is a smaller version of a drum brake and is serviced like any other drum brake.

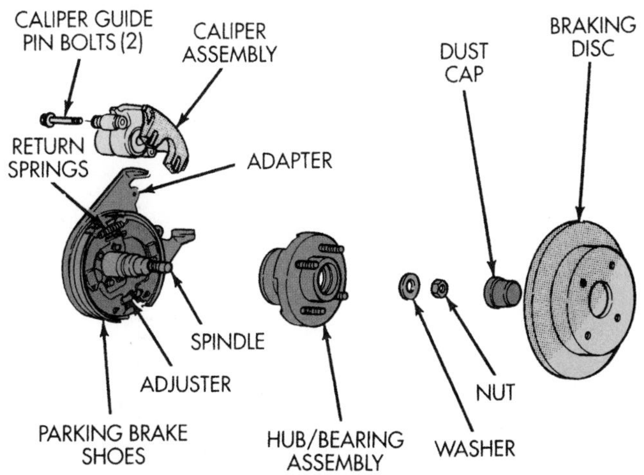

Figure 48-10 A typical parking brake that uses a drum built into the center of a rotor. *Courtesy of DaimlerChrysler Corporation*

Rear Disc Parking Brakes

Instead of using an auxiliary drum and shoes to hold the vehicle when parked, these brakes have a mechanism that forces the pads against the rotor mechanically. One method for doing this is the ball-and-ramp arrangement. Another method found on many vehicles has a threaded, spring-loaded pushrod **(Figure 48–11)**. As the parking brakes are applied, a mechanism rotates or unscrews the pushrod, which in turn pushes the piston out. Other types of actuating systems for rear disc brakes include the use of a variety of cams.

DISC BRAKE DIAGNOSIS

Many problems experienced on vehicles with disc brakes are the same as those evident with drum brake systems. Some problems occur only with disc brakes. Before covering the typical complaints, it is important to remind you to get as much information as possible about the complaint from the customer. Then road test the vehicle to verify the complaint. A complete inspection of the rotor, caliper, and pads **(Figure 48–12)** should be done anytime you are working on the brakes.

What follows is a brief discussion of common complaints and their typical causes.

Warning Lights

Today's vehicles are normally equipped with more than one brake warning light on the instrument panel. Regardless of what warning light is lit, it is an indication of warning to the driver. You need to understand what would cause the different lights to illuminate in order to take care of the problem. Keep in mind, a vehicle may have one, two, or all of these lights.

The red warning light indicates there is a problem in the regular brake system, such as low brake fluid levels or that the parking brake is on. A low fluid light may be present in addition to the red brake warning light. Whenever the fluid is low, you should suspect a leak or very worn brake pads.

The yellow or amber brake warning light is tied into the antilock brake system. This light turns on for two reasons: the ABS system is performing a self-test or there is a fault in the ABS system.

A blue or yellow warning light lets the driver know the wheels are slipping because of poor road conditions.

Pulsating Pedal

Customers will feel a vibration or pulsation in the brake pedal when the brakes are applied if a brake rotor is warped. If this symptom exists, check the rotors for runout and parallelism. A warped rotor should be

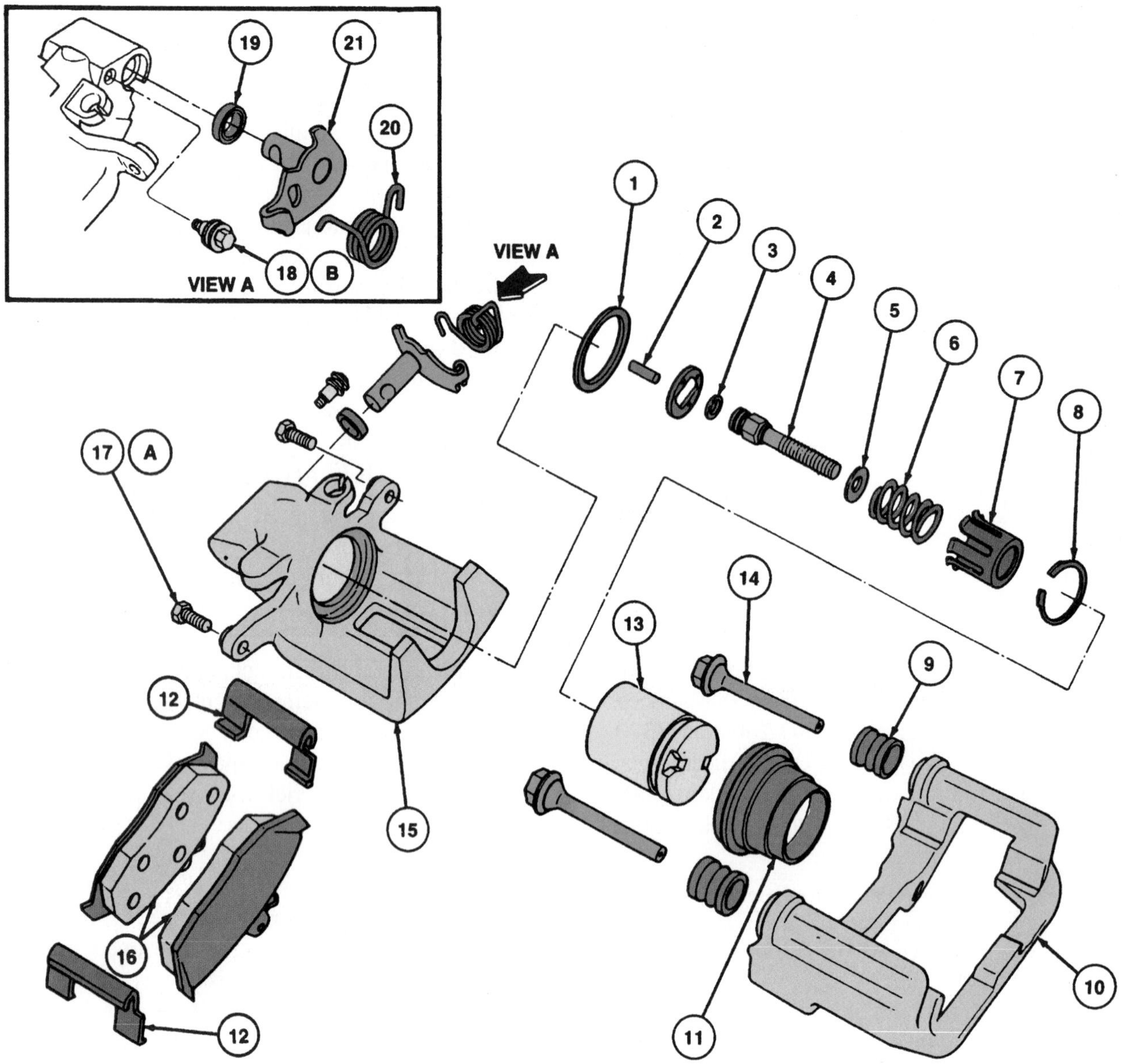

1. Piston seal
2. Pin
3. O-ring
4. Pushrod
5. Flat washer
6. Spring
7. Spring cage
8. Snapring
9. Slider pin boot seal
10. Rear support
11. Piston dust boot
12. Antirattle clip
13. Piston
14. Locating pin
15. Caliper
16. Brake pads
17. Pin retainer
18. Limiting bolt
19. Parking brake shaft seal
20. Parking brake lever return spring
21. Parking brake lever

Figure 48-11 A rear caliper with a threaded drive for the piston. *Courtesy of Ford Motor Company*

replaced and is often caused by improper tightening of the wheel lug nuts. In fact, uneven lug nut torque can cause a pulsating brake pedal. You should be aware that pedal pulsation is normal on vehicles with ABS when the antilock brake system is working.

Spongy Pedal

With a spongy pedal, the customer will probably feel the need to pump the brake pedal to get good stopping ability. The complaint may also be described as a soft pedal. This problem is caused by air in the hydraulic system. Al-

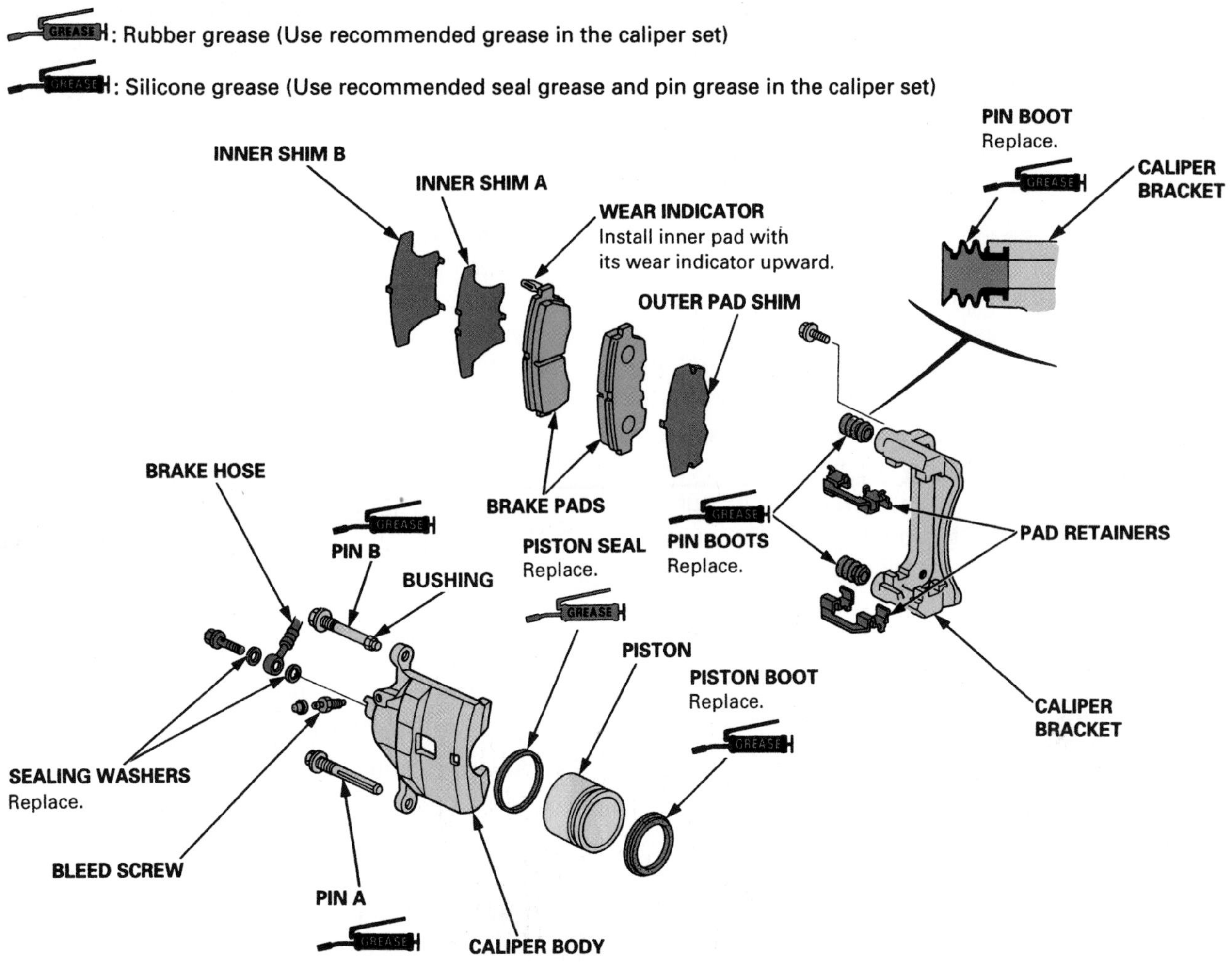

Figure 48-12 Whenever you are doing brake work, thoroughly inspect the entire brake assembly. *Courtesy of American Honda Motor Co., Inc.*

though bleeding the system may remove the air, you should always question how the air got in there. Check for leaks and for proper master cylinder operation.

Hard Pedal

The driver's complaint of a hard pedal normally indicates a problem with the power brake booster. However it can also be caused by a restricted brake line or hose. Carefully check the lines and hoses for damage. Feel the brake hoses. If they seem to have lost their rigidity, the hose may have collapsed on the inside and this is causing the restriction. Restrictions can also be caused by frozen caliper or wheel cylinder pistons.

Dragging Brakes

Dragging brakes make the vehicle feel as if it has lost or is losing power as it drives down the road. The problem also wastes a lot of fuel and generates destructive amounts of heat that can cause serious brake damage and brake failure. While trying to find the cause of this problem, check the parking brake first. Make sure it is off. Check the rear wheels to make sure the parking brakes are released when they should be. If the problem is not in the parking brakes, check for sticky or seized pistons at the calipers and wheel cylinders.

Grabbing Brakes

When the brakes seem to be overly sensitive to pedal pressure, they are grabbing. Normally this problem is caused by contaminated brake linings. If the linings are covered or saturated with oil, find the source of the oil and repair it. Then replace the pads and refinish or replace the rotor.

Noise

If the customer's complaint is noisy brakes, verify during the road test that the problem is in the brakes. If the noise is caused by the brakes, pay attention to the type of noise and let that lead to the source of the problem. Remember, some brake pads have wear sensors that are designed to make a high-pitched squeal when the pads are worn. Other causes could be the rotor rubbing against the splash shield or that something has become wedged between the rotor and another part of the vehicle. Noise may also be caused by failure to install all of the hardware when placing a caliper or brake pads in service.

Pulling

When a vehicle drifts or pulls to one side while cruising or when braking, the cause could be in the brake system or in the steering and suspension system. Check the inflation of the tires, the tires' tread condition, and verify that the tires on each axle are the same size. Check the operation of the brakes. If only one front wheel is actually doing the braking, the vehicle will seem to stumble or pivot on that one wheel. If no problems are found in the brake system, suspect an alignment or suspension problem.

SERVICE PRECAUTIONS

The following general service precautions apply to all disc brake systems and should always be followed:

1. Be sure the vehicle is properly centered and secured on stands or a hoist.
2. If the vehicle has antilock brakes, depressurize the system according to the procedures given in the service manual.
3. Disconnect the battery ground cable.
4. Before any service is performed, carefully check the following:
   - Tires for excessive wear or improper inflation
   - Wheels for bent or warped rims
   - Wheel bearings for looseness or wear
   - Suspension components to see if they are worn or broken
   - Brake fluid level
   - Master cylinder, brake lines or hoses, and each wheel for leaks
5. During servicing, grease, oil, brake fluid, or any other foreign material must be kept off the brake linings, caliper, surfaces of the disc, and external surfaces of the hub. Handle the brake disc and caliper in such a way as to avoid deformation of the disc and nicking or scratching of the brake linings.
6. When a hydraulic hose is disconnected, plug it to prevent any foreign material from entering.
7. Never permit the caliper assembly to hang with its weight on the brake hose. Support it on the suspension or hang it by a piece of wire.
8. Inspect the caliper for leaks. If leakage is present, the caliper must be overhauled.
9. When using compressed air to remove caliper pistons, avoid high pressures. A safe pressure to use is 30 psi (207 kPa).
10. Clean the brake components in either denatured alcohol or clean brake fluid. Do not use mineral-based cleaning solvent such as gasoline, kerosene, carbon tetrachloride, acetone, or paint thinner to clean the caliper. It causes rubber parts to become soft and swollen in an extremely short time.
11. Lubricate any moving member such as the caliper housing or mounting bracket to ensure a free-moving action. Use only recommended lubricant.
12. Before the brake pads are installed, apply a disc brake noise suppressor to the back of the pads to prevent brake squeal. For best results, follow the directions on the container.
13. Obtain a firm brake pedal after servicing the brakes and before moving the vehicle. Be sure to road test the vehicle.
14. Always torque the lug nuts when installing a wheel on a vehicle with disc brakes. Never use an impact gun to tighten the lug nuts. Warpage of the rotor could result if an impact gun is used.

Before beginning brake work, remove about two-thirds of the brake fluid from the master cylinder's reservoir. If this is not done, the fluid could overflow and spill when the pistons are forced back into the caliper bore, possibly damaging the painted surfaces. Replace the cover. Discard old brake fluid.

Another common procedure (and perhaps a better way) is to open the caliper bleeder screw and run a hose down to a container to catch the fluid that is expelled when the piston is forced back into its bore. This also makes it easier to move the piston.

If the bleeder screws are frozen tight with corrosion, it is sometimes possible to free them using a propane torch and penetrating oil. Of course, the caliper has to be removed from the car, taken to a bench, and worked on there. If the bleeder screws cannot be loosened, they can be drilled out and the caliper retapped for an insert, or the caliper can be replaced with a new or rebuilt unit. The bleeder screw should be removed when doing an overhaul.

WARNING!

When using the propane torch to loosen a bleeder screw, use it with extreme care.

GENERAL CALIPER INSPECTION AND SERVICING

Frequently, caliper service involves only the removal and installation of the brake pads. However, since the new pads are thicker than the worn-out set they replace, they locate the piston farther back in the bore where dirt and corrosion might cause the seals to leak. For this reason, it is often good practice to carefully inspect the calipers whenever installing new pads. Of course, it is also good practice to true-up or replace the rotors when replacing brake pads.

When bench working a caliper assembly, use a vise that is equipped with protector jaws. Excessive vise pressure causes bore and piston distortion.

Caliper Removal

To be able to replace brake pads, service the rotor, or to replace the caliper, the caliper must be removed. The procedure for doing this varies according to caliper design. Always follow the specific procedures given in a service manual. Use the following as an example of these procedures:

1. Remove the brake fluid from the master cylinder.
2. Raise the vehicle and remove the wheel and tire assembly.
3. On a sliding or floating caliper, install a C-clamp with the solid end of the clamp on the caliper housing and the screw end on the metal portion of the outboard brake pad. Tighten the clamp until the piston bottoms in the caliper bore **(Figure 48–13)**, then remove the clamp. Bottoming the piston allows room for the brake pad to slide over the ridge of rust that accumulates on the edge of the rotor.

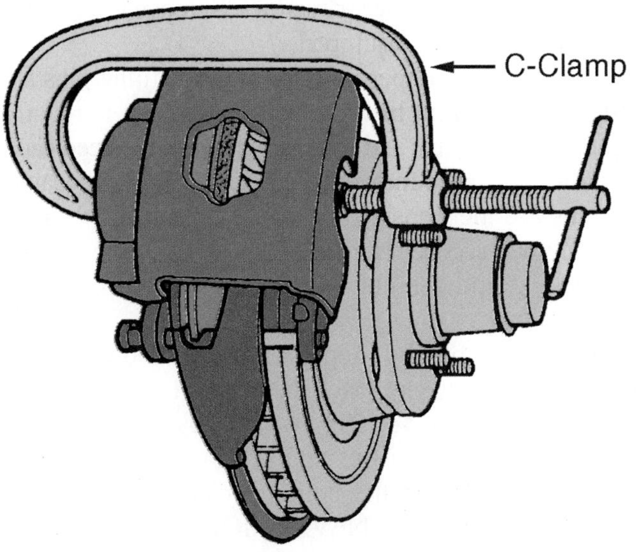

Figure 48-13 Bottoming the piston in the caliper's bore.

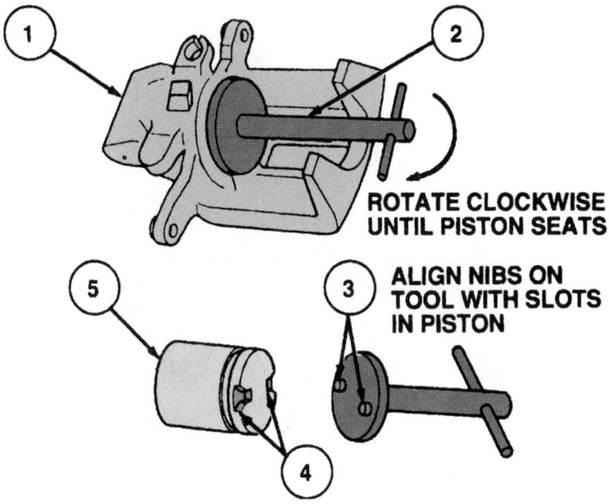

Item	Description
1	Caliper Housing
2	Rear Caliper Piston Adjuster Tool
3	Nibs
4	Slots
5	Rear Disc Brake Piston and Adjuster

Figure 48-14 A special tool is required to move a threaded piston into its bore. *Courtesy of Ford Motor Company*

4. On threaded-type rear calipers, the piston must be rotated to depress it, which requires a special tool **(Figure 48–14)**.
5. Disconnect the brake hose from the caliper and remove the copper gasket or washer and cap the end of the brake hose. If only the brake pads are to be replaced, do not disconnect the brake hose.
6. Remove the two mounting brackets to the steering knuckle bolts. Support the caliper when removing the second bolt to prevent the caliper from falling.
7. On a sliding caliper, remove the top bolts, retainer clip, and **antirattle springs (Figure 48–15)**. On a

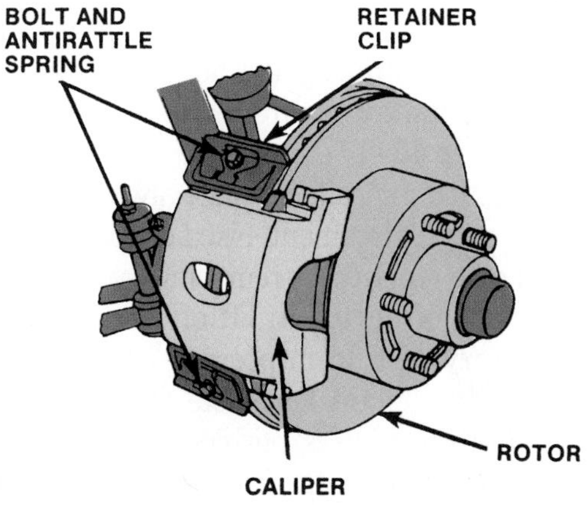

Figure 48-15 Sliding caliper removal.

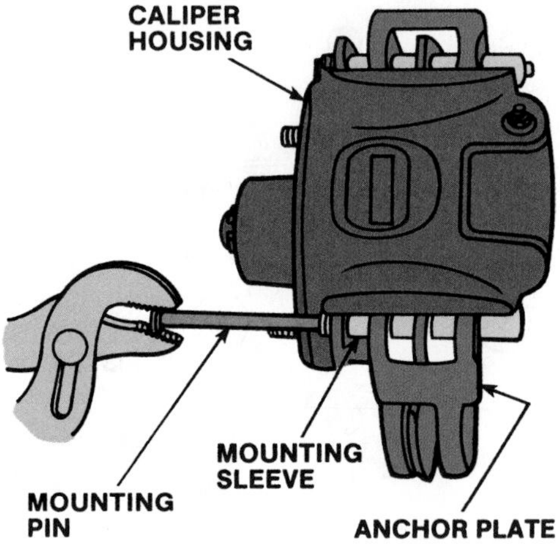

Figure 48–16 Floating caliper removal.

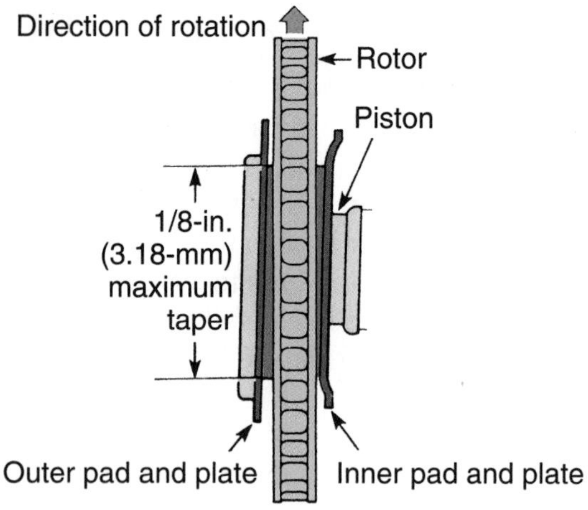

Figure 48–17 Normal pad wear pattern.

floating caliper, remove the two special pins that hold the caliper to the anchor plate **(Figure 48–16)**. On a fixed caliper, remove the bolts holding it to the steering knuckle. On all three types, get the caliper off by prying it straight up and lifting it clear of the rotor.

Brake Pad Removal

Disc brake linings should be checked periodically or whenever the wheels are removed. Some calipers have inspection holes in the caliper body. If they do not, the brake pads can be visually inspected from the outer ends of the caliper.

If you are not sure the pads are worn enough to warrant replacement, measure them at the thinnest part of the pad. Compare this measurement to the minimum brake pad lining thickness listed in the vehicle's service manual, and replace the pads if needed.

When a pad on one side of the rotor has worn more than on the other side, the condition is called uneven wear. Uneven pad wear often means the caliper is sticking and not giving equal pressure to both pads. On a sliding caliper, the problem could be caused by poor lubrication of or deformation of the machined sliding areas on the caliper and/or anchor plate. A slightly tapered wear pattern on the pads of certain models is caused by caliper twist during braking. It is normal if it does not exceed ⅛-inch (3-mm) taper from one end of the pad to the other **(Figure 48–17)**.

Sliding or floating calipers must always be lifted off the rotor for pad replacement. Fixed calipers might have pads that can be replaced by removing the retaining pins or clips instead of having to lift off the entire caliper. Brake pads may be held in position by retaining pins, guide pins, or a support key. Note the position of the shims, anti-rattle clips, keys, bushings, or pins during disassembly. A typical procedure for replacing brake pads is outlined in Photo Sequence 51, included in this chapter.

If only the pads are going to be replaced, lift the caliper off the rotor and hang it up by a wire. Remove the outer pad and inner pad. Remove the old sleeves and bushings and install new ones. Replace rusty pins on a floating caliper to provide for free movement. Transfer shoe retainers, which can be clips or springs, onto the new pads.

SHOP TALK

Most often, calipers are replaced rather than rebuilt. The old caliper is sent back to the manufacturer as a core. The following overview for rebuilding is intended only to give you an understanding of what may be done. Always refer to the appropriate service manual when rebuilding a caliper. ■

Caliper Disassembly

If the caliper must be rebuilt, it should be taken to the workbench for servicing. Drain any brake fluid from the caliper by way of bleeder screws. Remove the bleeder valve protector, if so equipped.

On a floating caliper, examine the mounting pins for rust that could limit travel. Most manufacturers recommend that these pins and their bushings be replaced each time the caliper is removed. This is a good idea because the pins are inexpensive and a good insurance against costly comebacks. On a fixed caliper, check the pistons for sticking and rebuild the caliper if this problem is found.

To disassemble the caliper, the piston and dust boot must first be removed. Place the caliper face down on a workbench **(Figure 48–18)**. Insert the used outer pad or a block of wood into the caliper. Place a folded shop towel on the face of the lining to cushion the piston. Apply low air pressure (NEVER MORE THAN 30 PSI) to the fluid inlet port of the caliper to force the piston from the caliper housing.

Figure 48–18 Using air to remove a piston.

CAUTION!

Wear safety glasses while disassembling the caliper to protect your eyes from spraying brake fluid.

WARNING!

Be careful to apply air pressure very gradually. Be sure there are enough cloths to catch the piston when it comes out of the bore. Never place your fingers in front of the piston for any reason when applying compressed air. Personal injury could occur if the piston is popped out of the bore.

If a piston is frozen, release air pressure and tap the piston into its bore with a soft-headed hammer or mallet. Reapply air pressure. Frozen phenolic (plastic) pistons can be broken into pieces with a chisel and hammer. Be careful not to damage the cylinder bore while doing this. Internal expanding pliers are sometimes used to remove pistons from caliper bores.

Inspect phenolic pistons for cracks, chips, or gouges. Replace the piston if any of these conditions are evident. If the plated surface of a steel piston is worn, pitted, scored, or corroded, it also should be replaced.

Dust boots vary in design depending on the type of piston and seal, but they all fit into one groove in the piston and another groove in the cylinder. One type comes out with the piston and peels off. Another type stays in place and the piston comes out through the boot, and then is removed from the cylinder **(Figure 48–19)**. In either case, peel the boot from its groove. In some cases it might be necessary to pry it out, but be careful not to scratch the cylinder bore while doing so. The old boot can be discarded since it must be replaced along with the seal.

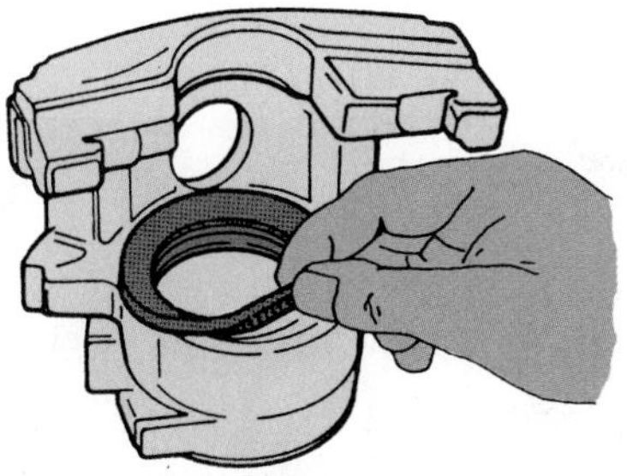

Figure 48–19 Peeling off the dust boot.

Remove the piston's and the cylinder's seal by prying them out with a wooden or plastic tool **(Figure 48–20)**. Do not use a screwdriver or other metal tool. Any of these could nick the metal in the caliper bore and cause a leak. Inspect the bore for pitting or scoring. A bore that shows light scratches or corrosion can usually be cleaned with crocus cloth. However, a bore that has deep

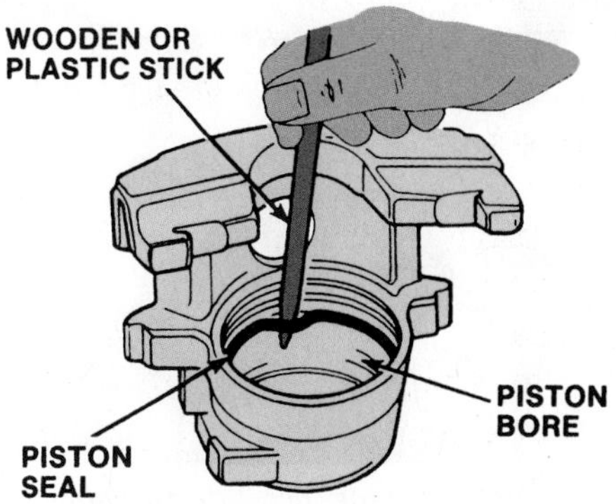

Figure 48–20 Removing a piston seal with a wooden or plastic stick.

PHOTO SEQUENCE 51

Removing and Replacing Brake Pads

P51-1 *Front brake pad replacement begins with removing brake fluid from the master cylinder reservoir.*

P51-2 *Raise the car. Make sure it is safely positioned on the lift. Remove its wheel assemblies.*

P51-3 *Inspect the brake assembly. Look for signs of fluid leaks, broken or cracked lines, or a damaged brake rotor. If any problem is found, correct it before installing the new brake pads.*

P51-4 *Loosen the bolts and remove the pad locator pins.*

P51-5 *Lift and rotate the caliper assembly from the rotor.*

P51-6 *Remove the brake pads from the caliper assembly.*

P51-7 *Fasten a piece of wire to the car's frame and support the caliper with the wire.*

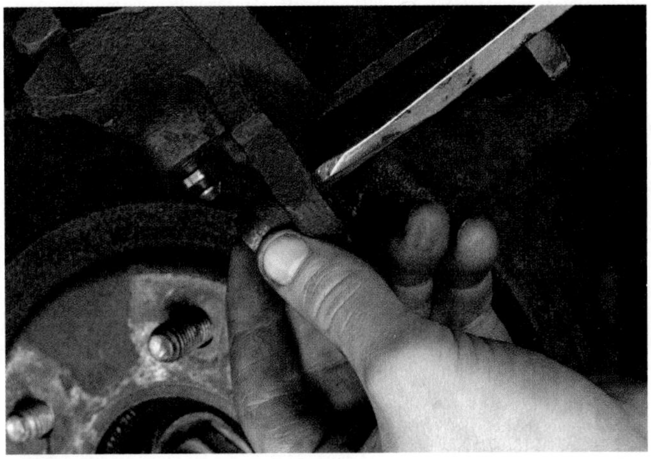

P51-8 *Check the condition of the locating pin insulators and sleeves.*

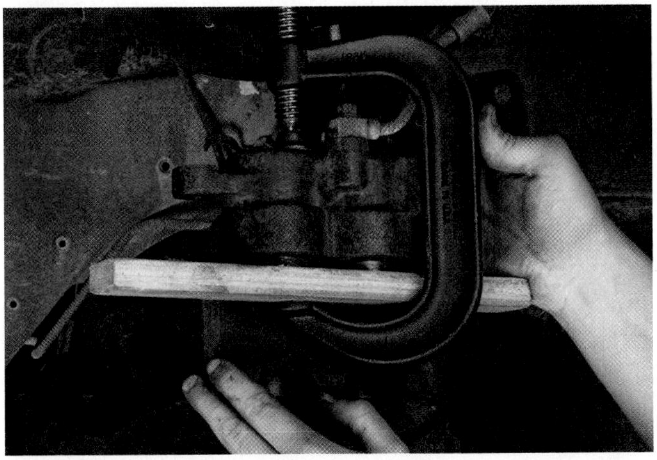

P51-9 *Place a piece of wood over the caliper's piston and install a C-clamp over the wood and caliper. Tighten the clamp to force the piston back into its bore.*

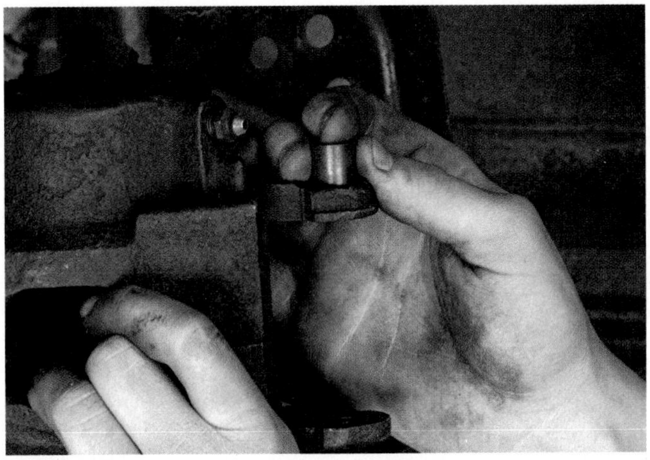

P51-10 *Remove the clamp and install new locating pin insulators and sleeves, if necessary.*

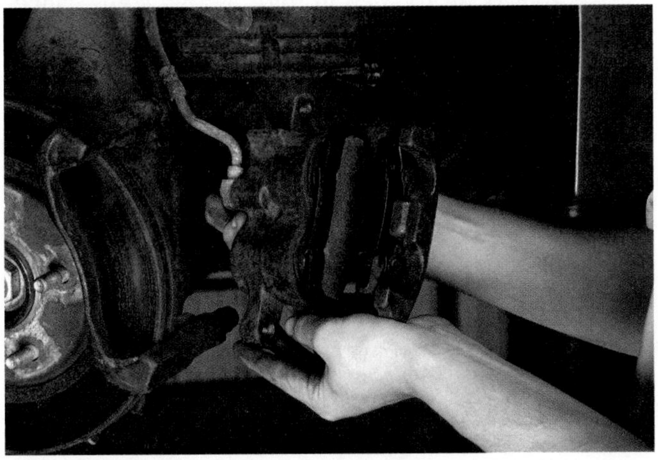

P51-11 *Install the new pads into the caliper.*

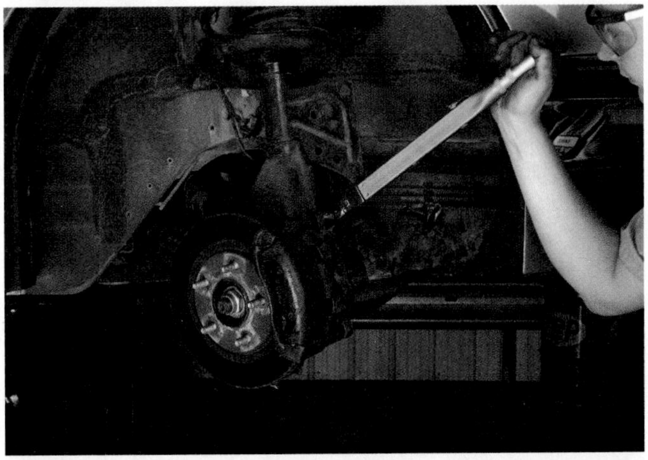

P51-12 *Set caliper with pads over the rotor and install the locating pins. After the assembly is in the proper position, torque the pins according to specifications.*

scratches or scoring normally indicates that the caliper should be replaced. In some cases, the cylinder can be honed. Check the service manual before doing this. If there is no mention of honing the bore, the manufacturer probably does not recommend it. Black stains on the bore walls are caused by piston seals. They do no harm.

When using a hone, be sure to install the hone baffle before honing the bore. The baffle protects the hone stones from damage. Use extreme care in cleaning the caliper after honing. Remove all dust and grit by flushing the caliper with alcohol. Wipe it dry with a clean lint-free cloth and then clean the caliper a second time in the same manner.

Loaded Calipers

There is now a trend toward installing loaded calipers **(Figure 48–21)** rather than overhauling calipers in the shop. Loaded calipers are completely assembled with friction pads and mounting hardware included. Besides the convenience and the savings of installation time, preassembled calipers also reduce the odds of errors during caliper overhaul.

Mistakes frequently made when replacing calipers include forgetting to bend pad-locating tabs that prevent pad vibration and noise, leaving off antirattle clips and pad insulators, and reusing corroded caliper mounting hardware that can cause a floating caliper to bind up and wear the pads unevenly.

One of the major causes of premature brake wear is rust. It causes improper slider and piston operation that leads to uneven pad wear. Tests have shown that when only the pads are replaced, the new pads can wear out in half the mileage as the originals when rust affects caliper operation. Installing a loaded caliper ensures that all parts requiring replacement are replaced.

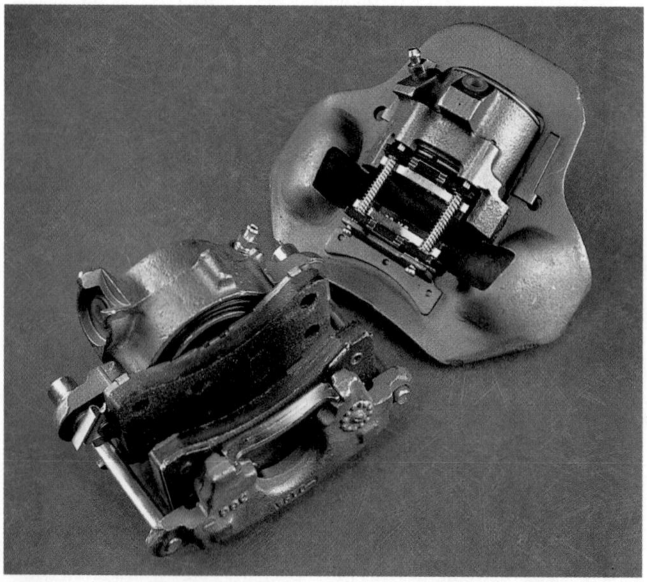

Figure 48-21 A pair of loaded brake calipers. *Courtesy of Bendix Brakes by AlliedSignal*

SHOP TALK

Avoid mismatching friction materials from side to side. When one caliper is bad, both calipers should be replaced using the same friction material on both sides. ■

Caliper Reassembly

Before assembling the caliper, clean the phenolic piston (if so equipped) and all metal parts to be reused in clean denatured alcohol or brake fluid. Then, clean out and dry the grooves and passageways with compressed air. Make sure that the caliper bore and component parts are thoroughly clean.

To replace a typical piston seal, dust boot, and piston, first lubricate the new piston seal with clean brake fluid or assembly lubricant (usually supplied with the caliper rebuild kit). Make sure the seal is not distorted. Insert it into the groove in the cylinder bore so it does not become twisted or rolled. Install a new dust boot by setting the flange squarely in the outer groove of the caliper bore. Next, coat the piston with brake fluid or assembly lubricant and install it in the cylinder bore. Be sure to use a wood block or other flat stock when installing the piston back into the piston bore. Never apply a C-clamp directly to a phenolic piston, and be sure the pistons are not cocked. Spread the dust boot over the piston as it is installed. Seat the dust boot in the piston groove.

With some types of boot/piston arrangements, the procedure of installation is slightly different from that already described. That is, the new dust boot is pulled over the end of the piston **(Figure 48–22)**. Lubricate the piston with brake fluid before installing it in the caliper. Then by hand, slip the piston carefully into the cylinder bore, pushing it straight, so the piston seal is not damaged during installation. Use an installation tool or wooden block to seat the new dust boot **(Figure 48–23)**.

Another point to keep in mind is that some caliper designs have a slot cut in the face of the pistons that must align with an **antisqueal shim**. Make sure that

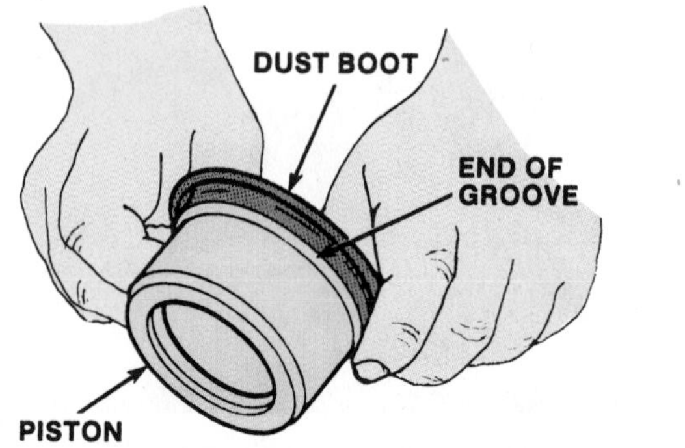

Figure 48-22 Some installation procedures require the dust boot to be pulled over the end of the piston.

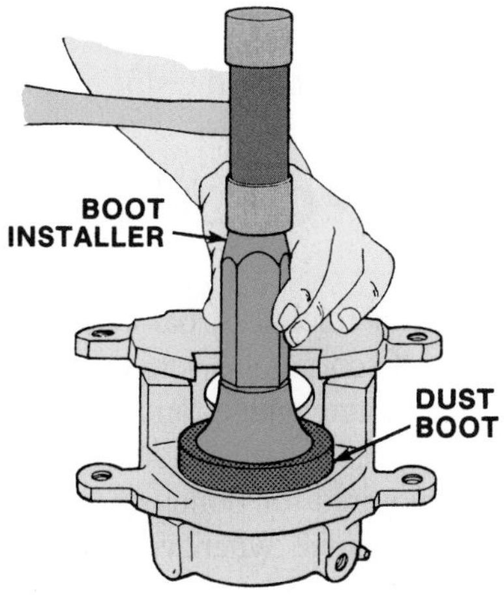

Figure 48-23 Seating a dust boot with a boot installer.

the piston and shim align. It might be necessary to turn the piston to achieve proper alignment. To complete the caliper assembly job, install the bleeder screw.

CAUTION!

On fixed calipers, bridge bolts are used to hold the two caliper halves together. These are high-tensile bolts ordered only by specific part number. They require accurate torque tightness to prevent leakage. Do not attempt to use standard bolts in place of bridge bolts.

Brake Pad Installation

It is a good practice to replace disc brake hardware **(Figure 48–24)** when replacing disc brake pads. Replacement of the hardware ensures proper caliper movement and brake pad retention. It also aids in preventing brake noise and uneven brake pad wear.

Fixed Caliper Brake Pads The designs of fixed caliper disc brakes vary slightly. Generally, to replace the pads, insert new pads and plates in the caliper with the metal plates against the end of the pistons. Be sure that the plates are properly seated in the caliper. Spread the pads apart and slide the caliper into position on the rotor. With some pads, mounting bolts are used to hold them in place. These bolts are usually tightened 80 to 90 foot-pounds (108 to 122 newton-meters). On some fixed disc brakes, the pads are held in place by retaining clips and/or retaining pins. Reinstall the antirattle spring/clips and other hardware (if so equipped).

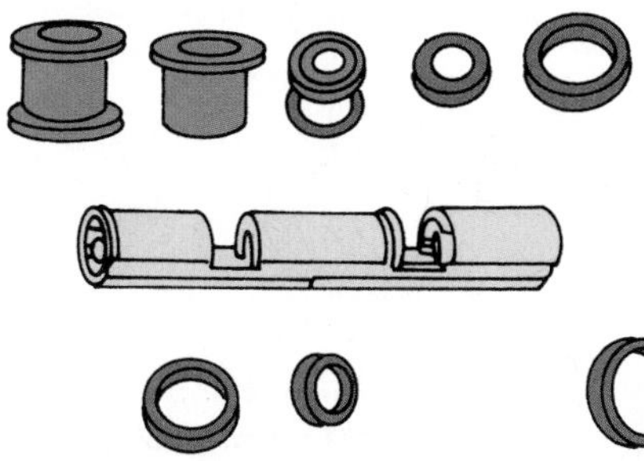

Figure 48-24 An assortment of caliper guide pin bushings and insulators. *Courtesy of Wagner Brake Products*

Sliding Caliper Brake Pads Push the piston carefully back into the bore until it bottoms. Lightly lubricate the sliding surfaces of the caliper and the caliper anchor. Slide a new outer pad into the recess of the caliper. No free play between the brake pad flanges and caliper fingers should exist. If free play is found, remove the pad from the caliper and bend the flanges to eliminate all vertical free play **(Figure 48–25)**. Install the pad.

Place the inner pad into position on the caliper anchor with the pad's flange on the machined sliding area. Fit the caliper over the rotor. Align the caliper to the anchor and slide it into position. Be careful not to pull the dust boot from its groove when the piston and boot slide over the inboard pad. Install the antirattle springs (if so equipped) on top of the retainer plate and tighten the retaining screws to specification.

On some calipers, especially those used as rear brakes, there is a notch or groove in the piston and a tab on the rear of the inner pad. During installation of the pad, the tab must fit into the groove in the piston **(Figure 48–26)**.

Floating Caliper Brake Pads For floating or pin caliper disc brakes, compress the flanges of the outer bushing in the caliper fingers and work them into position in the hole from the outer side of the caliper. Compress the flanges of the inner guide pin bushings and install them.

Slide the new pad and lining assemblies into position in the adapter and caliper. Be sure that the metal portion

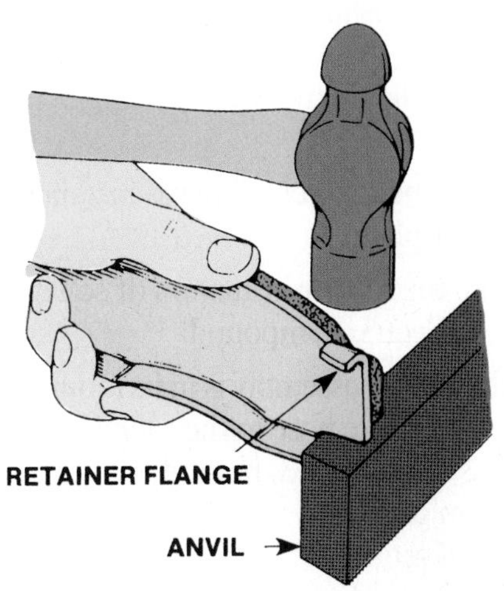

Figure 48-25 Bend the retaining flange if there is excessive free play.

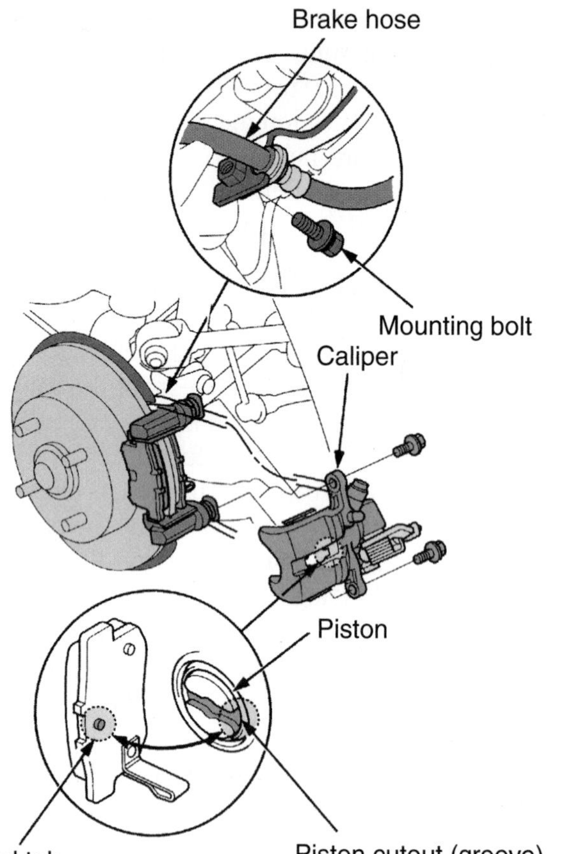

Figure 48-26 On some rear brake assemblies, there is a tab on the back of the pad that must line up with the groove in the piston. *Courtesy of American Honda Motor Co., Inc.*

of the pad is fully recessed in the caliper and adapter and that the proper pad is on the outer side of the caliper.

Hold the outer pad and carefully slide the caliper into position on the anchor and over the disc. Align the guide pin holes of the anchor with those of the inner and outer pads. Lightly lubricate and install the guide pins through the bushings, caliper, anchor, and inner and outer pads into the outer bushings in the caliper and antirattle spring.

When installing any type of caliper, follow these guidelines:

- Make sure the correct caliper is installed on the correct anchor plate.
- Lubricate the rubber insulators (if so equipped) with silicone dielectric compound.
- After the caliper assembly is in its mounting brackets, connect the brake hose to the caliper. If copper washers or gaskets are used, be sure to use new ones—the old ones might have taken a set and might not form a tight seal if reused.
- Fill the master cylinder reservoirs and bleed the hydraulic system.
- Check for fluid leaks under maximum pedal pressure.
- Lower the vehicle and road test it.

ROTOR INSPECTION AND SERVICING

Tolerances on rotor thickness, parallelism, runout, flatness, and depth of scoring are very critical and must be measured with exacting gauges and micrometers. Accurate measuring tools and up-to-date rotor resurfacing equipment are required for brake rotor service. The rotors should be inspected whenever brake pads are required and when the wheels are removed for other types of service. The following are typical disc brake rotor conditions that need careful inspection.

CAUTION!

Never turn the rotor on one side of the vehicle and not the other.

Lateral Runout

Excessive **lateral runout** is a wobbling of the rotor from side to side when it rotates. The excessive wobble knocks the pads farther back than normal, causing the pedal to pulse and vibrate during braking. Chatter can also result. It also causes excessive pedal travel because the pistons have farther to travel to reach the rotor. If runout exceeds specifications **(Figure 48–27)**, the rotor must be turned or replaced.

Lack of Parallelism

Parallelism refers to variations in thickness of the rotor **(Figure 48–28)**. If the rotor is out of parallel, it can cause

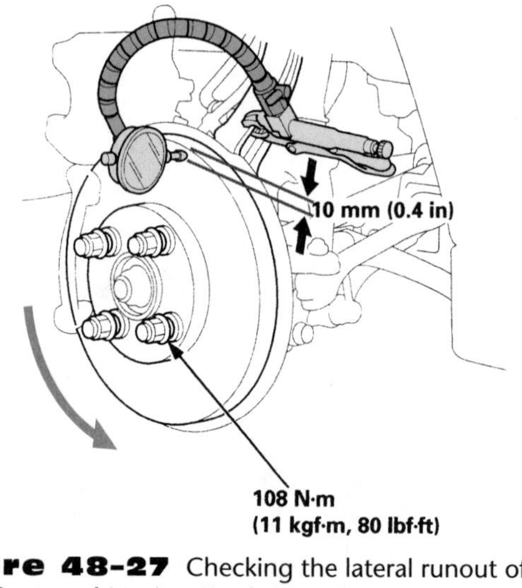

Figure 48-27 Checking the lateral runout of a brake rotor. *Courtesy of American Honda Motor Co., Inc.*

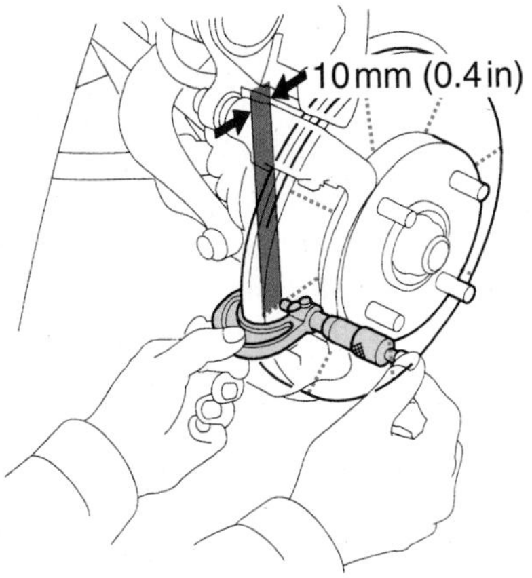

Figure 48–28 To check a rotor's parallelism, measure the thickness of the rotor at eight different spots. *Courtesy of American Honda Motor Co., Inc.*

excessive pedal travel, front-end vibration, pedal pulsation, chatter, and on occasion, grabbing of the brakes. It must be resurfaced or replaced.

Scoring

Rotor wear or scoring can be caused by linings that are worn through to the rivets or backing plate, or by friction material that is harsh or unkind to the mating surface. Rust, road dirt, and other contamination could also cause rotor scoring.

Light scoring (less than a depth of 0.015 inch [.381 mm]) of the disc braking surface can occur during normal brake use. This does not affect brake operation; however, it can result in a higher-than-average brake shoe lining wear rate. But, any rotor having score marks more than 0.15 inch (3.8 mm) should be refinished or replaced.

Bluing or Heat Checking

If the lining surface is charred, blued, or hard-ended with a heavy glaze, or if the rotor is severely heat checked, machine the rotor or replace it.

Hard or chill spots of steel in a rotor cast-iron surface usually result from a change in the metallurgy caused by braking heat. Pulling, rapid wear, hard pedal, and noise occur. These spots can be removed by grinding; however, only the raised surfaces are removed, and they could reappear when heat is again encountered. The rotor must be replaced.

Rusty Rotor

If the vehicle has not been driven for a period of time, the discs rust in the area not covered by the lining and cause noise and chatter. Excessive wear and scoring of the discs and lining result. Wear ridges on the discs can cause temporary improper pad lining contact if ridges are not removed before installation of new lining. Lining deposits on the disc can cause erratic friction characteristics if a new lining is installed without resurfacing or cleaning the disc.

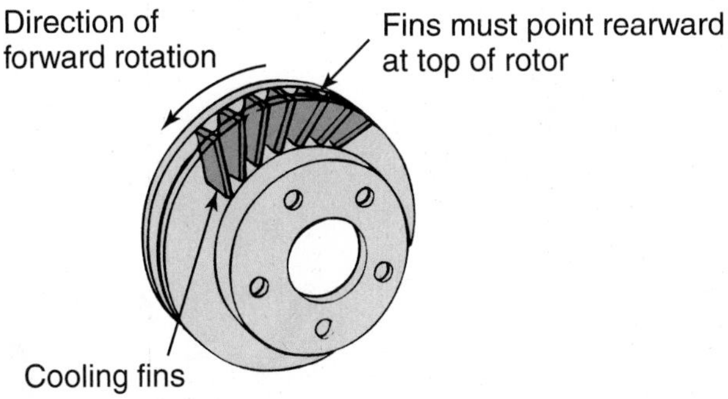

Figure 48–29 Some ventilated rotors have curved cooling fins and are directional. *Courtesy of EIS—Standard Motor Products*

Rotor Service

If the thickness of the rotor is below the minimal allowable thickness or is badly distorted, it must be replaced. If it has imperfections and has a thickness that is close to the minimum allowable thickness, it should also be replaced. If the thickness is well beyond minimum specifications, it can be trued and smoothed with a brake lathe. When replacing a rotor, make sure to tighten or secure it according to the manufacturer's recommendations.

On some vehicles the rotors are directional. These ventilated discs have curved cooling fins **(Figure 48–29)**. For maximum cooling, these fins must point rearward at the top of the rotor to allow centrifugal force to spin air outward from inside the rotor.

Rotors that have minor imperfections or are slightly unparallel can be turned true and smooth with a brake lathe. A brake lathe cuts metal away to achieve the desired surface finish. Basically two types of brake disc lathes are used by the industry. The first is one that has the capability of resurfacing brake drums and brake discs **(Figure 48–30)** after they have been removed from the vehicle. The lathe rotates the disc as cutting tools work their way across the braking surface of the disc. The second type is an on-vehicle brake lathe **(Figure 48–31)**. This type of brake lathe is a time saver because the rotor does not need to be removed from the vehicle. Special fixtures are used to straddle the rotor so the cutting tools can precisely cut both sides of the rotor. An electric motor is used to rotate the disc and hub assembly during cutting.

SHOP TALK

Special cutters are required to resurface composite rotors. Make sure you have the correct machine and cutting tools before attempting to true up a composite brake rotor. ■

Figure 48-30 A typical brake lathe. *Courtesy of Hunter Engineering Company*

Figure 48-31 An on-the-vehicle brake lathe. *Courtesy of Accu Industries*

CASE STUDY

An obviously embarrassed customer explains his problem to the brake shop technician. He just finished installing new disc pads on his light truck, but the disc brakes are dragging. The young man insists he performed the work correctly and that the rest of the braking system is in fine working order.

The technician inspects the front brakes and finds that the owner cleaned all components well and installed the pads correctly. The technician checks the action of the caliper. Its piston shows no signs of sticking, but the caliper appears to be binding. Removal and inspection of the caliper locates the problem. Although the owner had carefully cleaned the slides of the caliper, he had failed to apply any lubricant to the surface. This was causing the calipers to bind and the brakes to drag.

Omitting the simplest of tasks when performing repairs can often lead to failure and wasted time.

KEY TERMS

Antirattle spring	Lateral runout
Antisqueal shim	Parallelism
Audible sensor	Phenolic resin (piston)
Caliper	Rotor
Dust boot	Sliding caliper
Fixed caliper	Spindle anchor plate
Floating caliper	Splash shield

SUMMARY

- Disc brakes offer four major advantages over drum brakes: resistance to heat fade, resistance to water fade, increased straight-line stopping ability, and automatic adjustment.
- The typical rotor is attached to and rotates with the wheel hub assembly. Heavier vehicles generally use ventilated rotors. Splash shields protect the rotors and pads from road moisture and dirt.
- The caliper assembly includes cylinder bores and pistons, dust boots, and piston hydraulic seals.
- Brake pads are placed in each side of the caliper and together straddle the rotor. Some brake pads have wear sensors.

- Fixed caliper disc brakes do not move when the brakes are applied. Floating caliper disc brakes slide back and forth on pins or bolts. Sliding calipers slide on surfaces that have been machined smooth for this purpose.
- In a rear disc brake system the inside of each rear wheel hub and rotor assembly is used as the parking brake drum.
- Rear disc parking brakes have a mechanism that forces the pads against the rotor mechanically.
- The general procedures involved in a complete caliper overhaul include tasks such as: caliper removal, brake pad removal, caliper disassembly, caliper assembly, brake pad installation, and caliper installation.
- The first step in proper caliper service is to remove the caliper assembly from the vehicle.
- Disc brake pads should be checked periodically or whenever the wheels are removed. They should be replaced if they fail to exceed minimum lining thickness as listed in the service manual.
- To disassemble the caliper, the piston and dust boot must first be removed. Compressed air is used to pop the piston out of the bore.
- Before assembling the caliper, all metal parts and the phenolic piston are cleaned in denatured alcohol or brake fluid. The grooves and passageways of the caliper are cleaned out and dried with compressed air.
- It is a good practice to replace disc brake hardware when replacing disc brake pads.
- Disc brake rotor conditions that must be corrected include lateral runout, lack of parallelism, scoring, blueing or heat checking, and rusty rotors.

TECH MANUAL

The following procedures are included in Chapter 48 of the *Tech Manual* that accompanies this book:

1. Inspecting and servicing typical disc brakes
2. Servicing a single-piston rear disc brake caliper.
3. Servicing a floating rear disc brake caliper.
4. Removing and installing a typical hub and rotor assembly.
5. Inspecting and measuring disc brake rotors.

REVIEW QUESTIONS

1. Name four major advantages of disc brakes over drum brakes.
2. Name the three major assemblies that make up a disc brake.
3. Name the three types of calipers used on disc brakes.
4. What type of brake uses the inside of each rear wheel hub and rotor assembly as a parking brake drum?
5. How many cylinder bores and pistons can a caliper assembly contain?
6. *True or False?* Disc brakes are not as likely to fade during heavy braking as are drum brakes.
7. How do disc brakes self-adjust?
8. What are the two basic types of brake rotors used on today's vehicles?
9. What channels the flow of air over the exposed rotor surfaces?
   a. webs
   b. hub assembly
   c. splash shield
   d. caliper assembly
10. Technician A says a hard brake pedal on a disc brake vehicle can be caused by air in the hydraulic system. Technician B says a pulsating pedal on a disc brake vehicle can be caused by a restricted brake line. Who is correct?
    a. Technician A
    b. Technician B
    c. Both A and B
    d. Neither A nor B
11. What prevents moisture from entering the cylinder bore?
    a. phenolic piston
    b. drag caliper
    c. splash shield
    d. dust boot
12. When servicing disc brakes on a vehicle, Technician A works on one wheel before beginning work on another. Technician B uses a minimum of 50

psi (345 kPa) of air pressure to force the piston from the caliper housing. Who is correct?

a. Technician A **c.** Both A and B
b. Technician B **d.** Neither A nor B

13. When examining disc brakes, Technician A visually inspects the rotor and says the rotor can be reused if it is not damaged or scored. Technician B says it is normal for the inboard pad to be worn more than the outside pad. Who is correct?

a. Technician A **c.** Both A and B
b. Technician B **d.** Neither A nor B

14. What should be done to remove rust, corrosion, pitting, and scratches from the piston bore?

15. Which term refers to variations in thickness of the rotor?

a. torque **c.** parallelism
b. lateral runout **d.** pedal pulsation

16. Technician A cleans brake components in denatured alcohol. Technician B cleans brake components in clean brake fluid. Who is correct?

a. Technician A **c.** Both A and B
b. Technician B **d.** Neither A nor B

17. Technician A states that rotor wear can be caused by worn pad linings. Technician B states that hard spots in a rotor usually result from manufacturing defects. Who is correct?

a. Technician A **c.** Both A and B
b. Technician B **d.** Neither A nor B

18. Which of the following is not likely to cause a pulsating brake pedal?

a. loose wheel bearings
b. worn brake pad linings
c. excessive lateral runout
d. nonparallel rotors

19. While freeing up a stuck bleed screw from a caliper, Technician A carefully heats up the screw with a torch while it is mounted to its support at the wheel. Technician B removes the caliper and drills out the screw, then retaps the bore for a screw insert. Who is correct?

a. Technician A **c.** Both A and B
b. Technician B **d.** Neither A nor B

20. While discussing how to remove the piston from a brake caliper, Technician A says the dust boot should be removed, then a large dull screwdriver should be inserted into the piston groove to pry the piston out. Technician B says air pressure should be injected into the bleeder screw's bore to force the piston out of the caliper. Who is correct?

a. Technician A **c.** Both A and B
b. Technician B **d.** Neither A nor B

49

ANTILOCK BRAKE, TRACTION CONTROL, AND STABILITY CONTROL SYSTEMS

OBJECTIVES

■ Explain how antilock brake systems work to bring a vehicle to a controlled stop. ■ Describe the differences between an integrated and a nonintegrated antilock brake system. ■ Briefly describe the major components of a two-wheel antilock brake system. ■ Briefly describe the major components of a four-wheel antilock brake system. ■ Describe the operation of the major components of an antilock brake system. ■ Describe the operation of the major components of automatic traction and stability control systems. ■ Explain the best procedure for finding ABS faults. ■ List the precautions that should be followed whenever working on an antilock brake system.

Antilock brake systems (ABS) and traction and stability control systems are rapidly gaining popularity. ABS is now standard equipment on most vehicles. These systems add yet another group of electronically controlled systems to the increasingly complex modern vehicle.

ANTILOCK BRAKES

Modern antilock brake systems **(Figure 49–1)** can be thought of as electronic/hydraulic pumping of the brakes for straight-line stopping under panic conditions. Good drivers have always pumped the brake pedal during panic stops to avoid wheel lockup and the loss of steering control. Antilock brake systems simply get the pumping job done much faster and in a much more precise manner than the fastest human foot. Keep in mind that a tire on the verge of slipping produces more friction with respect to the road than one that is locked and skidding. Once a tire loses its grip, friction is reduced and the vehicle takes longer to stop.

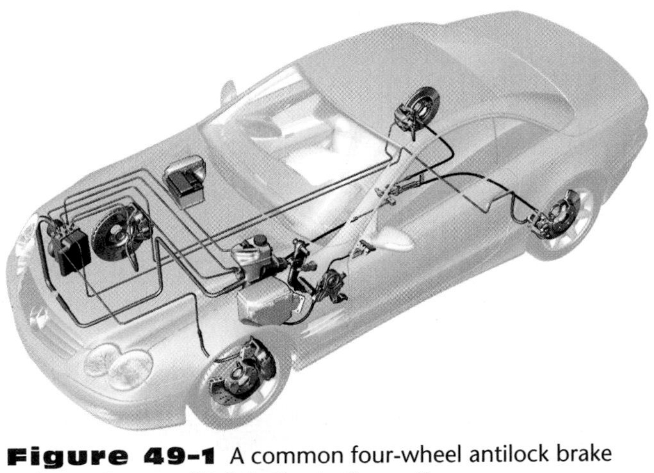

Figure 49-1 A common four-wheel antilock brake system. *Courtesy of DaimlerChrysler Corporation*

Pressure Modulation

When the driver quickly and firmly applies the brakes and holds the pedal down, the brakes of a vehicle not equipped with ABS will almost immediately lock the wheels. The vehicle slides rather than rolls to a stop. During this time, the driver also has a very difficult time keeping the vehicle straight and the vehicle will skid out of control. The skidding and lack of control was caused by the locking of the wheels. If the driver was able to release the brake pedal just before the wheels locked up then reapply the brakes, the skidding could be avoided.

This release and apply of the brake pedal is exactly what an antilock system does. When the brake pedal is pumped or pulsed, pressure is quickly applied and released at the wheels. This is called **pressure modulation**. Pressure modulation works to prevent wheel locking. Antilock brake systems can modulate the pressure to the brakes as often as fifteen times per second. By modulating the pressure to the brakes, friction between the tires and the road is maintained and the vehicle is able to come to a controllable stop.

The only time reduced friction aids in braking is when a tire is on loose snow. A locked tire allows a small wedge of snow to build up ahead of it, which allows it to stop in a shorter distance than a rolling tire.

Steering is another important consideration. As long as a tire does not slip, it goes only in the direction in which it is turned. But once it skids, it has little or no directional stability. One of the big advantages of ABS, therefore, is the ability to keep control of the vehicle under all conditions.

The maneuverability of the vehicle is reduced if the front wheels are locked, and the stability of the vehicle is reduced if the rear wheels are locked. Antilock brake systems precisely control the slip rate **(Figure 49–2)** of the wheels to ensure maximum grip force from the tires, and it thereby ensures maneuverability and stability of the vehicle. An ABS control module calculates the slip rate of the wheels based on the vehicle speed and the speed of the wheels, and then it controls the brake fluid pressure to attain the target slip rate.

Although ABS prevents complete wheel lockup, it allows some wheel slip in order to achieve the best braking possible. During ABS operation, the target slip rate can be from 10 to 30%. Zero (0)% slip means the wheel is rolling freely, while 100% means the wheel is fully locked. A slip rate of 25% means the velocity of a wheel is 25% less than that of a free rolling wheel at the same vehicle speed. Many things are considered when determining the target slip rate for a particular vehicle. For some the range is very low—5 to 10%—while on others it is high—20 to 30%.

CUSTOMER CARE

Remind your customers that pumping the brake pedal while stopping will prevent the ABS from activating. They should always keep firm steady pressure on the brake pedal during braking. ■

Pedal Feel

The brake pedal on a vehicle equipped with ABS has a different feel than that of a conventional braking system. When the ABS is activated, a small bump followed by rapid pedal pulsations will continue until the vehicle comes to a stop or the ABS turns off. These pulsations are the result of the modulation of pressure to the brakes and are felt more on some systems than on others. This is due to the use of damping valves in some modulation units. If pedal feel is of concern during diagnosis of a brake problem, compare the brake pedal feel with that of a similar vehicle with a normal operating antilock brake system. With ABS, the brake pedal effort and pedal feel during normal braking are similar to that of a conventional power brake system.

ABS COMPONENTS

Many different designs of antilock brake systems are found on today's vehicles. These designs vary in their basic layout, operation, and components. There are also variations based on the type of power-assist used and on whether the system is integral. The ABS components that may be found on a vehicle can be divided into two categories: hydraulic and electrical/electronic components. Keep in mind, no one system uses all of the parts discussed here. Normal or conventional brake parts are part of the overall brake system but are not in the following discussion.

Hydraulic Components

Accumulator An accumulator is used to store hydraulic fluid to maintain high pressure in the brake system and to provide residual pressure for power-assisted braking. Normally, the accumulator is charged with nitrogen gas **(Figure 49–3)** and is an integral part of the modulator unit. This unit is typically found on vehicles with a hydraulically assisted brake system.

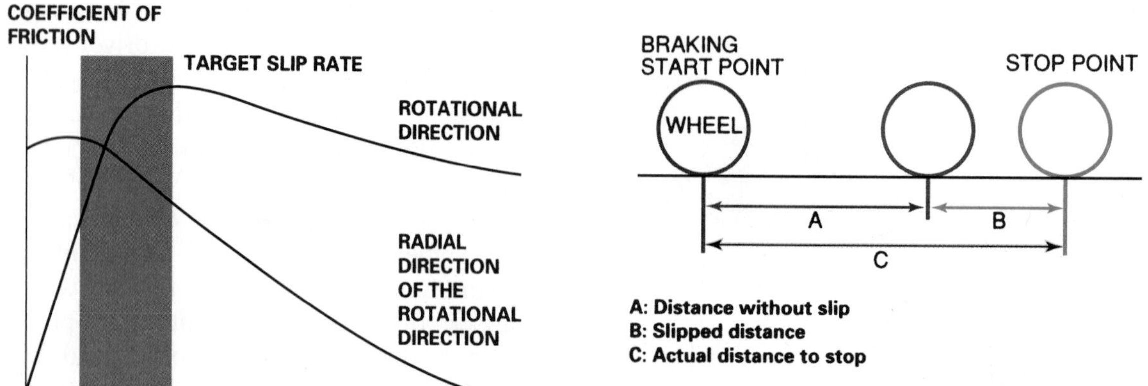

$$\text{SLIP RATE} = \frac{B}{C} = \frac{\text{VEHICLE SPEED} - \text{WHEEL SPEED}}{\text{VEHICLE SPEED}}$$

Figure 49-2 Defining slip rate. *Courtesy of American Honda Motor Co., Inc.*

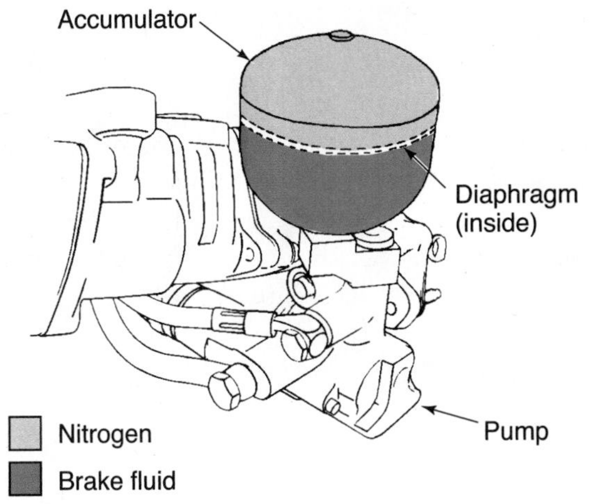

Figure 49-3 Pressure in an accumulator. *Courtesy of Wagner Brake Products*

Antilock Hydraulic Control Valve Assembly This assembly controls the release and application of brake system pressure to the wheel brake assemblies. It may be of the integral-type, meaning this unit is combined with the power-boost and master cylinder units into one assembly **(Figure 49–4)**. The nonintegral-type is mounted externally from the master cylinder/power booster unit and is located between the master cylinder and wheel brake assemblies. Both types generally contain solenoid valves that control the releasing, the holding, and the applying of brake system pressure.

Booster Pump The booster pump is an assembly of an electric motor and pump. The booster pump is used to provide pressurized hydraulic fluid for the ABS. The pump's motor is controlled by the system's control unit. The booster pump is also called the electric pump and motor assembly.

Booster/Master Cylinder Assembly The booster/master cylinder assembly **(Figure 49–5)**, sometimes referred to as the hydraulic unit, contains the valves and pistons needed to modulate hydraulic pressure in the wheel circuits during ABS operation. Power brake-assist is provided by pressurized brake fluid supplied by a hydraulic pump.

Fluid Accumulators Different than a pressure accumulator, fluid accumulators temporarily store brake fluid

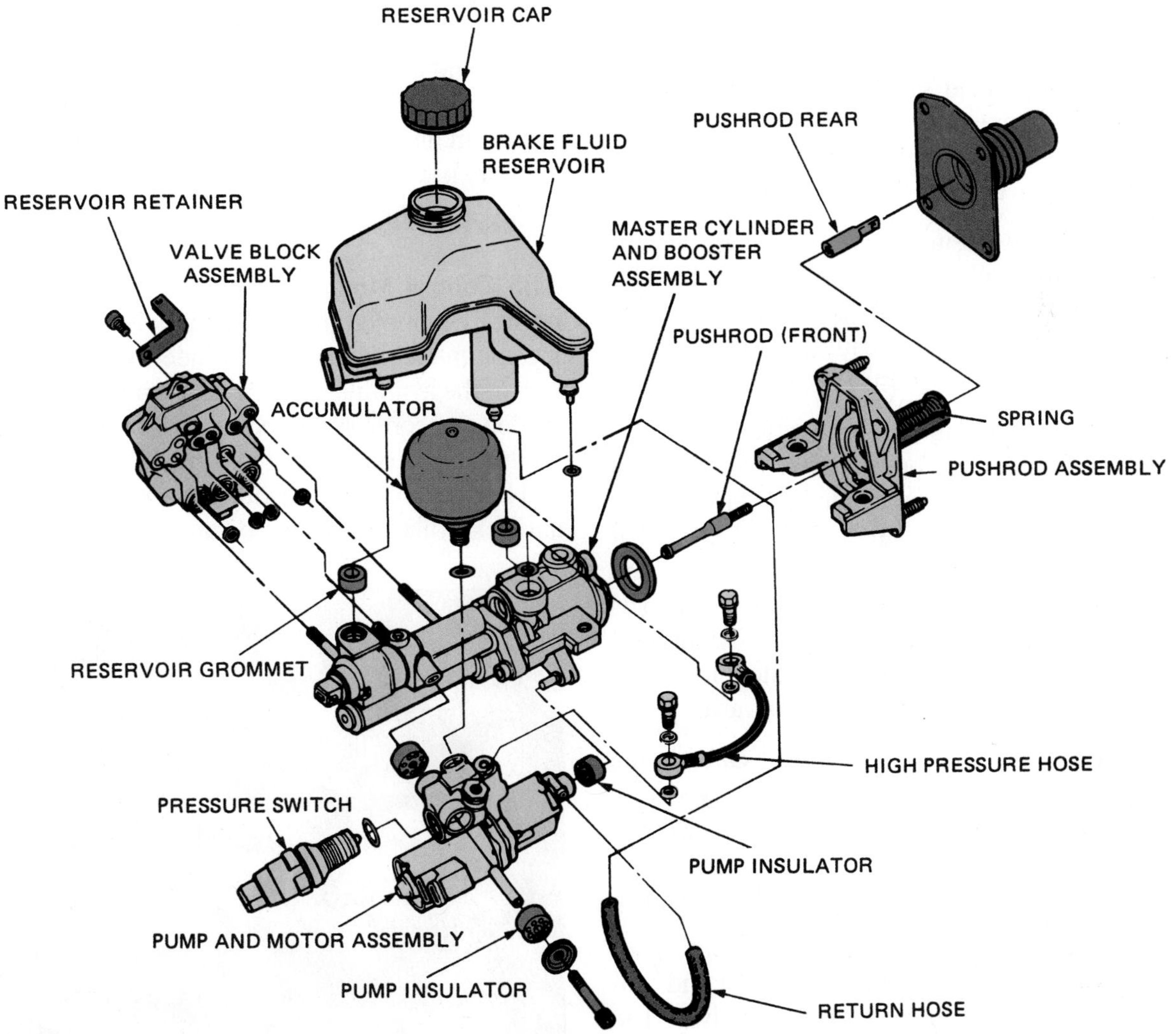

Figure 49-4 A hydraulic unit for a Teves antilock brake system. *Courtesy of Wagner Brake Products*

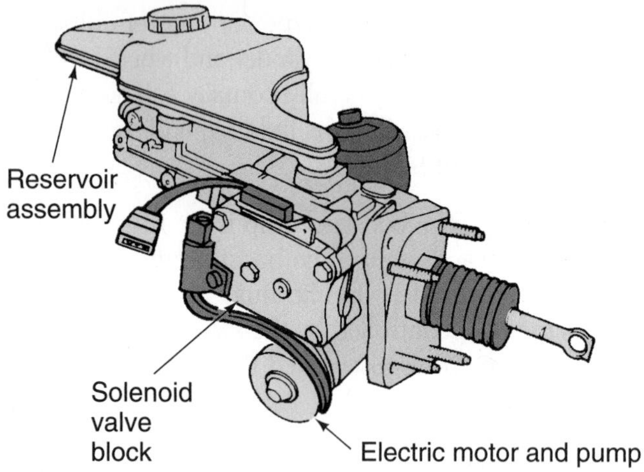

Figure 49-5 A master brake cylinder in a typical system uses an electric pump for power boost. *Courtesy of Ford Motor Company*

removed from the wheel brake units during an ABS cycle. This fluid is then used by the pump to build pressure for the brake hydraulic system. There are normally two fluid accumulators in a hydraulic control unit, one each for the primary and secondary hydraulic circuits.

Hydraulic Control Unit This assembly contains the solenoid valves, fluid accumulators, pump, and an electric motor. This is actually a combination unit of many individual components found separately in some systems. The unit may have one pump and one motor or it will have one motor and two pumps. One pump for half of the hydraulic system and the other for the other half.

Main Valve This two-position valve is also controlled by the ABS control module and is open only in the ABS mode. When open, pressurized brake fluid from the booster circuit is directed into the master cylinder (front brake) circuits to prevent excessive pedal travel.

Modulator Unit (Figure 49–6) The modulator unit controls the flow of pressurized brake fluid to the individual wheel circuits. Normally the modulator is made up of solenoids that open and close valves, several valves that control the flow of fluid to the wheel brake units, and electrical relays that activate or deactivate the solenoids through the commands of the control module. This unit may also be called the hydraulic actuator, hydraulic power unit, or the electrohydraulic control valve.

Solenoid Valves The solenoid valves are located in the modulator unit and are electrically operated by signals from the control module. The control module switches the solenoids on or off to increase, decrease, or maintain the hydraulic pressure to the individual wheel units.

Valve Block Assembly The valve block assembly attaches to the side of the booster/master cylinder and contains the hydraulic wheel circuit solenoid valves. The control module controls the position of these solenoid

Figure 49-6 A rear wheel ABS modulator assembly.

valves. The valve block is serviceable separate from the booster/master cylinder but should not be disassembled. An electrical connector links the valve block to the ABS control module.

Wheel Circuit Valves Two solenoid valves are used to control each circuit or channel. One controls the inlet valve of the circuit, the other controls the outlet valve. When inlet and outlet valves of a circuit are used in combination, pressure can be increased, decreased, or held steady in the circuit. The position of each valve is determined by the control module. Outlet valves are normally closed, and inlet valves are normally open. Valves are activated when the ABS control module switches 12 volts to the circuit solenoids. During normal driving, the circuits are not activated.

Electrical/Electronic Components

ABS Control Module This small control computer is normally mounted inside the trunk on the wheel housing, mounted to the master cylinder **(Figure 49–7)**, or is part of the hydraulic control unit. It monitors system operation and controls antilock function when needed. The module relies on inputs from the wheel speed sensors and feedback from the hydraulic unit to determine if the antilock brake system is operating correctly and to de-

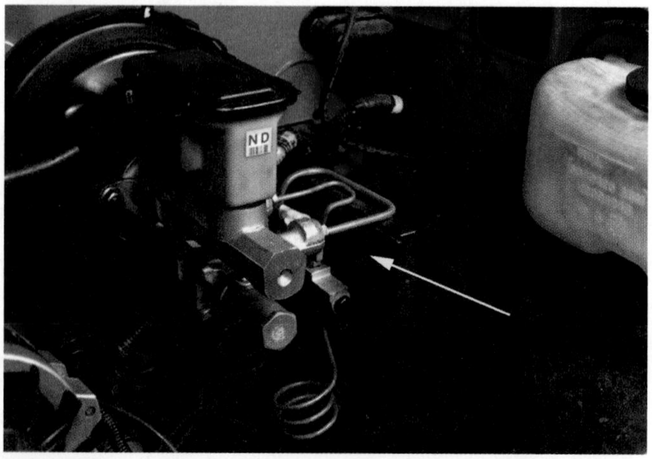

Figure 49-7 An ABS control module.

termine when the antilock mode is required. The module has a self-diagnostic function including numerous trouble codes. This module may also be called the ECU (electronic control unit), EBCM (electronic brake control module), the antilock brake controller, or the ECM (electronic control module). The name used depends on the manufacturer and year of the vehicle.

Brake Pedal Sensor The antilock brake pedal sensor switch is normally closed. When the brake pedal travel exceeds the antilock brake pedal sensor switch setting during an antilock stop, the antilock brake control module senses that the antilock brake pedal sensor switch is open and grounds the pump motor relay coil. This energizes the relay and turns the pump motor on. When the pump motor is running, the hydraulic reservoir is filled with high pressure brake fluid, and the brake pedal will be pushed up until the antilock brake pedal sensor switch closes. When the antilock brake pedal sensor switch closes, the pump motor is turned off and the brake pedal will drop some with each ABS control cycle until the antilock brake pedal sensor switch opens and the pump motor is turned on again. This minimizes pedal feedback during ABS cycling.

Data Link Connector (DLC). The DLC provides access and/or control of vehicle information, operating conditions, and diagnostic information.

Diagnostic Trouble Code (DTC). These trouble codes are numeric identifiers for fault conditions identified by the ABS's internal diagnostic system.

Indicator Lights Most ABS-equipped vehicles are fitted with two different warning lights. One of the warning lights is tied directly to the ABS, whereas the other lamp is part of the base brake system. All vehicles have a red warning light. This lamp is lit when there is a problem with the brake system or when the parking brake is on. An amber warning lamp lights when there is a fault in the ABS.

Lateral Acceleration Sensor Used on very few vehicles, this switch monitors the sideward movement of the vehicle while it is turning a corner. This information is sent to the control module to ensure proper braking during turns.

Pressure Switch This switch controls pump motor operation and the low pressure warning light circuit. The pressure switch grounds the pump motor relay coil circuit, activating the pump when accumulator pressure drops below 2,030 psi (14,000 kPa). The switch cuts off the motor when the pressure reaches 2,610 psi (18,000 kPa). The pressure switch also contains switches to activate the dash-mounted warning light if accumulator pressure drops below 1,500 psi (10,343 kPa). This unit is typically found on vehicles with a hydraulically assisted brake system.

Pressure Differential Switch The pressure differential switch is located in the modulator unit. This switch sends a signal to the control module whenever there is an undesirable difference in hydraulic pressures within the brake system.

Relays Relays are electromagnetic devices used to control a high-current circuit with a low-current switching circuit. In ABS, relays are used to switch motors and solenoids. A low-current signal from the control module energizes the relays that complete the electrical circuit for the motor or solenoid.

Toothed Ring The toothed ring can be located on an axle shaft, differential gear, or a wheel's hub. This ring is used in conjunction with the wheel speed sensor. The ring has a number of teeth around its circumference. The number of teeth varies by manufacturer and vehicle model. As the ring rotates and each tooth passes by the wheel speed sensor, an AC voltage signal is generated between the sensor and the tooth. As the tooth moves away from the sensor, the signal is broken until the next tooth comes close to the sensor. The end result is a pulsing signal that is sent to the control module. The control module translates the signal into wheel speed. The toothed ring may also be called the reluctor, tone ring, or gear pulser.

Wheel-Speed Sensor (Figure 49–8) The wheel-speed sensors are mounted near the different toothed

Figure 49-8 A wheel-speed sensor on a FWD drive axle. *Courtesy of Daimler Chrysler Corporation*

rings. As the ring's teeth rotate past the sensor, an AC voltage is generated. As the teeth move away from the sensor, the signal is broken until the next tooth comes close to the sensor. The end result is a pulsing signal that is sent to the control module. The control module translates the signal into wheel speed. The sensor is normally a small coil of wire with a permanent magnet in its center.

TYPES OF ANTILOCK BRAKE SYSTEMS

All antilock brake systems found on today's vehicles are manufactured by one of the following companies: Bendix, Bosch, Delco Moraine, ITT Teves, Kelsey-Hayes, or Lucas Girling. Each manufacturer has a unique way to accomplish the same thing—vehicle control during braking. When working with ABS, it is important that you identify the exact system you are working with and follow the specific service procedures for that system. Keep in mind, there have been nearly fifty different ABSs used by the industry in recent years.

The exact manner in which hydraulic pressure is controlled depends on the exact ABS design. The great majority of the earlier antilock brake systems were integrated or **integral antilock brake systems**. They combine the master cylinder, hydraulic booster, and ABS hydraulic circuitry into a single hydraulic assembly.

Other **antilock brake systems** are **nonintegral**. They use a conventional vacuum-assist booster and master cylinder. The ABS hydraulic control unit is a separate mechanism. In some nonintegrated systems, the master cylinder supplies brake fluid to the hydraulic unit. Although the hydraulic unit is a separate assembly, it still uses a high-pressure pump/motor, accumulator, and fast-acting solenoid valves to control hydraulic pressure to the wheels.

Both integral and nonintegral systems operate in much the same way, therefore an understanding of one system will lend itself to the understanding of the other systems.

General Motors' electromagnetic antilock brake system is a different type of nonintegral system that uses a conventional vacuum power booster and master brake cylinder. It does not use a high-pressure pump/motor and accumulator and fast-acting solenoid valves to control hydraulic pressure. Instead, it uses a hydraulic modulator.

In addition to being classified as integral and nonintegral antilock brake systems, systems can be divided into the level of control they provide. Antilock brake systems can be one-, two-, three-, or four-channel two- or four-wheel systems. A channel is merely a hydraulic circuit to the brakes.

Two-Wheel Systems

These basic systems offer antilock brake performance to the rear wheels only. They do not provide antilock performance to the steering wheels. Two-wheel systems are most often found on light trucks and some sport utility vehicles **(Figure 49–9)**.

These systems can be either one- or two-channel systems. In **one-channel systems**, the rear brakes on both sides of the vehicle are modulated at the same time to

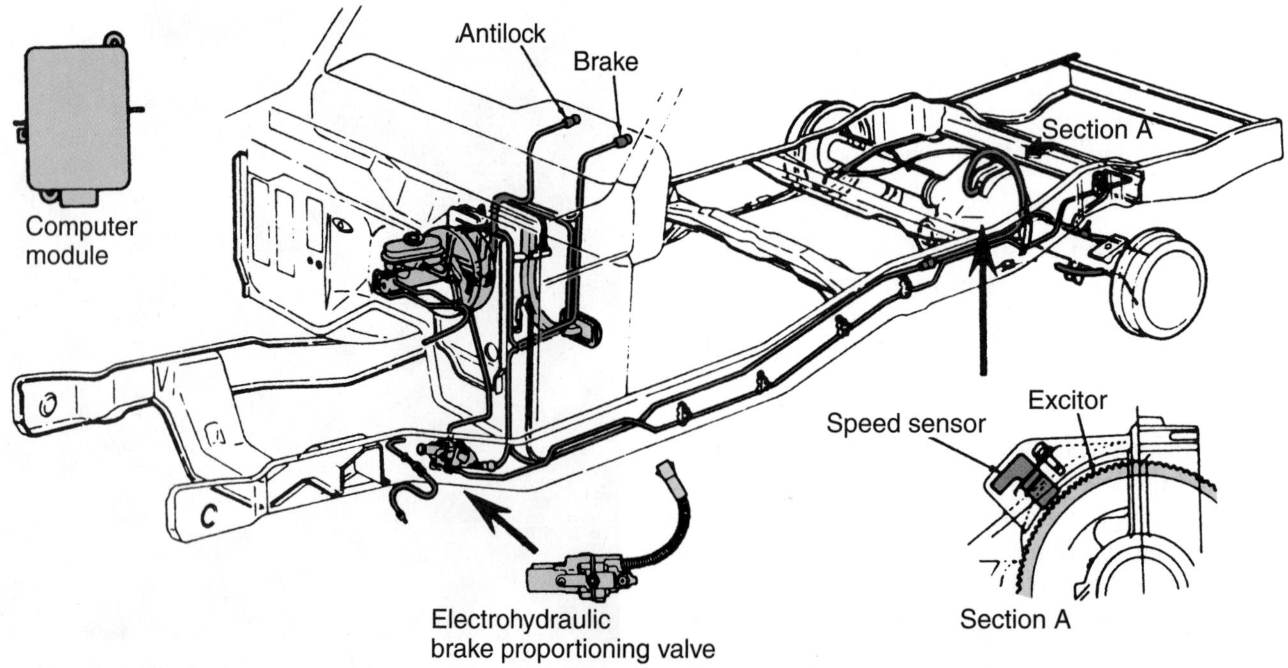

Figure 49-9 The main components of a rear-wheel ABS. *Courtesy of Wagner Brake Products*

control skidding. These systems rely on the input from a centrally located speed sensor. The speed sensor is normally positioned on the ring gear in the differential unit **(Figure 49–10)**, transmission, or transfer case.

Although not commonly found, a **two-channel system** can be used to modulate the pressure to each of the rear wheels independently of each other. Modulation is controlled by the speed variances recorded by speed sensors located at each wheel.

Two-channel systems may be found on some diagonally split brake systems. These systems use two speed sensors to provide wheel speed data for the regulation of all four wheels. One sensor has input that controls the right front wheel; the other sensor performs identically for the left front wheel.

Brake hydraulic pressure to the opposite rear wheel is controlled simultaneously with its diagonally located front wheel. For example, the right rear wheel receives the same pumping instructions as the left front wheel. This system is an upgrade from the two-wheel system since it does provide steering control. However, it can have shortcomings under certain operating conditions.

Full (Four-Wheel) Systems

Some hydraulic systems that are split from front to rear use a **three-channel system** and are called four-wheel antilock brake systems. These systems have individual hydraulic circuits to each of the two front wheels, and a single circuit to the two rear wheels.

The most effective and most common ABS available is a **four-channel system**, in which sensors monitor each of the four wheels. With this continuous information, the ABS control module ensures each wheel receives the exact braking force it needs to maintain both antilock and steering control.

ABS OPERATION

The exact operation of an antilock brake system depends on its design and manufacturer. It would take many pages to try to explain the operation of each, and as soon as you read the explanations there would be two or more new systems that would have to be explained. The exact operation of any system can be easily understood if you understand the basic operation of a few. The primary difference in operation between them all is based on the components used by the system. Therefore the following systems were chosen as examples of how certain systems operate with the components they have.

Two-Wheel Systems (Nonintegral)

These systems are used to prevent rear wheel lockup on pickup trucks and SUVs, especially under light payload conditions. They consist of a standard power brake

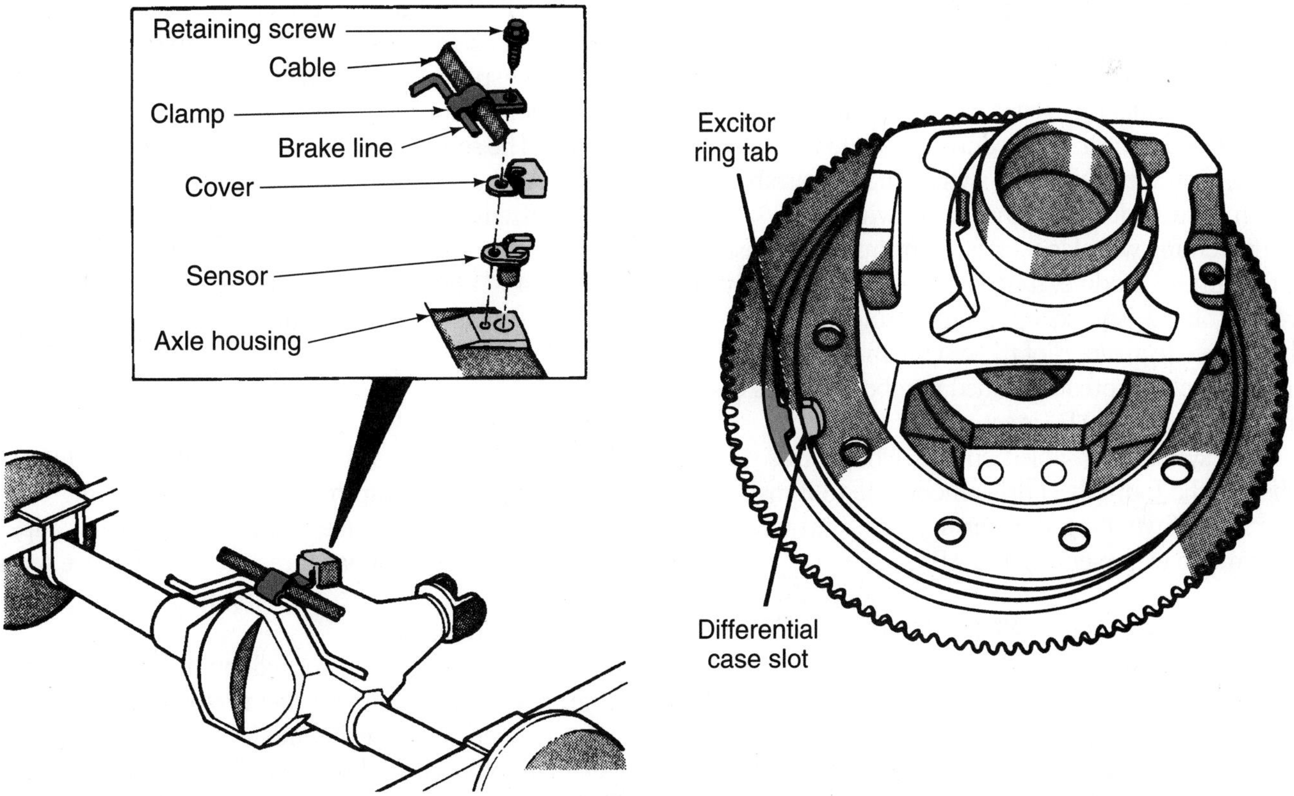

Figure 49–10 A speed sensor for both rear wheels located on the differential unit. *Courtesy of DaimlerChrysler Corporation*

system, an electronic control unit (control module), and an isolation/dump valve assembly. The valve assembly is attached to the master cylinder at the rear brake line. Both rear wheel brake assemblies are controlled by the valve assembly under ABS conditions.

Under normal braking, pressure will pass through the valve assembly. The control module receives a signal from the brake switch when brakes are applied and begins to monitor the vehicle speed sensor (VSS) signal at speeds over 5 miles per hour (8 km/h). If the control module detects a deceleration rate from the VSS that would indicate probable rear wheel lockup, it activates the isolation valve, which stops the buildup of pressure to the rear wheels. If further deceleration occurs that would indicate lockup, the control module will rapidly pulse the dump valve to release brake pressure into the accumulator. The control module continues to pulse the dump valve until rear wheel deceleration matches the vehicle's deceleration rate or the desired slip rate. When wheel speed picks up, the control module will turn off the isolation valve, allowing the fluid in the accumulator to return to the master cylinder and normal braking control to resume.

The control unit has three distinct functions: it performs self-test diagnostics, it monitors the ABS action and system, and it controls the ABS solenoid valves.

When the ignition switch is turned on, the control module checks its ROM and RAM. If an error is detected, a DTC is set in memory. A DTC will also be set if the control module senses a problem during ABS operation.

The control module continuously monitors the speed of the differential ring gear through signals from the rear wheel speed sensor. The control module also receives signals from the brake light switch, brake warning lamp switch, reset switch, and the 4WD switch.

Preventing wheel lockup is the primary responsibility of the control module. It does this by controlling the operation of the isolation and dump solenoid valves. To check the effectiveness of the system, the dump valve can only be cycled a predetermined number of times during one stop before a DTC is set.

This system is disabled on four-wheel drive vehicles when in the four-wheel drive mode due to transfer case operation. Switching the transfer case into two-wheel drive mode will re-enable the ABS.

Four-Wheel Systems (Nonintegral)

The hydraulic circuit for this type of system is an independent four-channel-type, one-channel for each wheel **(Figure 49–11)**. The hydraulic control unit is a separate unit. In the hydraulic control unit there are two valves per wheel, therefore a total of eight valves is used. Some systems have three channels, one for each of the front wheels and one for the rear axle. Obviously these system have only three pairs of solenoids **(Figure 49–12)**. Normal braking is accomplished by a conventional vacuum power-assist brake system.

The system prevents wheel lockup during an emergency stop by modulating brake pressure. It allows the driver to maintain steering control and stop the vehicle in the shortest possible distance under most conditions. During ABS operation, the driver will sense a pulsation in the brake pedal and a clicking sound.

Operation The ABS control module calculates the slip rate of the wheels and controls the brake fluid pressure to attain the target slip rate. If the control module senses that a wheel is about to lock based on input sensor data, it pulses the normally open inlet solenoid valve closed for that circuit. This prevents any more fluid from entering that circuit. The ABS control module then looks at the sensor signal from the affected wheel again. If that wheel is still decelerating faster than the other three wheels, it opens the normally closed outlet solenoid valve for that circuit. This dumps any pressure trapped between the closed inlet valve and the brake back to the master cylinder reservoir. Once the affected wheel returns to the same speed as the other wheels, the control module returns the valves to their normal condition allowing fluid flow to the affected brake.

Based on the inputs from the vehicle speed and wheel speed sensors, the control module calculates the slip rate of each wheel and transmits a control signal to the modulator unit solenoid valve when the slip rate is high.

Wheel speed at each wheel is measured by variable-reluctance sensors and sensor indicators (toothed ring or wheel). The sensors operate on magnetic induction principles. As the teeth on the brake sensor indicators rotate past the sensors, AC current is generated. The AC frequency changes in accordance with the wheel speed **(Figure 49–13)**. The ABS control unit detects the wheel sensor signal frequency and thereby detects the wheel speed.

Modulator Assembly The ABS modulator assembly consists of the inlet solenoid valve, outlet solenoid valve, reservoir, pump, pump motor, and the damping chamber. The modulator reduces fluid pressure to the calipers. The hydraulic control has three modes: pressure reduction (decrease), pressure retaining (hold), and pressure intensifying (increase).

While in the pressure reduction decrease mode **(Figure 49–14)**, the inlet valve is closed and the outlet valve is open. During this mode, fluid pressure to the caliper is blocked and the existing fluid in the caliper flows through the outlet valve back to the master cylinder reservoir. During the pressure-intensifying mode

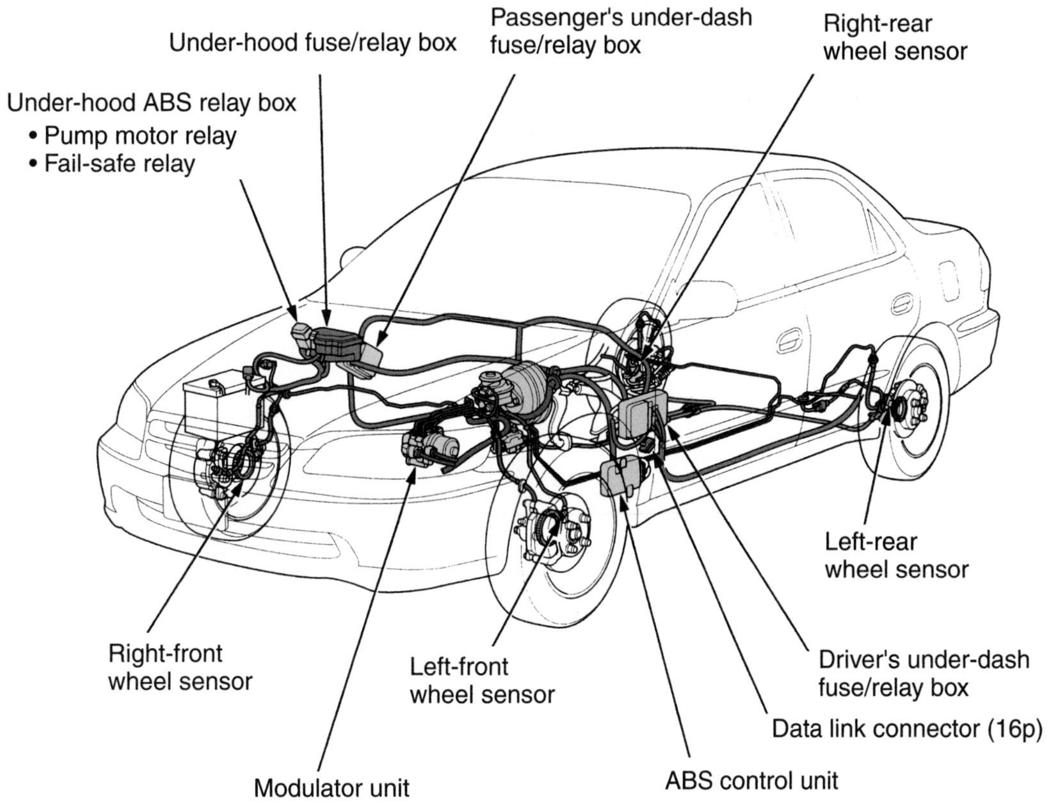

Figure 49-11 The basic electrical and hydraulic components of a four-wheel antilock brake system.
Courtesy of American Honda Motor Co., Inc.

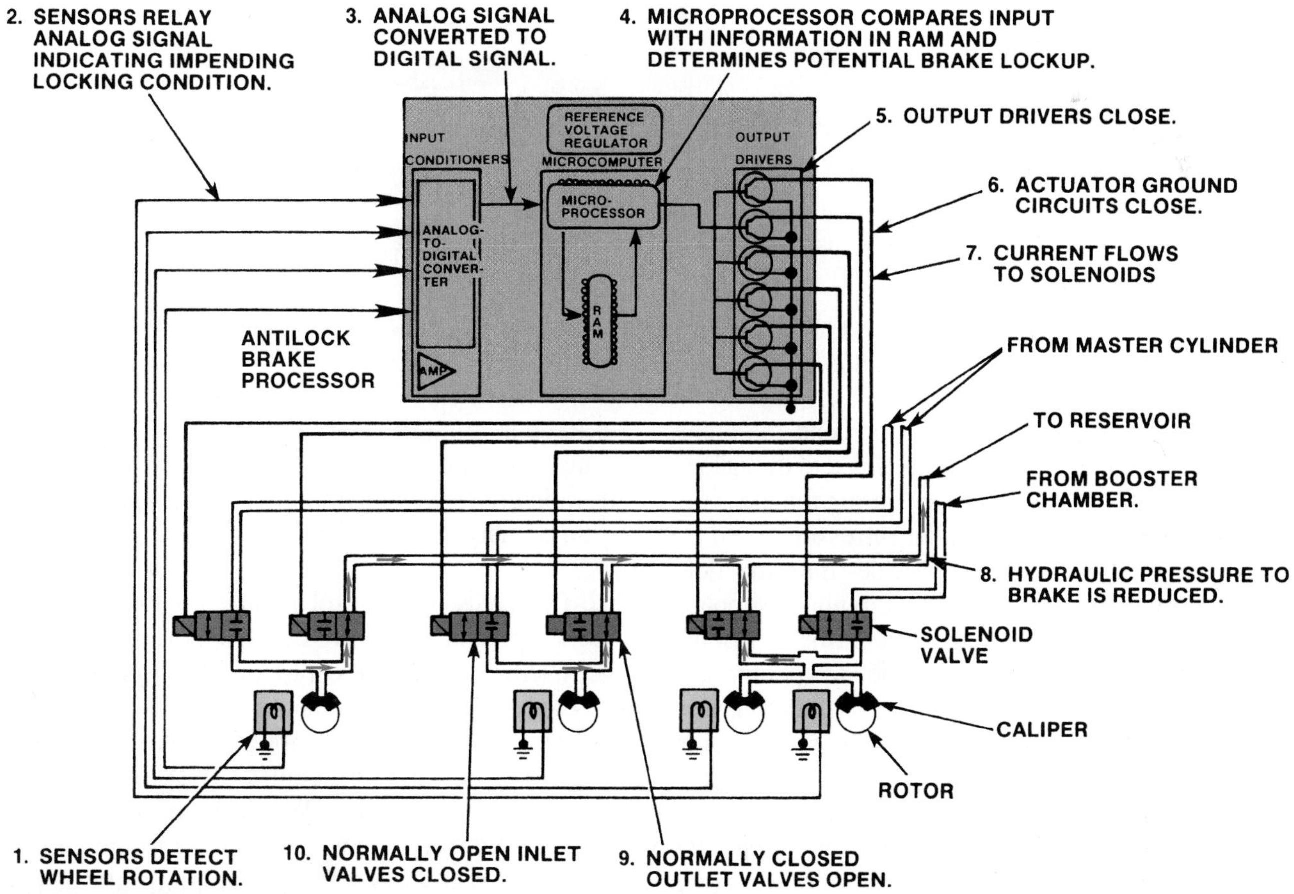

Figure 49-12 An ABS operation—potential brake lock condition.

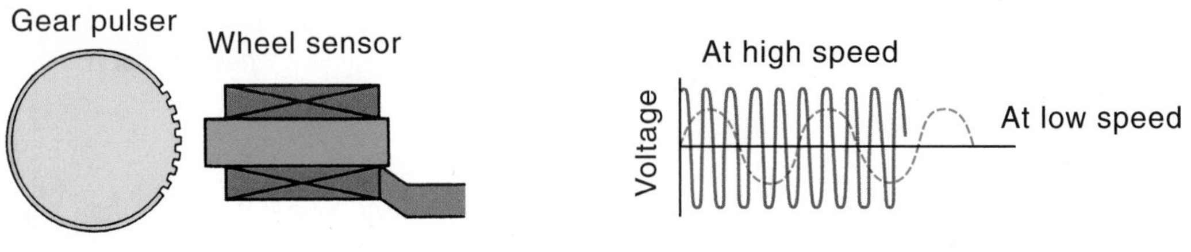

Figure 49–13 A wheel-speed sensor. *Courtesy of American Honda Motor Co., Inc.*

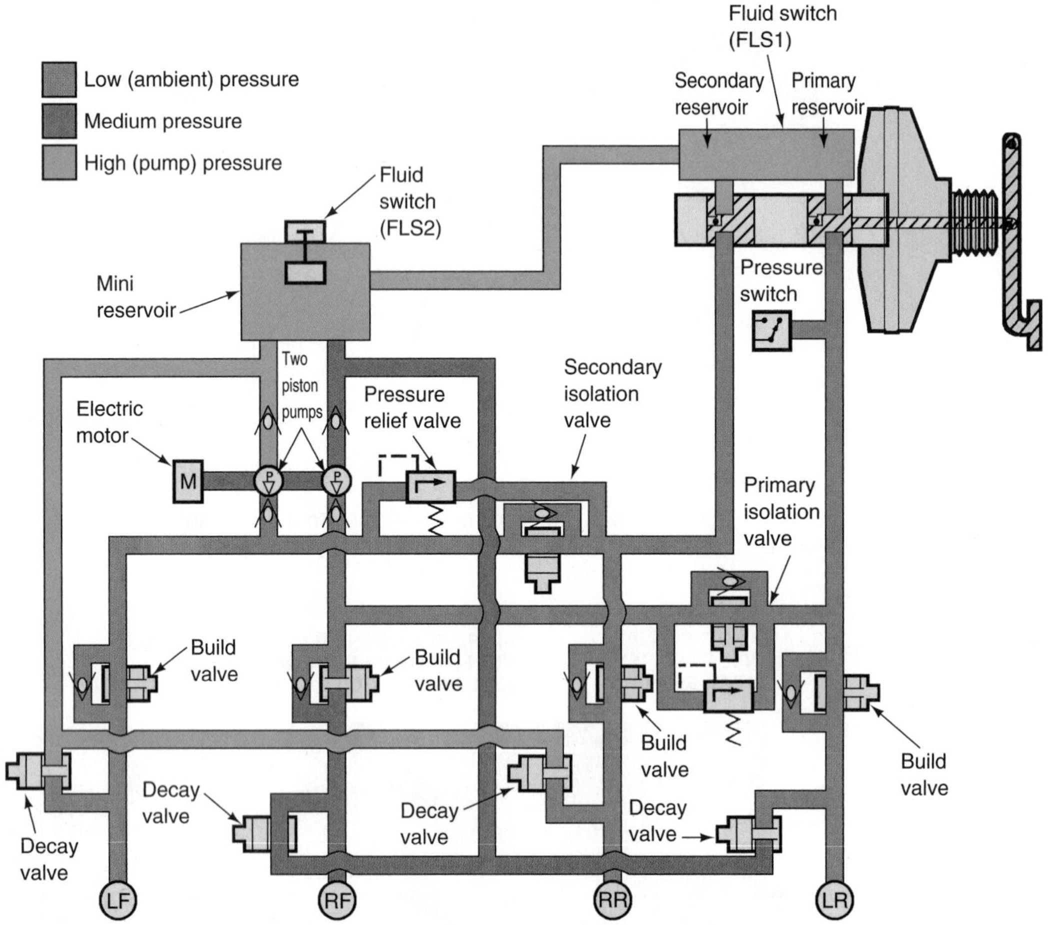

Figure 49–14 Antilock braking with pressure reduction. *Courtesy of DaimlerChrysler Corporation*

(Figure 49–15), the inlet valve is open and the outlet valve is closed. Pressurized fluid is pumped to the caliper. To keep the pressure at the caliper during the pressure-retaining mode, the inlet and outlet valves are closed.

The pump/motor assembly provides the extra fluid required during an ABS stop and the pump is supplied fluid that is released to the accumulators when the outlet valve is open. The accumulators provide temporary storage for the fluid during an ABS stop. The pump is also used to drain the accumulator circuits after the ABS stop is complete. The pump is run by an electric motor controlled by a relay that is controlled by the ABS control module. The pump is continuously on during an ABS stop and remains on for about 5 seconds after the stop is complete.

Keep in mind that the activity of the solenoid valves changes rapidly, several times each second. This means the fluid under pressure must be redirected quickly. This is the primary job of the pump.

Self-Diagnosis The control module monitors the electromechanical components of the system. A malfunction of the system will cause the control module to shut off or inhibit the system. However, normal power-assisted braking remains. Malfunctions are indicated by a warning indicator in the instrument cluster **(Figure 49–16)**. The system is self-monitoring. When the ignition switch is placed in the run position, the ABS control module will perform a preliminary self-check on its electrical system indicated by a second illumination of the amber ABS in-

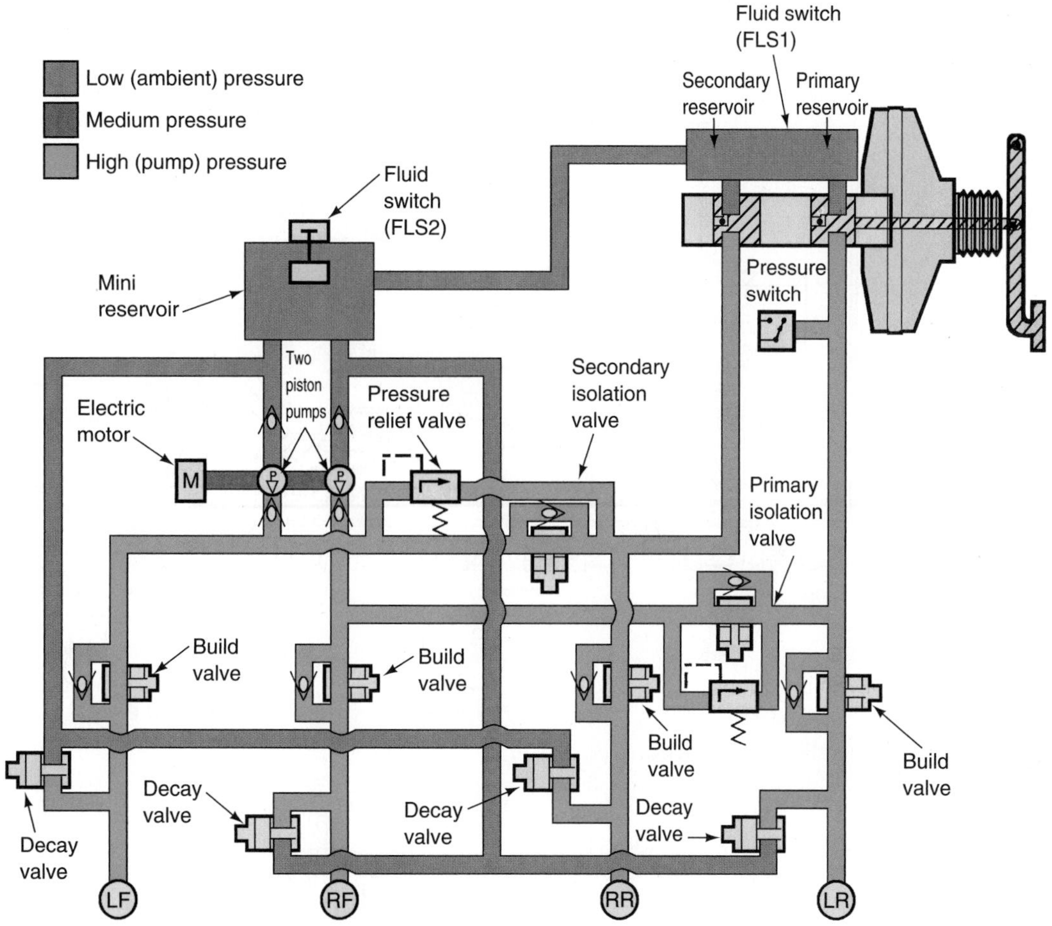

Figure 49-15 Antilock braking with pressure increase. *Courtesy of DaimlerChrysler Corporation*

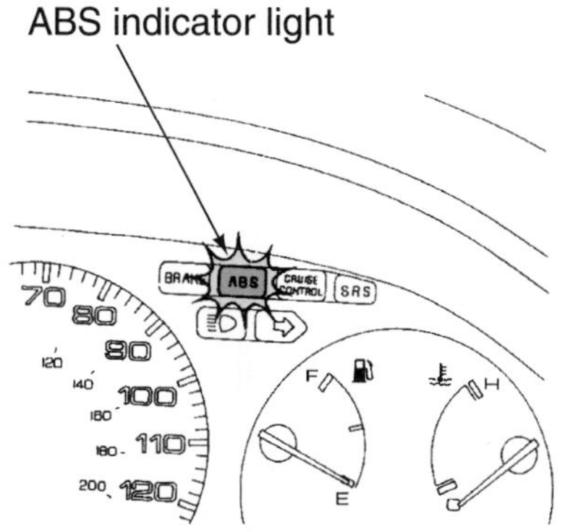

Figure 49-16 An ABS indicator light. *Courtesy of American Honda Motor Co., Inc.*

dicator in the instrument cluster. During vehicle operation and during normal and antilock braking, the control module monitors all electrical ABS functions and some hydraulic functions. With most malfunctions of the ABS, the amber ABS indicator is illuminated and a DTC recorded.

Four-Wheel Systems (Integral)

When a wheel speed sensor signals the ABS control module that a high rate of deceleration is taking place at its wheel (that is, that the wheel is likely to lock and skid), the control module initially signals the hydraulic unit to keep hydraulic pressure at the wheel constant. If the wheel continues to decelerate, the control module then signals the circuit valve solenoids to reduce hydraulic pressure to the affected wheel. This reduces braking at the wheel, reducing the risk of lockup.

The wheel accelerates again as a result of the reduced braking pressure. Once a specific limit has been reached, the control module registers the fact that the wheel is not being sufficiently braked. The wheel is decelerated again by increasing the pressure that was initially reduced. The control cycles may be executed every second depending upon the road conditions.

In operation, when the brakes are released, the piston in the master cylinder is retracted. The booster chamber is vented to the reservoir, and the fluid in the chamber is at the same low pressure as the reservoir **(Figure 49–17)**.

When the brakes are applied under normal conditions, the brake pedal actuates a pushrod **(Figure 49–18)**. This operates a scissors lever that moves a spool valve. When the spool valve moves, it closes the port from the booster chamber to the reservoir and partially opens the port from the accumulator in proportion to the

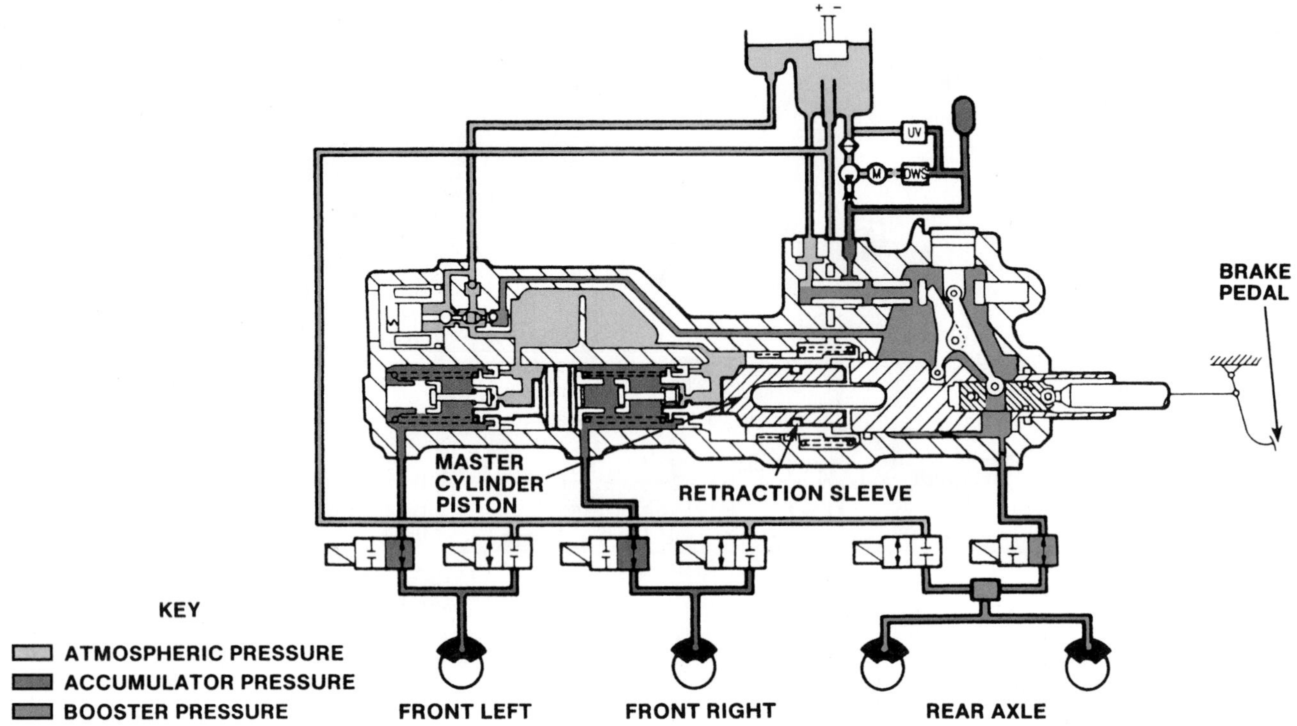

Figure 49-17 ABS operation—no braking.

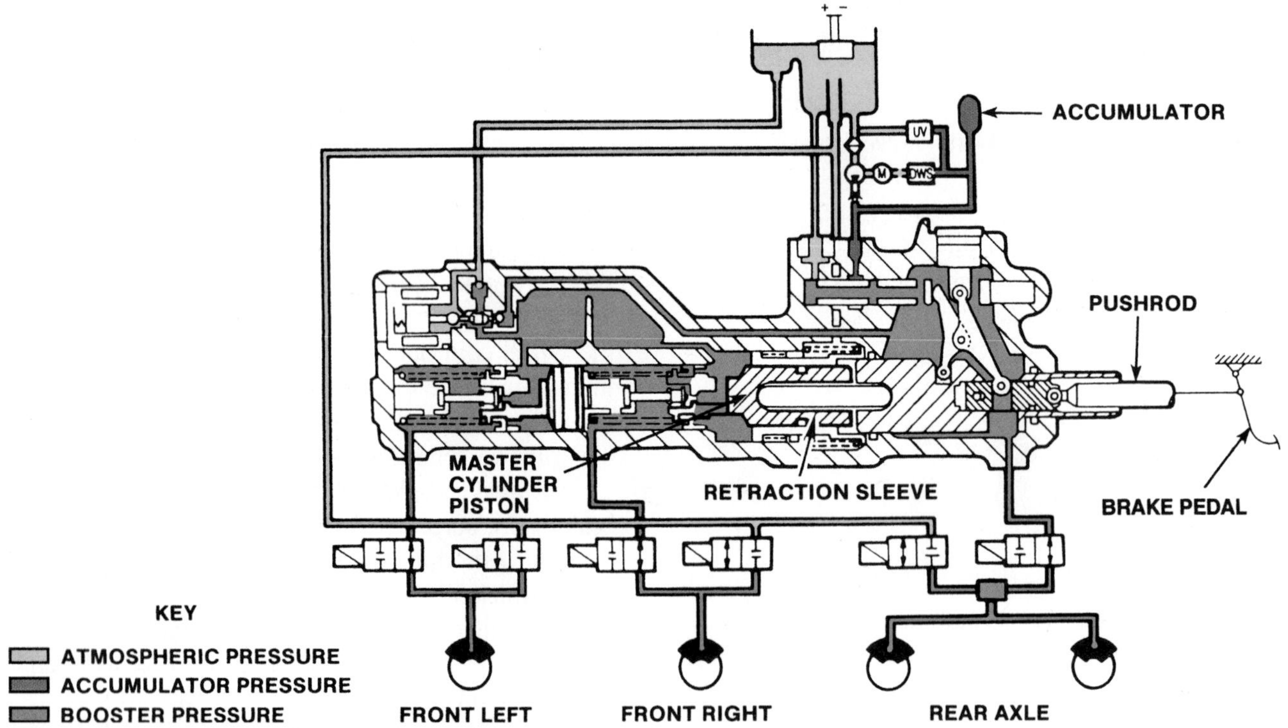

Figure 49-18 Normal braking with an antilock system.

pressure on the brake pedal. This allows hydraulic fluid under pressure from the accumulator to enter the booster chamber. As hydraulic pressure enters, it pushes the booster piston forward, providing hydraulic assist to the mechanical thrust from the pushrod.

When the control module determines the wheels are locking up, it opens a valve that supplies one chamber between the two master cylinder pistons and another chamber between the retraction sleeve and the first master cylinder piston **(Figure 49–19)**. The hydraulic pressure on the retraction sleeve retracts the pushrod, pushing back the brake pedal. In effect, the hydraulic pressure to the front wheels is now supplied by the accumulator, not by the brake pedal action. The control

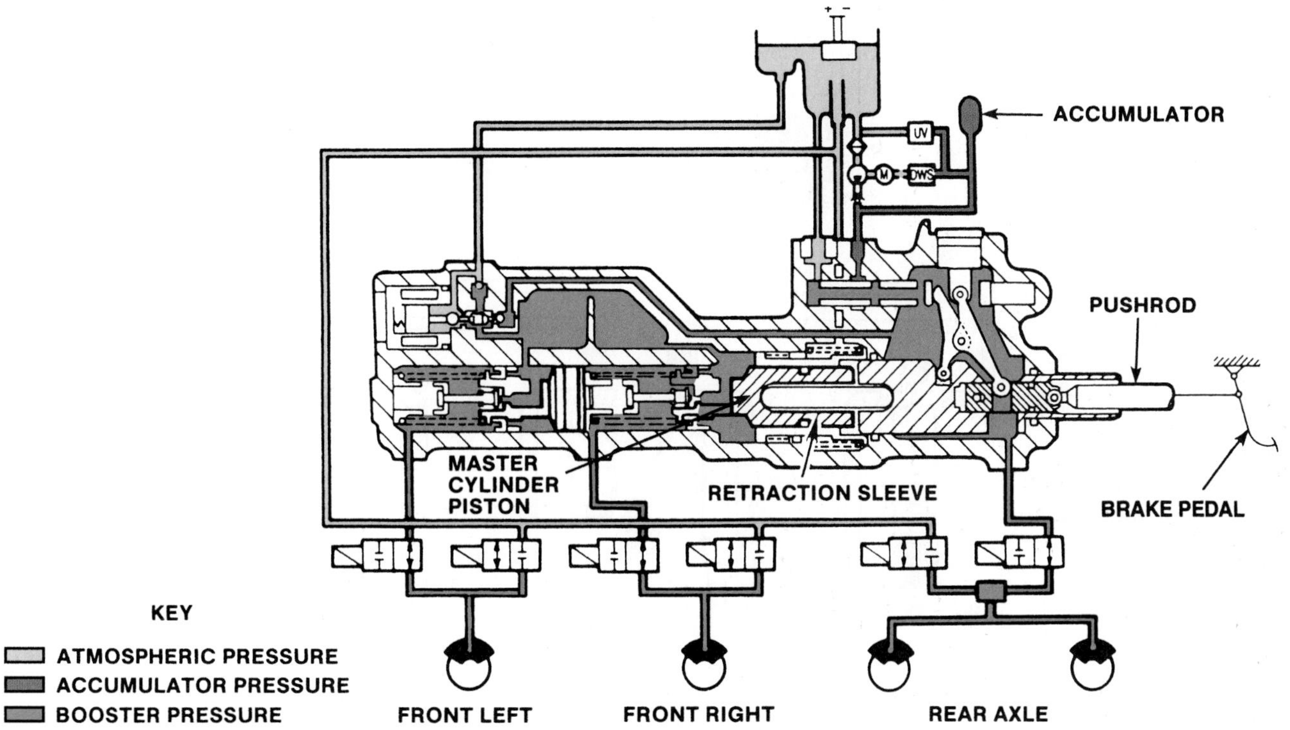

Figure 49-19 An antilock braking system in operation.

module also opens and closes the solenoid valves to cycle the brakes on the wheels that have been locking up.

When the solenoid valves are open, the master cylinder pistons supply hydraulic fluid to the front brakes and the boost pressure chamber provides hydraulic pressure to the rear. When the solenoid valves are closed, the hydraulic fluid from the master cylinder pistons and booster pressure chamber is cut off. The hydraulic fluid is returned from the brakes to the reservoir.

A few systems are equipped with a lateral acceleration switch to change the system's programming when hard braking while cornering. The switch may be no more than a left-hand and right-hand mercury switch located in a common housing that respond to forces generated by left-hand/right-hand turns and cornering movements. Late-model lateral acceleration sensors use a Hall-effect switch.

The electronic control module monitors the electromechanical components of the system. Malfunction of the ABS causes the electronic control module to shut off or inhibit the system. However, normal power-assisted braking remains. Malfunctions are indicated by one or two warning lights inside the vehicle. Loss of hydraulic fluid or power booster pressure disables the antilock brake system.

In most malfunctions in the antilock brake system, the "check antilock brakes" or brake light is illuminated. The sequence of illumination of these warning lights combined with the problem symptoms determine the appropriate diagnostic tests to perform. The diagnostic tests then pinpoint the exact component needing service.

General Motors' Electromagnetic Antilock Brake Systems

Beginning in 1991, General Motors began equipping certain small and midsize vehicles with an antilock braking system called the ABS-VI. This system is an add-on system that uses a conventional vacuum power booster and master brake cylinder. It does not use a high-pressure pump/motor and accumulator and fast-acting solenoid valves to control hydraulic pressure. Instead, it uses a hydraulic modulator **(Figure 49–20)** that operates using a principle called electromagnetic braking.

As in integrated systems, wheel speed is monitored using individual speed sensors. When one wheel begins to decelerate faster than the others while braking, the control module signals the hydraulic modulator assembly to reduce pressure to the affected brake.

The ABS-VI modulator contains three small screw plungers—one for each front brake circuit and one for the rear brake circuit **(Figure 49–21)**—that are driven by electric motors. At the top of each plunger cavity is a check ball that controls hydraulic pressure within the brake circuit.

The hydraulic valve body (modulator) assembly and motor pack is mounted to the master cylinder. The hydraulic brake circuit for each front wheel is controlled by a motor, gear-driven ball screw, solenoid, piston, and check valve. The rear wheel circuit is controlled by check balls and a single motor; therefore, both rear brakes are modulated together.

The motors are high-speed, bidirectional motors that quickly and precisely position the ball screws. Each

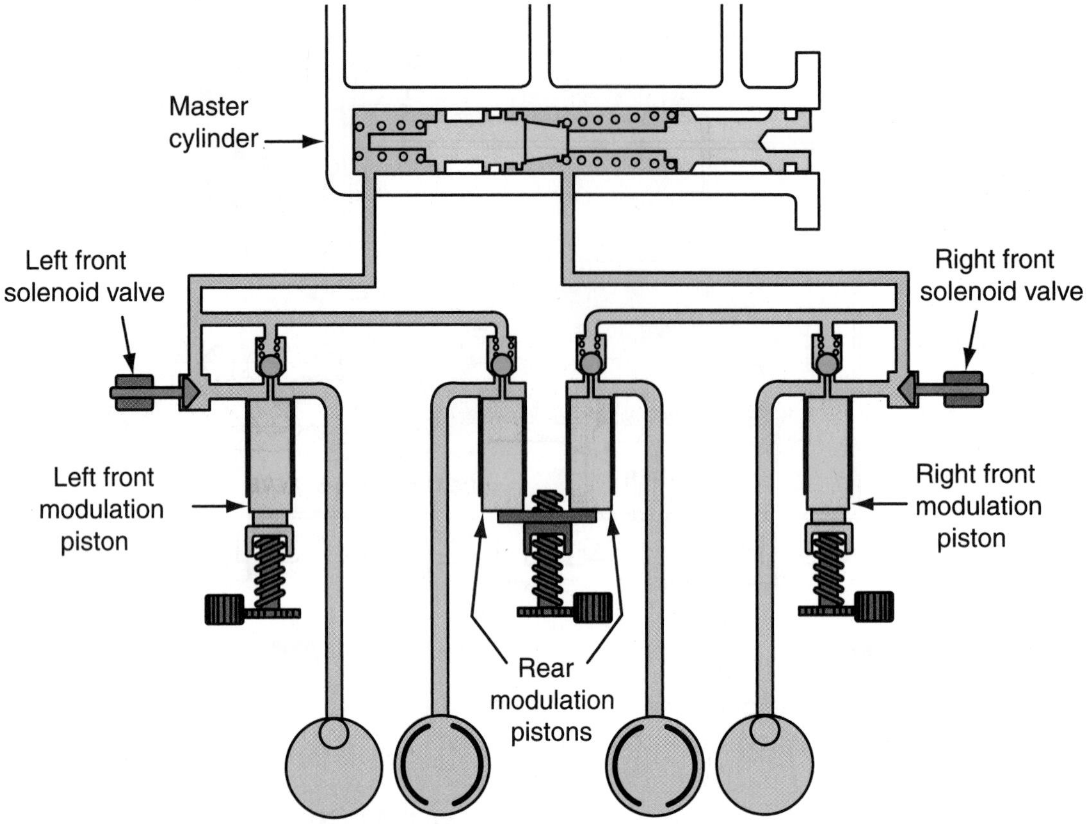

Figure 49-20 The hydraulic layout for GM's ABS-VI.

motor has a brake that allows its ball screw to maintain its position against hydraulic pressures. The front motors have an electromagnetic brake (EMB) and the rear motor uses an expansion spring brake (ESB).

During normal braking conditions, each plunger is all the way up. The check ball at the top is unseated and the bypass solenoid is normally open, which allows brake pressure from the master cylinder and vacuum power booster to apply the brakes during normal stopping.

During panic stops, the ABS mode operates. In this situation, brake pressure must be reduced to prevent wheel lockup. This is done by closing the normally open

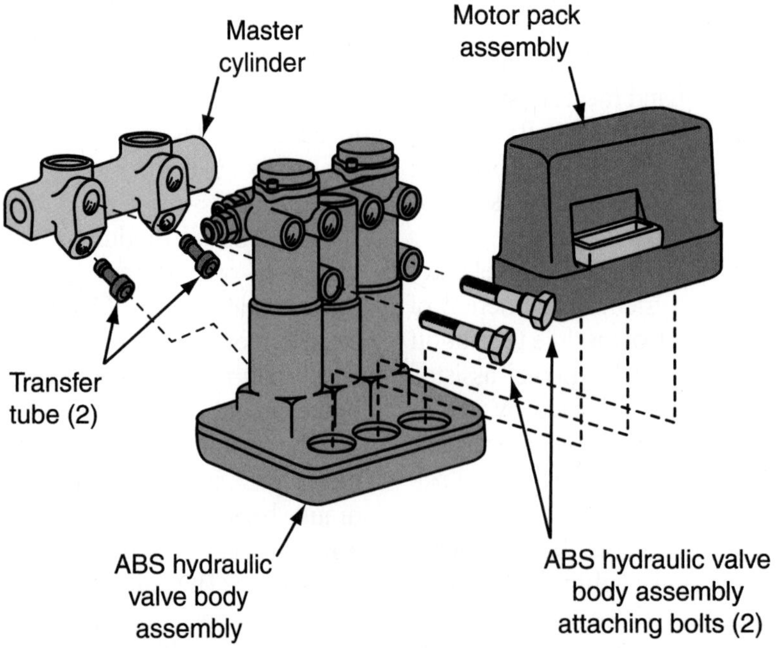

Figure 49-21 The hydraulic modulator is an add-on component to the conventional master cylinder. The electric motors can be replaced separately, as can the two solenoids.

solenoid to isolate the circuit and then turning the plunger down to reduce braking pressure. As the plunger turns down, it increases the volume within the brake circuit. This causes a drop in pressure that keeps the wheel from locking. The amount of pressure applied is controlled by running the plunger up and down as required. To decrease pressure further, the plunger is run down. To reapply pressure, the plunger moves back up.

The system can cycle the brakes seven times per second. Because the system does not have a high-pressure pump and accumulator, it cannot increase brake pressure above what can be provided by the master cylinder and vacuum-assist booster.

The ABS-VI control module has five separate diagnostic modes. Data available for troubleshooting includes wheel-speed sensor readings, vehicle speed, battery voltage, individual motor and solenoid command status, warning light status, and brake switch status. Numerous trouble codes are programmed into the control module to help pinpoint problems. Other diagnostic modes store past trouble codes. This data can help technicians determine if an earlier fault code, such as an intermittent wheel speed sensor, is linked to the present problem such as a completely failed wheel sensor. Another mode enables testing of individual system components.

Other Brake System Controls

A few automobiles are now equipped with an electronic brake-assist system that can recognize emergency braking and automatically applies full-power brake force for shorter stopping distances. This system is activated only in emergency braking situations and does not affect normal brake operation.

The system recognizes emergency braking by the speed at which it was depressed and the brakes are automatically applied under full power. The system is driver-adaptive as it learns the driver's braking habits by using sensors to monitor every movement of the brake pedal. When the sensors detect an emergency stop, an electronic valve at the power brake booster is turned on. This supplies full braking power to the wheels. The ABS prevents the wheels from locking in spite of the full-power braking force.

AUTOMATIC TRACTION CONTROL

Automakers use the technology and hardware of antilock braking systems to control tire traction and vehicle stability. As explained earlier, an ABS pumps the brakes when a braking wheel attempts to go into a locked condition. **Automatic traction control (ATC)** systems apply the brakes when a drive wheel attempts to spin and lose traction **(Figure 49–22)**.

The system works best when one drive wheel is working on a good traction surface and the other is not. The system also works well when the vehicle is accelerating on slippery road surfaces, especially when climbing hills.

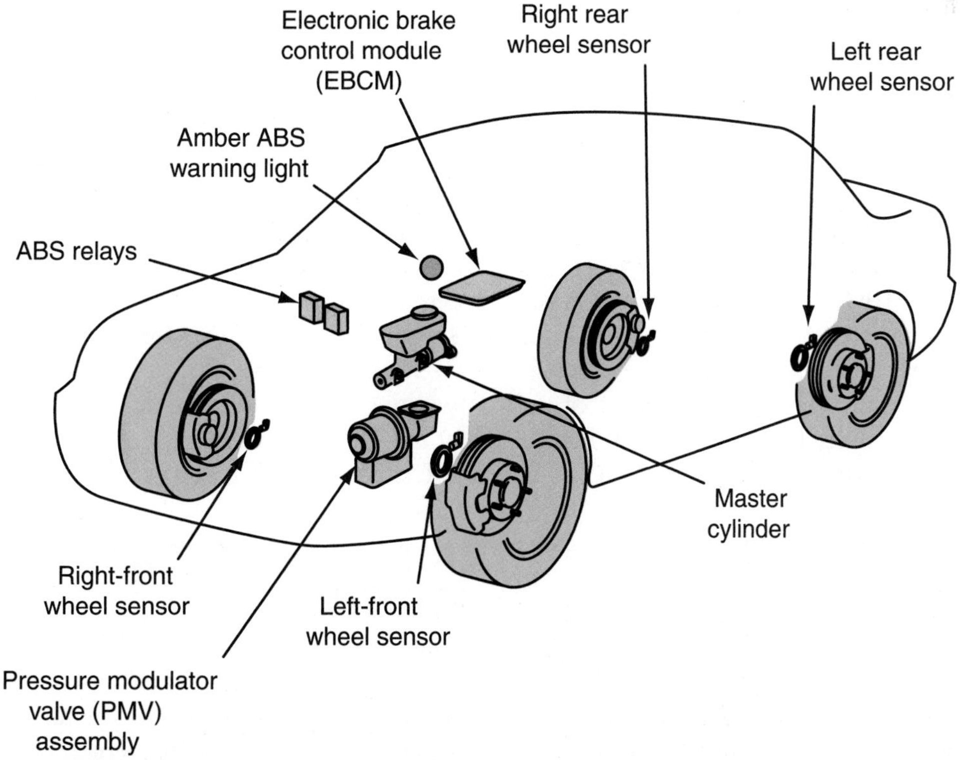

Figure 49–22 A typical ATC system.

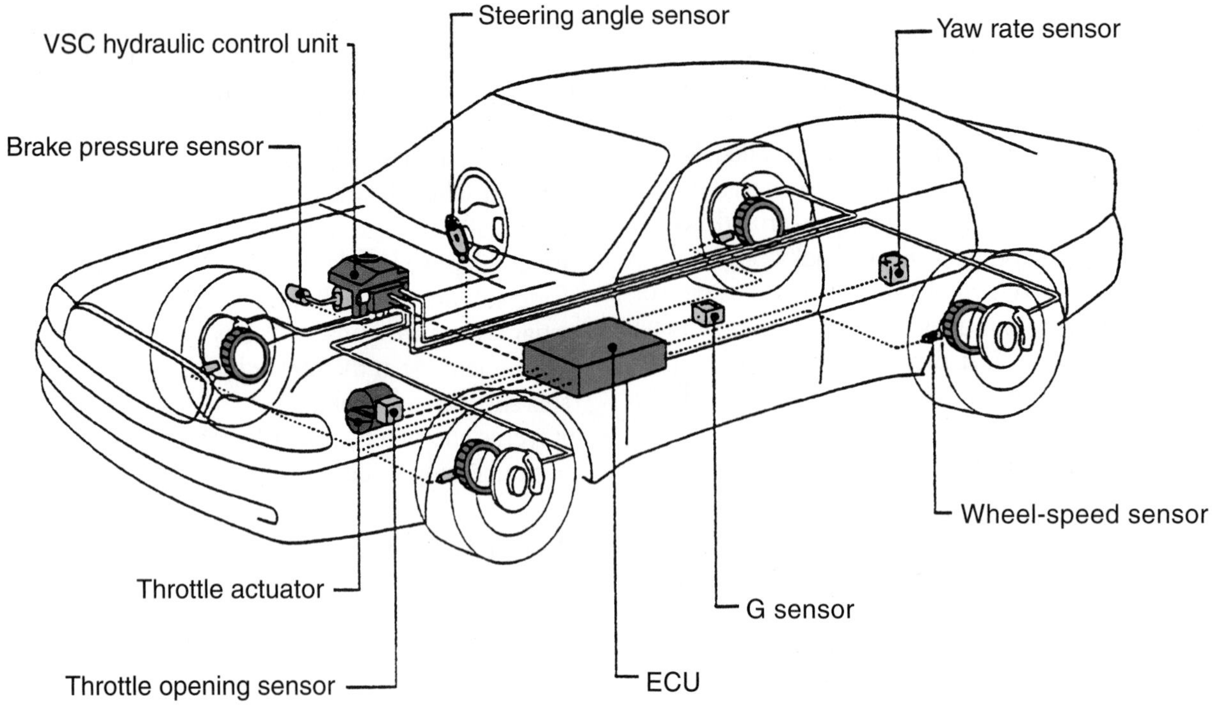

Figure 49–23 The components of a typical vehicle stability control system. *Courtesy of Lexus of America Co.*

ATC is most helpful on four-wheel- or all-wheel-drive vehicles in which loss of traction at one wheel could hamper driver control. It is also desirable on high-powered front-wheel-drive vehicles for the same reason. Often if traction control is fitted to a FWD vehicle, the modified ABS is a three-channel system since ATC is not needed at the rear wheels. On RWD and 4WD vehicles, the system is based on a four-channel ABS.

During road operation, the ATC system uses an electronic control module to monitor the wheel-speed sensors. If a wheel enters a loss-of-traction situation, the module applies braking force to the wheel in trouble. Loss of traction is identified by comparing the vehicle's speed to the speed of the wheel. If there is a loss of traction, the speed of the wheel will be greater than expected for the particular vehicle speed. Wheel spin is normally limited to a 10% slippage. Some traction control systems use separated hydraulic valve units and control modules for the ABS and ATC, while others integrate both systems into one hydraulic control unit and a single control module. The pulse rings and wheel-speed sensors remain unchanged from the ABS to the ATC.

Some ATC systems function only at low road speeds to 5 to 25 miles per hour (8 to 40 km/h). These systems are designed to reduce wheel slip and maintain traction at the drive wheels when the road is wet or snow covered. The control module monitors wheel speed. If during acceleration the module detects drive wheel slip and the brakes are not applied, the control module enters into the traction control mode. The inlet and outlet solenoid valves are pulsed and allow the brake to be quickly applied and released. The pump/motor assembly is turned on and supplies pressurized fluid to the slipping wheel's brake.

More advanced systems work at higher speeds and usually integrate some engine control functions into the control loop. For example, if the ATC system senses a loss of traction, it not only cycles the brakes but signals the engine control module to retard ignition timing and partially close the throttle as well. Timing and throttle reduce engine output. On these vehicles, there may be no mechanical connection between the gas pedal and the throttle plates. The pedal is nothing more than an electrical switch. This is sometimes referred to as "drive by wire." If wheel slippage continues, the ATC module may also cut fuel delivery to one or more engine cylinders. This action relieves engine stress and prevents the driver from overspeeding the engine in extremely slippery conditions.

Many systems are equipped with a dash-mounted warning light to alert the driver that the system is operating. There may also be a manual cutoff switch so the driver can turn off ATC operation.

AUTOMATIC STABILITY CONTROL

Various stability control systems are found on today's vehicles. Like traction control systems, stability controls are based on and linked to the antilock brake system **(Figure 49–23)**. On some vehicles the stability control system is also linked to the electronic suspension system

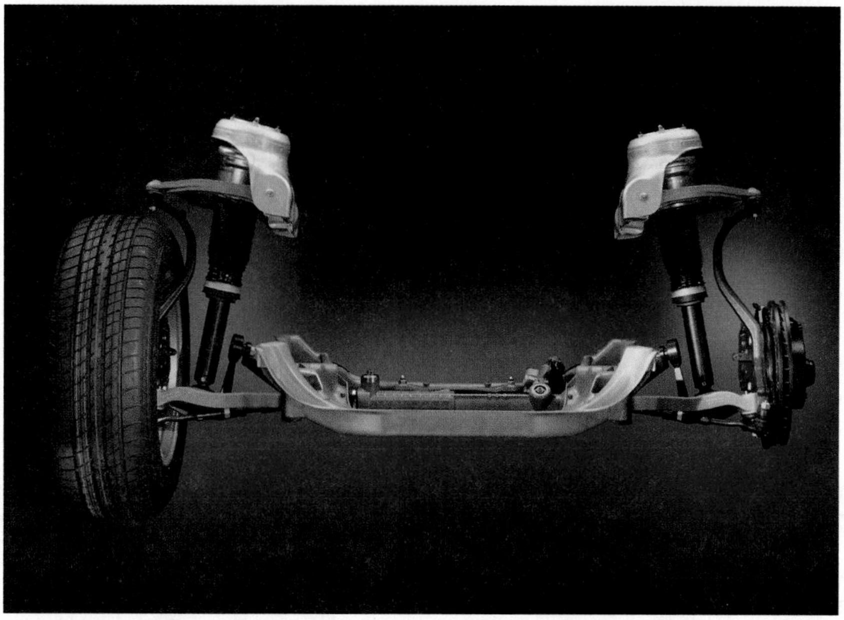

Figure 49–24 On some stability control systems, the suspension is also adjusted to provide for safe handling. *Courtesy of DaimlerChrysler Corporation*

(Figure 49–24). Also on a few cars, a switch allows the driver to disable the traction control but maintain the stability controls.

Stability control systems momentarily apply the brakes at any one wheel to correct oversteer or understeer **(Figure 49–25)**. The control unit receives signals from the typical sensors plus a yaw, lateral acceleration (G-force), and a steering angle sensor.

Understeer is a condition in which the vehicle is slow to respond to steering changes. Oversteer occurs when the rear wheels try to swing around or fishtail. When the system senses understeer in a turn, the brake at the inside

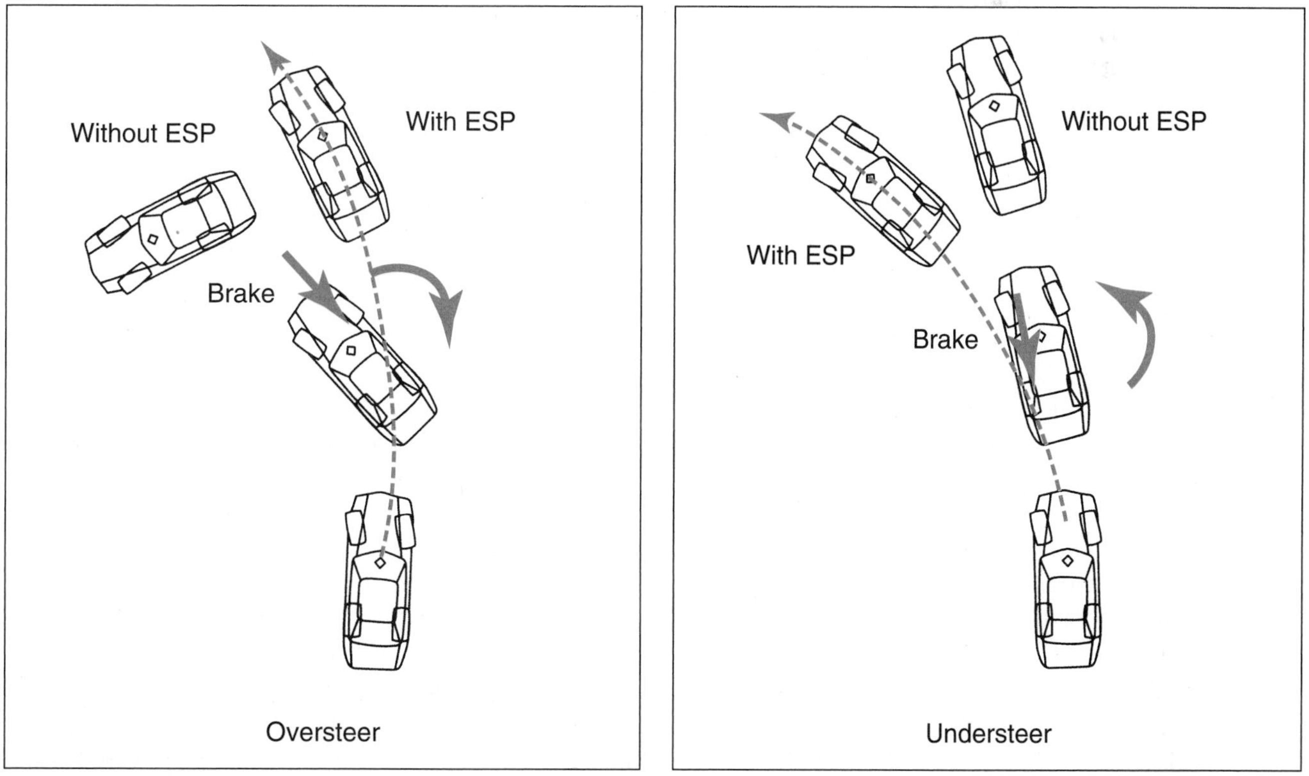

Figure 49–25 The effects of oversteer and understeer and how they are affected by stability control (ESP—Electronic Stability Program) systems. *Courtesy of DaimlerChrysler Corporation*

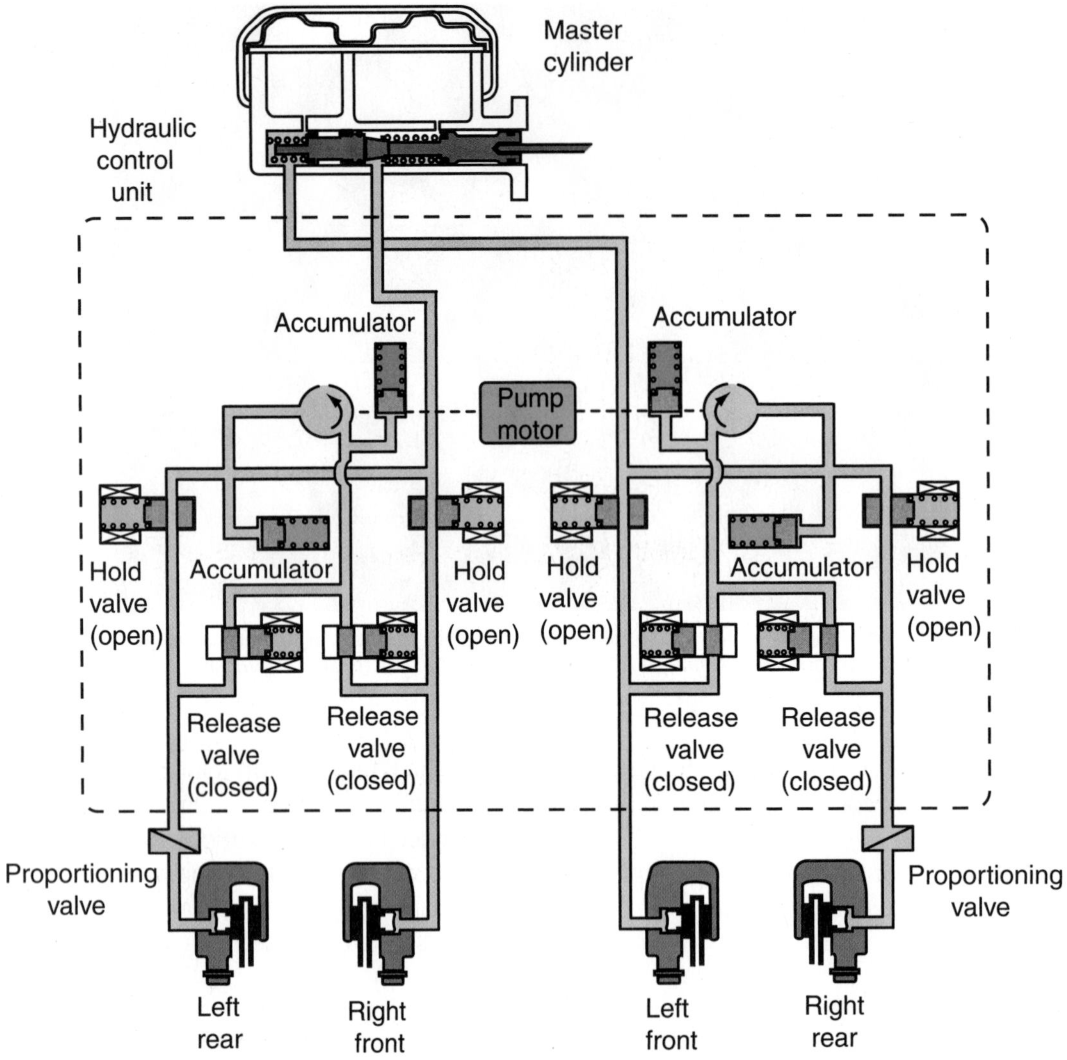

Figure 49-26 The hydraulic layout of a typical integrated antilock brake traction control and stability control system.

rear wheel is applied. During oversteer, the outside front brake is applied.

Relying on the input from the sensors and computer programming, the system calculates if the vehicle is going exactly in the direction in which it is being steered. If there is the slightest difference between what the driver is asking and what the vehicle is doing, the system corrects the situation by applying one of the right or left side brakes.

The system uses the angle of the steering wheel and the speed of the four wheels to calculate the path chosen by the driver. It then looks at lateral G-forces and vehicle yaw to measure where the vehicle is actually going. (Yaw is defined as the natural tendency for a vehicle to rotate on its vertical center axis.)

Stability control systems can control the vehicle during acceleration, braking, and coasting. If the brakes are already applied but oversteer or understeer is occurring, the fluid pressure to the appropriate brake is increased **(Figure 49–26)**.

ANTILOCK BRAKE SYSTEM SERVICE

Most of the service done to the brakes of an antilock brake system are identical to those in a conventional brake system. There are, however, some important differences. Always refer to the recommended procedures in the appropriate service manual before attempting to service the brakes on an ABS-equipped vehicle.

Many ABS components are simply remove-and-replace items. Normal brake repairs, such as replacing brake pads, caliper replacement, rotor machining or replacement, brake hose replacement, master cylinder or power booster replacement, or parking brake repair, can all be performed as usual. In other words, brake service on an ABS-equipped vehicle is similar to brake service on a conventional system with a few exceptions. Before beginning any service, check the service manual. It may be necessary to depressurize the accumulator to prevent personal injury from high-pressure fluid.

Safety Precautions

- When replacing brake lines and/or hoses, always use lines and hoses designed for and specifically labeled for use on ABS vehicles.
- Never use silicone brake fluids in ABS vehicles. Use only the brake fluid type recommended by the manufacturer (normally DOT-3).
- Never begin to bleed the hydraulic brake system on a vehicle equipped with ABS until you have checked a service manual for the proper procedure.
- Never open a bleeder screw or loosen a hydraulic brake line or hose while the ABS is pressurized.
- Disconnect all vehicle computers, including the ABS control unit, before doing any electrical welding on the vehicle.
- Never disconnect or reconnect electrical connectors while the ignition switch is on.
- Never install a telephone or CB antenna close to the ABS control unit or any other control module or computer.
- Check the wheel-speed sensor to sensor ring air gap after any parts of the wheel-speed circuit have been replaced.
- Keep the wheel sensor clean. Never cover the sensor with grease unless the manufacturer specifies doing so; in those cases, use only the recommended type.
- When replacing speed (toothed) rings, never beat them on with a hammer. These rings should be pressed onto their flange. Hammering may result in a loss of polarization or magnetization.

Relieving Accumulator Pressure

Some services require that brake tubing or hoses be disconnected. Many ABS use hydraulic pressures as high as 2,800 psi (19,300 kPa) and an accumulator to store this pressurized fluid. Before disconnecting any lines or fittings in many systems, the accumulator must be fully depressurized. A common method of depressurizing the ABS follows:

1. Turn the ignition switch to the off position.
2. Pump the brake pedal between twenty-five and fifty times.
3. The pedal should be noticeably harder when the accumulator is discharged.

Procedures for depressurizing the system vary with the design of the system. Always refer to the service manual.

DIAGNOSIS AND TESTING

Always follow the vehicle manufacturer's procedures when diagnosing an ABS. In general, ABS diagnostics requires three to five different types of testing that must be performed in the specified order listed in the service manual. Types of testing may include the following:

1. Prediagnostic inspections and test drive
2. Warning light symptom troubleshooting
3. On-board ABS control module testing (trouble code reading)
4. Individual trouble code or component troubleshooting

CAUTION!

Following the wrong sequence or bypassing steps may lead to unnecessary replacement of parts or incorrect resolution of the symptom. The information and procedures given in this chapter are typical of the various antilock systems on the market. For specific instructions, consult the vehicle's service manual.

Prediagnostic Inspection

Before undertaking any actual checks, take just a few minutes to talk with the customer about his or her ABS complaint. The customer is a very good source of information, especially when diagnosing intermittent problems. Make sure you find out what symptoms are present and under what conditions they occur.

All ABS have some sort of self-test **(Figure 49–27)**. This test is activated each time the ignition switch is turned on. You should begin all diagnosis with this simple test.

Place the ignition switch in the start position while observing both the red brake system light and amber ABS indicator lights. Both lights should turn on. Start the vehicle. The red brake system light should quickly turn off.

With the ignition switch in the run position, the antilock brake control module will perform a preliminary self-check on the antilock electrical system. The self-check takes three to six seconds, during which time the amber antilock indicator light remains on. Once the self-check is complete, the ABS indicator light should turn off. If any malfunction is detected during this test, the ABS warning light remains on and the ABS is shut down.

ON BOARD DIAGNOSTIC SYSTEM DESCRIPTION

[RE4F04B]

NOTE:

The CONSULT-II must be used in conjunction with a program card.
CONSULT-II does not require loading (Initialization) procedure.

Be sure the CONSULT-II is turned off before installing or removing a program card.

CONSULT-II Start Procedure

EAS0019C

1. Turn off the ignition switch.
2. Connect CONSULT-II and CONSULT-II CONVERTER to the data link connector.

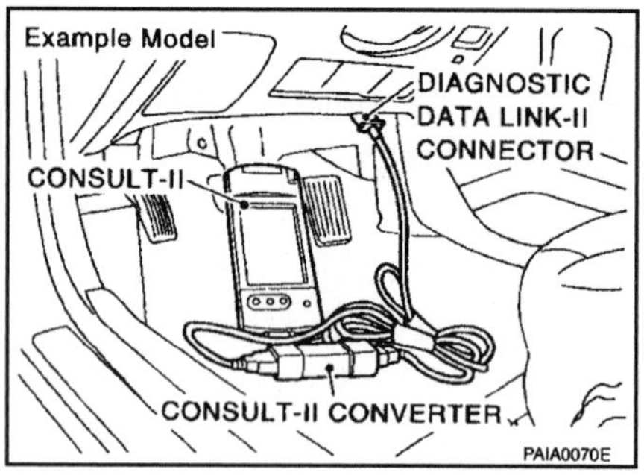

3. Turn on the ignition switch.
4. Touch "START (NISSAN BASED VECL)" or "System Shortcut" (eg: Engine) on the screen.

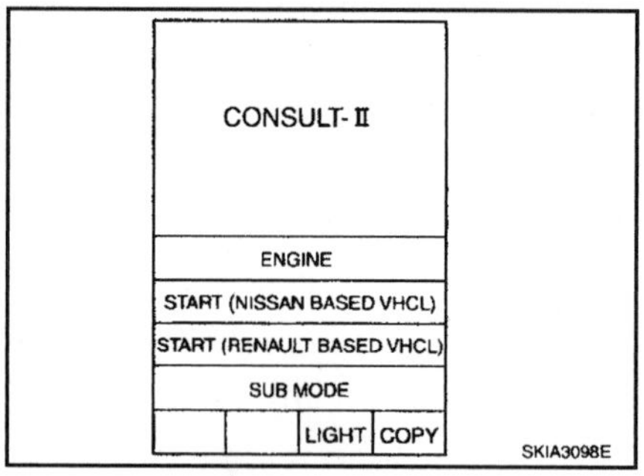

5. Touch "SELF DIAG RESULTS".
   Display shows malfunction experienced since the last erasing operation.
   CONSULT-II performs "Real Time Diagnosis".
   Also, any malfunction detected while in this mode will be displayed at real time.

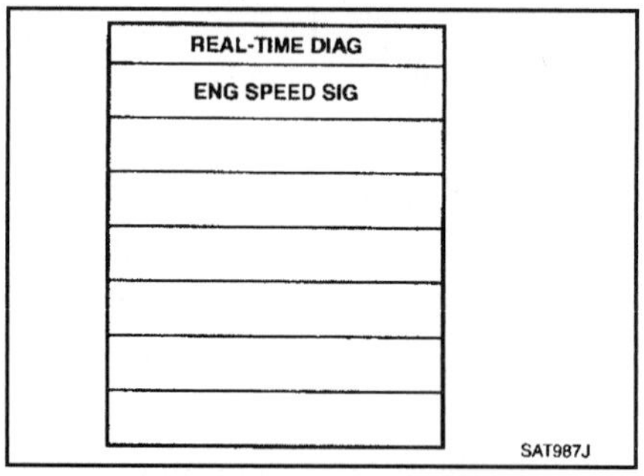

SELF-DIAGNOSTIC RESULT TEST MODE

Detected items (Screen terms for CONSULT-II, "SELF DIAGNOSIS" test mode)		Malfunction is detected when...	TCM self-diagnosis	OBD-II (DTC)
			Available by A/T check (position) indicator lamp or "A/T" on CONSULT-II	Available by malfunction indicator lamp*2, SERVICE ENGINE SOON "ENGINE" on CONSULT-II or GST
"A/T"	"ENGINE"			
Park/neutral position (PNP) switch circuit		TCM does not receive the correct voltage signal (based on the gear position) from the switch.	—	P0705
—	PNP SW/CIRC			
Revolution sensor		TCM does not receive the proper voltage signal from the sensor.	X	P0720
VHCL SPEED SEN·A/T	VEH SPD SEN/CIR AT			
Vehicle speed sensor (Meter)		TCM does not receive the proper voltage signal from the sensor.	X	—
VHCL SPEED SEN·MTR	—			

Figure 49-27 The self-diagnosis procedure for an Infiniti. *Courtesy of Infiniti Motor Company*

Visual Inspection

The prediagnosis inspection consists of a quick visual check of system components. Problems can often be spotted during this inspection, which can eliminate the need to conduct other more time-consuming procedures. This inspection should include the following:

1. Check the master cylinder fluid level.
2. Inspect all brake hoses, lines, and fittings for signs of damage, deterioration, and leakage. Inspect the hydraulic modulator unit for any leaks or wiring damage.
3. Inspect the brake components at all four wheels. Make sure that no brake drag exists and that all brakes react normally when they are applied.
4. Inspect for worn or damaged wheel bearings that may allow a wheel to wobble.
5. Check the alignment and operation of the outer CV joints.
6. Make sure the tires meet the legal tread depth requirements and that they are the correct size.
7. Inspect all electrical connections for signs of corrosion, damage, fraying, and disconnection.
8. Inspect the wheel-speed sensors and their wiring. Check the air gaps between the sensor and ring, and make sure these gaps are within the specified range. Also check the mounting of the sensors and the condition of the toothed ring and wiring to the sensor.

SHOP TALK

Remember that faulty base brake system components may cause the ABS to shut down. Do not condemn the ABS too quickly. ■

Test Drive

After the visual inspection is completed, test drive the vehicle to evaluate the performance of the entire brake system. Begin the test drive with a feel of the brake pedal while the vehicle is sitting still. Then accelerate to a speed of about 20 mph (32 km/h). Bring the vehicle to a stop using normal braking procedures. Look for any signs of swerving or improper operation. Next, accelerate the vehicle to about 25 mph (40 km/h) and apply the brakes with firm and constant pressure. You should feel the pedal pulsate if the antilock brake system is working properly.

During the test drive, both brake warning lights should remain off. If either of the lights turns on, take note of the condition that may have caused it. After you have stopped the vehicle, place the gear selector into park or neutral and observe the warning lights. They should both be off.

If the control module detects a problem with the system, the amber ABS indicator lamp will either flash or light continuously to alert the driver of the problem. In some systems, a flashing ABS indicator lamp indicates that the control unit detected a problem but has not suspended ABS operation. However, a flashing ABS indicator lamp is a signal that repairs must be made to the system as soon as possible.

A solid ABS indicator lamp indicates that a problem has been detected that affects the operation of ABS. No antilock braking will be available, but normal, nonantilock, brake performance will remain. In order to regain ABS braking ability, the ABS must be serviced.

The red brake warning lamp will be illuminated when the brake fluid level is low, the parking brake switch is closed, the bulb test switch section of the ignition switch is closed, or when certain ABS trouble codes are set.

Intermittent problems can often be identified during the test drive. If the scan tool is equipped with a "snapshot" feature, use it to capture system performance during normal acceleration, stopping, and turning maneuvers. If this does not reproduce the malfunction, perform an antilock stop on a low coefficient surface such as gravel, from approximately 30–50 mph (48–80 km/h) while triggering the snapshot mode on the scan tool.

Self-Diagnostics

The electronic control system of most ABSs include sophisticated on-board diagnostics that, when accessed with the proper scan tool, can identify the source of a problem within the system. Each of the codes represents a specific possible problem in the system. The service manual contains a detailed step-by-step troubleshooting chart for each of these trouble codes. The charts give specific instructions for confirming the suspect problem and performing the repair.

Testers and Scanning Tools

Different vehicle manufacturers provide ABS test and scan tools with varying capabilities **(Figure 49–28)**. Some testers are used simply to access the digital trouble codes. Others may also provide functional test modes for checking wheel sensor circuits, pump operation, solenoid testing, and so forth.

On some vehicles, the amber ABS light and red brake light flash out the digital trouble codes. As you can see, it is important to research the capabilities and proper use of the test equipment the vehicle manufacturer provides. Misuse of test equipment can be dangerous. For

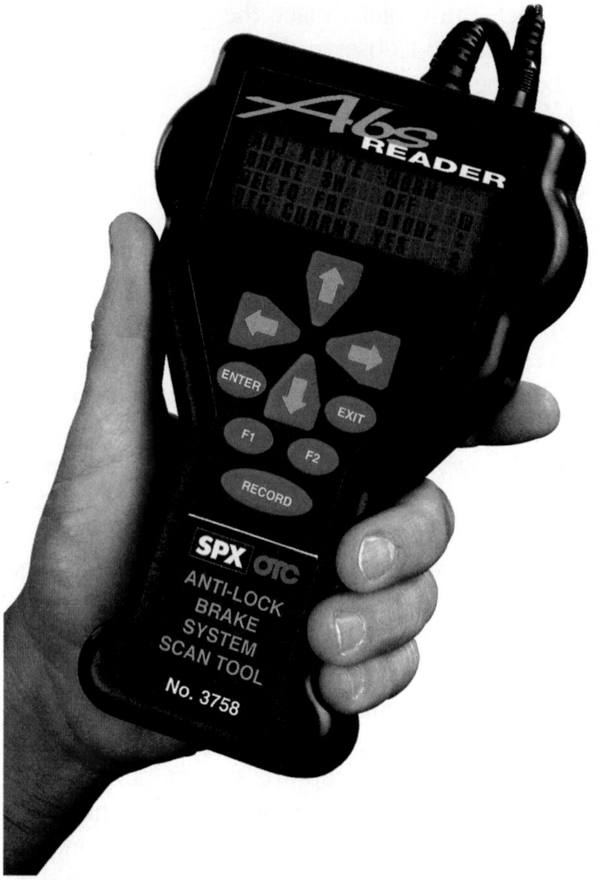

Figure 49-28 A hand-held antilock brake system scan tool. *Courtesy of OTC Tool and Equipment, a Division of SPX Corporation*

example, connecting test equipment during a test drive that is not designed for this use may lead to loss of braking ability.

Once all system malfunctions have been corrected, clear the ABS' DTCs. Codes cannot be erased until all codes have been retrieved, all faults have been corrected, and the vehicle has been driven above a set speed (usually 18 to 25 mph [30 to 40 km/h]). It may be necessary to disconnect a fuse for several seconds to clear the codes on some systems. After service work is performed on the ABS, repeat the previous test procedure to confirm that all codes have been erased.

Testing Components with ABS Scan Tools

ABS scan tools and testers can often be used to monitor and/or trigger input and output signals in the ABS. This allows you to confirm the presence of a suspected problem with an input sensor, switch, or output solenoid in the system. You can also check that the repair has been successful before driving the vehicle. Manual control of components and automated functional tests are also available when using many diagnostic testers. Details of typical functional tests follow.

Solenoid Leak Test

PROCEDURE

Checking for Solenoid Leaks in the Hydraulic Modulator

STEP 1 *Disconnect the inspection connector from the connector cover and connect the inspection connector to the tester.*

STEP 2 *Remove the modulator reservoir filter, then fill the reservoir to the max level with fresh fluid.*

STEP 3 *Bleed the high-pressure fluid from the maintenance bleeder connection. Often this procedure requires the use of a special tool.*

STEP 4 *Start the engine and release the parking brake.*

STEP 5 *Set the tester to the proper mode and press the start-test button.*

STEP 6 *While the ABS pump is running, place your finger over the top of the solenoid return tube in the modulator reservoir. If you can feel brake fluid coming from the return tube, one of the solenoids is leaking. Go to step 7. If you cannot feel brake fluid coming from the return tube, the solenoids are OK. Reinstall the modular reservoir filter and refill the reservoir to the max level.*

STEP 7 *Bleed the high-pressure fluid from the maintenance bleeder, then run through steps 3 and 6 with the tester. Repeat this procedure three or four times.*

STEP 8 *Repeat steps 5 and 6. If the solenoid leakage has stopped, reinstall the modulator reservoir filter and refill the reservoir to the max level. If one of the solenoids is leaking, the entire hydraulic modulator may require replacement. In some cases it is possible to remove, inspect, and replace individual solenoids.*

WARNING!

Certain components of the ABS are not intended to be serviced individually. Do not attempt to remove or disconnect these components. Only those components with approved removal and installation procedures in the manufacturer's service manual should be serviced.

Testing Components with a Lab Scope

Like most electrical/electronic systems, antilock brake traction control and stability control system components

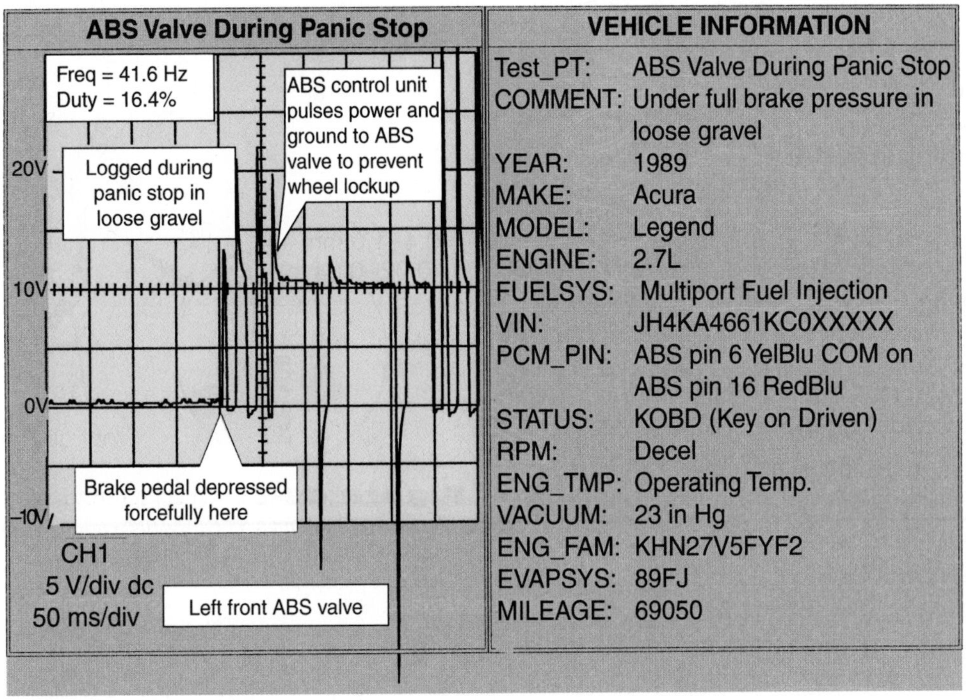

Figure 49–29 The waveform of a solenoid valve before and after the pedal was depressed hard and quickly. *Courtesy of Progressive Diagnostics—WaveFile AutoPro*

can be tested with a lab scope. A lab scope offers one distinct advantage over many other testing tools. You can watch the component's activity over time. For example, all antilock-based systems rely on the cycling of solenoid valves. **Figure 49–29** is a waveform of a solenoid valve showing the solenoid's activity immediately after the brake pedal was depressed hard. When looking at the solenoid on the lab scope you should see a change in the waveform as soon as ABS operation begins. When a wheel tries to slip, the ABS control module should begin pulsing the solenoid to that wheel.

A critical input to the antilock brake, traction control, and stability control systems is from the wheel sensors. These too can be monitored on a lab scope **(Figure 49–30)**. As the wheel begins to spin, the waveform of the sensor's output should begin to oscillate above and

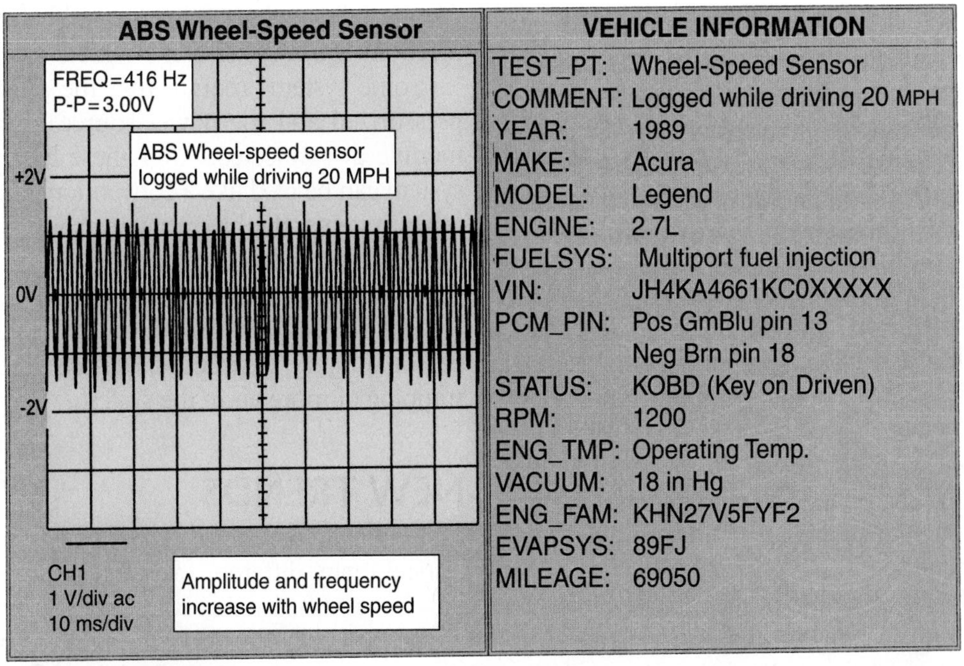

Figure 49–30 The waveform from a wheel-speed sensor. *Courtesy of Progressive Diagnostics—WaveFile AutoPro*

below zero volts. The oscillations should get taller as speed increases. If the wheel's speed is kept constant, the waveform should also stay constant.

Component Replacement

A typical antilock braking system consists of a conventional hydraulic brake system (the base system) plus a number of antilock components. The base brake system consists of a vacuum power booster, master cylinder, front disc brakes, rear drum or disc brakes, interconnecting hydraulic tubing and hoses, a low fluid sensor, and a red brake system warning light.

Antilock components are added to this base system to provide antilocking braking ability. Most ABS use the same operational principles, but the major components may be configured and/or named differently.

The electrical components of the ABS are generally very stable. Common electrical system failures are usually caused by poor or broken connections. Other common faults can be caused by malfunction of the wheel-speed sensors, pump and motor assembly, or the hydraulic module assembly.

Many of the components of ABS are simply remove-and-replace items. On some systems, wheel-speed sensors must be adjusted. Normal brake repairs, such as replacing brake pads, caliper replacement, rotor machining or replacement, brake hose replacement, master cylinder or power booster replacement, or parking brake repair, can all be performed as usual. In other words, brake service on an ABS-equipped vehicle is similar to service on a conventional system with a few exceptions.

Always refer to the service manual for the correct adjustment and replacements procedures for any and all brake parts in an ABS.

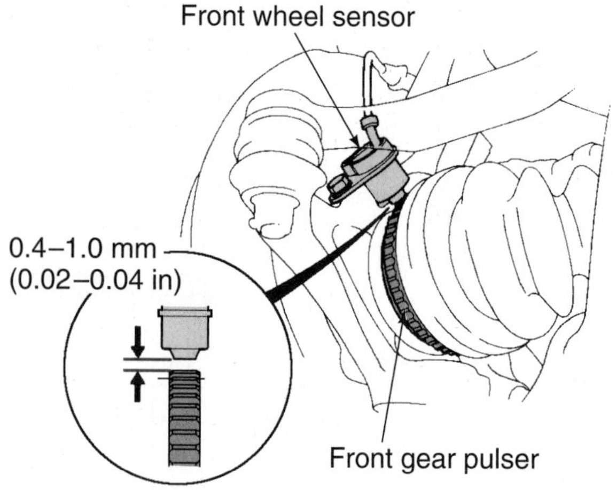

Figure 49–31 Wheel sensor gap measurement. *Courtesy of American Honda Motor Co., Inc.*

Wheel-Speed Sensor Service Visually inspect the wheel-speed sensor pulsers for chipped or damaged teeth. Use a feeler gauge to measure the air gap between the sensor and rotor **(Figure 49–31)** all the way around while rotating the drive shaft, wheel, or rear hub bearing unit by hand. If there is a specification on this gap, make certain it is within the required specification (typically 0.02 to 0.04 inch or 0.4 mm to 1.0 mm). If the gap exceeds specifications, the problem is likely a distorted knuckle that should be replaced.

Sensors are replaced by simply disconnecting the wiring at the sensor and unbolting the fasteners. Be careful not to twist the wiring cables or harness when installing the sensors.

On many vehicles a jumper harness made of highly flexible twisted pair wiring is located between each wheel-speed sensor and the main wiring harness. The rear jumper harnesses come together at a single connector at the rear body pass-through. This wiring exists because the main harness must connect to the suspension of the vehicle, thus the wiring in this area is subjected to the same motion as a spring or shock absorber. Consequently, any repair to this section of wiring will result in stiffening and eventual failure due to wire fracture. For this reason, the wheel-speed sensor jumper harnesses are not repairable and must be replaced. Do not attempt to solder, splice, or crimp these harnesses as eventual failure will likely result.

Brake System Bleeding This is one service area that is most likely to vary with the different designs of the ABS. Always refer to the service manual before attempting to bleed the brake system on a vehicle with ABS.

Some systems require that the accumulator be depressurized and the power source to the ABS control module disconnected. After these have been done, the system can be bled like a conventional brake system. On other systems, the bleeder screws are opened while the system is turned on, and they have one procedure for the front brakes and another for the rear brakes.

Failure to follow the correct procedure can result in personal injury, destruction of ABS components, or the trapping of more air in the system.

NEW TRENDS

The widespread use of ABS has allowed engineers to pursue many different features for an automobile. Not only have ABS, traction control, and stability control systems advanced, but the entire brake system is in a state of evolution.

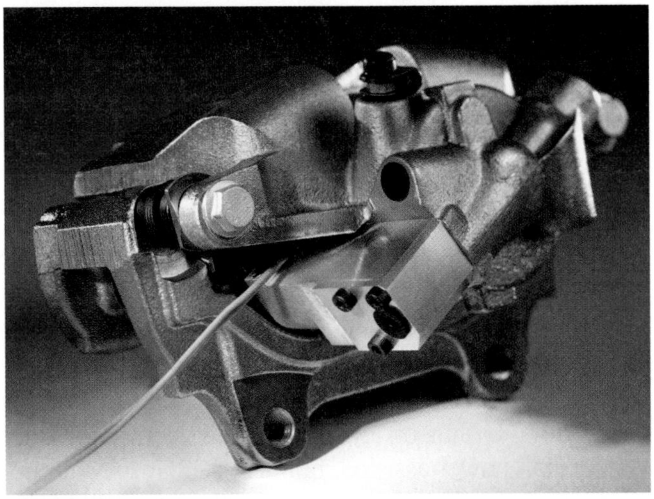

Figure 49-32 This caliper is fitted with an electronically controlled solenoid that controls the application of the parking brake. *Courtesy of Robert Bosch Corp.*

Automatic Parking Brake

The parking brake function of a disc brake can be activated by an electronically activated locking mechanism. This system is tied into the normal ABS system and uses hydraulic pressure to apply the parking brake. A solenoid at the caliper allows high fluid pressure to force the brake pads against the rotor **(Figure 49–32).**

Brake by Wire

In a brake-by-wire system, the braking command of the brake pedal is electronically processed. A control module receives sensor input from the brake pedal and activates the master cylinder **(Figure 49–33)**. For safety purposes, the normal linkage from the pedal to the master cylinder still exists. The advantage of brake by wire is simply that the pressure on the pedal can be immediately and directly relayed to the master cylinder.

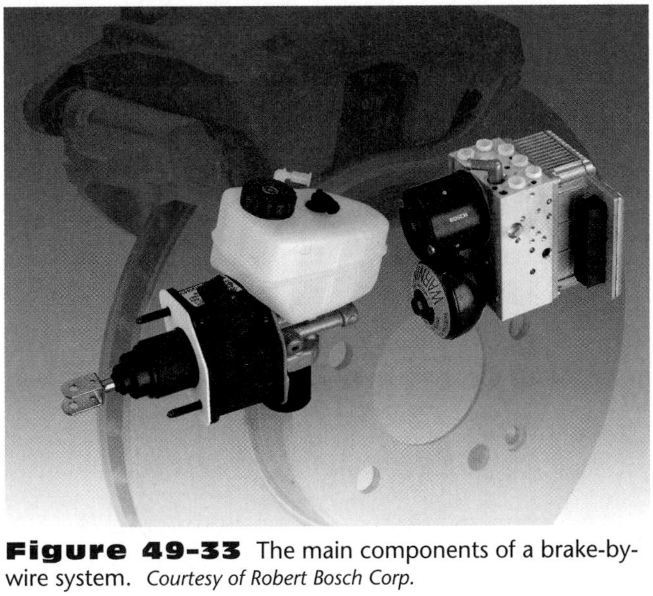

Figure 49-33 The main components of a brake-by-wire system. *Courtesy of Robert Bosch Corp.*

CASE STUDY

A customer brought in a late-model General Motors vehicle because her brake warning light came on and stayed on. Her expressed concern was she paid extra for ABS and is sure the system is not working, which is why the light is on. It seems she read her in the owner's manual that when the ABS light remains on, there is a problem with the system and the ABS is inoperative.

The technician verified the complaint and found the red brake warning lamp to be lit at all times. He then referred to the test procedures for that ABS. After the preliminary tests were completed, he used a scan tool to retrieve any trouble codes from the system. No DTCs were found and the brake warning light remained on.

Confused and frustrated, the technician sought help from a fellow certified technician. He summarized the problem and the tests he conducted. The certified technician went over to the car to see what he could find and immediately identified the problem. The red warning light was on, not the amber one. The amber lamp will light when there is an ABS problem. It will also indicate that the ABS is inoperative. The red light, however, will light when there is a problem of low brake fluid, a closed parking brake switch, or if there is a hydraulic problem in the base brake system.

Through a basic inspection, the technician found the brake fluid level to be low. Once the fluid level was corrected, the light went out. It seems the technician was sidetracked by the customer's concern over the light and her desire to have ABS. The moral of the story is simply to make sure you understand the function of all warning lights before you jump to any conclusions.

KEY TERMS

One-channel system
Two-channel system
Three-channel system
Four-channel system
Automatic traction control (ATC)
Integral antilock brake system
Nonintegral antilock brake system
Pressure modulation

SUMMARY

- Brake fluid is the lifeblood of any hydraulic brake system. DOT 3 is recommended for most antilock brake systems and some power brake systems.
- Modern antilock brake systems provide electronic/hydraulic pumping of the brakes for straight-line stopping under panic conditions.
- In addition to being classified as integral and nonintegral, antilock brake systems can be divided into the level of control they provide. They can be one-, two-, three-, or four-channel two- or four-wheel systems.
- Pressure modulation works to prevent wheel locking. Antilock brake systems can modulate the pressure to the brakes as often as fifteen times per second.
- Integrated antilock brake systems combine the master cylinder, hydraulic booster, and hydraulic circuitry into a single assembly. The great majority of antilock brake systems are of this type.
- On a nonintegrated ABS, the master cylinder and hydraulic valve unit are separate assemblies and a vacuum boost is used. In some nonintegrated systems, the master cylinder supplies brake fluid to the hydraulic unit.
- Automatic traction control (ATC) is a system that applies the brakes when a drive wheel attempts to spin and lose traction.
- Automatic stability systems correct oversteer and understeer by applying one wheel brake.
- Malfunction of the ABS causes the electronic control module to shut off or inhibit the system. However, normal power-assisted braking remains. Malfunctions are indicated by one or two warning lights inside the vehicle.
- Loss of hydraulic fluid or power booster pressure disables the antilock brake system,

TECH MANUAL

The following procedures are included in Chapter 49 of the *Tech Manual* that accompanies this book:

1. Performing ABS tests using a scan tool.
2. Monitoring the operation of the ABS solenoids using a lab scope.
3. Testing a wheel-speed sensor and adjusting its gap.

REVIEW QUESTIONS

1. Describe the ways antilock brake systems can be classified.
2. What is the primary difference between an automatic traction control system and a stability control system?
3. Explain the role of the wheel sensor(s) in the ABS.
4. Define the difference between an integrated and nonintegrated antilock braking system.
5. Briefly describe the proper steps and testing needed to accurately diagnose antilock braking systems.
6. An ABS modulates brake pressure. What does this mean?
7. Describe the normal pedal feel when the ABS is activated during hard braking.
8. What is the purpose of the hydraulic unit in an integrated ABS?
9. What is the basic difference between a Delco ABS VI and other typical systems?
10. Explain the difference between oversteer and understeer.
11. Technician A says in one-channel systems, the rear brakes on both sides of the vehicle are modulated at the same time to control skidding. Technician B says most single channel systems have a wheel-speed sensor at each rear wheel. Who is correct?
    a. Technician A
    b. Technician B
    c. Both A and B
    d. Neither A nor B
12. Technician A says a malfunction of the ABS causes the control module to shut off or inhibit the system. Technician B says a loss of hydraulic fluid or power booster pressure disables the antilock brake system. Who is correct?
    a. Technician A
    b. Technician B
    c. Both A and B
    d. Neither A nor B
13. Which of the following is a true statement?

a. In some antilock brake systems, the power brake assist is provided by pressurized brake fluid supplied by the hydraulic accumulator.
b. The accumulator is a small, sealed chamber mounted to the pump/motor assembly.
c. The accumulator holds a highly pressurized nitrogen gas that is used to generate a charging pressure.
d. All of the statements are true.

14. Technician A says some traction control systems have separate hydraulic valve units and control modules for an ABS and ATC. Technician B says some integrate both systems into one hydraulic control unit and a single control module. Who is correct?
    a. Technician A
    b. Technician B
    c. Both A and B
    d. Neither A nor B
15. While road testing a car equipped with an ABS, Technician A says that during heavy braking several pulses may be felt through the brake pedal. Technician B says a spongy pedal during normal braking is normal. Who is correct?
    a. Technician A
    b. Technician B
    c. Both A and B
    d. Neither A nor B
16. What is the name for the gas-filled pressure chamber that is part of the antilock braking system's pump and motor assembly?
    a. control module
    b. accumulator
    c. sensor
    d. reservoir
17. *True or False?* The ABS control module calculates the slip rate of the wheels and controls the brake fluid pressure throughout the entire system to attain the target slip rate.
18. Technician A says the accumulator in an ABS is used to store hydraulic fluid to provide residual pressure for power-assist braking. Technician B says the booster pump in an antilock system provides pressurized hydraulic fluid for the ABS. Who is correct?
    a. Technician A
    b. Technician B
    c. Both A and B
    d. Neither A nor B
19. Technician A says it is normal for the amber brake lamp to light when the ABS is activated during braking. Technician B says it is normal for the red brake lamp to be on whenever the ignition is on and the engine is off. Who is correct?
    a. Technician A
    b. Technician B
    c. Both A and B
    d. Neither A nor B
20. Technician A says some antilock brake systems use a lateral acceleration sensor in addition to the wheel-speed sensors. Technician B says traction control systems apply one wheel brake to get the vehicle going in the direction in which it is being steered. Who is correct?
    a. Technician A
    b. Technician B
    c. Both A and B
    d. Neither A nor B

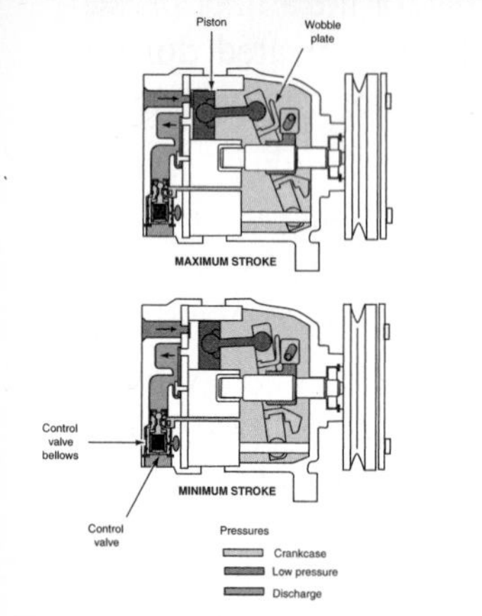

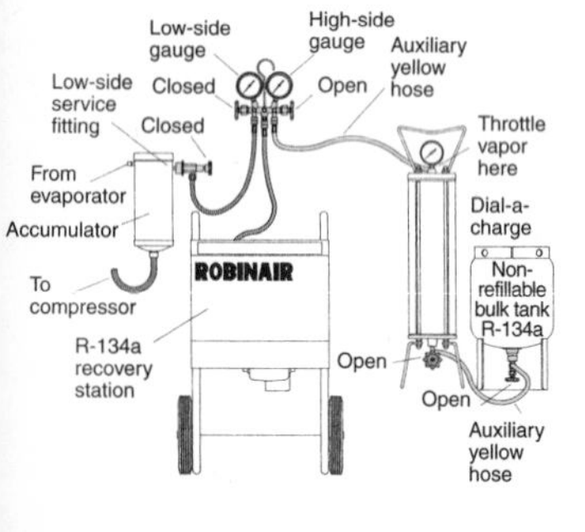

Delmar's Automotive Heating and Air Conditioning Video Series and CD-ROM Courseware

SECTION 9

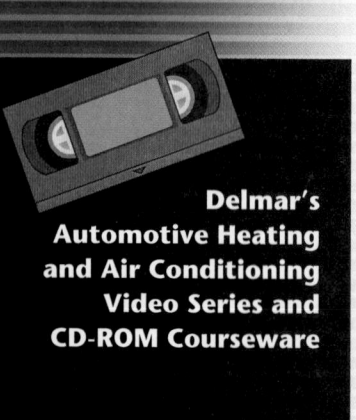

Passenger Comfort

Chapter 50 ■ *Heating and Air-Conditioning*

Chapter 51 ■ *Heating and Air-Conditioning Diagnosis and Service*

Modern vehicles are designed to provide passengers with comfort, convenience, and safety. Such accessory systems can be quite elaborate, adding electronic circuits and components to an already complex vehicle. One comfort system that few could live without is the heating and air-conditioning system. Concerns over the proper handling of air-conditioning refrigerant are particularly important, as are the environmentally based laws that dictate the current designs of these systems.

The information covered in Section 9 corresponds to materials found on the ASE certification tests on heating and air-conditioning systems.

WE ENCOURAGE PROFESSIONALISM

THROUGH TECHNICIAN CERTIFICATION

50

HEATING AND AIR-CONDITIONING

OBJECTIVES

■ **Identify the purpose of a ventilation system.** ■ **Identify the common parts of a heating system.** ■ **Compare the vacuum and mechanical controls of a heating system.** ■ **Describe how an automotive air-conditioning system operates.** ■ **Explain why R-134a is the current refrigerant of choice.** ■ **Locate, identify, and describe the function of the various air-conditioning components.** ■ **Describe the operation of the types of air-conditioning control systems.**

The discomforts of motoring's early days **(Figure 50–1)** would daunt the toughest motorcross racer. Paper-thin tires blew out if the vehicle ran over a good-size pebble. Headlights were so dim they could barely illuminate a locomotive in their path. There was no climate control. In winter, heavy coats and blankets were all that allowed passengers to survive behind a woefully inadequate windscreen. In summer, the breeze generated by 15 mph (25 km/h) travel was the only thing that retarded heat prostration.

Today, ventilation, heating, and air-conditioning systems are very important elements for providing passenger comfort. Ventilation and heating systems are standard equipment on all passenger vehicles and air-conditioning is standard on some and available for nearly all.

The large number of vehicles with air-conditioning plus recent changes in the methods used to cool a vehicle and to service the systems make a basic knowledge of air-conditioning systems a must for automotive technicians.

Figure 50–1 Early model cars did not offer the conveniences of today's vehicles.

VENTILATION SYSTEM

The ventilation system on most vehicles is designed to supply outside air to the passenger compartment through upper or lower vents or both. Several systems are used to vent air into the passenger compartment, the most common of which is the flow-through system **(Figure 50–2)**. In this arrangement, a supply of outside air, called ram air, flows into the car when it is moving. When the car is not moving, a steady flow of outside air can be produced by the heater fan. In operation, ram air is forced through an inlet grille. The pressurized air then circulates throughout the passenger and trunk compartment. From there the air is forced outside the vehicle through an exhaust area.

On certain older vehicles, air is admitted by opening or closing two vent knobs under the dashboard. The left knob controls air through the left inlet. The right knob controls air through the right inlet. The air is still considered ram air and is circulated through the passenger compartment.

Rather than using ram air (especially if the vehicle is stopped), a ventilation fan can be used. It can be accessible from under the dashboard or from inside the engine compartment. A blower assembly is placed inside the blower housing. As the squirrel cage blower rotates, it produces a strong suction on the intake. A pressure is also created on the output. When the fan motor is energized by using the temperature controls on the dashboard, air is moved through the passenger compartment.

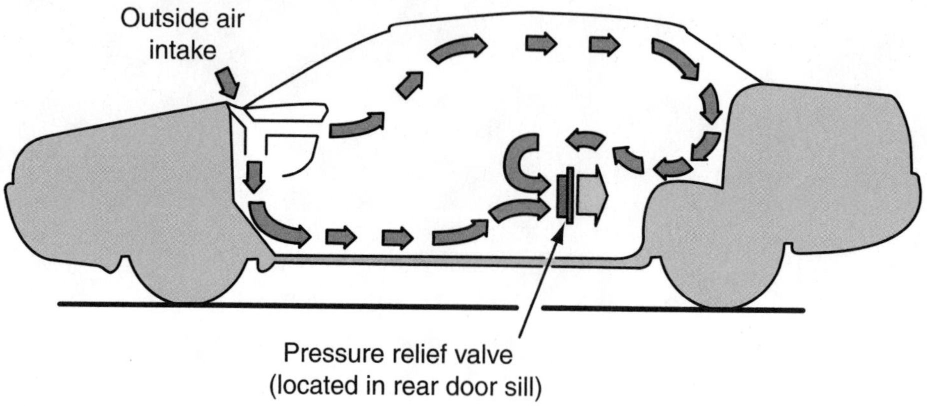

Figure 50-2 A flow-through ventilation system.

AUTOMOTIVE HEATING SYSTEMS

The automotive heating system has been designed to work hand in hand with the cooling system to maintain proper temperatures inside the car. The heating system's primary job is to provide a comfortable passenger compartment temperature and to keep car windows clear of fog or frost. The primary components of the automotive heating system are the heater core, the heater control valve, the blower motor and the fan, and the heater and defroster ducts **(Figure 50–3)**.

In the liquid-cooling system, heat from the coolant circulating inside the engine is converted to hot air, which is blown into the passenger compartment. Hot coolant from the engine is transferred by a heater hose to the heater control valve and then to the heater core inlet. As the coolant circulates through the core, heat is transferred from the coolant to the tubes and fins of the core. Air blown through the core by the blower motor and fan then picks up the heat from the surfaces of the core and transfers it into the passenger compartment of the car. After giving up its heat, the coolant is then pumped out through the heater core outlet, where it is returned to the engine to be recirculated by the water pump.

The **heater core** is generally designed and constructed much like a miniature radiator. It features an inlet or outlet tube and a tube and fin core to facilitate coolant flow between them.

Although all heater cores basically function in the same manner, several variations in design and materials are used by different automakers to achieve the same results. Most have an aluminum core with plastic tanks.

Like the radiator, heater core tanks, tubes, and fins can become clogged over time by rust, scale, and mineral deposits circulated by the coolant. Heater core failures are generally caused by leakage or clogging. Feel the heater inlet and outlet hoses while the engine is idling and warm with the heater temperature control on hot. If the hose downstream of the heater valve does not feel hot, the valve is not opening.

With cable-operated control valves, check the cable for sticking, slipping (loose mounting bracket) or misadjustment. With valves that are vacuum operated, there

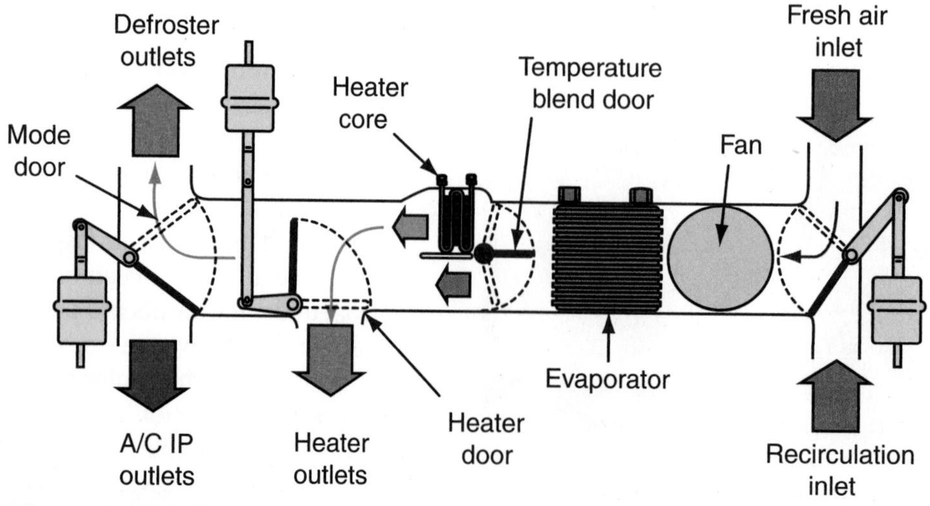

Figure 50-3 A typical HVAC airflow for a heater and air-conditioning.

should be no vacuum to the valve when the heater is on (except for those that are normally closed and need vacuum to open).

If the heater core appears to be plugged, the inlet hose may feel hot up to the core but the outlet hose remains cool. Reverse flushing the core with a power flusher may open up the blockage, but usually the core has to be removed for cleaning or replacement. Air pockets in the heater core can also interfere with proper coolant circulation. Air pockets form when the coolant level is low or when the cooling system is not properly filled after draining.

When the heater core leaks and must be repaired or replaced, it is a very difficult and time-consuming job primarily because of the core's location deep within the firewall of the car. For this reason always leak test a replacement heater core before installation. Also flush the cooling system and replace the coolant seasonally.

SHOP TALK

Engines with aluminum cylinder heads or blocks sometimes lose coolant through microscopic porosity leaks in the castings. Though the amount of coolant that is lost this way is not great, over time it can contaminate the oil enough to cause premature engine wear, especially in the cylinders and bearings. Pressure testing the cooling system usually does not reveal microscopic porosity leaks because the cracks are too small to allow a noticeable pressure drop. ■

Porosity leaks can be sealed by adding a hot melt sealer product to the cooling system. Some experts recommend adding stop-leak products as a preventive measure when changing the coolant in late-model engines with aluminum heads or blocks.

Heater Control Valve

The **heater control valve** (sometimes called the water flow valve) controls the flow of coolant into the heater core from the engine. In a closed position, the valve allows no flow of hot coolant to the heater core, keeping it cool. In an open position, the valve allows heated coolant to circulate through the heater core, maximizing heater efficiency. Heater control valves are operated in three basic ways: by cable, thermostat, or vacuum.

Cable-operated valves are controlled directly from the heater control lever on the dashboard **(Figure 50–4)**. Thermostatically controlled valves feature a liquid-filled capillary tube located in the discharge airstream off the heater core that senses air temperature and modulates the flow of water to maintain a constant temperature regardless of engine speed or temperature.

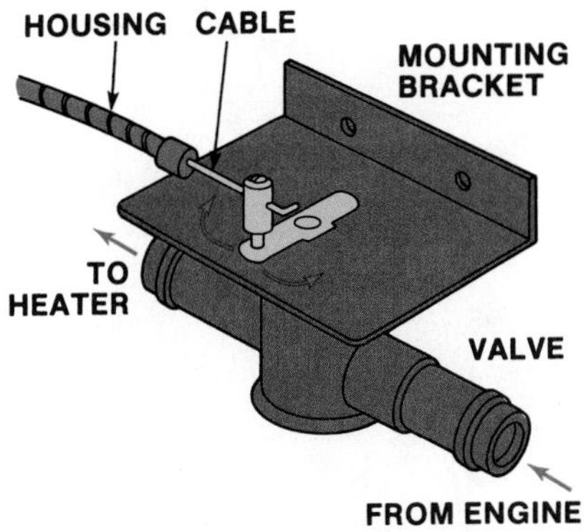

Figure 50-4 A cable-operated heater valve used to control the flow of water or coolant through the heater core.

Most heater valves utilized on today's car are vacuum operated. These valves are normally located in the heater hose line or mounted directly in the engine block. When a vacuum signal reaches the valve, a diaphragm inside the valve is raised, either opening or closing the valve against an opposing spring. When the temperature selection on the dashboard is changed, vacuum to the valve is vented and the valve returns to its original position. Vacuum-actuated heater control valves are either normally open or normally closed designs. Some vehicles do not use a heater control valve; rather, a heater door controls how much heat is released into the passenger compartment from the heater core.

On late-model vehicles, heater control valves are typically made of plastic for corrosion resistance and light weight. These valves feature few internal working parts and no external working parts. With the reduced weight of these valves, external mounting brackets are not required.

The thermostat, which helps regulate the coolant temperature in the cooling system, plays a large part in the heating system. A malfunctioning thermostat can cause the engine to overheat or not reach normal operating temperature, or it can be the cause of poor heater performance.

Blower Motor

The blower motor is usually located in the heater housing assembly. It ensures that air is circulated through the system **(Figure 50–5)**. Its speed is controlled by a multiposition switch in the control panel that works in conjunction with a resistor block that is usually located on the heater housing.

On some vehicles when the engine is running, the blower motor is in constant operation at low speed. On automatic temperature control systems, the blower motor is activated only when the engine reaches a predetermined temperature. The blower motor circuit is protected by a fuse located in the fuse panel. The fuse rating is usually 20 to 30 amperes.

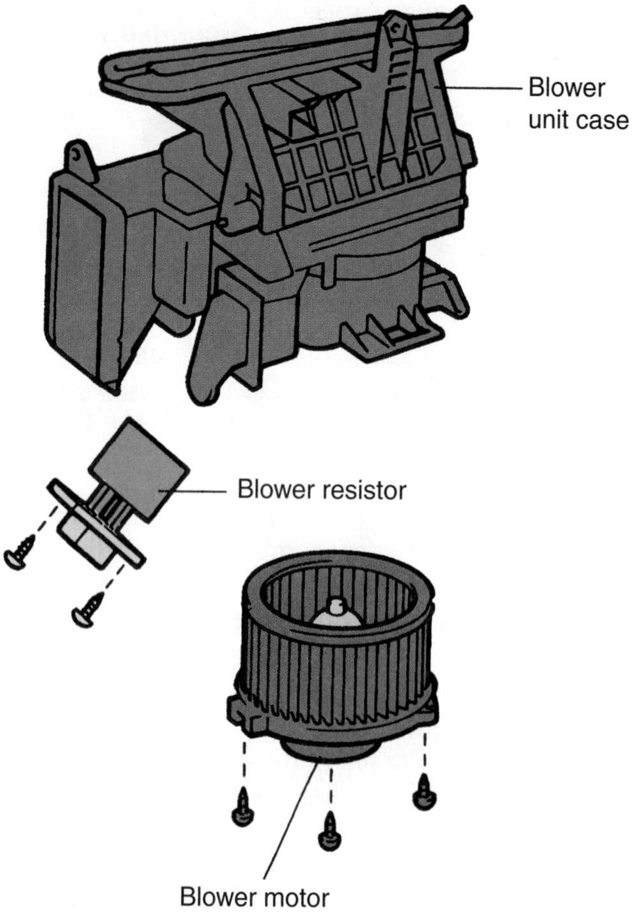

Figure 50-5 A blower motor assembly. *Reprinted with permission*

The blower motor resistor block is used to control the blower motor speed. The typical resistor block is composed of three or four wire resistors in series with the blower motor, which control its voltage and current. The speed of the motor is determined by the control panel switch, which puts the resistors in series. Increasing the resistance in the system slows the blower speed.

If the blower does not operate, use a test light to make sure there is voltage on both sides of the fuse. Then, check to see if current is arriving at the motor. On cars in which the blower motor is behind the inner fender shell, hunt out the wiring and check for current. If the blower motor is getting current, the problem is either a burned-out blower motor or a bad ground on the motor. In situations where no current is available at the motor, backtrack to check for an open resistor. Check also for burned or corroded connections in the blower relays or bulkhead connectors. In-line fuses are another potential problem.

Sliding the heat selector control switch to off on a vehicle without factory air-conditioning closes a door that blocks the flow of air into the heater core. Even if the blower is turned on, no heat comes out because the air inlet is closed. On vehicles with factory air-conditioning, a **blend door** usually routes some air through the heater core and some through the evaporator depending on the temperature range selected.

If the blower motor runs but no air comes out of the ducts, the problem is either a stuck or inoperative airflow control or blend door. This can also affect the operation of the defrosters. These doors may be cable or vacuum operated. Change the position of the temperature selector knob, sliding it from hot to cold. If the sound of doors opening and closing is not heard, the control cables can sometimes slip loose from the dash switch or door arm rendering the door inoperative. If this is the case, little or no resistance is felt when sliding the temperature control knob. A kinked or rusted cable can also prevent a door from working. If this is the case, resistance is felt trying to move the control knob. In either situation, it is necessary to get under the dash, find the cable, and replace, reroute, or reconnect it. Doors can also be jammed by objects that have fallen down the defroster ducts. Remove the obstruction from the plenum by fishing through the heater outlet with a coat hanger or magnet, or remove the plenum. Be careful when fishing for the obstruction. Some heater cores are easily pierced.

With vacuum-controlled doors, the most common reasons for failure are leaky or loose vacuum hoses or defective diaphragms in the little vacuum motors that move the doors. Check the vacuum by starting the engine and disconnecting the small hose that goes to the vacuum motor that works one of the doors. If you feel a vacuum or hear a hissing sound when trying different temperature settings, the vacuum source is good. If by applying vacuum to it with a hand-held pump, it moves and holds vacuum for about one minute then bleeds off, the problem is a bad vacuum motor. If it does not, check for leaky vacuum hose connections, a defective temperature control switch, or a leaky vacuum reservoir or check valve under the dash or in the engine compartment.

A low coolant level can starve the heater resulting in little or no heat output. Check the coolant level in the radiator (not the overflow tank) to see if it is low. If low, adding coolant does not cure the problem until the problem of low coolant is found and fixed.

Air pockets can form in long heater hoses or where the heater is mounted higher than the radiator. To vent the trapped air, some vehicles have bleeder valves on the hoses. Opening the valves allows air to escape as the system is filled. The valves are then closed when coolant reaches their level. On vehicles that lack these special bleeder valves, it may be necessary to temporarily loosen the heater outlet hose so air can bleed out as the system is filled.

Heater and Defroster Duct Hoses

Transferring heated air from the heater core to passenger compartment heater and defroster outlets is the job of the heater and defroster ducts. The ducts are typically part of a large plastic shell that connects to the necessary inside and outside vents. This ductwork also has mounting points for the evaporator and heater core assemblies. Contained inside the duct are also the doors required to direct air to the floor, dash, and/or windshield. Sometimes the duct is connected directly to the vents, others times hoses are used.

THEORY OF AUTOMOTIVE AIR-CONDITIONING

It is important for the technician to know the basic theory of automotive air-conditioning systems. Understanding how a system moves heat from the confined space of the passenger compartment and dissipates it into the atmosphere **(Figure 50–6)** will assist the technician in analyzing system failures and in performing the required maintenance and service.

All air-conditioning systems are based on three fundamental laws of nature, as follow:

Heat Flow

An air-conditioning system is designed to pump heat from one point to another. All materials or substances, as cold as –459°F (–273°C), have heat in them. Also, heat always flows from a warmer object to a colder one. For example, if one object is at 30°F (–1.1°C) and another object is 80°F (27°C), heat flows from the warmer object (80°F [27°C]) to the colder one (30°F [–1.1°C]). The greater the temperature difference between the objects, the greater the amount of heat flow.

Heat Absorption

Objects can be in one of three forms: solid, liquid, or gas. When objects change from one state to another, large amounts of heat can be transferred. For example, when water temperature goes below 32°F (0°C), water changes from a liquid to a solid (ice). If the temperature of water is raised to 212°F (100°C), the liquid turns into a gas (steam). But an interesting thing occurs when water, or any matter, changes from a solid to a liquid and then from a liquid to a gas. Additional heat is necessary to change the state of the substance, even though this heat does not register on a thermometer. For example, ice at 32°F (0°C) requires heat to change into water, which will also be at 32°F (0°C). Additional heat raises the temperature of the water until it reaches the boiling point of 212°F (100°C). More heat is required to change water into steam. But if the temperature of the steam were measured, it would also be 212°F (100°C). The amount of heat necessary to change the state of a substance is called **latent heat**—or hidden heat—because it cannot be measured with a thermometer. This hidden heat is the basic principle behind all air-conditioning systems.

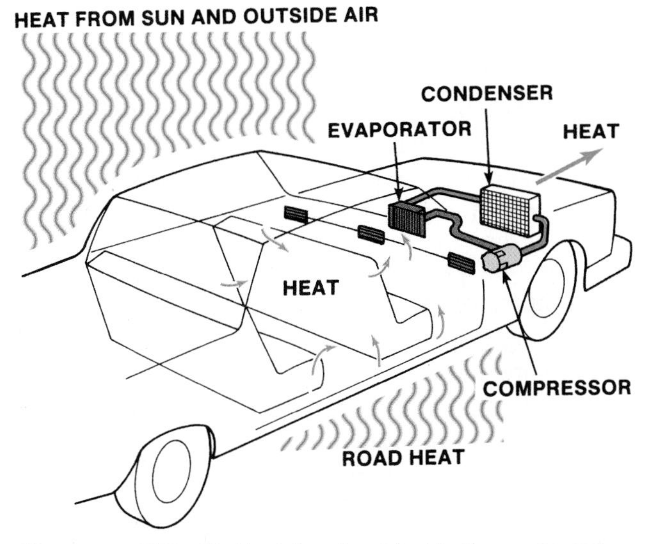

Figure 50-6 Heat flow from inside the car to outside. *Courtesy of Everco Industries, Inc.*

Pressure and Boiling Points

Pressure also plays an important part in air-conditioning. Pressure on a substance, such as a liquid, changes its boiling point. The greater the pressure on a liquid, the higher the boiling point. If pressure is placed on a vapor, the vapor condenses at a higher-than-normal temperature. In addition, as the pressure on a substance is reduced, the boiling point can also be reduced. For example, the boiling point of water is 212°F (100°C). The boiling point can be increased by increasing the pressure on the fluid. It can also be decreased by reducing the pressure or placing the fluid in a vacuum.

REFRIGERANTS IN AIR-CONDITIONING SYSTEMS

Types of Refrigerant

The substance used to remove heat from the inside of an air-conditioned vehicle is called the refrigerant. On older automotive air-conditioning systems, the refrigerant was Refrigerant-12 (commonly referred to as R-12 and Freon). R-12 is dichlorodifluoromethane (CCl_2F_2). An amendment to the United States' Clean Air Act and international agreements have mandated that R-12 be phased out as the refrigerant of choice. This came as a result of research that found the earth's ozone layer was being deteriorated by the chemicals found in R-12. The ozone layer is the earth's outermost shield of protection. This delicate layer protects against harmful effects of the sun's ultraviolet rays. The thinning of the ozone layer has become a worldwide concern. The ozone depletion is caused in part by release of **chlorofluorocarbons (CFCs)** into the atmosphere. R-12 is in the chemical family of CFCs. Since air-conditioning systems with R-12 are susceptible to leaks, further damage to the ozone layer could be avoided by not using R-12 in air-conditioning units.

Of the many chemicals that could be used in place of R-12, the automobile manufacturers have decided to use R-134a. This refrigerant may also be referred to as SUVA. R-134a is a **hydrofluorocarbon (HFC)** that causes less damage to the ozone layer when released into the atmosphere.

Although R-134a air conditioners operate in the same way and with the same basic components as R-12 systems, the two refrigerants are not interchangeable. Because it is less efficient than R-12, R-134a operates at higher pressures and is designed to handle this pressure. R-134a systems also require different service techniques and equipment. RAII R-134a systems are identified by an underhood decal and by the hoses and fittings used in the system.

The production phase-out for all CFCs, including R-12, was scheduled for December 1, 1995. To keep systems operating that were originally equipped with R-12; there are retrofit kits available for a changeover to R-134a. Although R-134a is less likely to have an adverse effect on the ozone layer, it still has the capability of contributing to the greenhouse effect when released into the atmosphere. Therefore, the recovery and recycling of R-12 and R-134a is mandatory by law.

Legislation has drastically affected the life of an air-conditioning technician. Laws have been passed that dictate the certification of equipment and technicians. The laws also define how older vehicles can be updated, through a conversion to R-134a or the use of an alternative refrigerant.

Basic Operation of an Air-Conditioning System

Refrigerants are used to carry heat from the inside of the vehicle to the outside of the vehicle. Automotive refrigerants have a low boiling point (the point at which **evaporation** occurs). For example, at any temperature above –21.6°F (–29.7°C), liquid R-12 can become a vapor. As a refrigerant changes state, it absorbs large amounts of heat. Since the heat that it absorbs is from the inside of the vehicle, passengers are cooler.

To understand how a refrigerant is used to cool the interior of a vehicle, the effects of pressure and temperature on it must first be understood. If the pressure of the refrigerant is high, so is its temperature; if the pressure is low, so is its temperature. Therefore, the temperature of R-12 can be changed by changing its pressure. **Figure 50–7** compares the pressures of R-12 and R-134a at various temperatures. Both refrigerants have an increase in pressure with an increase in temperature, and vice versa.

To absorb heat, the temperature and pressure of the refrigerant is kept low. To dissipate heat, the temperature and pressure are high. Remember that the heat from something hot always moves to something colder. The refrigerant absorbs heat from the inside of the vehicle and dissipates it to the cooler outside air. As the refrigerant absorbs heat, it changes from a liquid to a vapor. As it dissipates heat, it changes from a vapor to a liquid. The change from a vapor to a liquid is called **condensation**. These two changes of state—evaporation and condensation—occur continuously as the refrigerant circulates through the air-conditioning system.

THE AIR-CONDITIONING SYSTEM AND ITS COMPONENTS

An automotive air-conditioning system is a closed, pressurized system. It consists of a compressor, condenser, receiver/dryer or accumulator, expansion valve or orifice tube, and an evaporator.

In a basic air-conditioning system, the heat is absorbed and transferred in the following steps **(Figure 50–8)**:

1. Refrigerant leaves the compressor as a high-pressure, high-temperature vapor.
2. By removing heat via the condenser, the vapor becomes a high-pressure, high-temperature liquid.
3. Moisture and contaminants are removed by the receiver/dryer, where the cleaned refrigerant is stored until it is needed.
4. The expansion valve controls the flow of refrigerant into the evaporator.
5. Heat is absorbed from the air inside the passenger compartment by the low-pressure, warm refrigerant, causing the liquid to vaporize and greatly decrease its temperature.
6. The refrigerant returns to the compressor as a low-pressure, low-temperature vapor.

To understand the operation of the five major components, remember that an air-conditioning system is divided into two sides: the high side and the low side. **High side** refers to the side of the system that is under high pressure and high temperature. **Low side** refers to the low-pressure, low-temperature side of the system.

The Compressor

The compressor is the heart of the automotive air-conditioning system. It separates the high-pressure and low-pressure sides of the system. The primary purpose of the unit is to draw the low-pressure and low-temperature vapor from the **evaporator** and compress this vapor into high-temperature, high-pressure vapor. This action results in the refrigerant having a higher temperature than surrounding air, and enables the **condenser** to condense the vapor back to a liquid. The secondary purpose of the

STANDARD TEMPERATURE/ PRESSURE CHART FOR R-134a

°F	PSI	°F	PSI	°F	PSI	°F	PSI	°F	PSI
65	69	77	86	89	107	101	131	113	158
66	70	78	88	90	109	102	133	114	160
67	71	79	90	91	111	103	135	115	163
68	73	80	91	92	113	104	137	116	165
69	74	81	93	93	115	105	139	117	168
70	76	82	95	94	117	106	142	118	171
71	77	83	96	95	118	107	144	119	173
72	79	84	98	96	120	108	146	120	176
73	80	85	100	97	122	109	149		
74	82	86	102	98	125	110	151		
75	83	87	103	99	127	111	153		
76	85	88	105	100	129	112	156		

METRIC TEMPERATURE/ PRESSURE CHART FOR R-134a

°C	kPa	°C	kPa	°C	kPa
18	476	29	676	40	945
19	483	30	703	41	979
20	503	31	724	42	1007
21	524	32	752	43	1027
22	545	33	765	44	1055
23	552	34	793	45	1089
24	572	35	814	46	1124
25	593	36	841	47	1158
26	621	37	876	48	1179
27	642	38	889	49	1214
28	655	39	917		

STANDARD TEMPERATURE/ PRESSURE CHART FOR R-12

°F	PSI	°F	PSI	°F	PSI	°F	PSI	°F	PSI
65	74	75	87	85	102	95	118	105	136
66	75	76	88	86	103	96	120	106	138
67	76	77	90	87	105	97	122	107	140
68	78	78	92	88	107	98	124	108	142
69	79	79	94	89	108	99	125	109	144
70	80	80	96	90	110	100	127	110	146
71	82	81	98	91	111	101	129	111	148
72	83	82	99	92	113	102	130	112	150
73	84	83	100	93	115	103	132	113	152
74	86	84	101	94	116	104	134	114	154

METRIC TEMPERATURE/ PRESSURE CHART FOR R-12

°C	kg/cm^2	°C	kg/cm^2	°C	kg/cm^2
18	5.2	28	7.0	38	9.0
19	5.3	29	7.1	39	9.2
20	5.5	30	7.2	40	9.4
21	5.6	31	7.5	41	9.6
22	5.8	32	7.7	42	9.9
23	6.0	33	7.9	43	10.0
24	6.1	34	8.1	44	10.4
25	6.3	35	8.3	45	10.7
26	6.6	36	8.5	46	10.9
27	6.8	37	8.7	47	11.0

Figure 50-7 Refrigerant temperature and pressure charts. *© National Institute for Automotive Service Excellence*

compressor is to circulate or pump the refrigerant through the A/C system under the different pressures required for proper operation. The compressor is located on the engine and is driven by the engine's crankshaft via a drive belt.

Although there are numerous types of compressors in use today, they are usually one of three types, as follow:

Piston Compressor This type of compressor **(Figure 50–9)** can have its pistons arranged in an in-line, axial, radial, or V design. It is designed to have an intake stroke and a compression stroke for each cylinder. On the intake stroke, the refrigerant from the low side of the system (evaporator) is drawn into the compressor. The intake of refrigerant occurs through intake reed valves. These one-way valves control the flow of refrigerant vapors into the cylinder. During the compression stroke, the vaporous refrigerant is compressed. This increases both the pressure and the temperature of the heat-carrying refrigerant. The outlet or **discharge side** reed valves then open to allow the refrigerant to move to the condenser. The outlet reed valves are the beginning of the high side of the system. Reed valves are made of spring steel, which can be weakened or broken if improper charging procedures are used, such as liquid charging with the engine running.

A common variation of a piston-type compressor is the variable displacement compressor **(Figure 50–10)**. These compressors not only act as regular compressors but they also control the amount of refrigerant that

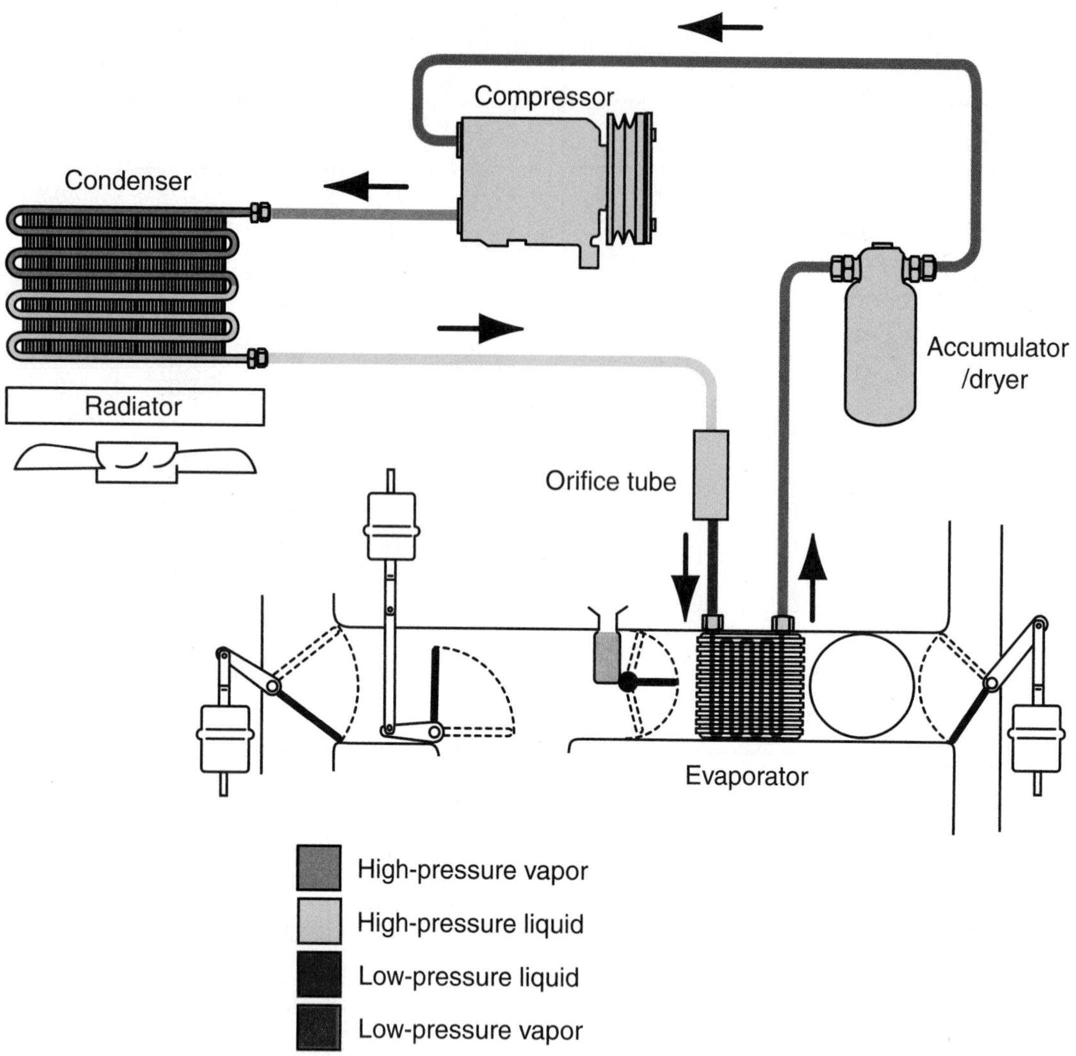

Figure 50-8 The basic refrigerant flow cycle.

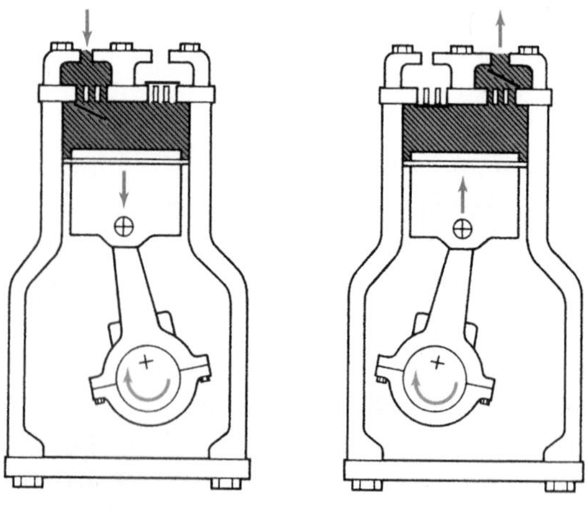

Figure 50-9 (A) A piston on the downstroke (intake) pulls low-pressure refrigerant into the cylinder cavity. The intake (suction) valve is open, and the discharge valve is closed. (B) A piston on the upstroke (discharge) compresses refrigerant vapor and forces it out through the discharge valve. The intake (suction) valve is closed, and the discharge valve is open.

passes through the evaporator. The pistons are connected to a wobble-plate. The angle of the wobble-plate determines the stroke of the pistons and is controlled by the difference in pressure between the outlet and inlet of the compressor. When the stroke of the pistons is increased, more refrigerant is being pumped and there is increased cooling.

Rotary Vane Compressor The rotary vane compressor does not have pistons. It consists of a rotor with several vanes and a carefully shaped housing. As the compressor shaft rotates, the vanes and housing form chambers. The refrigerant is drawn through the suction port into these chambers, which become smaller as the rotor turns. The discharge port is located at the point where the gas is completely compressed. No sealing rings are used in a vane compressor. The vanes are sealed against the housing by centrifugal force and lubricating oil. The oil sump is located on the discharge side, so the high pressure tends to force it around the vanes into the low-

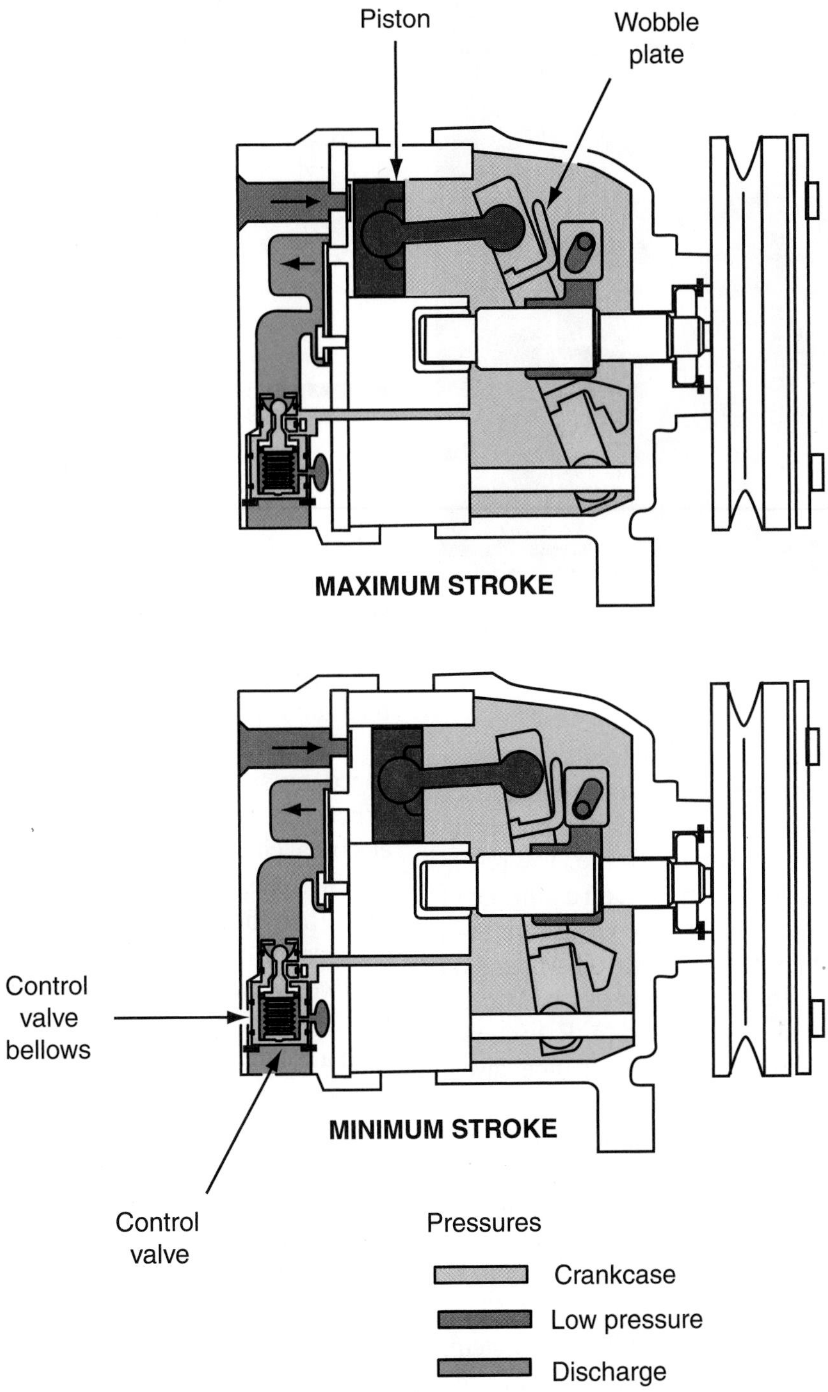

Figure 50–10 A V5 variable displacement compressor.

pressure side. This action ensures continuous lubrication. Because this type of compressor depends on a good oil supply, it is subject to damage if the system charge is lost. A protection device is used to disengage the clutch if pressure drops too low.

Scroll-Type Compressor The scroll-type compressor has a movable scroll and a fixed or nonmovable scroll that provide an eccentric-like motion. As the compressor's crankshaft rotates, the movable scroll forces the refrigerant against the fixed scroll and toward the center of the compressor. This motion pressurizes the refrigerant. The action of a scroll-type compressor can be compared to that of a tornado. The pressure of air moving in a circular pattern increases as it moves toward the center of the circle. A delivery port is positioned at the center of the compressor and allows the high-pressure refrigerant to flow into the air-conditioning system **(Figure 50–11)**. This type of compressor operates more smoothly than other designs.

Figure 50-11 A scroll-type compressor. *Courtesy of DaimlerChrysler Corporation*

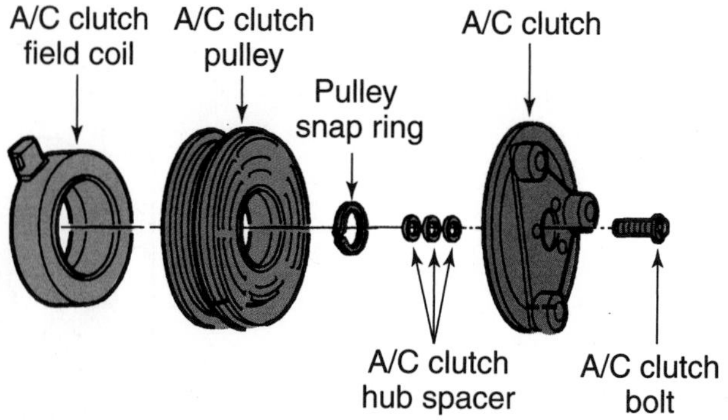

Figure 50-12 Parts of a compressor clutch assembly. *Courtesy of Ford Motor Company*

Refrigerant Oils

Normally the only source of lubrication for a compressor is the oil mixed with the refrigerant. Because of the loads and speeds at which the compressor operates, proper lubrication is a must for long compressor life. The refrigerant oil required by the system depends on a number of things, but it is primarily dictated by the refrigerant used in the system. R-12 systems use a mineral oil. Mineral oil mixes well with R-12 without breaking down. Mineral oil, however, cannot be used with R-134a. R-134a systems require that a synthetic oil, polyalkaline glycol (PAG) be used. There are a number of different blends of PAG oil. Always use the one recommended by the vehicle manufacturer or compressor manufacturer. Failure to use the correct oil will cause damage to the compressor.

Most often, when a system has been converted from R-12 to R-134a, ester oil will be recommended. Since there may some mineral oil and R-12 still in the system, it is important to use an oil that will mix well with the residue and not break down. Ester oil mixes well with mineral oil because it is hydrocarbon-based.

Compressor Clutches

Every compressor is equipped with an electromagnetic clutch as part of the compressor pulley assembly **(Figure 50–12)**. It is designed to engage the pulley to the compressor shaft when the clutch coil is energized. The purpose of the clutch is to transmit power from the engine to the compressor and to provide a means of engaging and disengaging the refrigeration system from engine operation.

The clutch is driven by power from the engine's crankshaft, which is transmitted through one or more belts (a few use gears) to the pulley, which is in operation whenever the engine is running. When the clutch is engaged, power is transmitted from the pulley to the compressor shaft by the clutch drive plate. When the clutch is not engaged, the compressor shaft does not rotate and the pulley freewheels.

The clutch is engaged by a magnetic field and disengaged by springs when the magnetic field is broken. When the controls call for compressor operation, the electrical circuit to the clutch is completed, the magnetic clutch is energized, and the clutch engages the compressor. When the electrical circuit is opened, the clutch disengages the compressor.

Two types of electromagnetic clutches have been in use for many years. Early-model air-conditioning systems used a rotating coil clutch. The magnetic coil, which engages or disengages the compressor, is mounted within the pulley and rotates with it. Electrical connections for the clutch operation are made through a stationary brush assembly and rotating slip rings, which are part of the field coil assembly. This older rotating coil clutch, now in limited use, has been largely replaced by the stationary coil clutch.

With the stationary coil, wear has been measurably reduced, efficiency increased, and serviceability made much easier. The clutch coil does not rotate. When the driver first energizes the air-conditioning system from the passenger compartment dashboard, the pulley assembly is magnetized by the stationary coil on the compressor body, thus engaging the clutch to the clutch hub attached to the compressor shaft. This activates the air-conditioning system. Depending on the system, the magnetic clutch is usually pressure controlled to cycle the operation of the compressor (depending on system temperature or pressure). In some system designs, the clutch might operate continually when the system is turned on. With stationary coil design, service is not usually necessary except for an occasional check on the electrical connections.

Condenser

The condenser **(Figure 50–13)** consists of coiled refrigerant tubing mounted in a series of thin cooling fins to provide maximum heat transfer in a minimum amount of space. The condenser is normally mounted just in front of the vehicle's radiator. It receives the full flow of ram air from the movement of the vehicle or airflow from the radiator fan when the vehicle is standing still.

The purpose of the condenser is to condense or liquefy the high-pressure, high-temperature vapor coming from the compressor. To do so, it must give up its heat. The condenser receives very hot (normally 200 to 400°F [93 to 204°C]), high-pressure refrigerant vapor from the compressor through its discharge hose. The refrigerant vapor enters the inlet at the top of the condenser and as the hot vapor passes down through the condenser coils, heat (following its natural tendencies) moves from the hot refrigerant into the cooler air as it flows across the condenser coils and fins. This process causes a large quantity of heat to be transferred to the outside air and the refrigerant to change from a high-pressure hot vapor to a high-pressure warm liquid. This high-pressure warm liquid flows from the outlet at the bottom of the condenser through a line to the receiver/dryer or to the refrigerant metering device if an accumulator instead of a dryer is used.

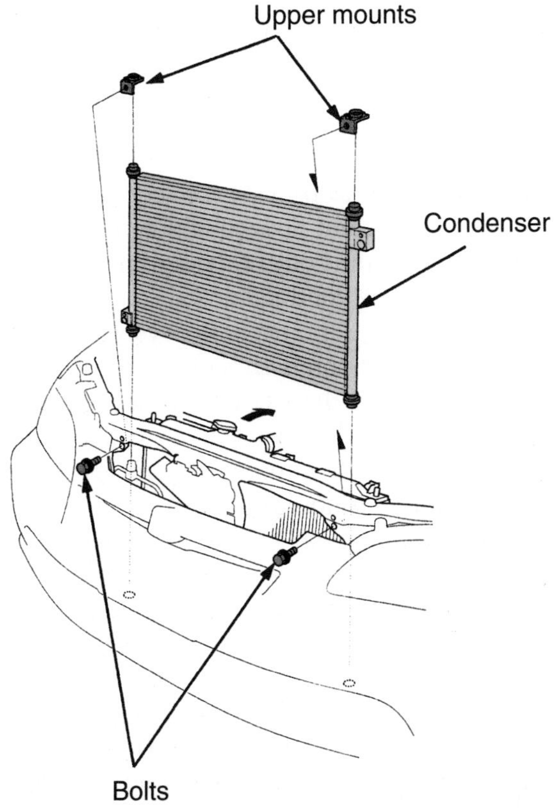

Figure 50–13 A typical condenser. *Courtesy of American Honda Motor Co., Inc.*

In an air-conditioning system operating under an average heat load, the condenser has a combination of hot refrigerant vapor in the upper two-thirds of its coils. The lower third of the coils contains the warm liquid refrigerant, which has condensed. This high-pressure, liquid refrigerant flows from the condenser and on toward the evaporator. In effect, the condenser is a true heat exchanger.

Receiver/Dryer

Used on many early systems, the **receiver/dryer** is a storage tank for the liquid refrigerant from the condenser, which flows into the upper portion of the receiver tank containing a bag of **desiccant** (moisture-absorbing material such as silica alumina or silica gel). As the refrigerant flows through an opening in the lower portion of the receiver, it is filtered through a mesh screen attached to a baffle at the bottom of the receiver. The desiccant in this unit absorbs any moisture present that might enter the system during assembly. These features of the unit prevent obstruction to the valves or damage to the compressor.

Depending on the manufacturer, the receiver/dryer may be known by such names as filter or dehydrator. Regardless of its name, the function is the same. Included in many receiver/dryers are additional features such as high-pressure fittings, a pressure relief valve, and a sight glass for determining the state and condition of the refrigerant in the system.

The receiver/dryer is often neglected when the air-conditioning system is serviced or repaired. Failure to replace it can lead to poor system performance or replacement part failure. It is recommended that the receiver/dryer and/or its desiccant be changed whenever a component is replaced, when the system has lost the refrigerant charge, or when the system has been open to the atmosphere for any length of time.

WARNING!

The desiccant draws moisture like a magnet and can become contaminated in less than five minutes if it is exposed to the atmosphere. Keep it sealed. Once opened, put it under a vacuum quickly.

Most late-model systems are not equipped with a receiver/dryer; rather, they use an accumulator to accomplish the same thing **(Figure 50–14)**. The accumulator is connected into the low side at the outlet of the evaporator. It also contains a desiccant and is designed to store, filter, and dry excess refrigerant. If liquid refrigerant flows out of the evaporator, it will be collected by and stored in the accumulator. The main purpose of an accumulator is to prevent liquid from entering the compressor.

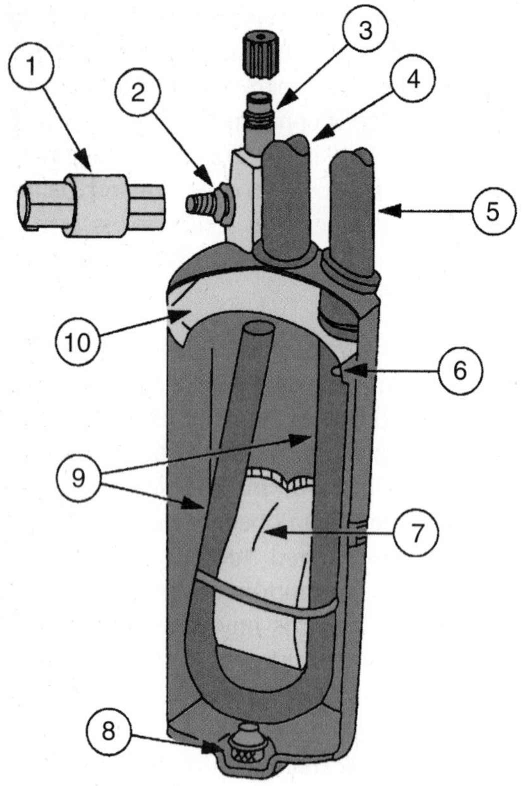

1. Cycling switch
2. O-ring seal
3. Low-side port
4. From evaporator
5. To compressor
6. Antisiphon hole
7. Desiccant bag
8. Oil return filter
9. Vapor return tube
10. Accumulator/dryer dome

Figure 50-14 An accumulator/dryer. *Courtesy of Ford Motor Company*

Thermostatic Expansion Valve/ Orifice Tube

The refrigerant flow to the evaporator must be controlled to obtain maximum cooling, while ensuring complete evaporation of the liquid refrigerant within the evaporator. This is accomplished by a thermostatic expansion valve (TEV or TXV) or a fixed **orifice tube (Figure 50–15).**

The TEV is mounted at the inlet to the evaporator and separates the high-pressure side of the system from the low-pressure side. The TEV regulates refrigerant flow to the evaporator to prevent evaporator flooding or starving. In operation, the TEV regulates the refrigerant flow to the evaporator by balancing the inlet flow to the outlet temperature.

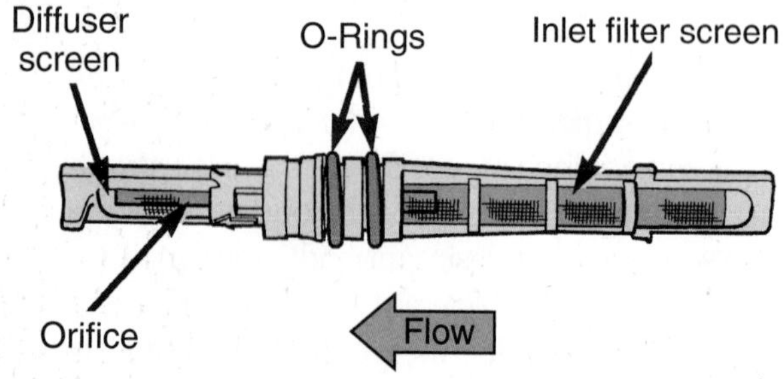

Figure 50-15 A typical orifice expansion tube. *Courtesy of DaimlerChrysler Corporation*

Both externally and internally equalized TEVs are used in air-conditioning systems. The only difference between the two valves is that the external TEV uses an equalizer line connected to the evaporator outlet line as a means of sensing evaporator outlet pressure. The internal TEV senses evaporator inlet pressure through an internal equalizer passage. Both valves have a capillary tube to sense evaporator outlet temperature. The tube is filled with a gas, which, if allowed to escape due to careless handling, will ruin the TEV.

During stabilized conditions, the pressure on the bottom of the expansion valve diaphragm becomes equal to the pressure on the top of the diaphragm, which allows the valve spring to close the valve. When the system is started, the pressure on the bottom of the diaphragm drops rapidly, allowing the valve to open and meter liquid refrigerant to the lower evaporator tubes where it begins to vaporize.

Compressor suction draws the vaporized refrigerant out of the top of the evaporator at the top tube where it passes by the sealed sensing bulb. The bottom of the valve diaphragm internally senses the evaporator pressure through the internal equalization passage around the sealed sensing bulb. As evaporator pressure is increased, the diaphragm flexes upward pulling the pushrod away from the ball seat of the expansion valve. The expansion valve spring forces the ball onto the tapered seat and the liquid refrigerant flow is reduced.

As the pressure is reduced due to restricted refrigerant flow, the diaphragm flexes downward again, opening the expansion valve to provide the required controlled pressure and refrigerant flow condition. As the cool refrigerant passes by the body of the sensing bulb, the gas above the diaphragm contracts and allows the expansion valve spring to close the expansion valve. When heat from the passenger compartment is absorbed by the refrigerant, it causes the gas to expand. The pushrod again forces the expansion valve to open, allowing more refrigerant to flow so that more heat can be absorbed.

Like the thermostatic expansion valve, the orifice tube is the dividing point between the high- and low-pressure parts of the system. However, its metering or flow rate control does not depend on comparing evaporator pressure and temperature. It is a fixed orifice. The flow rate is determined by pressure difference across the orifice and by subcooling. Subcooling is additional cooling of the refrigerant in the bottom of the condenser after it has changed from vapor to liquid. The flow rate through the orifice is more sensitive to subcooling than to pressure difference.

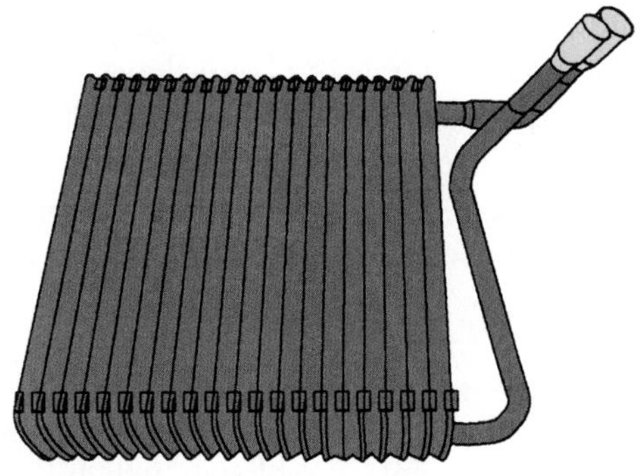

Figure 50–16 A typical evaporator. *Courtesy of Ford Motor Company*

Evaporator

The evaporator, like the condenser, consists of a refrigerant coil mounted in a series of thin cooling fins **(Figure 50–16)**. It provides a maximum amount of heat transfer in a minimum amount of space. The evaporator is usually located beneath the dashboard or instrument panel.

On receiving the low-pressure, low-temperature liquid refrigerant from the thermostatic expansion valve or orifice tube in the form of an atomized (or droplet) spray, the evaporator serves as a boiler or vaporizer. This regulated flow of refrigerant boils immediately. Heat from the core surface is lost to the boiling and vaporizing refrigerant, which is cooler than the core, thereby cooling the core. The air passing over the evaporator loses its heat to the cooler surface of the core, thereby cooling the air inside the car. As the process of heat loss from air to the evaporator core surface is taking place, any moisture (humidity) in the air condenses on the outside of the evaporator core and is drained off as water. A drain tube in the bottom of the evaporator housing leads the water outside the vehicle. This dehumidification of air is an added feature of the air-conditioning system that adds to passenger comfort. It can also be used as a means of controlling fogging of the vehicle windows. Under certain conditions, however, too much moisture can accumulate on the evaporator coils. An example would be when humidity is extremely high and the maximum cooling mode is selected. The evaporator temperature might become so low that moisture would freeze on the evaporator coils before it could drain off.

Through the metering, or controlling, action of the thermostatic expansion valve or orifice tube, greater or lesser amounts of refrigerant are provided in the evaporator to adequately cool the car under all heat load conditions. If too much refrigerant is allowed to enter, the evaporator floods. This results in poor cooling due to the higher pressure (and temperature) of the refrigerant. The refrigerant can neither boil away rapidly nor vaporize. Conversely, if too little refrigerant is metered, the evaporator starves. Poor cooling again results because the refrigerant boils away or vaporizes too quickly before passing through the evaporator.

The temperature of the refrigerant vapor at the evaporator outlet will be approximately 4 to 16°F (–15 to –8.8°C) higher than the temperature of the liquid refrigerant at the evaporator inlet. This temperature differential is the superheat that ensures that the vapor will not contain any droplets of liquid refrigerant that would be harmful to the compressor.

Refrigerant Lines

All the major components of the system have inlet and outlet connections that accommodate either flare or O-ring fittings. The refrigerant lines that connect between these units are made up of an appropriate length of hose or tubing with flare or O-ring fittings at each end as required **(Figure 50–17)**. In either case the hose or tube end of the fitting is constructed with sealing beads to accommodate a hose or tube clamp connection.

There are three major refrigerant lines: suction, liquid, and discharge. Suction lines are located between the outlet side of the evaporator and the inlet side or suction side of the compressor. They carry the low-pressure, low-temperature refrigerant vapor to the compressor where it is again recycled through the system. Suction lines are always distinguished from the discharge lines by touch and size. They are cold to the touch. The suction line is larger in diameter than the liquid line because refrigerant in a vapor state takes up more room than refrigerant in a liquid state.

Beginning at the discharge outlet on the compressor, the discharge or high-pressure line connects the compressor to the condenser. The liquid lines connect the condenser to the receiver/dryer and the receiver/dryer to the inlet side of the expansion valve. Through these lines, the refrigerant travels in its path from a gas state (compressor outlet) to a liquid state (condenser outlet)

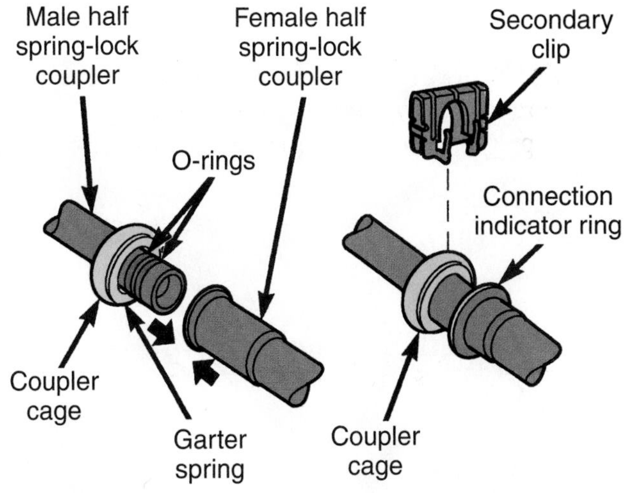

Figure 50–17 A spring-lock line coupler with O-rings. *Courtesy of DaimlerChrysler Corporation*

and then to the inlet side of the expansion valve where it vaporizes on entry to the evaporator. Discharge and liquid lines are always very warm to the touch and easily distinguishable from the suction lines.

Aluminum tubing is often used to connect air-conditioning components where flexibility is not required. Where the line is subjected to vibrations, special rubber hoses are used. Typically, the compressor outlet and inlet lines are rubber hoses with aluminum ends and fittings.

R-134a systems are required to be fitted with quick-disconnect fittings throughout the system. These systems also have hoses specially made for R-134a that have an additional layer of rubber that serves as a barrier to prevent the refrigerant from escaping through the pores of the hose. Some late-model R-12 systems also use these barrier hoses to prevent the loss of refrigerant through the walls of the hoses.

Sight Glass

The **sight glass** allows the technician to see the flow of refrigerant in the lines. A sight glass is normally found on systems using R-12 and a thermal expansion valve. It is not commonly found on R-134a systems. It can be located on the receiver/dryer or in-line between the receiver/dryer and the expansion valve or tube.

To check the refrigerant, open the windows and doors, set the controls for maximum cooling, and set the blower on its highest speed. Let the system run for about five minutes. Be sure the vehicle is in a well-ventilated area, or connect an exhaust gas ventilation system.

Use care to check the sight glass while the engine is running **(Figure 50–18)**. If oil streaking is seen, it indicates the system is empty. Bubbles, or foam, indicate the refrigerant is low. A sufficient level of refrigerant is indicated by what looks like a flow of clear water, with no bubbles. A clouded sight glass is an indication of desiccant contamination with subsequent infiltration and circulation through the system.

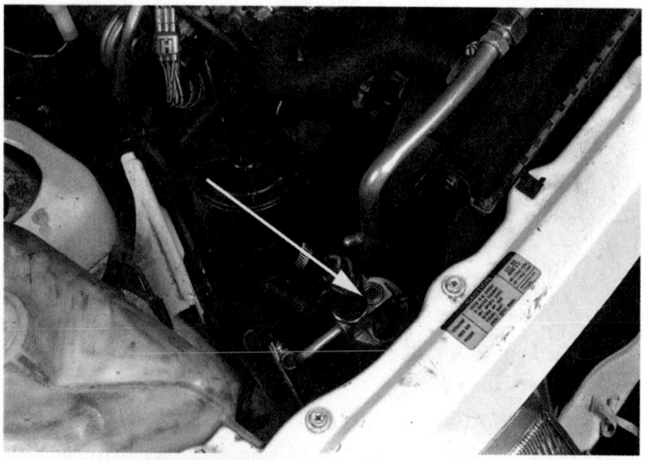

Figure 50-18 The location of a typical sight glass.

Accumulator systems do not have a receiver/drier to separate the gas from the liquid as it flows from the condenser. The liquid line will always have a certain amount of bubbles in it. Therefore it would be useless to have a sight glass in these systems. Pressure and performance testing are the only ways to identify low refrigerant levels.

Blower Motor/Fan

The blower motor/fan assembly is located in the evaporator housing. Its purpose is to increase airflow in the passenger compartment. The blower, which is basically the same type as that used in heater systems, draws warm air from the passenger compartment, forces it over the coils and fins of the evaporator, and blows the cooled, cleaned, and dehumidified air into the passenger compartment. The blower motor is controlled by a fan switch.

During cold weather, the blower motor/fan provides the airflow, and the heater core provides the heat for the passenger compartment.

AIR-CONDITIONING SYSTEMS AND CONTROLS

There are two basic types of automotive air-conditioning systems. They are classified according to the method used in obtaining temperature control and are known as cycling clutch systems or evaporator pressure or temperature control systems.

Evaporator Pressure Control System

Evaporator controls maintain a back pressure in the evaporator. Because of the refrigerant temperature/pressure relationship, the effect is to regulate evaporator temperature. The temperature is controlled to a point that provides effective air cooling but prevents the freezing of moisture that condenses on the evaporator.

In this type of system the compressor operates continually when dash controls are in the air-conditioning position. Evaporator outlet air temperature is automatically controlled by an evaporator pressure control valve. This type of valve throttles the flow of refrigerant out of the evaporator as required to establish a minimum evaporator pressure and thereby prevent freezing of condensation on the evaporator core.

Cycling Clutch System

In every **cycling clutch** system, the compressor is run intermittently by means of controlling the application and release of its clutch through a thermostatic or pressure switch. The thermostatic switch senses the evaporator's outlet air temperature through a capillary tube that is part of the switch assembly. With a high sensing temperature, the thermostatic switch is closed and the compressor clutch is energized. As the evaporator outlet temperature drops to a preset level, the thermostatic switch opens the circuit to the compressor clutch. The compressor then ceases to operate until such time as the evaporator temperature rises above the switch setting. From this on-and-off operation is derived the term cycling clutch. In effect, the thermostatic switch is calibrated to allow the lowest possible evaporator outlet temperature that would prevent the freezing of condensation that might form on the evaporator.

Variations of the cycling clutch system include a system with a thermostatic expansion valve and a system with an orifice tube.

Cycling Clutch System with Thermostatic Expansion Valve Some factory installations utilize a cycling clutch system that incorporates a TEV and receiver/dryer, as do some add-on units. The evaporator and control components are either in the engine compartment or an integral part of the cowl. In such cases there is a common blower and duct work for both heating and air-conditioning purposes. Also in these installations, the thermostatic switch has no temperature control knob and is usually mounted on the evaporator or its case. Temperature control is accomplished by using fresh or recirculating air and by reheating the cooled air in the heater core. The clutch cycles only to prevent evaporator icing.

A common form of the cycling clutch system is the field-installed (add-on) unit. With this installation, the evaporator, the thermostatic expansion valve, and thermostatic switch are self-contained in an underdash assembly. An installation of this type operates solely on passenger compartment recirculated air. Temperature control depends on intermittent operation of the compressor. The thermostatic switch has a control knob to cycle at higher temperatures when less cooling is desired.

Cycling Clutch System with Orifice Tube (CCOT) A typical **CCOT system** is illustrated in **Figure 50–19**. The system is factory installed and can use a thermostatic clutch cycling switch mounted on the evaporator case or a pressure cycling switch located on the accumulator. An expansion (orifice) tube is used in place of the TEV. Also, the system has an accumulator in the evaporator outlet. The accumulator is used primarily to separate vapor from liquid refrigerant before it enters the compressor. It also contains a drying agent or desiccant to remove moisture. The CCOT system has no receiver/dryer or sight glass. This system does have a special orifice, the oil bleed orifice, that allows refrigerant oil to return to the compressor rather than to collect in the accumulator.

Compressor Controls

Many controls are used to monitor and trigger the compressor during its operational cycle. Each of these represents the most common protective control devices designed to ensure safe and reliable operation of the compressor.

Ambient Temperature Switch This switch senses outside air temperature and is designed to prevent compressor clutch engagement when air-conditioning is not required or when compressor operation might cause internal damage to seals and other parts.

The switch is in series with the compressor clutch electrical circuit and closes at about 37°F (2.7°C). At all lower temperatures, the switch is open, preventing clutch engagement.

On some vehicles, the ambient switch is located in the air inlet duct of air-conditioning systems regulated by evaporator pressure controls; other makes have it installed near the radiator. It is not required on systems with a thermostatic or pressure switch.

Thermostatic Switch In cycling clutch systems, the thermostatic switch is placed in series with the compressor clutch circuit so it can turn the clutch on or off. It has two purposes. It de-energizes the clutch and stops the compressor if the evaporator is at the freezing point **(Figure 50–20)**. On hang-on units or systems without reheat temperature control, it also controls the air temperature by turning the compressor on and off intermittently. For this purpose, it has a control knob to change the switch setting.

When the temperature of the evaporator approaches the freezing point (or the low setting of the switch), the thermostatic switch opens the circuit and disengages the compressor clutch. The compressor remains inoperative until the evaporator temperature rises to the preset temperature, at which time the switch closes and compressor operation resumes.

Pressure Cycling Switch This switch is electrically connected in series with the compressor electromagnetic clutch. Like the thermostatic switch, the turning on and off of the pressure cycling switch controls the operation of the compressor.

Low-Pressure Cutoff or Discharge Pressure Switch This switch is located on the high side of the system and senses any low-pressure conditions. It is tied into the compressor clutch circuit, allowing it to immediately disengage the clutch when the pressure falls too low.

High-Pressure Cutout Switch This switch, normally located in the vicinity of the compressor or discharge (high side) muffler, is wired with the compressor clutch (in series). Designed to open (cut out) and disengage the clutch at 350 to 375 psi (2,400 to 2,600 kPa), it again closes and normally reengages the clutch when pressure returns to 250 psi (1700 kPa) (higher if the system uses R-134a).

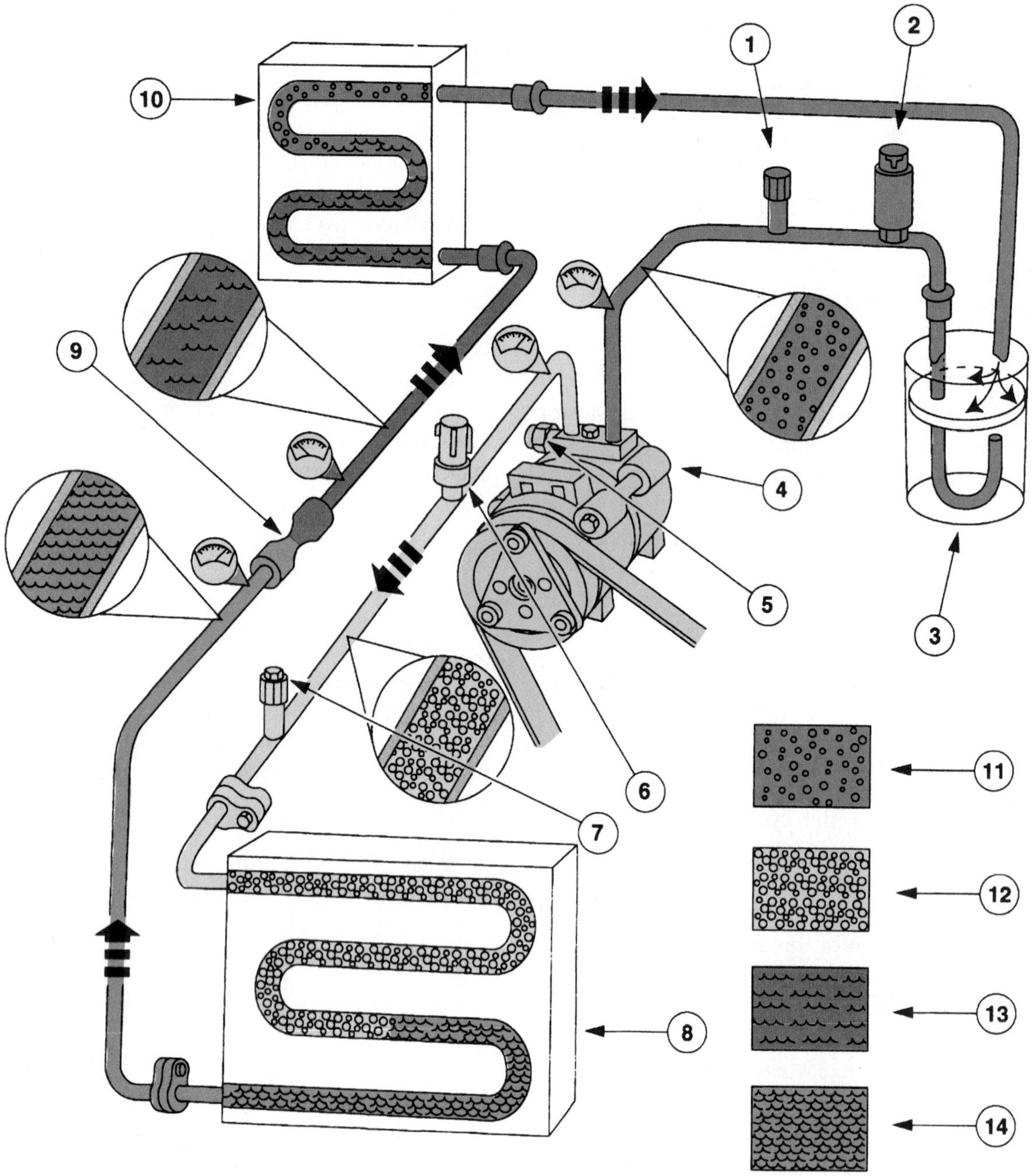

Item	Description	Item	Description
1	A/C charge valve port (low side)	8	A/C condenser core
2	A/C cycling switch	9	A/C evaporator core orifice
3	Suction accumulator/drier	10	A/C evaporator core
4	A/C compressor	11	Low-pressure vapor
5	A/C compressure pressure relief valve	12	High-pressure vapor
6	A/C pressure cutoff switch	13	Low-pressure liquid
7	A/C charge valve port (high side)	14	High-pressure liquid

Figure 50-19 A typical cycling clutch system with an expansion (orifice) tube. *Courtesy of Ford Motor Company*

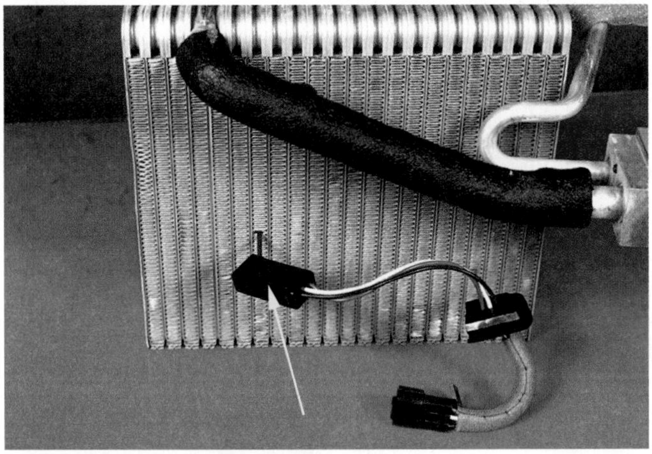

Figure 50-20 A thermostatic switch mounted onto an evaporator.

High-Pressure Relief Valve A high-pressure relief valve is incorporated into many air-conditioning systems. This valve may be installed on the receiver/dryer, compressor, or elsewhere in the high side of the system. It is a high-pressure protection device that opens (normally at 440 psi [3,033 kPa]) to bleed off excessive pressure that might occur in the system.

Compressor Control Valve This valve regulates the crankcase pressure in some compressors (commonly General Motors' V5 compressor). It has a pressure-sensitive bellows exposed to the suction side that acts on a ball and pin valve, which is exposed to high-side pressure. The bellows also controls a bleed port that is also exposed to the low side. The control valve is continuously modulating—changing the displacement of the compressor according to pressure or temperature.

Electronic Cycling Clutch Switch (ECCS) The ECCS prevents evaporator freeze-up by sending a signal to the engine control computer. The computer, in turn, cycles the compressor on and off by monitoring suction line temperature. If the temperature gets too low, the ECCS will open the input circuit to the computer, which causes the A/C clutch relay to open, disengaging the compressor clutch. Often this switch is the thermostatic switch at the evaporator.

TEMPERATURE CONTROL SYSTEMS

Temperature control systems for air conditioners usually are connected with heater controls. Most heater and air-conditioning systems use the same plenum chamber for air distribution. Two types of air conditioning controls are used: manual/semiautomatic and automatic.

Manual/Semiautomatic Temperature Controls

Air conditioner manual/semiautomatic temperature controls (MTC and SATC) operate in a manner similar to heater controls. Depending on the control setting, doors are opened and closed to direct airflow. The amount of cooling is controlled manually through the use of control settings and blower speed **(Figure 50–21)**.

Automatic Temperature Control

An automatic or electronic temperature control system **(Figure 50–22)** maintains a specific temperature automatically inside the passenger compartment. To maintain a selected temperature, heat sensors send signals to a computer unit that controls compressor, heater valve, blower, and plenum door operation. A typical electronic control system might contain a coolant temperature sensor, in-car temperature sensor **(Figure 50–23)**, outside temperature sensor, high-side temperature switch, low-side temperature switch, low-pressure switch, vehicle speed sensor, throttle position sensor, sunload sensor, and power-steering cutout switch.

The control panel is found in the instrument panel at a convenient location for both driver and front-seat passenger access. Three types of control panels may be found: manual, push-button, or touch pad. All serve the same purpose. They provide operator input control for the air-conditioning and heating system. Some control panels have features that other panels do not have, such as provisions to display in-car and outside air temperature in degrees **(Figure 50–24)**.

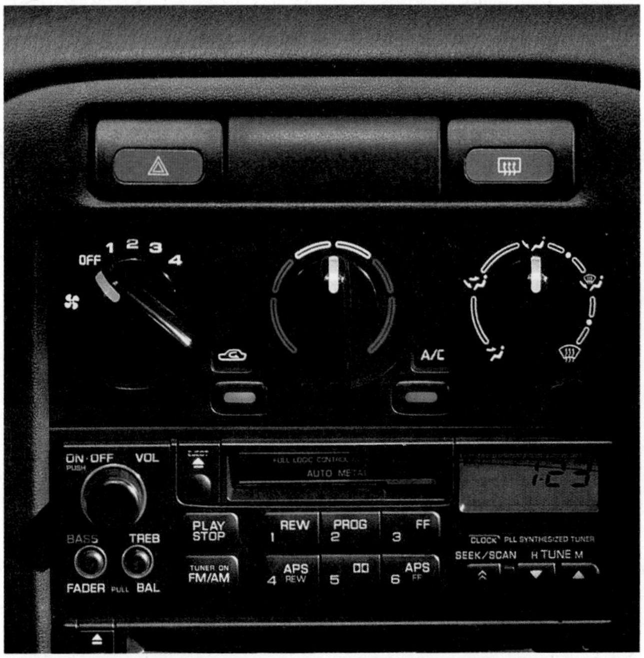

Figure 50-21 Typical air-conditioning controls. *Used with permission from Nissan North America, Inc.*

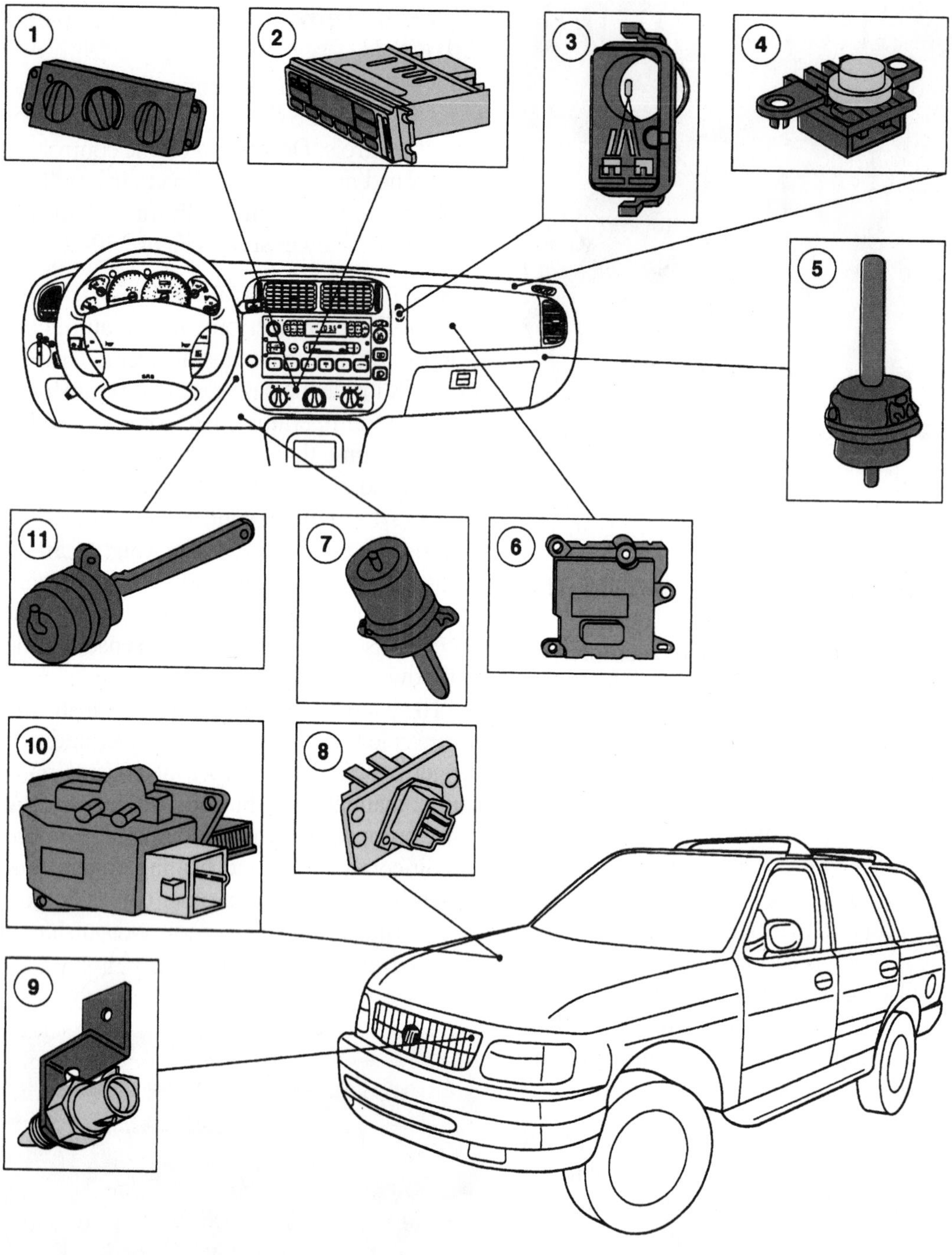

Item	Description
1	Manual A/C control assembly
2	Electronic automatic temperature control module
3	Automatic temperature control sensor hose and elbow (EATC only)
4	A/C sunload sensor (EATC only)
5	Vacuum control motor (air inlet door)
6	A/C electronic door actuator motor
7	Vacuum control motor (panel/defrost door)
8	Heater blower motor switch resistor
9	A/C ambient air temperature sensor and bracket (EATC only)
10	A/C blower motor speed control (EATC only)
11	Vacuum control motor (Floor/panel door)

Figure 50-22 Typical components in an automatic climate control system. *Courtesy of Ford Motor Company*

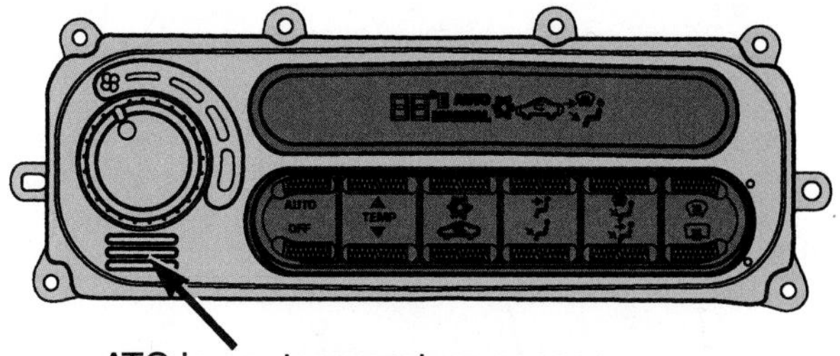

Figure 50-23 An in-car temperature sensor for ATC. *Courtesy of DaimlerChrysler Corporation*

Provisions are made on the control panel for operator selection of an in-car temperature between 65 and 85°F (18 and 29°C) in one-degree increments. Some have an override feature that provides for a setting of either 60 or 90°F (15 or 32°C). Either of these two settings overrides all in-car temperature control circuits to provide maximum cooling or heating conditions.

Usually, a microprocessor is located in the control head to input data to the programmer, based on operator-selected conditions. When the ignition switch is turned off, a memory circuit remembers the previous setting. These conditions are restored the next time the ignition switch is turned on. If the battery is disconnected, however, the memory circuit is cleared and must be reprogrammed.

Many automotive electronic temperature control systems have self-diagnostic test provisions in which an onboard microprocessor-controlled subsystem displays a code. This code (number, letter, or alphanumeric) is displayed to tell the technician the cause of the malfunction. Some systems also display a code to indicate which computer detected the malfunction. Manufacturers' specifications must be followed to identify the malfunction display codes, because they differ from car to car. For example, in some General Motors' car lines ".7,0" indicates no malfunction if "no trouble" codes are stored in the computer. In some Ford car lines the no trouble code is 888.

Case and Duct Systems

A typical automotive heater/air conditioner/case and duct system is shown in **Figure 50–25**. The purpose of the system is twofold: It is used to house the heater core and the air conditioner evaporator and to direct the selected supply air through these components into the passenger compartment of the vehicle. The supply air selected can be either fresh (outside) or recirculated air, depending on the system mode. After the air is heated or cooled, it is delivered to the floor outlet, dash panel outlets **(Figure 50–26)**, or the defrost outlets.

Figure 50-24 An ATC switch and message display assembly.

In domestic vehicles, there are two basic duct systems employed. In the stacked-core-reheat system the basic control is in the water valve. For maximum air, the water valve is completely closed. All air enters the vehicle compartment through the heater core.

The access door, which is activated by a cable, controls only fresh or recirculated air. Recirculated air is used during maximum cold operation. The air-conditioning unit is not operative and the evaporator will not be cold. The evaporator-only is used in the max air or maximum cold position. As the control level inside the car is moved, it controls the water valve by means of a vacuum or a cable to control the amount of hot water entering the heater core and the temperature of the air at the unit outlet.

The blend-air-reheat system is found on General Motors' and Ford vehicles and some truck units with factory-controlled heater system units. During heater-only operation, the air-conditioning unit is shut off, and the evaporator performs no function in air distribution or temperature control. During maximum air or extreme cold air, the air-conditioning system operates, the evaporator is cold, and the blend-air-door damper is completely closed. Only conditioned air enters the car.

As the control lever is moved in the vehicle from max air toward heat with the air conditioner on, the blend-air-door is moving. In maximum cold, it is completely shut. On maximum hot, it is completely open. The water valve on this unit is a vacuum on/off unit to regulate water flow. Normal position would be open. This type of blend-air system is extremely popular and can be used with or without a water valve.

To check the proper functioning of the duct work, move the temperature control lever to see if any change occurs. If it does not, shut off the air conditioner and turn on the heater. Move the temperature control arm again to see if any change occurs. If not, check the cable and the flap door connected to the temperature control lever. You might be able to reach under the dash to reconnect the cable or free a stuck flap.

If no substantial airflow is coming out of the registers, check the fuses in the blower circuit. Remove the fan switch and test it. Check the blower motor by hot-wiring it directly to the battery with jumper cables.

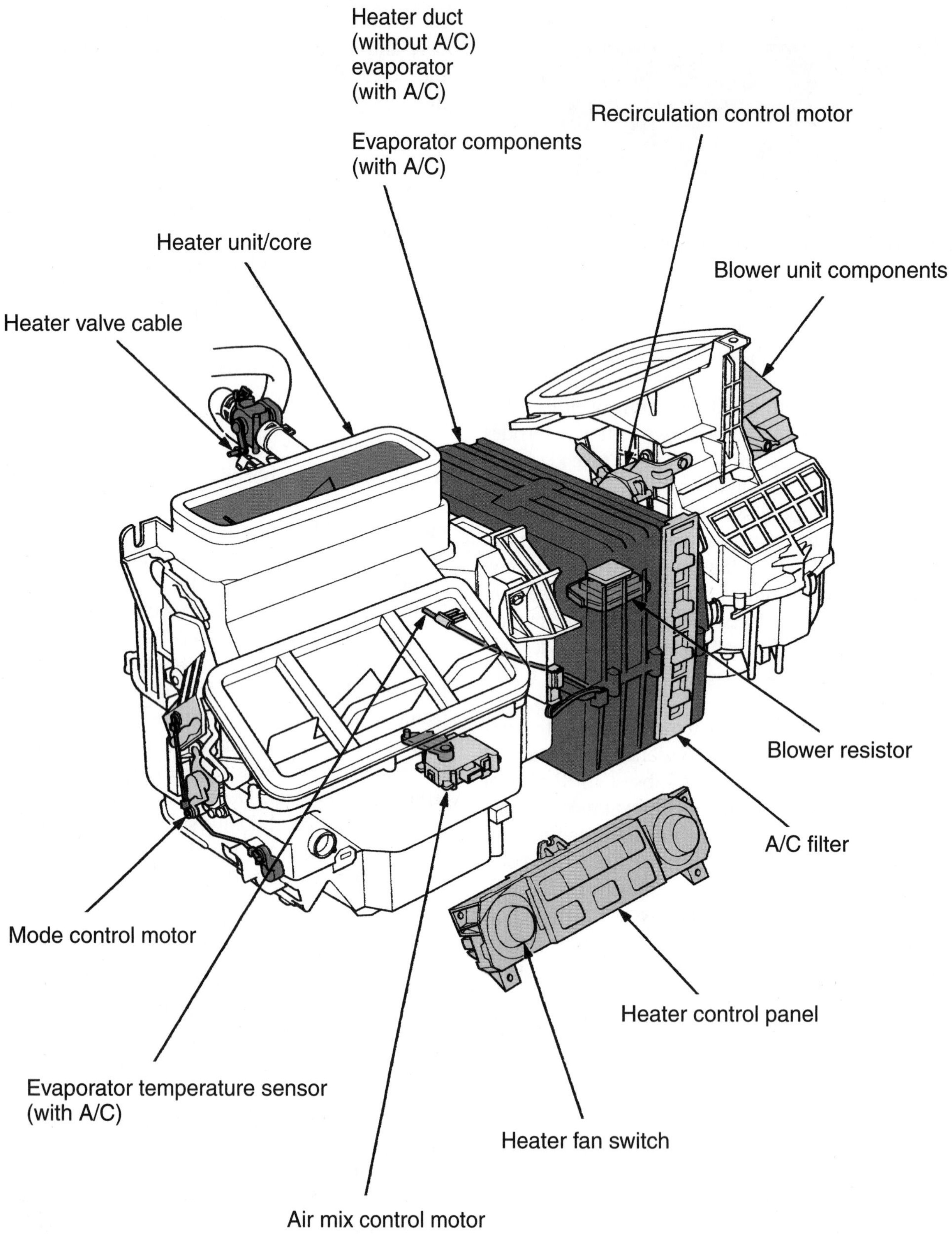

Figure 50-25 Typical heater/air conditioner ducts. *Courtesy of American Honda Motor Co., Inc.*

Figure 50-26 Air moving out from the dash vents to create comfort zones for the driver and passengers. *Courtesy of DaimlerChrysler Corporation*

CASE STUDY

A customer brings in a car, recently purchased from a used car dealer, for service with a complaint of a water leak. The situation is that water spills out from under the dash when she is turning right. She also complains that the passenger side floor mat seems to always be wet.

The technician verifies the wet floor mat and suspects that the heater core is leaking. However, there is no evidence of engine coolant on the floor mat. Further discussion with the customer verifies that the car is not overheating nor does it have a heater problem. While talking to the customer, the technician notices that there are not the familiar drips on the shop floor usually experienced when a car's air-conditioning is running.

He inspects the drain tube for the air conditioner's evaporator and notices that the car was recently undercoated. He further notices that the outside drain opening is plugged with undercoating. With no place to go, the moisture pulled out of the air in the car is backing up into the evaporator housing and spilling out. Cleaning the drain tube took care of the problem.

KEY TERMS

Blend door
CCOT system
Chlorofluorocarbon (CFC)
Condensation
Condenser
Cycling clutch
Desiccant
Discharge side
Evaporation
Evaporator
Heater control valve
Heater core
Hydrofluorocarbon (HFC)
High side
Latent heat
Low side
Orifice tube
Receiver/dryer
Sight glass

SUMMARY

- Ventilation, heating, and air-conditioning provide for the comfort of the vehicle's passengers.
- The ventilation system on most vehicles is designed to supply outside air to the passenger compartment through upper or lower vents or both. Several systems are used to vent air into the passenger compartment. The most common is the flow-through system. In this arrangement, a supply of outside air called ram air, flows into the car when it is moving.
- Automotive heating systems have been designed to work with the cooling system to maintain proper temperature inside the car. The heating system's primary job is to provide a comfortable passenger compartment temperature and to keep car windows clear of fog or frost.
- The main components of an automotive heating system are the heater control valve, the heater core, the blower motor and fan, and heater and defroster ducts.
- All air-conditioning systems are based on three fundamental laws of nature: heat flow, heat absorption, and pressure and boiling points.
- The major components of an air-conditioning system are compressor, condenser, receiver/dryer or accumulator, expansion valve or orifice tube, and evaporator.
- The compressor is the heart of an automotive air-conditioning system. It separates the high-pressure and low-pressure sides of the system. The primary purpose of the unit is to draw the low-pressure vapor from the evaporator and compress this vapor into high-temperature, high-pressure vapor. This action results in the refrigerant having a higher temperature than surrounding air, enabling the condenser to condense the vapor back to liquid.
- The secondary purpose of the compressor is to circulate or pump the refrigerant through the condenser under the different pressures required for proper operation. The compressor is located in the engine compartment.

- The condenser consists of a refrigerant coil tube mounted in a series of thin cooling fins to provide maximum heat transfer in a minimum amount of space. The condenser is normally mounted just in front of the vehicle's radiator.
- The receiver/dryer is a storage tank for the liquid refrigerant.
- The refrigerant flow to the evaporator must be controlled to obtain maximum cooling, while ensuring complete evaporation of the liquid refrigerant within the evaporator. This is accomplished by a thermostatic expansion valve or a fixed orifice tube.
- The evaporator, like the condenser, consists of a refrigerant coil mounted in a series of thin cooling fins. The evaporator is usually located beneath the dashboard or instrument panel.
- An amendment to the United States' Clean Air Act and international agreements have mandated that R-12 be phased out as the refrigerant of choice. This came as a result of research that found the earth's ozone layer was being deteriorated by the chemicals found in R-12.
- Of the many chemicals that could have been used in place of R-12, the automobile manufacturers have decided to use R-134a, a hydrofluorocarbon (HFC) that causes less damage to the ozone layer when released to the atmosphere.
- Although R-134a air conditioners operate in the same way and with the same basic components as R-12 systems, the two refrigerants are not interchangeable. Because it is less efficient than R-12, R-134a operates at higher pressures to make up for the loss of performance and requires new service techniques and system component designs. Basically, the higher system pressures of R-134a mean the system must be designed for those higher pressures.
- The production phase-out for all CFCs, including R-12, was scheduled for December 1, 1995. To keep systems operating that were originally equipped with R-12 there are retrofit kits available for a changeover to R-134a. The recovery and recycling of R-134a became mandatory on November 15, 1995.
- There are two basic types of automotive air-conditioning systems. They are classified according to the method used in obtaining temperature control and are known as cycling clutch systems or evaporator pressure control systems.
- Evaporator controls maintain a back pressure in the evaporator. Because of the refrigerant temperature/pressure relationship, the effect is to regulate evaporator temperature.
- Temperature control systems for air conditioners are usually connected with heater controls. Most heater and air conditioner systems use the same plenum chamber for air distribution.
- Two types of air conditioner controls are used: manual/semiautomatic and automatic.

TECH MANUAL

The following procedures are included in Chapters 50/51 of the *Tech Manual* that accompanies this book:

1. Identifying the type of air-conditioning system in a vehicle.
2. Identifying the components in an air-conditioning system.
3. Conducting a visual inspection of an air-conditioning system.
4. Conducting a system performance test on an air-cinditioning system.
5. Inspecting, testing, and servicing a compressor clutch.
6. Identifying and recovering refrigerant.
7. Recycling, labeling, and recovering refrigerant.

REVIEW QUESTIONS

1. What is the amount of heat necessary to change the state of a substance called?
2. What type of electromagnetic clutch is used in late-model vehicles?
3. Why is R-134a currently the preferred refrigerant for automotive air-conditioning systems?
4. What is the purpose of the thermostatic switch?
5. What does change of state mean? Why is it important to air-conditioning units?

6. What state is the refrigerant in when it leaves the condenser?
7. What causes condensed water to leak from the air-conditioning system?
8. On which of the following laws of nature is the air-conditioning system based?
   a. heat flow
   b. heat absorption
   c. pressure and boiling points
   d. all of the above
9. Technician A says a great amount of heat is transferred when a liquid boils or a vapor condenses. Technician B says a change in pressure does not change the boiling point of a substance. Who is correct?
   a. Technician A
   b. Technician B
   c. Both A and B
   d. Neither A nor B
10. Which of the following statements is true?
    a. Refrigerant leaves the compressor as a high-pressure, high-temperature liquid.
    b. Refrigerant leaves the condenser as a low-pressure, low-temperature liquid.
    c. Refrigerant returns to the compressor as a low-pressure, high-temperature vapor.
    d. None of the above.
11. Which of the following statements is false?
    a. The condenser is normally mounted just in front of the radiator.
    b. The receiver/dryer is a storage tank for the liquid refrigerant from the condenser.
    c. An accumulator is not used in any system with a receiver/dryer.
    d. All of the above.
12. Which of the following is true of R-12?
    a. It operates at higher pressures than R-134a.
    b. It has a high boiling point.
    c. It is destructive to the earth's ozone layer.
    d. All of the above.
13. Technician A says the thermostatic expansion valve and the fixed orifice tube perform the same function. Technician B says heat from the evaporator core surface is lost to the boiling and vaporizing refrigerant. Who is correct?
    a. Technician A
    b. Technician B
    c. Both A and B
    d. Neither A nor B
14. Technician A says suction lines are very warm to touch. Technician B says discharge lines connect the compressor to the condenser. Who is correct?
    a. Technician A
    b. Technician B
    c. Both A and B
    d. Neither A nor B
15. Technician A says the normal operating pressures of R-134a are slightly lower than the pressures of R-12. Technician B says R-134a is one of many refrigerants that are EPA approved for motor vehicles. Who is correct?
    a. Technician A
    b. Technician B
    c. Both A and B
    d. Neither A nor B
16. Which of the following is electrically connected in series with the compressor electromagnetic clutch?
    a. an ambient temperature switch
    b. a thermostatic switch
    c. a pressure cycling switch
    d. all of the above
17. How can you identify an R-134a air-conditioning system?
18. Which of the following oils is used in an original equipment R-134a system?
    a. Mineral oil
    b. CCOT
    c. Ester oil
    d. PAG
19. A heating system includes all of the following parts except a ______.
    a. heater core
    b. water control valve
    c. receiver/dryer
    d. radiator
20. All of the following statements are true except ______.
    a. The evaporator changes heated refrigerant to a liquid before it passes to the compressor.
    b. The automatic temperature control operates to maintain a set temperature.
    c. Air conditioner/heater systems can be computer controlled.
    d. Heaters and air conditioners use most of the same plenum chambers.

51 HEATING AND AIR-CONDITIONING DIAGNOSIS AND SERVICE

OBJECTIVES

■ Understand the special handling procedures for automotive refrigerants. ■ Explain the concerns and precautions regarding retrofitting an A/C system. ■ Describe how to connect a manifold gauge set to a system. ■ Describe methods used to check refrigerant leaks. ■ Use approved methods and equipment to discharge, reclaim/recycle, evacuate, and recharge an automotive air-conditioning system. ■ Perform a performance test on an A/C system. ■ Interpret pressure readings as an aid to diagnose A/C problems.

Air-conditioning service, in many ways, is different than service to other parts of the vehicle. Although there are few parts in the system **(Figure 51–1)**, each component has a specific purpose and service procedure. That in and of itself is not a big deal. The challenge with air-conditioning is how it operates. The system operates on changes of refrigerant pressure. Many things can cause the pressure to change; some are part of the system, some are part of the environment, and some are faults or bad components in the system.

MAINTENANCE PRECAUTIONS

Air-conditioning systems are extremely sensitive to moisture and dirt. Therefore, clean working conditions are extremely important. The smallest particle of foreign matter in an air-conditioning system contaminates the refrigerant, causing rust, ice, or damage to the compressor. For this reason, all replacement parts are sold in vacuum-sealed containers and should not be opened until they are ready to be installed in the system. If, for any reason, a part has been removed from its container for any length of time, the part must be completely flushed using only recommended flush solvent to remove any dust or moisture that might have accumulated during storage. When the system has been open for any length of time, the entire system must be purged completely and a new accumulator must be installed because the element of the existing unit will have become saturated and unable to remove any moisture from the system once the system is recharged. The following general practices should be observed to ensure chemical stability in the system.

SHOP TALK

It is important to remember that just one drop of water added to the refrigerant will start chemical changes that can result in corrosion and eventual breakdown of the chemicals in the system. Hydrochloric acid is the result of R-12 mixed with water. The smallest amount of moist air in the refrigerant system might start reactions that can cause malfunctions. ■

Whenever it becomes necessary to disconnect a refrigerant connection, wipe away any dirt or oil at and near the connection to eliminate the possibility of dirt entering the system. Both sides of the connection should be immediately capped or plugged to prevent the entrance of dirt, foreign material, and moisture. It must be remembered that all air contains moisture. Air that enters any part of the system carries moisture with it, and the exposed surface collects the moisture quickly.

Keep tools clean and dry, including the gauge set and replacement parts. Be careful not to overtighten any connection. Overtightening results in line and flare seat distortion and a system leak.

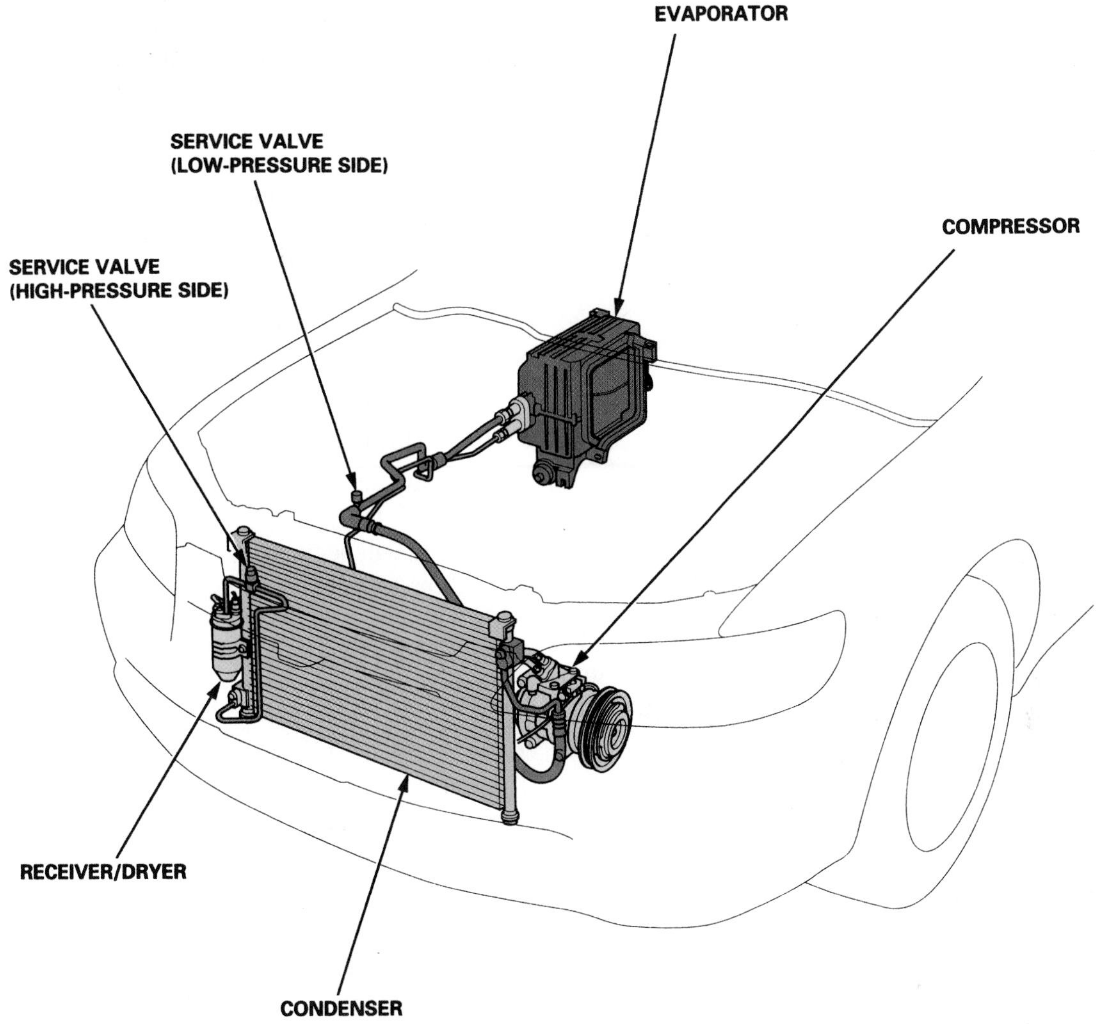

Figure 51-1 A late-model air-conditioning system. *Courtesy of American Honda Motor Co., Inc.*

When adding oil, the container and the transfer tube through which the oil will flow should be exceptionally clean and dry. Refrigerant oil is as moisture-free as it is possible to make it. Therefore, it quickly absorbs any moisture it contacts. For this reason, the oil container should not be opened until ready for use and then it should be capped immediately after use.

When it is necessary to open a system, have everything needed ready and handy so as little time as possible is required to perform the operation. Do not leave the system open any longer than is necessary.

Cleanliness is especially important when servicing compressors because of the very close tolerances used in these units. Consequently, repairs to the compressor itself should not be attempted unless all proper tools are at hand and a spotless work area is provided.

Any time the system has been opened and sealed again, it must be properly evacuated. Keep in mind that the complete and positive sealing of the entire system is vitally important and that this sealed condition is absolutely necessary to retain the chemicals and keep them in a pure and proper condition.

REFRIGERANT SAFETY PRECAUTIONS

- Always work in a well-ventilated and clean area. Refrigerant (R-12 and R-134a) is colorless and invisible as a gas. Refrigerant is heavier than oxygen and will displace oxygen in a confined area. Avoid breathing the refrigerant vapors. Exposure to refrigerant may irritate your eyes, nose, and throat.
- Refrigerant evaporates quickly when it is exposed to the atmosphere. Never rub your eyes or skin if you have been in contact with refrigerant. It will freeze anything it contacts. If liquid refrigerant gets in your eyes or on your skin it can cause frostbite. Immediately flush the exposed areas with cool water for fifteen minutes and seek medical help.

- An A/C system's high pressure can cause severe injury to your eyes and/or skin if a hose were to burst. Always wear eye protection when working around the A/C system and refrigerant. It is also advisable to wear protective gloves and clothing.
- Never use R-134a in combination with compressed air for leak testing. Pressurized R-134a in the presence of oxygen may form a combustible mixture. Never introduce compressed air into R-134a containers (empty or full ones), A/C systems, or A/C service equipment.
- Be careful when handling refrigerant containers. Never drop, strike, puncture, or burn the containers. Always use DOT-approved refrigerant containers.
- Never expose A/C system components to high temperatures. Heat will cause the refrigerant's pressure to increase. Never expose refrigerant to an open flame. A poisonous gas, **phosgene gas**, will result and if breathed, can cause severe respiratory irritation. This gas also occurs when an open flame leak detector is used. Phosgene fumes have an acrid (bitter) smell. If the refrigerant needs to be heated during service, the bottom of the refrigerant container should be placed in warm water (less than 125°F [52°C]).
- Never overfill refrigerant containers. The filling level of the container should never exceed 60% of the container's gross weight rating. Always store refrigerant containers in temperatures below 125°F (52°C) and keep them out of direct sunlight.
- Refrigerant comes in 30- and 50-pound (14- and 23-kg) cylinders. Remember that these drums are under considerable pressure and should be handled following precautions. Keep the drums in an upright position. Make sure that valves and safety plugs are protected by metal caps when the drums are not in use. Avoid dropping the drums. Handle them carefully. When transporting refrigerant, do not place containers in the vehicle's passenger compartment.
- R-12 should be stored and sold in white containers, whereas R-134a should be stored in light blue containers. R-12 and R-134a should never be mixed. Their oils and desiccants are not compatible. If the two refrigerants are mixed, contamination will occur and may result in A/C system failure. Separate service equipment, including recovery/recycling machines and service gauges should be used for the different refrigerants. Always read and follow the instructions from the equipment manufacturer when servicing A/C systems.
- To prevent cross-contamination, identify whether the A/C system being worked on uses R-12 or R-134a. Check the fittings in the system; all R-134a based-systems use ½-inch 16 ACME threaded fittings and quick-disconnect service couplings. Most R-134a systems can be identified by underhood labels that clearly state that R-134a is used. Most manufacturers identify the type of refrigerant used by labeling the compressor. Also look for a label with the words, *CAUTION—SYSTEM TO BE SERVICED BY QUALIFIED PERSONNEL.* This label or plate can be found under the hood, near a component of the system and will tell you what kind refrigerant is used, as well as the type of refrigerant oil. When servicing an A/C system, determine the type of refrigerant used before beginning.
- R-134a can also be identified by viewing the sight glass, if the system has one. The appearance of R-134a will be milky due to the mixture of refrigerant and the refrigerant oil.

GUIDELINES FOR CONVERTING (RETROFITTING) R-12 SYSTEMS TO R-134A

The following guidelines should be followed when converting an older A/C system to R-134a. These guidelines should allow you to provide the customer with a cool vehicle and also meet current legislative mandates. The guidelines are listed in order and reflect the necessary steps for making this conversion.

- Visually inspect all air-conditioning and heater system components.
- Use a refrigerant identifier to make sure the system only contains R-12.
- Check the system for leaks.
- Run a performance test and record the temperature and pressure readings.
- Remove all R-12 from the system with a recycling machine.
- If the system uses a compressor with an oil sump, remove the compressor and drain all the oil from it. Measure the amount of oil drained out.
- Remove and inspect the expansion valve or the orifice tube. Replace it if necessary. If either is contaminated, flush the condenser.
- Remove the filter/drier or accumulator, drain it and measure the amount of oil in it.
- Before converting, make all necessary system changes, such as hoses, gaskets, and seals.
- Install R-134a compatible oil into the system; put in the same amount you took out.
- Install a new filter/drier or accumulator with the correct desiccant.

- Permanently install conversion fittings using a thread locking chemical.
- Install a high-pressure cutoff switch if the system does not have one.
- Install conversion labels and remove the R-12 label.
- Connect R-134a system evacuation equipment and evacuate for at least thirty minutes.
- Recharge the system to approximately 80% of the original R-12 charge.
- Run a performance check and compare readings with those taken before the conversion.
- Leak check the system.

Refrigerants

Always use the same type of refrigerant and refrigerant oil that the system was designed for. If the system is being converted to accept an alternative refrigerant, make sure all old refrigerant and oil is removed from the system. Also make sure the vehicle is clearly labeled **(Figure 51–2)** as to the type of refrigerant it was converted to use. The EPA has approved a number of refrigerants other than R-134a **(Figure 51–3)**. These can be used, but the system must be modified to contain and effectively use these refrigerants. There are several disadvantages to using these alternative refrigerants. One is based on the future. If the system has been converted to use an alternative refrigerant and the vehicle needs service in the future, will the alternative refrigerant be available at that shop or at all? Another disadvantage is that many air-conditioning parts manufacturers will not warrant their parts if something other than R-12 or R-134a is used.

Keep in mind that all EPA-accepted refrigerants that substitute for R-12, except R-134a, are blends that contain ozone-depleting HCFCs. Therefore the venting of any of these refrigerants into the atmosphere is prohibited.

Figure 51–2 Air-conditioning systems should be labeled as to what refrigerant should be used in them.

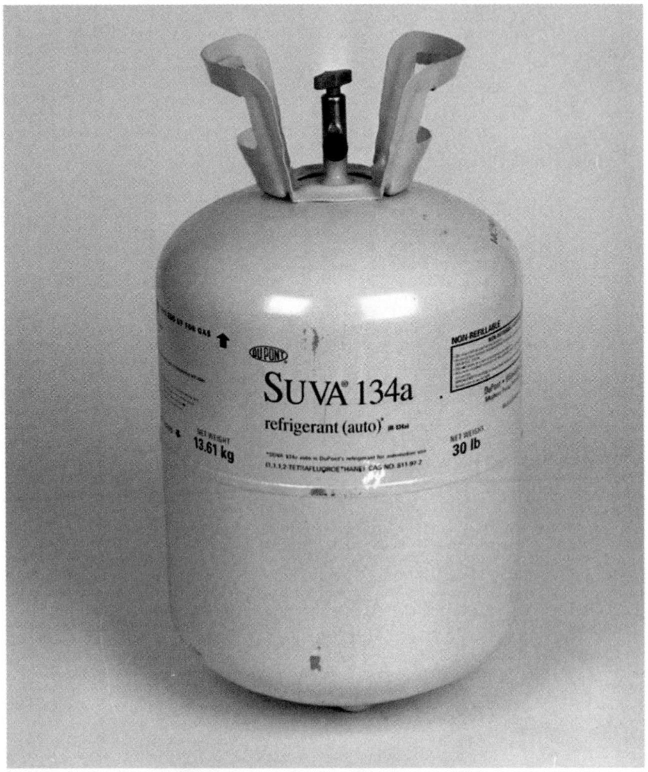

Figure 51–3 A container of R-134a.

WARNING!

Under no circumstances, because of environmental and health concerns, should refrigerants be mixed. Never add one type of refrigerant to a system that has or has had another type of refrigerant in it.

AIR-CONDITIONER TESTING AND SERVICING EQUIPMENT

Several specially designed pieces of equipment are required to perform test procedures. They include manifold gauge sets, thermometers, service valves, vacuum pumps, charging station, charging cylinders, recovery/recycling systems, and leak-detecting devices.

CAUTION!

Always wear safety goggles when working on or around air-conditioning systems.

Manifold Gauge Set

The **manifold gauge set** shown in **Figure 51–4** is one of the most valuable air-conditioning tools. It is used when discharging, charging, evacuating, and for diagnosing trouble in the system. With the new legislation on

Figure 51–4 A manifold gauge set for R-134a.

handling refrigerants, all gauge sets are required to have a valve device to close off the end of the hose so that the fitting not in use is automatically shut.

The low-pressure gauge is graduated into pounds of pressure from 1 to 120 (6 to 827 kPa) (with cushion to 250 [2,400 kPa]) in 1-pound graduations (10 kPa increments), and, in the opposite direction, in inches of vacuum from 0 to 30 (0 to 760 mm Hg). This is the gauge that should always be used in checking pressure on the low-pressure side of the system. The gauge in **Figure 51–4** is graduated from 0 to 500 pounds (0 to 3,400 kPa) of pressure in 10-pound graduations (50 kPa increments). This is the high-pressure gauge that is used for checking pressure on the high-pressure side of the system.

The center manifold fitting is common to both the low and the high side and is for evacuating or adding refrigerant to the system. When this fitting is not being used, it should be capped. A test hose connected to the fitting directly under the low-side gauge is used to connect the low side of the test manifold to the low side of the system. A similar connection is found on the high side.

The gauge manifold is designed to control refrigerant flow. When the manifold test set is connected into the system, pressure is registered on both gauges at all times. During all tests, both the low- and high-side hand valves are in the closed position (turned inward until the valve is seated).

Refrigerant flows around the valve stem to the respective gauges and registers the system's low-side pressure on the low-side gauge and the system's high-side pressure on the high-side gauge. The hand valves isolate the low and high side from the central portion of the manifold. When the gauges are first connected to the gauge fittings with the refrigeration system charged, the gauge lines should always be purged. Purging is done by cracking each valve on the gauge set to allow the pressure of the refrigerant in the refrigeration system to force the air to escape through the center gauge line. Failure to purge lines can result in air or other contaminants entering the refrigeration system.

WARNING!

Do not open the high-side hand valve with the system operating and a refrigerant can connected to the center hose. The refrigerant will flow out of the system under high pressure into the can. High-side pressure is between 150 and 300 psi (1,034 and 2,068 kPa) and will cause the refrigerant tank to burst. The only occasion for opening both hand valves at the same time would be when evacuating the system or when reclaiming refrigerant with the system off.

Because R-134a is not interchangeable with R-12, separate sets of hoses, gauges, and other equipment are required to service vehicles. All equipment used to service R-134a and R-12 systems must meet SAE standard J1991. The service hoses on the manifold gauge set must have manual or automatic backflow valves at the service port connector ends. This prevents the refrigerant from being released into the atmosphere during connection and disconnection. Manifold gauge sets for R-134a can be identified by one or all of the following: Labeled *FOR USE WITH R-134a*, labeled *HFC-134* or *R-134a*, and/or have a light blue color on the face of the gauges.

For identification purposes, R-134a service hoses must have a black stripe along their length and be clearly labeled *SAE J2196/R-134a*. The low-pressure hose is blue with a black stripe. The high-pressure hose is red with a black stripe and the center service hose is yellow with a black stripe. Service hoses for one type of refrigerant will not easily connect into the wrong system, as the fittings for an R-134a system are different than those used in an R-12 system.

Purity Test

When you are not sure of the refrigerant used in a system or if you suspect a mixing of refrigerants has occurred,

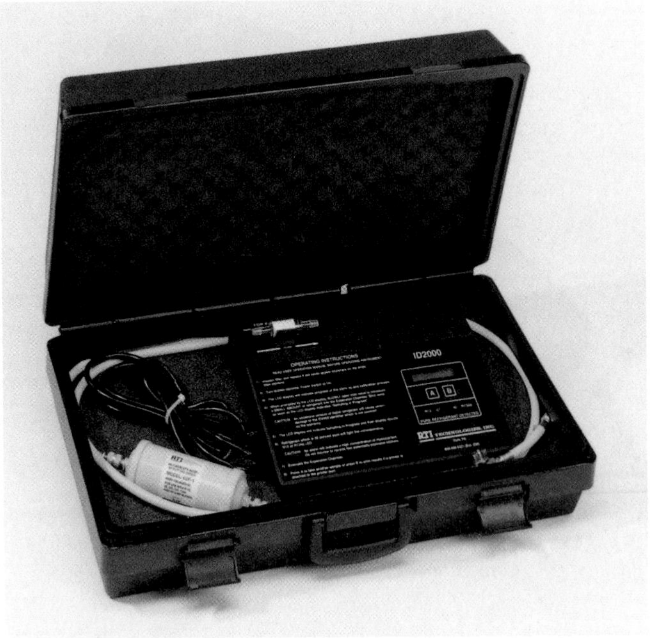

Figure 51–5 A refrigerant identifier. *Photo courtesy of RTI Technologies, Inc.*

you should run a purity test and/or use a refrigerant identifier **(Figure 51–5)**. Equipment is available to do both of these. Knowing what refrigerant is in the system, or what condition it is in, will help you determine what steps you need to take to properly service the system.

Service Valves

System service valves, which incorporate a service gauge port for manifold gauge set connections, are provided on the low- and high-sides of most air-conditioning systems **(Figure 51–6)**. When making gauge connections, purge the gauge lines first by cracking the charging valve and allowing a small amount of refrigerant to flow through the lines, then connect the lines immediately.

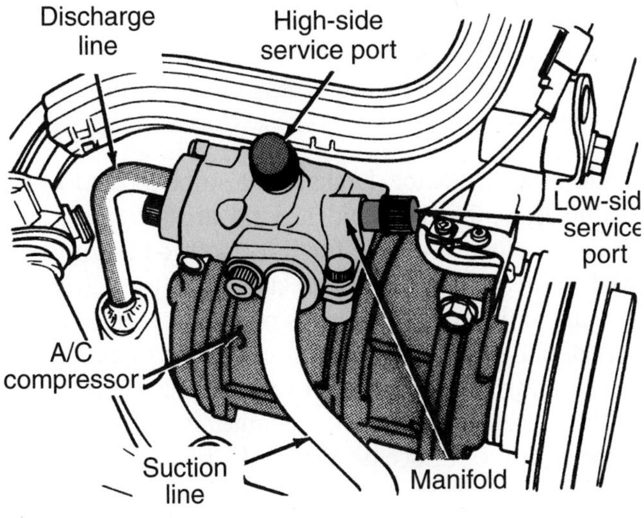

Figure 51–6 A/C service port locations. *Courtesy of DaimlerChrysler Corporation*

Two basic types of service valves are used: stem and Schrader.

Stem Service Valve The **stem valve** is sometimes used on two-cylinder reciprocating-piston compressors. Access to the high-pressure and low-pressure sides of the system is provided through service valves mounted on the compressor head. The low-pressure valve is mounted at the inlet to the compressor, while the high-pressure valve is mounted at the compressor outlet. Both of these valves can be used to shut off the rest of the air-conditioning system from the compressor when the compressor is being serviced. These valves have a stem under a cap **(Figure 51–7)** with the hose connection directly opposite.

A special wrench is used to open the valve to one of three positions. The back-seated position is the normal operating position with the valve stem rotated counterclockwise to seat the rear valve face and seal off the service gauge port. The midposition is the test position with the valve stem turned clockwise (inward) 1½ to 2 turns. This connects the service gauge port into the system so that gauge readings can be taken with the system operating. A service gauge hose must be connected with the valve completely back seated. In the front-seated position, the valve stem has been rotated clockwise to seat the front valve face and to isolate the compressor from the system. This position allows the compressor to

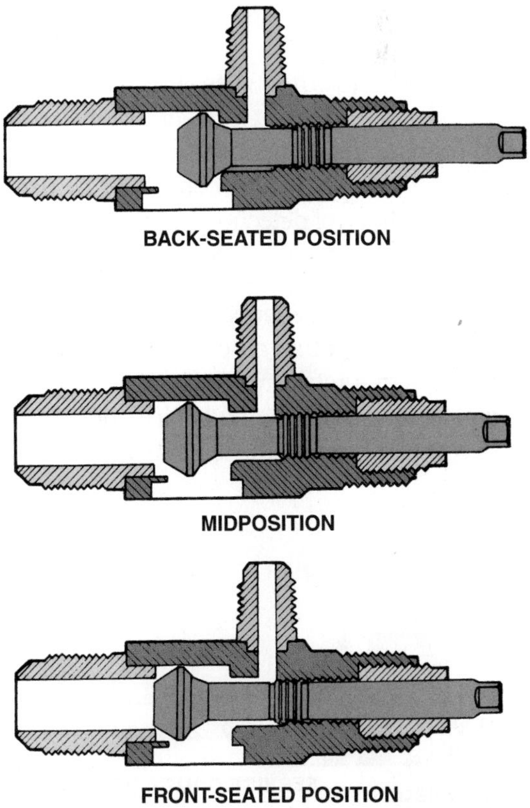

Figure 51–7 Stem service valve positions.

be serviced without discharging the entire system. The front-seated position is for service only. It is never used while the air conditioner is operating.

Schrader Service Valve Systems that do not utilize stem service have **Schrader service valves (Figure 51–8)** in both the high- and low-portions of the system for test purposes. Closely resembling a tire valve, Schrader valves are usually located in the high-pressure line (from compressor to condenser) and in the low-pressure line (from evaporator to condenser) to permit checking of the high side and low side of the system. All test hoses have a Schrader core depressor in them. As the hose is threaded into the service port, the pin in the center of the valve is depressed, allowing refrigerant to flow to the manifold gauge set. When the hose is removed, the valve closes automatically.

After disconnecting gauge lines, check the valve areas to be sure the service valves are correctly seated and the Schrader valves are not leaking.

Leak Testing a System

Testing the refrigerant system for leaks is one of the most important phases of troubleshooting. Over a period of time, all air conditioners lose or leak some refrigerant. In systems that are in good condition, R-12 losses up to ½ pound (.22 kg) per year are considered normal. Higher loss rates signal a need to locate and repair the leaks.

Leaks are most often found at the compressor hose connections and at the various fittings and joints in the system. Refrigerant can be lost through hose permeation. Leaks can also be traced to pinholes in the evaporator caused by acid that forms when water and refrigerant mix. Since oil and refrigerant leak out together, oily spots on hoses, fittings, and components mean there is a leak. Any suspected leak should be confirmed by using any one of these methods of detection. All leaks should be repaired before refrigerant is added to the system.

Visual Inspection As mentioned before, the presence of oil around the fitting of an air-conditioning line or hose is an indication of a refrigerant leak. Carefully check all system connections. This method of leak detection is the easiest to conduct but also the least effective.

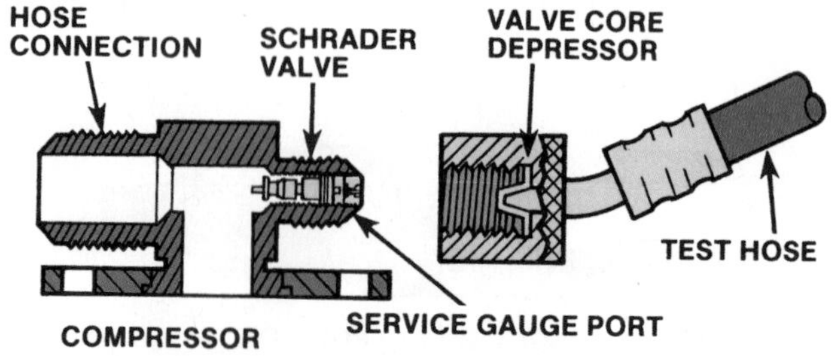

Figure 51–8 Schrader service valves.

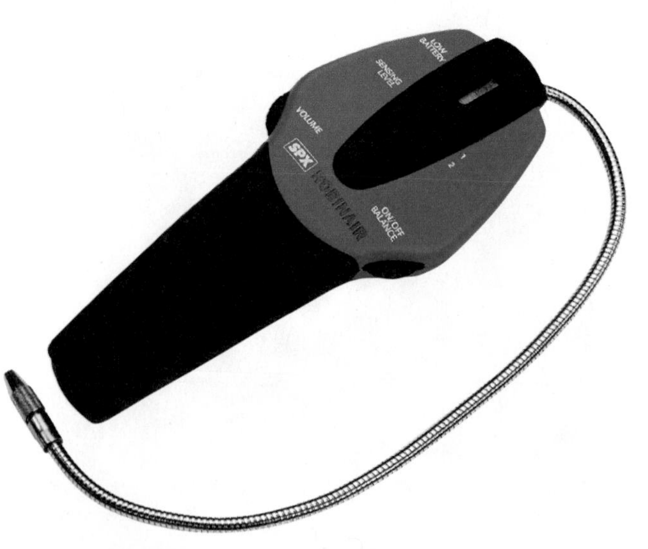

Figure 51–9 An electronic leak detector. *Courtesy of Robinair, SPX Corporation*

Electronic Leak Detector This is the preferred method of leak detection; it is safe, effective, and can be used with all types of refrigerants. The hand-held battery-operated electronic leak detector **(Figure 51–9)** contains a test probe that is moved about one inch per second in areas of suspected leaks. (Remember that refrigerant gas is heavier than air, thus the probe should be positioned below the test point.) An alarm or a buzzer on the detector indicates the presence of a leak. On some models, a light flashes to establish the leak. The electronic leak detector is the more sensitive of the two types.

Fluorescent Leak Tracer To find a refrigerant leak using the fluorescent tracer system, first introduce a fluorescent dye into the air-conditioning system by means of a special infuser. Run the air conditioner for a few minutes, giving the tracer dye fluid time to circulate and penetrate. Wear the tracer protective goggles and scan the system with a black-light glow gun. Leaks in the system will shine under the black light as a luminous yellow-green **(Figure 51–10)**. There are concerns about using this type of leak detection because it may create an oil viscosity problem.

Fluid Leak Detector Leaks can also be located by applying leak detector fluid around areas to be tested. There are dyes available for R-12 and R-134a systems. Make sure you use the correct one for the refrigerant you are looking for. If a leak is present, it will form clusters of bubbles around the source. A very small leak will cause a white foam to form around the leak source within several seconds to a minute. Adequate lighting over the entire surface being tested is necessary for an accurate diagnosis.

Regardless of which type of leak detector is used, the system should have a minimum of 70 psi (482 kPa) pres-

Figure 51–10 A fluorescent dye is added to the system before checking for leaks with the black light. *Courtesy of Tracer Products*

sure to accurately detect a leak. Be sure to check the entire system. If a leak is found at a connection, tighten the connection carefully and recheck. If the leak is still apparent, **discharge** the system. Replace the damaged components, evacuate, charge, and test the system for leaks.

SERVICE PROCEDURES

All refrigerant must be discharged from the system before repair or replacement of any component (except for compressors with stem service valves). This procedure is accomplished through the use of a manifold gauge set and a refrigerant reclaiming unit that makes it possible to control the rate of refrigerant discharge, preventing any loss of oil from the system.

Until recently, each time an air-conditioning system was serviced, the old refrigerant was vented into the atmosphere. Today, the refrigerant must be reclaimed and recycled. This means the refrigerant cannot be released into the atmosphere. According to Section 609 of the Federal Clean Air Act, no one repairing or servicing a motor vehicle's air-conditioning system can do so without the proper refrigerant recycling equipment. It further states that no one can do this service unless that person is properly trained and certified.

Certification

Before you can purchase R-12, you must be certified by an EPA-approved program. Actually you should be certified to work on the refrigerant part of any air-conditioning system. To become certified, a technician must have and use approved refrigerant recycling equipment and must pass an exam on refrigerant recovery and recycling. A common test for this certification is administered by ASE.

Recycling collects old refrigerant from the system being serviced and cleans it up and prepares it to be used in the future.

The SAE has set standards for recycled R-12 and R-134a to make sure the refrigerants provide proper system performance and longevity. J1991 and J2099 are purity standards for recycled refrigerant and specifies a limit, in ppm (parts per million) by weight. These limits are placed on three potential contaminants. There can be no more than 15 ppm of moisture in recycled R-12 and R-134a. The limit for the amount of refrigerant oil in R-12 is 4,000 ppm and 500 ppm for R-134a. The additional contaminant looked at by the SAE was air or noncondensable gases. There should be no more than 330 ppm in R-12 and no more than 150 ppm in R-134a.

Refrigerant Recovery

To minimize the amount of refrigerant released to the atmosphere when A/C systems are serviced, always follow these steps:

1. The recycling equipment **(Figure 51–11)** must have shutoff valves within 12 inches (305 mm) of the hoses' service ends. With the valves closed, connect the hoses to the vehicle's air-conditioning service fittings.
2. Always follow the equipment manufacturer's recommended procedures for use. Recover the refrigerant from the vehicle and continue the process until the vehicle's system shows vacuum instead of pressure. Turn off the recovery/recycling unit for at least five minutes. If the system

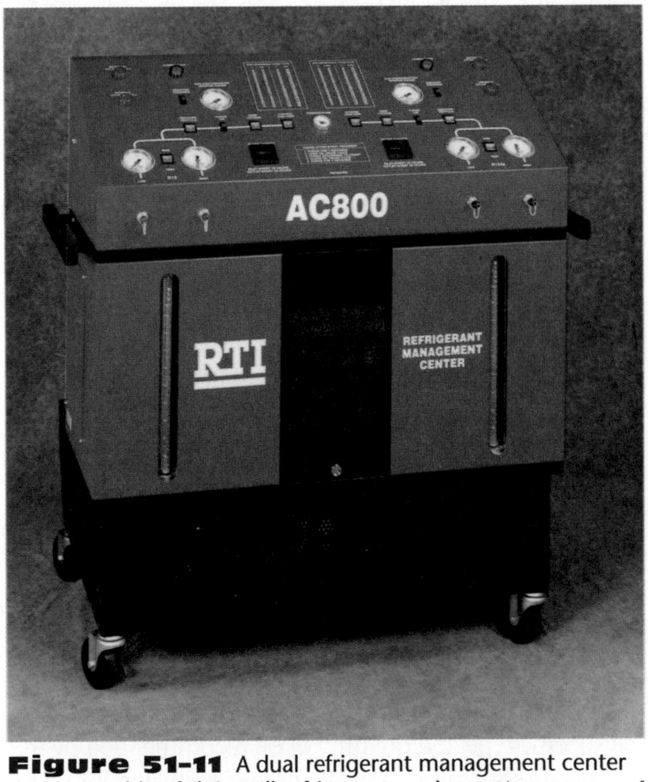

Figure 51–11 A dual refrigerant management center that is capable of doing all refrigerant services. *Photo courtesy of RTI Technologies, Inc.*

still has pressure, repeat the recovery process to remove any remaining refrigerant. Continue until the A/C system holds a stable vacuum for two minutes.

3. Close the valves in the recovery/recycling unit's service lines and disconnect them from the system's service fittings. On recovery/recycling stations with automatic shutoff valves, make sure they work properly.
4. You may now make repairs and/or replace parts in the system.

All recycled refrigerant must be safely stored in DOT CFR Title 49 or UL-approved containers. Containers specifically made for R-134a should be so marked. Before any container of recycled refrigerant can be used, it must be checked for noncondensable gases.

There are currently two types of refrigerant recovery machines, the single-pass and the multipass. Both have the ability to draw the refrigerant from the vehicle, filter and separate the oil from it, remove moisture and air from it, and store the refrigerant until it is reused.

In a single-pass system, the refrigerant goes through each stage before being stored. In multipass systems, the refrigerant may go through all stages or some of the stages before being stored. Either system is acceptable if it has the UL approved label.

System Flushing

Compressor failure causes foreign material to pass into the system. The condenser must be flushed and the receiver/dryer replaced. Filter screens are sometimes located in the suction side of the compressor and in the receiver/dryer. These screens confine foreign material to the compressor, condenser, receiver/dryer, and connecting hoses. Use only recommended flushing solvents. Never use CFCs or methylchloroform for flushing. Some recommend replacing the clogged components and installing a liquid line (in-line) filter just ahead of the expansion valve or orifice tube instead of flushing the system. If flushing is recommended by the manufacturer, use only the recommended flushing agent and follow the specified procedures. After the system has been flushed, be sure to oil all components that require it.

Compressor Oil Level Checks

Generally, compressor oil level is checked only where there is evidence of a major loss of system oil that could be caused by a broken refrigerant hose, severe hose fitting leak, badly leaking compressor seal, or collision damage to the system's components.

When replacing refrigerant oil, it is important to use the specific type and quantity of oil recommended by the compressor manufacturer. If there is a surplus of oil in the system, too much oil circulates with the refrigerant, causing the cooling capacity of the system to be reduced. Too little oil results in poor lubrication of the compressor. When there has been excessive leakage or it is necessary to replace a component of the refrigeration system, certain procedures must be followed to ensure that the total oil charge in the system is correct after leak repair or the new part is installed on the car.

When the compressor is operated, oil gradually leaves the compressor and is circulated through the system with the refrigerant. Eventually a balanced condition is reached in which a certain amount of oil is retained in the compressor and a certain amount is continually circulated. If a component of the system is replaced after the system has been operated, some oil goes with it. To maintain the original total oil charge, it is necessary to compensate for this by adding oil to the new replacement part. Because of the differences in compressor designs, be sure to follow the manufacturer's instructions when adding refrigerant oil to their unit.

R-12 based systems use mineral oil, while R-134a systems use synthetic polyalkylene gycol (PAG) oils. Using a mineral oil with R-134a will result in A/C compressor failure because of poor lubrication. Use only the oil specified for the system.

Evacuating and Charging the System

All of the refrigerant in the system must be recovered prior to evacuation. Evacuation is the name given to the process that pulls all traces of air, moisture, and refrigerant from the system. This is done by creating a vacuum in the system by connecting a vacuum pump to it **(Figure 51–12)**. The vacuum pump should remain on and con-

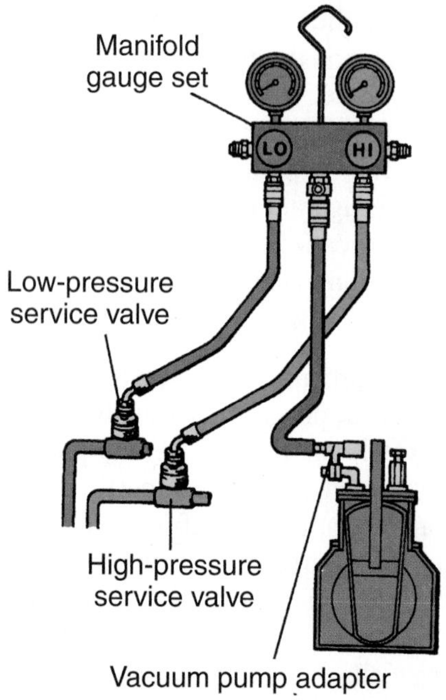

Figure 51-12 A vacuum pump connected to an A/C system through a gauge set. *Reprinted with permission*

nected to the system for at least thirty minutes after 26 to 29 inches of mercury (660 to 737 mm Hg) is reached.

CAUTION!

Always wear safety goggles when working with refrigerant containers or servicing air-conditioning systems.

Any air or moisture that is left inside an air-conditioning system reduces the system's efficiency and eventually leads to major problems, such as compressor failure.

Air causes excessive pressure within the system, restricting the refrigerant's ability to change its state from gas to liquid. Moisture, on the other hand, can cause freeze-up and restrict the refrigerant flow. Moisture also forms hydrochloric acid when mixed with refrigerant, which can cause corrosion inside the system. A vacuum pump removes the air and moisture from the system **(Figure 51–13)**. The vacuum pump reduces system pressure, vaporizes the moisture, and then exhausts it with the air.

Refilling the system with refrigerant is called charging the system. Refrigerant is added through the system's service ports. Depending on the method and source of new refrigerant, refrigerant is introduced through the low or high side of the system. When the system is operating, refrigerant is charged through the low side. On some vehicles when the system is off, charging is done through the high side. Always refer to the service manual for the correct procedure.

Thermistor Vacuum Gauge An electronic thermistor vacuum gauge is designed to work with the vacuum pump to measure the last, most critical inch of mercury during evacuation. It constantly monitors and visually indicates the vacuum level so you know when the system has a full vacuum and will be moisture free.

After the system is evacuated, it can be recharged. If the system will not pull down to a good vacuum, there is probably a leak somewhere in the system. Photo Sequence 52 illustrates a common way to evacuate and recharge a system.

SHOP TALK

The importance of the correct charge cannot be stressed enough. The efficient operation of the air-conditioning system depends greatly on the correct amount of refrigerant in the system. A low charge results in inadequate cooling under high heat loads due to a lack of reserve refrigerant and can cause the clutch cycling switch to cycle faster than normal. An overcharge can cause inadequate cooling because of a high liquid refrigerant level in the condenser. Refrigerant controls will not operate properly and compressor damage can result. In general, an overcharge of refrigerant will cause higher-than-normal gauge readings and noisy compressor operation. ■

Figure 51-13 A vacuum pump for evacuating the system. *Courtesy of Robinair, SPX Corporation*

Charging Cylinder With an increase in temperature in any cylinder filled with refrigerant, there is a corresponding increase in pressure and a change in the volume of liquid refrigerant in the cylinder. To measure an accurate charge by weight from a cylinder using the liquid level in a sight glass as a point of measurement, it is absolutely necessary to compensate for liquid volume variations caused by temperature differences. These temperature differences are directly related to pressure variations and accurate measurements by weight can be calibrated in relation to pressure.

The charging cylinder **(Figure 51–14)** is designed to meter out a desired amount of a specific refrigerant by weight. Compensation for temperature variations is accomplished by reading the pressure on the gauge of the cylinder and dialing the plastic shroud. The calibrated chart on the shroud contains corresponding pressure readings for the refrigerant being used.

When charging an air-conditioning system with refrigerant, the pressure in the system often reaches a point at which it is equal to the pressure in the cylinder from which the system is being charged. To get more refrigerant into the system to complete the charge, heat must be applied to the cylinder.

Performance Testing

Performance testing provides a measure of air-conditioning system operating efficiency. A manifold pressure gauge

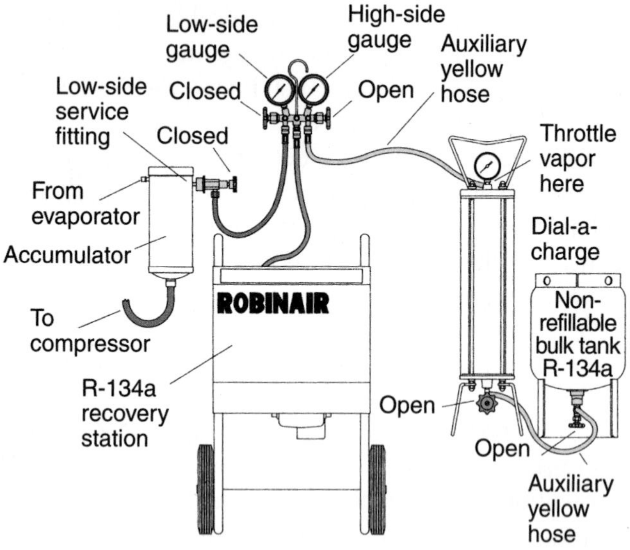

Figure 51-14 The connections for filling a system with refrigerant using a charging cylinder. *Courtesy of Robinair, SPX Corporation*

set is used to determine both high and low pressures in the refrigeration system. The desired pressure readings will vary according to temperature. Use temperature/pressure charts as a guide to determine the proper pressures. At the same time, a thermometer is used to determine air discharge temperature into the passenger compartment **(Figure 51–15)**.

Before making this test, it should be established that the air distribution (air door) portion of a factory-installed system is functioning properly. This ensures that all air passing through the evaporator is being routed directly to the air outlet nozzles.

Operating pressures vary with humidity as well as with outside air temperature. Accordingly, on more humid days, operating pressures will be on the high side of the range indicated in the service manual's performance chart. On less humid days, the operating pressures will read toward the lower side. If operating pressures are found to be within the normal range, the refrigeration portion of the air-conditioning system is functioning properly. This can be further confirmed with a check of evaporator outlet air temperatures.

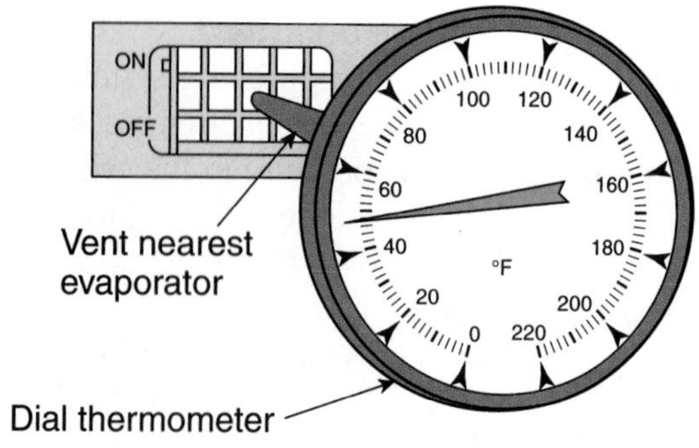

Figure 51-15 Checking outlet air temperature.

Always refer to the pressure charts given in the service manual when basing your diagnosis of the system on system pressures. Although the specifications will vary, here are some guidelines to help you interpret abnormal readings:

- If the high-side pressure is too high, suspect air in the system, too much refrigerant in the system, a restriction in the high side of the system, and poor airflow across the condenser.
- If the high-side pressure is too low, suspect a low refrigerant level or defective compressor.
- If the low-side pressure is higher than normal, suspect refrigerant overcharge, a defective compressor, or a faulty metering device.
- If the low-side pressure is lower than normal, suspect a faulty metering device, poor airflow across the evaporator, a restriction in the low side of the system, or a system undercharged with refrigerant.

PROCEDURE

Performance Testing the Air-Conditioning System

STEP 1 *Connect the manifold gauge set to the respective high- and low- pressure fittings. These fittings are found in various locations within the high- and low-pressure sides of the system.*

STEP 2 *Close the hood and all of the doors and windows of the vehicle.*

STEP 3 *Adjust the air-conditioning controls to maximum cooling and high blower position.*

STEP 4 *Idle the engine in neutral or park with the brake on. For the best results, place a high-volume fan in front of the radiator grille to ensure an adequate supply of airflow across the condenser.*

STEP 5 *Increase engine speed to 1,500–2,000 rpm.*

STEP 6 *Measure the temperature at the evaporator air outlet grille or air duct nozzle (35 to 40°F [1.6 to 4.4°C]).*

STEP 7 *Read the high and low pressures, and compare them to the normal range of the operating pressure given in the service manual.*

Evaporator outlet air temperature also varies according to outside (ambient) air and humidity conditions. Further variations can be found, depending on whether the system is controlled by a cycling clutch compressor or an evaporator pressure control valve. Because of these variations, it is difficult to pinpoint what evaporator outlet air temperature should be on all applications. In

Evacuating and Recharging an A/C System with Recycling and Charging Stations

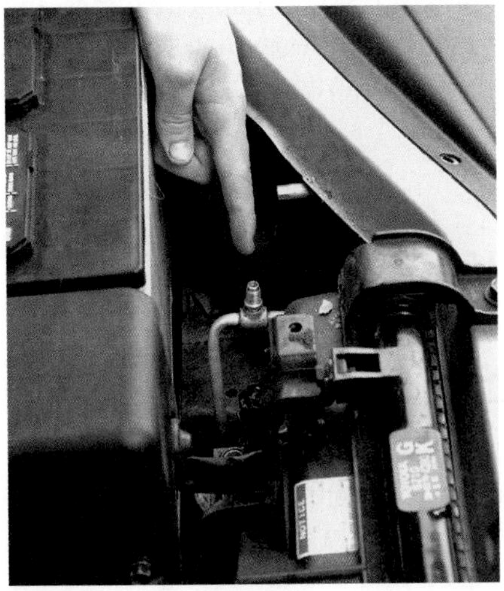

P52-1 *Physically locate the pressure fittings of the system you will be working with. Make sure you have the proper adapters for the fittings prior to starting any service work on air-conditioning systems.*

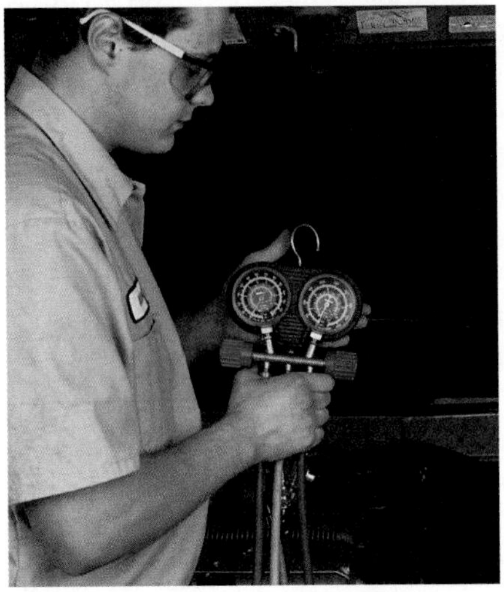

P52-2 *Connect a gauge set to the system.*

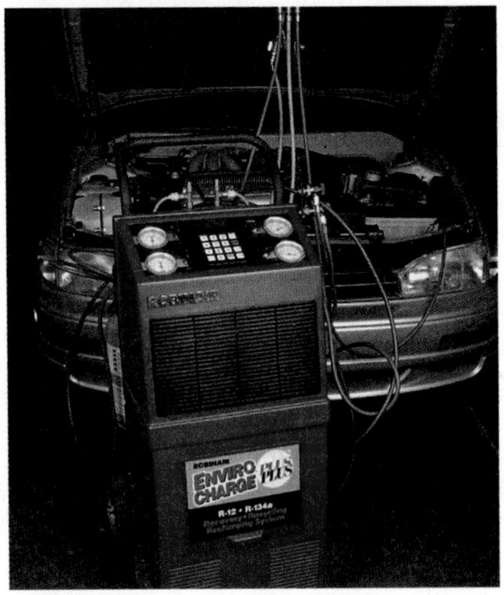

P52-3 *Typical refrigerant recycling/reclaiming machine.*

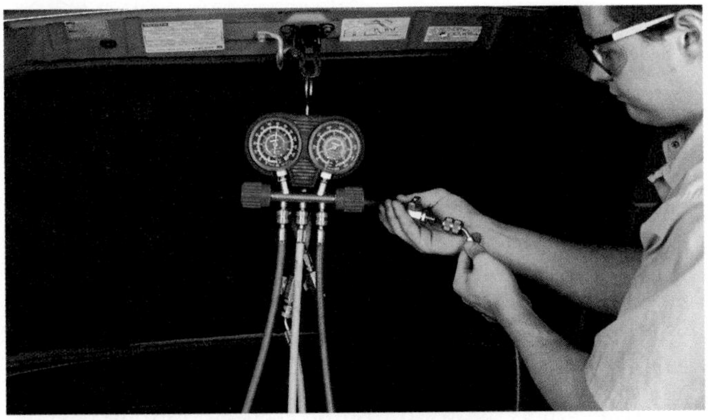

P52-4 *Connect the reclaiming unit to the gauge set.*

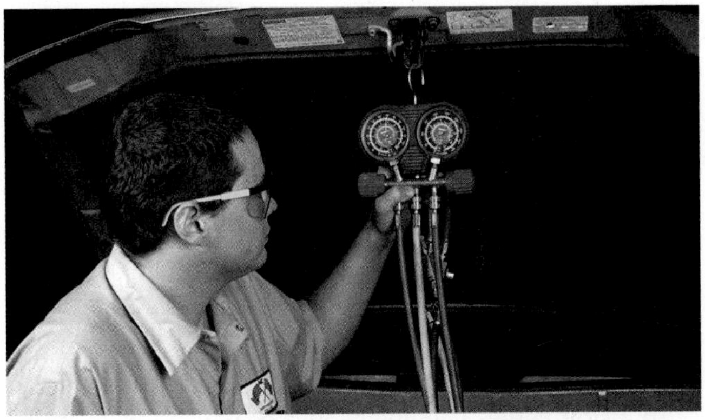

P52-5 *Activate the reclaimer to discharge the system, then begin evacuation. Allow enough time for the machine to draw a good vacuum (at least 29 in. Hg) in the system. Observe the gauge set to determine when a vacuum is created.*

P52-6 *Close the valves at the gauge set and disconnect the reclaiming machine. Connect the charging station and open the gauge set valves. Set the unit to deliver the required amount of refrigerant for that system.*

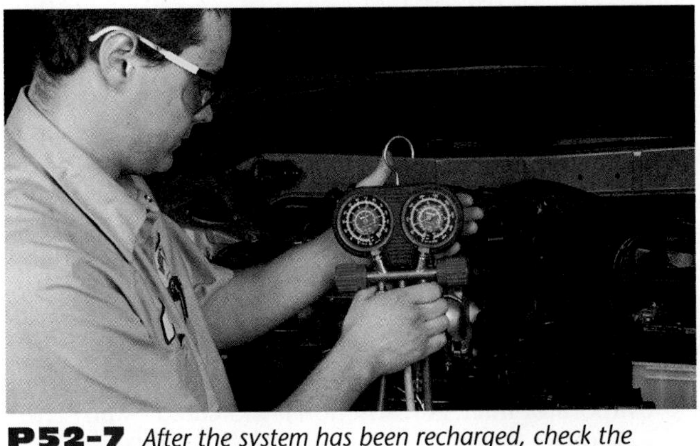

P52-7 *After the system has been recharged, check the gauge readings while the system is in operation to make sure it is working properly.*

general, with low-side air temperatures (70°F [21°C]) and humidity (20%), the evaporator outlet air temperature should be in the 35 to 40°F (2 to 4°C) range. On the other extreme of 80°F (27°C) outside air temperatures and 90% humidity condition, the evaporator air outlet temperature might be in the 55 to 60°F (13 to 60°C) range.

Since it is impractical to provide a specific performance chart for all the different types of air-conditioning systems, it is desirable to develop an experience factor for determining the correlation that can be anticipated between operating pressures and outlet air temperatures on the various systems. For example, feel the discharge line from the compressor to the condenser. The discharge line should be the same temperature along its full length. Any temperature change is a sign of restriction and the line should be flushed or replaced. Perform this test carefully because the discharge line will be hot.

Other tests can be performed with the engine running:

- Check the condenser by feeling up and down the face or along the return bends for a temperature change. There should be a gradual change from hot to warm as you go from the top to the bottom. Any abrupt change indicates a restriction, and the condenser has to be flushed or replaced.
- If the system has a receiver/dryer, check it. The inlet and outlet lines should be the same temperature. Any temperature difference or frost on the lines or receiver tank are signs of a restriction. The receiver/dryer must be replaced.
- If the system has a sight glass, check it as previously described.
- Feel the liquid line from the receiver/dryer to the expansion valve. The line should be warm for its entire length.
- The expansion valve should be free of frost and there should be a sharp temperature difference between its inlet and outlet.
- The suction line to the compressor should be cool to the touch from the evaporator to the compressor. If it is covered with thick frost, this might indicate that the expansion valve is flooding the evaporator.
- Typically, the formation of frost on the outside of a line or component means there is a restriction to the flow of refrigerant.
- On vehicles equipped with the orifice tube system, feel the liquid line from the condenser outlet to the evaporator inlet. A restriction is indicated by any temperature change in the liquid line before the crimp dimples the orifice tube in the evaporator inlet. Flush the liquid line or replace the orifice tube if restricted.
- The accumulator as well as the suction line must be cool to the touch from the evaporator outlet to the compressor.
- By combining the results of both the hands-on checks and an interpretation of pressure gauge readings, the technician has a good indication that some unit in the system is malfunctioning and that further diagnosis is needed.

DIAGNOSTIC AND TROUBLESHOOTING PROCEDURES

The first step in air-conditioning diagnosis, as in all automotive diagnostic work, is to get the customer's story. Is the problem no cold air, the air from the ducts never gets cold enough, or the unit only cools at certain times of the day? The complaint could also be that the system does not work at all, or that the air blows out of the wrong ducts. An accurate description of the customer's problem helps pinpoint whether the problem is refrigerant, mechanical, or electrical related. It also reduces diagnosis time and, most important, satisfies the customer.

Because of the many construction and operational variations that exist, there is no uniform or standard diagnostic procedure applicable to all automotive air-conditioning systems. However, the information given in the *Tech Manual* can be used as a general guide to common air-conditioning problems and their possible remedies. For complete specific diagnostic information on a given air conditioner, check the manufacturer's service manual.

ELECTRICAL SYSTEM INSPECTION

Because vehicle air-conditioning systems contain various switches, there will be times when one of them malfunctions. Sometimes this is not due to the switch itself but is caused instead by a bad connection or corrosion that has built up on the switch leads, leading to a broken contact. The switch can sometimes function intermittently because of this buildup. It is important to check the switches from time to time and clean off the leads. Check that each switch is making good contact and is free of dirt and corrosion.

The compressor clutch is an electromagnetic unit that can be checked with an ohmmeter **(Figure 51–16)**. Service manuals will give the acceptable resistance for the field coil of the clutch assembly.

Some late-model air-conditioning systems have a self-diagnostic capability. An indicator lamp on the control panel may be used to flash codes. On some vehicles, especially those with automatic climate control, part of the air-conditioning system can be checked with a scan tool

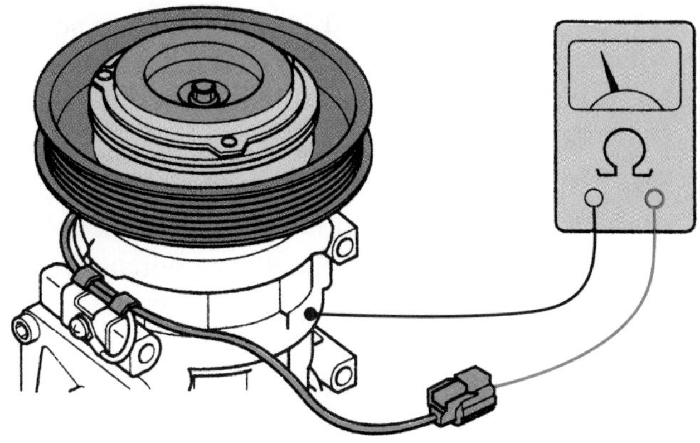

Figure 51–16 Testing a compressor clutch with an ohmmeter. *Reprinted with permission*

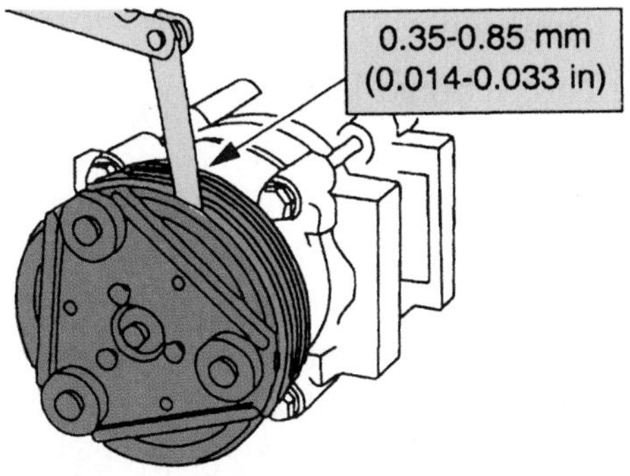

Figure 51–17 Checking the gap of a compressor clutch assembly. *Courtesy of Ford Motor Company*

connected to the vehicle's BCM. Normally the buttons on the control panel must be depressed in a certain sequence to activate the self-test.

One common repair to an air-conditioning system is the replacement of the compressor's clutch. Although the procedures for doing this vary with the manufacturer, one part of the procedure is consistent on all—measuring and adjusting the gap between the compressor pulley and the clutch assembly's pressure plate **(Figure 51–17)**.

HEATING SYSTEM SERVICE

When performing any checks and service procedures, always follow the procedures recommended in the manufacturer's service manual. In most cases, problems with the heating system are problems with the engine's cooling system. Therefore most service work and diagnosis is done to the cooling system. Problems that pertain specifically to the heater are few: the heater control valve and the heater. Most often if these two items are faulty, the engine's cooling system will be negatively affected. Both of these items are replaced rather than repaired. In some cases, it is possible to make repairs to vacuum hoses and electrical connections without removing the heater assembly. If it is necessary to remove the heater assembly, the cooling system must be drained before removing the heater core.

Basic Heater Inspection and Checks

When there is a problem of insufficient heat, begin your diagnosis with a visual inspection and a check of the coolant level. If the level is correct, turn the heater controls on, run the engine until it reaches normal operating temperature, then measure the temperature of the upper radiator hose. The temperature can be measured with a pyrometer. If one is not available, gently touch the hose. You should not be able to hold the hose long because of the heat. While doing this, make sure you stay clear of the area around the cooling fan. A spinning fan can chop off your hand. If the temperature of the hose is not within specifications, suspect a faulty thermostat.

CASE STUDY

A customer complained about his car's air-conditioning not working. Questioning the customer revealed that the system had not worked since the end of last summer. "It was going to get cool in a few weeks and I wouldn't need the air conditioner, so I put it off until spring."

A visual inspection of the system did not reveal any oil spots or ruptured lines and hoses, therefore no leaks were indicated. The technician then connected the manifold set to the system. He found the pressure to be equal on both sides. Further, the temperature/pressure chart indicated that the system pressure was within acceptable limits for the ambient temperature.

The technician noticed that the lead wire to the clutch coil had been disconnected. Assuming that it had been unintentionally disconnected, he reconnected it. Further questioning of the customer, however, revealed no knowledge of the disconnected wire.

Shortly after starting the engine and turning the A/C on, cool air was coming out of the dash vents. The manifold gauges showed normal readings. A thermometer was placed in the vents and the output temperature was within specifications.

Further discussion with the customer revealed that the car broke down while he was on vacation. While travelling, he had to replace a belt. What undoubtedly happened was that the technician who installed the belt must have unplugged the wire. Because the weather was cool, the customer had no reason to turn on the air conditioner and therefore could never link the problem to the belt replacement. Whenever you are working on a car, make sure everything is connected before returning it to the customer.

If the hose was the correct temperature, check the temperature of the two heater hoses. They should both be hot. If only one of the hoses is hot, suspect the heater control valve or a plugged heater core.

KEY TERMS

Discharge
Manifold gauge set
Phosgene gas
Schrader service valves
Stem valve

SUMMARY

- Air-conditioning systems are extremely sensitive to moisture and dirt. Therefore, clean working conditions are extremely important. The smallest particle of foreign matter in an air-conditioning system contaminates the refrigerant, causing rust, ice, or damage to the compressor.
- Air conditioner testing and servicing equipment includes a manifold gauge set, service valves, vacuum pumps, charging station, charging cylinder, recovery/recycling systems, and leak-detecting devices.
- To put air-conditioning systems in their best condition, the following service procedures should be performed: system discharging, system flushing, compressor oil level checks, evacuation, and system charging.
- Refrigerant evaporates quickly when it is exposed to the atmosphere. Never rub your eyes or skin if you have come in contact with refrigerant. It will freeze anything it contacts. If liquid refrigerant gets in your eyes or on your skin it can cause frostbite. Immediately flush the exposed areas with cool water for fifteen minutes and seek medical help.
- Never expose A/C system components to high temperatures. Heat will cause the refrigerant's pressure to increase. Never expose refrigerant to an open flame. A poisonous gas, phosgene gas, will result. If breathed, this gas can cause severe respiratory irritation.
- R-12 and R-134a should never be mixed. Their oils and desiccants are not compatible. If the two refrigerants are mixed, contamination will occur and may result in A/C system failure. Separate service equipment, including recovery/recycling machines and service gauges should be used for the different refrigerants.
- The manifold gauge set is used when discharging, charging, evacuating, and for diagnosing trouble in the system. Because R-134a is not interchangeable with R-12, separate sets of hoses, gauges, and other equipment are required to service vehicles.
- Performance testing provides a measure of air-conditioning system operating efficiency.
- A manifold pressure gauge set is used to determine both high and low pressures in the refrigeration system. The desired pressure readings will vary according to temperature.
- Using an electronic leak detector is the preferred method of leak detection.
- Fluorescent tracer systems use a fluorescent dye and a black-light glow gun to locate the source of leaks.
- All refrigerant must be discharged from the system before repair or replacement of any component, except for compressors with stem service valves.
- Before you can purchase R-12 or work with refrigerant, you must be certified by an EPA-approved program.
- Recycling collects old refrigerant from the system being serviced, cleans it up, and prepares it to be used in the future.
- When replacing refrigerant oil, it is important to use the specific type and quantity of oil recommended by the compressor manufacturer.
- R-12–based systems use mineral oil, while R-134a systems use synthetic polyalkylene gycol (PAG) oils. Using a mineral oil with R-134a will result in A/C compressor failure because of poor lubrication. Use only the oil specified for the system.
- While evacuating the system, the vacuum pump should remain on and connected to the system for at thirty minutes after 26 to 29 inches of mercury (660 to 737 mm Hg) is reached.
- When there is a problem of insufficient heat, begin your diagnosis with a visual inspection and a check of the coolant level. Then measure the temperature of the upper radiator hose.

TECH MANUAL

The following procedures are included in Chapters 50/51 of the *Tech Manual* that accompanies this book:

1. Identifying the type of air-conditioning system in a vehicle.
2. Identifying the components in an air-conditioning system.
3. Conducting a visual inspection of an air-conditioning system.
4. Conducting a system performance test on an air-cinditioning system.
5. Inspecting, testing, and servicing a compressor clutch.
6. Identifying and recovering refrigerant.
7. Recycling, labeling, and recovering refrigerant.

REVIEW QUESTIONS

1. Which of the following is not recommended while converting an R-12 system to R-134a?

a. Permanently installing conversion fittings using new hose clamps around the fittings.

b. Installing conversion labels and removing the R-12 label.

c. Recharging the system with R-134a to approximately 80% of the original R-12 charge.

d. Using a refrigerant identifier to make sure the system contains only R-12.

2. What should you do if you suspect that an alternative refrigerant had been used in the A/C system and it is not performing properly?

3. Which of the following statements best defines the term evacuation?

a. A process in which the refrigerant in a system is released.

b. A condition that exists when a compressor runs until there is no refrigerant left in the system.

c. The process that pulls all traces of air, moisture, and refrigerant from the system.

d. A process that uses air to force moisture and dirt out of the system.

4. Which of the following statements is not true about orifice tube systems?

a. The suction line must be cool to the touch from the evaporator outlet to the compressor.

b. The liquid line from the condenser outlet to the evaporator inlet should be cool.

c. The evaporator should feel very cold.

d. The accumulator must be cool to the touch.

5. Define vacuum.

6. What test instrument is commonly used to check the coil of a compressor clutch?

7. When is the manifold gauge set used?

a. when discharging

b. when charging

c. when evacuating

d. in all of the above

8. Which of the following does *not* cause excessive heat?

a. loose or damaged control cable

b. water flow control valve stuck open

c. flow control valve not opening fully

d. faulty thermostatic fan

9. To charge an air-conditioning system while it is running, the refrigerant should be added to ______.

a. the high side

b. the low side

c. both the high and low sides

d. either the high or the low side

10. Technician A says the inlet hose of the heater core should always be much hotter than the outlet hose. Technician B says the upper radiator hose should always be hotter than the lower radiator hose. Who is correct?

a. Technician A

b. Technician B

c. Both A and B

d. Neither A nor B

11. A heater does not supply enough heat. The coolant level and flow are correct. Technician A says a misadjusted heater control could be the cause. Technician B says a bad thermostat could be the cause. Who is correct?

a. Technician A

b. Technician B

c. Both A and B

d. Neither A nor B

12. Technician A says evacuating (pumping down) an air-conditioning system removes air and moisture from the system. Technician B says evacuating (pumping down) an air-conditioning system removes dirt particles from the system. Who is correct?

a. Technician A **c.** Both A and B
b. Technician B **d.** Neither A nor B

13. Technician A says some refrigerant leaks can also be traced to pinholes in the evaporator caused by acid that forms when water and refrigerant mix. Technician B says leaks are most often found at the compressor hose connections and at the various fittings and joints in the system. Who is correct?

a. Technician A **c.** Both A and B
b. Technician B **d.** Neither A nor B

14. Technician A says the presence of oil around a fitting of an air-conditioning line may indicate an oil leak but not a refrigerant leak. Technician B says using an electronic leak detector is the best way to find the source of a refrigerant leak. Who is correct?

a. Technician A **c.** Both A and B
b. Technician B **d.** Neither A nor B

15. Technician A says a faulty accumulator will allow foreign material to enter the A/C system. Technician B says a damaged compressor will allow foreign material to enter the A/C system. Who is correct?

a. Technician A **c.** Both A and B
b. Technician B **d.** Neither A nor B

16. Technician A says air in the system can cause excessive pressure within the system, restricting the refrigerant's ability to change its state from gas to liquid within the refrigeration cycle. Technician B says moisture can cause freeze-up at the cap tube or expansion valve that restricts refrigerant flow or blocks it completely. Who is correct?

a. Technician A **c.** Both A and B
b. Technician B **d.** Neither A nor B

17. Technician A feels up and down the face or along the return bends of the condenser for a temperature change and says there should be a gradual change from hot to warm as you go from the bottom to the top. Technician B says a restriction in the condenser will be evident by a sudden change in temperature. Who is correct?

a. Technician A **c.** Both A and B
b. Technician B **d.** Neither A nor B

18. Technician A says the oil level of the compressor should be checked every year. Technician B says the system should be charged only after it has been completely evacuated. Who is correct?

a. Technician A **c.** Both A and B
b. Technician B **d.** Neither A nor B

19. Technician A says if the suction line to the compressor is covered with thick frost, this might indicate that the expansion valve is flooding the evaporator. Technician B says the formation of frost on the outside of a line or component means there is a restriction to the flow of refrigerant. Who is correct?

a. Technician A **c.** Both A and B
b. Technician B **d.** Neither A nor B

20. Technician A says a manifold gauge set can be used during the evacuation of the system. Technician B says a manifold gauge set can be used to diagnose a problem in the A/C system. Who is correct?

a. Technician A **c.** Both A and B
b. Technician B **d.** Neither A nor B

APPENDIX A

DECIMAL AND METRIC EQUIVALENTS

DECIMAL AND METRIC EQUIVALENTS					
Fractions	**Decimal (in.)**	**Metric (mm)**	**Fractions**	**Decimal (in.)**	**Metric (mm)**
1/64	0.015625	0.397	33/64	0.515625	13.097
1/32	0.03125	0.794	17/32	0.53125	13.494
3/64	0.046875	1.191	35/64	0.546875	13.891
1/16	0.0625	1.588	9/16	0.5625	14.288
5/64	0.078125	1.984	37/64	0.578125	14.684
3/32	0.09375	2.381	19/32	0.59375	15.081
7/64	0.109375	2.778	39/64	0.609375	15.478
1/8	0.125	3.175	5/8	0.625	15.875
9/64	0.140625	3.572	41/64	0.640625	16.272
5/32	0.15625	3.969	21/32	0.65625	16.669
11/64	0.171875	4.366	43/64	0.671875	17.066
3/16	0.1875	4.763	11/16	0.6875	17.463
13/64	0.203125	5.159	45/64	0.703125	17.859
7/32	0.21875	5.556	23/32	0.71875	18.256
15/64	0.234275	5.953	47/64	0.734375	18.653
1/4	0.250	6.35	3/4	0.750	19.05
17/64	0.265625	6.747	49/64	0.765625	19.447
9/32	0.28125	7.144	25/32	0.78125	19.844
19/64	0.296875	7.54	51/64	0.796875	20.241
5/16	0.3125	7.938	13/16	0.8125	20.638
21/64	0.328125	8.334	53/64	0.828125	21.034
11/32	0.34375	8.731	27/32	0.84375	21.431
23/64	0.359375	9.128	55/64	0.859375	21.828
3/8	0.375	9.525	7/8	0.875	22.225
25/64	0.390625	9.922	57/64	0.890625	22.622
13/32	0.40625	10.319	29/32	0.90625	23.019
27/64	0.421875	10.716	59/64	0.921875	23.416
7/16	0.4375	11.113	15/16	0.9375	23.813
29/64	0.453125	11.509	61/64	0.953125	24.209
15/32	0.46875	11.906	31/32	0.96875	24.606
31/64	0.484375	12.303	63/64	0.984375	25.003
1/2	0.500	12.7	1	1.00	25.4

APPENDIX B

GENERAL TORQUE SPECIFICATIONS

The values in this chart should only be used when manufacturer's specifications are *not* available. Also, the values are only valid when SAE 10 oil is used to lubricate the threads of the bolt.

Bolt Diameter (in inches)	Torque (pounds-feet)		
	SAE 2	SAE 5	SAE 8
1/4	7	10	14
5/16	14	21	30
3/8	24	37	52
7/16	39	60	84
1/2	59	90	128
9/16	85	130	184
5/8	117	180	255
3/4	205	320	450
7/8	200	515	730
1	300	775	1090

Bolt Diameter (in millimeters)	TORQUE: kg. cm* kg. m Property Class:									
	4.6	4.8	5.6	5.8	6.6	6.8	6.9	8.8	10.9	12.9
6	49*	63*	61*	79*	74*	95*	103*	126*	172*	206*
8	119*	153*	148*	178*	178*	230*	250*	306*	417*	500*
10	235*	303*	294*	379*	353*	455*	495*	606*	8.2	10
12	411*	529*	427*	662*	616*	7.9	8.6	10.5	14	17
14	654*	8.4	8.2	10.5	10	12	13	17	23	27
16	10	13	12	16	15	20	21	26	36	43
18	14	18	17	23	21	27	30	36	49	59
22	27	35	34	44	41	52	57	70	95	114

APPENDIX C

CORRELATION OF CONTENTS IN CHAPTER 7 WITH OTHER CHAPTERS

The content in Chapter 7 covers all of the basic science standards and math for NATEF certification. *More importantly,* the theories and examples are provided to give you a better understanding of why things work the way they do. The following chart allows you to match the contents of the other chapters with the appropriate background information given in Chapter 7.

Chapter	Chapter 7 Contents	
Chapter 8	Matter	page 148
	Energy	page 149
	Volume	page 151
	Force	page 153
	Motion	page 155
	Work	page 158
	Gases	page 168
	Heat	page 169
	Chemical properties	page 171
Chapter 9	Matter	page 148
	Energy	page 149
	Force	page 153
	Chemical properties	page 171
Chapter 10	Matter	page 148
	Energy	page 149
	Volume	page 151
	Force	page 153
	Motion	page 155
	Work	page 158
	Waves and oscillations	page 162
	Gases	page 168
	Heat	page 169
	Chemical properties	page 171
Chapter 11	Matter	page 148
	Energy	page 149
	Volume	page 151
	Motion	page 155
	Work	page 158
	Gases	page 168
	Heat	page 169
	Chemical properties	page 171
Chapter 12	Matter	page 148
	Energy	page 149
	Volume	page 151
	Time	page 155
	Motion	page 155
	Liquids	page 165
	Heat	page 169
	Chemical properties	page 171
	Electricity and electromagnetism	page 175
Chapter 13	Matter	page 148
	Energy	page 149
	Volume	page 151
	Motion	page 155
	Liquids	page 165
	Gases	page 168
	Heat	page 169
	Chemical properties	page 171
Chapter 14	Matter	page 148

Chapter	Chapter 7 Contents	
	Energy	page 149
	Time	page 155
	Work	page 158
	Waves and oscillations	page 162
	Light	page 165
	Chemical properties	page 171
	Electricity and electromagnetism	page 175
Chapter 15	Matter	page 148
	Energy	page 149
	Waves and oscillations	page 162
	Heat	page 169
	Chemical properties	page 171
	Electricity and electromagnetism	page 175
Chapter 16	Matter	page 148
	Energy	page 149
	Time	page 155
	Waves and oscillations	page 162
	Light	page 165
	Heat	page 169
	Chemical properties	page 171
	Electricity and electromagnetism	page 175
Chapter 17	Matter	page 148
	Energy	page 149
	Volume	page 151
	Time	page 155
	Light	page 165
	Heat	page 169
	Chemical properties	page 171
	Electricity and electromagnetism	page 175
Chapter 18	Matter	page 148
	Energy	page 149
	Work	page 158
	Electricity and electromagnetism	page 175
Chapter 19	Matter	page 148
	Energy	page 149
	Work	page 158
	Waves and oscillations	page 162
	Electricity and electromagnetism	page 175
Chapter 20	Matter	page 148
	Energy	page 149
	Light	page 165

Chapter	Chapter 7 Contents	
	Heat	page 169
	Electricity and electromagnetism	page 175
Chapter 21	Matter	page 148
	Energy	page 149
	Time	page 155
	Motion	page 155
	Light	page 165
	Heat	page 169
	Electricity and electromagnetism	page 175
Chapter 22	Matter	page 148
	Energy	page 149
	Time	page 155
	Waves and oscillations	page 162
	Light	page 165
	Electricity and electromagnetism	page 175
Chapter 23	Matter	page 148
	Energy	page 149
	Force	page 153
	Motion	page 155
	Waves and oscillations	page 162
	Electricity and electromagnetism	page 175
Chapter 24	Matter	page 148
	Energy	page 149
	Time	page 155
	Light	page 165
	Electricity and electromagnetism	page 175
Chapter 25	Matter	page 148
	Energy	page 149
	Time	page 155
	Light	page 165
	Electricity and electromagnetism	page 175
Chapter 26	Matter	page 148
	Energy	page 149
	Volume	page 151
	Time	page 155
	Work	page 158
	Chemical properties	page 171
	Electricity and electromagnetism	page 175
Chapter 27	Matter	page 148
	Energy	page 149

Chapter	Chapter 7 Contents	
	Volume	page 151
	Time	page 155
	Liquids	page 165
	Gases	page 168
	Electricity and electromagnetism	page 175
Chapter 28	Matter	page 148
	Energy	page 149
	Volume	page 151
	Liquids	page 165
	Gases	page 168
	Heat	page 169
	Electricity and electromagnetism	page 175
Chapter 29	Matter	page 148
	Energy	page 149
	Volume	page 151
	Liquids	page 165
	Gases	page 168
	Heat	page 169
	Electricity and electromagnetism	page 175
Chapter 30	Matter	page 148
	Energy	page 149
	Volume	page 151
	Time	page 155
	Liquids	page 165
	Gases	page 168
	Heat	page 169
	Chemical properties	page 171
Chapter 31	Matter	page 148
	Energy	page 149
	Volume	page 151
	Gases	page 168
	Heat	page 169
	Chemical properties	page 171
	Electricity and electromagnetism	page 175
Chapter 32	Matter	page 148
	Energy	page 149
	Volume	page 151
	Gases	page 168
	Heat	page 169
	Chemical properties	page 171
	Electricity and electromagnetism	page 175
Chapter 33	Matter	page 148
	Energy	page 149
	Gases	page 168
	Heat	page 169
	Chemical properties	page 171
	Electricity and electromagnetism	page 175
Chapter 34	Matter	page 148
	Energy	page 149
	Volume	page 151
	Gases	page 168
	Heat	page 169
	Chemical properties	page 171
	Electricity and electromagnetism	page 175
Chapter 35	Matter	page 148
	Energy	page 149
	Motion	page 155
	Work	page 158
	Liquids	page 165
Chapter 36	Matter	page 148
	Energy	page 149
	Force	page 153
	Motion	page 155
	Work	page 158
Chapter 37	Matter	page 148
	Energy	page 149
	Force	page 153
	Motion	page 155
	Work	page 158
Chapter 38	Matter	page 148
	Energy	page 149
	Force	page 153
	Motion	page 155
	Work	page 158
Chapter 39	Matter	page 148
	Energy	page 149
	Volume	page 151
	Force	page 153
	Motion	page 155
	Work	page 158
	Liquids	page 165
	Gases	page 168
	Heat	page 169
	Chemical properties	page 171
Chapter 40	Matter	page 148
	Energy	page 149
	Volume	page 151

Chapter	Chapter 7 Contents	
	Force	page 153
	Motion	page 155
	Work	page 158
	Liquids	page 165
	Gases	page 168
	Heat	page 169
Chapter 41	Matter	page 148
	Energy	page 149
	Force	page 153
	Motion	page 155
	Work	page 158
	Liquids	page 165
Chapter 42	Matter	page 148
	Energy	page 149
	Force	page 153
	Motion	page 155
	Work	page 158
	Waves and oscillations	page 162
	Gases	page 168
	Heat	page 169
Chapter 43	Matter	page 148
	Energy	page 149
	Force	page 153
	Motion	page 155
	Work	page 158
	Waves and oscillations	page 162
	Chemical properties	page 171
	Electricity and electromagnetism	page 175
Chapter 44	Matter	page 148
	Energy	page 149
	Force	page 153
	Motion	page 155
	Work	page 158
	Waves and oscillations	page 162
	Electricity and electromagnetism	page 175
Chapter 45	Matter	page 148
	Energy	page 149
	Force	page 153
	Motion	page 155
	Work	page 158
	Waves and oscillations	page 162
Chapter 46	Matter	page 148
	Energy	page 149
	Volume	page 151
	Force	page 153
	Motion	page 155
	Work	page 158
	Heat	page 169
	Chemical properties	page 171
Chapter 47	Matter	page 148
	Energy	page 149
	Force	page 153
	Motion	page 155
	Heat	page 169
Chapter 48	Matter	page 148
	Energy	page 149
	Force	page 153
	Motion	page 155
	Heat	page 169
Chapter 49	Matter	page 148
	Energy	page 149
	Force	page 153
	Motion	page 155
	Electricity and electromagnetism	page 175
Chapter 50	Matter	page 148
	Energy	page 149
	Volume	page 151
	Work	page 158
	Gases	page 168
	Heat	page 169
	Chemical properties	page 171
	Electricity and electromagnetism	page 175
Chapter 51	Matter	page 148
	Energy	page 149
	Volume	page 151
	Work	page 158
	Gases	page 168
	Heat	page 169
	Chemical properties	page 171
	Electricity and electromagnetism	page 175

GLOSSARY

Note: Terms are highlighted in color, followed by Spanish translation in bold.

Abrasion Wearing or rubbing away of a part.

Abrasión El desgaste o consumo por rozamiento de una parte.

Abrasive cleaning Cleaning that relies on physical abrasion, such as wire brushing and glass bead blasting.

Limpieza abrasiva La limpieza que cuenta con un abrasión físico, tal como el limpiar con cepillo de alambre o con un chorro de perlitas de vidrio.

Acceleration The rate of increase in speed.

Aceleración El índice de incremento en velocidad.

Accessory An object or device not essential to the operation of the vehicle but adding to the beauty, convenience, or effectiveness of something else or some other system.

Accesorio Un objeto o dispositivo que no es esencial para el funcionamiento del vehículo, pero que agrega algo a la belleza, la conveniencia o la eficacia de alguna otra cosa u otro sistema.

Accumulator A device that cushions the motion of a clutch and servo action in an automatic transmission; a component used to store or hold liquid refrigerant in an air-conditioning system.

Acumulador Un dispositivo que suaviza los movimientos de un embrague y la acción del servo de una transmisión automática; un componente que se usa para guardar o reservar el refrigerante líquido de un sistema de un acondicionador del aire.

Acid A compound that has an excess of H ions and breaks into hydrogen (H^+) ions and another compound when placed in an aqueous (water) solution.

Ácido Un compuesto que tiene un exceso de iones de H. que se rompe en los iones de hidrógeno y un otro compuesto cuando está colocado en una solución acuosa (del agua).

Acidity In lubrication, acidity denotes the presence of acid-type chemicals, which are identified by the acid number. Acidity within oil causes corrosion, sludges, and varnish to increase.

Acidez En la lubricación, la acidez se refiere a la presencia de los químicos de tipo ácido, que se identifican por un número de ácido. La acidez en los aceites causa aumentos en la corrosión, la grasa y el barnís.

Ackerman principle The geometric principle used to provide toe-out on turns. The ends of the steering arms are angled so the inside wheel turns more than the outside wheel when a car is making a turn.

El principio Ackerman El principio geométrico que provee la divergencia en las vueltas. Las extremidades de los brazos de dirección estan en un ángulo para que el interior de la rueda gira más que el exterior de la rueda cuando un coche da la vuelta.

Active restraint A manual seat and shoulder belt assembly that must be activated by an occupant.

Sujeción activo Una asamblea manual de cinturón de seguridad que debe ser implementado por el ocupante.

Actuator A control device that delivers mechanical action in response to an electrical signal.

Solenoide Un dispositivo de control que entrega una acción mecánica como respuesta a un señal eléctrico.

A/D converter An electronic device that changes analog signals to digital signals.

Convertidor de A/D Un dispositivo electrónico que cambia los señales análogos a los señales digitales.

Adaptive learning An operational mode of a control computer that adjusts operation parameters according to various conditions.

Operación de adaptación Un modo de operación de una computadora de control que ajusta los parámetros de operación según varias condiciones.

Additive In automotive oils, a material added to the oil to give it certain properties; for example, a material added to engine oil to lessen its tendency to congeal or thicken at low temperatures.

Aditamento En los aceites automotrices, una materia añadida al aceite para que ésta tenga ciertas propriedades; por ejemplo, una materia agregada al aceite de motor para disminuir su tendencia a solidificarse o ponerse espeso en las temperaturas bajas.

Adhesion The property of lubricating oil that causes it to stick or cling to a bearing surface.

Adhesión La propriedad del aceite de lubricación que causa que se pega o se adhiera a una superficie portante.

Adhesive Substance that causes adjoining bodies to stick together.

Adhesivo Una substancia que causa que dos cuerpos contiguos se pegan.

Aeration The process of mixing air with a liquid. Aeration occurs in a shock absorber from rapid fluctuation in movement.

Aeración El proceso de mezclar el aire con un líquido. La aeración ocurre en un amortiguador de barquinazos por causa de una variación rápida en el movimiento.

Aerobic Typically refers to a sealant or adhesive that dries in the presence of oxygen.

Aeróbico Típicamente se refiere a un sellante o un adhesivo que se seca cuando esta presente el oxígeno.

Aerodynamics The study of the effects of air on a moving object.

Aerodinamica El estudio de los efectos del aire en un objeto móvil.

Air bag system System that uses impact sensors, vehicle's on-board computer, an inflation module, and a nylon bag in the steering column and dash to protect the driver and passenger during a head-on collision.

Sistema de bolsa de aire Un sistema que utiliza sensores de impacto, la computador del vehículo, un módulo inflador y bolsas de nilón en la columna de dirección y el tablero de instrumentos, para proteger al conductor y el pasajero en caso de un choque de frente.

Air cleaner A device connected to the carburetor in a manner that all incoming air must pass through it. Its purpose is to filter dirt and dust from the air before it passes into the engine.

Filtro (depurador) de aire Un dispositivo conectado al carburador de tal modo que todo el aire que llegue tenga que pasar por él. Su finalidad es la de retirar del aire la suciedad y el polvo antes de que pase al motor.

Air-conditioning The process of adjusting and regulating, by heating or refrigerating, the quality, quantity, temperature, humidity, and circulation of air in a space or enclosure; to condition the air.

Aire acondicionado El proceso de ajustar, regular calentando o enfriando, la calidad, la cantidad, la temperatura, la humedad, y la circulación de aire en un espacio o espacio cerrado; para acondicionar el aire.

Air-conditioning clutch An electromechanical device mounted on the air-conditioning compressor used to start and stop compressor action, thereby controlling refrigerant circulating through the system.

Embrague de climatizador (aire acondicionado) Dispositivo electromecánico montado en el compresor del equipo climatizador y que se utiliza para poner en marcha y detener dicho compresor, controlando de ese modo el refrigerante que circula por el sistema.

Air-conditioning compressor A vapor pump that pumps vapor (refrigerant or air) from one pressure level to a higher pressure level.

Compresor de aire acondicionado Dispositivo del climatizador que bombea vapor (refrigerante o aire) de un nivel de presión a otro más elevado.

Air ducts Tubes, channels, or other tubular structures used to carry air to a specific location.

Conductos de aire Los tubos, las canalizaciones, u otras estructuras tubulares que sirven para traer el aire a una posición específica.

Air filter A filter that removes dust, dirt, and particles from the air passing through it.

Filtro de aire Un filtro que retira polvo, suciedad y partículas del aire que pasa por él.

Air gap The space between spark plug electrodes, motor and generator armatures, and field shoes.

Entrehierro El espacio entre los electrodos de las bujías, el enducido del motor y del generador, y las piezas inductoras.

Air injection The introduction of fresh air into an exhaust manifold for additional burning and to provide oxygen to a catalytic converter.

Inyección por aire La introducción del aire fresco al múltiple de escape para mejor combustión y para proveer el oxígeno al convertidor catalítico.

Air lock A bubble of air trapped in a fluid circuit that interferes with normal circulation of the fluid.

Bolsa de vapor Una burbúja del aire atrapado en un circuito líquido que es un obstáculo en la circulación normal del líquido.

Air pump A device to produce a flow of air at higher-than-atmospheric pressure. Normally referred to as a thermactor air supply pump.

Bomba del aire Un dispositivo que produce un corriente de aire con una presión más alta que la del atmósfera. Por lo común se refiere como bomba de aire termactor.

Align boring A series of holes bored in a straight line.

Taladrado alineado Una seria de perforaciones que se han taladrado en una linea recta.

Alignment An adjustment to a line or to bring into a line.

Alineación Un ajuste que se efectúa en una linea o alinear.

Alloy A mixture of different metals such as solder, which is an alloy consisting of lead and tin.

Aleación Una mezcla de metales distinctos como la soldadura, que es una aleación que consiste del plomo y el estaño.

All-wheel drive System of driving all four wheels only when traction conditions dictate.

Tracción total (todoterreno) Un sistema de tracción que maneja las cuatro ruedas sólo cuando indican las condiciones de la tracción.

Alternating current Electrical current that changes direction between positive and negative.

Corriente alterna Corriente eléctrica que recorre un circuito ya sea en dirección positiva o en dirección negativa.

Alternator An AC (alternating current) generator that produces electrical current and forces it into the battery to recharge it.

Alternador Un generador de ca (corriente alterna) que produce corriente eléctrica y la dirige a una batería para cargarla.

Ambient temperature Temperature of air surrounding an object.

Temperatura del ambiente La temperatura en la atmósfera alrededor de un objeto.

Ammeter The instrument used to measure electrical current flow in a circuit.

Amperímetro El instrumento que se usa para medir los corrientes eléctricos de un circuito.

Ampere The unit for measuring electrical current; usually called an amp.

Amperio La unidad para medir el corriente eléctrico; por lo común se refiere como amp en inglés.

Amplifier A circuit or device used to increase the voltage or current of a signal.

Amplificador Un circuito o un dispositivo que se usa para aumentar el voltage o el corriente de un señal.

Amplify To enlarge or strengthen original characteristics; a term associated with electronics.

Amplificar Aumentar o fortalecer las características originales; un término que se asocia con lo electrónico.

Amplitude A measurement of a vibration's intensity.

Amplitud Una medida de intensidad de una vibración.

Anaerobic Typically refers to a sealant or adhesive that dries without being exposed to oxygen.

Anaerobio Típicamente se refiere a un sellante o un adhesivo que se seca sin ser expuesto al oxígeno.

Anaerobic sealer Liquid or gel that operates to bond two parts together in the absence of air.

Sellante anaerobio Un líquido o un gel cuyo propósito es adherir dos partes en la ausencia del aire.

Analog A nondigital measuring method that uses a needle to indicate readings. A typical dashboard gauge with a moving needle is an analog instrument.

Análogo Un método no numérico de medir que usa una aguja para indicar las lecturas. Un indicador típico de un tablero de intrumentos con una aguja que se mueve es un instrumento análogo.

AND gate A logic gate with at least two inputs and one output.

Puerta Y Una puerta lógica que tiene al menos dos entradas y una salida.

Aneroid bellows A device that responds to atmospheric pressure and can be used to move a metering needle within a jet to automatically compensate for altitude changes.

Fuelle aneroide Un dispositivo que se ajusta a la presión atmosférica y mueve una aguja de medición dentro de un chorro para compensar automáticamente por los cambios en el altitud.

Anodize An electrochemical process that coats and hardens the surface of aluminum.

Anodizar Un proceso electroquímico que cubre y endurece la superficie del aluminio.

Antifreeze A material, such as alcohol or glycerin, added to water to lower its freezing point.

Anticongelante Una materia, tal como el alcohol o la glicerina, que se agrega al agua para disminuir su punto de congelación.

Antilock braking system A series of sensing devices at each wheel that control braking action to prevent wheel lockup.

Sistema de frenos antideslizante Una seria de dispositivos sensibles en cada rueda que controla la acción del enfrenamiento y previene que las ruedas se bloquean.

Antiseize compounds A type of lubrication that prevents dissimilar metals from reacting with one another and seizing.

Compuestas antiagarrotamiento Un tipo de lubricación que previene que los metales desemejantes se reaccionan y se agarran.

Antisway bar A suspension geometry that resists a vehicle's tendency to drop or squat on the rear springs when accelerating.

Barra de anticabeceo Una geometría de suspensión que resiste la tendencia de un vehículo a desplomarse o apoyarse sobre los muelles posteriores al acelerar.

Apply side The side of a piston on which force or pressure is exerted to move the piston to do work.

Superficie de empuje El lado del pistón en el cual la fuerza o la presión se aplica para mover el pistón y hacerlo trabajar.

Aqueous A solution that is mostly water.

Acuosa Una solución acuosa es una solución que es agua en su mayor parte.

Arbor A shaft to which other parts may be attached.

Árbol Una flecha a la cual otras partes se pueden conectar.

Arcing Electrical energy jumping across a gap. Arcing across ignition points causes pitting and erosion.

Arco La energía eléctrica que salta por una holgura. Un arco entre los platinos del encendido causa las picaduras y la erosión.

Aspect ratio The height of a tire, from bead to tread, expressed as a percentage of the tire's section width.

Relación de aspecto La altura de un neumático, de la ceja a la banda, que se expresa como un porcentaje de lo ancho de una sección del neumático.

Asymmetric Unequal surfaces or sizes.

Asimétrico Desigual en las superficies o en los tamaños.

ATF Automatic transmission fluid.

ATF Fluido para transmisión automática.

Atmospheric pressure The weight of the air at sea level (about 14.7 pounds per square inch or less at higher altitudes).

Presión atmosférica El peso del aire a nivel del mar (aproximadamente unas 14.7 libras por pulgada cuadrada o menos en las altitudes más altas).

Atom The smallest particle of an element in which all the chemical characteristics of the element are present.

Átomo La partícula más pequeña de un elemento en el la cual todas las características químicas del elemento estén presentes.

Atomization The stage in which the metered air/fuel emulsion is drawn into the airstream in the form of tiny droplets.

Atomización La etapa en la cual la emulsión calibrada del aire/combustible se introduce al chorro del aire en la forma de pequeñas gotas.

Automatic transmission A transmission in which gear ratios are changed automatically.

Transmisión automática Una transmisión en la que las relaciones de engranajes cambian automáticamente.

Automotive aftermarket The network of businesses that supplies replacement parts and services to independent service shops, specialty repair shops, car and truck dealerships, fleet and industrial operations, and the general buying public.

Mercado de repuestas no originales La cadena de negocios que proveen las repuestas y los servicios a los talleres de servicio independientes, los talleres de refacción especializadas, los comerciantes de coches y camiones, las operaciones de flete e industrias, y las compras del público común.

Axial Having the same direction or being parallel to the axis or rotation.

Axial Teniendo el mismo dirección o siendo paralelo al eje o la rotación.

Axial load A type of load placed on a bearing that is parallel to the axis of the rotating shaft.

Carga axial Un tipo de carga puesto en un cojinete paralelo al eje de una flecha que gira.

Axial play Movement that is parallel to the axis or rotation.

Holgura axial El movimiento que es paralelo al eje o a la rotación.

Back pressure Pressure created by restriction in an exhaust system.

Contrapresión La presión creada por una restricción en un sistema de escape.

Backlash The clearance or play between two parts, such as meshed gears.

Juego La cantidad de holgura o juego entre dos partes, tal como los engranajes endentados.

Balancer shaft A weighted shaft used on some engines to reduce vibration.

Eje equilibrador Una flecha con un peso usado en algunos motores para disminuir la vibración.

Balancing coil gauge A type of gauge that utilizes coils to create magnetic fields instead of a permanent magnet.

Bobina equilibrada Un tipo de manómetro que utiliza las bobinas para crear un campo magnético en vez de un imán permanente.

Ball bearing An antifriction bearing that uses a series of steel balls held between inner and outer bearing races.

Cojinete de bolas Un cojinete de antifricción que usa una serie de bolas del acero sujetados entre pistas interiores e exteriores.

Ball joint A pivot point for turning a front wheel to the right or left. Ball joints can be considered either nonloaded or loaded when carrying the car's weight.

Unión esférica Un punto de pivote en el cual gira una rueda delantera hacia la derecha o a la izquierda. Las uniones esféricas se pueden considerar como sin carga o con carga al sostener el peso del coche.

Barometric pressure A sensor or its signal circuit that sends a varying frequency signal to the processor relating actual barometric pressure.

Presión barométrico Un sensor o su circuito de señal que manda un señal de frecuencia variada al procesor refiriendo la presión barométrico actual.

Base A solution that has an excess of OH ions, also called an alkali. Also the center layer of a bipolar transistor.

Base Una solución que tiene un exceso de iones del OH, también llamada un álcali. También la capa central de un transistor bipolar.

Battery A device for storing energy in chemical form so it can be released as electricity.

Batería Un dispositivo para almacenar energía en forma química para que se pueda liberar como electricidad.

Battery cable A heavy electrical conductor used to connect a vehicle's battery to the starter motor or chassis ground.

Cable de batería Un conductor eléctrico sólido que se usa para conectar la batería de un vehículo al motor de arranque o la masa del chasís (tierra).

Battery cell That part of a storage battery made from two dissimilar metals and an acid solution. A cell stores chemical energy for use later as electrical energy.

Célula de batería Aquella parte de una batería acumuladora hecha de dos metales desemejantes y una solución de ácido. Una célula almacena la energía química para usarse después como energía eléctrica.

BDC Bottom dead center, position of the piston.

BDC Punto muerto exterior, posición del pistón.

Bead The edge of a tire's sidewall, usually made of steel wires wrapped in rubber, used to hold the tire to the wheel.

Ceja La extremidad del refuerzo lateral de un neumático, suele ser hecho de alambres de acero cubiertos en caucho, que sujeta el neumático a la rueda.

Bearing Soft metallic shells used to reduce friction created by rotational forces.

Cojinete Una pieza hueca de metal blanda que sirve para reducir la fricción creada por las fuerzas giratorias.

Bearing clearance The amount of space left between a shaft and the bearing surface for lubricating oil to enter.

Holgura del cojinete La cantidad del espacio que se deja entre la flecha y la superficie portante por la cual entra el lubricante.

Bearing crush The process of compressing a bearing into place as the bearing cap is tightened.

Aplastamiento del cojinete El proceso de colocar un cojinete comprimiéndolo cuando se aprieta la tapa del cojinete.

Bearing race The machined circular surface of a bearing against which the roller or ball bearings ride.

Pista del cojinete La superficie circular maquinada del cojinete en la cual ruedan los rodillos o bolas.

Bearing spread The condition in which the distance across the outside parting edges of the bearing insert is slightly greater than the diameter of the housing bore.

Aplastamiento del cojinete La condición en la cual la distancia de las extremidades exteriores de la pieza inserta del cojinete es un poco más grande que el diámetro de la caja.

Bell housing A term often used for clutch housing.

Cárter del embrague Un término que se usa con frecuencia para la caja del embrague.

Bellows A movable cover or seal, usually of rubberlike material, that is pleated or folded like an accordion to allow for expansion and contraction.

Fuelle Una cubierta o un sello removible, normalmente de una materia parecida al caucho, que es plegado o doblado como un acordión para permitir la expansión y contracción.

Bias A diagonal line of direction. In relationship to tires, bias means that belts and plies are laid diagonally or crisscrossing each other.

Bies Una linea de dirección diagonal. Perteneciente a los neumáticos, bies quiere decir que las bandas y los pliegues se colocan diagonalmente y se cruzán entre sí.

Bimetal Two dissimilar metals with different expansion rates used to switch a circuit on and off.

Aleación bimetálica Dos metales desemejantes que se expanden en una tasa diferente que sirven para prender y apagar un circuito.

Binary code A series of numbers represented by 1's and 0's or offs and ons.

Código binario Una serie de números representados por los 1s y los 0s o por apagado y prendido.

Blowby The unburned fuel and products of combustion that leak past the piston rings and into the crankcase during the last part of the combustion stroke.

Fuga El combustible no consumido y los productos de la combustión que se escapan alrededor de los anillos de los pistones y entran al cárter en las últimas etapas de la carrera de combustión.

Blower A fan that pushes or blows air through a ventilation heater or air-conditioning system.

Ventilador Un ventilador que empuja o sopla el aire por un ventilador, un calentador o un sistema de aire acondicionado.

Body-over-frame Type of vehicle construction in which the frame is the foundation, with the body and all major parts of the vehicle attached to it.

Carrocería sobre armazón Un tipo de construcción de vehículos en el cual el armazón es la fundación, al cual se conectan la carrocería y los componentes principales del vehículo.

Bolt diameter The measurement across the major diameter of a bolt's threaded area or across the bolt shank.

Diámetro del perno La medida del diámetro mayor de la parte fileteada de un perno o atravéz del asta del perno.

Bolt head The part of a bolt that the socket or wrench fits over in order to torque or tighten the bolt.

Cabeza del perno La parte del perno sobre la cual se pone el dado o la llave para apretar o torcer al perno.

Boot Rubber protective cover with accordion pleats used to contain lubricants and exclude contaminating dirt, water, and grime. Located at each end of rack and pinion assembly and front-wheel-drive CV joints.

Protectores de caucho Una cubierta protectora hecho de caucho que tiene pliegues estilo acordión que sirve para contener las lubricantes y prevenir la entrada de los contaminantes como el lodo, el agua y el mugre. Se ubican en cada extremidad de la asamblea de piñon y cremallera.

Bore A dimension of cylinder size representing the diameter of the cylinder.

Diámetro del orificio Una dimensión de tamaño del cilindro que representa el diámetro del cilindro.

Bourdon tube A flexible hollow tube that uncoils and straightens as the pressure inside it is increased. A needle attached to the Bourdon tube moves over a scale to indicate the oil pressure.

Tubo bourdon Un tubo hueco y flexible que se desenreda y se endereza al aumentar el presión en su interior. Una aguja conectada al tubo bourdon mueve por encima de una escala para indicar la presión del aceite.

Bracket An attachment used to secure parts to the body or frame.

Soporte Una fijación que se usa para enganchar las parte a la carrocería o al armazón.

Brake band A circle-shaped part lined with friction material that acts as a brake or holding device to stop and hold a rotating drum that has a gear train member connected to it.

Banda de freno Una parte circular recubierta con materia fricativa que sirve para frenar o como un dispositivo de asir para detener y sostener un tambor giratorio al cual esta conectado un miembro del tren de engrenaje.

Brake drum A bowl-shaped cast-iron housing against which the brake shoes press to stop its rotation.

Tambor de freno Un alojamiento de hierro colado en forma de tazón contra el que se oprimen las zapatas de los frenos para detener su rotación.

Brake fade Occurs when friction surfaces become hot enough to cause the coefficient of friction to drop to a point where the application of severe pedal pressure results in little actual braking.

Amortiguamiento de frenar Ocurre cuando la superficie de fricción se sobrecalienta al punto de causar que caiga el coeficiente de fricción a tal punto que la aplicación de presión riguroso en el pedal de frenos resulta en muy poco enfrenamiento.

Brake fluid A hydraulic fluid used to transmit force through brake lines. Brake fluid must be noncorrosive to both the metal and rubber components of the brake system.

Líquido para frenos Un líquido hidráulico que se usa para transmitir potencia por las líneas de frenos. Dicho líquido debe ser no corrosivo para los componentes tanto metálicos como de caucho del sistema de frenos.

Brake pads The part of a disc brake system that holds the linings.

Tacos de presión Las piezas de un sistema de frenado que sostienen las guarniciones (balatas).

Brake rotor Disc-shaped component that revolves with hub and wheel. The lining pads are forced against the rotor to provide a friction surface for the brake system, so as to slow or stop a vehicle.

Rotor de freno Componente en forma de disco que gira con el cubo y la rueda.

Brake shoe The metal assembly onto which the frictional lining is attached for drum brake systems.

Zapata de freno El ensamblaje metálico sobre el que se sujeta la guarnición de fricción (balata) para los sistemas de freno de tambor.

Breaker points A set of contact points used in the ignition system to open and close the primary circuit. The points act like a switch operated from a distributor cam.

Contactos de interruptor Un conjunto de contactos utilizados en el sistema de encendido para abrir y cerrar el circuito primario. Los puntos actúan como un interruptor activado mediante una leva del distribuidor.

British thermal unit (Btu) A measurement of the amount of heat required to raise the temperature of 1 pound of water 1 degree Fahrenheit.

Unidad térmica Británica Una medida de la cantidad del calor que se requiere para aumentar la temperatura de 1 libra de agua por 1 grado fahrenheit.

BTDC Before top dead center, position of the piston.

BTDC Antes del punto muerto interior Una posición del pistón.

Buffer Any device used to reduce the shock or motion of opposing forces.

Parachoques Qualquier dispositivo que sirve para disminuir los choques o el movimiento de las fuerzas opuestas.

Bump steer Erratic steering that is caused from rolling over bumps, cornering, or heavy braking; same as orbital steer and roll steer.

Dirección de choques La dirección errática que se causa al pasar por encima de los topes, a dar la vuelta, o en el enfrenamiento violento; quiere decir lo mismo que la dirección orbital y la dirección de inclinación.

Burnish To smooth or polish with a sliding tool under pressure.

Bruñir Pulir o suavizar por medio de una herramienta deslizandola bajo presión.

CAFE standards Law requiring auto makers to not only manufacture clean-burning engines but also to equip vehicles with engines that burn gasoline efficiently.

Normas CAFE El ley requeriendo que los fabricantes de automóviles no sólo fabrican los motores de combustión limpia sino que tambien equipan los vehículos con los motores que consumen la gasolina en una manera eficiente.

Caliper Major component of a disc brake system. Houses the piston(s) and supports the brake pads.

Articulación Uno de los componentes principales de un sistema de freno de disco. Contiene el o los pistones y sostiene los tacos de presión.

Camber The attitude of a wheel and tire assembly when viewed from the front of a car. If it leans outward, away from the car at the top, the wheel is said to have positive camber. If it leans inward, it is said to have negative camber.

Camber (comba) La actitud de una asamblea de la rueda y el neumático al verse desde la frente del coche. Si se inclina hacia afuera, se dice que la rueda tiene una comba positiva. Si se inclina hacia el interior, se dice que tiene una comba negativa.

Camshaft The component in the engine that opens and closes the valves.

Árbol de levas El componente en el motor que abre y cierra las válvulas.

Candlepower A measurement of the brightness of light.

Unidad de intensidad luminosa Una medida del brillantez de la luz.

Capacitor A device for hólding and storing a surge of current.

Capacitor Un dispositivo que sostiene y almacena una sobretensión de corriente.

Cap screw A hex head bolt with a flange area above the threads.

Tornillo de cabeza (casquete) Un tornillo de cabeza hexagonal con una zona de reborde por encima de la rosca.

Carbon dioxide Compressed into solid form, this material is known as dry ice and remains at a temperature of –109 degrees. It goes directly from a solid to a vapor state.

Anhídrido carbónico Comprimido para formar un sólido, esta materia se conoce como hielo seco y sostiene una temperatura de –109 grados F. Va directamente del estado sólido a un vapor.

Carbon monoxide Poisonous gas formed in engine exhaust.

Óxido de carbono Un gas venenoso producido en el escape de un motor.

Carbonize The process of carbon formation within an engine, such as on the spark plugs and within the combustion chamber.

Carbonizar El proceso de la formación del carbono en un motor, tal como en las bujías y en el interior de la cámara de combustión.

Carburetor A fuel delivery device that mixes fuel and air to the proper ratio to produce a combustible gas.

Carburador Un dispositivo para entregar el combustible que mezcla el combustible y el aire en relaciones apropiadas correctas, para producir un gas combustible.

Carburizing A method used to surface-harden steel by heat or mechanical means to increase the hardness of the outer surface while leaving the core relatively soft.

Carburación Un método usado para endurecer la superficie del acero por el calor o medios mecánicos para aumentar la dureza de la superficie externa mientras que deja la base relativamente suave.

Case harden To harden the surface of steel.

Cementar Endurecer la superficie del acero.

Castellated nut A nut with slots through which a cotter pin can be passed to secure the nut to its bolt or stud.

Tuerca con entallas Una tuerca que tiene muescas por las cuales se puede insertar un pasador de chaveta para retener la tuerca al perno o a la espiga.

Caster Angle formed between the kingpin axis and a vertical axis as viewed from the side of the vehicle. Caster is considered positive when the top of the kingpin axis is behind the vertical axis.

Angulo de caster El ángulo formado entre la linea de pivote y un eje vertical al verse del lado del vehículo. El caster se considera positivo cuando la extremidad superior de la linea de pivote es atrás del eje vertical.

Catalyst A compound or substance that can speed up or slow down the reaction of other substances without being consumed itself. In an automatic catalytic converter, special metals (for example, platinum or palladium) are used to promote more complete combustion of unburned hydrocarbons and a reduction of carbon monoxide.

Agente catalítico Una compuesta o una substancia que puede acelerar o retardar la reacción de otras substancias sin ser consumida. En un convertidor catalítico automático, se usan metales especiales (por ejemplo, el platino y el paladio) para promover la combustión completa del hidrocarburo sobrante y para reducir el óxido de carbono.

Catalytic converter An emission device located in front of the muffler in the exhaust system. It looks very much like a heavy muffler and contains catalysts to clean up an engine's emissions before they leave the end of the exhaust pipe.

Convertidor catalítico Un dispositivo de emisiones ubicado en la parte delantera del silenciador en el sistema del escape. Se parece mucho a un silenciador de trabajo pesado y contiene loscatalizadores para limpiar las emisiones del motor antes de que salgan del tubo de escape.

Caustic Something that causes corrosion.

Cáustico Algo que causa la corrosión.

CC-ing The process of measuring a volume in cubic centimeters, commonly used to measure combustion chamber size.

Medida en CC El proceso de medir un volumen en centímetros cúbicos, suelen usarse al medir el tamaño de la cámara de combustión.

C-class oil Commercial engine oil designed for use in heavy-duty commercial applications.

Aceite tipo C El aceite del motor comercial que se diseñó para usarse en las aplicaciones comerciales de trabajo pesado.

Center link A steering linkage component connected between the pitman and idler arm.

Varilla central Un componente de enlace de dirección que conecta entre el brazo Pitman y el brazo loco.

Center of gravity The point about which the weight of a car is evenly distributed; the point of balance.

Centro de gravedad El punto en el cual el peso del coche es distribuido en una manera uniforme; el punto del equilibrio.

Centrifugal force A force tending to pull an object outward when it is rotating rapidly around a center.

Fuerza centrífuga Una fuerza que tiene la tendencia de jalar un objeto hacia afuera cuando éste gira rápidamente alrededor de un punto central.

Centripetal force A force that acts on something and keeps it in a circular motion by pulling it towards the center of the circle.

Fuerza centrípeta Una fuerza que actua sobre algo y lo mantiene moviendo circularmente tirandolo hacia el centro del círculo.

CFC Chlorofluorocarbon; CFCs are depleting the protective ozone layer through a chemical reaction.

Clorofluorocarburo Compuesto que está agotando la capa protectora de ozono mediante una reaccióin química.

Chaffing Damage caused by friction and rubbing.

Rozamiento Los daños incurridos por la fricción y el frotamiento.

Chamfer A bevel or taper at the edge of a hole.

Chaflán Un bisel o ahusamiento en la orilla de un hoyo.

Chamfering The process of removing the sharp edges around a bore or hole.

Achaflanar El proceso de quitar las orillas afiladas alrededor de un orificio o un hoyo.

Charcoal canister A small plastic or steel container filled with activated charcoal that can store gasoline vapors until the right time for them to be drawn into the engine and burned.

Bote de carbón vegetal Un pequeño recipiente de plástico o acero, lleno de carbón activado, que puede almacenar vapores de gasolina hasta el momento adecuado para su introducción al motor y su combustión.

Chase To straighten or repair damaged threads.

Embutir Enderezar o reparar los filetes dañados.

Check valve A gate or valve that allows passage of gas or fluid in one direction only.

Valvula de seguridad Una válvula que permite fluir la presión en un sólo sentido.

Chemical cleaning Cleaning that relies primarily on some type of chemical action to remove dirt, grease, scale, paint, or rust.

Limpieza química La limpieza que cuenta primariamente con la acción química para remover el lodo, la grasa, las incrustaciones, la pintura, o los óxidos.

Cherry picker A mobile crane that uses hydraulic power.

Burro Una grúa rodante que usa la energía hidráulica.

Chip A term commonly used for an integrated circuit that is small in size.

Pastilla Un término común que se refiere a circuito integrado que es de un tamaño pequeño.

Closed loop An electronic feedback system in which the sensors provide constant information on what is taking place in the engine.

Bucle cerrado Un sistema de reacción electrónico en el cual los detectores proveen información constante de lo que esta pasando en el interior del motor.

Closed-cooling systems Cooling systems with expansion tanks designed to hold any coolant that passes through the pressure cap when the engine is hot.

Sistema de enfriamiento cerrado Los sistemas de enfriamiento que tienen tanques de expansión diseñados para contener a cualquier flúido refrigerante que pasa por la tapa de presión al calentarse el motor.

Clutch An electromechanical device mounted on the air-conditioning compressor used to start and stop compressor action, thereby controlling refrigerant circulating through the system.

Embrague Un dispositivo electromecánico montado en el compresor del aire acondicionado, utilizado para comenzar y detener la acción del compresor, regulando así la circulación del refrigerante a través del sistema.

Clutch cover A term used by some manufacturers to describe a pressure plate.

Tapa del embrague Un término empleado por algunos fabricantes para describir una placa de presión.

Clutch disc The part of a clutch that receives the driving motion from the flywheel and pressure plate assembly and transmits that motion to the transmission input shaft.

Disco del embrague La parte de un embrague que recibe el movimiento propulsor del volante y la asamblea del plato opresor y transmite esa acción a la flecha de entrada de la transmisión.

Clutch fork A forked lever that moves the clutch release bearing and hub back and forth.

Horquilla del embrague Una palanca bifurcada que mueve al balero collarín del embrague hacia afrente y hacia atrás.

Clutch release bearing A sealed, prelubricated ball bearing that moves the pressure plate release levers or diaphragm spring through the engagement and disengagement process.

Collarín del embrague Un cojinete esférico, prelubricado y sellado, que mueve las palancas de liberación del plato opresor o el resorte diafragma por medio del proceso del enganche y el desenganche.

Coefficient of friction A relative measurement of the friction developed between two objects in contact with each other.

Coeficiente de la fricción La medida relativa de la fricción que ocurre entre dos superficies que estan en contacto la una con la otra.

Coil A term often used to describe a spring or ignition coil.

Bobina Término utilizado con frecuencia para describir un resorte o un devanado de encendido.

Coil pack The name commonly used to describe the assembly of multiple ignition coils on an engine with distributorless ignition.

Paquete de bobinas Término utilizado normalmente para describir el conjunto de varias bobinas en un motor con encendido sin distribuidor.

Coil spring pressure plate assembly An assembly that uses weighted levers to increase or decrease the force holding the clutch disc against the flywheel according to engine speed.

Asamblea de muelle y plato opresor Una asamblea que usa las palancas con pesos para incrementar o disminuir a la fuerza que aprieta el disco del embrague contra el volante según la velocidad del motor.

Collector The portion of a bipolar transistor that receives the majority of electrical current.

Colector La porción de un transistor bipolar que recibe la mayoría del corriente eléctrico.

Combination valve An H-valve, used in some early air-conditioning systems, combining a suction throttling valve and an expansion valve.

Válvula combinada Una válvula en H utilizada en algunos sistemas iniciales de climatización (aire acondicionado), que combina una válvula reguladora de succión y otra de expansión.

Combustion chamber The space between the top of a piston and the cylinder head where the engine's combustion takes place.

Cámara de combustión El espacio entre la parte superior del pistón y la cabeza del cilindro en donde se lleva acabo la combustión.

Compound A mixture of two or more ingredients.

Compuesta Una mezcla de dos o más ingredientes.

Compression The act of reducing volume by pressure.

Compresión La acción de reducir el volumen por efectos de la presión.

Compression ratio The ratio of the volume in the cylinder above the piston when the piston is at bottom dead center to the volume in the cylinder above the piston when the piston is at top dead center.

Relación índice de compresión La relación del volumen en el cilindro arriba del piston cuando el piston esta en el punto muerto inferior al volumen en el cilindro arriba del pistón cuando el piston esta en el punto muerto superior.

Compression stroke The second stroke of the four-stroke engine cycle, in which the piston moves from bottom dead center and the intake valve closes. This traps and compresses the air/fuel mixture in the cylinder.

Carrera de compresión La segunda carrera de un ciclo de un motor de cuatro tiempos, en la cual el pistón mueve del punto muerto inferior y se cierre la válvula de admisión. Esto encierre y comprime la mezcla de aire/combustible en el cilindro.

Concentric Two or more circles having a common center.

Concéntrico Dos círculos o más que comparten un centro común.

Condensation The process of a vapor becoming a liquid; the reverse of evaporation.

Condensación El proceso por el cual un vapor se convierte en un líquido; lo opuesto de la evaporación.

Condense To cool a vapor to below its boiling point. The vapor condenses into a liquid.

Condensar El proceso de enfriar un vapor a una temperatura más baja de su punto de ebullición. El vapor condensa, formando un líquido.

Condenser A capacitor made from two sheets of metal foil separated by an insulator.

Condensador Es un capacitador construido de dos láminas de metal separadas por un aislante.

Conduction The movement of heat through a material.

Conducción El movimiento de calor a través de un material.

Conductor A device that readily allows for current flow.

Conductor Un dispositivo que permite fácilmente el flujo del corriente.

Connecting rod The link between the piston and crankshaft.

Biela La acoplación entre el pistón y la cigüeñal.

Continuous injection system A system that uses fuel under pressure to modulate or change the fuel injection area.

Sistema de inyección continua Un sistema que usa al combustible bajo presión para moderar o cambiar el area del inyección de combustible.

Contraction A reduction in mass or dimension; the opposite of expansion.

Contracción Una reducción en la masa o en la dimensión; lo opuesto de la expansión.

Control arms Suspension parts that control coil spring action as a wheel is affected by road conditions.

Palanca de comando Las partes de suspención que controlan la acción del muelle de control al afectarse una rueda por las condiciones de la carretera.

Convection The transfer of heat by the movement of a heated object.

Conducción El transferimiento de calor a traves del movimiento de un objeto que ha sido calentado.

Convection oven Thermal cleaning oven that uses indirect heat to brake parts.

Horno de convección Un horno termal de limpieza que aplica el calor indirecto a las partes de los frenos.

Coolant A mixture of water and ethylene glycol-based antifreeze that circulates through the engine to help maintain proper temperatures.

Fluido refrigerante Una mezcla del agua con el anticongelante a base de glicol etileno que circula por el motor y ayuda en mantener las temperaturas indicadas.

Cords The inner materials running through the plies that produce strength in the tire. Common cord materials are fiberglass and steel.

Cuerdas Las materiales interiores recorriendo por los pliegues que producen la solidez del neumático. Las materiales más comunes para las cuerdas son la fibra de vidrio y el acero.

Core The center of the radiator, made of tubes and fins, used to transfer heat from the coolant to the air.

Núcleo El centro de un radiador, hecho de tubos y aletas, que sirve para transferir el calor del líquido refrigerante a la atmósfera.

Core plug A plug used to seal openings where necessary for casting the object in a cast structure.

Tapón del núcleo Un tapón diseñado para sellar las aperturas necesarias para fundar el objeto en un molde.

Corrosivity The characteristic of a material that enables it to dissolve metals and other materials or burn the skin.

Corrosidad La característica de una materia que permite que disuelva a los metales u otras materias o quema al piel.

Counterbore To enlarge a hole to a given depth.

Ensanchar con taladro Aumentar el tamaño de un hoyo a la profundidad indicada.

Countersink To cut or form a depression to allow the head of a screw to go below the surface.

Avellenar Cortar o formar una depresión bastante grande para permitir que la cabeza de un tornillo queda abajo de la superficie.

Counterweight Weight forged or cast into the crankshaft to reduce vibration.

Contrapeso Un peso que se ha forjado o colado al cigueñal para reducir la vibración.

Crank pin The machined, offset area of a crankshaft where the connecting rod journals are machined.

Espiga de manívela La superficie maquinada, descentrada de un cigueñal en donde los muñones de la biela se maquinan.

Crank throw The distance from the crankshaft main bearing centerline to the connecting rod journal centerline. The stroke of any engine is the crank throw.

Carrera La distancia del eje de quilla del cojinete principal del cigueñal al eje de quilla del muñon de la biela. Una carrera del pistón de cualquier motor es una carrera.

Crank web The unmachined portion of a crankshaft that lies between two crank pins or between a crank pin and main bearing journal.

Alma de cigueñal La porción no maquinada de un cigueñal que queda entre dos espigas de manívela o entre una espiga de manívela y el muñon del cojinete principal.

Crankshaft A rotating component mounted in the lower side of the block that changes vertical piston motion to rotary motion.

Cigüeñal Un componente giratorio montado en la parte inferior del bloque que convierte el movimiento vertical del pistón en movimiento giratorio.

Crimp The use of pressure to force a thin holding part to clamp to, or conform to the shape of, a part so it cannot move.

Engarzar El uso de la presión en forzar una parte delgada de sujeción para agarrar, o conformar a la forma de, una parte sujetada.

Crossflow radiator A radiator in which coolant enters on one side, travels through tubes, and collects on the opposite side.

Radiador de flujo horizontal Un radiador en el cual el líquido refrigerante entra de un lado, viaja por los tubos, y se colecta en el lado opuesto.

Crude oil The natural state of oil as it is pulled from the earth and before it is refined.

Aceite crudo El estado natural del aceite cuando sale de la tierra y antes de que sea refinado.

Current The number of electrons flowing past a given point in a given amount of time.

Corriente El número de los electronos que fluyen al través de un punto específico en un dado periódo del tiempo.

CV joint Constant velocity joint.

Junta CV Junta de velocidad continua.

Cylinder A circular tubelike opening in a compressor block or casting in which the piston moves up and down or back and forth; a circular drum used to store refrigerant.

Cilindro Una apertura circular parecida a un tubo en un bloque del compresor o una pieza en los que el pistón se mueve de arriba hacia abajo o de un lado a otro; un tambor circular utilizado para el almacenaje de anticongelante.

Cylinder bore dial gauge Tool used to determine cylinder bore size, out-of-round, and taper.

Calibrador del orificio del cilindro Una herramienta que se usa para determinar el tamaño, lo ovalado y lo cónico de un cilindro.

Cylinder head On most engines the cylinder head contains the valves, valve seats, valve guides, valve springs, and the upper portion of the combustion chamber.

Cabeza de los cilindros En la mayoría de los motores, la cabeza de los cilindros contiene las válvulas, asientos de las válvulas, sus guías, sus resortes y la parte superior de la cámara de combustión.

Dampen To slow or reduce oscillations or movement.

Amortiguar Retardar o reducir las vibraciones o el movimiento.

Dead axle An axle that does not rotate but merely forms a base on which to attach the wheels.

Eje muerto Un eje que no gira sino solamente forma una plataforma en la cual se puede conectar las ruedas.

Dead center The extreme upper or lower position of the crankshaft throw at which the piston is not moving in either direction.

Punto muerto La posición extrema superior o inferior de una carrera de cigueñal en la cual el pistón no mueve en ningúna dirección.

Deceleration The rate of decrease in speed.

Desaceleración El índice de la disminución de la velocidad.

Deck The top of the engine block where the cylinder head mounts.

Cubierta del monoblock La parte superior de un bloque del motor en donde se conecta la culata.

Deflection Bending or movement away from normal due to loading.

Desviación Curvación o movimiento fuera de lo normal debido a la carga.

Deflection angle The angle at which the oil is deflected inside the torque converter during operation. The greater the angle of deflection, the greater the amount of torque applied to the output shaft.

Ángulo de desviación El ángulo en el que se desvia el aceite en el convertidor de par durante su operación. Cuanto más el ángulo de desviación, más torsión se aplica en la flecha de potencia.

Density Compactness; relative mass of matter in a given volume.

Densidad La firmeza; la masa relativa de la materia en un volumen indicado.

Desiccant A special substance that absorbs moisture.

Desicante Una substancia especial que absorbe la humedad.

Detergent A compound of soaplike nature used in engine oil to remove engine deposits and hold them in suspension in the oil.

Detergente Una compuesta parecida al jabón que se usa en el aceite de motor para quitar los depósitos del motor y mantenerlos suspendidos en el aceite.

Detonation As used in an automobile, indicates a hasty burning or explosion of the mixture in the engine cylinders. It becomes audible through a vibration of the combustion chamber walls and is sometimes confused with a ping or spark knock.

Detonación Aplicado al automóvil, indica una combustión apresurada o una explosión de la mezcla en los cilindros del motor. Se oye por medio de las vibraciones de los paredes de la cámara de combustión y a veces se confunde con un golpeteo o una detonación de las bujías.

Dial caliper Versatile measuring instrument capable of taking inside, outside, depth, and step measurements.

Calibre de carátula (pie de rey) Un instrumento de medir versátil que es capable de tomar las medidas interiores, exteriores, de profundidad, y de paso a paso.

Dial indicator A measuring tool used to adjust small clearances up to 0.001 inch. The clearance is read on a dial.

Calibrador de carátula Una herramienta de medida que sirve para ajustar las pequeñas holguras hasta el 0.001 de una pulgada. La holgura se lee en un cuadrante.

Diaphragm A flexible, impermeable membrane on which pressure acts to produce mechanical movement.

Diafragma Una membrana flexible e impermeable sobre la cual tiene un efecto la presión para producir un movimiento mecánico.

Diaphragm spring pressure plate assembly An assembly that uses weighted levers to increase or decrease the force holding the clutch disc against the flywheel according to engine speed.

Asamblea del resorte diafragma del plato opresor Una asamblea que usa las palancas con pesos para incrementar o reducir la fuerza que aprieta el disco de embrague contra el volante según la velocidad del motor.

Dielectric An insulator material.

Dieléctrico Una material de aislador.

Differential A gear assembly that transmits power from the drive shaft to the wheels and allows two opposite wheels to turn at different speeds for cornering and traction.

Diferencial Una asamblea de engrenajes que transmite la potencia del árbol de mando a las ruedas y permite que dos ruedas opuestas giran a velocidades distinctos para ejecutar las vueltas y la tracción.

Diffusion The random movement of gas particles; also ensures that any two gases sharing the same container will totally mix.

Difusión El movimiento al azar de las partículas de gas también asegura que cualquier dos gases que comparten el mismo envase se mezclarán totalmente.

Digital A voltage signal that is pulsed on or off.

Digital Un señal del voltaje que pulsa prendida o apagada.

Dilution A thinner or weaker solution. Oil is diluted by the addition of fuel and water droplets.

Dilución Desleir o debilitar. El aceite se diluye por la agregación de las gotas del combustible o del agua.

Diode A simple semiconductor device that permits flow of electricity in one direction but not in the opposite direction.

Diodo Un dispositivo sencillo de semiconductor que permite el flujo de la electricidad en una dirección pero no en la dirección opuesta.

Direct current (DC) A type of electrical power used in mobile applications. A unidirectional current of substantially constant value.

Corriente continua Un tipo de potencia eléctrica utilizada en aplicaciones movibles. Una corriente que fluye en un solo sentido de un valor substancialmente constante.

Direct drive The downward gear engagement in which the input shaft and output shaft are locked together.

Transmisión directa El enganchamiento en descenso del engrenaje en el cual la flecha de entrada y la flecha de producción se enclavan.

Direct ignition A distributorless ignition system in which spark distribution is controlled by the vehicle's computer.

Encendido directo Un sistema del encendido sin distribuidor en el cual la distribución de la chispa se controla por medio de la computador del vehículo.

Direct injection A diesel fuel injection system in which fuel is injected directly onto the top of the piston.

Inyección directa Un sistema de inyección de diesel en el cual el combustible se inyecta directamente arriba del pistón.

Directional stability The ability of a car to travel in a straight line with a minimum of driver control.

Estabilidad direccional La abilidad de un coche de viajar en una linea recta con control mínimo del conductor.

Disable A type of microcomputer decision that results in an automotive system being deactivated and not permitted to operate.

Incapacitar Un tipo de decisión de la microcomputadora que resulta en que un sistema automotríz se deactiva y no permite que se opera.

Disc brakes Brakes in which the frictional forces act on the faces of a disc.

Frenos de disco Frenos en los que las fuerzas de fricción se aplican sobre las caras de un disco.

Discharge line Connects the compressor outlet to the condenser inlet.

Línea de descarga La que conecta la salida del compresor a la entrada del condensador.

Displacement The volume the cylinder holds between the top dead center and bottom dead center positions of the piston.

Desplazamiento (cilindrada) El volumen que contiene el cilindro entre la posiciones del punto muerto superior y el punto muerto inferior del pistón.

Distributor The mechanism within the ignition system that controls the primary circuit and directs the secondary voltage to the correct spark plug.

Distribuidor El mecanismo dentro del sistema de ignición que controla el circuito primario y dirige el voltaje secundario a las bujías correctas.

Distributor cap and rotor A mechanical device that distributes the ignition sparks from the coil secondary winding to the spark plugs.

Tapa y rotor del distribuidor Un dispositivo mecánico que distrubuye las chispas de la ignición desde el bobinado secundario hasta las bujías.

Distributor ignition (DI) system SAE J1930 terminology for an ignition system with a distributor.

Sistema de la ignición con distribuidor El término utilizado por la SAE J1930 para referirse a un sistema de la ignición que tiene un distribuidor.

DOHC Dual Overhead Camshaft.

DOHC Arbol de levas doble superior.

Dome Typically refers to the shape of the top of a piston.

Fondo hemisférico Típicamente se refiere al contorno de la parte superior de un pistón.

DOT U.S. Department of Transportation.

DOT Departamento de Transportes de los Estados Unidos de América.

Dowel A pin extending from one part to fit into a hole in an attached part; used for both location and retention.

Espiga Una clavija que extiende de una parte para quedarse en el hoyo de otra parte conjunta; se usa para la localización y la retención.

Downflow radiator A radiator in which coolant enters the top of the radiator and is drawn downward by gravity.

Radiador de flujo vertical Un radiador en el cual el líquido refrigerante entra en la parte superior del radiador y fluye hacia abajo atraído por la gravedad.

Drag link The steering component that connects the Pitman arm to the steering arm.

Contrabrazo Componente de dirección que conecta la biela a la columna de dirección.

Drive member A gear that drives, or provides power for, other gears in a planetary gearset.

Miembro de ataque Un engrenaje que impulsa, o provee la potencia para los otros engrenajes en un conjunto planetario.

Driveability The degree to which a vehicle operates properly. Includes starting, running smoothly, accelerating, and delivering reasonable fuel mileage.

Manejo El punto en el cual un vehículo opera correctamente. Incluye el encendido, marchar sin averías, la aceleración y la entregada de un kilometraje económico.

Driveshaft A hollow metal tube that has a universal joint at each end.

Eje motriz Un tubo metálico hueco que tiene una junta universal en cada extremo.

Dropping resistor A device that reduces the battery voltage.

Resistencia para bajar la tensión Un dispositivo que disminuye el voltaje de la bateria.

Duo-servo A drum brake design with increased stopping power due to the servo or self-energizing effect of the brake.

Frenos tipo servo Un diseño de freno de tambor que tiene más potencia de enfrenar debido al servo o al efecto autoenergético del freno.

Duty cycle The percentage of on-time to total cycle time of fuel injectors.

Ciclo del trabajo El porcentaje del tiempo de trabajar al tiempo del ciclo total de los inyectores del combustible.

Dwell time The degree of crankshaft rotation during which the primary circuit is on.

Ángulo de cierre El grado de rotación del cigueñal en el cual esta encendido el circuito primario.

Dynamic Refers to balance when the object is in motion.

Dinámico Refiere al balance mientras que un objeto esté en movimiento.

Dynamic pressure The pressure of a fluid while it is in motion.

Presión dinámica La presión de un líquido mientras que está en movimiento.

Eccentric The part of a camshaft that operates the fuel pump.

Eccéntrico La parte de un árbol de levas que opera la bomba del combustible.

Efficiency A ratio of the amount of energy put into an engine as compared to the amount of energy coming out of the engine; a measure of the quality of how well a particular machine works.

Eficiencia Un relación de la cantidad de energía que consume un motor comparado con la cantidad de energía que produce el motor; una medida de la calidad de la máquina y su marcha.

EGR The exhaust gas recirculation system that allows a small amount of exhaust gas to be routed into the incoming air/fuel mixture to reduce NO_x emissions.

EGR Sistema de recirculación de gas del escape Permite que se dirija una cantidad pequeña del gas del escape a la mezcla de admisión de aire y combustible para reducir descargas (emisiones) de NO_x.

EI The J1930 acronym for a distributorless ignition system.

EI La abreviación J1930 para un sistema de la ignición sin distrubuidor.

Elasticity The principle by which a bolt can be stretched a certain amount. Each time the stretching load is reduced, the bolt returns back to its exact, original, normal size.

Elasticidad El principio por el cual un perno puede extenderse por un cierta cantidad. Cada vez que la carga tensor se reduce, el perno regresa exactamente a su tamaño normal y original.

Electrochemical The chemical action of two dissimilar materials in a chemical solution.

Electroquímico El acción químico de dos materiales desemejantes en una solución química.

Electrode The firing terminals found in a spark plug.

Electrodo Los terminales de chispeo que se encuentran en una bujía.

Electrolysis A chemical and electrical decomposition process that can damage metals, such as brass, copper, and aluminum, in the cooling system.

Electrólisis (deposición) Un proceso de decomposición químico y eléctrico que puede dañar a los metales, tales como el latón, el cobre y el aluminio, en el sistema de enfriamiento.

Electrolyte A material that has atoms that become ionized, or electrically charged, in solution. Automobile battery electrolyte is a mixture of sulfuric acid and water.

Electrólito Una materia que tiene átomos que se han ionizado, o que tienen una carga eléctrica, en una solución. El electrólito para las baterías automotrices es una mezcla del ácido sulfúrico y el agua.

Electromagnet A magnet formed by electrical flow through a conductor.

Electroimán Un imán que se produce por un corriente eléctrico fluyendo por un conductor.

Electromagnetic induction Moving a wire through a magnetic field to create current flow in the wire.

Inducción electromagnético Crear un corriente eléctrico en un alambre al moverlo por un campo magnético.

Electromagnetism A form of magnetism that occurs when current flows through a conductor.

Electromagnetismo Una forma del magnetismo que ocurre al fluir un corriente por un conductor.

Electromechanical Refers to a device that incorporates both electronic and mechanical principles together in its operation.

Electromecánico Refiere a un dispositivo que incorpora ambos principios del electrónico y mecánico en su operación.

Electronic Pertaining to the control of systems or devices by the use of small electrical signals and various semiconductor devices and circuits.

Electrónico Perteneciente al control de los sistemas o de los dispositivos por medio de los pequeños señales eléctricos y por varios de los dispositivos semiconductores y los circuitos.

Electronic fuel injection (EFI) A generic term applied to various types of fuel injection systems.

Inyección electrónica de combustible Un término general aplicado a varios sistemas de inyección de combustible.

Element A substance with only one type of atom.

Elemento Una sustancia con solamente un tipo de átomo.

Embedability The ability of the bearing lining material to absorb dirt.

Incrustabilidad La abilidad que tiene una material de forro de un cojinete en absorber la suciedad.

Emitter A portion of a transistor from which electrons are emitted, or forced out.

Emisor Una porción de un transistor de la cual se emite o se expulsan los electrones.

End clearance play The extent of the possible forward and backward movement of the crankshaft rod bearing on the crank pin or the camshaft.

Holgura de las extremidades La cantidad del movimiento reciprocal del cigueñal, el cojinete del muñon o el árbol de levas.

End play The amount of axial or end-to-end movement in a shaft due to clearance in the bearings.

Juego axial La cantidad del movimiento axial o de una extremidad a otra de una flecha debido a la holgura de los cojinetes.

Energy The ability to do work.

Energía La abilidad de trabajar.

Engine The power plant that propels the vehicle.

Motor La energía central que propulsa el vehículo.

Engine block The main structure of the engine that houses the pistons and crankshaft. Most other engine components attach to the engine block.

Monobloque La estructura principal del motor que contiene los pistones y el cigüeñal. La mayoría de los otros componentes del motor se conectan al monobloque.

Engine efficiency A measure of the relationship between the amount of energy put into the engine and the amount of energy available from the engine.

Eficacia del motor Una medida de la relación entre la cantidad de energía puesta en el motor y la cantidad de energía disponible del motor.

EP toxicity The characteristic of a material that enables it to leach one or more of eight heavy metals in concentrations greater than 100 times standard drinking water concentrations.

Toxicidad EP La característica de una materia que permite que disuelva uno o más de los ocho metales pesados en concentraciones mayores de 100 veces la concentración del agua potable.

Equilibrium Exists when the applied forces on an object are balanced and there is no overall resultant force.

Equilibrio Existe cuando las fuerzas aplicadas en un objeto es equilibradas y no hay fuerza resultante total.

Ethanol A widely used gasoline additive known for its abilities as an octane enhancer.

Etanol Un aditivo muy común que se reconoce por su capacidad de elevar los niveles del octano.

Evacuate The process of applying vacuum to a closed refrigeration system to remove air and moisture.

Vaciar El proceso de aplicar un vacío a un sistema cerrado de refrigeración para quitarle el aire y la humedad.

Evaporate Action of atoms or molecules when they break free from the body of the liquid to become gas particles.

Evaporar Acción de los átomos o las moléculas cuando se escapan del cuerpo de un líquido para convertirse en partículas del gas.

Evaporation A natural process in which the moisture contained by an object leaves and enters the atmosphere.

Evaporación Un proceso natural en el cual la humedad deja un objeto y entra la atmósfera.

Evaporator The component of an air-conditioning system that conditions the air.

Evaporador El componente en un sistema de aire acondicionado que acondiciona el aire.

Excessive wear Wear caused by overloading an out-of-balance condition of a factor affecting wear, resulting in lower-than-normal life expectancy of the part or parts being subjected to the adverse operating condition.

Gasto excesivo El gasto causado por la condición de sobrecarga o de desequilibración de un factor que afecta al gasto, resultando en una durabilidad disminuida y menos de lo deseado de una parte o las partes que se exponen a las condiciones adversas.

Exhaust manifold A component that collects and then directs engine exhaust gases from the cylinders.

Múltiple de escape Un componente que colecciona y luego dirige los gases de escape del motor desde los cilindros.

Exhaust stroke The final stroke of the four-stroke engine cycle, in which the compressed fuel mixture is ignited in the combustion chamber.

Carrera de escape La última carrera de un ciclo de un motor de cuatro tiempos, en la cual se enciende la mezcla de aire/combustible en la cámara del cilindro.

Exhaust valve An engine part that controls the expulsion of spent gases and emissions out of the cylinder.

Válvula de escape Una parte del motor que controla la expulsión de los gases consumidos y las emisiones afuera del cilindro.

Expansion An increase in size. For example, when a metal rod is heated, it increases in length and perhaps also in diameter; expansion is the opposite of contraction.

Expansión Un aumento en el tamaño. Por ejemplo, al calentarse una varilla de metal, aumenta en longitud y quizá en su diámetro tambien; la expansión es lo contrario de la contracción.

Expansion tank A small tank sometimes located inside of the main gas tank that allows for expansion of fuel in a full tank on a hot day.

Tanque de dilatación Un tanque pequeño que suele estar al interior del depósito principal de gasolina y que permite que se dilate el combustible cuando el depósito esté lleno en días calurosos.

Expansion valve A component in an air conditioning system used to create a pressure on one side and reduce the pressure on the other side.

Válvula de expansión Un componente de un sistema de aire acondicionado, que sirve para crear presión en un lado y reducirla en el otro.

Fan clutch A device used on engine-driven fans to limit their terminal speed, reduce power requirements, and lower noise levels.

Embrague del ventilador Un dispositivo utilizado en los ventiladores accionados por motores para limitar su velocidad terminal, disminuir los requisitos de potencia, y bajar los niveles de ruido.

Fatigue Deterioration of a bearing metal under excessive intermittent loads or prolonged operation.

Fatiga La deterioración de un metal de apoyo bajo las cargas intermitentes o durante la operación prolongada.

Feedback Normally refers to the process in which computer commands and the results of such are monitored by the computer.

Retroalimentación Normalmente se refiere al proceso en el cual la computadora manda y los resultados se regulan por la computadora.

Ferrous metal A metal that contains iron or steel and is, therefore, subject to rust.

Metal férreo Un metal que contiene el hierro o el acero y que, por lo tanto, puede oxidarse.

Field coil A coil of wire on an alternator rotor or starter motor frame that produces a magnetic field when energized.

Bobina inductora Una bobina de alambre en un rotor de un alternador o un armazón de un motor de arranque que produce un campo magnético al establecer el corriente.

Filler neck A tube fitted to the fuel tank that allows fuel to be added from a remote location.

Boca de llenado Un tubo adaptado al depósito de combustible, que permite agregar éste último de un lugar remoto.

Fillets The smooth curve where the shank flows into the bolt head.

Filetes La curva lisa en donde el asta se convierte a la cabeza del tornillo.

Final drive The final set of reduction gears the engine's power passes through on its way to the drive wheels.

Engrane o velocidad final El ultimo juego de reductores por el cual pasa la fuerza del motor en camino a las ruedas motrices.

Firing order The order in which the cylinders of an engine move through the power stroke.

Orden de encendido El orden en el cual mueven los pistones de un motor en la carrera encendido-expansión.

Flange A projecting rim or collar on an object that keeps it in place.

Collarín Una orilla sobresaliente o un collar en un objeto que lo sujeta en su lugar.

Flare An expanded, shaped end on a metal tube or pipe.

Abocinado (abocardado) Una extremidad extendida o formada en un tubo de metal.

Flat-rate manuals Literature containing figures dealing with the length of time specific repairs are supposed to require. Flat-rate manuals often contain a parts list with prices as well.

Manuales de valuación La literatura que se trata de los detalles del tiempo de obra requirido por varias reparaciones específicas. Los manuales de valuación suelen incluir tambien una lista de repuestas con sus precios.

Flex plate A round, flywheel-like disc mounted to an engine's crankshaft when the vehicle is equipped with an automatic transmission.

Placa de flexión Una placa redonda parecida al volante montada al cigueñal de un motor de un vehículo que ha sido equipado con una transmisión automática.

Flooding A condition in which excess, unvaporized fuel in the intake manifold prevents the engine from starting.

Ahogo Una condición en la cual un exceso de combustible no vaporizado se queda en el múltiple de admisión y previene que arranque el motor.

Fluid Something that does not have a definite shape; therefore, liquids and gases are fluids.

Fluido Algo que no tiene una forma definida, por lo tanto los líquidos y los gases son fluidos.

Flutter (bounce) As applied to engine valves, refers to a condition in which the valve is not held tightly on its seat during the time the cam is not lifting it.

Titilación (rebote) Aplicado a las válvulas de los motores, se refiere a una condición en la cual la válvula no se asienta bien en cuando la leva no la tiene levantada.

Flux density The number of flux lines per square centimeter.

Densidad de flujo El número de lineas de flujo por centímetro cuadrado.

Flux field The magnetic field formed by magnetic lines of force.

Campo de flujo El campo magnético formado por lineas de fuerza magnética.

Flywheel A heavy circular component located on the rear of the crankshaft that keeps the crankshaft rotating during nonproductive strokes.

Volante del motor Un componente pesado redondo, ubicado en la parte trasera del cigüeñal que mantiene el cigüeñal girando durante las carreras no productivas.

Foot-pound A unit of measurement for torque. One foot-pound is the torque obtained by a force of 1 pound applied to a wrench handle 12 inches long.

Pies-libra Una unidad de medida de torsión. Un pie-libra es la torsión que se obtiene por la fuerza de una libra que se aplica a una llave de tuercas cuyo mango mide 12 pulgadas en longitud.

Force A push or pushing effort measured in pounds.

Fuerza Un esfuerzo de jalar o empujar que se mide en libras.

Forge The process of shaping metal by stamping it into a desired shape.

Forgar El proceso de formar el metal al imprimirlo en una forma requerida.

Forward bias A positive voltage that is applied to the P-material and a negative voltage that is applied to the N-material in a semiconductor.

Polarización directa Un voltaje positivo que se aplica al cristal P y un voltaje negativo que se aplica al cristal N de un semiconductor.

Free play Looseness in a linkage between the start of application and the actual movement of the device, such as the movement in the steering wheel before the wheels start to turn.

Juego libre Flojedad en una biela que aparece entre el tiempo en que se comienza usar a una aplicación y el movimiento actual del dispositivo, tal como el movimiento en un volante de dirección antes de que comienzan a girar las ruedas.

Free travel The distance a clutch pedal moves before it begins to take up slack in the clutch linkage.

Juego libre La distancia que mueve el pedal de embrague antes de que comienza accionar la biela del embrague.

Freewheeling A mechanical device that engages the driving member to impart motion to a driven member in one direction but not the other. Also known as an overrunning clutch.

Embrague de volante libre Un dispositivo mecánico que engrana el miembro de impulso con el miembro de tracción y confiere el movimiente en una dirección, pero no en la otra. Tambien se conoce como un embrague de sobremarcha.

Freeze plug A term used for core plug.

Tapón de extremo Término utilizado para el tapón de eje.

Frequency The rate at which something occurs.

Frecuencia La velocidad en la cual algo ocurre.

Friction The resistance to motion that occurs when two objects rub against each other.

Fricción La resistencia al movimiento que ocurre al frotarse dos objetos.

Fuel filter A device located in the fuel line to remove impurities from the fuel before it enters the carburetor or injector system.

Filtro de combustible Un dispositivo en la tubería de combustible para eliminar impurezas de la gasolina antes de que entre al carburador o el sistema de inyección.

Fuel gauge A gauge that indicates the amount of fuel remaining in the tank.

Medidor de combustible Un medidor que indica la cantidad de combustible que queda en el depósito.

Fuel injection Injecting fuel into the engine under pressure.

Inyección de combustible Inyección a presión de combustible al motor.

Fuel injector A carburetor mixes the air and fuel at a ratio of 14.7 to 1. However, today's vehicles are using fuel injectors. These injectors mix the fuel with the air just before the intake valve. The fuel is injected into the port under a low pressure. Computers control the amount of fuel injected.

Inyector de combustible Un carburador mezcla el aire y el combustible a razón de 14.7 a 1. Sin embargo, los vehículos actuales utilizan inyectores de combustible, que mezcla la gasolina con aire inmediatamente antes de la válvula de admisión. El combustible se inyecta en el orificio de entrada a baja presión. La cantidad inyectada se controla por medio de computadoras.

Fuel pressure regulator A device designed to limit the amount of pressure buildup in a fuel delivery system.

Regulador de presión del combustible Un dispositivo diseñado para limitar la cantidad de acumulación de presión en un sistema de suministro de combustible.

Fuel pump A mechanical or electrical device used to move fuel from the fuel tank to the carburetor or injectors.

Bomba de combustible Un dispositivo mecánico o eléctrico que se utiliza para llevar combustible del depósito al carburador o los inyectores.

Fuel rail A metal or plastic pipe in which the upper ends of the injectors are installed in port injection systems.

Carril del combustible Un tubo de metal o plástico en el cual se instalan las extremidades superiores de los inyectores en los sistemas de inyección de puerto/de lumbreras.

Fulcrum The support or point of rest on which a lever rests; also called the pivot point.

Punto de apoyo El soporte o punto de descanso en el cual descansa una palanca; tambien se llama el punto de pivote.

Full-floating A type of live axle arrangement in which the weight of the vehicle is not supported by the axle.

Flotante Un tipo de dispositivo de eje motor en el cual el peso del vehículo no se soporta por el eje.

Fuse An electrical device used to protect a circuit against accidental overload or unit malfunction.

Fusible Un dispositivo eléctrico utilizado para proteger un circuito contra una sobrecarga imprevista o un mal funcionamiento de la unidad.

Fusible link A type of fuse made of a special wire that melts to open a circuit when current draw is excessive.

Fusible térmico El tipo de fusible fabricado de un alambre especial que se funde para abrir un circuito cuando ocurre una sobrecarga del circuito.

Galling wear Uniting two solid surfaces that are in rubbing contact; used to describe the normal wear of valve lifters.

Desgaste por rozamiento Uniendo a dos superficies sólidos que estan en contacto frotativo; se usa para describir el gasto normal de las levantaválvulas.

Ganged A type of switch in which several circuits are controlled by moving one control or lever.

Acoplados Un tipo de interruptor en el cual varios circuitos se controlan al mover un control o una palanca.

Gas cap A removable cap that seals the fuel tank.

Tapón de gasolina Un tapón retirable que sella el depósito de combustible.

Gasket A thin layer of material or composition that is placed between two machined surfaces to provide a leakproof seal between them.

Junta Una capa delgada de material o compuesto que se coloca entre dos superficies rectificadas para proveer una junta hermética para evitar fugas entre ellas.

Gear A wheel with external or internal teeth that serves to transmit or change motion.

Engranaje Rueda con dientes externos o internos que sirve para transmitir o modificar un movimiento.

Gear pitch The number of teeth per given unit of pitch diameter. Gear pitch is determined by dividing the number of teeth by the pitch diameter of the gear.

Paso del engrenaje El número de los dientes por una unidad dada de un diámetro de paso. El paso del engrenaje se determine al dividir el número de los dientes por el diámetro del paso del engrenaje.

Gear pumps Positive displacement pumps that use two meshing external gears, one drive and one driven.

Bombas de engrenajes Las bombas de desplazamiento positivo que usan dos engrenajes externos engranados, uno propulsor y uno loco.

Gear ratio An expression of gear size and tooth count of gears that are meshed together. The ratio reflects torque multiplication.

Relación de engranes Una expresión del tamaño del engrenaje y el número de los dientes que se engranan. La relación refleja la multiplicación de par.

Generator An electrical device that produces alternating current that is rectified to DC current by the brushes and commutator.

Generador Un dispositivo eléctrico que produce corriente alterna que se rectifica como corriente continua mediante las escobillas y el conmutador.

Glaze A thin residue on the cylinder walls formed by a combination of heat, engine oil, and piston movement.

Porcelana (barniz) Un residuo en los paredes de los cilindros que se forma por una combinación del calor, el aceite del motor y el movimiento de los pistones.

Governor A device that limits the speed of a device.

Regulador Un dispositivo que limite la velocidad de un dispositivo.

Grade markings Marks on fasteners that indicate strength.

Marcos de grado Las marcas en las fijaciones que indican su fuerza.

Grease A combination of oil and a thickening agent.

Grasa Una combinación de aceite y un agente de espesamiento.

Ground The negatively charged side of a circuit. A ground can be a wire, the negative side of the battery, or even the vehicle chassis.

Tierra El lado de un circuito que tiene una carga negativa. Una tierra puede ser un alambre, el lado negativo de una batería, u hasta el chasis del vehículo.

Gum In automotive fuels, gum refers to oxidized petroleum products that accumulate in the fuel system, carburetor, or engine parts.

Sarro (goma) En los combustibles automotrices, el sarro se refiere a los productos petroléos oxidados que acumulan en el sistema de combustible, el carburador, o en las partes del motor.

Hairspring A fine wire spring.

Muelle resorte espiral Un resorte de alambre fino.

Half shaft Either of the two drive shafts that connect the transaxle to the wheel hubs in FWD cars. Half shafts have constant velocity joints attached to each end to allow for suspension motions and steering. The shafts may be of solid or tubular steel and may be of different lengths. Balance is not critical, as half shafts turn at roughly one-third the speed of RWD drive shafts.

Semieje cualquiera De los dos ejes motores que conectan el dispositivo de acoplamiento intermedio a los cubos de las ruedas en los vehículos de tracción en las cuatro ruedas. Los semiejes tienen juntas de velocidad constante fijas a cada extremo para permitir los movimientos de la suspención y la dirección. Los ejes pueden ser de acero tubular o sólido y tener longitudes diferentes. El equilibrio no es crítico, ya que los semiejes giran a aproximadamente un tercio de la velocidad de los ejes motores de los vehículos de tracción trasera.

Hall-effect The consequence of moving current through a thin conductor that is exposed to a magnetic field; as a result of this, voltage is produced.

Efecto Hall La consecuencia de pasar un corriente por un conductor delgado que se expone a un campo magnético; por consecuencia, se produce el voltaje.

Halogen The term used to define a group of chemically related nonmetallic elements.

Halógeno El termino para describir un grupo de elementos no metálicos del mismo género químico.

Hand tap A tool used for hand cutting internal threads.

Macho de roscar de mano Una herramiento que sirve para cortar a mano las roscas interiores.

Hand-threading dies Tools used to cut external threads on bolts, rods, and pipes.

Terraja Las herramientas que sirven para cortar las roscas exteriores en los pernos, las varillas, y los tubos.

Hardening A process that increases the hardness of a metal, deliberately or accidentally, by hammering, rolling, carburizing, heat treating, tempering, or other physical processes.

Endurecer Un proceso que aumenta la dureza de un metal, deliberadamente o accidentalmente, por el martilleo, rodar, la carburación, el calor, templar, o otros procesos físicos.

Hard spots Areas in the friction surface of a brake drum or rotor that have become harder than the surrounding metal; sometimes called islands of steel.

Regiones endurecidos Las averías en la superficie de fricción del tambor o el rotor de un freno que se han endurecido más que el metal que los rodea; a veces se llaman islas de acero.

Harmonics Periods of vibration that occur when valve springs are opened and closed rapidly.

Armónico Los períodos de la vibración que ocurren cuando los resortes de las válvulas abren y cierran rapidamente.

Hazardous waste Any material found on the Environmental Protection Agency list of known harmful materials. Waste is also considered hazardous if it has one or more of the following characteristics: ignitability, corrosivity, reactivity, and EP toxicity.

Desechos tóxicos Cualquier material que se nombra en la lista de materiales dañosos de la agencia de protección del medio ambiente (EPA). Los desechos tambien se consideran tóxicos si tiene uno o más de las características siguientes: la inflamabilidad, la calidad corrosiva, la reactividad y la toxicidad EP.

Head gasket Gasket used to prevent compression pressures, gases, and fluids from leaking. It is located on the connection between the cylinder head and the engine block.

Junta de la cabeza Una junta que se emplea para prevenir que se escapen las presiones, los gases y los flúidos de la compresión. Se ubica en la conexión entre la cabeza de los cilindros y el monobloque.

Headlight The lights at the front of a vehicle used to illuminate the road ahead.

Faros delanteros (Luces delanteras) Las luces del frente del vehículo que se utilizan para iluminar la carretera hacia adelante.

Heat A form of energy caused by the movement of atoms and molecules.

Calor Una forma de energía causada por el movimiento de átomos y de moléculas.

Heat dam The narrow groove cut into the top of the piston that restricts the flow of heat into the piston.

Borde hendido Una muesca estrecha cortada en la cabeza del pistón. Limita el flujo del calor al pistón.

Heat range A rating used to express how well a spark dissipates heat.

Rango de calor Una medida que expresa la abilidad de una chispa de dispersar el calor.

Heat shield An assembly designed to restrict the transfer of high heat from one component to another.

Pantalla térmica Un ensamblaje diseñado para restringir la transferencia de calor elevado de un componente a otro.

Heat sink A device used to dissipate heat and protect parts.

Fuente fría Un dispositivo que sirve para dispersar el calor y protejer las partes.

Heat treating The changing of the properties of a metal by using heat.

Tratamiento por calor El cambiar de las características de un metal usando calor.

Heater control valve A manual or automatic valve in the heater hose used for opening or closing, providing coolant flow control to the heater core.

Válvula de control del calentador Una válvula manual o automática en la manguera del calentador, que se usa para abrir o cerrar, proporcionando control de flujo de refrigerante al núcleo calentador.

Heater core A small heat exchanger in the passenger compartment through which engine coolant is circulated.

Núcleo del calentador Un pequeño intercambiador térmico en el compartimiento del pasajero por el que se hace circular el refrigerante del motor.

Heater hose The hose that connects the heater core to the engine's cooling system or heater control valve.

Manguera del calentador La manguera que conecta el núcleo del calentador al sistema de enfriamiento del motor o la válvula de control del calentador.

Helical gear A gear with teeth that are cut at an angle or are spiral to the gear's axis of rotation.

Engranaje helicoidal Un engranaje cuyos dientes han sido cortados en un ángulo o que son un espiral al eje de rotación del engrenaje.

Heptane A standard reference fuel with an octane number of zero, meaning that it knocks severely in an engine.

Eptano Un combustible de norma cuyo índice de octano es el cero, lo que significa que causa que el motor golpetea severamente.

Hertz (hz) The unit by which frequency is most often expressed, equal to one cycle per second.

Hertzio (hz) La unidad de medida que se utiliza para expresar frecuencia y es igual a un ciclo por segundo.

High tension High voltage. In automotive ignition systems, voltages (up to 40 kilovolts) in the secondary circuit of the system as contrasted with the low, primary circuit voltage (nominally 6 or 12 volts).

Alta tensión El voltaje alto. En los sistemas del encendido, los voltajes (hasta los 40 kilovoltios) en el circuito secundario del sistema en contraste del voltaje bajo del circuito principal (de 6 o 12 voltios nominalmente).

Hone An abrasive for correcting small irregularities of differences in diameter in a cylinder, such as an engine cylinder or brake cylinder.

Piedra de rectificar Un abrasivo que sirve para corregir las pequeñas irregularidades en los diámetros de un cilindro, tal como un cilindro de un motor o un cilindro de freno.

Horsepower The rate at which torque is produced.

Caballo de fuerza El índice de la cual se produce el esfuerzo de torsión.

Hose A flexible tube used to transfer a liquid or gas.

Manguera Un tubo flexible para la transferencia de un líquido o un gas.

Hose clamp A device used to attach hoses to the engine, heater core, radiator, or water pump. A popular type of replacement clamp is the high-torque, worm-gear clamp with a carbon steel screw and a stainless steel band.

Abrazadera de manguera Un dispositivo utilizado para sujetar mangueras al motor, el núcleo del calentador, el radiador o la bomba de gua. Uno de los tipos preferidos de de abrazaderas de reemplazo es el de las de par de torsión elevado y engranaje de gusano con un tornillo de acero al carbono y una banda de acero inoxidable.

Hot spot Refers to a comparatively thin section of area of the wall between the inlet and exhaust manifold of an engine, the purpose being to allow the hot exhaust gases to heat the comparatively cool incoming mixture. Also, used to designate local areas of the cooling system that have attained above average temperatures.

Región de calor Refiere a una sección delgada en el muro entre la admisión y el escape del múltiple de un motor, su propósito es permitir que los gases calientes del escape calientan a la mezcla entrando que es comparativamente fría. Este término tambien sirve para designar las areas locales del sistema del enfriamento, que han alcanzado las temperaturas más altas que las normales.

Hotchkiss drive A suspension design that uses leaf springs and the frame to control rear-end torque.

Propulsión del tipo Hotchkiss Un diseño de suspensión que usa los meulles de hojas y el armazón para controlar una torsión de la parte trasera del vehículo.

Hunting gearset A differential gearset in which one drive pinion gear tooth contacts every ring gear tooth after several rotations.

Conjunto de engranaje con diente suplementario Un conjunto de engranajes del diferencial en el cual cada diente del piñon de mando se engrena con cada diente del engranaje de corona después de varias rotaciones.

Hydraulic valve lifter A lifter that provides a rigid connection between the camshaft and the valves. The hydraulic valve lifter differs from the solid type in that it uses oil to absorb the shock that results from movement of the valve train.

Levantaválvulas hidráulica Una levantaválvulas que provee una conexión rígida entre el árbol de levas y las válvulas. La levantaválvulas hidráulica es distincta del tipo sólido en que usa el aceite para amortiguar los choques que resulten del movimiento del tren de válvulas.

Hydrocarbons Particles of gasoline present in the exhaust and in crankcase vapors that have not been fully burned.

Hidrocarburos Las partículas de la gasolina, presentes en los gases de escape y en los vapores de la caja de cigueñal, que no se han quemado completamente.

Hypoid gears A type of spiral, beveled ring and pinion gearset in a differential. Hypoid gears mesh below the ring gear centerline.

Engrenajes hipoíde Un conjunto de engrenajes de un diferencial de anillo y piñon de forma espiral y biselado. Los engrenajes hipoídes se engrenan abajo de la linea central de la corona.

Idler arm A steering linkage component fastened to the car frame that supports the right end of the center link.

Brazo intermedio Componente de enlace de dirección montado en el armazón del vehículo, que sostiene el extremo derecho del enlace central.

Idler pulley A pulley used to tension or torque the belt(s).

Polea tensora Una polea utilizada para proveer tensión o par de torsión a la(s) banda(s).

Ignitability The characteristic of a solid that enables it to spontaneously ignite. Any liquid with a flash point below 140°F (60°C) is also said to possess ignitability.

Inflamabilidad La característica de un sólido que lo permite encenderse espontáneamente. Cualquier líquido con una temperatura de inflamabilidad menos del 140°F (60°C) posea la caracteristica de la inflamabilidad.

Ignition coil A transformer containing a primary and secondary winding that acts to boost the battery voltage of 12 volts to as much as 30,000 volts to fire the spark plugs.

Bobina de encendido Un transformador contiene un devanado primario y otro secundario que actúan para reforzar el voltaje de la batería de 12 voltios a hasta 30,000 voltios para activar las bujías.

Ignition module An electronic device used to open and close the primary ignition circuit.

Módulo de la ignición Un dispositivo electrónico que sirve para abrir y cerrar el circuito primario de la ignición.

Ignition switch A five-position switch that is the power distribution point for most of the vehicle's primary electrical systems. The spring-loaded start position provides momentary contact and automatically moves to the run position when the key is released. The other switch detent positions are accessories, lock, and off.

Interruptor de encendido Un interruptor de cinco posiciones que es el punto de distribución de potencia de la mayoría de los sistemas eléctricos primarios del vehículo. La posición de arranque, de resorte, proporciona un contacto momentáneo y pasa automáticamente a la posición de marcha al soltar la llave. Las otras posiciones del interruptor son accesorios, cierre y apagado.

Ignition system The system that creates and distributes a timed spark to the cylinder.

Sistema de encendido Crea y distribuye una chispa sincronizada al cilindro.

Ignition timing The point at which ignition occurs in a cylinder.

Tiempo de encendido El punto en el que se produce el encendido en el cilindro.

Ignition wire set A kit containing coil wire and all of the spark plug wires for a particular engine.

Juego de cables de encendido Un juego que contiene un cable de la bobina y todos los cables de las bujías de un motor dado.

Impact sockets Heavier walled sockets made of hardened steel and designed for use with an impact wrench.

Dados de impacto Los dados cuyos muros gruesos son hechos de acero endurecido y se han diseñado para usarse con una llave neumática.

Impact wrench A portable hand-held reversible power wrench.

Llave neumático Una llave de potencia de mano portátil con acción invertible.

Impeller The part of a torque converter that is driven by the engine's crankshaft.

Impulsor Una parte de un convertidor del par que es accionada por el cigueñal del motor.

Impermeable Characteristic of materials that adsorb fluids.

Impermeable Una caracteristica de los materiales que fijan los líquidos por adsorcion.

Inboard joint The CV joint that connects the drive axle to the transaxle of the differential assembly.

Junta interior La junta de velocidad constante que conecta el eje motor al árbol intermedio del conjunto del diferencial.

Included angle The sum of the angle of camber and steering axis inclination.

Ángulo comprendido La suma del ángulo del camber y la inclinación del eje de la dirección.

Induction The process of producing electricity through magnetism rather than direct flow through a conductor.

Inducción El proceso de producción de la electricidad por el magnetismo en vez de un corriente directo por un conductor.

Inertia The constant moving force applied to carry the crankshaft from one firing stroke to the next.

Inercia La fuerza del movimiento constante aplicada a un cigueñal para progresar de una carrera a la próxima.

Inertia switch A switch that automatically shuts off the fuel pump if the vehicle is involved in a collision or rolls over.

Conmutador inercia Un interruptor que automáticamente apaga a la bomba de combustible si el vehículo se involucra en un choque o si se voltea.

Insert bearing A bearing made as a self-contained part and then inserted into the bearing housing.

Cojinete inserta Un cojinete fabricado como una parte autónoma y que se inserta en la caja del cojinete.

Installed spring height The distance from the valve spring seat to the underside of the retainer when it is assembled with keepers and held in place.

Altura del resorte instalado La distancia del asiento del resorte de la válvula a la parte inferior del retén cuando ha sido instalado con las cuñas y se fija en su lugar.

Installed stem height The distance from the valve spring seat and to the stem tip.

Altura del vástago instalado La distancia del asiento del muelle de la válvula al punto extremo del vástago.

Insulated circuit A circuit that includes all of the high-current cables and connections from the battery to the starter motor.

Circuito aislado Un circuito que incluye a todos los cables de alta tensión y las conexiones de la batería al motor de arranque.

Insulator A material that does not allow for good current flow.

Aislador Una materia que no permite un flujo del corriente.

Intake stroke The first stroke of the four-stroke engine cycle, in which the piston moves away from top dead center and the intake valve opens.

Carrera de admisión La primera carrera de un ciclo de un motor de cuatro tiempos, en la cual el pistón se aleja del punto muerto superior y se abre la válvula de admisión.

Intake valve The control passage of the air/fuel mixture entering the cylinder.

Válvula de admisión El pasaje de control para la mezcla de aire/combustible entrado en el cilindro.

Integral Make in one piece.

Integral Hecho en una sóla pieza.

Integrated circuit A large number of diodes, transistors, and other electronic components, all mounted on a single piece of semiconductor material and able to perform numerous functions.

Circuito integrado La cantidad de diodos, transistores u otros componentes electronicos, todos montados en una sola pieza de material semiconductor y capable de ejecutar varias funciones.

Intercooler A device used on some turbocharged engines to cool the compressed air.

Refrigerante intermedio Un dispositivo que se usa en algunos motores turbocargados para enfriar al aire comprimido.

Intermediate pipe Part of the exhaust system, used primarily to connect different major parts of the system.

Tubo intermedio Parte del sistema de escape, que se utiliza primordialmente para conectar entre sí diferentes piezas importantes del sistema.

Isooctane A standard reference fuel with an octane number of 100, meaning that it does not knock in an engine.

Iso-octano Un combustible de norma con un índice de octano de cien, lo que significa que resulta en que el motor no golpetea.

Jack stands Safety stands. Used to hold a vehicle up while using a floor jack.

Torres Los soportes de seguridad. Sirven para sostener un vehículo en alto mientras que se usa un gato de piso.

Jam nut A locknut; one nut is tightened against another nut.

Contra tuerca Una tuerca perturbador; una tuerca que se aprieta contra otra tuerca.

Jet A precisely sized, calibrated hole in a hollow passage through which fuel or air can pass.

Surtidor Un hoyo calibrado de medida precisa en un pasillo hueco por el cual puede pasar el combustible o el aire.

Jig Device that holds the work and guides the operation.

Plantilla Un dispositivo que sostiene al trabajo y guía a la operación.

Jounce Upward suspension movement.

Sacudo Un movimiento hacia arriba de la suspención.

Keep-alive memory A series of vehicle battery-powered memory locations in the microcomputer that allows the microcomputer to store information on input failure identified in normal operations for use in diagnostic routines. Keep-alive memory adopts some calibration parameters to compensate for changes in the vehicle system.

Memoria de retención Una seria de localizaciones de memoria mantenidas por la batería del vehículo que permiten que la microcomputador registra la información de fallos de entrada, que se identifican en las operaciones normales para usarse en las pruebas diagnósticos. La memoria de retención utiliza algunos de los parámetros calibrados para compensar por los cambios en el sistema del vehículo.

Key A small block inserted between the shaft and hub to prevent circumferential movement.

Chaveta Un tope pequeño que se meta entre la flecha y el cubo para prevenir un movimiento circunferencial.

Kickdown Normally refers to an automatic transmission shifting into a lower gear to meet the demands of the driver or a heavier load.

Disparo descendente Normalmente se refiere a una transmisión automática que efectúa un cambio descendente para acomodar al chofer o para acomodar a una carga más pesada.

Kinetic balance Balance of the radial forces on a spinning tire determined by an electronic wheel balancer.

Balanceo cinético El balanceo dinámico de las fuerzas radiales en una llanta que gira. Se determine por un balanceador de ruedas electrónico.

Kinetic energy Energy in motion.

Energía cinético La energía en movimiento.

Knurling A technique used for restoring the inside diameter dimensions of a worn valve guide by plowing tiny furrows through the surface of the metal.

Moletear Un método para restorar las dimensiones del diámetro interior de una guía de la válvula gastada partiendo al metal de la superficie en pequeños surcos.

Lambda sensor A form of oxygen sensor used on some European cars.

Sensor lambda Un tipo de detector del oxígeno que se usa en algunos coches europeos.

Lamination Thin layers of soft metal used as the core for a magnetic circuit.

Laminas Las hojas delgadas del metal blando que sirven de alma de un circuito magnético.

Land The areas on a piston between the grooves.

Meseta del pared Las areas de un pistón entre las muescas.

Lapping The process of fitting one surface to another by rubbing them together with an abrasive material between the two surfaces.

Pulido El proceso de ajustar a una superficie con otra por frotarlas juntas con una materia abrasiva entre las dos superficies.

Lash The operational gap between two objects.

Juego La holgura que permite operar dos objetos.

Latent heat The heat required to change a mass's state of matter.

Calor latente El calor requerido para cambiar una masa a un estado differente.

Leakdown The relative movement of the plunger with respect to the hydraulic valve lifter body after the check valve is seated by pressurized oil. A small amount of oil leakdown is necessary for proper hydraulic valve lifter operation.

Fuga por abajo El movimiento relativo del émbolo con respeto al cuerpo de la levantaválvula hidráulica después de que la válvula de un sólo paso se asienta por el aceite a presión. Una pequeña cantidad de fuga es necesario para asegurar una operación correcta de la levantaválvula hidráulica.

Lean An air/fuel mixture that has more air than is required for a stoichiometric mixture.

Mezcla pobre Una mezcla de combustible y aire que tiene más aire de lo que requiere una mezcla estoquiométrica.

Lever A device made up of a bar turning about a fixed point, called the fulcrum, that uses a force applied at one point to move a mass on the other end of the bar.

Palanca Un dispositivo compuesto de una barra que da la vuelta sobre un punto fijo, llamado el fulcro, que utiliza una fuerza aplicada en un punto para mover una masa en el otro extremo de la barra.

Lift Lift during acceleration and front end dive while braking.

Elevador de vehículos Un dispositivo que sirve para levantar un vehículo.

Light-emitting diode (LED) A type of digital electronic display used as either single indicator lights or grouped to show a set of letters or numbers.

Diodo emisor de luz (LED) Un tipo de indicador electrónico digital que tiene luces indicadores individuales o en grupos para formar un conjunto de letras o números.

Limited-slip A type of differential that allows more torque to be sent to the wheel with the most traction.

Deslizamiento limitado Un tipo de diferencial que permite que más energía del par se envía a la rueda con más tracción.

Line boring An engine block machining operation in which the main bearing housing bores are rebored to standard size and in perfect alignment.

Rectificación en linea Una operación maquinaria del bloque del motor en la cual los taladros principales de la caja se taladran de nuevo para asegurar un tamaño uniforme y una alineación perfecta.

Linear rate springs A coil spring with equal spacing between the coils, one basic shape, and constant wire diameter having a constant deflection rate regardless of load.

Resortes de variación lineal Resortes en espiral de espaciamiento igual entre espiras, una forma básica y un diámetro de alambre constante, con una tasa constante de desviación, sea cual sea la carga.

Liner Usually a thin section placed between two parts, such as a replaceable cylinder liner in an engine.

Forro Típicamente una sección delgada que se coloca entre dos partes, tal como un forro reemplazable del cilindro de un motor.

Lip-type seal An assembly consisting of a metal or plastic casing, a sealing element made of rubber, and a garter spring to help hold the seal against a turning shaft.

Sellos con borde Una asamblea que consiste de un cárter de metal o de plástico, un elemento sellante hecho de caucho, y un resorte de liga que sirve para apretar al sello contra una flecha que gira.

Liquid crystal diode (LCD) A type of digital electronic display made of special glass and liquid and requiring a separate light source.

Diodo de cristal líquido Un tipo de presentación electrónico hecho de un tipo especial de vidrio y líquido que requiere un suministro de luz independiente.

Live axle An axle on which the wheels are firmly affixed. The axle drives the wheels.

Eje motor Un eje en el cual las ruedas se montan rígidamente. El eje acciona las ruedas.

Load The work an engine must do, under which it operates more slowly and less efficiently. The load could be that of driving up a hill or pulling extra weight.

Carga El trabajo que debe ejecutar el motor, bajo el cual opera más lentamente y con menos eficiencia. La carga puede ser el de viajar cuesta arriba o de arrastrar el peso extra.

Lobe The part of the camshaft that raises the lifter.

Lóbulo La parte del árbol de levas que alza el buzo.

Lockup The point at which braking power overcomes the traction of the vehicle's tires and skidding occurs. The most efficient stopping occurs just before lockup is reached. Locked wheels cause loss of control, long stopping distances, and flat-spotting of the tires.

Enclavamiento El punto en el cual el enfrenamiento supera la tracción de los neumáticos del vehículo y ocurre el patinaje. El enfrenamiento más eficiente se efectúa justo antes de que ocurre el enclavamiento. Las ruedas enclavadas causan una pérdida del control, el enfrenamiento dilatorio y el gasto irregular de los neumáticos.

Lockup torque converter A converter with a friction disc that locks the impeller and turbine together.

Convertidor de par de torsión de bloqueo Un convertidor con un disco de fricción que bloquea al impulsor y la turbina.

Logic gate Electronic circuit that acts as a gate to output voltage signals depending on different combinations of input signals.

Compuerta lógico El circuito electrónico que sirve como una puerta a las señales del voltaje de salida según las combinaciones distinctas de señales de entrada.

Lookup tables The part of a microcomputer's memory that indicates in the form of calibrations and specifications how an engine should perform.

Obtención de datos de una tabla La parte de una memoria del microcomputador que indica en la forma de las calibraciones y las especificaciones como debe funcionar el motor.

Low and reverse band Transmission brake engaged in manual low and reverse gear ranges.

Banda de baja-reversa El freno de transmisión engranado en una posición de la banda baja-reversa.

Low and reverse clutch Transmission brake device engaged in manual low and reverse ranges.

Embrague baja-reversa Un dispositivo del frenado de transmisión engranado en la banda manual baja-reversa.

LP gas Liquified petroleum gas, often referred to as propane, which burns clean in the engine and can be precisely controlled.

Licuados del petróleo El gas de petroleo en forma líquido, típicamente llamado el propano, que se quema completamente en el motor y que puede ser controlado precisamente.

Lubrication The process of reducing friction between the moving parts of an engine.

Lubricación El proceso de reducir la fricción entre dos partes que mueven en un motor.

MacPherson strut suspension A suspension system in which the strut is connected from the steering knuckle to an upper strut mount, and the strut replaces the shock absorber.

Suspensión del puntal de tipo MacPherson Un sistema de suspensión en donde el puntal se conecta al muñón de la dirección a un montaje superior para el puntal, y el puntal reemplaza el amortiguador.

Magnet Any body with the property of attracting iron or steel.

Imán Cualquier cuerpo que tiene la propriedad de atraer al hierro o al acero.

Magnetic field The area surrounding the poles of a magnet that is affected by its attraction or repulsion forces.

Campo magnético El area alrededor de los polos de un imán que se afectan por las fuerzas de atracción o repulsión.

Magnetic gauges Electrical analog gauges that use magnetic forces to move the needle left or right.

Indicador magnético Las galgas análogas eléctricas que usan a las fuerzas magnéticas para mover la aguja indicadora a la izquierda o a la derecha.

Magnetic pulse generator An engine position sensor used to monitor the position of the crankshaft and control the flow of current to the center base terminal of the switching transistor.

Generador de pulsos magnéticos Un detector del posición de motor que sirve para regular la posición de la cigueñal y controlar al flujo del corriente al terminal del base central del transistor interruptor.

Magnitude Normally refers to the height of an electrical signal.

Magnitúd Normalmente se refiere a la intensidad de un señal eléctrico.

Main bearing A bearing used as a crankshaft support.

Cojinete principal del cigüeñal Un cojinete que se emplea como un soporte para el cigüeñal.

Main journal The central area of a crankshaft where the main bearings support the shaft in the block.

Muñon central El area central de un cigueñal en donde los cojinetes principales soportan al eje en el bloque.

Mainline pressure Pressure that is regulated in an automatic transmission hydraulic system.

Presión de la linea procedente de la bomba La presión que se regula en un sistema de transmisión hidráulica automática.

Major thrust face The side of the piston to which most force is applied by side thrust.

Cara de empuje principal El lado del pistón al cual la mayoría de la fuerza se aplica en un empuje lateral.

Malleable Able to be shaped.

Maleable Capable de ser formado.

Manifold absolute pressure A measure of the degree of vacuum or pressure within an intake manifold; used to measure air volume flow.

Presión absoluta en el múltiple Una medida del grado del vacío o de la presión dentro de un múltiple de admisión; sirve para medir el volumen del flujo del aire.

MAP sensor The sensor that measures changes in the intake manifold pressure that result from changes in engine load and speed.

Sensor MAP El detector que mide los cambios de presión en el múltiple de admisión que resultan según los cambios de la carga y la velocidad del motor.

Margin The area between the valve face and the head of the valve.

Margen El area entre la cara de la válvula y la cabeza de la válvula.

Mass The amount of matter in an object.

Masa La cantidad de materia en un objeto.

Mass airflow sensor An EFI air intake sensor that measures the mass, not the volume, of the air flowing into the intake manifold.

Sensor de la masa del aire Un detector del admisión del aire de inyección de combustible electrónico que mide la masa, en vez del volumen, del aire que fluye al múltiple de admisión.

Master cylinder The liquid-filled cylinder in the hydraulic brake system or clutch where hydraulic pressure is developed when the driver depresses a foot pedal.

Cilindro maestro El cilindro lleno de líquido del sistema de freno hidráulico o embrague, donde se desarrolla la presión hidráulica cuando el conductor oprime un pedal.

Material safety data sheets Information sheets containing chemical composition and precautionary information for all products that can present a health or safety hazard.

Folletos de información de materiales y precauciones El folleto que contiene la información pertinente a la composición química y la indicaciones precavidos para todos los productos que pueden presentar peligros a la salud o la seguridad.

Matter Anything that occupies space.

Materia Cualquier cosa que ocupa el espacio.

Mechanical advantage The result of a machine or system that increases the force available to do work.

Ventaja mecánica resultado de una máquina o de un sistema que aumente la fuerza disponible para hacer el trabajo.

Mechanical efficiency (engine) The ratio between the indicated horsepower and the brake horsepower of an engine.

Eficiencia mecánica (motor) La relación entre la potencia indicada en caballos y el caballo indicado al freno de un motor.

Memory The part of a computer that stores, or holds, the programs and other data.

Memoria La parte de una computador que archiva, o guarda, a las aplicaciones u otros datos.

Mercury switch A type of switch that uses the flow of liquid metal mercury to complete the electrical circuit. Mercury switches are frequently used to control trunk and underhood lights.

Interruptor de mercurio Un tipo de interruptor que usa el flujo del metal mercurio líquido para completar un circuito eléctrico. Los interruptores de mercurio se suelen usar en aplicaciones de los luces de la cajuela y bajo del cofre.

Mesh To fit together, as gear teeth.

Engrenar Ajustarse bien, tal como los dientes de un engrenaje.

Metered To control the amount of fuel passing into an injector. Fuel is metered to obtain the correct measured quantity.

Medir Controlar la cantidad del combustible que pasa al inyector. El combustible se mide para asegurar una cantidad medida correcta.

Metering The process of controlling something in response to quantities of something else.

Medida El proceso de controlar algo según las cantidades de otra cosa.

Metering valve A component that momentarily delays the application of front disc brakes until the rear drum brakes begin to move. Helps to provide balanced braking.

Válvula dosificadora Un componente que retrasa momentáneamente la aplicación de los frenos de disco delanteros hasta que comienzan a moverse los frenos de tambor traseros. Contribuye a asegurar un frenado más equilibrado.

Methanol The lightest and simplest of the alcohols; also known as wood alcohol.

Metano El más ligero y más sencillo del grupo de alcohol; tambien llamado el alcohol piroleñoso.

Microprocessor The portion of a microcomputer that receives sensor input and handles all calculations.

Microprocesor La parte de un microcomputador que recibe la información de los detectores y se encarga de todas las calculaciones.

Microsecond One millionth of a second.

Microsegundo Una millonésima de un segundo.

Milling The process of using blades to remove metal from an object.

Fresado El proceso de usar las cuchillas para quitar el metal de un objeto.

Millisecond One thousandth of a second.

Milisegundo Una milésima de un segundo.

Miscibility The ability of one chemical to mix well with another chemical.

Miscible La característica de una química a mezclarse bien con otra química.

Misfiring Failure of an explosion to occur in one or more cylinders while the engine is running; can be continuous or intermittent failure.

Fallo de encendido El fallo en que una detonación ocurre en uno o más cilindros mientras que este en marcha el motor; puede ser un fallo continuo o intermitente.

Missing A lack of power in one or more cylinders.

March irregular Una falta de potencia en uno o más cilindros.

Mode Manner or state of existence of a thing; e.g., heat or cool.

Modo Manera o estado de existencia de una cosa; calor o fresco.

Modulation Pulsing.

Modulación Pulsante.

Molecule The smallest particle of an element or compound that can exist in the free state and still retain the characteristics of the element or compound.

Molécula La partícula más pequeña de un elemento o un compuesto que puede existir en el estado libre y todavía conservar las características del elemento o del compuesto.

Momentum A type of mechanical energy that is the product of an object's weight times its speed.

Momento El momentu? es un tipo de energía mecánica que es el producto del peso de un objeto multiplicado por su velocidad.

Monolith A single body shaped like a pillar or long tubular structure used as a catalyst in a catalytic converter.

Monolito Un cuerpo sólo en forma de columna o una estructura larga y tubular que sirve de catalizador en un convertidor catalítico.

Muffler 1) A hollow, tubular device used in the lines of some air conditioners to minimize the compressor noise or surges transmitted to the inside of the car. 2) A device in the exhaust system used to reduce noise.

Silenciador 1) Un dispositivo tubular hueco que se usa en las líneas de algunos equipos de aire acondicionado para reducir al mínimo el ruido del compresor o las sobrecargas transmitidas al interior del vehículo. 2) Un dispositivo del sistema de escape que sirve para reducir el ruido.

Multimeter A tool that combines the voltmeter, ohmmeter, and ammeter together in one diagnostic instrument.

Multímetro Una herramienta que combina las funciones del voltiometro, del ohmímetro, y del amperímetro en un instrumento diagnóstico.

Multiplexing A means of transmitting information between computers. A system in which electrical signals are transmitted by a peripheral serial bus instead of common wires, allowing several devices to share signals on a common conductor.

Multiplexar Una manera de transferir la información entre las computadoras. Un sistema en el cual los señales eléctricos se transmiten por medio de un bus serial periférico en vez de por los alambres comunes, así permitiendo que varios dispositivos comparten los señales en un conductor común.

Multiviscosity oil A chemically modified oil that has been tested for viscosity at cold and hot temperatures.

Aceite de viscosidad de grado múltiple Un aceite modificado químicamente cuyo viscosidad se ha probado en temperaturas bajas y altas.

Neutral A solution that has a pH of 7. Also the condition in which work is not being transferred.

Neutral Una solución que tiene un pH de 7. También, neutro (punto muerto) describe una condición en la cual el trabajo no se esté transfiriendo.

Nodular iron A metal used in pressure plates that contains graphite, which acts as a lubricating agent.

Hierro nodular Un metal usado en los platos opresores que contiene el grafito, que funciona como un agente de lubricación.

Nonhunting gearset A differential gearset in which one drive pinion gear tooth contacts only three ring gear teeth after several rotations.

Conjunto de engrenaje sin diente suplementario Un conjunto de engrenaje en el cual un diente del engrenaje de piñon engrane solamente con tres dientes del engrenaje del anillo después de varias rotaciones.

Normal wear The average expected wear when operating under normal conditions.

Desgaste normal El desgaste que se espera de una parte al operar bajo condiciones normales.

Normally aspirated The method by which an internal combustion engine draws air into the combustion chamber. As the piston moves downward in the cylinder, it creates a vacuum that draws air into the combustion chamber through the intake manifold.

Aspirado normalmente El método por el cual un motor de combustión interno hace entrar al aire en la cámara de combustión. Al moverse hacia abajo el pistón en el cilindro, crea un vacío que hace entrar el aire a la cámara de combustión al travéz del múltiple de admisión.

Octane number A unit of measurement on a scale intended to indicate the tendency of a fuel to detonate or knock.

Índice de octano Una unidad de medida en una gama cuyo intención es de indicar la tendencia de un combustible de causar una detonación o los golpeteos.

OEM parts Parts made by the original vehicle manufacturer.

Partes EOF Las partes hechas por el fabricante original del vehículo.

Offset Placed off center. Also, the measurement between the center of the rim and the point where a wheel's center is mounted.

Descentrado Ubicado fuera del centro. Tambien, la medida entre el centro del ancho del rim y el punto en que se monta la rueda.

Ohm A unit of measured electrical resistance.

Ohmio Una unidad de medida de la resistencia eléctrica.

Ohm's law A basic law of electricity expressing the relationship between current, resistance, and voltage in any electrical circuit. It states that the voltage in the circuit is equal to the current (in amperes) multiplied by the resistance (in ohms).

Ley de los ohmios Una ley básica de la electricidad que describe la relación entre el corriente, la resistencia, y el voltaje en cualquier circuito eléctrico. Declare que el voltaje en un circuito egala al corriente (en amperios) multiplicado por la resistencia (en ohmios).

Oil filter A component located near the oil pump that removes abrasive particles from the motor oil by a straining process as the oil circulates through the lubrication system.

Filtro de aceite Un componente situado cerca de la bomba de aceite, que retira partículas abrasivas del aceite del motor mediante un proceso de cribado al pasar el aceite por el sistema de lubricación.

Oil seal A seal around a rotating shaft or other moving part that is used to prevent oil leakage.

Sello de aceite Un sello en torno a un eje giratorio u otra pieza móvil, que sirve para impedir que haya fugas de aceite.

Open circuit An electrical circuit that has a break in the wire.

Circuito abierto Un circuito eléctrico que tiene una quebradura en el alambre.

Open loop An electronic control system in which sensors provide information, the microcomputer gives orders, and the output actuators obey the orders without feedback to the microcomputer.

Bucle abierto Un sistema de control electrónico en el cual los detectores proveen la información, el microcomputador da los ordenes, y los actuadores de salida obedecen a éstas ordenes sin realimentación a la microcomputadora.

Orifice A precisely sized hole that controls fluid flow. Also, an opening.

Orificio Un hoyo de tamaño preciso que controla el flujo de los líquidos. Tambien, una apertura.

Oscillation Any single swing of an object back and forth between the extremes of its travel.

Oscilación Cualquier movimiento unico de un objeto hacia adelante y hacia atrás entre los extremos de su recorrido.

Out-of-round An inside or outside diameter, designed to be perfectly round, having varying diameters when measured at different points across its diameter.

Ovulado Un diámetro interior o exterior, por diseño perfectamente redondo, que tiene varios diámetros cuando se mide en puntos distinctos al través de su diámetro.

Outboard joint The CV joint that connects the drive axle to the wheel spindle.

Junta exterior La junta de velocidad constante que conecta el eje motor al husillo de la rueda.

Overbore The dimension by which a machined hole is larger than the standard size.

Rectificación sobremedida La dimensión por la cual un orificio maquinado es más grande del tamaño normal.

Overdrive Normally used to express a gear ratio that allows the driven gear to rotate faster than the drive gear.

Sobremarcha Normalemente se usa para describir una relación de engrenaje que permite que el engrenaje arrastrado gira más rapidamente que el engrenaje propulsor.

Overlap The period of time during which the intake and exhaust valves are open at the same time.

Temporización El periódo del tiempo en el cual las válvulas de admisión y de escape estan abiertas a la misma vez.

Override clutch The mechanism that disengages the starter from the engine as soon as the engine turns more rapidly than the starter has cranked it.

Embrague de invalidación El mecanismo que desembraga el motor de arranque en el momento que el motor gira más rapidamente que el motor de arranque.

Oversquare A term used to describe an engine that has a larger bore than stroke.

Oversquare Usado para describir un motor que tenga un cilindro con un diámetro interior más grande que la longitud del movimiento del pistón.

Oxidation The combination of a substance with oxygen to produce an oxygen-containing compound. Also, the chemical breakdown of a substance or compound caused by its combination with oxygen.

Oxidación La combinación de una sustancia con el oxígeno para producir una compuesta que tenga oxígeno. Tambien, una descomposición química de una sustancia o una compuesta causada por su combinación con el oxígeno.

Oxidation inhibitor Gasoline additives used to promote gasoline stability by controlling gum and deposit formation and staleness.

Inhibitor del oxígeno Los aditivos de la gasolina que promueban la estabilidad de la gasolina por medio de controlar la formación del sarro y los depósitos y el rancidez.

Oxidation rate Speed by which oxygen is taken into a substance.

Tasa de oxidación La velocidad en que el oxígeno se absorbe por una sustancia.

Oxides of nitrogen Various compounds of oxygen and nitrogen that are formed in the cylinders during combustion and are part of the exhaust gas.

Óxido de nitrógeno Varias compuesta del oxígeno y el nitrógeno que se forman en los cilindros durante la combustión y son parte del vapor del escape.

Oxygen (O_2) sensor An input sensor that sends a voltage signal to the computer in relation to the amount of oxygen in the exhaust stream.

Sensor de oxígeno (O_2) Es un sensor de entrada que le envía una señal de voltaje a la computadora referente a la cantidad de oxígeno en la pipa del escape.

Parallel circuit In this type of circuit, there is more than one path for the current to follow.

Circuito paralelo En este tipo de circuito, hay más que una senda en que puede viajar el corriente.

Parallelogram A four-sided figure in which opposite sides are parallel.

Paralelogramo Una figura de cuatro lados en la que los lados opuestos son paralelos.

Parasitic load An electrical load that is still present when the ignition switch is off.

Carga parásita Una carga eléctrica que siga presente cuando el interruptor de encendido esta apagado.

PCM Power control module.

PCM Módulo de control de la potencia.

PCV valve Positive crankcase ventilation valve. A valve that delivers crankcase vapors into the intake manifold rather than allowing them to escape to the atmosphere.

Válvula de ventilación positiva del cárter Una válvula que conduce vapores del cigüeñal al múltiple de admisión, en lugar de dejar que se descarguen a la atmósfera.

Peen To stretch or clinch over by pounding with the rounded end of a hammer.

Martillazo Estirar o remachar con la extremidad redondeado de un martillo de bola.

Periphery The external boundary of the torque converter.

Periferia El límite externo del convertidor de par.

Permeable A characteristic of materials that absorb fluids.

Permeable Una caracteristica de los materiales que absorben los líquidos.

pH scale A scale used to measure how acidic or basic a solution is.

Escala del pH Un sistema que mide el grado de acidez de una solución.

Phase Refers to the rotational positions of the various elements of a driveline.

Fase Se refiere a las posiciones giratorias de los varios elementos de una flecha motríz.

Phosgene gas A poisonous gas formed when R-12 is heated.

Gas fosgeno Un gas venenoso que forma al calentarse el R-12.

Pickup coil A weak permanent magnet and wire assembly that in combination with a reluctor forms a position sensor.

Bobina de captación Una asamblea de alambre y un imán débil permanente que en combinación con un reluctor forma un detector de posición.

Piezoresistive A characteristic of something that changes resistances in relationship to changes in pressure.

Piezoresistivo Una característica de algo que cambia la resistencia relativamente con los cambios de la presión.

Pilot bushing A plain bearing fitted in the end of a crankshaft. The primary purpose is to support the input shaft of the transmission.

Buje piloto Chumacera simple que se adapta al extremo de un cigüeñal. Su finalidad primordial es la de soportar al eje de entrada de la transmisión.

Pilot journal A mechanical journal that slides inside the pilot bearing and guides its movement.

Muñon piloto Un muñon mecánico que desliza dentro del cojinete piloto y guía su movimiento.

Pinion gear The smaller of two meshing gears.

Piñón diferencial El más pequeño de dos engranajes enganchados uno al otro.

Pinning Cold crack repair process involving the installation of tapered plugs in the crack or on either side of the crack.

Chavetear Un proceso de reparar las grietas sin calor que involucra la instalación de los espárragos cónicos en las grietas o a un lado de la grieta.

Pintle The center pin used to control a fluid passing through a hole; a small pin or pointed shaft used to open or close a passageway.

Clavija La aguja central que controla al flúido que pasa por un hoyo; una pequeña clavija o flecha puntiaguda que sirve para abrir o tapar un pasillo.

Piston An engine component in the form of a hollow cylinder that is enclosed at the top and open at the bottom. Combustion forces are applied to the top of the piston to force it down. The piston, when assembled to the connecting rod, is designed to transmit the power produced in the combustion chamber to the crankshaft.

Pistón Un componente del motor que consiste de un cilindro hueco cerrado en la parte de arriba y abierto en la parte de abajo. Las fuerzas de combustión se aplican en la parte superior del pistón para forzarlo hacia abajo. El pistón, al conectarse a la biela, es diseñado para transmitir la fuerza producida en la cámara de combustión al cigüeñal.

Piston rings Components that seal the compression and expansion gases and prevent oil from entering the combustion chamber.

Anillos (aros) del pistón Los componentes que sellan los gases de la compresión y la expansión y previenen que entre el aceite en la cámara de combustión.

Pitch The angle of the valve spring twist. A variable pitch valve spring has unevenly spaced coils.

Paso El ángulo de las espiras de un resorte de válvula. Un resorte de válvula con paso variable tiene separaciones inegales entre las espiras.

Pitch gauge A tool used to measure the thread pitch of a bolt.

Comprobador de paso de rosca Una herramiento que sirve para medir el paso de las roscas de un perno.

Pitman arm A steering linkage component that connects the steering gear to the linkage at the left end of the center link.

Brazo de dirección (biela de mando) Componente de enlace del sistema de dirección, que conecta el mecanismo al extremo izquierdo del enlace central.

Pitting Surface irregularities resulting from corrosion.

Picadura Las irregularidades de la superficie causadas por la corrosión.

Planetary gear set A group of gears named after the solar system because of their arrangement and action. This unit consists of a center (sun) gear around which pinion (planet) gears revolve. The assembly is placed inside a ring gear having internal teeth. All gears mesh constantly. Planetary gear sets may be used to increase or decrease torque and/or obtain neutral, low, intermediate, high, or reverse.

Juego de engranajes planetarios (piñones satélites) Un grupo de engranajes cuyo nombre se inspira en el sistema solar por su disposición y su modo de actuar. Esta unidad consiste en un engranaje central (sol) en torno al que giran piñones diferenciales (planetas). El conjunto se coloca al interior de un engranaje anular con dientes internos. Todos los engranajes están enganchados continuamente. Estos juegos se pueden utilizar para hacer que aumente o disminuya el par de torsión y/u obtener velocidades neutra, baja, intermedia, alta o de retroceso.

Plasma An ionized gas and the fourth state of matter after solid, liquid, and gas.

Plasma Un gas ionizado y el estado cuarto después de sólido, de líquido, y del gas.

Plastigage A special wax used to measure bearing clearances.

Plastigage (calibrador de plástico) Una cera especial que sirve para medir las holguras.

Play Movement between two parts.

Juego El movimiento entre dos partes.

Pneumatic Operated by compressed air.

Neumático Operado por el aire comprimido.

Polarity The particular state, either positive or negative, with reference to the two poles or to electrification.

Polaridad Un estado particular, sea positivo o negativo, en referencia a los dos polos o a la electrificación.

Poppet valve A valve consisting of a round head with a tapered face, an elongated stem that guides the valve, and a machined slot at the top of the stem for the valve spring retainer.

Válvula champiñón Una válvula que consiste de una cabeza redonda de cara cónica, un vástago alargado que guía a la válvula, y una muesca maquinada en la parte superior del vástago para el retenedor del resorte de la válvula.

Porosity Tiny holes in a casting caused by air bubbles.

Porosidad Los pequeños agujeros en una pieza moldeada causados por las burbujas del aire.

Port fuel injection A fuel injection system that uses one injector at each cylinder, thus making fuel distribution exactly equal among all of the cylinders.

Inyección de combustible por aperturas Un sistema de inyección de combustible que usa un inyector en cada cilindro, así proporcionando una distribución del combustible exactamente igual en todos los cilindros.

Positive displacement pumps Oil pump through which a fixed volume of oil passes with each revolution of its drive shaft.

Bombas de desplazamiento positivo Una bomba de aceite por el cual pasa un volumen fijo de aceite con cada revolución del eje propulsor.

Postcombustion control systems Emission control systems that clean up the exhaust gases after the fuel has been burned.

Controles de escape poscombustión Los sistemas de control de emisiones que purifican los vapores de gas después de que se haya quemado el combustible.

Potential energy Stored energy.

Energía potencial La energía almacenada.

Potentiometer A variable resistor that acts as a circuit divider to provide accurate voltage drop reading in response to the movement of an object.

Potenciómetro Un resistor variable que sirve de divisor de tensión para proveer una lectura precisa de una caída del voltaje según el movimiento de un objeto.

Power A measure of work being done.

Potencia Una medida del trabajo que se efectúa.

Power brake booster Used to increase pedal pressure applied to a brake master cylinder.

Reforzador de freno mecánico Se usa para incrementar la presión de pedal que se aplica al cilindro maestro del freno.

Power steering pump A hydraulic pump driven by a belt from a crankshaft pulley to provide up to 1,300 psi (8,964 kPa) "boost" pressure necessary to operate the power-steering system.

Bomba de dirección hidráulica (asistida) Una bomba hidráulica impulsada por una banda a partir de una polea de cigüeñal para proporcionar hasta 8,964 kPa (1,300 lbs/pulgada2) de presión "de refuerzo" necesaria para manejar el sistema de dirección hidráulica (asistida).

Power stroke The third stroke of the four-stroke engine cycle, in which the compressed fuel mixture is ignited in the combustion chamber.

Carrera encendido-expansión La tercera carrera de un ciclo de un motor de cuatro tiempos, en la cual se encienda la mezcla de aire/combustible en la cámara de combustión.

Precombustion control system Emission control systems that prevent emissions from being created in the engine, either during or before the combustion cycle.

Control de escape precombustión Los sistemas de control de emisión que previenen que se crean las emisiones en el motor, sea durante o antes del ciclo de combustión.

Preheating The application of heat as a preliminary step to some further thermal or mechanical treatment.

Precalentamiento La aplicación del calor como un paso preliminario de un tratamiento termal o mecánico.

Preignition The process of a glowing spark or deposit igniting the air/fuel mixture before the spark plug.

Detonación El proceso en que una chispa o un depósito caliente enciende la mezcla de aire/combustible antes de que lo haga la bujía.

Preload A thrust load applied to bearings that support a rotating part to eliminate axial play or movement.

Carga previa Una carga de empuje aplicada a los cojinetes que soportan una parte giratoria con el fin de eliminar el juego o movimiento axial.

Pressure The exertion of force upon a body in contact with it. Pressure is developed within the cooling system and is measured in pounds per square inch on a gauge.

Presión El esfuerzo de una fuerza sobre un cuerpo con el que esta en contacto. La presión se desarrolla dentro del sistema de enfriamiento y se mide en libras por pie cuadrado con un calibre.

Preventive maintenance Normally scheduled maintenance designed to prevent vehicle failures.

Mantenimiento preventivo El mantenimiento periódico con el fin de prevenir los fallos del vehículo.

Primary circuit The low-voltage circuit of an ignition system.

Circuito primario El circuito de bajo voltaje de un sistema de encendido.

Program A set of instructions or procedures that a computer must follow when controlling a system.

Programa Un conjunto de instrucciones o procedimientos que debe cumplir una computadora en el control de un sistema.

Proportioning valve A pressure reduction valve used in the rear brake circuit.

Válvula dosificadora Una válvula de reducción de presión que se usa en un circuito trasero de los frenos.

Pulley A wheel with a grooved rim in which a rope, belt, or chain runs to raise something by pulling on the other end of the rope, belt, or chain.

Polea Una rueda con un borde acanalado, en el cual mueve una cuerda, una correa, o una cadena. Es posible levantar un objeto pesado tirando en el otro extremo de la cuerda, de la correa, o de la cadena.

Pulse width The length of time in milliseconds that an injector is energized.

Anchura de impulso La cantidad del tiempo en milisegundos en la cual se acciona un inyector.

Purge To separate or clean by carrying off gasoline fumes. The carbon canister has a purge line to remove impurities.

Purgar Separar o limpiar al vaciar los vapores de la gasolina. El recipiente lleno de carbón tiene una linea de purgar para quitarle las impurezas.

Purge control valve A valve in some evaporative emission-control system charcoal canisters to limit the flow of vapor and air to the carburetor during idle.

Válvula de control de purga Una válvula de algunos recipientes de carbón del sistema de control de emisiones de evaporación para limitar el flujo de vapor y aire al carburador durante la marcha de vacío o en punto muerto.

Pushrod A connecting link between the lifter and rocker arm. Engines designed with the camshaft located in the block use pushrods to transfer motion from the lifters to the rocker arms.

Varilla de empuje Es una conexión entre los buzos y el balancín de las válvulas. Los motores diseñados con el árbol de levas en el bloque usan las varillas de empuje para transferir el movimiento de la buzos a los balancines.

Pushrod guide plates The devices used in some engines to limit side movement of the pushrods.

Placas de guía de varilla empujadora Los dispositivos que se usan en algunos motores para limitar el movimiento lateral de las varillas empujadoras.

Quad driver A group of transistors in a computer that controls specific outputs.

Ejecutor cuarteto Un grupo de transistores en una computadora que controla las salidas específicas.

Quenching The cooling of gases by pressing them into a thin area.

Templar El enfriamiento de los gases al oprimirlos en una area pequeña.

R-12 An air-conditioning system refrigerant that contains chlorofluorocarbons. When released into the atmosphere, this refrigerant is harmful to the earth's ozone layer.

R-12 Un refrigerante del sistema de aire acondicionado que contiene los clorofluorocarbonados. Al descargarse este refrigerante a la atmósfera, causa daños a la capa de la ozona de la tierra.

R-134a An air conditioning system refrigerant that replaces R-12. R-134a does not deplete the earth's ozone layer.

R-134a Un refrigerante del sistema del aire acondicionado que reemplaza el R-12. R-134a no daña la capa de ozona de la tierra.

Race A channel in the inner or outer ring of an antifriction bearing in which the balls or rollers operate.

Pista Un canal en el anillo interior o exterior de un cojinete antifricción en el cual operan las bolas o los rodillos.

Rack and pinion steering A steering system in which the end of the steering shaft has a pinion gear that meshes with a rack gear.

Dirección de piñón y cremallera El extremo del eje de dirección tiene un piñón que se endenta con un engranaje de cremallera.

Radial Perpendicular to the shaft or bearing bore.

Radial Perpendicular a la flecha o al taladro del cojinete.

Radial load A load that is applied at 90 degrees to an axis of rotation.

Carga Radial Una carga aplicada a 90 grados del eje de la rotación.

Radiation The transfer of heat by rays, such as heat from the sun.

Radiación La transferencia del calor por medio de los rayos, tal como el calor del sol.

Radiator A coolant-to-air heat exchanger. The device that removes heat from coolant passing through it.

Radiador Un intercambiador de calor hacia el aire fresco. El dispositivo que remueve el calor del refrigerante que pasa atraves de él.

Radiator cap A cap that seals in pressure from hot expanding coolant until a predetermined limit is reached, the valve opens, allowing excess pressure to escape, generally to a coolant-recovery tank.

Tapón del radiador Un tapón que sella al interior la presión de un refrigerante que se dilata al calentarse, hasta llegar a un límite predeterminado en el que se abre la válvula, permitiendo que escape la presión excesiva, por lo común a un depósito de recuperación de refrigerante.

Random-access memory (RAM) A type of memory used to store information temporarily.

Memoria de acceso aleatorio Un tipo de memoria que sirve para archivar la información temporalmente.

Ratio The relation of proportion that one number bears to another.

Relación La relación entre la proporción de un número a otro.

Reach A design feature of a spark plug. It is the length of the threaded portion of a spark plug.

Alcance Un rasgo del diseño de una bujía. Se trata de la longitud de la porción fileteada de una bujía.

Reactivity The characteristic of a material that enables it to react violently with water or other materials. Materials that release cyanide gas, hydrogen sulfide gas, or similar gases when exposed to low pH acid solutions are also said to possess reactivity, as are materials that generate toxic mists, fumes, vapors, and flammable gases.

Reactividad La característica de una materia que facilita que reacciona violentamente con el agua o con otras materias. Las materias que descargan el gas cianuro, el gas sulfhídrico, o los gases semejantes al exponerse a las soluciones de bajo pH se consideran tener la propriedad de reactividad, así como son las materiales que crean la toxicidad en forma de la llovizna, el humo y el vapor y los gases inflamables.

Read-only memory (ROM) A type of memory in an automotive microcomputer used to store information permanently.

Memoria de sólo lectura Un tipo de la memoria en una computador automotríz que sirve para archivar la información permanentemente.

Reaming A technique used to repair worn valve guides either by increasing the guide hole size to take an oversize valve stem or by restoring the guide to its original diameter.

Escariado Un método para reparar las guías de las válvulas gastadas sea por aumentar el tamaño del hoyo de la guía para que accepta un vástago más grande o por restaurar el diámetro original de la guía.

Rebound An expansion of a suspension spring after it has been compressed as the result of jounce.

Repercusión Una expansión de un muelle de suspensión después de que se haya comprimido por razón de un sacudo.

Receiver/dryer A refrigerant reservoir and dryer that supplies liquid refrigerant to the expansion valve.

Colector desecador Un depósito desecador de refrigerante líquido que suministra éste último a la válvula de dilatación (expansión).

Recess A shaped hollow space on a part.

Muesca Un espacio hueco formado en una parte.

Recirculating ball steering gear A popular low friction steering gear box. A sector gear meshes with a ball nut that rides on ball bearings on the worm shaft to provide a smooth steering feel.

Engranaje de dirección de rótula de recirculación Un cárter de dirección de baja fricción muy usado. Un sector dentado se engancha en una tuerca esférica que va sobre cojinetes de bolas (baleros) en el eje sinfín para proporcionar una sensación de manejo suave.

Reciprocating An up-and-down or back-and-forth motion.

Reciprocante Un movimiento de arriba y abajo o de un lado a otro.

Recovery tank See Expansion tank.

Depósito de recuperación Vea Depósito de expansión.

Rectifier An electrical device used to convert alternating current to direct current.

Convertidor estático (rectificador) Un dispositivo eléctrico que se usa para convertir la corriente alterna en continua.

Rectify To change one type of voltage to another.

Rectificar Cambiar un tipo del voltaje por otro.

Reduction The removal of oxygen from exhaust gases.

Reducción Remover el oxígeno de los gases de escape.

Reference voltage A voltage provided by a voltage regulator to operate potentiometers and other sensors at a constant level.

Voltaje de referencia Un voltaje proporcionado por un regulador de voltaje para operar a los potenciómetros u otros detectores en un nivel constante.

Refrigerant The chemical compound used in a refrigeration system to produce the desired cooling.

Refrigerante El compuesto químico utilizado en un sistema de refrigeración para producir el enfriamiento deseado.

Relay An electrical switching device that uses a low-current circuit to control a high-current circuit.

Relé Un dispositivo interruptor eléctrico que usa un circuito de bajo corriente para controlar un circuito de alta corriente.

Relief The amount one surface is set below or above another surface.

Relieve La cantidad por el cual una superficie queda arriba o abajo de otra superficie.

Reluctance A term used to indicate a material's resistance to the passage of magnetic lines of flux.

Reluctancia Un término que indica la resistencia de una materia al pasaje de las lineas de flujo magnético.

Residual Remaining or leftover pressure.

Residual La presión que queda o sobra.

Residue Surplus or what remains after a separation takes place.

Residuo Lo que sobra o se queda después de que se efectúa una separación.

Resilience Elastic or rebound action.

Resiliencia La acción elástica o de repercusión.

Resistance The opposition offered by a substance or body to the passage of electric current through it.

Resistencia La oposición que ofrece una sustancia o un cuerpo al pasaje de la corriente eléctrica que lo atraviesa.

Resonator A second muffler in line with the other muffler.

Resonador Un silenciador secundario en linea con el primario.

Reverse bias A positive voltage applied to N-material and a negative voltage applied to P-material in a semiconductor.

Polarización inversa Un voltaje positivo aplicado a un cristal N y un voltaje negativo aplicado al cristal P en un semiconductor.

Rheostat A two-terminal variable resistor used to regulate electrical current.

Reóstato Un resistor variable con dos terminales que sirve para regular al corriente eléctrico.

Rich An air/fuel mixture that has more fuel than is required for a stoichiometric mixture.

Rica Una mezcla de aire/combustible que contiene más combustible de lo que requiere una mezcla estoquiométrica.

Right-to-know laws Laws requiring employers to provide employees with a safe workplace as it relates to hazardous materials.

Leyes de derechos de los empleados Las leyes que requieren que los patrones proveen sus empleados con un ambiente del trabajo libre de los peligros asociados con las materiales tóxicas.

Ring gear 1) The gear around the edge of a flywheel. 2) A large, circular gear such as that in the final drive assembly.

Engranaje anular (anillo dentado) 1) El engranaje en torno al borde de un volante. 2) Un gran engranaje circular como el del ensamblaje de transmisión final.

Ring lands The high parts of a piston between the grooves.

Tierras de las coronas Las partes altas de un pistón entre las ranuras.

Rocker arm Pivots that transfer the motion of the pushrods or followers to the valve stem.

Balancín Un punto de pivote que transfiere el movimiento de los buzos o de los seguidores al vástago de la válvula.

Rolling resistance A term used to describe the amount of resistance a tire has to rolling on the road. Tires that have a lower rolling resistance usually get better gas mileage. Typically, radial tires have lower rolling resistance.

Resistencia a la rotación Un término que describe la cantidad de resistencia que tiene un neumático a rodar en el camino. Los neumáticos que tienen una resistencia a rodar más baja suelen proporcionar un kilometraje más económico.

Rotary A circular motion.

Rotativo Un movimiento circular.

Rotary flow Torque converter oil flow associated with the coupling stage of operation.

Flujo rotativo El flujo del aceite del convertidor de par que se asocia con la etapa de acoplamiento de su operación.

Rotor The rotating or freewheeling portion of a clutch; the belt slides on the rotor.

Rotor La parte giratoria o con marcha libre de un embrague; la correa/banda se desliza sobre el rotor.

Rotor pump A type of oil pump that utilizes two rotors, a four-lobe inner, and a five-lobe outer; output per revolution depends on rotor diameter and thickness.

Bomba de tipo rotor Un tipo de bomba de aceite que utiliza dos rotores, uno interior con cuatro resaltes, y uno exterior con cinco rebajes; la producción por revolución depende del diámetro y espesor del rotor.

RTV A formed-in-place gasket product used in place of conventional paper, cork, and cork/rubber gaskets.

Sellador RTV Un empaque que se forma en sitio que se usa en vez de los sellos más convencionales del papel, del caucho o de caucho con corcho.

Runner A case tube on an intake or exhaust manifold used to carry air in or out of the engine.

Tubería Un cárter tubular en un múltiple de admisión o escape que sirve para hacer entrar o sacar el aire del motor.

Runout Out-of-round or wobble.

Corrimiento Ovulado o con una oscilación irregular.

Sampling The act of periodically collecting information, as from a sensor. A microcomputer samples input from various sensors in the process of controlling a system.

Muestreo Coleccionar la información periódicamente, tal como de un detector. Una microcomputadora toma las muestras de varios detectores en el proceso de controlar a un sistema.

Saturation The point reached when current flowing through a coil or wire has built up the maximum magnetic field.

Saturación El punto en que el corriente que fluye por una bobina o un alambre llega a lo máximo de un campo magnético.

Scale A flaky deposit occurring on steel or iron. Ordinarily used to describe the accumulation of minerals and metals in an automobile cooling system.

Incrustación Un depósito que ocurre en el acero o el hierro. Típicamente se usa para describir la acumulación de los minerales y los metales en un sistema de enfriamiento automotríz.

Scan tool A microprocessor designed to communicate with a vehicle's on-board computer to perform diagnosis and troubleshooting.

Herramienta de exploración Un microprocesador diseñado para comunicar con una computadora a bordo de un vehículo para efectuar los prognósticos y localización de fallas.

Schematics Wiring diagrams used to show how circuits are constructed.

Esquemáticos Los dibujos de los conexiones que sirven para enseñar cómo se han construídos los circuitos.

S-class oil Standard engine oil designed for use in passenger cars and light trucks.

Aceite de aplicaciones S El aceite de motor común que se ha diseñado para usarse en los coches de pasajeros y en los camiones ligeros.

Score A scratch, ridge, or groove marring a finished surface.

Rayo Un rasguño, una arruga, o una muesca que echa a perder una superficie acabada.

Screw pitch gauge A tool used to check the threads per inch (pitch) of a fastener.

Calibrador de pasos de rosca Una herramienta que sirve para examinar las cuerdas por pulgada (el paso) de un elemento de fijación.

Scuffing Scraping and heavy wear from the piston on the cylinder walls.

Rozamiento La raspadura y el gasto extremo de un pistón en los paredes del cilíndro.

Seal Generally refers to a compressor shaft oil seal; matching shaft-mounted seal face and front head-mounted seal seat to prevent refrigerant and/or oil from escaping. May also refer to any gasket or O-ring used between two mating surfaces for the same purpose.

Junta hermética Generalmente se refiere a la junta hermética en el eje del compresor; el eje hace juego con el montaje de la cara del sello y el sello instalado con la cabeza mirando hacia adelante se asienta para prevenir que el refrigerante y/o aceite se escape. Puede referirse también a cualquier junta o sello de tipo anillo O utilizada entre dos superficies que hacen juego para el mismo propósito.

Sealant Material used as a seal between two mating objects.

Sellante Una material que sirve de empaque entre dos objetos de contacto.

Seat A surface, usually machined, on which another part rests or seats; for example, the surface on which a valve face rests.

Asiento Una superficie, típicamente maquinada, sobre la cual otra parte descanza o se asienta; por ejemplo, una superficie en la que descansa la cara de la válvula.

Secondary circuit A circuit in the ignition system that uses 20,000 or more volts to operate. It includes the secondary coil windings, the rotor, distributor cap, coil and spark plug wires, and spark plugs.

Circuito secundario Circuito del sistema de encendido que utiliza 20,000 voltios o más para funcionar. Incluye devanados secundarios de la bobina, rotor, tapa del distribuidor, bobina, bujías y sus cables.

Seize When one surface moving on another scratches. An example is a piston score or abrasion in a cylinder due to a lack of lubrication or overexpansion.

Agarrotar Cuando una superficie moviendo en otra la raya. Otro ejemplo sería un rayo o abrasión en un cilindro debido a la falta de lubricación o la sobreexpansión.

Semifloating A design of live axle in which only part of the vehicle's weight is supported by the axles.

Semiflotante Un diseño de un eje motor en el que sólo un porción del peso del vehículo se soporta por los ejes.

Sending unit A device that sends a signal to a gauge regarding oil pressure or coolant temperature.

Unidad emisora (detectora) Un dispositivo que envía una señal a un medidor sobre la temperatura del refrigerante o la presión del aceite.

Sensor Any device that provides an input to the computer.

Detector Cualquier dispositivo que provee una entrada para la computadora.

Serpentine belts Multiple-ribbed belts used to drive water pumps, power-steering pumps, air-conditioning compressors, alternators, and emission control pumps.

Correa serpentín Las bandas dentadas que sirven para propulsar las bombas de agua, las bombas de dirección hidráulica, los compresores acondicionadores, las alternadores y las bombas de control de emisiones.

Servo The part of a cruise control system that maintains the desired car speed by receiving a controlled amount of vacuum from the transducer.

Servo La parte de un sistema de control crucero que mantiene la velocidad deseada por medio de una cantidad de vacío que recibe y se controla por el transconductor.

Shift forks Semicircular castings connected to the shift rails that help control the movement of the synchronizer.

Horquillas de cambio de velocidades Partes moldeadas de forma semicircular conectadas a las palancas de cambio para ayudar en el control del movimiento del sincronizador.

Shift rails The parts of a transmission shift linkage that transfer motion from the driver-controlled gear shift lever to the shift forks.

Palancas de cambio Las partes de una biela de desembrague de una transmisión que transfieren el movimiento de la palanca del desembrague manual a las horquillas de desembrague.

Shim Thin sheets, usually metal, which are used as spacers between two parts, such as the two halves of a journal bearing.

Chapa Las hojas delgadas, por lo regular del metal, que sirven de arandelas entre dos partes, tal como las dos mitades de un cojinete de contacto plano.

Shock absorber A device that dampens spring oscillations by converting the energy from spring movement into heat energy.

Amortiguador Un dispositivo que absorbe las oscilaciones de los muelles, convirtiendo la energía del movimiento de estos últimos en energía térmica.

Short Of brief duration; for example, short cycling. Also refers to an intentional or unintentional grounding of an electrical circuit.

Breve/corto De una duración breve; funcionamiento cíclico breve. Se refiere también a una aplicación a tierra previsto o imprevista de un circuito eléctrico.

Short and long arm suspension (SLA) A suspension system using an upper and lower control arm. The upper arm is shorter than the lower arm. This is done to allow the wheel to deflect in a vertical direction with a minimum change in camber.

Suspención de brazo corto y largo Un sistema de suspencíon que usa un brazo de control superior e inferior. El brazo superior es más corto que el brazo inferior. Este arreglo permite que la rueda se desvía en una dirección vertical con un cambio mínimo del camber.

Shrink fit The shaft or part is slightly larger than the hole in which it is to be inserted. The outer part is heated above its normal operating temperature or the inner part chilled below its normal operating temperature and assembled in this condition. Upon cooling, an exceptionally tight fit is obtained.

Calado por contracción La flecha o la parte que se debe insertar es un poquito más grande que el orificio. La parte exterior se calienta a una temperatura más alta de lo normal al funcionar, o la parte interior se enfría a una temperatura más baja de lo normal al funcionar y se asamblea en estas condiciones. Al regresar a la temperatura normal, se obtiene un ajuste extremadamente apretado.

Shudder Momentary shake or quiver; can sometimes be severe.

Estremecimiento Un sacudo o temblor momentáneo; a veces puede ser severo.

Shunt More than one path for current to flow, such as a parallel part of a circuit.

Shunt (en derivación) Cuando hay más que una senda en que puede fluir un corriente, tal como una parte paralela de un circuito.

Shunt circuits The branches of the parallel circuit.

Circuitos en desviación Las ramas del circuito paralelo.

Signal flasher A device activated by the heat of the electricity traveling through it to cause the turn signal bulbs to flash.

Lámpara de señales Un dispositivo activado por el calor producido por la electricidad que lo alimenta para hacer que parpadeen (destellen) las luces direccionales.

Sine wave Basically, this wave is a circle drawn over time. It appears as a wave with equal positive and negative magnitudes and amplitudes.

Onda senoidal Basicamente, esta onda es un círculo que se dibuja al través del tiempo. Aparece como una onda con magnitudes y amplitudes positivos y negativos egales.

Sleeving A means of reconditioning an engine by boring the cylinder oversize and installing a thin metal liner called a sleeve. The inside diameter of the sleeve is then bored, usually to the original or standard piston size.

Restaurar con manguito Un metodo de recondicionar a un motor por medio de taladrar más grande al cilindro e instalar un forro de metal delgado llamado un manguito. Luego se taladra al diámetro interior del manguito al tamaño normal o indicado según el pistón.

Sliding fit Where sufficient clearance has been allowed between the shaft and journal to permit free running without overheating.

Ajuste deslizante Cuando una holgura suficiente grande se ha dejado entro la flecha y el muñon para permitir la operación sín que se sobrecalienta.

Slip A condition caused when a driving part rotates faster than a driven part.

Deslizamiento Una condición causada cuando una parte propulsor gira más rapidamente que la parte arrastrada.

Slip yoke A component having internal splines that slide on the transmission outputshaft external splines, allowing the driveline to adjust for variations in length as the rear axle assembly moves.

Horqueta deslizante Un componente con ranuras externas que se desliza sobre las acanaladuras externas del eje de salida de la transmisión, permitiendo que la línea de propulsión se adapte a las variaciones de longitud del conjunto del eje posterior al moverse.

Sludge As used in connection with automobile engines, it indicates a composition of oxidized petroleum products along with an emulsion formed by the mixture of oil and water. This forms a pasty substance and clogs oil lines and passages and interferes with engine lubrication.

Sebo (grasa) En referencia con los motores automotrices, indica una composición de productos petroléos oxidados junta con una emulsión formada de una mezcla del aceite y el agua. Esto forma una sustancia pastosa y causa las obstrucciones en las lineas y pasajes del aceite previniendo la lubricación.

Smog Air pollution created by the reaction of nitrogen oxides to sunlight.

Smog La contaminación del aire producido por la reacción de los óxidos de nitrógeno a la luz del sol.

Solenoid An electromagnetic switch with a movable core.

Solenoide Un interruptor electromagnético con un núcleo portátil.

Solution Formed when a solid dissolves into a liquid; its particles break away from this structure and mix evenly in the liquid.

Solución Se forma cuando un sólido disuelve en un líquido. Las partículas del sólido se separan y se mezclan uniformemente en el líquido.

Solvent The liquid in a solution.

Solvente El líquido en una solución.

Spark plug A device used to plug a hole in the combustion chamber while allowing a spark to enter the chamber.

Bujía Un dispositivo que sirve para tapar un hoyo en la cámara de combustión mientras que permite pasar una chispa a la cámara.

Spark plug reach The length of the spark plug shell from the contact surface at the seat to the bottom of the shell, including both threaded and nonthreaded sections.

Alcance de la bujía La longitud del casco de la bujía desde la superficie de contacto del asiento a la parte inferior del casco, incluyendo a las partes fileteadas y no fileteadas.

Specific gravity The weight of a given volume of a liquid divided by the weight of the same volume of water.

Gravedad específica El peso de un volumen dado de un líquido dividido por el peso del mismo volumen del agua.

Speed The distance an object travels in a set amount of time. Speed is the relationship between the distance traveled and the time it takes to travel it.

Velocidad La distancia que viaja un objeto dividido por la medida del tiempo usada para viajar esa distancia. La velocidad es la relación entre el espacio recorrido y el tiempo empleado en recorrerlo.

Speed ratio Comparison of the difference in speed between two moving parts, such as impeller speed and turbine speed.

Relación de la velocidad La comparación de la diferencia en velocidad entre dos partes en movimiento, tal como la velocidad del impulsor y la velocidad de la turbina.

Splay To spread or move outward from a central point.

Achaflanar Desplegar o moverse hacia afuera de un punto central.

Splice To join. Electrical wires can be joined by soldering or by using crimped connectors.

Empalmar Juntar. Los alambres eléctricos se pueden unir por soldado o por medio de los conectores de presión.

Splines External or internal teeth cut into a shaft that are used to keep a pulley or hub secured on a rotating shaft.

Espárragos Los dientes exteriores o interiores cortados en una flecha que sirven para fijar a una polea o un cubo en una flecha rotativa.

Sponginess A feel of a soft brake pedal.

Esponjoso Una sensación blanda en un pedal de freno.

Spontaneous combustion Process by which a combustible material ignites by itself and starts a fire.

Combustión espontánea Un proceso por el cual una material combustible se encienda sí mismo y causa un fuego.

Spring retainer The element that locates and provides reaction for the spring in a one-way roller clutch.

Retén de resorte El elemento que sitúa y proporciona reacción al resorte en un embrague de rodillo en un sentido.

Spur gear A gear with teeth that are cut straight across the gear.

Engrenaje recto Un engrenaje cuyos dientes atraviesan el engranaje en una linea recta.

Square wave Typically a digital waveform that shows the cycling of a circuit or device.

Onda cuadrada Típicamente una forma de onda digital que indica el ciclado de un circuito o un dispositivo.

Squirm To wiggle or twist about a body. When applied to tires, squirm is the wiggle or movement of the tread against the road surface. Squirm increases tire wear.

Encorvadura Torcerse o doblar alrededor de un cuerpo. Al aplicarse a los neumáticos, la encorvadura se refiere a la torción o el movimiento de la banda contra la superficie del camino. La encorvadura aumenta el desgaste de los neumáticos.

Stall speed With the engine operating at full throttle, the gear selection in D range, and the vehicle stationary, this speed can be read on a tachometer.

Velocidad de calar Al estar el motor en marcha a todo gas, el engrenaje selecionado en el rango D, el vehículo estacionario, este velocidad se lee en el taquímetro.

Stalling A condition in which the engine dies after starting.

Paro Una condición en la cual el motor para después de ser encendido.

Stamping A piece of sheet metal cut and formed into the desired shape with the use of dies.

Estampar Una pieza de metal de hoja que se corta y se forma a lo deseado con los matrices.

Star wheel adjuster A device used to adjust the clearance between the brake shoes and the brake drum.

Ajustador de rueda estrellada Un dispositivo utilizado para ajustar el franqueo (la holgura) entre las zapatas y el tambor del freno.

Starter drive The part of the starter motor that engages the armature to the engine flywheel ring gear.

Transmisión de arranque Pieza del motor de arranque que engrana el inducido al engranaje anular del volante del motor.

Starter motor A small electric motor used to turn the engine for starting.

Motor de arranque Un pequeño motor eléctrico utilizado para hacer que el motor gire para que arranque.

Starter relay A magnetic switch, generally operated by the ignition switch, that uses low current to close a circuit to control the flow of very high current to the starter.

Relé (relevador) del motor de arranque Un interruptor magnético, activado generalmente por el interruptor de encendido, que usa una corriente baja para cerrar un circuito y controlar el flujo de corriente muy alta al motor de arranque.

Static balance Balance at rest; still balance. It is the equal distribution of weight of the wheel and tire around the axis of rotation such that the wheel assembly has no tendency to rotate by itself regardless of its position.

Balanceo estático El equilibreo en descanso; el balanceo inmóvil. Es la distribución equilibrada del peso de la rueda y el neumático alrededor del eje de rotación para que la asamblea de la rueda no tenga tendencia a girar por sí mismo, sin que importa su posición.

Static pressure The pressure inside the hydraulic system.

Presión estática La presión dentro del sistema hydráulico.

Stator The name used to describe a part in a torque converter or the stationary windings of an AC generator.

Estátor El nombre que describe una parte en un convertidor de par o el devanado estático de un alternador.

Step-up transformer A transformer in which the voltage created in a secondary coil is greater than the voltage in the primary, or first, coil.

Transformador elevador Una transformador en la cual el voltaje creado en la bobina secundaria es más alta que el voltaje en la bobina primaria.

Stoichiometric Chemically correct. An air/fuel mixture is considered stoichiometric when it is neither too rich nor too lean; stoichiometric ratio is 14.7 parts of air for every part of fuel.

Estequiométrica Lo que es correcto químicamente. Una mezlca de aire/combustible se considera estequiométrica cuando no es demasiada rica o pobre; la relación estequiométrica es 14.7 partes del aire por cada parte del combustible.

Stress The force of strain to which a material is subjected.

Esfuerzo La fuerza de la tensión a la cual se somete una materia.

Stroke A term used to describe cylinder size that represents the amount of movement the piston has inside the bore.

Carrera Un término que describe el tamaño de un cilindro, representa la cantidad del movimiento que tiene el piston dentro del orificio.

Strut Components connected from the top of the steering knuckle to the upper strut mount that maintain the knuckle position and act as shock absorbers to control spring action in a vehicle's suspension system. Used on most front-wheel drive cars and some rear-wheel drive cars.

Riostras Componentes conectados de la parte superior de la charnela de dirección a la montura de tirante superior para mantener la posición de la charnela y actuar como amortiguadores para controlar la acción de resorte del sistema de suspensión de un vehículo. Se utiliza en la mayoría de los automóviles de tracción delantera y en muchos de tracción posterior.

Substrate A ceramic honeycomb grid structure coated with catalyst materials.

Sustrato Una estructura cerámica de rejilla agrietada cubierta con materiales de catalizador.

Suction Suction exists in a vessel when the pressure is lower than the atmospheric pressure.

Succión La succión existe en una vasija cuando la presión es más baja que la presión atmosférica.

Suction line The line connecting the evaporator outlet to the compressor inlet.

Línea de succión La línea que conecta la salida del evaporador a la entrada del compresor.

Supercharger A belt-driven pump, also called a blower.

Sobrealimentador Una bomba de banda, que se denomina también aventador.

Surface tension The result of forces of attraction, which pull on the particles of a liquid in all directions. These forces create liquid bubbles and drops in a spherical shape.

Tensión de superficie El resultado de las fuerzas de atracción, que halan las partículas de un líquido en todas las direcciones. Estas fuerzas crean burbujas y gotas líquidas en una forma esférica.

Surging A condition in which the engine speeds up and slows down with the throttle held steady.

Operación pulsatorio Una condición en la cual el motor acelera y desacelera mientras que la presión de admisión se mantiene fijo.

Sway bar Also called a stabilizer bar. It prevents the vehicle's body from diving into turns.

Barra de oscilación lateral Llamada también barra estabilizadora. Impide que la carrocería del vehículo se clave durante las curvas.

Swirl combustion A swirling of the air/fuel mixture in a corkscrew pattern. The swirling effect improves combustion.

Combustión remolino Un movimiento circular de la mezcla aire/combustible en el patrón de un remolino. El efecto de este movimiento es de amejorar la combustión.

Switch A device used to interrupt a circuit, providing a means of turning it on and off.

Interruptor Un dispositivo usado para interrumpir un circuito, proporcionando un medio de activarlo y desactivarlo.

Tail pipe The part of the exhaust system that allows the exhaust gases to leave the rear of the vehicle and go into the atmosphere.

Tubo de escape La parte del sistema de escape que permite que los gases se descarguen a la atmósfera por la parte trasera del vehículo.

Tang A projecting piece of metal placed on the end of the torque converter on automatic transmissions. The tangs are used to rotate the oil pump.

Lengueta Una pieza sobresaliente de metal ubicado en la extremidad del convertidor del par de una transmisión automática. Las lenguetas sirven para girar a la bomba de aceite.

Tap To cut threads in a hole with a tapered, fluted, threaded tool.

Roscar con macho Cortar las roscas en un agujero con una herramienta cónica, acanalada y fileteada.

Taper The difference in diameter between the cylinder bore at the bottom of the hole and the bore at the top of the hole, just below the ridge.

Ahuso La diferencia en el diámetro del orificio del cilindro en la parte inferior y del orificio en la parte superior de la apertura, justo abajo del reborde.

Tapered roller bearing A bearing containing tapered roller bearings mounted between the inner and outer races.

Balero de rodillos cónicos Los baleros que contienen rodillos cónicos montados entre las pistas interiores y exteriores.

Telescoping gauges Tools used for measuring bore diameters and other clearances; also known as snap gauges.

Calibradores telescópicos Las herramientas que sirven para medir los diámetros de los orificios y otras holguras; tambien se conocen como calibres de ensarte.

Temperature An indication of an object's kinetic energy.

Temperatura La temperatura es el grado de calor, relacionado con la energía cinética de las moléculas de los mismos.

Tempering The heat treatment of metal alloys, particularly steel, to result in specific properties.

Templar El tratar de las aleaciones del metal, particularmente de acero, con calor para dar ciertas características al metal.

Tensile strength The amount of pressure per square inch the bolt can withstand just before breaking when being pulled apart.

Resistencia a la tracción La cantidad de presión por pulgada cuadrada que puede resistir un perno al estirarse justo antes de que se quiebra.

Tension Effort that elongates or stretches a material.

Tracción Un esfuerzo que alarga o estira a una materia.

Thermal cleaning Cleaning that relies exclusively on heat to bake off or oxidize surface contaminants.

Limpieza termal La limpieza que cuenta exclusivamente con el calor para quemar u oxidar los contaminantes de una superficie.

Thermal contraction A decrease in the size of a mass due to the movement of atoms and molecules as heat moves out of a mass.

Contracción termal Una contracción termal es una disminución del volumen de una masa debido al movimiento de átomos y moléculas cuando el calor se mueve de una masa.

Thermal expansion An increase in the size of a mass due to the movement of atoms and molecules as heat moves into a mass.

Expansión termal Una expansión termal es un aumento en el volumen de una masa debido al movimiento de átomos y moléculas cuando el calor entra a una masa.

Thermal expansion valve (TXV) An air-conditioning control valve used to maintain evaporator pressure.

Válvula de dilatación térmica Una válvula de control del aire acondicionado que se usa para mantener la presión del evaporador.

Thermistor A solid-state variable resistor made from semiconductor material that changes resistance in response to changes in temperature.

Termistor Un resistor variable de estado sólido hecho de material semiconductor que cambia su resistencia según los cambios de la temperatura.

Thermo efficiency A gallon of fuel contains a certain amount of potential energy in the form of heat when burned in the combustion chamber. Some of this heat is lost and some is converted into power. The thermal efficiency is the ratio of work accomplished compared to the total quantity of heat contained in the fuel.

Rendimiento térmico Un galón del combustible contiene una cierta cantidad de energía potencial en la forma de calor cuando se quema en la cámara de combustión. Algo de éste calor se pierde y algo se convierte a la fuerza. El rendimiento térmico es la relación del trabajo que se efectúa comparado con la cantidad total del calor que contiene el combustible.

Thermostat A device used to cycle the clutch to control the rate of refrigerant flow as a means of temperature control. The driver has control of the temperature desired.

Termostato Un dispositivo utilizado para ciclar el embrague para regular la proporción del flujo de refrigerante como medio de regulación de temperatura. El accionador puede regular la temperatura deseada.

Thread pitch The number of threads in one inch of threaded bolt length. In the metric system, thread pitch is the distance in millimeters between two adjacent threads.

Paso de rosca El número de las cuerdas por pulgada de un perno con rosca. En el sistema métrico, el paso de las roscas es la medida de la distancia entre dos pasos contiguos en milímetros.

Throw With reference to an automobile engine, usually the distance from the center of the crankshaft main bearing to the center of the connecting rod journal.

Recorrido Al referirse a un motor de automóvil, típicamente es la distancia del cojinete principal del cigueñal al centro de los muñones de la biela.

Throw-out bearing In the clutch, the bearing that can be moved inward to the release levers and clutch-pedal action to cause declutching, which disengages the engine crankshaft from the transmission. A common name for a clutch release bearing.

Cojinete de desembrague En el embrague, el cojinete que puede moverse hacia adentro hasta las palancas de liberación, por medio de la acción del pedal del embrague, para logar el desembrague, desenganchando el cigüeñal del motor de la transmisión. Es un nombre común para designar una chumacera de liberación del embrague.

Thrust line A line that divides the total toe angle of the rear wheels.

Linea de empuje Una linea que divide el ángulo total de la convergencia de las ruedas traseras.

Thrust load Load placed on a part that is parallel to the center of the axis.

Carga de empuje Una carga en una parte que queda paralela al centro del eje.

Tie rod A rod connected from the steering arm to the rack or center link, depending on the type of steering linkage.

Barra de acoplamiento Una varilla conectada desde el brazo de la dirección a la cremallera o a la barra central, dependiendo del tipo de varilla de la dirección.

Time A measurement of the duration of an event that has happened, is happening, or will happen.

Tiempo El tiempo es una medida de la duración de un acontecimiento que ha sucedido, esta sucediendo, o sucedara en el futuro.

Tolerance A permissible variation between the two extremes of a specification or dimension.

Tolerancia Una variación permitible entre dos extremos de una especificación o una dimensión.

Torque A twisting force applied to a shaft or bolt.

Par de ajuste Una fuerza torcedura que se aplica a una flecha o un perno.

Torque converter A turbine device utilizing a rotary pump, one or more reactors (stators), and a driven circular turbine or vane whereby power is transmitted from a driving to a driven member by hydraulic action. It provides varying drive ratios; with a speed reduction, it increases torque.

Convertidor del par de torsión o de torque Una turbina que utiliza una bomba giratoria, uno o más reactores (estator) y una turbina circular accionada o paleta; la fuerza se transmite desde el mecanismo de accionamiento al mecanismo accionado mediante una acción hidráulica. Provee relaciones de accionamiento variadas; con una reducción de velocidad, aumenta el par de torsión.

Torque steer A twisting axle movement in front-wheel drive automobiles that causes a pulling action under acceleration toward the side with the longer driving axle.

Dirección de torsión Un movimiento de torsión del eje en un automóvil de tracción delantera que causa que jale mientras que accelera hacia el lado con un eje motor más largo.

Torsion bar A steel bar connected from the chassis to the lower control arm. As the vehicle's weight pushes the chassis downward, the torsion bar twists to support this weight. Torsion bars are used in place of coil springs.

Barra de torsión Una barra de acero conectada del chasis al brazo de control inferior de la suspensión. Mientras el peso del vehículo presiona el chasis hacia abajo, la barra de torsión se tuerce para soportar este peso. Las barras de torsión se utilizan en lugar de los resortes espirales.

Tracking The travel of the rear wheels in a parallel path with the front wheels.

Seguimiento El progreso de las ruedas traseras en una senda paralela con las ruedas delanteras.

Traction A tire's ability to hold or grip the road surface.

Tracción La abilidad de un neumático para agarrar o mantenerse en la superficie del camino.

Tractive effort Pushing force exerted by the vehicle's driving wheels against the road's surface.

Esfuerzo de tracción Una fuerza de empujo que resulta de tracción de las ruedas contra la superficie del camino.

Tramp Wheel hop caused by static balance.

Pisoteo Un brinquito de la rueda causado por el balanceo estático.

Transaxle A unit that combines a transmission and differential.

Eje de transmisión Una unidad que combina una transmisión y un diferencial.

Transfer case An additional component added to a four-wheel-drive vehicle in which the power from the engine is divided between both the front and rear wheels equally.

Caja de transferencia Componente adicional que se agrega a un vehículo de tracción en las cuatro ruedas, donde la potencia del motor se divide por igual entre las ruedas delanteras y las traseras.

Transistor An electronic device produced by joining three sections of semiconductor materials. A transistor is very useful as a switching device, functioning as either a conductor or an insulator.

Transistor Un dispositivo electrónico producido al unir tres secciones de materiales de semiconductor. Un transistor es muy útil como dispositivo interruptor, funciona como un conductor o como aislador.

Transverse Perpendicular, or at right angles, to a front-to-back centerline.

Transverso Perpendicular, o en ángulo recto, a una linea central de delantero a trasero.

Trimmer A resistor or capacitor that is adjusted to meet the needs of a circuit.

Trimer Un resistor o un condensador que se ajusta según los requerimientos de un circuito.

Trouble codes Output of the self-diagnostics program in the form of a numbered code that indicates faulty circuits or components. Trouble codes are two or three digit characters that are displayed in the diagnostic display if the testing and failure requirements are both met.

Códigos indicadores de fallas Los datos del programa autodiagnóstico en forma de código numerado que indica los circuitos o los componentes defectuosos. Los códigos de problemas son compuestos de dos o tres caracteres digitales que se muestran en el despliegue de diagnóstico si se llenan los requisitos de prueba y de falla.

Turbo boost The positive pressure increase created by a turbocharger.

Reforzar de turbo Un aumento del presión positivo creado por un turbocargador.

Turbocharger A small radial fan pump driven by the energy of the exhaust flow.

Turbosobrealimentador Un pequeño ventilador radial impulsado por la energía del flujo del escape.

Turning torque Amount of torque required to keep a shaft or gear rotating; measured with a torque wrench.

Torsión giratorio La cantidad de torsión requerida para mantener girando a una flecha o un engrenaje; se mide con una llave de torsión.

Type A fire A fire resulting from the burning of wood, paper, textiles, and clothing.

Fuego tipo A Un fuego que resulta de la combustión de la leña, el papel, los textiles, y la ropa.

Type B fire A fire resulting from the burning of gasoline, grease, oils, and other flammable liquids.

Fuego tipo B Un fuego que resulta de la combustión de la gasolina, la grasa, los aceites, y otros líquidos inflamables.

Type C fire A fire resulting from the burning of electrical equipment, motors, and switches.

Fuego tipo C Un fuego que resulta de la combustión del equipo eléctrico, los motores y los interruptores.

Ultrasonic cleaning Method of cleaning that utilizes high-frequency sound waves to create microscopic bubbles that work to loosen soil from parts.

Limpieza ultrasónica Un metodo de limpieza que usa las ondas de sonido de alta frecuencia para crear las burbújas microscópicas que hacen el trabajo de aflojar la suciedad de las partes.

Undersquare A term used to describe an engine that has a larger stroke than bore.

Undersquare Usado para describir un motor que tenga un cilindro con un diámetro interior más pequeño que la longitud del movimiento del pistón.

Unibody A stressed hull body structure that eliminates the need for a separate frame.

Monocasco Una estructura de casco arriostrada que elimina la necesidad de un chassis aparte.

Union A hydraulic coupling used to connect two brake lines.

Unión Un acoplador hidráulico que sirve para conectar dos lineas de freno.

Universal joint A joint that allows the driveshaft to transmit torque at different angles as the suspension moves up and down.

Junta universal Una junta que permite que la flecha motríz transmite el par en los ángulos diferentes mientras que la suspensión mueve hacia arriba y hacia abajo.

Vacuum The absence of atmospheric pressure; commonly used to refer to any pressure less than atmospheric pressure.

Vacio La ausencia de la presión atmosférica; palabra que se utiliza comúnmente para referir a cualquier presión que es menos que la presión atmosférica.

Vacuum fluorescent A type of digital electronic display utilizing glass tubes filled with argon or neon gas; when current is passed through the tubes, they glow brightly.

Flourescente en vacío Un tipo de muestra digital que utiliza los tubos de vidrio llenos del vapor de argo o neo; al pasar el corriente por estos tubos, brillan mucho.

Valve Device that controls the flow of gases into and out of the engine cylinder.

Válvula Un dispositivo que controla el flujo de los gases entrando y saliendo del cilindro del motor.

Valve body The hydraulic control assembly of the transmission.

Cuerpo de válvula El conjunto del control hidráulico de la transmisión.

Valve float A condition in which the valves fail to follow the actions of the valve train and remain open.

Acción flotante de la válvula Una condición en la cual las válvulas no siguen los movimientos del tren de válvulas y quedan abiertas.

Valve lifter A cylindrically shaped hydraulic or mechanical device in the valve train that rides on the camshaft lobe to lift the valve off its seat.

Filtro de válvula Un dispositivo hidráulico o mecánico de forma cilíndrica, en el tren de la válvula, que va sobre el lóbulo del árbol de levas para levantar la válvula a partir de su asiento.

Valve margin On a poppet valve, the space or rim between the surface of the head and the surface of the valve face.

Márgen de la válvula En una válvula de champiñones, el espacio o el borde entre la superficie de la cabeza y la superficie de la cara de la válvula.

Valve seat Machined surface of the cylinder head that provides the mating surface for the valve face. The valve seat can be either machined into the cylinder head or a separate component that is pressed into the cylinder head.

Asiento de la válvula La superficie rectificada a máquina en la cabeza del cilindro que provee una superficie de contacto para la cara de la válvula. El asiento puede ser rectificado a máquina en la cabeza del cilindro o puede ser un componente aparte para prensarlo en la cabeza de la válvula.

Valve spring A coil of specially constructed metal used to force the valve closed, providing a positive seal between the valve face and seat.

Resorte de la válvula Es un espiral de metal de construcción especial que provee la fuerza para mantener cerrada la válvula, así proveyendo un sello positivo entre la cara de la válvula y el asiento.

Vapor A substance in a gaseous state. Liquid becomes a vapor when brought above the boiling point.

Vapor Una substancia en un estado gaseoso. El líquido se convierte en un vapor al superar el punto de ebullición.

Vapor lock A condition wherein the fuel boils in the fuel system forming bubbles that retard or stop the flow of fuel to the carburetor.

Tapón de vapor Una condicion en la cual el combustible hierve en el sistema de combustible formando unas burbújas que retrasan o detienen al flujo del combustible al carburador.

Vapor/fuel separator A term used for a vapor/liquid separator. Part of the evaporative-emissions control system that prevents liquid (gasoline) from flowing through the vent line to the charcoal canister.

Separador de vapor y combustible Un término usado para el separador de vapor y líquido. Es parte del sistema de control de emisiones de evaporación que impide que el líquido (la gasolina) se desborde por la línea de ventilación al recipiente de carbón.

Vaporization The last stage of carburetion, in which a fine mist of fuel is created below the venturi in the bore.

Vaporización La última etapa del carburacíon, en la cual una llovizna muy fina del combustible se crea en el orificio abajo del venturi.

Variable-rate coil spring Rather than having a standard spring deflection rate, these springs have an average spring rate based on load at a predetermined deflection.

Resorte espiral de capacidad variable En vez de tener una capacidad de desviación de resorte estándar, estos resortes tienen un valor promedio de elasticidad basado en la carga a una desviación predeterminada.

Varnish A deposit in an engine lubrication system resulting from oxidation of the motor oil. Varnish is similar to, but softer than, lacquer.

Barníz Un depósito en el sistema de lubricación de un motor que resulta de una oxidación del aceite del motor. El barníz parece a la laca, pero es más blando.

V-belt A rubberlike continuous loop placed between the engine crankshaft pulley and accessories to transfer rotary motion of the crankshaft to the accessories.

Correa/banda en V Un ciclo de caucho continuo parecido a una liga de caucho ubicada entre la polea del cigüeñal del motor y los accesorios, para transferir el movimiento giratorio del cigüeñal a los accesorios.

Velocity The speed of an object in a particular direction.

Velocidad Velocidad es el índice de movimiento de un objeto en una dirección particular.

Ventilation The act of supplying fresh air to an enclosed space such as the inside of an automobile.

Ventilación El proceso de suministrar el aire fresco a un espacio cerrado, como por ejemplo al interior de un automóvil.

Viscosimeter An instrument for determining the viscosity of an oil by passing a certain quantity at a definite temperature through a standard-size orifice or port. The time required for the oil to pass through, which is expressed in seconds, gives the viscosity.

Viscosímetro Un instrumento que determina la viscosidad de un aceite al hacer que pasa una cierta cantidad en una temperatura definida por un orificio o una puerta de un tamaño normal. El tiempo requerido para que el aceite atraviesa, expresado en segundos, define la viscosidad.

VNT The variable nozzle turbine unit designed to allow a turbocharger to accelerate quickly, thus reducing lag time.

VNT La unidad de turbina con boquilla variante diseñada para permitir que un turbocargador se acelerara rápidamente, así disminuyendo el tiempo del retraso.

Volatile liquid A liquid that vaporizes very quickly.

Líquido volátil Un líquido que se vaporiza muy rápidamente.

Volatility The tendency for a fluid to evaporate rapidly or pass off in the form of vapor. For example, gasoline is more volatile than kerosene because it evaporates at a lower temperature.

Volatilidad La tendencia de un flúido a evaporarse rápidamente o escaparse en la forma de un vapor. Por ejemplo, la gasolina es más volátil que el kerosén porque se evapora en las temperatura más bajas.

Volt A unit of measurement of electromotive force. One volt of electromotive force applied steadily to a conductor of one-ohm resistance produces a current of one ampere.

Voltio Una unidad de medida de la fuerza electromotor. Un voltio de la fuerza electromotor aplicado constantemente a un conductor que tiene una resistencia de un ohmio produce una corriente de un amperio.

Voltage drop Voltage lost by the passage of electrical current through resistance.

Caída de voltaje El voltaje que se pierde al pasarse un corriente eléctrico por la resistencia.

Voltage regulator A device that controls the strength of the electromagnetic field in the rotor of an alternator.

Regulador del voltaje Un dispositivo que controla la fuerza del campo electromagnético en el rotor de un alternador.

Voltmeter A tool used to measure the voltage available at any point in an electrical system.

Voltímetro Una herramienta que sirve para medir el voltaje disponible en cualquier punto del sistema eléctrico.

Volume The measure of space expressed as cubic inches, cubic feet, and so forth.

Volumen La medida del espacio expresado como pulgadas cúbicas, pies cúbicos, etc.

Volumetric efficiency A measure of how well air flows in and out of an engine.

Rendimiento volumétrico Una medida de lo fácil que circula el aire en un motor.

Vortex flow A swirling, twisting motion of fluid.

Flujo vórtice Un movimiento como remolino, o giratorio de un flúido.

V-ribbed belt A belt that has multiple ribs on one side and is flat on the other side.

Banda encostillada en V Una banda que tiene varias nervaduras en un lado y esta plana en el otro.

Vulcanized A process of heating rubber under pressure to mold it into a special shape.

Vulcanizado Un proceso de calentar al caucho bajo presión para moldearlo en una forma especial.

Warpage Bending.
Alabeo Doblando.
Waste spark A spark occurring during the exhaust stroke on a computerized ignition system.
Chispa de desperdicio Chispa que se produce durante el ciclo de escape de un sistema de encendido computarizado.
Waste spark system A term that may be applied to any EI system in which a spark plug fires twice during one four-stroke cycle. The spark plug fires during the compression and the exhaust strokes. The firing during the exhaust stroke is wasted as it serves no purpose.
Sistema de la chispa perdida La frase sistema de la chispa perdida se puede aplicar a cualquier sistema de EI en el cual un enchufe de chispa encienda dos veces durante un ciclo de cuatro tiempos. El enchufe de chispa enciende durante el movimiento de la compresión y el movimiento del extractor. La ignición durante el movimiento del extractor se pierde porque no sirve por nada.
Water pump A device, usually located on the front of the engine and driven by one of the accessory drive belts, that circulates the coolant by causing it to move from the lower radiator-outlet section into the engine by centrifugal action of a finned impeller on the pump shaft.
Bomba de agua Un dispositivo, situado por lo común en la parte delantera del motor e impulsado por una de las bandas para accesorios, que hace circular el refrigerante, haciéndolo pasar de la sección de salida inferior del radiador al motor, mediante la acción centrífuga de un propulsor de aletas en el eje de la bomba.
Watt's law A basic law of electricity used to find the power of an electrical circuit expressed in watts. It states that power equals the voltage multiplied by the current (in amperes).
Ley de Watt Un ley básico de la electricidad que sirve para expresar la potencia de un circuito eléctrico en vatios. Declara que la potencia egala al voltaje multiplicado por el corriente (en ampérios).
Wave Any single swing of an object back and forth between the extremes of its travel through matter or space.
Onda Cualquier movimiento unico de un objeto hacia adelante y hacia atrás entre los extremos de su recorrido a través de materia o de espacio.
Wavelength The distance between each compression of a sound or electrical wave.
Longitud de onda La distancia entre cada compresión de un sonido o onda eléctrica.
Weight A force exerted on a mass by the gravitational force.
Peso El peso es una fuerza ejercida en una masa por la gravedad.
Wheatstone bridge A series-parallel arrangement of resistors between an input terminal and ground.
Puente Wheatstone Un arreglo en series-paralelo de los resistores entre un terminal de entrada y la tierra.
Wheel base The distance between the center of a front wheel and the center of a rear wheel.
Batalla (empate) La distancia entre el centro de una rueda delantera y el centro de una rueda trasera.
Wheel cylinder A device used to convert hydraulic fluid pressure to mechanical force for brake applications.
Cilindro de rueda Un dispositivo que sirve para convertir la presión de flúido hidráulico en una fuerza mecánica para aplicaciones de frenos.
Wheel spindle The short shaft on the front wheel upon which the wheel bearings ride and to which the wheel is attached.
Husillo de rueda El eje corto de la rueda delantera sobre el que se asientan los cojinetees y al que va acoplada la rueda.
Whip Flexing of a shaft that occurs during rotation.
Desviación (latigo) El doblar de una flecha que ocurre al girar.
Work What is accomplished when a force moves a certain mass a specific distance.
Trabajo Lo que se logra cuando una fuerza mueve una cierta masa una distancia específica.
Zener diode A diode that allows reverse current to flow above a set voltage limit.
Diodo zener Un diodo que permite que el corriente en reverso fluye en cantidades más altas del limite de voltaje predeterminado.

INDEX

IMPORTANT! READ CAREFULLY: This End User License Agreement ("Agreement") sets forth the conditions by which Delmar Cengage Learning will make electronic access to the Delmar Cengage Learning-owned licensed content and associated media, software, documentation, printed materials, and electronic documentation contained in this package and/or made available to you via this product (the "Licensed Content"), available to you (the "End User"). BY CLICKING THE "I ACCEPT" BUTTON AND/OR OPENING THIS PACKAGE, YOU ACKNOWLEDGE THAT YOU HAVE READ ALL OF THE TERMS AND CONDITIONS, AND THAT YOU AGREE TO BE BOUND BY ITS TERMS, CONDITIONS, AND ALL APPLICABLE LAWS AND REGULATIONS GOVERNING THE USE OF THE LICENSED CONTENT.

1.0 SCOPE OF LICENSE

1.1 Licensed Content. The Licensed Content may contain portions of modifiable content ("Modifiable Content") and content which may not be modified or otherwise altered by the End User ("Non-Modifiable Content"). For purposes of this Agreement, Modifiable Content and Non-Modifiable Content may be collectively referred to herein as the "Licensed Content." All Licensed Content shall be considered Non-Modifiable Content, unless such Licensed Content is presented to the End User in a modifiable format and it is clearly indicated that modification of the Licensed Content is permitted.

1.2 Subject to the End User's compliance with the terms and conditions of this Agreement, Delmar Cengage Learning hereby grants the End User, a nontransferable, nonexclusive, limited right to access and view a single copy of the Licensed Content on a single personal computer system for noncommercial, internal, personal use only. The End User shall not (i) reproduce, copy, modify (except in the case of Modifiable Content), distribute, display, transfer, sublicense, prepare derivative work(s) based on, sell, exchange, barter or transfer, rent, lease, loan, resell, or in any other manner exploit the Licensed Content; (ii) remove, obscure, or alter any notice of Delmar Cengage Learning's intellectual property rights present on or in the Licensed Content, including, but not limited to, copyright, trademark, and/or patent notices; or (iii) disassemble, decompile, translate, reverse engineer, or otherwise reduce the Licensed Content.

2.0 TERMINATION

2.1 Delmar Cengage Learning may at any time (without prejudice to its other rights or remedies) immediately terminate this Agreement and/or suspend access to some or all of the Licensed Content, in the event that the End User does not comply with any of the terms and conditions of this Agreement. In the event of such termination by Delmar Cengage Learning, the End User shall immediately return any and all copies of the Licensed Content to Delmar Cengage Learning.

3.0 PROPRIETARY RIGHTS

3.1 The End User acknowledges that Delmar Cengage Learning owns all rights, title and interest, including, but not limited to all copyright rights therein, in and to the Licensed Content, and that the End User shall not take any action inconsistent with such ownership. The Licensed Content is protected by U.S., Canadian and other applicable copyright laws and by international treaties, including the Berne Convention and the Universal Copyright Convention. Nothing contained in this Agreement shall be construed as granting the End User any ownership rights in or to the Licensed Content.

3.2 Delmar Cengage Learning reserves the right at any time to withdraw from the Licensed Content any item or part of an item for which it no longer retains the right to publish, or which it has reasonable grounds to believe infringes copyright or is defamatory, unlawful, or otherwise objectionable.

4.0 PROTECTION AND SECURITY

4.1 The End User shall use its best efforts and take all reasonable steps to safeguard its copy of the Licensed Content to ensure that no unauthorized reproduction, publication, disclosure, modification, or distribution of the Licensed Content, in whole or in part, is made. To the extent that the End User becomes aware of any such unauthorized use of the Licensed Content, the End User shall immediately notify Delmar Cengage Learning. Notification of such violations may be made by sending an e-mail to delmarhelp@cengage.com.

5.0 MISUSE OF THE LICENSED PRODUCT

5.1 In the event that the End User uses the Licensed Content in violation of this Agreement, Delmar Cengage Learning shall have the option of electing liquidated damages, which shall include all profits generated by the End User's use of the Licensed Content plus interest computed at the maximum rate permitted by law and all legal fees and other expenses incurred by Delmar Cengage Learning in enforcing its rights, plus penalties.

6.0 FEDERAL GOVERNMENT CLIENTS

6.1 Except as expressly authorized by Delmar Cengage Learning, Federal Government clients obtain only the rights specified in this Agreement and no other rights. The Government acknowledges that (i) all software and related documentation incorporated in the Licensed Content is existing commercial computer software within the meaning of FAR 27.405(b)(2); and (2) all other data delivered in whatever form, is limited rights data within the meaning of FAR 27.401. The restrictions in this section are acceptable as consistent with the Government's need for software and other data under this Agreement.

7.0 DISCLAIMER OF WARRANTIES AND LIABILITIES

7.1 Although Delmar Cengage Learning believes the Licensed Content to be reliable, Delmar Cengage Learning does not guarantee or warrant (i) any information or materials contained in or produced by the Licensed Content, (ii) the accuracy, completeness or reliability of the Licensed Content, or (iii) that the Licensed Content is free from errors or other material defects. THE LICENSED PRODUCT IS PROVIDED "AS IS," WITHOUT ANY WARRANTY OF ANY KIND AND DELMAR CENGAGE LEARNING DISCLAIMS ANY AND ALL WARRANTIES, EXPRESSED OR IMPLIED, INCLUDING, WITHOUT LIMITATION, WARRANTIES OF MERCHANTABILITY OR FITNESS FOR A PARTICULAR PURPOSE. IN NO EVENT SHALL DELMAR CENGAGE LEARNING BE LIABLE FOR: INDIRECT, SPECIAL, PUNITIVE OR CONSEQUENTIAL DAMAGES INCLUDING FOR LOST PROFITS, LOST DATA, OR OTHERWISE. IN NO EVENT SHALL DELMAR CENGAGE LEARNING'S AGGREGATE LIABILITY HEREUNDER, WHETHER ARISING IN CONTRACT, TORT, STRICT LIABILITY OR OTHERWISE, EXCEED THE AMOUNT OF FEES PAID BY THE END USER HEREUNDER FOR THE LICENSE OF THE LICENSED CONTENT.

8.0 GENERAL

8.1 Entire Agreement. This Agreement shall constitute the entire Agreement between the Parties and supercedes all prior Agreements and understandings oral or written relating to the subject matter hereof.

8.2 Enhancements/Modifications of Licensed Content. From time to time, and in Delmar Cengage Learning's sole discretion, Delmar Cengage Learning may advise the End User of updates, upgrades, enhancements and/or improvements to the Licensed Content, and may permit the End User to access and use, subject to the terms and conditions of this Agreement, such modifications, upon payment of prices as may be established by Delmar Cengage Learning.

8.3 No Export. The End User shall use the Licensed Content solely in the United States and shall not transfer or export, directly or indirectly, the Licensed Content outside the United States.

8.4 Severability. If any provision of this Agreement is invalid, illegal, or unenforceable under any applicable statute or rule of law, the provision shall be deemed omitted to the extent that it is invalid, illegal, or unenforceable. In such a case, the remainder of the Agreement shall be construed in a manner as to give greatest effect to the original intention of the parties hereto.

8.5 Waiver. The waiver of any right or failure of either party to exercise in any respect any right provided in this Agreement in any instance shall not be deemed to be a waiver of such right in the future or a waiver of any other right under this Agreement.

8.6 Choice of Law/Venue. This Agreement shall be interpreted, construed, and governed by and in accordance with the laws of the State of New York, applicable to contracts executed and to be wholly preformed therein, without regard to its principles governing conflicts of law. Each party agrees that any proceeding arising out of or relating to this Agreement or the breach or threatened breach of this Agreement may be commenced and prosecuted in a court in the State and County of New York. Each party consents and submits to the nonexclusive personal jurisdiction of any court in the State and County of New York in respect of any such proceeding.

8.7 Acknowledgment. By opening this package and/or by accessing the Licensed Content on this Web site, THE END USER ACKNOWLEDGES THAT IT HAS READ THIS AGREEMENT, UNDERSTANDS IT, AND AGREES TO BE BOUND BY ITS TERMS AND CONDITIONS. IF YOU DO NOT ACCEPT THESE TERMS AND CONDITIONS, YOU MUST NOT ACCESS THE LICENSED CONTENT AND RETURN THE LICENSED PRODUCT TO DELMAR CENGAGE LEARNING (WITHIN 30 CALENDAR DAYS OF THE END USER'S PURCHASE) WITH PROOF OF PAYMENT ACCEPTABLE TO DELMAR CENGAGE LEARNING, FOR A CREDIT OR A REFUND. Should the End User have any questions/comments regarding this Agreement, please contact Delmar Cengage Learning at delmarhelp@cemgage.com.